Be sure to keep up-to-date in 2010!

Register now and we'll notify you by email each quarter when the newest cost index is available free online.

Please provide your name, address, and email below and return this card by mail or Fax: 1-800-632-6732.

To register online: www.rsmeans411.com/register

Name ____

email ____

Company ____

Street ____

City/Town __________________ State/Prov. _____ Zip/Postal Code _______

RS**Means**

PLACE
STAMP
HERE

RSMeans
Quarterly Updates
63 Smiths Lane
Kingston, MA 02364-3008
USA

$169.95 per copy (in United States)
Price is subject to change without prior notice.

0010

RSMeans Building Construction Cost Data

68TH ANNUAL EDITION

Senior Editor
Phillip R. Waier, PE

Contributing Editors
Christopher Babbitt
Ted Baker
Barbara Balboni
Robert A. Bastoni
John H. Chiang, PE
Gary W. Christensen
David G. Drain, PE
Cheryl Elsmore
Robert J. Kuchta
Robert C. McNichols
Melville J. Mossman, PE
Jeannene D. Murphy
Stephen C. Plotner
Eugene R. Spencer
Marshall J. Stetson

President & CEO
Iain Melville

Vice President, Product & Development
Richard Remington

Vice President of Operations
David C. Walsh

Vice President of Sales
Thomas Kesler

Director, Cost Products
Daimon Bridge

Marketing Director
John M. Shea

Engineering Manager
Bob Mewis, CCC

Production Manager
Michael Kokernak

Production Coordinator
Jill Goodman

Production
Paula Reale-Camelio
Sheryl A. Rose
Jonathan Forgit
Debbie Panarelli
Mary Lou Geary

Technical Support
A.K. Dash
Tom Dion
Roger Hancock
Gary L. Hoitt
Genevieve Medeiros
Kathryn S. Rodriguez
Sharon Proulx

Book Design
Norman R. Forgit

Cover Design
Paul Robertson
Wayne Engebretson

2010

RSMeans
A division of Reed Construction Data
Construction Publishers & Consultants
63 Smiths Lane
Kingston, MA 02364-3008
(781) 422-5000

Printed in the United States of America
ISSN 0068-3531
ISBN 978-0-87629-746-9

This book is printed on recycled paper, using soy-based printing ink. This book is recyclable.

Foreword

Our Mission

Since 1942, RSMeans has been actively engaged in construction cost publishing and consulting throughout North America.

Today, more than 65 years after RSMeans began, our primary objective remains the same: to provide you, the construction and facilities professional, with the most current and comprehensive construction cost data possible.

Whether you are a contractor, owner, architect, engineer, facilities manager, or anyone else who needs a reliable construction cost estimate, you'll find this publication to be a highly useful and necessary tool.

With the constant flow of new construction methods and materials today, it's difficult to find the time to look at and evaluate all the different construction cost possibilities. In addition, because labor and material costs keep changing, last year's cost information is not a reliable basis for today's estimate or budget.

That's why so many construction professionals turn to RSMeans. We keep track of the costs for you, along with a wide range of other key information, from city cost indexes . . . to productivity rates . . . to crew composition . . . to contractor's overhead and profit rates.

RSMeans performs these functions by collecting data from all facets of the industry and organizing it in a format that is instantly accessible to you. From the preliminary budget to the detailed unit price estimate, you'll find the data in this book useful for all phases of construction cost determination.

The Staff, the Organization, and Our Services

When you purchase one of RSMeans' publications, you are, in effect, hiring the services of a full-time staff of construction and engineering professionals.

Our thoroughly experienced and highly qualified staff works daily at collecting, analyzing, and disseminating comprehensive cost information for your needs. These staff members have years of practical construction experience and engineering training prior to joining the firm. As a result, you can count on them not only for accurate cost figures, but also for additional background reference information that will help you create a realistic estimate.

The RSMeans organization is always prepared to help you solve construction problems through its variety of data solutions, including online, CD, and print book formats, as well as cost estimating expertise available via our business solutions, training, and seminars.

Besides a full array of construction cost estimating books, RSMeans also publishes a number of other reference works for the construction industry. Subjects include construction estimating and project and business management, special topics such as green building and job order contracting, and a library of facility management references.

In addition, you can access all of our construction cost data electronically using *Means CostWorks*® CD or on the Web using **MeansCostWorks.com.**

What's more, you can increase your knowledge and improve your construction estimating and management performance with an RSMeans construction seminar or in-house training program. These two-day seminar programs offer unparalleled opportunities for everyone in your organization to become updated on a wide variety of construction-related issues.

RSMeans is also a worldwide provider of construction cost management and analysis services for commercial and government owners.

In short, RSMeans can provide you with the tools and expertise for constructing accurate and dependable construction estimates and budgets in a variety of ways.

Robert Snow Means Established a Tradition of Quality That Continues Today

Robert Snow Means spent years building RSMeans, making certain he always delivered a quality product.

Today, at RSMeans, we do more than talk about the quality of our data and the usefulness of our books. We stand behind all of our data, from historical cost indexes to construction materials and techniques to current costs.

If you have any questions about our products or services, please call us toll-free at 1-800-334-3509. Our customer service representatives will be happy to assist you. You can also visit our Web site at **www.rsmeans.com.**

Table of Contents

Related RSMeans Products and Services

The engineers at RSMeans suggest the following products and services as companion information resources to *RSMeans Building Construction Cost Data:*

Construction Cost Data Books

Assemblies Cost Data 2010
Square Foot Costs 2010

Reference Books

ADA Compliance Pricing Guide
Estimating Building Costs
RSMeans Estimating Handbook
Green Building: Project Planning & Estimating
How to Estimate with Means Data and CostWorks
Plan Reading & Material Takeoff
Project Scheduling & Management for Construction
Repair & Remodeling Estimating Methods

Seminars and In-House Training

Unit Price Estimating
Means CostWorks® Training
Means Data for Job Order Contracting (JOC)
Practical Project Management
Scheduling & Project Management
Mechanical & Electrical Estimating

RSMeans on the Internet

Visit RSMeans at **www.rsmeans.com** for a one-stop portal for the most reliable and current resources available on the market. Here you can request or download **FREE** estimating software demos, visit our bookstore for convenient ordering, and learn more about new publications and companion products.

Also visit **www.reedconstructiondata.com** for RSMeans business solutions, seminars, training, and more.

RSMeans Electronic Data

Get the information found in RSMeans cost books electronically on *Means CostWorks®* CD or on the Web at **MeansCostWorks.com**.

RSMeans Business Solutions

Cost engineers and analysts conduct facility life cycle and benchmark research studies and predictive cost modeling, as well as offer consultation services for conceptual estimating, real property management, and job order contracting. Research studies are designed with the application of proprietary cost and project data from Reed Construction Data and RSMeans' extensive North American databases. Analysts offer building product manufacturers qualitative and quantitative market research, as well as time/motion studies for new products, and Web-based dashboards for market opportunity analysis. Clients are from both the public and private sectors, including federal, state, and municipal agencies, corporations, institutions, construction management firms, hospitals, and associations.

RSMeans for Job Order Contracting (JOC)

Best practice job order contracting (JOC) services support the owner in the development of their program. RSMeans develops all required contracting documents, works with stakeholders, and develops qualified and experienced contractor lists. Means JOCWorks™ software is the best of class in the functional tools that meet JOC specific requirements. RSMeans data is organized to meet JOC cost estimating requirements.

Construction Costs for Software Applications

Over 25 unit price and assemblies cost databases are available through a number of leading estimating and facilities management software providers (listed below). For more information see the "Other RSMeans Products" pages at the back of this publication.

MeansData™ is also available to federal, state, and local government agencies as multi-year, multi-seat licenses.

- 4Clicks-Solutions, LLC
- ArenaSoft Estimating
- Beck Technology
- BSD – Building Systems Design, Inc.
- CMS – Construction Management Software
- Corecon Technologies, Inc.
- CorVet Systems
- Estimating Systems, Inc.
- Hard Dollar Corporation
- MC²
- Parsons Corporation
- Sage Timberline Office
- UDA Technologies, Inc.
- US Cost, Inc.
- VFA – Vanderweil Facility Advisers
- WinEstimator, Inc.

How the Book is Built: An Overview

The Construction Specifications Institute (CSI) and Construction Specifications Canada (CSC) have produced the 2004 edition of MasterFormat™, a system of titles and numbers used extensively to organize construction information.

All unit price data in the RSMeans cost data books is now arranged in the 50-division MasterFormat™ 2004 system.

A Powerful Construction Tool

You have in your hands one of the most powerful construction tools available today. A successful project is built on the foundation of an accurate and dependable estimate. This book will enable you to construct just such an estimate.

For the casual user the book is designed to be:

- quickly and easily understood so you can get right to your estimate.
- filled with valuable information so you can understand the necessary factors that go into the cost estimate.

For the regular user, the book is designed to be:

- a handy desk reference that can be quickly referred to for key costs.
- a comprehensive, fully reliable source of current construction costs and productivity rates so you'll be prepared to estimate any project.
- a source book for preliminary project cost, product selections, and alternate materials and methods.

To meet all of these requirements, we have organized the book into the following clearly defined sections.

New: Quick Start

See our new "Quick Start" instructions on the following page to get started right away.

Estimating with RSMeans Unit Price Cost Data

Please refer to these steps for guidance on completing an estimate using RSMeans unit price cost data.

How To Use the Book: The Details

This section contains an in-depth explanation of how the book is arranged . . . and how you can use it to determine a reliable construction cost estimate. It includes information about how we develop our cost figures and how to completely prepare your estimate.

Unit Price Section

All cost data has been divided into the 50 divisions according to the MasterFormat system of classification and numbering.
For a listing of these divisions and an outline of their subdivisions, see the Unit Price Section Table of Contents.

Estimating tips are included at the beginning of each division.

Reference Section

This section includes information on Equipment Rental Costs, Crew Listings, Historical Cost Indexes, City Cost Indexes, Location Factors, Reference Tables, Change Orders, Square Foot Costs, and a listing of Abbreviations.

Equipment Rental Costs: This section contains the average costs to rent and operate hundreds of pieces of construction equipment.

Crew Listings: This section lists all of the crews referenced in the book. For the purposes of this book, a crew is composed of more than one trade classification and/or the addition of power equipment to any trade classification. Power equipment is included in the cost of the crew. Costs are shown both with bare labor rates and with the installing contractor's overhead and profit added. For each, the total crew cost per eight-hour day and the composite cost per labor-hour are listed.

Historical Cost Indexes: These indexes provide you with data to adjust construction costs over time.

City Cost Indexes: All costs in this book are U.S. national averages. Costs vary because of the regional economy. You can adjust costs by CSI Division to over 700 locations throughout the U.S. and Canada by using the data in this section.

Location Factors: You can adjust total project costs to over 900 locations throughout the U.S. and Canada by using the data in this section.

Reference Tables: At the beginning of selected major classifications in the Unit Price section are reference numbers shown in a shaded box. These numbers refer you to related information in the Reference Section. In this section, you'll find reference tables, explanations, estimating information that support how we develop the unit price data, technical data, and estimating procedures.

Change Orders: This section includes information on the factors that influence the pricing of change orders.

Square Foot Costs: This section contains costs for 59 different building types that allow you to make a rough estimate for the overall cost of a project or its major components.

Abbreviations: A listing of abbreviations used throughout this book, along with the terms they represent, is included in this section.

Index

A comprehensive listing of all terms and subjects in this book will help you quickly find what you need when you are not sure where it falls in MasterFormat.

The Scope of This Book

This book is designed to be as comprehensive and as easy to use as possible. To that end we have made certain assumptions and limited its scope in two key ways:

1. We have established material prices based on a national average.
2. We have computed labor costs based on a 30-city national average of union wage rates.

For a more detailed explanation of how the cost data is developed, see "How To Use the Book: The Details."

Project Size

This book is aimed primarily at commercial and industrial projects costing $1,000,000 and up, or large multi-family housing projects. Costs are primarily for new construction or major renovation of buildings rather than repairs or minor alterations.

***With reasonable exercise of judgment the figures can be used for any building work. However, for civil engineering structures such as bridges, dams, highways, or the like, please refer to* RSMeans Heavy Construction Cost Data.**

Absolute Essentials for a Quick Start

If you feel you are ready to use this book and don't think you will need the detailed instructions that begin on the following page, this Absolute Essentials for a Quick Start page is for you. These steps will allow you to get started estimating in a matter of minutes.

1 Scope

Think through the project that you will be estimating, and identify the many individual work tasks that will need to be covered in your estimate.

2 Quantify

Determine the number of units that will be required for each work task that you identified.

3 Pricing

Locate individual Unit Price line items that match the work tasks you identified. The Unit Price Section Table of Contents that begins on page 1 and the Index in the back of the book will help you find these line items.

4 Multiply

Multiply the Total Incl O&P cost for a Unit Price line item in the book by your quantity for that item. The price you calculate will be an estimate for a completed item of work performed by a subcontractor. Keep adding line items in this manner to build your estimate.

5 Project Overhead

Include project overhead items in your estimate. These items are needed to make the job run and are typically, but not always, provided by the General Contractor. They can be found in Division 1. An alternate method of estimating project overhead costs is to apply a percentage of the total project cost.

Include rented tools not included in crews, waste, rubbish handling, and cleanup.

6 Estimate Summary

Include General Contractor's markup on subcontractors, General Contractor's office overhead and profit, and sales tax on materials and equipment.

Adjust your estimate to the project's location by using the City Cost Indexes or Location Factors found in the Reference Section.

Editors' Note: We urge you to spend time reading and understanding the supporting material in the front of this book. An accurate estimate requires experience, knowledge, and careful calculation. The more you know about how we at RSMeans developed the data, the more accurate your estimate will be. In addition, it is important to take into consideration the reference material in the back of the book such as Equipment Listings, Crew Listings, City Cost Indexes, Location Factors, and Reference Numbers.

Estimating with RSMeans Unit Price Cost Data

Following these steps will allow you to complete an accurate estimate using RSMeans Unit Price cost data.

1 Scope out the project

- Identify the individual work tasks that will need to be covered in your estimate.
- The Unit Price data inside this book has been divided into 44 Divisions according to the CSI MasterFormat 2004—their titles are listed on the back cover of your book.
- Think through the project that you will be estimating, and identify those CSI Divisions that you will need to use in your estimate.
- The Unit Price Section Table of Contents that begins on page 1 may also be helpful when scoping out your project.
- Experienced estimators find it helpful to begin an estimate with Division 2, estimating Division 1 after the full scope of the project is known.

2 Quantify

- Determine the number of units that will be required for each work task that you previously identified.
- Experienced estimators will include an allowance for waste in their quantities. (Waste is not included in RSMeans Unit Price line items unless so stated.)

3 Price the quantities

- Use the Unit Price Table of Contents, and the Index, to locate the first individual Unit Price line item for your estimate.
- Reference Numbers indicated within a Unit Price section refer to additional information that you may find useful.
- The crew indicates who is performing the work for that task. Crew codes are expanded in the Crew Listings in the Reference Section to include all trades and equipment that comprise the crew.
- The Daily Output is the amount of work the crew is expected to do in one day.
- The Labor-Hours value is the amount of time it will take for the average crew member to install one unit of measure.
- The abbreviated Unit designation indicates the unit of measure upon which the crew, productivity, and prices are based.
- Bare Costs are shown for materials, labor, and equipment needed to complete the Unit Price line item. Bare costs do not include waste, project overhead, payroll insurance, payroll taxes, main office overhead, or profit.
- The Total Incl O&P cost is the billing rate or invoice amount of the installing contractor or subcontractor who performs the work for the Unit Price line item.

4 Multiply

- Multiply the total number of units needed for your project by the Total Incl O&P cost for the Unit Price line item.
- Be careful that the final unit of measure for your quantity of units matches the unit of measure in the Unit column in the book.
- The price you calculate will be an estimate for a completed item of work.
- Keep scoping individual tasks, determining the number of units required for those tasks, matching them up with individual Unit Price line items in the book, and multiplying quantities by Total Incl O&P costs. In this manner, keep building your estimate.
- An estimate completed to this point, in this manner, will be priced as if a subcontractor, or set of subcontractors, will perform the work. The estimate does not yet include Project Overhead or Estimate Summary components such as General Contractor markups on subcontracted and self-performed work, General Contractor office overhead and profit, contingency, and location factor.

5 Project Overhead

- Include project overhead items from Division 1 – General Requirements.
- These are items that will be needed to make the job run. These are typically, but not always, provided by the General Contractor. They include, but are not limited to, such items as field personnel, insurance, performance bond, permits, testing, temporary utilities, field office and storage facilities, temporary scaffolding and platforms, equipment mobilization and demobilization, temporary roads and sidewalks, winter protection, temporary barricades and fencing, temporary security, temporary signs, field engineering and layout, final cleaning and commissioning.
- These items should be scoped, quantified, matched to individual Unit Price line items in Division 1, priced and added to your estimate.
- An alternate method of estimating project overhead costs is to apply a percentage of the total project cost, usually 5% to 15% with an average of 10% (see General Conditions, page ix).
- Include other project related expenses in your estimate such as:
  - Rented equipment not itemized in the Crew Listings
  - Rubbish handling throughout the project (see 02 41 19.23)

6 Estimate Summary

- Includes sales tax on materials and equipment.
  Note: Sales tax must be added for materials in subcontracted work.
- Include the General Contractor's markup on self-performed work, usually 5% to 15% with an average of 10%.
- Include the General Contractor's markup on subcontracted work, usually 5% to 15% with an average of 10%.
- Include General Contractor's main office overhead and profit.
- RSMeans gives general guidelines on the General Contractor's main office overhead (see section 01 31 13.50 and Reference Number R013113-50).
- RSMeans gives no guidance on the General Contractor's profit.
- Markups will depend on the size of the General Contractor's operations, his projected annual revenue, the level of risk he is taking on, and on the level of competitiveness in the local area and for this project in particular.
- Include a contingency, usually 3% to 5%.
- Adjust your estimate to the project's location by using the City Cost Indexes or the Location Factors in the Reference Section.
- Look at the rules on the pages for How to Use the City Cost Indexes to see how to apply the Indexes for your location.
- When the proper Index or Factor has been identified for the project's location, convert it to a multiplier by dividing it by 100, then multiply that multiplier by your estimate total cost. Your original estimate total cost will now be adjusted up or down from the national average to a total that is appropriate for your location.

Editors' Notes:

1) *We urge you to spend time reading and understanding the supporting material in the front of this book. An accurate estimate requires experience, knowledge, and careful calculation. The more you know about how we at RSMeans developed the data, the more accurate your estimate will be. In addition, it is important to take into consideration the reference material in the back of the book such as Equipment Listings, Crew Listings, City Cost Indexes, Location Factors, and Reference Numbers.*
2) *Contractors who are bidding or are involved in JOC, DOC, SABER, or IDIQ type contracts are cautioned that workers' compensation insurance, federal and state payroll taxes, waste, project supervision, project overhead, main office overhead, and profit are not included in bare costs. Your coefficient or multiplier must cover these costs.*

How to Use the Book: The Details

What's Behind the Numbers? The Development of Cost Data

The staff at RSMeans continually monitors developments in the construction industry in order to ensure reliable, thorough, and up-to-date cost information.

While *overall* construction costs may vary relative to general economic conditions, price fluctuations within the industry are dependent upon many factors. Individual price variations may, in fact, be opposite to overall economic trends. Therefore, costs are continually monitored and complete updates are published yearly. Also, new items are frequently added in response to changes in materials and methods.

Costs—$ (U.S.)

All costs represent U.S. national averages and are given in U.S. dollars. The RSMeans City Cost Indexes can be used to adjust costs to a particular location. The City Cost Indexes for Canada can be used to adjust U.S. national averages to local costs in Canadian dollars. No exchange rate conversion is necessary.

G The processes or products identified by the green symbol in this publication have been determined to be environmentally responsible and/or resource-efficient solely by the RSMeans engineering staff. The inclusion of the green symbol does not represent compliance with any specific industry association or standard.

Material Costs

The RSMeans staff contacts manufacturers, dealers, distributors, and contractors all across the U.S. and Canada to determine national average material costs. If you have access to current material costs for your specific location, you may wish to make adjustments to reflect differences from the national average. Included within material costs are fasteners for a normal installation. RSMeans engineers use manufacturers' recommendations, written specifications, and/or standard construction practice for size and spacing of fasteners. Adjustments to material costs may be required for your specific application or location. Material costs do not include sales tax.

Labor Costs

Labor costs are based on the average of wage rates from 30 major U.S. cities. Rates are determined from labor union agreements or prevailing wages for construction trades for the current year. Rates, along with overhead and profit markups, are listed on the inside back cover of this book.

- If wage rates in your area vary from those used in this book, or if rate increases are expected within a given year, labor costs should be adjusted accordingly.

Labor costs reflect productivity based on actual working conditions. In addition to actual installation, these figures include time spent during a normal workday on tasks such as material receiving and handling, mobilization at site, site movement, breaks, and cleanup.

Productivity data is developed over an extended period so as not to be influenced by abnormal variations and reflects a typical average.

Equipment Costs

Equipment costs include not only rental, but also operating costs for equipment under normal use. The operating costs include parts and labor for routine servicing, such as repair and replacement of pumps, filters, and worn lines. Normal operating expendables, such as fuel, lubricants, tires, and electricity (where applicable), are also included. Extraordinary operating expendables with highly variable wear patterns, such as diamond bits and blades, are excluded. These costs are included under materials. Equipment rental rates are obtained from industry sources throughout North America—contractors, suppliers, dealers, manufacturers, and distributors.

Equipment costs do not include operators' wages; nor do they include the cost to move equipment to a job site (mobilization) or from a job site (demobilization).

Equipment Cost/Day—The cost of power equipment required for each crew is included in the Crew Listings in the Reference Section (small tools that are considered as essential everyday tools are not listed out separately). The Crew Listings itemize specialized tools and heavy equipment along with labor trades. The daily cost of itemized equipment included in a crew is based on dividing the weekly bare rental rate by 5 (number of working days per week) and then adding the hourly operating cost times 8 (the number of hours per day). This Equipment Cost/Day is shown in the last column of the Equipment Rental Cost pages in the Reference Section.

Mobilization/Demobilization—The cost to move construction equipment from an equipment yard or rental company to the job site and back again is not included in equipment costs. Mobilization (to the site) and demobilization (from the site) costs can be found in the Unit Price section. If a piece of equipment is already at the job site, it is not appropriate to utilize mobil./demob. costs again in an estimate.

Overhead and Profit

Total Cost including O&P for the *Installing Contractor* is shown in the last column on the Unit Price pages of this book. This figure is the sum of the bare material cost plus 10% for overhead and profit, the bare labor cost plus total overhead and profit, and the bare equipment cost plus 10% for overhead and profit. Details for the calculation of Overhead and Profit on labor are shown on the inside back cover of this book. (See the "How To Use the Unit Price Pages" for an example of this calculation.)

General Conditions

General Conditions, or General Requirements, of the contract should also be added to the Total Cost including O&P when applicable. Costs for General Conditions are listed in Division 1 and the Reference Section of this book. General Conditions for the *Installing Contractor* may range from 0% to 10% of the Total Cost including O&P. For the *General or Prime Contractor*, costs for General Conditions may range from 5% to 15% of the Total Cost including O&P, with a figure of 10% as the most typical allowance.

Factors Affecting Costs

Costs can vary depending upon a number of variables. Here's how we have handled the main factors affecting costs.

Quality—The prices for materials and the workmanship upon which productivity is based represent sound construction work. They are also in line with U.S. government specifications.

Overtime—We have made no allowance for overtime. If you anticipate premium time or work beyond normal working hours, be sure to make an appropriate adjustment to your labor costs.

Productivity—The productivity, daily output, and labor-hour figures for each line item are based on working an eight-hour day in daylight hours in moderate temperatures. For work that extends beyond normal work hours or is performed under adverse conditions, productivity may decrease. (See the section in "How To Use the Unit Price Pages" for more on productivity.)

Size of Project—The size, scope of work, and type of construction project will have a significant impact on cost. Economies of scale can reduce costs for large projects. Unit costs can often run higher for small projects. Costs in this book are intended for the size and type of project as previously described in "How the Book Is Built: An Overview." Costs for projects of a significantly different size or type should be adjusted accordingly.

Location—Material prices in this book are for metropolitan areas. However, in dense urban areas, traffic and site storage limitations may increase costs. Beyond a 20-mile radius of large cities, extra trucking or transportation charges may also increase the material costs slightly. On the other hand, lower wage rates may be in effect. Be sure to consider both of these factors when preparing an estimate, particularly if the job site is located in a central city or remote rural location.

In addition, highly specialized subcontract items may require travel and per-diem expenses for mechanics.

Other Factors—

- season of year
- contractor management
- weather conditions
- local union restrictions
- building code requirements
- availability of:
  - adequate energy
  - skilled labor
  - building materials
- owner's special requirements/restrictions
- safety requirements
- environmental considerations

Unpredictable Factors—General business conditions influence "in-place" costs of all items. Substitute materials and construction methods may have to be employed. These may affect the installed cost and/or life cycle costs. Such factors may be difficult to evaluate and cannot necessarily be predicted on the basis of the job's location in a particular section of the country. Thus, where these factors apply, you may find significant but unavoidable cost variations for which you will have to apply a measure of judgment to your estimate.

Rounding of Costs

In general, all unit prices in excess of $5.00 have been rounded to make them easier to use and still maintain adequate precision of the results. The rounding rules we have chosen are in the following table.

Prices from . . .	Rounded to the nearest . . .
$.01 to $5.00	$.01
$5.01 to $20.00	$.05
$20.01 to $100.00	$.50
$100.01 to $300.00	$1.00
$300.01 to $1,000.00	$5.00
$1,000.01 to $10,000.00	$25.00
$10,000.01 to $50,000.00	$100.00
$50,000.01 and above	$500.00

Final Checklist

Estimating can be a straightforward process provided you remember the basics. Here's a checklist of some of the steps you should remember to complete before finalizing your estimate.

Did you remember to . . .

- factor in the City Cost Index for your locale?
- take into consideration which items have been marked up and by how much?
- mark up the entire estimate sufficiently for your purposes?
- read the background information on techniques and technical matters that could impact your project time span and cost?
- include all components of your project in the final estimate?
- double check your figures for accuracy?
- call RSMeans if you have any questions about your estimate or the data you've found in our publications?

Remember, RSMeans stands behind its publications. If you have any questions about your estimate, about the costs you've used from our books, or even about the technical aspects of the job that may affect your estimate, feel free to call the RSMeans editors at 1-800-334-3509.

Unit Price Section

Table of Contents

Table of Contents (cont.)

Table of Contents (cont.)

How to Use the Unit Price Pages

The following is a detailed explanation of a sample entry in the Unit Price Section. Next to each bold number below is the described item with the appropriate component of the sample entry following in parentheses. Some prices are listed as bare costs; others as costs that include overhead and profit of the installing contractor. In most cases, if the work is to be subcontracted, the general contractor will need to add an additional markup (RSMeans suggests using 10%) to the figures in the column "Total Incl. O&P."

1 Division Number/Title (03 30/Cast-In-Place Concrete)

Use the Unit Price Section Table of Contents to locate specific items. The sections are classified according to the CSI MasterFormat 2004 system.

2 Line Numbers (03 30 53.40 3920)

Each unit price line item has been assigned a unique 12-digit code based on the CSI MasterFormat classification.

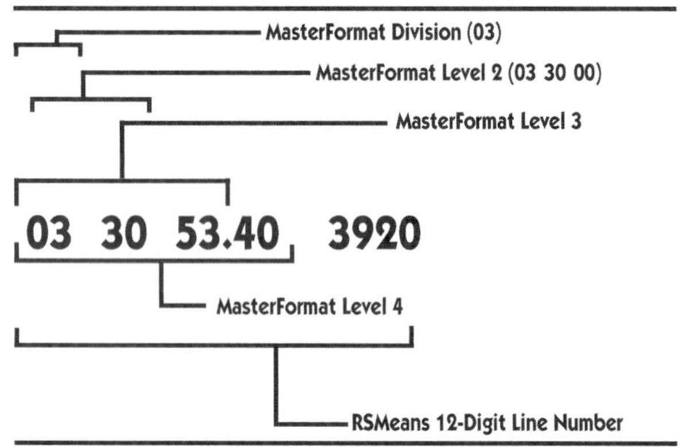

3 Description (Concrete In Place, etc.)

Each line item is described in detail. Sub-items and additional sizes are indented beneath the appropriate line items. The first line or two after the main item (in boldface) may contain descriptive information that pertains to all line items beneath this boldface listing.

Items which include the symbol **CN** are updated in The *RSMeans Quarterly Update Service* online. To obtain access to this service contact RSMeans customer service at 1-800-334-3509.

4 Reference Number Information

R033053 -50 You'll see reference numbers shown in shaded boxes at the beginning of some sections. These refer to related items in the Reference Section, visually identified by a vertical gray bar on the page edges.

The relation may be an estimating procedure that should be read before estimating, or technical information.

The "R" designates the Reference Section. The numbers refer to the MasterFormat 2004 classification system.

It is strongly recommended that you review all reference numbers that appear within the section in which you are working.

03 30 Cast-In-Place Concrete 1

03 30 53 – Miscellaneous Cast-In-Place Concrete

03 30 53.40 Concrete In Place 2		4	Crew	Daily Output	Labor-Hours	Unit	Material	2010 Bare Costs Labor	Equipment	Total	Total Incl O&P
0010	**CONCRETE IN PLACE**			6				8			9
0020	Including forms (4 uses), reinforcing steel, concrete, placement,										
0050	and finishing unless otherwise indicated	R033053-50									
0300	Beams (3500 psi), 5 kip per L.F., 10' span		C-14A	15.62	12.804	C.Y.	282	535	49.50	866.50	1,200
3850	Over 5 C.Y.			75	1.493		158	59.50	.40	217.90	266
3900	Footings, strip (3000 psi), 18" x 9", unreinforced 3		C-14L	40	2.400		112	93	.77	205.77	267
3920	18" x 9", reinforced		C-14C	35	3.200		131	128	.86	259.86	345
3925	20" x 10", unreinforced	5	C-14L	45	2.133		110	82.50	.68	193.18	249
3930	20" x 10", reinforced		C-14C	40	2.800	7	126	112	.76	238.76	310
3935	24" x 12", unreinforced		C-14L	55	1.745		110	67.50	.56	178.06	226
3940	24" x 12", reinforced		C-14C	48	2.333		126	93	.63	219.63	283

5

Crew (C-14C)

The "Crew" column designates the typical trade or crew used to install the item. If an installation can be accomplished by one trade and requires no power equipment, that trade and the number of workers are listed (for example, "2 Carpenters"). If an installation requires a composite crew, a crew code designation is listed (for example, "C-14C"). You'll find full details on all composite crews in the Crew Listings.

- For a complete list of all trades utilized in this book and their abbreviations, see the inside back cover.

Crews

Crew No.	Bare Costs		Incl. Subs O&P		Cost Per Labor-Hour	
Crew C-14C	Hr.	Daily	Hr.	Daily	Bare Costs	Incl. O&P
1 Carpenter Foreman (out)	$43.55	$348.40	$67.15	$537.20	$39.90	$61.74
6 Carpenters	41.55	1994.40	64.05	3074.40		
2 Rodmen (reinf.)	46.80	748.80	75.40	1206.40		
4 Laborers	33.10	1059.20	51.05	1633.60		
1 Cement Finisher	39.70	317.60	57.95	463.60		
1 Gas Engine Vibrator		30.40		33.44	.27	.30
112 L.H., Daily Totals		$4498.80		$6948.64	$40.17	$62.04

6

Productivity: Daily Output (35.0)/ Labor-Hours (3.20)

The "Daily Output" represents the typical number of units the designated crew will install in a normal 8-hour day. To find out the number of days the given crew would require to complete the installation, divide your quantity by the daily output. For example:

Quantity	÷	Daily Output	=	Duration
100 C.Y.	÷	35.0/ Crew Day	=	2.86 Crew Days

The "Labor-Hours" figure represents the number of labor-hours required to install one unit of work. To find out the number of labor-hours required for your particular task, multiply the quantity of the item times the number of labor-hours shown. For example:

Quantity	x	Productivity Rate	=	Duration
100 C.Y.	x	3.20 Labor-Hours/ C.Y.	=	320 Labor-Hours

7

Unit (C.Y.)

The abbreviated designation indicates the unit of measure upon which the price, production, and crew are based (C.Y. = Cubic Yard). For a complete listing of abbreviations, refer to the Abbreviations Listing in the Reference Section of this book.

Bare Costs:

Mat. (Bare Material Cost) (131)

The unit material cost is the "bare" material cost with no overhead or profit included. Costs shown reflect national average material prices for January of the current year and include delivery to the job site. No sales taxes are included.

Labor (128)

The unit labor cost is derived by multiplying bare labor-hour costs for Crew C-14C by labor-hour units. The bare labor-hour cost is found in the Crew Section under C-14C. (If a trade is listed, the hourly labor cost—the wage rate—is found on the inside back cover.)

Labor-Hour Cost Crew C-14C	x	Labor-Hour Units	=	Labor
$39.90	x	3.20		$128.00

Equip. (Equipment) (.86)

Costs for equipment included in each crew are listed in the Crew Listings in the Reference Section. Costs for small tools that are considered as essential everyday tools are covered in the "Overhead" column on the "Installing Contractor's Overhead and Profit" table that lists labor trades, base rates and markups. The unit equipment cost displayed here is derived by multiplying the bare equipment hourly cost in the Crew Listing by the unit Labor-Hours value.

Equipment Cost Crew C-14C	x	Labor-Hour Units	=	Equip.
$.27	x	3.20	=	$.86

Total (259.86)

The total of the bare costs is the arithmetic total of the three previous columns: mat., labor, and equip.

Material	+	Labor	+	Equip.	=	Total
$131	+	$128	+	$.86	=	$259.86

Total Costs Including O&P

This figure is the sum of the bare material cost plus 10% for profit; the bare labor cost plus total overhead and profit (per the inside back cover or, if a crew is listed, from the crew listings); and the bare equipment cost plus 10% for profit.

Material is Bare Material Cost + 10% = 131 + 13.10	=	$144.10
Labor for Crew C-14C = Labor-Hour Cost (61.74) x Labor-Hour Units (3.20)	=	$197.57
Equip. is Bare Equip. Cost + 10% = .86 + .09	=	$ .95
Total (Rounded)	=	$ 345

Estimating Tips

01 20 00 Price and Payment Procedures

- Allowances that should be added to estimates to cover contingencies and job conditions that are not included in the national average material and labor costs are shown in section 01 21.
- When estimating historic preservation projects (depending on the condition of the existing structure and the owner's requirements), a 15%-20% contingency or allowance is recommended, regardless of the stage of the drawings.

01 30 00 Administrative Requirements

- Before determining a final cost estimate, it is a good practice to review all the items listed in Subdivisions 01 31 and 01 32 to make final adjustments for items that may need customizing to specific job conditions.
- Requirements for initial and periodic submittals can represent a significant cost to the General Requirements of a job. Thoroughly check the submittal specifications when estimating a project to determine any costs that should be included.

01 40 00 Quality Requirements

- All projects will require some degree of Quality Control. This cost is not included in the unit cost of construction listed in each division. Depending upon the terms of the contract, the various costs of inspection and testing can be the responsibility of either the owner or the contractor. Be sure to include the required costs in your estimate.

01 50 00 Temporary Facilities and Controls

- Barricades, access roads, safety nets, scaffolding, security, and many more requirements for the execution of a safe project are elements of direct cost. These costs can easily be overlooked when preparing an estimate. When looking through the major classifications of this subdivision, determine which items apply to each division in your estimate.
- Construction Equipment Rental Costs can be found in the Reference Section in section 01 54 33. Operators' wages are not included in equipment rental costs.
- Equipment mobilization and demobilization costs are not included in equipment rental costs and must be considered separately in section 01 54 36.50.
- The cost of small tools provided by the installing contractor for his workers is covered in the "Overhead" column on the "Installing Contractor's Overhead and Profit" table that lists labor trades, base rates and markups and, therefore, is included in the "Total Incl. O&P" cost of any Unit Price line item. For those users who are constrained by contract terms to use only bare costs, there are two line items in section 01 54 39.70 for small tools as a percentage of bare labor cost. If some of those users are further constrained by contract terms to refrain from using line items from Division 1, you are advised to cover the cost of small tools within your coefficient or multiplier.

01 70 00 Execution and Closeout Requirements

- When preparing an estimate, thoroughly read the specifications to determine the requirements for Contract Closeout. Final cleaning, record documentation, operation and maintenance data, warranties and bonds, and spare parts and maintenance materials can all be elements of cost for the completion of a contract. Do not overlook these in your estimate.

Reference Numbers

Reference numbers are shown in shaded boxes at the beginning of some major classifications. These numbers refer to related items in the Reference Section. The reference information may be an estimating procedure, an alternate pricing method, or technical information.

Note: Not all subdivisions listed here necessarily appear in this publication.

01 11 Summary of Work

01 11 31 – Professional Consultants

01 11 31.10 Architectural Fees

	01 11 31.10 Architectural Fees	Crew	Daily Output	Labor-Hours	Unit	2010 Bare Costs Material	Labor	Equipment	Total	Total Incl O&P
0010	**ARCHITECTURAL FEES** R011110-10									
0020	For new construction									
0060	Minimum				Project					4.90%
0090	Maximum									16%
0100	For alteration work, to $500,000, add to new construction fee									50%
0150	Over $500,000, add to new construction fee									25%
2000	For "Greening" of building G									3%

01 11 31.20 Construction Management Fees

	01 11 31.20 Construction Management Fees	Crew	Daily Output	Labor-Hours	Unit	Material	Labor	Equipment	Total	Total Incl O&P
0010	**CONSTRUCTION MANAGEMENT FEES**									
0020	$1,000,000 job, minimum				Project					4.50%
0050	Maximum									7.50%
0300	$50,000,000 job, minimum									2.50%
0350	Maximum									4%

01 11 31.30 Engineering Fees

	01 11 31.30 Engineering Fees	Crew	Daily Output	Labor-Hours	Unit	Material	Labor	Equipment	Total	Total Incl O&P
0010	**ENGINEERING FEES** R011110-30									
0020	Educational planning consultant, minimum				Project					.50%
0100	Maximum				″					2.50%
0200	Electrical, minimum				Contrct					4.10%
0300	Maximum									10.10%
0400	Elevator & conveying systems, minimum									2.50%
0500	Maximum									5%
0600	Food service & kitchen equipment, minimum									8%
0700	Maximum									12%
0800	Landscaping & site development, minimum									2.50%
0900	Maximum									6%
1000	Mechanical (plumbing & HVAC), minimum									4.10%
1100	Maximum									10.10%
1200	Structural, minimum				Project					1%
1300	Maximum				″					2.50%
4000	Consultant, using DOE software energy analysis, small bldg, min G				SF Flr.					.25
4010	Maximum G									.45
4020	Medium building, minimum G									.15
4030	Maximum G									.35
4040	Large building, minimum G									.05
4050	Maximum G									.25

01 11 31.50 Models

	01 11 31.50 Models	Crew	Daily Output	Labor-Hours	Unit	Material	Labor	Equipment	Total	Total Incl O&P
0010	**MODELS**									
0020	Cardboard & paper, 1 building, minimum				Ea.	700			700	770
0050	Maximum					1,600			1,600	1,750
0100	2 buildings, minimum					935			935	1,025
0150	Maximum					2,125			2,125	2,325
0200	Plexiglass and metal, basic layout				SF Flr.	.06			.06	.07
0210	Including equipment and personnel				″	.31			.31	.34
0300	Site plan layout, minimum				Ea.	1,350			1,350	1,475
0350	Maximum				″	2,250			2,250	2,475

01 11 31.75 Renderings

	01 11 31.75 Renderings	Crew	Daily Output	Labor-Hours	Unit	Material	Labor	Equipment	Total	Total Incl O&P
0010	**RENDERINGS** Color, matted, 20″ x 30″, eye level,									
0020	1 building, minimum				Ea.	1,950			1,950	2,150
0050	Average					2,775			2,775	3,075
0100	Maximum					4,450			4,450	4,900
1000	5 buildings, minimum					3,900			3,900	4,300
1100	Maximum					7,800			7,800	8,575

01 11 Summary of Work

01 11 31 – Professional Consultants

01 11 31.75 Renderings		Crew	Daily Output	Labor-Hours	Unit	Material	2010 Bare Costs Labor	Equipment	Total	Total Incl O&P
2000	Aerial perspective, color, 1 building, minimum				Ea.	2,775			2,775	3,075
2100	Maximum					7,800			7,800	8,575
3000	5 buildings, minimum					5,600			5,600	6,175
3100	Maximum				↓	11,100			11,100	12,200

01 21 Allowances

01 21 16 – Contingency Allowances

01 21 16.50 Contingencies		Crew	Daily Output	Labor-Hours	Unit	Material	Labor	Equipment	Total	Total Incl O&P
0010	**CONTINGENCIES**, Add to estimate									
0020	Conceptual stage				Project					20%
0050	Schematic stage									15%
0100	Preliminary working drawing stage (Design Dev.)									10%
0150	Final working drawing stage				↓					3%

01 21 55 – Job Conditions Allowance

01 21 55.50 Job Conditions		Crew	Daily Output	Labor-Hours	Unit	Material	Labor	Equipment	Total	Total Incl O&P
0010	**JOB CONDITIONS** Modifications to applicable									
0020	cost summaries									
0100	Economic conditions, favorable, deduct				Project				2%	2%
0200	Unfavorable, add								5%	5%
0300	Hoisting conditions, favorable, deduct								2%	2%
0400	Unfavorable, add								5%	5%
0500	General Contractor management, experienced, deduct								2%	2%
0600	Inexperienced, add								10%	10%
0700	Labor availability, surplus, deduct								1%	1%
0800	Shortage, add								10%	10%
0900	Material storage area, available, deduct								1%	1%
1000	Not available, add								2%	2%
1100	Subcontractor availability, surplus, deduct								5%	5%
1200	Shortage, add								12%	12%
1300	Work space, available, deduct								2%	2%
1400	Not available, add				↓				5%	5%

01 21 57 – Overtime Allowance

01 21 57.50 Overtime		Crew	Daily Output	Labor-Hours	Unit	Material	Labor	Equipment	Total	Total Incl O&P
0010	**OVERTIME** for early completion of projects or where R012909-90									
0020	labor shortages exist, add to usual labor, up to				Costs		100%			

01 21 61 – Cost Indexes

01 21 61.10 Construction Cost Index		Crew	Daily Output	Labor-Hours	Unit	Material	Labor	Equipment	Total	Total Incl O&P
0010	**CONSTRUCTION COST INDEX** (Reference) over 930 zip code locations in									
0020	the U.S. and Canada, total bldg. cost, min. (Guymon, OK)				%					67.40%
0050	Average									100%
0100	Maximum (New York, NY)				↓					133.20%

01 21 61.20 Historical Cost Indexes		Crew	Daily Output	Labor-Hours	Unit	Material	Labor	Equipment	Total	Total Incl O&P
0010	**HISTORICAL COST INDEXES** (See Reference Section)									

01 21 61.30 Labor Index		Crew	Daily Output	Labor-Hours	Unit	Material	Labor	Equipment	Total	Total Incl O&P
0010	**LABOR INDEX** (Reference) For over 930 zip code locations in									
0020	the U.S. and Canada, minimum (Del Rio, TX)				%		33.70%			
0050	Average						100%			
0100	Maximum (New York, NY)				↓		166.30%			

01 21 Allowances

01 21 61 – Cost Indexes

01 21 61.50 Material Index

	01 21 61.50 Material Index	Crew	Daily Output	Labor-Hours	Unit	Material	2010 Bare Costs Labor	Equipment	Total	Total Incl O&P
0010	**MATERIAL INDEX** (Reference) For over 930 zip code locations in									
0020	the U.S. and Canada, minimum (Elizabethtown, KY)				%	90%				
0040	Average					100%				
0060	Maximum (Ketchikan, AK)					139.30%				

01 21 63 – Taxes

01 21 63.10 Taxes

	01 21 63.10 Taxes	Crew	Daily Output	Labor-Hours	Unit	Material	Labor	Equipment	Total	Total Incl O&P
0010	**TAXES** R012909-80									
0020	Sales tax, State, average				%	5.01%				
0050	Maximum R012909-85					8.25%				
0200	Social Security, on first $106,800 of wages						7.65%			
0300	Unemployment, combined Federal and State, minimum						.80%			
0350	Average						6.20%			
0400	Maximum						10.96%			

01 31 Project Management and Coordination

01 31 13 – Project Coordination

01 31 13.20 Field Personnel

	01 31 13.20 Field Personnel	Crew	Daily Output	Labor-Hours	Unit	Material	Labor	Equipment	Total	Total Incl O&P
0010	**FIELD PERSONNEL**									
0020	Clerk, average				Week		395		395	605
0100	Field engineer, minimum						935		935	1,425
0120	Average						1,215		1,215	1,875
0140	Maximum						1,400		1,400	2,150
0160	General purpose laborer, average						1,300		1,300	2,000
0180	Project manager, minimum						1,725		1,725	2,650
0200	Average						2,000		2,000	3,075
0220	Maximum						2,275		2,275	3,500
0240	Superintendent, minimum						1,675		1,675	2,575
0260	Average						1,850		1,850	2,850
0280	Maximum						2,100		2,100	3,225
0290	Timekeeper, average						1,085		1,085	1,675

01 31 13.30 Insurance

	01 31 13.30 Insurance	Crew	Daily Output	Labor-Hours	Unit	Material	Labor	Equipment	Total	Total Incl O&P
0010	**INSURANCE** R013113-40									
0020	Builders risk, standard, minimum				Job					.24%
0050	Maximum R013113-50									.64%
0200	All-risk type, minimum									.25%
0250	Maximum R013113-60									.62%
0400	Contractor's equipment floater, minimum				Value					.50%
0450	Maximum				"					1.50%
0600	Public liability, average				Job					2.02%
0800	Workers' compensation & employer's liability, average									
0850	by trade, carpentry, general				Payroll		16.88%			
0900	Clerical						.52%			
0950	Concrete						13.72%			
1000	Electrical						6.38%			
1050	Excavation						9.42%			
1100	Glazing						13.27%			
1150	Insulation						13.11%			
1200	Lathing						9.79%			
1250	Masonry						13.47%			
1300	Painting & decorating						11.46%			

01 31 13 – Project Coordination

01 31 13.30 Insurance		Crew	Daily Output	Labor-Hours	Unit	Material	2010 Bare Costs Labor	Equipment	Total	Total Incl O&P
1350	Pile driving				Payroll		18.54%			
1400	Plastering						11.92%			
1450	Plumbing						7.62%			
1500	Roofing						28.89%			
1550	Sheet metal work (HVAC)						9.14%			
1600	Steel erection, structural						35.99%			
1650	Tile work, interior ceramic						8.78%			
1700	Waterproofing, brush or hand caulking						6.48%			
1800	Wrecking						30.31%			
2000	Range of 35 trades in 50 states, excl. wrecking, min.						2.39%			
2100	Average						14.40%			
2200	Maximum						170.80%			

01 31 13.40 Main Office Expense

		Crew	Daily Output	Labor-Hours	Unit	Material	Labor	Equipment	Total	Total Incl O&P
0010	**MAIN OFFICE EXPENSE** Average for General Contractors R013113-50									
0020	As a percentage of their annual volume									
0125	Annual volume under 1 million dollars				% Vol.				17.50%	
0145	Up to 2.5 million dollars								8%	
0150	Up to 4.0 million dollars								6.80%	
0200	Up to 7.0 million dollars								5.60%	
0250	Up to 10 million dollars								5.10%	
0300	Over 10 million dollars								3.90%	

01 31 13.50 General Contractor's Mark-Up

		Crew	Daily Output	Labor-Hours	Unit	Material	Labor	Equipment	Total	Total Incl O&P
0010	**GENERAL CONTRACTOR'S MARK-UP** on Change Orders									
0200	Extra work, by subcontractors, add				%					10%
0250	By General Contractor, add									15%
0400	Omitted work, by subcontractors, deduct all but									5%
0450	By General Contractor, deduct all but									7.50%
0600	Overtime work, by subcontractors, add									15%
0650	By General Contractor, add									10%

01 31 13.60 Installing Contractor's Main Office Overhead

		Crew	Daily Output	Labor-Hours	Unit	Material	Labor	Equipment	Total	Total Incl O&P
0010	**INSTALLING CONTRACTOR'S MAIN OFFICE OVERHEAD** R013113-50									
0020	As percent of direct costs, minimum				%				5%	
0050	Average								13%	
0100	Maximum								30%	

01 31 13.80 Overhead and Profit

		Crew	Daily Output	Labor-Hours	Unit	Material	Labor	Equipment	Total	Total Incl O&P
0010	**OVERHEAD & PROFIT** Allowance to add to items in this									
0020	book that do not include Subs O&P, average				%				25%	
0100	Allowance to add to items in this book that									
0110	do include Subs O&P, minimum				%					5%
0150	Average									10%
0200	Maximum									15%
0300	Typical, by size of project, under $100,000								30%	
0350	$500,000 project								25%	
0400	$2,000,000 project								20%	
0450	Over $10,000,000 project								15%	

01 31 13.90 Performance Bond

		Crew	Daily Output	Labor-Hours	Unit	Material	Labor	Equipment	Total	Total Incl O&P
0010	**PERFORMANCE BOND** R013113-80									
0020	For buildings, minimum				Job					.60%
0100	Maximum				"					2.50%

01 32 Construction Progress Documentation

01 32 13 – Scheduling of work

01 32 13.50 Scheduling		Crew	Daily Output	Labor-Hours	Unit	Material	Labor	Equipment	Total	Total Incl O&P
						2010 Bare Costs				
0010	**SCHEDULING**									
0020	Critical path, as % of architectural fee, minimum				%					.50%
0100	Maximum				"					1%
0300	Computer-update, micro, no plots, minimum				Ea.				455	500
0400	Including plots, maximum				"				1,455	1,600
0600	Rule of thumb, CPM scheduling, small job ($10 Million)				Job					.05%
0650	Large job ($50 Million +)									.03%
0700	Including cost control, small job									.08%
0750	Large job									.04%

01 32 33 – Photographic Documentation

01 32 33.50 Photographs

		Crew	Daily Output	Labor-Hours	Unit	Material	Labor	Equipment	Total	Total Incl O&P
0010	**PHOTOGRAPHS**									
0020	8" x 10", 4 shots, 2 prints ea., std. mounting				Set	475			475	520
0100	Hinged linen mounts					530			530	580
0200	8" x 10", 4 shots, 2 prints each, in color					415			415	460
0300	For I.D. slugs, add to all above					5.30			5.30	5.85
0500	Aerial photos, initial fly-over, 6 shots, 1 print ea., 8" x 10"					825			825	905
0550	11" x 14" prints					1,000			1,000	1,100
0600	16" x 20" prints					1,175			1,175	1,300
0700	For full color prints, add					40%				40%
0750	Add for traffic control area					294			294	325
0900	For over 30 miles from airport, add per				Mile	5.30			5.30	5.85
1000	Vertical photography, 4 to 6 shots with									
1010	different scales, 1 print each				Set	1,100			1,100	1,200
1500	Time lapse equipment, camera and projector, buy					3,775			3,775	4,175
1550	Rent per month					565			565	620
1700	Cameraman and film, including processing, B.&W.				Day	1,375			1,375	1,525
1720	Color				"	1,375			1,375	1,525

01 41 Regulatory Requirements

01 41 26 – Permits

01 41 26.50 Permits

		Crew	Daily Output	Labor-Hours	Unit	Material	Labor	Equipment	Total	Total Incl O&P
0010	**PERMITS**									
0020	Rule of thumb, most cities, minimum				Job					.50%
0100	Maximum				"					2%

01 45 Quality Control

01 45 23 – Testing and Inspecting Services

01 45 23.50 Testing

		Crew	Daily Output	Labor-Hours	Unit	Material	Labor	Equipment	Total	Total Incl O&P
0010	**TESTING** and Inspecting Services									
0015	For concrete building costing $1,000,000, minimum				Project				4,725	5,200
0020	Maximum								38,000	41,800
0050	Steel building, minimum								4,727	5,200
0070	Maximum								14,818	16,300
0100	For building costing, $10,000,000, minimum								30,091	33,100
0150	Maximum								48,182	53,000
0200	Asphalt testing, compressive strength Marshall stability, set of 3				Ea.				145	165
0220	Density, set of 3								86	95
0250	Extraction, individual tests on sample								136	150

01 45 23 – Testing and Inspecting Services

01 45 23.50 Testing		Crew	Daily Output	Labor-Hours	Unit	Material	Labor	Equipment	Total	Total Incl O&P
						2010 Bare Costs				
0300	Penetration				Ea.				41	45
0350	Mix design, 5 specimens								182	200
0360	Additional specimen								36	40
0400	Specific gravity								41	45
0420	Swell test								64	70
0450	Water effect and cohesion, set of 6								182	200
0470	Water effect and plastic flow								64	70
0600	Concrete testing, aggregates, abrasion, ASTM C 131								136	150
0650	Absorption, ASTM C 127								42	46
0800	Petrographic analysis, ASTM C 295								773	850
0900	Specific gravity, ASTM C 127								50	55
1000	Sieve analysis, washed, ASTM C 136								59	65
1050	Unwashed								59	65
1200	Sulfate soundness								114	125
1300	Weight per cubic foot								36	40
1500	Cement, physical tests, ASTM C 150								318	350
1600	Chemical tests, ASTM C 150								245	270
1800	Compressive test, cylinder, delivered to lab, ASTM C 39								12	13
1900	Picked up by lab, minimum								14	15
1950	Average								18	20
2000	Maximum								27	30
2200	Compressive strength, cores (not incl. drilling), ASTM C 42				↓				36	40
2250	Core drilling, 4" diameter (plus technician)				Inch				23	25
2260	Technician for core drilling				Hr.				45	50
2300	Patching core holes				Ea.				22	24
2400	Drying shrinkage at 28 days								236	260
2500	Flexural test beams, ASTM C 78								59	65
2600	Mix design, one batch mix								259	285
2650	Added trial batches								120	132
2800	Modulus of elasticity, ASTM C 469								164	180
2900	Tensile test, cylinders, ASTM C 496								45	50
3000	Water-Cement ratio curve, 3 batches								141	155
3100	4 batches								186	205
3300	Masonry testing, absorption, per 5 brick, ASTM C 67								45	50
3350	Chemical resistance, per 2 brick								50	55
3400	Compressive strength, per 5 brick, ASTM C 67								68	75
3420	Efflorescence, per 5 brick, ASTM C 67								68	75
3440	Imperviousness, per 5 brick								87	96
3470	Modulus of rupture, per 5 brick								86	95
3500	Moisture, block only								32	35
3550	Mortar, compressive strength, set of 3								23	25
4100	Reinforcing steel, bend test								55	61
4200	Tensile test, up to #8 bar								36	40
4220	#9 to #11 bar								41	45
4240	#14 bar and larger								64	70
4400	Soil testing, Atterberg limits, liquid and plastic limits								59	65
4510	Hydrometer analysis								109	120
4530	Specific gravity, ASTM D 354								44	48
4600	Sieve analysis, washed, ASTM D 422								55	60
4700	Unwashed, ASTM D 422								59	65
4710	Consolidation test (ASTM D2435), minimum								250	275
4715	Maximum								432	475
4720	Density and classification of undisturbed sample				↓				73	80

01 45 Quality Control

01 45 23 – Testing and Inspecting Services

	01 45 23.50 Testing	Crew	Daily Output	Labor-Hours	Unit	2010 Bare Costs Material	Labor	Equipment	Total	Total Incl O&P
4735	Soil density, nuclear method, ASTM D2922				Ea.				35	38.67
4740	Sand cone method ASTM D1556								27	30.17
4750	Moisture content, ASTM D 2216								9	10
4780	Permeability test, double ring infiltrometer								500	550
4800	Permeability, var. or constant head, undist., ASTM D 2434								227	250
4850	Recompacted								250	275
4900	Proctor compaction, 4" standard mold, ASTM D 698								123	135
4950	6" modified mold								68	75
5100	Shear tests, triaxial, minimum								409	450
5150	Maximum								545	600
5300	Direct shear, minimum, ASTM D 3080								318	350
5350	Maximum								409	450
5550	Technician for inspection, per day, earthwork								210	231
5650	Bolting								268	295
5750	Roofing								244	256
5790	Welding								257	283
5820	Non-destructive metal testing, dye penetrant				Day				310	341
5840	Magnetic particle								310	341
5860	Radiography								450	495
5880	Ultrasonic								309	340
6000	Welding certification, minimum				Ea.				91	100
6100	Maximum				″				250	275
7000	Underground storage tank									
7500	Volumetric tightness test ,<=12,000 gal				Ea.				435	478
7510	<=30,000 gal				″				613	675
7600	Vadose zone (soil gas) sampling, 10-40 samples, min.				Day				1,364	1,500
7610	Maximum				″				2,273	2,500
7700	Ground water monitoring incl. drilling 3 wells, min.				Total				4,545	5,000
7710	Maximum				″				6,364	7,000
8000	X-ray concrete slabs				Ea.				182	200
9000	Thermographic testing, for bldg envelope heat loss, average 2,000 S.F. G				″					500

01 51 Temporary Utilities

01 51 13 – Temporary Electricity

	01 51 13.80 Temporary Utilities	Crew	Daily Output	Labor-Hours	Unit	Material	Labor	Equipment	Total	Total Incl O&P
0010	**TEMPORARY UTILITIES**									
0100	Heat, incl. fuel and operation, per week, 12 hrs. per day	1 Skwk	100	.080	CSF Flr	27	3.41		30.41	35
0200	24 hrs. per day	″	60	.133		52	5.70		57.70	66.50
0350	Lighting, incl. service lamps, wiring & outlets, minimum	1 Elec	34	.235		2.63	11.55		14.18	20
0360	Maximum	″	17	.471		5.70	23		28.70	41
0400	Power for temp lighting only, per month, min/month 6.6 KWH								.92	1.01
0430	Average/month 11.8 KWH								1.65	1.82
0450	Maximum/month 23.6 KWH								3.30	3.63
0600	Power for job duration incl. elevator, etc., minimum								47	51.70
0650	Maximum								110	121
1000	Toilet, portable, see Equip. Rental 01 54 33 in Reference Section									

01 52 Construction Facilities

01 52 13 – Field Offices and Sheds

01 52 13.20 Office and Storage Space		Crew	Daily Output	Labor-Hours	Unit	2010 Bare Costs Material	Labor	Equipment	Total	Total Incl O&P
0010	**OFFICE AND STORAGE SPACE**									
0020	Trailer, furnished, no hookups, 20′ x 8′, buy	2 Skwk	1	16	Ea.	8,550	680		9,230	10,500
0250	Rent per month					146			146	160
0300	32′ x 8′, buy	2 Skwk	.70	22.857		12,300	975		13,275	15,000
0350	Rent per month					193			193	213
0400	50′ x 10′, buy	2 Skwk	.60	26.667		23,300	1,125		24,425	27,500
0450	Rent per month					286			286	315
0500	50′ x 12′, buy	2 Skwk	.50	32		26,000	1,375		27,375	30,700
0550	Rent per month					360			360	395
0700	For air conditioning, rent per month, add				▼	41			41	45
0800	For delivery, add per mile				Mile	4.50			4.50	4.95
1000	Portable buildings, prefab, on skids, economy, 8′ x 8′	2 Carp	265	.060	S.F.	85	2.51		87.51	97.50
1100	Deluxe, 8′ x 12′	"	150	.107	"	95	4.43		99.43	112
1200	Storage boxes, 20′ x 8′, buy	2 Skwk	1.80	8.889	Ea.	3,225	380		3,605	4,125
1250	Rent per month					69.50			69.50	76.50
1300	40′ x 8′, buy	2 Skwk	1.40	11.429		4,225	485		4,710	5,375
1350	Rent per month				▼	92.50			92.50	102
5000	Air supported structures, see Div. 13 31 13.13									

01 52 13.40 Field Office Expense		Crew	Daily Output	Labor-Hours	Unit	Material	Labor	Equipment	Total	Total Incl O&P
0010	**FIELD OFFICE EXPENSE**									
0100	Office equipment rental average				Month	155			155	171
0120	Office supplies, average				"	85			85	93.50
0125	Office trailer rental, see Div. 01 52 13.20									
0140	Telephone bill; avg. bill/month incl. long dist.				Month	80			80	88
0160	Lights & HVAC				"	150			150	165

01 54 Construction Aids

01 54 09 – Protection Equipment

01 54 09.50 Personnel Protective Equipment		Crew	Daily Output	Labor-Hours	Unit	Material	Labor	Equipment	Total	Total Incl O&P
0010	**PERSONNEL PROTECTIVE EQUIPMENT**									
0015	Hazardous waste protection									
0020	Respirator mask only, full face, silicone				Ea.	229			229	251
0030	Half face, silicone					35			35	38.50
0040	Respirator cartridges, 2 req'd/mask, dust or asbestos					2.21			2.21	2.43
0050	Chemical vapor					2.49			2.49	2.74
0060	Combination vapor and dust					5.20			5.20	5.70
0100	Emergency escape breathing apparatus, 5 min					465			465	510
0110	10 min					500			500	550
0150	Self contained breathing apparatus with full face piece, 30 min					1,750			1,750	1,925
0160	60 min					2,925			2,925	3,225
0200	Encapsulating suits, limited use, level A					950			950	1,050
0210	Level B				▼	278			278	305
0300	Over boots, latex				Pr.	6.35			6.35	7
0310	PVC					21.50			21.50	24
0320	Neoprene					41.50			41.50	46
0400	Gloves, nitrile/PVC					21			21	23.50
0410	Neoprene coated				▼	24			24	26.50

01 54 09.60 Safety Nets		Crew	Daily Output	Labor-Hours	Unit	Material	Labor	Equipment	Total	Total Incl O&P
0010	**SAFETY NETS**									
0020	No supports, stock sizes, nylon, 4″ mesh				S.F.	1.10			1.10	1.21

01 54 Construction Aids

01 54 09 – Protection Equipment

01 54 09.60 Safety Nets		Crew	Daily Output	Labor-Hours	Unit	2010 Bare Costs Material	Labor	Equipment	Total	Total Incl O&P
0100	Polypropylene, 6" mesh				S.F.	1.59			1.59	1.75
0200	Small mesh debris nets, 1/4" & 3/4" mesh, stock sizes					.74			.74	.81
0220	Combined 4" mesh and 1/4" mesh, stock sizes					2.05			2.05	2.26
0300	Monthly rental, 4" mesh, stock sizes, 1st month					.50			.50	.55
0320	2nd month rental					.25			.25	.28
0340	Maximum rental/year					1.15			1.15	1.27

01 54 16 – Temporary Hoists

01 54 16.50 Weekly Forklift Crew

		Crew	Daily Output	Labor-Hours	Unit	Material	Labor	Equipment	Total	Total Incl O&P
0010	**WEEKLY FORKLIFT CREW**									
0100	All-terrain forklift, 45' lift, 35' reach, 9000 lb. capacity	A-3P	.20	40	Week		1,650	2,400	4,050	5,100

01 54 19 – Temporary Cranes

01 54 19.50 Daily Crane Crews

		Crew	Daily Output	Labor-Hours	Unit	Material	Labor	Equipment	Total	Total Incl O&P
0010	**DAILY CRANE CREWS** for small jobs, portal to portal									
0100	12-ton truck-mounted hydraulic crane	A-3H	1	8	Day		355	870	1,225	1,500
0200	25-ton	A-3I	1	8			355	1,025	1,380	1,650
0300	40-ton	A-3J	1	8			355	1,250	1,605	1,900
0400	55-ton	A-3K	1	16			660	1,775	2,435	2,950
0500	80-ton	A-3L	1	16			660	2,225	2,885	3,425
0600	100-ton	A-3M	1	16			660	2,425	3,085	3,675

01 54 19.60 Monthly Tower Crane Crew

		Crew	Daily Output	Labor-Hours	Unit	Material	Labor	Equipment	Total	Total Incl O&P
0010	**MONTHLY TOWER CRANE CREW**, excludes concrete footing									
0100	Static tower crane, 130' high, 106' jib, 6200 lb. capacity	A-3N	.05	176	Month		7,825	23,100	30,925	37,100

01 54 23 – Temporary Scaffolding and Platforms

01 54 23.60 Pump Staging

		Crew	Daily Output	Labor-Hours	Unit	Material	Labor	Equipment	Total	Total Incl O&P
0010	**PUMP STAGING**, Aluminum R015423-20									
0200	24' long pole section, buy				Ea.	415			415	460
0300	18' long pole section, buy					325			325	355
0400	12' long pole section, buy					218			218	240
0500	6' long pole section, buy					115			115	126
0600	6' long splice joint section, buy					85.50			85.50	94
0700	Pump jack, buy					139			139	153
0900	Foldable brace, buy					56.50			56.50	62.50
1000	Workbench/back safety rail support, buy					73.50			73.50	81
1100	Scaffolding planks/workbench, 14" wide x 24' long, buy					680			680	750
1200	Plank end safety rail, buy					370			370	405
1250	Safety net, 22' long, buy					335			335	365
1300	System in place, 50' working height, per use based on 50 uses	2 Carp	84.80	.189	C.S.F.	6.40	7.85		14.25	19.15
1400	100 uses		84.80	.189		3.21	7.85		11.06	15.65
1500	150 uses		84.80	.189		2.15	7.85		10	14.45

01 54 23.70 Scaffolding

		Crew	Daily Output	Labor-Hours	Unit	Material	Labor	Equipment	Total	Total Incl O&P
0010	**SCAFFOLDING** R015423-10									
0015	Steel tube, regular, no plank, labor only to erect & dismantle									
0090	Building exterior, wall face, 1 to 5 stories, 6'-4" x 5' frames	3 Carp	8	3	C.S.F.		125		125	192
0200	6 to 12 stories	4 Carp	8	4			166		166	256
0301	13 to 20 stories	5 Clab	8	5			166		166	255
0460	Building interior, wall face area, up to 16' high	3 Carp	12	2			83		83	128
0560	16' to 40' high		10	2.400			99.50		99.50	154
0800	Building interior floor area, up to 30' high		150	.160	C.C.F.		6.65		6.65	10.25
0900	Over 30' high	4 Carp	160	.200	"		8.30		8.30	12.80
0906	Complete system for face of walls, no plank, material only rent/mo				C.S.F.	36			36	39.50

01 54 23.70 Scaffolding		Crew	Daily Output	Labor-Hours	Unit	Material	2010 Bare Costs Labor	Equipment	Total	Total Incl O&P
0908	Interior spaces, no plank, material only rent/mo				C.C.F.	3.42			3.42	3.76
0910	Steel tubular, heavy duty shoring, buy									
0920	Frames 5′ high 2′ wide				Ea.	93			93	102
0925	5′ high 4′ wide					105			105	116
0930	6′ high 2′ wide					106			106	117
0935	6′ high 4′ wide					124			124	136
0940	Accessories									
0945	Cross braces				Ea.	18			18	19.80
0950	U-head, 8″ x 8″					21.50			21.50	24
0955	J-head, 4″ x 8″					15.80			15.80	17.40
0960	Base plate, 8″ x 8″					17.60			17.60	19.35
0965	Leveling jack					38			38	41.50
1000	Steel tubular, regular, buy									
1100	Frames 3′ high 5′ wide				Ea.	71			71	78
1150	5′ high 5′ wide					83			83	91.50
1200	6′-4″ high 5′ wide					104			104	114
1350	7′-6″ high 6′ wide					179			179	197
1500	Accessories cross braces					20.50			20.50	22.50
1550	Guardrail post					18.60			18.60	20.50
1600	Guardrail 7′ section					9.90			9.90	10.90
1650	Screw jacks & plates					27			27	29.50
1700	Sidearm brackets					43			43	47.50
1750	8″ casters					35.50			35.50	39
1800	Plank 2″ x 10″ x 16′-0″					51			51	56
1900	Stairway section					325			325	360
1910	Stairway starter bar					36.50			36.50	40
1920	Stairway inside handrail					66			66	72.50
1930	Stairway outside handrail					98			98	108
1940	Walk-thru frame guardrail					47			47	51.50
2000	Steel tubular, regular, rent/mo.									
2100	Frames 3′ high 5′ wide				Ea.	5			5	5.50
2150	5′ high 5′ wide					5			5	5.50
2200	6′-4″ high 5′ wide					5.15			5.15	5.65
2250	7′-6″ high 6′ wide					7			7	7.70
2500	Accessories, cross braces					1			1	1.10
2550	Guardrail post					1			1	1.10
2600	Guardrail 7′ section					1			1	1.10
2650	Screw jacks & plates					2			2	2.20
2700	Sidearm brackets					2			2	2.20
2750	8″ casters					8			8	8.80
2800	Outrigger for rolling tower					3			3	3.30
2850	Plank 2″ x 10″ x 16′-0″					6			6	6.60
2900	Stairway section					40			40	44
2940	Walk-thru frame guardrail					2.50			2.50	2.75
3000	Steel tubular, heavy duty shoring, rent/mo.									
3250	5′ high 2′ & 4′ wide				Ea.	5			5	5.50
3300	6′ high 2′ & 4′ wide					5			5	5.50
3500	Accessories, cross braces					1			1	1.10
3600	U - head, 8″ x 8″					1			1	1.10
3650	J - head, 4″ x 8″					1			1	1.10
3700	Base plate, 8″ x 8″					1			1	1.10
3750	Leveling jack					2			2	2.20
5700	Planks, 2″ x 10″ x 16′-0″, labor only to erect & remove to 50′ H	3 Carp	72	.333			13.85		13.85	21.50

01 54 Construction Aids

01 54 23 – Temporary Scaffolding and Platforms

01 54 23.70 Scaffolding		Crew	Daily Output	Labor-Hours	Unit	2010 Bare Costs Material	Labor	Equipment	Total	Total Incl O&P
5800	Over 50′ high	4 Carp	80	.400	Ea.		16.60		16.60	25.50
6000	Heavy duty shoring for elevated slab forms to 8′-2″ high, floor area									
6100	Labor only to erect & dismantle	4 Carp	16	2	C.S.F.		83		83	128
6110	Materials only, rent.mo				″	29.50			29.50	32.50
6500	To 14′-8″ high									
6600	Labor only to erect & dismantle	4 Carp	10	3.200	C.S.F.		133		133	205
6610	Materials only, rent/mo				″	43			43	47.50

01 54 23.75 Scaffolding Specialties

Line	Description	Crew	Daily Output	Labor-Hours	Unit	Material	Labor	Equipment	Total	Total Incl O&P
0010	**SCAFFOLDING SPECIALTIES**									
1200	Sidewalk bridge, heavy duty steel posts & beams, including									
1210	parapet protection & waterproofing									
1220	8′ to 10′ wide, 2 posts	3 Carp	15	1.600	L.F.	31.50	66.50		98	137
1230	3 posts	″	10	2.400	″	48.50	99.50		148	208
1500	Sidewalk bridge using tubular steel									
1510	scaffold frames, including planking	3 Carp	45	.533	L.F.	5.60	22		27.60	40
1600	For 2 uses per month, deduct from all above					50%				
1700	For 1 use every 2 months, add to all above					100%				
1900	Catwalks, 20″ wide, no guardrails, 7′ span, buy				Ea.	145			145	160
2000	10′ span, buy					203			203	223
3720	Putlog, standard, 8′ span, with hangers, buy					75			75	82.50
3730	Rent per month					10			10	11
3750	12′ span, buy					113			113	124
3755	Rent per month					15			15	16.50
3760	Trussed type, 16′ span, buy					260			260	286
3770	Rent per month					20			20	22
3790	22′ span, buy					310			310	345
3795	Rent per month					30			30	33
3800	Rolling ladders with handrails, 30″ wide, buy, 2 step					237			237	261
4000	7 step					725			725	800
4050	10 step					1,000			1,000	1,100
4100	Rolling towers, buy, 5′ wide, 7′ long, 10′ high					1,350			1,350	1,500
4200	For 5′ high added sections, to buy, add				↓	207			207	228
4300	Complete incl. wheels, railings, outriggers,									
4350	21′ high, to buy				Ea.	2,250			2,250	2,475
4400	Rent/month = 5% of purchase cost				″	113			113	124

01 54 23.80 Staging Aids

Line	Description	Crew	Daily Output	Labor-Hours	Unit	Material	Labor	Equipment	Total	Total Incl O&P
0010	**STAGING AIDS** and fall protection equipment									
0100	Sidewall staging bracket, tubular, buy				Ea.	46			46	50.50
0110	Cost each per day, based on 250 days use				Day	.18			.18	.20
0200	Guard post, buy				Ea.	22.50			22.50	25
0210	Cost each per day, based on 250 days use				Day	.09			.09	.10
0300	End guard chains, buy per pair				Pair	31			31	34
0310	Cost per set per day, based on 250 days use				Day	.18			.18	.20
1000	Roof shingling bracket, steel, buy				Ea.	9.85			9.85	10.85
1010	Cost each per day, based on 250 days use				Day	.04			.04	.04
1100	Wood bracket, buy				Ea.	16.60			16.60	18.25
1110	Cost each per day, based on 250 days use				Day	.07			.07	.07
2000	Ladder jack, aluminum, buy per pair				Pair	108			108	119
2010	Cost per pair per day, based on 250 days use				Day	.43			.43	.48
2100	Steel siderail jack, buy per pair				Pair	122			122	134
2110	Cost per pair per day, based on 250 days use				Day	.49			.49	.54
3000	Laminated wood plank, 2″ x 10″ x 16′, buy				Ea.	51			51	56

01 54 Construction Aids

01 54 23 – Temporary Scaffolding and Platforms

01 54 23.80 Staging Aids		Crew	Daily Output	Labor-Hours	Unit	Material	2010 Bare Costs Labor	Equipment	Total	Total Incl O&P
3010	Cost each per day, based on 250 days use				Day	.20			.20	.22
3100	Aluminum scaffolding plank, 20" wide x 24' long, buy				Ea.	890			890	975
3110	Cost each per day, based on 250 days use				Day	3.55			3.55	3.91
4000	Nylon full body harness, lanyard and rope grab				Ea.	215			215	236
4010	Cost each per day, based on 250 days use				Day	.86			.86	.95
4100	Rope for safety line, 5/8" x 100' nylon, buy				Ea.	70			70	77
4110	Cost each per day, based on 250 days use				Day	.28			.28	.31
4200	Permanent U-Bolt roof anchor, buy				Ea.	35			35	38.50
4300	Temporary (one use) roof ridge anchor, buy				"	28.50			28.50	31
5000	Installation (setup and removal) of staging aids									
5010	Sidewall staging bracket	2 Carp	64	.250	Ea.		10.40		10.40	16
5020	Guard post with 2 wood rails	"	64	.250			10.40		10.40	16
5030	End guard chains, set	1 Carp	64	.125			5.20		5.20	8
5100	Roof shingling bracket		96	.083			3.46		3.46	5.35
5200	Ladder jack		64	.125			5.20		5.20	8
5300	Wood plank, 2" x 10" x 16'	2 Carp	80	.200			8.30		8.30	12.80
5310	Aluminum scaffold plank, 20" x 24'	"	40	.400			16.60		16.60	25.50
5410	Safety rope	1 Carp	40	.200			8.30		8.30	12.80
5420	Permanent U-Bolt roof anchor (install only)	2 Carp	40	.400			16.60		16.60	25.50
5430	Temporary roof ridge anchor (install only)	1 Carp	64	.125			5.20		5.20	8

01 54 26 – Temporary Swing Staging

01 54 26.50 Swing Staging

Code	Description	Crew	Daily Output	Labor-Hours	Unit	Material	Labor	Equipment	Total	Total Incl O&P
0010	**SWING STAGING**, 500 lb cap., 2' wide to 24' long, hand operated									
0020	steel cable type, with 60' cables, buy				Ea.	4,050			4,050	4,450
0030	Rent per month				"	405			405	445
0600	Lightweight (not for masons) 24' long for 150' height,									
0610	manual type, buy				Ea.	4,150			4,150	4,575
0620	Rent per month					415			415	455
0700	Powered, electric or air, to 150' high, buy					13,300			13,300	14,700
0710	Rent per month					935			935	1,025
0780	To 300' high, buy					13,500			13,500	14,800
0800	Rent per month					945			945	1,025
1000	Bosun's chair or work basket 3' x 3.5', to 300' high, electric, buy					8,625			8,625	9,500
1010	Rent per month					605			605	665
2200	Move swing staging (setup and remove)	E-4	2	16	Move		760	73	833	1,400

01 54 36 – Equipment Mobilization

01 54 36.50 Mobilization

Code	Description	Crew	Daily Output	Labor-Hours	Unit	Material	Labor	Equipment	Total	Total Incl O&P
0010	**MOBILIZATION** (Use line item again for demobilization) R015433-10									
0015	Up to 25 mi haul dist (50 mi RT for mob/demob crew)									
0020	Dozer, loader, backhoe, excav., grader, paver, roller, 70 to 150 H.P.	B-34N	4	2	Ea.		66.50	136	202.50	251
0100	Above 150 HP	B-34K	3	2.667			88.50	305	393.50	475
0300	Scraper, towed type (incl. tractor), 6 C.Y. capacity		3	2.667			88.50	305	393.50	475
0400	10 C.Y.		2.50	3.200			106	370	476	565
0600	Self-propelled scraper, 15 C.Y.		2.50	3.200			106	370	476	565
0700	24 C.Y.		2	4			133	460	593	705
0900	Shovel or dragline, 3/4 C.Y.		3.60	2.222			73.50	256	329.50	395
1000	1-1/2 C.Y.		3	2.667			88.50	305	393.50	475
1100	Small equipment, placed in rear of, or towed by pickup truck	A-3A	8	1			32.50	18.90	51.40	70
1150	Equip up to 70 HP, on flatbed trailer behind pickup truck	A-3D	4	2			64.50	64	128.50	169
2000	Crane, truck-mounted, up to 75 ton, (driver only, one-way)	1 Eqhv	7.20	1.111			49.50		49.50	74
2100	Crane, truck-mounted, over 75 ton	A-3E	2.50	6.400			248	60.50	308.50	440
2200	Crawler-mounted, up to 75 ton	A-3F	2	8			310	475	785	990

01 54 Construction Aids

01 54 36 – Equipment Mobilization

01 54 36.50 Mobilization		Crew	Daily Output	Labor-Hours	Unit	Material	2010 Bare Costs Labor	Equipment	Total	Total Incl O&P
2300	Over 75 ton	A-3G	1.50	10.667	Ea.		415	715	1,130	1,400
2500	For each additional 5 miles haul distance, add						10%	10%		
3000	For large pieces of equipment, allow for assembly/knockdown									
3001	For mob/demob of vibrofloatation equip, see Div. 31 45 13.10									
3100	For mob/demob of micro-tunneling equip, see Div. 33 05 23.19									
3200	For mob/demob of pile driving equip, see Div. 31 62 19.10									
3300	For mob/demob of caisson drilling equip, see Div. 31 63 26.13									

01 54 39 – Construction Equipment

01 54 39.70 Small Tools

		Crew	Daily Output	Labor-Hours	Unit	Material	Labor	Equipment	Total	Total Incl O&P
0010	**SMALL TOOLS** R013113-50									
0020	As % of contractor's bare labor cost for project, minimum				Total		.50%			
0100	Maximum				"		2%			

01 55 Vehicular Access and Parking

01 55 23 – Temporary Roads

01 55 23.50 Roads and Sidewalks

		Crew	Daily Output	Labor-Hours	Unit	Material	Labor	Equipment	Total	Total Incl O&P
0010	**ROADS AND SIDEWALKS** Temporary									
0050	Roads, gravel fill, no surfacing, 4" gravel depth	B-14	715	.067	S.Y.	4	2.34	.47	6.81	8.50
0100	8" gravel depth	"	615	.078	"	8	2.72	.55	11.27	13.55
1000	Ramp, 3/4" plywood on 2" x 6" joists, 16" O.C.	2 Carp	300	.053	S.F.	.96	2.22		3.18	4.48
1100	On 2" x 10" joists, 16" O.C.	"	275	.058	"	1.30	2.42		3.72	5.15

01 56 Temporary Barriers and Enclosures

01 56 13 – Temporary Air Barriers

01 56 13.60 Tarpaulins

		Crew	Daily Output	Labor-Hours	Unit	Material	Labor	Equipment	Total	Total Incl O&P
0010	**TARPAULINS**									
0020	Cotton duck, 10 oz. to 13.13 oz. per S.Y., minimum				S.F.	.59			.59	.65
0050	Maximum					.53			.53	.58
0100	Polyvinyl coated nylon, 14 oz. to 18 oz., minimum					.48			.48	.53
0150	Maximum					.68			.68	.75
0200	Reinforced polyethylene 3 mils thick, white					.15			.15	.17
0300	4 mils thick, white, clear or black					.20			.20	.22
0400	5.5 mils thick, clear					.14			.14	.15
0500	White, fire retardant					.35			.35	.39
0600	7.5 mils, oil resistant, fire retardant					.40			.40	.44
0700	8.5 mils, black					.53			.53	.58
0710	Woven polyethylene, 6 mils thick					.35			.35	.39
0730	Polyester reinforced w/ integral fastening system 11 mils thick					1.07			1.07	1.18
0740	Mylar polyester, non-reinforced, 7 mils thick				↓	1.17			1.17	1.29

01 56 13.90 Winter Protection

		Crew	Daily Output	Labor-Hours	Unit	Material	Labor	Equipment	Total	Total Incl O&P
0010	**WINTER PROTECTION**									
0100	Framing to close openings	2 Clab	750	.021	S.F.	.39	.71		1.10	1.52
0200	Tarpaulins hung over scaffolding, 8 uses, not incl. scaffolding		1500	.011		.25	.35		.60	.82
0250	Tarpaulin polyester reinf. w/ integral fastening system 11 mils thick		1600	.010		.80	.33		1.13	1.39
0300	Prefab fiberglass panels, steel frame, 8 uses	↓	1200	.013	↓	.85	.44		1.29	1.62

01 56 23 – Temporary Barricades

01 56 23.10 Barricades		Crew	Daily Output	Labor-Hours	Unit	Material	2010 Bare Costs Labor	Equipment	Total	Total Incl O&P
0010	**BARRICADES**									
0020	5′ high, 3 rail @ 2″ x 8″, fixed	2 Carp	20	.800	L.F.	4.41	33		37.41	56
0150	Movable	″	30	.533	″	3.54	22		25.54	38
0300	Stock units, 6′ high, 8′ wide, plain, buy				Ea.	435			435	480
0350	With reflective tape, buy				″	525			525	580
0400	Break-a-way 3″ PVC pipe barricade									
0410	with 3 ea. 1′ x 4′ reflectorized panels, buy				Ea.	305			305	335
0500	Plywood with steel legs, 32″ wide					72			72	79
0600	Telescoping Christmas tree, 9′ high, 5 flags, buy					122			122	134
0800	Traffic cones, PVC, 18″ high					9.70			9.70	10.65
0850	28″ high					16.40			16.40	18.05
1000	Guardrail, wooden, 3′ high, 1″ x 6″, on 2″ x 4″ posts	2 Carp	200	.080	L.F.	.97	3.32		4.29	6.15
1100	2″ x 6″, on 4″ x 4″ posts	″	165	.097		2	4.03		6.03	8.40
1200	Portable metal with base pads, buy					19.30			19.30	21
1250	Typical installation, assume 10 reuses	2 Carp	600	.027		2.50	1.11		3.61	4.46
1300	Barricade tape, polyethylene, 7 mil, 3″ wide x 500′ long roll				Ea.	25			25	27.50
5000	Barricades, see Div. 01 54 33.40									

01 56 26 – Temporary Fencing

01 56 26.50 Temporary Fencing		Crew	Daily Output	Labor-Hours	Unit	Material	Labor	Equipment	Total	Total Incl O&P
0010	**TEMPORARY FENCING**									
0020	Chain link, 11 ga, 5′ high	2 Clab	400	.040	L.F.	2.75	1.32		4.07	5.05
0100	6′ high		300	.053		3.25	1.77		5.02	6.30
0200	Rented chain link, 6′ high, to 1000′ (up to 12 mo.)		400	.040		4.98	1.32		6.30	7.55
0250	Over 1000′ (up to 12 mo.)		300	.053		3.98	1.77		5.75	7.10
0350	Plywood, painted, 2″ x 4″ frame, 4′ high	A-4	135	.178		4.89	7.10		11.99	16.20
0400	4″ x 4″ frame, 8′ high	″	110	.218		9.45	8.70		18.15	23.50
0500	Wire mesh on 4″ x 4″ posts, 4′ high	2 Carp	100	.160		9.65	6.65		16.30	21
0550	8′ high	″	80	.200		14.75	8.30		23.05	29

01 56 29 – Temporary Protective Walkways

01 56 29.50 Protection		Crew	Daily Output	Labor-Hours	Unit	Material	Labor	Equipment	Total	Total Incl O&P
0010	**PROTECTION**									
0020	Stair tread, 2″ x 12″ planks, 1 use	1 Carp	75	.107	Tread	3.18	4.43		7.61	10.35
0100	Exterior plywood, 1/2″ thick, 1 use		65	.123		1.26	5.10		6.36	9.30
0200	3/4″ thick, 1 use		60	.133		1.75	5.55		7.30	10.45
2200	Sidewalks, 2″ x 12″ planks, 2 uses		350	.023	S.F.	.53	.95		1.48	2.04
2300	Exterior plywood, 2 uses, 1/2″ thick		750	.011		.21	.44		.65	.91
2400	5/8″ thick		650	.012		.25	.51		.76	1.07
2500	3/4″ thick		600	.013		.29	.55		.84	1.17

01 56 32 – Temporary Security

01 56 32.50 Watchman		Crew	Daily Output	Labor-Hours	Unit	Material	Labor	Equipment	Total	Total Incl O&P
0010	**WATCHMAN**									
0020	Service, monthly basis, uniformed person, minimum				Hr.				25	27.50
0100	Maximum								45.45	50
0200	Person and command dog, minimum								31	34
0300	Maximum								54.55	60
0500	Sentry dog, leased, with job patrol (yard dog), 1 dog				Week				290	319
0600	2 dogs				″				390	429
0800	Purchase, trained sentry dog, minimum				Ea.				1,364	1,500
0900	Maximum				″				2,727	3,000

01 58 Project Identification

01 58 13 – Temporary Project Signage

01 58 13.50 Signs		Crew	Daily Output	Labor-Hours	Unit	Material	2010 Bare Costs Labor	Equipment	Total	Total Incl O&P
0010	**SIGNS**									
0020	High intensity reflectorized, no posts, buy				S.F.	26.50			26.50	29.50

01 71 Examination and Preparation

01 71 23 – Field Engineering

01 71 23.13 Construction Layout

		Crew	Daily Output	Labor-Hours	Unit	Material	Labor	Equipment	Total	Total Incl O&P
0010	**CONSTRUCTION LAYOUT**									
1100	Crew for layout of building, trenching or pipe laying, 2 person crew	A-6	1	16	Day		660	70	730	1,075
1200	3 person crew	A-7	1	24			1,075	70	1,145	1,725
1400	Crew for roadway layout, 4 person crew	A-8	1	32	↓		1,400	70	1,470	2,200

01 71 23.19 Surveyor Stakes

		Crew	Daily Output	Labor-Hours	Unit	Material	Labor	Equipment	Total	Total Incl O&P
0010	**SURVEYOR STAKES**									
0020	Hardwood, 1" x 1" x 48" long				C	58			58	64
0100	2" x 2" x 18" long					66			66	72.50
0150	2" x 2" x 24" long					110			110	121
0200	2" x 2" x 30" long				↓	74			74	81.50

01 74 Cleaning and Waste Management

01 74 13 – Progress Cleaning

01 74 13.20 Cleaning Up

		Crew	Daily Output	Labor-Hours	Unit	Material	Labor	Equipment	Total	Total Incl O&P
0010	**CLEANING UP**									
0020	After job completion, allow, minimum				Job					.30%
0040	Maximum				"					1%
0050	Cleanup of floor area, continuous, per day, during const.	A-5	24	.750	M.S.F.	1.70	25	2.50	29.20	42.50
0100	Final by GC at end of job	"	11.50	1.565	"	2.71	51.50	5.20	59.41	88
0200	Rubbish removal, see Div. 02 41 19.23									

01 91 Commissioning

01 91 13 – General Commissioning Requirements

01 91 13.50 Commissioning

		Crew	Daily Output	Labor-Hours	Unit	Material	Labor	Equipment	Total	Total Incl O&P
0010	**COMMISSIONING** Including documentation of design intent									
0100	performance verification, O&M, training, minimum				Project					1%
0150	Maximum				"					1.25%

Estimating Tips

02 30 00 Subsurface Investigation

In preparing estimates on structures involving earthwork or foundations, all information concerning soil characteristics should be obtained. Look particularly for hazardous waste, evidence of prior dumping of debris, and previous stream beds.

02 41 00 Demolition and Structure Moving

The costs shown for selective demolition do not include rubbish handling or disposal. These items should be estimated separately using RSMeans data or other sources.

- Historic preservation often requires that the contractor remove materials from the existing structure, rehab them, and replace them. The estimator must be aware of any related measures and precautions that must be taken when doing selective demolition and cutting and patching. Requirements may include special handling and storage, as well as security.
- In addition to Subdivision 02 41 00, you can find selective demolition items in each division. Example: Roofing demolition is in Division 7.

02 42 10 Building Deconstruction

This is a new section on the careful demantling and recycling of most of low-rise building materials.

02 56 13 Containment of Hazardous Waste

This is a new section for disposal of hazardous waste on site.

02 80 00 Hazardous Material Disposal/ Remediation

This subdivision includes information on hazardous waste handling, asbestos remediation, lead remediation, and mold remediation. See reference R028213-20 and R028319-60 for further guidance in using these unit price lines.

02 91 10 Monitoring Chemical Sampling, Testing Analysis

This is a new section for on-site sampling and testing hazardous waste.

Reference Numbers

Reference numbers are shown in shaded boxes at the beginning of some major classifications. These numbers refer to related items in the Reference Section. The reference information may be an estimating procedure, an alternate pricing method, or technical information.

Note: Not all subdivisions listed here necessarily appear in this publication.

02 21 Surveys

02 21 13 – Site Surveys

02 21 13.09 Topographical Surveys		Crew	Daily Output	Labor-Hours	Unit	Material	2010 Bare Costs Labor	Equipment	Total	Total Incl O&P
0010	**TOPOGRAPHICAL SURVEYS**									
0020	Topographical surveying, conventional, minimum	A-7	3.30	7.273	Acre	18	325	21	364	545
0100	Maximum	A-8	.60	53.333	"	55	2,325	116	2,496	3,750

02 21 13.13 Boundary and Survey Markers

		Crew	Daily Output	Labor-Hours	Unit	Material	Labor	Equipment	Total	Total Incl O&P
0010	**BOUNDARY AND SURVEY MARKERS**									
0300	Lot location and lines, large quantities, minimum	A-7	2	12	Acre	32	540	35	607	900
0320	Average	"	1.25	19.200		51	865	56	972	1,450
0400	Small quantities, maximum	A-8	1	32	↓	68	1,400	70	1,538	2,275
0600	Monuments, 3′ long	A-7	10	2.400	Ea.	30	108	7	145	206
0800	Property lines, perimeter, cleared land	"	1000	.024	L.F.	.03	1.08	.07	1.18	1.76
0900	Wooded land	A-8	875	.037	"	.05	1.60	.08	1.73	2.59

02 21 13.16 Aerial Surveys

		Crew	Daily Output	Labor-Hours	Unit	Material	Labor	Equipment	Total	Total Incl O&P
0010	**AERIAL SURVEYS**									
1500	Aerial surveying, including ground control, minimum fee, 10 acres				Total					4,400
1510	100 acres									8,800
1550	From existing photography, deduct				↓					1,500
1600	2′ contours, 10 acres				Acre					160
1650	20 acres									80
1800	50 acres									35
1850	100 acres									28
2000	1000 acres									26
2050	10,000 acres				↓					24
2150	For 1′ contours and									
2160	dense urban areas, add to above				Acre					20%
3000	Inertial guidance system for									
3010	locating coordinates, rent per day				Ea.					4,400

02 32 Geotechnical Investigations

02 32 13 – Subsurface Drilling and Sampling

02 32 13.10 Boring and Exploratory Drilling

		Crew	Daily Output	Labor-Hours	Unit	Material	Labor	Equipment	Total	Total Incl O&P
0010	**BORING AND EXPLORATORY DRILLING**									
0020	Borings, initial field stake out & determination of elevations	A-6	1	16	Day		660	70	730	1,075
0100	Drawings showing boring details				Total		250		250	310
0200	Report and recommendations from P.E.						600		600	750
0300	Mobilization and demobilization, minimum	B-55	4	6	↓		197	261	458	595
0350	For over 100 miles, per added mile		450	.053	Mile		1.75	2.32	4.07	5.25
0600	Auger holes in earth, no samples, 2-1/2″ diameter		78.60	.305	L.F.		10	13.30	23.30	30
0650	4″ diameter		67.50	.356			11.65	15.50	27.15	35
0800	Cased borings in earth, with samples, 2-1/2″ diameter		55.50	.432		20.50	14.20	18.85	53.55	65
0850	4″ diameter	↓	32.60	.736		32.50	24	32	88.50	108
1000	Drilling in rock, "BX" core, no sampling	B-56	34.90	.458			17.05	42	59.05	72
1050	With casing & sampling		31.70	.505		20.50	18.80	46	85.30	102
1200	"NX" core, no sampling		25.92	.617			23	56.50	79.50	97
1250	With casing and sampling	↓	25	.640	↓	25	24	58.50	107.50	128
1400	Borings, earth, drill rig and crew with truck mounted auger	B-55	1	24	Day		790	1,050	1,840	2,350
1450	Rock using crawler type drill	B-56	1	16	"		595	1,475	2,070	2,500
1500	For inner city borings add, minimum									10%
1510	Maximum									20%

02 32 Geotechnical Investigations

02 32 19 – Exploratory Excavations

02 32 19.10 Test Pits		Crew	Daily Output	Labor-Hours	Unit	Material	2010 Bare Costs Labor	Equipment	Total	Total Incl O&P
0010	**TEST PITS**									
0020	Hand digging, light soil	1 Clab	4.50	1.778	C.Y.		59		59	91
0100	Heavy soil	"	2.50	3.200			106		106	163
0120	Loader-backhoe, light soil	B-11M	28	.571			21.50	13.85	35.35	48.50
0130	Heavy soil	"	20	.800	↓		30.50	19.40	49.90	67.50
1000	Subsurface exploration, mobilization				Mile				6	7.50
1010	Difficult access for rig, add				Hr.				160	190
1020	Auger borings, drill rig, incl. samples				L.F.				23	28
1030	Hand auger								29	36
1050	Drill and sample every 5′, split spoon				↓				29	36
1060	Extra samples				Ea.				35	44

02 41 Demolition

02 41 13 – Selective Site Demolition

02 41 13.15 Hydrodemolition

		Crew	Daily Output	Labor-Hours	Unit	Material	Labor	Equipment	Total	Total Incl O&P
0010	**HYDRODEMOLITION** R024119-10									
0015	Hydrodemolition, concrete pavement, 4000 PSI, 2″ depth	B-5	500	.112	S.F.		4.05	2.60	6.65	9.05
0120	4″ depth		450	.124			4.50	2.89	7.39	10.10
0130	6″ depth		400	.140			5.05	3.26	8.31	11.35
0410	6000 PSI, 2″ depth		410	.137			4.94	3.18	8.12	11.05
0420	4″ depth		350	.160			5.80	3.72	9.52	12.95
0430	6″ depth		300	.187			6.75	4.34	11.09	15.05
0510	8000 PSI, 2″ depth		330	.170			6.15	3.95	10.10	13.75
0520	4″ depth		280	.200			7.25	4.65	11.90	16.15
0530	6″ depth	↓	240	.233	↓		8.45	5.40	13.85	18.85

02 41 13.17 Demolish, Remove Pavement and Curb

		Crew	Daily Output	Labor-Hours	Unit	Material	Labor	Equipment	Total	Total Incl O&P
0010	**DEMOLISH, REMOVE PAVEMENT AND CURB** R024119-10									
5010	Pavement removal, bituminous roads, 3″ thick	B-38	690	.058	S.Y.		2.15	1.71	3.86	5.15
5050	4″ to 6″ thick		420	.095			3.53	2.80	6.33	8.50
5100	Bituminous driveways		640	.063			2.32	1.84	4.16	5.55
5200	Concrete to 6″ thick, hydraulic hammer, mesh reinforced		255	.157			5.80	4.62	10.42	13.95
5300	Rod reinforced		200	.200	↓		7.40	5.90	13.30	17.80
5400	Concrete, 7″ to 24″ thick, plain		33	1.212	C.Y.		45	35.50	80.50	108
5500	Reinforced	↓	24	1.667	"		62	49	111	148
5600	With hand held air equipment, bituminous, to 6″ thick	B-39	1900	.025	S.F.		.88	.12	1	1.48
5700	Concrete to 6″ thick, no reinforcing		1600	.030			1.04	.14	1.18	1.75
5800	Mesh reinforced		1400	.034			1.19	.16	1.35	2.01
5900	Rod reinforced	↓	765	.063	↓		2.18	.29	2.47	3.67
6000	Curbs, concrete, plain	B-6	360	.067	L.F.		2.39	.94	3.33	4.67
6100	Reinforced		275	.087			3.13	1.23	4.36	6.10
6200	Granite		360	.067			2.39	.94	3.33	4.67
6300	Bituminous	↓	528	.045	↓		1.63	.64	2.27	3.18

02 41 13.33 Minor Site Demolition

		Crew	Daily Output	Labor-Hours	Unit	Material	Labor	Equipment	Total	Total Incl O&P
0010	**MINOR SITE DEMOLITION** R024119-10									
0015	No hauling, abandon catch basin or manhole	B-6	7	3.429	Ea.		123	48.50	171.50	240
0020	Remove existing catch basin or manhole, masonry		4	6			215	84.50	299.50	425
0030	Catch basin or manhole frames and covers, stored		13	1.846			66	26	92	130
0040	Remove and reset	↓	7	3.429			123	48.50	171.50	240
0100	Roadside delineators, remove only	B-80	175	.183			6.50	4.26	10.76	14.60
0110	Remove and reset	"	100	.320	↓		11.35	7.45	18.80	25.50

02 41 Demolition

02 41 13 – Selective Site Demolition

02 41 13.33 Minor Site Demolition

	02 41 13.33 Minor Site Demolition	Crew	Daily Output	Labor-Hours	Unit	2010 Bare Costs Material	Labor	Equipment	Total	Total Incl O&P
0800	Guiderail, corrugated steel, remove only	B-80A	100	.240	L.F.		7.95	2.96	10.91	15.50
0850	Remove and reset	"	40	.600	"		19.85	7.40	27.25	38.50
0860	Guide posts, remove only	B-80B	120	.267	Ea.		9.35	2.02	11.37	16.55
0870	Remove and reset	B-55	50	.480			15.75	21	36.75	47
0900	Hydrants, fire, remove only	B-21A	5	8			330	131	461	645
0950	Remove and reset	"	2	20			825	330	1,155	1,600
1000	Masonry walls, block, solid	B-5	1800	.031	C.F.		1.13	.72	1.85	2.52
1200	Brick, solid		900	.062			2.25	1.45	3.70	5.05
1400	Stone, with mortar		900	.062			2.25	1.45	3.70	5.05
1500	Dry set		1500	.037			1.35	.87	2.22	3.01
1600	Median barrier, precast concrete, remove and store	B-3	430	.112	L.F.		3.92	5.60	9.52	12.15
1610	Remove and reset	"	390	.123			4.32	6.20	10.52	13.40
2900	Pipe removal, sewer/water, no excavation, 12" diameter	B-6	175	.137			4.91	1.93	6.84	9.60
2930	15"-18" diameter		150	.160			5.75	2.25	8	11.25
2960	21"-24" diameter		120	.200			7.15	2.82	9.97	14.05
3000	27"-36" diameter		90	.267			9.55	3.75	13.30	18.70
3200	Steel, welded connections, 4" diameter		160	.150			5.35	2.11	7.46	10.50
3300	10" diameter		80	.300			10.75	4.22	14.97	21
3500	Railroad track removal, ties and track	B-13	330	.170			6.05	2.27	8.32	11.75
3600	Ballast	B-14	500	.096	C.Y.		3.34	.68	4.02	5.85
3700	Remove and re-install, ties & track using new bolts & spikes		50	.960	L.F.		33.50	6.75	40.25	58.50
3800	Turnouts using new bolts and spikes		1	48	Ea.		1,675	340	2,015	2,925
4000	Sidewalk removal, bituminous, 2-1/2" thick	B-6	325	.074	S.Y.		2.65	1.04	3.69	5.20
4050	Brick, set in mortar		185	.130			4.65	1.83	6.48	9.10
4100	Concrete, plain, 4"		160	.150			5.35	2.11	7.46	10.50
4200	Mesh reinforced		150	.160			5.75	2.25	8	11.25
4300	Slab on grade removal, plain	B-5	45	1.244	C.Y.		45	29	74	101
4310	Mesh reinforced		33	1.697			61.50	39.50	101	138
4320	Rod reinforced		25	2.240			81	52	133	182
4400	For congested sites or small quantities, add up to								200%	200%
4450	For disposal on site, add	B-11A	232	.069			2.62	5.15	7.77	9.65
4500	To 5 miles, add	B-34D	76	.105			3.49	9.55	13.04	15.80

02 41 13.60 Selective Demolition Fencing

	02 41 13.60 Selective Demolition Fencing	Crew	Daily Output	Labor-Hours	Unit	Material	Labor	Equipment	Total	Total Incl O&P
0010	**SELECTIVE DEMOLITION FENCING** R024119-10									
1600	Fencing, barbed wire, 3 strand	2 Clab	430	.037	L.F.		1.23		1.23	1.90
1650	5 strand	"	280	.057			1.89		1.89	2.92
1700	Chain link, posts & fabric, 8' to 10' high, remove only	B-6	445	.054			1.93	.76	2.69	3.78
1750	Remove and reset	"	70	.343			12.30	4.83	17.13	24

02 41 16 – Structure Demolition

02 41 16.13 Building Demolition

	02 41 16.13 Building Demolition	Crew	Daily Output	Labor-Hours	Unit	Material	Labor	Equipment	Total	Total Incl O&P
0010	**BUILDING DEMOLITION** Large urban projects, incl. 20 mi. haul R024119-10									
0011	No foundation or dump fees, C.F. is vol. of building standing									
0020	Steel	B-8	21500	.003	C.F.		.11	.15	.26	.33
0050	Concrete		15300	.004			.15	.21	.36	.46
0080	Masonry		20100	.003			.12	.16	.28	.35
0100	Mixture of types, average		20100	.003			.12	.16	.28	.35
0500	Small bldgs, or single bldgs, no salvage included, steel	B-3	14800	.003			.11	.16	.27	.35
0600	Concrete		11300	.004			.15	.21	.36	.47
0650	Masonry		14800	.003			.11	.16	.27	.35
0700	Wood		14800	.003			.11	.16	.27	.35
0750	For buildings with no interior walls, deduct								50%	
1000	Single family, one story house, wood, minimum				Ea.				4,100	4,800

02 41 Demolition

02 41 16 – Structure Demolition

02 41 16.13 Building Demolition		Crew	Daily Output	Labor-Hours	Unit	Material	2010 Bare Costs Labor	Equipment	Total	Total Incl O&P
1020	Maximum				Ea.				8,200	9,600
1200	Two family, two story house, wood, minimum								6,100	7,200
1220	Maximum								10,900	12,800
1300	Three family, three story house, wood, minimum								8,200	9,600
1320	Maximum								13,600	16,000
5000	For buildings with no interior walls, deduct								50%	

02 41 16.15 Explosive/Implosive Demolition

		Crew	Daily Output	Labor-Hours	Unit	Material	Labor	Equipment	Total	Total Incl O&P
0010	**EXPLOSIVE/IMPLOSIVE DEMOLITION** R024119-10									
0011	Large projects,									
0020	No disposal fee based on building volume, steel building	B-5B	16900	.003	C.F.		.11	.16	.27	.34
0100	Concrete building		16900	.003			.11	.16	.27	.34
0200	Masonry building		16900	.003			.11	.16	.27	.34
0400	Disposal of material, minimum	B-3	445	.108	C.Y.		3.79	5.40	9.19	11.75
0500	Maximum	"	365	.132	"		4.61	6.60	11.21	14.30

02 41 16.17 Building Demolition Footings and Foundations

		Crew	Daily Output	Labor-Hours	Unit	Material	Labor	Equipment	Total	Total Incl O&P
0010	**BUILDING DEMOLITION FOOTINGS AND FOUNDATIONS** R024119-10									
0200	Floors, concrete slab on grade,									
0240	4" thick, plain concrete	B-9	500	.080	S.F.		2.68	.45	3.13	4.62
0280	Reinforced, wire mesh		470	.085			2.85	.48	3.33	4.93
0300	Rods		400	.100			3.35	.56	3.91	5.75
0400	6" thick, plain concrete		375	.107			3.57	.60	4.17	6.15
0420	Reinforced, wire mesh		340	.118			3.94	.66	4.60	6.85
0440	Rods		300	.133			4.47	.75	5.22	7.70
1000	Footings, concrete, 1' thick, 2' wide	B-5	300	.187	L.F.		6.75	4.34	11.09	15.05
1080	1'-6" thick, 2' wide		250	.224			8.10	5.20	13.30	18.15
1120	3' wide		200	.280			10.15	6.50	16.65	22.50
1140	2' thick, 3' wide		175	.320			11.60	7.45	19.05	26
1200	Average reinforcing, add								10%	10%
1220	Heavy reinforcing, add								20%	20%
2000	Walls, block, 4" thick	1 Clab	180	.044	S.F.		1.47		1.47	2.27
2040	6" thick		170	.047			1.56		1.56	2.40
2080	8" thick		150	.053			1.77		1.77	2.72
2100	12" thick		150	.053			1.77		1.77	2.72
2200	For horizontal reinforcing, add								10%	10%
2220	For vertical reinforcing, add								20%	20%
2400	Concrete, plain concrete, 6" thick	B-9	160	.250			8.40	1.41	9.81	14.45
2420	8" thick		140	.286			9.55	1.61	11.16	16.50
2440	10" thick		120	.333			11.15	1.87	13.02	19.25
2500	12" thick		100	.400			13.40	2.25	15.65	23
2600	For average reinforcing, add								10%	10%
2620	For heavy reinforcing, add								20%	20%
4000	For congested sites or small quantities, add up to								200%	200%
4200	Add for disposal, on site	B-11A	232	.069	C.Y.		2.62	5.15	7.77	9.65
4250	To five miles	B-30	220	.109	"		3.97	10.70	14.67	17.75

02 41 19 – Selective Structure Demolition

02 41 19.13 Selective Building Demolition

		Crew	Daily Output	Labor-Hours	Unit	Material	Labor	Equipment	Total	Total Incl O&P
0010	**SELECTIVE BUILDING DEMOLITION**									
0020	Costs related to selective demolition of specific building components									
0025	are included under Common Work Results (XX 05 00)									
0030	in the component's appropriate division.									

02 41 19 – Selective Structure Demolition

02 41 19.16 Selective Demolition, Cutout		Crew	Daily Output	Labor-Hours	Unit	Material	2010 Bare Costs Labor	Equipment	Total	Total Incl O&P
0010	**SELECTIVE DEMOLITION, CUTOUT** R024119-10									
0020	Concrete, elev. slab, light reinforcement, under 6 C.F.	B-9	65	.615	C.F.		20.50	3.46	23.96	36
0050	Light reinforcing, over 6 C.F.		75	.533	"		17.85	3	20.85	31
0200	Slab on grade to 6" thick, not reinforced, under 8 S.F.		85	.471	S.F.		15.75	2.64	18.39	27.50
0250	8 – 16 S.F.	↓	175	.229	"		7.65	1.28	8.93	13.20
0255	For over 16 S.F. see Div. 02 41 16.17 0400									
0600	Walls, not reinforced, under 6 C.F.	B-9	60	.667	C.F.		22.50	3.75	26.25	38.50
0650	6 – 12 C.F.	"	80	.500	"		16.75	2.81	19.56	29
0655	For over 12 C.F. see Div. 02 41 16.17 2500									
1000	Concrete, elevated slab, bar reinforced, under 6 C.F.	B-9	45	.889	C.F.		30	5	35	51.50
1050	Bar reinforced, over 6 C.F.		50	.800	"		27	4.50	31.50	46.50
1200	Slab on grade to 6" thick, bar reinforced, under 8 S.F.		75	.533	S.F.		17.85	3	20.85	31
1250	8 – 16 S.F.	↓	150	.267	"		8.95	1.50	10.45	15.45
1255	For over 16 S.F. see Div. 02 41 16.17 0440									
1400	Walls, bar reinforced, under 6 C.F.	B-9	50	.800	C.F.		27	4.50	31.50	46.50
1450	6 – 12 C.F.	"	70	.571	"		19.15	3.21	22.36	33
1455	For over 12 C.F. see Div. 02 41 16.17 2500 and 2600									
2000	Brick, to 4 S.F. opening, not including toothing									
2040	4" thick	B-9	30	1.333	Ea.		44.50	7.50	52	77.50
2060	8" thick		18	2.222			74.50	12.50	87	129
2080	12" thick		10	4			134	22.50	156.50	232
2400	Concrete block, to 4 S.F. opening, 2" thick		35	1.143			38.50	6.40	44.90	66
2420	4" thick		30	1.333			44.50	7.50	52	77.50
2440	8" thick		27	1.481			49.50	8.35	57.85	85.50
2460	12" thick		24	1.667			56	9.35	65.35	96.50
2600	Gypsum block, to 4 S.F. opening, 2" thick		80	.500			16.75	2.81	19.56	29
2620	4" thick		70	.571			19.15	3.21	22.36	33
2640	8" thick		55	.727			24.50	4.09	28.59	42
2800	Terra cotta, to 4 S.F. opening, 4" thick		70	.571			19.15	3.21	22.36	33
2840	8" thick		65	.615			20.50	3.46	23.96	36
2880	12" thick	↓	50	.800	↓		27	4.50	31.50	46.50
3000	Toothing masonry cutouts, brick, soft old mortar	1 Brhe	40	.200	V.L.F.		6.75		6.75	10.15
3100	Hard mortar		30	.267			8.95		8.95	13.55
3200	Block, soft old mortar		70	.114			3.85		3.85	5.80
3400	Hard mortar	↓	50	.160	↓		5.40		5.40	8.10
6000	Walls, interior, not including re-framing,									
6010	openings to 5 S.F.									
6100	Drywall to 5/8" thick	1 Clab	24	.333	Ea.		11.05		11.05	17
6200	Paneling to 3/4" thick		20	.400			13.25		13.25	20.50
6300	Plaster, on gypsum lath		20	.400			13.25		13.25	20.50
6340	On wire lath	↓	14	.571	↓		18.90		18.90	29
7000	Wood frame, not including re-framing, openings to 5 S.F.									
7200	Floors, sheathing and flooring to 2" thick	1 Clab	5	1.600	Ea.		53		53	81.50
7310	Roofs, sheathing to 1" thick, not including roofing		6	1.333			44		44	68
7410	Walls, sheathing to 1" thick, not including siding	↓	7	1.143	↓		38		38	58.50

02 41 19.18 Selective Demolition, Disposal Only		Crew	Daily Output	Labor-Hours	Unit	Material	Labor	Equipment	Total	Total Incl O&P
0010	**SELECTIVE DEMOLITION, DISPOSAL ONLY** R024119-10									
0015	Urban bldg w/salvage value allowed									
0020	Including loading and 5 mile haul to dump									
0200	Steel frame	B-3	430	.112	C.Y.		3.92	5.60	9.52	12.15
0300	Concrete frame		365	.132			4.61	6.60	11.21	14.30
0400	Masonry construction	↓	445	.108	↓		3.79	5.40	9.19	11.75

02 41 19 – Selective Structure Demolition

02 41 19.18 Selective Demolition, Disposal Only		Crew	Daily Output	Labor-Hours	Unit	Material	Labor	Equipment	Total	Total Incl O&P
						2010 Bare Costs				
0500	Wood frame	B-3	247	.194	C.Y.		6.80	9.75	16.55	21

02 41 19.19 Selective Demolition, Dump Charges

Line	Description	Crew	Daily Output	Labor-Hours	Unit	Material	Labor	Equipment	Total	Total Incl O&P
0010	**SELECTIVE DEMOLITION, DUMP CHARGES** R024119-10									
0020	Dump charges, typical urban city, tipping fees only									
0100	Building construction materials				Ton	90			90	99
0200	Trees, brush, lumber					75			75	83
0300	Rubbish only					80			80	88
0500	Reclamation station, usual charge					90			90	99

02 41 19.21 Selective Demolition, Gutting

Line	Description	Crew	Daily Output	Labor-Hours	Unit	Material	Labor	Equipment	Total	Total Incl O&P
0010	**SELECTIVE DEMOLITION, GUTTING** R024119-10									
0020	Building interior, including disposal, dumpster fees not included									
0500	Residential building									
0560	Minimum	B-16	400	.080	SF Flr.		2.69	1.67	4.36	6
0580	Maximum	"	360	.089	"		2.99	1.86	4.85	6.65
0900	Commercial building									
1000	Minimum	B-16	350	.091	SF Flr.		3.07	1.91	4.98	6.85
1020	Maximum	"	250	.128	"		4.30	2.67	6.97	9.55

02 41 19.23 Selective Demolition, Rubbish Handling

Line	Description	Crew	Daily Output	Labor-Hours	Unit	Material	Labor	Equipment	Total	Total Incl O&P
0010	**SELECTIVE DEMOLITION, RUBBISH HANDLING** R024119-10									
0020	The following are to be added to the demolition prices									
0400	Chute, circular, prefabricated steel, 18" diameter	B-1	40	.600	L.F.	48.50	20.50		69	84.50
0440	30" diameter	"	30	.800	"	50	27		77	96.50
0600	Dumpster, weekly rental, 1 dump/week, 6 C.Y. capacity (2 Tons)				Week	450			450	495
0700	10 C.Y. capacity (4 Tons)					525			525	578
0725	20 C.Y. capacity (8 Tons) R024119-20					700			700	770
0800	30 C.Y. capacity (10 Tons)					850			850	935
0840	40 C.Y. capacity (13 Tons)					1,000			1,000	1,100
1000	Dust partition, 6 mil polyethylene, 1" x 3" frame	2 Carp	2000	.008	S.F.	.45	.33		.78	1
1080	2" x 4" frame	"	2000	.008	"	.24	.33		.57	.77
2000	Load, haul, dump and return, 50' haul, hand carried	2 Clab	24	.667	C.Y.		22		22	34
2005	Wheeled		37	.432			14.30		14.30	22
2040	100' haul, hand carried		16.50	.970			32		32	49.50
2045	Wheeled		25	.640			21		21	32.50
2080	Over 100' haul, add per 100 L.F., hand carried		35.50	.451			14.90		14.90	23
2085	Wheeled		54	.296			9.80		9.80	15.15
2120	In elevators, per 10 floors, add		140	.114			3.78		3.78	5.85
2130	Load, haul, dump and return, 50' haul, incl. 5 riser stair, hand carried		23	.696			23		23	35.50
2135	Wheeled		35	.457			15.15		15.15	23.50
2140	6 – 10 riser stairs, hand carried		22	.727			24		24	37
2145	Wheeled		34	.471			15.60		15.60	24
2150	11 – 20 riser stairs, hand carried		20	.800			26.50		26.50	41
2155	Wheeled		31	.516			17.10		17.10	26.50
2160	21 – 40 riser stairs, hand carried		16	1			33		33	51
2165	Wheeled		24	.667			22		22	34
2170	100' haul, incl. 5 riser stair, hand carried		15	1.067			35.50		35.50	54.50
2175	Wheeled		23	.696			23		23	35.50
2180	6 – 10 riser stair, hand carried		14	1.143			38		38	58.50
2185	Wheeled		21	.762			25		25	39
2190	11 – 20 riser stair, hand carried		12	1.333			44		44	68
2195	Wheeled		18	.889			29.50		29.50	45.50
2200	21 – 40 riser stair, hand carried		8	2			66		66	102
2205	Wheeled		12	1.333			44		44	68

02 41 Demolition

02 41 19 – Selective Structure Demolition

02 41 19.23 Selective Demolition, Rubbish Handling		Crew	Daily Output	Labor-Hours	Unit	Material	2010 Bare Costs Labor	Equipment	Total	Total Incl O&P
2210	Over 100' haul, add per 100 L.F., hand carried	2 Clab	35.50	.451	C.Y.		14.90		14.90	23
2215	Wheeled		54	.296			9.80		9.80	15.15
2220	For each additional flight of stairs, 5 risers		550	.029	Flight		.96		.96	1.49
2225	6 – 10 risers		275	.058			1.93		1.93	2.97
2230	11-20 risers		138	.116			3.84		3.84	5.90
2235	21 – 40 risers		69	.232			7.70		7.70	11.85
3000	Loading & trucking, including 2 mile haul, chute loaded	B-16	45	.711	C.Y.		24	14.85	38.85	53.50
3040	Hand loading truck, 50' haul	"	48	.667			22.50	13.90	36.40	50
3080	Machine loading truck	B-17	120	.267			9.40	6.15	15.55	21
5000	Haul, per mile, up to 8 C.Y. truck	B-34B	1165	.007			.23	.57	.80	.98
5100	Over 8 C.Y. truck	"	1550	.005			.17	.43	.60	.73

02 41 19.25 Selective Demolition, Saw Cutting

		Crew	Daily Output	Labor-Hours	Unit	Material	Labor	Equipment	Total	Total Incl O&P
0010	SELECTIVE DEMOLITION, SAW CUTTING R024119-10									
0015	Asphalt, up to 3" deep	B-89	1050	.015	L.F.	.41	.56	.45	1.42	1.80
0020	Each additional inch of depth	"	1800	.009		.09	.33	.26	.68	.88
1200	Masonry walls, hydraulic saw, brick, per inch of depth	B-89B	300	.053		.43	1.96	2.81	5.20	6.50
1220	Block walls, solid, per inch of depth	"	250	.064		.43	2.35	3.37	6.15	7.75
2000	Brick or masonry w/hand held saw, per inch of depth	A-1	125	.064		.34	2.12	.59	3.05	4.30
5000	Wood sheathing to 1" thick, on walls	1 Carp	200	.040			1.66		1.66	2.56
5020	On roof	"	250	.032			1.33		1.33	2.05

02 41 19.27 Selective Demolition, Torch Cutting

		Crew	Daily Output	Labor-Hours	Unit	Material	Labor	Equipment	Total	Total Incl O&P
0010	SELECTIVE DEMOLITION, TORCH CUTTING R024119-10									
0020	Steel, 1" thick plate	1 Clab	360	.022	L.F.	.23	.74		.97	1.38
0040	1" diameter bar	"	210	.038	Ea.		1.26		1.26	1.95
1000	Oxygen lance cutting, reinforced concrete walls									
1040	12" to 16" thick walls	1 Clab	10	.800	L.F.		26.50		26.50	41
1080	24" thick walls	"	6	1.333	"		44		44	68

02 42 Removal and Salvage of Construction Material

02 42 10 – Building Deconstruction

02 42 10.10 Estimated Salvage Value or Savings

		Crew	Daily Output	Labor-Hours	Unit	Material	Labor	Equipment	Total	Total Incl O&P
0010	ESTIMATED SALVAGE VALUE OR SAVINGS									
0015	Excludes material handling, packaging, container costs and									
0020	transportation for salvage or disposal									
0050	All Items in section 02 42 10.10 are credit deducts and not costs									
0100	Copper Wire Salvage Value G				Lb.				1.30	1.30
0110	Disposal Savings G								.03	.03
0200	Copper Pipe Salvage Value G								1.34	1.34
0210	Disposal Savings G								.03	.03
0300	Steel Pipe Salvage Value G								.06	.06
0310	Disposal Savings G								.03	.03
0400	Cast Iron Pipe Salvage Value G								.06	.06
0410	Disposal Savings G								.03	.03
0500	Steel Doors or Windows Salvage Value G								.06	.06
0510	Aluminum G								.66	.66
0520	Disposal Savings G								.03	.03
0600	Aluminum Siding Salvage Value G								.49	.49
0630	Disposal Savings G								.03	.03
0640	Wood Siding (no lead or asbestos) G				C.Y.				10.92	10.92
0800	Clean Concrete Disposal Savings G				Ton				60	60

02 42 10 – Building Deconstruction

02 42 10.10 Estimated Salvage Value or Savings

	02 42 10.10 Estimated Salvage Value or Savings		Crew	Daily Output	Labor-Hours	Unit	Material	2010 Bare Costs Labor	Equipment	Total	Total Incl O&P
0850	Asphalt Shingles Disposal Savings	G				Ton				60	60
1000	Wood wall framing clean salvage value	G				M.B.F.				50	50
1010	Painted	G								40	40
1020	Floor framing	G								50	50
1030	Painted	G								40	40
1050	Roof framing	G								50	50
1060	Painted	G								40	40
1100	Wood beams salvage value	G								50	50
1200	Wood framing and beams disposal savings	G				Ton				60	60
1220	Wood sheating and sub-base flooring	G								60	60
1230	Wood wall paneling (1/4 inch thick)	G								60	60
1300	Wood panel 3/4-1 inch thick low salvage value	G				S.F.				.50	.50
1350	high salvage value	G				"				2	2
1400	Disposal savings	G				Ton				60	60
1500	Flooring tongue and groove 25/32 inch thick low salvage value	G				S.F.				.50	.50
1530	High salvage value	G				"				1	1
1560	Disposal savings	G				Ton				60	60
1600	Drywall or sheet rock salvage value	G								20	20
1650	Disposal savings	G								60	60

02 42 10.20 Deconstruction of Building Components

	02 42 10.20 Deconstruction of Building Components		Crew	Daily Output	Labor-Hours	Unit	Material	Labor	Equipment	Total	Total Incl O&P
0010	**DECONSTRUCTION OF BUILDING COMPONENTS**										
0012	Buildings one or two stories only										
0015	Excludes material handling, packaging, container costs and										
0020	transportation for salvage or disposal										
0050	Deconstruction of Plumbing Fixtures										
0100	Wall hung or countertop lavatory	G	2 Clab	16	1	Ea.		33		33	51
0110	Single or double compartment kitchen sink	G		14	1.143			38		38	58.50
0120	Wall hung urinal	G		14	1.143			38		38	58.50
0130	Floor mounted	G		8	2			66		66	102
0140	Floor mounted water closet	G		16	1			33		33	51
0150	Wall hung	G		14	1.143			38		38	58.50
0160	Water fountain, free standing	G		16	1			33		33	51
0170	Wall hung or deck mounted	G		12	1.333			44		44	68
0180	Bathtub, steel or fiberglass	G		10	1.600			53		53	81.50
0190	Cast iron	G		8	2			66		66	102
0200	Shower, single	G		6	2.667			88.50		88.50	136
0210	Group	G		7	2.286			75.50		75.50	117
0300	Deconstruction of Electrical Fixtures										
0310	Surface mount incandescent fixtures	G	2 Clab	48	.333	Ea.		11.05		11.05	17
0320	Fluorescent, 2 lamp	G		32	.500			16.55		16.55	25.50
0330	4 lamp	G		24	.667			22		22	34
0340	Strip Fluorescent, 1 lamp	G		40	.400			13.25		13.25	20.50
0350	2 lamp	G		32	.500			16.55		16.55	25.50
0400	Recessed drop-in fluorescent fixture, 2 lamp	G		27	.593			19.60		19.60	30.50
0410	4 lamp	G		18	.889			29.50		29.50	45.50
0500	Deconstruction of appliances										
0510	Cooking stoves	G	2 Clab	26	.615	Ea.		20.50		20.50	31.50
0520	Dishwashers	G	"	26	.615	"		20.50		20.50	31.50
0600	Deconstruction of millwork and trim										
0610	Cabinets, wood	G	2 Carp	40	.400	L.F.		16.60		16.60	25.50
0620	Countertops	G		100	.160	"		6.65		6.65	10.25
0630	Wall paneling, 1 inch thick	G		500	.032	S.F.		1.33		1.33	2.05

02 42 10.20 Deconstruction of Building Components			Crew	Daily Output	Labor-Hours	Unit	Material	2010 Bare Costs Labor	Equipment	Total	Total Incl O&P
0640	Ceiling trim	G	2 Carp	500	.032	L.F.		1.33		1.33	2.05
0650	Wainscoting	G		500	.032	S.F.		1.33		1.33	2.05
0660	Base, 3/4" to 1" thick	G	↓	600	.027	L.F.		1.11		1.11	1.71
0700	Deconstruction of doors and windows										
0710	Doors, wrap, interior, wood, single, no closers	G	2 Carp	21	.762	Ea.	3.50	31.50		35	53
0720	Double	G		13	1.231		7	51		58	86.50
0730	Solid core, single, exterior or interior	G		10	1.600		3.50	66.50		70	106
0740	Double	G		8	2		7	83		90	136
0810	Windows, wrap, wood, single		↓	21	.762	↓	3.50	31.50		35	53
0812	with no casement or cladding	G									
0820	with casement and/or cladding	G	2 Carp	18	.889	Ea.	3.50	37		40.50	61
0900	Deconstruction of interior finishes										
0910	Drywall for recycling	G	2 Clab	1775	.009	S.F.		.30		.30	.46
0920	Plaster wall, first floor	G		1775	.009			.30		.30	.46
0930	Second floor	G	↓	1330	.012	↓		.40		.40	.61
1000	Deconstruction of roofing and accessories										
1010	Built-up roofs	G	2 Clab	570	.028	S.F.		.93		.93	1.43
1020	Gutters, facia and rakes	G	"	1140	.014	L.F.		.46		.46	.72
2000	Deconstruction of wood components										
2010	Roof sheeting	G	2 Clab	570	.028	S.F.		.93		.93	1.43
2020	Main roof framing	G		760	.021	L.F.		.70		.70	1.07
2030	Porch roof framing	G	↓	445	.036			1.19		1.19	1.84
2040	Beams 4" x 8"	G	B-1	375	.064			2.16		2.16	3.33
2050	4" x 10"	G		300	.080			2.70		2.70	4.17
2055	4" x 12"	G		250	.096			3.24		3.24	5
2060	6" x 8"	G		250	.096			3.24		3.24	5
2065	6" x 10"	G		200	.120			4.05		4.05	6.25
2070	6" x 12"	G		170	.141			4.77		4.77	7.35
2075	8" x 12"	G		126	.190			6.45		6.45	9.90
2080	10" x 12"	G	↓	100	.240			8.10		8.10	12.50
2100	Ceiling joists	G	2 Clab	800	.020			.66		.66	1.02
2150	Wall framing, interior	G		1230	.013	↓		.43		.43	.66
2160	Sub-floor	G		2000	.008	S.F.		.26		.26	.41
2170	Floor joists	G		2000	.008	L.F.		.26		.26	.41
2200	Wood siding (no lead or asbestos)	G		1300	.012	S.F.		.41		.41	.63
2300	Wall framing, exterior	G		1600	.010	L.F.		.33		.33	.51
2400	Stair risers	G		53	.302	Ea.		10		10	15.40
2500	Posts	G	↓	800	.020	L.F.		.66		.66	1.02
3000	Deconstruction of exterior brick walls										
3010	Exterior brick walls, first floor	G	2 Clab	200	.080	S.F.		2.65		2.65	4.08
3020	Second floor	G		64	.250	"		8.30		8.30	12.75
3030	Brick chiminey	G	↓	100	.160	C.F.		5.30		5.30	8.15
4000	Deconstruction of concrete										
4010	Slab on grade, 4" thick, plain concrete	G	B-9	500	.080	S.F.		2.68	.45	3.13	4.62
4020	Wire mesh reinforced	G		470	.085			2.85	.48	3.33	4.93
4030	Rod reinforced	G		400	.100			3.35	.56	3.91	5.75
4110	Foundation wall, 6" thick, plain concrete	G		160	.250			8.40	1.41	9.81	14.45
4120	8" thick	G		140	.286			9.55	1.61	11.16	16.50
4130	10" thick	G	↓	120	.333	↓		11.15	1.87	13.02	19.25
9000	Deconstruction process, support equipment as needed										
9010	Daily use, portal to portal, 12-ton truck-mounted hydraulic crane crew	G	A-3H	1	8	Day		355	870	1,225	1,500
9020	Daily use, skid steer and operator	G	A-3C	1	8			330	291	621	815
9030	Daily use, backhoe 48 H. P., operator and labor	G	"	1	8	↓		330	291	621	815

02 42 Removal and Salvage of Construction Material

02 42 10 – Building Deconstruction

02 42 10.30 Deconstruction Material Handling		Crew	Daily Output	Labor-Hours	Unit	2010 Bare Costs Material	Labor	Equipment	Total	Total Incl O&P
0010	**DECONSTRUCTION MATERIAL HANDLING**									
0012	Buildings one or two stories only									
0100	Clean and stack brick on pallet G	2 Clab	1200	.013	Ea.		.44		.44	.68
0200	Haul 50′ and load rough lumber up to 2″ x 8″ size G		2000	.008	″		.26		.26	.41
0210	Lumber larger than 2″ x 8″ G		3200	.005	B.F.		.17		.17	.26
0300	Finish wood for recycling stack and wrap per pallet G		8	2	Ea.	28	66		94	133
1000	See section 02 41 19.23 for bulk material handling									

02 43 Structure Moving

02 43 13 – Structure Relocation

02 43 13.13 Building Relocation

		Crew	Daily Output	Labor-Hours	Unit	Material	Labor	Equipment	Total	Total Incl O&P
0010	**BUILDING RELOCATION**									
0011	One day move, up to 24′ wide									
0020	Reset on new foundation, patch & hook-up, average move				Total					11,220
0040	Wood or steel frame bldg., based on ground floor area G	B-4	185	.259	S.F.		8.70	2.69	11.39	16.30
0060	Masonry bldg., based on ground floor area G	″	137	.350			11.70	3.63	15.33	22
0200	For 24′ to 42′ wide, add									15%
0220	For each additional day on road, add G	B-4	1	48	Day		1,600	495	2,095	3,025
0240	Construct new basement, move building, 1 day									
0300	move, patch & hook-up, based on ground floor area G	B-3	155	.310	S.F.	9.10	10.85	15.55	35.50	43.50

02 56 Site containment

02 56 13 – Waste containment

02 56 13.10 Containment of Hazardous Waste

		Crew	Daily Output	Labor-Hours	Unit	Material	Labor	Equipment	Total	Total Incl O&P
0010	**CONTAINMENT OF HAZARDOUS WASTE**									
0020	OSHA hazard level C									
0030	OSHA Hazard level D decrease labor and equipment, deduct						-45%	-45%		
0035	OSHA Hazard level B increase labor and equipment, add						22%	22%		
0040	OSHA Hazard level A increase labor and equipment, add						71%	71%		
0100	Excavation of contaminated soil & waste									
0105	Includes one respirator filter and two disposable suits per work day									
0110	3/4 C.Y. excavator to 10 feet deep	B-12F	51	.314	B.C.Y.	1.64	12.15	13.30	27.09	35
0120	Labor crew to 6′ deep	B-2	19	2.105		11	70.50		81.50	121
0130	6′ - 12′ deep	″	12	3.333		17.40	112		129.40	191
0200	Move contaminated soil/waste upto 150′ on-site with 2.5 C.Y. loader	B-10T	300	.040	L.C.Y.	.28	1.59	1.45	3.32	4.31
0210	300′	″	186	.065	″	.45	2.56	2.34	5.35	6.95
0300	Secure burial cell construction									
0310	Various liner and cover materials									
0400	Very low density polyethelene (VLDPE)									
0410	50 mil top cover	B-47H	4000	.008	S.F.	.36	.34	.07	.77	1.01
0420	80 mil liner	″	4000	.008	″	.36	.34	.07	.77	1.01
0500	Chlorosulfunated polyethylene									
0510	36 mil hypalon top cover	B-47H	4000	.008	S.F.	1.17	.34	.07	1.58	1.90
0520	45 mil hypalon liner	″	4000	.008	″	1.23	.34	.07	1.64	1.96
0600	Polyvinyl chloride (PVC)									
0610	60 mil top cover	B-47H	4000	.008	S.F.	.64	.34	.07	1.05	1.32
0620	80 mil liner	″	4000	.008	″	.80	.34	.07	1.21	1.49
0700	Rough textured H.D. polyethylene (HDPE)									
0710	40 mil top cover	B-47H	4000	.008	S.F.	.35	.34	.07	.76	1

02 56 Site containment

02 56 13 – Waste containment

02 56 13.10 Containment of Hazardous Waste		Crew	Daily Output	Labor-Hours	Unit	Material	2010 Bare Costs Labor	Equipment	Total	Total Incl O&P
0720	60 mil top cover	B-47H	4000	.008	S.F.	.38	.34	.07	.79	1.03
0722	60 mil liner		4000	.008		.34	.34	.07	.75	.98
0730	80 mil liner		3800	.008		.41	.36	.08	.85	1.10
1000	3/4" crushed stone, 6" deep ballast around liner	B-6	30	.800	L.C.Y.	35	28.50	11.25	74.75	94.50
1100	Hazardous waste, ballast cover with common borrow material	B-63	56	.714		7.40	25	2.89	35.29	49.50
1110	Mixture of common borrow & topsoil		56	.714		15.20	25	2.89	43.09	58
1120	Bank sand		56	.714		10.55	25	2.89	38.44	53
1130	Medium priced clay		44	.909		26.50	31.50	3.67	61.67	81.50
1140	Mixture of common borrow & medium priced clay		56	.714		16.85	25	2.89	44.74	59.50

02 58 Snow Control

02 58 13 – Snow Fencing

02 58 13.10 Snow Fencing System

		Crew	Daily Output	Labor-Hours	Unit	Material	Labor	Equipment	Total	Total Incl O&P
0010	**SNOW FENCING SYSTEM**									
7001	Snow fence on steel posts 10' O.C., 4' high	B-1	500	.048	L.F.	2.62	1.62		4.24	5.40

02 65 Underground Storage Tank Removal

02 65 10 – Underground Tank and Contaminated Soil Removal

02 65 10.30 Removal of Underground Storage Tanks

			Crew	Daily Output	Labor-Hours	Unit	Material	Labor	Equipment	Total	Total Incl O&P
0010	**REMOVAL OF UNDERGROUND STORAGE TANKS**	R026510-20									
0011	Petroleum storage tanks, non-leaking										
0100	Excavate & load onto trailer										
0110	3000 gal. to 5000 gal. tank	G	B-14	4	12	Ea.		420	84.50	504.50	735
0120	6000 gal. to 8000 gal. tank	G	B-3A	3	13.333			470	340	810	1,100
0130	9000 gal. to 12000 gal. tank	G	"	2	20			700	510	1,210	1,625
0190	Known leaking tank, add					%				100%	100%
0200	Remove sludge, water and remaining product from tank bottom										
0201	of tank with vacuum truck										
0300	3000 gal. to 5000 gal. tank	G	A-13	5	1.600	Ea.		66	155	221	270
0310	6000 gal. to 8000 gal. tank	G		4	2			82.50	194	276.50	335
0320	9000 gal. to 12000 gal. tank	G		3	2.667			110	259	369	450
0390	Dispose of sludge off-site, average					Gal.				6	6.60
0400	Insert inert solid CO_2 "dry ice" into tank										
0401	For cleaning/transporting tanks (1.5 lbs./100 gal. cap)	G	1 Clab	500	.016	Lb.	1.72	.53		2.25	2.71
0403	Insert solid carbon dioxide, 1.5 lbs./100 gal.	G	"	400	.020	"	1.72	.66		2.38	2.91
0503	Disconnect and remove piping	G	1 Plum	160	.050	L.F.		2.60		2.60	3.90
0603	Transfer liquids, 10% of volume	G	"	1600	.005	Gal.		.26		.26	.39
0703	Cut accessway into underground storage tank	G	1 Clab	5.33	1.501	Ea.		49.50		49.50	76.50
0813	Remove sludge, wash and wipe tank, 500 gal.	G	1 Plum	8	1			52		52	78
0823	3,000 gal.	G		6.67	1.199			62.50		62.50	93.50
0833	5,000 gal.	G		6.15	1.301			67.50		67.50	101
0843	8,000 gal.	G		5.33	1.501			78		78	117
0853	10,000 gal.	G		4.57	1.751			91		91	137
0863	12,000 gal.	G		4.21	1.900			99		99	148
1020	Haul tank to certified salvage dump, 100 miles round trip										
1023	3000 gal. to 5000 gal. tank					Ea.				700	770
1026	6000 gal. to 8000 gal. tank									800	880
1029	9,000 gal. to 12,000 gal. tank									1,000	1,100
1100	Disposal of contaminated soil to landfill										

02 65 Underground Storage Tank Removal

02 65 10 – Underground Tank and Contaminated Soil Removal

02 65 10.30 Removal of Underground Storage Tanks		Crew	Daily Output	Labor-Hours	Unit	2010 Bare Costs Material	Labor	Equipment	Total	Total Incl O&P
1110	Minimum				C.Y.				120	132
1111	Maximum				"				380	418
1120	Disposal of contaminated soil to									
1121	bituminous concrete batch plant									
1130	Minimum				C.Y.				75	83
1131	Maximum				"				120	132
1203	Excavate, pull, & load tank, backfill hole, 8,000 gal. + G	B-12C	.50	32	Ea.		1,250	2,650	3,900	4,775
1213	Haul tank to certified dump, 100 miles rt, 8,000 gal. + G	B-34K	1	8			265	920	1,185	1,425
1223	Excavate, pull, & load tank, backfill hole, 500 gal. G	B-11C	1	16			610	340	950	1,300
1233	Excavate, pull, & load tank, backfill hole, 3,000 – 5,000 gal. G	B-11M	.50	32			1,225	775	2,000	2,700
1243	Haul tank to certified dump, 100 miles rt, 500 gal. G	B-34L	1	8	↓		330	240	570	760
2010	Decontamination of soil on site incl poly tarp on top/bottom									
2011	Soil containment berm, and chemical treatment									
2020	Minimum G	B-11C	100	.160	C.Y.	7.05	6.10	3.38	16.53	20.50
2021	Maximum G	"	100	.160		9.15	6.10	3.38	18.63	23
2050	Disposal of decontaminated soil, minimum								75	83
2055	Maximum				↓				150	165

02 81 Transportation and Disposal of Hazardous Materials

02 81 20 – Hazardous Waste Handling

02 81 20.10 Hazardous Waste Cleanup/Pickup/Disposal

		Crew	Daily Output	Labor-Hours	Unit	Material	Labor	Equipment	Total	Total Incl O&P
0010	**HAZARDOUS WASTE CLEANUP/PICKUP/DISPOSAL**									
0100	For contractor rental equipment, i.e., Dozer,									
0110	Front end loader, Dump truck, etc., see 01 54 33 Reference Section									
1000	Solid pickup									
1100	55 gal. drums				Ea.				230	253
1120	Bulk material, minimum				Ton				175	193
1130	Maximum				"				580	638
1200	Transportation to disposal site									
1220	Truckload = 80 drums or 25 C.Y. or 18 tons									
1260	Minimum				Mile				3.60	3.96
1270	Maximum				"				6.50	7.15
3000	Liquid pickup, vacuum truck, stainless steel tank									
3100	Minimum charge, 4 hours									
3110	1 compartment, 2200 gallon				Hr.				125	138
3120	2 compartment, 5000 gallon				"				175	193
3400	Transportation in 6900 gallon bulk truck				Mile				6.50	7.15
3410	In teflon lined truck				"				8	8.80
5000	Heavy sludge or dry vacuumable material				Hr.				125	138
6000	Dumpsite disposal charge, minimum				Ton				125	138
6020	Maximum				"				450	495

02 82 Asbestos Remediation

02 82 13 – Asbestos Abatement

02 82 13.41 Asbestos Abatement Equip.		Crew	Daily Output	Labor-Hours	Unit	2010 Bare Costs Material	Labor	Equipment	Total	Total Incl O&P
0010	**ASBESTOS ABATEMENT EQUIP.** R028213-20									
0011	Equipment and supplies, buy									
0200	Air filtration device, 2000 C.F.M.				Ea.	795			795	875
0250	Large volume air sampling pump, minimum					355			355	390
0260	Maximum					490			490	540
0300	Airless sprayer unit, 2 gun					4,800			4,800	5,300
0350	Light stand, 500 watt					52			52	57.50
0400	Personal respirators									
0410	Negative pressure, 1/2 face, dual operation, min.				Ea.	25.50			25.50	28
0420	Maximum					27.50			27.50	30.50
0450	P.A.P.R., full face, minimum					147			147	162
0460	Maximum					147			147	162
0470	Supplied air, full face, incl. air line, minimum					163			163	179
0480	Maximum					288			288	315
0500	Personnel sampling pump, minimum					221			221	244
1500	Power panel, 20 unit, incl. G.F.I.					570			570	625
1600	Shower unit, including pump and filters					1,175			1,175	1,275
1700	Supplied air system (type C)					9,800			9,800	10,800
1750	Vacuum cleaner, HEPA, 16 gal., stainless steel, wet/dry					1,050			1,050	1,150
1760	55 gallon					1,700			1,700	1,875
1800	Vacuum loader, 9 – 18 ton/hr					90,000			90,000	99,000
1900	Water atomizer unit, including 55 gal. drum					230			230	253
2000	Worker protection, whole body, foot, head cover & gloves, plastic					8.70			8.70	9.60
2500	Respirator, single use					28.50			28.50	31.50
2550	Cartridge for respirator					6.90			6.90	7.60
2570	Glove bag, 7 mil, 50" x 64"					17.25			17.25	19
2580	10 mil, 44" x 60"					74.50			74.50	82
2590	6 mil, 44" x 60"					74.50			74.50	82
3000	HEPA vacuum for work area, minimum					1,325			1,325	1,475
3050	Maximum					1,400			1,400	1,525
6000	Disposable polyethylene bags, 6 mil, 3 C.F.					2.26			2.26	2.49
6300	Disposable fiber drums, 3 C.F.					16.30			16.30	17.95
6400	Pressure sensitive caution labels, 3" x 5"					3.11			3.11	3.42
6450	11" x 17"					7.05			7.05	7.75
6500	Negative air machine, 1800 C.F.M.					810			810	890

02 82 13.42 Preparation of Asbestos Containment Area		Crew	Daily Output	Labor-Hours	Unit	Material	Labor	Equipment	Total	Total Incl O&P
0010	**PREPARATION OF ASBESTOS CONTAINMENT AREA**									
0100	Pre-cleaning, HEPA vacuum and wet wipe, flat surfaces	A-9	12000	.005	S.F.	.02	.24		.26	.40
0200	Protect carpeted area, 2 layers 6 mil poly on 3/4" plywood	"	1000	.064		2	2.92		4.92	6.75
0300	Separation barrier, 2" x 4" @ 16", 1/2" plywood ea. side, 8' high	2 Carp	400	.040		2	1.66		3.66	4.76
0310	12' high		320	.050		2	2.08		4.08	5.40
0320	16' high		200	.080		2	3.32		5.32	7.30
0400	Personnel decontam. chamber, 2" x 4" @ 16", 3/4" ply ea. side		280	.057		3	2.37		5.37	6.95
0450	Waste decontam. chamber, 2" x 4" studs @ 16", 3/4" ply ea. side		360	.044		3	1.85		4.85	6.15
0500	Cover surfaces with polyethylene sheeting									
0501	Including glue and tape									
0550	Floors, each layer, 6 mil	A-9	8000	.008	S.F.	.05	.36		.41	.63
0551	4 mil		9000	.007		.04	.32		.36	.54
0560	Walls, each layer, 6 mil		6000	.011		.05	.49		.54	.82
0561	4 mil		7000	.009		.04	.42		.46	.69
0570	For heights above 12', add						20%			
0575	For heights above 20', add						30%			

02 82 13.42 Preparation of Asbestos Containment Area		Crew	Daily Output	Labor-Hours	Unit	Material	2010 Bare Costs Labor	Equipment	Total	Total Incl O&P
0580	For fire retardant poly, add					100%				
0590	For large open areas, deduct					10%	20%			
0600	Seal floor penetrations with foam firestop to 36 Sq. In.	2 Carp	200	.080	Ea.	7.60	3.32		10.92	13.50
0610	36 Sq. In. to 72 Sq. In.		125	.128		15.25	5.30		20.55	25
0615	72 Sq. In. to 144 Sq. In.		80	.200		30.50	8.30		38.80	46.50
0620	Wall penetrations, to 36 square inches		180	.089		7.60	3.69		11.29	14.10
0630	36 Sq. In. to 72 Sq. In.		100	.160		15.25	6.65		21.90	27
0640	72 Sq. In. to 144 Sq. In.		60	.267		30.50	11.10		41.60	50.50
0800	Caulk seams with latex	1 Carp	230	.035	L.F.	.15	1.45		1.60	2.40
0900	Set up neg. air machine, 1-2k C.F.M. /25 M.C.F. volume	1 Asbe	4.30	1.860	Ea.		84.50		84.50	132

02 82 13.43 Bulk Asbestos Removal		Crew	Daily Output	Labor-Hours	Unit	Material	Labor	Equipment	Total	Total Incl O&P
0010	**BULK ASBESTOS REMOVAL**									
0020	Includes disposable tools and 2 suits and 1respirator filter/day/worker									
0100	Beams, W 10 x 19	A-9	235	.272	L.F.	.83	12.40		13.23	20
0110	W 12 x 22		210	.305		.93	13.90		14.83	22.50
0120	W 14 x 26		180	.356		1.08	16.20		17.28	26
0130	W 16 x 31		160	.400		1.22	18.25		19.47	30
0140	W 18 x 40		140	.457		1.39	21		22.39	34
0150	W 24 x 55		110	.582		1.77	26.50		28.27	43.50
0160	W 30 x 108		85	.753		2.29	34.50		36.79	56
0170	W 36 x 150		72	.889		2.70	40.50		43.20	66
0200	Boiler insulation		480	.133	S.F.	.48	6.10		6.58	9.95
0210	With metal lath, add				%				50%	
0300	Boiler breeching or flue insulation	A-9	520	.123	S.F.	.37	5.60		5.97	9.15
0310	For active boiler, add				%				100%	
0400	Duct or AHU insulation	A-10B	440	.073	S.F.	.22	3.32		3.54	5.40
0500	Duct vibration isolation joints, up to 24 Sq. In. duct	A-9	56	1.143	Ea.	3.48	52		55.48	85
0520	25 Sq. In. to 48 Sq. In. duct		48	1.333		4.06	61		65.06	99
0530	49 Sq. In. to 76 Sq. In. duct		40	1.600		4.87	73		77.87	118
0600	Pipe insulation, air cell type, up to 4" diameter pipe		900	.071	L.F.	.22	3.24		3.46	5.30
0610	4" to 8" diameter pipe		800	.080		.24	3.65		3.89	5.90
0620	10" to 12" diameter pipe		700	.091		.28	4.17		4.45	6.80
0630	14" to 16" diameter pipe		550	.116		.35	5.30		5.65	8.65
0650	Over 16" diameter pipe		650	.098	S.F.	.30	4.49		4.79	7.35
0700	With glove bag up to 3" diameter pipe		200	.320	L.F.	6.05	14.60		20.65	29
1000	Pipe fitting insulation up to 4" diameter pipe		320	.200	Ea.	.61	9.10		9.71	14.85
1100	6" to 8" diameter pipe		304	.211		.64	9.60		10.24	15.60
1110	10" to 12" diameter pipe		192	.333		1.01	15.20		16.21	24.50
1120	14" to 16" diameter pipe		128	.500		1.52	23		24.52	37
1130	Over 16" diameter pipe		176	.364	S.F.	1.11	16.60		17.71	27
1200	With glove bag, up to 8" diameter pipe		75	.853	L.F.	6.55	39		45.55	67.50
2000	Scrape foam fireproofing from flat surface		2400	.027	S.F.	.08	1.22		1.30	1.98
2100	Irregular surfaces		1200	.053		.16	2.43		2.59	3.96
3000	Remove cementitious material from flat surface		1800	.036		.11	1.62		1.73	2.64
3100	Irregular surface		1000	.064		.14	2.92		3.06	4.69
4000	Scrape acoustical coating/fireproofing, from ceiling		3200	.020		.06	.91		.97	1.49
5000	Remove VAT from floor by hand		2400	.027		.08	1.22		1.30	1.98
5100	By machine	A-11	4800	.013		.04	.61	.01	.66	1
5150	For 2 layers, add				%				50%	
6000	Remove contaminated soil from crawl space by hand	A-9	400	.160	C.F.	.49	7.30		7.79	11.90
6100	With large production vacuum loader	A-12	700	.091	"	.28	4.17	1.11	5.56	8.05
7000	Radiator backing, not including radiator removal	A-9	1200	.053	S.F.	.16	2.43		2.59	3.96

02 82 Asbestos Remediation

02 82 13 – Asbestos Abatement

02 82 13.43 Bulk Asbestos Removal

		Crew	Daily Output	Labor-Hours	Unit	Material	2010 Bare Costs Labor	Equipment	Total	Total Incl O&P
8000	Cement-asbestos transite board	2 Asbe	1000	.016	S.F.	.14	.73		.87	1.28
8100	Transite shingle siding	A-10B	750	.043		.21	1.95		2.16	3.26
8200	Shingle roofing	"	2000	.016		.08	.73		.81	1.23
8250	Built-up, no gravel, non-friable	B-2	1400	.029		.08	.96		1.04	1.57
8300	Asbestos millboard	2 Asbe	1000	.016	↓	.08	.73		.81	1.22
9000	For type B (supplied air) respirator equipment, add				%					10%

02 82 13.44 Demolition In Asbestos Contaminated Area

		Crew	Daily Output	Labor-Hours	Unit	Material	Labor	Equipment	Total	Total Incl O&P
0010	**DEMOLITION IN ASBESTOS CONTAMINATED AREA**									
0200	Ceiling, including suspension system, plaster and lath	A-9	2100	.030	S.F.	.09	1.39		1.48	2.26
0210	Finished plaster, leaving wire lath		585	.109		.33	4.99		5.32	8.10
0220	Suspended acoustical tile		3500	.018		.06	.83		.89	1.36
0230	Concealed tile grid system		3000	.021		.06	.97		1.03	1.58
0240	Metal pan grid system		1500	.043		.13	1.95		2.08	3.16
0250	Gypsum board		2500	.026	↓	.08	1.17		1.25	1.90
0260	Lighting fixtures up to 2′ x 4′	↓	72	.889	Ea.	2.70	40.50		43.20	66
0400	Partitions, non load bearing									
0410	Plaster, lath, and studs	A-9	690	.093	S.F.	.85	4.23		5.08	7.55
0450	Gypsum board and studs	"	1390	.046	"	.14	2.10		2.24	3.41
9000	For type B (supplied air) respirator equipment, add				%					10%

02 82 13.45 OSHA Testing

		Crew	Daily Output	Labor-Hours	Unit	Material	Labor	Equipment	Total	Total Incl O&P
0010	**OSHA TESTING**									
0100	Certified technician, minimum				Day				240	264
0110	Maximum				"				320	352
0200	Personal sampling, PCM analysis, NIOSH 7400, minimum	1 Asbe	8	1	Ea.	2.75	45.50		48.25	74
0210	Maximum	"	4	2	"	3	91		94	145
0300	Industrial hygienist, minimum				Day				320	352
0310	Maximum				"				480	528
1000	Cleaned area samples	1 Asbe	8	1	Ea.	2.63	45.50		48.13	74
1100	PCM air sample analysis, NIOSH 7400, minimum		8	1		30.50	45.50		76	105
1110	Maximum	↓	4	2		3.20	91		94.20	146
1200	TEM air sample analysis, NIOSH 7402, minimum								100	125
1210	Maximum				↓				400	500

02 82 13.46 Decontamination of Asbestos Containment Area

		Crew	Daily Output	Labor-Hours	Unit	Material	Labor	Equipment	Total	Total Incl O&P
0010	**DECONTAMINATION OF ASBESTOS CONTAINMENT AREA**									
0100	Spray exposed substrate with surfactant (bridging)									
0200	Flat surfaces	A-9	6000	.011	S.F.	.36	.49		.85	1.16
0250	Irregular surfaces		4000	.016	"	.31	.73		1.04	1.47
0300	Pipes, beams, and columns		2000	.032	L.F.	.56	1.46		2.02	2.89
1000	Spray encapsulate polyethylene sheeting		8000	.008	S.F.	.31	.36		.67	.91
1100	Roll down polyethylene sheeting		8000	.008	"		.36		.36	.57
1500	Bag polyethylene sheeting		400	.160	Ea.	.77	7.30		8.07	12.20
2000	Fine clean exposed substrate, with nylon brush		2400	.027	S.F.		1.22		1.22	1.89
2500	Wet wipe substrate		4800	.013			.61		.61	.95
2600	Vacuum surfaces, fine brush	↓	6400	.010	↓		.46		.46	.71
3000	Structural demolition									
3100	Wood stud walls	A-9	2800	.023	S.F.		1.04		1.04	1.62
3500	Window manifolds, not incl. window replacement		4200	.015			.70		.70	1.08
3600	Plywood carpet protection	↓	2000	.032	↓		1.46		1.46	2.27
4000	Remove custom decontamination facility	A-10A	8	3	Ea.	15	137		152	230
4100	Remove portable decontamination facility	3 Asbe	12	2	"	12.75	91		103.75	156
5000	HEPA vacuum, shampoo carpeting	A-9	4800	.013	S.F.	.05	.61		.66	1.01
9000	Final cleaning of protected surfaces	A-10A	8000	.003	"		.14		.14	.21

02 82 Asbestos Remediation

02 82 13 – Asbestos Abatement

02 82 13.47 Asbestos Waste Pkg., Handling, and Disp.		Crew	Daily Output	Labor-Hours	Unit	2010 Bare Costs Material	Labor	Equipment	Total	Total Incl O&P
0010	ASBESTOS WASTE PACKAGING, HANDLING, AND DISPOSAL									
0100	Collect and bag bulk material, 3 C.F. bags, by hand	A-9	400	.160	Ea.	2.26	7.30		9.56	13.85
0200	Large production vacuum loader	A-12	880	.073		.80	3.32	.88	5	7
1000	Double bag and decontaminate	A-9	960	.067		2.26	3.04		5.30	7.20
2000	Containerize bagged material in drums, per 3 C.F. drum	"	800	.080		16.30	3.65		19.95	23.50
3000	Cart bags 50' to dumpster	2 Asbe	400	.040	↓		1.82		1.82	2.83
5000	Disposal charges, not including haul, minimum				C.Y.				50	55
5020	Maximum				"				170	187
5100	Remove refrigerant from system	1 Stpi	40	.200	Lb.		10.40		10.40	15.55
9000	For type B (supplied air) respirator equipment, add				%					10%

02 82 13.48 Asbestos Encapsulation With Sealants		Crew	Daily Output	Labor-Hours	Unit	Material	Labor	Equipment	Total	Total Incl O&P
0010	ASBESTOS ENCAPSULATION WITH SEALANTS									
0100	Ceilings and walls, minimum	A-9	21000	.003	S.F.	.27	.14		.41	.52
0110	Maximum		10600	.006		.42	.28		.70	.89
0200	Columns and beams, minimum		13300	.005		.27	.22		.49	.64
0210	Maximum		5325	.012	↓	.47	.55		1.02	1.37
0300	Pipes to 12" diameter including minor repairs, minimum		800	.080	L.F.	.37	3.65		4.02	6.05
0310	Maximum	↓	400	.160	"	1.04	7.30		8.34	12.50

02 83 Lead Remediation

02 83 19 – Lead-Based Paint Remediation

02 83 19.23 Encapsulation of Lead-Based Paint		Crew	Daily Output	Labor-Hours	Unit	Material	Labor	Equipment	Total	Total Incl O&P
0010	ENCAPSULATION OF LEAD-BASED PAINT									
0020	Interior, brushwork, trim, under 6"	1 Pord	240	.033	L.F.	2.46	1.21		3.67	4.51
0030	6" to 12" wide		180	.044		3.28	1.62		4.90	6
0040	Balustrades		300	.027		1.98	.97		2.95	3.62
0050	Pipe to 4" diameter		500	.016		1.19	.58		1.77	2.18
0060	To 8" diameter		375	.021		1.57	.78		2.35	2.88
0070	To 12" diameter		250	.032		2.36	1.16		3.52	4.33
0080	To 16" diameter		170	.047	↓	3.47	1.71		5.18	6.35
0090	Cabinets, ornate design		200	.040	S.F.	2.97	1.45		4.42	5.45
0100	Simple design	↓	250	.032	"	2.36	1.16		3.52	4.33
0110	Doors, 3' x 7', both sides, incl. frame & trim									
0120	Flush	1 Pord	6	1.333	Ea.	30.50	48.50		79	106
0130	French, 10 – 15 lite		3	2.667		6.05	97		103.05	151
0140	Panel		4	2		36.50	72.50		109	148
0150	Louvered	↓	2.75	2.909	↓	33.50	106		139.50	194
0160	Windows, per interior side, per 15 S.F.									
0170	1 to 6 lite	1 Pord	14	.571	Ea.	21	21		42	54
0180	7 to 10 lite		7.50	1.067		23	39		62	83
0190	12 lite		5.75	1.391		31	50.50		81.50	110
0200	Radiators		8	1	↓	73.50	36.50		110	135
0210	Grilles, vents		275	.029	S.F.	2.15	1.06		3.21	3.94
0220	Walls, roller, drywall or plaster		1000	.008		.59	.29		.88	1.08
0230	With spunbonded reinforcing fabric		720	.011		.68	.40		1.08	1.35
0240	Wood		800	.010		.74	.36		1.10	1.35
0250	Ceilings, roller, drywall or plaster		900	.009		.68	.32		1	1.23
0260	Wood		700	.011	↓	.84	.42		1.26	1.54
0270	Exterior, brushwork, gutters and downspouts		300	.027	L.F.	1.98	.97		2.95	3.62
0280	Columns	↓	400	.020	S.F.	1.47	.73		2.20	2.70

02 83 Lead Remediation

02 83 19 – Lead-Based Paint Remediation

02 83 19.23 Encapsulation of Lead-Based Paint		Crew	Daily Output	Labor-Hours	Unit	2010 Bare Costs Material	Labor	Equipment	Total	Total Incl O&P
0290	Spray, siding	1 Pord	600	.013	S.F.	.99	.48		1.47	1.81
0300	Miscellaneous									
0310	Electrical conduit, brushwork, to 2" diameter	1 Pord	500	.016	L.F.	1.19	.58		1.77	2.18
0320	Brick, block or concrete, spray		500	.016	S.F.	1.19	.58		1.77	2.18
0330	Steel, flat surfaces and tanks to 12"		500	.016		1.19	.58		1.77	2.18
0340	Beams, brushwork		400	.020		1.47	.73		2.20	2.70
0350	Trusses		400	.020		1.47	.73		2.20	2.70

02 83 19.26 Removal of Lead-Based Paint

		Crew	Daily Output	Labor-Hours	Unit	Material	Labor	Equipment	Total	Total Incl O&P
0010	**REMOVAL OF LEAD-BASED PAINT** R028319-60									
0011	By chemicals, per application									
0050	Baseboard, to 6" wide	1 Pord	64	.125	L.F.	.66	4.54		5.20	7.50
0070	To 12" wide		32	.250	"	1.32	9.10		10.42	15
0200	Balustrades, one side		28	.286	S.F.	1.32	10.40		11.72	16.90
1400	Cabinets, simple design		32	.250		1.32	9.10		10.42	15
1420	Ornate design		25	.320		1.32	11.65		12.97	18.75
1600	Cornice, simple design		60	.133		1.32	4.85		6.17	8.65
1620	Ornate design		20	.400		5.05	14.55		19.60	27
2800	Doors, one side, flush		84	.095		1	3.46		4.46	6.25
2820	Two panel		80	.100		1.32	3.64		4.96	6.85
2840	Four panel		45	.178		1.32	6.45		7.77	11.05
2880	For trim, one side, add		64	.125	L.F.	.66	4.54		5.20	7.50
3000	Fence, picket, one side		30	.267	S.F.	1.32	9.70		11.02	15.90
3200	Grilles, one side, simple design		30	.267		1.32	9.70		11.02	15.90
3220	Ornate design		25	.320		1.32	11.65		12.97	18.75
4400	Pipes, to 4" diameter		90	.089	L.F.	1	3.23		4.23	5.90
4420	To 8" diameter		50	.160		2	5.80		7.80	10.85
4440	To 12" diameter		36	.222		3	8.10		11.10	15.30
4460	To 16" diameter		20	.400		4	14.55		18.55	26
4500	For hangers, add		40	.200	Ea.	2.54	7.25		9.79	13.60
4800	Siding		90	.089	S.F.	1.17	3.23		4.40	6.10
5000	Trusses, open		55	.145	SF Face	1.85	5.30		7.15	9.90
6200	Windows, one side only, double hung, 1/1 light, 24" x 48" high		4	2	Ea.	25.50	72.50		98	136
6220	30" x 60" high		3	2.667		34	97		131	182
6240	36" x 72" high		2.50	3.200		41	116		157	218
6280	40" x 80" high		2	4		51	145		196	272
6400	Colonial window, 6/6 light, 24" x 48" high		2	4		51	145		196	273
6420	30" x 60" high		1.50	5.333		68	194		262	365
6440	36" x 72" high		1	8		102	291		393	545
6480	40" x 80" high		1	8		102	291		393	545
6600	8/8 light, 24" x 48" high		2	4		51	145		196	273
6620	40" x 80" high		1	8		102	291		393	545
6800	12/12 light, 24" x 48" high		1	8		102	291		393	545
6820	40" x 80" high		.75	10.667		136	390		526	725
6840	Window frame & trim items, included in pricing above									

02 85 Mold Remediation

02 85 16 – Mold Remediation Preparation and Containment

02 85 16.40 Mold Remediation Plans and Methods		Crew	Daily Output	Labor-Hours	Unit	2010 Bare Costs Material	Labor	Equipment	Total	Total Incl O&P
0010	**MOLD REMEDIATION PLANS AND METHODS**									
0020	Initial inspection, average 3 bedroom home				Total				250	275
0030	Average 5 bedroom home								340	374
0040	Testing, air sample each								250	275
0050	Swab sample								150	165
0060	Tape sample								150	165
0070	Post remediation air test								250	275
0080	Mold abatement plan, average 3 bedroom home								1,300	1,430
0090	Average 5 bedroom home								1,580	1,740
0100	Packup & removal of contents, average 3 bedroom home, excl storage								10,500	11,550
0110	Average 5 bedroom home, excl storage								26,400	29,000
0120	For demolition in mold contaminated areas, see Div. 02 85 33.50									
0130	For personal protection equipment, see Div. 02 82 13.41									

02 85 16.50 Preparation of Mold Containment Area		Crew	Daily Output	Labor-Hours	Unit	Material	Labor	Equipment	Total	Total Incl O&P
0010	**PREPARATION OF MOLD CONTAINMENT AREA**									
0100	Pre-cleaning, HEPA vacuum and wet wipe, flat surfaces	A-9	12000	.005	S.F.	.02	.24		.26	.40
0300	Separation barrier, 2" x 4" @ 16", 1/2" plywood ea. side, 8' high	2 Carp	400	.040		2	1.66		3.66	4.76
0310	12' high		320	.050		2	2.08		4.08	5.40
0320	16' high		200	.080		2	3.32		5.32	7.30
0400	Personnel decontam. chamber, 2" x 4" @ 16", 3/4" ply ea. side		280	.057		3	2.37		5.37	6.95
0450	Waste decontam. chamber, 2" x 4" studs @ 16", 3/4" ply each side		360	.044		3	1.85		4.85	6.15
0500	Cover surfaces with polyethylene sheeting									
0501	Including glue and tape									
0550	Floors, each layer, 6 mil	A-9	8000	.008	S.F.	.06	.36		.42	.64
0551	4 mil		9000	.007		.04	.32		.36	.55
0560	Walls, each layer, 6 mil		6000	.011		.09	.49		.58	.86
0561	4 mil		7000	.009		.07	.42		.49	.73
0570	For heights above 12', add						20%			
0575	For heights above 20', add						30%			
0580	For fire retardant poly, add					100%				
0590	For large open areas, deduct					10%	20%			
0600	Seal floor penetrations with foam firestop to 36 sq. in.	2 Carp	200	.080	Ea.	7.60	3.32		10.92	13.50
0610	36 sq. in. to 72 sq. in.		125	.128		15.25	5.30		20.55	25
0615	72 sq. in. to 144 sq. in.		80	.200		30.50	8.30		38.80	46.50
0620	Wall penetrations, to 36 square inches		180	.089		7.60	3.69		11.29	14.10
0630	36 sq. in. to 72 sq. in.		100	.160		15.25	6.65		21.90	27
0640	72 Sq. in. to 144 sq. in.		60	.267		30.50	11.10		41.60	50.50
0800	Caulk seams with latex caulk	1 Carp	230	.035	L.F.	.15	1.45		1.60	2.40
0900	Set up neg. air machine, 1-2k C.F.M. /25 M.C.F. volume	1 Asbe	4.30	1.860	Ea.		84.50		84.50	132

02 85 33 – Removal and Disposal of Materials with Mold

02 85 33.50 Demolition in Mold Contaminated Area		Crew	Daily Output	Labor-Hours	Unit	Material	Labor	Equipment	Total	Total Incl O&P
0010	**DEMOLITION IN MOLD CONTAMINATED AREA**									
0200	Ceiling, including suspension system, plaster and lath	A-9	2100	.030	S.F.	.09	1.39		1.48	2.26
0210	Finished plaster, leaving wire lath		585	.109		.33	4.99		5.32	8.10
0220	Suspended acoustical tile		3500	.018		.06	.83		.89	1.36
0230	Concealed tile grid system		3000	.021		.06	.97		1.03	1.58
0240	Metal pan grid system		1500	.043		.13	1.95		2.08	3.16
0250	Gypsum board		2500	.026		.08	1.17		1.25	1.90
0255	Plywood		2500	.026		.08	1.17		1.25	1.90
0260	Lighting fixtures up to 2' x 4'		72	.889	Ea.	2.70	40.50		43.20	66
0400	Partitions, non load bearing									
0410	Plaster, lath, and studs	A-9	690	.093	S.F.	.85	4.23		5.08	7.55

02 85 Mold Remediation

02 85 33 – Removal and Disposal of Materials with Mold

02 85 33.50 Demolition in Mold Contaminated Area		Crew	Daily Output	Labor-Hours	Unit	Material	2010 Bare Costs Labor	Equipment	Total	Total Incl O&P
0450	Gypsum board and studs	A-9	1390	.046	S.F.	.14	2.10		2.24	3.41
0465	Carpet & pad		1390	.046		.14	2.10		2.24	3.41
0600	Pipe insulation, air cell type, up to 4" diameter pipe		900	.071	L.F.	.22	3.24		3.46	5.30
0610	4" to 8" diameter pipe		800	.080		.24	3.65		3.89	5.90
0620	10" to 12" diameter pipe		700	.091		.28	4.17		4.45	6.80
0630	14" to 16" diameter pipe		550	.116		.35	5.30		5.65	8.65
0650	Over 16" diameter pipe		650	.098	S.F.	.30	4.49		4.79	7.35
9000	For type B (supplied air) respirator equipment, add				%					10%

02 91 Chemical Sampling, Testing and Analysis

02 91 10 – MONITORING, SAMPLING, TESTING AND ANALYSIS

02 91 10.10 Monitoring, Chemical Sampling, Testing and Analysis		Crew	Daily Output	Labor-Hours	Unit	Material	Labor	Equipment	Total	Total Incl O&P
0010	**MONITORING, CHEMICAL SAMPLING, TESTING AND ANALYSIS**									
0015	Field Sampling of waste									
0100	Field samples, sample collection, sludge	1 Skwk	32	.250	Ea.		10.65		10.65	16.40
0110	Contaminated soils	"	32	.250	"		10.65		10.65	16.40
0200	Vials and bottles									
0210	32 Oz. clear wide mouth jar (case of 12)				Ea.	41.50			41.50	46
0220	32 Oz. Boston round bottle (case of 12)					29			29	32
0230	32 Oz. HDPE bottle (case of 12)					17.60			17.60	19.35
0300	Laboratory analytical services									
0310	Laboratory testing 13 metals				Ea.	190			190	209
0312	13 metals + mercury					223			223	245
0314	8 metals					160			160	176
0316	Mercury only					37.50			37.50	41.50
0318	Single metal (only Cs, Li, Sr, Ta)					20			20	22
0320	Single metal (excludes Hg, Cs, Li, Sr, Ta)					17.50			17.50	19.25
0400	Hydrocarbons standard					90			90	99
0410	Hydrocarbons fingerprint					163			163	179
0500	Radioactivity gross alpha					150			150	165
0510	Gross alpha & beta					150			150	165
0520	Radium 226					85			85	93.50
0530	Radium 228					125			125	138
0540	Radon					145			145	160
0550	Uranium					75			75	82.50
0600	Volatile organics without GC/MS					140			140	154
0610	Volatile organics including GC/MS					185			185	204
0630	Synthetic organic compounds					1,000			1,000	1,100
0640	Herbicides					213			213	234
0650	Pesticides					135			135	149
0660	PCB's					135			135	149

Estimating Tips

General

- Carefully check all the plans and specifications. Concrete often appears on drawings other than structural drawings, including mechanical and electrical drawings for equipment pads. The cost of cutting and patching is often difficult to estimate. See Subdivision 03 81 for Concrete Cutting, Subdivision 02 41 19.16 for Cutout Demolition, Subdivision 03 05 05.10 for Concrete Demolition, and Subdivision 02 41 19.23 for Rubbish Handling (handling, loading and hauling of debris).
- Always obtain concrete prices from suppliers near the job site. A volume discount can often be negotiated, depending upon competition in the area. Remember to add for waste, particularly for slabs and footings on grade.

03 10 00 Concrete Forming and Accessories

- A primary cost for concrete construction is forming. Most jobs today are constructed with prefabricated forms. The selection of the forms best suited for the job and the total square feet of forms required for efficient concrete forming and placing are key elements in estimating concrete construction. Enough forms must be available for erection to make efficient use of the concrete placing equipment and crew.
- Concrete accessories for forming and placing depend upon the systems used. Study the plans and specifications to ensure that all special accessory requirements have been included in the cost estimate, such as anchor bolts, inserts, and hangers.
- Included within costs for forms-in-place are all necessary bracing and shoring.

03 20 00 Concrete Reinforcing

- Ascertain that the reinforcing steel supplier has included all accessories, cutting, bending, and an allowance for lapping, splicing, and waste. A good rule of thumb is 10% for lapping, splicing, and waste. Also, 10% waste should be allowed for welded wire fabric.
- The unit price items in the subdivision for Reinforcing In Place include the labor to install accessories such as beam and slab bolsters, high chairs, and bar ties and tie wire. The material cost for these accessories is not included; they may be obtained from the Accessories Division.

03 30 00 Cast-In-Place Concrete

- When estimating structural concrete, pay particular attention to requirements for concrete additives, curing methods, and surface treatments. Special consideration for climate, hot or cold, must be included in your estimate. Be sure to include requirements for concrete placing equipment, and concrete finishing.
- For accurate concrete estimating, the estimator must consider each of the following major components individually: forms, reinforcing steel, ready-mix concrete, placement of the concrete, and finishing of the top surface. For faster estimating, Subdivision 03 30 53.40 for Concrete-In-Place can be used; here, various items of concrete work are presented that include the costs of all five major components (unless specifically stated otherwise).

03 40 00 Precast Concrete

03 50 00 Cast Decks and Underlayment

- The cost of hauling precast concrete structural members is often an important factor. For this reason, it is important to get a quote from the nearest supplier. It may become economically feasible to set up precasting beds on the site if the hauling costs are prohibitive.

Reference Numbers

Reference numbers are shown in shaded boxes at the beginning of some major classifications. These numbers refer to related items in the Reference Section. The reference information may be an estimating procedure, an alternate pricing method, or technical information.

Note: Not all subdivisions listed here necessarily appear in this publication.

Division 3 - Concrete

03 01 Maintenance of Concrete

03 01 30 – Maintenance of Cast-In-Place Concrete

03 01 30.62 Concrete Patching		Crew	Daily Output	Labor-Hours	Unit	Material	2010 Bare Costs Labor	Equipment	Total	Total Incl O&P
0010	**CONCRETE PATCHING**									
0100	Floors, 1/4" thick, small areas, regular grout	1 Cefi	170	.047	S.F.	1.18	1.87		3.05	4.03
0150	Epoxy grout	"	100	.080	"	7.20	3.18		10.38	12.55
2000	Walls, including chipping, cleaning and epoxy grout									
2100	1/4" deep	1 Cefi	65	.123	S.F.	7.50	4.89		12.39	15.40
2150	1/2" deep		50	.160		15.05	6.35		21.40	26
2200	3/4" deep	↓	40	.200	↓	22.50	7.95		30.45	36.50

03 05 Common Work Results for Concrete

03 05 05 – Selective Concrete Demolition

03 05 05.10 Selective Demolition, Concrete

		Crew	Daily Output	Labor-Hours	Unit	Material	Labor	Equipment	Total	Total Incl O&P
0010	**SELECTIVE DEMOLITION, CONCRETE** R024119-10									
0012	Excludes saw cutting, torch cutting, loading or hauling									
0050	Break up into small pieces, minimum reinforcing	B-9	24	1.667	C.Y.		56	9.35	65.35	96.50
0060	Average reinforcing		16	2.500			84	14.05	98.05	144
0070	Maximum reinforcing	↓	8	5	↓		168	28	196	289
0150	Remove whole pieces, up to 2 tons per piece	E-18	36	1.111	Ea.		51.50	28.50	80	120
0160	2 – 5 tons per piece		30	1.333			62	34.50	96.50	144
0170	5 – 10 tons per piece		24	1.667			77.50	43	120.50	180
0180	10 – 15 tons per piece	↓	18	2.222			103	57	160	240
0250	Precast unit embedded in masonry, up to 1 CF	D-1	16	1			37.50		37.50	57
0260	1 – 2 CF		12	1.333			50.50		50.50	76
0270	2 – 5 CF		10	1.600			60.50		60.50	91
0280	5 – 10 CF	↓	8	2	↓		75.50		75.50	114
0990	For hydrodemolition see 02 41 13.15									

03 05 13 – Basic Concrete Materials

03 05 13.20 Concrete Admixtures and Surface Treatments

		Crew	Daily Output	Labor-Hours	Unit	Material	Labor	Equipment	Total	Total Incl O&P
0010	**CONCRETE ADMIXTURES AND SURFACE TREATMENTS**									
0040	Abrasives, aluminum oxide, over 20 tons				Lb.	1.75			1.75	1.93
0050	1 to 20 tons					1.60			1.60	1.76
0070	Under 1 ton					1.67			1.67	1.84
0100	Silicon carbide, black, over 20 tons					2.44			2.44	2.68
0110	1 to 20 tons					2.57			2.57	2.83
0120	Under 1 ton				↓	2.68			2.68	2.95
0200	Air entraining agent, .7 to 1.5 oz. per bag, 55 gallon drum				Gal.	9.95			9.95	10.95
0220	5 gallon pail					13.75			13.75	15.15
0300	Bonding agent, acrylic latex, 250 S.F. per gallon, 5 gallon pail					19.60			19.60	21.50
0320	Epoxy resin, 80 S.F. per gallon, 3.5 gallon unit				↓	47.50			47.50	52.50
0400	Calcium chloride, 50 lb. bags, TL lots				Ton	580			580	635
0420	Less than truckload lots				Bag	17.95			17.95	19.75
0500	Carbon black, liquid, 2 to 8 lbs. per bag of cement				Lb.	6.45			6.45	7.10
0600	Colors, integral, 2 to 10 lb. per bag of cement, minimum					2.52			2.52	2.77
0610	Average					3.50			3.50	3.85
0620	Maximum				↓	4.88			4.88	5.35
0920	Dustproofing compound, 250 SF/gal, 5 gallon pail				Gal.	7.40			7.40	8.10
1010	Epoxy based, 125 SF/Gal, 5 gallon pail				"	50			50	55
1100	Hardeners, metallic, 55 lb. bags, natural (grey)				Lb.	.85			.85	.93
1200	Colors					1.51			1.51	1.66
1300	Non-metallic, 55 lb. bags, natural grey					.42			.42	.47
1320	Colors				↓	.71			.71	.78

03 05 13 – Basic Concrete Materials

03 05 13.20 Concrete Admixtures and Surface Treatments		Crew	Daily Output	Labor-Hours	Unit	Material	2010 Bare Costs Labor	Equipment	Total	Total Incl O&P
1550	Release agent, for tilt slabs, 5 gallon pail				Gal.	17			17	18.70
1570	For forms, 5 gallon pail					10.50			10.50	11.55
1590	Concrete release agent for forms, 100% biodegradeable, zero VOC, 5 gal pail G					14.75			14.75	16.25
1595	55 gallon drum G					11.75			11.75	12.95
1600	Sealer, hardener and dustproofer, epoxy-based, 125 SF/gal, 5 gallon, min					50			50	55
1620	5 gallon pail, maximum					55			55	60.50
1630	Sealer, solvent-based, 250 SF/gal, 55 gallon drum					17.30			17.30	19.05
1640	5 gallon pail					24.50			24.50	27
1650	Sealer, water based, 350 S.F., 55 gallon drum					20.50			20.50	22.50
1660	5 gallon pail					23.50			23.50	26
1900	Set retarder, 100 SF/gallon, 5 gallon pail				▼	19.70			19.70	21.50
2000	Waterproofing, integral 1 lb. per bag of cement				Lb.	1.52			1.52	1.67
2100	Powdered metallic, 40 lbs. per 100 S.F., minimum					1.79			1.79	1.97
2120	Maximum				▼	2.51			2.51	2.76
3000	For integral colors, 2500 psi (5 bag mix)									
3100	Red, yellow or brown, 1.8 lb. per bag, add				C.Y.	22.50			22.50	25
3200	9.4 lb. per bag, add					118			118	130
3400	Black, 1.8 lb. per bag, add					31.50			31.50	34.50
3500	7.5 lb. per bag, add					131			131	144
3700	Green, 1.8 lb. per bag, add					44			44	48.50
3800	7.5 lb. per bag, add				▼	183			183	201
6000	Concrete ready-mix additives, recycled coal fly ash, mixed at plant G				Ton	56			56	61.50
6010	Recycled blast furnace slag, mixed at plant G				"	87			87	95.50

03 05 13.25 Aggregate

		Crew	Daily Output	Labor-Hours	Unit	Material	Labor	Equipment	Total	Total Incl O&P
0010	**AGGREGATE** R033105-20									
0100	Lightweight vermiculite or perlite, 4 C.F. bag, C.L. lots G				Bag	18.90			18.90	21
0150	L.C.L. lots R033105-40				"	21			21	23
0250	Sand & stone, loaded at pit, crushed bank gravel				Ton	21.50			21.50	23.50
0350	Sand, washed, for concrete R033105-50					17.95			17.95	19.75
0400	For plaster or brick					17.95			17.95	19.75
0450	Stone, 3/4" to 1-1/2"					15.20			15.20	16.70
0470	Round, river stone					36.50			36.50	40
0500	3/8" roofing stone & 1/2" pea stone					23.50			23.50	26
0550	For trucking 10 miles (5 mile R/T), add to the above	B-34B	117	.068			2.27	5.70	7.97	9.75
0600	30 miles (15 mi R/T), add to the above	"	72	.111	▼		3.68	9.30	12.98	15.80
0850	Sand & stone, loaded at pit, crushed bank gravel				C.Y.	30			30	33
0950	Sand, washed, for concrete					25			25	27.50
1000	For plaster or brick					25			25	27.50
1050	Stone, 3/4" to 1-1/2"					29			29	32
1055	Round, river stone					33			33	36
1100	3/8" roofing stone & 1/2" pea stone					39			39	43
1150	For trucking 10 miles (5 mi R/T), add to the above	B-34B	78	.103			3.40	8.55	11.95	14.60
1200	30 miles (15 mi R/T), add to the above	"	48	.167	▼		5.55	13.90	19.45	24
1310	Quartz chips, 50 lb. bags				Cwt.	18.70			18.70	20.50
1330	Silica chips, 50 lb. bags					10.25			10.25	11.30
1410	White marble, 3/8" to 1/2", 50 lb. bags				▼	15.45			15.45	17
1430	3/4"				Ton	78			78	86

03 05 13.30 Cement

		Crew	Daily Output	Labor-Hours	Unit	Material	Labor	Equipment	Total	Total Incl O&P
0010	**CEMENT** R033105-20									
0240	Portland, Type I/II, TL lots, 94 lb bags				Bag	10.50			10.50	11.55
0250	LTL/LCL lots R033105-30				"	12.35			12.35	13.55
0300	Trucked in bulk, per cwt CN				Cwt.	7.80			7.80	8.55

03 05 Common Work Results for Concrete

03 05 13 – Basic Concrete Materials

03 05 13.30 Cement

03 05 13.30 Cement		Crew	Daily Output	Labor-Hours	Unit	Material	2010 Bare Costs Labor	Equipment	Total	Total Incl O&P
0400	Type III, high early strength, TL lots, 94 lb bags CN R033105-40				Bag	12.50			12.50	13.75
0420	L.T.L. or L.C.L. lots					12.40			12.40	13.65
0500	White, type III, high early strength, T.L. or C.L. lots, bags R033105-50					25			25	27.50
0520	L.T.L. or L.C.L. lots					25			25	27.50
0600	White, type I, T.L. or C.L. lots, bags					23.50			23.50	25.50
0620	L.T.L. or L.C.L. lots					24.50			24.50	27

03 05 13.80 Waterproofing and Dampproofing

03 05 13.80		Crew	Daily Output	Labor-Hours	Unit	Material	Labor	Equipment	Total	Total Incl O&P
0010	**WATERPROOFING AND DAMPPROOFING**									
0050	Integral waterproofing, add to cost of regular concrete				C.Y.	9.10			9.10	10.05

03 05 13.85 Winter Protection

03 05 13.85		Crew	Daily Output	Labor-Hours	Unit	Material	Labor	Equipment	Total	Total Incl O&P
0010	**WINTER PROTECTION**									
0012	For heated ready mix, add, minimum				C.Y.	4.25			4.25	4.68
0050	Maximum				"	5.30			5.30	5.85
0100	Temporary heat to protect concrete, 24 hours, minimum	2 Clab	50	.320	M.S.F.	520	10.60		530.60	590
0150	Maximum	"	25	.640	"	680	21		701	780
0200	Temporary shelter for slab on grade, wood frame/polyethylene sheeting									
0201	Build or remove, minimum	2 Carp	10	1.600	M.S.F.	215	66.50		281.50	340
0210	Maximum	"	3	5.333	"	261	222		483	625
0500	Electrically heated pads, 110 volts, 15 watts per S.F., buy				S.F.	6.65			6.65	7.30
0600	20 watts per S.F., buy					8.80			8.80	9.70
0710	Electrically, heated pads, 15 watts/S.F., 20 uses, minimum					.24			.24	.26
0800	Maximum					.39			.39	.43

03 11 Concrete Forming

03 11 13 – Structural Cast-In-Place Concrete Forming

03 11 13.20 Forms In Place, Beams and Girders

03 11 13.20		Crew	Daily Output	Labor-Hours	Unit	Material	Labor	Equipment	Total	Total Incl O&P
0010	**FORMS IN PLACE, BEAMS AND GIRDERS** R031113-40									
0500	Exterior spandrel, job-built plywood, 12" wide, 1 use	C-2	225	.213	SFCA	3.43	8.65		12.08	17.10
0550	2 use R031113-60		275	.175		1.84	7.05		8.89	12.95
0600	3 use		295	.163		1.37	6.60		7.97	11.65
0650	4 use		310	.155		1.12	6.25		7.37	10.90
1000	18" wide, 1 use		250	.192		2.77	7.75		10.52	15.05
1050	2 use		275	.175		1.53	7.05		8.58	12.60
1100	3 use		305	.157		1.11	6.35		7.46	11
1150	4 use		315	.152		.91	6.15		7.06	10.50
1500	24" wide, 1 use		265	.181		2.54	7.35		9.89	14.10
1550	2 use		290	.166		1.42	6.70		8.12	11.90
1600	3 use		315	.152		1.02	6.15		7.17	10.60
1650	4 use		325	.148		.83	6		6.83	10.10
2000	Interior beam, job-built plywood, 12" wide, 1 use		300	.160		4.03	6.50		10.53	14.45
2050	2 use		340	.141		2.01	5.70		7.71	11
2100	3 use		364	.132		1.61	5.35		6.96	10
2150	4 use		377	.127		1.31	5.15		6.46	9.40
2500	24" wide, 1 use		320	.150		2.57	6.05		8.62	12.20
2550	2 use		365	.132		1.44	5.30		6.74	9.80
2600	3 use		385	.125		1.02	5.05		6.07	8.95
2650	4 use		395	.122		.83	4.92		5.75	8.50
3000	Encasing steel beam, hung, job-built plywood, 1 use		325	.148		2.87	6		8.87	12.35
3050	2 use		390	.123		1.58	4.98		6.56	9.45
3100	3 use		415	.116		1.15	4.68		5.83	8.45

03 11 13 – Structural Cast-In-Place Concrete Forming

03 11 13.20 Forms In Place, Beams and Girders	Crew	Daily Output	Labor-Hours	Unit	Material	2010 Bare Costs Labor	Equipment	Total	Total Incl O&P
3150 4 use	C-2	430	.112	SFCA	.93	4.52		5.45	8
3500 Bottoms only, to 30" wide, job-built plywood, 1 use		230	.209		3.61	8.45		12.06	16.95
3550 2 use		265	.181		2.02	7.35		9.37	13.50
3600 3 use		280	.171		1.44	6.95		8.39	12.30
3650 4 use		290	.166		1.17	6.70		7.87	11.65
4000 Sides only, vertical, 36" high, job-built plywood, 1 use		335	.143		6.80	5.80		12.60	16.40
4050 2 use		405	.119		3.73	4.80		8.53	11.50
4100 3 use		430	.112		2.71	4.52		7.23	9.95
4150 4 use		445	.108		2.20	4.37		6.57	9.15
4500 Sloped sides, 36" high, 1 use		305	.157		6.15	6.35		12.50	16.55
4550 2 use		370	.130		3.41	5.25		8.66	11.85
4600 3 use		405	.119		2.46	4.80		7.26	10.10
4650 4 use		425	.113		2	4.57		6.57	9.25
5000 Upstanding beams, 36" high, 1 use		225	.213		9.10	8.65		17.75	23.50
5050 2 use		255	.188		5	7.60		12.60	17.25
5100 3 use		275	.175		3.65	7.05		10.70	14.90
5150 4 use	↓	280	.171	↓	2.96	6.95		9.91	13.95

03 11 13.25 Forms In Place, Columns	Crew	Daily Output	Labor-Hours	Unit	Material	Labor	Equipment	Total	Total Incl O&P
0010 **FORMS IN PLACE, COLUMNS** R031113-40									
0500 Round fiberglass, 4 use per mo., rent, 12" diameter	C-1	160	.200	L.F.	8.15	7.90		16.05	21
0550 16" diameter R031113-60		150	.213		9.70	8.40		18.10	23.50
0600 18" diameter		140	.229		10.85	9		19.85	26
0650 24" diameter		135	.237		13.50	9.35		22.85	29.50
0700 28" diameter		130	.246		15.10	9.70		24.80	31.50
0800 30" diameter		125	.256		15.75	10.10		25.85	33
0850 36" diameter		120	.267		21	10.50		31.50	39
1500 Round fiber tube, recycled paper, 1 use, 8" diameter G		155	.206		1.44	8.15		9.59	14.15
1550 10" diameter G		155	.206		1.99	8.15		10.14	14.75
1600 12" diameter G		150	.213		2.33	8.40		10.73	15.50
1650 14" diameter G		145	.221		3.31	8.70		12.01	17.05
1700 16" diameter G		140	.229		3.99	9		12.99	18.30
1720 18" diameter G		140	.229		4.67	9		13.67	19.05
1750 20" diameter G		135	.237		6.45	9.35		15.80	21.50
1800 24" diameter G		130	.246		8.15	9.70		17.85	24
1850 30" diameter G		125	.256		11.85	10.10		21.95	28.50
1900 36" diameter G		115	.278		14.90	10.95		25.85	33.50
1950 42" diameter G		100	.320		37	12.60		49.60	60
2000 48" diameter G	↓	85	.376	↓	41.50	14.85		56.35	69
2200 For seamless type, add					15%				
3000 Round, steel, 4 use per mo., rent, regular duty, 14" diameter G	C-1	145	.221	L.F.	15.15	8.70		23.85	30
3050 16" diameter G		125	.256		15.50	10.10		25.60	32.50
3100 Heavy duty, 20" diameter G		105	.305		17	12		29	37.50
3150 24" diameter G		85	.376		18.65	14.85		33.50	43.50
3200 30" diameter G		70	.457		21.50	18.05		39.55	51.50
3250 36" diameter G		60	.533		23	21		44	58
3300 48" diameter G		50	.640		68	25		93	114
3350 60" diameter G	↓	45	.711	↓	42	28		70	89
4500 For second and succeeding months, deduct					50%				
5000 Job-built plywood, 8" x 8" columns, 1 use	C-1	165	.194	SFCA	2.27	7.65		9.92	14.30
5050 2 use		195	.164		1.30	6.45		7.75	11.40
5100 3 use		210	.152		.91	6		6.91	10.25
5150 4 use	↓	215	.149	↓	.75	5.85		6.60	9.85

03 11 Concrete Forming

03 11 13 – Structural Cast-In-Place Concrete Forming

03 11 13.25 Forms In Place, Columns		Crew	Daily Output	Labor-Hours	Unit	Material	2010 Bare Costs Labor	Equipment	Total	Total Incl O&P
5500	12" x 12" columns, 1 use	C-1	180	.178	SFCA	2.11	7		9.11	13.10
5550	2 use		210	.152		1.16	6		7.16	10.55
5600	3 use		220	.145		.84	5.75		6.59	9.80
5650	4 use		225	.142		.69	5.60		6.29	9.40
6000	16" x 16" columns, 1 use		185	.173		2.06	6.80		8.86	12.75
6050	2 use		215	.149		1.11	5.85		6.96	10.25
6100	3 use		230	.139		.83	5.50		6.33	9.35
6150	4 use		235	.136		.67	5.35		6.02	9.05
6500	24" x 24" columns, 1 use		190	.168		2.28	6.65		8.93	12.75
6550	2 use		216	.148		1.25	5.85		7.10	10.40
6600	3 use		230	.139		.91	5.50		6.41	9.45
6650	4 use		238	.134		.74	5.30		6.04	8.95
7000	36" x 36" columns, 1 use		200	.160		1.65	6.30		7.95	11.55
7050	2 use		230	.139		.94	5.50		6.44	9.50
7100	3 use		245	.131		.66	5.15		5.81	8.70
7150	4 use	↓	250	.128	↓	.54	5.05		5.59	8.40
7400	Steel framed plywood, based on 50 uses of purchased									
7420	forms, and 4 uses of bracing lumber									
7500	8" x 8" column	C-1	340	.094	SFCA	4.51	3.71		8.22	10.65
7550	10" x 10"		350	.091		3.97	3.61		7.58	9.90
7600	12" x 12"		370	.086		3.36	3.41		6.77	8.95
7650	16" x 16"		400	.080		2.62	3.16		5.78	7.75
7700	20" x 20"		420	.076		2.34	3		5.34	7.20
7750	24" x 24"		440	.073		1.66	2.87		4.53	6.25
7755	30" x 30"		440	.073		2.13	2.87		5	6.75
7760	36" x 36"	↓	460	.070	↓	1.90	2.74		4.64	6.30

03 11 13.30 Forms In Place, Culvert

		Crew	Daily Output	Labor-Hours	Unit	Material	Labor	Equipment	Total	Total Incl O&P
0010	**FORMS IN PLACE, CULVERT** R031113-40									
0015	5' to 8' square or rectangular, 1 use	C-1	170	.188	SFCA	5.55	7.40		12.95	17.55
0050	2 use R031113-60		180	.178		3.42	7		10.42	14.55
0100	3 use		190	.168		2.71	6.65		9.36	13.25
0150	4 use	↓	200	.160	↓	2.36	6.30		8.66	12.35

03 11 13.35 Forms In Place, Elevated Slabs

		Crew	Daily Output	Labor-Hours	Unit	Material	Labor	Equipment	Total	Total Incl O&P
0010	**FORMS IN PLACE, ELEVATED SLABS** R031113-40									
1000	Flat plate, job-built plywood, to 15' high, 1 use R031113-60	C-2	470	.102	S.F.	4.06	4.13		8.19	10.80
1050	2 use		520	.092		2.23	3.74		5.97	8.20
1100	3 use		545	.088		1.63	3.57		5.20	7.30
1150	4 use		560	.086		1.32	3.47		4.79	6.80
1500	15' to 20' high ceilings, 4 use		495	.097		1.53	3.93		5.46	7.75
1600	21' to 35' high ceilings, 4 use		450	.107		2.41	4.32		6.73	9.30
2000	Flat slab, drop panels, job-built plywood, to 15' high, 1 use		449	.107		4.43	4.33		8.76	11.50
2050	2 use		509	.094		2.44	3.82		6.26	8.60
2100	3 use		532	.090		1.77	3.65		5.42	7.60
2150	4 use		544	.088		1.44	3.57		5.01	7.10
2250	15' to 20' high ceilings, 4 use		480	.100		5	4.05		9.05	11.80
2350	20' to 35' high ceilings, 4 use		435	.110		5.90	4.47		10.37	13.35
3000	Floor slab hung from steel beams, 1 use		485	.099		2.73	4.01		6.74	9.20
3050	2 use		535	.090		2.28	3.63		5.91	8.10
3100	3 use		550	.087		2.13	3.53		5.66	7.80
3150	4 use		565	.085		2.06	3.44		5.50	7.55
3500	Floor slab, with 1-way joist pans, 1 use		415	.116		6.20	4.68		10.88	14
3550	2 use	↓	445	.108	↓	4.19	4.37		8.56	11.35

03 11 13 – Structural Cast-In-Place Concrete Forming

03 11 13.35 Forms In Place, Elevated Slabs

	03 11 13.35 Forms In Place, Elevated Slabs		Crew	Daily Output	Labor-Hours	Unit	2010 Bare Costs Material	Labor	Equipment	Total	Total Incl O&P
3600	3 use		C-2	475	.101	S.F.	3.53	4.09		7.62	10.20
3650	4 use			500	.096		3.19	3.89		7.08	9.50
4500	With 2-way waffle domes, 1 use			405	.119		6.30	4.80		11.10	14.35
4520	2 use			450	.107		4.32	4.32		8.64	11.40
4530	3 use			460	.104		3.65	4.22		7.87	10.50
4550	4 use			470	.102		3.32	4.13		7.45	10
5000	Box out for slab openings, over 16" deep, 1 use			190	.253	SFCA	3.60	10.25		13.85	19.70
5050	2 use			240	.200	"	1.98	8.10		10.08	14.65
5500	Shallow slab box outs, to 10 S.F.			42	1.143	Ea.	7.20	46.50		53.70	79.50
5550	Over 10 S.F. (use perimeter)			600	.080	L.F.	.96	3.24		4.20	6.05
6000	Bulkhead forms for slab, with keyway, 1 use, 2 piece			500	.096		1.67	3.89		5.56	7.85
6100	3 piece (see also edge forms)			460	.104		1.79	4.22		6.01	8.45
6200	Slab bulkhead form, 4-1/2" high, exp metal, w/ keyway & stakes	G	C-1	1200	.027		1.94	1.05		2.99	3.75
6210	5-1/2" high	G		1100	.029		2.27	1.15		3.42	4.27
6215	7-1/2" high	G		960	.033		2.99	1.31		4.30	5.30
6220	9-1/2" high	G		840	.038		3.39	1.50		4.89	6.05
6500	Curb forms, wood, 6" to 12" high, on elevated slabs, 1 use			180	.178	SFCA	1.03	7		8.03	11.95
6550	2 use			205	.156		.56	6.15		6.71	10.10
6600	3 use			220	.145		.41	5.75		6.16	9.30
6650	4 use			225	.142		.33	5.60		5.93	9
7000	Edge forms to 6" high, on elevated slab, 4 use			500	.064	L.F.	.12	2.52		2.64	4.02
7500	Depressed area forms to 12" high, 4 use			300	.107		.80	4.21		5.01	7.40
7550	12" to 24" high, 4 use			175	.183		1.09	7.20		8.29	12.30
8000	Perimeter deck and rail for elevated slabs, straight			90	.356		12.40	14		26.40	35
8050	Curved			65	.492		17	19.40		36.40	48.50
8500	Void forms, round fiber, 3" diameter	G		450	.071		.97	2.80		3.77	5.40
8550	4" diameter	G		425	.075		1.44	2.97		4.41	6.15
8600	6" diameter	G		400	.080		2.25	3.16		5.41	7.35
8650	8" diameter	G		375	.085		3.85	3.37		7.22	9.45
8700	10" diameter	G		350	.091		2.40	3.61		6.01	8.20
8750	12" diameter	G		300	.107		3.05	4.21		7.26	9.85

03 11 13.40 Forms In Place, Equipment Foundations

			Crew	Daily Output	Labor-Hours	Unit	Material	Labor	Equipment	Total	Total Incl O&P
0010	**FORMS IN PLACE, EQUIPMENT FOUNDATIONS**	R031113-40									
0020	1 use		C-2	160	.300	SFCA	2.58	12.15		14.73	21.50
0050	2 use	R031113-60		190	.253		1.42	10.25		11.67	17.30
0100	3 use			200	.240		1.03	9.70		10.73	16.15
0150	4 use			205	.234		.84	9.50		10.34	15.55

03 11 13.45 Forms In Place, Footings

			Crew	Daily Output	Labor-Hours	Unit	Material	Labor	Equipment	Total	Total Incl O&P
0010	**FORMS IN PLACE, FOOTINGS**	R031113-40									
0020	Continuous wall, plywood, 1 use		C-1	375	.085	SFCA	5.25	3.37		8.62	10.95
0050	2 use	R031113-60		440	.073		2.88	2.87		5.75	7.60
0100	3 use			470	.068		2.09	2.69		4.78	6.45
0150	4 use			485	.066		1.70	2.60		4.30	5.90
0500	Dowel supports for footings or beams, 1 use			500	.064	L.F.	.56	2.52		3.08	4.51
1000	Integral starter wall, to 4" high, 1 use			400	.080		.57	3.16		3.73	5.50
1500	Keyway, 4 use, tapered wood, 2" x 4"		1 Carp	530	.015		.13	.63		.76	1.12
1550	2" x 6"			500	.016		.19	.66		.85	1.23
2000	Tapered plastic			530	.015		.58	.63		1.21	1.61
2250	For keyway hung from supports, add			150	.053		.56	2.22		2.78	4.04
3000	Pile cap, square or rectangular, job-built plywood, 1 use		C-1	290	.110	SFCA	2.15	4.35		6.50	9.05
3050	2 use			346	.092		1.18	3.65		4.83	6.90
3100	3 use			371	.086		.86	3.40		4.26	6.20

03 11 13.45 Forms In Place, Footings		Crew	Daily Output	Labor-Hours	Unit	2010 Bare Costs Material	Labor	Equipment	Total	Total Incl O&P
3150	4 use	C-1	383	.084	SFCA	.70	3.30		4	5.85
4000	Triangular or hexagonal, 1 use		225	.142		2.50	5.60		8.10	11.40
4050	2 use		280	.114		1.38	4.51		5.89	8.45
4100	3 use		305	.105		1	4.14		5.14	7.50
4150	4 use		315	.102		.81	4.01		4.82	7.10
5000	Spread footings, job-built lumber, 1 use		305	.105		1.54	4.14		5.68	8.10
5050	2 use		371	.086		.86	3.40		4.26	6.20
5100	3 use		401	.080		.62	3.15		3.77	5.55
5150	4 use		414	.077		.50	3.05		3.55	5.25
6000	Supports for dowels, plinths or templates, 2′ x 2′ footing		25	1.280	Ea.	4.92	50.50		55.42	83.50
6050	4′ x 4′ footing		22	1.455		9.85	57.50		67.35	99.50
6100	8′ x 8′ footing		20	1.600		19.70	63		82.70	119
6150	12′ x 12′ footing		17	1.882		24	74		98	140
7000	Plinths, job-built plywood, 1 use		250	.128	SFCA	2.48	5.05		7.53	10.55
7100	4 use		270	.119	″	.81	4.67		5.48	8.10

03 11 13.47 Forms In Place, Gas Station Forms		Crew	Daily Output	Labor-Hours	Unit	Material	Labor	Equipment	Total	Total Incl O&P
0010	**FORMS IN PLACE, GAS STATION FORMS**									
0050	Curb fascia, with template, 12 ga. steel, left in place, 9″ high G	1 Carp	50	.160	L.F.	11.55	6.65		18.20	23
1000	Sign or light bases, 18″ diameter, 9″ high G		9	.889	Ea.	72.50	37		109.50	137
1050	30″ diameter, 13″ high G		8	1		115	41.50		156.50	191
2000	Island forms, 10′ long, 9″ high, 3′- 6″ wide G	C-1	10	3.200		320	126		446	550
2050	4′ wide G		9	3.556		335	140		475	580
2500	20′ long, 9″ high, 4′ wide G		6	5.333		535	210		745	915
2550	5′ wide G		5	6.400		555	252		807	1,000

03 11 13.50 Forms In Place, Grade Beam		Crew	Daily Output	Labor-Hours	Unit	Material	Labor	Equipment	Total	Total Incl O&P
0010	**FORMS IN PLACE, GRADE BEAM** R031113-40									
0020	Job-built plywood, 1 use	C-2	530	.091	SFCA	2.30	3.67		5.97	8.20
0050	2 use R031113-60		580	.083		1.26	3.35		4.61	6.55
0100	3 use		600	.080		.92	3.24		4.16	6
0150	4 use		605	.079		.75	3.21		3.96	5.75

03 11 13.55 Forms In Place, Mat Foundation		Crew	Daily Output	Labor-Hours	Unit	Material	Labor	Equipment	Total	Total Incl O&P
0010	**FORMS IN PLACE, MAT FOUNDATION** R031113-40									
0020	Job-built plywood, 1 use	C-2	290	.166	SFCA	2.36	6.70		9.06	12.95
0050	2 use R031113-60		310	.155		.85	6.25		7.10	10.60
0100	3 use		330	.145		.54	5.90		6.44	9.70
0120	4 use		350	.137		.50	5.55		6.05	9.10

03 11 13.65 Forms In Place, Slab On Grade		Crew	Daily Output	Labor-Hours	Unit	Material	Labor	Equipment	Total	Total Incl O&P
0010	**FORMS IN PLACE, SLAB ON GRADE** R031113-40									
1000	Bulkhead forms w/keyway, wood, 6″ high, 1 use	C-1	510	.063	L.F.	.64	2.47		3.11	4.53
1050	2 uses R031113-60		400	.080		.35	3.16		3.51	5.25
1100	4 uses		350	.091		.21	3.61		3.82	5.80
1400	Bulkhead form for slab, 4-1/2″ high, exp metal, incl keyway & stakes G		1200	.027		1.94	1.05		2.99	3.75
1410	5-1/2″ high G		1100	.029		2.27	1.15		3.42	4.27
1420	7-1/2″ high G		960	.033		2.99	1.31		4.30	5.30
1430	9-1/2″ high G		840	.038		3.39	1.50		4.89	6.05
2000	Curb forms, wood, 6″ to 12″ high, on grade, 1 use		215	.149	SFCA	2.21	5.85		8.06	11.50
2050	2 use		250	.128		1.22	5.05		6.27	9.15
2100	3 use		265	.121		.88	4.76		5.64	8.30
2150	4 use		275	.116		.72	4.59		5.31	7.85
3000	Edge forms, wood, 4 use, on grade, to 6″ high		600	.053	L.F.	.27	2.10		2.37	3.54
3050	7″ to 12″ high		435	.074	SFCA	.59	2.90		3.49	5.10
3500	For depressed slabs, 4 use, to 12″ high		300	.107	L.F.	.46	4.21		4.67	7

03 11 13 – Structural Cast-In-Place Concrete Forming

03 11 13.65 Forms In Place, Slab On Grade		Crew	Daily Output	Labor-Hours	Unit	Material	2010 Bare Costs Labor	Equipment	Total	Total Incl O&P
3550	To 24" high	C-1	175	.183	L.F.	.60	7.20		7.80	11.75
4000	For slab blockouts, to 12" high, 1 use		200	.160		.49	6.30		6.79	10.30
4050	To 24" high, 1 use		120	.267		.62	10.50		11.12	16.90
4100	Plastic (extruded), to 6" high, multiple use, on grade		800	.040		7.90	1.58		9.48	11.15
5000	Screed, 24 ga. metal key joint, see Div. 03 15 05.25									
5020	Wood, incl. wood stakes, 1" x 3"	C-1	900	.036	L.F.	.68	1.40		2.08	2.91
5050	2" x 4"		900	.036	"	.59	1.40		1.99	2.81
6000	Trench forms in floor, wood, 1 use		160	.200	SFCA	1.13	7.90		9.03	13.40
6050	2 use		175	.183		.62	7.20		7.82	11.80
6100	3 use		180	.178		.45	7		7.45	11.30
6150	4 use		185	.173		.37	6.80		7.17	10.90
8760	Void form, corrugated fiberboard, 6" x 12", 10' long G		240	.133	S.F.	.95	5.25		6.20	9.15

03 11 13.85 Forms In Place, Walls		Crew	Daily Output	Labor-Hours	Unit	Material	Labor	Equipment	Total	Total Incl O&P
0010	**FORMS IN PLACE, WALLS** R031113-10									
0100	Box out for wall openings, to 16" thick, to 10 S.F.	C-2	24	2	Ea.	16.85	81		97.85	144
0150	Over 10 S.F. (use perimeter) R031113-40	"	280	.171	L.F.	1.44	6.95		8.39	12.30
0250	Brick shelf, 4" w, add to wall forms, use wall area abv shelf									
0260	1 use R031113-60	C-2	240	.200	SFCA	1.53	8.10		9.63	14.20
0300	2 use		275	.175		.84	7.05		7.89	11.85
0350	4 use		300	.160		.61	6.50		7.11	10.65
0500	Bulkhead, wood with keyway, 1 use, 2 piece		265	.181	L.F.	1.31	7.35		8.66	12.75
0600	Bulkhead forms with keyway, 1 piece expanded metal, 8" wall G	C-1	1000	.032		2.99	1.26		4.25	5.25
0610	10" wall G		800	.040		3.39	1.58		4.97	6.15
0620	12" wall G		525	.061		4.07	2.40		6.47	8.20
0700	Buttress, to 8' high, 1 use	C-2	350	.137	SFCA	5.25	5.55		10.80	14.35
0750	2 use		430	.112		2.90	4.52		7.42	10.15
0800	3 use		460	.104		2.11	4.22		6.33	8.80
0850	4 use		480	.100		1.74	4.05		5.79	8.15
1000	Corbel or haunch, to 12" wide, add to wall forms, 1 use		150	.320	L.F.	1.74	12.95		14.69	22
1050	2 use		170	.282		.96	11.45		12.41	18.65
1100	3 use		175	.274		.70	11.10		11.80	17.85
1150	4 use		180	.267		.57	10.80		11.37	17.25
2000	Wall, job-built plywood, to 8' high, 1 use		370	.130	SFCA	2.24	5.25		7.49	10.55
2050	2 use		435	.110		1.38	4.47		5.85	8.40
2100	3 use		495	.097		1	3.93		4.93	7.15
2150	4 use		505	.095		.81	3.85		4.66	6.85
2400	Over 8' to 16' high, 1 use		280	.171		2.51	6.95		9.46	13.45
2450	2 use		345	.139		1	5.65		6.65	9.80
2500	3 use		375	.128		.72	5.20		5.92	8.80
2550	4 use		395	.122		.59	4.92		5.51	8.25
2700	Over 16' high, 1 use		235	.204		2.05	8.25		10.30	15
2750	2 use		290	.166		1.13	6.70		7.83	11.60
2800	3 use		315	.152		.82	6.15		6.97	10.40
2850	4 use		330	.145		.67	5.90		6.57	9.85
3000	For architectural finish, add		1820	.026		5.45	1.07		6.52	7.65
4000	Radial, smooth curved, job-built plywood, 1 use		245	.196		1.84	7.95		9.79	14.25
4050	2 use		300	.160		1.01	6.50		7.51	11.10
4100	3 use		325	.148		.73	6		6.73	10
4150	4 use		335	.143		.60	5.80		6.40	9.60
4200	Below grade, job-built plywood, 1 use		225	.213		1.89	8.65		10.54	15.40
4210	2 use		225	.213		1.04	8.65		9.69	14.45
4220	3 use		225	.213		.85	8.65		9.50	14.25

03 11 13 – Structural Cast-In-Place Concrete Forming

03 11 13.85 Forms In Place, Walls		Crew	Daily Output	Labor-Hours	Unit	Material	2010 Bare Costs Labor	Equipment	Total	Total Incl O&P
4230	4 use	C-2	225	.213	SFCA	.62	8.65		9.27	14
4300	Curved, 2′ chords, job-built plywood, 1 use		290	.166		1.57	6.70		8.27	12.10
4350	2 use		355	.135		.86	5.45		6.31	9.40
4400	3 use		385	.125		.63	5.05		5.68	8.50
4450	4 use		400	.120		.51	4.86		5.37	8.05
4500	Over 8′ high, 1 use		290	.166		.58	6.70		7.28	11
4525	2 use		355	.135		.32	5.45		5.77	8.80
4550	3 use		385	.125		.23	5.05		5.28	8.05
4575	4 use		400	.120		.19	4.86		5.05	7.70
4600	Retaining wall, battered, job-built plyw'd, to 8′ high, 1 use		300	.160		1.49	6.50		7.99	11.65
4650	2 use		355	.135		.82	5.45		6.27	9.35
4700	3 use		375	.128		.60	5.20		5.80	8.65
4750	4 use		390	.123		.49	4.98		5.47	8.25
4900	Over 8′ to 16′ high, 1 use		240	.200		1.61	8.10		9.71	14.25
4950	2 use		295	.163		.88	6.60		7.48	11.10
5000	3 use		305	.157		.64	6.35		6.99	10.50
5050	4 use	↓	320	.150		.52	6.05		6.57	9.90
5500	For gang wall forming, 192 S.F. sections, deduct					10%	10%			
5550	384 S.F. sections, deduct				↓	20%	20%			
5750	Liners for forms (add to wall forms), A.B.S. plastic									
5800	Aged wood, 4″ wide, 1 use	1 Carp	256	.031	SFCA	3.79	1.30		5.09	6.15
5820	2 use		256	.031		2.08	1.30		3.38	4.29
5830	3 use		256	.031		1.52	1.30		2.82	3.67
5840	4 use		256	.031		1.23	1.30		2.53	3.35
5900	Fractured rope rib, 1 use		192	.042		5.45	1.73		7.18	8.65
5925	2 use		192	.042		3.01	1.73		4.74	6
5950	3 use		192	.042		2.19	1.73		3.92	5.10
6000	4 use		192	.042		1.78	1.73		3.51	4.63
6100	Ribbed, 3/4″ deep x 1-1/2″ O.C., 1 use		224	.036		5.45	1.48		6.93	8.30
6125	2 use		224	.036		3.01	1.48		4.49	5.60
6150	3 use		224	.036		2.19	1.48		3.67	4.70
6200	4 use		224	.036		1.78	1.48		3.26	4.25
6300	Rustic brick pattern, 1 use		224	.036		3.79	1.48		5.27	6.45
6325	2 use		224	.036		2.08	1.48		3.56	4.58
6350	3 use		224	.036		1.52	1.48		3	3.96
6400	4 use		224	.036		1.23	1.48		2.71	3.64
6500	3/8″ striated, random, 1 use		224	.036		3.79	1.48		5.27	6.45
6525	2 use		224	.036		2.08	1.48		3.56	4.58
6550	3 use		224	.036		1.52	1.48		3	3.96
6600	4 use		224	.036		1.23	1.48		2.71	3.64
6850	Random vertical rustication, 1 use		384	.021		7.20	.87		8.07	9.25
6900	2 use		384	.021		3.95	.87		4.82	5.65
6925	3 use		384	.021		2.87	.87		3.74	4.49
6950	4 use		384	.021	↓	2.33	.87		3.20	3.90
7050	Wood, beveled edge, 3/4″ deep, 1 use		384	.021	L.F.	.20	.87		1.07	1.55
7100	1″ deep, 1 use		384	.021	″	.34	.87		1.21	1.70
7200	4″ wide aged cedar, 1 use		256	.031	SFCA	3.79	1.30		5.09	6.15
7300	4″ variable depth rough cedar		224	.036		5.45	1.48		6.93	8.30
7500	Lintel or sill forms, 1 use		30	.267		2.29	11.10		13.39	19.60
7520	2 use		34	.235		1.26	9.80		11.06	16.45
7540	3 use		36	.222		.92	9.25		10.17	15.25
7560	4 use	↓	37	.216	↓	.74	9		9.74	14.65
7800	Modular prefabricated plywood, based on 20 uses of purchased									

03 11 Concrete Forming

03 11 13 – Structural Cast-In-Place Concrete Forming

03 11 13.85 Forms In Place, Walls		Crew	Daily Output	Labor-Hours	Unit	2010 Bare Costs Material	Labor	Equipment	Total	Total Incl O&P
7820	forms, and 4 uses of bracing lumber									
7860	To 8′ high	C-2	800	.060	SFCA	1.44	2.43		3.87	5.35
8060	Over 8′ to 16′ high		600	.080		1.47	3.24		4.71	6.60
8600	Pilasters, 1 use		270	.178		2.60	7.20		9.80	13.95
8620	2 use		330	.145		1.43	5.90		7.33	10.65
8640	3 use		370	.130		1.04	5.25		6.29	9.25
8660	4 use		385	.125		.85	5.05		5.90	8.75
9010	Steel framed plywood, based on 50 uses of purchased									
9020	forms, and 4 uses of bracing lumber									
9060	To 8′ high	C-2	600	.080	SFCA	.75	3.24		3.99	5.80
9260	Over 8′ to 16′ high		450	.107		.75	4.32		5.07	7.50
9460	Over 16′ to 20′ high		400	.120		.75	4.86		5.61	8.35
9475	For elevated walls, add						10%			
9480	For battered walls, 1 side battered, add					10%	10%			
9485	For battered walls, 2 sides battered, add					15%	15%			

03 11 19 – Insulating Concrete Forming

03 11 19.10 Insulating Forms, Left In Place

Line	Description	Crew	Daily Output	Labor-Hours	Unit	Material	Labor	Equipment	Total	Total Incl O&P
0010	**INSULATING FORMS, LEFT IN PLACE**									
0020	S.F. is for exterior face, but includes forms for both faces (total R22)									
2000	4″ wall, straight block, 16″ x 48″ (5.33 SF) G	2 Carp	90	.178	Ea.	15.60	7.40		23	28.50
2010	90 corner block, exterior 16″ x 38″ x 22″ (6.67 SF) G		75	.213		17.60	8.85		26.45	33
2020	45 corner block, exterior 16″ x 34″ x 18″ (5.78 SF) G		75	.213		17	8.85		25.85	32.50
2100	6″ wall, straight block, 16″ x 48″ (5.33 SF) G		90	.178		15.75	7.40		23.15	29
2110	90 corner block, exterior 16″ x 32″ x 24″ (6.22 SF) G		75	.213		18.05	8.85		26.90	33.50
2120	45 corner block, exterior 16″ x 26″ x 18″ (4.89 SF) G		75	.213		16.75	8.85		25.60	32
2130	Brick ledge block, 16″ x 48″ (5.33 SF) G		80	.200		19.45	8.30		27.75	34.50
2140	Taper top block, 16″ x 48″ (5.33 SF) G		80	.200		17.35	8.30		25.65	32
2200	8″ wall, straight block, 16″ x 48″ (5.33 SF) G		90	.178		16.55	7.40		23.95	29.50
2210	90 corner block, exterior 16″ x 34″ x 26″ (6.67 SF) G		75	.213		19.50	8.85		28.35	35
2220	45 corner block, exterior 16″ x 28″ x 20″ (5.33 SF) G		75	.213		17.55	8.85		26.40	33
2230	Brick ledge block, 16″ x 48″ (5.33 SF) G		80	.200		20	8.30		28.30	35
2240	Taper top block, 16″ x 48″ (5.33 SF) G		80	.200		18.10	8.30		26.40	32.50

03 11 19.60 Roof Deck Form Boards

Line	Description	Crew	Daily Output	Labor-Hours	Unit	Material	Labor	Equipment	Total	Total Incl O&P
0010	**ROOF DECK FORM BOARDS** R051223-50									
0050	Includes bulb tee sub-purlins @ 32-5/8″ O.C.									
0070	Non-asbestos fiber cement, 5/16″ thick	C-13	2950	.008	S.F.	2.41	.37	.05	2.83	3.32
0100	Fiberglass, 1″ thick		2700	.009		3.33	.40	.05	3.78	4.40
0500	Wood fiber, 1″ thick G		2700	.009		1.97	.40	.05	2.42	2.90

03 11 23 – Permanent Stair Forming

03 11 23.75 Forms In Place, Stairs

Line	Description	Crew	Daily Output	Labor-Hours	Unit	Material	Labor	Equipment	Total	Total Incl O&P
0010	**FORMS IN PLACE, STAIRS** R031113-40									
0015	(Slant length x width), 1 use	C-2	165	.291	S.F.	4.23	11.80		16.03	23
0050	2 use R031113-60		170	.282		2.63	11.45		14.08	20.50
0100	3 use		180	.267		2.10	10.80		12.90	18.95
0150	4 use		190	.253		1.83	10.25		12.08	17.75
1000	Alternate pricing method (1.0 L.F./S.F.), 1 use		100	.480	LF Rsr	4.23	19.45		23.68	34.50
1050	2 use		105	.457		2.63	18.50		21.13	31.50
1100	3 use		110	.436		2.10	17.65		19.75	29.50
1150	4 use		115	.417		1.83	16.90		18.73	28
2000	Stairs, cast on sloping ground (length x width), 1 use		220	.218	S.F.	1.43	8.85		10.28	15.15
2025	2 use		232	.207		.79	8.40		9.19	13.75

03 11 Concrete Forming

03 11 23 – Permanent Stair Forming

03 11 23.75 Forms In Place, Stairs			Crew	Daily Output	Labor-Hours	Unit	Material	2010 Bare Costs Labor	Equipment	Total	Total Incl O&P
2050	3 use		C-2	244	.197	S.F.	.57	7.95		8.52	12.95
2100	4 use		↓	256	.188	↓	.46	7.60		8.06	12.20

03 15 Concrete Accessories

03 15 05 – Concrete Forming Accessories

03 15 05.02 Anchor Bolts

			Crew	Daily Output	Labor-Hours	Unit	Material	Labor	Equipment	Total	Total Incl O&P
0010	**ANCHOR BOLTS**										
0015	J-type, plain, incl. nut and washer										
0020	1/2" diameter, 6" long	G	1 Carp	90	.089	Ea.	1.22	3.69		4.91	7.05
0050	10" long	G		85	.094		1.39	3.91		5.30	7.60
0100	12" long	G		85	.094		1.53	3.91		5.44	7.75
0200	5/8" diameter, 12" long	G		80	.100		2.72	4.16		6.88	9.40
0250	18" long	G		70	.114		3.20	4.75		7.95	10.80
0300	24" long	G		60	.133		3.68	5.55		9.23	12.60
0350	3/4" diameter, 8" long	G		80	.100		3.20	4.16		7.36	9.90
0400	12" long	G		70	.114		4	4.75		8.75	11.70
0450	18" long	G		60	.133		5.20	5.55		10.75	14.25
0500	24" long	G		50	.160		6.80	6.65		13.45	17.75
0600	7/8" diameter, 12" long	G		60	.133		5.80	5.55		11.35	14.95
0650	18" long	G		50	.160		7.80	6.65		14.45	18.80
0700	24" long	G		40	.200		8.05	8.30		16.35	21.50
0800	1" diameter, 12" long	G		55	.145		8.40	6.05		14.45	18.55
0850	18" long	G		45	.178		10.15	7.40		17.55	22.50
0900	24" long	G		35	.229		12.35	9.50		21.85	28.50
0950	36" long	G		25	.320		16.95	13.30		30.25	39
1200	1-1/2" diameter, 18" long	G		22	.364		28	15.10		43.10	54
1250	24" long	G		18	.444		33	18.45		51.45	65
1300	36" long	G	↓	12	.667	↓	41.50	27.50		69	88

03 15 05.12 Chamfer Strips

			Crew	Daily Output	Labor-Hours	Unit	Material	Labor	Equipment	Total	Total Incl O&P
0010	**CHAMFER STRIPS**										
2000	Polyvinyl chloride, 1/2" wide with leg		1 Carp	535	.015	L.F.	.43	.62		1.05	1.43
2200	3/4" wide with leg			525	.015		.60	.63		1.23	1.64
2400	1" radius with leg			515	.016		.27	.65		.92	1.29
2800	2" radius with leg			500	.016		.47	.66		1.13	1.54
5000	Wood, 1/2" wide			535	.015		.22	.62		.84	1.20
5200	3/4" wide			525	.015		.20	.63		.83	1.20
5400	1" wide		↓	515	.016	↓	.34	.65		.99	1.36

03 15 05.15 Column Form Accessories

			Crew	Daily Output	Labor-Hours	Unit	Material	Labor	Equipment	Total	Total Incl O&P
0010	**COLUMN FORM ACCESSORIES**										
1000	Column clamps, adjustable to 24" x 24", buy	G				Set	126			126	138
1100	Rent per month	G					11.80			11.80	13
1300	For sizes to 30" x 30", buy	G					142			142	157
1400	Rent per month	G					14			14	15.40
1600	For sizes to 36" x 36", buy	G					163			163	179
1700	Rent per month	G					16.70			16.70	18.40
2000	Bull winch (band iron) 36" x 36", buy	G					51.50			51.50	56.50
2100	Rent per month	G					5.25			5.25	5.80
2300	48" x 48", buy	G					52.50			52.50	58
2400	Rent per month	G					6.30			6.30	6.95
3000	Chain & wedge type 36" x 36", buy	G				↓	71.50			71.50	78.50

03 15 Concrete Accessories

03 15 05 – Concrete Forming Accessories

03 15 05.15 Column Form Accessories			Crew	Daily Output	Labor-Hours	Unit	2010 Bare Costs Material	Labor	Equipment	Total	Total Incl O&P
3100	Rent per month	G				Set	8.05			8.05	8.85
3300	60" x 60", buy	G					96.50			96.50	106
3400	Rent per month	G					11.55			11.55	12.70
4000	Friction collars 2'-6" diam., buy	G					730			730	800
4100	Rent per month	G					71.50			71.50	78.50
4300	4'-0" diam., buy	G					835			835	920
4400	Rent per month	G					92			92	101

03 15 05.20 Dovetail Anchor System

			Crew	Daily Output	Labor-Hours	Unit	Material	Labor	Equipment	Total	Total Incl O&P
0010	**DOVETAIL ANCHOR SYSTEM**										
0500	Anchor slot, galv., filled, 24 ga.	G	1 Carp	425	.019	L.F.	.74	.78		1.52	2.02
0600	20 ga.	G		400	.020		1.57	.83		2.40	3.01
0625	12 ga.	G		400	.020		1.77	.83		2.60	3.23
0900	26 ga. stainless steel, foam filled	G		375	.021		1.18	.89		2.07	2.67
1200	Brick anchor, corr., galv., 3-1/2" long, 16 ga.	G	1 Bric	10.50	.762	C	21	32		53	71
1300	12 ga.	G		10.50	.762		29.50	32		61.50	80.50
1500	Flat, galv., 3-1/2" long, 16 ga.	G		10.50	.762		38	32		70	89.50
1600	12 ga.	G		10.50	.762		53	32		85	107
2000	Cavity wall, corr., galv., 5" long, 16 ga.	G		10.50	.762		33	32		65	84.50
2100	12 ga.	G		10.50	.762		47	32		79	100
3000	Furring anchors, corr., galv., 1-1/2" long, 16 ga.	G		10.50	.762		12.80	32		44.80	62
3100	12 ga.	G		10.50	.762		22	32		54	72.50
6000	Stone anchors, 3-1/2" long, galv., 1/8" x 1" wide	G		10.50	.762		82	32		114	138
6100	1/4" x 1" wide	G		10.50	.762		94.50	32		126.50	152

03 15 05.25 Expansion Joints

			Crew	Daily Output	Labor-Hours	Unit	Material	Labor	Equipment	Total	Total Incl O&P
0010	**EXPANSION JOINTS**										
0020	Keyed, cold, 24 ga., incl. stakes, 3-1/2" high	G	1 Carp	200	.040	L.F.	1.71	1.66		3.37	4.44
0050	4-1/2" high	G		200	.040		1.94	1.66		3.60	4.69
0100	5-1/2" high	G		195	.041		2.27	1.70		3.97	5.15
0150	7-1/2" high	G		190	.042		2.99	1.75		4.74	6
0160	9-1/2" high	G		185	.043		3.39	1.80		5.19	6.50
0300	Poured asphalt, plain, 1/2" x 1"		1 Clab	450	.018		.66	.59		1.25	1.64
0350	1" x 2"			400	.020		2.64	.66		3.30	3.92
0500	Neoprene, liquid, cold applied, 1/2" x 1"			450	.018		2.22	.59		2.81	3.35
0550	1" x 2"			400	.020		8.85	.66		9.51	10.75
0700	Polyurethane, poured, 2 part, 1/2" x 1"			400	.020		1.30	.66		1.96	2.45
0750	1" x 2"			350	.023		5.20	.76		5.96	6.85
0900	Rubberized asphalt, hot or cold applied, 1/2" x 1"			450	.018		.36	.59		.95	1.31
0950	1" x 2"			400	.020		1.43	.66		2.09	2.59
1100	Hot applied, fuel resistant, 1/2" x 1"			450	.018		.54	.59		1.13	1.50
1150	1" x 2"			400	.020		2.15	.66		2.81	3.38
2000	Premolded, bituminous fiber, 1/2" x 6"		1 Carp	375	.021		.41	.89		1.30	1.82
2050	1" x 12"			300	.027		1.91	1.11		3.02	3.81
2140	Concrete expansion joint, recycled paper and fiber, 1/2" x 6"	G		390	.021		.36	.85		1.21	1.71
2150	1/2" x 12"	G		360	.022		.72	.92		1.64	2.22
2250	Cork with resin binder, 1/2" x 6"			375	.021		1.82	.89		2.71	3.37
2300	1" x 12"			300	.027		7.55	1.11		8.66	10
2500	Neoprene sponge, closed cell, 1/2" x 6"			375	.021		1.93	.89		2.82	3.49
2550	1" x 12"			300	.027		7.10	1.11		8.21	9.50
2750	Polyethylene foam, 1/2" x 6"			375	.021		.66	.89		1.55	2.10
2800	1" x 12"			300	.027		2.15	1.11		3.26	4.08
3000	Polyethylene backer rod, 3/8" diameter			460	.017		.05	.72		.77	1.17
3050	3/4" diameter			460	.017		.10	.72		.82	1.22

03 15 Concrete Accessories

03 15 05 – Concrete Forming Accessories

03 15 05.25 Expansion Joints			Crew	Daily Output	Labor-Hours	Unit	Material	2010 Bare Costs Labor	Equipment	Total	Total Incl O&P
3100	1" diameter		1 Carp	460	.017	L.F.	.19	.72		.91	1.32
3500	Polyurethane foam, with polybutylene, 1/2" x 1/2"			475	.017		.95	.70		1.65	2.13
3550	1" x 1"			450	.018		2.40	.74		3.14	3.78
3750	Polyurethane foam, regular, closed cell, 1/2" x 6"			375	.021		.70	.89		1.59	2.14
3800	1" x 12"			300	.027		2.50	1.11		3.61	4.46
4000	Polyvinyl chloride foam, closed cell, 1/2" x 6"			375	.021		1.90	.89		2.79	3.46
4050	1" x 12"			300	.027		6.55	1.11		7.66	8.90
4250	Rubber, gray sponge, 1/2" x 6"			375	.021		1.60	.89		2.49	3.13
4300	1" x 12"			300	.027		5.75	1.11		6.86	8.05
4500	Lead wool for joints					Lb.	10			10	11
5000	For installation in walls, add							75%			
5250	For installation in boxouts, add							25%			

03 15 05.30 Hangers			Crew	Daily Output	Labor-Hours	Unit	Material	Labor	Equipment	Total	Total Incl O&P
0010	**HANGERS**										
0020	Slab and beam form										
0500	Banding iron, 3/4" x 22 ga, 14 L.F. per lb or										
0550	1/2" x 14 ga, 7 L.F. per lb.	G				Lb.	1.29			1.29	1.42
1000	Fascia ties, coil type, to 24" long	G				C	560			560	620
1500	Frame ties to 8-1/8"	G					660			660	725
1550	8-1/8" to 10-1/8"	G					695			695	765
5000	Snap tie hanger, to 30" overall length, 3000 #	G					560			560	615
5050	To 36" overall length	G					610			610	670
5100	To 48" overall length	G					710			710	780
5500	Steel beam hanger										
5600	Flange to 8-1/8"	G				C	660			660	725
5650	8-1/8" to 10-1/8"	G					695			695	765
6000	Tie hangers to 30" overall length, 4000 #	G					635			635	700
6100	To 36" overall length	G					685			685	755
6500	Tie back hanger, up to 12-1/8" flange	G					1,775			1,775	1,950
8500	Wire, black annealed, 9 ga	G				Cwt.	79.50			79.50	87.50
8600	16 ga					"	122			122	134

03 15 05.35 Inserts			Crew	Daily Output	Labor-Hours	Unit	Material	Labor	Equipment	Total	Total Incl O&P
0010	**INSERTS**										
1000	Inserts, slotted nut type for 3/4" bolts, 4" long	G	1 Carp	84	.095	Ea.	17.60	3.96		21.56	25.50
2100	6" long	G		84	.095		19.65	3.96		23.61	27.50
2150	8" long	G		84	.095		24.50	3.96		28.46	33
2200	Slotted, strap type, 4" long	G		84	.095		17.10	3.96		21.06	25
2250	6" long	G		84	.095		19.65	3.96		23.61	27.50
2300	8" long	G		84	.095		24.50	3.96		28.46	33
2350	Strap for slotted insert, 4" long	G		84	.095		12.80	3.96		16.76	20
4100	6" long	G		84	.095		13.85	3.96		17.81	21.50
4150	8" long	G		84	.095		16.85	3.96		20.81	24.50
4200	10" long	G		84	.095		20.50	3.96		24.46	28.50
7000	Loop ferrule type										
7100	1/4" diameter bolt	G	1 Carp	84	.095	Ea.	3.38	3.96		7.34	9.80
7350	7/8" diameter bolt	G	"	84	.095	"	15.75	3.96		19.71	23.50
9000	Wedge type										
9100	For 3/4" diameter bolt	G	1 Carp	60	.133	Ea.	13.05	5.55		18.60	23
9800	Cut washers, black										
9900	3/4" bolt	G				Ea.	.84			.84	.92
9950	For galvanized inserts, add						30%				

03 15 Concrete Accessories

03 15 05 – Concrete Forming Accessories

03 15 05.70 Shores

	03 15 05.70 Shores		Crew	Daily Output	Labor-Hours	Unit	2010 Bare Costs Material	Labor	Equipment	Total	Total Incl O&P
0010	**SHORES**										
0020	Erect and strip, by hand, horizontal members										
0500	Aluminum joists and stringers	G	2 Carp	60	.267	Ea.		11.10		11.10	17.10
0600	Steel, adjustable beams	G		45	.356			14.75		14.75	23
0700	Wood joists			50	.320			13.30		13.30	20.50
0800	Wood stringers			30	.533			22		22	34
1000	Vertical members to 10′ high	G		55	.291			12.10		12.10	18.65
1050	To 13′ high	G		50	.320			13.30		13.30	20.50
1100	To 16′ high	G		45	.356			14.75		14.75	23
1500	Reshoring	G		1400	.011	S.F.	.50	.47		.97	1.28
1600	Flying truss system	G	C-17D	9600	.009	SFCA		.38	.08	.46	.67
1760	Horizontal, aluminum joists, 6-1/4″ high x 5′ to 21′ span, buy	G				L.F.	36			36	39.50
1770	Beams, 7-1/4″ high x 4′ to 30′ span	G				″	47.50			47.50	52.50
1810	Horizontal, steel beam, adjustable, 4′ to 7′ span, buy	G				Ea.	330			330	360
1830	6′ to 10′ span	G					330			330	360
1920	9′ to 15′ span	G					545			545	600
1940	12′ to 20′ span	G					650			650	715
1970	Steel stringer, W8x10, 4′ to 16′ span, buy	G				L.F.	25.50			25.50	28
3000	Rent for job duration, aluminum joist @ 2′ O.C., per mo	G				SF Flr.	.90			.90	.99
3050	Steel W8x10	G					.64			.64	.70
3060	Steel adjustable	G					.82			.82	.90
3500	#1 post shore, steel, 5′-7″ to 9′-6″ high, 10000# cap., buy	G				Ea.	410			410	450
3550	#2 post shore, 7′-3″ to 12′-10″ high, 7800# capacity	G					465			465	515
3600	#3 post shore, 8′-10″ to 16′-1″ high, 3800# capacity	G					540			540	595
5010	Frame shoring systems, steel, 12000#/leg, buy										
5040	Frame, 2′ wide x 6′ high	G				Ea.	106			106	117
5250	X-brace	G					18			18	19.80
5550	Base plate	G					17.60			17.60	19.35
5600	Screw jack	G					38			38	41.50
5650	U-head, 8″ x 8″	G					21.50			21.50	24

03 15 05.75 Sleeves and Chases

	03 15 05.75 Sleeves and Chases		Crew	Daily Output	Labor-Hours	Unit	Material	Labor	Equipment	Total	Total Incl O&P
0010	**SLEEVES AND CHASES**										
0100	Plastic, 1 use, 9″ long, 2″ diameter		1 Carp	100	.080	Ea.	1.36	3.32		4.68	6.60
0150	4″ diameter			90	.089		4.01	3.69		7.70	10.10
0200	6″ diameter			75	.107		7.05	4.43		11.48	14.65
0250	12″ diameter			60	.133		19.85	5.55		25.40	30.50

03 15 05.80 Snap Ties

	03 15 05.80 Snap Ties		Crew	Daily Output	Labor-Hours	Unit	Material	Labor	Equipment	Total	Total Incl O&P
0010	**SNAP TIES**, 8-1/4″ L&W (Lumber and wedge)										
0100	2250 lb., w/ flat washer, 8″ wall	G				C	178			178	195
0200	12″ wall	G					189			189	207
0250	16″ wall	G					211			211	232
0300	18″ wall	G					222			222	244
0500	With plastic cone, 8″ wall	G					191			191	210
0600	12″ wall	G					207			207	228
0650	16″ wall	G					227			227	249
0700	18″ wall	G					237			237	261
1000	3350 lb., w/ flat washer, 8″ wall	G					296			296	325
1150	12″ wall	G					335			335	370
1200	16″ wall	G					380			380	415
1250	18″ wall	G					395			395	435
1500	With plastic cone, 8″ wall	G					296			296	325
1600	12″ wall	G					335			335	370

03 15 Concrete Accessories

03 15 05 – Concrete Forming Accessories

03 15 05.80 Snap Ties			Crew	Daily Output	Labor-Hours	Unit	Material	Labor	Equipment	Total	Total Incl O&P
							2010 Bare Costs				
1650	16″ wall	G				C	385			385	420
1700	18″ wall	G					395			395	435

03 15 05.85 Stair Tread Inserts

			Crew	Daily Output	Labor-Hours	Unit	Material	Labor	Equipment	Total	Total Incl O&P
0010	**STAIR TREAD INSERTS**										
0015	Cast iron, abrasive, 3″ wide	G	1 Carp	90	.089	L.F.	10.10	3.69		13.79	16.80
0020	4″ wide	G		80	.100		13.45	4.16		17.61	21
0040	6″ wide	G		75	.107		20	4.43		24.43	29
0050	9″ wide	G		70	.114		30.50	4.75		35.25	41
0100	12″ wide	G		65	.123		40.50	5.10		45.60	52.50
0300	Cast aluminum, compared to cast iron, deduct						10%				
0500	Extruded aluminum safety tread, 3″ wide	G	1 Carp	75	.107		11.50	4.43		15.93	19.50
0550	4″ wide	G		75	.107		15.35	4.43		19.78	24
0600	6″ wide	G		75	.107		23	4.43		27.43	32.50
0650	9″ wide to resurface stairs	G		70	.114		34.50	4.75		39.25	45.50
1700	Cement fill for pan-type metal treads, plain		1 Cefi	115	.070	S.F.	1.80	2.76		4.56	6
1750	Non-slip		″	100	.080	″	1.98	3.18		5.16	6.80

03 15 05.95 Wall and Foundation Form Accessories

			Crew	Daily Output	Labor-Hours	Unit	Material	Labor	Equipment	Total	Total Incl O&P
0010	**WALL AND FOUNDATION FORM ACCESSORIES**										
2000	Footings, turnbuckle form aligner	G				Ea.	32.50			32.50	36
2050	Spreaders for footer, adjustable	G				″	19.60			19.60	21.50
3000	Form oil, coverage varies greatly, minimum					Gal.	9.25			9.25	10.20
3050	Maximum					″	11.90			11.90	13.10
3500	Form patches, 1-3/4″ diameter					C	22.50			22.50	25
3550	2-3/4″ diameter					″	38.50			38.50	42.50
4000	Nail stakes, 3/4″ diameter, 18″ long	G				Ea.	3.68			3.68	4.05
4050	24″ long	G					4.37			4.37	4.81
4200	30″ long	G					5.50			5.50	6.05
4250	36″ long	G					6.15			6.15	6.80

03 15 13 – Waterstops

03 15 13.50 Waterstops

			Crew	Daily Output	Labor-Hours	Unit	Material	Labor	Equipment	Total	Total Incl O&P
0010	**WATERSTOPS**, PVC and Rubber										
0020	PVC, ribbed 3/16″ thick, 4″ wide		1 Carp	155	.052	L.F.	1.08	2.14		3.22	4.50
0050	6″ wide			145	.055		1.98	2.29		4.27	5.70
0500	With center bulb, 6″ wide, 3/16″ thick			135	.059		1.75	2.46		4.21	5.75
0550	3/8″ thick			130	.062		3.16	2.56		5.72	7.40
0600	9″ wide x 3/8″ thick			125	.064		4.84	2.66		7.50	9.40
0800	Dumbbell type, 6″ wide, 3/16″ thick			150	.053		1.67	2.22		3.89	5.25
0850	3/8″ thick			145	.055		3.45	2.29		5.74	7.35
1000	9″ wide, 3/8″ thick, plain			130	.062		4.83	2.56		7.39	9.25
1050	Center bulb			130	.062		7.50	2.56		10.06	12.20
1250	Ribbed type, split, 3/16″ thick, 6″ wide			145	.055		1.99	2.29		4.28	5.70
1300	3/8″ thick			130	.062		3.83	2.56		6.39	8.15
2000	Rubber, flat dumbbell, 3/8″ thick, 6″ wide			145	.055		7.85	2.29		10.14	12.15
2050	9″ wide			135	.059		12.05	2.46		14.51	17.05
2500	Flat dumbbell split, 3/8″ thick, 6″ wide			145	.055		1.99	2.29		4.28	5.70
2550	9″ wide			135	.059		3.83	2.46		6.29	8
3000	Center bulb, 1/4″ thick, 6″ wide			145	.055		7.40	2.29		9.69	11.70
3050	9″ wide			135	.059		16.20	2.46		18.66	21.50
3500	Center bulb split, 3/8″ thick, 6″ wide			145	.055		11.45	2.29		13.74	16.15
3550	9″ wide			135	.059		19.75	2.46		22.21	25.50
5000	Waterstop fittings, rubber, flat										
5010	Dumbbell or center bulb, 3/8″ thick,										

03 15 Concrete Accessories

03 15 13 – Waterstops

03 15 13.50 Waterstops		Crew	Daily Output	Labor-Hours	Unit	Material	2010 Bare Costs Labor	Equipment	Total	Total Incl O&P
5200	Field union, 6" wide	1 Carp	50	.160	Ea.	18.30	6.65		24.95	30.50
5250	9" wide		50	.160		23.50	6.65		30.15	36.50
5500	Flat cross, 6" wide		30	.267		32.50	11.10		43.60	53
5550	9" wide		30	.267		47.50	11.10		58.60	69
6000	Flat tee, 6" wide		30	.267		27.50	11.10		38.60	47.50
6050	9" wide		30	.267		37.50	11.10		48.60	58
6500	Flat ell, 6" wide		40	.200		26.50	8.30		34.80	42
6550	9" wide		40	.200		33.50	8.30		41.80	50
7000	Vertical tee, 6" wide		25	.320		23	13.30		36.30	45.50
7050	9" wide		25	.320		32.50	13.30		45.80	56
7500	Vertical ell, 6" wide		35	.229		21.50	9.50		31	38
7550	9" wide		35	.229		28	9.50		37.50	45

03 21 Reinforcing Steel

03 21 05 – Reinforcing Steel Accessories

03 21 05.10 Rebar Accessories

Code	Description	Crew	Daily Output	Labor-Hours	Unit	Material	Labor	Equipment	Total	Total Incl O&P
0010	**REBAR ACCESSORIES**									
0030	Steel & plastic made from recycled materials G									
0100	Beam bolsters (BB), lower, 1-1/2" high, plain steel G				C.L.F.	70			70	77
0102	Galvanized G					84			84	92.50
0104	Stainless tipped legs G					360			360	395
0106	Plastic tipped legs G					121			121	133
0108	Epoxy dipped G					164			164	180
0110	2" high, plain G					88.50			88.50	97.50
0120	Galvanized G					106			106	117
0140	Stainless tipped legs G					350			350	385
0160	Plastic tipped legs G					123			123	135
0162	Epoxy dipped G					174			174	191
0200	Upper (BBU), 1-1/2" high, plain steel G					197			197	217
0210	3" high G					355			355	390
0300	Beam bolster with plate (BBP), 1-1/2" high, plain steel G					230			230	252
0310	3" high G					365			365	400
0500	Slab bolsters, continuous (SB), 1" high, plain steel G					60			60	66
0502	Galvanized G					67			67	73.50
0504	Stainless tipped legs G					350			350	385
0506	Plastic tipped legs G					82.50			82.50	91
0510	2" high, plain steel G					70			70	77
0515	Galvanized G					82.50			82.50	91
0520	Stainless tipped legs G					360			360	395
0525	Plastic tipped legs G					91.50			91.50	101
0530	For bolsters with wire runners (SBR), add G					101			101	111
0540	For bolsters with plates (SBP), add G					203			203	223
0700	Bag ties, 16 ga., plain, 4" long G				C	.95			.95	1.05
0710	5" long G					1			1	1.10
0720	6" long G					1			1	1.10
0730	7" long G					1.10			1.10	1.21
1200	High chairs, individual (HC), 3" high, plain steel G					84			84	92.50
1202	Galvanized G					119			119	130
1204	Stainless tipped legs G					375			375	410
1206	Plastic tipped legs G					112			112	123
1210	5" high, plain G					130			130	143

03 21 05.10 Rebar Accessories			Crew	Daily Output	Labor-Hours	Unit	Material	Labor	Equipment	Total	Total Incl O&P
							2010 Bare Costs				
1212	Galvanized	G				C	161			161	177
1214	Stainless tipped legs	G					420			420	460
1216	Plastic tipped legs	G					161			161	177
1220	8″ high, plain	G					280			280	310
1222	Galvanized	G					355			355	390
1224	Stainless tipped legs	G					570			570	625
1226	Plastic tipped legs	G					355			355	390
1230	12″ high, plain	G					585			585	640
1232	Galvanized	G					700			700	770
1234	Stainless tipped legs	G					875			875	960
1236	Plastic tipped legs	G					655			655	720
1400	Individual high chairs, with plate (HCP), 5″ high	G					335			335	370
1410	8″ high	G					510			510	560
1500	Bar chair (BC), 1-1/2″ high, plain steel	G					43			43	47.50
1520	Galvanized	G					50			50	55
1530	Stainless tipped legs	G					335			335	365
1540	Plastic tipped legs	G					88			88	97
1550	Joist chair (JC), 3/4″ x 5″, plain steel	G					58.50			58.50	64.50
1580	Galvanized	G					70			70	77
1600	Stainless tipped legs	G					350			350	385
1620	Plastic tipped legs	G					101			101	111
1630	Epoxy dipped	G					223			223	245
1700	Continuous high chairs (CHC), legs 8″ O.C., 4″ high, plain steel	G				C.L.F.	112			112	123
1705	Galvanized	G					131			131	144
1710	Stainless tipped legs	G					400			400	440
1715	Plastic tipped legs	G					139			139	153
1718	Epoxy dipped	G					198			198	218
1720	6″ high, plain	G					161			161	177
1725	Galvanized	G					222			222	244
1730	Stainless tipped legs	G					450			450	495
1735	Plastic tipped legs	G					222			222	244
1738	Epoxy dipped	G					269			269	295
1740	8″ high, plain	G					235			235	259
1745	Galvanized	G					282			282	310
1750	Stainless tipped legs	G					525			525	580
1755	Plastic tipped legs	G					305			305	340
1758	Epoxy dipped	G					410			410	450
1900	For continuous bottom wire runners, add	G					121			121	133
1940	For continuous bottom plate, add	G					235			235	259
2100	Paper tubing, 4′ lengths, for #4 bar	G					54.50			54.50	60
2120	For #6 bar	G					64.50			64.50	71
2200	Screed chair base, 1/2″ coil thread diam., 2-1/2″ high, plain steel	G				C	278			278	305
2210	Galvanized	G					295			295	325
2220	5-1/2″ high, plain	G					335			335	370
2250	Galvanized	G					355			355	390
2300	3/4″ coil thread diam., 2-1/2″ high, plain steel	G					355			355	390
2310	Galvanized	G					470			470	520
2320	5-1/2″ high, plain steel	G					430			430	475
2350	Galvanized	G					605			605	665
2400	Screed holder, 1/2″ coil thread diam. for pipe screed, plain steel, 6″ long	G					295			295	325
2420	12″ long	G					485			485	535
2500	3/4″ coil thread diam. for pipe screed, plain steel, 6″ long	G					530			530	580
2520	12″ long	G					860			860	945

03 21 05 – Reinforcing Steel Accessories

03 21 05.10 Rebar Accessories

	03 21 05.10 Rebar Accessories		Crew	Daily Output	Labor-Hours	Unit	Material	2010 Bare Costs Labor	Equipment	Total	Total Incl O&P
2700	Screw anchor for bolts, plain steel, 3/4" diameter x 4" long	G				C	335			335	365
2720	1" diameter x 6" long	G					570			570	630
2740	1-1/2" diameter x 8" long	G					955			955	1,050
2800	Screw anchor eye bolts, 3/4" x 7" long	G					4,275			4,275	4,700
2820	1" x 9" long	G					8,450			8,450	9,300
2840	1-1/2" x 14" long	G					21,400			21,400	23,500
2900	Screw anchor bolts, 3/4" x 9" long	G					1,500			1,500	1,650
2920	1" x 12" long	G					2,875			2,875	3,175
3001	Slab lifting inserts, single pickup, galv, 3/4" diam., 5" high	G					1,400			1,400	1,525
3010	6" high	G					1,525			1,525	1,675
3030	7" high	G					1,600			1,600	1,775
3100	1" diameter, 5-1/2" high	G					1,625			1,625	1,775
3120	7" high	G					1,700			1,700	1,875
3200	Double pickup lifting inserts, 1" diameter, 5-1/2" high	G					3,300			3,300	3,650
3220	7" high	G					3,625			3,625	4,000
3330	1-1/2" diameter, 5-1/2" high	G					4,350			4,350	4,775
3800	Subgrade chairs, #4 bar head, 3-1/2" high	G					41.50			41.50	45.50
3850	12" high	G					72			72	79
3900	#6 bar head, 3-1/2" high	G					41.50			41.50	45.50
3950	12" high	G					72			72	79
4200	Subgrade stakes, no nail holes, 3/4" diameter, 12" long	G					257			257	283
4250	24" long	G					395			395	435
4300	7/8" diameter, 12" long	G					370			370	405
4350	24" long	G				↓	525			525	575
4500	Tie wire, 16 ga. annealed steel	G				Cwt.	122			122	134
4550	Tie wire holder, plastic case	G				Ea.	23.50			23.50	26
4600	Aluminum case	G				"	36.50			36.50	40

03 21 05.75 Splicing Reinforcing Bars

	03 21 05.75 Splicing Reinforcing Bars		Crew	Daily Output	Labor-Hours	Unit	Material	Labor	Equipment	Total	Total Incl O&P
0010	**SPLICING REINFORCING BARS**	R032110-70									
0020	Including holding bars in place while splicing										
0100	Standard, self-aligning type, taper threaded, #4 bars	G	C-25	190	.168	Ea.	4.55	6.15		10.70	15
0105	#5 bars	G		170	.188		5.55	6.85		12.40	17.30
0110	#6 bars	G		150	.213		6.40	7.80		14.20	19.75
0120	#7 bars	G		130	.246		7.45	9		16.45	23
0300	#8 bars	G	↓	115	.278		12.90	10.15		23.05	31
0305	#9 bars	G	C-5	105	.533		14.10	24.50	7.10	45.70	62
0310	#10 bars	G		95	.589		15.70	27	7.85	50.55	68.50
0320	#11 bars	G		85	.659		16.70	30	8.80	55.50	75.50
0330	#14 bars	G		65	.862		25	39	11.50	75.50	102
0340	#18 bars	G		45	1.244		38	56.50	16.60	111.10	150
0500	Transition self-aligning, taper threaded, #18-14	G		45	1.244		40.50	56.50	16.60	113.60	152
0510	#18-11	G		45	1.244		41.50	56.50	16.60	114.60	154
0520	#14-11	G		65	.862		29.50	39	11.50	80	107
0540	#11-10	G		85	.659		23.50	30	8.80	62.30	83
0550	#10-9	G	↓	95	.589		22	27	7.85	56.85	75
0560	#9-8	G	C-25	105	.305		21	11.10		32.10	41
0580	#8-7	G		115	.278		19.55	10.15		29.70	38
0590	#7-6	G		130	.246		19.05	9		28.05	35.50
0600	Position coupler for curved bars, taper threaded, #4 bars	G		160	.200		22	7.30		29.30	36
0610	#5 bars	G		145	.221		23	8.05		31.05	38
0620	#6 bars	G		130	.246		26	9		35	43.50
0630	#7 bars	G	↓	110	.291	↓	38.50	10.60		49.10	60

03 21 05 – Reinforcing Steel Accessories

03 21 05.75 Splicing Reinforcing Bars			Crew	Daily Output	Labor-Hours	Unit	Material	Labor (2010 Bare Costs)	Equipment	Total	Total Incl O&P
0640	#8 bars	G	C-25	100	.320	Ea.	40	11.70		51.70	63
0650	#9 bars	G	C-5	90	.622		43.50	28.50	8.30	80.30	102
0660	#10 bars	G		80	.700		47	32	9.35	88.35	112
0670	#11 bars	G		70	.800		49	36.50	10.70	96.20	123
0680	#14 bars	G		55	1.018		61	46.50	13.60	121.10	156
0690	#18 bars	G		40	1.400		88	63.50	18.70	170.20	218
0700	Transition position coupler for curved bars, taper threaded, #18-14	G		40	1.400		111	63.50	18.70	193.20	244
0710	#18-11	G		40	1.400		112	63.50	18.70	194.20	245
0720	#14-11	G		55	1.018		95	46.50	13.60	155.10	192
0730	#11-10	G		70	.800		54.50	36.50	10.70	101.70	129
0740	#10-9	G		80	.700		52	32	9.35	93.35	118
0750	#9-8	G	C-25	90	.356		48	13		61	74
0760	#8-7	G		100	.320		44	11.70		55.70	67.50
0770	#7-6	G		110	.291		42.50	10.60		53.10	64.50
0800	Sleeve type w/ grout filler, for precast concrete, #6 bars	G		72	.444		23.50	16.20		39.70	52.50
0802	#7 bars	G		64	.500		25.50	18.25		43.75	57.50
0805	#8 bars	G		56	.571		29	21		50	66
0807	#9 bars	G		48	.667		31.50	24.50		56	74.50
0810	#10 bars	G	C-5	40	1.400		41	63.50	18.70	123.20	167
0900	#11 bars	G		32	1.750		49	79.50	23.50	152	206
0920	#14 bars	G		24	2.333		79.50	106	31	216.50	290
1000	Sleeve type w/ ferrous filler, for critical structures, #6 bars	G	C-25	72	.444		41	16.20		57.20	72
1210	#7 bars	G		64	.500		42	18.25		60.25	75.50
1220	#8 bars	G		56	.571		44	21		65	82.50
1230	#9 bars	G	C-5	48	1.167		45	53	15.60	113.60	151
1240	#10 bars	G		40	1.400		48	63.50	18.70	130.20	175
1250	#11 bars	G		32	1.750		58	79.50	23.50	161	216
1260	#14 bars	G		24	2.333		73	106	31	210	283
1270	#18 bars	G		16	3.500		106	159	46.50	311.50	420
2000	Weldable half coupler, taper threaded, #4 bars	G	E-16	120	.133		7.10	6.40	1.21	14.71	20.50
2100	#5 bars	G		112	.143		8.40	6.85	1.30	16.55	22.50
2200	#6 bars	G		104	.154		13.35	7.35	1.40	22.10	29
2300	#7 bars	G		96	.167		15.50	8	1.52	25.02	33
2400	#8 bars	G		88	.182		16.15	8.70	1.66	26.51	35
2500	#9 bars	G		80	.200		17.80	9.60	1.82	29.22	38.50
2600	#10 bars	G		72	.222		18.20	10.65	2.02	30.87	41
2700	#11 bars	G		64	.250		19.45	12	2.28	33.73	45
2800	#14 bars	G		56	.286		22.50	13.70	2.60	38.80	51.50
2900	#18 bars	G		48	.333		36.50	15.95	3.04	55.49	71.50

03 21 10 – Uncoated Reinforcing Steel

03 21 10.60 Reinforcing In Place			Crew	Daily Output	Labor-Hours	Unit	Material	Labor (2010 Bare Costs)	Equipment	Total	Total Incl O&P
0015	**REINFORCING IN PLACE**, 50-60 ton lots, A615 Grade 60 R032110-10										
0020	Includes labor, but not material cost, to install accessories										
0030	Made from recycled materials	G									
0100	Beams & Girders, #3 to #7 CN	G	4 Rodm	1.60	20	Ton	800	935		1,735	2,375
0150	#8 to #18 R032110-20	G		2.70	11.852		800	555		1,355	1,775
0200	Columns, #3 to #7	G		1.50	21.333		800	1,000		1,800	2,475
0250	#8 to #18	G		2.30	13.913		800	650		1,450	1,925
0300	Spirals, hot rolled, 8" to 15" diameter	G		2.20	14.545		1,175	680		1,855	2,375
0320	15" to 24" diameter R032110-40	G		2.20	14.545		1,100	680		1,780	2,325
0330	24" to 36" diameter	G		2.30	13.913		1,050	650		1,700	2,225
0340	36" to 48" diameter R032110-50	G		2.40	13.333		1,000	625		1,625	2,100

03 21 Reinforcing Steel

03 21 10 – Uncoated Reinforcing Steel

03 21 10.60 Reinforcing In Place			Crew	Daily Output	Labor-Hours	Unit	2010 Bare Costs Material	Labor	Equipment	Total	Total Incl O&P
0360	48" to 64" diameter	G	4 Rodm	2.50	12.800	Ton	1,100	600		1,700	2,200
0380	64" to 84" diameter R032110-70	G		2.60	12.308		1,175	575		1,750	2,200
0390	84" to 96" diameter	G		2.70	11.852		1,225	555		1,780	2,225
0400	Elevated slabs, #4 to #7 R032110-80	G		2.90	11.034		850	515		1,365	1,775
0500	Footings, #4 to #7	G		2.10	15.238		760	715		1,475	1,975
0550	#8 to #18	G		3.60	8.889		720	415		1,135	1,450
0600	Slab on grade, #3 to #7	G		2.30	13.913		760	650		1,410	1,875
0700	Walls, #3 to #7	G		3	10.667		760	500		1,260	1,650
0750	#8 to #18	G		4	8		760	375		1,135	1,450
0900	For other than 50-60 ton lots										
1000	Under 10 ton job, #3 to #7, add					Ton	25%	10%			
1010	#8 to #18, add						20%	10%			
1050	10 – 50 ton job, #3 to #7, add						10%				
1060	#8 to #18, add						5%				
1100	60 – 100 ton job, #3 - #7, deduct						5%				
1110	#8 to #18, deduct						10%				
1150	Over 100 ton job, #3 - #7, deduct						10%				
1160	#8 - #18, deduct						15%				
1200	High strength steel, Grade 75, #14 bars only, add	G					79.50			79.50	87.50
2000	Unloading & sorting, add to above		C-5	100	.560			25.50	7.50	33	49
2200	Crane cost for handling, add to above, minimum			135	.415			18.90	5.55	24.45	36
2210	Average			92	.609			27.50	8.15	35.65	53
2220	Maximum			35	1.600			73	21.50	94.50	139
2400	Dowels, 2 feet long, deformed, #3	G	2 Rodm	520	.031	Ea.	.33	1.44		1.77	2.68
2410	#4	G		480	.033		.59	1.56		2.15	3.16
2420	#5	G		435	.037		.92	1.72		2.64	3.78
2430	#6	G		360	.044		1.32	2.08		3.40	4.80
2450	Longer and heavier dowels, add	G		725	.022	Lb.	.44	1.03		1.47	2.14
2500	Smooth dowels, 12" long, 1/4" or 3/8" diameter	G		140	.114	Ea.	.74	5.35		6.09	9.40
2520	5/8" diameter	G		125	.128		1.29	6		7.29	11.05
2530	3/4" diameter	G		110	.145		1.60	6.80		8.40	12.70
2600	Dowel sleeves for CIP concrete, 2-part system										
2610	Sleeve base, plastic, for 5/8" smooth dowel sleeve, fasten to edge form		1 Rodm	200	.040	Ea.	.49	1.87		2.36	3.56
2615	Sleeve, plastic, 12" long, for 5/8" smooth dowel, snap onto base			400	.020		1.06	.94		2	2.68
2620	Sleeve base, for 3/4" smooth dowel sleeve			175	.046		.52	2.14		2.66	4.02
2625	Sleeve, 12" long, for 3/4" smooth dowel			350	.023		1.59	1.07		2.66	3.47
2630	Sleeve base, for 1" smooth dowel sleeve			150	.053		.63	2.50		3.13	4.71
2635	Sleeve, 12" long, for 1" smooth dowel			300	.027		2.26	1.25		3.51	4.50
2700	Dowel caps, visual warning only, plastic, #3 to #8		2 Rodm	800	.020		.42	.94		1.36	1.97
2720	#8 to #18			750	.021		1.05	1		2.05	2.77
2750	Impalement protective, plastic, #4 to #9			800	.020		1.62	.94		2.56	3.29

03 21 10.70 Glass Fiber Reinforced Polymer Bars

03 21 10.70		Crew	Daily Output	Labor-Hours	Unit	Material	Labor	Equipment	Total	Total Incl O&P
0010	**GLASS FIBER REINFORCED POLYMER BARS**									
0050	#2 bar, .043 lbs/ft.	4 Rodm	9500	.003	L.F.	.35	.16		.51	.64
0100	#3 bar, .092 lbs/ft.		9300	.003		.48	.16		.64	.79
0150	#4 bar, .160 lbs/ft.		9100	.004		.70	.16		.86	1.04
0200	#5 bar, .258 lbs/ft.		8700	.004		.98	.17		1.15	1.36
0250	#6 bar, .372 lbs/ft.		8300	.004		1.35	.18		1.53	1.78
0300	#7 bar, .497 lbs/ft.		7900	.004		1.70	.19		1.89	2.18
0350	#8 bar, .620 lbs/ft.		7400	.004		2.25	.20		2.45	2.81
0400	#9 bar, .800 lbs/ft.		6800	.005		2.90	.22		3.12	3.55
0450	#10 bar, 1.08 lbs/ft.		5800	.006		3.45	.26		3.71	4.22

03 21 Reinforcing Steel

03 21 10 – Uncoated Reinforcing Steel

03 21 10.70 Glass Fiber Reinforced Polymer Bars		Crew	Daily Output	Labor-Hours	Unit	Material	2010 Bare Costs Labor	Equipment	Total	Total Incl O&P
0500	For Bends, add per bend				Ea.	1			1	1.10

03 21 13 – Galvanized Reinforcing Steel

03 21 13.10 Galvanized Reinforcing		Crew	Daily Output	Labor-Hours	Unit	Material	Labor	Equipment	Total	Total Incl O&P
0010	**GALVANIZED REINFORCING**									
0150	Add to uncoated reinforcing price for galvanizing				Ton	610			610	670

03 21 16 – Epoxy-Coated Reinforcing Steel

03 21 16.10 Epoxy-Coated Reinforcing		Crew	Daily Output	Labor-Hours	Unit	Material	Labor	Equipment	Total	Total Incl O&P
0010	**EPOXY-COATED REINFORCING**									
0100	Add to uncoated reinforcing price for coating with epoxy				Ton	380			380	420

03 22 Welded Wire Fabric Reinforcing

03 22 05 – Uncoated Welded Wire Fabric

03 22 05.50 Welded Wire Fabric		Crew	Daily Output	Labor-Hours	Unit	Material	Labor	Equipment	Total	Total Incl O&P
0010	**WELDED WIRE FABRIC** ASTM A185 R032205-30									
0030	Made from recycled materials G									
0050	Sheets									
0100	6 x 6 - W1.4 x W1.4 (10 x 10) 21 lb. per C.S.F. **CN** G	2 Rodm	35	.457	C.S.F.	12	21.50		33.50	47.50
0200	6 x 6 - W2.1 x W2.1 (8 x 8) 30 lb. per C.S.F. G		31	.516		19.25	24		43.25	60
0300	6 x 6 - W2.9 x W2.9 (6 x 6) 42 lb. per C.S.F. G		29	.552		24	26		50	68
0400	6 x 6 - W4 x W4 (4 x 4) 58 lb. per C.S.F. G		27	.593		32	27.50		59.50	80
0500	4 x 4 - W1.4 x W1.4 (10 x 10) 31 lb. per C.S.F. G		31	.516		17.75	24		41.75	58.50
0600	4 x 4 - W2.1 x W2.1 (8 x 8) 44 lb. per C.S.F. G		29	.552		25	26		51	69
0650	4 x 4 - W2.9 x W2.9 (6 x 6) 61 lb. per C.S.F. G		27	.593		37	27.50		64.50	85.50
0700	4 x 4 - W4 x W4 (4 x 4) 85 lb. per C.S.F. G		25	.640		47	30		77	100
0750	Rolls									
0800	2 x 2 - #14 galv., 21 lb/C.S.F., beam & column wrap G	2 Rodm	6.50	2.462	C.S.F.	40	115		155	230
0900	2 x 2 - #12 galv. for gunite reinforcing G	"	6.50	2.462	"	35	115		150	225

03 23 Stressing Tendons

03 23 05 – Prestressing Tendons

03 23 05.50 Prestressing Steel		Crew	Daily Output	Labor-Hours	Unit	Material	Labor	Equipment	Total	Total Incl O&P
0010	**PRESTRESSING STEEL** R034136-90									
0100	Grouted strand, post-tensioned in field, 50′ span, 100 kip G	C-3	1200	.053	Lb.	2.29	2.29	.09	4.67	6.25
0150	300 kip G		2700	.024		1.04	1.02	.04	2.10	2.79
0300	100′ span, 100 kip G		1700	.038		2.30	1.62	.06	3.98	5.15
0350	300 kip G		3200	.020		1.99	.86	.03	2.88	3.59
0500	200′ span, 100 kip G		2700	.024		2.29	1.02	.04	3.35	4.17
0550	300 kip G		3500	.018		1.98	.79	.03	2.80	3.45
0800	Grouted bars, 50′ span, 42 kip G		2600	.025		.87	1.06	.04	1.97	2.68
0850	143 kip G		3200	.020		.83	.86	.03	1.72	2.32
1000	75′ span, 42 kip G		3200	.020		.88	.86	.03	1.77	2.37
1050	143 kip G		4200	.015		.74	.65	.03	1.42	1.88
1200	Ungrouted strand, 50′ span, 100 kip G	C-4	1275	.025		.61	1.19	.02	1.82	2.60
1250	300 kip G		1475	.022		.61	1.03	.02	1.66	2.34
1400	100′ span, 100 kip G		1500	.021		.61	1.01	.02	1.64	2.32
1450	300 kip G		1650	.019		.61	.92	.02	1.55	2.17
1600	200′ span, 100 kip G		1500	.021		.61	1.01	.02	1.64	2.32
1650	300 kip G		1700	.019		.61	.89	.02	1.52	2.12

03 23 Stressing Tendons

03 23 05 – Prestressing Tendons

03 23 05.50 Prestressing Steel			Crew	Daily Output	Labor-Hours	Unit	Material	2010 Bare Costs Labor	Equipment	Total	Total Incl O&P
1800	Ungrouted bars, 50′ span, 42 kip	G	C-4	1400	.023	Lb.	.40	1.08	.02	1.50	2.20
1850	143 kip	G		1700	.019		.40	.89	.02	1.31	1.89
2000	75′ span, 42 kip	G		1800	.018		.40	.84	.02	1.26	1.81
2050	143 kip	G		2200	.015		.40	.69	.01	1.10	1.56
2220	Ungrouted single strand, 100′ slab, 25 kip	G		1200	.027		.61	1.26	.02	1.89	2.73
2250	35 kip	G		1475	.022		.61	1.03	.02	1.66	2.34
3000	Slabs on grade, 0.5-inch diam. non-bonded strands, HDPE sheathed,										
3050	attached dead-end anchors, loose stressing-end anchors										
3100	25′ x 30′ slab, strands @ 36″ O.C., placing		2 Rodm	2940	.005	S.F.	.51	.25		.76	.97
3105	Stressing		C-4A	3750	.004			.20	.01	.21	.33
3110	42″ O.C., placing		2 Rodm	3200	.005		.45	.23		.68	.88
3115	Stressing		C-4A	4040	.004			.19	.01	.20	.31
3120	48″ O.C., placing		2 Rodm	3510	.005		.39	.21		.60	.77
3125	Stressing		C-4A	4390	.004			.17	.01	.18	.28
3150	25′ x 40′ slab, strands @ 36″ O.C., placing		2 Rodm	3370	.005		.50	.22		.72	.91
3155	Stressing		C-4A	4360	.004			.17	.01	.18	.29
3160	42″ O.C., placing		2 Rodm	3760	.004		.43	.20		.63	.80
3165	Stressing		C-4A	4820	.003			.16	.01	.17	.26
3170	48″ O.C., placing		2 Rodm	4090	.004		.38	.18		.56	.71
3175	Stressing		C-4A	5190	.003			.14	.01	.15	.24
3200	30′ x 30′ slab, strands @ 36″ O.C., placing		2 Rodm	3260	.005		.50	.23		.73	.92
3205	Stressing		C-4A	4190	.004			.18	.01	.19	.30
3210	42″ O.C., placing		2 Rodm	3530	.005		.45	.21		.66	.83
3215	Stressing		C-4A	4500	.004			.17	.01	.18	.28
3220	48″ O.C., placing		2 Rodm	3840	.004		.40	.20		.60	.75
3225	Stressing		C-4A	4850	.003			.15	.01	.16	.26
3230	30′ x 40′ slab, strands @ 36″ O.C., placing		2 Rodm	3780	.004		.49	.20		.69	.86
3235	Stressing		C-4A	4920	.003			.15	.01	.16	.26
3240	42″ O.C., placing		2 Rodm	4190	.004		.43	.18		.61	.76
3245	Stressing		C-4A	5410	.003			.14	.01	.15	.23
3250	48″ O.C., placing		2 Rodm	4520	.004		.39	.17		.56	.70
3255	Stressing		C-4A	5790	.003			.13	.01	.14	.22
3260	30′ x 50′ slab, strands @ 36″ O.C., placing		2 Rodm	4300	.004		.47	.17		.64	.80
3265	Stressing		C-4A	5650	.003			.13	.01	.14	.22
3270	42″ O.C., placing		2 Rodm	4720	.003		.42	.16		.58	.72
3275	Stressing		C-4A	6150	.003			.12	.01	.13	.21
3280	48″ O.C., placing		2 Rodm	5240	.003		.37	.14		.51	.63
3285	Stressing		C-4A	6760	.002			.11	.01	.12	.19

03 24 Fibrous Reinforcing

03 24 05 – Reinforcing Fibers

03 24 05.30 Synthetic Fibers

			Crew	Daily Output	Labor-Hours	Unit	Material	Labor	Equipment	Total	Total Incl O&P
0010	**SYNTHETIC FIBERS**										
0100	Synthetic fibers, add to concrete					Lb.	4.72			4.72	5.20
0110	1-1/2 lb. per C.Y.					C.Y.	7.30			7.30	8

03 24 05.70 Steel Fibers

			Crew	Daily Output	Labor-Hours	Unit	Material	Labor	Equipment	Total	Total Incl O&P
0010	**STEEL FIBERS**										
0150	Steel fibers, add to concrete	G				Lb.	.77			.77	.85
0155	25 lb. per C.Y.	G				C.Y.	19.25			19.25	21
0160	50 lb. per C.Y.	G					38.50			38.50	42.50
0170	75 lb. per C.Y.	G					59.50			59.50	65

03 24 Fibrous Reinforcing

03 24 05 – Reinforcing Fibers

03 24 05.70 Steel Fibers		Crew	Daily Output	Labor-Hours	Unit	Material	2010 Bare Costs Labor	Equipment	Total	Total Incl O&P
0180	100 lb. per C.Y. G				C.Y.	77			77	84.50

03 30 Cast-In-Place Concrete

03 30 53 – Miscellaneous Cast-In-Place Concrete

03 30 53.40 Concrete In Place

		Crew	Daily Output	Labor-Hours	Unit	Material	Labor	Equipment	Total	Total Incl O&P
0010	**CONCRETE IN PLACE**									
0020	Including forms (4 uses), reinforcing steel, concrete, placement,									
0050	and finishing unless otherwise indicated R033053-50									
0300	Beams (3500 psi), 5 kip per L.F., 10′ span	C-14A	15.62	12.804	C.Y.	282	535	49.50	866.50	1,200
0350	25′ span	"	18.55	10.782		290	450	41.50	781.50	1,075
0500	Chimney foundations (5000 psi), industrial, minimum	C-14C	32.22	3.476		140	139	.94	279.94	370
0510	Maximum	"	23.71	4.724		162	188	1.28	351.28	470
0700	Columns, square (4000 psi), 12″ x 12″, minimum reinforcing	C-14A	11.96	16.722		305	700	64.50	1,069.50	1,475
0720	Average reinforcing R033105-85		10.13	19.743		475	825	76	1,376	1,875
0740	Maximum reinforcing		9.03	22.148		700	925	85.50	1,710.50	2,300
0800	16″ x 16″, minimum reinforcing		16.22	12.330		246	515	47.50	808.50	1,125
0820	Average reinforcing		12.57	15.911		405	665	61.50	1,131.50	1,550
0840	Maximum reinforcing		10.25	19.512		615	815	75	1,505	2,050
0900	24″ x 24″, minimum reinforcing		23.66	8.453		211	355	32.50	598.50	815
0920	Average reinforcing		17.71	11.293		360	470	43.50	873.50	1,175
0940	Maximum reinforcing		14.15	14.134		565	590	54.50	1,209.50	1,600
1000	36″ x 36″, minimum reinforcing		33.69	5.936		188	248	23	459	615
1020	Average reinforcing		23.32	8.576		320	360	33	713	945
1040	Maximum reinforcing		17.82	11.223		530	470	43	1,043	1,350
1100	Columns, round (4000 psi), tied, 12″ diameter, minimum reinforcing		20.97	9.537		274	400	36.50	710.50	955
1120	Average reinforcing		15.27	13.098		440	545	50.50	1,035.50	1,400
1140	Maximum reinforcing		12.11	16.515		655	690	63.50	1,408.50	1,875
1200	16″ diameter, minimum reinforcing		31.49	6.351		251	265	24.50	540.50	715
1220	Average reinforcing		19.12	10.460		420	435	40.50	895.50	1,175
1240	Maximum reinforcing		13.77	14.524		620	605	56	1,281	1,675
1300	20″ diameter, minimum reinforcing		41.04	4.873		254	204	18.75	476.75	615
1320	Average reinforcing		24.05	8.316		405	345	32	782	1,025
1340	Maximum reinforcing		17.01	11.758		620	490	45.50	1,155.50	1,500
1400	24″ diameter, minimum reinforcing		51.85	3.857		238	161	14.85	413.85	530
1420	Average reinforcing		27.06	7.391		405	310	28.50	743.50	955
1440	Maximum reinforcing		18.29	10.935		610	455	42	1,107	1,425
1500	36″ diameter, minimum reinforcing		75.04	2.665		237	111	10.25	358.25	445
1520	Average reinforcing		37.49	5.335		385	223	20.50	628.50	795
1540	Maximum reinforcing		22.84	8.757		590	365	33.50	988.50	1,250
1900	Elevated slab (4000 psi), flat slab with drops, 125 psf Sup. Load, 20′ span	C-14B	38.45	5.410		244	226	20	490	640
1950	30′ span		50.99	4.079		252	170	15.10	437.10	555
2100	Flat plate, 125 psf Sup. Load, 15′ span		30.24	6.878		226	287	25.50	538.50	720
2150	25′ span		49.60	4.194		231	175	15.50	421.50	540
2300	Waffle const., 30″ domes, 125 psf Sup. Load, 20′ span		37.07	5.611		227	234	21	482	630
2350	30′ span		44.07	4.720		212	197	17.45	426.45	555
2500	One way joists, 30″ pans, 125 psf Sup. Load, 15′ span		27.38	7.597		274	315	28	617	820
2550	25′ span		31.15	6.677		258	278	24.50	560.50	740
2700	One way beam & slab, 125 psf Sup. Load, 15′ span		20.59	10.102		248	420	37.50	705.50	965
2750	25′ span		28.36	7.334		229	305	27	561	755
2900	Two way beam & slab, 125 psf Sup. Load, 15′ span		24.04	8.652		236	360	32	628	855
2950	25′ span		35.87	5.799		202	242	21.50	465.50	620

03 30 53 – Miscellaneous Cast-In-Place Concrete

03 30 53.40 Concrete In Place		Crew	Daily Output	Labor-Hours	Unit	2010 Bare Costs Material	Labor	Equipment	Total	Total Incl O&P
3100	Elevated slabs, flat plate, including finish, not									
3110	including forms or reinforcing									
3150	Regular concrete (4000 psi), 4″ slab	C-8	2613	.021	S.F.	1.33	.79	.28	2.40	2.96
3200	6″ slab		2585	.022		1.96	.79	.29	3.04	3.67
3250	2-1/2″ thick floor fill		2685	.021		.85	.77	.28	1.90	2.39
3300	Lightweight, 110# per C.F., 2-1/2″ thick floor fill		2585	.022		1.15	.79	.29	2.23	2.77
3400	Cellular concrete, 1-5/8″ fill, under 5000 S.F.		2000	.028		.77	1.03	.37	2.17	2.80
3450	Over 10,000 S.F.		2200	.025		.73	.93	.34	2	2.59
3500	Add per floor for 3 to 6 stories high		31800	.002			.06	.02	.08	.13
3520	For 7 to 20 stories high		21200	.003			.10	.03	.13	.19
3540	Equipment pad (3000 psi), 3′ x 3′ x 6″ thick	C-14H	45	1.067	Ea.	42.50	44	.67	87.17	115
3550	4′ x 4′ x 6″ thick		30	1.600		62.50	65.50	1.01	129.01	171
3560	5′ x 5′ x 8″ thick		18	2.667		108	109	1.68	218.68	290
3570	6′ x 6′ x 8″ thick		14	3.429		146	141	2.16	289.16	380
3580	8′ x 8′ x 10″ thick		8	6		310	246	3.78	559.78	725
3590	10′ x 10′ x 12″ thick		5	9.600		530	395	6.05	931.05	1,200
3800	Footings (3000 psi), spread under 1 C.Y.	C-14C	28	4	C.Y.	150	160	1.08	311.08	415
3825	1 C.Y to 5 C.Y.		43	2.605		173	104	.70	277.70	350
3850	Over 5 C.Y.		75	1.493		158	59.50	.40	217.90	266
3900	Footings, strip (3000 psi), 18″ x 9″, unreinforced	C-14L	40	2.400		112	93	.77	205.77	267
3920	18″ x 9″, reinforced	C-14C	35	3.200		131	128	.86	259.86	345
3925	20″ x 10″, unreinforced	C-14L	45	2.133		110	82.50	.68	193.18	249
3930	20″ x 10″, reinforced	C-14C	40	2.800		126	112	.76	238.76	310
3935	24″ x 12″, unreinforced	C-14L	55	1.745		110	67.50	.56	178.06	226
3940	24″ x 12″, reinforced	C-14C	48	2.333		126	93	.63	219.63	283
3945	36″ x 12″, unreinforced	C-14L	70	1.371		107	53	.44	160.44	200
3950	36″ x 12″, reinforced	C-14C	60	1.867		122	74.50	.50	197	250
4000	Foundation mat (3000 psi), under 10 C.Y.		38.67	2.896		180	116	.78	296.78	380
4050	Over 20 C.Y.		56.40	1.986		159	79	.54	238.54	299
4200	Wall, free-standing (3000 psi), 8″ thick, 8′ high	C-14D	45.83	4.364		175	180	16.80	371.80	490
4250	14′ high		27.26	7.337		204	305	28.50	537.50	725
4260	12″ thick, 8′ high		64.32	3.109		155	129	11.95	295.95	380
4270	14′ high		40.01	4.999		163	207	19.25	389.25	520
4300	15″ thick, 8′ high		80.02	2.499		145	103	9.60	257.60	330
4350	12′ high		51.26	3.902		145	161	15	321	425
4500	18′ high		48.85	4.094		160	169	15.75	344.75	455
4520	Handicap access ramp (4000 psi), railing both sides, 3′ wide	C-14H	14.58	3.292	L.F.	273	135	2.07	410.07	510
4525	5′ wide		12.22	3.928		283	161	2.47	446.47	560
4530	With 6″ curb and rails both sides, 3′ wide		8.55	5.614		282	230	3.54	515.54	670
4535	5′ wide		7.31	6.566		287	269	4.14	560.14	735
4650	Slab on grade (3500 psi), not including finish, 4″ thick	C-14E	60.75	1.449	C.Y.	117	59.50	.51	177.01	222
4700	6″ thick	″	92	.957	″	113	39.50	.33	152.83	186
4701	Thickened slab edge (3500 psi), for slab on grade poured									
4702	monolithically with slab; depth is in addition to slab thickness;									
4703	formed vertical outside edge, earthen bottom and inside slope									
4705	8″ deep x 8″ wide bottom, unreinforced	C-14L	2190	.044	L.F.	3.07	1.70	.01	4.78	6
4710	8″ x 8″, reinforced	C-14C	1670	.067		4.95	2.68	.02	7.65	9.60
4715	12″ deep x 12″ wide bottom, unreinforced	C-14L	1800	.053		6.40	2.07	.02	8.49	10.25
4720	12″ x 12″, reinforced	C-14C	1310	.086		9.80	3.41	.02	13.23	16.15
4725	16″ deep x 16″ wide bottom, unreinforced	C-14L	1440	.067		10.90	2.58	.02	13.50	16
4730	16″ x 16″, reinforced	C-14C	1120	.100		15	3.99	.03	19.02	22.50
4735	20″ deep x 20″ wide bottom, unreinforced	C-14L	1150	.083		16.65	3.23	.03	19.91	23.50
4740	20″ x 20″, reinforced	C-14C	920	.122		22	4.86	.03	26.89	31.50

03 30 Cast-In-Place Concrete

03 30 53 – Miscellaneous Cast-In-Place Concrete

03 30 53.40 Concrete In Place		Crew	Daily Output	Labor-Hours	Unit	2010 Bare Costs Material	Labor	Equipment	Total	Total Incl O&P
4745	24″ deep x 24″ wide bottom, unreinforced	C-14L	930	.103	L.F.	23.50	4	.03	27.53	32
4750	24″ x 24″, reinforced	C-14C	740	.151		30.50	6.05	.04	36.59	43
4751	Slab on grade (3500 psi), incl. troweled finish, not incl. forms									
4760	or reinforcing, over 10,000 S.F., 4″ thick	C-14F	3425	.021	S.F.	1.29	.79	.01	2.09	2.61
4820	6″ thick		3350	.021		1.89	.81	.01	2.71	3.29
4840	8″ thick		3184	.023		2.59	.85	.01	3.45	4.13
4900	12″ thick		2734	.026		3.88	.99	.01	4.88	5.75
4950	15″ thick		2505	.029		4.88	1.08	.01	5.97	6.95
5000	Slab on grade (3000 psi), incl. textured finish, not incl. forms									
5001	or reinforcing, 4″ thick	C-14G	2873	.019	S.F.	1.26	.72	.01	1.99	2.48
5010	6″ thick		2590	.022		1.97	.80	.01	2.78	3.38
5020	8″ thick		2320	.024		2.57	.90	.01	3.48	4.18
5200	Lift slab in place above the foundation, incl. forms,									
5210	reinforcing, concrete (4000 psi) and columns, minimum	C-14B	2113	.098	S.F.	6.10	4.10	.36	10.56	13.45
5250	Average		1650	.126		6.60	5.25	.47	12.32	15.95
5300	Maximum		1500	.139		7.15	5.80	.51	13.46	17.40
5500	Lightweight, ready mix, including screed finish only,									
5510	not including forms or reinforcing									
5550	1:4 (2500 psi) for structural roof decks	C-14B	260	.800	C.Y.	135	33.50	2.96	171.46	204
5600	1:6 (3000 psi) for ground slab with radiant heat	C-14F	92	.783		133	29.50	.33	162.83	191
5650	1:3:2 (2000 psi) with sand aggregate, roof deck	C-14B	260	.800		132	33.50	2.96	168.46	200
5700	Ground slab (2000 psi)	C-14F	107	.673		132	25.50	.28	157.78	183
5900	Pile caps (3000 psi), incl. forms and reinf., sq. or rect., under 10 C.Y.	C-14C	54.14	2.069		147	82.50	.56	230.06	291
5950	Over 10 C.Y.		75	1.493		140	59.50	.40	199.90	246
6000	Triangular or hexagonal, under 10 C.Y.		53	2.113		115	84.50	.57	200.07	257
6050	Over 10 C.Y.		85	1.318		127	52.50	.36	179.86	222
6200	Retaining walls (3000 psi), gravity, 4′ high see Div. 32 32	C-14D	66.20	3.021		133	125	11.65	269.65	355
6250	10′ high		125	1.600		126	66	6.15	198.15	247
6300	Cantilever, level backfill loading, 8′ high		70	2.857		153	118	11	282	365
6350	16′ high		91	2.198		140	91	8.45	239.45	305
6800	Stairs (3500 psi), not including safety treads, free standing, 3′-6″ wide	C-14H	83	.578	LF Nose	5.10	23.50	.36	28.96	42.50
6850	Cast on ground		125	.384	″	4.03	15.75	.24	20.02	29
7000	Stair landings, free standing		200	.240	S.F.	4.20	9.85	.15	14.20	20
7050	Cast on ground		475	.101	″	3.13	4.15	.06	7.34	9.90

03 31 Structural Concrete

03 31 05 – Normal Weight Structural Concrete

03 31 05.30 Concrete, Field Mix		Crew	Daily Output	Labor-Hours	Unit	Material	Labor	Equipment	Total	Total Incl O&P
0010	**CONCRETE, FIELD MIX** R033105-65									
0015	FOB forms 2250 psi				C.Y.	85			85	93.50
0020	3000 psi				″	90			90	99

03 31 05.35 Normal Weight Concrete, Ready Mix		Crew	Daily Output	Labor-Hours	Unit	Material	Labor	Equipment	Total	Total Incl O&P
0010	**NORMAL WEIGHT CONCRETE, READY MIX**, delivered R033105-10									
0012	Includes local aggregate, sand, Portland cement, and water									
0015	Excludes all additives and treatments R033105-20									
0020	2000 psi				C.Y.	91.50			91.50	101
0100	2500 psi R033105-30					94			94	103
0150	3000 psi *CN*					97			97	107
0200	3500 psi R033105-40					99.50			99.50	110
0300	4000 psi					103			103	113
0350	4500 psi R033105-50					106			106	116

03 31 05 – Normal Weight Structural Concrete

03 31 05.35 Normal Weight Concrete, Ready Mix		Crew	Daily Output	Labor-Hours	Unit	Material	2010 Bare Costs Labor	Equipment	Total	Total Incl O&P
0400	5000 psi CN				C.Y.	109			109	120
0411	6000 psi					124			124	137
0412	8000 psi					203			203	223
0413	10,000 psi					288			288	315
0414	12,000 psi					350			350	380
1000	For high early strength cement, add					10%				
1010	For structural lightweight with regular sand, add					25%				
1300	For winter concrete (hot water), add					4.25			4.25	4.68
1400	For hot weather concrete (ice), add					9.10			9.10	10
1410	For mid-range water reducer, add					4.13			4.13	4.54
1420	For high-range water reducer/superplasticizer, add					6.35			6.35	6.95
1430	For retarder, add					2.71			2.71	2.98
1440	For non-Chloride accelerator, add					4.83			4.83	5.30
1450	For Chloride accelerator, per 1%, add					3.28			3.28	3.61
1460	For fiber reinforcing, synthetic (1 Lb./C.Y.), add					6.65			6.65	7.30
1500	For Saturday delivery, add					8.85			8.85	9.70
1510	For truck holding/waiting time past 1st hour per load, add				Hr.	87.50			87.50	96.50
1520	For short load (less than 4 C.Y.), add per load				Ea.	112			112	124
2000	For all lightweight aggregate, add				C.Y.	45%				

03 31 05.70 Placing Concrete		Crew	Daily Output	Labor-Hours	Unit	Material	Labor	Equipment	Total	Total Incl O&P
0010	**PLACING CONCRETE** R033105-70									
0020	Includes labor and equipment to place, strike off and consolidate									
0050	Beams, elevated, small beams, pumped	C-20	60	1.067	C.Y.		38	13.30	51.30	72
0100	With crane and bucket	C-7	45	1.600			57	28.50	85.50	119
0200	Large beams, pumped	C-20	90	.711			25	8.90	33.90	48.50
0250	With crane and bucket	C-7	65	1.108			39.50	19.85	59.35	82
0400	Columns, square or round, 12" thick, pumped	C-20	60	1.067			38	13.30	51.30	72
0450	With crane and bucket	C-7	40	1.800			64.50	32.50	97	134
0600	18" thick, pumped	C-20	90	.711			25	8.90	33.90	48.50
0650	With crane and bucket	C-7	55	1.309			47	23.50	70.50	97
0800	24" thick, pumped	C-20	92	.696			24.50	8.70	33.20	47
0850	With crane and bucket	C-7	70	1.029			37	18.45	55.45	76.50
1000	36" thick, pumped	C-20	140	.457			16.20	5.70	21.90	31
1050	With crane and bucket	C-7	100	.720			25.50	12.90	38.40	53
1400	Elevated slabs, less than 6" thick, pumped	C-20	140	.457			16.20	5.70	21.90	31
1450	With crane and bucket	C-7	95	.758			27	13.60	40.60	56
1500	6" to 10" thick, pumped	C-20	160	.400			14.15	5	19.15	27
1550	With crane and bucket	C-7	110	.655			23.50	11.75	35.25	48.50
1600	Slabs over 10" thick, pumped	C-20	180	.356			12.60	4.44	17.04	24
1650	With crane and bucket	C-7	130	.554			19.80	9.95	29.75	41
1900	Footings, continuous, shallow, direct chute	C-6	120	.400			13.80	.51	14.31	21.50
1950	Pumped	C-20	150	.427			15.10	5.35	20.45	29
2000	With crane and bucket	C-7	90	.800			28.50	14.35	42.85	59.50
2100	Footings, continuous, deep, direct chute	C-6	140	.343			11.85	.44	12.29	18.55
2150	Pumped	C-20	160	.400			14.15	5	19.15	27
2200	With crane and bucket	C-7	110	.655			23.50	11.75	35.25	48.50
2400	Footings, spread, under 1 C.Y., direct chute	C-6	55	.873			30	1.11	31.11	47
2450	Pumped	C-20	65	.985			35	12.30	47.30	66.50
2500	With crane and bucket	C-7	45	1.600			57	28.50	85.50	119
2600	Over 5 C.Y., direct chute	C-6	120	.400			13.80	.51	14.31	21.50
2650	Pumped	C-20	150	.427			15.10	5.35	20.45	29
2700	With crane and bucket	C-7	100	.720			25.50	12.90	38.40	53

03 31 Structural Concrete

03 31 05 – Normal Weight Structural Concrete

03 31 05.70 Placing Concrete		Crew	Daily Output	Labor-Hours	Unit	Material	2010 Bare Costs Labor	Equipment	Total	Total Incl O&P
2900	Foundation mats, over 20 C.Y., direct chute	C-6	350	.137	C.Y.		4.74	.17	4.91	7.45
2950	Pumped	C-20	400	.160			5.65	2	7.65	10.85
3000	With crane and bucket	C-7	300	.240			8.60	4.30	12.90	17.80
3200	Grade beams, direct chute	C-6	150	.320			11.05	.41	11.46	17.30
3250	Pumped	C-20	180	.356			12.60	4.44	17.04	24
3300	With crane and bucket	C-7	120	.600			21.50	10.75	32.25	44.50
3500	High rise, for more than 5 stories, pumped, add per story	C-20	2100	.030			1.08	.38	1.46	2.06
3510	With crane and bucket, add per story	C-7	2100	.034			1.23	.61	1.84	2.54
3700	Pile caps, under 5 C.Y., direct chute	C-6	90	.533			18.40	.68	19.08	28.50
3750	Pumped	C-20	110	.582			20.50	7.25	27.75	39.50
3800	With crane and bucket	C-7	80	.900			32	16.15	48.15	67
3850	Pile cap, 5 C.Y. to 10 C.Y., direct chute	C-6	175	.274			9.45	.35	9.80	14.85
3900	Pumped	C-20	200	.320			11.35	4	15.35	21.50
3950	With crane and bucket	C-7	150	.480			17.15	8.60	25.75	35.50
4000	Over 10 C.Y., direct chute	C-6	215	.223			7.70	.28	7.98	12.05
4050	Pumped	C-20	240	.267			9.45	3.33	12.78	18.05
4100	With crane and bucket	C-7	185	.389			13.90	6.95	20.85	28.50
4300	Slab on grade, up to 6" thick, direct chute	C-6	110	.436			15.05	.55	15.60	23.50
4350	Pumped	C-20	130	.492			17.45	6.15	23.60	33.50
4400	With crane and bucket	C-7	110	.655			23.50	11.75	35.25	48.50
4600	Over 6" thick, direct chute	C-6	165	.291			10.05	.37	10.42	15.75
4650	Pumped	C-20	185	.346			12.25	4.32	16.57	23.50
4700	With crane and bucket	C-7	145	.497			17.75	8.90	26.65	37
4900	Walls, 8" thick, direct chute	C-6	90	.533			18.40	.68	19.08	28.50
4950	Pumped	C-20	100	.640			22.50	8	30.50	43.50
5000	With crane and bucket	C-7	80	.900			32	16.15	48.15	67
5050	12" thick, direct chute	C-6	100	.480			16.55	.61	17.16	26
5100	Pumped	C-20	110	.582			20.50	7.25	27.75	39.50
5200	With crane and bucket	C-7	90	.800			28.50	14.35	42.85	59.50
5300	15" thick, direct chute	C-6	105	.457			15.80	.58	16.38	24.50
5350	Pumped	C-20	120	.533			18.90	6.65	25.55	36.50
5400	With crane and bucket	C-7	95	.758			27	13.60	40.60	56
5600	Wheeled concrete dumping, add to placing costs above									
5610	Walking cart, 50' haul, add	C-18	32	.281	C.Y.		9.35	1.83	11.18	16.45
5620	150' haul, add		24	.375			12.50	2.43	14.93	22
5700	250' haul, add		18	.500			16.65	3.25	19.90	29
5800	Riding cart, 50' haul, add	C-19	80	.113			3.75	1.17	4.92	7.10
5810	150' haul, add		60	.150			5	1.57	6.57	9.40
5900	250' haul, add		45	.200			6.65	2.09	8.74	12.60

03 35 Concrete Finishing

03 35 29 – Tooled Concrete Finishing

03 35 29.30 Finishing Floors		Crew	Daily Output	Labor-Hours	Unit	Material	Labor	Equipment	Total	Total Incl O&P
0010	**FINISHING FLOORS**									
0012	Finishing requires that concrete first be placed, struck off & consolidated									
0015	Basic finishing for various unspecified flatwork									
0100	Bull float only	C-10	4000	.006	S.F.		.23		.23	.33
0125	Bull float & manual float		2000	.012			.45		.45	.67
0150	Bull float, manual float, & broom finish, w/edging & joints		1850	.013			.49		.49	.72
0200	Bull float, manual float & manual steel trowel		1265	.019			.71		.71	1.06
0210	For specified Random Access Floors in ACI Classes 1, 2, 3 and 4 to achieve									

03 35 Concrete Finishing

03 35 29 – Tooled Concrete Finishing

03 35 29.30 Finishing Floors		Crew	Daily Output	Labor-Hours	Unit	Material	2010 Bare Costs Labor	Equipment	Total	Total Incl O&P
0215	Composite Overall Floor Flatness and Levelness values up to F35/F25									
0250	Bull float, machine float & machine trowel (walk-behind)	C-10C	1715	.014	S.F.		.52	.03	.55	.81
0300	Power screed, bull float, machine float & trowel (walk-behind)	C-10D	2400	.010			.38	.05	.43	.61
0350	Power screed, bull float, machine float & trowel (ride-on)	C-10E	4000	.006			.23	.06	.29	.40
0352	For specified Random Access Floors in ACI Classes 5, 6, 7 and 8 to achieve									
0354	Composite Overall Floor Flatness and Levelness values up to F50/F50									
0356	Add for two-dimensional restraightening after power float	C-10	6000	.004	S.F.		.15		.15	.22
0358	For specified Random or Defined Access Floors in ACI Class 9 to achieve									
0360	Composite Overall Floor Flatness and Levelness values up to F100/F100									
0362	Add for two-dimensional restraightening after bull float & power float	C-10	3000	.008	S.F.		.30		.30	.45
0364	For specified Superflat Defined Access Floors in ACI Class 9 to achieve									
0366	Minimum Floor Flatness and Levelness values of F100/F100									
0368	Add for 2-dim'l restraightening after bull float, power float, power trowel	C-10	2000	.012	S.F.		.45		.45	.67
0400	Integral topping and finish, using 1:1:2 mix, 3/16" thick	C-10B	1000	.040		.10	1.43	.27	1.80	2.55
0450	1/2" thick		950	.042		.27	1.51	.28	2.06	2.88
0500	3/4" thick		850	.047		.41	1.68	.31	2.40	3.32
0600	1" thick		750	.053		.54	1.91	.36	2.81	3.86
0800	Granolithic topping, laid after, 1:1:1-1/2 mix, 1/2" thick		590	.068		.30	2.42	.45	3.17	4.48
0820	3/4" thick		580	.069		.45	2.47	.46	3.38	4.72
0850	1" thick		575	.070		.61	2.49	.46	3.56	4.92
0950	2" thick		500	.080		1.21	2.86	.53	4.60	6.20
1200	Heavy duty, 1:1:2, 3/4" thick, preshrunk, gray, 20 MSF		320	.125		.71	4.47	.83	6.01	8.45
1300	100 MSF		380	.105		.41	3.76	.70	4.87	6.85
1600	Exposed local aggregate finish, minimum	1 Cefi	625	.013		.23	.51		.74	.99
1650	Maximum		465	.017		.70	.68		1.38	1.77
1800	Floor abrasives, .25 psf, aluminum oxide		850	.009		.42	.37		.79	1.01
1850	Silicon carbide		850	.009		.67	.37		1.04	1.29
2000	Floor hardeners, metallic, light service, .50 psf, add		850	.009		.51	.37		.88	1.11
2050	Medium service, .75 psf		750	.011		.76	.42		1.18	1.45
2100	Heavy service, 1.0 psf		650	.012		1.01	.49		1.50	1.82
2150	Extra heavy, 1.5 psf		575	.014		1.52	.55		2.07	2.48
2300	Non-metallic, light service, .50 psf		850	.009		.23	.37		.60	.80
2350	Medium service, .75 psf		750	.011		.34	.42		.76	1
2400	Heavy service, 1.00 psf		650	.012		.45	.49		.94	1.21
2450	Extra heavy, 1.50 psf		575	.014		.68	.55		1.23	1.56
2800	Trap rock wearing surface for monolithic floors									
2810	2.0 psf	C-10B	1250	.032	S.F.	.02	1.14	.21	1.37	1.97
3000	Floor coloring, dusted on, minimum (0.6 psf), add to above	1 Cefi	1300	.006		.43	.24		.67	.83
3050	Maximum (1.0 psf), add to above	"	625	.013		.71	.51		1.22	1.52
3100	Colored powder only				Lb.	.71			.71	.78
3600	1/2" topping using 0.6 psf powdered color	C-10B	590	.068	S.F.	5.10	2.42	.45	7.97	9.75
3650	1/2" topping using 1.0 psf powdered color	"	590	.068		5.40	2.42	.45	8.27	10.10
3800	Dustproofing, solvent-based, 1 coat	1 Cefi	1900	.004		.17	.17		.34	.43
3850	2 coats		1300	.006		.61	.24		.85	1.04
4000	Epoxy-based, 1 coat		1500	.005		.12	.21		.33	.44
4050	2 coats		1500	.005		.23	.21		.44	.56
4400	Stair finish, float		275	.029			1.15		1.15	1.69
4500	Steel trowel finish		200	.040			1.59		1.59	2.32
4600	Silicon carbide finish, .25 psf		150	.053		.42	2.12		2.54	3.55
03 35 29.35 Control Joints, Saw Cut										
0010	**CONTROL JOINTS, SAW CUT**									
0100	Sawcut in green concrete									

03 35 Concrete Finishing

03 35 29 – Tooled Concrete Finishing

	03 35 29.35 Control Joints, Saw Cut	Crew	Daily Output	Labor-Hours	Unit	Material	2010 Bare Costs Labor	Equipment	Total	Total Incl O&P
0120	1" depth	C-27	2000	.008	L.F.	.06	.32	.08	.46	.62
0140	1-1/2" depth		1800	.009		.09	.35	.09	.53	.72
0160	2" depth		1600	.010		.12	.40	.10	.62	.82
0200	Blow construction debris out of control joint	C-28	6000	.001			.05		.05	.08
0300	Joint sealant									
0320	Backer rod, polyethylene, 1/4" diameter	1 Cefi	460	.017	L.F.	.03	.69		.72	1.04
0340	Sealant, polyurethane									
0360	1/4" x 1/4" (308 LF/Gal)	1 Cefi	270	.030	L.F.	.16	1.18		1.34	1.90
0380	1/4" x 1/2" (154 LF/Gal)	"	255	.031	"	.32	1.25		1.57	2.18

	03 35 29.60 Finishing Walls	Crew	Daily Output	Labor-Hours	Unit	Material	Labor	Equipment	Total	Total Incl O&P
0010	**FINISHING WALLS**									
0020	Break ties and patch voids	1 Cefi	540	.015	S.F.	.03	.59		.62	.90
0050	Burlap rub with grout		450	.018		.03	.71		.74	1.07
0100	Carborundum rub, dry		270	.030			1.18		1.18	1.72
0150	Wet rub		175	.046			1.81		1.81	2.65
0300	Bush hammer, green concrete	B-39	1000	.048			1.67	.22	1.89	2.81
0350	Cured concrete	"	650	.074			2.57	.35	2.92	4.32
0600	Float finish, 1/16" thick	1 Cefi	300	.027		.30	1.06		1.36	1.87
0700	Sandblast, light penetration	E-11	1100	.029		.76	1.09	.21	2.06	2.86
0750	Heavy penetration	"	375	.085		1.53	3.18	.61	5.32	7.60
0850	Grind form fins flush	1 Clab	700	.011	L.F.		.38		.38	.58

03 35 33 – Stamped Concrete Finishing

	03 35 33.50 Slab Texture Stamping	Crew	Daily Output	Labor-Hours	Unit	Material	Labor	Equipment	Total	Total Incl O&P
0010	**SLAB TEXTURE STAMPING**									
0050	Stamping requires that concrete first be placed, struck off, consolidated,									
0060	bull floated and free of bleed water. Decorative stamping tasks include:									
0100	Step 1 - first application of dry shake colored hardener	1 Cefi	6400	.001	S.F.	.38	.05		.43	.49
0110	Step 2 - bull float		6400	.001			.05		.05	.07
0130	Step 3 - second application of dry shake colored hardener		6400	.001		.19	.05		.24	.28
0140	Step 4 - bull float, manual float & steel trowel	3 Cefi	1280	.019			.74		.74	1.09
0150	Step 5 - application of dry shake colored release agent	1 Cefi	6400	.001		.09	.05		.14	.16
0160	Step 6 - place, tamp & remove mats	3 Cefi	2400	.010		1.41	.40		1.81	2.13
0170	Step 7 - touch up edges, mat joints & simulated grout lines	1 Cefi	1280	.006			.25		.25	.36
0300	Alternate stamping estimating method includes all tasks above	4 Cefi	800	.040		2.07	1.59		3.66	4.60
0400	Step 8 - pressure wash @ 3000 psi after 24 hours	1 Cefi	1600	.005			.20		.20	.29
0500	Step 9 - roll 2 coats cure/seal compound when dry	"	800	.010		.49	.40		.89	1.12

03 37 Specialty Placed Concrete

03 37 13 – Shotcrete

	03 37 13.30 Gunite (Dry-Mix)	Crew	Daily Output	Labor-Hours	Unit	Material	Labor	Equipment	Total	Total Incl O&P
0010	**GUNITE (DRY-MIX)**									
0020	Applied in 1" layers, no mesh included	C-8	2000	.028	S.F.	.33	1.03	.37	1.73	2.32
0100	Mesh for gunite 2 x 2, #12, to 3" thick	2 Rodm	800	.020		.35	.94		1.29	1.90
0150	Over 3" thick	"	500	.032		.35	1.50		1.85	2.80
0300	Typical in place, including mesh, 2" thick, minimum	C-16	1000	.072		1.01	2.80	.74	4.55	6.25
0350	Maximum		500	.144		1.01	5.60	1.48	8.09	11.35
0500	4" thick, minimum		750	.096		1.66	3.74	.99	6.39	8.65
0550	Maximum		350	.206		1.66	8	2.11	11.77	16.45
0900	Prepare old walls, no scaffolding, minimum	C-10	1000	.024			.90		.90	1.34
0950	Maximum	"	275	.087			3.27		3.27	4.86

03 37 Specialty Placed Concrete

03 37 13 – Shotcrete

03 37 13.30 Gunite (Dry-Mix)		Crew	Daily Output	Labor-Hours	Unit	2010 Bare Costs Material	Labor	Equipment	Total	Total Incl O&P
1100	For high finish requirement or close tolerance, add, minimum				S.F.		50%			
1150	Maximum				↓		110%			

03 39 Concrete Curing

03 39 13 – Water Concrete Curing

03 39 13.50 Water Curing

		Crew	Daily Output	Labor-Hours	Unit	Material	Labor	Equipment	Total	Total Incl O&P
0010	**WATER CURING**									
0015	With burlap, 4 uses assumed, 7.5 oz.	2 Clab	55	.291	C.S.F.	8.65	9.65		18.30	24.50
0100	10 oz.	"	55	.291	"	15.55	9.65		25.20	32
0400	Curing blankets, 1" to 2" thick, buy, minimum				S.F.	.33			.33	.36
0450	Maximum				"	.50			.50	.54

03 39 23 – Membrane Concrete Curing

03 39 23.13 Chemical Compound Membrane Concrete Curing

		Crew	Daily Output	Labor-Hours	Unit	Material	Labor	Equipment	Total	Total Incl O&P
0010	**CHEMICAL COMPOUND MEMBRANE CONCRETE CURING**									
0300	Sprayed membrane curing compound	2 Clab	95	.168	C.S.F.	6.10	5.55		11.65	15.35
0700	Curing compound, solvent based, 400 S.F./gal, 55 gallon lots				Gal.	17.30			17.30	19.05
0720	5 gallon lots					24.50			24.50	27
0800	Curing compound, water based, 250 S.F./gal, 55 gallon lots					20.50			20.50	22.50
0820	5 gallon lots				↓	23.50			23.50	26

03 39 23.23 Sheet Membrane Concrete Curing

		Crew	Daily Output	Labor-Hours	Unit	Material	Labor	Equipment	Total	Total Incl O&P
0010	**SHEET MEMBRANE CONCRETE CURING**									
0200	Curing blanket, burlap/poly, 2-ply	2 Clab	70	.229	C.S.F.	16	7.55		23.55	29.50

03 41 Precast Structural Concrete

03 41 05 – Precast Concrete Members

03 41 05.10 Precast Beams

		Crew	Daily Output	Labor-Hours	Unit	Material	Labor	Equipment	Total	Total Incl O&P
0010	**PRECAST BEAMS** R034105-30									
0011	L-shaped, 20' span, 12" x 20"	C-11	32	2.250	Ea.	2,000	103	62.50	2,165.50	2,450
1000	Inverted tee beams, add to above, small beams					15%				
1050	Large beams					20%				
1200	Rectangular, 20' span, 12" x 20"	C-11	32	2.250		1,500	103	62.50	1,665.50	1,900
1250	18" x 36"		24	3		2,725	138	83.50	2,946.50	3,325
1300	24" x 44"		22	3.273		3,775	150	91	4,016	4,525
1400	30' span, 12" x 36"		24	3		3,425	138	83.50	3,646.50	4,100
1450	18" x 44"		20	3.600		4,925	165	100	5,190	5,825
1500	24" x 52"		16	4.500		6,575	207	125	6,907	7,725
1600	40' span, 12" x 52"		20	3.600		6,375	165	100	6,640	7,400
1650	18" x 52"		16	4.500		7,575	207	125	7,907	8,825
1700	24" x 52"		12	6		8,775	275	167	9,217	10,300
2000	"T" shaped, 20' span, 12" x 20"		32	2.250		2,600	103	62.50	2,765.50	3,100
2050	18" x 36"		24	3		4,675	138	83.50	4,896.50	5,475
2100	24" x 44"		22	3.273		6,475	150	91	6,716	7,475
2200	30' span, 12" x 36"		24	3		5,825	138	83.50	6,046.50	6,725
2250	18" x 44"		20	3.600		8,375	165	100	8,640	9,600
2300	24" x 52"		16	4.500		11,200	207	125	11,532	12,800
2500	40' span, 12" x 52"		20	3.600		10,800	165	100	11,065	12,200
2550	18" x 52"		16	4.500		12,900	207	125	13,232	14,700
2600	24" x 52"	↓	12	6	↓	14,900	275	167	15,342	17,100

03 41 Precast Structural Concrete

03 41 05 – Precast Concrete Members

03 41 05.15 Precast Columns

	03 41 05.15 Precast Columns	Crew	Daily Output	Labor-Hours	Unit	Material	Labor	Equipment	Total	Total Incl O&P
						2010 Bare Costs				
0010	**PRECAST COLUMNS** R034105-30									
0020	Rectangular to 12′ high, small columns	C-11	120	.600	L.F.	106	27.50	16.70	150.20	181
0050	Large columns		96	.750		184	34.50	21	239.50	284
0300	24′ high, small columns		192	.375		106	17.20	10.45	133.65	157
0350	Large columns		144	.500		184	23	13.90	220.90	256

03 41 05.25 Precast Joists

	03 41 05.25 Precast Joists	Crew	Daily Output	Labor-Hours	Unit	Material	Labor	Equipment	Total	Total Incl O&P
0010	**PRECAST JOISTS** R034105-30									
0015	40 psf L.L., 6″ deep for 12′ spans	C-12	600	.080	L.F.	17.65	3.28	1.09	22.02	25.50
0050	8″ deep for 16′ spans		575	.083		29.50	3.42	1.14	34.06	39
0100	10″ deep for 20′ spans		550	.087		51.50	3.57	1.19	56.26	63.50
0150	12″ deep for 24′ spans		525	.091		70.50	3.74	1.25	75.49	84.50

03 41 13 – Precast Concrete Hollow Core Planks

03 41 13.50 Precast Slab Planks

	03 41 13.50 Precast Slab Planks	Crew	Daily Output	Labor-Hours	Unit	Material	Labor	Equipment	Total	Total Incl O&P
0010	**PRECAST SLAB PLANKS** R034105-30									
0020	Prestressed roof/floor members, grouted, solid, 4″ thick	C-11	2400	.030	S.F.	5.55	1.38	.83	7.76	9.35
0050	6″ thick		2800	.026		6.50	1.18	.72	8.40	9.95
0100	Hollow, 8″ thick **CN**		3200	.023		7.20	1.03	.63	8.86	10.35
0150	10″ thick		3600	.020		7.50	.92	.56	8.98	10.45
0200	12″ thick		4000	.018		8	.83	.50	9.33	10.75

03 41 16 – Precast Concrete Slabs

03 41 16.20 Precast Concrete Channel Slabs

	03 41 16.20 Precast Concrete Channel Slabs	Crew	Daily Output	Labor-Hours	Unit	Material	Labor	Equipment	Total	Total Incl O&P
0010	**PRECAST CONCRETE CHANNEL SLABS**									
0335	Lightweight concrete channel slab, long runs, 2-3/4″ thick	C-12	1575	.030	S.F.	7.25	1.25	.42	8.92	10.35
0375	3-3/4″ thick		1550	.031		7.55	1.27	.42	9.24	10.70
0475	4-3/4″ thick		1525	.031		8.25	1.29	.43	9.97	11.55
1275	Short pieces, 2-3/4″ thick		785	.061		10.90	2.50	.84	14.24	16.70
1375	3-3/4″ thick		770	.062		11.35	2.55	.85	14.75	17.30
1475	4-3/4″ thick		762	.063		12.40	2.58	.86	15.84	18.50

03 41 16.50 Precast Lightweight Concrete Plank

	03 41 16.50 Precast Lightweight Concrete Plank	Crew	Daily Output	Labor-Hours	Unit	Material	Labor	Equipment	Total	Total Incl O&P
0010	**PRECAST LIGHTWEIGHT CONCRETE PLANK**									
0015	Lightweight plank, nailable, T&G, 2″ thick	C-12	1800	.027	S.F.	6.65	1.09	.36	8.10	9.35
0150	For premium ceiling finish, add				″	10%				
0200	For sloping roofs, slope over 4 in 12, add						25%			
0250	Slope over 6 in 12, add						150%			

03 41 23 – Precast Concrete Stairs

03 41 23.50 Precast Stairs

	03 41 23.50 Precast Stairs	Crew	Daily Output	Labor-Hours	Unit	Material	Labor	Equipment	Total	Total Incl O&P
0010	**PRECAST STAIRS**									
0020	Precast concrete treads on steel stringers, 3′ wide	C-12	75	.640	Riser	105	26	8.75	139.75	166
0300	Front entrance, 5′ wide with 48″ platform, 2 risers		16	3	Flight	460	123	41	624	740
0350	5 risers		12	4		715	164	54.50	933.50	1,100
0500	6′ wide, 2 risers		15	3.200		510	131	43.50	684.50	810
0550	5 risers		11	4.364		790	179	59.50	1,028.50	1,200
0700	7′ wide, 2 risers		14	3.429		620	140	47	807	945
0750	5 risers		10	4.800		1,025	197	65.50	1,287.50	1,500
1200	Basement entrance stairs, steel bulkhead doors, minimum	B-51	22	2.182		1,350	72.50	10.90	1,433.40	1,600
1250	Maximum	″	11	4.364		2,225	145	22	2,392	2,700

03 41 33 – Precast Structural Pretensioned Concrete

03 41 33.60 Tees

	03 41 33.60 Tees	Crew	Daily Output	Labor-Hours	Unit	Material	Labor	Equipment	Total	Total Incl O&P
0010	**TEES** R034105-30									

03 41 Precast Structural Concrete

03 41 33 – Precast Structural Pretensioned Concrete

03 41 33.60 Tees		Crew	Daily Output	Labor-Hours	Unit	2010 Bare Costs Material	Labor	Equipment	Total	Total Incl O&P
0020	Quad tee, short spans, roof	C-11	7200	.010	S.F.	6.90	.46	.28	7.64	8.70
0050	Floor		7200	.010		6.90	.46	.28	7.64	8.70
0200	Double tee, floor members, 60′ span		8400	.009		8.60	.39	.24	9.23	10.40
0250	80′ span		8000	.009		10.10	.41	.25	10.76	12.10
0300	Roof members, 30′ span		4800	.015		7.30	.69	.42	8.41	9.70
0350	50′ span		6400	.011		8.35	.52	.31	9.18	10.40
0400	Wall members, up to 55′ high		3600	.020		11.60	.92	.56	13.08	15
0500	Single tee roof members, 40′ span		3200	.023		9.20	1.03	.63	10.86	12.60
0550	80′ span		5120	.014		11.30	.65	.39	12.34	13.95
0600	100′ span		6000	.012		17	.55	.33	17.88	20
0650	120′ span		6000	.012		18.55	.55	.33	19.43	22
1000	Double tees, floor members									
1100	Lightweight, 20″ x 8′ wide, 45′ span	C-11	20	3.600	Ea.	2,925	165	100	3,190	3,625
1150	24″ x 8′ wide, 50′ span		18	4		3,200	184	111	3,495	3,925
1200	32″ x 10′ wide, 60′ span		16	4.500		5,400	207	125	5,732	6,425
1250	Standard weight, 12″ x 8′ wide, 20′ span		22	3.273		1,100	150	91	1,341	1,550
1300	16″ x 8′ wide, 25′ span		20	3.600		1,475	165	100	1,740	2,000
1350	18″ x 8′ wide, 30′ span *CN*		20	3.600		1,850	165	100	2,115	2,450
1400	20″ x 8′ wide, 45′ span		18	4		2,100	184	111	2,395	2,725
1450	24″ x 8′ wide, 50′ span		16	4.500		2,700	207	125	3,032	3,475
1500	32″ x 10′ wide, 60′ span		14	5.143		4,950	236	143	5,329	6,000
2000	Roof members									
2050	Lightweight, 20″ x 8′ wide, 40′ span	C-11	20	3.600	Ea.	2,475	165	100	2,740	3,125
2100	24″ x 8′ wide, 50′ span		18	4		3,300	184	111	3,595	4,075
2150	32″ x 10′ wide, 60′ span		16	4.500		5,450	207	125	5,782	6,500
2200	Standard weight, 12″ x 8′ wide, 30′ span		22	3.273		1,650	150	91	1,891	2,175
2250	16″ x 8′ wide, 30′ span		20	3.600		1,725	165	100	1,990	2,300
2300	18″ x 8′ wide, 30′ span		20	3.600		1,925	165	100	2,190	2,525
2350	20″ x 8′ wide, 40′ span		18	4		1,975	184	111	2,270	2,600
2400	24″ x 8′ wide, 50′ span		16	4.500		2,650	207	125	2,982	3,400
2450	32″ x 10′ wide, 60′ span		14	5.143		4,625	236	143	5,004	5,650

03 41 36 – Precast Structural Post-Tensioned Concrete

03 41 36.50 Post-Tensioned Jobs

		Crew	Daily Output	Labor-Hours	Unit	Material	Labor	Equipment	Total	Total Incl O&P
0010	**POST-TENSIONED JOBS** R034105-30									
0100	Post-tensioned in place, small job R034136-90	C-17B	8.50	9.647	C.Y.	1,175	415	47	1,637	2,000
0200	Large job	″	10	8.200	″	885	355	40	1,280	1,550

03 45 Precast Architectural Concrete

03 45 13 – Faced Architectural Precast Concrete

03 45 13.50 Precast Wall Panels

		Crew	Daily Output	Labor-Hours	Unit	Material	Labor	Equipment	Total	Total Incl O&P
0010	**PRECAST WALL PANELS** R034513-10									
0050	Uninsulated 4″ thick, smooth gray									
0150	Low rise, 4′ x 8′ x 4″ thick	C-11	320	.225	S.F.	20.50	10.35	6.25	37.10	47
0210	8′ x 8′, 4″ thick		576	.125		20	5.75	3.48	29.23	35.50
0250	8′ x 16′ x 4″ thick		1024	.070		19.95	3.23	1.96	25.14	29.50
0600	High rise, 4′ x 8′ x 4″ thick		288	.250		20.50	11.45	6.95	38.90	49.50
0650	8′ x 8′ x 4″ thick		512	.141		20	6.45	3.91	30.36	37.50
0700	8′ x 16′ x 4″ thick		768	.094		19.95	4.30	2.61	26.86	32
0750	20′ x 10′, 6″ thick, smooth gray *CN*		1400	.051		31	2.36	1.43	34.79	39.50
0800	Insulated panel, 2″ polystyrene, add					1.04			1.04	1.14

03 45 Precast Architectural Concrete

03 45 13 – Faced Architectural Precast Concrete

03 45 13.50 Precast Wall Panels		Crew	Daily Output	Labor-Hours	Unit	Material	2010 Bare Costs Labor	Equipment	Total	Total Incl O&P
0850	2" urethane, add				S.F.	.84			.84	.92
1200	Finishes, white, add					2.73			2.73	3.01
1250	Exposed aggregate, add					1.69			1.69	1.86
1300	Granite faced, domestic, add				↓	28.50			28.50	31.50
2200	Fiberglass reinforced cement with urethane core									
2210	R20, 8′ x 8′, minimum	E-2	750	.075	S.F.	21	3.41	2.14	26.55	31
2220	Maximum	"	600	.093	"	31.50	4.26	2.68	38.44	44.50

03 47 Site-Cast Concrete

03 47 13 – Tilt-Up Concrete

03 47 13.50 Tilt-Up Wall Panels

		Crew	Daily Output	Labor-Hours	Unit	Material	Labor	Equipment	Total	Total Incl O&P
0010	**TILT-UP WALL PANELS** R034713-20									
0015	Wall panel construction, walls only, 5-1/2" thick	C-14	1600	.090	S.F.	6.80	3.66	.97	11.43	14.15
0100	7-1/2" thick		1550	.093		7.95	3.78	1	12.73	15.70
0500	Walls and columns, 5-1/2" thick walls, 12" x 12" columns		1565	.092		10.25	3.75	.99	14.99	18.15
0550	7-1/2" thick wall, 12" x 12" columns		1370	.105	↓	11.80	4.28	1.13	17.21	21
0800	Columns only, site precast, minimum		200	.720	L.F.	16.65	29.50	7.75	53.90	72.50
0850	Maximum	↓	105	1.371	"	19.30	56	14.70	90	124

03 48 Precast Concrete Specialties

03 48 43 – Precast Concrete Trim

03 48 43.40 Precast Lintels

		Crew	Daily Output	Labor-Hours	Unit	Material	Labor	Equipment	Total	Total Incl O&P
0010	**PRECAST LINTELS**, smooth gray, prestressed, stock units only									
0800	4" wide, 8" high, x 4′ long	D-10	28	1.143	Ea.	21	46.50	23.50	91	120
0850	8′ long		24	1.333		56.50	54.50	27.50	138.50	174
1000	6" wide, 8" high, x 4′ long		26	1.231		31	50.50	25	106.50	138
1050	10′ long		22	1.455		83	59.50	30	172.50	214
1200	8" wide, 8" high, x 4′ long		24	1.333		37.50	54.50	27.50	119.50	153
1250	12′ long	↓	20	1.600		130	65.50	33	228.50	278
1275	For custom sizes, types, colors, or finishes of precast lintels, add				↓	150%				

03 48 43.90 Precast Window Sills

		Crew	Daily Output	Labor-Hours	Unit	Material	Labor	Equipment	Total	Total Incl O&P
0010	**PRECAST WINDOW SILLS**									
0600	Precast concrete, 4" tapers to 3", 9" wide	D-1	70	.229	L.F.	12.25	8.60		20.85	26.50
0650	11" wide		60	.267		16	10.05		26.05	33
0700	13" wide, 3 1/2" tapers to 2 1/2", 12" wall	↓	50	.320	↓	16	12.05		28.05	36

03 51 Cast Roof Decks

03 51 13 – Cementitious Wood Fiber Decks

03 51 13.50 Cementitious/Wood Fiber Planks

		Crew	Daily Output	Labor-Hours	Unit	Material	Labor	Equipment	Total	Total Incl O&P
0010	**CEMENTITIOUS/WOOD FIBER PLANKS** R051223-50									
0050	Plank, beveled edge, 1" thick	2 Carp	1000	.016	S.F.	2.32	.66		2.98	3.57
0100	1-1/2" thick		975	.016		2.94	.68		3.62	4.28
0150	T & G, 2" thick		950	.017		3.54	.70		4.24	4.97
0200	2-1/2" thick		925	.017		4.14	.72		4.86	5.65
0250	3" thick	↓	900	.018		4.75	.74		5.49	6.40
1000	Bulb tee, sub-purlin and grout, 6′ span, add	E-1	5000	.005		2.07	.22	.03	2.32	2.67
1100	8′ span	"	4200	.006	↓	2.31	.26	.03	2.60	3.02

03 52 Lightweight Concrete Roof Insulation

03 52 16 – Lightweight Insulating Concrete

03 52 16.13 Lightweight Cellular Insulating Concrete		Crew	Daily Output	Labor-Hours	Unit	Material	2010 Bare Costs Labor	Equipment	Total	Total Incl O&P
0010	LIGHTWEIGHT CELLULAR INSULATING CONCRETE R035216-10									
0020	Portland cement and foaming agent G	C-8	50	1.120	C.Y.	144	41	14.75	199.75	236

03 52 16.16 Lightweight Aggregate Insulating Concrete		Crew	Daily Output	Labor-Hours	Unit	Material	Labor	Equipment	Total	Total Incl O&P
0010	LIGHTWEIGHT AGGREGATE INSULATING CONCRETE R035216-10									
0100	Poured vermiculite or perlite, field mix,									
0110	1:6 field mix G	C-8	50	1.120	C.Y.	219	41	14.75	274.75	320
0200	Ready mix, 1:6 mix, roof fill, 2" thick G		10000	.006	S.F.	1.22	.21	.07	1.50	1.73
0250	3" thick G		7700	.007		1.83	.27	.10	2.20	2.52
0400	Expanded volcanic glass rock, with binder, minimum G	2 Carp	1500	.011		.40	.44		.84	1.12
0450	Maximum G	"	1200	.013		1.19	.55		1.74	2.16

03 54 Cast Underlayment

03 54 13 – Gypsum Cement Underlayment

03 54 13.50 Gypsum Underlayment		Crew	Daily Output	Labor-Hours	Unit	Material	Labor	Equipment	Total	Total Incl O&P
0010	GYPSUM UNDERLAYMENT									
1000	Poured gypsum, 2" thick	C-8	6000	.009	S.F.	1.78	.34	.12	2.24	2.62
1100	3" thick	"	4800	.012	"	2.67	.43	.15	3.25	3.76

03 54 16 – Hydraulic Cement Underlayment

03 54 16.50 Cement Underlayment		Crew	Daily Output	Labor-Hours	Unit	Material	Labor	Equipment	Total	Total Incl O&P
0010	CEMENT UNDERLAYMENT									
2510	Underlayment, P.C based self-leveling, 4100 psi, pumped, 1/4"	C-8	20000	.003	S.F.	1.52	.10	.04	1.66	1.87
2520	1/2"		19000	.003		3.04	.11	.04	3.19	3.55
2530	3/4"		18000	.003		4.56	.11	.04	4.71	5.20
2540	1"		17000	.003		6.10	.12	.04	6.26	6.95
2550	1-1/2"		15000	.004		9.10	.14	.05	9.29	10.30
2560	Hand mix, 1/2"	C-18	4000	.002		3.04	.08	.01	3.13	3.49
2610	Topping, P.C. based self-level/dry 6100 psi, pumped, 1/4"	C-8	20000	.003		2.34	.10	.04	2.48	2.77
2620	1/2"		19000	.003		4.68	.11	.04	4.83	5.35
2630	3/4"		18000	.003		7	.11	.04	7.15	7.90
2660	1"		17000	.003		9.35	.12	.04	9.51	10.55
2670	1-1/2"		15000	.004		14.05	.14	.05	14.24	15.70
2680	Hand mix, 1/2"	C-18	4000	.002		4.68	.08	.01	4.77	5.30

03 62 Non-Shrink Grouting

03 62 13 – Non-Metallic Non-Shrink Grouting

03 62 13.50 Grout, Non-Metallic Non-shrink		Crew	Daily Output	Labor-Hours	Unit	Material	Labor	Equipment	Total	Total Incl O&P
0010	GROUT, NON-METALLIC NON-SHRINK									
0300	Non-shrink, non-metallic, 1" deep	1 Cefi	35	.229	S.F.	7.40	9.05		16.45	21.50
0350	2" deep	"	25	.320	"	14.75	12.70		27.45	35

03 62 16 – Metallic Non-Shrink Grouting

03 62 16.50 Grout, Metallic Non-Shrink		Crew	Daily Output	Labor-Hours	Unit	Material	Labor	Equipment	Total	Total Incl O&P
0010	GROUT, METALLIC NON-SHRINK									
0020	Column & machine bases, non-shrink, metallic, 1" deep	1 Cefi	35	.229	S.F.	8.90	9.05		17.95	23
0050	2" deep	"	25	.320	"	17.75	12.70		30.45	38

03 63 Epoxy Grouting

03 63 05 – Grouting of Dowels and Fasteners

03 63 05.10 Epoxy Only		Crew	Daily Output	Labor-Hours	Unit	2010 Bare Costs Material	Labor	Equipment	Total	Total Incl O&P
0010	**EPOXY ONLY**									
1500	Chemical anchoring, epoxy cartridge, excludes layout, drilling, fastener									
1530	For fastener 3/4" diam. x 6" embedment	2 Skwk	72	.222	Ea.	4.57	9.45		14.02	19.60
1535	1" diam. x 8" embedment		66	.242		6.85	10.35		17.20	23.50
1540	1-1/4" diam. x 10" embedment		60	.267		13.70	11.35		25.05	32.50
1545	1-3/4" diam. x 12" embedment		54	.296		23	12.60		35.60	44.50
1550	14" embedment		48	.333		27.50	14.20		41.70	52
1555	2" diam. x 12" embedment		42	.381		36.50	16.25		52.75	65
1560	18" embedment	↓	32	.500	↓	45.50	21.50		67	83.50

03 81 Concrete Cutting

03 81 13 – Flat Concrete Sawing

03 81 13.50 Concrete Floor/Slab Cutting		Crew	Daily Output	Labor-Hours	Unit	Material	Labor	Equipment	Total	Total Incl O&P
0010	**CONCRETE FLOOR/SLAB CUTTING**									
0400	Concrete slabs, mesh reinforcing, up to 3" deep	B-89	980	.016	L.F.	.50	.60	.49	1.59	1.99
0420	Each additional inch of depth	"	1600	.010	"	.17	.37	.30	.84	1.07

03 81 16 – Track Mounted Concrete Wall Sawing

03 81 16.50 Concrete Wall Cutting		Crew	Daily Output	Labor-Hours	Unit	Material	Labor	Equipment	Total	Total Incl O&P
0010	**CONCRETE WALL CUTTING**									
0800	Concrete walls, hydraulic saw, plain, per inch of depth	B-89B	250	.064	L.F.	.45	2.35	3.37	6.17	7.75
0820	Rod reinforcing, per inch of depth	"	150	.107	"	.64	3.92	5.60	10.16	12.80

03 82 Concrete Boring

03 82 13 – Concrete Core Drilling

03 82 13.10 Core Drilling		Crew	Daily Output	Labor-Hours	Unit	Material	Labor	Equipment	Total	Total Incl O&P
0010	**CORE DRILLING**									
0020	Reinf. conc slab, up to 6" thick, incl. bit, layout & set up									
0100	1" diameter core	B-89A	17	.941	Ea.	3.65	35.50	6.70	45.85	66.50
0150	Each added inch thick in same hole, add		1440	.011		.68	.42	.08	1.18	1.49
0300	3" diameter core		16	1		12.20	38	7.10	57.30	79.50
0350	Each added inch thick in same hole, add		720	.022		1.40	.84	.16	2.40	3.01
0500	4" diameter core		15	1.067		15.45	40.50	7.60	63.55	87.50
0550	Each added inch thick in same hole, add		480	.033		1.72	1.26	.24	3.22	4.09
0700	6" diameter core		14	1.143		20	43.50	8.15	71.65	97.50
0750	Each added inch thick in same hole, add		360	.044		2.73	1.68	.32	4.73	5.95
0900	8" diameter core		13	1.231		33	46.50	8.75	88.25	118
0950	Each added inch thick in same hole, add		288	.056		3.75	2.10	.40	6.25	7.80
1100	10" diameter core		12	1.333		26.50	50.50	9.50	86.50	117
1150	Each added inch thick in same hole, add		240	.067		3.75	2.52	.47	6.74	8.55
1300	12" diameter core		11	1.455		31	55	10.35	96.35	130
1350	Each added inch thick in same hole, add		206	.078		4.55	2.94	.55	8.04	10.15
1500	14" diameter core		10	1.600		37	60.50	11.40	108.90	147
1550	Each added inch thick in same hole, add		180	.089		6.20	3.36	.63	10.19	12.70
1700	18" diameter core		9	1.778		45.50	67.50	12.65	125.65	168
1750	Each added inch thick in same hole, add	↓	144	.111		8.80	4.21	.79	13.80	17.05
1760	For horizontal holes, add to above				↓		20%	20%		
1770	Prestressed hollow core plank, 8" thick									
1780	1" diameter core	B-89A	17.50	.914	Ea.	2.37	34.50	6.50	43.37	63.50
1790	Each added inch thick in same hole, add	↓	3840	.004	↓	.47	.16	.03	.66	.79

03 82 Concrete Boring

03 82 13 – Concrete Core Drilling

03 82 13.10 Core Drilling		Crew	Daily Output	Labor-Hours	Unit	Material	2010 Bare Costs Labor	Equipment	Total	Total Incl O&P
1800	3" diameter core	B-89A	17	.941	Ea.	4.95	35.50	6.70	47.15	68
1810	Each added inch thick in same hole, add		1920	.008		.85	.32	.06	1.23	1.50
1820	4" diameter core		16.50	.970		6.80	36.50	6.90	50.20	71.50
1830	Each added inch thick in same hole, add		1280	.013		1.25	.47	.09	1.81	2.21
1840	6" diameter core		15.50	1.032		8.35	39	7.35	54.70	77.50
1850	Each added inch thick in same hole, add		960	.017		1.42	.63	.12	2.17	2.66
1860	8" diameter core		15	1.067		11.85	40.50	7.60	59.95	83.50
1870	Each added inch thick in same hole, add		768	.021		1.99	.79	.15	2.93	3.56
1880	10" diameter core		14	1.143		15.70	43.50	8.15	67.35	92.50
1890	Each added inch thick in same hole, add		640	.025		2.07	.95	.18	3.20	3.94
1900	12" diameter core		13.50	1.185		18.35	45	8.45	71.80	98.50
1910	Each added inch thick in same hole, add		548	.029		3.32	1.11	.21	4.64	5.60
1999	Drilling, core, minimum labor/equipment charge		4	4	Job		151	28.50	179.50	265
3010	Bits for core drill, diamond, premium, 1" diameter				Ea.	108			108	119
3020	3" diameter					234			234	257
3040	4" diameter					260			260	286
3050	6" diameter					415			415	455
3080	8" diameter					585			585	645
3120	12" diameter					885			885	975
3180	18" diameter					1,725			1,725	1,875
3240	24" diameter					2,425			2,425	2,675

03 82 16 – Concrete Drilling

03 82 16.10 Concrete Impact Drilling		Crew	Daily Output	Labor-Hours	Unit	Material	Labor	Equipment	Total	Total Incl O&P
0010	**CONCRETE IMPACT DRILLING**									
0050	Up to 4" deep in conc/brick floor/wall, incl. bit & layout, no anchor									
0100	Holes, 1/4" diameter	1 Carp	75	.107	Ea.	.07	4.43		4.50	6.90
0150	For each additional inch of depth, add		430	.019		.02	.77		.79	1.21
0200	3/8" diameter		63	.127		.06	5.30		5.36	8.20
0250	For each additional inch of depth, add		340	.024		.02	.98		1	1.53
0300	1/2" diameter		50	.160		.06	6.65		6.71	10.30
0350	For each additional inch of depth, add		250	.032		.02	1.33		1.35	2.07
0400	5/8" diameter		48	.167		.09	6.95		7.04	10.80
0450	For each additional inch of depth, add		240	.033		.02	1.38		1.40	2.16
0500	3/4" diameter		45	.178		.11	7.40		7.51	11.50
0550	For each additional inch of depth, add		220	.036		.03	1.51		1.54	2.36
0600	7/8" diameter		43	.186		.13	7.75		7.88	12.05
0650	For each additional inch of depth, add		210	.038		.03	1.58		1.61	2.48
0700	1" diameter		40	.200		.15	8.30		8.45	12.95
0750	For each additional inch of depth, add		190	.042		.04	1.75		1.79	2.74
0800	1-1/4" diameter		38	.211		.22	8.75		8.97	13.75
0850	For each additional inch of depth, add		180	.044		.06	1.85		1.91	2.91
0900	1-1/2" diameter		35	.229		.32	9.50		9.82	15
0950	For each additional inch of depth, add		165	.048		.08	2.01		2.09	3.20
1000	For ceiling installations, add						40%			

Division Notes

	CREW	DAILY OUTPUT	LABOR-HOURS	UNIT	2010 BARE COSTS MAT.	LABOR	EQUIP.	TOTAL	TOTAL INCL O&P

Estimating Tips

04 05 00 Common Work Results for Masonry

- The terms *mortar* and *grout* are often used interchangeably, and incorrectly. Mortar is used to bed masonry units, seal the entry of air and moisture, provide architectural appearance, and allow for size variations in the units. Grout is used primarily in reinforced masonry construction and is used to bond the masonry to the reinforcing steel. Common mortar types are M(2500 psi), S(1800 psi), N(750 psi), and O(350 psi), and conform to ASTM C270. Grout is either fine or coarse and conforms to ASTM C476, and in-place strengths generally exceed 2500 psi. Mortar and grout are different components of masonry construction and are placed by entirely different methods. An estimator should be aware of their unique uses and costs.
- Waste, specifically the loss/droppings of mortar and the breakage of brick and block, is included in all masonry assemblies in this division. A factor of 25% is added for mortar and 3% for brick and concrete masonry units.
- Scaffolding or staging is not included in any of the Division 4 costs. Refer to Subdivision 01 54 23 for scaffolding and staging costs.

04 20 00 Unit Masonry

- The most common types of unit masonry are brick and concrete masonry. The major classifications of brick are building brick (ASTM C62), facing brick (ASTM C216), glazed brick, fire brick, and pavers. Many varieties of texture and appearance can exist within these classifications, and the estimator would be wise to check local custom and availability within the project area. For repair and remodeling jobs, matching the existing brick may be the most important criteria.
- Brick and concrete block are priced by the piece and then converted into a price per square foot of wall. Openings less than two square feet are generally ignored by the estimator because any savings in units used is offset by the cutting and trimming required.
- It is often difficult and expensive to find and purchase small lots of historic brick. Costs can vary widely. Many design issues affect costs, selection of mortar mix, and repairs or replacement of masonry materials. Cleaning techniques must be reflected in the estimate.
- All masonry walls, whether interior or exterior, require bracing. The cost of bracing walls during construction should be included by the estimator, and this bracing must remain in place until permanent bracing is complete. Permanent bracing of masonry walls is accomplished by masonry itself, in the form of pilasters or abutting wall corners, or by anchoring the walls to the structural frame. Accessories in the form of anchors, anchor slots, and ties are used, but their supply and installation can be by different trades. For instance, anchor slots on spandrel beams and columns are supplied and welded in place by the steel fabricator, but the ties from the slots into the masonry are installed by the bricklayer. Regardless of the installation method, the estimator must be certain that these accessories are accounted for in pricing.

Reference Numbers

Reference numbers are shown in shaded boxes at the beginning of some major classifications. These numbers refer to related items in the Reference Section. The reference information may be an estimating procedure, an alternate pricing method, or technical information.

Note: Not all subdivisions listed here necessarily appear in this publication.

04 01 Maintenance of Masonry

04 01 20 – Maintenance of Unit Masonry

04 01 20.20 Pointing Masonry		Crew	Daily Output	Labor-Hours	Unit	Material	2010 Bare Costs Labor	Equipment	Total	Total Incl O&P
0010	**POINTING MASONRY**									
0300	Cut and repoint brick, hard mortar, running bond	1 Bric	80	.100	S.F.	.53	4.18		4.71	6.90
0320	Common bond		77	.104		.53	4.34		4.87	7.15
0360	Flemish bond		70	.114		.56	4.77		5.33	7.80
0400	English bond		65	.123		.56	5.15		5.71	8.35
0600	Soft old mortar, running bond		100	.080		.53	3.34		3.87	5.65
0620	Common bond		96	.083		.53	3.48		4.01	5.85
0640	Flemish bond		90	.089		.56	3.71		4.27	6.20
0680	English bond		82	.098		.56	4.07		4.63	6.75
0700	Stonework, hard mortar		140	.057	L.F.	.70	2.39		3.09	4.37
0720	Soft old mortar		160	.050	"	.70	2.09		2.79	3.92
1000	Repoint, mask and grout method, running bond		95	.084	S.F.	.70	3.52		4.22	6.05
1020	Common bond		90	.089		.70	3.71		4.41	6.35
1040	Flemish bond		86	.093		.74	3.88		4.62	6.65
1060	English bond		77	.104		.74	4.34		5.08	7.35
2000	Scrub coat, sand grout on walls, minimum		120	.067		2.91	2.78		5.69	7.40
2020	Maximum		98	.082		4.04	3.41		7.45	9.60

04 01 20.30 Pointing CMU		Crew	Daily Output	Labor-Hours	Unit	Material	Labor	Equipment	Total	Total Incl O&P
0010	**POINTING CMU**									
0300	Cut and repoint block, hard mortar, running bond	1 Bric	190	.042	S.F.	.22	1.76		1.98	2.89
0310	Stacked bond		200	.040		.22	1.67		1.89	2.76
0600	Soft old mortar, running bond		230	.035		.22	1.45		1.67	2.43
0610	Stacked bond		245	.033		.22	1.36		1.58	2.30

04 01 20.40 Sawing Masonry		Crew	Daily Output	Labor-Hours	Unit	Material	Labor	Equipment	Total	Total Incl O&P
0010	**SAWING MASONRY**									
0050	Brick or block by hand, per inch depth	A-1	125	.064	L.F.		2.12	.59	2.71	3.92

04 01 30 – Unit Masonry Cleaning

04 01 30.20 Cleaning Masonry		Crew	Daily Output	Labor-Hours	Unit	Material	Labor	Equipment	Total	Total Incl O&P
0010	**CLEANING MASONRY**									
0200	Chemical cleaning, new construction, brush and wash, minimum	D-1	1000	.016	S.F.	.05	.60		.65	.96
0220	Average		800	.020		.07	.75		.82	1.22
0240	Maximum		600	.027		.09	1.01		1.10	1.62
0260	Light restoration, minimum		800	.020		.08	.75		.83	1.22
0270	Average		400	.040		.11	1.51		1.62	2.40
0280	Maximum		330	.048		.15	1.83		1.98	2.93
0300	Heavy restoration, minimum		600	.027		.09	1.01		1.10	1.62
0310	Average		400	.040		.13	1.51		1.64	2.42
0320	Maximum		250	.064		.18	2.41		2.59	3.83
0400	High pressure water only, minimum	B-9	2000	.020			.67	.11	.78	1.15
0420	Average		1500	.027			.89	.15	1.04	1.54
0440	Maximum		1000	.040			1.34	.22	1.56	2.32
2000	Steam cleaning, minimum		3000	.013			.45	.07	.52	.77
2020	Average		2500	.016			.54	.09	.63	.93
2040	Maximum		1500	.027			.89	.15	1.04	1.54
4000	Add for masking doors and windows	1 Clab	800	.010		.06	.33		.39	.58
4200	Add for pedestrian protection				Job					10%

04 01 30.40 Masonry Building Cleaning		Crew	Daily Output	Labor-Hours	Unit	Material	Labor	Equipment	Total	Total Incl O&P
0010	**MASONRY BUILDING CLEANING**									
0020	Steam clean building, not incl scaffolding, minimum	B-9	3000	.013	S.F.		.45	.07	.52	.77
0100	Maximum		1500	.027			.89	.15	1.04	1.54
0300	Common face brick		1750	.023			.77	.13	.90	1.32

04 01 Maintenance of Masonry

04 01 30 – Unit Masonry Cleaning

04 01 30.40 Masonry Building Cleaning		Crew	Daily Output	Labor-Hours	Unit	Material	Labor	Equipment	Total	Total Incl O&P
						2010 Bare Costs				
0400	Wire cut face brick	B-9	1250	.032	S.F.		1.07	.18	1.25	1.85

04 01 30.60 Brick Washing

Line	Description	Crew	Daily Output	Labor-Hours	Unit	Material	Labor	Equipment	Total	Total Incl O&P
0010	**BRICK WASHING** R040130-10									
0012	Acid cleanser, smooth brick surface	1 Bric	560	.014	S.F.	.04	.60		.64	.94
0050	Rough brick		400	.020		.05	.84		.89	1.32
0060	Stone, acid wash		600	.013		.06	.56		.62	.91
1000	Muriatic acid, price per gallon in 5 gallon lots				Gal.	7.55			7.55	8.30

04 05 Common Work Results for Masonry

04 05 05 – Selective Masonry Demolition

04 05 05.10 Selective Demolition

Line	Description	Crew	Daily Output	Labor-Hours	Unit	Material	Labor	Equipment	Total	Total Incl O&P
0010	**SELECTIVE DEMOLITION** R024119-10									
0200	Bond beams, 8" block with #4 bar	2 Clab	32	.500	L.F.		16.55		16.55	25.50
0300	Concrete block walls, unreinforced, 2" thick		1200	.013	S.F.		.44		.44	.68
0310	4" thick		1150	.014			.46		.46	.71
0320	6" thick		1100	.015			.48		.48	.74
0330	8" thick		1050	.015			.50		.50	.78
0340	10" thick		1000	.016			.53		.53	.82
0360	12" thick		950	.017			.56		.56	.86
0380	Reinforced alternate courses, 2" thick		1130	.014			.47		.47	.72
0390	4" thick		1080	.015			.49		.49	.76
0400	6" thick		1035	.015			.51		.51	.79
0410	8" thick		990	.016			.53		.53	.83
0420	10" thick		940	.017			.56		.56	.87
0430	12" thick		890	.018			.60		.60	.92
0440	Reinforced alternate courses & vertically 48" OC, 4" thick		900	.018			.59		.59	.91
0450	6" thick		850	.019			.62		.62	.96
0460	8" thick		800	.020			.66		.66	1.02
0480	10" thick		750	.021			.71		.71	1.09
0490	12" thick		700	.023			.76		.76	1.17
1000	Chimney, 16" x 16", soft old mortar	1 Clab	55	.145	C.F.		4.81		4.81	7.45
1020	Hard mortar		40	.200			6.60		6.60	10.20
1030	16" x 20", soft old mortar		55	.145			4.81		4.81	7.45
1040	Hard mortar		40	.200			6.60		6.60	10.20
1050	16" x 24", soft old mortar		55	.145			4.81		4.81	7.45
1060	Hard mortar		40	.200			6.60		6.60	10.20
1080	20" x 20", soft old mortar		55	.145			4.81		4.81	7.45
1100	Hard mortar		40	.200			6.60		6.60	10.20
1110	20" x 24", soft old mortar		55	.145			4.81		4.81	7.45
1120	Hard mortar		40	.200			6.60		6.60	10.20
1140	20" x 32", soft old mortar		55	.145			4.81		4.81	7.45
1160	Hard mortar		40	.200			6.60		6.60	10.20
1200	48" x 48", soft old mortar		55	.145			4.81		4.81	7.45
1220	Hard mortar		40	.200			6.60		6.60	10.20
1250	Metal, high temp steel jacket, 24" diameter	E-2	130	.431	V.L.F.		19.65	12.35	32	46.50
1260	60" diameter	"	60	.933			42.50	27	69.50	102
1280	Flue lining, up to 12" x 12"	1 Clab	200	.040			1.32		1.32	2.04
1282	Up to 24" x 24"		150	.053			1.77		1.77	2.72
2000	Columns, 8" x 8", soft old mortar		48	.167			5.50		5.50	8.50
2020	Hard mortar		40	.200			6.60		6.60	10.20
2060	16" x 16", soft old mortar		16	.500			16.55		16.55	25.50

04 05 Common Work Results for Masonry

04 05 05 – Selective Masonry Demolition

04 05 05.10 Selective Demolition		Crew	Daily Output	Labor-Hours	Unit	Material	2010 Bare Costs Labor	Equipment	Total	Total Incl O&P
2100	Hard mortar	1 Clab	14	.571	V.L.F.		18.90		18.90	29
2140	24" x 24", soft old mortar		8	1			33		33	51
2160	Hard mortar		6	1.333			44		44	68
2200	36" x 36", soft old mortar		4	2			66		66	102
2220	Hard mortar		3	2.667			88.50		88.50	136
2230	Alternate pricing method, soft old mortar		30	.267	C.F.		8.85		8.85	13.60
2240	Hard mortar		23	.348	"		11.50		11.50	17.75
3000	Copings, precast or masonry, to 8" wide									
3020	Soft old mortar	1 Clab	180	.044	L.F.		1.47		1.47	2.27
3040	Hard mortar	"	160	.050	"		1.66		1.66	2.55
3100	To 12" wide									
3120	Soft old mortar	1 Clab	160	.050	L.F.		1.66		1.66	2.55
3140	Hard mortar	"	140	.057	"		1.89		1.89	2.92
4000	Fireplace, brick, 30" x 24" opening									
4020	Soft old mortar	1 Clab	2	4	Ea.		132		132	204
4040	Hard mortar		1.25	6.400			212		212	325
4100	Stone, soft old mortar		1.50	5.333			177		177	272
4120	Hard mortar		1	8			265		265	410
5000	Veneers, brick, soft old mortar		140	.057	S.F.		1.89		1.89	2.92
5020	Hard mortar		125	.064			2.12		2.12	3.27
5100	Granite and marble, 2" thick		180	.044			1.47		1.47	2.27
5120	4" thick		170	.047			1.56		1.56	2.40
5140	Stone, 4" thick		180	.044			1.47		1.47	2.27
5160	8" thick		175	.046			1.51		1.51	2.33
5400	Alternate pricing method, stone, 4" thick		60	.133	C.F.		4.41		4.41	6.80
5420	8" thick		85	.094	"		3.12		3.12	4.80

04 05 13 – Masonry Mortaring

04 05 13.10 Cement

		Crew	Daily Output	Labor-Hours	Unit	Material	Labor	Equipment	Total	Total Incl O&P
0010	**CEMENT**									
0100	Masonry, 70 lb. bag, T.L. lots **CN**				Bag	8.80			8.80	9.70
0150	L.T.L. lots					9.35			9.35	10.30
0200	White, 70 lb. bag, T.L. lots					14.85			14.85	16.35
0250	L.T.L. lots					16.70			16.70	18.35

04 05 13.20 Lime

		Crew	Daily Output	Labor-Hours	Unit	Material	Labor	Equipment	Total	Total Incl O&P
0010	**LIME**									
0020	Masons, hydrated, 50 lb. bag, T.L. lots **CN**				Bag	8.80			8.80	9.70
0050	L.T.L. lots					9.30			9.30	10.25
0200	Finish, double hydrated, 50 lb. bag, T.L. lots					9.90			9.90	10.90
0250	L.T.L. lots					11			11	12.10

04 05 13.23 Surface Bonding Masonry Mortaring

		Crew	Daily Output	Labor-Hours	Unit	Material	Labor	Equipment	Total	Total Incl O&P
0010	**SURFACE BONDING MASONRY MORTARING**									
0020	Gray or white colors, not incl. block work	1 Bric	540	.015	S.F.	.10	.62		.72	1.04

04 05 13.30 Mortar

		Crew	Daily Output	Labor-Hours	Unit	Material	Labor	Equipment	Total	Total Incl O&P
0010	**MORTAR** R040513-10									
0020	With masonry cement									
0100	Type M, 1:1:6 mix	1 Brhe	143	.056	C.F.	4.83	1.88		6.71	8.15
0200	Type N, 1:3 mix	"	143	.056	"	4.22	1.88		6.10	7.50
2000	With portland cement and lime									
2100	Type M, 1:1/4:3 mix	1 Brhe	143	.056	C.F.	8.80	1.88		10.68	12.55
2200	Type N, 1:1:6 mix, 750 psi		143	.056		6.80	1.88		8.68	10.30
2300	Type O, 1:2:9 mix (Pointing Mortar)		143	.056		7.50	1.88		9.38	11.10

04 05 Common Work Results for Masonry

04 05 13 – Masonry Mortaring

04 05 13.30 Mortar		Crew	Daily Output	Labor-Hours	Unit	Material	2010 Bare Costs Labor	Equipment	Total	Total Incl O&P
2650	Pre-mixed, type S or N				C.F.	4.94			4.94	5.45
2700	Mortar for glass block	1 Brhe	143	.056		10	1.88		11.88	13.85
2900	Mortar for Fire Brick, 80 lb. bag, T.L. Lots				Bag	25.50			25.50	28

04 05 13.91 Masonry Restoration Mortaring

		Crew	Daily Output	Labor-Hours	Unit	Material	Labor	Equipment	Total	Total Incl O&P
0010	**MASONRY RESTORATION MORTARING**									
0020	Masonry restoration mix				Lb.	1.97			1.97	2.17
0050	White				"	2.27			2.27	2.50

04 05 13.93 Mortar Pigments

		Crew	Daily Output	Labor-Hours	Unit	Material	Labor	Equipment	Total	Total Incl O&P
0010	**MORTAR PIGMENTS**, 50 lb. bags (2 bags per M bricks) R040513-10									
0020	Color admixture, range 2 to 10 lb. per bag of cement, minimum				Lb.	5.25			5.25	5.80
0050	Average					7.80			7.80	8.60
0100	Maximum					11.10			11.10	12.20

04 05 13.95 Sand

		Crew	Daily Output	Labor-Hours	Unit	Material	Labor	Equipment	Total	Total Incl O&P
0010	**SAND**, screened and washed at pit									
0020	For mortar, per ton CN				Ton	17.95			17.95	19.75
0050	With 10 mile haul					28			28	30.50
0100	With 30 mile haul					36			36	39.50
0200	Screened and washed, at the pit				C.Y.	25			25	27.50
0250	With 10 mile haul					38.50			38.50	42.50
0300	With 30 mile haul					50			50	55

04 05 13.98 Mortar Admixtures

		Crew	Daily Output	Labor-Hours	Unit	Material	Labor	Equipment	Total	Total Incl O&P
0010	**MORTAR ADMIXTURES**									
0020	Waterproofing admixture, per quart (1 qt. to 2 bags of masonry cement)				Qt.	9.50			9.50	10.45

04 05 16 – Masonry Grouting

04 05 16.30 Grouting

		Crew	Daily Output	Labor-Hours	Unit	Material	Labor	Equipment	Total	Total Incl O&P
0010	**GROUTING** R040513-10									
0011	Bond beams & lintels, 8″ deep, 6″ thick, 0.15 C.F. per L.F.	D-4	1480	.022	L.F.	.64	.81	.09	1.54	2.02
0020	8″ thick, 0.2 C.F. per L.F.		1400	.023		.99	.86	.09	1.94	2.48
0050	10″ thick, 0.25 C.F. per L.F.		1200	.027		1.06	1	.11	2.17	2.80
0060	12″ thick, 0.3 C.F. per L.F.		1040	.031		1.28	1.16	.13	2.57	3.28
0200	Concrete block cores, solid, 4″ thk., by hand, 0.067 C.F./S.F. of wall	D-8	1100	.036	S.F.	.28	1.40		1.68	2.42
0210	6″ thick, pumped, 0.175 C.F. per S.F.	D-4	720	.044		.74	1.67	.18	2.59	3.53
0250	8″ thick, pumped, 0.258 C.F. per S.F.		680	.047		1.10	1.77	.19	3.06	4.08
0300	10″ thick, pumped, 0.340 C.F. per S.F.		660	.048		1.45	1.82	.20	3.47	4.55
0350	12″ thick, pumped, 0.422 C.F. per S.F.		640	.050		1.79	1.88	.21	3.88	5.05
0500	Cavity walls, 2″ space, pumped, 0.167 C.F./S.F. of wall		1700	.019		.71	.71	.08	1.50	1.93
0550	3″ space, 0.250 C.F./S.F.		1200	.027		1.06	1	.11	2.17	2.80
0600	4″ space, 0.333 C.F. per S.F.		1150	.028		1.42	1.05	.12	2.59	3.26
0700	6″ space, 0.500 C.F. per S.F.		800	.040		2.13	1.50	.17	3.80	4.78
0800	Door frames, 3′ x 7′ opening, 2.5 C.F. per opening		60	.533	Opng.	10.65	20	2.21	32.86	44
0850	6′ x 7′ opening, 3.5 C.F. per opening		45	.711	"	14.90	26.50	2.94	44.34	59.50
2000	Grout, C476, for bond beams, lintels and CMU cores		350	.091	C.F.	4.25	3.44	.38	8.07	10.25

04 05 19 – Masonry Anchorage and Reinforcing

04 05 19.05 Anchor Bolts

		Crew	Daily Output	Labor-Hours	Unit	Material	Labor	Equipment	Total	Total Incl O&P
0010	**ANCHOR BOLTS**									
0020	Hooked, with nut and washer, 1/2″ diam., 8″ long	1 Bric	200	.040	Ea.	.80	1.67		2.47	3.40
0030	12″ long		190	.042		1.53	1.76		3.29	4.33
0040	5/8″ diameter, 8″ long		180	.044		1.13	1.86		2.99	4.04
0050	12″ long		170	.047		1.24	1.96		3.20	4.33
0060	3/4″ diameter, 8″ long		160	.050		3.20	2.09		5.29	6.65

04 05 Common Work Results for Masonry

04 05 19 – Masonry Anchorage and Reinforcing

04 05 19.05 Anchor Bolts		Crew	Daily Output	Labor-Hours	Unit	Material	2010 Bare Costs Labor	Equipment	Total	Total Incl O&P
0070	12" long	1 Bric	150	.053	Ea.	4	2.23		6.23	7.75

04 05 19.16 Masonry Anchors

		Crew	Daily Output	Labor-Hours	Unit	Material	Labor	Equipment	Total	Total Incl O&P
0010	**MASONRY ANCHORS**									
0020	For brick veneer, galv., corrugated, 7/8" x 7", 22 Ga.	1 Bric	10.50	.762	C	9.60	32		41.60	58.50
0100	24 Ga.		10.50	.762		9	32		41	58
0150	16 Ga.		10.50	.762		23.50	32		55.50	74
0200	Buck anchors, galv., corrugated, 16 gauge, 2" bend, 8" x 2"		10.50	.762		148	32		180	211
0250	8" x 3"		10.50	.762		138	32		170	200
0670	3/16" diameter		10.50	.762		20	32		52	70.50
0680	1/4" diameter		10.50	.762		38	32		70	90
0850	8" long, 3/16" diameter		10.50	.762		23	32		55	73.50
0855	1/4" diameter		10.50	.762		45	32		77	97.50
1000	Rectangular type, galvanized, 1/4" diameter, 2" x 6"		10.50	.762		65.50	32		97.50	120
1050	4" x 6"		10.50	.762		81.50	32		113.50	138
1100	3/16" diameter, 2" x 6"		10.50	.762		35.50	32		67.50	87.50
1150	4" x 6"		10.50	.762		39	32		71	91
1500	Rigid partition anchors, plain, 8" long, 1" x 1/8"		10.50	.762		91	32		123	148
1550	1" x 1/4"		10.50	.762		150	32		182	213
1580	1-1/2" x 1/8"		10.50	.762		121	32		153	181
1600	1-1/2" x 1/4"		10.50	.762		246	32		278	320
1650	2" x 1/8"		10.50	.762		157	32		189	220
1700	2" x 1/4"		10.50	.762		288	32		320	365

04 05 19.26 Masonry Reinforcing Bars

		Crew	Daily Output	Labor-Hours	Unit	Material	Labor	Equipment	Total	Total Incl O&P
0010	**MASONRY REINFORCING BARS** R040519-50									
0015	Steel bars A615, placed horiz., #3 & #4 bars	1 Bric	450	.018	Lb.	.40	.74		1.14	1.56
0020	#5 & #6 bars		800	.010		.40	.42		.82	1.07
0050	Placed vertical, #3 & #4 bars		350	.023		.40	.95		1.35	1.88
0060	#5 & #6 bars		650	.012		.40	.51		.91	1.21
0200	Joint reinforcing, regular truss, to 6" wide, mill std galvanized		30	.267	C.L.F.	15.20	11.15		26.35	33.50
0250	12" wide		20	.400		16.95	16.70		33.65	43.50
0400	Cavity truss with drip section, to 6" wide		30	.267		13.45	11.15		24.60	31.50
0450	12" wide		20	.400		13.80	16.70		30.50	40

04 05 23 – Masonry Accessories

04 05 23.13 Control Joint

		Crew	Daily Output	Labor-Hours	Unit	Material	Labor	Equipment	Total	Total Incl O&P
0010	**CONTROL JOINT**									
0020	Rubber, for double wythe 8" minimum wall (Brick/CMU)	1 Bric	400	.020	L.F.	1.49	.84		2.33	2.90
0025	"T" shaped		320	.025		1.07	1.04		2.11	2.75
0030	Cross-shaped for CMU units		280	.029		1.44	1.19		2.63	3.38
0050	PVC, for double wythe 8" minimum wall (Brick/CMU)		400	.020		1.09	.84		1.93	2.46
0120	"T" shaped		320	.025		.64	1.04		1.68	2.27
0160	Cross-shaped for CMU units		280	.029		.88	1.19		2.07	2.77

04 05 23.19 Vent Box

		Crew	Daily Output	Labor-Hours	Unit	Material	Labor	Equipment	Total	Total Incl O&P
0010	**VENT BOX**									
0020	Extruded aluminum, 4" deep, 2-3/8" x 8-1/8"	1 Bric	30	.267	Ea.	30.50	11.15		41.65	50.50
0050	5" x 8-1/8"		25	.320		40	13.35		53.35	64
0100	2-1/4" x 25"		25	.320		69.50	13.35		82.85	96.50
0150	5" x 16-1/2"		22	.364		58.50	15.20		73.70	87
0200	6" x 16-1/2"		22	.364		81	15.20		96.20	112
0250	7-3/4" x 16-1/2"		20	.400		69	16.70		85.70	101
0400	For baked enamel finish, add					35%				
0500	For cast aluminum, painted, add					60%				

04 05 Common Work Results for Masonry

04 05 23 – Masonry Accessories

04 05 23.19 Vent Box		Crew	Daily Output	Labor-Hours	Unit	Material	Labor	Equipment	Total	Total Incl O&P
						2010 Bare Costs				
1000	Stainless steel ventilators, 6" x 6"	1 Bric	25	.320	Ea.	126	13.35		139.35	158
1050	8" x 8"		24	.333		132	13.90		145.90	167
1100	12" x 12"		23	.348		153	14.50		167.50	190
1150	12" x 6"		24	.333		134	13.90		147.90	168
1200	Foundation block vent, galv., 1-1/4" thk, 8" high, 16" long, no damper		30	.267		24.50	11.15		35.65	44
1250	For damper, add					8.20			8.20	9

04 05 23.95 Wall Plugs		Crew	Daily Output	Labor-Hours	Unit	Material	Labor	Equipment	Total	Total Incl O&P
0010	**WALL PLUGS** (for nailing to brickwork)									
0020	26 ga., galvanized, plain	1 Bric	10.50	.762	C	31	32		63	82
0050	Wood filled	"	10.50	.762	"	104	32		136	162

04 21 Clay Unit Masonry

04 21 13 – Brick Masonry

04 21 13.13 Brick Veneer Masonry		Crew	Daily Output	Labor-Hours	Unit	Material	Labor	Equipment	Total	Total Incl O&P
0010	**BRICK VENEER MASONRY**, T.L. lots, excl. scaff., grout & reinforcing R042110-20									
0015	Material costs incl. 3% brick and 25% mortar waste									
0020	Standard, select common, 4" x 2-2/3" x 8" (6.75/S.F.)	D-8	1.50	26.667	M	535	1,025		1,560	2,150
0050	Red, 4" x 2-2/3" x 8", running bond		1.50	26.667		690	1,025		1,715	2,300
0100	Full header every 6th course (7.88/S.F.) R042110-50		1.45	27.586		690	1,050		1,740	2,350
0150	English, full header every 2nd course (10.13/S.F.)		1.40	28.571		690	1,100		1,790	2,400
0200	Flemish, alternate header every course (9.00/S.F.)		1.40	28.571		690	1,100		1,790	2,400
0250	Flemish, alt. header every 6th course (7.13/S.F.)		1.45	27.586		690	1,050		1,740	2,350
0300	Full headers throughout (13.50/S.F.)		1.40	28.571		685	1,100		1,785	2,400
0350	Rowlock course (13.50/S.F.)		1.35	29.630		685	1,150		1,835	2,475
0400	Rowlock stretcher (4.50/S.F.)		1.40	28.571		695	1,100		1,795	2,425
0450	Soldier course (6.75/S.F.)		1.40	28.571		690	1,100		1,790	2,400
0500	Sailor course (4.50/S.F.)		1.30	30.769		695	1,175		1,870	2,550
0601	Buff or gray face, running bond, (6.75/S.F.)		1.50	26.667		690	1,025		1,715	2,300
0700	Glazed face, 4" x 2-2/3" x 8", running bond		1.40	28.571		1,850	1,100		2,950	3,675
0750	Full header every 6th course (7.88/S.F.)		1.35	29.630		1,775	1,150		2,925	3,675
1000	Jumbo, 6" x 4" x 12",(3.00/S.F.)		1.30	30.769		1,775	1,175		2,950	3,725
1051	Norman, 4" x 2-2/3" x 12" (4.50/S.F.)		1.45	27.586		1,100	1,050		2,150	2,825
1100	Norwegian, 4" x 3-1/5" x 12" (3.75/S.F.)		1.40	28.571		1,150	1,100		2,250	2,900
1150	Economy, 4" x 4" x 8" (4.50 per S.F.)		1.40	28.571		940	1,100		2,040	2,675
1201	Engineer, 4" x 3-1/5" x 8", (5.63/S.F.)		1.45	27.586		615	1,050		1,665	2,275
1251	Roman, 4" x 2" x 12", (6.00/S.F.)		1.50	26.667		985	1,025		2,010	2,625
1300	S.C.R. 6" x 2-2/3" x 12" (4.50/S.F.)		1.40	28.571		1,175	1,100		2,275	2,950
1350	Utility, 4" x 4" x 12" (3.00/S.F.)		1.35	29.630		1,550	1,150		2,700	3,425
1360	For less than truck load lots, add					15			15	16.50
1400	For battered walls, add						30%			
1450	For corbels, add						75%			
1500	For curved walls, add						30%			
1550	For pits and trenches, deduct						20%			
1999	Alternate method of figuring by square foot									
2000	Standard, sel. common, 4" x 2-2/3" x 8", (6.75/S.F.)	D-8	230	.174	S.F.	3.61	6.70		10.31	14.05
2020	Standard, red, 4" x 2-2/3" x 8", running bond (6.75/SF)		220	.182		4.67	7		11.67	15.70
2050	Full header every 6th course (7.88/S.F.)		185	.216		5.45	8.35		13.80	18.55
2100	English, full header every 2nd course (10.13/S.F.)		140	.286		6.95	11		17.95	24.50
2150	Flemish, alternate header every course (9.00/S.F.)		150	.267		6.20	10.25		16.45	22.50
2200	Flemish, alt. header every 6th course (7.13/S.F.)		205	.195		4.93	7.50		12.43	16.75
2250	Full headers throughout (13.50/S.F.)		105	.381		9.25	14.65		23.90	32

04 21 Clay Unit Masonry

04 21 13 – Brick Masonry

04 21 13.13 Brick Veneer Masonry		Crew	Daily Output	Labor-Hours	Unit	Material	Labor	Equipment	Total	Total Incl O&P
						2010 Bare Costs				
2300	Rowlock course (13.50/S.F.)	D-8	100	.400	S.F.	9.25	15.40		24.65	33
2350	Rowlock stretcher (4.50/S.F.)		310	.129		3.13	4.97		8.10	10.95
2400	Soldier course (6.75/S.F.)		200	.200		4.67	7.70		12.37	16.75
2450	Sailor course (4.50/S.F.)		290	.138		3.13	5.30		8.43	11.45
2600	Buff or gray face, running bond, (6.75/S.F.)		220	.182		4.94	7		11.94	16
2700	Glazed face brick, running bond		210	.190		11.95	7.35		19.30	24
2750	Full header every 6th course (7.88/S.F.)		170	.235		13.95	9.05		23	29
3000	Jumbo, 6" x 4" x 12" running bond (3.00/S.F.)		435	.092		4.87	3.54		8.41	10.70
3050	Norman, 4" x 2-2/3" x 12" running bond, (4.5/S.F.)		320	.125		5.70	4.81		10.51	13.50
3100	Norwegian, 4" x 3-1/5" x 12" (3.75/S.F.)		375	.107		4.15	4.11		8.26	10.75
3150	Economy, 4" x 4" x 8" (4.50/S.F.)		310	.129		4.19	4.97		9.16	12.10
3200	Engineer, 4" x 3-1/5" x 8" (5.63/S.F.)		260	.154		3.44	5.90		9.34	12.75
3250	Roman, 4" x 2" x 12" (6.00/S.F.)		250	.160		5.75	6.15		11.90	15.65
3300	SCR, 6" x 2-2/3" x 12" (4.50/S.F.)		310	.129		5.25	4.97		10.22	13.25
3350	Utility, 4" x 4" x 12" (3.00/S.F.)		450	.089		4.50	3.42		7.92	10.10
3400	For cavity wall construction, add						15%			
3450	For stacked bond, add						10%			
3500	For interior veneer construction, add						15%			
3550	For curved walls, add						30%			

04 21 13.14 Thin Brick Veneer

		Crew	Daily Output	Labor-Hours	Unit	Material	Labor	Equipment	Total	Total Incl O&P
0010	**THIN BRICK VENEER**									
0015	Material costs incl. 3% brick and 25% mortar waste									
0020	On & incl. metal panel support sys, modular, 2-2/3" x 5/8" x 8", red	D-7	92	.174	S.F.	8.75	6.15		14.90	18.60
0100	Closure, 4" x 5/8" x 8"		110	.145		8.75	5.15		13.90	17.15
0110	Norman, 2-2/3" x 5/8" x 12"		110	.145		8.95	5.15		14.10	17.35
0120	Utility, 4" x 5/8" x 12"		125	.128		8.95	4.52		13.47	16.45
0130	Emperor, 4" x 3/4" x 16"		175	.091		10.15	3.23		13.38	15.90
0140	Super emperor, 8" x 3/4" x 16"		195	.082		10.15	2.89		13.04	15.45
0150	For L shaped corners with 4" return, add				L.F.	9.25			9.25	10.20
0200	On masonry/plaster back-up, modular, 2-2/3" x 5/8" x 8", red	D-7	137	.117	S.F.	3.95	4.12		8.07	10.35
0210	Closure, 4" x 5/8" x 8"		165	.097		3.95	3.42		7.37	9.35
0220	Norman, 2-2/3" x 5/8" x 12"		165	.097		4.15	3.42		7.57	9.55
0230	Utility, 4" x 5/8" x 12"		185	.086		4.15	3.05		7.20	9.05
0240	Emperor, 4" x 3/4" x 16"		260	.062		5.35	2.17		7.52	9.05
0250	Super emperor, 8" x 3/4" x 16"		285	.056		5.35	1.98		7.33	8.80
0260	For L shaped corners with 4" return, add				L.F.	9.25			9.25	10.20
0270	For embedment into pre-cast concrete panels, add				S.F.	14.40			14.40	15.85

04 21 13.15 Chimney

		Crew	Daily Output	Labor-Hours	Unit	Material	Labor	Equipment	Total	Total Incl O&P
0010	**CHIMNEY**, excludes foundation, scaffolding, grout and reinforcing									
0100	Brick, 16" x 16", 8" flue	D-1	18.20	.879	V.L.F.	21	33		54	73.50
0150	16" x 20" with one 8" x 12" flue		16	1		33.50	37.50		71	94
0200	16" x 24" with two 8" x 8" flues		14	1.143		49	43		92	119
0250	20" x 20" with one 12" x 12" flue		13.70	1.168		38.50	44		82.50	109
0300	20" x 24" with two 8" x 12" flues		12	1.333		55.50	50.50		106	137
0350	20" x 32" with two 12" x 12" flues		10	1.600		68	60.50		128.50	166
1800	Metal, high temp. steel jacket, factory lining, 24" diam.	E-2	65	.862		210	39.50	24.50	274	325
1900	60" diameter	"	30	1.867		765	85	53.50	903.50	1,050
2100	Poured concrete, brick lining, 200' high x 10' diam.					7,275			7,275	8,000
2800	500' x 20' diameter					12,700			12,700	14,000

04 21 13.18 Columns

		Crew	Daily Output	Labor-Hours	Unit	Material	Labor	Equipment	Total	Total Incl O&P
0010	**COLUMNS**, solid, excludes scaffolding, grout and reinforcing R042110-10									
0050	Brick, 8" x 8", 9 brick per VLF	D-1	56	.286	V.L.F.	6.05	10.75		16.80	23

04 21 13 – Brick Masonry

04 21 13.18 Columns		Crew	Daily Output	Labor-Hours	Unit	Material	2010 Bare Costs Labor	Equipment	Total	Total Incl O&P
0100	12" x 8", 13.5 brick per VLF	D-1	37	.432	V.L.F.	9.10	16.30		25.40	34.50
0200	12" x 12", 20 brick per VLF		25	.640		13.50	24		37.50	51.50
0300	16" x 12", 27 brick per VLF		19	.842		18.20	32		50.20	68
0400	16" x 16", 36 brick per VLF		14	1.143		24.50	43		67.50	91.50
0500	20" x 16", 45 brick per VLF		11	1.455		30.50	55		85.50	116
0600	20" x 20", 56 brick per VLF		9	1.778		38	67		105	143
0700	24" x 20", 68 brick per VLF		7	2.286		46	86		132	181
0800	24" x 24", 81 brick per VLF		6	2.667		54.50	101		155.50	212
1000	36" x 36", 182 brick per VLF		3	5.333		123	201		324	440

04 21 13.30 Oversized Brick		Crew	Daily Output	Labor-Hours	Unit	Material	Labor	Equipment	Total	Total Incl O&P
0010	**OVERSIZED BRICK**, excludes scaffolding, grout and reinforcing									
0100	Veneer, 4" x 2.25" x 16"	D-8	387	.103	S.F.	6.40	3.98		10.38	13.05
0105	4" x 2.75" x 16"		412	.097		5.80	3.74		9.54	12
0110	4" x 4" x 16"		460	.087		5.30	3.35		8.65	10.90
0120	4" x 8" x 16"		533	.075		6.10	2.89		8.99	11.05
0125	Loadbearing, 6" x 4" x 16", grouted and reinforced		387	.103		8.95	3.98		12.93	15.85
0130	8" x 4" x 16", grouted and reinforced		327	.122		9.45	4.71		14.16	17.50
0135	6" x 8" x 16", grouted and reinforced		440	.091		8.70	3.50		12.20	14.85
0140	8" x 8" x 16", grouted and reinforced		400	.100		9.55	3.85		13.40	16.35
0145	Curtainwall/reinforced veneer, 6" x 4" x 16"		387	.103		13.25	3.98		17.23	20.50
0150	8" x 4" x 16"		327	.122		15.75	4.71		20.46	24.50
0155	6" x 8" x 16"		440	.091		13.35	3.50		16.85	20
0160	8" x 8" x 16"		400	.100		15.85	3.85		19.70	23.50
0200	For 1 to 3 slots in face, add					15%				
0210	For 4 to 7 slots in face, add					25%				
0220	For bond beams, add					20%				
0230	For bullnose shapes, add					20%				
0240	For open end knockout, add					10%				
0250	For white or gray color group, add					10%				
0260	For 135 degree corner, add					250%				

04 21 13.35 Common Building Brick		Crew	Daily Output	Labor-Hours	Unit	Material	Labor	Equipment	Total	Total Incl O&P
0010	**COMMON BUILDING BRICK**, C62, TL lots, material only R042110-20									
0020	Standard, minimum				M	400			400	440
0050	Average (select)				"	430			430	475

04 21 13.40 Structural Brick		Crew	Daily Output	Labor-Hours	Unit	Material	Labor	Equipment	Total	Total Incl O&P
0010	**STRUCTURAL BRICK** C652, Grade SW, incl. mortar, scaffolding not incl.									
0100	Standard unit, 4-5/8" x 2-3/4" x 9-5/8"	D-8	245	.163	S.F.	5.15	6.30		11.45	15.20
0120	Bond beam		225	.178		9.65	6.85		16.50	21
0140	V cut bond beam		225	.178		10.45	6.85		17.30	22
0160	Stretcher quoin, 5-5/8" x 2-3/4" x 9-5/8"		245	.163		10.10	6.30		16.40	20.50
0180	Corner quoin		245	.163		11.30	6.30		17.60	22
0200	Corner, 45 deg, 4-5/8" x 2-3/4" x 10-7/16"		235	.170		11.70	6.55		18.25	23

04 21 13.45 Face Brick		Crew	Daily Output	Labor-Hours	Unit	Material	Labor	Equipment	Total	Total Incl O&P
0010	**FACE BRICK** Material Only, C216, TL lots R042110-20									
0300	Standard modular, 4" x 2-2/3" x 8", minimum				M	585			585	640
0350	Maximum *CN*					700			700	770
0450	Economy, 4" x 4" x 8", minimum					815			815	895
0500	Maximum					1,050			1,050	1,150
0510	Economy, 4" x 4" x 12", minimum					1,200			1,200	1,325
0520	Maximum					1,550			1,550	1,725
0550	Jumbo, 6" x 4" x 12", minimum					1,475			1,475	1,625
0600	Maximum					1,875			1,875	2,075

04 21 Clay Unit Masonry

04 21 13 – Brick Masonry

04 21 13.45 Face Brick		Crew	Daily Output	Labor-Hours	Unit	Material	2010 Bare Costs Labor	Equipment	Total	Total Incl O&P
0610	Jumbo, 8" x 4" x 12", minimum				M	1,475			1,475	1,625
0620	Maximum					1,875			1,875	2,075
0650	Norwegian, 4" x 3-1/5" x 12", minimum					980			980	1,075
0700	Maximum					1,275			1,275	1,400
0710	Norwegian, 6" x 3-1/5" x 12", minimum					1,425			1,425	1,575
0720	Maximum					1,875			1,875	2,050
0850	Standard glazed, plain colors, 4" x 2-2/3" x 8", minimum					1,625			1,625	1,800
0900	Maximum					2,125			2,125	2,325
1000	Deep trim shades, 4" x 2-2/3" x 8", minimum					1,900			1,900	2,100
1050	Maximum					2,175			2,175	2,400
1080	Jumbo utility, 4" x 4" x 12"					1,350			1,350	1,500
1120	4" x 8" x 8"					1,675			1,675	1,850
1140	4" x 8" x 16"					4,100			4,100	4,500
1260	Engineer, 4" x 3-1/5" x 8", minimum					505			505	555
1270	Maximum					730			730	805
1350	King, 4" x 2-3/4" x 10", minimum					460			460	505
1360	Maximum					560			560	620
1400	Norman, 4" x 2-3/4" x 12"					460			460	505
1450	Roman, 4" x 2" x 12"					460			460	505
1500	SCR, 6" x 2-2/3" x 12"					460			460	505
1550	Double, 4" x 5-1/3" x 8"					460			460	505
1600	Triple, 4" x 5-1/3" x 12"					460			460	505
1770	Standard modular, double glazed, 4" x 2-2/3" x 8"					2,275			2,275	2,500
1850	Jumbo, colored glazed ceramic, 6" x 4" x 12"					2,400			2,400	2,650
2050	Jumbo utility, glazed, 4" x 4" x 12"					4,550			4,550	5,000
2100	4" x 8" x 8"					5,325			5,325	5,875
2150	4" x 16" x 8"					6,200			6,200	6,825
2170	For less than truck load lots, add					15			15	16.50
2180	For buff or gray brick, add				↓	16			16	17.60

04 21 26 – Glazed Structural Clay Tile Masonry

04 21 26.10 Structural Facing Tile		Crew	Daily Output	Labor-Hours	Unit	Material	Labor	Equipment	Total	Total Incl O&P
0010	**STRUCTURAL FACING TILE**, std. colors, excl. scaffolding, grout, reinforcing									
0020	6T series, 5-1/3" x 12", 2.3 pieces per S.F., glazed 1 side, 2" thick	D-8	225	.178	S.F.	7.95	6.85		14.80	19.05
0100	4" thick		220	.182		10.75	7		17.75	22.50
0150	Glazed 2 sides		195	.205		13.85	7.90		21.75	27
0250	6" thick		210	.190		15.75	7.35		23.10	28.50
0300	Glazed 2 sides		185	.216		18.70	8.35		27.05	33
0400	8" thick		180	.222	↓	21	8.55		29.55	36
0500	Special shapes, group 1		400	.100	Ea.	6.70	3.85		10.55	13.15
0550	Group 2		375	.107		10.90	4.11		15.01	18.20
0600	Group 3		350	.114		14.10	4.40		18.50	22
0650	Group 4		325	.123		29.50	4.74		34.24	39
0700	Group 5		300	.133		35	5.15		40.15	46.50
0750	Group 6		275	.145	↓	48	5.60		53.60	61
1000	Fire rated, 4" thick, 1 hr. rating		210	.190	S.F.	15.45	7.35		22.80	28
1300	Acoustic, 4" thick	↓	210	.190	"	34	7.35		41.35	48.50
2000	8W series, 8" x 16", 1.125 pieces per S.F.									
2050	2" thick, glazed 1 side	D-8	360	.111	S.F.	8.80	4.28		13.08	16.10
2100	4" thick, glazed 1 side		345	.116		11.10	4.46		15.56	18.95
2150	Glazed 2 sides		325	.123		13.45	4.74		18.19	22
2200	6" thick, glazed 1 side		330	.121		20.50	4.67		25.17	29.50
2250	8" thick, glazed 1 side	↓	310	.129	↓	21.50	4.97		26.47	31

04 21 26 – Glazed Structural Clay Tile Masonry

04 21 26.10 Structural Facing Tile		Crew	Daily Output	Labor-Hours	Unit	Material	Labor	Equipment	Total	Total Incl O&P
						2010 Bare Costs				
2500	Special shapes, group 1	D-8	300	.133	Ea.	12.75	5.15		17.90	22
2550	Group 2		280	.143		16.85	5.50		22.35	27
2600	Group 3		260	.154		18.10	5.90		24	29
2650	Group 4		250	.160		36	6.15		42.15	49
2700	Group 5		240	.167		38	6.40		44.40	51.50
2750	Group 6		230	.174		82	6.70		88.70	101
3000	4" thick, glazed 1 side		345	.116	S.F.	11.10	4.46		15.56	18.95
3100	Acoustic, 4" thick		345	.116	"	13.95	4.46		18.41	22
3120	4W series, 8" x 8", 2.25 pieces per S.F.									
3125	2" thick, glazed 1 side	D-8	360	.111	S.F.	8	4.28		12.28	15.25
3130	4" thick, glazed 1 side		345	.116		9.65	4.46		14.11	17.35
3135	Glazed 2 sides		325	.123		13.20	4.74		17.94	21.50
3140	6" thick, glazed 1 side		330	.121		13.80	4.67		18.47	22
3150	8" thick, glazed 1 side		310	.129		20	4.97		24.97	29.50
3155	Special shapes, group I		300	.133	Ea.	6.35	5.15		11.50	14.75
3160	Group II		280	.143	"	7.30	5.50		12.80	16.30
3200	For designer colors, add					25%				
3300	For epoxy mortar joints, add				S.F.	1.58			1.58	1.74

04 21 29 – Terra Cotta Masonry

04 21 29.10 Terra Cotta Masonry Components		Crew	Daily Output	Labor-Hours	Unit	Material	Labor	Equipment	Total	Total Incl O&P
0010	**TERRA COTTA MASONRY COMPONENTS**									
0020	Coping, split type, not glazed, 9" wide	D-1	90	.178	L.F.	10.40	6.70		17.10	21.50
0100	13" wide		80	.200		15.75	7.55		23.30	28.50
0200	Coping, split type, glazed, 9" wide		90	.178		17.40	6.70		24.10	29.50
0250	13" wide		80	.200		22.50	7.55		30.05	36.50
0500	Partition or back-up blocks, scored, in C.L. lots									
0700	Non-load bearing 12" x 12", 3" thick, special order	D-8	550	.073	S.F.	19.30	2.80		22.10	25
0750	4" thick, standard		500	.080		7.10	3.08		10.18	12.45
0800	6" thick		450	.089		8.70	3.42		12.12	14.70
0850	8" thick		400	.100		11.05	3.85		14.90	18
1000	Load bearing, 12" x 12", 4" thick, in walls		500	.080		6.95	3.08		10.03	12.30
1050	In floors		750	.053		6.95	2.05		9	10.75
1200	6" thick, in walls		450	.089		8.70	3.42		12.12	14.70
1250	In floors		675	.059		8.70	2.28		10.98	13
1400	8" thick, in walls		400	.100		11.05	3.85		14.90	18
1450	In floors		575	.070		11.05	2.68		13.73	16.25
1600	10" thick, in walls, special order		350	.114		14.50	4.40		18.90	22.50
1650	In floors, special order		500	.080		14.50	3.08		17.58	20.50
1800	12" thick, in walls, special order		300	.133		28	5.15		33.15	38.50
1850	In floors, special order		450	.089		28	3.42		31.42	35.50
2000	For reinforcing with steel rods, add to above					15%	5%			
2100	For smooth tile instead of scored, add					3.74			3.74	4.11
2200	For L.C.L. quantities, add					10%	10%			

04 21 29.20 Terra Cotta Tile		Crew	Daily Output	Labor-Hours	Unit	Material	Labor	Equipment	Total	Total Incl O&P
0010	**TERRA COTTA TILE**, on walls, dry set, 1/2" thick									
0100	Square, hexagonal or lattice shapes, unglazed	1 Tilf	135	.059	S.F.	6.35	2.33		8.68	10.35
0300	Glazed, plain colors		130	.062		8.75	2.42		11.17	13.15
0400	Intense colors		125	.064		10.20	2.52		12.72	14.90

04 22 Concrete Unit Masonry

04 22 10 – Concrete Masonry Units

04 22 10.11 Autoclave Aerated Concrete Block		Crew	Daily Output	Labor-Hours	Unit	Material	Labor	Equipment	Total	Total Incl O&P
						2010 Bare Costs				
0010	**AUTOCLAVE AERATED CONCRETE BLOCK**, excl. scaffolding, grout & reinforcing									
0050	Solid, 4" x 12" x 24", incl mortar G	D-8	600	.067	S.F.	1.36	2.57		3.93	5.35
0060	6" x 12" x 24" G		600	.067		2.03	2.57		4.60	6.10
0070	8" x 8" x 24" G		575	.070		2.72	2.68		5.40	7.05
0080	10" x 12" x 24" G		575	.070		3.32	2.68		6	7.70
0090	12" x 12" x 24" G	↓	550	.073	↓	4.07	2.80		6.87	8.70

04 22 10.12 Chimney Block

		Crew	Daily Output	Labor-Hours	Unit	Material	Labor	Equipment	Total	Total Incl O&P
0010	**CHIMNEY BLOCK**, excludes scaffolding, grout and reinforcing									
0220	1 piece, with 8" x 8" flue, 16" x 16"	D-1	28	.571	V.L.F.	17.20	21.50		38.70	51.50
0230	2 piece, 16" x 16"		26	.615		14.75	23		37.75	51
0240	2 piece, with 8" x 12" flue, 16" x 20"	↓	24	.667	↓	29	25		54	70

04 22 10.14 Concrete Block, Back-Up

		Crew	Daily Output	Labor-Hours	Unit	Material	Labor	Equipment	Total	Total Incl O&P
0010	**CONCRETE BLOCK, BACK-UP**, C90, 2000 psi R042210-20									
0020	Normal weight, 8" x 16" units, tooled joint 1 side									
0050	Not-reinforced, 2000 psi, 2" thick	D-8	475	.084	S.F.	1.39	3.24		4.63	6.40
0200	4" thick		460	.087		1.56	3.35		4.91	6.75
0300	6" thick		440	.091		2.17	3.50		5.67	7.70
0350	8" thick		400	.100		2.45	3.85		6.30	8.50
0400	10" thick	↓	330	.121		3.09	4.67		7.76	10.45
0450	12" thick	D-9	310	.155		3.51	5.85		9.36	12.65
1000	Reinforced, alternate courses, 4" thick	D-8	450	.089		1.67	3.42		5.09	7
1100	6" thick		430	.093		2.30	3.58		5.88	7.95
1150	8" thick		395	.101		2.58	3.90		6.48	8.75
1200	10" thick	↓	320	.125		3.23	4.81		8.04	10.80
1250	12" thick	D-9	300	.160	↓	3.66	6.05		9.71	13.10

04 22 10.16 Concrete Block, Bond Beam

		Crew	Daily Output	Labor-Hours	Unit	Material	Labor	Equipment	Total	Total Incl O&P
0010	**CONCRETE BLOCK, BOND BEAM**, C90, 2000 psi									
0020	Not including grout or reinforcing									
0125	Regular block, 6" thick	D-8	584	.068	L.F.	2.32	2.64		4.96	6.55
0130	8" high, 8" thick	"	565	.071		2.50	2.73		5.23	6.85
0150	12" thick	D-9	510	.094		3.50	3.55		7.05	9.20
0525	Lightweight, 6" thick	D-8	592	.068		2.55	2.60		5.15	6.75
0530	8" high, 8" thick	"	575	.070		3.09	2.68		5.77	7.45
0550	12" thick	D-9	520	.092	↓	4.16	3.48		7.64	9.80
2000	Including grout and 2 #5 bars									
2100	Regular block, 8" high, 8" thick	D-8	300	.133	L.F.	4.35	5.15		9.50	12.55
2150	12" thick	D-9	250	.192		5.90	7.25		13.15	17.40
2500	Lightweight, 8" high, 8" thick	D-8	305	.131		4.95	5.05		10	13.05
2550	12" thick	D-9	255	.188	↓	6.55	7.10		13.65	17.90

04 22 10.18 Concrete Block, Column

		Crew	Daily Output	Labor-Hours	Unit	Material	Labor	Equipment	Total	Total Incl O&P
0010	**CONCRETE BLOCK, COLUMN** or pilaster									
0050	Including vertical reinforcing (4-#4 bars) and grout									
0160	1 piece unit, 16" x 16"	D-1	26	.615	V.L.F.	17.95	23		40.95	55
0170	2 piece units, 16" x 20"		24	.667		23.50	25		48.50	64
0180	20" x 20"		22	.727		34.50	27.50		62	79.50
0190	22" x 24"		18	.889		48.50	33.50		82	104
0200	20" x 32"	↓	14	1.143	↓	53	43		96	124

04 22 10.19 Concrete Block, Insulation Inserts

		Crew	Daily Output	Labor-Hours	Unit	Material	Labor	Equipment	Total	Total Incl O&P
0010	**CONCRETE BLOCK, INSULATION INSERTS**									
0100	Styrofoam, plant installed, add to block prices									
0200	8" x 16" units, 6" thick				S.F.	1.81			1.81	1.99

04 22 Concrete Unit Masonry

04 22 10 – Concrete Masonry Units

04 22 10.19 Concrete Block, Insulation Inserts		Crew	Daily Output	Labor-Hours	Unit	Material	Labor	Equipment	Total	Total Incl O&P
						2010 Bare Costs				
0250	8″ thick				S.F.	1.81			1.81	1.99
0300	10″ thick					2.13			2.13	2.34
0350	12″ thick					2.24			2.24	2.46
0500	8″ x 8″ units, 8″ thick					1.49			1.49	1.64
0550	12″ thick					1.81			1.81	1.99

04 22 10.23 Concrete Block, Decorative		Crew	Daily Output	Labor-Hours	Unit	Material	Labor	Equipment	Total	Total Incl O&P
0010	**CONCRETE BLOCK, DECORATIVE**, C90, 2000 psi									
0020	Embossed, simulated brick face									
0100	8″ x 16″ units, 4″ thick	D-8	400	.100	S.F.	3.51	3.85		7.36	9.65
0200	8″ thick		340	.118		4.84	4.53		9.37	12.15
0250	12″ thick		300	.133		6.35	5.15		11.50	14.75
0400	Embossed both sides									
0500	8″ thick	D-8	300	.133	S.F.	5.45	5.15		10.60	13.70
0550	12″ thick	″	275	.145	″	6.85	5.60		12.45	16
1000	Fluted high strength									
1100	8″ x 16″ x 4″ thick, flutes 1 side,	D-8	345	.116	S.F.	4.17	4.46		8.63	11.35
1150	Flutes 2 sides		335	.119		5.05	4.60		9.65	12.50
1200	8″ thick		300	.133		6.60	5.15		11.75	15
1250	For special colors, add					.41			.41	.45
1400	Deep grooved, smooth face									
1450	8″ x 16″ x 4″ thick	D-8	345	.116	S.F.	2.72	4.46		7.18	9.75
1500	8″ thick	″	300	.133	″	4.69	5.15		9.84	12.90
2000	Formblock, incl. inserts & reinforcing									
2100	8″ x 16″ x 8″ thick	D-8	345	.116	S.F.	4.76	4.46		9.22	12
2150	12″ thick	″	310	.129	″	5.95	4.97		10.92	14.05
2500	Ground face									
2600	8″ x 16″ x 4″ thick	D-8	345	.116	S.F.	3.27	4.46		7.73	10.35
2650	6″ thick		325	.123		4.78	4.74		9.52	12.40
2700	8″ thick		300	.133		6.35	5.15		11.50	14.75
2750	12″ thick	D-9	265	.181		5.70	6.85		12.55	16.60
2900	For special colors, add, minimum					15%				
2950	For special colors, add, maximum					45%				
4000	Slump block									
4100	4″ face height x 16″ x 4″ thick	D-1	165	.097	S.F.	3.71	3.66		7.37	9.60
4150	6″ thick		160	.100		5	3.77		8.77	11.20
4200	8″ thick		155	.103		5.65	3.89		9.54	12.05
4250	10″ thick		140	.114		9.95	4.31		14.26	17.45
4300	12″ thick		130	.123		10.45	4.64		15.09	18.50
4400	6″ face height x 16″ x 6″ thick		155	.103		4.36	3.89		8.25	10.65
4450	8″ thick		150	.107		5.90	4.02		9.92	12.55
4500	10″ thick		130	.123		9.10	4.64		13.74	17
4550	12″ thick		120	.133		9.75	5.05		14.80	18.35
5000	Split rib profile units, 1″ deep ribs, 8 ribs									
5100	8″ x 16″ x 4″ thick	D-8	345	.116	S.F.	3.16	4.46		7.62	10.25
5150	6″ thick		325	.123		3.64	4.74		8.38	11.15
5200	8″ thick		300	.133		4.17	5.15		9.32	12.35
5250	12″ thick	D-9	275	.175		4.94	6.60		11.54	15.35
5400	For special deeper colors, 4″ thick, add					1.15			1.15	1.27
5450	12″ thick, add					1.19			1.19	1.31
5600	For white, 4″ thick, add					1.15			1.15	1.27
5650	6″ thick, add					1.14			1.14	1.26
5700	8″ thick, add					1.20			1.20	1.32

04 22 10 – Concrete Masonry Units

04 22 10.23 Concrete Block, Decorative		Crew	Daily Output	Labor-Hours	Unit	Material	Labor	Equipment	Total	Total Incl O&P
						2010 Bare Costs				
5750	12" thick, add				S.F.	1.23			1.23	1.35
6000	Split face									
6100	8" x 16" x 4" thick	D-8	350	.114	S.F.	2.82	4.40		7.22	9.75
6150	6" thick		325	.123		3.28	4.74		8.02	10.75
6200	8" thick		300	.133		3.73	5.15		8.88	11.85
6250	12" thick	D-9	270	.178		4.58	6.70		11.28	15.15
6300	For scored, add					.35			.35	.39
6400	For special deeper colors, 4" thick, add					.67			.67	.73
6450	6" thick, add					.66			.66	.72
6500	8" thick, add					.68			.68	.75
6550	12" thick, add					.70			.70	.77
6650	For white, 4" thick, add					1.14			1.14	1.26
6700	6" thick, add					1.13			1.13	1.24
6750	8" thick, add					1.16			1.16	1.28
6800	12" thick, add					1.19			1.19	1.31
7000	Scored ground face, 2 to 5 scores									
7100	8" x 16" x 4" thick	D-8	340	.118	S.F.	4.90	4.53		9.43	12.25
7150	6" thick		310	.129		5.60	4.97		10.57	13.70
7200	8" thick		290	.138		6.60	5.30		11.90	15.25
7250	12" thick	D-9	265	.181		8.20	6.85		15.05	19.35
8000	Hexagonal face profile units, 8" x 16" units									
8100	4" thick, hollow	D-8	340	.118	S.F.	3.12	4.53		7.65	10.30
8200	Solid		340	.118		4.03	4.53		8.56	11.30
8300	6" thick, hollow		310	.129		3.49	4.97		8.46	11.35
8350	8" thick, hollow		290	.138		4.74	5.30		10.04	13.20
8500	For stacked bond, add						26%			
8550	For high rise construction, add per story	D-8	67.80	.590	M.S.F.		22.50		22.50	34.50
8600	For scored block, add					10%				
8650	For honed or ground face, per face, add				Ea.	.38			.38	.42
8700	For honed or ground end, per end, add				"	2.98			2.98	3.28
8750	For bullnose block, add					10%				
8800	For special color, add					13%				

04 22 10.24 Concrete Block, Exterior		Crew	Daily Output	Labor-Hours	Unit	Material	Labor	Equipment	Total	Total Incl O&P
0010	**CONCRETE BLOCK, EXTERIOR**, C90, 2000 psi									
0020	Reinforced alt courses, tooled joints 2 sides									
0100	Normal weight, 8" x 16" x 6" thick	D-8	395	.101	S.F.	2.55	3.90		6.45	8.70
0200	8" thick		360	.111		3.67	4.28		7.95	10.50
0250	10" thick		290	.138		4.29	5.30		9.59	12.70
0300	12" thick	D-9	250	.192		4.53	7.25		11.78	15.90
0500	Lightweight, 8" x 16" x 6" thick	D-8	450	.089		2.72	3.42		6.14	8.15
0600	8" thick		430	.093		3.47	3.58		7.05	9.20
0650	10" thick		395	.101		4.32	3.90		8.22	10.65
0700	12" thick	D-9	350	.137		6.35	5.15		11.50	14.80

04 22 10.26 Concrete Block Foundation Wall		Crew	Daily Output	Labor-Hours	Unit	Material	Labor	Equipment	Total	Total Incl O&P
0010	**CONCRETE BLOCK FOUNDATION WALL**, C90/C145									
0050	Normal-weight, cut joints, horiz joint reinf, no vert reinf									
0200	Hollow, 8" x 16" x 6" thick	D-8	455	.088	S.F.	2.61	3.39		6	7.95
0250	8" thick		425	.094		2.90	3.62		6.52	8.65
0300	10" thick		350	.114		3.55	4.40		7.95	10.55
0350	12" thick	D-9	300	.160		3.98	6.05		10.03	13.50
0500	Solid, 8" x 16" block, 6" thick	D-8	440	.091		2.68	3.50		6.18	8.25
0550	8" thick	"	415	.096		3.78	3.71		7.49	9.75

04 22 10 – Concrete Masonry Units

04 22 10.26 Concrete Block Foundation Wall		Crew	Daily Output	Labor-Hours	Unit	Material	Labor	Equipment	Total	Total Incl O&P
						2010 Bare Costs				
0600	12" thick	D-9	350	.137	S.F.	5.45	5.15		10.60	13.80

04 22 10.28 Concrete Block, High Strength		Crew	Daily Output	Labor-Hours	Unit	Material	Labor	Equipment	Total	Total Incl O&P
0010	**CONCRETE BLOCK, HIGH STRENGTH**									
0050	Hollow, reinforced alternate courses, 8" x 16" units									
0200	3500 psi, 4" thick	D-8	440	.091	S.F.	2.22	3.50		5.72	7.75
0250	6" thick		395	.101		2.47	3.90		6.37	8.60
0300	8" thick		360	.111		3.58	4.28		7.86	10.40
0350	12" thick	D-9	250	.192		4.42	7.25		11.67	15.75
0500	5000 psi, 4" thick	D-8	440	.091		2.43	3.50		5.93	8
0550	6" thick		395	.101		3.05	3.90		6.95	9.25
0600	8" thick		360	.111		4.06	4.28		8.34	10.90
0650	12" thick	D-9	300	.160		6.05	6.05		12.10	15.75
1000	For 75% solid block, add					30%				
1050	For 100% solid block, add					50%				

04 22 10.30 Concrete Block, Interlocking		Crew	Daily Output	Labor-Hours	Unit	Material	Labor	Equipment	Total	Total Incl O&P
0010	**CONCRETE BLOCK, INTERLOCKING**									
0100	Not including grout or reinforcing									
0200	8" x 16" units, 2,000 psi, 8" thick	D-1	245	.065	S.F.	2.72	2.46		5.18	6.70
0300	12" thick		220	.073		4.03	2.74		6.77	8.55
0350	16" thick		185	.086		6.05	3.26		9.31	11.55
0400	Including grout & reinforcing, 8" thick	D-4	245	.131		7.30	4.91	.54	12.75	16.05
0450	12" thick		220	.145		8.80	5.45	.60	14.85	18.55
0500	16" thick		185	.173		10.95	6.50	.72	18.17	22.50

04 22 10.32 Concrete Block, Lintels		Crew	Daily Output	Labor-Hours	Unit	Material	Labor	Equipment	Total	Total Incl O&P
0010	**CONCRETE BLOCK, LINTELS**, C90, normal weight									
0100	Including grout and horizontal reinforcing									
0200	8" x 8" x 8", 1 #4 bar	D-4	300	.107	L.F.	4.83	4.01	.44	9.28	11.85
0250	2 #4 bars		295	.108		5	4.08	.45	9.53	12.15
0400	8" x 16" x 8", 1 #4 bar		275	.116		3.32	4.37	.48	8.17	10.80
0450	2 #4 bars		270	.119		3.50	4.46	.49	8.45	11.10
1000	12" x 8" x 8", 1 #4 bar		275	.116		6.80	4.37	.48	11.65	14.60
1100	2 #4 bars		270	.119		6.95	4.46	.49	11.90	14.90
1150	2 #5 bars		270	.119		7.15	4.46	.49	12.10	15.15
1200	2 #6 bars		265	.121		7.40	4.54	.50	12.44	15.55
1500	12" x 16" x 8", 1 #4 bar		250	.128		10.75	4.81	.53	16.09	19.70
1600	2 #3 bars		245	.131		10.80	4.91	.54	16.25	19.85
1650	2 #4 bars		245	.131		10.95	4.91	.54	16.40	20
1700	2 #5 bars		240	.133		11.15	5	.55	16.70	20.50

04 22 10.34 Concrete Block, Partitions		Crew	Daily Output	Labor-Hours	Unit	Material	Labor	Equipment	Total	Total Incl O&P
0010	**CONCRETE BLOCK, PARTITIONS**, excludes scaffolding R042210-20									
0100	Acoustical slotted block									
0200	4" thick, type A-1	D-8	315	.127	S.F.	3.74	4.89		8.63	11.45
0210	8" thick		275	.145		4.72	5.60		10.32	13.65
0250	8" thick, type Q		275	.145		8.30	5.60		13.90	17.60
0260	4" thick, type RSC		315	.127		5.80	4.89		10.69	13.75
0270	6" thick		295	.136		5.80	5.20		11	14.25
0280	8" thick		275	.145		5.80	5.60		11.40	14.85
0290	12" thick		250	.160		5.80	6.15		11.95	15.70
0300	8" thick, type RSR		275	.145		5.80	5.60		11.40	14.85
0400	8" thick, type RSC/RF		275	.145		6.55	5.60		12.15	15.65
0410	10" thick		260	.154		7	5.90		12.90	16.65
0420	12" thick		250	.160		7.55	6.15		13.70	17.60

04 22 10.34 Concrete Block, Partitions		Crew	Daily Output	Labor-Hours	Unit	Material	Labor	Equipment	Total	Total Incl O&P
						2010 Bare Costs				
0430	12" thick, type RSC/RF-4	D-8	250	.160	S.F.	8.95	6.15		15.10	19.15
0500	NRC .60 type R, 8" thick		265	.151		8	5.80		13.80	17.55
0600	NRC .65 type RR, 8" thick		265	.151		7.35	5.80		13.15	16.80
0700	NRC .65 type 4R-RF, 8" thick		265	.151		8.25	5.80		14.05	17.80
0710	NRC .70 type R, 12" thick		245	.163		8.65	6.30		14.95	19
1000	Lightweight block, tooled joints, 2 sides, hollow									
1100	Not reinforced, 8" x 16" x 4" thick	D-8	440	.091	S.F.	1.65	3.50		5.15	7.10
1150	6" thick		410	.098		2.37	3.76		6.13	8.25
1200	8" thick		385	.104		2.88	4		6.88	9.20
1250	10" thick		370	.108		3.49	4.16		7.65	10.15
1300	12" thick	D-9	350	.137		4.34	5.15		9.49	12.55
2000	Not reinforced, 8" x 24" x 4" thick, hollow		460	.104		1.21	3.93		5.14	7.30
2100	6" thick		440	.109		1.71	4.11		5.82	8.10
2150	8" thick		415	.116		2.12	4.36		6.48	8.95
2200	10" thick		385	.125		2.56	4.70		7.26	9.90
2250	12" thick		365	.132		3.17	4.96		8.13	11
2800	Solid, not reinforced, 8" x 16" x 2" thick	D-8	440	.091		1.46	3.50		4.96	6.90
2900	4" thick		420	.095		2.36	3.67		6.03	8.15
2950	6" thick		390	.103		2.63	3.95		6.58	8.85
3000	8" thick		365	.110		3.38	4.22		7.60	10.05
3050	10" thick		350	.114		4.22	4.40		8.62	11.30
3100	12" thick	D-9	330	.145		6.25	5.50		11.75	15.10
4000	Regular block, tooled joints, 2 sides, hollow									
4100	Not reinforced, 8" x 16" x 4" thick	D-8	430	.093	S.F.	1.47	3.58		5.05	7
4150	6" thick		400	.100		2.08	3.85		5.93	8.10
4200	8" thick **CN**		375	.107		2.36	4.11		6.47	8.80
4250	10" thick		360	.111		3	4.28		7.28	9.75
4300	12" thick	D-9	340	.141		3.42	5.30		8.72	11.80
4500	Reinforced alternate courses, 8" x 16" x 4" thick	D-8	425	.094		1.58	3.62		5.20	7.20
4550	6" thick		395	.101		2.21	3.90		6.11	8.35
4600	8" thick		370	.108		2.49	4.16		6.65	9.05
4650	10" thick		355	.113		3.43	4.34		7.77	10.30
4700	12" thick	D-9	335	.143		3.57	5.40		8.97	12.05
4900	Solid, not reinforced, 2" thick	D-8	435	.092		1.34	3.54		4.88	6.80
5000	3" thick		430	.093		1.60	3.58		5.18	7.15
5050	4" thick		415	.096		1.82	3.71		5.53	7.60
5100	6" thick		385	.104		2.15	4		6.15	8.40
5150	8" thick		360	.111		3.24	4.28		7.52	10
5200	12" thick	D-9	325	.148		4.88	5.55		10.43	13.75
5500	Solid, reinforced alternate courses, 4" thick	D-8	420	.095		1.89	3.67		5.56	7.65
5550	6" thick		380	.105		2.24	4.05		6.29	8.55
5600	8" thick		355	.113		3.34	4.34		7.68	10.20
5650	12" thick	D-9	320	.150		4.42	5.65		10.07	13.40

04 22 10.38 Concrete Brick		Crew	Daily Output	Labor-Hours	Unit	Material	Labor	Equipment	Total	Total Incl O&P
0010	**CONCRETE BRICK**, C55, grade N, type 1									
0100	Regular, 4 x 2-1/4 x 8	D-8	660	.061	Ea.	.45	2.33		2.78	4.02
0125	Rusticated, 4 x 2-1/4 x 8		660	.061		.50	2.33		2.83	4.07
0150	Frog, 4 x 2-1/4 x 8		660	.061		.49	2.33		2.82	4.05
0200	Double, 4 x 4-7/8 x 8		535	.075		.79	2.88		3.67	5.20

04 22 10.42 Concrete Block, Screen Block		Crew	Daily Output	Labor-Hours	Unit	Material	Labor	Equipment	Total	Total Incl O&P
0010	**CONCRETE BLOCK, SCREEN BLOCK**									
0200	8" x 16", 4" thick	D-8	330	.121	S.F.	1.97	4.67		6.64	9.20

04 22 Concrete Unit Masonry

04 22 10 – Concrete Masonry Units

04 22 10.42 Concrete Block, Screen Block		Crew	Daily Output	Labor-Hours	Unit	Material	Labor	Equipment	Total	Total Incl O&P
						2010 Bare Costs				
0300	8" thick	D-8	270	.148	S.F.	2.99	5.70		8.69	11.90
0350	12" x 12", 4" thick		290	.138		2.21	5.30		7.51	10.45
0500	8" thick		250	.160		3.06	6.15		9.21	12.65

04 22 10.44 Glazed Concrete Block		Crew	Daily Output	Labor-Hours	Unit	Material	Labor	Equipment	Total	Total Incl O&P
0010	**GLAZED CONCRETE BLOCK** C744									
0100	Single face, 8" x 16" units, 2" thick	D-8	360	.111	S.F.	8.25	4.28		12.53	15.55
0200	4" thick		345	.116		8.50	4.46		12.96	16.10
0250	6" thick		330	.121		9.15	4.67		13.82	17.10
0300	8" thick		310	.129		9.25	4.97		14.22	17.65
0350	10" thick		295	.136		11.10	5.20		16.30	20
0400	12" thick	D-9	280	.171		11.80	6.45		18.25	22.50
0700	Double face, 8" x 16" units, 4" thick	D-8	340	.118		12.20	4.53		16.73	20.50
0750	6" thick		320	.125		14.50	4.81		19.31	23
0800	8" thick		300	.133		15.20	5.15		20.35	24.50
1000	Jambs, bullnose or square, single face, 8" x 16", 2" thick		315	.127	Ea.	17.50	4.89		22.39	26.50
1050	4" thick		285	.140	"	20.50	5.40		25.90	30.50
1200	Caps, bullnose or square, 8" x 16", 2" thick		420	.095	L.F.	14.15	3.67		17.82	21
1250	4" thick		380	.105		15.95	4.05		20	23.50
1500	Cove base, 8" x 16", 2" thick		315	.127		11	4.89		15.89	19.45
1550	4" thick		285	.140		7.15	5.40		12.55	16
1600	6" thick		265	.151		7.70	5.80		13.50	17.20
1650	8" thick		245	.163		8.10	6.30		14.40	18.45

04 23 Glass Unit Masonry

04 23 13 – Vertical Glass Unit Masonry

04 23 13.10 Glass Block		Crew	Daily Output	Labor-Hours	Unit	Material	Labor	Equipment	Total	Total Incl O&P
0010	**GLASS BLOCK**									
0100	Plain, 4" thick, under 1,000 S.F., 6" x 6"	D-8	115	.348	S.F.	25.50	13.40		38.90	48
0150	8" x 8"		160	.250		15.55	9.65		25.20	31.50
0160	end block		160	.250		47	9.65		56.65	66
0170	90 deg corner		160	.250		46	9.65		55.65	65
0180	45 deg corner		160	.250		33.50	9.65		43.15	51.50
0200	12" x 12"		175	.229		20	8.80		28.80	35.50
0210	4" x 8"		160	.250		15.20	9.65		24.85	31
0220	6" x 8"		160	.250		14.60	9.65		24.25	30.50
0300	1,000 to 5,000 S.F., 6" x 6"		135	.296		25	11.40		36.40	44.50
0350	8" x 8"		190	.211		15.25	8.10		23.35	29
0400	12" x 12"		215	.186		19.70	7.15		26.85	32.50
0410	4" x 8"		215	.186		6	7.15		13.15	17.40
0420	6" x 8"		215	.186		5.70	7.15		12.85	17.05
0500	Over 5,000 S.F., 6" x 6"		145	.276		24	10.60		34.60	42.50
0550	8" x 8"		215	.186		14.75	7.15		21.90	27
0600	12" x 12"		240	.167		19.10	6.40		25.50	30.50
0610	4" x 8"		240	.167		5.60	6.40		12	15.85
0620	6" x 8"		240	.167		5.70	6.40		12.10	15.95
0700	For solar reflective blocks, add					100%				
1000	Thinline, plain, 3-1/8" thick, under 1,000 S.F., 6" x 6"	D-8	115	.348	S.F.	17.35	13.40		30.75	39
1050	8" x 8"		160	.250		9.70	9.65		19.35	25
1200	Over 5,000 S.F., 6" x 6"		145	.276		16.80	10.60		27.40	34.50
1250	8" x 8"		215	.186		9.45	7.15		16.60	21
1400	For cleaning block after installation (both sides), add		1000	.040		.11	1.54		1.65	2.44

04 23 Glass Unit Masonry

04 23 13 – Vertical Glass Unit Masonry

04 23 13.10 Glass Block		Crew	Daily Output	Labor-Hours	Unit	Material	2010 Bare Costs Labor	Equipment	Total	Total Incl O&P
4000	Accessories									
4100	Anchors, 20 ga. galv., 1-3/4″ wide x 24″ long				Ea.	3.09			3.09	3.40
4200	Emulsion asphalt				Gal.	8.10			8.10	8.90
4300	Expansion joint, fiberglass				L.F.	.67			.67	.74
4400	Steel mesh, double galvanized				″	.94			.94	1.03

04 24 Adobe Unit Masonry

04 24 16 – Manufactured Adobe Unit Masonry

04 24 16.06 Adobe Brick

		Crew	Daily Output	Labor-Hours	Unit	Material	Labor	Equipment	Total	Total Incl O&P
0010	**ADOBE BRICK**, Semi-stabilized, with cement mortar									
0060	Brick, 10″ x 4″ x 14″, 2.6/S.F. G	D-8	560	.071	S.F.	3.45	2.75		6.20	7.95
0080	12″ x 4″ x 16″, 2.3/S.F. G		580	.069		4.61	2.66		7.27	9.05
0100	10″ x 4″ x 16″, 2.3/S.F. G		590	.068		4.37	2.61		6.98	8.75
0120	8″ x 4″ x 16″, 2.3/S.F. G		560	.071		3.36	2.75		6.11	7.85
0140	4″ x 4″ x 16″, 2.3/S.F. G		540	.074		2.56	2.85		5.41	7.10
0160	6″ x 4″ x 16″, 2.3/S.F. G		540	.074		2.59	2.85		5.44	7.15
0180	4″ x 4″ x 12″, 3.0/S.F. G		520	.077		2.45	2.96		5.41	7.15
0200	8″ x 4″ x 12″, 3.0/S.F. G	↓	520	.077	↓	3.16	2.96		6.12	7.95

04 25 Unit Masonry Panels

04 25 20 – Pre-Fabricated Masonry Panels

04 25 20.10 Brick and Epoxy Mortar Panels

		Crew	Daily Output	Labor-Hours	Unit	Material	Labor	Equipment	Total	Total Incl O&P
0010	**BRICK AND EPOXY MORTAR PANELS**									
0020	Prefabricated brick & epoxy mortar, 4″ thick, minimum	C-11	775	.093	S.F.	9.25	4.26	2.58	16.09	20.50
0100	Maximum	″	500	.144		10.75	6.60	4.01	21.36	27.50
0200	For 2″ concrete back-up, add					50%				
0300	For 1″ urethane & 3″ concrete back-up, add				↓	70%				

04 27 Multiple-Wythe Unit Masonry

04 27 10 – Multiple-Wythe Masonry

04 27 10.20 Cavity Walls

		Crew	Daily Output	Labor-Hours	Unit	Material	Labor	Equipment	Total	Total Incl O&P
0010	**CAVITY WALLS**, brick and CMU, includes joint reinforcing and ties									
0200	4″ face brick, 4″ block	D-8	165	.242	S.F.	6.25	9.35		15.60	21
0400	6″ block		145	.276		6.75	10.60		17.35	23.50
0600	8″ block	↓	125	.320	↓	6.90	12.30		19.20	26

04 27 10.30 Brick Walls

		Crew	Daily Output	Labor-Hours	Unit	Material	Labor	Equipment	Total	Total Incl O&P
0010	**BRICK WALLS**, including mortar, excludes scaffolding R042110-20									
0020	Estimating by number of brick									
0140	Face brick, 4″ thick wall, 6.75 brick/S.F.	D-8	1.45	27.586	M	685	1,050		1,735	2,350
0150	Common brick, 4″ thick wall, 6.75 brick/S.F.		1.60	25		495	965		1,460	2,000
0204	8″ thick, 13.50 bricks per S.F.		1.80	22.222		510	855		1,365	1,850
0250	12″ thick, 20.25 bricks per S.F.		1.90	21.053		515	810		1,325	1,800
0304	16″ thick, 27.00 bricks per S.F.		2	20		520	770		1,290	1,725
0500	Reinforced, face brick, 4″ thick wall, 6.75 brick/S.F.		1.40	28.571		705	1,100		1,805	2,425
0520	Common brick, 4″ thick wall, 6.75 brick/S.F.		1.55	25.806		515	995		1,510	2,075
0550	8″ thick, 13.50 bricks per S.F.		1.75	22.857		530	880		1,410	1,900
0600	12″ thick, 20.25 bricks per S.F.	↓	1.85	21.622	↓	535	835		1,370	1,850

04 27 Multiple-Wythe Unit Masonry

04 27 10 – Multiple-Wythe Masonry

04 27 10.30 Brick Walls		Crew	Daily Output	Labor-Hours	Unit	Material	Labor	Equipment	Total	Total Incl O&P
						2010 Bare Costs				
0650	16" thick, 27.00 bricks per S.F.	D-8	1.95	20.513	M	540	790		1,330	1,800
0790	Alternate method of figuring by square foot									
0800	Face brick, 4" thick wall, 6.75 brick/S.F.	D-8	215	.186	S.F.	4.61	7.15		11.76	15.85
0850	Common brick, 4" thick wall, 6.75 brick/S.F.		240	.167		3.33	6.40		9.73	13.35
0900	8" thick, 13.50 bricks per S.F.		135	.296		6.90	11.40		18.30	25
1000	12" thick, 20.25 bricks per S.F.		95	.421		10.40	16.20		26.60	36
1050	16" thick, 27.00 bricks per S.F.		75	.533		14.05	20.50		34.55	46.50
1200	Reinforced, face brick, 4" thick wall, 6.75 brick/S.F.		210	.190		4.74	7.35		12.09	16.25
1220	Common brick, 4" thick wall, 6.75 brick/S.F.		235	.170		3.46	6.55		10.01	13.70
1250	8" thick, 13.50 bricks per S.F.		130	.308		7.15	11.85		19	26
1300	12" thick, 20.25 bricks per S.F.		90	.444		10.80	17.10		27.90	38
1350	16" thick, 27.00 bricks per S.F.	↓	70	.571	↓	14.60	22		36.60	49

04 27 10.40 Steps		Crew	Daily Output	Labor-Hours	Unit	Material	Labor	Equipment	Total	Total Incl O&P
0010	**STEPS**									
0012	Entry steps, select common brick	D-1	.30	53.333	M	430	2,000		2,430	3,500

04 41 Dry-Placed Stone

04 41 10 – Dry Placed Stone

04 41 10.10 Rough Stone Wall		Crew	Daily Output	Labor-Hours	Unit	Material	Labor	Equipment	Total	Total Incl O&P
0011	**ROUGH STONE WALL**, Dry									
0012	Dry laid (no mortar), under 18" thick	D-1	60	.267	C.F.	8.80	10.05		18.85	25
0100	Random fieldstone, under 18" thick G	D-12	60	.533		8.80	20.50		29.30	40
0150	Over 18" thick G	"	63	.508	↓	10.55	19.40		29.95	41
0500	Field stone veneer G	D-8	120	.333	S.F.	9.95	12.85		22.80	30.50
0510	Valley stone veneer G		120	.333		9.95	12.85		22.80	30.50
0520	River stone veneer G	↓	120	.333	↓	9.95	12.85		22.80	30.50
0600	Rubble stone walls, in mortar bed, up to 18" thick G	D-11	75	.320	C.F.	10.65	12.70		23.35	31

04 43 Stone Masonry

04 43 10 – Masonry with Natural and Processed Stone

04 43 10.05 Ashlar Veneer		Crew	Daily Output	Labor-Hours	Unit	Material	Labor	Equipment	Total	Total Incl O&P
0011	**ASHLAR VENEER** 4" + or - thk, random or random rectangular									
0150	Sawn face, split joints, low priced stone	D-8	140	.286	S.F.	9.35	11		20.35	27
0200	Medium priced stone		130	.308		11.15	11.85		23	30
0300	High priced stone		120	.333		14.35	12.85		27.20	35
0600	Seam face, split joints, medium price stone		125	.320		7.55	12.30		19.85	27
0700	High price stone		120	.333		15	12.85		27.85	36
1000	Split or rock face, split joints, medium price stone		125	.320		10.80	12.30		23.10	30.50
1100	High price stone	↓	120	.333	↓	15	12.85		27.85	36

04 43 10.10 Bluestone		Crew	Daily Output	Labor-Hours	Unit	Material	Labor	Equipment	Total	Total Incl O&P
0010	**BLUESTONE**, cut to size									
0500	Sills, natural cleft, 10" wide to 6' long, 1-1/2" thick	D-11	70	.343	L.F.	13.35	13.60		26.95	35
0550	2" thick	"	63	.381		14.60	15.15		29.75	39
1000	Stair treads, natural cleft, 12" wide, 6' long, 1-1/2" thick	D-10	115	.278		30	11.40	5.70	47.10	56.50
1050	2" thick		105	.305		30	12.45	6.25	48.70	58.50
1100	Smooth finish, 1-1/2" thick		115	.278		15	11.40	5.70	32.10	40
1300	Thermal finish, 1-1/2" thick		115	.278		15	11.40	5.70	32.10	40
1350	2" thick	↓	105	.305	↓	15	12.45	6.25	33.70	42

04 43 Stone Masonry

04 43 10 – Masonry with Natural and Processed Stone

04 43 10.45 Granite

04 43 10.45 Granite			Crew	Daily Output	Labor-Hours	Unit	Material	Labor	Equipment	Total	Total Incl O&P
							2010 Bare Costs				
0010	**GRANITE**, cut to size										
0050	Veneer, polished face, 3/4″ to 1-1/2″ thick										
0150	Low price, gray, light gray, etc.	CN	D-10	130	.246	S.F.	26	10.05	5.05	41.10	49
0220	High price, red, black, etc.	CN	″	130	.246	″	41	10.05	5.05	56.10	66
0300	1-1/2″ to 2-1/2″ thick, veneer										
0350	Low price, gray, light gray, etc.		D-10	130	.246	S.F.	28	10.05	5.05	43.10	51
0550	High price, red, black, etc.		″	130	.246	″	51	10.05	5.05	66.10	76.50
0700	2-1/2″ to 4″ thick, veneer										
0750	Low price, gray, light gray, etc.		D-10	110	.291	S.F.	37.50	11.90	5.95	55.35	66
0950	High price, red, black, etc.		″	110	.291		61.50	11.90	5.95	79.35	92
1000	For bush hammered finish, deduct						5%				
1050	Coarse rubbed finish, deduct						10%				
1100	Honed finish, deduct						5%				
1150	Thermal finish, deduct						18%				
2450	For radius under 5′, add					L.F.	100%				
2500	Steps, copings, etc., finished on more than one surface										
2550	Minimum		D-10	50	.640	C.F.	91	26	13.10	130.10	154
2600	Maximum		″	50	.640	″	146	26	13.10	185.10	214
2800	Pavers, 4″ x 4″ x 4″ blocks, split face and joints										
2850	Minimum		D-11	80	.300	S.F.	12.75	11.90		24.65	32
2900	Maximum		″	80	.300		28.50	11.90		40.40	49
4000	Soffits, 2″ thick, minimum		D-13	35	1.371		37.50	54.50	18.75	110.75	145
4100	Maximum			35	1.371		89	54.50	18.75	162.25	201
4200	4″ thick, minimum			35	1.371		61.50	54.50	18.75	134.75	171
4300	Maximum			35	1.371		116	54.50	18.75	189.25	230

04 43 10.50 Lightweight Natural Stone

04 43 10.50 Lightweight Natural Stone			Crew	Daily Output	Labor-Hours	Unit	Material	Labor	Equipment	Total	Total Incl O&P
0011	**LIGHTWEIGHT NATURAL STONE** Lava type										
0100	Veneer, rubble face, sawed back, irregular shapes	G	D-10	130	.246	S.F.	6.30	10.05	5.05	21.40	27.50
0200	Sawed face and back, irregular shapes	G	″	130	.246	″	6.30	10.05	5.05	21.40	27.50

04 43 10.55 Limestone

04 43 10.55 Limestone		Crew	Daily Output	Labor-Hours	Unit	Material	Labor	Equipment	Total	Total Incl O&P
0010	**LIMESTONE**, cut to size									
0020	Veneer facing panels									
0500	Texture finish, light stick, 4-1/2″ thick, 5′ x 12′	D-4	300	.107	S.F.	48	4.01	.44	52.45	59.50
0750	5″ thick, 5′ x 14′ panels	D-10	275	.116		52.50	4.76	2.38	59.64	67.50
1000	Sugarcube finish, 2″ Thick, 3′ x 5′ panels		275	.116		35	4.76	2.38	42.14	48.50
1050	3″ Thick, 4′ x 9′ panels		275	.116		38	4.76	2.38	45.14	52
1200	4″ Thick, 5′ x 11′ panels		275	.116		41	4.76	2.38	48.14	55
1400	Sugarcube, textured finish, 4-1/2″ thick, 5′ x 12′		275	.116		45	4.76	2.38	52.14	59.50
1450	5″ thick, 5′ x 14′ panels		275	.116		53.50	4.76	2.38	60.64	68.50
2000	Coping, sugarcube finish, top & 2 sides		30	1.067	C.F.	68	43.50	22	133.50	165
2100	Sills, lintels, jambs, trim, stops, sugarcube finish, average		20	1.600		68	65.50	33	166.50	210
2150	Detailed		20	1.600		68	65.50	33	166.50	210
2300	Steps, extra hard, 14″ wide, 6″ rise		50	.640	L.F.	41.50	26	13.10	80.60	100
3000	Quoins, plain finish, 6″ x 12″ x 12″	D-12	25	1.280	Ea.	38	49		87	116
3050	6″ x 16″ x 24″	″	25	1.280	″	50.50	49		99.50	129

04 43 10.60 Marble

04 43 10.60 Marble		Crew	Daily Output	Labor-Hours	Unit	Material	Labor	Equipment	Total	Total Incl O&P
0011	**MARBLE**, ashlar, split face, 4″ + or - thick, random									
0040	Lengths 1′ to 4′ & heights 2″ to 7-1/2″, average	D-8	175	.229	S.F.	17.75	8.80		26.55	33
0100	Base, polished, 3/4″ or 7/8″ thick, polished, 6″ high	D-10	65	.492	L.F.	16.15	20	10.10	46.25	59.50
0300	Carvings or bas relief, from templates, average		80	.400	S.F.	144	16.35	8.20	168.55	192
0350	Maximum		80	.400	″	335	16.35	8.20	359.55	405

04 43 Stone Masonry

04 43 10 – Masonry with Natural and Processed Stone

04 43 10.60 Marble		Crew	Daily Output	Labor-Hours	Unit	Material	2010 Bare Costs Labor	Equipment	Total	Total Incl O&P
0600	Columns, cornices, mouldings, etc.									
0650	Hand or special machine cut, average	D-10	35	.914	C.F.	144	37.50	18.75	200.25	235
0700	Maximum	"	35	.914	"	310	37.50	18.75	366.25	415
1000	Facing, polished finish, cut to size, 3/4" to 7/8" thick									
1050	Average	D-10	130	.246	S.F.	23.50	10.05	5.05	38.60	46.50
1100	Maximum		130	.246		55	10.05	5.05	70.10	81
1300	1-1/4" thick, average		125	.256		34.50	10.45	5.25	50.20	59
1350	Maximum		125	.256		68.50	10.45	5.25	84.20	97
1500	2" thick, average		120	.267		40	10.90	5.45	56.35	66.50
1550	Maximum		120	.267		69	10.90	5.45	85.35	98.50
2200	Window sills, 6" x 3/4" thick	D-1	85	.188	L.F.	8.95	7.10		16.05	20.50
2500	Flooring, polished tiles, 12" x 12" x 3/8" thick									
2510	Thin set, average	D-11	90	.267	S.F.	11.90	10.60		22.50	29
2600	Maximum		90	.267		183	10.60		193.60	217
2700	Mortar bed, average		65	.369		10.65	14.65		25.30	34
2740	Maximum		65	.369		183	14.65		197.65	223
2780	Travertine, 3/8" thick, average	D-10	130	.246		15.15	10.05	5.05	30.25	37.50
2790	Maximum	"	130	.246		37	10.05	5.05	52.10	61.50
2800	Patio tile, non-slip, 1/2" thick, flame finish	D-11	75	.320		15.45	12.70		28.15	36
2900	Shower or toilet partitions, 7/8" thick partitions									
3050	3/4" or 1-1/4" thick stiles, polished 2 sides, average	D-11	75	.320	S.F.	46	12.70		58.70	69.50
3201	Soffits, add to above prices				"	20%	100%			
3210	Stairs, risers, 7/8" thick x 6" high	D-10	115	.278	L.F.	15.55	11.40	5.70	32.65	40.50
3360	Treads, 12" wide x 1-1/4" thick	"	115	.278	"	22.50	11.40	5.70	39.60	48.50
3500	Thresholds, 3' long, 7/8" thick, 4" to 5" wide, plain	D-12	24	1.333	Ea.	17.15	51		68.15	96
3550	Beveled		24	1.333	"	19.90	51		70.90	99
3700	Window stools, polished, 7/8" thick, 5" wide		85	.376	L.F.	15.95	14.40		30.35	39

04 43 10.75 Sandstone or Brownstone		Crew	Daily Output	Labor-Hours	Unit	Material	Labor	Equipment	Total	Total Incl O&P
0011	**SANDSTONE OR BROWNSTONE**									
0100	Sawed face veneer, 2-1/2" thick, to 2' x 4' panels	D-10	130	.246	S.F.	20	10.05	5.05	35.10	43
0150	4" thick, to 3'-6" x 8' panels		100	.320		20	13.10	6.55	39.65	49.50
0300	Split face, random sizes		100	.320		13.55	13.10	6.55	33.20	42
0350	Cut stone trim (limestone)									
0360	Ribbon stone, 4" thick, 5' pieces	D-8	120	.333	Ea.	167	12.85		179.85	203
0370	Cove stone, 4" thick, 5' pieces		105	.381		168	14.65		182.65	207
0380	Cornice stone, 10" to 12" wide		90	.444		207	17.10		224.10	254
0390	Band stone, 4" thick, 5' pieces		145	.276		107	10.60		117.60	134
0410	Window and door trim, 3" to 4" wide		160	.250		91	9.65		100.65	115
0420	Key stone, 18" long		60	.667		95.50	25.50		121	144

04 43 10.80 Slate		Crew	Daily Output	Labor-Hours	Unit	Material	Labor	Equipment	Total	Total Incl O&P
0010	**SLATE**									
0040	Pennsylvania - blue gray to black									
0050	Vermont - unfading green, mottled green & purple, gray & purple									
0100	Virginia - blue black									
0200	Exterior paving, natural cleft, 1" thick									
0500	24" x 24", Pennsylvania	D-12	120	.267	S.F.	13	10.20		23.20	29.50
0550	Vermont		120	.267		31	10.20		41.20	49.50
0600	Virginia		120	.267		20.50	10.20		30.70	38
1000	Interior flooring, natural cleft, 1/2" thick									
1300	24" x 24" Pennsylvania	D-12	120	.267	S.F.	7.75	10.20		17.95	24
1350	Vermont		120	.267		25	10.20		35.20	43
1400	Virginia		120	.267		14.75	10.20		24.95	31.50

04 43 Stone Masonry

04 43 10 – Masonry with Natural and Processed Stone

04 43 10.80 Slate		Crew	Daily Output	Labor-Hours	Unit	Material	Labor	Equipment	Total	Total Incl O&P
						2010 Bare Costs				
2000	Facing panels, 1-1/4" thick, to 4' x 4' panels									
2100	Natural cleft finish, Pennsylvania	D-10	180	.178	S.F.	34	7.25	3.64	44.89	52
2110	Vermont		180	.178		27	7.25	3.64	37.89	45
2120	Virginia		180	.178		33	7.25	3.64	43.89	51.50
2150	Sand rubbed finish, surface, add					10.15			10.15	11.20
2200	Honed finish, add					7.35			7.35	8.10
2500	Ribbon, natural cleft finish, 1" thick, to 9 S.F.	D-10	80	.400		13.10	16.35	8.20	37.65	48
2700	1-1/2" thick		78	.410		17.05	16.80	8.40	42.25	53.50
2850	2" thick		76	.421		20.50	17.20	8.65	46.35	58
3500	Stair treads, sand finish, 1" thick x 12" wide									
3600	3 L.F. to 6 L.F.	D-10	120	.267	L.F.	24	10.90	5.45	40.35	48.50
3700	Ribbon, sand finish, 1" thick x 12" wide									
3750	To 6 L.F.	D-10	120	.267	L.F.	20	10.90	5.45	36.35	44.50
4000	Stools or sills, sand finish, 1" thick, 6" wide	D-12	160	.200		11.50	7.65		19.15	24
4200	10" wide		90	.356		17.75	13.60		31.35	40
4400	2" thick, 6" wide		140	.229		18.50	8.75		27.25	33.50
4600	10" wide		90	.356		29	13.60		42.60	52.50
4800	For lengths over 3', add					25%				

04 43 10.85 Window Sill		Crew	Daily Output	Labor-Hours	Unit	Material	Labor	Equipment	Total	Total Incl O&P
0010	**WINDOW SILL**									
0020	Bluestone, thermal top, 10" wide, 1-1/2" thick	D-1	85	.188	S.F.	16.05	7.10		23.15	28.50
0050	2" thick		75	.213	"	18.75	8.05		26.80	32.50
0100	Cut stone, 5" x 8" plain		48	.333	L.F.	11.90	12.55		24.45	32
0200	Face brick on edge, brick, 8" wide		80	.200		2.53	7.55		10.08	14.15
0400	Marble, 9" wide, 1" thick		85	.188		8.50	7.10		15.60	20
0900	Slate, colored, unfading, honed, 12" wide, 1" thick		85	.188		17.10	7.10		24.20	29.50
0950	2" thick		70	.229		24	8.60		32.60	39

04 51 Flue Liner Masonry

04 51 10 – Clay Flue Lining

04 51 10.10 Flue Lining		Crew	Daily Output	Labor-Hours	Unit	Material	Labor	Equipment	Total	Total Incl O&P
0010	**FLUE LINING**, including mortar									
0020	Clay, 8" x 8"	D-1	125	.128	V.L.F.	4.67	4.83		9.50	12.45
0100	8" x 12"		103	.155		7.20	5.85		13.05	16.75
0200	12" x 12"		93	.172		9.45	6.50		15.95	20
0300	12" x 18"		84	.190		17.70	7.20		24.90	30.50
0400	18" x 18"		75	.213		22	8.05		30.05	36
0500	20" x 20"		66	.242		34	9.15		43.15	51.50
0600	24" x 24"		56	.286		43.50	10.75		54.25	64.50
1000	Round, 18" diameter		66	.242		30	9.15		39.15	47
1100	24" diameter		47	.340		62.50	12.85		75.35	88.50

04 54 Refractory Brick Masonry

04 54 10 – Refractory Brick Work

04 54 10.10 Fire Brick		Crew	Daily Output	Labor-Hours	Unit	Material	Labor	Equipment	Total	Total Incl O&P
						2010 Bare Costs				
0010	**FIRE BRICK**									
0012	Low duty, 2000°F, 9" x 2-1/2" x 4-1/2"	D-1	.60	26.667	M	1,375	1,000		2,375	3,050
0050	High duty, 3000°F	"	.60	26.667	"	2,100	1,000		3,100	3,850
04 54 10.20 Fire Clay										
0010	**FIRE CLAY**									
0020	Gray, high duty, 100 lb. bag				Bag	59			59	65
0050	100 lb. drum, premixed (400 brick per drum)				Drum	73			73	80.50

04 57 Masonry Fireplaces

04 57 10 – Brick or Stone Fireplaces

04 57 10.10 Fireplace

		Crew	Daily Output	Labor-Hours	Unit	Material	Labor	Equipment	Total	Total Incl O&P
0010	**FIREPLACE**									
0100	Brick fireplace, not incl. foundations or chimneys									
0110	30" x 29" opening, incl. chamber, plain brickwork	D-1	.40	40	Ea.	535	1,500		2,035	2,875
0200	Fireplace box only (110 brick)	"	2	8	"	176	300		476	650
0300	For elaborate brickwork and details, add					35%	35%			
0400	For hearth, brick & stone, add	D-1	2	8	Ea.	197	300		497	670
0410	For steel angle, damper, cleanouts, add		4	4		138	151		289	380
0600	Plain brickwork, incl. metal circulator		.50	32	↓	1,025	1,200		2,225	2,950
0800	Face brick only, standard size, 8" x 2-2/3" x 4"	↓	.30	53.333	M	535	2,000		2,535	3,625

04 71 Manufactured Brick Masonry

04 71 10 – Simulated or Manufactured Brick

04 71 10.10 Simulated Brick

		Crew	Daily Output	Labor-Hours	Unit	Material	Labor	Equipment	Total	Total Incl O&P
0010	**SIMULATED BRICK**									
0020	Aluminum, baked on colors	1 Carp	200	.040	S.F.	3.53	1.66		5.19	6.45
0050	Fiberglass panels		200	.040		3.87	1.66		5.53	6.80
0100	Urethane pieces cemented in mastic		150	.053		6.70	2.22		8.92	10.75
0150	Vinyl siding panels	↓	200	.040		2.67	1.66		4.33	5.50
0160	Cement base, brick, incl. mastic	D-1	100	.160	↓	4.22	6.05		10.27	13.75
0170	Corner		50	.320	V.L.F.	10.55	12.05		22.60	30
0180	Stone face, incl. mastic		100	.160	S.F.	10.20	6.05		16.25	20.50
0190	Corner	↓	50	.320	V.L.F.	11.25	12.05		23.30	30.50

04 72 Cast Stone Masonry

04 72 10 – Cast Stone Masonry Features

04 72 10.10 Coping

		Crew	Daily Output	Labor-Hours	Unit	Material	Labor	Equipment	Total	Total Incl O&P
0010	**COPING**, stock units									
0050	Precast concrete, 10" wide, 4" tapers to 3-1/2", 8" wall	D-1	75	.213	L.F.	17.55	8.05		25.60	31.50
0100	12" wide, 3-1/2" tapers to 3", 10" wall		70	.229		17.55	8.60		26.15	32.50
0110	14" wide, 4" tapers to 3-1/2", 12" wall		65	.246		20.50	9.30		29.80	36.50
0150	16" wide, 4" tapers to 3-1/2", 14" wall		60	.267	↓	17.10	10.05		27.15	34
0250	Precast concrete corners		40	.400	Ea.	28.50	15.10		43.60	53.50
0300	Limestone for 12" wall, 4" thick		90	.178	L.F.	15.35	6.70		22.05	27
0350	6" thick		80	.200		23	7.55		30.55	37
0500	Marble, to 4" thick, no wash, 9" wide		90	.178		23	6.70		29.70	35.50
0550	12" wide	↓	80	.200	↓	35	7.55		42.55	50

04 72 Cast Stone Masonry

04 72 10 – Cast Stone Masonry Features

04 72 10.10 Coping		Crew	Daily Output	Labor-Hours	Unit	Material	2010 Bare Costs Labor	Equipment	Total	Total Incl O&P
0700	Terra cotta, 9" wide	D-1	90	.178	L.F.	5.60	6.70		12.30	16.25
0750	12" wide		80	.200		9.20	7.55		16.75	21.50
0800	Aluminum, for 12" wall		80	.200		13.30	7.55		20.85	26

04 72 20 – Cultured Stone Veneer

04 72 20.10 Cultured Stone Veneer Components		Crew	Daily Output	Labor-Hours	Unit	Material	Labor	Equipment	Total	Total Incl O&P
0010	**CULTURED STONE VENEER COMPONENTS**									
0110	On wood frame and sheathing substrate, random sized cobbles, corner stones	D-8	70	.571	V.L.F.	12.15	22		34.15	46.50
0120	Field stones		140	.286	S.F.	8.80	11		19.80	26.50
0130	Random sized flats, corner stones		70	.571	V.L.F.	11.95	22		33.95	46
0140	Field stones		140	.286	S.F.	10.10	11		21.10	27.50
0150	Horizontal lined ledgestones, corner stones		75	.533	V.L.F.	12.15	20.50		32.65	44.50
0160	Field stones		150	.267	S.F.	8.80	10.25		19.05	25
0170	Random shaped flats, corner stones		65	.615	V.L.F.	12.15	23.50		35.65	49
0180	Field stones		150	.267	S.F.	8.80	10.25		19.05	25
0190	Random shaped/textured face, corner stones		65	.615	V.L.F.	12.15	23.50		35.65	49
0200	Field stones		130	.308	S.F.	8.80	11.85		20.65	27.50
0210	Random shaped river rock, corner stones		65	.615	V.L.F.	12.15	23.50		35.65	49
0220	Field stones		130	.308	S.F.	8.80	11.85		20.65	27.50
0240	On concrete or CMU substrate, random sized cobbles, corner stones		70	.571	V.L.F.	11.50	22		33.50	45.50
0250	Field stones		140	.286	S.F.	8.50	11		19.50	26
0260	Random sized flats, corner stones		70	.571	V.L.F.	11.30	22		33.30	45.50
0270	Field stones		140	.286	S.F.	9.75	11		20.75	27.50
0280	Horizontal lined ledgestones, corner stones		75	.533	V.L.F.	11.50	20.50		32	43.50
0290	Field stones		150	.267	S.F.	8.50	10.25		18.75	25
0300	Random shaped flats, corner stones		70	.571	V.L.F.	11.50	22		33.50	45.50
0310	Field stones		140	.286	S.F.	8.50	11		19.50	26
0320	Random shaped/textured face, corner stones		65	.615	V.L.F.	11.50	23.50		35	48
0330	Field stones		130	.308	S.F.	8.50	11.85		20.35	27
0340	Random shaped river rock, corner stones		65	.615	V.L.F.	11.50	23.50		35	48
0350	Field stones		130	.308	S.F.	8.50	11.85		20.35	27
0360	Cultured stone veneer, #15 felt weather resistant barrier	1 Clab	3700	.002	Sq.	5.90	.07		5.97	6.55
0370	Expanded metal lath, diamond, 2.5 lb/sy, galvanized	1 Lath	85	.094	S.Y.	2.94	3.48		6.42	8.35
0390	Water table or window sill, 18" long	1 Bric	80	.100	Ea.	11.85	4.18		16.03	19.35

04 73 Manufactured Stone Masonry

04 73 20 – Simulated or Manufactured Stone

04 73 20.10 Simulated Stone		Crew	Daily Output	Labor-Hours	Unit	Material	Labor	Equipment	Total	Total Incl O&P
0010	**SIMULATED STONE**									
0100	Insulated fiberglass panels, 5/8" ply backer	L-4	200	.120	S.F.	11.40	4.67		16.07	19.70

Estimating Tips

05 05 00 Common Work Results for Metals

- Nuts, bolts, washers, connection angles, and plates can add a significant amount to both the tonnage of a structural steel job and the estimated cost. As a rule of thumb, add 10% to the total weight to account for these accessories.
- Type 2 steel construction, commonly referred to as "simple construction," consists generally of field-bolted connections with lateral bracing supplied by other elements of the building, such as masonry walls or x-bracing. The estimator should be aware, however, that shop connections may be accomplished by welding or bolting. The method may be particular to the fabrication shop and may have an impact on the estimated cost.

05 12 23 Structural Steel

- Steel items can be obtained from two sources: a fabrication shop or a metals service center. Fabrication shops can fabricate items under more controlled conditions than can crews in the field. They are also more efficient and can produce items more economically. Metal service centers serve as a source of long mill shapes to both fabrication shops and contractors.
- Most line items in this structural steel subdivision, and most items in 05 50 00 Metal Fabrications, are indicated as being shop fabricated. The bare material cost for these shop fabricated items is the "Invoice Cost" from the shop and includes the mill base price of steel plus mill extras, transportation to the shop, shop drawings and detailing where warranted, shop fabrication and handling, sandblasting and a shop coat of primer paint, all necessary structural bolts, and delivery to the job site. The bare labor cost and bare equipment cost for these shop fabricated items is for field installation or erection.
- Line items in Subdivision 05 12 23.40 Lightweight Framing, and other items scattered in Division 5, are indicated as being field fabricated. The bare material cost for these field fabricated items is the "Invoice Cost" from the metals service center and includes the mill base price of steel plus mill extras, transportation to the metals service center, material handling, and delivery of long lengths of mill shapes to the job site. Material costs for structural bolts and welding rods should be added to the estimate. The bare labor cost and bare equipment cost for these items is for both field fabrication and field installation or erection, and include time for cutting, welding and drilling in the fabricated metal items. Drilling into concrete and fasteners to fasten field fabricated items to other work are not included and should be added to the estimate.

05 20 00 Steel Joist Framing

- In any given project the total weight of open web steel joists is determined by the loads to be supported and the design. However, economies can be realized in minimizing the amount of labor used to place the joists. This is done by maximizing the joist spacing, and therefore minimizing the number of joists required to be installed on the job. Certain spacings and locations may be required by the design, but in other cases maximizing the spacing and keeping it as uniform as possible will keep the costs down.

05 30 00 Steel Decking

- The takeoff and estimating of metal deck involves more than simply the area of the floor or roof and the type of deck specified or shown on the drawings. Many different sizes and types of openings may exist. Small openings for individual pipes or conduits may be drilled after the floor/roof is installed, but larger openings may require special deck lengths as well as reinforcing or structural support. The estimator should determine who will be supplying this reinforcing. Additionally, some deck terminations are part of the deck package, such as screed angles and pour stops, and others will be part of the steel contract, such as angles attached to structural members and cast-in-place angles and plates. The estimator must ensure that all pieces are accounted for in the complete estimate.

05 50 00 Metal Fabrications

- The most economical steel stairs are those that use common materials, standard details, and most importantly, a uniform and relatively simple method of field assembly. Commonly available A36 channels and plates are very good choices for the main stringers of the stairs, as are angles and tees for the carrier members. Risers and treads are usually made by specialty shops, and it is most economical to use a typical detail in as many places as possible. The stairs should be pre-assembled and shipped directly to the site. The field connections should be simple and straightforward to be accomplished efficiently, and with minimum equipment and labor.

Reference Numbers

Reference numbers are shown in shaded boxes at the beginning of some major classifications. These numbers refer to related items in the Reference Section. The reference information may be an estimating procedure, an alternate pricing method, or technical information. *Note:* Not all subdivisions listed here necessarily appear in this publication.

05 01 Maintenance of Metals

05 01 10 – Maintenance of Structural Metal Framing

05 01 10.51 Cleaning of Structural Metal Framing		Crew	Daily Output	Labor-Hours	Unit	2010 Bare Costs Material	Labor	Equipment	Total	Total Incl O&P
0010	**CLEANING OF STRUCTURAL METAL FRAMING**									
6125	Steel surface treatments, PDCA guidelines									
6170	Wire brush, hand (SSPC-SP2)	1 Psst	400	.020	S.F.	.02	.75		.77	1.36
6180	Power tool (SSPC-SP3)	″	700	.011		.06	.43		.49	.83
6215	Pressure washing, 2800-6000 S.F./day	1 Pord	10000	.001			.03		.03	.04
6220	Steam cleaning, 2800-4000 S.F./day		2000	.004			.15		.15	.22
6225	Water blasting		2500	.003			.12		.12	.17
6230	Brush-off blast (SSPC-SP7)	E-11	1750	.018		.25	.68	.13	1.06	1.55
6235	Com'l blast (SSPC-SP6), loose scale, fine pwder rust, 2.0 #/S.F. sand		1200	.027		.51	.99	.19	1.69	2.42
6240	Tight mill scale, little/no rust, 3.0 #/S.F. sand		1000	.032		.76	1.19	.23	2.18	3.06
6245	Exist coat blistered/pitted, 4.0 #/S.F. sand		875	.037		1.02	1.36	.26	2.64	3.67
6250	Exist coat badly pitted/nodules, 6.7 #/S.F. sand		825	.039		1.71	1.45	.28	3.44	4.57
6255	Near white blast (SSPC-SP10), loose scale, fine rust, 5.6 #/S.F. sand		450	.071		1.43	2.65	.50	4.58	6.50
6260	Tight mill scale, little/no rust, 6.9 #/S.F. sand		325	.098		1.76	3.67	.70	6.13	8.75
6265	Exist coat blistered/pitted, 9.0 #/S.F. sand		225	.142		2.29	5.30	1.01	8.60	12.40
6270	Exist coat badly pitted/nodules, 11.3 #/S.F. sand		150	.213		2.88	7.95	1.51	12.34	17.95

05 05 Common Work Results for Metals

05 05 05 – Selective Metals Demolition

05 05 05.10 Selective Demolition, Metals

		Crew	Daily Output	Labor-Hours	Unit	Material	Labor	Equipment	Total	Total Incl O&P
0010	**SELECTIVE DEMOLITION, METALS** R024119-10									
0015	Excludes shores, bracing, cutting, loading, hauling, dumping									
0020	Remove nuts only up to 3/4″ diameter	1 Sswk	480	.017	Ea.		.78		.78	1.37
0030	7/8″ to 1-1/4″ diameter		240	.033			1.56		1.56	2.75
0040	1-3/8″ to 2″ diameter		160	.050			2.35		2.35	4.12
0060	Unbolt and remove structural bolts up to 3/4″ diameter		240	.033			1.56		1.56	2.75
0070	7/8″ to 2″ diameter		160	.050			2.35		2.35	4.12
0140	Light weight framing members, remove whole or cut up, up to 20 lb		240	.033			1.56		1.56	2.75
0150	21 – 40 lb	2 Sswk	210	.076			3.57		3.57	6.30
0160	41 – 80 lb	3 Sswk	180	.133			6.25		6.25	11
0170	81 – 120 lb	4 Sswk	150	.213			10		10	17.60
0230	Structural members, remove whole or cut up, up to 500 lb	E-19	48	.500			23	21.50	44.50	62
0240	1/4 – 2 tons	E-18	36	1.111			51.50	28.50	80	120
0250	2 – 5 tons	E-24	30	1.067			49	25	74	111
0260	5 – 10 tons	E-20	24	2.667			122	52.50	174.50	265
0270	10 – 15 tons	E-2	18	3.111			142	89.50	231.50	340
0340	Fabricated item, remove whole or cut up, up to 20 lb	1 Sswk	96	.083			3.91		3.91	6.85
0350	21 – 40 lb	2 Sswk	84	.190			8.95		8.95	15.70
0360	41 – 80 lb	3 Sswk	72	.333			15.65		15.65	27.50
0370	81 – 120 lb	4 Sswk	60	.533			25		25	44
0380	121 – 500 lb	E-19	48	.500			23	21.50	44.50	62
0390	501 – 1000 lb	″	36	.667			30.50	28.50	59	82.50

05 05 13 – Shop-Applied Coatings for Metal

05 05 13.50 Paints and Protective Coatings

		Crew	Daily Output	Labor-Hours	Unit	Material	Labor	Equipment	Total	Total Incl O&P
0010	**PAINTS AND PROTECTIVE COATINGS**									
5900	Galvanizing structural steel in shop, under 1 ton				Ton	485			485	535
5950	1 ton to 20 tons					425			425	470
6000	Over 20 tons					395			395	435

05 05 Common Work Results for Metals

05 05 21 – Fastening Methods for Metal

	05 05 21.10 Cutting Steel	Crew	Daily Output	Labor-Hours	Unit	Material	2010 Bare Costs Labor	Equipment	Total	Total Incl O&P
0010	**CUTTING STEEL**									
0020	Hand burning, incl. preparation, torch cutting & grinding, no staging									
0100	Steel to 1/2" thick	E-25	320	.025	L.F.		1.22	.31	1.53	2.49
0150	3/4" thick		260	.031			1.50	.38	1.88	3.06
0200	1" thick		200	.040			1.96	.49	2.45	3.98

05 05 21.15 Drilling Steel

	05 05 21.15 Drilling Steel	Crew	Daily Output	Labor-Hours	Unit	Material	Labor	Equipment	Total	Total Incl O&P
0010	**DRILLING STEEL**									
1910	Drilling & layout for steel, up to 1/4" deep, no anchor									
1920	Holes, 1/4" diameter	1 Sswk	112	.071	Ea.	.09	3.35		3.44	6
1925	For each additional 1/4" depth, add		336	.024		.09	1.12		1.21	2.06
1930	3/8" diameter		104	.077		.10	3.61		3.71	6.45
1935	For each additional 1/4" depth, add		312	.026		.10	1.20		1.30	2.22
1940	1/2" diameter		96	.083		.11	3.91		4.02	6.95
1945	For each additional 1/4" depth, add		288	.028		.11	1.30		1.41	2.41
1950	5/8" diameter		88	.091		.15	4.26		4.41	7.65
1955	For each additional 1/4" depth, add		264	.030		.15	1.42		1.57	2.66
1960	3/4" diameter		80	.100		.17	4.69		4.86	8.45
1965	For each additional 1/4" depth, add		240	.033		.17	1.56		1.73	2.93
1970	7/8" diameter		72	.111		.20	5.20		5.40	9.35
1975	For each additional 1/4" depth, add		216	.037		.20	1.74		1.94	3.26
1980	1" diameter		64	.125		.22	5.85		6.07	10.55
1985	For each additional 1/4" depth, add		192	.042		.22	1.95		2.17	3.69
1990	For drilling up, add						40%			

05 05 21.90 Welding Steel

	05 05 21.90 Welding Steel	Crew	Daily Output	Labor-Hours	Unit	Material	Labor	Equipment	Total	Total Incl O&P
0010	**WELDING STEEL**, Structural R050521-20									
0020	Field welding, 1/8" E6011, cost per welder, no oper. engr	E-14	8	1	Hr.	4.73	49	18.25	71.98	111
0200	With 1/2 operating engineer	E-13	8	1.500		4.73	69.50	18.25	92.48	142
0300	With 1 operating engineer	E-12	8	2		4.73	90	18.20	112.93	173
0500	With no operating engineer, 2# weld rod per ton	E-14	8	1	Ton	4.73	49	18.25	71.98	111
0600	8# E6011 per ton	"	2	4		18.90	196	73	287.90	445
0800	With one operating engineer per welder, 2# E6011 per ton	E-12	8	2		4.73	90	18.20	112.93	173
0900	8# E6011 per ton	"	2	8		18.90	360	73	451.90	690
1200	Continuous fillet, stick welding, incl. equipment									
1300	Single pass, 1/8" thick, 0.1#/L.F.	E-14	150	.053	L.F.	.24	2.61	.97	3.82	5.90
1400	3/16" thick, 0.2#/L.F.		75	.107		.47	5.20	1.94	7.61	11.80
1500	1/4" thick, 0.3#/L.F.		50	.160		.71	7.80	2.92	11.43	17.75
1610	5/16" thick, 0.4#/L.F.		38	.211		.95	10.30	3.84	15.09	23.50
1800	3 passes, 3/8" thick, 0.5#/L.F.		30	.267		1.18	13.05	4.86	19.09	29.50
2010	4 passes, 1/2" thick, 0.7#/L.F.		22	.364		1.66	17.80	6.65	26.11	40.50
2200	5 to 6 passes, 3/4" thick, 1.3#/L.F.		12	.667		3.07	32.50	12.15	47.72	74
2400	8 to 11 passes, 1" thick, 2.4#/L.F.		6	1.333		5.65	65	24.50	95.15	148
2600	For all position welding, add, minimum						20%			
2700	Maximum						300%			
2900	For semi-automatic welding, deduct, minimum						5%			
3000	Maximum						15%			
4000	Cleaning and welding plates, bars, or rods									
4010	to existing beams, columns, or trusses	E-14	12	.667	L.F.	1.18	32.50	12.15	45.83	72

05 05 23 – Metal Fastenings

05 05 23.05 Anchor Bolts

	05 05 23.05 Anchor Bolts	Crew	Daily Output	Labor-Hours	Unit	Material	Labor	Equipment	Total	Total Incl O&P
0010	**ANCHOR BOLTS**									
0015	Made from recycled materials [G]									

05 05 Common Work Results for Metals

05 05 23 – Metal Fastenings

05 05 23.05 Anchor Bolts			Crew	Daily Output	Labor-Hours	Unit	Material	Labor	Equipment	Total	Total Incl O&P
							2010 Bare Costs				
0100	J-type, incl. hex nut & washer, 1/2" diameter x 6" long	G	2 Carp	70	.229	Ea.	1.22	9.50		10.72	16
0110	12" long	G		65	.246		1.53	10.25		11.78	17.45
0120	18" long	G		60	.267		1.99	11.10		13.09	19.30
0130	3/4" diameter x 8" long	G		50	.320		3.20	13.30		16.50	24
0140	12" long	G		45	.356		4	14.75		18.75	27.50
0150	18" long	G		40	.400		5.20	16.60		21.80	31
0160	1" diameter x 12" long	G		35	.457		8.40	19		27.40	39
0170	18" long	G		30	.533		10.15	22		32.15	45
0180	24" long	G		25	.640		12.35	26.50		38.85	54.50
0190	36" long	G		20	.800		16.95	33		49.95	69.50
0200	1-1/2" diameter x 18" long	G		22	.727		28	30		58	77
0210	24" long	G		16	1		33	41.50		74.50	101
0300	L-type, incl. hex nut & washer, 3/4" diameter x 12" long	G		45	.356		3.72	14.75		18.47	27
0310	18" long	G		40	.400		4.77	16.60		21.37	31
0320	24" long	G		35	.457		5.80	19		24.80	36
0330	30" long	G		30	.533		7.40	22		29.40	42
0340	36" long	G		25	.640		8.45	26.50		34.95	50.50
0350	1" diameter x 12" long	G		35	.457		6.75	19		25.75	37
0360	18" long	G		30	.533		8.40	22		30.40	43
0370	24" long	G		25	.640		10.35	26.50		36.85	52.50
0380	30" long	G		23	.696		12.20	29		41.20	58
0390	36" long	G		20	.800		13.95	33		46.95	66.50
0400	42" long	G		18	.889		16.95	37		53.95	75.50
0410	48" long	G		15	1.067		19.05	44.50		63.55	89.50
0420	1-1/4" diameter x 18" long	G		25	.640		13.30	26.50		39.80	55.50
0430	24" long	G		22	.727		15.80	30		45.80	64
0440	30" long	G		20	.800		18.30	33		51.30	71
0450	36" long	G		18	.889		21	37		58	80
0460	42" long	G		16	1		23.50	41.50		65	90
0470	48" long	G		14	1.143		27	47.50		74.50	103
0480	54" long	G		12	1.333		31.50	55.50		87	121
0490	60" long	G		10	1.600		34.50	66.50		101	140
0500	1-1/2" diameter x 18" long	G		22	.727		19.80	30		49.80	68.50
0510	24" long	G		19	.842		23	35		58	79.50
0520	30" long	G		17	.941		26	39		65	89
0530	36" long	G		16	1		30	41.50		71.50	97
0540	42" long	G		15	1.067		34	44.50		78.50	106
0550	48" long	G		13	1.231		38.50	51		89.50	121
0560	54" long	G		11	1.455		47	60.50		107.50	145
0570	60" long	G		9	1.778		51.50	74		125.50	171
0580	1-3/4" diameter x 18" long	G		20	.800		28.50	33		61.50	82
0590	24" long	G		18	.889		33.50	37		70.50	93.50
0600	30" long	G		17	.941		38.50	39		77.50	103
0610	36" long	G		16	1		44	41.50		85.50	113
0620	42" long	G		14	1.143		49.50	47.50		97	128
0630	48" long	G		12	1.333		54.50	55.50		110	146
0640	54" long	G		10	1.600		67.50	66.50		134	177
0650	60" long	G		8	2		73	83		156	209
0660	2" diameter x 24" long	G		17	.941		45	39		84	110
0670	30" long	G		15	1.067		50.50	44.50		95	125
0680	36" long	G		13	1.231		55.50	51		106.50	141
0690	42" long	G		11	1.455		62	60.50		122.50	162
0700	48" long	G	↓	10	1.600	↓	71.50	66.50		138	181

05 05 23.05 Anchor Bolts

Line	Description		Crew	Daily Output	Labor-Hours	Unit	2010 Bare Costs Material	Labor	Equipment	Total	Total Incl O&P
0710	54″ long	G	2 Carp	9	1.778	Ea.	85	74		159	208
0720	60″ long	G		8	2		91.50	83		174.50	229
0730	66″ long	G		7	2.286		98	95		193	254
0740	72″ long	G	↓	6	2.667		107	111		218	289
0990	For galvanized, add					↓	75%				

05 05 23.10 Bolts and Hex Nuts

Line	Description		Crew	Daily Output	Labor-Hours	Unit	Material	Labor	Equipment	Total	Total Incl O&P
0010	**BOLTS & HEX NUTS**, Steel, A307										
0100	1/4″ diameter, 1/2″ long	G	1 Sswk	140	.057	Ea.	.10	2.68		2.78	4.82
0200	1″ long	G		140	.057		.12	2.68		2.80	4.84
0300	2″ long	G		130	.062		.15	2.89		3.04	5.20
0400	3″ long	G		130	.062		.22	2.89		3.11	5.30
0500	4″ long	G		120	.067		.24	3.13		3.37	5.75
0600	3/8″ diameter, 1″ long	G		130	.062		.22	2.89		3.11	5.30
0700	2″ long	G		130	.062		.27	2.89		3.16	5.35
0800	3″ long	G		120	.067		.36	3.13		3.49	5.90
0900	4″ long	G		120	.067		.45	3.13		3.58	6
1000	5″ long	G		115	.070		.56	3.26		3.82	6.35
1100	1/2″ diameter, 1-1/2″ long	G		120	.067		.50	3.13		3.63	6.05
1200	2″ long	G		120	.067		.57	3.13		3.70	6.10
1300	4″ long	G		115	.070		.86	3.26		4.12	6.70
1400	6″ long	G		110	.073		1.17	3.41		4.58	7.30
1500	8″ long	G		105	.076		1.51	3.57		5.08	7.95
1600	5/8″ diameter, 1-1/2″ long	G		120	.067		1.15	3.13		4.28	6.75
1700	2″ long	G		120	.067		1.26	3.13		4.39	6.90
1800	4″ long	G		115	.070		1.75	3.26		5.01	7.65
1900	6″ long	G		110	.073		2.20	3.41		5.61	8.40
2000	8″ long	G		105	.076		3.20	3.57		6.77	9.80
2100	10″ long	G		100	.080		3.99	3.75		7.74	11
2200	3/4″ diameter, 2″ long	G		120	.067		1.63	3.13		4.76	7.30
2300	4″ long	G		110	.073		2.30	3.41		5.71	8.55
2400	6″ long	G		105	.076		2.94	3.57		6.51	9.55
2500	8″ long	G		95	.084		4.39	3.95		8.34	11.80
2600	10″ long	G		85	.094		5.75	4.41		10.16	14.05
2700	12″ long	G		80	.100		6.70	4.69		11.39	15.60
2800	1″ diameter, 3″ long	G		105	.076		4.94	3.57		8.51	11.75
2900	6″ long	G		90	.089		7.75	4.17		11.92	15.85
3000	12″ long	G	↓	75	.107		14.85	5		19.85	25
3100	For galvanized, add						75%				
3200	For stainless, add					↓	350%				

05 05 23.15 Chemical Anchors

Line	Description	Crew	Daily Output	Labor-Hours	Unit	Material	Labor	Equipment	Total	Total Incl O&P
0010	**CHEMICAL ANCHORS**									
0020	Includes layout & drilling									
1430	Chemical anchor, w/rod & epoxy cartridge, 3/4″ diam. x 9-1/2″ long	B-89A	27	.593	Ea.	9.45	22.50	4.21	36.16	49.50
1435	1″ diameter x 11-3/4″ long		24	.667		16.55	25	4.74	46.29	62.50
1440	1-1/4″ diameter x 14″ long		21	.762		29	29	5.40	63.40	82
1445	1-3/4″ diameter x 15″ long		20	.800		50.50	30.50	5.70	86.70	108
1450	18″ long		17	.941		60.50	35.50	6.70	102.70	129
1455	2″ diameter x 18″ long		16	1		82.50	38	7.10	127.60	157
1460	24″ long	↓	15	1.067	↓	107	40.50	7.60	155.10	187

05 05 23.20 Expansion Anchors

Line	Description	Crew	Daily Output	Labor-Hours	Unit	Material	Labor	Equipment	Total	Total Incl O&P
0010	**EXPANSION ANCHORS**									
0100	Anchors for concrete, brick or stone, no layout and drilling									

05 05 23 – Metal Fastenings

05 05 23.20 Expansion Anchors			Crew	Daily Output	Labor-Hours	Unit	Material	2010 Bare Costs Labor	Equipment	Total	Total Incl O&P
0200	Expansion shields, zinc, 1/4" diameter, 1-5/16" long, single	G	1 Carp	90	.089	Ea.	.34	3.69		4.03	6.05
0300	1-3/8" long, double	G		85	.094		.47	3.91		4.38	6.55
0400	3/8" diameter, 1-1/2" long, single	G		85	.094		.63	3.91		4.54	6.75
0500	2" long, double	G		80	.100		1.17	4.16		5.33	7.70
0600	1/2" diameter, 2-1/16" long, single	G		80	.100		1.16	4.16		5.32	7.70
0700	2-1/2" long, double	G		75	.107		1.63	4.43		6.06	8.65
0800	5/8" diameter, 2-5/8" long, single	G		75	.107		2.26	4.43		6.69	9.35
0900	2-3/4" long, double	G		70	.114		2.33	4.75		7.08	9.85
1000	3/4" diameter, 2-3/4" long, single	G		70	.114		2.07	4.75		6.82	9.60
1100	3-15/16" long, double	G		65	.123		4.51	5.10		9.61	12.85
2100	Hollow wall anchors for gypsum wall board, plaster or tile										
2300	1/8" diameter, short	G	1 Carp	160	.050	Ea.	.18	2.08		2.26	3.40
2400	Long	G		150	.053		.21	2.22		2.43	3.65
2500	3/16" diameter, short	G		150	.053		.32	2.22		2.54	3.77
2600	Long	G		140	.057		.46	2.37		2.83	4.17
2700	1/4" diameter, short	G		140	.057		.54	2.37		2.91	4.25
2800	Long	G		130	.062		.48	2.56		3.04	4.47
3000	Toggle bolts, bright steel, 1/8" diameter, 2" long	G		85	.094		.18	3.91		4.09	6.25
3100	4" long	G		80	.100		.23	4.16		4.39	6.65
3200	3/16" diameter, 3" long	G		80	.100		.22	4.16		4.38	6.65
3300	6" long	G		75	.107		.31	4.43		4.74	7.20
3400	1/4" diameter, 3" long	G		75	.107		.29	4.43		4.72	7.15
3500	6" long	G		70	.114		.46	4.75		5.21	7.80
3600	3/8" diameter, 3" long	G		70	.114		.70	4.75		5.45	8.05
3700	6" long	G		60	.133		1.20	5.55		6.75	9.85
3800	1/2" diameter, 4" long	G		60	.133		1.69	5.55		7.24	10.40
3900	6" long	G		50	.160		2.02	6.65		8.67	12.45
4000	Nailing anchors										
4100	Nylon nailing anchor, 1/4" diameter, 1" long		1 Carp	3.20	2.500	C	11.30	104		115.30	172
4200	1-1/2" long			2.80	2.857		12.45	119		131.45	197
4300	2" long			2.40	3.333		17.50	139		156.50	233
4400	Metal nailing anchor, 1/4" diameter, 1" long	G		3.20	2.500		16.65	104		120.65	178
4500	1-1/2" long	G		2.80	2.857		22.50	119		141.50	208
4600	2" long	G		2.40	3.333		26	139		165	243
5000	Screw anchors for concrete, masonry,										
5100	stone & tile, no layout or drilling included										
5700	Lag screw shields, 1/4" diameter, short	G	1 Carp	90	.089	Ea.	.29	3.69		3.98	6
5800	Long	G		85	.094		.35	3.91		4.26	6.45
5900	3/8" diameter, short	G		85	.094		.59	3.91		4.50	6.70
6000	Long	G		80	.100		.72	4.16		4.88	7.20
6100	1/2" diameter, short	G		80	.100		.92	4.16		5.08	7.40
6200	Long	G		75	.107		1.15	4.43		5.58	8.10
6300	5/8" diameter, short	G		70	.114		1.45	4.75		6.20	8.90
6400	Long	G		65	.123		1.94	5.10		7.04	10.05
6600	Lead, #6 & #8, 3/4" long	G		260	.031		.12	1.28		1.40	2.10
6700	#10 - #14, 1-1/2" long	G		200	.040		.18	1.66		1.84	2.76
6800	#16 & #18, 1-1/2" long	G		160	.050		.29	2.08		2.37	3.52
6900	Plastic, #6 & #8, 3/4" long			260	.031		.03	1.28		1.31	2
7000	#8 & #10, 7/8" long			240	.033		.02	1.38		1.40	2.15
7100	#10 & #12, 1" long			220	.036		.03	1.51		1.54	2.36
7200	#14 & #16, 1-1/2" long			160	.050		.05	2.08		2.13	3.26
8000	Wedge anchors, not including layout or drilling										
8050	Carbon steel, 1/4" diameter, 1-3/4" long	G	1 Carp	150	.053	Ea.	.29	2.22		2.51	3.74

05 05 23 – Metal Fastenings

05 05 23.20 Expansion Anchors			Crew	Daily Output	Labor-Hours	Unit	Material	2010 Bare Costs Labor	Equipment	Total	Total Incl O&P
8100	3-1/4″ long	G	1 Carp	140	.057	Ea.	.38	2.37		2.75	4.08
8150	3/8″ diameter, 2-1/4″ long	G		145	.055		.34	2.29		2.63	3.91
8200	5″ long	G		140	.057		.60	2.37		2.97	4.32
8250	1/2″ diameter, 2-3/4″ long	G		140	.057		.73	2.37		3.10	4.46
8300	7″ long	G		125	.064		1.24	2.66		3.90	5.45
8350	5/8″ diameter, 3-1/2″ long	G		130	.062		1.15	2.56		3.71	5.20
8400	8-1/2″ long	G		115	.070		2.45	2.89		5.34	7.15
8450	3/4″ diameter, 4-1/4″ long	G		115	.070		2.02	2.89		4.91	6.70
8500	10″ long	G		95	.084		4.60	3.50		8.10	10.45
8550	1″ diameter, 6″ long	G		100	.080		7.70	3.32		11.02	13.60
8575	9″ long	G		85	.094		10	3.91		13.91	17.05
8600	12″ long	G		75	.107		10.80	4.43		15.23	18.75
8650	1-1/4″ diameter, 9″ long	G		70	.114		23	4.75		27.75	33
8700	12″ long	G	↓	60	.133	↓	29.50	5.55		35.05	41
8750	For type 303 stainless steel, add						350%				
8800	For type 316 stainless steel, add						450%				
8950	Self-drilling concrete screw, hex washer head, 3/16″ diam. x 1-3/4″ long	G	1 Carp	300	.027	Ea.	.19	1.11		1.30	1.92
8960	2-1/4″ long	G		250	.032		.20	1.33		1.53	2.27
8970	Phillips flat head, 3/16″ diam. x 1-3/4″ long	G		300	.027		.17	1.11		1.28	1.90
8980	2-1/4″ long	G	↓	250	.032	↓	.20	1.33		1.53	2.27

05 05 23.25 High Strength Bolts			Crew	Daily Output	Labor-Hours	Unit	Material	Labor	Equipment	Total	Total Incl O&P
0010	**HIGH STRENGTH BOLTS** R050523-10										
0020	A325 Type 1, structural steel, bolt-nut-washer set										
0100	1/2″ diameter x 1-1/2″ long	G	1 Sswk	130	.062	Ea.	.92	2.89		3.81	6.05
0120	2″ long	G		125	.064		.99	3		3.99	6.40
0150	3″ long	G		120	.067		1.37	3.13		4.50	7
0170	5/8″ diameter x 1-1/2″ long	G		125	.064		1.58	3		4.58	7.05
0180	2″ long	G		120	.067		1.70	3.13		4.83	7.35
0190	3″ long	G		115	.070		2.08	3.26		5.34	8.05
0200	3/4″ diameter x 2″ long	G		120	.067		2.47	3.13		5.60	8.20
0220	3″ long	G		115	.070		2.93	3.26		6.19	8.95
0250	4″ long	G		110	.073		3.56	3.41		6.97	9.90
0300	6″ long	G		105	.076		4.58	3.57		8.15	11.35
0350	8″ long	G		95	.084		8.90	3.95		12.85	16.75
0360	7/8″ diameter x 2″ long	G		115	.070		3.42	3.26		6.68	9.50
0365	3″ long	G		110	.073		4.02	3.41		7.43	10.40
0370	4″ long	G		105	.076		4.81	3.57		8.38	11.60
0380	6″ long	G		100	.080		6.10	3.75		9.85	13.30
0390	8″ long	G		90	.089		9.55	4.17		13.72	17.85
0400	1″ diameter x 2″ long	G		105	.076		4.70	3.57		8.27	11.45
0420	3″ long	G		100	.080		5.30	3.75		9.05	12.40
0450	4″ long	G		95	.084		5.95	3.95		9.90	13.50
0500	6″ long	G		90	.089		7.85	4.17		12.02	16
0550	8″ long	G		85	.094		13.50	4.41		17.91	22.50
0600	1-1/4″ diameter x 3″ long	G		85	.094		10.45	4.41		14.86	19.25
0650	4″ long	G		80	.100		2.93	4.69		7.62	11.50
0700	6″ long	G		75	.107		14.65	5		19.65	25
0750	8″ long	G	↓	70	.114	↓	18.55	5.35		23.90	30
1020	A490, bolt-nut-washer set										
1170	5/8″ diameter x 1-1/2″ long	G	1 Sswk	125	.064	Ea.	1.75	3		4.75	7.25
1180	2″ long	G		120	.067		2.06	3.13		5.19	7.75
1190	3″ long	G	↓	115	.070	↓	2.49	3.26		5.75	8.50

05 05 Common Work Results for Metals

05 05 23 – Metal Fastenings

05 05 23.25 High Strength Bolts			Crew	Daily Output	Labor-Hours	Unit	Material	2010 Bare Costs Labor	Equipment	Total	Total Incl O&P
1200	3/4" diameter x 2" long	G	1 Sswk	120	.067	Ea.	2.82	3.13		5.95	8.60
1220	3" long	G		115	.070		3.33	3.26		6.59	9.40
1250	4" long	G		110	.073		3.87	3.41		7.28	10.25
1300	6" long	G		105	.076		5.65	3.57		9.22	12.50
1350	8" long	G		95	.084		9.55	3.95		13.50	17.45
1360	7/8" diameter x 2" long	G		115	.070		4.48	3.26		7.74	10.70
1365	3" long	G		110	.073		5.25	3.41		8.66	11.80
1370	4" long	G		105	.076		6.55	3.57		10.12	13.50
1380	6" long	G		100	.080		9.20	3.75		12.95	16.70
1390	8" long	G		90	.089		13.40	4.17		17.57	22
1400	1" diameter x 2" long	G		105	.076		4.97	3.57		8.54	11.75
1420	3" long	G		100	.080		6.05	3.75		9.80	13.25
1450	4" long	G		95	.084		6.95	3.95		10.90	14.60
1500	6" long	G		90	.089		9.35	4.17		13.52	17.65
1550	8" long	G		85	.094		14.85	4.41		19.26	24
1600	1-1/4" diameter x 3" long	G		85	.094		12.05	4.41		16.46	21
1650	4" long	G		80	.100		13.75	4.69		18.44	23.50
1700	6" long	G		75	.107		18.75	5		23.75	29.50
1750	8" long	G		70	.114		24.50	5.35		29.85	36
05 05 23.30 Lag Screws											
0010	**LAG SCREWS**										
0020	Steel, 1/4" diameter, 2" long	G	1 Carp	200	.040	Ea.	.09	1.66		1.75	2.66
0100	3/8" diameter, 3" long	G		150	.053		.26	2.22		2.48	3.71
0200	1/2" diameter, 3" long	G		130	.062		.48	2.56		3.04	4.47
0300	5/8" diameter, 3" long	G		120	.067		1.11	2.77		3.88	5.50
05 05 23.35 Machine Screws											
0010	**MACHINE SCREWS**										
0020	Steel, round head, #8 x 1" long	G	1 Carp	4.80	1.667	C	3.45	69.50		72.95	111
0110	#8 x 2" long	G		2.40	3.333		4.34	139		143.34	219
0200	#10 x 1" long	G		4	2		5.90	83		88.90	135
0300	#10 x 2" long	G		2	4		7.25	166		173.25	264
05 05 23.40 Machinery Anchors											
0010	**MACHINERY ANCHORS**, heavy duty, incl. sleeve, floating base nut,										
0020	lower stud & coupling nut, fiber plug, connecting stud, washer & nut.										
0030	For flush mounted embedment in poured concrete heavy equip. pads.										
0200	Stud & bolt, 1/2" diameter	G	E-16	40	.400	Ea.	68.50	19.15	3.64	91.29	113
0300	5/8" diameter	G		35	.457		76	22	4.16	102.16	127
0500	3/4" diameter	G		30	.533		87.50	25.50	4.86	117.86	147
0600	7/8" diameter	G		25	.640		95.50	30.50	5.85	131.85	165
0800	1" diameter	G		20	.800		100	38.50	7.30	145.80	186
0900	1-1/4" diameter	G		15	1.067		133	51	9.70	193.70	248
05 05 23.50 Powder Actuated Tools and Fasteners											
0010	**POWDER ACTUATED TOOLS & FASTENERS**										
0020	Stud driver, .22 caliber, buy, minimum					Ea.	219			219	241
0100	Maximum					"	370			370	405
0300	Powder charges for above, low velocity					C	4.99			4.99	5.50
0400	Standard velocity						5.95			5.95	6.55
0600	Drive pins & studs, 1/4" & 3/8" diam., to 3" long, minimum	G	1 Carp	4.80	1.667		4.38	69.50		73.88	112
0700	Maximum	G	"	4	2		9.40	83		92.40	138

05 05 23.55 Rivets

	05 05 23.55 Rivets		Crew	Daily Output	Labor-Hours	Unit	Material	2010 Bare Costs Labor	Equipment	Total	Total Incl O&P
0010	**RIVETS**										
0100	Aluminum rivet & mandrel, 1/2" grip length x 1/8" diameter	G	1 Carp	4.80	1.667	C	6.35	69.50		75.85	114
0200	3/16" diameter	G		4	2		11.15	83		94.15	140
0300	Aluminum rivet, steel mandrel, 1/8" diameter	G		4.80	1.667		6.65	69.50		76.15	114
0400	3/16" diameter	G		4	2		10.55	83		93.55	140
0500	Copper rivet, steel mandrel, 1/8" diameter	G		4.80	1.667		7.75	69.50		77.25	116
0800	Stainless rivet & mandrel, 1/8" diameter	G		4.80	1.667		23.50	69.50		93	133
0900	3/16" diameter	G		4	2		38.50	83		121.50	171
1000	Stainless rivet, steel mandrel, 1/8" diameter	G		4.80	1.667		19.65	69.50		89.15	129
1100	3/16" diameter	G		4	2		27.50	83		110.50	158
1200	Steel rivet and mandrel, 1/8" diameter	G		4.80	1.667		6.50	69.50		76	114
1300	3/16" diameter	G		4	2		10.65	83		93.65	140
1400	Hand riveting tool, minimum					Ea.	40			40	44
1500	Maximum						227			227	250
1600	Power riveting tool, minimum						450			450	495
1700	Maximum						2,275			2,275	2,525

05 05 23.70 Structural Blind Bolts

	05 05 23.70 Structural Blind Bolts		Crew	Daily Output	Labor-Hours	Unit	Material	Labor	Equipment	Total	Total Incl O&P
0010	**STRUCTURAL BLIND BOLTS**										
0100	1/4" diameter x 1/4" grip	G	1 Sswk	240	.033	Ea.	1.24	1.56		2.80	4.11
0150	1/2" grip	G		216	.037		.95	1.74		2.69	4.10
0200	3/8" diameter x 1/2" grip	G		232	.034		1.75	1.62		3.37	4.77
0250	3/4" grip	G		208	.038		1.83	1.80		3.63	5.20
0300	1/2" diameter x 1/2" grip	G		224	.036		5.15	1.67		6.82	8.60
0350	3/4" grip	G		200	.040		5.60	1.88		7.48	9.45
0400	5/8" diameter x 3/4" grip	G		216	.037		8.25	1.74		9.99	12.15
0450	1" grip	G		192	.042		9.05	1.95		11	13.40

05 05 23.80 Vibration and Bearing Pads

	05 05 23.80 Vibration and Bearing Pads	Crew	Daily Output	Labor-Hours	Unit	Material	Labor	Equipment	Total	Total Incl O&P
0010	**VIBRATION & BEARING PADS**									
0300	Laminated synthetic rubber impregnated cotton duck, 1/2" thick	2 Sswk	24	.667	S.F.	76.50	31.50		108	139
0400	1" thick		20	.800		154	37.50		191.50	236
0600	Neoprene bearing pads, 1/2" thick		24	.667		29	31.50		60.50	86.50
0700	1" thick		20	.800		59	37.50		96.50	131
0900	Fabric reinforced neoprene, 5000 psi, 1/2" thick		24	.667		12.95	31.50		44.45	69.50
1000	1" thick		20	.800		26	37.50		63.50	94.50
1200	Felt surfaced vinyl pads, cork and sisal, 5/8" thick		24	.667		33.50	31.50		65	91.50
1300	1" thick		20	.800		61	37.50		98.50	133
1500	Teflon bonded to 10 ga. carbon steel, 1/32" layer		24	.667		57.50	31.50		89	118
1600	3/32" layer		24	.667		86	31.50		117.50	150
1800	Bonded to 10 ga. stainless steel, 1/32" layer		24	.667		102	31.50		133.50	167
1900	3/32" layer		24	.667		132	31.50		163.50	201
2100	Circular machine leveling pad & stud				Kip	8.35			8.35	9.15

05 05 23.85 Weld Shear Connectors

	05 05 23.85 Weld Shear Connectors		Crew	Daily Output	Labor-Hours	Unit	Material	Labor	Equipment	Total	Total Incl O&P
0010	**WELD SHEAR CONNECTORS**										
0020	3/4" diameter, 3-3/16" long	G	E-10	960	.017	Ea.	.51	.80	.46	1.77	2.47
0030	3-3/8" long	G		950	.017		.53	.81	.47	1.81	2.52
0200	3-7/8" long	G		945	.017		.57	.81	.47	1.85	2.57
0300	4-3/16" long	G		935	.017		.60	.82	.47	1.89	2.62
0500	4-7/8" long	G		930	.017		.67	.82	.47	1.96	2.71
0600	5-3/16" long	G		920	.017		.70	.83	.48	2.01	2.76
0800	5-3/8" long	G		910	.018		.70	.84	.49	2.03	2.78
0900	6-3/16" long	G		905	.018		.77	.85	.49	2.11	2.88

05 05 23.85 Weld Shear Connectors

	05 05 23.85 Weld Shear Connectors		Crew	Daily Output	Labor-Hours	Unit	Material	2010 Bare Costs Labor	Equipment	Total	Total Incl O&P
1000	7-3/16" long	G	E-10	895	.018	Ea.	.96	.86	.49	2.31	3.11
1100	8-3/16" long	G		890	.018		1.05	.86	.50	2.41	3.21
1500	7/8" diameter, 3-11/16" long	G		920	.017		.82	.83	.48	2.13	2.90
1600	4-3/16" long	G		910	.018		.88	.84	.49	2.21	2.98
1700	5-3/16" long	G		905	.018		1	.85	.49	2.34	3.13
1800	6-3/16" long	G		895	.018		1.12	.86	.49	2.47	3.28
1900	7-3/16" long	G		890	.018		1.24	.86	.50	2.60	3.43
2000	8-3/16" long	G		880	.018		1.36	.87	.50	2.73	3.57

05 05 23.87 Weld Studs

	05 05 23.87 Weld Studs		Crew	Daily Output	Labor-Hours	Unit	Material	Labor	Equipment	Total	Total Incl O&P
0010	**WELD STUDS**										
0020	1/4" diameter, 2-11/16" long	G	E-10	1120	.014	Ea.	.33	.68	.39	1.40	1.99
0100	4-1/8" long	G		1080	.015		.31	.71	.41	1.43	2.04
0200	3/8" diameter, 4-1/8" long	G		1080	.015		.36	.71	.41	1.48	2.10
0300	6-1/8" long	G		1040	.015		.47	.74	.42	1.63	2.28
0400	1/2" diameter, 2-1/8" long	G		1040	.015		.34	.74	.42	1.50	2.14
0500	3-1/8" long	G		1025	.016		.41	.75	.43	1.59	2.23
0600	4-1/8" long	G		1010	.016		.48	.76	.44	1.68	2.34
0700	5-5/16" long	G		990	.016		.59	.77	.45	1.81	2.50
0800	6-1/8" long	G		975	.016		.64	.79	.45	1.88	2.58
0900	8-1/8" long	G		960	.017		.90	.80	.46	2.16	2.90
1000	5/8" diameter, 2-11/16" long	G		1000	.016		.58	.77	.44	1.79	2.48
1010	4-3/16" long	G		990	.016		.72	.77	.45	1.94	2.64
1100	6-9/16" long	G		975	.016		.94	.79	.45	2.18	2.91
1200	8-3/16" long	G		960	.017		1.26	.80	.46	2.52	3.30

05 05 23.90 Welding Rod

	05 05 23.90 Welding Rod		Crew	Daily Output	Labor-Hours	Unit	Material	Labor	Equipment	Total	Total Incl O&P
0010	**WELDING ROD**										
0020	Steel, type 6011, 1/8" diam., less than 500#					Lb.	2.36			2.36	2.60
0100	500# to 2,000#						2.13			2.13	2.34
0200	2,000# to 5,000#						2			2	2.20
0300	5/32" diameter, less than 500#						2.28			2.28	2.50
0310	500# to 2,000#						2.05			2.05	2.26
0320	2,000# to 5,000#						1.93			1.93	2.12
0400	3/16" diam., less than 500#						2.26			2.26	2.49
0500	500# to 2,000#						2.04			2.04	2.24
0600	2,000# to 5,000#						1.92			1.92	2.11
0620	Steel, type 6010, 1/8" diam., less than 500#						2.26			2.26	2.49
0630	500# to 2,000#						2.04			2.04	2.24
0640	2,000# to 5,000#						1.92			1.92	2.11
0650	Steel, type 7018 Low Hydrogen, 1/8" diam., less than 500#						2.36			2.36	2.60
0660	500# to 2,000#	CN					2.13			2.13	2.34
0670	2,000# to 5,000#						2			2	2.20
0700	Steel, type 7024 Jet Weld, 1/8" diam., less than 500#						2.26			2.26	2.49
0710	500# to 2,000#						2.04			2.04	2.24
0720	2,000# to 5,000#						1.92			1.92	2.11
1550	Aluminum, type 4043 TIG, 1/8" diam., less than 10#						4.60			4.60	5.05
1560	10# to 60#						4.14			4.14	4.55
1570	Over 60#						3.89			3.89	4.28
1600	Aluminum, type 5356 TIG, 1/8" diam., less than 10#						4.77			4.77	5.25
1610	10# to 60#						4.30			4.30	4.73
1620	Over 60#						4.04			4.04	4.45
1900	Cast iron, type 8 Nickel, 1/8" diam., less than 500#						27.50			27.50	30
1910	500# to 1,000#						24.50			24.50	27

05 05 Common Work Results for Metals

05 05 23 – Metal Fastenings

05 05 23.90 Welding Rod			Crew	Daily Output	Labor-Hours	Unit	2010 Bare Costs Material	Labor	Equipment	Total	Total Incl O&P
1920	Over 1,000#					Lb.	23			23	25.50
2000	Stainless steel, type 316/316L, 1/8" diam., less than 500#						7.80			7.80	8.55
2100	500# to 1000#						7			7	7.70
2220	Over 1000#						6.60			6.60	7.25

05 12 Structural Steel Framing

05 12 23 – Structural Steel for Buildings

05 12 23.05 Canopy Framing

Line	Description		Crew	Daily Output	Labor-Hours	Unit	Material	Labor	Equipment	Total	Total Incl O&P
0010	CANOPY FRAMING										
0020	6" and 8" members, shop fabricated	G	E-4	3000	.011	Lb.	1.32	.51	.05	1.88	2.39

05 12 23.10 Ceiling Supports

Line	Description		Crew	Daily Output	Labor-Hours	Unit	Material	Labor	Equipment	Total	Total Incl O&P
0010	CEILING SUPPORTS										
1000	Entrance door/folding partition supports, shop fabricated	G	E-4	60	.533	L.F.	22	25.50	2.43	49.93	71
1100	Linear accelerator door supports	G		14	2.286		100	108	10.40	218.40	310
1200	Lintels or shelf angles, hung, exterior hot dipped galv.	G		267	.120		15	5.70	.55	21.25	27
1250	Two coats primer paint instead of galv.	G		267	.120		13	5.70	.55	19.25	25
1400	Monitor support, ceiling hung, expansion bolted	G		4	8	Ea.	350	380	36.50	766.50	1,100
1450	Hung from pre-set inserts	G		6	5.333		375	253	24.50	652.50	885
1600	Motor supports for overhead doors	G		4	8		177	380	36.50	593.50	900
1700	Partition support for heavy folding partitions, without pocket	G		24	1.333	L.F.	50	63	6.10	119.10	173
1750	Supports at pocket only	G		12	2.667		100	126	12.15	238.15	345
2000	Rolling grilles & fire door supports	G		34	.941		43	44.50	4.29	91.79	130
2100	Spider-leg light supports, expansion bolted to ceiling slab	G		8	4	Ea.	143	190	18.25	351.25	510
2150	Hung from pre-set inserts	G		12	2.667	"	154	126	12.15	292.15	405
2400	Toilet partition support	G		36	.889	L.F.	50	42	4.05	96.05	133
2500	X-ray travel gantry support	G		12	2.667	"	172	126	12.15	310.15	425

05 12 23.15 Columns, Lightweight

Line	Description		Crew	Daily Output	Labor-Hours	Unit	Material	Labor	Equipment	Total	Total Incl O&P
0010	COLUMNS, LIGHTWEIGHT										
1000	Lightweight units (lally), 3-1/2" diameter		E-2	780	.072	L.F.	6.60	3.27	2.06	11.93	15.05
1050	4" diameter		"	900	.062	"	7.65	2.84	1.79	12.28	15.15
5800	Adjustable jack post, 8' maximum height, 2-3/4" diameter	G				Ea.	33			33	36.50
5850	4" diameter	G				"	53			53	58

05 12 23.17 Columns, Structural

Line	Description		Crew	Daily Output	Labor-Hours	Unit	Material	Labor	Equipment	Total	Total Incl O&P
0010	COLUMNS, STRUCTURAL R051223-10										
0015	Made from recycled materials	G									
0020	Shop fab'd for 100-ton, 1-2 story project, bolted connections										
0800	Steel, concrete filled, extra strong pipe, 3-1/2" diameter		E-2	660	.085	L.F.	36.50	3.87	2.44	42.81	49
0830	4" diameter			780	.072		40.50	3.27	2.06	45.83	52.50
0890	5" diameter			1020	.055		48	2.50	1.58	52.08	59
0930	6" diameter			1200	.047		64	2.13	1.34	67.47	75
0940	8" diameter			1100	.051		64	2.32	1.46	67.78	75.50
1100	For galvanizing, add					Lb.	.21			.21	.23
1300	For web ties, angles, etc., add per added lb.		1 Sswk	945	.008		1.10	.40		1.50	1.91
1500	Steel pipe, extra strong, no concrete, 3" to 5" diameter	G	E-2	16000	.004		1.10	.16	.10	1.36	1.59
1600	6" to 12" diameter	G		14000	.004		1.10	.18	.11	1.39	1.65
1700	Steel pipe, extra strong, no concrete, 3" diameter x 12'-0"	G		60	.933	Ea.	135	42.50	27	204.50	251
1750	4" diameter x 12'-0"	G		58	.966		198	44	27.50	269.50	325
1800	6" diameter x 12'-0"	G		54	1.037		375	47.50	30	452.50	530
1850	8" diameter x 14'-0"	G		50	1.120		670	51	32	753	855
1900	10" diameter x 16'-0"	G		48	1.167		965	53	33.50	1,051.50	1,175

05 12 23.17 Columns, Structural

	05 12 23.17 Columns, Structural		Crew	Daily Output	Labor-Hours	Unit	Material	2010 Bare Costs Labor	Equipment	Total	Total Incl O&P
1950	12″ diameter x 18′-0″	G	E-2	45	1.244	Ea.	1,300	57	35.50	1,392.50	1,550
3300	Structural tubing, square, A500GrB, 4″ to 6″ square, light section	G		11270	.005	Lb.	1.10	.23	.14	1.47	1.75
3600	Heavy section	G		32000	.002	″	1.10	.08	.05	1.23	1.40
4000	Concrete filled, add					L.F.	4.04			4.04	4.45
4500	Structural tubing, sq, 4″ x 4″ x 1/4″ x 12′-0″	G	E-2	58	.966	Ea.	182	44	27.50	253.50	305
4550	6″ x 6″ x 1/4″ x 12′-0″	G		54	1.037		297	47.50	30	374.50	440
4600	8″ x 8″ x 3/8″ x 14′-0″	G		50	1.120		645	51	32	728	830
4650	10″ x 10″ x 1/2″ x 16′-0″	G		48	1.167		1,200	53	33.50	1,286.50	1,450
5100	Structural tubing, rect, 5″ to 6″ wide, light section	G		8000	.007	Lb.	1.10	.32	.20	1.62	1.97
5200	Heavy section	G		12000	.005		1.10	.21	.13	1.44	1.72
5300	7″ to 10″ wide, light section	G		15000	.004		1.10	.17	.11	1.38	1.62
5400	Heavy section	G		18000	.003		1.10	.14	.09	1.33	1.55
5500	Structural tubing, rect, 5″ x 3″ x 1/4″ x 12′-0″	G		58	.966	Ea.	176	44	27.50	247.50	299
5550	6″ x 4″ x 5/16″ x 12′-0″	G		54	1.037		275	47.50	30	352.50	420
5600	8″ x 4″ x 3/8″ x 12′-0″	G		54	1.037		400	47.50	30	477.50	555
5650	10″ x 6″ x 3/8″ x 14′-0″	G		50	1.120		645	51	32	728	830
5700	12″ x 8″ x 1/2″ x 16′-0″	G		48	1.167		1,200	53	33.50	1,286.50	1,425
6800	W Shape, A992 steel, 2 tier, W8 x 24	G		1080	.052	L.F.	29	2.36	1.49	32.85	37.50
6850	W8 x 31	G		1080	.052		37.50	2.36	1.49	41.35	47
6900	W8 x 48	G		1032	.054		58	2.47	1.56	62.03	70
6950	W8 x 67	G		984	.057		81	2.60	1.63	85.23	95
7000	W10 x 45	G		1032	.054		54.50	2.47	1.56	58.53	66
7050	W10 x 68	G		984	.057		82.50	2.60	1.63	86.73	96.50
7100	W10 x 112	G		960	.058		136	2.66	1.67	140.33	155
7150	W12 x 50	G		1032	.054		60.50	2.47	1.56	64.53	72.50
7200	W12 x 87	G		984	.057		105	2.60	1.63	109.23	122
7250	W12 x 120	G		960	.058		145	2.66	1.67	149.33	166
7300	W12 x 190	G		912	.061		230	2.80	1.76	234.56	260
7350	W14 x 74	G		984	.057		89.50	2.60	1.63	93.73	105
7400	W14 x 120	G		960	.058		145	2.66	1.67	149.33	166
7450	W14 x 176	G		912	.061		213	2.80	1.76	217.56	241
8090	For projects 75 to 99 tons, add					All	10%				
8092	50 to 74 tons, add						20%				
8094	25 to 49 tons, add						30%	10%			
8096	10 to 24 tons, add						50%	25%			
8098	2 to 9 tons, add						75%	50%			
8099	Less than 2 tons, add						100%	100%			

05 12 23.20 Curb Edging

	05 12 23.20 Curb Edging		Crew	Daily Output	Labor-Hours	Unit	Material	Labor	Equipment	Total	Total Incl O&P
0010	**CURB EDGING**										
0020	Steel angle w/anchors, shop fabricated, on forms, 1″ x 1″, 0.8#/L.F.	G	E-4	350	.091	L.F.	1.44	4.33	.42	6.19	9.65
0100	2″ x 2″ angles, 3.92#/L.F.	G		330	.097		5.55	4.60	.44	10.59	14.70
0200	3″ x 3″ angles, 6.1#/L.F.	G		300	.107		8.70	5.05	.49	14.24	19
0300	4″ x 4″ angles, 8.2#/L.F.	G		275	.116		11.45	5.50	.53	17.48	23
1000	6″ x 4″ angles, 12.3#/L.F.	G		250	.128		16.90	6.05	.58	23.53	30
1050	Steel channels with anchors, on forms, 3″ channel, 5#/L.F.	G		290	.110		7	5.25	.50	12.75	17.45
1100	4″ channel, 5.4#/L.F.	G		270	.119		7.50	5.60	.54	13.64	18.75
1200	6″ channel, 8.2#/L.F.	G		255	.125		11.45	5.95	.57	17.97	23.50
1300	8″ channel, 11.5#/L.F.	G		225	.142		15.80	6.75	.65	23.20	30
1400	10″ channel, 15.3#/L.F.	G		180	.178		21	8.45	.81	30.26	38.50
1500	12″ channel, 20.7#/L.F.	G		140	.229		28	10.85	1.04	39.89	51
2000	For curved edging, add						35%	10%			

05 12 23 – Structural Steel for Buildings

05 12 23.40 Lightweight Framing		Crew	Daily Output	Labor-Hours	Unit	Material	Labor	Equipment	Total	Total Incl O&P
						2010 Bare Costs				
0010	LIGHTWEIGHT FRAMING R051223-35									
0015	Made from recycled materials G									
0200	For load-bearing steel studs see Div. 05 41 13.30									
0400	Angle framing, field fabricated, 4" and larger R051223-45	E-3	440	.055	Lb.	.64	2.59	.33	3.56	5.60
0450	Less than 4" angles G		265	.091	"	.66	4.31	.55	5.52	8.90
0460	1/2" x 1/2" x 1/8" G		200	.120	L.F.	.13	5.70	.73	6.56	11
0462	3/4" x 3/4" x 1/8" G		160	.150		.37	7.15	.91	8.43	13.95
0464	1" x 1" x 1/8" G		135	.178		.53	8.45	1.08	10.06	16.60
0466	1-1/4" x 1-1/4" x 3/16" G		115	.209		.98	9.95	1.27	12.20	19.90
0468	1-1/2" x 1-1/2" x 3/16" G		100	.240		1.19	11.40	1.46	14.05	23
0470	2" x 2" x 1/4" G		90	.267		2.11	12.70	1.62	16.43	26.50
0472	2-1/2" x 2-1/2" x 1/4" G		72	.333		2.71	15.85	2.03	20.59	33
0474	3" x 2" x 3/8" G		65	.369		3.89	17.55	2.24	23.68	38
0476	3" x 3" x 3/8" G		57	.421	↓	4.75	20	2.56	27.31	43
0600	Channel framing, field fabricated, 8" and larger G		500	.048	Lb.	.66	2.28	.29	3.23	5.05
0650	Less than 8" channels G		335	.072	"	.66	3.41	.44	4.51	7.20
0660	C2 x 1.78 G		115	.209	L.F.	1.17	9.95	1.27	12.39	20
0662	C3 x 4.1 G		80	.300		2.71	14.25	1.82	18.78	30
0664	C4 x 5.4 G		66	.364		3.56	17.30	2.21	23.07	37
0666	C5 x 6.7 G		57	.421		4.42	20	2.56	26.98	42.50
0668	C6 x 8.2 G		55	.436		5.25	21	2.65	28.90	45
0670	C7 x 9.8 G		40	.600		6.45	28.50	3.65	38.60	61
0672	C8 x 11.5 G		36	.667		7.60	31.50	4.05	43.15	69
0710	Structural bar tee, field fabricated, 3/4" x 3/4" x 1/8" G		160	.150		.37	7.15	.91	8.43	13.95
0712	1" x 1" x 1/8" G		135	.178		.53	8.45	1.08	10.06	16.60
0714	1-1/2" x 1-1/2" x 1/4" G		114	.211		1.54	10	1.28	12.82	20.50
0716	2" x 2" x 1/4" G		89	.270		2.11	12.85	1.64	16.60	26.50
0718	2-1/2" x 2-1/2" x 3/8" G		72	.333		3.89	15.85	2.03	21.77	34.50
0720	3" x 3" x 3/8" G		57	.421		4.75	20	2.56	27.31	43
0730	Structural zee, field fabricated, 1-1/4" x 1-3/4" x 1-3/4" G		114	.211		.50	10	1.28	11.78	19.55
0732	2-11/16" x 3" x 2-11/16" G		114	.211		1.17	10	1.28	12.45	20.50
0734	3-1/16" x 4" x 3-1/16" G		133	.180		1.78	8.60	1.10	11.48	18.25
0736	3-1/4" x 5" x 3-1/4" G		133	.180		2.42	8.60	1.10	12.12	18.95
0738	3-1/2" x 6" x 3-1/2" G		160	.150		3.65	7.15	.91	11.71	17.55
0740	Junior beam, field fabricated, 3" G		80	.300		3.76	14.25	1.82	19.83	31
0742	4" G		72	.333		5.10	15.85	2.03	22.98	36
0744	5" G		67	.358		6.60	17.05	2.18	25.83	39.50
0746	6" G		62	.387		8.25	18.40	2.35	29	44
0748	7" G		57	.421		10.10	20	2.56	32.66	49
0750	8" G	↓	53	.453	↓	12.15	21.50	2.75	36.40	54.50
1000	Continuous slotted channel framing system, shop fab, minimum G	2 Sswk	2400	.007	Lb.	3.41	.31		3.72	4.30
1200	Maximum G	"	1600	.010		3.85	.47		4.32	5.05
1300	Cross bracing, rods, shop fabricated, 3/4" diameter G	E-3	700	.034		1.32	1.63	.21	3.16	4.55
1310	7/8" diameter G		850	.028		1.32	1.34	.17	2.83	4
1320	1" diameter G		1000	.024		1.32	1.14	.15	2.61	3.62
1330	Angle, 5" x 5" x 3/8" G		2800	.009		1.32	.41	.05	1.78	2.23
1350	Hanging lintels, shop fabricated, average G	↓	850	.028		1.32	1.34	.17	2.83	4
1380	Roof frames, shop fabricated, 3'-0" square, 5' span G	E-2	4200	.013		1.32	.61	.38	2.31	2.90
1400	Tie rod, not upset, 1-1/2" to 4" diameter, with turnbuckle G	2 Sswk	800	.020		1.43	.94		2.37	3.22
1420	No turnbuckle G		700	.023		1.38	1.07		2.45	3.39
1500	Upset, 1-3/4" to 4" diameter, with turnbuckle G		800	.020		1.43	.94		2.37	3.22
1520	No turnbuckle G	↓	700	.023	↓	1.38	1.07		2.45	3.39

05 12 23.45 Lintels

Line	05 12 23.45 Lintels		Crew	Daily Output	Labor-Hours	Unit	Material	2010 Bare Costs Labor	Equipment	Total	Total Incl O&P
0010	**LINTELS**										
0015	Made from recycled materials	G									
0020	Plain steel angles, shop fabricated, under 500 lb.	G	1 Bric	550	.015	Lb.	.85	.61		1.46	1.85
0100	500 to 1000 lb. *CN*	G		640	.013		.83	.52		1.35	1.70
0200	1,000 to 2,000 lb.	G		640	.013		.80	.52		1.32	1.67
0300	2,000 to 4,000 lb.	G		640	.013		.78	.52		1.30	1.65
0500	For built-up angles and plates, add to above	G					.28			.28	.30
0700	For engineering, add to above						.11			.11	.12
0900	For galvanizing, add to above, under 500 lb.						.27			.27	.29
0950	500 to 2,000 lb.						.24			.24	.27
1000	Over 2,000 lb.						.21			.21	.23
2000	Steel angles, 3-1/2" x 3", 1/4" thick, 2'-6" long	G	1 Bric	47	.170	Ea.	11.90	7.10		19	24
2100	4'-6" long	G		26	.308		21.50	12.85		34.35	43
2600	4" x 3-1/2", 1/4" thick, 5'-0" long	G		21	.381		27.50	15.90		43.40	54
2700	9'-0" long	G		12	.667		49	28		77	96

05 12 23.60 Pipe Support Framing

Line	05 12 23.60 Pipe Support Framing		Crew	Daily Output	Labor-Hours	Unit	Material	Labor	Equipment	Total	Total Incl O&P
0010	**PIPE SUPPORT FRAMING**										
0020	Under 10#/L.F., shop fabricated	G	E-4	3900	.008	Lb.	1.47	.39	.04	1.90	2.34
0200	10.1 to 15#/L.F.	G		4300	.007		1.45	.35	.03	1.83	2.26
0400	15.1 to 20#/L.F.	G		4800	.007		1.43	.32	.03	1.78	2.16
0600	Over 20#/L.F.	G		5400	.006		1.41	.28	.03	1.72	2.07

05 12 23.65 Plates

Line	05 12 23.65 Plates		Crew	Daily Output	Labor-Hours	Unit	Material	Labor	Equipment	Total	Total Incl O&P
0010	**PLATES** R051223-80										
0015	Made from recycled materials	G									
0020	For connections & stiffener plates, shop fabricated										
0050	1/8" thick (5.1 Lb./S.F.)	G				S.F.	5.60			5.60	6.15
0100	1/4" thick (10.2 Lb./S.F.)	G					11.20			11.20	12.35
0300	3/8" thick (15.3 Lb./S.F.)	G					16.85			16.85	18.50
0400	1/2" thick (20.4 Lb./S.F.)	G					22.50			22.50	24.50
0450	3/4" thick (30.6 Lb./S.F.)	G					33.50			33.50	37
0500	1" thick (40.8 Lb.S.F.)	G					45			45	49.50
2000	Steel plate, warehouse prices, no shop fabrication										
2100	1/4" thick (10.2 Lb./S.F.)	G				S.F.	8.40			8.40	9.25

05 12 23.70 Stressed Skin Steel Roof and Ceiling System

Line	05 12 23.70 Stressed Skin Steel Roof and Ceiling System		Crew	Daily Output	Labor-Hours	Unit	Material	Labor	Equipment	Total	Total Incl O&P
0010	**STRESSED SKIN STEEL ROOF & CEILING SYSTEM**										
0020	Double panel flat roof, spans to 100'	G	E-2	1150	.049	S.F.	8.80	2.22	1.40	12.42	15
0100	Double panel convex roof, spans to 200'	G		960	.058		14.30	2.66	1.67	18.63	22
0200	Double panel arched roof, spans to 300'	G		760	.074		22	3.36	2.11	27.47	32

05 12 23.75 Structural Steel Members

Line	05 12 23.75 Structural Steel Members		Crew	Daily Output	Labor-Hours	Unit	Material	Labor	Equipment	Total	Total Incl O&P
0010	**STRUCTURAL STEEL MEMBERS** R051223-10										
0015	Made from recycled materials	G									
0020	Shop fab'd for 100-ton, 1-2 story project, bolted connections										
0100	W 6 x 9	G	E-2	600	.093	L.F.	10.90	4.26	2.68	17.84	22
0120	x 15	G		600	.093		18.15	4.26	2.68	25.09	30
0140	x 20	G		600	.093		24	4.26	2.68	30.94	36.50
0300	W 8 x 10	G		600	.093		12.10	4.26	2.68	19.04	23.50
0320	x 15	G		600	.093		18.15	4.26	2.68	25.09	30
0350	x 21	G		600	.093		25.50	4.26	2.68	32.44	38
0360	x 24	G		550	.102		29	4.64	2.92	36.56	43
0370	x 28	G		550	.102		34	4.64	2.92	41.56	48.50
0500	x 31	G		550	.102		37.50	4.64	2.92	45.06	52.50

05 12 23 – Structural Steel for Buildings

05 12 23.75 Structural Steel Members			Crew	Daily Output	Labor-Hours	Unit	Material	2010 Bare Costs Labor	Equipment	Total	Total Incl O&P
0520	x 35	G	E-2	550	.102	L.F.	42.50	4.64	2.92	50.06	57.50
0540	x 48	G		550	.102		58	4.64	2.92	65.56	75
0600	W 10 x 12	G		600	.093		14.50	4.26	2.68	21.44	26
0620	x 15	G		600	.093		18.15	4.26	2.68	25.09	30
0700	x 22	G		600	.093		26.50	4.26	2.68	33.44	39.50
0720	x 26	G		600	.093		31.50	4.26	2.68	38.44	44.50
0740	x 33	G		550	.102		40	4.64	2.92	47.56	55
0900	x 49	G		550	.102		59.50	4.64	2.92	67.06	76
1100	W 12 x 16	G		880	.064		19.35	2.90	1.83	24.08	28.50
1300	x 22	G		880	.064		26.50	2.90	1.83	31.23	36.50
1500	x 26	G		880	.064		31.50	2.90	1.83	36.23	41.50
1520	x 35	G		810	.069		42.50	3.15	1.98	47.63	54
1560	x 50	G		750	.075		60.50	3.41	2.14	66.05	74.50
1580	x 58	G		750	.075		70	3.41	2.14	75.55	85
1700	x 72	G		640	.088		87	3.99	2.51	93.50	106
1740	x 87	G		640	.088		105	3.99	2.51	111.50	126
1900	W 14 x 26	G		990	.057		31.50	2.58	1.62	35.70	40.50
2100	x 30	G		900	.062		36.50	2.84	1.79	41.13	47
2300	x 34	G		810	.069		41	3.15	1.98	46.13	53
2320	x 43	G		810	.069		52	3.15	1.98	57.13	64.50
2340	x 53	G		800	.070		64	3.19	2.01	69.20	78
2360	x 74	G		760	.074		89.50	3.36	2.11	94.97	107
2380	x 90	G		740	.076		109	3.45	2.17	114.62	128
2500	x 120	G		720	.078		145	3.55	2.23	150.78	168
2700	W 16 x 26	G		1000	.056		31.50	2.55	1.61	35.66	40.50
2900	x 31	G		900	.062		37.50	2.84	1.79	42.13	48.50
3100	x 40	G		800	.070		48.50	3.19	2.01	53.70	60.50
3120	x 50	G		800	.070		60.50	3.19	2.01	65.70	74
3140	x 67	G		760	.074		81	3.36	2.11	86.47	97
3300	W 18 x 35	G	E-5	960	.083		42.50	3.85	1.83	48.18	55
3500	x 40	G		960	.083		48.50	3.85	1.83	54.18	61.50
3520	x 46	G		960	.083		55.50	3.85	1.83	61.18	69.50
3700	x 50	G		912	.088		60.50	4.05	1.92	66.47	75.50
3900	x 55	G		912	.088		66.50	4.05	1.92	72.47	82
3920	x 65	G		900	.089		78.50	4.11	1.95	84.56	95.50
3940	x 76	G		900	.089		92	4.11	1.95	98.06	110
3960	x 86	G		900	.089		104	4.11	1.95	110.06	123
3980	x 106	G		900	.089		128	4.11	1.95	134.06	150
4100	W 21 x 44	G		1064	.075		53	3.47	1.65	58.12	66.50
4300	x 50	G		1064	.075		60.50	3.47	1.65	65.62	74.50
4500	x 62	G		1036	.077		75	3.57	1.69	80.26	90.50
4700	x 68	G		1036	.077		82.50	3.57	1.69	87.76	98.50
4720	x 83	G		1000	.080		100	3.70	1.75	105.45	118
4740	x 93	G		1000	.080		113	3.70	1.75	118.45	132
4760	x 101	G		1000	.080		122	3.70	1.75	127.45	142
4780	x 122	G		1000	.080		148	3.70	1.75	153.45	170
4900	W 24 x 55	G		1110	.072		66.50	3.33	1.58	71.41	80.50
5100	x 62	G		1110	.072		75	3.33	1.58	79.91	90
5300	x 68	G		1110	.072		82.50	3.33	1.58	87.41	98
5500	x 76	G		1110	.072		92	3.33	1.58	96.91	108
5700	x 84	G		1080	.074		102	3.42	1.62	107.04	120
5720	x 94	G		1080	.074		114	3.42	1.62	119.04	133
5740	x 104	G		1050	.076		126	3.52	1.67	131.19	146

05 12 23 – Structural Steel for Buildings

05 12 23.75 Structural Steel Members

	05 12 23.75 Structural Steel Members		Crew	Daily Output	Labor-Hours	Unit	Material	2010 Bare Costs Labor	Equipment	Total	Total Incl O&P
5760	x 117	G	E-5	1050	.076	L.F.	142	3.52	1.67	147.19	164
5780	x 146	G		1050	.076		177	3.52	1.67	182.19	202
5800	W 27 x 84	G		1190	.067		102	3.11	1.47	106.58	119
5900	x 94	G		1190	.067		114	3.11	1.47	118.58	132
5920	x 114	G		1150	.070		138	3.21	1.52	142.73	159
5940	x 146	G		1150	.070		177	3.21	1.52	181.73	201
5960	x 161	G		1150	.070		195	3.21	1.52	199.73	221
6100	W 30 x 99	G		1200	.067		120	3.08	1.46	124.54	139
6300	x 108	G		1200	.067		131	3.08	1.46	135.54	151
6500	x 116	G		1160	.069		140	3.19	1.51	144.70	161
6520	x 132	G		1160	.069		160	3.19	1.51	164.70	183
6540	x 148	G		1160	.069		179	3.19	1.51	183.70	204
6560	x 173	G		1120	.071		209	3.30	1.57	213.87	237
6580	x 191	G		1120	.071		231	3.30	1.57	235.87	261
6700	W 33 x 118	G		1176	.068		143	3.14	1.49	147.63	164
6900	x 130	G		1134	.071		157	3.26	1.55	161.81	180
7100	x 141	G		1134	.071		171	3.26	1.55	175.81	195
7120	x 169	G		1100	.073		204	3.36	1.59	208.95	233
7140	x 201	G		1100	.073		243	3.36	1.59	247.95	276
7300	W 36 x 135	G		1170	.068		163	3.16	1.50	167.66	187
7500	x 150	G		1170	.068		182	3.16	1.50	186.66	207
7600	x 170	G		1150	.070		206	3.21	1.52	210.73	233
7700	x 194	G		1125	.071		235	3.28	1.56	239.84	265
7900	x 231	G		1125	.071		280	3.28	1.56	284.84	310
7920	x 262	G		1035	.077		315	3.57	1.69	320.26	360
8100	x 302	G	↓	1035	.077		365	3.57	1.69	370.26	410
8490	For projects 75 to 99 tons, add						10%				
8492	50 to 74 tons, add						20%				
8494	25 to 49 tons, add						30%	10%			
8496	10 to 24 tons, add						50%	25%			
8498	2 to 9 tons, add						75%	50%			
8499	Less than 2 tons, add					↓	100%	100%			

05 12 23.77 Structural Steel Projects

	05 12 23.77 Structural Steel Projects		Crew	Daily Output	Labor-Hours	Unit	Material	2010 Bare Costs Labor	Equipment	Total	Total Incl O&P
0010	**STRUCTURAL STEEL PROJECTS**	R050516-30									
0015	Made from recycled materials	G									
0020	Shop fab'd for 100-ton, 1-2 story project, bolted connections										
0200	Apartments, nursing homes, etc., 1 to 2 stories	R050523-10	E-5	10.30	7.767	Ton	2,200	360	170	2,730	3,225
0300	3 to 6 stories	G	"	10.10	7.921		2,250	365	174	2,789	3,300
0400	7 to 15 stories	R051223-10	E-6	14.20	9.014		2,300	415	137	2,852	3,400
0500	Over 15 stories	G	"	13.90	9.209		2,375	425	140	2,940	3,500
0700	Offices, hospitals, etc., steel bearing, 1 to 2 stories	R051223-20 G	E-5	10.30	7.767		2,200	360	170	2,730	3,225
0800	3 to 6 stories	G	E-6	14.40	8.889		2,250	410	135	2,795	3,325
0900	7 to 15 stories	R051223-25		14.20	9.014		2,300	415	137	2,852	3,400
1000	Over 15 stories	G	↓	13.90	9.209		2,375	425	140	2,940	3,500
1100	For multi-story masonry wall bearing construction, add	R051223-30						30%			
1300	Industrial bldgs., 1 story, beams & girders, steel bearing	G	E-5	12.90	6.202		2,200	286	136	2,622	3,075
1400	Masonry bearing	G	"	10	8	↓	2,200	370	175	2,745	3,250
1500	Industrial bldgs., 1 story, under 10 tons,										
1510	steel from warehouse, trucked	G	E-2	7.50	7.467	Ton	2,650	340	214	3,204	3,700
1600	1 story with roof trusses, steel bearing	G	E-5	10.60	7.547		2,600	350	165	3,115	3,625
1700	Masonry bearing	G	"	8.30	9.639		2,600	445	211	3,256	3,850
1900	Monumental structures, banks, stores, etc., minimum	G	E-6	13	9.846	↓	2,200	455	150	2,805	3,375

05 12 23.77 Structural Steel Projects			Crew	Daily Output	Labor-Hours	Unit	2010 Bare Costs Material	Labor	Equipment	Total	Total Incl O&P
2000	Maximum	G	E-6	9	14.222	Ton	3,650	655	216	4,521	5,400
2200	Churches, minimum	G	E-5	11.60	6.897		2,050	320	151	2,521	2,950
2300	Maximum	G	"	5.20	15.385		2,725	710	335	3,770	4,600
2800	Power stations, fossil fuels, minimum	G	E-6	11	11.636		2,200	540	177	2,917	3,550
2900	Maximum	G		5.70	22.456		3,300	1,050	340	4,690	5,775
2950	Nuclear fuels, non-safety steel, minimum	G		7	18.286		2,200	845	278	3,323	4,175
3000	Maximum	G		5.50	23.273		3,300	1,075	355	4,730	5,875
3040	Safety steel, minimum	G		2.50	51.200		3,200	2,375	780	6,355	8,425
3070	Maximum	G	↓	1.50	85.333		4,225	3,950	1,300	9,475	12,900
3100	Roof trusses, minimum	G	E-5	13	6.154		3,075	284	135	3,494	4,025
3200	Maximum	G		8.30	9.639		3,750	445	211	4,406	5,125
3210	Schools, minimum	G		14.50	5.517		2,200	255	121	2,576	3,000
3220	Maximum	G	↓	8.30	9.639		3,200	445	211	3,856	4,525
3400	Welded construction, simple commercial bldgs., 1 to 2 stories	G	E-7	7.60	10.526		2,250	485	250	2,985	3,575
3500	7 to 15 stories	G	E-9	8.30	15.422		2,600	715	281	3,596	4,375
3700	Welded rigid frame, 1 story, minimum	G	E-7	15.80	5.063		2,300	234	120	2,654	3,050
3800	Maximum	G	"	5.50	14.545	↓	2,975	670	345	3,990	4,800
3810	Fabrication shop costs (included in project material cost, above)										
3820	Mini mill base price, A992	G				Ton	770			770	845
3830	Mill extra for delivery to shop						235			235	259
3840	Shop extra for shop drawings and detailing						260			260	286
3850	Shop fabricating and handling						700			700	770
3860	Shop sandblasting and primer coat of paint						135			135	149
3870	Shop delivery to the job site						100			100	110
3880	Total material cost, shop fabricated, primed, delivered					↓	2,200			2,200	2,425
3900	High strength steel mill spec extras:										
3950	A529, A572 (50 ksi) and A36: same as A992 steel (no extra)										
4000	Add to A992 price for A572 (60, 65 ksi)	G				Ton	100			100	110
4100	A242 and A588 Weathering	G				"	90			90	99
4200	Mill size extras for W-Shapes: 0 to 30 plf: no extra charge										
4210	Member sizes 31 to 65 plf, deduct	G				Ton	.01			.01	.01
4220	Member sizes 66 to 100 plf, deduct	G					5.65			5.65	6.25
4230	Member sizes 101 to 387 plf, add	G				↓	57			57	62.50
4300	Column base plates, light, up to 150 lb	G	2 Sswk	2000	.008	Lb.	1.21	.38		1.59	1.99
4400	Heavy, over 150 lb	G	E-2	7500	.007	"	1.27	.34	.21	1.82	2.21
4600	Castellated beams, light sections, to 50#/L.F., minimum	G		10.70	5.234	Ton	2,300	239	150	2,689	3,125
4700	Maximum	G		7	8		2,525	365	230	3,120	3,650
4900	Heavy sections, over 50# per L.F., minimum	G		11.70	4.786		2,425	218	137	2,780	3,175
5000	Maximum	G	↓	7.80	7.179		2,650	325	206	3,181	3,675
5390	For projects 75 to 99 tons, add						10%				
5392	50 to 74 tons, add						20%				
5394	25 to 49 tons, add						30%	10%			
5396	10 to 24 tons, add						50%	25%			
5398	2 to 9 tons, add						75%	50%			
5399	Less than 2 tons, add					↓	100%	100%			

05 12 23.80 Subpurlins			Crew	Daily Output	Labor-Hours	Unit	Material	Labor	Equipment	Total	Total Incl O&P
0010	**SUBPURLINS** R051223-50										
0015	Made from recycled materials	G									
0020	Bulb tees, shop fabricated, painted, 32-5/8" O.C., 40 psf L.L.										
0100	Type 178, max 8'-9" span, 2.15 plf, 2" high x 1-5/8" wide	G	E-1	4200	.006	S.F.	1.53	.26	.03	1.82	2.16
0200	Type 218, max 10'-2" span, 3.19 plf, 2-1/8" high x 2-1/8" wide	G	"	3100	.008		1.77	.35	.05	2.17	2.59
1420	For 24-5/8" spacing, add					↓	33%	33%			

05 12 Structural Steel Framing

05 12 23 – Structural Steel for Buildings

05 12 23.80 Subpurlins			Crew	Daily Output	Labor-Hours	Unit	2010 Bare Costs Material	Labor	Equipment	Total	Total Incl O&P
1430	For 48-5/8" spacing, deduct					S.F.	50%	50%			

05 14 Structural Aluminum Framing

05 14 23 – Non-Exposed Structural Aluminum Framing

05 14 23.05 Aluminum Shapes

			Crew	Daily Output	Labor-Hours	Unit	Material	Labor	Equipment	Total	Total Incl O&P
0010	**ALUMINUM SHAPES**										
0015	Made from recycled materials	G									
0020	Structural shapes, 1" to 10" members, under 1 ton	G	E-2	1050	.053	Lb.	3.11	2.43	1.53	7.07	9.20
0050	1 to 5 tons	G		1330	.042		2.88	1.92	1.21	6.01	7.75
0100	Over 5 tons *CN*	G		1330	.042		2.90	1.92	1.21	6.03	7.75
0300	Extrusions, over 5 tons, stock shapes	G		1330	.042		2.51	1.92	1.21	5.64	7.35
0400	Custom shapes	G		1330	.042		2.49	1.92	1.21	5.62	7.30

05 15 Wire Rope Assemblies

05 15 16 – Steel Wire Rope Assemblies

05 15 16.05 Accessories for Steel Wire Rope

			Crew	Daily Output	Labor-Hours	Unit	Material	Labor	Equipment	Total	Total Incl O&P
0010	**ACCESSORIES FOR STEEL WIRE ROPE**										
0015	Made from recycled materials	G									
1500	Thimbles, heavy duty, 1/4"	G	E-17	160	.100	Ea.	.46	4.79		5.25	8.90
1510	1/2"	G		160	.100		2.01	4.79		6.80	10.60
1520	3/4"	G		105	.152		4.57	7.30		11.87	17.85
1530	1"	G		52	.308		9.15	14.75		23.90	36
1540	1-1/4"	G		38	.421		14.05	20		34.05	51
1550	1-1/2"	G		13	1.231		39.50	59		98.50	148
1560	1-3/4"	G		8	2		81.50	96		177.50	258
1570	2"	G		6	2.667		119	128		247	355
1580	2-1/4"	G		4	4		161	192		353	510
1600	Clips, 1/4" diameter	G		160	.100		2.73	4.79		7.52	11.40
1610	3/8" diameter	G		160	.100		2.99	4.79		7.78	11.70
1620	1/2" diameter	G		160	.100		4.80	4.79		9.59	13.70
1630	3/4" diameter	G		102	.157		7.80	7.50		15.30	22
1640	1" diameter	G		64	.250		13	12		25	35.50
1650	1-1/4" diameter	G		35	.457		21.50	22		43.50	62
1670	1-1/2" diameter	G		26	.615		28.50	29.50		58	83.50
1680	1-3/4" diameter	G		16	1		67	48		115	158
1690	2" diameter	G		12	1.333		74.50	64		138.50	194
1700	2-1/4" diameter	G		10	1.600		109	76.50		185.50	255
1800	Sockets, open swage, 1/4" diameter	G		160	.100		38	4.79		42.79	50.50
1810	1/2" diameter	G		77	.208		55	9.95		64.95	78
1820	3/4" diameter	G		19	.842		85.50	40.50		126	165
1830	1" diameter	G		9	1.778		152	85		237	320
1840	1-1/4" diameter	G		5	3.200		212	153		365	500
1850	1-1/2" diameter	G		3	5.333		465	255		720	960
1860	1-3/4" diameter	G		3	5.333		825	255		1,080	1,350
1870	2" diameter	G		1.50	10.667		1,250	510		1,760	2,275
1900	Closed swage, 1/4" diameter	G		160	.100		23.50	4.79		28.29	34.50
1910	1/2" diameter	G		104	.154		40.50	7.35		47.85	58
1920	3/4" diameter	G		32	.500		61	24		85	109
1930	1" diameter	G		15	1.067		107	51		158	208

05 15 16 – Steel Wire Rope Assemblies

05 15 16.05 Accessories for Steel Wire Rope		Crew	Daily Output	Labor-Hours	Unit	Material	Labor	Equipment	Total	Total Incl O&P
						2010 Bare Costs				
1940	1-1/4" diameter	E-17	7	2.286	Ea.	161	109		270	370
1950	1-1/2" diameter		4	4		291	192		483	655
1960	1-3/4" diameter		3	5.333		430	255		685	920
1970	2" diameter		2	8		835	385		1,220	1,600
2000	Open spelter, galv., 1/4" diameter		160	.100		53	4.79		57.79	66.50
2010	1/2" diameter		70	.229		55	10.95		65.95	80
2020	3/4" diameter		26	.615		83	29.50		112.50	143
2030	1" diameter		10	1.600		230	76.50		306.50	390
2040	1-1/4" diameter		5	3.200		330	153		483	630
2050	1-1/2" diameter		4	4		695	192		887	1,100
2060	1-3/4" diameter		2	8		1,225	385		1,610	2,025
2070	2" diameter		1.20	13.333		1,400	640		2,040	2,675
2080	2-1/2" diameter		1	16		2,575	765		3,340	4,175
2100	Closed spelter, galv., 1/4" diameter		160	.100		43	4.79		47.79	56
2110	1/2" diameter		88	.182		46	8.70		54.70	66
2120	3/4" diameter		30	.533		69.50	25.50		95	122
2130	1" diameter		13	1.231		148	59		207	267
2140	1-1/4" diameter		7	2.286		237	109		346	455
2150	1-1/2" diameter		6	2.667		510	128		638	785
2160	1-3/4" diameter		2.80	5.714		680	274		954	1,225
2170	2" diameter		2	8		840	385		1,225	1,600
2200	Jaw & jaw turnbuckles, 1/4" x 4"		160	.100		14.30	4.79		19.09	24
2250	1/2" x 6"		96	.167		18.05	8		26.05	34
2260	1/2" x 9"		77	.208		24	9.95		33.95	44
2270	1/2" x 12"		66	.242		27	11.60		38.60	50.50
2300	3/4" x 6"		38	.421		35.50	20		55.50	74.50
2310	3/4" x 9"		30	.533		39	25.50		64.50	88
2320	3/4" x 12"		28	.571		50.50	27.50		78	104
2330	3/4" x 18"		23	.696		60	33.50		93.50	125
2350	1" x 6"		17	.941		68.50	45		113.50	155
2360	1" x 12"		13	1.231		75	59		134	187
2370	1" x 18"		10	1.600		113	76.50		189.50	259
2380	1" x 24"		9	1.778		124	85		209	287
2400	1-1/4" x 12"		7	2.286		126	109		235	330
2410	1-1/4" x 18"		6.50	2.462		156	118		274	380
2420	1-1/4" x 24"		5.60	2.857		211	137		348	475
2450	1-1/2" x 12"		5.20	3.077		425	147		572	725
2460	1-1/2" x 18"		4	4		450	192		642	835
2470	1-1/2" x 24"		3.20	5		610	240		850	1,100
2500	1-3/4" x 18"		3.20	5		920	240		1,160	1,425
2510	1-3/4" x 24"		2.80	5.714		1,050	274		1,324	1,625
2550	2" x 24"		1.60	10		1,425	480		1,905	2,400

05 15 16.50 Steel Wire Rope

		Crew	Daily Output	Labor-Hours	Unit	Material	Labor	Equipment	Total	Total Incl O&P
0010	**STEEL WIRE ROPE**									
0015	Made from recycled materials									
0020	6 x 19, bright, fiber core, 5000' rolls, 1/2" diameter				L.F.	.60			.60	.66
0050	Steel core					.79			.79	.87
0100	Fiber core, 1" diameter					2.02			2.02	2.22
0150	Steel core					2.30			2.30	2.53
0300	6 x 19, galvanized, fiber core, 1/2" diameter					.88			.88	.97
0350	Steel core					1.01			1.01	1.11
0400	Fiber core, 1" diameter					2.58			2.58	2.84

05 15 Wire Rope Assemblies

05 15 16 – Steel Wire Rope Assemblies

05 15 16.50 Steel Wire Rope			Crew	Daily Output	Labor-Hours	Unit	2010 Bare Costs Material	Labor	Equipment	Total	Total Incl O&P
0450	Steel core	G				L.F.	2.71			2.71	2.98
0500	6 x 7, bright, IPS, fiber core, <500 L.F. w/acc., 1/4" diameter	G	E-17	6400	.003		1.36	.12		1.48	1.71
0510	1/2" diameter	G		2100	.008		3.32	.37		3.69	4.29
0520	3/4" diameter	G		960	.017		6	.80		6.80	8
0550	6 x 19, bright, IPS, IWRC, <500 L.F. w/acc., 1/4" diameter	G		5760	.003		.66	.13		.79	.96
0560	1/2" diameter	G		1730	.009		1.07	.44		1.51	1.96
0570	3/4" diameter	G		770	.021		1.86	1		2.86	3.79
0580	1" diameter	G		420	.038		3.15	1.83		4.98	6.70
0590	1-1/4" diameter	G		290	.055		5.25	2.64		7.89	10.40
0600	1-1/2" diameter	G		192	.083		6.45	3.99		10.44	14.05
0610	1-3/4" diameter	G	E-18	240	.167		10.25	7.75	4.29	22.29	29
0620	2" diameter	G		160	.250		13.15	11.65	6.45	31.25	41.50
0630	2-1/4" diameter	G		160	.250		17.60	11.65	6.45	35.70	46.50
0650	6 x 37, bright, IPS, IWRC, <500 L.F. w/acc., 1/4" diameter	G	E-17	6400	.003		.77	.12		.89	1.06
0660	1/2" diameter	G		1730	.009		1.30	.44		1.74	2.21
0670	3/4" diameter	G		770	.021		2.10	1		3.10	4.06
0680	1" diameter	G		430	.037		3.34	1.78		5.12	6.80
0690	1-1/4" diameter	G		290	.055		5.05	2.64		7.69	10.20
0700	1-1/2" diameter	G		190	.084		7.20	4.03		11.23	15.05
0710	1-3/4" diameter	G	E-18	260	.154		11.45	7.15	3.96	22.56	29
0720	2" diameter	G		200	.200		14.85	9.30	5.15	29.30	38
0730	2-1/4" diameter	G		160	.250		19.65	11.65	6.45	37.75	48.50
0800	6 x 19 & 6 x 37, swaged, 1/2" diameter	G	E-17	1220	.013		1.83	.63		2.46	3.11
0810	9/16" diameter	G		1120	.014		2.13	.68		2.81	3.54
0820	5/8" diameter	G		930	.017		2.52	.82		3.34	4.23
0830	3/4" diameter	G		640	.025		3.22	1.20		4.42	5.65
0840	7/8" diameter	G		480	.033		4.06	1.60		5.66	7.25
0850	1" diameter	G		350	.046		4.95	2.19		7.14	9.30
0860	1-1/8" diameter	G		288	.056		6.10	2.66		8.76	11.40
0870	1-1/4" diameter	G		230	.070		7.40	3.33		10.73	13.95
0880	1-3/8" diameter	G		192	.083		8.50	3.99		12.49	16.35
0890	1-1/2" diameter	G	E-18	300	.133		10.35	6.20	3.43	19.98	26

05 15 16.60 Galvanized Steel Wire Rope and Accessories

			Crew	Daily Output	Labor-Hours	Unit	Material	Labor	Equipment	Total	Total Incl O&P
0010	**GALVANIZED STEEL WIRE ROPE & ACCESSORIES**										
0015	Made from recycled materials	G									
3000	Aircraft cable, galvanized, 7 x 7 x 1/8"	G	E-17	5000	.003	L.F.	.18	.15		.33	.47
3100	Clamps, 1/8"	G	"	125	.128	Ea.	1.95	6.15		8.10	12.95

05 15 16.70 Temporary Cable Safety Railing

			Crew	Daily Output	Labor-Hours	Unit	Material	Labor	Equipment	Total	Total Incl O&P
0010	**TEMPORARY CABLE SAFETY RAILING**, Each 100' strand incl.										
0020	2 eyebolts, 1 turnbuckle, 100' cable, 2 thimbles, 6 clips										
0025	Made from recycled materials	G									
0100	One strand using 1/4" cable & accessories	G	2 Sswk	4	4	C.L.F.	189	188		377	540
0200	1/2" cable & accessories	G	"	2	8	"	400	375		775	1,100

05 21 13 – Deep Longspan Steel Joist Framing

05 21 13.50 Deep Longspan Joists			Crew	Daily Output	Labor-Hours	Unit	Material	Labor	Equipment	Total	Total Incl O&P
							2010 Bare Costs				
0010	DEEP LONGSPAN JOISTS										
3010	DLH series, 40-ton job lots, bolted cross bridging, shop primer										
3015	Made from recycled materials	G									
3020	Spans to 144' (shipped in 2 pieces), minimum	G	E-7	16	5	Ton	1,100	231	119	1,450	1,750
3040	Average	G		13	6.154		1,200	284	146	1,630	1,950
3100	Maximum	G		11	7.273		1,425	335	173	1,933	2,350
3200	52DLH11, 26 Lb/LF	G		2000	.040	L.F.	14.85	1.85	.95	17.65	20.50
3220	52DLH16, 45 Lb/LF	G		2000	.040		26.50	1.85	.95	29.30	33.50
3240	56DLH11, 26 Lb/LF	G		2000	.040		15.45	1.85	.95	18.25	21
3260	56DLH16, 46 Lb/LF	G		2000	.040		27.50	1.85	.95	30.30	34
3280	60DLH12, 29 Lb/LF	G		2000	.040		17.20	1.85	.95	20	23
3300	60DLH17, 52 Lb/LF	G		2000	.040		31	1.85	.95	33.80	38
3320	64DLH12, 31 Lb/LF	G		2200	.036		18.40	1.68	.86	20.94	24.50
3340	64DLH17, 52 Lb/LF	G		2200	.036		31	1.68	.86	33.54	38
3360	68DLH13, 37 Lb/LF	G		2200	.036		22	1.68	.86	24.54	28
3380	68DLH18, 61 Lb/LF	G		2200	.036		36	1.68	.86	38.54	44
3400	72DLH14, 41 Lb/LF	G		2200	.036		24.50	1.68	.86	27.04	31
3420	72DLH19, 70 Lb/LF	G		2200	.036		41.50	1.68	.86	44.04	49.50
3500	For less than 40-ton job lots										
3502	For 30 to 39 tons, add						10%				
3504	20 to 29 tons, add						20%				
3506	10 to 19 tons, add						30%				
3507	5 to 9 tons, add						50%	25%			
3508	1 to 4 tons, add						75%	50%			
3509	Less than 1 ton, add						100%	100%			
4010	SLH series, 40-ton job lots, bolted cross bridging, shop primer										
4020	Spans to 200' (shipped in 3 pieces), minimum	G	E-7	16	5	Ton	1,075	231	119	1,425	1,725
4040	Average	G		13	6.154		1,225	284	146	1,655	2,000
4060	Maximum	G		11	7.273		1,450	335	173	1,958	2,375
4200	80SLH15, 40 Lb/LF	G		1500	.053	L.F.	24.50	2.46	1.27	28.23	32.50
4220	80SLH20, 75 Lb/LF	G		1500	.053		46	2.46	1.27	49.73	56
4240	88SLH16, 46 Lb/LF	G		1500	.053		28	2.46	1.27	31.73	36.50
4260	88SLH21, 89 Lb/LF	G		1500	.053		54.50	2.46	1.27	58.23	65
4280	96SLH17, 52 Lb/LF	G		1500	.053		31.50	2.46	1.27	35.23	40.50
4300	96SLH22, 102 Lb/LF	G		1500	.053		62.50	2.46	1.27	66.23	74
4320	104SLH18, 59 Lb/LF	G		1800	.044		36	2.05	1.05	39.10	44
4340	104SLH23, 109 Lb/LF	G		1800	.044		66.50	2.05	1.05	69.60	77.50
4360	112SLH19, 67 Lb/LF	G		1800	.044		41	2.05	1.05	44.10	49.50
4380	112SLH24, 131 Lb/LF	G		1800	.044		80	2.05	1.05	83.10	92.50
4400	120SLH20, 77 Lb/LF	G		1800	.044		47	2.05	1.05	50.10	56
4420	120SLH25, 152 Lb/LF	G		1800	.044		93	2.05	1.05	96.10	107
6100	For less than 40-ton job lots										
6102	For 30 to 39 tons, add						10%				
6104	20 to 29 tons, add						20%				
6106	10 to 19 tons, add						30%				
6107	5 to 9 tons, add						50%	25%			
6108	1 to 4 tons, add						75%	50%			
6109	Less than 1 ton, add						100%	100%			

05 21 16 – Longspan Steel Joist Framing

05 21 16.50 Longspan Joists											
0010	LONGSPAN JOISTS										
2000	LH series, 40-ton job lots, bolted cross bridging, shop primer										

05 21 Steel Joist Framing

05 21 16 – Longspan Steel Joist Framing

05 21 16.50 Longspan Joists			Crew	Daily Output	Labor-Hours	Unit	2010 Bare Costs Material	Labor	Equipment	Total	Total Incl O&P
2015	Made from recycled materials	G									
2020	Spans to 96', minimum	G	E-7	16	5	Ton	1,025	231	119	1,375	1,650
2040	Average	G		13	6.154		1,125	284	146	1,555	1,900
2080	Maximum	G		11	7.273		1,325	335	173	1,833	2,250
2200	18LH04, 12 Lb/LF	G		1400	.057	L.F.	6.80	2.64	1.36	10.80	13.45
2220	18LH08, 19 Lb/LF	G		1400	.057		10.75	2.64	1.36	14.75	17.80
2240	20LH04, 12 Lb/LF	G		1400	.057		6.80	2.64	1.36	10.80	13.45
2260	20LH08, 19 Lb/LF	G		1400	.057		10.75	2.64	1.36	14.75	17.80
2280	24LH05, 13 Lb/LF	G		1400	.057		7.35	2.64	1.36	11.35	14.10
2300	24LH10, 23 Lb/LF	G		1400	.057		13	2.64	1.36	17	20.50
2320	28LH06, 16 Lb/LF	G		1800	.044		9.05	2.05	1.05	12.15	14.60
2340	28LH11, 25 Lb/LF	G		1800	.044		14.10	2.05	1.05	17.20	20
2360	32LH08, 17 Lb/LF	G		1800	.044		9.60	2.05	1.05	12.70	15.20
2380	32LH13, 30 Lb/LF	G		1800	.044		16.95	2.05	1.05	20.05	23.50
2400	36LH09, 21 Lb/LF	G		1800	.044		11.85	2.05	1.05	14.95	17.70
2420	36LH14, 36 Lb/LF	G		1800	.044		20.50	2.05	1.05	23.60	27
2440	40LH10, 21 Lb/LF	G		2200	.036		11.85	1.68	.86	14.39	16.85
2460	40LH15, 36 Lb/LF	G		2200	.036		20.50	1.68	.86	23.04	26.50
2480	44LH11, 22 Lb/LF	G		2200	.036		12.40	1.68	.86	14.94	17.45
2500	44LH16, 42 Lb/LF	G		2200	.036		23.50	1.68	.86	26.04	30
2520	48LH11, 22 Lb/LF	G		2200	.036		12.40	1.68	.86	14.94	17.45
2540	48LH16, 42 Lb/LF	G		2200	.036		23.50	1.68	.86	26.04	30
2600	For less than 40-ton job lots										
2602	For 30 to 39 tons, add						10%				
2604	20 to 29 tons, add						20%				
2606	10 to 19 tons, add						30%				
2607	5 to 9 tons, add						50%	25%			
2608	1 to 4 tons, add						75%	50%			
2609	Less than 1 ton, add						100%	100%			
6000	For welded cross bridging, add							30%			

05 21 19 – Open Web Steel Joist Framing

05 21 19.10 Open Web Joists			Crew	Daily Output	Labor-Hours	Unit	Material	Labor	Equipment	Total	Total Incl O&P
0010	**OPEN WEB JOISTS**										
0015	Made from recycled materials	G									
0020	K series, 40-ton lots, horiz. bridging, spans to 30', shop primer, minimum	G	E-7	15	5.333	Ton	910	246	127	1,283	1,550
0050	Average	G		12	6.667		1,025	310	158	1,493	1,825
0080	Maximum	G		9	8.889		1,225	410	211	1,846	2,275
0130	8K1, 5.1 Lb/LF	G		1200	.067	L.F.	2.60	3.08	1.58	7.26	9.85
0140	10K1, 5.0 Lb/LF	G		1200	.067		2.55	3.08	1.58	7.21	9.80
0160	12K3, 5.7 Lb/LF	G		1500	.053		2.91	2.46	1.27	6.64	8.80
0180	14K3, 6.0 Lb/LF	G		1500	.053		3.06	2.46	1.27	6.79	8.95
0200	16K3, 6.3 Lb/LF	G		1800	.044		3.21	2.05	1.05	6.31	8.20
0220	16K6, 8.1 Lb/LF	G		1800	.044		4.13	2.05	1.05	7.23	9.20
0240	18K5, 7.7 Lb/LF	G		2000	.040		3.93	1.85	.95	6.73	8.50
0260	18K9, 10.2 Lb/LF	G		2000	.040		5.20	1.85	.95	8	9.90
0410	Span 30' to 50', minimum	G		17	4.706	Ton	890	217	112	1,219	1,475
0440	Average **CN**	G		17	4.706		1,000	217	112	1,329	1,600
0460	Maximum	G		10	8		1,050	370	190	1,610	2,025
0500	20K5, 8.2 Lb/LF	G		2000	.040	L.F.	4.10	1.85	.95	6.90	8.70
0520	20K9, 10.8 Lb/LF	G		2000	.040		5.40	1.85	.95	8.20	10.15
0540	22K5, 8.8 Lb/LF	G		2000	.040		4.40	1.85	.95	7.20	9.05
0560	22K9, 11.3 Lb/LF	G		2000	.040		5.65	1.85	.95	8.45	10.40

05 21 Steel Joist Framing

05 21 19 – Open Web Steel Joist Framing

05 21 19.10 Open Web Joists			Crew	Daily Output	Labor-Hours	Unit	2010 Bare Costs Material	Labor	Equipment	Total	Total Incl O&P
0580	24K6, 9.7 Lb/LF	G	E-7	2200	.036	L.F.	4.85	1.68	.86	7.39	9.15
0600	24K10, 13.1 Lb/LF	G		2200	.036		6.55	1.68	.86	9.09	11
0620	26K6, 10.6 Lb/LF	G		2200	.036		5.30	1.68	.86	7.84	9.65
0640	26K10, 13.8 Lb/LF	G		2200	.036		6.90	1.68	.86	9.44	11.40
0660	28K8, 12.7 Lb/LF	G		2400	.033		6.35	1.54	.79	8.68	10.50
0680	28K12, 17.1 Lb/LF	G		2400	.033		8.55	1.54	.79	10.88	12.90
0700	30K8, 13.2 Lb/LF	G		2400	.033		6.60	1.54	.79	8.93	10.75
0720	30K12, 17.6 Lb/LF	G		2400	.033		8.80	1.54	.79	11.13	13.20
0800	For less than 40-ton job lots										
0802	For 30 to 39 tons, add						10%				
0804	20 to 29 tons, add						20%				
0806	10 to 19 tons, add						30%				
0807	5 to 9 tons, add						50%	25%			
0808	1 to 4 tons, add						75%	50%			
0809	Less than 1 ton, add						100%	100%			
1010	CS series, 40-ton job lots, horizontal bridging, shop primer										
1020	Spans to 30′, minimum	G	E-7	15	5.333	Ton	940	246	127	1,313	1,575
1040	Average	G		12	6.667		1,050	310	158	1,518	1,850
1060	Maximum	G		9	8.889		1,225	410	211	1,846	2,275
1100	10CS2, 7.5 Lb/LF	G		1200	.067	L.F.	3.93	3.08	1.58	8.59	11.30
1120	12CS2, 8.0 Lb/LF	G		1500	.053		4.19	2.46	1.27	7.92	10.20
1140	14CS2, 8.0Lb/LF	G		1500	.053		4.19	2.46	1.27	7.92	10.20
1160	16CS2, 8.5 Lb/LF	G		1800	.044		4.45	2.05	1.05	7.55	9.55
1180	16CS4, 14.5 Lb/LF	G		1800	.044		7.60	2.05	1.05	10.70	13
1200	18CS2, 9.0 Lb/LF	G		2000	.040		4.71	1.85	.95	7.51	9.40
1220	18CS4, 15.0 Lb/LF	G		2000	.040		7.85	1.85	.95	10.65	12.85
1240	20CS2, 9.5 Lb/LF	G		2000	.040		4.97	1.85	.95	7.77	9.65
1260	20CS4, 16.5 Lb/LF	G		2000	.040		8.65	1.85	.95	11.45	13.70
1280	22CS2, 10.0 Lb/LF	G		2000	.040		5.25	1.85	.95	8.05	9.95
1300	22CS4, 16.5 Lb/LF	G		2000	.040		8.65	1.85	.95	11.45	13.70
1320	24CS2, 10.0 Lb/LF	G		2200	.036		5.25	1.68	.86	7.79	9.55
1340	24CS4, 16.5 Lb/LF	G		2200	.036		8.65	1.68	.86	11.19	13.30
1360	26CS2, 10.0 Lb/LF	G		2200	.036		5.25	1.68	.86	7.79	9.55
1380	26CS4, 16.5 Lb/LF	G		2200	.036		8.65	1.68	.86	11.19	13.30
1400	28CS2, 10.5 Lb/LF	G		2400	.033		5.50	1.54	.79	7.83	9.55
1420	28CS4, 16.5 Lb/LF	G		2400	.033		8.65	1.54	.79	10.98	13
1440	30CS2, 11.0 Lb/LF	G		2400	.033		5.75	1.54	.79	8.08	9.85
1460	30CS4, 16.5 Lb/LF	G		2400	.033		8.65	1.54	.79	10.98	13
1500	For less than 40-ton job lots										
1502	For 30 to 39 tons, add						10%				
1504	20 to 29 tons, add						20%				
1506	10 to 19 tons, add						30%				
1507	5 to 9 tons, add						50%	25%			
1508	1 to 4 tons, add						75%	50%			
1509	Less than 1 ton, add						100%	100%			
6200	For shop prime paint other than mfrs. standard, add						20%				
6300	For bottom chord extensions, add per chord	G				Ea.	22			22	24.50
6400	Individual steel bearing plate, 6″ x 6″ x 1/4″ with J-hook	G	1 Bric	160	.050	″	6.60	2.09		8.69	10.40

05 21 23 – Steel Joist Girder Framing

05 21 23.50 Joist Girders			Crew	Daily Output	Labor-Hours	Unit	Material	Labor	Equipment	Total	Total Incl O&P
0010	**JOIST GIRDERS**										
0015	Made from recycled materials	G									

05 21 Steel Joist Framing

05 21 23 – Steel Joist Girder Framing

05 21 23.50 Joist Girders			Crew	Daily Output	Labor-Hours	Unit	Material	2010 Bare Costs Labor	Equipment	Total	Total Incl O&P
7000	Joist girders, 40-ton job lots, shop primer, minimum	G	E-5	15	5.333	Ton	930	246	117	1,293	1,575
7020	Average	G		13	6.154		1,025	284	135	1,444	1,750
7040	Maximum	G		11	7.273		1,075	335	159	1,569	1,925
7100	For less than 40-ton job lots										
7102	For 30 to 39 tons, add						10%				
7104	20 to 29 tons, add						20%				
7106	10 to 19 tons, add						30%				
7107	5 to 9 tons, add						50%	25%			
7108	1 to 4 tons, add						75%	50%			
7109	Less than 1 ton, add						100%	100%			
8000	Trusses, 40-ton job lots, shop fabricated WT chords, shop primer, average	G	E-5	11	7.273	Ton	3,350	335	159	3,844	4,425
8100	For less than 40-ton job lots										
8102	For 30 to 39 tons, add						10%				
8104	20 to 29 tons, add						20%				
8106	10 to 19 tons, add						30%				
8107	5 to 9 tons, add						50%	25%			
8108	1 to 4 tons, add						75%	50%			
8109	Less than 1 ton, add						100%	100%			

05 31 Steel Decking

05 31 13 – Steel Floor Decking

05 31 13.50 Floor Decking			Crew	Daily Output	Labor-Hours	Unit	Material	Labor	Equipment	Total	Total Incl O&P
0010	**FLOOR DECKING** R053100-10										
0015	Made from recycled materials	G									
5100	Non-cellular composite decking, galvanized, 1-1/2" deep, 16 gauge	G	E-4	3500	.009	S.F.	2.18	.43	.04	2.65	3.21
5120	18 gauge	G		3650	.009		1.73	.42	.04	2.19	2.67
5140	20 gauge	G		3800	.008		1.41	.40	.04	1.85	2.30
5200	2" deep, 22 gauge	G		3860	.008		1.22	.39	.04	1.65	2.07
5300	20 gauge	G		3600	.009		1.36	.42	.04	1.82	2.28
5400	18 gauge	G		3380	.009		1.69	.45	.04	2.18	2.70
5500	16 gauge	G		3200	.010		2.16	.47	.05	2.68	3.25
5700	3" deep, 22 gauge	G		3200	.010		1.33	.47	.05	1.85	2.34
5800	20 gauge	G		3000	.011		1.50	.51	.05	2.06	2.59
5900	18 gauge *CN*	G		2850	.011		1.80	.53	.05	2.38	2.98
6000	16 gauge	G		2700	.012		2.45	.56	.05	3.06	3.75

05 31 23 – Steel Roof Decking

05 31 23.50 Roof Decking			Crew	Daily Output	Labor-Hours	Unit	Material	Labor	Equipment	Total	Total Incl O&P
0010	**ROOF DECKING**										
0015	Made from recycled materials	G									
2100	Open type B wide rib, galvanized, 1-1/2" deep, 22 gauge, under 50 sq	G	E-4	4500	.007	S.F.	1.29	.34	.03	1.66	2.05
2200	50-500 squares *CN*	G		4900	.007		1	.31	.03	1.34	1.67
2400	Over 500 squares	G		5100	.006		.93	.30	.03	1.26	1.57
2600	20 gauge, under 50 squares	G		3865	.008		1.52	.39	.04	1.95	2.40
2650	50-500 squares	G		4170	.008		1.22	.36	.04	1.62	2.02
2700	Over 500 squares	G		4300	.007		1.09	.35	.03	1.47	1.86
2900	18 gauge, under 50 squares	G		3800	.008		1.96	.40	.04	2.40	2.89
2950	50-500 squares	G		4100	.008		1.57	.37	.04	1.98	2.41
3000	Over 500 squares	G		4300	.007		1.41	.35	.03	1.79	2.21
3050	16 gauge, under 50 squares	G		3700	.009		2.64	.41	.04	3.09	3.66
3060	50-500 squares	G		4000	.008		2.11	.38	.04	2.53	3.03

05 31 Steel Decking

05 31 23 – Steel Roof Decking

05 31 23.50 Roof Decking			Crew	Daily Output	Labor-Hours	Unit	Material	Labor	Equipment	Total	Total Incl O&P
							2010 Bare Costs				
3100	Over 500 squares	G	E-4	4200	.008	S.F.	1.90	.36	.03	2.29	2.77
3200	3" deep, 22 gauge, under 50 squares	G		3600	.009		1.91	.42	.04	2.37	2.88
3250	50-500 squares	G		3800	.008		1.52	.40	.04	1.96	2.42
3260	over 500 squares	G		4000	.008		1.37	.38	.04	1.79	2.22
3300	20 gauge, under 50 squares	G		3400	.009		2.06	.45	.04	2.55	3.10
3350	50-500 squares	G		3600	.009		1.65	.42	.04	2.11	2.59
3360	over 500 squares	G		3800	.008		1.48	.40	.04	1.92	2.37
3400	18 gauge, under 50 squares	G		3200	.010		2.67	.47	.05	3.19	3.81
3450	50-500 squares	G		3400	.009		2.13	.45	.04	2.62	3.18
3460	over 500 squares	G		3600	.009		1.92	.42	.04	2.38	2.89
3500	16 gauge, under 50 squares	G		3000	.011		3.51	.51	.05	4.07	4.81
3550	50-500 squares	G		3200	.010		2.81	.47	.05	3.33	3.97
3560	over 500 squares	G		3400	.009		2.53	.45	.04	3.02	3.61
3700	4-1/2" deep, 20 gauge, over 50 squares	G		2700	.012		3.31	.56	.05	3.92	4.69
3800	18 gauge	G		2460	.013		4.26	.62	.06	4.94	5.85
3900	16 gauge	G		2350	.014		3.19	.65	.06	3.90	4.71
4100	6" deep, 18 gauge, over 50 squares	G		2000	.016		6.10	.76	.07	6.93	8.10
4200	16 gauge	G		1930	.017		4.55	.79	.08	5.42	6.45
4300	14 gauge	G		1860	.017		5.85	.82	.08	6.75	7.95
4500	7-1/2" deep, 18 gauge, over 50 squares	G		1690	.019		6.70	.90	.09	7.69	9
4600	16 gauge	G		1590	.020		5	.95	.09	6.04	7.30
4700	14 gauge	G	↓	1490	.021	↓	6.45	1.02	.10	7.57	9
4800	For painted instead of galvanized, deduct						5%				
5000	For acoustical perforated with fiberglass insulation, add					S.F.	25%				
5100	For type F intermediate rib instead of type B wide rib, add	G					25%				
5150	For type A narrow rib instead of type B wide rib, add	G				↓	25%				

05 31 33 – Steel Form Decking

05 31 33.50 Form Decking			Crew	Daily Output	Labor-Hours	Unit	Material	Labor	Equipment	Total	Total Incl O&P
0010	**FORM DECKING**										
0015	Made from recycled materials	G									
6100	Slab form, steel, 28 gauge, 9/16" deep, uncoated	G	E-4	4000	.008	S.F.	.98	.38	.04	1.40	1.79
6200	Galvanized	G		4000	.008		.87	.38	.04	1.29	1.67
6220	24 gauge, 1" deep, uncoated	G		3900	.008		.94	.39	.04	1.37	1.76
6240	Galvanized	G		3900	.008		1.11	.39	.04	1.54	1.94
6300	24 gauge, 1-5/16" deep, uncoated	G		3800	.008		1	.40	.04	1.44	1.84
6400	Galvanized	G		3800	.008		1.18	.40	.04	1.62	2.04
6500	22 gauge, 1-5/16" deep, uncoated	G		3700	.009		1.25	.41	.04	1.70	2.14
6600	Galvanized	G		3700	.009		1.28	.41	.04	1.73	2.17
6700	22 gauge, 2" deep uncoated	G		3600	.009		1.65	.42	.04	2.11	2.60
6800	Galvanized	G	↓	3600	.009	↓	1.62	.42	.04	2.08	2.56
7000	Sheet metal edge closure form, 12" wide with 2 bends, galv										
7100	18 gauge	G	E-14	360	.022	L.F.	2.66	1.09	.41	4.16	5.30
7200	16 gauge	G	"	360	.022	"	3.60	1.09	.41	5.10	6.30

05 35 Raceway Decking Assemblies

05 35 13 – Steel Cellular Decking

05 35 13.50 Cellular Decking			Crew	Daily Output	Labor-Hours	Unit	Material	2010 Bare Costs Labor	Equipment	Total	Total Incl O&P
0010	**CELLULAR DECKING**										
0015	Made from recycled materials	G									
0200	Cellular units, galv, 2" deep, 20-20 gauge, over 15 squares	G	E-4	1460	.022	S.F.	5.10	1.04	.10	6.24	7.55
0250	18-20 gauge	G		1420	.023		5.80	1.07	.10	6.97	8.40
0300	18-18 gauge	G		1390	.023		5.95	1.09	.11	7.15	8.60
0320	16-18 gauge	G		1360	.024		7.10	1.12	.11	8.33	9.90
0340	16-16 gauge	G		1330	.024		7.90	1.14	.11	9.15	10.80
0400	3" deep, galvanized, 20-20 gauge	G		1375	.023		5.60	1.10	.11	6.81	8.25
0500	18-20 gauge	G		1350	.024		6.80	1.12	.11	8.03	9.60
0600	18-18 gauge	G		1290	.025		6.80	1.18	.11	8.09	9.65
0700	16-18 gauge	G		1230	.026		7.65	1.23	.12	9	10.70
0800	16-16 gauge	G		1150	.028		8.35	1.32	.13	9.80	11.60
1000	4-1/2" deep, galvanized, 20-18 gauge	G		1100	.029		7.85	1.38	.13	9.36	11.20
1100	18-18 gauge	G		1040	.031		7.80	1.46	.14	9.40	11.30
1200	16-18 gauge	G		980	.033		8.80	1.55	.15	10.50	12.55
1300	16-16 gauge	G		935	.034		9.60	1.62	.16	11.38	13.55
1500	For acoustical deck, add						15%				
1700	For cells used for ventilation, add						15%				
1900	For multi-story or congested site, add							50%			
8000	Metal deck and trench, 2" thick, 20 gauge, combination										
8010	60% cellular, 40% non-cellular, inserts and trench	G	R-4	1100	.036	S.F.	10.20	1.74	.13	12.07	14.35

05 41 Structural Metal Stud Framing

05 41 13 – Load-Bearing Metal Stud Framing

05 41 13.05 Bracing

05 41 13.05 Bracing			Crew	Daily Output	Labor-Hours	Unit	Material	Labor	Equipment	Total	Total Incl O&P
0010	**BRACING**, shear wall X-bracing, per 10' x 10' bay, one face										
0015	Made of recycled materials	G									
0120	Metal strap, 20 ga x 4" wide	G	2 Carp	18	.889	Ea.	19.35	37		56.35	78.50
0130	6" wide	G		18	.889		30	37		67	90
0160	18 ga x 4" wide	G		16	1		28	41.50		69.50	94.50
0170	6" wide	G		16	1		41	41.50		82.50	109
0410	Continuous strap bracing, per horizontal row on both faces										
0420	Metal strap, 20 ga x 2" wide, studs 12" O.C.	G	1 Carp	7	1.143	C.L.F.	48.50	47.50		96	126
0430	16" O.C.	G		8	1		48.50	41.50		90	117
0440	24" O.C.	G		10	.800		48.50	33		81.50	104
0450	18 ga x 2" wide, studs 12" O.C.	G		6	1.333		68	55.50		123.50	161
0460	16" O.C.	G		7	1.143		68	47.50		115.50	148
0470	24" O.C.	G		8	1		68	41.50		109.50	139

05 41 13.10 Bridging

05 41 13.10 Bridging			Crew	Daily Output	Labor-Hours	Unit	Material	Labor	Equipment	Total	Total Incl O&P
0010	**BRIDGING**, solid between studs w/ 1-1/4" leg track, per stud bay										
0015	Made from recycled materials	G									
0200	Studs 12" O.C., 18 ga x 2-1/2" wide	G	1 Carp	125	.064	Ea.	.82	2.66		3.48	5
0210	3-5/8" wide	G		120	.067		1	2.77		3.77	5.35
0220	4" wide	G		120	.067		1.07	2.77		3.84	5.45
0230	6" wide	G		115	.070		1.38	2.89		4.27	5.95
0240	8" wide	G		110	.073		1.75	3.02		4.77	6.60
0300	16 ga x 2-1/2" wide	G		115	.070		1.03	2.89		3.92	5.60
0310	3-5/8" wide	G		110	.073		1.26	3.02		4.28	6.05
0320	4" wide	G		110	.073		1.34	3.02		4.36	6.15
0330	6" wide	G		105	.076		1.72	3.17		4.89	6.80
0340	8" wide	G		100	.080		2.20	3.32		5.52	7.50

05 41 13 – Load-Bearing Metal Stud Framing

05 41 13.10 Bridging

	05 41 13.10 Bridging		Crew	Daily Output	Labor-Hours	Unit	Material	Labor	Equipment	Total	Total Incl O&P
							2010 Bare Costs				
1200	Studs 16" O.C., 18 ga x 2-1/2" wide	G	1 Carp	125	.064	Ea.	1.06	2.66		3.72	5.25
1210	3-5/8" wide	G		120	.067		1.29	2.77		4.06	5.70
1220	4" wide	G		120	.067		1.37	2.77		4.14	5.80
1230	6" wide	G		115	.070		1.77	2.89		4.66	6.40
1240	8" wide	G		110	.073		2.24	3.02		5.26	7.15
1300	16 ga x 2-1/2" wide	G		115	.070		1.32	2.89		4.21	5.90
1310	3-5/8" wide	G		110	.073		1.62	3.02		4.64	6.45
1320	4" wide	G		110	.073		1.72	3.02		4.74	6.55
1330	6" wide	G		105	.076		2.21	3.17		5.38	7.30
1340	8" wide	G		100	.080		2.82	3.32		6.14	8.20
2200	Studs 24" O.C., 18 ga x 2-1/2" wide	G		125	.064		1.53	2.66		4.19	5.80
2210	3-5/8" wide	G		120	.067		1.86	2.77		4.63	6.30
2220	4" wide	G		120	.067		1.98	2.77		4.75	6.45
2230	6" wide	G		115	.070		2.55	2.89		5.44	7.25
2240	8" wide	G		110	.073		3.25	3.02		6.27	8.25
2300	16 ga x 2-1/2" wide	G		115	.070		1.91	2.89		4.80	6.55
2310	3-5/8" wide	G		110	.073		2.34	3.02		5.36	7.25
2320	4" wide	G		110	.073		2.48	3.02		5.50	7.40
2330	6" wide	G		105	.076		3.20	3.17		6.37	8.40
2340	8" wide	G	↓	100	.080	↓	4.08	3.32		7.40	9.60
3000	Continuous bridging, per row										
3100	16 ga x 1-1/2" channel thru studs 12" O.C.	G	1 Carp	6	1.333	C.L.F.	45.50	55.50		101	136
3110	16" O.C.	G		7	1.143		45.50	47.50		93	123
3120	24" O.C.	G		8.80	.909		45.50	38		83.50	108
4100	2" x 2" angle x 18 ga, studs 12" O.C.	G		7	1.143		68	47.50		115.50	148
4110	16" O.C.	G		9	.889		68	37		105	132
4120	24" O.C.	G		12	.667		68	27.50		95.50	118
4200	16 ga, studs 12" O.C.	G		5	1.600		86	66.50		152.50	197
4210	16" O.C.	G		7	1.143		86	47.50		133.50	168
4220	24" O.C.	G	↓	10	.800	↓	86	33		119	146

05 41 13.25 Framing, Boxed Headers/Beams

	05 41 13.25 Framing, Boxed Headers/Beams		Crew	Daily Output	Labor-Hours	Unit	Material	Labor	Equipment	Total	Total Incl O&P
0010	**FRAMING, BOXED HEADERS/BEAMS**										
0015	Made from recycled materials	G									
0200	Double, 18 ga x 6" deep	G	2 Carp	220	.073	L.F.	4.81	3.02		7.83	9.95
0210	8" deep	G		210	.076		5.35	3.17		8.52	10.80
0220	10" deep	G		200	.080		6.45	3.32		9.77	12.20
0230	12" deep	G		190	.084		7.10	3.50		10.60	13.20
0300	16 ga x 8" deep	G		180	.089		6.15	3.69		9.84	12.45
0310	10" deep	G		170	.094		7.35	3.91		11.26	14.10
0320	12" deep	G		160	.100		8	4.16		12.16	15.20
0400	14 ga x 10" deep	G		140	.114		8.55	4.75		13.30	16.70
0410	12" deep	G		130	.123		9.40	5.10		14.50	18.20
1210	Triple, 18 ga x 8" deep	G		170	.094		7.70	3.91		11.61	14.55
1220	10" deep	G		165	.097		9.25	4.03		13.28	16.40
1230	12" deep	G		160	.100		10.20	4.16		14.36	17.65
1300	16 ga x 8" deep	G		145	.110		8.90	4.58		13.48	16.85
1310	10" deep	G		140	.114		10.60	4.75		15.35	18.95
1320	12" deep	G		135	.119		11.60	4.92		16.52	20.50
1400	14 ga x 10" deep	G		115	.139		11.65	5.80		17.45	22
1410	12" deep	G	↓	110	.145	↓	12.90	6.05		18.95	23.50

05 41 13 – Load-Bearing Metal Stud Framing

05 41 13.30 Framing, Stud Walls		Crew	Daily Output	Labor-Hours	Unit	Material	Labor	Equipment	Total	Total Incl O&P
0010	**FRAMING, STUD WALLS** w/ top & bottom track, no openings,									
0020	Headers, beams, bridging or bracing									
0025	Made from recycled materials G									
4100	8' high walls, 18 ga x 2-1/2" wide, studs 12" O.C. G	2 Carp	54	.296	L.F.	8.05	12.30		20.35	28
4110	16" O.C. G		77	.208		6.45	8.65		15.10	20.50
4120	24" O.C. G		107	.150		4.81	6.20		11.01	14.90
4130	3-5/8" wide, studs 12" O.C. G		53	.302		9.60	12.55		22.15	30
4140	16" O.C. G		76	.211		7.65	8.75		16.40	22
4150	24" O.C. G		105	.152		5.75	6.35		12.10	16.10
4160	4" wide, studs 12" O.C. G		52	.308		10.05	12.80		22.85	31
4170	16" O.C. G		74	.216		8.05	9		17.05	22.50
4180	24" O.C. G		103	.155		6.05	6.45		12.50	16.60
4190	6" wide, studs 12" O.C. G		51	.314		12.70	13.05		25.75	34
4200	16" O.C. G		73	.219		10.20	9.10		19.30	25.50
4210	24" O.C. G		101	.158		7.65	6.60		14.25	18.60
4220	8" wide, studs 12" O.C. G		50	.320		15.55	13.30		28.85	37.50
4230	16" O.C. G		72	.222		12.50	9.25		21.75	28
4240	24" O.C. G		100	.160		9.45	6.65		16.10	20.50
4300	16 ga x 2-1/2" wide, studs 12" O.C. G		47	.340		9.50	14.15		23.65	32.50
4310	16" O.C. G		68	.235		7.50	9.80		17.30	23.50
4320	24" O.C. G		94	.170		5.55	7.05		12.60	17
4330	3-5/8" wide, studs 12" O.C. G		46	.348		11.35	14.45		25.80	35
4340	16" O.C. G		66	.242		9	10.05		19.05	25.50
4350	24" O.C. G		92	.174		6.65	7.25		13.90	18.45
4360	4" wide, studs 12" O.C. G		45	.356		11.95	14.75		26.70	36
4370	16" O.C. G		65	.246		9.45	10.25		19.70	26
4380	24" O.C. G		90	.178		7	7.40		14.40	19.10
4390	6" wide, studs 12" O.C. G		44	.364		15	15.10		30.10	40
4400	16" O.C. G		64	.250		11.90	10.40		22.30	29
4410	24" O.C. G		88	.182		8.80	7.55		16.35	21.50
4420	8" wide, studs 12" O.C. G		43	.372		18.55	15.45		34	44.50
4430	16" O.C. G		63	.254		14.75	10.55		25.30	32.50
4440	24" O.C. G		86	.186		10.95	7.75		18.70	24
5100	10' high walls, 18 ga x 2-1/2" wide, studs 12" O.C. G		54	.296		9.65	12.30		21.95	29.50
5110	16" O.C. G		77	.208		7.65	8.65		16.30	21.50
5120	24" O.C. G		107	.150		5.60	6.20		11.80	15.80
5130	3-5/8" wide, studs 12" O.C. G		53	.302		11.50	12.55		24.05	32
5140	16" O.C. G		76	.211		9.10	8.75		17.85	23.50
5150	24" O.C. G		105	.152		6.70	6.35		13.05	17.15
5160	4" wide, studs 12" O.C. G		52	.308		12.05	12.80		24.85	33
5170	16" O.C. G		74	.216		9.55	9		18.55	24.50
5180	24" O.C. G		103	.155		7.05	6.45		13.50	17.70
5190	6" wide, studs 12" O.C. G		51	.314		15.20	13.05		28.25	37
5200	16" O.C. G		73	.219		12.05	9.10		21.15	27.50
5210	24" O.C. G		101	.158		8.90	6.60		15.50	19.95
5220	8" wide, studs 12" O.C. G		50	.320		18.65	13.30		31.95	41
5230	16" O.C. G		72	.222		14.80	9.25		24.05	30.50
5240	24" O.C. G		100	.160		11	6.65		17.65	22.50
5300	16 ga x 2-1/2" wide, studs 12" O.C. G		47	.340		11.45	14.15		25.60	34.50
5310	16" O.C. G		68	.235		9	9.80		18.80	25
5320	24" O.C. G		94	.170		6.50	7.05		13.55	18.05
5330	3-5/8" wide, studs 12" O.C. G	↓	46	.348	↓	13.70	14.45		28.15	37.50

Material, Labor, Equipment and Total are 2010 Bare Costs.

05 41 13 – Load-Bearing Metal Stud Framing

05 41 13.30 Framing, Stud Walls			Crew	Daily Output	Labor-Hours	Unit	Material	Labor	Equipment	Total	Total Incl O&P
							2010 Bare Costs				
5340	16" O.C.	G	2 Carp	66	.242	L.F.	10.75	10.05		20.80	27.50
5350	24" O.C.	G		92	.174		7.80	7.25		15.05	19.75
5360	4" wide, studs 12" O.C.	G		45	.356		14.45	14.75		29.20	39
5370	16" O.C.	G		65	.246		11.35	10.25		21.60	28
5380	24" O.C.	G		90	.178		8.25	7.40		15.65	20.50
5390	6" wide, studs 12" O.C.	G		44	.364		18.10	15.10		33.20	43.50
5400	16" O.C.	G		64	.250		14.25	10.40		24.65	31.50
5410	24" O.C.	G		88	.182		10.35	7.55		17.90	23
5420	8" wide, studs 12" O.C.	G		43	.372		22.50	15.45		37.95	48.50
5430	16" O.C.	G		63	.254		17.60	10.55		28.15	35.50
5440	24" O.C.	G		86	.186		12.85	7.75		20.60	26
6190	12' high walls, 18 ga x 6" wide, studs 12" O.C.	G		41	.390		17.75	16.20		33.95	44.50
6200	16" O.C.	G		58	.276		13.95	11.45		25.40	33
6210	24" O.C.	G		81	.198		10.20	8.20		18.40	24
6220	8" wide, studs 12" O.C.	G		40	.400		21.50	16.60		38.10	49.50
6230	16" O.C.	G		57	.281		17.10	11.65		28.75	37
6240	24" O.C.	G		80	.200		12.50	8.30		20.80	26.50
6390	16 ga x 6" wide, studs 12" O.C.	G		35	.457		21	19		40	53
6400	16" O.C.	G		51	.314		16.55	13.05		29.60	38.50
6410	24" O.C.	G		70	.229		11.90	9.50		21.40	28
6420	8" wide, studs 12" O.C.	G		34	.471		26	19.55		45.55	59
6430	16" O.C.	G		50	.320		20.50	13.30		33.80	43
6440	24" O.C.	G		69	.232		14.75	9.65		24.40	31
6530	14 ga x 3-5/8" wide, studs 12" O.C.	G		34	.471		20	19.55		39.55	52
6540	16" O.C.	G		48	.333		15.70	13.85		29.55	39
6550	24" O.C.	G		65	.246		11.30	10.25		21.55	28
6560	4" wide, studs 12" O.C.	G		33	.485		21.50	20		41.50	54.50
6570	16" O.C.	G		47	.340		16.60	14.15		30.75	40.50
6580	24" O.C.	G		64	.250		11.90	10.40		22.30	29
6730	12 ga x 3-5/8" wide, studs 12" O.C.	G		31	.516		28	21.50		49.50	63.50
6740	16" O.C.	G		43	.372		21.50	15.45		36.95	47.50
6750	24" O.C.	G		59	.271		15.10	11.25		26.35	34
6760	4" wide, studs 12" O.C.	G		30	.533		30	22		52	67
6770	16" O.C.	G		42	.381		23	15.85		38.85	50
6780	24" O.C.	G		58	.276		16.25	11.45		27.70	35.50
7390	16' high walls, 16 ga x 6" wide, studs 12" O.C.	G		33	.485		27.50	20		47.50	61
7400	16" O.C.	G		48	.333		21	13.85		34.85	45
7410	24" O.C.	G		67	.239		15	9.90		24.90	32
7420	8" wide, studs 12" O.C.	G		32	.500		33.50	21		54.50	69
7430	16" O.C.	G		47	.340		26	14.15		40.15	51
7440	24" O.C.	G		66	.242		18.55	10.05		28.60	36
7560	14 ga x 4" wide, studs 12" O.C.	G		31	.516		27.50	21.50		49	63.50
7570	16" O.C.	G		45	.356		21.50	14.75		36.25	46.50
7580	24" O.C.	G		61	.262		15.05	10.90		25.95	33.50
7590	6" wide, studs 12" O.C.	G		30	.533		34.50	22		56.50	72
7600	16" O.C.	G		44	.364		27	15.10		42.10	53
7610	24" O.C.	G		60	.267		18.95	11.10		30.05	38
7760	12 ga x 4" wide, studs 12" O.C.	G		29	.552		39	23		62	78.50
7770	16" O.C.	G		40	.400		30	16.60		46.60	58.50
7780	24" O.C.	G		55	.291		21	12.10		33.10	41.50
7790	6" wide, studs 12" O.C.	G		28	.571		49	23.50		72.50	90.50
7800	16" O.C.	G		39	.410		37.50	17.05		54.55	68
7810	24" O.C.	G		54	.296		26	12.30		38.30	48

05 41 Structural Metal Stud Framing

05 41 13 – Load-Bearing Metal Stud Framing

05 41 13.30 Framing, Stud Walls			Crew	Daily Output	Labor-Hours	Unit	Material	2010 Bare Costs Labor	Equipment	Total	Total Incl O&P
8590	20′ high walls, 14 ga x 6″ wide, studs 12″ O.C.	G	2 Carp	29	.552	L.F.	42.50	23		65.50	82
8600	16″ O.C.	G		42	.381		32.50	15.85		48.35	60.50
8610	24″ O.C.	G		57	.281		23	11.65		34.65	43
8620	8″ wide, studs 12″ O.C.	G		28	.571		52	23.50		75.50	93.50
8630	16″ O.C.	G		41	.390		40	16.20		56.20	69
8640	24″ O.C.	G		56	.286		28	11.85		39.85	49.50
8790	12 ga x 6″ wide, studs 12″ O.C.	G		27	.593		60.50	24.50		85	105
8800	16″ O.C.	G		37	.432		46.50	17.95		64.45	78.50
8810	24″ O.C.	G		51	.314		32	13.05		45.05	55
8820	8″ wide, studs 12″ O.C.	G		26	.615		74	25.50		99.50	121
8830	16″ O.C.	G		36	.444		56.50	18.45		74.95	90.50
8840	24″ O.C.	G	↓	50	.320	↓	39	13.30		52.30	63.50

05 42 Cold-Formed Metal Joist Framing

05 42 13 – Cold-Formed Metal Floor Joist Framing

05 42 13.05 Bracing

			Crew	Daily Output	Labor-Hours	Unit	Material	Labor	Equipment	Total	Total Incl O&P
0010	**BRACING**, continuous, per row, top & bottom										
0015	Made from recycled materials	G									
0120	Flat strap, 20 ga x 2″ wide, joists at 12″ O.C.	G	1 Carp	4.67	1.713	C.L.F.	50.50	71		121.50	166
0130	16″ O.C.	G		5.33	1.501		49	62.50		111.50	150
0140	24″ O.C.	G		6.66	1.201		47	50		97	129
0150	18 ga x 2″ wide, joists at 12″ O.C.	G		4	2		67.50	83		150.50	203
0160	16″ O.C.	G		4.67	1.713		66.50	71		137.50	183
0170	24″ O.C.	G	↓	5.33	1.501	↓	65	62.50		127.50	168

05 42 13.10 Bridging

			Crew	Daily Output	Labor-Hours	Unit	Material	Labor	Equipment	Total	Total Incl O&P
0010	**BRIDGING**, solid between joists w/ 1-1/4″ leg track, per joist bay										
0015	Made from recycled materials	G									
0230	Joists 12″ O.C., 18 ga track x 6″ wide	G	1 Carp	80	.100	Ea.	1.38	4.16		5.54	7.90
0240	8″ wide	G		75	.107		1.75	4.43		6.18	8.80
0250	10″ wide	G		70	.114		2.16	4.75		6.91	9.70
0260	12″ wide	G		65	.123		2.50	5.10		7.60	10.65
0330	16 ga track x 6″ wide	G		70	.114		1.72	4.75		6.47	9.20
0340	8″ wide	G		65	.123		2.20	5.10		7.30	10.30
0350	10″ wide	G		60	.133		2.70	5.55		8.25	11.50
0360	12″ wide	G		55	.145		3.11	6.05		9.16	12.75
0440	14 ga track x 8″ wide	G		60	.133		2.78	5.55		8.33	11.60
0450	10″ wide	G		55	.145		3.41	6.05		9.46	13.05
0460	12″ wide	G		50	.160		3.94	6.65		10.59	14.60
0550	12 ga track x 10″ wide	G		45	.178		5	7.40		12.40	16.90
0560	12″ wide	G		40	.200		5.65	8.30		13.95	19.05
1230	16″ O.C., 18 ga track x 6″ wide	G		80	.100		1.77	4.16		5.93	8.35
1240	8″ wide	G		75	.107		2.24	4.43		6.67	9.30
1250	10″ wide	G		70	.114		2.77	4.75		7.52	10.35
1260	12″ wide	G		65	.123		3.20	5.10		8.30	11.40
1330	16 ga track x 6″ wide	G		70	.114		2.21	4.75		6.96	9.75
1340	8″ wide	G		65	.123		2.82	5.10		7.92	11
1350	10″ wide	G		60	.133		3.47	5.55		9.02	12.35
1360	12″ wide	G		55	.145		3.99	6.05		10.04	13.70
1440	14 ga track x 8″ wide	G		60	.133		3.56	5.55		9.11	12.45
1450	10″ wide	G		55	.145		4.37	6.05		10.42	14.10
1460	12″ wide	G	↓	50	.160	↓	5.05	6.65		11.70	15.80

05 42 Cold-Formed Metal Joist Framing

05 42 13 – Cold-Formed Metal Floor Joist Framing

05 42 13.10 Bridging		Crew	Daily Output	Labor-Hours	Unit	Material	Labor	Equipment	Total	Total Incl O&P
						2010 Bare Costs				
1550 12 ga track x 10" wide	G	1 Carp	45	.178	Ea.	6.40	7.40		13.80	18.45
1560 12" wide	G		40	.200		7.25	8.30		15.55	21
2230 24" O.C., 18 ga track x 6" wide	G		80	.100		2.55	4.16		6.71	9.20
2240 8" wide	G		75	.107		3.25	4.43		7.68	10.40
2250 10" wide	G		70	.114		4.01	4.75		8.76	11.70
2260 12" wide	G		65	.123		4.63	5.10		9.73	13
2330 16 ga track x 6" wide	G		70	.114		3.20	4.75		7.95	10.80
2340 8" wide	G		65	.123		4.08	5.10		9.18	12.40
2350 10" wide	G		60	.133		5	5.55		10.55	14.05
2360 12" wide	G		55	.145		5.80	6.05		11.85	15.65
2440 14 ga track x 8" wide	G		60	.133		5.15	5.55		10.70	14.20
2450 10" wide	G		55	.145		6.35	6.05		12.40	16.25
2460 12" wide	G		50	.160		7.30	6.65		13.95	18.30
2550 12 ga track x 10" wide	G		45	.178		9.30	7.40		16.70	21.50
2560 12" wide	G	↓	40	.200	↓	10.50	8.30		18.80	24.50

05 42 13.25 Framing, Band Joist		Crew	Daily Output	Labor-Hours	Unit	Material	Labor	Equipment	Total	Total Incl O&P
0010 **FRAMING, BAND JOIST** (track) fastened to bearing wall										
0015 Made from recycled materials	G									
0220 18 ga track x 6" deep	G	2 Carp	1000	.016	L.F.	1.12	.66		1.78	2.26
0230 8" deep	G		920	.017		1.43	.72		2.15	2.68
0240 10" deep	G		860	.019		1.76	.77		2.53	3.13
0320 16 ga track x 6" deep	G		900	.018		1.41	.74		2.15	2.69
0330 8" deep	G		840	.019		1.80	.79		2.59	3.20
0340 10" deep	G		780	.021		2.21	.85		3.06	3.74
0350 12" deep	G		740	.022		2.54	.90		3.44	4.18
0430 14 ga track x 8" deep	G		750	.021		2.27	.89		3.16	3.86
0440 10" deep	G		720	.022		2.78	.92		3.70	4.48
0450 12" deep	G		700	.023		3.21	.95		4.16	4.99
0540 12 ga track x 10" deep	G		670	.024		4.08	.99		5.07	6
0550 12" deep	G	↓	650	.025	↓	4.62	1.02		5.64	6.70

05 42 13.30 Framing, Boxed Headers/Beams		Crew	Daily Output	Labor-Hours	Unit	Material	Labor	Equipment	Total	Total Incl O&P
0010 **FRAMING, BOXED HEADERS/BEAMS**										
0015 Made from recycled materials	G									
0200 Double, 18 ga x 6" deep	G	2 Carp	220	.073	L.F.	4.81	3.02		7.83	9.95
0210 8" deep	G		210	.076		5.35	3.17		8.52	10.80
0220 10" deep	G		200	.080		6.45	3.32		9.77	12.20
0230 12" deep	G		190	.084		7.10	3.50		10.60	13.20
0300 16 ga x 8" deep	G		180	.089		6.15	3.69		9.84	12.45
0310 10" deep	G		170	.094		7.35	3.91		11.26	14.10
0320 12" deep	G		160	.100		8	4.16		12.16	15.20
0400 14 ga x 10" deep	G		140	.114		8.55	4.75		13.30	16.70
0410 12" deep	G		130	.123		9.40	5.10		14.50	18.20
0500 12 ga x 10" deep	G		110	.145		11.30	6.05		17.35	22
0510 12" deep	G		100	.160		12.50	6.65		19.15	24
1210 Triple, 18 ga x 8" deep	G		170	.094		7.70	3.91		11.61	14.55
1220 10" deep	G		165	.097		9.25	4.03		13.28	16.40
1230 12" deep	G		160	.100		10.20	4.16		14.36	17.65
1300 16 ga x 8" deep	G		145	.110		8.90	4.58		13.48	16.85
1310 10" deep	G		140	.114		10.60	4.75		15.35	18.95
1320 12" deep	G		135	.119		11.60	4.92		16.52	20.50
1400 14 ga x 10" deep	G		115	.139		12.40	5.80		18.20	22.50
1410 12" deep	G	↓	110	.145	↓	13.65	6.05		19.70	24.50

05 42 Cold-Formed Metal Joist Framing

05 42 13 – Cold-Formed Metal Floor Joist Framing

05 42 13.30 Framing, Boxed Headers/Beams		Crew	Daily Output	Labor-Hours	Unit	Material	2010 Bare Costs Labor	Equipment	Total	Total Incl O&P
1500	12 ga x 10" deep G	2 Carp	90	.178	L.F.	16.55	7.40		23.95	29.50
1510	12" deep G		85	.188		18.40	7.80		26.20	32
05 42 13.40 Framing, Joists										
0010	**FRAMING, JOISTS**, no band joists (track), web stiffeners, headers,									
0020	Beams, bridging or bracing									
0025	Made from recycled materials G									
0030	Joists (2" flange) and fasteners, materials only									
0220	18 ga x 6" deep G				L.F.	1.48			1.48	1.63
0230	8" deep G					1.76			1.76	1.94
0240	10" deep G					2.07			2.07	2.28
0320	16 ga x 6" deep G					1.82			1.82	2
0330	8" deep G					2.18			2.18	2.40
0340	10" deep G					2.54			2.54	2.80
0350	12" deep G					2.89			2.89	3.18
0430	14 ga x 8" deep G					2.75			2.75	3.03
0440	10" deep G					3.17			3.17	3.49
0450	12" deep G					3.61			3.61	3.97
0540	12 ga x 10" deep G					4.62			4.62	5.10
0550	12" deep G					5.25			5.25	5.80
1010	Installation of joists to band joists, beams & headers, labor only									
1220	18 ga x 6" deep	2 Carp	110	.145	Ea.		6.05		6.05	9.30
1230	8" deep		90	.178			7.40		7.40	11.40
1240	10" deep		80	.200			8.30		8.30	12.80
1320	16 ga x 6" deep		95	.168			7		7	10.80
1330	8" deep		70	.229			9.50		9.50	14.65
1340	10" deep		60	.267			11.10		11.10	17.10
1350	12" deep		55	.291			12.10		12.10	18.65
1430	14 ga x 8" deep		65	.246			10.25		10.25	15.75
1440	10" deep		45	.356			14.75		14.75	23
1450	12" deep		35	.457			19		19	29.50
1540	12 ga x 10" deep		40	.400			16.60		16.60	25.50
1550	12" deep		30	.533			22		22	34
05 42 13.45 Framing, Web Stiffeners										
0010	**FRAMING, WEB STIFFENERS** at joist bearing, fabricated from									
0020	Stud piece (1-5/8" flange) to stiffen joist (2" flange)									
0025	Made from recycled materials G									
2120	For 6" deep joist, with 18 ga x 2-1/2" stud G	1 Carp	120	.067	Ea.	1.78	2.77		4.55	6.25
2130	3-5/8" stud G		110	.073		1.97	3.02		4.99	6.80
2140	4" stud G		105	.076		1.90	3.17		5.07	6.95
2150	6" stud G		100	.080		2.08	3.32		5.40	7.40
2160	8" stud G		95	.084		2.14	3.50		5.64	7.75
2220	8" deep joist, with 2-1/2" stud G		120	.067		1.95	2.77		4.72	6.40
2230	3-5/8" stud G		110	.073		2.12	3.02		5.14	7
2240	4" stud G		105	.076		2.08	3.17		5.25	7.15
2250	6" stud G		100	.080		2.28	3.32		5.60	7.60
2260	8" stud G		95	.084		2.46	3.50		5.96	8.10
2320	10" deep joist, with 2-1/2" stud G		110	.073		2.76	3.02		5.78	7.70
2330	3-5/8" stud G		100	.080		3.03	3.32		6.35	8.45
2340	4" stud G		95	.084		2.99	3.50		6.49	8.70
2350	6" stud G		90	.089		3.24	3.69		6.93	9.25
2360	8" stud G		85	.094		3.30	3.91		7.21	9.70
2420	12" deep joist, with 2-1/2" stud G		110	.073		2.92	3.02		5.94	7.85

05 42 Cold-Formed Metal Joist Framing

05 42 13 – Cold-Formed Metal Floor Joist Framing

05 42 13.45 Framing, Web Stiffeners			Crew	Daily Output	Labor-Hours	Unit	Material	2010 Bare Costs Labor	Equipment	Total	Total Incl O&P
2430	3-5/8″ stud	G	1 Carp	100	.080	Ea.	3.17	3.32		6.49	8.60
2440	4″ stud	G		95	.084		3.10	3.50		6.60	8.80
2450	6″ stud	G		90	.089		3.40	3.69		7.09	9.45
2460	8″ stud	G		85	.094		3.67	3.91		7.58	10.10
3130	For 6″ deep joist, with 16 ga x 3-5/8″ stud	G		100	.080		2.07	3.32		5.39	7.35
3140	4″ stud	G		95	.084		2.05	3.50		5.55	7.65
3150	6″ stud	G		90	.089		2.25	3.69		5.94	8.15
3160	8″ stud	G		85	.094		2.38	3.91		6.29	8.65
3230	8″ deep joist, with 3-5/8″ stud	G		100	.080		2.29	3.32		5.61	7.60
3240	4″ stud	G		95	.084		2.24	3.50		5.74	7.85
3250	6″ stud	G		90	.089		2.49	3.69		6.18	8.45
3260	8″ stud	G		85	.094		2.67	3.91		6.58	9
3330	10″ deep joist, with 3-5/8″ stud	G		85	.094		3.13	3.91		7.04	9.50
3340	4″ stud	G		80	.100		3.19	4.16		7.35	9.90
3350	6″ stud	G		75	.107		3.47	4.43		7.90	10.65
3360	8″ stud	G		70	.114		3.63	4.75		8.38	11.30
3430	12″ deep joist, with 3-5/8″ stud	G		85	.094		3.42	3.91		7.33	9.80
3440	4″ stud	G		80	.100		3.35	4.16		7.51	10.10
3450	6″ stud	G		75	.107		3.72	4.43		8.15	10.95
3460	8″ stud	G		70	.114		3.99	4.75		8.74	11.70
4230	For 8″ deep joist, with 14 ga x 3-5/8″ stud	G		90	.089		2.97	3.69		6.66	8.95
4240	4″ stud	G		85	.094		3.03	3.91		6.94	9.40
4250	6″ stud	G		80	.100		3.28	4.16		7.44	10
4260	8″ stud	G		75	.107		3.52	4.43		7.95	10.75
4330	10″ deep joist, with 3-5/8″ stud	G		75	.107		4.18	4.43		8.61	11.45
4340	4″ stud	G		70	.114		4.14	4.75		8.89	11.85
4350	6″ stud	G		65	.123		4.56	5.10		9.66	12.90
4360	8″ stud	G		60	.133		4.76	5.55		10.31	13.80
4430	12″ deep joist, with 3-5/8″ stud	G		75	.107		4.44	4.43		8.87	11.75
4440	4″ stud	G		70	.114		4.52	4.75		9.27	12.30
4450	6″ stud	G		65	.123		4.90	5.10		10	13.30
4460	8″ stud	G		60	.133		5.25	5.55		10.80	14.35
5330	For 10″ deep joist, with 12 ga x 3-5/8″ stud	G		65	.123		4.40	5.10		9.50	12.75
5340	4″ stud	G		60	.133		4.54	5.55		10.09	13.55
5350	6″ stud	G		55	.145		5	6.05		11.05	14.80
5360	8″ stud	G		50	.160		5.50	6.65		12.15	16.30
5430	12″ deep joist, with 3-5/8″ stud	G		65	.123		4.88	5.10		9.98	13.25
5440	4″ stud	G		60	.133		4.79	5.55		10.34	13.80
5450	6″ stud	G		55	.145		5.45	6.05		11.50	15.30
5460	8″ stud	G	↓	50	.160	↓	6.30	6.65		12.95	17.15

05 42 23 – Cold-Formed Metal Roof Joist Framing

05 42 23.05 Framing, Bracing			Crew	Daily Output	Labor-Hours	Unit	Material	Labor	Equipment	Total	Total Incl O&P
0010	**FRAMING, BRACING**										
0015	Made from recycled materials	G									
0020	Continuous bracing, per row										
0100	16 ga x 1-1/2″ channel thru rafters/trusses @ 16″ O.C.	G	1 Carp	4.50	1.778	C.L.F.	45.50	74		119.50	164
0120	24″ O.C.	G		6	1.333		45.50	55.50		101	136
0300	2″ x 2″ angle x 18 ga, rafters/trusses @ 16″ O.C.	G		6	1.333		68	55.50		123.50	161
0320	24″ O.C.	G		8	1		68	41.50		109.50	139
0400	16 ga, rafters/trusses @ 16″ O.C.	G		4.50	1.778		86	74		160	209
0420	24″ O.C.	G	↓	6.50	1.231	↓	86	51		137	174

05 42 Cold-Formed Metal Joist Framing

05 42 23 – Cold-Formed Metal Roof Joist Framing

05 42 23.10 Framing, Bridging

	05 42 23.10 Framing, Bridging		Crew	Daily Output	Labor-Hours	Unit	Material	2010 Bare Costs Labor	Equipment	Total	Total Incl O&P
0010	**FRAMING, BRIDGING**										
0015	Made from recycled materials	G									
0020	Solid, between rafters w/ 1-1/4" leg track, per rafter bay										
1200	Rafters 16" O.C., 18 ga x 4" deep	G	1 Carp	60	.133	Ea.	1.37	5.55		6.92	10.05
1210	6" deep	G		57	.140		1.77	5.85		7.62	10.95
1220	8" deep	G		55	.145		2.24	6.05		8.29	11.75
1230	10" deep	G		52	.154		2.77	6.40		9.17	12.90
1240	12" deep	G		50	.160		3.20	6.65		9.85	13.75
2200	24" O.C., 18 ga x 4" deep	G		60	.133		1.98	5.55		7.53	10.75
2210	6" deep	G		57	.140		2.55	5.85		8.40	11.80
2220	8" deep	G		55	.145		3.25	6.05		9.30	12.85
2230	10" deep	G		52	.154		4.01	6.40		10.41	14.25
2240	12" deep	G		50	.160		4.63	6.65		11.28	15.35

05 42 23.50 Framing, Parapets

	05 42 23.50 Framing, Parapets		Crew	Daily Output	Labor-Hours	Unit	Material	Labor	Equipment	Total	Total Incl O&P
0010	**FRAMING, PARAPETS**										
0015	Made from recycled materials	G									
0100	3' high installed on 1st story, 18 ga x 4" wide studs, 12" O.C.	G	2 Carp	100	.160	L.F.	5.05	6.65		11.70	15.80
0110	16" O.C.	G		150	.107		4.28	4.43		8.71	11.55
0120	24" O.C.	G		200	.080		3.53	3.32		6.85	9
0200	6" wide studs, 12" O.C.	G		100	.160		6.40	6.65		13.05	17.30
0210	16" O.C.	G		150	.107		5.45	4.43		9.88	12.85
0220	24" O.C.	G		200	.080		4.51	3.32		7.83	10.05
1100	Installed on 2nd story, 18 ga x 4" wide studs, 12" O.C.	G		95	.168		5.05	7		12.05	16.35
1110	16" O.C.	G		145	.110		4.28	4.58		8.86	11.75
1120	24" O.C.	G		190	.084		3.53	3.50		7.03	9.30
1200	6" wide studs, 12" O.C.	G		95	.168		6.40	7		13.40	17.85
1210	16" O.C.	G		145	.110		5.45	4.58		10.03	13.05
1220	24" O.C.	G		190	.084		4.51	3.50		8.01	10.35
2100	Installed on gable, 18 ga x 4" wide studs, 12" O.C.	G		85	.188		5.05	7.80		12.85	17.60
2110	16" O.C.	G		130	.123		4.28	5.10		9.38	12.60
2120	24" O.C.	G		170	.094		3.53	3.91		7.44	9.95
2200	6" wide studs, 12" O.C.	G		85	.188		6.40	7.80		14.20	19.10
2210	16" O.C.	G		130	.123		5.45	5.10		10.55	13.90
2220	24" O.C.	G		170	.094		4.51	3.91		8.42	11

05 42 23.60 Framing, Roof Rafters

	05 42 23.60 Framing, Roof Rafters		Crew	Daily Output	Labor-Hours	Unit	Material	Labor	Equipment	Total	Total Incl O&P
0010	**FRAMING, ROOF RAFTERS**										
0015	Made from recycled materials	G									
0100	Boxed ridge beam, double, 18 ga x 6" deep	G	2 Carp	160	.100	L.F.	4.81	4.16		8.97	11.70
0110	8" deep	G		150	.107		5.35	4.43		9.78	12.75
0120	10" deep	G		140	.114		6.45	4.75		11.20	14.40
0130	12" deep	G		130	.123		7.10	5.10		12.20	15.70
0200	16 ga x 6" deep	G		150	.107		5.45	4.43		9.88	12.85
0210	8" deep	G		140	.114		6.15	4.75		10.90	14.05
0220	10" deep	G		130	.123		7.35	5.10		12.45	15.95
0230	12" deep	G		120	.133		8	5.55		13.55	17.35
1100	Rafters, 2" flange, material only, 18 ga x 6" deep	G					1.48			1.48	1.63
1110	8" deep	G					1.76			1.76	1.94
1120	10" deep	G					2.07			2.07	2.28
1130	12" deep	G					2.40			2.40	2.64
1200	16 ga x 6" deep	G					1.82			1.82	2
1210	8" deep	G					2.18			2.18	2.40
1220	10" deep	G					2.54			2.54	2.80

05 42 Cold-Formed Metal Joist Framing

05 42 23 – Cold-Formed Metal Roof Joist Framing

05 42 23.60 Framing, Roof Rafters		Crew	Daily Output	Labor-Hours	Unit	Material	2010 Bare Costs Labor	Equipment	Total	Total Incl O&P
1230	12" deep	G			L.F.	2.89			2.89	3.18
2100	Installation only, ordinary rafter to 4:12 pitch, 18 ga x 6" deep		2 Carp	35	.457	Ea.		19		19	29.50
2110	8" deep			30	.533			22		22	34
2120	10" deep			25	.640			26.50		26.50	41
2130	12" deep			20	.800			33		33	51
2200	16 ga x 6" deep			30	.533			22		22	34
2210	8" deep			25	.640			26.50		26.50	41
2220	10" deep			20	.800			33		33	51
2230	12" deep			15	1.067			44.50		44.50	68.50
8100	Add to labor, ordinary rafters on steep roofs							25%			
8110	Dormers & complex roofs							50%			
8200	Hip & valley rafters to 4:12 pitch							25%			
8210	Steep roofs							50%			
8220	Dormers & complex roofs							75%			
8300	Hip & valley jack rafters to 4:12 pitch							50%			
8310	Steep roofs							75%			
8320	Dormers & complex roofs							100%			

05 42 23.70 Framing, Soffits and Canopies			Crew	Daily Output	Labor-Hours	Unit	Material	Labor	Equipment	Total	Total Incl O&P
0010	**FRAMING, SOFFITS & CANOPIES**										
0015	Made from recycled materials	G									
0130	Continuous ledger track @ wall, studs @ 16" O.C., 18 ga x 4" wide	G	2 Carp	535	.030	L.F.	.91	1.24		2.15	2.92
0140	6" wide	G		500	.032		1.18	1.33		2.51	3.34
0150	8" wide	G		465	.034		1.50	1.43		2.93	3.85
0160	10" wide	G		430	.037		1.85	1.55		3.40	4.41
0230	Studs @ 24" O.C., 18 ga x 4" wide	G		800	.020		.87	.83		1.70	2.24
0240	6" wide	G		750	.021		1.12	.89		2.01	2.61
0250	8" wide	G		700	.023		1.43	.95		2.38	3.03
0260	10" wide	G		650	.025		1.76	1.02		2.78	3.52
1000	Horizontal soffit and canopy members, material only										
1030	1-5/8" flange studs, 18 ga x 4" deep	G				L.F.	1.20			1.20	1.32
1040	6" deep	G					1.51			1.51	1.66
1050	8" deep	G					1.84			1.84	2.02
1140	2" flange joists, 18 ga x 6" deep	G					1.69			1.69	1.86
1150	8" deep	G					2.02			2.02	2.22
1160	10" deep	G					2.36			2.36	2.60
4030	Installation only, 18 ga, 1-5/8" flange x 4" deep		2 Carp	130	.123	Ea.		5.10		5.10	7.90
4040	6" deep			110	.145			6.05		6.05	9.30
4050	8" deep			90	.178			7.40		7.40	11.40
4140	2" flange, 18 ga x 6" deep			110	.145			6.05		6.05	9.30
4150	8" deep			90	.178			7.40		7.40	11.40
4160	10" deep			80	.200			8.30		8.30	12.80
6010	Clips to attach facia to rafter tails, 2" x 2" x 18 ga angle	G	1 Carp	120	.067		.81	2.77		3.58	5.15
6020	16 ga angle	G	"	100	.080		1.01	3.32		4.33	6.20

05 44 13 – Cold-Formed Metal Roof Trusses

05 44 13.60 Framing, Roof Trusses			Crew	Daily Output	Labor-Hours	Unit	Material	2010 Bare Costs Labor	Equipment	Total	Total Incl O&P
0010	**FRAMING, ROOF TRUSSES**										
0015	Made from recycled materials	G									
0020	Fabrication of trusses on ground, Fink (W) or King Post, to 4:12 pitch										
0120	18 ga x 4" chords, 16' span	G	2 Carp	12	1.333	Ea.	56	55.50		111.50	147
0130	20' span	G		11	1.455		70	60.50		130.50	170
0140	24' span	G		11	1.455		84	60.50		144.50	186
0150	28' span	G		10	1.600		98	66.50		164.50	210
0160	32' span	G		10	1.600		112	66.50		178.50	225
0250	6" chords, 28' span	G		9	1.778		123	74		197	250
0260	32' span	G		9	1.778		141	74		215	269
0270	36' span	G		8	2		159	83		242	305
0280	40' span	G		8	2		176	83		259	320
1120	5:12 to 8:12 pitch, 18 ga x 4" chords, 16' span	G		10	1.600		64	66.50		130.50	173
1130	20' span	G		9	1.778		80	74		154	202
1140	24' span	G		9	1.778		96	74		170	220
1150	28' span	G		8	2		112	83		195	251
1160	32' span	G		8	2		128	83		211	269
1250	6" chords, 28' span	G		7	2.286		141	95		236	300
1260	32' span	G		7	2.286		161	95		256	325
1270	36' span	G		6	2.667		181	111		292	370
1280	40' span	G		6	2.667		202	111		313	395
2120	9:12 to 12:12 pitch, 18 ga x 4" chords, 16' span	G		8	2		80	83		163	216
2130	20' span	G		7	2.286		100	95		195	256
2140	24' span	G		7	2.286		120	95		215	278
2150	28' span	G		6	2.667		140	111		251	325
2160	32' span	G		6	2.667		160	111		271	345
2250	6" chords, 28' span	G		5	3.200		176	133		309	400
2260	32' span	G		5	3.200		202	133		335	425
2270	36' span	G		4	4		227	166		393	505
2280	40' span	G		4	4		252	166		418	535
5120	Erection only of roof trusses, to 4:12 pitch, 16' span		F-6	48	.833			32.50	13.65	46.15	64.50
5130	20' span			46	.870			33.50	14.25	47.75	67
5140	24' span			44	.909			35	14.90	49.90	70.50
5150	28' span			42	.952			37	15.60	52.60	73.50
5160	32' span			40	1			38.50	16.40	54.90	77.50
5170	36' span			38	1.053			41	17.25	58.25	81.50
5180	40' span			36	1.111			43	18.20	61.20	86
5220	5:12 to 8:12 pitch, 16' span			42	.952			37	15.60	52.60	73.50
5230	20' span			40	1			38.50	16.40	54.90	77.50
5240	24' span			38	1.053			41	17.25	58.25	81.50
5250	28' span			36	1.111			43	18.20	61.20	86
5260	32' span			34	1.176			45.50	19.30	64.80	91
5270	36' span			32	1.250			48.50	20.50	69	96.50
5280	40' span			30	1.333			51.50	22	73.50	103
5320	9:12 to 12:12 pitch, 16' span			36	1.111			43	18.20	61.20	86
5330	20' span			34	1.176			45.50	19.30	64.80	91
5340	24' span			32	1.250			48.50	20.50	69	96.50
5350	28' span			30	1.333			51.50	22	73.50	103
5360	32' span			28	1.429			55.50	23.50	79	111
5370	36' span			26	1.538			59.50	25	84.50	119
5380	40' span			24	1.667			64.50	27.50	92	129

05 51 Metal Stairs

05 51 13 – Metal Pan Stairs

05 51 13.50 Pan Stairs

			Crew	Daily Output	Labor-Hours	Unit	Material	2010 Bare Costs Labor	Equipment	Total	Total Incl O&P
0010	**PAN STAIRS**, shop fabricated, steel stringers										
0015	Made from recycled materials	G									
0200	Cement fill metal pan, picket rail, 3′-6″ wide	G	E-4	35	.914	Riser	415	43.50	4.17	462.67	535
0300	4′-0″ wide	G		30	1.067		470	50.50	4.86	525.36	610
0350	Wall rail, both sides, 3′-6″ wide CN	G		53	.604		315	28.50	2.75	346.25	405
1500	Landing, steel pan, conventional	G		160	.200	S.F.	55	9.50	.91	65.41	78
1600	Pre-erected	G		255	.125	″	96.50	5.95	.57	103.02	117
1700	Pre-erected, steel pan tread, 3′-6″ wide, 2 line pipe rail	G	E-2	87	.644	Riser	455	29.50	18.45	502.95	570

05 51 16 – Metal Floor Plate Stairs

05 51 16.50 Floor Plate Stairs

			Crew	Daily Output	Labor-Hours	Unit	Material	Labor	Equipment	Total	Total Incl O&P
0010	**FLOOR PLATE STAIRS**, shop fabricated, steel stringers										
0015	Made from recycled materials	G									
0400	Cast iron tread and pipe rail, 3′-6″ wide	G	E-4	35	.914	Riser	440	43.50	4.17	487.67	565
0500	Checkered plate tread, industrial, 3′-6″ wide	G		28	1.143		275	54	5.20	334.20	405
0550	Circular, for tanks, 3′-0″ wide	G		33	.970		305	46	4.42	355.42	420
0600	For isolated stairs, add							100%			
0800	Custom steel stairs, 3′-6″ wide, minimum	G	E-4	35	.914		415	43.50	4.17	462.67	535
0810	Average	G		30	1.067		550	50.50	4.86	605.36	700
0900	Maximum	G		20	1.600		690	76	7.30	773.30	895
1100	For 4′ wide stairs, add						5%	5%			
1300	For 5′ wide stairs, add						10%	10%			

05 51 19 – Metal Grating Stairs

05 51 19.50 Grating Stairs

			Crew	Daily Output	Labor-Hours	Unit	Material	Labor	Equipment	Total	Total Incl O&P
0010	**GRATING STAIRS**, shop fabricated, steel stringers, safety nosing on treads										
0015	Made from recycled materials	G									
0020	Grating tread and pipe railing, 3′-6″ wide	G	E-4	35	.914	Riser	275	43.50	4.17	322.67	385
0100	4′-0″ wide	G	″	30	1.067	″	360	50.50	4.86	415.36	490

05 51 23 – Metal Fire Escapes

05 51 23.25 Fire Escapes

			Crew	Daily Output	Labor-Hours	Unit	Material	Labor	Equipment	Total	Total Incl O&P
0010	**FIRE ESCAPES**, shop fabricated										
0200	2′ wide balcony, 1″ x 1/4″ bars 1-1/2″ O.C.	G	1 Sswk	5	1.600	L.F.	48	75		123	185
0400	1st story cantilevered stair, standard	G		.09	88.889	Ea.	2,000	4,175		6,175	9,525
0700	Platform & fixed stair, 36″ x 40″	G		.17	47.059	Flight	890	2,200		3,090	4,850
0900	For 3′-6″ wide escapes, add to above						100%	150%			

05 51 23.50 Fire Escape Stairs

			Crew	Daily Output	Labor-Hours	Unit	Material	Labor	Equipment	Total	Total Incl O&P
0010	**FIRE ESCAPE STAIRS**, shop fabricated										
0020	One story, disappearing, stainless steel	G	2 Sswk	20	.800	V.L.F.	212	37.50		249.50	299
0100	Portable ladder					Ea.	99			99	109
1100	Fire escape, galvanized steel, 8′-0″ to 10′-4″ ceiling	G	2 Carp	1	16		1,525	665		2,190	2,700
1110	10′-6″ to 13′-6″ ceiling	G	″	1	16		2,000	665		2,665	3,225

05 51 33 – Metal Ladders

05 51 33.13 Vertical Metal Ladders

			Crew	Daily Output	Labor-Hours	Unit	Material	Labor	Equipment	Total	Total Incl O&P
0010	**VERTICAL METAL LADDERS**, shop fabricated										
0015	Made from recycled materials	G									
0020	Steel, 20″ wide, bolted to concrete, with cage	G	E-4	50	.640	V.L.F.	74	30.50	2.92	107.42	138
0100	Without cage	G		85	.376		38	17.85	1.72	57.57	75.50
0300	Aluminum, bolted to concrete, with cage	G		50	.640		120	30.50	2.92	153.42	189
0400	Without cage	G		85	.376		55.50	17.85	1.72	75.07	94.50

05 51 Metal Stairs

05 51 33 – Metal Ladders

05 51 33.16 Inclined Metal Ladders			Crew	Daily Output	Labor-Hours	Unit	Material	2010 Bare Costs Labor	Equipment	Total	Total Incl O&P
0010	**INCLINED METAL LADDERS**, shop fabricated										
3900	Industrial ships ladder, 3' W, grating treads, 2 line pipe rail	G	E-4	30	1.067	Riser	149	50.50	4.86	204.36	258
4000	Aluminum	G	"	30	1.067	"	229	50.50	4.86	284.36	345
05 51 33.23 Alternating Tread Ladders											
0010	**ALTERNATING TREAD LADDERS**, shop fabricated										
0015	Made from recycled materials	G									
1350	Alternating tread stair, 56/68°, steel, standard paint color	G	2 Sswk	50	.320	V.L.F.	202	15		217	249
1360	Non-standard paint color	G		50	.320		228	15		243	278
1370	Galvanized steel	G		50	.320		235	15		250	286
1380	Stainless steel	G		50	.320		340	15		355	395
1390	68°, aluminum	G	↓	50	.320	↓	248	15		263	299

05 52 Metal Railings

05 52 13 – Pipe and Tube Railings

05 52 13.50 Railings, Pipe

			Crew	Daily Output	Labor-Hours	Unit	Material	Labor	Equipment	Total	Total Incl O&P
0010	**RAILINGS, PIPE**, shop fab'd, 3'-6" high, posts @ 5' O.C.										
0015	Made from recycled materials	G									
0020	Aluminum, 2 rail, satin finish, 1-1/4" diameter	G	E-4	160	.200	L.F.	31.50	9.50	.91	41.91	52
0030	Clear anodized	G		160	.200		39	9.50	.91	49.41	60
0040	Dark anodized	G		160	.200		44	9.50	.91	54.41	65.50
0080	1-1/2" diameter, satin finish	G		160	.200		37.50	9.50	.91	47.91	59
0090	Clear anodized	G		160	.200		42	9.50	.91	52.41	63.50
0100	Dark anodized	G		160	.200		46.50	9.50	.91	56.91	68.50
0140	Aluminum, 3 rail, 1-1/4" diam., satin finish	G		137	.234		48	11.05	1.07	60.12	73.50
0150	Clear anodized	G		137	.234		60	11.05	1.07	72.12	86.50
0160	Dark anodized	G		137	.234		66.50	11.05	1.07	78.62	93.50
0200	1-1/2" diameter, satin finish	G		137	.234		57.50	11.05	1.07	69.62	84
0210	Clear anodized	G		137	.234		65.50	11.05	1.07	77.62	92.50
0220	Dark anodized	G		137	.234		71.50	11.05	1.07	83.62	99.50
0500	Steel, 2 rail, on stairs, primed, 1-1/4" diameter	G		160	.200		20.50	9.50	.91	30.91	40
0520	1-1/2" diameter	G		160	.200		22.50	9.50	.91	32.91	42.50
0540	Galvanized, 1-1/4" diameter	G		160	.200		28.50	9.50	.91	38.91	49
0560	1-1/2" diameter	G		160	.200		32	9.50	.91	42.41	52.50
0580	Steel, 3 rail, primed, 1-1/4" diameter	G		137	.234		30.50	11.05	1.07	42.62	54.50
0600	1-1/2" diameter	G		137	.234		32.50	11.05	1.07	44.62	56.50
0620	Galvanized, 1-1/4" diameter	G		137	.234		43	11.05	1.07	55.12	68
0640	1-1/2" diameter CN	G		137	.234		50.50	11.05	1.07	62.62	76.50
0700	Stainless steel, 2 rail, 1-1/4" diam. #4 finish	G		137	.234		90.50	11.05	1.07	102.62	121
0720	High polish	G		137	.234		146	11.05	1.07	158.12	182
0740	Mirror polish	G		137	.234		183	11.05	1.07	195.12	222
0760	Stainless steel, 3 rail, 1-1/2" diam., #4 finish	G		120	.267		137	12.65	1.22	150.87	174
0770	High polish	G		120	.267		226	12.65	1.22	239.87	272
0780	Mirror finish	G		120	.267		275	12.65	1.22	288.87	330
0900	Wall rail, alum. pipe, 1-1/4" diam., satin finish	G		213	.150		17.95	7.10	.69	25.74	33
0905	Clear anodized	G		213	.150		22	7.10	.69	29.79	37.50
0910	Dark anodized	G		213	.150		26.50	7.10	.69	34.29	42.50
0915	1-1/2" diameter, satin finish	G		213	.150		19.85	7.10	.69	27.64	35.50
0920	Clear anodized	G		213	.150		25	7.10	.69	32.79	41
0925	Dark anodized	G		213	.150		31	7.10	.69	38.79	47.50
0930	Steel pipe, 1-1/4" diameter, primed	G	↓	213	.150	↓	12.45	7.10	.69	20.24	27

05 52 Metal Railings

05 52 13 – Pipe and Tube Railings

05 52 13.50 Railings, Pipe			Crew	Daily Output	Labor-Hours	Unit	Material	Labor	Equipment	Total	Total Incl O&P
							2010 Bare Costs				
0935	Galvanized	G	E-4	213	.150	L.F.	18.05	7.10	.69	25.84	33
0940	1-1/2" diameter	G		176	.182		12.85	8.60	.83	22.28	30
0945	Galvanized	G		213	.150		18.15	7.10	.69	25.94	33
0955	Stainless steel pipe, 1-1/2" diam., #4 finish	G		107	.299		72.50	14.20	1.36	88.06	106
0960	High polish	G		107	.299		148	14.20	1.36	163.56	189
0965	Mirror polish	G		107	.299		174	14.20	1.36	189.56	218
2000	Aluminum pipe & picket railing, double top rail, pickets @ 4-1/2" OC										
2010	36" high, straight & level	G	2 Sswk	80	.200	L.F.	81	9.40		90.40	106
2020	Curved & level	G		60	.267		111	12.50		123.50	144
2030	Straight & sloped	G		40	.400		93	18.75		111.75	135

05 52 16 – Industrial Railings

05 52 16.50 Railings, Industrial			Crew	Daily Output	Labor-Hours	Unit	Material	Labor	Equipment	Total	Total Incl O&P
0010	**RAILINGS, INDUSTRIAL**, shop fab'd, 3'-6" high, posts @ 5' O.C.										
0020	2 rail, 3'-6" high, 1-1/2" pipe	G	E-4	255	.125	L.F.	22.50	5.95	.57	29.02	36
0100	2" angle rail	G	"	255	.125		20.50	5.95	.57	27.02	33.50
0200	For 4" high kick plate, 10 gauge, add	G					4.72			4.72	5.20
0300	1/4" thick, add	G					6.05			6.05	6.70
0500	For curved rails, add						30%	30%			

05 53 Metal Gratings

05 53 13 – Floor Grating Frame

05 53 13.50 Grating Frame			Crew	Daily Output	Labor-Hours	Unit	Material	Labor	Equipment	Total	Total Incl O&P
0010	**GRATING FRAME**, field fabricated										
0020	Aluminum, for gratings 1" to 1-1/2" deep	G	1 Sswk	70	.114	L.F.	3.19	5.35		8.54	12.90
0100	For each corner, add	G				Ea.	4.78			4.78	5.25

05 53 16 – Floor Grating Planks

05 53 16.50 Grating Planks			Crew	Daily Output	Labor-Hours	Unit	Material	Labor	Equipment	Total	Total Incl O&P
0010	**GRATING PLANKS**, field fabricated from planks										
0020	Aluminum, 9-1/2" wide, 14 ga., 2" rib	G	E-4	950	.034	L.F.	28	1.60	.15	29.75	34
0200	Galvanized steel, 9-1/2" wide, 14 ga., 2-1/2" rib	G		950	.034		19.50	1.60	.15	21.25	24.50
0300	4" rib	G		950	.034		23	1.60	.15	24.75	28
0500	12 gauge, 2-1/2" rib	G		950	.034		26	1.60	.15	27.75	31.50
0600	3" rib	G		950	.034		31	1.60	.15	32.75	37
0800	Stainless steel, type 304, 16 ga., 2" rib	G		950	.034		41	1.60	.15	42.75	48
0900	Type 316	G		950	.034		50.50	1.60	.15	52.25	58.50

05 53 19 – Aluminum Floor Grating

05 53 19.50 Floor Grating, Aluminum			Crew	Daily Output	Labor-Hours	Unit	Material	Labor	Equipment	Total	Total Incl O&P
0010	**FLOOR GRATING, ALUMINUM**, field fabricated from panels										
0015	Made from recycled materials	G									
0110	Bearing bars @ 1-3/16" O.C., cross bars @ 4" O.C.,										
0111	Up to 300 S.F., 1" x 1/8" bar	G	E-4	900	.036	S.F.	23.50	1.69	.16	25.35	29
0112	Over 300 S.F.	G		850	.038		21.50	1.78	.17	23.45	27
0113	1-1/4" x 1/8" bar, up to 300 S.F.	G		800	.040		28	1.90	.18	30.08	34
0114	Over 300 S.F.	G		1000	.032		25	1.52	.15	26.67	31
0122	1-1/4" x 3/16" bar, up to 300 S.F.	G		750	.043		36	2.02	.19	38.21	43.50
0124	Over 300 S.F.	G		1000	.032		32.50	1.52	.15	34.17	39
0132	1-1/2" x 3/16" bar, up to 300 S.F.	G		700	.046		41.50	2.17	.21	43.88	49.50
0134	Over 300 S.F.	G		1000	.032		37.50	1.52	.15	39.17	44.50
0136	1-3/4" x 3/16" bar, up to 300 S.F.	G		500	.064		50.50	3.03	.29	53.82	61

05 53 Metal Gratings

05 53 19 – Aluminum Floor Grating

05 53 19.50 Floor Grating, Aluminum			Crew	Daily Output	Labor-Hours	Unit	Material	2010 Bare Costs Labor	Equipment	Total	Total Incl O&P
0138	Over 300 S.F.	G	E-4	1000	.032	S.F.	46	1.52	.15	47.67	53.50
0146	2-1/4" x 3/16" bar, up to 300 S.F.	G		600	.053		69	2.53	.24	71.77	80.50
0148	Over 300 S.F.	G		1000	.032		62.50	1.52	.15	64.17	72
0162	Cross bars @ 2" O.C., 1" x 1/8", up to 300 S.F.	G		600	.053		30.50	2.53	.24	33.27	38
0164	Over 300 S.F.	G		1000	.032		27.50	1.52	.15	29.17	33.50
0172	1-1/4" x 3/16" bar, up to 300 S.F.	G		600	.053		41	2.53	.24	43.77	49.50
0174	Over 300 S.F.	G		1000	.032		37	1.52	.15	38.67	44
0182	1-1/2" x 3/16" bar, up to 300 S.F.	G		600	.053		46	2.53	.24	48.77	55
0184	Over 300 S.F.	G		1000	.032		42	1.52	.15	43.67	49
0186	1-3/4" x 3/16" bar, up to 300 S.F.	G		600	.053		69.50	2.53	.24	72.27	81
0188	Over 300 S.F.	G		1000	.032		63	1.52	.15	64.67	72.50
0200	For straight cuts, add					L.F.	3.60			3.60	3.96
0300	For curved cuts, add						4.40			4.40	4.84
0400	For straight banding, add	G					4.60			4.60	5.05
0500	For curved banding, add	G					5.60			5.60	6.15
0600	For aluminum checkered plate nosings, add	G					6			6	6.60
0700	For straight toe plate, add	G					9.10			9.10	10
0800	For curved toe plate, add	G					10.60			10.60	11.65
1000	For cast aluminum abrasive nosings, add	G					8.85			8.85	9.75
1200	Expanded aluminum, .65# per S.F.	G	E-4	1050	.030	S.F.	11.75	1.44	.14	13.33	15.65
1400	Extruded I bars are 10% less than 3/16" bars										
1600	Heavy duty, all extruded plank, 3/4" deep, 1.8 # per S.F.	G	E-4	1100	.029	S.F.	30	1.38	.13	31.51	35.50
1700	1-1/4" deep, 2.9# per S.F.	G		1000	.032		35	1.52	.15	36.67	41.50
1800	1-3/4" deep, 4.2# per S.F.	G		925	.035		42	1.64	.16	43.80	49
1900	2-1/4" deep, 5.0# per S.F.	G		875	.037		50	1.73	.17	51.90	58
2100	For safety serrated surface, add						15%				

05 53 21 – Steel Floor Grating

05 53 21.50 Floor Grating, Steel			Crew	Daily Output	Labor-Hours	Unit	Material	Labor	Equipment	Total	Total Incl O&P
0010	**FLOOR GRATING, STEEL**, field fabricated from panels										
0015	Made from recycled materials	G									
0300	Platforms, to 12' high, rectangular	G	E-4	3150	.010	Lb.	2.40	.48	.05	2.93	3.54
0400	Circular	G	"	2300	.014	"	3	.66	.06	3.72	4.53
0410	Painted bearing bars @ 1-3/16"										
0412	Cross bars @ 4" O.C., 3/4" x 1/8" bar, up to 300 S.F.	G	E-2	500	.112	S.F.	8.10	5.10	3.21	16.41	21
0414	Over 300 S.F.	G		750	.075		7.35	3.41	2.14	12.90	16.20
0422	1-1/4" x 3/16", up to 300 S.F.	G		400	.140		12.50	6.40	4.02	22.92	29
0424	Over 300 S.F.	G		600	.093		11.35	4.26	2.68	18.29	22.50
0432	1-1/2" x 3/16", up to 300 S.F.	G		400	.140		14.40	6.40	4.02	24.82	31
0434	Over 300 S.F.	G		600	.093		13.10	4.26	2.68	20.04	24.50
0436	1-3/4" x 3/16", up to 300 S.F.	G		400	.140		19.90	6.40	4.02	30.32	37
0438	Over 300 S.F.	G		600	.093		18.10	4.26	2.68	25.04	30
0452	2-1/4" x 3/16", up to 300 S.F.	G		300	.187		24	8.50	5.35	37.85	47
0454	Over 300 S.F.	G		450	.124		21.50	5.65	3.57	30.72	37.50
0462	Cross bars @ 2" O.C., 3/4" x 1/8", up to 300 S.F.	G		500	.112		13.65	5.10	3.21	21.96	27
0464	Over 300 S.F.	G		750	.075		11.35	3.41	2.14	16.90	20.50
0472	1-1/4" x 3/16", up to 300 S.F.	G		400	.140		17.85	6.40	4.02	28.27	35
0474	Over 300 S.F.	G		600	.093		14.85	4.26	2.68	21.79	26.50
0482	1-1/2" x 3/16", up to 300 S.F.	G		400	.140		20	6.40	4.02	30.42	37
0484	Over 300 S.F.	G		600	.093		16.75	4.26	2.68	23.69	28.50
0486	1-3/4" x 3/16", up to 300 S.F.	G		400	.140		30.50	6.40	4.02	40.92	48.50
0488	Over 300 S.F.	G		600	.093		25	4.26	2.68	31.94	37.50
0502	2-1/4" x 3/16", up to 300 S.F.	G		300	.187		29.50	8.50	5.35	43.35	53

05 53 Metal Gratings

05 53 21 – Steel Floor Grating

05 53 21.50 Floor Grating, Steel			Crew	Daily Output	Labor-Hours	Unit	2010 Bare Costs Material	Labor	Equipment	Total	Total Incl O&P
0504	Over 300 S.F.	G	E-2	450	.124	S.F.	24.50	5.65	3.57	33.72	40.50
0690	For galvanized grating, add						25%				
0800	For straight cuts, add					L.F.	5.20			5.20	5.70
0900	For curved cuts, add						6.55			6.55	7.20
1000	For straight banding, add	G					5.60			5.60	6.15
1100	For curved banding, add	G					7.30			7.30	8.05
1200	For checkered plate nosings, add	G					6.40			6.40	7.05
1300	For straight toe or kick plate, add	G					11.50			11.50	12.65
1400	For curved toe or kick plate, add	G					13			13	14.30
1500	For abrasive nosings, add	G					8.55			8.55	9.40
1510	For stair treads, see Div. 05 55 13.50										
1600	For safety serrated surface, minimum, add						15%				
1700	Maximum, add						25%				
2000	Stainless steel gratings, close spaced, 1" x 1/8" bars, up to 300 S.F.	G	E-4	450	.071	S.F.	105	3.37	.32	108.69	122
2100	Standard spacing, 3/4" x 1/8" bars	G		500	.064		75.50	3.03	.29	78.82	88.50
2200	1-1/4" x 3/16" bars	G		400	.080		116	3.79	.36	120.15	135
2400	Expanded steel grating, at ground, 3.0# per S.F.	G		900	.036		7.30	1.69	.16	9.15	11.20
2500	3.14# per S.F.	G		900	.036		7.85	1.69	.16	9.70	11.80
2600	4.0# per S.F.	G		850	.038		9.70	1.78	.17	11.65	14
2650	4.27# per S.F.	G		850	.038		10.40	1.78	.17	12.35	14.80
2700	5.0# per S.F.	G		800	.040		12.35	1.90	.18	14.43	17.15
2800	6.25# per S.F.	G		750	.043		16	2.02	.19	18.21	21.50
2900	7.0# per S.F.	G		700	.046		17.70	2.17	.21	20.08	23.50
3100	For flattened expanded steel grating, add						8%				
3300	For elevated installation above 15', add							15%			

05 54 Metal Floor Plates

05 54 13 – Floor Plates

05 54 13.20 Checkered Plates			Crew	Daily Output	Labor-Hours	Unit	Material	Labor	Equipment	Total	Total Incl O&P
0010	**CHECKERED PLATES**, steel, field fabricated										
0015	Made from recycled materials	G									
0020	1/4" & 3/8", 2000 to 5000 S.F., bolted	G	E-4	2900	.011	Lb.	.85	.52	.05	1.42	1.92
0100	Welded	G		4400	.007	"	.80	.34	.03	1.17	1.53
0300	Pit or trench cover and frame, 1/4" plate, 2' to 3' wide	G		100	.320	S.F.	10.10	15.15	1.46	26.71	39
0400	For galvanizing, add	G				Lb.	.30			.30	.33
0500	Platforms, 1/4" plate, no handrails included, rectangular	G	E-4	4200	.008		2.40	.36	.03	2.79	3.32
0600	Circular	G	"	2500	.013		3	.61	.06	3.67	4.43

05 54 13.70 Trench Covers			Crew	Daily Output	Labor-Hours	Unit	Material	Labor	Equipment	Total	Total Incl O&P
0010	**TRENCH COVERS**, field fabricated										
0020	Cast iron grating with bar stops and angle frame, to 18" wide	G	1 Sswk	20	.400	L.F.	58	18.75		76.75	97
0100	Frame only (both sides of trench), 1" grating	G		45	.178		1.50	8.35		9.85	16.30
0150	2" grating	G		35	.229		3.30	10.70		14	22.50
0200	Aluminum, stock units, including frames and										
0210	3/8" plain cover plate, 4" opening	G	E-4	205	.156	L.F.	38.50	7.40	.71	46.61	56.50
0300	6" opening	G		185	.173		48	8.20	.79	56.99	68
0400	10" opening	G		170	.188		65.50	8.90	.86	75.26	89
0500	16" opening	G		155	.206		90	9.80	.94	100.74	117
0700	Add per inch for additional widths to 24"	G					108			108	119
0900	For custom fabrication, add						50%				
1100	For 1/4" plain cover plate, deduct						12%				
1500	For cover recessed for tile, 1/4" thick, deduct						12%				

05 54 Metal Floor Plates

05 54 13 – Floor Plates

05 54 13.70 Trench Covers			Crew	Daily Output	Labor-Hours	Unit	Material	Labor	Equipment	Total	Total Incl O&P
							2010 Bare Costs				
1600	3/8″ thick, add					L.F.	5%				
1800	For checkered plate cover, 1/4″ thick, deduct						12%				
1900	3/8″ thick, add						2%				
2100	For slotted or round holes in cover, 1/4″ thick, add						3%				
2200	3/8″ thick, add						4%				
2300	For abrasive cover, add						12%				

05 55 Metal Stair Treads and Nosings

05 55 13 – Metal Stair Treads

05 55 13.50 Stair Treads			Crew	Daily Output	Labor-Hours	Unit	Material	Labor	Equipment	Total	Total Incl O&P
0010	**STAIR TREADS**										
0020	Aluminum grating, 3′ long, 1-1/2″ x 3/16″ rect. bars, 6″ wide	G	1 Sswk	24	.333	Ea.	62	15.65		77.65	96
0100	12″ wide	G		22	.364		91.50	17.05		108.55	131
0200	1-1/2″ x 3/16″ I-bars, 6″ wide	G		24	.333		52	15.65		67.65	85
0300	12″ wide	G		22	.364		84	17.05		101.05	123
0400	For abrasive nosings, add	G					16.30			16.30	17.90
0500	For narrow mesh, add						60%				
0600	Stair treads, not incl. stringers, see also Div. 03 15 05.85										
0700	Cast aluminum, abrasive, 3′ long x 12″ wide, 5/16″ thick	G	1 Sswk	15	.533	Ea.	138	25		163	196
0800	3/8″ thick	G		15	.533		147	25		172	205
0900	1/2″ thick	G		15	.533		161	25		186	221
1000	Cast bronze, abrasive, 3/8″ thick	G		8	1		625	47		672	775
1100	1/2″ thick	G		8	1		825	47		872	990
1200	Cast iron, abrasive, 3′ long x 12″ wide, 3/8″ thick	G		15	.533		121	25		146	177
1300	1/2″ thick	G		15	.533		140	25		165	198
1400	Fiberglass reinforced plastic with safety nosing,										
1500	1-1/2″ thick, 12″ wide, 24″ long		1 Sswk	22	.364	Ea.	152	17.05		169.05	197
1600	30″ long			22	.364		159	17.05		176.05	205
1700	36″ long			22	.364		173	17.05		190.05	221
2000	Steel grating, painted, 3′ long, 1-1/4″ x 3/16″ bars, 6″ wide	G		20	.400		47	18.75		65.75	85
2010	12″ wide	G		18	.444		54.50	21		75.50	96.50
2100	Add for abrasive nosing, 3′ long, painted	G					14.75			14.75	16.20
2200	Galvanized	G					19.75			19.75	22
2300	Painting 3′ treads, in shop, nonstandard paint						2.61			2.61	2.87
2400	Added coats, standard paint						1.12			1.12	1.23
2500	Expanded steel, 2-1/2″ deep, 9″ x 3′ long, 18 gauge	G	1 Sswk	20	.400		26	18.75		44.75	61.50
2600	14 gauge	G	″	20	.400		32	18.75		50.75	68.50

05 56 Metal Castings

05 56 13 – Metal Construction Castings

05 56 13.50 Construction Castings			Crew	Daily Output	Labor-Hours	Unit	Material	Labor	Equipment	Total	Total Incl O&P
0010	**CONSTRUCTION CASTINGS**										
0020	Manhole covers and frames, see Div. 33 44 13.13										
0100	Column bases, cast iron, 16″ x 16″, approx. 65 lb.	G	E-4	46	.696	Ea.	135	33	3.17	171.17	209
0200	32″ x 32″, approx. 256 lb.	G	″	23	1.391		505	66	6.35	577.35	680
0400	Cast aluminum for wood columns, 8″ x 8″	G	1 Carp	32	.250		43.50	10.40		53.90	63.50
0500	12″ x 12″	G	″	32	.250		93.50	10.40		103.90	119
0600	Miscellaneous C.I. castings, light sections, less than 150 lb.	G	E-4	3200	.010	Lb.	2.09	.47	.05	2.61	3.18
1100	Heavy sections, more than 150 lb	G		4200	.008		1.02	.36	.03	1.41	1.80

05 56 Metal Castings

05 56 13 – Metal Construction Castings

05 56 13.50 Construction Castings			Crew	Daily Output	Labor-Hours	Unit	Material	2010 Bare Costs Labor	Equipment	Total	Total Incl O&P
1300	Special low volume items	G	E-4	3200	.010	Lb.	3.63	.47	.05	4.15	4.87
1500	For ductile iron, add						100%				

05 58 Formed Metal Fabrications

05 58 21 – Formed Chain

05 58 21.05 Alloy Steel Chain

			Crew	Daily Output	Labor-Hours	Unit	Material	Labor	Equipment	Total	Total Incl O&P
0010	**ALLOY STEEL CHAIN**, Grade 80										
0015	Self-colored, cut lengths, w/accessories, 1/4"	G	E-17	4	4	C.L.F.	625	192		817	1,025
0020	3/8"	G		2	8		755	385		1,140	1,500
0030	1/2"	G		1.20	13.333		1,250	640		1,890	2,500
0040	5/8"	G		.72	22.222		2,025	1,075		3,100	4,100
0050	3/4"	G	E-18	.48	83.333		2,825	3,875	2,150	8,850	12,100
0060	7/8"	G		.40	100		4,675	4,650	2,575	11,900	15,900
0070	1"	G		.35	114		6,125	5,325	2,950	14,400	19,100
0080	1-1/4"	G		.24	166		16,200	7,750	4,300	28,250	35,800
0110	Clevis slip hook, 1/4"	G				Ea.	34			34	37.50
0120	3/8"	G					47			47	51.50
0130	1/2"	G					69			69	76
0140	5/8"	G					89			89	98
0150	3/4"	G					151			151	166
0160	Eye/sling hook w/ hammerlock coupling, 7/8"	G					283			283	310
0170	1"	G					650			650	715
0180	1-1/4"	G					2,300			2,300	2,525

05 58 23 – Formed Columns

05 58 23.10 Aluminum Columns

			Crew	Daily Output	Labor-Hours	Unit	Material	Labor	Equipment	Total	Total Incl O&P
0010	**ALUMINUM COLUMNS**										
0015	Made from recycled materials	G									
0020	Aluminum, extruded, stock units, no cap or base, 6" diameter	G	E-4	240	.133	L.F.	11.65	6.30	.61	18.56	24.50
0100	8" diameter	G		170	.188		15	8.90	.86	24.76	33
0200	10" diameter	G		150	.213		20.50	10.10	.97	31.57	41.50
0300	12" diameter	G		140	.229		37	10.85	1.04	48.89	60.50
0400	15" diameter	G		120	.267		44	12.65	1.22	57.87	72
0410	Caps and bases, plain, 6" diameter	G				Set	22			22	24
0420	8" diameter	G					29			29	31.50
0430	10" diameter	G					37			37	40.50
0440	12" diameter	G					50			50	55
0450	15" diameter	G					90			90	99
0460	Caps, ornamental, minimum	G					242			242	266
0470	Maximum	G					1,075			1,075	1,200
0500	For square columns, add to column prices above					L.F.	50%				
0700	Residential, flat, 8' high, plain	G	E-4	20	1.600	Ea.	95	76	7.30	178.30	245
0720	Fancy	G		20	1.600		181	76	7.30	264.30	340
0740	Corner type, plain	G		20	1.600		165	76	7.30	248.30	325
0760	Fancy	G		20	1.600		320	76	7.30	403.30	490

05 58 25 – Formed Lamp Posts

05 58 25.40 Lamp Posts

			Crew	Daily Output	Labor-Hours	Unit	Material	Labor	Equipment	Total	Total Incl O&P
0010	**LAMP POSTS**										
0020	Aluminum, 7' high, stock units, post only	G	1 Carp	16	.500	Ea.	31.50	21		52.50	66.50
0100	Mild steel, plain	G	"	16	.500	"	28	21		49	63

05 58 Formed Metal Fabrications

05 58 27 – Formed Guards

05 58 27.90 Window Guards		Crew	Daily Output	Labor-Hours	Unit	Material	2010 Bare Costs Labor	Equipment	Total	Total Incl O&P
0010	**WINDOW GUARDS**, shop fabricated									
0015	Expanded metal, steel angle frame, permanent G	E-4	350	.091	S.F.	19.10	4.33	.42	23.85	29
0025	Steel bars, 1/2" x 1/2", spaced 5" O.C. G	"	290	.110	"	13.25	5.25	.50	19	24.50
0030	Hinge mounted, add G				Opng.	38.50			38.50	42
0040	Removable type, add G				"	24.50			24.50	26.50
0050	For galvanized guards, add				S.F.	35%				
0070	For pivoted or projected type, add					105%	40%			
0100	Mild steel, stock units, economy G	E-4	405	.079		5.20	3.75	.36	9.31	12.70
0200	Deluxe G		405	.079		10.65	3.75	.36	14.76	18.75
0400	Woven wire, stock units, 3/8" channel frame, 3' x 5' opening G		40	.800	Opng.	140	38	3.65	181.65	225
0500	4' x 6' opening G		38	.842		224	40	3.84	267.84	320
0800	Basket guards for above, add G					192			192	211
1000	Swinging guards for above, add G					66			66	72.50

05 71 Decorative Metal Stairs

05 71 13 – Fabricated Metal Spiral Stairs

05 71 13.50 Spiral Stairs		Crew	Daily Output	Labor-Hours	Unit	Material	Labor	Equipment	Total	Total Incl O&P
0010	**SPIRAL STAIRS**, shop fabricated									
1810	Spiral aluminum, 5'-0" diameter, stock units G	E-4	45	.711	Riser	500	33.50	3.24	536.74	615
1820	Custom units G		45	.711		945	33.50	3.24	981.74	1,125
1900	Spiral, cast iron, 4'-0" diameter, ornamental, minimum G		45	.711		445	33.50	3.24	481.74	555
1920	Maximum G		25	1.280		605	60.50	5.85	671.35	785
2000	Spiral, steel, industrial checkered plate, 4' diameter G		45	.711		445	33.50	3.24	481.74	555
2200	Stock units, 6'-0" diameter G		40	.800		540	38	3.65	581.65	665
3110	Spiral steel, stock units, primed, flat metal tread, 3'-6" dia G	2 Carp	1.60	10	Flight	1,250	415		1,665	2,000
3120	4'-0" dia G		1.45	11.034		1,425	460		1,885	2,275
3130	4'-6" dia G		1.35	11.852		1,575	490		2,065	2,475
3140	5'-0" dia G		1.25	12.800		1,700	530		2,230	2,700
3210	Galvanized, 3'-6" dia G		1.60	10		1,575	415		1,990	2,400
3220	4'-0" dia G		1.45	11.034		1,800	460		2,260	2,700
3230	4'-6" dia G		1.35	11.852		2,000	490		2,490	2,950
3240	5'-0" dia G		1.25	12.800		2,150	530		2,680	3,200
3310	Checkered plate tread, 3'-6" dia G		1.45	11.034		1,450	460		1,910	2,300
3320	4'-0" dia G		1.35	11.852		1,650	490		2,140	2,550
3330	4'-6" dia G		1.25	12.800		1,800	530		2,330	2,800
3340	5'-0" dia G		1.15	13.913		1,950	580		2,530	3,050
3410	Galvanized, 3'-6" dia G		1.45	11.034		1,850	460		2,310	2,750
3420	4'-0" dia G		1.35	11.852		2,075	490		2,565	3,025
3430	4'-6" dia G		1.25	12.800		2,250	530		2,780	3,325
3440	5'-0" dia G		1.15	13.913		2,425	580		3,005	3,575
3510	Red oak tread on flat metal, 3'-6" dia		1.35	11.852		2,550	490		3,040	3,575
3520	4'-0" dia		1.25	12.800		2,825	530		3,355	3,925
3530	4'-6" dia		1.15	13.913		3,025	580		3,605	4,250
3540	5'-0" dia		1.05	15.238		3,325	635		3,960	4,625

05 73 Decorative Metal Railings

05 73 16 – Wire Rope Decorative Metal Railings

05 73 16.10 Cable Railings			Crew	Daily Output	Labor-Hours	Unit	Material	2010 Bare Costs Labor	Equipment	Total	Total Incl O&P
0010	**CABLE RAILINGS**, with 316 stainless steel 1 x 19 cable, 3/16" diameter										
0015	Made from recycled materials	G									
0100	1-3/4" diameter stainless steel posts x 42" high, cables 4" OC	G	2 Sswk	25	.640	L.F.	56.50	30		86.50	115

05 73 23 – Ornamental Railings

05 73 23.50 Railings, Ornamental			Crew	Daily Output	Labor-Hours	Unit	Material	Labor	Equipment	Total	Total Incl O&P
0010	**RAILINGS, ORNAMENTAL**, shop fab'd, 3'-6" high, posts @ 5' O.C.										
0020	Bronze or stainless, minimum	G	1 Sswk	24	.333	L.F.	35	15.65		50.65	66
0100	Maximum	G		9	.889		345	41.50		386.50	455
0200	Aluminum ornamental rail, minimum	G		15	.533		35	25		60	82.50
0300	Maximum	G		8	1		105	47		152	198
0400	Hand-forged wrought iron, minimum	G		12	.667		109	31.50		140.50	175
0500	Maximum	G		8	1		325	47		372	445
0550	Steel, minimum	G		12	.667		26	31.50		57.50	83.50
0560	Maximum	G		8	1		78.50	47		125.50	169
0600	Composite metal/wood/glass, minimum			6	1.333		202	62.50		264.50	335
0700	Maximum		↓	5	1.600	↓	405	75		480	575

05 75 Decorative Formed Metal

05 75 13 – Columns

05 75 13.20 Columns, Ornamental			Crew	Daily Output	Labor-Hours	Unit	Material	Labor	Equipment	Total	Total Incl O&P
0010	**COLUMNS, ORNAMENTAL**, shop fabricated	R051223-10									
6400	Mild steel, flat, 9" wide, stock units, painted, plain	G	E-4	160	.200	V.L.F.	8.90	9.50	.91	19.31	27.50
6450	Fancy	G		160	.200		17	9.50	.91	27.41	36.50
6500	Corner columns, painted, plain	G		160	.200		15.50	9.50	.91	25.91	34.50
6550	Fancy	G	↓	160	.200	↓	30	9.50	.91	40.41	50.50

Division Notes

		CREW	DAILY OUTPUT	LABOR-HOURS	UNIT	2010 BARE COSTS MAT.	LABOR	EQUIP.	TOTAL	TOTAL INCL O&P

Estimating Tips

06 05 00 Common Work Results for Wood, Plastics, and Composites

- Common to any wood-framed structure are the accessory connector items such as screws, nails, adhesives, hangers, connector plates, straps, angles, and hold-downs. For typical wood-framed buildings, such as residential projects, the aggregate total for these items can be significant, especially in areas where seismic loading is a concern. For floor and wall framing, the material cost is based on 10 to 25 lbs. per MBF. Hold-downs, hangers, and other connectors should be taken off by the piece.

06 10 00 Carpentry

- Lumber is a traded commodity and therefore sensitive to supply and demand in the marketplace. Even in "budgetary" estimating of wood-framed projects, it is advisable to call local suppliers for the latest market pricing.
- Common quantity units for wood-framed projects are "thousand board feet" (MBF). A board foot is a volume of wood, 1″ x 1′ x 1′, or 144 cubic inches. Board-foot quantities are generally calculated using nominal material dimensions—dressed sizes are ignored. Board foot per lineal foot of any stick of lumber can be calculated by dividing the nominal cross-sectional area by 12. As an example, 2,000 lineal feet of 2 x 12 equates to 4 MBF by dividing the nominal area, 2 x 12, by 12, which equals 2, and multiplying by 2,000 to give 4,000 board feet. This simple rule applies to all nominal dimensioned lumber.
- Waste is an issue of concern at the quantity takeoff for any area of construction. Framing lumber is sold in even foot lengths, i.e., 10′, 12′, 14′, 16′, and depending on spans, wall heights, and the grade of lumber, waste is inevitable. A rule of thumb for lumber waste is 5%–10% depending on material quality and the complexity of the framing.
- Wood in various forms and shapes is used in many projects, even where the main structural framing is steel, concrete, or masonry. Plywood as a back-up partition material and 2x boards used as blocking and cant strips around roof edges are two common examples. The estimator should ensure that the costs of all wood materials are included in the final estimate.

06 20 00 Finish Carpentry

- It is necessary to consider the grade of workmanship when estimating labor costs for erecting millwork and interior finish. In practice, there are three grades: premium, custom, and economy. The RSMeans daily output for base and case moldings is in the range of 200 to 250 L.F. per carpenter per day. This is appropriate for most average custom-grade projects. For premium projects, an adjustment to productivity of 25%–50% should be made, depending on the complexity of the job.

Reference Numbers

Reference numbers are shown in shaded boxes at the beginning of some major classifications. These numbers refer to related items in the Reference Section. The reference information may be an estimating procedure, an alternate pricing method, or technical information.

Note: Not all subdivisions listed here necessarily appear in this publication.

06 05 05 – Selective Wood and Plastics Demolition

06 05 05.10 Selective Demolition Wood Framing		Crew	Daily Output	Labor-Hours	Unit	Material	2010 Bare Costs Labor	Equipment	Total	Total Incl O&P
0010	SELECTIVE DEMOLITION WOOD FRAMING R024119-10									
0100	Timber connector, nailed, small	1 Clab	96	.083	Ea.		2.76		2.76	4.25
0110	Medium		60	.133			4.41		4.41	6.80
0120	Large		48	.167			5.50		5.50	8.50
0130	Bolted, small		48	.167			5.50		5.50	8.50
0140	Medium		32	.250			8.30		8.30	12.75
0150	Large		24	.333			11.05		11.05	17
2958	Beams, 2" x 6"	2 Clab	1100	.015	L.F.		.48		.48	.74
2960	2" x 8"		825	.019			.64		.64	.99
2965	2" x 10"		665	.024			.80		.80	1.23
2970	2" x 12"		550	.029			.96		.96	1.49
2972	2" x 14"		470	.034			1.13		1.13	1.74
2975	4" x 8"	B-1	413	.058			1.96		1.96	3.03
2980	4" x 10"		330	.073			2.46		2.46	3.79
2985	4" x 12"		275	.087			2.95		2.95	4.54
3000	6" x 8"		275	.087			2.95		2.95	4.54
3040	6" x 10"		220	.109			3.68		3.68	5.70
3080	6" x 12"		185	.130			4.38		4.38	6.75
3120	8" x 12"		140	.171			5.80		5.80	8.95
3160	10" x 12"		110	.218			7.35		7.35	11.35
3162	Alternate pricing method		1.10	21.818	M.B.F.		735		735	1,125
3170	Blocking, in 16" OC wall framing, 2" x 4"	1 Clab	600	.013	L.F.		.44		.44	.68
3172	2" x 6"		400	.020			.66		.66	1.02
3174	In 24" OC wall framing, 2" x 4"		600	.013			.44		.44	.68
3176	2" x 6"		400	.020			.66		.66	1.02
3178	Alt method, wood blocking removal from wood framing		.40	20	M.B.F.		660		660	1,025
3179	Wood blocking removal from steel framing		.36	22.222	"		735		735	1,125
3180	Bracing, let in, 1" x 3", studs 16" OC		1050	.008	L.F.		.25		.25	.39
3181	Studs 24" OC		1080	.007			.25		.25	.38
3182	1" x 4", studs 16" OC		1050	.008			.25		.25	.39
3183	Studs 24" OC		1080	.007			.25		.25	.38
3184	1" x 6", studs 16" OC		1050	.008			.25		.25	.39
3185	Studs 24" OC		1080	.007			.25		.25	.38
3186	2" x 3", studs 16" OC		800	.010			.33		.33	.51
3187	Studs 24" OC		830	.010			.32		.32	.49
3188	2" x 4", studs 16" OC		800	.010			.33		.33	.51
3189	Studs 24" OC		830	.010			.32		.32	.49
3190	2" x 6", studs 16" OC		800	.010			.33		.33	.51
3191	Studs 24" OC		830	.010			.32		.32	.49
3192	2" x 8", studs 16" OC		800	.010			.33		.33	.51
3193	Studs 24" OC		830	.010			.32		.32	.49
3194	"T" shaped metal bracing, studs at 16" OC		1060	.008			.25		.25	.39
3195	Studs at 24" OC		1200	.007			.22		.22	.34
3196	Metal straps, studs at 16" OC		1200	.007			.22		.22	.34
3197	Studs at 24" OC		1240	.006			.21		.21	.33
3200	Columns, round, 8' to 14' tall		40	.200	Ea.		6.60		6.60	10.20
3202	Dimensional lumber sizes	2 Clab	1.10	14.545	M.B.F.		480		480	745
3250	Blocking, between joists	1 Clab	320	.025	Ea.		.83		.83	1.28
3252	Bridging, metal strap, between joists		320	.025	Pr.		.83		.83	1.28
3254	Wood, between joists		320	.025	"		.83		.83	1.28
3260	Door buck, studs, header & access., 8' high 2" x 4" wall, 3' wide		32	.250	Ea.		8.30		8.30	12.75
3261	4' wide		32	.250			8.30		8.30	12.75
3262	5' wide		32	.250			8.30		8.30	12.75

06 05 05 – Selective Wood and Plastics Demolition

06 05 05.10 Selective Demolition Wood Framing		Crew	Daily Output	Labor-Hours	Unit	Material	2010 Bare Costs Labor	Equipment	Total	Total Incl O&P
3263	6′ wide	1 Clab	32	.250	Ea.		8.30		8.30	12.75
3264	8′ wide		30	.267			8.85		8.85	13.60
3265	10′ wide		30	.267			8.85		8.85	13.60
3266	12′ wide		30	.267			8.85		8.85	13.60
3267	2″ x 6″ wall, 3′ wide		32	.250			8.30		8.30	12.75
3268	4′ wide		32	.250			8.30		8.30	12.75
3269	5′ wide		32	.250			8.30		8.30	12.75
3270	6′ wide		32	.250			8.30		8.30	12.75
3271	8′ wide		30	.267			8.85		8.85	13.60
3272	10′ wide		30	.267			8.85		8.85	13.60
3273	12′ wide		30	.267			8.85		8.85	13.60
3274	Window buck, studs, header & access, 8′ high 2″ x 4″ wall, 2′ wide		24	.333			11.05		11.05	17
3275	3′ wide		24	.333			11.05		11.05	17
3276	4′ wide		24	.333			11.05		11.05	17
3277	5′ wide		24	.333			11.05		11.05	17
3278	6′ wide		24	.333			11.05		11.05	17
3279	7′ wide		24	.333			11.05		11.05	17
3280	8′ wide		22	.364			12.05		12.05	18.55
3281	10′ wide		22	.364			12.05		12.05	18.55
3282	12′ wide		22	.364			12.05		12.05	18.55
3283	2″ x 6″ wall, 2′ wide		24	.333			11.05		11.05	17
3284	3′ wide		24	.333			11.05		11.05	17
3285	4′ wide		24	.333			11.05		11.05	17
3286	5′ wide		24	.333			11.05		11.05	17
3287	6′ wide		24	.333			11.05		11.05	17
3288	7′ wide		24	.333			11.05		11.05	17
3289	8′ wide		22	.364			12.05		12.05	18.55
3290	10′ wide		22	.364			12.05		12.05	18.55
3291	12′ wide		22	.364			12.05		12.05	18.55
3400	Fascia boards, 1″ x 6″		500	.016	L.F.		.53		.53	.82
3440	1″ x 8″		450	.018			.59		.59	.91
3480	1″ x 10″		400	.020			.66		.66	1.02
3490	2″ x 6″		450	.018			.59		.59	.91
3500	2″ x 8″		400	.020			.66		.66	1.02
3510	2″ x 10″		350	.023			.76		.76	1.17
3610	Furring, on wood walls or ceiling		4000	.002	S.F.		.07		.07	.10
3620	On masonry or concrete walls or ceiling		1200	.007	″		.22		.22	.34
3800	Headers over openings, 2 @ 2″ x 6″		110	.073	L.F.		2.41		2.41	3.71
3840	2 @ 2″ x 8″		100	.080			2.65		2.65	4.08
3880	2 @ 2″ x 10″		90	.089			2.94		2.94	4.54
3885	Alternate pricing method		.26	30.651	M.B.F.		1,025		1,025	1,575
3920	Joists, 1″ x 4″		1250	.006	L.F.		.21		.21	.33
3930	1″ x 6″		1135	.007			.23		.23	.36
3940	1″ x 8″		1000	.008			.26		.26	.41
3950	1″ x 10″		895	.009			.30		.30	.46
3960	1″ x 12″		765	.010			.35		.35	.53
4200	2″ x 4″	2 Clab	1000	.016			.53		.53	.82
4230	2″ x 6″		970	.016			.55		.55	.84
4240	2″ x 8″		940	.017			.56		.56	.87
4250	2″ x 10″		910	.018			.58		.58	.90
4280	2″ x 12″		880	.018			.60		.60	.93
4281	2″ x 14″		850	.019			.62		.62	.96
4282	Composite joists, 9-1/2″		960	.017			.55		.55	.85

06 05 05 – Selective Wood and Plastics Demolition

06 05 05.10 Selective Demolition Wood Framing		Crew	Daily Output	Labor-Hours	Unit	Material	2010 Bare Costs Labor	Equipment	Total	Total Incl O&P
4283	11-7/8"	2 Clab	930	.017	L.F.		.57		.57	.88
4284	14"		897	.018			.59		.59	.91
4285	16"		865	.019			.61		.61	.94
4290	Wood joists, alternate pricing method		1.50	10.667	M.B.F.		355		355	545
4500	Open web joist, 12" deep		500	.032	L.F.		1.06		1.06	1.63
4505	14" deep		475	.034			1.11		1.11	1.72
4510	16" deep		450	.036			1.18		1.18	1.82
4520	18" deep		425	.038			1.25		1.25	1.92
4530	24" deep		400	.040			1.32		1.32	2.04
4550	Ledger strips, 1" x 2"	1 Clab	1200	.007			.22		.22	.34
4560	1" x 3"		1200	.007			.22		.22	.34
4570	1" x 4"		1200	.007			.22		.22	.34
4580	2" x 2"		1100	.007			.24		.24	.37
4590	2" x 4"		1000	.008			.26		.26	.41
4600	2" x 6"		1000	.008			.26		.26	.41
4601	2" x 8 or 2" x 10"		800	.010			.33		.33	.51
4602	4" x 6"		600	.013			.44		.44	.68
4604	4" x 8"		450	.018			.59		.59	.91
5400	Posts, 4" x 4"	2 Clab	800	.020			.66		.66	1.02
5405	4" x 6"		550	.029			.96		.96	1.49
5410	4" x 8"		440	.036			1.20		1.20	1.86
5425	4" x 10"		390	.041			1.36		1.36	2.09
5430	4" x 12"		350	.046			1.51		1.51	2.33
5440	6" x 6"		400	.040			1.32		1.32	2.04
5445	6" x 8"		350	.046			1.51		1.51	2.33
5450	6" x 10"		320	.050			1.66		1.66	2.55
5455	6" x 12"		290	.055			1.83		1.83	2.82
5480	8" x 8"		300	.053			1.77		1.77	2.72
5500	10" x 10"		240	.067			2.21		2.21	3.40
5660	Tongue and groove floor planks		2	8	M.B.F.		265		265	410
5750	Rafters, ordinary, 16" OC, 2" x 4"		880	.018	S.F.		.60		.60	.93
5755	2" x 6"		840	.019			.63		.63	.97
5760	2" x 8"		820	.020			.65		.65	1
5770	2" x 10"		820	.020			.65		.65	1
5780	2" x 12"		810	.020			.65		.65	1.01
5785	24" OC, 2" x 4"		1170	.014			.45		.45	.70
5786	2" x 6"		1117	.014			.47		.47	.73
5787	2" x 8"		1091	.015			.49		.49	.75
5788	2" x 10"		1091	.015			.49		.49	.75
5789	2" x 12"		1077	.015			.49		.49	.76
5795	Rafters, ordinary, 2" x 4" (alternate method)		862	.019	L.F.		.61		.61	.95
5800	2" x 6" (alternate method)		850	.019			.62		.62	.96
5840	2" x 8" (alternate method)		837	.019			.63		.63	.98
5855	2" x 10" (alternate method)		825	.019			.64		.64	.99
5865	2" x 12" (alternate method)		812	.020			.65		.65	1.01
5870	Sill plate, 2" x 4"	1 Clab	1170	.007			.23		.23	.35
5871	2" x 6"		780	.010			.34		.34	.52
5872	2" x 8"		586	.014			.45		.45	.70
5873	Alternate pricing method		.78	10.256	M.B.F.		340		340	525
5885	Ridge board, 1" x 4"	2 Clab	900	.018	L.F.		.59		.59	.91
5886	1" x 6"		875	.018			.61		.61	.93
5887	1" x 8"		850	.019			.62		.62	.96
5888	1" x 10"		825	.019			.64		.64	.99

06 05 Common Work Results for Wood, Plastics, and Composites

06 05 05 – Selective Wood and Plastics Demolition

06 05 05.10 Selective Demolition Wood Framing		Crew	Daily Output	Labor-Hours	Unit	Material	2010 Bare Costs Labor	Equipment	Total	Total Incl O&P
5889	1" x 12"	2 Clab	800	.020	L.F.		.66		.66	1.02
5890	2" x 4"		900	.018			.59		.59	.91
5892	2" x 6"		875	.018			.61		.61	.93
5894	2" x 8"		850	.019			.62		.62	.96
5896	2" x 10"		825	.019			.64		.64	.99
5898	2" x 12"		800	.020			.66		.66	1.02
6050	Rafter tie, 1" x 4"		1250	.013			.42		.42	.65
6052	1" x 6"		1135	.014			.47		.47	.72
6054	2" x 4"		1000	.016			.53		.53	.82
6056	2" x 6"		970	.016			.55		.55	.84
6070	Sleepers, on concrete, 1" x 2"	1 Clab	4700	.002			.06		.06	.09
6075	1" x 3"		4000	.002			.07		.07	.10
6080	2" x 4"		3000	.003			.09		.09	.14
6085	2" x 6"		2600	.003			.10		.10	.16
6086	Sheathing from roof, 5/16"	2 Clab	1600	.010	S.F.		.33		.33	.51
6088	3/8"		1525	.010			.35		.35	.54
6090	1/2"		1400	.011			.38		.38	.58
6092	5/8"		1300	.012			.41		.41	.63
6094	3/4"		1200	.013			.44		.44	.68
6096	Board sheathing from roof		1400	.011			.38		.38	.58
6100	Sheathing, from walls, 1/4"		1200	.013			.44		.44	.68
6110	5/16"		1175	.014			.45		.45	.70
6120	3/8"		1150	.014			.46		.46	.71
6130	1/2"		1125	.014			.47		.47	.73
6140	5/8"		1100	.015			.48		.48	.74
6150	3/4"		1075	.015			.49		.49	.76
6152	Board sheathing from walls		1500	.011			.35		.35	.54
6158	Subfloor, with boards		1050	.015			.50		.50	.78
6160	Plywood, 1/2" thick		768	.021			.69		.69	1.06
6162	5/8" thick		760	.021			.70		.70	1.07
6164	3/4" thick		750	.021			.71		.71	1.09
6165	1-1/8" thick		720	.022			.74		.74	1.13
6166	Underlayment, particle board, 3/8" thick	1 Clab	780	.010			.34		.34	.52
6168	1/2" thick		768	.010			.34		.34	.53
6170	5/8" thick		760	.011			.35		.35	.54
6172	3/4" thick		750	.011			.35		.35	.54
6200	Stairs and stringers, minimum	2 Clab	40	.400	Riser		13.25		13.25	20.50
6240	Maximum	"	26	.615	"		20.50		20.50	31.50
6300	Components, tread	1 Clab	110	.073	Ea.		2.41		2.41	3.71
6320	Riser		80	.100	"		3.31		3.31	5.10
6390	Stringer, 2" x 10"		260	.031	L.F.		1.02		1.02	1.57
6400	2" x 12"		260	.031			1.02		1.02	1.57
6410	3" x 10"		250	.032			1.06		1.06	1.63
6420	3" x 12"		250	.032			1.06		1.06	1.63
6590	Wood studs, 2" x 3"	2 Clab	3076	.005			.17		.17	.27
6600	2" x 4"		2000	.008			.26		.26	.41
6640	2" x 6"		1600	.010			.33		.33	.51
6720	Wall framing, including studs, plates and blocking, 2" x 4"	1 Clab	600	.013	S.F.		.44		.44	.68
6740	2" x 6"		480	.017	"		.55		.55	.85
6750	Headers, 2" x 4"		1125	.007	L.F.		.24		.24	.36
6755	2" x 6"		1125	.007			.24		.24	.36
6760	2" x 8"		1050	.008			.25		.25	.39
6765	2" x 10"		1050	.008			.25		.25	.39

06 05 Common Work Results for Wood, Plastics, and Composites

06 05 05 – Selective Wood and Plastics Demolition

06 05 05.10 Selective Demolition Wood Framing

	06 05 05.10 Selective Demolition Wood Framing	Crew	Daily Output	Labor-Hours	Unit	Material	2010 Bare Costs Labor	Equipment	Total	Total Incl O&P
6770	2" x 12"	1 Clab	1000	.008	L.F.		.26		.26	.41
6780	4" x 10"		525	.015			.50		.50	.78
6785	4" x 12"		500	.016			.53		.53	.82
6790	6" x 8"		560	.014			.47		.47	.73
6795	6" x 10"		525	.015			.50		.50	.78
6797	6" x 12"		500	.016			.53		.53	.82
8000	Soffit, T & G wood		520	.015	S.F.		.51		.51	.79
8010	Hardboard, vinyl or aluminum		640	.013			.41		.41	.64
8030	Plywood	2 Carp	315	.051			2.11		2.11	3.25
9500	See Div. 02 41 19.23 for rubbish handling									

06 05 05.20 Selective Demolition Millwork and Trim

	06 05 05.20 Selective Demolition Millwork and Trim	Crew	Daily Output	Labor-Hours	Unit	Material	Labor	Equipment	Total	Total Incl O&P
0010	**SELECTIVE DEMOLITION MILLWORK AND TRIM** R024119-10									
1000	Cabinets, wood, base cabinets, per L.F.	2 Clab	80	.200	L.F.		6.60		6.60	10.20
1020	Wall cabinets, per L.F.	"	80	.200	"		6.60		6.60	10.20
1060	Remove and reset, base cabinets	2 Carp	18	.889	Ea.		37		37	57
1070	Wall cabinets	"	20	.800	"		33		33	51
1100	Steel, painted, base cabinets	2 Clab	60	.267	L.F.		8.85		8.85	13.60
1120	Wall cabinets		60	.267	"		8.85		8.85	13.60
1200	Casework, large area		320	.050	S.F.		1.66		1.66	2.55
1220	Selective		200	.080	"		2.65		2.65	4.08
1500	Counter top, minimum		200	.080	L.F.		2.65		2.65	4.08
1510	Maximum		120	.133			4.41		4.41	6.80
1550	Remove and reset, minimum	2 Carp	50	.320			13.30		13.30	20.50
1560	Maximum	"	40	.400			16.60		16.60	25.50
2000	Paneling, 4' x 8' sheets	2 Clab	2000	.008	S.F.		.26		.26	.41
2100	Boards, 1" x 4"		700	.023			.76		.76	1.17
2120	1" x 6"		750	.021			.71		.71	1.09
2140	1" x 8"		800	.020			.66		.66	1.02
3000	Trim, baseboard, to 6" wide		1200	.013	L.F.		.44		.44	.68
3040	Greater than 6" and up to 12" wide		1000	.016			.53		.53	.82
3080	Remove and reset, minimum	2 Carp	400	.040			1.66		1.66	2.56
3090	Maximum	"	300	.053			2.22		2.22	3.42
3100	Ceiling trim	2 Clab	1000	.016			.53		.53	.82
3120	Chair rail		1200	.013			.44		.44	.68
3140	Railings with balusters		240	.067			2.21		2.21	3.40
3160	Wainscoting		700	.023	S.F.		.76		.76	1.17

06 05 23 – Wood, Plastic, and Composite Fastenings

06 05 23.10 Nails

	06 05 23.10 Nails	Crew	Daily Output	Labor-Hours	Unit	Material	Labor	Equipment	Total	Total Incl O&P
0010	**NAILS**, material only, based upon 50# box purchase									
0020	Copper nails, plain				Lb.	10			10	11
0400	Stainless steel, plain					7.55			7.55	8.30
0500	Box, 3d to 20d, bright					1.26			1.26	1.39
0520	Galvanized					1.77			1.77	1.95
0600	Common, 3d to 60d, plain					1.30			1.30	1.43
0700	Galvanized					1.56			1.56	1.72
0800	Aluminum					4.55			4.55	5
1000	Annular or spiral thread, 4d to 60d, plain					1.80			1.80	1.98
1200	Galvanized					3			3	3.30
1400	Drywall nails, plain					.85			.85	.94
1600	Galvanized					1.75			1.75	1.93
1800	Finish nails, 4d to 10d, plain					1.40			1.40	1.54
2000	Galvanized					1.79			1.79	1.97

06 05 Common Work Results for Wood, Plastics, and Composites

06 05 23 – Wood, Plastic, and Composite Fastenings

06 05 23.10 Nails		Crew	Daily Output	Labor-Hours	Unit	Material	Labor (2010 Bare Costs)	Equipment	Total	Total Incl O&P
2100	Aluminum				Lb.	4.10			4.10	4.51
2300	Flooring nails, hardened steel, 2d to 10d, plain					2.55			2.55	2.81
2400	Galvanized					3.88			3.88	4.27
2500	Gypsum lath nails, 1-1/8", 13 ga. flathead, blued					2.68			2.68	2.95
2600	Masonry nails, hardened steel, 3/4" to 3" long, plain					2.45			2.45	2.70
2700	Galvanized					3.72			3.72	4.09
2900	Roofing nails, threaded, galvanized					1.63			1.63	1.79
3100	Aluminum					4.80			4.80	5.30
3300	Compressed lead head, threaded, galvanized					2.47			2.47	2.72
3600	Siding nails, plain shank, galvanized					2			2	2.20
3800	Aluminum					4.11			4.11	4.52
5000	Add to prices above for cement coating					.11			.11	.12
5200	Zinc or tin plating					.14			.14	.15
5500	Vinyl coated sinkers, 8d to 16d					.62			.62	.68

06 05 23.40 Sheet Metal Screws		Crew	Daily Output	Labor-Hours	Unit	Material	Labor	Equipment	Total	Total Incl O&P
0010	**SHEET METAL SCREWS**									
0020	Steel, standard, #8 x 3/4", plain				C	3.42			3.42	3.76
0100	Galvanized					3.42			3.42	3.76
0300	#10 x 1", plain					4.70			4.70	5.15
0400	Galvanized					4.70			4.70	5.15
0600	With washers, #14 x 1", plain					16.05			16.05	17.65
0700	Galvanized					16.05			16.05	17.65
0900	#14 x 2", plain					21			21	23
1000	Galvanized					21			21	23
1500	Self-drilling, with washers, (pinch point) #8 x 3/4", plain					12.25			12.25	13.45
1600	Galvanized					12.25			12.25	13.45
1800	#10 x 3/4", plain					14.50			14.50	15.95
1900	Galvanized					14.50			14.50	15.95
3000	Stainless steel w/aluminum or neoprene washers, #14 x 1", plain					35			35	38.50
3100	#14 x 2", plain					44			44	48.50

06 05 23.50 Wood Screws		Crew	Daily Output	Labor-Hours	Unit	Material	Labor	Equipment	Total	Total Incl O&P
0010	**WOOD SCREWS**									
0020	#8 x 1" long, steel				C	5.10			5.10	5.60
0100	Brass					10.30			10.30	11.35
0200	#8, 2" long, steel					5.10			5.10	5.60
0300	Brass					17.40			17.40	19.10
0400	#10, 1" long, steel					3.63			3.63	3.99
0500	Brass					15.80			15.80	17.40
0600	#10, 2" long, steel					6.15			6.15	6.80
0700	Brass					24			24	26.50
0800	#10, 3" long, steel					10.30			10.30	11.30
1000	#12, 2" long, steel					9.05			9.05	9.95
1100	Brass					31.50			31.50	34.50
1500	#12, 3" long, steel					12.10			12.10	13.30
2000	#12, 4" long, steel					18.15			18.15	19.95

06 05 23.60 Timber Connectors		Crew	Daily Output	Labor-Hours	Unit	Material	Labor	Equipment	Total	Total Incl O&P
0010	**TIMBER CONNECTORS**									
0020	Add up cost of each part for total cost of connection									
0100	Connector plates, steel, with bolts, straight	2 Carp	75	.213	Ea.	25.50	8.85		34.35	41.50
0110	Tee, 7 gauge		50	.320		29.50	13.30		42.80	52.50
0120	T- Strap, 14 gauge, 12" x 8" x 2"		50	.320		29.50	13.30		42.80	52.50
0150	Anchor plates, 7 ga, 9" x 7"		75	.213		25.50	8.85		34.35	41.50

06 05 Common Work Results for Wood, Plastics, and Composites

06 05 23 – Wood, Plastic, and Composite Fastenings

06 05 23.60 Timber Connectors		Crew	Daily Output	Labor-Hours	Unit	Material	2010 Bare Costs Labor	Equipment	Total	Total Incl O&P
0200	Bolts, machine, sq. hd. with nut & washer, 1/2" diameter, 4" long	1 Carp	140	.057	Ea.	.86	2.37		3.23	4.61
0300	7-1/2" long		130	.062		1.50	2.56		4.06	5.60
0500	3/4" diameter, 7-1/2" long		130	.062		4.39	2.56		6.95	8.75
0610	Machine bolts, w/ nut, washer, 3/4" diam., 15" L, HD's & beam hangers		95	.084		8.10	3.50		11.60	14.35
0800	Drilling bolt holes in timber, 1/2" diameter		450	.018	Inch		.74		.74	1.14
0900	1" diameter		350	.023	"		.95		.95	1.46
1100	Framing anchor, angle, 3" x 3" x 1-1/2", 12 ga		175	.046	Ea.	2.20	1.90		4.10	5.35
1150	Framing anchors, 18 gauge, 4-1/2" x 2-3/4"		175	.046		2.20	1.90		4.10	5.35
1160	Framing anchors, 18 gauge, 4-1/2" x 3"		175	.046		2.20	1.90		4.10	5.35
1170	Clip anchors plates, 18 gauge, 12" x 1-1/8"		175	.046		2.20	1.90		4.10	5.35
1250	Holdowns, 3 gauge base, 10 gauge body		8	1		21	41.50		62.50	87
1260	Holdowns, 7 gauge 11-1/16" x 3-1/4"		8	1		21	41.50		62.50	87
1270	Holdowns, 7 gauge 14-3/8" x 3-1/8"		8	1		21	41.50		62.50	87
1275	Holdowns, 12 gauge 8" x 2-1/2"		8	1		21	41.50		62.50	87
1300	Joist and beam hangers, 18 ga. galv., for 2" x 4" joist		175	.046		.61	1.90		2.51	3.60
1400	2" x 6" to 2" x 10" joist		165	.048		1.17	2.01		3.18	4.40
1600	16 ga. galv., 3" x 6" to 3" x 10" joist		160	.050		2.49	2.08		4.57	5.95
1700	3" x 10" to 3" x 14" joist		160	.050		4.14	2.08		6.22	7.75
1800	4" x 6" to 4" x 10" joist		155	.052		2.58	2.14		4.72	6.15
1900	4" x 10" to 4" x 14" joist		155	.052		4.23	2.14		6.37	7.95
2000	Two-2" x 6" to two-2" x 10" joists		150	.053		3.52	2.22		5.74	7.30
2100	Two-2" x 10" to two-2" x 14" joists		150	.053		3.94	2.22		6.16	7.75
2300	3/16" thick, 6" x 8" joist		145	.055		54.50	2.29		56.79	63.50
2400	6" x 10" joist		140	.057		57	2.37		59.37	66
2500	6" x 12" joist		135	.059		59.50	2.46		61.96	69
2700	1/4" thick, 6" x 14" joist		130	.062		61.50	2.56		64.06	72
2900	Plywood clips, extruded aluminum H clip, for 3/4" panels					.20			.20	.22
3000	Galvanized 18 ga. back-up clip					.16			.16	.18
3200	Post framing, 16 ga. galv. for 4" x 4" base, 2 piece	1 Carp	130	.062		13.80	2.56		16.36	19.10
3300	Cap		130	.062		19.15	2.56		21.71	25
3500	Rafter anchors, 18 ga. galv., 1-1/2" wide, 5-1/4" long		145	.055		.41	2.29		2.70	3.98
3600	10-3/4" long		145	.055		1.23	2.29		3.52	4.88
3800	Shear plates, 2-5/8" diameter		120	.067		2	2.77		4.77	6.45
3900	4" diameter		115	.070		4.75	2.89		7.64	9.70
4000	Sill anchors, embedded in concrete or block, 25-1/2" long		115	.070		10.70	2.89		13.59	16.20
4100	Spike grids, 3" x 6"		120	.067		.81	2.77		3.58	5.15
4400	Split rings, 2-1/2" diameter		120	.067		1.65	2.77		4.42	6.10
4500	4" diameter		110	.073		2.50	3.02		5.52	7.40
4550	Tie plate, 20 gauge, 7" x 3 1/8"		110	.073		2.50	3.02		5.52	7.40
4560	Tie plate, 20 gauge, 5" x 4 1/8"		110	.073		2.50	3.02		5.52	7.40
4575	Twist straps, 18 gauge, 12" x 1 1/4"		110	.073		2.50	3.02		5.52	7.40
4580	Twist straps, 18 gauge, 16" x 1 1/4"		110	.073		2.50	3.02		5.52	7.40
4600	Strap ties, 20 ga., 2 -1/16" wide, 12 13/16" long		180	.044		.80	1.85		2.65	3.73
4700	Strap ties, 16 ga., 1-3/8" wide, 12" long		180	.044		.80	1.85		2.65	3.73
4800	21-5/8" x 1-1/4"		160	.050		2.49	2.08		4.57	5.95
5000	Toothed rings, 2-5/8" or 4" diameter		90	.089		1.49	3.69		5.18	7.35
5200	Truss plates, nailed, 20 gauge, up to 32' span		17	.471	Truss	10.85	19.55		30.40	42
5400	Washers, 2" x 2" x 1/8"				Ea.	.34			.34	.37
5500	3" x 3" x 3/16"				"	.89			.89	.98
6000	Angles and gussets, painted									
6012	7 ga., 3-1/4" x 3-1/4" x 2-1/2" long	1 Carp	1.90	4.211	C	930	175		1,105	1,300
6014	3-1/4" x 3-1/4" x 5" long		1.90	4.211		1,825	175		2,000	2,275
6016	3-1/4" x 3-1/4" x 7-1/2" long		1.85	4.324		3,425	180		3,605	4,050

06 05 23 – Wood, Plastic, and Composite Fastenings

06 05 23.60 Timber Connectors		Crew	Daily Output	Labor-Hours	Unit	Material	2010 Bare Costs Labor	Equipment	Total	Total Incl O&P
6018	5-3/4" x 5-3/4" x 2-1/2" long	1 Carp	1.85	4.324	C	2,225	180		2,405	2,700
6020	5-3/4" x 5-3/4" x 5" long		1.85	4.324		3,525	180		3,705	4,150
6022	5-3/4" x 5-3/4" x 7-1/2" long		1.80	4.444		5,200	185		5,385	6,000
6024	3 ga., 4-1/4" x 4-1/4" x 3" long		1.85	4.324		2,350	180		2,530	2,850
6026	4-1/4" x 4-1/4" x 6" long		1.85	4.324		5,050	180		5,230	5,825
6028	4-1/4" x 4-1/4" x 9" long		1.80	4.444		5,675	185		5,860	6,525
6030	7-1/4" x 7-1/4" x 3" long		1.80	4.444		4,075	185		4,260	4,750
6032	7-1/4" x 7-1/4" x 6" long		1.80	4.444		5,475	185		5,660	6,300
6034	7-1/4" x 7-1/4" x 9" long	↓	1.75	4.571	↓	12,300	190		12,490	13,800
6036	Gussets									
6038	7 ga., 8-1/8" x 8-1/8" x 2-3/4" long	1 Carp	1.80	4.444	C	3,875	185		4,060	4,550
6040	3 ga., 9-3/4" x 9-3/4" x 3-1/4" long	"	1.80	4.444	"	5,400	185		5,585	6,200
6101	Beam hangers, polymer painted									
6102	Bolted, 3 ga., (W x H x L)									
6104	3-1/4" x 9" x 12" top flange	1 Carp	1	8	C	16,500	330		16,830	18,700
6106	5-1/4" x 9" x 12" top flange		1	8		17,200	330		17,530	19,400
6108	5-1/4" x 11" x 11-3/4" top flange		1	8		19,600	330		19,930	22,000
6110	6-7/8" x 9" x 12" top flange		1	8		17,800	330		18,130	20,100
6112	6-7/8" x 11" x 13-1/2" top flange		1	8		20,600	330		20,930	23,100
6114	8-7/8" x 11" x 15-1/2" top flange	↓	1	8	↓	22,000	330		22,330	24,700
6116	Nailed, 3 ga., (W x H x L)									
6118	3-1/4" x 10-1/2" x 10" top flange	1 Carp	1.80	4.444	C	17,200	185		17,385	19,200
6120	3-1/4" x 10-1/2" x 12" top flange		1.80	4.444		17,200	185		17,385	19,200
6122	5-1/4" x 9-1/2" x 10" top flange		1.80	4.444		17,600	185		17,785	19,600
6124	5-1/4" x 9-1/2" x 12" top flange		1.80	4.444		17,600	185		17,785	19,600
6128	6-7/8" x 8-1/2" x 12" top flange	↓	1.80	4.444	↓	18,000	185		18,185	20,100
6134	Saddle hangers, glu-lam (W x H x L)									
6136	3-1/4" x 10-1/2" x 5-1/4" x 6" saddle	1 Carp	.50	16	C	12,900	665		13,565	15,200
6138	3-1/4" x 10-1/2" x 6-7/8" x 6" saddle		.50	16		13,600	665		14,265	15,900
6140	3-1/4" x 10-1/2" x 8-7/8" x 6" saddle		.50	16		14,200	665		14,865	16,700
6142	3-1/4" x 19-1/2" x 5-1/4" x 10-1/8" saddle		.40	20		12,900	830		13,730	15,500
6144	3-1/4" x 19-1/2" x 6-7/8" x 10-1/8" saddle		.40	20		13,600	830		14,430	16,200
6146	3-1/4" x 19-1/2" x 8-7/8" x 10-1/8" saddle		.40	20		14,200	830		15,030	17,000
6148	5-1/4" x 9-1/2" x 5-1/4" x 12" saddle		.50	16		15,400	665		16,065	17,900
6150	5-1/4" x 9-1/2" x 6-7/8" x 9" saddle		.50	16		16,800	665		17,465	19,500
6152	5-1/4" x 10-1/2" x spec x 12" saddle		.50	16		18,300	665		18,965	21,100
6154	5-1/4" x 18" x 5-1/4" x 12-1/8" saddle		.40	20		15,400	830		16,230	18,200
6156	5-1/4" x 18" x 6-7/8" x 12-1/8" saddle		.40	20		16,800	830		17,630	19,800
6158	5-1/4" x 18" x spec x 12-1/8" saddle		.40	20		18,300	830		19,130	21,400
6160	6-7/8" x 8-1/2" x 6-7/8" x 12" saddle		.50	16		18,400	665		19,065	21,200
6162	6-7/8" x 8-1/2" x 8-7/8" x 12" saddle		.50	16		19,000	665		19,665	21,900
6164	6-7/8" x 10-1/2" x spec x 12" saddle		.50	16		18,400	665		19,065	21,200
6166	6-7/8" x 18" x 6-7/8" x 13-3/4" saddle		.40	20		18,400	830		19,230	21,500
6168	6-7/8" x 18" x 8-7/8" x 13-3/4" saddle		.40	20		19,000	830		19,830	22,200
6170	6-7/8" x 18" x spec x 13-3/4" saddle		.40	20		20,500	830		21,330	23,800
6172	8-7/8" x 18" x spec x 15-3/4" saddle	↓	.40	20	↓	32,000	830		32,830	36,500
6201	Beam and purlin hangers, galvanized, 12 ga.									
6202	Purlin or joist size, 3" x 8"	1 Carp	1.70	4.706	C	1,700	196		1,896	2,175
6204	3" x 10"		1.70	4.706		1,850	196		2,046	2,325
6206	3" x 12"		1.65	4.848		2,100	201		2,301	2,625
6208	3" x 14"		1.65	4.848		2,250	201		2,451	2,775
6210	3" x 16"		1.65	4.848		2,375	201		2,576	2,925
6212	4" x 8"	↓	1.65	4.848	↓	1,725	201		1,926	2,175

06 05 23.60	Timber Connectors	Crew	Daily Output	Labor-Hours	Unit	2010 Bare Costs Material	Labor	Equipment	Total	Total Incl O&P
6214	4" x 10"	1 Carp	1.65	4.848	C	1,850	201		2,051	2,325
6216	4" x 12"		1.60	5		2,200	208		2,408	2,725
6218	4" x 14"		1.60	5		2,325	208		2,533	2,875
6220	4" x 16"		1.60	5		2,450	208		2,658	3,025
6222	6" x 8"		1.60	5		2,200	208		2,408	2,750
6224	6" x 10"		1.55	5.161		2,250	214		2,464	2,800
6226	6" x 12"		1.55	5.161		3,850	214		4,064	4,575
6228	6" x 14"		1.50	5.333		4,075	222		4,297	4,850
6230	6" x 16"		1.50	5.333		4,325	222		4,547	5,100
6250	Beam seats									
6252	Beam size, 5-1/4" wide									
6254	5" x 7" x 1/4"	1 Carp	1.80	4.444	C	6,500	185		6,685	7,425
6256	6" x 7" x 3/8"		1.80	4.444		7,275	185		7,460	8,275
6258	7" x 7" x 3/8"		1.80	4.444		7,775	185		7,960	8,825
6260	8" x 7" x 3/8"		1.80	4.444		9,175	185		9,360	10,400
6262	Beam size, 6-7/8" wide									
6264	5" x 9" x 1/4"	1 Carp	1.80	4.444	C	7,750	185		7,935	8,800
6266	6" x 9" x 3/8"		1.80	4.444		10,100	185		10,285	11,400
6268	7" x 9" x 3/8"		1.80	4.444		10,200	185		10,385	11,600
6270	8" x 9" x 3/8"		1.80	4.444		12,100	185		12,285	13,600
6272	Special beams, over 6-7/8" wide									
6274	5" x 10" x 3/8"	1 Carp	1.80	4.444	C	10,500	185		10,685	11,900
6276	6" x 10" x 3/8"		1.80	4.444		12,200	185		12,385	13,800
6278	7" x 10" x 3/8"		1.80	4.444		12,800	185		12,985	14,400
6280	8" x 10" x 3/8"		1.75	4.571		13,700	190		13,890	15,400
6282	5-1/4" x 12" x 5/16"		1.75	4.571		10,600	190		10,790	12,000
6284	6-1/2" x 12" x 3/8"		1.75	4.571		17,600	190		17,790	19,700
6286	5-1/4" x 16" x 5/16"		1.70	4.706		15,600	196		15,796	17,500
6288	6-1/2" x 16" x 3/8"		1.70	4.706		20,400	196		20,596	22,700
6290	5-1/4" x 20" x 5/16"		1.70	4.706		18,400	196		18,596	20,500
6292	6-1/2" x 20" x 3/8"		1.65	4.848		24,000	201		24,201	26,700
6300	Column bases									
6302	4 x 4, 16 ga.	1 Carp	1.80	4.444	C	655	185		840	1,000
6306	7 ga.		1.80	4.444		2,400	185		2,585	2,925
6308	4 x 6, 16 ga.		1.80	4.444		1,525	185		1,710	1,950
6312	7 ga.		1.80	4.444		2,500	185		2,685	3,025
6314	6 x 6, 16 ga.		1.75	4.571		1,725	190		1,915	2,200
6318	7 ga.		1.75	4.571		3,425	190		3,615	4,050
6320	6 x 8, 7 ga.		1.70	4.706		2,675	196		2,871	3,225
6322	6 x 10, 7 ga.		1.70	4.706		2,850	196		3,046	3,450
6324	6 x 12, 7 ga.		1.70	4.706		3,100	196		3,296	3,700
6326	8 x 8, 7 ga.		1.65	4.848		5,225	201		5,426	6,050
6330	8 x 10, 7 ga.		1.65	4.848		6,250	201		6,451	7,175
6332	8 x 12, 7 ga.		1.60	5		6,800	208		7,008	7,800
6334	10 x 10, 3 ga.		1.60	5		6,925	208		7,133	7,950
6336	10 x 12, 3 ga.		1.60	5		7,975	208		8,183	9,075
6338	12 x 12, 3 ga.		1.55	5.161		8,650	214		8,864	9,825
6350	Column caps, painted, 3 ga.									
6352	3-1/4" x 3-5/8"	1 Carp	1.80	4.444	C	8,800	185		8,985	9,950
6354	3-1/4" x 5-1/2"		1.80	4.444		8,800	185		8,985	9,950
6356	3-5/8" x 3-5/8"		1.80	4.444		7,200	185		7,385	8,175
6358	3-5/8" x 5-1/2"		1.80	4.444		7,200	185		7,385	8,175
6360	5-1/4" x 5-1/2"		1.75	4.571		9,400	190		9,590	10,600

06 05 23.60 Timber Connectors		Crew	Daily Output	Labor-Hours	Unit	2010 Bare Costs Material	Labor	Equipment	Total	Total Incl O&P
6362	5-1/4" x 7-1/2"	1 Carp	1.75	4.571	C	9,400	190		9,590	10,600
6364	5-1/2" x 3-5/8"		1.75	4.571		10,100	190		10,290	11,500
6366	5-1/2" x 5-1/2"		1.75	4.571		10,100	190		10,290	11,500
6368	5-1/2" x 7-1/2"		1.70	4.706		10,100	196		10,296	11,500
6370	6-7/8" x 5-1/2"		1.70	4.706		10,600	196		10,796	11,900
6372	6-7/8" x 6-7/8"		1.70	4.706		10,600	196		10,796	11,900
6374	6-7/8" x 7-1/2"		1.70	4.706		10,600	196		10,796	11,900
6376	7-1/2" x 5-1/2"		1.65	4.848		11,100	201		11,301	12,500
6378	7-1/2" x 7-1/2"		1.65	4.848		11,100	201		11,301	12,500
6380	8-7/8" x 5-1/2"		1.60	5		11,700	208		11,908	13,200
6382	8-7/8" x 7-1/2"		1.60	5		11,700	208		11,908	13,200
6384	9-1/2" x 5-1/2"	↓	1.60	5	↓	15,800	208		16,008	17,700
6400	Floor tie anchors, polymer paint									
6402	10 ga., 3" x 37-1/2"	1 Carp	1.80	4.444	C	4,200	185		4,385	4,900
6404	3-1/2" x 45-1/2"		1.75	4.571		4,400	190		4,590	5,150
6406	3 ga., 3-1/2" x 56"	↓	1.70	4.706	↓	7,775	196		7,971	8,850
6410	Girder hangers									
6412	6" wall thickness, 4" x 6"	1 Carp	1.80	4.444	C	2,375	185		2,560	2,875
6414	4" x 8"		1.80	4.444		2,650	185		2,835	3,175
6416	8" wall thickness, 4" x 6"		1.80	4.444		2,725	185		2,910	3,275
6418	4" x 8"	↓	1.80	4.444	↓	2,725	185		2,910	3,275
6420	Hinge connections, polymer painted									
6422	3/4" thick top plate									
6424	5-1/4" x 12" w/ 5" x 5" top	1 Carp	1	8	C	30,200	330		30,530	33,700
6426	5-1/4" x 15" w/ 6" x 6" top		.80	10		32,100	415		32,515	35,900
6428	5-1/4" x 18" w/ 7" x 7" top		.70	11.429		33,700	475		34,175	37,800
6430	5-1/4" x 26" w/ 9" x 9" top	↓	.60	13.333	↓	35,900	555		36,455	40,400
6432	1" thick top plate									
6434	6-7/8" x 14" w/ 5" x 5" top	1 Carp	.80	10	C	36,800	415		37,215	41,000
6436	6-7/8" x 17" w/ 6" x 6" top		.80	10		40,900	415		41,315	45,600
6438	6-7/8" x 21" w/ 7" x 7" top		.70	11.429		44,600	475		45,075	49,700
6440	6-7/8" x 31" w/ 9" x 9" top	↓	.60	13.333	↓	48,800	555		49,355	54,500
6442	1-1/4" thick top plate									
6444	8-7/8" x 16" w/ 5" x 5" top	1 Carp	.60	13.333	C	45,800	555		46,355	51,500
6446	8-7/8" x 21" w/ 6" x 6" top		.50	16		50,500	665		51,165	56,500
6448	8-7/8" x 26" w/ 7" x 7" top		.40	20		57,000	830		57,830	64,500
6450	8-7/8" x 39" w/ 9" x 9" top	↓	.30	26.667	↓	71,000	1,100		72,100	80,000
6460	Holddowns									
6462	Embedded along edge									
6464	26" long, 12 ga.	1 Carp	.90	8.889	C	1,150	370		1,520	1,850
6466	35" long, 12 ga.		.85	9.412		1,525	390		1,915	2,275
6468	35" long, 10 ga.	↓	.85	9.412	↓	1,200	390		1,590	1,925
6470	Embedded away from edge									
6472	Medium duty, 12 ga.									
6474	18-1/2" long	1 Carp	.95	8.421	C	695	350		1,045	1,300
6476	23-3/4" long		.90	8.889		810	370		1,180	1,450
6478	28" long		.85	9.412		825	390		1,215	1,525
6480	35" long	↓	.85	9.412	↓	1,125	390		1,515	1,850
6482	Heavy duty, 10 ga.									
6484	28" long	1 Carp	.85	9.412	C	1,475	390		1,865	2,225
6486	35" long	"	.85	9.412	"	1,625	390		2,015	2,375
6490	Surface mounted (W x H)									
6492	2-1/2" x 5-3/4", 7 ga.	1 Carp	1	8	C	1,675	330		2,005	2,350

06 05 23 – Wood, Plastic, and Composite Fastenings

06 05 23.60 Timber Connectors		Crew	Daily Output	Labor-Hours	Unit	Material	Labor	Equipment	Total	Total Incl O&P
						2010 Bare Costs				
6494	2-1/2" x 8", 12 ga.	1 Carp	1	8	C	1,025	330		1,355	1,650
6496	2-7/8" x 6-3/8", 7 ga.		1	8		3,725	330		4,055	4,600
6498	2-7/8" x 12-1/2", 3 ga.		1	8		3,825	330		4,155	4,725
6500	3-3/16" x 9-3/8", 10 ga.		1	8		2,550	330		2,880	3,300
6502	3-1/2" x 11-5/8", 3 ga.		1	8		4,650	330		4,980	5,625
6504	3-1/2" x 14-3/4", 3 ga.		1	8		3,100	330		3,430	3,925
6506	3-1/2" x 16-1/2", 3 ga.		1	8		7,125	330		7,455	8,325
6508	3-1/2" x 20-1/2", 3 ga.		.90	8.889		7,275	370		7,645	8,600
6510	3-1/2" x 24-1/2", 3 ga.		.90	8.889		9,200	370		9,570	10,700
6512	4-1/4" x 20-3/4", 3 ga.	↓	.90	8.889	↓	6,275	370		6,645	7,475
6520	Joist hangers									
6522	Sloped, field adjustable, 18 ga.									
6524	2" x 6"	1 Carp	1.65	4.848	C	460	201		661	815
6526	2" x 8"		1.65	4.848		825	201		1,026	1,225
6528	2" x 10" and up		1.65	4.848		1,375	201		1,576	1,825
6530	3" x 10" and up		1.60	5		1,025	208		1,233	1,450
6532	4" x 10" and up	↓	1.55	5.161	↓	1,250	214		1,464	1,700
6536	Skewed 45°, 16 ga.									
6538	2" x 4"	1 Carp	1.75	4.571	C	735	190		925	1,100
6540	2" x 6" or 2" x 8"		1.65	4.848		745	201		946	1,125
6542	2" x 10" or 2" x 12"		1.65	4.848		855	201		1,056	1,250
6544	2" x 14" or 2" x 16"		1.60	5		1,525	208		1,733	2,000
6546	(2) 2" x 6" or (2) 2" x 8"		1.60	5		1,375	208		1,583	1,850
6548	(2) 2" x 10" or (2) 2" x 12"		1.55	5.161		1,475	214		1,689	1,950
6550	(2) 2" x 14" or (2) 2" x 16"		1.50	5.333		2,325	222		2,547	2,925
6552	4" x 6" or 4" x 8"		1.60	5		1,150	208		1,358	1,600
6554	4" x 10" or 4" x 12"		1.55	5.161		1,350	214		1,564	1,800
6556	4" x 14" or 4" x 16"	↓	1.55	5.161	↓	2,100	214		2,314	2,625
6560	Skewed 45°, 14 ga.									
6562	(2) 2" x 6" or (2) 2" x 8"	1 Carp	1.60	5	C	1,600	208		1,808	2,075
6564	(2) 2" x 10" or (2) 2" x 12"		1.55	5.161		2,200	214		2,414	2,750
6566	(2) 2" x 14" or (2) 2" x 16"		1.50	5.333		3,225	222		3,447	3,900
6568	4" x 6" or 4" x 8"		1.60	5		1,900	208		2,108	2,400
6570	4" x 10" or 4" x 12"		1.55	5.161		2,000	214		2,214	2,525
6572	4" x 14" or 4" x 16"	↓	1.55	5.161	↓	2,675	214		2,889	3,275
6590	Joist hangers, heavy duty 12 ga., galvanized									
6592	2" x 4"	1 Carp	1.75	4.571	C	1,075	190		1,265	1,475
6594	2" x 6"		1.65	4.848		1,175	201		1,376	1,600
6595	2" x 6", 16 gauge		1.65	4.848		1,175	201		1,376	1,600
6596	2" x 8"		1.65	4.848		1,775	201		1,976	2,250
6597	2" x 8", 16 gauge		1.65	4.848		1,775	201		1,976	2,250
6598	2" x 10"		1.65	4.848		1,825	201		2,026	2,300
6600	2" x 12"		1.65	4.848		2,250	201		2,451	2,775
6602	2" x 14"		1.65	4.848		2,350	201		2,551	2,875
6604	2" x 16"		1.65	4.848		2,475	201		2,676	3,025
6606	3" x 4"		1.65	4.848		1,475	201		1,676	1,925
6608	3" x 6"		1.65	4.848		2,000	201		2,201	2,500
6610	3" x 8"		1.65	4.848		2,025	201		2,226	2,525
6612	3" x 10"		1.60	5		2,350	208		2,558	2,900
6614	3" x 12"		1.60	5		2,800	208		3,008	3,425
6616	3" x 14"		1.60	5		3,300	208		3,508	3,950
6618	3" x 16"		1.60	5		3,625	208		3,833	4,325
6620	(2) 2" x 4"	↓	1.75	4.571	↓	1,750	190		1,940	2,225

06 05 23 – Wood, Plastic, and Composite Fastenings

06 05 23.60 Timber Connectors		Crew	Daily Output	Labor-Hours	Unit	Material	Labor	Equipment	Total	Total Incl O&P
						2010 Bare Costs				
6622	(2) 2" x 6"	1 Carp	1.60	5	C	2,100	208		2,308	2,625
6624	(2) 2" x 8"		1.60	5		2,150	208		2,358	2,675
6626	(2) 2" x 10"		1.55	5.161		2,325	214		2,539	2,875
6628	(2) 2" x 12"		1.55	5.161		2,975	214		3,189	3,600
6630	(2) 2" x 14"		1.50	5.333		3,000	222		3,222	3,650
6632	(2) 2" x 16"		1.50	5.333		3,025	222		3,247	3,675
6634	4" x 4"		1.65	4.848		1,275	201		1,476	1,700
6636	4" x 6"		1.60	5		1,375	208		1,583	1,850
6638	4" x 8"		1.60	5		1,600	208		1,808	2,100
6640	4" x 10"		1.55	5.161		1,975	214		2,189	2,500
6642	4" x 12"		1.55	5.161		2,100	214		2,314	2,625
6644	4" x 14"		1.55	5.161		2,500	214		2,714	3,075
6646	4" x 16"		1.55	5.161		2,750	214		2,964	3,350
6648	(3) 2" x 10"		1.50	5.333		3,050	222		3,272	3,700
6650	(3) 2" x 12"		1.50	5.333		3,400	222		3,622	4,075
6652	(3) 2" x 14"		1.45	5.517		3,650	229		3,879	4,350
6654	(3) 2" x 16"		1.45	5.517		3,700	229		3,929	4,425
6656	6" x 6"		1.60	5		1,625	208		1,833	2,125
6658	6" x 8"		1.60	5		1,675	208		1,883	2,175
6660	6" x 10"		1.55	5.161		2,025	214		2,239	2,550
6662	6" x 12"		1.55	5.161		2,300	214		2,514	2,850
6664	6" x 14"		1.50	5.333		2,875	222		3,097	3,525
6666	6" x 16"	↓	1.50	5.333	↓	3,400	222		3,622	4,075
6690	Knee braces, galvanized, 12 ga.									
6692	Beam depth, 10" x 15" x 5' long	1 Carp	1.80	4.444	C	4,400	185		4,585	5,125
6694	15" x 22-1/2" x 7' long		1.70	4.706		5,050	196		5,246	5,850
6696	22-1/2" x 28-1/2" x 8' long		1.60	5		5,425	208		5,633	6,300
6698	28-1/2" x 36" x 10' long		1.55	5.161		5,650	214		5,864	6,550
6700	36" x 42" x 12' long	↓	1.50	5.333	↓	6,250	222		6,472	7,225
6710	Mudsill anchors									
6714	2" x 4" or 3" x 4"	1 Carp	115	.070	C	1,125	2.89		1,127.89	1,250
6716	2" x 6" or 3" x 6"		115	.070		1,125	2.89		1,127.89	1,250
6718	Block wall, 13-1/4" long		115	.070		780	2.89		782.89	865
6720	21-1/4" long	↓	115	.070	↓	116	2.89		118.89	132
6730	Post bases, 12 ga. galvanized									
6732	Adjustable, 3-9/16" x 3-9/16"	1 Carp	1.30	6.154	C	900	256		1,156	1,375
6734	3-9/16" x 5-1/2"		1.30	6.154		1,175	256		1,431	1,675
6736	4" x 4"		1.30	6.154		655	256		911	1,125
6738	4" x 6"		1.30	6.154		1,525	256		1,781	2,075
6740	5-1/2" x 5-1/2"		1.30	6.154		2,250	256		2,506	2,900
6742	6" x 6"		1.30	6.154		1,725	256		1,981	2,300
6744	Elevated, 3-9/16" x 3-1/4"		1.30	6.154		975	256		1,231	1,475
6746	5-1/2" x 3-5/16"		1.30	6.154		1,375	256		1,631	1,925
6748	5-1/2" x 5"		1.30	6.154		2,075	256		2,331	2,675
6750	Regular, 3-9/16" x 3-3/8"		1.30	6.154		745	256		1,001	1,225
6752	4" x 3-3/8"		1.30	6.154		1,050	256		1,306	1,550
6754	18 gauge, 5-1/4" x 3-1/8"		1.30	6.154		1,100	256		1,356	1,600
6755	5-1/2" x 3-3/8"		1.30	6.154		1,100	256		1,356	1,600
6756	5-1/2" x 5-3/8"		1.30	6.154		1,575	256		1,831	2,125
6758	6" x 3-3/8"		1.30	6.154		1,900	256		2,156	2,500
6760	6" x 5-3/8"	↓	1.30	6.154	↓	2,125	256		2,381	2,725
6762	Post combination cap/bases									
6764	3-9/16" x 3-9/16"	1 Carp	1.20	6.667	C	425	277		702	895

06 05 23 – Wood, Plastic, and Composite Fastenings

06 05 23.60 Timber Connectors		Crew	Daily Output	Labor-Hours	Unit	Material	2010 Bare Costs Labor	Equipment	Total	Total Incl O&P
6766	3-9/16" x 5-1/2"	1 Carp	1.20	6.667	C	965	277		1,242	1,475
6768	4" x 4"		1.20	6.667		1,925	277		2,202	2,525
6770	5-1/2" x 5-1/2"		1.20	6.667		1,075	277		1,352	1,600
6772	6" x 6"		1.20	6.667		3,500	277		3,777	4,275
6774	7-1/2" x 7-1/2"		1.20	6.667		3,850	277		4,127	4,650
6776	8" x 8"		1.20	6.667		3,975	277		4,252	4,800
6790	Post-beam connection caps									
6792	Beam size 3-9/16"									
6794	12 ga. post, 4" x 4"	1 Carp	1	8	C	2,425	330		2,755	3,175
6796	4" x 6"		1	8		3,250	330		3,580	4,075
6798	4" x 8"		1	8		4,800	330		5,130	5,775
6800	16 ga. post, 4" x 4"		1	8		1,100	330		1,430	1,700
6802	4" x 6"		1	8		1,825	330		2,155	2,525
6804	4" x 8"		1	8		3,100	330		3,430	3,900
6805	18 ga. post, 2-7/8" x 3"		1	8		3,100	330		3,430	3,900
6806	Beam size 5-1/2"									
6808	12 ga. post, 6" x 4"	1 Carp	1	8	C	3,175	330		3,505	4,000
6810	6" x 6"		1	8		5,000	330		5,330	6,000
6812	6" x 8"		1	8		3,450	330		3,780	4,300
6816	16 ga. post, 6" x 4"		1	8		1,725	330		2,055	2,400
6818	6" x 6"		1	8		1,825	330		2,155	2,525
6820	Beam size 7-1/2"									
6822	12 ga. post, 8" x 4"	1 Carp	1	8	C	4,400	330		4,730	5,350
6824	8" x 6"		1	8		4,600	330		4,930	5,575
6826	8" x 8"		1	8		6,950	330		7,280	8,125
6840	Purlin anchors, embedded									
6842	Heavy duty, 10 ga.									
6844	Straight, 28" long	1 Carp	1.60	5	C	1,200	208		1,408	1,650
6846	35" long		1.50	5.333		1,500	222		1,722	2,000
6848	Twisted, 28" long		1.60	5		1,200	208		1,408	1,650
6850	35" long		1.50	5.333		1,500	222		1,722	2,000
6852	Regular duty, 12 ga.									
6854	Straight, 18-1/2" long	1 Carp	1.80	4.444	C	695	185		880	1,050
6856	23-3/4" long		1.70	4.706		870	196		1,066	1,250
6858	29" long		1.60	5		890	208		1,098	1,300
6860	35" long		1.50	5.333		1,225	222		1,447	1,700
6862	Twisted, 18" long		1.80	4.444		695	185		880	1,050
6866	28" long		1.60	5		825	208		1,033	1,225
6868	35" long		1.50	5.333		1,225	222		1,447	1,700
6870	Straight, plastic coated									
6872	23-1/2" long	1 Carp	1.60	5	C	1,700	208		1,908	2,200
6874	26-7/8" long		1.60	5		2,000	208		2,208	2,525
6876	32-1/2" long		1.50	5.333		2,125	222		2,347	2,700
6878	35-7/8" long		1.50	5.333		2,225	222		2,447	2,775
6890	Purlin hangers, painted									
6892	12 ga., 2" x 6"	1 Carp	1.80	4.444	C	1,575	185		1,760	2,000
6894	2" x 8"		1.80	4.444		1,700	185		1,885	2,150
6896	2" x 10"		1.80	4.444		1,825	185		2,010	2,300
6898	2" x 12"		1.75	4.571		1,975	190		2,165	2,475
6900	2" x 14"		1.75	4.571		2,100	190		2,290	2,600
6902	2" x 16"		1.75	4.571		2,225	190		2,415	2,750
6904	3" x 6"		1.70	4.706		1,575	196		1,771	2,025
6906	3" x 8"		1.70	4.706		1,700	196		1,896	2,175

06 05 23 – Wood, Plastic, and Composite Fastenings

06 05 23.60 Timber Connectors		Crew	Daily Output	Labor-Hours	Unit	Material	2010 Bare Costs Labor	Equipment	Total	Total Incl O&P
6908	3" x 10"	1 Carp	1.70	4.706	C	1,850	196		2,046	2,325
6910	3" x 12"		1.65	4.848		2,100	201		2,301	2,625
6912	3" x 14"		1.65	4.848		2,250	201		2,451	2,775
6914	3" x 16"		1.65	4.848		2,375	201		2,576	2,925
6916	4" x 6"		1.65	4.848		1,575	201		1,776	2,025
6918	4" x 8"		1.65	4.848		1,725	201		1,926	2,175
6920	4" x 10"		1.65	4.848		1,850	201		2,051	2,325
6922	4" x 12"		1.60	5		2,200	208		2,408	2,725
6924	4" x 14"		1.60	5		2,325	208		2,533	2,875
6926	4" x 16"		1.60	5		2,450	208		2,658	3,025
6928	6" x 6"		1.60	5		2,075	208		2,283	2,625
6930	6" x 8"		1.60	5		2,200	208		2,408	2,750
6932	6" x 10"		1.55	5.161		2,250	214		2,464	2,800
6934	double 2" x 6"		1.70	4.706		1,700	196		1,896	2,175
6936	double 2" x 8"		1.70	4.706		1,850	196		2,046	2,325
6938	double 2" x 10"		1.70	4.706		1,975	196		2,171	2,475
6940	double 2" x 12"		1.65	4.848		2,100	201		2,301	2,625
6942	double 2" x 14"		1.65	4.848		2,250	201		2,451	2,775
6944	double 2" x 16"		1.65	4.848		2,375	201		2,576	2,925
6960	11 ga., 4" x 6"		1.65	4.848		3,125	201		3,326	3,725
6962	4" x 8"		1.65	4.848		3,350	201		3,551	4,000
6964	4" x 10"		1.65	4.848		3,575	201		3,776	4,250
6966	6" x 6"		1.60	5		3,150	208		3,358	3,800
6968	6" x 8"		1.60	5		3,400	208		3,608	4,050
6970	6" x 10"		1.55	5.161		3,625	214		3,839	4,300
6972	6" x 12"		1.55	5.161		3,850	214		4,064	4,575
6974	6" x 14"		1.55	5.161		4,075	214		4,289	4,825
6976	6" x 16"		1.50	5.333		4,325	222		4,547	5,100
6978	7 ga., 8" x 6"		1.60	5		3,425	208		3,633	4,100
6980	8" x 8"		1.60	5		3,650	208		3,858	4,350
6982	8" x 10"		1.55	5.161		3,900	214		4,114	4,600
6984	8" x 12"		1.55	5.161		4,125	214		4,339	4,850
6986	8" x 14"		1.50	5.333		4,350	222		4,572	5,150
6988	8" x 16"	↓	1.50	5.333	↓	4,575	222		4,797	5,400
7000	Strap connectors, galvanized									
7002	12 ga., 2-1/16" x 36"	1 Carp	1.55	5.161	C	995	214		1,209	1,425
7004	2-1/16" x 47"		1.50	5.333		1,375	222		1,597	1,875
7005	10 ga., 2-1/16" x 72"		1.50	5.333		1,375	222		1,597	1,875
7006	7 ga., 2-1/16" x 34"		1.55	5.161		2,475	214		2,689	3,050
7008	2-1/16" x 45"		1.50	5.333		3,200	222		3,422	3,875
7010	3 ga., 3" x 32"		1.55	5.161		4,150	214		4,364	4,900
7012	3" x 41"		1.55	5.161		4,325	214		4,539	5,075
7014	3" x 50"		1.50	5.333		6,575	222		6,797	7,575
7016	3" x 59"		1.50	5.333		8,000	222		8,222	9,150
7018	3-1/2" x 68"	↓	1.45	5.517	↓	8,125	229		8,354	9,300
7030	Tension ties									
7032	19-1/8" long, 16 ga., 3/4" anchor bolt	1 Carp	1.80	4.444	C	1,150	185		1,335	1,525
7034	20" long, 12 ga., 1/2" anchor bolt		1.80	4.444		1,475	185		1,660	1,900
7036	20" long, 12 ga., 3/4" anchor bolt		1.80	4.444		1,475	185		1,660	1,900
7038	27-3/4" long, 12 ga., 3/4" anchor bolt	↓	1.75	4.571	↓	2,625	190		2,815	3,175
7050	Truss connectors, galvanized									
7052	Adjustable hanger									
7054	18 ga., 2" x 6"	1 Carp	1.65	4.848	C	445	201		646	800

06 05 Common Work Results for Wood, Plastics, and Composites

06 05 23 – Wood, Plastic, and Composite Fastenings

06 05 23.60 Timber Connectors		Crew	Daily Output	Labor-Hours	Unit	Material	2010 Bare Costs Labor	Equipment	Total	Total Incl O&P
7056	4" x 6"	1 Carp	1.65	4.848	C	585	201		786	955
7058	16 ga., 4" x 10"		1.60	5		860	208		1,068	1,275
7060	(2) 2" x 10"	↓	1.60	5	↓	860	208		1,068	1,275
7062	Connectors to plate									
7064	16 ga., 2" x 4" plate	1 Carp	1.80	4.444	C	440	185		625	770
7066	2" x 6" plate	"	1.80	4.444	"	570	185		755	910
7068	Hip jack connector									
7070	14 ga.	1 Carp	1.50	5.333	C	2,325	222		2,547	2,900

06 05 23.70 Rough Hardware		Crew	Daily Output	Labor-Hours	Unit	Material	Labor	Equipment	Total	Total Incl O&P
0010	**ROUGH HARDWARE**, average percent of carpentry material									
0020	Minimum				Job	.50%				
0200	Maximum					1.50%				
0210	In seismic or hurricane areas, up to				↓	10%				

06 05 23.80 Metal Bracing		Crew	Daily Output	Labor-Hours	Unit	Material	Labor	Equipment	Total	Total Incl O&P
0010	**METAL BRACING**									
0302	Let-in, "T" shaped, 22 ga. galv. steel, studs at 16" O.C.	1 Carp	580	.014	L.F.	.75	.57		1.32	1.71
0402	Studs at 24" O.C.		600	.013		.75	.55		1.30	1.68
0502	16 ga. galv. steel straps, studs at 16" O.C.		600	.013		.97	.55		1.52	1.92
0602	Studs at 24" O.C.	↓	620	.013	↓	.97	.54		1.51	1.90

06 11 Wood Framing

06 11 10 – Framing with Dimensional, Engineered or Composite Lumber

06 11 10.02 Blocking		Crew	Daily Output	Labor-Hours	Unit	Material	Labor	Equipment	Total	Total Incl O&P
0010	**BLOCKING**									
2600	Miscellaneous, to wood construction									
2620	2" x 4"	1 Carp	.17	47.059	M.B.F.	375	1,950		2,325	3,425
2625	Pneumatic nailed		.21	38.095		375	1,575		1,950	2,850
2660	2" x 8"		.27	29.630		420	1,225		1,645	2,350
2665	Pneumatic nailed	↓	.33	24.242	↓	420	1,000		1,420	2,000
2720	To steel construction									
2740	2" x 4"	1 Carp	.14	57.143	M.B.F.	375	2,375		2,750	4,050
2780	2" x 8"	"	.21	38.095	"	420	1,575		1,995	2,900

06 11 10.04 Wood Bracing		Crew	Daily Output	Labor-Hours	Unit	Material	Labor	Equipment	Total	Total Incl O&P
0010	**WOOD BRACING**									
0011	Let-in, with 1" x 6" boards, studs 16" O.C.	1 Carp	1.50	5.333	C.L.F.	64	222		286	410
0200	Studs @ 24" O.C.	"	2.30	3.478	"	64	145		209	294

06 11 10.06 Bridging		Crew	Daily Output	Labor-Hours	Unit	Material	Labor	Equipment	Total	Total Incl O&P
0010	**BRIDGING**									
0011	Wood, for joists 16" O.C., 1" x 3"	1 Carp	1.30	6.154	C.Pr.	51	256		307	450
0015	Pneumatic nailed		1.70	4.706		51	196		247	355
0100	2" x 3" bridging		1.30	6.154		40	256		296	440
0105	Pneumatic nailed		1.70	4.706		40	196		236	345
0300	Steel, galvanized, 18 ga., for 2" x 10" joists at 12" O.C.		1.30	6.154		158	256		414	570
0400	24" O.C.		1.40	5.714		169	237		406	550
0600	For 2" x 14" joists at 16" O.C.		1.30	6.154		170	256		426	580
0700	24" O.C.		1.40	5.714		260	237		497	650
0900	Compression type, 16" O.C., 2" x 8" joists		2	4		164	166		330	435
1000	2" x 12" joists	↓	2	4	↓	164	166		330	435

06 11 10 – Framing with Dimensional, Engineered or Composite Lumber

06 11 10.10 Beam and Girder Framing

06 11 10.10 Beam and Girder Framing		Crew	Daily Output	Labor-Hours	Unit	Material	Labor	Equipment	Total	Total Incl O&P
0010	**BEAM AND GIRDER FRAMING** R061110-30									
3500	Single, 2" x 6"	2 Carp	.70	22.857	M.B.F.	375	950		1,325	1,900
3505	Pneumatic nailed		.81	19.704		375	820		1,195	1,675
3520	2" x 8"		.86	18.605		420	775		1,195	1,650
3525	Pneumatic nailed		1	16.048		420	665		1,085	1,475
3540	2" x 10"		1	16		435	665		1,100	1,500
3545	Pneumatic nailed		1.16	13.793		435	575		1,010	1,350
3560	2" x 12"		1.10	14.545		530	605		1,135	1,525
3565	Pneumatic nailed		1.28	12.539		530	520		1,050	1,400
3580	2" x 14"		1.17	13.675		855	570		1,425	1,825
3585	Pneumatic nailed		1.36	11.791		855	490		1,345	1,700
3600	3" x 8"		1.10	14.545		1,050	605		1,655	2,075
3620	3" x 10"		1.25	12.800		1,100	530		1,630	2,050
3640	3" x 12"		1.35	11.852		1,175	490		1,665	2,050
3660	3" x 14"		1.40	11.429		1,275	475		1,750	2,125
3680	4" x 8"	F-3	2.66	15.038		1,075	635	246	1,956	2,450
3700	4" x 10"		3.16	12.658		1,100	535	207	1,842	2,250
3720	4" x 12"		3.60	11.111		1,525	470	182	2,177	2,600
3740	4" x 14"		3.96	10.101		1,525	425	166	2,116	2,500
4000	Double, 2" x 6"	2 Carp	1.25	12.800		375	530		905	1,225
4005	Pneumatic nailed		1.45	11.034		375	460		835	1,125
4020	2" x 8"		1.60	10		420	415		835	1,100
4025	Pneumatic nailed		1.86	8.621		420	360		780	1,000
4040	2" x 10"		1.92	8.333		435	345		780	1,000
4045	Pneumatic nailed		2.23	7.185		435	299		734	935
4060	2" x 12"		2.20	7.273		530	300		830	1,050
4065	Pneumatic nailed		2.55	6.275		530	261		791	985
4080	2" x 14"		2.45	6.531		855	271		1,126	1,350
4085	Pneumatic nailed		2.84	5.634		855	234		1,089	1,300
5000	Triple, 2" x 6"		1.65	9.697		375	405		780	1,025
5005	Pneumatic nailed		1.91	8.377		375	350		725	950
5020	2" x 8"		2.10	7.619		420	315		735	950
5025	Pneumatic nailed		2.44	6.568		420	273		693	880
5040	2" x 10"		2.50	6.400		435	266		701	885
5045	Pneumatic nailed		2.90	5.517		435	229		664	830
5060	2" x 12"		2.85	5.614		530	233		763	945
5065	Pneumatic nailed		3.31	4.840		530	201		731	895
5080	2" x 14"		3.15	5.079		855	211		1,066	1,275
5085	Pneumatic nailed		3.35	4.770		855	198		1,053	1,250

06 11 10.12 Ceiling Framing

		Crew	Daily Output	Labor-Hours	Unit	Material	Labor	Equipment	Total	Total Incl O&P
0010	**CEILING FRAMING**									
6400	Suspended, 2" x 3"	2 Carp	.50	32	M.B.F.	535	1,325		1,860	2,650
6450	2" x 4"		.59	27.119		375	1,125		1,500	2,125
6500	2" x 6"		.80	20		375	830		1,205	1,700
6550	2" x 8"		.86	18.605		420	775		1,195	1,650

06 11 10.14 Posts and Columns

		Crew	Daily Output	Labor-Hours	Unit	Material	Labor	Equipment	Total	Total Incl O&P
0010	**POSTS AND COLUMNS**									
0400	4" x 4"	2 Carp	.52	30.769	M.B.F.	1,150	1,275		2,425	3,250
0420	4" x 6"		.55	29.091		1,200	1,200		2,400	3,200
0440	4" x 8"		.59	27.119		1,075	1,125		2,200	2,925
0460	6" x 6"		.65	24.615		1,075	1,025		2,100	2,750
0480	6" x 8"		.70	22.857		1,075	950		2,025	2,650

Material, Labor, Equipment and Total columns: 2010 Bare Costs.

06 11 10 – Framing with Dimensional, Engineered or Composite Lumber

06 11 10.14 Posts and Columns		Crew	Daily Output	Labor-Hours	Unit	Material	2010 Bare Costs Labor	Equipment	Total	Total Incl O&P
0500	6" x 10"	2 Carp	.75	21.333	M.B.F.	1,325	885		2,210	2,825

06 11 10.18 Joist Framing		Crew	Daily Output	Labor-Hours	Unit	Material	Labor	Equipment	Total	Total Incl O&P
0010	**JOIST FRAMING** R061110-30									
2650	Joists, 2" x 4"	2 Carp	.83	19.277	M.B.F.	375	800		1,175	1,625
2655	Pneumatic nailed		.96	16.667		375	695		1,070	1,475
2680	2" x 6"		1.25	12.800		375	530		905	1,225
2685	Pneumatic nailed		1.44	11.111		375	460		835	1,125
2700	2" x 8"		1.46	10.959		420	455		875	1,150
2705	Pneumatic nailed		1.68	9.524		420	395		815	1,075
2720	2" x 10"		1.49	10.738		435	445		880	1,175
2725	Pneumatic nailed		1.71	9.357		435	390		825	1,075
2740	2" x 12"		1.75	9.143		530	380		910	1,175
2745	Pneumatic nailed		2.01	7.960		530	330		860	1,100
2760	2" x 14"		1.79	8.939		855	370		1,225	1,525
2765	Pneumatic nailed		2.06	7.767		855	325		1,180	1,425
2780	3" x 6"		1.39	11.511		930	480		1,410	1,750
2790	3" x 8"		1.90	8.421		1,050	350		1,400	1,700
2800	3" x 10"		1.95	8.205		1,100	340		1,440	1,750
2820	3" x 12"		1.80	8.889		1,175	370		1,545	1,875
2840	4" x 6"		1.60	10		1,200	415		1,615	1,975
2850	4" x 8"		4.15	3.855		1,075	160		1,235	1,450
2860	4" x 10"		2	8		1,100	330		1,430	1,700
2880	4" x 12"		1.80	8.889		1,525	370		1,895	2,250
3000	Composite wood joist 9-1/2" deep		.90	17.778	M.L.F.	1,575	740		2,315	2,875
3010	11-1/2" deep		.88	18.182		1,700	755		2,455	3,025
3020	14" deep		.82	19.512		2,325	810		3,135	3,825
3030	16" deep		.78	20.513		3,075	850		3,925	4,700
4000	Open web joist 12" deep		.88	18.182		2,775	755		3,530	4,225
4010	14" deep		.82	19.512		2,975	810		3,785	4,525
4020	16" deep		.78	20.513		2,950	850		3,800	4,575
4030	18" deep		.74	21.622		3,175	900		4,075	4,875
6000	Composite rim joist, 1-1/4" x 9-1/2"		.90	17.778		1,700	740		2,440	3,025
6010	1-1/4" x 11-1/2"		.88	18.182		2,125	755		2,880	3,500
6020	1-1/4" x 14-1/2"		.82	19.512		2,425	810		3,235	3,925
6030	1-1/4" x 16-1/2"		.78	20.513		3,100	850		3,950	4,750

06 11 10.24 Miscellaneous Framing		Crew	Daily Output	Labor-Hours	Unit	Material	Labor	Equipment	Total	Total Incl O&P
0010	**MISCELLANEOUS FRAMING**									
8500	Firestops, 2" x 4"	2 Carp	.51	31.373	M.B.F.	375	1,300		1,675	2,400
8505	Pneumatic nailed		.62	25.806		375	1,075		1,450	2,050
8520	2" x 6"		.60	26.667		375	1,100		1,475	2,125
8525	Pneumatic nailed		.73	21.858		375	910		1,285	1,825
8540	2" x 8"		.60	26.667		420	1,100		1,520	2,150
8560	2" x 12"		.70	22.857		530	950		1,480	2,050
8600	Nailers, treated, wood construction, 2" x 4"		.53	30.189		520	1,250		1,770	2,500
8605	Pneumatic nailed		.64	25.157		520	1,050		1,570	2,175
8620	2" x 6"		.75	21.333		555	885		1,440	2,000
8625	Pneumatic nailed		.90	17.778		555	740		1,295	1,775
8640	2" x 8"		.93	17.204		775	715		1,490	1,950
8645	Pneumatic nailed		1.12	14.337		775	595		1,370	1,775
8660	Steel construction, 2" x 4"		.50	32		520	1,325		1,845	2,625
8680	2" x 6"		.70	22.857		555	950		1,505	2,100
8700	2" x 8"		.87	18.391		775	765		1,540	2,025

06 11 10 – Framing with Dimensional, Engineered or Composite Lumber

06 11 10.24 Miscellaneous Framing		Crew	Daily Output	Labor-Hours	Unit	2010 Bare Costs Material	Labor	Equipment	Total	Total Incl O&P
8760	Rough bucks, treated, for doors or windows, 2" x 6"	2 Carp	.40	40	M.B.F.	555	1,650		2,205	3,175
8765	Pneumatic nailed		.48	33.333		555	1,375		1,930	2,750
8780	2" x 8"		.51	31.373		775	1,300		2,075	2,850
8785	Pneumatic nailed		.61	26.144		775	1,075		1,850	2,525
8800	Stair stringers, 2" x 10"		.22	72.727		435	3,025		3,460	5,125
8820	2" x 12"		.26	61.538		530	2,550		3,080	4,525
8840	3" x 10"		.31	51.613		1,100	2,150		3,250	4,525
8860	3" x 12"		.38	42.105		1,175	1,750		2,925	4,000
8870	Laminated structural lumber, 1-1/4" x 11-1/2"		130	.123	L.F.	2.11	5.10		7.21	10.20
8880	1-1/4" x 14-1/2"		130	.123	"	2.43	5.10		7.53	10.60

06 11 10.26 Partitions

Line	Description	Crew	Daily Output	Labor-Hours	Unit	Material	Labor	Equipment	Total	Total Incl O&P
0010	**PARTITIONS**									
0020	Single bottom and double top plate, no waste, std. & better lumber									
0180	2" x 4" studs, 8' high, studs 12" O.C.	2 Carp	80	.200	L.F.	2.74	8.30		11.04	15.80
0185	12" O.C., pneumatic nailed		96	.167		2.74	6.95		9.69	13.70
0200	16" O.C.		100	.160		2.24	6.65		8.89	12.70
0205	16" O.C., pneumatic nailed		120	.133		2.24	5.55		7.79	11
0300	24" O.C.		125	.128		1.74	5.30		7.04	10.10
0305	24" O.C., pneumatic nailed		150	.107		1.74	4.43		6.17	8.75
0380	10' high, studs 12" O.C.		80	.200		3.23	8.30		11.53	16.35
0385	12" O.C., pneumatic nailed		96	.167		3.23	6.95		10.18	14.25
0400	16" O.C.		100	.160		2.61	6.65		9.26	13.10
0405	16" O.C., pneumatic nailed		120	.133		2.61	5.55		8.16	11.40
0500	24" O.C.		125	.128		1.99	5.30		7.29	10.40
0505	24" O.C., pneumatic nailed		150	.107		1.99	4.43		6.42	9.05
0580	12' high, studs 12" O.C.		65	.246		3.73	10.25		13.98	19.85
0585	12" O.C., pneumatic nailed		78	.205		3.73	8.50		12.23	17.25
0600	16" O.C.		80	.200		2.98	8.30		11.28	16.10
0605	16" O.C., pneumatic nailed		96	.167		2.98	6.95		9.93	14
0700	24" O.C.		100	.160		2.24	6.65		8.89	12.70
0705	24" O.C., pneumatic nailed		120	.133		2.24	5.55		7.79	11
0780	2" x 6" studs, 8' high, studs 12" O.C.		70	.229		4.15	9.50		13.65	19.20
0785	12" O.C., pneumatic nailed		84	.190		4.15	7.90		12.05	16.75
0800	16" O.C.		90	.178		3.39	7.40		10.79	15.15
0805	16" O.C., pneumatic nailed		108	.148		3.39	6.15		9.54	13.25
0900	24" O.C.		115	.139		2.64	5.80		8.44	11.80
0905	24" O.C., pneumatic nailed		138	.116		2.64	4.82		7.46	10.35
0980	10' high, studs 12" O.C.		70	.229		4.90	9.50		14.40	20
0985	12" O.C., pneumatic nailed		84	.190		4.90	7.90		12.80	17.60
1000	16" O.C.		90	.178		3.96	7.40		11.36	15.75
1005	16" O.C., pneumatic nailed		108	.148		3.96	6.15		10.11	13.85
1100	24" O.C.		115	.139		3.02	5.80		8.82	12.20
1105	24" O.C., pneumatic nailed		138	.116		3.02	4.82		7.84	10.75
1180	12' high, studs 12" O.C.		55	.291		5.65	12.10		17.75	25
1185	12" O.C., pneumatic nailed		66	.242		5.65	10.05		15.70	22
1200	16" O.C.		70	.229		4.52	9.50		14.02	19.65
1205	16" O.C., pneumatic nailed		84	.190		4.52	7.90		12.42	17.20
1300	24" O.C.		90	.178		3.39	7.40		10.79	15.15
1305	24" O.C., pneumatic nailed		108	.148		3.39	6.15		9.54	13.25
1400	For horizontal blocking, 2" x 4", add		600	.027		.25	1.11		1.36	1.98
1500	2" x 6", add		600	.027		.38	1.11		1.49	2.12
1600	For openings, add		250	.064			2.66		2.66	4.10

06 11 Wood Framing

06 11 10 – Framing with Dimensional, Engineered or Composite Lumber

06 11 10.26 Partitions		Crew	Daily Output	Labor-Hours	Unit	2010 Bare Costs Material	Labor	Equipment	Total	Total Incl O&P
1700	Headers for above openings, material only, add				M.B.F.	420			420	460

06 11 10.28 Porch or Deck Framing

		Crew	Daily Output	Labor-Hours	Unit	Material	Labor	Equipment	Total	Total Incl O&P
0010	**PORCH OR DECK FRAMING**									
0100	Treated lumber, posts or columns, 4" x 4"	2 Carp	390	.041	L.F.	1.56	1.70		3.26	4.34
0110	4" x 6"		275	.058		2.30	2.42		4.72	6.25
0120	4" x 8"		220	.073		4.74	3.02		7.76	9.85
0130	Girder, single, 4" x 4"		675	.024		1.56	.98		2.54	3.23
0140	4" x 6"		600	.027		2.30	1.11		3.41	4.24
0150	4" x 8"		525	.030		4.74	1.27		6.01	7.15
0160	Double, 2" x 4"		625	.026		.72	1.06		1.78	2.44
0170	2" x 6"		600	.027		1.16	1.11		2.27	2.99
0180	2" x 8"		575	.028		2.13	1.16		3.29	4.12
0190	2" x 10"		550	.029		2.78	1.21		3.99	4.92
0200	2" x 12"		525	.030		3.69	1.27		4.96	6
0210	Triple, 2" x 4"		575	.028		1.09	1.16		2.25	2.97
0220	2" x 6"		550	.029		1.74	1.21		2.95	3.78
0230	2" x 8"		525	.030		3.19	1.27		4.46	5.45
0240	2" x 10"		500	.032		4.17	1.33		5.50	6.65
0250	2" x 12"		475	.034		5.55	1.40		6.95	8.25
0260	Ledger, bolted 4' O.C., 2" x 4"		400	.040		.47	1.66		2.13	3.07
0270	2" x 6"		395	.041		.68	1.68		2.36	3.33
0280	2" x 8"		390	.041		1.15	1.70		2.85	3.90
0300	2" x 12"		380	.042		1.92	1.75		3.67	4.81
0310	Joists, 2" x 4"		1250	.013		.36	.53		.89	1.22
0320	2" x 6"		1250	.013		.58	.53		1.11	1.46
0330	2" x 8"		1100	.015		1.07	.60		1.67	2.11
0340	2" x 10"		900	.018		1.40	.74		2.14	2.68
0350	2" x 12"		875	.018		1.68	.76		2.44	3.01
0360	Railings and trim , 1" x 4"	1 Carp	300	.027		.44	1.11		1.55	2.19
0370	2" x 2"		300	.027		.32	1.11		1.43	2.06
0380	2" x 4"		300	.027		.35	1.11		1.46	2.09
0390	2" x 6"		300	.027		.56	1.11		1.67	2.32
0400	Decking, 1" x 4"		275	.029	S.F.	1.55	1.21		2.76	3.57
0410	2" x 4"		300	.027		1.19	1.11		2.30	3.01
0420	2" x 6"		320	.025		1.21	1.04		2.25	2.94
0430	5/4" x 6"		320	.025		2.35	1.04		3.39	4.19
0440	Balusters, square, 2" x 2"	2 Carp	660	.024	L.F.	.33	1.01		1.34	1.91
0450	Turned, 2" x 2"		420	.038		.44	1.58		2.02	2.92
0460	Stair stringer, 2" x 10"		130	.123		1.40	5.10		6.50	9.45
0470	2" x 12"		130	.123		1.68	5.10		6.78	9.75
0480	Stair treads, 1" x 4"		140	.114		1.56	4.75		6.31	9
0490	2" x 4"		140	.114		.36	4.75		5.11	7.70
0500	2" x 6"		160	.100		.55	4.16		4.71	7
0510	5/4" x 6"		160	.100		1.10	4.16		5.26	7.60
0520	Turned handrail post, 4" x 4"		64	.250	Ea.	32	10.40		42.40	51
0530	Lattice panel, 4' x 8', 1/2"		1600	.010	S.F.	.68	.42		1.10	1.39
0535	3/4"		1600	.010	"	1.02	.42		1.44	1.76
0540	Cedar, posts or columns, 4" x 4"		390	.041	L.F.	3.01	1.70		4.71	5.95
0550	4" x 6"		275	.058		5.75	2.42		8.17	10.05
0560	4" x 8"		220	.073		9.70	3.02		12.72	15.35
0800	Decking, 1" x 4"		550	.029		1.65	1.21		2.86	3.67
0810	2" x 4"		600	.027		3.40	1.11		4.51	5.45

06 11 10 – Framing with Dimensional, Engineered or Composite Lumber

06 11 10.28 Porch or Deck Framing		Crew	Daily Output	Labor-Hours	Unit	2010 Bare Costs Material	Labor	Equipment	Total	Total Incl O&P
0820	2" x 6"	2 Carp	640	.025	L.F.	5.55	1.04		6.59	7.70
0830	5/4" x 6"		640	.025		3.73	1.04		4.77	5.70
0840	Railings and trim, 1" x 4"		600	.027		1.65	1.11		2.76	3.52
0860	2" x 4"		600	.027		3.40	1.11		4.51	5.45
0870	2" x 6"		600	.027		5.55	1.11		6.66	7.80
0920	Stair treads, 1" x 4"		140	.114		1.65	4.75		6.40	9.10
0930	2" x 4"		140	.114		3.40	4.75		8.15	11.05
0940	2" x 6"		160	.100		5.55	4.16		9.71	12.50
0950	5/4" x 6"		160	.100		3.73	4.16		7.89	10.50
0980	Redwood, posts or columns, 4" x 4"		390	.041		6.40	1.70		8.10	9.70
0990	4" x 6"		275	.058		11.20	2.42		13.62	16.05
1000	4" x 8"		220	.073		20.50	3.02		23.52	27
1240	Redwood decking, 1" x 4"	1 Carp	275	.029	S.F.	3.97	1.21		5.18	6.20
1260	2" x 6"		340	.024		7.50	.98		8.48	9.75
1270	5/4" x 6"		320	.025		4.74	1.04		5.78	6.80
1280	Railings and trim, 1" x 4"	2 Carp	600	.027	L.F.	1.17	1.11		2.28	2.99
1310	2" x 6"		600	.027		7.50	1.11		8.61	9.95
1420	Alternative decking, wood / plastic composite, 5/4" x 6" G		640	.025		3.01	1.04		4.05	4.92
1430	Vinyl, 1-1/2" x 5-1/2"		640	.025		4.15	1.04		5.19	6.15
1440	1" x 4" square edge fir		550	.029		1.57	1.21		2.78	3.58
1450	1" x 4" tongue and groove fir		450	.036		1.41	1.48		2.89	3.83
1460	1" x 4" mahogany		550	.029		2.51	1.21		3.72	4.62
1462	5/4" x 6" PVC		550	.029		2.72	1.21		3.93	4.85
1470	Accessories, joist hangers, 2" x 4"	1 Carp	160	.050	Ea.	.61	2.08		2.69	3.87
1480	2" x 6" through 2" x 12"	"	150	.053		1.17	2.22		3.39	4.71
1530	Post footing, incl excav, backfill, tube form & concrete, 4' deep, 8" dia	F-7	12	2.667		10.60	99.50		110.10	165
1540	10" diameter		11	2.909		15.70	109		124.70	184
1550	12" diameter		10	3.200		21	119		140	207

06 11 10.30 Roof Framing		Crew	Daily Output	Labor-Hours	Unit	Material	Labor	Equipment	Total	Total Incl O&P
0010	**ROOF FRAMING**									
5250	Composite rafter, 9-1/2" deep	2 Carp	575	.028	L.F.	1.56	1.16		2.72	3.50
5260	11-1/2" deep		575	.028	"	1.69	1.16		2.85	3.64
6070	Fascia boards, 2" x 8"		.30	53.333	M.B.F.	420	2,225		2,645	3,875
6080	2" x 10"		.30	53.333		435	2,225		2,660	3,900
7000	Rafters, to 4 in 12 pitch, 2" x 6"		1	16		375	665		1,040	1,450
7060	2" x 8"		1.26	12.698		420	530		950	1,275
7300	Hip and valley rafters, 2" x 6"		.76	21.053		375	875		1,250	1,775
7360	2" x 8"		.96	16.667		420	695		1,115	1,525
7540	Hip and valley jacks, 2" x 6"		.60	26.667		375	1,100		1,475	2,125
7600	2" x 8"		.65	24.615		420	1,025		1,445	2,025
7780	For slopes steeper than 4 in 12, add						30%			
7790	For dormers or complex roofs, add						50%			
7800	Rafter tie, 1" x 4", #3	2 Carp	.27	59.259	M.B.F.	1,325	2,450		3,775	5,250
7820	Ridge board, #2 or better, 1" x 6" CN		.30	53.333		1,275	2,225		3,500	4,825
7840	1" x 8"		.37	43.243		1,250	1,800		3,050	4,150
7860	1" x 10"		.42	38.095		1,400	1,575		2,975	4,000
7880	2" x 6"		.50	32		375	1,325		1,700	2,475
7900	2" x 8"		.60	26.667		420	1,100		1,520	2,150
7920	2" x 10"		.66	24.242		435	1,000		1,435	2,025
7940	Roof cants, split, 4" x 4"		.86	18.605		1,150	775		1,925	2,475
7960	6" x 6"		1.80	8.889		1,075	370		1,445	1,750
7980	Roof curbs, untreated, 2" x 6"		.52	30.769		375	1,275		1,650	2,400

06 11 Wood Framing

06 11 10 – Framing with Dimensional, Engineered or Composite Lumber

06 11 10.30 Roof Framing		Crew	Daily Output	Labor-Hours	Unit	Material	2010 Bare Costs Labor	Equipment	Total	Total Incl O&P
8000	2" x 12"	2 Carp	.80	20	M.B.F.	530	830		1,360	1,850

06 11 10.32 Sill and Ledger Framing

		Crew	Daily Output	Labor-Hours	Unit	Material	Labor	Equipment	Total	Total Incl O&P
0010	SILL AND LEDGER FRAMING									
4482	Ledgers, nailed, 2" x 4"	2 Carp	.50	32	M.B.F.	375	1,325		1,700	2,450
4484	2" x 6"		.60	26.667		375	1,100		1,475	2,125
4486	Bolted, not including bolts, 3" x 8"		.65	24.615		1,050	1,025		2,075	2,725
4488	3" x 12"		.70	22.857		1,175	950		2,125	2,775
4490	Mud sills, redwood, construction grade, 2" x 4"		.59	27.119		3,400	1,125		4,525	5,450
4492	2" x 6"		.78	20.513		3,400	850		4,250	5,075
4500	Sills, 2" x 4"		.40	40		375	1,650		2,025	2,950
4520	2" x 6"		.55	29.091		375	1,200		1,575	2,300
4540	2" x 8"		.67	23.881		420	990		1,410	1,975
4600	Treated, 2" x 4"		.36	44.444		520	1,850		2,370	3,425
4620	2" x 6"		.50	32		555	1,325		1,880	2,675
4640	2" x 8"		.60	26.667		775	1,100		1,875	2,550
4700	4" x 4"		.60	26.667		1,150	1,100		2,250	2,950
4720	4" x 6"		.70	22.857		1,125	950		2,075	2,725
4740	4" x 8"		.80	20		1,750	830		2,580	3,200
4760	4" x 10"	↓	.87	18.391	↓	1,750	765		2,515	3,100

06 11 10.34 Sleepers

		Crew	Daily Output	Labor-Hours	Unit	Material	Labor	Equipment	Total	Total Incl O&P
0010	SLEEPERS									
0300	On concrete, treated, 1" x 2"	2 Carp	.39	41.026	M.B.F.	1,975	1,700		3,675	4,800
0320	1" x 3"		.50	32		2,100	1,325		3,425	4,350
0340	2" x 4"		.99	16.162		520	670		1,190	1,600
0360	2" x 6"	↓	1.30	12.308	↓	555	510		1,065	1,400

06 11 10.36 Soffit and Canopy Framing

		Crew	Daily Output	Labor-Hours	Unit	Material	Labor	Equipment	Total	Total Incl O&P
0010	SOFFIT AND CANOPY FRAMING									
1300	Canopy or soffit framing, 1" x 4"	2 Carp	.30	53.333	M.B.F.	1,325	2,225		3,550	4,875
1340	1" x 8"		.50	32		1,250	1,325		2,575	3,425
1360	2" x 4"		.41	39.024		375	1,625		2,000	2,900
1400	2" x 8"		.67	23.881		420	990		1,410	1,975
1420	3" x 4"		.50	32		825	1,325		2,150	2,950
1460	3" x 8"	↓	.60	26.667	↓	1,050	1,100		2,150	2,850

06 11 10.38 Treated Lumber Framing Material

		Crew	Daily Output	Labor-Hours	Unit	Material	Labor	Equipment	Total	Total Incl O&P
0010	TREATED LUMBER FRAMING MATERIAL									
0100	2" x 4"				M.B.F.	520			520	570
0110	2" x 6"					555			555	615
0120	2" x 8"					775			775	855
0130	2" x 10"					810			810	890
0140	2" x 12"					900			900	990
0200	4" x 4"					1,150			1,150	1,250
0210	4" x 6"					1,125			1,125	1,250
0220	4" x 8"				↓	1,750			1,750	1,925

06 11 10.40 Wall Framing

		Crew	Daily Output	Labor-Hours	Unit	Material	Labor	Equipment	Total	Total Incl O&P
0010	WALL FRAMING R061110-30									
5860	Headers over openings, 2" x 6"	2 Carp	.36	44.444	M.B.F.	375	1,850		2,225	3,275
5865	2" x 6", pneumatic nailed		.43	37.209		375	1,550		1,925	2,800
5880	2" x 8"		.45	35.556		420	1,475		1,895	2,725
5885	2" x 8", pneumatic nailed		.54	29.630		420	1,225		1,645	2,350
5900	2" x 10"		.53	30.189		435	1,250		1,685	2,400
5905	2" x 10", pneumatic nailed	↓	.67	23.881	↓	435	990		1,425	2,000

06 11 Wood Framing

06 11 10 – Framing with Dimensional, Engineered or Composite Lumber

06 11 10.40 Wall Framing		Crew	Daily Output	Labor-Hours	Unit	2010 Bare Costs Material	Labor	Equipment	Total	Total Incl O&P
5920	2" x 12"	2 Carp	.60	26.667	M.B.F.	530	1,100		1,630	2,275
5925	2" x 12", pneumatic nailed		.72	22.222		530	925		1,455	2,000
5940	4" x 12"		.76	21.053		1,525	875		2,400	3,025
5945	4" x 12", pneumatic nailed		.92	17.391		1,525	725		2,250	2,800
5960	6" x 12"		.84	19.048		1,700	790		2,490	3,100
5965	6" x 12", pneumatic nailed		1.01	15.873		1,700	660		2,360	2,900
6000	Plates, untreated, 2" x 3"		.43	37.209		535	1,550		2,085	2,975
6005	2" x 3", pneumatic nailed		.52	30.769		535	1,275		1,810	2,575
6020	2" x 4" **CN**		.53	30.189		375	1,250		1,625	2,325
6025	2" x 4", pneumatic nailed		.67	23.881		375	990		1,365	1,925
6040	2" x 6"		.75	21.333		375	885		1,260	1,800
6045	2" x 6", pneumatic nailed		.90	17.778		375	740		1,115	1,575
6120	Studs, 8' high wall, 2" x 3"		.60	26.667		535	1,100		1,635	2,300
6125	2" x 3", pneumatic nailed		.72	22.222		535	925		1,460	2,025
6140	2" x 4"		.92	17.391		375	725		1,100	1,525
6145	2" x 4", pneumatic nailed		1.10	14.493		375	600		975	1,350
6160	2" x 6"		1	16		375	665		1,040	1,450
6165	2" x 6", pneumatic nailed		1.20	13.333		375	555		930	1,275
6180	3" x 4"		.80	20		825	830		1,655	2,175
6185	3" x 4", pneumatic nailed	▼	.96	16.667	▼	825	695		1,520	1,975
8200	For 12' high walls, deduct						5%			
8220	For stub wall, 6' high, add						20%			
8240	3' high, add						40%			
8250	For second story & above, add						5%			
8300	For dormer & gable, add						15%			

06 11 10.42 Furring		Crew	Daily Output	Labor-Hours	Unit	Material	Labor	Equipment	Total	Total Incl O&P
0010	**FURRING**									
0012	Wood strips, 1" x 2", on walls, on wood	1 Carp	550	.015	L.F.	.25	.60		.85	1.20
0015	On wood, pneumatic nailed		710	.011		.25	.47		.72	.99
0300	On masonry		495	.016		.25	.67		.92	1.31
0400	On concrete		260	.031		.25	1.28		1.53	2.24
0600	1" x 3", on walls, on wood		550	.015		.34	.60		.94	1.31
0605	On wood, pneumatic nailed		710	.011		.34	.47		.81	1.10
0700	On masonry		495	.016		.34	.67		1.01	1.42
0800	On concrete		260	.031		.34	1.28		1.62	2.35
0850	On ceilings, on wood		350	.023		.34	.95		1.29	1.84
0855	On wood, pneumatic nailed		450	.018		.34	.74		1.08	1.52
0900	On masonry		320	.025		.34	1.04		1.38	1.98
0950	On concrete	▼	210	.038	▼	.34	1.58		1.92	2.82

06 11 10.44 Grounds		Crew	Daily Output	Labor-Hours	Unit	Material	Labor	Equipment	Total	Total Incl O&P
0010	**GROUNDS**									
0020	For casework, 1" x 2" wood strips, on wood	1 Carp	330	.024	L.F.	.25	1.01		1.26	1.82
0100	On masonry		285	.028		.25	1.17		1.42	2.07
0200	On concrete		250	.032		.25	1.33		1.58	2.32
0400	For plaster, 3/4" deep, on wood		450	.018		.25	.74		.99	1.41
0500	On masonry		225	.036		.25	1.48		1.73	2.55
0600	On concrete		175	.046		.25	1.90		2.15	3.20
0700	On metal lath	▼	200	.040	▼	.25	1.66		1.91	2.83

06 12 Structural Panels

06 12 10 – Structural Insulated Panels

06 12 10.10 OSB Faced Panels			Crew	Daily Output	Labor-Hours	Unit	Material	2010 Bare Costs Labor	Equipment	Total	Total Incl O&P
0010	**OSB FACED PANELS**										
0100	Structural insul. panels, 7/16" OSB both faces, EPS insul, 3-5/8" T	G	F-3	2075	.019	S.F.	2.91	.81	.32	4.04	4.79
0110	5-5/8" thick	G		1725	.023		3.33	.98	.38	4.69	5.60
0120	7-3/8" thick	G		1425	.028		3.69	1.18	.46	5.33	6.40
0130	9-3/8" thick	G		1125	.036		4.13	1.50	.58	6.21	7.45
0140	7/16" OSB one face, EPS insul, 3-5/8" thick	G		2175	.018		2.77	.77	.30	3.84	4.57
0150	5-5/8" thick	G		1825	.022		3.05	.92	.36	4.33	5.15
0160	7-3/8" thick	G		1525	.026		3.33	1.10	.43	4.86	5.80
0170	9-3/8" thick	G		1225	.033		3.64	1.38	.54	5.56	6.70
0190	7/16" OSB - 1/2" GWB faces , EPS insul, 3-5/8" T	G		2075	.019		2.87	.81	.32	4	4.75
0200	5-5/8" thick	G		1725	.023		3.29	.98	.38	4.65	5.55
0210	7-3/8" thick	G		1425	.028		3.45	1.18	.46	5.09	6.10
0220	9-3/8" thick	G		1125	.036		3.87	1.50	.58	5.95	7.20
0240	7/16" OSB - 1/2" MRGWB faces , EPS insul, 3-5/8" T	G		2075	.019		2.87	.81	.32	4	4.75
0250	5-5/8" thick	G		1725	.023		3.29	.98	.38	4.65	5.55
0260	7-3/8" thick	G		1425	.028		3.45	1.18	.46	5.09	6.10
0270	9-3/8" thick	G	↓	1125	.036		3.87	1.50	.58	5.95	7.20
0300	For 1/2" GWB added to OSB skin, add	G					.60			.60	.66
0310	For 1/2" MRGWB added to OSB skin, add	G					.75			.75	.83
0320	For one T1-11 skin, add to OSB-OSB	G					.74			.74	.81
0330	For one 19/32" CDX skin, add to OSB-OSB	G				↓	.69			.69	.76
0500	Structural insulated panel, 7/16" OSB both sides, straw core										
0510	4-3/8" T, walls (w/sill, splines, plates)	G	F-6	2400	.017	S.F.	7.60	.65	.27	8.52	9.65
0520	Floors (w/splines)	G		2400	.017		7.60	.65	.27	8.52	9.65
0530	Roof (w/ splines)	G		2400	.017		7.60	.65	.27	8.52	9.65
0550	7-7/8" T, walls (w/sill, splines, plates)	G		2400	.017		10.95	.65	.27	11.87	13.35
0560	Floors (w/ splines)	G		2400	.017		10.95	.65	.27	11.87	13.35
0570	Roof (w/ splines)	G	↓	2400	.017	↓	10.95	.65	.27	11.87	13.35

06 13 Heavy Timber

06 13 23 – Heavy Timber Construction

06 13 23.10 Heavy Framing

		Crew	Daily Output	Labor-Hours	Unit	Material	Labor	Equipment	Total	Total Incl O&P
0010	**HEAVY FRAMING**									
0020	Beams, single 6" x 10"	2 Carp	1.10	14.545	M.B.F.	1,600	605		2,205	2,675
0100	Single 8" x 16"		1.20	13.333		1,975	555		2,530	3,025
0200	Built from 2" lumber, multiple 2" x 14"		.90	17.778		855	740		1,595	2,100
0210	Built from 3" lumber, multiple 3" x 6"		.70	22.857		930	950		1,880	2,500
0220	Multiple 3" x 8"		.80	20		1,050	830		1,880	2,425
0230	Multiple 3" x 10"		.90	17.778		1,100	740		1,840	2,375
0240	Multiple 3" x 12"		1	16		1,175	665		1,840	2,325
0250	Built from 4" lumber, multiple 4" x 6"		.80	20		1,200	830		2,030	2,600
0260	Multiple 4" x 8"		.90	17.778		1,075	740		1,815	2,350
0270	Multiple 4" x 10"		1	16		1,100	665		1,765	2,225
0280	Multiple 4" x 12"		1.10	14.545		1,525	605		2,130	2,600
0290	Columns, structural grade, 1500f, 4" x 4"		.60	26.667		1,275	1,100		2,375	3,100
0300	6" x 6"		.65	24.615		1,325	1,025		2,350	3,025
0400	8" x 8"		.70	22.857		1,350	950		2,300	2,950
0500	10" x 10"		.75	21.333		1,450	885		2,335	2,975
0600	12" x 12"		.80	20		1,525	830		2,355	2,950
0800	Floor planks, 2" thick, T & G, 2" x 6"		1.05	15.238		1,050	635		1,685	2,125
0900	2" x 10"	↓	1.10	14.545	↓	1,050	605		1,655	2,075

06 13 Heavy Timber

06 13 23 – Heavy Timber Construction

06 13 23.10 Heavy Framing		Crew	Daily Output	Labor-Hours	Unit	Material	2010 Bare Costs Labor	Equipment	Total	Total Incl O&P
1100	3" thick, 3" x 6"	2 Carp	1.05	15.238	M.B.F.	1,250	635		1,885	2,350
1200	3" x 10"		1.10	14.545		1,250	605		1,855	2,300
1400	Girders, structural grade, 12" x 12"		.80	20		1,525	830		2,355	2,950
1500	10" x 16"		1	16		2,425	665		3,090	3,700
2300	Roof purlins, 4" thick, structural grade		1.05	15.238		1,325	635		1,960	2,425
2500	Roof trusses, add timber connectors, Division 06 05 23.60		.45	35.556		1,225	1,475		2,700	3,625

06 15 Wood Decking

06 15 16 – Wood Roof Decking

06 15 16.10 Solid Wood Roof Decking

		Crew	Daily Output	Labor-Hours	Unit	Material	Labor	Equipment	Total	Total Incl O&P
0010	**SOLID WOOD ROOF DECKING**									
0400	Cedar planks, 3" thick	2 Carp	320	.050	S.F.	7.70	2.08		9.78	11.70
0500	4" thick		250	.064		10.30	2.66		12.96	15.45
0700	Douglas fir, 3" thick		320	.050		3.28	2.08		5.36	6.80
0800	4" thick		250	.064		4.36	2.66		7.02	8.90
1000	Hemlock, 3" thick		320	.050		3.28	2.08		5.36	6.80
1100	4" thick		250	.064		4.36	2.66		7.02	8.90
1300	Western white spruce, 3" thick		320	.050		3.28	2.08		5.36	6.80
1400	4" thick		250	.064		4.36	2.66		7.02	8.90

06 15 23 – Laminated Wood Decking

06 15 23.10 Laminated Roof Deck

		Crew	Daily Output	Labor-Hours	Unit	Material	Labor	Equipment	Total	Total Incl O&P
0010	**LAMINATED ROOF DECK**									
0020	Pine or hemlock, 3" thick	2 Carp	425	.038	S.F.	3.36	1.56		4.92	6.10
0100	4" thick		325	.049		4.47	2.05		6.52	8.05
0300	Cedar, 3" thick		425	.038		4.11	1.56		5.67	6.95
0400	4" thick		325	.049		5.25	2.05		7.30	8.90
0600	Fir, 3" thick		425	.038		3.43	1.56		4.99	6.20
0700	4" thick		325	.049		4.29	2.05		6.34	7.85

06 16 Sheathing

06 16 23 – Subflooring

06 16 23.10 Subfloor

		Crew	Daily Output	Labor-Hours	Unit	Material	Labor	Equipment	Total	Total Incl O&P
0010	**SUBFLOOR** R061636-20									
0011	Plywood, CDX, 1/2" thick	2 Carp	1500	.011	SF Flr.	.42	.44		.86	1.14
0015	Pneumatic nailed		1860	.009		.42	.36		.78	1.01
0100	5/8" thick		1350	.012		.51	.49		1	1.32
0105	Pneumatic nailed		1674	.010		.51	.40		.91	1.17
0200	3/4" thick		1250	.013		.58	.53		1.11	1.46
0205	Pneumatic nailed		1550	.010		.58	.43		1.01	1.30
0300	1-1/8" thick, 2-4-1 including underlayment		1050	.015		1.50	.63		2.13	2.63
0440	With boards, 1" x 6", S4S, laid regular		900	.018		1.40	.74		2.14	2.67
0450	1" x 8", laid regular		1000	.016		1.34	.66		2	2.49
0460	Laid diagonal		850	.019		1.34	.78		2.12	2.68
0500	1" x 10", laid regular		1100	.015		1.47	.60		2.07	2.55
0600	Laid diagonal		900	.018		1.47	.74		2.21	2.76
8990	Subfloor adhesive, 3/8" bead	1 Carp	2300	.003	L.F.	.09	.14		.23	.32

06 16 26 – Underlayment

06 16 26.10 Wood Product Underlayment			Crew	Daily Output	Labor-Hours	Unit	Material	Labor	Equipment	Total	Total Incl O&P
							2010 Bare Costs				
0010	**WOOD PRODUCT UNDERLAYMENT**	R061636-20									
0030	Plywood, underlayment grade, 3/8" thick		2 Carp	1500	.011	SF Flr.	.77	.44		1.21	1.53
0070	Pneumatic nailed			1860	.009		.77	.36		1.13	1.40
0100	1/2" thick			1450	.011		.80	.46		1.26	1.59
0105	Pneumatic nailed			1798	.009		.80	.37		1.17	1.45
0200	5/8" thick			1400	.011		1.17	.47		1.64	2.02
0205	Pneumatic nailed			1736	.009		1.17	.38		1.55	1.88
0300	3/4" thick			1300	.012		1.23	.51		1.74	2.14
0305	Pneumatic nailed			1612	.010		1.23	.41		1.64	1.99
0500	Particle board, 3/8" thick	G		1500	.011		.36	.44		.80	1.08
0505	Pneumatic nailed	G		1860	.009		.36	.36		.72	.95
0600	1/2" thick	G		1450	.011		.39	.46		.85	1.14
0605	Pneumatic nailed	G		1798	.009		.39	.37		.76	1
0800	5/8" thick	G		1400	.011		.50	.47		.97	1.28
0805	Pneumatic nailed	G		1736	.009		.50	.38		.88	1.14
0900	3/4" thick	G		1300	.012		.65	.51		1.16	1.51
0905	Pneumatic nailed	G		1612	.010		.65	.41		1.06	1.36
0955	Particleboard, 100% recycled straw/wheat, 4' x 8' x 1/4"	G		1450	.011	S.F.	.26	.46		.72	1
0960	4' x 8' x 3/8"	G		1450	.011		.39	.46		.85	1.14
0965	4' x 8' x 1/2"	G		1350	.012		.53	.49		1.02	1.34
0970	4' x 8' x 5/8"	G		1300	.012		.63	.51		1.14	1.48
0975	4' x 8' x 3/4"	G		1250	.013		.73	.53		1.26	1.62
0980	4' x 8' x 1"	G		1150	.014		.98	.58		1.56	1.97
0985	4' x 8' x 1-1/4"	G		1100	.015		1.15	.60		1.75	2.20
1100	Hardboard, underlayment grade, 4' x 4', .215" thick	G		1500	.011	SF Flr.	.57	.44		1.01	1.31

06 16 36 – Wood Panel Product Sheathing

06 16 36.10 Sheathing			Crew	Daily Output	Labor-Hours	Unit	Material	Labor	Equipment	Total	Total Incl O&P
0010	**SHEATHING**	R061636-20									
0012	Plywood on roofs, CDX										
0030	5/16" thick		2 Carp	1600	.010	S.F.	.53	.42		.95	1.22
0035	Pneumatic nailed	R061110-30		1952	.008		.53	.34		.87	1.11
0050	3/8" thick			1525	.010		.36	.44		.80	1.07
0055	Pneumatic nailed			1860	.009		.36	.36		.72	.95
0100	1/2" thick	***CN***		1400	.011		.42	.47		.89	1.19
0105	Pneumatic nailed			1708	.009		.42	.39		.81	1.06
0200	5/8" thick			1300	.012		.51	.51		1.02	1.35
0205	Pneumatic nailed			1586	.010		.51	.42		.93	1.21
0300	3/4" thick			1200	.013		.58	.55		1.13	1.49
0305	Pneumatic nailed			1464	.011		.58	.45		1.03	1.34
0500	Plywood on walls with exterior CDX, 3/8" thick			1200	.013		.36	.55		.91	1.25
0505	Pneumatic nailed			1488	.011		.36	.45		.81	1.09
0600	1/2" thick			1125	.014		.42	.59		1.01	1.37
0605	Pneumatic nailed			1395	.011		.42	.48		.90	1.19
0700	5/8" thick			1050	.015		.51	.63		1.14	1.54
0705	Pneumatic nailed			1302	.012		.51	.51		1.02	1.35
0800	3/4" thick			975	.016		.58	.68		1.26	1.69
0805	Pneumatic nailed			1209	.013		.58	.55		1.13	1.49
0840	Oriented strand board, 7/16" thick	G		1400	.011		.22	.47		.69	.97
0845	Pneumatic nailed	G		1736	.009		.22	.38		.60	.83
0846	1/2" thick	G		1325	.012		.22	.50		.72	1.01
0847	Pneumatic nailed	G		1643	.010		.22	.40		.62	.86
0852	5/8" thick	G		1250	.013		.34	.53		.87	1.19

06 16 Sheathing

06 16 36 – Wood Panel Product Sheathing

06 16 36.10 Sheathing		Crew	Daily Output	Labor-Hours	Unit	Material	2010 Bare Costs Labor	Equipment	Total	Total Incl O&P
0857	Pneumatic nailed G	2 Carp	1550	.010	S.F.	.34	.43		.77	1.03
1000	For shear wall construction, add						20%			
1200	For structural 1 exterior plywood, add				S.F.	10%				
1400	With boards, on roof 1" x 6" boards, laid horizontal	2 Carp	725	.022		1.40	.92		2.32	2.94
1500	Laid diagonal		650	.025		1.40	1.02		2.42	3.11
1700	1" x 8" boards, laid horizontal		875	.018		1.34	.76		2.10	2.64
1800	Laid diagonal		725	.022		1.34	.92		2.26	2.88
2000	For steep roofs, add						40%			
2200	For dormers, hips and valleys, add					5%	50%			
2400	Boards on walls, 1" x 6" boards, laid regular	2 Carp	650	.025		1.40	1.02		2.42	3.11
2500	Laid diagonal		585	.027		1.40	1.14		2.54	3.28
2700	1" x 8" boards, laid regular		765	.021		1.34	.87		2.21	2.81
2800	Laid diagonal		650	.025		1.34	1.02		2.36	3.05
2850	Gypsum, weatherproof, 1/2" thick		1125	.014		.38	.59		.97	1.33
2900	With embedded glass mats		1100	.015		.49	.60		1.09	1.47
3000	Wood fiber, regular, no vapor barrier, 1/2" thick		1200	.013		.56	.55		1.11	1.47
3100	5/8" thick		1200	.013		.71	.55		1.26	1.63
3300	No vapor barrier, in colors, 1/2" thick		1200	.013		.75	.55		1.30	1.68
3400	5/8" thick		1200	.013		.80	.55		1.35	1.73
3600	With vapor barrier one side, white, 1/2" thick		1200	.013		.55	.55		1.10	1.46
3700	Vapor barrier 2 sides, 1/2" thick		1200	.013		.77	.55		1.32	1.70
3800	Asphalt impregnated, 25/32" thick		1200	.013		.28	.55		.83	1.16
3850	Intermediate, 1/2" thick		1200	.013		.20	.55		.75	1.07

06 17 Shop-Fabricated Structural Wood

06 17 33 – Wood I-Joists

06 17 33.10 Wood and Composite I-Joists		Crew	Daily Output	Labor-Hours	Unit	Material	Labor	Equipment	Total	Total Incl O&P
0010	**WOOD AND COMPOSITE I-JOISTS**									
0100	Plywood webs, incl. bridging & blocking, panels 24" O.C.									
1200	15' to 24' span, 50 psf live load	F-5	2400	.013	SF Flr.	1.78	.56		2.34	2.82
1300	55 psf live load		2250	.014		1.93	.60		2.53	3.04
1400	24' to 30' span, 45 psf live load		2600	.012		2.66	.52		3.18	3.72
1500	55 psf live load		2400	.013		3.50	.56		4.06	4.71
1600	Tubular steel open webs, 45 psf, 24" O.C., 40' span	F-3	6250	.006		2.24	.27	.10	2.61	2.99
1700	55' span		7750	.005		2.17	.22	.08	2.47	2.81
1800	70' span		9250	.004		2.82	.18	.07	3.07	3.46
1900	85 psf live load, 26' span		2300	.017		2.63	.73	.29	3.65	4.32

06 17 53 – Shop-Fabricated Wood Trusses

06 17 53.10 Roof Trusses		Crew	Daily Output	Labor-Hours	Unit	Material	Labor	Equipment	Total	Total Incl O&P
0010	**ROOF TRUSSES**									
0100	Fink (W) or King post type, 2'-0" O.C.									
0200	Metal plate connected, 4 in 12 slope									
0210	24' to 29' span	F-3	3000	.013	SF Flr.	2.10	.56	.22	2.88	3.41
0300	30' to 43' span		3000	.013		2.28	.56	.22	3.06	3.61
0400	44' to 60' span		3000	.013		2.52	.56	.22	3.30	3.87
0700	Glued and nailed, add					50%				

06 18 13.20 Laminated Framing		Crew	Daily Output	Labor-Hours	Unit	Material	2010 Bare Costs Labor	Equipment	Total	Total Incl O&P
0010	**LAMINATED FRAMING**									
0020	30 lb., short term live load, 15 lb. dead load									
0200	Straight roof beams, 20′ clear span, beams 8′ O.C.	F-3	2560	.016	SF Flr.	1.76	.66	.26	2.68	3.23
0300	Beams 16′ O.C.		3200	.013		1.27	.53	.20	2	2.44
0500	40′ clear span, beams 8′ O.C.		3200	.013		3.36	.53	.20	4.09	4.74
0600	Beams 16′ O.C.		3840	.010		2.75	.44	.17	3.36	3.89
0800	60′ clear span, beams 8′ O.C.	F-4	2880	.017		5.80	.69	.42	6.91	7.85
0900	Beams 16′ O.C.	″	3840	.013		4.31	.52	.31	5.14	5.90
1100	Tudor arches, 30′ to 40′ clear span, frames 8′ O.C.	F-3	1680	.024		7.55	1	.39	8.94	10.25
1200	Frames 16′ O.C.	″	2240	.018		5.90	.75	.29	6.94	7.95
1400	50′ to 60′ clear span, frames 8′ O.C.	F-4	2200	.022		8.15	.91	.55	9.61	10.95
1500	Frames 16′ O.C.		2640	.018		6.90	.75	.46	8.11	9.25
1700	Radial arches, 60′ clear span, frames 8′ O.C.		1920	.025		7.60	1.04	.63	9.27	10.60
1800	Frames 16′ O.C.		2880	.017		5.85	.69	.42	6.96	7.95
2000	100′ clear span, frames 8′ O.C.		1600	.030		7.90	1.24	.76	9.90	11.40
2100	Frames 16′ O.C.		2400	.020		6.90	.83	.50	8.23	9.40
2300	120′ clear span, frames 8′ O.C.		1440	.033		10.45	1.38	.84	12.67	14.55
2400	Frames 16′ O.C.		1920	.025		9.55	1.04	.63	11.22	12.75
2600	Bowstring trusses, 20′ O.C., 40′ clear span	F-3	2400	.017		4.72	.70	.27	5.69	6.60
2700	60′ clear span	F-4	3600	.013		4.24	.55	.34	5.13	5.85
2800	100′ clear span		4000	.012		6	.50	.30	6.80	7.70
2900	120′ clear span		3600	.013		6.45	.55	.34	7.34	8.25
3000	For less than 1000 B.F., add					20%				
3050	For over 5000 B.F., deduct					10%				
3100	For premium appearance, add to S.F. prices					5%				
3300	For industrial type, deduct					15%				
3500	For stain and varnish, add					5%				
3900	For 3/4″ laminations, add to straight					25%				
4100	Add to curved					15%				
4300	Alternate pricing method: (use nominal footage of									
4310	components). Straight beams, camber less than 6″	F-3	3.50	11.429	M.B.F.	2,600	480	187	3,267	3,825
4400	Columns, including hardware		2	20		2,800	840	330	3,970	4,725
4600	Curved members, radius over 32′		2.50	16		2,875	675	262	3,812	4,475
4700	Radius 10′ to 32′		3	13.333		2,850	560	219	3,629	4,225
4900	For complicated shapes, add maximum					100%				
5100	For pressure treating, add to straight					35%				
5200	Add to curved					45%				
6000	Laminated veneer members, southern pine or western species									
6050	1-3/4″ wide x 5-1/2″ deep	2 Carp	480	.033	L.F.	2.39	1.38		3.77	4.76
6100	9-1/2″ deep		480	.033		3.64	1.38		5.02	6.15
6150	14″ deep		450	.036		5.60	1.48		7.08	8.45
6200	18″ deep		450	.036		7.25	1.48		8.73	10.30
6300	Parallel strand members, southern pine or western species									
6350	1-3/4″ wide x 9-1/4″ deep	2 Carp	480	.033	L.F.	4.50	1.38		5.88	7.10
6400	11-1/4″ deep		450	.036		5.60	1.48		7.08	8.50
6450	14″ deep		400	.040		6.65	1.66		8.31	9.85
6500	3-1/2″ wide x 9-1/4″ deep		480	.033		15.15	1.38		16.53	18.85
6550	11-1/4″ deep		450	.036		18.10	1.48		19.58	22
6600	14″ deep		400	.040		22.50	1.66		24.16	27
6650	7″ wide x 9-1/4″ deep		450	.036		30.50	1.48		31.98	36
6700	11-1/4″ deep		420	.038		38	1.58		39.58	44
6750	14″ deep		400	.040		44.50	1.66		46.16	51.50
8000	Straight beams									

06 18 13 – Glued-Laminated Beams

06 18 13.20 Laminated Framing		Crew	Daily Output	Labor-Hours	Unit	2010 Bare Costs Material	Labor	Equipment	Total	Total Incl O&P
8102	20' span									
8104	3-1/8" x 9"	F-3	30	1.333	Ea.	122	56	22	200	245
8106	X 10-1/2"		30	1.333		143	56	22	221	267
8108	X 12"		30	1.333		163	56	22	241	289
8110	X 13-1/2"		30	1.333		184	56	22	262	310
8112	X 15"		29	1.379		204	58	22.50	284.50	340
8114	5-1/8" x 10-1/2"		30	1.333		234	56	22	312	370
8116	X 12"		30	1.333		268	56	22	346	405
8118	X 13-1/2"		30	1.333		300	56	22	378	440
8120	X 15"		29	1.379		335	58	22.50	415.50	485
8122	X 16-1/2"		29	1.379		370	58	22.50	450.50	520
8124	X 18"		29	1.379		400	58	22.50	480.50	555
8126	X 19-1/2"		29	1.379		435	58	22.50	515.50	595
8128	X 21"		28	1.429		470	60	23.50	553.50	635
8130	X 22-1/2"		28	1.429		500	60	23.50	583.50	670
8132	X 24"		28	1.429		535	60	23.50	618.50	710
8134	6-3/4" x 12"		29	1.379		350	58	22.50	430.50	505
8136	X 13-1/2"		29	1.379		395	58	22.50	475.50	550
8138	X 15"		29	1.379		440	58	22.50	520.50	600
8140	X 16-1/2"		28	1.429		485	60	23.50	568.50	655
8142	X 18"		28	1.429		530	60	23.50	613.50	700
8144	X 19-1/2"		28	1.429		575	60	23.50	658.50	750
8146	X 21"		27	1.481		615	62.50	24.50	702	800
8148	X 22-1/2"		27	1.481		660	62.50	24.50	747	845
8150	X 24"		27	1.481		705	62.50	24.50	792	895
8152	X 25-1/2"		27	1.481		750	62.50	24.50	837	945
8154	X 27"		26	1.538		795	65	25	885	995
8156	X 28-1/2"		26	1.538		835	65	25	925	1,050
8158	X 30"	↓	26	1.538	↓	880	65	25	970	1,100
8200	30' span									
8250	3-1/8" x 9"	F-3	30	1.333	Ea.	184	56	22	262	310
8252	X 10-1/2"		30	1.333		214	56	22	292	345
8254	X 12"		30	1.333		245	56	22	323	380
8256	X 13-1/2"		30	1.333		275	56	22	353	415
8258	X 15"		29	1.379		305	58	22.50	385.50	450
8260	5-1/8" x 10-1/2"		30	1.333		350	56	22	428	495
8262	X 12"		30	1.333		400	56	22	478	550
8264	X 13-1/2"		30	1.333		450	56	22	528	605
8266	X 15"		29	1.379		500	58	22.50	580.50	665
8268	X 16-1/2"		29	1.379		550	58	22.50	630.50	720
8270	X 18"		29	1.379		600	58	22.50	680.50	775
8272	X 19-1/2"		29	1.379		650	58	22.50	730.50	835
8274	X 21"		28	1.429		700	60	23.50	783.50	895
8276	X 22-1/2"		28	1.429		755	60	23.50	838.50	950
8278	X 24"		28	1.429		805	60	23.50	888.50	1,000
8280	6-3/4" x 12"		29	1.379		530	58	22.50	610.50	695
8282	X 13-1/2"		29	1.379		595	58	22.50	675.50	770
8284	X 15"		29	1.379		660	58	22.50	740.50	840
8286	X 16-1/2"		28	1.429		725	60	23.50	808.50	920
8288	X 18"		28	1.429		795	60	23.50	878.50	990
8290	X 19-1/2"		28	1.429		860	60	23.50	943.50	1,075
8292	X 21"		27	1.481		925	62.50	24.50	1,012	1,150
8294	X 22-1/2"	↓	27	1.481	↓	990	62.50	24.50	1,077	1,225

06 18 Glued-Laminated Construction

06 18 13 – Glued-Laminated Beams

06 18 13.20 Laminated Framing		Crew	Daily Output	Labor-Hours	Unit	Material	2010 Bare Costs Labor	Equipment	Total	Total Incl O&P
8296	X 24"	F-3	27	1.481	Ea.	1,050	62.50	24.50	1,137	1,300
8298	X 25-1/2"		27	1.481		1,125	62.50	24.50	1,212	1,350
8300	X 27"		26	1.538		1,200	65	25	1,290	1,425
8302	X 28-1/2"		26	1.538		1,250	65	25	1,340	1,500
8304	X 30"	↓	26	1.538	↓	1,325	65	25	1,415	1,575
8400	40' span									
8402	3-1/8" x 9"	F-3	30	1.333	Ea.	245	56	22	323	380
8404	X 10-1/2"		30	1.333		286	56	22	364	425
8406	X 12"		30	1.333		325	56	22	403	470
8408	X 13-1/2"		30	1.333		365	56	22	443	515
8410	X 15"		29	1.379		410	58	22.50	490.50	565
8412	5-1/8" x 10-1/2"		30	1.333		470	56	22	548	625
8414	X 12"		30	1.333		535	56	22	613	700
8416	X 13-1/2"		30	1.333		600	56	22	678	770
8418	X 15"		29	1.379		670	58	22.50	750.50	850
8420	X 16-1/2"		29	1.379		735	58	22.50	815.50	925
8422	X 18"		29	1.379		805	58	22.50	885.50	1,000
8424	X 19-1/2"		29	1.379		870	58	22.50	950.50	1,075
8426	X 21"		28	1.429		935	60	23.50	1,018.50	1,150
8428	X 22-1/2"		28	1.429		1,000	60	23.50	1,083.50	1,225
8430	X 24"		28	1.429		1,075	60	23.50	1,158.50	1,300
8432	6-3/4" x 12"		29	1.379		705	58	22.50	785.50	890
8434	X 13-1/2"		29	1.379		795	58	22.50	875.50	985
8436	X 15"		29	1.379		880	58	22.50	960.50	1,075
8438	X 16-1/2"		28	1.429		970	60	23.50	1,053.50	1,200
8440	X 18"		28	1.429		1,050	60	23.50	1,133.50	1,300
8442	X 19-1/2"		28	1.429		1,150	60	23.50	1,233.50	1,375
8444	X 21"		27	1.481		1,225	62.50	24.50	1,312	1,475
8446	X 22-1/2"		27	1.481		1,325	62.50	24.50	1,412	1,575
8448	X 24"		27	1.481		1,400	62.50	24.50	1,487	1,675
8450	X 25-1/2"		27	1.481		1,500	62.50	24.50	1,587	1,775
8452	X 27"		26	1.538		1,575	65	25	1,665	1,875
8454	X 28-1/2"		26	1.538		1,675	65	25	1,765	1,975
8456	X 30"	↓	26	1.538	↓	1,750	65	25	1,840	2,075

06 22 Millwork

06 22 13 – Standard Pattern Wood Trim

06 22 13.15 Moldings, Base

		Crew	Daily Output	Labor-Hours	Unit	Material	Labor	Equipment	Total	Total Incl O&P
0010	**MOLDINGS, BASE**									
0500	Base, stock pine, 9/16" x 3-1/2"	1 Carp	240	.033	L.F.	2.22	1.38		3.60	4.57
0550	9/16" x 4-1/2"		200	.040		2.46	1.66		4.12	5.25
0561	Base shoe, oak, 3/4" x 1"	↓	240	.033	↓	1.13	1.38		2.51	3.37

06 22 13.30 Moldings, Casings

		Crew	Daily Output	Labor-Hours	Unit	Material	Labor	Equipment	Total	Total Incl O&P
0010	**MOLDINGS, CASINGS**									
0090	Apron, stock pine, 5/8" x 2"	1 Carp	250	.032	L.F.	1.22	1.33		2.55	3.39
0110	5/8" x 3-1/2"		220	.036		1.62	1.51		3.13	4.11
0300	Band, stock pine, 11/16" x 1-1/8"		270	.030		.82	1.23		2.05	2.80
0350	11/16" x 1-3/4"		250	.032		1.06	1.33		2.39	3.22
0700	Casing, stock pine, 11/16" x 2-1/2"		240	.033		1.33	1.38		2.71	3.59
0750	11/16" x 3-1/2"	↓	215	.037	↓	1.62	1.55		3.17	4.16

06 22 Millwork

06 22 13 – Standard Pattern Wood Trim

06 22 13.35 Moldings, Ceilings

	06 22 13.35 Moldings, Ceilings	Crew	Daily Output	Labor-Hours	Unit	2010 Bare Costs Material	Labor	Equipment	Total	Total Incl O&P
0010	**MOLDINGS, CEILINGS**									
0600	Bed, stock pine, 9/16" x 1-3/4"	1 Carp	270	.030	L.F.	1.02	1.23		2.25	3.02
0650	9/16" x 2"		240	.033		1.22	1.38		2.60	3.47
1200	Cornice molding, stock pine, 9/16" x 1-3/4"		330	.024		1.02	1.01		2.03	2.67
1300	9/16" x 2-1/4"		300	.027		1.24	1.11		2.35	3.07
2400	Cove scotia, stock pine, 9/16" x 1-3/4"		270	.030		.67	1.23		1.90	2.64
2500	11/16" x 2-3/4"		255	.031		1.86	1.30		3.16	4.06
2600	Crown, stock pine, 9/16" x 3-5/8"		250	.032		2.30	1.33		3.63	4.58
2700	11/16" x 4-5/8"	↓	220	.036	↓	3.87	1.51		5.38	6.60

06 22 13.40 Moldings, Exterior

	06 22 13.40 Moldings, Exterior	Crew	Daily Output	Labor-Hours	Unit	Material	Labor	Equipment	Total	Total Incl O&P
0010	**MOLDINGS, EXTERIOR**									
1500	Cornice, boards, pine, 1" x 2"	1 Carp	330	.024	L.F.	.25	1.01		1.26	1.82
1700	1" x 6"		250	.032		.64	1.33		1.97	2.75
2000	1" x 12"		180	.044		1.42	1.85		3.27	4.41
2200	Three piece, built-up, pine, minimum		80	.100		1.33	4.16		5.49	7.85
2300	Maximum		65	.123		3.42	5.10		8.52	11.65
3000	Corner board, sterling pine, 1" x 4"		200	.040		.79	1.66		2.45	3.43
3100	1" x 6"		200	.040		1.05	1.66		2.71	3.72
3350	Fascia, sterling pine, 1" x 6"		250	.032		1.05	1.33		2.38	3.21
3370	1" x 8"		225	.036		1.43	1.48		2.91	3.85
3400	Trim, back band, 11/16" x 1-1/16"		250	.032		.83	1.33		2.16	2.96
3500	Casing, 11/16" x 4-1/4"		250	.032		2.74	1.33		4.07	5.05
3600	Crown, 11/16" x 4-1/4"		250	.032		3.85	1.33		5.18	6.30
3700	1" x 4" and 1" x 6" railing w/ balusters 4" O.C.		22	.364		22.50	15.10		37.60	48
3800	Insect screen framing stock, 1-1/16" x 1-3/4"		395	.020		2.03	.84		2.87	3.53
4100	Verge board, sterling pine, 1" x 4"		200	.040		.79	1.66		2.45	3.43
4200	1" x 6"		200	.040		1.05	1.66		2.71	3.72
4300	2" x 6"		165	.048		2.10	2.01		4.11	5.40
4400	2" x 8"	↓	165	.048	↓	2.86	2.01		4.87	6.25
4700	For redwood trim, add					200%				

06 22 13.45 Moldings, Trim

	06 22 13.45 Moldings, Trim	Crew	Daily Output	Labor-Hours	Unit	Material	Labor	Equipment	Total	Total Incl O&P
0010	**MOLDINGS, TRIM**									
0200	Astragal, stock pine, 11/16" x 1-3/4"	1 Carp	255	.031	L.F.	1.22	1.30		2.52	3.35
0250	1-5/16" x 2-3/16"		240	.033		2.57	1.38		3.95	4.96
0800	Chair rail, stock pine, 5/8" x 2-1/2"		270	.030		1.43	1.23		2.66	3.47
0900	5/8" x 3-1/2"		240	.033		2.26	1.38		3.64	4.62
1000	Closet pole, stock pine, 1-1/8" diameter		200	.040		1.05	1.66		2.71	3.72
1100	Fir, 1-5/8" diameter		200	.040		1.93	1.66		3.59	4.68
3300	Half round, stock pine, 1/4" x 1/2"		270	.030		.24	1.23		1.47	2.16
3350	1/2" x 1"	↓	255	.031	↓	.62	1.30		1.92	2.69
3400	Handrail, fir, single piece, stock, hardware not included									
3450	1-1/2" x 1-3/4"	1 Carp	80	.100	L.F.	2.03	4.16		6.19	8.65
3470	Pine, 1-1/2" x 1-3/4"		80	.100		2.03	4.16		6.19	8.65
3500	1-1/2" x 2-1/2"		76	.105		3.04	4.37		7.41	10.10
3600	Lattice, stock pine, 1/4" x 1-1/8"		270	.030		.42	1.23		1.65	2.36
3700	1/4" x 1-3/4"		250	.032		.65	1.33		1.98	2.77
3800	Miscellaneous, custom, pine, 1" x 1"		270	.030		.35	1.23		1.58	2.28
3900	1" x 3"		240	.033		1.05	1.38		2.43	3.28
4100	Birch or oak, nominal 1" x 1"		240	.033		.45	1.38		1.83	2.62
4200	Nominal 1" x 3"		215	.037		1.35	1.55		2.90	3.86
4400	Walnut, nominal 1" x 1"		215	.037		.58	1.55		2.13	3.02
4500	Nominal 1" x 3"	↓	200	.040	↓	1.74	1.66		3.40	4.47

06 22 Millwork

06 22 13 – Standard Pattern Wood Trim

06 22 13.45 Moldings, Trim		Crew	Daily Output	Labor-Hours	Unit	Material	Labor	Equipment	Total	Total Incl O&P
						2010 Bare Costs				
4700	Teak, nominal 1" x 1"	1 Carp	215	.037	L.F.	1.60	1.55		3.15	4.14
4800	Nominal 1" x 3"		200	.040		4.79	1.66		6.45	7.80
4900	Quarter round, stock pine, 1/4" x 1/4"		275	.029		.22	1.21		1.43	2.10
4950	3/4" x 3/4"		255	.031		.60	1.30		1.90	2.67
5600	Wainscot moldings, 1-1/8" x 9/16", 2' high, minimum		76	.105	S.F.	11.15	4.37		15.52	19.05
5700	Maximum		65	.123	"	20	5.10		25.10	30.50

06 22 13.50 Moldings, Window and Door

		Crew	Daily Output	Labor-Hours	Unit	Material	Labor	Equipment	Total	Total Incl O&P
0010	**MOLDINGS, WINDOW AND DOOR**									
2800	Door moldings, stock, decorative, 1-1/8" wide, plain	1 Carp	17	.471	Set	45	19.55		64.55	79.50
2900	Detailed		17	.471	"	87.50	19.55		107.05	127
2960	Clear pine door jamb, no stops, 11/16" x 4-9/16"		240	.033	L.F.	5.15	1.38		6.53	7.80
3150	Door trim set, 1 head and 2 sides, pine, 2-1/2 wide		12	.667	Opng.	22.50	27.50		50	67.50
3170	3-1/2" wide		11	.727	"	27.50	30		57.50	77
3250	Glass beads, stock pine, 3/8" x 1/2"		275	.029	L.F.	.31	1.21		1.52	2.20
3270	3/8" x 7/8"		270	.030		.41	1.23		1.64	2.35
4850	Parting bead, stock pine, 3/8" x 3/4"		275	.029		.36	1.21		1.57	2.26
4870	1/2" x 3/4"		255	.031		.45	1.30		1.75	2.51
5000	Stool caps, stock pine, 11/16" x 3-1/2"		200	.040		2.27	1.66		3.93	5.05
5100	1-1/16" x 3-1/4"		150	.053		3.56	2.22		5.78	7.35
5300	Threshold, oak, 3' long, inside, 5/8" x 3-5/8"		32	.250	Ea.	9	10.40		19.40	26
5400	Outside, 1-1/2" x 7-5/8"		16	.500	"	38	21		59	74
5900	Window trim sets, including casings, header, stops,									
5910	stool and apron, 2-1/2" wide, minimum	1 Carp	13	.615	Opng.	32.50	25.50		58	75
5950	Average		10	.800		38	33		71	93
6000	Maximum		6	1.333		62.50	55.50		118	154

06 22 13.60 Moldings, Soffits

		Crew	Daily Output	Labor-Hours	Unit	Material	Labor	Equipment	Total	Total Incl O&P
0010	**MOLDINGS, SOFFITS**									
0200	Soffits, pine, 1" x 4"	2 Carp	420	.038	L.F.	.44	1.58		2.02	2.92
0210	1" x 6"		420	.038		.64	1.58		2.22	3.14
0220	1" x 8"		420	.038		.84	1.58		2.42	3.36
0230	1" x 10"		400	.040		1.17	1.66		2.83	3.84
0240	1" x 12"		400	.040		1.42	1.66		3.08	4.12
0250	STK cedar, 1" x 4"		420	.038		.52	1.58		2.10	3.01
0260	1" x 6"		420	.038		.84	1.58		2.42	3.36
0270	1" x 8"		420	.038		1.31	1.58		2.89	3.88
0280	1" x 10"		400	.040		1.89	1.66		3.55	4.64
0290	1" x 12"		400	.040		2.63	1.66		4.29	5.45
1000	Exterior AC plywood, 1/4" thick		420	.038	S.F.	.68	1.58		2.26	3.19
1050	3/8" thick		420	.038		.77	1.58		2.35	3.29
1100	1/2" thick		420	.038		.80	1.58		2.38	3.32
1150	Polyvinyl chloride, white, solid	1 Carp	230	.035		1.09	1.45		2.54	3.43
1160	Perforated	"	230	.035		1.09	1.45		2.54	3.43

06 25 Prefinished Paneling

06 25 13 – Prefinished Hardboard Paneling

06 25 13.10 Paneling, Hardboard		Crew	Daily Output	Labor-Hours	Unit	Material	Labor	Equipment	Total	Total Incl O&P
						2010 Bare Costs				
0010	**PANELING, HARDBOARD**									
0050	Not incl. furring or trim, hardboard, tempered, 1/8" thick G	2 Carp	500	.032	S.F.	.39	1.33		1.72	2.48
0100	1/4" thick G		500	.032		.53	1.33		1.86	2.63
0300	Tempered pegboard, 1/8" thick G		500	.032		.44	1.33		1.77	2.53
0400	1/4" thick G		500	.032		.67	1.33		2	2.79
0600	Untempered hardboard, natural finish, 1/8" thick G		500	.032		.38	1.33		1.71	2.47
0700	1/4" thick G		500	.032		.46	1.33		1.79	2.56
0900	Untempered pegboard, 1/8" thick G		500	.032		.38	1.33		1.71	2.47
1000	1/4" thick G		500	.032		.42	1.33		1.75	2.51
1200	Plastic faced hardboard, 1/8" thick G		500	.032		.67	1.33		2	2.79
1300	1/4" thick G		500	.032		.85	1.33		2.18	2.99
1500	Plastic faced pegboard, 1/8" thick G		500	.032		.61	1.33		1.94	2.72
1600	1/4" thick G		500	.032		.75	1.33		2.08	2.88
1800	Wood grained, plain or grooved, 1/4" thick, minimum G		500	.032		.57	1.33		1.90	2.68
1900	Maximum G		425	.038		1.19	1.56		2.75	3.72
2100	Moldings for hardboard, wood or aluminum, minimum		500	.032	L.F.	.38	1.33		1.71	2.47
2200	Maximum		425	.038	"	1.08	1.56		2.64	3.60

06 25 16 – Prefinished Plywood Paneling

06 25 16.10 Paneling, Plywood		Crew	Daily Output	Labor-Hours	Unit	Material	Labor	Equipment	Total	Total Incl O&P
0010	**PANELING, PLYWOOD** R061636-20									
2400	Plywood, prefinished, 1/4" thick, 4' x 8' sheets									
2410	with vertical grooves. Birch faced, minimum	2 Carp	500	.032	S.F.	.89	1.33		2.22	3.03
2420	Average		420	.038		1.35	1.58		2.93	3.93
2430	Maximum		350	.046		1.97	1.90		3.87	5.10
2600	Mahogany, African		400	.040		2.51	1.66		4.17	5.30
2700	Philippine (Lauan)		500	.032		1.08	1.33		2.41	3.24
2900	Oak or Cherry, minimum		500	.032		2.10	1.33		3.43	4.36
3000	Maximum		400	.040		3.22	1.66		4.88	6.10
3200	Rosewood		320	.050		4.58	2.08		6.66	8.25
3400	Teak		400	.040		3.22	1.66		4.88	6.10
3600	Chestnut		375	.043		4.77	1.77		6.54	8
3800	Pecan		400	.040		2.06	1.66		3.72	4.83
3900	Walnut, minimum		500	.032		2.75	1.33		4.08	5.10
3950	Maximum		400	.040		5.20	1.66		6.86	8.30
4000	Plywood, prefinished, 3/4" thick, stock grades, minimum		320	.050		1.24	2.08		3.32	4.56
4100	Maximum		224	.071		5.40	2.97		8.37	10.50
4300	Architectural grade, minimum		224	.071		3.98	2.97		6.95	8.95
4400	Maximum		160	.100		6.10	4.16		10.26	13.10
4600	Plywood, "A" face, birch, V.C., 1/2" thick, natural		450	.036		1.89	1.48		3.37	4.36
4700	Select		450	.036		2.06	1.48		3.54	4.55
4900	Veneer core, 3/4" thick, natural		320	.050		1.99	2.08		4.07	5.40
5000	Select		320	.050		2.23	2.08		4.31	5.65
5200	Lumber core, 3/4" thick, natural		320	.050		2.99	2.08		5.07	6.50
5500	Plywood, knotty pine, 1/4" thick, A2 grade		450	.036		1.63	1.48		3.11	4.07
5600	A3 grade		450	.036		2.06	1.48		3.54	4.55
5800	3/4" thick, veneer core, A2 grade		320	.050		2.11	2.08		4.19	5.50
5900	A3 grade		320	.050		2.38	2.08		4.46	5.80
6100	Aromatic cedar, 1/4" thick, plywood		400	.040		2.08	1.66		3.74	4.85
6200	1/4" thick, particle board		400	.040		1.01	1.66		2.67	3.67

06 25 Prefinished Paneling

06 25 26 – Panel System

06 25 26.10 Panel Systems		Crew	Daily Output	Labor-Hours	Unit	Material	Labor	Equipment	Total	Total Incl O&P
						2010 Bare Costs				
0010	PANEL SYSTEMS									
0100	Raised panel, eng. wood core w/ wood veneer, std., paint grade	2 Carp	300	.053	S.F.	11.90	2.22		14.12	16.50
0110	Oak veneer		300	.053		19.70	2.22		21.92	25
0120	Maple veneer		300	.053		25	2.22		27.22	31
0130	Cherry veneer		300	.053		32	2.22		34.22	38.50
0300	Class I fire rated, paint grade		300	.053		14.25	2.22		16.47	19.05
0310	Oak veneer		300	.053		23.50	2.22		25.72	29.50
0320	Maple veneer		300	.053		30	2.22		32.22	36.50
0330	Cherry veneer		300	.053		38.50	2.22		40.72	45.50
0510	Beadboard, 5/8" MDF, standard, primed		300	.053		8.40	2.22		10.62	12.65
0520	Oak veneer, unfinished		300	.053		14.60	2.22		16.82	19.45
0530	Maple veneer, unfinished		300	.053		17.25	2.22		19.47	22.50
0610	Rustic paneling, 5/8" MDF, standard, maple veneer, unfinished	↓	300	.053	↓	22	2.22		24.22	28

06 26 Board Paneling

06 26 13 – Profile Board Paneling

06 26 13.10 Paneling, Boards

Line	Description	Crew	Daily Output	Labor-Hours	Unit	Material	Labor	Equipment	Total	Total Incl O&P
0010	PANELING, BOARDS									
6400	Wood board paneling, 3/4" thick, knotty pine	2 Carp	300	.053	S.F.	1.44	2.22		3.66	5
6500	Rough sawn cedar		300	.053		1.85	2.22		4.07	5.45
6700	Redwood, clear, 1" x 4" boards		300	.053		4.96	2.22		7.18	8.85
6900	Aromatic cedar, closet lining, boards	↓	275	.058	↓	3.32	2.42		5.74	7.40

06 43 Wood Stairs and Railings

06 43 13 – Wood Stairs

06 43 13.20 Prefabricated Wood Stairs

Line	Description	Crew	Daily Output	Labor-Hours	Unit	Material	Labor	Equipment	Total	Total Incl O&P
0010	PREFABRICATED WOOD STAIRS									
0100	Box stairs, prefabricated, 3'-0" wide									
0110	Oak treads, up to 14 risers	2 Carp	39	.410	Riser	85	17.05		102.05	120
0600	With pine treads for carpet, up to 14 risers	"	39	.410	"	55	17.05		72.05	87
1100	For 4' wide stairs, add				Flight	25%				
1550	Stairs, prefabricated stair handrail with balusters	1 Carp	30	.267	L.F.	75	11.10		86.10	99
1700	Basement stairs, prefabricated, pine treads									
1710	Pine risers, 3' wide, up to 14 risers	2 Carp	52	.308	Riser	55	12.80		67.80	80
4000	Residential, wood, oak treads, prefabricated		1.50	10.667	Flight	1,100	445		1,545	1,875
4200	Built in place	↓	.44	36.364	"	1,750	1,500		3,250	4,250
4400	Spiral, oak, 4'-6" diameter, unfinished, prefabricated,									
4500	incl. railing, 9' high	2 Carp	1.50	10.667	Flight	3,625	445		4,070	4,650

06 43 13.40 Wood Stair Parts

Line	Description	Crew	Daily Output	Labor-Hours	Unit	Material	Labor	Equipment	Total	Total Incl O&P
0010	WOOD STAIR PARTS									
0020	Pin top balusters, 1-1/4", oak, 34"	1 Carp	96	.083	Ea.	7.80	3.46		11.26	13.95
0030	38"		96	.083		9.85	3.46		13.31	16.20
0040	42"		96	.083		12.10	3.46		15.56	18.65
0050	Poplar, 34"		96	.083		3.05	3.46		6.51	8.70
0060	38"		96	.083		3.62	3.46		7.08	9.35
0070	42"		96	.083		8.55	3.46		12.01	14.80
0080	Maple, 34"		96	.083		4.90	3.46		8.36	10.75
0090	38"	↓	96	.083	↓	6.20	3.46		9.66	12.15

06 43 13 – Wood Stairs

06 43 13.40 Wood Stair Parts		Crew	Daily Output	Labor-Hours	Unit	Material	Labor	Equipment	Total	Total Incl O&P
						2010 Bare Costs				
0100	42"	1 Carp	96	.083	Ea.	7.60	3.46		11.06	13.70
0130	Primed, 34"		96	.083		2.85	3.46		6.31	8.50
0140	38"		96	.083		3.62	3.46		7.08	9.35
0150	42"		96	.083		4.42	3.46		7.88	10.20
0180	Box top balusters, 1-1/4", oak, 34"		60	.133		10.85	5.55		16.40	20.50
0190	38"		60	.133		12.50	5.55		18.05	22.50
0200	42"		60	.133		12.95	5.55		18.50	23
0210	Poplar, 34"		60	.133		8.65	5.55		14.20	18.05
0220	38"		60	.133		9.50	5.55		15.05	19
0230	42"		60	.133		10	5.55		15.55	19.55
0240	Maple, 34"		60	.133		11.90	5.55		17.45	21.50
0250	38"		60	.133		13.80	5.55		19.35	24
0260	42"		60	.133		14.20	5.55		19.75	24
0290	Primed, 34"		60	.133		6.95	5.55		12.50	16.20
0300	38"		60	.133		8	5.55		13.55	17.35
0310	42"		60	.133		8.35	5.55		13.90	17.75
0340	Square balusters, cut from lineal stock, pine, 1-1/16" x 1-1/16"		180	.044	L.F.	1.32	1.85		3.17	4.30
0350	1-5/16" x 1-5/16"		180	.044		1.70	1.85		3.55	4.72
0360	1-5/8" x 1-5/8"		180	.044		2.82	1.85		4.67	5.95
0370	Turned newel, oak, 3-1/2" square, 48" high		8	1	Ea.	94	41.50		135.50	167
0380	62" high		8	1		138	41.50		179.50	216
0390	Poplar, 3-1/2" square, 48" high		8	1		52	41.50		93.50	121
0400	62" high		8	1		75	41.50		116.50	147
0410	Maple, 3-1/2" square, 48" high		8	1		66	41.50		107.50	137
0420	62" high		8	1		93	41.50		134.50	166
0430	Square newel, oak, 3-1/2" square, 48" high		8	1		74	41.50		115.50	146
0440	58" high		8	1		78	41.50		119.50	150
0450	Poplar, 3-1/2" square, 48" high		8	1		49	41.50		90.50	118
0460	58" high		8	1		54	41.50		95.50	124
0470	Maple, 3" square, 48" high		8	1		66	41.50		107.50	137
0480	58" high		8	1		70	41.50		111.50	141
0490	Railings, oak, minimum		96	.083	L.F.	11	3.46		14.46	17.45
0500	Average		96	.083		14	3.46		17.46	21
0510	Maximum		96	.083		14.50	3.46		17.96	21.50
0520	Maple, minimum		96	.083		15	3.46		18.46	22
0530	Average		96	.083		15.25	3.46		18.71	22
0540	Maximum		96	.083		15.50	3.46		18.96	22.50
0550	Oak, for bending rail, minimum		48	.167		21	6.95		27.95	33.50
0560	Average		48	.167		22	6.95		28.95	34.50
0570	Maximum		48	.167		23	6.95		29.95	36
0580	Maple, for bending rail, minimum		48	.167		23	6.95		29.95	35.50
0590	Average		48	.167		25	6.95		31.95	37.50
0600	Maximum		48	.167		27	6.95		33.95	40
0610	Risers, oak, 3/4" x 8", 36" long		80	.100	Ea.	18	4.16		22.16	26
0620	42" long		80	.100		21	4.16		25.16	29.50
0630	48" long		80	.100		24	4.16		28.16	33
0640	54" long		80	.100		27	4.16		31.16	36
0650	60" long		80	.100		30	4.16		34.16	39.50
0660	72" long		80	.100		36	4.16		40.16	46
0670	Poplar, 3/4" x 8", 36" long		80	.100		11.10	4.16		15.26	18.60
0680	42" long		80	.100		12.95	4.16		17.11	20.50
0690	48" long		80	.100		14.80	4.16		18.96	22.50
0700	54" long		80	.100		16.65	4.16		20.81	24.50

06 43 Wood Stairs and Railings

06 43 13 – Wood Stairs

06 43 13.40 Wood Stair Parts		Crew	Daily Output	Labor-Hours	Unit	Material	Labor	Equipment	Total	Total Incl O&P
						2010 Bare Costs				
0710	60″ long	1 Carp	80	.100	Ea.	18.50	4.16		22.66	27
0720	72″ long		80	.100		22	4.16		26.16	31
0730	Pine, 1″ x 8″, 36″ long		80	.100		2.51	4.16		6.67	9.15
0740	42″ long		80	.100		2.93	4.16		7.09	9.60
0750	48″ long		80	.100		3.34	4.16		7.50	10.10
0760	54″ long		80	.100		3.76	4.16		7.92	10.55
0770	60″ long		80	.100		4.18	4.16		8.34	11
0780	72″ long		80	.100		5	4.16		9.16	11.90
0790	Treads, oak, no returns, 1-1/32″ x 11-1/2″ x 36″ long		32	.250		32	10.40		42.40	51
0800	42″ long		32	.250		37.50	10.40		47.90	57
0810	48″ long		32	.250		42.50	10.40		52.90	63
0820	54″ long		32	.250		48	10.40		58.40	69
0830	60″ long		32	.250		53.50	10.40		63.90	74.50
0840	72″ long		32	.250		64	10.40		74.40	86.50
0850	Mitred return one end, 1-1/32″ x 11-1/2″ x 36″ long		24	.333		45	13.85		58.85	71
0860	42″ long		24	.333		52.50	13.85		66.35	79.50
0870	48″ long		24	.333		60	13.85		73.85	87.50
0880	54″ long		24	.333		67.50	13.85		81.35	96
0890	60″ long		24	.333		75	13.85		88.85	104
0900	72″ long		24	.333		90	13.85		103.85	121
0910	Mitred return two ends, 1-1/32″ x 11-1/2″ x 36″ long		12	.667		49	27.50		76.50	96.50
0920	42″ long		12	.667		57	27.50		84.50	106
0930	48″ long		12	.667		65.50	27.50		93	115
0940	54″ long		12	.667		73.50	27.50		101	124
0950	60″ long		12	.667		81.50	27.50		109	133
0960	72″ long		12	.667		98	27.50		125.50	151
0970	Starting step, oak, 48″, bullnose		8	1		176	41.50		217.50	258
0980	Double end bullnose		8	1		275	41.50		316.50	370
1030	Skirt board, pine, 1″ x 10″		55	.145	L.F.	1.17	6.05		7.22	10.60
1040	1″ x 12″		52	.154	″	1.42	6.40		7.82	11.40
1050	Oak landing tread, 1-1/16″ thick		54	.148	S.F.	8.25	6.15		14.40	18.60
1060	Oak cove molding		96	.083	L.F.	1.45	3.46		4.91	6.95
1070	Oak stringer molding		96	.083	″	2.75	3.46		6.21	8.40
1090	Rail bolt, 5/16″ x 3-1/2″		48	.167	Ea.	2.75	6.95		9.70	13.75
1100	5/16″ x 4-1/2″		48	.167		2.75	6.95		9.70	13.75
1120	Newel post anchor		16	.500		13.50	21		34.50	47
1130	Tapered plug, 1/2″		240	.033		.30	1.38		1.68	2.46
1140	1″		240	.033		.40	1.38		1.78	2.57

06 43 16 – Wood Railings

06 43 16.10 Wood Handrails and Railings		Crew	Daily Output	Labor-Hours	Unit	Material	Labor	Equipment	Total	Total Incl O&P
0010	**WOOD HANDRAILS AND RAILINGS**									
0020	Custom design, architectural grade, hardwood, minimum	1 Carp	38	.211	L.F.	9.50	8.75		18.25	24
0100	Maximum		30	.267		50	11.10		61.10	72
0300	Stock interior railing with spindles 4″ O.C., 4′ long		40	.200		37	8.30		45.30	53.50
0400	8′ long		48	.167		37	6.95		43.95	51

06 44 Ornamental Woodwork

06 44 19 – Wood Grilles

06 44 19.10 Grilles		Crew	Daily Output	Labor-Hours	Unit	Material	2010 Bare Costs Labor	Equipment	Total	Total Incl O&P
0010	**GRILLES** and panels, hardwood, sanded									
0020	2′ x 4′ to 4′ x 8′, custom designs, unfinished, minimum	1 Carp	38	.211	S.F.	54	8.75		62.75	73
0050	Average		30	.267		62.50	11.10		73.60	86
0100	Maximum		19	.421		71	17.50		88.50	105

06 44 33 – Wood Mantels

06 44 33.10 Fireplace Mantels		Crew	Daily Output	Labor-Hours	Unit	Material	Labor	Equipment	Total	Total Incl O&P
0010	**FIREPLACE MANTELS**									
0015	6″ molding, 6′ x 3′-6″ opening, minimum	1 Carp	5	1.600	Opng.	228	66.50		294.50	355
0100	Maximum		5	1.600		410	66.50		476.50	555
0300	Prefabricated pine, colonial type, stock, deluxe		2	4		1,250	166		1,416	1,650
0400	Economy		3	2.667		530	111		641	755

06 44 33.20 Fireplace Mantel Beam		Crew	Daily Output	Labor-Hours	Unit	Material	Labor	Equipment	Total	Total Incl O&P
0010	**FIREPLACE MANTEL BEAM**									
0020	Rough texture wood, 4″ x 8″	1 Carp	36	.222	L.F.	6.90	9.25		16.15	22
0100	4″ x 10″		35	.229	″	12.80	9.50		22.30	29
0300	Laminated hardwood, 2-1/4″ x 10-1/2″ wide, 6′ long		5	1.600	Ea.	110	66.50		176.50	223
0400	8′ long		5	1.600	″	150	66.50		216.50	267
0600	Brackets for above, rough sawn		12	.667	Pr.	10	27.50		37.50	53.50
0700	Laminated		12	.667	″	15	27.50		42.50	59

06 44 39 – Wood Posts and Columns

06 44 39.10 Decorative Beams		Crew	Daily Output	Labor-Hours	Unit	Material	Labor	Equipment	Total	Total Incl O&P
0010	**DECORATIVE BEAMS**									
0020	Rough sawn cedar, non-load bearing, 4″ x 4″	2 Carp	180	.089	L.F.	1	3.69		4.69	6.80
0100	4″ x 6″		170	.094		1.42	3.91		5.33	7.60
0200	4″ x 8″		160	.100		1.89	4.16		6.05	8.50
0300	4″ x 10″		150	.107		2.37	4.43		6.80	9.45
0400	4″ x 12″		140	.114		2.84	4.75		7.59	10.40
0500	8″ x 8″		130	.123		4.96	5.10		10.06	13.35

06 44 39.20 Columns		Crew	Daily Output	Labor-Hours	Unit	Material	Labor	Equipment	Total	Total Incl O&P
0010	**COLUMNS**									
0050	Aluminum, round colonial, 6″ diameter	2 Carp	80	.200	V.L.F.	16	8.30		24.30	30.50
0100	8″ diameter		62.25	.257		19.30	10.70		30	37.50
0200	10″ diameter		55	.291		23.50	12.10		35.60	44.50
0250	Fir, stock units, hollow round, 6″ diameter		80	.200		26.50	8.30		34.80	42
0300	8″ diameter		80	.200		31	8.30		39.30	47.50
0350	10″ diameter		70	.229		39.50	9.50		49	58
0400	Solid turned, to 8′ high, 3-1/2″ diameter		80	.200		9	8.30		17.30	22.50
0500	4-1/2″ diameter		75	.213		14	8.85		22.85	29
0600	5-1/2″ diameter		70	.229		18.40	9.50		27.90	34.50
0800	Square columns, built-up, 5″ x 5″		65	.246		15.50	10.25		25.75	33
0900	Solid, 3-1/2″ x 3-1/2″		130	.123		9	5.10		14.10	17.80
1600	Hemlock, tapered, T & G, 12″ diam, 10′ high		100	.160		53	6.65		59.65	68.50
1700	16′ high		65	.246		79.50	10.25		89.75	103
1900	10′ high, 14″ diameter		100	.160		121	6.65		127.65	143
2000	18′ high		65	.246		107	10.25		117.25	134
2200	18″ diameter, 12′ high		65	.246		153	10.25		163.25	185
2300	20′ high		50	.320		148	13.30		161.30	184
2500	20″ diameter, 14′ high		40	.400		159	16.60		175.60	201
2600	20′ high		35	.457		149	19		168	193
2800	For flat pilasters, deduct					33%				
3000	For splitting into halves, add				Ea.	116			116	128

06 44 Ornamental Woodwork

06 44 39 – Wood Posts and Columns

06 44 39.20 Columns		Crew	Daily Output	Labor-Hours	Unit	Material	2010 Bare Costs Labor	Equipment	Total	Total Incl O&P
4000	Rough sawn cedar posts, 4" x 4"	2 Carp	250	.064	V.L.F.	6.30	2.66		8.96	11.05
4100	4" x 6"		235	.068		15	2.83		17.83	21
4200	6" x 6"		220	.073		26.50	3.02		29.52	33.50
4300	8" x 8"		200	.080		60	3.32		63.32	71

06 48 Wood Frames

06 48 13 – Exterior Wood Door Frames

06 48 13.10 Exterior Wood Door Frames and Accessories

		Crew	Daily Output	Labor-Hours	Unit	Material	Labor	Equipment	Total	Total Incl O&P
0010	**EXTERIOR WOOD DOOR FRAMES AND ACCESSORIES**									
0400	Exterior frame, incl. ext. trim, pine, 5/4 x 4-9/16" deep	2 Carp	375	.043	L.F.	5.55	1.77		7.32	8.85
0420	5-3/16" deep		375	.043		11.35	1.77		13.12	15.25
0440	6-9/16" deep		375	.043		8.50	1.77		10.27	12.10
0600	Oak, 5/4 x 4-9/16" deep		350	.046		10.70	1.90		12.60	14.75
0620	5-3/16" deep		350	.046		11.80	1.90		13.70	15.90
0640	6-9/16" deep		350	.046		14.50	1.90		16.40	18.90
1000	Sills, 8/4 x 8" deep, oak, no horns		100	.160		17	6.65		23.65	29
1020	2" horns		100	.160		17.10	6.65		23.75	29
1040	3" horns		100	.160		17.10	6.65		23.75	29
1100	8/4 x 10" deep, oak, no horns		90	.178		20.50	7.40		27.90	34
1120	2" horns		90	.178		21	7.40		28.40	35
1140	3" horns		90	.178		21	7.40		28.40	35
2000	Exterior, colonial, frame & trim, 3' opng., in-swing, minimum		22	.727	Ea.	315	30		345	395
2010	Average		21	.762		470	31.50		501.50	565
2020	Maximum		20	.800		1,075	33		1,108	1,225
2100	5'-4" opening, in-swing, minimum		17	.941		360	39		399	455
2120	Maximum		15	1.067		1,075	44.50		1,119.50	1,250
2140	Out-swing, minimum		17	.941		370	39		409	465
2160	Maximum		15	1.067		1,100	44.50		1,144.50	1,300
2400	6'-0" opening, in-swing, minimum		16	1		345	41.50		386.50	445
2420	Maximum		10	1.600		1,100	66.50		1,166.50	1,325
2460	Out-swing, minimum		16	1		375	41.50		416.50	480
2480	Maximum		10	1.600		1,300	66.50		1,366.50	1,525
2600	For two sidelights, add, minimum		30	.533	Opng.	355	22		377	425
2620	Maximum		20	.800	"	1,125	33		1,158	1,300
2700	Custom birch frame, 3'-0" opening		16	1	Ea.	205	41.50		246.50	290
2750	6'-0" opening		16	1		305	41.50		346.50	400
2900	Exterior, modern, plain trim, 3' opng., in-swing, minimum		26	.615		35.50	25.50		61	78.50
2920	Average		24	.667		42	27.50		69.50	88.50
2940	Maximum		22	.727		50.50	30		80.50	102

06 48 16 – Interior Wood Door Frames

06 48 16.10 Interior Wood Door Jamb and Frames

		Crew	Daily Output	Labor-Hours	Unit	Material	Labor	Equipment	Total	Total Incl O&P
0010	**INTERIOR WOOD DOOR JAMB AND FRAMES**									
3000	Interior frame, pine, 11/16" x 3-5/8" deep	2 Carp	375	.043	L.F.	5.55	1.77		7.32	8.85
3020	4-9/16" deep		375	.043		6.50	1.77		8.27	9.90
3200	Oak, 11/16" x 3-5/8" deep		350	.046		4.35	1.90		6.25	7.70
3220	4-9/16" deep		350	.046		4.50	1.90		6.40	7.90
3240	5-3/16" deep		350	.046		4.58	1.90		6.48	8
3400	Walnut, 11/16" x 3-5/8" deep		350	.046		7.30	1.90		9.20	11
3420	4-9/16" deep		350	.046		7.50	1.90		9.40	11.20
3440	5-3/16" deep		350	.046		7.90	1.90		9.80	11.60

06 48 Wood Frames

06 48 16 – Interior Wood Door Frames

06 48 16.10 Interior Wood Door Jamb and Frames		Crew	Daily Output	Labor-Hours	Unit	2010 Bare Costs Material	Labor	Equipment	Total	Total Incl O&P
3600	Pocket door frame	2 Carp	16	1	Ea.	74.50	41.50		116	146
3800	Threshold, oak, 5/8" x 3-5/8" deep		200	.080	L.F.	3.83	3.32		7.15	9.30
3820	4-5/8" deep		190	.084		4.58	3.50		8.08	10.45
3840	5-5/8" deep	↓	180	.089	↓	7.15	3.69		10.84	13.55

06 49 Wood Screens and Exterior Wood Shutters

06 49 19 – Exterior Wood Shutters

06 49 19.10 Shutters, Exterior

Code	Description	Crew	Daily Output	Labor-Hours	Unit	Material	Labor	Equipment	Total	Total Incl O&P
0010	**SHUTTERS, EXTERIOR**									
0012	Aluminum, louvered, 1'-4" wide, 3'-0" long	1 Carp	10	.800	Pr.	60.50	33		93.50	118
0400	6'-8" long		9	.889		122	37		159	191
1000	Pine, louvered, primed, each 1'-2" wide, 3'-3" long		10	.800		105	33		138	167
1100	4'-7" long		10	.800		142	33		175	207
1500	Each 1'-6" wide, 3'-3" long		10	.800		112	33		145	174
1600	4'-7" long		10	.800		157	33		190	224
1620	Hemlock, louvered, 1'-2" wide, 5'-7" long		10	.800		172	33		205	240
1630	Each 1'-4" wide, 2'-2" long		10	.800		107	33		140	169
1670	4'-3" long		10	.800		128	33		161	192
1690	5'-11" long		10	.800		181	33		214	250
1700	Door blinds, 6'-9" long, each 1'-3" wide		9	.889		182	37		219	257
1710	1'-6" wide		9	.889		196	37		233	272
1720	Hemlock, solid raised panel, each 1'-4" wide, 3'-3" long		10	.800		174	33		207	242
1740	4'-3" long		10	.800		220	33		253	293
1770	5'-11" long		10	.800		294	33		327	375
1800	Door blinds, 6'-9" long, each 1'-3" wide		9	.889		330	37		367	420
1900	1'-6" wide		9	.889		360	37		397	450
2500	Polystyrene, solid raised panel, each 1'-4" wide, 3'-3" long		10	.800		40.50	33		73.50	95.50
2700	4'-7" long		10	.800		51	33		84	107
4500	Polystyrene, louvered, each 1'-2" wide, 3'-3" long		10	.800		27.50	33		60.50	81.50
4750	5'-3" long		10	.800		39	33		72	94
6000	Vinyl, louvered, each 1'-2" x 4'-7" long		10	.800		34	33		67	88.50
6200	Each 1'-4" x 6'-8" long	↓	9	.889	↓	47.50	37		84.50	110

06 51 Structural Plastic Shapes and Plates

06 51 13 – Plastic Lumber

06 51 13.10 Recycled Plastic Lumber

Code	Description		Crew	Daily Output	Labor-Hours	Unit	Material	Labor	Equipment	Total	Total Incl O&P
0010	**RECYCLED PLASTIC LUMBER**										
4000	Sheeting, recycled plastic, black or white, 4' x 8' x 1/8"	G	2 Carp	1100	.015	S.F.	1.56	.60		2.16	2.65
4010	4' x 8' x 3/16"	G		1100	.015		2.33	.60		2.93	3.49
4020	4' x 8' x 1/4"	G		950	.017		3.11	.70		3.81	4.50
4030	4' x 8' x 3/8"	G		950	.017		4.67	.70		5.37	6.25
4040	4' x 8' x 1/2"	G		900	.018		6.20	.74		6.94	8
4050	4' x 8' x 5/8"	G		900	.018		7.70	.74		8.44	9.60
4060	4' x 8' x 3/4"	G	↓	850	.019	↓	9.15	.78		9.93	11.30
4070	Add for colors	G				Ea.	5%			5%	6%
8500	100% recycled plastic, var colors, NLB, 2" x 2"	G				L.F.	1.41			1.41	1.55
8510	2" x 4"	G					2.95			2.95	3.25
8520	2" x 6"	G					4.63			4.63	5.10
8530	2" x 8"	G				↓	6.35			6.35	6.95

06 51 Structural Plastic Shapes and Plates

06 51 13 – Plastic Lumber

06 51 13.10 Recycled Plastic Lumber			Crew	Daily Output	Labor-Hours	Unit	Material	2010 Bare Costs Labor	Equipment	Total	Total Incl O&P
8540	2" x 10"	G				L.F.	9.25			9.25	10.15
8550	5/4" x 4"	G					3.02			3.02	3.32
8560	5/4" x 6"	G					3			3	3.30
8570	1" x 6"	G					2.28			2.28	2.51
8580	1/2" x 8"	G					2.46			2.46	2.71
8590	2" x 10" T & G	G					8.60			8.60	9.45
8600	3" x 10" T & G	G					14.25			14.25	15.70
8610	Add for premium colors	G				↓	20%			20%	22%

06 52 Plastic Structural Assemblies

06 52 10 – Fiberglass Structural Assemblies

06 52 10.10 Castings, Fiberglass

		Crew	Daily Output	Labor-Hours	Unit	Material	Labor	Equipment	Total	Total Incl O&P
0010	**CASTINGS, FIBERGLASS**									
0100	Angle, 1" x 1" x 1/8" thick	2 Sswk	240	.067	L.F.	1.44	3.13		4.57	7.10
0120	3" x 3" x 1/4" thick		200	.080		6	3.75		9.75	13.20
0140	4" x 4" x 1/4" thick		200	.080		7.70	3.75		11.45	15.10
0160	4" x 4" x 3/8" thick		200	.080		11.85	3.75		15.60	19.65
0180	6" x 6" x 1/2" thick		160	.100	↓	23.50	4.69		28.19	34.50
1000	Flat sheet, 1/8" thick		140	.114	S.F.	5.05	5.35		10.40	14.95
1020	1/4" thick		120	.133		9.90	6.25		16.15	22
1040	3/8" thick		100	.160		16.85	7.50		24.35	31.50
1060	1/2" thick		80	.200	↓	22	9.40		31.40	41
2000	Handrail, 42" high, 2" diam. rails pickets 5' O.C.		32	.500	L.F.	51	23.50		74.50	97
3000	Round bar, 1/4" diam.		240	.067		.58	3.13		3.71	6.15
3020	1/2" diam.		200	.080		1.01	3.75		4.76	7.70
3040	3/4" diam.		200	.080		2.05	3.75		5.80	8.85
3060	1" diam.		160	.100		3.62	4.69		8.31	12.25
3080	1-1/4" diam.		160	.100		4.13	4.69		8.82	12.80
3100	1-1/2" diam.		140	.114		5.05	5.35		10.40	14.95
3500	Round tube, 1" diam. x 1/8" thick		240	.067		1.97	3.13		5.10	7.65
3520	2" diam. x 1/4" thick		200	.080		6.35	3.75		10.10	13.60
3540	3" diam. x 1/4" thick		160	.100		10.10	4.69		14.79	19.35
4000	Square bar, 1/2" square		240	.067		3.98	3.13		7.11	9.90
4020	1" square		200	.080		5.10	3.75		8.85	12.20
4040	1-1/2" square		160	.100		10.40	4.69		15.09	19.70
4500	Square tube, 1" x 1" x 1/8" thick		240	.067		2.24	3.13		5.37	7.95
4520	2" x 2" x 1/8" thick		200	.080		4.34	3.75		8.09	11.35
4540	3" x 3" x 1/4" thick		160	.100		11.85	4.69		16.54	21.50
5000	Threaded rod, 3/8" diam.		320	.050		3.45	2.35		5.80	7.90
5020	1/2" diam.		320	.050		4.06	2.35		6.41	8.60
5040	5/8" diam.		280	.057		4.49	2.68		7.17	9.65
5060	3/4" diam.		280	.057		5.10	2.68		7.78	10.30
6000	Wide flange beam, 4" x 4" x 1/4" thick		120	.133		10.55	6.25		16.80	22.50
6020	6" x 6" x 1/4" thick		100	.160		19.95	7.50		27.45	35
6040	8" x 8" x 3/8" thick	↓	80	.200	↓	35	9.40		44.40	55

06 52 10.20 Fiberglass Stair Treads

		Crew	Daily Output	Labor-Hours	Unit	Material	Labor	Equipment	Total	Total Incl O&P
0010	**FIBERGLASS STAIR TREADS**									
0100	24" wide	2 Sswk	52	.308	Ea.	39.50	14.45		53.95	69
0140	30" wide		52	.308		47	14.45		61.45	77.50
0180	36" wide		52	.308		55.50	14.45		69.95	86.50
0220	42" wide	↓	52	.308	↓	65.50	14.45		79.95	97.50

06 52 Plastic Structural Assemblies

06 52 10 – Fiberglass Structural Assemblies

06 52 10.30 Fiberglass Grating

	06 52 10.30 Fiberglass Grating	Crew	Daily Output	Labor-Hours	Unit	Material	Labor	Equipment	Total	Total Incl O&P
						2010 Bare Costs				
0010	**FIBERGLASS GRATING**									
0100	Molded, green (for mod. corrosive environment)									
0140	1" x 4" mesh, 1" thick	2 Sswk	400	.040	S.F.	11.15	1.88		13.03	15.60
0180	1-1/2" square mesh, 1" thick		400	.040		16.60	1.88		18.48	21.50
0220	1-1/4" thick		400	.040		13.25	1.88		15.13	17.90
0260	1-1/2" thick		400	.040		24	1.88		25.88	30
0300	2" square mesh, 2" thick		320	.050		22.50	2.35		24.85	29
1000	Orange (for highly corrosive environment)									
1040	1" x 4" mesh, 1" thick	2 Sswk	400	.040	S.F.	14.05	1.88		15.93	18.75
1080	1-1/2" square mesh, 1" thick		400	.040		18.60	1.88		20.48	24
1120	1-1/4" thick		400	.040		19.30	1.88		21.18	25
1160	1-1/2" thick		400	.040		21	1.88		22.88	26.50
1200	2" square mesh, 2" thick		320	.050		25	2.35		27.35	31.50
3000	Pultruded, green (for mod. corrosive environment)									
3040	1" O.C. bar spacing, 1" thick	2 Sswk	400	.040	S.F.	16.65	1.88		18.53	21.50
3080	1-1/2" thick		320	.050		18.50	2.35		20.85	24.50
3120	1-1/2" O.C. bar spacing, 1" thick		400	.040		13.50	1.88		15.38	18.15
3160	1-1/2" thick		400	.040		15.15	1.88		17.03	19.95
4000	Grating support legs, fixed height, no base				Ea.	43.50			43.50	48
4040	With base					39			39	43
4080	Adjustable to 60"					59			59	65

06 52 10.40 Fiberglass Floor Grating

	06 52 10.40 Fiberglass Floor Grating	Crew	Daily Output	Labor-Hours	Unit	Material	Labor	Equipment	Total	Total Incl O&P
0010	**FIBERGLASS FLOOR GRATING**									
0100	Reinforced polyester, fire retardant, 1" x 4" grid, 1" thick	E-4	510	.063	S.F.	14.25	2.97	.29	17.51	21
0200	1-1/2" x 6" mesh, 1-1/2" thick		500	.064		16.65	3.03	.29	19.97	24
0300	With grit surface, 1-1/2" x 6" grid, 1-1/2" thick		500	.064		17	3.03	.29	20.32	24.50

06 63 Plastic Railings

06 63 10 – Plastic (PVC) Railings

06 63 10.10 Plastic Railings

	06 63 10.10 Plastic Railings	Crew	Daily Output	Labor-Hours	Unit	Material	Labor	Equipment	Total	Total Incl O&P
0010	**PLASTIC RAILINGS**									
0100	Horizontal PVC handrail with balusters, 3-1/2" wide, 36" high	1 Carp	96	.083	L.F.	24	3.46		27.46	32
0150	42" high		96	.083		27	3.46		30.46	35
0200	Angled PVC handrail with balusters, 3-1/2" wide, 36" high		72	.111		28	4.62		32.62	37.50
0250	42" high		72	.111		30.50	4.62		35.12	40.50
0300	Post sleeve for 4 x 4 post		96	.083		12	3.46		15.46	18.55
0400	Post cap for 4 x 4 post, flat profile		48	.167	Ea.	11.40	6.95		18.35	23.50
0450	Newel post style profile		48	.167		21	6.95		27.95	33.50
0500	Raised corbeled profile		48	.167		23	6.95		29.95	35.50
0550	Post base trim for 4 x 4 post		96	.083		12.80	3.46		16.26	19.45

06 65 10 – PVC Trim

06 65 10.10 PVC Trim, Exterior		Crew	Daily Output	Labor-Hours	Unit	2010 Bare Costs Material	Labor	Equipment	Total	Total Incl O&P
0010	**PVC TRIM, EXTERIOR**									
0100	Cornerboards, 5/4" x 6" x 6"	1 Carp	240	.033	L.F.	5.70	1.38		7.08	8.45
0110	Door / window casing, 1" x 4"		200	.040		1.30	1.66		2.96	3.99
0120	1" x 6"		200	.040		2.05	1.66		3.71	4.82
0130	1" x 8"		195	.041		2.70	1.70		4.40	5.60
0140	1" x 10"		195	.041		3.42	1.70		5.12	6.40
0150	1" x 12"		190	.042		4.19	1.75		5.94	7.30
0160	5/4" x 4"		195	.041		1.73	1.70		3.43	4.53
0170	5/4" x 6"		195	.041		2.70	1.70		4.40	5.60
0180	5/4" x 8"		190	.042		3.55	1.75		5.30	6.60
0190	5/4" x 10"		190	.042		4.53	1.75		6.28	7.70
0200	5/4" x 12"		185	.043		5.50	1.80		7.30	8.80
0210	Fascia, 1" x 4"		250	.032		1.30	1.33		2.63	3.48
0220	1" x 6"		250	.032		2.05	1.33		3.38	4.31
0230	1" x 8"		225	.036		2.70	1.48		4.18	5.25
0240	1" x 10"		225	.036		3.42	1.48		4.90	6.05
0250	1" x 12"		200	.040		4.19	1.66		5.85	7.15
0260	5/4" x 4"		240	.033		1.73	1.38		3.11	4.03
0270	5/4" x 6"		240	.033		2.70	1.38		4.08	5.10
0280	5/4" x 8"		215	.037		3.55	1.55		5.10	6.30
0290	5/4" x 10"		215	.037		4.53	1.55		6.08	7.35
0300	5/4" x 12"		190	.042		5.50	1.75		7.25	8.75
0310	Frieze, 1" x 4"		250	.032		1.30	1.33		2.63	3.48
0320	1" x 6"		250	.032		2.05	1.33		3.38	4.31
0330	1" x 8"		225	.036		2.70	1.48		4.18	5.25
0340	1" x 10"		225	.036		3.42	1.48		4.90	6.05
0350	1" x 12"		200	.040		4.19	1.66		5.85	7.15
0360	5/4" x 4"		240	.033		1.73	1.38		3.11	4.03
0370	5/4" x 6"		240	.033		2.70	1.38		4.08	5.10
0380	5/4" x 8"		215	.037		3.55	1.55		5.10	6.30
0390	5/4" x 10"		215	.037		4.53	1.55		6.08	7.35
0400	5/4" x 12"		190	.042		5.50	1.75		7.25	8.75
0410	Rake, 1" x 4"		200	.040		1.30	1.66		2.96	3.99
0420	1" x 6"		200	.040		2.05	1.66		3.71	4.82
0430	1" x 8"		190	.042		2.70	1.75		4.45	5.65
0440	1" x 10"		190	.042		3.42	1.75		5.17	6.45
0450	1" x 12"		180	.044		4.19	1.85		6.04	7.45
0460	5/4" x 4"		195	.041		1.73	1.70		3.43	4.53
0470	5/4" x 6"		195	.041		2.70	1.70		4.40	5.60
0480	5/4" x 8"		185	.043		3.55	1.80		5.35	6.70
0490	5/4" x 10"		185	.043		4.53	1.80		6.33	7.75
0500	5/4" x 12"		175	.046		5.50	1.90		7.40	9
0510	Rake trim, 1" x 4"		225	.036		1.30	1.48		2.78	3.71
0520	1" x 6"		225	.036		2.05	1.48		3.53	4.54
0560	5/4" x 4"		220	.036		1.73	1.51		3.24	4.23
0570	5/4" x 6"		220	.036		2.70	1.51		4.21	5.30
0610	Soffit, 1" x 4"	2 Carp	420	.038		1.30	1.58		2.88	3.87
0620	1" x 6"		420	.038		2.05	1.58		3.63	4.70
0630	1" x 8"		420	.038		2.70	1.58		4.28	5.40
0640	1" x 10"		400	.040		3.42	1.66		5.08	6.30
0650	1" x 12"		400	.040		4.19	1.66		5.85	7.15
0660	5/4" x 4"		410	.039		1.73	1.62		3.35	4.40
0670	5/4" x 6"		410	.039	L.F.	2.70	1.62		4.32	5.45

06 65 Plastic Simulated Wood Trim

06 65 10 – PVC Trim

06 65 10.10 PVC Trim, Exterior		Crew	Daily Output	Labor-Hours	Unit	2010 Bare Costs Material	Labor	Equipment	Total	Total Incl O&P
0680	5/4" x 8"	2 Carp	410	.039	L.F.	3.55	1.62		5.17	6.40
0690	5/4" x 10"	↓	390	.041	↓	4.53	1.70		6.23	7.60
0700	5/4" x 12"	↓	390	.041	↓	5.50	1.70		7.20	8.70

Division Notes

		CREW	DAILY OUTPUT	LABOR-HOURS	UNIT	2010 BARE COSTS				TOTAL INCL O&P
						MAT.	LABOR	EQUIP.	TOTAL	

Estimating Tips

07 10 00 Dampproofing and Waterproofing

- Be sure of the job specifications before pricing this subdivision. The difference in cost between waterproofing and dampproofing can be great. Waterproofing will hold back standing water. Dampproofing prevents the transmission of water vapor. Also included in this section are vapor retarding membranes.

07 20 00 Thermal Protection

- Insulation and fireproofing products are measured by area, thickness, volume or R-value. Specifications may give only what the specific R-value should be in a certain situation. The estimator may need to choose the type of insulation to meet that R-value.

07 30 00 Steep Slope Roofing

07 40 00 Roofing and Siding Panels

- Many roofing and siding products are bought and sold by the square. One square is equal to an area that measures 100 square feet.

This simple change in unit of measure could create a large error if the estimator is not observant. Accessories necessary for a complete installation must be figured into any calculations for both material and labor.

07 50 00 Membrane Roofing

07 60 00 Flashing and Sheet Metal

07 70 00 Roofing and Wall Specialties and Accessories

- The items in these subdivisions compose a roofing system. No one component completes the installation, and all must be estimated. Built-up or single-ply membrane roofing systems are made up of many products and installation trades. Wood blocking at roof perimeters or penetrations, parapet coverings, reglets, roof drains, gutters, downspouts, sheet metal flashing, skylights, smoke vents, and roof hatches all need to be considered along with the roofing material. Several different installation trades will need to work together on the roofing system. Inherent difficulties in the scheduling and coordination of various trades must be accounted for when estimating labor costs.

07 90 00 Joint Protection

- To complete the weather-tight shell, the sealants and caulkings must be estimated. Where different materials meet—at expansion joints, at flashing penetrations, and at hundreds of other locations throughout a construction project—they provide another line of defense against water penetration. Often, an entire system is based on the proper location and placement of caulking or sealants. The detailed drawings that are included as part of a set of architectural plans show typical locations for these materials. When caulking or sealants are shown at typical locations, this means the estimator must include them for all the locations where this detail is applicable. Be careful to keep different types of sealants separate, and remember to consider backer rods and primers if necessary.

Reference Numbers

Reference numbers are shown in shaded boxes at the beginning of some major classifications. These numbers refer to related items in the Reference Section. The reference information may be an estimating procedure, an alternate pricing method, or technical information.

Note: Not all subdivisions listed here necessarily appear in this publication.

07 01 Operation and Maint. of Thermal and Moisture Protection

07 01 50 – Maintenance of Membrane Roofing

07 01 50.10 Roof Coatings		Crew	Daily Output	Labor-Hours	Unit	Material	2010 Bare Costs Labor	Equipment	Total	Total Incl O&P
0010	**ROOF COATINGS**									
0012	Asphalt, brush grade, material only **CN**				Gal.	8.05			8.05	8.85
0200	Asphalt base, fibered aluminum coating G					9.75			9.75	10.75
0300	Asphalt primer, 5 gallon					7.80			7.80	8.60
0600	Coal tar pitch, 200 lb. barrels				Ton	1,025			1,025	1,125
0700	Tar roof cement, 5 gal. lots				Gal.	13.20			13.20	14.50
0800	Glass fibered roof & patching cement, 5 gallon				"	8			8	8.80
0900	Reinforcing glass membrane, 450 S.F./roll				Ea.	51			51	56
1000	Neoprene roof coating, 5 gal, 2 gal/sq				Gal.	25.50			25.50	28
1100	Roof patch & flashing cement, 5 gallon					9.70			9.70	10.65
1200	Roof resaturant, glass fibered, 3 gal/sq					7.90			7.90	8.70

07 05 Common Work Results for Thermal and Moisture Protection

07 05 05 – Selective Demolition

07 05 05.10 Selective Demo., Thermal and Moist. Protection		Crew	Daily Output	Labor-Hours	Unit	Material	Labor	Equipment	Total	Total Incl O&P
0010	**SELECTIVE DEMOLITION, THERMAL AND MOISTURE PROTECTION**									
0020	Caulking/sealant, to 1" x 1" joint R024119-10	1 Clab	600	.013	L.F.		.44		.44	.68
0120	Downspouts, including hangers		350	.023	"		.76		.76	1.17
0220	Flashing, sheet metal		290	.028	S.F.		.91		.91	1.41
0420	Gutters, aluminum or wood, edge hung		240	.033	L.F.		1.10		1.10	1.70
0520	Built-in		100	.080	"		2.65		2.65	4.08
0620	Insulation, air/vapor barrier		3500	.002	S.F.		.08		.08	.12
0670	Batts or blankets		1400	.006	C.F.		.19		.19	.29
0720	Foamed or sprayed in place	2 Clab	1000	.016	B.F.		.53		.53	.82
0770	Loose fitting	1 Clab	3000	.003	C.F.		.09		.09	.14
0870	Rigid board		3450	.002	B.F.		.08		.08	.12
1120	Roll roofing, cold adhesive		12	.667	Sq.		22		22	34
1170	Roof accessories, adjustable metal chimney flashing		9	.889	Ea.		29.50		29.50	45.50
1325	Plumbing vent flashing		32	.250	"		8.30		8.30	12.75
1375	Ridge vent strip, aluminum		310	.026	L.F.		.85		.85	1.32
1620	Skylight to 10 S.F.		8	1	Ea.		33		33	51
2120	Roof edge, aluminum soffit and fascia		570	.014	L.F.		.46		.46	.72
2170	Concrete coping, up to 12" wide	2 Clab	160	.100			3.31		3.31	5.10
2220	Drip edge	1 Clab	1650	.005			.16		.16	.25
2270	Gravel stop		1650	.005			.16		.16	.25
2370	Sheet metal coping, up to 12" wide		240	.033			1.10		1.10	1.70
2470	Roof insulation board, over 2" thick	B-2	7800	.005	B.F.		.17		.17	.27
2520	Up to 2" thick	"	3900	.010	S.F.		.34		.34	.53
2620	Roof ventilation, louvered gable vent	1 Clab	16	.500	Ea.		16.55		16.55	25.50
2670	Remove, roof hatch	G-3	15	2.133			87.50		87.50	134
2675	Rafter vents	1 Clab	960	.008			.28		.28	.43
2720	Soffit vent and/or fascia vent		575	.014	L.F.		.46		.46	.71
2775	Soffit vent strip, aluminum, 3" to 4" wide		160	.050			1.66		1.66	2.55
2820	Roofing accessories, shingle moulding, to 1" x 4"		1600	.005			.17		.17	.26
2870	Cant strip	B-2	2000	.020			.67		.67	1.03
2920	Concrete block walkway	1 Clab	230	.035			1.15		1.15	1.78
3070	Roofing, felt paper, 15#		70	.114	Sq.		3.78		3.78	5.85
3125	#30 felt		30	.267	"		8.85		8.85	13.60
3170	Asphalt shingles, 1 layer	B-2	3500	.011	S.F.		.38		.38	.59
3180	2 layers		1750	.023	"		.77		.77	1.18
3370	Modified bitumen		26	1.538	Sq.		51.50		51.50	79.50

07 05 Common Work Results for Thermal and Moisture Protection

07 05 05 – Selective Demolition

07 05 05.10 Selective Demo., Thermal and Moist. Protection		Crew	Daily Output	Labor-Hours	Unit	Material	2010 Bare Costs Labor	Equipment	Total	Total Incl O&P
3420	Built-up, no gravel, 3 ply	B-2	25	1.600	Sq.		53.50		53.50	82.50
3470	4 ply		21	1.905			64		64	98.50
3620	5 ply		1600	.025	S.F.		.84		.84	1.29
3720	5 ply, with gravel		890	.045			1.51		1.51	2.32
3725	Gravel removal, minimum		5000	.008			.27		.27	.41
3730	Maximum		2000	.020			.67		.67	1.03
3870	Fiberglass sheet		1200	.033			1.12		1.12	1.72
4120	Slate shingles		1900	.021			.71		.71	1.09
4170	Ridge shingles, clay or slate		2000	.020	L.F.		.67		.67	1.03
4320	Single ply membrane, attached at seams		52	.769	Sq.		26		26	39.50
4370	Ballasted		75	.533			17.85		17.85	27.50
4420	Fully adhered		39	1.026			34.50		34.50	53
4550	Roof hatch, 2′-6″ x 3′-0″	1 Clab	10	.800	Ea.		26.50		26.50	41
4670	Wood shingles	B-2	2200	.018	S.F.		.61		.61	.94
4820	Sheet metal roofing	″	2150	.019			.62		.62	.96
4970	Siding, horizontal wood clapboards	1 Clab	380	.021			.70		.70	1.07
5025	Exterior insulation finish system	″	120	.067			2.21		2.21	3.40
5070	Tempered hardboard, remove and reset	1 Carp	380	.021			.87		.87	1.35
5120	Tempered hardboard sheet siding	″	375	.021			.89		.89	1.37
5170	Metal, corner strips	1 Clab	850	.009	L.F.		.31		.31	.48
5225	Horizontal strips		444	.018	S.F.		.60		.60	.92
5320	Vertical strips		400	.020			.66		.66	1.02
5520	Wood shingles		350	.023			.76		.76	1.17
5620	Stucco siding		360	.022			.74		.74	1.13
5670	Textured plywood		725	.011			.37		.37	.56
5720	Vinyl siding		510	.016			.52		.52	.80
5770	Corner strips		900	.009	L.F.		.29		.29	.45
5870	Wood, boards, vertical		400	.020	S.F.		.66		.66	1.02
5920	Waterproofing, protection / drain board	2 Clab	3900	.004	B.F.		.14		.14	.21
5970	Over 1/2″ thick		1750	.009	S.F.		.30		.30	.47
6020	To 1/2″ thick		2000	.008	″		.26		.26	.41

07 11 Dampproofing

07 11 13 – Bituminous Dampproofing

07 11 13.10 Bituminous Asphalt Coating

		Crew	Daily Output	Labor-Hours	Unit	Material	Labor	Equipment	Total	Total Incl O&P
0010	**BITUMINOUS ASPHALT COATING**									
0030	Brushed on, below grade, 1 coat	1 Rofc	665	.012	S.F.	.20	.43		.63	.93
0100	2 coat		500	.016		.40	.57		.97	1.38
0300	Sprayed on, below grade, 1 coat, 25.6 S.F./gal.		830	.010		.20	.34		.54	.79
0400	2 coat, 20.5 S.F./gal.		500	.016		.39	.57		.96	1.37
0500	Asphalt coating, with fibers				Gal.	9.75			9.75	10.75
0600	Troweled on, asphalt with fibers, 1/16″ thick	1 Rofc	500	.016	S.F.	.42	.57		.99	1.41
0700	1/8″ thick		400	.020		.75	.71		1.46	2.01
1000	1/2″ thick		350	.023		2.44	.81		3.25	4.03

07 11 16 – Cementitious Dampproofing

07 11 16.20 Cementitious Parging

		Crew	Daily Output	Labor-Hours	Unit	Material	Labor	Equipment	Total	Total Incl O&P
0010	**CEMENTITIOUS PARGING**									
0020	Portland cement, 2 coats, 1/2″ thick	D-1	250	.064	S.F.	.29	2.41		2.70	3.96
0100	Waterproofed Portland cement, 1/2″ thick, 2 coats	″	250	.064	″	2.87	2.41		5.28	6.80

07 12 Built-up Bituminous Waterproofing

07 12 13 – Built-Up Asphalt Waterproofing

07 12 13.20 Membrane Waterproofing		Crew	Daily Output	Labor-Hours	Unit	2010 Bare Costs Material	Labor	Equipment	Total	Total Incl O&P
0010	**MEMBRANE WATERPROOFING**									
0012	On slabs, 1 ply, felt, mopped	G-1	3000	.019	S.F.	.38	.62	.16	1.16	1.62
0100	On slabs, 1 ply, glass fiber fabric, mopped		2100	.027		.39	.88	.23	1.50	2.15
0300	On slabs, 2 ply, felt, mopped		2500	.022		.77	.74	.19	1.70	2.28
0400	On slabs, 2 ply, glass fiber fabric, mopped		1650	.034		.88	1.12	.29	2.29	3.16
0600	On slabs, 3 ply, felt, mopped		2100	.027		1.15	.88	.23	2.26	2.99
0700	On slabs, 3 ply, glass fiber fabric, mopped		1550	.036		1.17	1.19	.31	2.67	3.62
0710	Asphaltic hardboard protection board, 1/8" thick	2 Rofc	500	.032		.40	1.13		1.53	2.32
1000	1/4" EPS membrane protection board		3500	.005		.29	.16		.45	.58
1050	3/8" thick		3500	.005		.33	.16		.49	.63
1060	1/2" thick		3500	.005		.38	.16		.54	.68
1070	Fiberglass fabric, black, 20/10 mesh		116	.138	Sq.	20	4.88		24.88	30
1080	White, 20/10 mesh		116	.138	"	20	4.88		24.88	30

07 13 Sheet Waterproofing

07 13 53 – Elastomeric Sheet Waterproofing

07 13 53.10 Elastomeric Sheet Waterproofing and Access.		Crew	Daily Output	Labor-Hours	Unit	Material	Labor	Equipment	Total	Total Incl O&P
0010	**ELASTOMERIC SHEET WATERPROOFING AND ACCESS.**									
0090	EPDM, plain, 45 mils thick	2 Rofc	580	.028	S.F.	1.31	.98		2.29	3.06
0100	60 mils thick		570	.028		1.36	.99		2.35	3.15
0300	Nylon reinforced sheets, 45 mils thick		580	.028		1.27	.98		2.25	3.02
0400	60 mils thick		570	.028		1.58	.99		2.57	3.39
0600	Vulcanizing splicing tape for above, 2" wide				C.L.F.	48			48	53
0700	4" wide				"	110			110	121
0900	Adhesive, bonding, 60 SF per gal				Gal.	21			21	23
1000	Splicing, 75 SF per gal				"	31			31	34
1200	Neoprene sheets, plain, 45 mils thick	2 Rofc	580	.028	S.F.	1.68	.98		2.66	3.47
1300	60 mils thick		570	.028		2.87	.99		3.86	4.81
1500	Nylon reinforced, 45 mils thick		580	.028		1.75	.98		2.73	3.55
1600	60 mils thick		570	.028		2.25	.99		3.24	4.13
1800	120 mils thick		500	.032		4	1.13		5.13	6.30
1900	Adhesive, splicing, 150 S.F. per gal. per coat				Gal.	20			20	22
2100	Fiberglass reinforced, fluid applied, 1/8" thick	2 Rofc	500	.032	S.F.	1.90	1.13		3.03	3.97
2200	Polyethylene and rubberized asphalt sheets, 1/8" thick		550	.029		.76	1.03		1.79	2.55
2400	Polyvinyl chloride sheets, plain, 10 mils thick		580	.028		.17	.98		1.15	1.81
2500	20 mils thick		570	.028		.28	.99		1.27	1.96
2700	30 mils thick		560	.029		.39	1.01		1.40	2.11
3000	Adhesives, trowel grade, 40-100 SF per gal				Gal.	22			22	24
3100	Brush grade, 100-250 SF per gal.				"	22			22	24
3300	Bitumen modified polyurethane, fluid applied, 55 mils thick	2 Rofc	665	.024	S.F.	.78	.85		1.63	2.28
3600	Vinyl plastic, sprayed on, 25 to 40 mils thick	"	475	.034	"	1.21	1.19		2.40	3.31

07 16 Cementitious and Reactive Waterproofing

07 16 16 – Crystalline Waterproofing

07 16 16.20 Cementitious Waterproofing		Crew	Daily Output	Labor-Hours	Unit	2010 Bare Costs Material	Labor	Equipment	Total	Total Incl O&P
0010	**CEMENTITIOUS WATERPROOFING**									
0020	1/8" application, sprayed on	G-2A	1000	.024	S.F.	1.64	.76	.69	3.09	3.79
0030	2 coat, cementitious/metallic slurry, troweled, 1/4" thick	1 Cefi	2.48	3.226	C.S.F.	24	128		152	214
0040	3 coat, 3/8" thick	↓	1.84	4.348	↓	42.50	173		215.50	299
0050	4 coat, 1/2" thick	↓	1.20	6.667	↓	57	265		322	450

07 17 Bentonite Waterproofing

07 17 13 – Bentonite Panel Waterproofing

07 17 13.10 Bentonite		Crew	Daily Output	Labor-Hours	Unit	Material	Labor	Equipment	Total	Total Incl O&P
0010	**BENTONITE**									
0020	Panels, 4' x 4', 3/16" thick	1 Rofc	625	.013	S.F.	1.08	.45		1.53	1.94
0100	Rolls, 3/8" thick, with geotextile fabric both sides	"	550	.015	"	1.26	.52		1.78	2.25
0300	Granular bentonite, 50 lb. bags (.625 C.F.)				Bag	9.80			9.80	10.80
0400	3/8" thick, troweled on	1 Rofc	475	.017	S.F.	.49	.60		1.09	1.53
0500	Drain board, expanded polystyrene, 1-1/2" thick	1 Rohe	1600	.005		.54	.13		.67	.81
0510	2" thick		1600	.005		.72	.13		.85	1.01
0520	3" thick		1600	.005		1.08	.13		1.21	1.41
0530	4" thick		1600	.005		1.44	.13		1.57	1.80
0600	With filter fabric, 1-1/2" thick		1600	.005		.61	.13		.74	.90
0625	2" thick		1600	.005		.79	.13		.92	1.09
0650	3" thick		1600	.005		1.15	.13		1.28	1.49
0675	4" thick	↓	1600	.005	↓	1.51	.13		1.64	1.89

07 19 Water Repellents

07 19 19 – Silicone Water Repellents

07 19 19.10 Silicone Based Water Repellents		Crew	Daily Output	Labor-Hours	Unit	Material	Labor	Equipment	Total	Total Incl O&P
0010	**SILICONE BASED WATER REPELLENTS**									
0020	Water base liquid, roller applied	2 Rofc	7000	.002	S.F.	.65	.08		.73	.85
0200	Silicone or stearate, sprayed on CMU, 1 coat	1 Rofc	4000	.002		.36	.07		.43	.52
0300	2 coats	"	3000	.003	↓	.73	.09		.82	.96

07 21 Thermal Insulation

07 21 13 – Board Insulation

07 21 13.10 Rigid Insulation			Crew	Daily Output	Labor-Hours	Unit	Material	Labor	Equipment	Total	Total Incl O&P
0010	**RIGID INSULATION**, for walls										
0040	Fiberglass, 1.5#/CF, unfaced, 1" thick, R4.1	G	1 Carp	1000	.008	S.F.	.44	.33		.77	.99
0060	1-1/2" thick, R6.2	G		1000	.008		.64	.33		.97	1.21
0080	2" thick, R8.3	G		1000	.008		.69	.33		1.02	1.27
0120	3" thick, R12.4	G		800	.010		.80	.42		1.22	1.52
0370	3#/CF, unfaced, 1" thick, R4.3	G		1000	.008		.49	.33		.82	1.05
0390	1-1/2" thick, R6.5	G		1000	.008		.95	.33		1.28	1.56
0400	2" thick, R8.7	G		890	.009		1.15	.37		1.52	1.85
0420	2-1/2" thick, R10.9	G		800	.010		1.01	.42		1.43	1.75
0440	3" thick, R13	G		800	.010		1.01	.42		1.43	1.75
0520	Foil faced, 1" thick, R4.3	G		1000	.008		.92	.33		1.25	1.52
0540	1-1/2" thick, R6.5	G		1000	.008		1.36	.33		1.69	2.01
0560	2" thick, R8.7	G	↓	890	.009	↓	1.71	.37		2.08	2.46

07 21 Thermal Insulation

07 21 13 – Board Insulation

07 21 13.10 Rigid Insulation			Crew	Daily Output	Labor-Hours	Unit	Material	2010 Bare Costs Labor	Equipment	Total	Total Incl O&P
0580	2-1/2" thick, R10.9	G	1 Carp	800	.010	S.F.	2.02	.42		2.44	2.86
0600	3" thick, R13	G		800	.010		2.19	.42		2.61	3.05
1500	Foamglass, 1-1/2" thick, R4.5	G		800	.010		1.40	.42		1.82	2.18
1550	3" thick, R9	G		730	.011		3.45	.46		3.91	4.50
1600	Isocyanurate, 4' x 8' sheet, foil faced, both sides										
1610	1/2" thick	G	1 Carp	800	.010	S.F.	.31	.42		.73	.98
1620	5/8" thick	G		800	.010		.34	.42		.76	1.01
1630	3/4" thick	G		800	.010		.36	.42		.78	1.04
1640	1" thick	G		800	.010		.55	.42		.97	1.25
1650	1-1/2" thick	G		730	.011		.67	.46		1.13	1.44
1660	2" thick *CN*	G		730	.011		.84	.46		1.30	1.62
1670	3" thick	G		730	.011		1.90	.46		2.36	2.79
1680	4" thick	G		730	.011		2.14	.46		2.60	3.05
1700	Perlite, 1" thick, R2.77	G		800	.010		.30	.42		.72	.97
1750	2" thick, R5.55	G		730	.011		.60	.46		1.06	1.36
1900	Extruded polystyrene, 25 PSI compressive strength, 1" thick, R5	G		800	.010		.51	.42		.93	1.20
1940	2" thick R10	G		730	.011		1.04	.46		1.50	1.84
1960	3" thick, R15	G		730	.011		1.40	.46		1.86	2.24
2100	Expanded polystyrene, 1" thick, R3.85	G		800	.010		.36	.42		.78	1.04
2120	2" thick, R7.69	G		730	.011		.72	.46		1.18	1.49
2140	3" thick, R11.49	G		730	.011		1.08	.46		1.54	1.89

07 21 13.13 Foam Board Insulation			Crew	Daily Output	Labor-Hours	Unit	Material	Labor	Equipment	Total	Total Incl O&P
0010	**FOAM BOARD INSULATION**										
0600	Polystyrene, expanded, 1" thick, R4	G	1 Carp	680	.012	S.F.	.36	.49		.85	1.15
0700	2" thick, R8	G	"	675	.012	"	.72	.49		1.21	1.55

07 21 16 – Blanket Insulation

07 21 16.10 Blanket Insulation for Floors/Ceilings			Crew	Daily Output	Labor-Hours	Unit	Material	Labor	Equipment	Total	Total Incl O&P
0010	**BLANKET INSULATION FOR FLOORS/CEILINGS**										
0020	Including spring type wire fasteners										
2000	Fiberglass, blankets or batts, paper or foil backing										
2100	3-1/2" thick, R13	G	1 Carp	700	.011	S.F.	.38	.47		.85	1.15
2150	6-1/4" thick, R19	G		600	.013		.48	.55		1.03	1.38
2210	9-1/2" thick, R 30	G		500	.016		.69	.66		1.35	1.78
2220	12" thick, R 38	G		475	.017		1	.70		1.70	2.18
3000	Unfaced, 3-1/2" thick, R 13	G		600	.013		.30	.55		.85	1.18
3010	6-1/4" thick, R 19	G		500	.016		.45	.66		1.11	1.52
3020	9-1/2" thick, R30	G		450	.018		.69	.74		1.43	1.90
3030	12" thick, R 38	G		425	.019		.81	.78		1.59	2.10

07 21 16.20 Blanket Insulation for Walls			Crew	Daily Output	Labor-Hours	Unit	Material	Labor	Equipment	Total	Total Incl O&P
0010	**BLANKET INSULATION FOR WALLS**										
0020	Kraft faced fiberglass, 3-1/2" thick, R11, 15" wide	G	1 Carp	1350	.006	S.F.	.28	.25		.53	.69
0030	23" wide	G		1600	.005		.28	.21		.49	.63
0060	R13, 11" wide	G		1150	.007		.33	.29		.62	.81
0080	15" wide *CN*	G		1350	.006		.33	.25		.58	.74
0100	23" wide	G		1600	.005		.33	.21		.54	.68
0110	R-15, 11" wide	G		1150	.007		.45	.29		.74	.95
0120	15" wide	G		1350	.006		.45	.25		.70	.88
0130	23" wide	G		1600	.005		.45	.21		.66	.82
0140	6" thick, R19, 11" wide	G		1150	.007		.43	.29		.72	.92
0160	15" wide	G		1350	.006		.43	.25		.68	.85
0180	23" wide	G		1600	.005		.43	.21		.64	.79
0182	R21, 11" wide	G		1150	.007		.65	.29		.94	1.17

07 21 16 – Blanket Insulation

07 21 16.20 Blanket Insulation for Walls			Crew	Daily Output	Labor-Hours	Unit	Material	2010 Bare Costs Labor	Equipment	Total	Total Incl O&P
0184	15" wide	G	1 Carp	1350	.006	S.F.	.65	.25		.90	1.10
0186	23" wide	G		1600	.005		.65	.21		.86	1.04
0188	9" thick, R-30, 11" wide	G		985	.008		.70	.34		1.04	1.29
0200	15" wide	G		1150	.007		.70	.29		.99	1.22
0220	23" wide	G		1350	.006		.70	.25		.95	1.15
0230	12" thick, R38, 11" wide	G		985	.008		1	.34		1.34	1.62
0240	15" wide	G		1150	.007		1	.29		1.29	1.55
0260	23" wide	G		1350	.006		1	.25		1.25	1.48
0410	Foil faced fiberglass, 3-1/2" thick, R13, 11" wide	G		1150	.007		.54	.29		.83	1.04
0420	15" wide	G		1350	.006		.54	.25		.79	.97
0440	23" wide	G		1600	.005		.54	.21		.75	.91
0442	R15, 11" wide	G		1150	.007		.56	.29		.85	1.07
0444	15" wide	G		1350	.006		.56	.25		.81	1
0446	23" wide	G		1600	.005		.56	.21		.77	.94
0448	6" thick, R19, 11" wide	G		1150	.007		.77	.29		1.06	1.30
0460	15" wide	G		1350	.006		.77	.25		1.02	1.23
0480	23" wide	G		1600	.005		.77	.21		.98	1.17
0482	R-21, 11" wide	G		1150	.007		.82	.29		1.11	1.35
0484	15" wide	G		1350	.006		.82	.25		1.07	1.28
0486	23" wide	G		1600	.005		.82	.21		1.03	1.22
0488	9" thick, R-30, 11" wide	G		985	.008		.89	.34		1.23	1.50
0500	9" thick, R30, 15" wide	G		1150	.007		.89	.29		1.18	1.43
0550	23" wide	G		1350	.006		.89	.25		1.14	1.36
0560	12" thick, R-38, 11" wide	G		985	.008		.90	.34		1.24	1.51
0570	15" wide	G		1150	.007		.90	.29		1.19	1.44
0580	23" wide	G		1350	.006		.90	.25		1.15	1.37
0620	Unfaced fiberglass, 3-1/2" thick, R-13, 11" wide	G		1150	.007		.30	.29		.59	.78
0820	15" wide	G		1350	.006		.30	.25		.55	.71
0830	23" wide	G		1600	.005		.30	.21		.51	.65
0832	R15, 11" wide	G		1150	.007		.44	.29		.73	.93
0834	15" wide	G		1350	.006		.44	.25		.69	.86
0836	23" wide	G		1600	.005		.44	.21		.65	.80
0838	6" thick, R19, 11" wide	G		1150	.007		.45	.29		.74	.95
0860	15" wide	G		1150	.007		.45	.29		.74	.95
0880	23" wide	G		1350	.006		.45	.25		.70	.88
0882	R-21, 11" wide	G		1150	.007		.46	.29		.75	.96
0886	15" wide	G		1350	.006		.46	.25		.71	.89
0888	23" wide	G		1600	.005		.46	.21		.67	.83
0890	9" thick, R30, 11" wide	G		985	.008		.64	.34		.98	1.22
0900	15" wide	G		1150	.007		.64	.29		.93	1.15
0920	23" wide	G		1350	.006		.64	.25		.89	1.08
0930	12" thick, R38, 11" wide	G		985	.008		.81	.34		1.15	1.41
0940	15" wide	G		1000	.008		.81	.33		1.14	1.40
0960	23" wide	G	↓	1150	.007	↓	.81	.29		1.10	1.34
1300	Mineral fiber batts, kraft faced										
1320	3-1/2" thick, R12	G	1 Carp	1600	.005	S.F.	.38	.21		.59	.74
1340	6" thick, R19	G		1600	.005		.48	.21		.69	.85
1380	10" thick, R30	G		1350	.006	↓	.73	.25		.98	1.18
1850	Friction fit wire insulation supports, 16" O.C.		↓	960	.008	Ea.	.05	.35		.40	.59

07 21 Thermal Insulation

07 21 23 – Loose-Fill Insulation

07 21 23.10 Poured Loose-Fill Insulation

			Crew	Daily Output	Labor-Hours	Unit	Material	2010 Bare Costs Labor	Equipment	Total	Total Incl O&P
0010	**POURED LOOSE-FILL INSULATION**										
0020	Cellulose fiber, R3.8 per inch	G	1 Carp	200	.040	C.F.	.60	1.66		2.26	3.22
0040	Ceramic type (perlite), R3.2 per inch	G		200	.040		1.75	1.66		3.41	4.49
0080	Fiberglass wool, R4 per inch	G		200	.040		.46	1.66		2.12	3.07
0100	Mineral wool, R3 per inch	G		200	.040		.39	1.66		2.05	2.99
0300	Polystyrene, R4 per inch	G		200	.040		3.09	1.66		4.75	5.95
0400	Vermiculite or perlite, R2.7 per inch	G		200	.040		1.75	1.66		3.41	4.49
0700	Wood fiber, R3.85 per inch	G		200	.040		.70	1.66		2.36	3.33

07 21 23.20 Masonry Loose-Fill Insulation

			Crew	Daily Output	Labor-Hours	Unit	Material	Labor	Equipment	Total	Total Incl O&P
0010	**MASONRY LOOSE-FILL INSULATION**, vermiculite or perlite										
0100	In cores of concrete block, 4" thick wall, .115 CF/SF	G	D-1	4800	.003	S.F.	.20	.13		.33	.41
0200	6" thick wall, .175 CF/SF	G		3000	.005		.31	.20		.51	.64
0300	8" thick wall, .258 CF/SF	G		2400	.007		.45	.25		.70	.88
0400	10" thick wall, .340 CF/SF	G		1850	.009		.60	.33		.93	1.14
0500	12" thick wall, .422 CF/SF	G		1200	.013		.74	.50		1.24	1.57
0550	For sand fill, deduct from above	G					70%				
0600	Poured cavity wall, vermiculite or perlite, water repellant	G	D-1	250	.064	C.F.	1.75	2.41		4.16	5.55
0700	Foamed in place, urethane in 2-5/8" cavity	G	G-2A	1035	.023	S.F.	.42	.73	.67	1.82	2.39
0800	For each 1" added thickness, add	G	"	2372	.010	"	.12	.32	.29	.73	.97

07 21 26 – Blown Insulation

07 21 26.10 Blown Insulation

			Crew	Daily Output	Labor-Hours	Unit	Material	Labor	Equipment	Total	Total Incl O&P
0010	**BLOWN INSULATION** Ceilings, with open access										
0020	Cellulose, 3-1/2" thick, R13	G	G-4	5000	.005	S.F.	.19	.16	.08	.43	.55
0030	5-3/16" thick, R19	G		3800	.006		.28	.21	.10	.59	.76
0050	6-1/2" thick, R22	G		3000	.008		.36	.27	.13	.76	.97
0100	8-11/16" thick, R30	G		2600	.009		.48	.31	.15	.94	1.18
0120	10-7/8" thick, R38	G		1800	.013		.62	.45	.22	1.29	1.61
1000	Fiberglass, 5.5" thick, R11	G		3800	.006		.19	.21	.10	.50	.66
1050	6" thick, R12	G		3000	.008		.23	.27	.13	.63	.82
1100	8.8" thick, R19	G		2200	.011		.33	.37	.18	.88	1.13
1200	10" thick, R22	G		1800	.013		.38	.45	.22	1.05	1.35
1300	11.5" thick, R26	G		1500	.016		.46	.54	.27	1.27	1.63
1350	13" thick, R30	G		1400	.017		.50	.58	.28	1.36	1.75
1450	16" thick, R38	G		1145	.021		.61	.71	.35	1.67	2.14
1500	20" thick, R49	G		920	.026		.77	.88	.43	2.08	2.68
2000	Mineral wool, 4" thick, R12	G		3500	.007		.23	.23	.11	.57	.74
2050	6" thick, R17	G		2500	.010		.26	.32	.16	.74	.97
2100	9" thick, R23	G		1750	.014		.34	.46	.23	1.03	1.33

07 21 27 – Reflective Insulation

07 21 27.10 Reflective Insulation Options

			Crew	Daily Output	Labor-Hours	Unit	Material	Labor	Equipment	Total	Total Incl O&P
0010	**REFLECTIVE INSULATION OPTIONS**										
0020	Aluminum foil on reinforced scrim	G	1 Carp	19	.421	C.S.F.	14.20	17.50		31.70	42.50
0100	Reinforced with woven polyolefin	G		19	.421		17.20	17.50		34.70	46
0500	With single bubble air space, R8.8	G		15	.533		25.50	22		47.50	62
0600	With double bubble air space, R9.8	G		15	.533		26.50	22		48.50	63

07 21 29 – Sprayed Insulation

07 21 29.10 Sprayed-On Insulation

			Crew	Daily Output	Labor-Hours	Unit	Material	Labor	Equipment	Total	Total Incl O&P
0010	**SPRAYED-ON INSULATION**										
0020	Fibrous/cementitious, finished wall, 1" thick, R3.7	G	G-2	2050	.012	S.F.	.26	.41	.06	.73	.97
0100	Attic, 5.2" thick, R19	G		1550	.015	"	.42	.54	.09	1.05	1.36

07 21 Thermal Insulation

07 21 29 – Sprayed Insulation

07 21 29.10 Sprayed-On Insulation			Crew	Daily Output	Labor-Hours	Unit	2010 Bare Costs Material	Labor	Equipment	Total	Total Incl O&P
0200	Fiberglass, R4 per inch, vertical	G	G-2	2050	.012	B.F.	.15	.41	.06	.62	.85
0210	Horizontal	G	↓	1550	.015	"	.15	.54	.09	.78	1.07
0300	Closed cell, spray polyurethane foam, 2 pounds per cubic foot density										
0310	1" thick	G	G-2A	6000	.004	S.F.	.41	.13	.12	.66	.78
0320	2" thick	G		3000	.008		.82	.25	.23	1.30	1.56
0330	3" thick	G		2000	.012		1.23	.38	.35	1.96	2.35
0335	3-1/2" thick	G		1715	.014		1.44	.44	.40	2.28	2.74
0340	4" thick	G		1500	.016		1.64	.51	.46	2.61	3.14
0350	5" thick	G		1200	.020		2.05	.63	.58	3.26	3.91
0355	5-1/2" thick	G		1090	.022		2.26	.70	.63	3.59	4.31
0360	6" thick	G	↓	1000	.024	↓	2.46	.76	.69	3.91	4.70

07 22 Roof and Deck Insulation

07 22 16 – Roof Board Insulation

07 22 16.10 Roof Deck Insulation			Crew	Daily Output	Labor-Hours	Unit	Material	Labor	Equipment	Total	Total Incl O&P
0010	**ROOF DECK INSULATION**										
0020	Fiberboard low density, 1/2" thick R1.39	G	1 Rofc	1000	.008	S.F.	.24	.28		.52	.73
0030	1" thick R2.78	G		800	.010		.42	.35		.77	1.05
0080	1 1/2" thick R4.17	G		800	.010		.64	.35		.99	1.29
0100	2" thick R5.56	G		800	.010		.85	.35		1.20	1.53
0110	Fiberboard high density, 1/2" thick R1.3	G		1000	.008		.22	.28		.50	.71
0120	1" thick R2.5	G		800	.010		.44	.35		.79	1.07
0130	1-1/2" thick R3.8	G		800	.010		.66	.35		1.01	1.32
0200	Fiberglass, 3/4" thick R2.78	G		1000	.008		.56	.28		.84	1.09
0400	15/16" thick R3.70	G		1000	.008		.74	.28		1.02	1.28
0460	1-1/16" thick R4.17	G		1000	.008		.93	.28		1.21	1.49
0600	1-5/16" thick R5.26	G		1000	.008		1.26	.28		1.54	1.86
0650	2-1/16" thick R8.33	G		800	.010		1.35	.35		1.70	2.08
0700	2-7/16" thick R10	G		800	.010		1.55	.35		1.90	2.30
1500	Foamglass, 1-1/2" thick R4.5	G		800	.010		1.57	.35		1.92	2.32
1530	3" thick R9	G		700	.011	↓	2.89	.40		3.29	3.85
1600	Tapered for drainage	G		600	.013	B.F.	1.16	.47		1.63	2.06
1650	Perlite, 1/2" thick R1.32	G		1050	.008	S.F.	.34	.27		.61	.82
1655	3/4" thick R2.08	G		800	.010		.37	.35		.72	1
1660	1" thick R2.78	G		800	.010		.46	.35		.81	1.10
1670	1-1/2" thick R4.17	G		800	.010		.48	.35		.83	1.12
1680	2" thick R5.56	G		700	.011		.80	.40		1.20	1.55
1685	2-1/2" thick R6.67	G		700	.011	↓	.95	.40		1.35	1.72
1690	Tapered for drainage	G		800	.010	B.F.	.73	.35		1.08	1.39
1700	Polyisocyanurate, 2#/CF density, 3/4" thick	G		1500	.005	S.F.	.44	.19		.63	.79
1705	1" thick	G		1400	.006		.51	.20		.71	.90
1715	1-1/2" thick	G		1250	.006		.62	.23		.85	1.06
1725	2" thick	G		1100	.007		.86	.26		1.12	1.38
1735	2-1/2" thick	G		1050	.008		1.01	.27		1.28	1.56
1745	3" thick	G		1000	.008		1.24	.28		1.52	1.83
1755	3-1/2" thick	G		1000	.008	↓	1.95	.28		2.23	2.62
1765	Tapered for drainage	G	↓	1400	.006	B.F.	1.95	.20		2.15	2.49
1900	Extruded Polystyrene										
1910	15 PSI compressive strength, 1" thick, R5	G	1 Rofc	1500	.005	S.F.	.49	.19		.68	.85
1920	2" thick, R10	G		1250	.006		.62	.23		.85	1.06
1930	3" thick R15	G	↓	1000	.008	↓	1.27	.28		1.55	1.87

07 22 Roof and Deck Insulation

07 22 16 – Roof Board Insulation

07 22 16.10 Roof Deck Insulation			Crew	Daily Output	Labor-Hours	Unit	2010 Bare Costs Material	Labor	Equipment	Total	Total Incl O&P
1932	4" thick R20	G	1 Rofc	1000	.008	S.F.	1.64	.28		1.92	2.27
1934	Tapered for drainage	G		1500	.005	B.F.	.53	.19		.72	.89
1940	25 PSI compressive strength, 1" thick R5	G		1500	.005	S.F.	.67	.19		.86	1.05
1942	2" thick R10	G		1250	.006		1.27	.23		1.50	1.78
1944	3" thick R15	G		1000	.008		1.93	.28		2.21	2.59
1946	4" thick R20	G		1000	.008		2.72	.28		3	3.46
1948	Tapered for drainage	G		1500	.005	B.F.	.56	.19		.75	.93
1950	40 psi compressive strength, 1" thick R5	G		1500	.005	S.F.	.51	.19		.70	.87
1952	2" thick R10	G		1250	.006		.96	.23		1.19	1.44
1954	3" thick R15	G		1000	.008		1.40	.28		1.68	2.01
1956	4" thick R20	G		1000	.008		1.88	.28		2.16	2.54
1958	Tapered for drainage	G		1400	.006	B.F.	.71	.20		.91	1.12
1960	60 PSI compressive strength, 1" thick R5	G		1450	.006	S.F.	.59	.20		.79	.97
1962	2" thick R10	G		1200	.007		1.05	.24		1.29	1.55
1964	3" thick R15	G		975	.008		1.55	.29		1.84	2.19
1966	4" thick R20	G		950	.008		2.17	.30		2.47	2.89
1968	Tapered for drainage	G		1400	.006	B.F.	.85	.20		1.05	1.28
2010	Expanded polystyrene, 1#/CF density, 3/4" thick R2.89	G		1500	.005	S.F.	.27	.19		.46	.61
2020	1" thick R3.85	G		1500	.005		.36	.19		.55	.71
2100	2" thick R7.69	G		1250	.006		.72	.23		.95	1.17
2110	3" thick R11.49	G		1250	.006		1.08	.23		1.31	1.57
2120	4" thick R15.38	G		1200	.007		1.44	.24		1.68	1.97
2130	5" thick R19.23	G		1150	.007		1.80	.25		2.05	2.39
2140	6" thick R23.26	G		1150	.007		2.16	.25		2.41	2.79
2150	Tapered for drainage	G		1500	.005	B.F.	.54	.19		.73	.90
2400	Composites with 2" EPS										
2410	1" fiberboard	G	1 Rofc	950	.008	S.F.	1.24	.30		1.54	1.86
2420	7/16" oriented strand board	G		800	.010		.94	.35		1.29	1.62
2430	1/2" plywood	G		800	.010		1.14	.35		1.49	1.84
2440	1" perlite	G		800	.010		1.18	.35		1.53	1.89
2450	Composites with 1-1/2" polyisocyanurate										
2460	1" fiberboard	G	1 Rofc	800	.010	S.F.	1.14	.35		1.49	1.84
2470	1" perlite	G		850	.009		1.08	.33		1.41	1.74
2480	7/16" oriented strand board	G		800	.010		.84	.35		1.19	1.51

07 24 Exterior Insulation and Finish Systems

07 24 13 – Polymer Based Exterior Insulation and Finish Systems

07 24 13.10 Exterior Insulation and Finish Systems

			Crew	Daily Output	Labor-Hours	Unit	Material	Labor	Equipment	Total	Total Incl O&P
0010	**EXTERIOR INSULATION AND FINISH SYSTEMS**										
0095	Field applied, 1" EPS insulation	G	J-1	390	.103	S.F.	2.30	3.69	.35	6.34	8.40
0100	With 1/2" cement board sheathing	G		268	.149		2.95	5.35	.51	8.81	11.80
0105	2" EPS insulation	G		390	.103		2.66	3.69	.35	6.70	8.80
0110	With 1/2" cement board sheathing	G		268	.149		3.31	5.35	.51	9.17	12.20
0115	3" EPS insulation	G		390	.103		3.02	3.69	.35	7.06	9.20
0120	With 1/2" cement board sheathing	G		268	.149		3.67	5.35	.51	9.53	12.60
0125	4" EPS insulation	G		390	.103		3.38	3.69	.35	7.42	9.60
0130	With 1/2" cement board sheathing	G		268	.149		4.67	5.35	.51	10.53	13.70
0140	Premium finish add			1265	.032		.31	1.14	.11	1.56	2.16
0150	Heavy duty reinforcement add			914	.044		1.08	1.57	.15	2.80	3.70
0160	2.5#/S.Y. metal lath substrate add		1 Lath	75	.107	S.Y.	2.37	3.94		6.31	8.40
0170	3.4#/S.Y. metal lath substrate add		"	75	.107	"	2.58	3.94		6.52	8.65

07 24 Exterior Insulation and Finish Systems

07 24 13 – Polymer Based Exterior Insulation and Finish Systems

07 24 13.10 Exterior Insulation and Finish Systems		Crew	Daily Output	Labor-Hours	Unit	Material	2010 Bare Costs Labor	Equipment	Total	Total Incl O&P
0180	Color or texture change,	J-1	1265	.032	S.F.	.83	1.14	.11	2.08	2.73
0190	With substrate leveling base coat	1 Plas	530	.015		.83	.56		1.39	1.75
0210	With substrate sealing base coat	1 Pord	1224	.007	↓	.08	.24		.32	.44
0370	V groove shape in panel face				L.F.	.60			.60	.66
0380	U groove shape in panel face				"	.79			.79	.87
0440	For higher than one story, add						25%			

07 25 Weather Barriers

07 25 10 – Weather Barriers or Wraps

07 25 10.10 Weather Barriers		Crew	Daily Output	Labor-Hours	Unit	Material	Labor	Equipment	Total	Total Incl O&P
0010	**WEATHER BARRIERS**									
0400	Asphalt felt paper, 15#	1 Carp	37	.216	Sq.	5.90	9		14.90	20.50
0401	Per square foot	"	3700	.002	S.F.	.06	.09		.15	.20
0450	Housewrap, exterior, spun bonded polypropylene									
0470	Small roll	1 Carp	3800	.002	S.F.	.23	.09		.32	.39
0480	Large roll	"	4000	.002	"	.13	.08		.21	.27
2100	Asphalt felt roof deck vapor barrier, class 1 metal decks	1 Rofc	37	.216	Sq.	21.50	7.65		29.15	36.50
2200	For all other decks	"	37	.216		15.85	7.65		23.50	30
2800	Asphalt felt, 50% recycled content, 15 lb, 4 sq per roll	1 Carp	36	.222		4.77	9.25		14.02	19.50
2810	30 lb, 2 sq per roll	"	36	.222	↓	9.55	9.25		18.80	25
3000	Building wrap, spunbonded polyethylene	2 Carp	8000	.002	S.F.	.13	.08		.21	.27

07 26 Vapor Retarders

07 26 10 – Above-Grade Vapor Retarders

07 26 10.10 Vapor Retarders		Crew	Daily Output	Labor-Hours	Unit	Material	Labor	Equipment	Total	Total Incl O&P
0010	**VAPOR RETARDERS**									
0020	Aluminum and kraft laminated, foil 1 side G	1 Carp	37	.216	Sq.	8.80	9		17.80	23.50
0100	Foil 2 sides G		37	.216		9.50	9		18.50	24.50
0600	Polyethylene vapor barrier, standard, .002" thick G		37	.216		1.09	9		10.09	15.05
0700	.004" thick G		37	.216		3.39	9		12.39	17.60
0900	.006" thick G		37	.216		5.20	9		14.20	19.60
1200	.010" thick G		37	.216		6.70	9		15.70	21
1300	Clear reinforced, fire retardant, .008" thick G		37	.216		10.15	9		19.15	25
1350	Cross laminated type, .003" thick G		37	.216		7.10	9		16.10	21.50
1400	.004" thick G		37	.216		7.80	9		16.80	22.50
1800	Reinf. waterproof, .002" polyethylene backing, 1 side		37	.216		5.65	9		14.65	20
1900	2 sides		37	.216		7.45	9		16.45	22
2400	Waterproofed kraft with sisal or fiberglass fibers, minimum		37	.216		6.10	9		15.10	20.50
2500	Maximum	↓	37	.216	↓	15.10	9		24.10	30.50

07 31 Shingles and Shakes

07 31 13 – Asphalt Shingles

07 31 13.10 Asphalt Roof Shingles

	07 31 13.10 Asphalt Roof Shingles	Crew	Daily Output	Labor-Hours	Unit	2010 Bare Costs Material	Labor	Equipment	Total	Total Incl O&P
0010	**ASPHALT ROOF SHINGLES**									
0100	Standard strip shingles									
0150	Inorganic, class A, 210-235 lb/sq **CN**	1 Rofc	5.50	1.455	Sq.	73.50	51.50		125	166
0155	Pneumatic nailed		7	1.143		73.50	40.50		114	148
0200	Organic, class C, 235-240 lb/sq		5	1.600		64	56.50		120.50	164
0205	Pneumatic nailed		6.25	1.280		64	45.50		109.50	146
0250	Standard, laminated multi-layered shingles									
0300	Class A, 240-260 lb/sq	1 Rofc	4.50	1.778	Sq.	86.50	63		149.50	201
0305	Pneumatic nailed		5.63	1.422		86.50	50.50		137	179
0350	Class C, 260-300 lb/square, 4 bundles/square		4	2		84	71		155	211
0355	Pneumatic nailed		5	1.600		84	56.50		140.50	187
0400	Premium, laminated multi-layered shingles									
0450	Class A, 260-300 lb, 4 bundles/sq	1 Rofc	3.50	2.286	Sq.	100	81		181	245
0455	Pneumatic nailed		4.37	1.831		100	65		165	218
0500	Class C, 300-385 lb/square, 5 bundles/square		3	2.667		132	94.50		226.50	300
0505	Pneumatic nailed		3.75	2.133		132	75.50		207.50	271
0800	#15 felt underlayment		64	.125		5.90	4.43		10.33	13.80
0825	#30 felt underlayment		58	.138		11.95	4.88		16.83	21.50
0850	Self adhering polyethylene and rubberized asphalt underlayment		22	.364		60	12.85		72.85	87.50
0900	Ridge shingles		330	.024	L.F.	1.90	.86		2.76	3.52
0905	Pneumatic nailed		412.50	.019	"	1.90	.69		2.59	3.23
1000	For steep roofs (7 to 12 pitch or greater), add						50%			

07 31 16 – Metal Shingles

07 31 16.10 Aluminum Shingles

	07 31 16.10 Aluminum Shingles	Crew	Daily Output	Labor-Hours	Unit	Material	Labor	Equipment	Total	Total Incl O&P
0010	**ALUMINUM SHINGLES**									
0020	Mill finish, .019 thick	1 Carp	5	1.600	Sq.	195	66.50		261.50	315
0100	.020" thick	"	5	1.600		188	66.50		254.50	310
0300	For colors, add					16			16	17.60
0600	Ridge cap, .024" thick	1 Carp	170	.047	L.F.	2.10	1.96		4.06	5.30
0700	End wall flashing, .024" thick		170	.047		1.71	1.96		3.67	4.89
0900	Valley section, .024" thick		170	.047		2.81	1.96		4.77	6.10
1000	Starter strip, .024" thick		400	.020		1.36	.83		2.19	2.78
1200	Side wall flashing, .024" thick		170	.047		1.71	1.96		3.67	4.89
1500	Gable flashing, .024" thick		400	.020		1.36	.83		2.19	2.78

07 31 16.20 Steel Shingles

	07 31 16.20 Steel Shingles	Crew	Daily Output	Labor-Hours	Unit	Material	Labor	Equipment	Total	Total Incl O&P
0010	**STEEL SHINGLES**									
0012	Galvanized, 26 gauge	1 Rots	2.20	3.636	Sq.	232	129		361	470
0200	24 gauge	"	2.20	3.636		244	129		373	485
0300	For colored galvanized shingles, add					55			55	60.50
0500	For 1" factory applied polystyrene insulation, add					40			40	44

07 31 19 – Mineral-Fiber Cement Shingles

07 31 19.10 Fiber Cement Shingles

	07 31 19.10 Fiber Cement Shingles	Crew	Daily Output	Labor-Hours	Unit	Material	Labor	Equipment	Total	Total Incl O&P
0010	**FIBER CEMENT SHINGLES**									
0012	Field shingles, 16" x 9.35", 500 lb per square	1 Rofc	2.20	3.636	Sq.	335	129		464	585
0110	Starters, 16" x 9.35"		3	2.667	C.L.F.	115	94.50		209.50	284
0120	Hip & ridge, 4.75" x 14"		1	8	"	830	283		1,113	1,375
0200	Shakes, 16" x 9.35", 550 lb per square		2.20	3.636	Sq.	305	129		434	550
0300	Hip & ridge, 4.75 x 14"		1	8	C.L.F.	830	283		1,113	1,375
0400	Hexagonal, 16" x 16"		3	2.667	Sq.	230	94.50		324.50	410
0500	Square, 16" x 16"		3	2.667	"	205	94.50		299.50	385

07 31 Shingles and Shakes

07 31 26 – Slate Shingles

07 31 26.10 Slate Roof Shingles		Crew	Daily Output	Labor-Hours	Unit	Material	2010 Bare Costs Labor	Equipment	Total	Total Incl O&P
0010	SLATE ROOF SHINGLES R073126-20									
0100	Buckingham Virginia black, 3/16" - 1/4" thick G	1 Rots	1.75	4.571	Sq.	475	163		638	795
0200	1/4" thick G		1.75	4.571		475	163		638	795
0900	Pennsylvania black, Bangor, #1 clear G		1.75	4.571		490	163		653	810
1200	Vermont, unfading, green, mottled green G		1.75	4.571		480	163		643	795
1300	Semi-weathering green & gray G		1.75	4.571		350	163		513	650
1400	Purple G		1.75	4.571		425	163		588	735
1500	Black or gray G		1.75	4.571		460	163		623	775
1600	Red G		1.75	4.571		1,175	163		1,338	1,550
1700	Variegated purple		1.75	4.571		415	163		578	725

07 31 29 – Wood Shingles and Shakes

07 31 29.13 Wood Shingles

		Crew	Daily Output	Labor-Hours	Unit	Material	Labor	Equipment	Total	Total Incl O&P
0010	WOOD SHINGLES									
0012	16" No. 1 red cedar shingles, 5" exposure, on roof	1 Carp	2.50	3.200	Sq.	310	133		443	545
0015	Pneumatic nailed		3.25	2.462		310	102		412	500
0200	7-1/2" exposure, on walls		2.05	3.902		207	162		369	475
0205	Pneumatic nailed		2.67	2.996		207	124		331	420
0300	18" No. 1 red cedar perfections, 5-1/2" exposure, on roof		2.75	2.909		265	121		386	480
0305	Pneumatic nailed		3.57	2.241		265	93		358	435
0500	7-1/2" exposure, on walls		2.25	3.556		195	148		343	440
0505	Pneumatic nailed		2.92	2.740		195	114		309	390
0600	Resquared, and rebutted, 5-1/2" exposure, on roof		3	2.667		236	111		347	430
0605	Pneumatic nailed		3.90	2.051		236	85		321	390
0900	7-1/2" exposure, on walls		2.45	3.265		173	136		309	400
0905	Pneumatic nailed		3.18	2.516		173	105		278	350
1000	Add to above for fire retardant shingles					55			55	60.50
1050	18" long					55			55	60.50
1060	Preformed ridge shingles	1 Carp	400	.020	L.F.	3.60	.83		4.43	5.25
2000	White cedar shingles, 16" long, extras, 5" exposure, on roof		2.40	3.333	Sq.	174	139		313	405
2005	Pneumatic nailed		3.12	2.564		174	107		281	355
2050	5" exposure on walls		2	4		174	166		340	445
2055	Pneumatic nailed		2.60	3.077		174	128		302	390
2100	7-1/2" exposure, on walls		2	4		124	166		290	395
2105	Pneumatic nailed		2.60	3.077		124	128		252	335
2150	"B" grade, 5" exposure on walls		2	4		150	166		316	420
2155	Pneumatic nailed		2.60	3.077		150	128		278	360
2300	For 15# organic felt underlayment on roof, 1 layer, add		64	.125		5.90	5.20		11.10	14.45
2400	2 layers, add		32	.250		11.75	10.40		22.15	29
2600	For steep roofs (7/12 pitch or greater), add to above						50%			
2700	Panelized systems, No.1 cedar shingles on 5/16" CDX plywood									
2800	On walls, 8' strips, 7" or 14" exposure	2 Carp	700	.023	S.F.	5.55	.95		6.50	7.55
3500	On roofs, 8' strips, 7" or 14" exposure	1 Carp	3	2.667	Sq.	555	111		666	780
3505	Pneumatic nailed	"	4	2	"	555	83		638	740

07 31 29.16 Wood Shakes

		Crew	Daily Output	Labor-Hours	Unit	Material	Labor	Equipment	Total	Total Incl O&P
0010	WOOD SHAKES									
1100	Hand-split red cedar shakes, 1/2" thick x 24" long, 10" exp. on roof	1 Carp	2.50	3.200	Sq.	165	133		298	385
1105	Pneumatic nailed		3.25	2.462		165	102		267	340
1110	3/4" thick x 24" long, 10" exp. on roof		2.25	3.556		165	148		313	410
1115	Pneumatic nailed		2.92	2.740		165	114		279	355
1200	1/2" thick, 18" long, 8-1/2" exp. on roof		2	4		146	166		312	415
1205	Pneumatic nailed		2.60	3.077		146	128		274	360

07 31 Shingles and Shakes

07 31 29 – Wood Shingles and Shakes

07 31 29.16 Wood Shakes		Crew	Daily Output	Labor-Hours	Unit	Material	Labor	Equipment	Total	Total Incl O&P
						2010 Bare Costs				
1210	3/4" thick x 18" long, 8 1/2" exp. on roof	1 Carp	1.80	4.444	Sq.	146	185		331	445
1215	Pneumatic nailed		2.34	3.419		146	142		288	380
1255	10" exp. on walls		2	4		141	166		307	410
1260	10" exposure on walls, pneumatic nailed		2.60	3.077		141	128		269	350
1700	Add to above for fire retardant shakes, 24" long					55			55	60.50
1800	18" long					55			55	60.50
1810	Ridge shakes	1 Carp	350	.023	L.F.	3.60	.95		4.55	5.40

07 32 Roof Tiles

07 32 13 – Clay Roof Tiles

07 32 13.10 Clay Tiles

Code	Description		Crew	Daily Output	Labor-Hours	Unit	Material	Labor	Equipment	Total	Total Incl O&P
0010	**CLAY TILES**										
0200	Lanai tile or Classic tile, 158 pc per sq	G	1 Rots	1.65	4.848	Sq.	500	173		673	835
0300	Americana, 158 pc per sq, most colors	G		1.65	4.848		660	173		833	1,025
0350	Green, gray or brown	G		1.65	4.848		625	173		798	970
0400	Blue	G		1.65	4.848		625	173		798	970
0600	Spanish tile, 171 pc per sq, red	G		1.80	4.444		325	158		483	620
0800	Buff, green, gray, brown	G		1.80	4.444		600	158		758	925
0900	Glazed white			1.80	4.444		665	158		823	995
1100	Mission tile, 192 pc per sq, machine scored finish, red	G		1.15	6.957		740	248		988	1,225
1700	French tile, 133 pc per sq, smooth finish, red	G		1.35	5.926		675	211		886	1,100
1750	Blue or green	G		1.35	5.926		870	211		1,081	1,300
1800	Norman black 317 pc per sq	G		1	8		1,050	285		1,335	1,625
2200	Williamsburg tile, 158 pc per sq, aged cedar	G		1.35	5.926		730	211		941	1,150
2250	Gray or green	G		1.35	5.926		595	211		806	1,000
2510	One piece mission tile, natural red, 75 pc per square	G		1.65	4.848		231	173		404	540
2530	Mission Tile, 134 pc per square	G		1.15	6.957		263	248		511	700
3010	Clay tile, #15 felt underlayment		1 Rofc	64	.125		5.90	4.43		10.33	13.80
3020	Clay tile, #30 felt underlayment			58	.138		11.95	4.88		16.83	21.50
3040	Clay tile, polyethylene and rubberized asph. underlayment			22	.364		60	12.85		72.85	87.50

07 32 16 – Concrete Roof Tiles

07 32 16.10 Concrete Tiles

Code	Description	Crew	Daily Output	Labor-Hours	Unit	Material	Labor	Equipment	Total	Total Incl O&P
0010	**CONCRETE TILES**									
0020	Corrugated, 13" x 16-1/2", 90 per sq, 950 lb per sq									
0050	Earthtone colors, nailed to wood deck	1 Rots	1.35	5.926	Sq.	103	211		314	465
0150	Blues		1.35	5.926		104	211		315	465
0200	Greens		1.35	5.926		104	211		315	465
0250	Premium colors		1.35	5.926		229	211		440	600
0500	Shakes, 13" x 16-1/2", 90 per sq, 950 lb per sq									
0600	All colors, nailed to wood deck	1 Rots	1.50	5.333	Sq.	270	190		460	610
1500	Accessory pieces, ridge & hip, 10" x 16-1/2", 8 lbs. each	"	120	.067	Ea.	3.64	2.37		6.01	7.95
1700	Rake, 6-1/2" x 16-3/4", 9 lbs. each					3.64			3.64	4
1800	Mansard hip, 10" x 16-1/2", 9.2 lbs. each					3.64			3.64	4
1900	Hip starter, 10" x 16-1/2", 10.5 lbs. each					11.45			11.45	12.60
2000	3 or 4 way apex, 10" each side, 11.5 lbs. each					13.25			13.25	14.55

07 32 19 – Metal Roof Tiles

07 32 19.10 Metal Roof Tiles

Code	Description	Crew	Daily Output	Labor-Hours	Unit	Material	Labor	Equipment	Total	Total Incl O&P
0010	**METAL ROOF TILES**									
0020	Accessories included, .032" thick aluminum, mission tile	1 Carp	2.50	3.200	Sq.	815	133		948	1,100
0200	Spanish tiles	"	3	2.667	"	560	111		671	785

07 33 Natural Roof Coverings

07 33 63 – Vegetated Roofing

07 33 63.10 Green Roof Systems			Crew	Daily Output	Labor-Hours	Unit	Material	Labor	Equipment	Total	Total Incl O&P
							2010 Bare Costs				
0010	**GREEN ROOF SYSTEMS**										
0020	Soil mixture for green roof 30% sand, 55% gravel, 15% soil	G									
0100	Hoist and spread soil mixture 4 inch depth up to five stories tall roof	G	B-13B	4000	.014	S.F.	.22	.50	.30	1.02	1.33
0150	6 inch depth	G		2667	.021		.33	.75	.45	1.53	2.01
0200	8 inch depth	G		2000	.028		.44	1	.60	2.04	2.68
0250	10 inch depth	G		1600	.035		.55	1.25	.76	2.56	3.35
0300	12 inch depth	G		1335	.042		.66	1.50	.91	3.07	4.02
0350	Mobilization 55 ton crane to site	G	1 Eqhv	3.60	2.222	Ea.		98.50		98.50	148
0355	Hoisting cost to five stories per day (Avg. 28 picks per day)	G	B-13B	1	56	Day		2,000	1,200	3,200	4,375
0360	Mobilization or demobilization, 100 ton crane to site driver & escort	G	A-3E	2.50	6.400	Ea.		248	60.50	308.50	440
0365	Hoisting cost six to ten stories per day (Avg. 21 picks per day)	G	B-13C	1	56	Day		2,000	1,725	3,725	4,950
0370	Hoist and spread soil mixture 4 inch depth six to ten stories tall roof	G		4000	.014	S.F.	.22	.50	.43	1.15	1.48
0375	6 inch depth	G		2667	.021		.33	.75	.65	1.73	2.22
0380	8 inch depth	G		2000	.028		.44	1	.87	2.31	2.96
0385	10 inch depth	G		1600	.035		.55	1.25	1.08	2.88	3.71
0390	12 inch depth	G		1335	.042		.66	1.50	1.30	3.46	4.45
0400	Green roof edging treated lumber 4″ x4″ no hoisting included	G	2 Carp	400	.040	L.F.	1.56	1.66		3.22	4.27
0410	4″ x 6″	G		400	.040		2.30	1.66		3.96	5.10
0420	4″ x 8″	G		360	.044		4.74	1.85		6.59	8.05
0430	4″ x 6″ double stacked	G		300	.053		4.60	2.22		6.82	8.45
0500	Green roof edging redwood lumber 4″ x4″ no hoisting included	G		400	.040		6.40	1.66		8.06	9.60
0510	4″ x 6″	G		400	.040		11.20	1.66		12.86	14.85
0520	4″ x 8″	G		360	.044		20.50	1.85		22.35	25.50
0530	4″ x 6″ double stacked	G		300	.053		11.20	2.22		13.42	15.75
0600	Planting sedum, light soil, potted, 2-1/4″ diameter, two per SF	G	1 Clab	420	.019	S.F.	4.98	.63		5.61	6.45
0610	one per SF	G	″	840	.010		2.49	.32		2.81	3.23
0630	Planting sedum mat per SF including shipping (4000 SF Minimum)	G	4 Clab	4000	.008		6	.26		6.26	6.95
0640	Installation sedum mat system (no soil required) per SF (4000 SF minimum)	G	″	4000	.008		8.30	.26		8.56	9.55
0645	Note: pricing of sedum mats shipped in full truck loads (4000-5000 SF)	G									

07 41 Roof Panels

07 41 13 – Metal Roof Panels

07 41 13.10 Aluminum Roof Panels		Crew	Daily Output	Labor-Hours	Unit	Material	Labor	Equipment	Total	Total Incl O&P
0010	**ALUMINUM ROOF PANELS**									
0020	Corrugated or ribbed, .0155″ thick, natural	G-3	1200	.027	S.F.	.96	1.10		2.06	2.73
0300	Painted		1200	.027		1.40	1.10		2.50	3.21
0400	Corrugated, .018″ thick, on steel frame, natural finish		1200	.027		1.27	1.10		2.37	3.07
0600	Painted		1200	.027		1.55	1.10		2.65	3.38
0700	Corrugated, on steel frame, natural, .024″ thick		1200	.027		1.80	1.10		2.90	3.65
0800	Painted		1200	.027		2.19	1.10		3.29	4.08
0900	.032″ thick, natural		1200	.027		2.32	1.10		3.42	4.22
1200	Painted		1200	.027		2.97	1.10		4.07	4.94
1300	V-Beam, on steel frame construction, .032″ thick, natural		1200	.027		2.45	1.10		3.55	4.37
1500	Painted		1200	.027		3.01	1.10		4.11	4.98
1600	.040″ thick, natural		1200	.027		2.99	1.10		4.09	4.96
1800	Painted		1200	.027		3.59	1.10		4.69	5.60
1900	.050″ thick, natural		1200	.027		3.58	1.10		4.68	5.60
2100	Painted		1200	.027		4.33	1.10		5.43	6.45
2200	For roofing on wood frame, deduct		4600	.007		.08	.29		.37	.53
2400	Ridge cap, .032″ thick, natural		800	.040	L.F.	2.83	1.64		4.47	5.60

07 41 Roof Panels

07 41 13 – Metal Roof Panels

07 41 13.20 Steel Roofing Panels		Crew	Daily Output	Labor-Hours	Unit	Material	2010 Bare Costs Labor	Equipment	Total	Total Incl O&P
0010	**STEEL ROOFING PANELS**									
0012	Corrugated or ribbed, on steel framing, 30 ga galv	G-3	1100	.029	S.F.	1.68	1.20		2.88	3.67
0100	28 ga		1050	.030		1.79	1.25		3.04	3.88
0300	26 ga		1000	.032		1.85	1.32		3.17	4.05
0400	24 ga		950	.034		2.91	1.38		4.29	5.30
0600	Colored, 28 ga		1050	.030		1.68	1.25		2.93	3.76
0700	26 ga		1000	.032		1.99	1.32		3.31	4.20
0710	Flat profile, 1-3/4" standing seams, 10" wide, standard finish, 26 ga		1000	.032		3.89	1.32		5.21	6.30
0715	24 ga		950	.034		4.51	1.38		5.89	7.05
0720	22 ga		900	.036		5.55	1.46		7.01	8.40
0725	Zinc aluminum alloy finish, 26 ga		1000	.032		3.05	1.32		4.37	5.35
0730	24 ga		950	.034		3.64	1.38		5.02	6.10
0735	22 ga		900	.036		4.17	1.46		5.63	6.80
0740	12" wide, standard finish, 26 ga		1000	.032		3.88	1.32		5.20	6.30
0745	24 ga		950	.034		5.10	1.38		6.48	7.70
0750	Zinc aluminum alloy finish, 26 ga		1000	.032		4.41	1.32		5.73	6.85
0755	24 ga		950	.034		3.63	1.38		5.01	6.10
0840	Flat profile, 1" x 3/8" batten, 12" wide, standard finish, 26 ga		1000	.032		3.42	1.32		4.74	5.75
0845	24 ga		950	.034		4.01	1.38		5.39	6.50
0850	22 ga		900	.036		4.80	1.46		6.26	7.55
0855	Zinc aluminum alloy finish, 26 ga		1000	.032		3.28	1.32		4.60	5.60
0860	24 ga		950	.034		3.66	1.38		5.04	6.15
0865	22 ga		900	.036		4.23	1.46		5.69	6.90
0870	16-1/2" wide, standard finish, 24 ga		950	.034		3.95	1.38		5.33	6.45
0875	22 ga		900	.036		4.42	1.46		5.88	7.10
0880	Zinc aluminum alloy finish, 24 ga		950	.034		3.45	1.38		4.83	5.90
0885	22 ga		900	.036		3.85	1.46		5.31	6.45
0890	Flat profile, 2" x 2" batten, 12" wide, standard finish, 26 ga		1000	.032		3.93	1.32		5.25	6.35
0895	24 ga		950	.034		4.69	1.38		6.07	7.25
0900	22 ga		900	.036		5.75	1.46		7.21	8.60
0905	Zinc aluminum alloy finish, 26 ga		1000	.032		3.66	1.32		4.98	6.05
0910	24 ga		950	.034		4.18	1.38		5.56	6.70
0915	22 ga		900	.036		4.87	1.46		6.33	7.60
0920	16-1/2" wide, standard finish, 24 ga		950	.034		4.32	1.38		5.70	6.85
0925	22 ga		900	.036		5.05	1.46		6.51	7.80
0930	Zinc aluminum alloy finish, 24 ga		950	.034		3.91	1.38		5.29	6.40
0935	22 ga		900	.036		4.45	1.46		5.91	7.15
1200	Ridge, galvanized, 10" wide		800	.040	L.F.	3.18	1.64		4.82	6
1210	20" wide		750	.043	"	4.93	1.75		6.68	8.10

07 41 33 – Plastic Roof Panels

07 41 33.10 Fiberglass Panels		Crew	Daily Output	Labor-Hours	Unit	Material	Labor	Equipment	Total	Total Incl O&P
0010	**FIBERGLASS PANELS**									
0012	Corrugated panels, roofing, 8 oz per S.F.	G-3	1000	.032	S.F.	1.58	1.32		2.90	3.75
0100	12 oz per S.F.		1000	.032		3.43	1.32		4.75	5.80
0300	Corrugated siding, 6 oz per S.F.		880	.036		1.58	1.49		3.07	4.02
0400	8 oz per S.F.		880	.036		1.58	1.49		3.07	4.02
0500	Fire retardant		880	.036		3.21	1.49		4.70	5.80
0600	12 oz. siding, textured		880	.036		3.32	1.49		4.81	5.95
0700	Fire retardant		880	.036		4.39	1.49		5.88	7.10
0900	Flat panels, 6 oz per S.F., clear or colors		880	.036		1.83	1.49		3.32	4.29
1100	Fire retardant, class A		880	.036		3.18	1.49		4.67	5.80
1300	8 oz per S.F., clear or colors		880	.036		2.38	1.49		3.87	4.90

07 41 Roof Panels

07 41 33 – Plastic Roof Panels

07 41 33.10 Fiberglass Panels		Crew	Daily Output	Labor-Hours	Unit	Material	2010 Bare Costs Labor	Equipment	Total	Total Incl O&P
1700	Sandwich panels, fiberglass, 1-9/16" thick, panels to 20 S.F.	G-3	180	.178	S.F.	23	7.30		30.30	36.50
1900	As above, but 2-3/4" thick, panels to 100 S.F.	↓	265	.121	↓	17.05	4.96		22.01	26.50

07 42 Wall Panels

07 42 13 – Metal Wall Panels

07 42 13.10 Mansard Panels

		Crew	Daily Output	Labor-Hours	Unit	Material	Labor	Equipment	Total	Total Incl O&P
0010	**MANSARD PANELS**									
0600	Aluminum, stock units, straight surfaces	1 Shee	115	.070	S.F.	3.35	3.42		6.77	8.85
0700	Concave or convex surfaces		75	.107	"	3.67	5.25		8.92	12
0800	For framing, to 5' high, add		115	.070	L.F.	3.67	3.42		7.09	9.20
0900	Soffits, to 1' wide	↓	125	.064	S.F.	1.82	3.14		4.96	6.75

07 42 13.20 Aluminum Siding Panels

		Crew	Daily Output	Labor-Hours	Unit	Material	Labor	Equipment	Total	Total Incl O&P
0010	**ALUMINUM SIDING PANELS**									
0012	Corrugated, on steel framing, .019 thick, natural finish	G-3	775	.041	S.F.	1.47	1.70		3.17	4.21
0100	Painted		775	.041		1.58	1.70		3.28	4.33
0400	Farm type, .021" thick on steel frame, natural		775	.041		1.48	1.70		3.18	4.22
0600	Painted		775	.041		1.58	1.70		3.28	4.33
0700	Industrial type, corrugated, on steel, .024" thick, mill		775	.041		2.06	1.70		3.76	4.86
0900	Painted		775	.041		2.22	1.70		3.92	5.05
1000	.032" thick, mill		775	.041		2.37	1.70		4.07	5.20
1200	Painted		775	.041		2.89	1.70		4.59	5.75
1300	V-Beam, on steel frame, .032" thick, mill		775	.041		2.67	1.70		4.37	5.55
1500	Painted		775	.041		3.06	1.70		4.76	5.95
1600	.040" thick, mill		775	.041		3.24	1.70		4.94	6.15
1800	Painted		775	.041		3.80	1.70		5.50	6.75
1900	.050" thick, mill		775	.041		3.83	1.70		5.53	6.80
2100	Painted		775	.041		4.57	1.70		6.27	7.65
2200	Ribbed, 3" profile, on steel frame, .032" thick, natural		775	.041		2.31	1.70		4.01	5.15
2400	Painted		775	.041		2.93	1.70		4.63	5.80
2500	.040" thick, natural		775	.041		2.64	1.70		4.34	5.50
2700	Painted		775	.041		3.08	1.70		4.78	6
2750	.050" thick, natural		775	.041		3.04	1.70		4.74	5.95
2760	Painted		775	.041		3.49	1.70		5.19	6.45
3300	For siding on wood frame, deduct from above	↓	2800	.011	↓	.09	.47		.56	.82
3400	Screw fasteners, aluminum, self tapping, neoprene washer, 1"				M	200			200	220
3600	Stitch screws, self tapping, with neoprene washer, 5/8"				"	150			150	165
3630	Flashing, sidewall, .032" thick	G-3	800	.040	L.F.	2.77	1.64		4.41	5.55
3650	End wall, .040" thick		800	.040		3.24	1.64		4.88	6.05
3670	Closure strips, corrugated, .032" thick		800	.040		.82	1.64		2.46	3.41
3680	Ribbed, 4" or 8", .032" thick		800	.040		.82	1.64		2.46	3.41
3690	V-beam, .040" thick	↓	800	.040	↓	1.11	1.64		2.75	3.73
3800	Horizontal, colored clapboard, 8" wide, plain	2 Carp	515	.031	S.F.	1.73	1.29		3.02	3.89
3810	Insulated		515	.031		1.86	1.29		3.15	4.04
4000	Vertical board & batten, colored, non-insulated	↓	515	.031		1.53	1.29		2.82	3.67
4200	For simulated wood design, add				↓	.10			.10	.11
4300	Corners for above, outside	2 Carp	515	.031	V.L.F.	3.05	1.29		4.34	5.35
4500	Inside corners	"	515	.031	"	1.39	1.29		2.68	3.52
4600	Sandwich panels, 1" insulation, single story	G-3	395	.081	S.F.	7.35	3.33		10.68	13.15
4900	Multi-story	"	345	.093		10.30	3.81		14.11	17.15
5100	For baked enamel finish 1 side, add				↓	.40			.40	.44

07 42 Wall Panels

07 42 13 – Metal Wall Panels

07 42 13.30 Steel Siding		Crew	Daily Output	Labor-Hours	Unit	Material	2010 Bare Costs Labor	Equipment	Total	Total Incl O&P
0010	**STEEL SIDING**									
0020	Beveled, vinyl coated, 8" wide	1 Carp	265	.030	S.F.	1.97	1.25		3.22	4.10
0050	10" wide	"	275	.029		2.11	1.21		3.32	4.18
0080	Galv, corrugated or ribbed, on steel frame, 30 gauge	G-3	800	.040		1.32	1.64		2.96	3.96
0100	28 gauge		795	.040		1.38	1.65		3.03	4.04
0300	26 gauge		790	.041		1.94	1.67		3.61	4.67
0400	24 gauge		785	.041		1.95	1.68		3.63	4.71
0600	22 gauge		770	.042		2.25	1.71		3.96	5.10
0700	Colored, corrugated/ribbed, on steel frame, 10 yr fnsh, 28 ga.		800	.040		2.05	1.64		3.69	4.77
0900	26 gauge		795	.040		2.14	1.65		3.79	4.87
1000	24 gauge		790	.041		2.49	1.67		4.16	5.30
1020	20 gauge		785	.041		3.16	1.68		4.84	6.05
1200	Factory sandwich panel, 26 ga., 1" insulation, galvanized		380	.084		5.25	3.46		8.71	11.10
1300	Colored 1 side		380	.084		6.45	3.46		9.91	12.35
1500	Galvanized 2 sides		380	.084		8	3.46		11.46	14.10
1600	Colored 2 sides		380	.084		8.25	3.46		11.71	14.40
1800	Acrylic paint face, regular paint liner	↓	380	.084		6.25	3.46		9.71	12.15
1900	For 2" thick polystyrene, add					.96			.96	1.06
2000	22 ga, galv, 2" insulation, baked enamel exterior	G-3	360	.089		12.20	3.65		15.85	19
2100	Polyvinylidene exterior finish	"	360	.089	↓	12.85	3.65		16.50	19.70

07 44 Faced Panels

07 44 73 – Metal Faced Panels

07 44 73.10 Metal Faced Panels and Accessories		Crew	Daily Output	Labor-Hours	Unit	Material	2010 Bare Costs Labor	Equipment	Total	Total Incl O&P
0010	**METAL FACED PANELS AND ACCESSORIES**									
0400	Textured aluminum, 4' x 8' x 5/16" plywood backing, single face	2 Shee	375	.043	S.F.	3.67	2.10		5.77	7.20
0600	Double face		375	.043		4.99	2.10		7.09	8.65
0700	4' x 10' x 5/16" plywood backing, single face		375	.043		3.93	2.10		6.03	7.50
0900	Double face		375	.043		5.25	2.10		7.35	8.95
1000	4' x 12' x 5/16" plywood backing, single face		375	.043		3.33	2.10		5.43	6.85
1300	Smooth aluminum, 1/4" plywood panel, fluoropolymer finish, double face		375	.043		5.45	2.10		7.55	9.10
1350	Clear anodized finish, double face		375	.043		9.55	2.10		11.65	13.65
1400	Double face textured aluminum, structural panel, 1" EPS insulation	↓	375	.043	↓	4.92	2.10		7.02	8.55
1500	Accessories, outside corner	1 Shee	175	.046	L.F.	1.90	2.24		4.14	5.50
1600	Inside corner		175	.046		1.34	2.24		3.58	4.87
1800	Batten mounting clip		200	.040		.49	1.96		2.45	3.51
1900	Low profile batten		480	.017		.61	.82		1.43	1.91
2100	High profile batten		480	.017		1.41	.82		2.23	2.79
2200	Water table		200	.040		2.12	1.96		4.08	5.30
2400	Horizontal joint connector		200	.040		1.67	1.96		3.63	4.81
2500	Corner cap		200	.040		1.84	1.96		3.80	4.99
2700	H - moulding	↓	480	.017	↓	1.25	.82		2.07	2.62

07 46 Siding

07 46 23 – Wood Siding

07 46 23.10 Wood Board Siding		Crew	Daily Output	Labor-Hours	Unit	Material	Labor	Equipment	Total	Total Incl O&P
						2010 Bare Costs				
0010	**WOOD BOARD SIDING**									
3200	Wood, cedar bevel, A grade, 1/2" x 6"	1 Carp	295	.027	S.F.	4.73	1.13		5.86	6.95
3300	1/2" x 8"		330	.024		5.55	1.01		6.56	7.65
3500	3/4" x 10", clear grade		375	.021		6.10	.89		6.99	8.05
3600	"B" grade		375	.021		5.20	.89		6.09	7.05
3800	Cedar, rough sawn, 1" x 4", A grade, natural		220	.036		3.77	1.51		5.28	6.50
3900	Stained		220	.036		4.17	1.51		5.68	6.90
4100	1" x 12", board & batten, #3 & Btr., natural		420	.019		3.94	.79		4.73	5.55
4200	Stained		420	.019		4.20	.79		4.99	5.85
4400	1" x 8" channel siding, #3 & Btr., natural		330	.024		2.34	1.01		3.35	4.12
4500	Stained		330	.024		2.50	1.01		3.51	4.30
4700	Redwood, clear, beveled, vertical grain, 1/2" x 4"		220	.036		3.10	1.51		4.61	5.75
4750	1/2" x 6"		295	.027		3.10	1.13		4.23	5.15
4800	1/2" x 8"		330	.024		3.24	1.01		4.25	5.10
5000	3/4" x 10"		375	.021		3.32	.89		4.21	5
5200	Channel siding, 1" x 10", B grade		375	.021		3.32	.89		4.21	5
5250	Redwood, T&G boards, B grade, 1" x 4"		220	.036		3.10	1.51		4.61	5.75
5270	1" x 8"		330	.024		3.24	1.01		4.25	5.10
5400	White pine, rough sawn, 1" x 8", natural		330	.024		2.20	1.01		3.21	3.97
5500	Stained		330	.024		2.60	1.01		3.61	4.41

07 46 29 – Plywood Siding

07 46 29.10 Plywood Siding Options		Crew	Daily Output	Labor-Hours	Unit	Material	Labor	Equipment	Total	Total Incl O&P
0010	**PLYWOOD SIDING OPTIONS**									
0900	Plywood, medium density overlaid, 3/8" thick	2 Carp	750	.021	S.F.	1.14	.89		2.03	2.62
1000	1/2" thick		700	.023		1.22	.95		2.17	2.80
1100	3/4" thick		650	.025		1.55	1.02		2.57	3.29
1600	Texture 1-11, cedar, 5/8" thick, natural		675	.024		2.43	.98		3.41	4.19
1700	Factory stained		675	.024		1.95	.98		2.93	3.67
1900	Texture 1-11, fir, 5/8" thick, natural		675	.024		1.37	.98		2.35	3.03
2000	Factory stained		675	.024		1.73	.98		2.71	3.42
2050	Texture 1-11, S.Y.P., 5/8" thick, natural		675	.024		1.19	.98		2.17	2.83
2100	Factory stained		675	.024		1.24	.98		2.22	2.88
2200	Rough sawn cedar, 3/8" thick, natural		675	.024		1.20	.98		2.18	2.84
2300	Factory stained		675	.024		.96	.98		1.94	2.58
2500	Rough sawn fir, 3/8" thick, natural		675	.024		.76	.98		1.74	2.36
2600	Factory stained		675	.024		.96	.98		1.94	2.58
2800	Redwood, textured siding, 5/8" thick		675	.024		1.92	.98		2.90	3.63
3000	Polyvinyl chloride coated, 3/8" thick		750	.021		1.11	.89		2	2.59

07 46 33 – Plastic Siding

07 46 33.10 Vinyl Siding		Crew	Daily Output	Labor-Hours	Unit	Material	Labor	Equipment	Total	Total Incl O&P
0010	**VINYL SIDING**									
3995	Clapboard profile, woodgrain texture, .048 thick, double 4	2 Carp	495	.032	S.F.	.92	1.34		2.26	3.08
4000	Double 5		550	.029		.92	1.21		2.13	2.87
4005	Single 8		495	.032		.96	1.34		2.30	3.13
4010	Single 10		550	.029		1.15	1.21		2.36	3.13
4015	.044 thick, double 4		495	.032		.84	1.34		2.18	2.99
4020	Double 5		550	.029		.84	1.21		2.05	2.78
4025	.042 thick, double 4		495	.032		.84	1.34		2.18	2.99
4030	Double 5		550	.029		.84	1.21		2.05	2.78
4035	Cross sawn texture, .040 thick, double 4		495	.032		.66	1.34		2	2.80
4040	Double 5		550	.029		.66	1.21		1.87	2.59
4045	Smooth texture, .042 thick, double 4		495	.032		.75	1.34		2.09	2.90

07 46 33 – Plastic Siding

07 46 33.10 Vinyl Siding		Crew	Daily Output	Labor-Hours	Unit	Material	Labor	Equipment	Total	Total Incl O&P
						2010 Bare Costs				
4050	Double 5	2 Carp	550	.029	S.F.	.75	1.21		1.96	2.69
4055	Single 8		495	.032		.96	1.34		2.30	3.13
4060	Cedar texture, .044 thick, double 4		495	.032		.92	1.34		2.26	3.08
4065	Double 6		600	.027		.92	1.11		2.03	2.72
4070	Dutch lap profile, woodgrain texture, .048 thick, double 5		550	.029		.92	1.21		2.13	2.87
4075	.044 thick, double 4.5		525	.030		.92	1.27		2.19	2.96
4080	.042 thick, double 4.5		525	.030		.75	1.27		2.02	2.78
4085	.040 thick, double 4.5		525	.030		.66	1.27		1.93	2.68
4090	Shingle profile, random grooves, double 7		400	.040		2.57	1.66		4.23	5.40
4095	Triple 5		400	.040		2.57	1.66		4.23	5.40
4100	Shake profile, 10" wide		400	.040		2.19	1.66		3.85	4.97
4105	Vertical pattern, .046 thick, double 5		550	.029		1.22	1.21		2.43	3.20
4110	.044 thick, triple 3		550	.029		1.37	1.21		2.58	3.37
4115	.040 thick, triple 4		550	.029		1.22	1.21		2.43	3.20
4120	.040 thick, triple 2.66		550	.029		1.47	1.21		2.68	3.48
4125	Insulation, fan folded extruded polystyrene, 1/4"		2000	.008		.25	.33		.58	.79
4130	3/8"		2000	.008	↓	.27	.33		.60	.81
4135	Accessories, J channel, 5/8" pocket		700	.023	L.F.	.43	.95		1.38	1.93
4140	3/4" pocket		695	.023		.48	.96		1.44	1.99
4145	1-1/4" pocket		680	.024		.61	.98		1.59	2.18
4150	Flexible, 3/4" pocket		600	.027		1.81	1.11		2.92	3.70
4155	Under sill finish trim		500	.032		.45	1.33		1.78	2.54
4160	Vinyl starter strip		700	.023		.36	.95		1.31	1.85
4165	Aluminum starter strip		700	.023		.27	.95		1.22	1.75
4170	Window casing, 2-1/2" wide, 3/4" pocket		510	.031		1.14	1.30		2.44	3.26
4175	Outside corner, woodgrain finish, 4" face, 3/4" pocket		700	.023		1.75	.95		2.70	3.39
4180	5/8" pocket		700	.023		1.67	.95		2.62	3.30
4185	Smooth finish, 4" face, 3/4" pocket		700	.023		1.84	.95		2.79	3.48
4190	7/8" pocket		690	.023		1.78	.96		2.74	3.45
4195	1-1/4" pocket		700	.023		1.23	.95		2.18	2.81
4200	Soffit and fascia, 1' overhang, solid		120	.133		3.37	5.55		8.92	12.25
4205	Vented		120	.133		3.37	5.55		8.92	12.25
4207	18" overhang, solid		110	.145		3.98	6.05		10.03	13.65
4208	Vented		110	.145		3.98	6.05		10.03	13.65
4210	2' overhang, solid		100	.160		4.58	6.65		11.23	15.30
4215	Vented		100	.160		4.58	6.65		11.23	15.30
4217	3' overhang, solid		100	.160		5.80	6.65		12.45	16.60
4218	Vented	↓	100	.160	↓	5.80	6.65		12.45	16.60
4220	Colors for siding and soffits, add				S.F.	.16			.16	.18
4225	Colors for accessories and trim, add				L.F.	.32			.32	.35

07 46 46 – Mineral-Fiber Cement Siding

07 46 46.10 Fiber Cement Siding		Crew	Daily Output	Labor-Hours	Unit	Material	Labor	Equipment	Total	Total Incl O&P
0010	**FIBER CEMENT SIDING**									
0020	Lap siding, 5/16" thick, 6" wide, 4-3/4" exposure, smooth texture	2 Carp	415	.039	S.F.	1.29	1.60		2.89	3.88
0025	Woodgrain texture		415	.039		1.29	1.60		2.89	3.88
0030	7-1/2" wide, 6-1/4" exposure, smooth texture		425	.038		1.36	1.56		2.92	3.91
0035	Woodgrain texture		425	.038		1.36	1.56		2.92	3.91
0040	8" wide, 6-3/4" exposure, smooth texture		425	.038		1.51	1.56		3.07	4.07
0045	Roughsawn texture		425	.038		1.51	1.56		3.07	4.07
0050	9-1/2" wide, 8-1/4" exposure, smooth texture		440	.036		1.45	1.51		2.96	3.93
0055	Woodgrain texture		440	.036		1.45	1.51		2.96	3.93
0060	12" wide, 10-3/8" exposure, smooth texture	↓	455	.035	↓	1.34	1.46		2.80	3.72

07 46 Siding

07 46 46 – Mineral-Fiber Cement Siding

07 46 46.10 Fiber Cement Siding		Crew	Daily Output	Labor-Hours	Unit	Material	Labor	Equipment	Total	Total Incl O&P
						2010 Bare Costs				
0065	Woodgrain texture	2 Carp	455	.035	S.F.	1.34	1.46		2.80	3.72
0070	Panel siding, 5/16" thick, smooth texture		750	.021		1.12	.89		2.01	2.61
0075	Stucco texture		750	.021		1.12	.89		2.01	2.61
0080	Grooved woodgrain texture		750	.021		1.12	.89		2.01	2.61
0085	V - grooved woodgrain texture		750	.021		1.12	.89		2.01	2.61
0090	Wood starter strip		400	.040	L.F.	.53	1.66		2.19	3.14

07 46 73 – Soffit

07 46 73.10 Soffit Options		Crew	Daily Output	Labor-Hours	Unit	Material	Labor	Equipment	Total	Total Incl O&P
0010	**SOFFIT OPTIONS**									
0012	Aluminum, residential, .020" thick	1 Carp	210	.038	S.F.	1.62	1.58		3.20	4.22
0100	Baked enamel on steel, 16 or 18 gauge		105	.076		5.35	3.17		8.52	10.80
0300	Polyvinyl chloride, white, solid		230	.035		1.09	1.45		2.54	3.43
0400	Perforated		230	.035		1.09	1.45		2.54	3.43
0500	For colors, add					.13			.13	.14

07 51 Built-Up Bituminous Roofing

07 51 13 – Built-Up Asphalt Roofing

07 51 13.10 Built-Up Roofing Components		Crew	Daily Output	Labor-Hours	Unit	Material	Labor	Equipment	Total	Total Incl O&P
0010	**BUILT-UP ROOFING COMPONENTS**									
0012	Asphalt saturated felt, #30, 2 square per roll	1 Rofc	58	.138	Sq.	11.95	4.88		16.83	21.50
0200	#15, 4 sq per roll, plain or perforated, not mopped		58	.138		5.90	4.88		10.78	14.55
0300	Roll roofing, smooth, #65		15	.533		9.65	18.90		28.55	42
0500	#90		15	.533		34.50	18.90		53.40	69.50
0520	Mineralized		15	.533		34.50	18.90		53.40	69.50
0540	D.C. (Double coverage), 19" selvage edge		10	.800		44	28.50		72.50	95
0580	Adhesive (lap cement)				Gal.	4.79			4.79	5.25
0600	Steep, flat or dead level asphalt, 10 ton lots, bulk				Ton	490			490	540
0800	Packaged				"	830			830	915

07 51 13.13 Cold-Applied Built-Up Asphalt Roofing		Crew	Daily Output	Labor-Hours	Unit	Material	Labor	Equipment	Total	Total Incl O&P
0010	**COLD-APPLIED BUILT-UP ASPHALT ROOFING**									
0020	3 ply system, installation only (components listed below)	G-5	50	.800	Sq.		25.50	3.77	29.27	46.50
0100	Spunbond poly. fabric, 1.35 oz/SY, 36"W, 10.8 Sq/roll				Ea.	175			175	193
0200	49" wide, 14.6 Sq./roll					244			244	268
0300	2.10 oz./S.Y., 36" wide, 10.8 Sq./roll					266			266	293
0400	49" wide, 14.6 Sq./roll					360			360	395
0500	Base & finish coat, 3 gal./Sq., 5 gal./can				Gal.	5			5	5.50
0600	Coating, ceramic granules, 1/2 Sq./bag				Ea.	16.75			16.75	18.45
0700	Aluminum, 2 gal./Sq.				Gal.	13.50			13.50	14.85
0800	Emulsion, fibered or non-fibered, 4 gal./Sq.				"	4.70			4.70	5.15

07 51 13.20 Built-Up Roofing Systems		Crew	Daily Output	Labor-Hours	Unit	Material	Labor	Equipment	Total	Total Incl O&P
0010	**BUILT-UP ROOFING SYSTEMS** R075113-20									
0120	Asphalt flood coat with gravel/slag surfacing, not including									
0140	Insulation, flashing or wood nailers									
0200	Asphalt base sheet, 3 plies #15 asphalt felt, mopped	G-1	22	2.545	Sq.	94	84	21.50	199.50	267
0350	On nailable decks		21	2.667		98	88	22.50	208.50	280
0500	4 plies #15 asphalt felt, mopped		20	2.800		130	92.50	24	246.50	325
0550	On nailable decks		19	2.947		115	97.50	25	237.50	315
0700	Coated glass base sheet, 2 plies glass (type IV), mopped		22	2.545		91.50	84	21.50	197	265
0850	3 plies glass, mopped		20	2.800		110	92.50	24	226.50	300
0950	On nailable decks		19	2.947		103	97.50	25	225.50	305

07 51 13 – Built-Up Asphalt Roofing

07 51 13.20 Built-Up Roofing Systems		Crew	Daily Output	Labor-Hours	Unit	Material	Labor	Equipment	Total	Total Incl O&P
						2010 Bare Costs				
1100	4 plies glass fiber felt (type IV), mopped	G-1	20	2.800	Sq.	135	92.50	24	251.50	330
1150	On nailable decks		19	2.947		121	97.50	25	243.50	325
1200	Coated & saturated base sheet, 3 plies #15 asph. felt, mopped		20	2.800		104	92.50	24	220.50	294
1250	On nailable decks		19	2.947		96.50	97.50	25	219	296
1300	4 plies #15 asphalt felt, mopped		22	2.545		121	84	21.50	226.50	297
2000	Asphalt flood coat, smooth surface									
2200	Asphalt base sheet & 3 plies #15 asphalt felt, mopped	G-1	24	2.333	Sq.	100	77	19.85	196.85	260
2400	On nailable decks		23	2.435		92.50	80.50	20.50	193.50	259
2600	4 plies #15 asphalt felt, mopped		24	2.333		117	77	19.85	213.85	279
2700	On nailable decks		23	2.435		110	80.50	20.50	211	278
2900	Coated glass fiber base sheet, mopped, and 2 plies of									
2910	glass fiber felt (type IV)	G-1	25	2.240	Sq.	86.50	74	19.05	179.55	239
3100	On nailable decks		24	2.333		81	77	19.85	177.85	240
3200	3 plies, mopped		23	2.435		104	80.50	20.50	205	272
3300	On nailable decks		22	2.545		97.50	84	21.50	203	271
3800	4 plies glass fiber felt (type IV), mopped		23	2.435		123	80.50	20.50	224	292
3900	On nailable decks		22	2.545		116	84	21.50	221.50	291
4000	Coated & saturated base sheet, 3 plies #15 asph. felt, mopped		24	2.333		98.50	77	19.85	195.35	258
4200	On nailable decks		23	2.435		91.50	80.50	20.50	192.50	257
4300	4 plies #15 organic felt, mopped		22	2.545		116	84	21.50	221.50	291
4500	Coal tar pitch with gravel/slag surfacing									
4600	4 plies #15 tarred felt, mopped	G-1	21	2.667	Sq.	183	88	22.50	293.50	375
4800	3 plies glass fiber felt (type IV), mopped	"	19	2.947	"	150	97.50	25	272.50	355
5000	Coated glass fiber base sheet, and 2 plies of									
5010	glass fiber felt, (type IV), mopped	G-1	19	2.947	Sq.	148	97.50	25	270.50	355
5300	On nailable decks		18	3.111		130	103	26.50	259.50	345
5600	4 plies glass fiber felt (type IV), mopped		21	2.667		206	88	22.50	316.50	400
5800	On nailable decks		20	2.800		188	92.50	24	304.50	385

07 51 13.30 Cants		Crew	Daily Output	Labor-Hours	Unit	Material	Labor	Equipment	Total	Total Incl O&P
0010	**CANTS**									
0012	Lumber, treated, 4" x 4" cut diagonally	1 Rofc	325	.025	L.F.	1.52	.87		2.39	3.12
0100	Foamglass		325	.025		2.23	.87		3.10	3.90
0300	Mineral or fiber, trapezoidal, 1" x 4" x 48"		325	.025		.19	.87		1.06	1.66
0400	1-1/2" x 5-5/8" x 48"		325	.025		.31	.87		1.18	1.79

07 51 13.40 Felts		Crew	Daily Output	Labor-Hours	Unit	Material	Labor	Equipment	Total	Total Incl O&P
0010	**FELTS**									
0012	Glass fibered roofing felt, #15, not mopped	1 Rofc	58	.138	Sq.	6.55	4.88		11.43	15.30
0300	Base sheet, #45, channel vented		58	.138		26	4.88		30.88	36.50
0400	#50, coated		58	.138		14	4.88		18.88	23.50
0500	Cap, mineral surfaced		58	.138		34.50	4.88		39.38	46
0600	Flashing membrane, #65		16	.500		9.65	17.70		27.35	40
0800	Coal tar fibered, #15, no mopping		58	.138		8	4.88		12.88	16.90
0900	Asphalt felt, #15, 4 sq per roll, no mopping **CN**		58	.138		5.90	4.88		10.78	14.55
1100	#30, 2 sq per roll		58	.138		11.95	4.88		16.83	21.50
1200	Double coated, #33		58	.138		9.90	4.88		14.78	19
1400	#40, base sheet		58	.138		11.30	4.88		16.18	20.50
1450	Coated and saturated		58	.138		9.90	4.88		14.78	18.95
1500	Tarred felt, organic, #15, 4 sq rolls		58	.138		15.60	4.88		20.48	25.50
1550	#30, 2 sq roll		58	.138		22.50	4.88		27.38	33
1700	Add for mopping above felts, per ply, asphalt, 24 lb per sq	G-1	192	.292		9.95	9.65	2.48	22.08	29.50
1800	Coal tar mopping, 30 lb per sq		186	.301		15.30	9.95	2.56	27.81	36
1900	Flood coat, with asphalt, 60 lb per sq		60	.933		25	31	7.95	63.95	88

07 51 Built-Up Bituminous Roofing

07 51 13 – Built-Up Asphalt Roofing

07 51 13.40 Felts		Crew	Daily Output	Labor-Hours	Unit	Material	Labor (2010 Bare Costs)	Equipment	Total	Total Incl O&P
2000	With coal tar, 75 lb per sq	G-1	56	1	Sq.	38	33	8.50	79.50	106

07 51 13.50 Walkways for Built-Up Roofs

		Crew	Daily Output	Labor-Hours	Unit	Material	Labor	Equipment	Total	Total Incl O&P
0010	**WALKWAYS FOR BUILT-UP ROOFS**									
0020	Asphalt impregnated, 3′ x 6′ x 1/2″ thick	1 Rofc	400	.020	S.F.	1.70	.71		2.41	3.05
0100	3′ x 3′ x 3/4″ thick	"	400	.020		3.13	.71		3.84	4.62
0300	Concrete patio blocks, 2″ thick, natural	1 Clab	115	.070		3.71	2.30		6.01	7.65
0400	Colors	"	115	.070		3.75	2.30		6.05	7.70

07 52 Modified Bituminous Membrane Roofing

07 52 13 – Atactic-Polypropylene-Modified Bituminous Membrane Roofing

07 52 13.10 APP Modified Bituminous Membrane

		Crew	Daily Output	Labor-Hours	Unit	Material	Labor	Equipment	Total	Total Incl O&P
0010	**APP MODIFIED BITUMINOUS MEMBRANE** R075213-30									
0020	Base sheet, #15 glass fiber felt, nailed to deck	1 Rofc	58	.138	Sq.	7.45	4.88		12.33	16.30
0030	Spot mopped to deck	G-1	295	.190		11.55	6.30	1.61	19.46	25
0040	Fully mopped to deck	"	192	.292		16.50	9.65	2.48	28.63	37
0050	#15 organic felt, nailed to deck	1 Rofc	58	.138		6.80	4.88		11.68	15.60
0060	Spot mopped to deck	G-1	295	.190		10.85	6.30	1.61	18.76	24
0070	Fully mopped to deck	"	192	.292		15.85	9.65	2.48	27.98	36
2100	APP mod., smooth surf. cap sheet, poly. reinf., torched, 160 mils	G-5	2100	.019	S.F.	.48	.61	.09	1.18	1.65
2150	170 mils		2100	.019		.55	.61	.09	1.25	1.73
2200	Granule surface cap sheet, poly. reinf., torched, 180 mils		2000	.020		.59	.64	.09	1.32	1.82
2250	Smooth surface flashing, torched, 160 mils		1260	.032		.48	1.02	.15	1.65	2.39
2300	170 mils		1260	.032		.55	1.02	.15	1.72	2.47
2350	Granule surface flashing, torched, 180 mils		1260	.032		.59	1.02	.15	1.76	2.51
2400	Fibrated aluminum coating	1 Rofc	3800	.002		.10	.07		.17	.23

07 52 16 – Styrene-Butadiene-Styrene Modified Bituminous Membrane Roofing

07 52 16.10 SBS Modified Bituminous Membrane

		Crew	Daily Output	Labor-Hours	Unit	Material	Labor	Equipment	Total	Total Incl O&P
0010	**SBS MODIFIED BITUMINOUS MEMBRANE**									
0080	SBS modified, granule surf cap sheet, polyester rein., mopped									
0600	150 mils	G-1	2000	.028	S.F.	.58	.93	.24	1.75	2.44
1100	160 mils		2000	.028		.77	.93	.24	1.94	2.65
1500	Glass fiber reinforced, mopped, 160 mils		2000	.028		.49	.93	.24	1.66	2.34
1600	Smooth surface cap sheet, mopped, 145 mils		2100	.027		.49	.88	.23	1.60	2.26
1700	Smooth surface flashing, 145 mils		1260	.044		.49	1.47	.38	2.34	3.40
1800	150 mils		1260	.044		.48	1.47	.38	2.33	3.39
1900	Granular surface flashing, 150 mils		1260	.044		.58	1.47	.38	2.43	3.50
2000	160 mils		1260	.044		.77	1.47	.38	2.62	3.71

07 53 Elastomeric Membrane Roofing

07 53 16 – Chlorosulfonate Polyethylene Roofing

07 53 16.10 Chlorosulfonated Polyethylene Roofing

		Crew	Daily Output	Labor-Hours	Unit	Material	Labor	Equipment	Total	Total Incl O&P
0010	**CHLOROSULFONATED POLYETHYLENE ROOFING**									
0800	Chlorosulfonated polyethylene-hypalon (CSPE), 45 mils,									
0900	0.29 P.S.F., fully adhered	G-5	26	1.538	Sq.	185	49.50	7.25	241.75	294
1100	Loose-laid & ballasted with stone (10 P.S.F.)		51	.784		193	25	3.69	221.69	258
1200	Mechanically attached		35	1.143		186	36.50	5.40	227.90	272
1300	Plates with adhesive attachment		35	1.143		183	36.50	5.40	224.90	268

07 53 Elastomeric Membrane Roofing

07 53 23 – Ethylene-Propylene-Diene-Monomer Roofing

07 53 23.20 Ethylene-Propylene-Diene-Monomer Roofing		Crew	Daily Output	Labor-Hours	Unit	2010 Bare Costs Material	Labor	Equipment	Total	Total Incl O&P
0010	**ETHYLENE-PROPYLENE-DIENE-MONOMER ROOFING (E.P.D.M.)**									
3500	Ethylene-propylene-diene-monomer (EPDM), 45 mils, 0.28 P.S.F.									
3600	Loose-laid & ballasted with stone (10 P.S.F.)	G-5	51	.784	Sq.	76	25	3.69	104.69	130
3700	Mechanically attached		35	1.143		68.50	36.50	5.40	110.40	142
3800	Fully adhered with adhesive		26	1.538		104	49.50	7.25	160.75	204
4500	60 mils, 0.40 P.S.F.									
4600	Loose-laid & ballasted with stone (10 P.S.F.)	G-5	51	.784	Sq.	90	25	3.69	118.69	145
4700	Mechanically attached		35	1.143		81.50	36.50	5.40	123.40	156
4800	Fully adhered with adhesive		26	1.538		117	49.50	7.25	173.75	218
4810	45 mil, .28 PSF, membrane only **CN**					43			43	47.50
4820	60 mil, .40 PSF, membrane only					55			55	60.50
4850	Seam tape for membrane, 3" x 100' roll				Ea.	48			48	53
4900	Batten strips, 10' sections					2.80			2.80	3.08
4910	Cover tape for batten strips, 6" x 100' roll					110			110	121
4930	Plate anchors				M	105			105	116
4970	Adhesive for fully adhered systems, 60 S.F./gal.				Gal.	21			21	23

07 53 29 – Polyisobutylene Roofing

07 53 29.10 Polyisobutylene Roofing

		Crew	Daily Output	Labor-Hours	Unit	Material	Labor	Equipment	Total	Total Incl O&P
0010	**POLYISOBUTYLENE ROOFING**									
7500	Polyisobutylene (PIB), 100 mils, 0.57 P.S.F.									
7600	Loose-laid & ballasted with stone/gravel (10 P.S.F.)	G-5	51	.784	Sq.	181	25	3.69	209.69	245
7700	Partially adhered with adhesive		35	1.143		227	36.50	5.40	268.90	315
7800	Hot asphalt attachment		35	1.143		217	36.50	5.40	258.90	305
7900	Fully adhered with contact cement		26	1.538		234	49.50	7.25	290.75	350

07 54 Thermoplastic Membrane Roofing

07 54 19 – Polyvinyl-Chloride Roofing

07 54 19.10 Polyvinyl-Chloride Roofing (P.V.C.)

		Crew	Daily Output	Labor-Hours	Unit	Material	Labor	Equipment	Total	Total Incl O&P
0010	**POLYVINYL-CHLORIDE ROOFING (P.V.C.)**									
8200	Heat welded seams									
8700	Reinforced, 48 mils, 0.33 P.S.F.									
8750	Loose-laid & ballasted with stone/gravel (12 P.S.F.)	G-5	51	.784	Sq.	97	25	3.69	125.69	153
8800	Mechanically attached		35	1.143		90	36.50	5.40	131.90	166
8850	Fully adhered with adhesive		26	1.538		118	49.50	7.25	174.75	220
8860	Reinforced, 60 mils, .40 P.S.F.									
8870	Loose-laid & ballasted with stone/gravel (12 P.S.F.)	G-5	51	.784	Sq.	96.50	25	3.69	125.19	152
8880	Mechanically attached		35	1.143		89.50	36.50	5.40	131.40	165
8890	Fully adhered with adhesive		26	1.538		118	49.50	7.25	174.75	219

07 54 23 – Thermoplastic Polyolefin Roofing

07 54 23.10 Thermoplastic Polyolefin Roofing (T.P.O)

		Crew	Daily Output	Labor-Hours	Unit	Material	Labor	Equipment	Total	Total Incl O&P
0010	**THERMOPLASTIC POLYOLEFIN ROOFING (T.P.O.)**									
0100	45 mils, loose laid & ballasted with stone(1/2 ton/sq.)	G-5	51	.784	Sq.	90.50	25	3.69	119.19	146
0120	Fully adhered		25	1.600		80.50	51.50	7.55	139.55	182
0140	Mechanically attached		34	1.176		80	38	5.55	123.55	157
0160	Self adhered		35	1.143		88	36.50	5.40	129.90	164
0180	60 mil membrane, heat welded seams, ballasted		50	.800		107	25.50	3.77	136.27	165
0200	Fully adhered		25	1.600		96	51.50	7.55	155.05	200
0220	Mechanically attached		34	1.176		99	38	5.55	142.55	178
0240	Self adhered		35	1.143		109	36.50	5.40	150.90	187

07 56 Fluid-Applied Roofing

07 56 10 – Fluid-Applied Roofing Elastomers

07 56 10.10 Elastomeric Roofing		Crew	Daily Output	Labor-Hours	Unit	Material	2010 Bare Costs Labor	Equipment	Total	Total Incl O&P
0010	**ELASTOMERIC ROOFING**									
0110	Acrylic rubber, fluid applied, 20 mils thick	G-5	2000	.020	S.F.	2.04	.64	.09	2.77	3.41
0120	50 mils, reinforced		1200	.033		3.16	1.07	.16	4.39	5.45
0130	For walking surface, add		900	.044		.96	1.43	.21	2.60	3.66
0300	Hypalon neoprene, fluid applied, 20 mil thick, not-reinforced	G-1	1135	.049		2.57	1.63	.42	4.62	6
0600	Non-woven polyester, reinforced		960	.058		2.61	1.93	.50	5.04	6.65
0700	5 coat neoprene deck, 60 mil thick, under 10,000 SF		325	.172		5.75	5.70	1.46	12.91	17.35
0900	Over 10,000 SF		625	.090		5.35	2.96	.76	9.07	11.65
1300	Vinyl plastic traffic deck, sprayed, 2 to 4 mils thick		625	.090		1.68	2.96	.76	5.40	7.60
1500	Vinyl and neoprene membrane traffic deck		1550	.036		1.77	1.19	.31	3.27	4.28

07 57 Coated Foamed Roofing

07 57 13 – Sprayed Polyurethane Foam Roofing

07 57 13.10 Sprayed Polyurethane Foam Roofing (S.P.F.)		Crew	Daily Output	Labor-Hours	Unit	Material	Labor	Equipment	Total	Total Incl O&P
0010	**SPRAYED POLYURETHANE FOAM ROOFING (S.P.F.)**									
0100	Primer for metal substrate (when required)	G-2A	3000	.008	S.F.	.42	.25	.23	.90	1.12
0200	Primer for non-metal substrate (when required)		3000	.008		.16	.25	.23	.64	.84
0300	Closed cell spray, polyurethane foam, 3 lbs per CF density, 1", R6.7		15000	.002		.55	.05	.05	.65	.74
0400	2", R13.4		13125	.002		1.10	.06	.05	1.21	1.36
0500	3", R18.6		11485	.002		1.66	.07	.06	1.79	2
0550	4", R24.8		10080	.002		2.21	.08	.07	2.36	2.63
0700	Spray-on silicone coating		2500	.010		.92	.30	.28	1.50	1.80
0800	Warranty 5-20 year manufacturer's									.15
0900	Warranty 20 year, no dollar limit									.20

07 58 Roll Roofing

07 58 10 – Asphalt Roll Roofing

07 58 10.10 Roll Roofing		Crew	Daily Output	Labor-Hours	Unit	Material	Labor	Equipment	Total	Total Incl O&P
0010	**ROLL ROOFING**									
0100	Asphalt, mineral surface									
0200	1 ply #15 organic felt, 1 ply mineral surfaced									
0300	Selvage roofing, lap 19", nailed & mopped	G-1	27	2.074	Sq.	64	68.50	17.65	150.15	203
0400	3 plies glass fiber felt (type IV), 1 ply mineral surfaced									
0500	Selvage roofing, lapped 19", mopped	G-1	25	2.240	Sq.	103	74	19.05	196.05	258
0600	Coated glass fiber base sheet, 2 plies of glass fiber									
0700	Felt (type IV), 1 ply mineral surfaced selvage									
0800	Roofing, lapped 19", mopped	G-1	25	2.240	Sq.	111	74	19.05	204.05	266
0900	On nailable decks	"	24	2.333	"	101	77	19.85	197.85	261
1000	3 plies glass fiber felt (type III), 1 ply mineral surfaced									
1100	Selvage roofing, lapped 19", mopped	G-1	25	2.240	Sq.	103	74	19.05	196.05	258

07 61 Sheet Metal Roofing

07 61 13 – Standing Seam Sheet Metal Roofing

07 61 13.10 Standing Seam Sheet Metal Roofing, Field Fab.		Crew	Daily Output	Labor-Hours	Unit	2010 Bare Costs Material	Labor	Equipment	Total	Total Incl O&P
0010	**STANDING SEAM SHEET METAL ROOFING, FIELD FABRICATED**									
0400	Copper standing seam roofing, over 10 squares, 16 oz, 125 lb per sq	1 Shee	1.30	6.154	Sq.	1,075	300		1,375	1,650
0600	18 oz, 140 lb per sq		1.20	6.667		1,200	325		1,525	1,800
0700	20 oz, 150 lb per sq		1.10	7.273		1,300	355		1,655	1,975
1200	For abnormal conditions or small areas, add					25%	100%			
1300	For lead-coated copper, add					25%				

07 61 16 – Batten Seam Sheet Metal Roofing

07 61 16.10 Batten Seam Sheet Metal Roofing, Field Fabricated		Crew	Daily Output	Labor-Hours	Unit	Material	Labor	Equipment	Total	Total Incl O&P
0010	**BATTEN SEAM SHEET METAL ROOFING, FIELD FABRICATED**									
0012	Copper batten seam roofing, over 10 sq, 16 oz, 130 lb per sq	1 Shee	1.10	7.273	Sq.	1,075	355		1,430	1,750
0020	Lead batten seam roofing, 5 lb. per SF		1.20	6.667		600	325		925	1,150
0100	Zinc/copper alloy batten seam roofing, .020 thick		1.20	6.667		1,075	325		1,400	1,700
0200	Copper roofing, batten seam, over 10 sq, 18 oz, 145 lb per sq		1	8		1,200	395		1,595	1,900
0300	20 oz, 160 lb per sq		1	8		1,300	395		1,695	2,025
0500	Stainless steel batten seam roofing, type 304, 28 gauge		1.20	6.667		690	325		1,015	1,250
0600	26 gauge		1.15	6.957		695	340		1,035	1,275
0800	Zinc, copper alloy roofing, batten seam, .027" thick		1.15	6.957		1,550	340		1,890	2,225
0900	.032" thick		1.10	7.273		1,825	355		2,180	2,550
1000	.040" thick		1.05	7.619		2,200	375		2,575	2,975

07 61 19 – Flat Seam Sheet Metal Roofing

07 61 19.10 Flat Seam Sheet Metal Roofing, Field Fabricated		Crew	Daily Output	Labor-Hours	Unit	Material	Labor	Equipment	Total	Total Incl O&P
0010	**FLAT SEAM SHEET METAL ROOFING, FIELD FABRICATED**									
0900	Copper flat seam roofing, over 10 squares, 16 oz, 115 lb per sq	1 Shee	1.20	6.667	Sq.	1,075	325		1,400	1,700
1000	20 oz, 145 lb per sq		1.10	7.273		1,300	355		1,655	1,975
1100	Lead flat seam roofing, 5 lb per SF		1.30	6.154		600	300		900	1,125

07 65 Flexible Flashing

07 65 10 – Sheet Metal Flashing

07 65 10.10 Sheet Metal Flashing and Counter Flashing		Crew	Daily Output	Labor-Hours	Unit	Material	Labor	Equipment	Total	Total Incl O&P
0010	**SHEET METAL FLASHING AND COUNTER FLASHING**									
0011	Including up to 4 bends									
0020	Aluminum, mill finish, .013" thick	1 Rofc	145	.055	S.F.	.67	1.95		2.62	3.99
0030	.016" thick		145	.055		.79	1.95		2.74	4.12
0060	.019" thick		145	.055		1.02	1.95		2.97	4.37
0100	.032" thick		145	.055		1.48	1.95		3.43	4.88
0200	.040" thick		145	.055		2.21	1.95		4.16	5.70
0300	.050" thick		145	.055		2.52	1.95		4.47	6
0325	Mill finish 5" x 7" step flashing, .016" thick		1920	.004	Ea.	.16	.15		.31	.43
0350	Mill finish 12" x 12" step flashing, .016" thick		1600	.005	"	.61	.18		.79	.96
0400	Painted finish, add				S.F.	.37			.37	.41
1000	Mastic-coated 2 sides, .005" thick	1 Rofc	330	.024		1.67	.86		2.53	3.27
1100	.016" thick		330	.024		1.84	.86		2.70	3.45
1600	Copper, 16 oz, sheets, under 1000 lbs.		115	.070		7.55	2.46		10.01	12.40
1700	Over 4000 lbs.		155	.052		6.75	1.83		8.58	10.50
1900	20 oz sheets, under 1000 lbs.		110	.073		9.05	2.57		11.62	14.25
2000	Over 4000 lbs.		145	.055		8.15	1.95		10.10	12.20
2200	24 oz sheets, under 1000 lbs.		105	.076		11.30	2.70		14	16.95
2300	Over 4000 lbs.		135	.059		10.15	2.10		12.25	14.70
2500	32 oz sheets, under 1000 lbs.		100	.080		14.10	2.83		16.93	20
2600	Over 4000 lbs.		130	.062		12.70	2.18		14.88	17.55

07 65 Flexible Flashing

07 65 10 – Sheet Metal Flashing

07 65 10.10 Sheet Metal Flashing and Counter Flashing		Crew	Daily Output	Labor-Hours	Unit	2010 Bare Costs Material	Labor	Equipment	Total	Total Incl O&P
2700	W shape for valleys, 16 oz, 24" wide	1 Rofc	100	.080	L.F.	13.50	2.83		16.33	19.55
5800	Lead, 2.5 lb. per SF, up to 12" wide		135	.059	S.F.	4.55	2.10		6.65	8.50
5900	Over 12" wide		135	.059		4.55	2.10		6.65	8.50
8650	Copper, 16 oz		100	.080		6	2.83		8.83	11.30
8900	Stainless steel sheets, 32 ga, .010" thick		155	.052		3.79	1.83		5.62	7.20
9000	28 ga, .015" thick		155	.052		4.70	1.83		6.53	8.20
9100	26 ga, .018" thick		155	.052		5.70	1.83		7.53	9.30
9200	24 ga, .025" thick		155	.052		7.40	1.83		9.23	11.20
9290	For mechanically keyed flashing, add					40%				
9320	Steel sheets, galvanized, 20 gauge	1 Rofc	130	.062	S.F.	1.19	2.18		3.37	4.93
9340	30 gauge		160	.050		.50	1.77		2.27	3.49
9400	Terne coated stainless steel, .015" thick, 28 ga		155	.052		7.35	1.83		9.18	11.15
9500	.018" thick, 26 ga		155	.052		8.30	1.83		10.13	12.15
9600	Zinc and copper alloy (brass), .020" thick		155	.052		5.15	1.83		6.98	8.70
9700	.027" thick		155	.052		6.90	1.83		8.73	10.65
9800	.032" thick		155	.052		8.05	1.83		9.88	11.90
9900	.040" thick		155	.052		9.85	1.83		11.68	13.85

07 65 12 – Fabric and Mastic Flashings

07 65 12.10 Fabric and Mastic Flashing and Counter Flashing		Crew	Daily Output	Labor-Hours	Unit	Material	Labor	Equipment	Total	Total Incl O&P
0010	**FABRIC AND MASTIC FLASHING AND COUNTER FLASHING**									
1300	Asphalt flashing cement, 5 gallon				Gal.	10.80			10.80	11.85
4900	Fabric, asphalt-saturated cotton, specification grade	1 Rofc	35	.229	S.Y.	2.37	8.10		10.47	16.05
5000	Utility grade		35	.229		1.50	8.10		9.60	15.10
5200	Open-mesh fabric, saturated, 40 oz per S.Y.		35	.229		1.67	8.10		9.77	15.30
5300	Close-mesh fabric, saturated, 17 oz per S.Y.		35	.229		1.75	8.10		9.85	15.40
5500	Fiberglass, resin-coated		35	.229		1.43	8.10		9.53	15
5600	Asphalt-coated, 40 oz per S.Y.		35	.229		9.75	8.10		17.85	24
8500	Shower pan, bituminous membrane, 7 oz		155	.052	S.F.	2	1.83		3.83	5.25

07 65 13 – Laminated Sheet Flashing

07 65 13.10 Laminated Sheet Flashing		Crew	Daily Output	Labor-Hours	Unit	Material	Labor	Equipment	Total	Total Incl O&P
0010	**LAMINATED SHEET FLASHING**, Including up to 4 bends									
0500	Fabric-backed 2 sides, .004" thick	1 Rofc	330	.024	S.F.	1.41	.86		2.27	2.98
0700	.005" thick		330	.024		1.67	.86		2.53	3.27
0750	Mastic-backed, self adhesive		460	.017		3.46	.62		4.08	4.83
0800	Mastic-coated 2 sides, .004" thick		330	.024		1.41	.86		2.27	2.98
2800	Copper, paperbacked 1 side, 2 oz		330	.024		1.42	.86		2.28	2.99
2900	3 oz		330	.024		1.85	.86		2.71	3.47
3100	Paperbacked 2 sides, 2 oz		330	.024		1.43	.86		2.29	3
3150	3 oz		330	.024		1.84	.86		2.70	3.45
3200	5 oz		330	.024		2.76	.86		3.62	4.47
3250	7 oz		330	.024		4.50	.86		5.36	6.40
3400	Mastic-backed 2 sides, copper, 2 oz		330	.024		1.69	.86		2.55	3.29
3500	3 oz		330	.024		2.08	.86		2.94	3.72
3700	5 oz		330	.024		3.05	.86		3.91	4.79
3800	Fabric-backed 2 sides, copper, 2 oz		330	.024		1.79	.86		2.65	3.40
4000	3 oz		330	.024		2.31	.86		3.17	3.97
4100	5 oz		330	.024		3.14	.86		4	4.88
4300	Copper-clad stainless steel, .015" thick, under 500 lbs.		115	.070		5.25	2.46		7.71	9.85
4400	Over 2000 lbs.		155	.052		5.05	1.83		6.88	8.60
4600	.018" thick, under 500 lbs.		100	.080		6.90	2.83		9.73	12.30
4700	Over 2000 lbs.		145	.055		5.05	1.95		7	8.80
6100	Lead-coated copper, fabric-backed, 2 oz		330	.024		2.42	.86		3.28	4.09

07 65 Flexible Flashing

07 65 13 – Laminated Sheet Flashing

07 65 13.10 Laminated Sheet Flashing		Crew	Daily Output	Labor-Hours	Unit	Material	2010 Bare Costs Labor	Equipment	Total	Total Incl O&P
6200	5 oz	1 Rofc	330	.024	S.F.	2.78	.86		3.64	4.49
6400	Mastic-backed 2 sides, 2 oz		330	.024		1.89	.86		2.75	3.51
6500	5 oz		330	.024		2.34	.86		3.20	4
6700	Paperbacked 1 side, 2 oz		330	.024		1.63	.86		2.49	3.22
6800	3 oz		330	.024		1.92	.86		2.78	3.54
7000	Paperbacked 2 sides, 2 oz		330	.024		1.69	.86		2.55	3.29
7100	5 oz		330	.024		2.75	.86		3.61	4.46
8550	3 ply copper and fabric, 3 oz		155	.052		2.31	1.83		4.14	5.60
8600	7 oz		155	.052		3.25	1.83		5.08	6.60
8700	Lead on copper and fabric, 5 oz		155	.052		2.78	1.83		4.61	6.10
8800	7 oz		155	.052		5.25	1.83		7.08	8.85
9300	Stainless steel, paperbacked 2 sides, .005" thick	↓	330	.024	↓	3.34	.86		4.20	5.10

07 65 19 – Plastic Sheet Flashing

07 65 19.10 Plastic Sheet Flashing and Counter Flashing		Crew	Daily Output	Labor-Hours	Unit	Material	Labor	Equipment	Total	Total Incl O&P
0010	**PLASTIC SHEET FLASHING AND COUNTER FLASHING**									
7300	Polyvinyl chloride, black, .010" thick	1 Rofc	285	.028	S.F.	.23	.99		1.22	1.90
7400	.020" thick		285	.028		.33	.99		1.32	2.01
7600	.030" thick		285	.028		.42	.99		1.41	2.11
7700	.056" thick		285	.028		1.03	.99		2.02	2.78
7900	Black or white for exposed roofs, .060" thick	↓	285	.028	↓	2.25	.99		3.24	4.13
8060	PVC tape, 5" x 45 mils, for joint covers, 100 L.F./roll				Ea.	120			120	132
8850	Polyvinyl chloride, .030" thick	1 Rofc	160	.050	S.F.	.46	1.77		2.23	3.45

07 65 23 – Rubber Sheet Flashing

07 65 23.10 Rubber Sheet Flashing and Counterflashing		Crew	Daily Output	Labor-Hours	Unit	Material	Labor	Equipment	Total	Total Incl O&P
0010	**RUBBER SHEET FLASHING AND COUNTERFLASHING**									
8100	Rubber, butyl, 1/32" thick	1 Rofc	285	.028	S.F.	1.07	.99		2.06	2.83
8200	1/16" thick		285	.028		1.57	.99		2.56	3.38
8300	Neoprene, cured, 1/16" thick		285	.028		2.03	.99		3.02	3.88
8400	1/8" thick	↓	285	.028	↓	4.09	.99		5.08	6.15

07 71 Roof Specialties

07 71 19 – Manufactured Gravel Stops and Fascias

07 71 19.10 Gravel Stop		Crew	Daily Output	Labor-Hours	Unit	Material	Labor	Equipment	Total	Total Incl O&P
0010	**GRAVEL STOP**									
0020	Aluminum, .050" thick, 4" face height, mill finish	1 Shee	145	.055	L.F.	6.10	2.71		8.81	10.85
0080	Duranodic finish		145	.055		5.90	2.71		8.61	10.60
0100	Painted		145	.055		6.85	2.71		9.56	11.60
0300	6" face height		135	.059		6.55	2.91		9.46	11.65
0350	Duranodic finish		135	.059		6.95	2.91		9.86	12.05
0400	Painted		135	.059		8.10	2.91		11.01	13.30
0600	8" face height		125	.064		8.20	3.14		11.34	13.80
0650	Duranodic finish		125	.064		8	3.14		11.14	13.55
0700	Painted		125	.064		8.15	3.14		11.29	13.70
0900	12" face height, .080 thick, 2 piece		100	.080		10.95	3.93		14.88	18
0950	Duranodic finish		100	.080		10.05	3.93		13.98	17
1000	Painted		100	.080		11.90	3.93		15.83	19.05
1350	Galv steel, 24 ga., 4" leg, plain, with continuous cleat, 4" face		145	.055		2.41	2.71		5.12	6.75
1360	6" face height		145	.055		3.64	2.71		6.35	8.10
1500	Polyvinyl chloride, 6" face height		135	.059		4.45	2.91		7.36	9.30
1600	9" face height	↓	125	.064	↓	5.25	3.14		8.39	10.55

07 71 Roof Specialties

07 71 19 – Manufactured Gravel Stops and Fascias

07 71 19.10 Gravel Stop		Crew	Daily Output	Labor-Hours	Unit	Material	Labor	Equipment	Total	Total Incl O&P
						2010 Bare Costs				
1800	Stainless steel, 24 ga., 6″ face height	1 Shee	135	.059	L.F.	11.45	2.91		14.36	17
1900	12″ face height		100	.080		24	3.93		27.93	32.50
2100	20 ga., 6″ face height		135	.059		13	2.91		15.91	18.70
2200	12″ face height		100	.080		27.50	3.93		31.43	36

07 71 19.30 Fascia

Code	Description	Crew	Daily Output	Labor-Hours	Unit	Material	Labor	Equipment	Total	Total Incl O&P
0010	**FASCIA**									
0100	Aluminum, reverse board and batten, .032″ thick, colored, no furring incl	1 Shee	145	.055	S.F.	5.95	2.71		8.66	10.65
0300	Steel, galv and enameled, stock, no furring, long panels		145	.055		4	2.71		6.71	8.50
0600	Short panels		115	.070		5.45	3.42		8.87	11.15

07 71 23 – Manufactured Gutters and Downspouts

07 71 23.10 Downspouts

Code	Description	Crew	Daily Output	Labor-Hours	Unit	Material	Labor	Equipment	Total	Total Incl O&P
0010	**DOWNSPOUTS**									
0020	Aluminum 2″ x 3″, .020″ thick, embossed	1 Shee	190	.042	L.F.	1.04	2.07		3.11	4.27
0100	Enameled		190	.042		1.59	2.07		3.66	4.88
0300	Enameled, .024″ thick, 2″ x 3″		180	.044		1.84	2.18		4.02	5.30
0400	3″ x 4″		140	.057		2.16	2.81		4.97	6.65
0600	Round, corrugated aluminum, 3″ diameter, .020″ thick		190	.042		1.42	2.07		3.49	4.69
0700	4″ diameter, .025″ thick		140	.057		2.39	2.81		5.20	6.90
0900	Wire strainer, round, 2″ diameter		155	.052	Ea.	3.32	2.53		5.85	7.50
1000	4″ diameter		155	.052		6.10	2.53		8.63	10.60
1200	Rectangular, perforated, 2″ x 3″		145	.055		2.30	2.71		5.01	6.65
1300	3″ x 4″		145	.055		3.32	2.71		6.03	7.75
1500	Copper, round, 16 oz., stock, 2″ diameter		190	.042	L.F.	6.65	2.07		8.72	10.50
1600	3″ diameter		190	.042		6.25	2.07		8.32	10
1800	4″ diameter		145	.055		9	2.71		11.71	14
1900	5″ diameter		130	.062		14.20	3.02		17.22	20
2100	Rectangular, corrugated copper, stock, 2″ x 3″		190	.042		7.35	2.07		9.42	11.25
2200	3″ x 4″		145	.055		8.60	2.71		11.31	13.55
2400	Rectangular, plain copper, stock, 2″ x 3″		190	.042		7	2.07		9.07	10.85
2500	3″ x 4″		145	.055		9.10	2.71		11.81	14.10
2700	Wire strainers, rectangular, 2″ x 3″		145	.055	Ea.	5.50	2.71		8.21	10.15
2800	3″ x 4″		145	.055		6.60	2.71		9.31	11.35
3000	Round, 2″ diameter		145	.055		6.90	2.71		9.61	11.70
3100	3″ diameter		145	.055		6.90	2.71		9.61	11.70
3300	4″ diameter		145	.055		12.70	2.71		15.41	18.10
3400	5″ diameter		115	.070		16.90	3.42		20.32	24
3600	Lead-coated copper, round, stock, 2″ diameter		190	.042	L.F.	22.50	2.07		24.57	27.50
3700	3″ diameter		190	.042		22	2.07		24.07	27
3900	4″ diameter		145	.055		23.50	2.71		26.21	29.50
4200	6″ diameter, corrugated		105	.076		30	3.74		33.74	38.50
4300	Rectangular, corrugated, stock, 2″ x 3″		190	.042		12.70	2.07		14.77	17.10
4500	Plain, stock, 2″ x 3″		190	.042		11	2.07		13.07	15.25
4600	3″ x 4″		145	.055		12	2.71		14.71	17.30
4800	Steel, galvanized, round, corrugated, 2″ or 3″ diameter, 28 gauge		190	.042		2.24	2.07		4.31	5.60
4900	4″ diameter, 28 gauge		145	.055		2.73	2.71		5.44	7.10
5100	5″ diameter, 28 gauge		130	.062		3.53	3.02		6.55	8.45
5200	26 gauge		130	.062		3.75	3.02		6.77	8.70
5400	6″ diameter, 28 gauge		105	.076		4	3.74		7.74	10.05
5500	26 gauge		105	.076		5.45	3.74		9.19	11.65
5700	Rectangular, corrugated, 28 gauge, 2″ x 3″		190	.042		1.28	2.07		3.35	4.54
5800	3″ x 4″		145	.055		2.16	2.71		4.87	6.50
6000	Rectangular, plain, 28 gauge, galvanized, 2″ x 3″		190	.042		1.28	2.07		3.35	4.54

07 71 23 – Manufactured Gutters and Downspouts

07 71 23.10 Downspouts		Crew	Daily Output	Labor-Hours	Unit	Material	2010 Bare Costs Labor	Equipment	Total	Total Incl O&P
6100	3" x 4"	1 Shee	145	.055	L.F.	2.64	2.71		5.35	7
6300	Epoxy painted, 24 gauge, corrugated, 2" x 3"		190	.042		2	2.07		4.07	5.35
6400	3" x 4"		145	.055		2.50	2.71		5.21	6.85
6600	Wire strainers, rectangular, 2" x 3"		145	.055	Ea.	12	2.71		14.71	17.30
6700	3" x 4"		145	.055		14	2.71		16.71	19.50
6900	Round strainers, 2" or 3" diameter		145	.055		3	2.71		5.71	7.40
7000	4" diameter		145	.055		5.60	2.71		8.31	10.25
7800	Stainless steel tubing, schedule 5, 2" x 3" or 3" diameter		190	.042	L.F.	49.50	2.07		51.57	57
7900	3" x 4" or 4" diameter		145	.055		62.50	2.71		65.21	73
8100	4" x 5" or 5" diameter		135	.059		129	2.91		131.91	146
8200	Vinyl, rectangular, 2" x 3"		210	.038		1.29	1.87		3.16	4.25
8300	Round, 2-1/2"		220	.036		1.28	1.79		3.07	4.11

07 71 23.20 Downspout Elbows		Crew	Daily Output	Labor-Hours	Unit	Material	Labor	Equipment	Total	Total Incl O&P
0010	**DOWNSPOUT ELBOWS**									
0020	Aluminum, 2" x 3", embossed	1 Shee	100	.080	Ea.	1.31	3.93		5.24	7.40
0100	Enameled		100	.080		1.31	3.93		5.24	7.40
0200	3" x 4", .025" thick, embossed		100	.080		3.99	3.93		7.92	10.35
0300	Enameled		100	.080		5.15	3.93		9.08	11.60
0400	Round corrugated, 3", embossed, .020" thick		100	.080		2.27	3.93		6.20	8.45
0500	4", .025" thick		100	.080		6.50	3.93		10.43	13.10
0600	Copper, 16 oz. round, 2" diameter		100	.080		13.40	3.93		17.33	20.50
0700	3" diameter		100	.080		11.60	3.93		15.53	18.70
0800	4" diameter		100	.080		16.95	3.93		20.88	24.50
1000	2" x 3" corrugated		100	.080		6.50	3.93		10.43	13.10
1100	3" x 4" corrugated		100	.080		13.15	3.93		17.08	20.50
1300	Vinyl, 2-1/2" diameter, 45° or 75°		100	.080		3.88	3.93		7.81	10.20
1400	Tee Y junction		75	.107		12.90	5.25		18.15	22

07 71 23.30 Gutters		Crew	Daily Output	Labor-Hours	Unit	Material	Labor	Equipment	Total	Total Incl O&P
0010	**GUTTERS**									
0012	Aluminum, stock units, 5" K type, .027" thick, plain	1 Shee	120	.067	L.F.	2.75	3.27		6.02	8
0100	Enameled		120	.067		2.44	3.27		5.71	7.65
0300	5" K type type, .032" thick, plain		120	.067		2.74	3.27		6.01	7.95
0400	Enameled		120	.067		2.90	3.27		6.17	8.15
0700	Copper, half round, 16 oz, stock units, 4" wide		120	.067		6.95	3.27		10.22	12.60
0900	5" wide		120	.067		7.35	3.27		10.62	13
1000	6" wide		115	.070		9.95	3.42		13.37	16.10
1200	K type, 16 oz, stock, 4" wide		120	.067		6.80	3.27		10.07	12.45
1300	5" wide		120	.067		8.20	3.27		11.47	13.95
1500	Lead coated copper, half round, stock, 4" wide		120	.067		9.45	3.27		12.72	15.35
1600	6" wide		115	.070		15.60	3.42		19.02	22.50
1800	K type, stock, 4" wide		120	.067		11.35	3.27		14.62	17.45
1900	5" wide		120	.067		16.35	3.27		19.62	23
2100	Stainless steel, half round or box, stock, 4" wide		120	.067		5.90	3.27		9.17	11.40
2200	5" wide		120	.067		6.35	3.27		9.62	11.90
2400	Steel, galv, half round or box, 28 ga, 5" wide, plain		120	.067		1.46	3.27		4.73	6.55
2500	Enameled		120	.067		1.51	3.27		4.78	6.60
2700	26 ga, stock, 5" wide		120	.067		1.59	3.27		4.86	6.70
2800	6" wide		120	.067		2.99	3.27		6.26	8.25
3000	Vinyl, O.G., 4" wide	1 Carp	110	.073		1.08	3.02		4.10	5.85
3100	5" wide		110	.073		1.38	3.02		4.40	6.20
3200	4" half round, stock units		110	.073		1.08	3.02		4.10	5.85
3250	Joint connectors				Ea.	2.88			2.88	3.17

07 71 23 – Manufactured Gutters and Downspouts

07 71 23.30 Gutters		Crew	Daily Output	Labor-Hours	Unit	Material	2010 Bare Costs Labor	Equipment	Total	Total Incl O&P
3300	Wood, clear treated cedar, fir or hemlock, 3" x 4"	1 Carp	100	.080	L.F.	7.95	3.32		11.27	13.85
3400	4" x 5"	"	100	.080	"	9.45	3.32		12.77	15.50

07 71 23.35 Gutter Guard		Crew	Daily Output	Labor-Hours	Unit	Material	Labor	Equipment	Total	Total Incl O&P
0010	**GUTTER GUARD**									
0020	6" wide strip, aluminum mesh	1 Carp	500	.016	L.F.	2.18	.66		2.84	3.42
0100	Vinyl mesh	"	500	.016	"	2.58	.66		3.24	3.86

07 71 26 – Reglets

07 71 26.10 Reglets and Accessories		Crew	Daily Output	Labor-Hours	Unit	Material	Labor	Equipment	Total	Total Incl O&P
0010	**REGLETS AND ACCESSORIES**									
0020	Aluminum, .025" thick, in concrete parapet	1 Carp	225	.036	L.F.	1.39	1.48		2.87	3.81
0100	Copper, 10 oz.		225	.036		2.62	1.48		4.10	5.15
0300	16 oz.		225	.036		5.10	1.48		6.58	7.95
0400	Galvanized steel, 24 gauge		225	.036		.94	1.48		2.42	3.31
0600	Stainless steel, .020" thick		225	.036		3.15	1.48		4.63	5.75
0700	Zinc and copper alloy, 20 oz.		225	.036		2.62	1.48		4.10	5.15
0900	Counter flashing for above, 12" wide, .032" aluminum	1 Shee	150	.053		1.63	2.62		4.25	5.75
1000	Copper, 10 oz.		150	.053		4.95	2.62		7.57	9.40
1200	16 oz.		150	.053		4.80	2.62		7.42	9.25
1300	Galvanized steel, .020" thick		150	.053		1.17	2.62		3.79	5.25
1500	Stainless steel, .020" thick		150	.053		5.10	2.62		7.72	9.55
1600	Zinc and copper alloy, 20 oz.		150	.053		4.78	2.62		7.40	9.20

07 71 29 – Manufactured Roof Expansion Joints

07 71 29.10 Expansion Joints		Crew	Daily Output	Labor-Hours	Unit	Material	Labor	Equipment	Total	Total Incl O&P
0010	**EXPANSION JOINTS**									
0300	Butyl or neoprene center with foam insulation, metal flanges									
0400	Aluminum, .032" thick for openings to 2-1/2"	1 Rofc	165	.048	L.F.	11.65	1.72		13.37	15.65
0600	For joint openings to 3-1/2"		165	.048		13.60	1.72		15.32	17.80
0610	For joint openings to 5"		165	.048		15.50	1.72		17.22	19.90
0620	For joint openings to 8"		165	.048		26	1.72		27.72	31.50
0700	Copper, 16 oz. for openings to 2-1/2"		165	.048		18.95	1.72		20.67	24
0900	For joint openings to 3-1/2"		165	.048		22	1.72		23.72	27
0910	For joint openings to 5"		165	.048		25.50	1.72		27.22	31
0920	For joint openings to 8"		165	.048		38	1.72		39.72	45
1000	Galvanized steel, 26 ga. for openings to 2-1/2"		165	.048		9.55	1.72		11.27	13.40
1200	For joint openings to 3-1/2"		165	.048		11.30	1.72		13.02	15.30
1210	For joint openings to 5"		165	.048		14.15	1.72		15.87	18.45
1220	For joint openings to 8"		165	.048		24	1.72		25.72	29.50
1300	Lead-coated copper, 16 oz. for openings to 2-1/2"		165	.048		33.50	1.72		35.22	40
1500	For joint openings to 3-1/2"		165	.048		40	1.72		41.72	47
1600	Stainless steel, .018", for openings to 2-1/2"		165	.048		14.55	1.72		16.27	18.90
1800	For joint openings to 3-1/2"		165	.048		16.60	1.72		18.32	21
1810	For joint openings to 5"		165	.048		20.50	1.72		22.22	25.50
1820	For joint openings to 8"		165	.048		31	1.72		32.72	37.50
1900	Neoprene, double-seal type with thick center, 4-1/2" wide		125	.064		13.60	2.27		15.87	18.75
1950	Polyethylene bellows, with galv steel flat flanges		100	.080		5.25	2.83		8.08	10.50
1960	With galvanized angle flanges		100	.080		5.80	2.83		8.63	11.10
2000	Roof joint with extruded aluminum cover, 2"	1 Shee	115	.070		36.50	3.42		39.92	45
2100	Roof joint, plastic curbs, foam center, standard	1 Rofc	100	.080		13.30	2.83		16.13	19.35
2200	Large		100	.080		17.75	2.83		20.58	24.50
2300	Transitions, regular, minimum		10	.800	Ea.	115	28.50		143.50	173
2350	Maximum		4	2		146	71		217	279

07 71 Roof Specialties

07 71 29 – Manufactured Roof Expansion Joints

07 71 29.10 Expansion Joints		Crew	Daily Output	Labor-Hours	Unit	Material	2010 Bare Costs Labor	Equipment	Total	Total Incl O&P
2400	Large, minimum	1 Rofc	9	.889	Ea.	168	31.50		199.50	238
2450	Maximum		3	2.667		176	94.50		270.50	350
2500	Roof to wall joint with extruded aluminum cover	1 Shee	115	.070	L.F.	31	3.42		34.42	39
2700	Wall joint, closed cell foam on PVC cover, 9" wide	1 Rofc	125	.064		4.74	2.27		7.01	8.95
2800	12" wide	"	115	.070		5.35	2.46		7.81	10

07 71 43 – Drip Edge

07 71 43.10 Drip Edge, Rake Edge, Ice Belts

		Crew	Daily Output	Labor-Hours	Unit	Material	Labor	Equipment	Total	Total Incl O&P
0010	**DRIP EDGE, RAKE EDGE, ICE BELTS**									
0020	Aluminum, .016" thick, 5" wide, mill finish	1 Carp	400	.020	L.F.	.38	.83		1.21	1.70
0100	White finish		400	.020		.38	.83		1.21	1.70
0200	8" wide, mill finish		400	.020		.46	.83		1.29	1.79
0300	Ice belt, 28" wide, mill finish		100	.080		3.85	3.32		7.17	9.35
0310	Vented, mill finish		400	.020		2.76	.83		3.59	4.32
0320	Painted finish		400	.020		2.86	.83		3.69	4.43
0400	Galvanized, 5" wide		400	.020		.48	.83		1.31	1.81
0500	8" wide, mill finish		400	.020		.69	.83		1.52	2.04
0510	Rake edge, aluminum, 1-1/2" x 1-1/2"		400	.020		.26	.83		1.09	1.57
0520	3-1/2" x 1-1/2"		400	.020		.43	.83		1.26	1.75

07 72 Roof Accessories

07 72 23 – Relief Vents

07 72 23.10 Roof Vents

		Crew	Daily Output	Labor-Hours	Unit	Material	Labor	Equipment	Total	Total Incl O&P
0010	**ROOF VENTS**									
0020	Mushroom shape, for built-up roofs, aluminum	1 Rofc	30	.267	Ea.	68.50	9.45		77.95	91
0100	PVC, 6" high	"	30	.267	"	28.50	9.45		37.95	47

07 72 26 – Ridge Vents

07 72 26.10 Ridge Vents and Accessories

		Crew	Daily Output	Labor-Hours	Unit	Material	Labor	Equipment	Total	Total Incl O&P
0010	**RIDGE VENTS AND ACCESSORIES**									
0100	Aluminum strips, mill finish	1 Rofc	160	.050	L.F.	2.62	1.77		4.39	5.80
0150	Painted finish		160	.050	"	3.74	1.77		5.51	7.05
0200	Connectors		48	.167	Ea.	3.68	5.90		9.58	13.85
0300	End caps		48	.167	"	1.37	5.90		7.27	11.30
0400	Galvanized strips		160	.050	L.F.	2.26	1.77		4.03	5.45
0430	Molded polyethylene, shingles not included		160	.050	"	3.05	1.77		4.82	6.30
0440	End plugs		48	.167	Ea.	1.37	5.90		7.27	11.30
0450	Flexible roll, shingles not included		160	.050	L.F.	2.25	1.77		4.02	5.40
2300	Ridge vent strip, mill finish	1 Shee	155	.052	"	2.73	2.53		5.26	6.85

07 72 33 – Roof Hatches

07 72 33.10 Roof Hatch Options

		Crew	Daily Output	Labor-Hours	Unit	Material	Labor	Equipment	Total	Total Incl O&P
0010	**ROOF HATCH OPTIONS**									
0500	2'-6" x 3', aluminum curb and cover	G-3	10	3.200	Ea.	985	132		1,117	1,275
0520	Galvanized steel curb and aluminum cover ***CN***		10	3.200		595	132		727	855
0540	Galvanized steel curb and cover		10	3.200		500	132		632	750
0600	2'-6" x 4'-6", aluminum curb and cover		9	3.556		850	146		996	1,150
0800	Galvanized steel curb and aluminum cover		9	3.556		600	146		746	885
0900	Galvanized steel curb and cover		9	3.556		825	146		971	1,125
1100	4' x 4' aluminum curb and cover		8	4		1,675	164		1,839	2,075
1120	Galvanized steel curb and aluminum cover		8	4		1,675	164		1,839	2,075
1140	Galvanized steel curb and cover		8	4		1,600	164		1,764	2,025

07 72 Roof Accessories

07 72 33 – Roof Hatches

07 72 33.10 Roof Hatch Options		Crew	Daily Output	Labor-Hours	Unit	2010 Bare Costs Material	Labor	Equipment	Total	Total Incl O&P
1200	2'-6" x 8'-0", aluminum curb and cover	G-3	6.60	4.848	Ea.	1,675	199		1,874	2,125
1400	Galvanized steel curb and aluminum cover		6.60	4.848		1,600	199		1,799	2,050
1500	Galvanized steel curb and cover		6.60	4.848		1,200	199		1,399	1,625
1800	For plexiglass panels, 2'-6" x 3'-0", add to above					400			400	440

07 72 36 – Smoke Vents

07 72 36.10 Smoke Hatches		Crew	Daily Output	Labor-Hours	Unit	Material	Labor	Equipment	Total	Total Incl O&P
0010	**SMOKE HATCHES**									
0200	For 3'-0" long, add to roof hatches from Division 07 72 33.10				Ea.	25%	5%			
0250	For 4'-0" long, add to roof hatches from Division 07 72 33.10					20%	5%			
0300	For 8'-0" long, add to roof hatches from Division 07 72 33.10					10%	5%			

07 72 36.20 Smoke Vent Options		Crew	Daily Output	Labor-Hours	Unit	Material	Labor	Equipment	Total	Total Incl O&P
0010	**SMOKE VENT OPTIONS**									
0100	4' x 4' aluminum cover and frame	G-3	13	2.462	Ea.	1,900	101		2,001	2,250
0200	Galvanized steel cover and frame		13	2.462		1,675	101		1,776	1,975
0300	4' x 8' aluminum cover and frame		8	4		2,600	164		2,764	3,100
0400	Galvanized steel cover and frame		8	4		2,200	164		2,364	2,675

07 72 53 – Snow Guards

07 72 53.10 Snow Guard Options		Crew	Daily Output	Labor-Hours	Unit	Material	Labor	Equipment	Total	Total Incl O&P
0010	**SNOW GUARD OPTIONS**									
0100	Slate & asphalt shingle roofs, fastened with nails	1 Rofc	160	.050	Ea.	9.40	1.77		11.17	13.30
0200	Standing seam metal roofs, fastened with set screws		48	.167		14.60	5.90		20.50	26
0300	Surface mount for metal roofs, fastened with solder		48	.167		7.70	5.90		13.60	18.25
0400	Double rail pipe type, including pipe		130	.062	L.F.	22.50	2.18		24.68	28.50

07 72 73 – Pitch Pockets

07 72 73.10 Pitch Pockets, Variable Sizes		Crew	Daily Output	Labor-Hours	Unit	Material	Labor	Equipment	Total	Total Incl O&P
0010	**PITCH POCKETS, VARIABLE SIZES**									
0100	Adjustable, 4" to 7", welded corners, 4" deep	1 Rofc	48	.167	Ea.	12.90	5.90		18.80	24
0200	Side extenders, 6"	"	240	.033	"	1.97	1.18		3.15	4.13

07 72 80 – Vents

07 72 80.30 Vent Options		Crew	Daily Output	Labor-Hours	Unit	Material	Labor	Equipment	Total	Total Incl O&P
0010	**VENT OPTIONS**									
0020	Plastic, for insulated decks, 1 per M.S.F., minimum	1 Rofc	40	.200	Ea.	19	7.10		26.10	33
0100	Maximum		20	.400		38	14.15		52.15	65.50
0300	Aluminum		30	.267		19	9.45		28.45	36.50
0800	Polystyrene baffles, 12" wide for 16" O.C. rafter spacing	1 Carp	90	.089		.40	3.69		4.09	6.15
0900	For 24" O.C. rafter spacing	"	110	.073		.75	3.02		3.77	5.50

07 81 Applied Fireproofing

07 81 16 – Cementitious Fireproofing

07 81 16.10 Sprayed Cementitious Fireproofing		Crew	Daily Output	Labor-Hours	Unit	Material	Labor	Equipment	Total	Total Incl O&P
0010	**SPRAYED CEMENTITIOUS FIREPROOFING**									
0050	Not incl tamping or canvas protection									
0100	1" thick, on flat plate steel	G-2	3000	.008	S.F.	.53	.28	.04	.85	1.05
0200	Flat decking		2400	.010		.53	.35	.06	.94	1.16
0400	Beams		1500	.016		.53	.56	.09	1.18	1.52
0500	Corrugated or fluted decks		1250	.019		.79	.67	.11	1.57	2
0700	Columns, 1-1/8" thick		1100	.022		.59	.76	.12	1.47	1.92
0800	2-3/16" thick		700	.034		1.13	1.19	.19	2.51	3.25

07 81 Applied Fireproofing

07 81 16 – Cementitious Fireproofing

07 81 16.10 Sprayed Cementitious Fireproofing		Crew	Daily Output	Labor-Hours	Unit	Material	2010 Bare Costs Labor	Equipment	Total	Total Incl O&P
0850	For tamping, add						10%			
0900	For canvas protection, add	G-2	5000	.005	S.F.	.07	.17	.03	.27	.36
1500	Intumescent epoxy fireproofing on wire mesh, 3/16" thick									
1550	1 hour rating, exterior use	G-2	136	.176	S.F.	6.90	6.15	.98	14.03	17.85
1600	Magnesium oxychloride, 35# to 40# density, 1/4" thick		3000	.008		1.44	.28	.04	1.76	2.05
1650	1/2" thick		2000	.012		2.90	.42	.07	3.39	3.89
1700	60# to 70# density, 1/4" thick		3000	.008		1.92	.28	.04	2.24	2.58
1750	1/2" thick		2000	.012		3.85	.42	.07	4.34	4.94
2000	Vermiculite cement, troweled or sprayed, 1/4" thick		3000	.008		1.31	.28	.04	1.63	1.91
2050	1/2" thick	↓	2000	.012	↓	2.60	.42	.07	3.09	3.56

07 84 Firestopping

07 84 13 – Penetration Firestopping

07 84 13.10 Firestopping		Crew	Daily Output	Labor-Hours	Unit	Material	Labor	Equipment	Total	Total Incl O&P
0010	**FIRESTOPPING** R078413-30									
0100	Metallic piping, non insulated									
0110	Through walls, 2" diameter	1 Carp	16	.500	Ea.	12.40	21		33.40	45.50
0120	4" diameter		14	.571		18.90	23.50		42.40	57.50
0130	6" diameter		12	.667		25.50	27.50		53	70.50
0140	12" diameter		10	.800		45	33		78	101
0150	Through floors, 2" diameter		32	.250		7.50	10.40		17.90	24.50
0160	4" diameter		28	.286		10.80	11.85		22.65	30
0170	6" diameter		24	.333		14.20	13.85		28.05	37
0180	12" diameter	↓	20	.400	↓	24	16.60		40.60	52
0190	Metallic piping, insulated									
0200	Through walls, 2" diameter	1 Carp	16	.500	Ea.	17.55	21		38.55	51.50
0210	4" diameter		14	.571		24	23.50		47.50	63
0220	6" diameter		12	.667		30.50	27.50		58	76
0230	12" diameter		10	.800		50	33		83	106
0240	Through floors, 2" diameter		32	.250		12.70	10.40		23.10	30
0250	4" diameter		28	.286		15.95	11.85		27.80	36
0260	6" diameter		24	.333		19.35	13.85		33.20	43
0270	12" diameter	↓	20	.400	↓	24	16.60		40.60	52
0280	Non metallic piping, non insulated									
0290	Through walls, 2" diameter	1 Carp	12	.667	Ea.	51	27.50		78.50	98.50
0300	4" diameter		10	.800		64	33		97	122
0310	6" diameter		8	1		89	41.50		130.50	162
0330	Through floors, 2" diameter		16	.500		40	21		61	76
0340	4" diameter		6	1.333		49.50	55.50		105	140
0350	6" diameter	↓	6	1.333	↓	59.50	55.50		115	151
0370	Ductwork, insulated & non insulated, round									
0380	Through walls, 6" diameter	1 Carp	12	.667	Ea.	26	27.50		53.50	71
0390	12" diameter		10	.800		51.50	33		84.50	108
0400	18" diameter		8	1		84	41.50		125.50	157
0410	Through floors, 6" diameter		16	.500		14.20	21		35.20	47.50
0420	12" diameter		14	.571		26	23.50		49.50	65
0430	18" diameter	↓	12	.667	↓	45	27.50		72.50	92
0440	Ductwork, insulated & non insulated, rectangular									
0450	With stiffener/closure angle, through walls, 6" x 12"	1 Carp	8	1	Ea.	21.50	41.50		63	87.50
0460	12" x 24"		6	1.333		28.50	55.50		84	117
0470	24" x 48"	↓	4	2	↓	81.50	83		164.50	218

07 84 13.10 Firestopping		Crew	Daily Output	Labor-Hours	Unit	Material	2010 Bare Costs Labor	Equipment	Total	Total Incl O&P
0480	With stiffener/closure angle, through floors, 6" x 12"	1 Carp	10	.800	Ea.	11.65	33		44.65	64
0490	12" x 24"		8	1		21	41.50		62.50	87
0500	24" x 48"		6	1.333		41	55.50		96.50	131
0510	Multi trade openings									
0520	Through walls, 6" x 12"	1 Carp	2	4	Ea.	45	166		211	305
0530	12" x 24"	"	1	8		182	330		512	710
0540	24" x 48"	2 Carp	1	16		730	665		1,395	1,825
0550	48" x 96"	"	.75	21.333		2,925	885		3,810	4,600
0560	Through floors, 6" x 12"	1 Carp	2	4		45	166		211	305
0570	12" x 24"	"	1	8		182	330		512	710
0580	24" x 48"	2 Carp	.75	21.333		730	885		1,615	2,175
0590	48" x 96"	"	.50	32		2,925	1,325		4,250	5,275
0600	Structural penetrations, through walls									
0610	Steel beams, W8 x 10	1 Carp	8	1	Ea.	28.50	41.50		70	95.50
0620	W12 x 14		6	1.333		45	55.50		100.50	135
0630	W21 x 44		5	1.600		90.50	66.50		157	202
0640	W36 x 135		3	2.667		220	111		331	415
0650	Bar joists, 18" deep		6	1.333		41.50	55.50		97	131
0660	24" deep		6	1.333		51.50	55.50		107	143
0670	36" deep		5	1.600		77.50	66.50		144	188
0680	48" deep		4	2		90.50	83		173.50	228
0690	Construction joints, floor slab at exterior wall									
0700	Precast, brick, block or drywall exterior									
0710	2" wide joint	1 Carp	125	.064	L.F.	6.45	2.66		9.11	11.20
0720	4" wide joint	"	75	.107	"	12.90	4.43		17.33	21
0730	Metal panel, glass or curtain wall exterior									
0740	2" wide joint	1 Carp	40	.200	L.F.	15.30	8.30		23.60	29.50
0750	4" wide joint	"	25	.320	"	21	13.30		34.30	43.50
0760	Floor slab to drywall partition									
0770	Flat joint	1 Carp	100	.080	L.F.	6.35	3.32		9.67	12.05
0780	Fluted joint		50	.160		12.90	6.65		19.55	24.50
0790	Etched fluted joint		75	.107		8.40	4.43		12.83	16.10
0800	Floor slab to concrete/masonry partition									
0810	Flat joint	1 Carp	75	.107	L.F.	14.20	4.43		18.63	22.50
0820	Fluted joint	"	50	.160	"	16.80	6.65		23.45	29
0830	Concrete/CMU wall joints									
0840	1" wide	1 Carp	100	.080	L.F.	7.75	3.32		11.07	13.65
0850	2" wide		75	.107		14.20	4.43		18.63	22.50
0860	4" wide		50	.160		27	6.65		33.65	40.50
0870	Concrete/CMU floor joints									
0880	1" wide	1 Carp	200	.040	L.F.	3.88	1.66		5.54	6.85
0890	2" wide		150	.053		7.10	2.22		9.32	11.20
0900	4" wide		100	.080		13.55	3.32		16.87	20

07 91 Pre-formed Joint Seals

07 91 13 – Compression Seals

07 91 13.10 Compression Seals		Crew	Daily Output	Labor-Hours	Unit	Material	Labor	Equipment	Total	Total Incl O&P
						2010 Bare Costs				
0010	**COMPRESSION SEALS**									
4900	Compression seals, O-ring type cord, 1/4"	1 Bric	472	.017	L.F.	.47	.71		1.18	1.59
4910	1/2"		440	.018		1.78	.76		2.54	3.10
4920	3/4"		424	.019		3.61	.79		4.40	5.15
4930	1"		408	.020		6.95	.82		7.77	8.90
4940	1-1/4"		384	.021		13.60	.87		14.47	16.25
4950	1-1/2"		368	.022		17.20	.91		18.11	20.50
4960	1-3/4"		352	.023		36	.95		36.95	41
4970	2"	↓	344	.023	↓	63.50	.97		64.47	71.50

07 91 16 – Joint Gaskets

07 91 16.10 Joint Gaskets		Crew	Daily Output	Labor-Hours	Unit	Material	Labor	Equipment	Total	Total Incl O&P
0010	**JOINT GASKETS**									
4400	Joint gaskets, neoprene, closed cell w/adh, 1/8" x 3/8"	1 Bric	240	.033	L.F.	.26	1.39		1.65	2.39
4500	1/4" x 3/4"		215	.037		.58	1.55		2.13	2.98
4700	1/2" x 1"		200	.040		1.30	1.67		2.97	3.95
4800	3/4" x 1-1/2"	↓	165	.048	↓	1.44	2.02		3.46	4.63

07 91 23 – Backer Rods

07 91 23.10 Backer Rods		Crew	Daily Output	Labor-Hours	Unit	Material	Labor	Equipment	Total	Total Incl O&P
0010	**BACKER RODS**									
0030	Backer rod, polyethylene, 1/4" diameter	1 Bric	4.60	1.739	C.L.F.	2.82	72.50		75.32	112
0050	1/2" diameter		4.60	1.739		6.60	72.50		79.10	116
0070	3/4" diameter		4.60	1.739		10.10	72.50		82.60	120
0090	1" diameter	↓	4.60	1.739	↓	18.70	72.50		91.20	130

07 91 26 – Joint Fillers

07 91 26.10 Joint Fillers		Crew	Daily Output	Labor-Hours	Unit	Material	Labor	Equipment	Total	Total Incl O&P
0010	**JOINT FILLERS**									
4360	Butyl rubber filler, 1/4" x 1/4"	1 Bric	290	.028	L.F.	.14	1.15		1.29	1.90
4365	1/2" x 1/2"		250	.032		.57	1.34		1.91	2.63
4370	1/2" x 3/4"		210	.038		.85	1.59		2.44	3.34
4375	3/4" x 3/4"		230	.035		1.28	1.45		2.73	3.60
4380	1" x 1"	↓	180	.044	↓	1.71	1.86		3.57	4.68
4390	For coloring, add					12%				
4980	Polyethylene joint backing, 1/4" x 2"	1 Bric	2.08	3.846	C.L.F.	12	161		173	255
4990	1/4" x 6"		1.28	6.250	"	36	261		297	435
5600	Silicone, room temp vulcanizing foam seal, 1/4" x 1/2"		1312	.006	L.F.	.29	.25		.54	.70
5610	1/2" x 1/2"		656	.012		.57	.51		1.08	1.40
5620	1/2" x 3/4"		442	.018		.86	.76		1.62	2.09
5630	3/4" x 3/4"		328	.024		1.29	1.02		2.31	2.96
5640	1/8" x 1"		1312	.006		.29	.25		.54	.70
5650	1/8" x 3"		442	.018		.86	.76		1.62	2.09
5670	1/4" x 3"		295	.027		1.72	1.13		2.85	3.60
5680	1/4" x 6"		148	.054		3.44	2.26		5.70	7.20
5690	1/2" x 6"		82	.098		6.90	4.07		10.97	13.70
5700	1/2" x 9"		52.50	.152		10.30	6.35		16.65	21
5710	1/2" x 12"	↓	33	.242	↓	13.75	10.10		23.85	30.50

07 92 Joint Sealants

07 92 13 – Elastomeric Joint Sealants

07 92 13.20 Caulking and Sealant Options		Crew	Daily Output	Labor-Hours	Unit	Material	Labor	Equipment	Total	Total Incl O&P
						2010 Bare Costs				
0010	**CAULKING AND SEALANT OPTIONS**									
0050	Latex acrylic based, bulk				Gal.	27			27	29.50
0055	Bulk in place 1/4" x 1/4" bead	1 Bric	300	.027	L.F.	.09	1.11		1.20	1.77
0060	1/4" x 3/8"		294	.027		.14	1.14		1.28	1.87
0065	1/4" x 1/2"		288	.028		.19	1.16		1.35	1.96
0075	3/8" x 3/8"		284	.028		.21	1.18		1.39	2
0080	3/8" x 1/2"		280	.029		.28	1.19		1.47	2.11
0085	3/8" x 5/8"		276	.029		.35	1.21		1.56	2.21
0095	3/8" x 3/4"		272	.029		.42	1.23		1.65	2.32
0100	1/2" x 1/2"		275	.029		.38	1.21		1.59	2.25
0105	1/2" x 5/8"		269	.030		.47	1.24		1.71	2.39
0110	1/2" x 3/4"		263	.030		.57	1.27		1.84	2.54
0115	1/2" x 7/8"		256	.031		.66	1.30		1.96	2.70
0120	1/2" x 1"		250	.032		.76	1.34		2.10	2.84
0125	3/4" x 3/4"		244	.033		.85	1.37		2.22	3
0130	3/4" x 1"		225	.036		1.14	1.48		2.62	3.49
0135	1" x 1"	↓	200	.040	↓	1.52	1.67		3.19	4.19
0190	Cartridges				Gal.	22.50			22.50	24.50
0200	11 fl. oz cartridge				Ea.	1.93			1.93	2.12
0500	1/4" x 1/2"	1 Bric	288	.028	L.F.	.16	1.16		1.32	1.92
0600	1/2" x 1/2"		275	.029		.32	1.21		1.53	2.18
0800	3/4" x 3/4"		244	.033		.71	1.37		2.08	2.84
0900	3/4" x 1"		225	.036		.95	1.48		2.43	3.28
1000	1" x 1"	↓	200	.040	↓	1.18	1.67		2.85	3.82
1400	Butyl based, bulk				Gal.	26			26	28.50
1500	Cartridges				"	32			32	35
1700	1/4" x 1/2", 154 L.F./gal.	1 Bric	288	.028	L.F.	.17	1.16		1.33	1.94
1800	1/2" x 1/2", 77 L.F./gal.	"	275	.029	"	.34	1.21		1.55	2.20
2300	Polysulfide compounds, 1 component, bulk				Gal.	51			51	56
2600	1 or 2 component, in place, 1/4" x 1/4", 308 L.F./gal.	1 Bric	300	.027	L.F.	.17	1.11		1.28	1.86
2700	1/2" x 1/4", 154 L.F./gal.		288	.028		.33	1.16		1.49	2.12
2900	3/4" x 3/8", 68 L.F./gal.		272	.029		.75	1.23		1.98	2.68
3000	1" x 1/2", 38 L.F./gal.	↓	250	.032	↓	1.35	1.34		2.69	3.49
3200	Polyurethane, 1 or 2 component				Gal.	50			50	55
3500	Bulk, in place, 1/4" x 1/4"	1 Bric	300	.027	L.F.	.16	1.11		1.27	1.86
3655	1/2" x 1/4"		288	.028		.32	1.16		1.48	2.11
3800	3/4" x 3/8", 68 L.F./gal.		272	.029		.74	1.23		1.97	2.66
3900	1" x 1/2"	↓	250	.032	↓	1.30	1.34		2.64	3.44
4100	Silicone rubber, bulk				Gal.	41			41	45
4200	Cartridges				"	41			41	45

07 92 16 – Rigid Joint Sealants

07 92 16.10 Rigid Joint Sealants		Crew	Daily Output	Labor-Hours	Unit	Material	Labor	Equipment	Total	Total Incl O&P
0010	**RIGID JOINT SEALANTS**									
5800	Tapes, sealant, P.V.C. foam adhesive, 1/16" x 1/4"				C.L.F.	5.65			5.65	6.20
5900	1/16" x 1/2"					8.35			8.35	9.20
5950	1/16" x 1"					13.85			13.85	15.25
6000	1/8" x 1/2"				↓	9.35			9.35	10.30

07 92 19 – Acoustical Joint Sealants

07 92 19.10 Acoustical Sealant		Crew	Daily Output	Labor-Hours	Unit	Material	Labor	Equipment	Total	Total Incl O&P
0010	**ACOUSTICAL SEALANT**									
0020	Acoustical sealant, elastomeric, cartridges				Ea.	2.25			2.25	2.48
0025	In place, 1/4" x 1/4"	1 Bric	300	.027	L.F.	.09	1.11		1.20	1.78

07 92 Joint Sealants

07 92 19 – Acoustical Joint Sealants

07 92 19.10 Acoustical Sealant		Crew	Daily Output	Labor-Hours	Unit	Material	2010 Bare Costs Labor	Equipment	Total	Total Incl O&P
0030	1/4" x 1/2"	1 Bric	288	.028	L.F.	.18	1.16		1.34	1.95
0035	1/2" x 1/2"		275	.029		.37	1.21		1.58	2.23
0040	1/2" x 3/4"		263	.030		.55	1.27		1.82	2.52
0045	3/4" x 3/4"		244	.033		.83	1.37		2.20	2.97
0050	1" x 1"	↓	200	.040	↓	1.47	1.67		3.14	4.14

07 95 Expansion Control

07 95 13 – Expansion Joint Cover Assemblies

07 95 13.50 Expansion Joint Assemblies		Crew	Daily Output	Labor-Hours	Unit	Material	Labor	Equipment	Total	Total Incl O&P
0010	**EXPANSION JOINT ASSEMBLIES**									
0200	Floor cover assemblies, 1" space, aluminum	1 Sswk	38	.211	L.F.	20.50	9.85		30.35	40
0300	Bronze		38	.211		41.50	9.85		51.35	63
0500	2" space, aluminum		38	.211		24.50	9.85		34.35	44.50
0600	Bronze		38	.211		44.50	9.85		54.35	66.50
0800	Wall and ceiling assemblies, 1" space, aluminum		38	.211		12.15	9.85		22	31
0900	Bronze		38	.211		38.50	9.85		48.35	60
1100	2" space, aluminum		38	.211		20.50	9.85		30.35	40
1200	Bronze		38	.211		47.50	9.85		57.35	70
1400	Floor to wall assemblies, 1" space, aluminum		38	.211		18.55	9.85		28.40	38
1500	Bronze or stainless		38	.211		47	9.85		56.85	69
1700	Gym floor angle covers, aluminum, 3" x 3" angle		46	.174		16.05	8.15		24.20	32
1800	3" x 4" angle		46	.174		18.95	8.15		27.10	35.50
2000	Roof closures, aluminum, flat roof, low profile, 1" space		57	.140		37	6.60		43.60	52.50
2100	High profile		57	.140		45.50	6.60		52.10	61.50
2300	Roof to wall, low profile, 1" space		57	.140		20.50	6.60		27.10	34
2400	High profile	↓	57	.140	↓	26.50	6.60		33.10	40.50

Estimating Tips

08 10 00 Doors and Frames

All exterior doors should be addressed for their energy conservation.

- Most metal doors and frames look alike, but there may be significant differences among them. When estimating these items, be sure to choose the line item that most closely compares to the specification or door schedule requirements regarding:
  - — type of metal
  - — metal gauge
  - — door core material
  - — fire rating
  - — finish
- Wood and plastic doors vary considerably in price. The primary determinant is the veneer material. Lauan, birch, and oak are the most common veneers. Other variables include the following:
  - — hollow or solid core
  - — fire rating
  - — flush or raised panel
  - — finish

08 30 00 Specialty Doors and Frames

- There are many varieties of special doors, and they are usually priced per each. Add frames, hardware, or operators required for a complete installation.

08 40 00 Entrances, Storefronts, and Curtain Walls

- Glazed curtain walls consist of the metal tube framing and the glazing material. The cost data in this subdivision is presented for the metal tube framing alone or the composite wall. If your estimate requires a detailed takeoff of the framing, be sure to add the glazing cost.

08 50 00 Windows

- Most metal windows are delivered preglazed. However, some metal windows are priced without glass. Refer to 08 80 00 Glazing for glass pricing. The grade C indicates commercial grade windows, usually ASTM C-35.
- All wood windows and vinyl are priced preglazed. The glazing is insulating glass. Some wood windows may have single pane float glass. Add the cost of screens and grills if required, and not already included.

08 70 00 Hardware

- Hardware costs add considerably to the cost of a door. The most efficient method to determine the hardware requirements for a project is to review the door schedule.
- Door hinges are priced by the pair, with most doors requiring 1-1/2 pairs per door. The hinge prices do not include installation labor because it is included in door installation. Hinges are classified according to the frequency of use.

08 80 00 Glazing

- Different openings require different types of glass. The three most common types are:
  - — float
  - — tempered
  - — insulating
- Most exterior windows are glazed with insulating glass. Entrance doors and window walls, where the glass is less than 18″ from the floor, are generally glazed with tempered glass. Interior windows and some residential windows are glazed with float glass.
- Coastal communities are starting to require the use of impact-resistant glass.
- The insulation or 'u' value is a strong consideration, along with solar heat gain.

Reference Numbers

Reference numbers are shown in shaded boxes at the beginning of some major classifications. These numbers refer to related items in the Reference Section. The reference information may be an estimating procedure, an alternate pricing method, or technical information.

Note: Not all subdivisions listed here necessarily appear in this publication.

08 05 05 – Selective Windows and Doors Demolition

08 05 05.10 Selective Demolition Doors		Crew	Daily Output	Labor-Hours	Unit	Material	2010 Bare Costs Labor	Equipment	Total	Total Incl O&P
0010	SELECTIVE DEMOLITION DOORS R024119-10									
0200	Doors, exterior, 1-3/4" thick, single, 3' x 7' high	1 Clab	16	.500	Ea.		16.55		16.55	25.50
0220	Double, 6' x 7' high		12	.667			22		22	34
0500	Interior, 1-3/8" thick, single, 3' x 7' high		20	.400			13.25		13.25	20.50
0520	Double, 6' x 7' high		16	.500			16.55		16.55	25.50
0700	Bi-folding, 3' x 6'-8" high		20	.400			13.25		13.25	20.50
0720	6' x 6'-8" high		18	.444			14.70		14.70	22.50
0900	Bi-passing, 3' x 6'-8" high		16	.500			16.55		16.55	25.50
0940	6' x 6'-8" high		14	.571			18.90		18.90	29
1500	Remove and reset, minimum	1 Carp	8	1			41.50		41.50	64
1520	Maximum		6	1.333			55.50		55.50	85.50
2000	Frames, including trim, metal		8	1			41.50		41.50	64
2200	Wood	2 Carp	32	.500			21		21	32
3000	Special doors, counter doors		6	2.667			111		111	171
3100	Double acting		10	1.600			66.50		66.50	102
3200	Floor door (trap type), or access type		8	2			83		83	128
3300	Glass, sliding, including frames		12	1.333			55.50		55.50	85.50
3400	Overhead, commercial, 12' x 12' high		4	4			166		166	256
3440	up to 20' x 16' high		3	5.333			222		222	340
3445	up to 35' x 30' high		1	16			665		665	1,025
3500	Residential, 9' x 7' high		8	2			83		83	128
3540	16' x 7' high		7	2.286			95		95	146
3600	Remove and reset, minimum		4	4			166		166	256
3620	Maximum		2.50	6.400			266		266	410
3700	Roll-up grille		5	3.200			133		133	205
3800	Revolving door		2	8			330		330	510
3900	Storefront swing door		3	5.333			222		222	340
6600	Demo flexible transparent strip entrance	3 Shee	115	.209	SF Surf		10.25		10.25	15.50
7100	Remove double swing pneumatic doors, openers and sensors	2 Skwk	.50	32	Opng.		1,375		1,375	2,100
7110	Remove automatic operators, industrial, sliding doors, to 12' wide	"	.40	40	"		1,700		1,700	2,625
7550	Hangar door demo	2 Sswk	330	.048	S.F.		2.27		2.27	4
7570	Remove shock absorbing door	"	1.90	8.421	Opng.		395		395	695

08 05 05.20 Selective Demolition of Windows		Crew	Daily Output	Labor-Hours	Unit	Material	2010 Bare Costs Labor	Equipment	Total	Total Incl O&P
0010	SELECTIVE DEMOLITION OF WINDOWS R024119-10									
0200	Aluminum, including trim, to 12 S.F.	1 Clab	16	.500	Ea.		16.55		16.55	25.50
0240	To 25 S.F.		11	.727			24		24	37
0280	To 50 S.F.		5	1.600			53		53	81.50
0320	Storm windows/screens, to 12 S.F.		27	.296			9.80		9.80	15.15
0360	To 25 S.F.		21	.381			12.60		12.60	19.45
0400	To 50 S.F.		16	.500			16.55		16.55	25.50
0600	Glass, minimum		200	.040	S.F.		1.32		1.32	2.04
0620	Maximum		150	.053	"		1.77		1.77	2.72
1000	Steel, including trim, to 12 S.F.		13	.615	Ea.		20.50		20.50	31.50
1020	To 25 S.F.		9	.889			29.50		29.50	45.50
1040	To 50 S.F.		4	2			66		66	102
2000	Wood, including trim, to 12 S.F.		22	.364			12.05		12.05	18.55
2020	To 25 S.F.		18	.444			14.70		14.70	22.50
2060	To 50 S.F.		13	.615			20.50		20.50	31.50
2065	To 180 S.F.		8	1			33		33	51
4300	Remove bay/bow window	2 Carp	6	2.667			111		111	171
4410	Remove skylight, plstc domes, flush/curb mtd	G-3	395	.081	S.F.		3.33		3.33	5.10
5020	Remove and reset window, minimum	1 Carp	6	1.333	Ea.		55.50		55.50	85.50

08 05 Common Work Results for Openings

08 05 05 – Selective Windows and Doors Demolition

08 05 05.20 Selective Demolition of Windows		Crew	Daily Output	Labor-Hours	Unit	Material	2010 Bare Costs Labor	Equipment	Total	Total Incl O&P
5040	Average	1 Carp	4	2	Ea.		83		83	128
5080	Maximum		2	4			166		166	256

08 11 Metal Doors and Frames

08 11 16 – Aluminum Doors and Frames

08 11 16.10 Entrance Doors and Frames

		Crew	Daily Output	Labor-Hours	Unit	Material	Labor	Equipment	Total	Total Incl O&P
0010	**ENTRANCE DOORS AND FRAMES** Aluminum, narrow stile									
0011	Including standard hardware, clear finish, no glass									
0020	3′-0″ x 7′-0″ opening	2 Sswk	2	8	Ea.	745	375		1,120	1,475
0030	3′-6″ x 7′-0″ opening		2	8		695	375		1,070	1,425
0100	3′-0″ x 10′-0″ opening, 3′ high transom		1.80	8.889		1,175	415		1,590	2,025
0200	3′-6″ x 10′-0″ opening, 3′ high transom		1.80	8.889		1,225	415		1,640	2,075
0280	5′-0″ x 7′-0″ opening		2	8		1,175	375		1,550	1,925
0300	6′-0″ x 7′-0″ opening		1.30	12.308		1,075	575		1,650	2,200
0400	6′-0″ x 10′-0″ opening, 3′ high transom		1.10	14.545	Pr.	1,525	680		2,205	2,875
0420	7′-0″ x 7′-0″ opening		1	16	″	1,250	750		2,000	2,700
0520	3′-0″ x 7′-0″ opening, wide stile		2	8	Ea.	890	375		1,265	1,650
0540	3′-6″ x 7′-0″ opening		2	8		1,000	375		1,375	1,775
0560	5′-0″ x 7′-0″ opening		2	8		1,425	375		1,800	2,200
0580	6′-0″ x 7′-0″ opening		1.30	12.308	Pr.	1,450	575		2,025	2,600
0600	7′-0″ x 7′-0″ opening		1	16	″	1,600	750		2,350	3,075
1100	For full vision doors, with 1/2″ glass, add				Leaf	55%				
1200	For non-standard size, add					67%				
1300	Light bronze finish, add					36%				
1400	Dark bronze finish, add					18%				
1500	For black finish, add					36%				
1600	Concealed panic device, add					910			910	1,000
1700	Electric striker release, add				Opng.	252			252	277
1800	Floor check, add				Leaf	630			630	695
1900	Concealed closer, add				″	505			505	555
2000	Flush 3′ x 7′ Insulated, 12″x 12″ lite, clear finish	2 Sswk	2	8	Ea.	1,275	375		1,650	2,050

08 11 63 – Metal Screen and Storm Doors and Frames

08 11 63.23 Aluminum Screen and Storm Doors and Frames

		Crew	Daily Output	Labor-Hours	Unit	Material	Labor	Equipment	Total	Total Incl O&P
0010	**ALUMINUM SCREEN AND STORM DOORS AND FRAMES**									
0020	Combination storm and screen									
0400	Clear anodic coating, 6′-8″ x 2′-6″ wide	2 Carp	15	1.067	Ea.	172	44.50		216.50	258
0420	2′-8″ wide		14	1.143		156	47.50		203.50	244
0440	3′-0″ wide		14	1.143		156	47.50		203.50	244
0500	For 7′ door height, add					5%				
1000	Mill finish, 6′-8″ x 2′-6″ wide	2 Carp	15	1.067	Ea.	225	44.50		269.50	315
1020	2′-8″ wide		14	1.143		225	47.50		272.50	320
1040	3′-0″ wide		14	1.143		246	47.50		293.50	345
1100	For 7′-0″ door, add					5%				
1500	White painted, 6′-8″ x 2′-6″ wide	2 Carp	15	1.067		282	44.50		326.50	380
1520	2′-8″ wide		14	1.143		278	47.50		325.50	380
1540	3′-0″ wide		14	1.143		292	47.50		339.50	395
1600	For 7′-0″ door, add					5%				
2000	Wood door & screen, see Div. 08 14 33.20									

08 11 Metal Doors and Frames

08 11 74 – Sliding Metal Grilles

08 11 74.10 Rolling Grille Supports		Crew	Daily Output	Labor-Hours	Unit	Material	2010 Bare Costs Labor	Equipment	Total	Total Incl O&P
0010	ROLLING GRILLE SUPPORTS									
0020	Rolling grille supports, overhead framed	E-4	36	.889	L.F.	23.50	42	4.05	69.55	104

08 12 Metal Frames

08 12 13 – Hollow Metal Frames

08 12 13.13 Standard Hollow Metal Frames

		Crew	Daily Output	Labor-Hours	Unit	Material	Labor	Equipment	Total	Total Incl O&P
0010	STANDARD HOLLOW METAL FRAMES									
0020	16 ga., up to 5-3/4" jamb depth									
0025	6'-8" high, 3'-0" wide, single	2 Carp	16	1	Ea.	136	41.50		177.50	213
0028	3'-6" wide, single		16	1		142	41.50		183.50	221
0030	4'-0" wide, single		16	1		142	41.50		183.50	221
0040	6'-0" wide, double		14	1.143		195	47.50		242.50	287
0045	8'-0" wide, double		14	1.143		203	47.50		250.50	296
0100	7'-0" high, 3'-0" wide, single		16	1		140	41.50		181.50	218
0110	3'-6" wide, single		16	1		147	41.50		188.50	226
0112	4'-0" wide, single		16	1		147	41.50		188.50	226
0140	6'-0" wide, double		14	1.143		185	47.50		232.50	277
0145	8'-0" wide, double		14	1.143		219	47.50		266.50	315
0160	6'-0" wide, double		14	1.143		185	47.50		232.50	277
1000	16 ga., up to 4-7/8" deep, 7'-0" H, 3'-0" W, single **CN**		16	1		154	41.50		195.50	234
1140	6'-0" wide, double		14	1.143		171	47.50		218.50	261
1200	16 ga., 8-3/4" deep, 7'-0" H, 3'-0" W, single		16	1		163	41.50		204.50	243
2800	14 ga., up to 3-7/8" deep, 7'-0" high, 3'-0" wide, single		16	1		149	41.50		190.50	227
2840	6'-0" wide, double		14	1.143		174	47.50		221.50	264
3000	14 ga., up to 5-3/4" deep, 6'-8" high, 3'-0" wide, single		16	1		140	41.50		181.50	218
3002	3'-6" wide, single		16	1		183	41.50		224.50	265
3005	4'-0" wide, single		16	1		183	41.50		224.50	265
3600	up to 5-3/4" jamb depth, 7'-0" high, 4'-0" wide, single		15	1.067		140	44.50		184.50	223
3620	6'-0" wide, double		12	1.333		182	55.50		237.50	286
3640	8'-0" wide, double		12	1.333		179	55.50		234.50	283
3700	8'-0" high, 4'-0" wide, single		15	1.067		169	44.50		213.50	255
3740	8'-0" wide, double		12	1.333		260	55.50		315.50	370
4000	6-3/4" deep, 7'-0" high, 4'-0" wide, single		15	1.067		198	44.50		242.50	287
4020	6'-0" wide, double		12	1.333		243	55.50		298.50	355
4040	8'-0", wide double		12	1.333		251	55.50		306.50	360
4100	8'-0" high, 4'-0" wide, single		15	1.067		208	44.50		252.50	298
4140	8'-0" wide, double		12	1.333		208	55.50		263.50	315
4400	8-3/4" deep, 7'-0" high, 4'-0" wide, single		15	1.067		176	44.50		220.50	262
4440	8'-0" wide, double		12	1.333		237	55.50		292.50	345
4500	8'-0" high, 4'-0" wide, single		15	1.067		177	44.50		221.50	263
4540	8'-0" wide, double		12	1.333		208	55.50		263.50	315
4900	For welded frames, add					60			60	66
5400	14 ga., "B" label, up to 5-3/4" deep, 7'-0" high, 4'-0" wide, single	2 Carp	15	1.067		194	44.50		238.50	282
5440	8'-0" wide, double		12	1.333		252	55.50		307.50	365
5800	6-3/4" deep, 7'-0" high, 4'-0" wide, single		15	1.067		208	44.50		252.50	298
5840	8'-0" wide, double		12	1.333		360	55.50		415.50	480
6200	8-3/4" deep, 7'-0" high, 4'-0" wide, single		15	1.067		289	44.50		333.50	390
6240	8'-0" wide, double		12	1.333		335	55.50		390.50	455
6300	For "A" label use same price as "B" label									
6400	For baked enamel finish, add					30%	15%			

08 12 Metal Frames

08 12 13 – Hollow Metal Frames

08 12 13.13 Standard Hollow Metal Frames		Crew	Daily Output	Labor-Hours	Unit	Material	2010 Bare Costs Labor	Equipment	Total	Total Incl O&P
6500	For galvanizing, add					15%				
6600	For hospital stop, add				Ea.	282			282	310
7900	Transom lite frames, fixed, add	2 Carp	155	.103	S.F.	52	4.29		56.29	63.50
8000	Movable, add	"	130	.123	"	65	5.10		70.10	79.50

08 12 13.25 Channel Metal Frames		Crew	Daily Output	Labor-Hours	Unit	Material	Labor	Equipment	Total	Total Incl O&P
0010	**CHANNEL METAL FRAMES**									
0020	Steel channels with anchors and bar stops									
0100	6" channel @ 8.2#/L.F., 3' x 7' door, weighs 150#	E-4	13	2.462	Ea.	198	117	11.20	326.20	435
0200	8" channel @ 11.5#/L.F., 6' x 8' door, weighs 275#		9	3.556		365	169	16.20	550.20	715
0300	8' x 12' door, weighs 400#		6.50	4.923		530	233	22.50	785.50	1,025
0400	10" channel @ 15.3#/L.F., 10' x 10' door, weighs 500# CN		6	5.333		660	253	24.50	937.50	1,200
0500	12' x 12' door, weighs 600#		5.50	5.818		790	276	26.50	1,092.50	1,375
0600	12" channel @ 20.7#/L.F., 12' x 12' door, weighs 825#		4.50	7.111		1,100	335	32.50	1,467.50	1,825
0700	12' x 16' door, weighs 1000#	↓	4	8		1,325	380	36.50	1,741.50	2,150
0800	For frames without bar stops, light sections, deduct					15%				
0900	Heavy sections, deduct				↓	10%				

08 13 Metal Doors

08 13 13 – Hollow Metal Doors

08 13 13.13 Standard Hollow Metal Doors		Crew	Daily Output	Labor-Hours	Unit	Material	Labor	Equipment	Total	Total Incl O&P
0010	**STANDARD HOLLOW METAL DOORS** R081313-20									
0015	Flush, full panel, hollow core									
0020	1-3/8" thick, 20 ga., 2'-0"x 6'-8" G	2 Carp	20	.800	Ea.	320	33		353	400
0040	2'-8" x 6'-8" G		18	.889		330	37		367	415
0060	3'-0" x 6'-8" G		17	.941		330	39		369	425
0100	3'-0" x 7'-0" G	↓	17	.941		345	39		384	435
0120	For vision lite, add					85			85	93.50
0140	For narrow lite, add					92			92	101
0320	Half glass, 20 ga., 2'-0" x 6'-8" G	2 Carp	20	.800		470	33		503	565
0340	2'-8" x 6'-8" G		18	.889		480	37		517	580
0360	3'-0" x 6'-8" G		17	.941		480	39		519	590
0400	3'-0" x 7'-0" G		17	.941		560	39		599	675
0410	1-3/8" thick, 18 ga., 2'-0"x 6'-8" G		20	.800		380	33		413	470
0420	3'-0" x 6'-8" G		17	.941		385	39		424	485
0425	3'-0" x 7'-0" G	↓	17	.941		400	39		439	500
0450	For vision lite, add					85			85	93.50
0452	For narrow lite, add					92			92	101
0460	Half glass, 18 ga., 2'-0" x 6'-8" G	2 Carp	20	.800		530	33		563	635
0465	2'-8" x 6'-8" G		18	.889		550	37		587	660
0470	3'-0" x 6'-8" G		17	.941		535	39		574	650
0475	3'-0" x 7'-0" G		17	.941		550	39		589	665
0500	Hollow core, 1-3/4" thick, full panel, 20 ga., 2'-8" x 6'-8" G		18	.889		375	37		412	465
0520	3'-0" x 6'-8" G		17	.941		335	39		374	430
0640	3'-0" x 7'-0" G		17	.941		390	39		429	490
0680	4'-0" x 7'-0" G		15	1.067		535	44.50		579.50	660
0700	4'-0" x 8'-0" G		13	1.231		610	51		661	750
1000	18 ga., 2'-8" x 6'-8" G		17	.941		400	39		439	500
1020	3'-0" x 6'-8" G		16	1		385	41.50		426.50	490
1120	3'-0" x 7'-0" CN G		17	.941		445	39		484	550
1180	4'-0" x 7'-0" G		14	1.143		535	47.50		582.50	665
1200	4'-0" x 8'-0" G	↓	17	.941	↓	540	39		579	655

08 13 Metal Doors

08 13 13 – Hollow Metal Doors

08 13 13.13 Standard Hollow Metal Doors			Crew	Daily Output	Labor-Hours	Unit	2010 Bare Costs Material	Labor	Equipment	Total	Total Incl O&P
1212	For vision lite, add					Ea.	85			85	93.50
1214	For narrow lite, add						92			92	101
1230	Half glass, 20 ga., 2′-8″ x 6′-8″	G	2 Carp	20	.800		430	33		463	525
1240	3′-0″ x 6′-8″	G		18	.889		420	37		457	520
1260	3′-0″ x 7′-0″	G		18	.889		425	37		462	520
1320	18 ga., 2′-8″ x 6′-8″	G		18	.889		460	37		497	560
1340	3′-0″ x 6′-8″	G		17	.941		455	39		494	560
1360	3′-0″ x 7′-0″	G		17	.941		465	39		504	570
1380	4′-0″ x 7′-0″	G		15	1.067		580	44.50		624.50	705
1400	4′-0″ x 8′-0″	G		14	1.143		660	47.50		707.50	805
1720	Insulated, 1-3/4″ thick, full panel, 18 ga., 3′-0″ x 6′-8″	G		15	1.067		400	44.50		444.50	515
1740	2′-8″ x 7′-0″	G		16	1		420	41.50		461.50	525
1760	3′-0″ x 7′-0″	G		15	1.067		415	44.50		459.50	525
1800	4′-0″ x 8′-0″	G		13	1.231		605	51		656	745
1805	For vision lite, add						85			85	93.50
1810	For narrow lite, add						92			92	101
1820	Half glass, 18 ga., 3′-0″ x 6′-8″	G	2 Carp	16	1		535	41.50		576.50	655
1840	2′-8″ x 7′-0″	G		17	.941		545	39		584	660
1860	3′-0″ x 7′-0″	G		16	1		585	41.50		626.50	710
1900	4′-0″ x 8′-0″	G		14	1.143		655	47.50		702.50	795
2000	For bottom louver, add						245			245	269
2020	For baked enamel finish, add						30%	15%			
2040	For galvanizing, add						15%				

08 13 13.15 Metal Fire Doors			Crew	Daily Output	Labor-Hours	Unit	Material	Labor	Equipment	Total	Total Incl O&P
0010	**METAL FIRE DOORS**	R081313-20									
0015	Steel, flush, "B" label, 90 minute										
0020	Full panel, 20 ga., 2′-0″ x 6′-8″		2 Carp	20	.800	Ea.	340	33		373	425
0040	2′-8″ x 6′-8″			18	.889		345	37		382	435
0060	3′-0″ x 6′-8″			17	.941		345	39		384	440
0080	3′-0″ x 7′-0″			17	.941		360	39		399	455
0140	18 ga., 3′-0″ x 6′-8″			16	1		395	41.50		436.50	500
0160	2′-8″ x 7′-0″			17	.941		415	39		454	515
0180	3′-0″ x 7′-0″			16	1		395	41.50		436.50	500
0200	4′-0″ x 7′-0″			15	1.067		520	44.50		564.50	640
0220	For "A" label, 3 hour, 18 ga., use same price as "B" label										
0240	For vision lite, add					Ea.	130			130	143
0520	Flush, "B" label 90 min., composite, 20 ga., 2′-0″ x 6′-8″		2 Carp	18	.889		430	37		467	530
0540	2′-8″ x 6′-8″			17	.941		435	39		474	540
0560	3′-0″ x 6′-8″			16	1		440	41.50		481.50	545
0580	3′-0″ x 7′-0″			16	1		450	41.50		491.50	560
0640	Flush, "A" label 3 hour, composite, 18 ga., 3′-0″ x 6′-8″			15	1.067		495	44.50		539.50	615
0660	2′-8″ x 7′-0″			16	1		515	41.50		556.50	635
0680	3′-0″ x 7′-0″			15	1.067		505	44.50		549.50	625
0700	4′-0″ x 7′-0″			14	1.143		630	47.50		677.50	770

08 13 13.20 Residential Steel Doors			Crew	Daily Output	Labor-Hours	Unit	Material	Labor	Equipment	Total	Total Incl O&P
0010	**RESIDENTIAL STEEL DOORS**										
0020	Prehung, insulated, exterior										
0030	Embossed, full panel, 2′-8″ x 6′-8″	G	2 Carp	17	.941	Ea.	293	39		332	380
0040	3′-0″ x 6′-8″	G		15	1.067		272	44.50		316.50	370
0060	3′-0″ x 7′-0″	G		15	1.067		320	44.50		364.50	420
0070	5′-4″ x 6′-8″, double	G		8	2		575	83		658	765
0220	Half glass, 2′-8″ x 6′-8″			17	.941		290	39		329	380

08 13 Metal Doors

08 13 13 – Hollow Metal Doors

08 13 13.20 Residential Steel Doors		Crew	Daily Output	Labor-Hours	Unit	2010 Bare Costs Material	Labor	Equipment	Total	Total Incl O&P
0240	3′-0″ x 6′-8″	2 Carp	16	1	Ea.	290	41.50		331.50	385
0260	3′-0″ x 7′-0″		16	1		340	41.50		381.50	440
0270	5′-4″ x 6′-8″, double		8	2		585	83		668	770
1320	Flush face, full panel, 2′-8″ x 6′-8″		16	1		230	41.50		271.50	315
1340	3′-0″ x 6′-8″		15	1.067		230	44.50		274.50	320
1360	3′-0″ x 7′-0″		15	1.067		290	44.50		334.50	390
1380	5′-4″ x 6′-8″, double		8	2		470	83		553	645
1420	Half glass, 2′-8″ x 6′-8″		17	.941		289	39		328	380
1440	3′-0″ x 6′-8″		16	1		289	41.50		330.50	385
1460	3′-0″ x 7′-0″		16	1		335	41.50		376.50	435
1480	5′-4″ x 6′-8″, double	↓	8	2		555	83		638	740
1500	Sidelight, full lite, 1′-0″ x 6′-8″with grille					236			236	260
1510	1′-0″ x 6′-8″, low e					257			257	282
1520	1′-0″ x 6′-8″, half lite					254			254	280
1530	1′-0″ x 6′-8″, half lite, low e					262			262	289
2300	Interior, residential, closet, bi-fold, 6′-8″ x 2′-0″ wide	2 Carp	16	1		158	41.50		199.50	238
2330	3′-0″ wide		16	1		197	41.50		238.50	281
2360	4′-0″ wide		15	1.067		255	44.50		299.50	350
2400	5′-0″ wide		14	1.143		310	47.50		357.50	415
2420	6′-0″ wide	↓	13	1.231	↓	330	51		381	445

08 13 16 – Aluminum Doors

08 13 16.10 Commercial Aluminum Doors		Crew	Daily Output	Labor-Hours	Unit	Material	Labor	Equipment	Total	Total Incl O&P
0010	**COMMERCIAL ALUMINUM DOORS**, no glazing									
0020	Incl. hinges, push/pull, deadlock, cyl., threshold									
1000	Narrow stile, no glazing, standard hardware, 3′-0″ x 7′-0″, single	2 Carp	3	5.333	Ea.	590	222		812	985
1200	Pair of 3′-0″ x 7′-0″		1.70	9.412	Pr.	1,175	390		1,565	1,875
1500	3′-6″ x 7′-0″, single		3	5.333	Ea.	685	222		907	1,100
2100	Medium stile, 3′-0″ x 7′-0″, single		3	5.333	″	720	222		942	1,125
2200	Pair of 3′-0″ x 7′-0″		1.70	9.412	Pr.	1,450	390		1,840	2,175
2300	3′-6″ x 7′-0″, single	↓	3	5.333	Ea.	930	222		1,152	1,375
5000	Flush panel doors, pair of 2′-6″ x 7′-0″	2 Sswk	2	8	Pr.	1,325	375		1,700	2,100
5050	3′-0″ x 7′-0″, single		2.50	6.400	Ea.	755	300		1,055	1,350
5100	Pair of 3′-0″ x 7′-0″		2	8	Pr.	1,425	375		1,800	2,200
5150	3′-6″ x 7′-0″, single	↓	2.50	6.400	Ea.	900	300		1,200	1,525

08 13 73 – Sliding Metal Doors

08 13 73.10 Steel Sliding Doors		Crew	Daily Output	Labor-Hours	Unit	Material	Labor	Equipment	Total	Total Incl O&P
0010	**STEEL SLIDING DOORS**									
0020	Up to 50′ x 18′, electric, standard duty, minimum	L-5	360	.156	S.F.	28	7.30	2.08	37.38	46
0100	Maximum		340	.165		44	7.70	2.20	53.90	64
0500	Heavy duty, minimum		297	.189		36	8.85	2.52	47.37	57.50
0600	Maximum	↓	277	.202	↓	90	9.45	2.70	102.15	118

08 14 Wood Doors

08 14 13 – Carved Wood Doors

08 14 13.10 Types of Wood Doors, Carved		Crew	Daily Output	Labor-Hours	Unit	Material	2010 Bare Costs Labor	Equipment	Total	Total Incl O&P
0010	**TYPES OF WOOD DOORS, CARVED**									
3000	Solid wood, 1-3/4" thick stile and rail									
3020	Mahogany, 3'-0" x 7'-0", minimum	2 Carp	14	1.143	Ea.	945	47.50		992.50	1,125
3030	Maximum		10	1.600		1,500	66.50		1,566.50	1,750
3040	3'-6" x 8'-0", minimum		10	1.600		1,250	66.50		1,316.50	1,475
3050	Maximum		8	2		1,850	83		1,933	2,175
3100	Pine, 3'-0" x 7'-0", minimum		14	1.143		450	47.50		497.50	570
3110	Maximum		10	1.600		750	66.50		816.50	925
3120	3'-6" x 8'-0", minimum		10	1.600		885	66.50		951.50	1,075
3130	Maximum		8	2		1,800	83		1,883	2,100
3200	Red oak, 3'-0" x 7'-0", minimum		14	1.143		1,700	47.50		1,747.50	1,950
3210	Maximum		10	1.600		2,025	66.50		2,091.50	2,325
3220	3'-6" x 8'-0", minimum		10	1.600		2,200	66.50		2,266.50	2,525
3230	Maximum	↓	8	2	↓	3,200	83		3,283	3,650
4000	Hand carved door, mahogany									
4020	3'-0" x 7'-0", minimum	2 Carp	14	1.143	Ea.	1,600	47.50		1,647.50	1,825
4030	Maximum		11	1.455		3,600	60.50		3,660.50	4,050
4040	3'-6" x 8'-0", minimum		10	1.600		2,500	66.50		2,566.50	2,850
4050	Maximum		8	2		3,500	83		3,583	3,975
4200	Rose wood, 3'-0" x 7'-0", minimum		14	1.143		5,000	47.50		5,047.50	5,575
4210	Maximum		11	1.455		13,600	60.50		13,660.50	15,100
4220	3'-6" x 8'-0", minimum	↓	10	1.600	↓	5,700	66.50		5,766.50	6,375
4280	For 6'-8" high door, deduct from 7'-0" door					34			34	37.50
4400	For custom finish, add					400			400	440
4600	Side light, mahogany, 7'-0" x 1'-6" wide, minimum	2 Carp	18	.889		900	37		937	1,050
4610	Maximum		14	1.143		2,625	47.50		2,672.50	2,975
4620	8'-0" x 1'-6" wide, minimum		14	1.143		1,700	47.50		1,747.50	1,950
4630	Maximum		10	1.600		1,950	66.50		2,016.50	2,250
4640	Side light, oak, 7'-0" x 1'-6" wide, minimum		18	.889		1,100	37		1,137	1,250
4650	Maximum		14	1.143		1,900	47.50		1,947.50	2,175
4660	8'-0" x 1'-6" wide, minimum		14	1.143		1,000	47.50		1,047.50	1,175
4670	Maximum	↓	10	1.600	↓	1,900	66.50		1,966.50	2,200

08 14 16 – Flush Wood Doors

08 14 16.09 Smooth Wood Doors		Crew	Daily Output	Labor-Hours	Unit	Material	Labor	Equipment	Total	Total Incl O&P
0010	**SMOOTH WOOD DOORS**									
0015	Flush, interior, 1-3/8", 7 ply, hollow core									
0020	Lauan face, 2'-0" x 6'-8"	2 Carp	17	.941	Ea.	32	39		71	96
0080	3'-0" x 6'-8"		17	.941		52	39		91	118
0100	4'-0" x 6'-8"		16	1		126	41.50		167.50	203
0108	1-3/4", 3'-0" x 7'-0"		16	1	↓	103	41.50		144.50	177
0112	Pair of 3'-0" x 7'-0"		9	1.778	Pr.	182	74		256	315
0120	Birch face, 2'-0" x 6'-8"		17	.941	Ea.	45	39		84	110
0140	2'-6" x 6'-8"		17	.941		82.50	39		121.50	152
0180	3'-0" x 6'-8"		17	.941		94	39		133	164
0200	4'-0" x 6'-8"		16	1		158	41.50		199.50	238
0210	1-3/4", 3'-0" x 7'-0"		16	1	↓	103	41.50		144.50	177
0214	Pair of 3'-0" x 7'-0"		9	1.778	Pr.	197	74		271	330
0220	Oak face, 2'-0" x 6'-8"		17	.941	Ea.	86.50	39		125.50	156
0240	2'-6" x 6'-8"		17	.941		94	39		133	164
0280	3'-0" x 6'-8"		17	.941		101	39		140	172
0300	4'-0" x 6'-8"		16	1		125	41.50		166.50	201
0320	Walnut face, 2'-0" x 6'-8"	↓	17	.941	↓	171	39		210	249

08 14 16.09 Smooth Wood Doors		Crew	Daily Output	Labor-Hours	Unit	Material	2010 Bare Costs Labor	Equipment	Total	Total Incl O&P
0340	2′-6″ x 6′-8″	2 Carp	17	.941	Ea.	175	39		214	253
0380	3′-0″ x 6′-8″		17	.941		182	39		221	262
0400	4′-0″ x 6′-8″		16	1		205	41.50		246.50	290
0430	For 7′-0″ high, add					16			16	17.60
0440	For 8′-0″ high, add					30			30	33
0480	For prefinishing, clear, add					35			35	38.50
0500	For prefinishing, stain, add					45			45	49.50
1320	M.D. overlay on hardboard, 2′-0″ x 6′-8″	2 Carp	17	.941		94	39		133	164
1340	2′-6″ x 6′-8″		17	.941		94	39		133	164
1380	3′-0″ x 6′-8″		17	.941		112	39		151	184
1400	4′-0″ x 6′-8″		16	1		160	41.50		201.50	240
1420	For 7′-0″ high, add					10			10	11
1440	For 8′-0″ high, add					25			25	27.50
1720	H.P. plastic laminate, 2′-0″ x 6′-8″	2 Carp	16	1		224	41.50		265.50	310
1740	2′-6″ x 6′-8″		16	1		224	41.50		265.50	310
1780	3′-0″ x 6′-8″		15	1.067		278	44.50		322.50	375
1800	4′-0″ x 6′-8″		14	1.143		355	47.50		402.50	470
1820	For 7′-0″ high, add					10			10	11
1840	For 8′-0″ high, add					25			25	27.50
2020	5 ply particle core, lauan face, 2′-6″ x 6′-8″	2 Carp	15	1.067		82	44.50		126.50	159
2040	3′-0″ x 6′-8″		14	1.143		85	47.50		132.50	167
2080	3′-0″ x 7′-0″		13	1.231		93	51		144	181
2100	4′-0″ x 7′-0″		12	1.333		110	55.50		165.50	207
2120	Birch face, 2′-6″ x 6′-8″		15	1.067		94	44.50		138.50	172
2140	3′-0″ x 6′-8″		14	1.143		103	47.50		150.50	186
2180	3′-0″ x 7′-0″		13	1.231		104	51		155	193
2200	4′-0″ x 7′-0″		12	1.333		127	55.50		182.50	226
2220	Oak face, 2′-6″ x 6′-8″		15	1.067		103	44.50		147.50	182
2240	3′-0″ x 6′-8″		14	1.143		114	47.50		161.50	198
2280	3′-0″ x 7′-0″		13	1.231		117	51		168	208
2300	4′-0″ x 7′-0″		12	1.333		143	55.50		198.50	243
2320	Walnut face, 2′-0″ x 6′-8″		15	1.067		114	44.50		158.50	194
2340	2′-6″ x 6′-8″		14	1.143		130	47.50		177.50	216
2380	3′-0″ x 6′-8″		13	1.231		147	51		198	241
2400	4′-0″ x 6′-8″		12	1.333		192	55.50		247.50	297
2440	For 8′-0″ high, add					30			30	33
2460	For 8′-0″ high walnut, add					16			16	17.60
2480	For solid wood core, add					35			35	38.50
2720	For prefinishing, clear, add					23			23	25.50
2740	For prefinishing, stain, add					50			50	55
3320	M.D. overlay on hardboard, 2′-6″ x 6′-8″	2 Carp	14	1.143		104	47.50		151.50	187
3340	3′-0″ x 6′-8″		13	1.231		110	51		161	200
3380	3′-0″ x 7′-0″		12	1.333		112	55.50		167.50	209
3400	4′-0″ x 7′-0″		10	1.600		145	66.50		211.50	262
3440	For 8′-0″ height, add					32			32	35
3460	For solid wood core, add					40			40	44
3720	H.P. plastic laminate, 2′-6″ x 6′-8″	2 Carp	13	1.231		152	51		203	246
3740	3′-0″ x 6′-8″		12	1.333		175	55.50		230.50	279
3780	3′-0″ x 7′-0″		11	1.455		180	60.50		240.50	291
3800	4′-0″ x 7′-0″		8	2		218	83		301	370
3840	For 8′-0″ height, add					32			32	35
3860	For solid wood core, add					38			38	42
4000	Exterior, flush, solid wood stave core, birch, 1-3/4″ x 7′-0″ x 2′-6″	2 Carp	15	1.067		162	44.50		206.50	247

08 14 Wood Doors

08 14 16 – Flush Wood Doors

08 14 16.09 Smooth Wood Doors		Crew	Daily Output	Labor-Hours	Unit	Material	2010 Bare Costs Labor	Equipment	Total	Total Incl O&P
4020	2'-8" wide	2 Carp	15	1.067	Ea.	170	44.50		214.50	256
4040	3'-0" wide		14	1.143		185	47.50		232.50	277
4100	Oak faced 1-3/4" x 7'-0" x 2'-6" wide		15	1.067		180	44.50		224.50	267
4120	2'-8" wide		15	1.067		195	44.50		239.50	284
4140	3'-0" wide		14	1.143		206	47.50		253.50	300
4200	Walnut faced, 1-3/4" x 7'-0" x 2'-6" wide		15	1.067		266	44.50		310.50	360
4220	2'-8" wide		15	1.067		280	44.50		324.50	380
4240	3'-0" wide		14	1.143		290	47.50		337.50	395
4300	For 6'-8" high door, deduct from 7'-0" door					16			16	17.60
5000	Wood doors, for vision lite, add					85			85	93.50
5010	Wood doors, for narrow lite, add					92			92	101
5015	Wood doors, for bottom (or top) louver, add					245			245	269

08 14 16.10 Wood Doors Decorator		Crew	Daily Output	Labor-Hours	Unit	Material	Labor	Equipment	Total	Total Incl O&P
0010	**WOOD DOORS DECORATOR**									
0040	7 ply hollow core lauan face, 2'-6" x 6'-8"	2 Carp	17	.941	Ea.	37.50	39		76.50	102

08 14 16.20 Wood Fire Doors		Crew	Daily Output	Labor-Hours	Unit	Material	Labor	Equipment	Total	Total Incl O&P
0010	**WOOD FIRE DOORS**									
0020	Particle core, 7 face plys, "B" label,									
0040	1 hour, birch face, 1-3/4" x 2'-6" x 6'-8"	2 Carp	14	1.143	Ea.	380	47.50		427.50	495
0080	3'-0" x 6'-8"		13	1.231		390	51		441	510
0090	3'-0" x 7'-0"		12	1.333		410	55.50		465.50	535
0100	4'-0" x 7'-0"		12	1.333		510	55.50		565.50	645
0140	Oak face, 2'-6" x 6'-8"		14	1.143		335	47.50		382.50	445
0180	3'-0" x 6'-8"		13	1.231		350	51		401	465
0190	3'-0" x 7'-0"		12	1.333		365	55.50		420.50	485
0200	4'-0" x 7'-0"		12	1.333		480	55.50		535.50	615
0240	Walnut face, 2'-6" x 6'-8"		14	1.143		450	47.50		497.50	570
0280	3'-0" x 6'-8"		13	1.231		470	51		521	595
0290	3'-0" x 7'-0"		12	1.333		490	55.50		545.50	625
0300	4'-0" x 7'-0"		12	1.333		610	55.50		665.50	755
0440	M.D. overlay on hardboard, 2'-6" x 6'-8"		15	1.067		300	44.50		344.50	400
0480	3'-0" x 6'-8"		14	1.143		315	47.50		362.50	420
0490	3'-0" x 7'-0"		13	1.231		330	51		381	445
0500	4'-0" x 7'-0"		12	1.333		400	55.50		455.50	525
0740	90 minutes, birch face, 1-3/4" x 2'-6" x 6'-8"		14	1.143		320	47.50		367.50	425
0780	3'-0" x 6'-8"		13	1.231		305	51		356	415
0790	3'-0" x 7'-0"		12	1.333		375	55.50		430.50	500
0800	4'-0" x 7'-0"		12	1.333		470	55.50		525.50	605
0840	Oak face, 2'-6" x 6'-8"		14	1.143		425	47.50		472.50	540
0880	3'-0" x 6'-8"		13	1.231		435	51		486	555
0890	3'-0" x 7'-0"		12	1.333		450	55.50		505.50	580
0900	4'-0" x 7'-0"		12	1.333		580	55.50		635.50	725
0940	Walnut face, 2'-6" x 6'-8"		14	1.143		390	47.50		437.50	505
0980	3'-0" x 6'-8"		13	1.231		400	51		451	520
0990	3'-0" x 7'-0"		12	1.333		460	55.50		515.50	590
1000	4'-0" x 7'-0"		12	1.333		610	55.50		665.50	755
1140	M.D. overlay on hardboard, 2'-6" x 6'-8"		15	1.067		350	44.50		394.50	455
1180	3'-0" x 6'-8"		14	1.143		370	47.50		417.50	485
1190	3'-0" x 7'-0"		13	1.231		400	51		451	520
1200	4'-0" x 7'-0"		12	1.333		450	55.50		505.50	580
1240	For 8'-0" height, add					65			65	71.50
1260	For 8'-0" height walnut, add					80			80	88

08 14 Wood Doors

08 14 16 – Flush Wood Doors

08 14 16.20 Wood Fire Doors		Crew	Daily Output	Labor-Hours	Unit	Material	2010 Bare Costs Labor	Equipment	Total	Total Incl O&P
2200	Custom architectural "B" label, flush, 1-3/4" thick, birch,									
2210	Solid core									
2220	2'-6" x 7'-0"	2 Carp	15	1.067	Ea.	296	44.50		340.50	395
2260	3'-0" x 7'-0"		14	1.143		305	47.50		352.50	410
2300	4'-0" x 7'-0"		13	1.231		400	51		451	515
2420	4'-0" x 8'-0"		11	1.455		430	60.50		490.50	565
2480	For oak veneer, add					50%				
2500	For walnut veneer, add					75%				

08 14 23 – Clad Wood Doors

08 14 23.13 Metal-Faced Wood Doors		Crew	Daily Output	Labor-Hours	Unit	Material	Labor	Equipment	Total	Total Incl O&P
0010	**METAL-FACED WOOD DOORS**									
0020	Interior, flush type, 3' x 7'	2 Carp	4.30	3.721	Opng.	240	155		395	500

08 14 23.20 Tin Clad Wood Doors		Crew	Daily Output	Labor-Hours	Unit	Material	Labor	Equipment	Total	Total Incl O&P
0010	**TIN CLAD WOOD DOORS**									
0020	3 ply, 6' x 7', double sliding, doors only	2 Carp	1	16	Opng.	1,900	665		2,565	3,125
1000	For electric operator, add	1 Elec	2	4	"	3,250	196		3,446	3,875

08 14 33 – Stile and Rail Wood Doors

08 14 33.10 Wood Doors Paneled		Crew	Daily Output	Labor-Hours	Unit	Material	Labor	Equipment	Total	Total Incl O&P
0010	**WOOD DOORS PANELED**									
0020	Interior, six panel, hollow core, 1-3/8" thick									
0040	Molded hardboard, 2'-0" x 6'-8"	2 Carp	17	.941	Ea.	56	39		95	122
0060	2'-6" x 6'-8"		17	.941		60	39		99	127
0070	2'-8" x 6'-8"		17	.941		63	39		102	130
0080	3'-0" x 6'-8"		17	.941		66	39		105	133
0140	Embossed print, molded hardboard, 2'-0" x 6'-8"		17	.941		60	39		99	127
0160	2'-6" x 6'-8"		17	.941		60	39		99	127
0180	3'-0" x 6'-8"		17	.941		66	39		105	133
0540	Six panel, solid, 1-3/8" thick, pine, 2'-0" x 6'-8"		15	1.067		145	44.50		189.50	229
0560	2'-6" x 6'-8"		14	1.143		162	47.50		209.50	251
0580	3'-0" x 6'-8"		13	1.231		136	51		187	229
1020	Two panel, bored rail, solid, 1-3/8" thick, pine, 1'-6" x 6'-8"		16	1		263	41.50		304.50	355
1040	2'-0" x 6'-8"		15	1.067		345	44.50		389.50	450
1060	2'-6" x 6'-8"		14	1.143		395	47.50		442.50	510
1340	Two panel, solid, 1-3/8" thick, fir, 2'-0" x 6'-8"		15	1.067		150	44.50		194.50	234
1360	2'-6" x 6'-8"		14	1.143		200	47.50		247.50	293
1380	3'-0" x 6'-8"		13	1.231		405	51		456	525
1740	Five panel, solid, 1-3/8" thick, fir, 2'-0" x 6'-8"		15	1.067		270	44.50		314.50	365
1760	2'-6" x 6'-8"		14	1.143		415	47.50		462.50	530
1780	3'-0" x 6'-8"		13	1.231		415	51		466	535

08 14 33.20 Wood Doors Residential		Crew	Daily Output	Labor-Hours	Unit	Material	Labor	Equipment	Total	Total Incl O&P
0010	**WOOD DOORS RESIDENTIAL**									
0200	Exterior, combination storm & screen, pine									
0260	2'-8" wide	2 Carp	10	1.600	Ea.	286	66.50		352.50	415
0280	3'-0" wide		9	1.778		295	74		369	440
0300	7'-1" x 3'-0" wide		9	1.778		335	74		409	485
0400	Full lite, 6'-9" x 2'-6" wide		11	1.455		296	60.50		356.50	420
0420	2'-8" wide		10	1.600		296	66.50		362.50	425
0440	3'-0" wide		9	1.778		300	74		374	445
0500	7'-1" x 3'-0" wide		9	1.778		325	74		399	470
0700	Dutch door, pine, 1-3/4" x 6'-8" x 2'-8" wide, minimum		12	1.333		675	55.50		730.50	830
0720	Maximum		10	1.600		825	66.50		891.50	1,000

08 14 Wood Doors

08 14 33 – Stile and Rail Wood Doors

	08 14 33.20 Wood Doors Residential	Crew	Daily Output	Labor-Hours	Unit	2010 Bare Costs Material	Labor	Equipment	Total	Total Incl O&P
0800	3′-0″ wide, minimum	2 Carp	12	1.333	Ea.	690	55.50		745.50	840
0820	Maximum		10	1.600		900	66.50		966.50	1,100
1000	Entrance door, colonial, 1-3/4″ x 6′-8″ x 2′-8″ wide		16	1		460	41.50		501.50	570
1020	6 panel pine, 3′-0″ wide		15	1.067		430	44.50		474.50	545
1100	8 panel pine, 2′-8″ wide		16	1		570	41.50		611.50	695
1120	3′-0″ wide		15	1.067		555	44.50		599.50	680
1200	For tempered safety glass lites, (min of 2) add					74.50			74.50	82
1300	Flush, birch, solid core, 1-3/4″ x 6′-8″ x 2′-8″ wide	2 Carp	16	1		104	41.50		145.50	178
1320	3′-0″ wide		15	1.067		114	44.50		158.50	194
1350	7′-0″ x 2′-8″ wide		16	1		113	41.50		154.50	188
1360	3′-0″ wide		15	1.067		121	44.50		165.50	202
1380	For tempered safety glass lites, add					102			102	112
2700	Interior, closet, bi-fold, w/hardware, no frame or trim incl.									
2720	Flush, birch, 6′-6″ or 6′-8″ x 2′-6″ wide	2 Carp	13	1.231	Ea.	51	51		102	135
2740	3′-0″ wide		13	1.231		55	51		106	139
2760	4′-0″ wide		12	1.333		97.50	55.50		153	193
2780	5′-0″ wide		11	1.455		100	60.50		160.50	203
2800	6′-0″ wide		10	1.600		126	66.50		192.50	241
3000	Raised panel pine, 6′-6″ or 6′-8″ x 2′-6″ wide		13	1.231		178	51		229	275
3020	3′-0″ wide		13	1.231		273	51		324	380
3040	4′-0″ wide		12	1.333		305	55.50		360.50	420
3060	5′-0″ wide		11	1.455		365	60.50		425.50	495
3080	6′-0″ wide		10	1.600		400	66.50		466.50	540
3200	Louvered, pine 6′-6″ or 6′-8″ x 2′-6″ wide		13	1.231		137	51		188	230
3220	3′-0″ wide		13	1.231		197	51		248	296
3240	4′-0″ wide		12	1.333		222	55.50		277.50	330
3260	5′-0″ wide		11	1.455		256	60.50		316.50	375
3280	6′-0″ wide		10	1.600		280	66.50		346.50	405
4400	Bi-passing closet, incl. hardware and frame, no trim incl.									
4420	Flush, lauan, 6′-8″ x 4′-0″ wide	2 Carp	12	1.333	Opng.	169	55.50		224.50	271
4440	5′-0″ wide		11	1.455		184	60.50		244.50	295
4460	6′-0″ wide		10	1.600		198	66.50		264.50	320
4600	Flush, birch, 6′-8″ x 4′-0″ wide		12	1.333		217	55.50		272.50	325
4620	5′-0″ wide		11	1.455		209	60.50		269.50	325
4640	6′-0″ wide		10	1.600		254	66.50		320.50	380
4800	Louvered, pine, 6′-8″ x 4′-0″ wide		12	1.333		400	55.50		455.50	525
4820	5′-0″ wide		11	1.455		390	60.50		450.50	520
4840	6′-0″ wide		10	1.600		515	66.50		581.50	670
5000	Paneled, pine, 6′-8″ x 4′-0″ wide		12	1.333		500	55.50		555.50	635
5020	5′-0″ wide		11	1.455		405	60.50		465.50	540
5040	6′-0″ wide		10	1.600		595	66.50		661.50	755
6100	Folding accordion, closet, including track and frame									
6120	Vinyl, 2 layer, stock see Div. 10 22 26.13	2 Carp	10	1.600	Ea.	61	66.50		127.50	169
6140	Woven mahogany and vinyl, stock		10	1.600		45	66.50		111.50	152
6160	Wood slats with vinyl overlay, stock		10	1.600		139	66.50		205.50	255
6180	Economy vinyl, stock		10	1.600		28.50	66.50		95	134
6200	Rigid PVC		10	1.600		33.50	66.50		100	139
6220	For custom partition, add					25%	10%			
7310	Passage doors, flush, no frame included									
7320	Hardboard, hollow core, 1-3/8″ x 6′-8″ x 1′-6″ wide	2 Carp	18	.889	Ea.	42.50	37		79.50	104
7330	2′-0″ wide		18	.889		42	37		79	103
7340	2′-6″ wide		18	.889		46.50	37		83.50	108
7350	2′-8″ wide		18	.889		48	37		85	110

08 14 Wood Doors

08 14 33 – Stile and Rail Wood Doors

08 14 33.20 Wood Doors Residential		Crew	Daily Output	Labor-Hours	Unit	Material	2010 Bare Costs Labor	Equipment	Total	Total Incl O&P
7360	3'-0" wide	2 Carp	17	.941	Ea.	51	39		90	117
7420	Lauan, hollow core, 1-3/8" x 6'-8" x 1'-6" wide		18	.889		31	37		68	91
7440	2'-0" wide		18	.889		31	37		68	91
7450	2'-4" wide		18	.889		34.50	37		71.50	95
7460	2'-6" wide		18	.889		34.50	37		71.50	95
7480	2'-8" wide		18	.889		36	37		73	96.50
7500	3'-0" wide		17	.941		37	39		76	101
7700	Birch, hollow core, 1-3/8" x 6'-8" x 1'-6" wide		18	.889		38	37		75	99
7720	2'-0" wide		18	.889		41	37		78	102
7740	2'-6" wide		18	.889		49	37		86	111
7760	2'-8" wide		18	.889		49.50	37		86.50	112
7780	3'-0" wide CN		17	.941		51	39		90	117
8000	Pine louvered, 1-3/8" x 6'-8" x 1'-6" wide		19	.842		101	35		136	165
8020	2'-0" wide		18	.889		128	37		165	197
8040	2'-6" wide		18	.889		139	37		176	210
8060	2'-8" wide		18	.889		146	37		183	218
8080	3'-0" wide		17	.941		167	39		206	244
8300	Pine paneled, 1-3/8" x 6'-8" x 1'-6" wide		19	.842		111	35		146	176
8320	2'-0" wide		18	.889		130	37		167	200
8330	2'-4" wide		18	.889		150	37		187	222
8340	2'-6" wide		18	.889		158	37		195	230
8360	2'-8" wide		18	.889		161	37		198	234
8380	3'-0" wide	↓	17	.941	↓	167	39		206	244

08 14 40 – Interior Cafe Doors

08 14 40.10 Cafe Style Doors		Crew	Daily Output	Labor-Hours	Unit	Material	Labor	Equipment	Total	Total Incl O&P
0010	**CAFE STYLE DOORS**									
6520	Interior cafe doors, 2'-6" opening, stock, panel pine	2 Carp	16	1	Ea.	188	41.50		229.50	270
6540	3'-0" opening	"	16	1	"	195	41.50		236.50	279
6550	Louvered pine									
6560	2'-6" opening	2 Carp	16	1	Ea.	164	41.50		205.50	244
8000	3'-0" opening		16	1		175	41.50		216.50	256
8010	2'-6" opening, hardwood		16	1		238	41.50		279.50	325
8020	3'-0" opening	↓	16	1	↓	264	41.50		305.50	355

08 16 Composite Doors

08 16 13 – Fiberglass Doors

08 16 13.10 Entrance Doors, Fiberous Glass		Crew	Daily Output	Labor-Hours	Unit	Material	Labor	Equipment	Total	Total Incl O&P
0010	**ENTRANCE DOORS, FIBEROUS GLASS**									
0020	Exterior, fiberglass, door, 2'-8" wide x 6'-8" high G	2 Carp	15	1.067	Ea.	252	44.50		296.50	345
0040	3'-0" wide x 6'-8" high G		15	1.067		252	44.50		296.50	345
0060	3'-0" wide x 7'-0" high G		15	1.067		450	44.50		494.50	565
0080	3'-0" wide x 6'-8" high, with two lites G		15	1.067		294	44.50		338.50	395
0100	3'-0" wide x 8'-0" high, with two lites G		15	1.067		490	44.50		534.50	605
0110	Half glass, 3'-0" wide x 6'-8" high G		15	1.067		294	44.50		338.50	395
0120	3'-0" wide x 6'-8" high, low e G		15	1.067		325	44.50		369.50	425
0130	3'-0" wide x 8'-0" high G		15	1.067		580	44.50		624.50	705
0140	3'-0" wide x 8'-0" high, low e G	↓	15	1.067		640	44.50		684.50	770
0150	Side lights, 1'-0" wide x 6'-8" high, G					247			247	271
0160	1'-0" wide x 6'-8" high, low e G					256			256	281
0180	1'-0" wide x 6'-8" high, full glass G				↓	275			275	305

08 16 Composite Doors

08 16 13 – Fiberglass Doors

08 16 13.10 Entrance Doors, Fiberous Glass		Crew	Daily Output	Labor-Hours	Unit	Material	2010 Bare Costs Labor	Equipment	Total	Total Incl O&P
0190	1'-0" wide x 6'-8" high, low e G				Ea.	284			284	315

08 16 14 – French Doors

08 16 14.10 Exterior Doors With Glass Lites		Crew	Daily Output	Labor-Hours	Unit	Material	Labor	Equipment	Total	Total Incl O&P
0010	**EXTERIOR DOORS WITH GLASS LITES**									
0020	French, Fir, 1-3/4", 3'-0"wide x 6'-8" high	2 Carp	12	1.333	Ea.	460	55.50		515.50	590
0025	double		12	1.333		925	55.50		980.50	1,100
0030	Maple, 1-3/4", 3'-0"wide x 6'-8" high		12	1.333		545	55.50		600.50	685
0035	double		12	1.333		1,100	55.50		1,155.50	1,275
0040	Cherry, 1-3/4", 3'-0"wide x 6'-8" high		12	1.333		625	55.50		680.50	770
0045	double		12	1.333		1,250	55.50		1,305.50	1,450
0100	Mahogany, 1-3/4", 3'-0"wide x 8'-0" high		10	1.600		510	66.50		576.50	660
0105	double		10	1.600		1,025	66.50		1,091.50	1,225
0110	Fir, 1-3/4", 3'-0"wide x 8'-0" high		10	1.600		870	66.50		936.50	1,050
0115	double		10	1.600		1,750	66.50		1,816.50	2,025
0120	Oak, 1-3/4", 3'-0"wide x 8'-0" high		10	1.600		1,800	66.50		1,866.50	2,075
0125	double	↓	10	1.600	↓	3,600	66.50		3,666.50	4,050

08 17 Integrated Door Opening Assemblies

08 17 13 – Integrated Metal Door Opening Assemblies

08 17 13.20 Tubular Steel Swing Doors		Crew	Daily Output	Labor-Hours	Unit	Material	Labor	Equipment	Total	Total Incl O&P
0010	**TUBULAR STEEL SWING DOORS**									
0020	Tubular steel, 7' high, single, 3'-4" opening	2 Sswk	2.50	6.400	Ea.	875	300		1,175	1,500
0100	Double, 6'-0" opening	"	2	8	Pr.	1,100	375		1,475	1,850

08 17 23 – Integrated Wood Door Opening Assemblies

08 17 23.10 Pre-Hung Doors		Crew	Daily Output	Labor-Hours	Unit	Material	Labor	Equipment	Total	Total Incl O&P
0010	**PRE-HUNG DOORS**									
0300	Exterior, wood, comb. storm & screen, 6'-9" x 2'-6" wide	2 Carp	15	1.067	Ea.	289	44.50		333.50	390
0320	2'-8" wide		15	1.067		289	44.50		333.50	390
0340	3'-0" wide	↓	15	1.067		296	44.50		340.50	395
0360	For 7'-0" high door, add				↓	24			24	26.50
1600	Entrance door, flush, birch, solid core									
1620	4-5/8" solid jamb, 1-3/4" x 6'-8" x 2'-8" wide	2 Carp	16	1	Ea.	285	41.50		326.50	380
1640	3'-0" wide		16	1		350	41.50		391.50	450
1642	5-5/8" jamb	↓	16	1		325	41.50		366.50	420
1680	For 7'-0" high door, add				↓	21			21	23
2000	Entrance door, colonial, 6 panel pine									
2020	4-5/8" solid jamb, 1-3/4" x 6'-8" x 2'-8" wide	2 Carp	16	1	Ea.	530	41.50		571.50	645
2040	3'-0" wide	"	16	1		530	41.50		571.50	645
2060	For 7'-0" high door, add					52.50			52.50	58
2200	For 5-5/8" solid jamb, add				↓	40.50			40.50	44.50
4000	Interior, passage door, 4-5/8" solid jamb									
4400	Lauan, flush, solid core, 1-3/8" x 6'-8" x 2'-6" wide	2 Carp	17	.941	Ea.	182	39		221	261
4420	2'-8" wide		17	.941		182	39		221	261
4440	3'-0" wide		16	1		195	41.50		236.50	279
4600	Hollow core, 1-3/8" x 6'-8" x 2'-6" wide		17	.941		120	39		159	193
4620	2'-8" wide		17	.941		120	39		159	193
4640	3'-0" wide	↓	16	1		135	41.50		176.50	213
4700	For 7'-0" high door, add					25			25	27.50
5000	Birch, flush, solid core, 1-3/8" x 6'-8" x 2'-6" wide	2 Carp	17	.941		187	39		226	267
5020	2'-8" wide	↓	17	.941	↓	195	39		234	276

08 17 Integrated Door Opening Assemblies

08 17 23 – Integrated Wood Door Opening Assemblies

08 17 23.10 Pre-Hung Doors		Crew	Daily Output	Labor-Hours	Unit	Material	2010 Bare Costs Labor	Equipment	Total	Total Incl O&P
5040	3'-0" wide	2 Carp	16	1	Ea.	217	41.50		258.50	305
5200	Hollow core, 1-3/8" x 6'-8" x 2'-6" wide		17	.941		147	39		186	223
5220	2'-8" wide		17	.941		153	39		192	229
5240	3'-0" wide		16	1		152	41.50		193.50	232
5280	For 7'-0" high door, add					22			22	24
5500	Hardboard paneled, 1-3/8" x 6'-8" x 2'-6" wide	2 Carp	17	.941		142	39		181	217
5520	2'-8" wide		17	.941		150	39		189	226
5540	3'-0" wide		16	1		147	41.50		188.50	226
6000	Pine paneled, 1-3/8" x 6'-8" x 2'-6" wide		17	.941		245	39		284	330
6020	2'-8" wide		17	.941		265	39		304	355
6040	3'-0" wide		16	1		270	41.50		311.50	360
6500	For 5-5/8" solid jamb, add					19			19	21
6520	For split jamb, deduct					21			21	23

08 31 Access Doors and Panels

08 31 13 – Access Doors and Frames

08 31 13.10 Types of Framed Access Doors		Crew	Daily Output	Labor-Hours	Unit	Material	Labor	Equipment	Total	Total Incl O&P
0010	TYPES OF FRAMED ACCESS DOORS									
1000	Fire rated door with lock									
1100	Metal, 12" x 12"	1 Carp	10	.800	Ea.	158	33		191	225
1150	18" x 18"		9	.889		205	37		242	283
1200	24" x 24"		9	.889		325	37		362	415
1250	24" x 36"		8	1		335	41.50		376.50	430
1300	24" x 48"		8	1		410	41.50		451.50	520
1350	36" x 36"		7.50	1.067		495	44.50		539.50	615
1400	48" x 48"		7.50	1.067		635	44.50		679.50	765
1600	Stainless steel, 12" x 12"		10	.800		282	33		315	360
1650	18" x 18"		9	.889		410	37		447	505
1700	24" x 24"		9	.889		500	37		537	605
1750	24" x 36"		8	1		640	41.50		681.50	770
2000	Flush door for finishing									
2100	Metal 8" x 8"	1 Carp	10	.800	Ea.	45	33		78	101
2150	12" x 12"	"	10	.800	"	50	33		83	106
3000	Recessed door for acoustic tile									
3100	Metal, 12" x 12"	1 Carp	4.50	1.778	Ea.	69	74		143	190
3150	12" x 24"		4.50	1.778		90	74		164	213
3200	24" x 24"		4	2		121	83		204	261
3250	24" x 36"		4	2		155	83		238	299
4000	Recessed door for drywall									
4100	Metal 12" x 12"	1 Carp	6	1.333	Ea.	77	55.50		132.50	170
4150	12" x 24"		5.50	1.455		114	60.50		174.50	218
4200	24" x 36"		5	1.600		182	66.50		248.50	300
6000	Standard door									
6100	Metal, 8" x 8"	1 Carp	10	.800	Ea.	40	33		73	95
6150	12" x 12"		10	.800		45	33		78	101
6200	18" x 18"		9	.889		62	37		99	125
6250	24" x 24"		9	.889		82	37		119	147
6300	24" x 36"		8	1		120	41.50		161.50	196
6350	36" x 36"		8	1		146	41.50		187.50	225
6500	Stainless steel, 8" x 8"		10	.800		80	33		113	139
6550	12" x 12"		10	.800		107	33		140	169

08 31 Access Doors and Panels

08 31 13 – Access Doors and Frames

08 31 13.10 Types of Framed Access Doors		Crew	Daily Output	Labor-Hours	Unit	Material	2010 Bare Costs Labor	Equipment	Total	Total Incl O&P
6600	18" x 18"	1 Carp	9	.889	Ea.	198	37		235	275
6650	24" x 24"	↓	9	.889		260	37		297	345
7010	Aluminum cover, Ceiling hatches, 2'-6" x 2'-6", single leaf, st fr	G-3	11	2.909	↓	540	120		660	775

08 31 13.20 Bulkhead/Cellar Doors

		Crew	Daily Output	Labor-Hours	Unit	Material	Labor	Equipment	Total	Total Incl O&P
0010	**BULKHEAD/CELLAR DOORS**									
0020	Steel, not incl. sides, 44" x 62"	1 Carp	5.50	1.455	Ea.	490	60.50		550.50	635
0100	52" x 73"		5.10	1.569		660	65		725	825
0500	With sides and foundation plates, 57" x 45" x 24"		4.70	1.702		700	70.50		770.50	880
0600	42" x 49" x 51"	↓	4.30	1.860	↓	790	77.50		867.50	990

08 31 13.30 Commercial Floor Doors

		Crew	Daily Output	Labor-Hours	Unit	Material	Labor	Equipment	Total	Total Incl O&P
0010	**COMMERCIAL FLOOR DOORS**									
0020	Aluminum tile, steel frame, one leaf, 2' x 2' opng.	2 Sswk	3.50	4.571	Opng.	775	214		989	1,225
0050	3'-6" x 3'-6" opening		3.50	4.571		1,475	214		1,689	2,000
0500	Double leaf, 4' x 4' opening		3	5.333		1,575	250		1,825	2,175
0550	5' x 5' opening	↓	3	5.333	↓	2,800	250		3,050	3,525

08 31 13.35 Industrial Floor Doors

		Crew	Daily Output	Labor-Hours	Unit	Material	Labor	Equipment	Total	Total Incl O&P
0010	**INDUSTRIAL FLOOR DOORS**									
0020	Steel 300 psf L.L., single leaf, 2' x 2', 175#	2 Sswk	6	2.667	Opng.	735	125		860	1,025
0050	3' x 3' opening, 300#		5.50	2.909		1,000	136		1,136	1,350
0300	Double leaf, 4' x 4' opening, 455#		5	3.200		1,625	150		1,775	2,075
0350	5' x 5' opening, 645#		4.50	3.556		2,450	167		2,617	3,000
1000	Aluminum, 300 psf L.L., single leaf, 2' x 2', 60#		6	2.667		740	125		865	1,025
1050	3' x 3' opening, 100#		5.50	2.909		1,100	136		1,236	1,450
1500	Double leaf, 4' x 4' opening, 160#		5	3.200		1,950	150		2,100	2,425
1550	5' x 5' opening, 235#		4.50	3.556		2,675	167		2,842	3,250
2000	Aluminum, 150 psf L.L., single leaf, 2' x 2', 60#		6	2.667		610	125		735	890
2050	3' x 3' opening, 95#		5.50	2.909		920	136		1,056	1,250
2500	Double leaf, 4' x 4' opening, 150#		5	3.200		1,350	150		1,500	1,750
2550	5' x 5' opening, 230#	↓	4.50	3.556	↓	1,825	167		1,992	2,325

08 31 13.40 Kennel Doors

		Crew	Daily Output	Labor-Hours	Unit	Material	Labor	Equipment	Total	Total Incl O&P
0010	**KENNEL DOORS**									
0020	2 way, swinging type, 13" x 19" opening	2 Carp	11	1.455	Opng.	100	60.50		160.50	203
0100	17" x 29" opening	"	11	1.455	"	224	60.50		284.50	340

08 32 Sliding Glass Doors

08 32 13 – Sliding Aluminum-Framed Glass Doors

08 32 13.10 Sliding Alumimum Doors

		Crew	Daily Output	Labor-Hours	Unit	Material	Labor	Equipment	Total	Total Incl O&P
0010	**SLIDING ALUMINUM DOORS**									
0350	Aluminum, 5/8" tempered insulated glass, 6' wide									
0400	Premium	2 Carp	4	4	Ea.	1,550	166		1,716	1,975
0450	Economy		4	4		870	166		1,036	1,225
0500	8' wide, premium		3	5.333		1,625	222		1,847	2,125
0550	Economy		3	5.333		1,400	222		1,622	1,875
0600	12' wide, premium		2.50	6.400		2,950	266		3,216	3,650
0650	Economy		2.50	6.400		1,500	266		1,766	2,050
0700	Non-insulated, 6' wide		4	4		1,075	166		1,241	1,425
4000	Aluminum, baked on enamel, temp glass, 6'-8" x 10'-0" wide		4	4		970	166		1,136	1,325
4020	Insulating glass, 6'-8" x 6'-0" wide		4	4		860	166		1,026	1,200
4040	8'-0" wide		3	5.333		1,000	222		1,222	1,450
4060	10'-0" wide	↓	2	8	↓	1,275	330		1,605	1,900

08 32 Sliding Glass Doors

08 32 13 – Sliding Aluminum-Framed Glass Doors

08 32 13.10 Sliding Alumimum Doors		Crew	Daily Output	Labor-Hours	Unit	Material	2010 Bare Costs Labor	Equipment	Total	Total Incl O&P
4080	Anodized, temp glass, 6′-8″ x 6′-0″ wide	2 Carp	4	4	Ea.	440	166		606	740
4100	8′-0″ wide		3	5.333		560	222		782	955
4120	10′-0″ wide		2	8		630	330		960	1,200
4200	Vinyl clad, anodized, temp glass, 6′-8″ x 6′-0″ wide		4	4		410	166		576	705
4240	8′-0″ wide		3	5.333		570	222		792	965
4260	10′-0″ wide		2	8		620	330		950	1,200
4280	Insulating glass, 6′-8″ x 6′-0′ wide		4	4		995	166		1,161	1,350
4300	8′-0″ wide		3	5.333		920	222		1,142	1,350
4320	10′-0″ wide	↓	4	4	↓	1,500	166		1,666	1,900

08 32 19 – Sliding Wood-Framed Glass Doors

08 32 19.15 Sliding Glass Vinyl-Clad Wood Doors		Crew	Daily Output	Labor-Hours	Unit	Material	Labor	Equipment	Total	Total Incl O&P
0010	**SLIDING GLASS VINYL-CLAD WOOD DOORS**									
0012	Vinyl clad, 1″ insul. glass, 6′-0″ x 6′-10″ high G	2 Carp	4	4	Opng.	1,575	166		1,741	2,000
0020	Glass, sliding vinyl clad, insul. glass, 6′-0″ x 6′-8″, G		4	4	Ea.	1,525	166		1,691	1,925
0030	6′-0″ x 8′-0″ high G		4	4		1,950	166		2,116	2,400
0050	5′-0″ x 6′-8″ high G		4	4	↓	1,475	166		1,641	1,875
0100	8′-0″ x 6′-10″ high G		4	4	Opng.	2,000	166		2,166	2,450
0500	4 leaf, 9′-0″ x 6′-10″ high G		3	5.333		3,175	222		3,397	3,850
0600	12′-0″ x 6′-10″ high G	↓	3	5.333	↓	3,800	222		4,022	4,525

08 33 Coiling Doors and Grilles

08 33 13 – Coiling Counter Doors

08 33 13.10 Counter Doors, Coiling Type		Crew	Daily Output	Labor-Hours	Unit	Material	Labor	Equipment	Total	Total Incl O&P
0010	**COUNTER DOORS, COILING TYPE**									
0020	Manual, incl. frm and hdwe, galv. stl., 4′ roll-up, 6′ long	2 Carp	2	8	Opng.	1,150	330		1,480	1,750
0300	Galvanized steel, UL label		1.80	8.889		1,200	370		1,570	1,900
0600	Stainless steel, 4′ high roll-up, 6′ long		2	8		2,100	330		2,430	2,800
0700	10′ long		1.80	8.889		2,450	370		2,820	3,275
2000	Aluminum, 4′ high, 4′ long		2.20	7.273		1,275	300		1,575	1,875
2020	6′ long		2	8		1,425	330		1,755	2,050
2040	8′ long		1.90	8.421		1,650	350		2,000	2,350
2060	10′ long		1.80	8.889		1,725	370		2,095	2,475
2080	14′ long		1.40	11.429		2,500	475		2,975	3,475
2100	6′ high, 4′ long		2	8		1,425	330		1,755	2,075
2120	6′ long		1.60	10		1,450	415		1,865	2,250
2140	10′ long	↓	1.40	11.429	↓	1,975	475		2,450	2,900

08 33 16 – Coiling Counter Grilles

08 33 16.10 Coiling Grilles		Crew	Daily Output	Labor-Hours	Unit	Material	Labor	Equipment	Total	Total Incl O&P
0010	**COILING GRILLES**									
0015	Aluminum, manual, incl. frame, mill finish									
0020	Top coiling, 4′ high, 4′ long	2 Sswk	3.20	5	Opng.	1,425	235		1,660	1,975
0030	6′ long		3.20	5		1,675	235		1,910	2,225
0040	8′ long		2.40	6.667		1,850	315		2,165	2,600
0050	12′ long		2.40	6.667		2,250	315		2,565	3,025
0060	16′ long		1.60	10		2,425	470		2,895	3,475
0070	6′ high, 4′ long		3.20	5		1,675	235		1,910	2,225
0080	6′ long		3.20	5		1,675	235		1,910	2,250
0090	8′ long		2.40	6.667		1,675	315		1,990	2,400
0100	12′ long		1.60	10		2,225	470		2,695	3,250
0110	16′ long	↓	1.20	13.333	↓	2,650	625		3,275	4,025

08 33 Coiling Doors and Grilles

08 33 16 – Coiling Counter Grilles

08 33 16.10 Coiling Grilles		Crew	Daily Output	Labor-Hours	Unit	Material	2010 Bare Costs Labor	Equipment	Total	Total Incl O&P
0200	Side coiling, 8′ high, 12′ long	2 Sswk	.60	26.667	Opng.	2,175	1,250		3,425	4,600
0220	18′ long		.50	32		3,475	1,500		4,975	6,475
0240	24′ long		.40	40		3,450	1,875		5,325	7,100
0260	12′ high, 12′ long		.50	32		3,175	1,500		4,675	6,150
0280	18′ long		.40	40		4,700	1,875		6,575	8,475
0300	24′ long	↓	.28	57.143	↓	6,300	2,675		8,975	11,600

08 33 23 – Overhead Coiling Doors

08 33 23.10 Rolling Service Doors

		Crew	Daily Output	Labor-Hours	Unit	Material	Labor	Equipment	Total	Total Incl O&P
0010	**ROLLING SERVICE DOORS** Steel, manual, 20 ga., incl. hardware									
0050	8′ x 8′ high	2 Sswk	1.60	10	Ea.	1,050	470		1,520	2,000
0100	10′ x 10′ high		1.40	11.429		1,775	535		2,310	2,900
0200	20′ x 10′ high		1	16		2,950	750		3,700	4,575
0300	12′ x 12′ high		1.20	13.333		1,775	625		2,400	3,075
0400	20′ x 12′ high		.90	17.778		2,050	835		2,885	3,725
0500	14′ x 14′ high		.80	20		2,900	940		3,840	4,850
0600	20′ x 16′ high		.60	26.667		3,250	1,250		4,500	5,775
0700	10′ x 20′ high		.50	32		2,300	1,500		3,800	5,175
1000	12′ x 12′, crank operated, crank on door side		.80	20		1,575	940		2,515	3,375
1100	Crank thru wall	↓	.70	22.857		1,800	1,075		2,875	3,850
1300	For vision panel, add				↓	310			310	340
1400	For 22 ga., deduct				S.F.	1.10			1.10	1.21
1600	3′ x 7′ pass door within rolling steel door, new construction				Ea.	1,700			1,700	1,850
1700	Existing construction	2 Sswk	2	8		1,775	375		2,150	2,625
2000	Class A fire doors, manual, 20 ga., 8′ x 8′ high		1.40	11.429		1,400	535		1,935	2,500
2100	10′ x 10′ high		1.10	14.545		1,900	680		2,580	3,300
2200	20′ x 10′ high		.80	20		3,900	940		4,840	5,950
2300	12′ x 12′ high		1	16		3,000	750		3,750	4,625
2400	20′ x 12′ high		.80	20		4,225	940		5,165	6,300
2500	14′ x 14′ high		.60	26.667		3,300	1,250		4,550	5,825
2600	20′ x 16′ high		.50	32		5,300	1,500		6,800	8,475
2700	10′ x 20′ high	↓	.40	40	↓	4,100	1,875		5,975	7,800
3000	For 18 ga. doors, add				S.F.	1.35			1.35	1.49
3300	For enamel finish, add				"	1.60			1.60	1.76
3600	For safety edge bottom bar, pneumatic, add				L.F.	19.90			19.90	22
3700	Electric, add					37.50			37.50	41.50
4000	For weatherstripping, extruded rubber, jambs, add					12.90			12.90	14.20
4100	Hood, add					7.80			7.80	8.55
4200	Sill, add				↓	4.46			4.46	4.91
4500	Motor operators, to 14′ x 14′ opening	2 Sswk	5	3.200	Ea.	1,025	150		1,175	1,400
4600	Over 14′ x 14′, jack shaft type	"	5	3.200		1,075	150		1,225	1,450
4700	For fire door, additional fusible link, add				↓	24.50			24.50	27

08 34 Special Function Doors

08 34 13 – Cold Storage Doors

08 34 13.10 Doors for Cold Area Storage		Crew	Daily Output	Labor-Hours	Unit	Material	2010 Bare Costs Labor	Equipment	Total	Total Incl O&P
0010	**DOORS FOR COLD AREA STORAGE**									
0020	Single, 20 ga. galvanized steel									
0300	Horizontal sliding, 5′ x 7′, manual operation, 3.5″ thick	2 Carp	2	8	Ea.	3,150	330		3,480	3,950
0400	4″ thick		2	8		3,825	330		4,155	4,725
0500	6″ thick		2	8		3,200	330		3,530	4,025
0800	5′ x 7′, power operation, 2″ thick		1.90	8.421		5,225	350		5,575	6,300
0900	4″ thick		1.90	8.421		5,350	350		5,700	6,425
1000	6″ thick		1.90	8.421		6,100	350		6,450	7,250
1300	9′ x 10′, manual operation, 2″ insulation		1.70	9.412		4,250	390		4,640	5,275
1400	4″ insulation		1.70	9.412		4,350	390		4,740	5,375
1500	6″ insulation		1.70	9.412		5,225	390		5,615	6,350
1800	Power operation, 2″ insulation		1.60	10		7,200	415		7,615	8,575
1900	4″ insulation		1.60	10		7,425	415		7,840	8,800
2000	6″ insulation		1.70	9.412		8,400	390		8,790	9,850
2300	For stainless steel face, add					22%				
3000	Hinged, lightweight, 3′ x 7′-0″, galvanized 1 face, 2″ thick	2 Carp	2	8	Ea.	1,350	330		1,680	1,975
3050	4″ thick		1.90	8.421		1,700	350		2,050	2,400
3300	Aluminum doors, 3′ x 7′-0″, 4″ thick		1.90	8.421		1,250	350		1,600	1,925
3350	6″ thick		1.40	11.429		2,200	475		2,675	3,150
3600	Stainless steel, 3′ x 7′-0″, 4″ thick		1.90	8.421		1,625	350		1,975	2,325
3650	6″ thick		1.40	11.429		2,775	475		3,250	3,775
3900	Painted, 3′ x 7′-0″, 4″ thick		1.90	8.421		1,200	350		1,550	1,875
3950	6″ thick		1.40	11.429		2,200	475		2,675	3,150
5000	Bi-parting, electric operated									
5010	6′ x 8′ opening, galv. faces, 4″ thick for cooler	2 Carp	.80	20	Opng.	6,800	830		7,630	8,750
5050	For freezer, 4″ thick		.80	20		7,500	830		8,330	9,525
5300	For door buck framing and door protection, add		2.50	6.400		600	266		866	1,075
6000	Galvanized batten door, galvanized hinges, 4′ x 7′		2	8		1,700	330		2,030	2,375
6050	6′ x 8′		1.80	8.889		2,325	370		2,695	3,125
6500	Fire door, 3 hr., 6′ x 8′, single slide		.80	20		7,800	830		8,630	9,850
6550	Double, bi-parting		.70	22.857		12,200	950		13,150	14,900

08 34 16 – Hangar Doors

08 34 16.10 Aircraft Hangar Doors

		Crew	Daily Output	Labor-Hours	Unit	Material	Labor	Equipment	Total	Total Incl O&P
0010	**AIRCRAFT HANGAR DOORS**									
0020	Bi-fold, ovhd., 20 PSF wind load, incl. elec. oper.									
0100	12′ high x 40′	2 Sswk	240	.067	S.F.	15	3.13		18.13	22
0200	16′ high x 60′		230	.070		18	3.26		21.26	25.50
0300	20′ high x 80′		220	.073		19.40	3.41		22.81	27.50

08 34 36 – Darkroom Doors

08 34 36.10 Various Types of Darkroom Doors

		Crew	Daily Output	Labor-Hours	Unit	Material	Labor	Equipment	Total	Total Incl O&P
0010	**VARIOUS TYPES OF DARKROOM DOORS**									
0015	Revolving, standard, 2 way, 36″ diameter	2 Carp	3.10	5.161	Opng.	2,250	214		2,464	2,800
0020	41″ diameter		3.10	5.161		2,425	214		2,639	3,000
0050	3 way, 51″ diameter		1.40	11.429		3,075	475		3,550	4,100
1000	4 way, 49″ diameter		1.40	11.429		3,625	475		4,100	4,725
2000	Hinged safety, 2 way, 41″ diameter		2.30	6.957		2,825	289		3,114	3,575
2500	3 way, 51″ diameter		1.40	11.429		3,625	475		4,100	4,725
3000	Pop out safety, 2 way, 41″ diameter		3.10	5.161		3,600	214		3,814	4,275
4000	3 way, 51″ diameter		1.40	11.429		3,550	475		4,025	4,625
5000	Wheelchair-type, pop out, 51″ diameter		1.40	11.429		3,525	475		4,000	4,600
5020	72″ diameter		.90	17.778		7,225	740		7,965	9,100
9300	For complete darkrooms, see Div. 13 21 53.50									

08 34 Special Function Doors

08 34 53 – Security Doors and Frames

08 34 53.20 Steel Door		Crew	Daily Output	Labor-Hours	Unit	Material	Labor	Equipment	Total	Total Incl O&P
0010	**STEEL DOOR** with ballistic core and welded frame both 14 gauge									
0050	Flush, UL 752 Level 3, 1-3/4", 3'-0" x 6'-8"	2 Carp	1.50	10.667	Opng.	2,075	445		2,520	2,975
0055	1-3/4", 3'-6" x 6'-8"		1.50	10.667		2,175	445		2,620	3,075
0060	1-3/4", 4'-0" x 6'-8"		1.20	13.333		2,350	555		2,905	3,450
0100	UL 752 Level 8, 1-3/4", 3'-0" x 6'-8"		1.50	10.667		9,200	445		9,645	10,800
0105	1-3/4", 3'-6" x 6'-8"		1.50	10.667		9,300	445		9,745	10,900
0110	1-3/4", 4'-0" x 6'-8"		1.20	13.333		10,600	555		11,155	12,500
0120	UL 752 Level 3, 1-3/4", 3'-0" x 7'-0"		1.50	10.667		2,100	445		2,545	3,000
0125	1-3/4", 3'-6" x 7'-0"		1.50	10.667		2,200	445		2,645	3,100
0130	1-3/4", 4'-0" x 7'-0"		1.20	13.333		2,400	555		2,955	3,475
0150	UL 752 Level 8, 1-3/4", 3'-0" x 7'-0"		1.50	10.667		9,225	445		9,670	10,900
0155	1-3/4", 3'-6" x 7'-0"		1.50	10.667		9,325	445		9,770	11,000
0160	1-3/4", 4'-0" x 7'-0"		1.20	13.333		10,600	555		11,155	12,600
1000	Safe Room sliding door and hardware, 1-3/4", 3'-0" x 7'-0" UL 752 Level 3		.50	32		22,000	1,325		23,325	26,300
1050	Safe Room swinging door and hardware, 1-3/4", 3'-0" x 7'-0" UL 752 Level 3	↓	.50	32	↓	26,000	1,325		27,325	30,700

08 34 53.30 Wood Ballistic Doors		Crew	Daily Output	Labor-Hours	Unit	Material	Labor	Equipment	Total	Total Incl O&P
0010	**WOOD BALLISTIC DOORS** with frames and hardware									
0050	Wood, 1-3/4", 3'-0" x 7'-0" UL 752 Level 3	2 Carp	1.50	10.667	Opng.	2,000	445		2,445	2,875

08 34 56 – Security Gates

08 34 56.10 For Information On Security Gates		Crew	Daily Output	Labor-Hours	Unit	Material	Labor	Equipment	Total	Total Incl O&P
0010	**FOR INFORMATION ON SECURITY GATES** see Div.10 22 16.10									

08 34 59 – Vault Doors and Day Gates

08 34 59.10 Secure Storage Doors		Crew	Daily Output	Labor-Hours	Unit	Material	Labor	Equipment	Total	Total Incl O&P
0010	**SECURE STORAGE DOORS**									
0020	Door and frame, 32" x 78", clear opening									
0100	1 hour test, 32" door, weighs 750 lbs.	2 Sswk	1.50	10.667	Opng.	4,100	500		4,600	5,375
0200	2 hour test, 32" door, weighs 950 lbs.		1.30	12.308		4,350	575		4,925	5,825
0250	40" door, weighs 1130 lbs.		1	16		4,950	750		5,700	6,775
0300	4 hour test, 32" door, weighs 1025 lbs.		1.20	13.333		4,725	625		5,350	6,300
0350	40" door, weighs 1140 lbs.	↓	.90	17.778		5,500	835		6,335	7,525
0500	For stainless steel front, including frame, add to above					1,900			1,900	2,100
0550	Back, add				↓	1,900			1,900	2,100
0600	For time lock, two movement, add	1 Elec	2	4	Ea.	1,450	196		1,646	1,875
0650	Three movement, add	"	2	4		2,200	196		2,396	2,725
0800	Day gate, painted, steel, 32" wide	2 Sswk	1.50	10.667		1,850	500		2,350	2,925
0850	40" wide		1.40	11.429		2,050	535		2,585	3,200
0900	Aluminum, 32" wide		1.50	10.667		2,950	500		3,450	4,125
0950	40" wide	↓	1.40	11.429	↓	3,275	535		3,810	4,550
2050	Security vault door, class I, 3' wide, 3 1/2" thick	E-24	.19	166	Opng.	15,200	7,650	3,900	26,750	34,000
2100	Class II, 3' wide, 7" thick		.19	166		17,600	7,650	3,900	29,150	36,600
2150	Class III, 9R, 3' wide, 10" thick, minimum		.13	250	↓	21,800	11,500	5,850	39,150	49,900
2160	Class V, type 1, 40" door		2.48	12.903	Ea.	6,325	590	300	7,215	8,300
2170	Class V, type 2, 40" door	↓	2.48	12.903		6,300	590	300	7,190	8,250
2180	Day gate for class V vault	2 Sswk	2	8	↓	1,600	375		1,975	2,400

08 34 63 – Detention Doors and Frames

08 34 63.13 Steel Detention Doors and Frames		Crew	Daily Output	Labor-Hours	Unit	Material	Labor	Equipment	Total	Total Incl O&P
0010	**STEEL DETENTION DOORS AND FRAMES**									
0500	Rolling cell door, bar front, 7/8" bars, 4" O.C., 7' H, 5' W, with hardware G	E-4	2	16	Ea.	5,025	760	73	5,858	6,925
0550	Actuator for rolling cell door, bar front	2 Skwk	2	8		4,200	340		4,540	5,150
1000	Doors & frames, 3' x 7', complete, with hardware, single plate	E-4	4	8	↓	4,200	380	36.50	4,616.50	5,325

08 34 Special Function Doors

08 34 63 – Detention Doors and Frames

08 34 63.13 Steel Detention Doors and Frames		Crew	Daily Output	Labor-Hours	Unit	Material	2010 Bare Costs Labor	Equipment	Total	Total Incl O&P
1650	Double plate	E-4	4	8	Ea.	5,175	380	36.50	5,591.50	6,400

08 34 73 – Sound Control Door Assemblies

08 34 73.10 Acoustical Doors

		Crew	Daily Output	Labor-Hours	Unit	Material	Labor	Equipment	Total	Total Incl O&P
0010	**ACOUSTICAL DOORS**									
0020	Including framed seals, 3′ x 7′, wood, 27 STC rating	2 Carp	1.50	10.667	Ea.	800	445		1,245	1,575
0100	Steel, 40 STC rating		1.50	10.667		2,800	445		3,245	3,750
0200	45 STC rating		1.50	10.667		3,300	445		3,745	4,300
0300	48 STC rating		1.50	10.667		3,800	445		4,245	4,850
0400	52 STC rating		1.50	10.667		4,300	445		4,745	5,400

08 36 Panel Doors

08 36 13 – Sectional Doors

08 36 13.10 Overhead Commercial Doors

		Crew	Daily Output	Labor-Hours	Unit	Material	Labor	Equipment	Total	Total Incl O&P
0010	**OVERHEAD COMMERCIAL DOORS**									
1000	Stock, sectional, heavy duty, wood, 1-3/4″ thick, 8′ x 8′ high	2 Carp	2	8	Ea.	775	330		1,105	1,350
1100	10′ x 10′ high		1.80	8.889		1,150	370		1,520	1,825
1200	12′ x 12′ high		1.50	10.667		1,675	445		2,120	2,500
1300	Chain hoist, 14′ x 14′ high		1.30	12.308		2,550	510		3,060	3,600
1400	12′ x 16′ high		1	16		2,525	665		3,190	3,800
1500	20′ x 8′ high		1.30	12.270		2,275	510		2,785	3,275
1600	20′ x 16′ high		.65	24.615		5,975	1,025		7,000	8,150
1800	Center mullion openings, 8′ high		4	4		1,100	166		1,266	1,450
1900	20′ high		2	8		1,900	330		2,230	2,575
2100	For medium duty custom door, deduct					5%	5%			
2150	For medium duty stock doors, deduct					10%	5%			
2300	Fiberglass and aluminum, heavy duty, sectional, 12′ x 12′ high	2 Carp	1.50	10.667	Ea.	2,575	445		3,020	3,525
2450	Chain hoist, 20′ x 20′ high		.50	32		6,500	1,325		7,825	9,200
2600	Steel, 24 ga. sectional, manual, 8′ x 8′ high		2	8		765	330		1,095	1,350
2650	10′ x 10′ high		1.80	8.889		985	370		1,355	1,650
2700	12′ x 12′ high		1.50	10.667		1,200	445		1,645	2,000
2800	Chain hoist, 20′ x 14′ high		.70	22.857		3,425	950		4,375	5,250
2850	For 1-1/4″ rigid insulation and 26 ga. galv.									
2860	back panel, add				S.F.	4			4	4.40
2900	For electric trolley operator, 1/3 H.P., to 12′ x 12′, add	1 Carp	2	4	Ea.	900	166		1,066	1,250
2950	Over 12′ x 12′, 1/2 H.P., add	″	1	8	″	1,025	330		1,355	1,625

08 36 13.20 Residential Garage Doors

		Crew	Daily Output	Labor-Hours	Unit	Material	Labor	Equipment	Total	Total Incl O&P
0010	**RESIDENTIAL GARAGE DOORS**									
0050	Hinged, wood, custom, double door, 9′ x 7′	2 Carp	4	4	Ea.	585	166		751	900
0070	16′ x 7′		3	5.333		965	222		1,187	1,400
0200	Overhead, sectional, incl. hardware, fiberglass, 9′ x 7′, standard		5.28	3.030		735	126		861	1,000
0220	Deluxe		5.28	3.030		935	126		1,061	1,225
0300	16′ x 7′, standard		6	2.667		1,550	111		1,661	1,875
0320	Deluxe		6	2.667		1,975	111		2,086	2,350
0500	Hardboard, 9′ x 7′, standard		8	2		550	83		633	735
0520	Deluxe		8	2		775	83		858	985
0600	16′ x 7′, standard		6	2.667		1,125	111		1,236	1,425
0620	Deluxe		6	2.667		1,350	111		1,461	1,650
0700	Metal, 9′ x 7′, standard		5.28	3.030		580	126		706	835
0720	Deluxe		8	2		855	83		938	1,075
0800	16′ x 7′, standard		3	5.333		730	222		952	1,150

08 36 Panel Doors

08 36 13 – Sectional Doors

08 36 13.20 Residential Garage Doors		Crew	Daily Output	Labor-Hours	Unit	Material	2010 Bare Costs Labor	Equipment	Total	Total Incl O&P
0820	Deluxe	2 Carp	6	2.667	Ea.	1,375	111		1,486	1,675
0900	Wood, 9′ x 7′, standard		8	2		720	83		803	920
0920	Deluxe		8	2		2,050	83		2,133	2,375
1000	16′ x 7′, standard		6	2.667		1,425	111		1,536	1,750
1020	Deluxe		6	2.667		2,975	111		3,086	3,450
1800	Door hardware, sectional	1 Carp	4	2		320	83		403	480
1810	Door tracks only		4	2		138	83		221	280
1820	One side only		7	1.143		110	47.50		157.50	194
3000	Swing-up, including hardware, fiberglass, 9′ x 7′, standard	2 Carp	8	2		980	83		1,063	1,200
3020	Deluxe		8	2		790	83		873	1,000
3100	16′ x 7′, standard		6	2.667		1,225	111		1,336	1,525
3120	Deluxe		6	2.667		980	111		1,091	1,250
3200	Hardboard, 9′ x 7′, standard		8	2		345	83		428	510
3220	Deluxe		8	2		455	83		538	635
3300	16′ x 7′, standard		6	2.667		480	111		591	700
3320	Deluxe		6	2.667		715	111		826	955
3400	Metal, 9′ x 7′, standard		8	2		380	83		463	550
3420	Deluxe		8	2		835	83		918	1,050
3500	16′ x 7′, standard		6	2.667		600	111		711	830
3520	Deluxe		6	2.667		960	111		1,071	1,225
3600	Wood, 9′ x 7′, standard		8	2		450	83		533	625
3620	Deluxe		8	2		935	83		1,018	1,150
3700	16′ x 7′, standard		6	2.667		725	111		836	970
3720	Deluxe		6	2.667		1,025	111		1,136	1,300
3900	Door hardware only, swing up	1 Carp	4	2		145	83		228	288
3920	One side only		7	1.143		70	47.50		117.50	150
4000	For electric operator, economy, add		8	1		410	41.50		451.50	515
4100	Deluxe, including remote control		8	1		600	41.50		641.50	725
4500	For transmitter/receiver control , add to operator				Total	105			105	116
4600	Transmitters, additional				"	50			50	55

08 36 19 – Multi-Leaf Vertical Lift Doors

08 36 19.10 Sectional Vertical Lift Doors		Crew	Daily Output	Labor-Hours	Unit	Material	Labor	Equipment	Total	Total Incl O&P
0010	**SECTIONAL VERTICAL LIFT DOORS**									
0020	Motorized, 14 ga. stl, incl., frm and ctrl pnl									
0050	16′ x 16′ high	L-10	.50	48	Ea.	21,300	2,250	1,300	24,850	28,600
0100	10′ x 20′ high		1.30	18.462		31,000	865	505	32,370	36,100
0120	15′ x 20′ high		1.30	18.462		38,300	865	505	39,670	44,200
0140	20′ x 20′ high		1	24		44,700	1,125	655	46,480	51,500
0160	25′ x 20′ high		1	24		51,000	1,125	655	52,780	58,500
0170	32′ x 24′ high		.75	32		46,000	1,500	875	48,375	54,000
0180	20′ x 25′ high		1	24		52,500	1,125	655	54,280	60,000
0200	25′ x 25′ high		.70	34.286		61,000	1,600	935	63,535	70,500
0220	25′ x 30′ high		.70	34.286		66,000	1,600	935	68,535	76,000
0240	30′ x 30′ high		.70	34.286		77,000	1,600	935	79,535	88,000
0260	35′ x 30′ high		.70	34.286		87,000	1,600	935	89,535	99,000

08 36 23 – Telescoping Vertical Lift Doors

08 36 23.10 Telescoping Steel Doors		Crew	Daily Output	Labor-Hours	Unit	Material	Labor	Equipment	Total	Total Incl O&P
0010	**TELESCOPING STEEL DOORS**									
1000	Overhead, .03" thick, electric operated, 10′ x 10′	E-3	.80	30	Ea.	14,000	1,425	182	15,607	18,100
2000	20′ x 10′		.60	40		18,000	1,900	243	20,143	23,400
3000	20′ x 16′		.40	60		18,800	2,850	365	22,015	26,100

08 38 Traffic Doors

08 38 13 – Flexible Strip Doors

08 38 13.10 Flexible Transparent Strip Doors		Crew	Daily Output	Labor-Hours	Unit	Material	2010 Bare Costs Labor	Equipment	Total	Total Incl O&P
0010	**FLEXIBLE TRANSPARENT STRIP DOORS**									
0100	12" strip width, 2/3 overlap	3 Shee	135	.178	SF Surf	7.50	8.75		16.25	21.50
0200	Full overlap		115	.209		9.40	10.25		19.65	26
0220	8" strip width, 1/2 overlap		140	.171		6	8.40		14.40	19.35
0240	Full overlap		120	.200		7.60	9.80		17.40	23
0300	Add for suspension system, header mount				L.F.	8.90			8.90	9.80
0400	Wall mount				"	9			9	9.90

08 38 19 – Rigid Traffic Doors

08 38 19.20 Double Acting Swing Doors		Crew	Daily Output	Labor-Hours	Unit	Material	Labor	Equipment	Total	Total Incl O&P
0010	**DOUBLE ACTING SWING DOORS**									
1000	.063" aluminum, 7'-0" high, 4'-0" wide	2 Carp	4.20	3.810	Pr.	1,900	158		2,058	2,350
1025	6'-0" wide		4	4		2,100	166		2,266	2,550
1050	6'-8" wide		4	4		2,300	166		2,466	2,775
2000	Solid core wood, 3/4" thick, metal frame, stainless steel									
2010	base plate, 7' high opening, 4' wide	2 Carp	4	4	Pr.	2,200	166		2,366	2,675
2050	7' wide	"	3.80	4.211	"	2,500	175		2,675	3,025

08 38 19.30 Shock Absorbing Doors		Crew	Daily Output	Labor-Hours	Unit	Material	Labor	Equipment	Total	Total Incl O&P
0010	**SHOCK ABSORBING DOORS**									
0020	Rigid, no frame, 1-1/2" thick, 5' x 7'	2 Sswk	1.90	8.421	Opng.	1,400	395		1,795	2,250
0100	8' x 8'		1.80	8.889		2,000	415		2,415	2,925
0500	Flexible, no frame, insulated, .16" thick, economy, 5' x 7'		2	8		1,750	375		2,125	2,575
0600	Deluxe		1.90	8.421		2,625	395		3,020	3,575
1000	8' x 8' opening, economy		2	8		2,750	375		3,125	3,675
1100	Deluxe		1.90	8.421		3,500	395		3,895	4,550

08 41 Entrances and Storefronts

08 41 13 – Aluminum-Framed Entrances and Storefronts

08 41 13.20 Tube Framing		Crew	Daily Output	Labor-Hours	Unit	Material	Labor	Equipment	Total	Total Incl O&P
0010	**TUBE FRAMING**, For window walls and store fronts, aluminum stock									
0050	Plain tube frame, mill finish, 1-3/4" x 1-3/4"	2 Glaz	103	.155	L.F.	8.70	6.25		14.95	19
0150	1-3/4" x 4"		98	.163		11.90	6.55		18.45	23
0200	1-3/4" x 4-1/2" **CN**		95	.168		13.80	6.75		20.55	25.50
0250	2" x 6"		89	.180		20.50	7.25		27.75	33.50
0350	4" x 4"		87	.184		23	7.40		30.40	36.50
0400	4-1/2" x 4-1/2"		85	.188		25	7.55		32.55	39
0450	Glass bead		240	.067		2.60	2.68		5.28	6.90
1000	Flush tube frame, mill finish, 1/4" glass, 1-3/4" x 4", open header		80	.200		11.45	8.05		19.50	24.50
1050	Open sill		82	.195		9.30	7.85		17.15	22
1100	Closed back header		83	.193		16.35	7.75		24.10	29.50
1150	Closed back sill		85	.188		15.60	7.55		23.15	28.50
1160	Tube fmg, spandrel cover both sides, alum 1" wide	1 Sswk	85	.094	S.F.	86	4.41		90.41	102
1170	Tube fmg, spandrel cover both sides, alum 2" wide	"	85	.094	"	33	4.41		37.41	44.50
1200	Vertical mullion, one piece	2 Glaz	75	.213	L.F.	17.15	8.60		25.75	32
1250	Two piece		73	.219		18.20	8.80		27	33.50
1300	90° or 180° vertical corner post		75	.213		27.50	8.60		36.10	43.50
1400	1-3/4" x 4-1/2", open header		80	.200		13.90	8.05		21.95	27.50
1450	Open sill		82	.195		11.60	7.85		19.45	24.50
1500	Closed back header		83	.193		16.90	7.75		24.65	30.50
1550	Closed back sill		85	.188		16.35	7.55		23.90	29.50
1600	Vertical mullion, one piece		75	.213		18.30	8.60		26.90	33

08 41 Entrances and Storefronts

08 41 13 – Aluminum-Framed Entrances and Storefronts

08 41 13.20 Tube Framing		Crew	Daily Output	Labor-Hours	Unit	2010 Bare Costs Material	Labor	Equipment	Total	Total Incl O&P
1650	Two piece	2 Glaz	73	.219	L.F.	19.30	8.80		28.10	34.50
1700	90° or 180° vertical corner post		75	.213		19.70	8.60		28.30	34.50
2000	Flush tube frame, mil fin.,ins. glass w/thml brk, 2" x 4-1/2", open header		75	.213		14.50	8.60		23.10	29
2050	Open sill		77	.208		12.25	8.35		20.60	26
2100	Closed back header		78	.205		13.75	8.25		22	27.50
2150	Closed back sill		80	.200		14.70	8.05		22.75	28.50
2200	Vertical mullion, one piece		70	.229		15.25	9.20		24.45	30.50
2250	Two piece		68	.235		16.45	9.45		25.90	32.50
2300	90° or 180° vertical corner post		70	.229		15.70	9.20		24.90	31
5000	Flush tube frame, mill fin., thermal brk., 2-1/4"x 4-1/2", open header		74	.216		15.15	8.70		23.85	30
5050	Open sill		75	.213		13.45	8.60		22.05	27.50
5100	Vertical mullion, one piece		69	.232		16.50	9.30		25.80	32
5150	Two piece		67	.239		19.15	9.60		28.75	35.50
5200	90° or 180° vertical corner post		69	.232		16.85	9.30		26.15	32.50
6980	Door stop (snap in)	↓	380	.042	↓	3	1.69		4.69	5.85
7000	For joints, 90°, clip type, add				Ea.	23			23	25
7050	Screw spline joint, add					20.50			20.50	22.50
7100	For joint other than 90°, add				↓	43			43	47
8000	For bronze anodized aluminum, add					15%				
8020	For black finish, add					27%				
8050	For stainless steel materials, add					350%				
8100	For monumental grade, add					52%				
8150	For steel stiffener, add	2 Glaz	200	.080	L.F.	10.40	3.22		13.62	16.30
8200	For 2 to 5 stories, add per story				Story		6%			

08 41 19 – Stainless-Steel-Framed Entrances and Storefronts

08 41 19.10 Stainless-Steel and Glass Entrance Unit		Crew	Daily Output	Labor-Hours	Unit	Material	Labor	Equipment	Total	Total Incl O&P
0010	**STAINLESS-STEEL AND GLASS ENTRANCE UNIT**, narrow stiles									
0020	3' x 7' opening, including hardware, minimum	2 Sswk	1.60	10	Opng.	6,100	470		6,570	7,525
0050	Average		1.40	11.429		6,500	535		7,035	8,100
0100	Maximum	↓	1.20	13.333		7,000	625		7,625	8,800
1000	For solid bronze entrance units, statuary finish, add					62%				
1100	Without statuary finish, add				↓	45%				
2000	Balanced doors, 3' x 7', economy	2 Sswk	.90	17.778	Ea.	8,200	835		9,035	10,500
2100	Premium	"	.70	22.857	"	14,100	1,075		15,175	17,400

08 41 26 – All-Glass Entrances and Storefronts

08 41 26.10 Window Walls Aluminum, Stock		Crew	Daily Output	Labor-Hours	Unit	Material	Labor	Equipment	Total	Total Incl O&P
0010	**WINDOW WALLS ALUMINUM, STOCK**, including glazing									
0020	Minimum	H-2	160	.150	S.F.	40	5.65		45.65	52.50
0050	Average		140	.171		55	6.50		61.50	70.50
0100	Maximum	↓	110	.218	↓	155	8.25		163.25	183
0500	For translucent sandwich wall systems, see Div. 07 41 33.10									
0850	Cost of the above walls depends on material,									
0860	finish, repetition, and size of units.									
0870	The larger the opening, the lower the S.F. cost									
1200	Double glazed acoustical window wall for airports,									
1220	including 1" thick glass with 2" x 4-1/2" tube frame	H-2	40	.600	S.F.	112	22.50		134.50	158

08 42 Entrances

08 42 26 – All-Glass Entrances

08 42 26.10 Swinging Glass Doors		Crew	Daily Output	Labor-Hours	Unit	Material	Labor	Equipment	Total	Total Incl O&P
						2010 Bare Costs				
0010	**SWINGING GLASS DOORS**									
0020	Including hardware, 1/2" thick, tempered, 3' x 7' opening	2 Glaz	2	8	Opng.	2,075	320		2,395	2,750
0100	6' x 7' opening		1.40	11.429	"	3,900	460		4,360	5,000
9000	Minimum labor/equipment charge		2	8	Job		320		320	485

08 42 29 – Automatic Entrances

08 42 29.23 Sliding Automatic Entrances		Crew	Daily Output	Labor-Hours	Unit	Material	Labor	Equipment	Total	Total Incl O&P
0010	**SLIDING AUTOMATIC ENTRANCES** 12' x 7'-6" opng., 5' x 7' door, 2 way traffic									
0020	Mat or electronic activated, panic pushout, incl. operator & hardware,									
0030	not including glass or glazing	2 Glaz	.70	22.857	Opng.	7,800	920		8,720	9,950
9000	Minimum labor/equipment charge	"	.70	22.857	Job		920		920	1,375

08 42 33 – Revolving Door Entrances

08 42 33.10 Circular Rotating Entrance Doors		Crew	Daily Output	Labor-Hours	Unit	Material	Labor	Equipment	Total	Total Incl O&P
0010	**CIRCULAR ROTATING ENTRANCE DOORS**, Aluminum									
0020	6'-10" to 7' high, stock units, minimum	4 Sswk	.75	42.667	Opng.	19,000	2,000		21,000	24,400
0050	Average		.60	53.333		22,800	2,500		25,300	29,500
0100	Maximum		.45	71.111		39,300	3,325		42,625	49,100
1000	Stainless steel		.30	105		37,800	4,975		42,775	50,500
1100	Solid bronze		.15	213		44,100	10,000		54,100	66,000
1500	For automatic controls, add	2 Elec	2	8		14,400	390		14,790	16,400

08 42 36 – Balanced Door Entrances

08 42 36.10 Balanced Entrance Doors		Crew	Daily Output	Labor-Hours	Unit	Material	Labor	Equipment	Total	Total Incl O&P
0010	**BALANCED ENTRANCE DOORS**									
0020	Hardware & frame, alum. & glass, 3' x 7', econ.	2 Sswk	.90	17.778	Ea.	5,900	835		6,735	7,950
0150	Premium	"	.70	22.857	"	7,300	1,075		8,375	9,900

08 43 Storefronts

08 43 13 – Aluminum-Framed Storefronts

08 43 13.10 Aluminum-Framed Entrance Doors and Frames		Crew	Daily Output	Labor-Hours	Unit	Material	Labor	Equipment	Total	Total Incl O&P
0010	**ALUMINUM-FRAMED ENTRANCE DOORS AND FRAMES**									
0015	Standard hardware and glass stops but no glazing									
0020	Entrance door, 3' x 7' opening, clear anodized finish	2 Sswk	7	2.286	Opng.	355	107		462	585
0040	Bronze finish		7	2.286		375	107		482	605
0060	Black finish		7	2.286		420	107		527	650
0500	6' x 7' opening, clear finish		6	2.667		715	125		840	1,000
0520	Bronze finish		6	2.667		750	125		875	1,050
0540	Black finish		6	2.667		835	125		960	1,150
1000	With 3' high transom above, 3' x 7' opening, clear finish		5.50	2.909		410	136		546	690
1050	Bronze finish		5.50	2.909		435	136		571	715
1100	Black finish		5.50	2.909		465	136		601	750
1300	3'-6" x 7'-0" opening, clear finish		5.50	2.909		320	136		456	595
1320	Bronze finish		5.50	2.909		330	136		466	605
1340	Black finish		5.50	2.909		350	136		486	625
1500	6' x 7' opening, clear finish		5.50	2.909		500	136		636	790
1550	Bronze finish		5.50	2.909		520	136		656	815
1600	Black finish		5.50	2.909		580	136		716	875

08 43 13.20 Storefront Systems		Crew	Daily Output	Labor-Hours	Unit	Material	Labor	Equipment	Total	Total Incl O&P
0010	**STOREFRONT SYSTEMS**, aluminum frame clear 3/8" plate glass									
0020	incl. 3' x 7' door with hardware (400 sq. ft. max. wall)									
0500	Wall height to 12' high, commercial grade	2 Glaz	150	.107	S.F.	19.30	4.29		23.59	27.50

08 43 Storefronts

08 43 13 – Aluminum-Framed Storefronts

08 43 13.20 Storefront Systems		Crew	Daily Output	Labor-Hours	Unit	Material	2010 Bare Costs Labor	Equipment	Total	Total Incl O&P
0600	Institutional grade	2 Glaz	130	.123	S.F.	24	4.95		28.95	34
0700	Monumental grade		115	.139		36.50	5.60		42.10	48.50
1000	6′ x 7′ door with hardware, commercial grade		135	.119		19.50	4.76		24.26	28.50
1100	Institutional grade		115	.139		27	5.60		32.60	38
1200	Monumental grade		100	.160		49.50	6.45		55.95	63.50
1500	For bronze anodized finish, add					15%				
1600	For black anodized finish, add					35%				
1700	For stainless steel framing, add to monumental					76%				

08 43 29 – Sliding Storefronts

08 43 29.10 Sliding Panels		Crew	Daily Output	Labor-Hours	Unit	Material	Labor	Equipment	Total	Total Incl O&P
0010	**SLIDING PANELS**									
0020	Mall fronts, aluminum & glass, 15′ x 9′ high	2 Glaz	1.30	12.308	Opng.	3,100	495		3,595	4,150
0100	24′ x 9′ high		.70	22.857		4,400	920		5,320	6,225
0200	48′ x 9′ high, with fixed panels		.90	17.778		8,200	715		8,915	10,100
0500	For bronze finish, add					17%				

08 44 Curtain Wall and Glazed Assemblies

08 44 13 – Glazed Aluminum Curtain Walls

08 44 13.10 Glazed Curtain Walls		Crew	Daily Output	Labor-Hours	Unit	Material	Labor	Equipment	Total	Total Incl O&P
0010	**GLAZED CURTAIN WALLS**, aluminum, stock, including glazing									
0020	Minimum	H-1	205	.156	S.F.	32	6.80		38.80	46.50
0050	Average, single glazed		195	.164		45.50	7.15		52.65	62.50
0053	Replace average, single glazed		150	.213		45.50	9.30		54.80	66
0150	Average, double glazed		180	.178		59	7.75		66.75	77.50
0200	Maximum		160	.200		161	8.70		169.70	191

08 45 Translucent Wall and Roof Assemblies

08 45 10 – Translucent Roof Assemblies

08 45 10.10 Skyroofs		Crew	Daily Output	Labor-Hours	Unit	Material	Labor	Equipment	Total	Total Incl O&P
0010	**SKYROOFS**, Translucent panels, 2-3/4″ thick									
0020	Under 5000 S.F.	G-3	395	.081	SF Hor.	32	3.33		35.33	40.50
0100	Over 5000 S.F.		465	.069		30	2.83		32.83	37.50
0300	Continuous vaulted, semi-circular, to 8′ wide, double glazed G		145	.221		50	9.05		59.05	69
0400	Single glazed		160	.200		34	8.20		42.20	50
0600	To 20′ wide, single glazed		175	.183		37.50	7.50		45	53
0700	Over 20′ wide, single glazed		200	.160		43.50	6.60		50.10	57.50
0900	Motorized opening type, single glazed, 1/3 opening		145	.221		44.50	9.05		53.55	63
1000	Full opening		130	.246		51	10.10		61.10	71.50
1200	Pyramid type units, self-supporting, to 30′ clear opening,									
1300	square or circular, single glazed, minimum	G-3	200	.160	SF Hor.	26	6.60		32.60	38.50
1310	Average		165	.194		35.50	7.95		43.45	51
1400	Maximum		130	.246		51	10.10		61.10	71.50
1500	Grid type, 4′ to 10′ modules, single glass glazed, minimum		200	.160		34	6.60		40.60	47.50
1550	Maximum		128	.250		53.50	10.30		63.80	74
1600	Preformed acrylic, minimum		300	.107		41	4.38		45.38	51.50
1650	Maximum		175	.183		56	7.50		63.50	73
1800	Skyroofs, dome type, self-supporting, to 100′ clear opening, circular									
1900	Rise to span ratio = 0.20									
1920	Minimum	G-3	197	.162	SF Hor.	15.30	6.70		22	27

08 45 Translucent Wall and Roof Assemblies

08 45 10 – Translucent Roof Assemblies

08 45 10.10 Skyroofs		Crew	Daily Output	Labor-Hours	Unit	Material	2010 Bare Costs Labor	Equipment	Total	Total Incl O&P
1950	Maximum	G-3	113	.283	SF Hor.	51	11.65		62.65	74
2100	Rise to span ratio = 0.33, minimum		169	.189		30.50	7.80		38.30	45.50
2150	Maximum		101	.317		60.50	13		73.50	86.50
2200	Rise to span ratio = 0.50, minimum		148	.216		46.50	8.90		55.40	64.50
2250	Maximum		87	.368		67	15.10		82.10	97
2400	Ridge units, continuous, to 8′ wide, double (G)		130	.246		117	10.10		127.10	144
2500	Single (G)		200	.160		80.50	6.60		87.10	98.50
2700	Ridge and furrow units, over 4′ O.C., double, minimum (G)		200	.160		25.50	6.60		32.10	38
2750	Maximum (G)		120	.267		47	10.95		57.95	68
2800	Single, minimum		214	.150		21.50	6.15		27.65	33
2850	Maximum		153	.209		45.50	8.60		54.10	63
3000	Rolling roof, translucent panels, flat roof, minimum		253	.126	S.F.	19.50	5.20		24.70	29.50
3030	Maximum (G)		160	.200		39	8.20		47.20	55.50
3100	Lean-to skyroof, long span, double, minimum (G)		197	.162		26	6.70		32.70	38.50
3150	Maximum (G)		101	.317		52	13		65	77.50
3300	Single, minimum		321	.100		19.80	4.10		23.90	28.50
3350	Maximum		160	.200		32.50	8.20		40.70	48.50

08 51 Metal Windows

08 51 13 – Aluminum Windows

08 51 13.10 Aluminum Sash

		Crew	Daily Output	Labor-Hours	Unit	Material	Labor	Equipment	Total	Total Incl O&P
0010	**ALUMINUM SASH**									
0020	Stock, grade C, glaze & trim not incl., casement	2 Sswk	200	.080	S.F.	36	3.75		39.75	46
0050	Double hung		200	.080		36.50	3.75		40.25	46.50
0100	Fixed casement		200	.080		15.65	3.75		19.40	24
0150	Picture window		200	.080		16.65	3.75		20.40	25
0200	Projected window		200	.080		33	3.75		36.75	42.50
0250	Single hung		200	.080		15.80	3.75		19.55	24
0300	Sliding		200	.080		20.50	3.75		24.25	29
1000	Mullions for above, tubular		240	.067	L.F.	5.45	3.13		8.58	11.45
2000	Custom aluminum sash, grade HC, glazing not included, minimum		200	.080	S.F.	38	3.75		41.75	48.50
2100	Maximum		85	.188	"	49.50	8.85		58.35	70

08 51 13.20 Aluminum Windows

		Crew	Daily Output	Labor-Hours	Unit	Material	Labor	Equipment	Total	Total Incl O&P
0010	**ALUMINUM WINDOWS**, incl. frame and glazing, commercial grade									
0020	See Div. 05 12 23.40									
1000	Stock units, casement, 3′-1″ x 3′-2″ opening	2 Sswk	10	1.600	Ea.	355	75		430	520
1050	Add for storms					115			115	126
1600	Projected, with screen, 3′-1″ x 3′-2″ opening	2 Sswk	10	1.600		335	75		410	500
1700	Add for storms					112			112	123
2000	4′-5″ x 5′-3″ opening	2 Sswk	8	2		380	94		474	585
2100	Add for storms					120			120	132
2500	Enamel finish windows, 3′-1″ x 3′-2″	2 Sswk	10	1.600		340	75		415	505
2600	4′-5″ x 5′-3″		8	2		385	94		479	590
3000	Single hung, 2′ x 3′ opening, enameled, standard glazed		10	1.600		198	75		273	350
3100	Insulating glass		10	1.600		240	75		315	395
3300	2′-8″ x 6′-8″ opening, standard glazed		8	2		350	94		444	550
3400	Insulating glass		8	2		450	94		544	660
3700	3′-4″ x 5′-0″ opening, standard glazed		9	1.778		286	83.50		369.50	460
3800	Insulating glass		9	1.778		315	83.50		398.50	495
3890	Awning type, 3′ x 3′ opening standard glass		14	1.143		405	53.50		458.50	540
3900	Insulating glass		14	1.143		430	53.50		483.50	570

08 51 Metal Windows

08 51 13 – Aluminum Windows

08 51 13.20 Aluminum Windows		Crew	Daily Output	Labor-Hours	Unit	Material	2010 Bare Costs Labor	Equipment	Total	Total Incl O&P
3910	3' x 4' opening, standard glass	2 Sswk	10	1.600	Ea.	470	75		545	650
3920	Insulating glass		10	1.600		545	75		620	725
3930	3' x 5'-4" opening, standard glass		10	1.600		570	75		645	755
3940	Insulating glass		10	1.600		670	75		745	865
3950	4' x 5'-4" opening, standard glass		9	1.778		625	83.50		708.50	830
3960	Insulating glass		9	1.778		750	83.50		833.50	970
4000	Sliding aluminum, 3' x 2' opening, standard glazed		10	1.600		209	75		284	360
4100	Insulating glass		10	1.600		224	75		299	380
4300	5' x 3' opening, standard glazed		9	1.778		320	83.50		403.50	495
4400	Insulating glass		9	1.778		370	83.50		453.50	555
4600	8' x 4' opening, standard glazed		6	2.667		335	125		460	590
4700	Insulating glass		6	2.667		545	125		670	815
5000	9' x 5' opening, standard glazed		4	4		510	188		698	890
5100	Insulating glass		4	4		820	188		1,008	1,225
5500	Sliding, with thermal barrier and screen, 6' x 4', 2 track		8	2		695	94		789	930
5700	4 track	↓	8	2		880	94		974	1,125
6000	For above units with bronze finish, add					12%				
6200	For installation in concrete openings, add				↓	6%				

08 51 23 – Steel Windows

08 51 23.10 Steel Sash		Crew	Daily Output	Labor-Hours	Unit	Material	Labor	Equipment	Total	Total Incl O&P
0010	**STEEL SASH** Custom units, glazing and trim not included R085123-10									
0100	Casement, 100% vented	2 Sswk	200	.080	S.F.	48	3.75		51.75	59
0200	50% vented		200	.080		44	3.75		47.75	55
0300	Fixed		200	.080		30	3.75		33.75	39.50
1000	Projected, commercial, 40% vented		200	.080		48.50	3.75		52.25	59.50
1100	Intermediate, 50% vented		200	.080		54.50	3.75		58.25	66.50
1500	Industrial, horizontally pivoted		200	.080		50.50	3.75		54.25	62
1600	Fixed		200	.080		29	3.75		32.75	38.50
2000	Industrial security sash, 50% vented		200	.080		54	3.75		57.75	66
2100	Fixed		200	.080		44.50	3.75		48.25	55.50
2500	Picture window		200	.080		28.50	3.75		32.25	38
3000	Double hung		200	.080	↓	56	3.75		59.75	68
5000	Mullions for above, open interior face		240	.067	L.F.	9.80	3.13		12.93	16.25
5100	With interior cover	↓	240	.067	"	16.35	3.13		19.48	23.50

08 51 23.20 Steel Windows		Crew	Daily Output	Labor-Hours	Unit	Material	Labor	Equipment	Total	Total Incl O&P
0010	**STEEL WINDOWS** Stock, including frame, trim and insul. glass									
0020	See Div. 13 34 19.50									
1000	Custom units, double hung, 2'-8" x 4'-6" opening R085123-10	2 Sswk	12	1.333	Ea.	680	62.50		742.50	860
1100	2'-4" x 3'-9" opening		12	1.333		565	62.50		627.50	730
1500	Commercial projected, 3'-9" x 5'-5" opening		10	1.600		1,200	75		1,275	1,425
1600	6'-9" x 4'-1" opening		7	2.286		1,575	107		1,682	1,925
2000	Intermediate projected, 2'-9" x 4'-1" opening		12	1.333		670	62.50		732.50	845
2100	4'-1" x 5'-5" opening	↓	10	1.600	↓	1,350	75		1,425	1,600
9000	Minimum labor/equipment charge	1 Sswk	3	2.667	Job		125		125	220

08 51 66 – Metal Window Screens

08 51 66.10 Screens		Crew	Daily Output	Labor-Hours	Unit	Material	Labor	Equipment	Total	Total Incl O&P
0010	**SCREENS**									
0020	For metal sash, aluminum or bronze mesh, flat screen	2 Sswk	1200	.013	S.F.	4.10	.63		4.73	5.60
0500	Wicket screen, inside window		1000	.016		6.30	.75		7.05	8.25
0800	Security screen, aluminum frame with stainless steel cloth		1200	.013		22.50	.63		23.13	26
0900	Steel grate, painted, on steel frame	↓	1600	.010	↓	12.50	.47		12.97	14.55

08 51 Metal Windows

08 51 66 – Metal Window Screens

08 51 66.10 Screens		Crew	Daily Output	Labor-Hours	Unit	Material	2010 Bare Costs Labor	Equipment	Total	Total Incl O&P
1000	For solar louvers, add	2 Sswk	160	.100	S.F.	23.50	4.69		28.19	34
4000	See Div. 05 58 27.90									

08 52 Wood Windows

08 52 10 – Plain Wood Windows

08 52 10.20 Awning Window

		Crew	Daily Output	Labor-Hours	Unit	Material	Labor	Equipment	Total	Total Incl O&P
0010	**AWNING WINDOW**, Including frame, screens and grilles									
0100	Average quality, builders model, 34" x 22", double insulated glass	1 Carp	10	.800	Ea.	253	33		286	330
0200	Low E glass		10	.800		257	33		290	335
0300	40" x 28", double insulated glass		9	.889		310	37		347	395
0400	Low E Glass		9	.889		330	37		367	420
0500	48" x 36", double insulated glass		8	1		450	41.50		491.50	560
0600	Low E glass		8	1		475	41.50		516.50	585
1000	Vinyl clad, 34" x 22"		10	.800		259	33		292	335
1100	40" x 22"		10	.800		283	33		316	360
1200	36" x 28"		9	.889		300	37		337	385
1300	36" x 36"		9	.889		335	37		372	425
1400	48" x 28"		8	1		360	41.50		401.50	465
1500	60" x 36"		8	1		505	41.50		546.50	620
2200	Metal clad, 36" x 25"		9	.889		269	37		306	355
2300	40" x 30"		9	.889		335	37		372	425
2400	48" x 28"		8	1		345	41.50		386.50	445
2500	60" x 36"	↓	8	1		365	41.50		406.50	470
4000	Impact windows, minimum, add					60%				
4010	Impact windows, maximum, add				↓	160%				

08 52 10.30 Wood Windows

		Crew	Daily Output	Labor-Hours	Unit	Material	Labor	Equipment	Total	Total Incl O&P
0010	**WOOD WINDOWS**, double hung									
3200	20 S.F. and over	1 Carp	106	.075	S.F.	16.85	3.14		19.99	23.50
3800	Triple glazing for above, add					8.35			8.35	9.15
6000	Replacement sash, double hung, double glazing, to 12 S.F.	1 Carp	64	.125		25	5.20		30.20	36
6100	12 S.F. to 20 S.F.		94	.085	↓	19.50	3.54		23.04	27
7000	Sash, single lite, 2'-0" x 2'-0" high		20	.400	Ea.	49	16.60		65.60	79
7050	2'-6" x 2'-0" high		19	.421		60	17.50		77.50	93
7100	2'-6" x 2'-6" high		18	.444		67.50	18.45		85.95	103
7150	3'-0" x 2'-0" high	↓	17	.471	↓	88	19.55		107.55	127

08 52 10.40 Casement Window

		Crew	Daily Output	Labor-Hours	Unit	Material	Labor	Equipment	Total	Total Incl O&P
0010	**CASEMENT WINDOW**, including frame, screen and grilles									
0100	Avg. quality, bldrs. model, 2'-0" x 3'-0" H, dbl. insulated glass G	1 Carp	10	.800	Ea.	276	33		309	355
0150	Low E glass CN G		10	.800		225	33		258	299
0200	2'-0" x 4'-6" high, double insulated glass G		9	.889		345	37		382	435
0250	Low E glass G		9	.889		263	37		300	345
0260	Casement 4'-2" x 4'-2" double insulated glass G		11	.727		855	30		885	985
0270	4'-0" x 4'-0" Low E glass G		11	.727		465	30		495	560
0290	6'-4" x 5'-7" Low E glass G		9	.889		1,100	37		1,137	1,250
0300	2'-4" x 6'-0" high, double insulated glass G		8	1		400	41.50		441.50	505
0350	Low E glass G		8	1		465	41.50		506.50	580
0522	Vinyl clad, premium, double insulated glass, 2'-0" x 3'-0" G		10	.800		266	33		299	345
0524	2'-0" x 4'-0" G		9	.889		310	37		347	395
0525	2'-0" x 5'-0" G		8	1		355	41.50		396.50	455
0528	2'-0" x 6'-0" G	↓	8	1	↓	355	41.50		396.50	455

08 52 Wood Windows

08 52 10 – Plain Wood Windows

08 52 10.40 Casement Window			Crew	Daily Output	Labor-Hours	Unit	Material	2010 Bare Costs Labor	Equipment	Total	Total Incl O&P
0600	3′-0″ x 5′-0″		1 Carp	8	1	Ea.	650	41.50		691.50	780
0700	4′-0″ x 3′-0″			8	1		715	41.50		756.50	855
0710	4′-0″ x 4′-0″	G		8	1		610	41.50		651.50	740
0720	4′-8″ x 4′-0″			8	1		675	41.50		716.50	810
0730	4′-8″ x 5′-0″			6	1.333		770	55.50		825.50	935
0740	4′-8″ x 6′-0″			6	1.333		870	55.50		925.50	1,050
0750	6′-0″ x 4′-0″			6	1.333		790	55.50		845.50	950
0800	6′-0″ x 5′-0″			6	1.333		885	55.50		940.50	1,050
0900	5′-6″ x 5′-6″	G	2 Carp	15	1.067		1,400	44.50		1,444.50	1,600
2000	Bay, casement units, 8′ x 5′, w/screens, dbl insul. glass			2.50	6.400	Opng.	1,575	266		1,841	2,125
2100	Low E glass			2.50	6.400	″	1,650	266		1,916	2,200
8100	Metal clad, deluxe, dbl. insul. glass, 2′-0″ x 3′-0″ high	G	1 Carp	10	.800	Ea.	272	33		305	350
8120	2′-0″ x 4′-0″ high	G		9	.889		305	37		342	390
8140	2′-0″ x 5′-0″ high	G		8	1		320	41.50		361.50	415
8160	2′-0″ x 6′-0″ high	G		8	1		355	41.50		396.50	455
8190	For installation, add per leaf							15%			
8200	For multiple leaf units, deduct for stationary sash										
8220	2′ high					Ea.	22.50			22.50	24.50
8240	4′-6″ high						25.50			25.50	28
8260	6′ high						33.50			33.50	37
8300	Impact windows, minimum, add						60%				
8310	Impact windows, maximum, add						160%				

08 52 10.50 Double Hung

Line	Description		Crew	Daily Output	Labor-Hours	Unit	Material	Labor	Equipment	Total	Total Incl O&P
0010	**DOUBLE HUNG**, Including frame, screens and grilles R085216-10										
0100	Avg. quality, bldrs. model, 2′-0″ x 3′-0″ high, dbl insul. glass	G	1 Carp	10	.800	Ea.	184	33		217	253
0150	Low E glass	G		10	.800		208	33		241	280
0200	3′-0″ x 4′-0″ high, double insulated glass	G		9	.889		257	37		294	340
0250	Low E glass	G		9	.889		278	37		315	360
0300	4′-0″ x 4′-6″ high, double insulated glass	G		8	1		315	41.50		356.50	410
0350	Low E glass	G		8	1		340	41.50		381.50	440
1000	Vinyl clad, premium, double insulated glass, 2′-6″ x 3′-0″	G		10	.800		225	33		258	299
1005	2′-6″ x 4′-0″	G		10	.800		276	33		309	355
1100	3′-0″ x 3′-6″	G		10	.800		261	33		294	340
1200	3′-0″ x 4′-0″	G		9	.889		370	37		407	460
1300	3′-0″ x 4′-6″	G		9	.889		340	37		377	425
1400	3′-0″ x 5′-0″	G		8	1		315	41.50		356.50	410
1490	3′-4″ x 5′-0″	G		8	1		355	41.50		396.50	455
1500	3′-6″ x 6′-0″	G		8	1		370	41.50		411.50	470
1520	4′-0″ x 5′-0″	G		7	1.143		550	47.50		597.50	680
1530	4′-0″ x 6′-0″	G		7	1.143		685	47.50		732.50	830
2000	Metal clad, deluxe, dbl. insul. glass, 2′-6″ x 3′-0″ high	G		10	.800		265	33		298	345
2100	3′-0″ x 3′-6″ high	G		10	.800		305	33		338	385
2200	3′-0″ x 4′-0″ high	G		9	.889		320	37		357	405
2300	3′-0″ x 4′-6″ high	G		9	.889		335	37		372	425
2400	3′-0″ x 5′-0″ high	G		8	1		365	41.50		406.50	465
2500	3′-6″ x 6′-0″ high	G		8	1		440	41.50		481.50	545
8000	Impact windows, minimum, add						60%				
8010	Impact windows, maximum, add						160%				

08 52 10.55 Picture Window

Line	Description		Crew	Daily Output	Labor-Hours	Unit	Material	Labor	Equipment	Total	Total Incl O&P
0010	**PICTURE WINDOW**, Including frame and grilles										
0100	Average quality, bldrs. model, 3′-6″ x 4′-0″ high, dbl insulated glass		2 Carp	12	1.333	Ea.	345	55.50		400.50	465
0150	Low E glass			12	1.333		410	55.50		465.50	535

08 52 Wood Windows

08 52 10 – Plain Wood Windows

08 52 10.55 Picture Window

08 52 10.55 Picture Window		Crew	Daily Output	Labor-Hours	Unit	Material	2010 Bare Costs Labor	Equipment	Total	Total Incl O&P
0200	4'-0" x 4'-6" high, double insulated glass	2 Carp	11	1.455	Ea.	490	60.50		550.50	635
0250	Low E glass		11	1.455		520	60.50		580.50	665
0300	5'-0" x 4'-0" high, double insulated glass		11	1.455		570	60.50		630.50	720
0350	Low E glass		11	1.455		590	60.50		650.50	745
0400	6'-0" x 4'-6" high, double insulated glass		10	1.600		610	66.50		676.50	775
0450	Low E glass		10	1.600		620	66.50		686.50	785

08 52 10.65 Wood Sash

		Crew	Daily Output	Labor-Hours	Unit	Material	Labor	Equipment	Total	Total Incl O&P
0010	**WOOD SASH**, Including glazing but not trim									
0050	Custom, 5'-0" x 4'-0", 1" dbl. glazed, 3/16" thick lites	2 Carp	3.20	5	Ea.	167	208		375	505
0100	1/4" thick lites		5	3.200		172	133		305	395
0200	1" thick, triple glazed		5	3.200		395	133		528	635
0300	7'-0" x 4'-6" high, 1" double glazed, 3/16" thick lites		4.30	3.721		400	155		555	680
0400	1/4" thick lites		4.30	3.721		450	155		605	735
0500	1" thick, triple glazed		4.30	3.721		515	155		670	805
0600	8'-6" x 5'-0" high, 1" double glazed, 3/16" thick lites		3.50	4.571		540	190		730	885
0700	1/4" thick lites		3.50	4.571		590	190		780	945
0800	1" thick, triple glazed		3.50	4.571		595	190		785	950
0900	Window frames only, based on perimeter length				L.F.	3.83			3.83	4.21
1200	Window sill, stock, per lineal foot					8.25			8.25	9.10
1250	Casing, stock					3.15			3.15	3.47

08 52 10.70 Sliding Windows

		Crew	Daily Output	Labor-Hours	Unit	Material	Labor	Equipment	Total	Total Incl O&P
0010	**SLIDING WINDOWS**									
0100	Average quality, bldrs. model, 3'-0" x 3'-0" high, double insulated G	1 Carp	10	.800	Ea.	272	33		305	350
0120	Low E glass G		10	.800		293	33		326	370
0200	4'-0" x 3'-6" high, double insulated G		9	.889		325	37		362	415
0220	Low E glass G		9	.889		350	37		387	440
0300	6'-0" x 5'-0" high, double insulated G		8	1		460	41.50		501.50	575
0320	Low E glass G		8	1		500	41.50		541.50	615

08 52 13 – Metal-Clad Wood Windows

08 52 13.10 Awning Windows, Metal-Clad

		Crew	Daily Output	Labor-Hours	Unit	Material	Labor	Equipment	Total	Total Incl O&P
0010	**AWNING WINDOWS, METAL-CLAD**									
2000	Metal clad, awning deluxe, double insulated glass, 34" x 22"	1 Carp	10	.800	Ea.	242	33		275	315
2100	40" x 22"	"	10	.800	"	285	33		318	365

08 52 13.35 Picture and Sliding Windows Metal-Clad

		Crew	Daily Output	Labor-Hours	Unit	Material	Labor	Equipment	Total	Total Incl O&P
0010	**PICTURE AND SLIDING WINDOWS METAL-CLAD**									
2000	Metal clad, dlx picture, dbl. insul. glass, 4'-0" x 4'-0" high	2 Carp	12	1.333	Ea.	365	55.50		420.50	490
2100	4'-0" x 6'-0" high		11	1.455		535	60.50		595.50	685
2200	5'-0" x 6'-0" high		10	1.600		595	66.50		661.50	755
2300	6'-0" x 6'-0" high		10	1.600		680	66.50		746.50	850
2400	Metal clad, dlx sliding, double insulated glass, 3'-0" x 3'-0" high G	1 Carp	10	.800		330	33		363	410
2420	4'-0" x 3'-6" high G		9	.889		400	37		437	495
2440	5'-0" x 4'-0" high G		9	.889		480	37		517	580
2460	6'-0" x 5'-0" high G		8	1		705	41.50		746.50	840

08 52 16 – Plastic-Clad Wood Windows

08 52 16.10 Bow Window

		Crew	Daily Output	Labor-Hours	Unit	Material	Labor	Equipment	Total	Total Incl O&P
0010	**BOW WINDOW** Including frames, screens, and grilles									
0020	End panels operable									
1000	Bow type, casement, wood, bldrs mdl, 8' x 5' dbl insltd glass, 4 panel	2 Carp	10	1.600	Ea.	1,500	66.50		1,566.50	1,750
1050	Low E glass		10	1.600		1,325	66.50		1,391.50	1,550
1100	10'-0" x 5'-0", double insulated glass, 6 panels		6	2.667		1,350	111		1,461	1,675
1200	Low E glass, 6 panels		6	2.667		1,450	111		1,561	1,775

08 52 16 – Plastic-Clad Wood Windows

08 52 16.10 Bow Window		Crew	Daily Output	Labor-Hours	Unit	Material	2010 Bare Costs Labor	Equipment	Total	Total Incl O&P
1300	Vinyl clad, bldrs model, double insulated glass, 6'-0" x 4'-0", 3 panel	2 Carp	10	1.600	Ea.	1,025	66.50		1,091.50	1,225
1340	9'-0" x 4'-0", 4 panel		8	2		1,350	83		1,433	1,625
1380	10'-0" x 6'-0", 5 panels		7	2.286		2,250	95		2,345	2,625
1420	12'-0" x 6'-0", 6 panels		6	2.667		2,925	111		3,036	3,400
1600	Metal clad, casement, bldrs mdl, 6'-0" x 4'-0", dbl insltd gls, 3 panels		10	1.600		920	66.50		986.50	1,100
1640	9'-0" x 4'-0", 4 panels		8	2		1,250	83		1,333	1,500
1680	10'-0" x 5'-0", 5 panels		7	2.286		1,725	95		1,820	2,050
1720	12'-0" x 6'-0", 6 panels		6	2.667		2,400	111		2,511	2,825
2000	Bay window, builders model, 8' x 5' dbl insul glass,		10	1.600		1,825	66.50		1,891.50	2,125
2050	Low E glass,		10	1.600		2,225	66.50		2,291.50	2,550
2100	12'-0" x 6'-0", double insulated glass, 6 panels		6	2.667		2,300	111		2,411	2,700
2200	Low E glass		6	2.667		2,350	111		2,461	2,775
2280	6'-0" x 4'-0"		11	1.455		1,225	60.50		1,285.50	1,450
2300	Vinyl clad, premium, double insulated glass, 8'-0" x 5'-0"		10	1.600		1,750	66.50		1,816.50	2,025
2340	10'-0" x 5'-0"		8	2		2,275	83		2,358	2,625
2380	10'-0" x 6'-0"		7	2.286		2,525	95		2,620	2,925
2420	12'-0" x 6'-0"		6	2.667		3,200	111		3,311	3,700
2600	Metal clad, deluxe, dbl insul. glass, 8'-0" x 5'-0" high, 4 panels		10	1.600		1,625	66.50		1,691.50	1,875
2640	10'-0" x 5'-0" high, 5 panels		8	2		1,750	83		1,833	2,050
2680	10'-0" x 6'-0" high, 5 panels		7	2.286		2,050	95		2,145	2,425
2720	12'-0" x 6'-0" high, 6 panels		6	2.667		2,850	111		2,961	3,325
3000	Double hung, bldrs. model, bay, 8' x 4' high,dbl insulated glass		10	1.600		1,300	66.50		1,366.50	1,525
3050	Low E glass		10	1.600		1,375	66.50		1,441.50	1,625
3100	9'-0" x 5'-0" high, double insulated glass		6	2.667		1,400	111		1,511	1,700
3200	Low E glass		6	2.667		1,475	111		1,586	1,800
3300	Vinyl clad, premium, double insulated glass, 7'-0" x 4'-6"		10	1.600		1,325	66.50		1,391.50	1,575
3340	8'-0" x 4'-6"		8	2		1,375	83		1,458	1,625
3380	8'-0" x 5'-0"		7	2.286		1,425	95		1,520	1,725
3420	9'-0" x 5'-0"		6	2.667		1,475	111		1,586	1,800
3600	Metal clad, deluxe, dbl insul. glass, 7'-0" x 4'-0" high		10	1.600		1,225	66.50		1,291.50	1,450
3640	8'-0" x 4'-0" high		8	2		1,275	83		1,358	1,525
3680	8'-0" x 5'-0" high		7	2.286		1,325	95		1,420	1,600
3720	9'-0" x 5'-0" high	↓	6	2.667	↓	1,400	111		1,511	1,725

08 52 16.30 Palladian Windows		Crew	Daily Output	Labor-Hours	Unit	Material	Labor	Equipment	Total	Total Incl O&P
0010	**PALLADIAN WINDOWS**									
0020	Vinyl clad, double insulated glass, including frame and grilles									
0040	3'-2" x 2'-6" high	2 Carp	11	1.455	Ea.	1,250	60.50		1,310.50	1,475
0060	3'-2" x 4'-10"		11	1.455		1,750	60.50		1,810.50	2,000
0080	3'-2" x 6'-4"		10	1.600		1,700	66.50		1,766.50	1,975
0100	4'-0" x 4'-0"	↓	10	1.600		1,525	66.50		1,591.50	1,775
0120	4'-0" x 5'-4"	3 Carp	10	2.400		1,875	99.50		1,974.50	2,200
0140	4'-0" x 6'-0"		9	2.667		1,950	111		2,061	2,325
0160	4'-0" x 7'-4"		9	2.667		2,125	111		2,236	2,500
0180	5'-5" x 4'-10"		9	2.667		2,275	111		2,386	2,675
0200	5'-5" x 6'-10"		9	2.667		2,575	111		2,686	3,025
0220	5'-5" x 7'-9"		9	2.667		2,800	111		2,911	3,250
0240	6'-0" x 7'-11"		8	3		3,475	125		3,600	4,025
0260	8'-0" x 6'-0"	↓	8	3	↓	3,075	125		3,200	3,575

08 52 16.40 Transom Windows		Crew	Daily Output	Labor-Hours	Unit	Material	Labor	Equipment	Total	Total Incl O&P
0010	**TRANSOM WINDOWS**									
1000	Vinyl clad, premium, dbl. insul. glass, 4'-0" x 4'-0"	2 Carp	12	1.333	Ea.	495	55.50		550.50	630
1100	4'-0" x 6'-0"	↓	11	1.455	↓	910	60.50		970.50	1,100

08 52 Wood Windows

08 52 16 – Plastic-Clad Wood Windows

08 52 16.40 Transom Windows		Crew	Daily Output	Labor-Hours	Unit	Material	2010 Bare Costs Labor	Equipment	Total	Total Incl O&P
1200	5′-0″ x 6′-0″	2 Carp	10	1.600	Ea.	1,000	66.50		1,066.50	1,225
1300	6′-0″ x 6′-0″	↓	10	1.600	↓	1,025	66.50		1,091.50	1,225

08 52 16.70 Vinyl Clad, Premium, Dbl. Insulated Glass		Crew	Daily Output	Labor-Hours	Unit	Material	Labor	Equipment	Total	Total Incl O&P
0010	**VINYL CLAD, PREMIUM, DBL. INSULATED GLASS**									
1000	Sliding, 3′-0″ x 3′-0″ G	1 Carp	10	.800	Ea.	575	33		608	680
1050	4′-0″ x 3′-6″ G		9	.889		670	37		707	790
1100	5′-0″ x 4′-0″ G		9	.889		860	37		897	1,000
1150	6′-0″ x 5′-0″ G	↓	8	1	↓	1,075	41.50		1,116.50	1,275

08 52 50 – Window Accessories

08 52 50.10 Window Grille or Muntin		Crew	Daily Output	Labor-Hours	Unit	Material	Labor	Equipment	Total	Total Incl O&P
0010	**WINDOW GRILLE OR MUNTIN**, snap in type									
0020	Standard pattern interior grilles									
2000	Wood, awning window, glass size 28″ x 16″ high	1 Carp	30	.267	Ea.	24	11.10		35.10	43.50
2060	44″ x 24″ high		32	.250		36	10.40		46.40	55.50
2100	Casement, glass size, 20″ x 36″ high		30	.267		29	11.10		40.10	48.50
2180	20″ x 56″ high		32	.250	↓	40	10.40		50.40	60
2200	Double hung, glass size, 16″ x 24″ high		24	.333	Set	48	13.85		61.85	74.50
2280	32″ x 32″ high		34	.235	″	128	9.80		137.80	156
2500	Picture, glass size, 48″ x 48″ high		30	.267	Ea.	114	11.10		125.10	142
2580	60″ x 68″ high		28	.286	″	160	11.85		171.85	194
2600	Sliding, glass size, 14″ x 36″ high		24	.333	Set	26.50	13.85		40.35	50.50
2680	36″ x 36″ high	↓	22	.364	″	39	15.10		54.10	66.50

08 52 66 – Wood Window Screens

08 52 66.10 Wood Screens		Crew	Daily Output	Labor-Hours	Unit	Material	Labor	Equipment	Total	Total Incl O&P
0010	**WOOD SCREENS**									
0020	Over 3 S.F., 3/4″ frames	2 Carp	375	.043	S.F.	4.37	1.77		6.14	7.55
0100	1-1/8″ frames	″	375	.043	″	7.70	1.77		9.47	11.25

08 52 69 – Wood Storm Windows

08 52 69.10 Storm Windows		Crew	Daily Output	Labor-Hours	Unit	Material	Labor	Equipment	Total	Total Incl O&P
0010	**STORM WINDOWS**, aluminum residential									
0300	Basement, mill finish, incl. fiberglass screen									
0320	1′-10″ x 1′-0″ high	2 Carp	30	.533	Ea.	35	22		57	72.50
0340	2′-9″ x 1′-6″ high	″	30	.533	″	38	22		60	76
1600	Double-hung, combination, storm & screen									
2000	Average quality, clear anodic coating, 2′-0″ x 3′-5″ high	2 Carp	30	.533	Ea.	85	22		107	128
2020	2′-6″ x 5′-0″ high		28	.571		107	23.50		130.50	154
2040	4′-0″ x 6′-0″ high		25	.640		126	26.50		152.50	180
2400	White painted, 2′-0″ x 3′-5″ high		30	.533		85	22		107	128
2420	2′-6″ x 5′-0″ high		28	.571		90	23.50		113.50	136
2440	4′-0″ x 6′-0″ high		25	.640		98	26.50		124.50	149
2600	Mill finish, 2′-0″ x 3′-5″ high		30	.533		75	22		97	117
2620	2′-6″ x 5′-0″ high		28	.571		85	23.50		108.50	130
2640	4′-0″ x 6′-8″ high	↓	25	.640	↓	99	26.50		125.50	150

08 53 13.20 Vinyl Single Hung Windows

	08 53 13.20 Vinyl Single Hung Windows		Crew	Daily Output	Labor-Hours	Unit	Material	Labor	Equipment	Total	Total Incl O&P
							2010 Bare Costs				
0010	**VINYL SINGLE HUNG WINDOWS**										
0100	Grids, low E, J fin, ext. jambs, 21" x 53"	G	2 Carp	18	.889	Ea.	162	37		199	235
0110	21" x 57"	G		17	.941		166	39		205	244
0120	21" x 65"	G		16	1		173	41.50		214.50	254
0130	25" x 41"	G		20	.800		157	33		190	224
0140	25" x 49"	G		18	.889		169	37		206	243
0150	25" x 57"	G		17	.941		173	39		212	251
0160	25" x 65"	G		16	1		180	41.50		221.50	262
0170	29" x 41"	G		18	.889		163	37		200	236
0180	29" x 53"	G		18	.889		175	37		212	250
0190	29" x 57"	G		17	.941		179	39		218	258
0200	29" x 65"	G		16	1		185	41.50		226.50	268
0210	33" x 41"	G		20	.800		169	33		202	237
0220	33" x 53"	G		18	.889		182	37		219	257
0230	33" x 57"	G		17	.941		185	39		224	265
0240	33" x 65"	G		16	1		196	41.50		237.50	280
0250	37" x 41"	G		20	.800		178	33		211	247
0260	37" x 53"	G		18	.889		191	37		228	267
0270	37" x 57"	G		17	.941		195	39		234	276
0280	37" x 65"	G		16	1		203	41.50		244.50	287

08 53 13.30 Vinyl Double Hung Windows

	08 53 13.30 Vinyl Double Hung Windows		Crew	Daily Output	Labor-Hours	Unit	Material	Labor	Equipment	Total	Total Incl O&P
0010	**VINYL DOUBLE HUNG WINDOWS**										
0100	Grids, low E, J fin, ext. jambs, 21" x 53"	G	2 Carp	18	.889	Ea.	186	37		223	262
0102	21" x 37"	G		18	.889		165	37		202	239
0104	21" x 41"	G		18	.889		170	37		207	244
0106	21" x 49"	G		18	.889		178	37		215	253
0110	21" x 57"	G		17	.941		190	39		229	270
0120	21" x 65"	G		16	1		198	41.50		239.50	282
0128	25" x 37"	G		20	.800		175	33		208	244
0130	25" x 41"	G		20	.800		179	33		212	248
0140	25" x 49"	G		18	.889		184	37		221	259
0145	25" x 53"	G		18	.889		192	37		229	268
0150	25" x 57"	G		17	.941		192	39		231	272
0160	25" x 65"	G		16	1		204	41.50		245.50	288
0162	25" x 69"	G		16	1		212	41.50		253.50	297
0164	25" x 77"	G		16	1		225	41.50		266.50	310
0168	29" x 37"	G		18	.889		180	37		217	255
0170	29" x 41"	G		18	.889		184	37		221	259
0172	29" x 49"	G		18	.889		193	37		230	269
0180	29" x 53"	G		18	.889		197	37		234	274
0190	29" x 57"	G		17	.941		202	39		241	283
0200	29" x 65"	G		16	1		217	41.50		258.50	305
0202	29" x 69"	G		16	1		217	41.50		258.50	305
0205	29" x 77"	G		16	1		230	41.50		271.50	315
0208	33" x 37"	G		20	.800		185	33		218	255
0210	33" x 41"	G		20	.800		189	33		222	259
0215	33" x 49"	G		20	.800		199	33		232	270
0220	33" x 53"	G		18	.889		204	37		241	281
0230	33" x 57"	G		17	.941		208	39		247	290
0240	33" x 65"	G		16	1		214	41.50		255.50	299
0242	33" x 69"	G		16	1		227	41.50		268.50	315
0246	33" x 77"	G		16	1		238	41.50		279.50	325

08 53 13.30 Vinyl Double Hung Windows			Crew	Daily Output	Labor-Hours	Unit	Material	2010 Bare Costs Labor	Equipment	Total	Total Incl O&P
0250	37" x 41"	G	2 Carp	20	.800	Ea.	191	33		224	261
0255	37" x 49"	G		20	.800		204	33		237	275
0260	37" x 53"	G		18	.889		212	37		249	290
0270	37" x 57"	G		17	.941		217	39		256	300
0280	37" x 65"	G		16	1		222	41.50		263.50	310
0282	37" x 69"	G		16	1		227	41.50		268.50	315
0286	37" x 77"	G		16	1		248	41.50		289.50	335
0300	Solid vinyl, average quality, double insulated glass, 2'-0" x 3'-0"	G	1 Carp	10	.800		279	33		312	355
0310	3'-0" x 4'-0"	G		9	.889		195	37		232	272
0320	4'-0" x 4'-6"	G		8	1		293	41.50		334.50	385
0330	Premium, double insulated glass, 2'-6" x 3'-0"	G		10	.800		214	33		247	287
0340	3'-0" x 3'-6"	G		9	.889		245	37		282	325
0350	3'-0" x 4"-0"	G		9	.889		256	37		293	340
0360	3'-0" x 4'-6"	G		9	.889		248	37		285	330
0370	3'-0" x 5'-0"	G		8	1		268	41.50		309.50	360
0380	3'-6" x 6"-0"	G		8	1		300	41.50		341.50	395

08 53 13.40 Vinyl Casement Windows			Crew	Daily Output	Labor-Hours	Unit	Material	Labor	Equipment	Total	Total Incl O&P
0010	**VINYL CASEMENT WINDOWS**										
0100	Grids, low E, J fin, ext. jambs, 1 lt, 21" x 41"	G	2 Carp	20	.800	Ea.	242	33		275	315
0110	21" x 47"	G		20	.800		264	33		297	340
0120	21" x 53"	G		20	.800		286	33		319	365
0128	24" x 35"	G		19	.842		233	35		268	310
0130	24" x 41"	G		19	.842		253	35		288	330
0140	24" x 47"	G		19	.842		274	35		309	355
0150	24" x 53"	G		19	.842		296	35		331	380
0158	28" x 35"	G		19	.842		248	35		283	325
0160	28" x 41"	G		19	.842		268	35		303	350
0170	28" x 47"	G		19	.842		290	35		325	375
0180	28" x 53"	G		19	.842		320	35		355	405
0184	28" x 59"	G		19	.842		325	35		360	415
0188	Two lites, 33" x 35"	G		18	.889		400	37		437	495
0190	33" x 41"	G		18	.889		425	37		462	525
0200	33" x 47"	G		18	.889		460	37		497	560
0210	33" x 53"	G		18	.889		490	37		527	595
0212	33" x 59"	G		18	.889		520	37		557	625
0215	33" x 72"	G		18	.889		540	37		577	650
0220	41" x 41"	G		18	.889		465	37		502	570
0230	41" x 47"	G		18	.889		500	37		537	605
0240	41" x 53"	G		17	.941		530	39		569	640
0242	41" x 59"	G		17	.941		560	39		599	675
0246	41" x 72"	G		17	.941		585	39		624	700
0250	47" x 41"	G		17	.941		470	39		509	580
0260	47" x 47"	G		17	.941		500	39		539	610
0270	47" x 53"	G		17	.941		530	39		569	645
0272	47" x 59"	G		17	.941		585	39		624	700
0280	56" x 41"	G		15	1.067		505	44.50		549.50	625
0290	56" x 47"	G		15	1.067		530	44.50		574.50	655
0300	56" x 53"	G		15	1.067		580	44.50		624.50	710
0302	56" x 59"	G		15	1.067		605	44.50		649.50	740
0310	56" x 72"	G		15	1.067		660	44.50		704.50	795
0340	Solid vinyl, premium, double insulated glass, 2'-0" x 3'-0" high	G	1 Carp	10	.800		252	33		285	330
0360	2'-0" x 4'-0" high	G		9	.889		280	37		317	365

08 53 Plastic Windows

08 53 13 – Vinyl Windows

08 53 13.40 Vinyl Casement Windows			Crew	Daily Output	Labor-Hours	Unit	Material	2010 Bare Costs Labor	Equipment	Total	Total Incl O&P
0380	2'-0" x 5'-0" high	G	1 Carp	8	1	Ea.	291	41.50		332.50	385

08 53 13.50 Vinyl Picture Windows

		Crew	Daily Output	Labor-Hours	Unit	Material	Labor	Equipment	Total	Total Incl O&P
0010	**VINYL PICTURE WINDOWS**									
0100	Grids, low E, J fin, ext. jambs, 33" x 47"	2 Carp	12	1.333	Ea.	238	55.50		293.50	350
0110	35" x 71"		12	1.333		310	55.50		365.50	425
0120	41" x 47"		12	1.333		252	55.50		307.50	365
0130	41" x 71"		12	1.333		320	55.50		375.50	440
0140	47" x 47"		12	1.333		277	55.50		332.50	390
0150	47" x 71"		11	1.455		350	60.50		410.50	480
0160	53" x 47"		11	1.455		300	60.50		360.50	425
0170	53" x 71"		11	1.455		315	60.50		375.50	440
0180	59" x 47"		11	1.455		340	60.50		400.50	470
0190	59" x 71"		11	1.455		365	60.50		425.50	495
0200	71" x 47"		10	1.600		375	66.50		441.50	515
0210	71" x 71"	↓	10	1.600	↓	395	66.50		461.50	535

08 54 Composite Windows

08 54 13 – Fiberglass Windows

08 54 13.10 Fiberglass Single Hung Windows

			Crew	Daily Output	Labor-Hours	Unit	Material	Labor	Equipment	Total	Total Incl O&P
0010	**FIBERGLASS SINGLE HUNG WINDOWS**										
0100	Grids, low E, 18" x 24"	G	2 Carp	18	.889	Ea.	335	37		372	425
0110	18" x 40"	G		17	.941		300	39		339	390
0130	24" x 40"	G		20	.800		330	33		363	415
0230	36" x 36"	G		17	.941		370	39		409	465
0250	36" x 48"	G		20	.800		405	33		438	495
0260	36" x 60"	G		18	.889		445	37		482	545
0280	36" x 72"	G		16	1		470	41.50		511.50	580
0290	48" x 40"	G	↓	16	1	↓	470	41.50		511.50	580

08 56 Special Function Windows

08 56 63 – Detention Windows

08 56 63.13 Visitor Cubicle Windows

		Crew	Daily Output	Labor-Hours	Unit	Material	Labor	Equipment	Total	Total Incl O&P
0010	**VISITOR CUBICLE WINDOWS**									
4000	Visitor cubicle, vision panel, no intercom	E-4	2	16	Ea.	2,925	760	73	3,758	4,625

08 62 Unit Skylights

08 62 13 – Domed Unit Skylights

08 62 13.20 Skylights

			Crew	Daily Output	Labor-Hours	Unit	Material	Labor	Equipment	Total	Total Incl O&P
0010	**SKYLIGHTS**, Plastic domes, flush or curb mounted ten or										
0100	more units										
0300	Nominal size under 10 S.F., double	G	G-3	130	.246	S.F.	28	10.10		38.10	46.50
0400	Single			160	.200		24.50	8.20		32.70	39.50
0600	10 S.F. to 20 S.F., double	G		315	.102		24	4.18		28.18	33
0700	Single			395	.081		24	3.33		27.33	31
0900	20 S.F. to 30 S.F., double	G		395	.081		22	3.33		25.33	29.50
1000	Single			465	.069		19.70	2.83		22.53	26
1200	30 S.F. to 65 S.F., double	G	↓	465	.069	↓	22	2.83		24.83	29

08 62 Unit Skylights

08 62 13 – Domed Unit Skylights

08 62 13.20 Skylights

	08 62 13.20 Skylights		Crew	Daily Output	Labor-Hours	Unit	Material	2010 Bare Costs Labor	Equipment	Total	Total Incl O&P
1300	Single		G-3	610	.052	S.F.	17	2.16		19.16	22
1500	For insulated 4″ curbs, double, add						27%				
1600	Single, add						30%				
1800	For integral insulated 9″ curbs, double, add						30%				
1900	Single, add						40%				
2120	Ventilating insulated plexiglass dome with										
2130	curb mounting, 36″ x 36″	G	G-3	12	2.667	Ea.	460	110		570	670
2150	52″ x 52″	G		12	2.667		640	110		750	870
2160	28″ x 52″	G		10	3.200		470	132		602	715
2170	36″ x 52″	G		10	3.200		520	132		652	770
2180	For electric opening system, add	G					300			300	330
2200	Field fabricated, factory type, aluminum and wire glass	G	G-3	120	.267	S.F.	16	10.95		26.95	34.50
2300	Insulated safety glass with aluminum frame	G		160	.200		90	8.20		98.20	112
2400	Sandwich panels, fiberglass, for walls, 1-9/16″ thick, to 250 SF	G		200	.160		16.20	6.60		22.80	28
2500	250 SF and up	G		265	.121		14.60	4.96		19.56	23.50
2700	As above, but for roofs, 2-3/4″ thick, to 250 SF	G		295	.108		23.50	4.46		27.96	32.50
2800	250 SF and up	G		330	.097		19	3.99		22.99	27

08 63 Metal-Framed Skylights

08 63 23 – Ridge Metal-Framed Skylights

08 63 23.10 Prefabricated

	08 63 23.10 Prefabricated		Crew	Daily Output	Labor-Hours	Unit	Material	Labor	Equipment	Total	Total Incl O&P
0010	**PREFABRICATED** glass block with metal frame										
0020	Minimum	G	G-3	265	.121	S.F.	55	4.96		59.96	68
0100	Maximum	G	″	160	.200	″	107	8.20		115.20	131

08 71 Door Hardware

08 71 13 – Automatic Door Operators

08 71 13.10 Automatic Openers Commercial

	08 71 13.10 Automatic Openers Commercial	Crew	Daily Output	Labor-Hours	Unit	Material	Labor	Equipment	Total	Total Incl O&P
0010	**AUTOMATIC OPENERS COMMERCIAL**									
0020	Pneumatic, incl opener, motion sens, control box, tubing, compressor									
0050	For single swing door, per opening	2 Skwk	.80	20	Ea.	4,200	850		5,050	5,925
0100	Pair, per opening		.50	32	Opng.	6,900	1,375		8,275	9,700
1000	For single sliding door, per opening		.60	26.667		4,600	1,125		5,725	6,800
1300	Bi-parting pair		.50	32		7,000	1,375		8,375	9,800
1420	Electronic door opener incl motion sens, 12V control box, motor									
1450	For single swing door, per opening	2 Skwk	.80	20	Opng.	3,500	850		4,350	5,150
1500	Pair, per opening		.50	32		5,700	1,375		7,075	8,375
1600	For single sliding door, per opening		.60	26.667		3,800	1,125		4,925	5,925
1700	Bi-parting pair		.50	32		5,800	1,375		7,175	8,475
1750	Handicap actuator buttons, 2, including 12V DC wiring, add	1 Carp	1.50	5.333	Pr.	415	222		637	795

08 71 13.20 Automatic Openers Industrial

	08 71 13.20 Automatic Openers Industrial	Crew	Daily Output	Labor-Hours	Unit	Material	Labor	Equipment	Total	Total Incl O&P
0010	**AUTOMATIC OPENERS INDUSTRIAL**									
0015	Sliding doors up to 6′ wide	2 Skwk	.60	26.667	Opng.	5,400	1,125		6,525	7,700
0200	To 12′ wide	″	.40	40	″	6,600	1,700		8,300	9,875
0400	Over 12′ wide, add per L.F. of excess				L.F.	750			750	825
1000	Swing doors, to 5′ wide	2 Skwk	.80	20	Ea.	3,200	850		4,050	4,825
1860	Add for controls, wall pushbutton, 3 button		4	4		200	170		370	480
1870	Ceiling pull cord		4.30	3.721		170	159		329	430

08 71 20.10 Bolts, Flush		Crew	Daily Output	Labor-Hours	Unit	Material	2010 Bare Costs Labor	Equipment	Total	Total Incl O&P
0010	**BOLTS, FLUSH**									
0020	Standard, concealed	1 Carp	7	1.143	Ea.	22.50	47.50		70	98
0800	Automatic fire exit	"	5	1.600		299	66.50		365.50	430
1600	For electric release, add	1 Elec	3	2.667		125	131		256	330
3000	Barrel, brass, 2" long	1 Carp	40	.200		5.20	8.30		13.50	18.50
3020	4" long		40	.200		5.15	8.30		13.45	18.45
3060	6" long		40	.200		10.40	8.30		18.70	24.50
08 71 20.15 Hardware										
0009	**HARDWARE**									
0010	Average percentage for hardware, total job cost									
0025	Minimum				Job					1%
0050	Maximum									4%
0500	Total hardware for building, average distribution					85%	15%			
1000	Door hardware, apartment, interior				Door	490			490	540
1300	Average, door hardware, motel/hotel interior					635			635	700
1500	Hospital bedroom, minimum					595			595	650
2000	Maximum					710			710	780
2100	Pocket door				Ea.	134			134	147
2250	School, single exterior, incl. lever, incl. panic device				Door	1,075			1,075	1,200
2500	Single interior, regular use, lever included					600			600	660
2550	Average, door hdwe, school, classroom, ANSI F84, lever handle					735			735	810
2600	Average, door hdwe, school, classroom, ANSI F88, incl. lever					780			780	860
2850	Stairway, single interior					515			515	570
3100	Double exterior, with panic device				Pr.	2,025			2,025	2,250
3600	Toilet, public, single interior				Door	210			210	231
6020	Add for fire alram door holder, electro-magnetic	1 Elec	4	2	Ea.	99	98		197	255
08 71 20.20 Door Protectors										
0010	**DOOR PROTECTORS**									
0020	1-3/4" x 3/4" U channel	2 Carp	80	.200	L.F.	31	8.30		39.30	47
0021	1-3/4" x 1-1/4" U channel		80	.200	"	15.30	8.30		23.60	29.50
1000	Tear drop, spring-stl, 8" high x 19" long		15	1.067	Ea.	92	44.50		136.50	170
1010	8" high x 32" long		15	1.067		114	44.50		158.50	194
1100	Tear drop, stainless stl, 8" high x 19" long		15	1.067		240	44.50		284.50	335
1200	8" high x 32" long		15	1.067		295	44.50		339.50	395
08 71 20.30 Door Closers										
0010	**DOOR CLOSERS**									
0020	Adjustable backcheck, 3 way mount, all sizes, regular arm	1 Carp	6	1.333	Ea.	166	55.50		221.50	268
0040	Hold open arm		6	1.333		210	55.50		265.50	315
0100	Fusible link		6.50	1.231		140	51		191	233
0200	Non sized, regular arm		6	1.333		160	55.50		215.50	262
0240	Hold open arm		6	1.333		195	55.50		250.50	300
0400	4 way mount, non sized, regular arm		6	1.333		210	55.50		265.50	315
0440	Hold open arm		6	1.333		220	55.50		275.50	330
2000	Backcheck and adjustable power, hinge face mount									
2010	All sizes, regular arm	1 Carp	6.50	1.231	Ea.	200	51		251	299
2040	Hold open arm		6.50	1.231		240	51		291	345
2400	Top jamb mount, all sizes, regular arm		6	1.333		195	55.50		250.50	300
2440	Hold open arm		6	1.333		205	55.50		260.50	310
2800	Top face mount, all sizes, regular arm		6.50	1.231		190	51		241	288
2840	Hold open arm		6.50	1.231		208	51		259	310
4000	Backcheck, overhead concealed, all sizes, regular arm		5.50	1.455		210	60.50		270.50	325

08 71 20.30 Door Closers		Crew	Daily Output	Labor-Hours	Unit	Material	2010 Bare Costs Labor	Equipment	Total	Total Incl O&P
4040	Concealed arm	1 Carp	5	1.600	Ea.	225	66.50		291.50	350
4400	Compact overhead, concealed, all sizes, regular arm		5.50	1.455		340	60.50		400.50	470
4440	Concealed arm		5	1.600		380	66.50		446.50	520
4800	Concealed in door, all sizes, regular arm		5.50	1.455		350	60.50		410.50	480
4840	Concealed arm		5	1.600		370	66.50		436.50	505
4900	Floor concealed, all sizes, single acting		2.20	3.636		375	151		526	650
4940	Double acting		2.20	3.636		300	151		451	565
5000	For cast aluminum cylinder, deduct					30			30	33
5040	For delayed action, add					40			40	44
5080	For fusible link arm, add					25			25	27.50
5120	For shock absorbing arm, add					45			45	49.50
5160	For spring power adjustment, add					35			35	38.50
6000	Closer-holder, hinge face mount, all sizes, exposed arm	1 Carp	6.50	1.231		160	51		211	255
7000	Electronic closer-holder, hinge facemount, concealed arm		5	1.600		260	66.50		326.50	390
7400	With built-in detector		5	1.600		660	66.50		726.50	825
8000	Surface mounted, stand. duty, parallel arm, primed, traditional		6	1.333		184	55.50		239.50	288
8030	Light duty		6	1.333		119	55.50		174.50	217
8050	Heavy duty		6	1.333		211	55.50		266.50	320
8100	Standard duty, parallel arm, modern		6	1.333		204	55.50		259.50	310
8150	Heavy duty		6	1.333		221	55.50		276.50	330

08 71 20.31 Door Closers		Crew	Daily Output	Labor-Hours	Unit	Material	Labor	Equipment	Total	Total Incl O&P
0010	**DOOR CLOSERS**									
0015	Door closer, rack and pinion	1 Carp	6.50	1.231	Ea.	155	51		206	250
1520	Door, single acting, standard arm		1	8		195	330		525	725
1522	Hold open arm		1	8		211	330		541	745
1526	Frame, single acting, standard arm		1	8		330	330		660	875
1530	Hold open arm		1	8		350	330		680	895
1534	Double acting, standard arm		1	8		445	330		775	1,000
1536	Hold open arm		1	8		455	330		785	1,000
1554	Floor, center hung, single acting, bottom arm		1	8		285	330		615	825
1558	Double acting		1	8		315	330		645	855
1560	Offset hung, single acting, bottom arm		1	8		330	330		660	870
6500	Electro magnetic closer/holder									
6510	Single point, no detector	1 Carp	1	8	Ea.	370	330		700	915
6515	Including detector		1	8		665	330		995	1,250
6520	Multi-point, no detector		1	8		665	330		995	1,250
6524	Including detector		1	8		770	330		1,100	1,350
6550	Electric automatic operators									
6555	Operator	1 Carp	1	8	Ea.	2,375	330		2,705	3,100
6570	Wall plate actuator		1	8		193	330		523	725
8010	Light duty, regular arm		1	8		108	330		438	630
8032	Parallel arm		1	8		111	330		441	630
8034	Hold open arm		1	8		117	330		447	640
8036	Fusible link arm		1	8		172	330		502	700
8040	Medium duty, regular arm		1	8		134	330		464	660
8042	Extra duty parallel arm		1	8		147	330		477	670
8044	Hold open arm		1	8		155	330		485	680
8046	Positive stop arm		1	8		168	330		498	695
8052	Heavy duty, regular arm		1	8		164	330		494	690
8054	Top jamb mount		1	8		164	330		494	690
8056	Extra duty parallel arm		1	8		164	330		494	690
8058	Hold open arm		1	8		183	330		513	710

08 71 Door Hardware

08 71 20 – Hardware

08 71 20.31 Door Closers		Crew	Daily Output	Labor-Hours	Unit	2010 Bare Costs Material	Labor	Equipment	Total	Total Incl O&P
8060	Positive stop arm	1 Carp	1	8	Ea.	186	330		516	715
8062	Fusible link arm		1	8		213	330		543	745
8080	Universal heavy duty, regular arm		1	8		183	330		513	710
8084	Parallel arm		1	8		183	330		513	710
8088	Extra duty, parallel arm		1	8		186	330		516	715
8090	Hold open arm		1	8		268	330		598	805
8094	Positive stop arm		1	8		194	330		524	725
8098	For delayed action add	↓	1	8	↓	26	330		356	540

08 71 20.35 Panic Devices

Line	Description	Crew	Daily Output	Labor-Hours	Unit	Material	Labor	Equipment	Total	Total Incl O&P
0010	**PANIC DEVICES**									
0015	For rim locks, single door exit only	1 Carp	6	1.333	Ea.	495	55.50		550.50	630
0020	Outside key and pull		5	1.600		540	66.50		606.50	690
0200	Bar and vertical rod, exit only		5	1.600		715	66.50		781.50	885
0210	Outside key and pull		4	2		790	83		873	1,000
0400	Bar and concealed rod		4	2		590	83		673	780
0600	Touch bar, exit only		6	1.333		500	55.50		555.50	635
0610	Outside key and pull		5	1.600		570	66.50		636.50	730
0700	Touch bar and vertical rod, exit only		5	1.600		645	66.50		711.50	810
0710	Outside key and pull		4	2		745	83		828	950
1000	Mortise, bar, exit only		4	2		530	83		613	710
1600	Touch bar, exit only		4	2		605	83		688	795
2000	Narrow stile, rim mounted, bar, exit only		6	1.333		635	55.50		690.50	785
2010	Outside key and pull		5	1.600		690	66.50		756.50	855
2200	Bar and vertical rod, exit only		5	1.600		655	66.50		721.50	820
2210	Outside key and pull		4	2		655	83		738	850
2400	Bar and concealed rod, exit only		3	2.667		760	111		871	1,000
3000	Mortise, bar, exit only		4	2		575	83		658	760
3600	Touch bar, exit only	↓	4	2	↓	830	83		913	1,050

08 71 20.40 Lockset

Line	Description	Crew	Daily Output	Labor-Hours	Unit	Material	Labor	Equipment	Total	Total Incl O&P
0010	**LOCKSET**, Standard duty									
0020	Non-keyed, passage, w/sect.trim	1 Carp	12	.667	Ea.	53.50	27.50		81	102
0100	Privacy		12	.667		62	27.50		89.50	111
0400	Keyed, single cylinder function **CN**		10	.800		86	33		119	146
0420	Hotel		8	1		130	41.50		171.50	207
0500	Lever handled, keyed, single cylinder function		10	.800		156	33		189	223
1000	Heavy duty with sectional trim, non-keyed, passages		12	.667		140	27.50		167.50	197
1100	Privacy		12	.667		180	27.50		207.50	241
1400	Keyed, single cylinder function		10	.800		215	33		248	288
1420	Hotel		8	1		320	41.50		361.50	415
1600	Communicating	↓	10	.800		240	33		273	315
1690	For re-core cylinder, add					38			38	42
1700	Residential, interior door, minimum	1 Carp	16	.500		22	21		43	56
1720	Maximum		8	1		55	41.50		96.50	125
1800	Exterior, minimum		14	.571		40	23.50		63.50	80.50
1810	Average		8	1		72	41.50		113.50	144
1820	Maximum	↓	8	1	↓	160	41.50		201.50	240

08 71 20.41 Dead Locks

Line	Description	Crew	Daily Output	Labor-Hours	Unit	Material	Labor	Equipment	Total	Total Incl O&P
0010	**DEAD LOCKS**									
0011	Mortise heavy duty outside key (security item)	1 Carp	9	.889	Ea.	260	37		297	345
0020	Double cylinder		9	.889		265	37		302	350
0100	Medium duty, outside key		10	.800		140	33		173	205
0110	Double cylinder	↓	10	.800	↓	155	33		188	222

08 71 20.41 Dead Locks		Crew	Daily Output	Labor-Hours	Unit	Material	2010 Bare Costs Labor	Equipment	Total	Total Incl O&P
1000	Tubular, standard duty, outside key	1 Carp	10	.800	Ea.	65	33		98	123
1010	Double cylinder		10	.800		82	33		115	141
1200	Night latch, outside key		10	.800		81	33		114	140

08 71 20.42 Mortise Locksets		Crew	Daily Output	Labor-Hours	Unit	Material	Labor	Equipment	Total	Total Incl O&P
0010	**MORTISE LOCKSETS**, Comm., wrought knobs & full escutcheon trim									
0020	Non-keyed, passage, minimum	1 Carp	9	.889	Ea.	182	37		219	257
0030	Maximum		8	1		294	41.50		335.50	390
0040	Privacy, minimum		9	.889		195	37		232	272
0050	Maximum		8	1		320	41.50		361.50	415
0100	Keyed, office/entrance/apartment, minimum		8	1		226	41.50		267.50	315
0110	Maximum		7	1.143		380	47.50		427.50	495
0120	Single cylinder, typical, minimum		8	1		194	41.50		235.50	277
0130	Maximum		7	1.143		350	47.50		397.50	460
0200	Hotel, minimum		7	1.143		225	47.50		272.50	320
0210	Maximum		6	1.333		370	55.50		425.50	490
0300	Communication, double cylinder, minimum		8	1		220	41.50		261.50	305
0310	Maximum		7	1.143		290	47.50		337.50	395
1000	Wrought knobs and sectional trim, non-keyed, passage, minimum		10	.800		120	33		153	183
1010	Maximum		9	.889		235	37		272	315
1040	Privacy, minimum		10	.800		141	33		174	206
1050	Maximum		9	.889		256	37		293	340
1100	Keyed, entrance, office/apartment, minimum		9	.889		216	37		253	295
1110	Maximum		8	1		300	41.50		341.50	395
1120	Single cylinder, typical, minimum		9	.889		210	37		247	288
1130	Maximum		8	1		291	41.50		332.50	385
2000	Cast knobs and full escutcheon trim									
2010	Non-keyed, passage, minimum	1 Carp	9	.889	Ea.	256	37		293	340
2020	Maximum		8	1		410	41.50		451.50	515
2040	Privacy, minimum		9	.889		310	37		347	395
2050	Maximum		8	1		440	41.50		481.50	550
2120	Keyed, single cylinder, typical, minimum		8	1		315	41.50		356.50	410
2130	Maximum		7	1.143		490	47.50		537.50	615
2200	Hotel, minimum		7	1.143		335	47.50		382.50	445
2210	Maximum		6	1.333		595	55.50		650.50	740
3000	Cast knob and sectional trim, non-keyed, passage, minimum		10	.800		200	33		233	271
3010	Maximum		10	.800		390	33		423	480
3040	Privacy, minimum		10	.800		225	33		258	299
3050	Maximum		10	.800		400	33		433	490
3100	Keyed, office/entrance/apartment, minimum		9	.889		255	37		292	340
3110	Maximum		9	.889		400	37		437	495
3120	Single cylinder, typical, minimum		9	.889		255	37		292	340
3130	Maximum		9	.889		490	37		527	595
3190	For re-core cylinder, add					35			35	38.50
3800	Cipher lockset w/key pad (security item)	1 Carp	13	.615		810	25.50		835.50	935
3900	Cipher lockset, with dial for swinging doors (security item)		13	.615		1,725	25.50		1,750.50	1,950
3920	with dial for swinging doors & drill resistant plate (security item)		12	.667		2,100	27.50		2,127.50	2,350
3950	Cipher lockset with dial for safe/vault door (security item)		12	.667		1,400	27.50		1,427.50	1,600
3980	Keyless, pushbutton type									
4000	Residential/light commercial, deadbolt, standard	1 Carp	9	.889	Ea.	125	37		162	195
4010	Heavy duty		9	.889		142	37		179	213
4020	Industrial, heavy duty, with deadbolt		9	.889		284	37		321	370
4030	Key override		9	.889		320	37		357	405

08 71 Door Hardware

08 71 20 – Hardware

08 71 20.42 Mortise Locksets		Crew	Daily Output	Labor-Hours	Unit	Material	2010 Bare Costs Labor	Equipment	Total	Total Incl O&P
4040	Lever activated handle	1 Carp	9	.889	Ea.	335	37		372	425
4050	Key override		9	.889		375	37		412	470
4060	Double sided pushbutton type		8	1		695	41.50		736.50	825
4070	Key override		8	1		690	41.50		731.50	820
08 71 20.45 Peepholes										
0010	**PEEPHOLES**									
2010	Peephole	1 Carp	32	.250	Ea.	15.80	10.40		26.20	33.50
08 71 20.50 Door Stops										
0010	**DOOR STOPS**									
0020	Holder & bumper, floor or wall	1 Carp	32	.250	Ea.	34.50	10.40		44.90	54
1300	Wall bumper, 4″ diameter, with rubber pad, aluminum		32	.250		10.95	10.40		21.35	28
1600	Door bumper, floor type, aluminum		32	.250		7	10.40		17.40	23.50
1900	Plunger type, door mounted		32	.250		28	10.40		38.40	46.50
08 71 20.55 Push-Pull Plates										
0010	**PUSH-PULL PLATES**									
0100	Push plate, .050 thick, 4″ x 16″, aluminum	1 Carp	12	.667	Ea.	9.15	27.50		36.65	52.50
0500	Bronze		12	.667		21	27.50		48.50	66
1500	Pull handle and push bar, aluminum		11	.727		131	30		161	191
2000	Bronze		10	.800		169	33		202	236
3000	Push plate both sides, aluminum		14	.571		18.25	23.50		41.75	56.50
3500	Bronze		13	.615		43	25.50		68.50	87
4000	Door pull, designer style, cast aluminum, minimum		12	.667		71	27.50		98.50	121
5000	Maximum		8	1		370	41.50		411.50	470
6000	Cast bronze, minimum		12	.667		87	27.50		114.50	138
7000	Maximum		8	1		405	41.50		446.50	510
8000	Walnut, minimum		12	.667		67	27.50		94.50	116
9000	Maximum		8	1		370	41.50		411.50	470
08 71 20.60 Entrance Locks										
0010	**ENTRANCE LOCKS**									
0015	Cylinder, grip handle deadlocking latch	1 Carp	9	.889	Ea.	132	37		169	202
0020	Deadbolt		8	1		160	41.50		201.50	240
0100	Push and pull plate, dead bolt		8	1		150	41.50		191.50	229
0900	For handicapped lever, add					170			170	187
08 71 20.65 Thresholds										
0010	**THRESHOLDS**									
0011	Threshold 3′ long saddles aluminum	1 Carp	48	.167	L.F.	4.16	6.95		11.11	15.30
0100	Aluminum, 8″ wide, 1/2″ thick		12	.667	Ea.	39	27.50		66.50	85.50
0500	Bronze		60	.133	L.F.	40	5.55		45.55	52.50
0600	Bronze, panic threshold, 5″ wide, 1/2″ thick		12	.667	Ea.	66	27.50		93.50	115
0700	Rubber, 1/2″ thick, 5-1/2″ wide		20	.400		38	16.60		54.60	67.50
0800	2-3/4″ wide		20	.400		16	16.60		32.60	43
08 71 20.70 Floor Checks										
0010	**FLOOR CHECKS**									
0020	For over 3′ wide doors single acting	1 Carp	2.50	3.200	Ea.	520	133		653	775
0500	Double acting	″	2.50	3.200	″	670	133		803	940
08 71 20.75 Door Hardware Accessories										
0010	**DOOR HARDWARE ACCESSORIES**									
0050	Door closing coordinator, 36″ (for paired openings up to 56″)	1 Carp	8	1	Ea.	94	41.50		135.50	167
0060	48″ (for paired openings up to 84″)		8	1		101	41.50		142.50	175
0070	56″ (for paired openings up to 96″)		8	1		111	41.50		152.50	186

08 71 20.80 Hasps		Crew	Daily Output	Labor-Hours	Unit	2010 Bare Costs Material	Labor	Equipment	Total	Total Incl O&P
0010	**HASPS**, steel assembly									
0015	3"	1 Carp	26	.308	Ea.	4.04	12.80		16.84	24
0020	4-1/2"		13	.615		5.15	25.50		30.65	45
0040	6"		12.50	.640		8.60	26.50		35.10	50.50
08 71 20.90 Hinges										
0010	**HINGES** R087120-10									
0012	Full mortise, avg. freq., steel base, USP, 4-1/2" x 4-1/2"				Pr.	25.50			25.50	28
0100	5" x 5", USP					42.50			42.50	46.50
0200	6" x 6", USP					90			90	99
0400	Brass base, 4-1/2" x 4-1/2", US10					52.50			52.50	57.50
0500	5" x 5", US10					76			76	84
0600	6" x 6", US10					130			130	143
0800	Stainless steel base, 4-1/2" x 4-1/2", US32					77			77	85
0900	For non removable pin, add(security item)				Ea.	4.47			4.47	4.92
0910	For floating pin, driven tips, add					3.15			3.15	3.47
0930	For hospital type tip on pin, add					13.80			13.80	15.20
0940	For steeple type tip on pin, add					12			12	13.20
0950	Full mortise, high frequency, steel base, 3-1/2" x 3-1/2", US26D				Pr.	27			27	29.50
1000	4-1/2" x 4-1/2", USP					61			61	67.50
1100	5" x 5", USP					57			57	62.50
1200	6" x 6", USP					140			140	154
1400	Brass base, 3-1/2" x 3-1/2", US4					46.50			46.50	51.50
1430	4-1/2" x 4-1/2", US10					82			82	90.50
1500	5" x 5", US10					121			121	133
1600	6" x 6", US10					175			175	192
1800	Stainless steel base, 4-1/2" x 4-1/2", US32					127			127	140
1810	5" x 4-1/2", US32					180			180	198
1930	For hospital type tip on pin, add				Ea.	12.30			12.30	13.50
1950	Full mortise, low frequency, steel base, 3-1/2" x 3-1/2", US26D				Pr.	11.20			11.20	12.35
2000	4-1/2" x 4-1/2", USP					12.55			12.55	13.85
2100	5" x 5", USP					31			31	34
2200	6" x 6", USP					62			62	68
2300	4-1/2" x 4-1/2", US3					18.10			18.10	19.90
2310	5" x 5", US3					43.50			43.50	48
2400	Brass bass, 4-1/2" x 4-1/2", US10					43.50			43.50	48
2500	5" x 5", US10					66			66	72.50
2800	Stainless steel base, 4-1/2" x 4-1/2", US32					75			75	82.50
08 71 20.91 Special Hinges										
0010	**SPECIAL HINGES**									
0015	Paumelle, high frequency									
0020	Steel base, 6" x 4-1/2", US10				Pr.	145			145	160
0100	Bronze base, 5" x 4-1/2", US10					181			181	199
0200	Paumelle, average frequency, steel base, 4-1/2" x 3-1/2", US10					98.50			98.50	108
0400	Olive knuckle, low frequency, brass base, 6" x 4-1/2", US10					165			165	181
1000	Electric hinge with concealed conductor, average frequency									
1010	Steel base, 4-1/2" x 4-1/2", US26D				Pr.	315			315	345
1100	Bronze base, 4-1/2" x 4-1/2", US26D				"	320			320	350
1200	Electric hinge with concealed conductor, high frequency									
1210	Steel base, 4-1/2" x 4-1/2", US26D				Pr.	240			240	264
1600	Double weight, 800 lb., steel base, removable pin, 5" x 6", USP					132			132	145
1700	Steel base-welded pin, 5" x 6", USP					162			162	178
1800	Triple weight, 2000 lb., steel base, welded pin, 5" x 6", USP					151			151	166

08 71 20.91 Special Hinges		Crew	Daily Output	Labor-Hours	Unit	Material	2010 Bare Costs Labor	Equipment	Total	Total Incl O&P
2000	Pivot reinf., high frequency, steel base, 7-3/4" door plate, USP				Pr.	182			182	200
2200	Bronze base, 7-3/4" door plate, US10				↓	229			229	252
3000	Swing clear, full mortise, full or half surface, high frequency,									
3010	Steel base, 5" high, USP				Pr.	159			159	175
3200	Swing clear, full mortise, average frequency									
3210	Steel base, 4-1/2" high, USP				Pr.	127			127	139
4000	Wide throw, average frequency, steel base, 4-1/2" x 6", USP					96.50			96.50	106
4200	High frequency, steel base, 4-1/2" x 6", USP				↓	148			148	163
4600	Spring hinge, single acting, 6" flange, steel				Ea.	54			54	59.50
4700	Brass					95.50			95.50	105
4900	Double acting, 6" flange, steel					96.50			96.50	106
4950	Brass				↓	157			157	173
9000	Continuous hinge, steel, full mortise, heavy duty	2 Carp	64	.250	L.F.	22.50	10.40		32.90	41

08 71 20.95 Kick Plates		Crew	Daily Output	Labor-Hours	Unit	Material	Labor	Equipment	Total	Total Incl O&P
0010	**KICK PLATES**									
0020	Stainless steel, .050, 16 ga, 8" x 28", US32	1 Carp	15	.533	Ea.	34	22		56	71.50
0030	8" x 30"		15	.533		35	22		57	72.50
0040	8" x 34"		15	.533		39.50	22		61.50	77.50
0050	10" x 28"		15	.533		66.50	22		88.50	107
0060	10" x 30"		15	.533		71	22		93	112
0070	10" x 34"		15	.533		80.50	22		102.50	123
0080	Mop/Kick, 4" x 28"		15	.533		29.50	22		51.50	66
0090	4" x 30"		15	.533		31.50	22		53.50	68.50
0100	4" x 34"		15	.533		35.50	22		57.50	73
0110	6" x 28"		15	.533		38	22		60	76
0120	6" x 30"		15	.533		40.50	22		62.50	79
0130	6" x 34"		15	.533		46	22		68	85
0500	Bronze, .050", 8" x 28"		15	.533		46.50	22		68.50	85
0510	8" x 30"		15	.533		62.50	22		84.50	103
0520	8" x 34"		15	.533		71	22		93	112
0530	10" x 28"		15	.533		66.50	22		88.50	107
0540	10" x 30"		15	.533		71	22		93	112
0550	10" x 34"		15	.533		80.50	22		102.50	123
0560	Mop/Kick, 4" x 28"		15	.533		31	22		53	68
0570	4" x 30"		15	.533		33	22		55	70.50
0580	4" x 34"		15	.533		35	22		57	72.50
0590	6" x 28"		15	.533		43.50	22		65.50	82
0600	6" x 30"		15	.533		50	22		72	89
0610	6" x 34"		15	.533		50	22		72	89
1000	Acrylic, .125", 8" x 26"		15	.533		26	22		48	62.50
1010	8" x 36"		15	.533		36	22		58	73.50
1020	8" x 42"		15	.533		42	22		64	80
1030	10" x 26"		15	.533		32.50	22		54.50	70
1040	10" x 36"		15	.533		45	22		67	83.50
1050	10" x 42"		15	.533		68.50	22		90.50	110
1060	Mop/Kick, 4" x 26"		15	.533		15.60	22		37.60	51
1070	4" x 23"		15	.533		21.50	22		43.50	57.50
1080	4" x 42"		15	.533		24.50	22		46.50	61
1090	6" x 26"		15	.533		23	22		45	59.50
1100	6" x 36"		15	.533		31.50	22		53.50	69
1110	6" x 42"		15	.533		37	22		59	74.50
2000	Aluminum, .050, with 3 beveled edges, 10" x 28"	↓	15	.533	↓	23	22		45	59.50

08 71 20 – Hardware

08 71 20.95 Kick Plates		Crew	Daily Output	Labor-Hours	Unit	Material	2010 Bare Costs Labor	Equipment	Total	Total Incl O&P
2010	10" x 30"	1 Carp	15	.533	Ea.	25	22		47	61.50
2020	10" x 34"		15	.533		26	22		48	62.50
2040	10" x 38"		15	.533		28	22		50	64.50

08 71 21 – Astragals

08 71 21.10 Exterior Mouldings, Astragals

		Crew	Daily Output	Labor-Hours	Unit	Material	Labor	Equipment	Total	Total Incl O&P
0010	**EXTERIOR MOULDINGS, ASTRAGALS**									
0400	One piece, overlapping cadmium plated steel, flat, 3/16" x 2"	1 Carp	90	.089	L.F.	3.50	3.69		7.19	9.55
0600	Prime coated steel, flat, 1/8" x 3"		90	.089		4.70	3.69		8.39	10.85
0800	Stainless steel, flat, 3/32" x 1-5/8"		90	.089		17.90	3.69		21.59	25.50
1000	Aluminum, flat, 1/8" x 2"		90	.089		3.55	3.69		7.24	9.60
1200	Nail on, "T" extrusion		120	.067		.90	2.77		3.67	5.25
1300	Vinyl bulb insert		105	.076		1.30	3.17		4.47	6.30
1600	Screw on, "T" extrusion		90	.089		5.15	3.69		8.84	11.35
1700	Vinyl insert		75	.107		3.45	4.43		7.88	10.65
2000	"L" extrusion, neoprene bulbs		75	.107		2.05	4.43		6.48	9.10
2100	Neoprene sponge insert		75	.107		6	4.43		10.43	13.45
2200	Magnetic		75	.107		9.90	4.43		14.33	17.75
2400	Spring hinged security seal, with cam		75	.107		6.40	4.43		10.83	13.90
2600	Spring loaded locking bolt, vinyl insert		45	.178		8.65	7.40		16.05	21
2800	Neoprene sponge strip, "Z" shaped, aluminum		60	.133		4.20	5.55		9.75	13.15
2900	Solid neoprene strip, nail on aluminum strip		90	.089		3.45	3.69		7.14	9.50
3000	One piece stile protection									
3020	Neoprene fabric loop, nail on aluminum strips	1 Carp	60	.133	L.F.	.80	5.55		6.35	9.45
3110	Flush mounted aluminum extrusion, 1/2" x 1-1/4"		60	.133		3.32	5.55		8.87	12.20
3140	3/4" x 1-3/8"		60	.133		4.22	5.55		9.77	13.20
3160	1-1/8" x 1-3/4"		60	.133		6.65	5.55		12.20	15.85
3300	Mortise, 9/16" x 3/4"		60	.133		3.60	5.55		9.15	12.50
3320	13/16" x 1-3/8"		60	.133		3.85	5.55		9.40	12.80
3600	Spring bronze strip, nail on type		105	.076		3.10	3.17		6.27	8.30
3620	Screw on, with retainer		75	.107		2.50	4.43		6.93	9.60
3800	Flexible stainless steel housing, pile insert, 1/2" door		105	.076		6.80	3.17		9.97	12.40
3820	3/4" door		105	.076		7.65	3.17		10.82	13.30
4000	Extruded aluminum retainer, flush mount, pile insert		105	.076		2.41	3.17		5.58	7.55
4080	Mortise, felt insert		90	.089		4.28	3.69		7.97	10.40
4160	Mortise with spring, pile insert		90	.089		3.35	3.69		7.04	9.40
4400	Rigid vinyl retainer, mortise, pile insert		105	.076		2.40	3.17		5.57	7.50
4600	Wool pile filler strip, aluminum backing		105	.076		2.45	3.17		5.62	7.60
5000	Two piece overlapping astragal, extruded aluminum retainer									
5010	Pile insert	1 Carp	60	.133	L.F.	3.10	5.55		8.65	11.95
5020	Vinyl bulb insert		60	.133		2.05	5.55		7.60	10.80
5040	Vinyl flap insert		60	.133		6.20	5.55		11.75	15.40
5060	Solid neoprene flap insert		60	.133		6.15	5.55		11.70	15.30
5080	Hypalon rubber flap insert		60	.133		6.30	5.55		11.85	15.45
5090	Snap on cover, pile insert		60	.133		7.15	5.55		12.70	16.40
5400	Magnetic aluminum, surface mounted		60	.133		24.50	5.55		30.05	35.50
5500	Interlocking aluminum, 5/8" x 1" neoprene bulb insert		45	.178		3.91	7.40		11.31	15.70
5600	Adjustable aluminum, 9/16" x 21/32", pile insert		45	.178		18.20	7.40		25.60	31.50
5790	For vinyl bulb, deduct					.60			.60	.66
5800	Magnetic, adjustable, 9/16" x 21/32"	1 Carp	45	.178		23.50	7.40		30.90	37
6000	Two piece stile protection									
6010	Cloth backed rubber loop, 1" gap, nail on aluminum strips	1 Carp	45	.178	L.F.	4	7.40		11.40	15.80
6040	Screw on aluminum strips		45	.178		6.15	7.40		13.55	18.15

08 71 Door Hardware

08 71 21 – Astragals

	08 71 21.10 Exterior Mouldings, Astragals	Crew	Daily Output	Labor-Hours	Unit	Material	2010 Bare Costs Labor	Equipment	Total	Total Incl O&P
6100	1-1/2" gap, screw on aluminum extrusion	1 Carp	45	.178	L.F.	5.50	7.40		12.90	17.45
6240	Vinyl fabric loop, slotted aluminum extrusion, 1" gap		45	.178		2	7.40		9.40	13.60
6300	1-1/4" gap		45	.178		5.80	7.40		13.20	17.80

08 71 25 – Weatherstripping

	08 71 25.10 Mechanical Seals, Weatherstripping	Crew	Daily Output	Labor-Hours	Unit	Material	Labor	Equipment	Total	Total Incl O&P
0010	**MECHANICAL SEALS, WEATHERSTRIPPING**									
1000	Doors, wood frame, interlocking, for 3' x 7' door, zinc	1 Carp	3	2.667	Opng.	15.70	111		126.70	188
1100	Bronze		3	2.667		25	111		136	199
1300	6' x 7' opening, zinc		2	4		17	166		183	275
1400	Bronze		2	4		32	166		198	292
1700	Wood frame, spring type, bronze									
1800	3' x 7' door	1 Carp	7.60	1.053	Opng.	22	43.50		65.50	91.50
1900	6' x 7' door	"	7	1.143	"	28	47.50		75.50	104
2200	Metal frame, spring type, bronze									
2300	3' x 7' door	1 Carp	3	2.667	Opng.	36.50	111		147.50	212
2400	6' x 7' door	"	2.50	3.200	"	46.50	133		179.50	256
2500	For stainless steel, spring type, add					133%				
2700	Metal frame, extruded sections, 3' x 7' door, aluminum	1 Carp	3	2.667	Opng.	45	111		156	221
2800	Bronze		3	2.667		114	111		225	296
3100	6' x 7' door, aluminum		1.50	5.333		57.50	222		279.50	405
3200	Bronze		1.50	5.333		136	222		358	490
3500	Threshold weatherstripping									
3650	Door sweep, flush mounted, aluminum	1 Carp	25	.320	Ea.	13.50	13.30		26.80	35.50
3700	Vinyl		25	.320		15.90	13.30		29.20	38
5000	Garage door bottom weatherstrip, 12' aluminum, clear		14	.571		21.50	23.50		45	60
5010	Bronze		14	.571		83	23.50		106.50	128
5050	Bottom protection, Rubber		14	.571		22	23.50		45.50	61
5100	Threshold		14	.571		90.50	23.50		114	136

08 74 Access Control Hardware

08 74 13 – Card Key Access Control Hardware

	08 74 13.50 Card Key Access	Crew	Daily Output	Labor-Hours	Unit	Material	Labor	Equipment	Total	Total Incl O&P
0010	**CARD KEY ACCESS**, does not include door hardware									
0020	Card type, 1 time zone, minimum				Ea.	750			750	820
0040	Maximum					1,775			1,775	1,950
0060	3 time zones, minimum					2,200			2,200	2,425
0080	Maximum					2,450			2,450	2,700
0100	System with printer, and control console, 3 zones				Total	9,150			9,150	10,100
0120	6 zones				"	12,000			12,000	13,200
0140	For each door, minimum, add				Ea.	1,400			1,400	1,550
0160	Maximum, add				"	2,000			2,000	2,200

08 74 19 – Biometric Identity Access Control Hardware

	08 74 19.50 Biometric Identity Access	Crew	Daily Output	Labor-Hours	Unit	Material	Labor	Equipment	Total	Total Incl O&P
0010	**BIOMETRIC IDENTITY ACCESS**									
0220	Hand geometry scanner, mem of 512 users, excl striker/powr	1 Elec	3	2.667	Ea.	2,025	131		2,156	2,425
0230	Memory upgrade for, adds 9,700 user profiles		8	1		264	49		313	365
0240	Adds 32,500 user profiles		8	1		580	49		629	715
0250	Prison type, memory of 256 users, excl striker, power		3	2.667		2,600	131		2,731	3,050
0260	Memory upgrade for, adds 3,300 user profiles		8	1		211	49		260	305
0270	Adds 9,700 user profiles		8	1		425	49		474	540

08 74 Access Control Hardware

08 74 19 – Biometric Identity Access Control Hardware

08 74 19.50 Biometric Identity Access		Crew	Daily Output	Labor-Hours	Unit	Material	2010 Bare Costs Labor	Equipment	Total	Total Incl O&P
0280	Adds 27,900 user profiles	1 Elec	8	1	Ea.	580	49		629	715
0290	All weather, mem of 512 users, excl striker/pwr		3	2.667		3,900	131		4,031	4,475
0300	Facial & fingerprint scanner, combination unit, excl striker/power		3	2.667		4,200	131		4,331	4,825
0310	Access for, for initial setup, excl striker/power	↓	3	2.667	↓	1,000	131		1,131	1,300

08 75 Window Hardware

08 75 30 – Weatherstripping

08 75 30.10 Mechanical Weather Seals		Crew	Daily Output	Labor-Hours	Unit	Material	Labor	Equipment	Total	Total Incl O&P
0010	**MECHANICAL WEATHER SEALS**, Window, double hung, 3′ X 5′									
0020	Zinc	1 Carp	7.20	1.111	Opng.	12.80	46		58.80	85
0100	Bronze		7.20	1.111		27	46		73	101
0500	As above but heavy duty, zinc		4.60	1.739		17.60	72.50		90.10	130
0600	Bronze	↓	4.60	1.739	↓	30	72.50		102.50	144
9000	Minimum labor/equipment charge	1 Clab	4.60	1.739	Job		57.50		57.50	89

08 79 Hardware Accessories

08 79 13 – Key Storage Equipment

08 79 13.10 Key Cabinets		Crew	Daily Output	Labor-Hours	Unit	Material	Labor	Equipment	Total	Total Incl O&P
0010	**KEY CABINETS**									
0020	Wall mounted, 60 key capacity	1 Carp	20	.400	Ea.	95.50	16.60		112.10	131
0200	Drawer type, 600 key capacity	1 Clab	15	.533		725	17.65		742.65	825
0300	2,400 key capacity		20	.400		4,200	13.25		4,213.25	4,650
0400	Tray type, 20 key capacity		50	.160		41.50	5.30		46.80	53.50
0500	50 key capacity	↓	40	.200	↓	92.50	6.60		99.10	112

08 79 20 – Door Accessories

08 79 20.10 Door Hardware Accessories		Crew	Daily Output	Labor-Hours	Unit	Material	Labor	Equipment	Total	Total Incl O&P
0010	**DOOR HARDWARE ACCESSORIES**									
0140	Door bolt, surface, 4″	1 Carp	32	.250	Ea.	9	10.40		19.40	26
0160	Door latch	″	12	.667	″	8.35	27.50		35.85	51.50
0200	Sliding closet door									
0220	Track and hanger, single	1 Carp	10	.800	Ea.	54	33		87	111
0240	Double		8	1		73	41.50		114.50	145
0260	Door guide, single		48	.167		24	6.95		30.95	37
0280	Double		48	.167		33	6.95		39.95	47
0600	Deadbolt and lock cover plate, brass or stainless steel		30	.267		27	11.10		38.10	46.50
0620	Hole cover plate, brass or chrome		35	.229		7.15	9.50		16.65	22.50
2240	Mortise lockset, passage, lever handle		9	.889		194	37		231	270
4000	Security chain, standard	↓	18	.444	↓	7.90	18.45		26.35	37

08 81 Glass Glazing

08 81 10 – Float Glass

08 81 10.10 Various Types and Thickness of Float Glass		Crew	Daily Output	Labor-Hours	Unit	2010 Bare Costs Material	Labor	Equipment	Total	Total Incl O&P
0010	VARIOUS TYPES AND THICKNESS OF FLOAT GLASS R088110-10									
0020	3/16" Plain	2 Glaz	130	.123	S.F.	4.58	4.95		9.53	12.50
0200	Tempered, clear		130	.123		6.15	4.95		11.10	14.25
0300	Tinted		130	.123		7.25	4.95		12.20	15.40
0600	1/4" thick, clear, plain CN		120	.133		5.35	5.35		10.70	13.95
0700	Tinted		120	.133		7.40	5.35		12.75	16.20
0800	Tempered, clear		120	.133		7.60	5.35		12.95	16.40
0900	Tinted		120	.133		9.90	5.35		15.25	18.95
1600	3/8" thick, clear, plain		75	.213		9.35	8.60		17.95	23
1700	Tinted		75	.213		14.80	8.60		23.40	29
1800	Tempered, clear		75	.213		15.40	8.60		24	30
1900	Tinted		75	.213		17.15	8.60		25.75	32
2200	1/2" thick, clear, plain		55	.291		18.50	11.70		30.20	38
2300	Tinted		55	.291		26	11.70		37.70	46
2400	Tempered, clear		55	.291		22.50	11.70		34.20	42.50
2500	Tinted		55	.291		26.50	11.70		38.20	46.50
2800	5/8" thick, clear, plain		45	.356		26	14.30		40.30	50
2900	Tempered, clear		45	.356		29.50	14.30		43.80	54
3200	3/4" thick, clear, plain		35	.457		33.50	18.40		51.90	64
3300	Tempered, clear		35	.457		39	18.40		57.40	70.50
3600	1" thick, clear, plain	↓	30	.533		55.50	21.50		77	93.50
8900	For low emissivity coating for 3/16" & 1/4" only, add to above				↓	16%				

08 81 13 – Decorative Glass Glazing

08 81 13.10 Beveled Glass

		Crew	Daily Output	Labor-Hours	Unit	Material	Labor	Equipment	Total	Total Incl O&P
0010	**BEVELED GLASS**, with design patterns									
0020	Minimum	2 Glaz	150	.107	S.F.	55	4.29		59.29	67
0050	Average		125	.128		129	5.15		134.15	150
0100	Maximum	↓	100	.160	↓	229	6.45		235.45	261

08 81 13.20 Faceted Glass

		Crew	Daily Output	Labor-Hours	Unit	Material	Labor	Equipment	Total	Total Incl O&P
0010	**FACETED GLASS**, Color tinted 3/4" thick									
0020	Minimum	2 Glaz	95	.168	S.F.	46.50	6.75		53.25	61
0100	Maximum	"	75	.213	"	81.50	8.60		90.10	102

08 81 13.30 Sandblasted Glass

		Crew	Daily Output	Labor-Hours	Unit	Material	Labor	Equipment	Total	Total Incl O&P
0010	**SANDBLASTED GLASS**, float glass									
0020	1/8" thick	2 Glaz	160	.100	S.F.	10.25	4.02		14.27	17.35
0100	3/16" thick		130	.123		11.40	4.95		16.35	19.95
0500	1/4" thick		120	.133		11.80	5.35		17.15	21
0600	3/8" thick	↓	75	.213	↓	12.85	8.60		21.45	27

08 81 20 – Vision Panels

08 81 20.10 Full Vision

		Crew	Daily Output	Labor-Hours	Unit	Material	Labor	Equipment	Total	Total Incl O&P
0010	**FULL VISION**, window system with 3/4" glass mullions									
0020	Up to 10' high	H-2	130	.185	S.F.	59	7		66	75
0100	10' to 20' high, minimum		110	.218		62	8.25		70.25	81
0150	Average		100	.240		67	9.10		76.10	88
0200	Maximum	↓	80	.300	↓	75.50	11.35		86.85	100

08 81 25 – Glazing Variables

08 81 25.10 Applications of Glazing

		Crew	Daily Output	Labor-Hours	Unit	Material	Labor	Equipment	Total	Total Incl O&P
0010	APPLICATIONS OF GLAZING R088110-10									
0500	For high rise glazing, exterior, add per S.F. per story				S.F.					.45
0600	For glass replacement, add				"		100%			

08 81 Glass Glazing

08 81 25 – Glazing Variables

08 81 25.10 Applications of Glazing		Crew	Daily Output	Labor-Hours	Unit	Material	Labor	Equipment	Total	Total Incl O&P
0700	For gasket settings, add				L.F.	5.55			5.55	6.10
0900	For sloped glazing, add				S.F.		25%			
2000	Fabrication, polished edges, 1/4" thick				Inch	.48			.48	.53
2100	1/2" thick					1.21			1.21	1.33
2500	Mitered edges, 1/4" thick					1.21			1.21	1.33
2600	1/2" thick				↓	1.96			1.96	2.16

(2010 Bare Costs: Material, Labor, Equipment, Total)

08 81 30 – Insulating Glass

08 81 30.10 Reduce Heat Transfer Glass

		Crew	Daily Output	Labor-Hours	Unit	Material	Labor	Equipment	Total	Total Incl O&P
0010	REDUCE HEAT TRANSFER GLASS R088110-10									
0015	2 lites 1/8" float, /2" thk under 15 S.F.									
0020	Clear G	2 Glaz	95	.168	S.F.	9.25	6.75		16	20.50
0100	Tinted G		95	.168		12.90	6.75		19.65	24.50
0200	2 lites 3/16" float, for 5/8" thk unit, 15 to 30 S.F., clear G		90	.178		12.95	7.15		20.10	25
0300	Tinted G		90	.178		13.30	7.15		20.45	25.50
0400	1" thk, dbl. glazed, 1/4" float, 30-70 S.F., clear CN G		75	.213		15.50	8.60		24.10	30
0500	Tinted G		75	.213		22	8.60		30.60	37.50
0600	1" thick double glazed, 1/4" float, 1/4" wire		75	.213		21.50	8.60		30.10	36.50
0700	1/4" float, 1/4" tempered		75	.213		21	8.60		29.60	36
0800	1/4" wire, 1/4" tempered		75	.213		29.50	8.60		38.10	45.50
0900	Both lites, 1/4" wire		75	.213		35	8.60		43.60	51.50
2000	Both lites, light & heat reflective G		85	.188		29.50	7.55		37.05	44
2500	Heat reflective, film inside, 1" thick unit, clear G		85	.188		26	7.55		33.55	40
2600	Tinted G		85	.188		28	7.55		35.55	42.50
3000	Film on weatherside, clear, 1/2" thick unit G		95	.168		18.50	6.75		25.25	30.50
3100	5/8" thick unit G		90	.178		18.50	7.15		25.65	31.50
3200	1" thick unit G	↓	85	.188	↓	25.50	7.55		33.05	39.50

08 81 35 – Translucent Glass

08 81 35.10 Obscure Glass

		Crew	Daily Output	Labor-Hours	Unit	Material	Labor	Equipment	Total	Total Incl O&P
0010	OBSCURE GLASS									
0020	1/8" thick, Minimum	2 Glaz	140	.114	S.F.	10.45	4.59		15.04	18.40
0100	Maximum		125	.128		12.55	5.15		17.70	21.50
0300	7/32" thick, minimum		120	.133		11.60	5.35		16.95	21
0400	Maximum	↓	105	.152	↓	14.80	6.15		20.95	25.50

08 81 35.20 Patterned Glass

		Crew	Daily Output	Labor-Hours	Unit	Material	Labor	Equipment	Total	Total Incl O&P
0010	PATTERNED GLASS, colored									
0020	1/8" thick, Minimum	2 Glaz	140	.114	S.F.	8.70	4.59		13.29	16.45
0100	Maximum		125	.128		14.40	5.15		19.55	23.50
0300	7/32" thick, minimum		120	.133		10.75	5.35		16.10	19.85
0400	Maximum	↓	105	.152	↓	15.40	6.15		21.55	26

08 81 45 – Sheet Glass

08 81 45.10 Window Glass, Sheet

		Crew	Daily Output	Labor-Hours	Unit	Material	Labor	Equipment	Total	Total Incl O&P
0010	WINDOW GLASS, SHEET gray									
0020	1/8" thick	2 Glaz	160	.100	S.F.	5.30	4.02		9.32	11.90
0200	1/4" thick	"	130	.123	"	6.55	4.95		11.50	14.65

08 81 50 – Spandrel Glass

08 81 50.10 Glass for Non Vision Areas

		Crew	Daily Output	Labor-Hours	Unit	Material	Labor	Equipment	Total	Total Incl O&P
0010	GLASS FOR NON VISION AREAS, 1/4" thick standard colors									
0020	Up to 1000 S.F.	2 Glaz	110	.145	S.F.	15.35	5.85		21.20	25.50
0200	1,000 to 2,000 S.F.	"	120	.133	"	14.30	5.35		19.65	24
0300	For custom colors, add				Total	10%				

08 81 Glass Glazing

08 81 50 – Spandrel Glass

08 81 50.10 Glass for Non Vision Areas		Crew	Daily Output	Labor-Hours	Unit	Material	2010 Bare Costs Labor	Equipment	Total	Total Incl O&P
0500	For 3/8" thick, add				S.F.	10.65			10.65	11.70
1000	For double coated, 1/4" thick, add					3.97			3.97	4.37
1200	For insulation on panels, add					6.55			6.55	7.20
2000	Panels, insulated, with aluminum backed fiberglass, 1" thick	2 Glaz	120	.133		15.85	5.35		21.20	25.50
2100	2" thick	"	120	.133		18.45	5.35		23.80	28.50
2500	With galvanized steel backing, add				↓	5.50			5.50	6.05

08 81 55 – Window Glass

08 81 55.10 Sheet Glass		Crew	Daily Output	Labor-Hours	Unit	Material	Labor	Equipment	Total	Total Incl O&P
0010	**SHEET GLASS** (window), clear float, stops, putty bed									
0015	1/8" thick, clear float	2 Glaz	480	.033	S.F.	4.35	1.34		5.69	6.80
0500	3/16" thick, clear		480	.033		5.35	1.34		6.69	7.90
0600	Tinted		480	.033		7.15	1.34		8.49	9.85
0700	Tempered	↓	480	.033	↓	8.65	1.34		9.99	11.50

08 81 65 – Wire Glass

08 81 65.10 Glass Reinforced With Wire		Crew	Daily Output	Labor-Hours	Unit	Material	Labor	Equipment	Total	Total Incl O&P
0010	**GLASS REINFORCED WITH WIRE**									
0012	1/4" thick rough obscure	2 Glaz	135	.119	S.F.	14.10	4.76		18.86	22.50
1000	Polished wire, 1/4" thick, diamond, clear		135	.119		19.40	4.76		24.16	28.50
1500	Pinstripe, obscure	↓	135	.119	↓	21	4.76		25.76	30

08 83 Mirrors

08 83 13 – Mirrored Glass Glazing

08 83 13.10 Mirrors		Crew	Daily Output	Labor-Hours	Unit	Material	Labor	Equipment	Total	Total Incl O&P
0010	**MIRRORS**, No frames, wall type, 1/4" plate glass, polished edge									
0100	Up to 5 S.F.	2 Glaz	125	.128	S.F.	9.10	5.15		14.25	17.75
0200	Over 5 S.F.		160	.100		8.80	4.02		12.82	15.75
0500	Door type, 1/4" plate glass, up to 12 S.F.		160	.100		8	4.02		12.02	14.85
1000	Float glass, up to 10 S.F., 1/8" thick		160	.100		5.25	4.02		9.27	11.80
1100	3/16" thick		150	.107		6.35	4.29		10.64	13.45
1500	12" x 12" wall tiles, square edge, clear		195	.082		1.90	3.30		5.20	7.05
1600	Veined		195	.082		4.86	3.30		8.16	10.30
2000	1/4" thick, stock sizes, one way transparent		125	.128		18	5.15		23.15	27.50
2010	Bathroom, unframed, laminated	↓	160	.100	↓	15	4.02		19.02	22.50

08 83 13.15 Reflective Glass		Crew	Daily Output	Labor-Hours	Unit	Material	Labor	Equipment	Total	Total Incl O&P
0010	**REFLECTIVE GLASS**									
0100	1/4" float with fused metallic oxide fixed [G]	2 Glaz	115	.139	S.F.	15.40	5.60		21	25.50
0500	1/4" float glass with reflective applied coating [G]	"	115	.139	"	12.85	5.60		18.45	22.50

08 84 Plastic Glazing

08 84 10 – Plexiglass Glazing

08 84 10.10 Plexiglass Acrylic		Crew	Daily Output	Labor-Hours	Unit	Material	Labor (2010 Bare Costs)	Equipment	Total	Total Incl O&P
0010	**PLEXIGLASS ACRYLIC**, clear, masked,									
0020	1/8" thick, cut sheets	2 Glaz	170	.094	S.F.	5	3.78		8.78	11.20
0200	Full sheets		195	.082		2.60	3.30		5.90	7.85
0500	1/4" thick, cut sheets		165	.097		8.70	3.90		12.60	15.40
0600	Full sheets		185	.086		4.75	3.48		8.23	10.50
0900	3/8" thick, cut sheets		155	.103		15.90	4.15		20.05	24
1000	Full sheets		180	.089		8.60	3.57		12.17	14.85
1300	1/2" thick, cut sheets		135	.119		18.30	4.76		23.06	27
1400	Full sheets		150	.107		17.80	4.29		22.09	26
1700	3/4" thick, cut sheets		115	.139		65	5.60		70.60	80
1800	Full sheets		130	.123		37.50	4.95		42.45	49
2100	1" thick, cut sheets		105	.152		73.50	6.15		79.65	90
2200	Full sheets		125	.128		45.50	5.15		50.65	58
3000	Colored, 1/8" thick, cut sheets		170	.094		15.10	3.78		18.88	22.50
3200	Full sheets		195	.082		9.80	3.30		13.10	15.75
3500	1/4" thick, cut sheets		165	.097		17	3.90		20.90	24.50
3600	Full sheets		185	.086		11.55	3.48		15.03	17.95
4000	Mirrors, untinted, cut sheets, 1/8" thick		185	.086		7.10	3.48		10.58	13.05
4200	1/4" thick		180	.089		11.50	3.57		15.07	18.05

08 84 20 – Polycarbonate

08 84 20.10 Thermoplastic		Crew	Daily Output	Labor-Hours	Unit	Material	Labor	Equipment	Total	Total Incl O&P
0010	**THERMOPLASTIC**, clear, masked, cut sheets									
0020	1/8" thick	2 Glaz	170	.094	S.F.	9.70	3.78		13.48	16.35
0500	3/16" thick		165	.097		11.10	3.90		15	18.05
1000	1/4" thick		155	.103		12.50	4.15		16.65	20
1500	3/8" thick		150	.107		22	4.29		26.29	31

08 84 30 – Vinyl Glass

08 84 30.10 Vinyl Sheets		Crew	Daily Output	Labor-Hours	Unit	Material	Labor	Equipment	Total	Total Incl O&P
0010	**VINYL SHEETS**, Steel mesh reinforced, stock sizes									
0020	.090" thick	2 Glaz	170	.094	S.F.	12.50	3.78		16.28	19.45
0500	.120" thick		170	.094		15.30	3.78		19.08	22.50
1000	.250" thick		155	.103		20.50	4.15		24.65	29
1500	For non-standard sizes, add					18%				

08 87 Glazing Surface Films

08 87 13 – Solar Control Films

08 87 13.10 Solar Films On Glass		Crew	Daily Output	Labor-Hours	Unit	Material	Labor	Equipment	Total	Total Incl O&P
0010	**SOLAR FILMS ON GLASS** (glass not included)									
2000	Minimum G	2 Glaz	180	.089	S.F.	6.20	3.57		9.77	12.25
2050	Maximum G	"	225	.071	"	14.40	2.86		17.26	20

08 87 16 – Safety Films

08 87 16.10 Window Protection Film		Crew	Daily Output	Labor-Hours	Unit	Material	Labor	Equipment	Total	Total Incl O&P
0010	**WINDOW PROTECTION FILM**									
2010	Window protection film, controls blast damage G	2 Glaz	80	.200	S.F.	6.20	8.05		14.25	18.95

08 87 53 – Security Films

08 87 53.10 Security Film		Crew	Daily Output	Labor-Hours	Unit	Material	Labor	Equipment	Total	Total Incl O&P
0010	**SECURITY FILM**, clear, 32000psi tensile strength, adhered to glass									
0100	.002" thick, daylight installation	H-2	950	.025	S.F.	1.85	.96		2.81	3.49
0150	.004" thick, daylight installation R088110-10		800	.030		2	1.13		3.13	3.92

08 87 Glazing Surface Films

08 87 53 – Security Films

08 87 53.10 Security Film		Crew	Daily Output	Labor-Hours	Unit	Material	Labor	Equipment	Total	Total Incl O&P
						2010 Bare Costs				
0200	.006" thick, daylight installation	H-2	700	.034	S.F.	2.15	1.30		3.45	4.34
0210	Install for anchorage		600	.040		2.39	1.51		3.90	4.93
0400	.007" thick, daylight installation		600	.040		2.25	1.51		3.76	4.78
0410	Install for anchorage		500	.048		2.50	1.82		4.32	5.50
0500	.008" thick, daylight installation		500	.048		2.65	1.82		4.47	5.65
0510	Install for anchorage		500	.048		2.94	1.82		4.76	6
0600	.015" thick, daylight installation		400	.060		4.40	2.27		6.67	8.30
0610	Install for anchorage		400	.060		2.94	2.27		5.21	6.70
0900	Security film anchorage, mechanical attachment and cover plate	H-3	370	.043	L.F.	8.70	1.55		10.25	11.95
0950	Security film anchorage, wet glaze structural caulking	1 Glaz	225	.036	"	.83	1.43		2.26	3.06
1000	Adhered security film removal	1 Clab	275	.029	S.F.		.96		.96	1.49

08 88 Special Function Glazing

08 88 40 – Acoustical Glass Units

08 88 40.10 Sound Reduction Units		Crew	Daily Output	Labor-Hours	Unit	Material	Labor	Equipment	Total	Total Incl O&P
0010	**SOUND REDUCTION UNITS**, 1lite at 3/8", 1 lite at 3/16"									
0020	For 1" thick	2 Glaz	100	.160	S.F.	32	6.45		38.45	45
0100	For 4" thick	"	80	.200	"	55	8.05		63.05	72.50

08 88 56 – Ballistics-Resistant Glazing

08 88 56.10 Laminated Glass		Crew	Daily Output	Labor-Hours	Unit	Material	Labor	Equipment	Total	Total Incl O&P
0010	**LAMINATED GLASS**									
0020	Clear float .03" vinyl 1/4"	2 Glaz	90	.178	S.F.	11.35	7.15		18.50	23.50
0100	3/8" thick		78	.205		20	8.25		28.25	35
0200	.06" vinyl, 1/2" thick		65	.246		23.50	9.90		33.40	41
1000	5/8" thick		90	.178		28	7.15		35.15	41.50
2000	Bullet-resisting, 1-3/16" thick, to 15 S.F.		16	1		77	40		117	146
2100	Over 15 S.F.		16	1		94	40		134	164
2500	2-1/4" thick, to 15 S.F.		12	1.333		91	53.50		144.50	181
2600	Over 15 S.F.		12	1.333		81	53.50		134.50	170
2700	Level 2 (.357 magnum)		12	1.333		63	53.50		116.50	150
2750	Level 3A (.44 magnum)		12	1.333		68	53.50		121.50	156
2800	Level 4 (AK-47)		12	1.333		89	53.50		142.50	179
2850	Level 5 (M-16)		12	1.333		99	53.50		152.50	190
2900	Level 3 (7.62 Armor Piercing)		12	1.333		118	53.50		171.50	211

08 91 Louvers

08 91 19 – Fixed Louvers

08 91 19.10 Aluminum Louvers		Crew	Daily Output	Labor-Hours	Unit	Material	Labor	Equipment	Total	Total Incl O&P
0010	**ALUMINUM LOUVERS**									
0020	Aluminum with screen, residential, 8" x 8"	1 Carp	38	.211	Ea.	10.95	8.75		19.70	25.50
0100	12" x 12"		38	.211		14	8.75		22.75	29
0200	12" x 18"		35	.229		13.95	9.50		23.45	30
0250	14" x 24"		30	.267		25	11.10		36.10	44.50
0300	18" x 24"		27	.296		27	12.30		39.30	48.50
0500	24" x 30"		24	.333		32	13.85		45.85	56.50
0700	Triangle, adjustable, small		20	.400		32.50	16.60		49.10	61.50
0800	Large		15	.533		40.50	22		62.50	78.50
1200	Extruded aluminum, see Div. 23 37 15.40									
2100	Midget, aluminum, 3/4" deep, 1" diameter	1 Carp	85	.094	Ea.	.73	3.91		4.64	6.85

08 91 Louvers

08 91 19 – Fixed Louvers

08 91 19.10 Aluminum Louvers		Crew	Daily Output	Labor-Hours	Unit	Material	2010 Bare Costs Labor	Equipment	Total	Total Incl O&P
2150	3″ diameter	1 Carp	60	.133	Ea.	1.78	5.55		7.33	10.50
2200	4″ diameter		50	.160		3.64	6.65		10.29	14.25
2250	6″ diameter	↓	30	.267	↓	3.74	11.10		14.84	21

08 95 Vents

08 95 13 – Soffit Vents

08 95 13.10 Wall Louvers		Crew	Daily Output	Labor-Hours	Unit	Material	Labor	Equipment	Total	Total Incl O&P
0010	**WALL LOUVERS**									
2330	Soffit vent, continuous, 3″ wide, aluminum, mill finish	1 Carp	200	.040	L.F.	.48	1.66		2.14	3.09
2340	Baked enamel finish		200	.040	″	.52	1.66		2.18	3.13
2400	Under eaves vent, aluminum, mill finish, 16″ x 4″		48	.167	Ea.	1.33	6.95		8.28	12.15
2500	16″ x 8″	↓	48	.167	″	1.54	6.95		8.49	12.40

08 95 16 – Wall Vents

08 95 16.10 Louvers		Crew	Daily Output	Labor-Hours	Unit	Material	Labor	Equipment	Total	Total Incl O&P
0010	**LOUVERS**									
0020	Redwood, 2′-0″ diameter, full circle	1 Carp	16	.500	Ea.	169	21		190	218
0100	Half circle		16	.500		189	21		210	240
0200	Octagonal		16	.500		129	21		150	174
0300	Triangular, 5/12 pitch, 5′-0″ at base		16	.500		189	21		210	240
7000	Vinyl gable vent, 8″ x 8″		38	.211		11.55	8.75		20.30	26
7020	12″ x 12″		38	.211		24	8.75		32.75	40
7080	12″ x 18″		35	.229		31	9.50		40.50	48.50
7200	18″ x 24″	↓	30	.267	↓	37.50	11.10		48.60	58.50

Division Notes

	CREW	DAILY OUTPUT	LABOR-HOURS	UNIT	2010 BARE COSTS MAT.	LABOR	EQUIP.	TOTAL	TOTAL INCL O&P

Estimating Tips

General

- Room Finish Schedule: A complete set of plans should contain a room finish schedule. If one is not available, it would be well worth the time and effort to obtain one.

09 20 00 Plaster and Gypsum Board

- Lath is estimated by the square yard plus a 5% allowance for waste. Furring, channels, and accessories are measured by the linear foot. An extra foot should be allowed for each accessory miter or stop.
- Plaster is also estimated by the square yard. Deductions for openings vary by preference, from zero deduction to 50% of all openings over 2 feet in width. The estimator should allow one extra square foot for each linear foot of horizontal interior or exterior angle located below the ceiling level. Also, double the areas of small radius work.
- Drywall accessories, studs, track, and acoustical caulking are all measured by the linear foot. Drywall taping is figured by the square foot. Gypsum wallboard is estimated by the square foot. No material deductions should be made for door or window openings under 32 S.F.

09 60 00 Flooring

- Tile and terrazzo areas are taken off on a square foot basis. Trim and base materials are measured by the linear foot. Accent tiles are listed per each. Two basic methods of installation are used. Mud set is approximately 30% more expensive than thin set. In terrazzo work, be sure to include the linear footage of embedded decorative strips, grounds, machine rubbing, and power cleanup.
- Wood flooring is available in strip, parquet, or block configuration. The latter two types are set in adhesives with quantities estimated by the square foot. The laying pattern will influence labor costs and material waste. In addition to the material and labor for laying wood floors, the estimator must make allowances for sanding and finishing these areas unless the flooring is prefinished.
- Sheet flooring is measured by the square yard. Roll widths vary, so consideration should be given to use the most economical width, as waste must be figured into the total quantity. Consider also the installation methods available, direct glue down or stretched.

09 70 00 Wall Finishes

- Wall coverings are estimated by the square foot. The area to be covered is measured, length by height of wall above baseboards, to calculate the square footage of each wall. This figure is divided by the number of square feet in the single roll which is being used. Deduct, in full, the areas of openings such as doors and windows. Where a pattern match is required allow 25%–30% waste.

09 80 00 Acoustic Treatment

- Acoustical systems fall into several categories. The takeoff of these materials should be by the square foot of area with a 5% allowance for waste. Do not forget about scaffolding, if applicable, when estimating these systems.

09 90 00 Painting and Coating

- A major portion of the work in painting involves surface preparation. Be sure to include cleaning, sanding, filling, and masking costs in the estimate.
- Protection of adjacent surfaces is not included in painting costs. When considering the method of paint application, an important factor is the amount of protection and masking required. These must be estimated separately and may be the determining factor in choosing the method of application.

Reference Numbers

Reference numbers are shown in shaded boxes at the beginning of some major classifications. These numbers refer to related items in the Reference Section. The reference information may be an estimating procedure, an alternate pricing method, or technical information.

Note: Not all subdivisions listed here necessarily appear in this publication.

09 01 Maintenance of Finishes

09 01 60 – Maintenance of Flooring

09 01 60.10 Carpet Maintenance		Crew	Daily Output	Labor-Hours	Unit	Material	2010 Bare Costs Labor	Equipment	Total	Total Incl O&P
0010	**CARPET MAINTENANCE**									
0020	Steam clean, per cleaning, minimum	1 Clab	3000	.003	S.F.	.07	.09		.16	.22
0500	Maximum	"	2000	.004	"	.12	.13		.25	.33

09 01 70 – Maintenance of Wall Finishes

09 01 70.10 Gypsum Wallboard Repairs		Crew	Daily Output	Labor-Hours	Unit	Material	Labor	Equipment	Total	Total Incl O&P
0010	**GYPSUM WALLBOARD REPAIRS**									
0100	Fill and sand, pin / nail holes	1 Carp	960	.008	Ea.		.35		.35	.53
0110	Screw head pops		480	.017			.69		.69	1.07
0120	Dents, up to 2" square		48	.167		.01	6.95		6.96	10.70
0130	2" to 4" square		24	.333		.03	13.85		13.88	21.50
0140	Cut square, patch, sand and finish, holes, up to 2" square		12	.667		.04	27.50		27.54	42.50
0150	2" to 4" square		11	.727		.09	30		30.09	46.50
0160	4" to 8" square		10	.800		.22	33		33.22	51
0170	8" to 12" square		8	1		.42	41.50		41.92	64.50
0180	12" to 32" square		6	1.333		2.62	55.50		58.12	88.50
0210	16" by 48"		5	1.600		2.01	66.50		68.51	104
0220	32" by 48"		4	2		3.36	83		86.36	132
0230	48" square		3.50	2.286		4.73	95		99.73	151
0240	60" square		3.20	2.500		8.30	104		112.30	169
0500	Skim coat surface with joint compound		1600	.005	S.F.	.03	.21		.24	.35

09 05 Common Work Results for Finishes

09 05 05 – Selective Finishes Demolition

09 05 05.10 Selective Demolition, Ceilings		Crew	Daily Output	Labor-Hours	Unit	Material	Labor	Equipment	Total	Total Incl O&P
0010	**SELECTIVE DEMOLITION, CEILINGS** R024119-10									
0200	Ceiling, drywall, furred and nailed or screwed	2 Clab	800	.020	S.F.		.66		.66	1.02
0220	On metal frame		760	.021			.70		.70	1.07
0240	On suspension system, including system		720	.022			.74		.74	1.13
1000	Plaster, lime and horse hair, on wood lath, incl. lath		700	.023			.76		.76	1.17
1020	On metal lath		570	.028			.93		.93	1.43
1100	Gypsum, on gypsum lath		720	.022			.74		.74	1.13
1120	On metal lath		500	.032			1.06		1.06	1.63
1200	Suspended ceiling, mineral fiber, 2' x 2' or 2' x 4'		1500	.011			.35		.35	.54
1250	On suspension system, incl. system		1200	.013			.44		.44	.68
1500	Tile, wood fiber, 12" x 12", glued		900	.018			.59		.59	.91
1540	Stapled		1500	.011			.35		.35	.54
1580	On suspension system, incl. system		760	.021			.70		.70	1.07
2000	Wood, tongue and groove, 1" x 4"		1000	.016			.53		.53	.82
2040	1" x 8"		1100	.015			.48		.48	.74
2400	Plywood or wood fiberboard, 4' x 8' sheets		1200	.013			.44		.44	.68

09 05 05.20 Selective Demolition, Flooring		Crew	Daily Output	Labor-Hours	Unit	Material	Labor	Equipment	Total	Total Incl O&P
0010	**SELECTIVE DEMOLITION, FLOORING** R024119-10									
0200	Brick with mortar	2 Clab	475	.034	S.F.		1.11		1.11	1.72
0400	Carpet, bonded, including surface scraping		2000	.008			.26		.26	.41
0440	Scrim applied		8000	.002			.07		.07	.10
0480	Tackless		9000	.002			.06		.06	.09
0550	Carpet tile, releasable adhesive		5000	.003			.11		.11	.16
0560	Permanent adhesive		1850	.009			.29		.29	.44
0600	Composition, acrylic or epoxy		400	.040			1.32		1.32	2.04
0700	Concrete, scarify skin	A-1A	225	.036			1.51	.96	2.47	3.39

09 05 05 – Selective Finishes Demolition

09 05 05.20 Selective Demolition, Flooring

	09 05 05.20 Selective Demolition, Flooring	Crew	Daily Output	Labor-Hours	Unit	Material	2010 Bare Costs Labor	Equipment	Total	Total Incl O&P
0800	Resilient, sheet goods	2 Clab	1400	.011	S.F.		.38		.38	.58
0820	For gym floors	"	900	.018	↓		.59		.59	.91
0850	Vinyl or rubber cove base	1 Clab	1000	.008	L.F.		.26		.26	.41
0860	Vinyl or rubber cove base, molded corner	"	1000	.008	Ea.		.26		.26	.41
0870	For glued and caulked installation, add to labor						50%			
0900	Vinyl composition tile, 12" x 12"	2 Clab	1000	.016	S.F.		.53		.53	.82
2000	Tile, ceramic, thin set		675	.024			.78		.78	1.21
2020	Mud set		625	.026			.85		.85	1.31
2200	Marble, slate, thin set		675	.024			.78		.78	1.21
2220	Mud set		625	.026			.85		.85	1.31
2600	Terrazzo, thin set		450	.036			1.18		1.18	1.82
2620	Mud set		425	.038			1.25		1.25	1.92
2640	Terrazzo, cast in place	↓	300	.053			1.77		1.77	2.72
3000	Wood, block, on end	1 Carp	400	.020			.83		.83	1.28
3200	Parquet		450	.018			.74		.74	1.14
3400	Strip flooring, interior, 2-1/4" x 25/32" thick		325	.025			1.02		1.02	1.58
3500	Exterior, porch flooring, 1" x 4"		220	.036			1.51		1.51	2.33
3800	Subfloor, tongue and groove, 1" x 6"		325	.025			1.02		1.02	1.58
3820	1" x 8"		430	.019			.77		.77	1.19
3840	1" x 10"		520	.015			.64		.64	.99
4000	Plywood, nailed		600	.013			.55		.55	.85
4100	Glued and nailed		400	.020			.83		.83	1.28
4200	Hardboard, 1/4" thick	↓	760	.011			.44		.44	.67
8000	Remove flooring, bead blast, minimum	A-1A	1000	.008			.34	.22	.56	.76
8100	Maximum		400	.020			.85	.54	1.39	1.90
8150	Mastic only	↓	1500	.005	↓		.23	.14	.37	.51

09 05 05.30 Selective Demolition, Walls and Partitions

	09 05 05.30 Selective Demolition, Walls and Partitions	Crew	Daily Output	Labor-Hours	Unit	Material	2010 Bare Costs Labor	Equipment	Total	Total Incl O&P
0010	**SELECTIVE DEMOLITION, WALLS AND PARTITIONS** R024119-10									
0020	Walls, concrete, reinforced	B-39	120	.400	C.F.		13.90	1.87	15.77	23.50
0025	Plain	"	160	.300			10.45	1.40	11.85	17.55
0100	Brick, 4" to 12" thick	B-9	220	.182	↓		6.10	1.02	7.12	10.50
0200	Concrete block, 4" thick		1000	.040	S.F.		1.34	.22	1.56	2.32
0280	8" thick		810	.049			1.65	.28	1.93	2.86
0300	Exterior stucco 1" thick over mesh	↓	3200	.013			.42	.07	.49	.73
1000	Drywall, nailed or screwed	1 Clab	1000	.008			.26		.26	.41
1010	2 layers		400	.020			.66		.66	1.02
1020	Glued and nailed		900	.009			.29		.29	.45
1500	Fiberboard, nailed		900	.009			.29		.29	.45
1520	Glued and nailed		800	.010			.33		.33	.51
1568	Plenum barrier, sheet lead		300	.027			.88		.88	1.36
2000	Movable walls, metal, 5' high		300	.027			.88		.88	1.36
2020	8' high	↓	400	.020			.66		.66	1.02
2200	Metal or wood studs, finish 2 sides, fiberboard	B-1	520	.046			1.56		1.56	2.40
2250	Lath and plaster		260	.092			3.12		3.12	4.81
2300	Plasterboard (drywall)		520	.046			1.56		1.56	2.40
2350	Plywood	↓	450	.053			1.80		1.80	2.78
2800	Paneling, 4' x 8' sheets	1 Clab	475	.017			.56		.56	.86
3000	Plaster, lime and horsehair, on wood lath		400	.020			.66		.66	1.02
3020	On metal lath		335	.024			.79		.79	1.22
3400	Gypsum or perlite, on gypsum lath		410	.020			.65		.65	1
3420	On metal lath		300	.027	↓		.88		.88	1.36
3450	Plaster, interior gypsum, acoustic, or cement	↓	60	.133	S.Y.		4.41		4.41	6.80

09 05 Common Work Results for Finishes

09 05 05 – Selective Finishes Demolition

09 05 05.30 Selective Demolition, Walls and Partitions		Crew	Daily Output	Labor-Hours	Unit	2010 Bare Costs Material	Labor	Equipment	Total	Total Incl O&P
3500	Stucco, on masonry	1 Clab	145	.055	S.Y.		1.83		1.83	2.82
3510	Commercial 3-coat		80	.100			3.31		3.31	5.10
3520	Interior stucco		25	.320			10.60		10.60	16.35
3600	Plywood, one side	B-1	1500	.016	S.F.		.54		.54	.83
3750	Terra cotta block and plaster, to 6″ thick	″	175	.137			4.63		4.63	7.15
3760	Tile, ceramic, on walls, thin set	1 Clab	300	.027			.88		.88	1.36
3765	Mud set		250	.032			1.06		1.06	1.63
3800	Toilet partitions, slate or marble		5	1.600	Ea.		53		53	81.50
3820	Metal or plastic		8	1	″		33		33	51
5000	Wallcovering, vinyl	1 Pape	700	.011	S.F.		.42		.42	.62
5010	With release agent		1500	.005			.19		.19	.29
5025	Wallpaper, 2 layers or less, by hand		250	.032			1.17		1.17	1.74
5035	3 layers or more		165	.048			1.77		1.77	2.64
5040	Designer		480	.017			.61		.61	.91

09 21 Plaster and Gypsum Board Assemblies

09 21 13 – Plaster Assemblies

09 21 13.10 Plaster Partition Wall

		Crew	Daily Output	Labor-Hours	Unit	Material	Labor	Equipment	Total	Total Incl O&P
0010	**PLASTER PARTITION WALL**									
0400	Stud walls, 3.4 lb. metal lath, 3 coat gypsum plaster, 2 sides									
0600	2″ x 4″ wood studs, 16″ O.C.	J-2	315	.152	S.F.	3.31	5.50	.43	9.24	12.30
0700	2-1/2″ metal studs, 25 ga., 12″ O.C.		325	.148		3.06	5.35	.42	8.83	11.80
0800	3-5/8″ metal studs, 25 ga., 16″ O.C.		320	.150		3.08	5.40	.43	8.91	11.90
0900	Gypsum lath, 2 coat vermiculite plaster, 2 sides									
1000	2″ x 4″ wood studs, 16″ O.C.	J-2	355	.135	S.F.	3.69	4.88	.39	8.96	11.75
1200	2-1/2″ metal studs, 25 ga., 12″ O.C.		365	.132		3.27	4.75	.37	8.39	11.05
1300	3-5/8″ metal studs, 25 ga., 16″ O.C.		360	.133		3.36	4.81	.38	8.55	11.25

09 21 16 – Gypsum Board Assemblies

09 21 16.23 Gypsum Board Shaft Wall Assemblies

		Crew	Daily Output	Labor-Hours	Unit	Material	Labor	Equipment	Total	Total Incl O&P
0010	**GYPSUM BOARD SHAFT WALL ASSEMBLIES**									
0020	Cavity type on 25 ga. J-track & C-H studs, 24″ O.C.									
0030	1″ thick coreboard wall liner on shaft side									
0040	2-hour assembly with double layer									
0060	5/8″ fire rated gypsum board on room side	2 Carp	220	.073	S.F.	1.67	3.02		4.69	6.50
0100	3-hour assembly with triple layer									
0300	5/8″ fire rated gypsum board on room side	2 Carp	180	.089	S.F.	1.39	3.69		5.08	7.25
0400	4-hour assembly, 1″ coreboard, 5/8″ fire rated gypsum board									
0600	and 3/4″ galv. metal furring channels, 24″ O.C., with									
0700	Double layer 5/8″ fire rated gypsum board on room side	2 Carp	110	.145	S.F.	1.30	6.05		7.35	10.75
0900	For taping & finishing, add per side	1 Carp	1050	.008	″	.04	.32		.36	.54
1000	For insulation, see Div. 07 21									
5200	For work over 8′ high, add	2 Carp	3060	.005	S.F.		.22		.22	.34
5300	For distribution cost over 3 stories high, add per story	″	6100	.003	″		.11		.11	.17

09 21 16.33 Partition Wall

		Crew	Daily Output	Labor-Hours	Unit	Material	Labor	Equipment	Total	Total Incl O&P
0010	**PARTITION WALL** Stud wall, 8′ to 12′ high									
0050	1/2″, interior, gypsum board, std, tape & finish 2 sides									
0500	Installed on and incl., 2″ x 4″ wood studs, 16″ O.C.	2 Carp	310	.052	S.F.	.84	2.14		2.98	4.24
1000	Metal studs, NLB, 25 ga., 16″ O.C., 3-5/8″ wide		350	.046		.89	1.90		2.79	3.91
1200	6″ wide		330	.048		1.05	2.01		3.06	4.27
1400	Water resistant, on 2″ x 4″ wood studs, 16″ O.C.		310	.052		1.06	2.14		3.20	4.48

09 21 16 – Gypsum Board Assemblies

09 21 16.33 Partition Wall		Crew	Daily Output	Labor-Hours	Unit	Material	Labor	Equipment	Total	Total Incl O&P
						2010 Bare Costs				
1600	Metal studs, NLB, 25 ga., 16" O.C., 3-5/8" wide	2 Carp	350	.046	S.F.	1.11	1.90		3.01	4.15
1800	6" wide		330	.048		1.27	2.01		3.28	4.51
2000	Fire res., 2 layers, 1-1/2 hr., on 2" x 4" wood studs, 16" O.C.		210	.076		1.48	3.17		4.65	6.50
2200	Metal studs, NLB, 25 ga., 16" O.C., 3-5/8" wide		250	.064		1.53	2.66		4.19	5.80
2400	6" wide		230	.070		1.69	2.89		4.58	6.30
2600	Fire & water res., 2 layers, 1-1/2 hr., 2" x 4" studs, 16" O.C.		210	.076		1.48	3.17		4.65	6.50
2800	Metal studs, NLB, 25 ga., 16" O.C., 3-5/8" wide		250	.064		1.53	2.66		4.19	5.80
3000	6" wide	↓	230	.070	↓	1.69	2.89		4.58	6.30
3200	5/8", interior, gypsum board, std, tape & finish 2 sides									
3400	Installed on and including 2" x 4" wood studs, 16" O.C.	2 Carp	300	.053	S.F.	.92	2.22		3.14	4.44
3600	24" O.C.		330	.048		.86	2.01		2.87	4.06
3800	Metal studs, NLB, 25 ga., 16" O.C., 3-5/8" wide		340	.047		.97	1.96		2.93	4.07
4000	6" wide		320	.050		1.13	2.08		3.21	4.45
4200	24" O.C., 3-5/8" wide		360	.044		.88	1.85		2.73	3.82
4400	6" wide		340	.047		1	1.96		2.96	4.11
4800	Water resistant, on 2" x 4" wood studs, 16" O.C.		300	.053		1.22	2.22		3.44	4.77
5000	24" O.C.		330	.048		1.16	2.01		3.17	4.39
5200	Metal studs, NLB, 25 ga. 16" O.C., 3-5/8" wide		340	.047		1.27	1.96		3.23	4.40
5400	6" wide		320	.050		1.43	2.08		3.51	4.78
5600	24" O.C., 3-5/8" wide		360	.044		1.18	1.85		3.03	4.15
5800	6" wide		340	.047		1.30	1.96		3.26	4.44
6000	Fire resistant, 2 layers, 2 hr., on 2" x 4" wood studs, 16" O.C.		205	.078		1.38	3.24		4.62	6.50
6200	24" O.C.		235	.068		1.38	2.83		4.21	5.90
6400	Metal studs, NLB, 25 ga., 16" O.C., 3-5/8" wide		245	.065		1.49	2.71		4.20	5.80
6600	6" wide		225	.071		1.65	2.95		4.60	6.35
6800	24" O.C., 3-5/8" wide		265	.060		1.40	2.51		3.91	5.40
7000	6" wide		245	.065		1.52	2.71		4.23	5.85
7200	Fire & water resistant, 2 layers, 2 hr., 2" x 4" studs, 16" O.C.		205	.078		1.44	3.24		4.68	6.60
7400	24" O.C.		235	.068		1.38	2.83		4.21	5.90
7600	Metal studs, NLB, 25 ga., 16" O.C., 3-5/8" wide		245	.065		1.49	2.71		4.20	5.80
7800	6" wide		225	.071		1.65	2.95		4.60	6.35
8000	24" O.C., 3-5/8" wide		265	.060		1.40	2.51		3.91	5.40
8200	6" wide	↓	245	.065	↓	1.52	2.71		4.23	5.85
8600	1/2" blueboard, mesh tape both sides									
8620	Installed on and including 2" x 4" wood studs, 16" O.C.	2 Carp	300	.053	S.F.	.92	2.22		3.14	4.44
8640	Metal studs, NLB, 25 ga., 16" O.C., 3-5/8" wide		340	.047		.97	1.96		2.93	4.07
8660	6" wide	↓	320	.050	↓	1.13	2.08		3.21	4.45
9000	Exterior, 1/2" gypsum sheathing, 1/2" gypsum finished, interior,									
9100	including foil faced insulation, metal studs, 20 ga.									
9200	16" O.C., 3-5/8" wide	2 Carp	290	.055	S.F.	1.61	2.29		3.90	5.30
9400	6" wide		270	.059		1.77	2.46		4.23	5.75
9600	Partitions, for work over 8' high, add	↓	1530	.010	↓		.43		.43	.67

09 22 03 – Fastening Methods for Finishes

09 22 03.20 Drilling Plaster/Drywall		Crew	Daily Output	Labor-Hours	Unit	Material	2010 Bare Costs Labor	Equipment	Total	Total Incl O&P
0010	**DRILLING PLASTER/DRYWALL**									
1100	Drilling & layout for drywall/plaster walls, up to 1" deep, no anchor									
1200	Holes, 1/4" diameter	1 Carp	150	.053	Ea.	.01	2.22		2.23	3.43
1300	3/8" diameter		140	.057		.01	2.37		2.38	3.67
1400	1/2" diameter		130	.062		.01	2.56		2.57	3.95
1500	3/4" diameter		120	.067		.01	2.77		2.78	4.29
1600	1" diameter		110	.073		.02	3.02		3.04	4.68
1700	1-1/4" diameter		100	.080		.03	3.32		3.35	5.15
1800	1-1/2" diameter		90	.089		.04	3.69		3.73	5.75
1900	For ceiling installations, add						40%			

09 22 13 – Metal Furring

09 22 13.13 Metal Channel Furring		Crew	Daily Output	Labor-Hours	Unit	Material	Labor	Equipment	Total	Total Incl O&P
0010	**METAL CHANNEL FURRING**									
0030	Beams and columns, 7/8" channels, galvanized, 12" O.C.	1 Lath	155	.052	S.F.	.34	1.91		2.25	3.18
0050	16" O.C.		170	.047		.28	1.74		2.02	2.87
0070	24" O.C.		185	.043		.19	1.60		1.79	2.55
0100	Ceilings, on steel, 7/8" channels, galvanized, 12" O.C.		210	.038		.31	1.41		1.72	2.41
0300	16" O.C.		290	.028		.28	1.02		1.30	1.81
0400	24" O.C.		420	.019		.19	.70		.89	1.24
0600	1-5/8" channels, galvanized, 12" O.C.		190	.042		.41	1.56		1.97	2.75
0700	16" O.C.		260	.031		.37	1.14		1.51	2.08
0900	24" O.C.		390	.021		.25	.76		1.01	1.38
0930	7/8" channels with sound isolation clips, 12" O.C.		120	.067		1.37	2.46		3.83	5.15
0940	16" O.C.		100	.080		1.91	2.96		4.87	6.45
0950	24" O.C.		165	.048		1.25	1.79		3.04	4
0960	1-5/8" channels, galvanized, 12" O.C.		110	.073		1.48	2.69		4.17	5.60
0970	16" O.C.		100	.080		2	2.96		4.96	6.55
0980	24" O.C.		155	.052		1.31	1.91		3.22	4.25
1000	Walls, 7/8" channels, galvanized, 12" O.C.		235	.034		.31	1.26		1.57	2.19
1200	16" O.C.		265	.030		.28	1.12		1.40	1.95
1300	24" O.C.		350	.023		.19	.84		1.03	1.44
1500	1-5/8" channels, galvanized, 12" O.C.		210	.038		.41	1.41		1.82	2.53
1600	16" O.C.		240	.033		.37	1.23		1.60	2.22
1800	24" O.C.		305	.026		.25	.97		1.22	1.70
1920	7/8" channels with sound isolation clips, 12" O.C.		125	.064		1.37	2.36		3.73	4.99
1940	16" O.C.		100	.080		1.91	2.96		4.87	6.45
1950	24" O.C.		150	.053		1.25	1.97		3.22	4.27
1960	1-5/8" channels, galvanized, 12" O.C.		115	.070		1.48	2.57		4.05	5.40
1970	16" O.C.		95	.084		2	3.11		5.11	6.80
1980	24" O.C.		140	.057		1.31	2.11		3.42	4.55

09 22 16 – Non-Structural Metal Framing

09 22 16.13 Metal Studs and Track		Crew	Daily Output	Labor-Hours	Unit	Material	Labor	Equipment	Total	Total Incl O&P
0010	**METAL STUDS AND TRACK**									
1600	Non-load bearing, galv, 8' high, 25 ga. 1-5/8" wide, 16" O.C.	1 Carp	619	.013	S.F.	.25	.54		.79	1.10
1610	24" O.C.		950	.008		.18	.35		.53	.74
1620	2-1/2" wide, 16" O.C.		613	.013		.30	.54		.84	1.17
1630	24" O.C.		938	.009		.22	.35		.57	.80
1640	3-5/8" wide, 16" O.C.		600	.013		.34	.55		.89	1.23
1650	24" O.C.		925	.009		.26	.36		.62	.83
1660	4" wide, 16" O.C.		594	.013		.35	.56		.91	1.24
1670	24" O.C.		925	.009		.26	.36		.62	.83
1680	6" wide, 16" O.C.		588	.014		.51	.57		1.08	1.44

09 22 16.13 Metal Studs and Track		Crew	Daily Output	Labor-Hours	Unit	Material	2010 Bare Costs Labor	Equipment	Total	Total Incl O&P
1690	24" O.C.	1 Carp	906	.009	S.F.	.39	.37		.76	.99
1700	20 ga. studs, 1-5/8" wide, 16" O.C.		494	.016		.34	.67		1.01	1.42
1710	24" O.C.		763	.010		.26	.44		.70	.95
1720	2-1/2" wide, 16" O.C.		488	.016		.39	.68		1.07	1.48
1730	24" O.C.		750	.011		.30	.44		.74	1.01
1740	3-5/8" wide, 16" O.C.		481	.017		.43	.69		1.12	1.54
1750	24" O.C.		738	.011		.32	.45		.77	1.04
1760	4" wide, 16" O.C.		475	.017		.50	.70		1.20	1.63
1770	24" O.C.		738	.011		.38	.45		.83	1.11
1780	6" wide, 16" O.C.		469	.017		.60	.71		1.31	1.74
1790	24" O.C.		725	.011		.45	.46		.91	1.20
2000	Non-load bearing, galv, 10' high, 25 ga. 1-5/8" wide, 16" O.C.		495	.016		.23	.67		.90	1.30
2100	24" O.C.		760	.011		.17	.44		.61	.86
2200	2-1/2" wide, 16" O.C.		490	.016		.28	.68		.96	1.36
2250	24" O.C.		750	.011		.21	.44		.65	.91
2300	3-5/8" wide, 16" O.C.		480	.017		.32	.69		1.01	1.43
2350	24" O.C.		740	.011		.24	.45		.69	.95
2400	4" wide, 16" O.C.		475	.017		.33	.70		1.03	1.44
2450	24" O.C.		740	.011		.24	.45		.69	.96
2500	6" wide, 16" O.C.		470	.017		.49	.71		1.20	1.63
2550	24" O.C.		725	.011		.36	.46		.82	1.11
2600	20 ga. studs, 1-5/8" wide, 16" O.C.		395	.020		.33	.84		1.17	1.66
2650	24" O.C.		610	.013		.24	.54		.78	1.10
2700	2-1/2" wide, 16" O.C.		390	.021		.37	.85		1.22	1.72
2750	24" O.C.		600	.013		.28	.55		.83	1.15
2800	3-5/8" wide, 16" OC **CN**		385	.021		.40	.86		1.26	1.77
2850	24" O.C.		590	.014		.30	.56		.86	1.20
2900	4" wide, 16" O.C.		380	.021		.48	.87		1.35	1.88
2950	24" O.C.		590	.014		.35	.56		.91	1.26
3000	6" wide, 16" O.C.		375	.021		.56	.89		1.45	1.99
3050	24" O.C.		580	.014		.42	.57		.99	1.34
3060	Non-load bearing, galv, 12' high, 25 ga. 1-5/8" wide, 16" O.C.		413	.019		.22	.80		1.02	1.48
3070	24" O.C.		633	.013		.16	.53		.69	.99
3080	2-1/2" wide, 16" O.C.		408	.020		.27	.81		1.08	1.56
3090	24" O.C.		625	.013		.20	.53		.73	1.04
3100	3-5/8" wide, 16" O.C.		400	.020		.31	.83		1.14	1.62
3110	24" O.C.		617	.013		.23	.54		.77	1.08
3120	4" wide, 16" O.C.		396	.020		.31	.84		1.15	1.64
3130	24" O.C.		617	.013		.23	.54		.77	1.08
3140	6" wide, 16" O.C.		392	.020		.47	.85		1.32	1.82
3150	24" O.C.		604	.013		.34	.55		.89	1.22
3160	20 ga. studs, 1-5/8" wide, 16" O.C.		329	.024		.31	1.01		1.32	1.90
3170	24" O.C.		508	.016		.23	.65		.88	1.26
3180	2-1/2" wide, 16" O.C.		325	.025		.36	1.02		1.38	1.97
3190	24" O.C.		500	.016		.26	.66		.92	1.31
3200	3-5/8" wide, 16" O.C.		321	.025		.39	1.04		1.43	2.02
3210	24" O.C.		492	.016		.28	.68		.96	1.35
3220	4" wide, 16" O.C.		317	.025		.46	1.05		1.51	2.12
3230	24" O.C.		492	.016		.33	.68		1.01	1.41
3240	6" wide, 16" O.C.		313	.026		.54	1.06		1.60	2.23
3250	24" O.C.	↓	483	.017	↓	.39	.69		1.08	1.49
5000	Load bearing studs, see Div. 05 41 13.30									

09 22 Supports for Plaster and Gypsum Board

09 22 26 – Suspension Systems

09 22 26.13 Ceiling Suspension Systems		Crew	Daily Output	Labor-Hours	Unit	2010 Bare Costs Material	Labor	Equipment	Total	Total Incl O&P
0010	**CEILING SUSPENSION SYSTEMS** For gypsum board or plaster									
8000	Suspended ceilings, including carriers									
8200	1-1/2" carriers, 24" O.C. with:									
8300	7/8" channels, 16" O.C.	1 Lath	275	.029	S.F.	.47	1.07		1.54	2.10
8320	24" O.C.		310	.026		.38	.95		1.33	1.82
8400	1-5/8" channels, 16" O.C.		205	.039		.57	1.44		2.01	2.74
8420	24" O.C.		250	.032		.44	1.18		1.62	2.23
8600	2" carriers, 24" O.C. with:									
8700	7/8" channels, 16" O.C.	1 Lath	250	.032	S.F.	.53	1.18		1.71	2.32
8720	24" O.C.		285	.028		.44	1.04		1.48	2.01
8800	1-5/8" channels, 16" O.C.		190	.042		.62	1.56		2.18	2.97
8820	24" O.C.		225	.036		.50	1.31		1.81	2.48

09 22 36 – Lath

09 22 36.13 Gypsum Lath

		Crew	Daily Output	Labor-Hours	Unit	Material	Labor	Equipment	Total	Total Incl O&P
0010	**GYPSUM LATH** R092000-50									
0020	Plain or perforated, nailed, 3/8" thick	1 Lath	85	.094	S.Y.	5.05	3.48		8.53	10.65
0100	1/2" thick		80	.100		4.95	3.70		8.65	10.90
0300	Clipped to steel studs, 3/8" thick		75	.107		5.05	3.94		8.99	11.35
0400	1/2" thick		70	.114		4.95	4.22		9.17	11.65
1500	For ceiling installations, add		216	.037			1.37		1.37	2.01
1600	For columns and beams, add		170	.047			1.74		1.74	2.56

09 22 36.23 Metal Lath

		Crew	Daily Output	Labor-Hours	Unit	Material	Labor	Equipment	Total	Total Incl O&P
0010	**METAL LATH** R092000-50									
0020	Diamond, expanded, 2.5 lb. per S.Y., painted				S.Y.	4.52			4.52	4.97
0100	Galvanized					2.94			2.94	3.23
0300	3.4 lb. per S.Y., painted					3.99			3.99	4.39
0400	Galvanized					4.53			4.53	4.98
0600	For 15# asphalt sheathing paper, add					.53			.53	.58
0900	Flat rib, 1/8" high, 2.75 lb., painted					3.69			3.69	4.06
1000	Foil backed					3.60			3.60	3.96
1200	3.4 lb. per S.Y., painted					4.71			4.71	5.20
1300	Galvanized					4.19			4.19	4.61
1500	For 15# asphalt sheathing paper, add					.53			.53	.58
1800	High rib, 3/8" high, 3.4 lb. per S.Y., painted					5.70			5.70	6.25
1900	Galvanized					5.70			5.70	6.25
2400	3/4" high, painted, .60 lb. per S.F.				S.F.	.59			.59	.65
2500	.75 lb. per S.F.				"	1.27			1.27	1.40
2800	Stucco mesh, painted, 3.6 lb.				S.Y.	3.32			3.32	3.65
3000	K-lath, perforated, absorbent paper, regular					4.43			4.43	4.87
3100	Heavy duty					5.25			5.25	5.75
3300	Waterproof, heavy duty, grade B backing					5.10			5.10	5.65
3400	Fire resistant backing					5.65			5.65	6.25
3600	2.5 lb. diamond painted, on wood framing, on walls	1 Lath	85	.094		4.52	3.48		8	10.05
3700	On ceilings		75	.107		4.52	3.94		8.46	10.75
3900	3.4 lb. diamond painted, on wood framing, on walls		80	.100		4.71	3.70		8.41	10.65
4000	On ceilings		70	.114		4.71	4.22		8.93	11.40
4200	3.4 lb. diamond painted, wired to steel framing		75	.107		4.71	3.94		8.65	11
4300	On ceilings		60	.133		4.71	4.93		9.64	12.45
4500	Columns and beams, wired to steel		40	.200		4.71	7.40		12.11	16.05
4600	Cornices, wired to steel		35	.229		4.71	8.45		13.16	17.60
4800	Screwed to steel studs, 2.5 lb.		80	.100		4.52	3.70		8.22	10.40

09 22 Supports for Plaster and Gypsum Board

09 22 36 – Lath

09 22 36.23 Metal Lath		Crew	Daily Output	Labor-Hours	Unit	Material	2010 Bare Costs Labor	Equipment	Total	Total Incl O&P
4900	3.4 lb.	1 Lath	75	.107	S.Y.	3.99	3.94		7.93	10.20
5100	Rib lath, painted, wired to steel, on walls, 2.5 lb.		75	.107		3.69	3.94		7.63	9.85
5200	3.4 lb.		70	.114		5.70	4.22		9.92	12.45
5400	4.0 lb.	↓	65	.123		5.70	4.55		10.25	13
5500	For self-furring lath, add					.10			.10	.11
5700	Suspended ceiling system, incl. 3.4 lb. diamond lath, painted	1 Lath	15	.533		4.28	19.70		23.98	33.50
5800	Galvanized	"	15	.533	↓	7.30	19.70		27	37
6000	Hollow metal stud partitions, 3.4 lb. painted lath both sides									
6010	Non-load bearing, 25 ga., w/rib lath 2-1/2" studs, 12" O.C.	1 Lath	20.30	.394	S.Y.	14.60	14.55		29.15	37.50
6300	16" O.C.		21.10	.379		13.90	14		27.90	36
6350	24" O.C.		22.70	.352		13.25	13		26.25	34
6400	3-5/8" studs, 16" O.C.		19.50	.410		14.30	15.15		29.45	38
6600	24" O.C.		20.40	.392		13.50	14.50		28	36.50
6700	4" studs, 16" O.C.		20.40	.392		14.35	14.50		28.85	37.50
6900	24" O.C.		21.60	.370		13.55	13.70		27.25	35
7000	6" studs, 16" O.C.		19.50	.410		15.75	15.15		30.90	40
7100	24" O.C.		21.10	.379		14.60	14		28.60	36.50
7200	L.B. partitions, 16 ga., w/rib lath, 2-1/2" studs, 16" O.C.		20	.400		13.85	14.80		28.65	37
7300	3-5/8" studs, 16 ga.		19.70	.406		15.50	15		30.50	39
7500	4" studs, 16 ga.		19.50	.410		16	15.15		31.15	40
7600	6" studs, 16 ga.	↓	18.70	.428	↓	18.60	15.80		34.40	44

09 22 36.43 Security Mesh		Crew	Daily Output	Labor-Hours	Unit	Material	Labor	Equipment	Total	Total Incl O&P
0010	**SECURITY MESH**, expanded metal, flat, screwed to framing									
0100	On walls, 3/4", 1.76 lb/S.F.	2 Carp	1500	.011	S.F.	1.61	.44		2.05	2.45
0110	1-1/2", 1.14 lb/S.F.		1600	.010		1.30	.42		1.72	2.07
0200	On ceilings, 3/4", 1.76 lb/S.F.		1350	.012		1.61	.49		2.10	2.53
0210	1-1/2", 1.14 lb/S.F.	↓	1450	.011	↓	1.30	.46		1.76	2.14

09 22 36.83 Accessories, Plaster		Crew	Daily Output	Labor-Hours	Unit	Material	Labor	Equipment	Total	Total Incl O&P
0010	**ACCESSORIES, PLASTER**									
0020	Casing bead, expanded flange, galvanized	1 Lath	2.70	2.963	C.L.F.	46.50	109		155.50	212
0900	Channels, cold rolled, 16 ga., 3/4" deep, galvanized					31			31	34
1200	1-1/2" deep, 16 ga., galvanized					41.50			41.50	45.50
1620	Corner bead, expanded bullnose, 3/4" radius, #10, galvanized	1 Lath	2.60	3.077		33.50	114		147.50	204
1650	#1, galvanized		2.55	3.137		56	116		172	233
1670	Expanded wing, 2-3/4" wide, #1, galvanized		2.65	3.019		34	112		146	201
1700	Inside corner (corner rite), 3" x 3", painted		2.60	3.077		30	114		144	200
1750	Strip-ex, 4" wide, painted		2.55	3.137		30	116		146	204
1800	Expansion joint, 3/4" grounds, limited expansion, galv., 1 piece		2.70	2.963		108	109		217	280
2100	Extreme expansion, galvanized, 2 piece	↓	2.60	3.077	↓	174	114		288	360

09 23 Gypsum Plastering

09 23 13 – Acoustical Gypsum Plastering

09 23 13.10 Perlite or Vermiculite Plaster		Crew	Daily Output	Labor-Hours	Unit	Material	Labor	Equipment	Total	Total Incl O&P
0010	**PERLITE OR VERMICULITE PLASTER** R092000-50									
0020	In 100 lb. bags, under 200 bags				Bag	14.10			14.10	15.50
0100	Over 200 bags				"	12.80			12.80	14.05
0300	2 coats, no lath included, on walls	J-1	92	.435	S.Y.	3.78	15.65	1.49	20.92	29.50
0400	On ceilings	"	79	.506		3.78	18.20	1.73	23.71	33
0600	On and incl. 3/8" gypsum lath, on metal studs	J-2	84	.571		9.45	20.50	1.63	31.58	42.50
0700	On ceilings	"	70	.686	↓	9.45	25	1.95	36.40	49.50

09 23 Gypsum Plastering

09 23 13 – Acoustical Gypsum Plastering

09 23 13.10 Perlite or Vermiculite Plaster		Crew	Daily Output	Labor-Hours	Unit	Material	Labor	Equipment	Total	Total Incl O&P
						2010 Bare Costs				
0900	3 coats, no lath included, on walls	J-1	74	.541	S.Y.	6.30	19.45	1.85	27.60	38
1000	On ceilings	"	63	.635		6.30	23	2.17	31.47	43.50
1200	On and incl. painted metal lath, on metal studs	J-2	72	.667		12	24	1.90	37.90	51.50
1300	On ceilings		61	.787		12	28.50	2.24	42.74	58
1500	On and incl. suspended metal lath ceiling	↓	37	1.297		10.60	47	3.70	61.30	85
1700	For irregular or curved surfaces, add to above						30%			
1800	For columns and beams, add to above						50%			
1900	For soffits, add to ceiling prices				↓		40%			

09 23 20 – Gypsum Plaster

09 23 20.10 Gypsum Plaster On Walls and Ceilings

		Crew	Daily Output	Labor-Hours	Unit	Material	Labor	Equipment	Total	Total Incl O&P
0010	**GYPSUM PLASTER ON WALLS AND CEILINGS** R092000-50									
0020	80# bag, less than 1 ton				Bag	15			15	16.50
0100	Over 1 ton				"	14.50			14.50	15.95
0300	2 coats, no lath included, on walls	J-1	105	.381	S.Y.	3.74	13.70	1.30	18.74	26
0400	On ceilings	"	92	.435		3.74	15.65	1.49	20.88	29
0600	On and incl. 3/8" gypsum lath on steel, on walls	J-2	97	.495		8.80	17.85	1.41	28.06	37.50
0700	On ceilings	"	83	.578		8.80	21	1.65	31.45	42.50
0900	3 coats, no lath included, on walls	J-1	87	.460		5.35	16.50	1.57	23.42	32
1000	On ceilings	"	78	.513		5.35	18.45	1.75	25.55	35.50
1200	On and including painted metal lath, on wood studs	J-2	86	.558		10.80	20	1.59	32.39	43.50
1300	On ceilings	"	76.50	.627	↓	10.80	22.50	1.79	35.09	47.50
1600	For irregular or curved surfaces, add						30%			
1800	For columns & beams, add						50%			

09 23 20.20 Gauging Plaster

		Crew	Daily Output	Labor-Hours	Unit	Material	Labor	Equipment	Total	Total Incl O&P
0010	**GAUGING PLASTER** R092000-50									
0020	100 lb. bags, less than 1 ton				Bag	19.45			19.45	21.50
0100	Over 1 ton				"	16.55			16.55	18.20

09 23 20.30 Keenes Cement

		Crew	Daily Output	Labor-Hours	Unit	Material	Labor	Equipment	Total	Total Incl O&P
0010	**KEENES CEMENT** R092000-50									
0020	In 100 lb. bags, less than 1 ton				Bag	26.50			26.50	29.50
0100	Over 1 ton				"	25.50			25.50	28
0300	Finish only, add to plaster prices, standard	J-1	215	.186	S.Y.	2.45	6.70	.64	9.79	13.40
0400	High quality	"	144	.278	"	2.47	10	.95	13.42	18.65

09 24 Portland Cement Plastering

09 24 23 – Portland Cement Stucco

09 24 23.40 Stucco

		Crew	Daily Output	Labor-Hours	Unit	Material	Labor	Equipment	Total	Total Incl O&P
0010	**STUCCO** R092000-50									
0015	3 coats 1" thick, float finish, with mesh, on wood frame	J-2	63	.762	S.Y.	5.60	27.50	2.17	35.27	49.50
0100	On masonry construction, no mesh incl.	J-1	67	.597		2.28	21.50	2.04	25.82	37
0300	For trowel finish, add	1 Plas	170	.047			1.76		1.76	2.62
0400	For 3/4" thick, on masonry, deduct	J-1	880	.045		.61	1.63	.16	2.40	3.28
0600	For coloring and special finish, add, minimum		685	.058		.39	2.10	.20	2.69	3.78
0700	Maximum	↓	200	.200		1.36	7.20	.68	9.24	13
0900	For soffits, add	J-2	155	.310		2.11	11.20	.88	14.19	19.95
1000	Exterior stucco, with bonding agent, 3 coats, on walls, no mesh incl.	J-1	200	.200		3.54	7.20	.68	11.42	15.40
1200	Ceilings		180	.222		3.54	8	.76	12.30	16.65
1300	Beams		80	.500		3.54	17.95	1.71	23.20	33
1500	Columns	↓	100	.400		3.54	14.40	1.37	19.31	27
1600	Mesh, painted, nailed to wood, 1.8 lb.	1 Lath	60	.133	↓	5.90	4.93		10.83	13.70

09 24 Portland Cement Plastering

09 24 23 – Portland Cement Stucco

09 24 23.40 Stucco		Crew	Daily Output	Labor-Hours	Unit	Material	2010 Bare Costs Labor	Equipment	Total	Total Incl O&P
1800	3.6 lb.	1 Lath	55	.145	S.Y.	3.32	5.35		8.67	11.55
1900	Wired to steel, painted, 1.8 lb.		53	.151		5.90	5.60		11.50	14.65
2100	3.6 lb.	↓	50	.160	↓	3.32	5.90		9.22	12.35

09 26 Veneer Plastering

09 26 13 – Gypsum Veneer Plastering

09 26 13.20 Blueboard

		Crew	Daily Output	Labor-Hours	Unit	Material	Labor	Equipment	Total	Total Incl O&P
0010	**BLUEBOARD** For use with thin coat									
0100	plaster application (see Div. 09 26 13.80)									
1000	3/8" thick, on walls or ceilings, standard, no finish included	2 Carp	1900	.008	S.F.	.26	.35		.61	.83
1100	With thin coat plaster finish		875	.018		.34	.76		1.10	1.55
1400	On beams, columns, or soffits, standard, no finish included		675	.024		.30	.98		1.28	1.85
1450	With thin coat plaster finish		475	.034		.38	1.40		1.78	2.58
3000	1/2" thick, on walls or ceilings, standard, no finish included		1900	.008		.29	.35		.64	.86
3100	With thin coat plaster finish		875	.018		.37	.76		1.13	1.58
3300	Fire resistant, no finish included		1900	.008		.29	.35		.64	.86
3400	With thin coat plaster finish		875	.018		.37	.76		1.13	1.58
3450	On beams, columns, or soffits, standard, no finish included		675	.024		.33	.98		1.31	1.89
3500	With thin coat plaster finish		475	.034		.42	1.40		1.82	2.62
3700	Fire resistant, no finish included		675	.024		.33	.98		1.31	1.89
3800	With thin coat plaster finish		475	.034		.42	1.40		1.82	2.62
5000	5/8" thick, on walls or ceilings, fire resistant, no finish included		1900	.008		.29	.35		.64	.86
5100	With thin coat plaster finish		875	.018		.37	.76		1.13	1.58
5500	On beams, columns, or soffits, no finish included		675	.024		.33	.98		1.31	1.89
5600	With thin coat plaster finish		475	.034		.42	1.40		1.82	2.62
6000	For high ceilings, over 8' high, add		3060	.005			.22		.22	.34
6500	For over 3 stories high, add per story	↓	6100	.003	↓		.11		.11	.17

09 26 13.80 Thin Coat Plaster

		Crew	Daily Output	Labor-Hours	Unit	Material	Labor	Equipment	Total	Total Incl O&P
0010	**THIN COAT PLASTER**									
0012	1 coat veneer, not incl. lath	J-1	3600	.011	S.F.	.08	.40	.04	.52	.73
1000	In 50 lb. bags				Bag	11.30			11.30	12.45

09 28 Backing Boards and Underlayments

09 28 13 – Cementitious Backing Boards

09 28 13.10 Cementitious Backerboard

		Crew	Daily Output	Labor-Hours	Unit	Material	Labor	Equipment	Total	Total Incl O&P
0010	**CEMENTITIOUS BACKERBOARD**									
0070	Cementitious backerboard, on floor, 3' x 4' x 1/2" sheets	2 Carp	525	.030	S.F.	.69	1.27		1.96	2.71
0080	3' x 5' x 1/2" sheets		525	.030		.72	1.27		1.99	2.74
0090	3' x 6' x 1/2" sheets		525	.030		.64	1.27		1.91	2.66
0100	3' x 4' x 5/8" sheets		525	.030		.93	1.27		2.20	2.97
0110	3' x 5' x 5/8" sheets		525	.030		.92	1.27		2.19	2.96
0120	3' x 6' x 5/8" sheets		525	.030		.75	1.27		2.02	2.77
0150	On wall, 3' x 4' x 1/2" sheets		350	.046		.69	1.90		2.59	3.69
0160	3' x 5' x 1/2" sheets		350	.046		.72	1.90		2.62	3.72
0170	3' x 6' x 1/2" sheets		350	.046		.64	1.90		2.54	3.64
0180	3' x 4' x 5/8" sheets		350	.046		.93	1.90		2.83	3.95
0190	3' x 5' x 5/8" sheets		350	.046		.92	1.90		2.82	3.94
0200	3' x 6' x 5/8" sheets		350	.046		.75	1.90		2.65	3.75
0250	On counter, 3' x 4' x 1/2" sheets	↓	180	.089	↓	.69	3.69		4.38	6.45

09 28 Backing Boards and Underlayments

09 28 13 – Cementitious Backing Boards

09 28 13.10 Cementitious Backerboard		Crew	Daily Output	Labor-Hours	Unit	Material	2010 Bare Costs Labor	Equipment	Total	Total Incl O&P
0260	3′ x 5′ x 1/2″ sheets	2 Carp	180	.089	S.F.	.72	3.69		4.41	6.50
0270	3′ x 6′ x 1/2″ sheets		180	.089		.64	3.69		4.33	6.40
0300	3′ x 4′ x 5/8″ sheets		180	.089		.93	3.69		4.62	6.70
0310	3′ x 5′ x 5/8″ sheets		180	.089		.92	3.69		4.61	6.70
0320	3′ x 6′ x 5/8″ sheets		180	.089		.75	3.69		4.44	6.50

09 29 Gypsum Board

09 29 10 – Gypsum Board Panels

09 29 10.30 Gypsum Board

		Crew	Daily Output	Labor-Hours	Unit	Material	Labor	Equipment	Total	Total Incl O&P
0010	**GYPSUM BOARD** on walls & ceilings R092910-10									
0100	Nailed or screwed to studs unless otherwise noted									
0150	3/8″ thick, on walls, standard, no finish included	2 Carp	2000	.008	S.F.	.24	.33		.57	.77
0200	On ceilings, standard, no finish included		1800	.009		.24	.37		.61	.83
0250	On beams, columns, or soffits, no finish included		675	.024		.24	.98		1.22	1.78
0300	1/2″ thick, on walls, standard, no finish included ***CN***		2000	.008		.24	.33		.57	.77
0350	Taped and finished (level 4 finish)		965	.017		.28	.69		.97	1.37
0390	With compound skim coat (level 5 finish)		775	.021		.33	.86		1.19	1.68
0400	Fire resistant, no finish included		2000	.008		.28	.33		.61	.82
0450	Taped and finished (level 4 finish)		965	.017		.32	.69		1.01	1.41
0490	With compound skim coat (level 5 finish)		775	.021		.37	.86		1.23	1.72
0500	Water resistant, no finish included		2000	.008		.35	.33		.68	.90
0550	Taped and finished (level 4 finish)		965	.017		.39	.69		1.08	1.49
0590	With compound skim coat (level 5 finish)		775	.021		.44	.86		1.30	1.80
0600	Prefinished, vinyl, clipped to studs		900	.018		.60	.74		1.34	1.80
0700	Mold resistant, no finish included		2000	.008		.42	.33		.75	.97
0710	Taped and finished (level 4 finish)		965	.017		.46	.69		1.15	1.57
0720	With compound skim coat (level 5 finish)		775	.021		.51	.86		1.37	1.88
1000	On ceilings, standard, no finish included		1800	.009		.24	.37		.61	.83
1050	Taped and finished (level 4 finish)		765	.021		.28	.87		1.15	1.65
1090	With compound skim coat (level 5 finish)		610	.026		.33	1.09		1.42	2.04
1100	Fire resistant, no finish included		1800	.009		.28	.37		.65	.88
1150	Taped and finished (level 4 finish)		765	.021		.32	.87		1.19	1.69
1195	With compound skim coat (level 5 finish)		610	.026		.37	1.09		1.46	2.08
1200	Water resistant, no finish included		1800	.009		.35	.37		.72	.96
1250	Taped and finished (level 4 finish)		765	.021		.39	.87		1.26	1.77
1290	With compound skim coat (level 5 finish)		610	.026		.44	1.09		1.53	2.16
1310	Mold resistant, no finish included		1800	.009		.42	.37		.79	1.03
1320	Taped and finished (level 4 finish)		765	.021		.46	.87		1.33	1.85
1330	With compound skim coat (level 5 finish)		610	.026		.51	1.09		1.60	2.24
1500	On beams, columns, or soffits, standard, no finish included		675	.024		.28	.98		1.26	1.82
1550	Taped and finished (level 4 finish)		475	.034		.28	1.40		1.68	2.47
1590	With compound skim coat (level 5 finish)		540	.030		.33	1.23		1.56	2.26
1600	Fire resistant, no finish included		675	.024		.28	.98		1.26	1.83
1650	Taped and finished (level 4 finish)		475	.034		.32	1.40		1.72	2.51
1690	With compound skim coat (level 5 finish)		540	.030		.37	1.23		1.60	2.30
1700	Water resistant, no finish included		675	.024		.40	.98		1.38	1.96
1750	Taped and finished (level 4 finish)		475	.034		.39	1.40		1.79	2.59
1790	With compound skim coat (level 5 finish)		540	.030		.44	1.23		1.67	2.38
1800	Mold resistant, no finish included		675	.024		.48	.98		1.46	2.05
1810	Taped and finished (level 4 finish)		475	.034		.46	1.40		1.86	2.67
1820	With compound skim coat (level 5 finish)		540	.030		.51	1.23		1.74	2.46

09 29 10 – Gypsum Board Panels

09 29 10.30 Gypsum Board		Crew	Daily Output	Labor-Hours	Unit	Material	Labor	Equipment	Total	Total Incl O&P
						2010 Bare Costs				
2000	5/8" thick, on walls, standard, no finish included	2 Carp	2000	.008	S.F.	.28	.33		.61	.82
2050	Taped and finished (level 4 finish)		965	.017		.32	.69		1.01	1.41
2090	With compound skim coat (level 5 finish)		775	.021		.37	.86		1.23	1.72
2100	Fire resistant, no finish included		2000	.008		.27	.33		.60	.81
2150	Taped and finished (level 4 finish)		965	.017		.31	.69		1	1.40
2195	With compound skim coat (level 5 finish)		775	.021		.36	.86		1.22	1.71
2200	Water resistant, no finish included		2000	.008		.43	.33		.76	.98
2250	Taped and finished (level 4 finish)		965	.017		.47	.69		1.16	1.58
2290	With compound skim coat (level 5 finish)		775	.021		.52	.86		1.38	1.89
2300	Prefinished, vinyl, clipped to studs		900	.018		.71	.74		1.45	1.92
2510	Mold resistant, no finish included		2000	.008		.49	.33		.82	1.05
2520	Taped and finished (level 4 finish)		965	.017		.53	.69		1.22	1.65
2530	With compound skim coat (level 5 finish)		775	.021		.58	.86		1.44	1.95
3000	On ceilings, standard, no finish included		1800	.009		.28	.37		.65	.88
3050	Taped and finished (level 4 finish)		765	.021		.32	.87		1.19	1.69
3090	With compound skim coat (level 5 finish)		615	.026		.37	1.08		1.45	2.07
3100	Fire resistant, no finish included		1800	.009		.27	.37		.64	.87
3150	Taped and finished (level 4 finish)		765	.021		.31	.87		1.18	1.68
3190	With compound skim coat (level 5 finish)		615	.026		.36	1.08		1.44	2.06
3200	Water resistant, no finish included		1800	.009		.43	.37		.80	1.04
3250	Taped and finished (level 4 finish)		765	.021		.47	.87		1.34	1.86
3290	With compound skim coat (level 5 finish)		615	.026		.52	1.08		1.60	2.24
3300	Mold resistant, no finish included		1800	.009		.49	.37		.86	1.11
3310	Taped and finished (level 4 finish)		765	.021		.53	.87		1.40	1.93
3320	With compound skim coat (level 5 finish)		615	.026		.58	1.08		1.66	2.30
3500	On beams, columns, or soffits, no finish included		675	.024		.32	.98		1.30	1.87
3550	Taped and finished (level 4 finish)		475	.034		.37	1.40		1.77	2.57
3590	With compound skim coat (level 5 finish)		380	.042		.42	1.75		2.17	3.16
3600	Fire resistant, no finish included		675	.024		.31	.98		1.29	1.86
3650	Taped and finished (level 4 finish)		475	.034		.36	1.40		1.76	2.55
3690	With compound skim coat (level 5 finish)		380	.042		.36	1.75		2.11	3.09
3700	Water resistant, no finish included		675	.024		.49	.98		1.47	2.06
3750	Taped and finished (level 4 finish)		475	.034		.54	1.40		1.94	2.76
3790	With compound skim coat (level 5 finish)		380	.042		.52	1.75		2.27	3.27
3800	Mold resistant, no finish included		675	.024		.56	.98		1.54	2.14
3810	Taped and finished (level 4 finish)		475	.034		.61	1.40		2.01	2.83
3820	With compound skim coat (level 5 finish)		380	.042		.58	1.75		2.33	3.33
4000	Fireproofing, beams or columns, 2 layers, 1/2" thick, incl finish		330	.048		.60	2.01		2.61	3.77
4010	Mold resistant		330	.048		.88	2.01		2.89	4.08
4050	5/8" thick		300	.053		.62	2.22		2.84	4.11
4060	Mold resistant		300	.053		1.06	2.22		3.28	4.59
4100	3 layers, 1/2" thick		225	.071		.88	2.95		3.83	5.50
4110	Mold resistant		225	.071		1.30	2.95		4.25	6
4150	5/8" thick		210	.076		.94	3.17		4.11	5.90
4160	Mold resistant		210	.076		1.60	3.17		4.77	6.65
5050	For 1" thick coreboard on columns		480	.033		.55	1.38		1.93	2.74
5100	For foil-backed board, add					.14			.14	.15
5200	For work over 8' high, add	2 Carp	3060	.005			.22		.22	.34
5270	For textured spray, add	2 Lath	1600	.010		.05	.37		.42	.60
5300	For distribution cost over 3 stories high, add per story	2 Carp	6100	.003			.11		.11	.17
5350	For finishing inner corners, add		950	.017	L.F.	.09	.70		.79	1.18
5355	For finishing outer corners, add		1250	.013		.21	.53		.74	1.05
5500	For acoustical sealant, add per bead	1 Carp	500	.016		.04	.66		.70	1.07

09 29 Gypsum Board

09 29 10 – Gypsum Board Panels

09 29 10.30 Gypsum Board		Crew	Daily Output	Labor-Hours	Unit	Material	Labor	Equipment	Total	Total Incl O&P
						2010 Bare Costs				
5550	Sealant, 1 quart tube				Ea.	6.85			6.85	7.55
5600	Sound deadening board, 1/4" gypsum	2 Carp	1800	.009	S.F.	.33	.37		.70	.93
5650	1/2" wood fiber	"	1800	.009	"	.82	.37		1.19	1.47

09 29 10.50 High Abuse Gypsum Board

Line	Description	Crew	Daily Output	Labor-Hours	Unit	Material	Labor	Equipment	Total	Total Incl O&P
0010	**HIGH ABUSE GYPSUM BOARD**, fiber reinforced, nailed or									
0100	screwed to studs unless otherwise noted									
0110	1/2" thick, on walls, no finish included	2 Carp	1800	.009	S.F.	.84	.37		1.21	1.49
0120	Taped and finished (level 4 finish)		870	.018		.88	.76		1.64	2.15
0130	With compound skim coat (level 5 finish)		700	.023		.93	.95		1.88	2.48
0150	On ceilings, no finish included		1620	.010		.84	.41		1.25	1.55
0160	Taped and finished (level 4 finish)		690	.023		.88	.96		1.84	2.46
0170	With compound skim coat (level 5 finish)		550	.029		.93	1.21		2.14	2.88
0210	5/8" thick, on walls, no finish included		1800	.009		.89	.37		1.26	1.55
0220	Taped and finished (level 4 finish)		870	.018		.93	.76		1.69	2.21
0230	With compound skim coat (level 5 finish)		700	.023		.98	.95		1.93	2.53
0250	On ceilings, no finish included		1620	.010		.89	.41		1.30	1.61
0260	Taped and finished (level 4 finish)		690	.023		.93	.96		1.89	2.52
0270	With compound skim coat (level 5 finish)		550	.029		.98	1.21		2.19	2.93
0310	5/8" thick, on walls, very high impact, no finish included		1800	.009		.95	.37		1.32	1.62
0320	Taped and finished (level 4 finish)		870	.018		.99	.76		1.75	2.27
0330	With compound skim coat (level 5 finish)		700	.023		1.04	.95		1.99	2.60
0350	On ceilings, no finish included		1620	.010		.95	.41		1.36	1.68
0360	Taped and finished (level 4 finish)		690	.023		.99	.96		1.95	2.58
0370	With compound skim coat (level 5 finish)	↓	550	.029	↓	1.04	1.21		2.25	3
0400	High abuse, gypsum core, paper face									
0410	1/2" thick, on walls, no finish included	2 Carp	1800	.009	S.F.	.76	.37		1.13	1.41
0420	Taped and finished (level 4 finish)		870	.018		.80	.76		1.56	2.06
0430	With compound skim coat (level 5 finish)		700	.023		.85	.95		1.80	2.39
0450	On ceilings, no finish included		1620	.010		.76	.41		1.17	1.47
0460	Taped and finished (level 4 finish)		690	.023		.80	.96		1.76	2.37
0470	With compound skim coat (level 5 finish)		550	.029		.85	1.21		2.06	2.79
0510	5/8" thick, on walls, no finish included		1800	.009		.77	.37		1.14	1.42
0520	Taped and finished (level 4 finish)		870	.018		.81	.76		1.57	2.07
0530	With compound skim coat (level 5 finish)		700	.023		.86	.95		1.81	2.40
0550	On ceilings, no finish included		1620	.010		.77	.41		1.18	1.48
0560	Taped and finished (level 4 finish)		690	.023		.81	.96		1.77	2.38
0570	With compound skim coat (level 5 finish)		550	.029		.86	1.21		2.07	2.80
1000	For high ceilings, over 8' high, add		2750	.006			.24		.24	.37
1010	For distribution cost over 3 stories high, add per story	↓	5500	.003	↓		.12		.12	.19

09 29 15 – Gypsum Board Accessories

09 29 15.10 Accessories, Gypsum Board

Line	Description	Crew	Daily Output	Labor-Hours	Unit	Material	Labor	Equipment	Total	Total Incl O&P
0010	**ACCESSORIES, GYPSUM BOARD**									
0020	Casing bead, galvanized steel	1 Carp	2.90	2.759	C.L.F.	23.50	115		138.50	203
0100	Vinyl		3	2.667		22.50	111		133.50	196
0300	Corner bead, galvanized steel, 1" x 1"		4	2		14.85	83		97.85	144
0400	1-1/4" x 1-1/4"		3.50	2.286		16.60	95		111.60	164
0600	Vinyl		4	2		17.80	83		100.80	148
0900	Furring channel, galv. steel, 7/8" deep, standard		2.60	3.077		25	128		153	225
1000	Resilient		2.55	3.137		25.50	130		155.50	229
1100	J trim, galvanized steel, 1/2" wide		3	2.667		25	111		136	199
1120	5/8" wide		2.95	2.712		24	113		137	201
1140	L trim, galvanized	↓	3	2.667	↓	20.50	111		131.50	194

09 29 Gypsum Board

09 29 15 – Gypsum Board Accessories

09 29 15.10 Accessories, Gypsum Board		Crew	Daily Output	Labor-Hours	Unit	Material	Labor	Equipment	Total	Total Incl O&P
						2010 Bare Costs				
1150	U trim, galvanized	1 Carp	2.95	2.712	C.L.F.	19.55	113		132.55	196
1160	Screws #6 x 1" A				M	7.35			7.35	8.10
1170	#6 x 1-5/8" A				"	10.80			10.80	11.85
1200	For stud partitions, see Div. 05 41 13.30 and 09 22 16.13									
1500	Z stud, galvanized steel, 1-1/2" wide	1 Carp	2.60	3.077	C.L.F.	36	128		164	237
1600	2" wide	"	2.55	3.137	"	55.50	130		185.50	262

09 30 Tiling

09 30 13 – Ceramic Tiling

09 30 13.10 Ceramic Tile

		Crew	Daily Output	Labor-Hours	Unit	Material	Labor	Equipment	Total	Total Incl O&P
0010	**CERAMIC TILE**									
0050	Base, using 1' x 4" high pc. with 1" x 1" tiles, mud set	D-7	82	.195	L.F.	4.95	6.90		11.85	15.50
0100	Thin set	"	128	.125		4.45	4.41		8.86	11.35
0300	For 6" high base, 1" x 1" tile face, add					.72			.72	.79
0400	For 2" x 2" tile face, add to above					.38			.38	.42
0600	Cove base, 4-1/4" x 4-1/4" high, mud set	D-7	91	.176		3.75	6.20		9.95	13.15
0700	Thin set		128	.125		3.64	4.41		8.05	10.45
0900	6" x 4-1/4" high, mud set		100	.160		3.34	5.65		8.99	11.90
1000	Thin set		137	.117		3.06	4.12		7.18	9.35
1200	Sanitary cove base, 6" x 4-1/4" high, mud set		93	.172		3.84	6.05		9.89	13.05
1300	Thin set		124	.129		3.59	4.55		8.14	10.60
1500	6" x 6" high, mud set		84	.190		4.48	6.70		11.18	14.75
1600	Thin set		117	.137		4.20	4.82		9.02	11.65
1800	Bathroom accessories, average		82	.195	Ea.	10.70	6.90		17.60	22
1900	Bathtub, 5', rec. 4-1/4" x 4-1/4" tile wainscot, adhesive set 6' high		2.90	5.517		156	195		351	455
2100	7' high wainscot		2.50	6.400		179	226		405	525
2200	8' high wainscot		2.20	7.273		190	257		447	585
2400	Bullnose trim, 4-1/4" x 4-1/4", mud set		82	.195	L.F.	3.55	6.90		10.45	13.95
2500	Thin set		128	.125		3.11	4.41		7.52	9.85
2700	6" x 4-1/4" bullnose trim, mud set		84	.190		2.70	6.70		9.40	12.75
2800	Thin set		124	.129		2.50	4.55		7.05	9.40
3000	Floors, natural clay, random or uniform, thin set, color group 1		183	.087	S.F.	4.19	3.08		7.27	9.10
3100	Color group 2		183	.087		4.51	3.08		7.59	9.45
3255	Floors, glazed, thin set, 6" x 6", color group 1		200	.080		3.58	2.82		6.40	8.05
3260	8" x 8" tile		250	.064		3.58	2.26		5.84	7.25
3270	12" x 12" tile		325	.049		4.47	1.74		6.21	7.45
3280	16" x 16" tile		550	.029		6.15	1.03		7.18	8.25
3285	Border, 6" x 12" tile		275	.058		11.55	2.05		13.60	15.70
3290	3" x 12" tile		200	.080		34	2.82		36.82	41
3300	Porcelain type, 1 color, color group 2, 1" x 1"		183	.087		4.71	3.08		7.79	9.70
3310	2" x 2" or 2" x 1", thin set		190	.084		4.98	2.97		7.95	9.85
3350	For random blend, 2 colors, add					.88			.88	.97
3360	4 colors, add					1.24			1.24	1.36
3370	For color group 3, add					.50			.50	.55
3380	For abrasive non-slip tile, add					.49			.49	.54
4300	Specialty tile, 4-1/4" x 4-1/4" x 1/2", decorator finish	D-7	183	.087		10	3.08		13.08	15.50
4500	Add for epoxy grout, 1/16" joint, 1" x 1" tile		800	.020		.62	.71		1.33	1.71
4600	2" x 2" tile		820	.020		.59	.69		1.28	1.66
4800	Pregrouted sheets, walls, 4-1/4" x 4-1/4", 6" x 4-1/4"									
4810	and 8-1/2" x 4-1/4", 4 S.F. sheets, silicone grout	D-7	240	.067	S.F.	4.73	2.35		7.08	8.65
5100	Floors, unglazed, 2 S.F. sheets,									

09 30 Tiling

09 30 13 – Ceramic Tiling

09 30 13.10 Ceramic Tile		Crew	Daily Output	Labor-Hours	Unit	2010 Bare Costs Material	Labor	Equipment	Total	Total Incl O&P
5110	Urethane adhesive	D-7	180	.089	S.F.	4.71	3.14		7.85	9.80
5400	Walls, interior, thin set, 4-1/4" x 4-1/4" tile CN		190	.084		2.34	2.97		5.31	6.90
5500	6" x 4-1/4" tile		190	.084		2.71	2.97		5.68	7.30
5700	8-1/2" x 4-1/4" tile		190	.084		3.78	2.97		6.75	8.50
5800	6" x 6" tile		200	.080		3.04	2.82		5.86	7.45
5810	8" x 8" tile		225	.071		4.06	2.51		6.57	8.15
5820	12" x 12" tile		300	.053		3.47	1.88		5.35	6.55
5830	16" x 16" tile		500	.032		3.75	1.13		4.88	5.80
6000	Decorated wall tile, 4-1/4" x 4-1/4", minimum		270	.059		3.51	2.09		5.60	6.90
6100	Maximum		180	.089		41.50	3.14		44.64	50.50
6300	Exterior walls, frostproof, mud set, 4-1/4" x 4-1/4"		102	.157		7.30	5.55		12.85	16.10
6400	1-3/8" x 1-3/8"		93	.172		4.93	6.05		10.98	14.25
6600	Crystalline glazed, 4-1/4" x 4-1/4", mud set, plain		100	.160		4.28	5.65		9.93	12.95
6700	4-1/4" x 4-1/4", scored tile		100	.160		5.25	5.65		10.90	14
6900	6" x 6" plain		93	.172		5.30	6.05		11.35	14.65
7000	For epoxy grout, 1/16" joints, 4-1/4" tile, add		800	.020		.39	.71		1.10	1.46
7200	For tile set in dry mortar, add		1735	.009			.33		.33	.48
7300	For tile set in Portland cement mortar, add		290	.055		.27	1.95		2.22	3.14
9000	Regrout tile 4-1/2 x 4-1/2, or larger, wall	1 Tilf	100	.080		.14	3.15		3.29	4.75
9220	Floor	"	125	.064		.15	2.52		2.67	3.85
9300	Ceramic tiles, recycled glass, standard colors, 2" x 2" thru 6" x 6" G	D-7	190	.084		17	2.97		19.97	23
9310	6" x 6" G		200	.080		17	2.82		19.82	23
9320	8" x 8" G		225	.071		18.45	2.51		20.96	24
9330	12"x12" G		300	.053		18.45	1.88		20.33	23.50
9340	Earthtones, 2" x 2" to 4" x 8" G		190	.084		21.50	2.97		24.47	28
9350	6" x 6" G		200	.080		21.50	2.82		24.32	27.50
9360	8" x 8" G		225	.071		23	2.51		25.51	28.50
9370	12" x 12" G		300	.053		23	1.88		24.88	28
9380	Deep colors, 2" x 2" to 4" x 8" G		190	.084		31	2.97		33.97	38.50
9390	6" x 6" G		200	.080		31	2.82		33.82	38
9400	8" x 8" G		225	.071		32	2.51		34.51	39
9410	12" x 12" G		300	.053		32	1.88		33.88	38.50

09 30 16 – Quarry Tiling

09 30 16.10 Quarry Tile		Crew	Daily Output	Labor-Hours	Unit	Material	Labor	Equipment	Total	Total Incl O&P
0010	**QUARRY TILE**									
0100	Base, cove or sanitary, mud set, to 5" high, 1/2" thick	D-7	110	.145	L.F.	5.30	5.15		10.45	13.35
0300	Bullnose trim, red, mud set, 6" x 6" x 1/2" thick		120	.133		4.36	4.70		9.06	11.65
0400	4" x 4" x 1/2" thick		110	.145		4.78	5.15		9.93	12.75
0600	4" x 8" x 1/2" thick, using 8" as edge		130	.123		4.24	4.34		8.58	11
0700	Floors, mud set, 1,000 S.F. lots, red, 4" x 4" x 1/2" thick		120	.133	S.F.	7.50	4.70		12.20	15.10
0900	6" x 6" x 1/2" thick		140	.114		6.85	4.03		10.88	13.40
1000	4" x 8" x 1/2" thick		130	.123		7.50	4.34		11.84	14.60
1300	For waxed coating, add					.67			.67	.74
1500	For non-standard colors, add					.40			.40	.44
1600	For abrasive surface, add					.47			.47	.52
1800	Brown tile, imported, 6" x 6" x 3/4"	D-7	120	.133		8.40	4.70		13.10	16.10
1900	8" x 8" x 1"		110	.145		9.05	5.15		14.20	17.45
2100	For thin set mortar application, deduct		700	.023			.81		.81	1.18
2200	For epoxy grout & mortar, 6" x 6" x 1/2", add		350	.046		1.91	1.61		3.52	4.46
2700	Stair tread, 6" x 6" x 3/4", plain		50	.320		5.05	11.30		16.35	22
2800	Abrasive		47	.340		5.30	12		17.30	23.50
3000	Wainscot, 6" x 6" x 1/2", thin set, red		105	.152		4.09	5.40		9.49	12.35

09 30 Tiling

09 30 16 – Quarry Tiling

09 30 16.10 Quarry Tile		Crew	Daily Output	Labor-Hours	Unit	2010 Bare Costs Material	Labor	Equipment	Total	Total Incl O&P
3100	Non-standard colors	D-7	105	.152	S.F.	4.54	5.40		9.94	12.85
3300	Window sill, 6" wide, 3/4" thick		90	.178	L.F.	5.10	6.25		11.35	14.75
3400	Corners		80	.200	Ea.	5.70	7.05		12.75	16.55

09 30 23 – Glass Mosaic Tiling

09 30 23.10 Glass Mosaics		Crew	Daily Output	Labor-Hours	Unit	Material	Labor	Equipment	Total	Total Incl O&P
0010	**GLASS MOSAICS** 3/4" tile on 12" sheets, standard grout									
0300	Color group 1 & 2	D-7	73	.219	S.F.	17.35	7.75		25.10	30.50
0350	Color group 3		73	.219		18.25	7.75		26	31.50
0400	Color group 4		73	.219		25	7.75		32.75	39
0450	Color group 5		73	.219		28	7.75		35.75	42.50
0500	Color group 6		73	.219		34	7.75		41.75	48.50
0600	Color group 7		73	.219		39	7.75		46.75	54.50
0700	Color group 8, golds, silvers & specialties		64	.250		56.50	8.80		65.30	75.50

09 30 29 – Metal Tiling

09 30 29.10 Metal Tile		Crew	Daily Output	Labor-Hours	Unit	Material	Labor	Equipment	Total	Total Incl O&P
0010	**METAL TILE** 4' x 4' sheet, 24 ga., tile pattern, nailed									
0200	Stainless steel	2 Carp	512	.031	S.F.	24.50	1.30		25.80	29
0400	Aluminized steel	"	512	.031	"	13.15	1.30		14.45	16.45

09 51 Acoustical Ceilings

09 51 23 – Acoustical Tile Ceilings

09 51 23.10 Suspended Acoustic Ceiling Tiles		Crew	Daily Output	Labor-Hours	Unit	Material	Labor	Equipment	Total	Total Incl O&P
0010	**SUSPENDED ACOUSTIC CEILING TILES**, not including									
0100	suspension system									
0300	Fiberglass boards, film faced, 2' x 2' or 2' x 4', 5/8" thick	1 Carp	625	.013	S.F.	.70	.53		1.23	1.59
0400	3/4" thick		600	.013		1.61	.55		2.16	2.62
0500	3" thick, thermal, R11		450	.018		1.53	.74		2.27	2.82
0600	Glass cloth faced fiberglass, 3/4" thick		500	.016		2.87	.66		3.53	4.18
0700	1" thick		485	.016		2.18	.69		2.87	3.46
0820	1-1/2" thick, nubby face		475	.017		2.66	.70		3.36	4.01
1110	Mineral fiber tile, lay-in, 2' x 2' or 2' x 4', 5/8" thick, fine texture		625	.013		.55	.53		1.08	1.43
1115	Rough textured		625	.013		1.23	.53		1.76	2.17
1125	3/4" thick, fine textured		600	.013		1.50	.55		2.05	2.50
1130	Rough textured		600	.013		1.79	.55		2.34	2.82
1135	Fissured		600	.013		2.45	.55		3	3.55
1150	Tegular, 5/8" thick, fine textured		470	.017		1.45	.71		2.16	2.69
1155	Rough textured		470	.017		1.90	.71		2.61	3.18
1165	3/4" thick, fine textured		450	.018		2.08	.74		2.82	3.43
1170	Rough textured		450	.018		2.35	.74		3.09	3.73
1175	Fissured		450	.018		3.63	.74		4.37	5.15
1180	For aluminum face, add					6.05			6.05	6.65
1185	For plastic film face, add					.94			.94	1.03
1190	For fire rating, add					.45			.45	.50
1300	Mirror tile, acrylic, 2' x 2'	1 Carp	500	.016		8.70	.66		9.36	10.55
1900	Eggcrate, acrylic, 1/2" x 1/2" x 1/2" cubes		500	.016		1.82	.66		2.48	3.02
2100	Polystyrene eggcrate, 3/8" x 3/8" x 1/2" cubes		510	.016		1.53	.65		2.18	2.68
2200	1/2" x 1/2" x 1/2" cubes		500	.016		1.75	.66		2.41	2.95
2400	Luminous panels, prismatic, acrylic		400	.020		2.22	.83		3.05	3.72
2500	Polystyrene		400	.020		1.14	.83		1.97	2.53
2700	Flat white acrylic		400	.020		3.86	.83		4.69	5.55

09 51 Acoustical Ceilings

09 51 23 – Acoustical Tile Ceilings

09 51 23.10 Suspended Acoustic Ceiling Tiles		Crew	Daily Output	Labor-Hours	Unit	2010 Bare Costs Material	Labor	Equipment	Total	Total Incl O&P
2800	Polystyrene	1 Carp	400	.020	S.F.	2.65	.83		3.48	4.20
3000	Drop pan, white, acrylic		400	.020		5.65	.83		6.48	7.55
3100	Polystyrene		400	.020		4.73	.83		5.56	6.50
3600	Perforated aluminum sheets, .024" thick, corrugated, painted		490	.016		2.25	.68		2.93	3.53
3700	Plain		500	.016		3.76	.66		4.42	5.15
3720	Mineral fiber, 24" x 24" or 48", reveal edge, painted, 5/8" thick		600	.013		1.38	.55		1.93	2.37
3740	3/4" thick		575	.014		2.24	.58		2.82	3.35
5020	66 – 78% recycled content, 3/4" thick G		600	.013		1.82	.55		2.37	2.85
5040	Mylar, 42% recycled content, 3/4" thick G		600	.013		4.28	.55		4.83	5.55

09 51 23.30 Suspended Ceilings, Complete

		Crew	Daily Output	Labor-Hours	Unit	Material	Labor	Equipment	Total	Total Incl O&P
0010	**SUSPENDED CEILINGS, COMPLETE** Including standard									
0100	suspension system but not incl. 1-1/2" carrier channels									
0600	Fiberglass ceiling board, 2' x 4' x 5/8", plain faced	1 Carp	500	.016	S.F.	1.38	.66		2.04	2.54
0700	Offices, 2' x 4' x 3/4"		380	.021		2.29	.87		3.16	3.87
0800	Mineral fiber, on 15/16" T bar susp. 2' x 2' x 3/4" lay-in board		345	.023		2.35	.96		3.31	4.07
0810	2' x 4' x 5/8" tile		380	.021		1.23	.87		2.10	2.70
0820	Tegular, 2' x 2' x 5/8" tile on 9/16" grid		250	.032		2.46	1.33		3.79	4.75
0830	2' x 4' x 3/4" tile		275	.029		2.92	1.21		4.13	5.05
0900	Luminous panels, prismatic, acrylic		255	.031		2.90	1.30		4.20	5.20
1200	Metal pan with acoustic pad, steel		75	.107		4.21	4.43		8.64	11.50
1300	Painted aluminum		75	.107		2.93	4.43		7.36	10.05
1500	Aluminum, degreased finish		75	.107		4.85	4.43		9.28	12.20
1600	Stainless steel		75	.107		9.10	4.43		13.53	16.85
1800	Tile, Z bar suspension, 5/8" mineral fiber tile		150	.053		2.16	2.22		4.38	5.80
1900	3/4" mineral fiber tile		150	.053		2.31	2.22		4.53	5.95
2400	For strip lighting, see Div. 26 51 13.50									
2500	For rooms under 500 S.F., add				S.F.		25%			

09 51 53 – Direct-Applied Acoustical Ceilings

09 51 53.10 Ceiling Tile

		Crew	Daily Output	Labor-Hours	Unit	Material	Labor	Equipment	Total	Total Incl O&P
0010	**CEILING TILE**, Stapled or cemented									
0100	12" x 12" or 12" x 24", not including furring									
0600	Mineral fiber, vinyl coated, 5/8" thick *CN*	1 Carp	300	.027	S.F.	1.79	1.11		2.90	3.68
0700	3/4" thick		300	.027		1.91	1.11		3.02	3.81
0900	Fire rated, 3/4" thick, plain faced		300	.027		1.40	1.11		2.51	3.25
1000	Plastic coated face		300	.027		1.18	1.11		2.29	3.01
1200	Aluminum faced, 5/8" thick, plain		300	.027		1.51	1.11		2.62	3.37
3700	Wall application of above, add		1000	.008			.33		.33	.51
3900	For ceiling primer, add					.13			.13	.14
4000	For ceiling cement, add					.37			.37	.41

09 53 Acoustical Ceiling Suspension Assemblies

09 53 23 – Metal Acoustical Ceiling Suspension Assemblies

09 53 23.30 Ceiling Suspension Systems			Crew	Daily Output	Labor-Hours	Unit	2010 Bare Costs Material	Labor	Equipment	Total	Total Incl O&P
0010	**CEILING SUSPENSION SYSTEMS** for boards and tile										
0050	Class A suspension system, 15/16" T bar, 2' x 4' grid	CN	1 Carp	800	.010	S.F.	.68	.42		1.10	1.39
0300	2' x 2' grid			650	.012		.85	.51		1.36	1.72
0310	25% recycled steel, 2' x 4' grid	G		800	.010		.79	.42		1.21	1.51
0320	2' x 2' grid	G		650	.012		.99	.51		1.50	1.88
0350	For 9/16" grid, add						.16			.16	.18
0360	For fire rated grid, add						.09			.09	.10
0370	For colored grid, add						.21			.21	.23
0400	Concealed Z bar suspension system, 12" module		1 Carp	520	.015		.69	.64		1.33	1.75
0600	1-1/2" carrier channels, 4' O.C., add		"	470	.017		.11	.71		.82	1.21
0700	Carrier channels for ceilings with										
0900	recessed lighting fixtures, add		1 Carp	460	.017	S.F.	.19	.72		.91	1.32
1040	Hanging wire, 12 ga., 4' long			65	.123	C.S.F.	8.10	5.10		13.20	16.85
1080	8' long			65	.123	"	16.25	5.10		21.35	26
1200	Seismic bracing, 2-1/2" compression post with four bracing wires			50	.160	Ea.	2.74	6.65		9.39	13.25

09 54 Specialty Ceilings

09 54 33 – Decorative Panel Ceilings

09 54 33.20 Metal Panel Ceilings		Crew	Daily Output	Labor-Hours	Unit	Material	Labor	Equipment	Total	Total Incl O&P
0010	**METAL PANEL CEILINGS**									
0020	Lay-in or screwed to furring, not including grid									
0100	Tin ceilings, 2' x 2' or 2' x 4', bare steel finish	2 Carp	300	.053	S.F.	2.37	2.22		4.59	6.05
0120	Painted white finish		300	.053	"	3.59	2.22		5.81	7.35
0140	Copper, chrome or brass finish		300	.053	L.F.	5.95	2.22		8.17	9.95
0200	Cornice molding, 2-1/2" to 3-1/2" wide, 4' long, bare steel finish		200	.080	S.F.	2.25	3.32		5.57	7.55
0220	Painted white finish		200	.080		3.27	3.32		6.59	8.70
0240	Copper, chrome or brass finish		200	.080		4.51	3.32		7.83	10.05
0320	5" to 6-1/2" wide, 4' long, bare steel finish		150	.107		3.12	4.43		7.55	10.30
0340	Painted white finish		150	.107		4.51	4.43		8.94	11.80
0360	Copper, chrome or brass finish		150	.107		7	4.43		11.43	14.55
0420	Flat molding, 3-1/2" to 5" wide, 4' long, bare steel finish		250	.064		3.44	2.66		6.10	7.90
0440	Painted white finish		250	.064		4.86	2.66		7.52	9.45
0460	Copper, chrome or brass finish		250	.064		10.05	2.66		12.71	15.15

09 62 Specialty Flooring

09 62 19 – Laminate Flooring

09 62 19.10 Floating Floor		Crew	Daily Output	Labor-Hours	Unit	Material	Labor	Equipment	Total	Total Incl O&P
0010	**FLOATING FLOOR**									
8300	Floating floor, laminate, wood pattern strip, complete	1 Clab	133	.060	S.F.	4.25	1.99		6.24	7.75
8310	Components, T & G wood composite strips					3.66			3.66	4.03
8320	Film					.15			.15	.16
8330	Foam					.30			.30	.33
8340	Adhesive					.17			.17	.18
8350	Installation kit					.17			.17	.19
8360	Trim, 2" wide x 3' long				L.F.	2.79			2.79	3.07
8370	Reducer moulding				"	5.85			5.85	6.40

09 62 Specialty Flooring

09 62 23 – Bamboo Flooring

09 62 23.10 Flooring, Bamboo			Crew	Daily Output	Labor-Hours	Unit	2010 Bare Costs Material	Labor	Equipment	Total	Total Incl O&P
0010	**FLOORING, BAMBOO**										
8600	Flooring, wood, bamboo strips, unfinished, 5/8" x 4" x 3'	G	1 Carp	255	.031	S.F.	4.44	1.30		5.74	6.90
8610	5/8" x 4" x 4'	G		275	.029		4.61	1.21		5.82	6.90
8620	5/8" x 4" x 6'	G		295	.027		5.05	1.13		6.18	7.30
8630	Finished, 5/8" x 4" x 3'	G		255	.031		4.89	1.30		6.19	7.40
8640	5/8" x 4" x 4'	G		275	.029		5.10	1.21		6.31	7.50
8650	5/8" x 4" x 6'	G		295	.027		4.96	1.13		6.09	7.20
8660	Stair treads, unfinished, 1-1/16" x 11-1/2" x 4'	G		18	.444	Ea.	42.50	18.45		60.95	75.50
8670	Finished, 1-1/16" x 11-1/2" x 4'	G		18	.444		75.50	18.45		93.95	112
8680	Stair risers, unfinished, 5/8" x 7-1/2" x 4'	G		18	.444		15.75	18.45		34.20	46
8690	Finished, 5/8" x 7-1/2" x 4'	G		18	.444		30	18.45		48.45	61.50
8700	Stair nosing, unfinished, 6' long	G		16	.500		35	21		56	70.50
8710	Finished, 6' long	G		16	.500		40	21		61	76

09 63 Masonry Flooring

09 63 13 – Brick Flooring

09 63 13.10 Miscellaneous Brick Flooring			Crew	Daily Output	Labor-Hours	Unit	Material	Labor	Equipment	Total	Total Incl O&P
0010	**MISCELLANEOUS BRICK FLOORING**										
0020	Acid-proof shales, red, 8" x 3-3/4" x 1-1/4" thick		D-7	.43	37.209	M	860	1,325		2,185	2,875
0050	2-1/4" thick		D-1	.40	40		1,000	1,500		2,500	3,375
0200	Acid-proof clay brick, 8" x 3-3/4" x 2-1/4" thick	G		.40	40		895	1,500		2,395	3,250
0250	9" x 4-1/2" x 3"	G		95	.168	S.F.	4.12	6.35		10.47	14.10
0260	Cast ceramic, pressed, 4" x 8" x 1/2", unglazed		D-7	100	.160		6.25	5.65		11.90	15.15
0270	Glazed			100	.160		8.35	5.65		14	17.40
0280	Hand molded flooring, 4" x 8" x 3/4", unglazed			95	.168		8.25	5.95		14.20	17.80
0290	Glazed			95	.168		10.35	5.95		16.30	20
0300	8" hexagonal, 3/4" thick, unglazed			85	.188		9.05	6.65		15.70	19.65
0310	Glazed			85	.188		16.35	6.65		23	27.50
0400	Heavy duty industrial, cement mortar bed, 2" thick, not incl. brick		D-1	80	.200		.83	7.55		8.38	12.25
0450	Acid-proof joints, 1/4" wide		"	65	.246		1.44	9.30		10.74	15.60
0500	Pavers, 8" x 4", 1" to 1-1/4" thick, red		D-7	95	.168		3.64	5.95		9.59	12.70
0510	Ironspot		"	95	.168		5.15	5.95		11.10	14.35
0540	1-3/8" to 1-3/4" thick, red		D-1	95	.168		3.51	6.35		9.86	13.40
0560	Ironspot			95	.168		5.10	6.35		11.45	15.15
0580	2-1/4" thick, red			90	.178		3.57	6.70		10.27	14.05
0590	Ironspot			90	.178		5.55	6.70		12.25	16.20
0700	Paver, adobe brick, 6" x 12", 1/2" joint	G		42	.381		.90	14.35		15.25	22.50
0710	Mexican red, 12" x 12"	G	1 Tilf	48	.167		1.61	6.55		8.16	11.35
0720	Saltillo, 12" x 12"	G	"	48	.167		1.13	6.55		7.68	10.85
0800	For sidewalks and patios with pavers, see Div. 32 14 16.10										
0870	For epoxy joints, add		D-1	600	.027	S.F.	2.72	1.01		3.73	4.51
0880	For Furan underlayment, add		"	600	.027		2.25	1.01		3.26	4
0890	For waxed surface, steam cleaned, add		A-1H	1000	.008		.20	.26	.06	.52	.70

09 63 40 – Stone Flooring

09 63 40.10 Marble			Crew	Daily Output	Labor-Hours	Unit	Material	Labor	Equipment	Total	Total Incl O&P
0010	**MARBLE**										
0020	Thin gauge tile, 12" x 6", 3/8", white Carara		D-7	60	.267	S.F.	11.15	9.40		20.55	26
0100	Travertine			60	.267		12.20	9.40		21.60	27
0200	12" x 12" x 3/8", thin set, floors			60	.267		6.05	9.40		15.45	20.50
0300	On walls			52	.308		9.35	10.85		20.20	26

09 63 Masonry Flooring

09 63 40 – Stone Flooring

09 63 40.10 Marble		Crew	Daily Output	Labor-Hours	Unit	2010 Bare Costs Material	Labor	Equipment	Total	Total Incl O&P
1000	Marble threshold, 4" wide x 36" long x 5/8" thick, white	D-7	60	.267	Ea.	6.05	9.40		15.45	20.50

09 63 40.20 Slate Tile		Crew	Daily Output	Labor-Hours	Unit	Material	Labor	Equipment	Total	Total Incl O&P
0010	**SLATE TILE**									
0020	Vermont, 6" x 6" x 1/4" thick, thin set	D-7	180	.089	S.F.	6.80	3.14		9.94	12.05
0200	See also Div. 32 14 40.10									

09 64 Wood Flooring

09 64 16 – Wood Block Flooring

09 64 16.10 End Grain Block Flooring		Crew	Daily Output	Labor-Hours	Unit	Material	Labor	Equipment	Total	Total Incl O&P
0010	**END GRAIN BLOCK FLOORING**									
0020	End grain flooring, coated, 2" thick	1 Carp	295	.027	S.F.	3.29	1.13		4.42	5.35
0400	Natural finish, 1" thick, fir		125	.064		3.40	2.66		6.06	7.85
0600	1-1/2" thick, pine		125	.064		3.34	2.66		6	7.75
0700	2" thick, pine	↓	125	.064	↓	4.64	2.66		7.30	9.20

09 64 19 – Wood Composition Flooring

09 64 19.10 Wood Composition		Crew	Daily Output	Labor-Hours	Unit	Material	Labor	Equipment	Total	Total Incl O&P
0010	**WOOD COMPOSITION** Gym floors									
0100	2-1/4" x 6-7/8" x 3/8", on 2" grout setting bed	D-7	150	.107	S.F.	5.85	3.76		9.61	11.95
0200	Thin set, on concrete	"	250	.064		5.35	2.26		7.61	9.20
0300	Sanding and finishing, add	1 Carp	200	.040	↓	.79	1.66		2.45	3.43

09 64 23 – Wood Parquet Flooring

09 64 23.10 Wood Parquet		Crew	Daily Output	Labor-Hours	Unit	Material	Labor	Equipment	Total	Total Incl O&P
0010	**WOOD PARQUET** flooring									
5200	Parquetry, standard, 5/16" thick, not incl. finish, oak, minimum	1 Carp	160	.050	S.F.	4.41	2.08		6.49	8.05
5300	Maximum		100	.080		4.95	3.32		8.27	10.55
5500	Teak, minimum		160	.050		4.79	2.08		6.87	8.45
5600	Maximum		100	.080		8.35	3.32		11.67	14.30
5650	13/16" thick, select grade oak, minimum		160	.050		9.20	2.08		11.28	13.35
5700	Maximum		100	.080		14	3.32		17.32	20.50
5800	Custom parquetry, including finish, minimum		100	.080		15.45	3.32		18.77	22
5900	Maximum		50	.160		20.50	6.65		27.15	33
6700	Parquetry, prefinished white oak, 5/16" thick, minimum		160	.050		3.06	2.08		5.14	6.55
6800	Maximum		100	.080		6.55	3.32		9.87	12.30
7000	Walnut or teak, parquetry, minimum		160	.050		5.35	2.08		7.43	9.10
7100	Maximum	↓	100	.080	↓	9.35	3.32		12.67	15.40
7200	Acrylic wood parquet blocks, 12" x 12" x 5/16",									
7210	Irradiated, set in epoxy	1 Carp	160	.050	S.F.	7.80	2.08		9.88	11.80

09 64 29 – Wood Strip and Plank Flooring

09 64 29.10 Wood		Crew	Daily Output	Labor-Hours	Unit	Material	Labor	Equipment	Total	Total Incl O&P
0010	**WOOD**									
0020	Fir, vertical grain, 1" x 4", not incl. finish, grade B & better	1 Carp	255	.031	S.F.	2.74	1.30		4.04	5
0100	C grade & better		255	.031		2.58	1.30		3.88	4.85
4000	Maple, strip, 25/32" x 2-1/4", not incl. finish, select		170	.047		5.25	1.96		7.21	8.80
4100	#2 & better		170	.047		3.32	1.96		5.28	6.65
4300	33/32" x 3-1/4", not incl. finish, #1 grade		170	.047		4.19	1.96		6.15	7.60
4400	#2 & better	↓	170	.047	↓	3.73	1.96		5.69	7.10
4600	Oak, white or red, 25/32" x 2-1/4", not incl. finish									
4700	#1 common **CN**	1 Carp	170	.047	S.F.	2.40	1.96		4.36	5.65
4900	Select quartered, 2-1/4" wide	↓	170	.047	↓	2.94	1.96		4.90	6.25

09 64 Wood Flooring

09 64 29 – Wood Strip and Plank Flooring

09 64 29.10 Wood		Crew	Daily Output	Labor-Hours	Unit	Material	2010 Bare Costs Labor	Equipment	Total	Total Incl O&P
5000	Clear	1 Carp	170	.047	S.F.	3.83	1.96		5.79	7.20
6100	Prefinished, white oak, prime grade, 2-1/4" wide		170	.047		4.79	1.96		6.75	8.25
6200	3-1/4" wide		185	.043		6.15	1.80		7.95	9.50
6400	Ranch plank		145	.055		6.75	2.29		9.04	10.95
6500	Hardwood blocks, 9" x 9", 25/32" thick		160	.050		5.75	2.08		7.83	9.55
7400	Yellow pine, 3/4" x 3-1/8", T & G, C & better, not incl. finish		200	.040		1.99	1.66		3.65	4.75
7500	Refinish wood floor, sand, 2 cts poly, wax, soft wood, min.	1 Clab	400	.020		.78	.66		1.44	1.88
7600	Hard wood, max		130	.062		1.16	2.04		3.20	4.42
7800	Sanding and finishing, 2 coats polyurethane		295	.027		.78	.90		1.68	2.24
7900	Subfloor and underlayment, see Div. 06 16									
8015	Transition molding, 2 1/4" wide, 5' long	1 Carp	19.20	.417	Ea.	15	17.30		32.30	43

09 64 66 – Wood Athletic Flooring

09 64 66.10 Gymnasium Flooring		Crew	Daily Output	Labor-Hours	Unit	Material	Labor	Equipment	Total	Total Incl O&P
0010	**GYMNASIUM FLOORING**									
0600	Gym floor, in mastic, over 2 ply felt, #2 & better									
0700	25/32" thick maple	1 Carp	100	.080	S.F.	4.03	3.32		7.35	9.55
0900	33/32" thick maple		98	.082		4.65	3.39		8.04	10.35
1000	For 1/2" corkboard underlayment, add		750	.011		.90	.44		1.34	1.67
1300	For #1 grade maple, add					.47			.47	.52
1600	Maple flooring, over sleepers, #2 & better									
1700	25/32" thick	1 Carp	85	.094	S.F.	4.32	3.91		8.23	10.80
1900	33/32" thick	"	83	.096		5.05	4.01		9.06	11.70
2000	For #1 grade, add					.50			.50	.55
2200	For 3/4" subfloor, add	1 Carp	350	.023		1.12	.95		2.07	2.69
2300	With two 1/2" subfloors, 25/32" thick	"	69	.116		5.40	4.82		10.22	13.40
2500	Maple, incl. finish, #2 & btr., 25/32" thick, on rubber									
2600	Sleepers, with two 1/2" subfloors	1 Carp	76	.105	S.F.	5.80	4.37		10.17	13.15
2800	With steel spline, double connection to channels	"	73	.110		6.20	4.55		10.75	13.80
2900	For 33/32" maple, add					.67			.67	.74
3100	For #1 grade maple, add					.50			.50	.55
3500	For termite proofing all of the above, add					.27			.27	.30
3700	Portable hardwood, prefinished panels	1 Carp	83	.096		7.75	4.01		11.76	14.65
3720	Insulated with polystyrene, 1" thick, add		165	.048		.67	2.01		2.68	3.85
3750	Running tracks, Sitka spruce surface, 25/32" x 2-1/4"		62	.129		14.15	5.35		19.50	24
3770	3/4" plywood surface, finished		100	.080		3.36	3.32		6.68	8.80

09 65 Resilient Flooring

09 65 10 – Resilient Tile Underlayment

09 65 10.10 Latex Underlayment		Crew	Daily Output	Labor-Hours	Unit	Material	Labor	Equipment	Total	Total Incl O&P
0010	**LATEX UNDERLAYMENT**									
3600	Latex underlayment, 1/8" thk., cementitious for resilient flooring	1 Tilf	160	.050	S.F.	.79	1.97		2.76	3.75
4000	Liquid, fortified				Gal.	38.50			38.50	42

09 65 13 – Resilient Base and Accessories

09 65 13.13 Resilient Base		Crew	Daily Output	Labor-Hours	Unit	Material	Labor	Equipment	Total	Total Incl O&P
0010	**RESILIENT BASE**									
0800	Base, cove, rubber or vinyl									
1100	Standard colors, 0.080" thick, 2-1/2" high	1 Tilf	315	.025	L.F.	.80	1		1.80	2.34
1150	4" high		315	.025		.94	1		1.94	2.49
1200	6" high		315	.025		1.30	1		2.30	2.89
1450	1/8" thick, 2-1/2" high		315	.025		.80	1		1.80	2.34

09 65 Resilient Flooring

09 65 13 – Resilient Base and Accessories

09 65 13.13 Resilient Base		Crew	Daily Output	Labor-Hours	Unit	Material	2010 Bare Costs Labor	Equipment	Total	Total Incl O&P
1500	4" high CN	1 Tilf	315	.025	L.F.	.73	1		1.73	2.26
1600	Corners, 2-1/2" high		315	.025	Ea.	1.50	1		2.50	3.11
1630	4" high		315	.025		2.85	1		3.85	4.60
1660	6" high		315	.025		2.95	1		3.95	4.71

09 65 13.23 Resilient Stair Treads and Risers		Crew	Daily Output	Labor-Hours	Unit	Material	Labor	Equipment	Total	Total Incl O&P
0010	RESILIENT STAIR TREADS AND RISERS									
0300	Rubber, molded tread, 12" wide, 5/16" thick, black	1 Tilf	115	.070	L.F.	12	2.74		14.74	17.20
0400	Colors		115	.070		12.60	2.74		15.34	17.85
0600	1/4" thick, black		115	.070		11.85	2.74		14.59	17.05
0700	Colors		115	.070		12.60	2.74		15.34	17.85
0900	Grip strip safety tread, colors, 5/16" thick		115	.070		16.20	2.74		18.94	22
1000	3/16" thick		120	.067		11.30	2.62		13.92	16.30
1200	Landings, smooth sheet rubber, 1/8" thick		120	.067	S.F.	5.95	2.62		8.57	10.40
1300	3/16" thick		120	.067	"	7.50	2.62		10.12	12.10
1500	Nosings, 3" wide, 3/16" thick, black		140	.057	L.F.	3.20	2.25		5.45	6.80
1600	Colors		140	.057		3.15	2.25		5.40	6.75
1800	Risers, 7" high, 1/8" thick, flat		250	.032		6.20	1.26		7.46	8.70
1900	Coved		250	.032		5.10	1.26		6.36	7.45
2100	Vinyl, molded tread, 12" wide, colors, 1/8" thick		115	.070		5.80	2.74		8.54	10.40
2200	1/4" thick		115	.070		7.70	2.74		10.44	12.45
2300	Landing material, 1/8" thick		200	.040	S.F.	5.80	1.57		7.37	8.70
2400	Riser, 7" high, 1/8" thick, coved		175	.046	L.F.	3.35	1.80		5.15	6.30
2500	Tread and riser combined, 1/8" thick		80	.100	"	9	3.94		12.94	15.65

09 65 16 – Resilient Sheet Flooring

09 65 16.10 Rubber and Vinyl Sheet Flooring		Crew	Daily Output	Labor-Hours	Unit	Material	Labor	Equipment	Total	Total Incl O&P
0010	RUBBER AND VINYL SHEET FLOORING									
5500	Linoleum, sheet goods G	1 Tilf	360	.022	S.F.	4.55	.87		5.42	6.30
5900	Rubber, sheet goods, 36" wide, 1/8" thick		120	.067		6.45	2.62		9.07	10.95
5950	3/16" thick		100	.080		8.75	3.15		11.90	14.25
6000	1/4" thick		90	.089		10.30	3.50		13.80	16.45
8000	Vinyl sheet goods, backed, .065" thick, minimum		250	.032		3.50	1.26		4.76	5.70
8050	Maximum		200	.040		3.61	1.57		5.18	6.25
8100	.080" thick, minimum		230	.035		3.60	1.37		4.97	5.95
8150	Maximum		200	.040		5.10	1.57		6.67	7.90
8200	.125" thick, minimum		230	.035		4.10	1.37		5.47	6.50
8250	Maximum		200	.040		6.10	1.57		7.67	9
8700	Adhesive cement, 1 gallon per 200 to 300 S.F.				Gal.	25			25	27.50
8800	Asphalt primer, 1 gallon per 300 S.F.					12			12	13.20
8900	Emulsion, 1 gallon per 140 S.F.					15.50			15.50	17.05

09 65 19 – Resilient Tile Flooring

09 65 19.10 Miscellaneous Resilient Tile Flooring		Crew	Daily Output	Labor-Hours	Unit	Material	Labor	Equipment	Total	Total Incl O&P
0010	MISCELLANEOUS RESILIENT TILE FLOORING									
2200	Cork tile, standard finish, 1/8" thick G	1 Tilf	315	.025	S.F.	6	1		7	8.05
2250	3/16" thick G		315	.025		6	1		7	8.05
2300	5/16" thick G		315	.025		7.45	1		8.45	9.65
2350	1/2" thick G		315	.025		8.80	1		9.80	11.10
2500	Urethane finish, 1/8" thick G		315	.025		7.10	1		8.10	9.30
2550	3/16" thick G		315	.025		7.45	1		8.45	9.65
2600	5/16" thick G		315	.025		9.55	1		10.55	11.95
2650	1/2" thick G		315	.025		12.85	1		13.85	15.60
6050	Rubber tile, marbleized colors, 12" x 12", 1/8" thick		400	.020		6.45	.79		7.24	8.20

09 65 Resilient Flooring

09 65 19 – Resilient Tile Flooring

09 65 19.10 Miscellaneous Resilient Tile Flooring		Crew	Daily Output	Labor-Hours	Unit	2010 Bare Costs Material	Labor	Equipment	Total	Total Incl O&P
6100	3/16" thick	1 Tilf	400	.020	S.F.	9.50	.79		10.29	11.60
6300	Special tile, plain colors, 1/8" thick		400	.020		9.50	.79		10.29	11.60
6350	3/16" thick		400	.020		9.65	.79		10.44	11.75
6410	Raised, radial or square, minimum		400	.020		8.75	.79		9.54	10.80
6430	Maximum		400	.020		11.85	.79		12.64	14.20
6450	For golf course, skating rink, etc., 1/4" thick		275	.029		11.85	1.14		12.99	14.70
6700	Synthetic turf, 3/8" thick		90	.089		5.80	3.50		9.30	11.50
6750	Interlocking 2' x 2' squares, 1/2" thick, not									
6810	cemented, for playgrounds, minimum	1 Tilf	210	.038	S.F.	4.65	1.50		6.15	7.30
6850	Maximum		190	.042		11.50	1.66		13.16	15.05
7000	Vinyl composition tile, 12" x 12", 1/16" thick		500	.016		.92	.63		1.55	1.93
7050	Embossed		500	.016		2.17	.63		2.80	3.31
7100	Marbleized		500	.016		2.17	.63		2.80	3.31
7150	Solid		500	.016		2.79	.63		3.42	3.99
7200	3/32" thick, embossed		500	.016		1.38	.63		2.01	2.44
7250	Marbleized		500	.016		2.49	.63		3.12	3.66
7300	Solid		500	.016		2.31	.63		2.94	3.46
7350	1/8" thick, marbleized **CN**		500	.016		1.64	.63		2.27	2.72
7400	Solid		500	.016		1.80	.63		2.43	2.90
7450	Conductive		500	.016		5.45	.63		6.08	6.85
7500	Vinyl tile, 12" x 12", .050" thick, minimum		500	.016		3	.63		3.63	4.22
7550	Maximum		500	.016		6	.63		6.63	7.50
7600	1/8" thick, minimum		500	.016		4.35	.63		4.98	5.70
7650	Solid colors		500	.016		4.95	.63		5.58	6.35
7700	Marbleized or Travertine pattern		500	.016		5.65	.63		6.28	7.10
7750	Florentine pattern		500	.016		6	.63		6.63	7.50
7800	Maximum		500	.016		12.70	.63		13.33	14.85

09 65 33 – Conductive Resilient Flooring

09 65 33.10 Conductive Rubber and Vinyl Flooring		Crew	Daily Output	Labor-Hours	Unit	Material	Labor	Equipment	Total	Total Incl O&P
0010	**CONDUCTIVE RUBBER AND VINYL FLOORING**									
1700	Conductive flooring, rubber tile, 1/8" thick	1 Tilf	315	.025	S.F.	5.10	1		6.10	7.05
1800	Homogeneous vinyl tile, 1/8" thick	"	315	.025	"	7.15	1		8.15	9.30

09 66 Terrazzo Flooring

09 66 13 – Portland Cement Terrazzo Flooring

09 66 13.10 Portland Cement Terrazzo		Crew	Daily Output	Labor-Hours	Unit	Material	Labor	Equipment	Total	Total Incl O&P
0010	**PORTLAND CEMENT TERRAZZO**, cast-in-place R096613-10									
0020	Cove base, 6" high, 16ga. zinc	1 Mstz	20	.400	L.F.	3.32	15.60		18.92	26.50
0100	Curb, 6" high and 6" wide		6	1.333		5.25	52		57.25	82
0300	Divider strip for floors, 14 ga., 1-1/4" deep, zinc		375	.021		1.37	.83		2.20	2.73
0400	Brass		375	.021		2.69	.83		3.52	4.18
0600	Heavy top strip 1/4" thick, 1-1/4" deep, zinc		300	.027		1.88	1.04		2.92	3.59
0900	Galv. bottoms, brass		300	.027		3.53	1.04		4.57	5.40
1200	For thin set floors, 16 ga., 1/2" x 1/2", zinc		350	.023		.80	.89		1.69	2.18
1300	Brass		350	.023		2.24	.89		3.13	3.76
1500	Floor, bonded to concrete, 1-3/4" thick, gray cement	J-3	75	.213	S.F.	2.87	7.55	3.56	13.98	18.15
1600	White cement, mud set		75	.213		3.27	7.55	3.56	14.38	18.55
1800	Not bonded, 3" total thickness, gray cement		70	.229		3.59	8.10	3.81	15.50	20
1900	White cement, mud set		70	.229		3.92	8.10	3.81	15.83	20.50
2100	For Venetian terrazzo, 1" topping, add					50%	50%			

09 66 Terrazzo Flooring

09 66 13 – Portland Cement Terrazzo Flooring

09 66 13.10 Portland Cement Terrazzo		Crew	Daily Output	Labor-Hours	Unit	2010 Bare Costs Material	Labor	Equipment	Total	Total Incl O&P
2200	For heavy duty abrasive terrazzo, add					50%	50%			
2700	Monolithic terrazzo, 1/2″ thick									
2710	10′ panels	J-3	125	.128	S.F.	2.61	4.54	2.14	9.29	11.85
3000	Stairs, cast in place, pan filled treads		30	.533	L.F.	2.61	18.90	8.90	30.41	40
3100	Treads and risers		14	1.143	″	5.40	40.50	19.05	64.95	86
3300	For stair landings, add to floor prices						50%			
3400	Stair stringers and fascia	J-3	30	.533	S.F.	4.37	18.90	8.90	32.17	42
3600	For abrasive metal nosings on stairs, add		150	.107	L.F.	8.40	3.78	1.78	13.96	16.75
3700	For abrasive surface finish, add		600	.027	S.F.	1.03	.95	.45	2.43	3
3900	For raised abrasive strips, add		150	.107	L.F.	.98	3.78	1.78	6.54	8.60
4000	Wainscot, bonded, 1-1/2″ thick		30	.533	S.F.	3.32	18.90	8.90	31.12	41
4200	Epoxy terrazzo, 1/4″ thick		40	.400	″	4.88	14.20	6.70	25.78	33
4300	Stone chips, onyx gemstone, per 50 lb. bag				Bag	16.30			16.30	17.95

09 66 13.30 Terrazzo, Precast		Crew	Daily Output	Labor-Hours	Unit	Material	Labor	Equipment	Total	Total Incl O&P
0010	**TERRAZZO, PRECAST**									
0020	Base, 6″ high, straight	1 Mstz	70	.114	L.F.	10.40	4.46		14.86	17.95
0100	Cove		60	.133		12.15	5.20		17.35	21
0300	8″ high, straight		60	.133		10.85	5.20		16.05	19.50
0400	Cove		50	.160		15.95	6.25		22.20	26.50
0600	For white cement, add					.43			.43	.47
0700	For 16 ga. zinc toe strip, add					1.63			1.63	1.79
0900	Curbs, 4″ x 4″ high	1 Mstz	40	.200		30.50	7.80		38.30	45
1000	8″ x 8″ high	″	30	.267		35.50	10.40		45.90	54
1200	Floor tiles, non-slip, 1″ thick, 12″ x 12″	D-1	60	.267	S.F.	18.30	10.05		28.35	35
1300	1-1/4″ thick, 12″ x 12″		60	.267		19.60	10.05		29.65	36.50
1500	16″ x 16″		50	.320		21.50	12.05		33.55	41.50
1600	1-1/2″ thick, 16″ x 16″		45	.356		19.45	13.40		32.85	41.50
1800	For Venetian terrazzo, add					5.85			5.85	6.40
1900	For white cement, add					.55			.55	.61
2400	Stair treads, 1-1/2″ thick, non-slip, three line pattern	2 Mstz	70	.229	L.F.	39.50	8.90		48.40	56.50
2500	Nosing and two lines		70	.229		39.50	8.90		48.40	56.50
2700	2″ thick treads, straight		60	.267		43	10.40		53.40	62
2800	Curved		50	.320		56	12.50		68.50	80
3000	Stair risers, 1″ thick, to 6″ high, straight sections		60	.267		9.80	10.40		20.20	26
3100	Cove		50	.320		14.35	12.50		26.85	34
3300	Curved, 1″ thick, to 6″ high, vertical		48	.333		19.55	13		32.55	40.50
3400	Cove		38	.421		37	16.40		53.40	65
3600	Stair tread and riser, single piece, straight, minimum		60	.267		50	10.40		60.40	70
3700	Maximum		40	.400		65	15.60		80.60	94
3900	Curved tread and riser, minimum		40	.400		70	15.60		85.60	100
4000	Maximum		32	.500		88.50	19.50		108	126
4200	Stair stringers, notched, 1″ thick		25	.640		28.50	25		53.50	68
4300	2″ thick		22	.727		34	28.50		62.50	78.50
4500	Stair landings, structural, non-slip, 1-1/2″ thick		85	.188	S.F.	31	7.35		38.35	45.50
4600	3″ thick		75	.213		44.50	8.30		52.80	60.50
4800	Wainscot, 12″ x 12″ x 1″ tiles	1 Mstz	12	.667		6.50	26		32.50	45
4900	16″ x 16″ x 1-1/2″ tiles	″	8	1		13.70	39		52.70	72

09 66 16 – Terrazzo Floor Tile

09 66 16.10 Tile or Terrazzo Base		Crew	Daily Output	Labor-Hours	Unit	Material	Labor	Equipment	Total	Total Incl O&P
0010	**TILE OR TERRAZZO BASE**									
0020	Scratch coat only	1 Mstz	150	.053	S.F.	.41	2.08		2.49	3.49
0500	Scratch and brown coat only	″	75	.107	″	.78	4.16		4.94	6.95

09 66 Terrazzo Flooring

09 66 33 – Conductive Terrazzo Flooring

	09 66 33.10 Conductive Terrazzo	Crew	Daily Output	Labor-Hours	Unit	Material	2010 Bare Costs Labor	Equipment	Total	Total Incl O&P
0010	**CONDUCTIVE TERRAZZO**									
2400	Bonded conductive floor for hospitals	J-3	90	.178	S.F.	3.92	6.30	2.97	13.19	16.75
2500	Epoxy terrazzo, 1/4" thick, minimum		100	.160		4.43	5.65	2.67	12.75	16.10
2550	Average		75	.213		4.88	7.55	3.56	15.99	20.50
2600	Maximum		60	.267		5.35	9.45	4.45	19.25	24.50

09 67 Fluid-Applied Flooring

09 67 20 – Epoxy-Marble Chip Flooring

09 67 20.13 Elastomeric Liquid Flooring

		Crew	Daily Output	Labor-Hours	Unit	Material	Labor	Equipment	Total	Total Incl O&P
0010	**ELASTOMERIC LIQUID FLOORING**									
0020	Cementitious acrylic, 1/4" thick	C-6	520	.092	S.F.	1.45	3.19	.12	4.76	6.60
0100	3/8" thick	"	450	.107		1.85	3.68	.14	5.67	7.80
0200	Methyl methachrylate, 1/4" thick	C-8A	3000	.016		5.40	.57		5.97	6.80
0210	1/8" thick	"	3000	.016		2.97	.57		3.54	4.13
0300	Cupric oxychloride, on bond coat, minimum	C-6	480	.100		3.15	3.45	.13	6.73	8.85
0400	Maximum		420	.114		5.25	3.95	.15	9.35	11.95
2400	Mastic, hot laid, 2 coat, 1-1/2" thick, standard, minimum		690	.070		3.65	2.40	.09	6.14	7.80
2500	Maximum		520	.092		4.69	3.19	.12	8	10.15
2700	Acid-proof, minimum		605	.079		4.69	2.74	.10	7.53	9.45
2800	Maximum		350	.137		6.50	4.74	.17	11.41	14.60
3000	Neoprene, trowelled on, 1/4" thick, minimum		545	.088		3.59	3.04	.11	6.74	8.70
3100	Maximum		430	.112		4.94	3.85	.14	8.93	11.50
4300	Polyurethane, with suspended vinyl chips, minimum		1065	.045		7	1.56	.06	8.62	10.15
4500	Maximum		860	.056		10.05	1.93	.07	12.05	14.10

09 67 20.16 Epoxy Terrazzo

		Crew	Daily Output	Labor-Hours	Unit	Material	Labor	Equipment	Total	Total Incl O&P
0010	**EPOXY TERRAZZO**									
1800	Epoxy terrazzo, 1/4" thick, chemical resistant, minimum	J-3	200	.080	S.F.	5.25	2.84	1.34	9.43	11.40
1900	Maximum		150	.107		7.70	3.78	1.78	13.26	16
2100	Conductive, minimum		210	.076		6.90	2.70	1.27	10.87	12.95
2200	Maximum		160	.100		8.95	3.55	1.67	14.17	16.90

09 67 20.19 Polyacrylate Terrazzo

		Crew	Daily Output	Labor-Hours	Unit	Material	Labor	Equipment	Total	Total Incl O&P
0010	**POLYACRYLATE TERRAZZO**									
3150	Polyacrylate, 1/4" thick, minimum	C-6	735	.065	S.F.	2.98	2.26	.08	5.32	6.80
3170	Maximum		480	.100		3.68	3.45	.13	7.26	9.45
3200	3/8" thick, minimum		620	.077		3.88	2.67	.10	6.65	8.45
3220	Maximum		480	.100		5.15	3.45	.13	8.73	11.05
3300	Conductive, 1/4" thick, minimum		450	.107		6.25	3.68	.14	10.07	12.60
3330	Maximum		305	.157		7.75	5.45	.20	13.40	17
3350	3/8" thick, minimum		365	.132		8.20	4.54	.17	12.91	16.15
3370	Maximum		255	.188		9.70	6.50	.24	16.44	21
3450	Granite, conductive, 1/4" thick, minimum		695	.069		7.90	2.38	.09	10.37	12.45
3470	Maximum		420	.114		10.20	3.95	.15	14.30	17.35
3500	3/8" thick, minimum		695	.069		11.55	2.38	.09	14.02	16.45
3520	Maximum		380	.126		13.95	4.36	.16	18.47	22

09 67 20.26 Quartz Flooring

		Crew	Daily Output	Labor-Hours	Unit	Material	Labor	Equipment	Total	Total Incl O&P
0010	**QUARTZ FLOORING**									
0600	Epoxy, with colored quartz chips, broadcast, minimum	C-6	675	.071	S.F.	2.49	2.46	.09	5.04	6.60
0700	Maximum		490	.098		3.02	3.38	.12	6.52	8.60
0900	Trowelled, minimum		560	.086		3.21	2.96	.11	6.28	8.15
1000	Maximum		480	.100		4.69	3.45	.13	8.27	10.55

09 67 Fluid-Applied Flooring

09 67 20 – Epoxy-Marble Chip Flooring

09 67 20.26 Quartz Flooring		Crew	Daily Output	Labor-Hours	Unit	2010 Bare Costs Material	Labor	Equipment	Total	Total Incl O&P
1200	Heavy duty epoxy topping, 1/4" thick,									
1300	500 to 1,000 S.F.	C-6	420	.114	S.F.	4.67	3.95	.15	8.77	11.30
1500	1,000 to 2,000 S.F.		450	.107		3.87	3.68	.14	7.69	10
1600	Over 10,000 S.F.		480	.100		3.62	3.45	.13	7.20	9.35
3600	Polyester, with colored quartz chips, 1/16" thick, minimum		1065	.045		2.67	1.56	.06	4.29	5.40
3700	Maximum		560	.086		3.63	2.96	.11	6.70	8.65
3900	1/8" thick, minimum		810	.059		3.12	2.05	.08	5.25	6.65
4000	Maximum		675	.071		4.22	2.46	.09	6.77	8.50
4200	Polyester, heavy duty, compared to epoxy, add	↓	2590	.019	↓	1.27	.64	.02	1.93	2.41

09 68 Carpeting

09 68 05 – Carpet Accessories

09 68 05.11 Flooring Transition Strip		Crew	Daily Output	Labor-Hours	Unit	Material	Labor	Equipment	Total	Total Incl O&P
0010	**FLOORING TRANSITION STRIP**									
0107	Clamp down brass divider, 12' strip, vinyl to carpet	1 Tilf	31.25	.256	Ea.	20.50	10.05		30.55	37
0117	Vinyl to hard surface	"	31.25	.256	"	20.50	10.05		30.55	37

09 68 10 – Carpet Pad

09 68 10.10 Commercial Grade Carpet Pad		Crew	Daily Output	Labor-Hours	Unit	Material	Labor	Equipment	Total	Total Incl O&P
0010	**COMMERCIAL GRADE CARPET PAD**									
9000	Sponge rubber pad, minimum	1 Tilf	150	.053	S.Y.	3.97	2.10		6.07	7.45
9100	Maximum		150	.053		10	2.10		12.10	14.05
9200	Felt pad, minimum		150	.053		4.14	2.10		6.24	7.60
9300	Maximum		150	.053		7.05	2.10		9.15	10.80
9400	Bonded urethane pad, minimum		150	.053		4.57	2.10		6.67	8.10
9500	Maximum		150	.053		7.80	2.10		9.90	11.65
9600	Prime urethane pad, minimum		150	.053		2.62	2.10		4.72	5.95
9700	Maximum	↓	150	.053	↓	4.84	2.10		6.94	8.35

09 68 13 – Tile Carpeting

09 68 13.10 Carpet Tile		Crew	Daily Output	Labor-Hours	Unit	Material	Labor	Equipment	Total	Total Incl O&P
0010	**CARPET TILE**									
0100	Tufted nylon, 18" x 18", hard back, 20 oz.	1 Tilf	150	.053	S.Y.	22	2.10		24.10	27.50
0110	26 oz.		150	.053		38	2.10		40.10	44.50
0200	Cushion back, 20 oz.		150	.053		28	2.10		30.10	33.50
0210	26 oz.		150	.053		43.50	2.10		45.60	51
1100	Tufted, 24" x 24", hard back, 24 oz. nylon		80	.100		29	3.94		32.94	37.50
1180	35 oz.		80	.100		34	3.94		37.94	43.50
5060	42 oz.	↓	80	.100	↓	45.50	3.94		49.44	56

09 68 16 – Sheet Carpeting

09 68 16.10 Sheet Carpet		Crew	Daily Output	Labor-Hours	Unit	Material	Labor	Equipment	Total	Total Incl O&P
0010	**SHEET CARPET**									
0700	Nylon, level loop, 26 oz., light to medium traffic **CN**	1 Tilf	75	.107	S.Y.	27	4.20		31.20	35.50
0720	28 oz., light to medium traffic		75	.107		28	4.20		32.20	36.50
0900	32 oz., medium traffic		75	.107		33	4.20		37.20	42.50
1100	40 oz., medium to heavy traffic		75	.107		49.50	4.20		53.70	60.50
2920	Nylon plush, 30 oz., medium traffic		57	.140		25.50	5.50		31	36
3000	36 oz., medium traffic		75	.107		34	4.20		38.20	43.50
3100	42 oz., medium to heavy traffic		70	.114		38	4.50		42.50	48.50
3200	46 oz., medium to heavy traffic		70	.114		44.50	4.50		49	55.50
3300	54 oz., heavy traffic	↓	70	.114	↓	49	4.50		53.50	60

09 68 Carpeting

09 68 16 – Sheet Carpeting

09 68 16.10 Sheet Carpet		Crew	Daily Output	Labor-Hours	Unit	2010 Bare Costs Material	Labor	Equipment	Total	Total Incl O&P
3340	60 oz., heavy traffic	1 Tilf	70	.114	S.Y.	68	4.50		72.50	81
3665	Olefin, 24 oz., light to medium traffic		75	.107		15.35	4.20		19.55	23
3670	26 oz., medium traffic		75	.107		16.50	4.20		20.70	24.50
3680	28 oz., medium to heavy traffic		75	.107		18	4.20		22.20	26
3700	32 oz., medium to heavy traffic		75	.107		23	4.20		27.20	31.50
3730	42 oz., heavy traffic		70	.114		27.50	4.50		32	36.50
4110	Wool, level loop, 40 oz., medium traffic		75	.107		114	4.20		118.20	131
4500	50 oz., medium to heavy traffic		75	.107		116	4.20		120.20	134
4700	Patterned, 32 oz., medium to heavy traffic		70	.114		115	4.50		119.50	134
4900	48 oz., heavy traffic		70	.114		118	4.50		122.50	137
5000	For less than full roll (approx. 1500 S.F.), add					25%				
5100	For small rooms, less than 12′ wide, add						25%			
5200	For large open areas (no cuts), deduct						25%			
5600	For bound carpet baseboard, add	1 Tilf	300	.027	L.F.	2.40	1.05		3.45	4.17
5610	For stairs, not incl. price of carpet, add	"	30	.267	Riser		10.50		10.50	15.35
5620	For borders and patterns, add to labor						18%			
8950	For tackless, stretched installation, add padding to above									
9850	For brand-named specific fiber, add				S.Y.	25%				

09 68 20 – Athletic Carpet

09 68 20.10 Indoor Athletic Carpet		Crew	Daily Output	Labor-Hours	Unit	Material	Labor	Equipment	Total	Total Incl O&P
0010	**INDOOR ATHLETIC CARPET**									
3700	Polyethylene, in rolls, no base incl., landscape surfaces	1 Tilf	275	.029	S.F.	3.20	1.14		4.34	5.20
3800	Nylon action surface, 1/8″ thick		275	.029		3.36	1.14		4.50	5.35
3900	1/4″ thick		275	.029		4.84	1.14		5.98	6.95
4000	3/8″ thick		275	.029		6.10	1.14		7.24	8.35
4100	Golf tee surface with foam back		235	.034		6	1.34		7.34	8.55
4200	Practice putting, knitted nylon surface		235	.034		5.10	1.34		6.44	7.55
4400	Polyurethane, thermoset, prefabricated in place, indoor									
4500	3/8″ thick for basketball, gyms, etc.	1 Tilf	100	.080	S.F.	4.81	3.15		7.96	9.90
4600	1/2″ thick for professional sports		95	.084		6.45	3.31		9.76	11.95
4700	Outdoor, 1/4″ thick, smooth, for tennis		100	.080		5	3.15		8.15	10.10
4800	Rough, for track, 3/8″ thick		95	.084		5.80	3.31		9.11	11.20
5000	Poured in place, indoor, with finish, 1/4″ thick		80	.100		3.55	3.94		7.49	9.65
5050	3/8″ thick		65	.123		4.32	4.84		9.16	11.85
5100	1/2″ thick		50	.160		5.50	6.30		11.80	15.25
5500	Polyvinyl chloride, sheet goods for gyms, 1/4″ thick		80	.100		7.75	3.94		11.69	14.30
5600	3/8″ thick		60	.133		8.75	5.25		14	17.30

09 69 Access Flooring

09 69 13 – Rigid-Grid Access Flooring

09 69 13.10 Access Floors		Crew	Daily Output	Labor-Hours	Unit	Material	Labor	Equipment	Total	Total Incl O&P
0010	**ACCESS FLOORS**									
0015	Access floor package including panel, pedestal, stringers & laminate cover									
0100	Computer room, greater than 6,000 S.F.	4 Carp	750	.043	S.F.	8.65	1.77		10.42	12.25
0110	Less than 6,000 S.F.	2 Carp	375	.043		9.95	1.77		11.72	13.70
0120	Office, greater than 6,000 S.F.	4 Carp	1050	.030		4.56	1.27		5.83	6.95
0250	Panels, particle board or steel, 1250# load, no covering, under 6,000 S.F.	2 Carp	600	.027		4.24	1.11		5.35	6.35
0300	Over 6,000 S.F.		640	.025		3.72	1.04		4.76	5.70
0400	Aluminum, 24″ panels		500	.032		31.50	1.33		32.83	37
0600	For carpet covering, add					8.55			8.55	9.40

09 69 Access Flooring

09 69 13 – Rigid-Grid Access Flooring

09 69 13.10 Access Floors		Crew	Daily Output	Labor-Hours	Unit	2010 Bare Costs Material	Labor	Equipment	Total	Total Incl O&P
0700	For vinyl floor covering, add				S.F.	6.70			6.70	7.35
0900	For high pressure laminate covering, add					5.45			5.45	6
0910	For snap on stringer system, add	2 Carp	1000	.016	↓	1.44	.66		2.10	2.60
0950	Office applications, steel or concrete panels,									
0960	no covering, over 6,000 S.F.	2 Carp	960	.017	S.F.	10.10	.69		10.79	12.15
1000	Machine cutouts after initial installation	1 Carp	50	.160	Ea.	5.05	6.65		11.70	15.80
1050	Pedestals, 6" to 12"	2 Carp	85	.188		7.95	7.80		15.75	21
1100	Air conditioning grilles, 4" x 12"	1 Carp	17	.471		66	19.55		85.55	103
1150	4" x 18"	"	14	.571	↓	90.50	23.50		114	136
1200	Approach ramps, minimum	2 Carp	60	.267	S.F.	24.50	11.10		35.60	44
1300	Maximum	"	40	.400	"	33.50	16.60		50.10	62
1500	Handrail, 2 rail, aluminum	1 Carp	15	.533	L.F.	94.50	22		116.50	138

09 72 Wall Coverings

09 72 23 – Wallpapering

09 72 23.10 Wallpaper

09 72 23.10 Wallpaper		Crew	Daily Output	Labor-Hours	Unit	Material	Labor	Equipment	Total	Total Incl O&P
0010	**WALLPAPER** including sizing; add 10-30 percent waste @ takeoff R097223-10									
0050	Aluminum foil	1 Pape	275	.029	S.F.	1.02	1.06		2.08	2.70
0100	Copper sheets, .025" thick, vinyl backing		240	.033		5.45	1.22		6.67	7.80
0300	Phenolic backing		240	.033		7.05	1.22		8.27	9.60
0600	Cork tiles, light or dark, 12" x 12" x 3/16"		240	.033		4.47	1.22		5.69	6.75
0700	5/16" thick		235	.034		3.80	1.24		5.04	6.05
0900	1/4" basketweave		240	.033		5.85	1.22		7.07	8.25
1000	1/2" natural, non-directional pattern		240	.033		7.85	1.22		9.07	10.45
1100	3/4" natural, non-directional pattern		240	.033		10.80	1.22		12.02	13.65
1200	Granular surface, 12" x 36", 1/2" thick		385	.021		1.26	.76		2.02	2.52
1300	1" thick		370	.022		1.63	.79		2.42	2.97
1500	Polyurethane coated, 12" x 12" x 3/16" thick		240	.033		3.95	1.22		5.17	6.15
1600	5/16" thick		235	.034		5.65	1.24		6.89	8.05
1800	Cork wallpaper, paperbacked, natural		480	.017		1.99	.61		2.60	3.10
1900	Colors		480	.017		2.78	.61		3.39	3.97
2100	Flexible wood veneer, 1/32" thick, plain woods		100	.080		2.36	2.92		5.28	6.95
2200	Exotic woods	↓	95	.084	↓	3.58	3.08		6.66	8.50
2400	Gypsum-based, fabric-backed, fire resistant									
2500	for masonry walls, minimum, 21 oz./S.Y.	1 Pape	800	.010	S.F.	.82	.37		1.19	1.44
2700	Maximum (small quantities)	"	640	.013		1.37	.46		1.83	2.19
2750	Acrylic, modified, semi-rigid PVC, .028" thick	2 Carp	330	.048		1.13	2.01		3.14	4.35
2800	.040" thick	"	320	.050		1.49	2.08		3.57	4.84
3000	Vinyl wall covering, fabric-backed, lightweight, type 1 (12-15 oz./S.Y.)	1 Pape	640	.013		.69	.46		1.15	1.44
3300	Medium weight, type 2 (20-24 oz./S.Y.)		480	.017		.82	.61		1.43	1.81
3400	Heavy weight, type 3 (28 oz./S.Y.)	↓	435	.018	↓	1.64	.67		2.31	2.80
3600	Adhesive, 5 gal. lots (18SY/Gal.)				Gal.	10.50			10.50	11.55
3700	Wallpaper, average workmanship, solid pattern, low cost paper	1 Pape	640	.013	S.F.	.37	.46		.83	1.09
3900	basic patterns (matching required), avg. cost paper		535	.015		.71	.55		1.26	1.59
4000	Paper at $85 per double roll, quality workmanship	↓	435	.018	↓	2.47	.67		3.14	3.72
4100	Linen wall covering, paper backed									
4150	Flame treatment, minimum				S.F.	1.05			1.05	1.16
4180	Maximum					1.55			1.55	1.71
4200	Grass cloths with lining paper, minimum	1 Pape	400	.020		.78	.73		1.51	1.95
4300	Maximum	"	350	.023	↓	2.49	.84		3.33	3.98

09 77 Special Wall Surfacing

09 77 33 – Fiberglass Reinforced Panels

09 77 33.10 Fiberglass Reinforced Plastic Panels		Crew	Daily Output	Labor-Hours	Unit	Material	2010 Bare Costs Labor	Equipment	Total	Total Incl O&P
0010	**FIBERGLASS REINFORCED PLASTIC PANELS**, .090" thick									
0020	On walls, adhesive mounted, embossed surface	2 Carp	640	.025	S.F.	1.23	1.04		2.27	2.95
0030	Smooth surface		640	.025		1.51	1.04		2.55	3.26
0040	Fire rated, embossed surface		640	.025		2.19	1.04		3.23	4.01
0050	Nylon rivet mounted, on drywall, embossed surface		480	.033		1.15	1.38		2.53	3.40
0060	Smooth surface		480	.033		1.43	1.38		2.81	3.70
0070	Fire rated, embossed surface		480	.033		2.11	1.38		3.49	4.45
0080	On masonry, embossed surface		320	.050		1.15	2.08		3.23	4.47
0090	Smooth surface		320	.050		1.43	2.08		3.51	4.77
0100	Fire rated, embossed surface		320	.050		2.11	2.08		4.19	5.50
0110	Nylon rivet and adhesive mounted, on drywall, embossed surface		240	.067		1.30	2.77		4.07	5.70
0120	Smooth surface		240	.067		1.58	2.77		4.35	6
0130	Fire rated, embossed surface		240	.067		2.26	2.77		5.03	6.75
0140	On masonry, embossed surface		190	.084		1.30	3.50		4.80	6.85
0150	Smooth surface		190	.084		1.58	3.50		5.08	7.15
0160	Fire rated, embossed surface		190	.084		2.26	3.50		5.76	7.90
0170	For moldings add	1 Carp	250	.032	L.F.	.28	1.33		1.61	2.36
0180	On ceilings, for lay in grid system, embossed surface		400	.020	S.F.	1.23	.83		2.06	2.63
0190	Smooth surface		400	.020		1.51	.83		2.34	2.94
0200	Fire rated, embossed surface		400	.020		2.19	.83		3.02	3.69

09 77 43 – Panel Systems

09 77 43.20 Slatwall Panels and Accessories		Crew	Daily Output	Labor-Hours	Unit	Material	Labor	Equipment	Total	Total Incl O&P
0010	**SLATWALL PANELS AND ACCESSORIES**									
0100	Slatwall panel, 4' x 8' x 3/4" T, MDF, paint grade	1 Carp	500	.016	S.F.	1.46	.66		2.12	2.63
0110	Melamine finish		500	.016		2.09	.66		2.75	3.32
0120	High pressure plastic laminate finish		500	.016		3.41	.66		4.07	4.77
0130	Aluminum channel inserts, add					3.34			3.34	3.67
0200	Accessories, corner forms, 8' L				L.F.	4.49			4.49	4.94
0210	T-connector, 8' L					5.50			5.50	6.05
0220	J-mold, 8' L					1.50			1.50	1.65
0230	Edge cap, 8' L					1.08			1.08	1.19
0240	Finish end cap, 8' L					3.39			3.39	3.73
0300	Display hook, metal, 4" L				Ea.	1			1	1.10
0310	6" L					1.09			1.09	1.20
0320	8" L					1.16			1.16	1.28
0330	10" L					1.31			1.31	1.44
0340	12" L					1.45			1.45	1.60
0350	Acrylic, 4" L					.91			.91	1
0360	6" L					1.05			1.05	1.16
0370	8" L					1.10			1.10	1.21
0380	10" L					1.12			1.12	1.23
0400	Waterfall hanger, metal, 12" - 16"					5.85			5.85	6.40
0410	Acrylic					9.85			9.85	10.85
0500	Shelf bracket, metal, 8"					6.35			6.35	6.95
0510	10"					6.55			6.55	7.20
0520	12"					6.75			6.75	7.45
0530	14"					7.15			7.15	7.85
0540	16"					8.60			8.60	9.45
0550	Acrylic, 8"					3.16			3.16	3.48
0560	10"					3.53			3.53	3.88
0570	12"					3.91			3.91	4.30
0580	14"					4.43			4.43	4.87

09 77 Special Wall Surfacing

09 77 43 – Panel Systems

09 77 43.20 Slatwall Panels and Accessories		Crew	Daily Output	Labor-Hours	Unit	Material	Labor	Equipment	Total	Total Incl O&P
0600	Shelf, acrylic, 12" x 16" x 1/4"				Ea.	26			26	28.50
0610	12" x 24" x 1/4"				↓	43.50			43.50	47.50

09 81 Acoustic Insulation

09 81 16 – Acoustic Blanket Insulation

09 81 16.10 Sound Attenuation Blanket

		Crew	Daily Output	Labor-Hours	Unit	Material	Labor	Equipment	Total	Total Incl O&P
0010	**SOUND ATTENUATION BLANKET**									
0020	Blanket, 1" thick	1 Carp	925	.009	S.F.	.25	.36		.61	.83
0500	1-1/2" thick		920	.009		.25	.36		.61	.84
1000	2" thick		915	.009		.34	.36		.70	.93
1500	3" thick		910	.009		.50	.37		.87	1.11
3000	Thermal or acoustical batt above ceiling, 2" thick		900	.009		.49	.37		.86	1.11
3100	3" thick		900	.009		.74	.37		1.11	1.38
3200	4" thick	↓	900	.009		.92	.37		1.29	1.58
3400	Urethane plastic foam, open cell, on wall, 2" thick	2 Carp	2050	.008		3.05	.32		3.37	3.86
3500	3" thick		1550	.010		4.05	.43		4.48	5.10
3600	4" thick		1050	.015		5.70	.63		6.33	7.25
3700	On ceiling, 2" thick		1700	.009		3.04	.39		3.43	3.94
3800	3" thick		1300	.012		4.05	.51		4.56	5.25
3900	4" thick	↓	900	.018	↓	5.70	.74		6.44	7.40
4000	Nylon matting 0.4" thick, with carbon black spinerette									
4010	plus polyester fabric, on floor	D-7	4000	.004	S.F.	2.26	.14		2.40	2.70
4200	Fiberglass reinf. backer board underlayment, 7/16" thick, on floor	"	800	.020	"	1.90	.71		2.61	3.12

09 84 Acoustic Room Components

09 84 13 – Fixed Sound-Absorptive Panels

09 84 13.10 Fixed Panels

		Crew	Daily Output	Labor-Hours	Unit	Material	Labor	Equipment	Total	Total Incl O&P
0010	**FIXED PANELS** Perforated steel facing, painted with									
0100	Fiberglass or mineral filler, no backs, 2-1/4" thick, modular									
0200	space units, ceiling or wall hung, white or colored	1 Carp	100	.080	S.F.	10	3.32		13.32	16.10
0300	Fiberboard sound deadening panels, 1/2" thick	"	600	.013	"	.32	.55		.87	1.20
0500	Fiberglass panels, 4' x 8' x 1" thick, with									
0600	glass cloth face for walls, cemented	1 Carp	155	.052	S.F.	7	2.14		9.14	11
0700	1-1/2" thick, dacron covered, inner aluminum frame,									
0710	wall mounted	1 Carp	300	.027	S.F.	8.75	1.11		9.86	11.30
0900	Mineral fiberboard panels, fabric covered, 30" x 108",									
1000	3/4" thick, concealed spline, wall mounted	1 Carp	150	.053	S.F.	6.25	2.22		8.47	10.30

09 84 36 – Sound-Absorbing Ceiling Units

09 84 36.10 Barriers

		Crew	Daily Output	Labor-Hours	Unit	Material	Labor	Equipment	Total	Total Incl O&P
0010	**BARRIERS** Plenum									
0600	Aluminum foil, fiberglass reinf., parallel with joists	1 Carp	275	.029	S.F.	.85	1.21		2.06	2.80
0700	Perpendicular to joists		180	.044		.85	1.85		2.70	3.79
0900	Aluminum mesh, kraft paperbacked		275	.029		.77	1.21		1.98	2.71
0970	Fiberglass batts, kraft faced, 3-1/2" thick		1400	.006		.36	.24		.60	.77
0980	6" thick		1300	.006		.65	.26		.91	1.11
1000	Sheet lead, 1 lb., 1/64" thick, perpendicular to joists		150	.053		8.40	2.22		10.62	12.60
1100	Vinyl foam reinforced, 1/8" thick, 1.0 lb. per S.F.	↓	150	.053	↓	9.80	2.22		12.02	14.15

09 91 03 – Paint Restoration

09 91 03.20 Sanding

	09 91 03.20 Sanding	Crew	Daily Output	Labor-Hours	Unit	Material	2010 Bare Costs Labor	Equipment	Total	Total Incl O&P
0010	**SANDING** and puttying interior trim, compared to R099100-10									
0100	Painting 1 coat, on quality work				L.F.		100%			
0300	Medium work						50%			
0400	Industrial grade						25%			
0500	Surface protection, placement and removal									
0510	Basic drop cloths	1 Pord	6400	.001	S.F.		.05		.05	.07
0520	Masking with paper		800	.010		.06	.36		.42	.61
0530	Volume cover up (using plastic sheathing, or building paper)		16000	.001			.02		.02	.03

09 91 03.30 Exterior Surface Preparation

	09 91 03.30 Exterior Surface Preparation	Crew	Daily Output	Labor-Hours	Unit	Material	Labor	Equipment	Total	Total Incl O&P
0010	**EXTERIOR SURFACE PREPARATION** R099100-10									
0015	Doors, per side, not incl. frames or trim									
0020	Scrape & sand									
0030	Wood, flush	1 Pord	616	.013	S.F.		.47		.47	.70
0040	Wood, detail		496	.016			.59		.59	.87
0050	Wood, louvered		280	.029			1.04		1.04	1.55
0060	Wood, overhead		616	.013			.47		.47	.70
0070	Wire brush									
0080	Metal, flush	1 Pord	640	.013	S.F.		.45		.45	.68
0090	Metal, detail		520	.015			.56		.56	.83
0100	Metal, louvered		360	.022			.81		.81	1.20
0110	Metal or fibr., overhead		640	.013			.45		.45	.68
0120	Metal, roll up		560	.014			.52		.52	.77
0130	Metal, bulkhead		640	.013			.45		.45	.68
0140	Power wash, based on 2500 lb. operating pressure									
0150	Metal, flush	A-1H	2240	.004	S.F.		.12	.03	.15	.21
0160	Metal, detail		2120	.004			.12	.03	.15	.22
0170	Metal, louvered		2000	.004			.13	.03	.16	.23
0180	Metal or fibr., overhead		2400	.003			.11	.03	.14	.20
0190	Metal, roll up		2400	.003			.11	.03	.14	.20
0200	Metal, bulkhead		2200	.004			.12	.03	.15	.22
0400	Windows, per side, not incl. trim									
0410	Scrape & sand									
0420	Wood, 1-2 lite	1 Pord	320	.025	S.F.		.91		.91	1.35
0430	Wood, 3-6 lite		280	.029			1.04		1.04	1.55
0440	Wood, 7-10 lite		240	.033			1.21		1.21	1.80
0450	Wood, 12 lite		200	.040			1.45		1.45	2.16
0460	Wood, Bay / Bow		320	.025			.91		.91	1.35
0470	Wire brush									
0480	Metal, 1-2 lite	1 Pord	480	.017	S.F.		.61		.61	.90
0490	Metal, 3-6 lite		400	.020			.73		.73	1.08
0500	Metal, Bay / Bow		480	.017			.61		.61	.90
0510	Power wash, based on 2500 lb. operating pressure									
0520	1-2 lite	A-1H	4400	.002	S.F.		.06	.01	.07	.11
0530	3-6 lite		4320	.002			.06	.01	.07	.11
0540	7-10 lite		4240	.002			.06	.01	.07	.12
0550	12 lite		4160	.002			.06	.01	.07	.12
0560	Bay / Bow		4400	.002			.06	.01	.07	.11
0600	Siding, scrape and sand, light=10-30%, med.=30-70%									
0610	Heavy=70-100% of surface to sand									
0650	Texture 1-11, light	1 Pord	480	.017	S.F.		.61		.61	.90
0660	Med.		440	.018			.66		.66	.98
0670	Heavy		360	.022			.81		.81	1.20

09 91 03 – Paint Restoration

09 91 03.30 Exterior Surface Preparation		Crew	Daily Output	Labor-Hours	Unit	Material	2010 Bare Costs Labor	Equipment	Total	Total Incl O&P
0680	Wood shingles, shakes, light	1 Pord	440	.018	S.F.		.66		.66	.98
0690	Med.		360	.022			.81		.81	1.20
0700	Heavy		280	.029			1.04		1.04	1.55
0710	Clapboard, light		520	.015			.56		.56	.83
0720	Med.		480	.017			.61		.61	.90
0730	Heavy		400	.020			.73		.73	1.08
0740	Wire brush									
0750	Aluminum, light	1 Pord	600	.013	S.F.		.48		.48	.72
0760	Med.		520	.015			.56		.56	.83
0770	Heavy		440	.018			.66		.66	.98
0780	Pressure wash, based on 2500 lb. operating pressure									
0790	Stucco	A-1H	3080	.003	S.F.		.09	.02	.11	.15
0800	Aluminum or vinyl		3200	.003			.08	.02	.10	.15
0810	Siding, masonry, brick & block		2400	.003			.11	.03	.14	.20
1300	Miscellaneous, wire brush									
1310	Metal, pedestrian gate	1 Pord	100	.080	S.F.		2.91		2.91	4.33
8000	For chemical washing, see Div. 04 01 30									
8010	For steam cleaning, see Div. 04 01 30.20									
8020	For sand blasting, see Div. 05 01 10.51 and 03 35 29.60									

09 91 03.40 Interior Surface Preparation		Crew	Daily Output	Labor-Hours	Unit	Material	Labor	Equipment	Total	Total Incl O&P
0010	**INTERIOR SURFACE PREPARATION** R099100-10									
0020	Doors, per side, not incl. frames or trim									
0030	Scrape & sand									
0040	Wood, flush	1 Pord	616	.013	S.F.		.47		.47	.70
0050	Wood, detail		496	.016			.59		.59	.87
0060	Wood, louvered		280	.029			1.04		1.04	1.55
0070	Wire brush									
0080	Metal, flush	1 Pord	640	.013	S.F.		.45		.45	.68
0090	Metal, detail		520	.015			.56		.56	.83
0100	Metal, louvered		360	.022			.81		.81	1.20
0110	Hand wash									
0120	Wood, flush	1 Pord	2160	.004	S.F.		.13		.13	.20
0130	Wood, detailed		2000	.004			.15		.15	.22
0140	Wood, louvered		1360	.006			.21		.21	.32
0150	Metal, flush		2160	.004			.13		.13	.20
0160	Metal, detail		2000	.004			.15		.15	.22
0170	Metal, louvered		1360	.006			.21		.21	.32
0400	Windows, per side, not incl. trim									
0410	Scrape & sand									
0420	Wood, 1-2 lite	1 Pord	360	.022	S.F.		.81		.81	1.20
0430	Wood, 3-6 lite		320	.025			.91		.91	1.35
0440	Wood, 7-10 lite		280	.029			1.04		1.04	1.55
0450	Wood, 12 lite		240	.033			1.21		1.21	1.80
0460	Wood, Bay / Bow		360	.022			.81		.81	1.20
0470	Wire brush									
0480	Metal, 1-2 lite	1 Pord	520	.015	S.F.		.56		.56	.83
0490	Metal, 3-6 lite		440	.018			.66		.66	.98
0500	Metal, Bay / Bow		520	.015			.56		.56	.83
0600	Walls, sanding, light=10-30%, medium - 30-70%,									
0610	heavy=70-100% of surface to sand									
0650	Walls, sand									
0660	Gypsum board or plaster, light	1 Pord	3077	.003	S.F.		.09		.09	.14

09 91 Painting

09 91 03 – Paint Restoration

09 91 03.40 Interior Surface Preparation		Crew	Daily Output	Labor-Hours	Unit	Material	Labor (2010 Bare Costs)	Equipment	Total	Total Incl O&P
0670	Gypsum board or plaster, medium	1 Pord	2160	.004	S.F.		.13		.13	.20
0680	Gypsum board or plaster, heavy		923	.009			.32		.32	.47
0690	Wood, T&G, light		2400	.003			.12		.12	.18
0700	Wood, T&G, med.		1600	.005			.18		.18	.27
0710	Wood, T&G, heavy		800	.010			.36		.36	.54
0720	Walls, wash									
0730	Gypsum board or plaster	1 Pord	3200	.003	S.F.		.09		.09	.14
0740	Wood, T&G		3200	.003			.09		.09	.14
0750	Masonry, brick & block, smooth		2800	.003			.10		.10	.15
0760	Masonry, brick & block, coarse		2000	.004			.15		.15	.22
8000	For chemical washing, see Div. 04 01 30									
8010	For steam cleaning, see Div. 04 01 30.20									
8020	For sand blasting, see Div. 03 35 29.60 and 05 01 10.51									

09 91 13 – Exterior Painting

09 91 13.30 Fences

		Crew	Daily Output	Labor-Hours	Unit	Material	Labor	Equipment	Total	Total Incl O&P
0010	**FENCES** R099100-20									
0100	Chain link or wire metal, one side, water base									
0110	Roll & brush, first coat	1 Pord	960	.008	S.F.	.05	.30		.35	.51
0120	Second coat		1280	.006		.05	.23		.28	.40
0130	Spray, first coat		2275	.004		.05	.13		.18	.25
0140	Second coat		2600	.003		.05	.11		.16	.23
0150	Picket, water base									
0160	Roll & brush, first coat	1 Pord	865	.009	S.F.	.06	.34		.40	.56
0170	Second coat		1050	.008		.06	.28		.34	.47
0180	Spray, first coat		2275	.004		.06	.13		.19	.25
0190	Second coat		2600	.003		.06	.11		.17	.23
0200	Stockade, water base									
0210	Roll & brush, first coat	1 Pord	1040	.008	S.F.	.06	.28		.34	.48
0220	Second coat		1200	.007		.06	.24		.30	.42
0230	Spray, first coat		2275	.004		.06	.13		.19	.25
0240	Second coat		2600	.003		.06	.11		.17	.23

09 91 13.42 Miscellaneous, Exterior

		Crew	Daily Output	Labor-Hours	Unit	Material	Labor	Equipment	Total	Total Incl O&P
0010	**MISCELLANEOUS, EXTERIOR** R099100-20									
0015	For painting metals, see Div. 09 97 13.23									
0100	Railing, ext., decorative wood, incl. cap & baluster									
0110	Newels & spindles @ 12″ O.C.									
0120	Brushwork, stain, sand, seal & varnish									
0130	First coat	1 Pord	90	.089	L.F.	.69	3.23		3.92	5.55
0140	Second coat	″	120	.067	″	.69	2.42		3.11	4.37
0150	Rough sawn wood, 42″ high, 2″ x 2″ verticals, 6″ O.C.									
0160	Brushwork, stain, each coat	1 Pord	90	.089	L.F.	.22	3.23		3.45	5.05
0170	Wrought iron, 1″ rail, 1/2″ sq. verticals									
0180	Brushwork, zinc chromate, 60″ high, bars 6″ O.C.									
0190	Primer	1 Pord	130	.062	L.F.	.59	2.24		2.83	3.98
0200	Finish coat		130	.062		.18	2.24		2.42	3.53
0210	Additional coat		190	.042		.21	1.53		1.74	2.51
0220	Shutters or blinds, single panel, 2′ x 4′, paint all sides									
0230	Brushwork, primer	1 Pord	20	.400	Ea.	.60	14.55		15.15	22
0240	Finish coat, exterior latex		20	.400		.43	14.55		14.98	22
0250	Primer & 1 coat, exterior latex		13	.615		.89	22.50		23.39	34.50
0260	Spray, primer		35	.229		.88	8.30		9.18	13.30
0270	Finish coat, exterior latex		35	.229		.90	8.30		9.20	13.35

09 91 Painting

09 91 13 – Exterior Painting

09 91 13.42 Miscellaneous, Exterior

Line	Description	Crew	Daily Output	Labor-Hours	Unit	Material (2010 Bare Costs)	Labor (2010 Bare Costs)	Equipment (2010 Bare Costs)	Total (2010 Bare Costs)	Total Incl O&P
0280	Primer & 1 coat, exterior latex	1 Pord	20	.400	Ea.	.93	14.55		15.48	22.50
0290	For louvered shutters, add				S.F.	10%				
0300	Stair stringers, exterior, metal									
0310	Roll & brush, zinc chromate, to 14", each coat	1 Pord	320	.025	L.F.	.06	.91		.97	1.42
0320	Rough sawn wood, 4" x 12"									
0330	Roll & brush, exterior latex, each coat	1 Pord	215	.037	L.F.	.06	1.35		1.41	2.08
0340	Trellis/lattice, 2" x 2" @ 3" O.C. with 2" x 8" supports									
0350	Spray, latex, per side, each coat	1 Pord	475	.017	S.F.	.06	.61		.67	.98
0450	Decking, ext., sealer, alkyd, brushwork, sealer coat		1140	.007		.09	.26		.35	.48
0460	1st coat		1140	.007		.09	.26		.35	.47
0470	2nd coat		1300	.006		.06	.22		.28	.40
0500	Paint, alkyd, brushwork, primer coat		1140	.007		.07	.26		.33	.46
0510	1st coat		1140	.007		.16	.26		.42	.55
0520	2nd coat		1300	.006		.12	.22		.34	.46
0600	Sand paint, alkyd, brushwork, 1 coat		150	.053		.11	1.94		2.05	3.01

09 91 13.60 Siding Exterior

Line	Description	Crew	Daily Output	Labor-Hours	Unit	Material	Labor	Equipment	Total	Total Incl O&P
0010	**SIDING EXTERIOR**, Alkyd (oil base) R099100-10									
0450	Steel siding, oil base, paint 1 coat, brushwork	2 Pord	2015	.008	S.F.	.13	.29		.42	.57
0500	Spray R099100-20		4550	.004		.19	.13		.32	.40
0800	Paint 2 coats, brushwork		1300	.012		.25	.45		.70	.95
1000	Spray		2750	.006		.17	.21		.38	.50
1200	Stucco, rough, oil base, paint 2 coats, brushwork		1300	.012		.25	.45		.70	.95
1400	Roller		1625	.010		.27	.36		.63	.82
1600	Spray		2925	.005		.28	.20		.48	.61
1800	Texture 1-11 or clapboard, oil base, primer coat, brushwork		1300	.012		.10	.45		.55	.78
2000	Spray		4550	.004		.10	.13		.23	.30
2400	Paint 2 coats, brushwork		810	.020		.37	.72		1.09	1.47
2600	Spray		2600	.006		.41	.22		.63	.78
3400	Stain 2 coats, brushwork		950	.017		.15	.61		.76	1.07
4000	Spray		3050	.005		.17	.19		.36	.46
4200	Wood shingles, oil base primer coat, brushwork		1300	.012		.09	.45		.54	.77
4400	Spray		3900	.004		.08	.15		.23	.31
5000	Paint 2 coats, brushwork		810	.020		.31	.72		1.03	1.41
5200	Spray		2275	.007		.29	.26		.55	.70
6500	Stain 2 coats, brushwork		950	.017		.15	.61		.76	1.07
7000	Spray		2660	.006		.21	.22		.43	.56
8000	For latex paint, deduct					10%				
8100	For work over 12' H, from pipe scaffolding, add						15%			
8200	For work over 12' H, from extension ladder, add						25%			
8300	For work over 12' H, from swing staging, add						35%			

09 91 13.62 Siding, Misc.

Line	Description	Crew	Daily Output	Labor-Hours	Unit	Material	Labor	Equipment	Total	Total Incl O&P
0010	**SIDING, MISC.**, latex paint R099100-10									
0100	Aluminum siding									
0110	Brushwork, primer R099100-20	2 Pord	2275	.007	S.F.	.05	.26		.31	.44
0120	Finish coat, exterior latex		2275	.007		.04	.26		.30	.43
0130	Primer & 1 coat exterior latex		1300	.012		.11	.45		.56	.79
0140	Primer & 2 coats exterior latex		975	.016		.15	.60		.75	1.06
0150	Mineral fiber shingles									
0160	Brushwork, primer	2 Pord	1495	.011	S.F.	.10	.39		.49	.69
0170	Finish coat, industrial enamel		1495	.011		.15	.39		.54	.74
0180	Primer & 1 coat enamel		810	.020		.25	.72		.97	1.34
0190	Primer & 2 coats enamel		540	.030		.39	1.08		1.47	2.03

09 91 13.62 Siding, Misc.

	09 91 13.62 Siding, Misc.	Crew	Daily Output	Labor-Hours	Unit	2010 Bare Costs Material	Labor	Equipment	Total	Total Incl O&P
0200	Roll, primer	2 Pord	1625	.010	S.F.	.11	.36		.47	.65
0210	Finish coat, industrial enamel		1625	.010		.16	.36		.52	.71
0220	Primer & 1 coat enamel		975	.016		.27	.60		.87	1.18
0230	Primer & 2 coats enamel		650	.025		.43	.89		1.32	1.80
0240	Spray, primer		3900	.004		.08	.15		.23	.31
0250	Finish coat, industrial enamel		3900	.004		.13	.15		.28	.37
0260	Primer & 1 coat enamel		2275	.007		.22	.26		.48	.62
0270	Primer & 2 coats enamel		1625	.010		.35	.36		.71	.92
0280	Waterproof sealer, first coat		4485	.004		.07	.13		.20	.26
0290	Second coat		5235	.003		.06	.11		.17	.24
0300	Rough wood incl. shingles, shakes or rough sawn siding									
0310	Brushwork, primer	2 Pord	1280	.013	S.F.	.12	.45		.57	.81
0320	Finish coat, exterior latex		1280	.013		.08	.45		.53	.77
0330	Primer & 1 coat exterior latex		960	.017		.20	.61		.81	1.12
0340	Primer & 2 coats exterior latex		700	.023		.28	.83		1.11	1.55
0350	Roll, primer		2925	.005		.16	.20		.36	.48
0360	Finish coat, exterior latex		2925	.005		.09	.20		.29	.40
0370	Primer & 1 coat exterior latex		1790	.009		.26	.33		.59	.76
0380	Primer & 2 coats exterior latex		1300	.012		.35	.45		.80	1.06
0390	Spray, primer		3900	.004		.14	.15		.29	.37
0400	Finish coat, exterior latex		3900	.004		.07	.15		.22	.30
0410	Primer & 1 coat exterior latex		2600	.006		.21	.22		.43	.56
0420	Primer & 2 coats exterior latex		2080	.008		.28	.28		.56	.73
0430	Waterproof sealer, first coat		4485	.004		.12	.13		.25	.32
0440	Second coat		4485	.004		.07	.13		.20	.26
0450	Smooth wood incl. butt, T&G, beveled, drop or B&B siding									
0460	Brushwork, primer	2 Pord	2325	.007	S.F.	.09	.25		.34	.47
0470	Finish coat, exterior latex		1280	.013		.08	.45		.53	.77
0480	Primer & 1 coat exterior latex		800	.020		.17	.73		.90	1.26
0490	Primer & 2 coats exterior latex		630	.025		.24	.92		1.16	1.64
0500	Roll, primer		2275	.007		.10	.26		.36	.49
0510	Finish coat, exterior latex		2275	.007		.08	.26		.34	.47
0520	Primer & 1 coat exterior latex		1300	.012		.18	.45		.63	.87
0530	Primer & 2 coats exterior latex		975	.016		.27	.60		.87	1.18
0540	Spray, primer		4550	.004		.08	.13		.21	.27
0550	Finish coat, exterior latex		4550	.004		.07	.13		.20	.27
0560	Primer & 1 coat exterior latex		2600	.006		.15	.22		.37	.49
0570	Primer & 2 coats exterior latex		1950	.008		.22	.30		.52	.68
0580	Waterproof sealer, first coat		5230	.003		.07	.11		.18	.24
0590	Second coat		5980	.003		.07	.10		.17	.22
0600	For oil base paint, add					10%				

09 91 13.70 Doors and Windows, Exterior

	09 91 13.70 Doors and Windows, Exterior	Crew	Daily Output	Labor-Hours	Unit	Material	Labor	Equipment	Total	Total Incl O&P
0010	**DOORS AND WINDOWS, EXTERIOR** R099100-10									
0100	Door frames & trim, only									
0110	Brushwork, primer R099100-20	1 Pord	512	.016	L.F.	.05	.57		.62	.91
0120	Finish coat, exterior latex		512	.016		.05	.57		.62	.91
0130	Primer & 1 coat, exterior latex		300	.027		.11	.97		1.08	1.56
0140	Primer & 2 coats, exterior latex		265	.030		.16	1.10		1.26	1.81
0150	Doors, flush, both sides, incl. frame & trim									
0160	Roll & brush, primer	1 Pord	10	.800	Ea.	4.16	29		33.16	48
0170	Finish coat, exterior latex		10	.800		4.04	29		33.04	48
0180	Primer & 1 coat, exterior latex		7	1.143		8.20	41.50		49.70	71

09 91 13 – Exterior Painting

09 91 13.70 Doors and Windows, Exterior		Crew	Daily Output	Labor-Hours	Unit	2010 Bare Costs Material	Labor	Equipment	Total	Total Incl O&P
0190	Primer & 2 coats, exterior latex	1 Pord	5	1.600	Ea.	12.25	58		70.25	100
0200	Brushwork, stain, sealer & 2 coats polyurethane	↓	4	2	↓	24	72.50		96.50	135
0210	Doors, French, both sides, 10-15 lite, incl. frame & trim									
0220	Brushwork, primer	1 Pord	6	1.333	Ea.	2.08	48.50		50.58	74.50
0230	Finish coat, exterior latex		6	1.333		2.02	48.50		50.52	74
0240	Primer & 1 coat, exterior latex		3	2.667		4.10	97		101.10	149
0250	Primer & 2 coats, exterior latex		2	4		6	145		151	223
0260	Brushwork, stain, sealer & 2 coats polyurethane	↓	2.50	3.200	↓	8.80	116		124.80	183
0270	Doors, louvered, both sides, incl. frame & trim									
0280	Brushwork, primer	1 Pord	7	1.143	Ea.	4.16	41.50		45.66	66.50
0290	Finish coat, exterior latex		7	1.143		4.04	41.50		45.54	66.50
0300	Primer & 1 coat, exterior latex		4	2		8.20	72.50		80.70	117
0310	Primer & 2 coats, exterior latex		3	2.667		12	97		109	157
0320	Brushwork, stain, sealer & 2 coats polyurethane	↓	4.50	1.778	↓	24	64.50		88.50	123
0330	Doors, panel, both sides, incl. frame & trim									
0340	Roll & brush, primer	1 Pord	6	1.333	Ea.	4.16	48.50		52.66	76.50
0350	Finish coat, exterior latex		6	1.333		4.04	48.50		52.54	76.50
0360	Primer & 1 coat, exterior latex		3	2.667		8.20	97		105.20	153
0370	Primer & 2 coats, exterior latex		2.50	3.200		12	116		128	186
0380	Brushwork, stain, sealer & 2 coats polyurethane	↓	3	2.667	↓	24	97		121	171
0400	Windows, per ext. side, based on 15 SF									
0410	1 to 6 lite									
0420	Brushwork, primer	1 Pord	13	.615	Ea.	.82	22.50		23.32	34.50
0430	Finish coat, exterior latex		13	.615		.80	22.50		23.30	34.50
0440	Primer & 1 coat, exterior latex		8	1		1.62	36.50		38.12	56
0450	Primer & 2 coats, exterior latex		6	1.333		2.37	48.50		50.87	74.50
0460	Stain, sealer & 1 coat varnish	↓	7	1.143	↓	3.47	41.50		44.97	66
0470	7 to 10 lite									
0480	Brushwork, primer	1 Pord	11	.727	Ea.	.82	26.50		27.32	40.50
0490	Finish coat, exterior latex		11	.727		.80	26.50		27.30	40.50
0500	Primer & 1 coat, exterior latex		7	1.143		1.62	41.50		43.12	64
0510	Primer & 2 coats, exterior latex		5	1.600		2.37	58		60.37	89
0520	Stain, sealer & 1 coat varnish	↓	6	1.333	↓	3.47	48.50		51.97	76
0530	12 lite									
0540	Brushwork, primer	1 Pord	10	.800	Ea.	.82	29		29.82	44.50
0550	Finish coat, exterior latex		10	.800		.80	29		29.80	44.50
0560	Primer & 1 coat, exterior latex		6	1.333		1.62	48.50		50.12	74
0570	Primer & 2 coats, exterior latex		5	1.600		2.37	58		60.37	89
0580	Stain, sealer & 1 coat varnish	↓	6	1.333		3.40	48.50		51.90	75.50
0590	For oil base paint, add				↓	10%				

09 91 13.80 Trim, Exterior			Crew	Daily Output	Labor-Hours	Unit	Material	Labor	Equipment	Total	Total Incl O&P
0010	**TRIM, EXTERIOR**	R099100-10									
0100	Door frames & trim (see Doors, interior or exterior)										
0110	Fascia, latex paint, one coat coverage	R099100-20									
0120	1" x 4", brushwork		1 Pord	640	.013	L.F.	.02	.45		.47	.70
0130	Roll			1280	.006		.02	.23		.25	.36
0140	Spray			2080	.004		.01	.14		.15	.23
0150	1" x 6" to 1" x 10", brushwork			640	.013		.06	.45		.51	.74
0160	Roll			1230	.007		.06	.24		.30	.42
0170	Spray			2100	.004		.05	.14		.19	.26
0180	1" x 12", brushwork			640	.013		.06	.45		.51	.74
0190	Roll		↓	1050	.008	↓	.06	.28		.34	.48

09 91 13.80 Trim, Exterior		Crew	Daily Output	Labor-Hours	Unit	Material	2010 Bare Costs Labor	Equipment	Total	Total Incl O&P
0200	Spray	1 Pord	2200	.004	L.F.	.05	.13		.18	.25
0210	Gutters & downspouts, metal, zinc chromate paint									
0220	Brushwork, gutters, 5", first coat	1 Pord	640	.013	L.F.	.06	.45		.51	.75
0230	Second coat		960	.008		.06	.30		.36	.52
0240	Third coat		1280	.006		.05	.23		.28	.39
0250	Downspouts, 4", first coat		640	.013		.06	.45		.51	.75
0260	Second coat		960	.008		.06	.30		.36	.52
0270	Third coat		1280	.006		.05	.23		.28	.39
0280	Gutters & downspouts, wood									
0290	Brushwork, gutters, 5", primer	1 Pord	640	.013	L.F.	.05	.45		.50	.74
0300	Finish coat, exterior latex		640	.013		.05	.45		.50	.73
0310	Primer & 1 coat exterior latex		400	.020		.11	.73		.84	1.20
0320	Primer & 2 coats exterior latex		325	.025		.16	.89		1.05	1.51
0330	Downspouts, 4", primer		640	.013		.05	.45		.50	.74
0340	Finish coat, exterior latex		640	.013		.05	.45		.50	.73
0350	Primer & 1 coat exterior latex		400	.020		.11	.73		.84	1.20
0360	Primer & 2 coats exterior latex		325	.025		.08	.89		.97	1.42
0370	Molding, exterior, up to 14" wide									
0380	Brushwork, primer	1 Pord	640	.013	L.F.	.07	.45		.52	.75
0390	Finish coat, exterior latex		640	.013		.06	.45		.51	.74
0400	Primer & 1 coat exterior latex		400	.020		.13	.73		.86	1.22
0410	Primer & 2 coats exterior latex		315	.025		.13	.92		1.05	1.51
0420	Stain & fill		1050	.008		.09	.28		.37	.51
0430	Shellac		1850	.004		.09	.16		.25	.33
0440	Varnish		1275	.006		.10	.23		.33	.45

09 91 13.90 Walls, Masonry (CMU), Exterior		Crew	Daily Output	Labor-Hours	Unit	Material	Labor	Equipment	Total	Total Incl O&P
0350	**WALLS, MASONRY (CMU), EXTERIOR** R099100-10									
0360	Concrete masonry units (CMU), smooth surface									
0370	Brushwork, latex, first coat	1 Pord	640	.013	S.F.	.06	.45		.51	.74
0380	Second coat		960	.008		.05	.30		.35	.50
0390	Waterproof sealer, first coat		736	.011		.24	.40		.64	.86
0400	Second coat		1104	.007		.24	.26		.50	.66
0410	Roll, latex, paint, first coat		1465	.005		.07	.20		.27	.38
0420	Second coat		1790	.004		.05	.16		.21	.30
0430	Waterproof sealer, first coat		1680	.005		.24	.17		.41	.53
0440	Second coat		2060	.004		.24	.14		.38	.48
0450	Spray, latex, paint, first coat		1950	.004		.05	.15		.20	.28
0460	Second coat		2600	.003		.04	.11		.15	.22
0470	Waterproof sealer, first coat		2245	.004		.24	.13		.37	.46
0480	Second coat		2990	.003		.24	.10		.34	.42
0490	Concrete masonry unit (CMU), porous									
0500	Brushwork, latex, first coat	1 Pord	640	.013	S.F.	.12	.45		.57	.81
0510	Second coat		960	.008		.06	.30		.36	.51
0520	Waterproof sealer, first coat		736	.011		.24	.40		.64	.86
0530	Second coat		1104	.007		.24	.26		.50	.66
0540	Roll latex, first coat		1465	.005		.09	.20		.29	.40
0550	Second coat		1790	.004		.06	.16		.22	.30
0560	Waterproof sealer, first coat		1680	.005		.24	.17		.41	.53
0570	Second coat		2060	.004		.24	.14		.38	.48
0580	Spray latex, first coat		1950	.004		.06	.15		.21	.29
0590	Second coat		2600	.003		.04	.11		.15	.22
0600	Waterproof sealer, first coat		2245	.004		.24	.13		.37	.46

09 91 Painting

09 91 13 – Exterior Painting

09 91 13.90 Walls, Masonry (CMU), Exterior			Crew	Daily Output	Labor-Hours	Unit	2010 Bare Costs Material	Labor	Equipment	Total	Total Incl O&P
0610	Second coat		1 Pord	2990	.003	S.F.	.24	.10		.34	.42

09 91 23 – Interior Painting

09 91 23.20 Cabinets and Casework

			Crew	Daily Output	Labor-Hours	Unit	Material	Labor	Equipment	Total	Total Incl O&P
0010	**CABINETS AND CASEWORK**	R099100-10									
1000	Primer coat, oil base, brushwork		1 Pord	650	.012	S.F.	.07	.45		.52	.75
2000	Paint, oil base, brushwork, 1 coat	R099100-20		650	.012		.10	.45		.55	.79
3000	Stain, brushwork, wipe off			650	.012		.07	.45		.52	.75
4000	Shellac, 1 coat, brushwork			650	.012		.07	.45		.52	.75
4500	Varnish, 3 coats, brushwork, sand after 1st coat			325	.025		.24	.89		1.13	1.59
5000	For latex paint, deduct						10%				

09 91 23.33 Doors and Windows, Interior Alkyd (Oil Base)

			Crew	Daily Output	Labor-Hours	Unit	Material	Labor	Equipment	Total	Total Incl O&P
0010	**DOORS AND WINDOWS, INTERIOR ALKYD (OIL BASE)**	R099100-10									
0500	Flush door & frame, 3′ x 7′, oil, primer, brushwork		1 Pord	10	.800	Ea.	2.90	29		31.90	46.50
1000	Paint, 1 coat	R099100-20		10	.800		4.03	29		33.03	48
1400	Stain, brushwork, wipe off			18	.444		1.57	16.15		17.72	25.50
1600	Shellac, 1 coat, brushwork			25	.320		1.52	11.65		13.17	18.95
1800	Varnish, 3 coats, brushwork, sand after 1st coat			9	.889		5	32.50		37.50	53.50
2000	Panel door & frame, 3′ x 7′, oil, primer, brushwork			6	1.333		2.72	48.50		51.22	75
2200	Paint, 1 coat			6	1.333		4.03	48.50		52.53	76.50
2600	Stain, brushwork, panel door, 3′ x 7′, not incl. frame			16	.500		1.57	18.20		19.77	28.50
2800	Shellac, 1 coat, brushwork			22	.364		1.52	13.20		14.72	21.50
3000	Varnish, 3 coats, brushwork, sand after 1st coat			7.50	1.067		5	39		44	63
4400	Windows, including frame and trim, per side										
4600	Colonial type, 6/6 lites, 2′ x 3′, oil, primer, brushwork		1 Pord	14	.571	Ea.	.43	21		21.43	31.50
5800	Paint, 1 coat			14	.571		.64	21		21.64	31.50
6200	3′ x 5′ opening, 6/6 lites, primer coat, brushwork			12	.667		1.07	24		25.07	37
6400	Paint, 1 coat			12	.667		1.59	24		25.59	38
6800	4′ x 8′ opening, 6/6 lites, primer coat, brushwork			8	1		2.29	36.50		38.79	56.50
7000	Paint, 1 coat			8	1		3.39	36.50		39.89	57.50
8000	Single lite type, 2′ x 3′, oil base, primer coat, brushwork			33	.242		.43	8.80		9.23	13.55
8200	Paint, 1 coat			33	.242		.64	8.80		9.44	13.80
8600	3′ x 5′ opening, primer coat, brushwork			20	.400		1.07	14.55		15.62	22.50
8800	Paint, 1 coat			20	.400		1.59	14.55		16.14	23.50
9200	4′ x 8′ opening, primer coat, brushwork			14	.571		2.29	21		23.29	33.50
9400	Paint, 1 coat			14	.571		3.39	21		24.39	34.50

09 91 23.35 Doors and Windows, Interior Latex

			Crew	Daily Output	Labor-Hours	Unit	Material	Labor	Equipment	Total	Total Incl O&P
0010	**DOORS & WINDOWS, INTERIOR LATEX**	R099100-10									
0100	Doors, flush, both sides, incl. frame & trim										
0110	Roll & brush, primer	R099100-20	1 Pord	10	.800	Ea.	3.78	29		32.78	47.50
0120	Finish coat, latex			10	.800		4.09	29		33.09	48
0130	Primer & 1 coat latex			7	1.143		7.85	41.50		49.35	70.50
0140	Primer & 2 coats latex			5	1.600		11.75	58		69.75	99.50
0160	Spray, both sides, primer			20	.400		3.98	14.55		18.53	26
0170	Finish coat, latex			20	.400		4.30	14.55		18.85	26
0180	Primer & 1 coat latex			11	.727		8.35	26.50		34.85	48.50
0190	Primer & 2 coats latex			8	1		12.40	36.50		48.90	67.50
0200	Doors, French, both sides, 10-15 lite, incl. frame & trim										
0210	Roll & brush, primer		1 Pord	6	1.333	Ea.	1.89	48.50		50.39	74
0220	Finish coat, latex			6	1.333		2.05	48.50		50.55	74.50
0230	Primer & 1 coat latex			3	2.667		3.94	97		100.94	148
0240	Primer & 2 coats latex			2	4		5.85	145		150.85	222
0260	Doors, louvered, both sides, incl. frame & trim										

09 91 23.35 Doors and Windows, Interior Latex		Crew	Daily Output	Labor-Hours	Unit	Material	2010 Bare Costs Labor	Equipment	Total	Total Incl O&P
0270	Roll & brush, primer	1 Pord	7	1.143	Ea.	3.78	41.50		45.28	66
0280	Finish coat, latex		7	1.143		4.09	41.50		45.59	66.50
0290	Primer & 1 coat, latex		4	2		7.65	72.50		80.15	116
0300	Primer & 2 coats, latex		3	2.667		11.95	97		108.95	157
0320	Spray, both sides, primer		20	.400		3.98	14.55		18.53	26
0330	Finish coat, latex		20	.400		4.30	14.55		18.85	26
0340	Primer & 1 coat, latex		11	.727		8.35	26.50		34.85	48.50
0350	Primer & 2 coats, latex		8	1		12.70	36.50		49.20	68
0360	Doors, panel, both sides, incl. frame & trim									
0370	Roll & brush, primer	1 Pord	6	1.333	Ea.	3.98	48.50		52.48	76.50
0380	Finish coat, latex		6	1.333		4.09	48.50		52.59	76.50
0390	Primer & 1 coat, latex		3	2.667		7.85	97		104.85	153
0400	Primer & 2 coats, latex		2.50	3.200		11.95	116		127.95	186
0420	Spray, both sides, primer		10	.800		3.98	29		32.98	48
0430	Finish coat, latex		10	.800		4.30	29		33.30	48
0440	Primer & 1 coat, latex		5	1.600		8.35	58		66.35	95.50
0450	Primer & 2 coats, latex		4	2		12.70	72.50		85.20	122
0460	Windows, per interior side, based on 15 SF									
0470	1 to 6 lite									
0480	Brushwork, primer	1 Pord	13	.615	Ea.	.75	22.50		23.25	34.50
0490	Finish coat, enamel		13	.615		.81	22.50		23.31	34.50
0500	Primer & 1 coat enamel		8	1		1.55	36.50		38.05	55.50
0510	Primer & 2 coats enamel		6	1.333		2.36	48.50		50.86	74.50
0530	7 to 10 lite									
0540	Brushwork, primer	1 Pord	11	.727	Ea.	.75	26.50		27.25	40.50
0550	Finish coat, enamel		11	.727		.81	26.50		27.31	40.50
0560	Primer & 1 coat enamel		7	1.143		1.55	41.50		43.05	63.50
0570	Primer & 2 coats enamel		5	1.600		2.36	58		60.36	89
0590	12 lite									
0600	Brushwork, primer	1 Pord	10	.800	Ea.	.75	29		29.75	44.50
0610	Finish coat, enamel		10	.800		.81	29		29.81	44.50
0620	Primer & 1 coat enamel		6	1.333		1.55	48.50		50.05	73.50
0630	Primer & 2 coats enamel		5	1.600		2.36	58		60.36	89
0650	For oil base paint, add					10%				

09 91 23.39 Doors and Windows, Interior Latex, Zero VOC		Crew	Daily Output	Labor-Hours	Unit	Material	Labor	Equipment	Total	Total Incl O&P
0010	**DOORS & WINDOWS, INTERIOR LATEX, ZERO VOC**									
0100	Doors flush, both sides, incl. frame & trim									
0110	Roll & brush, primer G	1 Pord	10	.800	Ea.	5.05	29		34.05	49
0120	Finish coat, latex G		10	.800		5.25	29		34.25	49.50
0130	Primer & 1 coat latex G		7	1.143		10.30	41.50		51.80	73.50
0140	Primer & 2 coats latex G		5	1.600		15.25	58		73.25	103
0160	Spray, both sides, primer G		20	.400		5.35	14.55		19.90	27.50
0170	Finish coat, latex G		20	.400		5.50	14.55		20.05	27.50
0180	Primer & 1 coat latex G		11	.727		10.90	26.50		37.40	51.50
0190	Primer & 2 coats latex G		8	1		16.15	36.50		52.65	72
0200	Doors, French, both sides, 10-15 lite, incl. frame & trim									
0210	Roll & brush, primer G	1 Pord	6	1.333	Ea.	2.53	48.50		51.03	75
0220	Finish coat, latex G		6	1.333		2.62	48.50		51.12	75
0230	Primer & 1 coat latex G		3	2.667		5.15	97		102.15	150
0240	Primer & 2 coats latex G		2	4		7.60	145		152.60	224
0360	Doors, panel, both sides, incl. frame & trim									
0370	Roll & brush, primer G	1 Pord	6	1.333	Ea.	5.35	48.50		53.85	78

JOCWorks™ Software

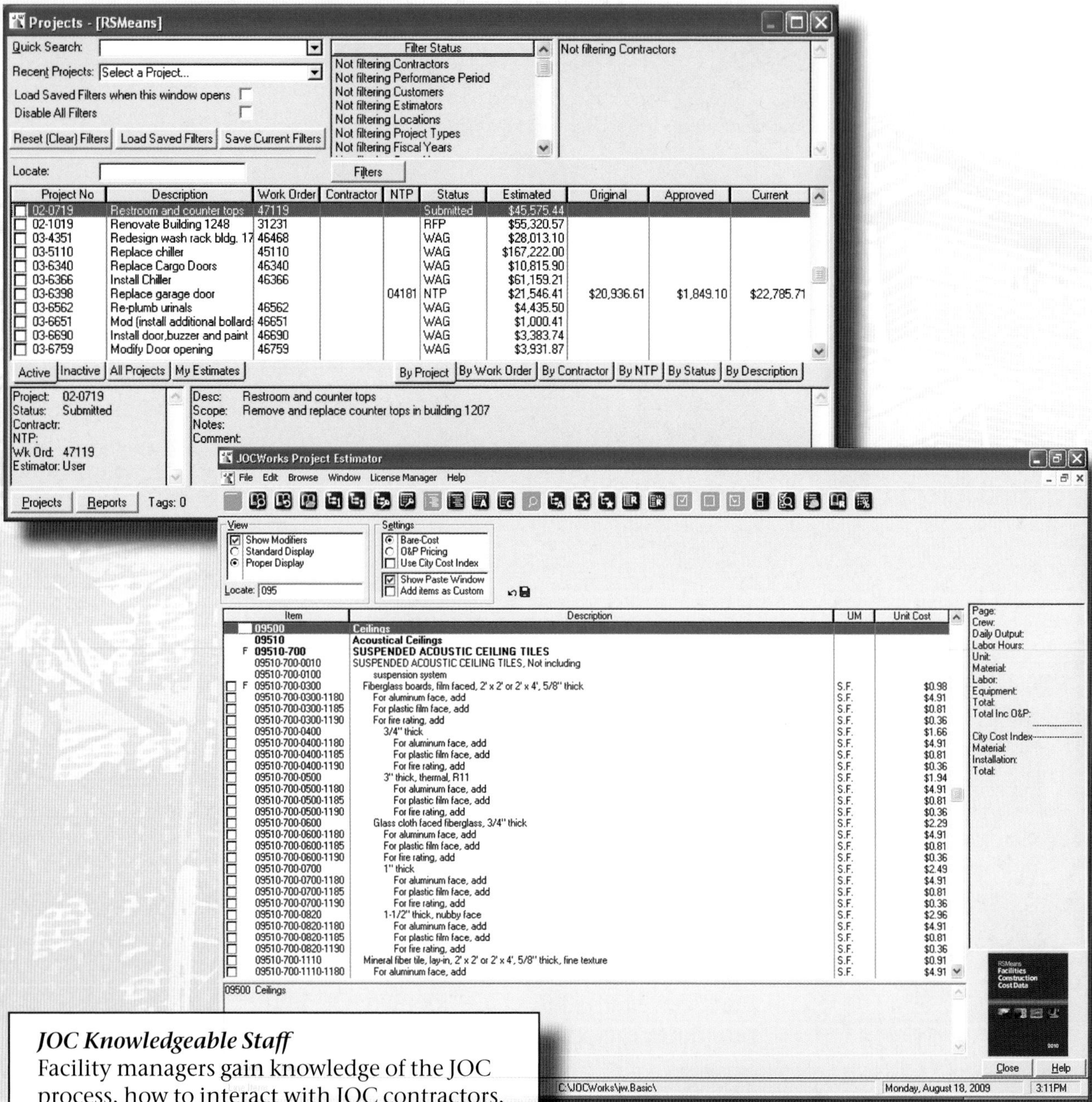

Projects - [RSMeans]

Quick Search:
Recent Projects: Select a Project...
Load Saved Filters when this window opens
Disable All Filters
Reset (Clear) Filters | Load Saved Filters | Save Current Filters

Filter Status:
Not filtering Contractors
Not filtering Performance Period
Not filtering Customers
Not filtering Estimators
Not filtering Locations
Not filtering Project Types
Not filtering Fiscal Years

Not filtering Contractors

Locate: | Filters

Project No	Description	Work Order	Contractor	NTP	Status	Estimated	Original	Approved	Current
02-0719	Restroom and counter tops	47119			Submitted	$45,575.44			
02-1019	Renovate Building 1248	31231			RFP	$55,320.57			
03-4351	Redesign wash rack bldg. 17	46468			WAG	$28,013.10			
03-5110	Replace chiller	45110			WAG	$167,222.00			
03-6340	Replace Cargo Doors	46340			WAG	$10,815.90			
03-6366	Install Chiller	46366			WAG	$61,159.21			
03-6398	Replace garage door			04181	NTP	$21,546.41	$20,936.61	$1,849.10	$22,785.71
03-6562	Re-plumb urinals	46562			WAG	$4,435.50			
03-6651	Mod (install additional bollard	46651			WAG	$1,000.41			
03-6690	Install door,buzzer and paint	46690			WAG	$3,383.74			
03-6759	Modify Door opening	46759			WAG	$3,931.87			

Active | Inactive | All Projects | My Estimates — By Project | By Work Order | By Contractor | By NTP | By Status | By Description

Project: 02-0719
Status: Submitted
Contractr:
NTP:
Wk Ord: 47119
Estimator: User

Desc: Restroom and counter tops
Scope: Remove and replace counter tops in building 1207
Notes:
Comment:

Projects | Reports | Tags: 0

JOCWorks Project Estimator

File Edit Browse Window License Manager Help

View: Show Modifiers / Standard Display / Proper Display
Settings: Bare Cost / O&P Pricing / Use City Cost Index / Show Paste Window / Add items as Custom
Locate: 095

Item	Description	UM	Unit Cost
09500	Ceilings		
09510	**Acoustical Ceilings**		
F 09510-700	**SUSPENDED ACOUSTIC CEILING TILES**		
09510-700-0010	SUSPENDED ACOUSTIC CEILING TILES, Not including		
09510-700-0100	suspension system		
F 09510-700-0300	Fiberglass boards, film faced, 2' x 2' or 2' x 4', 5/8" thick	S.F.	$0.98
09510-700-0300-1180	For aluminum face, add	S.F.	$4.91
09510-700-0300-1185	For plastic film face, add	S.F.	$0.81
09510-700-0300-1190	For fire rating, add	S.F.	$0.36
09510-700-0400	3/4" thick	S.F.	$1.66
09510-700-0400-1180	For aluminum face, add	S.F.	$4.91
09510-700-0400-1185	For plastic film face, add	S.F.	$0.81
09510-700-0400-1190	For fire rating, add	S.F.	$0.36
09510-700-0500	3" thick, thermal, R11	S.F.	$1.94
09510-700-0500-1180	For aluminum face, add	S.F.	$4.91
09510-700-0500-1185	For plastic film face, add	S.F.	$0.81
09510-700-0500-1190	For fire rating, add	S.F.	$0.36
09510-700-0600	Glass cloth faced fiberglass, 3/4" thick	S.F.	$2.29
09510-700-0600-1180	For aluminum face, add	S.F.	$4.91
09510-700-0600-1185	For plastic film face, add	S.F.	$0.81
09510-700-0600-1190	For fire rating, add	S.F.	$0.36
09510-700-0700	1" thick	S.F.	$2.49
09510-700-0700-1180	For aluminum face, add	S.F.	$4.91
09510-700-0700-1185	For plastic film face, add	S.F.	$0.81
09510-700-0700-1190	For fire rating, add	S.F.	$0.36
09510-700-0820	1-1/2" thick, nubby face	S.F.	$2.96
09510-700-0820-1180	For aluminum face, add	S.F.	$4.91
09510-700-0820-1185	For plastic film face, add	S.F.	$0.81
09510-700-0820-1190	For fire rating, add	S.F.	$0.36
09510-700-1110	Mineral fiber tile, lay-in, 2' x 2' or 2' x 4', 5/8" thick, fine texture	S.F.	$0.91
09510-700-1110-1180	For aluminum face, add	S.F.	$4.91

Page:
Crew:
Daily Output:
Labor Hours:
Unit:
Material:
Labor:
Equipment:
Total:
Total Inc O&P:

City Cost Index:
Material:
Installation:
Total:

09500 Ceilings

Close | Help

C:\JOCWorks\jw.Basic\ | Monday, August 18, 2009 | 3:11PM

JOC Knowledgeable Staff
Facility managers gain knowledge of the JOC process, how to interact with JOC contractors, an understanding of the master bidding document (i.e., Unit Price Book) and the development of JOC project specifications and contracts.

RSMeans Unit Price Book (UPB)
A line item analysis of JOC tasks and the components of those tasks. When both the JOC client and contractor agree to this document, negotiations are reduced to project-specific peculiarities.

JOC Training
Designed for working facilities professionals, training increases understanding of how to use JOCWorks™, as well as the Unit Price Book.

JOCWorks™ – a leading project management and document management software program with specialized features for renovation and repair projects.

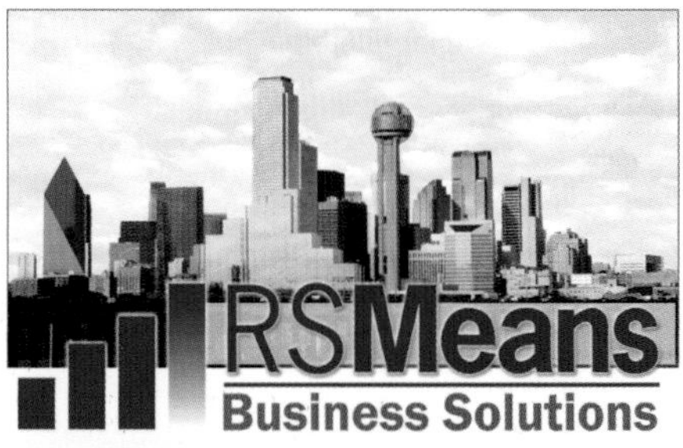

09 91 Painting

09 91 23 – Interior Painting

09 91 23.39 Doors and Windows, Interior Latex, Zero VOC		Crew	Daily Output	Labor-Hours	Unit	Material	Labor	Equipment	Total	Total Incl O&P
						2010 Bare Costs				
0380	Finish coat, latex G	1 Pord	6	1.333	Ea.	5.25	48.50		53.75	78
0390	Primer & 1 coat, latex G		3	2.667		10.30	97		107.30	155
0400	Primer & 2 coats, latex G		2.50	3.200		15.55	116		131.55	190
0420	Spray, both sides, primer G		10	.800		5.35	29		34.35	49.50
0430	Finish coat, latex G		10	.800		5.50	29		34.50	49.50
0440	Primer & 1 coat, latex G		5	1.600		10.90	58		68.90	98.50
0450	Primer & 2 coats, latex G		4	2		16.45	72.50		88.95	126
0460	Windows, per interior side, based on 15 SF									
0470	1 to 6 lite									
0480	Brushwork, primer G	1 Pord	13	.615	Ea.	1	22.50		23.50	34.50
0490	Finish coat, enamel G		13	.615		1.03	22.50		23.53	34.50
0500	Primer & 1 coat enamel G		8	1		2.03	36.50		38.53	56
0510	Primer & 2 coats enamel G		6	1.333		3.07	48.50		51.57	75.50

09 91 23.40 Floors, Interior		Crew	Daily Output	Labor-Hours	Unit	Material	Labor	Equipment	Total	Total Incl O&P
0010	**FLOORS, INTERIOR** R099100-10									
0100	Concrete paint, latex									
0110	Brushwork									
0120	1st coat	1 Pord	975	.008	S.F.	.15	.30		.45	.61
0130	2nd coat		1150	.007		.10	.25		.35	.49
0140	3rd coat		1300	.006		.08	.22		.30	.42
0150	Roll									
0160	1st coat	1 Pord	2600	.003	S.F.	.20	.11		.31	.39
0170	2nd coat		3250	.002		.12	.09		.21	.26
0180	3rd coat		3900	.002		.09	.07		.16	.21
0190	Spray									
0200	1st coat	1 Pord	2600	.003	S.F.	.17	.11		.28	.36
0210	2nd coat		3250	.002		.09	.09		.18	.23
0220	3rd coat		3900	.002		.08	.07		.15	.19
0300	Acid stain and sealer									
0310	Stain, one coat	1 Pord	650	.012	S.F.	.11	.45		.56	.79
0320	Two coats		570	.014		.22	.51		.73	1
0330	Acrylic sealer, one coat		2600	.003		.15	.11		.26	.34
0340	Two coats		1400	.006		.30	.21		.51	.64

09 91 23.52 Miscellaneous, Interior		Crew	Daily Output	Labor-Hours	Unit	Material	Labor	Equipment	Total	Total Incl O&P
0010	**MISCELLANEOUS, INTERIOR** R099100-10									
2400	Floors, conc./wood, oil base, primer/sealer coat, brushwork	2 Pord	1950	.008	S.F.	.06	.30		.36	.51
2450	Roller		5200	.003		.06	.11		.17	.24
2600	Spray		6000	.003		.06	.10		.16	.21
2650	Paint 1 coat, brushwork		1950	.008		.10	.30		.40	.55
2800	Roller		5200	.003		.10	.11		.21	.28
2850	Spray		6000	.003		.11	.10		.21	.26
3000	Stain, wood floor, brushwork, 1 coat		4550	.004		.07	.13		.20	.27
3200	Roller		5200	.003		.08	.11		.19	.26
3250	Spray		6000	.003		.08	.10		.18	.23
3400	Varnish, wood floor, brushwork		4550	.004		.08	.13		.21	.28
3450	Roller		5200	.003		.08	.11		.19	.26
3600	Spray		6000	.003		.09	.10		.19	.24
3650	For dust proofing or anti skid, see Div. 03 35 29.30									
3800	Grilles, per side, oil base, primer coat, brushwork	1 Pord	520	.015	S.F.	.14	.56		.70	.99
3850	Spray		1140	.007		.15	.26		.41	.54
3920	Paint 2 coats, brushwork		325	.025		.41	.89		1.30	1.78
3940	Spray		650	.012		.47	.45		.92	1.19

09 91 Painting

09 91 23 – Interior Painting

09 91 23.52 Miscellaneous, Interior		Crew	Daily Output	Labor-Hours	Unit	Material	Labor	Equipment	Total	Total Incl O&P
						2010 Bare Costs				
5000	Pipe, 1" - 4" diameter, primer or sealer coat, oil base, brushwork	2 Pord	1250	.013	L.F.	.06	.47		.53	.76
5100	Spray		2165	.007		.06	.27		.33	.47
5350	Paint 2 coats, brushwork		775	.021		.15	.75		.90	1.28
5400	Spray		1240	.013		.16	.47		.63	.88
6300	13" - 16" diameter, primer or sealer coat, brushwork		310	.052		.25	1.88		2.13	3.07
6350	Spray		540	.030		.28	1.08		1.36	1.91
6500	Paint 2 coats, brushwork		195	.082		.58	2.98		3.56	5.10
6550	Spray		310	.052		.65	1.88		2.53	3.50
7000	Trim, wood, incl. puttying, under 6" wide									
7200	Primer coat, oil base, brushwork	1 Pord	650	.012	L.F.	.04	.45		.49	.71
7250	Paint, 1 coat, brushwork		650	.012		.05	.45		.50	.73
7450	3 coats		325	.025		.15	.89		1.04	1.50
7500	Over 6" wide, primer coat, brushwork		650	.012		.07	.45		.52	.75
7550	Paint, 1 coat, brushwork		650	.012		.11	.45		.56	.79
7650	3 coats		325	.025		.31	.89		1.20	1.67
8000	Cornice, simple design, primer coat, oil base, brushwork		650	.012	S.F.	.07	.45		.52	.75
8250	Paint, 1 coat		650	.012		.11	.45		.56	.79
8350	Ornate design, primer coat		350	.023		.07	.83		.90	1.32
8400	Paint, 1 coat		350	.023		.11	.83		.94	1.36
8600	Balustrades, primer coat, oil base, brushwork		520	.015		.07	.56		.63	.91
8650	Paint, 1 coat		520	.015		.11	.56		.67	.95
8900	Trusses and wood frames, primer coat, oil base, brushwork		800	.010		.07	.36		.43	.62
8950	Spray		1200	.007		.07	.24		.31	.44
9220	Paint 2 coats, brushwork		500	.016		.21	.58		.79	1.10
9240	Spray		600	.013		.23	.48		.71	.97
9260	Stain, brushwork, wipe off		600	.013		.07	.48		.55	.80
9280	Varnish, 3 coats, brushwork		275	.029		.24	1.06		1.30	1.83
9350	For latex paint, deduct					10%				

09 91 23.62 Electrostatic Painting		Crew	Daily Output	Labor-Hours	Unit	Material	Labor	Equipment	Total	Total Incl O&P
0010	**ELECTROSTATIC PAINTING**									
0100	In Shop									
0200	Flat Surfaces (lockers, casework, elev doors. etc)									
0300	One coat	1 Pord	200	.040	S.F.	.46	1.45		1.91	2.67
0400	Two coats	"	120	.067	"	.67	2.42		3.09	4.35
0500	Irregular Surfaces (furniture, door frames, etc)									
0600	One coat	1 Pord	150	.053	S.F.	.46	1.94		2.40	3.40
0700	Two coats	"	100	.080	"	.67	2.91		3.58	5.05
0800	On Site									
0900	Flat Surfaces (lockers, casework, elev doors. etc)									
1000	One coat	1 Pord	150	.053	S.F.	.46	1.94		2.40	3.40
1100	Two coats	"	100	.080	"	.67	2.91		3.58	5.05
1200	Irregular Surfaces (furniture, door frames, etc)									
1300	One coat	1 Pord	115	.070	S.F.	.46	2.53		2.99	4.27
1400	Two coats	"	70	.114	"	.67	4.15		4.82	6.95

09 91 23.72 Walls and Ceilings, Interior		Crew	Daily Output	Labor-Hours	Unit	Material	Labor	Equipment	Total	Total Incl O&P
0010	**WALLS AND CEILINGS, INTERIOR** R099100-10									
0100	Concrete, drywall or plaster, oil base, primer or sealer coat R099100-20									
0200	Smooth finish, brushwork	1 Pord	1150	.007	S.F.	.07	.25		.32	.46
0240	Roller		1350	.006		.07	.22		.29	.39
0280	Spray		2750	.003		.05	.11		.16	.22
0300	Sand finish, brushwork		975	.008		.07	.30		.37	.51
0340	Roller		1150	.007		.08	.25		.33	.46

09 91 23.72 Walls and Ceilings, Interior		Crew	Daily Output	Labor-Hours	Unit	Material	Labor	Equipment	Total	Total Incl O&P
0380	Spray	1 Pord	2275	.004	S.F.	.06	.13		.19	.25
0800	Paint 2 coats, smooth finish, brushwork		680	.012		.18	.43		.61	.84
0840	Roller		800	.010		.19	.36		.55	.75
0880	Spray		1625	.005		.16	.18		.34	.45
0900	Sand finish, brushwork		605	.013		.18	.48		.66	.92
0940	Roller		1020	.008		.19	.29		.48	.63
0980	Spray		1700	.005		.16	.17		.33	.43
1200	Paint 3 coats, smooth finish, brushwork		510	.016		.27	.57		.84	1.14
1240	Roller		650	.012		.28	.45		.73	.98
1280	Spray		1625	.005		.24	.18		.42	.54
1600	Glaze coating, 2 coats, spray, clear		1200	.007		.49	.24		.73	.90
1640	Multicolor		1200	.007		.95	.24		1.19	1.41
1660	Painting walls, complete, including surface prep, primer &									
1670	2 coats finish, on drywall or plaster, with roller	1 Pord	325	.025	S.F.	.18	.89		1.07	1.53
1700	For latex paint, deduct					10%				
1800	For ceiling installations, add						25%			
2000	Masonry or concrete block, oil base, primer or sealer coat									
2100	Smooth finish, brushwork	1 Pord	1224	.007	S.F.	.06	.24		.30	.42
2180	Spray		2400	.003		.09	.12		.21	.28
2200	Sand finish, brushwork		1089	.007		.10	.27		.37	.51
2280	Spray		2400	.003		.09	.12		.21	.28
2800	Paint 2 coats, smooth finish, brushwork		756	.011		.20	.38		.58	.79
2880	Spray		1360	.006		.18	.21		.39	.52
2900	Sand finish, brushwork		672	.012		.20	.43		.63	.86
2980	Spray		1360	.006		.18	.21		.39	.52
3600	Glaze coating, 3 coats, spray, clear		900	.009		.70	.32		1.02	1.25
3620	Multicolor		900	.009		1.10	.32		1.42	1.69
4000	Block filler, 1 coat, brushwork		425	.019		.13	.68		.81	1.17
4100	Silicone, water repellent, 2 coats, spray		2000	.004		.26	.15		.41	.50
4120	For latex paint, deduct					10%				
8200	For work 8' - 15' H, add						10%			
8300	For work over 15' H, add						20%			
8400	For light textured surfaces, add						10%			
8410	Heavy textured, add						25%			

09 91 23.74 Walls and Ceilings, Interior, Zero VOC Latex			Crew	Daily Output	Labor-Hours	Unit	Material	Labor	Equipment	Total	Total Incl O&P
0010	**WALLS AND CEILINGS, INTERIOR, ZERO VOC LATEX**										
0100	Concrete, dry wall or plaster, latex, primer or sealer coat										
0200	Smooth finish, brushwork	G	1 Pord	1150	.007	S.F.	.07	.25		.32	.45
0240	Roller	G		1350	.006		.06	.22		.28	.39
0280	Spray	G		2750	.003		.05	.11		.16	.21
0300	Sand finish, brushwork	G		975	.008		.06	.30		.36	.51
0340	Roller	G		1150	.007		.07	.25		.32	.46
0380	Spray	G		2275	.004		.05	.13		.18	.25
0800	Paint 2 coats, smooth finish, brushwork	G		680	.012		.13	.43		.56	.78
0840	Roller	G		800	.010		.13	.36		.49	.68
0880	Spray	G		1625	.005		.11	.18		.29	.39
0900	Sand finish, brushwork	G		605	.013		.12	.48		.60	.85
0940	Roller	G		1020	.008		.13	.29		.42	.56
0980	Spray	G		1700	.005		.11	.17		.28	.37
1200	Paint 3 coats, smooth finish, brushwork	G		510	.016		.18	.57		.75	1.05
1240	Roller	G		650	.012		.19	.45		.64	.88
1280	Spray	G		1625	.005		.17	.18		.35	.45

09 91 Painting

09 91 23 – Interior Painting

09 91 23.74 Walls and Ceilings, Interior, Zero VOC Latex		Crew	Daily Output	Labor-Hours	Unit	Material	2010 Bare Costs Labor	Equipment	Total	Total Incl O&P
1800	For ceiling installations, add G				S.F.		25%			
8200	For work 8' - 15' H, add						10%			
8300	For work over 15' H, add				↓		20%			

09 91 23.75 Dry Fall Painting		Crew	Daily Output	Labor-Hours	Unit	Material	Labor	Equipment	Total	Total Incl O&P
0010	**DRY FALL PAINTING** R099100-10									
0100	Sprayed on walls, gypsum board or plaster									
0220	One coat R099100-20	1 Pord	2600	.003	S.F.	.05	.11		.16	.22
0250	Two coats		1560	.005		.10	.19		.29	.39
0280	Concrete or textured plaster, one coat		1560	.005		.05	.19		.24	.33
0310	Two coats		1300	.006		.10	.22		.32	.44
0340	Concrete block, one coat		1560	.005		.05	.19		.24	.33
0370	Two coats		1300	.006		.10	.22		.32	.44
0400	Wood, one coat		877	.009		.05	.33		.38	.54
0430	Two coats	↓	650	.012	↓	.10	.45		.55	.78
0440	On ceilings, gypsum board or plaster									
0470	One coat	1 Pord	1560	.005	S.F.	.05	.19		.24	.33
0500	Two coats		1300	.006		.10	.22		.32	.44
0530	Concrete or textured plaster, one coat		1560	.005		.05	.19		.24	.33
0560	Two coats		1300	.006		.10	.22		.32	.44
0570	Structural steel, bar joists or metal deck, one coat		1560	.005		.05	.19		.24	.33
0580	Two coats	↓	1040	.008	↓	.10	.28		.38	.53

09 93 Staining and Transparent Finishing

09 93 23 – Interior Staining and Finishing

09 93 23.10 Varnish		Crew	Daily Output	Labor-Hours	Unit	Material	Labor	Equipment	Total	Total Incl O&P
0010	**VARNISH**									
0012	1 coat + sealer, on wood trim, brush, no sanding included	1 Pord	400	.020	S.F.	.07	.73		.80	1.16
0100	Hardwood floors, 2 coats, no sanding included, roller	"	1890	.004	"	.15	.15		.30	.40

09 96 High-Performance Coatings

09 96 23 – Graffiti-Resistant Coatings

09 96 23.10 Graffiti Resistant Treatments		Crew	Daily Output	Labor-Hours	Unit	Material	Labor	Equipment	Total	Total Incl O&P
0010	**GRAFFITI RESISTANT TREATMENTS**, sprayed on walls									
0100	Non-sacrificial, permanent non-stick coating, clear, on metals	1 Pord	2000	.004	S.F.	2.06	.15		2.21	2.49
0200	Concrete		2000	.004		2.35	.15		2.50	2.80
0300	Concrete block		2000	.004		3.03	.15		3.18	3.56
0400	Brick		2000	.004		3.44	.15		3.59	4
0500	Stone		2000	.004		3.44	.15		3.59	4
0600	Unpainted wood		2000	.004		3.97	.15		4.12	4.59
2000	Semi-permanent cross linking polymer primer, on metals		2000	.004		.40	.15		.55	.66
2100	Concrete		2000	.004		.48	.15		.63	.75
2200	Concrete block		2000	.004		.60	.15		.75	.88
2300	Brick		2000	.004		.48	.15		.63	.75
2400	Stone		2000	.004		.48	.15		.63	.75
2500	Unpainted wood		2000	.004		.67	.15		.82	.95
3000	Top coat, on metals		2000	.004		.43	.15		.58	.69
3100	Concrete		2000	.004		.49	.15		.64	.75
3200	Concrete block		2000	.004		.68	.15		.83	.97
3300	Brick	↓	2000	.004	↓	.57	.15		.72	.84

09 96 High-Performance Coatings

09 96 23 – Graffiti-Resistant Coatings

09 96 23.10 Graffiti Resistant Treatments		Crew	Daily Output	Labor-Hours	Unit	Material	2010 Bare Costs Labor	Equipment	Total	Total Incl O&P
3400	Stone	1 Pord	2000	.004	S.F.	.57	.15		.72	.84
3500	Unpainted wood		2000	.004		.68	.15		.83	.97
5000	Sacrificial, water based, on metal		2000	.004		.24	.15		.39	.49
5100	Concrete		2000	.004		.24	.15		.39	.49
5200	Concrete block		2000	.004		.24	.15		.39	.49
5300	Brick		2000	.004		.24	.15		.39	.49
5400	Stone		2000	.004		.24	.15		.39	.49
5500	Unpainted wood	↓	2000	.004	↓	.24	.15		.39	.49
8000	Cleaner for use after treatment									
8100	Towels or wipes, per package of 30				Ea.	.63			.63	.70
8200	Aerosol spray, 24 oz. can				"	18			18	19.80

09 96 56 – Epoxy Coatings

09 96 56.20 Wall Coatings		Crew	Daily Output	Labor-Hours	Unit	Material	Labor	Equipment	Total	Total Incl O&P
0010	**WALL COATINGS**									
0100	Acrylic glazed coatings, minimum	1 Pord	525	.015	S.F.	.28	.55		.83	1.13
0200	Maximum		305	.026		.60	.95		1.55	2.08
0300	Epoxy coatings, minimum		525	.015		.37	.55		.92	1.23
0400	Maximum		170	.047		1.12	1.71		2.83	3.78
0600	Exposed aggregate, troweled on, 1/16" to 1/4", minimum		235	.034		.56	1.24		1.80	2.46
0700	Maximum (epoxy or polyacrylate)		130	.062		1.20	2.24		3.44	4.65
0900	1/2" to 5/8" aggregate, minimum		130	.062		1.11	2.24		3.35	4.55
1000	Maximum		80	.100		1.90	3.64		5.54	7.50
1500	Exposed aggregate, sprayed on, 1/8" aggregate, minimum		295	.027		.51	.99		1.50	2.03
1600	Maximum		145	.055		.96	2.01		2.97	4.04
1800	High build epoxy, 50 mil, minimum		390	.021		.62	.75		1.37	1.79
1900	Maximum		95	.084		1.06	3.06		4.12	5.75
2100	Laminated epoxy with fiberglass, minimum		295	.027		.67	.99		1.66	2.21
2200	Maximum		145	.055		1.19	2.01		3.20	4.29
2400	Sprayed perlite or vermiculite, 1/16" thick, minimum		2935	.003		.24	.10		.34	.41
2500	Maximum		640	.013		.68	.45		1.13	1.43
2700	Vinyl plastic wall coating, minimum		735	.011		.30	.40		.70	.92
2800	Maximum		240	.033		.75	1.21		1.96	2.63
3000	Urethane on smooth surface, 2 coats, minimum		1135	.007		.24	.26		.50	.64
3100	Maximum		665	.012		.53	.44		.97	1.23
3600	Ceramic-like glazed coating, cementitious, minimum		440	.018		.44	.66		1.10	1.46
3700	Maximum		345	.023		.74	.84		1.58	2.06
3900	Resin base, minimum		640	.013		.30	.45		.75	1.01
4000	Maximum	↓	330	.024	↓	.49	.88		1.37	1.85

09 97 Special Coatings

09 97 13 – Steel Coatings

09 97 13.23 Exterior Steel Coatings		Crew	Daily Output	Labor-Hours	Unit	Material	Labor	Equipment	Total	Total Incl O&P
0010	**EXTERIOR STEEL COATINGS** R050516-30									
6100	Cold galvanizing, brush in field	1 Psst	1100	.007	S.F.	.07	.27		.34	.56
6510	Paints & protective coatings, sprayed in field									
6520	Alkyds, primer	2 Psst	3600	.004	S.F.	.06	.17		.23	.37
6540	Gloss topcoats		3200	.005		.06	.19		.25	.40
6560	Silicone alkyd		3200	.005		.12	.19		.31	.46
6610	Epoxy, primer		3000	.005		.24	.20		.44	.62
6630	Intermediate or topcoat	↓	2800	.006	↓	.21	.21		.42	.62

09 97 Special Coatings

09 97 13 – Steel Coatings

09 97 13.23 Exterior Steel Coatings		Crew	Daily Output	Labor-Hours	Unit	2010 Bare Costs Material	Labor	Equipment	Total	Total Incl O&P
6650	Enamel coat	2 Psst	2800	.006	S.F.	.27	.21		.48	.67
6700	Epoxy ester, primer		2800	.006		.52	.21		.73	.95
6720	Topcoats		2800	.006		.18	.21		.39	.58
6810	Latex primer		3600	.004		.05	.17		.22	.36
6830	Topcoats		3200	.005		.06	.19		.25	.39
6910	Universal primers, one part, phenolic, modified alkyd		2000	.008		.11	.30		.41	.66
6940	Two part, epoxy spray		2000	.008		.28	.30		.58	.85
7000	Zinc rich primers, self cure, spray, inorganic		1800	.009		.59	.33		.92	1.24
7010	Epoxy, spray, organic	↓	1800	.009	↓	.21	.33		.54	.83
7020	Above one story, spray painting simple structures, add						25%			
7030	Intricate structures, add						50%			

Estimating Tips

General

- The items in this division are usually priced per square foot or each.
- Many items in Division 10 require some type of support system or special anchors that are not usually furnished with the item. The required anchors must be added to the estimate in the appropriate division.
- Some items in Division 10, such as lockers, may require assembly before installation. Verify the amount of assembly required. Assembly can often exceed installation time.

10 20 00 Interior Specialties

- Support angles and blocking are not included in the installation of toilet compartments, shower/dressing compartments, or cubicles. Appropriate line items from Divisions 5 or 6 may need to be added to support the installations.
- Toilet partitions are priced by the stall. A stall consists of a side wall, pilaster, and door with hardware. Toilet tissue holders and grab bars are extra.
- The required acoustical rating of a folding partition can have a significant impact on costs. Verify the sound transmission coefficient rating of the panel priced to the specification requirements.
- Grab bar installation does not include supplemental blocking or backing to support the required load. When grab bars are installed at an existing facility, provisions must be made to attach the grab bars to solid structure.

Reference Numbers

Reference numbers are shown in shaded boxes at the beginning of some major classifications. These numbers refer to related items in the Reference Section. The reference information may be an estimating procedure, an alternate pricing method, or technical information.

Note: Not all subdivisions listed here necessarily appear in this publication.

10 05 Common Work Results for Specialties

10 05 05 – Selective Specialties Demolition

10 05 05.10 Selective Demolition, Specialties		Crew	Daily Output	Labor-Hours	Unit	Material	2010 Bare Costs Labor	Equipment	Total	Total Incl O&P
0010	**SELECTIVE DEMOLITION, SPECIALTIES**									
1100	Boards and panels, wall mounted	2 Clab	15	1.067	Ea.		35.50		35.50	54.50
1200	Cases, for directory and/or bulletin boards, incl doors		24	.667			22		22	34
1850	Shower partitions, cabinet or stall, incl base and door		8	2			66		66	102
1855	Shower receptor, terrazzo or concrete	1 Clab	14	.571			18.90		18.90	29
1900	Curtain track or rod, hospital type, ceiling mounted or suspended	"	220	.036	L.F.		1.20		1.20	1.86
1910	Toilet Cubicles, remove	2 Clab	8	2	Ea.		66		66	102
1930	Urinal Screen, remove	1 Clab	12	.667	"		22		22	34
2650	Wall guard, misc. wall or corner protection	"	320	.025	L.F.		.83		.83	1.28
2750	Access floor, metal panel system, incl pedestals, covering	2 Clab	850	.019	S.F.		.62		.62	.96
3050	Fireplace, prefab, freestanding or wall hung, incl hood and screen	1 Clab	2	4	Ea.		132		132	204
3054	Chimney top, simulated brick, 4′ high	"	15	.533			17.65		17.65	27
3200	Stove, woodburning, cast iron	2 Clab	2	8			265		265	410
3440	Weathervane, residential	1 Clab	12	.667			22		22	34
3500	Flagpole, groundset, to 70′ high, excl base/fndn	K-1	1	16			590	296	886	1,225
3555	To 30′ high	"	2.50	6.400			236	118	354	490
4300	Letter, signs or plaques, exterior on wall	1 Clab	20	.400			13.25		13.25	20.50
4310	Signs, street, reflective aluminum, incl post and bracket		60	.133			4.41		4.41	6.80
4320	Door signs interior on door 6″ x 6″, selective demolition		20	.400			13.25		13.25	20.50
4550	Turnstiles, manual or electric	2 Clab	2	8			265		265	410
5050	Lockers	1 Clab	15	.533	Opng.		17.65		17.65	27
5250	Cabinets, recessed	Q-12	12	1.333	Ea.		60.50		60.50	90.50
5260	Mail boxes, Horiz., Key Lock, front loading, Remove	1 Carp	34	.235	"		9.80		9.80	15.05
5350	Awning, fabric, including frame	2 Clab	100	.160	S.F.		5.30		5.30	8.15
6050	Partition, woven wire		1400	.011			.38		.38	.58
6100	Folding gate, security, door or window		500	.032			1.06		1.06	1.63
6580	Acoustic air wall		650	.025			.81		.81	1.26
7550	Telephone enclosure, exterior, post mounted		3	5.333	Ea.		177		177	272
8850	Scale, platform, excludes foundation or pit		.25	64	"		2,125		2,125	3,275

10 11 Visual Display Surfaces

10 11 13 – Chalkboards

10 11 13.13 Fixed Chalkboards		Crew	Daily Output	Labor-Hours	Unit	Material	Labor	Equipment	Total	Total Incl O&P
0010	**FIXED CHALKBOARDS** Porcelain enamel steel									
3900	Wall hung									
4000	Aluminum frame and chalktrough									
4200	3′ x 4′	2 Carp	16	1	Ea.	185	41.50		226.50	268
4300	3′ x 5′		15	1.067		248	44.50		292.50	340
4500	4′ x 8′		14	1.143		355	47.50		402.50	465
4600	4′ x 12′		13	1.231		515	51		566	645
4700	Wood frame and chalktrough									
4800	3′ x 4′	2 Carp	16	1	Ea.	188	41.50		229.50	271
5000	3′ x 5′		15	1.067		233	44.50		277.50	325
5100	4′ x 5′		14	1.143		248	47.50		295.50	345
5300	4′ x 8′		13	1.231		350	51		401	465
5400	Liquid chalk, white porcelain enamel, wall hung									
5420	Deluxe units, aluminum trim and chalktrough									
5450	4′ x 4′	2 Carp	16	1	Ea.	216	41.50		257.50	300
5500	4′ x 8′		14	1.143		335	47.50		382.50	445
5550	4′ x 12′		12	1.333		500	55.50		555.50	635
5700	Wood trim and chalktrough									

10 11 13 – Chalkboards

10 11 13.13 Fixed Chalkboards		Crew	Daily Output	Labor-Hours	Unit	Material	2010 Bare Costs Labor	Equipment	Total	Total Incl O&P
5900	4′ x 4′	2 Carp	16	1	Ea.	550	41.50		591.50	670
6000	4′ x 6′		15	1.067		655	44.50		699.50	790
6200	4′ x 8′		14	1.143		760	47.50		807.50	910
6300	Liquid chalk, felt tip markers					1.52			1.52	1.67
6500	Erasers					1.54			1.54	1.69
6600	Board cleaner, 8 oz. bottle					4.49			4.49	4.94

10 11 13.23 Modular-Support-Mounted Chalkboards		Crew	Daily Output	Labor-Hours	Unit	Material	Labor	Equipment	Total	Total Incl O&P
0010	**MODULAR-SUPPORT-MOUNTED CHALKBOARDS**									
0400	Sliding chalkboards									
0450	Vertical, one sliding board with back panel, wall mounted									
0500	8′ x 4′	2 Carp	8	2	Ea.	2,050	83		2,133	2,375
0520	8′ x 8′		7.50	2.133		2,875	88.50		2,963.50	3,300
0540	8′ x 12′		7	2.286		3,500	95		3,595	4,000
0600	Two sliding boards, with back panel									
0620	8′ x 4′	2 Carp	8	2	Ea.	3,300	83		3,383	3,750
0640	8′ x 8′		7.50	2.133		4,900	88.50		4,988.50	5,500
0660	8′ x 12′		7	2.286		7,925	95		8,020	8,875
0700	Horizontal, two track									
0800	4′ x 8′, 2 sliding panels	2 Carp	8	2	Ea.	1,825	83		1,908	2,150
0820	4′ x 12′, 2 sliding panels		7.50	2.133		2,400	88.50		2,488.50	2,750
0840	4′ x 16′, 4 sliding panels		7	2.286		3,225	95		3,320	3,700
0900	Four track, four sliding panels									
0920	4′ x 8′	2 Carp	8	2	Ea.	2,975	83		3,058	3,375
0940	4′ x 12′		7.50	2.133		3,875	88.50		3,963.50	4,375
0960	4′ x 16′		7	2.286		5,050	95		5,145	5,700
1200	Vertical, motor operated									
1400	One sliding panel with back panel									
1450	10′ x 4′	2 Carp	4	4	Ea.	5,325	166		5,491	6,125
1500	10′ x 10′		3.75	4.267		6,425	177		6,602	7,350
1550	10′ x 16′		3.50	4.571		7,575	190		7,765	8,650
1700	Two sliding panels with back panel									
1750	10′ x 4′	2 Carp	4	4	Ea.	9,475	166		9,641	10,700
1800	10′ x 10′		3.75	4.267		11,000	177		11,177	12,400
1850	10′ x 16′		3.50	4.571		12,600	190		12,790	14,200
2000	Three sliding panels with back panel									
2100	10′ x 4′	2 Carp	4	4	Ea.	13,200	166		13,366	14,800
2150	10′ x 10′		3.75	4.267		14,500	177		14,677	16,200
2200	10′ x 16′		3.50	4.571		17,500	190		17,690	19,500
2400	For projection screen, glass beaded, add				S.F.	7.65			7.65	8.40
2500	For remote control, 1 panel control, add				Ea.	360			360	395
2600	2 panel control, add				"	620			620	680
2800	For units without back panels, deduct				S.F.	4.84			4.84	5.30
2850	For liquid chalk porcelain panels, add				"	5.15			5.15	5.70
3000	Swing leaf, any comb. of chalkboard & cork, aluminum frame									
3100	Floor style, 6 panels									
3150	30″ x 40″ panels				Ea.	1,525			1,525	1,675
3200	48″ x 40″ panels				"	2,550			2,550	2,800
3300	Wall mounted, 6 panels									
3400	30″ x 40″ panels	2 Carp	16	1	Ea.	1,425	41.50		1,466.50	1,650
3450	48″ x 40″ panels	"	16	1	"	1,775	41.50		1,816.50	2,025
3600	Extra panels for swing leaf units									
3700	30″ x 40″ panels				Ea.	287			287	315

10 11 Visual Display Surfaces

10 11 13 – Chalkboards

10 11 13.23 Modular-Support-Mounted Chalkboards		Crew	Daily Output	Labor-Hours	Unit	2010 Bare Costs Material	Labor	Equipment	Total	Total Incl O&P
3750	48″ x 40″ panels				Ea.	355			355	390

10 11 13.43 Portable Chalkboards

		Crew	Daily Output	Labor-Hours	Unit	Material	Labor	Equipment	Total	Total Incl O&P
0010	**PORTABLE CHALKBOARDS**									
0100	Freestanding, reversible									
0120	Economy, wood frame, 4′ x 6′									
0140	Chalkboard both sides				Ea.	525			525	575
0160	Chalkboard one side, cork other side				″	500			500	555
0200	Standard, lightweight satin finished aluminum, 4′ x 6′									
0220	Chalkboard both sides				Ea.	565			565	620
0240	Chalkboard one side, cork other side				″	595			595	655
0300	Deluxe, heavy duty extruded aluminum, 4′ x 6′									
0320	Chalkboard both sides				Ea.	990			990	1,100
0340	Chalkboard one side, cork other side				″	890			890	980

10 11 23 – Tackboards

10 11 23.10 Fixed Tackboards

		Crew	Daily Output	Labor-Hours	Unit	Material	Labor	Equipment	Total	Total Incl O&P
0010	**FIXED TACKBOARDS**									
0020	Cork sheets, unbacked, no frame, 1/4″ thick	2 Carp	290	.055	S.F.	3.24	2.29		5.53	7.10
0100	1/2″ thick		290	.055		5.70	2.29		7.99	9.80
0300	Fabric-face, no frame, on 7/32″ cork underlay		290	.055		6	2.29		8.29	10.15
0400	On 1/4″ cork on 1/4″ hardboard		290	.055		7.85	2.29		10.14	12.20
0600	With edges wrapped		290	.055		8.30	2.29		10.59	12.65
0700	On 7/16″ fire retardant core		290	.055		7.60	2.29		9.89	11.90
0900	With edges wrapped	↓	290	.055		12.30	2.29		14.59	17.05
1000	Designer fabric only, cut to size					1.83			1.83	2.01
1200	1/4″ vinyl cork, on 1/4″ hardboard, no frame	2 Carp	290	.055		8.40	2.29		10.69	12.75
1300	On 1/4″ coreboard		290	.055	↓	7.95	2.29		10.24	12.30
2000	For map and display rail, economy, add		385	.042	L.F.	2.40	1.73		4.13	5.30
2100	Deluxe, add		350	.046	″	4.20	1.90		6.10	7.55
2120	Prefabricated, 1/4″ cork, 3′ x 5′ with aluminum frame		16	1	Ea.	113	41.50		154.50	188
2140	Wood frame		16	1		143	41.50		184.50	221
2160	4′ x 4′ with aluminum frame		16	1		98.50	41.50		140	172
2180	Wood frame		16	1		128	41.50		169.50	205
2200	4′ x 8′ with aluminum frame		14	1.143		177	47.50		224.50	267
2210	With wood frame		14	1.143		130	47.50		177.50	216
2220	4′ x 12′ with aluminum frame	↓	12	1.333	↓	263	55.50		318.50	375
2230	Bulletin board case, single glass door, with lock									
2240	36″ x 24″, economy	2 Carp	12	1.333	Ea.	297	55.50		352.50	410
2250	Deluxe		12	1.333		520	55.50		575.50	655
2260	42″ x 30″, economy		12	1.333		340	55.50		395.50	460
2270	Deluxe	↓	12	1.333	↓	200	55.50		255.50	305
2300	Glass enclosed cabinets, alum., cork panel, hinged doors									
2400	3′ x 3′, 1 door	2 Carp	12	1.333	Ea.	570	55.50		625.50	715
2500	4′ x 4′, 2 door		11	1.455		955	60.50		1,015.50	1,150
2600	4′ x 7′, 3 door		10	1.600		1,650	66.50		1,716.50	1,925
2800	4′ x 10′, 4 door	↓	8	2		2,200	83		2,283	2,525
2900	For lights, add per door opening	1 Elec	13	.615		131	30		161	190
3100	Horizontal sliding units, 4 doors, 4′ x 8′, 8′ x 4′	2 Carp	9	1.778		1,825	74		1,899	2,150
3200	4′ x 12′		7	2.286		2,400	95		2,495	2,775
3400	8 doors, 4′ x 16′		5	3.200		3,250	133		3,383	3,775
3500	4′ x 24′	↓	4	4	↓	4,375	166		4,541	5,075

10 11 Visual Display Surfaces

10 11 23 – Tackboards

10 11 23.20 Control Boards		Crew	Daily Output	Labor-Hours	Unit	Material	2010 Bare Costs Labor	Equipment	Total	Total Incl O&P
0010	**CONTROL BOARDS**									
0020	Magnetic, porcelain finish, 18" x 24", framed	2 Carp	8	2	Ea.	199	83		282	345
0100	24" x 36"		7.50	2.133		255	88.50		343.50	420
0200	36" x 48"		7	2.286		390	95		485	575
0300	48" x 72"		6	2.667		860	111		971	1,125
0400	48" x 96"		5	3.200		1,075	133		1,208	1,375

10 13 Directories

10 13 10 – Building Directories

10 13 10.10 Directory Boards		Crew	Daily Output	Labor-Hours	Unit	Material	Labor	Equipment	Total	Total Incl O&P
0010	**DIRECTORY BOARDS**									
0050	Plastic, glass covered, 30" x 20"	2 Carp	3	5.333	Ea.	243	222		465	605
0100	36" x 48"		2	8		745	330		1,075	1,325
0300	Grooved cork, 30" x 20"		3	5.333		375	222		597	750
0400	36" x 48"		2	8		505	330		835	1,075
0600	Black felt, 30" x 20"		3	5.333		193	222		415	550
0700	36" x 48"		2	8		375	330		705	925
0900	Outdoor, weatherproof, black plastic, 36" x 24"		2	8		720	330		1,050	1,300
1000	36" x 36"		1.50	10.667		830	445		1,275	1,600
1800	Indoor, economy, open face, 18" x 24"		7	2.286		128	95		223	287
1900	24" x 36"		7	2.286		124	95		219	282
2000	36" x 24"		6	2.667		141	111		252	325
2100	36" x 48"		6	2.667		218	111		329	410
2400	Building directory, alum., black felt panels, 1 door, 24" x 18"		4	4		300	166		466	585
2500	36" x 24"		3.50	4.571		355	190		545	685
2600	48" x 32"		3	5.333		575	222		797	975
2700	2 door, 36" x 48"		2.50	6.400		630	266		896	1,100
2800	36" x 60"		2	8		825	330		1,155	1,425
2900	48" x 60"		1	16		915	665		1,580	2,025
3100	For bronze enamel finish, add					15%				
3200	For bronze anodized finish, add					25%				
3400	For illuminated directory, single door unit, add					110			110	120
3500	For 6" header panel, 6 letters per foot, add				L.F.	20.50			20.50	22.50

10 14 Signage

10 14 19 – Dimensional Letter Signage

10 14 19.10 Exterior Signs		Crew	Daily Output	Labor-Hours	Unit	Material	Labor	Equipment	Total	Total Incl O&P
0010	**EXTERIOR SIGNS**									
0020	Letters, 2" high, 3/8" deep, cast bronze	1 Carp	24	.333	Ea.	22	13.85		35.85	46
0140	1/2" deep, cast aluminum		18	.444		22	18.45		40.45	53
0160	Cast bronze		32	.250		30	10.40		40.40	49
0300	6" high, 5/8" deep, cast aluminum		24	.333		27.50	13.85		41.35	52
0400	Cast bronze		24	.333		56.50	13.85		70.35	83.50
0600	8" high, 3/4" deep, cast aluminum		14	.571		33.50	23.50		57	73.50
0700	Cast bronze		20	.400		80.50	16.60		97.10	114
0900	10" high, 1" deep, cast aluminum		18	.444		49.50	18.45		67.95	83
1000	Bronze		18	.444		110	18.45		128.45	150
1200	12" high, 1-1/4" deep, cast aluminum		12	.667		54	27.50		81.50	102
1500	Cast bronze		18	.444		129	18.45		147.45	171

10 14 19 – Dimensional Letter Signage

10 14 19.10 Exterior Signs		Crew	Daily Output	Labor-Hours	Unit	Material	Labor	Equipment	Total	Total Incl O&P
						2010 Bare Costs				
1600	14" high, 2-5/16" deep, cast aluminum	1 Carp	12	.667	Ea.	75	27.50		102.50	125
1800	Fabricated stainless steel, 6" high, 2" deep		20	.400		53	16.60		69.60	83.50
1900	12" high, 3" deep		18	.444		118	18.45		136.45	158
2100	18" high, 3" deep		12	.667		228	27.50		255.50	293
2200	24" high, 4" deep		10	.800		385	33		418	470
2700	Acrylic, on high density foam, 12" high, 2" deep		20	.400		20.50	16.60		37.10	48
2800	18" high, 2" deep		18	.444		51	18.45		69.45	84.50
3900	Plaques, custom, 20" x 30", for up to 450 letters, cast aluminum	2 Carp	4	4		1,050	166		1,216	1,400
4000	Cast bronze		4	4		1,550	166		1,716	1,975
4200	30" x 36", up to 900 letters cast aluminum		3	5.333		1,925	222		2,147	2,475
4300	Cast bronze		3	5.333		3,700	222		3,922	4,400
4500	36" x 48", for up to 1300 letters, cast bronze		2	8		4,225	330		4,555	5,150
4800	Signs, reflective alum. directional signs, dbl. face, 2-way, w/bracket		30	.533		86	22		108	129
4900	4-way		30	.533		172	22		194	224
5100	Exit signs, 24 ga. alum., 14" x 12" surface mounted	1 Carp	30	.267		41.50	11.10		52.60	62.50
5200	10" x 7"		20	.400		25	16.60		41.60	53
5400	Bracket mounted, double face, 12" x 10"		30	.267		41.50	11.10		52.60	62.50
5500	Sticky back, stock decals, 14" x 10"	1 Clab	50	.160		11.40	5.30		16.70	20.50
6000	Interior elec., wall mount, fiberglass panels, 2 lamps, 6"	1 Elec	8	1		111	49		160	195
6100	8"	"	8	1		89	49		138	171
6400	Replacement sign faces, 6" or 8"	1 Clab	50	.160		48	5.30		53.30	61

10 14 23 – Panel Signage

10 14 23.13 Engraved Interior Panel Signage		Crew	Daily Output	Labor-Hours	Unit	Material	Labor	Equipment	Total	Total Incl O&P
0010	ENGRAVED INTERIOR PANEL SIGNAGE									
1010	Flexible door sign, adhesive back, w/Braille, 5/8" letters, 4" x 4"	1 Clab	32	.250	Ea.	26.50	8.30		34.80	42.50
1050	6" x 6"		32	.250		39.50	8.30		47.80	56.50
1100	8" x 2"		32	.250		27.50	8.30		35.80	43
1150	8" x 4"		32	.250		35	8.30		43.30	51.50
1200	8" x 8"		32	.250		48	8.30		56.30	66
1250	12" x 2"		32	.250		35	8.30		43.30	51.50
1300	12" x 6"		32	.250		50	8.30		58.30	68
1350	12" x 12"		32	.250		99	8.30		107.30	122
1500	Graphic symbols, 2" x 2"		32	.250		12	8.30		20.30	26
1550	6" x 6"		32	.250		30	8.30		38.30	46
1600	8" x 8"		32	.250		30	8.30		38.30	46

10 14 53 – Traffic Signage

10 14 53.20 Traffic Signs		Crew	Daily Output	Labor-Hours	Unit	Material	Labor	Equipment	Total	Total Incl O&P
0010	TRAFFIC SIGNS									
0012	Stock, 24" x 24", no posts, .080" alum. reflectorized	B-80	70	.457	Ea.	71.50	16.20	10.65	98.35	115
0100	High intensity		70	.457		71.50	16.20	10.65	98.35	115
0300	30" x 30", reflectorized		70	.457		146	16.20	10.65	172.85	196
0400	High intensity		70	.457		146	16.20	10.65	172.85	196
0600	Guide and directional signs, 12" x 18", reflectorized		70	.457		41.50	16.20	10.65	68.35	81.50
0700	High intensity		70	.457		43.50	16.20	10.65	70.35	84
0900	18" x 24", stock signs, reflectorized		70	.457		43.50	16.20	10.65	70.35	84
1000	High intensity		70	.457		43.50	16.20	10.65	70.35	84
1200	24" x 24", stock signs, reflectorized		70	.457		53.50	16.20	10.65	80.35	94.50
1300	High intensity		70	.457		59.50	16.20	10.65	86.35	102
1500	Add to above for steel posts, galvanized, 10'-0" upright, bolted		200	.160		32.50	5.65	3.73	41.88	49
1600	12'-0" upright, bolted		140	.229		51	8.10	5.30	64.40	74
1800	Highway road signs, aluminum, over 20 S. F., reflectorized		350	.091	S.F.	28	3.24	2.13	33.37	38.50
2000	High intensity		350	.091		28	3.24	2.13	33.37	38.50

10 14 Signage

10 14 53 – Traffic Signage

10 14 53.20 Traffic Signs		Crew	Daily Output	Labor-Hours	Unit	Material	2010 Bare Costs Labor	Equipment	Total	Total Incl O&P
2200	Highway, suspended over road, 80 S.F. min., reflectorized	B-80	165	.194	S.F.	28	6.85	4.52	39.37	46.50
2300	High intensity		165	.194		28	6.85	4.52	39.37	46.50

10 17 Telephone Specialties

10 17 16 – Telephone Enclosures

10 17 16.10 Commercial Telephone Enclosures		Crew	Daily Output	Labor-Hours	Unit	Material	2010 Bare Costs Labor	Equipment	Total	Total Incl O&P
0010	**COMMERCIAL TELEPHONE ENCLOSURES**									
0300	Shelf type, wall hung, minimum	2 Carp	5	3.200	Ea.	1,050	133		1,183	1,350
0400	Maximum		5	3.200		2,350	133		2,483	2,775
0600	Booth type, painted steel, indoor or outdoor, minimum		1.50	10.667		3,475	445		3,920	4,500
0700	Maximum (stainless steel)		1.50	10.667		11,600	445		12,045	13,400
1300	Outdoor, acoustical, on post		3	5.333		1,525	222		1,747	2,025
1400	Phone carousel, pedestal mounted with dividers		.60	26.667		5,800	1,100		6,900	8,075
1900	Outdoor, drive-up type, wall mounted		4	4		930	166		1,096	1,275
2000	Post mounted, stainless steel posts		3	5.333		1,425	222		1,647	1,925
2200	Directory shelf, wall mounted, stainless steel									
2300	3 binders	2 Carp	8	2	Ea.	1,100	83		1,183	1,350
2500	4 binders		7	2.286		1,275	95		1,370	1,550
2600	5 binders		6	2.667		1,625	111		1,736	1,975
2800	Table type, stainless steel, 4 binders		8	2		1,300	83		1,383	1,575
2900	7 binders		7	2.286		1,625	95		1,720	1,950

10 21 Compartments and Cubicles

10 21 13 – Toilet Compartments

10 21 13.13 Metal Toilet Compartments		Crew	Daily Output	Labor-Hours	Unit	Material	2010 Bare Costs Labor	Equipment	Total	Total Incl O&P
0010	**METAL TOILET COMPARTMENTS**									
0110	Cubicles, ceiling hung									
0200	Powder coated steel	2 Carp	4	4	Ea.	445	166		611	745
0500	Stainless steel	"	4	4		1,400	166		1,566	1,800
0600	For handicap units, incl. 52" grab bars, add					405			405	445
0900	Floor and ceiling anchored									
1000	Powder coated steel	2 Carp	5	3.200	Ea.	535	133		668	795
1300	Stainless steel	"	5	3.200		1,525	133		1,658	1,875
1400	For handicap units, incl. 52" grab bars, add					287			287	315
1610	Floor anchored									
1700	Powder coated steel	2 Carp	7	2.286	Ea.	535	95		630	735
2000	Stainless steel	"	7	2.286		1,475	95		1,570	1,775
2100	For handicap units, incl. 52" grab bars, add					285			285	315
2200	For juvenile units, deduct					41.50			41.50	45.50
2450	Floor anchored, headrail braced									
2500	Powder coated steel	2 Carp	6	2.667	Ea.	470	111		581	685
2804	Stainless steel	"	4.60	3.478		1,175	145		1,320	1,525
2900	For handicap units, incl. 52" grab bars, add					330			330	365
3000	Wall hung partitions, powder coated steel	2 Carp	7	2.286		625	95		720	830
3300	Stainless steel	"	7	2.286		1,650	95		1,745	1,975
3400	For handicap units, incl. 52" grab bars, add					330			330	365
4000	Screens, entrance, floor mounted, 58" high, 48" wide									
4200	Powder coated steel	2 Carp	15	1.067	Ea.	233	44.50		277.50	325
4500	Stainless steel	"	15	1.067	"	895	44.50		939.50	1,050

10 21 Compartments and Cubicles

10 21 13 – Toilet Compartments

10 21 13.13 Metal Toilet Compartments		Crew	Daily Output	Labor-Hours	Unit	Material	2010 Bare Costs Labor	Equipment	Total	Total Incl O&P
4650	Urinal screen, 18" wide									
4704	Powder coated steel	2 Carp	6.15	2.602	Ea.	180	108		288	365
5004	Stainless steel	"	6.15	2.602	"	580	108		688	805
5100	Floor mounted, head rail braced									
5300	Powder coated steel	2 Carp	8	2	Ea.	209	83		292	360
5600	Stainless steel	"	8	2	"	575	83		658	760
5750	Pilaster, flush									
5800	Powder coated steel	2 Carp	10	1.600	Ea.	274	66.50		340.50	400
6100	Stainless steel		10	1.600		670	66.50		736.50	835
6300	Post braced, powder coated steel		10	1.600		175	66.50		241.50	295
6600	Stainless steel	↓	10	1.600	↓	450	66.50		516.50	595
6700	Wall hung, bracket supported									
6800	Powder coated steel	2 Carp	10	1.600	Ea.	305	66.50		371.50	435
7100	Stainless steel		10	1.600		269	66.50		335.50	400
7400	Flange supported, powder coated steel		10	1.600		104	66.50		170.50	217
7700	Stainless steel		10	1.600		305	66.50		371.50	435
7800	Wedge type, powder coated steel		10	1.600		134	66.50		200.50	250
8100	Stainless steel	↓	10	1.600	↓	565	66.50		631.50	720

10 21 13.14 Metal Toilet Compartment Components		Crew	Daily Output	Labor-Hours	Unit	Material	Labor	Equipment	Total	Total Incl O&P
0010	**METAL TOILET COMPARTMENT COMPONENTS**									
0100	Pilasters									
0110	Overhead braced, powder coated steel, 7" wide x 82" high	2 Carp	22.20	.721	Ea.	75	30		105	129
0120	Stainless steel		22.20	.721		189	30		219	254
0130	Floor braced, powder coated steel, 7" wide x 70" high		23.30	.687		116	28.50		144.50	172
0140	Stainless steel		23.30	.687		277	28.50		305.50	350
0150	Ceiling hung, powder coated steel, 7" wide x 83" high		13.30	1.203		123	50		173	212
0160	Stainless steel		13.30	1.203		300	50		350	410
0170	Wall hung, powder coated steel, 3" wide x 58" high		18.90	.847		115	35		150	181
0180	Stainless steel	↓	18.90	.847	↓	166	35		201	236
0200	Panels									
0210	Powder coated steel, 31" wide x 58" high	2 Carp	18.90	.847	Ea.	128	35		163	194
0220	Stainless steel		18.90	.847		400	35		435	495
0230	Powder coated steel, 53" wide x 58" high		18.90	.847		163	35		198	233
0240	Stainless steel		18.90	.847		530	35		565	635
0250	Powder coated steel, 63" wide x 58" high		18.90	.847		195	35		230	269
0260	Stainless steel	↓	18.90	.847	↓	645	35		680	765
0300	Doors									
0310	Powder coated steel, 24" wide x 58" high	2 Carp	14.10	1.135	Ea.	141	47		188	228
0320	Stainless steel		14.10	1.135		385	47		432	495
0330	Powder coated steel, 26" wide x 58" high		14.10	1.135		144	47		191	231
0340	Stainless steel		14.10	1.135		390	47		437	505
0350	Powder coated steel, 28" wide x 58" high		14.10	1.135		152	47		199	241
0360	Stainless steel		14.10	1.135		410	47		457	525
0370	Powder coated steel, 36" wide x 58" high		14.10	1.135		175	47		222	266
0380	Stainless steel	↓	14.10	1.135	↓	470	47		517	595
0400	Headrails									
0410	For powder coated steel, 62" long	2 Carp	65	.246	Ea.	23.50	10.25		33.75	42
0420	Stainless steel		65	.246		23.50	10.25		33.75	42
0430	For powder coated steel, 84" long		50	.320		32	13.30		45.30	55.50
0440	Stainless steel		50	.320		30.50	13.30		43.80	54
0450	For powder coated steel, 120" long		30	.533		44	22		66	82
0460	Stainless steel	↓	30	.533	↓	44	22		66	82

10 21 Compartments and Cubicles

10 21 13 – Toilet Compartments

10 21 13.16 Plastic-Laminate-Clad Toilet Compartments

	10 21 13.16 Plastic-Laminate-Clad Toilet Compartments	Crew	Daily Output	Labor-Hours	Unit	2010 Bare Costs Material	Labor	Equipment	Total	Total Incl O&P
0010	**PLASTIC-LAMINATE-CLAD TOILET COMPARTMENTS**									
0110	Cubicles, ceiling hung									
0300	Plastic laminate on particle board	2 Carp	4	4	Ea.	560	166		726	870
0600	For handicap units, incl. 52" grab bars, add				"	405			405	445
0900	Floor and ceiling anchored									
1100	Plastic laminate on particle board	2 Carp	5	3.200	Ea.	765	133		898	1,050
1400	For handicap units, incl. 52" grab bars, add				"	287			287	315
1610	Floor mounted									
1800	Plastic laminate on particle board	2 Carp	7	2.286	Ea.	555	95		650	755
2450	Floor mounted, headrail braced									
2600	Plastic laminate on particle board CN	2 Carp	6	2.667	Ea.	800	111		911	1,050
3400	For handicap units, incl. 52" grab bars, add					330			330	365
4300	Entrance screen, floor mtd., plas. lam., 58" high, 48" wide	2 Carp	15	1.067		580	44.50		624.50	710
4800	Urinal screen, 18" wide, ceiling braced, plastic laminate		8	2		174	83		257	320
5400	Floor mounted, headrail braced		8	2		220	83		303	370
5900	Pilaster, flush, plastic laminate		10	1.600		485	66.50		551.50	630
6400	Post braced, plastic laminate		10	1.600		285	66.50		351.50	415
6700	Wall hung, bracket supported									
6900	Plastic laminate on particle board	2 Carp	10	1.600	Ea.	91	66.50		157.50	202
7450	Flange supported									
7500	Plastic laminate on particle board	2 Carp	10	1.600	Ea.	220	66.50		286.50	345

10 21 13.17 Plastic-Laminate Clad Toilet Compartment Components

	10 21 13.17 Plastic-Laminate Clad Toilet Compartment Components	Crew	Daily Output	Labor-Hours	Unit	Material	Labor	Equipment	Total	Total Incl O&P
0010	**PLASTIC-LAMINATE CLAD TOILET COMPARTMENT COMPONENTS**									
0100	Pilasters									
0110	Overhead braced, 7" wide x 82" high	2 Carp	22.20	.721	Ea.	100	30		130	156
0130	Floor anchored, 7" wide x 70" high		23.30	.687		98.50	28.50		127	152
0150	Ceiling hung, 7" wide x 83" high		13.30	1.203		97	50		147	184
0180	Wall hung, 3" wide x 58" high		18.90	.847		87.50	35		122.50	151
0200	Panels									
0210	31" wide x 58" high	2 Carp	18.90	.847	Ea.	135	35		170	202
0230	51" wide x 58" high		18.90	.847		188	35		223	261
0250	63" wide x 58" high		18.90	.847		223	35		258	300
0300	Doors									
0310	24" wide x 58" high	2 Carp	14.10	1.135	Ea.	135	47		182	222
0330	26" wide x 58" high		14.10	1.135		140	47		187	226
0350	28" wide x 58" high		14.10	1.135		145	47		192	232
0370	36" wide x 58" high		14.10	1.135		172	47		219	262
0400	Headrails									
0410	62" long	2 Carp	65	.246	Ea.	25.50	10.25		35.75	44
0430	84" long		60	.267		30.50	11.10		41.60	50.50
0450	120" long		30	.533		44	22		66	82

10 21 13.19 Plastic Toilet Compartments

	10 21 13.19 Plastic Toilet Compartments	Crew	Daily Output	Labor-Hours	Unit	Material	Labor	Equipment	Total	Total Incl O&P
0010	**PLASTIC TOILET COMPARTMENTS**									
0110	Cubicles, ceiling hung									
0250	Phenolic	2 Carp	4	4	Ea.	905	166		1,071	1,250
0600	For handicap units, incl. 52" grab bars, add				"	405			405	445
0900	Floor and ceiling anchored									
1050	Phenolic	2 Carp	5	3.200	Ea.	940	133		1,073	1,225
1400	For handicap units, incl. 52" grab bars, add				"	287			287	315
1610	Floor mounted									
1750	Phenolic	2 Carp	7	2.286	Ea.	1,350	95		1,445	1,625
2100	For handicap units, incl. 52" grab bars, add					285			285	315

10 21 Compartments and Cubicles

10 21 13 – Toilet Compartments

10 21 13.19 Plastic Toilet Compartments		Crew	Daily Output	Labor-Hours	Unit	Material	2010 Bare Costs Labor	Equipment	Total	Total Incl O&P
2200	For juvenile units, deduct				Ea.	41.50			41.50	45.50
2450	Floor mounted, headrail braced									
2550	Phenolic	2 Carp	6	2.667	Ea.	895	111		1,006	1,150

10 21 13.20 Plastic Toilet Compartment Components		Crew	Daily Output	Labor-Hours	Unit	Material	Labor	Equipment	Total	Total Incl O&P
0010	**PLASTIC TOILET COMPARTMENT COMPONENTS**									
0100	Pilasters									
0110	Overhead braced, polymer plastic, 7" wide x 82" high	2 Carp	22.20	.721	Ea.	158	30		188	220
0120	Phenolic		22.20	.721		153	30		183	214
0130	Floor braced, polymer plastic, 7" wide x 70" high		23.30	.687		165	28.50		193.50	225
0140	Phenolic		23.30	.687		142	28.50		170.50	200
0150	Ceiling hung, polymer plastic, 7" wide x 83" high		13.30	1.203		141	50		191	232
0160	Phenolic		13.30	1.203		161	50		211	254
0180	Wall hung, phenolic, 3" wide x 58" high	↓	18.90	.847	↓	90.50	35		125.50	154
0200	Panels									
0210	Polymer plastic, 31" high x 55" high	2 Carp	18.90	.847	Ea.	330	35		365	420
0220	Phenolic, 31" wide x 58" high		18.90	.847		259	35		294	340
0230	Polymer plastic, 51" wide x 55" high		18.90	.847		505	35		540	610
0240	Phenolic, 51" wide x 58" high		18.90	.847		395	35		430	485
0250	Polymer plastic, 63" wide x 55" high		18.90	.847		610	35		645	725
0260	Phenolic, 63" wide x 58" high	↓	18.90	.847	↓	465	35		500	570
0300	Doors									
0310	Polymer plastic, 24" wide x 55" high	2 Carp	14.10	1.135	Ea.	263	47		310	360
0320	Phenolic, 24" wide x 58" high		14.10	1.135		289	47		336	395
0330	Polymer plastic, 26" high x 55" high		14.10	1.135		279	47		326	380
0340	Phenolic, 26" wide x 58" high		14.10	1.135		305	47		352	410
0350	Polymer plastic, 28" wide x 55" high		14.10	1.135		295	47		342	400
0360	Phenolic, 28" wide x 58" high		14.10	1.135		325	47		372	430
0370	Polymer plastic, 36" wide x 55" high		14.10	1.135		370	47		417	485
0380	Phenolic, 36" wide x 58" high	↓	14.10	1.135	↓	405	47		452	520
0400	Headrails									
0410	For polymer plastic, 62" long	2 Carp	65	.246	Ea.	32	10.25		42.25	51.50
0420	Phenolic		65	.246		23	10.25		33.25	41.50
0430	For polymer plastic, 84" long		50	.320		35.50	13.30		48.80	59.50
0440	Phenolic		50	.320		28.50	13.30		41.80	52
0450	For polymer plastic, 120" long		30	.533		50.50	22		72.50	89.50
0460	Phenolic	↓	30	.533	↓	38	22		60	76

10 21 13.40 Stone Toilet Compartments		Crew	Daily Output	Labor-Hours	Unit	Material	Labor	Equipment	Total	Total Incl O&P
0010	**STONE TOILET COMPARTMENTS**									
0100	Cubicles, ceiling hung, marble	2 Marb	2	8	Ea.	1,450	320		1,770	2,075
0600	For handicap units, incl. 52" grab bars, add					405			405	445
0800	Floor & ceiling anchored, marble	2 Marb	2.50	6.400		1,825	255		2,080	2,375
1400	For handicap units, incl. 52" grab bars, add					287			287	315
1600	Floor mounted, marble	2 Marb	3	5.333		1,075	213		1,288	1,500
2400	Floor mounted, headrail braced, marble	"	3	5.333		1,025	213		1,238	1,450
2900	For handicap units, incl. 52" grab bars, add					330			330	365
4100	Entrance screen, floor mounted marble, 58" high, 48" wide	2 Marb	9	1.778		675	71		746	850
4600	Urinal screen, 18" wide, ceiling braced, marble	D-1	6	2.667	↓	675	101		776	895
5100	Floor mounted, head rail braced									
5200	Marble	D-1	6	2.667	Ea.	575	101		676	780
5700	Pilaster, flush, marble		9	1.778		750	67		817	925
6200	Post braced, marble	↓	9	1.778	↓	745	67		812	920

10 21 16 – Shower and Dressing Compartments

10 21 16.10 Partitions, Shower		Crew	Daily Output	Labor-Hours	Unit	Material	Labor	Equipment	Total	Total Incl O&P
						2010 Bare Costs				
0010	**PARTIONS, SHOWER** Floor mounted, no plumbing									
0400	Cabinet, one piece, fiberglass, 32" x 32"	2 Carp	5	3.200	Ea.	475	133		608	725
0420	36" x 36"		5	3.200		530	133		663	785
0440	36" x 48"		5	3.200		1,375	133		1,508	1,725
0460	Acrylic, 32" x 32"		5	3.200		325	133		458	560
0480	36" x 36"		5	3.200		980	133		1,113	1,275
0500	36" x 48"		5	3.200		2,100	133		2,233	2,525
0520	Shower door for above, clear plastic, 24" wide	1 Carp	8	1		162	41.50		203.50	242
0540	28" wide		8	1		184	41.50		225.50	266
0560	Tempered glass, 24" wide		8	1		175	41.50		216.50	256
0580	28" wide		8	1		197	41.50		238.50	281
2400	Glass stalls, with doors, no receptors, chrome on brass	2 Shee	3	5.333		1,525	262		1,787	2,075
2700	Anodized aluminum	"	4	4		1,050	196		1,246	1,450
2900	Marble shower stall, stock design, with shower door	2 Marb	1.20	13.333		2,275	530		2,805	3,300
3000	With curtain		1.30	12.308		2,000	490		2,490	2,950
3200	Receptors, precast terrazzo, 32" x 32"		14	1.143		147	45.50		192.50	231
3300	48" x 34"		9.50	1.684		203	67		270	325
3500	Plastic, simulated terrazzo receptor, 32" x 32"		14	1.143		120	45.50		165.50	201
3600	32" x 48"		12	1.333		204	53		257	305
3800	Precast concrete, colors, 32" x 32"		14	1.143		221	45.50		266.50	310
3900	48" x 48"		8	2		395	80		475	555
4100	Shower doors, economy plastic, 24" wide	1 Shee	9	.889		132	43.50		175.50	211
4200	Tempered glass door, economy		8	1		219	49		268	315
4400	Folding, tempered glass, aluminum frame		6	1.333		395	65.50		460.50	535
4500	Sliding, tempered glass, 48" opening		6	1.333		263	65.50		328.50	390
4700	Deluxe, tempered glass, chrome on brass frame, minimum		8	1		235	49		284	335
4800	Maximum		1	8		910	395		1,305	1,600
4850	On anodized aluminum frame, minimum		2	4		148	196		344	460
4900	Maximum		1	8		535	395		930	1,175
5100	Shower enclosure, tempered glass, anodized alum. frame									
5120	2 panel & door, corner unit, 32" x 32"	1 Shee	2	4	Ea.	520	196		716	870
5140	Neo-angle corner unit, 16" x 24" x 16"	"	2	4		960	196		1,156	1,350
5200	Shower surround, 3 wall, polypropylene, 32" x 32"	1 Carp	4	2		385	83		468	555
5220	PVC, 32" x 32"		4	2		350	83		433	515
5240	Fiberglass		4	2		385	83		468	550
5250	2 wall, polypropylene, 32" x 32"		4	2		290	83		373	450
5270	PVC		4	2		360	83		443	525
5290	Fiberglass		4	2		365	83		448	535
5300	Tub doors, tempered glass & frame, minimum	1 Shee	8	1		210	49		259	305
5400	Maximum		6	1.333		490	65.50		555.50	640
5600	Chrome plated, brass frame, minimum		8	1		278	49		327	380
5700	Maximum		6	1.333		680	65.50		745.50	850
5900	Tub/shower enclosure, temp. glass, alum. frame, minimum		2	4		375	196		571	710
6200	Maximum		1.50	5.333		785	262		1,047	1,250
6500	On chrome-plated brass frame, minimum		2	4		520	196		716	865
6600	Maximum		1.50	5.333		1,100	262		1,362	1,625
6800	Tub surround, 3 wall, polypropylene	1 Carp	4	2		235	83		318	385
6900	PVC		4	2		355	83		438	525
7000	Fiberglass, minimum		4	2		370	83		453	535
7100	Maximum		3	2.667		625	111		736	855

10 21 Compartments and Cubicles

10 21 23 – Cubicles

10 21 23.16 Cubicle Track and Hardware		Crew	Daily Output	Labor-Hours	Unit	Material	Labor	Equipment	Total	Total Incl O&P
						2010 Bare Costs				
0010	**CUBICLE TRACK AND HARDWARE**									
0020	Curtain track, box channel, ceiling mounted	1 Carp	135	.059	L.F.	5.80	2.46		8.26	10.15
0100	Suspended	"	100	.080	"	7.05	3.32		10.37	12.85
0300	Curtains, nylon mesh tops, fire resistant, 11 oz. per lineal yard									
0310	Polyester oxford cloth, 9′ ceiling height	1 Carp	425	.019	L.F.	12	.78		12.78	14.40
0500	8′ ceiling height		425	.019		10.15	.78		10.93	12.35
0700	Designer oxford cloth		425	.019		25.50	.78		26.28	29.50
0800	I.V. track systems									
0820	I.V. track, oval	1 Carp	135	.059	L.F.	5.25	2.46		7.71	9.60
0830	I.V. trolley		32	.250	Ea.	36	10.40		46.40	56
0840	I.V. pendent, (tree, 5 hook)		32	.250	"	129	10.40		139.40	158

10 22 Partitions

10 22 13 – Wire Mesh Partitions

10 22 13.10 Partitions, Woven Wire		Crew	Daily Output	Labor-Hours	Unit	Material	Labor	Equipment	Total	Total Incl O&P
0010	**PARTITIONS, WOVEN WIRE** For tool or stockroom enclosures									
0100	Channel frame, 1-1/2" diamond mesh, 10 ga. wire, painted									
0300	Wall panels, 4′-0" wide, 7′ high	2 Carp	25	.640	Ea.	126	26.50		152.50	180
0400	8′ high		23	.696		130	29		159	188
0600	10′ high		18	.889		171	37		208	246
0700	For 5′ wide panels, add					5%				
0900	Ceiling panels, 10′ long, 2′ wide	2 Carp	25	.640		130	26.50		156.50	184
1000	4′ wide		15	1.067		165	44.50		209.50	250
1200	Panel with service window & shelf, 5′ wide, 7′ high		20	.800		405	33		438	495
1300	8′ high		15	1.067		405	44.50		449.50	515
1500	Sliding doors, full height, 3′ wide, 7′ high		6	2.667		420	111		531	630
1600	10′ high		5	3.200		385	133		518	630
1800	6′ wide sliding door, 7′ full height		5	3.200		585	133		718	850
1900	10′ high		4	4		720	166		886	1,050
2100	Swinging doors, 3′ wide, 7′ high, no transom		6	2.667		273	111		384	470
2200	7′ high, 3′ transom		5	3.200		325	133		458	560

10 22 16 – Folding Gates

10 22 16.10 Security Gates		Crew	Daily Output	Labor-Hours	Unit	Material	Labor	Equipment	Total	Total Incl O&P
0010	**SECURITY GATES** For roll up type, see Div. 08 33 13.10									
0300	Scissors type folding gate, ptd. steel, single, 6-1/2′ high, 5-1/2′wide	2 Sswk	4	4	Opng.	214	188		402	565
0350	6-1/2′ wide		4	4		213	188		401	565
0400	7-1/2′ wide		4	4		231	188		419	585
0600	Double gate, 8′ high, 8′ wide		2.50	6.400		340	300		640	905
0650	10′ wide		2.50	6.400		400	300		700	970
0700	12′ wide		2	8		530	375		905	1,250
0750	14′ wide		2	8		605	375		980	1,325
0900	Door gate, folding steel, 4′ wide, 61" high		4	4		115	188		303	455
1000	71" high		4	4		105	188		293	445
1200	81" high		4	4		114	188		302	455
1300	Window gates, 2′ to 4′ wide, 31" high		4	4		58	188		246	395
1500	55" high		3.75	4.267		90	200		290	450
1600	79" high		3.50	4.571		105	214		319	490

10 22 19 – Demountable Partitions

10 22 19.43 Demountable Composite Partitions		Crew	Daily Output	Labor-Hours	Unit	Material	2010 Bare Costs Labor	Equipment	Total	Total Incl O&P
0010	**DEMOUNTABLE COMPOSITE PARTITIONS**, add for doors									
0100	Do not deduct door openings from total L.F.									
0900	Demountable gypsum system on 2″ to 2-1/2″									
1000	steel studs, 9′ high, 3″ to 3-3/4″ thick									
1200	Vinyl clad gypsum **CN**	2 Carp	48	.333	L.F.	58	13.85		71.85	85
1300	Fabric clad gypsum		44	.364		144	15.10		159.10	182
1500	Steel clad gypsum		40	.400		156	16.60		172.60	198
1600	1.75 system, aluminum framing, vinyl clad hardboard,									
1800	paper honeycomb core panel, 1-3/4″ to 2-1/2″ thick									
1900	9′ high	2 Carp	48	.333	L.F.	97	13.85		110.85	129
2100	7′ high		60	.267		87	11.10		98.10	113
2200	5′ high		80	.200		73.50	8.30		81.80	94
2250	Unitized gypsum system									
2300	Unitized panel, 9′ high, 2″ to 2-1/2″ thick									
2350	Vinyl clad gypsum	2 Carp	48	.333	L.F.	125	13.85		138.85	159
2400	Fabric clad gypsum	″	44	.364	″	206	15.10		221.10	250
2500	Unitized mineral fiber system									
2510	Unitized panel, 9′ high, 2-1/4″ thick, aluminum frame									
2550	Vinyl clad mineral fiber	2 Carp	48	.333	L.F.	124	13.85		137.85	159
2600	Fabric clad mineral fiber	″	44	.364	″	185	15.10		200.10	228
2800	Movable steel walls, modular system									
2900	Unitized panels, 9′ high, 48″ wide									
3100	Baked enamel, pre-finished	2 Carp	60	.267	L.F.	141	11.10		152.10	172
3200	Fabric clad steel		56	.286	″	203	11.85		214.85	242
5310	Trackless wall, cork finish, semi-acoustic, 1-5/8″ thick, minimum		325	.049	S.F.	38.50	2.05		40.55	45.50
5320	Maximum		190	.084		36.50	3.50		40	45.50
5330	Acoustic, 2″ thick, minimum		305	.052		31	2.18		33.18	37.50
5340	Maximum		225	.071		53.50	2.95		56.45	63.50
5500	For acoustical partitions, add, minimum					2.22			2.22	2.44
5550	Maximum					10.35			10.35	11.40
5700	For doors, see Div. 08 11 & 08 16									
5800	For door hardware, see Div. 08 71									
6100	In-plant modular office system, w/prehung hollow core door									
6200	3″ thick polystyrene core panels									
6250	12′ x 12′, 2 wall	2 Clab	3.80	4.211	Ea.	3,650	139		3,789	4,250
6300	4 wall		1.90	8.421		4,650	279		4,929	5,525
6350	16′ x 16′, 2 wall		3.60	4.444		4,850	147		4,997	5,550
6400	4 wall		1.80	8.889		6,525	294		6,819	7,625

10 22 23 – Portable Partitions, Screens, and Panels

10 22 23.13 Wall Screens		Crew	Daily Output	Labor-Hours	Unit	Material	Labor	Equipment	Total	Total Incl O&P
0010	**WALL SCREENS**, divider panels, free standing, fiber core									
0020	Fabric face straight									
0100	3′-0″ long, 4′-0″ high	2 Carp	100	.160	L.F.	131	6.65		137.65	154
0200	5′-0″ high		90	.178		115	7.40		122.40	137
0500	6′-0″ high		75	.213		117	8.85		125.85	143
0900	5′-0″ long, 4′-0″ high		175	.091		79	3.80		82.80	93
1000	5′-0″ high		150	.107		90.50	4.43		94.93	106
1500	6″-0″ high		125	.128		93.50	5.30		98.80	111
1600	6′-0″ long, 5′-0″ high		162	.099		78	4.10		82.10	92.50
3200	Economical panels, fabric face, 4′-0″ long, 5′-0″ high		132	.121		50.50	5.05		55.55	63.50
3250	6′-0″ high		112	.143		54	5.95		59.95	68
3300	5′-0″ long, 5′-0″ high		150	.107		42	4.43		46.43	53

10 22 Partitions

10 22 23 – Portable Partitions, Screens, and Panels

10 22 23.13 Wall Screens		Crew	Daily Output	Labor-Hours	Unit	Material	2010 Bare Costs Labor	Equipment	Total	Total Incl O&P
3350	6'-0" high	2 Carp	125	.128	L.F.	46	5.30		51.30	58.50
3450	Acoustical panels, 60 to 90 NRC, 3'-0" long, 5'-0" high		90	.178		89.50	7.40		96.90	110
3550	6'-0" high		75	.213		99.50	8.85		108.35	123
3600	5'-0" long, 5'-0" high		150	.107		70	4.43		74.43	84
3650	6'-0" high		125	.128		68	5.30		73.30	82.50
3700	6'-0" long, 5'-0" high		162	.099		61.50	4.10		65.60	74
3750	6'-0" high		138	.116		56	4.82		60.82	69
3800	Economy acoustical panels, 40 N.R.C., 4'-0" long, 5'-0" high		132	.121		50.50	5.05		55.55	63.50
3850	6'-0" high		112	.143		54	5.95		59.95	68
3900	5'-0" long, 6'-0" high		125	.128		46	5.30		51.30	58.50
3950	6'-0" long, 5'-0" high		162	.099		36	4.10		40.10	46
4000	Metal chalkboard, 6'-6" high, chalkboard, 1 side		125	.128		104	5.30		109.30	122
4100	Metal chalkboard, 2 sides		120	.133		118	5.55		123.55	139
4300	Tackboard, both sides	↓	123	.130	↓	94	5.40		99.40	111

10 22 26 – Operable Partitions

10 22 26.13 Accordion Folding Partitions		Crew	Daily Output	Labor-Hours	Unit	Material	Labor	Equipment	Total	Total Incl O&P
0010	**ACCORDION FOLDING PARTITIONS**									
0100	Vinyl covered, over 150 S.F., frame not included									
0300	Residential, 1.25 lb. per S.F., 8' maximum height	2 Carp	300	.053	S.F.	19.35	2.22		21.57	25
0400	Commercial, 1.75 lb. per S.F., 8' maximum height		225	.071		22	2.95		24.95	29
0600	2 lb. per S.F., 17' maximum height		150	.107		23	4.43		27.43	32
0700	Industrial, 4 lb. per S.F., 20' maximum height		75	.213		32.50	8.85		41.35	49.50
0900	Acoustical, 3 lb. per S.F., 17' maximum height		100	.160		25.50	6.65		32.15	38.50
1200	5 lb. per S.F., 20' maximum height		95	.168		35.50	7		42.50	50
1300	5.5 lb. per S.F., 17' maximum height		90	.178		41.50	7.40		48.90	57.50
1400	Fire rated, 4.5 psf, 20' maximum height		160	.100		41.50	4.16		45.66	52.50
1500	Vinyl clad wood or steel, electric operation, 5.0 psf		160	.100		48.50	4.16		52.66	59.50
1900	Wood, non-acoustic, birch or mahogany, to 10' high	↓	300	.053	↓	26	2.22		28.22	32

10 22 26.33 Folding Panel Partitions		Crew	Daily Output	Labor-Hours	Unit	Material	Labor	Equipment	Total	Total Incl O&P
0010	**FOLDING PANEL PARTITIONS**, acoustic, wood									
0100	Vinyl faced, to 18' high, 6 psf, minimum	2 Carp	60	.267	S.F.	51.50	11.10		62.60	73.50
0150	Average		45	.356		61.50	14.75		76.25	90.50
0200	Maximum		30	.533		79.50	22		101.50	122
0400	Formica or hardwood finish, minimum		60	.267		53	11.10		64.10	75.50
0500	Maximum		30	.533		56.50	22		78.50	96.50
0600	Wood, low acoustical type, 4.5 psf, to 14' high		50	.320		38.50	13.30		51.80	63
1100	Steel, acoustical, 9 to 12 lb. per S.F., vinyl faced, minimum		60	.267		55	11.10		66.10	77.50
1200	Maximum		30	.533		67	22		89	108
1700	Aluminum framed, acoustical, to 12' high, 5.5 psf, minimum		60	.267		37	11.10		48.10	57.50
1800	Maximum		30	.533		44.50	22		66.50	83
2000	6.5 lb. per S.F., minimum		60	.267		39	11.10		50.10	60
2100	Maximum	↓	30	.533	↓	48	22		70	87

10 22 26.43 Sliding Partitions		Crew	Daily Output	Labor-Hours	Unit	Material	Labor	Equipment	Total	Total Incl O&P
0010	**SLIDING PARTITIONS**									
0020	Acoustic air wall, 1-5/8" thick, minimum	2 Carp	375	.043	S.F.	31.50	1.77		33.27	37
0100	Maximum		365	.044		54	1.82		55.82	62.50
0300	2-1/4" thick, minimum		360	.044		35.50	1.85		37.35	42
0400	Maximum	↓	330	.048	↓	62	2.01		64.01	71
0600	For track type, add to above				L.F.	115			115	127
0700	Overhead track type, acoustical, 3" thick, 11 psf, minimum	2 Carp	350	.046	S.F.	79.50	1.90		81.40	90.50
0800	Maximum	"	300	.053	"	95.50	2.22		97.72	108

10 26 Wall and Door Protection

10 26 13 – Corner Guards

10 26 13.10 Metal Corner Guards		Crew	Daily Output	Labor-Hours	Unit	Material	Labor	Equipment	Total	Total Incl O&P
						2010 Bare Costs				
0010	**METAL CORNER GUARDS**									
0020	Steel angle w/anchors, 1" x 1" x 1/4", 1.5#/L.F.	2 Carp	160	.100	L.F.	5.05	4.16		9.21	11.95
0100	2" x 2" x 1/4" angles, 3.2#/L.F.		150	.107		8.95	4.43		13.38	16.65
0200	3" x 3" x 5/16" angles, 6.1#/L.F.		140	.114		12.45	4.75		17.20	21
0300	4" x 4" x 5/16" angles, 8.2#/L.F.		120	.133		13.25	5.55		18.80	23
0350	For angles drilled and anchored to masonry, add					15%	120%			
0370	Drilled and anchored to concrete, add					20%	170%			
0400	For galvanized angles, add					35%				
0450	For stainless steel angles, add					100%				
0500	Steel door track/wheel guards, 4' - 0" high	E-4	22	1.455	Ea.	156	69	6.65	231.65	299
0800	Pipe bumper for truck doors, 8' long, 6" diameter, filled		20	1.600		425	76	7.30	508.30	605
0900	8" diameter		20	1.600		650	76	7.30	733.30	850

10 26 13.20 Corner Protection		Crew	Daily Output	Labor-Hours	Unit	Material	Labor	Equipment	Total	Total Incl O&P
0010	**CORNER PROTECTION**									
0100	Stainless steel, 16 ga., adhesive mount, 3-1/2" leg	1 Sswk	80	.100	L.F.	18.70	4.69		23.39	29
0200	12 ga. stainless, adhesive mount	"	80	.100		21	4.69		25.69	31.50
0300	For screw mount, add						10%			
0500	Vinyl acrylic, adhesive mount, 3" leg	1 Carp	128	.063		7.45	2.60		10.05	12.20
0550	1-1/2" leg		160	.050		5	2.08		7.08	8.70
0600	Screw mounted, 3" leg		80	.100		7.90	4.16		12.06	15.10
0650	1-1/2" leg		100	.080		4.25	3.32		7.57	9.80
0700	Clear plastic, screw mounted, 2-1/2"		60	.133		3.77	5.55		9.32	12.70
1000	Vinyl cover, alum. retainer, surface mount, 3" x 3"		48	.167		9.15	6.95		16.10	21
1050	2" x 2"		48	.167		9.30	6.95		16.25	21
1100	Flush mounted, 3" x 3"		32	.250		19.25	10.40		29.65	37
1150	2" x 2"		32	.250		15.65	10.40		26.05	33.50

10 26 16 – Bumper Guards

10 26 16.10 Wallguard		Crew	Daily Output	Labor-Hours	Unit	Material	Labor	Equipment	Total	Total Incl O&P
0010	**WALLGUARD**									
0400	Rub rail, vinyl, adhesive mounted	1 Carp	185	.043	L.F.	7.70	1.80		9.50	11.20
0500	Neoprene, aluminum backing, 1-1/2" x 2"		110	.073		8.65	3.02		11.67	14.15
1000	Trolley rail, PVC, clipped to wall, 5" high		185	.043		7.80	1.80		9.60	11.30
1050	8" high		180	.044		10.70	1.85		12.55	14.65
1200	Bed bumper, vinyl acrylic, alum. retainer, 21" long		10	.800	Ea.	39.50	33		72.50	94.50
1300	53" long with aligner		9	.889	"	100	37		137	167
1400	Bumper, vinyl cover, alum. retain., cush. mnt., 1-1/2" x 2-3/4"		80	.100	L.F.	13.95	4.16		18.11	21.50
1500	2" x 4-1/4"		80	.100		20.50	4.16		24.66	29
1600	Surface mounted, 1-3/4" x 3-5/8"		80	.100		11.70	4.16		15.86	19.30
2000	Crash rail, vinyl cover, alum. retainer, 1" x 4"		110	.073		10.80	3.02		13.82	16.55
2100	1" x 8"		90	.089		16.90	3.69		20.59	24.50
2150	Vinyl inserts, aluminum plate, 1" x 2-1/2"		110	.073		14.60	3.02		17.62	20.50
2200	1" x 5"		90	.089		22.50	3.69		26.19	30.50
3000	Handrail/bumper, vinyl cover, alum. retainer									
3010	Bracket mounted, flat rail, 5-1/2"	1 Carp	80	.100	L.F.	17.55	4.16		21.71	26
3100	6-1/2"		80	.100		22	4.16		26.16	30.50
3200	Bronze bracket, 1-3/4" diam. rail		80	.100		15.85	4.16		20.01	24

10 28 13 – Toilet Accessories

10 28 13.13 Commercial Toilet Accessories		Crew	Daily Output	Labor-Hours	Unit	2010 Bare Costs Material	Labor	Equipment	Total	Total Incl O&P
0010	**COMMERCIAL TOILET ACCESSORIES**									
0200	Curtain rod, stainless steel, 5′ long, 1″ diameter	1 Carp	13	.615	Ea.	34.50	25.50		60	77.50
0300	1-1/4″ diameter		13	.615		24	25.50		49.50	66
0400	Diaper changing station, horizontal, wall mounted, plastic		10	.800		217	33		250	289
0500	Dispenser units, combined soap & towel dispensers,									
0510	mirror and shelf, flush mounted	1 Carp	10	.800	Ea.	310	33		343	390
0600	Towel dispenser and waste receptacle,									
0610	18 gallon capacity	1 Carp	10	.800	Ea.	263	33		296	340
0800	Grab bar, straight, 1-1/4″ diameter, stainless steel, 18″ long		24	.333		30	13.85		43.85	55
0900	24″ long		23	.348		31.50	14.45		45.95	57
1000	30″ long		22	.364		32.50	15.10		47.60	59
1100	36″ long		20	.400		34.50	16.60		51.10	63.50
1105	42″ long		20	.400		37.50	16.60		54.10	67
1200	1-1/2″ diameter, 24″ long		23	.348		28.50	14.45		42.95	54
1300	36″ long		20	.400		32.50	16.60		49.10	61.50
1310	42″ long		18	.444		43	18.45		61.45	76
1500	Tub bar, 1-1/4″ diameter, 24″ x 36″		14	.571		75	23.50		98.50	119
1600	Plus vertical arm		12	.667		85.50	27.50		113	137
1900	End tub bar, 1″ diameter, 90° angle, 16″ x 32″		12	.667		96	27.50		123.50	148
2010	Tub/shower/toilet, 2-wall, 36″ x 24″		12	.667		96	27.50		123.50	149
2300	Hand dryer, surface mounted, electric, 115 volt, 20 amp		4	2		390	83		473	555
2400	230 volt, 10 amp		4	2		655	83		738	850
2600	Hat and coat strip, stainless steel, 4 hook, 36″ long		24	.333		71	13.85		84.85	100
2700	6 hook, 60″ long		20	.400		112	16.60		128.60	149
3000	Mirror, with stainless steel 3/4″ square frame, 18″ x 24″		20	.400		56.50	16.60		73.10	88
3100	36″ x 24″		15	.533		105	22		127	150
3200	48″ x 24″		10	.800		152	33		185	219
3300	72″ x 24″		6	1.333		210	55.50		265.50	315
3500	With 5″ stainless steel shelf, 18″ x 24″		20	.400		189	16.60		205.60	234
3600	36″ x 24″		15	.533		202	22		224	256
3700	48″ x 24″		10	.800		214	33		247	287
3800	72″ x 24″		6	1.333		264	55.50		319.50	375
4100	Mop holder strip, stainless steel, 5 holders, 48″ long		20	.400		78	16.60		94.60	112
4200	Napkin/tampon dispenser, recessed		15	.533		540	22		562	630
4250	Napkin receptacle, recessed		6.50	1.231		142	51		193	235
4300	Robe hook, single, regular		36	.222		11.65	9.25		20.90	27
4400	Heavy duty, concealed mounting		36	.222		13.35	9.25		22.60	29
4600	Soap dispenser, chrome, surface mounted, liquid		20	.400		42.50	16.60		59.10	72.50
4700	Powder		20	.400		50.50	16.60		67.10	81
5000	Recessed stainless steel, liquid		10	.800		126	33		159	190
5100	Powder		10	.800		200	33		233	271
5300	Soap tank, stainless steel, 1 gallon		10	.800		200	33		233	271
5400	5 gallon		5	1.600		275	66.50		341.50	400
5600	Shelf, stainless steel, 5″ wide, 18 ga., 24″ long		24	.333		57	13.85		70.85	84.50
5700	48″ long		16	.500		134	21		155	179
5800	8″ wide shelf, 18 ga., 24″ long		22	.364		63	15.10		78.10	93
5900	48″ long		14	.571		116	23.50		139.50	165
6000	Toilet seat cover dispenser, stainless steel, recessed		20	.400		162	16.60		178.60	204
6050	Surface mounted		15	.533		30.50	22		52.50	67.50
6100	Toilet tissue dispenser, surface mounted, SS, single roll		30	.267		17.40	11.10		28.50	36.50
6200	Double roll		24	.333		24	13.85		37.85	48
6400	Towel bar, stainless steel, 18″ long		23	.348		39.50	14.45		53.95	66
6500	30″ long		21	.381		111	15.85		126.85	148

10 28 Toilet, Bath, and Laundry Accessories

10 28 13 – Toilet Accessories

10 28 13.13 Commercial Toilet Accessories		Crew	Daily Output	Labor-Hours	Unit	Material	Labor	Equipment	Total	Total Incl O&P
						2010 Bare Costs				
6700	Towel dispenser, stainless steel, surface mounted	1 Carp	16	.500	Ea.	35	21		56	70.50
6800	Flush mounted, recessed		10	.800		237	33		270	310
7000	Towel holder, hotel type, 2 guest size		20	.400		43.50	16.60		60.10	73.50
7200	Towel shelf, stainless steel, 24" long, 8" wide		20	.400		56	16.60		72.60	87
7400	Tumbler holder, tumbler only		30	.267		40	11.10		51.10	61
7500	Soap, tumbler & toothbrush		30	.267		18.30	11.10		29.40	37
7700	Wall urn ash receiver, surface mount, 11" long		12	.667		72	27.50		99.50	122
7800	7-1/2", long		18	.444		104	18.45		122.45	143
8000	Waste receptacles, stainless steel, with top, 13 gallon		10	.800		295	33		328	375
8100	36 gallon	↓	8	1	↓	325	41.50		366.50	425

10 28 16 – Bath Accessories

10 28 16.20 Medicine Cabinets		Crew	Daily Output	Labor-Hours	Unit	Material	Labor	Equipment	Total	Total Incl O&P
0010	**MEDICINE CABINETS**									
0020	With mirror, st. st. frame, 16" x 22", unlighted	1 Carp	14	.571	Ea.	66.50	23.50		90	110
0100	Wood frame		14	.571		119	23.50		142.50	168
0300	Sliding mirror doors, 20" x 16" x 4-3/4", unlighted		7	1.143		110	47.50		157.50	194
0400	24" x 19" x 8-1/2", lighted		5	1.600		159	66.50		225.50	277
0600	Triple door, 30" x 32", unlighted, plywood body		7	1.143		279	47.50		326.50	380
0700	Steel body		7	1.143		310	47.50		357.50	415
0900	Oak door, wood body, beveled mirror, single door		7	1.143		160	47.50		207.50	249
1000	Double door		6	1.333		375	55.50		430.50	495
1200	Hotel cabinets, stainless, with lower shelf, unlighted		10	.800		202	33		235	273
1300	Lighted	↓	5	1.600	↓	300	66.50		366.50	430

10 28 23 – Laundry Accessories

10 28 23.13 Built-In Ironing Boards		Crew	Daily Output	Labor-Hours	Unit	Material	Labor	Equipment	Total	Total Incl O&P
0010	**BUILT-IN IRONING BOARDS**									
0020	Including cabinet, board & light, minimum	1 Carp	2	4	Ea.	315	166		481	600
0100	Maximum, see also Div. 11 23 13.13	"	1.50	5.333	"	525	222		747	920

10 31 Manufactured Fireplaces

10 31 13 – Manufactured Fireplace Chimneys

10 31 13.10 Fireplace Chimneys		Crew	Daily Output	Labor-Hours	Unit	Material	Labor	Equipment	Total	Total Incl O&P
0010	**FIREPLACE CHIMNEYS**									
0500	Chimney dbl. wall, all stainless, over 8'-6", 7" diam., add to fireplace	1 Carp	33	.242	V.L.F.	71.50	10.05		81.55	94
0600	10" diameter, add to fireplace		32	.250		109	10.40		119.40	136
0700	12" diameter, add to fireplace		31	.258		136	10.70		146.70	166
0800	14" diameter, add to fireplace		30	.267	↓	183	11.10		194.10	218
1000	Simulated brick chimney top, 4' high, 16" x 16"		10	.800	Ea.	258	33		291	335
1100	24" x 24"	↓	7	1.143	"	480	47.50		527.50	605

10 31 13.20 Chimney Accessories		Crew	Daily Output	Labor-Hours	Unit	Material	Labor	Equipment	Total	Total Incl O&P
0010	**CHIMNEY ACCESSORIES**									
0020	Chimney screens, galv., 13" x 13" flue	1 Bric	8	1	Ea.	54	42		96	122
0050	24" x 24" flue		5	1.600		124	67		191	238
0200	Stainless steel, 13" x 13" flue		8	1		330	42		372	425
0250	20" x 20" flue		5	1.600		450	67		517	595
2400	Squirrel and bird screens, galvanized, 8" x 8" flue		16	.500		46.50	21		67.50	83
2450	13" x 13" flue	↓	12	.667	↓	50.50	28		78.50	97.50

10 31 Manufactured Fireplaces

10 31 16 – Manufactured Fireplace Forms

10 31 16.10 Fireplace Forms		Crew	Daily Output	Labor-Hours	Unit	Material	Labor	Equipment	Total	Total Incl O&P
						2010 Bare Costs				
0010	**FIREPLACE FORMS**									
1800	Fireplace forms, no accessories, 32" opening	1 Bric	3	2.667	Ea.	650	111		761	885
1900	36" opening		2.50	3.200		830	134		964	1,100
2000	40" opening		2	4		1,100	167		1,267	1,450
2100	78" opening	↓	1.50	5.333	↓	1,600	223		1,823	2,075

10 31 23 – Prefabricated Fireplaces

10 31 23.10 Fireplace, Prefabricated

		Crew	Daily Output	Labor-Hours	Unit	Material	Labor	Equipment	Total	Total Incl O&P
0010	**FIREPLACE, PREFABRICATED**, free standing or wall hung									
0100	With hood & screen, minimum	1 Carp	1.30	6.154	Ea.	1,200	256		1,456	1,725
0150	Average		1	8		1,700	330		2,030	2,350
0200	Maximum		.90	8.889	↓	3,375	370		3,745	4,275
1500	Simulated logs, gas fired, 40,000 BTU, 2' long, minimum		7	1.143	Set	450	47.50		497.50	570
1600	Maximum		6	1.333		980	55.50		1,035.50	1,150
1700	Electric, 1,500 BTU, 1'-6" long, minimum		7	1.143		175	47.50		222.50	266
1800	11,500 BTU, maximum		6	1.333	↓	395	55.50		450.50	515
2000	Fireplace, built-in, 36" hearth, radiant		1.30	6.154	Ea.	595	256		851	1,050
2100	Recirculating, small fan		1	8		925	330		1,255	1,525
2150	Large fan		.90	8.889		1,925	370		2,295	2,675
2200	42" hearth, radiant		1.20	6.667		835	277		1,112	1,350
2300	Recirculating, small fan		.90	8.889		1,075	370		1,445	1,775
2350	Large fan		.80	10		1,925	415		2,340	2,750
2400	48" hearth, radiant		1.10	7.273		1,800	300		2,100	2,450
2500	Recirculating, small fan		.80	10		2,225	415		2,640	3,100
2550	Large fan		.70	11.429		3,450	475		3,925	4,525
3000	See through, including doors		.80	10		2,925	415		3,340	3,875
3200	Corner (2 wall)	↓	1	8	↓	2,950	330		3,280	3,725

10 32 Fireplace Specialties

10 32 13 – Fireplace Dampers

10 32 13.10 Dampers

		Crew	Daily Output	Labor-Hours	Unit	Material	Labor	Equipment	Total	Total Incl O&P
0010	**DAMPERS**									
0800	Damper, rotary control, steel, 30" opening	1 Bric	6	1.333	Ea.	79.50	55.50		135	172
0850	Cast iron, 30" opening		6	1.333		88.50	55.50		144	182
0880	36" opening		6	1.333		96.50	55.50		152	190
0900	48" opening		6	1.333		137	55.50		192.50	235
0920	60" opening		6	1.333		310	55.50		365.50	425
0950	72" opening		5	1.600		370	67		437	505
1000	84" opening, special order		5	1.600		795	67		862	970
1050	96" opening, special order		4	2		805	83.50		888.50	1,000
1200	Steel plate, poker control, 60" opening		8	1		281	42		323	375
1250	84" opening, special order		5	1.600		510	67		577	665
1400	"Universal" type, chain operated, 32" x 20" opening		8	1		218	42		260	305
1450	48" x 24" opening	↓	5	1.600	↓	325	67		392	460

10 32 23 – Fireplace Doors

10 32 23.10 Doors

		Crew	Daily Output	Labor-Hours	Unit	Material	Labor	Equipment	Total	Total Incl O&P
0010	**DOORS**									
0400	Cleanout doors and frames, cast iron, 8" x 8"	1 Bric	12	.667	Ea.	37.50	28		65.50	83
0450	12" x 12"		10	.800		45.50	33.50		79	101
0500	18" x 24"	↓	8	1	↓	131	42		173	207

10 32 Fireplace Specialties

10 32 23 – Fireplace Doors

10 32 23.10 Doors		Crew	Daily Output	Labor-Hours	Unit	Material	2010 Bare Costs Labor	Equipment	Total	Total Incl O&P
0550	Cast iron frame, steel door, 24" x 30"	1 Bric	5	1.600	Ea.	283	67		350	410
1600	Dutch Oven door and frame, cast iron, 12" x 15" opening	↓	13	.615	↓	115	25.50		140.50	165
1650	Copper plated, 12" x 15" opening	↓	13	.615	↓	224	25.50		249.50	286

10 35 Stoves

10 35 13 – Heating Stoves

10 35 13.10 Woodburning Stoves

		Crew	Daily Output	Labor-Hours	Unit	Material	Labor	Equipment	Total	Total Incl O&P
0010	**WOODBURNING STOVES**									
0015	Cast iron, minimum	2 Carp	1.30	12.308	Ea.	1,175	510		1,685	2,100
0020	Average	↓	1	16	↓	1,600	665		2,265	2,775
0030	Maximum	↓	.80	20	↓	2,850	830		3,680	4,425
0050	For gas log lighter, add				↓	42			42	46.50

10 44 Fire Protection Specialties

10 44 13 – Fire Extinguisher Cabinets

10 44 13.53 Fire Equipment Cabinets

		Crew	Daily Output	Labor-Hours	Unit	Material	Labor	Equipment	Total	Total Incl O&P
0010	**FIRE EQUIPMENT CABINETS**, not equipped, 20 ga. steel box									
0040	recessed, D.S. glass in door, box size given									
1000	Portable extinguisher, single, 8" x 12" x 27", alum. door & frame	Q-12	8	2	Ea.	124	90.50		214.50	273
1100	Steel door and frame	"	8	2	"	93	90.50		183.50	238
3000	Hose rack assy., 1-1/2" valve & 100' hose, 24" x 40" x 5-1/2"									
3100	Aluminum door and frame	Q-12	6	2.667	Ea.	305	121		426	515
3200	Steel door and frame	↓	6	2.667	↓	291	121		412	500
3300	Stainless steel door and frame	↓	6	2.667	↓	700	121		821	950
4000	Hose rack assy., 2-1/2" x 1-1/2" valve, 100' hose, 24" x 40" x 8"									
4100	Aluminum door and frame	Q-12	6	2.667	Ea.	305	121		426	515
4200	Steel door and frame	↓	6	2.667	↓	208	121		329	410
4300	Stainless steel door and frame	↓	6	2.667	↓	410	121		531	630
5000	Hose rack assy., 2-1/2" x 1-1/2" valve, 100' hose									
5010	and extinguisher, 30" x 40" x 8"									
5100	Aluminum door and frame	Q-12	5	3.200	Ea.	390	145		535	645
5200	Steel door and frame	↓	5	3.200	↓	229	145		374	470
5300	Stainless steel door and frame	↓	5	3.200	↓	465	145		610	730
8000	Valve cabinet for 2-1/2" FD angle valve, 18" x 18" x 8"									
8100	Aluminum door and frame	Q-12	12	1.333	Ea.	135	60.50		195.50	240
8200	Steel door and frame	↓	12	1.333	↓	111	60.50		171.50	214
8300	Stainless steel door and frame	↓	12	1.333	↓	182	60.50		242.50	291

10 44 16 – Fire Extinguishers

10 44 16.13 Portable Fire Extinguishers

		Crew	Daily Output	Labor-Hours	Unit	Material	Labor	Equipment	Total	Total Incl O&P
0010	**PORTABLE FIRE EXTINGUISHERS**									
0140	CO_2, with hose and "H" horn, 10 lb.				Ea.	191			191	210
0160	15 lb.				↓	250			250	276
0180	20 lb.				↓	277			277	305
1000	Dry chemical, pressurized									
1040	Standard type, portable, painted, 2-1/2 lb.				Ea.	31.50			31.50	34.50
1060	5 lb.				↓	44			44	48.50
1080	10 lb.				↓	69.50			69.50	76.50
1100	20 lb.				↓	113			113	124
1120	30 lb.				↓	375			375	415

10 44 Fire Protection Specialties

10 44 16 – Fire Extinguishers

10 44 16.13 Portable Fire Extinguishers		Crew	Daily Output	Labor-Hours	Unit	Material	2010 Bare Costs Labor	Equipment	Total	Total Incl O&P
1300	Standard type, wheeled, 150 lb.				Ea.	1,875			1,875	2,075
2000	ABC all purpose type, portable, 2-1/2 lb.					21			21	23.50
2060	5 lb.					26.50			26.50	29
2080	9-1/2 lb.					44			44	48
2100	20 lb.					79.50			79.50	87.50
5000	Pressurized water, 2-1/2 gallon, stainless steel					97.50			97.50	107
5060	With anti-freeze					106			106	116
9400	Installation of extinguishers, 12 or more, on wood	1 Carp	30	.267			11.10		11.10	17.10
9420	On masonry or concrete	"	15	.533			22		22	34

10 44 16.16 Wheeled Fire Extinguisher Units		Crew	Daily Output	Labor-Hours	Unit	Material	Labor	Equipment	Total	Total Incl O&P
0010	**WHEELED FIRE EXTINGUISHER UNITS**									
0350	CO_2, portable, with swivel horn									
0360	Wheeled type, cart mounted, 50 lb.				Ea.	940			940	1,025
0400	100 lb.				"	2,750			2,750	3,025
2200	ABC all purpose type									
2300	Wheeled, 45 lb.				Ea.	735			735	810
2360	150 lb.				"	1,850			1,850	2,025

10 51 Lockers

10 51 13 – Metal Lockers

10 51 13.10 Lockers		Crew	Daily Output	Labor-Hours	Unit	Material	Labor	Equipment	Total	Total Incl O&P
0011	**LOCKERS** Steel, baked enamel, pre-assembled									
0110	Single tier box locker, 12" x 15" x 72"	1 Shee	20	.400	Ea.	221	19.65		240.65	273
0120	18" x 15" x 72"		20	.400		240	19.65		259.65	293
0130	12" x 18" x 72"		20	.400		255	19.65		274.65	310
0140	18" x 18" x 72"		20	.400		251	19.65		270.65	305
0410	Double tier, 12" x 15" x 36"		30	.267		233	13.10		246.10	276
0420	18" x 15" x 36"		30	.267		269	13.10		282.10	315
0430	12" x 18" x 36"		30	.267		251	13.10		264.10	296
0440	18" x 18" x 36"		30	.267		249	13.10		262.10	294
0500	Two person, 18" x 15" x 72"		20	.400		273	19.65		292.65	330
0510	18" x 18" x 72"		20	.400		285	19.65		304.65	345
0520	Duplex, 15" x 15" x 72"		20	.400		299	19.65		318.65	360
0530	15" x 21" x 72"		20	.400		325	19.65		344.65	385
0600	5 tier box lockers, minimum		30	.267	Opng.	45.50	13.10		58.60	70
0700	Maximum		24	.333		52.50	16.35		68.85	82.50
0900	6 tier box lockers, minimum		36	.222		37	10.90		47.90	57
1000	Maximum		30	.267		47	13.10		60.10	72
1100	Wire meshed wardrobe, floor. mtd., open front varsity type		7.50	1.067	Ea.	234	52.50		286.50	340
2400	16-person locker unit with clothing rack									
2500	72 wide x 15" deep x 72" high	1 Shee	15	.533	Ea.	465	26		491	550
2550	18" deep	"	15	.533	"	610	26		636	715
3000	Wall mounted lockers, 4 person, with coat bar									
3100	48" wide x 18" deep x 12" high	1 Shee	20	.400	Ea.	310	19.65		329.65	370
3250	Rack w/24 wire mesh baskets		1.50	5.333	Set	345	262		607	775
3260	30 baskets		1.25	6.400		267	315		582	770
3270	36 baskets		.95	8.421		350	415		765	1,000
3280	42 baskets		.80	10		385	490		875	1,175
3300	For built-in lock with 2 keys, add				Ea.	8.70			8.70	9.55
3600	For hanger rods, add					1.86			1.86	2.05
3650	For number plate kit, 100 plates #1 - #100, add	1 Shee	4	2		60	98		158	215

10 51 Lockers

10 51 13 – Metal Lockers

10 51 13.10 Lockers		Crew	Daily Output	Labor-Hours	Unit	Material	2010 Bare Costs Labor	Equipment	Total	Total Incl O&P
3700	For locker base, closed front panel	1 Shee	90	.089	Ea.	6.45	4.36		10.81	13.70
3710	End panel, bolted		36	.222		8.20	10.90		19.10	25.50
3800	For sloping top, 12″ wide		24	.333		22.50	16.35		38.85	50
3810	15″ wide		24	.333		23.50	16.35		39.85	50.50
3820	18″ wide		24	.333		25	16.35		41.35	52.50
3850	Sloping top end panel, 12″ deep		72	.111		8.55	5.45		14	17.65
3860	15″ deep		72	.111		9.50	5.45		14.95	18.70
3870	18″ deep		72	.111		10.10	5.45		15.55	19.35
3900	For finish end panels, steel, 60″ high, 15″ deep		12	.667		50	32.50		82.50	105
3910	72″ high, 12″ deep		12	.667		50	32.50		82.50	105
3920	18″ deep	↓	12	.667	↓	59.50	32.50		92	115
5000	For 'ready to assemble' lockers,									
5010	Add to labor						75%			
5020	Deduct from material					20%				

10 51 26 – Plastic Lockers

10 51 26.13 Lockers, Plastic		Crew	Daily Output	Labor-Hours	Unit	Material	Labor	Equipment	Total	Total Incl O&P
0011	**LOCKERS, PLASTIC,** 30% recycled									
0110	Single tier box locker, 12″ x 12″ x 72″ G	1 Shee	8	1	Ea.	390	49		439	505
0120	12″ x 15″ x 72″ G		8	1		420	49		469	535
0130	12″ x 18″ x 72″ G		8	1		350	49		399	460
0410	Double tier, 12″ x 12″ x 72″ G		21	.381		400	18.70		418.70	470
0420	12″ x 15″ x 72″ G		21	.381		400	18.70		418.70	470
0430	12″ x 18″ x 72″ G	↓	21	.381	↓	400	18.70		418.70	470

10 51 53 – Locker Room Benches

10 51 53.10 Benches		Crew	Daily Output	Labor-Hours	Unit	Material	Labor	Equipment	Total	Total Incl O&P
0010	**BENCHES**									
2100	Locker bench, laminated maple, top only	1 Shee	100	.080	L.F.	14.35	3.93		18.28	22
2200	Pedestals, steel pipe	″	25	.320	Ea.	36	15.70		51.70	64

10 55 Postal Specialties

10 55 23 – Mail Boxes

10 55 23.10 Commercial Mail Boxes		Crew	Daily Output	Labor-Hours	Unit	Material	Labor	Equipment	Total	Total Incl O&P
0010	**COMMERCIAL MAIL BOXES**									
0020	Horiz., key lock, 5″H x 6″W x 15″D, alum., rear load	1 Carp	34	.235	Ea.	35	9.80		44.80	53.50
0100	Front loading		34	.235		35	9.80		44.80	53.50
0200	Double, 5″H x 12″W x 15″D, rear loading		26	.308		71.50	12.80		84.30	98.50
0300	Front loading		26	.308		66	12.80		78.80	92
0500	Quadruple, 10″H x 12″W x 15″D, rear loading		20	.400		91	16.60		107.60	126
0600	Front loading		20	.400		108	16.60		124.60	145
0800	Vertical, front load, 15″H x 5″W x 6″D, alum., per compartment		34	.235		30	9.80		39.80	48
0900	Bronze, duranodic finish		34	.235		40	9.80		49.80	59
1000	Steel, enameled		34	.235		30	9.80		39.80	48
1700	Alphabetical directories, 120 names		10	.800		127	33		160	191
1800	Letter collection box	↓	6	1.333		710	55.50		765.50	865
1830	Lobby collection boxes, aluminum	2 Shee	5	3.200		2,025	157		2,182	2,475
1840	Bronze or stainless	″	4.50	3.556		2,500	175		2,675	3,025
1900	Letter slot, residential	1 Carp	20	.400		88	16.60		104.60	123
2000	Post office type		8	1		105	41.50		146.50	180
2200	Post office counter window, with grille		2	4		550	166		716	860
2250	Key keeper, single key, aluminum	↓	26	.308	↓	37	12.80		49.80	60.50

10 55 Postal Specialties

10 55 23 – Mail Boxes

10 55 23.10 Commercial Mail Boxes		Crew	Daily Output	Labor-Hours	Unit	Material	2010 Bare Costs Labor	Equipment	Total	Total Incl O&P
2300	Steel, enameled	1 Carp	26	.308	Ea.	107	12.80		119.80	138

10 56 Storage Assemblies

10 56 13 – Metal Storage Shelving

10 56 13.10 Shelving

		Crew	Daily Output	Labor-Hours	Unit	Material	Labor	Equipment	Total	Total Incl O&P
0010	**SHELVING**									
0020	Metal, industrial, cross-braced, 3′ wide, 12″ deep	1 Sswk	175	.046	SF Shlf	6.60	2.14		8.74	11
0100	24″ deep		330	.024		4.71	1.14		5.85	7.20
0300	4′ wide, 12″ deep		185	.043		5.95	2.03		7.98	10.10
0400	24″ deep		380	.021		3.69	.99		4.68	5.80
1200	Enclosed sides, cross-braced back, 3′ wide, 12″ deep		175	.046		11.25	2.14		13.39	16.10
1300	24″ deep		290	.028		8.95	1.29		10.24	12.10
1500	Fully enclosed, sides and back, 3′ wide, 12″ deep		150	.053		12.15	2.50		14.65	17.75
1600	24″ deep		255	.031		8.35	1.47		9.82	11.80
1800	4′ wide, 12″ deep		150	.053		9.35	2.50		11.85	14.70
1900	24″ deep		290	.028		7	1.29		8.29	9.95
2200	Wide span, 1600 lb. capacity per shelf, 6′ wide, 24″ deep		380	.021		13.35	.99		14.34	16.40
2400	36″ deep		440	.018		8.25	.85		9.10	10.55
2600	8′ wide, 24″ deep		440	.018		8.60	.85		9.45	10.95
2800	36″ deep	↓	520	.015		6.95	.72		7.67	8.90
4000	Pallet racks, steel frame 5,000 lb. capacity, 8′ long, 36″ deep	2 Sswk	450	.036		8.10	1.67		9.77	11.90
4200	42″ deep		500	.032		6.75	1.50		8.25	10.10
4400	48″ deep	↓	520	.031	↓	6.55	1.44		7.99	9.80

10 56 13.20 Parts Bins

		Crew	Daily Output	Labor-Hours	Unit	Material	Labor	Equipment	Total	Total Incl O&P
0010	**PARTS BINS** Metal, gray baked enamel finish									
0100	6′-3″ high, 3′ wide									
0300	12 bins, 18″ wide x 12″ high, 12″ deep	2 Clab	10	1.600	Ea.	325	53		378	440
0400	24″ deep		10	1.600		430	53		483	550
0600	72 bins, 6″ wide x 6″ high, 12″ deep		8	2		505	66		571	655
0700	18″ deep	↓	8	2	↓	780	66		846	960
1000	7′-3″ high, 3′ wide									
1200	14 bins, 18″ wide x 12″ high, 12″ deep	2 Clab	10	1.600	Ea.	310	53		363	420
1300	24″ deep		10	1.600		410	53		463	530
1500	84 bins, 6″ wide x 6″ high, 12″ deep		8	2		835	66		901	1,025
1600	24″ deep	↓	8	2	↓	1,050	66		1,116	1,250

10 57 Wardrobe and Closet Specialties

10 57 13 – Hat and Coat Racks

10 57 13.10 Coat Racks and Wardrobes

		Crew	Daily Output	Labor-Hours	Unit	Material	Labor	Equipment	Total	Total Incl O&P
0010	**COAT RACKS AND WARDROBES**									
0020	Hat & coat rack, floor model, 6 hangers									
0050	Standing, beech wood, 21″ x 21″ x 72″, chrome				Ea.	237			237	260
0100	18 gauge tubular steel, 21″ x 21″ x 69″, wood walnut				″	275			275	305
0500	16 gauge steel frame, 22 gauge steel shelves									
0650	Single pedestal, 30″ x 18″ x 63″				Ea.	245			245	270
0800	Single face rack, 29″ x 18-1/2″ x 62″					246			246	271
0900	51″ x 18-1/2″ x 70″					375			375	415
0910	Double face rack, 39″ x 26″ x 70″					350			350	385
0920	63″ x 26″ x 70″				↓	425			425	465

10 57 Wardrobe and Closet Specialties

10 57 13 – Hat and Coat Racks

10 57 13.10 Coat Racks and Wardrobes		Crew	Daily Output	Labor-Hours	Unit	2010 Bare Costs Material	Labor	Equipment	Total	Total Incl O&P
0940	For 2" ball casters, add				Set	88			88	97
1400	Utility hook strips, 3/8" x 2-1/2" x 18", 6 hooks	1 Carp	48	.167	Ea.	56.50	6.95		63.45	73
1500	34" long, 12 hooks	"	48	.167	"	52.50	6.95		59.45	68.50
1650	Wall mounted racks, 16 gauge steel frame, 22 gauge steel shelves									
1850	12" x 15" x 26", 6 hangers	1 Carp	32	.250	Ea.	155	10.40		165.40	186
2000	12" x 15" x 50", 12 hangers	"	32	.250	"	181	10.40		191.40	215
2150	Wardrobe cabinet, steel, baked enamel finish									
2300	36" x 21" x 78", incl. top shelf & hanger rod				Ea.	282			282	310

10 57 23 – Closet and Utility Shelving

10 57 23.19 Wood Closet and Utility Shelving		Crew	Daily Output	Labor-Hours	Unit	Material	Labor	Equipment	Total	Total Incl O&P
0010	**WOOD CLOSET AND UTILITY SHELVING**									
0020	Pine, clear grade, no edge band, 1" x 8"	1 Carp	115	.070	L.F.	2.79	2.89		5.68	7.55
0100	1" x 10"		110	.073		3.48	3.02		6.50	8.50
0200	1" x 12"		105	.076		4.19	3.17		7.36	9.50
0600	Plywood, 3/4" thick with lumber edge, 12" wide		75	.107		1.58	4.43		6.01	8.60
0700	24" wide		70	.114		2.81	4.75		7.56	10.40
0900	Bookcase, clear grade pine, shelves 12" O.C., 8" deep, per SF shelf		70	.114	S.F.	9.10	4.75		13.85	17.30
1000	12" deep shelves		65	.123	"	13.60	5.10		18.70	23
1200	Adjustable closet rod and shelf, 12" wide, 3' long		20	.400	Ea.	7.80	16.60		24.40	34
1300	8' long		15	.533	"	16.40	22		38.40	52
1500	Prefinished shelves with supports, stock, 8" wide		75	.107	L.F.	3.14	4.43		7.57	10.30
1600	10" wide		70	.114	"	4.27	4.75		9.02	12

10 71 Exterior Protection

10 71 13 – Exterior Sun Control Devices

10 71 13.19 Rolling Exterior Shutters		Crew	Daily Output	Labor-Hours	Unit	Material	Labor	Equipment	Total	Total Incl O&P
0010	**ROLLING EXTERIOR SHUTTERS**									
0020	Roll-up, manual operation, aluminum, 3' x 4', incl. frame	2 Carp	8	2	Ea.	780	83		863	990
0030	6' x 7'	"	8	2	"	1,725	83		1,808	2,025

10 73 Protective Covers

10 73 13 – Awnings

10 73 13.10 Awnings, Fabric		Crew	Daily Output	Labor-Hours	Unit	Material	Labor	Equipment	Total	Total Incl O&P
0010	**AWNINGS, FABRIC**									
0020	Including acrylic canvas and frame, standard design									
0100	Door and window, slope, 3' high, 4' wide	1 Carp	4.50	1.778	Ea.	690	74		764	875
0110	6' wide		3.50	2.286		890	95		985	1,125
0120	8' wide		3	2.667		1,100	111		1,211	1,375
0200	Quarter round convex, 4' wide		3	2.667		1,075	111		1,186	1,350
0210	6' wide		2.25	3.556		1,400	148		1,548	1,750
0220	8' wide		1.80	4.444		1,700	185		1,885	2,150
0300	Dome, 4' wide		7.50	1.067		415	44.50		459.50	525
0310	6' wide		3.50	2.286		935	95		1,030	1,175
0320	8' wide		2	4		1,650	166		1,816	2,075
0350	Elongated dome, 4' wide		1.33	6.015		1,550	250		1,800	2,075
0360	6' wide		1.11	7.207		1,850	299		2,149	2,500
0370	8' wide		1	8		2,175	330		2,505	2,900
1000	Entry or walkway, peak, 12' long, 4' wide	2 Carp	.90	17.778		4,925	740		5,665	6,550
1010	6' wide		.60	26.667		7,575	1,100		8,675	10,000

10 73 Protective Covers

10 73 13 – Awnings

10 73 13.10 Awnings, Fabric		Crew	Daily Output	Labor-Hours	Unit	Material	2010 Bare Costs Labor	Equipment	Total	Total Incl O&P
1020	8' wide	2 Carp	.40	40	Ea.	10,500	1,650		12,150	14,100
1100	Radius with dome end, 4' wide		1.10	14.545		3,725	605		4,330	5,025
1110	6' wide		.70	22.857		6,000	950		6,950	8,075
1120	8' wide	↓	.50	32	↓	8,525	1,325		9,850	11,400
2000	Retractable lateral arm awning, manual									
2010	To 12' wide, 8' - 6" projection	2 Carp	1.70	9.412	Ea.	1,100	390		1,490	1,825
2020	To 14' wide, 8' - 6" projection		1.10	14.545		1,300	605		1,905	2,350
2030	To 19' wide, 8' - 6" projection		.85	18.824		1,750	780		2,530	3,125
2040	To 24' wide, 8' - 6" projection	↓	.67	23.881		2,225	990		3,215	3,975
2050	Motor for above, add	1 Carp	2.67	3	↓	955	125		1,080	1,250
3000	Patio/deck canopy with frame									
3010	12' wide, 12' projection	2 Carp	2	8	Ea.	1,575	330		1,905	2,225
3020	16' wide, 14' projection	"	1.20	13.333		2,450	555		3,005	3,525
9000	For fire retardant canvas, add					7%				
9010	For lettering or graphics, add					35%				
9020	For painted or coated acrylic canvas, deduct					8%				
9030	For translucent or opaque vinyl canvas, add					10%				
9040	For 6 or more units, deduct				↓	20%	15%			

10 73 16 – Canopies

10 73 16.20 Metal Canopies		Crew	Daily Output	Labor-Hours	Unit	Material	Labor	Equipment	Total	Total Incl O&P
0010	**METAL CANOPIES**									
0020	Wall hung, .032", aluminum, prefinished, 8' x 10'	K-2	1.30	18.462	Ea.	2,150	790	228	3,168	3,950
0300	8' x 20'		1.10	21.818		4,275	930	269	5,474	6,575
0500	10' x 10'		1.30	18.462		2,400	790	228	3,418	4,225
0700	10' x 20'		1.10	21.818		4,450	930	269	5,649	6,775
1000	12' x 20'		1	24		5,475	1,025	296	6,796	8,100
1360	12' x 30'		.80	30		8,050	1,275	370	9,695	11,400
1700	12' x 40'	↓	.60	40		9,800	1,700	495	11,995	14,200
1900	For free standing units, add				↓	20%	10%			
2300	Aluminum entrance canopies, flat soffit, .032"									
2500	3'-6" x 4'-0", clear anodized	2 Carp	4	4	Ea.	900	166		1,066	1,250
2700	Bronze anodized		4	4		1,575	166		1,741	2,000
3000	Polyurethane painted		4	4		1,275	166		1,441	1,650
3300	4'-6" x 10'-0", clear anodized		2	8		2,475	330		2,805	3,225
3500	Bronze anodized		2	8		3,175	330		3,505	3,975
3700	Polyurethane painted	↓	2	8		2,650	330		2,980	3,425
4000	Wall downspout, 10 L.F., clear anodized	1 Carp	7	1.143		154	47.50		201.50	243
4300	Bronze anodized		7	1.143		269	47.50		316.50	370
4500	Polyurethane painted	↓	7	1.143	↓	231	47.50		278.50	325
7000	Carport, baked vinyl finish, .032", 20' x 10', no foundations, min.	K-2	4	6	Car	3,850	256	74	4,180	4,750
7250	Maximum		2	12	"	7,000	510	148	7,658	8,725
7500	Walkway cover, to 12' wide, stl., vinyl finish, .032",no fndtns., min.		250	.096	S.F.	22.50	4.10	1.18	27.78	33.50
7750	Maximum	↓	200	.120	"	24.50	5.10	1.48	31.08	37.50

10 74 Manufactured Exterior Specialties

10 74 23 – Cupolas

10 74 23.10 Wood Cupolas		Crew	Daily Output	Labor-Hours	Unit	Material	2010 Bare Costs Labor	Equipment	Total	Total Incl O&P
0010	WOOD CUPOLAS									
0020	Stock units, pine, painted, 18" sq., 28" high, alum. roof	1 Carp	4.10	1.951	Ea.	162	81		243	305
0100	Copper roof		3.80	2.105		208	87.50		295.50	365
0300	23" square, 33" high, aluminum roof		3.70	2.162		345	90		435	520
0400	Copper roof		3.30	2.424		465	101		566	665
0600	30" square, 37" high, aluminum roof		3.70	2.162		495	90		585	685
0700	Copper roof		3.30	2.424		590	101		691	805
0900	Hexagonal, 31" wide, 46" high, copper roof		4	2		730	83		813	935
1000	36" wide, 50" high, copper roof		3.50	2.286		1,275	95		1,370	1,550
1200	For deluxe stock units, add to above					25%				
1400	For custom built units, add to above					50%	50%			

10 74 29 – Steeples

10 74 29.10 Prefabricated Steeples		Crew	Daily Output	Labor-Hours	Unit	Material	Labor	Equipment	Total	Total Incl O&P
0010	PREFABRICATED STEEPLES									
4000	Steeples, translucent fiberglass, 30" square, 15' high	F-3	2	20	Ea.	5,450	840	330	6,620	7,650
4150	25' high		1.80	22.222		6,350	935	365	7,650	8,825
4350	Opaque fiberglass, 24" square, 14' high		2	20		4,525	840	330	5,695	6,650
4500	28' high		1.80	22.222		4,775	935	365	6,075	7,100
4600	Aluminum, baked finish, 16" square, 14' high					4,200			4,200	4,625
4620	20' high, 3'-6" base					7,075			7,075	7,800
4640	35' high, 8' base					27,400			27,400	30,200
4660	60' high, 14' base					59,000			59,000	65,000
4680	152' high, custom					495,000			495,000	544,500
4700	Porcelain enamel steeples, custom, 40' high	F-3	.50	80		14,100	3,375	1,300	18,775	22,200
4800	60' high	"	.30	133		24,500	5,625	2,175	32,300	38,000

10 74 46 – Window Wells

10 74 46.10 Area Window Wells		Crew	Daily Output	Labor-Hours	Unit	Material	Labor	Equipment	Total	Total Incl O&P
0010	AREA WINDOW WELLS, Galvanized steel									
0020	20 ga., 3'-2" wide, 1' deep	1 Sswk	29	.276	Ea.	14	12.95		26.95	38
0100	2' deep		23	.348		25	16.30		41.30	56
0300	16 ga., 3'-2" wide, 1' deep		29	.276		20	12.95		32.95	44.50
0400	3' deep		23	.348		39	16.30		55.30	71.50
0600	Welded grating for above, 15 lbs., painted		45	.178		76	8.35		84.35	98
0700	Galvanized		45	.178		112	8.35		120.35	138
0900	Translucent plastic cap for above		60	.133		16	6.25		22.25	28.50

10 75 Flagpoles

10 75 16 – Ground-Set Flagpoles

10 75 16.10 Flagpoles		Crew	Daily Output	Labor-Hours	Unit	Material	Labor	Equipment	Total	Total Incl O&P
0010	FLAGPOLES, Ground set									
0050	Not including base or foundation									
0100	Aluminum, tapered, ground set 20' high	K-1	2	8	Ea.	990	295	148	1,433	1,700
0200	25' high		1.70	9.412		1,275	345	174	1,794	2,125
0300	30' high		1.50	10.667		1,250	395	197	1,842	2,200
0400	35' high		1.40	11.429		1,675	420	211	2,306	2,725
0500	40' high		1.20	13.333		2,350	490	247	3,087	3,600
0600	50' high		1	16		2,800	590	296	3,686	4,300
0700	60' high		.90	17.778		5,175	655	330	6,160	7,025
0800	70' high		.80	20		7,975	740	370	9,085	10,300
1100	Counterbalanced, internal halyard, 20' high		1.80	8.889		2,350	330	164	2,844	3,250

10 75 Flagpoles

10 75 16 – Ground-Set Flagpoles

10 75 16.10 Flagpoles		Crew	Daily Output	Labor-Hours	Unit	2010 Bare Costs Material	Labor	Equipment	Total	Total Incl O&P
1200	30′ high	K-1	1.50	10.667	Ea.	2,600	395	197	3,192	3,675
1300	40′ high		1.30	12.308		5,675	455	228	6,358	7,200
1400	50′ high		1	16		7,900	590	296	8,786	9,900
2820	Aluminum, electronically operated, 30′ high		1.40	11.429		4,300	420	211	4,931	5,600
2840	35′ high		1.30	12.308		4,625	455	228	5,308	6,050
2860	39′ high		1.10	14.545		5,050	535	269	5,854	6,675
2880	45′ high		1	16		6,450	590	296	7,336	8,325
2900	50′ high		.90	17.778		6,875	655	330	7,860	8,900
3000	Fiberglass, tapered, ground set, 23′ high		2	8		1,100	295	148	1,543	1,825
3100	29′-7″ high		1.50	10.667		1,450	395	197	2,042	2,400
3200	36′-1″ high		1.40	11.429		1,900	420	211	2,531	2,975
3300	39′-5″ high		1.20	13.333		2,225	490	247	2,962	3,475
3400	49′-2″ high		1	16		3,975	590	296	4,861	5,600
3500	59′ high	↓	.90	17.778	↓	4,800	655	330	5,785	6,625
4300	Steel, direct imbedded installation									
4400	Internal halyard, 20′ high	K-1	2.50	6.400	Ea.	1,550	236	118	1,904	2,225
4500	25′ high		2.50	6.400		1,750	236	118	2,104	2,400
4600	30′ high		2.30	6.957		2,050	257	129	2,436	2,775
4700	40′ high		2.10	7.619		3,050	281	141	3,472	3,925
4800	50′ high		1.90	8.421		4,500	310	156	4,966	5,600
5000	60′ high		1.80	8.889		6,800	330	164	7,294	8,150
5100	70′ high		1.60	10		8,300	370	185	8,855	9,900
5200	80′ high		1.40	11.429		10,900	420	211	11,531	12,900
5300	90′ high		1.20	13.333		15,000	490	247	15,737	17,500
5500	100′ high	↓	1	16	↓	15,500	590	296	16,386	18,300
6400	Wood poles, tapered, clear vertical grain fir with tilting									
6410	base, not incl. foundation, 4″ butt, 25′ high	K-1	1.90	8.421	Ea.	1,375	310	156	1,841	2,175
6800	6″ butt, 30′ high	″	1.30	12.308	″	1,925	455	228	2,608	3,075
7300	Foundations for flagpoles, including									
7400	excavation and concrete, to 35′ high poles	C-1	10	3.200	Ea.	540	126		666	790
7600	40′ to 50′ high		3.50	9.143		985	360		1,345	1,625
7700	Over 60′ high	↓	2	16	↓	1,100	630		1,730	2,175

10 75 23 – Wall-Mounted Flagpoles

10 75 23.10 Flagpoles		Crew	Daily Output	Labor-Hours	Unit	Material	Labor	Equipment	Total	Total Incl O&P
0010	**FLAGPOLES**, Structure mounted									
0100	Fiberglass, vertical wall set, 19′-8″ long	K-1	1.50	10.667	Ea.	1,300	395	197	1,892	2,250
0200	23′ long		1.40	11.429		1,375	420	211	2,006	2,375
0300	26′-3″ long		1.30	12.308		1,775	455	228	2,458	2,900
0800	19′-8″ long outrigger		1.30	12.308		1,325	455	228	2,008	2,400
1300	Aluminum, vertical wall set, tapered, with base, 20′ high		1.20	13.333		980	490	247	1,717	2,100
1400	29′-6″ high		1	16		2,250	590	296	3,136	3,700
2400	Outrigger poles with base, 12′ long		1.30	12.308		930	455	228	1,613	1,975
2500	14′ long	↓	1	16	↓	1,425	590	296	2,311	2,800

10 81 Pest Control Devices

10 81 13 – Bird Control Devices

10 81 13.10 Bird Control Netting		Crew	Daily Output	Labor-Hours	Unit	Material	Labor	Equipment	Total	Total Incl O&P
						2010 Bare Costs				
0010	**BIRD CONTROL NETTING**									
0020	1/8" square mesh	4 Clab	4000	.008	S.F.	.06	.26		.32	.48
0100	1/4" square mesh		4000	.008		.07	.26		.33	.49
0120	1/2" square mesh		4000	.008		.08	.26		.34	.50
0140	5/8" x 3/4" mesh		4000	.008		.08	.26		.34	.50
0160	1-1/4" x 1-1/2" mesh		4000	.008		.10	.26		.36	.52
0200	4" square mesh		4000	.008		.11	.26		.37	.53
1000	Poly clips				Ea.	.02			.02	.02

10 88 Scales

10 88 05 – Commercial Scales

10 88 05.10 Scales		Crew	Daily Output	Labor-Hours	Unit	Material	Labor	Equipment	Total	Total Incl O&P
0010	**SCALES** Built-in floor scale, not incl. foundations									
0100	Dial type, 5 ton capacity, 8' x 6' platform	3 Carp	.50	48	Ea.	7,275	2,000		9,275	11,100
0300	9' x 7' platform		.40	60		9,450	2,500		11,950	14,300
0400	10 ton capacity, steel, 8' x 6' platform		.40	60		11,400	2,500		13,900	16,500
0600	9' x 7' platform		.35	68.571		10,700	2,850		13,550	16,200
0700	Truck scales, incl. steel weigh bridge,									
0800	not including foundation, pits									
1550	Digital, electronic, 100 ton capacity, steel deck 12' x 10' platform	3 Carp	.20	120	Ea.	12,100	4,975		17,075	21,000
1600	40' x 10' platform		.14	171		24,600	7,125		31,725	38,100
1640	60' x 10' platform		.13	184		30,400	7,675		38,075	45,300
1680	70' x 10' platform		.12	200		34,000	8,300		42,300	50,000
2000	For standard automatic printing device, add					2,450			2,450	2,700
2100	For remote reading electronic system, add					2,225			2,225	2,450
2300	Concrete foundation pits for above, 8' x 6', 5 C.Y. required	C-1	.50	64		925	2,525		3,450	4,925
2400	14' x 6' platform, 10 C.Y. required		.35	91.429		1,350	3,600		4,950	7,050
2600	50' x 10' platform, 30 C.Y. required		.25	128		1,825	5,050		6,875	9,775
2700	70' x 10' platform, 40 C.Y. required		.15	213		4,000	8,425		12,425	17,400
2750	Crane scales, dial, 1 ton capacity					1,000			1,000	1,100
2780	5 ton capacity					1,375			1,375	1,500
2800	Digital, 1 ton capacity					1,875			1,875	2,050
2850	10 ton capacity					4,375			4,375	4,825
2900	Low profile electronic warehouse scale,									
3000	not incl. printer, 4' x 4' platform, 10,000 lb. capacity	2 Carp	.30	53.333	Ea.	1,425	2,225		3,650	4,975
3300	5' x 7' platform, 10,000 lb. capacity		.25	64		4,000	2,650		6,650	8,500
3400	20,000 lb. capacity		.20	80		5,150	3,325		8,475	10,800
3500	For printers, incl. time, date & numbering, add					1,275			1,275	1,400
3800	Portable, beam type, capacity 1000#, platform 18" x 24"					745			745	820
3900	Dial type, capacity 2000#, platform 24" x 24"					1,275			1,275	1,425
4000	Digital type, capacity 1000#, platform 24" x 30"					2,050			2,050	2,275
4100	Portable contractor truck scales, 50 ton cap., 40' x 10' platform					32,000			32,000	35,200
4200	60' x 10' platform					28,600			28,600	31,500

Division Notes

		CREW	DAILY OUTPUT	LABOR-HOURS	UNIT	2010 BARE COSTS MAT.	LABOR	EQUIP.	TOTAL	TOTAL INCL O&P

Estimating Tips

General

- The items in this division are usually priced per square foot or each. Many of these items are purchased by the owner for installation by the contractor. Check the specifications for responsibilities and include time for receiving, storage, installation, and mechanical and electrical hookups in the appropriate divisions.
- Many items in Division 11 require some type of support system that is not usually furnished with the item. Examples of these systems include blocking for the attachment of casework and support angles for ceiling-hung projection screens. The required blocking or supports must be added to the estimate in the appropriate division.
- Some items in Division 11 may require assembly or electrical hookups. Verify the amount of assembly required or the need for a hard electrical connection and add the appropriate costs.

Reference Numbers

Reference numbers are shown in shaded boxes at the beginning of some major classifications. These numbers refer to related items in the Reference Section. The reference information may be an estimating procedure, an alternate pricing method, or technical information.

Note: Not all subdivisions listed here necessarily appear in this publication.

11 05 05 – Selective Equipment Demolition

11 05 05.10 Selective Demolition		Crew	Daily Output	Labor-Hours	Unit	Material	2010 Bare Costs Labor	Equipment	Total	Total Incl O&P
0010	**SELECTIVE DEMOLITION**									
0130	Central vacuum, motor unit, residential or commercial	1 Clab	2	4	Ea.		132		132	204
0210	Vault door and frame	2 Skwk	2	8			340		340	525
0215	Day gate, for vault	″	3	5.333			227		227	350
0380	Bank equipment, teller window, bullet resistant	1 Clab	1.20	6.667			221		221	340
0381	Counter	2 Clab	1.50	10.667	Station		355		355	545
0382	Drive-up window, including drawer and glass		1.50	10.667	″		355		355	545
0383	Thru-wall boxes and chests, selective demolition		2.50	6.400	Ea.		212		212	325
0384	Bullet resistant partitions		20	.800	L.F.		26.50		26.50	41
0385	Pneumatic tube system, 2 lane drive-up	L-3	.45	35.556	Ea.		1,600		1,600	2,450
0386	Safety deposit box	1 Clab	50	.160	Opng.		5.30		5.30	8.15
0387	Surveillance system, video, complete	2 Elec	2	8	Ea.		390		390	585
0410	Church equipment, misc moveable fixtures	2 Clab	1	16			530		530	815
0412	Steeple, to 28′ high	F-3	3	13.333			560	219	779	1,100
0414	40′ to 60′ high	″	.80	50			2,100	820	2,920	4,125
0510	Library equipment, bookshelves, wood, to 90″ high	1 Clab	20	.400	L.F.		13.25		13.25	20.50
0515	Carrels, hardwood, 36″ x 24″	″	9	.889	Ea.		29.50		29.50	45.50
0630	Stage equipment, light control panel	1 Elec	1	8	″		390		390	585
0632	Border lights		40	.200	L.F.		9.80		9.80	14.55
0634	Spotlights		8	1	Ea.		49		49	73
0636	Telescoping platforms and risers	2 Clab	175	.091	SF Stg.		3.03		3.03	4.67
1020	Barber equipment, hydraulic chair	1 Clab	40	.200	Ea.		6.60		6.60	10.20
1030	Checkout counter, supermarket or warehouse conveyor	2 Clab	18	.889			29.50		29.50	45.50
1040	Food cases, refrigerated or frozen	Q-5	6	2.667			125		125	187
1190	Laundry equipment, commercial	L-6	3	4			204		204	305
1360	Movie equipment, lamphouse, to 4000 watt, incl rectifier	1 Elec	4	2			98		98	146
1365	Sound system, incl amplifier	″	1.25	6.400			315		315	465
1410	Air compressor, to 5 H.P.	2 Clab	2.50	6.400			212		212	325
1412	Lubrication equipment, automotive, 3 reel type, incl pump, excl piping	L-4	1	24	Set		935		935	1,425
1414	Booth, spray paint, complete, to 26′ long	″	.80	30	″		1,175		1,175	1,800
1560	Parking equipment, cashier booth	B-22	2	15	Ea.		575	104	679	995
1600	Loading dock equipment, dock bumpers, rubber	1 Clab	50	.160	″		5.30		5.30	8.15
1610	Door seal for door perimeter	″	50	.160	L.F.		5.30		5.30	8.15
1620	Platform lifter, fixed, 6′ x 8′, 5000 lb capacity	E-16	1.50	10.667	Ea.		510	97	607	1,000
1630	Dock leveller	″	2	8			385	73	458	755
1640	Lights, single or double arm	1 Elec	8	1			49		49	73
1650	Shelter, fabric, truck or train	1 Clab	1.50	5.333			177		177	272
1790	Waste handling equipment, commercial compactor	L-4	2	12			465		465	720
1792	Commercial or municipal incinerator, gas	″	2	12			465		465	720
1795	Crematory, excluding building	Q-3	.25	128			6,350		6,350	9,500
1910	Detection equipment, cell bar front	E-4	4	8			380	36.50	416.50	705
1912	Cell door and frame		8	4			190	18.25	208.25	355
1914	Prefab cell, 4′ to 5′ wide, 7′ to 8′ high, 7′ deep		8	4			190	18.25	208.25	355
1916	Cot, bolted, single		40	.800			38	3.65	41.65	70.50
1918	Visitor cubicle		4	8			380	36.50	416.50	705
2850	Hydraulic gates, canal, flap, knife, slide or sluice, to 18″ diameter	L-5A	8	4			187	82	269	405
2852	19″ to 36″ diameter		6	5.333			249	109	358	545
2854	37″ to 48″ diameter		2	16			750	330	1,080	1,625
2856	49″ to 60″ diameter		1	32			1,500	655	2,155	3,275
2858	Over 60″ diameter		.30	106			5,000	2,175	7,175	10,900
3100	Sewage pumping system, prefabricated, to 1000 GPM	C-17D	.20	420			18,100	3,875	21,975	32,100
3110	Sewage treatment, holding tank for recirc chemical water closet	1 Plum	8	1			52		52	78
3900	Wastewater treatment system, to 1500 gallons	B-21	2	14			530	69	599	890

11 05 Common Work Results for Equipment

11 05 05 – Selective Equipment Demolition

11 05 05.10 Selective Demolition		Crew	Daily Output	Labor-Hours	Unit	Material	2010 Bare Costs Labor	Equipment	Total	Total Incl O&P
4050	Food storage equipment, walk-in refrigerator/freezer	2 Clab	64	.250	S.F.		8.30		8.30	12.75
4052	Shelving, stainless steel, 4 tier or dunnage rack	1 Clab	12	.667	Ea.		22		22	34
4100	Food preparation equipment, small countertop		18	.444			14.70		14.70	22.50
4150	Food delivery carts, heated cabinets		18	.444			14.70		14.70	22.50
4200	Cooking equipment, commercial range	Q-1	12	1.333			62.50		62.50	93.50
4250	Hood and ventilation equipment, kitchen exhaust hood, excl fire prot	1 Clab	3	2.667			88.50		88.50	136
4255	Fire protection system	Q-1	3	5.333			250		250	375
4300	Food dispensing equipment, countertop items	1 Clab	15	.533			17.65		17.65	27
4310	Serving counter	"	65	.123	L.F.		4.07		4.07	6.30
4350	Ice machine, ice cube maker, flakers and storage bins, to 2000 lb/day	Q-1	1.60	10	Ea.		470		470	700
4400	Cleaning and disposal, commercial dishwasher, to 50 racks per hour	L-6	1	12			610		610	915
4405	To 275 racks per hour	L-4	1	24			935		935	1,425
4410	Dishwasher hood	2 Clab	5	3.200			106		106	163
4420	Garbage disposal, commercial, to 5 H.P.	L-1	8	2			101		101	151
4540	Water heater, residential, to 80 gal/day	"	5	3.200			162		162	241
4542	Water softener, automatic	2 Plum	10	1.600			83.50		83.50	125
4544	Disappearing stairway, to 15' floor height	2 Clab	6	2.667			88.50		88.50	136
4710	Darkroom equipment, light	L-7	10	2.800			113		113	172
4712	Heavy	"	1.50	18.667			750		750	1,150
4720	Doors	2 Clab	3.50	4.571	Opng.		151		151	233
4830	Bowling alley, complete, incl pinsetter, scorer, counters, misc supplies	4 Clab	.40	80	Lane		2,650		2,650	4,075
4840	Health club equipment, circuit training apparatus	2 Clab	2	8	Set		265		265	410
4842	Squat racks	"	10	1.600	Ea.		53		53	81.50
4860	School equipment, basketball backstop	L-2	2	8			293		293	450
4862	Table and benches, folding, in wall, 14' long	L-4	4	6			234		234	360
4864	Bleachers, telescoping, to 30 tier	F-5	120	.267	Seat		11.20		11.20	17.30
4866	Boxing ring, elevated	L-4	.20	120	Ea.		4,675		4,675	7,175
4867	Boxing ring, floor level	"	2	12			465		465	720
4868	Exercise equipment	1 Clab	6	1.333			44		44	68
4870	Gym divider	L-4	1000	.024	S.F.		.93		.93	1.44
4875	Scoreboard	R-3	2	10	Ea.		485	69	554	795
4880	Shooting range, incl bullet traps, targets, excl structure	L-9	1	36	Point		1,375		1,375	2,175
5200	Vocational shop equipment	2 Clab	8	2	Ea.		66		66	102
6200	Fume hood, incl countertop, excl HVAC	"	6	2.667	L.F.		88.50		88.50	136
7100	Medical sterilizing, distiller, water, steam heated, 50 gal capacity	1 Plum	2.80	2.857	Ea.		149		149	223
7200	Medical equipment, surgery table, minor	1 Clab	1	8			265		265	410
7210	Surgical lights, doctors office, single or double arm	2 Elec	3	5.333			261		261	390
7300	Physical therapy, table	2 Clab	4	4			132		132	204
7310	Whirlpool bath, fixed, incl mixing valves	1 Plum	4	2			104		104	156
7400	Dental equipment, chair, electric or hydraulic	1 Clab	.75	10.667			355		355	545
7410	Central suction system	1 Plum	2	4			208		208	310
7420	Drill console with accessories	1 Clab	3.20	2.500			83		83	128
7430	X-ray unit	"	4	2			66		66	102
7440	X-ray developer	1 Plum	10	.800			41.50		41.50	62.50

11 05 10 – Equipment Installation

11 05 10.10 Industrial Equipment Installation		Crew	Daily Output	Labor-Hours	Unit	Material	Labor	Equipment	Total	Total Incl O&P
0010	**INDUSTRIAL EQUIPMENT INSTALLATION**									
0020	Industrial equipment, minimum	E-2	12	4.667	Ton		213	134	347	505
0200	Maximum	"	2	28	"		1,275	805	2,080	3,025

11 11 Vehicle Service Equipment

11 11 13 – Compressed-Air Vehicle Service Equipment

11 11 13.10 Compressed Air Equipment		Crew	Daily Output	Labor-Hours	Unit	2010 Bare Costs Material	Labor	Equipment	Total	Total Incl O&P
0010	**COMPRESSED AIR EQUIPMENT**									
0030	Compressors, electric, 1-1/2 H.P., standard controls	L-4	1.50	16	Ea.	400	625		1,025	1,400
0550	Dual controls		1.50	16		715	625		1,340	1,750
0600	5 H.P., 115/230 volt, standard controls		1	24		2,050	935		2,985	3,675
0650	Dual controls	↓	1	24	↓	2,175	935		3,110	3,825

11 11 19 – Vehicle Lubrication Equipment

11 11 19.10 Lubrication Equipment		Crew	Daily Output	Labor-Hours	Unit	Material	Labor	Equipment	Total	Total Incl O&P
0010	**LUBRICATION EQUIPMENT**									
3000	Lube equipment, 3 reel type, with pumps, not including piping	L-4	.50	48	Set	7,900	1,875		9,775	11,600

11 11 33 – Vehicle Spray Painting Equipment

11 11 33.10 Spray Painting Equipment		Crew	Daily Output	Labor-Hours	Unit	Material	Labor	Equipment	Total	Total Incl O&P
0010	**SPRAY PAINTING EQUIPMENT**									
4000	Spray painting booth, 26′ long, complete	L-4	.40	60	Ea.	15,900	2,325		18,225	21,000

11 12 Parking Control Equipment

11 12 13 – Parking Key and Card Control Units

11 12 13.10 Parking Control Units		Crew	Daily Output	Labor-Hours	Unit	Material	Labor	Equipment	Total	Total Incl O&P
0010	**PARKING CONTROL UNITS**									
5100	Card reader	1 Elec	2	4	Ea.	1,825	196		2,021	2,300
5120	Proximity with customer display	2 Elec	1	16		5,500	785		6,285	7,225
6000	Parking control software, minimum	1 Elec	.50	16		21,900	785		22,685	25,300
6020	Maximum	"	.20	40	↓	96,000	1,950		97,950	108,500

11 12 16 – Parking Ticket Dispensers

11 12 16.10 Ticket Dispensers		Crew	Daily Output	Labor-Hours	Unit	Material	Labor	Equipment	Total	Total Incl O&P
0010	**TICKET DISPENSERS**									
5900	Ticket spitter with time/date stamp, standard	2 Elec	2	8	Ea.	5,975	390		6,365	7,125
5920	Mag stripe encoding	"	2	8	"	18,600	390		18,990	21,000

11 12 26 – Parking Fee Collection Equipment

11 12 26.13 Parking Fee Coin Collection Equipment		Crew	Daily Output	Labor-Hours	Unit	Material	Labor	Equipment	Total	Total Incl O&P
0010	**PARKING FEE COIN COLLECTION EQUIPMENT**									
5200	Cashier booth, average	B-22	1	30	Ea.	9,650	1,150	207	11,007	12,600
5300	Collector station, pay on foot	2 Elec	.20	80		109,500	3,925		113,425	126,500
5320	Credit card only	"	.50	32	↓	20,200	1,575		21,775	24,600

11 12 26.23 Fee Equipment		Crew	Daily Output	Labor-Hours	Unit	Material	Labor	Equipment	Total	Total Incl O&P
0010	**FEE EQUIPMENT**									
5600	Fee computer	1 Elec	1.50	5.333	Ea.	13,400	261		13,661	15,100

11 12 33 – Parking Gates

11 12 33.13 Parking Gate Equipment		Crew	Daily Output	Labor-Hours	Unit	Material	Labor	Equipment	Total	Total Incl O&P
0010	**PARKING GATE EQUIPMENT**									
5000	Barrier gate with programmable controller	2 Elec	3	5.333	Ea.	3,350	261		3,611	4,075
5020	Industrial		3	5.333		4,575	261		4,836	5,425
5500	Exit verifier	↓	1	16		17,400	785		18,185	20,400
5700	Full sign, 4″ letters	1 Elec	2	4		1,225	196		1,421	1,625
5800	Inductive loop	2 Elec	4	4		169	196		365	475
5950	Vehicle detector, microprocessor based	1 Elec	3	2.667	↓	390	131		521	625

11 13 Loading Dock Equipment

11 13 13 – Loading Dock Bumpers

11 13 13.10 Dock Bumpers		Crew	Daily Output	Labor-Hours	Unit	2010 Bare Costs Material	Labor	Equipment	Total	Total Incl O&P
0010	**DOCK BUMPERS** Bolts not included									
0020	2" x 6" to 4" x 8", average	1 Carp	.30	26.667	M.B.F.	1,200	1,100		2,300	3,025
0050	Bumpers, rubber blocks 4-1/2" thick, 10" high, 14" long		26	.308	Ea.	49	12.80		61.80	73
0200	24" long		22	.364		86.50	15.10		101.60	119
0300	36" long		17	.471		91.50	19.55		111.05	131
0500	12" high, 14" long		25	.320		91	13.30		104.30	121
0550	24" long		20	.400		101	16.60		117.60	137
0600	36" long		15	.533		112	22		134	157
0800	Rubber blocks 6" thick, 10" high, 14" long		22	.364		87.50	15.10		102.60	120
0850	24" long		18	.444		110	18.45		128.45	150
0900	36" long		13	.615		140	25.50		165.50	194
0910	20" high, 11" long		13	.615		133	25.50		158.50	187
0920	Extruded rubber bumpers, T section, 22" x 22" x 3" thick		41	.195		51.50	8.10		59.60	69.50
0940	Molded rubber bumpers, 24" x 12" x 3" thick		20	.400		52	16.60		68.60	83
1000	Welded installation of above bumpers	E-14	8	1		3.40	49	18.25	70.65	110
1100	For drilled anchors, add per anchor	1 Carp	36	.222		6.10	9.25		15.35	21
1300	Steel bumpers, see Div. 10 26 13.10									

11 13 16 – Loading Dock Seals and Shelters

11 13 16.10 Dock Seals and Shelters		Crew	Daily Output	Labor-Hours	Unit	Material	Labor	Equipment	Total	Total Incl O&P
0010	**DOCK SEALS AND SHELTERS**									
3600	Door seal for door perimeter, 12" x 12", vinyl covered	1 Carp	26	.308	L.F.	29	12.80		41.80	51
3900	Folding gates, see Div. 10 22 16.10									
6200	Shelters, fabric, for truck or train, scissor arms, minimum	1 Carp	1	8	Ea.	1,500	330		1,830	2,150
6300	Maximum	"	.50	16	"	1,950	665		2,615	3,175

11 13 19 – Stationary Loading Dock Equipment

11 13 19.10 Dock Equipment		Crew	Daily Output	Labor-Hours	Unit	Material	Labor	Equipment	Total	Total Incl O&P
0010	**DOCK EQUIPMENT**									
2200	Dock boards, heavy duty, 60" x 60", aluminum, 5,000 lb. capacity				Ea.	1,175			1,175	1,275
2700	9,000 lb. capacity					1,425			1,425	1,575
3200	15,000 lb. capacity					1,575			1,575	1,725
4200	Platform lifter, 6' x 6', portable, 3,000 lb. capacity					8,300			8,300	9,150
4250	4,000 lb. capacity					10,200			10,200	11,300
4400	Fixed, 6' x 8', 5,000 lb. capacity	E-16	.70	22.857		8,950	1,100	208	10,258	12,000
4500	Levelers, hinged for trucks, 10 ton capacity, 6' x 8'		1.08	14.815		4,525	710	135	5,370	6,375
4650	7' x 8'		1.08	14.815		5,075	710	135	5,920	6,975
4670	Air bag power operated, 10 ton cap., 6'x8'		1.08	14.815		5,250	710	135	6,095	7,175
4680	7' x 8'		1.08	14.815		5,275	710	135	6,120	7,200
4700	Hydraulic, 10 ton capacity, 6' x 8' **CN**		1.08	14.815		7,875	710	135	8,720	10,000
4800	7' x 8'		1.08	14.815		8,475	710	135	9,320	10,700
5800	Loading dock safety restraints, manual style		1.08	14.815		2,875	710	135	3,720	4,575
5900	Automatic style		1.08	14.815		4,850	710	135	5,695	6,725

11 13 26 – Loading Dock Lights

11 13 26.10 Dock Lights		Crew	Daily Output	Labor-Hours	Unit	Material	Labor	Equipment	Total	Total Incl O&P
0010	**DOCK LIGHTS**									
5000	Lights for loading docks, single arm, 24" long	1 Elec	3.80	2.105	Ea.	139	103		242	305
5700	Double arm, 60" long	"	3.80	2.105	"	175	103		278	345

11 14 Pedestrian Control Equipment

11 14 13 – Pedestrian Gates

11 14 13.13 Portable Posts and Railings		Crew	Daily Output	Labor-Hours	Unit	Material	2010 Bare Costs Labor	Equipment	Total	Total Incl O&P
0010	PORTABLE POSTS AND RAILINGS									
0020	Portable for pedestrian traffic control, standard, minimum				Ea.	126			126	139
0100	Maximum					186			186	204
0300	Deluxe posts, minimum					179			179	197
0400	Maximum					325			325	355
0600	Ropes for above posts, plastic covered, 1-1/2" diameter				L.F.	15.55			15.55	17.10
0700	Chain core				"	10.90			10.90	12
1500	Portable security or safety barrier, black with 7' yellow strap				Ea.	250			250	275
1510	12' yellow strap					290			290	320
1550	Sign holder, standard design					130			130	143

11 14 13.19 Turnstiles

		Crew	Daily Output	Labor-Hours	Unit	Material	Labor	Equipment	Total	Total Incl O&P
0010	TURNSTILES									
0020	One way, 4 arm, 46" diameter, economy, manual	2 Carp	5	3.200	Ea.	805	133		938	1,100
0100	Electric		1.20	13.333		1,300	555		1,855	2,300
0300	High security, galv., 5'-5" diameter, 7' high, manual		1	16		3,975	665		4,640	5,400
0350	Electric		.60	26.667		4,500	1,100		5,600	6,650
0420	Three arm, 24" opening, light duty, manual		2	8		1,250	330		1,580	1,900
0450	Heavy duty		1.50	10.667		3,125	445		3,570	4,125
0460	Manual, with registering & controls, light duty		2	8		2,725	330		3,055	3,500
0470	Heavy duty		1.50	10.667		3,300	445		3,745	4,300
0480	Electric, heavy duty		1.10	14.545		3,825	605		4,430	5,125
0500	For coin or token operating, add					610			610	670
1200	One way gate with horizontal bars, 5'-5" diameter									
1300	7' high, recreation or transit type	2 Carp	.80	20	Ea.	4,425	830		5,255	6,125
1500	For electronic counter, add				"	210			210	231

11 16 Vault Equipment

11 16 16 – Safes

11 16 16.10 Commercial Safes

		Crew	Daily Output	Labor-Hours	Unit	Material	Labor	Equipment	Total	Total Incl O&P
0010	COMMERCIAL SAFES									
0200	Office, 1 hr. rating, 30" x 18" x 18"				Ea.	2,175			2,175	2,400
0250	40" x 18" x 18"					4,100			4,100	4,525
0300	60" x 36" x 18", double door					8,150			8,150	8,975
0600	Data, 1 hr. rating, 27" x 19" x 16"					7,600			7,600	8,350
0700	63" x 34" x 16"					14,500			14,500	15,900
0750	Diskette, 1 hr., 14" x 12" x 11", inside					4,200			4,200	4,625
0800	Money, "B" label, 9" x 14" x 14"					525			525	580
0900	Tool resistive, 24" x 24" x 20"					3,575			3,575	3,925
1050	Tool and torch resistive, 24" x 24" x 20"					9,150			9,150	10,100
1150	Jewelers, 23" x 20" x 18"					11,700			11,700	12,800
1200	63" x 25" x 18"					22,400			22,400	24,700
1300	For handling into building, add, minimum	A-2	8.50	2.824			92.50	28	120.50	173
1400	Maximum	"	.78	30.769			1,000	305	1,305	1,900

11 17 Teller and Service Equipment

11 17 13 – Teller Equipment Systems

11 17 13.10 Bank Equipment		Crew	Daily Output	Labor-Hours	Unit	Material	2010 Bare Costs Labor	Equipment	Total	Total Incl O&P
0010	**BANK EQUIPMENT**									
0020	Alarm system, police	2 Elec	1.60	10	Ea.	4,675	490		5,165	5,875
0100	With vault alarm	"	.40	40		18,600	1,950		20,550	23,300
0400	Bullet resistant teller window, 44" x 60"	1 Glaz	.60	13.333		3,275	535		3,810	4,400
0500	48" x 60"	"	.60	13.333	↓	4,175	535		4,710	5,375
3000	Counters for banks, frontal only	2 Carp	1	16	Station	1,750	665		2,415	2,950
3100	Complete with steel undercounter	"	.50	32	"	3,400	1,325		4,725	5,800
4600	Door and frame, bullet-resistant, with vision panel, minimum	2 Sswk	1.10	14.545	Ea.	5,000	680		5,680	6,700
4700	Maximum		1.10	14.545		6,775	680		7,455	8,675
4800	Drive-up window, drawer & mike, not incl. glass, minimum		1	16		5,575	750		6,325	7,475
4900	Maximum		.50	32		8,875	1,500		10,375	12,400
5000	Night depository, with chest, minimum		1	16		7,250	750		8,000	9,300
5100	Maximum		.50	32		10,300	1,500		11,800	14,000
5200	Package receiver, painted		3.20	5		1,300	235		1,535	1,825
5300	Stainless steel	↓	3.20	5	↓	2,225	235		2,460	2,825
5400	Partitions, bullet-resistant, 1-3/16" glass, 8' high	2 Carp	10	1.600	L.F.	189	66.50		255.50	310
5450	Acrylic	"	10	1.600	"	360	66.50		426.50	495
5500	Pneumatic tube systems, 2 lane drive-up, complete	L-3	.25	64	Total	24,600	2,900		27,500	31,500
5550	With T.V. viewer	"	.20	80	"	47,600	3,625		51,225	58,000
5570	Safety deposit boxes, minimum	1 Sswk	44	.182	Opng.	54.50	8.55		63.05	75
5580	Maximum, 10" x 15" opening		19	.421		116	19.75		135.75	162
5590	Teller locker, average	↓	15	.533	↓	1,500	25		1,525	1,700
5600	Pass thru, bullet-res. window, painted steel, 24" x 36"	2 Sswk	1.60	10	Ea.	2,300	470		2,770	3,350
5700	48" x 48"		1.20	13.333		2,550	625		3,175	3,900
5800	72" x 40"	↓	.80	20	↓	3,550	940		4,490	5,550
5900	For stainless steel frames, add					20%				
6100	Surveillance system, video camera, complete	2 Elec	1	16	Ea.	10,400	785		11,185	12,700
6110	For each additional camera, add				"	945			945	1,050
6120	CCTV system, see Div. 27 41 19.10									
6200	Twenty-four hour teller, single unit,									
6300	automated deposit, cash and memo	L-3	.25	64	Ea.	42,800	2,900		45,700	51,500
7000	Vault front, see Div. 08 34 59.10									

11 19 Detention Equipment

11 19 30 – Detention Cell Equipment

11 19 30.10 Cell Equipment		Crew	Daily Output	Labor-Hours	Unit	Material	Labor	Equipment	Total	Total Incl O&P
0010	**CELL EQUIPMENT**									
3000	Toilet apparatus including wash basin, average	L-8	1.50	13.333	Ea.	3,350	580		3,930	4,600

11 21 Mercantile and Service Equipment

11 21 13 – Cash Registers and Checking Equipment

11 21 13.10 Checkout Counter		Crew	Daily Output	Labor-Hours	Unit	Material	Labor	Equipment	Total	Total Incl O&P
0010	**CHECKOUT COUNTER**									
0020	Supermarket conveyor, single belt	2 Clab	10	1.600	Ea.	2,900	53		2,953	3,275
0100	Double belt, power take-away		9	1.778		4,175	59		4,234	4,700
0400	Double belt, power take-away, incl. side scanning		7	2.286		4,900	75.50		4,975.50	5,525
0800	Warehouse or bulk type	↓	6	2.667	↓	5,800	88.50		5,888.50	6,500
1000	Scanning system, 2 lanes, w/registers, scan gun & memory				System	15,200			15,200	16,700
1100	10 lanes, single processor, full scan, with scales				"	144,500			144,500	158,500

11 21 Mercantile and Service Equipment

11 21 13 – Cash Registers and Checking Equipment

11 21 13.10 Checkout Counter		Crew	Daily Output	Labor-Hours	Unit	Material	2010 Bare Costs Labor	Equipment	Total	Total Incl O&P
2000	Register, restaurant, minimum				Ea.	615			615	675
2100	Maximum					2,700			2,700	2,975
2150	Store, minimum					615			615	675
2200	Maximum				↓	2,700			2,700	2,975

11 21 53 – Barber and Beauty Shop Equipment

11 21 53.10 Barber Equipment

		Crew	Daily Output	Labor-Hours	Unit	Material	Labor	Equipment	Total	Total Incl O&P
0010	**BARBER EQUIPMENT**									
0020	Chair, hydraulic, movable, minimum	1 Carp	24	.333	Ea.	420	13.85		433.85	480
0050	Maximum	"	16	.500		2,975	21		2,996	3,300
0200	Wall hung styling station with mirrors, minimum	L-2	8	2		370	73		443	520
0300	Maximum	"	4	4		1,975	146		2,121	2,400
0500	Sink, hair washing basin, rough plumbing not incl.	1 Plum	8	1		425	52		477	545
1000	Sterilizer, liquid solution for tools					138			138	151
1100	Total equipment, rule of thumb, per chair, minimum	L-8	1	20		1,575	875		2,450	3,050
1150	Maximum	"	1	20	↓	4,400	875		5,275	6,175

11 23 Commercial Laundry and Dry Cleaning Equipment

11 23 13 – Dry Cleaning Equipment

11 23 13.13 Dry Cleaners

		Crew	Daily Output	Labor-Hours	Unit	Material	Labor	Equipment	Total	Total Incl O&P
0010	**DRY CLEANERS** Not incl. rough-in									
2000	Dry cleaners, electric, 20 lb. capacity	L-1	.20	80	Ea.	33,300	4,050		37,350	42,600
2050	25 lb. capacity		.17	94.118		45,100	4,750		49,850	56,500
2100	30 lb. capacity		.15	106		47,600	5,400		53,000	60,500
2150	60 lb. capacity	↓	.09	177	↓	73,000	8,975		81,975	93,500

11 23 16 – Drying and Conditioning Equipment

11 23 16.13 Dryers

		Crew	Daily Output	Labor-Hours	Unit	Material	Labor	Equipment	Total	Total Incl O&P
0010	**DRYERS**, Not including rough-in									
1500	Industrial, 30 lb. capacity	1 Plum	2	4	Ea.	2,975	208		3,183	3,575
1600	50 lb. capacity	"	1.70	4.706		3,200	245		3,445	3,900
4700	Lint collector, ductwork not included, 8,000 to 10,000 C.F.M.	Q-10	.30	80	↓	8,575	3,675		12,250	15,000

11 23 19 – Finishing Equipment

11 23 19.13 Folders and Spreaders

		Crew	Daily Output	Labor-Hours	Unit	Material	Labor	Equipment	Total	Total Incl O&P
0010	**FOLDERS AND SPREADERS**									
3500	Folders, blankets & sheets, minimum	1 Elec	.17	47.059	Ea.	30,600	2,300		32,900	37,100
3700	King size with automatic stacker		.10	80		55,500	3,925		59,425	67,000
3800	For conveyor delivery, add	↓	.45	17.778		6,000	870		6,870	7,900
4900	Spreader feeders, 240V, 2 station	L-6	.70	17.143		51,000	875		51,875	57,500
4920	4 station	"	.35	34.286	↓	61,500	1,750		63,250	70,000

11 23 23 – Commercial Ironing Equipment

11 23 23.13 Irons and Pressers

		Crew	Daily Output	Labor-Hours	Unit	Material	Labor	Equipment	Total	Total Incl O&P
0010	**IRONS AND PRESSERS**									
4500	Ironers, institutional, 110", single roll	1 Elec	.20	40	Ea.	29,700	1,950		31,650	35,600
4800	Pressers, low capacity air operated	L-6	1.75	6.857		8,775	350		9,125	10,200
4820	Hand operated		1.75	6.857		8,125	350		8,475	9,475
4840	Ironer 48", 240V		3.50	3.429		103,000	175		103,175	114,000
6600	Hand operated presser		.70	17.143		5,525	875		6,400	7,375
6620	Mushroom press 115V	↓	.70	17.143	↓	6,925	875		7,800	8,900

11 23 Commercial Laundry and Dry Cleaning Equipment

11 23 26 – Commercial Washers and Extractors

11 23 26.13 Washers and Extractors		Crew	Daily Output	Labor-Hours	Unit	Material	2010 Bare Costs Labor	Equipment	Total	Total Incl O&P
0010	**WASHERS AND EXTRACTORS**, not including rough-in									
6000	Combination washer/extractor, 20 lb. capacity	L-6	1.50	8	Ea.	5,350	410		5,760	6,500
6100	30 lb. capacity		.80	15		8,500	765		9,265	10,500
6200	50 lb. capacity		.68	17.647		9,925	900		10,825	12,300
6300	75 lb. capacity		.30	40		18,600	2,050		20,650	23,600
6350	125 lb. capacity		.16	75		24,900	3,825		28,725	33,100
6380	Washer extractor/dryer, 110 lb., 240V		1	12		8,150	610		8,760	9,900
6400	Washer extractor, 135 lb, 240V		1	12		23,600	610		24,210	26,900
6450	Pass through		1	12		64,000	610		64,610	71,500
6500	200 lb. washer extractor		1	12		66,500	610		67,110	74,500
6550	Pass through		1	12		69,500	610		70,110	77,500
6600	Extractor, low capacity	↓	1.75	6.857	↓	6,625	350		6,975	7,800

11 23 33 – Coin-Operated Laundry Equipment

11 23 33.13 Coin Operated Washers and Dryers		Crew	Daily Output	Labor-Hours	Unit	Material	Labor	Equipment	Total	Total Incl O&P
0010	**COIN OPERATED WASHERS AND DRYERS**									
0990	Dryer, gas fired									
1000	Commercial, 30 lb. capacity, coin operated, single	1 Plum	3	2.667	Ea.	3,100	139		3,239	3,600
1100	Double stacked	"	2	4		6,275	208		6,483	7,200
4860	Coin dry cleaner 20 lb.	L-6	1.75	6.857	↓	25,900	350		26,250	29,000
5290	Clothes washer									
5300	Commercial, coin operated, average	1 Plum	3	2.667	Ea.	1,125	139		1,264	1,450

11 24 Maintenance Equipment

11 24 19 – Vacuum Cleaning Systems

11 24 19.10 Vacuum Cleaning		Crew	Daily Output	Labor-Hours	Unit	Material	Labor	Equipment	Total	Total Incl O&P
0010	**VACUUM CLEANING**									
0020	Central, 3 inlet, residential	1 Skwk	.90	8.889	Total	1,025	380		1,405	1,725
0200	Commercial		.70	11.429		1,225	485		1,710	2,075
0400	5 inlet system, residential		.50	16		1,450	680		2,130	2,650
0600	7 inlet system, commercial		.40	20		1,625	850		2,475	3,075
0800	9 inlet system, residential	↓	.30	26.667		3,600	1,125		4,725	5,700
4010	Rule of thumb: First 1200 S.F., installed				↓				1,325	1,450
4020	For each additional S.F., add				S.F.					.24

11 26 Unit Kitchens

11 26 13 – Metal Unit Kitchens

11 26 13.10 Commercial Unit Kitchens		Crew	Daily Output	Labor-Hours	Unit	Material	Labor	Equipment	Total	Total Incl O&P
0010	**COMMERCIAL UNIT KITCHENS**									
1500	Combination range, refrigerator and sink, 30" wide, minimum	L-1	2	8	Ea.	925	405		1,330	1,625
1550	Maximum		1	16		3,675	810		4,485	5,250
1570	60" wide, average		1.40	11.429		3,225	575		3,800	4,375
1590	72" wide, average		1.20	13.333		4,200	675		4,875	5,625
1600	Office model, 48" wide **CN**		2	8		2,650	405		3,055	3,500
1620	Refrigerator and sink only	↓	2.40	6.667	↓	2,725	335		3,060	3,475
1640	Combination range, refrigerator, sink, microwave									
1660	Oven and ice maker	L-1	.80	20	Ea.	4,450	1,000		5,450	6,400

11 27 Photographic Processing Equipment

11 27 13 – Darkroom Processing Equipment

11 27 13.10 Darkroom Equipment		Crew	Daily Output	Labor-Hours	Unit	Material	2010 Bare Costs Labor	Equipment	Total	Total Incl O&P
0010	**DARKROOM EQUIPMENT**									
0020	Developing sink, 5" deep, 24" x 48"	Q-1	2	8	Ea.	5,625	375		6,000	6,725
0050	48" x 52"		1.70	9.412		5,675	440		6,115	6,875
0200	10" deep, 24" x 48"		1.70	9.412		7,775	440		8,215	9,200
0250	24" x 108"		1.50	10.667		10,200	500		10,700	12,100
0500	Dryers, dehumidified filtered air, 36" x 25" x 68" high	L-7	6	4.667		5,075	188		5,263	5,850
0550	48" x 25" x 68" high		5	5.600		10,000	225		10,225	11,300
2000	Processors, automatic, color print, minimum		4	7		15,700	281		15,981	17,700
2050	Maximum		.60	46.667		24,700	1,875		26,575	30,100
2300	Black and white print, minimum		2	14		11,900	565		12,465	14,000
2350	Maximum		.80	35		61,500	1,400		62,900	69,500
2600	Manual processor, 16" x 20" maximum print size		2	14		9,450	565		10,015	11,300
2650	20" x 24" maximum print size		1	28		8,975	1,125		10,100	11,600
3000	Viewing lights, 20" x 24"		6	4.667		380	188		568	700
3100	20" x 24" with color correction		6	4.667		535	188		723	875
3500	Washers, round, minimum sheet 11" x 14"	Q-1	2	8		3,400	375		3,775	4,300
3550	Maximum sheet 20" x 24"		1	16		3,475	750		4,225	4,950
3800	Square, minimum sheet 20" x 24"		1	16		3,125	750		3,875	4,550
3900	Maximum sheet 50" x 56"		.80	20		4,850	935		5,785	6,725
4500	Combination tank sink, tray sink, washers, with									
4510	Dry side tables, average	Q-1	.45	35.556	Ea.	10,600	1,675		12,275	14,200

11 31 Residential Appliances

11 31 13 – Residential Kitchen Appliances

11 31 13.13 Cooking Equipment

		Crew	Daily Output	Labor-Hours	Unit	Material	Labor	Equipment	Total	Total Incl O&P
0010	**COOKING EQUIPMENT**									
0020	Cooking range, 30" free standing, 1 oven, minimum	2 Clab	10	1.600	Ea.	345	53		398	460
0050	Maximum		4	4		1,900	132		2,032	2,300
0150	2 oven, minimum		10	1.600		1,400	53		1,453	1,600
0200	Maximum		10	1.600		1,600	53		1,653	1,850
0350	Built-in, 30" wide, 1 oven, minimum	1 Elec	6	1.333		680	65.50		745.50	845
0400	Maximum	2 Carp	2	8		1,500	330		1,830	2,150
0500	2 oven, conventional, minimum		4	4		1,550	166		1,716	1,950
0550	1 conventional, 1 microwave, maximum		2	8		1,375	330		1,705	2,025
0700	Free-standing, 1 oven, 21" wide range, minimum	2 Clab	10	1.600		330	53		383	440
0750	21" wide, maximum	"	4	4		335	132		467	575
0900	Countertop cooktops, 4 burner, standard, minimum	1 Elec	6	1.333		300	65.50		365.50	425
0950	Maximum		3	2.667		600	131		731	855
1050	As above, but with grill and griddle attachment, minimum		6	1.333		1,375	65.50		1,440.50	1,625
1100	Maximum		3	2.667		3,350	131		3,481	3,875
1200	Induction cooktop, 30" wide		3	2.667		1,125	131		1,256	1,450
1250	Microwave oven, minimum		4	2		110	98		208	267
1300	Maximum		2	4		420	196		616	750

11 31 13.23 Refrigeration Equipment

		Crew	Daily Output	Labor-Hours	Unit	Material	Labor	Equipment	Total	Total Incl O&P
0010	**REFRIGERATION EQUIPMENT**									
2000	Deep freeze, 15 to 23 C.F., minimum	2 Clab	10	1.600	Ea.	530	53		583	665
2050	Maximum		5	3.200		660	106		766	895
2200	30 C.F., minimum		8	2		830	66		896	1,025
2250	Maximum		3	5.333		940	177		1,117	1,300
5200	Icemaker, automatic, 20 lb. per day	1 Plum	7	1.143		880	59.50		939.50	1,050
5350	51 lb. per day	"	2	4		1,250	208		1,458	1,675

11 31 Residential Appliances

11 31 13 – Residential Kitchen Appliances

11 31 13.23 Refrigeration Equipment			Crew	Daily Output	Labor-Hours	Unit	2010 Bare Costs Material	Labor	Equipment	Total	Total Incl O&P
5500	Refrigerator, no frost, 10 C.F. to 12 C.F. minimum		2 Clab	10	1.600	Ea.	445	53		498	570
5600	Maximum			6	2.667		450	88.50		538.50	630
5750	14 C.F. to 16 C.F., minimum			9	1.778		475	59		534	610
5800	Maximum			5	3.200		710	106		816	945
5950	18 C.F. to 20 C.F., minimum			8	2		590	66		656	750
6000	Maximum			4	4		1,100	132		1,232	1,400
6150	21 C.F. to 29 C.F., minimum			7	2.286		940	75.50		1,015.50	1,150
6200	Maximum			3	5.333		3,075	177		3,252	3,650
6790	Energy-star qualified, 18 CF, minimum	G	2 Carp	4	4		495	166		661	800
6795	Maximum	G		2	8		1,300	330		1,630	1,925
6797	21.7 CF, minimum	G		4	4		815	166		981	1,150
6799	Maximum	G		4	4		1,300	166		1,466	1,675

11 31 13.33 Kitchen Cleaning Equipment			Crew	Daily Output	Labor-Hours	Unit	Material	Labor	Equipment	Total	Total Incl O&P
0010	KITCHEN CLEANING EQUIPMENT										
2750	Dishwasher, built-in, 2 cycles, minimum		L-1	4	4	Ea.	212	202		414	535
2800	Maximum			2	8		320	405		725	955
2950	4 or more cycles, minimum			4	4		279	202		481	605
2960	Average			4	4		370	202		572	710
3000	Maximum			2	8		1,050	405		1,455	1,750
3100	Energy-star qualified, minimum	G		4	4		340	202		542	675
3110	Maximum	G		2	8		1,250	405		1,655	1,975

11 31 13.43 Waste Disposal Equipment			Crew	Daily Output	Labor-Hours	Unit	Material	Labor	Equipment	Total	Total Incl O&P
0010	WASTE DISPOSAL EQUIPMENT										
1750	Compactor, residential size, 4 to 1 compaction, minimum		1 Carp	5	1.600	Ea.	535	66.50		601.50	685
1800	Maximum		"	3	2.667		830	111		941	1,075
3300	Garbage disposal, sink type, minimum		L-1	10	1.600		72.50	81		153.50	201
3350	Maximum		"	10	1.600		191	81		272	330

11 31 13.53 Kitchen Ventilation Equipment			Crew	Daily Output	Labor-Hours	Unit	Material	Labor	Equipment	Total	Total Incl O&P
0010	KITCHEN VENTILATION EQUIPMENT										
4150	Hood for range, 2 speed, vented, 30" wide, minimum		L-3	5	3.200	Ea.	53.50	145		198.50	279
4200	Maximum			3	5.333		820	242		1,062	1,275
4300	42" wide, minimum			5	3.200		234	145		379	475
4330	Custom			5	3.200		1,500	145		1,645	1,900
4350	Maximum			3	5.333		1,850	242		2,092	2,400
4500	For ventless hood, 2 speed, add						17.05			17.05	18.75
4650	For vented 1 speed, deduct from maximum						45.50			45.50	50

11 31 23 – Residential Laundry Appliances

11 31 23.13 Washers			Crew	Daily Output	Labor-Hours	Unit	Material	Labor	Equipment	Total	Total Incl O&P
0010	WASHERS										
5000	Residential, 4 cycle, average		1 Plum	3	2.667	Ea.	800	139		939	1,100
6650	Washing machine, automatic, minimum			3	2.667		355	139		494	600
6700	Maximum			1	8		1,000	415		1,415	1,750
6750	Energy star qualified, front loading, minimum	G		3	2.667		480	139		619	740
6760	Maximum	G		1	8		1,700	415		2,115	2,500
6764	Top loading, minimum	G		3	2.667		575	139		714	845
6766	Maximum	G		3	2.667		1,050	139		1,189	1,350

11 31 23.23 Dryers			Crew	Daily Output	Labor-Hours	Unit	Material	Labor	Equipment	Total	Total Incl O&P
0010	DRYERS										
0500	Gas fired residential, 16 lb. capacity, average		1 Plum	3	2.667	Ea.	625	139		764	900
6770	Electric, front loading, energy-star qualified, minimum	G	L-2	3	5.333		360	195		555	695
6780	Maximum	G	"	2	8		1,300	293		1,593	1,875

11 31 Residential Appliances

11 31 23 – Residential Laundry Appliances

11 31 23.23 Dryers		Crew	Daily Output	Labor-Hours	Unit	Material	2010 Bare Costs Labor	Equipment	Total	Total Incl O&P
7450	Vent kits for dryers	1 Carp	10	.800	Ea.	25.50	33		58.50	79

11 31 33 – Miscellaneous Residential Appliances

11 31 33.13 Sump Pumps

		Crew	Daily Output	Labor-Hours	Unit	Material	Labor	Equipment	Total	Total Incl O&P
0010	**SUMP PUMPS**									
6400	Cellar drainer, pedestal, 1/3 H.P., molded PVC base	1 Plum	3	2.667	Ea.	119	139		258	340
6450	Solid brass	"	2	4	"	263	208		471	600
6460	Sump pump, see also Div. 22 14 29.16									

11 31 33.23 Water Heaters

		Crew	Daily Output	Labor-Hours	Unit	Material	Labor	Equipment	Total	Total Incl O&P
0010	**WATER HEATERS**									
6900	Electric, glass lined, 30 gallon, minimum	L-1	5	3.200	Ea.	520	162		682	810
6950	Maximum		3	5.333		720	269		989	1,200
7100	80 gallon, minimum		2	8		935	405		1,340	1,625
7150	Maximum	↓	1	16		1,300	810		2,110	2,625
7180	Gas, glass lined, 30 gallon, minimum	2 Plum	5	3.200		805	167		972	1,125
7220	Maximum		3	5.333		1,125	278		1,403	1,650
7260	50 gallon, minimum		2.50	6.400		845	335		1,180	1,425
7300	Maximum	↓	1.50	10.667	↓	1,175	555		1,730	2,100
7310	Water heater, see also Div. 22 33 30.13									

11 31 33.43 Air Quality

		Crew	Daily Output	Labor-Hours	Unit	Material	Labor	Equipment	Total	Total Incl O&P
0010	**AIR QUALITY**									
2450	Dehumidifier, portable, automatic, 15 pint				Ea.	154			154	170
2550	40 pint					171			171	188
3550	Heater, electric, built-in, 1250 watt, ceiling type, minimum	1 Elec	4	2		96	98		194	251
3600	Maximum		3	2.667		156	131		287	365
3700	Wall type, minimum		4	2		154	98		252	315
3750	Maximum		3	2.667		166	131		297	375
3900	1500 watt wall type, with blower		4	2		154	98		252	315
3950	3000 watt	↓	3	2.667		315	131		446	540
4850	Humidifier, portable, 8 gallons per day					165			165	182
5000	15 gallons per day				↓	198			198	218

11 33 Retractable Stairs

11 33 10 – Disappearing Stairs

11 33 10.10 Disappearing Stairway

		Crew	Daily Output	Labor-Hours	Unit	Material	Labor	Equipment	Total	Total Incl O&P
0010	**DISAPPEARING STAIRWAY** No trim included									
0100	Custom grade, pine, 8′-6″ ceiling, minimum	1 Carp	4	2	Ea.	138	83		221	279
0150	Average		3.50	2.286		174	95		269	335
0200	Maximum		3	2.667		257	111		368	455
0500	Heavy duty, pivoted, from 7′-7″ to 12′-10″ floor to floor		3	2.667		650	111		761	885
0600	16′-0″ ceiling		2	4		1,325	166		1,491	1,700
0800	Economy folding, pine, 8′-6″ ceiling		4	2		121	83		204	261
0900	9′-6″ ceiling	↓	4	2		130	83		213	271
1100	Automatic electric, aluminum, floor to floor height, 8′ to 9′	2 Carp	1	16		8,250	665		8,915	10,100
1400	11′ to 12′		.90	17.778		8,875	740		9,615	10,900
1700	14′ to 15′	↓	.70	22.857	↓	9,475	950		10,425	11,900

11 41 13 – Refrigerated Food Storage Cases

11 41 13.10 Refrigerated Food Cases

	11 41 13.10 Refrigerated Food Cases	Crew	Daily Output	Labor-Hours	Unit	2010 Bare Costs Material	Labor	Equipment	Total	Total Incl O&P
0010	REFRIGERATED FOOD CASES									
0030	Dairy, multi-deck, 12′ long	Q-5	3	5.333	Ea.	10,400	249		10,649	11,900
0100	For rear sliding doors, add					1,500			1,500	1,650
0200	Delicatessen case, service deli, 12′ long, single deck	Q-5	3.90	4.103		7,025	192		7,217	8,000
0300	Multi-deck, 18 S.F. shelf display		3	5.333		6,450	249		6,699	7,450
0400	Freezer, self-contained, chest-type, 30 C.F.		3.90	4.103		7,550	192		7,742	8,575
0500	Glass door, upright, 78 C.F.		3.30	4.848		9,750	226		9,976	11,000
0600	Frozen food, chest type, 12′ long		3.30	4.848		7,050	226		7,276	8,100
0700	Glass door, reach-in, 5 door		3	5.333		13,500	249		13,749	15,200
0800	Island case, 12′ long, single deck		3.30	4.848		7,975	226		8,201	9,125
0900	Multi-deck		3	5.333		16,900	249		17,149	19,000
1000	Meat case, 12′ long, single deck		3.30	4.848		5,800	226		6,026	6,725
1050	Multi-deck		3.10	5.161		9,975	241		10,216	11,400
1100	Produce, 12′ long, single deck		3.30	4.848		7,675	226		7,901	8,775
1200	Multi-deck		3.10	5.161		8,375	241		8,616	9,550

11 41 13.20 Refrigerated Food Storage Equipment

	11 41 13.20 Refrigerated Food Storage Equipment	Crew	Daily Output	Labor-Hours	Unit	Material	Labor	Equipment	Total	Total Incl O&P
0010	REFRIGERATED FOOD STORAGE EQUIPMENT									
2350	Cooler, reach-in, beverage, 6′ long	Q-1	6	2.667	Ea.	4,525	125		4,650	5,150
4300	Freezers, reach-in, 44 C.F.		4	4		2,700	187		2,887	3,250
4500	68 C.F.		3	5.333		3,650	250		3,900	4,375
4600	Freezer, pre-fab, 8′ x 8′ w/refrigeration	2 Carp	.45	35.556		10,900	1,475		12,375	14,300
4620	8′ x 12′		.35	45.714		12,900	1,900		14,800	17,100
4640	8′ x 16′		.25	64		15,600	2,650		18,250	21,200
4660	8′ x 20′		.17	94.118		18,700	3,900		22,600	26,600
4680	Reach-in, 1 compartment	Q-1	4	4		1,650	187		1,837	2,100
4685	Energy star rated G	R-18	7.80	3.333		2,250	128		2,378	2,650
4700	2 compartment	Q-1	3	5.333		3,425	250		3,675	4,150
4705	Energy star rated G	R-18	6.20	4.194		3,200	161		3,361	3,775
4710	3 compartment	Q-1	3	5.333		4,850	250		5,100	5,725
4715	Energy star rated G	R-18	5.60	4.643		4,325	178		4,503	5,025
8320	Refrigerator, reach-in, 1 compartment		7.80	3.333		2,100	128		2,228	2,525
8325	Energy star rated G		7.80	3.333		2,250	128		2,378	2,650
8330	2 compartment		6.20	4.194		3,125	161		3,286	3,675
8335	Energy star rated G		6.20	4.194		3,200	161		3,361	3,775
8340	3 compartment		5.60	4.643		4,175	178		4,353	4,875
8345	Energy star rated G		5.60	4.643		4,325	178		4,503	5,025
8350	Pre-fab, with refrigeration, 8′ x 8′	2 Carp	.45	35.556		6,675	1,475		8,150	9,625
8360	8′ x 12′		.35	45.714		7,975	1,900		9,875	11,700
8370	8′ x 16′		.25	64		11,400	2,650		14,050	16,600
8380	8′ x 20′		.17	94.118		14,400	3,900		18,300	21,900
8390	Pass-thru/roll-in, 1 compartment	R-18	7.80	3.333		4,000	128		4,128	4,600
8400	2 compartment		6.24	4.167		5,875	160		6,035	6,725
8410	3 compartment		5.60	4.643		7,525	178		7,703	8,550
8420	Walk-in, alum, door & floor only, no refrig, 6′ x 6′ x 7′-6″	2 Carp	1.40	11.429		8,675	475		9,150	10,300
8430	10′ x 6′ x 7′-6″		.55	29.091		12,400	1,200		13,600	15,500
8440	12′ x 14′ x 7′-6″		.25	64		17,600	2,650		20,250	23,400
8450	12′ x 20′ x 7′-6″		.17	94.118		21,400	3,900		25,300	29,600
8460	Refrigerated cabinets, mobile					3,575			3,575	3,925
8470	Refrigerator/freezer, reach-in, 1 compartment	R-18	5.60	4.643		4,700	178		4,878	5,450
8480	2 compartment	″	4.80	5.417		6,625	208		6,833	7,625

11 41 Food Storage Equipment

11 41 13 – Refrigerated Food Storage Cases

11 41 13.30 Wine Cellar		Crew	Daily Output	Labor-Hours	Unit	2010 Bare Costs Material	Labor	Equipment	Total	Total Incl O&P
0010	**WINE CELLAR**, refrigerated, Redwood interior, carpeted, walk-in type									
0020	6'-8" high, including racks									
0200	80" W x 48" D for 900 bottles	2 Carp	1.50	10.667	Ea.	3,350	445		3,795	4,350
0250	80" W x 72" D for 1300 bottles		1.33	12.030		4,425	500		4,925	5,625
0300	80" W x 94" D for 1900 bottles		1.17	13.675		5,725	570		6,295	7,175
0400	80" W x 124" D for 2500 bottles		1	16		6,925	665		7,590	8,650
0600	Portable cabinets, red oak, reach-in temp.& humidity controlled									
0650	26-5/8"W x 26-1/2"D x 68"H for 235 bottles				Ea.	2,450			2,450	2,700
0660	32"W x 21-1/2"D x 73-1/2"H for 144 bottles					2,625			2,625	2,900
0670	32"W x 29-1/2"D x 73-1/2"H for 288 bottles					2,700			2,700	2,975
0680	39-1/2"W x 29-1/2"D x 86-1/2"H for 440 bottles					3,125			3,125	3,450
0690	52-1/2"W x 29-1/2"D x 73-1/2"H for 468 bottles					3,725			3,725	4,075
0700	52-1/2"W x 29-1/2"D x 86-1/2"H for 572 bottles					3,825			3,825	4,225
0730	Portable, red oak, can be built-in with glass door									
0750	23-7/8"W x 24"D x 34-1/2"H for 50 bottles				Ea.	965			965	1,050

11 41 33 – Food Storage Shelving

11 41 33.20 Metal Food Storage Shelving		Crew	Daily Output	Labor-Hours	Unit	Material	Labor	Equipment	Total	Total Incl O&P
0010	**METAL FOOD STORAGE SHELVING**									
8600	Stainless steel shelving, louvered 4-tier, 20" x 3'	1 Clab	6	1.333	Ea.	1,225	44		1,269	1,425
8605	20" x 4'		6	1.333		1,750	44		1,794	2,000
8610	20" x 6'		6	1.333		2,550	44		2,594	2,875
8615	24" x 3'		6	1.333		1,600	44		1,644	1,850
8620	24" x 4'		6	1.333		1,875	44		1,919	2,125
8625	24" x 6'		6	1.333		2,725	44		2,769	3,050
8630	Flat 4-tier, 20" x 3'		6	1.333		1,100	44		1,144	1,275
8635	20" x 4'		6	1.333		1,325	44		1,369	1,525
8640	20" x 5'		6	1.333		1,525	44		1,569	1,750
8645	24" x 3'		6	1.333		1,250	44		1,294	1,450
8650	24" x 4'		6	1.333		2,000	44		2,044	2,275
8655	24" x 6'		6	1.333		2,425	44		2,469	2,725
8700	Galvanized shelving, louvered 4-tier, 20" x 3'		6	1.333		625	44		669	755
8705	20" x 4'		6	1.333		695	44		739	835
8710	20" x 6'		6	1.333		870	44		914	1,025
8715	24" x 3'		6	1.333		555	44		599	680
8720	24" x 4'		6	1.333		745	44		789	890
8725	24" x 6'		6	1.333		1,075	44		1,119	1,250
8730	Flat 4-tier, 20" x 3'		6	1.333		570	44		614	695
8735	20" x 4'		6	1.333		625	44		669	760
8740	20" x 6'		6	1.333		875	44		919	1,025
8745	24" x 3'		6	1.333		600	44		644	730
8750	24" x 4'		6	1.333		635	44		679	770
8755	24" x 6'		6	1.333		910	44		954	1,075
8760	Stainless steel dunnage rack, 24" x 3'		8	1		415	33		448	505
8765	24" x 4'		8	1		800	33		833	930
8770	Galvanized dunnage rack, 24" x 3'		8	1		157	33		190	224
8775	24" x 4'		8	1		184	33		217	254

11 42 Food Preparation Equipment

11 42 10 – Commercial Food Preparation Equipment

11 42 10.10 Choppers, Mixers and Misc. Equipment		Crew	Daily Output	Labor-Hours	Unit	Material	2010 Bare Costs Labor	Equipment	Total	Total Incl O&P
0010	**CHOPPERS, MIXERS AND MISC. EQUIPMENT**									
1700	Choppers, 5 pounds	R-18	7	3.714	Ea.	1,725	142		1,867	2,125
1720	16 pounds		5	5.200		2,125	199		2,324	2,650
1740	35 to 40 pounds		4	6.500		3,150	249		3,399	3,825
1840	Coffee brewer, 5 burners	1 Plum	3	2.667		1,225	139		1,364	1,550
1850	Coffee urn, twin 6 gallon urns		2	4		2,475	208		2,683	3,025
1860	Single, 3 gallon		3	2.667		1,825	139		1,964	2,225
3000	Fast food equipment, total package, minimum	6 Skwk	.08	600		184,500	25,600		210,100	242,000
3100	Maximum	"	.07	685		251,000	29,200		280,200	321,000
3800	Food mixers, 20 quarts	L-7	7	4		2,775	161		2,936	3,300
3850	40 quarts		5.40	5.185		6,950	208		7,158	7,975
3900	60 quarts		5	5.600		8,725	225		8,950	9,950
4040	80 quarts		3.90	7.179		12,400	289		12,689	14,000
4080	130 quarts		2.20	12.727		17,100	510		17,610	19,600
4100	Floor type, 20 quarts		15	1.867		3,175	75		3,250	3,600
4120	60 quarts		14	2		7,925	80.50		8,005.50	8,850
4140	80 quarts		12	2.333		10,600	94		10,694	11,800
4160	140 quarts		8.60	3.256		18,700	131		18,831	20,800
6700	Peelers, small	R-18	8	3.250		1,525	125		1,650	1,875
6720	Large	"	6	4.333		4,625	166		4,791	5,325
6800	Pulper/extractor, close coupled, 5 HP	1 Plum	1.90	4.211		3,525	219		3,744	4,200
8580	Slicer with table	R-18	9	2.889		4,850	111		4,961	5,500

11 43 Food Delivery Carts and Conveyors

11 43 13 – Food Delivery Carts

11 43 13.10 Mobile Carts, Racks and Trays		Crew	Daily Output	Labor-Hours	Unit	Material	Labor	Equipment	Total	Total Incl O&P
0010	**MOBILE CARTS, RACKS AND TRAYS**									
1650	Cabinet, heated, 1 compartment, reach-in	R-18	5.60	4.643	Ea.	3,475	178		3,653	4,100
1655	Pass-thru roll-in		5.60	4.643		5,125	178		5,303	5,900
1660	2 compartment, reach-in		4.80	5.417		7,450	208		7,658	8,500
1670	Mobile					3,175			3,175	3,500
6850	Mobile rack w/pan slide					1,275			1,275	1,400
9180	Tray and silver dispenser, mobile	1 Clab	16	.500		840	16.55		856.55	950

11 44 Food Cooking Equipment

11 44 13 – Commercial Ranges

11 44 13.10 Cooking Equipment		Crew	Daily Output	Labor-Hours	Unit	Material	Labor	Equipment	Total	Total Incl O&P
0010	**COOKING EQUIPMENT**									
0020	Bake oven, gas, one section	Q-1	8	2	Ea.	4,650	93.50		4,743.50	5,275
0300	Two sections		7	2.286		10,500	107		10,607	11,700
0600	Three sections		6	2.667		13,500	125		13,625	15,000
0900	Electric convection, single deck	L-7	4	7		6,425	281		6,706	7,475
1300	Broiler, without oven, standard	Q-1	8	2		3,550	93.50		3,643.50	4,050
1550	Infrared	L-7	4	7		9,325	281		9,606	10,700
4750	Fryer, with twin baskets, modular model	Q-1	7	2.286		1,500	107		1,607	1,800
5000	Floor model, on 6" legs	"	5	3.200		2,275	150		2,425	2,725
5100	Extra single basket, large					96			96	105
5170	Energy star rated, 50 lb. capacity G	R-18	4	6.500		3,775	249		4,024	4,525
5175	85 lb. capacity G	"	4	6.500		7,675	249		7,924	8,825

11 44 Food Cooking Equipment

11 44 13 – Commercial Ranges

11 44 13.10 Cooking Equipment		Crew	Daily Output	Labor-Hours	Unit	Material	2010 Bare Costs Labor	Equipment	Total	Total Incl O&P
5300	Griddle, SS, 24" plate, w/4" legs, elec, 208V, 3 phase, 3' long	Q-1	7	2.286	Ea.	1,125	107		1,232	1,400
5550	4' long	"	6	2.667		1,525	125		1,650	1,850
6200	Iced tea brewer	1 Plum	3.44	2.326		675	121		796	920
6350	Kettle, w/steam jacket, tilting, w/positive lock, SS, 20 gallons	L-7	7	4		7,125	161		7,286	8,075
6600	60 gallons	"	6	4.667		9,225	188		9,413	10,500
6900	Range, restaurant type, 6 burners and 1 standard oven, 36" wide	Q-1	7	2.286		2,325	107		2,432	2,700
6950	Convection		7	2.286		5,125	107		5,232	5,800
7150	2 standard ovens, 24" griddle, 60" wide		6	2.667		7,700	125		7,825	8,650
7200	1 standard, 1 convection oven		6	2.667		6,800	125		6,925	7,650
7450	Heavy duty, single 34" standard oven, open top		5	3.200		5,575	150		5,725	6,350
7500	Convection oven		5	3.200		4,650	150		4,800	5,325
7700	Griddle top		6	2.667		3,250	125		3,375	3,750
7750	Convection oven	↓	6	2.667		6,800	125		6,925	7,650
8850	Steamer, electric 27 KW	L-7	7	4		7,875	161		8,036	8,925
9100	Electric, 10 KW or gas 100,000 BTU	"	5	5.600		5,725	225		5,950	6,625
9150	Toaster, conveyor type, 16-22 slices per minute					1,200			1,200	1,325
9160	Pop-up, 2 slot				↓	570			570	630
9200	For deluxe models of above equipment, add					75%				
9400	Rule of thumb: Equipment cost based									
9410	on kitchen work area									
9420	Office buildings, minimum	L-7	77	.364	S.F.	82	14.60		96.60	113
9450	Maximum		58	.483		138	19.40		157.40	182
9550	Public eating facilities, minimum		77	.364		107	14.60		121.60	141
9600	Maximum		46	.609		175	24.50		199.50	230
9750	Hospitals, minimum		58	.483		110	19.40		129.40	152
9800	Maximum	↓	39	.718	↓	203	29		232	268

11 46 Food Dispensing Equipment

11 46 16 – Service Line Equipment

11 46 16.10 Commercial Food Dispensing Equipment		Crew	Daily Output	Labor-Hours	Unit	Material	2010 Bare Costs Labor	Equipment	Total	Total Incl O&P
0010	**COMMERCIAL FOOD DISPENSING EQUIPMENT**									
1050	Butter pat dispenser	1 Clab	13	.615	Ea.	910	20.50		930.50	1,025
1100	Bread dispenser, counter top		13	.615		865	20.50		885.50	985
1900	Cup and glass dispenser, drop in		4	2		975	66		1,041	1,175
1920	Disposable cup, drop in		16	.500		425	16.55		441.55	495
2650	Dish dispenser, drop in, 12"		11	.727		1,975	24		1,999	2,175
2660	Mobile	↓	10	.800		2,375	26.50		2,401.50	2,650
3300	Food warmer, counter, 1.2 KW					670			670	740
3550	1.6 KW					1,950			1,950	2,150
3600	Well, hot food, built-in, rectangular, 12" x 20"	R-30	10	2.600		630	102		732	845
3610	Circular, 7 qt		10	2.600		335	102		437	525
3620	Refrigerated, 2 compartments		10	2.600		2,675	102		2,777	3,100
3630	3 compartments		9	2.889		3,325	114		3,439	3,850
3640	4 compartments		8	3.250		3,850	128		3,978	4,450
4720	Frost cold plate	↓	9	2.889		16,700	114		16,814	18,500
5700	Hot chocolate dispenser	1 Plum	4	2		1,075	104		1,179	1,325
6250	Jet spray dispenser	R-18	4.50	5.778		3,100	221		3,321	3,725
6300	Juice dispenser, concentrate	"	4.50	5.778		2,100	221		2,321	2,650
6690	Milk dispenser, bulk, 2 flavor	R-30	8	3.250		1,450	128		1,578	1,800
6695	3 flavor	"	8	3.250	↓	2,000	128		2,128	2,400
8800	Serving counter, straight	1 Carp	40	.200	L.F.	785	8.30		793.30	875

11 46 Food Dispensing Equipment

11 46 16 – Service Line Equipment

11 46 16.10 Commercial Food Dispensing Equipment		Crew	Daily Output	Labor-Hours	Unit	Material	2010 Bare Costs Labor	Equipment	Total	Total Incl O&P
8820	Curved section	1 Carp	30	.267	L.F.	980	11.10		991.10	1,100
8825	Solid surface, see Div. 12 36 61.16									
8830	Soft serve ice cream machine, medium	R-18	11	2.364	Ea.	10,300	90.50		10,390.50	11,500
8840	Large	"	9	2.889	"	18,000	111		18,111	20,000

11 47 Ice Machines

11 47 10 – Commercial Ice Machines

11 47 10.10 Commercial Ice Equipment		Crew	Daily Output	Labor-Hours	Unit	Material	Labor	Equipment	Total	Total Incl O&P
0010	**COMMERCIAL ICE EQUIPMENT**									
5800	Ice cube maker, 50 pounds per day	Q-1	6	2.667	Ea.	1,550	125		1,675	1,875
5900	250 pounds per day		1.20	13.333		2,550	625		3,175	3,725
6050	500 pounds per day		4	4		3,125	187		3,312	3,700
6060	With bin		1.20	13.333		2,850	625		3,475	4,050
6090	1000 pounds per day, with bin		1	16		4,975	750		5,725	6,600
6100	Ice flakers, 300 pounds per day		1.60	10		3,150	470		3,620	4,175
6120	600 pounds per day		.95	16.842		4,000	790		4,790	5,550
6130	1000 pounds per day		.75	21.333		4,400	1,000		5,400	6,325
6140	2000 pounds per day		.65	24.615		19,500	1,150		20,650	23,200
6160	Ice storage bin, 500 pound capacity	Q-5	1	16		1,000	745		1,745	2,225
6180	1000 pound	"	.56	28.571		1,950	1,325		3,275	4,150

11 48 Cleaning and Disposal Equipment

11 48 13 – Commercial Dishwashers

11 48 13.10 Dishwashers		Crew	Daily Output	Labor-Hours	Unit	Material	Labor	Equipment	Total	Total Incl O&P
0010	**DISHWASHERS**									
2700	Dishwasher, commercial, rack type									
2720	10 to 12 racks per hour	Q-1	3.20	5	Ea.	4,300	234		4,534	5,075
2730	Energy star rated, 35-40 racks/hour G		1.30	12.308		4,650	575		5,225	5,975
2740	50-60 racks/hour G		1.30	12.308		8,525	575		9,100	10,300
2800	Automatic, 190 to 230 racks per hour	L-6	.35	34.286		13,100	1,750		14,850	17,000
2820	235 to 275 racks per hour		.25	48		30,300	2,450		32,750	37,000
2840	8,750 to 12,500 dishes per hour		.10	120		49,500	6,125		55,625	63,500
2950	Dishwasher hood, canopy type	L-3A	10	1.200	L.F.	800	54.50		854.50	965
2960	Pant leg type	"	2.50	4.800	Ea.	7,875	218		8,093	9,000
5200	Garbage disposal 1.5 HP, 100 GPH	L-1	4.80	3.333		1,800	168		1,968	2,225
5210	3 HP, 120 GPH		4.60	3.478		2,175	176		2,351	2,650
5220	5 HP, 250 GPH		4.50	3.556		3,550	180		3,730	4,175
6750	Pot sink, 3 compartment	1 Plum	7.25	1.103	L.F.	925	57.50		982.50	1,100
6760	Pot washer, small		1.60	5	Ea.	8,850	260		9,110	10,100
6770	Large		1.20	6.667		15,100	345		15,445	17,100
9170	Trash compactor, small, up to 125 lb. compacted weight	L-4	4	6		21,900	234		22,134	24,400
9175	Large, up to 175 lb. compacted weight	"	3	8		26,700	310		27,010	29,900

11 52 Audio-Visual Equipment

11 52 13 – Projection Screens

11 52 13.10 Projection Screens, Wall or Ceiling Hung		Crew	Daily Output	Labor-Hours	Unit	2010 Bare Costs Material	Labor	Equipment	Total	Total Incl O&P
0010	**PROJECTION SCREENS, WALL OR CEILING HUNG**, matte white									
0100	Manually operated, economy	2 Carp	500	.032	S.F.	5.85	1.33		7.18	8.50
0300	Intermediate		450	.036		6.80	1.48		8.28	9.80
0400	Deluxe		400	.040		9.45	1.66		11.11	12.90
0600	Electric operated, matte white, 25 S.F., economy		5	3.200	Ea.	855	133		988	1,150
0700	Deluxe		4	4		1,775	166		1,941	2,200
0900	50 S.F., economy		3	5.333		750	222		972	1,175
1000	Deluxe		2	8		1,975	330		2,305	2,675
1200	Heavy duty, electric operated, 200 S.F.		1.50	10.667		3,900	445		4,345	4,975
1300	400 S.F.		1	16		4,800	665		5,465	6,325
1500	Rigid acrylic in wall, for rear projection, 1/4" thick	2 Glaz	30	.533	S.F.	47	21.50		68.50	84.50
1600	1/2" thick (maximum size 10' x 20')	"	25	.640	"	83.50	25.50		109	131

11 52 16 – Projectors

11 52 16.10 Movie Equipment

11 52 16.10 Movie Equipment		Crew	Daily Output	Labor-Hours	Unit	Material	Labor	Equipment	Total	Total Incl O&P
0010	**MOVIE EQUIPMENT**									
0020	Changeover, minimum				Ea.	465			465	510
0100	Maximum					905			905	995
0400	Film transport, incl. platters and autowind, minimum					5,025			5,025	5,525
0500	Maximum					14,300			14,300	15,700
0800	Lamphouses, incl. rectifiers, xenon, 1,000 watt	1 Elec	2	4		6,650	196		6,846	7,600
0900	1,600 watt		2	4		7,100	196		7,296	8,100
1000	2,000 watt		1.50	5.333		7,625	261		7,886	8,775
1100	4,000 watt		1.50	5.333		9,425	261		9,686	10,800
1400	Lenses, anamorphic, minimum					1,275			1,275	1,400
1500	Maximum					2,875			2,875	3,150
1800	Flat 35 mm, minimum					1,100			1,100	1,225
1900	Maximum					1,725			1,725	1,900
2200	Pedestals, for projectors					1,475			1,475	1,625
2300	Console type					10,700			10,700	11,800
2600	Projector mechanisms, incl. soundhead, 35 mm, minimum					11,100			11,100	12,200
2700	Maximum					15,200			15,200	16,700
3000	Projection screens, rigid, in wall, acrylic, 1/4" thick	2 Glaz	195	.082	S.F.	42	3.30		45.30	51
3100	1/2" thick	"	130	.123	"	48.50	4.95		53.45	61
3300	Electric operated, heavy duty, 400 S.F.	2 Carp	1	16	Ea.	2,950	665		3,615	4,275
3320	Theater projection screens, matte white, including frames	"	200	.080	S.F.	6.65	3.32		9.97	12.40
3400	Also see Div. 11 52 13.10									
3700	Sound systems, incl. amplifier, mono, minimum	1 Elec	.90	8.889	Ea.	3,300	435		3,735	4,275
3800	Dolby/Super Sound, maximum		.40	20		18,100	980		19,080	21,400
4100	Dual system, 2 channel, front surround, minimum		.70	11.429		4,625	560		5,185	5,925
4200	Dolby/Super Sound, 4 channel, maximum		.40	20		16,500	980		17,480	19,700
4500	Sound heads, 35 mm					5,300			5,300	5,825
4900	Splicer, tape system, minimum					740			740	815
5000	Tape type, maximum					1,325			1,325	1,450
5300	Speakers, recessed behind screen, minimum	1 Elec	2	4		1,050	196		1,246	1,475
5400	Maximum	"	1	8		3,100	390		3,490	3,975
5700	Seating, painted steel, upholstered, minimum	2 Carp	35	.457		132	19		151	175
5800	Maximum	"	28	.571		420	23.50		443.50	500
6100	Rewind tables, minimum					2,650			2,650	2,900
6200	Maximum					4,725			4,725	5,175
7000	For automation, varying sophistication, minimum	1 Elec	1	8	System	2,375	390		2,765	3,200
7100	Maximum	2 Elec	.30	53.333	"	5,575	2,625		8,200	10,000

11 53 Laboratory Equipment

11 53 03 – Laboratory Test Equipment

11 53 03.13 Test Equipment		Crew	Daily Output	Labor-Hours	Unit	2010 Bare Costs Material	Labor	Equipment	Total	Total Incl O&P
0010	**TEST EQUIPMENT**									
1700	Thermometer, electric, portable				Ea.	360			360	400
1800	Titration unit, four 2000 ml reservoirs				"	5,500			5,500	6,050

11 53 13 – Laboratory Fume Hoods

11 53 13.13 Recirculating Laboratory Fume Hoods

		Crew	Daily Output	Labor-Hours	Unit	Material	Labor	Equipment	Total	Total Incl O&P
0010	**RECIRCULATING LABORATORY FUME HOODS**									
0600	Fume hood, with countertop & base, not including HVAC									
0610	Simple, minimum	2 Carp	5.40	2.963	L.F.	605	123		728	855
0620	Complex, including fixtures		2.40	6.667		1,800	277		2,077	2,400
0630	Special, maximum		1.70	9.412		1,825	390		2,215	2,600
0670	Service fixtures, average				Ea.	273			273	300
0680	For sink assembly with hot and cold water, add	1 Plum	1.40	5.714		705	297		1,002	1,225
0750	Glove box, fiberglass, bacteriological					16,900			16,900	18,600
0760	Controlled atmosphere					18,700			18,700	20,500
0770	Radioisotope					16,900			16,900	18,600
0780	Carcinogenic					16,900			16,900	18,600

11 53 13.23 Exhaust Hoods

		Crew	Daily Output	Labor-Hours	Unit	Material	Labor	Equipment	Total	Total Incl O&P
0010	**EXHAUST HOODS**									
0650	Ductwork, minimum	2 Shee	1	16	Hood	3,575	785		4,360	5,125
0660	Maximum	"	.50	32	"	5,500	1,575		7,075	8,425

11 53 16 – Laboratory Incubators

11 53 16.13 Incubators

		Crew	Daily Output	Labor-Hours	Unit	Material	Labor	Equipment	Total	Total Incl O&P
0010	**INCUBATORS**									
1000	Incubators, minimum				Ea.	3,400			3,400	3,750
1010	Maximum				"	20,100			20,100	22,100

11 53 19 – Laboratory Sterilizers

11 53 19.13 Sterilizers

		Crew	Daily Output	Labor-Hours	Unit	Material	Labor	Equipment	Total	Total Incl O&P
0010	**STERILIZERS**									
0700	Glassware washer, undercounter, minimum	L-1	1.80	8.889	Ea.	5,650	450		6,100	6,900
0710	Maximum	"	1	16		11,800	810		12,610	14,200
1850	Utensil washer-sanitizer	1 Plum	2	4		10,900	208		11,108	12,300

11 53 23 – Laboratory Refrigerators

11 53 23.13 Refrigerators

		Crew	Daily Output	Labor-Hours	Unit	Material	Labor	Equipment	Total	Total Incl O&P
0010	**REFRIGERATORS**									
1200	Blood bank, 28.6 C.F. emergency signal				Ea.	7,700			7,700	8,475
1210	Reach-in, 16.9 C.F.				"	7,750			7,750	8,525

11 53 33 – Emergency Safety Appliances

11 53 33.13 Emergency Equipment

		Crew	Daily Output	Labor-Hours	Unit	Material	Labor	Equipment	Total	Total Incl O&P
0010	**EMERGENCY EQUIPMENT**									
1400	Safety equipment, eye wash, hand held				Ea.	405			405	445
1450	Deluge shower				"	740			740	815

11 53 43 – Service Fittings and Accessories

11 53 43.13 Fittings

		Crew	Daily Output	Labor-Hours	Unit	Material	Labor	Equipment	Total	Total Incl O&P
0010	**FITTINGS**									
1600	Sink, one piece plastic, flask wash, hose, free standing	1 Plum	1.60	5	Ea.	1,800	260		2,060	2,375
1610	Epoxy resin sink, 25" x 16" x 10"	"	2	4	"	198	208		406	530
1950	Utility table, acid resistant top with drawers	2 Carp	30	.533	L.F.	147	22		169	195
8000	Alternate pricing method: as percent of lab furniture									
8050	Installation, not incl. plumbing & duct work				% Furn.				20%	22%

11 53 Laboratory Equipment

11 53 43 – Service Fittings and Accessories

11 53 43.13 Fittings		Crew	Daily Output	Labor-Hours	Unit	Material	2010 Bare Costs Labor	Equipment	Total	Total Incl O&P
8100	Plumbing, final connections, simple system				% Furn.				9.09%	10%
8110	Moderately complex system								13.64%	15%
8120	Complex system								18.18%	20%
8150	Electrical, simple system								9.09%	10%
8160	Moderately complex system								18.18%	20%
8170	Complex system								31.80%	35%

11 57 Vocational Shop Equipment

11 57 10 – Shop Equipment

11 57 10.10 Vocational School Shop Equipment		Crew	Daily Output	Labor-Hours	Unit	Material	Labor	Equipment	Total	Total Incl O&P
0010	**VOCATIONAL SCHOOL SHOP EQUIPMENT**									
0020	Benches, work, wood, average	2 Carp	5	3.200	Ea.	490	133		623	745
0100	Metal, average		5	3.200		450	133		583	700
0400	Combination belt & disc sander, 6"		4	4		1,375	166		1,541	1,775
0700	Drill press, floor mounted, 12", 1/2 H.P.		4	4		405	166		571	700
0800	Dust collector, not incl. ductwork, 6" diameter	1 Shee	1.10	7.273		4,125	355		4,480	5,100
0810	Dust collector bag, 20" diameter	"	5	1.600		310	78.50		388.50	460
1000	Grinders, double wheel, 1/2 H.P.	2 Carp	5	3.200		247	133		380	475
1300	Jointer, 4", 3/4 H.P.		4	4		1,250	166		1,416	1,625
1600	Kilns, 16 C.F., to 2000°		4	4		2,300	166		2,466	2,775
1900	Lathe, woodworking, 10", 1/2 H.P.		4	4		735	166		901	1,075
2200	Planer, 13" x 6"		4	4		970	166		1,136	1,325
2500	Potter's wheel, motorized		4	4		1,175	166		1,341	1,550
2800	Saws, band, 14", 3/4 H.P.		4	4		950	166		1,116	1,300
3100	Metal cutting band saw, 14"		4	4		2,175	166		2,341	2,625
3400	Radial arm saw, 10", 2 H.P.		4	4		1,250	166		1,416	1,625
3700	Scroll saw, 24"		4	4		580	166		746	895
4000	Table saw, 10", 3 H.P.		4	4		2,100	166		2,266	2,550
4300	Welder AC arc, 30 amp capacity		4	4		2,100	166		2,266	2,575

11 61 Theater and Stage Equipment

11 61 23 – Folding and Portable Stages

11 61 23.10 Portable Stages		Crew	Daily Output	Labor-Hours	Unit	Material	Labor	Equipment	Total	Total Incl O&P
0010	**PORTABLE STAGES**									
1500	Flooring, portable oak parquet, 3' x 3' sections				S.F.	13.40			13.40	14.70
1600	Cart to carry 225 S.F. of flooring				Ea.	400			400	440
5000	Stages, portable with steps, folding legs, stock, 8" high				SF Stg.	27.50			27.50	30.50
5100	16" high					23.50			23.50	26
5200	32" high					38			38	41.50
5300	40" high					59.50			59.50	65
6000	Telescoping platforms, extruded alum., straight, minimum	4 Carp	157	.204		30	8.45		38.45	46
6100	Maximum		77	.416		40.50	17.25		57.75	71
6500	Pie-shaped, minimum		150	.213		64.50	8.85		73.35	84.50
6600	Maximum		70	.457		72	19		91	109
6800	For 3/4" plywood covered deck, deduct					3.85			3.85	4.24
7000	Band risers, steel frame, plywood deck, minimum	4 Carp	275	.116		30.50	4.83		35.33	41
7100	Maximum	"	138	.232		63.50	9.65		73.15	84.50
7500	Chairs for above, self-storing, minimum	2 Carp	43	.372	Ea.	103	15.45		118.45	137
7600	Maximum	"	40	.400	"	179	16.60		195.60	222

11 61 Theater and Stage Equipment

11 61 33 – Rigging Systems and Controls

11 61 33.10 Controls		Crew	Daily Output	Labor-Hours	Unit	Material	2010 Bare Costs Labor	Equipment	Total	Total Incl O&P
0010	CONTROLS									
0050	Control boards with dimmers and breakers, minimum	1 Elec	1	8	Ea.	12,300	390		12,690	14,100
0100	Average		.50	16		38,600	785		39,385	43,600
0150	Maximum		.20	40		126,000	1,950		127,950	141,500
8000	Rule of thumb: total stage equipment, minimum	4 Carp	100	.320	SF Stg.	97	13.30		110.30	128
8100	Maximum	"	25	1.280	"	545	53		598	680

11 61 43 – Stage Curtains

11 61 43.10 Curtains		Crew	Daily Output	Labor-Hours	Unit	Material	Labor	Equipment	Total	Total Incl O&P
0010	CURTAINS									
0500	Curtain track, straight, light duty	2 Carp	20	.800	L.F.	27	33		60	80.50
0600	Heavy duty		18	.889		61	37		98	124
0700	Curved sections		12	1.333		173	55.50		228.50	276
1000	Curtains, velour, medium weight		600	.027	S.F.	7.90	1.11		9.01	10.40
1150	Silica based yarn, inherently fire retardant		50	.320	"	15.20	13.30		28.50	37

11 62 Musical Equipment

11 62 16 – Carillons

11 62 16.10 Bell Tower Equipment		Crew	Daily Output	Labor-Hours	Unit	Material	Labor	Equipment	Total	Total Incl O&P
0010	BELL TOWER EQUIPMENT									
0300	Carillon, 4 octave (48 bells), with keyboard				System	937,000			937,000	1,031,000
0320	2 octave (24 bells)					441,000			441,000	485,000
0340	3 to 4 bell peal, minimum					110,500			110,500	121,500
0360	Maximum					661,500			661,500	727,500
0380	Cast bronze bell, average				Ea.	99,000			99,000	109,000
0400	Electronic, digital, minimum					16,500			16,500	18,200
0410	With keyboard, maximum					82,500			82,500	91,000

11 66 Athletic Equipment

11 66 13 – Exercise Equipment

11 66 13.10 Physical Training Equipment		Crew	Daily Output	Labor-Hours	Unit	Material	Labor	Equipment	Total	Total Incl O&P
0010	PHYSICAL TRAINING EQUIPMENT									
0020	Abdominal rack, 2 board capacity				Ea.	470			470	515
0050	Abdominal board, upholstered					525			525	575
0200	Bicycle trainer, minimum					695			695	765
0300	Deluxe, electric					4,150			4,150	4,575
0400	Barbell set, chrome plated steel, 25 lbs.					244			244	269
0420	100 lbs.					370			370	405
0450	200 lbs.					715			715	785
0500	Weight plates, cast iron, per lb.				Lb.	5			5	5.50
0520	Storage rack, 10 station				Ea.	790			790	870
0600	Circuit training apparatus, 12 machines minimum	2 Clab	1.25	12.800	Set	27,600	425		28,025	31,000
0700	Average		1	16		33,800	530		34,330	38,000
0800	Maximum		.75	21.333		40,100	705		40,805	45,200
0820	Dumbbell set, cast iron, with rack and 5 pair					580			580	640
0900	Squat racks	2 Clab	5	3.200	Ea.	765	106		871	1,000
1200	Multi-station gym machine, 5 station					7,950			7,950	8,750
1250	9 station					11,200			11,200	12,300
1280	Rowing machine, hydraulic					1,475			1,475	1,625

11 66 Athletic Equipment

11 66 13 – Exercise Equipment

11 66 13.10 Physical Training Equipment		Crew	Daily Output	Labor-Hours	Unit	Material	2010 Bare Costs Labor	Equipment	Total	Total Incl O&P
1300	Treadmill, manual				Ea.	1,450			1,450	1,575
1320	Motorized					3,500			3,500	3,850
1340	Electronic					5,450			5,450	6,000
1360	Cardio-testing					7,775			7,775	8,550
1400	Treatment/massage tables, minimum					535			535	590
1420	Deluxe, with accessories					685			685	755
4150	Exercise equipment, bicycle trainer					595			595	655
4180	Chinning bar, adjustable, wall mounted	1 Carp	5	1.600		425	66.50		491.50	570
4200	Exercise ladder, 16' x 1'-7", suspended	L-2	3	5.333		1,475	195		1,670	1,925
4210	High bar, floor plate attached	1 Carp	4	2		1,100	83		1,183	1,350
4240	Parallel bars, adjustable		4	2		2,100	83		2,183	2,425
4270	Uneven parallel bars, adjustable		4	2		2,675	83		2,758	3,050
4280	Wall mounted, adjustable	L-2	1.50	10.667	Set	770	390		1,160	1,450
4300	Rope, ceiling mounted, 18' long	1 Carp	3.66	2.186	Ea.	257	91		348	425
4330	Side horse, vaulting		5	1.600		1,700	66.50		1,766.50	1,975
4360	Treadmill, motorized, deluxe, training type		5	1.600		3,500	66.50		3,566.50	3,950
4390	Weight lifting multi-station, minimum	2 Clab	1	16		395	530		925	1,250
4450	Maximum	"	.50	32		14,700	1,050		15,750	17,800

11 66 23 – Gymnasium Equipment

11 66 23.13 Basketball Equipment		Crew	Daily Output	Labor-Hours	Unit	Material	Labor	Equipment	Total	Total Incl O&P
0010	**BASKETBALL EQUIPMENT**									
1000	Backstops, wall mtd., 6' extended, fixed, minimum	L-2	1	16	Ea.	1,150	585		1,735	2,150
1100	Maximum		1	16		1,675	585		2,260	2,750
1200	Swing up, minimum		1	16		1,400	585		1,985	2,425
1250	Maximum		1	16		5,350	585		5,935	6,800
1300	Portable, manual, heavy duty, spring operated		1.90	8.421		13,300	310		13,610	15,100
1400	Ceiling suspended, stationary, minimum		.78	20.513		2,850	750		3,600	4,300
1450	Fold up, with accessories, maximum		.40	40		6,100	1,475		7,575	8,975
1600	For electrically operated, add	1 Elec	1	8		2,000	390		2,390	2,775
5800	Wall pads, 1-1/2" thick, standard (not fire rated)	2 Carp	640	.025	S.F.	6.80	1.04		7.84	9.10

11 66 23.19 Boxing Ring		Crew	Daily Output	Labor-Hours	Unit	Material	Labor	Equipment	Total	Total Incl O&P
0010	**BOXING RING**									
4100	Elevated, 22' x 22'	L-4	.10	240	Ea.	7,225	9,350		16,575	22,400
4110	For cellular plastic foam padding, add		.10	240		850	9,350		10,200	15,300
4120	Floor level, including posts and ropes only, 20' x 20'		.80	30		3,000	1,175		4,175	5,100
4130	Canvas, 30' x 30'		5	4.800		2,575	187		2,762	3,100

11 66 23.47 Gym Mats		Crew	Daily Output	Labor-Hours	Unit	Material	Labor	Equipment	Total	Total Incl O&P
0010	**GYM MATS**									
5500	2" thick, naugahyde covered				S.F.	3.59			3.59	3.95
5600	Vinyl/nylon covered					6.20			6.20	6.85
6000	Wrestling mats, 1" thick, heavy duty					5.50			5.50	6.05

11 66 43 – Interior Scoreboards

11 66 43.10 Scoreboards		Crew	Daily Output	Labor-Hours	Unit	Material	Labor	Equipment	Total	Total Incl O&P
0010	**SCOREBOARDS**									
7000	Baseball, minimum	R-3	1.30	15.385	Ea.	4,000	745	106	4,851	5,625
7200	Maximum		.05	400		17,500	19,300	2,775	39,575	51,000
7300	Football, minimum		.86	23.256		5,175	1,125	161	6,461	7,525
7400	Maximum		.20	100		15,400	4,825	690	20,915	24,900
7500	Basketball (one side), minimum		2.07	9.662		2,325	465	67	2,857	3,325
7600	Maximum		.30	66.667		3,400	3,225	460	7,085	9,050
7700	Hockey-basketball (four sides), minimum		.25	80		5,500	3,850	555	9,905	12,400

11 66 Athletic Equipment

11 66 43 – Interior Scoreboards

	11 66 43.10 Scoreboards	Crew	Daily Output	Labor-Hours	Unit	2010 Bare Costs Material	Labor	Equipment	Total	Total Incl O&P
7800	Maximum	R-3	.15	133	Ea.	5,575	6,425	920	12,920	16,700

11 66 53 – Gymnasium Dividers

11 66 53.10 Divider Curtains

	11 66 53.10 Divider Curtains	Crew	Daily Output	Labor-Hours	Unit	Material	Labor	Equipment	Total	Total Incl O&P
0010	**DIVIDER CURTAINS**									
4500	Gym divider curtain, mesh top, vinyl bottom, manual	L-4	500	.048	S.F.	8.15	1.87		10.02	11.85
4700	Electric roll up	L-7	400	.070	"	9.95	2.81		12.76	15.25

11 67 Recreational Equipment

11 67 13 – Bowling Alley Equipment

11 67 13.10 Bowling Alleys

	11 67 13.10 Bowling Alleys	Crew	Daily Output	Labor-Hours	Unit	Material	Labor	Equipment	Total	Total Incl O&P
0010	**BOWLING ALLEYS** Including alley, pinsetter, scorer,									
0020	Counters and misc. supplies, minimum	4 Carp	.20	160	Lane	41,300	6,650		47,950	55,500
0150	Average		.19	168		45,400	7,000		52,400	60,500
0300	Maximum		.18	177		52,000	7,375		59,375	69,000
0400	Combo table ball rack, add					1,150			1,150	1,250
0600	For automatic scorer, add, minimum					8,225			8,225	9,050
0700	Maximum					9,875			9,875	10,900

11 67 23 – Shooting Range Equipment

11 67 23.10 Shooting Range

	11 67 23.10 Shooting Range	Crew	Daily Output	Labor-Hours	Unit	Material	Labor	Equipment	Total	Total Incl O&P
0010	**SHOOTING RANGE** Incl. bullet traps, target provisions, controls,									
0100	Separators, ceiling system, etc. Not incl. structural shell									
0200	Commercial	L-9	.64	56.250	Point	25,000	2,150		27,150	30,900
0300	Law enforcement		.28	128		30,500	4,900		35,400	41,400
0400	National Guard armories		.71	50.704		18,000	1,925		19,925	22,900
0500	Reserve training centers		.71	50.704		12,000	1,925		13,925	16,300
0600	Schools and colleges		.32	112		28,000	4,275		32,275	37,600
0700	Major academies		.19	189		45,000	7,200		52,200	61,000
0800	For acoustical treatment, add					10%	10%			
0900	For lighting, add					28%	25%			
1000	For plumbing, add					5%	5%			
1100	For ventilating system, add, minimum					40%	40%			
1200	Add, average					25%	25%			
1300	Add, maximum					35%	35%			

11 68 Play Field Equipment and Structures

11 68 13 – Playground Equipment

11 68 13.10 Free-Standing Playground Equipment

	11 68 13.10 Free-Standing Playground Equipment	Crew	Daily Output	Labor-Hours	Unit	Material	Labor	Equipment	Total	Total Incl O&P
0010	**FREE-STANDING PLAYGROUND EQUIPMENT** See also individual items									
0200	Bike rack, 10′ long, permanent G	B-1	12	2	Ea.	560	67.50		627.50	720
0400	Horizontal monkey ladder, 14′ long, 6′ high		4	6		860	203		1,063	1,250
0590	Parallel bars, 10′ long		4	6		330	203		533	675
0600	Posts, tether ball set, 2-3/8″ O.D.		12	2		281	67.50		348.50	415
0800	Poles, multiple purpose, 10′-6″ long		12	2	Pr.	158	67.50		225.50	278
1000	Ground socket for movable posts, 2-3/8″ post		10	2.400		140	81		221	279
1100	3-1/2″ post		10	2.400		141	81		222	280
1300	See-saw, spring, steel, 2 units		6	4	Ea.	815	135		950	1,100
1400	4 units		4	6		1,250	203		1,453	1,675
1500	6 units		3	8		1,225	270		1,495	1,775

11 68 Play Field Equipment and Structures

11 68 13 – Playground Equipment

11 68 13.10 Free-Standing Playground Equipment		Crew	Daily Output	Labor-Hours	Unit	2010 Bare Costs Material	Labor	Equipment	Total	Total Incl O&P
1700	Shelter, fiberglass golf tee, 3 person	B-1	4.60	5.217	Ea.	3,300	176		3,476	3,900
1900	Slides, stainless steel bed, 12′ long, 6′ high		3	8		4,850	270		5,120	5,750
2000	20′ long, 10′ high		2	12		2,500	405		2,905	3,375
2200	Swings, plain seats, 8′ high, 4 seats		2	12		1,275	405		1,680	2,025
2300	8 seats		1.30	18.462		2,125	625		2,750	3,275
2500	12′ high, 4 seats		2	12		1,925	405		2,330	2,750
2600	8 seats		1.30	18.462		2,975	625		3,600	4,225
2800	Whirlers, 8′ diameter		3	8		3,075	270		3,345	3,800
2900	10′ diameter		3	8		4,450	270		4,720	5,325

11 68 13.20 Modular Playground

Line	Description	Crew	Daily Output	Labor-Hours	Unit	Material	Labor	Equipment	Total	Total Incl O&P
0010	**MODULAR PLAYGROUND** Basic components									
0100	Deck, square, steel, 48″ x 48″	B-1	1	24	Ea.	890	810		1,700	2,225
0110	Recycled polyurethane		1	24		880	810		1,690	2,225
0120	Triangular, steel, 48″ side		1	24		635	810		1,445	1,950
0130	Post, steel, 5″ square		18	1.333	L.F.	31.50	45		76.50	104
0140	Aluminum, 2-3/8″ square		20	1.200		26.50	40.50		67	91.50
0150	5″ square		18	1.333		34.50	45		79.50	108
0160	Roof, square poly, 54″ side		18	1.333	Ea.	1,075	45		1,120	1,250
0170	Wheelchair transfer module, for 3′ high deck		3	8	″	2,825	270		3,095	3,525
0180	Guardrail, pipe, 36″ high		60	.400	L.F.	130	13.50		143.50	164
0190	Steps, deck-to-deck, 3 – 8″ steps		8	3	Ea.	895	101		996	1,150
0200	Activity panel, crawl through panel		2	12		365	405		770	1,025
0210	Alphabet/spelling panel		2	12		525	405		930	1,200
0360	With guardrails		3	8		1,575	270		1,845	2,175
0370	Crawl tunnel, straight, 56″ long		4	6		1,500	203		1,703	1,950
0380	90°, 4′ long		4	6		1,650	203		1,853	2,125
1200	Slide, tunnel, for 56″ high deck		8	3		1,575	101		1,676	1,875
1210	Straight, poly		8	3		2,250	101		2,351	2,625
1220	Stainless steel, 54″ high deck		6	4		435	135		570	690
1230	Curved, poly, 40″ high deck		6	4		585	135		720	855
1240	Spiral slide, 56″ - 72″ high		5	4.800		3,500	162		3,662	4,100
1300	Ladder, vertical, for 24″ - 72″ high deck		5	4.800		645	162		807	955
1310	Horizontal, 8′ long		5	4.800		740	162		902	1,075
1320	Corkscrew climber, 6′ high		3	8		550	270		820	1,025
1330	Fire pole for 72″ high deck		6	4		470	135		605	725
1340	Bridge, ring climber, 8′ long		4	6		1,425	203		1,628	1,875
1350	Suspension		4	6	L.F.	610	203		813	980

11 68 16 – Play Structures

11 68 16.10 Handball/Squash Court

Line	Description	Crew	Daily Output	Labor-Hours	Unit	Material	Labor	Equipment	Total	Total Incl O&P
0010	**HANDBALL/SQUASH COURT**, outdoor									
0900	Handball or squash court, outdoor, wood	2 Carp	.50	32	Ea.	4,475	1,325		5,800	6,975
1000	Masonry handball/squash court	D-1	.30	53.333	″	24,500	2,000		26,500	29,900

11 68 16.30 Platform/Paddle Tennis Court

Line	Description	Crew	Daily Output	Labor-Hours	Unit	Material	Labor	Equipment	Total	Total Incl O&P
0010	**PLATFORM/PADDLE TENNIS COURT** Complete with lighting, etc.									
0100	Aluminum slat deck with aluminum frame	B-1	.08	300	Court	49,500	10,100		59,600	70,000
0500	Aluminum slat deck with wood frame	C-1	.12	266		49,500	10,500		60,000	70,500
0800	Aluminum deck heater, add	B-1	1.18	20.339		4,900	685		5,585	6,450
0900	Douglas fir planking with wood frame 2″ x 6″ x 30′	C-1	.12	266		46,600	10,500		57,100	67,000
1000	Plywood deck with steel frame		.12	266		46,600	10,500		57,100	67,000
1100	Steel slat deck with wood frame		.12	266		33,000	10,500		43,500	52,500

11 68 Play Field Equipment and Structures

11 68 33 – Athletic Field Equipment

11 68 33.13 Football Field Equipment		Crew	Daily Output	Labor-Hours	Unit	Material	2010 Bare Costs Labor	Equipment	Total	Total Incl O&P
0010	**FOOTBALL FIELD EQUIPMENT**									
0020	Goal posts, steel, football, double post	B-1	1.50	16	Pr.	2,700	540		3,240	3,800
0100	Deluxe, single post		1.50	16		4,575	540		5,115	5,875
0300	Football, convertible to soccer		1.50	16		3,600	540		4,140	4,775
0500	Soccer, regulation	↓	2	12	↓	2,750	405		3,155	3,650

11 71 Medical Sterilizing Equipment

11 71 10 – Medical Sterilizers & Distillers

11 71 10.10 Sterilizers and Distillers		Crew	Daily Output	Labor-Hours	Unit	Material	Labor	Equipment	Total	Total Incl O&P
0010	**STERILIZERS AND DISTILLERS**									
0700	Distiller, water, steam heated, 50 gal. capacity	1 Plum	1.40	5.714	Ea.	18,500	297		18,797	20,700
5600	Sterilizers, floor loading, 26" x 62" x 42", single door, steam					158,500			158,500	174,500
5650	Double door, steam					204,000			204,000	224,000
5800	General purpose, 20" x 20" x 38", single door					15,300			15,300	16,800
6000	Portable, counter top, steam, minimum					3,800			3,800	4,175
6020	Maximum					5,950			5,950	6,550
6050	Portable, counter top, gas, 17" x 15" x 32-1/2"					39,300			39,300	43,200
6150	Manual washer/sterilizer, 16" x 16" x 26"	1 Plum	2	4	↓	54,000	208		54,208	59,500
6200	Steam generators, electric 10 kW to 180 kW, freestanding									
6250	Minimum	1 Elec	3	2.667	Ea.	8,000	131		8,131	9,000
6300	Maximum	"	.70	11.429		28,800	560		29,360	32,500
8200	Bed pan washer-sanitizer	1 Plum	2	4	↓	7,125	208		7,333	8,150

11 72 Examination and Treatment Equipment

11 72 13 – Examination Equipment

11 72 13.13 Examination Equipment		Crew	Daily Output	Labor-Hours	Unit	Material	Labor	Equipment	Total	Total Incl O&P
0010	**EXAMINATION EQUIPMENT**									
0300	Blood pressure unit, mercurial, wall				Ea.	198			198	218
0400	Diagnostic set, wall					650			650	715
4400	Scale, physician's, with height rod				↓	292			292	320

11 72 53 – Treatment Equipment

11 72 53.13 Medical Treatment Equipment		Crew	Daily Output	Labor-Hours	Unit	Material	Labor	Equipment	Total	Total Incl O&P
0010	**MEDICAL TREATMENT EQUIPMENT**									
6500	Surgery table, minor minimum	1 Sswk	.70	11.429	Ea.	13,900	535		14,435	16,100
6520	Maximum	"	.50	16		26,100	750		26,850	30,000
6700	Surgical lights, doctor's office, single arm	2 Elec	2	8		1,225	390		1,615	1,925
6750	Dual arm	"	1	16	↓	5,800	785		6,585	7,575

11 73 Patient Care Equipment

11 73 10 – Patient Treatment Equipment

11 73 10.10 Treatment Equipment		Crew	Daily Output	Labor-Hours	Unit	2010 Bare Costs Material	Labor	Equipment	Total	Total Incl O&P
0010	**TREATMENT EQUIPMENT**									
0750	Exam room furnishings, average per room				Ea.	6,450			6,450	7,100
1800	Heat therapy unit, humidified, 26" x 78" x 28"				"	3,400			3,400	3,725
2100	Hubbard tank with accessories, stainless steel,									
2110	125 GPM at 45 psi water pressure				Ea.	26,300			26,300	29,000
2150	For electric overhead hoist, add					2,875			2,875	3,150
2900	K-Module for heat therapy, 20 oz. capacity, 75°F to 110°F					405			405	445
3600	Paraffin bath, 126°F, auto controlled					1,475			1,475	1,600
3900	Parallel bars for walking training, 12'-0"					1,200			1,200	1,325
4600	Station, dietary, medium, with ice					16,000			16,000	17,600
4700	Medicine					7,250			7,250	7,975
7000	Tables, physical therapy, walk off, electric	2 Carp	3	5.333		3,200	222		3,422	3,875
7150	Standard, vinyl top with base cabinets, minimum		3	5.333		1,300	222		1,522	1,775
7200	Maximum		2	8		3,850	330		4,180	4,750
8400	Whirlpool bath, mobile, sst, 18" x 24" x 60"					4,500			4,500	4,950
8450	Fixed, incl. mixing valves	1 Plum	2	4		9,200	208		9,408	10,400

11 74 Dental Equipment

11 74 10 – Dental Office Equipment

11 74 10.10 Diagnostic and Treatment Equipment		Crew	Daily Output	Labor-Hours	Unit	Material	Labor	Equipment	Total	Total Incl O&P
0010	**DIAGNOSTIC AND TREATMENT EQUIPMENT**									
0020	Central suction system, minimum	1 Plum	1.20	6.667	Ea.	1,550	345		1,895	2,225
0100	Maximum	"	.90	8.889		4,250	465		4,715	5,375
0300	Air compressor, minimum	1 Skwk	.80	10		2,975	425		3,400	3,925
0400	Maximum		.50	16		7,200	680		7,880	8,975
0600	Chair, electric or hydraulic, minimum		.50	16		2,550	680		3,230	3,875
0700	Maximum		.25	32		14,400	1,375		15,775	18,000
0800	Doctor's/assistant's stool, minimum					119			119	131
0850	Maximum					810			810	890
1000	Drill console with accessories, minimum	1 Skwk	1.60	5		1,575	213		1,788	2,050
1100	Maximum		1.60	5		5,375	213		5,588	6,250
2000	Light, ceiling mounted, minimum		8	1		1,225	42.50		1,267.50	1,425
2100	Maximum		8	1		1,625	42.50		1,667.50	1,875
2200	Unit light, minimum	2 Skwk	5.33	3.002		600	128		728	855
2210	Maximum		5.33	3.002		1,625	128		1,753	2,000
2220	Track light, minimum		3.20	5		1,525	213		1,738	2,000
2230	Maximum		3.20	5		3,725	213		3,938	4,425
2300	Sterilizers, steam portable, minimum					1,850			1,850	2,050
2350	Maximum					10,400			10,400	11,400
2600	Steam, institutional					4,250			4,250	4,675
2650	Dry heat, electric, portable, 3 trays					1,225			1,225	1,325
2700	Ultra-sonic cleaner, portable, minimum					239			239	263
2750	Maximum (institutional)					1,525			1,525	1,675
3000	X-ray unit, wall, minimum	1 Skwk	4	2		2,200	85		2,285	2,525
3010	Maximum		4	2		5,125	85		5,210	5,775
3100	Panoramic unit		.60	13.333		14,300	570		14,870	16,600
3105	Deluxe, minimum	2 Skwk	1.60	10		16,000	425		16,425	18,300
3110	Maximum	"	1.60	10		54,000	425		54,425	60,000
3500	Developers, X-ray, average	1 Plum	5.33	1.501		4,950	78		5,028	5,575
3600	Maximum	"	5.33	1.501		8,100	78		8,178	9,025

11 76 Operating Room Equipment

11 76 10 – Operating Room Equipment

11 76 10.10 Surgical Equipment		Crew	Daily Output	Labor-Hours	Unit	Material	2010 Bare Costs Labor	Equipment	Total	Total Incl O&P
0010	**SURGICAL EQUIPMENT**									
5000	Scrub, surgical, stainless steel, single station, minimum	1 Plum	3	2.667	Ea.	4,100	139		4,239	4,700
5100	Maximum					8,300			8,300	9,125
6550	Major surgery table, minimum	1 Sswk	.50	16		31,400	750		32,150	35,900
6570	Maximum	"	.50	16		103,500	750		104,250	115,500
6800	Surgical lights, major operating room, dual head, minimum	2 Elec	1	16		17,500	785		18,285	20,400
6850	Maximum	"	1	16	↓	29,100	785		29,885	33,200

11 77 Radiology Equipment

11 77 10 – Radiology Equipment

11 77 10.10 X-Ray Equipment		Crew	Daily Output	Labor-Hours	Unit	Material	Labor	Equipment	Total	Total Incl O&P
0010	**X-RAY EQUIPMENT**									
8700	X-ray, mobile, minimum				Ea.	13,400			13,400	14,700
8750	Maximum					75,500			75,500	83,000
8900	Stationary, minimum					41,800			41,800	46,000
8950	Maximum					217,500			217,500	239,500
9150	Developing processors, minimum					11,600			11,600	12,800
9200	Maximum				↓	30,600			30,600	33,600

11 78 Mortuary Equipment

11 78 13 – Mortuary Refrigerators

11 78 13.10 Mortuary and Autopsy Equipment		Crew	Daily Output	Labor-Hours	Unit	Material	Labor	Equipment	Total	Total Incl O&P
0010	**MORTUARY AND AUTOPSY EQUIPMENT**									
0015	Autopsy table, standard	1 Plum	1	8	Ea.	8,850	415		9,265	10,400
0020	Deluxe	"	.60	13.333		17,700	695		18,395	20,500
3200	Mortuary refrigerator, end operated, 2 capacity					12,800			12,800	14,100
3300	6 capacity				↓	23,900			23,900	26,300

11 78 16 – Crematorium Equipment

11 78 16.10 Crematory		Crew	Daily Output	Labor-Hours	Unit	Material	Labor	Equipment	Total	Total Incl O&P
0010	**CREMATORY**									
1500	Crematory, not including building, 1 place	Q-3	.20	160	Ea.	61,500	7,925		69,425	79,500
1750	2 place	"	.10	320	"	87,500	15,900		103,400	120,500

11 82 Solid Waste Handling Equipment

11 82 19 – Packaged Incinerators

11 82 19.10 Packaged Gas Fired Incinerators		Crew	Daily Output	Labor-Hours	Unit	Material	Labor	Equipment	Total	Total Incl O&P
0010	**PACKAGED GAS FIRED INCINERATORS**									
4400	Incinerator, gas, not incl. chimney, elec. or pipe, 50#/hr., minimum	Q-3	.80	40	Ea.	11,200	1,975		13,175	15,300
4420	Maximum		.70	45.714		36,100	2,275		38,375	43,100
4440	200 lb. per hr., minimum (batch type)		.60	53.333		62,000	2,650		64,650	72,000
4460	Maximum (with feeder)		.50	64		87,500	3,175		90,675	101,500
4480	400 lb. per hr., minimum (batch type)		.30	106		68,500	5,300		73,800	83,500
4500	Maximum (with feeder)		.25	128		92,500	6,350		98,850	111,500
4520	800 lb. per hr., with feeder, minimum		.20	160		112,000	7,925		119,925	135,000
4540	Maximum		.17	188		154,500	9,325		163,825	184,000
4560	1,200 lb. per hr., with feeder, minimum		.15	213		150,500	10,600		161,100	181,500
4580	Maximum	↓	.11	290	↓	179,500	14,400		193,900	219,000

11 82 Solid Waste Handling Equipment

11 82 19 – Packaged Incinerators

11 82 19.10 Packaged Gas Fired Incinerators		Crew	Daily Output	Labor-Hours	Unit	2010 Bare Costs Material	Labor	Equipment	Total	Total Incl O&P
4600	2,000 lb. per hr., with feeder, minimum	Q-3	.10	320	Ea.	368,000	15,900		383,900	429,000
4620	Maximum		.05	640		504,000	31,700		535,700	602,000
4700	For heat recovery system, add, minimum		.25	128		71,000	6,350		77,350	87,500
4710	Add, maximum		.11	290		227,000	14,400		241,400	271,500
4720	For automatic ash conveyer, add		.50	64		29,800	3,175		32,975	37,600
4750	Large municipal incinerators, incl. stack, minimum		.25	128	Ton/day	18,100	6,350		24,450	29,500
4850	Maximum		.10	320	"	48,200	15,900		64,100	77,000

11 82 26 – Waste Compactors and Destructors

11 82 26.10 Compactors		Crew	Daily Output	Labor-Hours	Unit	Material	Labor	Equipment	Total	Total Incl O&P
0010	**COMPACTORS**									
0020	Compactors, 115 volt, 250#/hr., chute fed	L-4	1	24	Ea.	11,200	935		12,135	13,800
0100	Hand fed		2.40	10		8,225	390		8,615	9,625
0300	Multi-bag, 230 volt, 600#/hr, chute fed		1	24		10,200	935		11,135	12,600
0400	Hand fed		1	24		8,750	935		9,685	11,100
0500	Containerized, hand fed, 2 to 6 C.Y. containers, 250#/hr.		1	24		10,800	935		11,735	13,300
0550	For chute fed, add per floor		1	24		1,250	935		2,185	2,800
1000	Heavy duty industrial compactor, 0.5 C.Y. capacity		1	24		7,300	935		8,235	9,475
1050	1.0 C.Y. capacity		1	24		11,100	935		12,035	13,700
1100	3.0 C.Y. capacity		.50	48		15,200	1,875		17,075	19,600
1150	5.0 C.Y. capacity		.50	48		18,800	1,875		20,675	23,600
1200	Combination shredder/compactor (5,000 lbs./hr.)		.50	48		37,200	1,875		39,075	43,800
1400	For handling hazardous waste materials, 55 gallon drum packer, std.					17,300			17,300	19,000
1410	55 gallon drum packer w/HEPA filter					21,600			21,600	23,700
1420	55 gallon drum packer w/charcoal & HEPA filter					28,800			28,800	31,700
1430	All of the above made explosion proof, add					13,100			13,100	14,500
5500	Shredder, municipal use, 35 ton per hour					270,000			270,000	297,000
5600	60 ton per hour					575,000			575,000	632,500
5750	Shredder & baler, 50 ton per day					539,500			539,500	593,500
5800	Shredder, industrial, minimum					21,200			21,200	23,400
5850	Maximum					114,000			114,000	125,500
5900	Baler, industrial, minimum					8,500			8,500	9,350
5950	Maximum					497,000			497,000	547,000
6000	Transfer station compactor, with power unit									
6050	and pedestal, not including pit, 50 ton per hour				Ea.	170,000			170,000	187,000

11 91 Religious Equipment

11 91 13 – Baptistries

11 91 13.10 Baptistry		Crew	Daily Output	Labor-Hours	Unit	Material	Labor	Equipment	Total	Total Incl O&P
0010	**BAPTISTRY**									
0150	Fiberglass, 3'-6" deep, x 13'-7" long,									
0160	steps at both ends, incl. plumbing, minimum	L-8	1	20	Ea.	3,825	875		4,700	5,525
0200	Maximum	"	.70	28.571		6,250	1,250		7,500	8,775
0250	Add for filter, heater and lights					1,375			1,375	1,500

11 91 23 – Sanctuary Equipment

11 91 23.10 Sanctuary Furnishings		Crew	Daily Output	Labor-Hours	Unit	Material	Labor	Equipment	Total	Total Incl O&P
0010	**SANCTUARY FURNISHINGS**									
0020	Altar, wood, custom design, plain	1 Carp	1.40	5.714	Ea.	1,925	237		2,162	2,475
0050	Deluxe	"	.20	40		9,225	1,650		10,875	12,800
0070	Granite or marble, average	2 Marb	.50	32		10,700	1,275		11,975	13,600
0090	Deluxe	"	.20	80		30,400	3,200		33,600	38,200

11 91 23 – Sanctuary Equipment

11 91 23.10 Sanctuary Furnishings		Crew	Daily Output	Labor-Hours	Unit	2010 Bare Costs Material	Labor	Equipment	Total	Total Incl O&P
0100	Arks, prefabricated, plain	2 Carp	.80	20	Ea.	8,025	830		8,855	10,100
0130	Deluxe, maximum	"	.20	80		114,000	3,325		117,325	130,500
0500	Reconciliation room, wood, prefabricated, single, plain	1 Carp	.60	13.333		2,425	555		2,980	3,525
0550	Deluxe		.40	20		6,675	830		7,505	8,625
0650	Double, plain		.40	20		4,850	830		5,680	6,625
0700	Deluxe		.20	40		15,200	1,650		16,850	19,300
1000	Lecterns, wood, plain		5	1.600		750	66.50		816.50	925
1100	Deluxe		2	4		5,125	166		5,291	5,900
2000	Pulpits, hardwood, prefabricated, plain		2	4		1,225	166		1,391	1,600
2100	Deluxe		1.60	5		8,325	208		8,533	9,475
2500	Railing, hardwood, average		25	.320	L.F.	166	13.30		179.30	204
3000	Seating, individual, oak, contour, laminated		21	.381	Person	147	15.85		162.85	186
3100	Cushion seat		21	.381		133	15.85		148.85	171
3200	Fully upholstered		21	.381		130	15.85		145.85	168
3300	Combination, self-rising		21	.381		335	15.85		350.85	390
3500	For cherry, add					30%				
5000	Wall cross, aluminum, extruded, 2" x 2" section	1 Carp	34	.235	L.F.	132	9.80		141.80	160
5150	4" x 4" section		29	.276		190	11.45		201.45	227
5300	Bronze, extruded, 1" x 2" section		31	.258		260	10.70		270.70	305
5350	2-1/2" x 2-1/2" section		34	.235		395	9.80		404.80	450
5450	Solid bar stock, 1/2" x 3" section		29	.276		515	11.45		526.45	590
5600	Fiberglass, stock		34	.235		107	9.80		116.80	132
5700	Stainless steel, 4" deep, channel section		29	.276		420	11.45		431.45	480
5800	4" deep box section		29	.276		570	11.45		581.45	645

Division Notes

		CREW	DAILY OUTPUT	LABOR-HOURS	UNIT	2010 BARE COSTS MAT.	LABOR	EQUIP.	TOTAL	TOTAL INCL O&P

Estimating Tips

General

- The items in this division are usually priced per square foot or each. Most of these items are purchased by the owner and placed by the contractor. Do not assume the items in Division 12 will be purchased and installed by the contractor. Check the specifications for responsibilities and include receiving, storage, installation, and mechanical and electrical hookups in the appropriate divisions.
- Some items in this division require some type of support system that is not usually furnished with the item. Examples of these systems include blocking for the attachment of casework and heavy drapery rods. The required blocking must be added to the estimate in the appropriate division.

Reference Numbers

Reference numbers are shown in shaded boxes at the beginning of some major classifications. These numbers refer to related items in the Reference Section. The reference information may be an estimating procedure, an alternate pricing method, or technical information.

Note: Not all subdivisions listed here necessarily appear in this publication.

12 21 Window Blinds

12 21 13 – Horizontal Louver Blinds

12 21 13.13 Metal Horizontal Louver Blinds		Crew	Daily Output	Labor-Hours	Unit	2010 Bare Costs Material	Labor	Equipment	Total	Total Incl O&P
0010	**METAL HORIZONTAL LOUVER BLINDS**									
0020	Horizontal, 1" aluminum slats, solid color, stock	1 Carp	590	.014	S.F.	4.75	.56		5.31	6.10
0090	Custom, minimum		590	.014		5.25	.56		5.81	6.65
0100	Maximum		440	.018		7.15	.76		7.91	9
0250	2" aluminum slats, custom, minimum		590	.014		8.10	.56		8.66	9.80
0350	Maximum		440	.018		13.15	.76		13.91	15.65
0450	Stock, minimum		590	.014		4.86	.56		5.42	6.20
0500	Maximum		440	.018		7.90	.76		8.66	9.85
0600	2" steel slats, stock, minimum		590	.014		1.84	.56		2.40	2.89
0630	Maximum		440	.018		5.15	.76		5.91	6.85
0750	Custom, minimum		590	.014		1.67	.56		2.23	2.71
0850	Maximum	↓	400	.020	↓	8.25	.83		9.08	10.40

12 21 16 – Vertical Louver Blinds

12 21 16.13 Metal Vertical Louver Blinds		Crew	Daily Output	Labor-Hours	Unit	Material	Labor	Equipment	Total	Total Incl O&P
0010	**METAL VERTICAL LOUVER BLINDS**									
1500	Vertical, 3" to 5" PVC or cloth strips, minimum **CN**	1 Carp	460	.017	S.F.	3.61	.72		4.33	5.10
1600	Maximum		400	.020		12.65	.83		13.48	15.20
1800	4" aluminum slats, minimum		460	.017		7.65	.72		8.37	9.50
1900	Maximum		400	.020		9.10	.83		9.93	11.30
1950	Mylar mirror-finish strips, to 8" wide, minimum		460	.017		4.62	.72		5.34	6.20
1970	Maximum	↓	400	.020	↓	10.50	.83		11.33	12.85

12 22 Curtains and Drapes

12 22 16 – Drapery Track and Accessories

12 22 16.10 Drapery Hardware		Crew	Daily Output	Labor-Hours	Unit	Material	Labor	Equipment	Total	Total Incl O&P
0010	**DRAPERY HARDWARE**									
0030	Standard traverse, per foot, minimum	1 Carp	59	.136	L.F.	3.42	5.65		9.07	12.45
0100	Maximum		51	.157	"	8.05	6.50		14.55	18.90
4000	Traverse rods, adjustable, 28" to 48"		22	.364	Ea.	21	15.10		36.10	46.50
4020	48" to 84"		20	.400		29.50	16.60		46.10	58
4040	66" to 120"		18	.444		31.50	18.45		49.95	63
4060	84" to 156"		16	.500		33	21		54	68
4080	100" to 180"		14	.571		42.50	23.50		66	83
4090	156" to 228"		13	.615		47.50	25.50		73	92
4100	228" to 312"		13	.615		59.50	25.50		85	105
4200	Double rods, adjustable, 30" to 48"		9	.889		38	37		75	99
4220	48" to 86"		9	.889		52.50	37		89.50	115
4240	86" to 150"		8	1		57.50	41.50		99	128
4260	100" to 180"		7	1.143		60.50	47.50		108	140
4300	Tray and curtain rod, adjustable, 30" to 48"		9	.889		23.50	37		60.50	83
4320	48" to 86"		9	.889		35	37		72	95.50
4340	86" to 150"		8	1		41.50	41.50		83	110
4360	100" to 180"	↓	7	1.143		47	47.50		94.50	125
4600	Valance, pinch pleated fabric, 12" deep, up to 54" long, minimum					36.50			36.50	40.50
4610	Maximum					91.50			91.50	101
4620	Up to 77" long, minimum					56.50			56.50	62
4630	Maximum					148			148	163
5000	Stationary rods, first 2'				↓	9.35			9.35	10.25
5020	Each additional foot, add				L.F.	3.64			3.64	4

12 22 Curtains and Drapes

12 22 16 – Drapery Track and Accessories

	12 22 16.20 Blast Curtains	Crew	Daily Output	Labor-Hours	Unit	Material	2010 Bare Costs Labor	Equipment	Total	Total Incl O&P
0010	**BLAST CURTAINS** per L.F. horizontal opening width, off-white or gray fabric									
0100	Blast curtains, drapery system, complete, including hardware, minimum				L.F.				130	142
0120	Average								140	153
0140	Maximum				↓				149	165

12 23 Interior Shutters

12 23 10 – Wood Interior Shutters

	12 23 10.10 Wood Interior Shutters	Crew	Daily Output	Labor-Hours	Unit	Material	Labor	Equipment	Total	Total Incl O&P
0010	**WOOD INTERIOR SHUTTERS**, louvered									
0200	Two panel, 27" wide, 36" high	1 Carp	5	1.600	Set	124	66.50		190.50	238
0300	33" wide, 36" high		5	1.600		162	66.50		228.50	281
0500	47" wide, 36" high		5	1.600		216	66.50		282.50	340
1000	Four panel, 27" wide, 36" high		5	1.600		147	66.50		213.50	263
1100	33" wide, 36" high		5	1.600		180	66.50		246.50	300
1300	47" wide, 36" high	↓	5	1.600	↓	255	66.50		321.50	385

	12 23 10.13 Wood Panels	Crew	Daily Output	Labor-Hours	Unit	Material	Labor	Equipment	Total	Total Incl O&P
0010	**WOOD PANELS**									
3000	Wood folding panels with movable louvers, 7" x 20" each	1 Carp	17	.471	Pr.	47	19.55		66.55	81.50
3300	8" x 28" each		17	.471		61	19.55		80.55	97.50
3450	9" x 36" each		17	.471		75.50	19.55		95.05	113
3600	10" x 40" each		17	.471		81.50	19.55		101.05	120
4000	Fixed louver type, stock units, 8" x 20" each		17	.471		84	19.55		103.55	123
4150	10" x 28" each		17	.471		90	19.55		109.55	129
4300	12" x 36" each		17	.471		106	19.55		125.55	147
4450	18" x 40" each		17	.471		118	19.55		137.55	160
5000	Insert panel type, stock, 7" x 20" each		17	.471		19.65	19.55		39.20	51.50
5150	8" x 28" each		17	.471		36	19.55		55.55	69.50
5300	9" x 36" each		17	.471		46	19.55		65.55	80.50
5450	10" x 40" each		17	.471		49	19.55		68.55	84
5600	Raised panel type, stock, 10" x 24" each		17	.471		116	19.55		135.55	158
5650	12" x 26" each		17	.471		116	19.55		135.55	158
5700	14" x 30" each		17	.471		214	19.55		233.55	266
5750	16" x 36" each	↓	17	.471		294	19.55		313.55	355
6000	For custom built pine, add					22%				
6500	For custom built hardwood blinds, add				↓	42%				

12 24 Window Shades

12 24 13 – Roller Window Shades

	12 24 13.10 Shades	Crew	Daily Output	Labor-Hours	Unit	Material	Labor	Equipment	Total	Total Incl O&P
0010	**SHADES**									
0020	Basswood, roll-up, stain finish, 3/8" slats	1 Carp	300	.027	S.F.	13.95	1.11		15.06	17.05
0200	7/8" slats		300	.027		13.15	1.11		14.26	16.20
0300	Vertical side slide, stain finish, 3/8" slats		300	.027		18.05	1.11		19.16	21.50
0400	7/8" slats	↓	300	.027		18.05	1.11		19.16	21.50
0500	For fire retardant finishes, add					16%				
0600	For "B" rated finishes, add					20%				
0900	Mylar, single layer, non-heat reflective	1 Carp	685	.012		6.70	.49		7.19	8.10
1000	Double layered, heat reflective	↓	685	.012	↓	9.25	.49		9.74	10.95

12 24 Window Shades

12 24 13 – Roller Window Shades

12 24 13.10 Shades		Crew	Daily Output	Labor-Hours	Unit	Material	Labor	Equipment	Total	Total Incl O&P
						2010 Bare Costs				
1100	Triple layered, heat reflective	1 Carp	685	.012	S.F.	10.80	.49		11.29	12.60
1200	For metal roller instead of wood, add per				Shade	4.20			4.20	4.62
1300	Vinyl coated cotton, standard	1 Carp	685	.012	S.F.	3.43	.49		3.92	4.52
1400	Lightproof decorator shades		685	.012		2.15	.49		2.64	3.12
1500	Vinyl, lightweight, 4 gauge		685	.012		.56	.49		1.05	1.37
1600	Heavyweight, 6 gauge		685	.012		1.70	.49		2.19	2.62
1700	Vinyl laminated fiberglass, 6 ga., translucent		685	.012		2.41	.49		2.90	3.40
1800	Lightproof		685	.012		3.91	.49		4.40	5.05
3000	Woven aluminum, 3/8" thick, lightproof and fireproof		350	.023		5.50	.95		6.45	7.50

12 32 Manufactured Wood Casework

12 32 16 – Manufactured Plastic-Laminate-Clad Casework

12 32 16.20 Plastic Laminate Casework Doors

Code	Description	Crew	Daily Output	Labor-Hours	Unit	Material	Labor	Equipment	Total	Total Incl O&P
0010	**PLASTIC LAMINATE CASEWORK DOORS**									
1000	For casework frames, see Div. 12 32 23.15									
1100	For casework hardware, see Div. 12 32 23.35									
6000	Plastic laminate on particle board									
6100	12" wide, 18" high	1 Carp	25	.320	Ea.	10.50	13.30		23.80	32
6140	30" high		23	.348		17.50	14.45		31.95	42
6500	18" wide, 18" high		24	.333		15.75	13.85		29.60	39
6600	30" high		22	.364		26.50	15.10		41.60	52.50

12 32 16.25 Plastic Laminate Drawer Fronts

Code	Description	Crew	Daily Output	Labor-Hours	Unit	Material	Labor	Equipment	Total	Total Incl O&P
0010	**PLASTIC LAMINATE DRAWER FRONTS**									
2800	Plastic laminate on particle board front									
3000	4" high, 12" wide	1 Carp	17	.471	Ea.	3.51	19.55		23.06	34
3200	18" wide	"	16	.500	"	5.25	21		26.25	38

12 32 23 – Hardwood Casework

12 32 23.10 Manufactured Wood Casework, Stock Units

Code	Description	Crew	Daily Output	Labor-Hours	Unit	Material	Labor	Equipment	Total	Total Incl O&P
0010	**MANUFACTURED WOOD CASEWORK, STOCK UNITS**									
0300	Built-in drawer units, pine, 18" deep, 32" high, unfinished									
0400	Minimum	2 Carp	53	.302	L.F.	145	12.55		157.55	179
0500	Maximum	"	40	.400	"	180	16.60		196.60	224
0700	Kitchen base cabinets, hardwood, not incl. counter tops,									
0710	24" deep, 35" high, prefinished									
0800	One top drawer, one door below, 12" wide	2 Carp	24.80	.645	Ea.	229	27		256	294
0840	18" wide		23.30	.687		259	28.50		287.50	330
0880	24" wide		22.30	.717		310	30		340	390
1000	Four drawers, 12" wide		24.80	.645		240	27		267	305
1040	18" wide		23.30	.687		268	28.50		296.50	340
1060	24" wide		22.30	.717		305	30		335	380
1200	Two top drawers, two doors below, 27" wide		22	.727		330	30		360	410
1260	36" wide		20.30	.788		400	33		433	490
1300	48" wide		18.90	.847		460	35		495	560
1500	Range or sink base, two doors below, 30" wide		21.40	.748		305	31		336	385
1540	36" wide		20.30	.788		345	33		378	430
1580	48" wide		18.90	.847		385	35		420	480
1800	For sink front units, deduct					145			145	160
2000	Corner base cabinets, 36" wide, standard	2 Carp	18	.889		540	37		577	650
2100	Lazy Susan with revolving door	"	16.50	.970		725	40.50		765.50	855
4000	Kitchen wall cabinets, hardwood, 12" deep with two doors									

12 32 23 – Hardwood Casework

12 32 23.10 Manufactured Wood Casework, Stock Units		Crew	Daily Output	Labor-Hours	Unit	2010 Bare Costs Material	Labor	Equipment	Total	Total Incl O&P
4050	12" high, 30" wide	2 Carp	24.80	.645	Ea.	205	27		232	267
4100	36" wide		24	.667		240	27.50		267.50	305
4400	15" high, 30" wide		24	.667		165	27.50		192.50	224
4440	36" wide		22.70	.705		240	29.50		269.50	310
4700	24" high, 30" wide		23.30	.687		275	28.50		303.50	350
4720	36" wide		22.70	.705		305	29.50		334.50	380
5000	30" high, one door, 12" wide		22	.727		185	30		215	251
5040	18" wide		20.90	.766		229	32		261	300
5060	24" wide		20.30	.788		267	33		300	345
5300	Two doors, 27" wide		19.80	.808		294	33.50		327.50	375
5340	36" wide		18.80	.851		350	35.50		385.50	440
5380	48" wide		18.40	.870		430	36		466	530
6000	Corner wall, 30" high, 24" wide		18	.889		233	37		270	315
6050	30" wide		17.20	.930		264	38.50		302.50	350
6100	36" wide		16.50	.970		295	40.50		335.50	385
6500	Revolving Lazy Susan		15.20	1.053		415	43.50		458.50	525
7000	Broom cabinet, 84" high, 24" deep, 18" wide		10	1.600		575	66.50		641.50	735
7500	Oven cabinets, 84" high, 24" deep, 27" wide		8	2		860	83		943	1,075
7750	Valance board trim		396	.040	L.F.	12.40	1.68		14.08	16.25
9000	For deluxe models of all cabinets, add					40%				
9500	For custom built in place, add					25%	10%			
9550	Rule of thumb, kitchen cabinets not including									
9560	appliances & counter top, minimum	2 Carp	30	.533	L.F.	146	22		168	195
9600	Maximum	"	25	.640	"	310	26.50		336.50	385
9610	For metal cabinets, see Div. 12 35 70.13									

12 32 23.15 Manufactured Wood Casework Frames		Crew	Daily Output	Labor-Hours	Unit	Material	Labor	Equipment	Total	Total Incl O&P
0010	**MANUFACTURED WOOD CASEWORK FRAMES**									
0050	Base cabinets, counter storage, 36" high									
0100	One bay, 18" wide	1 Carp	2.70	2.963	Ea.	134	123		257	335
0400	Two bay, 36" wide		2.20	3.636		204	151		355	455
1100	Three bay, 54" wide		1.50	5.333		243	222		465	605
2800	Bookcases, one bay, 7' high, 18" wide		2.40	3.333		157	139		296	385
3500	Two bay, 36" wide		1.60	5		228	208		436	570
4100	Three bay, 54" wide		1.20	6.667		375	277		652	840
5100	Coat racks, one bay, 7' high, 24" wide		4.50	1.778		157	74		231	287
5300	Two bay, 48" wide		2.75	2.909		219	121		340	425
5800	Three bay, 72" wide		2.10	3.810		320	158		478	600
6100	Wall mounted cabinet, one bay, 24" high, 18" wide		3.60	2.222		86.50	92.50		179	237
6800	Two bay, 36" wide		2.20	3.636		126	151		277	370
7400	Three bay, 54" wide		1.70	4.706		157	196		353	475
8400	30" high, one bay, 18" wide		3.60	2.222		94	92.50		186.50	245
9000	Two bay, 36" wide		2.15	3.721		125	155		280	375
9400	Three bay, 54" wide		1.60	5		156	208		364	490
9800	Wardrobe, 7' high, single, 24" wide		2.70	2.963		173	123		296	380
9880	Partition & adjustable shelves, 48" wide		1.70	4.706		220	196		416	540
9950	Partition, adjustable shelves & drawers, 48" wide		1.40	5.714		330	237		567	730

12 32 23.20 Manufactured Hardwood Casework Doors		Crew	Daily Output	Labor-Hours	Unit	Material	Labor	Equipment	Total	Total Incl O&P
0010	**MANUFACTURED HARDWOOD CASEWORK DOORS**									
2000	Glass panel, hardwood frame									
2200	12" wide, 18" high	1 Carp	34	.235	Ea.	18.75	9.80		28.55	35.50
2600	30" high		32	.250		31.50	10.40		41.90	50.50
4450	18" wide, 18" high		32	.250		28	10.40		38.40	47

12 32 Manufactured Wood Casework

12 32 23 – Hardwood Casework

12 32 23.20 Manufactured Hardwood Casework Doors		Crew	Daily Output	Labor-Hours	Unit	2010 Bare Costs Material	Labor	Equipment	Total	Total Incl O&P
4550	30" high	1 Carp	29	.276	Ea.	47	11.45		58.45	69
5000	Hardwood, raised panel									
5100	12" wide, 18" high	1 Carp	16	.500	Ea.	24	21		45	58.50
5200	30" high		15	.533		40	22		62	78
5500	18" wide, 18" high		15	.533		36	22		58	73.50
5600	30" high	↓	14	.571	↓	60	23.50		83.50	103

12 32 23.25 Manufactured Wood Casework Drawer Fronts		Crew	Daily Output	Labor-Hours	Unit	Material	Labor	Equipment	Total	Total Incl O&P
0010	**MANUFACTURED WOOD CASEWORK DRAWER FRONTS**									
0100	Solid hardwood front									
1000	4" high, 12" wide	1 Carp	17	.471	Ea.	2.83	19.55		22.38	33
1200	18" wide	"	16	.500	"	4.25	21		25.25	36.50

12 32 23.30 Manufactured Wood Casework Vanities		Crew	Daily Output	Labor-Hours	Unit	Material	Labor	Equipment	Total	Total Incl O&P
0010	**MANUFACTURED WOOD CASEWORK VANITIES**									
8000	Vanity bases, 2 doors, 30" high, 21" deep, 24" wide	2 Carp	20	.800	Ea.	227	33		260	300
8050	30" wide		16	1		262	41.50		303.50	350
8100	36" wide		13.33	1.200		350	50		400	460
8150	48" wide	↓	11.43	1.400		415	58		473	550
9000	For deluxe models of all vanities, add to above					40%				
9500	For custom built in place, add to above				↓	25%	10%			

12 32 23.35 Manufactured Wood Casework Hardware		Crew	Daily Output	Labor-Hours	Unit	Material	Labor	Equipment	Total	Total Incl O&P
0010	**MANUFACTURED WOOD CASEWORK HARDWARE**									
1000	Catches, minimum	1 Carp	235	.034	Ea.	1	1.41		2.41	3.28
1040	Maximum	"	80	.100	"	6.25	4.16		10.41	13.25
2000	Door/drawer pulls, handles									
2200	Handles and pulls, projecting, metal, minimum	1 Carp	160	.050	Ea.	4.52	2.08		6.60	8.15
2240	Maximum		68	.118		9.55	4.89		14.44	18.10
2300	Wood, minimum		160	.050		4.76	2.08		6.84	8.45
2340	Maximum		68	.118		8.75	4.89		13.64	17.15
2600	Flush, metal, minimum		160	.050		4.76	2.08		6.84	8.45
2640	Maximum		68	.118	↓	8.75	4.89		13.64	17.15
3000	Drawer tracks/glides, minimum		48	.167	Pr.	8.05	6.95		15	19.60
3040	Maximum		24	.333		23.50	13.85		37.35	47.50
4000	Cabinet hinges, minimum		160	.050		2.74	2.08		4.82	6.20
4040	Maximum	↓	68	.118	↓	10.35	4.89		15.24	18.95

12 35 Specialty Casework

12 35 50 – Educational/Library Casework

12 35 50.13 Educational Casework		Crew	Daily Output	Labor-Hours	Unit	Material	Labor	Equipment	Total	Total Incl O&P
0010	**EDUCATIONAL CASEWORK**									
5000	School, 24" deep, metal, 84" high units	2 Carp	15	1.067	L.F.	385	44.50		429.50	495
5150	Counter height units		20	.800		258	33		291	335
5450	Wood, custom fabricated, 32" high counter		20	.800		215	33		248	287
5600	Add for counter top		56	.286		23	11.85		34.85	43.50
5800	84" high wall units	↓	15	1.067	↓	420	44.50		464.50	530
6000	Laminated plastic finish is same price as wood									

12 35 53 – Laboratory Casework

12 35 53.13 Metal Laboratory Casework		Crew	Daily Output	Labor-Hours	Unit	Material	Labor	Equipment	Total	Total Incl O&P
0010	**METAL LABORATORY CASEWORK**									
0020	Cabinets, base, door units, metal	2 Carp	18	.889	L.F.	192	37		229	268
0300	Drawer units	↓	18	.889	↓	430	37		467	525

12 35 Specialty Casework

12 35 53 – Laboratory Casework

12 35 53.13 Metal Laboratory Casework		Crew	Daily Output	Labor-Hours	Unit	2010 Bare Costs Material	Labor	Equipment	Total	Total Incl O&P
0700	Tall storage cabinets, open, 7′ high	2 Carp	20	.800	L.F.	410	33		443	500
0900	With glazed doors		20	.800		490	33		523	590
1300	Wall cabinets, metal, 12-1/2″ deep, open		20	.800		135	33		168	200
1500	With doors		20	.800		281	33		314	360
6300	Rule of thumb: lab furniture including installation & connection									
6320	High school				S.F.				33.50	37
6340	College								49.50	54.60
6360	Clinical, health care								42.75	46.85
6380	Industrial								69	75.70

12 35 59 – Display Casework

12 35 59.10 Display Cases		Crew	Daily Output	Labor-Hours	Unit	Material	Labor	Equipment	Total	Total Incl O&P
0010	**DISPLAY CASES** Free standing, all glass									
0020	Aluminum frame, 42″ high x 36″ x 12″ deep	2 Carp	8	2	Ea.	1,250	83		1,333	1,475
0100	70″ high x 48″ x 18″ deep	″	6	2.667		3,075	111		3,186	3,550
0500	For wood bases, add					9%				
0600	For hardwood frames, deduct					8%				
0700	For bronze, baked enamel finish, add					10%				
2000	Wall mounted, glass front, aluminum frame									
2010	Non-illuminated, one section 3′ x 4′ x 1′-4″	2 Carp	5	3.200	Ea.	2,150	133		2,283	2,550
2100	5′ x 4′ x 1′-4″		5	3.200		2,475	133		2,608	2,925
2200	6′ x 4′ x 1′-4″		4	4		2,975	166		3,141	3,525
2500	Two sections, 8′ x 4′ x 1′-4″		2	8		1,900	330		2,230	2,600
2600	10′ x 4′ x 1′-4″		2	8		2,250	330		2,580	2,975
3000	Three sections, 16′ x 4′ x 1′-4″		1.50	10.667		3,900	445		4,345	4,975
3500	For fluorescent lights, add				Section	330			330	365
4000	Table exhibit cases, 2′ wide, 3′ high, 4′ long, flat top	2 Carp	5	3.200	Ea.	1,150	133		1,283	1,475
4100	3′ wide, 3′ high, 4′ long, sloping top	″	3	5.333	″	1,850	222		2,072	2,375

12 35 70 – Healthcare Casework

12 35 70.13 Hospital Casework		Crew	Daily Output	Labor-Hours	Unit	Material	Labor	Equipment	Total	Total Incl O&P
0010	**HOSPITAL CASEWORK**									
0500	Base cabinets, laminated plastic	2 Carp	10	1.600	L.F.	246	66.50		312.50	375
1000	Stainless steel	″	10	1.600		455	66.50		521.50	600
1200	For all drawers, add					26			26	28.50
1300	Cabinet base trim, 4″ high, enameled steel	2 Carp	200	.080		41.50	3.32		44.82	50.50
1400	Stainless steel		200	.080		82.50	3.32		85.82	96
1450	Countertop, laminated plastic, no backsplash		40	.400		42.50	16.60		59.10	72.50
1650	With backsplash		40	.400		53	16.60		69.60	83.50
1800	For sink cutout, add		12.20	1.311	Ea.		54.50		54.50	84
1900	Stainless steel counter top		40	.400	L.F.	138	16.60		154.60	177
2000	For drop-in stainless 43″ x 21″ sink, add				Ea.	905			905	995
2500	Wall cabinets, laminated plastic	2 Carp	15	1.067	L.F.	184	44.50		228.50	272
2600	Enameled steel		15	1.067		227	44.50		271.50	320
2700	Stainless steel		15	1.067		450	44.50		494.50	565
2800	For glass doors, add					31			31	34
12 35 70.16 Nurse Station Casework										
0010	**NURSE STATION CASEWORK**									
2100	Door type, laminated plastic	2 Carp	10	1.600	L.F.	286	66.50		352.50	415
2200	Enameled steel		10	1.600		273	66.50		339.50	400
2300	Stainless steel		10	1.600		545	66.50		611.50	700
2400	For drawer type, add					232			232	255

12 35 Specialty Casework

12 35 80 – Commercial Kitchen Casework

12 35 80.13 Metal Kitchen Casework		Crew	Daily Output	Labor-Hours	Unit	Material	2010 Bare Costs Labor	Equipment	Total	Total Incl O&P
0010	**METAL KITCHEN CASEWORK**									
3500	Base cabinets, metal, minimum	2 Carp	30	.533	L.F.	66.50	22		88.50	107
3600	Maximum		25	.640		169	26.50		195.50	227
3700	Wall cabinets, metal, minimum		30	.533		66.50	22		88.50	107
3800	Maximum	↓	25	.640	↓	153	26.50		179.50	209

12 36 Countertops

12 36 16 – Metal Countertops

12 36 16.10 Stainless Steel Countertops

		Crew	Daily Output	Labor-Hours	Unit	Material	Labor	Equipment	Total	Total Incl O&P
0010	**STAINLESS STEEL COUNTERTOPS**									
3200	Stainless steel, custom	1 Carp	24	.333	S.F.	143	13.85		156.85	179

12 36 19 – Wood Countertops

12 36 19.10 Maple Countertops

		Crew	Daily Output	Labor-Hours	Unit	Material	Labor	Equipment	Total	Total Incl O&P
0010	**MAPLE COUNTERTOPS**									
2900	Solid, laminated, 1-1/2" thick, no splash	1 Carp	28	.286	L.F.	62.50	11.85		74.35	87
3000	With square splash		28	.286	"	74	11.85		85.85	100
3400	Recessed cutting block with trim, 16" x 20" x 1"	↓	8	1	Ea.	72.50	41.50		114	144

12 36 23 – Plastic Countertops

12 36 23.13 Plastic-Laminate-Clad Countertops

		Crew	Daily Output	Labor-Hours	Unit	Material	Labor	Equipment	Total	Total Incl O&P
0010	**PLASTIC-LAMINATE-CLAD COUNTERTOPS**									
0020	Stock, 24" wide w/backsplash, minimum	1 Carp	30	.267	L.F.	9.10	11.10		20.20	27
0100	Maximum		25	.320		19.50	13.30		32.80	42
0300	Custom plastic, 7/8" thick, aluminum molding, no splash		30	.267		21	11.10		32.10	40
0400	Cove splash		30	.267		27.50	11.10		38.60	47
0600	1-1/4" thick, no splash		28	.286		31	11.85		42.85	52.50
0700	Square splash		28	.286		30	11.85		41.85	51.50
0900	Square edge, plastic face, 7/8" thick, no splash		30	.267		26	11.10		37.10	46
1000	With splash	↓	30	.267		33.50	11.10		44.60	53.50
1200	For stainless channel edge, 7/8" thick, add					2.75			2.75	3.03
1300	1-1/4" thick, add					3.22			3.22	3.54
1500	For solid color suede finish, add				↓	2.68			2.68	2.95
1700	For end splash, add				Ea.	17.85			17.85	19.65
1900	For cut outs, standard, add, minimum	1 Carp	32	.250		3.57	10.40		13.97	19.95
2000	Maximum		8	1	↓	5.95	41.50		47.45	70.50
2100	Postformed, including backsplash and front edge		30	.267	L.F.	10.70	11.10		21.80	29
2110	Mitred, add		12	.667	Ea.		27.50		27.50	42.50
2200	Built-in place, 25" wide, plastic laminate	↓	25	.320	L.F.	14.30	13.30		27.60	36

12 36 33 – Tile Countertops

12 36 33.10 Ceramic Tile Countertops

		Crew	Daily Output	Labor-Hours	Unit	Material	Labor	Equipment	Total	Total Incl O&P
0010	**CERAMIC TILE COUNTERTOPS**									
2300	Ceramic tile mosaic	1 Carp	25	.320	L.F.	31	13.30		44.30	54.50

12 36 40 – Stone Countertops

12 36 40.10 Natural Stone Countertops

		Crew	Daily Output	Labor-Hours	Unit	Material	Labor	Equipment	Total	Total Incl O&P
0010	**NATURAL STONE COUNTERTOPS**									
2500	Marble, stock, with splash, 1/2" thick, minimum	1 Bric	17	.471	L.F.	38	19.65		57.65	71.50
2700	3/4" thick, maximum		13	.615		95.50	25.50		121	144
2800	Granite, average, 1-1/4" thick, 24" wide, no splash	↓	13.01	.615	↓	131	25.50		156.50	183

12 36 53 – Laboratory Countertops

	12 36 53.10 Laboratory Countertops and Sinks	Crew	Daily Output	Labor-Hours	Unit	Material	Labor	Equipment	Total	Total Incl O&P
						2010 Bare Costs				
0010	**LABORATORY COUNTERTOPS AND SINKS**									
0020	Countertops, not incl. base cabinets, acid-proof, minimum	2 Carp	82	.195	S.F.	33	8.10		41.10	48.50
0030	Maximum		70	.229		42.50	9.50		52	61
0040	Stainless steel		82	.195		98	8.10		106.10	121

12 36 61 – Simulated Stone Countertops

	12 36 61.16 Solid Surface Countertops	Crew	Daily Output	Labor-Hours	Unit	Material	Labor	Equipment	Total	Total Incl O&P
0010	**SOLID SURFACE COUNTERTOPS**, Acrylic polymer									
0020	Pricing for orders of 100 L.F. or greater									
0100	25″ wide, solid colors	2 Carp	28	.571	L.F.	44.50	23.50		68	85.50
0200	Patterned colors		28	.571		56.50	23.50		80	98.50
0300	Premium patterned colors		28	.571		70.50	23.50		94	115
0400	With silicone attached 4″ backsplash, solid colors		27	.593		49	24.50		73.50	92
0500	Patterned colors		27	.593		62	24.50		86.50	106
0600	Premium patterned colors		27	.593		77	24.50		101.50	123
0700	With hard seam attached 4″ backsplash, solid colors		23	.696		49	29		78	98.50
0800	Patterned colors		23	.696		62	29		91	113
0900	Premium patterned colors		23	.696		77	29		106	130
1000	Pricing for order of 51 – 99 L.F.									
1100	25″ wide, solid colors	2 Carp	24	.667	L.F.	51.50	27.50		79	99
1200	Patterned colors		24	.667		65	27.50		92.50	114
1300	Premium patterned colors		24	.667		81.50	27.50		109	132
1400	With silicone attached 4″ backsplash, solid colors		23	.696		56.50	29		85.50	107
1500	Patterned colors		23	.696		71.50	29		100.50	123
1600	Premium patterned colors		23	.696		89	29		118	142
1700	With hard seam attached 4″ backsplash, solid colors		20	.800		56.50	33		89.50	113
1800	Patterned colors		20	.800		71.50	33		104.50	130
1900	Premium patterned colors		20	.800		89	33		122	149
2000	Pricing for order of 1 – 50 L.F.									
2100	25″ wide, solid colors	2 Carp	20	.800	L.F.	60	33		93	117
2200	Patterned colors		20	.800		76.50	33		109.50	135
2300	Premium patterned colors		20	.800		95.50	33		128.50	156
2400	With silicone attached 4″ backsplash, solid colors		19	.842		66	35		101	127
2500	Patterned colors		19	.842		83.50	35		118.50	146
2600	Premium patterned colors		19	.842		104	35		139	169
2700	With hard seam attached 4″ backsplash, solid colors		15	1.067		66	44.50		110.50	141
2800	Patterned colors		15	1.067		83.50	44.50		128	161
2900	Premium patterned colors		15	1.067		104	44.50		148.50	184
3000	Sinks, pricing for order of 100 or greater units									
3100	Single bowl, hard seamed, solid colors, 13″ x 17″	1 Carp	3	2.667	Ea.	300	111		411	500
3200	10″ x 15″		7	1.143		139	47.50		186.50	226
3300	Cutouts for sinks		8	1			41.50		41.50	64
3400	Sinks, pricing for order of 51 – 99 units									
3500	Single bowl, hard seamed, solid colors, 13″ x 17″	1 Carp	2.55	3.137	Ea.	345	130		475	580
3600	10″ x 15″		6	1.333		160	55.50		215.50	262
3700	Cutouts for sinks		7	1.143			47.50		47.50	73
3800	Sinks, pricing for order of 1 – 50 units									
3900	Single bowl, hard seamed, solid colors, 13″ x 17″	1 Carp	2	4	Ea.	405	166		571	700
4000	10″ x 15″		4.55	1.758		188	73		261	320
4100	Cutouts for sinks		5.25	1.524			63.50		63.50	97.50
4200	Cooktop cutouts, pricing for 100 or greater units		4	2		22	83		105	153
4300	51 – 99 units		3.40	2.353		25.50	98		123.50	179
4400	1 – 50 units		3	2.667		30	111		141	204

12 36 Countertops

12 36 61 – Simulated Stone Countertops

12 36 61.17 Solid Surface Vanity Tops		Crew	Daily Output	Labor-Hours	Unit	2010 Bare Costs Material	Labor	Equipment	Total	Total Incl O&P
0010	**SOLID SURFACE VANITY TOPS**									
0015	Solid surface, center bowl, 17" x 19"	1 Carp	12	.667	Ea.	189	27.50		216.50	251
0020	19" x 25"		12	.667		229	27.50		256.50	295
0030	19" x 31"		12	.667		278	27.50		305.50	350
0040	19" x 37"		12	.667		325	27.50		352.50	400
0050	22" x 25"		10	.800		202	33		235	273
0060	22" x 31"		10	.800		236	33		269	310
0070	22" x 37"		10	.800		274	33		307	350
0080	22" x 43"		10	.800		315	33		348	395
0090	22" x 49"		10	.800		345	33		378	430
0110	22" x 55"		8	1		395	41.50		436.50	500
0120	22" x 61"		8	1		450	41.50		491.50	560
0220	Double bowl, 22" x 61"		8	1		505	41.50		546.50	625
0230	Double bowl, 22" x 73"	↓	8	1	↓	705	41.50		746.50	840
0240	For aggregate colors, add					35%				
0250	For faucets and fittings, see Div. 22 41 39.10									

12 36 61.19 Engineered Stone Countertops		Crew	Daily Output	Labor-Hours	Unit	Material	Labor	Equipment	Total	Total Incl O&P
0010	**ENGINEERED STONE COUNTERTOPS**									
0100	25" wide, 4" backsplash, color group A, minimum	2 Carp	15	1.067	L.F.	17	44.50		61.50	87
0110	Maximum		15	1.067		42	44.50		86.50	115
0120	Color group B, minimum		15	1.067		22	44.50		66.50	92.50
0130	Maximum		15	1.067		49	44.50		93.50	123
0140	Color group C, minimum		15	1.067		29.50	44.50		74	101
0150	Maximum		15	1.067		60	44.50		104.50	135
0160	Color group D, minimum		15	1.067		37	44.50		81.50	109
0170	Maximum	↓	15	1.067	↓	70	44.50		114.50	146

12 46 Furnishing Accessories

12 46 13 – Ash Receptacles

12 46 13.10 Ash/Trash Receivers		Crew	Daily Output	Labor-Hours	Unit	Material	Labor	Equipment	Total	Total Incl O&P
0010	**ASH/TRASH RECEIVERS**									
1000	Ash urn, cylindrical metal									
1020	8" diameter, 20" high	1 Clab	60	.133	Ea.	85	4.41		89.41	100
1060	10" diameter, 26" high	"	60	.133	"	100	4.41		104.41	117
2000	Combination ash/trash urn, metal									
2020	8" diameter, 20" high	1 Clab	60	.133	Ea.	85	4.41		89.41	100
2050	10" diameter, 26" high	"	60	.133	"	105	4.41		109.41	122

12 46 19 – Clocks

12 46 19.50 Wall Clocks		Crew	Daily Output	Labor-Hours	Unit	Material	Labor	Equipment	Total	Total Incl O&P
0010	**WALL CLOCKS**									
0080	12" diameter, single face	1 Elec	8	1	Ea.	139	49		188	225
0100	Double face	"	6.20	1.290	"	350	63		413	480

12 46 33 – Waste Receptacles

12 46 33.13 Trash Receptacles		Crew	Daily Output	Labor-Hours	Unit	Material	Labor	Equipment	Total	Total Incl O&P
0010	**TRASH RECEPTACLES**									
4000	Trash receptacle, metal									
4020	8" diameter, 15" high	1 Clab	60	.133	Ea.	67	4.41		71.41	80.50
4040	10" diameter, 18" high		60	.133		135	4.41		139.41	155
5040	16" x 8" x 14" high	↓	60	.133	↓	21	4.41		25.41	30

12 46 Furnishing Accessories

12 46 33 – Waste Receptacles

12 46 33.13 Trash Receptacles		Crew	Daily Output	Labor-Hours	Unit	Material	Labor	Equipment	Total	Total Incl O&P
						2010 Bare Costs				
5500	Plastic, with lid									
5520	35 gallon	1 Clab	60	.133	Ea.	135	4.41		139.41	155
5540	45 gallon		60	.133		220	4.41		224.41	249
5550	Plastic recycling barrel, w/lid & wheels, 32 gal. G		60	.133		75	4.41		79.41	89.50
5560	65 gal. G		60	.133		135	4.41		139.41	155
5570	95 gal. G		60	.133		220	4.41		224.41	249

12 48 Rugs and Mats

12 48 13 – Entrance Floor Mats and Frames

12 48 13.13 Entrance Floor Mats		Crew	Daily Output	Labor-Hours	Unit	Material	Labor	Equipment	Total	Total Incl O&P
0010	**ENTRANCE FLOOR MATS**									
0020	Recessed, in-laid black rubber, 3/8" thick, solid CN	1 Clab	155	.052	S.F.	19.85	1.71		21.56	24.50
0050	Perforated		155	.052		13.80	1.71		15.51	17.85
0100	1/2" thick, solid		155	.052		16.60	1.71		18.31	21
0150	Perforated		155	.052		22	1.71		23.71	26.50
0200	In colors, 3/8" thick, solid		155	.052		17.85	1.71		19.56	22.50
0250	Perforated		155	.052		18.45	1.71		20.16	23
0300	1/2" thick, solid		155	.052		23	1.71		24.71	28
0350	Perforated		155	.052		23.50	1.71		25.21	28.50
2000	Recycled rubber tire tile, 12" x 12" x 3/8" thick G		125	.064		8.50	2.12		10.62	12.60
2510	Natural cocoa fiber, 1/2" thick G		125	.064		8.10	2.12		10.22	12.20
2520	3/4" thick G		125	.064		6.25	2.12		8.37	10.10
2530	1" thick G		125	.064		6.90	2.12		9.02	10.85

12 51 Office Furniture

12 51 16 – Case Goods

12 51 16.13 Metal Case Goods		Crew	Daily Output	Labor-Hours	Unit	Material	Labor	Equipment	Total	Total Incl O&P
0010	**METAL CASE GOODS**									
0020	Desks, 29" high, double pedestal, 30" x 60", metal, minimum				Ea.	500			500	550
0030	Maximum					1,200			1,200	1,325
0600	Desks, single pedestal, 30" x 60", metal, minimum					490			490	535
0620	Maximum					970			970	1,075
0720	Desks, secretarial, 30" x 60", metal, minimum					430			430	470
0730	Maximum					730			730	800
0740	Return, 20" x 42", minimum					315			315	350
0750	Maximum					480			480	525
0940	59" x 12" x 23" high, steel, minimum					265			265	292
0960	Maximum					340			340	370

12 51 16.16 Wood Case Goods		Crew	Daily Output	Labor-Hours	Unit	Material	Labor	Equipment	Total	Total Incl O&P
0010	**WOOD CASE GOODS**									
0150	Desk, 29" high, double pedestal, 30" x 60"									
0160	Wood, minimum				Ea.	330			330	365
0180	Maximum				"	2,325			2,325	2,550
0630	Single pedestal, 30" x 60"									
0640	Wood, minimum				Ea.	560			560	615
0650	Maximum					730			730	800
0670	Executive return, 24" x 42", with box, file, wood, minimum					380			380	420
0680	Maximum					730			730	800
0790	Desk, 29" high, secretarial, 30" x 60"									

12 51 Office Furniture

12 51 16 – Case Goods

12 51 16.16 Wood Case Goods		Crew	Daily Output	Labor-Hours	Unit	2010 Bare Costs Material	Labor	Equipment	Total	Total Incl O&P
0800	Wood, minimum				Ea.	455			455	500
0810	Maximum					2,350			2,350	2,600
0820	Return, 20" x 42", minimum					275			275	300
0830	Maximum					955			955	1,050
0900	Desktop organizer, 72" x 14" x 36" high, wood, minimum					164			164	181
0920	Maximum				↓	380			380	420
1110	Furniture, credenza, 29" high, 18" to 22" x 60" to 72"									
1120	Wood, minimum				Ea.	615			615	675
1140	Maximum				"	2,925			2,925	3,200

12 51 23 – Office Tables

12 51 23.33 Conference Tables		Crew	Daily Output	Labor-Hours	Unit	Material	Labor	Equipment	Total	Total Incl O&P
0010	**CONFERENCE TABLES**									
6050	Boat, 96" x 42", minimum				Ea.	775			775	855
6150	Maximum					3,075			3,075	3,375
6720	Rectangle, 96" x 42", minimum					1,075			1,075	1,175
6740	Maximum				↓	3,075			3,075	3,375

12 52 Seating

12 52 23 – Office Seating

12 52 23.13 Office Chairs		Crew	Daily Output	Labor-Hours	Unit	Material	Labor	Equipment	Total	Total Incl O&P
0010	**OFFICE CHAIRS**									
2000	Standard office chair, executive, minimum				Ea.	330			330	365
2150	Maximum					1,800			1,800	2,000
2200	Management, minimum					197			197	216
2250	Maximum					2,300			2,300	2,525
2280	Task, minimum					130			130	143
2290	Maximum					575			575	630
2300	Arm kit, minimum					58			58	64
2320	Maximum				↓	87			87	95.50

12 54 Hospitality Furniture

12 54 13 – Hotel and Motel Furniture

12 54 13.10 Hotel Furniture		Crew	Daily Output	Labor-Hours	Unit	Material	Labor	Equipment	Total	Total Incl O&P
0010	**HOTEL FURNITURE**									
0020	Standard quality set, minimum				Room	2,325			2,325	2,550
0200	Maximum				"	8,300			8,300	9,125

12 54 16 – Restaurant Furniture

12 54 16.10 Tables, Folding		Crew	Daily Output	Labor-Hours	Unit	Material	Labor	Equipment	Total	Total Incl O&P
0010	**TABLES, FOLDING** Laminated plastic tops									
1000	Tubular steel legs with glides									
1020	18" x 60", minimum				Ea.	226			226	249
1040	Maximum					1,300			1,300	1,425
1840	36" x 96", minimum					275			275	305
1860	Maximum					2,600			2,600	2,875
2000	Round, wood stained, plywood top, 60" diameter, minimum					385			385	420
2020	Maximum				↓	1,400			1,400	1,525

12 54 Hospitality Furniture

12 54 16 – Restaurant Furniture

12 54 16.20 Furniture, Restaurant		Crew	Daily Output	Labor-Hours	Unit	Material	2010 Bare Costs Labor	Equipment	Total	Total Incl O&P
0010	**FURNITURE, RESTAURANT**									
0020	Bars, built-in, front bar	1 Carp	5	1.600	L.F.	250	66.50		316.50	375
0200	Back bar	"	5	1.600	"	182	66.50		248.50	300
0300	Booth seating, see Div. 12 54 16.70									
2000	Chair, bentwood side chair, metal, minimum				Ea.	83.50			83.50	92
2020	Maximum					98			98	108
2600	Upholstered seat & back, arms, minimum					136			136	150
2620	Maximum				↓	385			385	425

12 54 16.70 Booths		Crew	Daily Output	Labor-Hours	Unit	Material	Labor	Equipment	Total	Total Incl O&P
0010	**BOOTHS**									
1000	Banquet, upholstered seat and back, custom									
1500	Straight, minimum	2 Carp	40	.400	L.F.	169	16.60		185.60	212
1520	Maximum		36	.444		330	18.45		348.45	390
1600	"L" or "U" shape, minimum		35	.457		172	19		191	220
1620	Maximum	↓	30	.533	↓	305	22		327	370
1800	Upholstered outside finished backs for									
1810	single booths and custom banquets									
1820	Minimum	2 Carp	44	.364	L.F.	19.50	15.10		34.60	45
1840	Maximum	"	40	.400	"	58.50	16.60		75.10	90
3000	Fixed seating, one piece plastic chair and									
3010	plastic laminate table top									
3100	Two seat, 24" x 24" table, minimum	F-7	30	1.067	Ea.	680	40		720	810
3120	Maximum		26	1.231		970	46		1,016	1,150
3200	Four seat, 24" x 48" table, minimum		28	1.143		675	42.50		717.50	810
3220	Maximum	↓	24	1.333	↓	1,150	50		1,200	1,350
5000	Mount in floor, wood fiber core with									
5010	plastic laminate face, single booth									
5050	24" wide	F-7	30	1.067	Ea.	261	40		301	350
5100	48" wide	"	28	1.143	"	330	42.50		372.50	430

12 55 Detention Furniture

12 55 13 – Detention Bunks

12 55 13.13 Cots		Crew	Daily Output	Labor-Hours	Unit	Material	Labor	Equipment	Total	Total Incl O&P
0010	**COTS**									
2500	Bolted, single, painted steel	E-4	20	1.600	Ea.	330	76	7.30	413.30	505
2700	Stainless steel	"	20	1.600	"	960	76	7.30	1,043.30	1,200

12 56 Institutional Furniture

12 56 33 – Classroom Furniture

12 56 33.10 Furniture, School		Crew	Daily Output	Labor-Hours	Unit	Material	Labor	Equipment	Total	Total Incl O&P
0010	**FURNITURE, SCHOOL**									
0500	Classroom, movable chair & desk type, minimum				Set				77.25	85
0600	Maximum				"				141	155
1000	Chair, molded plastic									
1100	Integral tablet arm, minimum				Ea.	74			74	81.50
1150	Maximum					151			151	166
2000	Desk, single pedestal, top book compartment, minimum					56			56	61.50
2020	Maximum				↓	69			69	76

12 56 Institutional Furniture

12 56 33 – Classroom Furniture

12 56 33.10 Furniture, School		Crew	Daily Output	Labor-Hours	Unit	Material	2010 Bare Costs Labor	Equipment	Total	Total Incl O&P
2200	Flip top, minimum				Ea.	95.50			95.50	105
2220	Maximum				↓	122			122	134

12 56 43 – Dormitory Furniture

12 56 43.10 Dormitory Furnishings		Crew	Daily Output	Labor-Hours	Unit	Material	Labor	Equipment	Total	Total Incl O&P
0010	**DORMITORY FURNISHINGS**									
0300	Bunkable bed, twin, minimum				Ea.	300			300	330
0320	Maximum					685			685	750
1000	Chest, four drawer, minimum					440			440	485
1020	Maximum				↓	580			580	635
1050	Built-in, minimum	2 Carp	13	1.231	L.F.	145	51		196	239
1150	Maximum		10	1.600		268	66.50		334.50	395
1200	Desk top, built-in, laminated plastic, 24" deep, minimum		50	.320		39	13.30		52.30	63
1300	Maximum		40	.400		116	16.60		132.60	154
1450	30" deep, minimum		50	.320		49.50	13.30		62.80	75
1550	Maximum		40	.400		217	16.60		233.60	265
1750	Dressing unit, built-in, minimum		12	1.333		218	55.50		273.50	325
1850	Maximum	↓	8	2	↓	655	83		738	850
8000	Rule of thumb: total cost for furniture, minimum				Student				2,225	2,450
8050	Maximum				"				4,300	4,675

12 56 51 – Library Furniture

12 56 51.10 Library Furnishings		Crew	Daily Output	Labor-Hours	Unit	Material	Labor	Equipment	Total	Total Incl O&P
0010	**LIBRARY FURNISHINGS**									
0100	Attendant desk, 36" x 62" x 29" high	1 Carp	16	.500	Ea.	2,700	21		2,721	3,000
0200	Book display, "A" frame display, both sides, 42" x 42" x 60" high		16	.500		1,500	21		1,521	1,675
0220	Table with bulletin board, 42" x 24" x 49" high		16	.500		1,325	21		1,346	1,475
0800	Card catalogue, 30 tray unit		16	.500		2,850	21		2,871	3,150
0840	60 tray unit	↓	16	.500		5,225	21		5,246	5,775
0880	72 tray unit	2 Carp	16	1		6,400	41.50		6,441.50	7,125
1000	Carrels, single face, initial unit	1 Carp	16	.500		720	21		741	825
1500	Double face, initial unit	2 Carp	16	1		1,150	41.50		1,191.50	1,325
1710	Carrels, hardwood, 36" x 24", minimum	1 Carp	5	1.600		460	66.50		526.50	605
1720	Maximum	"	4	2		930	83		1,013	1,150
2700	Card catalog file, 60 trays, complete					5,050			5,050	5,550
2720	Alternate method: each tray				↓	84			84	92.50
3800	Charging desk, built-in, with counter, plastic laminated top	1 Carp	7	1.143	L.F.	410	47.50		457.50	530
4000	Dictionary stand, stationary		16	.500	Ea.	880	21		901	1,000
4020	Revolving		16	.500		345	21		366	410
4200	Exhibit case, table style, 60" x 28" x 36"		11	.727	↓	2,800	30		2,830	3,150
6010	Bookshelf, metal, 90" high, 10" shelf, double face		11.50	.696	L.F.	118	29		147	175
6020	Single face	↓	12	.667	"	113	27.50		140.50	167
6050	For 8" shelving, subtract from above					10%				
6060	For 12" shelving, add to above					10%				
6070	For 42" high with countertop, subtract from above					20%				
6100	Mobile compacted shelving, hand crank, 9'-0" high									
6110	Double face, including track, 3' section				Ea.	1,125			1,125	1,250
6150	For electrical operation, add					25%				
6200	Magazine shelving, 82" high, 12" deep, single face	1 Carp	11.50	.696	L.F.	92.50	29		121.50	147
6210	Double face		11.50	.696	"	105	29		134	160
7000	Tables, card catalog reference, 24" x 60" x 42"	↓	16	.500	Ea.	975	21		996	1,100
7200	Reading table, laminated top, 60" x 36"				"	655			655	720

12 56 Institutional Furniture

12 56 70 – Healthcare Furniture

12 56 70.10 Furniture, Hospital		Crew	Daily Output	Labor-Hours	Unit	2010 Bare Costs Material	Labor	Equipment	Total	Total Incl O&P
0010	**FURNITURE, HOSPITAL**									
0020	Beds, manual, minimum				Ea.	755			755	830
0100	Maximum					2,575			2,575	2,825
0300	Manual and electric beds, minimum					1,050			1,050	1,150
0400	Maximum					2,975			2,975	3,275
0600	All electric hospital beds, minimum					1,325			1,325	1,450
0700	Maximum					4,300			4,300	4,750
0900	Manual, nursing home beds, minimum					850			850	935
1000	Maximum					1,775			1,775	1,950
1020	Overbed table, laminated top, minimum					360			360	400
1040	Maximum					900			900	990
1100	Patient wall systems, not incl. plumbing, minimum				Room	1,075			1,075	1,200
1200	Maximum				"	2,000			2,000	2,200
2000	Geriatric chairs, minimum				Ea.	420			420	465
2020	Maximum				"	720			720	790

12 61 Fixed Audience Seating

12 61 13 – Upholstered Audience Seating

12 61 13.13 Auditorium Chairs		Crew	Daily Output	Labor-Hours	Unit	Material	Labor	Equipment	Total	Total Incl O&P
0010	**AUDITORIUM CHAIRS**									
2000	All veneer construction	2 Carp	22	.727	Ea.	175	30		205	240
2200	Veneer back, padded seat		22	.727		221	30		251	290
2350	Fully upholstered, spring seat		22	.727		211	30		241	279
2450	For tablet arms, add					58			58	63.50
2500	For fire retardancy, CATB-133, add					25			25	27.50
12 61 13.23 Lecture Hall Seating										
0010	**LECTURE HALL SEATING**									
1000	Pedestal type, minimum	2 Carp	22	.727	Ea.	165	30		195	229
1200	Maximum	"	14.50	1.103	"	555	46		601	680

12 63 Stadium and Arena Seating

12 63 13 – Stadium and Arena Bench Seating

12 63 13.13 Bleachers		Crew	Daily Output	Labor-Hours	Unit	Material	Labor	Equipment	Total	Total Incl O&P
0010	**BLEACHERS**									
3000	Telescoping, manual to 15 tier, minimum	F-5	65	.492	Seat	78.50	20.50		99	118
3100	Maximum		60	.533		118	22.50		140.50	164
3300	16 to 20 tier, minimum		60	.533		188	22.50		210.50	242
3400	Maximum		55	.582		235	24.50		259.50	297
3600	21 to 30 tier, minimum		50	.640		196	27		223	257
3700	Maximum		40	.800		235	33.50		268.50	310
3900	For integral power operation, add, minimum	2 Elec	300	.053		39	2.61		41.61	47
4000	Maximum	"	250	.064		62.50	3.14		65.64	73.50
5000	Benches, folding, in wall, 14' table, 2 benches	L-4	2	12	Set	665	465		1,130	1,450

12 67 Pews and Benches

12 67 13 – Pews

12 67 13.13 Sanctuary Pews		Crew	Daily Output	Labor-Hours	Unit	Material	2010 Bare Costs Labor	Equipment	Total	Total Incl O&P
0010	**SANCTUARY PEWS**									
1500	Bench type, hardwood, minimum	1 Carp	20	.400	L.F.	74	16.60		90.60	107
1550	Maximum	"	15	.533		147	22		169	195
1570	For kneeler, add				↓	18.15			18.15	19.95

12 92 Interior Planters and Artificial Plants

12 92 33 – Interior Planters

12 92 33.10 Planters

		Crew	Daily Output	Labor-Hours	Unit	Material	Labor	Equipment	Total	Total Incl O&P
0010	**PLANTERS**									
1000	Fiberglass, hanging, 12" diameter, 7" high				Ea.	105			105	115
1500	Rectangular, 48" long, 16" high x 15" wide					690			690	760
1650	60" long, 30" high, 28" wide					1,050			1,050	1,150
2000	Round, 12" diameter, 13" high					119			119	131
2050	25" high					135			135	148
5000	Square, 10" side, 20" high					158			158	173
5100	14" side, 15" high					190			190	209
6000	Metal bowl, 32" diameter, 8" high, minimum					480			480	525
6050	Maximum				↓	620			620	680
8750	Wood, fiberglass liner, square									
8780	14" square, 15" high, minimum				Ea.	300			300	330
8800	Maximum					370			370	410
9400	Plastic cylinder, molded, 10" diameter, 10" high					12.25			12.25	13.45
9500	11" diameter, 11" high				↓	39			39	42.50

12 93 Site Furnishings

12 93 23 – Trash and Litter Receptors

12 93 23.10 Trash Receptacles

		Crew	Daily Output	Labor-Hours	Unit	Material	Labor	Equipment	Total	Total Incl O&P
0010	**TRASH RECEPTACLES**									
0020	Fiberglass, 2' square, 18" high	2 Clab	30	.533	Ea.	310	17.65		327.65	370
0100	2' square, 2'-6" high		30	.533		435	17.65		452.65	505
0300	Circular , 2' diameter, 18" high		30	.533		280	17.65		297.65	335
0400	2' diameter, 2'-6" high		30	.533		375	17.65		392.65	435
0500	Recycled plastic, var colors, round, 32 gal., 28" x 38" H G		5	3.200		291	106		397	485
0510	32 gal., 31" x 32" H G	↓	5	3.200	↓	360	106		466	560

12 93 23.20 Trash Closure

		Crew	Daily Output	Labor-Hours	Unit	Material	Labor	Equipment	Total	Total Incl O&P
0010	**TRASH CLOSURE**									
0020	Steel with pullover cover, 2'-3" wide, 4'-7" high, 6'-2" long	2 Clab	5	3.200	Ea.	1,150	106		1,256	1,425
0100	10'-1" long		4	4		1,525	132		1,657	1,875
0300	Wood, 10' wide, 6' high, 10' long	↓	1.20	13.333	↓	1,100	440		1,540	1,875

12 93 33 – Manufactured Planters

12 93 33.10 Planters

		Crew	Daily Output	Labor-Hours	Unit	Material	Labor	Equipment	Total	Total Incl O&P
0010	**PLANTERS**									
0012	Concrete, sandblasted, precast, 48" diameter, 24" high	2 Clab	15	1.067	Ea.	655	35.50		690.50	775
0100	Fluted, precast, 7' diameter, 36" high		10	1.600		1,100	53		1,153	1,300
0300	Fiberglass, circular, 36" diameter, 24" high		15	1.067		475	35.50		510.50	575
0320	36" diameter, 27" high		12	1.333		505	44		549	625
0330	33" high		15	1.067		515	35.50		550.50	620
0335	24" diameter, 36" high	↓	15	1.067	↓	430	35.50		465.50	530

12 93 Site Furnishings

12 93 33 – Manufactured Planters

12 93 33.10 Planters		Crew	Daily Output	Labor-Hours	Unit	Material	2010 Bare Costs Labor	Equipment	Total	Total Incl O&P
0340	60" diameter, 39" high	2 Clab	8	2	Ea.	1,125	66		1,191	1,350
0400	60" diameter, 24" high		10	1.600		800	53		853	960
0600	Square, 24" side, 36" high		15	1.067		460	35.50		495.50	560
0610	24" side, 27" high		12	1.333		305	44		349	405
0620	24" side, 16" high		20	.800		335	26.50		361.50	410
0700	48" side, 36" high		15	1.067		815	35.50		850.50	950
0900	Planter/bench, 72" square, 36" high		5	3.200		1,275	106		1,381	1,575
1000	96" square, 27" high		5	3.200		1,950	106		2,056	2,325
1200	Wood, square, 48" side, 24" high		15	1.067		1,000	35.50		1,035.50	1,150
1300	Circular, 48" diameter, 30" high		10	1.600		825	53		878	990
1500	72" diameter, 30" high		10	1.600		1,450	53		1,503	1,675
1600	Planter/bench, 72"	↓	5	3.200	↓	3,225	106		3,331	3,725

12 93 43 – Site Seating and Tables

12 93 43.13 Site Seating		Crew	Daily Output	Labor-Hours	Unit	Material	2010 Bare Costs Labor	Equipment	Total	Total Incl O&P
0010	**SITE SEATING**									
0012	Seating, benches, park, precast conc, w/backs, wood rails, 4' long	2 Clab	5	3.200	Ea.	420	106		526	630
0100	8' long		4	4		800	132		932	1,075
0300	Fiberglass, without back, one piece, 4' long		10	1.600		610	53		663	750
0400	8' long		7	2.286		1,225	75.50		1,300.50	1,475
0500	Steel barstock pedestals w/backs, 2" x 3" wood rails, 4' long		10	1.600		1,050	53		1,103	1,225
0510	8' long		7	2.286		1,250	75.50		1,325.50	1,500
0520	3" x 8" wood plank, 4' long		10	1.600		1,050	53		1,103	1,250
0530	8' long		7	2.286		1,225	75.50		1,300.50	1,475
0540	Backless, 4" x 4" wood plank, 4' square		10	1.600		1,025	53		1,078	1,200
0550	8' long		7	2.286		975	75.50		1,050.50	1,200
0600	Aluminum pedestals, with backs, aluminum slats, 8' long		8	2		335	66		401	470
0610	15' long		5	3.200		585	106		691	810
0620	Portable, aluminum slats, 8' long		8	2		335	66		401	470
0630	15' long		5	3.200		735	106		841	975
0800	Cast iron pedestals, back & arms, wood slats, 4' long		8	2		295	66		361	425
0820	8' long		5	3.200		980	106		1,086	1,250
0840	Backless, wood slats, 4' long		8	2		545	66		611	700
0860	8' long		5	3.200		930	106		1,036	1,200
1700	Steel frame, fir seat, 10' long	↓	10	1.600	↓	240	53		293	345

Division Notes

		CREW	DAILY OUTPUT	LABOR-HOURS	UNIT	2010 BARE COSTS MAT.	LABOR	EQUIP.	TOTAL	TOTAL INCL O&P

Estimating Tips

General

- The items and systems in this division are usually estimated, purchased, supplied, and installed as a unit by one or more subcontractors. The estimator must ensure that all parties are operating from the same set of specifications and assumptions, and that all necessary items are estimated and will be provided. Many times the complex items and systems are covered, but the more common ones, such as excavation or a crane, are overlooked for the very reason that everyone assumes nobody could miss them. The estimator should be the central focus and be able to ensure that all systems are complete.
- Another area where problems can develop in this division is at the interface between systems. The estimator must ensure, for instance, that anchor bolts, nuts, and washers are estimated and included for the air-supported structures and pre-engineered buildings to be bolted to their foundations. Utility supply is a common area where essential items or pieces of equipment can be missed or overlooked due to the fact that each subcontractor may feel it is another's responsibility. The estimator should also be aware of certain items which may be supplied as part of a package but installed by others, and ensure that the installing contractor's estimate includes the cost of installation. Conversely, the estimator must also ensure that items are not costed by two different subcontractors, resulting in an inflated overall estimate.

13 30 00 Special Structures

- The foundations and floor slab, as well as rough mechanical and electrical, should be estimated, as this work is required for the assembly and erection of the structure. Generally, as noted in the book, the pre-engineered building comes as a shell. Pricing is based on the size and structural design parameters stated in the reference section. Additional features, such as windows and doors, must also be included by the estimator. Here again, the estimator must have a clear understanding of the scope of each portion of the work and all the necessary interfaces.

Reference Numbers

Reference numbers are shown in shaded boxes at the beginning of some major classifications. These numbers refer to related items in the Reference Section. The reference information may be an estimating procedure, an alternate pricing method, or technical information.

Note: Not all subdivisions listed here necessarily appear in this publication.

13 05 05 – Selective Special Construction Demolition

13 05 05.10 Selective Demolition, Air Supported Structures		Crew	Daily Output	Labor-Hours	Unit	Material	2010 Bare Costs Labor	Equipment	Total	Total Incl O&P
0010	**SELECTIVE DEMOLITION, AIR SUPPORTED STRUCTURES**									
0020	Tank covers, scrim, dbl layer, vinyl poly w/ hdw, blower & controls									
0050	Round and rectangular R024119-10	B-2	9000	.004	S.F.		.15		.15	.23
0100	Warehouse structures									
0120	Poly/vinyl fabric, 28 oz, incl tension cables & inflation system	4 Clab	9000	.004	SF Flr.		.12		.12	.18
0150	Reinforced vinyl, 12 oz., 3000 S.F.	″	5000	.006			.21		.21	.33
0200	12,000 to 24,000 S.F.	8 Clab	20000	.003			.11		.11	.16
0250	Tedlar vinyl fabric, 28 oz. w/liner, to 3000 S.F.	4 Clab	5000	.006			.21		.21	.33
0300	12,000 to 24,000 S.F.	8 Clab	20000	.003			.11		.11	.16
0350	Greenhouse/shelter, woven polyethylene with liner									
0400	3000 S.F.	4 Clab	5000	.006	SF Flr.		.21		.21	.33
0450	12,000 to 24,000 S.F.	8 Clab	20000	.003			.11		.11	.16
0500	Tennis/gymnasium, poly/vinyl fabric, 28 oz., incl thermal liner	4 Clab	9000	.004			.12		.12	.18
0600	Stadium/convention center, teflon coated fiberglass, incl thermal liner	9 Clab	40000	.002			.06		.06	.09
0700	Doors, air lock, 15′ long, 10′ x 10′	2 Carp	1.50	10.667	Ea.		445		445	685
0720	15′ x 15′		.80	20			830		830	1,275
0750	Revolving personnel door, 6′ dia. x 6′-6″ high		1.50	10.667			445		445	685

13 05 05.20 Selective Demolition, Garden Houses		Crew	Daily Output	Labor-Hours	Unit	Material	Labor	Equipment	Total	Total Incl O&P
0010	**SELECTIVE DEMOLITION, GARDEN HOUSES** R024119-10									
0020	Prefab, wood, excl foundation, average	2 Clab	400	.040	SF Flr.		1.32		1.32	2.04

13 05 05.25 Selective Demolition, Geodesic Domes		Crew	Daily Output	Labor-Hours	Unit	Material	Labor	Equipment	Total	Total Incl O&P
0010	**SELECTIVE DEMOLITION, GEODESIC DOMES**									
0050	Shell only, interlocking plywood panels, 30′ diameter	F-5	3.20	10	Ea.		420		420	650
0060	34′ diameter		2.30	13.913			585		585	900
0070	39′ diameter		2	16			675		675	1,025
0080	45′ diameter	F-3	2.20	18.182			765	298	1,063	1,500
0090	55′ diameter		2	20			840	330	1,170	1,650
0100	60′ diameter		2	20			840	330	1,170	1,650
0110	65′ diameter		1.60	25			1,050	410	1,460	2,075

13 05 05.30 Selective Demolition, Greenhouses		Crew	Daily Output	Labor-Hours	Unit	Material	Labor	Equipment	Total	Total Incl O&P
0010	**SELECTIVE DEMOLITION, GREENHOUSES** R024119-10									
0020	Resi-type, free standing, excl foundations, 9′ long x 8′ wide	2 Clab	160	.100	SF Flr.		3.31		3.31	5.10
0030	9′ long x 11′ wide		170	.094			3.12		3.12	4.80
0040	9′ long x 14′ wide		220	.073			2.41		2.41	3.71
0050	9′ long x 17′ wide		320	.050			1.66		1.66	2.55
0060	Lean-to type, 4′ wide		64	.250			8.30		8.30	12.75
0070	7′ wide		120	.133			4.41		4.41	6.80
0080	Geodesic hemisphere, 1/8″ plexiglass glazing, 8′ diam.		4	4	Ea.		132		132	204
0090	24′ dia		.80	20			660		660	1,025
0100	48′ dia		.40	40			1,325		1,325	2,050

13 05 05.35 Selective Demolition, Hangars		Crew	Daily Output	Labor-Hours	Unit	Material	Labor	Equipment	Total	Total Incl O&P
0010	**SELECTIVE DEMOLITION, HANGARS**									
0020	T type hangars, prefab, steel , galv roof & walls, incl doors, excl fndtn	E-2	2550	.022	SF Flr.		1	.63	1.63	2.38
0030	Circular type, prefab, steel frame, plastic skin, incl foundation, 80′ diam	″	.50	112	Total		5,100	3,225	8,325	12,200

13 05 05.45 Selective Demolition, Lightning Protection		Crew	Daily Output	Labor-Hours	Unit	Material	Labor	Equipment	Total	Total Incl O&P
0010	**SELECTIVE DEMOLITION, LIGHTNING PROTECTION**									
0020	Air terminal & base, copper, 3/8″ diam. x 10″, to 75′ h	1 Clab	16	.500	Ea.		16.55		16.55	25.50
0030	1/2″ diam. x 12″, over 75′ h		16	.500			16.55		16.55	25.50
0050	Aluminum, 1/2″ diam. x 12″, to 75′ h		16	.500			16.55		16.55	25.50
0060	5/8″ diam. x 12″, over 75′ h		16	.500			16.55		16.55	25.50
0070	Cable, copper, 220 lb per thousand feet, to 75′ high		640	.013	L.F.		.41		.41	.64

13 05 Common Work Results for Special Construction

13 05 05 – Selective Special Construction Demolition

13 05 05.45 Selective Demolition, Lightning Protection		Crew	Daily Output	Labor-Hours	Unit	Material	2010 Bare Costs Labor	Equipment	Total	Total Incl O&P
0080	375 lb per thousand feet, over 75′ high	1 Clab	460	.017	L.F.		.58		.58	.89
0090	Aluminum, 101 lb per thousand feet, to 75′ high		560	.014			.47		.47	.73
0100	199 lb per thousand feet, over 75′ high		480	.017			.55		.55	.85
0110	Arrester, 175 V AC, to ground		16	.500	Ea.		16.55		16.55	25.50
0120	650 V AC, to ground		13	.615	"		20.50		20.50	31.50

13 05 05.50 Selective Demolition, Pre-Engineered Steel Buildings		Crew	Daily Output	Labor-Hours	Unit	Material	Labor	Equipment	Total	Total Incl O&P
0010	**SELECTIVE DEMOLITION, PRE-ENGINEERED STEEL BUILDINGS**									
0500	Pre-engd steel bldgs, rigid frame, clear span & multi post, excl salvage									
0550	3,500 to 7,500 S.F	L-10	1000	.024	SF Flr.		1.12	.66	1.78	2.60
0600	7,501 to 12,500 S.F		1500	.016			.75	.44	1.19	1.73
0650	12,500 S.F or greater		1650	.015			.68	.40	1.08	1.58
0700	Pre-engd steel building components									
0710	Entrance canopy, including frame 4′ x 4′	E-24	8	4	Ea.		184	93.50	277.50	415
0720	4′ x 8′	"	7	4.571			210	107	317	470
0730	H.M doors, self framing, single leaf	2 Skwk	8	2			85		85	131
0740	Double leaf		5	3.200			136		136	210
0760	Gutter, eave type		600	.027	L.F.		1.14		1.14	1.75
0770	Sash, single slide, double slide or fixed		24	.667	Ea.		28.50		28.50	43.50
0780	Skylight, fiberglass, to 30 S.F.		16	1			42.50		42.50	65.50
0785	Roof vents, circular, 12″ to 24″ diameter		12	1.333			57		57	87.50
0790	Continuous, 10′ long		8	2			85		85	131
0900	Shelters, aluminum frame									
0910	Acrylic glazing, 3′ x 9′ x 8′ high	2 Skwk	2	8	Ea.		340		340	525
0920	9′ x 12′ x 8′ high	"	1.50	10.667	"		455		455	700

13 05 05.60 Selective Demolition, Silos		Crew	Daily Output	Labor-Hours	Unit	Material	Labor	Equipment	Total	Total Incl O&P
0010	**SELECTIVE DEMOLITION, SILOS**									
0020	Conc stave, indstrl, conical/sloping bott, excl fndtn, 12′ diam., 35′ h	E-24	.18	177	Ea.		8,150	4,150	12,300	18,500
0030	16′ diam., 45′ h		.12	266			12,200	6,225	18,425	27,700
0040	25′ dia, 75′ h		.08	400			18,400	9,350	27,750	41,500
0050	Steel, factory fabricated, 30,000 gal cap, painted or epoxy lined	L-5	2	28			1,300	375	1,675	2,650

13 05 05.65 Selective Demolition, Sound Control		Crew	Daily Output	Labor-Hours	Unit	Material	Labor	Equipment	Total	Total Incl O&P
0010	**SELECTIVE DEMOLITION, SOUND CONTROL** R024119-10									
0120	Acoustical enclosure, 4″ thick walls & ceiling panels, 8 lbs/SF	3 Carp	144	.167	SF Surf		6.95		6.95	10.70
0130	10.5 lbs/SF		128	.188			7.80		7.80	12
0140	Reverb chamber, parallel walls, 4″ thick		120	.200			8.30		8.30	12.80
0150	Skewed walls, parallel roof, 4″ thick		110	.218			9.05		9.05	13.95
0160	Skewed walls/roof, 4″ layer/air space		96	.250			10.40		10.40	16
0170	Sound-absorbing panels, painted metal, 2′-6″ x 8′, under 1,000 SF		430	.056			2.32		2.32	3.57
0180	Over 1,000 SF		480	.050			2.08		2.08	3.20
0190	Flexible transparent curtain, clear	3 Shee	430	.056			2.74		2.74	4.15
0192	50% clear, 50% foam		430	.056			2.74		2.74	4.15
0194	25% clear, 75% foam		430	.056			2.74		2.74	4.15
0196	100% foam		430	.056			2.74		2.74	4.15
0200	Audio-masking sys, incl speakers, amplfr, signal gnrtr, clng mntd									
0205	Ceiling mounted, 5,000 SF	2 Elec	4800	.003	S.F.		.16		.16	.24
0210	10,000 SF		5600	.003			.14		.14	.21
0220	Plenum mounted, 5,000 SF		7600	.002			.10		.10	.15
0230	10,000 SF		8800	.002			.09		.09	.13

13 05 05.70 Selective Demolition, Special Purpose Rooms		Crew	Daily Output	Labor-Hours	Unit	Material	Labor	Equipment	Total	Total Incl O&P
0010	**SELECTIVE DEMOLITION, SPECIAL PURPOSE ROOMS** R024119-10									
0100	Audiometric rooms, under 500 S.F. surface	4 Carp	200	.160	SF Surf		6.65		6.65	10.25
0110	Over 500 S.F. surface	"	240	.133	"		5.55		5.55	8.55

13 05 05 – Selective Special Construction Demolition

13 05 05.70 Selective Demolition, Special Purpose Rooms		Crew	Daily Output	Labor-Hours	Unit	Material	2010 Bare Costs Labor	Equipment	Total	Total Incl O&P
0200	Clean rooms, 12' x 12' soft wall, class 100	1 Carp	.30	26.667	Ea.		1,100		1,100	1,700
0210	Class 1000		.30	26.667			1,100		1,100	1,700
0220	Class 10,000		.35	22.857			950		950	1,475
0230	Class 100,000		.35	22.857			950		950	1,475
0300	Darkrooms, shell complete, 8' high	2 Carp	220	.073	SF Flr.		3.02		3.02	4.66
0310	12' high		110	.145	"		6.05		6.05	9.30
0350	Darkrooms doors, mini-cylindrical, revolving		4	4	Ea.		166		166	256
0400	Music room, practice modular		140	.114	SF Surf		4.75		4.75	7.30
0500	Refrigeration structures and finishes									
0510	Wall finish, 2 coat portland cement plaster, 1/2" thick	1 Clab	200	.040	S.F.		1.32		1.32	2.04
0520	Fiberglass panels, 1/8" thick		400	.020			.66		.66	1.02
0530	Ceiling finish, polystyrene plastic, 1" to 2" thick		500	.016			.53		.53	.82
0540	4" thick		450	.018			.59		.59	.91
0550	Refrigerator, prefab aluminum walk-in, 7'-6" high, 6' x 6' OD	2 Carp	100	.160	SF Flr.		6.65		6.65	10.25
0560	10' x 10' OD		160	.100			4.16		4.16	6.40
0570	Over 150 S.F.		200	.080			3.32		3.32	5.10
0600	Sauna, prefabricated, including heater & controls, 7' high, to 30 S.F.		120	.133			5.55		5.55	8.55
0610	To 40 S.F.		140	.114			4.75		4.75	7.30
0620	To 60 S.F.		175	.091			3.80		3.80	5.85
0630	To 100 S.F.		220	.073			3.02		3.02	4.66
0640	To 130 S.F.		250	.064			2.66		2.66	4.10
0650	Steam bath, heater, timer, head, single, to 140 CF	1 Plum	2.20	3.636	Ea.		189		189	284
0660	To 300 CF		2.20	3.636			189		189	284
0670	Steam bath, comm. size, w/blow-down assembly, to 800 CF		1.80	4.444			231		231	345
0680	To 2500 CF		1.60	5			260		260	390
0690	Steam bath, comm. size, multiple, for motels, apts, 500 CF, 2 baths		2	4			208		208	310
0700	1,000 CF, 4 baths		1.40	5.714			297		297	445

13 05 05.75 Selective Demolition, Storage Tanks		Crew	Daily Output	Labor-Hours	Unit	Material	Labor	Equipment	Total	Total Incl O&P
0010	**SELECTIVE DEMOLITION, STORAGE TANKS**									
0500	Steel tank, single wall, above ground, not incl fdn, pumps or piping									
0510	Single wall, 275 gallon R024119-10	Q-1	3	5.333	Ea.		250		250	375
0520	550 thru 2,000 gallon	B-34P	2	12			510	315	825	1,125
0530	5,000 thru 10,000 gallon	B-34Q	2	12			515	625	1,140	1,475
0540	15,000 thru 30,000 gallon	B-34S	2	16			725	1,800	2,525	3,075
0600	Steel tank, double wall, above ground not incl fdn, pumps & piping									
0620	500 thru 2,000 gallon	B-34P	2	12	Ea.		510	315	825	1,125

13 05 05.85 Selective Demolition, Swimming Pool Equip		Crew	Daily Output	Labor-Hours	Unit	Material	Labor	Equipment	Total	Total Incl O&P
0010	**SELECTIVE DEMOLITION, SWIMMING POOL EQUIP**									
0020	Diving stand, stainless steel, 3 meter	2 Clab	3	5.333	Ea.		177		177	272
0030	1 meter		5	3.200			106		106	163
0040	Diving board, 16' long, aluminum		5.40	2.963			98		98	151
0050	Fiberglass		5.40	2.963			98		98	151
0070	Ladders, heavy duty, stainless steel, 2 tread		14	1.143			38		38	58.50
0080	4 tread		12	1.333			44		44	68
0090	Lifeguard chair, stainless steel, fixed		5	3.200			106		106	163
0100	Slide, tubular, fiberglass, aluminum handrails & ladder, 5', straight		4	4			132		132	204
0110	8', curved		6	2.667			88.50		88.50	136
0120	10', curved		3	5.333			177		177	272
0130	12' straight, with platform		2.50	6.400			212		212	325
0140	Removable access ramp, stainless steel		4	4			132		132	204
0150	Removable stairs, stainless steel, collapsible		4	4			132		132	204

13 05 Common Work Results for Special Construction

13 05 05 – Selective Special Construction Demolition

13 05 05.90 Selective Demolition, Tension Structures		Crew	Daily Output	Labor-Hours	Unit	Material	2010 Bare Costs Labor	Equipment	Total	Total Incl O&P
0010	**SELECTIVE DEMOLITION, TENSION STRUCTURES**									
0020	Tension structure, steel/alum frame, fabric shell, 60′ clear span, 6,000 SF	B-41	2000	.022	SF Flr.		.75	.15	.90	1.32
0030	12,000 SF		2200	.020			.68	.14	.82	1.20
0040	80′ clear span, 20,800 SF		2440	.018			.62	.12	.74	1.08
0050	100′ clear span, 10,000 SF	L-5	4350	.013			.60	.17	.77	1.23
0060	26,000 SF	″	4600	.012			.57	.16	.73	1.16

13 05 05.95 Selective Demo, X-Ray/Radio Freq Protection		Crew	Daily Output	Labor-Hours	Unit	Material	Labor	Equipment	Total	Total Incl O&P
0010	**SELECTIVE DEMO, X-RAY/RADIO FREQ PROTECTION**									
0020	Shielding lead, lined door frame, excl hdwe, 1/16″ thick	1 Clab	4.80	1.667	Ea.		55		55	85
0030	Lead sheets, 1/16″ thick R024119-10	2 Clab	270	.059	S.F.		1.96		1.96	3.03
0040	1/8″ thick		240	.067			2.21		2.21	3.40
0050	Lead shielding, 1/4″ thick		270	.059			1.96		1.96	3.03
0060	1/2″ thick		240	.067			2.21		2.21	3.40
0070	Lead glass, 1/4″ thick, 2.0 mm LE, 12″ x 16″	2 Glaz	16	1	Ea.		40		40	60.50
0080	24″ x 36″		8	2			80.50		80.50	121
0090	36″ x 60″		4	4			161		161	242
0100	Lead glass window frame, with 1/16″ lead & voice passage, 36″ x 60″		4	4			161		161	242
0110	Lead glass window frame, 24″ x 36″		8	2			80.50		80.50	121
0120	Lead gypsum board, 5/8″ thick with 1/16″ lead	2 Clab	320	.050	S.F.		1.66		1.66	2.55
0130	1/8″ lead		280	.057			1.89		1.89	2.92
0140	1/32″ lead		400	.040			1.32		1.32	2.04
0150	Butt joints, 1/8″ lead or thicker, 2″ x 7′ long batten strip		480	.033	Ea.		1.10		1.10	1.70
0160	X-ray protection, average radiography room, up to 300 SF, 1/16″ lead, min		.50	32	Total		1,050		1,050	1,625
0170	Maximum		.30	53.333			1,775		1,775	2,725
0180	Deep therapy X-ray room, 250 KV cap, up to 300 SF, 1/4″ lead, min		.20	80			2,650		2,650	4,075
0190	Maximum		.12	133			4,425		4,425	6,800
0880	Radio frequency shielding, prefab or screen-type copper or steel, minimum		360	.044	SF Surf		1.47		1.47	2.27
0890	Average		310	.052			1.71		1.71	2.63
0895	Maximum		290	.055			1.83		1.83	2.82

13 11 Swimming Pools

13 11 13 – Below-Grade Swimming Pools

13 11 13.50 Swimming Pools		Crew	Daily Output	Labor-Hours	Unit	Material	Labor	Equipment	Total	Total Incl O&P
0010	**SWIMMING POOLS** Residential in-ground, vinyl lined, concrete									
0020	Sides including equipment, sand bottom	B-52	300	.187	SF Surf	13.90	7.15	1.80	22.85	28
0100	Metal or polystyrene sides R131113-20	B-14	410	.117		11.60	4.07	.82	16.49	19.90
0200	Add for vermiculite bottom					.89			.89	.98
0500	Gunite bottom and sides, white plaster finish									
0600	12′ x 30′ pool	B-52	145	.386	SF Surf	26	14.85	3.72	44.57	55
0720	16′ x 32′ pool		155	.361		23.50	13.90	3.48	40.88	51
0750	20′ x 40′ pool		250	.224		21	8.60	2.15	31.75	38.50
0810	Concrete bottom and sides, tile finish									
0820	12′ x 30′ pool	B-52	80	.700	SF Surf	26	27	6.75	59.75	77
0830	16′ x 32′ pool		95	.589		21.50	22.50	5.65	49.65	64.50
0840	20′ x 40′ pool		130	.431		17.15	16.55	4.14	37.84	49
1100	Motel, gunite with plaster finish, incl. medium									
1150	capacity filtration & chlorination	B-52	115	.487	SF Surf	32	18.70	4.68	55.38	68.50
1200	Municipal, gunite with plaster finish, incl. high									
1250	capacity filtration & chlorination	B-52	100	.560	SF Surf	41	21.50	5.40	67.90	84.50
1350	Add for formed gutters				L.F.	60.50			60.50	66.50

13 11 Swimming Pools

13 11 13 – Below-Grade Swimming Pools

13 11 13.50 Swimming Pools		Crew	Daily Output	Labor-Hours	Unit	2010 Bare Costs Material	Labor	Equipment	Total	Total Incl O&P
1360	Add for stainless steel gutters				L.F.	179			179	197
1600	For water heating system, see Div. 23 52 28.10									
1700	Filtration and deck equipment only, as % of total				Total				20%	20%
1800	Deck equipment, rule of thumb, 20′ x 40′ pool				SF Pool				1.18	1.30
1900	5000 S.F. pool				″				1.73	1.90
3000	Painting pools, preparation + 3 coats, 20′ x 40′ pool, epoxy	2 Pord	.33	48.485	Total	1,625	1,750		3,375	4,400
3100	Rubber base paint, 18 gallons	″	.33	48.485		1,100	1,750		2,850	3,825
3500	42′ x 82′ pool, 75 gallons, epoxy paint	3 Pord	.14	171		6,850	6,225		13,075	16,800
3600	Rubber base paint	″	.14	171		4,500	6,225		10,725	14,200

13 11 46 – Swimming Pool Accessories

13 11 46.50 Swimming Pool Equipment		Crew	Daily Output	Labor-Hours	Unit	Material	Labor	Equipment	Total	Total Incl O&P
0010	**SWIMMING POOL EQUIPMENT**									
0020	Diving stand, stainless steel, 3 meter	2 Carp	.40	40	Ea.	10,500	1,650		12,150	14,100
0300	1 meter		2.70	5.926		6,175	246		6,421	7,175
0600	Diving boards, 16′ long, aluminum		2.70	5.926		3,375	246		3,621	4,075
0700	Fiberglass		2.70	5.926		2,550	246		2,796	3,175
1100	Gutter system, stainless steel, with grating, stock,									
1110	contains supply and drainage system	E-1	20	1.200	L.F.	268	55	7.30	330.30	395
1120	Integral gutter and 5′ high wall system, stainless steel	″	10	2.400	″	360	110	14.60	484.60	595
1200	Ladders, heavy duty, stainless steel, 2 tread	2 Carp	7	2.286	Ea.	585	95		680	790
1500	4 tread		6	2.667		775	111		886	1,025
1800	Lifeguard chair, stainless steel, fixed		2.70	5.926		2,975	246		3,221	3,650
1900	Portable					2,725			2,725	3,000
2100	Lights, underwater, 12 volt, with transformer, 300 watt	1 Elec	1	8		263	390		653	875
2200	110 volt, 500 watt, standard		1	8		209	390		599	815
2400	Low water cutoff type		1	8		254	390		644	865
2800	Heaters, see Div. 23 52 28.10									
3000	Pool covers, reinforced vinyl	3 Clab	1800	.013	S.F.	.36	.44		.80	1.08
3050	Automatic, electric								8.77	9.65
3100	Vinyl water tube	3 Clab	3200	.008		.50	.25		.75	.93
3200	Maximum	″	3000	.008		.63	.26		.89	1.10
3250	Sealed air bubble polyethylene solar blanket, 16 mils					.36			.36	.40
3300	Slides, tubular, fiberglass, aluminum handrails & ladder, 5′-0″, straight	2 Carp	1.60	10	Ea.	3,625	415		4,040	4,650
3320	8′-0″, curved		3	5.333		6,375	222		6,597	7,375
3400	10′-0″, curved		1	16		26,300	665		26,965	29,900
3420	12′-0″, straight with platform		1.20	13.333		12,400	555		12,955	14,600
4500	Hydraulic lift, movable pool bottom, single ram									
4520	Under 1,000 S.F. area	L-9	72	.500	S.F.	130	19		149	174
4600	Four ram lift, over 1,000 S.F.	″	109	.330	″	105	12.55		117.55	136
5000	Removable access ramp, stainless steel	2 Clab	2	8	Ea.	10,200	265		10,465	11,600
5500	Removable stairs, stainless steel, collapsible	″	2	8	″	4,450	265		4,715	5,275

13 17 Tubs and Pools

13 17 33 – Whirlpool Tubs

13 17 33.10 Whirlpool Bath		Crew	Daily Output	Labor-Hours	Unit	2010 Bare Costs Material	Labor	Equipment	Total	Total Incl O&P
0010	**WHIRLPOOL BATH**									
6000	Whirlpool, bath with vented overflow, molded fiberglass									
6100	66" x 48" x 24"	Q-1	1	16	Ea.	3,225	750		3,975	4,675

13 18 Ice Rinks

13 18 13 – Ice Rink Floor Systems

13 18 13.50 Ice Skating		Crew	Daily Output	Labor-Hours	Unit	Material	Labor	Equipment	Total	Total Incl O&P
0010	**ICE SKATING** Equipment incl. refrigeration, plumbing & cooling									
0020	coils & concrete slab, 85' x 200' rink									
0300	55° system, 5 mos., 100 ton				Total	575,000			575,000	632,500
0700	90° system, 12 mos., 135 ton				"	600,000			600,000	660,000
1200	Subsoil heating system (recycled from compressor), 85' x 200'	Q-7	.27	118	Ea.	25,000	5,850		30,850	36,300
1300	Subsoil insulation, 2 lb. polystyrene with vapor barrier, 85' x 200'	2 Carp	.14	114	"	30,000	4,750		34,750	40,300

13 18 16 – Ice Rink Dasher Boards

13 18 16.50 Ice Rink Dasher Boards		Crew	Daily Output	Labor-Hours	Unit	Material	Labor	Equipment	Total	Total Incl O&P
0010	**ICE RINK DASHER BOARDS**									
1000	Dasher boards, 1/2" H.D. polyethylene faced steel frame, 3' acrylic									
1020	screen at sides, 5' acrylic ends, 85' x 200'	F-5	.06	533	Ea.	125,000	22,400		147,400	172,000
1100	Fiberglass & aluminum construction, same sides and ends	"	.06	533	"	125,000	22,400		147,400	172,000

13 21 Controlled Environment Rooms

13 21 13 – Clean Rooms

13 21 13.50 Clean Room Components		Crew	Daily Output	Labor-Hours	Unit	Material	Labor	Equipment	Total	Total Incl O&P
0010	**CLEAN ROOM COMPONENTS**									
1100	Clean room, soft wall, 12' x 12', Class 100	1 Carp	.18	44.444	Ea.	15,500	1,850		17,350	20,000
1110	Class 1,000		.18	44.444		12,600	1,850		14,450	16,800
1120	Class 10,000		.21	38.095		11,100	1,575		12,675	14,700
1130	Class 100,000		.21	38.095		10,400	1,575		11,975	14,000
2800	Ceiling grid support, slotted channel struts 4'-0" O.C., ea. way				S.F.				5.91	6.50
3000	Ceiling panel, vinyl coated foil on mineral substrate									
3020	Sealed, non-perforated				S.F.				1.27	1.40
4000	Ceiling panel seal, silicone sealant, 150 L.F./gal.	1 Carp	150	.053	L.F.	.28	2.22		2.50	3.73
4100	Two sided adhesive tape	"	240	.033	"	.13	1.38		1.51	2.27
4200	Clips, one per panel				Ea.	1.09			1.09	1.20
6000	HEPA filter, 2' x 4', 99.97% eff., 3" dp beveled frame (silicone seal)					365			365	400
6040	6" deep skirted frame (channel seal)					390			390	430
6100	99.99% efficient, 3" deep beveled frame (silicone seal)					380			380	420
6140	6" deep skirted frame (channel seal)					405			405	445
6200	99.999% efficient, 3" deep beveled frame (silicone seal)					395			395	435
6240	6" deep skirted frame (channel seal)					420			420	465
7000	Wall panel systems, including channel strut framing									
7020	Polyester coated aluminum, particle board				S.F.				18.18	20
7100	Porcelain coated aluminum, particle board								31.82	35
7400	Wall panel support, slotted channel struts, to 12' high								16.36	18

13 21 26 – Cold Storage Rooms

13 21 26.50 Refrigeration		Crew	Daily Output	Labor-Hours	Unit	Material	Labor	Equipment	Total	Total Incl O&P
0010	**REFRIGERATION**									
0020	Curbs, 12" high, 4" thick, concrete	2 Carp	58	.276	L.F.	4.37	11.45		15.82	22.50

13 21 Controlled Environment Rooms

13 21 26 – Cold Storage Rooms

13 21 26.50 Refrigeration		Crew	Daily Output	Labor-Hours	Unit	Material	Labor	Equipment	Total	Total Incl O&P
						2010 Bare Costs				
1000	Doors, see Div. 08 34 13.10									
2400	Finishes, 2 coat portland cement plaster, 1/2" thick	1 Plas	48	.167	S.F.	1.20	6.20		7.40	10.60
2500	For galvanized reinforcing mesh, add	1 Lath	335	.024		.80	.88		1.68	2.18
2700	3/16" thick latex cement	1 Plas	88	.091		2.05	3.39		5.44	7.30
2900	For glass cloth reinforced ceilings, add	"	450	.018		.49	.66		1.15	1.53
3100	Fiberglass panels, 1/8" thick	1 Carp	149.45	.054		2.71	2.22		4.93	6.40
3200	Polystyrene, plastic finish ceiling, 1" thick		274	.029		2.50	1.21		3.71	4.62
3400	2" thick		274	.029		2.85	1.21		4.06	5
3500	4" thick		219	.037		3.17	1.52		4.69	5.85
3800	Floors, concrete, 4" thick	1 Cefi	93	.086		1.19	3.42		4.61	6.30
3900	6" thick	"	85	.094		1.77	3.74		5.51	7.40
4000	Insulation, 1" to 6" thick, cork				B.F.	1.16			1.16	1.28
4100	Urethane					1.14			1.14	1.25
4300	Polystyrene, regular					.77			.77	.85
4400	Bead board					.58			.58	.64
4600	Installation of above, add per layer	2 Carp	657.60	.024	S.F.	.38	1.01		1.39	1.98
4700	Wall and ceiling juncture		298.90	.054	L.F.	1.87	2.22		4.09	5.50
4900	Partitions, galvanized sandwich panels, 4" thick, stock		219.20	.073	S.F.	7.85	3.03		10.88	13.30
5000	Aluminum or fiberglass		219.20	.073	"	8.55	3.03		11.58	14.15
5200	Prefab walk-in, 7'-6" high, aluminum, incl. door & floors,									
5210	not incl. partitions or refrigeration, 6' x 6' O.D. nominal	2 Carp	54.80	.292	SF Flr.	140	12.15		152.15	173
5500	10' x 10' O.D. nominal		82.20	.195		113	8.10		121.10	136
5700	12' x 14' O.D. nominal		109.60	.146		101	6.05		107.05	120
5800	12' x 20' O.D. nominal		109.60	.146		88	6.05		94.05	106
6100	For 8'-6" high, add					5%				
6300	Rule of thumb for complete units, w/o doors & refrigeration, cooler	2 Carp	146	.110		127	4.55		131.55	147
6400	Freezer		109.60	.146		150	6.05		156.05	174
6600	Shelving, plated or galvanized, steel wire type		360	.044	SF Hor.	11.10	1.85		12.95	15.10
6700	Slat shelf type		375	.043		13.70	1.77		15.47	17.85
6900	For stainless steel shelving, add					300%				
7000	Vapor barrier, on wood walls	2 Carp	1644	.010	S.F.	.16	.40		.56	.80
7200	On masonry walls	"	1315	.012	"	.42	.51		.93	1.24
7500	For air curtain doors, see Div. 23 34 33.10									

13 21 48 – Sound-Conditioned Rooms

13 21 48.10 Anechoic Chambers

Line	Description	Crew	Daily Output	Labor-Hours	Unit	Material	Labor	Equipment	Total	Total Incl O&P
0010	**ANECHOIC CHAMBERS** Standard units, 7' ceiling heights									
0100	Area for pricing is net inside dimensions									
0300	200 cycles per second cutoff, 25 S.F. floor area				SF Flr.	1,600			1,600	1,750
0400	50 S.F.								1,045	1,150
0600	75 S.F.								1,000	1,100
0700	100 S.F.					1,200			1,200	1,325
0900	For 150 cycles per second cutoff, add to 100 S.F. room									30%
1000	For 100 cycles per second cutoff, add to 100 S.F. room									45%

13 21 48.15 Audiometric Rooms

Line	Description	Crew	Daily Output	Labor-Hours	Unit	Material	Labor	Equipment	Total	Total Incl O&P
0010	**AUDIOMETRIC ROOMS**									
0020	Under 500 S.F. surface	4 Carp	98	.327	SF Surf	51.50	13.55		65.05	77.50
0100	Over 500 S.F. surface	"	120	.267	"	49	11.10		60.10	70.50

13 21 53 – Darkrooms

13 21 53.50 Darkrooms

Line	Description	Crew	Daily Output	Labor-Hours	Unit	Material	Labor	Equipment	Total	Total Incl O&P
0010	**DARKROOMS**									
0020	Shell, complete except for door, 64 S.F., 8' high	2 Carp	128	.125	SF Flr.	53.50	5.20		58.70	66.50

13 21 Controlled Environment Rooms

13 21 53 – Darkrooms

13 21 53.50 Darkrooms		Crew	Daily Output	Labor-Hours	Unit	Material	2010 Bare Costs Labor	Equipment	Total	Total Incl O&P
0100	12′ high	2 Carp	64	.250	SF Flr.	68.50	10.40		78.90	91
0500	120 S.F. floor, 8′ high		120	.133		37.50	5.55		43.05	49.50
0600	12′ high		60	.267		51.50	11.10		62.60	73.50
0800	240 S.F. floor, 8′ high		120	.133		27	5.55		32.55	38
0900	12′ high		60	.267		38	11.10		49.10	58.50
1200	Mini-cylindrical, revolving, unlined, 4′ diameter		3.50	4.571	Ea.	2,825	190		3,015	3,400
1400	5′-6″ diameter		2.50	6.400		3,850	266		4,116	4,625
1600	Add for lead lining, inner cylinder, 1/32″ thick					1,475			1,475	1,625
1700	1/16″ thick					3,975			3,975	4,375
1800	Add for lead lining, inner and outer cylinder, 1/32″ thick					2,725			2,725	3,000
1900	1/16″ thick					6,100			6,100	6,700
2000	For darkroom door, see Div. 08 34 36.10									

13 21 56 – Music Rooms

13 21 56.50 Music Rooms		Crew	Daily Output	Labor-Hours	Unit	Material	Labor	Equipment	Total	Total Incl O&P
0010	**MUSIC ROOMS**									
0020	Practice room, modular, perforated steel, under 500 S.F.	2 Carp	70	.229	SF Surf	31.50	9.50		41	49
0100	Over 500 S.F.	"	80	.200	"	26.50	8.30		34.80	42.50

13 24 Special Activity Rooms

13 24 16 – Saunas

13 24 16.50 Saunas and Heaters		Crew	Daily Output	Labor-Hours	Unit	Material	Labor	Equipment	Total	Total Incl O&P
0010	**SAUNAS AND HEATERS**									
0020	Prefabricated, incl. heater & controls, 7′ high, 6′ x 4′, C/C	L-7	2.20	12.727	Ea.	4,625	510		5,135	5,875
0050	6′ x 4′, C/P		2	14		4,275	565		4,840	5,550
0400	6′ x 5′, C/C		2	14		5,175	565		5,740	6,525
0450	6′ x 5′, C/P		2	14		4,800	565		5,365	6,150
0600	6′ x 6′, C/C		1.80	15.556		5,475	625		6,100	6,975
0650	6′ x 6′, C/P		1.80	15.556		5,125	625		5,750	6,575
0800	6′ x 9′, C/C		1.60	17.500		6,800	705		7,505	8,550
0850	6′ x 9′, C/P		1.60	17.500		6,300	705		7,005	8,025
1000	8′ x 12′, C/C		1.10	25.455		10,600	1,025		11,625	13,300
1050	8′ x 12′, C/P		1.10	25.455		9,725	1,025		10,750	12,300
1200	8′ x 8′, C/C		1.40	20		8,100	805		8,905	10,100
1250	8′ x 8′, C/P		1.40	20		7,600	805		8,405	9,600
1400	8′ x 10′, C/C		1.20	23.333		8,975	940		9,915	11,300
1450	8′ x 10′, C/P		1.20	23.333		8,350	940		9,290	10,600
1600	10′ x 12′, C/C		1	28		11,200	1,125		12,325	14,000
1650	10′ x 12′, C/P		1	28		10,100	1,125		11,225	12,800
1700	Door only, cedar, 2′ x 6′, with tempered insulated glass window	2 Carp	3.40	4.706		610	196		806	970
1800	Prehung, incl. jambs, pulls & hardware	"	12	1.333		615	55.50		670.50	760
2500	Heaters only (incl. above), wall mounted, to 200 C.F.					605			605	665
2750	To 300 C.F.					770			770	845
3000	Floor standing, to 720 C.F., 10,000 watts, w/controls	1 Elec	3	2.667		2,025	131		2,156	2,425
3250	To 1,000 C.F., 16,000 watts	"	3	2.667		2,525	131		2,656	2,975

13 24 26 – Steam Baths

13 24 26.50 Steam Baths and Components		Crew	Daily Output	Labor-Hours	Unit	Material	Labor	Equipment	Total	Total Incl O&P
0010	**STEAM BATHS AND COMPONENTS**									
0020	Heater, timer & head, single, to 140 C.F.	1 Plum	1.20	6.667	Ea.	1,400	345		1,745	2,075
0500	To 300 C.F.		1.10	7.273		1,625	380		2,005	2,350
1000	Commercial size, with blow-down assembly, to 800 C.F.		.90	8.889		5,200	465		5,665	6,425

13 24 Special Activity Rooms

13 24 26 – Steam Baths

13 24 26.50 Steam Baths and Components		Crew	Daily Output	Labor-Hours	Unit	2010 Bare Costs Material	Labor	Equipment	Total	Total Incl O&P
1500	To 2500 C.F.	1 Plum	.80	10	Ea.	8,025	520		8,545	9,600
2000	Multiple, motels, apts., 2 baths, w/ blow-down assm., 500 C.F.	Q-1	1.30	12.308		5,900	575		6,475	7,375
2500	4 baths	"	.70	22.857		7,450	1,075		8,525	9,800
2700	Conversion unit for residential tub, including door					4,100			4,100	4,500

13 28 Athletic and Recreational Special Construction

13 28 33 – Athletic and Recreational Court Walls

13 28 33.50 Sport Court

Line	Description	Crew	Daily Output	Labor-Hours	Unit	Material	Labor	Equipment	Total	Total Incl O&P
0010	**SPORT COURT**									
0020	Floors, No. 2 & better maple, 25/32" thick				SF Flr.				6.05	6.65
0100	Walls, laminated plastic bonded to galv. steel studs				SF Wall				7.68	8.45
0150	Laminated fiberglass surfacing, minimum								1.91	2.10
0180	Maximum								2.15	2.37
0300	Squash, regulation court in existing building, minimum				Court	36,800			36,800	40,400
0400	Maximum				"	41,000			41,000	45,000
0450	Rule of thumb for components:									
0470	Walls	3 Carp	.15	160	Court	11,000	6,650		17,650	22,300
0500	Floor	"	.25	96		8,725	4,000		12,725	15,700
0550	Lighting	2 Elec	.60	26.667		2,100	1,300		3,400	4,250
0600	Handball, racquetball court in existing building, minimum	C-1	.20	160		39,800	6,300		46,100	53,500
0800	Maximum	"	.10	320		43,100	12,600		55,700	67,000
0900	Rule of thumb for components: walls	3 Carp	.12	200		12,600	8,300		20,900	26,700
1000	Floor		.25	96		8,725	4,000		12,725	15,700
1100	Ceiling		.33	72.727		4,200	3,025		7,225	9,275
1200	Lighting	2 Elec	.60	26.667		2,200	1,300		3,500	4,375

13 31 Fabric Structures

13 31 13 – Air-Supported Fabric Structures

13 31 13.09 Air Supported Tank Covers

Line	Description	Crew	Daily Output	Labor-Hours	Unit	Material	Labor	Equipment	Total	Total Incl O&P
0010	**AIR SUPPORTED TANK COVERS**, vinyl polyester									
0100	Scrim, double layer, with hardware, blower, standby & controls									
0200	Round, 75' diameter	B-2	4500	.009	S.F.	10	.30		10.30	11.45
0300	100' diameter		5000	.008		9.10	.27		9.37	10.40
0400	150' diameter		5000	.008		7.20	.27		7.47	8.35
0500	Rectangular, 20' x 20'		4500	.009		23	.30		23.30	25.50
0600	30' x 40'		4500	.009		23	.30		23.30	25.50
0700	50' x 60'		4500	.009		23	.30		23.30	25.50
0800	For single wall construction, deduct, minimum					.77			.77	.85
0900	Maximum					2.29			2.29	2.52
1000	For maximum resistance to atmosphere or cold, add					1.12			1.12	1.23
1100	For average shipping charges, add				Total	1,925			1,925	2,125

13 31 13.13 Single-Walled Air-Supported Structures

Line	Description	Crew	Daily Output	Labor-Hours	Unit	Material	Labor	Equipment	Total	Total Incl O&P
0010	**SINGLE-WALLED AIR-SUPPORTED STRUCTURES** R133113-10									
0020	Site preparation, incl. anchor placement and utilities	B-11B	1000	.016	SF Flr.	1.14	.60	.29	2.03	2.47
0030	For concrete curb, see Div. 03 30 53.40									
0050	Warehouse, polyester/vinyl fabric, 28 oz., over 10 yr. life, welded									
0060	Seams, tension cables, primary & auxiliary inflation system,									
0070	airlock, personnel doors and liner									
0100	5,000 S.F.	4 Clab	5000	.006	SF Flr.	24.50	.21		24.71	27.50

13 31 Fabric Structures

13 31 13 – Air-Supported Fabric Structures

13 31 13.13 Single-Walled Air-Supported Structures		Crew	Daily Output	Labor-Hours	Unit	2010 Bare Costs Material	Labor	Equipment	Total	Total Incl O&P
0250	12,000 S.F.	4 Clab	6000	.005	SF Flr.	17.65	.18		17.83	19.65
0400	24,000 S.F.	8 Clab	12000	.005		12.40	.18		12.58	13.90
0500	50,000 S.F.	"	12500	.005		11.50	.17		11.67	12.90
0700	12 oz. reinforced vinyl fabric, 5 yr. life, sewn seams,									
0710	accordian door, including liner									
0750	3000 S.F.	4 Clab	3000	.011	SF Flr.	11.75	.35		12.10	13.45
0800	12,000 S.F.	"	6000	.005		10	.18		10.18	11.25
0850	24,000 S.F.	8 Clab	12000	.005		8.45	.18		8.63	9.55
0950	Deduct for single layer					1.01			1.01	1.11
1000	Add for welded seams					1.01			1.01	1.11
1050	Add for double layer, welded seams included					2.08			2.08	2.29
1250	Tedlar/vinyl fabric, 28 oz., with liner, over 10 yr. life,									
1260	incl. overhead and personnel doors									
1300	3000 S.F.	4 Clab	3000	.011	SF Flr.	24	.35		24.35	27
1450	12,000 S.F.	"	6000	.005		16.85	.18		17.03	18.80
1550	24,000 S.F.	8 Clab	12000	.005		13.05	.18		13.23	14.60
1700	Deduct for single layer					1.44			1.44	1.58
2250	Greenhouse/shelter, woven polyethylene with liner, 2 yr. life,									
2260	sewn seams, including doors									
2300	3000 S.F.	4 Clab	3000	.011	SF Flr.	6.60	.35		6.95	7.80
2350	12,000 S.F.	"	6000	.005		6.70	.18		6.88	7.65
2450	24,000 S.F.	8 Clab	12000	.005		5.60	.18		5.78	6.40
2550	Deduct for single layer					.63			.63	.69
2600	Tennis/gymnasium, polyester/vinyl fabric, 28 oz., over 10 yr. life,									
2610	including thermal liner, heat and lights									
2650	7,200 S.F.	4 Clab	6000	.005	SF Flr.	23	.18		23.18	26
2750	13,000 S.F.	"	6500	.005		17.65	.16		17.81	19.65
2850	Over 24,000 S.F.	8 Clab	12000	.005		16.10	.18		16.28	18
2860	For low temperature conditions, add					1.12			1.12	1.23
2870	For average shipping charges, add				Total	5,475			5,475	6,000
2900	Thermal liner, translucent reinforced vinyl				SF Flr.	1.12			1.12	1.23
2950	Metalized mylar fabric and mesh, double liner				"	2.29			2.29	2.52
3050	Stadium/convention center, teflon coated fiberglass, heavy weight,									
3060	over 20 yr. life, incl. thermal liner and heating system									
3100	Minimum	9 Clab	26000	.003	SF Flr.	56	.09		56.09	62
3110	Maximum	"	19000	.004	"	67	.13		67.13	73.50
3400	Doors, air lock, 15′ long, 10′ x 10′	2 Carp	.80	20	Ea.	21,100	830		21,930	24,500
3600	15′ x 15′	"	.50	32		29,700	1,325		31,025	34,800
3700	For each added 5′ length, add					5,300			5,300	5,850
3900	Revolving personnel door, 6′ diameter, 6′-6″ high	2 Carp	.80	20		14,900	830		15,730	17,700
4200	Double wall, self supporting, shell only, minimum				SF Flr.				19.09	21
4300	Maximum				"				35.45	39

13 31 23 – Tensioned Fabric Structures

13 31 23.50 Tension Structures		Crew	Daily Output	Labor-Hours	Unit	Material	Labor	Equipment	Total	Total Incl O&P
0010	**TENSION STRUCTURES** Rigid steel/alum. frame, vinyl coated poly									
0100	Fabric shell, 60′ clear span, not incl. foundations or floors									
0200	6,000 S.F.	B-41	1000	.044	SF Flr.	14.65	1.51	.30	16.46	18.75
0300	12,000 S.F.		1100	.040		13.55	1.37	.27	15.19	17.30
0400	80′ clear span, 20,800 S.F.		1220	.036		13.45	1.23	.24	14.92	16.90
0410	100′ clear span, 10,000 S.F.	L-5	2175	.026		14.80	1.21	.34	16.35	18.75
0430	26,000 S.F.		2300	.024		13.95	1.14	.33	15.42	17.65
0450	36,000 S.F.		2500	.022		13.60	1.05	.30	14.95	17.10

13 31 Fabric Structures

13 31 23 – Tensioned Fabric Structures

	13 31 23.50 Tension Structures	Crew	Daily Output	Labor-Hours	Unit	Material	2010 Bare Costs Labor	Equipment	Total	Total Incl O&P
0460	120′ clear span, 24,000 S.F.	L-5	3000	.019	SF Flr.	15.55	.87	.25	16.67	18.90
0470	150′ clear span, 30,000 S.F.		6000	.009		16.70	.44	.12	17.26	19.25
0480	200′ clear span, 40,000 S.F.	E-6	8000	.016		19.05	.74	.24	20.03	22.50
0500	For roll-up door, 12′ x 14′, add	L-2	1	16	Ea.	6,075	585		6,660	7,575
0600	For personnel doors, add, minimum				SF Flr.	5%				
0700	Add, maximum					15%				
0800	For site work, simple foundation, etc., add, minimum								1.25	1.95
0900	Add, maximum								2.75	3.05

13 34 Fabricated Engineered Structures

13 34 13 – Glazed Structures

	13 34 13.13 Greenhouses	Crew	Daily Output	Labor-Hours	Unit	Material	2010 Bare Costs Labor	Equipment	Total	Total Incl O&P
0010	**GREENHOUSES**, Shell only, stock units, not incl. 2′ stub walls,									
0020	foundation, floors, heat or compartments									
0300	Residential type, free standing, 8′-6″ long x 7′-6″ wide	2 Carp	59	.271	SF Flr.	38	11.25		49.25	59
0400	10′-6″ wide		85	.188		27	7.80		34.80	41.50
0600	13′-6″ wide		108	.148		23	6.15		29.15	35
0700	17′-0″ wide		160	.100		27	4.16		31.16	36.50
0900	Lean-to type, 3′-10″ wide		34	.471		33.50	19.55		53.05	67
1000	6′-10″ wide		58	.276		38	11.45		49.45	59.50
1500	Commercial, custom, truss frame, incl. equip., plumbing, elec.,									
1510	benches and controls, under 2,000 S.F., minimum				SF Flr.				27.27	30
1550	Maximum					30.50			30.50	33.50
1700	Over 5,000 S.F., minimum					21			21	23.50
1750	Maximum								27.27	30
2000	Institutional, custom, rigid frame, including compartments and									
2010	multi-controls, under 500 S.F., minimum				SF Flr.	45			45	49.50
2050	Maximum					76			76	84
2150	Over 2,000 S.F., minimum					29.50			29.50	32.50
2200	Maximum					59			59	65
2400	Concealed rigid frame, under 500 S.F., minimum								81.82	90
2450	Maximum								100	110
2550	Over 2,000 S.F., minimum								61.82	68
2600	Maximum								71.82	79
2800	Lean-to type, under 500 S.F., minimum								72.73	80
2850	Maximum								109.09	120
3000	Over 2,000 S.F., minimum								40	44
3050	Maximum								66.36	73
3600	For 1/4″ clear plate glass, add				SF Surf	1.86			1.86	2.05
3700	For 1/4″ tempered glass, add				″	4.21			4.21	4.63
3900	For cooling, add, minimum				SF Flr.	2.81			2.81	3.09
4000	Maximum					6.95			6.95	7.65
4200	For heaters, 13.6 MBH, add					5.35			5.35	5.85
4300	60 MBH, add					2			2	2.20
4500	For benches, 2′ x 3′-6″, add				SF Hor.	24			24	26.50
4600	3′ x 10′, add				S.F.	12.95			12.95	14.25
4800	For controls, add, minimum				Total	2,400			2,400	2,650
4900	Maximum				″	14,300			14,300	15,700
5100	For humidification equipment, add				M.C.F.	6.15			6.15	6.75
5200	For vinyl shading, add				S.F.	1.29			1.29	1.42
6000	Geodesic hemisphere, 1/8″ plexiglass glazing									

13 34 13 – Glazed Structures

13 34 13.13 Greenhouses		Crew	Daily Output	Labor-Hours	Unit	Material	2010 Bare Costs Labor	Equipment	Total	Total Incl O&P
6050	8′ diameter	2 Carp	2	8	Ea.	2,700	330		3,030	3,475
6150	24′ diameter		.35	45.714		13,900	1,900		15,800	18,200
6250	48′ diameter		.20	80		37,600	3,325		40,925	46,500

13 34 13.19 Swimming Pool Enclosures		Crew	Daily Output	Labor-Hours	Unit	Material	Labor	Equipment	Total	Total Incl O&P
0010	**SWIMMING POOL ENCLOSURES** Translucent, free standing									
0020	not including foundations, heat or light									
0200	Economy, minimum	2 Carp	200	.080	SF Hor.	14.80	3.32		18.12	21.50
0300	Maximum		100	.160		43	6.65		49.65	58
0400	Deluxe, minimum		100	.160		50.50	6.65		57.15	66
0600	Maximum		70	.229		227	9.50		236.50	265
0700	For motorized roof, 40% opening, solid roof, add					8.30			8.30	9.15
0800	Skylight type roof, add					8.80			8.80	9.70
0900	Air-inflated, including blowers and heaters, minimum								3.18	3.50
1000	Maximum								6.09	6.70

13 34 16 – Grandstands and Bleachers

13 34 16.13 Grandstands		Crew	Daily Output	Labor-Hours	Unit	Material	Labor	Equipment	Total	Total Incl O&P
0010	**GRANDSTANDS** Permanent, municipal, including foundation									
0050	Steel understructure w/aluminum closed deck, minimum				Seat				118.18	130
0100	Maximum								205	225
0300	Steel, minimum					55			55	60.50
0400	Maximum					110			110	121
0600	Aluminum, extruded, stock design, minimum								63.64	70
0700	Maximum								104.55	115
0900	Composite, steel, wood and plastic, stock design, minimum					130			130	143
1000	Maximum					260			260	286

13 34 16.53 Bleachers		Crew	Daily Output	Labor-Hours	Unit	Material	Labor	Equipment	Total	Total Incl O&P
0010	**BLEACHERS**									
0020	Bleachers, outdoor, portable, 3 to 5 tiers, to 300′ long, min	2 Sswk	120	.133	Seat	48	6.25		54.25	63.50
0100	Maximum, less than 15′ long, prefabricated		80	.200		68.50	9.40		77.90	92
0200	6 to 20 tiers, minimum, up to 300′ long		120	.133		86	6.25		92.25	106
0300	Max., under 15′, (highly prefabricated, on wheels)		80	.200		83.50	9.40		92.90	109
0500	Permanent grandstands, wood seat, steel frame, 24″ row									
0600	3 to 15 tiers, minimum	2 Sswk	60	.267	Seat	122	12.50		134.50	156
0700	Maximum		48	.333		137	15.65		152.65	178
0900	16 to 30 tiers, minimum		60	.267		187	12.50		199.50	227
0950	Average		55	.291		210	13.65		223.65	255
1000	Maximum		48	.333		226	15.65		241.65	276
1200	Seat backs only, 30″ row, fiberglass		160	.100		26.50	4.69		31.19	38
1300	Steel and wood		160	.100		31.50	4.69		36.19	43
1400	NOTE: average seating is 1.5′ in width									

13 34 19 – Metal Building Systems

13 34 19.50 Pre-Engineered Steel Buildings		Crew	Daily Output	Labor-Hours	Unit	Material	Labor	Equipment	Total	Total Incl O&P
0010	**PRE-ENGINEERED STEEL BUILDINGS** R133419-10									
0100	Clear span rigid frame, 26 ga. colored roofing and siding									
0150	20′ wide, 10′ eave height	E-2	425	.132	SF Flr.	8.70	6	3.78	18.48	24
0160	14′ eave height		350	.160		9.25	7.30	4.59	21.14	27.50
0170	16′ eave height		320	.175		9.70	8	5	22.70	29.50
0180	20′ eave height		275	.204		10.70	9.30	5.85	25.85	34
0190	24′ eave height		240	.233		12.25	10.65	6.70	29.60	39
0200	30′ to 40′ wide, 10′ eave height		535	.105		7.10	4.77	3	14.87	19.20
0300	14′ eave height		450	.124		7.50	5.65	3.57	16.72	22

13 34 19.50 Pre-Engineered Steel Buildings		Crew	Daily Output	Labor-Hours	Unit	Material	2010 Bare Costs Labor	Equipment	Total	Total Incl O&P
0400	16′ eave height	E-2	415	.135	SF Flr.	7.85	6.15	3.87	17.87	23.50
0500	20′ eave height		360	.156		8.60	7.10	4.46	20.16	26.50
0600	24′ eave height		320	.175		9.70	8	5	22.70	29.50
0700	50′ to 100′ wide, 10′ eave height		865	.065		6.10	2.95	1.86	10.91	13.80
0800	14′ eave height		770	.073		10.30	3.32	2.09	15.71	19.25
0900	16′ eave height		730	.077		6.85	3.50	2.20	12.55	15.85
1000	20′ eave height		660	.085		7.40	3.87	2.44	13.71	17.35
1100	24′ eave height		605	.093		8.10	4.22	2.66	14.98	18.95
1200	Clear span tapered beam frame, 26 ga. colored roofing/siding									
1300	30′ wide, 10′ eave height	E-2	535	.105	SF Flr.	7.80	4.77	3	15.57	19.95
1400	14′ eave height		450	.124		8.60	5.65	3.57	17.82	23
1500	16′ eave height		415	.135		9.25	6.15	3.87	19.27	25
1600	20′ eave height		360	.156		10.20	7.10	4.46	21.76	28
1700	40′ wide, 10′ eave height		600	.093		7.20	4.26	2.68	14.14	18.05
1800	14′ eave height		510	.110		7.80	5	3.15	15.95	20.50
1900	16′ eave height		475	.118		8.15	5.40	3.38	16.93	22
2000	20′ eave height		415	.135		8.90	6.15	3.87	18.92	24.50
2100	50′ to 80′ wide, 10′ eave height		770	.073		6.65	3.32	2.09	12.06	15.20
2200	14′ eave height		675	.083		7.20	3.78	2.38	13.36	16.95
2300	16′ eave height		635	.088		7.50	4.02	2.53	14.05	17.85
2400	20′ eave height		490	.114		11.35	5.20	3.28	19.83	25
2500	Single post 2-span frame, 26 ga. colored roofing and siding									
2600	80′ wide, 14′ eave height	E-2	740	.076	SF Flr.	5.85	3.45	2.17	11.47	14.70
2700	16′ eave height		695	.081		6.20	3.67	2.31	12.18	15.60
2800	20′ eave height		625	.090		6.75	4.09	2.57	13.41	17.20
2900	24′ eave height		570	.098		7.30	4.48	2.82	14.60	18.70
3000	100′ wide, 14′ eave height		835	.067		5.65	3.06	1.92	10.63	13.50
3100	16′ eave height		795	.070		5.55	3.21	2.02	10.78	13.80
3200	20′ eave height		730	.077		6.55	3.50	2.20	12.25	15.50
3300	24′ eave height		670	.084		7.05	3.81	2.40	13.26	16.85
3400	120′ wide, 14′ eave height		870	.064		5.60	2.94	1.85	10.39	13.15
3500	16′ eave height		830	.067		5.85	3.08	1.94	10.87	13.80
3600	20′ eave height		765	.073		6.30	3.34	2.10	11.74	14.90
3700	24′ eave height		705	.079		6.90	3.62	2.28	12.80	16.20
3800	Double post 3-span frame, 26 ga. colored roofing and siding									
3900	150′ wide, 14′ eave height	E-2	925	.061	SF Flr.	4.83	2.76	1.74	9.33	11.90
4000	16′ eave height		890	.063		4.98	2.87	1.81	9.66	12.35
4100	20′ eave height		820	.068		5.25	3.11	1.96	10.32	13.20
4200	24′ eave height		765	.073		6.25	3.34	2.10	11.69	14.85
4300	Triple post 4-span frame, 26 ga. colored roofing and siding									
4400	160′ wide, 14′ eave height	E-2	970	.058	SF Flr.	4.68	2.63	1.66	8.97	11.40
4500	16′ eave height		930	.060		4.90	2.75	1.73	9.38	11.95
4600	20′ eave height		870	.064		4.89	2.94	1.85	9.68	12.40
4700	24′ eave height		815	.069		5.85	3.13	1.97	10.95	13.85
4800	200′ wide, 14′ eave height		1030	.054		4.49	2.48	1.56	8.53	10.85
4900	16′ eave height		995	.056		4.61	2.57	1.62	8.80	11.15
5000	20′ eave height		935	.060		4.98	2.73	1.72	9.43	12
5100	24′ eave height		885	.063		5.85	2.89	1.82	10.56	13.35
5200	Accessory items: add to the basic building cost above									
5250	Eave overhang, 2′ wide, 26 ga., with soffit	E-2	360	.156	L.F.	24.50	7.10	4.46	36.06	44
5300	4′ wide, without soffit		300	.187		24	8.50	5.35	37.85	47
5350	With soffit		250	.224		33.50	10.20	6.45	50.15	61.50
5400	6′ wide, without soffit		250	.224		33	10.20	6.45	49.65	61

13 34 19.50 Pre-Engineered Steel Buildings		Crew	Daily Output	Labor-Hours	Unit	Material	2010 Bare Costs Labor	Equipment	Total	Total Incl O&P
5450	With soffit	E-2	200	.280	L.F.	43	12.75	8.05	63.80	77.50
5500	Entrance canopy, incl. frame, 4' x 4'		25	2.240	Ea.	350	102	64.50	516.50	630
5550	4' x 8'		19	2.947	"	480	134	84.50	698.50	850
5600	End wall roof overhang, 4' wide, without soffit		850	.066	L.F.	16	3	1.89	20.89	25
5650	With soffit		500	.112	"	23.50	5.10	3.21	31.81	38
5700	Doors, H.M. self-framing, incl. butts, lockset and trim									
5750	Single leaf, 3070 (3' x 7'), economy	2 Sswk	5	3.200	Opng.	530	150		680	850
5800	Deluxe		4	4		555	188		743	940
5825	Glazed		4	4		565	188		753	950
5850	3670 (3'-6" x 7')		4	4		675	188		863	1,075
5900	4070 (4' x 7')		3	5.333		710	250		960	1,225
5950	Double leaf, 6070 (6' x 7')		2	8		960	375		1,335	1,700
6000	Glazed		2	8		1,025	375		1,400	1,775
6050	Framing only, for openings, 3' x 7'	E-2	25	2.240		153	102	64.50	319.50	415
6100	10' x 10'		21	2.667		505	122	76.50	703.50	845
6150	For windows below, 2020 (2' x 2')		25	2.240		162	102	64.50	328.50	425
6200	4030 (4' x 3')		22	2.545		198	116	73	387	495
6250	Flashings, 26 ga., corner or eave, painted	2 Sswk	240	.067	L.F.	4.01	3.13		7.14	9.90
6300	Galvanized		240	.067		3.50	3.13		6.63	9.35
6350	Rake flashing, painted		240	.067		4.29	3.13		7.42	10.20
6400	Galvanized		240	.067		3.84	3.13		6.97	9.70
6450	Ridge flashing, 18" wide, painted		240	.067		5.25	3.13		8.38	11.30
6500	Galvanized		240	.067		4.95	3.13		8.08	10.95
6550	Gutter, eave type, 26 ga., painted		320	.050		5.50	2.35		7.85	10.15
6600	Galvanized		320	.050		3.54	2.35		5.89	8
6650	Valley type, between buildings, painted		120	.133		9.35	6.25		15.60	21.50
6710	Insulation, rated .6 lb density, unfaced 4" thick, R13	2 Carp	2300	.007	S.F.	.30	.29		.59	.78
6720	6" thick, R-19		2300	.007		.42	.29		.71	.91
6730	10" thick, R-30		2300	.007		.72	.29		1.01	1.24
6750	Insulation, rated .6 lb density, poly/scrim/foil (PSF) faced									
6760	4" thick R-13	2 Carp	2300	.007	S.F.	.40	.29		.69	.89
6770	6" thick, R-19		2300	.007		.53	.29		.82	1.03
6780	10" thick, R-30		2300	.007		.72	.29		1.01	1.24
6800	Insulation, rated .6 lb density, vinyl faced 1-1/2" thick, R5		2300	.007		.27	.29		.56	.75
6850	3" thick, R10		2300	.007		.29	.29		.58	.77
6900	4" thick, R13		2300	.007		.41	.29		.70	.90
6920	6" thick, R19		2300	.007		.58	.29		.87	1.09
6930	10" thick, R30		2300	.007		.92	.29		1.21	1.46
6950	Foil/scrim/kraft (FSK) faced, 1-1/2" thick, R5		2300	.007		.29	.29		.58	.77
7000	2" thick, R6		2300	.007		.37	.29		.66	.86
7050	3" thick, R10		2300	.007		.37	.29		.66	.86
7100	4" thick, R13		2300	.007		.47	.29		.76	.97
7110	6" thick, R19		2300	.007		.55	.29		.84	1.06
7120	10" thick, R-30		2300	.007		.82	.29		1.11	1.35
7150	Met polyester/scrim/kraft (PSK) facing,1-1/2" thk,R5		2300	.007		.47	.29		.76	.97
7200	2" thick, R6		2300	.007		.55	.29		.84	1.06
7250	3" thick, R11		2300	.007		.52	.29		.81	1.02
7300	4" thick, R13		2300	.007		.63	.29		.92	1.14
7310	6" thick, R19		2300	.007		.81	.29		1.10	1.34
7320	10" thick, R-30		2300	.007		.96	.29		1.25	1.51
7350	Vinyl/scrim/foil (VSF), 1-1/2" thick, R5		2300	.007		.41	.29		.70	.90
7400	2" thick, R6		2300	.007		.52	.29		.81	1.02
7450	3" thick, R10		2300	.007		.58	.29		.87	1.09

13 34 19 – Metal Building Systems

13 34 19.50 Pre-Engineered Steel Buildings		Crew	Daily Output	Labor-Hours	Unit	2010 Bare Costs Material	Labor	Equipment	Total	Total Incl O&P
7500	4" thick, R13	2 Carp	2300	.007	S.F.	.67	.29		.96	1.19
7510	Vinyl/scrim/vinyl (VSV) 4" thick, R-13		2300	.007		.45	.29		.74	.95
7520	6" thick R-19		2300	.007		.59	.29		.88	1.10
7530	10" thick, R-30		2300	.007		.80	.29		1.09	1.33
7540	Polyprop/scrim/polyester (PSP), 4",R-13		2300	.007		.40	.29		.69	.89
7550	6" thick, R-19		2300	.007		.54	.29		.83	1.04
7555	10" thick, R-30		2300	.007		.99	.29		1.28	1.54
7560	Polyprop/scrim/kraft/polyester (PSKP), 4" thick, R-13		2300	.007		.42	.29		.71	.91
7570	6" thick, R-19		2300	.007		.55	.29		.84	1.06
7580	10" thick, R-19		2300	.007		1.05	.29		1.34	1.61
7585	Vinyl/scrim/polyester (VSP), 4" thick, R-13		2300	.007		.44	.29		.73	.93
7590	6" thick, R-19		2300	.007		.60	.29		.89	1.11
7600	10" thick, R-30		2300	.007		1.08	.29		1.37	1.64
7635	Insulation installation, over the purlin, second layer, up to 4" thick, add						90%			
7640	Insulation installation, between the purlins, up to 4" thick, add						100%			
7650	Sash, single slide, glazed, with screens, 2020 (2'x 2')	E-1	22	1.091	Opng.	117	50	6.65	173.65	220
7700	3030 (3' x 3')		14	1.714		264	78.50	10.40	352.90	435
7750	4030 (4' x 3')		13	1.846		350	84.50	11.20	445.70	545
7800	6040 (6' x 4')		12	2		705	91.50	12.15	808.65	940
7850	Double slide sash, 3030 (3' x 3')		14	1.714		177	78.50	10.40	265.90	340
7900	6040 (6' x 4')		12	2		475	91.50	12.15	578.65	685
7950	Fixed glass, no screens, 3030 (3' x 3')		14	1.714		218	78.50	10.40	306.90	385
8000	6040 (6' x 4')		12	2		580	91.50	12.15	683.65	805
8050	Prefinished storm sash, 3030 (3' x 3')		70	.343		72	15.65	2.08	89.73	108
8100	Siding and roofing, see Div. 07 41 13.00 & 07 42 13.00									
8200	Skylight, fiberglass panels, to 30 S.F.	E-1	10	2.400	Ea.	103	110	14.60	227.60	315
8250	Larger sizes, add for excess over 30 S.F.	"	300	.080	S.F.	3.43	3.66	.49	7.58	10.45
8300	Roof vents, circular with damper, birdscreen									
8350	and operator hardware, painted									
8400	26 ga., 12" diameter	1 Sswk	4	2	Ea.	69	94		163	241
8450	20" diameter		3	2.667		142	125		267	375
8500	24 ga., 24" diameter		2	4		282	188		470	640
8550	Galvanized		2	4		236	188		424	590
8600	Continuous, 26 ga., 10' long, 9" wide	2 Sswk	4	4		22.50	188		210.50	355
8650	12" wide	"	4	4		27	188		215	360

13 34 23 – Fabricated Structures

13 34 23.10 Comfort Stations

		Crew	Daily Output	Labor-Hours	Unit	Material	Labor	Equipment	Total	Total Incl O&P
0010	**COMFORT STATIONS** Prefab., stock, w/doors, windows & fixt.									
0100	Not incl. interior finish or electrical									
0300	Mobile, on steel frame, minimum				S.F.	46.50			46.50	51
0350	Maximum					75			75	82.50
0400	Permanent, including concrete slab, minimum	B-12J	50	.320		199	12.40	16.25	227.65	256
0500	Maximum	"	43	.372		288	14.40	18.90	321.30	360
0600	Alternate pricing method, mobile, minimum				Fixture	2,150			2,150	2,375
0650	Maximum					3,175			3,175	3,500
0700	Permanent, minimum	B-12J	.70	22.857		12,200	885	1,150	14,235	16,100
0750	Maximum	"	.50	32		20,400	1,250	1,625	23,275	26,100

13 34 23.15 Domes

		Crew	Daily Output	Labor-Hours	Unit	Material	Labor	Equipment	Total	Total Incl O&P
0010	**DOMES**									
0020	Domes, rev alum, elec drive, for astronomy obsv shell only, stock units									
0600	10'-0" diameter, 800#, dome	2 Carp	.25	64	Ea.	11,000	2,650		13,650	16,200
0700	Base		.67	23.881		3,800	990		4,790	5,700

13 34 Fabricated Engineered Structures

13 34 23 – Fabricated Structures

13 34 23.15 Domes		Crew	Daily Output	Labor-Hours	Unit	Material	2010 Bare Costs Labor	Equipment	Total	Total Incl O&P
0900	18'-0" diameter, 2,500#, dome	2 Carp	.17	94.118	Ea.	32,100	3,900		36,000	41,300
1000	Base		.33	48.485		10,600	2,025		12,625	14,800
1200	24'-0" diameter, 4,500#, dome		.08	200		59,500	8,300		67,800	78,500
1300	Base		.25	64		19,200	2,650		21,850	25,200
1500	Domes, bulk storage, shell only, dual radius hemisphere, arch, steel									
1600	framing, corrugated steel covering, 150' diameter	E-2	550	.102	SF Flr.	28	4.64	2.92	35.56	41.50
1700	400' diameter	"	720	.078		22.50	3.55	2.23	28.28	33.50
1800	Wood framing, wood decking, to 400' diameter	F-4	400	.120		20.50	4.98	3.02	28.50	33.50
1900	Radial framed wood (2" x 6"), 1/2" thick									
2000	plywood, asphalt shingles, 50' diameter	F-3	2000	.020	SF Flr.	38.50	.84	.33	39.67	44
2100	60' diameter		1900	.021		29.50	.89	.35	30.74	34
2200	72' diameter		1800	.022		26.50	.94	.36	27.80	31
2300	116' diameter		1730	.023		23	.97	.38	24.35	27.50
2400	150' diameter		1500	.027		20	1.12	.44	21.56	24

13 34 23.16 Fabricated Control Booths		Crew	Daily Output	Labor-Hours	Unit	Material	Labor	Equipment	Total	Total Incl O&P
0010	**FABRICATED CONTROL BOOTHS**									
0100	Guard House, prefab conc w/bullet resistant doors & windows, roof & wiring									
0110	8' x 8', Level III	L-10	1	24	Ea.	22,000	1,125	655	23,780	26,800
0120	8' x 8', Level IV	"	1	24	"	27,500	1,125	655	29,280	32,900

13 34 23.25 Garage Costs		Crew	Daily Output	Labor-Hours	Unit	Material	Labor	Equipment	Total	Total Incl O&P
0010	**GARAGE COSTS**									
0020	Public parking, average				Car				16,500	18,200
0100	See also Square Foot Costs in Reference Section									
0300	Residential, prefab shell, stock, wood, single car, minimum	2 Carp	1	16	Total	5,100	665		5,765	6,625
0350	Maximum		.67	23.881		5,875	990		6,865	8,000
0400	Two car, minimum		.67	23.881		10,200	990		11,190	12,700
0450	Maximum		.50	32		11,800	1,325		13,125	15,000

13 34 23.30 Garden House		Crew	Daily Output	Labor-Hours	Unit	Material	Labor	Equipment	Total	Total Incl O&P
0010	**GARDEN HOUSE** Prefab wood, no floors or foundations									
0100	32 to 200 S.F., minimum	2 Carp	200	.080	SF Flr.	30	3.32		33.32	38
0300	Maximum	"	48	.333	"	43.50	13.85		57.35	69.50

13 34 23.35 Geodesic Domes		Crew	Daily Output	Labor-Hours	Unit	Material	Labor	Equipment	Total	Total Incl O&P
0010	**GEODESIC DOMES** Shell only, interlocking plywood panels R133423-30									
0400	30' diameter	F-5	1.60	20	Ea.	15,800	840		16,640	18,700
0500	34' diameter		1.14	28.070		19,400	1,175		20,575	23,200
0600	39' diameter		1	32		22,600	1,350		23,950	27,000
0700	45' diameter	F-3	1.13	35.556		27,200	1,500	585	29,285	32,900
0750	55' diameter		1	40		40,500	1,675	655	42,830	47,900
0800	60' diameter		1	40		53,000	1,675	655	55,330	61,500
0850	65' diameter		.80	50		62,500	2,100	820	65,420	73,000
1100	Aluminum panel, with 6" insulation									
1200	100' diameter				SF Flr.	41.50			41.50	45.50
1300	500' diameter				"	35.50			35.50	39
1600	Aluminum framed, plexiglass closure panels									
1700	40' diameter				SF Flr.	105			105	116
1800	200' diameter				"	84			84	92.50
2100	Aluminum framed, aluminum closure panels									
2200	40' diameter				SF Flr.	33.50			33.50	36.50
2300	100' diameter					25			25	27.50
2400	200' diameter					24			24	26
2500	For VRP faced bonded fiberglass insulation, add									10
2700	Aluminum framed, fiberglass sandwich panel closure									

13 34 Fabricated Engineered Structures

13 34 23 – Fabricated Structures

13 34 23.35 Geodesic Domes		Crew	Daily Output	Labor-Hours	Unit	Material	2010 Bare Costs Labor	Equipment	Total	Total Incl O&P
2800	6′ diameter	2 Carp	150	.107	SF Flr.	37	4.43		41.43	47.50
2900	28′ diameter	"	350	.046	"	33.50	1.90		35.40	40
13 34 23.45 Kiosks										
0010	**KIOSKS**									
0020	Round, 5′ diameter, 8′ high, 1/4″ fiberglass wall				Ea.	5,800			5,800	6,375
0100	1″ insulated double wall, fiberglass					6,600			6,600	7,250
0500	Rectangular, 5′ x 9′, 7′-6″ high, 1/4″ fiberglass wall					8,450			8,450	9,300
0600	1″ insulated double wall, fiberglass					10,000			10,000	11,000
13 34 23.60 Portable Booths										
0010	**PORTABLE BOOTHS** Prefab aluminum with doors, windows, ext. roof									
0100	lights wiring & insulation, 15 S.F. building, O.D., painted, minimum				S.F.	310			310	340
0300	30 S.F. building, minimum					212			212	233
0400	50 S.F. building, minimum					157			157	173
0600	80 S.F. building, minimum					128			128	141
0700	100 S.F. building, minimum					120			120	132
0900	Acoustical booth, 27 Db @ 1,000 Hz, 15 S.F. floor				Ea.	3,625			3,625	3,975
1000	7′ x 7′-6″, including light & ventilation					7,425			7,425	8,175
1200	Ticket booth, galv. steel, not incl. foundations., 4′ x 4′					4,950			4,950	5,425
1300	4′ x 6′					6,400			6,400	7,050
13 34 23.70 Shelters										
0010	**SHELTERS**									
0020	Aluminum frame, acrylic glazing, 3′ x 9′ x 8′ high	2 Sswk	1.14	14.035	Ea.	3,350	660		4,010	4,825
0100	9′ x 12′ x 8′ high	"	.73	21.918	"	7,900	1,025		8,925	10,500

13 34 43 – Aircraft Hangars

13 34 43.50 Hangars		Crew	Daily Output	Labor-Hours	Unit	Material	2010 Bare Costs Labor	Equipment	Total	Total Incl O&P
0010	**HANGARS** Prefabricated steel T hangars, Galv. steel roof &									
0100	walls, incl. electric bi-folding doors, 4 or more units,									
0110	not including floors or foundations, minimum	E-2	1275	.044	SF Flr.	13.20	2	1.26	16.46	19.35
0130	Maximum		1063	.053		14.65	2.40	1.51	18.56	22
0900	With bottom rolling doors, minimum		1386	.040		12.85	1.84	1.16	15.85	18.55
1000	Maximum		966	.058		14.25	2.64	1.66	18.55	22
1200	Alternate pricing method:									
1300	Galv. roof and walls, electric bi-folding doors, minimum	E-2	1.06	52.830	Plane	17,100	2,400	1,525	21,025	24,600
1500	Maximum		.91	61.538		20,400	2,800	1,775	24,975	29,200
1600	With bottom rolling doors, minimum		1.25	44.800		12,900	2,050	1,275	16,225	19,100
1800	Maximum		.97	57.732		15,000	2,625	1,650	19,275	22,800
2000	Circular type, prefab., steel frame, plastic skin, electric									
2010	door, including foundations, 80′ diameter,									
2020	for up to 5 light planes, minimum	E-2	.50	112	Total	72,500	5,100	3,225	80,825	92,000
2200	Maximum	"	.25	224	"	82,000	10,200	6,425	98,625	114,500

13 34 53 – Agricultural Structures

13 34 53.50 Silos		Crew	Daily Output	Labor-Hours	Unit	Material	2010 Bare Costs Labor	Equipment	Total	Total Incl O&P
0010	**SILOS** Concrete stave industrial, conical or sloping bottom									
0100	excluding foundation, 12′ diameter, 35′ high	E-24	.09	355	Ea.	42,400	16,300	8,300	67,000	83,500
0200	16′ diameter, 45′ high		.06	533		69,000	24,500	12,500	106,000	131,500
0400	25′ diameter, 75′ high		.04	800		159,000	36,700	18,700	214,400	258,000
0500	Steel, factory fab., 30,000 gallon cap., painted, minimum	L-5	1	56		21,000	2,625	750	24,375	28,500
0700	Maximum		.50	112		33,400	5,250	1,500	40,150	47,500
0800	Epoxy lined, minimum		1	56		34,300	2,625	750	37,675	43,200
1000	Maximum		.50	112		43,600	5,250	1,500	50,350	58,500

13 34 Fabricated Engineered Structures

13 34 63 – Natural Fiber Construction

13 34 63.50 Straw Bale Construction		Crew	Daily Output	Labor-Hours	Unit	Material	2010 Bare Costs Labor	Equipment	Total	Total Incl O&P
0010	**STRAW BALE CONSTRUCTION**									
2000	Straw bales wall, incl labor & material complete, minimum G				SF Flr.					130
2010	Maximum G				"					160
2020	Straw bales in walls w/ modified post and beam frame G	2 Carp	320	.050	S.F.	4.11	2.08		6.19	7.70

13 36 Towers

13 36 13 – Metal Towers

13 36 13.50 Control Towers		Crew	Daily Output	Labor-Hours	Unit	Material	Labor	Equipment	Total	Total Incl O&P
0010	**CONTROL TOWERS**									
0020	Modular 12′ x 10′, incl. instruments, min.				Ea.	670,000			670,000	737,000
0100	Maximum								927,273	1,020,000
0500	With standard 40′ tower, average				↓	935,000			935,000	1,028,500
1000	Temporary portable control towers, 8′ x 12′,									
1010	complete with one position communications, minimum				Ea.				266,000	293,000
2000	For fixed facilities, depending on height, minimum								57,200	63,000
2010	Maximum				↓				115,000	127,000

13 42 Building Modules

13 42 63 – Detention Cell Modules

13 42 63.16 Steel Detention Cell Modules		Crew	Daily Output	Labor-Hours	Unit	Material	Labor	Equipment	Total	Total Incl O&P
0010	**STEEL DETENTION CELL MODULES**									
2000	Cells, prefab., 5′ to 6′ wide, 7′ to 8′ high, 7′ to 8′ deep,									
2010	bar front, cot, not incl. plumbing	E-4	1.50	21.333	Ea.	9,650	1,000	97.50	10,747.50	12,500

13 48 Sound, Vibration, and Seismic Control

13 48 13 – Manufactured Sound and Vibration Control Components

13 48 13.50 Audio Masking		Crew	Daily Output	Labor-Hours	Unit	Material	Labor	Equipment	Total	Total Incl O&P
0010	**AUDIO MASKING**, acoustical enclosure, 4″ thick wall and ceiling									
0020	8# per S.F., up to 12′ span	3 Carp	72	.333	SF Surf	32	13.85		45.85	57
0300	Better quality panels, 10.5# per S.F.		64	.375		36.50	15.60		52.10	64
0400	Reverb-chamber, 4″ thick, parallel walls		60	.400		45.50	16.60		62.10	75.50
0600	Skewed wall, parallel roof, 4″ thick panels		55	.436		52	18.15		70.15	85
0700	Skewed walls, skewed roof, 4″ layers, 4″ air space		48	.500		58.50	21		79.50	96.50
0900	Sound-absorbing panels, pntd mtl, 2′-6″ x 8′, under 1,000 S.F.		215	.112		12.10	4.64		16.74	20.50
1100	Over 1000 S.F.		240	.100		11.65	4.16		15.81	19.20
1200	Fabric faced	↓	240	.100		9.45	4.16		13.61	16.75
1500	Flexible transparent curtain, clear	3 Shee	215	.112		7.30	5.50		12.80	16.30
1600	50% foam		215	.112		10.20	5.50		15.70	19.50
1700	75% foam		215	.112		10.20	5.50		15.70	19.50
1800	100% foam	↓	215	.112	↓	10.20	5.50		15.70	19.50
3100	Audio masking system, including speakers, amplification									
3110	and signal generator									
3200	Ceiling mounted, 5,000 S.F.	2 Elec	2400	.007	S.F.	1.25	.33		1.58	1.87
3300	10,000 S.F.		2800	.006		1.01	.28		1.29	1.53
3400	Plenum mounted, 5,000 S.F.		3800	.004		1.06	.21		1.27	1.48
3500	10,000 S.F.	↓	4400	.004	↓	.72	.18		.90	1.06

13 49 Radiation Protection

13 49 13 – Lead Sheet

13 49 13.50 Lead Sheets		Crew	Daily Output	Labor-Hours	Unit	Material	2010 Bare Costs Labor	Equipment	Total	Total Incl O&P
0010	**LEAD SHEETS**									
0300	Lead sheets, 1/16″ thick	2 Lath	135	.119	S.F.	7.15	4.38		11.53	14.35
0400	1/8″ thick		120	.133		14.45	4.93		19.38	23
0500	Lead shielding, 1/4″ thick		135	.119		34	4.38		38.38	44
0550	1/2″ thick		120	.133		59.50	4.93		64.43	73
0950	Lead headed nails (average 1 lb. per sheet)				Lb.	7.15			7.15	7.90
1000	Butt joints in 1/8″ lead or thicker, 2″ batten strip x 7′ long	2 Lath	240	.067	Ea.	16.95	2.46		19.41	22.50
1200	X-ray protection, average radiography or fluoroscopy									
1210	room, up to 300 S.F. floor, 1/16″ lead, minimum	2 Lath	.25	64	Total	8,075	2,375		10,450	12,400
1500	Maximum, 7′-0″ walls	″	.15	106	″	9,700	3,950		13,650	16,500
1600	Deep therapy X-ray room, 250 kV capacity,									
1800	up to 300 S.F. floor, 1/4″ lead, minimum	2 Lath	.08	200	Total	22,600	7,400		30,000	35,700
1900	Maximum, 7′-0″ walls	″	.06	266	″	27,800	9,850		37,650	45,100

13 49 19 – Lead-Lined Materials

13 49 19.50 Shielding Lead		Crew	Daily Output	Labor-Hours	Unit	Material	Labor	Equipment	Total	Total Incl O&P
0010	**SHIELDING LEAD**									
0100	Laminated lead in wood doors, 1/16″ thick, no hardware				S.F.	45			45	49.50
0200	Lead lined door frame, not incl. hardware,									
0210	1/16″ thick lead, butt prepared for hardware	1 Lath	2.40	3.333	Ea.	575	123		698	810
0850	Window frame with 1/16″ lead and voice passage, 36″ x 60″	2 Glaz	2	8		1,125	320		1,445	1,700
0870	24″ x 36″ frame		4	4		900	161		1,061	1,225
0900	Lead gypsum board, 5/8″ thick with 1/16″ lead		160	.100	S.F.	7.25	4.02		11.27	14.05
0910	1/8″ lead *CN*		140	.114		14.45	4.59		19.04	23
0930	1/32″ lead	2 Lath	200	.080		4.43	2.96		7.39	9.20

13 49 21 – Lead Glazing

13 49 21.50 Lead Glazing		Crew	Daily Output	Labor-Hours	Unit	Material	Labor	Equipment	Total	Total Incl O&P
0010	**LEAD GLAZING**									
0600	Lead glass, 1/4″ thick, 2.0 mm LE, 12″ x 16″	2 Glaz	13	1.231	Ea.	265	49.50		314.50	365
0700	24″ x 36″		8	2		980	80.50		1,060.50	1,200
0800	36″ x 60″		2	8		3,850	320		4,170	4,725
2000	X-ray viewing panels, clear lead plastic									
2010	7 mm thick, 0.3 mm LE, 2.3 lbs/S.F.	H-3	139	.115	S.F.	143	4.13		147.13	164
2020	12 mm thick, 0.5 mm LE, 3.9 lbs/S.F.		82	.195		194	7		201	224
2030	18 mm thick, 0.8 mm LE, 5.9 lbs/S.F.		54	.296		210	10.65		220.65	247
2040	22 mm thick, 1.0 mm LE, 7.2 lbs/S.F.		44	.364		215	13.05		228.05	256
2050	35 mm thick, 1.5 mm LE, 11.5 lbs/S.F.		28	.571		241	20.50		261.50	296
2060	46 mm thick, 2.0 mm LE, 15.0 lbs/S.F.		21	.762		315	27.50		342.50	390
2090	For panels 12 S.F. to 48 S.F., add crating charge				Ea.					50

13 49 23 – Modular Shielding Partitions

13 49 23.50 Modular Shielding Partitions		Crew	Daily Output	Labor-Hours	Unit	Material	Labor	Equipment	Total	Total Incl O&P
0010	**MODULAR SHIELDING PARTITIONS**									
4000	X-ray barriers, modular, panels mounted within framework for									
4002	attaching to floor, wall or ceiling, upper portion is clear lead									
4005	plastic window panels 48″H, lower portion is opaque leaded									
4008	steel panels 36″H, structural supports not incl.									
4010	1-section barrier, 36″W x 84″H overall									
4020	0.5 mm LE panels	H-3	6.40	2.500	Ea.	4,525	90		4,615	5,100
4030	0.8 mm LE panels		6.40	2.500		4,825	90		4,915	5,425
4040	1.0 mm LE panels		5.33	3.002		4,900	108		5,008	5,575
4050	1.5 mm LE panels		5.33	3.002		5,225	108		5,333	5,925
4060	2-section barrier, 72″W x 84″H overall									

13 49 Radiation Protection

13 49 23 – Modular Shielding Partitions

13 49 23.50 Modular Shielding Partitions		Crew	Daily Output	Labor-Hours	Unit	2010 Bare Costs Material	Labor	Equipment	Total	Total Incl O&P
4070	0.5 mm LE panels	H-3	4	4	Ea.	9,250	144		9,394	10,400
4080	0.8 mm LE panels		4	4		9,775	144		9,919	11,000
4090	1.0 mm LE panels		3.56	4.494		9,975	161		10,136	11,200
5000	1.5 mm LE panels	↓	3.20	5	↓	11,300	180		11,480	12,800
5010	3-section barrier, 108"W x 84"H overall									
5020	0.5 mm LE panels	H-3	3.20	5	Ea.	13,900	180		14,080	15,600
5030	0.8 mm LE panels		3.20	5		14,700	180		14,880	16,400
5040	1.0 mm LE panels		2.67	5.993		14,900	215		15,115	16,700
5050	1.5 mm LE panels	↓	2.46	6.504	↓	16,200	234		16,434	18,200
7000	X-ray barriers, mobile, mounted within framework w/casters on									
7005	bottom, clear lead plastic window panels on upper portion,									
7010	opaque on lower, 30"W x 75"H overall, incl. framework									
7020	24"H upper w/0.5 mm LE, 48"H lower w/0.8 mm LE	1 Carp	16	.500	Ea.	2,400	21		2,421	2,675
7030	48"W x 75"H overall, incl. framework									
7040	36"H upper w/0.5 mm LE, 36"H lower w/0.8 mm LE	1 Carp	16	.500	Ea.	4,425	21		4,446	4,875
7050	36"H upper w/1.0 mm LE, 36"H lower w/1.5 mm LE	"	16	.500	"	5,250	21		5,271	5,800
7060	72"W x 75"H overall, incl. framework									
7070	36"H upper w/0.5 mm LE, 36"H lower w/0.8 mm LE	1 Carp	16	.500	Ea.	5,225	21		5,246	5,775
7080	36"H upper w/1.0 mm LE, 36"H lower w/1.5 mm LE	"	16	.500	"	6,550	21		6,571	7,225

13 49 33 – Radio Frequency Shielding

13 49 33.50 Shielding, Radio Frequency		Crew	Daily Output	Labor-Hours	Unit	Material	Labor	Equipment	Total	Total Incl O&P
0010	**SHIELDING, RADIO FREQUENCY**									
0020	Prefabricated or screen-type copper or steel, minimum	2 Carp	180	.089	SF Surf	31.50	3.69		35.19	40
0100	Average		155	.103		33.50	4.29		37.79	43.50
0150	Maximum	↓	145	.110	↓	35.50	4.58		40.08	46

13 53 Meteorological Instrumentation

13 53 09 – Weather Instrumentation

13 53 09.50 Weather Station		Crew	Daily Output	Labor-Hours	Unit	Material	Labor	Equipment	Total	Total Incl O&P
0010	**WEATHER STATION**									
0020	Remote recording				Ea.	675			675	745
0100	1 mile range				"	995			995	1,100

Division Notes

		CREW	DAILY OUTPUT	LABOR-HOURS	UNIT	2010 BARE COSTS MAT.	LABOR	EQUIP.	TOTAL	TOTAL INCL O&P

Estimating Tips

General

- Many products in Division 14 will require some type of support or blocking for installation not included with the item itself. Examples are supports for conveyors or tube systems, attachment points for lifts, and footings for hoists or cranes. Add these supports in the appropriate division.

14 10 00 Dumbwaiters 14 20 00 Elevators

- Dumbwaiters and elevators are estimated and purchased in a method similar to buying a car. The manufacturer has a base unit with standard features. Added to this base unit price will be whatever options the owner or specifications require. Increased load capacity, additional vertical travel, additional stops, higher speed, and cab finish options are items to be considered. When developing an estimate for dumbwaiters and elevators, remember that some items needed by the installers may have to be included as part of the general contract.

  Examples are:
  - — shaftway
  - — rail support brackets
  - — machine room
  - — electrical supply
  - — sill angles
  - — electrical connections
  - — pits
  - — roof penthouses
  - — pit ladders

Check the job specifications and drawings before pricing.

- Installation of elevators and handicapped lifts in historic structures can require significant additional costs. The associated structural requirements may involve cutting into and repairing finishes, mouldings, flooring, etc. The estimator must account for these special conditions.

14 30 00 Escalators and Moving Walks

- Escalators and moving walks are specialty items installed by specialty contractors. There are numerous options associated with these items. For specific options, contact a manufacturer or contractor. In a method similar to estimating dumbwaiters and elevators, you should verify the extent of general contract work and add items as necessary.

14 40 00 Lifts 14 90 00 Other Conveying Equipment

- Products such as correspondence lifts, chutes, and pneumatic tube systems, as well as other items specified in this subdivision, may require trained installers. The general contractor might not have any choice as to who will perform the installation or when it will be performed. Long lead times are often required for these products, making early decisions in scheduling necessary.

Reference Numbers

Reference numbers are shown in shaded boxes at the beginning of some major classifications. These numbers refer to related items in the Reference Section. The reference information may be an estimating procedure, an alternate pricing method, or technical information.

Note: Not all subdivisions listed here necessarily appear in this publication.

14 11 Manual Dumbwaiters

14 11 10 – Hand Operated Dumbwaiters

14 11 10.20 Manual Dumbwaiters		Crew	Daily Output	Labor-Hours	Unit	Material	2010 Bare Costs Labor	Equipment	Total	Total Incl O&P
0010	**MANUAL DUMBWAITERS**									
0020	2 stop, minimum	2 Elev	.75	21.333	Ea.	2,925	1,325		4,250	5,175
0100	Maximum		.50	32	"	6,650	1,975		8,625	10,300
0300	For each additional stop, add		.75	21.333	Stop	1,050	1,325		2,375	3,125

14 12 Electric Dumbwaiters

14 12 10 – Dumbwaiters

14 12 10.10 Electric Dumbwaiters		Crew	Daily Output	Labor-Hours	Unit	Material	2010 Bare Costs Labor	Equipment	Total	Total Incl O&P
0010	**ELECTRIC DUMBWAITERS**									
0020	2 stop, minimum	2 Elev	.13	123	Ea.	7,375	7,600		14,975	19,400
0100	Maximum		.11	145	"	22,200	8,975		31,175	37,700
0600	For each additional stop, add		.54	29.630	Stop	3,250	1,825		5,075	6,300
0750	Correspondence lift, 1 floor, 2 stop, 45 lb capacity	2 Elec	.20	80	Ea.	8,350	3,925		12,275	15,000

14 21 Electric Traction Elevators

14 21 13 – Electric Traction Freight Elevators

14 21 13.10 Electric Traction Freight Elevators and Options		Crew	Daily Output	Labor-Hours	Unit	Material	2010 Bare Costs Labor	Equipment	Total	Total Incl O&P
0010	**ELECTRIC TRACTION FREIGHT ELEVATORS AND OPTIONS** R142000-10									
0425	Electric freight, base unit, 4000 lb, 200 fpm, 4 stop, std. fin.	2 Elev	.05	320	Ea.	98,000	19,700		117,700	137,000
0450	For 5000 lb capacity, add					5,525			5,525	6,100
0500	For 6000 lb capacity, add					17,300			17,300	19,100
0525	For 7000 lb capacity, add					21,300			21,300	23,400
0550	For 8000 lb capacity, add					28,900			28,900	31,800
0575	For 10000 lb capacity, add					32,200			32,200	35,400
0600	For 12000 lb capacity, add					39,000			39,000	42,900
0625	For 16000 lb capacity, add					49,700			49,700	54,500
0650	For 20000 lb capacity, add					59,000			59,000	65,000
0675	For increased speed, 250 fpm, add					22,200			22,200	24,500
0700	300 fpm, geared electric, add					28,200			28,200	31,000
0725	350 fpm, geared electric, add					34,100			34,100	37,500
0750	400 fpm, geared electric, add					39,600			39,600	43,600
0775	500 fpm, gearless electric, add					47,200			47,200	52,000
0800	600 fpm, gearless electric, add					57,000			57,000	63,000
0825	700 fpm, gearless electric, add					68,500			68,500	75,000
0850	800 fpm, gearless electric, add					79,000			79,000	87,000
0875	For class "B" loading, add					6,500			6,500	7,150
0900	For class "C-1" loading, add					9,000			9,000	9,900
0925	For class "C-2" loading, add					10,200			10,200	11,300
0950	For class "C-3" loading, add					12,900			12,900	14,200
0975	For travel over 40 V.L.F., add	2 Elev	7.25	2.207	V.L.F.	540	136		676	795
1000	For number of stops over 4, add	"	.27	59.259	Stop	3,975	3,650		7,625	9,800

14 21 23 – Electric Traction Passenger Elevators

14 21 23.10 Electric Traction Passenger Elevators and Options		Crew	Daily Output	Labor-Hours	Unit	Material	2010 Bare Costs Labor	Equipment	Total	Total Incl O&P
0010	**ELECTRIC TRACTION PASSENGER ELEVATORS AND OPTIONS**									
1625	Electric pass., base unit, 2000 lb, 200 fpm, 4 stop, std. fin.	2 Elev	.05	320	Ea.	88,000	19,700		107,700	126,500
1650	For 2500 lb capacity, add					3,675			3,675	4,025
1675	For 3000 lb capacity, add					4,375			4,375	4,825
1700	For 3500 lb capacity, add					7,325			7,325	8,050
1725	For 4000 lb capacity, add					6,875			6,875	7,575

14 21 Electric Traction Elevators

14 21 23 – Electric Traction Passenger Elevators

14 21 23.10 Electric Traction Passenger Elevators and Options		Crew	Daily Output	Labor-Hours	Unit	2010 Bare Costs Material	Labor	Equipment	Total	Total Incl O&P
1750	For 4500 lb capacity, add				Ea.	8,700			8,700	9,575
1775	For 5000 lb capacity, add					10,500			10,500	11,500
1800	For increased speed, 250 fpm, geared electric, add					6,950			6,950	7,650
1825	300 fpm, geared electric, add					10,500			10,500	11,600
1850	350 fpm, geared electric, add					11,900			11,900	13,100
1875	400 fpm, geared electric, add					15,200			15,200	16,700
1900	500 fpm, gearless electric, add					31,200			31,200	34,400
1925	600 fpm, gearless electric, add					34,600			34,600	38,100
1950	700 fpm, gearless electric, add					39,700			39,700	43,700
1975	800 fpm, gearless electric, add				↓	45,400			45,400	49,900
2000	For travel over 40 V.L.F., add	2 Elev	7.25	2.207	V.L.F.	650	136		786	915
2025	For number of stops over 4, add		.27	59.259	Stop	7,775	3,650		11,425	14,000
2400	Electric hospital, base unit, 4000 lb, 200 fpm, 4 stop, std fin.	↓	.05	320	Ea.	76,000	19,700		95,700	113,500
2425	For 4500 lb capacity, add					5,450			5,450	5,975
2450	For 5000 lb capacity, add					6,525			6,525	7,175
2475	For increased speed, 250 fpm, geared electric, add					7,025			7,025	7,725
2500	300 fpm, geared electric, add					10,400			10,400	11,500
2525	350 fpm, geared electric, add					11,900			11,900	13,100
2550	400 fpm, geared electric, add					15,000			15,000	16,500
2575	500 fpm, gearless electric, add					30,900			30,900	33,900
2600	600 fpm, gearless electric, add					35,300			35,300	38,800
2625	700 fpm, gearless electric, add					40,000			40,000	44,000
2650	800 fpm, gearless electric, add				↓	46,100			46,100	50,500
2675	For travel over 40 V.L.F., add	2 Elev	7.25	2.207	V.L.F.	665	136		801	935
2700	For number of stops over 4, add	"	.27	59.259	Stop	4,050	3,650		7,700	9,900

14 21 33 – Electric Traction Residential Elevators

14 21 33.20 Residential Elevators		Crew	Daily Output	Labor-Hours	Unit	Material	Labor	Equipment	Total	Total Incl O&P
0010	**RESIDENTIAL ELEVATORS**									
7000	Residential, cab type, 1 floor, 2 stop, minimum	2 Elev	.20	80	Ea.	11,000	4,925		15,925	19,400
7100	Maximum		.10	160		18,600	9,875		28,475	35,200
7200	2 floor, 3 stop, minimum		.12	133		16,400	8,225		24,625	30,200
7300	Maximum	↓	.06	266	↓	26,700	16,500		43,200	54,000

14 24 Hydraulic Elevators

14 24 13 – Hydraulic Freight Elevators

14 24 13.10 Hydraulic Freight Elevators and Options		Crew	Daily Output	Labor-Hours	Unit	Material	Labor	Equipment	Total	Total Incl O&P
0010	**HYDRAULIC FREIGHT ELEVATORS AND OPTIONS**									
1025	Hydraulic freight, base unit, 2000 lb, 50 fpm, 2 stop, std. fin.	2 Elev	.10	160	Ea.	78,500	9,875		88,375	100,500
1050	For 2500 lb capacity, add					4,925			4,925	5,425
1075	For 3000 lb capacity, add					8,725			8,725	9,600
1100	For 3500 lb capacity, add					15,800			15,800	17,400
1125	For 4000 lb capacity, add					16,900			16,900	18,600
1150	For 4500 lb capacity, add					19,800			19,800	21,800
1175	For 5000 lb capacity, add					24,900			24,900	27,400
1200	For 6000 lb capacity, add					31,100			31,100	34,200
1225	For 7000 lb capacity, add					37,400			37,400	41,100
1250	For 8000 lb capacity, add					41,100			41,100	45,200
1275	For 10000 lb capacity, add					52,000			52,000	57,000
1300	For 12000 lb capacity, add					68,000			68,000	75,000
1325	For 16000 lb capacity, add				↓	93,000			93,000	102,500

14 24 Hydraulic Elevators

14 24 13 – Hydraulic Freight Elevators

14 24 13.10 Hydraulic Freight Elevators and Options		Crew	Daily Output	Labor-Hours	Unit	Material	2010 Bare Costs Labor	Equipment	Total	Total Incl O&P
1350	For 20000 lb capacity, add				Ea.	113,500			113,500	125,000
1375	For increased speed, 100 fpm, add					2,150			2,150	2,350
1400	125 fpm, add					4,625			4,625	5,075
1425	150 fpm, add					5,500			5,500	6,050
1450	175 fpm, add					9,375			9,375	10,300
1475	For class "B" loading, add					6,475			6,475	7,125
1500	For class "C-1" loading, add					8,975			8,975	9,850
1525	For class "C-2" loading, add					10,200			10,200	11,200
1550	For class "C-3" loading, add					12,900			12,900	14,200
1575	For travel over 20 V.L.F., add	2 Elev	7.25	2.207	V.L.F.	1,675	136		1,811	2,050
1600	For number of stops over 2, add	"	.27	59.259	Stop	4,550	3,650		8,200	10,400

14 24 23 – Hydraulic Passenger Elevators

14 24 23.10 Hydraulic Passenger Elevators and Options		Crew	Daily Output	Labor-Hours	Unit	Material	Labor	Equipment	Total	Total Incl O&P
0010	**HYDRAULIC PASSENGER ELEVATORS AND OPTIONS**									
2050	Hyd. pass., base unit, 1500 lb, 100 fpm, 2 stop, std. fin. **CN**	2 Elev	.10	160	Ea.	36,900	9,875		46,775	55,500
2075	For 2000 lb capacity, add					2,000			2,000	2,200
2100	For 2500 lb. capacity, add					3,125			3,125	3,425
2125	For 3000 lb capacity, add					4,100			4,100	4,525
2150	For 3500 lb capacity, add					5,950			5,950	6,550
2175	For 4000 lb capacity, add					7,300			7,300	8,050
2200	For 4500 lb capacity, add					8,600			8,600	9,475
2225	For 5000 lb capacity, add					11,300			11,300	12,500
2250	For increased speed, 125 fpm, add					2,950			2,950	3,250
2275	150 fpm, add					3,700			3,700	4,075
2300	175 fpm, add					5,525			5,525	6,075
2325	200 fpm, add					7,650			7,650	8,400
2350	For travel over 12 V.L.F., add	2 Elev	7.25	2.207	V.L.F.	890	136		1,026	1,175
2375	For number of stops over 2, add		.27	59.259	Stop	2,550	3,650		6,200	8,250
2725	Hydraulic hospital, base unit, 4000 lb, 100 fpm, 2 stop, std. fin.		.10	160	Ea.	59,500	9,875		69,375	80,000
2775	For 4500 lb capacity, add					6,775			6,775	7,450
2800	For 5000 lb capacity, add					8,925			8,925	9,825
2825	For increased speed, 125 fpm, add					3,425			3,425	3,750
2850	150 fpm, add					4,100			4,100	4,525
2875	175 fpm, add					5,975			5,975	6,550
2900	200 fpm, add					7,175			7,175	7,900
2925	For travel over 12 V.L.F., add	2 Elev	7.25	2.207	V.L.F.	745	136		881	1,025
2950	For number of stops over 2, add	"	.27	59.259	Stop	4,075	3,650		7,725	9,900

14 27 Custom Elevator Cabs

14 27 13 – Custom Elevator Cab Finishes

14 27 13.10 Cab Finishes		Crew	Daily Output	Labor-Hours	Unit	Material	Labor	Equipment	Total	Total Incl O&P
0010	**CAB FINISHES**									
3325	Passenger elevator cab finishes (based on 3500 lb cab size)									
3350	Acrylic panel ceiling				Ea.	735			735	810
3375	Aluminum eggcrate ceiling					1,200			1,200	1,325
3400	Stainless steel doors					2,925			2,925	3,225
3425	Carpet flooring					605			605	665
3450	Epoxy flooring					460			460	505
3475	Quarry tile flooring					825			825	910
3500	Slate flooring					1,450			1,450	1,600

14 27 Custom Elevator Cabs

14 27 13 – Custom Elevator Cab Finishes

14 27 13.10 Cab Finishes		Crew	Daily Output	Labor-Hours	Unit	Material (2010 Bare Costs)	Labor	Equipment	Total	Total Incl O&P
3525	Textured rubber flooring				Ea.	200			200	220
3550	Stainless steel walls					5,125			5,125	5,650
3575	Stainless steel returns at door				↓	1,200			1,200	1,300
4450	Hospital elevator cab finishes (based on 3500 lb cab size)									
4475	Aluminum eggcrate ceiling				Ea.	1,200			1,200	1,300
4500	Stainless steel doors					2,675			2,675	2,950
4525	Epoxy flooring					370			370	405
4550	Quarry tile flooring					845			845	930
4575	Textured rubber flooring					660			660	730
4600	Stainless steel walls					5,125			5,125	5,650
4625	Stainless steel returns at door				↓	1,200			1,200	1,300

14 28 Elevator Equipment and Controls

14 28 10 – Elevator Equipment and Control Options

14 28 10.10 Elevator Controls and Doors		Crew	Daily Output	Labor-Hours	Unit	Material	Labor	Equipment	Total	Total Incl O&P
0010	**ELEVATOR CONTROLS AND DOORS**									
2975	Passenger elevator options									
3000	2 car group automatic controls	2 Elev	.66	24.242	Ea.	6,600	1,500		8,100	9,500
3025	3 car group automatic controls		.44	36.364		12,800	2,250		15,050	17,400
3050	4 car group automatic controls		.33	48.485		26,100	3,000		29,100	33,300
3075	5 car group automatic controls		.26	61.538		47,900	3,800		51,700	58,000
3100	6 car group automatic controls		.22	72.727		98,500	4,475		102,975	114,500
3125	Intercom service		3	5.333		1,375	330		1,705	2,000
3150	Duplex car selective collective		.66	24.242		12,100	1,500		13,600	15,500
3175	Center opening 1 speed doors		2	8		1,800	495		2,295	2,700
3200	Center opening 2 speed doors		2	8		2,525	495		3,020	3,500
3225	Rear opening doors (opposite front)		2	8		3,000	495		3,495	4,050
3250	Side opening 2 speed doors		2	8		4,725	495		5,220	5,925
3275	Automatic emergency power switching		.66	24.242		1,150	1,500		2,650	3,500
3300	Manual emergency power switching	↓	8	2		480	123		603	715
3625	Hall finishes, stainless steel doors					2,000			2,000	2,200
3650	Stainless steel frames					1,450			1,450	1,600
3675	12 month maintenance contract								3,600	3,960
3700	Signal devices, hall lanterns	2 Elev	8	2		535	123		658	775
3725	Position indicators, up to 3		9.40	1.702		640	105		745	860
3750	Position indicators, per each over 3	↓	32	.500		620	31		651	725
3775	High speed heavy duty door opener					4,200			4,200	4,600
3800	Variable voltage, O.H. gearless machine, min.	2 Elev	.16	100		69,000	6,175		75,175	85,000
3815	Maximum		.07	228		102,000	14,100		116,100	133,000
3825	Basement installed geared machine	↓	.33	48.485	↓	71,500	3,000		74,500	83,000
3850	Freight elevator options									
3875	Doors, bi-parting	2 Elev	.66	24.242	Ea.	9,825	1,500		11,325	13,000
3900	Power operated door and gate	"	.66	24.242		22,800	1,500		24,300	27,200
3925	Finishes, steel plate floor					2,450			2,450	2,700
3950	14 ga. 1/4" x 4' steel plate walls					1,850			1,850	2,050
3975	12 month maintenance contract								3,600	3,960
4000	Signal devices, hall lanterns	2 Elev	8	2		535	123		658	775
4025	Position indicators, up to 3		9.40	1.702		645	105		750	865
4050	Position indicators, per each over 3		32	.500		620	31		651	725
4075	Variable voltage basement installed geared machine	↓	.66	24.242	↓	18,800	1,500		20,300	22,900
4100	Hospital elevator options									

14 28 Elevator Equipment and Controls

14 28 10 – Elevator Equipment and Control Options

14 28 10.10 Elevator Controls and Doors		Crew	Daily Output	Labor-Hours	Unit	Material	Labor	Equipment	Total	Total Incl O&P
						2010 Bare Costs				
4125	2 car group automatic controls	2 Elev	.66	24.242	Ea.	6,550	1,500		8,050	9,425
4150	3 car group automatic controls		.44	36.364		12,900	2,250		15,150	17,500
4175	4 car group automatic controls		.33	48.485		26,300	3,000		29,300	33,400
4200	5 car group automatic controls		.26	61.538		48,000	3,800		51,800	58,000
4225	6 car group automatic controls		.22	72.727		98,500	4,475		102,975	114,500
4250	Intercom service		3	5.333		1,375	330		1,705	2,000
4275	Duplex car selective collective		.66	24.242		12,100	1,500		13,600	15,500
4300	Center opening 1 speed doors		2	8		1,800	495		2,295	2,700
4325	Center opening 2 speed doors		2	8		2,450	495		2,945	3,425
4350	Rear opening doors (opposite front)		2	8		3,000	495		3,495	4,025
4375	Side opening 2 speed doors		2	8		3,775	495		4,270	4,875
4400	Automatic emergency power switching		.66	24.242		1,125	1,500		2,625	3,450
4425	Manual emergency power switching	↓	8	2		470	123		593	700
4675	Hall finishes, stainless steel doors					2,000			2,000	2,200
4700	Stainless steel frames					1,450			1,450	1,600
4725	12 month maintenance contract								3,600	3,960
4750	Signal devices, hall lanterns	2 Elev	8	2		535	123		658	775
4775	Position indicators, up to 3		9.40	1.702		640	105		745	860
4800	Position indicators, per each over 3	↓	32	.500		620	31		651	725
4825	High speed heavy duty door opener					4,200			4,200	4,600
4850	Variable voltage, O.H. gearless machine, min.	2 Elev	.16	100		69,000	6,175		75,175	85,000
4865	Maximum		.07	228		102,000	14,100		116,100	133,000
4875	Basement installed geared machine	↓	.33	48.485	↓	26,100	3,000		29,100	33,200
5000	Drilling for piston, casing included, 18" diameter	B-48	80	.700	V.L.F.	46	26	37	109	131

14 31 Escalators

14 31 10 – Glass and Steel Escalators

14 31 10.10 Escalators		Crew	Daily Output	Labor-Hours	Unit	Material	Labor	Equipment	Total	Total Incl O&P
0010	**ESCALATORS** R143110-10									
1000	Glass, 32" wide x 10' floor to floor height	M-1	.07	457	Ea.	84,500	26,800	825	112,125	133,000
1010	48" wide x 10' floor to floor height		.07	457		91,000	26,800	825	118,625	140,500
1020	32" wide x 15' floor to floor height **CN**		.06	533		89,500	31,300	965	121,765	146,000
1030	48" wide x 15' floor to floor height		.06	533		94,500	31,300	965	126,765	151,500
1040	32" wide x 20' floor to floor height		.05	653		93,000	38,300	1,175	132,475	161,000
1050	48" wide x 20' floor to floor height		.05	653		101,000	38,300	1,175	140,475	170,000
1060	32" wide x 25' floor to floor height		.04	800		100,500	46,900	1,450	148,850	181,500
1070	48" wide x 25' floor to floor height		.04	800		116,500	46,900	1,450	164,850	199,500
1080	Enameled steel, 32" wide x 10' floor to floor height		.07	457		90,000	26,800	825	117,625	139,500
1090	48" wide x 10' floor to floor height		.07	457		98,000	26,800	825	125,625	148,000
1110	32" wide x 15' floor to floor height		.06	533		98,000	31,300	965	130,265	155,000
1120	48" wide x 15' floor to floor height		.06	533		98,000	31,300	965	130,265	155,000
1130	32" wide x 20' floor to floor height		.05	653		99,500	38,300	1,175	138,975	168,000
1140	48" wide x 20' floor to floor height		.05	653		107,000	38,300	1,175	146,475	176,000
1150	32" wide x 25' floor to floor height		.04	800		107,000	46,900	1,450	155,350	188,500
1160	48" wide x 25' floor to floor height		.04	800		107,000	46,900	1,450	155,350	188,500
1170	Stainless steel, 32" wide x 10' floor to floor height		.07	457		107,000	26,800	825	134,625	158,000
1180	48" wide x 10' floor to floor height		.07	457		107,000	26,800	825	134,625	158,000
1500	32" wide x 15' floor to floor height		.06	533		98,500	31,300	965	130,765	156,000
1700	48" wide x 15' floor to floor height		.06	533		100,000	31,300	965	132,265	157,500
2300	32" wide x 25' floor to floor height		.04	800		106,500	46,900	1,450	154,850	188,000
2500	48" wide x 25' floor to floor height	↓	.04	800	↓	128,500	46,900	1,450	176,850	212,500

14 32 Moving Walks

14 32 10 – Moving Walkways

14 32 10.10 Moving Walks		Crew	Daily Output	Labor-Hours	Unit	Material	2010 Bare Costs Labor	Equipment	Total	Total Incl O&P
0010	**MOVING WALKS** R143210-20									
0020	Walk, 27" tread width, minimum	M-1	6.50	4.923	L.F.	850	289	8.90	1,147.90	1,375
0100	300' to 500', maximum		4.43	7.223		1,175	425	13.05	1,613.05	1,950
0300	48" tread width walk, minimum		4.43	7.223		1,925	425	13.05	2,363.05	2,775
0400	100' to 350', maximum		3.82	8.377		2,250	490	15.15	2,755.15	3,225
0600	Ramp, 12° incline, 36" tread width, minimum		5.27	6.072		1,575	355	11	1,941	2,275
0700	70' to 90' maximum		3.82	8.377		2,250	490	15.15	2,755.15	3,225
0900	48" tread width, minimum		3.57	8.964		2,300	525	16.20	2,841.20	3,325
1000	40' to 70', maximum	↓	2.91	10.997	↓	2,900	645	19.90	3,564.90	4,175

14 42 Wheelchair Lifts

14 42 13 – Inclined Wheelchair Lifts

14 42 13.10 Inclined Wheelchair Lifts and Stairclimbers		Crew	Daily Output	Labor-Hours	Unit	Material	Labor	Equipment	Total	Total Incl O&P
0010	**INCLINED WHEELCHAIR LIFTS AND STAIRCLIMBERS**									
7700	Stair climber (chair lift), single seat, minimum	2 Elev	1	16	Ea.	5,325	985		6,310	7,350
7800	Maximum		.20	80		7,350	4,925		12,275	15,400
8700	Stair lift, minimum		1	16		14,500	985		15,485	17,400
8900	Maximum	↓	.20	80	↓	22,900	4,925		27,825	32,500

14 42 16 – Vertical Wheelchair Lifts

14 42 16.10 Wheelchair Lifts		Crew	Daily Output	Labor-Hours	Unit	Material	Labor	Equipment	Total	Total Incl O&P
0010	**WHEELCHAIR LIFTS**									
8000	Wheelchair lift, minimum	2 Elev	1	16	Ea.	7,325	985		8,310	9,525
8500	Maximum	"	.50	32	"	17,300	1,975		19,275	21,900

14 45 Vehicle Lifts

14 45 10 – Hydraulic Vehicle Lifts

14 45 10.10 Hydraulic Lifts		Crew	Daily Output	Labor-Hours	Unit	Material	Labor	Equipment	Total	Total Incl O&P
0010	**HYDRAULIC LIFTS**									
0200	Double post, 9000 lb frame	L-4	1.15	20.870	Ea.	5,425	810		6,235	7,200
0400	26,000 lb frame		.22	109		25,400	4,250		29,650	34,500
2200	Hoists, single post, 8000 lb capacity, swivel arms		.40	60		4,975	2,325		7,300	9,075
2400	Two posts, adjustable frames, 11,000 lb capacity		.25	96		6,400	3,725		10,125	12,800
2500	24,000 lb capacity		.15	160		8,550	6,225		14,775	19,000
2700	7500 lb capacity, frame supports		.50	48		7,125	1,875		9,000	10,700
2800	Four post, roll on ramp		.50	48		6,400	1,875		8,275	9,925
2810	Hydraulic lifts, above ground, 2 post, clear floor, 6000 lb cap		2.67	8.989		7,025	350		7,375	8,275
2815	9000 lb capacity		2.29	10.480		16,700	410		17,110	18,900
2820	15,000 lb capacity		2	12		18,800	465		19,265	21,400
2825	30,000 lb capacity		1.60	15		41,600	585		42,185	46,700
2830	4 post, ramp style, 25,000 lb capacity		2	12		16,300	465		16,765	18,600
2835	35,000 lb capacity		1	24		76,000	935		76,935	85,000
2840	50,000 lb capacity		1	24		85,000	935		85,935	95,000
2845	75,000 lb capacity	↓	1	24		98,500	935		99,435	110,000
2850	For drive thru tracks, add, minimum					1,050			1,050	1,150
2855	Maximum					1,800			1,800	1,975
2860	Ramp extensions, 3' (set of 2)					860			860	945
2865	Rolling jack platform					3,000			3,000	3,300
2870	Elec/hyd jacking beam					8,000			8,000	8,800
2880	Scissor lift, portable, 6000 lb capacity				↓	7,850			7,850	8,625

14 51 Correspondence and Parcel Lifts

14 51 10 – Electric Correspondence and Parcel Lifts

14 51 10.10 Correspondence Lifts		Crew	Daily Output	Labor-Hours	Unit	2010 Bare Costs Material	Labor	Equipment	Total	Total Incl O&P
0010	**CORRESPONDENCE LIFTS**									
0020	1 floor, 2 stop, 25 lb capacity, electric	2 Elev	.20	80	Ea.	5,975	4,925		10,900	13,900
0100	Hand, 5 lb capacity	"	.20	80	"	2,250	4,925		7,175	9,825
14 51 10.20 Parcel Lifts										
0010	**PARCEL LIFTS**									
0020	20" x 20", 100 lb capacity, electric, per floor	2 Mill	.25	64	Ea.	9,350	2,750		12,100	14,300

14 91 Facility Chutes

14 91 33 – Laundry and Linen Chutes

14 91 33.10 Chutes		Crew	Daily Output	Labor-Hours	Unit	Material	Labor	Equipment	Total	Total Incl O&P
0011	**CHUTES**, linen, trash or refuse									
0050	Aluminized steel, 16 ga., 18" diameter	2 Shee	3.50	4.571	Floor	1,275	224		1,499	1,775
0100	24" diameter		3.20	5		1,350	246		1,596	1,875
0200	30" diameter		3	5.333		1,550	262		1,812	2,100
0300	36" diameter		2.80	5.714		1,725	281		2,006	2,325
0400	Galvanized steel, 16 ga., 18" diameter		3.50	4.571		1,000	224		1,224	1,450
0500	24" diameter		3.20	5		1,125	246		1,371	1,625
0600	30" diameter		3	5.333		1,275	262		1,537	1,800
0700	36" diameter		2.80	5.714		1,500	281		1,781	2,075
0800	Stainless steel, 18" diameter		3.50	4.571		2,650	224		2,874	3,250
0900	24" diameter		3.20	5		2,875	246		3,121	3,525
1000	30" diameter		3	5.333		3,550	262		3,812	4,300
1005	36" diameter		2.80	5.714		3,975	281		4,256	4,800
1200	Linen chute bottom collector, aluminized steel		4	4	Ea.	1,400	196		1,596	1,850
1300	Stainless steel		4	4		1,800	196		1,996	2,275
1500	Refuse, bottom hopper, aluminized steel, 18" diameter		3	5.333		985	262		1,247	1,475
1600	24" diameter		3	5.333		1,200	262		1,462	1,700
1800	36" diameter		3	5.333		2,050	262		2,312	2,650

14 91 82 – Trash Chutes

14 91 82.10 Trash Chutes and Accessories		Crew	Daily Output	Labor-Hours	Unit	Material	Labor	Equipment	Total	Total Incl O&P
0010	**TRASH CHUTES AND ACCESSORIES**									
2900	Package chutes, spiral type, minimum	2 Shee	4.50	3.556	Floor	2,425	175		2,600	2,950
3000	Maximum	"	1.50	10.667	"	6,325	525		6,850	7,775

14 92 Pneumatic Tube Systems

14 92 10 – Conventional, Automatic and Computer Controlled Pneumatic Tube Systems

14 92 10.10 Pneumatic Tube Systems		Crew	Daily Output	Labor-Hours	Unit	Material	Labor	Equipment	Total	Total Incl O&P
0010	**PNEUMATIC TUBE SYSTEMS**									
0020	100' long, single tube, 2 stations, stock									
0100	3" diameter	2 Stpi	.12	133	Total	6,600	6,925		13,525	17,700
0300	4" diameter	"	.09	177	"	7,450	9,225		16,675	22,000
0400	Twin tube, two stations or more, conventional system									
0600	2-1/2" round	2 Stpi	62.50	.256	L.F.	13.30	13.30		26.60	34.50
0700	3" round		46	.348		15.25	18.05		33.30	44
0900	4" round		49.60	.323		16.90	16.75		33.65	43.50
1000	4" x 7" oval		37.60	.426		24.50	22		46.50	60
1050	Add for blower		2	8	System	5,200	415		5,615	6,350
1110	Plus for each round station, add		7.50	2.133	Ea.	585	111		696	810

14 92 Pneumatic Tube Systems

14 92 10 – Conventional, Automatic and Computer Controlled Pneumatic Tube Systems

14 92 10.10 Pneumatic Tube Systems		Crew	Daily Output	Labor-Hours	Unit	Material	2010 Bare Costs Labor	Equipment	Total	Total Incl O&P
1150	Plus for each oval station, add	2 Stpi	7.50	2.133	Ea.	585	111		696	810
1200	Alternate pricing method: base cost, minimum		.75	21.333	Total	6,375	1,100		7,475	8,675
1300	Maximum		.25	64	"	12,700	3,325		16,025	18,900
1500	Plus total system length, add, minimum		93.40	.171	L.F.	8.20	8.90		17.10	22.50
1600	Maximum		37.60	.426	"	24.50	22		46.50	59.50
1800	Completely automatic system, 4" round, 15 to 50 stations		.29	55.172	Station	20,100	2,875		22,975	26,400
2200	51 to 144 stations		.32	50		15,600	2,600		18,200	21,100
2400	6" round or 4" x 7" oval, 15 to 50 stations		.24	66.667		25,200	3,450		28,650	32,900
2800	51 to 144 stations	↓	.23	69.565	↓	21,100	3,600		24,700	28,600

Division Notes

	CREW	DAILY OUTPUT	LABOR-HOURS	UNIT	2010 BARE COSTS MAT.	LABOR	EQUIP.	TOTAL	TOTAL INCL O&P

Estimating Tips

Pipe for fire protection and all uses is located in Subdivision 22 11 13.

The labor adjustment factors listed in Subdivision 22 01 02.20 also apply to Division 21.

Many, but not all, areas in the U.S. require backflow protection in the fire system. It is advisable to check local building codes for specific requirements.

For your reference, the following is a list of the most applicable Fire Codes and Standards which may be purchased from the NFPA, 1 Batterymarch Park, Quincy, MA 02169-7471.

NFPA 1: Uniform Fire Code
NFPA 10: Portable Fire Extinguishers
NFPA 11: Low-, Medium-, and High-Expansion Foam
NFPA 12: Carbon Dioxide Extinguishing Systems (Also companion 12A)
NFPA 13: Installation of Sprinkler Systems (Also companion 13D, 13E, and 13R)
NFPA 14: Installation of Standpipe and Hose Systems
NFPA 15: Water Spray Fixed Systems for Fire Protection
NFPA 16: Installation of Foam-Water Sprinkler and Foam-Water Spray Systems
NFPA 17: Dry Chemical Extinguishing Systems (Also companion 17A)
NFPA 18: Wetting Agents
NFPA 20: Installation of Stationary Pumps for Fire Protection
NFPA 22: Water Tanks for Private Fire Protection
NFPA 24: Installation of Private Fire Service Mains and their Appurtenances
NFPA 25: Inspection, Testing and Maintenance of Water-Based Fire Protection

Reference Numbers

Reference numbers are shown in shaded boxes at the beginning of some major classifications. These numbers refer to related items in the Reference Section. The reference information may be an estimating procedure, an alternate pricing method, or technical information.

Note: Not all subdivisions listed here necessarily appear in this publication.

Note: **Trade Service,** *in part, has been used as a reference source for some of the material prices used in Division 21.*

21 05 Common Work Results for Fire Suppression

21 05 23 – General-Duty Valves for Water-Based Fire-Suppression Piping

21 05 23.50 General-Duty Valves for Water-Based Fire Supp.		Crew	Daily Output	Labor-Hours	Unit	2010 Bare Costs Material	Labor	Equipment	Total	Total Incl O&P
0010	GENERAL-DUTY VALVES FOR WATER-BASED FIRE SUPPRESSION PIPING									
6200	Valves and components									
6500	Check, swing, C.I. body, brass fittings, auto. ball drip									
6520	4" size	Q-12	3	5.333	Ea.	255	242		497	640
6800	Check, wafer, butterfly type, C.I. body, bronze fittings									
6820	4" size	Q-12	4	4	Ea.	283	181		464	580

21 11 Facility Fire-Suppression Water-Service Piping

21 11 16 – Facility Fire Hydrants

21 11 16.50 Fire Hydrants for Buildings

		Crew	Daily Output	Labor-Hours	Unit	Material	Labor	Equipment	Total	Total Incl O&P
0010	FIRE HYDRANTS FOR BUILDINGS									
3750	Hydrants, wall, w/caps, single, flush, polished brass									
3800	2-1/2" x 2-1/2"	Q-12	5	3.200	Ea.	175	145		320	410
3840	2-1/2" x 3"	"	5	3.200	"	355	145		500	605
3950	Double, flush, polished brass									
4000	2-1/2" x 2-1/2" x 4"	Q-12	5	3.200	Ea.	470	145		615	735
4040	2-1/2" x 2-1/2" x 6"	"	4.60	3.478	↓	675	158		833	980
4200	For polished chrome, add					10%				
4350	Double, projecting, polished brass									
4400	2-1/2" x 2-1/2" x 4"	Q-12	5	3.200	Ea.	209	145		354	445
4450	2-1/2" x 2-1/2" x 6"	"	4.60	3.478	"	425	158		583	705
4460	Valve control, dbl. flush/projecting hydrant, cap &									
4470	chain, extension rod & cplg., escutcheon, polished brass	Q-12	8	2	Ea.	249	90.50		339.50	410

21 11 19 – Fire-Department Connections

21 11 19.50 Connections for the Fire-Department

		Crew	Daily Output	Labor-Hours	Unit	Material	Labor	Equipment	Total	Total Incl O&P
0010	CONNECTIONS FOR THE FIRE-DEPARTMENT									
7140	Standpipe connections, wall, w/plugs & chains									
7160	Single, flush, brass, 2-1/2" x 2-1/2"	Q-12	5	3.200	Ea.	123	145		268	350
7180	2-1/2" x 3"	"	5	3.200	"	135	145		280	365
7240	For polished chrome, add					15%				
7280	Double, flush, polished brass									
7300	2-1/2" x 2-1/2" x 4"	Q-12	5	3.200	Ea.	425	145		570	685
7330	2-1/2" x 2-1/2" x 6"	"	4.60	3.478	"	595	158		753	890
7400	For polished chrome, add					15%				
7440	For sill cock combination, add				Ea.	50.50			50.50	56
7900	Three way, flush, polished brass									
7920	2-1/2" (3) x 4"	Q-12	4.80	3.333	Ea.	1,225	151		1,376	1,575
7930	2-1/2" (3) x 6"	"	4.80	3.333	↓	1,150	151		1,301	1,475
8000	For polished chrome, add					9%				
8020	Three way, projecting, polished brass									
8040	2-1/2"(3) x 4"	Q-12	4.80	3.333	Ea.	985	151		1,136	1,300

21 12 Fire-Suppression Standpipes

21 12 13 – Fire-Suppression Hoses and Nozzles

21 12 13.50 Fire Hoses and Nozzles		Crew	Daily Output	Labor-Hours	Unit	Material	Labor	Equipment	Total	Total Incl O&P
						2010 Bare Costs				
0010	**FIRE HOSES AND NOZZLES**									
0200	Adapters, rough brass, straight hose threads									
0220	One piece, female to male, rocker lugs									
0240	1" x 1"				Ea.	39.50			39.50	43.50
2200	Hose, less couplings									
2260	Synthetic jacket, lined, 300 lb. test, 1-1/2" diameter	Q-12	2600	.006	L.F.	2.39	.28		2.67	3.05
2280	2-1/2" diameter		2200	.007		3.98	.33		4.31	4.87
2360	High strength, 500 lb. test, 1-1/2" diameter		2600	.006		2.47	.28		2.75	3.14
2380	2-1/2" diameter	↓	2200	.007	↓	4.29	.33		4.62	5.20
5600	Nozzles, brass									
5620	Adjustable fog, 3/4" booster line				Ea.	91			91	100
5630	1" booster line					109			109	119
5640	1-1/2" leader line					66			66	72.50
5660	2-1/2" direct connection				↓	172			172	189
5780	For chrome plated, add					8%				
5850	Electrical fire, adjustable fog, no shock									
5900	1-1/2"				Ea.	340			340	375
5920	2-1/2"					455			455	505
5980	For polished chrome, add				↓	6%				
6200	Heavy duty, comb. adj. fog and str. stream, with handle									
6210	1" booster line				Ea.	325			325	360

21 12 19 – Fire-Suppression Hose Racks

21 12 19.50 Fire Hose Racks		Crew	Daily Output	Labor-Hours	Unit	Material	Labor	Equipment	Total	Total Incl O&P
0010	**FIRE HOSE RACKS**									
2600	Hose rack, swinging, for 1-1/2" diameter hose,									
2620	Enameled steel, 50' & 75' lengths of hose	Q-12	20	.800	Ea.	48.50	36.50		85	108
2640	100' and 125' lengths of hose	"	20	.800	"	49	36.50		85.50	109

21 12 23 – Fire-Suppression Hose Valves

21 12 23.70 Fire Hose Valves		Crew	Daily Output	Labor-Hours	Unit	Material	Labor	Equipment	Total	Total Incl O&P
0010	**FIRE HOSE VALVES**									
0080	Wheel handle, 300 lb., 1-1/2"	1 Spri	12	.667	Ea.	74.50	33.50		108	133
0090	2-1/2"	"	7	1.143	"	60	57.50		117.50	152
0100	For polished brass, add					35%				
0110	For polished chrome, add					50%				

21 13 Fire-Suppression Sprinkler Systems

21 13 13 – Wet-Pipe Sprinkler Systems

21 13 13.50 Wet-Pipe Sprinkler System Components		Crew	Daily Output	Labor-Hours	Unit	Material	Labor	Equipment	Total	Total Incl O&P
0010	**WET-PIPE SPRINKLER SYSTEM COMPONENTS**									
2600	Sprinkler heads, not including supply piping									
3700	Standard spray, pendent or upright, brass, 135°F to 286°F									
3730	1/2" NPT, 7/16" orifice	1 Spri	16	.500	Ea.	11.35	25		36.35	50
3740	1/2" NPT, 1/2" orifice **CN**	"	16	.500		7.35	25		32.35	45.50
3860	For wax and lead coating, add					24.50			24.50	26.50
3880	For wax coating, add					14.50			14.50	15.95
3900	For lead coating, add				↓	15.65			15.65	17.20
3920	For 360°F, same cost									
3930	For 400°F, add				Ea.	48.50			48.50	53
3940	For 500°F, add				"	48.50			48.50	53
4500	Sidewall, horizontal, brass, 135°F to 286°F									

21 13 Fire-Suppression Sprinkler Systems

21 13 13 – Wet-Pipe Sprinkler Systems

21 13 13.50 Wet-Pipe Sprinkler System Components		Crew	Daily Output	Labor-Hours	Unit	Material	2010 Bare Costs Labor	Equipment	Total	Total Incl O&P
4520	1/2" NPT, 1/2" orifice	1 Spri	16	.500	Ea.	15	25		40	54
4540	For 360°F, same cost									
4800	Recessed pendent, brass, 135°F to 286°F									
4820	1/2" NPT, 3/8" orifice	1 Spri	10	.800	Ea.	40	40.50		80.50	105
4830	1/2" NPT, 7/16" orifice	↓	10	.800	↓	18.75	40.50		59.25	81
4840	1/2" NPT, 1/2" orifice	↓	10	.800	↓	13.60	40.50		54.10	75.50

21 13 16 – Dry-Pipe Sprinkler Systems

21 13 16.50 Dry-Pipe Sprinkler System Components		Crew	Daily Output	Labor-Hours	Unit	Material	Labor	Equipment	Total	Total Incl O&P
0010	**DRY-PIPE SPRINKLER SYSTEM COMPONENTS**									
0600	Accelerator	1 Spri	8	1	Ea.	560	50.50		610.50	690
2600	Sprinkler heads, not including supply piping									
2640	Dry, pendent, 1/2" orifice, 3/4" or 1" NPT									
2700	15-1/4" to 18" length	1 Spri	14	.571	Ea.	106	29		135	160
2710	18-1/4" to 21" length	↓	13	.615	↓	110	31		141	168
2720	21-1/4" to 24" length	↓	13	.615	↓	114	31		145	173
2730	24-1/4" to 27" length	↓	13	.615	↓	119	31		150	177

21 13 26 – Deluge Fire-Suppression Sprinkler Systems

21 13 26.50 Deluge Fire-Suppression Sprinkler System Components		Crew	Daily Output	Labor-Hours	Unit	Material	Labor	Equipment	Total	Total Incl O&P
0010	**DELUGE FIRE-SUPPRESSION SPRINKLER SYSTEM COMPONENTS**									
6200	Valves and components									
7000	Deluge, assembly, incl. trim, pressure									
7020	operated relief, emergency release, gauges									
7040	2" size	Q-12	2	8	Ea.	2,000	365		2,365	2,750
7060	3" size	"	1.50	10.667	"	2,325	485		2,810	3,275

21 13 39 – Foam-Water Systems

21 13 39.50 Foam-Water System Components		Crew	Daily Output	Labor-Hours	Unit	Material	Labor	Equipment	Total	Total Incl O&P
0010	**FOAM-WATER SYSTEM COMPONENTS**									
2600	Sprinkler heads, not including supply piping									
3600	Foam-water, pendent or upright, 1/2" NPT	1 Spri	12	.667	Ea.	130	33.50		163.50	193

21 21 Carbon-Dioxide Fire-Extinguishing Systems

21 21 16 – Carbon-Dioxide Fire-Extinguishing Equipment

21 21 16.50 C02 Fire Extinguishing System		Crew	Daily Output	Labor-Hours	Unit	Material	Labor	Equipment	Total	Total Incl O&P
0010	**CO² FIRE EXTINGUISHING SYSTEM**									
0042	For detectors and control stations, see Div. 28 31 23.50									
1000	Dispersion nozzle, CO_2, 3" x 5"	1 Plum	18	.444	Ea.	56	23		79	96.50
2000	Extinguisher, CO_2 system, high pressure, 75 lb. cylinder	Q-1	6	2.667	↓	1,075	125		1,200	1,350
3000	Electro/mechanical release	L-1	4	4	↓	140	202		342	455
3400	Manual pull station	1 Plum	6	1.333	↓	50.50	69.50		120	160

21 22 Clean-Agent Fire-Extinguishing Systems

21 22 16 – Clean-Agent Fire-Extinguishing Equipment

21 22 16.50 FM200 Fire Extinguishing System		Crew	Daily Output	Labor-Hours	Unit	Material	2010 Bare Costs Labor	Equipment	Total	Total Incl O&P
0010	**FM200 FIRE EXTINGUISHING SYSTEM**									
1100	Dispersion nozzle FM200, 1-1/2"	1 Plum	14	.571	Ea.	56	29.50		85.50	107
2400	Extinguisher, FM 200 system, filled, with mounting bracket									
2540	196 lb. container	Q-1	4	4	Ea.	6,525	187		6,712	7,450
6000	Average FM200 system, minimum				C.F.	1.48			1.48	1.63
6020	Maximum				"	2.94			2.94	3.23

21 31 Centrifugal Fire Pumps

21 31 13 – Electric-Drive, Centrifugal Fire Pumps

21 31 13.50 Electric-Drive Fire Pumps		Crew	Daily Output	Labor-Hours	Unit	Material	Labor	Equipment	Total	Total Incl O&P
0010	**ELECTRIC-DRIVE FIRE PUMPS** Including controller, fittings and relief valve									
3100	250 GPM, 55 psi, 15 HP, 3,550 RPM, 2" pump	Q-13	.70	45.714	Ea.	13,300	2,200		15,500	18,000
3200	500 GPM, 50 psi, 27 HP, 1,770 RPM, 4" pump		.68	47.059		15,000	2,250		17,250	19,900
3350	750 GPM, 50 psi, 44 HP, 1,770 RPM, 5" pump		.64	50		15,200	2,400		17,600	20,300
3400	750 GPM, 100 psi, 66 HP, 3,550 RPM, 4" pump	↓	.58	55.172		17,200	2,650		19,850	23,000
5000	For jockey pump 1", 3 HP, with control, add	Q-12	2	8	↓	2,000	365		2,365	2,750

21 31 16 – Diesel-Drive, Centrifugal Fire Pumps

21 31 16.50 Diesel-Drive Fire Pumps		Crew	Daily Output	Labor-Hours	Unit	Material	Labor	Equipment	Total	Total Incl O&P
0010	**DIESEL-DRIVE FIRE PUMPS** Including controller, fittings and relief valve									
0050	500 GPM, 50 psi, 27 HP, 4" pump	Q-13	.64	50	Ea.	25,600	2,400		28,000	31,800
0200	750 GPM, 50 psi, 44 HP, 5" pump		.60	53.333		26,400	2,550		28,950	32,900
0400	1,000 GPM, 100 psi, 89 HP, 4" pump		.56	57.143		67,000	2,750		69,750	77,500
0700	2,000 GPM, 100 psi, 167 HP, 6" pump		.34	94.118		33,100	4,525		37,625	43,200
0950	3,500 GPM, 100 psi, 300 HP, 10" pump	↓	.24	133	↓	61,000	6,400		67,400	76,500

Division Notes

	CREW	DAILY OUTPUT	LABOR-HOURS	UNIT	2010 BARE COSTS MAT.	LABOR	EQUIP.	TOTAL	TOTAL INCL O&P

Estimating Tips

22 10 00 Plumbing Piping and Pumps

This subdivision is primarily basic pipe and related materials. The pipe may be used by any of the mechanical disciplines, i.e., plumbing, fire protection, heating, and air conditioning.

- The labor adjustment factors listed in Subdivision 22 01 02.20 apply throughout Divisions 21, 22, and 23. CAUTION: the correct percentage may vary for the same items. For example, the percentage add for the basic pipe installation should be based on the maximum height that the craftsman must install for that particular section. If the pipe is to be located 14′ above the floor but it is suspended on threaded rod from beams, the bottom flange of which is 18′ high (4′ rods), then the height is actually 18′ and the add is 20%. The pipe coverer, however, does not have to go above the 14′, and so his or her add should be 10%.
- Most pipe is priced first as straight pipe with a joint (coupling, weld, etc.) every 10′ and a hanger usually every 10′. There are exceptions with hanger spacing such as for cast iron pipe (5′) and plastic pipe (3 per 10′). Following each type of pipe there are several lines listing sizes and the amount to be subtracted to delete couplings and hangers. This is for pipe that is to be buried or supported together on trapeze hangers. The reason that the couplings are deleted is that these runs are usually long, and frequently longer lengths of pipe are used. By deleting the couplings, the estimator is expected to look up and add back the correct reduced number of couplings.
- When preparing an estimate, it may be necessary to approximate the fittings. Fittings usually run between 25% and 50% of the cost of the pipe. The lower percentage is for simpler runs, and the higher number is for complex areas, such as mechanical rooms.
- For historic restoration projects, the systems must be as invisible as possible, and pathways must be sought for pipes, conduit, and ductwork. While installations in accessible spaces (such as basements and attics) are relatively straightforward to estimate, labor costs may be more difficult to determine when delivery systems must be concealed.

22 40 00 Plumbing Fixtures

- Plumbing fixture costs usually require two lines: the fixture itself and its "rough-in, supply, and waste."
- In the Assemblies Section (Plumbing D2010) for the desired fixture, the System Components Group at the center of the page shows the fixture on the first line. The rest of the list (fittings, pipe, tubing, etc.) will total up to what we refer to in the Unit Price section as "Rough-in, supply, waste, and vent." Note that for most fixtures we allow a nominal 5′ of tubing to reach from the fixture to a main or riser.
- Remember that gas- and oil-fired units need venting.

Reference Numbers

Reference numbers are shown in shaded boxes at the beginning of some major classifications. These numbers refer to related items in the Reference Section. The reference information may be an estimating procedure, an alternate pricing method, or technical information.

Note: Not all subdivisions listed here necessarily appear in this publication.

Note: **Trade Service,** *in part, has been used as a reference source for some of the material prices used in Division 22.*

22 01 Operation and Maintenance of Plumbing

22 01 02 – Labor Adjustments

22 01 02.10 Boilers, General		Crew	Daily Output	Labor-Hours	Unit	Material	2010 Bare Costs Labor	Equipment	Total	Total Incl O&P
0010	**BOILERS, GENERAL**, Prices do not include flue piping, elec. wiring,									
0020	gas or oil piping, boiler base, pad, or tankless unless noted									
0100	Boiler H.P.: 10 KW = 34 lbs/steam/hr = 33,475 BTU/hr. R235000-50									
0150	To convert SFR to BTU rating: Hot water, 150 x SFR;									
0160	Forced hot water, 180 x SFR; steam, 240 x SFR									

22 01 02.20 Labor Adjustment Factors		Crew	Daily Output	Labor-Hours	Unit	Material	Labor	Equipment	Total	Total Incl O&P
0010	**LABOR ADJUSTMENT FACTORS**, (For Div. 21,22 and 23)									
1000	Add to labor for elevated installation (Above floor level)									
1080	10′ to 14.5′ high						10%			
1100	15′ to 19.5′ high						20%			
1120	20′ to 24.5′ high						25%			
1140	25′ to 29.5′ high						35%			
1160	30′ to 34.5′ high						40%			
1180	35′ to 39.5′ high						50%			
1200	40′ and higher						55%			
2000	Add to labor for crawl space									
2100	3′ high						40%			
2140	4′ high						30%			
3000	Add to labor for multi-story building									
3100	Add per floor for floors 3 thru 19						2%			
3140	Add per floor for floors 20 and up						4%			
4000	Add to labor for working in existing occupied buildings									
4100	Hospital						35%			
4140	Office building						25%			
4180	School						20%			
4220	Factory or warehouse						15%			
4260	Multi dwelling						15%			
5000	Add to labor, miscellaneous									
5100	Cramped shaft						35%			
5140	Congested area						15%			
5180	Excessive heat or cold						30%			
9000	Labor factors, The above are reasonable suggestions, however									
9010	each project should be evaluated for its own peculiarities.									
9100	Other factors to be considered are:									
9140	Movement of material and equipment through finished areas									
9180	Equipment room									
9220	Attic space									
9260	No service road									
9300	Poor unloading/storage area									
9340	Congested site area/heavy traffic									

22 05 Common Work Results for Plumbing

22 05 05 – Selective Plumbing Demolition

22 05 05.10 Plumbing Demolition		Crew	Daily Output	Labor-Hours	Unit	Material	2010 Bare Costs Labor	Equipment	Total	Total Incl O&P
0010	**PLUMBING DEMOLITION**									
1020	Fixtures, including 10′ piping									
1100	Bathtubs, cast iron	1 Plum	4	2	Ea.		104		104	156
1120	Fiberglass		6	1.333			69.50		69.50	104
1140	Steel		5	1.600			83.50		83.50	125
1200	Lavatory, wall hung		10	.800			41.50		41.50	62.50
1220	Counter top		8	1			52		52	78
1300	Sink, single compartment		8	1			52		52	78
1320	Double compartment		7	1.143			59.50		59.50	89
1400	Water closet, floor mounted		8	1			52		52	78
1420	Wall mounted		7	1.143			59.50		59.50	89
1500	Urinal, floor mounted		4	2			104		104	156
1520	Wall mounted		7	1.143			59.50		59.50	89
1600	Water fountains, free standing		8	1			52		52	78
1620	Wall or deck mounted		6	1.333			69.50		69.50	104
2000	Piping, metal, up thru 1-1/2″ diameter		200	.040	L.F.		2.08		2.08	3.12
2050	2″ thru 3-1/2″ diameter		150	.053			2.78		2.78	4.16
2100	4″ thru 6″ diameter	2 Plum	100	.160			8.35		8.35	12.50
2150	8″ thru 14″ diameter	″	60	.267			13.90		13.90	21
2153	16″ thru 20″ diameter	Q-18	70	.343			16.60	.84	17.44	26
2155	24″ thru 26″ diameter		55	.436			21	1.06	22.06	32.50
2156	30″ thru 36″ diameter		40	.600			29	1.46	30.46	45
2212	Deduct for salvage, aluminum scrap				Ton				728	800
2214	Brass scrap								2,000	2,200
2216	Copper scrap								2,820	3,100
2218	Lead scrap								635	700
2220	Steel scrap								180	200
2250	Water heater, 40 gal.	1 Plum	6	1.333	Ea.		69.50		69.50	104
9470	Water softener	Q-1	2	8	″		375		375	560

22 05 23 – General-Duty Valves for Plumbing Piping

22 05 23.10 Valves, Brass		Crew	Daily Output	Labor-Hours	Unit	Material	Labor	Equipment	Total	Total Incl O&P
0010	**VALVES, BRASS**									
0500	Gas cocks, threaded									
0530	1/2″	1 Plum	24	.333	Ea.	8.90	17.35		26.25	36
0540	3/4″		22	.364		11.20	18.95		30.15	41
0550	1″		19	.421		17.65	22		39.65	52.50
0560	1-1/4″		15	.533		38	28		66	83.50

22 05 23.20 Valves, Bronze		Crew	Daily Output	Labor-Hours	Unit	Material	Labor	Equipment	Total	Total Incl O&P
0010	**VALVES, BRONZE**									
1020	Angle, 150 lb., rising stem, threaded									
1030	1/8″	1 Plum	24	.333	Ea.	89.50	17.35		106.85	125
1040	1/4″		24	.333		92.50	17.35		109.85	128
1050	3/8″		24	.333		92.50	17.35		109.85	128
1060	1/2″		22	.364		92.50	18.95		111.45	131
1070	3/4″		20	.400		127	21		148	170
1080	1″		19	.421		182	22		204	234
1100	1-1/2″		13	.615		305	32		337	385
1110	2″		11	.727		490	38		528	595
1380	Ball, 150 psi, threaded									
1400	1/4″	1 Plum	24	.333	Ea.	10.80	17.35		28.15	38
1430	3/8″		24	.333		10.80	17.35		28.15	38
1450	1/2″		22	.364		10.80	18.95		29.75	40.50

22 05 23.20 Valves, Bronze		Crew	Daily Output	Labor-Hours	Unit	Material	Labor	Equipment	Total	Total Incl O&P
						2010 Bare Costs				
1460	3/4"	1 Plum	20	.400	Ea.	17.80	21		38.80	50.50
1470	1"		19	.421		22.50	22		44.50	57.50
1480	1-1/4"		15	.533		42	28		70	87.50
1490	1-1/2"		13	.615		54	32		86	108
1500	2"		11	.727		65.50	38		103.50	129
1750	Check, swing, class 150, regrinding disc, threaded									
1800	1/8"	1 Plum	24	.333	Ea.	43.50	17.35		60.85	73.50
1830	1/4"		24	.333		43.50	17.35		60.85	73.50
1840	3/8"		24	.333		45.50	17.35		62.85	76
1850	1/2"		24	.333		45.50	17.35		62.85	76
1860	3/4"		20	.400		65.50	21		86.50	103
1870	1"		19	.421		94.50	22		116.50	137
1880	1-1/4"		15	.533		136	28		164	192
1890	1-1/2"		13	.615		158	32		190	222
1900	2"		11	.727		233	38		271	315
1910	2-1/2"	Q-1	15	1.067		520	50		570	650
2000	For 200 lb, add					5%	10%			
2040	For 300 lb, add					15%	15%			
2850	Gate, N.R.S., soldered, 125 psi									
2900	3/8"	1 Plum	24	.333	Ea.	38	17.35		55.35	67.50
2920	1/2"		24	.333		32	17.35		49.35	61
2940	3/4"		20	.400		36.50	21		57.50	71
2950	1"		19	.421		52	22		74	90
2960	1-1/4"		15	.533		82.50	28		110.50	133
2970	1-1/2"		13	.615		88.50	32		120.50	145
2980	2"		11	.727		125	38		163	195
2990	2-1/2"	Q-1	15	1.067		315	50		365	420
3000	3"	"	13	1.231		410	57.50		467.50	535
3850	Rising stem, soldered, 300 psi									
3950	1"	1 Plum	19	.421	Ea.	136	22		158	183
3980	2"	"	11	.727		360	38		398	450
4000	3"	Q-1	13	1.231		1,150	57.50		1,207.50	1,325
4250	Threaded, class 150									
4310	1/4"	1 Plum	24	.333	Ea.	48	17.35		65.35	79
4320	3/8"		24	.333		48	17.35		65.35	79
4330	1/2"		24	.333		45	17.35		62.35	76
4340	3/4"		20	.400		52.50	21		73.50	89
4350	1"		19	.421		70.50	22		92.50	111
4360	1-1/4"		15	.533		95	28		123	146
4370	1-1/2"		13	.615		120	32		152	180
4380	2"		11	.727		162	38		200	236
4390	2-1/2"	Q-1	15	1.067		375	50		425	490
4400	3"	"	13	1.231		525	57.50		582.50	665
4500	For 300 psi, threaded, add					100%	15%			
4540	For chain operated type, add					15%				
4850	Globe, class 150, rising stem, threaded									
4920	1/4"	1 Plum	24	.333	Ea.	69	17.35		86.35	102
4940	3/8" size		24	.333		69	17.35		86.35	102
4950	1/2"		24	.333		69	17.35		86.35	102
4960	3/4"		20	.400		92.50	21		113.50	133
4970	1"		19	.421		145	22		167	193
4980	1-1/4"		15	.533		231	28		259	296
4990	1-1/2"		13	.615		281	32		313	360

22 05 23 – General-Duty Valves for Plumbing Piping

22 05 23.20 Valves, Bronze		Crew	Daily Output	Labor-Hours	Unit	Material	2010 Bare Costs Labor	Equipment	Total	Total Incl O&P
5000	2"	1 Plum	11	.727	Ea.	420	38		458	520
5010	2-1/2"	Q-1	15	1.067		850	50		900	1,000
5020	3"	"	13	1.231	↓	1,200	57.50		1,257.50	1,400
5120	For 300 lb threaded, add					50%	15%			
5600	Relief, pressure & temperature, self-closing, ASME, threaded									
5640	3/4"	1 Plum	28	.286	Ea.	124	14.85		138.85	160
5650	1"		24	.333		190	17.35		207.35	234
5660	1-1/4"		20	.400		360	21		381	425
5670	1-1/2"		18	.444		735	23		758	845
5680	2"	↓	16	.500	↓	755	26		781	870
5950	Pressure, poppet type, threaded									
6000	1/2"	1 Plum	30	.267	Ea.	33	13.90		46.90	57
6040	3/4"	"	28	.286	"	35	14.85		49.85	61
6400	Pressure, water, ASME, threaded									
6440	3/4"	1 Plum	28	.286	Ea.	70	14.85		84.85	99.50
6450	1"		24	.333		149	17.35		166.35	190
6460	1-1/4"		20	.400		236	21		257	290
6470	1-1/2"		18	.444		325	23		348	395
6480	2"		16	.500		470	26		496	560
6490	2-1/2"	↓	15	.533	↓	2,375	28		2,403	2,650
6900	Reducing, water pressure									
6920	300 psi to 25-75 psi, threaded or sweat									
6940	1/2"	1 Plum	24	.333	Ea.	225	17.35		242.35	273
6950	3/4"		20	.400		225	21		246	278
6960	1"		19	.421		350	22		372	420
6970	1-1/4"		15	.533		625	28		653	725
6980	1-1/2"	↓	13	.615	↓	940	32		972	1,075
8350	Tempering, water, sweat connections									
8400	1/2"	1 Plum	24	.333	Ea.	74.50	17.35		91.85	108
8440	3/4"	"	20	.400	"	91	21		112	131
8650	Threaded connections									
8700	1/2"	1 Plum	24	.333	Ea.	112	17.35		129.35	149
8740	3/4"		20	.400		390	21		411	460
8750	1"		19	.421		430	22		452	510
8760	1-1/4"		15	.533		685	28		713	790
8770	1-1/2"		13	.615		745	32		777	870
8780	2"	↓	11	.727	↓	1,125	38		1,163	1,275

22 05 23.60 Valves, Plastic		Crew	Daily Output	Labor-Hours	Unit	Material	Labor	Equipment	Total	Total Incl O&P
0010	**VALVES, PLASTIC**									
1100	Angle, PVC, threaded									
1110	1/4"	1 Plum	26	.308	Ea.	61	16		77	91
1120	1/2"		26	.308		61	16		77	91
1130	3/4"		25	.320		72.50	16.65		89.15	105
1140	1"	↓	23	.348	↓	88	18.10		106.10	124
1150	Ball, PVC, socket or threaded, single union									
1230	1/2"	1 Plum	26	.308	Ea.	21	16		37	47
1240	3/4"		25	.320		24	16.65		40.65	51.50
1250	1"		23	.348		30.50	18.10		48.60	60.50
1260	1-1/4"		21	.381		41	19.85		60.85	74.50
1270	1-1/2"		20	.400		48.50	21		69.50	84.50
1280	2"	↓	17	.471		70.50	24.50		95	114
1360	For PVC, flanged, add				↓	100%	15%			

22 05 23 – General-Duty Valves for Plumbing Piping

22 05 23.60 Valves, Plastic		Crew	Daily Output	Labor-Hours	Unit	Material	2010 Bare Costs Labor	Equipment	Total	Total Incl O&P
1650	CPVC, socket or threaded, single union									
1700	1/2"	1 Plum	26	.308	Ea.	40	16		56	68
1720	3/4"		25	.320		50.50	16.65		67.15	80.50
1730	1"		23	.348		60	18.10		78.10	93
1750	1-1/4"		21	.381		95.50	19.85		115.35	135
1760	1-1/2"		20	.400		95.50	21		116.50	136
1840	For CPVC, flanged, add					65%	15%			
1880	For true union, socket or threaded, add					50%	5%			
2050	Polypropylene, threaded									
2100	1/4"	1 Plum	26	.308	Ea.	42.50	16		58.50	70.50
2120	3/8"		26	.308		42.50	16		58.50	70.50
2130	1/2"		26	.308		42.50	16		58.50	70.50
2140	3/4"		25	.320		53	16.65		69.65	83
2150	1"		23	.348		61.50	18.10		79.60	95
2160	1-1/4"		21	.381		90.50	19.85		110.35	129
2170	1-1/2"		20	.400		104	21		125	145
2180	2"		17	.471		141	24.50		165.50	192
4850	Foot valve, PVC, socket or threaded									
4900	1/2"	1 Plum	34	.235	Ea.	61.50	12.25		73.75	86.50
4930	3/4"		32	.250		69.50	13		82.50	96
4940	1"		28	.286		90.50	14.85		105.35	122
4950	1-1/4"		27	.296		174	15.40		189.40	214
4960	1-1/2"		26	.308		174	16		190	215
6350	Y sediment strainer, PVC, socket or threaded									
6400	1/2"	1 Plum	26	.308	Ea.	41	16		57	69
6440	3/4"		24	.333		44	17.35		61.35	74.50
6450	1"		23	.348		53	18.10		71.10	85
6460	1-1/4"		21	.381		87	19.85		106.85	126
6470	1-1/2"		20	.400		87	21		108	127

22 05 48 – Vibration and Seismic Controls for Plumbing Piping and Equipment

22 05 48.10 Seismic Bracing Supports		Crew	Daily Output	Labor-Hours	Unit	Material	Labor	Equipment	Total	Total Incl O&P
0010	**SEISMIC BRACING SUPPORTS**									
0020	Clamps									
0030	C-clamp, for mounting on steel beam									
0040	3/8" threaded rod	1 Skwk	160	.050	Ea.	2.83	2.13		4.96	6.40
0050	1/2" threaded rod		160	.050		3.50	2.13		5.63	7.15
0060	5/8" threaded rod		160	.050		5.35	2.13		7.48	9.20
0070	3/4" threaded rod		160	.050		7.25	2.13		9.38	11.30
0100	Brackets									
0110	Beam side or wall mallable iron									
0120	3/8" threaded rod	1 Skwk	48	.167	Ea.	3.01	7.10		10.11	14.20
0130	1/2" threaded rod		48	.167		5.35	7.10		12.45	16.75
0140	5/8" threaded rod		48	.167		8.20	7.10		15.30	19.95
0150	3/4" threaded rod		48	.167		9.85	7.10		16.95	22
0160	7/8" threaded rod		48	.167		11.50	7.10		18.60	23.50
0170	For concrete installation, add						30%			
0180	Wall, welded steel									
0190	0 size 12" wide 18" deep	1 Skwk	34	.235	Ea.	204	10		214	239
0200	1 size 18" wide 24" deep		34	.235		242	10		252	282
0210	2 size 24" wide 30" deep		34	.235		320	10		330	370
0300	Rod, carbon steel									
0310	Continuous thread									

22 05 48 – Vibration and Seismic Controls for Plumbing Piping and Equipment

22 05 48.10 Seismic Bracing Supports		Crew	Daily Output	Labor-Hours	Unit	Material	Labor	Equipment	Total	Total Incl O&P
						2010 Bare Costs				
0320	1/4" thread	1 Skwk	144	.056	L.F.	.16	2.37		2.53	3.82
0330	3/8" thread		144	.056		.25	2.37		2.62	3.92
0340	1/2" thread		144	.056		.45	2.37		2.82	4.14
0350	5/8" thread		144	.056		.82	2.37		3.19	4.54
0360	3/4" thread		144	.056		1.20	2.37		3.57	4.96
0370	7/8" thread	↓	144	.056	↓	1.73	2.37		4.10	5.55
0380	For galvanized, add					30%				
0400	Channel, steel									
0410	3/4" x 1-1/2"	1 Skwk	80	.100	L.F.	8	4.26		12.26	15.35
0420	1-1/2" x 1-1/2"		70	.114		10.70	4.87		15.57	19.25
0430	1-7/8" x 1-1/2"		60	.133		26	5.70		31.70	37.50
0440	3" x 1-1/2"	↓	50	.160	↓	30	6.80		36.80	43.50
0450	Spring nuts									
0460	3/8"	1 Skwk	100	.080	Ea.	3	3.41		6.41	8.55
0470	1/2"	"	80	.100	"	3.70	4.26		7.96	10.60
0500	Welding, field									
0510	Cleaning and welding plates, bars, or rods									
0520	To existing beams, columns, or trusses									
0530	1" weld	1 Skwk	144	.056	Ea.	.19	2.37		2.56	3.85
0540	2" weld		72	.111		.34	4.73		5.07	7.65
0550	3" weld		54	.148		.52	6.30		6.82	10.25
0560	4" weld		36	.222		.73	9.45		10.18	15.35
0570	5" weld		30	.267		.91	11.35		12.26	18.45
0580	6" weld	↓	24	.333	↓	1.04	14.20		15.24	23
0600	Vibration absorbers									
0610	Hangers, neoprene flex									
0620	10-120 lb. capacity	1 Skwk	8	1	Ea.	22.50	42.50		65	90
0630	75-550 lb. capacity		8	1		29	42.50		71.50	97.50
0640	250-1100 lb. capacity		6	1.333		59	57		116	153
0650	1000-4000 lb. capacity	↓	6	1.333	↓	93.50	57		150.50	191
0660	Spring flex									
0670	60-450 lb. capacity	1 Skwk	8	1	Ea.	61	42.50		103.50	133
0680	85-450 lb. capacity		8	1		90	42.50		132.50	165
0690	600-900 lb. capacity		8	1		93	42.50		135.50	168
0700	1100-1300 lb. capacity	↓	6	1.333	↓	88.50	57		145.50	185
0710	Mounts, neoprene									
0720	135-380 lb. capacity	1 Skwk	7	1.143	Ea.	14.85	48.50		63.35	91.50
0730	250-1100 lb. capacity		7	1.143		39	48.50		87.50	118
0740	1000-4000 lb. capacity	↓	5	1.600	↓	87	68		155	201
0750	Swivel hangers, rod									
0760	1/4" thread size	1 Skwk	24	.333	Ea.	2.50	14.20		16.70	25
0770	3/8" thread size		24	.333		3.07	14.20		17.27	25.50
0780	1/2" thread size		24	.333		4.53	14.20		18.73	27
0790	5/8" thread size		24	.333		6.55	14.20		20.75	29
0800	3/4" thread size		24	.333		9.35	14.20		23.55	32.50
0810	7/8" thread size	↓	24	.333	↓	13.55	14.20		27.75	37
0900	Wire support									
0910	Light fixture, recessed (pair)	1 Skwk	96	.083	Ea.	3.07	3.55		6.62	8.85

22 05 76 – Facility Drainage Piping Cleanouts

22 05 76.10 Cleanouts		Crew	Daily Output	Labor-Hours	Unit	Material	Labor	Equipment	Total	Total Incl O&P
0010	**CLEANOUTS**									
0060	Floor type									

22 05 Common Work Results for Plumbing

22 05 76 – Facility Drainage Piping Cleanouts

22 05 76.10 Cleanouts		Crew	Daily Output	Labor-Hours	Unit	2010 Bare Costs Material	Labor	Equipment	Total	Total Incl O&P
0080	Round or square, scoriated nickel bronze top									
0100	2" pipe size	1 Plum	10	.800	Ea.	155	41.50		196.50	233
0120	3" pipe size		8	1		232	52		284	335
0140	4" pipe size		6	1.333		232	69.50		301.50	360
0980	Round top, recessed for terrazzo									
1000	2" pipe size	1 Plum	9	.889	Ea.	155	46.50		201.50	240
1080	3" pipe size		6	1.333		232	69.50		301.50	360
1100	4" pipe size		4	2		232	104		336	410
1120	5" pipe size	Q-1	6	2.667		295	125		420	510

22 05 76.20 Cleanout Tees		Crew	Daily Output	Labor-Hours	Unit	Material	Labor	Equipment	Total	Total Incl O&P
0010	**CLEANOUT TEES**									
0100	Cast iron, B&S, with countersunk plug									
0200	2" pipe size	1 Plum	4	2	Ea.	211	104		315	390
0220	3" pipe size		3.60	2.222		230	116		346	425
0240	4" pipe size		3.30	2.424		286	126		412	505
0280	6" pipe size	Q-1	5	3.200		770	150		920	1,075
0500	For round smooth access cover, same price									
4000	Plastic, tees and adapters. Add plugs									
4010	ABS, DWV									
4020	Cleanout tee, 1-1/2" pipe size	1 Plum	15	.533	Ea.	4.62	28		32.62	46.50
4030	2" pipe size	Q-1	27	.593		5.40	28		33.40	47.50
4040	3" pipe size		21	.762		10.85	35.50		46.35	65.50
4050	4" pipe size		16	1		18.95	47		65.95	91
4100	Cleanout plug, 1-1/2" pipe size	1 Plum	32	.250		.84	13		13.84	20.50
4110	2" pipe size	Q-1	56	.286		1.11	13.40		14.51	21
4120	3" pipe size		36	.444		1.79	21		22.79	33
4130	4" pipe size		30	.533		3.48	25		28.48	41.50
4180	Cleanout adapter fitting, 1-1/2" pipe size	1 Plum	32	.250		2.48	13		15.48	22
4190	2" pipe size	Q-1	56	.286		1.87	13.40		15.27	22
4200	3" pipe size		36	.444		5.25	21		26.25	37
4210	4" pipe size		30	.533		8.45	25		33.45	47
5000	PVC, DWV									
5010	Cleanout tee, 1-1/2" pipe size	1 Plum	15	.533	Ea.	3.63	28		31.63	45.50
5020	2" pipe size	Q-1	27	.593		4.46	28		32.46	46.50
5030	3" pipe size		21	.762		8.90	35.50		44.40	63.50
5040	4" pipe size		16	1		16.55	47		63.55	88
5090	Cleanout plug, 1-1/2" pipe size	1 Plum	32	.250		.97	13		13.97	20.50
5100	2" pipe size	Q-1	56	.286		.99	13.40		14.39	21
5110	3" pipe size		36	.444		1.93	21		22.93	33
5120	4" pipe size		30	.533		2.85	25		27.85	40.50
5130	6" pipe size		24	.667		9.45	31		40.45	57.50
5170	Cleanout adapter fitting, 1-1/2" pipe size	1 Plum	32	.250		1.19	13		14.19	21
5180	2" pipe size	Q-1	56	.286		1.67	13.40		15.07	22
5190	3" pipe size		36	.444		4.70	21		25.70	36
5200	4" pipe size		30	.533		7.75	25		32.75	46
5210	6" pipe size		24	.667		20.50	31		51.50	70

22 07 19.10 Piping Insulation		Crew	Daily Output	Labor-Hours	Unit	Material	2010 Bare Costs Labor	Equipment	Total	Total Incl O&P
0010	**PIPING INSULATION**									
0100	Rule of thumb, as a percentage of total mechanical costs				Job				10%	
0110	Insulation req'd is based on the surface size/area to be covered									
4000	Pipe covering (price copper tube one size less than IPS)									
6600	Fiberglass, with all service jacket									
6840	1" wall, 1/2" iron pipe size G	Q-14	240	.067	L.F.	.95	2.73		3.68	5.30
6870	1" iron pipe size G		220	.073		1.12	2.98		4.10	5.85
6900	2" iron pipe size G		200	.080		1.41	3.28		4.69	6.65
6920	3" iron pipe size G		180	.089		1.72	3.64		5.36	7.55
6940	4" iron pipe size G		150	.107		2.27	4.37		6.64	9.30
7320	2" wall, 1/2" iron pipe size G		220	.073		2.81	2.98		5.79	7.75
7440	6" iron pipe size G		100	.160		5.65	6.55		12.20	16.40
7460	8" iron pipe size G		80	.200		6.85	8.20		15.05	20.50
7480	10" iron pipe size G		70	.229		8.20	9.35		17.55	23.50
7490	12" iron pipe size G	↓	65	.246	↓	9.20	10.10		19.30	26
7800	For fiberglass with standard canvas jacket, deduct					5%				
7802	For fittings, add 3 L.F. for each fitting									
7804	plus 4 L.F. for each flange of the fitting									
7810	Finishes									
7811	Finishes, for .010" aluminum jacket, add G	Q-14	200	.080	S.F.	.77	3.28		4.05	5.95
7812	For .016" aluminum jacket, add G		200	.080		1.12	3.28		4.40	6.35
7813	For .010" stainless steel, add G	↓	160	.100	↓	3.11	4.10		7.21	9.75
7814	For single layer of felt, add					10%	10%			
7816	For roofing paper, 45 lb to 55 lb, add					25%	10%			
7879	Rubber tubing, flexible closed cell foam									
7880	3/8" wall, 1/4" iron pipe size G	1 Asbe	120	.067	L.F.	.55	3.04		3.59	5.35
7910	1/2" iron pipe size G		115	.070		.70	3.17		3.87	5.70
7920	3/4" iron pipe size G		115	.070		.85	3.17		4.02	5.85
7930	1" iron pipe size G		110	.073		.98	3.31		4.29	6.25
7950	1-1/2" iron pipe size G		110	.073		1.30	3.31		4.61	6.60
8100	1/2" wall, 1/4" iron pipe size G		90	.089		.50	4.05		4.55	6.85
8130	1/2" iron pipe size G		89	.090		.61	4.09		4.70	7
8140	3/4" iron pipe size G		89	.090		.68	4.09		4.77	7.10
8150	1" iron pipe size G		88	.091		.75	4.14		4.89	7.30
8170	1-1/2" iron pipe size G		87	.092		1.06	4.19		5.25	7.65
8180	2" iron pipe size G		86	.093		1.38	4.24		5.62	8.10
8200	3" iron pipe size G		85	.094		1.95	4.29		6.24	8.80
8300	3/4" wall, 1/4" iron pipe size G		90	.089		.77	4.05		4.82	7.15
8330	1/2" iron pipe size G		89	.090		1.01	4.09		5.10	7.45
8340	3/4" iron pipe size G		89	.090		1.23	4.09		5.32	7.70
8350	1" iron pipe size G		88	.091		1.41	4.14		5.55	8
8370	1-1/2" iron pipe size G		87	.092		2.15	4.19		6.34	8.85
8380	2" iron pipe size G		86	.093		2.52	4.24		6.76	9.35
8400	3" iron pipe size G		85	.094		3.85	4.29		8.14	10.90
8444	1" wall, 1/2" iron pipe size G		86	.093		1.93	4.24		6.17	8.70
8445	3/4" iron pipe size G		84	.095		2.34	4.34		6.68	9.30
8446	1" iron pipe size G		84	.095		2.72	4.34		7.06	9.75
8447	1-1/4" iron pipe size G		82	.098		3.08	4.44		7.52	10.30
8448	1-1/2" iron pipe size G		82	.098		3.58	4.44		8.02	10.85
8449	2" iron pipe size G		80	.100		4.78	4.56		9.34	12.35
8450	2-1/2" iron pipe size G	↓	80	.100	↓	6.25	4.56		10.81	13.95
8456	Rubber insulation tape, 1/8" x 2" x 30' G				Ea.	11.40			11.40	12.50

22 11 Facility Water Distribution

22 11 13 – Facility Water Distribution Piping

22 11 13.14 Pipe, Brass		Crew	Daily Output	Labor-Hours	Unit	Material	Labor	Equipment	Total	Total Incl O&P
						2010 Bare Costs				
0010	**PIPE, BRASS**, Plain end									
0900	Field threaded, coupling & clevis hanger assembly 10′ O.C.									
0920	Regular weight									
1120	1/2″ diameter	1 Plum	48	.167	L.F.	5.80	8.70		14.50	19.40
1140	3/4″ diameter		46	.174		7.85	9.05		16.90	22
1160	1″ diameter		43	.186		11.45	9.70		21.15	27
1180	1-1/4″ diameter	Q-1	72	.222		17.50	10.40		27.90	35
1200	1-1/2″ diameter		65	.246		21	11.55		32.55	40.50
1220	2″ diameter		53	.302		29	14.15		43.15	53

22 11 13.23 Pipe/Tube, Copper		Crew	Daily Output	Labor-Hours	Unit	Material	Labor	Equipment	Total	Total Incl O&P
0010	**PIPE/TUBE, COPPER**, Solder joints									
1000	Type K tubing, couplings & clevis hanger assemblies 10′ O.C.									
1100	1/4″ diameter	1 Plum	84	.095	L.F.	2.80	4.96		7.76	10.55
1200	1″ diameter		66	.121		7.95	6.30		14.25	18.20
1260	2″ diameter		40	.200		20	10.40		30.40	37.50
2000	Type L tubing, couplings & clevis hanger assemblies 10′ O.C.									
2100	1/4″ diameter	1 Plum	88	.091	L.F.	1.60	4.73		6.33	8.85
2120	3/8″ diameter		84	.095		2.27	4.96		7.23	9.95
2140	1/2″ diameter **CN**		81	.099		2.49	5.15		7.64	10.45
2160	5/8″ diameter		79	.101		3.87	5.25		9.12	12.15
2180	3/4″ diameter		76	.105		3.83	5.50		9.33	12.40
2200	1″ diameter		68	.118		5.95	6.10		12.05	15.75
2220	1-1/4″ diameter		58	.138		8.35	7.20		15.55	19.95
2240	1-1/2″ diameter		52	.154		10.75	8		18.75	24
2260	2″ diameter		42	.190		16.70	9.90		26.60	33.50
2280	2-1/2″ diameter	Q-1	62	.258		26	12.10		38.10	46.50
2300	3″ diameter		56	.286		35.50	13.40		48.90	59
2320	3-1/2″ diameter		43	.372		49	17.45		66.45	80
2340	4″ diameter		39	.410		60.50	19.20		79.70	95.50
2360	5″ diameter		34	.471		130	22		152	176
2380	6″ diameter	Q-2	40	.600		182	29		211	244
2400	8″ diameter	″	36	.667		305	32.50		337.50	385
2410	For other than full hard temper, add					21%				
2590	For silver solder, add						15%			
4000	Type DWV tubing, couplings & clevis hanger assemblies 10′ O.C.									
4100	1-1/4″ diameter	1 Plum	60	.133	L.F.	9.80	6.95		16.75	21
4120	1-1/2″ diameter		54	.148		9.25	7.70		16.95	22
4140	2″ diameter		44	.182		12.55	9.45		22	28
4160	3″ diameter	Q-1	58	.276		23.50	12.90		36.40	45.50
4180	4″ diameter		40	.400		42	18.75		60.75	74
4200	5″ diameter		36	.444		117	21		138	160
4220	6″ diameter	Q-2	42	.571		169	28		197	228

22 11 13.44 Pipe, Steel		Crew	Daily Output	Labor-Hours	Unit	Material	Labor	Equipment	Total	Total Incl O&P
0010	**PIPE, STEEL**									
0012	The steel pipe in this section does not include fittings such as ells, tees									
0014	For fittings either add a % (usually 25 to 35%) or see									
0015	the Mechanical or Plumbing Cost Data									
0020	All pipe sizes are to Spec. A-53 unless noted otherwise R221113-50									
0050	Schedule 40, threaded, with couplings, and clevis hanger									
0060	assemblies sized for covering, 10′ O.C.									
0540	Black, 1/4″ diameter	1 Plum	66	.121	L.F.	2.78	6.30		9.08	12.50
0550	3/8″ diameter		65	.123		3.16	6.40		9.56	13.10

22 11 13 – Facility Water Distribution Piping

22 11 13.44 Pipe, Steel		Crew	Daily Output	Labor-Hours	Unit	Material	2010 Bare Costs Labor	Equipment	Total	Total Incl O&P
0560	1/2" diameter	1 Plum	63	.127	L.F.	3.61	6.60		10.21	13.85
0570	3/4" diameter		61	.131		4.28	6.85		11.13	14.95
0580	1" diameter		53	.151		6.25	7.85		14.10	18.65
0590	1-1/4" diameter	Q-1	89	.180		7.95	8.40		16.35	21.50
0600	1-1/2" diameter		80	.200		9.45	9.35		18.80	24.50
0610	2" diameter CN		64	.250		12.60	11.70		24.30	31.50
0620	2-1/2" diameter		50	.320		19.45	15		34.45	44
0630	3" diameter		43	.372		25	17.45		42.45	53.50
0640	3-1/2" diameter		40	.400		34	18.75		52.75	65.50
0650	4" diameter		36	.444		36.50	21		57.50	71.50
1280	All pipe sizes are to Spec. A-53 unless noted otherwise									
1281	Schedule 40, threaded, with couplings and clevis hanger									
1282	assemblies sized for covering, 10' O. C.									
1290	Galvanized, 1/4" diameter	1 Plum	66	.121	L.F.	3.80	6.30		10.10	13.65
1300	3/8" diameter		65	.123		4.22	6.40		10.62	14.25
1310	1/2" diameter		63	.127		5.15	6.60		11.75	15.55
1320	3/4" diameter		61	.131		5	6.85		11.85	15.75
1330	1" diameter		53	.151		8.45	7.85		16.30	21
1340	1-1/4" diameter	Q-1	89	.180		10.90	8.40		19.30	24.50
1350	1-1/2" diameter		80	.200		12.90	9.35		22.25	28.50
1360	2" diameter		64	.250		17.25	11.70		28.95	36.50
1370	2-1/2" diameter		50	.320		29	15		44	54
1380	3" diameter		43	.372		37	17.45		54.45	66.50
1390	3-1/2" diameter		40	.400		45.50	18.75		64.25	78
1400	4" diameter		36	.444		52	21		73	88.50
2000	Welded, sch. 40, on yoke & roll hanger assy's, sized for covering, 10' O.C.									
2040	Black, 1" diameter	Q-15	93	.172	L.F.	3.75	8.05	.63	12.43	16.90
2070	2" diameter		61	.262		6.55	12.30	.96	19.81	26.50
2090	3" diameter		43	.372		11.40	17.45	1.36	30.21	40
2110	4" diameter		37	.432		15.95	20.50	1.58	38.03	50
2120	5" diameter		32	.500		21.50	23.50	1.83	46.83	61
2130	6" diameter	Q-16	36	.667		26.50	32.50	1.63	60.63	79.50
2140	8" diameter		29	.828		39	40	2.02	81.02	105
2150	10" diameter		24	1		55	48.50	2.44	105.94	136
2160	12" diameter		19	1.263		83	61.50	3.08	147.58	187

22 11 13.48 Pipe, Fittings and Valves, Steel, Grooved-Joint

Line	Description	Crew	Daily Output	Labor-Hours	Unit	Material	Labor	Equipment	Total	Total Incl O&P
0010	**PIPE, FITTINGS AND VALVES, STEEL, GROOVED-JOINT**									
0012	Fittings are ductile iron. Steel fittings noted.									
0020	Pipe includes coupling & clevis type hanger assemblies, 10' O.C.									
1000	Schedule 40, black									
1040	3/4" diameter	1 Plum	71	.113	L.F.	5.30	5.85		11.15	14.60
1050	1" diameter		63	.127		3.38	6.60		9.98	13.60
1060	1-1/4" diameter		58	.138		4.43	7.20		11.63	15.60
1070	1-1/2" diameter		51	.157		5.10	8.15		13.25	17.85
1080	2" diameter		40	.200		6.30	10.40		16.70	22.50
1090	2-1/2" diameter	Q-1	57	.281		9.15	13.15		22.30	30
1100	3" diameter		50	.320		11.50	15		26.50	35
1110	4" diameter		45	.356		16.35	16.65		33	43
1120	5" diameter		37	.432		22.50	20.50		43	55.50
1130	6" diameter	Q-2	42	.571		28.50	28		56.50	73
1800	Galvanized									
1840	3/4" diameter	1 Plum	71	.113	L.F.	6	5.85		11.85	15.40

22 11 13 – Facility Water Distribution Piping

22 11 13.48 Pipe, Fittings and Valves, Steel, Grooved-Joint		Crew	Daily Output	Labor-Hours	Unit	Material	2010 Bare Costs Labor	Equipment	Total	Total Incl O&P
1850	1" diameter	1 Plum	63	.127	L.F.	8.35	6.60		14.95	19.10
1860	1-1/4" diameter		58	.138		10.80	7.20		18	22.50
1870	1-1/2" diameter		51	.157		12.70	8.15		20.85	26
1880	2" diameter		40	.200		16.50	10.40		26.90	34
1890	2-1/2" diameter	Q-1	57	.281		25	13.15		38.15	47
1900	3" diameter		50	.320		32.50	15		47.50	58
1910	4" diameter		45	.356		45	16.65		61.65	74.50
1920	5" diameter		37	.432		51.50	20.50		72	87
1930	6" diameter	Q-2	42	.571		60	28		88	108
3990	Fittings: coupling material required at joints not incl. in fitting price.									
3994	Add 1 selected coupling, material only, per joint for installed price.									
4000	Elbow, 90° or 45°, painted									
4030	3/4" diameter	1 Plum	50	.160	Ea.	40	8.35		48.35	56.50
4040	1" diameter		50	.160		21.50	8.35		29.85	36
4050	1-1/4" diameter		40	.200		21.50	10.40		31.90	39
4060	1-1/2" diameter		33	.242		21.50	12.60		34.10	42.50
4070	2" diameter		25	.320		21.50	16.65		38.15	48.50
4080	2-1/2" diameter	Q-1	40	.400		21.50	18.75		40.25	51.50
4090	3" diameter		33	.485		38	22.50		60.50	75.50
4100	4" diameter		25	.640		41.50	30		71.50	90.50
4110	5" diameter		20	.800		99	37.50		136.50	165
4120	6" diameter	Q-2	25	.960		116	46.50		162.50	198
4250	For galvanized elbows, add					26%				
4690	Tee, painted									
4700	3/4" diameter	1 Plum	38	.211	Ea.	43.50	10.95		54.45	64
4740	1" diameter		33	.242		33	12.60		45.60	55.50
4750	1-1/4" diameter		27	.296		33	15.40		48.40	59.50
4760	1-1/2" diameter		22	.364		33	18.95		51.95	65
4770	2" diameter		17	.471		33	24.50		57.50	73
4780	2-1/2" diameter	Q-1	27	.593		33	28		61	78
4790	3" diameter		22	.727		46	34		80	102
4800	4" diameter		17	.941		70	44		114	143
4810	5" diameter		13	1.231		163	57.50		220.50	267
4820	6" diameter	Q-2	17	1.412		189	68.50		257.50	310
4900	For galvanized tees, add					24%				
4906	Couplings, rigid style, painted									
4908	1" diameter	1 Plum	100	.080	Ea.	16	4.16		20.16	24
4909	1-1/4" diameter		100	.080		16	4.16		20.16	24
4910	1-1/2" diameter		67	.119		16	6.20		22.20	27
4912	2" diameter		50	.160		16.40	8.35		24.75	30.50
4914	2-1/2" diameter	Q-1	80	.200		18.90	9.35		28.25	35
4916	3" diameter		67	.239		22	11.20		33.20	41
4918	4" diameter		50	.320		31	15		46	56.50
4920	5" diameter		40	.400		40	18.75		58.75	72
4922	6" diameter	Q-2	50	.480		53	23.50		76.50	93.50
4940	Flexible, standard, painted									
4950	3/4" diameter	1 Plum	100	.080	Ea.	11.50	4.16		15.66	18.90
4960	1" diameter		100	.080		11.50	4.16		15.66	18.90
4970	1-1/4" diameter		80	.100		15.30	5.20		20.50	24.50
4980	1-1/2" diameter		67	.119		16.80	6.20		23	28
4990	2" diameter		50	.160		17.80	8.35		26.15	32
5000	2-1/2" diameter	Q-1	80	.200		21	9.35		30.35	37
5010	3" diameter		67	.239		23	11.20		34.20	42.50

22 11 13 – Facility Water Distribution Piping

22 11 13.48 Pipe, Fittings and Valves, Steel, Grooved-Joint		Crew	Daily Output	Labor-Hours	Unit	Material (2010 Bare Costs)	Labor (2010 Bare Costs)	Equipment (2010 Bare Costs)	Total (2010 Bare Costs)	Total Incl O&P
5020	3-1/2" diameter	Q-1	57	.281	Ea.	33.50	13.15		46.65	56.50
5030	4" diameter		50	.320		33.50	15		48.50	59.50
5040	5" diameter	↓	40	.400		51	18.75		69.75	84
5050	6" diameter	Q-2	50	.480	↓	60	23.50		83.50	101

22 11 13.64 Pipe, Stainless Steel

		Crew	Daily Output	Labor-Hours	Unit	Material	Labor	Equipment	Total	Total Incl O&P
0010	**PIPE, STAINLESS STEEL**									
3500	Threaded, couplings and clevis hanger assemblies, 10' O.C.									
3520	Schedule 40, type 304									
3540	1/4" diameter	1 Plum	54	.148	L.F.	7.35	7.70		15.05	19.65
3550	3/8" diameter		53	.151		7.45	7.85		15.30	19.95
3560	1/2" diameter		52	.154		8.25	8		16.25	21
3580	1" diameter	↓	45	.178		14.70	9.25		23.95	30
3610	2" diameter	Q-1	57	.281		32.50	13.15		45.65	55
3640	4" diameter	Q-2	51	.471		102	23		125	147
3740	For small quantities, add				↓	10%				
4250	Schedule 40, type 316									
4290	1/4" diameter	1 Plum	54	.148	L.F.	8.90	7.70		16.60	21.50
4300	3/8" diameter		53	.151		9.60	7.85		17.45	22.50
4310	1/2" diameter		52	.154		13.20	8		21.20	26.50
4320	3/4" diameter		51	.157		17.20	8.15		25.35	31
4330	1" diameter	↓	45	.178		25	9.25		34.25	41.50
4360	2" diameter	Q-1	57	.281		55.50	13.15		68.65	81
4390	4" diameter	Q-2	51	.471		167	23		190	218
4490	For small quantities, add				↓	10%				

22 11 13.74 Pipe, Plastic

		Crew	Daily Output	Labor-Hours	Unit	Material	Labor	Equipment	Total	Total Incl O&P
0010	**PIPE, PLASTIC**									
1800	PVC, couplings 10' O.C., clevis hanger assemblies, 3 per 10'									
1820	Schedule 40									
1860	1/2" diameter	1 Plum	54	.148	L.F.	1.65	7.70		9.35	13.35
1870	3/4" diameter		51	.157		1.89	8.15		10.04	14.35
1880	1" diameter		46	.174		2.18	9.05		11.23	15.95
1890	1-1/4" diameter		42	.190		2.72	9.90		12.62	17.85
1900	1-1/2" diameter	↓	36	.222		2.92	11.55		14.47	20.50
1910	2" diameter **CN**	Q-1	59	.271		3.51	12.70		16.21	23
1920	2-1/2" diameter		56	.286		5.40	13.40		18.80	26
1930	3" diameter		53	.302		6.95	14.15		21.10	28.50
1940	4" diameter		48	.333		9.60	15.60		25.20	34
1950	5" diameter		43	.372		13.95	17.45		31.40	41.50
1960	6" diameter	↓	39	.410	↓	17.80	19.20		37	48.50
4100	DWV type, schedule 40, couplings 10' O.C., clevis hanger assy's, 3 per 10'									
4210	ABS, schedule 40, foam core type									
4212	Plain end black									
4214	1-1/2" diameter	1 Plum	39	.205	L.F.	1.93	10.70		12.63	18.15
4216	2" diameter	Q-1	62	.258		2.35	12.10		14.45	20.50
4218	3" diameter		56	.286		4.42	13.40		17.82	25
4220	4" diameter		51	.314		6.10	14.70		20.80	28.50
4222	6" diameter	↓	42	.381	↓	12.35	17.85		30.20	40.50
4240	To delete coupling & hangers, subtract									
4244	1-1/2" diam. to 6" diam.					43%	48%			
4400	PVC									
4410	1-1/4" diameter	1 Plum	42	.190	L.F.	1.99	9.90		11.89	17.05
4420	1-1/2" diameter	"	36	.222	↓	2	11.55		13.55	19.55

22 11 13 – Facility Water Distribution Piping

22 11 13.74 Pipe, Plastic		Crew	Daily Output	Labor-Hours	Unit	Material	2010 Bare Costs Labor	Equipment	Total	Total Incl O&P
4460	2" diameter	Q-1	59	.271	L.F.	2.41	12.70		15.11	21.50
4470	3" diameter		53	.302		4.51	14.15		18.66	26
4480	4" diameter		48	.333		6.10	15.60		21.70	30.50
4490	6" diameter		39	.410		11.10	19.20		30.30	41
5300	CPVC, socket joint, couplings 10' O.C., clevis hanger assemblies, 3 per 10'									
5302	Schedule 40									
5304	1/2" diameter	1 Plum	54	.148	L.F.	2.35	7.70		10.05	14.15
5305	3/4" diameter		51	.157		2.85	8.15		11	15.40
5306	1" diameter		46	.174		3.76	9.05		12.81	17.70
5307	1-1/4" diameter		42	.190		4.88	9.90		14.78	20
5308	1-1/2" diameter		36	.222		5.75	11.55		17.30	23.50
5309	2" diameter	Q-1	59	.271		7.15	12.70		19.85	27
5310	2-1/2" diameter		56	.286		12.25	13.40		25.65	33.50
5311	3" diameter		53	.302		14.90	14.15		29.05	37.50
5360	CPVC, threaded, couplings 10' O.C., clevis hanger assemblies, 3 per 10'									
5380	Schedule 40									
5460	1/2" diameter	1 Plum	54	.148	L.F.	3.28	7.70		10.98	15.15
5470	3/4" diameter		51	.157		3.88	8.15		12.03	16.50
5480	1" diameter		46	.174		4.76	9.05		13.81	18.80
5490	1-1/4" diameter		42	.190		5.85	9.90		15.75	21.50
5500	1-1/2" diameter		36	.222		6.60	11.55		18.15	24.50
5510	2" diameter	Q-1	59	.271		8.20	12.70		20.90	28
5520	2-1/2" diameter		56	.286		13.35	13.40		26.75	34.50
5530	3" diameter		53	.302		16.50	14.15		30.65	39
7280	PEX, flexible, no couplings or hangers									
7282	Note: For labor costs add 25% to the couplings and fittings labor total.									
7300	Non-barrier type, hot/cold tubing rolls									
7310	1/4" diameter x 100'				L.F.	.44			.44	.48
7350	3/8" diameter x 100'					.49			.49	.54
7360	1/2" diameter x 100'					.55			.55	.61
7370	1/2" diameter x 500'					.55			.55	.61
7380	1/2" diameter x 1000'					.55			.55	.61
7400	3/4" diameter x 100'					1			1	1.10
7410	3/4" diameter x 500'					1			1	1.10
7420	3/4" diameter x 1000'					1			1	1.10
7460	1" diameter x 100'					1.71			1.71	1.88
7470	1" diameter x 300'					1.71			1.71	1.88
7480	1" diameter x 500'					1.71			1.71	1.88
7500	1-1/4" diameter x 100'					2.90			2.90	3.19
7510	1-1/4" diameter x 300'					2.90			2.90	3.19
7540	1-1/2" diameter x 100'					3.64			3.64	4
7550	1-1/2" diameter x 300'					3.40			3.40	3.74
7596	Most sizes available in red or blue									
7700	Non-barrier type, hot/cold tubing straight lengths									
7710	1/2" diameter x 20'				L.F.	.55			.55	.61
7750	3/4" diameter x 20'					.99			.99	1.09
7760	1" diameter x 20'					1.71			1.71	1.88
7770	1-1/4" diameter x 20'					2.90			2.90	3.19
7780	1-1/2" diameter x 20'					3.64			3.64	4
7790	2" diameter					5.05			5.05	5.55
7796	Most sizes available in red or blue									

22 11 19 – Domestic Water Piping Specialties

22 11 19.10 Flexible Connectors

	22 11 19.10 Flexible Connectors	Crew	Daily Output	Labor-Hours	Unit	2010 Bare Costs Material	Labor	Equipment	Total	Total Incl O&P
0010	**FLEXIBLE CONNECTORS**, Corrugated, 7/8″ O.D., 1/2″ I.D.									
0050	Gas, seamless brass, steel fittings									
0200	12″ long	1 Plum	36	.222	Ea.	15.10	11.55		26.65	34
0220	18″ long		36	.222		18.70	11.55		30.25	38
0240	24″ long		34	.235		22	12.25		34.25	43
0280	36″ long		32	.250		26.50	13		39.50	48.50
0340	60″ long		30	.267		40	13.90		53.90	65
2000	Water, copper tubing, dielectric separators									
2100	12″ long	1 Plum	36	.222	Ea.	14	11.55		25.55	33
2260	24″ long	″	34	.235	″	20.50	12.25		32.75	41

22 11 19.14 Flexible Metal Hose

	22 11 19.14 Flexible Metal Hose	Crew	Daily Output	Labor-Hours	Unit	Material	Labor	Equipment	Total	Total Incl O&P
0010	**FLEXIBLE METAL HOSE**, Connectors, standard lengths									
0100	Bronze braided, bronze ends									
0120	3/8″ diameter x 12″	1 Stpi	26	.308	Ea.	18.30	15.95		34.25	44
0160	3/4″ diameter x 12″		20	.400		25.50	21		46.50	59
0180	1″ diameter x 18″		19	.421		33	22		55	69
0200	1-1/2″ diameter x 18″		13	.615		52	32		84	106
0220	2″ diameter x 18″		11	.727		63	38		101	126

22 11 19.26 Pressure Regulators

	22 11 19.26 Pressure Regulators	Crew	Daily Output	Labor-Hours	Unit	Material	Labor	Equipment	Total	Total Incl O&P
0010	**PRESSURE REGULATORS**									
3000	Steam, high capacity, bronze body, stainless steel trim									
3020	Threaded, 1/2″ diameter	1 Stpi	24	.333	Ea.	1,375	17.30		1,392.30	1,525
3030	3/4″ diameter		24	.333		1,375	17.30		1,392.30	1,525
3040	1″ diameter		19	.421		1,525	22		1,547	1,700
3060	1-1/4″ diameter		15	.533		1,675	27.50		1,702.50	1,900
3080	1-1/2″ diameter		13	.615		1,925	32		1,957	2,175
3100	2″ diameter		11	.727		2,350	38		2,388	2,650
3120	2-1/2″ diameter	Q-5	12	1.333		2,950	62.50		3,012.50	3,350
3140	3″ diameter	″	11	1.455		3,375	68		3,443	3,800
3500	Flanged connection, iron body, 125 lb. W.S.P.									
3520	3″ diameter	Q-5	11	1.455	Ea.	3,700	68		3,768	4,150
3540	4″ diameter	″	5	3.200	″	4,650	149		4,799	5,350

22 11 19.38 Water Supply Meters

	22 11 19.38 Water Supply Meters	Crew	Daily Output	Labor-Hours	Unit	Material	Labor	Equipment	Total	Total Incl O&P
0010	**WATER SUPPLY METERS**									
2000	Domestic/commercial, bronze									
2020	Threaded									
2060	5/8″ diameter, to 20 GPM	1 Plum	16	.500	Ea.	42	26		68	85
2080	3/4″ diameter, to 30 GPM		14	.571		76.50	29.50		106	129
2100	1″ diameter, to 50 GPM		12	.667		116	34.50		150.50	180
2300	Threaded/flanged									
2340	1-1/2″ diameter, to 100 GPM	1 Plum	8	1	Ea.	284	52		336	390
2360	2″ diameter, to 160 GPM	″	6	1.333	″	385	69.50		454.50	530
2600	Flanged, compound									
2640	3″ diameter, 320 GPM	Q-1	3	5.333	Ea.	2,625	250		2,875	3,250
2660	4″ diameter, to 500 GPM		1.50	10.667		4,200	500		4,700	5,350
2680	6″ diameter, to 1,000 GPM		1	16		6,675	750		7,425	8,475
2700	8″ diameter, to 1,800 GPM		.80	20		10,500	935		11,435	12,900

22 11 19.42 Backflow Preventers

	22 11 19.42 Backflow Preventers	Crew	Daily Output	Labor-Hours	Unit	Material	Labor	Equipment	Total	Total Incl O&P
0010	**BACKFLOW PREVENTERS**, Includes valves									
0020	and four test cocks, corrosion resistant, automatic operation									
4000	Reduced pressure principle									

22 11 19 – Domestic Water Piping Specialties

22 11 19.42 Backflow Preventers		Crew	Daily Output	Labor-Hours	Unit	Material	Labor	Equipment	Total	Total Incl O&P
						2010 Bare Costs				
4100	Threaded, bronze, valves are ball									
4120	3/4" pipe size	1 Plum	16	.500	Ea.	365	26		391	440
4140	1" pipe size		14	.571		390	29.50		419.50	475
4150	1-1/4" pipe size		12	.667		670	34.50		704.50	790
4160	1-1/2" pipe size		10	.800		735	41.50		776.50	875
4180	2" pipe size	↓	7	1.143	↓	825	59.50		884.50	1,000
5000	Flanged, bronze, valves are OS&Y									
5060	2-1/2" pipe size	Q-1	5	3.200	Ea.	3,525	150		3,675	4,100
5080	3" pipe size		4.50	3.556		3,700	167		3,867	4,325
5100	4" pipe size	↓	3	5.333		4,650	250		4,900	5,475
5120	6" pipe size	Q-2	3	8	↓	6,725	390		7,115	7,975
5600	Flanged, iron, valves are OS&Y									
5660	2-1/2" pipe size	Q-1	5	3.200	Ea.	1,925	150		2,075	2,350
5680	3" pipe size		4.50	3.556		2,950	167		3,117	3,500
5700	4" pipe size	↓	3	5.333		3,800	250		4,050	4,550
5720	6" pipe size	Q-2	3	8		5,375	390		5,765	6,475
5740	8" pipe size		2	12		9,450	585		10,035	11,300
5760	10" pipe size	↓	1	24	↓	12,700	1,175		13,875	15,700

22 11 19.50 Vacuum Breakers		Crew	Daily Output	Labor-Hours	Unit	Material	Labor	Equipment	Total	Total Incl O&P
0010	**VACUUM BREAKERS**									
0013	See also backflow preventers Div. 22 11 19.42									
1000	Anti-siphon continuous pressure type									
1010	Max. 150 PSI - 210°F									
1020	Bronze body									
1030	1/2" size	1 Stpi	24	.333	Ea.	172	17.30		189.30	215
1040	3/4" size		20	.400		172	21		193	220
1050	1" size		19	.421		178	22		200	228
1060	1-1/4" size		15	.533		355	27.50		382.50	430
1070	1-1/2" size		13	.615		425	32		457	520
1080	2" size	↓	11	.727	↓	440	38		478	540
1200	Max. 125 PSI with atmospheric vent									
1210	Brass, in-line construction									
1220	1/4" size	1 Stpi	24	.333	Ea.	59.50	17.30		76.80	91.50
1230	3/8" size	"	24	.333		59.50	17.30		76.80	91.50
1260	For polished chrome finish, add				↓	13%				
2000	Anti-siphon, non-continuous pressure type									
2010	Hot or cold water 125 PSI - 210°F									
2020	Bronze body									
2030	1/4" size	1 Stpi	24	.333	Ea.	38	17.30		55.30	67.50
2040	3/8" size		24	.333		38	17.30		55.30	67.50
2050	1/2" size		24	.333		43	17.30		60.30	73.50
2060	3/4" size		20	.400		52.50	21		73.50	88.50
2070	1" size		19	.421		79.50	22		101.50	121
2080	1-1/4" size		15	.533		139	27.50		166.50	195
2090	1-1/2" size		13	.615		164	32		196	228
2100	2" size		11	.727		255	38		293	335
2110	2-1/2" size		8	1		730	52		782	885
2120	3" size	↓	6	1.333	↓	975	69		1,044	1,175
2150	For polished chrome finish, add					50%				

22 11 19 – Domestic Water Piping Specialties

22 11 19.54 Water Hammer Arresters/Shock Absorbers

	22 11 19.54 Water Hammer Arresters/Shock Absorbers	Crew	Daily Output	Labor-Hours	Unit	Material	2010 Bare Costs Labor	Equipment	Total	Total Incl O&P
0010	**WATER HAMMER ARRESTERS/SHOCK ABSORBERS**									
0490	Copper									
0500	3/4" male I.P.S. For 1 to 11 fixtures	1 Plum	12	.667	Ea.	20.50	34.50		55	74.50
0600	1" male I.P.S., For 12 to 32 fixtures		8	1		41.50	52		93.50	124
0700	1-1/4" male I.P.S. For 33 to 60 fixtures		8	1		44	52		96	127
0800	1-1/2" male I.P.S. For 61 to 113 fixtures		8	1		63	52		115	148
0900	2" male I.P.S.For 114 to 154 fixtures		8	1		92	52		144	179
1000	2-1/2" male I.P.S. For 155 to 330 fixtures	↓	4	2	↓	285	104		389	470

22 11 19.64 Hydrants

	22 11 19.64 Hydrants	Crew	Daily Output	Labor-Hours	Unit	Material	Labor	Equipment	Total	Total Incl O&P
0010	**HYDRANTS**									
0050	Wall type, moderate climate, bronze, encased									
0200	3/4" IPS connection	1 Plum	16	.500	Ea.	600	26		626	700
0300	1" IPS connection	"	14	.571		690	29.50		719.50	805
0500	Anti-siphon type, 3/4" connection				↓	520			520	570
1000	Non-freeze, bronze, exposed									
1100	3/4" IPS connection, 4" to 9" thick wall	1 Plum	14	.571	Ea.	405	29.50		434.50	490
1120	10" to 14" thick wall		12	.667		520	34.50		554.50	620
1140	15" to 19" thick wall		12	.667		520	34.50		554.50	620
1160	20" to 24" thick wall	↓	10	.800	↓	530	41.50		571.50	650
1200	For 1" IPS connection, add					15%	10%			
1240	For 3/4" adapter type vacuum breaker, add				Ea.	51			51	56
1280	For anti-siphon type, add				"	106			106	117
2000	Non-freeze bronze, encased, anti-siphon type									
2100	3/4" IPS connection, 5" to 9" thick wall	1 Plum	14	.571	Ea.	1,025	29.50		1,054.50	1,175
2140	15" to 19" thick wall	"	12	.667	"	1,100	34.50		1,134.50	1,275
3000	Ground box type, bronze frame, 3/4" IPS connection									
3080	Non-freeze, all bronze, polished face, set flush									
3100	2 feet depth of bury	1 Plum	8	1	Ea.	765	52		817	925
3180	6 feet depth of bury		7	1.143		1,000	59.50		1,059.50	1,200
3220	8 feet depth of bury	↓	5	1.600		1,125	83.50		1,208.50	1,350
3400	For 1" IPS connection, add					15%	10%			
3550	For 2" connection, add					445%	24%			
3600	For tapped drain port in box, add				↓	69.50			69.50	76
5000	Moderate climate, all bronze, polished face									
5020	and scoriated cover, set flush									
5100	3/4" IPS connection	1 Plum	16	.500	Ea.	530	26		556	625
5120	1" IPS connection	"	14	.571		655	29.50		684.50	765
5200	For tapped drain port in box, add				↓	69.50			69.50	76

22 11 23 – Domestic Water Pumps

22 11 23.10 General Utility Pumps

	22 11 23.10 General Utility Pumps	Crew	Daily Output	Labor-Hours	Unit	Material	Labor	Equipment	Total	Total Incl O&P
0010	**GENERAL UTILITY PUMPS**									
2000	Single stage									
3000	Double suction,									
3190	75 HP, to 2500 GPM	Q-3	.28	114	Ea.	16,400	5,675		22,075	26,500
3220	100 HP, to 3000 GPM		.26	123		20,600	6,100		26,700	31,800
3240	150 HP, to 4000 GPM	↓	.24	133	↓	31,700	6,600		38,300	44,800

22 13 Facility Sanitary Sewerage

22 13 16 – Sanitary Waste and Vent Piping

22 13 16.20 Pipe, Cast Iron

	22 13 16.20 Pipe, Cast Iron	Crew	Daily Output	Labor-Hours	Unit	2010 Bare Costs Material	Labor	Equipment	Total	Total Incl O&P
0010	**PIPE, CAST IRON**, Soil, on clevis hanger assemblies, 5′ O.C.									
0020	Single hub, service wt., lead & oakum joints 10′ O.C.									
2120	2″ diameter	Q-1	63	.254	L.F.	7	11.90		18.90	25.50
2140	3″ diameter		60	.267		9.80	12.50		22.30	29.50
2160	4″ diameter **CN**		55	.291		12.70	13.65		26.35	34.50
2180	5″ diameter	Q-2	76	.316		17.65	15.35		33	42.50
2200	6″ diameter	″	73	.329		21.50	15.95		37.45	47.50
2220	8″ diameter	Q-3	59	.542		33.50	27		60.50	77.50
2240	10″ diameter		54	.593		55.50	29.50		85	105
2260	12″ diameter		48	.667		79.50	33		112.50	137
2320	For service weight, double hub, add					10%				
2340	For extra heavy, single hub, add					48%	4%			
2360	For extra heavy, double hub, add					71%	4%			
2400	Lead for caulking, (1#/dia in)	Q-1	160	.100	Lb.	1.04	4.69		5.73	8.15
2420	Oakum for caulking, (1/8#/dia in)	″	40	.400	″	4.20	18.75		22.95	32.50
4000	No hub, couplings 10′ O.C.									
4100	1-1/2″ diameter	Q-1	71	.225	L.F.	7.05	10.55		17.60	23.50
4120	2″ diameter		67	.239		7.35	11.20		18.55	25
4160	4″ diameter		58	.276		13.20	12.90		26.10	34

22 13 16.50 Shower Drains

	22 13 16.50 Shower Drains	Crew	Daily Output	Labor-Hours	Unit	Material	Labor	Equipment	Total	Total Incl O&P
0010	**SHOWER DRAINS**									
2780	Shower, with strainer, uniform diam. trap, bronze top									
2800	2″ and 3″ pipe size	Q-1	8	2	Ea.	335	93.50		428.50	505
2820	4″ pipe size	″	7	2.286		370	107		477	565
2840	For galvanized body, add					148			148	163

22 13 16.60 Traps

	22 13 16.60 Traps	Crew	Daily Output	Labor-Hours	Unit	Material	Labor	Equipment	Total	Total Incl O&P
0010	**TRAPS**									
0030	Cast iron, service weight									
0050	Running P trap, without vent									
1100	2″	Q-1	16	1	Ea.	107	47		154	188
1140	3″		14	1.143		107	53.50		160.50	199
1150	4″		13	1.231		107	57.50		164.50	205
1160	6″	Q-2	17	1.412		465	68.50		533.50	620
1180	Running trap, single hub, with vent									
2080	3″ pipe size, 3″ vent	Q-1	14	1.143	Ea.	85	53.50		138.50	174
2120	4″ pipe size, 4″ vent	″	13	1.231		115	57.50		172.50	214
2300	For double hub, vent, add					10%	20%			
3000	P trap, B&S, 2″ pipe size	Q-1	16	1		25.50	47		72.50	98
3040	3″ pipe size	″	14	1.143		38	53.50		91.50	122
3350	Deep seal trap, B&S									
3400	1-1/4″ pipe size	Q-1	14	1.143	Ea.	40.50	53.50		94	125
3410	1-1/2″ pipe size		14	1.143		40.50	53.50		94	125
3420	2″ pipe size		14	1.143		34.50	53.50		88	118
3440	3″ pipe size		12	1.333		46.50	62.50		109	145
4700	Copper, drainage, drum trap									
4800	3″ x 5″ solid, 1-1/2″ pipe size	1 Plum	16	.500	Ea.	127	26		153	178
4840	3″ x 6″ swivel, 1-1/2″ pipe size	″	16	.500	″	201	26		227	260
5100	P trap, standard pattern									
5200	1-1/4″ pipe size	1 Plum	18	.444	Ea.	93.50	23		116.50	138
5240	1-1/2″ pipe size		17	.471		90	24.50		114.50	136
5260	2″ pipe size		15	.533		194	28		222	255
5280	3″ pipe size		11	.727		335	38		373	425

22 13 Facility Sanitary Sewerage

22 13 16 – Sanitary Waste and Vent Piping

22 13 16.60 Traps		Crew	Daily Output	Labor-Hours	Unit	Material	2010 Bare Costs Labor	Equipment	Total	Total Incl O&P
5340	With cleanout and slip joint									
5360	1-1/4" pipe size	1 Plum	18	.444	Ea.	68.50	23		91.50	110
5400	1-1/2" pipe size	"	17	.471	"	138	24.50		162.50	188

22 13 16.80 Vent Flashing and Caps

		Crew	Daily Output	Labor-Hours	Unit	Material	Labor	Equipment	Total	Total Incl O&P
0010	**VENT FLASHING AND CAPS**									
0120	Vent caps									
0140	Cast iron									
0180	2-1/2" - 3-5/8" pipe	1 Plum	21	.381	Ea.	40	19.85		59.85	73.50
0190	4" - 4-1/8" pipe	"	19	.421	"	48	22		70	86
0900	Vent flashing									
1000	Aluminum with lead ring									
1020	1-1/4" pipe	1 Plum	20	.400	Ea.	10.35	21		31.35	42.50
1030	1-1/2" pipe		20	.400		10.60	21		31.60	42.50
1040	2" pipe		18	.444		11.10	23		34.10	47
1050	3" pipe		17	.471		12.30	24.50		36.80	50
1060	4" pipe	↓	16	.500	↓	14.85	26		40.85	55.50
1350	Copper with neoprene ring									
1400	1-1/4" pipe	1 Plum	20	.400	Ea.	19.95	21		40.95	53
1430	1-1/2" pipe		20	.400		19.95	21		40.95	53
1440	2" pipe		18	.444		21	23		44	57.50
1450	3" pipe		17	.471		24.50	24.50		49	63.50
1460	4" pipe	↓	16	.500	↓	27	26		53	69

22 13 19 – Sanitary Waste Piping Specialties

22 13 19.13 Sanitary Drains

		Crew	Daily Output	Labor-Hours	Unit	Material	Labor	Equipment	Total	Total Incl O&P
0010	**SANITARY DRAINS**									
0400	Deck, auto park, C.I., 13" top									
0440	3", 4", 5", and 6" pipe size	Q-1	8	2	Ea.	1,125	93.50		1,218.50	1,400
0480	For galvanized body, add				"	610			610	675
2000	Floor, medium duty, C.I., deep flange, 7" dia top									
2040	2" and 3" pipe size	Q-1	12	1.333	Ea.	158	62.50		220.50	268
2080	For galvanized body, add					75.50			75.50	83
2120	For polished bronze top, add				↓	89			89	98
2400	Heavy duty, with sediment bucket, C.I., 12" dia. loose grate									
2420	2", 3", 4", 5", and 6" pipe size	Q-1	9	1.778	Ea.	545	83.50		628.50	725
2460	For polished bronze top, add				"	225			225	248
2500	Heavy duty, cleanout & trap w/bucket, C.I., 15" top									
2540	2", 3", and 4" pipe size	Q-1	6	2.667	Ea.	5,175	125		5,300	5,850
2560	For galvanized body, add					1,325			1,325	1,450
2580	For polished bronze top, add				↓	570			570	630

22 13 23 – Sanitary Waste Interceptors

22 13 23.10 Interceptors

		Crew	Daily Output	Labor-Hours	Unit	Material	Labor	Equipment	Total	Total Incl O&P
0010	**INTERCEPTORS**									
0150	Grease, cast iron, 4 GPM, 8 lb. fat capacity	1 Plum	4	2	Ea.	935	104		1,039	1,175
0200	7 GPM, 14 lb. fat capacity		4	2		1,300	104		1,404	1,575
1000	10 GPM, 20 lb. fat capacity		4	2		1,525	104		1,629	1,825
1040	15 GPM, 30 lb. fat capacity		4	2		2,250	104		2,354	2,625
1060	20 GPM, 40 lb. fat capacity	↓	3	2.667		2,775	139		2,914	3,250
1120	Fabricated steel, 50 GPM, 100 lb. fat capacity	Q-1	2	8		8,150	375		8,525	9,500
1160	100 GPM, 200 lb. fat capacity	"	2	8	↓	11,500	375		11,875	13,300
1580	For seepage pan, add					7%				
3000	Hair, cast iron, 1-1/4" and 1-1/2" pipe connection	1 Plum	8	1	Ea.	350	52		402	465

22 13 Facility Sanitary Sewerage

22 13 23 – Sanitary Waste Interceptors

22 13 23.10 Interceptors		Crew	Daily Output	Labor-Hours	Unit	2010 Bare Costs Material	Labor	Equipment	Total	Total Incl O&P
3100	For chrome-plated cast iron, add				Ea.	215			215	236
4000	Oil, fabricated steel, 10 GPM, 2" pipe size	1 Plum	4	2		2,075	104		2,179	2,450
4100	15 GPM, 2" or 3" pipe size		4	2		2,850	104		2,954	3,300
4120	20 GPM, 2" or 3" pipe size		3	2.667		3,450	139		3,589	4,000
4220	100 GPM, 3" pipe size	Q-1	2	8		11,500	375		11,875	13,300
6000	Solids, precious metals recovery, C.I., 1-1/4" to 2" pipe	1 Plum	4	2		530	104		634	735
6100	Dental Lab., large, C.I., 1-1/2" to 2" pipe	"	3	2.667		1,850	139		1,989	2,225

22 13 29 – Sanitary Sewerage Pumps

22 13 29.13 Wet-Pit-Mounted, Vertical Sewerage Pumps

Code	Description	Crew	Daily Output	Labor-Hours	Unit	Material	Labor	Equipment	Total	Total Incl O&P
0010	**WET-PIT-MOUNTED, VERTICAL SEWERAGE PUMPS**									
0020	Controls incl. alarm/disconnect panel w/wire. Excavation not included									
0260	Simplex, 9 GPM at 60 PSIG, 91 gal. tank				Ea.	2,975			2,975	3,275
0300	Unit with manway, 26" I.D., 18" high					3,375			3,375	3,700
0340	26" I.D., 36" high					3,500			3,500	3,850
0380	43" I.D., 4' high					3,575			3,575	3,925
3000	Indoor residential type installation									
3020	Simplex, 9 GPM at 60 PSIG, 91 gal. HDPE tank				Ea.	2,950			2,950	3,225

22 13 29.14 Sewage Ejector Pumps

Code	Description	Crew	Daily Output	Labor-Hours	Unit	Material	Labor	Equipment	Total	Total Incl O&P
0010	**SEWAGE EJECTOR PUMPS**, With operating and level controls									
0100	Simplex system incl. tank, cover, pump 15' head									
0500	37 gal PE tank, 12 GPM, 1/2 HP, 2" discharge	Q-1	3.20	5	Ea.	440	234		674	835
0510	3" discharge		3.10	5.161		480	242		722	885
0530	87 GPM, .7 HP, 2" discharge		3.20	5		670	234		904	1,075
0540	3" discharge		3.10	5.161		725	242		967	1,150
0600	45 gal. coated stl tank, 12 GPM, 1/2 HP, 2" discharge		3	5.333		780	250		1,030	1,225
0610	3" discharge		2.90	5.517		810	258		1,068	1,275
0630	87 GPM, .7 HP, 2" discharge		3	5.333		1,000	250		1,250	1,475
0640	3" discharge		2.90	5.517		1,050	258		1,308	1,525
0660	134 GPM, 1 HP, 2" discharge		2.80	5.714		1,075	268		1,343	1,600
0680	3" discharge		2.70	5.926		1,150	278		1,428	1,675
0700	70 gal. PE tank, 12 GPM, 1/2 HP, 2" discharge		2.60	6.154		855	288		1,143	1,375
0710	3" discharge		2.40	6.667		910	310		1,220	1,475
0730	87 GPM, 0.7 HP, 2" discharge		2.50	6.400		1,100	300		1,400	1,650
0740	3" discharge		2.30	6.957		1,150	325		1,475	1,775
0760	134 GPM, 1 HP, 2" discharge		2.20	7.273		1,200	340		1,540	1,800
0770	3" discharge		2	8		1,275	375		1,650	1,950

22 14 Facility Storm Drainage

22 14 23 – Storm Drainage Piping Specialties

22 14 23.33 Backwater Valves

Code	Description	Crew	Daily Output	Labor-Hours	Unit	Material	Labor	Equipment	Total	Total Incl O&P
0010	**BACKWATER VALVES**, C.I. Body									
6980	Bronze gate and automatic flapper valves									
7000	3" and 4" pipe size	Q-1	13	1.231	Ea.	1,600	57.50		1,657.50	1,850
7100	5" and 6" pipe size	"	13	1.231	"	2,475	57.50		2,532.50	2,775
7240	Bronze flapper valve, bolted cover									
7260	2" pipe size	Q-1	16	1	Ea.	470	47		517	585
7300	4" pipe size	"	13	1.231		905	57.50		962.50	1,075
7340	6" pipe size	Q-2	17	1.412		1,300	68.50		1,368.50	1,525

22 14 Facility Storm Drainage

22 14 26 – Facility Storm Drains

	22 14 26.13 Roof Drains	Crew	Daily Output	Labor-Hours	Unit	Material	2010 Bare Costs Labor	Equipment	Total	Total Incl O&P
0010	**ROOF DRAINS**									
0140	Cornice, C.I., 45° or 90° outlet									
0200	3" and 4" pipe size	Q-1	12	1.333	Ea.	259	62.50		321.50	380
0260	For galvanized body, add					59.50			59.50	65.50
0280	For polished bronze dome, add				↓	46			46	51
3860	Roof, flat metal deck, C.I. body, 12" C.I. dome									
3890	3" pipe size	Q-1	14	1.143	Ea.	310	53.50		363.50	420
3920	6" pipe size	"	10	1.600	"	540	75		615	700
4620	Main, all aluminum, 12" low profile dome									
4640	2", 3" and 4" pipe size	Q-1	14	1.143	Ea.	345	53.50		398.50	460

22 14 26.16 Facility Area Drains

		Crew	Daily Output	Labor-Hours	Unit	Material	Labor	Equipment	Total	Total Incl O&P
0010	**FACILITY AREA DRAINS**									
4980	Scupper floor, oblique strainer, C.I.									
5000	6" x 7" top, 2", 3" and 4" pipe size	Q-1	16	1	Ea.	223	47		270	315
5100	8" x 12" top, 5" and 6" pipe size	"	14	1.143		435	53.50		488.50	555
5160	For galvanized body, add					40%				
5200	For polished bronze strainer, add				↓	85%				

22 14 26.19 Facility Trench Drains

		Crew	Daily Output	Labor-Hours	Unit	Material	Labor	Equipment	Total	Total Incl O&P
0010	**FACILITY TRENCH DRAINS**									
5980	Trench, floor, hvy duty, modular, C.I., 12" x 12" top									
6000	2", 3", 4", 5", & 6" pipe size	Q-1	8	2	Ea.	705	93.50		798.50	915
6100	For unit with polished bronze top				"	1,050			1,050	1,150
6600	Trench, floor, for cement concrete encasement									
6610	Not including trenching or concrete									
6640	Polyester polymer concrete									
6650	4" internal width, with grate									
6660	Light duty steel grate	Q-1	120	.133	L.F.	26.50	6.25		32.75	38.50
6670	Medium duty steel grate		115	.139		33	6.50		39.50	46.50
6680	Heavy duty iron grate	↓	110	.145	↓	51	6.80		57.80	66
6700	12" internal width, with grate									
6770	Heavy duty galvanized grate	Q-1	80	.200	L.F.	128	9.35		137.35	155
6800	Fiberglass									
6810	8" internal width, with grate									
6820	Medium duty galvanized grate	Q-1	115	.139	L.F.	89	6.50		95.50	107
6830	Heavy duty iron grate	"	110	.145	"	86	6.80		92.80	105

22 14 29 – Sump Pumps

22 14 29.13 Wet-Pit-Mounted, Vertical Sump Pumps

		Crew	Daily Output	Labor-Hours	Unit	Material	Labor	Equipment	Total	Total Incl O&P
0010	**WET-PIT-MOUNTED, VERTICAL SUMP PUMPS**									
0400	Molded PVC base, 21 GPM at 15' head, 1/3 HP	1 Plum	5	1.600	Ea.	119	83.50		202.50	256
0800	Iron base, 21 GPM at 15' head, 1/3 HP		5	1.600		150	83.50		233.50	290
1200	Solid brass, 21 GPM at 15' head, 1/3 HP	↓	5	1.600	↓	263	83.50		346.50	415

22 14 29.16 Submersible Sump Pumps

		Crew	Daily Output	Labor-Hours	Unit	Material	Labor	Equipment	Total	Total Incl O&P
0010	**SUBMERSIBLE SUMP PUMPS**									
7000	Sump pump, automatic									
7100	Plastic, 1-1/4" discharge, 1/4 HP	1 Plum	6	1.333	Ea.	215	69.50		284.50	340
7140	1/3 HP		5	1.600		175	83.50		258.50	315
7160	1/2 HP		5	1.600		215	83.50		298.50	360
7180	1-1/2" discharge, 1/2 HP		4	2		238	104		342	420
7500	Cast iron, 1-1/4" discharge, 1/4 HP		6	1.333		166	69.50		235.50	286
7540	1/3 HP		6	1.333		195	69.50		264.50	320
7560	1/2 HP	↓	5	1.600	↓	236	83.50		319.50	385

22 31 Domestic Water Softeners

22 31 13 – Residential Domestic Water Softeners

	22 31 13.10 Residential Water Softeners	Crew	Daily Output	Labor-Hours	Unit	Material	2010 Bare Costs Labor	Equipment	Total	Total Incl O&P
0010	**RESIDENTIAL WATER SOFTENERS**									
7350	Water softener, automatic, to 30 grains per gallon	2 Plum	5	3.200	Ea.	380	167		547	670
7400	To 100 grains per gallon	″	4	4	″	750	208		958	1,125

22 31 16 – Commercial Domestic Water Softeners

	22 31 16.10 Water Softeners	Crew	Daily Output	Labor-Hours	Unit	Material	Labor	Equipment	Total	Total Incl O&P
0010	**WATER SOFTENERS**									
5800	Softener systems, automatic, intermediate sizes									
5820	available, may be used in multiples.									
6000	Hardness capacity between regenerations and flow									
6100	150,000 grains, 37 GPM cont., 51 GPM peak	Q-1	1.20	13.333	Ea.	3,900	625		4,525	5,200
6200	300,000 grains, 81 GPM cont., 113 GPM peak		1	16		8,275	750		9,025	10,200
6300	750,000 grains, 160 GPM cont., 230 GPM peak		.80	20		10,800	935		11,735	13,300
6400	900,000 grains, 185 GPM cont., 270 GPM peak		.70	22.857		17,300	1,075		18,375	20,700

22 33 Electric Domestic Water Heaters

22 33 13 – Instantaneous Electric Domestic Water Heaters

	22 33 13.10 Hot Water Dispensers	Crew	Daily Output	Labor-Hours	Unit	Material	Labor	Equipment	Total	Total Incl O&P
0010	**HOT WATER DISPENSERS**									
0160	Commercial, 100 cup, 11.3 amp	1 Plum	14	.571	Ea.	400	29.50		429.50	485
3180	Household, 60 cup	″	14	.571	″	210	29.50		239.50	276

22 33 30 – Residential, Electric Domestic Water Heaters

	22 33 30.13 Residential, Small-Capacity Elec. Water Heaters	Crew	Daily Output	Labor-Hours	Unit	Material	Labor	Equipment	Total	Total Incl O&P
0010	**RESIDENTIAL, SMALL-CAPACITY ELECTRIC DOMESTIC WATER HEATERS**									
1000	Residential, electric, glass lined tank, 5 yr, 10 gal., single element	1 Plum	2.30	3.478	Ea.	385	181		566	695
1040	20 gallon, single element		2.20	3.636		440	189		629	770
1060	30 gallon, double element		2.20	3.636		575	189		764	920
1080	40 gallon, double element		2	4		615	208		823	985
1100	52 gallon, double element		2	4		690	208		898	1,075
1180	120 gallon, double element		1.40	5.714		1,450	297		1,747	2,050

22 33 33 – Light-Commercial Electric Domestic Water Heaters

	22 33 33.10 Commercial Electric Water Heaters	Crew	Daily Output	Labor-Hours	Unit	Material	Labor	Equipment	Total	Total Incl O&P
0010	**COMMERCIAL ELECTRIC WATER HEATERS**									
4000	Commercial, 100° rise. NOTE: for each size tank, a range of									
4010	heaters between the ones shown are available									
4020	Electric									
4100	5 gal., 3 kW, 12 GPH, 208V	1 Plum	2	4	Ea.	2,725	208		2,933	3,300
4120	10 gal., 6 kW, 25 GPH, 208V		2	4		3,025	208		3,233	3,625
4140	50 gal., 9 kW, 37 GPH, 208V		1.80	4.444		4,150	231		4,381	4,925
4160	50 gal., 36 kW, 148 GPH, 208V		1.80	4.444		6,350	231		6,581	7,350
4300	200 gal., 15 kW, 61 GPH, 480V	Q-1	1.70	9.412		19,900	440		20,340	22,500
4320	200 gal., 120 kW , 490 GPH, 480V		1.70	9.412		27,100	440		27,540	30,500
4460	400 gal., 30 kW, 123 GPH, 480V		1	16		27,400	750		28,150	31,200
5400	Modulating step control, 2-5 steps	1 Elec	5.30	1.509		2,000	74		2,074	2,300
5440	6-10 steps		3.20	2.500		2,575	123		2,698	3,025
5460	11-15 steps		2.70	2.963		3,125	145		3,270	3,675
5480	16-20 steps		1.60	5		3,625	245		3,870	4,375

22 34 Fuel-Fired Domestic Water Heaters

22 34 13 – Instantaneous, Tankless, Gas Domestic Water Heaters

22 34 13.10 Instantaneous, Tankless, Gas Water Heaters		Crew	Daily Output	Labor-Hours	Unit	Material	2010 Bare Costs Labor	Equipment	Total	Total Incl O&P
0010	**INSTANTANEOUS, TANKLESS, GAS WATER HEATERS**									
9410	Natural gas/propane, incl vent, 3.2 GPM G	1 Plum	2	4	Ea.	535	208		743	900

22 34 30 – Residential Gas Domestic Water Heaters

22 34 30.13 Residential, Atmos, Gas Domestic Wtr Heaters		Crew	Daily Output	Labor-Hours	Unit	Material	Labor	Equipment	Total	Total Incl O&P
0010	**RESIDENTIAL, ATMOSPHERIC, GAS DOMESTIC WATER HEATERS**									
2000	Gas fired, foam lined tank, 10 yr, vent not incl.									
2040	30 gallon	1 Plum	2	4	Ea.	895	208		1,103	1,300
2100	75 gallon		1.50	5.333		1,225	278		1,503	1,775
2120	100 gallon	↓	1.30	6.154	↓	1,525	320		1,845	2,150

22 34 36 – Commercial Gas Domestic Water Heaters

22 34 36.13 Commercial, Atmos., Gas Domestic Water Htrs.		Crew	Daily Output	Labor-Hours	Unit	Material	Labor	Equipment	Total	Total Incl O&P
0010	**COMMERCIAL, ATMOSPHERIC, GAS DOMESTIC WATER HEATERS**									
6000	Gas fired, flush jacket, std. controls, vent not incl.									
6040	75 MBH input, 73 GPH	1 Plum	1.40	5.714	Ea.	1,575	297		1,872	2,175
6060	98 MBH input, 95 GPH		1.40	5.714		4,575	297		4,872	5,500
6080	120 MBH input, 110 GPH CN		1.20	6.667		4,975	345		5,320	5,975
6180	200 MBH input, 192 GPH		.60	13.333		7,650	695		8,345	9,450
6200	250 MBH input, 245 GPH		.50	16		7,975	835		8,810	10,000
6900	For low water cutoff, add		8	1		275	52		327	380
6960	For bronze body hot water circulator, add	↓	4	2	↓	1,750	104		1,854	2,075

22 34 46 – Oil-Fired Domestic Water Heaters

22 34 46.10 Residential Oil-Fired Water Heaters		Crew	Daily Output	Labor-Hours	Unit	Material	Labor	Equipment	Total	Total Incl O&P
0010	**RESIDENTIAL OIL-FIRED WATER HEATERS**									
3000	Oil fired, glass lined tank, 5 yr, vent not included, 30 gallon	1 Plum	2	4	Ea.	665	208		873	1,050
3040	50 gallon		1.80	4.444		1,050	231		1,281	1,500
3060	70 gallon		1.50	5.333		1,725	278		2,003	2,325
3080	85 gallon	↓	1.40	5.714	↓	4,750	297		5,047	5,675

22 34 46.20 Commercial Oil-Fired Water Heaters		Crew	Daily Output	Labor-Hours	Unit	Material	Labor	Equipment	Total	Total Incl O&P
0010	**COMMERCIAL OIL-FIRED WATER HEATERS**									
8000	Oil fired, glass lined, UL listed, std. controls, vent not incl.									
8060	140 gal., 140 MBH input, 134 GPH	Q-1	2.13	7.512	Ea.	14,900	350		15,250	16,900
8080	140 gal., 199 MBH input, 191 GPH		2	8		15,400	375		15,775	17,500
8100	140 gal., 255 MBH input, 247 GPH		1.60	10		15,900	470		16,370	18,100
8160	140 gal., 540 MBH input, 519 GPH		.96	16.667		21,000	780		21,780	24,300
8180	140 gal., 720 MBH input, 691 GPH	↓	.92	17.391		21,400	815		22,215	24,700
8280	201 gal., 1250 MBH input, 1200 GPH	Q-2	1.22	19.672		32,900	955		33,855	37,500
8300	201 gal., 1500 MBH input, 1441 GPH	"	1.16	20.690		35,700	1,000		36,700	40,800
8900	For low water cutoff, add	1 Plum	8	1		275	52		327	380
8960	For bronze body hot water circulator, add	"	4	2	↓	710	104		814	935

22 35 Domestic Water Heat Exchangers

22 35 30 – Water Heating by Steam

22 35 30.10 Water Heating Transfer Package		Crew	Daily Output	Labor-Hours	Unit	Material	2010 Bare Costs Labor	Equipment	Total	Total Incl O&P
0010	**WATER HEATING TRANSFER PACKAGE**, Complete controls,									
0020	expansion tank, converter, air separator									
1000	Hot water, 180°F enter, 200°F leaving, 15# steam									
1010	One pump system, 28 GPM	Q-6	.75	32	Ea.	15,800	1,550		17,350	19,700
1020	35 GPM		.70	34.286		17,200	1,650		18,850	21,400
1040	55 GPM		.65	36.923		20,300	1,800		22,100	25,000
1060	130 GPM		.55	43.636		25,000	2,125		27,125	30,700
1080	255 GPM		.40	60		33,800	2,900		36,700	41,600
1100	550 GPM	↓	.30	80	↓	42,700	3,875		46,575	53,000

22 41 Residential Plumbing Fixtures

22 41 06 – Plumbing Fixtures General

22 41 06.10 Plumbing Fixture Notes		Crew	Daily Output	Labor-Hours	Unit	Material	Labor	Equipment	Total	Total Incl O&P
0010	**PLUMBING FIXTURE NOTES**, Incl. trim fittings unless otherwise noted R224000-40									
0080	For rough-in, supply, waste, and vent, see add for each type									
0122	For electric water coolers, see Div. 22 47 16.10									
0160	For color, unless otherwise noted, add				Ea.	20%				

22 41 13 – Residential Water Closets, Urinals, and Bidets

22 41 13.40 Water Closets		Crew	Daily Output	Labor-Hours	Unit	Material	Labor	Equipment	Total	Total Incl O&P
0010	**WATER CLOSETS** R224000-40									
0032	For automatic flush, see Div. 22 42 39.10 0972									
0150	Tank type, vitreous china, incl. seat, supply pipe w/stop									
0200	Wall hung									
0400	Two piece, close coupled	Q-1	5.30	3.019	Ea.	595	141		736	865
0960	For rough-in, supply, waste, vent and carrier		2.73	5.861		910	275		1,185	1,400
1000	Floor mounted		5.30	3.019		540	141		681	800
1100	Two piece, close coupled CN		5.30	3.019		258	141		399	495
1110	Two piece, close coupled, dual flush		5.30	3.019		256	141		397	495
1140	Two piece, close coupled, 1.28 gpf, ADA G	↓	5.30	3.019	↓	295	141		436	535
1960	For color, add					30%				
1980	For rough-in, supply, waste and vent	Q-1	3.05	5.246	Ea.	283	246		529	680

22 41 16 – Residential Lavatories and Sinks

22 41 16.10 Lavatories		Crew	Daily Output	Labor-Hours	Unit	Material	Labor	Equipment	Total	Total Incl O&P
0010	**LAVATORIES**, With trim, white unless noted otherwise R224000-40									
0500	Vanity top, porcelain enamel on cast iron									
0600	20" x 18"	Q-1	6.40	2.500	Ea.	260	117		377	460
0640	33" x 19" oval		6.40	2.500		610	117		727	850
0720	19" round	↓	6.40	2.500	↓	246	117		363	445
0860	For color, add					25%				
1000	Cultured marble, 19" x 17", single bowl	Q-1	6.40	2.500	Ea.	190	117		307	385
1040	25" x 19", single bowl	"	6.40	2.500	"	222	117		339	420
1580	For color, same price									
1900	Stainless steel, self-rimming, 25" x 22", single bowl, ledge	Q-1	6.40	2.500	Ea.	345	117		462	555
1960	17" x 22", single bowl		6.40	2.500		340	117		457	545
2600	Steel, enameled, 20" x 17", single bowl		5.80	2.759		184	129		313	395
2660	19" round		5.80	2.759		179	129		308	390
2900	Vitreous china, 20" x 16", single bowl		5.40	2.963		278	139		417	515
2960	20" x 17", single bowl		5.40	2.963		194	139		333	420
3020	19" round, single bowl		5.40	2.963		192	139		331	420
3200	22" x 13", single bowl	↓	5.40	2.963	↓	285	139		424	525

22 41 16 – Residential Lavatories and Sinks

22 41 16.10 Lavatories		Crew	Daily Output	Labor-Hours	Unit	2010 Bare Costs Material	Labor	Equipment	Total	Total Incl O&P
3560	For color, add					50%				
3580	Rough-in, supply, waste and vent for all above lavatories	Q-1	2.30	6.957	Ea.	252	325		577	765
4000	Wall hung									
4040	Porcelain enamel on cast iron, 16" x 14", single bowl	Q-1	8	2	Ea.	455	93.50		548.50	640
4180	20" x 18", single bowl		8	2		355	93.50		448.50	530
4240	22" x 19", single bowl		8	2		605	93.50		698.50	805
4580	For color, add					30%				
6000	Vitreous china, 18" x 15", single bowl with backsplash	Q-1	7	2.286	Ea.	249	107		356	435
6500	For color, add					30%				
6960	Rough-in, supply, waste and vent for above lavatories	Q-1	1.66	9.639	Ea.	385	450		835	1,100
7000	Pedestal type									
7600	Vitreous china, 27" x 21", white	Q-1	6.60	2.424	Ea.	565	114		679	790
7610	27" x 21", colored		6.60	2.424		675	114		789	910
7620	27" x 21", premium color		6.60	2.424		740	114		854	980
7660	26" x 20", white		6.60	2.424		515	114		629	735
7670	26" x 20", colored		6.60	2.424		610	114		724	840
7680	26" x 20", premium color		6.60	2.424		670	114		784	905
7700	24" x 18", white		6.60	2.424		480	114		594	700
7710	24" x 18", colored		6.60	2.424		570	114		684	795
7720	24" x 18", premium color		6.60	2.424		625	114		739	855
7760	21" x 18", white		6.60	2.424		233	114		347	425
7770	21" x 18", colored		6.60	2.424		250	114		364	445
7990	Rough-in, supply, waste and vent for pedestal lavatories		1.66	9.639		385	450		835	1,100

22 41 16.30 Sinks		Crew	Daily Output	Labor-Hours	Unit	Material	Labor	Equipment	Total	Total Incl O&P
0010	**SINKS**, With faucets and drain R224000-40									
2000	Kitchen, counter top style, P.E. on C.I., 24" x 21" single bowl	Q-1	5.60	2.857	Ea.	257	134		391	485
2100	31" x 22" single bowl		5.60	2.857		300	134		434	530
2200	32" x 21" double bowl		4.80	3.333		385	156		541	660
2300	42" x 21" double bowl		4.80	3.333		540	156		696	830
3000	Stainless steel, self rimming, 19" x 18" single bowl		5.60	2.857		555	134		689	810
3100	25" x 22" single bowl		5.60	2.857		615	134		749	880
4000	Steel, enameled, with ledge, 24" x 21" single bowl		5.60	2.857		180	134		314	400
4100	32" x 21" double bowl		4.80	3.333		207	156		363	460
4960	For color sinks except stainless steel, add					10%				
4980	For rough-in, supply, waste and vent, counter top sinks	Q-1	2.14	7.477		272	350		622	825
5000	Kitchen, raised deck, P.E. on C.I.									
5100	32" x 21", dual level, double bowl	Q-1	2.60	6.154	Ea.	325	288		613	790
5700	For color, add					20%				
5790	For rough-in, supply, waste & vent, sinks	Q-1	1.85	8.649		272	405		677	905

22 41 19 – Residential Bathtubs

22 41 19.10 Baths		Crew	Daily Output	Labor-Hours	Unit	Material	Labor	Equipment	Total	Total Incl O&P
0010	**BATHS**									
0100	Tubs, recessed porcelain enamel on cast iron, with trim									
0180	48" x 42"	Q-1	4	4	Ea.	1,650	187		1,837	2,100
0220	72" x 36"		3	5.333		2,225	250		2,475	2,825
2000	Enameled formed steel, 4'-6" long		5.80	2.759		385	129		514	620
4000	Soaking, acrylic with pop-up drain, 60" x 32" x 21" deep		5.50	2.909		1,000	136		1,136	1,300
4100	60" x 48" x 18-1/2" deep		5	3.200		920	150		1,070	1,225
9600	Rough-in, supply, waste and vent, for all above tubs, add		2.07	7.729		315	360		675	890

22 41 Residential Plumbing Fixtures

22 41 23 – Residential Shower Receptors and Basins

	22 41 23.20 Showers	Crew	Daily Output	Labor-Hours	Unit	2010 Bare Costs Material	Labor	Equipment	Total	Total Incl O&P
0010	**SHOWERS**									
1500	Stall, with drain only. Add for valve and door/curtain									
3000	Fiberglass, one piece, with 3 walls, 32" x 32" square	Q-1	5.50	2.909	Ea.	515	136		651	775
3100	36" x 36" square		5.50	2.909		530	136		666	790
3250	64" x 65-3/4" x 81-1/2" fold. seat, whlchr. ♿		3.80	4.211		2,025	197		2,222	2,525
4000	Polypropylene, stall only, w/ molded-stone floor, 30" x 30"		2	8		605	375		980	1,225
4200	Rough-in, supply, waste and vent for above showers		2.05	7.805		390	365		755	980

	22 41 23.40 Shower System Components	Crew	Daily Output	Labor-Hours	Unit	Material	Labor	Equipment	Total	Total Incl O&P
0010	**SHOWER SYSTEM COMPONENTS**									
4500	Receptor only									
4510	For tile, 36" x 36"	1 Plum	4	2	Ea.	365	104		469	560
4520	Fiberglass receptor only, 32" x 32"		8	1		102	52		154	190
4530	34" x 34"		7.80	1.026		120	53.50		173.50	212
4540	36" x 36"		7.60	1.053		133	55		188	228
4600	Rectangular									
4620	32" x 48"	1 Plum	7.40	1.081	Ea.	141	56.50		197.50	240
4630	32" x 54"		7.20	1.111		186	58		244	292
4640	32" x 60"		7	1.143		197	59.50		256.50	305
5000	Built-in, head, arm, 2.5 GPM valve		4	2		80	104		184	244
5200	Head, arm, by-pass, integral stops, handles		3.60	2.222		226	116		342	420

22 41 36 – Residential Laundry Trays

	22 41 36.10 Laundry Sinks	Crew	Daily Output	Labor-Hours	Unit	Material	Labor	Equipment	Total	Total Incl O&P
0010	**LAUNDRY SINKS**, With trim									
0020	Porcelain enamel on cast iron, black iron frame									
0050	24" x 21", single compartment	Q-1	6	2.667	Ea.	465	125		590	695
0100	26" x 21", single compartment	"	6	2.667	"	450	125		575	680
2000	Molded stone, on wall hanger or legs									
2020	22" x 23", single compartment	Q-1	6	2.667	Ea.	146	125		271	350
2100	45" x 21", double compartment	"	5	3.200	"	259	150		409	510
3000	Plastic, on wall hanger or legs									
3020	18" x 23", single compartment	Q-1	6.50	2.462	Ea.	141	115		256	330
3300	40" x 24", double compartment		5.50	2.909		248	136		384	475
5000	Stainless steel, counter top, 22" x 17" single compartment		6	2.667		61.50	125		186.50	255
5200	33" x 22", double compartment		5	3.200		73.50	150		223.50	305
9600	Rough-in, supply, waste and vent, for all laundry sinks		2.14	7.477		272	350		622	825

22 41 39 – Residential Faucets, Supplies and Trim

	22 41 39.10 Faucets and Fittings	Crew	Daily Output	Labor-Hours	Unit	Material	Labor	Equipment	Total	Total Incl O&P
0010	**FAUCETS AND FITTINGS**									
0150	Bath, faucets, diverter spout combination, sweat	1 Plum	8	1	Ea.	86.50	52		138.50	173
0200	For integral stops, IPS unions, add					109			109	120
0420	Bath, press-bal mix valve w/diverter, spout, shower hd, arm/flange	1 Plum	8	1		155	52		207	249
0810	Bidet									
0812	Fitting, over the rim, swivel spray/pop-up drain	1 Plum	8	1	Ea.	212	52		264	310
1000	Kitchen sink faucets, top mount, cast spout		10	.800		61.50	41.50		103	130
1100	For spray, add		24	.333		16.15	17.35		33.50	44
1300	Single control lever handle									
1310	With pull out spray									
1320	Polished chrome	1 Plum	10	.800	Ea.	157	41.50		198.50	235
2000	Laundry faucets, shelf type, IPS or copper unions		12	.667		49.50	34.50		84	107
2100	Lavatory faucet, centerset, without drain		10	.800		44.50	41.50		86	112
2210	Porcelain cross handles and pop-up drain									

22 41 Residential Plumbing Fixtures

22 41 39 – Residential Faucets, Supplies and Trim

22 41 39.10 Faucets and Fittings			Crew	Daily Output	Labor-Hours	Unit	Material	Labor (2010 Bare Costs)	Equipment	Total	Total Incl O&P
2220	Polished chrome		1 Plum	6.66	1.201	Ea.	126	62.50		188.50	233
2230	Polished brass		"	6.66	1.201	"	189	62.50		251.50	300
2260	Single lever handle and pop-up drain										
2270	Black nickel		1 Plum	6.66	1.201	Ea.	220	62.50		282.50	335
2280	Polished brass			6.66	1.201		220	62.50		282.50	335
2290	Polished chrome			6.66	1.201		172	62.50		234.50	283
2810	Automatic sensor and operator, with faucet head	G		6.15	1.301		370	67.50		437.50	510
4000	Shower by-pass valve with union			18	.444		85.50	23		108.50	129
4200	Shower thermostatic mixing valve, concealed		↓	8	1	↓	365	52		417	485
4220	Shower pressure balancing mixing valve,										
4230	With shower head, arm, flange and diverter tub spout										
4240	Chrome		1 Plum	6.14	1.303	Ea.	222	68		290	345
4250	Polished brass			6.14	1.303		279	68		347	405
4260	Satin			6.14	1.303		279	68		347	405
4270	Polished chrome/brass			6.14	1.303		229	68		297	355
5000	Sillcock, compact, brass, IPS or copper to hose		↓	24	.333	↓	8.05	17.35		25.40	35

22 42 Commercial Plumbing Fixtures

22 42 13 – Commercial Water Closets, Urinals, and Bidets

22 42 13.30 Urinals			Crew	Daily Output	Labor-Hours	Unit	Material	Labor	Equipment	Total	Total Incl O&P
0010	**URINALS**										
3000	Wall hung, vitreous china, with hanger & self-closing valve										
3100	Siphon jet type		Q-1	3	5.333	Ea.	293	250		543	700
3120	Blowout type			3	5.333		435	250		685	855
3140	Water saving .5 gpf	G		3	5.333		715	250		965	1,150
3300	Rough-in, supply, waste & vent			2.83	5.654		240	265		505	660
5000	Stall type, vitreous china, includes valve			2.50	6.400		665	300		965	1,175
6980	Rough-in, supply, waste and vent		↓	1.99	8.040	↓	435	375		810	1,050
8000	Waterless (no flush) urinal										
8010	Wall hung										
8014	Fiberglass reinforced polyester										
8020	Standard unit	G	Q-1	21.30	.751	Ea.	385	35		420	475
8030	ADA compliant unit	G	"	21.30	.751		400	35		435	495
8070	For solid color, add	G					60			60	66
8080	For 2" brass flange, (new const.), add	G	Q-1	96	.167		19.20	7.80		27	32.50
8090	Rough-in, supply, waste & vent	G	"	3.86	4.145	↓	211	194		405	525
8200	Vitreous china										
8220	Standard unit	G	Q-1	21.30	.751	Ea.	280	35		315	365
8240	ADA compliant unit	G		21.30	.751		320	35		355	405
8250	ADA compliant unit, compact		↓	21.30	.751		272	35		307	350
8270	For solid color, add	G					60			60	66
8290	Rough-in, supply, waste & vent	G	Q-1	2.92	5.479	↓	470	257		727	900
8400	Trap liquid										
8410	1 quart	G				Ea.	14.35			14.35	15.80
8420	1 gallon	G				"	56			56	61.50

22 42 13.40 Water Closets			Crew	Daily Output	Labor-Hours	Unit	Material	Labor	Equipment	Total	Total Incl O&P
0010	**WATER CLOSETS**										
3000	Bowl only, with flush valve, seat, 1.6 gpf unless noted										
3100	Wall hung		Q-1	5.80	2.759	Ea.	305	129		434	530
3200	For rough-in, supply, waste and vent, single WC			2.56	6.250		950	293		1,243	1,500
3300	Floor mounted		↓	5.80	2.759	↓	325	129		454	555

22 42 Commercial Plumbing Fixtures

22 42 13 – Commercial Water Closets, Urinals, and Bidets

22 42 13.40 Water Closets		Crew	Daily Output	Labor-Hours	Unit	Material	Labor	Equipment	Total	Total Incl O&P
						2010 Bare Costs				
3350	With wall outlet	Q-1	5.80	2.759	Ea.	480	129		609	720
3360	With floor outlet, 1.28 gpf G		5.80	2.759		565	129		694	815
3362	With floor outlet, 1.28 gpf, ADA G		5.80	2.759		595	129		724	850
3400	For rough-in, supply, waste and vent, single WC		2.84	5.634		325	264		589	755

22 42 16 – Commercial Lavatories and Sinks

22 42 16.10 Handwasher-Dryer Module		Crew	Daily Output	Labor-Hours	Unit	Material	Labor	Equipment	Total	Total Incl O&P
0010	**HANDWASHER-DRYER MODULE**									
0030	Wall mounted									
0040	With electric dryer									
0050	Sensor operated	Q-1	8	2	Ea.	3,650	93.50		3,743.50	4,150
0110	Sensor operated (ADA)		8	2		3,650	93.50		3,743.50	4,175
0140	Sensor operated (ADA), surface mounted		8	2		4,400	93.50		4,493.50	5,000
0150	With paper towels									
0180	Sensor operated	Q-1	8	2	Ea.	3,550	93.50		3,643.50	4,050

22 42 16.14 Lavatories		Crew	Daily Output	Labor-Hours	Unit	Material	Labor	Equipment	Total	Total Incl O&P
0010	**LAVATORIES**, With trim, white unless noted otherwise									
0020	Commercial lavatories same as residential. See Div. 22 41 16.10									

22 42 16.40 Service Sinks		Crew	Daily Output	Labor-Hours	Unit	Material	Labor	Equipment	Total	Total Incl O&P
0010	**SERVICE SINKS**									
6650	Service, floor, corner, P.E. on C.I., 28" x 28"	Q-1	4.40	3.636	Ea.	695	170		865	1,025
6790	For rough-in, supply, waste & vent, floor service sinks		1.64	9.756		730	455		1,185	1,475
7000	Service, wall, P.E. on C.I., roll rim, 22" x 18"		4	4		640	187		827	985
7100	24" x 20"		4	4		700	187		887	1,050
8600	Vitreous china, 22" x 20"		4	4		520	187		707	855
8960	For stainless steel rim guard, front or side, add					37			37	41
8980	For rough-in, supply, waste & vent, wall service sinks	Q-1	1.30	12.308		1,025	575		1,600	2,000

22 42 23 – Commercial Shower Receptors and Basins

22 42 23.30 Group Showers		Crew	Daily Output	Labor-Hours	Unit	Material	Labor	Equipment	Total	Total Incl O&P
0010	**GROUP SHOWERS**									
6000	Group, w/pressure balancing valve, rough-in and rigging not included									
6800	Column, 6 heads, no receptors, less partitions	Q-1	3	5.333	Ea.	3,000	250		3,250	3,675
6900	With stainless steel partitions		1	16		9,250	750		10,000	11,300
7600	5 heads, no receptors, less partitions		3	5.333		2,475	250		2,725	3,100
7620	4 heads (1 handicap) no receptors, less partitions		3	5.333		3,975	250		4,225	4,750
7700	With stainless steel partitions		1	16		7,600	750		8,350	9,475
8000	Wall, 2 heads, no receptors, less partitions		4	4		1,375	187		1,562	1,800
8100	With stainless steel partitions		2	8		3,225	375		3,600	4,100

22 42 33 – Wash Fountains

22 42 33.20 Commercial Wash Fountains		Crew	Daily Output	Labor-Hours	Unit	Material	Labor	Equipment	Total	Total Incl O&P
0010	**COMMERCIAL WASH FOUNTAINS**									
1900	Group, foot control									
2000	Precast terrazzo, circular, 36" diam., 5 or 6 persons	Q-2	3	8	Ea.	4,200	390		4,590	5,175
2100	54" diameter for 8 or 10 persons		2.50	9.600		5,225	465		5,690	6,425
2400	Semi-circular, 36" diam. for 3 persons		3	8		3,850	390		4,240	4,825
2500	54" diam. for 4 or 5 persons		2.50	9.600		4,650	465		5,115	5,825
2700	Quarter circle (corner), 54" for 3 persons		3.50	6.857		4,525	335		4,860	5,475
3000	Stainless steel, circular, 36" diameter		3.50	6.857		4,100	335		4,435	5,025
3100	54" diameter		2.80	8.571		5,300	415		5,715	6,450
3400	Semi-circular, 36" diameter		3.50	6.857		3,575	335		3,910	4,425
3500	54" diameter		2.80	8.571		4,575	415		4,990	5,650

22 42 33 – Wash Fountains

22 42 33.20 Commercial Wash Fountains		Crew	Daily Output	Labor-Hours	Unit	Material	2010 Bare Costs Labor	Equipment	Total	Total Incl O&P
5610	Group, infrared control, barrier free									
5614	Precast terrazzo									
5620	Semi-circular 36" diam. for 3 persons	Q-2	3	8	Ea.	5,875	390		6,265	7,050
5630	46" diam. for 4 persons		2.80	8.571		6,425	415		6,840	7,675
5640	Circular, 54" diam. for 8 persons, button control		2.50	9.600		7,700	465		8,165	9,175
5700	Rough-in, supply, waste and vent for above wash fountains	Q-1	1.82	8.791		420	410		830	1,075
6200	Duo for small washrooms, stainless steel		2	8		2,725	375		3,100	3,525
6500	Rough-in, supply, waste & vent for duo fountains		2.02	7.921		215	370		585	790

22 42 39 – Commercial Faucets, Supplies, and Trim

22 42 39.10 Faucets and Fittings		Crew	Daily Output	Labor-Hours	Unit	Material	Labor	Equipment	Total	Total Incl O&P
0010	**FAUCETS AND FITTINGS**									
0840	Flush valves, with vacuum breaker									
0850	Water closet									
0860	Exposed, rear spud	1 Plum	8	1	Ea.	172	52		224	268
0870	Top spud		8	1		161	52		213	255
0880	Concealed, rear spud		8	1		180	52		232	276
0890	Top spud		8	1		149	52		201	242
0900	Wall hung		8	1		159	52		211	253
0920	Urinal									
0930	Exposed, stall	1 Plum	8	1	Ea.	161	52		213	255
0940	Wall, (washout)		8	1		161	52		213	255
0950	Pedestal, top spud		8	1		161	52		213	255
0960	Concealed, stall		8	1		145	52		197	237
0970	Wall (washout)		8	1		156	52		208	250
0971	Automatic flush sensor and operator for G									
0972	urinals or water closets, standard G	1 Plum	8	1	Ea.	415	52		467	540
0980	High efficiency water saving									
0984	Water closets, 1.28 gpf G	1 Plum	8	1	Ea.	395	52		447	515
0988	Urinals, .5 gpf G	"	8	1	"	395	52		447	515
2790	Faucets for lavatories									
2800	Self-closing, center set	1 Plum	10	.800	Ea.	131	41.50		172.50	207
2810	Automatic sensor and operator, with faucet head		6.15	1.301		370	67.50		437.50	510
3000	Service sink faucet, cast spout, pail hook, hose end		14	.571		80	29.50		109.50	133

22 42 39.30 Carriers and Supports		Crew	Daily Output	Labor-Hours	Unit	Material	Labor	Equipment	Total	Total Incl O&P
0010	**CARRIERS AND SUPPORTS**, For plumbing fixtures									
0500	Drinking fountain, wall mounted									
0600	Plate type with studs, top back plate	1 Plum	7	1.143	Ea.	70.50	59.50		130	167
0700	Top front and back plate		7	1.143		87	59.50		146.50	185
0800	Top & bottom, front & back plates, w/bearing jacks		7	1.143		126	59.50		185.50	228
3000	Lavatory, concealed arm									
3050	Floor mounted, single									
3100	High back fixture	1 Plum	6	1.333	Ea.	395	69.50		464.50	535
3200	Flat slab fixture		6	1.333		340	69.50		409.50	480
3220	Paraplegic		6	1.333		440	69.50		509.50	590
3250	Floor mounted, back to back									
3300	High back fixtures	1 Plum	5	1.600	Ea.	560	83.50		643.50	740
3400	Flat slab fixtures		5	1.600		690	83.50		773.50	885
3430	Paraplegic		5	1.600		640	83.50		723.50	830
3500	Wall mounted, in stud or masonry									
3600	High back fixture	1 Plum	6	1.333	Ea.	232	69.50		301.50	360
3700	Flat slab fixture	"	6	1.333	"	201	69.50		270.50	325
4600	Sink, floor mounted									

22 42 39.30 Carriers and Supports		Crew	Daily Output	Labor-Hours	Unit	2010 Bare Costs Material	Labor	Equipment	Total	Total Incl O&P
4650	Exposed arm system									
4700	Single heavy fixture	1 Plum	5	1.600	Ea.	655	83.50		738.50	850
4750	Single heavy sink with slab		5	1.600		810	83.50		893.50	1,025
4800	Back to back, standard fixtures		5	1.600		480	83.50		563.50	655
4850	Back to back, heavy fixtures		5	1.600		690	83.50		773.50	885
4900	Back to back, heavy sink with slab		5	1.600		690	83.50		773.50	885
4950	Exposed offset arm system									
5000	Single heavy deep fixture	1 Plum	5	1.600	Ea.	625	83.50		708.50	815
5100	Plate type system									
5200	With bearing jacks, single fixture	1 Plum	5	1.600	Ea.	720	83.50		803.50	915
5300	With exposed arms, single heavy fixture		5	1.600		920	83.50		1,003.50	1,150
5400	Wall mounted, exposed arms, single heavy fixture		5	1.600		335	83.50		418.50	495
6000	Urinal, floor mounted, 2" or 3" coupling, blowout type		6	1.333		415	69.50		484.50	565
6100	With fixture or hanger bolts, blowout or washout		6	1.333		298	69.50		367.50	435
6200	With bearing plate		6	1.333		330	69.50		399.50	470
6300	Wall mounted, plate type system		6	1.333		257	69.50		326.50	385
6980	Water closet, siphon jet									
7000	Horizontal, adjustable, caulk									
7040	Single, 4" pipe size	1 Plum	5.33	1.501	Ea.	755	78		833	945
7050	4" pipe size, paraplegic		5.33	1.501		755	78		833	945
7060	5" pipe size		5.33	1.501		955	78		1,033	1,175
7100	Double, 4" pipe size		5	1.600		1,100	83.50		1,183.50	1,350
7110	4" pipe size, paraplegic		5	1.600		1,100	83.50		1,183.50	1,350
7120	5" pipe size		5	1.600		1,600	83.50		1,683.50	1,875
7160	Horizontal, adjustable, extended, caulk									
7180	Single, 4" pipe size	1 Plum	5.33	1.501	Ea.	755	78		833	945
7200	5" pipe size		5.33	1.501		955	78		1,033	1,175
7240	Double, 4" pipe size		5	1.600		1,375	83.50		1,458.50	1,625
7260	5" pipe size		5	1.600		1,600	83.50		1,683.50	1,875
7400	Vertical, adjustable, caulk or thread									
7440	Single, 4" pipe size	1 Plum	5.33	1.501	Ea.	765	78		843	960
7460	5" pipe size		5.33	1.501		945	78		1,023	1,175
7480	6" pipe size		5	1.600		1,075	83.50		1,158.50	1,325
7520	Double, 4" pipe size		5	1.600		1,150	83.50		1,233.50	1,375
7540	5" pipe size		5	1.600		1,325	83.50		1,408.50	1,575
7560	6" pipe size		4	2		1,450	104		1,554	1,750
7600	Vertical, adjustable, extended, caulk									
7620	Single, 4" pipe size	1 Plum	5.33	1.501	Ea.	765	78		843	960
7640	5" pipe size		5.33	1.501		945	78		1,023	1,175
7680	6" pipe size		5	1.600		1,075	83.50		1,158.50	1,325
7720	Double, 4" pipe size		5	1.600		1,150	83.50		1,233.50	1,375
7740	5" pipe size		5	1.600		1,325	83.50		1,408.50	1,575
7760	6" pipe size		4	2		1,450	104		1,554	1,750
7780	Water closet, blow out									
7800	Vertical offset, caulk or thread									
7820	Single, 4" pipe size	1 Plum	5.33	1.501	Ea.	565	78		643	740
7840	Double, 4" pipe size	"	5	1.600	"	970	83.50		1,053.50	1,200
7880	Vertical offset, extended, caulk									
7900	Single, 4" pipe size	1 Plum	5.33	1.501	Ea.	710	78		788	895
7920	Double, 4" pipe size	"	5	1.600	"	1,125	83.50		1,208.50	1,350
7960	Vertical, for floor mounted back-outlet									
7980	Single, 4" thread, 2" vent	1 Plum	5.33	1.501	Ea.	500	78		578	665
8000	Double, 4" thread, 2" vent	"	6	1.333	"	1,475	69.50		1,544.50	1,700

22 42 Commercial Plumbing Fixtures

22 42 39 – Commercial Faucets, Supplies, and Trim

22 42 39.30 Carriers and Supports		Crew	Daily Output	Labor-Hours	Unit	2010 Bare Costs Material	Labor	Equipment	Total	Total Incl O&P
8040	Vertical, for floor mounted back-outlet, extended									
8060	Single, 4" caulk, 2" vent	1 Plum	6	1.333	Ea.	500	69.50		569.50	655
8080	Double, 4" caulk, 2" vent	"	6	1.333	"	1,475	69.50		1,544.50	1,700
8200	Water closet, residential									
8220	Vertical centerline, floor mount									
8240	Single, 3" caulk, 2" or 3" vent	1 Plum	6	1.333	Ea.	455	69.50		524.50	605
8260	4" caulk, 2" or 4" vent		6	1.333		575	69.50		644.50	740
8280	3" copper sweat, 3" vent		6	1.333		405	69.50		474.50	550
8300	4" copper sweat, 4" vent		6	1.333		490	69.50		559.50	645
8400	Vertical offset, floor mount									
8420	Single, 3" or 4" caulk, vent	1 Plum	4	2	Ea.	565	104		669	780
8440	3" or 4" copper sweat, vent		5	1.600		565	83.50		648.50	750
8460	Double, 3" or 4" caulk, vent		4	2		970	104		1,074	1,225
8480	3" or 4" copper sweat, vent		5	1.600		970	83.50		1,053.50	1,200
9000	Water cooler (electric), floor mounted									
9100	Plate type with bearing plate, single	1 Plum	6	1.333	Ea.	286	69.50		355.50	420

22 45 Emergency Plumbing Fixtures

22 45 13 – Emergency Showers

22 45 13.10 Emergency Showers		Crew	Daily Output	Labor-Hours	Unit	Material	Labor	Equipment	Total	Total Incl O&P
0010	**EMERGENCY SHOWERS**, Rough-in not included									
5000	Shower, single head, drench, ball valve, pull, freestanding	Q-1	4	4	Ea.	340	187		527	655
5200	Horizontal or vertical supply		4	4		490	187		677	820
6000	Multi-nozzle, eye/face wash combination		4	4		615	187		802	955
6400	Multi-nozzle, 12 spray, shower only		4	4		1,600	187		1,787	2,025
6600	For freeze-proof, add		6	2.667		365	125		490	590

22 45 16 – Eyewash Equipment

22 45 16.10 Eyewash Safety Equipment		Crew	Daily Output	Labor-Hours	Unit	Material	Labor	Equipment	Total	Total Incl O&P
0010	**EYEWASH SAFETY EQUIPMENT**, Rough-in not included									
1000	Eye wash fountain									
1400	Plastic bowl, pedestal mounted	Q-1	4	4	Ea.	262	187		449	570
1600	Unmounted		4	4		198	187		385	500
1800	Wall mounted		4	4		380	187		567	695
2000	Stainless steel, pedestal mounted		4	4		330	187		517	645
2200	Unmounted		4	4		248	187		435	555
2400	Wall mounted		4	4		244	187		431	550

22 45 19 – Self-Contained Eyewash Equipment

22 45 19.10 Self-Contained Eyewash Safety Equipment		Crew	Daily Output	Labor-Hours	Unit	Material	Labor	Equipment	Total	Total Incl O&P
0010	**SELF-CONTAINED EYEWASH SAFETY EQUIPMENT**									
3000	Eye wash, portable, self-contained				Ea.	945			945	1,050

22 45 26 – Eye/Face Wash Equipment

22 45 26.10 Eye/Face Wash Safety Equipment		Crew	Daily Output	Labor-Hours	Unit	Material	Labor	Equipment	Total	Total Incl O&P
0010	**EYE/FACE WASH SAFETY EQUIPMENT**, Rough-in not included									
4000	Eye and face wash, combination fountain									
4200	Stainless steel, pedestal mounted	Q-1	4	4	Ea.	490	187		677	820
4400	Unmounted		4	4		345	187		532	660
4600	Wall mounted		4	4		395	187		582	715

22 47 Drinking Fountains and Water Coolers

22 47 13 – Drinking Fountains

22 47 13.10 Drinking Water Fountains		Crew	Daily Output	Labor-Hours	Unit	Material	Labor	Equipment	Total	Total Incl O&P
						2010 Bare Costs				
0010	DRINKING WATER FOUNTAINS, For connection to cold water supply R224000-40									
1000	Wall mounted, non-recessed									
1400	Bronze, with no back	1 Plum	4	2	Ea.	1,000	104		1,104	1,250
1800	Cast aluminum, enameled, for correctional institutions		4	2		1,375	104		1,479	1,650
2000	Fiberglass, 12" back, single bubbler unit		4	2		1,750	104		1,854	2,075
2040	Dual bubbler		3.20	2.500		1,850	130		1,980	2,250
2400	Precast stone, no back		4	2		725	104		829	950
2700	Stainless steel, single bubbler, no back		4	2		920	104		1,024	1,150
2740	With back		4	2		970	104		1,074	1,225
2780	Dual handle & wheelchair projection type		4	2		670	104		774	895
2820	Dual level for handicapped type	↓	3.20	2.500	↓	1,350	130		1,480	1,700
3300	Vitreous china									
3340	7" back	1 Plum	4	2	Ea.	570	104		674	780
3940	For vandal-resistant bottom plate, add					59.50			59.50	65.50
3960	For freeze-proof valve system, add	1 Plum	2	4		510	208		718	870
3980	For rough-in, supply and waste, add	"	2.21	3.620	↓	190	188		378	490
4000	Wall mounted, semi-recessed									
4200	Poly-marble, single bubbler	1 Plum	4	2	Ea.	920	104		1,024	1,175
4600	Stainless steel, satin finish, single bubbler		4	2		730	104		834	955
4900	Vitreous china, single bubbler		4	2		645	104		749	860
5980	For rough-in, supply and waste, add	↓	1.83	4.372	↓	190	228		418	550
6000	Wall mounted, fully recessed									
6400	Poly-marble, single bubbler	1 Plum	4	2	Ea.	1,525	104		1,629	1,825
6800	Stainless steel, single bubbler		4	2		1,250	104		1,354	1,525
7560	For freeze-proof valve system, add		2	4		655	208		863	1,025
7580	For rough-in, supply and waste, add	↓	1.83	4.372	↓	190	228		418	550
7600	Floor mounted, pedestal type									
7700	Aluminum, architectural style, C.I. base	1 Plum	2	4	Ea.	2,075	208		2,283	2,600
7780	Wheelchair handicap unit		2	4		1,400	208		1,608	1,850
8400	Stainless steel, architectural style		2	4		1,600	208		1,808	2,075
8600	Enameled iron, heavy duty service, 2 bubblers		2	4		2,400	208		2,608	2,925
8660	4 bubblers		2	4		3,650	208		3,858	4,300
8880	For freeze-proof valve system, add		2	4		385	208		593	735
8900	For rough-in, supply and waste, add	↓	1.83	4.372	↓	190	228		418	550
9100	Deck mounted									
9500	Stainless steel, circular receptor	1 Plum	4	2	Ea.	350	104		454	540
9760	White enameled steel, 14" x 9" receptor		4	2		300	104		404	485
9860	White enameled cast iron, 24" x 16" receptor		3	2.667		315	139		454	555
9980	For rough-in, supply and waste, add	↓	1.83	4.372	↓	190	228		418	550

22 47 16 – Pressure Water Coolers

22 47 16.10 Electric Water Coolers		Crew	Daily Output	Labor-Hours	Unit	Material	Labor	Equipment	Total	Total Incl O&P
0010	ELECTRIC WATER COOLERS R224000-50									
0100	Wall mounted, non-recessed									
0140	4 GPH	Q-1	4	4	Ea.	575	187		762	910
0160	8 GPH, barrier free, sensor operated		4	4		830	187		1,017	1,200
0180	8.2 GPH	↓	4	4		650	187		837	995
0600	For hot and cold water, add					141			141	155
0640	For stainless steel cabinet, add					84			84	92
1000	Dual height, 8.2 GPH	Q-1	3.80	4.211		1,975	197		2,172	2,475
1040	14.3 GPH	"	3.80	4.211		1,075	197		1,272	1,475
1240	For stainless steel cabinet, add					131			131	144
2600	Wheelchair type, 8 GPH	Q-1	4	4	↓	1,650	187		1,837	2,075

22 47 Drinking Fountains and Water Coolers

22 47 16 – Pressure Water Coolers

22 47 16.10 Electric Water Coolers		Crew	Daily Output	Labor-Hours	Unit	Material	Labor	Equipment	Total	Total Incl O&P
						2010 Bare Costs				
3300	Semi-recessed, 8.1 GPH	Q-1	4	4	Ea.	600	187		787	940
3320	12 GPH	↓	4	4	↓	810	187		997	1,175
4600	Floor mounted, flush-to-wall									
4640	4 GPH	1 Plum	3	2.667	Ea.	715	139		854	995
4680	8.2 GPH		3	2.667		745	139		884	1,025
4720	14.3 GPH		3	2.667		715	139		854	995
4960	14 GPH hot and cold water	↓	3	2.667		1,200	139		1,339	1,525
4980	For stainless steel cabinet, add					128			128	141
5000	Dual height, 8.2 GPH	1 Plum	2	4		1,000	208		1,208	1,400
5040	14.3 GPH	"	2	4		1,050	208		1,258	1,450
5120	For stainless steel cabinet, add					168			168	185
9800	For supply, waste & vent, all coolers	1 Plum	2.21	3.620	↓	190	188		378	490

22 51 Swimming Pool Plumbing Systems

22 51 19 – Swimming Pool Water Treatment Equipment

22 51 19.50 Swimming Pool Filtration Equipment

		Crew	Daily Output	Labor-Hours	Unit	Material	Labor	Equipment	Total	Total Incl O&P
0010	**SWIMMING POOL FILTRATION EQUIPMENT**									
0900	Filter system, sand or diatomite type, incl. pump, 6,000 gal./hr.	2 Plum	1.80	8.889	Total	1,700	465		2,165	2,575
1020	Add for chlorination system, 800 S.F. pool		3	5.333	Ea.	103	278		381	530
1040	5,000 S.F. pool	↓	3	5.333	"	2,050	278		2,328	2,675

22 52 Fountain Plumbing Systems

22 52 16 – Fountain Pumps

22 52 16.10 Fountain Water Pumps

		Crew	Daily Output	Labor-Hours	Unit	Material	Labor	Equipment	Total	Total Incl O&P
0010	**FOUNTAIN WATER PUMPS**									
0100	Pump w/controls									
0200	Single phase, 100′ cord, 1/2 H.P. pump	2 Skwk	4.40	3.636	Ea.	1,825	155		1,980	2,250
0300	3/4 H.P. pump		4.30	3.721		2,075	159		2,234	2,525
0400	1 H.P. pump		4.20	3.810		3,850	162		4,012	4,475
0500	1-1/2 H.P. pump		4.10	3.902		3,525	166		3,691	4,150
0600	2 H.P. pump		4	4		4,775	170		4,945	5,500
0700	Three phase, 200′ cord, 5 H.P. pump		3.90	4.103		5,050	175		5,225	5,825
0800	7-1/2 H.P. pump		3.80	4.211		8,575	179		8,754	9,725
0900	10 H.P. pump		3.70	4.324		9,375	184		9,559	10,600
1000	15 H.P. pump	↓	3.60	4.444	↓	11,400	189		11,589	12,900
2000	DESIGN NOTE: Use two horsepower per surface acre									

22 52 33 – Fountain Ancillary

22 52 33.10 Fountain Miscellaneous

		Crew	Daily Output	Labor-Hours	Unit	Material	Labor	Equipment	Total	Total Incl O&P
0010	**FOUNTAIN MISCELLANEOUS**									
1100	Nozzles, minimum	2 Skwk	8	2	Ea.	151	85		236	298
1200	Maximum		8	2		305	85		390	465
1300	Lights w/mounting kits, 200 watt		18	.889		1,000	38		1,038	1,150
1400	300 watt		18	.889		1,200	38		1,238	1,375
1500	500 watt		18	.889		1,325	38		1,363	1,500
1600	Color blender	↓	12	1.333	↓	535	57		592	680

22 66 Chemical-Waste Systems for Lab. and Healthcare Facilities

22 66 53 – Laboratory Chemical-Waste and Vent Piping

22 66 53.30 Glass Pipe		Crew	Daily Output	Labor-Hours	Unit	Material	2010 Bare Costs Labor	Equipment	Total	Total Incl O&P
0010	**GLASS PIPE**, Borosilicate, couplings & clevis hanger assemblies, 10' O.C.									
0020	Drainage									
1100	1-1/2" diameter	Q-1	52	.308	L.F.	13.20	14.40		27.60	36
1120	2" diameter		44	.364		17.45	17.05		34.50	44.50
1140	3" diameter		39	.410		23.50	19.20		42.70	55
1160	4" diameter		30	.533		42.50	25		67.50	84.50
1180	6" diameter	↓	26	.615	↓	78	29		107	129
22 66 53.60 Corrosion Resistant Pipe										
0010	**CORROSION RESISTANT PIPE**, No couplings or hangers									
0020	Iron alloy, drain, mechanical joint									
1000	1-1/2" diameter	Q-1	70	.229	L.F.	28.50	10.70		39.20	47.50
1100	2" diameter		66	.242		29	11.35		40.35	49
1120	3" diameter		60	.267		37.50	12.50		50	60.50
1140	4" diameter	↓	52	.308	↓	58.50	14.40		72.90	86
2980	Plastic, epoxy, fiberglass filament wound, B&S joint									
3000	2" diameter	Q-1	62	.258	L.F.	9.60	12.10		21.70	28.50
3100	3" diameter		51	.314		11.35	14.70		26.05	34.50
3120	4" diameter		45	.356		16.45	16.65		33.10	43
3140	6" diameter	↓	32	.500	↓	23.50	23.50		47	60.50
3980	Polyester, fiberglass filament wound, B&S joint									
4000	2" diameter	Q-1	62	.258	L.F.	10.50	12.10		22.60	29.50
4100	3" diameter		51	.314		14.35	14.70		29.05	38
4120	4" diameter		45	.356		20.50	16.65		37.15	47.50
4140	6" diameter	↓	32	.500	↓	30	23.50		53.50	68
4980	Polypropylene, acid resistant, fire retardant, schedule 40									
5000	1-1/2" diameter	Q-1	68	.235	L.F.	6.05	11		17.05	23
5100	2" diameter		62	.258		8.20	12.10		20.30	27
5120	3" diameter		51	.314		16.80	14.70		31.50	40.50
5140	4" diameter	↓	45	.356	↓	21.50	16.65		38.15	48.50
5980	Proxylene, fire retardant, Schedule 40									
6000	1-1/2" diameter	Q-1	68	.235	L.F.	9.40	11		20.40	27
6100	2" diameter		62	.258		12.90	12.10		25	32.50
6120	3" diameter		51	.314		23.50	14.70		38.20	48
6140	4" diameter	↓	45	.356	↓	33	16.65		49.65	61.50

Estimating Tips

The labor adjustment factors listed in Subdivision 22 01 02.20 also apply to Division 23.

23 10 00 Facility Fuel Systems

- The prices in this subdivision for above- and below-ground storage tanks do not include foundations or hold-down slabs, unless noted. The estimator should refer to Divisions 3 and 31 for foundation system pricing. In addition to the foundations, required tank accessories, such as tank gauges, leak detection devices, and additional manholes and piping, must be added to the tank prices.

23 50 00 Central Heating Equipment

- When estimating the cost of an HVAC system, check to see who is responsible for providing and installing the temperature control system. It is possible to overlook controls, assuming that they would be included in the electrical estimate.
- When looking up a boiler, be careful on specified capacity. Some manufacturers rate their products on output while others use input.
- Include HVAC insulation for pipe, boiler, and duct (wrap and liner).
- Be careful when looking up mechanical items to get the correct pressure rating and connection type (thread, weld, flange).

23 70 00 Central HVAC Equipment

- Combination heating and cooling units are sized by the air conditioning requirements. (See Reference No. R236000-20 for preliminary sizing guide.)
- A ton of air conditioning is nominally 400 CFM.
- Rectangular duct is taken off by the linear foot for each size, but its cost is usually estimated by the pound. Remember that SMACNA standards now base duct on internal pressure.
- Prefabricated duct is estimated and purchased like pipe: straight sections and fittings.
- Note that cranes or other lifting equipment are not included on any lines in Division 23. For example, if a crane is required to lift a heavy piece of pipe into place high above a gym floor, or to put a rooftop unit on the roof of a four-story building, etc., it must be added. Due to the potential for extreme variation—from nothing additional required to a major crane or helicopter—we feel that including a nominal amount for "lifting contingency" would be useless and detract from the accuracy of the estimate. When using equipment rental cost data from RSMeans, do not forget to include the cost of the operator(s).

Reference Numbers

Reference numbers are shown in shaded boxes at the beginning of some major classifications. These numbers refer to related items in the Reference Section. The reference information may be an estimating procedure, an alternate pricing method, or technical information.

Note: Not all subdivisions listed here necessarily appear in this publication.

Note: **Trade Service,** *in part, has been used as a reference source for some of the material prices used in Division 23.*

23 05 Common Work Results for HVAC

23 05 05 – Selective HVAC Demolition

23 05 05.10 HVAC Demolition		Crew	Daily Output	Labor-Hours	Unit	Material	2010 Bare Costs Labor	Equipment	Total	Total Incl O&P
0010	**HVAC DEMOLITION**									
0100	Air conditioner, split unit, 3 ton	Q-5	2	8	Ea.		375		375	560
0150	Package unit, 3 ton	Q-6	3	8	"		385		385	580
0298	Boilers									
0300	Electric, up thru 148 kW	Q-19	2	12	Ea.		570		570	850
0310	150 thru 518 kW	"	1	24			1,150		1,150	1,700
0320	550 thru 2000 kW	Q-21	.40	80			3,875		3,875	5,825
0330	2070 kW and up	"	.30	106			5,175		5,175	7,750
0340	Gas and/or oil, up thru 150 MBH	Q-7	2.20	14.545			720		720	1,075
0350	160 thru 2000 MBH		.80	40			1,975		1,975	2,975
0360	2100 thru 4500 MBH		.50	64			3,175		3,175	4,750
0370	4600 thru 7000 MBH		.30	106			5,275		5,275	7,900
0380	7100 thru 12,000 MBH		.16	200			9,875		9,875	14,800
0390	12,200 thru 25,000 MBH		.12	266			13,200		13,200	19,800
1000	Ductwork, 4" high, 8" wide	1 Clab	200	.040	L.F.		1.32		1.32	2.04
1100	6" high, 8" wide		165	.048			1.60		1.60	2.47
1200	10" high, 12" wide		125	.064			2.12		2.12	3.27
1300	12"-14" high, 16"-18" wide		85	.094			3.12		3.12	4.80
1400	18" high, 24" wide		67	.119			3.95		3.95	6.10
1500	30" high, 36" wide		56	.143			4.73		4.73	7.30
1540	72" wide		50	.160			5.30		5.30	8.15
3000	Mechanical equipment, light items. Unit is weight, not cooling.	Q-5	.90	17.778	Ton		830		830	1,250
3600	Heavy items	"	1.10	14.545	"		680		680	1,025
3700	Deduct for salvage (when applicable), minimum				Job				73	80
3710	Maximum				"				455	500

23 05 23 – General-Duty Valves for HVAC Piping

23 05 23.30 Valves, Iron Body		Crew	Daily Output	Labor-Hours	Unit	Material	Labor	Equipment	Total	Total Incl O&P
0010	**VALVES, IRON BODY**									
1020	Butterfly, wafer type, gear actuator, 200 lb.									
1030	2"	1 Plum	14	.571	Ea.	167	29.50		196.50	228
1040	2-1/2"	Q-1	9	1.778		172	83.50		255.50	315
1050	3"		8	2		178	93.50		271.50	335
1060	4"		5	3.200		202	150		352	450
1070	5"	Q-2	5	4.800		269	233		502	645
1080	6"	"	5	4.800		305	233		538	685
1650	Gate, 125 lb., N.R.S.									
2150	Flanged									
2200	2"	1 Plum	5	1.600	Ea.	525	83.50		608.50	705
2240	2-1/2"	Q-1	5	3.200		540	150		690	820
2260	3"		4.50	3.556		605	167		772	915
2280	4"		3	5.333		865	250		1,115	1,325
2300	6"	Q-2	3	8		1,475	390		1,865	2,200
3550	OS&Y, 125 lb., flanged									
3600	2"	1 Plum	5	1.600	Ea.	320	83.50		403.50	475
3660	3"	Q-1	4.50	3.556		370	167		537	655
3680	4"	"	3	5.333		525	250		775	950
3700	6"	Q-2	3	8		865	390		1,255	1,550
3900	For 175 lb, flanged, add					200%	10%			
5450	Swing check, 125 lb., threaded									
5470	1"	1 Plum	13	.615	Ea.	575	32		607	685
5500	2"	"	11	.727		675	38		713	795
5540	2-1/2"	Q-1	15	1.067		660	50		710	805

23 05 23 – General-Duty Valves for HVAC Piping

23 05 23.30 Valves, Iron Body		Crew	Daily Output	Labor-Hours	Unit	2010 Bare Costs Material	Labor	Equipment	Total	Total Incl O&P
5550	3"	Q-1	13	1.231	Ea.	735	57.50		792.50	895
5560	4"	↓	10	1.600	↓	1,250	75		1,325	1,475
5950	Flanged									
6000	2"	1 Plum	5	1.600	Ea.	237	83.50		320.50	385
6040	2-1/2"	Q-1	5	3.200		265	150		415	515
6050	3"		4.50	3.556		285	167		452	565
6060	4"	↓	3	5.333		445	250		695	865
6070	6"	Q-2	3	8	↓	765	390		1,155	1,425

23 05 23.80 Valves, Steel		Crew	Daily Output	Labor-Hours	Unit	Material	Labor	Equipment	Total	Total Incl O&P
0010	**VALVES, STEEL**									
0800	Cast									
1350	Check valve, swing type, 150 lb., flanged									
1370	1"	1 Plum	10	.800	Ea.	525	41.50		566.50	640
1400	2"	"	8	1		660	52		712	805
1440	2-1/2"	Q-1	5	3.200		765	150		915	1,075
1450	3"		4.50	3.556		780	167		947	1,100
1460	4"	↓	3	5.333		1,125	250		1,375	1,600
1540	For 300 lb., flanged, add					50%	15%			
1548	For 600 lb., flanged, add				↓	110%	20%			
1950	Gate valve, 150 lb., flanged									
2000	2"	1 Plum	8	1	Ea.	710	52		762	860
2040	2-1/2"	Q-1	5	3.200		1,000	150		1,150	1,325
2050	3"		4.50	3.556		1,000	167		1,167	1,350
2060	4"	↓	3	5.333		1,250	250		1,500	1,750
2070	6"	Q-2	3	8	↓	1,850	390		2,240	2,600
3650	Globe valve, 150 lb., flanged									
3700	2"	1 Plum	8	1	Ea.	890	52		942	1,050
3740	2-1/2"	Q-1	5	3.200		1,125	150		1,275	1,475
3750	3"		4.50	3.556		1,125	167		1,292	1,500
3760	4"	↓	3	5.333		1,650	250		1,900	2,200
3770	6"	Q-2	3	8	↓	2,600	390		2,990	3,425
5150	Forged									
5650	Check valve, class 800, horizontal, socket									
5698	Threaded									
5700	1/4"	1 Plum	24	.333	Ea.	88	17.35		105.35	123
5720	3/8"		24	.333		88	17.35		105.35	123
5730	1/2"		24	.333		88	17.35		105.35	123
5740	3/4"		20	.400		94.50	21		115.50	135
5750	1"		19	.421		111	22		133	155
5760	1-1/4"	↓	15	.533	↓	217	28		245	281

23 05 93 – Testing, Adjusting, and Balancing for HVAC

23 05 93.10 Balancing, Air		Crew	Daily Output	Labor-Hours	Unit	Material	Labor	Equipment	Total	Total Incl O&P
0010	**BALANCING, AIR** (Subcontractor's quote incl. material and labor)									
0900	Heating and ventilating equipment									
1000	Centrifugal fans, utility sets				Ea.				325	355
1100	Heating and ventilating unit								485	535
1200	In-line fan								485	535
1300	Propeller and wall fan								92	101
1400	Roof exhaust fan								217	238
2000	Air conditioning equipment, central station								705	775
2100	Built-up low pressure unit								650	715
2200	Built-up high pressure unit				↓				760	835

23 05 Common Work Results for HVAC

23 05 93 – Testing, Adjusting, and Balancing for HVAC

23 05 93.10 Balancing, Air		Crew	Daily Output	Labor-Hours	Unit	Material (2010 Bare Costs)	Labor (2010 Bare Costs)	Equipment (2010 Bare Costs)	Total (2010 Bare Costs)	Total Incl O&P
2500	Multi-zone A.C. and heating unit				Ea.				485	535
2600	For each zone over one, add								108	119
2700	Package A.C. unit								270	297
2800	Rooftop heating and cooling unit								379	415
3000	Supply, return, exhaust, registers & diffusers, avg. height ceiling								65	71
3100	High ceiling								97	107
3200	Floor height				↓				54	59

23 05 93.20 Balancing, Water		Crew	Daily Output	Labor-Hours	Unit	Material	Labor	Equipment	Total	Total Incl O&P
0010	**BALANCING, WATER** (Subcontractor's quote incl. material and labor)									
0050	Air cooled condenser				Ea.				198	218
0080	Boiler								400	440
0100	Cabinet unit heater								68	74.50
0200	Chiller								480	530
0300	Convector								56.50	62
0500	Cooling tower								370	405
0600	Fan coil unit, unit ventilator								101	112
0700	Fin tube and radiant panels								113	124
0800	Main and duct re-heat coils								105	115
0810	Heat exchanger								105	115
1000	Pumps								249	274
1100	Unit heater				↓				79.50	87

23 07 HVAC Insulation

23 07 13 – Duct Insulation

23 07 13.10 Duct Thermal Insulation		Crew	Daily Output	Labor-Hours	Unit	Material	Labor	Equipment	Total	Total Incl O&P
0010	**DUCT THERMAL INSULATION**									
0100	Rule of thumb, as a percentage of total mechanical costs				Job				10%	
0110	Insulation req'd is based on the surface size/area to be covered									
3000	Ductwork									
3020	Blanket type, fiberglass, flexible									
3140	FSK vapor barrier wrap, .75 lb. density									
3160	1" thick G	Q-14	350	.046	S.F.	.20	1.87		2.07	3.13
3170	1-1/2" thick G		320	.050		.23	2.05		2.28	3.44
3180	2" thick G		300	.053		.28	2.19		2.47	3.71
3190	3" thick G		260	.062		.40	2.52		2.92	4.36
3200	4" thick G		242	.066		.60	2.71		3.31	4.87
3212	Vinyl Jacket, .75 lb. density, 1-1/2" thick G	↓	320	.050	↓	.23	2.05		2.28	3.44
3280	Unfaced, 1 lb. density									
3310	1" thick G	Q-14	360	.044	S.F.	.19	1.82		2.01	3.04
3320	1-1/2" thick G		330	.048		.23	1.99		2.22	3.34
3330	2" thick G	↓	310	.052	↓	.31	2.12		2.43	3.63
3400	FSK facing, 1 lb. density									
3420	1-1/2" thick G	Q-14	310	.052	S.F.	.18	2.12		2.30	3.49
3430	2" thick G	"	300	.053	"	.20	2.19		2.39	3.62
3450	FSK facing, 1.5 lb. density									
3470	1-1/2" thick G	Q-14	300	.053	S.F.	.27	2.19		2.46	3.70
3480	2" thick G	"	290	.055	"	.33	2.26		2.59	3.88
3795	Finishes									
3800	Stainless steel woven mesh	Q-14	100	.160	S.F.	.84	6.55		7.39	11.10
3820	18 oz. fiberglass cloth, pasted on		170	.094		.77	3.86		4.63	6.85
3900	8 oz. canvas, pasted on	↓	180	.089	↓	.49	3.64		4.13	6.20

23 07 HVAC Insulation

23 07 13 – Duct Insulation

23 07 13.10 Duct Thermal Insulation		Crew	Daily Output	Labor-Hours	Unit	Material	2010 Bare Costs Labor	Equipment	Total	Total Incl O&P
9600	Minimum labor/equipment charge	1 Stpi	4	2	Job		104		104	156

23 07 16 – HVAC Equipment Insulation

23 07 16.10 HVAC Equipment Thermal Insulation

		Crew	Daily Output	Labor-Hours	Unit	Material	Labor	Equipment	Total	Total Incl O&P
0010	**HVAC EQUIPMENT THERMAL INSULATION**									
0100	Rule of thumb, as a percentage of total mechanical costs				Job				10%	
0110	Insulation req'd is based on the surface size/area to be covered									
1000	Boiler, 1-1/2" calcium silicate only G	Q-14	110	.145	S.F.	2.92	5.95		8.87	12.45
1020	Plus 2" fiberglass G	"	80	.200	"	3.82	8.20		12.02	16.95
2000	Breeching, 2" calcium silicate									
2020	Rectangular G	Q-14	42	.381	S.F.	5.70	15.60		21.30	31
2040	Round G	"	38.70	.413	"	5.95	16.95		22.90	33

23 09 Instrumentation and Control for HVAC

23 09 33 – Electric and Electronic Control System for HVAC

23 09 33.10 Electronic Control Systems

		Crew	Daily Output	Labor-Hours	Unit	Material	Labor	Equipment	Total	Total Incl O&P
0010	**ELECTRONIC CONTROL SYSTEMS**									
0020	For electronic costs, add to Div. 23 09 43.10				Ea.					15%

23 09 43 – Pneumatic Control System for HVAC

23 09 43.10 Pneumatic Control Systems

		Crew	Daily Output	Labor-Hours	Unit	Material	Labor	Equipment	Total	Total Incl O&P
0010	**PNEUMATIC CONTROL SYSTEMS**									
0011	Including a nominal 50 Ft. of tubing. Add control panelboard if req'd.									
0100	Heating and ventilating, split system									
0200	Mixed air control, economizer cycle, panel readout, tubing									
0220	Up to 10 tons G	Q-19	.68	35.294	Ea.	3,575	1,675		5,250	6,425
0240	For 10 to 20 tons G		.63	37.915		3,825	1,800		5,625	6,900
0260	For over 20 tons G	↓	.58	41.096	↓	4,150	1,950		6,100	7,475
0300	Heating coil, hot water, 3 way valve,									
0320	Freezestat, limit control on discharge, readout	Q-5	.69	23.088	Ea.	2,650	1,075		3,725	4,550
0500	Cooling coil, chilled water, room									
0520	Thermostat, 3 way valve	Q-5	2	8	Ea.	1,175	375		1,550	1,850
0600	Cooling tower, fan cycle, damper control,									
0620	Control system including water readout in/out at panel	Q-19	.67	35.821	Ea.	4,700	1,700		6,400	7,725
1000	Unit ventilator, day/night operation,									
1100	freezestat, ASHRAE, cycle 2	Q-19	.91	26.374	Ea.	2,600	1,250		3,850	4,725
2000	Compensated hot water from boiler, valve control,									
2100	readout and reset at panel, up to 60 GPM	Q-19	.55	43.956	Ea.	4,875	2,075		6,950	8,475
2120	For 120 GPM		.51	47.059		5,200	2,225		7,425	9,075
2140	For 240 GPM		.49	49.180		5,450	2,325		7,775	9,500
3000	Boiler room combustion air, damper to 5 SF, controls		1.37	17.582		2,350	835		3,185	3,825
3500	Fan coil, heating and cooling valves, 4 pipe control system		3	8		1,050	380		1,430	1,750
3600	Heat exchanger system controls	↓	.86	27.907		2,275	1,325		3,600	4,475
4000	Pneumatic thermostat, including controlling room radiator valve	Q-5	2.43	6.593		705	310		1,015	1,225
4060	Pump control system	Q-19	3	8	↓	1,075	380		1,455	1,775
4500	Air supply for pneumatic control system									
4600	Tank mounted duplex compressor, starter, alternator,									
4620	piping, dryer, PRV station and filter									
4630	1/2 HP	Q-19	.68	35.139	Ea.	8,725	1,675		10,400	12,100
4660	1-1/2 HP		.58	41.739		10,700	1,975		12,675	14,700
4690	5 HP	↓	.42	57.143	↓	25,300	2,725		28,025	32,000

23 12 Facility Fuel Pumps

23 12 16 – Facility Gasoline Dispensing Pumps

23 12 16.20 Fuel Dispensing Equipment		Crew	Daily Output	Labor-Hours	Unit	2010 Bare Costs Material	Labor	Equipment	Total	Total Incl O&P
0010	**FUEL DISPENSING EQUIPMENT**									
1100	Product dispenser with vapor recovery for 6 nozzles, installed, not									
1110	including piping to storage tanks				Ea.	22,500			22,500	24,800

23 13 Facility Fuel-Storage Tanks

23 13 13 – Facility Underground Fuel-Oil, Storage Tanks

23 13 13.09 Single-Wall Steel Fuel-Oil Tanks

		Crew	Daily Output	Labor-Hours	Unit	Material	Labor	Equipment	Total	Total Incl O&P
0010	**SINGLE-WALL STEEL FUEL-OIL TANKS**									
5000	Steel underground, sti-P3, set in place, not incl. hold-down bars.									
5500	Excavation, pad, pumps and piping not included									
5510	Single wall, 500 gallon capacity, 7 gauge shell	Q-5	2.70	5.926	Ea.	1,975	277		2,252	2,600
5520	1,000 gallon capacity, 7 gauge shell	"	2.50	6.400		3,150	299		3,449	3,925
5530	2,000 gallon capacity, 1/4" thick shell	Q-7	4.60	6.957		5,125	345		5,470	6,150
5535	2,500 gallon capacity, 7 gauge shell	Q-5	3	5.333		5,650	249		5,899	6,600
5540	5,000 gallon capacity, 1/4" thick shell	Q-7	3.20	10		8,250	495		8,745	9,825
5580	15,000 gallon capacity, 5/16" thick shell		1.70	18.824		12,700	930		13,630	15,300
5600	20,000 gallon capacity, 5/16" thick shell		1.50	21.333		21,100	1,050		22,150	24,800
5610	25,000 gallon capacity, 3/8" thick shell		1.30	24.615		25,700	1,225		26,925	30,100
5620	30,000 gallon capacity, 3/8" thick shell		1.10	29.091		31,700	1,450		33,150	37,100
5630	40,000 gallon capacity, 3/8" thick shell		.90	35.556		34,600	1,750		36,350	40,600
5640	50,000 gallon capacity, 3/8" thick shell		.80	40		38,400	1,975		40,375	45,200

23 13 13.23 Glass-Fiber-Reinfcd-Plastic, Fuel-Oil, Storage

		Crew	Daily Output	Labor-Hours	Unit	Material	Labor	Equipment	Total	Total Incl O&P
0010	**GLASS-FIBER-REINFCD-PLASTIC, UNDERGRND FUEL-OIL, STORAGE**									
0210	Fiberglass, underground, single wall, U.L. listed, not including									
0220	manway or hold-down strap									
0240	2,000 gallon capacity	Q-7	4.57	7.002	Ea.	5,750	345		6,095	6,850
0250	4,000 gallon capacity		3.55	9.014		8,150	445		8,595	9,650
0260	6,000 gallon capacity		2.67	11.985		9,050	590		9,640	10,900
0280	10,000 gallon capacity		2	16		12,200	790		12,990	14,600
0284	15,000 gallon capacity		1.68	19.048		22,600	940		23,540	26,200
0290	20,000 gallon capacity		1.45	22.069		24,400	1,100		25,500	28,400
0500	For manway, fittings and hold-downs, add					20%	15%			
1020	Fiberglass, underground, double wall, U.L. listed									
1030	includes manways, not incl. hold-down straps									
1040	600 gallon capacity	Q-5	2.42	6.612	Ea.	6,700	310		7,010	7,850
1050	1,000 gallon capacity	"	2.25	7.111		8,475	330		8,805	9,825
1060	2,500 gallon capacity	Q-7	4.16	7.692		12,900	380		13,280	14,800
1070	3,000 gallon capacity		3.90	8.205		13,900	405		14,305	15,900
1080	4,000 gallon capacity		3.64	8.791		15,700	435		16,135	18,000
1090	6,000 gallon capacity		2.42	13.223		15,700	655		16,355	18,300
1100	8,000 gallon capacity **CN**		2.08	15.385		20,100	760		20,860	23,300
1110	10,000 gallon capacity		1.82	17.582		23,300	870		24,170	26,900
1120	12,000 gallon capacity		1.70	18.824		28,700	930		29,630	32,900
2210	Fiberglass, underground, single wall, U.L. listed, including									
2220	hold-down straps, no manways									
2240	2,000 gallon capacity	Q-7	3.55	9.014	Ea.	6,175	445		6,620	7,450
2250	4,000 gallon capacity		2.90	11.034		8,575	545		9,120	10,300
2260	6,000 gallon capacity		2	16		9,900	790		10,690	12,100
2280	10,000 gallon capacity		1.60	20		13,000	990		13,990	15,800
2284	15,000 gallon capacity		1.39	23.022		23,400	1,150		24,550	27,500
2290	20,000 gallon capacity		1.14	28.070		25,600	1,400		27,000	30,300

23 13 Facility Fuel-Storage Tanks

23 13 13 – Facility Underground Fuel-Oil, Storage Tanks

23 13 13.23 Glass-Fiber-Reinfcd-Plastic, Fuel-Oil, Storage		Crew	Daily Output	Labor-Hours	Unit	Material	2010 Bare Costs Labor	Equipment	Total	Total Incl O&P
3020	Fiberglass, underground, double wall, U.L. listed									
3030	includes manways and hold-down straps									
3040	600 gallon capacity	Q-5	1.86	8.602	Ea.	7,125	400		7,525	8,450
3050	1,000 gallon capacity	"	1.70	9.412		8,900	440		9,340	10,500
3060	2,500 gallon capacity	Q-7	3.29	9.726		13,300	480		13,780	15,300
3070	3,000 gallon capacity		3.13	10.224		14,300	505		14,805	16,600
3080	4,000 gallon capacity		2.93	10.922		16,100	540		16,640	18,600
3090	6,000 gallon capacity		1.86	17.204		16,500	850		17,350	19,500
3100	8,000 gallon capacity		1.65	19.394		21,000	960		21,960	24,500
3110	10,000 gallon capacity		1.48	21.622		24,100	1,075		25,175	28,100
3120	12,000 gallon capacity	↓	1.40	22.857	↓	29,500	1,125		30,625	34,200

23 13 23 – Facility Aboveground Fuel-Oil, Storage Tanks

23 13 23.13 Vertical, Steel, Aboveground Fuel-Oil, Storage Tanks		Crew	Daily Output	Labor-Hours	Unit	Material	Labor	Equipment	Total	Total Incl O&P
0010	**VERTICAL, STEEL, ABOVEGROUND FUEL-OIL, STORAGE TANKS**									
4000	Fixed roof oil storage tanks, steel, (1 BBL=42 GAL w/foundation 3'D x 1'W)									
4200	5,000 barrels				Ea.				194,000	213,500
4300	24,000 barrels								333,500	367,000
4500	56,000 barrels								729,000	802,000
4600	110,000 barrels								1,060,000	1,166,000
4800	143,000 barrels								1,250,000	1,375,000
4900	224,000 barrels								1,360,000	1,496,000
5100	Floating roof gasoline tanks, steel, 5,000 barrels (w/foundation 3'D x 1'W)								204,000	225,000
5200	25,000 barrels								381,000	419,000
5400	55,000 barrels								839,000	923,000
5500	100,000 barrels								1,253,000	1,379,000
5700	150,000 barrels								1,532,000	1,685,000
5800	225,000 barrels				↓				2,300,000	2,783,000

23 13 23.16 Horizontal, Stl, Abvgrd Fuel-Oil, Storage Tanks		Crew	Daily Output	Labor-Hours	Unit	Material	Labor	Equipment	Total	Total Incl O&P
0010	**HORIZONTAL, STEEL, ABOVEGROUND FUEL-OIL, STORAGE TANKS**									
3000	Steel, storage, above ground, including cradles, coating,									
3020	fittings, not including foundation, pumps or piping									
3040	Single wall, 275 gallon	Q-5	5	3.200	Ea.	375	149		524	635
3060	550 gallon	"	2.70	5.926		2,250	277		2,527	2,900
3080	1,000 gallon	Q-7	5	6.400		2,825	315		3,140	3,600
3100	1,500 gallon		4.75	6.737		4,275	335		4,610	5,200
3120	2,000 gallon		4.60	6.957		4,875	345		5,220	5,875
3140	5,000 gallon		3.20	10		6,075	495		6,570	7,450
3150	10,000 gallon		2	16		18,200	790		18,990	21,200
3160	15,000 gallon		1.70	18.824		20,700	930		21,630	24,200
3170	20,000 gallon		1.45	22.069		26,000	1,100		27,100	30,200
3180	25,000 gallon		1.30	24.615		31,900	1,225		33,125	36,900
3190	30,000 gallon	↓	1.10	29.091		35,700	1,450		37,150	41,500
3320	Double wall, 500 gallon capacity	Q-5	2.40	6.667		2,375	310		2,685	3,075
3330	2000 gallon capacity	Q-7	4.15	7.711		8,250	380		8,630	9,650
3340	4000 gallon capacity		3.60	8.889		14,700	440		15,140	16,900
3350	6000 gallon capacity		2.40	13.333		17,400	660		18,060	20,100
3360	8000 gallon capacity		2	16		22,300	790		23,090	25,700
3370	10000 gallon capacity		1.80	17.778		25,000	880		25,880	28,800
3380	15000 gallon capacity		1.50	21.333		37,900	1,050		38,950	43,300
3390	20000 gallon capacity		1.30	24.615		43,200	1,225		44,425	49,300
3400	25000 gallon capacity		1.15	27.826		52,500	1,375		53,875	59,500
3410	30000 gallon capacity	↓	1	32	↓	57,500	1,575		59,075	66,000

23 21 20 – Hydronic HVAC Piping Specialties

23 21 20.10 Air Control		Crew	Daily Output	Labor-Hours	Unit	Material	2010 Bare Costs Labor	Equipment	Total	Total Incl O&P
0010	**AIR CONTROL**									
0030	Air separator, with strainer									
0040	2" diameter	Q-5	6	2.667	Ea.	835	125		960	1,100
0080	2-1/2" diameter		5	3.200		945	149		1,094	1,275
0100	3" diameter		4	4		1,450	187		1,637	1,850
0120	4" diameter		3	5.333		2,075	249		2,324	2,675
0130	5" diameter	Q-6	3.60	6.667		2,650	325		2,975	3,400
0140	6" diameter	"	3.40	7.059		3,175	340		3,515	4,000

23 21 20.18 Automatic Air Vent

Line	Description	Crew	Daily Output	Labor-Hours	Unit	Material	Labor	Equipment	Total	Total Incl O&P
0010	**AUTOMATIC AIR VENT**									
0020	Cast iron body, stainless steel internals, float type									
0060	1/2" NPT inlet, 300 psi	1 Stpi	12	.667	Ea.	88	34.50		122.50	149
0220	3/4" NPT inlet, 250 psi	"	10	.800		275	41.50		316.50	360
0340	1-1/2" NPT inlet, 250 psi	Q-5	12	1.333		870	62.50		932.50	1,050

23 21 20.42 Expansion Joints

Line	Description	Crew	Daily Output	Labor-Hours	Unit	Material	Labor	Equipment	Total	Total Incl O&P
0010	**EXPANSION JOINTS**									
0100	Bellows type, neoprene cover, flanged spool									
0140	6" face to face, 1-1/4" diameter	1 Stpi	11	.727	Ea.	248	38		286	330
0160	1-1/2" diameter	"	10.60	.755		248	39		287	330
0180	2" diameter	Q-5	13.30	1.203		251	56		307	360
0190	2-1/2" diameter		12.40	1.290		260	60.50		320.50	375
0200	3" diameter		11.40	1.404		291	65.50		356.50	420
0480	10" face to face, 2" diameter		13	1.231		360	57.50		417.50	480
0500	2-1/2" diameter		12	1.333		380	62.50		442.50	515
0520	3" diameter		11	1.455		390	68		458	525
0540	4" diameter		8	2		440	93.50		533.50	625
0560	5" diameter		7	2.286		525	107		632	740
0580	6" diameter		6	2.667		545	125		670	785

23 21 20.46 Expansion Tanks

Line	Description	Crew	Daily Output	Labor-Hours	Unit	Material	Labor	Equipment	Total	Total Incl O&P
0010	**EXPANSION TANKS**									
1507	Fiberglass and steel single/double wall storage, see Div. 23 13 13									
1512	Tank leak detection systems, see Div. 28 33 33.50									
2000	Steel, liquid expansion, ASME, painted, 15 gallon capacity	Q-5	17	.941	Ea.	520	44		564	640
2020	24 gallon capacity		14	1.143		555	53.50		608.50	690
2040	30 gallon capacity		12	1.333		580	62.50		642.50	735
2060	40 gallon capacity		10	1.600		680	74.50		754.50	860
2080	60 gallon capacity		8	2		815	93.50		908.50	1,050
2100	80 gallon capacity		7	2.286		875	107		982	1,125
2120	100 gallon capacity		6	2.667		1,175	125		1,300	1,475
3000	Steel ASME expansion, rubber diaphragm, 19 gal. cap. accept.		12	1.333		2,150	62.50		2,212.50	2,475
3020	31 gallon capacity		8	2		2,375	93.50		2,468.50	2,775
3040	61 gallon capacity		6	2.667		3,350	125		3,475	3,875
3080	119 gallon capacity		4	4		3,625	187		3,812	4,250
3100	158 gallon capacity		3.80	4.211		5,025	197		5,222	5,825
3140	317 gallon capacity		2.80	5.714		7,600	267		7,867	8,750
3180	528 gallon capacity		2.40	6.667		12,300	310		12,610	14,100

23 21 20.58 Hydronic Heating Control Valves

Line	Description	Crew	Daily Output	Labor-Hours	Unit	Material	Labor	Equipment	Total	Total Incl O&P
0010	**HYDRONIC HEATING CONTROL VALVES**									
0100	Radiator supply, 1/2" diameter	1 Stpi	24	.333	Ea.	53.50	17.30		70.80	85
0120	3/4" diameter		20	.400		57	21		78	94
0140	1" diameter		19	.421		72	22		94	112
0160	1-1/4" diameter		15	.533		103	27.50		130.50	155

23 21 20 – Hydronic HVAC Piping Specialties

23 21 20.58 Hydronic Heating Control Valves		Crew	Daily Output	Labor-Hours	Unit	Material	2010 Bare Costs Labor	Equipment	Total	Total Incl O&P
0500	For low pressure steam, add					25%				

23 21 20.70 Steam Traps		Crew	Daily Output	Labor-Hours	Unit	Material	Labor	Equipment	Total	Total Incl O&P
0010	**STEAM TRAPS**									
0030	Cast iron body, threaded									
0040	Inverted bucket									
0050	1/2" pipe size	1 Stpi	12	.667	Ea.	131	34.50		165.50	196
0070	3/4" pipe size		10	.800		228	41.50		269.50	310
0100	1" pipe size		9	.889		350	46		396	455
0120	1-1/4" pipe size	↓	8	1	↓	530	52		582	660
1000	Float & thermostatic, 15 psi									
1010	3/4" pipe size	1 Stpi	16	.500	Ea.	105	26		131	155
1020	1" pipe size		15	.533		127	27.50		154.50	181
1040	1-1/2" pipe size		9	.889		223	46		269	315
1060	2" pipe size	↓	6	1.333	↓	410	69		479	555

23 21 20.76 Strainers, Y Type, Bronze Body		Crew	Daily Output	Labor-Hours	Unit	Material	Labor	Equipment	Total	Total Incl O&P
0010	**STRAINERS, Y TYPE, BRONZE BODY**									
0050	Screwed, 150 lb., 1/4" pipe size	1 Stpi	24	.333	Ea.	17.30	17.30		34.60	45
0070	3/8" pipe size		24	.333		23	17.30		40.30	51
0100	1/2" pipe size		20	.400		23	21		44	56
0140	1" pipe size		17	.471		30.50	24.50		55	70.50
0160	1-1/2" pipe size		14	.571		66	29.50		95.50	118
0180	2" pipe size		13	.615		88.50	32		120.50	145
0182	3" pipe size	↓	12	.667		670	34.50		704.50	785
0200	300 lb., 2-1/2" pipe size	Q-5	17	.941		420	44		464	530
0220	3" pipe size		16	1		835	46.50		881.50	990
0240	4" pipe size	↓	15	1.067	↓	1,900	50		1,950	2,175
0500	For 300 lb rating 1/4" thru 2", add					15%				
1000	Flanged, 150 lb., 1-1/2" pipe size	1 Stpi	11	.727	Ea.	435	38		473	535
1020	2" pipe size	"	8	1		590	52		642	730
1030	2-1/2" pipe size	Q-5	5	3.200		875	149		1,024	1,175
1040	3" pipe size		4.50	3.556		1,075	166		1,241	1,425
1060	4" pipe size	↓	3	5.333		1,625	249		1,874	2,175
1100	6" pipe size	Q-6	3	8		3,125	385		3,510	4,000
1106	8" pipe size	"	2.60	9.231	↓	3,425	445		3,870	4,450
1500	For 300 lb rating, add					40%				

23 21 20.78 Strainers, Y Type, Iron Body		Crew	Daily Output	Labor-Hours	Unit	Material	Labor	Equipment	Total	Total Incl O&P
0010	**STRAINERS, Y TYPE, IRON BODY**									
0050	Screwed, 250 lb., 1/4" pipe size	1 Stpi	20	.400	Ea.	9.60	21		30.60	41.50
0070	3/8" pipe size		20	.400		9.60	21		30.60	41.50
0100	1/2" pipe size		20	.400		9.60	21		30.60	41.50
0140	1" pipe size		16	.500		15.85	26		41.85	56.50
0160	1-1/2" pipe size		12	.667		26	34.50		60.50	80.50
0180	2" pipe size	↓	8	1		39	52		91	121
0220	3" pipe size	Q-5	11	1.455		251	68		319	380
0240	4" pipe size	"	5	3.200	↓	425	149		574	690
0500	For galvanized body, add					50%				
1000	Flanged, 125 lb., 1-1/2" pipe size	1 Stpi	11	.727	Ea.	135	38		173	205
1020	2" pipe size	"	8	1		142	52		194	234
1040	3" pipe size	Q-5	4.50	3.556		133	166		299	395
1060	4" pipe size	"	3	5.333		242	249		491	640
1080	5" pipe size	Q-6	3.40	7.059		380	340		720	925
1100	6" pipe size	"	3	8	↓	460	385		845	1,100

23 21 20 – Hydronic HVAC Piping Specialties

23 21 20.78 Strainers, Y Type, Iron Body	Crew	Daily Output	Labor-Hours	Unit	Material	2010 Bare Costs Labor	Equipment	Total	Total Incl O&P
1500 For 250 lb rating, add					20%				
2000 For galvanized body, add					50%				
2500 For steel body, add					40%				
23 21 20.88 Venturi Flow									
0010 **VENTURI FLOW**, Measuring device									
0050 1/2" diameter	1 Stpi	24	.333	Ea.	281	17.30		298.30	335
0120 1" diameter		19	.421		276	22		298	340
0140 1-1/4" diameter		15	.533		340	27.50		367.50	415
0160 1-1/2" diameter		13	.615		355	32		387	440
0180 2" diameter		11	.727		365	38		403	460
0220 3" diameter	Q-5	14	1.143		515	53.50		568.50	650
0240 4" diameter	"	11	1.455		775	68		843	950
0280 6" diameter	Q-6	3.50	6.857		1,125	330		1,455	1,750
0500 For meter, add					2,125			2,125	2,350

23 21 23 – Hydronic Pumps

23 21 23.13 In-Line Centrifugal Hydronic Pumps	Crew	Daily Output	Labor-Hours	Unit	Material	Labor	Equipment	Total	Total Incl O&P
0010 **IN-LINE CENTRIFUGAL HYDRONIC PUMPS**									
0600 Bronze, sweat connections, 1/40 HP, in line									
0640 3/4" size	Q-1	16	1	Ea.	192	47		239	281
1000 Flange connection, 3/4" to 1-1/2" size									
1040 1/12 HP	Q-1	6	2.667	Ea.	495	125		620	730
1060 1/8 HP		6	2.667		855	125		980	1,125
1100 1/3 HP		6	2.667		955	125		1,080	1,225
1140 2" size, 1/6 HP		5	3.200		1,025	150		1,175	1,350
1180 2-1/2" size, 1/4 HP		5	3.200		1,525	150		1,675	1,925
2000 Cast iron, flange connection									
2040 3/4" to 1-1/2" size, in line, 1/12 HP	Q-1	6	2.667	Ea.	320	125		445	535
2100 1/3 HP		6	2.667		595	125		720	840
2140 2" size, 1/6 HP		5	3.200		655	150		805	945
2180 2-1/2" size, 1/4 HP		5	3.200		830	150		980	1,125
2220 3" size, 1/4 HP		4	4		840	187		1,027	1,200
2600 For non-ferrous impeller, add					3%				

23 21 29 – Automatic Condensate Pump Units

23 21 29.10 Condensate Removal Pump System	Crew	Daily Output	Labor-Hours	Unit	Material	Labor	Equipment	Total	Total Incl O&P
0010 **CONDENSATE REMOVAL PUMP SYSTEM**									
0020 Pump with 1 Gal. ABS tank G									
0100 115 V G									
0120 1/50 HP, 200 GPH G	1 Stpi	12	.667	Ea.	148	34.50		182.50	215
0140 1/18 HP, 270 GPH G		10	.800		164	41.50		205.50	242
0160 1/5 HP, 450 GPH G		8	1		355	52		407	470
0200 230 V G									
0240 1/18 HP, 270 GPH G	1 Stpi	10	.800	Ea.	164	41.50		205.50	242
0260 1/5 HP, 450 GPH G	"	8	1	"	390	52		442	510

23 22 Steam and Condensate Piping and Pumps

23 22 13 – Steam and Condensate Heating Piping

23 22 13.23 Aboveground Steam and Condensate Piping		Crew	Daily Output	Labor-Hours	Unit	2010 Bare Costs Material	Labor	Equipment	Total	Total Incl O&P
0010	**ABOVEGROUND STEAM AND CONDENSATE HEATING PIPING**									
0020	Condensate meter									
0100	500 lb. per hour	1 Stpi	14	.571	Ea.	2,350	29.50		2,379.50	2,650
0140	1500 lb. per hour	"	7	1.143	"	2,550	59.50		2,609.50	2,900

23 22 23 – Steam Condensate Pumps

23 22 23.10 Condensate Return System		Crew	Daily Output	Labor-Hours	Unit	Material	Labor	Equipment	Total	Total Incl O&P
0010	**CONDENSATE RETURN SYSTEM**									
0200	Simplex, 3/4 H.P. mtr, float switch, controls, 10 Gal. C.I. rcvr, 6-15GPM	Q-1	1	16	Ea.	2,625	750		3,375	4,000
1000	Duplex, 2 pumps, 3/4 H.P. motors, float switch,									
1060	alternator assembly, 15 Gal. C.I. receiver	Q-1	.50	32	Ea.	4,300	1,500		5,800	6,975

23 31 HVAC Ducts and Casings

23 31 13 – Metal Ducts

23 31 13.13 Rectangular Metal Ducts		Crew	Daily Output	Labor-Hours	Unit	Material	Labor	Equipment	Total	Total Incl O&P
0010	**RECTANGULAR METAL DUCTS** R233100-20									
0020	Fabricated rectangular, includes fittings, joints, supports,									
0030	allowance for flexible connections, no insulation									
0031	NOTE: Fabrication and installation are combined									
0040	as LABOR cost. Approx. 25% fittings assumed.									
0050	Add to labor for elevated installation									
0051	of fabricated ductwork									
0052	10′ to 15′ high						6%			
0053	15′ to 20′ high						12%			
0054	20′ to 25′ high						15%			
0055	25′ to 30′ high						21%			
0056	30′ to 35′ high						24%			
0057	35′ to 40′ high						30%			
0058	Over 40′ high						33%			
0072	For duct insulation see Div. 23 07 13.10 3000									
0100	Aluminum, alloy 3003-H14, under 100 lb.	Q-10	75	.320	Lb.	3.72	14.65		18.37	26
0110	100 to 500 lb.		80	.300		2.48	13.75		16.23	23.50
0120	500 to 1,000 lb.		95	.253		2.31	11.60		13.91	20
0140	1,000 to 2,000 lb.		120	.200		2.15	9.15		11.30	16.25
0150	2,000 to 5,000 lb. **CN**		130	.185		2.15	8.45		10.60	15.15
0160	Over 5,000 lb.		145	.166		2.15	7.60		9.75	13.85
0500	Galvanized steel, under 200 lb.		235	.102		.47	4.68		5.15	7.60
0520	200 to 500 lb.		245	.098		.46	4.49		4.95	7.30
0540	500 to 1,000 lb.		255	.094		.45	4.31		4.76	7.05
0560	1,000 to 2,000 lb.		265	.091		.45	4.15		4.60	6.80
0570	2,000 to 5,000 lb.		275	.087		.44	4		4.44	6.55
0580	Over 5,000 lb. **CN**		285	.084		.43	3.86		4.29	6.30
1000	Stainless steel, type 304, under 100 lb.		165	.145		4.35	6.65		11	14.90
1020	100 to 500 lb.		175	.137		4.02	6.30		10.32	13.90
1030	500 to 1,000 lb.		190	.126		3.68	5.80		9.48	12.80
1040	1,000 to 2,000 lb.		200	.120		3.65	5.50		9.15	12.35
1050	2,000 to 5,000 lb.		225	.107		3.31	4.89		8.20	11.05
1060	Over 5,000 lb.	↓	235	.102		3.24	4.68		7.92	10.65
1100	For medium pressure ductwork, add						15%			
1200	For high pressure ductwork, add						40%			
1210	For welded ductwork, add				↓		85%			

23 31 HVAC Ducts and Casings

23 31 13 – Metal Ducts

23 31 13.13 Rectangular Metal Ducts		Crew	Daily Output	Labor-Hours	Unit	Material	2010 Bare Costs Labor	Equipment	Total	Total Incl O&P
1220	For 30% fittings, add				Lb.		11%			
1224	For 40% fittings, add						34%			
1228	For 50% fittings, add						56%			
1232	For 60% fittings, add						79%			
1236	For 70% fittings, add						101%			
1240	For 80% fittings, add						124%			
1244	For 90% fittings, add						147%			
1248	For 100% fittings, add						169%			
1252	Note: Fittings add includes time for detailing and installation.									

23 31 13.19 Metal Duct Fittings		Crew	Daily Output	Labor-Hours	Unit	Material	Labor	Equipment	Total	Total Incl O&P
0010	**METAL DUCT FITTINGS**									
0050	Air extractors, 12" x 4"	1 Shee	24	.333	Ea.	16.10	16.35		32.45	43
0100	8" x 6"		22	.364		16.10	17.85		33.95	45
0200	20" x 8"		16	.500		36.50	24.50		61	77
0280	24" x 12"		10	.800		49.50	39.50		89	114
2000	Fabrics for flexible connections, with metal edge		100	.080	L.F.	2.13	3.93		6.06	8.30
2100	Without metal edge		160	.050	"	1.53	2.46		3.99	5.40

23 31 16 – Nonmetal Ducts

23 31 16.13 Fibrous-Glass Ducts		Crew	Daily Output	Labor-Hours	Unit	Material	Labor	Equipment	Total	Total Incl O&P
0010	**FIBROUS-GLASS DUCTS** R233100-20									
3490	Rigid fiberglass duct board, foil reinf. kraft facing									
3500	Rectangular, 1" thick, alum. faced, (FRK), std. weight	Q-10	350	.069	SF Surf	.77	3.14		3.91	5.60

23 33 Air Duct Accessories

23 33 13 – Dampers

23 33 13.13 Volume-Control Dampers		Crew	Daily Output	Labor-Hours	Unit	Material	Labor	Equipment	Total	Total Incl O&P
0010	**VOLUME-CONTROL DAMPERS**									
5990	Multi-blade dampers, opposed blade, 8" x 6"	1 Shee	24	.333	Ea.	20	16.35		36.35	47
5994	8" x 8"		22	.364		21	17.85		38.85	50
5996	10" x 10"		21	.381		24.50	18.70		43.20	55.50
6000	12" x 12"		21	.381		27.50	18.70		46.20	58.50
6020	12" x 18"		18	.444		36.50	22		58.50	73.50
6030	14" x 10"		20	.400		26.50	19.65		46.15	59
6031	14" x 14"		17	.471		32.50	23		55.50	70.50
6033	16" x 12"		17	.471		32.50	23		55.50	70.50
6035	16" x 16"		16	.500		40.50	24.50		65	81.50
6037	18" x 16"		15	.533		44.50	26		70.50	88.50
6038	18" x 18"		15	.533		48	26		74	92.50
6070	20" x 16"		14	.571		48	28		76	95.50
6072	20" x 20"		13	.615		57.50	30		87.50	110
6074	22" x 18"		14	.571		57.50	28		85.50	106
6076	24" x 16"		11	.727		57	35.50		92.50	117
6078	24" x 20"		8	1		67	49		116	148
6080	24" x 24"		8	1		78.50	49		127.50	161
6110	26" x 26"		6	1.333		86.50	65.50		152	194
6133	30" x 30"	Q-9	6.60	2.424		124	107		231	298
6135	32" x 32"		6.40	2.500		144	111		255	325
6180	48" x 36"		5.60	2.857		237	126		363	450
8000	Multi-blade dampers, parallel blade									
8100	8" x 8"	1 Shee	24	.333	Ea.	73	16.35		89.35	106

23 33 13 – Dampers

23 33 13.13 Volume-Control Dampers		Crew	Daily Output	Labor-Hours	Unit	Material	Labor	Equipment	Total	Total Incl O&P
						2010 Bare Costs				
8140	16" x 10"	1 Shee	20	.400	Ea.	94	19.65		113.65	133
8200	24" x 16"		11	.727		120	35.50		155.50	185
8260	30" x 18"	↓	7	1.143	↓	164	56		220	265

23 33 13.16 Fire Dampers

		Crew	Daily Output	Labor-Hours	Unit	Material	Labor	Equipment	Total	Total Incl O&P
0010	**FIRE DAMPERS**									
3000	Fire damper, curtain type, 1-1/2 hr rated, vertical, 6" x 6"	1 Shee	24	.333	Ea.	21.50	16.35		37.85	49
3020	8" x 6"		22	.364		21.50	17.85		39.35	51
3240	16" x 14"		18	.444		29.50	22		51.50	65.50
3400	24" x 20"	↓	8	1	↓	42	49		91	121

23 33 13.28 Splitter Damper Assembly

		Crew	Daily Output	Labor-Hours	Unit	Material	Labor	Equipment	Total	Total Incl O&P
0009	**SPLITTER DAMPER ASSEMBLY**									
0010	Self locking, 1' rod	1 Shee	24	.333	Ea.	21.50	16.35		37.85	49
7020	3' rod		22	.364		28	17.85		45.85	58
7040	4' rod		20	.400		31.50	19.65		51.15	64
7060	6' rod	↓	18	.444	↓	38	22		60	75

23 33 19 – Duct Silencers

23 33 19.10 Duct Silencers

		Crew	Daily Output	Labor-Hours	Unit	Material	Labor	Equipment	Total	Total Incl O&P
0009	**DUCT SILENCERS**									
0010	Silencers, noise control for air flow, duct				MCFM	55.50			55.50	61

23 33 33 – Duct-Mounting Access Doors

23 33 33.13 Duct Access Doors

		Crew	Daily Output	Labor-Hours	Unit	Material	Labor	Equipment	Total	Total Incl O&P
0010	**DUCT ACCESS DOORS**									
1000	Duct access door, insulated, 6" x 6"	1 Shee	14	.571	Ea.	15	28		43	59
1020	10" x 10"		11	.727		18.15	35.50		53.65	74
1040	12" x 12"		10	.800		19.40	39.50		58.90	81
1050	12" x 18"		9	.889		34.50	43.50		78	104
1070	18" x 18"		8	1		34.50	49		83.50	113
1074	24" x 18"	↓	8	1	↓	41	49		90	120

23 33 46 – Flexible Ducts

23 33 46.10 Flexible Air Ducts

		Crew	Daily Output	Labor-Hours	Unit	Material	Labor	Equipment	Total	Total Incl O&P
0010	**FLEXIBLE AIR DUCTS** R233100-20									
1280	Add to labor for elevated installation									
1282	of prefabricated (purchased) ductwork									
1283	10' to 15' high						10%			
1284	15' to 20' high						20%			
1285	20' to 25' high						25%			
1286	25' to 30' high						35%			
1287	30' to 35' high						40%			
1288	35' to 40' high						50%			
1289	Over 40' high						55%			
1300	Flexible, coated fiberglass fabric on corr. resist. metal helix									
1400	pressure to 12" (WG) UL-181									
1500	Non-insulated, 3" diameter	Q-9	400	.040	L.F.	1.02	1.77		2.79	3.80
1540	5" diameter		320	.050		1.25	2.21		3.46	4.73
1560	6" diameter		280	.057		1.53	2.53		4.06	5.50
1580	7" diameter		240	.067		1.82	2.95		4.77	6.45
1600	8" diameter		200	.080		2.10	3.54		5.64	7.65
1640	10" diameter		160	.100		2.61	4.42		7.03	9.55
1660	12" diameter		120	.133		3.17	5.90		9.07	12.40
1900	Insulated, 1" thick, PE jacket, 3" diameter G	↓	380	.042	↓	2.15	1.86		4.01	5.20

23 33 Air Duct Accessories

23 33 46 – Flexible Ducts

23 33 46.10 Flexible Air Ducts		Crew	Daily Output	Labor-Hours	Unit	Material	2010 Bare Costs Labor	Equipment	Total	Total Incl O&P
1910	4″ diameter	G	Q-9	340	.047	L.F.	2.15	2.08		4.23	5.50
1920	5″ diameter	G		300	.053		2.15	2.36		4.51	5.95
1940	6″ diameter	G		260	.062		2.38	2.72		5.10	6.75
1960	7″ diameter	G		220	.073		2.78	3.21		5.99	7.95
1980	8″ diameter	G		180	.089		3	3.93		6.93	9.25
2020	10″ diameter	G		140	.114		3.51	5.05		8.56	11.50
2040	12″ diameter	G	↓	100	.160	↓	4.36	7.05		11.41	15.50

23 33 53 – Duct Liners

23 33 53.10 Duct Liner Board			Crew	Daily Output	Labor-Hours	Unit	Material	Labor	Equipment	Total	Total Incl O&P
0010	**DUCT LINER BOARD**										
3490	Board type, fiberglass liner, 3 lb. density										
3500	Fire resistant, black pigmented, 1 side										
3520	1″ thick	G	Q-14	150	.107	S.F.	.77	4.37		5.14	7.65
3540	1-1/2″ thick	G		130	.123		.97	5.05		6.02	8.90
3560	2″ thick	G	↓	120	.133	↓	1.18	5.45		6.63	9.80
3600	FSK vapor barrier										
3620	1″ thick	G	Q-14	150	.107	S.F.	.77	4.37		5.14	7.65
3630	1-1/2″ thick	G		130	.123		.97	5.05		6.02	8.90
3640	2″ thick	G	↓	120	.133	↓	1.18	5.45		6.63	9.80
3680	No finish										
3700	1″ thick	G	Q-14	170	.094	S.F.	.40	3.86		4.26	6.45
3710	1-1/2″ thick	G		140	.114		.60	4.69		5.29	7.95
3720	2″ thick	G	↓	130	.123	↓	.81	5.05		5.86	8.75
3940	Board type, non-fibrous foam										
3950	Temperature, bacteria and fungi resistant										
3960	1″ thick	G	Q-14	150	.107	S.F.	2.25	4.37		6.62	9.30
3970	1-1/2″ thick	G		130	.123		3	5.05		8.05	11.15
3980	2″ thick	G	↓	120	.133	↓	3.60	5.45		9.05	12.45

23 34 HVAC Fans

23 34 13 – Axial HVAC Fans

23 34 13.10 Axial Flow HVAC Fans		Crew	Daily Output	Labor-Hours	Unit	Material	Labor	Equipment	Total	Total Incl O&P
0010	**AXIAL FLOW HVAC FANS** R233400-10									
0020	Air conditioning and process air handling									
0030	Axial flow, compact, low sound, 2.5″ S.P.									
0050	3,800 CFM, 5 HP	Q-20	3.40	5.882	Ea.	4,500	266		4,766	5,350
0080	6,400 CFM, 5 HP		2.80	7.143		5,025	325		5,350	6,000
0100	10,500 CFM, 7-1/2 HP		2.40	8.333		6,250	375		6,625	7,450
0120	15,600 CFM, 10 HP		1.60	12.500		7,875	565		8,440	9,525
1500	Vaneaxial, low pressure, 2000 CFM, 1/2 HP		3.60	5.556		1,950	251		2,201	2,525
1520	4,000 CFM, 1 HP		3.20	6.250		2,250	282		2,532	2,900
1540	8,000 CFM, 2 HP	↓	2.80	7.143	↓	2,900	325		3,225	3,675

23 34 14 – Blower HVAC Fans

23 34 14.10 Blower Type HVAC Fans		Crew	Daily Output	Labor-Hours	Unit	Material	Labor	Equipment	Total	Total Incl O&P
0010	**BLOWER TYPE HVAC FANS**									
2500	Ceiling fan, right angle, extra quiet, 0.10″ S.P.									
2520	95 CFM	Q-20	20	1	Ea.	240	45		285	330
2540	210 CFM		19	1.053		283	47.50		330.50	380
2560	385 CFM		18	1.111		360	50		410	470
2580	885 CFM	↓	16	1.250	↓	710	56.50		766.50	865

23 34 14 – Blower HVAC Fans

23 34 14.10 Blower Type HVAC Fans		Crew	Daily Output	Labor-Hours	Unit	2010 Bare Costs Material	Labor	Equipment	Total	Total Incl O&P
2600	1,650 CFM	Q-20	13	1.538	Ea.	980	69.50		1,049.50	1,175
2620	2,960 CFM		11	1.818		1,300	82		1,382	1,575
2640	For wall or roof cap, add	1 Shee	16	.500		240	24.50		264.50	300
2660	For straight thru fan, add					10%				
2680	For speed control switch, add	1 Elec	16	.500		131	24.50		155.50	181
7500	Utility set, steel construction, pedestal, 1/4" S.P.									
7520	Direct drive, 150 CFM, 1/8 HP	Q-20	6.40	3.125	Ea.	755	141		896	1,050
7540	485 CFM, 1/6 HP		5.80	3.448		950	156		1,106	1,275
7560	1950 CFM, 1/2 HP		4.80	4.167		1,125	188		1,313	1,500
7580	2410 CFM, 3/4 HP		4.40	4.545		2,050	205		2,255	2,550
7600	3328 CFM, 1-1/2 HP		3	6.667		2,300	300		2,600	2,975
7680	V-belt drive, drive cover, 3 phase									
7700	800 CFM, 1/4 HP	Q-20	6	3.333	Ea.	775	151		926	1,075
7720	1,300 CFM, 1/3 HP		5	4		815	181		996	1,175
7740	2,000 CFM, 1 HP		4.60	4.348		965	196		1,161	1,350
7760	2,900 CFM, 3/4 HP		4.20	4.762		1,300	215		1,515	1,750

23 34 16 – Centrifugal HVAC Fans

23 34 16.10 Centrifugal Type HVAC Fans		Crew	Daily Output	Labor-Hours	Unit	Material	Labor	Equipment	Total	Total Incl O&P
0010	**CENTRIFUGAL TYPE HVAC FANS**									
0200	In-line centrifugal, supply/exhaust booster									
0220	aluminum wheel/hub, disconnect switch, 1/4" S.P.									
0240	500 CFM, 10" diameter connection	Q-20	3	6.667	Ea.	1,150	300		1,450	1,725
0260	1,380 CFM, 12" diameter connection		2	10		1,225	450		1,675	2,025
0280	1,520 CFM, 16" diameter connection		2	10		1,325	450		1,775	2,125
0300	2,560 CFM, 18" diameter connection		1	20		1,450	905		2,355	2,950
0320	3,480 CFM, 20" diameter connection		.80	25		1,725	1,125		2,850	3,575
0326	5,080 CFM, 20" diameter connection		.75	26.667		1,875	1,200		3,075	3,875
3500	Centrifugal, airfoil, motor and drive, complete									
3520	1000 CFM, 1/2 HP	Q-20	2.50	8	Ea.	1,650	360		2,010	2,375
3540	2,000 CFM, 1 HP		2	10		1,850	450		2,300	2,700
3560	4,000 CFM, 3 HP		1.80	11.111		2,300	500		2,800	3,275
3580	8,000 CFM, 7-1/2 HP		1.40	14.286		3,575	645		4,220	4,900
3600	12,000 CFM, 10 HP		1	20		4,300	905		5,205	6,075
4500	Corrosive fume resistant, plastic									
4600	roof ventilators, centrifugal, V belt drive, motor									
4620	1/4" S.P., 250 CFM, 1/4 HP	Q-20	6	3.333	Ea.	3,875	151		4,026	4,475
4640	895 CFM, 1/3 HP		5	4		4,200	181		4,381	4,900
4660	1630 CFM, 1/2 HP		4	5		4,975	226		5,201	5,825
4680	2240 CFM, 1 HP		3	6.667		5,175	300		5,475	6,150
5000	Utility set, centrifugal, V belt drive, motor									
5020	1/4" S.P., 1200 CFM, 1/4 HP	Q-20	6	3.333	Ea.	4,275	151		4,426	4,925
5040	1520 CFM, 1/3 HP		5	4		1,650	181		1,831	2,075
5060	1850 CFM, 1/2 HP		4	5		1,700	226		1,926	2,225
5080	2180 CFM, 3/4 HP		3	6.667		1,950	300		2,250	2,600
5100	1/2" S.P., 3600 CFM, 1 HP		2	10		2,225	450		2,675	3,125
5120	4250 CFM, 1-1/2 HP		1.60	12.500		2,475	565		3,040	3,575
5140	4800 CFM, 2 HP		1.40	14.286		2,725	645		3,370	3,975
7000	Roof exhauster, centrifugal, aluminum housing, 12" galvanized									
7020	curb, bird screen, back draft damper, 1/4" S.P.									
7100	Direct drive, 320 CFM, 11" sq. damper	Q-20	7	2.857	Ea.	600	129		729	855
7120	600 CFM, 11" sq. damper		6	3.333		750	151		901	1,050
7140	815 CFM, 13" sq. damper		5	4		750	181		931	1,100

23 34 16 – Centrifugal HVAC Fans

23 34 16.10 Centrifugal Type HVAC Fans		Crew	Daily Output	Labor-Hours	Unit	Material	Labor	Equipment	Total	Total Incl O&P
						2010 Bare Costs				
7160	1450 CFM, 13″ sq. damper	Q-20	4.20	4.762	Ea.	1,125	215		1,340	1,550
7180	2050 CFM, 16″ sq. damper		4	5		1,450	226		1,676	1,925
7200	V-belt drive, 1650 CFM, 12″ sq. damper		6	3.333		1,075	151		1,226	1,425
7220	2750 CFM, 21″ sq. damper		5	4		1,300	181		1,481	1,700
7230	3500 CFM, 21″ sq. damper		4.50	4.444		1,450	201		1,651	1,900
7240	4910 CFM, 23″ sq. damper		4	5		1,775	226		2,001	2,300
7260	8525 CFM, 28″ sq. damper		3	6.667		2,400	300		2,700	3,075
7280	13,760 CFM, 35″ sq. damper		2	10		3,200	450		3,650	4,200
7300	20,558 CFM, 43″ sq. damper		1	20		6,625	905		7,530	8,650
7320	For 2 speed winding, add					15%				
7340	For explosionproof motor, add					495			495	545
7360	For belt driven, top discharge, add					15%				
8500	Wall exhausters, centrifugal, auto damper, 1/8″ S.P.									
8520	Direct drive, 610 CFM, 1/20 HP	Q-20	14	1.429	Ea.	350	64.50		414.50	485
8540	796 CFM, 1/12 HP		13	1.538		735	69.50		804.50	910
8560	822 CFM, 1/6 HP		12	1.667		885	75.50		960.50	1,075
8580	1,320 CFM, 1/4 HP		12	1.667		975	75.50		1,050.50	1,200
9500	V-belt drive, 3 phase									
9520	2,800 CFM, 1/4 HP	Q-20	9	2.222	Ea.	1,575	100		1,675	1,900
9540	3,740 CFM, 1/2 HP	″	8	2.500	″	1,650	113		1,763	1,975

23 34 23 – HVAC Power Ventilators

23 34 23.10 HVAC Power Circulators and Ventilators		Crew	Daily Output	Labor-Hours	Unit	Material	Labor	Equipment	Total	Total Incl O&P
0010	**HVAC POWER CIRCULATORS AND VENTILATORS**									
3000	Paddle blade air circulator, 3 speed switch									
3020	42″, 5,000 CFM high, 3000 CFM low G	1 Elec	2.40	3.333	Ea.	130	163		293	385
3040	52″, 6,500 CFM high, 4000 CFM low G	″	2.20	3.636	″	142	178		320	420
3100	For antique white motor, same cost									
3200	For brass plated motor, same cost									
3300	For light adaptor kit, add G				Ea.	38.50			38.50	42.50
6000	Propeller exhaust, wall shutter, 1/4″ S.P.									
6020	Direct drive, two speed									
6100	375 CFM, 1/10 HP	Q-20	10	2	Ea.	430	90.50		520.50	610
6120	730 CFM, 1/7 HP		9	2.222		480	100		580	675
6140	1000 CFM, 1/8 HP		8	2.500		665	113		778	905
6200	4720 CFM, 1 HP		5	4		1,075	181		1,256	1,450
6300	V-belt drive, 3 phase									
6320	6175 CFM, 3/4 HP	Q-20	5	4	Ea.	870	181		1,051	1,225
6340	7500 CFM, 3/4 HP		5	4		935	181		1,116	1,300
6360	10,100 CFM, 1 HP		4.50	4.444		1,000	201		1,201	1,400
6380	14,300 CFM, 1-1/2 HP		4	5		1,075	226		1,301	1,525
6650	Residential, bath exhaust, grille, back draft damper									
6660	50 CFM	Q-20	24	.833	Ea.	44.50	37.50		82	106
6670	110 CFM		22	.909		71.50	41		112.50	141
6680	Light combination, squirrel cage, 100 watt, 70 CFM		24	.833		80.50	37.50		118	146
6700	Light/heater combination, ceiling mounted									
6710	70 CFM, 1450 watt	Q-20	24	.833	Ea.	114	37.50		151.50	182
6800	Heater combination, recessed, 70 CFM		24	.833		46.50	37.50		84	108
6820	With 2 infrared bulbs		23	.870		70	39.50		109.50	136
6900	Kitchen exhaust, grille, complete, 160 CFM		22	.909		83.50	41		124.50	154
6910	180 CFM		20	1		70.50	45		115.50	146
6920	270 CFM		18	1.111		128	50		178	217
6930	350 CFM		16	1.250		99.50	56.50		156	195

23 34 HVAC Fans

23 34 23 – HVAC Power Ventilators

23 34 23.10 HVAC Power Circulators and Ventilators		Crew	Daily Output	Labor-Hours	Unit	Material	2010 Bare Costs Labor	Equipment	Total	Total Incl O&P
6940	Residential roof jacks and wall caps									
6944	Wall cap with back draft damper									
6946	3" & 4" dia. round duct	1 Shee	11	.727	Ea.	19.05	35.50		54.55	75
6948	6" dia. round duct	"	11	.727	"	46	35.50		81.50	105
6958	Roof jack with bird screen and back draft damper									
6960	3" & 4" dia. round duct	1 Shee	11	.727	Ea.	18.30	35.50		53.80	74
6962	3-1/4" x 10" rectangular duct	"	10	.800	"	34	39.50		73.50	97
6980	Transition									
6982	3-1/4" x 10" to 6" dia. round	1 Shee	20	.400	Ea.	19.95	19.65		39.60	51.50

23 34 33 – Air Curtains

23 34 33.10 Air Barrier Curtains		Crew	Daily Output	Labor-Hours	Unit	Material	Labor	Equipment	Total	Total Incl O&P
0010	**AIR BARRIER CURTAINS**, Incl. motor starters, transformers,									
0050	door switches & temperature controls									
0100	Shipping and receiving doors, unheated, minimal wind stoppage									
0150	8' high, multiples of 3' wide	2 Shee	6	2.667	L.F.	335	131		466	570
0160	5' wide		10	1.600		320	78.50		398.50	475
0210	10' high, multiples of 4' wide		8	2		193	98		291	360
0250	12' high, 3'-6" wide		7	2.286		214	112		326	405
0260	12' wide		6	2.667		223	131		354	445
0350	16' high, 3'-6" wide		7	2.286		213	112		325	405
0360	12' wide	↓	6	2.667	↓	223	131		354	445
0500	Maximum wind stoppage									
0550	10' high, multiples of 4' wide	2 Shee	8	2	L.F.	253	98		351	430
0650	14' high, multiples of 4' wide		8	2		375	98		473	565
0750	20' high, multiples of 8' wide	↓	6	2.667	↓	1,025	131		1,156	1,325
1100	Heated, maximum wind stoppage, steam heat									
1150	10' high, multiples of 4' wide	2 Shee	6	2.667	L.F.	895	131		1,026	1,175
1250	14' high, multiples of 4' wide		6	2.667		375	131		506	615
1350	20' high, multiples of 8' wide	↓	4	4	↓	1,025	196		1,221	1,425
1500	Customer entrance doors, unheated, minimal wind stoppage									
1550	10' high, multiples of 3' wide	2 Shee	6	2.667	L.F.	335	131		466	570
1560	5' wide		10	1.600		300	78.50		378.50	450
1650	Maximum wind stoppage, 12' high, multiples of 4' wide	↓	8	2	↓	350	98		448	535
1700	Heated, minimal wind stoppage, electric heat									
1750	8' high, multiples of 3' wide	2 Shee	6	2.667	L.F.	242	131		373	465
1760	Multiples of 5' wide		10	1.600		233	78.50		311.50	375
1850	10' high, multiples of 3' wide		6	2.667		855	131		986	1,150
1860	Multiples of 5' wide	↓	10	1.600	↓	560	78.50		638.50	735
1950	Maximum wind stoppage, steam heat									
1960	12' high, multiples of 4' wide	2 Shee	8	2	L.F.	630	98		728	840
2000	Walk-in coolers and freezers, ambient air, minimal wind stoppage									
2050	8' high, multiples of 3' wide	2 Shee	6	2.667	L.F.	365	131		496	600
2060	Multiples of 5' wide		10	1.600		400	78.50		478.50	560
2250	Maximum wind stoppage, 12' high, multiples of 3' wide		6	2.667		445	131		576	690
2450	Conveyor openings or service windows, unheated, 5' high		5	3.200		171	157		328	425
2460	Heated, electric, 5' high, 2'-6" wide	↓	5	3.200	↓	146	157		303	400

23 37 Air Outlets and Inlets

23 37 13 – Diffusers, Registers, and Grilles

23 37 13.10 Diffusers		Crew	Daily Output	Labor-Hours	Unit	Material	2010 Bare Costs Labor	Equipment	Total	Total Incl O&P
0010	**DIFFUSERS**, Aluminum, opposed blade damper unless noted									
0100	Ceiling, linear, also for sidewall									
0500	Perforated, 24" x 24" lay-in panel size, 6" x 6"	1 Shee	16	.500	Ea.	57.50	24.50		82	101
0520	8" x 8"		15	.533		59	26		85	105
0530	9" x 9"		14	.571		63	28		91	112
0540	10" x 10"		14	.571		65	28		93	114
0560	12" x 12"		12	.667		66	32.50		98.50	122
0590	16" x 16"		11	.727		109	35.50		144.50	174
0600	18" x 18"		10	.800		115	39.50		154.50	187
0610	20" x 20"		10	.800		133	39.50		172.50	206
0620	24" x 24"		9	.889		165	43.50		208.50	248
1000	Rectangular, 1 to 4 way blow, 6" x 6"		16	.500		46.50	24.50		71	88
1010	8" x 8"		15	.533		55.50	26		81.50	101
1014	9" x 9"		15	.533		61	26		87	107
1016	10" x 10"		15	.533		65	26		91	111
1020	12" x 6"		15	.533		67.50	26		93.50	114
1040	12" x 9"		14	.571		73.50	28		101.50	124
1060	12" x 12"		12	.667		73	32.50		105.50	130
1070	14" x 6"		13	.615		73	30		103	127
1074	14" x 14"		12	.667		103	32.50		135.50	163
1150	18" x 18"		9	.889		128	43.50		171.50	207
1160	21" x 21"		8	1		183	49		232	276
1170	24" x 12"		10	.800		136	39.50		175.50	210
1500	Round, butterfly damper, steel, diffuser size, 6" diameter		18	.444		23	22		45	58.50
1520	8" diameter		16	.500		25	24.50		49.50	64.50
1540	10" diameter		14	.571		30.50	28		58.50	76
1560	12" diameter		12	.667		40.50	32.50		73	94
1580	14" diameter		10	.800		50.50	39.50		90	115
2000	T bar mounting, 24" x 24" lay-in frame, 6" x 6"		16	.500		42.50	24.50		67	84
2020	8" x 8"		14	.571		44.50	28		72.50	91
2040	12" x 12"		12	.667		52	32.50		84.50	107
2060	16" x 16"		11	.727		70	35.50		105.50	131
2080	18" x 18"		10	.800		78	39.50		117.50	146
6000	For steel diffusers instead of aluminum, deduct					10%				

23 37 13.30 Grilles		Crew	Daily Output	Labor-Hours	Unit	Material	Labor	Equipment	Total	Total Incl O&P
0010	**GRILLES**									
0020	Aluminum									
1000	Air return, 6" x 6"	1 Shee	26	.308	Ea.	17.55	15.10		32.65	42.50
1020	10" x 6"		24	.333		17.55	16.35		33.90	44.50
1080	16" x 8"		22	.364		22	17.85		39.85	51.50
1100	12" x 12"		22	.364		24	17.85		41.85	53
1120	24" x 12"		18	.444		33	22		55	69
1220	24" x 18"		16	.500		41.50	24.50		66	82.50
1280	36" x 24"		14	.571		71	28		99	121
3000	Filter grille with filter, 12" x 12"		24	.333		47	16.35		63.35	76.50
3020	18" x 12"		20	.400		65.50	19.65		85.15	102
3040	24" x 18"		18	.444		77	22		99	118
3060	24" x 24"		16	.500		89	24.50		113.50	135
6000	For steel grilles instead of aluminum in above, deduct					10%				

23 37 13.60 Registers		Crew	Daily Output	Labor-Hours	Unit	Material	Labor	Equipment	Total	Total Incl O&P
0010	**REGISTERS**									
0980	Air supply									

23 37 13 – Diffusers, Registers, and Grilles

23 37 13.60 Registers

	23 37 13.60 Registers	Crew	Daily Output	Labor-Hours	Unit	2010 Bare Costs Material	Labor	Equipment	Total	Total Incl O&P
1000	Ceiling/wall, O.B. damper, anodized aluminum									
1010	One or two way deflection, adj. curved face bars									
1020	8" x 4"	1 Shee	26	.308	Ea.	32.50	15.10		47.60	59
1120	12" x 12"		18	.444		59.50	22		81.50	98.50
1240	20" x 6"		18	.444		53	22		75	91.50
1340	24" x 8"		13	.615		71.50	30		101.50	125
1350	24" x 18"		12	.667		133	32.50		165.50	196
2700	Above registers in steel instead of aluminum, deduct					10%				
4000	Floor, toe operated damper, enameled steel									
4020	4" x 8"	1 Shee	32	.250	Ea.	25.50	12.30		37.80	47
4100	8" x 10"		22	.364		31.50	17.85		49.35	61.50
4140	10" x 10"		20	.400		37.50	19.65		57.15	71
4220	14" x 14"		16	.500		132	24.50		156.50	182
4240	14" x 20"		15	.533		176	26		202	233
4980	Air return									
5000	Ceiling or wall, fixed 45° face blades									
5010	Adjustable O.B. damper, anodized aluminum									
5020	4" x 8"	1 Shee	26	.308	Ea.	24	15.10		39.10	49
5060	6" x 10"		19	.421		27	20.50		47.50	61.50
5280	24" x 24"		11	.727		92	35.50		127.50	155
5300	24" x 36"		8	1		144	49		193	234
6000	For steel construction instead of aluminum, deduct					10%				

23 37 15 – Louvers

23 37 15.40 HVAC Louvers

	23 37 15.40 HVAC Louvers	Crew	Daily Output	Labor-Hours	Unit	Material	Labor	Equipment	Total	Total Incl O&P
0010	**HVAC LOUVERS**									
0100	Aluminum, extruded, with screen, mill finish									
1002	Brick vent, see also Div. 04 05 23.19									
1100	Standard, 4" deep, 8" wide, 5" high	1 Shee	24	.333	Ea.	34.50	16.35		50.85	62.50
1200	Modular, 4" deep, 7-3/4" wide, 5" high		24	.333		36	16.35		52.35	64.50
1300	Speed brick, 4" deep, 11-5/8" wide, 3-7/8" high		24	.333		36	16.35		52.35	64.50
1400	Fuel oil brick, 4" deep, 8" wide, 5" high		24	.333		61.50	16.35		77.85	93
2000	Cooling tower and mechanical equip., screens, light weight		40	.200	S.F.	15.40	9.80		25.20	32
2020	Standard weight		35	.229		41	11.20		52.20	62
2500	Dual combination, automatic, intake or exhaust		20	.400		56	19.65		75.65	91
2520	Manual operation		20	.400		41.50	19.65		61.15	75.50
2540	Electric or pneumatic operation		20	.400		41.50	19.65		61.15	75.50
2560	Motor, for electric or pneumatic		14	.571	Ea.	480	28		508	575
3000	Fixed blade, continuous line									
3100	Mullion type, stormproof	1 Shee	28	.286	S.F.	41.50	14.05		55.55	67
3200	Stormproof		28	.286		41.50	14.05		55.55	67
3300	Vertical line		28	.286		49.50	14.05		63.55	75
3500	For damper to use with above, add					50%	30%			
3520	Motor, for damper, electric or pneumatic	1 Shee	14	.571	Ea.	480	28		508	575
4000	Operating, 45°, manual, electric or pneumatic		24	.333	S.F.	50	16.35		66.35	80
4100	Motor, for electric or pneumatic		14	.571	Ea.	480	28		508	575
4200	Penthouse, roof		56	.143	S.F.	24.50	7		31.50	37.50
4300	Walls		40	.200		58	9.80		67.80	79
5000	Thinline, under 4" thick, fixed blade		40	.200		24	9.80		33.80	41.50
5010	Finishes, applied by mfr. at additional cost, available in colors									
5020	Prime coat only, add				S.F.	3.30			3.30	3.63
5040	Baked enamel finish coating, add					6.10			6.10	6.70
5060	Anodized finish, add					6.60			6.60	7.25

23 37 Air Outlets and Inlets

23 37 15 – Louvers

23 37 15.40 HVAC Louvers		Crew	Daily Output	Labor-Hours	Unit	2010 Bare Costs Material	Labor	Equipment	Total	Total Incl O&P
5080	Duranodic finish, add				S.F.	12			12	13.20
5100	Fluoropolymer finish coating, add					18.90			18.90	21
9980	For small orders (under 10 pieces), add				↓	25%				

23 37 23 – HVAC Gravity Ventilators

23 37 23.10 HVAC Gravity Air Ventilators		Crew	Daily Output	Labor-Hours	Unit	Material	Labor	Equipment	Total	Total Incl O&P
0010	**HVAC GRAVITY AIR VENTILATORS**, Incl. base and damper									
1280	Rotary ventilators, wind driven, galvanized									
1300	4″ neck diameter	Q-9	20	.800	Ea.	50	35.50		85.50	109
1340	6″ neck diameter		16	1		56	44		100	129
1400	12″ neck diameter		10	1.600		81	70.50		151.50	197
1500	24″ neck diameter, 3,100 CFM		8	2		236	88.50		324.50	395
1540	36″ neck diameter, 5,500 CFM	↓	6	2.667	↓	695	118		813	945
2000	Stationary, gravity, syphon, galvanized									
2160	6″ neck diameter, 66 CFM	Q-9	16	1	Ea.	53	44		97	126
2240	12″ neck diameter, 160 CFM		10	1.600		91	70.50		161.50	207
2340	24″ neck diameter, 900 CFM		8	2		248	88.50		336.50	405
2380	36″ neck diameter, 2,000 CFM		6	2.667		775	118		893	1,025
4200	Stationary mushroom, aluminum, 16″ orifice diameter		10	1.600		535	70.50		605.50	695
4220	26″ orifice diameter		6.15	2.602		790	115		905	1,050
4230	30″ orifice diameter		5.71	2.802		1,150	124		1,274	1,475
4240	38″ orifice diameter		5	3.200		1,650	141		1,791	2,050
4250	42″ orifice diameter		4.70	3.404		2,200	150		2,350	2,650
4260	50″ orifice diameter	↓	4.44	3.604	↓	2,625	159		2,784	3,125
5000	Relief vent									
5500	Rectangular, aluminum, galvanized curb									
5510	intake/exhaust, 0.033″ SP									
5580	500 CFM, 12″ x 12″	Q-9	8.60	1.860	Ea.	610	82		692	795
5600	600 CFM, 12″ x 16″		8	2		685	88.50		773.50	890
5640	1000 CFM, 12″ x 24″		6.60	2.424		775	107		882	1,000
5680	3000 CFM, 20″ x 42″	↓	4	4	↓	1,350	177		1,527	1,775
5880	Size is throat area, volume is at 500 FPM									

23 38 Ventilation Hoods

23 38 13 – Commercial-Kitchen Hoods

23 38 13.10 Hood and Ventilation Equipment		Crew	Daily Output	Labor-Hours	Unit	Material	Labor	Equipment	Total	Total Incl O&P
0010	**HOOD AND VENTILATION EQUIPMENT**									
2970	Exhaust hood, sst, gutter on all sides, 4′ x 4′ x 2′	1 Carp	1.80	4.444	Ea.	4,325	185		4,510	5,050
2980	4′ x 4′ x 7′	″	1.60	5		6,800	208		7,008	7,800
7950	Hood fire protection system, minimum	Q-1	3	5.333		4,300	250		4,550	5,100
8050	Maximum	″	1	16	↓	33,800	750		34,550	38,300

23 41 Particulate Air Filtration

23 41 13 – Panel Air Filters

23 41 13.10 Panel Type Air Filters		Crew	Daily Output	Labor-Hours	Unit	2010 Bare Costs Material	Labor	Equipment	Total	Total Incl O&P
0010	**PANEL TYPE AIR FILTERS**									
2950	Mechanical media filtration units									
3000	High efficiency type, with frame, non-supported G				MCFM	45			45	49.50
3100	Supported type G				"	60			60	66
5500	Throwaway glass or paper media type				Ea.	3.35			3.35	3.69

23 41 16 – Renewable-Media Air Filters

23 41 16.10 Disposable Media Air Filters

		Crew	Daily Output	Labor-Hours	Unit	Material	Labor	Equipment	Total	Total Incl O&P
0010	**DISPOSABLE MEDIA AIR FILTERS**									
5000	Renewable disposable roll				MCFM	250			250	275

23 41 19 – Washable Air Filters

23 41 19.10 Permanent Air Filters

		Crew	Daily Output	Labor-Hours	Unit	Material	Labor	Equipment	Total	Total Incl O&P
0010	**PERMANENT AIR FILTERS**									
4500	Permanent washable G				MCFM	20			20	22

23 41 23 – Extended Surface Filters

23 41 23.10 Expanded Surface Filters

		Crew	Daily Output	Labor-Hours	Unit	Material	Labor	Equipment	Total	Total Incl O&P
0010	**EXPANDED SURFACE FILTERS**									
4000	Medium efficiency, extended surface G				MCFM	5.50			5.50	6.05

23 42 Gas-Phase Air Filtration

23 42 13 – Activated-Carbon Air Filtration

23 42 13.10 Charcoal Type Air Filtration

		Crew	Daily Output	Labor-Hours	Unit	Material	Labor	Equipment	Total	Total Incl O&P
0010	**CHARCOAL TYPE AIR FILTRATION**									
0050	Activated charcoal type, full flow				MCFM	600			600	660
0060	Full flow, impregnated media 12" deep				↓	225			225	248
0070	HEPA filter & frame for field erection				↓	300			300	330
0080	HEPA filter-diffuser, ceiling install.				↓	275			275	305

23 43 Electronic Air Cleaners

23 43 13 – Washable Electronic Air Cleaners

23 43 13.10 Electronic Air Cleaners

		Crew	Daily Output	Labor-Hours	Unit	Material	Labor	Equipment	Total	Total Incl O&P
0010	**ELECTRONIC AIR CLEANERS**									
2000	Electronic air cleaner, duct mounted									
2150	400 – 1000 CFM	1 Shee	2.30	3.478	Ea.	1,675	171		1,846	2,100
2200	1000 – 1400 CFM	↓	2.20	3.636	↓	1,525	179		1,704	1,950
2250	1400 – 2000 CFM	↓	2.10	3.810	↓	1,525	187		1,712	1,950

23 51 Breechings, Chimneys, and Stacks

23 51 13 – Draft Control Devices

23 51 13.13 Draft-Induction Fans		Crew	Daily Output	Labor-Hours	Unit	Material	2010 Bare Costs Labor	Equipment	Total	Total Incl O&P
0010	**DRAFT-INDUCTION FANS**									
1000	Breeching installation									
1800	Hot gas, 600°F, variable pitch pulley and motor									
1860	8" diam. inlet, 1/4 H.P., 1phase, 1120 CFM	Q-9	4	4	Ea.	3,100	177		3,277	3,700
1900	12" diam. inlet, 3/4 H.P., 3 phase, 2960 CFM		3	5.333		4,250	236		4,486	5,025
1940	18" diam. inlet, 3 H.P., 3 phase, 9120 CFM		2	8		6,600	355		6,955	7,800
1980	24" diam. inlet, 7-1/2 H.P., 3 phase, 17,760 CFM		.80	20		10,400	885		11,285	12,800
2300	For multi-blade damper at fan inlet, add					20%				

23 51 23 – Gas Vents

23 51 23.10 Gas Chimney Vents		Crew	Daily Output	Labor-Hours	Unit	Material	Labor	Equipment	Total	Total Incl O&P
0010	**GAS CHIMNEY VENTS,** Prefab metal, U.L. listed									
0020	Gas, double wall, galvanized steel									
0080	3" diameter	Q-9	72	.222	V.L.F.	5.35	9.80		15.15	21
0100	4" diameter		68	.235		6.70	10.40		17.10	23
0120	5" diameter		64	.250		7.80	11.05		18.85	25.50
0140	6" diameter		60	.267		9.10	11.80		20.90	28
0160	7" diameter		56	.286		13.35	12.65		26	34
0180	8" diameter		52	.308		14.85	13.60		28.45	37
0200	10" diameter		48	.333		34	14.75		48.75	59.50
0220	12" diameter		44	.364		45	16.05		61.05	74
0260	16" diameter		40	.400		103	17.70		120.70	140
0300	20" diameter	Q-10	36	.667		139	30.50		169.50	200
0480	38" diameter	"	24	1		500	46		546	620

23 51 26 – All-Fuel Vent Chimneys

23 51 26.30 All-Fuel Vent Chimneys, Double Wall, St. Stl.		Crew	Daily Output	Labor-Hours	Unit	Material	Labor	Equipment	Total	Total Incl O&P
0010	**ALL-FUEL VENT CHIMNEYS, DOUBLE WALL, STAINLESS STEEL**									
7800	All fuel, double wall, stainless steel, 6" diameter	Q-9	60	.267	V.L.F.	56.50	11.80		68.30	80
7802	7" diameter		56	.286		73	12.65		85.65	99.50
7804	8" diameter		52	.308		86.50	13.60		100.10	116
7806	10" diameter		48	.333		125	14.75		139.75	160
7808	12" diameter		44	.364		167	16.05		183.05	209
7810	14" diameter		42	.381		219	16.85		235.85	267

23 51 33 – Insulated Sectional Chimneys

23 51 33.10 Prefabricated Insulated Sectional Chimneys		Crew	Daily Output	Labor-Hours	Unit	Material	Labor	Equipment	Total	Total Incl O&P
0010	**PREFABRICATED INSULATED SECTIONAL CHIMNEYS**									
9000	High temp. (2000°F), steel jacket, acid resistant refractory lining									
9010	11 ga. galvanized jacket, U.L. listed									
9020	Straight section, 48" long, 10" diameter	Q-10	13.30	1.805	Ea.	400	82.50		482.50	565
9040	18" diameter		7.40	3.243		615	149		764	900
9070	36" diameter		2.70	8.889		1,325	405		1,730	2,075
9090	48" diameter	Q-11	2.70	11.852		1,925	555		2,480	2,950
9120	Tee section, 10" diameter	Q-10	4.40	5.455		855	250		1,105	1,325
9140	18" diameter		2.40	10		1,225	460		1,685	2,050
9170	36" diameter		.80	30		3,975	1,375		5,350	6,450
9190	48" diameter	Q-11	.90	35.556		6,775	1,675		8,450	9,975
9220	Cleanout pier section, 10" diameter	Q-10	3.50	6.857		655	315		970	1,200
9240	18" diameter		1.90	12.632		885	580		1,465	1,850
9270	36" diameter		.75	32		1,625	1,475		3,100	4,000
9290	48" diameter	Q-11	.75	42.667		2,475	2,000		4,475	5,750
9320	For drain, add					53%				
9330	Elbow, 30° and 45°, 10" diameter	Q-10	6.60	3.636		555	167		722	860

23 51 Breechings, Chimneys, and Stacks

23 51 33 – Insulated Sectional Chimneys

23 51 33.10 Prefabricated Insulated Sectional Chimneys		Crew	Daily Output	Labor-Hours	Unit	Material	2010 Bare Costs Labor	Equipment	Total	Total Incl O&P
9350	18" diameter	Q-10	3.70	6.486	Ea.	900	297		1,197	1,450
9380	36" diameter		1.30	18.462		2,225	845		3,070	3,725
9400	48" diameter	Q-11	1.35	23.704		3,900	1,100		5,000	5,950
9430	For 60° and 90° elbow, add					112%				
9440	End cap, 10" diameter	Q-10	26	.923		955	42.50		997.50	1,125
9460	18" diameter		15	1.600		1,350	73.50		1,423.50	1,575
9490	36" diameter		5	4.800		2,350	220		2,570	2,925
9510	48" diameter	Q-11	5.30	6.038		3,475	282		3,757	4,225
9540	Increaser (1 diameter), 10" diameter	Q-10	6.60	3.636		475	167		642	770
9560	18" diameter		3.70	6.486		740	297		1,037	1,275
9590	36" diameter		1.30	18.462		1,450	845		2,295	2,875
9592	42" diameter	Q-11	1.60	20		1,600	935		2,535	3,175
9594	48" diameter		1.35	23.704		1,925	1,100		3,025	3,775
9596	54" diameter		1.10	29.091		2,325	1,350		3,675	4,600
9600	For expansion joints, add to straight section					7%				
9610	For 1/4" hot rolled steel jacket, add					157%				
9620	For 2950°F very high temperature, add					89%				
9630	26 ga. aluminized jacket, straight section, 48" long									
9640	10" diameter	Q-10	15.30	1.569	V.L.F.	237	72		309	370
9660	18" diameter		8.50	2.824		365	129		494	600
9690	36" diameter		3.10	7.742		845	355		1,200	1,450
9810	Draw band, galv. stl., 11 gauge, 10" diameter		32	.750	Ea.	65	34.50		99.50	124
9820	12" diameter		30	.800		69	36.50		105.50	131
9830	18" diameter		26	.923		86	42.50		128.50	159
9860	36" diameter		18	1.333		146	61		207	253
9880	48" diameter	Q-11	20	1.600		251	75		326	390

23 52 Heating Boilers

23 52 13 – Electric Boilers

23 52 13.10 Electric Boilers, ASME

		Crew	Daily Output	Labor-Hours	Unit	Material	Labor	Equipment	Total	Total Incl O&P
0010	**ELECTRIC BOILERS, ASME**, Standard controls and trim.									
1000	Steam, 6 KW, 20.5 MBH	Q-19	1.20	20	Ea.	3,500	950		4,450	5,275
1160	60 KW, 205 MBH		1	24		6,025	1,150		7,175	8,325
1220	112 KW, 382 MBH		.75	32		8,250	1,525		9,775	11,400
1280	222 KW, 758 MBH		.55	43.636		13,100	2,075		15,175	17,500
1380	518 KW, 1768 MBH	Q-21	.36	88.889		22,200	4,325		26,525	30,900
1480	814 KW, 2778 MBH		.25	128		31,000	6,225		37,225	43,400
1600	2,340 KW, 7984 MBH		.16	200		63,500	9,725		73,225	84,500
2000	Hot water, 7.5 KW, 25.6 MBH	Q-19	1.30	18.462		3,750	875		4,625	5,425
2100	90 KW, 307 MBH		1.10	21.818		5,475	1,025		6,500	7,575
2220	296 KW, 1010 MBH		.55	43.636		11,600	2,075		13,675	15,800
2500	1036 KW, 3536 MBH	Q-21	.34	94.118		26,900	4,575		31,475	36,500
2680	2400 KW, 8191 MBH		.25	128		56,000	6,225		62,225	71,000
2820	3600 KW, 12,283 MBH		.16	200		79,000	9,725		88,725	101,500

23 52 23 – Cast-Iron Boilers

23 52 23.20 Gas-Fired Boilers

		Crew	Daily Output	Labor-Hours	Unit	Material	Labor	Equipment	Total	Total Incl O&P
0010	**GAS-FIRED BOILERS**, Natural or propane, standard controls, packaged									
1000	Cast iron, with insulated jacket									
2000	Steam, gross output, 81 MBH	Q-7	1.40	22.857	Ea.	1,950	1,125		3,075	3,850
2080	203 MBH		.90	35.556		3,100	1,750		4,850	6,050

23 52 Heating Boilers

23 52 23 – Cast-Iron Boilers

23 52 23.20 Gas-Fired Boilers

Line	23 52 23.20 Gas-Fired Boilers	Crew	Daily Output	Labor-Hours	Unit	2010 Bare Costs Material	Labor	Equipment	Total	Total Incl O&P
2180	400 MBH	Q-7	.56	56.838	Ea.	5,200	2,800		8,000	9,925
2240	765 MBH		.43	74.419		10,200	3,675		13,875	16,800
2320	1,875 MBH		.30	106		18,300	5,275		23,575	28,000
2440	4,720 MBH		.15	207		30,700	10,300		41,000	49,200
2480	6,100 MBH		.13	246		77,000	12,200		89,200	102,500
2540	6,970 MBH		.10	320		86,500	15,800		102,300	118,500
3000	Hot water, gross output, 80 MBH		1.46	21.918		1,775	1,075		2,850	3,600
3140	320 MBH *CN*		.80	40		4,200	1,975		6,175	7,600
3260	1,088 MBH		.40	80		12,400	3,950		16,350	19,500
3360	2,856 MBH		.20	160		21,500	7,900		29,400	35,600
3380	3,264 MBH		.18	179		22,900	8,875		31,775	38,500
3480	6,100 MBH		.13	250		89,000	12,400		101,400	116,000
3540	6,970 MBH		.09	359		94,000	17,800		111,800	130,000
7000	For tankless water heater, add					10%				

23 52 23.30 Gas/Oil Fired Boilers

Line	23 52 23.30 Gas/Oil Fired Boilers	Crew	Daily Output	Labor-Hours	Unit	Material	Labor	Equipment	Total	Total Incl O&P
0010	**GAS/OIL FIRED BOILERS,** Combination with burners and controls, packaged									
1000	Cast iron with insulated jacket									
2000	Steam, gross output, 720 MBH	Q-7	.43	74.074	Ea.	13,900	3,650		17,550	20,800
2080	1,600 MBH		.30	107		17,000	5,300		22,300	26,600
2140	2,700 MBH		.19	165		27,500	8,200		35,700	42,600
2280	5,520 MBH		.14	235		84,000	11,600		95,600	109,500
2340	6,390 MBH		.11	296		90,000	14,600		104,600	120,500
2380	6,970 MBH		.09	372		95,000	18,400		113,400	132,000
2900	Hot water, gross output									
2910	200 MBH	Q-6	.62	39.024	Ea.	7,775	1,900		9,675	11,400
2920	300 MBH		.49	49.080		7,775	2,375		10,150	12,100
2930	400 MBH		.41	57.971		9,100	2,800		11,900	14,200
2940	500 MBH		.36	67.039		9,800	3,250		13,050	15,700
3000	584 MBH	Q-7	.44	72.072		8,125	3,575		11,700	14,300
3060	1,460 MBH		.28	113		22,900	5,600		28,500	33,600
3160	4,088 MBH		.16	195		45,500	9,650		55,150	64,500
3300	13,500 MBH, 403.3 BHP		.04	727		140,500	35,900		176,400	208,500

23 52 23.40 Oil-Fired Boilers

Line	23 52 23.40 Oil-Fired Boilers	Crew	Daily Output	Labor-Hours	Unit	Material	Labor	Equipment	Total	Total Incl O&P
0010	**OIL-FIRED BOILERS,** Standard controls, flame retention burner, packaged									
1000	Cast iron, with insulated flush jacket									
2000	Steam, gross output, 109 MBH	Q-7	1.20	26.667	Ea.	1,975	1,325		3,300	4,150
2060	207 MBH		.90	35.556		2,700	1,750		4,450	5,600
2180	1,084 MBH		.38	85.106		9,400	4,200		13,600	16,600
2280	3,000 MBH		.19	170		21,600	8,425		30,025	36,300
2380	5,520 MBH		.14	235		68,500	11,600		80,100	93,000
2460	6,970 MBH		.09	363		88,000	18,000		106,000	124,000
3000	Hot water, same price as steam									
4000	For tankless coil in smaller sizes, add				Ea.	15%				

23 52 26 – Steel Boilers

23 52 26.40 Oil-Fired Boilers

Line	23 52 26.40 Oil-Fired Boilers	Crew	Daily Output	Labor-Hours	Unit	Material	Labor	Equipment	Total	Total Incl O&P
0010	**OIL-FIRED BOILERS,** Standard controls, flame retention burner									
5000	Steel, with insulated flush jacket									
7000	Hot water, gross output, 103 MBH	Q-6	1.60	15	Ea.	1,725	725		2,450	3,000
7120	420 MBH		.70	34.483		5,975	1,675		7,650	9,075
7320	3,150 MBH		.13	184		31,600	8,950		40,550	48,200
7340	For tankless coil in steam or hot water, add					7%				

23 52 Heating Boilers

23 52 28 – Swimming Pool Boilers

23 52 28.10 Swimming Pool Heaters		Crew	Daily Output	Labor-Hours	Unit	Material	2010 Bare Costs Labor	Equipment	Total	Total Incl O&P
0010	**SWIMMING POOL HEATERS**, Not including wiring, external									
0020	piping, base or pad,									
0160	Gas fired, input, 155 MBH	Q-6	1.50	16	Ea.	1,575	775		2,350	2,900
0200	199 MBH		1	24		1,700	1,150		2,850	3,625
0280	500 MBH		.40	60		7,225	2,900		10,125	12,300
0400	1,800 MBH	↓	.14	171		15,700	8,300		24,000	29,700
2000	Electric, 12 KW, 4,800 gallon pool	Q-19	3	8		2,025	380		2,405	2,800
2020	15 KW, 7,200 gallon pool		2.80	8.571		2,050	405		2,455	2,850
2040	24 KW, 9,600 gallon pool		2.40	10		2,375	475		2,850	3,300
2100	57 KW, 24,000 gallon pool	↓	1.20	20	↓	3,475	950		4,425	5,250

23 54 Furnaces

23 54 13 – Electric-Resistance Furnaces

23 54 13.10 Electric Furnaces

		Crew	Daily Output	Labor-Hours	Unit	Material	Labor	Equipment	Total	Total Incl O&P
0010	**ELECTRIC FURNACES**, Hot air, blowers, std. controls									
0011	not including gas, oil or flue piping									
1000	Electric, UL listed									
1020	10.2 MBH	Q-20	5	4	Ea.	475	181		656	790
1040	17.1 MBH		4.80	4.167		495	188		683	830
1060	27.3 MBH		4.60	4.348		600	196		796	955
1100	34.1 MBH	↓	4.40	4.545	↓	615	205		820	985

23 54 16 – Fuel-Fired Furnaces

23 54 16.13 Gas-Fired Furnaces

		Crew	Daily Output	Labor-Hours	Unit	Material	Labor	Equipment	Total	Total Incl O&P
0010	**GAS-FIRED FURNACES**									
3000	Gas, AGA certified, upflow, direct drive models									
3020	45 MBH input	Q-9	4	4	Ea.	575	177		752	905
3040	60 MBH input		3.80	4.211		615	186		801	955
3060	75 MBH input		3.60	4.444		685	196		881	1,050
3100	100 MBH input	↓	3.20	5	↓	695	221		916	1,100

23 54 16.16 Oil-Fired Furnaces

		Crew	Daily Output	Labor-Hours	Unit	Material	Labor	Equipment	Total	Total Incl O&P
0010	**OIL-FIRED FURNACES**									
6000	Oil, UL listed, atomizing gun type burner									
6020	56 MBH output	Q-9	3.60	4.444	Ea.	1,850	196		2,046	2,325
6030	84 MBH output		3.50	4.571		1,850	202		2,052	2,350
6040	95 MBH output		3.40	4.706		1,875	208		2,083	2,400
6060	134 MBH output		3.20	5		1,950	221		2,171	2,475
6080	151 MBH output	↓	3	5.333	↓	2,275	236		2,511	2,850

23 55 Fuel-Fired Heaters

23 55 13 – Fuel-Fired Duct Heaters

23 55 13.16 Gas-Fired Duct Heaters		Crew	Daily Output	Labor-Hours	Unit	Material	2010 Bare Costs Labor	Equipment	Total	Total Incl O&P
0010	**GAS-FIRED DUCT HEATERS**, Includes burner, controls, stainless steel									
0020	heat exchanger. Gas fired, electric ignition									
0030	Indoor installation									
0100	120 MBH output	Q-5	4	4	Ea.	2,125	187		2,312	2,625
0130	200 MBH output		2.70	5.926		2,850	277		3,127	3,550
0140	240 MBH output		2.30	6.957		2,975	325		3,300	3,725
0180	320 MBH output		1.60	10		3,550	465		4,015	4,625
0300	For powered venter and adapter, add					460			460	510
0502	For required flue pipe, see Div. 23 51 23.10									
1000	Outdoor installation, with power venter									
1020	75 MBH output	Q-5	4	4	Ea.	2,775	187		2,962	3,325
1060	120 MBH output		4	4		3,525	187		3,712	4,175
1100	187 MBH output		3	5.333		5,050	249		5,299	5,925
1140	300 MBH output		1.80	8.889		6,100	415		6,515	7,325
1180	450 MBH output		1.40	11.429		7,450	535		7,985	9,000

23 55 33 – Fuel-Fired Unit Heaters

23 55 33.13 Oil-Fired Unit Heaters

		Crew	Daily Output	Labor-Hours	Unit	Material	Labor	Equipment	Total	Total Incl O&P
0010	**OIL-FIRED UNIT HEATERS**, Cabinet, grilles, fan, ctrl, burner, no piping									
6000	Oil fired, suspension mounted, 94 MBH output	Q-5	4	4	Ea.	3,400	187		3,587	4,000
6040	140 MBH output		3	5.333		3,625	249		3,874	4,350
6060	184 MBH output		3	5.333		3,900	249		4,149	4,675

23 55 33.16 Gas-Fired Unit Heaters

		Crew	Daily Output	Labor-Hours	Unit	Material	Labor	Equipment	Total	Total Incl O&P
0010	**GAS-FIRED UNIT HEATERS**, Cabinet, grilles, fan, ctrls., burner, no piping									
0022	thermostat, no piping. For flue see Div. 23 51 23.10									
1000	Gas fired, floor mounted									
1100	60 MBH output	Q-5	10	1.600	Ea.	805	74.50		879.50	995
1140	100 MBH output		8	2		890	93.50		983.50	1,125
1180	180 MBH output		6	2.667		1,275	125		1,400	1,575
2000	Suspension mounted, propeller fan, 20 MBH output		8.50	1.882		985	88		1,073	1,200
2040	60 MBH output		7	2.286		1,200	107		1,307	1,475
2060	80 MBH output		6	2.667		1,225	125		1,350	1,525
2100	130 MBH output		5	3.200		1,500	149		1,649	1,875
2240	320 MBH output		2	8		2,650	375		3,025	3,450
2500	For powered venter and adapter, add					355			355	390
5000	Wall furnace, 17.5 MBH output	Q-5	6	2.667		715	125		840	970
5020	24 MBH output		5	3.200		740	149		889	1,050
5040	35 MBH output		4	4		895	187		1,082	1,275

23 56 Solar Energy Heating Equipment

23 56 16 – Packaged Solar Heating Equipment

23 56 16.40 Solar Heating Systems

		Crew	Daily Output	Labor-Hours	Unit	Material	Labor	Equipment	Total	Total Incl O&P
0010	**SOLAR HEATING SYSTEMS** R235616-60									
0020	System/Package prices, not including connecting									
0030	pipe, insulation, or special heating/plumbing fixtures G									
0500	Hot water, standard package, low temperature G									
0540	1 collector, circulator, fittings, 65 gal. tank G	Q-1	.50	32	Ea.	1,500	1,500		3,000	3,900
0580	2 collectors, circulator, fittings, 120 gal. tank G		.40	40		2,325	1,875		4,200	5,350
0620	3 collectors, circulator, fittings, 120 gal. tank G		.34	47.059		3,175	2,200		5,375	6,775
0700	Medium temperature package G									
0720	1 collector, circulator, fittings, 80 gal. tank G	Q-1	.50	32	Ea.	2,350	1,500		3,850	4,825

23 56 Solar Energy Heating Equipment

23 56 16 – Packaged Solar Heating Equipment

23 56 16.40 Solar Heating Systems			Crew	Daily Output	Labor-Hours	Unit	2010 Bare Costs Material	Labor	Equipment	Total	Total Incl O&P
0740	2 collectors, circulator, fittings, 120 gal. tank	G	Q-1	.40	40	Ea.	3,300	1,875		5,175	6,450
0780	3 collectors, circulator, fittings, 120 gal. tank	G		.30	53.333		4,300	2,500		6,800	8,475
0980	For each additional 120 gal. tank, add	G					1,475			1,475	1,625

23 56 19 – Solar Heating Components

23 56 19.50 Solar Heating Ancillary			Crew	Daily Output	Labor-Hours	Unit	2010 Bare Costs Material	Labor	Equipment	Total	Total Incl O&P
0010	**SOLAR HEATING ANCILLARY**										
2300	Circulators, air	G									
2310	Blowers										
2400	Reversible fan, 20″ diameter, 2 speed	G	Q-9	18	.889	Ea.	108	39.50		147.50	179
2870	1/12 HP, 30 GPM	G	Q-1	10	1.600	″	330	75		405	475
3000	Collector panels, air with aluminum absorber plate										
3010	Wall or roof mount										
3040	Flat black, plastic glazing										
3080	4′ x 8′	G	Q-9	6	2.667	Ea.	635	118		753	875
3200	Flush roof mount, 10′ to 16′ x 22″ wide	G	″	96	.167	L.F.	450	7.35		457.35	505
3300	Collector panels, liquid with copper absorber plate										
3330	Alum. frame, 4′ x 8′, 5/32″ single glazing	G	Q-1	9.50	1.684	Ea.	945	79		1,024	1,175
3390	Alum. frame, 4′ x 10′, 5/32″ single glazing	G		6	2.667		1,075	125		1,200	1,375
3450	Flat black, alum. frame, 3.5′ x 7.5′	G		9	1.778		690	83.50		773.50	885
3500	4′ x 8′	G		5.50	2.909		865	136		1,001	1,150
3520	4′ x 10′	G		10	1.600		970	75		1,045	1,175
3540	4′ x 12.5′	G		5	3.200		1,175	150		1,325	1,500
3600	Liquid, full wetted, plastic, alum. frame, 3′ x 10′	G		5	3.200		234	150		384	480
3650	Collector panel mounting, flat roof or ground rack	G		7	2.286		217	107		324	400
3670	Roof clamps	G		70	.229	Set	2.35	10.70		13.05	18.65
3700	Roof strap, teflon	G	1 Plum	205	.039	L.F.	19.95	2.03		21.98	25
3900	Differential controller with two sensors										
3930	Thermostat, hard wired	G	1 Plum	8	1	Ea.	79.50	52		131.50	165
4100	Five station with digital read-out	G	″	3	2.667	″	218	139		357	445
4300	Heat exchanger										
4580	Fluid to fluid package includes two circulating pumps										
4590	expansion tank, check valve, relief valve										
4600	controller, high temperature cutoff and sensors	G	Q-1	2.50	6.400	Ea.	695	300		995	1,225
4650	Heat transfer fluid										
4700	Propylene glycol, inhibited anti-freeze	G	1 Plum	28	.286	Gal.	12.85	14.85		27.70	36.50
8250	Water storage tank with heat exchanger and electric element										
8300	80 gal. with 2″ x 2 lb. density insulation	G	1 Plum	1.60	5	Ea.	1,100	260		1,360	1,600
8380	120 gal. with 2″ x 2 lb. density insulation	G		1.40	5.714		1,225	297		1,522	1,800
8400	120 gal. with 2″ x 2 lb. density insul., 40 S.F. heat coil	G		1.40	5.714		1,575	297		1,872	2,175

23 57 Heat Exchangers for HVAC

23 57 16 – Steam-to-Water Heat Exchangers

23 57 16.10 Shell/Tube Type Steam-to-Water Heat Exch.		Crew	Daily Output	Labor-Hours	Unit	Material	Labor	Equipment	Total	Total Incl O&P
0010	**SHELL AND TUBE TYPE STEAM-TO-WATER HEAT EXCHANGERS**									
0016	Shell & tube type, 2 or 4 pass, 3/4″ O.D. copper tubes,									
0020	C.I. heads, C.I. tube sheet, steel shell									
0100	Hot water 40°F to 180°F, by steam at 10 PSI									
0120	8 GPM	Q-5	6	2.667	Ea.	1,725	125		1,850	2,075
0140	10 GPM		5	3.200		2,600	149		2,749	3,100
0160	40 GPM		4	4		4,050	187		4,237	4,725

23 57 Heat Exchangers for HVAC

23 57 16 – Steam-to-Water Heat Exchangers

23 57 16.10 Shell/Tube Type Steam-to-Water Heat Exch.		Crew	Daily Output	Labor-Hours	Unit	Material	2010 Bare Costs Labor	Equipment	Total	Total Incl O&P
0180	64 GPM	Q-5	2	8	Ea.	6,200	375		6,575	7,375
0200	96 GPM	↓	1	16		8,300	745		9,045	10,300
0220	120 GPM	Q-6	1.50	16	↓	10,900	775		11,675	13,200

23 57 19 – Liquid-to-Liquid Heat Exchangers

23 57 19.13 Plate-Type, Liquid-to-Liquid Heat Exchangers		Crew	Daily Output	Labor-Hours	Unit	Material	Labor	Equipment	Total	Total Incl O&P
0010	**PLATE-TYPE, LIQUID-TO-LIQUID HEAT EXCHANGERS**									
3000	Plate type,									
3100	400 GPM	Q-6	.80	30	Ea.	31,400	1,450		32,850	36,700
3120	800 GPM	"	.50	48		54,000	2,325		56,325	63,000
3140	1200 GPM	Q-7	.34	94.118		80,500	4,650		85,150	95,500
3160	1800 GPM	"	.24	133	↓	106,500	6,600		113,100	127,000

23 57 19.16 Shell-Type, Liquid-to-Liquid Heat Exchangers		Crew	Daily Output	Labor-Hours	Unit	Material	Labor	Equipment	Total	Total Incl O&P
0010	**SHELL-TYPE, LIQUID-TO-LIQUID HEAT EXCHANGERS**									
1000	Hot water 40°F to 140°F, by water at 200°F									
1020	7 GPM	Q-5	6	2.667	Ea.	2,125	125		2,250	2,525
1040	16 GPM		5	3.200		3,025	149		3,174	3,550
1060	34 GPM		4	4		4,575	187		4,762	5,300
1100	74 GPM	↓	1.50	10.667	↓	8,275	500		8,775	9,850

23 62 Packaged Compressor and Condenser Units

23 62 13 – Packaged Air-Cooled Refrigerant Compressor and Condenser Units

23 62 13.10 Packaged Air-Cooled Refrig. Condensing Units		Crew	Daily Output	Labor-Hours	Unit	Material	Labor	Equipment	Total	Total Incl O&P
0010	**PACKAGED AIR-COOLED REFRIGERANT CONDENSING UNITS**									
0020	Condensing unit									
0030	Air cooled, compressor, standard controls									
0050	1.5 ton	Q-5	2.50	6.400	Ea.	1,175	299		1,474	1,750
0500	5 ton		.60	26.667		2,125	1,250		3,375	4,225
0600	10 ton	↓	.50	32		4,750	1,500		6,250	7,475
0700	20 ton	Q-6	.40	60	↓	10,900	2,900		13,800	16,400

23 63 Refrigerant Condensers

23 63 13 – Air-Cooled Refrigerant Condensers

23 63 13.10 Air-Cooled Refrig. Condensers		Crew	Daily Output	Labor-Hours	Unit	Material	Labor	Equipment	Total	Total Incl O&P
0010	**AIR-COOLED REFRIG. CONDENSERS**, Ratings for 30°F TD, R-22.									
0080	Air cooled, belt drive, propeller fan									
0240	50 ton	Q-6	.69	34.985	Ea.	12,700	1,700		14,400	16,600
0280	59 ton		.58	41.308		14,200	2,000		16,200	18,600
0320	73 ton		.47	51.173		16,700	2,475		19,175	22,100
0360	86 ton		.40	60.302		19,800	2,925		22,725	26,200
0380	88 ton	↓	.39	61.697	↓	21,200	3,000		24,200	27,800
1550	Air cooled, direct drive, propeller fan									
1590	1 ton	Q-5	3.80	4.211	Ea.	685	197		882	1,050
1600	1-1/2 ton		3.60	4.444		840	208		1,048	1,225
1620	2 ton		3.20	5		900	234		1,134	1,350
1640	5 ton		2	8		1,825	375		2,200	2,575
1660	10 ton		1.40	11.429		3,025	535		3,560	4,125
1690	16 ton		1.10	14.545		4,150	680		4,830	5,600
1720	26 ton	↓	.84	19.002	↓	5,400	885		6,285	7,275

23 63 Refrigerant Condensers

23 63 13 – Air-Cooled Refrigerant Condensers

23 63 13.10 Air-Cooled Refrig. Condensers		Crew	Daily Output	Labor-Hours	Unit	2010 Bare Costs Material	Labor	Equipment	Total	Total Incl O&P
1760	41 ton	Q-6	.77	31.008	Ea.	9,700	1,500		11,200	13,000
1800	63 ton	"	.55	44.037	↓	15,300	2,125		17,425	20,000

23 64 Packaged Water Chillers

23 64 13 – Absorption Water Chillers

23 64 13.16 Indirect-Fired Absorption Water Chillers

Line	Description	Crew	Daily Output	Labor-Hours	Unit	Material	Labor	Equipment	Total	Total Incl O&P
0010	**INDIRECT-FIRED ABSORPTION WATER CHILLERS**									
0020	Steam or hot water, water cooled									
0050	100 ton	Q-7	.13	240	Ea.	125,000	11,900		136,900	155,500
0400	420 ton	"	.10	323	"	325,500	16,000		341,500	382,000

23 64 16 – Centrifugal Water Chillers

23 64 16.10 Centrifugal Type Water Chillers

Line	Description	Crew	Daily Output	Labor-Hours	Unit	Material	Labor	Equipment	Total	Total Incl O&P
0010	**CENTRIFUGAL TYPE WATER CHILLERS**, With standard controls									
0020	Centrifugal liquid chiller, water cooled									
0030	not including water tower									
0100	2000 ton (twin 1000 ton units)	Q-7	.07	477	Ea.	832,500	23,600		856,100	951,500

23 64 19 – Reciprocating Water Chillers

23 64 19.10 Reciprocating Type Water Chillers

Line	Description	Crew	Daily Output	Labor-Hours	Unit	Material	Labor	Equipment	Total	Total Incl O&P
0010	**RECIPROCATING TYPE WATER CHILLERS**, With standard controls									
0494	Water chillers, integral air cooled condenser									
0600	100 ton cooling	Q-7	.25	129	Ea.	71,500	6,375		77,875	88,000
0980	Water cooled, multiple compressor, semi-hermetic, tower not incl.									
1000	15 ton cooling	Q-6	.36	65.934	Ea.	16,000	3,200		19,200	22,400
1020	25 ton cooling	Q-7	.41	78.049		17,900	3,850		21,750	25,500
1060	35 ton cooling		.31	101		20,400	5,025		25,425	30,000
1100	50 ton cooling		.28	113		30,500	5,625		36,125	41,900
1160	100 ton cooling		.18	179		58,500	8,875		67,375	78,000
1180	125 ton cooling		.16	196		66,500	9,700		76,200	87,500
1200	145 ton cooling	↓	.16	202	↓	77,500	10,000		87,500	100,500
1451	Water cooled, dual compressors, semi-hermetic, tower not incl.									
1500	80 ton cooling	Q-7	.14	222	Ea.	27,400	11,000		38,400	46,600
1520	100 ton cooling		.14	228		37,000	11,300		48,300	57,500
1540	120 ton cooling	↓	.14	231	↓	44,800	11,500		56,300	66,500

23 64 23 – Scroll Water Chillers

23 64 23.10 Scroll Water Chillers

Line	Description	Crew	Daily Output	Labor-Hours	Unit	Material	Labor	Equipment	Total	Total Incl O&P
0010	**SCROLL WATER CHILLERS,** With standard controls									
0490	Packaged w/integral air cooled condenser, 15 ton cool	Q-7	.37	86.486	Ea.	17,400	4,275		21,675	25,500
0500	20 ton cooling		.34	94.118		23,600	4,650		28,250	32,900
0520	40 ton cooling	↓	.30	108	↓	31,700	5,350		37,050	42,800
0680	Scroll water cooled, single compressor, hermetic, tower not incl.									
0700	2 ton cooling	Q-5	.57	28.070	Ea.	3,225	1,300		4,525	5,525
0710	5 ton cooling		.57	28.070		3,850	1,300		5,150	6,225
0740	8 ton cooling	↓	.31	52.117		5,325	2,425		7,750	9,500
0760	10 ton cooling	Q-6	.36	67.039		6,125	3,250		9,375	11,600
0800	20 ton cooling	Q-7	.38	83.990		10,900	4,150		15,050	18,200
0820	30 ton cooling	"	.33	96.096	↓	12,200	4,750		16,950	20,600

23 64 26 – Rotary-Screw Water Chillers

23 64 26.10 Rotary-Screw Type Water Chillers

Line	Description	Crew	Daily Output	Labor-Hours	Unit	Material	Labor	Equipment	Total	Total Incl O&P
0010	**ROTARY-SCREW TYPE WATER CHILLERS**, With standard controls									

23 64 Packaged Water Chillers

23 64 26 – Rotary-Screw Water Chillers

23 64 26.10 Rotary-Screw Type Water Chillers		Crew	Daily Output	Labor-Hours	Unit	Material	2010 Bare Costs Labor	Equipment	Total	Total Incl O&P
0110	Screw, liquid chiller, air cooled, insulated evaporator									
0120	130 ton	Q-7	.14	228	Ea.	81,000	11,300		92,300	106,000
0124	160 ton		.13	246		100,000	12,200		112,200	128,000
0128	180 ton		.13	250		112,500	12,400		124,900	142,000
0132	210 ton		.12	258		123,500	12,800		136,300	155,000
0136	270 ton		.12	266		141,500	13,200		154,700	175,500
0140	320 ton		.12	275		177,500	13,600		191,100	215,500
0200	Packaged unit, water cooled, not incl. tower									
0210	80 ton	Q-7	.14	223	Ea.	42,100	11,100		53,200	63,000
0240	200 ton		.13	251		82,000	12,500		94,500	108,500
0270	350 ton		.12	275		156,000	13,600		169,600	192,000
1450	Water cooled, tower not included									
1580	150 ton cooling, screw compressors	Q-7	.13	240	Ea.	63,500	11,900		75,400	88,000
1620	200 ton cooling, screw compressors		.13	250		87,500	12,400		99,900	115,000
1660	291 ton cooling, screw compressors		.12	260		91,500	12,900		104,400	120,500

23 65 Cooling Towers

23 65 13 – Forced-Draft Cooling Towers

23 65 13.10 Forced-Draft Type Cooling Towers		Crew	Daily Output	Labor-Hours	Unit	Material	Labor	Equipment	Total	Total Incl O&P
0010	**FORCED-DRAFT TYPE COOLING TOWERS**, Packaged units									
0070	Galvanized steel									
0080	Induced draft, crossflow									
0100	Vertical, belt drive, 61 tons	Q-6	90	.267	TonAC	163	12.90		175.90	198
0150	100 ton		100	.240		156	11.60		167.60	188
0200	115 ton		109	.220		145	10.65		155.65	176
0250	131 ton		120	.200		144	9.70		153.70	173
0260	162 ton		132	.182		142	8.80		150.80	169
1500	Induced air, double flow									
1900	Vertical, gear drive, 167 ton	Q-6	126	.190	TonAC	138	9.20		147.20	166
2000	297 ton		129	.186		86	9		95	109
2100	582 ton		132	.182		69.50	8.80		78.30	89.50
2150	849 ton		142	.169		67.50	8.20		75.70	86.50
2200	1016 ton		150	.160		62.50	7.75		70.25	80.50
3000	For higher capacities, use multiples									
3500	For pumps and piping, add	Q-6	38	.632	TonAC	81.50	30.50		112	136
4000	For absorption systems, add				"	75%	75%			
5000	Fiberglass tower on galvanized steel support structure									
5010	Draw thru									
5100	100 ton	Q-6	1.40	17.143	Ea.	10,100	830		10,930	12,400
5120	120 ton		1.20	20		11,700	970		12,670	14,400
5140	140 ton		1	24		12,700	1,150		13,850	15,800
5160	160 ton		.80	30		14,100	1,450		15,550	17,700
5180	180 ton		.65	36.923		16,000	1,800		17,800	20,300
5200	200 ton		.48	50		18,100	2,425		20,525	23,500
5300	For stainless steel support structure, add					30%				
5360	For higher capacities, use multiples of each size									
6000	Stainless steel									
6010	Induced draft, crossflow, horizontal, belt drive									
6100	57 ton	Q-6	1.50	16	Ea.	21,200	775		21,975	24,600
6120	91 ton		.99	24.242		25,700	1,175		26,875	30,000
6140	111 ton		.43	55.814		38,500	2,700		41,200	46,500

23 65 Cooling Towers

23 65 13 – Forced-Draft Cooling Towers

23 65 13.10 Forced-Draft Type Cooling Towers		Crew	Daily Output	Labor-Hours	Unit	Material	2010 Bare Costs Labor	Equipment	Total	Total Incl O&P
6160	126 ton	Q-6	.22	109	Ea.	38,500	5,275		43,775	50,500

23 73 Indoor Central-Station Air-Handling Units

23 73 13 – Modular Indoor Central-Station Air-Handling Units

23 73 13.10 Air Handling Units		Crew	Daily Output	Labor-Hours	Unit	Material	Labor	Equipment	Total	Total Incl O&P
0010	**AIR HANDLING UNITS**, Built-Up									
0100	With cooling/heating coil section, filters, mixing box									
0880	Single zone, horizontal / vertical									
0890	Constant volume									
0900	1600 CFM	Q-5	1.20	13.333	Ea.	3,225	625		3,850	4,475
0920	5000 CFM	Q-6	1.40	17.143		8,375	830		9,205	10,500
0940	11,500 CFM		1	24		17,000	1,150		18,150	20,500
0970	22,000 CFM		.60	40		32,500	1,925		34,425	38,600
1000	40,000 CFM	↓	.30	80		49,700	3,875		53,575	60,500
1140	60,000 CFM	Q-7	.18	177	↓	74,000	8,800		82,800	94,500

23 73 39 – Indoor, Direct Gas-Fired Heating and Ventilating Units

23 73 39.10 Make-Up Air Unit		Crew	Daily Output	Labor-Hours	Unit	Material	Labor	Equipment	Total	Total Incl O&P
0010	**MAKE-UP AIR UNIT**									
0020	Indoor suspension, natural/LP gas, direct fired,									
0032	standard control. For flue see Div. 23 51 23.10									
0040	70°F temperature rise, MBH is input									
0100	75 MBH input	Q-6	3.60	6.667	Ea.	2,400	325		2,725	3,125
0160	150 MBH input		3	8		2,850	385		3,235	3,700
0220	225 MBH input		2.40	10		3,300	485		3,785	4,350
0300	400 MBH input	↓	1.60	15		7,450	725		8,175	9,275
0600	For discharge louver assembly, add					5%				
0700	For filters, add					10%				
0800	For air shut-off damper section, add				↓	30%				

23 74 Packaged Outdoor HVAC Equipment

23 74 33 – Packaged, Outdoor, Heating and Cooling Makeup Air-Conditioners

23 74 33.10 Roof Top Air Conditioners			Crew	Daily Output	Labor-Hours	Unit	Material	Labor	Equipment	Total	Total Incl O&P
0010	**ROOF TOP AIR CONDITIONERS**, Standard controls, curb, economizer										
1000	Single zone, electric cool, gas heat										
1100	3 ton cooling, 60 MBH heating	R236000-20	Q-5	.70	22.857	Ea.	3,475	1,075		4,550	5,425
1120	4 ton cooling, 95 MBH heating			.61	26.403		4,225	1,225		5,450	6,500
1140	5 ton cooling, 112 MBH heating			.56	28.521		4,850	1,325		6,175	7,350
1145	6 ton cooling, 140 MBH heating			.52	30.769		5,625	1,425		7,050	8,350
1150	7.5 ton cooling, 170 MBH heating		↓	.50	32.258		7,575	1,500		9,075	10,600
1160	10 ton cooling, 200 MBH heating		Q-6	.67	35.982		10,600	1,750		12,350	14,200
1170	12.5 ton cooling, 230 MBH heating			.63	37.975		12,100	1,850		13,950	16,100
1190	17.5 ton cooling, 330 MBH heating		↓	.52	45.889		17,900	2,225		20,125	23,000
1200	20 ton cooling, 360 MBH heating	**CN**	Q-7	.67	47.976		21,900	2,375		24,275	27,700
1210	25 ton cooling, 450 MBH heating			.56	57.554		25,000	2,850		27,850	31,800
1220	30 ton cooling, 540 MBH heating			.47	68.376		25,500	3,375		28,875	33,200
1240	40 ton cooling, 675 MBH heating		↓	.35	91.168	↓	33,400	4,500		37,900	43,500
2000	Multizone, electric cool, gas heat, economizer										
2100	15 ton cooling, 360 MBH heating		Q-7	.61	52.545	Ea.	57,000	2,600		59,600	66,500
2120	20 ton cooling, 360 MBH heating		↓	.53	60.038	↓	61,000	2,975		63,975	72,000

23 74 Packaged Outdoor HVAC Equipment

23 74 33 – Packaged, Outdoor, Heating and Cooling Makeup Air-Conditioners

23 74 33.10 Roof Top Air Conditioners		Crew	Daily Output	Labor-Hours	Unit	2010 Bare Costs Material	Labor	Equipment	Total	Total Incl O&P
2200	40 ton cooling, 540 MBH heating	Q-7	.28	113	Ea.	107,000	5,625		112,625	126,500
2210	50 ton cooling, 540 MBH heating		.23	142		133,500	7,025		140,525	157,000
2220	70 ton cooling, 1500 MBH heating		.16	198		144,500	9,825		154,325	173,500
2240	80 ton cooling, 1500 MBH heating		.14	228		165,000	11,300		176,300	198,500
2260	90 ton cooling, 1500 MBH heating		.13	256		173,000	12,700		185,700	209,000
2280	105 ton cooling, 1500 MBH heating		.11	290		190,500	14,400		204,900	231,000
2400	For hot water heat coil, deduct					5%				
2500	For steam heat coil, deduct					2%				
2600	For electric heat, deduct					3%	5%			

23 81 Decentralized Unitary HVAC Equipment

23 81 13 – Packaged Terminal Air-Conditioners

23 81 13.10 Packaged Cabinet Type Air-Conditioners		Crew	Daily Output	Labor-Hours	Unit	Material	Labor	Equipment	Total	Total Incl O&P
0010	**PACKAGED CABINET TYPE AIR-CONDITIONERS**, Cabinet, wall sleeve,									
0100	louver, electric heat, thermostat, manual changeover, 208 V									
0200	6,000 BTUH cooling, 8800 BTU heat	Q-5	6	2.667	Ea.	680	125		805	935
0220	9,000 BTUH cooling, 13,900 BTU heat		5	3.200		790	149		939	1,100
0240	12,000 BTUH cooling, 13,900 BTU heat		4	4		850	187		1,037	1,225
0260	15,000 BTUH cooling, 13,900 BTU heat		3	5.333		915	249		1,164	1,375
0500	For hot water coil, increase heat by 10%, add					5%	10%			
1000	For steam, increase heat output by 30%, add					8%	10%			

23 81 19 – Self-Contained Air-Conditioners

23 81 19.20 Self-Contained Single Package		Crew	Daily Output	Labor-Hours	Unit	Material	Labor	Equipment	Total	Total Incl O&P
0010	**SELF-CONTAINED SINGLE PACKAGE**									
0100	Air cooled, for free blow or duct, not incl. remote condenser									
0200	3 ton cooling	Q-5	1	16	Ea.	3,300	745		4,045	4,750
0220	5 ton cooling	Q-6	1.20	20		3,925	970		4,895	5,775
0240	10 ton cooling	Q-7	1	32		6,775	1,575		8,350	9,825
0260	20 ton cooling		.90	35.556		13,500	1,750		15,250	17,500
0280	30 ton cooling		.80	40		21,600	1,975		23,575	26,800
0340	60 ton cooling	Q-8	.40	80		50,500	3,950	146	54,596	61,500
0490	For duct mounting no price change									
0500	For steam heating coils, add				Ea.	10%	10%			
1000	Water cooled for free blow or duct, not including tower									
1010	Constant volume									
1100	3 ton cooling	Q-6	1	24	Ea.	3,250	1,150		4,400	5,325
1120	5 ton cooling	″	1	24		4,275	1,150		5,425	6,450
1140	10 ton cooling	Q-7	.90	35.556		8,300	1,750		10,050	11,800
1160	20 ton cooling		.80	40		25,800	1,975		27,775	31,300
1180	30 ton cooling		.70	45.714		34,200	2,250		36,450	41,000

23 81 23 – Computer-Room Air-Conditioners

23 81 23.10 Computer Room Units		Crew	Daily Output	Labor-Hours	Unit	Material	Labor	Equipment	Total	Total Incl O&P
0010	**COMPUTER ROOM UNITS**									
1000	Air cooled, includes remote condenser but not									
1020	interconnecting tubing or refrigerant									
1080	3 ton	Q-5	.50	32	Ea.	16,100	1,500		17,600	20,000
1120	5 ton		.45	35.556		17,200	1,650		18,850	21,400
1160	6 ton		.30	53.333		30,400	2,500		32,900	37,100
1200	8 ton		.27	59.259		32,300	2,775		35,075	39,800
1240	10 ton		.25	64		33,800	3,000		36,800	41,700

23 81 Decentralized Unitary HVAC Equipment

23 81 23 – Computer-Room Air-Conditioners

23 81 23.10 Computer Room Units		Crew	Daily Output	Labor-Hours	Unit	Material	2010 Bare Costs Labor	Equipment	Total	Total Incl O&P
1280	15 ton	Q-5	.22	72.727	Ea.	37,200	3,400		40,600	46,000
1290	18 ton	↓	.20	80		41,400	3,725		45,125	51,000
1300	20 ton	Q-6	.26	92.308		44,500	4,475		48,975	55,500
1320	22 ton		.24	100		45,100	4,850		49,950	57,000
1360	30 ton	↓	.21	114	↓	55,500	5,525		61,025	69,500
2200	Chilled water, for connection to									
2220	existing chiller system of adequate capacity									
2260	5 ton	Q-5	.74	21.622	Ea.	12,300	1,000		13,300	15,100

23 81 43 – Air-Source Unitary Heat Pumps

23 81 43.10 Air-Source Heat Pumps		Crew	Daily Output	Labor-Hours	Unit	Material	Labor	Equipment	Total	Total Incl O&P
0010	**AIR-SOURCE HEAT PUMPS**, Not including interconnecting tubing									
1000	Air to air, split system, not including curbs, pads, fan coil and ductwork									
1012	Outside condensing unit only, for fan coil see Div. 23 82 19.10									
1020	2 ton cooling, 8.5 MBH heat @ 0°F	Q-5	2	8	Ea.	2,125	375		2,500	2,875
1060	5 ton cooling, 27 MBH heat @ 0°F		.50	32		3,425	1,500		4,925	6,025
1080	7.5 ton cooling, 33 MBH heat @ 0°F	↓	.45	35.556		5,625	1,650		7,275	8,675
1100	10 ton cooling, 50 MBH heat @ 0°F	Q-6	.64	37.500		7,900	1,825		9,725	11,400
1120	15 ton cooling, 64 MBH heat @ 0°F		.50	48		10,400	2,325		12,725	15,000
1130	20 ton cooling, 85 MBH heat @ 0°F		.35	68.571		15,600	3,325		18,925	22,200
1140	25 ton cooling, 119 MBH heat @ 0°F	↓	.25	96	↓	18,800	4,650		23,450	27,700
1500	Single package, not including curbs, pads, or plenums									
1520	2 ton cooling, 6.5 MBH heat @ 0°F	Q-5	1.50	10.667	Ea.	2,750	500		3,250	3,775
1580	4 ton cooling, 13 MBH heat @ 0°F		.96	16.667		3,850	780		4,630	5,425
1640	7.5 ton cooling, 35 MBH heat @ 0°F	↓	.40	40	↓	6,550	1,875		8,425	10,000

23 81 46 – Water-Source Unitary Heat Pumps

23 81 46.10 Water Source Heat Pumps		Crew	Daily Output	Labor-Hours	Unit	Material	Labor	Equipment	Total	Total Incl O&P
0010	**WATER SOURCE HEAT PUMPS**, Not incl. connecting tubing or water source									
2000	Water source to air, single package									
2100	1 ton cooling, 13 MBH heat @ 75°F	Q-5	2	8	Ea.	1,475	375		1,850	2,175
2140	2 ton cooling, 19 MBH heat @ 75°F		1.70	9.412		1,600	440		2,040	2,425
2220	5 ton cooling, 29 MBH heat @ 75°F	↓	.90	17.778		2,625	830		3,455	4,125
3960	For supplementary heat coil, add				↓	10%				
4000	For increase in capacity thru use									
4020	of solar collector, size boiler at 60%									

23 82 Convection Heating and Cooling Units

23 82 16 – Air Coils

23 82 16.10 Flanged Coils		Crew	Daily Output	Labor-Hours	Unit	Material	Labor	Equipment	Total	Total Incl O&P
0010	**FLANGED COILS**									
0500	Chilled water cooling, 6 rows, 24" x 48"	Q-5	3.20	5	Ea.	3,650	234		3,884	4,350
1000	Direct expansion cooling, 6 rows, 24" x 48"		2.80	5.714		3,950	267		4,217	4,750
1500	Hot water heating, 1 row, 24" x 48"		4	4		1,450	187		1,637	1,875
2000	Steam heating, 1 row, 24" x 48"	↓	3.06	5.229	↓	2,050	244		2,294	2,650

23 82 16.20 Duct Heaters		Crew	Daily Output	Labor-Hours	Unit	Material	Labor	Equipment	Total	Total Incl O&P
0010	**DUCT HEATERS**, Electric, 480 V, 3 Ph.									
0020	Finned tubular insert, 500°F									
0100	8" wide x 6" high, 4.0 kW	Q-20	16	1.250	Ea.	705	56.50		761.50	860
0120	12" high, 8.0 kW		15	1.333		1,175	60		1,235	1,375
0140	18" high, 12.0 kW		14	1.429		1,650	64.50		1,714.50	1,900
0160	24" high, 16.0 kW	↓	13	1.538	↓	2,100	69.50		2,169.50	2,425

23 82 Convection Heating and Cooling Units

23 82 16 – Air Coils

23 82 16.20 Duct Heaters		Crew	Daily Output	Labor-Hours	Unit	2010 Bare Costs Material	Labor	Equipment	Total	Total Incl O&P
0180	30" high, 20.0 kW	Q-20	12	1.667	Ea.	2,575	75.50		2,650.50	2,950
0300	12" wide x 6" high, 6.7 kW		15	1.333		750	60		810	915
0360	24" high, 26.7 kW		12	1.667		2,175	75.50		2,250.50	2,525
0700	24" wide x 6" high, 17.8 kW		13	1.538		890	69.50		959.50	1,075
0760	24" high, 71.1 kW	↓	10	2	↓	2,725	90.50		2,815.50	3,125
8000	To obtain BTU multiply kW by 3413									

23 82 19 – Fan Coil Units

23 82 19.10 Fan Coil Air Conditioning		Crew	Daily Output	Labor-Hours	Unit	Material	Labor	Equipment	Total	Total Incl O&P
0010	**FAN COIL AIR CONDITIONING** Cabinet mounted, filters, controls									
0100	Chilled water, 1/2 ton cooling	Q-5	8	2	Ea.	710	93.50		803.50	920
0120	1 ton cooling		6	2.667		815	125		940	1,075
0140	1.5 ton cooling		5.50	2.909		920	136		1,056	1,200
0150	2 ton cooling		5.25	3.048		1,150	142		1,292	1,500
0180	3 ton cooling	↓	4	4	↓	1,725	187		1,912	2,175
0262	For hot water coil, add					40%	10%			
0940	Direct expansion, for use w/air cooled condensing unit, 1.5 ton cooling	Q-5	5	3.200	Ea.	690	149		839	985
1000	5 ton cooling	"	3	5.333		1,325	249		1,574	1,825
1040	10 ton cooling	Q-6	2.60	9.231		2,775	445		3,220	3,725
1060	20 ton cooling	"	.70	34.286	↓	5,375	1,650		7,025	8,400

23 82 19.20 Heating and Ventilating Units		Crew	Daily Output	Labor-Hours	Unit	Material	Labor	Equipment	Total	Total Incl O&P
0010	**HEATING AND VENTILATING UNITS,** Classroom units									
0020	Includes filter, heating/cooling coils, standard controls									
0080	750 CFM, 2 tons cooling	Q-6	2	12	Ea.	3,625	580		4,205	4,850
0120	1250 CFM, 3 tons cooling		1.40	17.143		4,425	830		5,255	6,100
0140	1500 CFM, 4 tons cooling	↓	.80	30		4,725	1,450		6,175	7,375
0500	For electric heat, add					35%				
1000	For no cooling, deduct				↓	25%	10%			

23 82 27 – Infrared Units

23 82 27.10 Infrared Type Heating Units		Crew	Daily Output	Labor-Hours	Unit	Material	Labor	Equipment	Total	Total Incl O&P
0010	**INFRARED TYPE HEATING UNITS**									
0020	Gas fired, unvented, electric ignition, 100% shutoff.									
0030	Piping and wiring not included									
0060	Input, 15 MBH	Q-5	7	2.286	Ea.	435	107		542	640
0120	45 MBH		5	3.200		500	149		649	775
0160	60 MBH		4	4		585	187		772	925
0240	120 MBH	↓	2	8	↓	970	375		1,345	1,625

23 82 29 – Radiators

23 82 29.10 Hydronic Heating		Crew	Daily Output	Labor-Hours	Unit	Material	Labor	Equipment	Total	Total Incl O&P
0010	**HYDRONIC HEATING,** Terminal units, not incl. main supply pipe									
1000	Radiation									
1100	Panel, baseboard, C.I., including supports, no covers	Q-5	46	.348	L.F.	32.50	16.25		48.75	60
3000	Radiators, cast iron									
3100	Free standing or wall hung, 6 tube, 25" high	Q-5	96	.167	Section	36	7.80		43.80	51
3200	4 tube, 19" high	"	96	.167	"	23.50	7.80		31.30	37.50
3250	Adj. brackets, 2 per wall radiator up to 30 sections	1 Stpi	32	.250	Ea.	27	13		40	49.50
9500	To convert SFR to BTU rating: Hot water, 150 x SFR									
9510	Forced hot water, 180 x SFR; steam, 240 x SFR									

23 82 33 – Convectors

23 82 33.10 Convector Units		Crew	Daily Output	Labor-Hours	Unit	Material	2010 Bare Costs Labor	Equipment	Total	Total Incl O&P
0010	**CONVECTOR UNITS,** Terminal units, not incl. main supply pipe									
2204	Convector, multifin, 2 pipe w/ cabinet									
2210	17" H x 24" L	Q-5	10	1.600	Ea.	98.50	74.50		173	220
2214	17" H x 36" L		8.60	1.860		148	87		235	292
2218	17" H x 48" L		7.40	2.162		197	101		298	365
2222	21" H x 24" L		9	1.778		96.50	83		179.50	230
2226	21" H x 36" L		8.20	1.951		145	91		236	296
2228	21" H x 48" L	↓	6.80	2.353	↓	193	110		303	375
2240	For knob operated damper, add					140%				
2241	For metal trim strips, add	Q-5	64	.250	Ea.	19.25	11.70		30.95	38.50
2243	For snap-on inlet grille, add					10%	10%			
2245	For hinged access door, add	Q-5	64	.250	Ea.	35.50	11.70		47.20	56.50
2246	For air chamber, auto-venting, add	"	58	.276	"	6.35	12.90		19.25	26.50

23 82 36 – Finned-Tube Radiation Heaters

23 82 36.10 Finned Tube Radiation		Crew	Daily Output	Labor-Hours	Unit	Material	Labor	Equipment	Total	Total Incl O&P
0010	**FINNED TUBE RADIATION,** Terminal units, not incl. main supply pipe									
1150	Fin tube, wall hung, 14" slope top cover, with damper									
1200	1-1/4" copper tube, 4-1/4" alum. fin	Q-5	38	.421	L.F.	40.50	19.65		60.15	74.50
1250	1-1/4" steel tube, 4-1/4" steel fin	"	36	.444	"	38.50	21		59.50	73.50
1500	Note: fin tube may also require corners, caps, etc.									

23 82 39 – Unit Heaters

23 82 39.16 Propeller Unit Heaters		Crew	Daily Output	Labor-Hours	Unit	Material	Labor	Equipment	Total	Total Incl O&P
0010	**PROPELLER UNIT HEATERS**									
3950	Unit heaters, propeller, 115 V 2 psi steam, 60°F entering air									
4000	Horizontal, 12 MBH	Q-5	12	1.333	Ea.	335	62.50		397.50	465
4060	43.9 MBH		8	2		505	93.50		598.50	695
4140	96.8 MBH		6	2.667		735	125		860	995
4180	157.6 MBH		4	4		980	187		1,167	1,350
4240	286.9 MBH		2	8		1,600	375		1,975	2,300
4250	326.0 MBH		1.90	8.421		1,825	395		2,220	2,600
4260	364 MBH		1.80	8.889		1,900	415		2,315	2,725
4270	404 MBH	↓	1.60	10		2,500	465		2,965	3,450
4300	For vertical diffuser, add					174			174	191
4310	Vertical flow, 40 MBH	Q-5	11	1.455		500	68		568	650
4314	58.5 MBH		8	2		520	93.50		613.50	710
4326	131.0 MBH		4	4		780	187		967	1,150
4346	297.0 MBH	↓	1.80	8.889		1,500	415		1,915	2,275
4354	420 MBH,(460 V)	Q-6	1.80	13.333		2,000	645		2,645	3,175
4358	500 MBH, (460 V)		1.71	14.035		2,650	680		3,330	3,925
4362	570 MBH, (460 V)		1.40	17.143		3,725	830		4,555	5,350
4366	620 MBH, (460 V)		1.30	18.462		4,275	895		5,170	6,050
4370	960 MBH, (460 V)	↓	1.10	21.818	↓	7,200	1,050		8,250	9,475

23 83 16.10 Radiant Floor Heating		Crew	Daily Output	Labor-Hours	Unit	2010 Bare Costs Material	Labor	Equipment	Total	Total Incl O&P
0010	**RADIANT FLOOR HEATING**									
0100	Tubing, PEX (cross-linked polyethylene)									
0110	Oxygen barrier type for systems with ferrous materials									
0120	1/2″	Q-5	800	.020	L.F.	1.01	.93		1.94	2.51
0130	3/4″		535	.030		1.29	1.40		2.69	3.51
0140	1″		400	.040		2.21	1.87		4.08	5.25
0200	Non barrier type for ferrous free systems									
0210	1/2″	Q-5	800	.020	L.F.	.55	.93		1.48	2.01
0220	3/4″		535	.030		1	1.40		2.40	3.19
0230	1″		400	.040		1.71	1.87		3.58	4.68
1000	Manifolds									
1110	Brass									
1120	With supply and return valves, flow meter, thermometer,									
1122	auto air vent and drain/fill valve.									
1130	1″, 2 circuit	Q-5	14	1.143	Ea.	181	53.50		234.50	279
1140	1″, 3 circuit		13.50	1.185		207	55.50		262.50	310
1150	1″, 4 circuit		13	1.231		225	57.50		282.50	335
1154	1″, 5 circuit		12.50	1.280		268	60		328	385
1158	1″, 6 circuit		12	1.333		292	62.50		354.50	415
1162	1″, 7 circuit		11.50	1.391		320	65		385	450
1166	1″, 8 circuit		11	1.455		355	68		423	490
1172	1″, 9 circuit		10.50	1.524		380	71		451	525
1174	1″, 10 circuit		10	1.600		410	74.50		484.50	560
1178	1″, 11 circuit		9.50	1.684		425	78.50		503.50	590
1182	1″, 12 circuit		9	1.778		470	83		553	640
1610	Copper manifold header, (cut to size)									
1620	1″ header, 12 – 1/2″ sweat outlets	Q-5	3.33	4.805	Ea.	72.50	224		296.50	415
1630	1-1/4″ header, 12 – 1/2″ sweat outlets		3.20	5		84	234		318	445
1640	1-1/4″ header, 12 – 3/4″ sweat outlets		3	5.333		90.50	249		339.50	475
1650	1-1/2″ header, 12 – 3/4″ sweat outlets		3.10	5.161		109	241		350	480
1660	2″ header, 12 – 3/4″ sweat outlets		2.90	5.517		160	258		418	560
3000	Valves									
3110	Thermostatic zone valve actuator with end switch	Q-5	40	.400	Ea.	36.50	18.70		55.20	68
3114	Thermostatic zone valve actuator	″	36	.444	″	81	21		102	120
3120	Motorized straight zone valve with operator complete									
3130	3/4″	Q-5	35	.457	Ea.	125	21.50		146.50	170
3140	1″		32	.500		137	23.50		160.50	186
3150	1-1/4″		29.60	.541		165	25		190	220
3500	4 Way mixing valve, manual, brass									
3530	1″	Q-5	13.30	1.203	Ea.	126	56		182	223
3540	1-1/4″		11.40	1.404		137	65.50		202.50	250
3550	1-1/2″		11	1.455		203	68		271	325
3560	2″		10.60	1.509		288	70.50		358.50	420
3800	Mixing valve motor, 4 way for valves, 1″ and 1-1/4″		34	.471		310	22		332	375
3810	Mixing valve motor, 4 way for valves, 1-1/2″ and 2″		30	.533		355	25		380	430
5000	Radiant floor heating, zone control panel									
5120	4 Zone actuator valve control, expandable	Q-5	20	.800	Ea.	163	37.50		200.50	235
5130	6 Zone actuator valve control, expandable		18	.889		199	41.50		240.50	281
6070	Thermal track, straight panel for long continuous runs, 5.333 S.F.		40	.400		16.60	18.70		35.30	46.50
6080	Thermal track, utility panel, for direction reverse at run end, 5.333 S.F.		40	.400		16.60	18.70		35.30	46.50
6090	Combination panel, for direction reverse plus straight run, 5.333 S.F.		40	.400		16.60	18.70		35.30	46.50
7000	PEX tubing fittings									
7100	Compression type									

23 83 Radiant Heating Units

23 83 16 – Radiant-Heating Hydronic Piping

23 83 16.10 Radiant Floor Heating		Crew	Daily Output	Labor-Hours	Unit	Material	Labor	Equipment	Total	Total Incl O&P
						2010 Bare Costs				
7116	Coupling									
7120	1/2" x 1/2"	1 Stpi	27	.296	Ea.	6.40	15.40		21.80	30
7124	3/4" x 3/4"	"	23	.348	"	10.10	18.05		28.15	38
7130	Adapter									
7132	1/2" x female sweat 1/2"	1 Stpi	27	.296	Ea.	4.15	15.40		19.55	27.50
7134	1/2" x female sweat 3/4"		26	.308		4.88	15.95		20.83	29.50
7136	5/8" x female sweat 3/4"	↓	24	.333	↓	6.65	17.30		23.95	33.50
7140	Elbow									
7142	1/2" x female sweat 1/2"	1 Stpi	27	.296	Ea.	6.35	15.40		21.75	30
7144	1/2" x female sweat 3/4"		26	.308		7.40	15.95		23.35	32
7146	5/8" x female sweat 3/4"	↓	24	.333	↓	8.75	17.30		26.05	35.50
7200	Insert type									
7206	PEX x male NPT									
7210	1/2" x 1/2"	1 Stpi	29	.276	Ea.	1.43	14.30		15.73	23
7220	3/4" x 3/4"		27	.296		2.10	15.40		17.50	25.50
7230	1" x 1"	↓	26	.308	↓	2.66	15.95		18.61	27
7300	PEX coupling									
7310	1/2" x 1/2"	1 Stpi	30	.267	Ea.	.58	13.85		14.43	21.50
7320	3/4" x 3/4"		29	.276		.71	14.30		15.01	22.50
7330	1" x 1"	↓	28	.286	↓	1.41	14.85		16.26	23.50
7400	PEX stainless crimp ring									
7410	1/2"	1 Stpi	86	.093	Ea.	.25	4.83		5.08	7.55
7420	3/4" x 3/4"		84	.095		.34	4.94		5.28	7.75
7430	1" x 1"	↓	82	.098	↓	.40	5.05		5.45	8.05

23 83 33 – Electric Radiant Heaters

23 83 33.10 Electric Heating		Crew	Daily Output	Labor-Hours	Unit	Material	Labor	Equipment	Total	Total Incl O&P
0010	**ELECTRIC HEATING**, not incl. conduit or feed wiring									
1100	Rule of thumb: Baseboard units, including control	1 Elec	4.40	1.818	kW	92.50	89		181.50	234
1300	Baseboard heaters, 2' long, 375 watt		8	1	Ea.	40	49		89	117
1400	3' long, 500 watt		8	1		46	49		95	124
1600	4' long, 750 watt		6.70	1.194		54	58.50		112.50	147
1800	5' long, 935 watt		5.70	1.404		68	69		137	177
2000	6' long, 1125 watt		5	1.600		74	78.50		152.50	199
2400	8' long, 1500 watt	↓	4	2	↓	92	98		190	247
2950	Wall heaters with fan, 120 to 277 volt									
3600	Thermostats, integral	1 Elec	16	.500	Ea.	30	24.50		54.50	69.50
3800	Line voltage, 1 pole	"	8	1	"	31	49		80	107

23 84 Humidity Control Equipment

23 84 13 – Humidifiers

23 84 13.10 Humidifier Units		Crew	Daily Output	Labor-Hours	Unit	Material	Labor	Equipment	Total	Total Incl O&P
0010	**HUMIDIFIER UNITS**									
0520	Steam, room or duct, filter, regulators, auto. controls, 220 V									
0540	11 lb. per hour	Q-5	6	2.667	Ea.	2,350	125		2,475	2,750
0560	22 lb. per hour		5	3.200		2,600	149		2,749	3,075
0580	33 lb. per hour		4	4		2,650	187		2,837	3,200
0600	50 lb. per hour		4	4		3,250	187		3,437	3,850
0620	100 lb. per hour	↓	3	5.333	↓	3,900	249		4,149	4,675

Division Notes

		CREW	DAILY OUTPUT	LABOR-HOURS	UNIT	2010 BARE COSTS MAT.	LABOR	EQUIP.	TOTAL	TOTAL INCL O&P

Estimating Tips

26 05 00 Common Work Results for Electrical

- Conduit should be taken off in three main categories—power distribution, branch power, and branch lighting—so the estimator can concentrate on systems and components, therefore making it easier to ensure all items have been accounted for.
- For cost modifications for elevated conduit installation, add the percentages to labor according to the height of installation, and only to the quantities exceeding the different height levels, not to the total conduit quantities.
- Remember that aluminum wiring of equal ampacity is larger in diameter than copper and may require larger conduit.
- If more than three wires at a time are being pulled, deduct percentages from the labor hours of that grouping of wires.
- When taking off grounding systems, identify separately the type and size of wire, and list each unique type of ground connection.
- The estimator should take the weights of materials into consideration when completing a takeoff. Topics to consider include: How will the materials be supported? What methods of support are available? How high will the support structure have to reach? Will the final support structure be able to withstand the total burden? Is the support material included or separate from the fixture, equipment, and material specified?
- Do not overlook the costs for equipment used in the installation. If scaffolding or highlifts are available in the field, contractors may use them in lieu of the proposed ladders and rolling staging.

26 20 00 Low-Voltage Electrical Transmission

- Supports and concrete pads may be shown on drawings for the larger equipment, or the support system may be only a piece of plywood for the back of a panelboard. In either case, it must be included in the costs.

26 40 00 Electrical and Cathodic Protection

- When taking off cathodic protections systems, identify the type and size of cable, and list each unique type of anode connection.

26 50 00 Lighting

- Fixtures should be taken off room by room, using the fixture schedule, specifications, and the ceiling plan. For large concentrations of lighting fixtures in the same area, deduct the percentages from labor hours.

Reference Numbers

Reference numbers are shown in shaded boxes at the beginning of some major classifications. These numbers refer to related items in the Reference Section. The reference information may be an estimating procedure, an alternate pricing method, or technical information.

Note: Not all subdivisions listed here necessarily appear in this publication.

Note: **Trade Service,** *in part, has been used as a reference source for some of the material prices used in Division 26.*

26 05 05 – Selective Electrical Demolition

26 05 05.10 Electrical Demolition		Crew	Daily Output	Labor-Hours	Unit	Material	2010 Bare Costs Labor	Equipment	Total	Total Incl O&P
0010	**ELECTRICAL DEMOLITION**									
0020	Conduit to 15′ high, including fittings & hangers									
0100	Rigid galvanized steel, 1/2″ to 1″ diameter	1 Elec	242	.033	L.F.		1.62		1.62	2.41
0120	1-1/4″ to 2″	″	200	.040			1.96		1.96	2.91
0140	2-1/2″ to 3-1/2″	2 Elec	302	.053			2.60		2.60	3.86
0160	4″ to 6″	″	160	.100			4.90		4.90	7.30
0200	Electric metallic tubing (EMT), 1/2″ to 1″	1 Elec	394	.020			.99		.99	1.48
0220	1-1/4″ to 1-1/2″	″	326	.025			1.20		1.20	1.79
0260	3-1/2″ to 4″	2 Elec	310	.052	↓		2.53		2.53	3.76
0270	Armored cable, (BX) avg. 50′ runs									
0280	#14, 2 wire	1 Elec	690	.012	L.F.		.57		.57	.84
0290	#14, 3 wire		571	.014			.69		.69	1.02
0300	#12, 2 wire		605	.013			.65		.65	.96
0310	#12, 3 wire		514	.016			.76		.76	1.13
0320	#10, 2 wire		514	.016			.76		.76	1.13
0330	#10, 3 wire		425	.019			.92		.92	1.37
0340	#8, 3 wire	↓	342	.023	↓		1.15		1.15	1.70
0350	Non metallic sheathed cable (Romex)									
0360	#14, 2 wire	1 Elec	720	.011	L.F.		.54		.54	.81
0370	#14, 3 wire		657	.012			.60		.60	.89
0380	#12, 2 wire		629	.013			.62		.62	.93
0390	#10, 3 wire	↓	450	.018	↓		.87		.87	1.30
0400	Wiremold raceway, including fittings & hangers									
0420	No. 3000	1 Elec	250	.032	L.F.		1.57		1.57	2.33
0440	No. 4000		217	.037			1.81		1.81	2.69
0460	No. 6000		166	.048	↓		2.36		2.36	3.51
0465	Telephone/power pole		12	.667	Ea.		32.50		32.50	48.50
0470	Non-metallic, straight section	↓	480	.017	L.F.		.82		.82	1.21
0500	Channels, steel, including fittings & hangers									
0520	3/4″ x 1-1/2″	1 Elec	308	.026	L.F.		1.27		1.27	1.89
0540	1-1/2″ x 1-1/2″		269	.030			1.46		1.46	2.17
0560	1-1/2″ x 1-7/8″	↓	229	.035	↓		1.71		1.71	2.54
0600	Copper bus duct, indoor, 3 phase									
0610	Including hangers & supports									
0620	225 amp	2 Elec	135	.119	L.F.		5.80		5.80	8.65
0640	400 amp		106	.151			7.40		7.40	11
0660	600 amp		86	.186			9.10		9.10	13.55
0680	1000 amp		60	.267			13.05		13.05	19.45
0700	1600 amp		40	.400			19.60		19.60	29
0720	3000 amp	↓	10	1.600	↓		78.50		78.50	117
1300	Transformer, dry type, 1 phase, incl. removal of									
1320	supports, wire & conduit terminations									
1340	1 kVA	1 Elec	7.70	1.039	Ea.		51		51	75.50
1420	75 kVA	2 Elec	2.50	6.400	″		315		315	465
1440	3 phase to 600V, primary									
1460	3 kVA	1 Elec	3.85	2.078	Ea.		102		102	151
1520	75 kVA	2 Elec	2.70	5.926			290		290	430
1550	300 kVA	R-3	1.80	11.111			535	77	612	885
1570	750 kVA	″	1.10	18.182	↓		880	126	1,006	1,450
1800	Wire, THW-THWN-THHN, removed from									
1810	in place conduit, to 15′ high									
1830	#14	1 Elec	65	.123	C.L.F.		6.05		6.05	8.95
1840	#12	↓	55	.145	↓		7.15		7.15	10.60

26 05 Common Work Results for Electrical

26 05 05 – Selective Electrical Demolition

26 05 05.10 Electrical Demolition		Crew	Daily Output	Labor-Hours	Unit	Material	2010 Bare Costs Labor	Equipment	Total	Total Incl O&P
1850	#10	1 Elec	45.50	.176	C.L.F.		8.60		8.60	12.80
1860	#8		40.40	.198			9.70		9.70	14.45
1870	#6		32.60	.245			12		12	17.90
1880	#4	2 Elec	53	.302			14.80		14.80	22
1890	#3		50	.320			15.70		15.70	23.50
1900	#2		44.60	.359			17.60		17.60	26
1910	1/0		33.20	.482			23.50		23.50	35
1920	2/0		29.20	.548			27		27	40
1930	3/0		25	.640			31.50		31.50	46.50
1940	4/0		22	.727			35.50		35.50	53
1950	250 kcmil		20	.800			39		39	58.50
1960	300 kcmil		19	.842			41.50		41.50	61.50
1970	350 kcmil		18	.889			43.50		43.50	65
1980	400 kcmil		17	.941			46		46	68.50
1990	500 kcmil		16.20	.988			48.50		48.50	72
2000	Interior fluorescent fixtures, incl. supports									
2010	& whips, to 15′ high									
2100	Recessed drop-in 2′ x 2′, 2 lamp	2 Elec	35	.457	Ea.		22.50		22.50	33.50
2120	2′ x 4′, 2 lamp		33	.485			24		24	35.50
2140	2′ x 4′, 4 lamp		30	.533			26		26	39
2160	4′ x 4′, 4 lamp		20	.800			39		39	58.50
2180	Surface mount, acrylic lens & hinged frame									
2200	1′ x 4′, 2 lamp	2 Elec	44	.364	Ea.		17.80		17.80	26.50
2220	2′ x 2′, 2 lamp		44	.364			17.80		17.80	26.50
2260	2′ x 4′, 4 lamp		33	.485			24		24	35.50
2280	4′ x 4′, 4 lamp		23	.696			34		34	50.50
2300	Strip fixtures, surface mount									
2320	4′ long, 1 lamp	2 Elec	53	.302	Ea.		14.80		14.80	22
2340	4′ long, 2 lamp		50	.320			15.70		15.70	23.50
2360	8′ long, 1 lamp		42	.381			18.65		18.65	28
2380	8′ long, 2 lamp		40	.400			19.60		19.60	29
2400	Pendant mount, industrial, incl. removal									
2410	of chain or rod hangers, to 15′ high									
2420	4′ long, 2 lamp	2 Elec	35	.457	Ea.		22.50		22.50	33.50
2440	8′ long, 2 lamp	"	27	.593	"		29		29	43

26 05 13 – Medium-Voltage Cables

26 05 13.16 Medium-Voltage, Single Cable		Crew	Daily Output	Labor-Hours	Unit	Material	Labor	Equipment	Total	Total Incl O&P
0010	**MEDIUM-VOLTAGE, SINGLE CABLE** Splicing & terminations not included									
0040	Copper, XLP shielding, 5 kV, #6	2 Elec	4.40	3.636	C.L.F.	129	178		307	405
0050	#4		4.40	3.636		168	178		346	450
0100	#2		4	4		199	196		395	510
0200	#1		4	4		228	196		424	540
0400	1/0		3.80	4.211		251	206		457	580
0600	2/0		3.60	4.444		315	218		533	670
0800	4/0		3.20	5		410	245		655	815
1000	250 kcmil	3 Elec	4.50	5.333		510	261		771	950
1200	350 kcmil		3.90	6.154		655	300		955	1,175
1400	500 kcmil		3.60	6.667		800	325		1,125	1,375
1600	15 kV, ungrounded neutral, #1	2 Elec	4	4		277	196		473	595
1800	1/0		3.80	4.211		330	206		536	670
2000	2/0		3.60	4.444		375	218		593	740
2200	4/0		3.20	5		500	245		745	915

26 05 Common Work Results for Electrical

26 05 13 – Medium-Voltage Cables

26 05 13.16 Medium-Voltage, Single Cable		Crew	Daily Output	Labor-Hours	Unit	Material	2010 Bare Costs Labor	Equipment	Total	Total Incl O&P
2400	250 kcmil	3 Elec	4.50	5.333	C.L.F.	555	261		816	1,000
2600	350 kcmil		3.90	6.154		700	300		1,000	1,225
2800	500 kcmil		3.60	6.667		860	325		1,185	1,425

26 05 19 – Low-Voltage Electrical Power Conductors and Cables

26 05 19.20 Armored Cable		Crew	Daily Output	Labor-Hours	Unit	Material	Labor	Equipment	Total	Total Incl O&P
0010	**ARMORED CABLE**									
0050	600 volt, copper (BX), #14, 2 conductor, solid	1 Elec	2.40	3.333	C.L.F.	70.50	163		233.50	320
0100	3 conductor, solid		2.20	3.636		111	178		289	385
0150	#12, 2 conductor, solid		2.30	3.478		71.50	170		241.50	330
0200	3 conductor, solid		2	4		114	196		310	415
0250	#10, 2 conductor, solid		2	4		130	196		326	435
0300	3 conductor, solid		1.60	5		179	245		424	560
0350	#8, 3 conductor, solid		1.30	6.154		335	300		635	820
0400	3 conductor with PVC jacket, in cable tray, #6		3.10	2.581		375	126		501	600
0450	#4	2 Elec	5.40	2.963		480	145		625	745
0500	#2		4.60	3.478		640	170		810	960
0550	#1		4	4		870	196		1,066	1,250
0600	1/0		3.60	4.444		1,025	218		1,243	1,450
0650	2/0		3.40	4.706		1,250	231		1,481	1,725
0700	3/0		3.20	5		1,450	245		1,695	1,975
0750	4/0		3	5.333		1,650	261		1,911	2,225
0800	250 kcmil	3 Elec	3.60	6.667		1,875	325		2,200	2,525
0850	350 kcmil		3.30	7.273		2,500	355		2,855	3,275
0900	500 kcmil		3	8		3,250	390		3,640	4,150
1050	5 kV, copper, 3 conductor with PVC jacket,									
1060	non-shielded, in cable tray, #4	2 Elec	380	.042	L.F.	7.35	2.06		9.41	11.15
1100	#2		360	.044		9.60	2.18		11.78	13.80
1200	#1		300	.053		12.20	2.61		14.81	17.30
1400	1/0		290	.055		14.05	2.70		16.75	19.50
1600	2/0		260	.062		16.30	3.02		19.32	22.50
2000	4/0		240	.067		21.50	3.27		24.77	29
2100	250 kcmil	3 Elec	330	.073		30	3.56		33.56	38
2150	350 kcmil		315	.076		36.50	3.73		40.23	46
2200	500 kcmil		270	.089		49	4.36		53.36	60
2400	15 kV, copper, 3 conductor with PVC jacket galv steel armored									
2500	grounded neutral, in cable tray, #2	2 Elec	300	.053	L.F.	15.50	2.61		18.11	21
2600	#1		280	.057		16.50	2.80		19.30	22.50
2800	1/0		260	.062		18.95	3.02		21.97	25.50
2900	2/0		220	.073		25	3.56		28.56	33
3000	4/0		190	.084		28.50	4.13		32.63	37.50
3100	250 kcmil	3 Elec	270	.089		31.50	4.36		35.86	41.50
3150	350 kcmil		240	.100		37.50	4.90		42.40	48.50
3200	500 kcmil		210	.114		50	5.60		55.60	63.50
3400	15 kV, copper, 3 conductor with PVC jacket,									
3450	ungrounded neutral, in cable tray, #2	2 Elec	260	.062	L.F.	16.70	3.02		19.72	23
3500	#1		230	.070		18.50	3.41		21.91	25.50
3600	1/0		200	.080		21	3.92		24.92	29.50
3700	2/0		190	.084		26	4.13		30.13	34.50
3800	4/0		160	.100		31.50	4.90		36.40	42
4000	250 kcmil	3 Elec	210	.114		36.50	5.60		42.10	48.50
4050	350 kcmil		195	.123		48	6.05		54.05	62
4100	500 kcmil		180	.133		58.50	6.55		65.05	74

26 05 19 – Low-Voltage Electrical Power Conductors and Cables

26 05 19.20 Armored Cable

	26 05 19.20 Armored Cable	Crew	Daily Output	Labor-Hours	Unit	Material	2010 Bare Costs Labor	Equipment	Total	Total Incl O&P
9010	600 volt, copper (MC) steel clad, #14, 2 wire	1 Elec	2.40	3.333	C.L.F.	74	163		237	325
9020	3 wire		2.20	3.636		114	178		292	390
9040	#12, 2 wire		2.30	3.478		75	170		245	335
9050	3 wire		2	4		126	196		322	430
9070	#10, 2 wire		2	4		158	196		354	465
9080	3 wire		1.60	5		219	245		464	605
9100	#8, 2 wire, stranded		1.80	4.444		282	218		500	635
9110	3 wire, stranded		1.30	6.154		420	300		720	910
9200	600 volt, copper (MC) aluminum clad, #14, 2 wire		2.65	3.019		71.50	148		219.50	299
9210	3 wire		2.45	3.265		111	160		271	360
9220	4 wire		2.20	3.636		150	178		328	430
9230	#12, 2 wire		2.55	3.137		73	154		227	310
9240	3 wire		2.20	3.636		122	178		300	400
9250	4 wire		2	4		164	196		360	470
9260	#10, 2 wire		2.20	3.636		154	178		332	435
9270	3 wire		1.80	4.444		213	218		431	560
9280	4 wire	↓	1.55	5.161	↓	330	253		583	740

26 05 19.35 Cable Terminations

	26 05 19.35 Cable Terminations	Crew	Daily Output	Labor-Hours	Unit	Material	Labor	Equipment	Total	Total Incl O&P
0010	**CABLE TERMINATIONS**									
0015	Wire connectors, screw type, #22 to #14	1 Elec	260	.031	Ea.	.08	1.51		1.59	2.33
0020	#18 to #12		240	.033		.10	1.63		1.73	2.54
0025	#18 to #10		240	.033		.16	1.63		1.79	2.61
0030	Screw-on connectors, insulated, #18 to #12		240	.033		.12	1.63		1.75	2.56
0035	#16 to #10		230	.035		.16	1.70		1.86	2.71
0040	#14 to #8		210	.038		.35	1.87		2.22	3.17
0045	#12 to #6		180	.044		.58	2.18		2.76	3.88
0050	Terminal lugs, solderless, #16 to #10		50	.160		.56	7.85		8.41	12.25
0100	#8 to #4		30	.267		.87	13.05		13.92	20.50
0150	#2 to #1		22	.364		1.17	17.80		18.97	28
0200	1/0 to 2/0		16	.500		2.22	24.50		26.72	39
0250	3/0		12	.667		4	32.50		36.50	53
0300	4/0		11	.727		4	35.50		39.50	57.50
0350	250 kcmil		9	.889		4	43.50		47.50	69.50
0400	350 kcmil		7	1.143		5.15	56		61.15	89
0450	500 kcmil	↓	6	1.333	↓	10.05	65.50		75.55	108
1600	Crimp 1 hole lugs, copper or aluminum, 600 volt									
1620	#14	1 Elec	60	.133	Ea.	.43	6.55		6.98	10.15
1630	#12		50	.160		.58	7.85		8.43	12.30
1640	#10		45	.178		.58	8.70		9.28	13.60
1780	#8		36	.222		2.30	10.90		13.20	18.75
1800	#6		30	.267		2.92	13.05		15.97	22.50
2000	#4		27	.296		3.50	14.50		18	25.50
2200	#2		24	.333		6.45	16.35		22.80	31.50
2400	#1		20	.400		6.60	19.60		26.20	36.50
2600	2/0		15	.533		8.25	26		34.25	48
2800	3/0		12	.667		9.70	32.50		42.20	59
3000	4/0		11	.727		10.95	35.50		46.45	65
3200	250 kcmil		9	.889		12.70	43.50		56.20	79
3400	300 kcmil		8	1		15.30	49		64.30	90
3500	350 kcmil		7	1.143		15.05	56		71.05	100
3600	400 kcmil		6.50	1.231		19.10	60.50		79.60	111
3800	500 kcmil	↓	6	1.333	↓	21	65.50		86.50	121

26 05 Common Work Results for Electrical

26 05 19 – Low-Voltage Electrical Power Conductors and Cables

26 05 19.50 Mineral Insulated Cable

	26 05 19.50 Mineral Insulated Cable	Crew	Daily Output	Labor-Hours	Unit	Material	2010 Bare Costs Labor	Equipment	Total	Total Incl O&P
0010	**MINERAL INSULATED CABLE** 600 volt									
0100	1 conductor, #12	1 Elec	1.60	5	C.L.F.	350	245		595	750
0200	#10		1.60	5		450	245		695	860
0400	#8		1.50	5.333		540	261		801	985
0500	#6		1.40	5.714		605	280		885	1,075
0600	#4	2 Elec	2.40	6.667		840	325		1,165	1,400
0800	#2		2.20	7.273		1,125	355		1,480	1,750
0900	#1		2.10	7.619		1,275	375		1,650	1,950
1000	1/0		2	8		1,475	390		1,865	2,200
1100	2/0		1.90	8.421		1,750	415		2,165	2,550
1200	3/0		1.80	8.889		2,075	435		2,510	2,925
1400	4/0		1.60	10		2,400	490		2,890	3,350
1410	250 kcmil	3 Elec	2.40	10		2,700	490		3,190	3,700
1420	350 kcmil		1.95	12.308		3,025	605		3,630	4,225
1430	500 kcmil		1.95	12.308		3,800	605		4,405	5,075

26 05 19.55 Non-Metallic Sheathed Cable

	26 05 19.55 Non-Metallic Sheathed Cable	Crew	Daily Output	Labor-Hours	Unit	Material	2010 Bare Costs Labor	Equipment	Total	Total Incl O&P
0010	**NON-METALLIC SHEATHED CABLE** 600 volt									
0100	Copper with ground wire, (Romex)									
0150	#14, 2 conductor	1 Elec	2.70	2.963	C.L.F.	23	145		168	241
0200	3 conductor		2.40	3.333		32	163		195	278
0250	#12, 2 conductor		2.50	3.200		34	157		191	271
0300	3 conductor		2.20	3.636		48.50	178		226.50	320
0350	#10, 2 conductor		2.20	3.636		54	178		232	325
0400	3 conductor		1.80	4.444		77.50	218		295.50	410
0450	#8, 3 conductor		1.50	5.333		121	261		382	525
0500	#6, 3 conductor		1.40	5.714		196	280		476	630
0550	SE type SER aluminum cable, 3 RHW and									
0600	1 bare neutral, 3 #8 & 1 #8	1 Elec	1.60	5	C.L.F.	120	245		365	495
0650	3 #6 & 1 #6	"	1.40	5.714		136	280		416	565
0700	3 #4 & 1 #6	2 Elec	2.40	6.667		152	325		477	650
0750	3 #2 & 1 #4		2.20	7.273		224	355		579	775
0800	3 #1/0 & 1 #2		2	8		340	390		730	960
0850	3 #2/0 & 1 #1		1.80	8.889		400	435		835	1,100
0900	3 #4/0 & 1 #2/0		1.60	10		570	490		1,060	1,350

26 05 19.90 Wire

	26 05 19.90 Wire	Crew	Daily Output	Labor-Hours	Unit	Material	2010 Bare Costs Labor	Equipment	Total	Total Incl O&P
0010	**WIRE** R260519-92									
0020	600 volt type THW, copper, solid, #14	1 Elec	13	.615	C.L.F.	7.80	30		37.80	53.50
0030	#12		11	.727		11.65	35.50		47.15	66
0040	#10		10	.800		18.25	39		57.25	78.50
0050	Stranded, #14 R260533-22		13	.615		8.80	30		38.80	54.50
0100	#12		11	.727		12.65	35.50		48.15	67
0120	#10		10	.800		19.15	39		58.15	79.50
0140	#8		8	1		33	49		82	110
0160	#6		6.50	1.231		54.50	60.50		115	150
0180	#4	2 Elec	10.60	1.509		86.50	74		160.50	205
0200	#3		10	1.600		107	78.50		185.50	234
0220	#2		9	1.778		134	87		221	278
0240	#1		8	2		174	98		272	335
0260	1/0		6.60	2.424		211	119		330	410
0280	2/0		5.80	2.759		265	135		400	495
0300	3/0		5	3.200		335	157		492	600
0350	4/0		4.40	3.636		420	178		598	725

26 05 19 – Low-Voltage Electrical Power Conductors and Cables

26 05 19.90 Wire		Crew	Daily Output	Labor-Hours	Unit	2010 Bare Costs Material	Labor	Equipment	Total	Total Incl O&P
0400	250 kcmil	3 Elec	6	4	C.L.F.	495	196		691	835
0420	300 kcmil		5.70	4.211		595	206		801	955
0450	350 kcmil		5.40	4.444		695	218		913	1,100
0480	400 kcmil		5.10	4.706		790	231		1,021	1,225
0490	500 kcmil		4.80	5		960	245		1,205	1,425
0540	Aluminum, stranded, #6	1 Elec	8	1		22	49		71	97
0560	#4	2 Elec	13	1.231		28	60.50		88.50	121
0580	#2		10.60	1.509		37.50	74		111.50	152
0600	#1		9	1.778		54.50	87		141.50	190
0620	1/0		8	2		65.50	98		163.50	218
0640	2/0		7.20	2.222		77.50	109		186.50	248
0680	3/0		6.60	2.424		96	119		215	283
0700	4/0		6.20	2.581		107	126		233	305
0720	250 kcmil	3 Elec	8.70	2.759		131	135		266	345
0740	300 kcmil		8.10	2.963		180	145		325	415
0760	350 kcmil		7.50	3.200		183	157		340	435
0780	400 kcmil		6.90	3.478		214	170		384	490
0800	500 kcmil		6	4		236	196		432	550
0850	600 kcmil		5.70	4.211		299	206		505	635
0880	700 kcmil		5.10	4.706		345	231		576	725
0900	750 kcmil		4.80	5		350	245		595	750
0920	Type THWN-THHN, copper, solid, #14	1 Elec	13	.615		7.80	30		37.80	53.50
0940	#12		11	.727		11.65	35.50		47.15	66
0960	#10		10	.800		18.25	39		57.25	78.50
1000	Stranded, #14		13	.615		8.80	30		38.80	54.50
1200	#12		11	.727		12.65	35.50		48.15	67
1250	#10 **CN**		10	.800		19.15	39		58.15	79.50
1300	#8		8	1		33	49		82	110
1350	#6		6.50	1.231		54.50	60.50		115	150
1400	#4	2 Elec	10.60	1.509		86.50	74		160.50	205

26 05 23 – Control-Voltage Electrical Power Cables

26 05 23.10 Control Cable		Crew	Daily Output	Labor-Hours	Unit	Material	Labor	Equipment	Total	Total Incl O&P
0010	**CONTROL CABLE**									
0020	600 volt, copper, #14 THWN wire with PVC jacket, 2 wires	1 Elec	9	.889	C.L.F.	18	43.50		61.50	85
0100	4 wires		7	1.143		30.50	56		86.50	117
0200	6 wires		6	1.333		54	65.50		119.50	157
0300	8 wires		5.30	1.509		67	74		141	184
0400	10 wires		4.80	1.667		79	81.50		160.50	208
0500	12 wires		4.30	1.860		89	91		180	234
0600	14 wires		3.80	2.105		109	103		212	273
0700	16 wires		3.50	2.286		112	112		224	290
0800	18 wires		3.30	2.424		126	119		245	315
0900	20 wires		3	2.667		146	131		277	355
1000	22 wires		2.80	2.857		150	140		290	375

26 05 26 – Grounding and Bonding for Electrical Systems

26 05 26.80 Grounding		Crew	Daily Output	Labor-Hours	Unit	Material	Labor	Equipment	Total	Total Incl O&P
0010	**GROUNDING**									
0030	Rod, copper clad, 8′ long, 1/2″ diameter	1 Elec	5.50	1.455	Ea.	16.75	71.50		88.25	124
0050	3/4″ diameter		5.30	1.509		30.50	74		104.50	144
0080	10′ long, 1/2″ diameter		4.80	1.667		19.65	81.50		101.15	143
0100	3/4″ diameter		4.40	1.818		36.50	89		125.50	172
0130	15′ long, 3/4″ diameter		4	2		92	98		190	247

26 05 26 – Grounding and Bonding for Electrical Systems

26 05 26.80 Grounding		Crew	Daily Output	Labor-Hours	Unit	Material	2010 Bare Costs Labor	Equipment	Total	Total Incl O&P
0390	Bare copper wire, stranded, #8	1 Elec	11	.727	C.L.F.	30.50	35.50		66	86.50
0400	#6		10	.800		53.50	39		92.50	117
0600	#2	2 Elec	10	1.600		134	78.50		212.50	265
0800	3/0		6.60	2.424		305	119		424	510
1000	4/0		5.70	2.807		385	138		523	630
1200	250 kcmil	3 Elec	7.20	3.333		455	163		618	745
1800	Water pipe ground clamps, heavy duty									
2000	Bronze, 1/2" to 1" diameter	1 Elec	8	1	Ea.	26	49		75	102
2100	1-1/4" to 2" diameter		8	1		34	49		83	111
2200	2-1/2" to 3" diameter		6	1.333		57	65.50		122.50	160
2800	Brazed connections, #6 wire		12	.667		13.45	32.50		45.95	63.50
3000	#2 wire		10	.800		18	39		57	78.50
3100	3/0 wire		8	1		27	49		76	103
3200	4/0 wire		7	1.143		31	56		87	118
3400	250 kcmil wire		5	1.600		36	78.50		114.50	157
3600	500 kcmil wire		4	2		44.50	98		142.50	195

26 05 33 – Raceway and Boxes for Electrical Systems

26 05 33.05 Conduit

		Crew	Daily Output	Labor-Hours	Unit	Material	Labor	Equipment	Total	Total Incl O&P
0010	**CONDUIT** To 15' high, includes 2 terminations, 2 elbows, R260533-22									
0020	11 beam clamps, and 11 couplings per 100 L.F.									
0300	Aluminum, 1/2" diameter	1 Elec	100	.080	L.F.	2.44	3.92		6.36	8.55
0500	3/4" diameter		90	.089		3.32	4.36		7.68	10.15
0700	1" diameter		80	.100		4.38	4.90		9.28	12.10
1000	1-1/4" diameter		70	.114		5.90	5.60		11.50	14.85
1030	1-1/2" diameter		65	.123		7.10	6.05		13.15	16.75
1050	2" diameter		60	.133		9.70	6.55		16.25	20.50
1070	2-1/2" diameter		50	.160		16.10	7.85		23.95	29.50
1100	3" diameter	2 Elec	90	.178		22.50	8.70		31.20	37.50
1130	3-1/2" diameter		80	.200		29.50	9.80		39.30	47
1140	4" diameter		70	.229		36	11.20		47.20	56
1750	Rigid galvanized steel, 1/2" diameter	1 Elec	90	.089		2.62	4.36		6.98	9.40
1770	3/4" diameter		80	.100		2.86	4.90		7.76	10.45
1800	1" diameter		65	.123		3.90	6.05		9.95	13.25
1830	1-1/4" diameter		60	.133		5.35	6.55		11.90	15.60
1850	1-1/2" diameter		55	.145		6.25	7.15		13.40	17.50
1870	2" diameter		45	.178		8.15	8.70		16.85	22
1900	2-1/2" diameter		35	.229		16.10	11.20		27.30	34.50
1930	3" diameter	2 Elec	50	.320		19.50	15.70		35.20	45
1950	3-1/2" diameter		44	.364		24	17.80		41.80	53
1970	4" diameter		40	.400		28.50	19.60		48.10	60.50
2500	Steel, intermediate conduit (IMC), 1/2" diameter	1 Elec	100	.080		1.90	3.92		5.82	7.95
2530	3/4" diameter		90	.089		2.26	4.36		6.62	9
2550	1" diameter		70	.114		3.13	5.60		8.73	11.80
2570	1-1/4" diameter		65	.123		4.13	6.05		10.18	13.50
2600	1-1/2" diameter		60	.133		4.95	6.55		11.50	15.15
2630	2" diameter		50	.160		6.30	7.85		14.15	18.60
2650	2-1/2" diameter		40	.200		12.35	9.80		22.15	28
2670	3" diameter	2 Elec	60	.267		16.70	13.05		29.75	38
2700	3-1/2" diameter		54	.296		21.50	14.50		36	45.50
2730	4" diameter		50	.320		24	15.70		39.70	50
5000	Electric metallic tubing (EMT), 1/2" diameter	1 Elec	170	.047		.62	2.31		2.93	4.11
5020	3/4" diameter CN		130	.062		.96	3.02		3.98	5.55

26 05 33 – Raceway and Boxes for Electrical Systems

26 05 33.05 Conduit		Crew	Daily Output	Labor-Hours	Unit	2010 Bare Costs Material	Labor	Equipment	Total	Total Incl O&P
5040	1" diameter	1 Elec	115	.070	L.F.	1.70	3.41		5.11	6.90
5060	1-1/4" diameter		100	.080		2.77	3.92		6.69	8.90
5080	1-1/2" diameter		90	.089		3.59	4.36		7.95	10.45
5100	2" diameter		80	.100		4.67	4.90		9.57	12.45
5120	2-1/2" diameter	↓	60	.133		11.15	6.55		17.70	22
5140	3" diameter	2 Elec	100	.160		13.25	7.85		21.10	26.50
5160	3-1/2" diameter		90	.178		16.80	8.70		25.50	31.50
5180	4" diameter	↓	80	.200	↓	18.60	9.80		28.40	35
9900	Add to labor for higher elevated installation									
9910	15' to 20' high, add						10%			
9920	20' to 25' high, add						20%			
9930	25' to 30' high, add						25%			
9940	30' to 35' high, add						30%			
9950	35' to 40' high, add						35%			
9960	Over 40' high, add						40%			
9980	Allow. for cond. ftngs., 5% min.-20% max.									

26 05 33.45 Wireway		Crew	Daily Output	Labor-Hours	Unit	Material	Labor	Equipment	Total	Total Incl O&P
0010	**WIREWAY** to 15' high									
0100	Screw cover, NEMA 1 w/fittings and supports, 2-1/2" x 2-1/2"	1 Elec	45	.178	L.F.	11.45	8.70		20.15	25.50
0200	4" x 4"	"	40	.200		12.65	9.80		22.45	28.50
0400	6" x 6"	2 Elec	60	.267		19.70	13.05		32.75	41
0600	8" x 8"	"	40	.400		33	19.60		52.60	65.50
4475	Screw cover, NEMA 3R w/fittings and supports, 4" x 4"	1 Elec	36	.222		27	10.90		37.90	45.50
4480	6" x 6"	2 Elec	55	.291		34	14.25		48.25	58.50
4485	8" x 8"		36	.444		55	22		77	93
4490	12" x 12"	↓	18	.889	↓	77	43.50		120.50	150
9980	Allow. for wireway ftngs., 5% min.-20% max.									

26 05 33.50 Outlet Boxes		Crew	Daily Output	Labor-Hours	Unit	Material	Labor	Equipment	Total	Total Incl O&P
0010	**OUTLET BOXES**									
0020	Pressed steel, octagon, 4"	1 Elec	20	.400	Ea.	3.36	19.60		22.96	32.50
0060	Covers, blank		64	.125		1.40	6.15		7.55	10.65
0100	Extension rings		40	.200		5.60	9.80		15.40	20.50
0150	Square, 4"		20	.400		2.38	19.60		21.98	31.50
0200	Extension rings		40	.200		5.65	9.80		15.45	21
0250	Covers, blank		64	.125		1.53	6.15		7.68	10.80
0300	Plaster rings		64	.125		3.09	6.15		9.24	12.50
0650	Switchbox		27	.296		5.40	14.50		19.90	27.50
1100	Concrete, floor, 1 gang		5.30	1.509		89.50	74		163.50	209
2000	Poke-thru fitting, fire rated, for 3-3/4" floor		6.80	1.176		119	57.50		176.50	217
2040	For 7" floor		6.80	1.176		119	57.50		176.50	217
2100	Pedestal, 15 amp, duplex receptacle & blank plate		5.25	1.524		131	74.50		205.50	256
2120	Duplex receptacle and telephone plate		5.25	1.524		131	74.50		205.50	256
2140	Pedestal, 20 amp, duplex recept. & phone plate		5	1.600		132	78.50		210.50	263
2200	Abandonment plate	↓	32	.250	↓	37.50	12.25		49.75	59

26 05 33.65 Pull Boxes		Crew	Daily Output	Labor-Hours	Unit	Material	Labor	Equipment	Total	Total Incl O&P
0010	**PULL BOXES**									
0100	Sheet metal, pull box, NEMA 1, type SC, 6" W x 6" H x 4" D	1 Elec	8	1	Ea.	13.65	49		62.65	88
0200	8" W x 8" H x 4" D		8	1		18.70	49		67.70	93.50
0300	10" W x 12" H x 6" D		5.30	1.509		33	74		107	147
0400	16" W x 20" H x 8" D		4	2		124	98		222	282
0500	20" W x 24" H x 8" D		3.20	2.500		134	123		257	330
0600	24" W x 36" H x 8" D	↓	2.70	2.963	↓	207	145		352	445

26 05 33 – Raceway and Boxes for Electrical Systems

26 05 33.65 Pull Boxes		Crew	Daily Output	Labor-Hours	Unit	Material	2010 Bare Costs Labor	Equipment	Total	Total Incl O&P
0650	Pull box, hinged , NEMA 1, 6″ W x 6″ H x 4″ D	1 Elec	8	1	Ea.	14	49		63	88.50
0800	12″ W x 16″ H x 6″ D		4.70	1.702		46	83.50		129.50	175
1000	20″ W x 20″ H x 6″ D		3.60	2.222		92.50	109		201.50	264
1200	20″ W x 20″ H x 8″ D		3.20	2.500		185	123		308	385
1400	24″ W x 36″ H x 8″ D		2.70	2.963		296	145		441	540
1600	24″ W x 42″ H x 8″ D		2	4		450	196		646	785
2100	Pull box, NEMA 3R, type SC, raintight & weatherproof									
2150	6″ L x 6″ W x 6″ D	1 Elec	10	.800	Ea.	23.50	39		62.50	84.50
2200	8″ L x 6″ W x 6″ D		8	1		29.50	49		78.50	106
2250	10″ L x 6″ W x 6″ D		7	1.143		39	56		95	127
2300	12″ L x 12″ W x 6″ D		5	1.600		55.50	78.50		134	178
2350	16″ L x 16″ W x 6″ D		4.50	1.778		110	87		197	251
2400	20″ L x 20″ W x 6″ D		4	2		150	98		248	310
2450	24″ L x 18″ W x 8″ D		3	2.667		166	131		297	375
2500	24″ L x 24″ W x 10″ D		2.50	3.200		200	157		357	455
2550	30″ L x 24″ W x 12″ D		2	4		405	196		601	735
2600	36″ L x 36″ W x 12″ D		1.50	5.333		530	261		791	975
2800	Cast iron, pull boxes for surface mounting									
3000	NEMA 4, watertight & dust tight									
3050	6″ L x 6″ W x 6″ D	1 Elec	4	2	Ea.	233	98		331	400
3100	8″ L x 6″ W x 6″ D		3.20	2.500		253	123		376	460
3150	10″ L x 6″ W x 6″ D		2.50	3.200		315	157		472	580
3200	12″ L x 12″ W x 6″ D		2.30	3.478		555	170		725	870
3250	16″ L x 16″ W x 6″ D		1.30	6.154		1,325	300		1,625	1,900
3300	20″ L x 20″ W x 6″ D		.80	10		2,450	490		2,940	3,425
3350	24″ L x 18″ W x 8″ D		.70	11.429		2,575	560		3,135	3,650
3400	24″ L x 24″ W x 10″ D		.50	16		3,950	785		4,735	5,525
3450	30″ L x 24″ W x 12″ D		.40	20		6,625	980		7,605	8,725
3500	36″ L x 36″ W x 12″ D		.20	40		8,175	1,950		10,125	11,900
6000	J.I.C. wiring boxes, NEMA 12, dust tight & drip tight									
6050	6″ L x 8″ W x 4″ D	1 Elec	10	.800	Ea.	56.50	39		95.50	121
6100	8″ L x 10″ W x 4″ D		8	1		70.50	49		119.50	151
6150	12″ L x 14″ W x 6″ D		5.30	1.509		118	74		192	240
6200	14″ L x 16″ W x 6″ D		4.70	1.702		148	83.50		231.50	287
6250	16″ L x 20″ W x 6″ D		4.40	1.818		291	89		380	450
6300	24″ L x 30″ W x 6″ D		3.20	2.500		430	123		553	650
6350	24″ L x 30″ W x 8″ D		2.90	2.759		460	135		595	705
6400	24″ L x 36″ W x 8″ D		2.70	2.963		510	145		655	775
6450	24″ L x 42″ W x 8″ D		2.30	3.478		555	170		725	865
6500	24″ L x 48″ W x 8″ D		2	4		605	196		801	955

26 05 36 – Cable Trays for Electrical Systems

26 05 36.10 Cable Tray Ladder Type		Crew	Daily Output	Labor-Hours	Unit	Material	Labor	Equipment	Total	Total Incl O&P
0010	**CABLE TRAY LADDER TYPE** w/ftngs. & supports, 4″ dp., to 15′ elev.									
0160	Galvanized steel tray									
0170	4″ rung spacing, 6″ wide	2 Elec	98	.163	L.F.	13.40	8		21.40	26.50
0200	12″ wide		86	.186		16.10	9.10		25.20	31.50
0400	18″ wide		82	.195		18.70	9.55		28.25	34.50
0600	24″ wide		78	.205		21.50	10.05		31.55	38.50
3200	Aluminum tray, 4″ deep, 6″ rung spacing, 6″ wide		134	.119		16.80	5.85		22.65	27
3220	12″ wide		124	.129		18.85	6.30		25.15	30.50
3230	18″ wide		114	.140		21	6.90		27.90	33
3240	24″ wide		106	.151		24.50	7.40		31.90	38

26 05 Common Work Results for Electrical

26 05 36 – Cable Trays for Electrical Systems

26 05 36.10 Cable Tray Ladder Type		Crew	Daily Output	Labor-Hours	Unit	Material	2010 Bare Costs Labor	Equipment	Total	Total Incl O&P
9980	Allow. for tray ftngs., 5% min.-20% max.									

26 05 39 – Underfloor Raceways for Electrical Systems

26 05 39.30 Conduit In Concrete Slab

		Crew	Daily Output	Labor-Hours	Unit	Material	Labor	Equipment	Total	Total Incl O&P
0010	**CONDUIT IN CONCRETE SLAB** Including terminations,									
0020	fittings and supports									
3230	PVC, schedule 40, 1/2" diameter	1 Elec	270	.030	L.F.	.67	1.45		2.12	2.90
3250	3/4" diameter		230	.035		.81	1.70		2.51	3.42
3270	1" diameter		200	.040		1.11	1.96		3.07	4.13
3300	1-1/4" diameter		170	.047		1.55	2.31		3.86	5.15
3330	1-1/2" diameter		140	.057		1.85	2.80		4.65	6.20
3350	2" diameter		120	.067		2.38	3.27		5.65	7.50
4350	Rigid galvanized steel, 1/2" diameter		200	.040		2.26	1.96		4.22	5.40
4400	3/4" diameter		170	.047		2.50	2.31		4.81	6.20
4450	1" diameter		130	.062		3.54	3.02		6.56	8.35
4500	1-1/4" diameter		110	.073		4.73	3.56		8.29	10.50
4600	1-1/2" diameter		100	.080		5.65	3.92		9.57	12.05
4800	2" diameter	↓	90	.089	↓	7.30	4.36		11.66	14.50

26 05 39.40 Conduit In Trench

		Crew	Daily Output	Labor-Hours	Unit	Material	Labor	Equipment	Total	Total Incl O&P
0010	**CONDUIT IN TRENCH** Includes terminations and fittings									
0020	Does not include excavation or backfill, see Div. 31 23 16.00									
0200	Rigid galvanized steel, 2" diameter	1 Elec	150	.053	L.F.	7.05	2.61		9.66	11.65
0400	2-1/2" diameter	"	100	.080		14.05	3.92		17.97	21.50
0600	3" diameter	2 Elec	160	.100		17.35	4.90		22.25	26.50
0800	3-1/2" diameter		140	.114		22.50	5.60		28.10	33.50
1000	4" diameter		100	.160		25.50	7.85		33.35	39.50
1200	5" diameter		80	.200		53.50	9.80		63.30	73.50
1400	6" diameter	↓	60	.267	↓	78.50	13.05		91.55	106

26 05 43 – Underground Ducts and Raceways for Electrical Systems

26 05 43.10 Trench Duct

		Crew	Daily Output	Labor-Hours	Unit	Material	Labor	Equipment	Total	Total Incl O&P
0010	**TRENCH DUCT** Steel with cover									
0020	Standard adjustable, depths to 4"									
0100	Straight, single compartment, 9" wide	2 Elec	40	.400	L.F.	107	19.60		126.60	146
0200	12" wide		32	.500		121	24.50		145.50	170
0400	18" wide		26	.615		161	30		191	222
0600	24" wide		22	.727		207	35.50		242.50	280
0800	30" wide		20	.800		250	39		289	335
1000	36" wide		16	1	↓	300	49		349	405
1200	Horizontal elbow, 9" wide		5.40	2.963	Ea.	395	145		540	650
1400	12" wide		4.60	3.478		455	170		625	755
1600	18" wide		4	4		585	196		781	930
1800	24" wide		3.20	5		820	245		1,065	1,275
2000	30" wide		2.60	6.154		1,100	300		1,400	1,650
2200	36" wide		2.40	6.667		1,425	325		1,750	2,050
2400	Vertical elbow, 9" wide		5.40	2.963		138	145		283	370
2600	12" wide		4.60	3.478		150	170		320	415
2800	18" wide		4	4		172	196		368	480
3000	24" wide		3.20	5		213	245		458	600
3200	30" wide		2.60	6.154		235	300		535	710
3400	36" wide		2.40	6.667		259	325		584	770
3600	Cross, 9" wide		4	4		645	196		841	1,000
3800	12" wide	↓	3.20	5	↓	685	245		930	1,125

26 05 Common Work Results for Electrical

26 05 43 – Underground Ducts and Raceways for Electrical Systems

26 05 43.10 Trench Duct		Crew	Daily Output	Labor-Hours	Unit	2010 Bare Costs Material	Labor	Equipment	Total	Total Incl O&P
4000	18" wide	2 Elec	2.60	6.154	Ea.	820	300		1,120	1,350
4200	24" wide		2.20	7.273		1,025	355		1,380	1,675
4400	30" wide		2	8		1,350	390		1,740	2,050
4600	36" wide		1.80	8.889		1,675	435		2,110	2,500
4800	End closure, 9" wide		14.40	1.111		40.50	54.50		95	126
5000	12" wide		12	1.333		46.50	65.50		112	148
5200	18" wide		10	1.600		71	78.50		149.50	195
5400	24" wide		8	2		93.50	98		191.50	249
5600	30" wide		6.60	2.424		117	119		236	305
5800	36" wide		5.80	2.759		139	135		274	355
6000	Tees, 9" wide		4	4		395	196		591	720
6200	12" wide		3.60	4.444		455	218		673	825
6400	18" wide		3.20	5		585	245		830	1,000
6600	24" wide		3	5.333		840	261		1,101	1,300
6800	30" wide		2.60	6.154		1,100	300		1,400	1,650
7000	36" wide		2	8		1,425	390		1,815	2,150
7200	Riser, and cabinet connector, 9" wide		5.40	2.963		172	145		317	405
7400	12" wide		4.60	3.478		200	170		370	475
7600	18" wide		4	4		246	196		442	560
7800	24" wide		3.20	5		298	245		543	690
8000	30" wide		2.60	6.154		340	300		640	825
8200	36" wide	↓	2	8		400	390		790	1,025
8400	Insert assembly, cell to conduit adapter, 1-1/4"	1 Elec	16	.500	↓	68	24.50		92.50	112

26 05 43.20 Underfloor Duct		Crew	Daily Output	Labor-Hours	Unit	Material	Labor	Equipment	Total	Total Incl O&P
0010	**UNDERFLOOR DUCT**									
0100	Duct, 1-3/8" x 3-1/8" blank, standard	2 Elec	160	.100	L.F.	13	4.90		17.90	21.50
0200	1-3/8" x 7-1/4" blank, super duct		120	.133		26	6.55		32.55	38
0400	7/8" or 1-3/8" insert type, 24" O.C., 1-3/8" x 3-1/8", std.		140	.114		17.40	5.60		23	27.50
0600	1-3/8" x 7-1/4", super duct	↓	100	.160	↓	30.50	7.85		38.35	45
0800	Junction box, single duct, 1 level, 3-1/8"	1 Elec	4	2	Ea.	390	98		488	575
1000	Junction box, single duct, 1 level, 7-1/4"		2.70	2.963		455	145		600	715
1200	1 level, 2 duct, 3-1/8"		3.20	2.500		520	123		643	755
1400	Junction box, 1 level, 2 duct, 7-1/4"		2.30	3.478		1,350	170		1,520	1,725
1600	Triple duct, 3-1/8"		2.30	3.478		885	170		1,055	1,225
1800	Insert to conduit adapter, 3/4" & 1"		32	.250		32	12.25		44.25	53
2000	Support, single cell		27	.296		48	14.50		62.50	74.50
2200	Super duct		16	.500		48	24.50		72.50	89.50
2400	Double cell		16	.500		48	24.50		72.50	89.50
2600	Triple cell		11	.727		48	35.50		83.50	106
2800	Vertical elbow, standard duct		10	.800		86	39		125	154
3000	Super duct		8	1		86	49		135	168
3200	Cabinet connector, standard duct		32	.250		65	12.25		77.25	89.50
3400	Super duct		27	.296		65	14.50		79.50	93
3600	Conduit adapter, 1" to 1-1/4"		32	.250		65	12.25		77.25	89.50
3800	2" to 1-1/4"		27	.296		78	14.50		92.50	107
4000	Outlet, low tension (tele, computer, etc.)		8	1		91	49		140	173
4200	High tension, receptacle (120 volt)	↓	8	1	↓	91	49		140	173

26 05 80 – Wiring Connections

26 05 80.10 Motor Connections		Crew	Daily Output	Labor-Hours	Unit	Material	Labor	Equipment	Total	Total Incl O&P
0010	**MOTOR CONNECTIONS**									
0020	Flexible conduit and fittings, 115 volt, 1 phase, up to 1 HP motor	1 Elec	8	1	Ea.	9.55	49		58.55	83.50
0120	230 volt, 10 HP motor, 3 phase	↓	4.20	1.905	↓	15.90	93.50		109.40	156

26 05 80 – Wiring Connections

26 05 80.10 Motor Connections		Crew	Daily Output	Labor-Hours	Unit	Material	2010 Bare Costs Labor	Equipment	Total	Total Incl O&P
0200	25 HP motor	1 Elec	2.70	2.963	Ea.	31.50	145		176.50	251
0400	50 HP motor		2.20	3.636		80.50	178		258.50	355
0600	100 HP motor		1.50	5.333		197	261		458	605

26 05 90 – Residential Applications

26 05 90.10 Residential Wiring		Crew	Daily Output	Labor-Hours	Unit	Material	Labor	Equipment	Total	Total Incl O&P
0010	**RESIDENTIAL WIRING**									
0020	20' avg. runs and #14/2 wiring incl. unless otherwise noted									
1000	Service & panel, includes 24' SE-AL cable, service eye, meter,									
1010	Socket, panel board, main bkr., ground rod, 15 or 20 amp									
1020	1-pole circuit breakers, and misc. hardware									
1100	100 amp, with 10 branch breakers	1 Elec	1.19	6.723	Ea.	565	330		895	1,125
1110	With PVC conduit and wire		.92	8.696		630	425		1,055	1,325
1120	With RGS conduit and wire		.73	10.959		805	535		1,340	1,700
1150	150 amp, with 14 branch breakers		1.03	7.767		880	380		1,260	1,525
1170	With PVC conduit and wire		.82	9.756		1,025	480		1,505	1,825
1180	With RGS conduit and wire		.67	11.940		1,350	585		1,935	2,375
1200	200 amp, with 18 branch breakers	2 Elec	1.80	8.889		1,150	435		1,585	1,925
1220	With PVC conduit and wire		1.46	10.959		1,300	535		1,835	2,225
1230	With RGS conduit and wire		1.24	12.903		1,750	630		2,380	2,875
1800	Lightning surge suppressor for above services, add	1 Elec	32	.250		51.50	12.25		63.75	74.50
2000	Switch devices									
2100	Single pole, 15 amp, Ivory, with a 1-gang box, cover plate,									
2110	Type NM (Romex) cable	1 Elec	17.10	.468	Ea.	10.05	23		33.05	45
2120	Type MC (BX) cable		14.30	.559		27.50	27.50		55	71.50
2130	EMT & wire		5.71	1.401		32.50	68.50		101	138
2150	3-way, #14/3, type NM cable		14.55	.550		14.05	27		41.05	55.50
2170	Type MC cable		12.31	.650		38	32		70	89
2180	EMT & wire		5	1.600		36	78.50		114.50	157
2200	4-way, #14/3, type NM cable		14.55	.550		30.50	27		57.50	73.50
2220	Type MC cable		12.31	.650		54	32		86	107
2230	EMT & wire		5	1.600		52.50	78.50		131	175
2250	S.P., 20 amp, #12/2, type NM cable		13.33	.600		16.85	29.50		46.35	62
2270	Type MC cable		11.43	.700		32	34.50		66.50	86.50
2280	EMT & wire		4.85	1.649		41	81		122	165
2290	S.P. rotary dimmer, 600W, no wiring		17	.471		20.50	23		43.50	57.50
2300	S.P. rotary dimmer, 600W, type NM cable		14.55	.550		25.50	27		52.50	68
2320	Type MC cable		12.31	.650		43	32		75	94.50
2330	EMT & wire		5	1.600		49	78.50		127.50	171
2350	3-way rotary dimmer, type NM cable		13.33	.600		32.50	29.50		62	79.50
2370	Type MC cable		11.43	.700		50	34.50		84.50	106
2380	EMT & wire		4.85	1.649		56.50	81		137.50	182
2400	Interval timer wall switch, 20 amp, 1-30 min., #12/2									
2410	Type NM cable	1 Elec	14.55	.550	Ea.	50	27		77	95
2420	Type MC cable		12.31	.650		58.50	32		90.50	112
2430	EMT & wire		5	1.600		74	78.50		152.50	199
2500	Decorator style									
2510	S.P., 15 amp, type NM cable	1 Elec	17.10	.468	Ea.	14.90	23		37.90	50.50
2520	Type MC cable		14.30	.559		32.50	27.50		60	76.50
2530	EMT & wire		5.71	1.401		37.50	68.50		106	143
2550	3-way, #14/3, type NM cable		14.55	.550		18.95	27		45.95	61
2570	Type MC cable		12.31	.650		43	32		75	94.50
2580	EMT & wire		5	1.600		41	78.50		119.50	162

26 05 90.10 Residential Wiring		Crew	Daily Output	Labor-Hours	Unit	Material	Labor	Equipment	Total	Total Incl O&P
						2010 Bare Costs				
2600	4-way, #14/3, type NM cable	1 Elec	14.55	.550	Ea.	35	27		62	78.50
2620	Type MC cable		12.31	.650		59	32		91	113
2630	EMT & wire		5	1.600		57.50	78.50		136	180
2650	S.P., 20 amp, #12/2, type NM cable		13.33	.600		21.50	29.50		51	67.50
2670	Type MC cable		11.43	.700		37	34.50		71.50	92
2680	EMT & wire		4.85	1.649		45.50	81		126.50	170
2700	S.P., slide dimmer, type NM cable		17.10	.468		33	23		56	70.50
2720	Type MC cable		14.30	.559		50.50	27.50		78	96.50
2730	EMT & wire		5.71	1.401		57	68.50		125.50	165
2750	S.P., touch dimmer, type NM cable		17.10	.468		27	23		50	63.50
2770	Type MC cable		14.30	.559		44.50	27.50		72	90
2780	EMT & wire		5.71	1.401		51	68.50		119.50	158
2800	3-way touch dimmer, type NM cable		13.33	.600		45.50	29.50		75	93.50
2820	Type MC cable		11.43	.700		63	34.50		97.50	120
2830	EMT & wire	↓	4.85	1.649	↓	69.50	81		150.50	196
3000	Combination devices									
3100	S.P. switch/15 amp recpt., Ivory, 1-gang box, plate									
3110	Type NM cable	1 Elec	11.43	.700	Ea.	21.50	34.50		56	74.50
3120	Type MC cable		10	.800		39	39		78	102
3130	EMT & wire		4.40	1.818		45.50	89		134.50	182
3150	S.P. switch/pilot light, type NM cable		11.43	.700		22	34.50		56.50	75.50
3170	Type MC cable		10	.800		39.50	39		78.50	102
3180	EMT & wire		4.43	1.806		46	88.50		134.50	183
3190	2-S.P. switches, 2-#14/2, no wiring		14	.571		8.25	28		36.25	50.50
3200	2-S.P. switches, 2-#14/2, type NM cables		10	.800		25	39		64	86
3220	Type MC cable		8.89	.900		52	44		96	123
3230	EMT & wire		4.10	1.951		49	95.50		144.50	196
3250	3-way switch/15 amp recpt., #14/3, type NM cable		10	.800		29	39		68	90.50
3270	Type MC cable		8.89	.900		53	44		97	124
3280	EMT & wire		4.10	1.951		51	95.50		146.50	199
3300	2-3 way switches, 2-#14/3, type NM cables		8.89	.900		38	44		82	108
3320	Type MC cable		8	1		78	49		127	159
3330	EMT & wire		4	2		58.50	98		156.50	211
3350	S.P. switch/20 amp recpt., #12/2, type NM cable		10	.800		33.50	39		72.50	95.50
3370	Type MC cable		8.89	.900		42.50	44		86.50	112
3380	EMT & wire	↓	4.10	1.951	↓	57.50	95.50		153	206
3400	Decorator style									
3410	S.P. switch/15 amp recpt., type NM cable	1 Elec	11.43	.700	Ea.	26.50	34.50		61	80
3420	Type MC cable		10	.800		44	39		83	107
3430	EMT & wire		4.40	1.818		50	89		139	187
3450	S.P. switch/pilot light, type NM cable		11.43	.700		27	34.50		61.50	81
3470	Type MC cable		10	.800		44.50	39		83.50	108
3480	EMT & wire		4.40	1.818		51	89		140	188
3500	2-S.P. switches, 2-#14/2, type NM cables		10	.800		30	39		69	91.50
3520	Type MC cable		8.89	.900		57	44		101	128
3530	EMT & wire		4.10	1.951		54	95.50		149.50	202
3550	3-way/15 amp recpt., #14/3, type NM cable		10	.800		34	39		73	96
3570	Type MC cable		8.89	.900		58	44		102	129
3580	EMT & wire		4.10	1.951		56	95.50		151.50	204
3650	2-3 way switches, 2-#14/3, type NM cables		8.89	.900		43	44		87	113
3670	Type MC cable		8	1		83	49		132	164
3680	EMT & wire		4	2		63.50	98		161.50	216
3700	S.P. switch/20 amp recpt., #12/2, type NM cable	↓	10	.800	↓	38.50	39		77.50	101

26 05 90 – Residential Applications

26 05 90.10 Residential Wiring		Crew	Daily Output	Labor-Hours	Unit	Material	Labor	Equipment	Total	Total Incl O&P
						2010 Bare Costs				
3720	Type MC cable	1 Elec	8.89	.900	Ea.	47	44		91	118
3730	EMT & wire	↓	4.10	1.951	↓	62.50	95.50		158	211
4000	Receptacle devices									
4010	Duplex outlet, 15 amp recpt., Ivory, 1-gang box, plate									
4015	Type NM cable	1 Elec	14.55	.550	Ea.	8.55	27		35.55	49.50
4020	Type MC cable		12.31	.650		26	32		58	76
4030	EMT & wire		5.33	1.501		31	73.50		104.50	143
4050	With #12/2, type NM cable		12.31	.650		10.80	32		42.80	59.50
4070	Type MC cable		10.67	.750		26	36.50		62.50	83.50
4080	EMT & wire		4.71	1.699		34.50	83		117.50	162
4100	20 amp recpt., #12/2, type NM cable		12.31	.650		19.85	32		51.85	69.50
4120	Type MC cable		10.67	.750		35	36.50		71.50	93.50
4130	EMT & wire	↓	4.71	1.699	↓	44	83		127	172
4140	For GFI see Div. 26 05 90.10 line 4300 below									
4150	Decorator style, 15 amp recpt., type NM cable	1 Elec	14.55	.550	Ea.	13.40	27		40.40	55
4170	Type MC cable		12.31	.650		31	32		63	81.50
4180	EMT & wire		5.33	1.501		36	73.50		109.50	149
4200	With #12/2, type NM cable		12.31	.650		15.70	32		47.70	65
4220	Type MC cable		10.67	.750		31	36.50		67.50	88.50
4230	EMT & wire		4.71	1.699		39.50	83		122.50	168
4250	20 amp recpt. #12/2, type NM cable		12.31	.650		24.50	32		56.50	74.50
4270	Type MC cable		10.67	.750		40	36.50		76.50	98.50
4280	EMT & wire		4.71	1.699		48.50	83		131.50	178
4300	GFI, 15 amp recpt., type NM cable		12.31	.650		43	32		75	95
4320	Type MC cable		10.67	.750		60.50	36.50		97	121
4330	EMT & wire		4.71	1.699		65.50	83		148.50	196
4350	GFI with #12/2, type NM cable		10.67	.750		45.50	36.50		82	105
4370	Type MC cable		9.20	.870		61	42.50		103.50	131
4380	EMT & wire		4.21	1.900		69.50	93		162.50	214
4400	20 amp recpt., #12/2 type NM cable		10.67	.750		47.50	36.50		84	107
4420	Type MC cable		9.20	.870		63	42.50		105.50	133
4430	EMT & wire		4.21	1.900		71.50	93		164.50	217
4500	Weather-proof cover for above receptacles, add	↓	32	.250	↓	5.25	12.25		17.50	24
4550	Air conditioner outlet, 20 amp-240 volt recpt.									
4560	30′ of #12/2, 2 pole circuit breaker									
4570	Type NM cable	1 Elec	10	.800	Ea.	60.50	39		99.50	125
4580	Type MC cable		9	.889		79.50	43.50		123	153
4590	EMT & wire		4	2		84.50	98		182.50	239
4600	Decorator style, type NM cable		10	.800		65	39		104	130
4620	Type MC cable		9	.889		84.50	43.50		128	158
4630	EMT & wire	↓	4	2	↓	89	98		187	244
4650	Dryer outlet, 30 amp-240 volt recpt., 20′ of #10/3									
4660	2 pole circuit breaker									
4670	Type NM cable	1 Elec	6.41	1.248	Ea.	66	61		127	164
4680	Type MC cable		5.71	1.401		80	68.50		148.50	190
4690	EMT & wire	↓	3.48	2.299	↓	81	113		194	256
4700	Range outlet, 50 amp-240 volt recpt., 30′ of #8/3									
4710	Type NM cable	1 Elec	4.21	1.900	Ea.	93	93		186	240
4720	Type MC cable		4	2		163	98		261	325
4730	EMT & wire		2.96	2.703		111	132		243	320
4750	Central vacuum outlet, Type NM cable		6.40	1.250		62.50	61.50		124	160
4770	Type MC cable		5.71	1.401		90	68.50		158.50	201
4780	EMT & wire	↓	3.48	2.299	↓	88.50	113		201.50	265

26 05 90 – Residential Applications

26 05 90.10 Residential Wiring		Crew	Daily Output	Labor-Hours	Unit	Material	2010 Bare Costs Labor	Equipment	Total	Total Incl O&P
4800	30 amp-110 volt locking recpt., #10/2 circ. bkr.									
4810	Type NM cable	1 Elec	6.20	1.290	Ea.	72.50	63		135.50	174
4820	Type MC cable		5.40	1.481		104	72.50		176.50	223
4830	EMT & wire	↓	3.20	2.500	↓	99	123		222	291
4900	Low voltage outlets									
4910	Telephone recpt., 20′ of 4/C phone wire	1 Elec	26	.308	Ea.	10.35	15.10		25.45	34
4920	TV recpt., 20′ of RG59U coax wire, F type connector	"	16	.500	"	19.30	24.50		43.80	57.50
4950	Door bell chime, transformer, 2 buttons, 60′ of bellwire									
4970	Economy model	1 Elec	11.50	.696	Ea.	75	34		109	133
4980	Custom model		11.50	.696		119	34		153	182
4990	Luxury model, 3 buttons	↓	9.50	.842	↓	320	41.50		361.50	415
6000	Lighting outlets									
6050	Wire only (for fixture), type NM cable	1 Elec	32	.250	Ea.	6.45	12.25		18.70	25.50
6070	Type MC cable		24	.333		17.05	16.35		33.40	43.50
6080	EMT & wire		10	.800		20.50	39		59.50	81
6100	Box (4″), and wire (for fixture), type NM cable		25	.320		16.25	15.70		31.95	41.50
6120	Type MC cable		20	.400		27	19.60		46.60	58.50
6130	EMT & wire	↓	11	.727	↓	30.50	35.50		66	86.50
6200	Fixtures (use with lines 6050 or 6100 above)									
6210	Canopy style, economy grade	1 Elec	40	.200	Ea.	34	9.80		43.80	52
6220	Custom grade		40	.200		52.50	9.80		62.30	72.50
6250	Dining room chandelier, economy grade		19	.421		85	20.50		105.50	124
6260	Custom grade		19	.421		250	20.50		270.50	305
6270	Luxury grade		15	.533		565	26		591	660
6310	Kitchen fixture (fluorescent), economy grade		30	.267		63.50	13.05		76.55	89.50
6320	Custom grade		25	.320		173	15.70		188.70	214
6350	Outdoor, wall mounted, economy grade		30	.267		32	13.05		45.05	54.50
6360	Custom grade		30	.267		111	13.05		124.05	141
6370	Luxury grade		25	.320		250	15.70		265.70	299
6410	Outdoor PAR floodlights, 1 lamp, 150 watt		20	.400		35	19.60		54.60	67.50
6420	2 lamp, 150 watt each		20	.400		58	19.60		77.60	93
6425	Motion sensing, 2 lamp, 150 watt each		20	.400		70	19.60		89.60	106
6430	For infrared security sensor, add		32	.250		117	12.25		129.25	147
6450	Outdoor, quartz-halogen, 300 watt flood		20	.400		42	19.60		61.60	75.50
6600	Recessed downlight, round, pre-wired, 50 or 75 watt trim		30	.267		43.50	13.05		56.55	67
6610	With shower light trim		30	.267		53.50	13.05		66.55	78.50
6620	With wall washer trim		28	.286		64.50	14		78.50	92
6630	With eye-ball trim	↓	28	.286		64.50	14		78.50	92
6640	For direct contact with insulation, add					2.20			2.20	2.42
6700	Porcelain lamp holder	1 Elec	40	.200		4.20	9.80		14	19.15
6710	With pull switch		40	.200		5.05	9.80		14.85	20
6750	Fluorescent strip, 1-20 watt tube, wrap around diffuser, 24″		24	.333		54	16.35		70.35	84
6760	1-34 watt tube, 48″		24	.333		69.50	16.35		85.85	101
6770	2-34 watt tubes, 48″		20	.400		82.50	19.60		102.10	120
6780	With residential ballast		20	.400		92.50	19.60		112.10	131
6800	Bathroom heat lamp, 1-250 watt		28	.286		46	14		60	71.50
6810	2-250 watt lamps	↓	28	.286	↓	71.50	14		85.50	99.50
6820	For timer switch, see Div. 26 05 90.10 line 2400									
6900	Outdoor post lamp, incl. post, fixture, 35′ of #14/2									
6910	Type NMC cable	1 Elec	3.50	2.286	Ea.	202	112		314	390
6920	Photo-eye, add		27	.296		35	14.50		49.50	60
6950	Clock dial time switch, 24 hr., w/enclosure, type NM cable		11.43	.700		61	34.50		95.50	118
6970	Type MC cable	↓	11	.727	↓	78.50	35.50		114	140

26 05 90.10 Residential Wiring		Crew	Daily Output	Labor-Hours	Unit	2010 Bare Costs Material	Labor	Equipment	Total	Total Incl O&P
6980	EMT & wire	1 Elec	4.85	1.649	Ea.	83.50	81		164.50	212
7000	Alarm systems									
7050	Smoke detectors, box, #14/3, type NM cable	1 Elec	14.55	.550	Ea.	34	27		61	77.50
7070	Type MC cable		12.31	.650		51.50	32		83.50	104
7080	EMT & wire		5	1.600		49.50	78.50		128	172
7090	For relay output to security system, add					13			13	14.30
8000	Residential equipment									
8050	Disposal hook-up, incl. switch, outlet box, 3′ of flex									
8060	20 amp-1 pole circ. bkr., and 25′ of #12/2									
8070	Type NM cable	1 Elec	10	.800	Ea.	30.50	39		69.50	92
8080	Type MC cable		8	1		48	49		97	126
8090	EMT & wire		5	1.600		56.50	78.50		135	180
8100	Trash compactor or dishwasher hook-up, incl. outlet box,									
8110	3′ of flex, 15 amp-1 pole circ. bkr., and 25′ of #14/2									
8120	Type NM cable	1 Elec	10	.800	Ea.	22.50	39		61.50	83.50
8130	Type MC cable		8	1		43	49		92	120
8140	EMT & wire		5	1.600		49	78.50		127.50	171
8150	Hot water sink dispensor hook-up, use line 8100									
8200	Vent/exhaust fan hook-up, type NM cable	1 Elec	32	.250	Ea.	6.45	12.25		18.70	25.50
8220	Type MC cable		24	.333		17.05	16.35		33.40	43.50
8230	EMT & wire		10	.800		20.50	39		59.50	81
8250	Bathroom vent fan, 50 CFM (use with above hook-up)									
8260	Economy model	1 Elec	15	.533	Ea.	25	26		51	66.50
8270	Low noise model		15	.533		32	26		58	74
8280	Custom model		12	.667		117	32.50		149.50	178
8300	Bathroom or kitchen vent fan, 110 CFM									
8310	Economy model	1 Elec	15	.533	Ea.	69.50	26		95.50	116
8320	Low noise model	″	15	.533	″	86	26		112	134
8350	Paddle fan, variable speed (w/o lights)									
8360	Economy model (AC motor)	1 Elec	10	.800	Ea.	106	39		145	176
8370	Custom model (AC motor)		10	.800		182	39		221	259
8380	Luxury model (DC motor)		8	1		360	49		409	470
8390	Remote speed switch for above, add		12	.667		30	32.50		62.50	81.50
8500	Whole house exhaust fan, ceiling mount, 36″, variable speed									
8510	Remote switch, incl. shutters, 20 amp-1 pole circ. bkr.									
8520	30′ of #12/2, type NM cable	1 Elec	4	2	Ea.	1,025	98		1,123	1,275
8530	Type MC cable		3.50	2.286		1,050	112		1,162	1,325
8540	EMT & wire		3	2.667		1,050	131		1,181	1,375
8600	Whirlpool tub hook-up, incl. timer switch, outlet box									
8610	3′ of flex, 20 amp-1 pole GFI circ. bkr.									
8620	30′ of #12/2, type NM cable	1 Elec	5	1.600	Ea.	124	78.50		202.50	253
8630	Type MC cable		4.20	1.905		136	93.50		229.50	288
8640	EMT & wire		3.40	2.353		144	115		259	330
8650	Hot water heater hook-up, incl. 1-2 pole circ. bkr., box;									
8660	3′ of flex, 20′ of #10/2, type NM cable	1 Elec	5	1.600	Ea.	32	78.50		110.50	153
8670	Type MC cable		4.20	1.905		55.50	93.50		149	200
8680	EMT & wire		3.40	2.353		52	115		167	229
9000	Heating/air conditioning									
9050	Furnace/boiler hook-up, incl. firestat, local on-off switch									
9060	Emergency switch, and 40′ of type NM cable	1 Elec	4	2	Ea.	54.50	98		152.50	206
9070	Type MC cable		3.50	2.286		81.50	112		193.50	257
9080	EMT & wire		1.50	5.333		89.50	261		350.50	490
9100	Air conditioner hook-up, incl. local 60 amp disc. switch									

26 05 Common Work Results for Electrical

26 05 90 – Residential Applications

26 05 90.10 Residential Wiring		Crew	Daily Output	Labor-Hours	Unit	Material	2010 Bare Costs Labor	Equipment	Total	Total Incl O&P
9110	3' sealtite, 40 amp, 2 pole circuit breaker									
9130	40' of #8/2, type NM cable	1 Elec	3.50	2.286	Ea.	216	112		328	405
9140	Type MC cable		3	2.667		315	131		446	540
9150	EMT & wire	↓	1.30	6.154	↓	252	300		552	730
9200	Heat pump hook-up, 1-40 & 1-100 amp 2 pole circ. bkr.									
9210	Local disconnect switch, 3' sealtite									
9220	40' of #8/2 & 30' of #3/2									
9230	Type NM cable	1 Elec	1.30	6.154	Ea.	545	300		845	1,050
9240	Type MC cable		1.08	7.407		745	365		1,110	1,350
9250	EMT & wire	↓	.94	8.511	↓	615	415		1,030	1,300
9500	Thermostat hook-up, using low voltage wire									
9520	Heating only, 25' of #18-3	1 Elec	24	.333	Ea.	9.85	16.35		26.20	35.50
9530	Heating/cooling, 25' of #18-4	"	20	.400	"	12.30	19.60		31.90	42.50

26 09 Instrumentation and Control for Electrical Systems

26 09 13 – Electrical Power Monitoring and Control

26 09 13.10 Switchboard Instruments		Crew	Daily Output	Labor-Hours	Unit	Material	Labor	Equipment	Total	Total Incl O&P
0010	**SWITCHBOARD INSTRUMENTS** 3 phase, 4 wire									
0100	AC indicating, ammeter & switch	1 Elec	8	1	Ea.	2,075	49		2,124	2,350
0200	Voltmeter & switch		8	1		2,075	49		2,124	2,350
0300	Wattmeter		8	1		4,100	49		4,149	4,575
0400	AC recording, ammeter		4	2		7,300	98		7,398	8,200
0500	Voltmeter		4	2		7,300	98		7,398	8,200
0600	Ground fault protection, zero sequence		2.70	2.963		6,450	145		6,595	7,325
0700	Ground return path		2.70	2.963		6,450	145		6,595	7,325
0800	3 current transformers, 5 to 800 amp		2	4		3,000	196		3,196	3,600
0900	1000 to 1500 amp		1.30	6.154		4,325	300		4,625	5,200
1200	2000 to 4000 amp		1	8		5,100	390		5,490	6,175
1300	Fused potential transformer, maximum 600 volt	↓	8	1	↓	1,125	49		1,174	1,325

26 12 Medium-Voltage Transformers

26 12 19 – Pad-Mounted, Liquid-Filled, Medium-Voltage Transformers

26 12 19.10 Transformer, Oil-Filled		Crew	Daily Output	Labor-Hours	Unit	Material	Labor	Equipment	Total	Total Incl O&P
0010	**TRANSFORMER, OIL-FILLED** primary delta or Y,									
0050	Pad mounted 5 kV or 15 kV, with taps, 277/480 V secondary, 3 phase									
0100	150 kVA	R-3	.65	30.769	Ea.	9,675	1,475	213	11,363	13,000
0200	300 kVA		.45	44.444		12,400	2,150	305	14,855	17,100
0300	500 kVA		.40	50		19,300	2,425	345	22,070	25,300
0400	750 kVA		.38	52.632		22,300	2,550	365	25,215	28,700
0500	1000 kVA		.26	76.923		26,400	3,725	530	30,655	35,200
0600	1500 kVA		.23	86.957		31,400	4,200	600	36,200	41,400
0700	2000 kVA		.20	100		39,600	4,825	690	45,115	51,500
0800	3750 kVA	↓	.16	125	↓	74,500	6,025	865	81,390	91,500

26 22 Low-Voltage Transformers

26 22 13 – Low-Voltage Distribution Transformers

26 22 13.10 Transformer, Dry-Type		Crew	Daily Output	Labor-Hours	Unit	Material	2010 Bare Costs Labor	Equipment	Total	Total Incl O&P
0010	**TRANSFORMER, DRY-TYPE**									
0050	Single phase, 240/480 volt primary, 120/240 volt secondary									
0100	1 kVA	1 Elec	2	4	Ea.	272	196		468	590
0300	2 kVA		1.60	5		410	245		655	815
0500	3 kVA		1.40	5.714		505	280		785	970
0700	5 kVA		1.20	6.667		700	325		1,025	1,250
0900	7.5 kVA	2 Elec	2.20	7.273		970	355		1,325	1,600
1100	10 kVA		1.60	10		1,250	490		1,740	2,100
1300	15 kVA		1.20	13.333		1,700	655		2,355	2,825
1500	25 kVA **CN**		1	16		2,300	785		3,085	3,725
1700	37.5 kVA		.80	20		3,000	980		3,980	4,750
1900	50 kVA		.70	22.857		3,550	1,125		4,675	5,575
2100	75 kVA		.65	24.615		4,700	1,200		5,900	6,975
2190	480V primary 120/240V secondary, nonvent., 15 kVA		1.20	13.333		1,700	655		2,355	2,850
2200	25 kVA		.90	17.778		2,525	870		3,395	4,075
2210	37 kVA		.75	21.333		3,000	1,050		4,050	4,850
2220	50 kVA		.65	24.615		3,550	1,200		4,750	5,700
2300	3 phase, 480 volt primary 120/208 volt secondary									
2310	Ventilated, 3 kVA	1 Elec	1	8	Ea.	885	390		1,275	1,550
2700	6 kVA		.80	10		1,225	490		1,715	2,075
2900	9 kVA		.70	11.429		1,375	560		1,935	2,350
3100	15 kVA	2 Elec	1.10	14.545		1,850	715		2,565	3,075
3300	30 kVA		.90	17.778		2,150	870		3,020	3,675
3500	45 kVA		.80	20		2,600	980		3,580	4,300
3700	75 kVA		.70	22.857		3,900	1,125		5,025	5,975
3900	112.5 kVA	R-3	.90	22.222		5,200	1,075	154	6,429	7,500
4100	150 kVA		.85	23.529		6,775	1,125	163	8,063	9,325
4300	225 kVA		.65	30.769		9,200	1,475	213	10,888	12,500
4500	300 kVA		.55	36.364		11,600	1,750	251	13,601	15,700
4700	500 kVA		.45	44.444		19,200	2,150	305	21,655	24,600
4800	750 kVA		.35	57.143		33,700	2,750	395	36,845	41,600

26 24 Switchboards and Panelboards

26 24 13 – Switchboards

26 24 13.10 Incoming Switchboards		Crew	Daily Output	Labor-Hours	Unit	Material	Labor	Equipment	Total	Total Incl O&P
0010	**INCOMING SWITCHBOARDS** main service section									
0100	Aluminum bus bars, not including CT's or PT's									
0200	No main disconnect, includes CT compartment									
0300	120/208 volt, 4 wire, 600 amp	2 Elec	1	16	Ea.	4,350	785		5,135	5,950
0400	800 amp		.88	18.182		4,350	890		5,240	6,100
0500	1000 amp		.80	20		5,225	980		6,205	7,200
0600	1200 amp		.72	22.222		5,225	1,100		6,325	7,375
0700	1600 amp		.66	24.242		5,225	1,200		6,425	7,525
0800	2000 amp		.62	25.806		5,625	1,275		6,900	8,075
1000	3000 amp		.56	28.571		7,425	1,400		8,825	10,300
2000	Fused switch & CT compartment									
2100	120/208 volt, 4 wire, 400 amp	2 Elec	1.12	14.286	Ea.	4,175	700		4,875	5,625
2200	600 amp		.94	17.021		4,950	835		5,785	6,700
2300	800 amp		.84	19.048		11,800	935		12,735	14,300
2400	1200 amp		.68	23.529		15,300	1,150		16,450	18,500
2900	Pressure switch & CT compartment									

26 24 Switchboards and Panelboards

26 24 13 – Switchboards

	26 24 13.10 Incoming Switchboards	Crew	Daily Output	Labor-Hours	Unit	Material	2010 Bare Costs Labor	Equipment	Total	Total Incl O&P
3000	120/208 volt, 4 wire, 800 amp	2 Elec	.80	20	Ea.	10,600	980		11,580	13,100
3100	1200 amp		.66	24.242		20,400	1,200		21,600	24,300
3200	1600 amp		.62	25.806		21,700	1,275		22,975	25,800
3300	2000 amp		.56	28.571		23,100	1,400		24,500	27,500
4400	Circuit breaker, molded case & CT compartment									
4600	3 pole, 4 wire, 600 amp	2 Elec	.94	17.021	Ea.	9,100	835		9,935	11,300
4800	800 amp		.84	19.048		10,900	935		11,835	13,400
5000	1200 amp		.68	23.529		14,800	1,150		15,950	18,000
5100	Copper bus bars, not incl. CT's or PT's, add, minimum					15%				

	26 24 13.30 Distribution Switchboards Section	Crew	Daily Output	Labor-Hours	Unit	Material	Labor	Equipment	Total	Total Incl O&P
0010	**DISTRIBUTION SWITCHBOARDS SECTION**									
0100	Aluminum bus bars, not including breakers									
0200	120/208 or 277/480 volt, 4 wire, 600 amp	2 Elec	1	16	Ea.	2,275	785		3,060	3,675
0300	800 amp		.88	18.182		3,000	890		3,890	4,625
0400	1000 amp		.80	20		3,000	980		3,980	4,750
0500	1200 amp		.72	22.222		3,725	1,100		4,825	5,725
0600	1600 amp		.66	24.242		4,275	1,200		5,475	6,475
0700	2000 amp		.62	25.806		5,000	1,275		6,275	7,375
0800	2500 amp		.60	26.667		5,650	1,300		6,950	8,175
0900	3000 amp		.56	28.571		6,850	1,400		8,250	9,600
0950	4000 amp		.52	30.769		10,000	1,500		11,500	13,300

	26 24 13.40 Switchboards Feeder Section	Crew	Daily Output	Labor-Hours	Unit	Material	Labor	Equipment	Total	Total Incl O&P
0010	**SWITCHBOARDS FEEDER SECTION** group mounted devices									
0030	Circuit breakers									
0160	FA frame, 15 to 60 amp, 240 volt, 1 pole	1 Elec	8	1	Ea.	114	49		163	198
0280	FA frame, 70 to 100 amp, 240 volt, 1 pole		7	1.143		150	56		206	249
0420	KA frame, 70 to 225 amp		3.20	2.500		1,250	123		1,373	1,550
0430	LA frame, 125 to 400 amp		2.30	3.478		2,825	170		2,995	3,375
0460	MA frame, 450 to 600 amp		1.60	5		4,675	245		4,920	5,525
0470	700 to 800 amp		1.30	6.154		6,075	300		6,375	7,150
0480	MAL frame, 1000 amp		1	8		6,325	390		6,715	7,525
0490	PA frame, 1200 amp		.80	10		12,800	490		13,290	14,800
0500	Branch circuit, fusible switch, 600 volt, double 30/30 amp		4	2		1,250	98		1,348	1,500
0550	60/60 amp		3.20	2.500		1,250	123		1,373	1,550
0600	100/100 amp		2.70	2.963		1,750	145		1,895	2,150
0650	Single, 30 amp		5.30	1.509		795	74		869	985
0700	60 amp		4.70	1.702		795	83.50		878.50	1,000
0750	100 amp		4	2		1,175	98		1,273	1,450
0800	200 amp		2.70	2.963		1,750	145		1,895	2,150
0850	400 amp		2.30	3.478		3,300	170		3,470	3,875
0900	600 amp		1.80	4.444		3,900	218		4,118	4,625
0950	800 amp		1.30	6.154		6,400	300		6,700	7,475
1000	1200 amp		.80	10		7,625	490		8,115	9,100

26 24 16 – Panelboards

	26 24 16.20 Panelboard and Load Center Circuit Breakers	Crew	Daily Output	Labor-Hours	Unit	Material	Labor	Equipment	Total	Total Incl O&P
0010	**PANELBOARD AND LOAD CENTER CIRCUIT BREAKERS**									
0050	Bolt-on, 10,000 amp IC, 120 volt, 1 pole									
0100	15 to 50 amp	1 Elec	10	.800	Ea.	16	39		55	76
0200	60 amp		8	1		16	49		65	90.50
0300	70 amp		8	1		30.50	49		79.50	107
0350	240 volt, 2 pole									
0400	15 to 50 amp	1 Elec	8	1	Ea.	35	49		84	112

26 24 16 – Panelboards

26 24 16.20 Panelboard and Load Center Circuit Breakers		Crew	Daily Output	Labor-Hours	Unit	Material	2010 Bare Costs Labor	Equipment	Total	Total Incl O&P
0500	60 amp	1 Elec	7.50	1.067	Ea.	35	52.50		87.50	116
0600	80 to 100 amp		5	1.600		90.50	78.50		169	217
0700	3 pole, 15 to 60 amp		6.20	1.290		111	63		174	216
0800	70 amp		5	1.600		140	78.50		218.50	271
0900	80 to 100 amp		3.60	2.222		159	109		268	335
1000	22,000 amp I.C., 240 volt, 2 pole, 70 – 225 amp		2.70	2.963		685	145		830	965
1100	3 pole, 70 – 225 amp		2.30	3.478		760	170		930	1,100
1200	14,000 amp I.C., 277 volts, 1 pole, 15 – 30 amp		8	1		42.50	49		91.50	120
1300	22,000 amp I.C., 480 volts, 2 pole, 70 – 225 amp		2.70	2.963		685	145		830	965
1400	3 pole, 70 – 225 amp	↓	2.30	3.478	↓	845	170		1,015	1,175

26 24 16.30 Panelboards Commercial Applications		Crew	Daily Output	Labor-Hours	Unit	Material	Labor	Equipment	Total	Total Incl O&P
0010	**PANELBOARDS COMMERCIAL APPLICATIONS**									
0050	NQOD, w/20 amp 1 pole bolt-on circuit breakers									
0100	3 wire, 120/240 volts, 100 amp main lugs									
0150	10 circuits	1 Elec	1	8	Ea.	510	390		900	1,150
0200	14 circuits		.88	9.091		595	445		1,040	1,325
0250	18 circuits		.75	10.667		650	525		1,175	1,500
0300	20 circuits	↓	.65	12.308		735	605		1,340	1,700
0350	225 amp main lugs, 24 circuits	2 Elec	1.20	13.333		830	655		1,485	1,875
0400	30 circuits		.90	17.778		970	870		1,840	2,375
0450	36 circuits		.80	20		1,100	980		2,080	2,675
0500	38 circuits		.72	22.222		1,200	1,100		2,300	2,925
0550	42 circuits	↓	.66	24.242		1,250	1,200		2,450	3,150
0600	4 wire, 120/208 volts, 100 amp main lugs, 12 circuits	1 Elec	1	8		575	390		965	1,225
0650	16 circuits		.75	10.667		660	525		1,185	1,500
0700	20 circuits		.65	12.308		770	605		1,375	1,750
0750	24 circuits *CN*		.60	13.333		835	655		1,490	1,900
0800	30 circuits	↓	.53	15.094		965	740		1,705	2,150
0850	225 amp main lugs, 32 circuits	2 Elec	.90	17.778		1,075	870		1,945	2,500
0900	34 circuits		.84	19.048		1,125	935		2,060	2,625
0950	36 circuits		.80	20		1,150	980		2,130	2,700
1000	42 circuits	↓	.68	23.529	↓	1,275	1,150		2,425	3,125
1200	NEHB,w/20 amp, 1 pole bolt-on circuit breakers									
1250	4 wire, 277/480 volts, 100 amp main lugs, 12 circuits	1 Elec	.88	9.091	Ea.	1,100	445		1,545	1,850
1300	20 circuits	"	.60	13.333		1,625	655		2,280	2,750
1350	225 amp main lugs, 24 circuits	2 Elec	.90	17.778		1,875	870		2,745	3,350
1400	30 circuits		.80	20		2,250	980		3,230	3,925
1450	36 circuits	↓	.72	22.222	↓	2,625	1,100		3,725	4,500
1600	NQOD panel, w/20 amp, 1 pole, circuit breakers									
1650	3 wire, 120/240 volt with main circuit breaker									
1700	100 amp main, 12 circuits	1 Elec	.80	10	Ea.	705	490		1,195	1,500
1750	20 circuits	"	.60	13.333		900	655		1,555	1,950
1800	225 amp main, 30 circuits	2 Elec	.68	23.529		1,725	1,150		2,875	3,625
1850	42 circuits		.52	30.769		2,000	1,500		3,500	4,450
1900	400 amp main, 30 circuits		.54	29.630		2,400	1,450		3,850	4,775
1950	42 circuits	↓	.50	32	↓	2,675	1,575		4,250	5,250
2000	4 wire, 120/208 volts with main circuit breaker									
2050	100 amp main, 24 circuits	1 Elec	.47	17.021	Ea.	1,050	835		1,885	2,400
2100	30 circuits	"	.40	20		1,200	980		2,180	2,750
2200	225 amp main, 32 circuits	2 Elec	.72	22.222		2,000	1,100		3,100	3,825
2250	42 circuits		.56	28.571		2,200	1,400		3,600	4,500
2300	400 amp main, 42 circuits	↓	.48	33.333	↓	2,975	1,625		4,600	5,700

26 24 Switchboards and Panelboards

26 24 16 – Panelboards

26 24 16.30 Panelboards Commercial Applications		Crew	Daily Output	Labor-Hours	Unit	2010 Bare Costs Material	Labor	Equipment	Total	Total Incl O&P
2350	600 amp main, 42 circuits	2 Elec	.40	40	Ea.	4,425	1,950		6,375	7,775
2400	NEHB, with 20 amp, 1 pole circuit breaker									
2450	4 wire, 277/480 volts with main circuit breaker									
2500	100 amp main, 24 circuits	1 Elec	.42	19.048	Ea.	2,175	935		3,110	3,775
2550	30 circuits	"	.38	21.053		2,550	1,025		3,575	4,325
2600	225 amp main, 30 circuits	2 Elec	.72	22.222		3,200	1,100		4,300	5,150
2650	42 circuits	"	.56	28.571		3,950	1,400		5,350	6,425

26 24 19 – Motor-Control Centers

26 24 19.40 Motor Starters and Controls

		Crew	Daily Output	Labor-Hours	Unit	Material	Labor	Equipment	Total	Total Incl O&P
0010	**MOTOR STARTERS AND CONTROLS**									
0050	Magnetic, FVNR, with enclosure and heaters, 480 volt									
0100	5 HP, size 0	1 Elec	2.30	3.478	Ea.	260	170		430	540
0200	10 HP, size 1	"	1.60	5		292	245		537	685
0300	25 HP, size 2	2 Elec	2.20	7.273		550	355		905	1,125
0400	50 HP, size 3		1.80	8.889		895	435		1,330	1,625
0500	100 HP, size 4		1.20	13.333		1,975	655		2,630	3,150
0600	200 HP, size 5		.90	17.778		4,650	870		5,520	6,425
0700	Combination, with motor circuit protectors, 5 HP, size 0	1 Elec	1.80	4.444		845	218		1,063	1,250
0800	10 HP, size 1	"	1.30	6.154		875	300		1,175	1,425
0900	25 HP, size 2	2 Elec	2	8		1,225	390		1,615	1,925
1000	50 HP, size 3		1.32	12.121		1,775	595		2,370	2,825
1200	100 HP, size 4		.80	20		3,850	980		4,830	5,700
1400	Combination, with fused switch, 5 HP, size 0	1 Elec	1.80	4.444		645	218		863	1,025
1600	10 HP, size 1	"	1.30	6.154		690	300		990	1,200
1800	25 HP, size 2	2 Elec	2	8		1,125	390		1,515	1,800
2000	50 HP, size 3		1.32	12.121		1,900	595		2,495	2,950
2200	100 HP, size 4		.80	20		3,300	980		4,280	5,075

26 25 Enclosed Bus Assemblies

26 25 13 – Bus Duct/Busway and Fittings

26 25 13.40 Copper Bus Duct

		Crew	Daily Output	Labor-Hours	Unit	Material	Labor	Equipment	Total	Total Incl O&P
0010	**COPPER BUS DUCT** 10 ft. long									
0050	Indoor 3 pole 4 wire, plug-in, straight section, 225 amp	2 Elec	40	.400	L.F.	186	19.60		205.60	233
1000	400 amp		32	.500		186	24.50		210.50	241
1500	600 amp		26	.615		186	30		216	249
2400	800 amp		20	.800		220	39		259	300
2450	1000 amp		18	.889		244	43.50		287.50	335
2500	1350 amp		16	1		350	49		399	460
2510	1600 amp		12	1.333		395	65.50		460.50	530
2520	2000 amp		10	1.600		500	78.50		578.50	665
2550	Feeder, 600 amp		28	.571		162	28		190	221
2600	800 amp		22	.727		197	35.50		232.50	270
2700	1000 amp		20	.800		220	39		259	300
2800	1350 amp		18	.889		325	43.50		368.50	420
2900	1600 amp		14	1.143		370	56		426	495
3000	2000 amp		12	1.333		475	65.50		540.50	620
3100	Elbows, 225 amp		4	4	Ea.	1,100	196		1,296	1,500
3200	400 amp		3.60	4.444		1,100	218		1,318	1,525
3300	600 amp		3.20	5		1,100	245		1,345	1,575
3400	800 amp		2.80	5.714		1,200	280		1,480	1,725

26 25 13 – Bus Duct/Busway and Fittings

26 25 13.40 Copper Bus Duct		Crew	Daily Output	Labor-Hours	Unit	Material	2010 Bare Costs Labor	Equipment	Total	Total Incl O&P
3500	1000 amp	2 Elec	2.60	6.154	Ea.	1,325	300		1,625	1,925
3600	1350 amp		2.40	6.667		1,575	325		1,900	2,200
3700	1600 amp		2.20	7.273		1,700	355		2,055	2,400
3800	2000 amp		1.80	8.889		2,100	435		2,535	2,950
4000	End box, 225 amp		34	.471		148	23		171	198
4100	400 amp		32	.500		148	24.50		172.50	200
4200	600 amp		28	.571		148	28		176	205
4300	800 amp		26	.615		148	30		178	208
4400	1000 amp		24	.667		148	32.50		180.50	212
4500	1350 amp		22	.727		148	35.50		183.50	216
4600	1600 amp		20	.800		148	39		187	222
4700	2000 amp		18	.889		182	43.50		225.50	265
4800	Cable tap box end, 225 amp		3.20	5		1,200	245		1,445	1,700
5000	400 amp		2.60	6.154		1,200	300		1,500	1,775
5100	600 amp		2.20	7.273		1,200	355		1,555	1,850
5200	800 amp		2	8		1,375	390		1,765	2,075
5300	1000 amp		1.60	10		1,475	490		1,965	2,350
5400	1350 amp		1.40	11.429		1,825	560		2,385	2,850
5500	1600 amp		1.20	13.333		2,075	655		2,730	3,250
5600	2000 amp		1	16		2,325	785		3,110	3,725
5700	Switchboard stub, 225 amp		5.40	2.963		1,125	145		1,270	1,450
5800	400 amp		4.60	3.478		1,125	170		1,295	1,475
5900	600 amp		4	4		1,125	196		1,321	1,525
6000	800 amp		3.20	5		1,350	245		1,595	1,850
6100	1000 amp		3	5.333		1,550	261		1,811	2,125
6200	1350 amp		2.60	6.154		2,000	300		2,300	2,650
6300	1600 amp		2.40	6.667		2,275	325		2,600	2,975
6400	2000 amp		2	8		2,750	390		3,140	3,600
6490	Tee fittings, 225 amp		2.40	6.667		1,525	325		1,850	2,150
6500	400 amp		2	8		1,525	390		1,915	2,250
6600	600 amp		1.80	8.889		1,525	435		1,960	2,325
6700	800 amp		1.60	10		1,750	490		2,240	2,650
6800	1350 amp		1.20	13.333		2,500	655		3,155	3,750
7000	1600 amp		1	16		2,850	785		3,635	4,325
7100	2000 amp		.80	20		3,400	980		4,380	5,175
7200	Plug-in fusible switches w/3 fuses, 600 volt, 3 pole, 30 amp	1 Elec	4	2		675	98		773	885
7300	60 amp		3.60	2.222		755	109		864	990
7400	100 amp		2.70	2.963		1,150	145		1,295	1,500
7500	200 amp	2 Elec	3.20	5		2,050	245		2,295	2,650
7600	400 amp		1.40	11.429		6,025	560		6,585	7,450
7700	600 amp		.90	17.778		6,850	870		7,720	8,825
7800	800 amp		.66	24.242		9,600	1,200		10,800	12,400
7900	1200 amp		.50	32		18,000	1,575		19,575	22,100
8000	Plug-in circuit breakers, molded case, 15 to 50 amp	1 Elec	4.40	1.818		635	89		724	830
8100	70 to 100 amp	"	3.10	2.581		705	126		831	970
8200	150 to 225 amp	2 Elec	3.40	4.706		1,925	231		2,156	2,450
8300	250 to 400 amp		1.40	11.429		3,375	560		3,935	4,525
8400	500 to 600 amp		1	16		4,550	785		5,335	6,175
8500	700 to 800 amp		.64	25		5,600	1,225		6,825	8,000
8600	900 to 1000 amp		.56	28.571		8,025	1,400		9,425	10,900
8700	1200 amp		.44	36.364		9,625	1,775		11,400	13,300

26 27 Low-Voltage Distribution Equipment

26 27 16 – Electrical Cabinets and Enclosures

26 27 16.10 Cabinets		Crew	Daily Output	Labor-Hours	Unit	Material	2010 Bare Costs Labor	Equipment	Total	Total Incl O&P
0010	**CABINETS**									
7000	Cabinets, current transformer									
7050	Single door, 24" H x 24" W x 10" D	1 Elec	1.60	5	Ea.	206	245		451	590
7100	30" H x 24" W x 10" D		1.30	6.154		210	300		510	680
7150	36" H x 24" W x 10" D		1.10	7.273		215	355		570	765
7200	30" H x 30" W x 10" D		1	8		225	390		615	835
7250	36" H x 30" W x 10" D		.90	8.889		305	435		740	985
7300	36" H x 36" W x 10" D		.80	10		325	490		815	1,100
7500	Double door, 48" H x 36" W x 10" D		.60	13.333		670	655		1,325	1,700
7550	24" H x 24" W x 12" D	↓	1	8	↓	274	390		664	885

26 27 23 – Indoor Service Poles

26 27 23.40 Surface Raceway		Crew	Daily Output	Labor-Hours	Unit	Material	Labor	Equipment	Total	Total Incl O&P
0010	**SURFACE RACEWAY**									
0090	Metal, straight section									
0100	No. 500	1 Elec	100	.080	L.F.	1.03	3.92		4.95	7
0110	No. 700		100	.080		1.16	3.92		5.08	7.15
0400	No. 1500, small pancake		90	.089		2.10	4.36		6.46	8.80
0600	No. 2000, base & cover, blank		90	.089		2.15	4.36		6.51	8.85
0800	No. 3000, base & cover, blank		75	.107		4.10	5.25		9.35	12.25
1000	No. 4000, base & cover, blank		65	.123		6.70	6.05		12.75	16.30
1200	No. 6000, base & cover, blank		50	.160	↓	11.10	7.85		18.95	24
2400	Fittings, elbows, No. 500		40	.200	Ea.	1.85	9.80		11.65	16.60
2800	Elbow cover, No. 2000		40	.200		3.50	9.80		13.30	18.40
2880	Tee, No. 500		42	.190		3.60	9.35		12.95	17.85
2900	No. 2000		27	.296		11.70	14.50		26.20	34.50
3000	Switch box, No. 500		16	.500		12.40	24.50		36.90	50
3400	Telephone outlet, No. 1500		16	.500		14.50	24.50		39	52.50
3600	Junction box, No. 1500	↓	16	.500	↓	9.90	24.50		34.40	47.50
3800	Plugmold wired sections, No. 2000									
4000	1 circuit, 6 outlets, 3 ft. long	1 Elec	8	1	Ea.	36	49		85	113
4100	2 circuits, 8 outlets, 6 ft. long	"	5.30	1.509	"	59.50	74		133.50	176

26 27 26 – Wiring Devices

26 27 26.10 Low Voltage Switching		Crew	Daily Output	Labor-Hours	Unit	Material	Labor	Equipment	Total	Total Incl O&P
0010	**LOW VOLTAGE SWITCHING**									
3600	Relays, 120 V or 277 V standard	1 Elec	12	.667	Ea.	37.50	32.50		70	89.50
3800	Flush switch, standard		40	.200		11.30	9.80		21.10	27
4000	Interchangeable		40	.200		14.75	9.80		24.55	31
4100	Surface switch, standard		40	.200		8	9.80		17.80	23.50
4200	Transformer 115 V to 25 V		12	.667		128	32.50		160.50	190
4400	Master control, 12 circuit, manual		4	2		118	98		216	276
4500	25 circuit, motorized		4	2		131	98		229	290
4600	Rectifier, silicon		12	.667		38.50	32.50		71	90.50
4800	Switchplates, 1 gang, 1, 2 or 3 switch, plastic		80	.100		3.60	4.90		8.50	11.25
5000	Stainless steel		80	.100		9.65	4.90		14.55	17.90
5400	2 gang, 3 switch, stainless steel		53	.151		18.60	7.40		26	31.50
5500	4 switch, plastic		53	.151		8	7.40		15.40	19.80
5800	3 gang, 9 switch, stainless steel	↓	32	.250	↓	59.50	12.25		71.75	83.50

26 27 26.20 Wiring Devices Elements		Crew	Daily Output	Labor-Hours	Unit	Material	Labor	Equipment	Total	Total Incl O&P
0010	**WIRING DEVICES ELEMENTS**									
0200	Toggle switch, quiet type, single pole, 15 amp **CN**	1 Elec	40	.200	Ea.	5.30	9.80		15.10	20.50
0600	3 way, 15 amp	↓	23	.348	↓	7.65	17.05		24.70	34

26 27 Low-Voltage Distribution Equipment

26 27 26 – Wiring Devices

26 27 26.20 Wiring Devices Elements		Crew	Daily Output	Labor-Hours	Unit	Material	2010 Bare Costs Labor	Equipment	Total	Total Incl O&P
0900	4 way, 15 amp	1 Elec	15	.533	Ea.	25	26		51	66.50
1650	Dimmer switch, 120 volt, incandescent, 600 watt, 1 pole [G]		16	.500		13.25	24.50		37.75	51
2460	Receptacle, duplex, 120 volt, grounded, 15 amp		40	.200		1.37	9.80		11.17	16.05
2470	20 amp		27	.296		10.40	14.50		24.90	33
2490	Dryer, 30 amp		15	.533		6.25	26		32.25	46
2500	Range, 50 amp		11	.727		14.60	35.50		50.10	69
2600	Wall plates, stainless steel, 1 gang		80	.100		2.50	4.90		7.40	10.05
2800	2 gang		53	.151		4.25	7.40		11.65	15.70
3200	Lampholder, keyless		26	.308		12.80	15.10		27.90	36.50
3400	Pullchain with receptacle	↓	22	.364	↓	13.60	17.80		31.40	41.50

26 27 73 – Door Chimes

26 27 73.10 Doorbell System		Crew	Daily Output	Labor-Hours	Unit	Material	Labor	Equipment	Total	Total Incl O&P
0010	**DOORBELL SYSTEM**, incl. transformer, button & signal									
0100	6" bell	1 Elec	4	2	Ea.	110	98		208	268
0200	Buzzer	"	4	2	"	87.50	98		185.50	243

26 28 Low-Voltage Circuit Protective Devices

26 28 16 – Enclosed Switches and Circuit Breakers

26 28 16.10 Circuit Breakers		Crew	Daily Output	Labor-Hours	Unit	Material	Labor	Equipment	Total	Total Incl O&P
0010	**CIRCUIT BREAKERS** (in enclosure)									
0100	Enclosed (NEMA 1), 600 volt, 3 pole, 30 amp	1 Elec	3.20	2.500	Ea.	630	123		753	875
0200	60 amp		2.80	2.857		630	140		770	905
0400	100 amp		2.30	3.478		720	170		890	1,050
0600	225 amp	↓	1.50	5.333		1,675	261		1,936	2,225
0700	400 amp	2 Elec	1.60	10		2,850	490		3,340	3,850
0800	600 amp		1.20	13.333		4,125	655		4,780	5,525
1000	800 amp	↓	.94	17.021	↓	5,375	835		6,210	7,150

26 28 16.20 Safety Switches		Crew	Daily Output	Labor-Hours	Unit	Material	Labor	Equipment	Total	Total Incl O&P
0010	**SAFETY SWITCHES**									
0100	General duty 240 volt, 3 pole NEMA 1, fusible, 30 amp	1 Elec	3.20	2.500	Ea.	113	123		236	305
0200	60 amp		2.30	3.478		192	170		362	465
0300	100 amp		1.90	4.211		330	206		536	665
0400	200 amp	↓	1.30	6.154		710	300		1,010	1,225
0500	400 amp	2 Elec	1.80	8.889		1,775	435		2,210	2,625
0600	600 amp	"	1.20	13.333	↓	3,325	655		3,980	4,650
2900	Heavy duty, 240 volt, 3 pole NEMA 1 fusible									
2910	30 amp	1 Elec	3.20	2.500	Ea.	183	123		306	385
3000	60 amp		2.30	3.478		310	170		480	595
3300	100 amp		1.90	4.211		490	206		696	840
3500	200 amp	↓	1.30	6.154		840	300		1,140	1,375
3700	400 amp	2 Elec	1.80	8.889		2,150	435		2,585	3,025
3900	600 amp	"	1.20	13.333	↓	3,700	655		4,355	5,050

26 29 Low-Voltage Controllers

26 29 13 – Enclosed Controllers

26 29 13.20 Control Stations		Crew	Daily Output	Labor-Hours	Unit	Material	2010 Bare Costs Labor	Equipment	Total	Total Incl O&P
0010	**CONTROL STATIONS**									
0050	NEMA 1, heavy duty, stop/start	1 Elec	8	1	Ea.	148	49		197	235
0100	Stop/start, pilot light		6.20	1.290		201	63		264	315
0200	Hand/off/automatic		6.20	1.290		109	63		172	214
0400	Stop/start/reverse	↓	5.30	1.509	↓	199	74		273	330

26 32 Packaged Generator Assemblies

26 32 13 – Engine Generators

26 32 13.13 Diesel-Engine-Driven Generator Sets

		Crew	Daily Output	Labor-Hours	Unit	Material	Labor	Equipment	Total	Total Incl O&P
0010	**DIESEL-ENGINE-DRIVEN GENERATOR SETS**									
2000	Diesel engine, including battery, charger,									
2010	muffler, automatic transfer switch & day tank, 30 kW	R-3	.55	36.364	Ea.	20,200	1,750	251	22,201	25,100
2100	50 kW		.42	47.619		24,900	2,300	330	27,530	31,200
2200	75 kW		.35	57.143		32,300	2,750	395	35,445	40,100
2300	100 kW		.31	64.516		36,000	3,125	445	39,570	44,600
2400	125 kW		.29	68.966		37,900	3,325	475	41,700	47,200
2500	150 kW		.26	76.923		43,500	3,725	530	47,755	54,000
2600	175 kW		.25	80		47,300	3,850	555	51,705	58,500
2700	200 kW		.24	83.333		48,800	4,025	575	53,400	60,000
2800	250 kW		.23	86.957		57,500	4,200	600	62,300	70,000
2900	300 kW		.22	90.909		62,000	4,400	630	67,030	75,500
3000	350 kW		.20	100		70,000	4,825	690	75,515	85,000
3100	400 kW		.19	105		86,500	5,075	725	92,300	104,000
3200	500 kW	↓	.18	111	↓	109,000	5,375	770	115,145	129,000

26 32 13.16 Gas-Engine-Driven Generator Sets

		Crew	Daily Output	Labor-Hours	Unit	Material	Labor	Equipment	Total	Total Incl O&P
0010	**GAS-ENGINE-DRIVEN GENERATOR SETS**									
0020	Gas or gasoline operated, includes battery,									
0050	charger, muffler & transfer switch									
0200	3 phase 4 wire, 277/480 volt, 7.5 kW	R-3	.83	24.096	Ea.	7,925	1,175	167	9,267	10,600
0300	11.5 kW		.71	28.169		11,200	1,350	195	12,745	14,600
0400	20 kW		.63	31.746		13,200	1,525	219	14,944	17,100
0500	35 kW		.55	36.364		15,800	1,750	251	17,801	20,200
0600	80 kW		.40	50		25,900	2,425	345	28,670	32,400
0700	100 kW		.33	60.606		28,300	2,925	420	31,645	36,000
0800	125 kW		.28	71.429		58,000	3,450	495	61,945	69,500
0900	185 kW	↓	.25	80	↓	76,500	3,850	555	80,905	91,000

26 35 Power Filters and Conditioners

26 35 13 – Capacitors

26 35 13.10 Capacitors Indoor

		Crew	Daily Output	Labor-Hours	Unit	Material	Labor	Equipment	Total	Total Incl O&P
0010	**CAPACITORS INDOOR**									
0020	240 volts, single & 3 phase, 0.5 kVAR	1 Elec	2.70	2.963	Ea.	385	145		530	640
0100	1.0 kVAR		2.70	2.963		465	145		610	730
0150	2.5 kVAR		2	4		525	196		721	870
0200	5.0 kVAR		1.80	4.444		630	218		848	1,025
0250	7.5 kVAR		1.60	5		730	245		975	1,175
0300	10 kVAR		1.50	5.333		880	261		1,141	1,350
0350	15 kVAR		1.30	6.154		1,225	300		1,525	1,775
0400	20 kVAR	↓	1.10	7.273	↓	1,500	355		1,855	2,175

26 35 Power Filters and Conditioners

26 35 13 – Capacitors

26 35 13.10 Capacitors Indoor		Crew	Daily Output	Labor-Hours	Unit	Material	Labor	Equipment	Total	Total Incl O&P
						2010 Bare Costs				
0450	25 kVAR	1 Elec	1	8	Ea.	1,775	390		2,165	2,525
1000	480 volts, single & 3 phase, 1 kVAR		2.70	2.963		355	145		500	605
1050	2 kVAR		2.70	2.963		405	145		550	660
1100	5 kVAR		2	4		510	196		706	855
1150	7.5 kVAR		2	4		550	196		746	895
1200	10 kVAR		2	4		615	196		811	970
1250	15 kVAR		2	4		760	196		956	1,125
1300	20 kVAR		1.60	5		850	245		1,095	1,300
1350	30 kVAR		1.50	5.333		1,050	261		1,311	1,575
1400	40 kVAR		1.20	6.667		1,350	325		1,675	1,950
1450	50 kVAR	↓	1.10	7.273	↓	1,525	355		1,880	2,225

26 42 Cathodic Protection

26 42 16 – Passive Cathodic Protection for Underground Storage Tank

26 42 16.50 Cathodic Protection Wiring Methods

		Crew	Daily Output	Labor-Hours	Unit	Material	Labor	Equipment	Total	Total Incl O&P
0010	**CATHODIC PROTECTION WIRING METHODS**									
1000	Anodes, magnesium type, 9 #	R-15	18.50	2.595	Ea.	45	124	18.80	187.80	255
1010	17 #		13	3.692		83	176	26.50	285.50	385
1020	32 #		10	4.800		150	229	35	414	545
1030	48 #	↓	7.20	6.667		228	320	48.50	596.50	780
1100	Graphite type w/epoxy cap, 3" x 60" (32 #)	R-22	8.40	4.438		117	185		302	405
1110	4" x 80" (68 #)		6	6.213		222	258		480	635
1120	6" x 72" (80 #)		5.20	7.169		1,475	298		1,773	2,075
1130	6" x 36" (45 #)	↓	9.60	3.883		745	162		907	1,075
2000	Rectifiers, silicon type, air cooled, 28 V/10 A	R-19	3.50	5.714		1,950	281		2,231	2,575
2010	20 V/20 A		3.50	5.714		2,025	281		2,306	2,650
2100	Oil immersed, 28 V/10 A		3	6.667		2,750	325		3,075	3,500
2110	20 V/20 A	↓	3	6.667	↓	2,850	325		3,175	3,625
3000	Anode backfill, coke breeze	R-22	3850	.010	Lb.	.14	.40		.54	.75
4000	Cable, HMWPE, No. 8		2.40	15.533	M.L.F.	282	645		927	1,275
4010	No. 6		2.40	15.533		405	645		1,050	1,425
4020	No. 4		2.40	15.533		580	645		1,225	1,600
4030	No. 2		2.40	15.533		925	645		1,570	2,000
4040	No. 1		2.20	16.945		1,225	705		1,930	2,400
4050	No. 1/0		2.20	16.945		1,500	705		2,205	2,700
4060	No. 2/0		2.20	16.945		1,825	705		2,530	3,050
4070	No. 4/0	↓	2	18.640	↓	3,475	775		4,250	5,000
5000	Test station, 7 terminal box, flush curb type w/lockable cover	R-19	12	1.667	Ea.	70.50	82		152.50	200
5010	Reference cell, 2" dia PVC conduit, cplg, plug, set flush	"	4.80	4.167	"	148	205		353	470

26 51 Interior Lighting

26 51 13 – Interior Lighting Fixtures, Lamps, and Ballasts

26 51 13.50 Interior Lighting Fixtures		Crew	Daily Output	Labor-Hours	Unit	Material	2010 Bare Costs Labor	Equipment	Total	Total Incl O&P
0010	**INTERIOR LIGHTING FIXTURES** Including lamps, mounting									
0030	hardware and connections									
0100	Fluorescent, C.W. lamps, troffer, recess mounted in grid, RS									
0200	Acrylic lens, 1'W x 4'L, two 40 watt	1 Elec	5.70	1.404	Ea.	53.50	69		122.50	161
0210	1'W x 4'L, three 40 watt		5.40	1.481		62.50	72.50		135	177
0300	2'W x 2'L, two U40 watt		5.70	1.404		57	69		126	165
0400	2'W x 4'L, two 40 watt		5.30	1.509		57	74		131	173
0500	2'W x 4'L, three 40 watt		5	1.600		61.50	78.50		140	185
0600	2'W x 4'L, four 40 watt *CN*		4.70	1.702		65.50	83.50		149	196
0700	4'W x 4'L, four 40 watt	2 Elec	6.40	2.500		340	123		463	555
0800	4'W x 4'L, six 40 watt		6.20	2.581		350	126		476	575
0900	4'W x 4'L, eight 40 watt		5.80	2.759		365	135		500	600
0910	Acrylic lens, 1'W x 4'L, two 32 watt T-8 G	1 Elec	5.70	1.404		67	69		136	176
0930	2'W x 2'L, two U32 watt T-8 G		5.70	1.404		80	69		149	190
0940	2'W x 4'L, two 32 watt T-8 G		5.30	1.509		72	74		146	190
0950	2'W x 4'L, three 32 watt T-8 G		5	1.600		76.50	78.50		155	201
0960	2'W x 4'L, four 32 watt T-8 G		4.70	1.702		78.50	83.50		162	211
1000	Surface mounted, RS									
1030	Acrylic lens with hinged & latched door frame									
1100	1'W x 4'L, two 40 watt	1 Elec	7	1.143	Ea.	74	56		130	165
1110	1'W x 4'L, three 40 watt		6.70	1.194		76.50	58.50		135	171
1200	2'W x 2'L, two U40 watt		7	1.143		79.50	56		135.50	171
1300	2'W x 4'L, two 40 watt		6.20	1.290		91	63		154	194
1400	2'W x 4'L, three 40 watt		5.70	1.404		92	69		161	203
1500	2'W x 4'L, four 40 watt		5.30	1.509		94.50	74		168.50	214
1600	4'W x 4'L, four 40 watt	2 Elec	7.20	2.222		465	109		574	670
1700	4'W x 4'L, six 40 watt		6.60	2.424		505	119		624	730
1800	4'W x 4'L, eight 40 watt		6.20	2.581		525	126		651	770
1900	2'W x 8'L, four 40 watt		6.40	2.500		185	123		308	385
2000	2'W x 8'L, eight 40 watt		6.20	2.581		199	126		325	405
2100	Strip fixture									
2130	Surface mounted									
2200	4' long, one 40 watt, RS	1 Elec	8.50	.941	Ea.	31.50	46		77.50	103
2300	4' long, two 40 watt, RS		8	1		34.50	49		83.50	111
2400	4' long, one 40 watt, SL		8	1		46	49		95	124
2500	4' long, two 40 watt, SL		7	1.143		62.50	56		118.50	153
2600	8' long, one 75 watt, SL	2 Elec	13.40	1.194		47	58.50		105.50	139
2700	8' long, two 75 watt, SL	"	12.40	1.290		57	63		120	157
2800	4' long, two 60 watt, HO	1 Elec	6.70	1.194		91.50	58.50		150	188
2900	8' long, two 110 watt, HO	2 Elec	10.60	1.509		96.50	74		170.50	216
2950	High bay pendent mounted, 16" W x 4' L, four 54 watt, T5HO G		8.90	1.798		223	88		311	375
2952	2' W x 4' L, six 54 watt, T5HO G		8.50	1.882		295	92		387	460
2954	2' W x 4' L, six 32 watt, T8 G		8.50	1.882		170	92		262	325
3000	Strip, pendent mounted, industrial, white porcelain enamel									
3100	4' long, two 40 watt, RS	1 Elec	5.70	1.404	Ea.	52	69		121	159
3200	4' long, two 60 watt, HO	"	5	1.600		81	78.50		159.50	206
3300	8' long, two 75 watt, SL	2 Elec	8.80	1.818		96	89		185	238
3400	8' long, two 110 watt, HO	"	8	2		123	98		221	281
3470	Troffer, air handling, 2'W x 4'L with four 32 watt T8 G	1 Elec	4	2		111	98		209	268
3480	2'W x 2'L with two U32 watt T8 G		5.50	1.455		106	71.50		177.50	223
3490	Air connector insulated, 5" diameter		20	.400		66	19.60		85.60	102
3500	6" diameter		20	.400		67.50	19.60		87.10	103
3510	Troffer parabolic lay-in, 1'W x 4'L with one 32 W T-8 G		5.70	1.404		112	69		181	225

26 51 Interior Lighting

26 51 13 – Interior Lighting Fixtures, Lamps, and Ballasts

26 51 13.50 Interior Lighting Fixtures		Crew	Daily Output	Labor-Hours	Unit	2010 Bare Costs Material	Labor	Equipment	Total	Total Incl O&P
3520	1'W x 4'L with two 32 W T-8 G	1 Elec	5.30	1.509	Ea.	123	74		197	245
3525	2'W x 2'L with two U32 W T-8 G		5.70	1.404		113	69		182	226
3530	2'W x 4'L with three 32 W T-8 G		5	1.600		127	78.50		205.50	257
4450	Incandescent, high hat can, round alzak reflector, prewired									
4470	100 watt	1 Elec	8	1	Ea.	70	49		119	150
4480	150 watt		8	1		102	49		151	185
4500	300 watt		6.70	1.194		236	58.50		294.50	345
4600	Square glass lens with metal trim, prewired									
4630	100 watt	1 Elec	6.70	1.194	Ea.	54	58.50		112.50	147
4700	200 watt		6.70	1.194		95	58.50		153.50	192
4800	300 watt		5.70	1.404		142	69		211	258
4900	Ceiling/wall, surface mounted, metal cylinder, 75 watt		10	.800		55	39		94	119
4920	150 watt		10	.800		79	39		118	146
5200	Ceiling, surface mounted, opal glass drum									
5300	8", one 60 watt lamp	1 Elec	10	.800	Ea.	43.50	39		82.50	107
5400	10", two 60 watt lamps		8	1		49	49		98	127
5500	12", four 60 watt lamps		6.70	1.194		69	58.50		127.50	163
6010	Vapor tight, incandescent, ceiling mounted, 200 watt		6.20	1.290		70.50	63		133.50	172
6100	Fluorescent, surface mounted, 2 lamps, 4'L, RS, 40 watt		3.20	2.500		113	123		236	305
6850	Vandalproof, surface mounted, fluorescent, two 32 watt T-8 G		3.20	2.500		247	123		370	455
6860	Incandescent, one 150 watt		8	1		73	49		122	153
7500	Ballast replacement, by weight of ballast, to 15' high									
7520	Indoor fluorescent, less than 2 lbs.	1 Elec	10	.800	Ea.	25	39		64	86
7540	Two 40W, watt reducer, 2 to 5 lbs.		9.40	.851		40	41.50		81.50	106
7560	Two F96 slimline, over 5 lbs.		8	1		75.50	49		124.50	156
7580	Vaportite ballast, less than 2 lbs.		9.40	.851		25	41.50		66.50	89.50
7600	2 lbs. to 5 lbs.		8.90	.899		40	44		84	110
7620	Over 5 lbs.		7.60	1.053		75.50	51.50		127	160
7630	Electronic ballast for two tubes		8	1		36.50	49		85.50	113
7640	Dimmable ballast one lamp G		8	1		108	49		157	192
7650	Dimmable ballast two-lamp G		7.60	1.053		108	51.50		159.50	196

26 52 Emergency Lighting

26 52 13 – Emergency Lighting Equipments

26 52 13.10 Emergency Lighting and Battery Units		Crew	Daily Output	Labor-Hours	Unit	Material	Labor	Equipment	Total	Total Incl O&P
0010	**EMERGENCY LIGHTING AND BATTERY UNITS**									
0300	Emergency light units, battery operated									
0350	Twin sealed beam light, 25 watt, 6 volt each									
0500	Lead battery operated	1 Elec	4	2	Ea.	129	98		227	287
0700	Nickel cadmium battery operated		4	2		635	98		733	845
0900	Self-contained fluorescent lamp pack		10	.800		150	39		189	224

26 53 Exit Signs

26 53 13 – Exit Lighting

26 53 13.10 Exit Lighting Fixtures		Crew	Daily Output	Labor-Hours	Unit	Material	2010 Bare Costs Labor	Equipment	Total	Total Incl O&P
0010	**EXIT LIGHTING FIXTURES**									
0080	Exit light ceiling or wall mount, incandescent, single face	1 Elec	8	1	Ea.	42.50	49		91.50	120
0100	Double face		6.70	1.194		49	58.50		107.50	141
0200	L.E.D. standard, single face G		8	1		73	49		122	154
0220	Double face G		6.70	1.194		73	58.50		131.50	168
0230	L.E.D. vandal-resistant, single face G		7.27	1.100		198	54		252	298
0240	L.E.D. w/battery unit, single face G		4.40	1.818		116	89		205	260
0260	Double face G		4	2		131	98		229	290
0262	L.E.D. w/battery unit, vandal-resistant, single face G		4.40	1.818		330	89		419	495
0270	Combination emergency light units and exit sign	↓	4	2	↓	180	98		278	345

26 54 Classified Location Lighting

26 54 13 – Classified Lighting

26 54 13.20 Explosionproof		Crew	Daily Output	Labor-Hours	Unit	Material	Labor	Equipment	Total	Total Incl O&P
0010	**EXPLOSIONPROOF**, incl lamps, mounting hardware and connections									
6510	Incandescent, ceiling mounted, 200 watt	1 Elec	4	2	Ea.	835	98		933	1,075
6600	Fluorescent, RS, 4′ long, ceiling mounted, two 40 watt	"	2.70	2.963	"	2,200	145		2,345	2,650

26 55 Special Purpose Lighting

26 55 61 – Theatrical Lighting

26 55 61.10 Lights		Crew	Daily Output	Labor-Hours	Unit	Material	Labor	Equipment	Total	Total Incl O&P
0010	**LIGHTS**									
2000	Lights, border, quartz, reflector, vented,									
2100	colored or white	1 Elec	20	.400	L.F.	173	19.60		192.60	219
2500	Spotlight, follow spot, with transformer, 2,100 watt	"	4	2	Ea.	3,075	98		3,173	3,525
2600	For no transformer, deduct					910			910	1,000
3000	Stationary spot, fresnel quartz, 6″ lens	1 Elec	4	2		131	98		229	290
3100	8″ lens		4	2		231	98		329	400
3500	Ellipsoidal quartz, 1,000W, 6″ lens		4	2		335	98		433	515
3600	12″ lens		4	2		600	98		698	805
4000	Strobe light, 1 to 15 flashes per second, quartz		3	2.667		750	131		881	1,025
4500	Color wheel, portable, five hole, motorized	↓	4	2	↓	194	98		292	360

26 56 Exterior Lighting

26 56 13 – Lighting Poles and Standards

26 56 13.10 Lighting Poles		Crew	Daily Output	Labor-Hours	Unit	Material	Labor	Equipment	Total	Total Incl O&P
0010	**LIGHTING POLES**									
2800	Light poles, anchor base									
2820	not including concrete bases									
2840	Aluminum pole, 8′ high	1 Elec	4	2	Ea.	675	98		773	885
3000	20′ high	R-3	2.90	6.897		880	335	47.50	1,262.50	1,525
3200	30′ high		2.60	7.692		1,725	370	53	2,148	2,500
3400	35′ high		2.30	8.696		1,850	420	60	2,330	2,725
3600	40′ high	↓	2	10		2,125	485	69	2,679	3,125
3800	Bracket arms, 1 arm	1 Elec	8	1		115	49		164	200
4000	2 arms		8	1		228	49		277	325
4200	3 arms	↓	5.30	1.509	↓	345	74		419	485

26 56 Exterior Lighting

26 56 13 – Lighting Poles and Standards

26 56 13.10 Lighting Poles		Crew	Daily Output	Labor-Hours	Unit	Material	2010 Bare Costs Labor	Equipment	Total	Total Incl O&P
4400	4 arms	1 Elec	5.30	1.509	Ea.	455	74		529	615
4500	Steel pole, galvanized, 8′ high		3.80	2.105		590	103		693	800
4600	20′ high	R-3	2.60	7.692		1,050	370	53	1,473	1,775
4800	30′ high		2.30	8.696		1,225	420	60	1,705	2,050
5000	35′ high		2.20	9.091		1,350	440	63	1,853	2,200
5200	40′ high		1.70	11.765		1,650	570	81.50	2,301.50	2,750
5400	Bracket arms, 1 arm	1 Elec	8	1		171	49		220	261
5600	2 arms		8	1		265	49		314	365
5800	3 arms		5.30	1.509		287	74		361	425
6000	4 arms		5.30	1.509		400	74		474	550

26 56 19 – Roadway Lighting

26 56 19.20 Roadway Area Luminaire		Crew	Daily Output	Labor-Hours	Unit	Material	Labor	Equipment	Total	Total Incl O&P
0010	**ROADWAY AREA LUMINAIRE**									
2650	Roadway area luminaire, low pressure sodium, 135 watt	1 Elec	2	4	Ea.	620	196		816	975
2700	180 watt	"	2	4		655	196		851	1,000
2750	Metal halide, 400 watt	2 Elec	4.40	3.636		510	178		688	825
2760	1000 watt		4	4		575	196		771	925
2780	High pressure sodium, 400 watt		4.40	3.636		530	178		708	845
2790	1000 watt		4	4		605	196		801	955

26 56 23 – Area Lighting

26 56 23.10 Exterior Fixtures		Crew	Daily Output	Labor-Hours	Unit	Material	Labor	Equipment	Total	Total Incl O&P
0010	**EXTERIOR FIXTURES** With lamps									
0200	Wall mounted, incandescent, 100 watt	1 Elec	8	1	Ea.	31.50	49		80.50	108
0400	Quartz, 500 watt		5.30	1.509		56	74		130	172
1100	Wall pack, low pressure sodium, 35 watt		4	2		234	98		332	405
1150	55 watt		4	2		278	98		376	450

26 56 36 – Flood Lighting

26 56 36.20 Floodlights		Crew	Daily Output	Labor-Hours	Unit	Material	Labor	Equipment	Total	Total Incl O&P
0010	**FLOODLIGHTS** with ballast and lamp,									
1400	pole mounted, pole not included									
1950	Metal halide, 175 watt	1 Elec	2.70	2.963	Ea.	325	145		470	575
2000	400 watt	2 Elec	4.40	3.636		405	178		583	710
2200	1000 watt	"	4	4		555	196		751	900
2340	High pressure sodium, 70 watt	1 Elec	2.70	2.963		235	145		380	475
2400	400 watt	2 Elec	4.40	3.636		365	178		543	665
2600	1000 watt	"	4	4		625	196		821	980

26 61 Lighting Systems and Accessories

26 61 23 – Lamps Applications

26 61 23.10 Lamps		Crew	Daily Output	Labor-Hours	Unit	Material	Labor	Equipment	Total	Total Incl O&P
0010	**LAMPS**									
0080	Fluorescent, rapid start, cool white, 2′ long, 20 watt	1 Elec	1	8	C	355	390		745	975
0100	4′ long, 40 watt		.90	8.889		300	435		735	980
0200	Slimline, 4′ long, 40 watt		.90	8.889		895	435		1,330	1,625
0210	4′ long, 30 watt energy saver G		.90	8.889		895	435		1,330	1,625
0400	High output, 4′ long, 60 watt		.90	8.889		1,025	435		1,460	1,775
0410	8′ long, 95 watt energy saver G		.80	10		1,025	490		1,515	1,875
0500	8′ long, 110 watt		.80	10		1,025	490		1,515	1,875
0512	2′ long, T5, 14 watt energy saver G		1	8		1,400	390		1,790	2,125

26 61 Lighting Systems and Accessories

26 61 23 – Lamps Applications

26 61 23.10 Lamps			Crew	Daily Output	Labor-Hours	Unit	2010 Bare Costs Material	Labor	Equipment	Total	Total Incl O&P
0514	3′ long, T5, 21 watt energy saver	G	1 Elec	.90	8.889	C	1,400	435		1,835	2,200
0516	4′ long, T5, 28 watt energy saver	G		.90	8.889		1,225	435		1,660	1,975
0517	4′ long, T5, 54 watt energy saver	G		.90	8.889		1,575	435		2,010	2,400
0560	Twin tube compact lamp	G		.90	8.889		585	435		1,020	1,300
0570	Double twin tube compact lamp	G		.80	10		1,300	490		1,790	2,175
0600	Mercury vapor, mogul base, deluxe white, 100 watt			.30	26.667		3,100	1,300		4,400	5,375
0700	250 watt			.30	26.667		4,800	1,300		6,100	7,250
0800	400 watt			.30	26.667		3,875	1,300		5,175	6,200
0900	1000 watt			.20	40		9,025	1,950		10,975	12,900
1000	Metal halide, mogul base, 175 watt			.30	26.667		4,125	1,300		5,425	6,500
1200	400 watt			.30	26.667		4,425	1,300		5,725	6,825
1300	1000 watt			.20	40		10,700	1,950		12,650	14,700
1350	High pressure sodium, 70 watt			.30	26.667		4,800	1,300		6,100	7,225
1380	250 watt			.30	26.667		5,475	1,300		6,775	7,975
1400	400 watt			.30	26.667		5,625	1,300		6,925	8,125
1450	1000 watt			.20	40		15,500	1,950		17,450	20,000
3000	Guards, fluorescent lamp, 4′ long			1	8		1,175	390		1,565	1,850
3200	8′ long		↓	.90	8.889	↓	2,325	435		2,760	3,225

26 71 Electrical Machines

26 71 13 – Motors Applications

26 71 13.20 Motors

			Crew	Daily Output	Labor-Hours	Unit	Material	Labor	Equipment	Total	Total Incl O&P
0010	**MOTORS** 230/460 volts, 60 HZ										
0050	Dripproof, premium efficiency, 1.15 service factor										
0060	1800 RPM, 1/4 HP		1 Elec	5.33	1.501	Ea.	173	73.50		246.50	299
0070	1/3 HP			5.33	1.501		176	73.50		249.50	305
0080	1/2 HP			5.33	1.501		266	73.50		339.50	400
0090	3/4 HP			5.33	1.501		276	73.50		349.50	415
0100	1 HP			4.50	1.778		285	87		372	445
0250	5 HP			4.50	1.778		430	87		517	605
0350	10 HP		↓	4	2		660	98		758	870
0450	20 HP	CN	2 Elec	5.20	3.077	↓	1,100	151		1,251	1,425

Estimating Tips

27 20 00 Data Communications
27 30 00 Voice Communications
27 40 00 Audio-Video Communications

When estimating material costs for special systems, it is always prudent to obtain manufacturers' quotations for equipment prices and special installation requirements which will affect the total costs.

Reference Numbers

Reference numbers are shown in shaded boxes at the beginning of some major classifications. These numbers refer to related items in the Reference Section. The reference information may be an estimating procedure, an alternate pricing method, or technical information.

Note: Not all subdivisions listed here necessarily appear in this publication.

Note: **Trade Service,** *in part, has been used as a reference source for some of the material prices used in Division 27.*

27 13 Communications Backbone Cabling

27 13 23 – Communications Optical Fiber Backbone Cabling

27 13 23.13 Communications Optical Fiber		Crew	Daily Output	Labor-Hours	Unit	2010 Bare Costs Material	Labor	Equipment	Total	Total Incl O&P
0010	**COMMUNICATIONS OPTICAL FIBER**									
0040	Specialized tools & techniques cause installation costs to vary.									
0070	Fiber optic, cable, bulk simplex, minimum	1 Elec	8	1	C.L.F.	23	49		72	98.50
0080	Bulk plenum quad, maximum	″	2.29	3.493	″	66	171		237	330
0150	Jumper				Ea.	55.50			55.50	61
0200	Pigtail					30			30	33
0300	Connector	1 Elec	24	.333		18.45	16.35		34.80	45
0350	Finger splice		32	.250		32	12.25		44.25	53
0400	Transceiver (low cost bi-directional)		8	1		300	49		349	405
0450	Rack housing, 4 rack spaces, 12 panels (144 fibers)		2	4		375	196		571	705
0500	Patch panel, 12 ports		6	1.333		242	65.50		307.50	365

27 41 Audio-Video Systems

27 41 19 – Portable Audio-Video Equipment

27 41 19.10 T.V. Systems

Line	Description	Crew	Daily Output	Labor-Hours	Unit	Material	Labor	Equipment	Total	Total Incl O&P
0010	**T.V. SYSTEMS**, not including rough-in wires, cables & conduits									
0100	Master TV antenna system									
0200	VHF reception & distribution, 12 outlets	1 Elec	6	1.333	Outlet	205	65.50		270.50	320
0400	30 outlets		10	.800		135	39		174	207
0600	100 outlets		13	.615		140	30		170	199
0800	VHF & UHF reception & distribution, 12 outlets		6	1.333		203	65.50		268.50	320
1000	30 outlets		10	.800		135	39		174	207
1200	100 outlets		13	.615		137	30		167	196
1400	School and deluxe systems, 12 outlets		2.40	3.333		268	163		431	540
1600	30 outlets		4	2		235	98		333	405
1800	80 outlets		5.30	1.509		226	74		300	360

27 51 Distributed Audio-Video Communications Systems

27 51 16 – Public Address and Mass Notification Systems

27 51 16.10 Public Address System

Line	Description	Crew	Daily Output	Labor-Hours	Unit	Material	Labor	Equipment	Total	Total Incl O&P
0010	**PUBLIC ADDRESS SYSTEM**									
0100	Conventional, office	1 Elec	5.33	1.501	Speaker	129	73.50		202.50	250
0200	Industrial	″	2.70	2.963	″	248	145		393	490

27 51 19 – Sound Masking Systems

27 51 19.10 Sound System

Line	Description	Crew	Daily Output	Labor-Hours	Unit	Material	Labor	Equipment	Total	Total Incl O&P
0010	**SOUND SYSTEM**, not including rough-in wires, cables & conduits									
0100	Components, projector outlet	1 Elec	8	1	Ea.	59.50	49		108.50	139
0200	Microphone		4	2		66	98		164	219
0400	Speakers, ceiling or wall		8	1		112	49		161	196
0600	Trumpets		4	2		209	98		307	375
0800	Privacy switch		8	1		83	49		132	165
1000	Monitor panel		4	2		370	98		468	555
1200	Antenna, AM/FM		4	2		207	98		305	375
1400	Volume control		8	1		83	49		132	165
1600	Amplifier, 250 watts		1	8		1,650	390		2,040	2,400
1800	Cabinets		1	8		810	390		1,200	1,475
2000	Intercom, 25 station capacity, master station	2 Elec	2	8		1,950	390		2,340	2,725
2200	Remote station	1 Elec	8	1		156	49		205	244
2400	Intercom outlets		8	1		91.50	49		140.50	174

27 51 Distributed Audio-Video Communications Systems

27 51 19 – Sound Masking Systems

27 51 19.10 Sound System		Crew	Daily Output	Labor-Hours	Unit	Material	Labor	Equipment	Total	Total Incl O&P
						2010 Bare Costs				
2600	Handset	1 Elec	4	2	Ea.	305	98		403	480
2800	Emergency call system, 12 zones, annunciator		1.30	6.154		915	300		1,215	1,450
3000	Bell		5.30	1.509		94.50	74		168.50	214
3200	Light or relay		8	1		47	49		96	125
3400	Transformer		4	2		207	98		305	375
3600	House telephone, talking station		1.60	5		445	245		690	855
3800	Press to talk, release to listen		5.30	1.509		104	74		178	224
4000	System-on button					62			62	68
4200	Door release	1 Elec	4	2		111	98		209	268
4400	Combination speaker and microphone		8	1		189	49		238	281
4600	Termination box		3.20	2.500		59.50	123		182.50	248
4800	Amplifier or power supply		5.30	1.509		680	74		754	860
5000	Vestibule door unit		16	.500	Name	125	24.50		149.50	175
5200	Strip cabinet		27	.296	Ea.	237	14.50		251.50	283
5400	Directory		16	.500	"	112	24.50		136.50	160

27 52 Healthcare Communications and Monitoring Systems

27 52 23 – Nurse Call/Code Blue Systems

27 52 23.10 Nurse Call Systems		Crew	Daily Output	Labor-Hours	Unit	Material	Labor	Equipment	Total	Total Incl O&P
0010	**NURSE CALL SYSTEMS**									
0100	Single bedside call station	1 Elec	8	1	Ea.	222	49		271	320
0200	Ceiling speaker station		8	1		67.50	49		116.50	147
0400	Emergency call station		8	1		114	49		163	198
0600	Pillow speaker		8	1		209	49		258	305
0800	Double bedside call station		4	2		223	98		321	390
1000	Duty station		4	2		173	98		271	335
1200	Standard call button		8	1		106	49		155	190
1400	Lights, corridor, dome or zone indicator		8	1		61	49		110	140
1600	Master control station for 20 stations	2 Elec	.65	24.615	Total	3,950	1,200		5,150	6,125

27 53 Distributed Systems

27 53 13 – Clock Systems

27 53 13.50 Clock Equipments		Crew	Daily Output	Labor-Hours	Unit	Material	Labor	Equipment	Total	Total Incl O&P
0010	**CLOCK EQUIPMENTS**, not including wires & conduits									
0100	Time system components, master controller	1 Elec	.33	24.242	Ea.	1,825	1,200		3,025	3,775
0200	Program bell		8	1		85	49		134	167
0400	Combination clock & speaker		3.20	2.500		210	123		333	415
0600	Frequency generator		2	4		2,350	196		2,546	2,900
0800	Job time automatic stamp recorder		4	2		460	98		558	650
2000	Time clock, 100 cards in & out, 1 color		3.20	2.500		1,325	123		1,448	1,625
2200	2 colors		3.20	2.500		1,250	123		1,373	1,575
2800	Metal rack for 25 cards		7	1.143		38.50	56		94.50	126

Division Notes

	CREW	DAILY OUTPUT	LABOR-HOURS	UNIT	2010 BARE COSTS MAT.	LABOR	EQUIP.	TOTAL	TOTAL INCL O&P

Estimating Tips

- When estimating material costs for electronic safety and security systems, it is always prudent to obtain manufacturers' quotations for equipment prices and special installation requirements that affect the total cost.
- Fire alarm systems consist of control panels, annunciator panels, battery with rack, charger, and fire alarm actuating and indicating devices. Some fire alarm systems include speakers, telephone lines, door closer controls, and other components. Be careful not to overlook the costs related to installation for these items. Also be aware of costs for integrated automation instrumentation and terminal devices, control equipment, control wiring, and programming.
- Security equipment includes items such as CCTV, access control, and other detection and identification systems to perform alert and alarm functions. Be sure to consider the costs related to installation for this security equipment, such as for integrated automation instrumentation and terminal devices, control equipment, control wiring, and programming.

Reference Numbers

Reference numbers are shown in shaded boxes at the beginning of some major classifications. These numbers refer to related items in the Reference Section. The reference information may be an estimating procedure, an alternate pricing method, or technical information.

Note: Not all subdivisions listed here necessarily appear in this publication.

28 13 Access Control

28 13 53 – Security Access Detection

28 13 53.13 Security Access Metal Detectors		Crew	Daily Output	Labor-Hours	Unit	2010 Bare Costs Material	Labor	Equipment	Total	Total Incl O&P
0010	**SECURITY ACCESS METAL DETECTORS**									
0240	Metal detector, hand-held, wand type, unit only				Ea.	81.50			81.50	90
0250	Metal detector, walk through portal type, single zone	1 Elec	2	4		2,750	196		2,946	3,325
0260	Multi-zone	"	2	4		3,500	196		3,696	4,150

28 13 53.16 Security Access X-Ray Equipment		Crew	Daily Output	Labor-Hours	Unit	Material	Labor	Equipment	Total	Total Incl O&P
0010	**SECURITY ACCESS X-RAY EQUIPMENT**									
0290	X-ray machine, desk top, for mail/small packages/letters	1 Elec	4	2	Ea.	3,000	98		3,098	3,450
0300	Conveyor type, incl monitor, minimum		2	4		14,000	196		14,196	15,700
0310	Maximum		2	4		25,000	196		25,196	27,800
0320	X-ray machine, large unit, for airports, incl monitor, min	2 Elec	1	16		35,000	785		35,785	39,700
0330	Maximum	"	.50	32		60,000	1,575		61,575	68,500

28 13 53.23 Security Access Explosive Detection Equipment		Crew	Daily Output	Labor-Hours	Unit	Material	Labor	Equipment	Total	Total Incl O&P
0010	**SECURITY ACCESS EXPLOSIVE DETECTION EQUIPMENT**									
0270	Explosives detector, walk through portal type	1 Elec	2	4	Ea.	3,500	196		3,696	4,150
0280	Hand-held, battery operated				"				25,500	28,100

28 16 Intrusion Detection

28 16 16 – Intrusion Detection Systems Infrastructure

28 16 16.50 Intrusion Detection		Crew	Daily Output	Labor-Hours	Unit	Material	Labor	Equipment	Total	Total Incl O&P
0010	**INTRUSION DETECTION,** not including wires & conduits									
0100	Burglar alarm, battery operated, mechanical trigger	1 Elec	4	2	Ea.	272	98		370	445
0200	Electrical trigger		4	2		325	98		423	500
0400	For outside key control, add		8	1		82	49		131	164
0600	For remote signaling circuitry, add		8	1		122	49		171	208
0800	Card reader, flush type, standard		2.70	2.963		910	145		1,055	1,225
1000	Multi-code		2.70	2.963		1,175	145		1,320	1,525
1200	Door switches, hinge switch		5.30	1.509		57.50	74		131.50	173
1400	Magnetic switch		5.30	1.509		67.50	74		141.50	185
1600	Exit control locks, horn alarm		4	2		340	98		438	520
1800	Flashing light alarm		4	2		380	98		478	565
2000	Indicating panels, 1 channel		2.70	2.963		360	145		505	610
2200	10 channel	2 Elec	3.20	5		1,225	245		1,470	1,725
2400	20 channel		2	8		2,400	390		2,790	3,225
2600	40 channel		1.14	14.035		4,375	690		5,065	5,825
2800	Ultrasonic motion detector, 12 volt	1 Elec	2.30	3.478		225	170		395	500
3000	Infrared photoelectric detector	"	2.30	3.478		186	170		356	455

28 23 Video Surveillance

28 23 13 – Video Surveillance Control and Management Systems

28 23 13.10 Closed Circuit Television System		Crew	Daily Output	Labor-Hours	Unit	Material	Labor	Equipment	Total	Total Incl O&P
0010	**CLOSED CIRCUIT TELEVISION SYSTEM**									
2000	Surveillance, one station (camera & monitor)	2 Elec	2.60	6.154	Total	1,300	300		1,600	1,875
2200	For additional camera stations, add	1 Elec	2.70	2.963	Ea.	725	145		870	1,025
2400	Industrial quality, one station (camera & monitor)	2 Elec	2.60	6.154	Total	2,700	300		3,000	3,400
2600	For additional camera stations, add	1 Elec	2.70	2.963	Ea.	1,650	145		1,795	2,050
2610	For low light, add		2.70	2.963		1,325	145		1,470	1,675
2620	For very low light, add		2.70	2.963		9,750	145		9,895	10,900
2800	For weatherproof camera station, add		1.30	6.154		1,025	300		1,325	1,575
3000	For pan and tilt, add		1.30	6.154		2,625	300		2,925	3,350

28 23 Video Surveillance

28 23 13 – Video Surveillance Control and Management Systems

28 23 13.10 Closed Circuit Television System		Crew	Daily Output	Labor-Hours	Unit	Material	2010 Bare Costs Labor	Equipment	Total	Total Incl O&P
3200	For zoom lens - remote control, add, minimum	1 Elec	2	4	Ea.	2,425	196		2,621	2,975
3400	Maximum		2	4		8,850	196		9,046	10,000
3410	For automatic iris for low light, add	↓	2	4	↓	2,125	196		2,321	2,625
3600	Educational T.V. studio, basic 3 camera system, black & white,									
3800	electrical & electronic equip. only, minimum	4 Elec	.80	40	Total	12,600	1,950		14,550	16,700
4000	Maximum (full console)		.28	114		53,500	5,600		59,100	67,500
4100	As above, but color system, minimum		.28	114		71,000	5,600		76,600	86,500
4120	Maximum	↓	.12	266	↓	308,500	13,100		321,600	358,500
4200	For film chain, black & white, add	1 Elec	1	8	Ea.	14,400	390		14,790	16,400
4250	Color, add		.25	32		17,500	1,575		19,075	21,600
4400	For video tape recorders, add, minimum	↓	1	8		3,025	390		3,415	3,900
4600	Maximum	4 Elec	.40	80	↓	25,200	3,925		29,125	33,500

28 23 23 – Video Surveillance Systems Infrastructure

28 23 23.50 Video Surveillance Equipments		Crew	Daily Output	Labor-Hours	Unit	Material	2010 Bare Costs Labor	Equipment	Total	Total Incl O&P
0010	**VIDEO SURVEILLANCE EQUIPMENTS**									
0200	Video cameras, wireless, hidden in exit signs, clocks, etc, incl receiver	1 Elec	3	2.667	Ea.	149	131		280	360
0210	Accessories for VCR, single camera		3	2.667		570	131		701	820
0220	For multiple cameras		3	2.667		1,225	131		1,356	1,550
0230	Video cameras, wireless, for under vehicle searching, complete	↓	2	4	↓	9,975	196		10,171	11,300

28 31 Fire Detection and Alarm

28 31 23 – Fire Detection and Alarm Annunciation Panels and Fire Stations

28 31 23.50 Alarm Panels and Devices		Crew	Daily Output	Labor-Hours	Unit	Material	2010 Bare Costs Labor	Equipment	Total	Total Incl O&P
0010	**ALARM PANELS AND DEVICES**, not including wires & conduits									
3594	Fire, alarm control panel									
3600	4 zone	2 Elec	2	8	Ea.	730	390		1,120	1,400
3800	8 zone		1	16		760	785		1,545	2,000
4000	12 zone	↓	.67	23.988		2,300	1,175		3,475	4,275
4020	Alarm device	1 Elec	8	1		229	49		278	325
4050	Actuating device		8	1		325	49		374	430
4200	Battery and rack		4	2		395	98		493	580
4400	Automatic charger		8	1		560	49		609	695
4600	Signal bell		8	1		70.50	49		119.50	151
4800	Trouble buzzer or manual station		8	1		80.50	49		129.50	162
5600	Strobe and horn		5.30	1.509		147	74		221	271
5800	Fire alarm horn		6.70	1.194		58.50	58.50		117	152
6000	Door holder, electro-magnetic		4	2		99	98		197	255
6200	Combination holder and closer		3.20	2.500		119	123		242	315
6600	Drill switch		8	1		360	49		409	470
6800	Master box		2.70	2.963		6,175	145		6,320	7,025
7000	Break glass station		8	1		65	49		114	145
7800	Remote annunciator, 8 zone lamp	↓	1.80	4.444		201	218		419	545
8000	12 zone lamp	2 Elec	2.60	6.154		345	300		645	830
8200	16 zone lamp	"	2.20	7.273	↓	345	355		700	910

28 31 43 – Fire Detection Sensors

28 31 43.50 Fire and Heat Detectors		Crew	Daily Output	Labor-Hours	Unit	Material	2010 Bare Costs Labor	Equipment	Total	Total Incl O&P
0010	**FIRE & HEAT DETECTORS**									
5000	Detector, rate of rise	1 Elec	8	1	Ea.	49	49		98	127

28 31 Fire Detection and Alarm

28 31 46 – Smoke Detection Sensors

28 31 46.50 Smoke Detectors		Crew	Daily Output	Labor-Hours	Unit	Material	2010 Bare Costs Labor	Equipment	Total	Total Incl O&P
0010	**SMOKE DETECTORS**									
5200	Smoke detector, ceiling type	1 Elec	6.20	1.290	Ea.	126	63		189	233
5400	Duct type	"	3.20	2.500	"	315	123		438	525

28 33 Fuel-Gas Detection and Alarm

28 33 33 – Fuel-Gas Detection Sensors

28 33 33.50 Tank Leak Detection Systems		Crew	Daily Output	Labor-Hours	Unit	Material	Labor	Equipment	Total	Total Incl O&P
0010	**TANK LEAK DETECTION SYSTEMS** Liquid and vapor									
0100	For hydrocarbons and hazardous liquids/vapors									
0120	Controller, data acquisition, incl. printer, modem, RS232 port									
0140	24 channel, for use with all probes				Ea.	3,700			3,700	4,075
0160	9 channel, for external monitoring				"	805			805	885
0200	Probes									
0210	Well monitoring									
0220	Liquid phase detection				Ea.	425			425	470
0230	Hydrocarbon vapor, fixed position					425			425	470
0240	Hydrocarbon vapor, float mounted					425			425	470
0250	Both liquid and vapor hydrocarbon				↓	425			425	470
0300	Secondary containment, liquid phase									
0310	Pipe trench/manway sump				Ea.	740			740	810
0320	Double wall pipe and manual sump					740			740	810
0330	Double wall fiberglass annular space					290			290	320
0340	Double wall steel tank annular space				↓	300			300	330
0500	Accessories									
0510	Modem, non-dedicated phone line				Ea.	272			272	299
0600	Monitoring, internal									
0610	Automatic tank gauge, incl. overfill				Ea.	1,025			1,025	1,125
0620	Product line				"	1,025			1,025	1,125
0700	Monitoring, special									
0710	Cathodic protection				Ea.	650			650	715
0720	Annular space chemical monitor				"	880			880	970

Estimating Tips

31 05 00 Common Work Results for Earthwork

- Estimating the actual cost of performing earthwork requires careful consideration of the variables involved. This includes items such as type of soil, whether water will be encountered, dewatering, whether banks need bracing, disposal of excavated earth, and length of haul to fill or spoil sites, etc. If the project has large quantities of cut or fill, consider raising or lowering the site to reduce costs, while paying close attention to the effect on site drainage and utilities.
- If the project has large quantities of fill, creating a borrow pit on the site can significantly lower the costs.
- It is very important to consider what time of year the project is scheduled for completion. Bad weather can create large cost overruns from dewatering, site repair, and lost productivity from cold weather.

Reference Numbers

Reference numbers are shown in shaded boxes at the beginning of some major classifications. These numbers refer to related items in the Reference Section. The reference information may be an estimating procedure, an alternate pricing method, or technical information.

Note: Not all subdivisions listed here necessarily appear in this publication.

31 05 Common Work Results for Earthwork

31 05 13 – Soils for Earthwork

31 05 13.10 Borrow		Crew	Daily Output	Labor-Hours	Unit	Material	Labor	Equipment	Total	Total Incl O&P
						2010 Bare Costs				
0010	BORROW R312316-40									
0020	Spread, 200 H.P. dozer, no compaction, 2 mi. RT haul									
0200	Common borrow	B-15	600	.047	C.Y.	7.40	1.68	4.21	13.29	15.35
0700	Screened loam CN		600	.047		30	1.68	4.21	35.89	40
0800	Topsoil, weed free		600	.047		23	1.68	4.21	28.89	32.50
0900	For 5 mile haul, add	B-34B	200	.040			1.33	3.34	4.67	5.70

31 05 16 – Aggregates for Earthwork

31 05 16.10 Borrow		Crew	Daily Output	Labor-Hours	Unit	Material	Labor	Equipment	Total	Total Incl O&P
0010	BORROW R312316-40									
0020	Spread, with 200 H.P. dozer, no compaction, 2 mi. RT haul									
0100	Bank run gravel CN	B-15	600	.047	C.Y.	22	1.68	4.21	27.89	31
0300	Crushed stone (1.40 tons per CY) , 1-1/2"		600	.047		35	1.68	4.21	40.89	45.50
0320	3/4" CN		600	.047		35	1.68	4.21	40.89	45.50
0340	1/2"		600	.047		30	1.68	4.21	35.89	40
0360	3/8"		600	.047		29.50	1.68	4.21	35.39	39.50
0400	Sand, washed, concrete		600	.047		33	1.68	4.21	38.89	43
0500	Dead or bank sand		600	.047		10.55	1.68	4.21	16.44	18.80
0600	Select structural fill		600	.047		12.65	1.68	4.21	18.54	21
0900	For 5 mile haul, add	B-34B	200	.040			1.33	3.34	4.67	5.70

31 05 23 – Cement and Concrete for Earthwork

31 05 23.30 Plant Mixed Bituminous Concrete		Crew	Daily Output	Labor-Hours	Unit	Material	Labor	Equipment	Total	Total Incl O&P
0010	PLANT MIXED BITUMINOUS CONCRETE									
0020	Asphaltic concrete plant mix (145 LB per C.F.) CN				Ton	65			65	71.50
0040	Asphaltic concrete less than 300 tons add trucking costs									
0050	See Div. 31 23 23.20 for hauling costs									
0200	All weather patching mix, hot				Ton	68			68	75
0250	Cold patch					68			68	75
0300	Berm mix					68			68	75

31 06 Schedules for Earthwork

31 06 60 – Schedules for Special Foundations and Load Bearing Elements

31 06 60.14 Piling Special Costs		Crew	Daily Output	Labor-Hours	Unit	Material	Labor	Equipment	Total	Total Incl O&P
0010	PILING SPECIAL COSTS									
0011	Piling special costs, pile caps, see Div. 03 30 53.40									
0500	Cutoffs, concrete piles, plain	1 Pile	5.50	1.455	Ea.		58.50		58.50	94.50
0600	With steel thin shell, add		38	.211			8.50		8.50	13.65
0700	Steel pile or "H" piles		19	.421			16.95		16.95	27.50
0800	Wood piles		38	.211			8.50		8.50	13.65
0900	Pre-augering up to 30' deep, average soil, 24" diameter	B-43	180	.267	L.F.		9.65	14.40	24.05	30.50
0920	36" diameter		115	.417			15.10	22.50	37.60	48
0960	48" diameter		70	.686			25	37	62	78.50
0980	60" diameter		50	.960			34.50	52	86.50	110
1000	Testing, any type piles, test load is twice the design load									
1050	50 ton design load, 100 ton test				Ea.				16,000	17,600
1100	100 ton design load, 200 ton test								21,000	23,100
1150	150 ton design load, 300 ton test								26,500	29,100
1200	200 ton design load, 400 ton test								28,500	31,300
1250	400 ton design load, 800 ton test								33,000	36,300
1500	Wet conditions, soft damp ground									
1600	Requiring mats for crane, add								40%	40%

31 06 Schedules for Earthwork

31 06 60 – Schedules for Special Foundations and Load Bearing Elements

31 06 60.14 Piling Special Costs		Crew	Daily Output	Labor-Hours	Unit	Material	2010 Bare Costs Labor	Equipment	Total	Total Incl O&P
1700	Barge mounted driving rig, add								30%	30%

31 06 60.15 Mobilization

		Crew	Daily Output	Labor-Hours	Unit	Material	Labor	Equipment	Total	Total Incl O&P
0010	**MOBILIZATION**									
0020	Set up & remove, air compressor, 600 C.F.M.	A-5	3.30	5.455	Ea.		180	18.15	198.15	297
0100	1200 C.F.M.	"	2.20	8.182			270	27.50	297.50	445
0200	Crane, with pile leads and pile hammer, 75 ton	B-19	.60	106			4,400	3,225	7,625	10,500
0300	150 ton	"	.36	177			7,350	5,375	12,725	17,400
0500	Drill rig, for caissons, to 36", minimum	B-43	2	24			870	1,300	2,170	2,750
0600	Up to 84"	"	1	48			1,725	2,600	4,325	5,500
0800	Auxiliary boiler, for steam small	A-5	1.66	10.843			360	36	396	590
0900	Large	"	.83	21.687			715	72	787	1,175
1100	Rule of thumb: complete pile driving set up, small	B-19	.45	142			5,875	4,300	10,175	13,900
1200	Large	"	.27	237	↓		9,800	7,150	16,950	23,200
1500	Mobilization, barge, by tug boat	B-83	25	.640	Mile		24.50	33.50	58	74

31 11 Clearing and Grubbing Land

31 11 10 – Clearing and Grubbing Land

31 11 10.10 Clear and Grub Site

		Crew	Daily Output	Labor-Hours	Unit	Material	Labor	Equipment	Total	Total Incl O&P
0010	**CLEAR AND GRUB SITE**									
0020	Cut & chip light trees to 6" diam.	B-7	1	48	Acre		1,675	1,400	3,075	4,125
0150	Grub stumps and remove	B-30	2	12			435	1,175	1,610	1,950
0200	Cut & chip medium, trees to 12" diam.	B-7	.70	68.571			2,400	2,025	4,425	5,900
0250	Grub stumps and remove	B-30	1	24			875	2,350	3,225	3,925
0300	Cut & chip heavy, trees to 24" diam.	B-7	.30	160			5,625	4,700	10,325	13,800
0350	Grub stumps and remove	B-30	.50	48			1,750	4,700	6,450	7,825
0400	If burning is allowed, reduce cut & chip				↓					40%
3000	Chipping stumps, to 18" deep, 12" diam.	B-86	20	.400	Ea.		17.20	9.55	26.75	36
3040	18" diameter		16	.500			21.50	11.95	33.45	45
3080	24" diameter		14	.571			24.50	13.65	38.15	51.50
3100	30" diameter		12	.667			28.50	15.90	44.40	60.50
3120	36" diameter		10	.800			34.50	19.10	53.60	72.50
3160	48" diameter	↓	8	1	↓		43	24	67	91
5000	Tree thinning, feller buncher, conifer									
5080	Up to 8" diameter	B-93	240	.033	Ea.		1.43	2.54	3.97	4.93
5120	12" diameter		160	.050			2.15	3.81	5.96	7.40
5240	Hardwood, up to 4" diameter		240	.033			1.43	2.54	3.97	4.93
5280	8" diameter		180	.044			1.91	3.39	5.30	6.60
5320	12" diameter	↓	120	.067	↓		2.86	5.10	7.96	9.90
7000	Tree removal, congested area, aerial lift truck									
7040	8" diameter	B-85	7	5.714	Ea.		200	132	332	450
7080	12" diameter		6	6.667			234	154	388	525
7120	18" diameter		5	8			281	185	466	635
7160	24" diameter		4	10			350	231	581	790
7240	36" diameter		3	13.333			470	310	780	1,050
7280	48" diameter	↓	2	20	↓		700	465	1,165	1,575

31 13 13 – Selective Tree and Shrub Removal

31 13 13.10 Selective Clearing		Crew	Daily Output	Labor-Hours	Unit	Material	2010 Bare Costs Labor	Equipment	Total	Total Incl O&P
0010	**SELECTIVE CLEARING**									
0020	Clearing brush with brush saw	A-1C	.25	32	Acre		1,050	113	1,163	1,750
0100	By hand	1 Clab	.12	66.667			2,200		2,200	3,400
0300	With dozer, ball and chain, light clearing	B-11A	2	8			305	595	900	1,125
0400	Medium clearing		1.50	10.667			405	795	1,200	1,500
0500	With dozer and brush rake, light		10	1.600			61	119	180	224
0550	Medium brush to 4" diameter		8	2			76	149	225	279
0600	Heavy brush to 4" diameter		6.40	2.500			95	186	281	350
1000	Brush mowing, tractor w/rotary mower, no removal									
1020	Light density	B-84	2	4	Acre		172	151	323	425
1040	Medium density		1.50	5.333			229	201	430	565
1080	Heavy density		1	8			345	300	645	845

31 13 13.20 Selective Tree Removal		Crew	Daily Output	Labor-Hours	Unit	Material	2010 Bare Costs Labor	Equipment	Total	Total Incl O&P
0010	**SELECTIVE TREE REMOVAL**									
0011	With tractor, large tract, firm									
0020	level terrain, no boulders, less than 12" diam. trees									
0300	300 HP dozer, up to 400 trees/acre, 0 to 25% hardwoods	B-10M	.75	16	Acre		635	2,125	2,760	3,275
0340	25% to 50% hardwoods		.60	20			795	2,650	3,445	4,125
0370	75% to 100% hardwoods		.45	26.667			1,050	3,550	4,600	5,500
0400	500 trees/acre, 0% to 25% hardwoods		.60	20			795	2,650	3,445	4,125
0440	25% to 50% hardwoods		.48	25			990	3,325	4,315	5,150
0470	75% to 100% hardwoods		.36	33.333			1,325	4,425	5,750	6,875
0500	More than 600 trees/acre, 0 to 25% hardwoods		.52	23.077			915	3,050	3,965	4,750
0540	25% to 50% hardwoods		.42	28.571			1,125	3,800	4,925	5,875
0570	75% to 100% hardwoods		.31	38.710			1,525	5,125	6,650	7,975
0900	Large tract clearing per tree									
1500	300 HP dozer, to 12" diameter, softwood	B-10M	320	.038	Ea.		1.49	4.98	6.47	7.70
1550	Hardwood		100	.120			4.76	15.90	20.66	24.50
1600	12" to 24" diameter, softwood		200	.060			2.38	7.95	10.33	12.35
1650	Hardwood		80	.150			5.95	19.90	25.85	31
1700	24" to 36" diameter, softwood		100	.120			4.76	15.90	20.66	24.50
1750	Hardwood		50	.240			9.50	32	41.50	49.50
1800	36" to 48" diameter, softwood		70	.171			6.80	22.50	29.30	35.50
1850	Hardwood		35	.343			13.60	45.50	59.10	70.50
2000	Stump removal on site by hydraulic backhoe, 1-1/2 C.Y.									
2040	4" to 6" diameter	B-17	60	.533	Ea.		18.75	12.35	31.10	42
2050	8" to 12" diameter	B-30	33	.727			26.50	71.50	98	119
2100	14" to 24" diameter		25	.960			35	94	129	157
2150	26" to 36" diameter		16	1.500			54.50	147	201.50	245
3000	Remove selective trees, on site using chain saws and chipper,									
3050	not incl. stumps, up to 6" diameter	B-7	18	2.667	Ea.		93.50	78.50	172	229
3100	8" to 12" diameter		12	4			140	117	257	345
3150	14" to 24" diameter		10	4.800			168	141	309	415
3200	26" to 36" diameter		8	6			210	176	386	520
3300	Machine load, 2 mile haul to dump, 12" diam. tree	A-3B	8	2			76	138	214	267

31 14 Earth Stripping and Stockpiling

31 14 13 – Soil Stripping and Stockpiling

31 14 13.23 Topsoil Stripping and Stockpiling		Crew	Daily Output	Labor-Hours	Unit	Material	2010 Bare Costs Labor	Equipment	Total	Total Incl O&P
0010	**TOPSOIL STRIPPING AND STOCKPILING**									
0020	200 H.P. dozer, ideal conditions	B-10B	2300	.005	C.Y.		.21	.52	.73	.88
0100	Adverse conditions	"	1150	.010			.41	1.04	1.45	1.76
0200	300 H.P. dozer, ideal conditions	B-10M	3000	.004			.16	.53	.69	.82
0300	Adverse conditions	"	1650	.007			.29	.96	1.25	1.50
0400	400 H.P. dozer, ideal conditions	B-10X	3900	.003			.12	.53	.65	.77
0500	Adverse conditions	"	2000	.006			.24	1.04	1.28	1.50
0600	Clay, dry and soft, 200 H.P. dozer, ideal conditions	B-10B	1600	.008			.30	.75	1.05	1.27
0700	Adverse conditions	"	800	.015			.60	1.49	2.09	2.54
1000	Medium hard, 300 H.P. dozer, ideal conditions	B-10M	2000	.006			.24	.80	1.04	1.24
1100	Adverse conditions	"	1100	.011			.43	1.45	1.88	2.24
1200	Very hard, 400 H.P. dozer, ideal conditions	B-10X	2600	.005			.18	.80	.98	1.16
1300	Adverse conditions	"	1340	.009	↓		.36	1.55	1.91	2.25
1400	Loam or topsoil, remove and stockpile on site									
1420	6" deep, 200' haul	B-10B	865	.014	C.Y.		.55	1.38	1.93	2.35
1430	300' haul		520	.023			.92	2.29	3.21	3.90
1440	500' haul		225	.053	↓		2.12	5.30	7.42	9.05
1450	Alternate method: 6" deep, 200' haul		5090	.002	S.Y.		.09	.23	.32	.40
1460	500' haul	↓	1325	.009	"		.36	.90	1.26	1.53

31 22 Grading

31 22 16 – Fine Grading

31 22 16.10 Finish Grading

		Crew	Daily Output	Labor-Hours	Unit	Material	Labor	Equipment	Total	Total Incl O&P
0010	**FINISH GRADING**									
0012	Finish grading area to be paved with grader, small area	B-11L	400	.040	S.Y.		1.52	1.56	3.08	4.02
0100	Large area		2000	.008			.30	.31	.61	.80
1100	Fine grade for slab on grade, machine	↓	1040	.015			.58	.60	1.18	1.55
1150	Hand grading	B-18	700	.034	↓		1.16	.06	1.22	1.86
3500	Finish grading lagoon bottoms	B-11L	4	4	M.S.F.		152	156	308	400

31 23 Excavation and Fill

31 23 16 – Excavation

31 23 16.13 Excavating, Trench

		Crew	Daily Output	Labor-Hours	Unit	Material	Labor	Equipment	Total	Total Incl O&P
0010	**EXCAVATING, TRENCH**									
0011	Or continuous footing									
0020	Common earth with no sheeting or dewatering included									
0050	1' to 4' deep, 3/8 C.Y. excavator	B-11C	150	.107	B.C.Y.		4.06	2.25	6.31	8.65
0060	1/2 C.Y. excavator	B-11M	200	.080			3.04	1.94	4.98	6.75
0090	4' to 6' deep, 1/2 C.Y. excavator	"	200	.080			3.04	1.94	4.98	6.75
0100	5/8 C.Y. excavator	B-12Q	250	.064			2.48	2.26	4.74	6.25
0110	3/4 C.Y. excavator	B-12F	300	.053			2.07	2.26	4.33	5.60
0300	1/2 C.Y. excavator, truck mounted	B-12J	200	.080			3.10	4.06	7.16	9.15
0500	6' to 10' deep, 3/4 C.Y. excavator	B-12F	225	.071			2.76	3.01	5.77	7.50
0510	1 C.Y. excavator	B-12A	400	.040			1.55	2	3.55	4.55
0600	1 C.Y. excavator, truck mounted	B-12K	400	.040			1.55	2.40	3.95	4.99
0610	1-1/2 C.Y.excavator	B-12B	600	.027			1.03	1.70	2.73	3.43
0900	10' to 14' deep, 3/4 C.Y. excavator	B-12F	200	.080			3.10	3.39	6.49	8.45
0910	1 C.Y. excavator	B-12A	360	.044			1.72	2.23	3.95	5.05
1000	1-1/2 C.Y. excavator	B-12B	540	.030	↓		1.15	1.88	3.03	3.81

31 23 16.13 Excavating, Trench		Crew	Daily Output	Labor-Hours	Unit	Material	2010 Bare Costs Labor	Equipment	Total	Total Incl O&P
1300	14′ to 20′ deep, 1 C.Y. excavator	B-12A	320	.050	B.C.Y.		1.94	2.51	4.45	5.70
1310	1-1/2 C.Y. excavator	B-12B	480	.033			1.29	2.12	3.41	4.29
1320	2-1/2 C.Y. excavator	B-12S	850	.019			.73	2.10	2.83	3.42
1340	20′ to 24′ deep, 1 C.Y. excavator	B-12A	288	.056			2.15	2.78	4.93	6.30
1342	1-1/2 C.Y. excavator	B-12B	432	.037			1.44	2.35	3.79	4.77
1344	2-1/2 C.Y. excavator	B-12S	765	.021			.81	2.33	3.14	3.80
1352	4′ to 6′ deep, 1/2 C.Y. excavator w/trench box	B-13H	188	.085			3.30	4.93	8.23	10.40
1354	5/8 C.Y. excavator	″	235	.068			2.64	3.94	6.58	8.35
1356	3/4 C.Y. excavator	B-13G	282	.057			2.20	2.81	5.01	6.40
1362	6′ to 10′ deep, 3/4 C.Y. excavator w/trench box	″	212	.075			2.92	3.73	6.65	8.55
1370	1 C.Y. excavator	B-13D	376	.043			1.65	2.43	4.08	5.20
1371	1-1/2 C.Y. excavator	B-13E	564	.028			1.10	2	3.10	3.87
1374	10′ to 14′ deep, 3/4 C.Y. excavator w/trench box	B-13G	188	.085			3.30	4.21	7.51	9.65
1375	1 C.Y. excavator	B-13D	338	.047			1.83	2.71	4.54	5.75
1376	1-1/2 C.Y. excavator	B-13E	508	.032			1.22	2.23	3.45	4.30
1381	14′ to 20′ deep, 1 C.Y. excavator w/trench box	B-13D	301	.053			2.06	3.04	5.10	6.45
1382	1-1/2 C.Y. excavator	B-13E	451	.035			1.37	2.51	3.88	4.84
1383	2-1/2 C.Y. excavator	B-13J	799	.020			.78	2.38	3.16	3.79
1386	20′ to 24′ deep, 1 C.Y. excavator w/trench box	B-13D	271	.059			2.29	3.38	5.67	7.20
1387	1-1/2 C.Y. excavator	B-13E	406	.039			1.53	2.78	4.31	5.40
1388	2-1/2 C.Y. excavator	B-13J	719	.022			.86	2.64	3.50	4.21
1391	Shoring by SF/day trench wall protected loose mat., 4′ W	B-6	3200	.008	SF Wall	.39	.27	.11	.77	.96
1392	Rent shoring per week per SF wall protected, loose mat., 4′ W					1.07			1.07	1.18
1395	Hydraulic shoring, SF trench wall protected stable mat., 4′ W	2 Clab	2700	.006		.14	.20		.34	.45
1397	semi-stable material, 4′ W	″	2400	.007		.18	.22		.40	.54
1398	Rent hydraulic shoring per day/SF wall, stable mat., 4′ W					.24			.24	.26
1399	semi-stable material					.28			.28	.31
1400	By hand with pick and shovel 2′ to 6′ deep, light soil	1 Clab	8	1	B.C.Y.		33		33	51
1500	Heavy soil	″	4	2	″		66		66	102
1700	For tamping backfilled trenches, air tamp, add	A-1G	100	.080	E.C.Y.		2.65	.52	3.17	4.65
1900	Vibrating plate, add	B-18	180	.133	″		4.50	.24	4.74	7.20
2100	Trim sides and bottom for concrete pours, common earth		1500	.016	S.F.		.54	.03	.57	.86
2300	Hardpan		600	.040	″		1.35	.07	1.42	2.16
2400	Pier and spread footing excavation, add to above				B.C.Y.				30%	30%
3000	Backfill trench, F.E. loader, wheel mtd., 1 C.Y. bucket									
3020	Minimal haul	B-10R	400	.030	L.C.Y.		1.19	.67	1.86	2.54
3040	100′ haul	″	200	.060			2.38	1.35	3.73	5.05
3080	2-1/4 C.Y. bucket, minimum haul	B-10T	600	.020			.79	.73	1.52	2
3090	100′ haul	″	300	.040			1.59	1.45	3.04	4
5020	Loam & Sandy clay with no sheeting or dewatering included									
5050	1′ to 4′ deep, 3/8 C.Y. tractor loader/backhoe	B-11C	162	.099	B.C.Y.		3.76	2.09	5.85	8
5060	1/2 C.Y. excavator	B-11M	216	.074			2.82	1.79	4.61	6.25
5080	4′ to 6′ deep, 1/2 C.Y. excavator	″	216	.074			2.82	1.79	4.61	6.25
5090	5/8 C.Y. excavator	B-12Q	276	.058			2.25	2.05	4.30	5.65
5100	3/4 C.Y. excavator	B-12F	324	.049			1.91	2.09	4	5.20
5130	1/2 C.Y. excavator, truck mounted	B-12J	216	.074			2.87	3.76	6.63	8.50
5140	6′ to 10′ deep, 3/4 C.Y. excavator	B-12F	243	.066			2.55	2.79	5.34	6.95
5150	1 C.Y. excavator	B-12A	432	.037			1.44	1.86	3.30	4.22
5160	1 C.Y. excavator, truck mounted	B-12K	432	.037			1.44	2.22	3.66	4.62
5170	1-1/2 C.Y. excavator	B-12B	648	.025			.96	1.57	2.53	3.18
5190	10′ to 14′ deep, 3/4 C.Y. excavator	B-12F	216	.074			2.87	3.14	6.01	7.80
5200	1 C.Y. excavator	B-12A	389	.041			1.59	2.06	3.65	4.69
5210	1-1/2 C.Y. excavator	B-12B	583	.027			1.06	1.74	2.80	3.53

31 23 16.13 Excavating, Trench		Crew	Daily Output	Labor-Hours	Unit	Material	2010 Bare Costs Labor	Equipment	Total	Total Incl O&P
5250	14′ to 20′ deep, 1 C.Y. excavator	B-12A	346	.046	B.C.Y.		1.79	2.32	4.11	5.25
5260	1-1/2 C.Y. excavator	B-12B	518	.031			1.20	1.96	3.16	3.97
5270	2-1/2 C.Y. excavator	B-12S	918	.017			.68	1.94	2.62	3.16
5300	20′ to 24′ deep, 1 C.Y. excavator	B-12A	311	.051			1.99	2.58	4.57	5.85
5310	1-1/2 C.Y. excavator	B-12B	467	.034			1.33	2.18	3.51	4.41
5320	2-1/2 C.Y. excavator	B-12S	826	.019			.75	2.16	2.91	3.52
5352	4′ to 6′ deep, 1/2 C.Y. excavator w/trench box	B-13H	205	.078			3.02	4.52	7.54	9.55
5354	5/8 C.Y. excavator	"	257	.062			2.41	3.60	6.01	7.60
5356	3/4 C.Y. excavator	B-13G	308	.052			2.01	2.57	4.58	5.90
5362	6′ to 10′ deep, 3/4 C.Y. excavator w/trench box	"	231	.069			2.68	3.42	6.10	7.85
5364	1 C.Y. excavator	B-13D	410	.039			1.51	2.23	3.74	4.74
5366	1-1/2 C.Y. excavator	B-13E	616	.026			1.01	1.83	2.84	3.55
5370	10′ to 14′ deep, 3/4 C.Y. excavator w/trench box	B-13G	205	.078			3.02	3.86	6.88	8.85
5372	1 C.Y. excavator	B-13D	370	.043			1.68	2.47	4.15	5.25
5374	1-1/2 C.Y. excavator	B-13E	554	.029			1.12	2.04	3.16	3.94
5382	14′ to 20′ deep, 1 C.Y. excavator w/trench box	B-13D	329	.049			1.88	2.78	4.66	5.90
5384	1-1/2 C.Y. excavator	B-13E	492	.033			1.26	2.30	3.56	4.44
5386	2-1/2 C.Y. excavator	B-13J	872	.018			.71	2.18	2.89	3.47
5392	20′ to 24′ deep, 1 C.Y. excavator w/trench box	B-13D	295	.054			2.10	3.10	5.20	6.60
5394	1-1/2 C.Y. excavator	B-13E	444	.036			1.40	2.55	3.95	4.92
5396	2-1/2 C.Y. excavator	B-13J	785	.020	↓		.79	2.42	3.21	3.86
6020	Sand & gravel with no sheeting or dewatering included									
6050	1′ to 4′ deep, 3/8 C.Y. excavator	B-11C	165	.097	B.C.Y.		3.69	2.05	5.74	7.85
6060	1/2 C.Y. excavator	B-11M	220	.073			2.77	1.76	4.53	6.15
6080	4′ to 6′ deep, 1/2 C.Y. excavator	"	220	.073			2.77	1.76	4.53	6.15
6090	5/8 C.Y. excavator	B-12Q	275	.058			2.25	2.05	4.30	5.70
6100	3/4 C.Y. excavator	B-12F	330	.048			1.88	2.05	3.93	5.10
6130	1/2 C.Y. excavator, truck mounted	B-12J	220	.073			2.82	3.69	6.51	8.35
6140	6′ to 10′ deep, 3/4 C.Y. excavator	B-12F	248	.065			2.50	2.73	5.23	6.80
6150	1 C.Y. excavator	B-12A	440	.036			1.41	1.82	3.23	4.14
6160	1 C.Y. excavator, truck mounted	B-12K	440	.036			1.41	2.18	3.59	4.54
6170	1-1/2 C.Y. excavator	B-12B	660	.024			.94	1.54	2.48	3.11
6190	10′ to 14′ deep, 3/4 C.Y. excavator	B-12F	220	.073			2.82	3.08	5.90	7.65
6200	1 C.Y. excavator	B-12A	396	.040			1.57	2.02	3.59	4.60
6210	1-1/2 C.Y. excavator	B-12B	594	.027			1.04	1.71	2.75	3.46
6250	14′ to 20′ deep, 1 C.Y. excavator	B-12A	352	.045			1.76	2.28	4.04	5.15
6260	1-1/2 C.Y. excavator	B-12B	528	.030			1.17	1.93	3.10	3.90
6270	2-1/2 C.Y. excavator	B-12S	935	.017			.66	1.91	2.57	3.11
6300	20′ to 24′ deep, 1 C.Y. excavator	B-12A	317	.050			1.96	2.53	4.49	5.75
6310	1-1/2 C.Y. excavator	B-12B	475	.034			1.31	2.14	3.45	4.33
6320	2-1/2 C.Y. excavator	B-12S	842	.019			.74	2.12	2.86	3.45
6352	4′ to 6′ deep, 1/2 C.Y. excavator w/trench box	B-13H	209	.077			2.97	4.43	7.40	9.40
6354	5/8 C.Y. excavator	"	261	.061			2.38	3.55	5.93	7.50
6356	3/4 C.Y. excavator	B-13G	314	.051			1.97	2.52	4.49	5.75
6362	6′ to 10′ deep, 3/4 C.Y. excavator w/trench box	"	236	.068			2.63	3.35	5.98	7.65
6364	1 C.Y. excavator	B-13D	418	.038			1.48	2.19	3.67	4.66
6366	1-1/2 C.Y. excavator	B-13E	627	.026			.99	1.80	2.79	3.48
6370	10′ to 14′ deep, 3/4 C.Y. excavator w/trench box	B-13G	209	.077			2.97	3.79	6.76	8.65
6372	1 C.Y. excavator	B-13D	376	.043			1.65	2.43	4.08	5.20
6374	1-1/2 C.Y. excavator	B-13E	564	.028			1.10	2	3.10	3.87
6382	14′ to 20′ deep, 1 C.Y. excavator w/trench box	B-13D	334	.048			1.86	2.74	4.60	5.80
6384	1-1/2 C.Y. excavator	B-13E	502	.032			1.24	2.25	3.49	4.35
6386	2-1/2 C.Y. excavator	B-13J	888	.018	↓		.70	2.14	2.84	3.41

31 23 16.13 Excavating, Trench

	31 23 16.13 Excavating, Trench	Crew	Daily Output	Labor-Hours	Unit	Material	2010 Bare Costs Labor	Equipment	Total	Total Incl O&P
6392	20′ to 24′ deep, 1 C.Y. excavator w/trench box	B-13D	301	.053	B.C.Y.		2.06	3.04	5.10	6.45
6394	1-1/2 C.Y. excavator	B-13E	452	.035			1.37	2.50	3.87	4.83
6396	2-1/2 C.Y. excavator	B-13J	800	.020			.78	2.37	3.15	3.79
7020	Dense hard clay with no sheeting or dewatering included									
7050	1′ to 4′ deep, 3/8 C.Y. excavator	B-11C	132	.121	B.C.Y.		4.61	2.56	7.17	9.80
7060	1/2 C.Y. excavator	B-11M	176	.091			3.46	2.20	5.66	7.65
7080	4′ to 6′ deep, 1/2 C.Y. excavator	"	176	.091			3.46	2.20	5.66	7.65
7090	5/8 C.Y. excavator	B-12Q	220	.073			2.82	2.57	5.39	7.10
7100	3/4 C.Y. excavator	B-12F	264	.061			2.35	2.57	4.92	6.40
7130	1/2 C.Y. excavator, truck mounted	B-12J	176	.091			3.52	4.62	8.14	10.45
7140	6′ to 10′ deep, 3/4 C.Y. excavator	B-12F	198	.081			3.13	3.42	6.55	8.50
7150	1 C.Y. excavator	B-12A	352	.045			1.76	2.28	4.04	5.15
7160	1 C.Y. excavator, truck mounted	B-12K	352	.045			1.76	2.72	4.48	5.65
7170	1-1/2 C.Y. excavator	B-12B	528	.030			1.17	1.93	3.10	3.90
7190	10′ to 14′ deep, 3/4 C.Y. excavator	B-12F	176	.091			3.52	3.85	7.37	9.60
7200	1 C.Y. excavator	B-12A	317	.050			1.96	2.53	4.49	5.75
7210	1-1/2 C.Y. excavator	B-12B	475	.034			1.31	2.14	3.45	4.33
7250	14′ to 20′ deep, 1 C.Y. excavator	B-12A	282	.057			2.20	2.84	5.04	6.45
7260	1-1/2 C.Y. excavator	B-12B	422	.038			1.47	2.41	3.88	4.88
7270	2-1/2 C.Y. excavator	B-12S	748	.021			.83	2.39	3.22	3.88
7300	20′ to 24′ deep, 1 C.Y. excavator	B-12A	254	.063			2.44	3.16	5.60	7.15
7310	1-1/2 C.Y. excavator	B-12B	380	.042			1.63	2.68	4.31	5.40
7320	2-1/2 C.Y. excavator	B-12S	673	.024			.92	2.65	3.57	4.32

31 23 16.14 Excavating, Utility Trench

	31 23 16.14 Excavating, Utility Trench	Crew	Daily Output	Labor-Hours	Unit	Material	Labor	Equipment	Total	Total Incl O&P
0010	**EXCAVATING, UTILITY TRENCH**									
0011	Common earth									
0050	Trenching with chain trencher, 12 H.P., operator walking									
0100	4″ wide trench, 12″ deep	B-53	800	.010	L.F.		.41	.08	.49	.71
0150	18″ deep		750	.011			.44	.09	.53	.76
0200	24″ deep		700	.011			.47	.09	.56	.81
0300	6″ wide trench, 12″ deep		650	.012			.51	.10	.61	.87
0350	18″ deep		600	.013			.55	.11	.66	.94
0400	24″ deep		550	.015			.60	.12	.72	1.03
0450	36″ deep		450	.018			.73	.15	.88	1.26
0600	8″ wide trench, 12″ deep		475	.017			.70	.14	.84	1.19
0650	18″ deep		400	.020			.83	.17	1	1.42
0700	24″ deep		350	.023			.94	.19	1.13	1.62
0750	36″ deep		300	.027			1.10	.22	1.32	1.89
1000	Backfill by hand including compaction, add									
1050	4″ wide trench, 12″ deep	A-1G	800	.010	L.F.		.33	.06	.39	.58
1100	18″ deep		530	.015			.50	.10	.60	.88
1150	24″ deep		400	.020			.66	.13	.79	1.16
1300	6″ wide trench, 12″ deep		540	.015			.49	.10	.59	.87
1350	18″ deep		405	.020			.65	.13	.78	1.15
1400	24″ deep		270	.030			.98	.19	1.17	1.72
1450	36″ deep		180	.044			1.47	.29	1.76	2.59
1600	8″ wide trench, 12″ deep		400	.020			.66	.13	.79	1.16
1650	18″ deep		265	.030			1	.20	1.20	1.76
1700	24″ deep		200	.040			1.32	.26	1.58	2.32
1750	36″ deep		135	.059			1.96	.38	2.34	3.45
2000	Chain trencher, 40 H.P. operator riding									
2050	6″ wide trench and backfill, 12″ deep	B-54	1200	.007	L.F.		.28	.28	.56	.72

31 23 16.14 Excavating, Utility Trench		Crew	Daily Output	Labor-Hours	Unit	Material	2010 Bare Costs Labor	Equipment	Total	Total Incl O&P
2100	18" deep	B-54	1000	.008	L.F.		.33	.33	.66	.86
2150	24" deep		975	.008			.34	.34	.68	.89
2200	36" deep		900	.009			.37	.37	.74	.96
2250	48" deep		750	.011			.44	.45	.89	1.15
2300	60" deep		650	.012			.51	.51	1.02	1.33
2400	8" wide trench and backfill, 12" deep		1000	.008			.33	.33	.66	.86
2450	18" deep		950	.008			.35	.35	.70	.91
2500	24" deep		900	.009			.37	.37	.74	.96
2550	36" deep		800	.010			.41	.42	.83	1.08
2600	48" deep		650	.012			.51	.51	1.02	1.33
2700	12" wide trench and backfill, 12" deep		975	.008			.34	.34	.68	.89
2750	18" deep		860	.009			.38	.39	.77	1.01
2800	24" deep		800	.010			.41	.42	.83	1.08
2850	36" deep		725	.011			.46	.46	.92	1.19
3000	16" wide trench and backfill, 12" deep		835	.010			.40	.40	.80	1.03
3050	18" deep		750	.011			.44	.45	.89	1.15
3100	24" deep		700	.011			.47	.48	.95	1.24
3200	Compaction with vibratory plate, add								35%	35%
5100	Hand excavate and trim for pipe bells after trench excavation									
5200	8" pipe	1 Clab	155	.052	L.F.		1.71		1.71	2.63
5300	18" pipe	"	130	.062	"		2.04		2.04	3.14

31 23 16.16 Structural Excavation for Minor Structures		Crew	Daily Output	Labor-Hours	Unit	Material	2010 Bare Costs Labor	Equipment	Total	Total Incl O&P
0010	**STRUCTURAL EXCAVATION FOR MINOR STRUCTURES** R312316-40									
0015	Hand, pits to 6' deep, sandy soil	1 Clab	8	1	B.C.Y.		33		33	51
0100	Heavy soil or clay		4	2			66		66	102
0300	Pits 6' to 12' deep, sandy soil		5	1.600			53		53	81.50
0500	Heavy soil or clay		3	2.667			88.50		88.50	136
0700	Pits 12' to 18' deep, sandy soil		4	2			66		66	102
0900	Heavy soil or clay		2	4			132		132	204
1100	Hand loading trucks from stock pile, sandy soil		12	.667			22		22	34
1300	Heavy soil or clay		8	1			33		33	51
1500	For wet or muck hand excavation, add to above				%				50%	50%
6000	Machine excavation, for spread and mat footings, elevator pits,									
6001	and small building foundations									
6030	Common earth, hydraulic backhoe, 1/2 C.Y. bucket	B-12E	55	.291	B.C.Y.		11.25	7.05	18.30	25
6035	3/4 C.Y. bucket	B-12F	90	.178			6.90	7.55	14.45	18.75
6040	1 C.Y. bucket	B-12A	108	.148			5.75	7.40	13.15	16.85
6050	1-1/2 C.Y. bucket	B-12B	144	.111			4.31	7.05	11.36	14.30
6060	2 C.Y. bucket	B-12C	200	.080			3.10	6.60	9.70	11.95
6070	Sand and gravel, 3/4 C.Y. bucket	B-12F	100	.160			6.20	6.80	13	16.85
6080	1 C.Y. bucket	B-12A	120	.133			5.15	6.70	11.85	15.20
6090	1-1/2 C.Y. bucket	B-12B	160	.100			3.88	6.35	10.23	12.90
6100	2 C.Y. bucket	B-12C	220	.073			2.82	6	8.82	10.85
6110	Clay, till, or blasted rock, 3/4 C.Y. bucket	B-12F	80	.200			7.75	8.45	16.20	21
6120	1 C.Y. bucket	B-12A	95	.168			6.55	8.45	15	19.20
6130	1-1/2 C.Y. bucket	B-12B	130	.123			4.77	7.80	12.57	15.85
6140	2 C.Y. bucket	B-12C	175	.091			3.54	7.55	11.09	13.65
6230	Sandy clay & loam, hydraulic backhoe, 1/2 C.Y. bucket	B-12E	60	.267			10.35	6.45	16.80	23
6235	3/4 C.Y. bucket	B-12F	98	.163			6.35	6.90	13.25	17.20
6240	1 C.Y. bucket	B-12A	116	.138			5.35	6.90	12.25	15.70
6250	1-1/2 C.Y. bucket	B-12B	156	.103			3.97	6.50	10.47	13.20
9010	For mobilization or demobilization, see Div. 01 54 36.50									

31 23 Excavation and Fill

31 23 16 – Excavation

31 23 16.16 Structural Excavation for Minor Structures

	31 23 16.16 Structural Excavation for Minor Structures	Crew	Daily Output	Labor-Hours	Unit	Material	2010 Bare Costs Labor	Equipment	Total	Total Incl O&P
9020	For dewatering, see Div. 31 23 19.20									
9022	For larger structures, see Bulk Excavation, Div. 31 23 16.42									
9024	For loading onto trucks, add								15%	
9026	For hauling, see Div. 31 23 23.20									
9030	For sheeting or soldier bms/lagging, see Div. 31 52 16.10									
9040	For trench excavation of strip ftgs, see Div. 31 23 16.13									

31 23 16.26 Rock Removal

		Crew	Daily Output	Labor-Hours	Unit	Material	Labor	Equipment	Total	Total Incl O&P
0010	**ROCK REMOVAL** R312316-40									
0015	Drilling only rock, 2" hole for rock bolts	B-47	316	.076	L.F.		2.77	4.69	7.46	9.40
0800	2-1/2" hole for pre-splitting		600	.040			1.46	2.47	3.93	4.94
4600	Quarry operations, 2-1/2" to 3-1/2" diameter		715	.034			1.23	2.07	3.30	4.15

31 23 16.30 Drilling and Blasting Rock

		Crew	Daily Output	Labor-Hours	Unit	Material	Labor	Equipment	Total	Total Incl O&P
0010	**DRILLING AND BLASTING ROCK**									
0020	Rock, open face, under 1500 C.Y.	B-47	225	.107	B.C.Y.	2.81	3.89	6.60	13.30	16.30
0100	Over 1500 C.Y.		300	.080		2.81	2.92	4.94	10.67	13
0200	Areas where blasting mats are required, under 1500 C.Y.		175	.137		2.81	5	8.45	16.26	20
0250	Over 1500 C.Y.		250	.096		2.81	3.50	5.90	12.21	14.95
0300	Bulk drilling and blasting, can vary greatly, average								10	11
0500	Pits, average								40	44
1300	Deep hole method, up to 1500 C.Y.	B-47	50	.480		2.81	17.50	29.50	49.81	62
1400	Over 1500 C.Y.		66	.364		2.81	13.25	22.50	38.56	47.50
1900	Restricted areas, up to 1500 C.Y.		13	1.846		2.81	67.50	114	184.31	231
2000	Over 1500 C.Y.		20	1.200		2.81	44	74	120.81	152
2200	Trenches, up to 1500 C.Y.		22	1.091		8.15	40	67.50	115.65	143
2300	Over 1500 C.Y.		26	.923		8.15	33.50	57	98.65	123
2500	Pier holes, up to 1500 C.Y.		22	1.091		2.81	40	67.50	110.31	138
2600	Over 1500 C.Y.		31	.774		2.81	28.50	48	79.31	98.50
2800	Boulders under 1/2 C.Y., loaded on truck, no hauling	B-100	80	.150			5.95	10.50	16.45	20.50
2900	Boulders, drilled, blasted	B-47	100	.240		2.81	8.75	14.80	26.36	32.50
3100	Jackhammer operators with foreman compressor, air tools	B-9	1	40	Day		1,350	225	1,575	2,325
3300	Track drill, compressor, operator and foreman	B-47	1	24	"		875	1,475	2,350	2,950
3500	Blasting caps				Ea.	5.20			5.20	5.75
3700	Explosives					.39			.39	.43
3900	Blasting mats, rent, for first day					122			122	134
4000	Per added day					40.50			40.50	45
4200	Preblast survey for 6 room house, individual lot, minimum	A-6	2.40	6.667			276	29	305	450
4300	Maximum	"	1.35	11.852			490	52	542	800
4500	City block within zone of influence, minimum	A-8	25200	.001	S.F.		.06		.06	.08
4600	Maximum	"	15100	.002	"		.09		.09	.15

31 23 16.42 Excavating, Bulk Bank Measure

		Crew	Daily Output	Labor-Hours	Unit	Material	Labor	Equipment	Total	Total Incl O&P
0010	**EXCAVATING, BULK BANK MEASURE** R312316-40									
0011	Common earth piled									
0020	For loading onto trucks, add								15%	15%
0050	For mobilization and demobilization, see Div. 01 54 36.50 R312316-45									
0100	For hauling, see Div. 31 23 23.20									
0200	Excavator, hydraulic, crawler mtd., 1 C.Y. cap. = 100 C.Y./hr.	B-12A	800	.020	B.C.Y.		.78	1	1.78	2.28
0250	1-1/2 C.Y. cap. = 125 C.Y./hr.	B-12B	1000	.016			.62	1.02	1.64	2.06
0260	2 C.Y. cap. = 165 C.Y./hr.	B-12C	1320	.012			.47	1	1.47	1.81
0300	3 C.Y. cap. = 260 C.Y./hr.	B-12D	2080	.008			.30	1.13	1.43	1.70
0305	3.5 C.Y. cap. = 300 C.Y./hr.	"	2400	.007			.26	.98	1.24	1.47
0310	Wheel mounted, 1/2 C.Y. cap. = 40 C.Y./hr.	B-12E	320	.050			1.94	1.21	3.15	4.27
0360	3/4 C.Y. cap. = 60 C.Y./hr.	B-12F	480	.033			1.29	1.41	2.70	3.51

31 23 16.42 Excavating, Bulk Bank Measure		Crew	Daily Output	Labor-Hours	Unit	Material	2010 Bare Costs Labor	Equipment	Total	Total Incl O&P
0500	Clamshell, 1/2 C.Y. cap. = 20 C.Y./hr.	B-12G	160	.100	B.C.Y.		3.88	4.57	8.45	10.90
0550	1 C.Y. cap. = 35 C.Y./hr.	B-12H	280	.057			2.21	4.41	6.62	8.20
0950	Dragline, 1/2 C.Y. cap. = 30 C.Y./hr.	B-12I	240	.067			2.58	3.80	6.38	8.10
1000	3/4 C.Y. cap. = 35 C.Y./hr.	"	280	.057			2.21	3.26	5.47	6.95
1050	1-1/2 C.Y. cap. = 65 C.Y./hr.	B-12P	520	.031			1.19	2.36	3.55	4.40
1200	Front end loader, track mtd., 1-1/2 C.Y. cap. = 70 C.Y./hr.	B-10N	560	.021			.85	.77	1.62	2.13
1250	2-1/2 C.Y. cap. = 95 C.Y./hr.	B-10O	760	.016			.63	1.10	1.73	2.17
1300	3 C.Y. cap. = 130 C.Y./hr.	B-10P	1040	.012			.46	1.04	1.50	1.83
1350	5 C.Y. cap. = 160 C.Y./hr.	B-10Q	1280	.009			.37	1.21	1.58	1.89
1500	Wheel mounted, 3/4 C.Y. cap. = 45 C.Y./hr.	B-10R	360	.033			1.32	.75	2.07	2.82
1550	1-1/2 C.Y. cap. = 80 C.Y./hr.	B-10S	640	.019			.74	.57	1.31	1.75
1600	2-1/4 C.Y. cap. = 100 C.Y./hr.	B-10T	800	.015			.60	.54	1.14	1.50
1650	5 C.Y. cap. = 185 C.Y./hr.	B-10U	1480	.008			.32	.67	.99	1.23
1800	Hydraulic excavator, truck mtd. 1/2 C.Y. = 30 C.Y./hr.	B-12J	240	.067			2.58	3.39	5.97	7.65
1850	48 inch bucket, 1 C.Y. = 45 C.Y./hr.	B-12K	360	.044			1.72	2.66	4.38	5.55
3700	Shovel, 1/2 C.Y. capacity = 55 C.Y./hr.	B-12L	440	.036			1.41	1.70	3.11	4.01
3750	3/4 C.Y. capacity = 85 C.Y./hr.	B-12M	680	.024			.91	1.40	2.31	2.92
3800	1 C.Y. capacity = 120 C.Y./hr.	B-12N	960	.017			.65	1.31	1.96	2.42
3850	1-1/2 C.Y. capacity = 160 C.Y./hr.	B-12O	1280	.013			.48	.99	1.47	1.82
3900	3 C.Y. cap. = 250 C.Y./hr.	B-12T	2000	.008			.31	.81	1.12	1.36
4000	For soft soil or sand, deduct								15%	15%
4100	For heavy soil or stiff clay, add								60%	60%
4200	For wet excavation with clamshell or dragline, add								100%	100%
4250	All other equipment, add								50%	50%
4400	Clamshell in sheeting or cofferdam, minimum	B-12H	160	.100			3.88	7.70	11.58	14.40
4450	Maximum	"	60	.267			10.35	20.50	30.85	38
5000	Excavating, bulk bank measure, sandy clay & loam piled									
5020	For loading onto trucks, add								15%	15%
5100	Excavator, hydraulic, crawler mtd., 1 C.Y. cap. = 120 C.Y./hr.	B-12A	960	.017	B.C.Y.		.65	.84	1.49	1.90
5150	1-1/2 C.Y. cap. = 150 C.Y./hr.	B-12B	1200	.013			.52	.85	1.37	1.71
5300	2 C.Y. cap. = 195 C.Y./hr.	B-12C	1560	.010			.40	.85	1.25	1.53
5400	3 C.Y. cap. = 300 C.Y./hr.	B-12D	2400	.007			.26	.98	1.24	1.47
5500	3.5 C.Y. cap. = 350 C.Y./hr.	"	2800	.006			.22	.84	1.06	1.27
5610	Wheel mounted, 1/2 C.Y. cap. = 44 C.Y./hr.	B-12E	352	.045			1.76	1.10	2.86	3.88
5660	3/4 C.Y. cap. = 66 C.Y./hr.	B-12F	528	.030			1.17	1.28	2.45	3.19
8000	For hauling excavated material, see Div. 31 23 23.20									

31 23 16.46 Excavating, Bulk, Dozer		Crew	Daily Output	Labor-Hours	Unit	Material	Labor	Equipment	Total	Total Incl O&P
0010	**EXCAVATING, BULK, DOZER**									
0011	Open site									
2000	80 H.P., 50' haul, sand & gravel	B-10L	460	.026	B.C.Y.		1.04	.95	1.99	2.60
2010	Sandy clay & loam		440	.027			1.08	.99	2.07	2.72
2020	Common earth		400	.030			1.19	1.09	2.28	3
2040	Clay		250	.048			1.90	1.75	3.65	4.79
2200	150' haul, sand & gravel		230	.052			2.07	1.90	3.97	5.20
2210	Sandy clay & loam		220	.055			2.16	1.98	4.14	5.45
2220	Common earth		200	.060			2.38	2.18	4.56	6
2240	Clay		125	.096			3.81	3.49	7.30	9.60
2400	300' haul, sand & gravel		120	.100			3.97	3.64	7.61	10
2410	Sandy clay & loam		115	.104			4.14	3.80	7.94	10.45
2420	Common earth		100	.120			4.76	4.37	9.13	12
2440	Clay		65	.185			7.30	6.70	14	18.45
3000	105 H.P., 50' haul, sand & gravel	B-10W	700	.017			.68	.94	1.62	2.07

31 23 16.46 Excavating, Bulk, Dozer		Crew	Daily Output	Labor-Hours	Unit	Material	2010 Bare Costs Labor	Equipment	Total	Total Incl O&P
3010	Sandy clay & loam	B-10W	680	.018	B.C.Y.		.70	.97	1.67	2.13
3020	Common earth		610	.020			.78	1.08	1.86	2.37
3040	Clay		385	.031			1.24	1.72	2.96	3.76
3200	150' haul, sand & gravel		310	.039			1.54	2.13	3.67	4.66
3210	Sandy clay & loam		300	.040			1.59	2.20	3.79	4.82
3220	Common earth		270	.044			1.76	2.45	4.21	5.35
3240	Clay		170	.071			2.80	3.88	6.68	8.50
3300	300' haul, sand & gravel		140	.086			3.40	4.72	8.12	10.35
3310	Sandy clay & loam		135	.089			3.53	4.89	8.42	10.70
3320	Common earth		120	.100			3.97	5.50	9.47	12.05
3340	Clay		100	.120			4.76	6.60	11.36	14.45
4000	200 H.P., 50' haul, sand & gravel	B-10B	1400	.009			.34	.85	1.19	1.45
4010	Sandy clay & loam		1360	.009			.35	.88	1.23	1.49
4020	Common earth		1230	.010			.39	.97	1.36	1.65
4040	Clay		770	.016			.62	1.55	2.17	2.63
4200	150' haul, sand & gravel		595	.020			.80	2	2.80	3.41
4210	Sandy clay & loam		580	.021			.82	2.06	2.88	3.50
4220	Common earth		516	.023			.92	2.31	3.23	3.93
4240	Clay		325	.037			1.46	3.67	5.13	6.25
4400	300' haul, sand & gravel		310	.039			1.54	3.85	5.39	6.55
4410	Sandy clay & loam		300	.040			1.59	3.97	5.56	6.75
4420	Common earth		270	.044			1.76	4.41	6.17	7.50
4440	Clay		170	.071			2.80	7	9.80	11.95
5000	300 H.P., 50' haul, sand & gravel	B-10M	1900	.006			.25	.84	1.09	1.30
5010	Sandy clay & loam		1850	.006			.26	.86	1.12	1.34
5020	Common earth		1650	.007			.29	.96	1.25	1.50
5040	Clay		1025	.012			.46	1.55	2.01	2.41
5200	150' haul, sand & gravel		920	.013			.52	1.73	2.25	2.68
5210	Sandy clay & loam		895	.013			.53	1.78	2.31	2.76
5220	Common earth		800	.015			.60	1.99	2.59	3.09
5240	Clay		500	.024			.95	3.18	4.13	4.94
5400	300' haul, sand & gravel		470	.026			1.01	3.39	4.40	5.25
5410	Sandy clay & loam		455	.026			1.05	3.50	4.55	5.45
5420	Common earth		410	.029			1.16	3.88	5.04	6
5440	Clay		250	.048			1.90	6.35	8.25	9.85
5500	460 H.P., 50' haul, sand & gravel	B-10X	1930	.006			.25	1.08	1.33	1.56
5506	Sandy clay & loam		1880	.006			.25	1.11	1.36	1.60
5510	Common earth		1680	.007			.28	1.24	1.52	1.79
5520	Clay		1050	.011			.45	1.98	2.43	2.86
5530	150' haul, sand & gravel		1290	.009			.37	1.61	1.98	2.33
5535	Sandy clay & loam		1250	.010			.38	1.66	2.04	2.40
5540	Common earth		1120	.011			.42	1.86	2.28	2.68
5550	Clay		700	.017			.68	2.97	3.65	4.30
5560	300' haul, sand & gravel		660	.018			.72	3.15	3.87	4.55
5565	Sandy clay & loam		640	.019			.74	3.25	3.99	4.69
5570	Common earth		575	.021			.83	3.62	4.45	5.25
5580	Clay		350	.034			1.36	5.95	7.31	8.60
6000	700 H.P., 50' haul, sand & gravel	B-10V	3500	.003			.14	1.31	1.45	1.66
6006	Sandy clay & loam		3400	.004			.14	1.35	1.49	1.70
6010	Common earth		3035	.004			.16	1.51	1.67	1.90
6020	Clay		1925	.006			.25	2.39	2.64	3
6030	150' haul, sand & gravel		2025	.006			.24	2.27	2.51	2.86
6035	Sandy clay & loam		1960	.006			.24	2.35	2.59	2.95

31 23 16 – Excavation

31 23 16.46 Excavating, Bulk, Dozer		Crew	Daily Output	Labor-Hours	Unit	Material	2010 Bare Costs Labor	Equipment	Total	Total Incl O&P
6040	Common earth	B-10V	1750	.007	B.C.Y.		.27	2.63	2.90	3.30
6050	Clay		1100	.011			.43	4.18	4.61	5.25
6060	300' haul, sand & gravel		1030	.012			.46	4.46	4.92	5.60
6065	Sandy clay & loam		1005	.012			.47	4.58	5.05	5.75
6070	Common earth		900	.013			.53	5.10	5.63	6.40
6080	Clay	↓	550	.022	↓		.87	8.35	9.22	10.50

31 23 16.50 Excavation, Bulk, Scrapers		Crew	Daily Output	Labor-Hours	Unit	Material	Labor	Equipment	Total	Total Incl O&P
0010	**EXCAVATION, BULK, SCRAPERS** R312316-40									
0100	Elev. scraper 11 C.Y., sand & gravel 1500' haul, 1/4 dozer	B-33F	690	.020	B.C.Y.		.81	2.29	3.10	3.75
0150	3000' haul		610	.023			.92	2.60	3.52	4.24
0200	5000' haul		505	.028			1.11	3.13	4.24	5.15
0300	Common earth, 1500' haul		600	.023			.94	2.64	3.58	4.31
0350	3000' haul		530	.026			1.06	2.99	4.05	4.89
0400	5000' haul		440	.032			1.28	3.60	4.88	5.90
0410	Sandy clay & loam, 1500' haul		648	.022			.87	2.44	3.31	4
0420	3000' haul		572	.024			.98	2.77	3.75	4.52
0430	5000' haul		475	.029			1.18	3.33	4.51	5.45
0500	Clay, 1500' haul		375	.037			1.50	4.22	5.72	6.90
0550	3000' haul		330	.042			1.70	4.80	6.50	7.85
0600	5000' haul	↓	275	.051	↓		2.04	5.75	7.79	9.45
1000	Self propelled scraper, 14 C.Y. 1/4 push dozer, sand									
1050	Sand and gravel, 1500' haul	B-33D	920	.015	B.C.Y.		.61	2.30	2.91	3.45
1100	3000' haul		805	.017			.70	2.63	3.33	3.94
1200	5000' haul		645	.022			.87	3.28	4.15	4.92
1300	Common earth, 1500' haul		800	.018			.70	2.64	3.34	3.97
1350	3000' haul		700	.020			.80	3.02	3.82	4.53
1400	5000' haul		560	.025			1	3.78	4.78	5.65
1420	Sandy clay & loam, 1500' haul		864	.016			.65	2.45	3.10	3.67
1430	3000' haul		786	.018			.71	2.69	3.40	4.04
1440	5000' haul		605	.023			.93	3.50	4.43	5.25
1500	Clay, 1500' haul		500	.028			1.12	4.23	5.35	6.35
1550	3000' haul		440	.032			1.28	4.81	6.09	7.25
1600	5000' haul	↓	350	.040			1.61	6.05	7.66	9.05
2000	21 C.Y., 1/4 push dozer, sand & gravel, 1500' haul	B-33E	1180	.012			.48	2.24	2.72	3.18
2100	3000' haul		910	.015			.62	2.90	3.52	4.12
2200	5000' haul		750	.019			.75	3.52	4.27	5
2300	Common earth, 1500' haul		1030	.014			.55	2.56	3.11	3.64
2350	3000' haul		790	.018			.71	3.34	4.05	4.75
2400	5000' haul		650	.022			.86	4.06	4.92	5.75
2420	Sandy clay & loam, 1500' haul		1112	.013			.51	2.37	2.88	3.37
2430	3000' haul		854	.016			.66	3.09	3.75	4.39
2440	5000' haul		702	.020			.80	3.76	4.56	5.35
2500	Clay, 1500' haul		645	.022			.87	4.09	4.96	5.80
2550	3000' haul		495	.028			1.14	5.35	6.49	7.55
2600	5000' haul	↓	405	.035			1.39	6.50	7.89	9.25
2700	Towed, 10 C.Y., 1/4 push dozer, sand & gravel, 1500' haul	B-33B	560	.025			1	3.83	4.83	5.75
2720	3000' haul		450	.031			1.25	4.77	6.02	7.15
2730	5000' haul		365	.038			1.54	5.90	7.44	8.75
2750	Common earth, 1500' haul		420	.033			1.34	5.10	6.44	7.60
2770	3000' haul		400	.035			1.40	5.35	6.75	8
2780	5000' haul		310	.045			1.81	6.90	8.71	10.35
2785	Sandy clay & Loam, 1500' haul	↓	454	.031	↓		1.24	4.73	5.97	7.05

31 23 Excavation and Fill

31 23 16 – Excavation

31 23 16.50 Excavation, Bulk, Scrapers		Crew	Daily Output	Labor-Hours	Unit	Material	2010 Bare Costs Labor	Equipment	Total	Total Incl O&P
2790	3000′ haul	B-33B	432	.032	B.C.Y.		1.30	4.97	6.27	7.40
2795	5000′ haul		340	.041			1.65	6.30	7.95	9.45
2800	Clay, 1500′ haul		315	.044			1.78	6.80	8.58	10.20
2820	3000′ haul		300	.047			1.87	7.15	9.02	10.65
2840	5000′ haul		225	.062			2.50	9.55	12.05	14.25
2900	15 C.Y., 1/4 push dozer, sand & gravel, 1500′ haul	B-33C	800	.018			.70	2.70	3.40	4.03
2920	3000′ haul		640	.022			.88	3.38	4.26	5.05
2940	5000′ haul		520	.027			1.08	4.16	5.24	6.20
2960	Common earth, 1500′ haul		600	.023			.94	3.60	4.54	5.35
2980	3000′ haul		560	.025			1	3.86	4.86	5.75
3000	5000′ haul		440	.032			1.28	4.91	6.19	7.35
3005	Sandy clay & Loam, 1500′ haul		648	.022			.87	3.34	4.21	4.98
3010	3000′ haul		605	.023			.93	3.57	4.50	5.35
3015	5000′ haul		475	.029			1.18	4.55	5.73	6.80
3020	Clay, 1500′ haul		450	.031			1.25	4.80	6.05	7.20
3040	3000′ haul		420	.033			1.34	5.15	6.49	7.65
3060	5000′ haul		320	.044			1.76	6.75	8.51	10.10

31 23 19 – Dewatering

31 23 19.20 Dewatering Systems		Crew	Daily Output	Labor-Hours	Unit	Material	Labor	Equipment	Total	Total Incl O&P
0010	**DEWATERING SYSTEMS**									
0020	Excavate drainage trench, 2′ wide, 2′ deep	B-11C	90	.178	C.Y.		6.75	3.75	10.50	14.40
0100	2′ wide, 3′ deep, with backhoe loader	"	135	.119			4.51	2.50	7.01	9.60
0200	Excavate sump pits by hand, light soil	1 Clab	7.10	1.127			37.50		37.50	57.50
0300	Heavy soil	"	3.50	2.286			75.50		75.50	117
0500	Pumping 8 hr., attended 2 hrs. per day, including 20 L.F.									
0550	of suction hose & 100 L.F. discharge hose									
0600	2″ diaphragm pump used for 8 hours	B-10H	4	3	Day		119	17.95	136.95	200
0650	4″ diaphragm pump used for 8 hours	B-10I	4	3			119	25	144	207
0800	8 hrs. attended, 2″ diaphragm pump	B-10H	1	12			475	72	547	800
0900	3″ centrifugal pump	B-10J	1	12			475	81	556	810
1000	4″ diaphragm pump	B-10I	1	12			475	99	574	830
1100	6″ centrifugal pump	B-10K	1	12			475	365	840	1,125
1300	CMP, incl. excavation 3′ deep, 12″ diameter	B-6	115	.209	L.F.	13.25	7.50	2.94	23.69	29
1400	18″ diameter		100	.240	"	16.50	8.60	3.38	28.48	35
1600	Sump hole construction, incl. excavation and gravel, pit		1250	.019	C.F.	.98	.69	.27	1.94	2.42
1700	With 12″ gravel collar, 12″ pipe, corrugated, 16 ga.		70	.343	L.F.	23	12.30	4.83	40.13	49.50
1800	15″ pipe, corrugated, 16 ga.		55	.436		29.50	15.65	6.15	51.30	63.50
1900	18″ pipe, corrugated, 16 ga.		50	.480		34.50	17.20	6.75	58.45	71.50
2000	24″ pipe, corrugated, 14 ga.		40	.600		41.50	21.50	8.45	71.45	88
2200	Wood lining, up to 4′ x 4′, add		300	.080	SFCA	17	2.87	1.13	21	24.50
9950	See Div. 31 23 19.40 for wellpoints									
9960	See Div. 31 23 19.30 for deep well systems									

31 23 19.30 Wells		Crew	Daily Output	Labor-Hours	Unit	Material	Labor	Equipment	Total	Total Incl O&P
0010	**WELLS**									
0011	For dewatering 10′ to 20′ deep, 2′ diameter									
0020	with steel casing, minimum	B-6	165	.145	V.L.F.	40	5.20	2.05	47.25	54
0050	Average		98	.245		40	8.75	3.45	52.20	61
0100	Maximum		49	.490		40	17.55	6.90	64.45	78.50
0300	For dewatering pumps see 01 54 33 in Reference Section									
0500	For domestic water wells, see Div. 33 21 13.10									

31 23 19 – Dewatering

31 23 19.40 Wellpoints		Crew	Daily Output	Labor-Hours	Unit	Material	2010 Bare Costs Labor	Equipment	Total	Total Incl O&P
0010	**WELLPOINTS** R312319-90									
0011	For equipment rental, see 01 54 33 in Reference Section									
0100	Installation and removal of single stage system									
0110	Labor only, .75 labor-hours per L.F., minimum	1 Clab	10.70	.748	LF Hdr		25		25	38
0200	2.0 labor-hours per L.F., maximum	"	4	2	"		66		66	102
0400	Pump operation, 4 @ 6 hr. shifts									
0410	Per 24 hour day	4 Eqlt	1.27	25.197	Day		1,050		1,050	1,550
0500	Per 168 hour week, 160 hr. straight, 8 hr. double time		.18	177	Week		7,350		7,350	11,000
0550	Per 4.3 week month		.04	800	Month		33,000		33,000	49,500
0600	Complete installation, operation, equipment rental, fuel &									
0610	removal of system with 2" wellpoints 5' O.C.									
0700	100' long header, 6" diameter, first month	4 Eqlt	3.23	9.907	LF Hdr	157	410		567	785
0800	Thereafter, per month		4.13	7.748		125	320		445	620
1000	200' long header, 8" diameter, first month		6	5.333		142	220		362	485
1100	Thereafter, per month		8.39	3.814		70.50	158		228.50	315
1300	500' long header, 8" diameter, first month		10.63	3.010		55	124		179	247
1400	Thereafter, per month		20.91	1.530		39	63		102	138
1600	1,000' long header, 10" diameter, first month		11.62	2.754		47	114		161	222
1700	Thereafter, per month		41.81	.765		23.50	31.50		55	73.50
1900	Note: above figures include pumping 168 hrs. per week									
1910	and include the pump operator and one stand-by pump.									

31 23 23 – Fill

31 23 23.13 Backfill

		Crew	Daily Output	Labor-Hours	Unit	Material	Labor	Equipment	Total	Total Incl O&P
0010	**BACKFILL** R312323-30									
0015	By hand, no compaction, light soil	1 Clab	14	.571	L.C.Y.		18.90		18.90	29
0100	Heavy soil		11	.727	"		24		24	37
0300	Compaction in 6" layers, hand tamp, add to above		20.60	.388	E.C.Y.		12.85		12.85	19.85
0400	Roller compaction operator walking, add	B-10A	100	.120			4.76	1.50	6.26	8.85
0500	Air tamp, add	B-9D	190	.211			7.05	1.35	8.40	12.40
0600	Vibrating plate, add	A-1D	60	.133			4.41	.57	4.98	7.45
0800	Compaction in 12" layers, hand tamp, add to above	1 Clab	34	.235			7.80		7.80	12
0900	Roller compaction operator walking, add	B-10A	150	.080			3.17	1	4.17	5.90
1000	Air tamp, add	B-9	285	.140			4.70	.79	5.49	8.10
1100	Vibrating plate, add	A-1E	90	.089			2.94	.48	3.42	5.05
1300	Dozer backfilling, bulk, up to 300' haul, no compaction	B-10B	1200	.010	L.C.Y.		.40	.99	1.39	1.69
1400	Air tamped, add	B-11B	80	.200	E.C.Y.		7.45	3.67	11.12	15.35
1600	Compacting backfill, 6" to 12" lifts, vibrating roller	B-10C	800	.015			.60	2.03	2.63	3.14
1700	Sheepsfoot roller	B-10D	750	.016			.63	2.24	2.87	3.43
1900	Dozer backfilling, trench, up to 300' haul, no compaction	B-10B	900	.013	L.C.Y.		.53	1.32	1.85	2.26
2000	Air tamped, add	B-11B	80	.200	E.C.Y.		7.45	3.67	11.12	15.35
2200	Compacting backfill, 6" to 12" lifts, vibrating roller	B-10C	700	.017			.68	2.32	3	3.59
2300	Sheepsfoot roller	B-10D	650	.018			.73	2.59	3.32	3.95
2350	Spreading in 8" layers, small dozer	B-10B	1060	.011	L.C.Y.		.45	1.12	1.57	1.92

31 23 23.14 Backfill, Structural

		Crew	Daily Output	Labor-Hours	Unit	Material	Labor	Equipment	Total	Total Incl O&P
0010	**BACKFILL, STRUCTURAL**									
0011	Dozer or F.E. loader									
0020	From existing stockpile, no compaction									
2000	80 H.P., 50' haul, sand & gravel	B-10L	1100	.011	L.C.Y.		.43	.40	.83	1.09
2010	Sandy clay & loam		1070	.011			.44	.41	.85	1.12
2020	Common earth		975	.012			.49	.45	.94	1.23
2040	Clay		850	.014			.56	.51	1.07	1.42

31 23 23.14 Backfill, Structural		Crew	Daily Output	Labor-Hours	Unit	Material	2010 Bare Costs Labor	Equipment	Total	Total Incl O&P
2400	300' haul, sand & gravel	B-10L	370	.032	L.C.Y.		1.29	1.18	2.47	3.24
2410	Sandy clay & loam		360	.033			1.32	1.21	2.53	3.33
2420	Common earth		330	.036			1.44	1.32	2.76	3.64
2440	Clay		290	.041			1.64	1.51	3.15	4.14
3000	105 H.P., 50' haul, sand & gravel	B-10W	1350	.009			.35	.49	.84	1.07
3010	Sandy clay & loam		1325	.009			.36	.50	.86	1.09
3020	Common earth		1225	.010			.39	.54	.93	1.18
3040	Clay		1100	.011			.43	.60	1.03	1.31
3300	300' haul, sand & gravel		465	.026			1.02	1.42	2.44	3.11
3310	Sandy clay & loam		455	.026			1.05	1.45	2.50	3.18
3320	Common earth		415	.029			1.15	1.59	2.74	3.48
3340	Clay		370	.032			1.29	1.78	3.07	3.90
4000	200 H.P., 50' haul, sand & gravel	B-10B	2500	.005			.19	.48	.67	.81
4010	Sandy clay & loam		2435	.005			.20	.49	.69	.84
4020	Common earth		2200	.005			.22	.54	.76	.93
4040	Clay		1950	.006			.24	.61	.85	1.04
4400	300' haul, sand & gravel		805	.015			.59	1.48	2.07	2.52
4410	Sandy clay & loam		790	.015			.60	1.51	2.11	2.57
4420	Common earth		735	.016			.65	1.62	2.27	2.76
4440	Clay		660	.018			.72	1.81	2.53	3.08
5000	300 H.P., 50' haul, sand & gravel	B-10M	3170	.004			.15	.50	.65	.78
5010	Sandy clay & loam		3110	.004			.15	.51	.66	.79
5020	Common earth		2900	.004			.16	.55	.71	.85
5040	Clay		2700	.004			.18	.59	.77	.92
5400	300' haul, sand & gravel		1500	.008			.32	1.06	1.38	1.65
5410	Sandy clay & loam		1470	.008			.32	1.08	1.40	1.68
5420	Common earth		1350	.009			.35	1.18	1.53	1.83
5440	Clay		1225	.010			.39	1.30	1.69	2.02
6010	For trench backfill, see Div. 31 23 16.13 and 31 23 16.14									
6100	For compaction, see Div. 31 23 23.24									

31 23 23.16 Fill By Borrow and Utility Bedding		Crew	Daily Output	Labor-Hours	Unit	Material	Labor	Equipment	Total	Total Incl O&P
0010	**FILL BY BORROW AND UTILITY BEDDING**									
0015	Fill by borrow, load, 1 mile haul, spread with dozer									
0020	for embankments	B-15	1200	.023	L.C.Y.	7.40	.84	2.11	10.35	11.75
0035	Select fill for shoulders & embankments	"	1200	.023	"	12.65	.84	2.11	15.60	17.50
0040	Fill,for hauling over 1 mile,add to above per C.Y., see Div. 31 23 23.20				Mile				1.29	1.55
0049	Utility bedding, for pipe & conduit, not incl. compaction									
0050	Crushed or screened bank run gravel	B-6	150	.160	L.C.Y.	25	5.75	2.25	33	38.50
0100	Crushed stone 3/4" to 1/2"		150	.160		35	5.75	2.25	43	49.50
0200	Sand, dead or bank		150	.160		10.55	5.75	2.25	18.55	23
0500	Compacting bedding in trench	A-1D	90	.089	E.C.Y.		2.94	.38	3.32	4.96
0600	If material source exceeds 2 miles, add for extra mileage.									
0610	See Div. 31 23 23 .20 for hauling mileage add.									

31 23 23.17 General Fill		Crew	Daily Output	Labor-Hours	Unit	Material	Labor	Equipment	Total	Total Incl O&P
0010	**GENERAL FILL**									
0011	Spread dumped material, no compaction									
0020	By dozer, no compaction	B-10B	1000	.012	L.C.Y.		.48	1.19	1.67	2.03
0100	By hand	1 Clab	12	.667	"		22		22	34
0500	Gravel fill, compacted, under floor slabs, 4" deep	B-37	10000	.005	S.F.	.25	.17	.02	.44	.56
0600	6" deep		8600	.006		.38	.19	.02	.59	.74
0700	9" deep		7200	.007		.63	.23	.02	.88	1.08
0800	12" deep		6000	.008		.89	.28	.03	1.20	1.43

31 23 23.17 General Fill		Crew	Daily Output	Labor-Hours	Unit	Material	2010 Bare Costs Labor	Equipment	Total	Total Incl O&P
1000	Alternate pricing method, 4″ deep	B-37	120	.400	E.C.Y.	19	13.90	1.27	34.17	44
1100	6″ deep		160	.300		19	10.45	.95	30.40	38
1200	9″ deep		200	.240		19	8.35	.76	28.11	34.50
1300	12″ deep		220	.218		19	7.60	.69	27.29	33.50
1500	For fill under exterior paving, see Div. 32 11 23.23									
1600	For flowable fill, see Div. 03 31 05.35									
31 23 23.20 Hauling										
0010	**HAULING**									
0011	Excavated or borrow, loose cubic yards									
0012	no loading equipment, including hauling,waiting, loading/dumping									
0013	time per cycle (wait, load, travel, unload or dump & return)									
0014	8 CY truck, 15 MPH ave, cycle 0.5 miles, 10 min. wait/Ld./Uld.	B-34A	320	.025	L.C.Y.		.83	1.26	2.09	2.64
0016	cycle 1 mile		272	.029			.97	1.48	2.45	3.11
0018	cycle 2 miles		208	.038			1.27	1.93	3.20	4.06
0020	cycle 4 miles		144	.056			1.84	2.79	4.63	5.90
0022	cycle 6 miles		112	.071			2.37	3.59	5.96	7.55
0024	cycle 8 miles		88	.091			3.01	4.57	7.58	9.60
0026	20 MPH ave,cycle 0.5 mile		336	.024			.79	1.20	1.99	2.52
0028	cycle 1 mile		296	.027			.90	1.36	2.26	2.86
0030	cycle 2 miles		240	.033			1.10	1.67	2.77	3.52
0032	cycle 4 miles		176	.045			1.51	2.28	3.79	4.81
0034	cycle 6 miles		136	.059			1.95	2.95	4.90	6.20
0036	cycle 8 miles		112	.071			2.37	3.59	5.96	7.55
0044	25 MPH ave, cycle 4 miles		192	.042			1.38	2.09	3.47	4.41
0046	cycle 6 miles		160	.050			1.66	2.51	4.17	5.30
0048	cycle 8 miles		128	.063			2.07	3.14	5.21	6.60
0050	30 MPH ave, cycle 4 miles		216	.037			1.23	1.86	3.09	3.92
0052	cycle 6 miles		176	.045			1.51	2.28	3.79	4.81
0054	cycle 8 miles		144	.056			1.84	2.79	4.63	5.90
0114	15 MPH ave, cycle 0.5 mile, 15 min. wait/Ld./Uld.		224	.036			1.18	1.79	2.97	3.78
0116	cycle 1 mile		200	.040			1.33	2.01	3.34	4.23
0118	cycle 2 miles		168	.048			1.58	2.39	3.97	5.05
0120	cycle 4 miles		120	.067			2.21	3.35	5.56	7.05
0122	cycle 6 miles		96	.083			2.76	4.19	6.95	8.80
0124	cycle 8 miles		80	.100			3.32	5	8.32	10.60
0126	20 MPH ave, cycle 0.5 mile		232	.034			1.14	1.73	2.87	3.65
0128	cycle 1 mile		208	.038			1.27	1.93	3.20	4.06
0130	cycle 2 miles		184	.043			1.44	2.18	3.62	4.60
0132	cycle 4 miles		144	.056			1.84	2.79	4.63	5.90
0134	cycle 6 miles		112	.071			2.37	3.59	5.96	7.55
0136	cycle 8 miles		96	.083			2.76	4.19	6.95	8.80
0144	25 MPH ave, cycle 4 miles		152	.053			1.74	2.64	4.38	5.55
0146	cycle 6 miles		128	.063			2.07	3.14	5.21	6.60
0148	cycle 8 miles		112	.071			2.37	3.59	5.96	7.55
0150	30 MPH ave, cycle 4 miles		168	.048			1.58	2.39	3.97	5.05
0152	cycle 6 miles		144	.056			1.84	2.79	4.63	5.90
0154	cycle 8 miles		120	.067			2.21	3.35	5.56	7.05
0214	15 MPH ave, cycle 0.5 mile, 20 min wait/Ld./Uld.		176	.045			1.51	2.28	3.79	4.81
0216	cycle 1 mile		160	.050			1.66	2.51	4.17	5.30
0218	cycle 2 miles		136	.059			1.95	2.95	4.90	6.20
0220	cycle 4 miles		104	.077			2.55	3.86	6.41	8.15
0222	cycle 6 miles		88	.091			3.01	4.57	7.58	9.60

31 23 23.20 Hauling		Crew	Daily Output	Labor-Hours	Unit	Material	2010 Bare Costs Labor	Equipment	Total	Total Incl O&P
0224	cycle 8 miles	B-34A	72	.111	L.C.Y.		3.68	5.60	9.28	11.75
0226	20 MPH ave, cycle 0.5 mile		176	.045			1.51	2.28	3.79	4.81
0228	cycle 1 mile		168	.048			1.58	2.39	3.97	5.05
0230	cycle 2 miles		144	.056			1.84	2.79	4.63	5.90
0232	cycle 4 miles		120	.067			2.21	3.35	5.56	7.05
0234	cycle 6 miles		96	.083			2.76	4.19	6.95	8.80
0236	cycle 8 miles		88	.091			3.01	4.57	7.58	9.60
0244	25 MPH ave, cycle 4 miles		128	.063			2.07	3.14	5.21	6.60
0246	cycle 6 miles		112	.071			2.37	3.59	5.96	7.55
0248	cycle 8 miles		96	.083			2.76	4.19	6.95	8.80
0250	30 MPH ave, cycle 4 miles		136	.059			1.95	2.95	4.90	6.20
0252	cycle 6 miles		120	.067			2.21	3.35	5.56	7.05
0254	cycle 8 miles		104	.077			2.55	3.86	6.41	8.15
0314	15 MPH ave, cycle 0.5 mile, 25 min wait/Ld./Uld.		144	.056			1.84	2.79	4.63	5.90
0316	cycle 1 mile		128	.063			2.07	3.14	5.21	6.60
0318	cycle 2 miles		112	.071			2.37	3.59	5.96	7.55
0320	cycle 4 miles		96	.083			2.76	4.19	6.95	8.80
0322	cycle 6 miles		80	.100			3.32	5	8.32	10.60
0324	cycle 8 miles		64	.125			4.14	6.30	10.44	13.20
0326	20 MPH ave, cycle 0.5 mile		144	.056			1.84	2.79	4.63	5.90
0328	cycle 1 mile		136	.059			1.95	2.95	4.90	6.20
0330	cycle 2 miles		120	.067			2.21	3.35	5.56	7.05
0332	cycle 4 miles		104	.077			2.55	3.86	6.41	8.15
0334	cycle 6 miles		88	.091			3.01	4.57	7.58	9.60
0336	cycle 8 miles		80	.100			3.32	5	8.32	10.60
0344	25 MPH ave, cycle 4 miles		112	.071			2.37	3.59	5.96	7.55
0346	cycle 6 miles		96	.083			2.76	4.19	6.95	8.80
0348	cycle 8 miles		88	.091			3.01	4.57	7.58	9.60
0350	30 MPH ave, cycle 4 miles		112	.071			2.37	3.59	5.96	7.55
0352	cycle 6 miles		104	.077			2.55	3.86	6.41	8.15
0354	cycle 8 miles		96	.083			2.76	4.19	6.95	8.80
0414	15 MPH ave, cycle 0.5 mile, 30 min wait/Ld./Uld.		120	.067			2.21	3.35	5.56	7.05
0416	cycle 1 mile		112	.071			2.37	3.59	5.96	7.55
0418	cycle 2 miles		96	.083			2.76	4.19	6.95	8.80
0420	cycle 4 miles		80	.100			3.32	5	8.32	10.60
0422	cycle 6 miles		72	.111			3.68	5.60	9.28	11.75
0424	cycle 8 miles		64	.125			4.14	6.30	10.44	13.20
0426	20 MPH ave, cycle 0.5 mile		120	.067			2.21	3.35	5.56	7.05
0428	cycle 1 mile		112	.071			2.37	3.59	5.96	7.55
0430	cycle 2 miles		104	.077			2.55	3.86	6.41	8.15
0432	cycle 4 miles		88	.091			3.01	4.57	7.58	9.60
0434	cycle 6 miles		80	.100			3.32	5	8.32	10.60
0436	cycle 8 miles		72	.111			3.68	5.60	9.28	11.75
0444	25 MPH ave, cycle 4 miles		96	.083			2.76	4.19	6.95	8.80
0446	cycle 6 miles		88	.091			3.01	4.57	7.58	9.60
0448	cycle 8 miles		80	.100			3.32	5	8.32	10.60
0450	30 MPH ave, cycle 4 miles		96	.083			2.76	4.19	6.95	8.80
0452	cycle 6 miles		88	.091			3.01	4.57	7.58	9.60
0454	cycle 8 miles		80	.100			3.32	5	8.32	10.60
0514	15 MPH ave, cycle 0.5 mile, 35 min wait/Ld./Uld.		104	.077			2.55	3.86	6.41	8.15
0516	cycle 1 mile		96	.083			2.76	4.19	6.95	8.80
0518	cycle 2 miles		88	.091			3.01	4.57	7.58	9.60
0520	cycle 4 miles		72	.111			3.68	5.60	9.28	11.75

31 23 23.20 Hauling		Crew	Daily Output	Labor-Hours	Unit	Material	2010 Bare Costs Labor	Equipment	Total	Total Incl O&P
0522	cycle 6 miles	B-34A	64	.125	L.C.Y.		4.14	6.30	10.44	13.20
0524	cycle 8 miles		56	.143			4.74	7.20	11.94	15.10
0526	20 MPH ave, cycle 0.5 mile		104	.077			2.55	3.86	6.41	8.15
0528	cycle 1 mile		96	.083			2.76	4.19	6.95	8.80
0530	cycle 2 miles		96	.083			2.76	4.19	6.95	8.80
0532	cycle 4 miles		80	.100			3.32	5	8.32	10.60
0534	cycle 6 miles		72	.111			3.68	5.60	9.28	11.75
0536	cycle 8 miles		64	.125			4.14	6.30	10.44	13.20
0544	25 MPH ave, cycle 4 miles		88	.091			3.01	4.57	7.58	9.60
0546	cycle 6 miles		80	.100			3.32	5	8.32	10.60
0548	cycle 8 miles		72	.111			3.68	5.60	9.28	11.75
0550	30 MPH ave, cycle 4 miles		88	.091			3.01	4.57	7.58	9.60
0552	cycle 6 miles		80	.100			3.32	5	8.32	10.60
0554	cycle 8 miles		72	.111			3.68	5.60	9.28	11.75
1014	12 CY truck, cycle 0.5 mile, 15 MPH ave, 15 min. wait/Ld./Uld.	B-34B	336	.024			.79	1.99	2.78	3.39
1016	cycle 1 mile		300	.027			.88	2.23	3.11	3.80
1018	cycle 2 miles		252	.032			1.05	2.65	3.70	4.53
1020	cycle 4 miles		180	.044			1.47	3.71	5.18	6.35
1022	cycle 6 miles		144	.056			1.84	4.64	6.48	7.90
1024	cycle 8 miles		120	.067			2.21	5.55	7.76	9.45
1025	cycle 10 miles		96	.083			2.76	6.95	9.71	11.85
1026	20 MPH ave, cycle 0.5 mile		348	.023			.76	1.92	2.68	3.27
1028	cycle 1 mile		312	.026			.85	2.14	2.99	3.65
1030	cycle 2 miles		276	.029			.96	2.42	3.38	4.13
1032	cycle 4 miles		216	.037			1.23	3.09	4.32	5.25
1034	cycle 6 miles		168	.048			1.58	3.98	5.56	6.80
1036	cycle 8 miles		144	.056			1.84	4.64	6.48	7.90
1038	cycle 10 miles		120	.067			2.21	5.55	7.76	9.45
1040	25 MPH ave, cycle 4 miles		228	.035			1.16	2.93	4.09	4.99
1042	cycle 6 miles		192	.042			1.38	3.48	4.86	5.95
1044	cycle 8 miles		168	.048			1.58	3.98	5.56	6.80
1046	cycle 10 miles		144	.056			1.84	4.64	6.48	7.90
1050	30 MPH ave, cycle 4 miles		252	.032			1.05	2.65	3.70	4.53
1052	cycle 6 miles		216	.037			1.23	3.09	4.32	5.25
1054	cycle 8 miles		180	.044			1.47	3.71	5.18	6.35
1056	cycle 10 miles		156	.051			1.70	4.28	5.98	7.30
1060	35 MPH ave, cycle 4 miles		264	.030			1	2.53	3.53	4.31
1062	cycle 6 miles		228	.035			1.16	2.93	4.09	4.99
1064	cycle 8 miles		204	.039			1.30	3.27	4.57	5.60
1066	cycle 10 miles		180	.044			1.47	3.71	5.18	6.35
1068	cycle 20 miles		120	.067			2.21	5.55	7.76	9.45
1069	cycle 30 miles		84	.095			3.16	7.95	11.11	13.55
1070	cycle 40 miles		72	.111			3.68	9.30	12.98	15.80
1072	40 MPH ave, cycle 6 miles		240	.033			1.10	2.78	3.88	4.74
1074	cycle 8 miles		216	.037			1.23	3.09	4.32	5.25
1076	cycle 10 miles		192	.042			1.38	3.48	4.86	5.95
1078	cycle 20 miles		120	.067			2.21	5.55	7.76	9.45
1080	cycle 30 miles		96	.083			2.76	6.95	9.71	11.85
1082	cycle 40 miles		72	.111			3.68	9.30	12.98	15.80
1084	cycle 50 miles		60	.133			4.42	11.15	15.57	19
1094	45 MPH ave, cycle 8 miles		216	.037			1.23	3.09	4.32	5.25
1096	cycle 10 miles		204	.039			1.30	3.27	4.57	5.60
1098	cycle 20 miles		132	.061			2.01	5.05	7.06	8.60

31 23 23.20 Hauling		Crew	Daily Output	Labor-Hours	Unit	Material	2010 Bare Costs Labor	Equipment	Total	Total Incl O&P
1100	cycle 30 miles	B-34B	108	.074	L.C.Y.		2.46	6.20	8.66	10.55
1102	cycle 40 miles		84	.095			3.16	7.95	11.11	13.55
1104	cycle 50 miles		72	.111			3.68	9.30	12.98	15.80
1106	50 MPH ave, cycle 10 miles		216	.037			1.23	3.09	4.32	5.25
1108	cycle 20 miles		144	.056			1.84	4.64	6.48	7.90
1110	cycle 30 miles		108	.074			2.46	6.20	8.66	10.55
1112	cycle 40 miles		84	.095			3.16	7.95	11.11	13.55
1114	cycle 50 miles		72	.111			3.68	9.30	12.98	15.80
1214	15 MPH ave, cycle 0.5 mile, 20 min. wait/Ld./Uld.		264	.030			1	2.53	3.53	4.31
1216	cycle 1 mile		240	.033			1.10	2.78	3.88	4.74
1218	cycle 2 miles		204	.039			1.30	3.27	4.57	5.60
1220	cycle 4 miles		156	.051			1.70	4.28	5.98	7.30
1222	cycle 6 miles		132	.061			2.01	5.05	7.06	8.60
1224	cycle 8 miles		108	.074			2.46	6.20	8.66	10.55
1225	cycle 10 miles		96	.083			2.76	6.95	9.71	11.85
1226	20 MPH ave, cycle 0.5 mile		264	.030			1	2.53	3.53	4.31
1228	cycle 1 mile		252	.032			1.05	2.65	3.70	4.53
1230	cycle 2 miles		216	.037			1.23	3.09	4.32	5.25
1232	cycle 4 miles		180	.044			1.47	3.71	5.18	6.35
1234	cycle 6 miles		144	.056			1.84	4.64	6.48	7.90
1236	cycle 8 miles		132	.061			2.01	5.05	7.06	8.60
1238	cycle 10 miles		108	.074			2.46	6.20	8.66	10.55
1240	25 MPH ave, cycle 4 miles		192	.042			1.38	3.48	4.86	5.95
1242	cycle 6 miles		168	.048			1.58	3.98	5.56	6.80
1244	cycle 8 miles		144	.056			1.84	4.64	6.48	7.90
1246	cycle 10 miles		132	.061			2.01	5.05	7.06	8.60
1250	30 MPH ave, cycle 4 miles		204	.039			1.30	3.27	4.57	5.60
1252	cycle 6 miles		180	.044			1.47	3.71	5.18	6.35
1254	cycle 8 miles		156	.051			1.70	4.28	5.98	7.30
1256	cycle 10 miles		144	.056			1.84	4.64	6.48	7.90
1260	35 MPH ave, cycle 4 miles		216	.037			1.23	3.09	4.32	5.25
1262	cycle 6 miles		192	.042			1.38	3.48	4.86	5.95
1264	cycle 8 miles		168	.048			1.58	3.98	5.56	6.80
1266	cycle 10 miles		156	.051			1.70	4.28	5.98	7.30
1268	cycle 20 miles		108	.074			2.46	6.20	8.66	10.55
1269	cycle 30 miles		72	.111			3.68	9.30	12.98	15.80
1270	cycle 40 miles		60	.133			4.42	11.15	15.57	19
1272	40 MPH ave, cycle 6 miles		192	.042			1.38	3.48	4.86	5.95
1274	cycle 8 miles		180	.044			1.47	3.71	5.18	6.35
1276	cycle 10 miles		156	.051			1.70	4.28	5.98	7.30
1278	cycle 20 miles		108	.074			2.46	6.20	8.66	10.55
1280	cycle 30 miles		84	.095			3.16	7.95	11.11	13.55
1282	cycle 40 miles		72	.111			3.68	9.30	12.98	15.80
1284	cycle 50 miles		60	.133			4.42	11.15	15.57	19
1294	45 MPH ave, cycle 8 miles		180	.044			1.47	3.71	5.18	6.35
1296	cycle 10 miles		168	.048			1.58	3.98	5.56	6.80
1298	cycle 20 miles		120	.067			2.21	5.55	7.76	9.45
1300	cycle 30 miles		96	.083			2.76	6.95	9.71	11.85
1302	cycle 40 miles		72	.111			3.68	9.30	12.98	15.80
1304	cycle 50 miles		60	.133			4.42	11.15	15.57	19
1306	50 MPH ave, cycle 10 miles		180	.044			1.47	3.71	5.18	6.35
1308	cycle 20 miles		132	.061			2.01	5.05	7.06	8.60
1310	cycle 30 miles		96	.083			2.76	6.95	9.71	11.85

31 23 23.20 Hauling		Crew	Daily Output	Labor-Hours	Unit	Material	2010 Bare Costs Labor	Equipment	Total	Total Incl O&P
1312	cycle 40 miles	B-34B	84	.095	L.C.Y.		3.16	7.95	11.11	13.55
1314	cycle 50 miles		72	.111			3.68	9.30	12.98	15.80
1414	15 MPH ave, cycle 0.5 mile, 25 min. wait/Ld./Uld.		204	.039			1.30	3.27	4.57	5.60
1416	cycle 1 mile		192	.042			1.38	3.48	4.86	5.95
1418	cycle 2 miles		168	.048			1.58	3.98	5.56	6.80
1420	cycle 4 miles		132	.061			2.01	5.05	7.06	8.60
1422	cycle 6 miles		120	.067			2.21	5.55	7.76	9.45
1424	cycle 8 miles		96	.083			2.76	6.95	9.71	11.85
1425	cycle 10 miles		84	.095			3.16	7.95	11.11	13.55
1426	20 MPH ave, cycle 0.5 mile		216	.037			1.23	3.09	4.32	5.25
1428	cycle 1 mile		204	.039			1.30	3.27	4.57	5.60
1430	cycle 2 miles		180	.044			1.47	3.71	5.18	6.35
1432	cycle 4 miles		156	.051			1.70	4.28	5.98	7.30
1434	cycle 6 miles		132	.061			2.01	5.05	7.06	8.60
1436	cycle 8 miles		120	.067			2.21	5.55	7.76	9.45
1438	cycle 10 miles		96	.083			2.76	6.95	9.71	11.85
1440	25 MPH ave, cycle 4 miles		168	.048			1.58	3.98	5.56	6.80
1442	cycle 6 miles		144	.056			1.84	4.64	6.48	7.90
1444	cycle 8 miles		132	.061			2.01	5.05	7.06	8.60
1446	cycle 10 miles		108	.074			2.46	6.20	8.66	10.55
1450	30 MPH ave, cycle 4 miles		168	.048			1.58	3.98	5.56	6.80
1452	cycle 6 miles		156	.051			1.70	4.28	5.98	7.30
1454	cycle 8 miles		132	.061			2.01	5.05	7.06	8.60
1456	cycle 10 miles		120	.067			2.21	5.55	7.76	9.45
1460	35 MPH ave, cycle 4 miles		180	.044			1.47	3.71	5.18	6.35
1462	cycle 6 miles		156	.051			1.70	4.28	5.98	7.30
1464	cycle 8 miles		144	.056			1.84	4.64	6.48	7.90
1466	cycle 10 miles		132	.061			2.01	5.05	7.06	8.60
1468	cycle 20 miles		96	.083			2.76	6.95	9.71	11.85
1469	cycle 30 miles		72	.111			3.68	9.30	12.98	15.80
1470	cycle 40 miles		60	.133			4.42	11.15	15.57	19
1472	40 MPH ave, cycle 6 miles		168	.048			1.58	3.98	5.56	6.80
1474	cycle 8 miles		156	.051			1.70	4.28	5.98	7.30
1476	cycle 10 miles		144	.056			1.84	4.64	6.48	7.90
1478	cycle 20 miles		96	.083			2.76	6.95	9.71	11.85
1480	cycle 30 miles		84	.095			3.16	7.95	11.11	13.55
1482	cycle 40 miles		60	.133			4.42	11.15	15.57	19
1484	cycle 50 miles		60	.133			4.42	11.15	15.57	19
1494	45 MPH ave, cycle 8 miles		156	.051			1.70	4.28	5.98	7.30
1496	cycle 10 miles		144	.056			1.84	4.64	6.48	7.90
1498	cycle 20 miles		108	.074			2.46	6.20	8.66	10.55
1500	cycle 30 miles		84	.095			3.16	7.95	11.11	13.55
1502	cycle 40 miles		72	.111			3.68	9.30	12.98	15.80
1504	cycle 50 miles		60	.133			4.42	11.15	15.57	19
1506	50 MPH ave, cycle 10 miles		156	.051			1.70	4.28	5.98	7.30
1508	cycle 20 miles		120	.067			2.21	5.55	7.76	9.45
1510	cycle 30 miles		96	.083			2.76	6.95	9.71	11.85
1512	cycle 40 miles		72	.111			3.68	9.30	12.98	15.80
1514	cycle 50 miles		60	.133			4.42	11.15	15.57	19
1614	15 MPH, cycle 0.5 mile, 30 min. wait/Ld./Uld.		180	.044			1.47	3.71	5.18	6.35
1616	cycle 1 mile		168	.048			1.58	3.98	5.56	6.80
1618	cycle 2 miles		144	.056			1.84	4.64	6.48	7.90
1620	cycle 4 miles		120	.067			2.21	5.55	7.76	9.45

31 23 23.20 Hauling		Crew	Daily Output	Labor-Hours	Unit	Material	2010 Bare Costs Labor	Equipment	Total	Total Incl O&P
1622	cycle 6 miles	B-34B	108	.074	L.C.Y.		2.46	6.20	8.66	10.55
1624	cycle 8 miles		84	.095			3.16	7.95	11.11	13.55
1625	cycle 10 miles		84	.095			3.16	7.95	11.11	13.55
1626	20 MPH ave, cycle 0.5 mile		180	.044			1.47	3.71	5.18	6.35
1628	cycle 1 mile		168	.048			1.58	3.98	5.56	6.80
1630	cycle 2 miles		156	.051			1.70	4.28	5.98	7.30
1632	cycle 4 miles		132	.061			2.01	5.05	7.06	8.60
1634	cycle 6 miles		120	.067			2.21	5.55	7.76	9.45
1636	cycle 8 miles		108	.074			2.46	6.20	8.66	10.55
1638	cycle 10 miles		96	.083			2.76	6.95	9.71	11.85
1640	25 MPH ave, cycle 4 miles		144	.056			1.84	4.64	6.48	7.90
1642	cycle 6 miles		132	.061			2.01	5.05	7.06	8.60
1644	cycle 8 miles		108	.074			2.46	6.20	8.66	10.55
1646	cycle 10 miles		108	.074			2.46	6.20	8.66	10.55
1650	30 MPH ave, cycle 4 miles		144	.056			1.84	4.64	6.48	7.90
1652	cycle 6 miles		132	.061			2.01	5.05	7.06	8.60
1654	cycle 8 miles		120	.067			2.21	5.55	7.76	9.45
1656	cycle 10 miles		108	.074			2.46	6.20	8.66	10.55
1660	35 MPH ave, cycle 4 miles		156	.051			1.70	4.28	5.98	7.30
1662	cycle 6 miles		144	.056			1.84	4.64	6.48	7.90
1664	cycle 8 miles		132	.061			2.01	5.05	7.06	8.60
1666	cycle 10 miles		120	.067			2.21	5.55	7.76	9.45
1668	cycle 20 miles		84	.095			3.16	7.95	11.11	13.55
1669	cycle 30 miles		72	.111			3.68	9.30	12.98	15.80
1670	cycle 40 miles		60	.133			4.42	11.15	15.57	19
1672	40 MPH, cycle 6 miles		144	.056			1.84	4.64	6.48	7.90
1674	cycle 8 miles		132	.061			2.01	5.05	7.06	8.60
1676	cycle 10 miles		120	.067			2.21	5.55	7.76	9.45
1678	cycle 20 miles		96	.083			2.76	6.95	9.71	11.85
1680	cycle 30 miles		72	.111			3.68	9.30	12.98	15.80
1682	cycle 40 miles		60	.133			4.42	11.15	15.57	19
1684	cycle 50 miles		48	.167			5.55	13.90	19.45	24
1694	45 MPH ave, cycle 8 miles		144	.056			1.84	4.64	6.48	7.90
1696	cycle 10 miles		132	.061			2.01	5.05	7.06	8.60
1698	cycle 20 miles		96	.083			2.76	6.95	9.71	11.85
1700	cycle 30 miles		84	.095			3.16	7.95	11.11	13.55
1702	cycle 40 miles		60	.133			4.42	11.15	15.57	19
1704	cycle 50 miles		60	.133			4.42	11.15	15.57	19
1706	50 MPH ave, cycle 10 miles		132	.061			2.01	5.05	7.06	8.60
1708	cycle 20 miles		108	.074			2.46	6.20	8.66	10.55
1710	cycle 30 miles		84	.095			3.16	7.95	11.11	13.55
1712	cycle 40 miles		72	.111			3.68	9.30	12.98	15.80
1714	cycle 50 miles		60	.133			4.42	11.15	15.57	19
2000	Hauling, 8 CY truck, small project cost per hour	B-34A	8	1	Hr.		33	50	83	106
2100	12 CY Truck	B-34B	8	1			33	83.50	116.50	143
2150	16.5 CY Truck	B-34C	8	1			33	89	122	149
2175	18 CY 8 wheel Truck	B-34I	8	1			33	105	138	166
2200	20 CY Truck	B-34D	8	1			33	91	124	151
2300	Grading at dump, or embankment if required, by dozer	B-10B	1000	.012	L.C.Y.		.48	1.19	1.67	2.03
2310	Spotter at fill or cut, if required	1 Clab	8	1	Hr.		33		33	51
9014	18 CY truck, 8 wheels,15 min. wait/Ld./Uld.,15 MPH, cycle 0.5 mi.	B-34I	504	.016	L.C.Y.		.53	1.66	2.19	2.63
9016	cycle 1 mile		450	.018			.59	1.86	2.45	2.95
9018	cycle 2 miles		378	.021			.70	2.22	2.92	3.51

31 23 23.20 Hauling		Crew	Daily Output	Labor-Hours	Unit	Material	2010 Bare Costs Labor	Equipment	Total	Total Incl O&P
9020	cycle 4 miles	B-34I	270	.030	L.C.Y.		.98	3.11	4.09	4.92
9022	cycle 6 miles		216	.037			1.23	3.88	5.11	6.15
9024	cycle 8 miles		180	.044			1.47	4.66	6.13	7.35
9025	cycle 10 miles		144	.056			1.84	5.80	7.64	9.20
9026	20 MPH ave, cycle 0.5 mile		522	.015			.51	1.61	2.12	2.54
9028	cycle 1 mile		468	.017			.57	1.79	2.36	2.83
9030	cycle 2 miles		414	.019			.64	2.02	2.66	3.21
9032	cycle 4 miles		324	.025			.82	2.59	3.41	4.10
9034	cycle 6 miles		252	.032			1.05	3.33	4.38	5.25
9036	cycle 8 miles		216	.037			1.23	3.88	5.11	6.15
9038	cycle 10 miles		180	.044			1.47	4.66	6.13	7.35
9040	25 MPH ave, cycle 4 miles		342	.023			.78	2.45	3.23	3.88
9042	cycle 6 miles		288	.028			.92	2.91	3.83	4.60
9044	cycle 8 miles		252	.032			1.05	3.33	4.38	5.25
9046	cycle 10 miles		216	.037			1.23	3.88	5.11	6.15
9050	30 MPH ave, cycle 4 miles		378	.021			.70	2.22	2.92	3.51
9052	cycle 6 miles		324	.025			.82	2.59	3.41	4.10
9054	cycle 8 miles		270	.030			.98	3.11	4.09	4.92
9056	cycle 10 miles		234	.034			1.13	3.58	4.71	5.65
9060	35 MPH ave, cycle 4 miles		396	.020			.67	2.12	2.79	3.35
9062	cycle 6 miles		342	.023			.78	2.45	3.23	3.88
9064	cycle 8 miles		288	.028			.92	2.91	3.83	4.60
9066	cycle 10 miles		270	.030			.98	3.11	4.09	4.92
9068	cycle 20 miles		162	.049			1.64	5.20	6.84	8.20
9070	cycle 30 miles		126	.063			2.10	6.65	8.75	10.50
9072	cycle 40 miles		90	.089			2.95	9.30	12.25	14.75
9074	40 MPH ave, cycle 6 miles		360	.022			.74	2.33	3.07	3.68
9076	cycle 8 miles		324	.025			.82	2.59	3.41	4.10
9078	cycle 10 miles		288	.028			.92	2.91	3.83	4.60
9080	cycle 20 miles		180	.044			1.47	4.66	6.13	7.35
9082	cycle 30 miles		144	.056			1.84	5.80	7.64	9.20
9084	cycle 40 miles		108	.074			2.46	7.75	10.21	12.30
9086	cycle 50 miles		90	.089			2.95	9.30	12.25	14.75
9094	45 MPH ave, cycle 8 miles		324	.025			.82	2.59	3.41	4.10
9096	cycle 10 miles		306	.026			.87	2.74	3.61	4.33
9098	cycle 20 miles		198	.040			1.34	4.23	5.57	6.70
9100	cycle 30 miles		144	.056			1.84	5.80	7.64	9.20
9102	cycle 40 miles		126	.063			2.10	6.65	8.75	10.50
9104	cycle 50 miles		108	.074			2.46	7.75	10.21	12.30
9106	50 MPH ave, cycle 10 miles		324	.025			.82	2.59	3.41	4.10
9108	cycle 20 miles		216	.037			1.23	3.88	5.11	6.15
9110	cycle 30 miles		162	.049			1.64	5.20	6.84	8.20
9112	cycle 40 miles		126	.063			2.10	6.65	8.75	10.50
9114	cycle 50 miles		108	.074			2.46	7.75	10.21	12.30
9214	20 min. wait/Ld./Uld.,15 MPH, cycle 0.5 mi.		396	.020			.67	2.12	2.79	3.35
9216	cycle 1 mile		360	.022			.74	2.33	3.07	3.68
9218	cycle 2 miles		306	.026			.87	2.74	3.61	4.33
9220	cycle 4 miles		234	.034			1.13	3.58	4.71	5.65
9222	cycle 6 miles		198	.040			1.34	4.23	5.57	6.70
9224	cycle 8 miles		162	.049			1.64	5.20	6.84	8.20
9225	cycle 10 miles		144	.056			1.84	5.80	7.64	9.20
9226	20 MPH ave, cycle 0.5 mile		396	.020			.67	2.12	2.79	3.35
9228	cycle 1 mile		378	.021			.70	2.22	2.92	3.51

31 23 23.20 Hauling		Crew	Daily Output	Labor-Hours	Unit	Material	2010 Bare Costs Labor	Equipment	Total	Total Incl O&P
9230	cycle 2 miles	B-34I	324	.025	L.C.Y.		.82	2.59	3.41	4.10
9232	cycle 4 miles		270	.030			.98	3.11	4.09	4.92
9234	cycle 6 miles		216	.037			1.23	3.88	5.11	6.15
9236	cycle 8 miles		198	.040			1.34	4.23	5.57	6.70
9238	cycle 10 miles		162	.049			1.64	5.20	6.84	8.20
9240	25 MPH ave, cycle 4 miles		288	.028			.92	2.91	3.83	4.60
9242	cycle 6 miles		252	.032			1.05	3.33	4.38	5.25
9244	cycle 8 miles		216	.037			1.23	3.88	5.11	6.15
9246	cycle 10 miles		198	.040			1.34	4.23	5.57	6.70
9250	30 MPH ave, cycle 4 miles		306	.026			.87	2.74	3.61	4.33
9252	cycle 6 miles		270	.030			.98	3.11	4.09	4.92
9254	cycle 8 miles		234	.034			1.13	3.58	4.71	5.65
9256	cycle 10 miles		216	.037			1.23	3.88	5.11	6.15
9260	35 MPH ave, cycle 4 miles		324	.025			.82	2.59	3.41	4.10
9262	cycle 6 miles		288	.028			.92	2.91	3.83	4.60
9264	cycle 8 miles		252	.032			1.05	3.33	4.38	5.25
9266	cycle 10 miles		234	.034			1.13	3.58	4.71	5.65
9268	cycle 20 miles		162	.049			1.64	5.20	6.84	8.20
9270	cycle 30 miles		108	.074			2.46	7.75	10.21	12.30
9272	cycle 40 miles		90	.089			2.95	9.30	12.25	14.75
9274	40 MPH ave, cycle 6 miles		288	.028			.92	2.91	3.83	4.60
9276	cycle 8 miles		270	.030			.98	3.11	4.09	4.92
9278	cycle 10 miles		234	.034			1.13	3.58	4.71	5.65
9280	cycle 20 miles		162	.049			1.64	5.20	6.84	8.20
9282	cycle 30 miles		126	.063			2.10	6.65	8.75	10.50
9284	cycle 40 miles		108	.074			2.46	7.75	10.21	12.30
9286	cycle 50 miles		90	.089			2.95	9.30	12.25	14.75
9294	45 MPH ave, cycle 8 miles		270	.030			.98	3.11	4.09	4.92
9296	cycle 10 miles		252	.032			1.05	3.33	4.38	5.25
9298	cycle 20 miles		180	.044			1.47	4.66	6.13	7.35
9300	cycle 30 miles		144	.056			1.84	5.80	7.64	9.20
9302	cycle 40 miles		108	.074			2.46	7.75	10.21	12.30
9304	cycle 50 miles		90	.089			2.95	9.30	12.25	14.75
9306	50 MPH ave, cycle 10 miles		270	.030			.98	3.11	4.09	4.92
9308	cycle 20 miles		198	.040			1.34	4.23	5.57	6.70
9310	cycle 30 miles		144	.056			1.84	5.80	7.64	9.20
9312	cycle 40 miles		126	.063			2.10	6.65	8.75	10.50
9314	cycle 50 miles		108	.074			2.46	7.75	10.21	12.30
9414	25 min. wait/Ld./Uld.,15 MPH, cycle 0.5 mi.		306	.026			.87	2.74	3.61	4.33
9416	cycle 1 mile		288	.028			.92	2.91	3.83	4.60
9418	cycle 2 miles		252	.032			1.05	3.33	4.38	5.25
9420	cycle 4 miles		198	.040			1.34	4.23	5.57	6.70
9422	cycle 6 miles		180	.044			1.47	4.66	6.13	7.35
9424	cycle 8 miles		144	.056			1.84	5.80	7.64	9.20
9425	cycle 10 miles		126	.063			2.10	6.65	8.75	10.50
9426	20 MPH ave, cycle 0.5 mile		324	.025			.82	2.59	3.41	4.10
9428	cycle 1 mile		306	.026			.87	2.74	3.61	4.33
9430	cycle 2 miles		270	.030			.98	3.11	4.09	4.92
9432	cycle 4 miles		234	.034			1.13	3.58	4.71	5.65
9434	cycle 6 miles		198	.040			1.34	4.23	5.57	6.70
9436	cycle 8 miles		180	.044			1.47	4.66	6.13	7.35
9438	cycle 10 miles		144	.056			1.84	5.80	7.64	9.20
9440	25 MPH ave, cycle 4 miles	↓	252	.032	↓		1.05	3.33	4.38	5.25

31 23 23.20 Hauling		Crew	Daily Output	Labor-Hours	Unit	Material	2010 Bare Costs Labor	Equipment	Total	Total Incl O&P
9442	cycle 6 miles	B-34I	216	.037	L.C.Y.		1.23	3.88	5.11	6.15
9444	cycle 8 miles		198	.040			1.34	4.23	5.57	6.70
9446	cycle 10 miles		180	.044			1.47	4.66	6.13	7.35
9450	30 MPH ave, cycle 4 miles		252	.032			1.05	3.33	4.38	5.25
9452	cycle 6 miles		234	.034			1.13	3.58	4.71	5.65
9454	cycle 8 miles		198	.040			1.34	4.23	5.57	6.70
9456	cycle 10 miles		180	.044			1.47	4.66	6.13	7.35
9460	35 MPH ave, cycle 4 miles		270	.030			.98	3.11	4.09	4.92
9462	cycle 6 miles		234	.034			1.13	3.58	4.71	5.65
9464	cycle 8 miles		216	.037			1.23	3.88	5.11	6.15
9466	cycle 10 miles		198	.040			1.34	4.23	5.57	6.70
9468	cycle 20 miles		144	.056			1.84	5.80	7.64	9.20
9470	cycle 30 miles		108	.074			2.46	7.75	10.21	12.30
9472	cycle 40 miles		90	.089			2.95	9.30	12.25	14.75
9474	40 MPH ave, cycle 6 miles		252	.032			1.05	3.33	4.38	5.25
9476	cycle 8 miles		234	.034			1.13	3.58	4.71	5.65
9478	cycle 10 miles		216	.037			1.23	3.88	5.11	6.15
9480	cycle 20 miles		144	.056			1.84	5.80	7.64	9.20
9482	cycle 30 miles		126	.063			2.10	6.65	8.75	10.50
9484	cycle 40 miles		90	.089			2.95	9.30	12.25	14.75
9486	cycle 50 miles		90	.089			2.95	9.30	12.25	14.75
9494	45 MPH ave, cycle 8 miles		234	.034			1.13	3.58	4.71	5.65
9496	cycle 10 miles		216	.037			1.23	3.88	5.11	6.15
9498	cycle 20 miles		162	.049			1.64	5.20	6.84	8.20
9500	cycle 30 miles		126	.063			2.10	6.65	8.75	10.50
9502	cycle 40 miles		108	.074			2.46	7.75	10.21	12.30
9504	cycle 50 miles		90	.089			2.95	9.30	12.25	14.75
9506	50 MPH ave, cycle 10 miles		234	.034			1.13	3.58	4.71	5.65
9508	cycle 20 miles		180	.044			1.47	4.66	6.13	7.35
9510	cycle 30 miles		144	.056			1.84	5.80	7.64	9.20
9512	cycle 40 miles		108	.074			2.46	7.75	10.21	12.30
9514	cycle 50 miles		90	.089			2.95	9.30	12.25	14.75
9614	25 min. wait/Ld./Uld.,15 MPH, cycle 0.5 mi.		270	.030			.98	3.11	4.09	4.92
9616	cycle 1 mile		252	.032			1.05	3.33	4.38	5.25
9618	cycle 2 miles		216	.037			1.23	3.88	5.11	6.15
9620	cycle 4 miles		180	.044			1.47	4.66	6.13	7.35
9622	cycle 6 miles		162	.049			1.64	5.20	6.84	8.20
9624	cycle 8 miles		126	.063			2.10	6.65	8.75	10.50
9625	cycle 10 miles		126	.063			2.10	6.65	8.75	10.50
9626	20 MPH ave, cycle 0.5 mile		270	.030			.98	3.11	4.09	4.92
9628	cycle 1 mile		252	.032			1.05	3.33	4.38	5.25
9630	cycle 2 miles		234	.034			1.13	3.58	4.71	5.65
9632	cycle 4 miles		198	.040			1.34	4.23	5.57	6.70
9634	cycle 6 miles		180	.044			1.47	4.66	6.13	7.35
9636	cycle 8 miles		162	.049			1.64	5.20	6.84	8.20
9638	cycle 10 miles		144	.056			1.84	5.80	7.64	9.20
9640	25 MPH ave, cycle 4 miles		216	.037			1.23	3.88	5.11	6.15
9642	cycle 6 miles		198	.040			1.34	4.23	5.57	6.70
9644	cycle 8 miles		180	.044			1.47	4.66	6.13	7.35
9646	cycle 10 miles		162	.049			1.64	5.20	6.84	8.20
9650	30 MPH ave, cycle 4 miles		216	.037			1.23	3.88	5.11	6.15
9652	cycle 6 miles		198	.040			1.34	4.23	5.57	6.70
9654	cycle 8 miles		180	.044			1.47	4.66	6.13	7.35

31 23 23 – Fill

31 23 23.20 Hauling

	31 23 23.20 Hauling	Crew	Daily Output	Labor-Hours	Unit	Material	2010 Bare Costs Labor	Equipment	Total	Total Incl O&P
9656	cycle 10 miles	B-34I	162	.049	L.C.Y.		1.64	5.20	6.84	8.20
9660	35 MPH ave, cycle 4 miles		234	.034			1.13	3.58	4.71	5.65
9662	cycle 6 miles		216	.037			1.23	3.88	5.11	6.15
9664	cycle 8 miles		198	.040			1.34	4.23	5.57	6.70
9666	cycle 10 miles		180	.044			1.47	4.66	6.13	7.35
9668	cycle 20 miles		126	.063			2.10	6.65	8.75	10.50
9670	cycle 30 miles		108	.074			2.46	7.75	10.21	12.30
9672	cycle 40 miles		90	.089			2.95	9.30	12.25	14.75
9674	40 MPH ave, cycle 6 miles		216	.037			1.23	3.88	5.11	6.15
9676	cycle 8 miles		198	.040			1.34	4.23	5.57	6.70
9678	cycle 10 miles		180	.044			1.47	4.66	6.13	7.35
9680	cycle 20 miles		144	.056			1.84	5.80	7.64	9.20
9682	cycle 30 miles		108	.074			2.46	7.75	10.21	12.30
9684	cycle 40 miles		90	.089			2.95	9.30	12.25	14.75
9686	cycle 50 miles		72	.111			3.68	11.65	15.33	18.40
9694	45 MPH ave, cycle 8 miles		216	.037			1.23	3.88	5.11	6.15
9696	cycle 10 miles		198	.040			1.34	4.23	5.57	6.70
9698	cycle 20 miles		144	.056			1.84	5.80	7.64	9.20
9700	cycle 30 miles		126	.063			2.10	6.65	8.75	10.50
9702	cycle 40 miles		108	.074			2.46	7.75	10.21	12.30
9704	cycle 50 miles		90	.089			2.95	9.30	12.25	14.75
9706	50 MPH ave, cycle 10 miles		198	.040			1.34	4.23	5.57	6.70
9708	cycle 20 miles		162	.049			1.64	5.20	6.84	8.20
9710	cycle 30 miles		126	.063			2.10	6.65	8.75	10.50
9712	cycle 40 miles		108	.074			2.46	7.75	10.21	12.30
9714	cycle 50 miles		90	.089			2.95	9.30	12.25	14.75

31 23 23.24 Compaction, Structural

	31 23 23.24 Compaction, Structural	Crew	Daily Output	Labor-Hours	Unit	Material	Labor	Equipment	Total	Total Incl O&P
0010	**COMPACTION, STRUCTURAL** R312323-30									
0020	Steel wheel tandem roller, 5 tons	B-10E	8	1.500	Hr.		59.50	19	78.50	111
0100	10 tons	B-10F	8	1.500	"		59.50	29.50	89	123
0300	Sheepsfoot or wobbly wheel roller, 8" lifts, common fill	B-10G	1300	.009	E.C.Y.		.37	.96	1.33	1.60
0400	Select fill	"	1500	.008			.32	.83	1.15	1.39
0600	Vibratory plate, 8" lifts, common fill	A-1D	200	.040			1.32	.17	1.49	2.23
0700	Select fill	"	216	.037			1.23	.16	1.39	2.06

31 25 Erosion and Sedimentation Controls

31 25 13 – Erosion Controls

31 25 13.10 Synthetic Erosion Control

	31 25 13.10 Synthetic Erosion Control		Crew	Daily Output	Labor-Hours	Unit	Material	Labor	Equipment	Total	Total Incl O&P
0010	**SYNTHETIC EROSION CONTROL**										
0020	Jute mesh, 100 SY per roll, 4' wide, stapled	G	B-80A	2400	.010	S.Y.	1.04	.33	.12	1.49	1.79
0070	Paper biodegradable mesh	G	B-1	2500	.010		.11	.32		.43	.62
0080	Paper mulch	G	B-64	20000	.001		.07	.03	.02	.12	.14
0100	Plastic netting, stapled, 2" x 1" mesh, 20 mil	G	B-1	2500	.010		.79	.32		1.11	1.37
0200	Polypropylene mesh, stapled, 6.5 oz./S.Y.	G		2500	.010		1.70	.32		2.02	2.37
0300	Tobacco netting, or jute mesh #2, stapled	G		2500	.010		.10	.32		.42	.61
1000	Silt fence, polypropylene, 3' high, ideal conditions	G	2 Clab	1600	.010	L.F.	.40	.33		.73	.95
1100	Adverse conditions	G	"	950	.017	"	.40	.56		.96	1.30
1200	Place and remove hay bales	G	A-2	3	8	Ton	164	263	80	507	675
1250	Hay bales, staked	G	"	2500	.010	L.F.	6.60	.32	.10	7.02	7.85

31 31 Soil Treatment

31 31 16 – Termite Control

31 31 16.13 Chemical Termite Control		Crew	Daily Output	Labor-Hours	Unit	Material	Labor	Equipment	Total	Total Incl O&P
0010	**CHEMICAL TERMITE CONTROL**									
0020	Slab and walls, residential	1 Skwk	1200	.007	SF Flr.	.31	.28		.59	.78
0100	Commercial, minimum		2496	.003		.33	.14		.47	.57
0200	Maximum		1645	.005		.50	.21		.71	.86
0400	Insecticides for termite control, minimum		14.20	.563	Gal.	12.75	24		36.75	51
0500	Maximum		11	.727	"	22	31		53	71.50

31 32 Soil Stabilization

31 32 13 – Soil Mixing Stabilization

31 32 13.30 Calcium Chloride		Crew	Daily Output	Labor-Hours	Unit	Material	Labor	Equipment	Total	Total Incl O&P
0010	**CALCIUM CHLORIDE**									
0020	Calcium chloride delivered, 100 LB bags, truckload lots				Ton	550			550	605
0030	Solution, 4 lb. flake per gallon, tank truck delivery				Gal.	1.18			1.18	1.30

31 33 Rock Stabilization

31 33 13 – Rock Bolting and Grouting

31 33 13.10 Rock Bolting		Crew	Daily Output	Labor-Hours	Unit	Material	Labor	Equipment	Total	Total Incl O&P
0010	**ROCK BOLTING**									
2020	Hollow core, prestressable anchor, 1" diameter, 5' long	2 Skwk	32	.500	Ea.	94	21.50		115.50	137
2025	10' long		24	.667		188	28.50		216.50	251
2060	2" diameter, 5' long		32	.500		350	21.50		371.50	420
2065	10' long		24	.667		705	28.50		733.50	820
2100	Super high-tensile, 3/4" diameter, 5' long		32	.500		21.50	21.50		43	56.50
2105	10' long		24	.667		42.50	28.50		71	90.50
2160	2" diameter, 5' long		32	.500		190	21.50		211.50	242
2165	10' long		24	.667		380	28.50		408.50	465
4400	Drill hole for rock bolt, 1-3/4" diam., 5' long (for 3/4" bolt)	B-56	17	.941			35	86	121	148
4405	10' long		9	1.778			66	163	229	279
4420	2" diameter, 5' long (for 1" bolt)		13	1.231			46	113	159	194
4425	10' long		7	2.286			85	209	294	360
4460	3-1/2" diameter, 5' long (for 2" bolt)		10	1.600			59.50	146	205.50	252
4465	10' long		5	3.200			119	293	412	500

31 36 Gabions

31 36 13 – Gabion Boxes

31 36 13.10 Gabion Box Systems		Crew	Daily Output	Labor-Hours	Unit	Material	Labor	Equipment	Total	Total Incl O&P
0010	**GABION BOX SYSTEMS**									
0400	Gabions, galvanized steel mesh mats or boxes, stone filled, 6" deep	B-13	200	.280	S.Y.	23.50	10	3.74	37.24	45.50
0500	9" deep		163	.344		36	12.30	4.59	52.89	63.50
0600	12" deep		153	.366		40	13.10	4.89	57.99	69.50
0700	18" deep		102	.549		56.50	19.60	7.35	83.45	100
0800	36" deep		60	.933		80.50	33.50	12.45	126.45	153

31 37 Riprap

31 37 13 – Machined Riprap

31 37 13.10 Riprap and Rock Lining		Crew	Daily Output	Labor-Hours	Unit	Material	2010 Bare Costs Labor	Equipment	Total	Total Incl O&P
0010	**RIPRAP AND ROCK LINING**									
0011	Random, broken stone									
0100	Machine placed for slope protection	B-12G	62	.258	L.C.Y.	30	10	11.80	51.80	61
0110	3/8 to 1/4 C.Y. pieces, grouted	B-13	80	.700	S.Y.	69	25	9.35	103.35	124
0200	18" minimum thickness, not grouted	"	53	1.057	"	18.85	38	14.10	70.95	93.50
0300	Dumped, 50 lb. average	B-11A	800	.020	Ton	27.50	.76	1.49	29.75	33
0350	100 lb. average		700	.023		39	.87	1.70	41.57	46
0370	300 lb. average	↓	600	.027	↓	45.50	1.01	1.99	48.50	53.50

31 41 Shoring

31 41 13 – Timber Shoring

31 41 13.10 Building Shoring

Code	Description	Crew	Daily Output	Labor-Hours	Unit	Material	Labor	Equipment	Total	Total Incl O&P
0010	**BUILDING SHORING**									
0020	Shoring, existing building, with timber, no salvage allowance	B-51	2.20	21.818	M.B.F.	530	725	109	1,364	1,825
1000	On cribbing with 35 ton screw jacks, per box and jack	"	3.60	13.333	Jack	65	445	66.50	576.50	830

31 41 16 – Sheet Piling

31 41 16.10 Sheet Piling Systems

Code	Description	Crew	Daily Output	Labor-Hours	Unit	Material	Labor	Equipment	Total	Total Incl O&P
0010	**SHEET PILING SYSTEMS**									
0020	Sheet piling steel, not incl. wales, 22 psf, 15' excav., left in place	B-40	10.81	5.920	Ton	1,250	245	355	1,850	2,150
0100	Drive, extract & salvage R314116-40		6	10.667		515	440	640	1,595	1,950
0300	20' deep excavation, 27 psf, left in place		12.95	4.942		1,250	204	296	1,750	2,025
0400	Drive, extract & salvage R314116-45		6.55	9.771		515	405	585	1,505	1,850
0600	25' deep excavation, 38 psf, left in place		19	3.368		1,250	139	202	1,591	1,825
0700	Drive, extract & salvage		10.50	6.095		515	252	365	1,132	1,350
0900	40' deep excavation, 38 psf, left in place ***CN***		21.20	3.019		1,250	125	181	1,556	1,775
1000	Drive, extract & salvage		12.25	5.224	↓	515	216	315	1,046	1,250
1200	15' deep excavation, 22 psf, left in place		983	.065	S.F.	14.65	2.69	3.90	21.24	24.50
1300	Drive, extract & salvage		545	.117		5.75	4.85	7.05	17.65	21.50
1500	20' deep excavation, 27 psf, left in place		960	.067		18.40	2.76	3.99	25.15	28.50
1600	Drive, extract & salvage		485	.132		7.50	5.45	7.90	20.85	25.50
1800	25' deep excavation, 38 psf, left in place		1000	.064		27	2.65	3.83	33.48	38.50
1900	Drive, extract & salvage	↓	553	.116	↓	10.25	4.78	6.95	21.98	26.50
2100	Rent steel sheet piling and wales, first month				Ton	268			268	295
2200	Per added month					27			27	29.50
2300	Rental piling left in place, add to rental					1,000			1,000	1,100
2500	Wales, connections & struts, 2/3 salvage					280			280	310
2700	High strength piling, 50,000 psi, add					68			68	74.50
2800	55,000 psi, add					72			72	79.50
3000	Tie rod, not upset, 1-1/2" to 4" diameter with turnbuckle					2,225			2,225	2,450
3100	No turnbuckle					1,725			1,725	1,875
3300	Upset, 1-3/4" to 4" diameter with turnbuckle					2,500			2,500	2,750
3400	No turnbuckle				↓	2,175			2,175	2,400
3600	Lightweight, 18" to 28" wide, 7 ga., 9.22 psf, and									
3610	9 ga., 8.6 psf, minimum				Lb.	.92			.92	1.01
3700	Average					1			1	1.10
3750	Maximum				↓	1.14			1.14	1.25
3900	Wood, solid sheeting, incl. wales, braces and spacers, R314116-40									
3910	drive, extract & salvage, 8' deep excavation	B-31	330	.121	S.F.	1.16	4.27	.64	6.07	8.60
4000	10' deep, 50 S.F./hr. in & 150 S.F./hr. out		300	.133		1.20	4.69	.71	6.60	9.35
4100	12' deep, 45 S.F./hr. in & 135 S.F./hr. out	↓	270	.148	↓	1.23	5.20	.78	7.21	10.25

31 41 Shoring

31 41 16 – Sheet Piling

31 41 16.10 Sheet Piling Systems		Crew	Daily Output	Labor-Hours	Unit	Material	2010 Bare Costs Labor	Equipment	Total	Total Incl O&P
4200	14' deep, 42 S.F./hr. in & 126 S.F./hr. out	B-31	250	.160	S.F.	1.27	5.65	.85	7.77	11
4300	16' deep, 40 S.F./hr. in & 120 S.F./hr. out		240	.167		1.31	5.85	.88	8.04	11.45
4400	18' deep, 38 S.F./hr. in & 114 S.F./hr. out		230	.174		1.35	6.10	.92	8.37	11.95
4500	20' deep, 35 S.F./hr. in & 105 S.F./hr. out		210	.190		1.39	6.70	1.01	9.10	13
4520	Left in place, 8' deep, 55 S.F./hr.		440	.091		2.09	3.20	.48	5.77	7.75
4540	10' deep, 50 S.F./hr.		400	.100		2.20	3.52	.53	6.25	8.45
4560	12' deep, 45 S.F./hr.		360	.111		2.32	3.91	.59	6.82	9.25
4565	14' deep, 42 S.F./hr.		335	.119		2.46	4.20	.63	7.29	9.90
4570	16' deep, 40 S.F./hr.		320	.125		2.61	4.40	.66	7.67	10.40
4580	18' deep, 38 S.F./hr.		305	.131		2.79	4.62	.69	8.10	10.95
4590	20' deep, 35 S.F./hr.		280	.143		2.99	5.05	.76	8.80	11.85
4700	Alternate pricing, left in place, 8' deep		1.76	22.727	M.B.F.	470	800	120	1,390	1,875
4800	Drive, extract and salvage, 8' deep		1.32	30.303	"	420	1,075	160	1,655	2,275
5000	For treated lumber add cost of treatment to lumber									

31 43 Concrete Raising

31 43 13 – Pressure Grouting

31 43 13.13 Concrete Pressure Grouting		Crew	Daily Output	Labor-Hours	Unit	Material	Labor	Equipment	Total	Total Incl O&P
0010	**CONCRETE PRESSURE GROUTING**									
0020	Grouting, pressure, cement & sand, 1:1 mix, minimum	B-61	124	.323	Bag	12.40	11.35	2.80	26.55	34
0100	Maximum		51	.784	"	12.40	27.50	6.80	46.70	63
0200	Cement and sand, 1:1 mix, minimum		250	.160	C.F.	25	5.60	1.39	31.99	37.50
0300	Maximum		100	.400		37	14.05	3.48	54.53	66.50
0400	Epoxy cement grout, minimum		137	.292		450	10.25	2.54	462.79	515
0500	Maximum		57	.702		450	24.50	6.10	480.60	540
0700	Alternate pricing method: (Add for materials)									
0710	5 person crew and equipment	B-61	1	40	Day		1,400	350	1,750	2,525

31 45 Vibroflotation and Densification

31 45 13 – Vibroflotation

31 45 13.10 Vibroflotation Densification		Crew	Daily Output	Labor-Hours	Unit	Material	Labor	Equipment	Total	Total Incl O&P
0010	**VIBROFLOTATION DENSIFICATION** R314513-90									
0900	Vibroflotation compacted sand cylinder, minimum	B-60	750	.075	V.L.F.		2.84	2.91	5.75	7.50
0950	Maximum		325	.172			6.55	6.70	13.25	17.35
1100	Vibro replacement compacted stone cylinder, minimum		500	.112			4.27	4.36	8.63	11.25
1150	Maximum		250	.224			8.55	8.70	17.25	22.50
1300	Mobilization and demobilization, minimum		.47	119	Total		4,550	4,625	9,175	12,000
1400	Maximum		.14	400	"		15,200	15,600	30,800	40,200

31 46 Needle Beams

31 46 13 – Cantilever Needle Beams

31 46 13.10 Needle Beams		Crew	Daily Output	Labor-Hours	Unit	Material	2010 Bare Costs Labor	Equipment	Total	Total Incl O&P
0010	**NEEDLE BEAMS**									
0011	Incl. wood shoring 10' x 10' opening									
0400	Block, concrete, 8" thick	B-9	7.10	5.634	Ea.	45	189	31.50	265.50	375
0420	12" thick		6.70	5.970		53	200	33.50	286.50	405
0800	Brick, 4" thick with 8" backup block		5.70	7.018		53	235	39.50	327.50	465
1000	Brick, solid, 8" thick		6.20	6.452		45	216	36.50	297.50	425
1040	12" thick		4.90	8.163		53	273	46	372	530
1080	16" thick		4.50	8.889		68	298	50	416	590
2000	Add for additional floors of shoring	B-1	6	4		45	135		180	258

31 48 Underpinning

31 48 13 – Underpinning Piers

31 48 13.10 Underpinning Foundations

Line	Description	Crew	Daily Output	Labor-Hours	Unit	Material	Labor	Equipment	Total	Total Incl O&P
0010	**UNDERPINNING FOUNDATIONS**									
0011	Including excavation,									
0020	forming, reinforcing, concrete and equipment									
0100	5' to 16' below grade, 100 to 500 C.Y.	B-52	2.30	24.348	C.Y.	273	935	234	1,442	1,975
0200	Over 500 C.Y.		2.50	22.400		245	860	215	1,320	1,825
0400	16' to 25' below grade, 100 to 500 C.Y.		2	28		300	1,075	269	1,644	2,275
0500	Over 500 C.Y.		2.10	26.667		283	1,025	257	1,565	2,175
0700	26' to 40' below grade, 100 to 500 C.Y.		1.60	35		325	1,350	335	2,010	2,775
0800	Over 500 C.Y.		1.80	31.111		300	1,200	299	1,799	2,475
0900	For under 50 C.Y., add					10%	40%			
1000	For 50 C.Y. to 100 C.Y., add					5%	20%			

31 52 Cofferdams

31 52 16 – Timber Cofferdams

31 52 16.10 Cofferdams

Line	Description	Crew	Daily Output	Labor-Hours	Unit	Material	Labor	Equipment	Total	Total Incl O&P
0010	**COFFERDAMS**									
0011	Incl. mobilization and temporary sheeting									
0080	Soldier beams & lagging H piles with 3" wood sheeting									
0090	horizontal between piles, including removal of wales & braces									
0100	No hydrostatic head, 15' deep, 1 line of braces, minimum	B-50	545	.206	S.F.	5.30	8.10	4.48	17.88	23.50
0200	Maximum		495	.226		5.90	8.95	4.94	19.79	26
0400	15' to 22' deep with 2 lines of braces, 10" H, minimum		360	.311		6.25	12.30	6.80	25.35	33.50
0500	Maximum		330	.339		7.05	13.40	7.40	27.85	37
0700	23' to 35' deep with 3 lines of braces, 12" H, minimum		325	.345		8.15	13.60	7.50	29.25	38.50
0800	Maximum		295	.380		8.85	15	8.30	32.15	42.50
1000	36' to 45' deep with 4 lines of braces, 14" H, minimum		290	.386		9.15	15.25	8.45	32.85	43.50
1100	Maximum		265	.423		9.65	16.70	9.20	35.55	47
1300	No hydrostatic head, left in place, 15' dp., 1 line of braces, min.		635	.176		7.05	6.95	3.85	17.85	23
1400	Maximum		575	.195		7.55	7.70	4.25	19.50	25
1600	15' to 22' deep with 2 lines of braces, minimum		455	.246		10.60	9.70	5.35	25.65	33
1700	Maximum		415	.270		11.80	10.65	5.90	28.35	36
1900	23' to 35' deep with 3 lines of braces, minimum		420	.267		12.60	10.55	5.80	28.95	37
2000	Maximum		380	.295		13.95	11.65	6.45	32.05	40.50
2200	36' to 45' deep with 4 lines of braces, minimum		385	.291		15.15	11.50	6.35	33	41.50
2300	Maximum		350	.320		17.65	12.65	7	37.30	47
2350	Lagging only, 3" thick wood between piles 8' O.C., minimum	B-46	400	.120		1.18	4.44	.10	5.72	8.40

31 52 Cofferdams

31 52 16 – Timber Cofferdams

31 52 16.10 Cofferdams		Crew	Daily Output	Labor-Hours	Unit	Material	2010 Bare Costs Labor	Equipment	Total	Total Incl O&P
2370	Maximum	B-46	250	.192	S.F.	1.77	7.10	.16	9.03	13.30
2400	Open sheeting no bracing, for trenches to 10′ deep, min.		1736	.028		.53	1.02	.02	1.57	2.22
2450	Maximum	↓	1510	.032		.59	1.18	.03	1.80	2.54
2500	Tie-back method, add to open sheeting, add, minimum								20%	20%
2550	Maximum				↓				60%	60%
2700	Tie-backs only, based on tie-backs total length, minimum	B-46	86.80	.553	L.F.	13.20	20.50	.45	34.15	47.50
2750	Maximum		38.50	1.247	″	23	46	1.02	70.02	99.50
3500	Tie-backs only, typical average, 25′ long		2	24	Ea.	580	890	19.70	1,489.70	2,050
3600	35′ long	↓	1.58	30.380	″	775	1,125	25	1,925	2,650

31 56 Slurry Walls

31 56 23 – Lean Concrete Slurry Walls

31 56 23.20 Slurry Trench		Crew	Daily Output	Labor-Hours	Unit	Material	Labor	Equipment	Total	Total Incl O&P
0010	**SLURRY TRENCH**									
0011	Excavated slurry trench in wet soils									
0020	backfilled with 3000 PSI concrete, no reinforcing steel									
0050	Minimum	C-7	333	.216	C.F.	7.20	7.75	3.87	18.82	24
0100	Maximum		200	.360	″	12.05	12.85	6.45	31.35	40
0200	Alternate pricing method, minimum		150	.480	S.F.	14.35	17.15	8.60	40.10	51.50
0300	Maximum	↓	120	.600		21.50	21.50	10.75	53.75	68
0500	Reinforced slurry trench, minimum	B-48	177	.316		10.80	11.70	16.65	39.15	48
0600	Maximum	″	69	.812	↓	36	30	42.50	108.50	132
0800	Haul for disposal, 2 mile haul, excavated material, add	B-34B	99	.081	C.Y.		2.68	6.75	9.43	11.50
0900	Haul bentonite castings for disposal, add	″	40	.200	″		6.65	16.70	23.35	28.50

31 62 Driven Piles

31 62 13 – Concrete Piles

31 62 13.23 Prestressed Concrete Piles		Crew	Daily Output	Labor-Hours	Unit	Material	Labor	Equipment	Total	Total Incl O&P
0010	**PRESTRESSED CONCRETE PILES**, 200 piles									
0020	Unless specified otherwise, not incl. pile caps or mobilization									
2200	Precast, prestressed, 50′ long, 12″ diam., 2-3/8″ wall	B-19	720	.089	V.L.F.	20	3.67	2.68	26.35	30.50
2300	14″ diameter, 2-1/2″ wall		680	.094		27.50	3.89	2.84	34.23	39
2500	16″ diameter, 3″ wall	↓	640	.100		39	4.13	3.02	46.15	52.50
2600	18″ diameter, 3″ wall	B-19A	600	.107		49	4.41	3.93	57.34	65
2800	20″ diameter, 3-1/2″ wall		560	.114		57	4.72	4.21	65.93	75
2900	24″ diameter, 3-1/2″ wall	↓	520	.123		70.50	5.10	4.54	80.14	90.50
3100	Precast, prestressed, 40′ long, 10″ thick, square	B-19	700	.091		14.85	3.78	2.76	21.39	25
3200	12″ thick, square		680	.094		18.70	3.89	2.84	25.43	29.50
3400	14″ thick, square		600	.107		22	4.41	3.22	29.63	35
3500	Octagonal		640	.100		30.50	4.13	3.02	37.65	43.50
3700	16″ thick, square		560	.114		35	4.72	3.45	43.17	49.50
3800	Octagonal	↓	600	.107		36.50	4.41	3.22	44.13	51
4000	18″ thick, square	B-19A	520	.123		42	5.10	4.54	51.64	59
4100	Octagonal	B-19	560	.114		43.50	4.72	3.45	51.67	58.50
4300	20″ thick, square	B-19A	480	.133		52	5.50	4.91	62.41	71
4400	Octagonal	B-19	520	.123		48	5.10	3.72	56.82	64.50
4600	24″ thick, square	B-19A	440	.145		74	6	5.35	85.35	97
4700	Octagonal	B-19	480	.133		69	5.50	4.03	78.53	89
4730	Precast, prestressed, 60′ long, 10″ thick, square	↓	700	.091	↓	15.60	3.78	2.76	22.14	26

31 62 Driven Piles

31 62 13 – Concrete Piles

31 62 13.23 Prestressed Concrete Piles		Crew	Daily Output	Labor-Hours	Unit	2010 Bare Costs Material	Labor	Equipment	Total	Total Incl O&P
4740	12" thick, square (60' long)	B-19	680	.094	V.L.F.	19.45	3.89	2.84	26.18	30.50
4750	Mobilization for 10,000 L.F. pile job, add		3300	.019			.80	.59	1.39	1.89
4800	25,000 L.F. pile job, add		8500	.008			.31	.23	.54	.74

31 62 16 – Steel Piles

31 62 16.13 Sheet Steel Piles		Crew	Daily Output	Labor-Hours	Unit	Material	Labor	Equipment	Total	Total Incl O&P
0010	**SHEET STEEL PILES**									
0100	Step tapered, round, concrete filled									
0110	8" tip, 60 ton capacity, 30' depth	B-19	760	.084	V.L.F.	9.95	3.48	2.54	15.97	19.15
0120	60' depth		740	.086		11.20	3.57	2.61	17.38	21
0130	80' depth		700	.091		11.60	3.78	2.76	18.14	21.50
0150	10" tip, 90 ton capacity, 30' depth		700	.091		12.20	3.78	2.76	18.74	22.50
0160	60' depth		690	.093		12.60	3.83	2.80	19.23	23
0170	80' depth		670	.096		13.55	3.95	2.88	20.38	24.50
0190	12" tip, 120 ton capacity, 30' depth		660	.097		16.20	4.01	2.93	23.14	27.50
0200	60' depth, 12" diameter		630	.102		16.25	4.20	3.07	23.52	28
0210	80' depth		590	.108		14.90	4.48	3.28	22.66	27
0250	"H" Sections, 50' long, HP8 x 36		640	.100		17.05	4.13	3.02	24.20	28.50
0400	HP10 X 42		610	.105		19.90	4.34	3.17	27.41	32.50
0500	HP10 X 57		610	.105		27	4.34	3.17	34.51	40
0700	HP12 X 53		590	.108		25.50	4.48	3.28	33.26	38.50
0800	HP12 X 74	B-19A	590	.108		35.50	4.48	4	43.98	50.50
1000	HP14 X 73		540	.119		35	4.90	4.37	44.27	51
1100	HP14 X 89		540	.119		43	4.90	4.37	52.27	60
1300	HP14 X 102		510	.125		49	5.20	4.63	58.83	67
1400	HP14 X 117		510	.125		56.50	5.20	4.63	66.33	75
1600	Splice on standard points, not in leads, 8" or 10"	1 Sswl	5	1.600	Ea.	112	75		187	255
1700	12" or 14"		4	2		163	94		257	345
1900	Heavy duty points, not in leads, 10" wide		4	2		173	94		267	355
2100	14" wide		3.50	2.286		224	107		331	435

31 62 19 – Timber Piles

31 62 19.10 Wood Piles		Crew	Daily Output	Labor-Hours	Unit	Material	Labor	Equipment	Total	Total Incl O&P
0010	**WOOD PILES**									
0011	Friction or end bearing, not including									
0050	mobilization or demobilization									
0100	Untreated piles, up to 30' long, 12" butts, 8" points	B-19	625	.102	V.L.F.	7.40	4.23	3.09	14.72	18.10
0200	30' to 39' long, 12" butts, 8" points		700	.091		7.40	3.78	2.76	13.94	17.05
0300	40' to 49' long, 12" butts, 7" points		720	.089		7.40	3.67	2.68	13.75	16.80
0400	50' to 59' long, 13"butts, 7" points		800	.080		7.45	3.31	2.42	13.18	16
0500	60' to 69' long, 13" butts, 7" points		840	.076		8.40	3.15	2.30	13.85	16.70
0600	70' to 80' long, 13" butts, 6" points		840	.076		9.35	3.15	2.30	14.80	17.70
0800	Treated piles, 12 lb. per C.F.,									
0810	friction or end bearing, ASTM class B									
1000	Up to 30' long, 12" butts, 8" points	B-19	625	.102	V.L.F.	10.90	4.23	3.09	18.22	22
1100	30' to 39' long, 12" butts, 8" points		700	.091		10.65	3.78	2.76	17.19	20.50
1200	40' to 49' long, 12" butts, 7" points		720	.089		11.40	3.67	2.68	17.75	21.50
1300	50' to 59' long, 13" butts, 7" points		800	.080		12.60	3.31	2.42	18.33	21.50
1400	60' to 69' long, 13" butts, 6" points	B-19A	840	.076		17.20	3.15	2.81	23.16	27
1500	70' to 80' long, 13" butts, 6" points	"	840	.076		21.50	3.15	2.81	27.46	32
1600	Treated piles, C.C.A., 2.5# per C.F.									
1610	8" butts, 10' long	B-19	400	.160	V.L.F.	7.95	6.60	4.83	19.38	24.50
1620	11' to 16' long		500	.128		7.95	5.30	3.87	17.12	21.50
1630	17' to 20' long		575	.111		7.95	4.60	3.36	15.91	19.60

31 62 19 – Timber Piles

31 62 19.10 Wood Piles		Crew	Daily Output	Labor-Hours	Unit	Material	Labor	Equipment	Total	Total Incl O&P
						2010 Bare Costs				
1640	10" butts, 10' to 16' long	B-19	500	.128	V.L.F.	9.05	5.30	3.87	18.22	22.50
1650	17' to 20' long		575	.111		9.05	4.60	3.36	17.01	21
1660	21' to 40' long		700	.091		9.05	3.78	2.76	15.59	18.90
1670	12" butts, 10' to 20' long		575	.111		9.75	4.60	3.36	17.71	21.50
1680	21' to 35' long		650	.098		9.75	4.07	2.97	16.79	20.50
1690	36' to 40' long		700	.091		9.75	3.78	2.76	16.29	19.70
1695	14" butts. to 40' long		700	.091		14	3.78	2.76	20.54	24.50
1700	Boot for pile tip, minimum	1 Pile	27	.296	Ea.	22.50	11.95		34.45	43.50
1800	Maximum		21	.381		67	15.35		82.35	98
2000	Point for pile tip, minimum		20	.400		22.50	16.10		38.60	50.50
2100	Maximum		15	.533		80	21.50		101.50	123
2300	Splice for piles over 50' long, minimum	B-46	35	1.371		56	51	1.12	108.12	143
2400	Maximum		20	2.400		67	89	1.97	157.97	216
2600	Concrete encasement with wire mesh and tube		331	.145	V.L.F.	10.50	5.35	.12	15.97	20
2700	Mobilization for 10,000 L.F. pile job, add	B-19	3300	.019			.80	.59	1.39	1.89
2800	25,000 L.F. pile job, add	"	8500	.008			.31	.23	.54	.74

31 62 23 – Composite Piles

31 62 23.13 Concrete Filled Steel Piles		Crew	Daily Output	Labor-Hours	Unit	Material	Labor	Equipment	Total	Total Incl O&P
0010	**CONCRETE FILLED STEEL PILES** no mobilization or demobilization									
2600	Pipe piles, 50' lg. 8" diam., 29 lb. per L.F., no concrete	B-19	500	.128	V.L.F.	18.45	5.30	3.87	27.62	33
2700	Concrete filled		460	.139		19.40	5.75	4.20	29.35	35
2900	10" diameter, 34 lb. per L.F., no concrete		500	.128		23	5.30	3.87	32.17	38
3000	Concrete filled		450	.142		25	5.90	4.30	35.20	41.50
3200	12" diameter, 44 lb. per L.F., no concrete		475	.135		28.50	5.55	4.07	38.12	44.50
3300	Concrete filled		415	.154		29	6.35	4.66	40.01	47
3500	14" diameter, 46 lb. per L.F., no concrete		430	.149		30.50	6.15	4.50	41.15	48
3600	Concrete filled		355	.180		32	7.45	5.45	44.90	52.50
3800	16" diameter, 52 lb. per L.F., no concrete		385	.166		34	6.85	5	45.85	54
3900	Concrete filled		335	.191		36.50	7.90	5.75	50.15	59
4100	18" diameter, 59 lb. per L.F., no concrete		355	.180		44.50	7.45	5.45	57.40	66.50
4200	Concrete filled		310	.206		50.50	8.55	6.25	65.30	75.50
4400	Splices for pipe piles, not in leads, 8" diameter	1 Sswl	4.67	1.713	Ea.	83	80.50		163.50	232
4500	14" diameter		3.79	2.111		108	99		207	293
4600	16" diameter		3.03	2.640		134	124		258	365
4800	Points, standard, 8" diameter		4.61	1.735		92.50	81.50		174	245
4900	14" diameter		4.05	1.975		129	92.50		221.50	305
5000	16" diameter		3.37	2.374		157	111		268	370
5200	Points, heavy duty, 10" diameter		2.89	2.768		64.50	130		194.50	299
5300	14" or 16" diameter		2.02	3.960		102	186		288	440
5500	For reinforcing steel, add		1150	.007	Lb.	.76	.33		1.09	1.41
5700	For thick wall sections, add				"	.87			.87	.96

31 63 26.13 Fixed End Cassion Piles		Crew	Daily Output	Labor-Hours	Unit	Material	2010 Bare Costs Labor	Equipment	Total	Total Incl O&P
0010	**FIXED END CASSION PILES** R316326-60									
0015	Including excavation, concrete, 50 lbs reinforcing									
0020	per C.Y., not incl. mobilization, boulder removal, disposal									
0100	Open style, machine drilled, to 50′ deep, in stable ground, no									
0110	casings or ground water, 18″ diam., 0.065 C.Y./L.F.	B-43	200	.240	V.L.F.	7.60	8.70	12.95	29.25	36
0200	24″ diameter, 0.116 C.Y./L.F.		190	.253		13.55	9.15	13.65	36.35	44
0300	30″ diameter, 0.182 C.Y./L.F.		150	.320		21.50	11.60	17.25	50.35	60
0400	36″ diameter, 0.262 C.Y./L.F.		125	.384		30.50	13.90	20.50	64.90	77.50
0500	48″ diameter, 0.465 C.Y./L.F.		100	.480		54.50	17.35	26	97.85	115
0600	60″ diameter, 0.727 C.Y./L.F.		90	.533		85	19.30	29	133.30	155
0700	72″ diameter, 1.05 C.Y./L.F.		80	.600		123	21.50	32.50	177	204
0800	84″ diameter, 1.43 C.Y./L.F.	↓	75	.640	↓	167	23	34.50	224.50	258
1000	For bell excavation and concrete, add									
1020	4′ bell diameter, 24″ shaft, 0.444 C.Y.	B-43	20	2.400	Ea.	43	87	130	260	325
1040	6′ bell diameter, 30″ shaft, 1.57 C.Y.		5.70	8.421		152	305	455	912	1,125
1060	8′ bell diameter, 36″ shaft, 3.72 C.Y.		2.40	20		360	725	1,075	2,160	2,700
1080	9′ bell diameter, 48″ shaft, 4.48 C.Y.		2	24		435	870	1,300	2,605	3,225
1100	10′ bell diameter, 60″ shaft, 5.24 C.Y.		1.70	28.235		510	1,025	1,525	3,060	3,775
1120	12′ bell diameter, 72″ shaft, 8.74 C.Y.		1	48		850	1,725	2,600	5,175	6,425
1140	14′ bell diameter, 84″ shaft, 13.6 C.Y.	↓	.70	68.571	↓	1,325	2,475	3,700	7,500	9,300
1200	Open style, machine drilled, to 50′ deep, in wet ground, pulled									
1300	casing and pumping, 18″ diameter, 0.065 C.Y./L.F.	B-48	160	.350	V.L.F.	7.60	12.90	18.45	38.95	48.50
1400	24″ diameter, 0.116 C.Y./L.F.		125	.448		13.55	16.55	23.50	53.60	66
1500	30″ diameter, 0.182 C.Y./L.F.		85	.659		21.50	24.50	34.50	80.50	98.50
1600	36″ diameter, 0.262 C.Y./L.F.	↓	60	.933		30.50	34.50	49	114	140
1700	48″ diameter, 0.465 C.Y./L.F.	B-49	55	1.600		54.50	61.50	67	183	228
1800	60″ diameter, 0.727 C.Y./L.F.		35	2.514		85	96.50	106	287.50	360
1900	72″ diameter, 1.05 C.Y./L.F.		30	2.933		123	112	123	358	445
2000	84″ diameter, 1.43 C.Y./L.F.	↓	25	3.520	↓	167	135	148	450	555
2100	For bell excavation and concrete, add									
2120	4′ bell diameter, 24″ shaft, 0.444 C.Y.	B-48	19.80	2.828	Ea.	43	104	149	296	370
2140	6′ bell diameter, 30″ shaft, 1.57 C.Y.		5.70	9.825		152	365	515	1,032	1,300
2160	8′ bell diameter, 36″ shaft, 3.72 C.Y.	↓	2.40	23.333		360	860	1,225	2,445	3,050
2180	9′ bell diameter, 48″ shaft, 4.48 C.Y.	B-49	3.30	26.667		435	1,025	1,125	2,585	3,275
2200	10′ bell diameter, 60″ shaft, 5.24 C.Y.		2.80	31.429		510	1,200	1,325	3,035	3,850
2220	12′ bell diameter, 72″ shaft, 8.74 C.Y.		1.60	55		850	2,100	2,300	5,250	6,700
2240	14′ bell diameter, 84″ shaft, 13.6 C.Y.	↓	1	88	↓	1,325	3,375	3,700	8,400	10,700
2300	Open style, machine drilled, to 50′ deep, in soft rocks and									
2400	medium hard shales, 18″ diameter, 0.065 C.Y./L.F.	B-49	50	1.760	V.L.F.	7.60	67.50	74	149.10	193
2500	24″ diameter, 0.116 C.Y./L.F.		30	2.933		13.55	112	123	248.55	325
2600	30″ diameter, 0.182 C.Y./L.F.		20	4.400		21.50	169	185	375.50	485
2700	36″ diameter, 0.262 C.Y./L.F.		15	5.867		30.50	225	246	501.50	650
2800	48″ diameter, 0.465 C.Y./L.F.		10	8.800		54.50	335	370	759.50	980
2900	60″ diameter, 0.727 C.Y./L.F.		7	12.571		85	480	530	1,095	1,425
3000	72″ diameter, 1.05 C.Y./L.F.		6	14.667		123	560	615	1,298	1,675
3100	84″ diameter, 1.43 C.Y./L.F.	↓	5	17.600	↓	167	675	740	1,582	2,025
3200	For bell excavation and concrete, add									
3220	4′ bell diameter, 24″ shaft, 0.444 C.Y.	B-49	10.90	8.073	Ea.	43	310	340	693	900
3240	6′ bell diameter, 30″ shaft, 1.57 C.Y.		3.10	28.387		152	1,100	1,200	2,452	3,150
3260	8′ bell diameter, 36″ shaft, 3.72 C.Y.		1.30	67.692		360	2,600	2,850	5,810	7,500
3280	9′ bell diameter, 48″ shaft, 4.48 C.Y.		1.10	80		435	3,075	3,350	6,860	8,875
3300	10′ bell diameter, 60″ shaft, 5.24 C.Y.		.90	97.778		510	3,750	4,100	8,360	10,800
3320	12′ bell diameter, 72″ shaft, 8.74 C.Y.	↓	.60	146	↓	850	5,625	6,150	12,625	16,300

31 63 26 – Drilled Caissons

31 63 26.13 Fixed End Cassion Piles		Crew	Daily Output	Labor-Hours	Unit	Material	Labor	Equipment	Total	Total Incl O&P
						2010 Bare Costs				
3340	14′ bell diameter, 84″ shaft, 13.6 C.Y.	B-49	.40	220	Ea.	1,325	8,425	9,250	19,000	24,600
3600	For rock excavation, sockets, add, minimum		120	.733	C.F.		28	31	59	77
3650	Average		95	.926			35.50	39	74.50	97.50
3700	Maximum		48	1.833			70.50	77	147.50	193
3900	For 50′ to 100′ deep, add				V.L.F.				7%	7%
4000	For 100′ to 150′ deep, add								25%	25%
4100	For 150′ to 200′ deep, add								30%	30%
4200	For casings left in place, add				Lb.	.97			.97	1.07
4300	For other than 50 lb. reinf. per C.Y., add or deduct				"	1.03			1.03	1.13
4400	For steel "I" beam cores, add	B-49	8.30	10.602	Ton	1,750	405	445	2,600	3,050
4500	Load and haul excess excavation, 2 miles	B-34B	178	.045	L.C.Y.		1.49	3.75	5.24	6.40
4600	For mobilization, 50 mile radius, rig to 36″	B-43	2	24	Ea.		870	1,300	2,170	2,750
4650	Rig to 84″	B-48	1.75	32			1,175	1,675	2,850	3,650
4700	For low headroom, add								50%	
5000	Bottom inspection	1 Skwk	1.20	6.667			284		284	435

31 63 26.16 Concrete Caissons for Marine Construction		Crew	Daily Output	Labor-Hours	Unit	Material	Labor	Equipment	Total	Total Incl O&P
0010	**CONCRETE CAISSONS FOR MARINE CONSTRUCTION**									
0100	Caissons, incl. mobilization and demobilization, up to 50 miles									
0200	Uncased shafts, 30 to 80 tons cap., 17″ diam., 10′ depth	B-44	88	.727	V.L.F.	19.40	29.50	21	69.90	91
0300	25′ depth		165	.388		13.85	15.80	11.15	40.80	52.50
0400	80-150 ton capacity, 22″ diameter, 10′ depth		80	.800		24.50	32.50	23	80	103
0500	20′ depth		130	.492		19.40	20	14.15	53.55	68.50
0700	Cased shafts, 10 to 30 ton capacity, 10-5/8″ diam., 20′ depth		175	.366		13.85	14.90	10.50	39.25	50.50
0800	30′ depth		240	.267		12.95	10.85	7.65	31.45	40
0850	30 to 60 ton capacity, 12″ diameter, 20′ depth		160	.400		19.40	16.25	11.50	47.15	59.50
0900	40′ depth		230	.278		14.90	11.30	8	34.20	43
1000	80 to 100 ton capacity, 16″ diameter, 20′ depth		160	.400		27.50	16.25	11.50	55.25	68.50
1100	40′ depth		230	.278		26	11.30	8	45.30	55
1200	110 to 140 ton capacity, 17-5/8″ diameter, 20′ depth		160	.400		30	16.25	11.50	57.75	71
1300	40′ depth		230	.278		27.50	11.30	8	46.80	57
1400	140 to 175 ton capacity, 19″ diameter, 20′ depth		130	.492		32.50	20	14.15	66.65	82.50
1500	40′ depth		210	.305		30	12.40	8.75	51.15	62
1700	Over 30′ long, L.F. cost tends to be lower									
1900	Maximum depth is about 90′									

31 63 29 – Drilled Concrete Piers and Shafts

31 63 29.13 Uncased Drilled Concrete Piers		Crew	Daily Output	Labor-Hours	Unit	Material	Labor	Equipment	Total	Total Incl O&P
0010	**UNCASED DRILLED CONCRETE PIERS**									
0020	Unless specified otherwise, not incl. pile caps or mobilization									
0050	Cast in place augered piles, no casing or reinforcing									
0060	8″ diameter	B-43	540	.089	V.L.F.	3.91	3.22	4.80	11.93	14.50
0065	10″ diameter		480	.100		6.20	3.62	5.40	15.22	18.30
0070	12″ diameter		420	.114		8.75	4.14	6.15	19.04	23
0075	14″ diameter		360	.133		11.80	4.82	7.20	23.82	28.50
0080	16″ diameter		300	.160		15.90	5.80	8.65	30.35	36
0085	18″ diameter		240	.200		19.70	7.25	10.80	37.75	44.50
0100	Cast in place, thin wall shell pile, straight sided,									
0110	not incl. reinforcing, 8″ diam., 16 ga., 5.8 lb./L.F.	B-19	700	.091	V.L.F.	9.10	3.78	2.76	15.64	18.95
0200	10″ diameter, 16 ga. corrugated, 7.3 lb./L.F.		650	.098		11.90	4.07	2.97	18.94	22.50
0300	12″ diameter, 16 ga. corrugated, 8.7 lb./L.F.		600	.107		15.45	4.41	3.22	23.08	27.50
0400	14″ diameter, 16 ga. corrugated, 10.0 lb./L.F.		550	.116		18.15	4.81	3.51	26.47	31.50
0500	16″ diameter, 16 ga. corrugated, 11.6 lb./L.F.		500	.128		22.50	5.30	3.87	31.67	37
0800	Cast in place friction pile, 50′ long, fluted,									

31 63 Bored Piles

31 63 29 – Drilled Concrete Piers and Shafts

31 63 29.13 Uncased Drilled Concrete Piers		Crew	Daily Output	Labor-Hours	Unit	2010 Bare Costs Material	Labor	Equipment	Total	Total Incl O&P
0810	tapered steel, 4000 psi concrete, no reinforcing									
0900	12" diameter, 7 ga.	B-19	600	.107	V.L.F.	27.50	4.41	3.22	35.13	40.50
1000	14" diameter, 7 ga.		560	.114		29.50	4.72	3.45	37.67	43.50
1100	16" diameter, 7 ga.		520	.123		35	5.10	3.72	43.82	50.50
1200	18" diameter, 7 ga.	↓	480	.133	↓	41	5.50	4.03	50.53	58
1300	End bearing, fluted, constant diameter,									
1320	4000 psi concrete, no reinforcing									
1340	12" diameter, 7 ga.	B-19	600	.107	V.L.F.	28.50	4.41	3.22	36.13	42
1360	14" diameter, 7 ga.		560	.114		35.50	4.72	3.45	43.67	50
1380	16" diameter, 7 ga.		520	.123		41.50	5.10	3.72	50.32	57.50
1400	18" diameter, 7 ga.	↓	480	.133	↓	45.50	5.50	4.03	55.03	63
31 63 29.20 Cast In Place Piles, Adds										
0010	**CAST IN PLACE PILES, ADDS**									
1500	For reinforcing steel, add				Lb.	.80			.80	.88
1700	For ball or pedestal end, add	B-19	11	5.818	C.Y.	146	240	176	562	730
1900	For lengths above 60', concrete, add	"	11	5.818	"	152	240	176	568	735
2000	For steel thin shell, pipe only				Lb.	1.08			1.08	1.19

Estimating Tips

32 01 00 Operations and Maintenance of Exterior Improvements

- Recycling of asphalt pavement is becoming very popular and is an alternative to removal and replacement of asphalt pavement. It can be a good value engineering proposal if removed pavement can be recycled either at the site or another site that is reasonably close to the project site.

32 10 00 Bases, Ballasts, and Paving

- When estimating paving, keep in mind the project schedule. If an asphaltic paving project is in a colder climate and runs through to the spring, consider placing the base course in the autumn and then topping it in the spring, just prior to completion. This could save considerable costs in spring repair. Keep in mind that prices for asphalt and concrete are generally higher in the cold seasons.

32 90 00 Planting

- The timing of planting and guarantee specifications often dictate the costs for establishing tree and shrub growth and a stand of grass or ground cover. Establish the work performance schedule to coincide with the local planting season. Maintenance and growth guarantees can add from 20%–100% to the total landscaping cost. The cost to replace trees and shrubs can be as high as 5% of the total cost, depending on the planting zone, soil conditions, and time of year.

Reference Numbers

Reference numbers are shown in shaded boxes at the beginning of some major classifications. These numbers refer to related items in the Reference Section. The reference information may be an estimating procedure, an alternate pricing method, or technical information.

Note: Not all subdivisions listed here necessarily appear in this publication.

32 01 Operation and Maintenance of Exterior Improvements

32 01 13 – Flexible Paving Surface Treatment

32 01 13.61 Slurry Seal (Latex Modified)		Crew	Daily Output	Labor-Hours	Unit	2010 Bare Costs Material	Labor	Equipment	Total	Total Incl O&P
0010	**SLURRY SEAL (LATEX MODIFIED)**									
3780	Rubberized asphalt (latex) seal	B-45	5000	.003	S.Y.	2.31	.12	.18	2.61	2.92

32 01 13.64 Sand Seal

Code	Description	Crew	Daily Output	Labor-Hours	Unit	Material	Labor	Equipment	Total	Total Incl O&P
0010	**SAND SEAL**									
2080	Sand sealing, sharp sand, asphalt emulsion, small area	B-91	10000	.006	S.Y.	1.29	.25	.23	1.77	2.04
2120	Roadway or large area	″	18000	.004	″	1.11	.14	.13	1.38	1.57

32 01 13.66 Fog Seal

Code	Description	Crew	Daily Output	Labor-Hours	Unit	Material	Labor	Equipment	Total	Total Incl O&P
0010	**FOG SEAL**									
0012	Sealcoating, 2 coat coal tar pitch emulsion over 10,000 SY	B-45	5000	.003	S.Y.	1.05	.12	.18	1.35	1.54
0030	1000 to 10,000 S.Y.	″	3000	.005		1.05	.20	.30	1.55	1.80
0100	Under 1000 S.Y.	B-1	1050	.023		1.05	.77		1.82	2.35
0300	Petroleum resistant, over 10,000 S.Y.	B-45	5000	.003		1.30	.12	.18	1.60	1.81
0320	1000 to 10,000 S.Y.	″	3000	.005		1.30	.20	.30	1.80	2.07
0400	Under 1000 S.Y.	B-1	1050	.023		1.30	.77		2.07	2.62
0600	Non-skid pavement renewal, over 10,000 S.Y.	B-45	5000	.003		1.50	.12	.18	1.80	2.03
0620	1000 to 10,000 S.Y.	″	3000	.005		1.50	.20	.30	2	2.29
0700	Under 1000 S.Y.	B-1	1050	.023		1.50	.77		2.27	2.84
0800	Prepare and clean surface for above	A-2	8545	.003			.09	.03	.12	.17
1000	Hand seal asphalt curbing	B-1	4420	.005	L.F.	.80	.18		.98	1.16
1900	Asphalt surface treatment, single course, small area									
1901	0.30 gal/S.Y. asphalt material, 20#/S.Y. aggregate	B-91	5000	.013	S.Y.	1.75	.49	.46	2.70	3.17
1910	Roadway or large area		10000	.006		1.61	.25	.23	2.09	2.39
1950	Asphalt surface treatment, dbl. course for small area		3000	.021		3.20	.82	.76	4.78	5.60
1960	Roadway or large area		6000	.011		2.88	.41	.38	3.67	4.21
1980	Asphalt surface treatment, single course, for shoulders		7500	.009		1.86	.33	.31	2.50	2.88

32 06 Schedules for Exterior Improvements

32 06 10 – Schedules for Bases, Ballasts, and Paving

32 06 10.10 Sidewalks, Driveways and Patios

Code	Description	Crew	Daily Output	Labor-Hours	Unit	Material	Labor	Equipment	Total	Total Incl O&P
0010	**SIDEWALKS, DRIVEWAYS AND PATIOS** No base									
0020	Asphaltic concrete, 2″ thick	B-37	720	.067	S.Y.	6.90	2.32	.21	9.43	11.35
0100	2-1/2″ thick	″	660	.073	″	8.75	2.53	.23	11.51	13.75
0300	Concrete, 3000 psi, CIP, 6 x 6 - W1.4 x W1.4 mesh,									
0310	broomed finish, no base, 4″ thick	B-24	600	.040	S.F.	1.62	1.52		3.14	4.09
0350	5″ thick		545	.044		2.16	1.68		3.84	4.91
0400	6″ thick		510	.047		2.51	1.79		4.30	5.50
0450	For bank run gravel base, 4″ thick, add	B-18	2500	.010		.51	.32	.02	.85	1.08
0520	8″ thick, add	″	1600	.015		1.03	.51	.03	1.57	1.95
0550	Exposed aggregate finish, add to above, minimum	B-24	1875	.013		.12	.49		.61	.87
0600	Maximum	″	455	.053		.39	2.01		2.40	3.47
1000	Crushed stone, 1″ thick, white marble	2 Clab	1700	.009		.22	.31		.53	.72
1050	Bluestone	″	1700	.009		.24	.31		.55	.74
1700	Redwood, prefabricated, 4′ x 4′ sections	2 Carp	316	.051		4.74	2.10		6.84	8.45
1750	Redwood planks, 1″ thick, on sleepers	″	240	.067		4.74	2.77		7.51	9.45
2250	Stone dust, 4″ thick	B-62	900	.027	S.Y.	3.27	.96	.18	4.41	5.25

32 06 10.20 Steps

Code	Description	Crew	Daily Output	Labor-Hours	Unit	Material	Labor	Equipment	Total	Total Incl O&P
0010	**STEPS**									
0011	Incl. excav., borrow & concrete base as required									
0100	Brick steps	B-24	35	.686	LF Riser	12.85	26		38.85	53.50
0200	Railroad ties	2 Clab	25	.640		3.17	21		24.17	36

32 06 Schedules for Exterior Improvements

32 06 10 – Schedules for Bases, Ballasts, and Paving

32 06 10.20 Steps		Crew	Daily Output	Labor-Hours	Unit	Material	2010 Bare Costs Labor	Equipment	Total	Total Incl O&P
0300	Bluestone treads, 12″ x 2″ or 12″ x 1-1/2″	B-24	30	.800	LF Riser	28	30.50		58.50	77
0500	Concrete, cast in place, see Div. 03 30 53.40									
0600	Precast concrete, see Div. 03 41 23.50									
4025	Steel edge strips, incl. stakes, 1/4″ x 5″	B-1	390	.062	L.F.	4.20	2.08		6.28	7.80
4050	Edging, landscape timber or railroad ties, 6″ x 8″	2 Carp	170	.094	″	3.15	3.91		7.06	9.50

32 11 Base Courses

32 11 23 – Aggregate Base Courses

32 11 23.23 Base Course Drainage Layers

		Crew	Daily Output	Labor-Hours	Unit	Material	Labor	Equipment	Total	Total Incl O&P
0010	**BASE COURSE DRAINAGE LAYERS**									
0011	For roadways and large areas									
0050	Crushed 3/4″ stone base, compacted, 3″ deep	B-36C	5200	.008	S.Y.	3.38	.30	.71	4.39	4.96
0100	6″ deep		5000	.008		6.80	.32	.74	7.86	8.75
0200	9″ deep		4600	.009		10.15	.34	.81	11.30	12.60
0300	12″ deep		4200	.010		13.55	.38	.88	14.81	16.45
0301	Crushed 1-1/2″ stone base, compacted to 4″ deep	B-36B	6000	.011		4.66	.41	.71	5.78	6.55
0302	6″ deep		5400	.012		7	.45	.78	8.23	9.25
0303	8″ deep		4500	.014		9.30	.54	.94	10.78	12.10
0304	12″ deep		3800	.017		14	.64	1.12	15.76	17.60
0350	Bank run gravel, spread and compacted									
0370	6″ deep	B-32	6000	.005	S.Y.	4.29	.22	.34	4.85	5.40
0390	9″ deep		4900	.007		6.45	.26	.42	7.13	7.90
0400	12″ deep		4200	.008		8.55	.31	.49	9.35	10.45
6000	Stabilization fabric, polypropylene, 6 oz./S.Y.	B-6	10000	.002		1.50	.09	.03	1.62	1.82
6900	For small and irregular areas, add						50%	50%		
7000	Prepare and roll sub-base, small areas to 2500 S.Y.	B-32A	1500	.016	S.Y.		.63	.84	1.47	1.88
8000	Large areas over 2500 S.Y.	″	3500	.007			.27	.36	.63	.80
8050	For roadways	B-32	4000	.008			.32	.51	.83	1.05

32 11 26 – Asphaltic Base Courses

32 11 26.19 Bituminous-Stabilized Base Courses

		Crew	Daily Output	Labor-Hours	Unit	Material	Labor	Equipment	Total	Total Incl O&P
0010	**BITUMINOUS-STABILIZED BASE COURSES**									
0020	And large paved areas									
0700	Liquid application to gravel base, asphalt emulsion	B-45	6000	.003	Gal.	6	.10	.15	6.25	6.90
0800	Prime and seal, cut back asphalt		6000	.003	″	7.10	.10	.15	7.35	8.10
1000	Macadam penetration crushed stone, 2 gal. per S.Y., 4″ thick		6000	.003	S.Y.	12	.10	.15	12.25	13.50
1100	6″ thick, 3 gal. per S.Y.		4000	.004		18	.15	.22	18.37	20.50
1200	8″ thick, 4 gal. per S.Y.		3000	.005		24	.20	.30	24.50	27
8900	For small and irregular areas, add						50%	50%		

32 12 16.13 Plant-Mix Asphalt Paving		Crew	Daily Output	Labor-Hours	Unit	2010 Bare Costs Material	Labor	Equipment	Total	Total Incl O&P
0010	**PLANT-MIX ASPHALT PAVING**									
0020	And large paved areas with no hauling included									
0025	See Div. 31 23 23.20 for hauling costs									
0080	Binder course, 1-1/2" thick	B-25	7725	.011	S.Y.	5.30	.41	.33	6.04	6.85
0120	2" thick		6345	.014		7.05	.50	.40	7.95	9
0130	2-1/2" thick		5620	.016		8.85	.56	.46	9.87	11.05
0160	3" thick		4905	.018		10.60	.65	.52	11.77	13.20
0170	3-1/2 " thick		4520	.019		12.35	.70	.57	13.62	15.30
0200	4" thick		4140	.021		14.15	.76	.62	15.53	17.40
0300	Wearing course, 1" thick	B-25B	10575	.009		3.51	.33	.27	4.11	4.66
0340	1-1/2" thick		7725	.012		5.90	.45	.36	6.71	7.60
0380	2" thick		6345	.015		7.90	.55	.44	8.89	10.05
0420	2-1/2" thick		5480	.018		9.75	.64	.51	10.90	12.30
0460	3" thick		4900	.020		11.65	.72	.57	12.94	14.50
0470	3-1/2" thick		4520	.021		13.65	.78	.62	15.05	16.85
0480	4 " thick		4140	.023		15.60	.85	.68	17.13	19.20
0500	Open graded friction course	B-25C	5000	.010		3.30	.35	.45	4.10	4.66
0800	Alternate method of figuring paving costs									
0810	Binder course, 1-1/2" thick	B-25	630	.140	Ton	65	5	4.07	74.07	83.50
0811	2" thick		690	.128		65	4.59	3.72	73.31	82.50
0812	3" thick		800	.110		65	3.96	3.21	72.17	81
0813	4" thick		900	.098		65	3.52	2.85	71.37	80
0850	Wearing course, 1" thick	B-25B	575	.167		68	6.10	4.87	78.97	89.50
0851	1-1/2" thick		630	.152		68	5.55	4.45	78	88.50
0852	2" thick		690	.139		68	5.10	4.06	77.16	87
0853	2-1/2" thick		765	.125		68	4.59	3.66	76.25	86
0854	3" thick		800	.120		68	4.39	3.50	75.89	85.50
1000	Pavement replacement over trench, 2" thick	B-37	90	.533	S.Y.	7.30	18.55	1.69	27.54	38.50
1050	4" thick		70	.686		14.45	24	2.17	40.62	55
1080	6" thick		55	.873		23	30.50	2.77	56.27	75

32 12 16.14 Asphaltic Concrete Paving		Crew	Daily Output	Labor-Hours	Unit	Material	Labor	Equipment	Total	Total Incl O&P
0011	**ASPHALTIC CONCRETE PAVING**, parking lots & driveways									
0015	No asphalt hauling included									
0018	Use 6.05 C. Y. per inch per MSF for hauling									
0020	6" stone base, 2" binder course, 1" topping	B-25C	9000	.005	S.F.	1.82	.20	.25	2.27	2.58
0025	2" binder course, 2" topping		9000	.005		2.24	.20	.25	2.69	3.03
0030	3" binder course, 2" topping		9000	.005		2.63	.20	.25	3.08	3.46
0035	4" binder course, 2" topping		9000	.005		3.02	.20	.25	3.47	3.89
0040	1.5" binder course, 1" topping		9000	.005		1.63	.20	.25	2.08	2.36
0042	3" binder course, 1" topping		9000	.005		2.22	.20	.25	2.67	3.01
0045	3" binder course, 3" topping		9000	.005		3.05	.20	.25	3.50	3.92
0050	4" binder course, 3" topping		9000	.005		3.44	.20	.25	3.89	4.35
0055	4" binder course, 4" topping		9000	.005		3.84	.20	.25	4.29	4.80
0300	Binder course, 1-1/2" thick		35000	.001		.59	.05	.06	.70	.80
0400	2" thick		25000	.002		.76	.07	.09	.92	1.05
0500	3" thick		15000	.003		1.18	.12	.15	1.45	1.64
0600	4" thick		10800	.004		1.55	.16	.21	1.92	2.18
0800	Sand finish course, 3/4" thick		41000	.001		.32	.04	.05	.41	.49
0900	1" thick		34000	.001		.40	.05	.07	.52	.59
1000	Fill pot holes, hot mix, 2" thick	B-16	4200	.008		.84	.26	.16	1.26	1.49
1100	4" thick		3500	.009		1.23	.31	.19	1.73	2.03
1120	6" thick		3100	.010		1.65	.35	.22	2.22	2.59

32 12 Flexible Paving

32 12 16 – Asphalt Paving

32 12 16.14 Asphaltic Concrete Paving		Crew	Daily Output	Labor-Hours	Unit	2010 Bare Costs Material	Labor	Equipment	Total	Total Incl O&P
1140	Cold patch, 2" thick	B-51	3000	.016	S.F.	.85	.53	.08	1.46	1.85
1160	4" thick		2700	.018		1.62	.59	.09	2.30	2.79
1180	6" thick		1900	.025		2.52	.84	.13	3.49	4.20

32 13 Rigid Paving

32 13 13 – Concrete Paving

32 13 13.23 Concrete Paving Surface Treatment		Crew	Daily Output	Labor-Hours	Unit	Material	Labor	Equipment	Total	Total Incl O&P
0010	**CONCRETE PAVING SURFACE TREATMENT**									
0015	Including joints, finishing and curing									
0020	Fixed form, 12' pass, unreinforced, 6" thick	B-26	3000	.029	S.Y.	21	1.08	1.10	23.18	26
0100	8" thick		2750	.032		31.50	1.18	1.20	33.88	38
0110	8" thick, small area		1375	.064		31.50	2.36	2.40	36.26	41.50
0200	9" thick		2500	.035		35.50	1.30	1.32	38.12	42.50
0300	10" thick		2100	.042		39	1.55	1.57	42.12	46.50
0310	10" thick, small area		1050	.084		39	3.09	3.14	45.23	50.50
0400	12" thick		1800	.049		44	1.81	1.83	47.64	53.50
0410	Conc pavement, w/jt,fnsh&curing,fix form,24' pass,unreinforced,6"T		6000	.015		19.45	.54	.55	20.54	23
0430	8" thick		5500	.016		28.50	.59	.60	29.69	33
0440	9" thick		5000	.018		32.50	.65	.66	33.81	37
0450	10" thick		4200	.021		35.50	.77	.79	37.06	41
0460	12" thick		3600	.024		41	.90	.92	42.82	47.50
0470	15" thick		3000	.029		53	1.08	1.10	55.18	61.50
0500	Fixed form 12' pass 15" thick		1500	.059		57	2.17	2.20	61.37	68
0510	For small irregular areas, add				%	10%	100%	100%		
0520	Welded wire fabric, sheets for rigid paving 2.33 lbs/SY	2 Rodm	389	.041	S.Y.	1.08	1.92		3	4.29
0530	Reinforcing steel for rigid paving 12 lbs/SY		666	.024		5.05	1.12		6.17	7.35
0540	Reinforcing steel for rigid paving 18 lbs/SY		444	.036		7.55	1.69		9.24	11
0620	Slip form, 12' pass, unreinforced, 6" thick	B-26A	5600	.016		20.50	.58	.62	21.70	24
0624	8" thick		5300	.017		31	.61	.65	32.26	35.50
0626	9" thick		4820	.018		34.50	.67	.72	35.89	40
0628	10" thick		4050	.022		38	.80	.85	39.65	43.50
0630	12" thick		3470	.025		43	.94	1	44.94	50
0632	15" thick		2890	.030		55.50	1.12	1.20	57.82	64
0640	Slip form, 24' pass, unreinforced, 6" thick		11200	.008		19.70	.29	.31	20.30	22.50
0644	8" thick		10600	.008		28	.31	.33	28.64	32
0646	9" thick		9640	.009		32	.34	.36	32.70	36
0648	10" thick		8100	.011		35	.40	.43	35.83	39.50
0650	12" thick		6940	.013		40.50	.47	.50	41.47	46
0652	15" thick		5780	.015		51.50	.56	.60	52.66	58
0700	Finishing, broom finish small areas	2 Cefi	120	.133			5.30		5.30	7.75
1000	Curing, with sprayed membrane by hand	2 Clab	1500	.011		.55	.35		.90	1.15
1650	For integral coloring, see Div. 03 05 13.20									

32 14 13 – Precast Concrete Unit Paving

32 14 13.13 Interlocking Precast Concrete Unit Paving		Crew	Daily Output	Labor-Hours	Unit	Material	2010 Bare Costs Labor	Equipment	Total	Total Incl O&P
0010	INTERLOCKING PRECAST CONCRETE UNIT PAVING									
0020	"V" blocks for retaining soil	D-1	205	.078	S.F.	8.60	2.94		11.54	13.95

32 14 13.16 Precast Concrete Unit Paving Slabs		Crew	Daily Output	Labor-Hours	Unit	Material	Labor	Equipment	Total	Total Incl O&P
0010	PRECAST CONCRETE UNIT PAVING SLABS									
0710	Precast concrete patio blocks, 2-3/8" thick, colors, 8" x 16"	D-1	265	.060	S.F.	1.51	2.28		3.79	5.10
0750	Exposed local aggregate, natural	2 Bric	250	.064		6.50	2.67		9.17	11.20
0800	Colors		250	.064		7.05	2.67		9.72	11.80
0850	Exposed granite or limestone aggregate		250	.064		7.85	2.67		10.52	12.65
0900	Exposed white tumblestone aggregate		250	.064		4.64	2.67		7.31	9.15

32 14 16 – Brick Unit Paving

32 14 16.10 Brick Paving		Crew	Daily Output	Labor-Hours	Unit	Material	Labor	Equipment	Total	Total Incl O&P
0010	BRICK PAVING									
0012	4" x 8" x 1-1/2", without joints (4.5 brick/S.F.)	D-1	110	.145	S.F.	3.27	5.50		8.77	11.85
0100	Grouted, 3/8" joint (3.9 brick/S.F.)		90	.178		3.38	6.70		10.08	13.80
0200	4" x 8" x 2-1/4", without joints (4.5 bricks/S.F.)		110	.145		3.74	5.50		9.24	12.35
0300	Grouted, 3/8" joint (3.9 brick/S.F.)		90	.178		3.45	6.70		10.15	13.90
0500	Bedding, asphalt, 3/4" thick	B-25	5130	.017		.65	.62	.50	1.77	2.21
0540	Course washed sand bed, 1" thick	B-18	5000	.005		.30	.16	.01	.47	.59
0580	Mortar, 1" thick	D-1	300	.053		.66	2.01		2.67	3.75
0620	2" thick		200	.080		1.31	3.02		4.33	6
1500	Brick on 1" thick sand bed laid flat, 4.5 per S.F.		100	.160		3	6.05		9.05	12.40
2000	Brick pavers, laid on edge, 7.2 per S.F.		70	.229		3.18	8.60		11.78	16.50
2500	For 4" thick concrete bed and joints, add		595	.027		1.12	1.01		2.13	2.77
2800	For steam cleaning, add	A-1H	950	.008		.07	.28	.06	.41	.58

32 14 23 – Asphalt Unit Paving

32 14 23.10 Asphalt Blocks		Crew	Daily Output	Labor-Hours	Unit	Material	Labor	Equipment	Total	Total Incl O&P
0010	ASPHALT BLOCKS									
0020	Rectangular, 6" x 12" x 1-1/4", w/bed & neopr. adhesive	D-1	135	.119	S.F.	7.80	4.47		12.27	15.35
0100	3" thick		130	.123		10.95	4.64		15.59	19.05
0300	Hexagonal tile, 8" wide, 1-1/4" thick		135	.119		7.80	4.47		12.27	15.35
0400	2" thick		130	.123		10.95	4.64		15.59	19.05
0500	Square, 8" x 8", 1-1/4" thick		135	.119		7.80	4.47		12.27	15.35
0600	2" thick		130	.123		10.95	4.64		15.59	19.05
0900	For exposed aggregate (ground finish) add					.54			.54	.59
0910	For colors, add					.41			.41	.45

32 14 40 – Stone Paving

32 14 40.10 Stone Pavers		Crew	Daily Output	Labor-Hours	Unit	Material	Labor	Equipment	Total	Total Incl O&P
0010	STONE PAVERS									
1100	Flagging, bluestone, irregular, 1" thick,	D-1	81	.198	S.F.	5.95	7.45		13.40	17.80
1150	Snapped random rectangular, 1" thick		92	.174		9.05	6.55		15.60	19.85
1200	1-1/2" thick		85	.188		10.85	7.10		17.95	22.50
1250	2" thick		83	.193		12.65	7.25		19.90	25
1300	Slate, natural cleft, irregular, 3/4" thick		92	.174		7	6.55		13.55	17.60
1350	Random rectangular, gauged, 1/2" thick		105	.152		15.15	5.75		20.90	25.50
1400	Random rectangular, butt joint, gauged, 1/4" thick		150	.107		16.30	4.02		20.32	24
1500	For interior setting, add								25%	25%
1550	Granite blocks, 3-1/2" x 3-1/2" x 3-1/2"	D-1	92	.174	S.F.	10.05	6.55		16.60	21
1600	4" to 12" long, 3" to 5" wide, 3" to 5" thick		98	.163		8.40	6.15		14.55	18.50
1650	6" to 15" long, 3" to 6" wide, 3" to 5" thick		105	.152		4.47	5.75		10.22	13.55

32 16 Curbs and Gutters

32 16 13 – Concrete Curbs and Gutters

32 16 13.13 Cast-in-Place Concrete Curbs and Gutters

	32 16 13.13 Cast-in-Place Concrete Curbs and Gutters	Crew	Daily Output	Labor-Hours	Unit	2010 Bare Costs Material	Labor	Equipment	Total	Total Incl O&P
0010	**CAST-IN-PLACE CONCRETE CURBS AND GUTTERS**									
0290	Forms only, no concrete									
0300	Concrete, wood forms, 6″ x 18″, straight	C-2	500	.096	L.F.	3.63	3.89		7.52	10
0400	6″ x 18″, radius	″	200	.240	″	3.70	9.70		13.40	19.05
0402	Forms and concrete complete									
0404	Concrete, wood forms, 6″ x 18″, straight & concrete	C-2A	500	.096	L.F.	6.35	3.86		10.21	12.90
0406	6″ x 18″, radius		200	.240		6.40	9.65		16.05	22
0410	Steel forms, 6″ x 18″, straight		700	.069		4	2.75		6.75	8.60
0411	6″ x 18″, radius		400	.120		6.70	4.82		11.52	14.75
0415	Machine formed, 6″ x 18″, straight	B-69A	2000	.024		6.20	.87	.41	7.48	8.60
0416	6″ x 18″, radius	″	900	.053		6.35	1.93	.92	9.20	10.95
0421	Curb and gutter, straight									
0422	with 6″ high curb and 6″ thick gutter, wood forms									
0430	24″ wide, .055 C.Y. per L.F.	C-2A	375	.128	L.F.	13.90	5.15		19.05	23
0435	30″ wide, .066 C.Y. per L.F.		340	.141		15.30	5.65		20.95	25.50
0440	Steel forms, 24″ wide, straight		700	.069		11.15	2.75		13.90	16.45
0441	Radius		300	.160		11.15	6.45		17.60	22
0442	30″ wide, straight		700	.069		13.15	2.75		15.90	18.70
0443	Radius		300	.160		13.15	6.45		19.60	24.50
0445	Machine formed, 24″ wide, straight	B-69A	2000	.024		5.45	.87	.41	6.73	7.75
0446	Radius		900	.053		5.45	1.93	.92	8.30	9.95
0447	30″ wide, straight		2000	.024		6.30	.87	.41	7.58	8.70
0448	Radius		900	.053		6.30	1.93	.92	9.15	10.90

32 16 13.26 Precast Concrete Curbs

	32 16 13.26 Precast Concrete Curbs	Crew	Daily Output	Labor-Hours	Unit	Material	Labor	Equipment	Total	Total Incl O&P
0010	**PRECAST CONCRETE CURBS**									
0550	Precast, 6″ x 18″, straight	B-29	700	.080	L.F.	10.85	2.86	1.16	14.87	17.55
0600	6″ x 18″, radius	″	325	.172	″	12.45	6.15	2.50	21.10	26

32 16 19 – Asphalt Curbs

32 16 19.10 Bituminous Concrete Curbs

	32 16 19.10 Bituminous Concrete Curbs	Crew	Daily Output	Labor-Hours	Unit	Material	Labor	Equipment	Total	Total Incl O&P
0010	**BITUMINOUS CONCRETE CURBS**									
0012	Curbs, asphaltic, machine formed, 8″ wide, 6″ high, 40 L.F./ton	B-27	1000	.032	L.F.	1.70	1.08	.29	3.07	3.84
0100	8″ wide, 8″ high, 30 L.F. per ton		900	.036		2.27	1.19	.32	3.78	4.69
0150	Asphaltic berm, 12″ W, 3″-6″ H, 35 L.F./ton, before pavement		700	.046		.04	1.54	.41	1.99	2.86
0200	12″ W, 1-1/2″ to 4″ H, 60 L.F. per ton, laid with pavement	B-2	1050	.038		.02	1.28		1.30	2

32 16 40 – Stone Curbs

32 16 40.13 Cut Stone Curbs

	32 16 40.13 Cut Stone Curbs	Crew	Daily Output	Labor-Hours	Unit	Material	Labor	Equipment	Total	Total Incl O&P
0010	**CUT STONE CURBS**									
1000	Granite, split face, straight, 5″ x 16″	D-13	500	.096	L.F.	12.05	3.82	1.31	17.18	20.50
1100	6″ x 18″	″	450	.107		15.85	4.24	1.46	21.55	25.50
1300	Radius curbing, 6″ x 18″, over 10′ radius	B-29	260	.215		19.40	7.70	3.13	30.23	36.50
1400	Corners, 2′ radius		80	.700	Ea.	65	25	10.15	100.15	121
1600	Edging, 4-1/2″ x 12″, straight		300	.187	L.F.	6.05	6.65	2.71	15.41	19.85
1800	Curb inlets, (guttermouth) straight		41	1.366	Ea.	145	49	19.80	213.80	256
2000	Indian granite (belgian block)									
2100	Jumbo, 10-1/2″ x 7-1/2″ x 4″, grey	D-1	150	.107	L.F.	2.21	4.02		6.23	8.50
2150	Pink		150	.107		2.88	4.02		6.90	9.20
2200	Regular, 9″ x 4-1/2″ x 4-1/2″, grey		160	.100		2	3.77		5.77	7.90
2250	Pink		160	.100		2.76	3.77		6.53	8.75
2300	Cubes, 4″ x 4″ x 4″, grey		175	.091		1.91	3.45		5.36	7.30
2350	Pink		175	.091		2	3.45		5.45	7.40
2400	6″ x 6″ x 6″, pink		155	.103		4.96	3.89		8.85	11.30

32 16 Curbs and Gutters

32 16 40 – Stone Curbs

32 16 40.13 Cut Stone Curbs		Crew	Daily Output	Labor-Hours	Unit	Material	2010 Bare Costs Labor	Equipment	Total	Total Incl O&P
2500	Alternate pricing method for indian granite									
2550	Jumbo, 10-1/2" x 7-1/2" x 4" (30 lb), grey				Ton	126			126	139
2600	Pink					167			167	184
2650	Regular, 9" x 4-1/2" x 4-1/2" (20 lb), grey					141			141	156
2700	Pink					193			193	212
2750	Cubes, 4" x 4" x 4" (5 lb), grey					231			231	255
2800	Pink					257			257	283
2850	6" x 6" x 6" (25 lb), pink					193			193	212
2900	For pallets, add					20			20	22

32 17 Paving Specialties

32 17 13 – Parking Bumpers

32 17 13.13 Metal Parking Bumpers		Crew	Daily Output	Labor-Hours	Unit	Material	Labor	Equipment	Total	Total Incl O&P
0010	**METAL PARKING BUMPERS**									
0015	Bumper rails for garages, 12 Ga. rail, 6" wide, with steel									
0020	posts 12'-6" O.C., minimum	E-4	190	.168	L.F.	17.35	8	.77	26.12	34
0030	Average		165	.194		21.50	9.20	.88	31.58	41
0100	Maximum		140	.229		26	10.85	1.04	37.89	48.50
0300	12" channel rail, minimum		160	.200		21.50	9.50	.91	31.91	41.50
0400	Maximum		120	.267		32.50	12.65	1.22	46.37	59.50
1300	Pipe bollards, conc filled/paint, 8' L x 4' D hole, 6" diam.	B-6	20	1.200	Ea.	400	43	16.90	459.90	525
1400	8" diam.		15	1.600		610	57.50	22.50	690	785
1500	12" diam.		12	2		790	71.50	28	889.50	1,000
1592	Bollards, steel, 3' H, retractable, incl hydraulic controls, min		4	6		39,800	215	84.50	40,099.50	44,200
1594	Max		2	12		44,400	430	169	44,999	49,600
2030	Folding with individual padlocks	B-2	50	.800		970	27		997	1,125
8000	Parking lot control, see Div. 11 12 13.10									

32 17 13.16 Plastic Parking Bumpers		Crew	Daily Output	Labor-Hours	Unit	Material	Labor	Equipment	Total	Total Incl O&P
0010	**PLASTIC PARKING BUMPERS**									
1200	Thermoplastic, 6" x 10" x 6'-0"	B-2	120	.333	Ea.	106	11.15		117.15	133

32 17 13.19 Precast Concrete Parking Bumpers		Crew	Daily Output	Labor-Hours	Unit	Material	Labor	Equipment	Total	Total Incl O&P
0010	**PRECAST CONCRETE PARKING BUMPERS**									
1000	Wheel stops, precast concrete incl. dowels, 6" x 10" x 6'-0"	B-2	120	.333	Ea.	57.50	11.15		68.65	80
1100	8" x 13" x 6'-0"	"	120	.333	"	66.50	11.15		77.65	90

32 17 13.26 Wood Parking Bumpers		Crew	Daily Output	Labor-Hours	Unit	Material	Labor	Equipment	Total	Total Incl O&P
0010	**WOOD PARKING BUMPERS**									
0020	Parking barriers, timber w/saddles, treated type									
0100	4" x 4" for cars	B-2	520	.077	L.F.	1.93	2.58		4.51	6.10
0200	6" x 6" for trucks		520	.077	"	3.96	2.58		6.54	8.35
0600	Flexible fixed stanchion, 2' high, 3" diameter		100	.400	Ea.	34	13.40		47.40	58

32 17 23 – Pavement Markings

32 17 23.13 Painted Pavement Markings		Crew	Daily Output	Labor-Hours	Unit	Material	Labor	Equipment	Total	Total Incl O&P
0010	**PAINTED PAVEMENT MARKINGS**									
0020	Acrylic waterborne, white or yellow, 4" wide	B-78	20000	.002	L.F.	.26	.08	.03	.37	.44
0200	6" wide		11000	.004		.20	.15	.05	.40	.50
0500	8" wide		10000	.005		.34	.16	.06	.56	.68
0600	12" wide		4000	.012		.62	.40	.15	1.17	1.47
0620	Arrows or gore lines		2300	.021	S.F.	.69	.69	.26	1.64	2.11
0640	Temporary paint, white or yellow		15000	.003	L.F.	.24	.11	.04	.39	.46
0660	Removal	1 Clab	300	.027			.88		.88	1.36

32 17 Paving Specialties

32 17 23 – Pavement Markings

32 17 23.13 Painted Pavement Markings		Crew	Daily Output	Labor-Hours	Unit	Material	2010 Bare Costs Labor	Equipment	Total	Total Incl O&P
0680	Temporary tape	2 Clab	1500	.011	L.F.	1.78	.35		2.13	2.50
0710	Thermoplastic, white or yellow, 4" wide	B-79	15000	.003		.89	.09	.09	1.07	1.22
0730	6" wide		14000	.003		1.29	.10	.10	1.49	1.68
0740	8" wide		12000	.003		1.74	.11	.11	1.96	2.21
0750	12" wide		6000	.007		2.58	.22	.23	3.03	3.43
0760	Arrows		660	.061	S.F.	1.86	2.02	2.09	5.97	7.45
0770	Gore lines		2500	.016		1.26	.53	.55	2.34	2.82
0780	Letters		660	.061		1.55	2.02	2.09	5.66	7.10
1000	Airport painted markings									
1100	Painting, white or yellow, taxiway markings	B-79	4000	.010	S.F.	.26	.33	.34	.93	1.18
1200	Runway markings		3500	.011		.26	.38	.39	1.03	1.31
1300	Pavement location or direction signs		2500	.016		.26	.53	.55	1.34	1.72
1350	Mobilization airport pavement painting		4	10	Ea.		335	345	680	895
1400	Paint markings or pavement signs removal daytime	B-78B	400	.045	S.F.		1.53	.90	2.43	3.35
1500	Removal nighttime		335	.054	"		1.83	1.08	2.91	4
1600	Mobilization pavement paint removal		4	4.500	Ea.		153	90.50	243.50	335

32 17 23.14 Pavement Parking Markings

		Crew	Daily Output	Labor-Hours	Unit	Material	Labor	Equipment	Total	Total Incl O&P
0010	**PAVEMENT PARKING MARKINGS**									
0790	Layout of pavement marking	A-2	25000	.001	L.F.		.03	.01	.04	.06
0800	Lines on pvmt, parking stall, paint, white, 4" wide	B-78	440	.109	Stall	6.05	3.63	1.34	11.02	13.70
1000	Street letters and numbers	"	1600	.030	S.F.	.64	1	.37	2.01	2.64

32 18 Athletic and Recreational Surfacing

32 18 13 – Synthetic Grass Surfacing

32 18 13.10 Artificial Grass Surfacing

		Crew	Daily Output	Labor-Hours	Unit	Material	Labor	Equipment	Total	Total Incl O&P
0010	**ARTIFICIAL GRASS SURFACING**									
0015	Not including asphalt base or drainage,									
0020	but including cushion pad, over 50,000 S.F.									
0200	1/2" pile and 5/16" cushion pad, standard	C-17	3200	.025	S.F.	9.45	1.08		10.53	12.05
0300	Deluxe		2560	.031		11.20	1.34		12.54	14.40
0500	1/2" pile and 5/8" cushion pad, standard		2844	.028		13.80	1.21		15.01	17
0600	Deluxe		2327	.034		15.10	1.48		16.58	18.85
0800	For asphaltic concrete base, 2-1/2" thick,									
0900	with 6" crushed stone sub-base, add	B-25	12000	.007	S.F.	1.73	.26	.21	2.20	2.55

32 18 23 – Athletic Surfacing

32 18 23.33 Running Track Surfacing

		Crew	Daily Output	Labor-Hours	Unit	Material	Labor	Equipment	Total	Total Incl O&P
0010	**RUNNING TRACK SURFACING**									
0020	Running track, asphalt, incl base, 3" thick	B-37	300	.160	S.Y.	20.50	5.55	.51	26.56	32
0100	Surface, latex rubber system, 3/8" thick, black	B-20	125	.192		9.30	7.10		16.40	21
0150	Colors		125	.192		16.60	7.10		23.70	29
0300	Urethane rubber system, 3/8" thick, black		120	.200		25	7.40		32.40	39
0400	Color coating		115	.209		30.50	7.70		38.20	45.50

32 18 23.53 Tennis Court Surfacing

		Crew	Daily Output	Labor-Hours	Unit	Material	Labor	Equipment	Total	Total Incl O&P
0010	**TENNIS COURT SURFACING**									
0020	Tennis court, asphalt, incl. base, 2-1/2" thick, one court	B-37	450	.107	S.Y.	33.50	3.71	.34	37.55	43
0200	Two courts		675	.071		15.05	2.47	.23	17.75	20.50
0300	Clay courts		360	.133		37.50	4.64	.42	42.56	48.50
0400	Pulverized natural greenstone with 4" base, fast dry		250	.192		36	6.70	.61	43.31	50.50
0800	Rubber-acrylic base resilient pavement		600	.080		51	2.78	.25	54.03	60.50
1000	Colored sealer, acrylic emulsion, 3 coats	2 Clab	800	.020		6.15	.66		6.81	7.75

32 18 Athletic and Recreational Surfacing

32 18 23 – Athletic Surfacing

32 18 23.53 Tennis Court Surfacing		Crew	Daily Output	Labor-Hours	Unit	2010 Bare Costs Material	Labor	Equipment	Total	Total Incl O&P
1100	3 coat, 2 colors	2 Clab	900	.018	S.Y.	8.50	.59		9.09	10.25
1200	For preparing old courts, add	1 Clab	825	.010			.32		.32	.50
1400	Posts for nets, 3-1/2" diameter with eye bolts	B-1	3.40	7.059	Pr.	211	238		449	600
1500	With pulley & reel		3.40	7.059	"	300	238		538	700
1700	Net, 42' long, nylon thread with binder		50	.480	Ea.	279	16.20		295.20	330
1800	All metal		6.50	3.692	"	490	125		615	730
2000	Paint markings on asphalt, 2 coats	1 Pord	1.78	4.494	Court	130	163		293	385
2200	Complete court with fence, etc., asphaltic conc., minimum	B-37	.20	240		25,000	8,350	760	34,110	41,100
2300	Maximum		.16	300		50,000	10,400	950	61,350	72,000
2800	Clay courts, minimum		.20	240		25,000	8,350	760	34,110	41,100
2900	Maximum		.16	300		40,000	10,400	950	51,350	61,000

32 31 Fences and Gates

32 31 13 – Chain Link Fences and Gates

32 31 13.20 Fence, Chain Link Industrial		Crew	Daily Output	Labor-Hours	Unit	Material	Labor	Equipment	Total	Total Incl O&P
0010	**FENCE, CHAIN LINK INDUSTRIAL**									
0011	Schedule 40, including concrete									
0020	3 strands barb wire, 2" post @ 10' O.C., set in concrete, 6' H									
0200	9 ga. wire, galv. steel, in concrete	B-80C	240	.100	L.F.	14.80	3.28	1.03	19.11	22.50
0248	Fence, add for vinyl coated fabric				S.F.	.16			.16	.18
0300	Aluminized steel	B-80C	240	.100	L.F.	18.95	3.28	1.03	23.26	27
0500	6 ga. wire, galv. steel		240	.100		23	3.28	1.03	27.31	31.50
0600	Aluminized steel		240	.100		26.50	3.28	1.03	30.81	35
0800	6 ga. wire, 6' high but omit barbed wire, galv. steel		250	.096		22.50	3.15	.99	26.64	31
0900	Aluminized steel, in concrete		250	.096		31.50	3.15	.99	35.64	41
0920	8' H, 6 ga. wire, 2-1/2" line post, galv. steel, in concrete		180	.133		36	4.38	1.38	41.76	47.50
0940	Aluminized steel, in concrete		180	.133		44	4.38	1.38	49.76	56.50
1100	Add for corner posts, 3" diam., galv. steel, in concrete		40	.600	Ea.	105	19.70	6.20	130.90	152
1200	Aluminized steel, in concrete		40	.600		125	19.70	6.20	150.90	175
1300	Add for braces, galv. steel		80	.300		28	9.85	3.10	40.95	49.50
1350	Aluminized steel		80	.300		37.50	9.85	3.10	50.45	60
1400	Gate for 6' high fence, 1-5/8" frame, 3' wide, galv. steel		10	2.400		172	79	25	276	335
1500	Aluminized steel, in concrete		10	2.400		211	79	25	315	380
2000	5'-0" high fence, 9 ga., no barbed wire, 2" line post, in concrete									
2010	10' O.C., 1-5/8" top rail, in concrete									
2100	Galvanized steel, in concrete	B-80C	300	.080	L.F.	12.60	2.63	.83	16.06	18.80
2200	Aluminized steel, in concrete		300	.080	"	15.15	2.63	.83	18.61	21.50
2400	Gate, 4' wide, 5' high, 2" frame, galv. steel, in concrete		10	2.400	Ea.	185	79	25	289	350
2500	Aluminized steel, in concrete		10	2.400	"	208	79	25	312	375
3100	Overhead slide gate, chain link, 6' high, to 18' wide, in concrete		38	.632	L.F.	162	20.50	6.50	189	217
3110	Cantilever type, in concrete	B-80	48	.667		75	23.50	15.55	114.05	136
3120	8' high, in concrete		24	1.333		108	47.50	31	186.50	225
3130	10' high, in concrete		18	1.778		127	63	41.50	231.50	282
5000	Double swing gates, incl. posts & hardware, in concrete									
5010	5' high, 12' opening, in concrete	B-80C	3.40	7.059	Opng.	480	232	73	785	965
5020	20' opening, in concrete		2.80	8.571		655	281	88.50	1,024.50	1,250
5060	6' high, 12' opening, in concrete		3.20	7.500		810	246	77.50	1,133.50	1,350
5070	20' opening, in concrete		2.60	9.231		1,125	305	95.50	1,525.50	1,800
5080	8' high, 12' opening, in concrete	B-80	2.13	15.002		1,250	530	350	2,130	2,600
5090	20' opening, in concrete		1.45	22.069		1,650	780	515	2,945	3,600
5100	10' high, 12' opening, in concrete		1.31	24.427		1,575	865	570	3,010	3,675

32 31 Fences and Gates

32 31 13 – Chain Link Fences and Gates

32 31 13.20 Fence, Chain Link Industrial		Crew	Daily Output	Labor-Hours	Unit	2010 Bare Costs Material	Labor	Equipment	Total	Total Incl O&P
5110	20' opening, in concrete	B-80	1.03	31.068	Opng.	2,375	1,100	725	4,200	5,075
5120	12' high, 12' opening, in concrete		1.05	30.476		2,300	1,075	710	4,085	4,950
5130	20' opening, in concrete		.85	37.647		2,950	1,325	875	5,150	6,250
5190	For aluminized steel add					20%				
7075	Fence, for small jobs 100 LF or less fence w/or wo gate, add				S.F.	20%			20%	

32 31 13.25 Fence, Chain Link Residential		Crew	Daily Output	Labor-Hours	Unit	Material	Labor	Equipment	Total	Total Incl O&P
0010	**FENCE, CHAIN LINK RESIDENTIAL**									
0011	Schedule 20, 11 gauge wire, 1-5/8" post									
0020	10' O.C., 1-3/8" top rail, 2" corner post, galv. stl. 3' high	B-80C	500	.048	L.F.	4.29	1.58	.50	6.37	7.70
0050	4' high		400	.060		4.95	1.97	.62	7.54	9.15
0100	6' high		200	.120		6	3.94	1.24	11.18	14
0150	Add for gate 3' wide, 1-3/8" frame, 3' high		12	2	Ea.	61.50	65.50	20.50	147.50	191
0170	4' high		10	2.400		69	79	25	173	224
0190	6' high		10	2.400		96.50	79	25	200.50	254
0200	Add for gate 4' wide, 1-3/8" frame, 3' high		9	2.667		73	87.50	27.50	188	245
0220	4' high		9	2.667		77.50	87.50	27.50	192.50	250
0240	6' high		8	3		92	98.50	31	221.50	286
0350	Aluminized steel, 11 ga. wire, 3' high		500	.048	L.F.	5.90	1.58	.50	7.98	9.40
0380	4' high		400	.060		7.50	1.97	.62	10.09	11.95
0400	6' high		200	.120		10.60	3.94	1.24	15.78	19.05
0450	Add for gate 3' wide, 1-3/8" frame, 3' high		12	2	Ea.	80.50	65.50	20.50	166.50	213
0470	4' high		10	2.400		132	79	25	236	293
0490	6' high		10	2.400		166	79	25	270	330
0500	Add for gate 4' wide, 1-3/8" frame, 3' high		10	2.400		91.50	79	25	195.50	249
0520	4' high		9	2.667		101	87.50	27.50	216	277
0540	6' high		8	3		118	98.50	31	247.50	315
0620	Vinyl covered, 9 ga. wire, 3' high		500	.048	L.F.	4.66	1.58	.50	6.74	8.10
0640	4' high		400	.060		5.40	1.97	.62	7.99	9.65
0660	6' high		200	.120		6.90	3.94	1.24	12.08	15
0720	Add for gate 3' wide, 1-3/8" frame, 3' high		12	2	Ea.	90	65.50	20.50	176	223
0740	4' high		10	2.400		102	79	25	206	260
0760	6' high		10	2.400		159	79	25	263	325
0780	Add for gate 4' wide, 1-3/8" frame, 3' high		10	2.400		104	79	25	208	262
0800	4' high		9	2.667		118	87.50	27.50	233	295
0820	6' high		8	3		128	98.50	31	257.50	325
7076	Fence, for small jobs 100 LF fence or less w/ or wo gate, add				S.F.	20%			20%	

32 31 13.26 Tennis Court Fences and Gates		Crew	Daily Output	Labor-Hours	Unit	Material	Labor	Equipment	Total	Total Incl O&P
0010	**TENNIS COURT FENCES AND GATES**									
0860	Tennis courts, 11 ga. wire, 2-1/2" post set									
0870	in concrete, 10' O.C., 1-5/8" top rail									
0900	10' high	B-80	190	.168	L.F.	13.85	5.95	3.92	23.72	28.50
0920	12' high		170	.188	"	16.35	6.65	4.38	27.38	33
1000	Add for gate 4' wide, 1-5/8" frame 7' high		10	3.200	Ea.	132	113	74.50	319.50	400
1040	Aluminized steel, 11 ga. wire 10' high		190	.168	L.F.	17.20	5.95	3.92	27.07	32.50
1100	12' high		170	.188	"	21.50	6.65	4.38	32.53	38.50
1140	Add for gate 4' wide, 1-5/8" frame, 7' high		10	3.200	Ea.	155	113	74.50	342.50	425
1250	Vinyl covered, 9 ga. wire, 10' high		190	.168	L.F.	16.30	5.95	3.92	26.17	31.50
1300	12' high		170	.188	"	27	6.65	4.38	38.03	45
1310	Fence, CL, tennis ct, transom gate, single, galv, 4' x 7' x 3'	B-80A	8.72	2.752	Ea.	294	91	34	419	505
1400	Add for gate 4' wide, 1-5/8" frame, 7' high	B-80	10	3.200	"	155	113	74.50	342.50	425

32 31 Fences and Gates

32 31 13 – Chain Link Fences and Gates

32 31 13.33 Chain Link Backstops		Crew	Daily Output	Labor-Hours	Unit	Material	2010 Bare Costs Labor	Equipment	Total	Total Incl O&P
0010	**CHAIN LINK BACKSTOPS**									
0015	Backstops, baseball, prefabricated, 30′ wide, 12′ high & 1 overhang	B-1	1	24	Ea.	3,150	810		3,960	4,725
0100	40′ wide, 12′ high & 2 overhangs	"	.75	32		5,450	1,075		6,525	7,675
0300	Basketball, steel, single goal	B-13	3.04	18.421		1,250	660	246	2,156	2,650
0400	Double goal	"	1.92	29.167		820	1,050	390	2,260	2,925
0600	Tennis, wire mesh with pair of ends	B-1	2.48	9.677	Set	1,925	325		2,250	2,600
0700	Enclosed court	"	1.30	18.462	Ea.	5,450	625		6,075	6,950

32 31 13.53 High-Security Chain Link Fences, Gates and Systems

		Crew	Daily Output	Labor-Hours	Unit	Material	Labor	Equipment	Total	Total Incl O&P
0010	**HIGH-SECURITY CHAIN LINK FENCES, GATES AND SYSTEMS**									
0100	Fence, chain link, security, 7′ H, standard FE-7, incl excavation & posts	B-80C	480	.050	L.F.	41	1.64	.52	43.16	48
0200	Fence, barbed wire, security, 7′ high, with 3 wire barbed wire arm	"	400	.060	"	8.75	1.97	.62	11.34	13.30
0300	Complete systems, including material and installation									
0310	Taunt wire fence detection system				M.L.F.				22,300	24,500
0410	Microwave fence detection system								36,800	40,500
0510	Passive magnetic fence detection system								17,400	19,100
0610	Infrared fence detection system								11,600	12,800
0710	Strain relief fence detection system								22,300	24,500
0810	Electro-shock fence detection system								32,000	35,200
0910	Photo-electric fence detection system								14,500	15,900

32 31 19 – Decorative Metal Fences and Gates

32 31 19.10 Decorative Fence

		Crew	Daily Output	Labor-Hours	Unit	Material	Labor	Equipment	Total	Total Incl O&P
0010	**DECORATIVE FENCE**									
5300	Tubular picket, steel, 6′ sections, 1-9/16″ posts, 4′ high	B-80C	300	.080	L.F.	30	2.63	.83	33.46	38.50
5400	2″ posts, 5′ high		240	.100		42	3.28	1.03	46.31	52
5600	2″ posts, 6′ high		200	.120		47.50	3.94	1.24	52.68	59.50
5700	Staggered picket 1-9/16″ posts, 4′ high		300	.080		27.50	2.63	.83	30.96	35
5800	2″ posts, 5′ high		240	.100		45	3.28	1.03	49.31	55.50
5900	2″ posts, 6′ high		200	.120		46.50	3.94	1.24	51.68	58.50
6200	Gates, 4′ high, 3′ wide	B-1	10	2.400	Ea.	263	81		344	415
6300	5′ high, 3′ wide		10	2.400		340	81		421	500
6400	6′ high, 3′ wide		10	2.400		350	81		431	510
6500	4′ wide		10	2.400		410	81		491	575

32 31 26 – Wire Fences and Gates

32 31 26.10 Fences, Misc. Metal

		Crew	Daily Output	Labor-Hours	Unit	Material	Labor	Equipment	Total	Total Incl O&P
0010	**FENCES, MISC. METAL**									
0012	Chicken wire, posts @ 4′, 1″ mesh, 4′ high	B-80C	410	.059	L.F.	2.10	1.92	.60	4.62	5.90
0100	2″ mesh, 6′ high		350	.069		1.89	2.25	.71	4.85	6.30
0200	Galv. steel, 12 ga., 2″ x 4″ mesh, posts 5′ O.C., 3′ high		300	.080		3.03	2.63	.83	6.49	8.25
0300	5′ high		300	.080		4.04	2.63	.83	7.50	9.40
0400	14 ga., 1″ x 2″ mesh, 3′ high		300	.080		3.21	2.63	.83	6.67	8.45
0500	5′ high		300	.080		4.44	2.63	.83	7.90	9.80
1000	Kennel fencing, 1-1/2″ mesh, 6′ long, 3′-6″ wide, 6′-2″ high	2 Clab	4	4	Ea.	505	132		637	760
1050	12′ long		4	4		605	132		737	870
1200	Top covers, 1-1/2″ mesh, 6′ long		15	1.067		102	35.50		137.50	168
1250	12′ long		12	1.333		164	44		208	248
1300	For kennel doors, see Div. 08 31 13.40									
4500	Security fence, prison grade, set in concrete, 12′ high	B-80	25	1.280	L.F.	44.50	45.50	30	120	151
4600	16′ high	"	20	1.600	"	53.50	56.50	37.50	147.50	186

32 31 26 – Wire Fences and Gates

	32 31 26.20 Wire Fencing, General	Crew	Daily Output	Labor-Hours	Unit	Material	2010 Bare Costs Labor	Equipment	Total	Total Incl O&P
0010	**WIRE FENCING, GENERAL**									
0015	Barbed wire, galvanized, domestic steel, hi-tensile 15-1/2 ga.				M.L.F.	35			35	38.50
0020	Standard, 12-3/4 ga.					46.50			46.50	51.50
0210	Barbless wire, 2-strand galvanized, 12-1/2 ga.					46.50			46.50	51.50
0500	Helical razor ribbon, stainless steel, 18" dia x 18" spacing				C.L.F.	136			136	149
0600	Hardware cloth galv., 1/4" mesh, 23 ga., 2' wide				C.S.F.	61			61	67
0700	3' wide					59			59	65
0900	1/2" mesh, 19 ga., 2' wide					53.50			53.50	59
1000	4' wide					52.50			52.50	58
1200	Chain link fabric, steel, 2" mesh, 6 ga, galvanized					201			201	221
1300	9 ga, galvanized					99			99	109
1350	Vinyl coated					83			83	91.50
1360	Aluminized					129			129	142
1400	2-1/4" mesh, 11.5 ga, galvanized					67			67	73.50
1600	1-3/4" mesh (tennis courts), 11.5 ga (core), vinyl coated					95			95	104
1700	9 ga, galvanized					85.50			85.50	94
2100	Welded wire fabric, galvanized, 1" x 2", 14 ga.					57			57	62.50
2200	2" x 4", 12-1/2 ga.					37.50			37.50	41.50

32 31 29 – Wood Fences and Gates

	32 31 29.20 Fence, Wood Rail	Crew	Daily Output	Labor-Hours	Unit	Material	Labor	Equipment	Total	Total Incl O&P
0010	**FENCE, WOOD RAIL**									
0012	Picket, No. 2 cedar, Gothic, 2 rail, 3' high	B-1	160	.150	L.F.	6.05	5.05		11.10	14.45
0050	Gate, 3'-6" wide	B-80C	9	2.667	Ea.	52	87.50	27.50	167	222
0400	3 rail, 4' high		150	.160	L.F.	6.95	5.25	1.65	13.85	17.45
0500	Gate, 3'-6" wide		9	2.667	Ea.	62.50	87.50	27.50	177.50	233
0600	Open rail, rustic, No. 1 cedar, 2 rail, 3' high		160	.150	L.F.	5.40	4.92	1.55	11.87	15.20
0650	Gate, 3' wide		9	2.667	Ea.	60	87.50	27.50	175	231
0700	3 rail, 4' high		150	.160	L.F.	6.15	5.25	1.65	13.05	16.65
0900	Gate, 3' wide		9	2.667	Ea.	77.50	87.50	27.50	192.50	250
1200	Stockade, No. 2 cedar, treated wood rails, 6' high		160	.150	L.F.	7.50	4.92	1.55	13.97	17.50
1250	Gate, 3' wide		9	2.667	Ea.	61.50	87.50	27.50	176.50	233
1300	No. 1 cedar, 3-1/4" cedar rails, 6' high		160	.150	L.F.	18.60	4.92	1.55	25.07	30
1500	Gate, 3' wide		9	2.667	Ea.	156	87.50	27.50	271	335
1520	Open rail, split, No. 1 cedar, 2 rail, 3' high		160	.150	L.F.	5.40	4.92	1.55	11.87	15.20
1540	3 rail, 4'-0" high		150	.160		6.55	5.25	1.65	13.45	17.05
3300	Board, shadow box, 1" x 6", treated pine, 6' high		160	.150		10.20	4.92	1.55	16.67	20.50
3400	No. 1 cedar, 6' high		150	.160		20	5.25	1.65	26.90	32
3900	Basket weave, No. 1 cedar, 6' high		160	.150		19.95	4.92	1.55	26.42	31.50
3950	Gate, 3'-6" wide	B-1	8	3	Ea.	136	101		237	305
4000	Treated pine, 6' high		150	.160	L.F.	13.05	5.40		18.45	23
4200	Gate, 3'-6" wide		9	2.667	Ea.	63.50	90		153.50	209
5000	Fence rail, redwood, 2" x 4", merch grade 8'		2400	.010	L.F.	1.17	.34		1.51	1.81
5050	Select grade, 8'		2400	.010	"	3.52	.34		3.86	4.39
6000	Fence post, select redwood, earthpacked & treated, 4" x 4" x 6'		96	.250	Ea.	11.25	8.45		19.70	25.50
6010	4" x 4" x 8'		96	.250		13.35	8.45		21.80	27.50
6020	Set in concrete, 4" x 4" x 6'		50	.480		14.50	16.20		30.70	41
6030	4" x 4" x 8'		50	.480		17.30	16.20		33.50	44
6040	Wood post, 4' high, set in concrete, incl. concrete		50	.480		8.05	16.20		24.25	34
6050	Earth packed		96	.250		5.55	8.45		14	19.10
6060	6' high, set in concrete, incl. concrete		50	.480		10.35	16.20		26.55	36.50
6070	Earth packed		96	.250		7.50	8.45		15.95	21.50

32 32 Retaining Walls

32 32 13 – Cast-in-Place Concrete Retaining Walls

32 32 13.10 Retaining Walls, Cast Concrete		Crew	Daily Output	Labor-Hours	Unit	2010 Bare Costs Material	Labor	Equipment	Total	Total Incl O&P
0010	RETAINING WALLS, CAST CONCRETE									
1800	Concrete gravity wall with vertical face including excavation & backfill									
1850	No reinforcing									
1900	6' high, level embankment	C-17C	36	2.306	L.F.	70.50	99.50	16.10	186.10	247
2000	33° slope embankment		32	2.594		64.50	112	18.10	194.60	263
2200	8' high, no surcharge		27	3.074		88	132	21.50	241.50	325
2300	33° slope embankment		24	3.458		107	149	24	280	375
2500	10' high, level embankment		19	4.368		126	188	30.50	344.50	460
2600	33° slope embankment		18	4.611		175	199	32	406	535
2800	Reinforced concrete cantilever, incl. excavation, backfill & reinf.									
2900	6' high, 33° slope embankment	C-17C	35	2.371	L.F.	64.50	102	16.55	183.05	246
3000	8' high, 33° slope embankment		29	2.862		74.50	123	20	217.50	293
3100	10' high, 33° slope embankment		20	4.150		97	179	29	305	415
3200	20' high, 500 lb. per L.F. surcharge		7.50	11.067		291	475	77.50	843.50	1,125
3500	Concrete cribbing, incl. excavation and backfill									
3700	12' high, open face	B-13	210	.267	S.F.	32.50	9.55	3.56	45.61	54
3900	Closed face	"	210	.267	"	30.50	9.55	3.56	43.61	52
4100	Concrete filled slurry trench, see Div. 31 56 23.20									

32 32 23 – Segmental Retaining Walls

32 32 23.13 Segmental Conc. Unit Masonry Retaining Walls		Crew	Daily Output	Labor-Hours	Unit	Material	Labor	Equipment	Total	Total Incl O&P
0010	SEGMENTAL CONC. UNIT MASONRY RETAINING WALLS									
7100	Segmental Retaining Wall system, incl pins, and void fill									
7120	base not included									
7140	Large unit, 8" high x 18" wide x 20" deep, 3 plane split	B-62	300	.080	S.F.	10.10	2.87	.54	13.51	16.05
7150	Straight split		300	.080		10.10	2.87	.54	13.51	16.05
7160	Medium, ltwt, 8" high x 18" wide x 12" deep, 3 plane split		400	.060		11.30	2.15	.40	13.85	16.10
7170	Straight split		400	.060		8.50	2.15	.40	11.05	13.05
7180	Small unit, 4" x 18" x 10" deep, 3 plane split		400	.060		12	2.15	.40	14.55	16.90
7190	Straight split		400	.060		12.70	2.15	.40	15.25	17.65
7200	Cap unit, 3 plane split		300	.080		14.20	2.87	.54	17.61	20.50
7210	Cap unit, straight split		300	.080		14.20	2.87	.54	17.61	20.50
7260	For reinforcing, add								4.25	4.68
8000	For higher walls, add components as necessary									

32 32 26 – Metal Crib Retaining Walls

32 32 26.10 Metal Bin Retaining Walls		Crew	Daily Output	Labor-Hours	Unit	Material	Labor	Equipment	Total	Total Incl O&P
0010	METAL BIN RETAINING WALLS									
0011	Aluminized steel bin, excavation									
0020	and backfill not included, 10' wide									
0100	4' high, 5.5' deep	B-13	650	.086	S.F.	23	3.08	1.15	27.23	31.50
0200	8' high, 5.5' deep		615	.091		26.50	3.25	1.22	30.97	35.50
0300	10' high, 7.7' deep		580	.097		28	3.45	1.29	32.74	37
0400	12' high, 7.7' deep		530	.106		30	3.78	1.41	35.19	40.50
0500	16' high, 7.7' deep		515	.109		31.50	3.89	1.45	36.84	42.50
0600	16' high, 9.9' deep		500	.112		33.50	4	1.50	39	44.50
0700	20' high, 9.9' deep		470	.119		37.50	4.26	1.59	43.35	50
0800	20' high, 12.1' deep		460	.122		40	4.35	1.63	45.98	52.50
0900	24' high, 12.1' deep		455	.123		43	4.40	1.64	49.04	55.50
1000	24' high, 14.3' deep		450	.124		44.50	4.45	1.66	50.61	57.50
1100	28' high, 14.3' deep		440	.127		46	4.55	1.70	52.25	60
1300	For plain galvanized bin type walls, deduct					10%				

32 32 Retaining Walls

32 32 36 – Gabion Retaining Walls

32 32 36.10 Stone Gabion Retaining Walls		Crew	Daily Output	Labor-Hours	Unit	Material	2010 Bare Costs Labor	Equipment	Total	Total Incl O&P
0010	**STONE GABION RETAINING WALLS**									
4300	Stone filled gabions, not incl. excavation,									
4310	Stone, delivered, 3′ wide									
4350	Galvanized, 6′ high, 33° slope embankment	B-13	49	1.143	L.F.	22	41	15.25	78.25	103
4500	Highway surcharge		27	2.074		46.50	74	27.50	148	195
4600	9′ high, up to 33° slope embankment		24	2.333		49.50	83.50	31	164	216
4700	Highway surcharge		16	3.500		81.50	125	46.50	253	335
4900	12′ high, up to 33° slope embankment		14	4		77	143	53.50	273.50	360
5000	Highway surcharge	↓	11	5.091		117	182	68	367	480
5950	For PVC coating, add				↓	20%				

32 32 60 – Stone Retaining Walls

32 32 60.10 Retaining Walls, Stone		Crew	Daily Output	Labor-Hours	Unit	Material	Labor	Equipment	Total	Total Incl O&P
0010	**RETAINING WALLS, STONE**									
0015	Including excavation, concrete footing and									
0020	stone 3′ below grade. Price is exposed face area.									
0200	Decorative random stone, to 6′ high, 1′-6″ thick, dry set	D-1	35	.457	S.F.	46	17.25		63.25	77
0300	Mortar set		40	.400		46	15.10		61.10	73.50
0500	Cut stone, to 6′ high, 1′-6″ thick, dry set		35	.457		46	17.25		63.25	77
0600	Mortar set		40	.400		46	15.10		61.10	73.50
0800	Random stone, 6′ to 10′ high, 2′ thick, dry set		45	.356		46	13.40		59.40	71
0900	Mortar set		50	.320		46	12.05		58.05	69
1100	Cut stone, 6′ to 10′ high, 2′ thick, dry set		45	.356		46	13.40		59.40	71
1200	Mortar set	↓	50	.320	↓	46	12.05		58.05	69

32 34 Fabricated Bridges

32 34 20 – Fabricated Pedestrian Bridges

32 34 20.10 Bridges, Pedestrian		Crew	Daily Output	Labor-Hours	Unit	Material	Labor	Equipment	Total	Total Incl O&P
0010	**BRIDGES, PEDESTRIAN**									
0011	Spans over streams, roadways, etc.									
0020	including erection, not including foundations									
0050	Precast concrete, complete in place, 8′ wide, 60′ span	E-2	215	.260	S.F.	67	11.90	7.50	86.40	102
0100	100′ span		185	.303		73	13.80	8.70	95.50	114
0150	120′ span		160	.350		79.50	15.95	10.05	105.50	126
0200	150′ span		145	.386		83	17.60	11.10	111.70	133
0300	Steel, trussed or arch spans, compl. in place, 8′ wide, 40′ span		320	.175		87	8	5	100	115
0400	50′ span		395	.142		78	6.45	4.07	88.52	101
0500	60′ span		465	.120		78	5.50	3.46	86.96	99
0600	80′ span		570	.098		93	4.48	2.82	100.30	113
0700	100′ span		465	.120		130	5.50	3.46	138.96	157
0800	120′ span		365	.153		165	7	4.40	176.40	198
0900	150′ span		310	.181		176	8.25	5.20	189.45	213
1000	160′ span		255	.220		176	10	6.30	192.30	217
1100	10′ wide, 80′ span		640	.088		96.50	3.99	2.51	103	116
1200	120′ span		415	.135		125	6.15	3.87	135.02	153
1300	150′ span		445	.126		141	5.75	3.61	150.36	169
1400	200′ span	↓	205	.273		150	12.45	7.85	170.30	195
1600	Wood, laminated type, complete in place, 80′ span	C-12	203	.236		57.50	9.70	3.23	70.43	81.50
1700	130′ span	″	153	.314	↓	59.50	12.85	4.29	76.64	90

32 35 Screening Devices

32 35 16 – Sound Barriers

32 35 16.10 Traffic Barriers, Highway Sound Barriers		Crew	Daily Output	Labor-Hours	Unit	2010 Bare Costs Material	Labor	Equipment	Total	Total Incl O&P
0010	**TRAFFIC BARRIERS, HIGHWAY SOUND BARRIERS**									
0020	Highway sound barriers, not including footing									
0100	Precast concrete, concrete columns @ 30′ OC, 8″ T, 8′ H	C-12	400	.120	L.F.	125	4.91	1.64	131.55	147
0110	12′ H		265	.181		188	7.40	2.47	197.87	220
0120	16′ H		200	.240		250	9.85	3.28	263.13	294
0130	20′ H	↓	160	.300		315	12.30	4.10	331.40	370
0400	Lt. Wt. composite panel, cementitious face, St. posts @ 12′ OC, 8′ H	B-80B	190	.168		140	5.90	1.28	147.18	164
0410	12′ H		125	.256		209	9	1.94	219.94	246
0420	16′ H		95	.337		279	11.85	2.55	293.40	325
0430	20′ H	↓	75	.427	↓	350	15	3.23	368.23	410

32 84 Planting Irrigation

32 84 23 – Underground Sprinklers

32 84 23.10 Sprinkler Irrigation System

		Crew	Daily Output	Labor-Hours	Unit	Material	Labor	Equipment	Total	Total Incl O&P
0010	**SPRINKLER IRRIGATION SYSTEM**									
0011	For lawns									
0100	Golf course with fully automatic system	C-17	.05	1600	9 holes	99,000	69,000		168,000	215,000
0200	24′ diam. head at 15′ O.C incl. piping, auto oper., minimum	B-20	70	.343	Head	20	12.65		32.65	41.50
0300	Maximum		40	.600		50	22		72	89
0500	60′ diam. head at 40′ O.C. incl. piping, auto oper., minimum		28	.857		65	31.50		96.50	121
0600	Maximum		23	1.043	↓	180	38.50		218.50	258
0800	Residential system, custom, 1″ supply		2000	.012	S.F.	.30	.44		.74	1.01
0900	1-1/2″ supply	↓	1800	.013	″	.35	.49		.84	1.15
1020	Pop up spray head w/risers, hi-pop, full circle pattern, 4″	2 Skwk	76	.211	Ea.	4.28	8.95		13.23	18.50
1030	1/2 circle pattern, 4″		76	.211		4.54	8.95		13.49	18.80
1040	6″, full circle pattern		76	.211		10.25	8.95		19.20	25
1050	1/2 circle pattern, 6″		76	.211		10.40	8.95		19.35	25
1060	12″, full circle pattern		76	.211		15.20	8.95		24.15	30.50
1070	1/2 circle pattern, 12″		76	.211		14.95	8.95		23.90	30.50
1080	Pop up bubbler head w/risers, hi-pop bubbler head, 4″		76	.211		4.64	8.95		13.59	18.90
1090	6″		76	.211		11	8.95		19.95	26
1100	12″		76	.211		15.20	8.95		24.15	30.50
1110	Impact full/part circle sprinklers, 28′-54′ 25-60 PSI		37	.432		14.80	18.40		33.20	45
1120	Spaced 37′-49′ @ 25-50 PSI		37	.432		31.50	18.40		49.90	63
1130	Spaced 43′-61′ @ 30-60 PSI		37	.432		62.50	18.40		80.90	97
1140	Spaced 54′-78′ @ 40-80 PSI	↓	37	.432	↓	129	18.40		147.40	171
1145	Impact rotor pop-up full/part commercial circle sprinklers									
1150	Spaced 42′-65′ 35-80 PSI	2 Skwk	25	.640	Ea.	24.50	27.50		52	69
1160	Spaced 48′-76′ 45-85 PSI	″	25	.640	″	12.15	27.50		39.65	55.50
1165	Impact rotor pop-up part. circle comm., 53′-75′, 55-100 PSI, w/accessories									
1170	Plastic case, metal cover	2 Skwk	25	.640	Ea.	174	27.50		201.50	233
1180	Rubber cover		25	.640		132	27.50		159.50	187
1190	Iron case, metal cover		22	.727		163	31		194	227
1200	Rubber cover		22	.727		182	31		213	248
1250	Plastic case, 2 nozzle, metal cover		25	.640		153	27.50		180.50	211
1260	Rubber cover		25	.640		163	27.50		190.50	221
1270	Iron case, 2 nozzle, metal cover		22	.727		220	31		251	289
1280	Rubber cover	↓	22	.727	↓	221	31		252	291
1282	Impact rotor pop-up full circle commercial, 39′-99′, 30-100 PSI									
1284	Plastic case, metal cover	2 Skwk	25	.640	Ea.	154	27.50		181.50	211
1286	Rubber cover	↓	25	.640	↓	171	27.50		198.50	230

32 84 Planting Irrigation

32 84 23 – Underground Sprinklers

32 84 23.10 Sprinkler Irrigation System		Crew	Daily Output	Labor-Hours	Unit	2010 Bare Costs Material	Labor	Equipment	Total	Total Incl O&P
1288	Iron case, metal cover	2 Skwk	22	.727	Ea.	223	31		254	293
1290	Rubber cover		22	.727		233	31		264	305
1292	Plastic case, 2 nozzle, metal cover		22	.727		166	31		197	230
1294	Rubber cover		22	.727		172	31		203	237
1296	Iron case, 2 nozzle, metal cover		20	.800		213	34		247	287
1298	Rubber cover		20	.800		228	34		262	305
1305	Electric remote control valve, plastic, 3/4"		18	.889		18.65	38		56.65	78.50
1310	1"		18	.889		21.50	38		59.50	81.50
1320	1-1/2"		18	.889		112	38		150	181
1330	2"	↓	18	.889	↓	130	38		168	201
1335	Quick coupling valves, brass, locking cover									
1340	Inlet coupling valve, 3/4"	2 Skwk	18.75	.853	Ea.	50	36.50		86.50	111
1350	1"		18.75	.853		57	36.50		93.50	119
1360	Controller valve boxes, 6" round boxes		18.75	.853		5.10	36.50		41.60	61.50
1370	10" round boxes		14.25	1.123		12.55	48		60.55	87.50
1380	12" square box	↓	9.75	1.641	↓	27.50	70		97.50	138
1388	Electromech. control, 14 day 3-60 min, auto start to 23/day									
1390	4 station	2 Skwk	1.04	15.385	Ea.	204	655		859	1,225
1400	7 station		.64	25		320	1,075		1,395	2,000
1410	12 station		.40	40		375	1,700		2,075	3,050
1420	Dual programs, 18 station		.24	66.667		1,225	2,850		4,075	5,725
1430	23 station	↓	.16	100	↓	1,975	4,250		6,225	8,725
1435	Backflow preventer, bronze, 0-175 PSI, w/valves, test cocks									
1440	3/4"	2 Skwk	2	8	Ea.	119	340		459	655
1450	1"		2	8		137	340		477	675
1460	1-1/2"		2	8		274	340		614	825
1470	2"	↓	2	8	↓	310	340		650	865
1475	Pressure vacuum breaker, brass, 15-150 PSI									
1480	3/4"	2 Skwk	2	8	Ea.	74	340		414	605
1490	1"		2	8		106	340		446	640
1500	1-1/2"		2	8		201	340		541	745
1510	2"	↓	2	8	↓	258	340		598	810

32 91 Planting Preparation

32 91 13 – Soil Preparation

32 91 13.16 Mulching		Crew	Daily Output	Labor-Hours	Unit	Material	Labor	Equipment	Total	Total Incl O&P
0010	**MULCHING**									
0100	Aged barks, 3" deep, hand spread	1 Clab	100	.080	S.Y.	2.93	2.65		5.58	7.30
0150	Skid steer loader	B-63	13.50	2.963	M.S.F.	325	103	11.95	439.95	530
0200	Hay, 1" deep, hand spread	1 Clab	475	.017	S.Y.	.52	.56		1.08	1.43
0250	Power mulcher, small	B-64	180	.089	M.S.F.	58	2.90	2.17	63.07	70.50
0350	Large	B-65	530	.030	"	58	.99	1.05	60.04	66
0400	Humus peat, 1" deep, hand spread	1 Clab	700	.011	S.Y.	4.55	.38		4.93	5.60
0450	Push spreader	"	2500	.003	"	4.55	.11		4.66	5.15
0550	Tractor spreader	B-66	700	.011	M.S.F.	505	.47	.31	505.78	555
0600	Oat straw, 1" deep, hand spread	1 Clab	475	.017	S.Y.	.49	.56		1.05	1.40
0650	Power mulcher, small	B-64	180	.089	M.S.F.	54.50	2.90	2.17	59.57	67
0700	Large	B-65	530	.030	"	54.50	.99	1.05	56.54	62.50
0750	Add for asphaltic emulsion	B-45	1770	.009	Gal.	2.28	.34	.51	3.13	3.59
0800	Peat moss, 1" deep, hand spread	1 Clab	900	.009	S.Y.	1.90	.29		2.19	2.54
0850	Push spreader	"	2500	.003	"	1.90	.11		2.01	2.25

32 91 Planting Preparation

32 91 13 – Soil Preparation

32 91 13.16 Mulching

	32 91 13.16 Mulching	Crew	Daily Output	Labor-Hours	Unit	Material	2010 Bare Costs Labor	Equipment	Total	Total Incl O&P
0950	Tractor spreader	B-66	700	.011	M.S.F.	211	.47	.31	211.78	233
1000	Polyethylene film, 6 mil.	2 Clab	2000	.008	S.Y.	.22	.26		.48	.65
1100	Redwood nuggets, 3" deep, hand spread	1 Clab	150	.053	"	4.70	1.77		6.47	7.85
1150	Skid steer loader	B-63	13.50	2.963	M.S.F.	520	103	11.95	634.95	745
1200	Stone mulch, hand spread, ceramic chips, economy	1 Clab	125	.064	S.Y.	6.95	2.12		9.07	10.90
1250	Deluxe	"	95	.084	"	10.60	2.79		13.39	16
1300	Granite chips	B-1	10	2.400	C.Y.	34.50	81		115.50	163
1400	Marble chips		10	2.400		129	81		210	267
1600	Pea gravel		28	.857		68.50	29		97.50	120
1700	Quartz		10	2.400		166	81		247	305
1800	Tar paper, 15 Lb. felt	1 Clab	800	.010	S.Y.	.53	.33		.86	1.09
1900	Wood chips, 2" deep, hand spread	"	220	.036	"	1.93	1.20		3.13	3.98
1950	Skid steer loader	B-63	20.30	1.970	M.S.F.	214	68.50	7.95	290.45	350

32 91 13.26 Planting Beds

	32 91 13.26 Planting Beds	Crew	Daily Output	Labor-Hours	Unit	Material	Labor	Equipment	Total	Total Incl O&P
0010	**PLANTING BEDS**									
0100	Backfill planting pit, by hand, on site topsoil	2 Clab	18	.889	C.Y.		29.50		29.50	45.50
0200	Prepared planting mix, by hand	"	24	.667			22		22	34
0300	Skid steer loader, on site topsoil	B-62	340	.071			2.53	.48	3.01	4.38
0400	Prepared planting mix	"	410	.059			2.10	.39	2.49	3.63
1000	Excavate planting pit, by hand, sandy soil	2 Clab	16	1			33		33	51
1100	Heavy soil or clay	"	8	2			66		66	102
1200	1/2 C.Y. backhoe, sandy soil	B-11C	150	.107			4.06	2.25	6.31	8.65
1300	Heavy soil or clay	"	115	.139			5.30	2.94	8.24	11.30
2000	Mix planting soil, incl. loam, manure, peat, by hand	2 Clab	60	.267		40.50	8.85		49.35	58
2100	Skid steer loader	B-62	150	.160		40.50	5.75	1.08	47.33	54.50
3000	Pile sod, skid steer loader	"	2800	.009	S.Y.		.31	.06	.37	.53
3100	By hand	2 Clab	400	.040			1.32		1.32	2.04
4000	Remove sod, F.E. loader	B-10S	2000	.006			.24	.18	.42	.56
4100	Sod cutter	B-12K	3200	.005			.19	.30	.49	.62
4200	By hand	2 Clab	240	.067			2.21		2.21	3.40

32 91 19 – Landscape Grading

32 91 19.13 Topsoil Placement and Grading

	32 91 19.13 Topsoil Placement and Grading	Crew	Daily Output	Labor-Hours	Unit	Material	Labor	Equipment	Total	Total Incl O&P
0010	**TOPSOIL PLACEMENT AND GRADING**									
0400	Spread from pile to rough finish grade, F.E. loader, 1.5 CY	B-10S	200	.060	C.Y.		2.38	1.84	4.22	5.60
0500	Up to 200' radius, by hand	1 Clab	14	.571			18.90		18.90	29
0600	Top dress by hand, 1 C.Y. for 600 S.F.	"	11.50	.696		30	23		53	68.50
0700	Furnish and place, truck dumped, screened, 4" deep	B-10S	1300	.009	S.Y.	3.75	.37	.28	4.40	4.99
0800	6" deep	"	820	.015	"	4.80	.58	.45	5.83	6.65

32 92 Turf and Grasses

32 92 19 – Seeding

32 92 19.13 Mechanical Seeding

	32 92 19.13 Mechanical Seeding	Crew	Daily Output	Labor-Hours	Unit	Material	Labor	Equipment	Total	Total Incl O&P
0010	**MECHANICAL SEEDING** R329219-50									
0020	Mechanical seeding, 215 lb./acre	B-66	1.50	5.333	Acre	565	220	146	931	1,125
0100	44 lb./M.S.Y.	"	2500	.003	S.Y.	.18	.13	.09	.40	.50
0300	Fine grading and seeding incl. lime, fertilizer & seed,									
0310	with equipment	B-14	1000	.048	S.Y.	.20	1.67	.34	2.21	3.15
0600	Limestone hand push spreader, 50 lbs. per M.S.F.	1 Clab	180	.044	M.S.F.	4.02	1.47		5.49	6.70
0800	Grass seed hand push spreader, 4.5 lbs. per M.S.F.	"	180	.044	"	20	1.47		21.47	24.50
1000	Hydro or air seeding for large areas, incl. seed and fertilizer	B-81	8900	.003	S.Y.	.19	.10	.08	.37	.45

32 92 Turf and Grasses

32 92 19 – Seeding

32 92 19.13 Mechanical Seeding		Crew	Daily Output	Labor-Hours	Unit	Material	Labor	Equipment	Total	Total Incl O&P
						2010 Bare Costs				
1100	With wood fiber mulch added	B-81	8900	.003	S.Y.	.33	.10	.08	.51	.60
1300	Seed only, over 100 lbs., field seed, minimum				Lb.	1.28			1.28	1.41
1400	Maximum					4.15			4.15	4.57
1500	Lawn seed, minimum					1.95			1.95	2.15
1600	Maximum					5			5	5.50
1800	Aerial operations, seeding only, field seed	B-58	50	.480	Acre	415	17.20	57	489.20	550
1900	Lawn seed		50	.480		635	17.20	57	709.20	785
2100	Seed and liquid fertilizer, field seed		50	.480		505	17.20	57	579.20	645
2200	Lawn seed		50	.480		725	17.20	57	799.20	885

32 92 23 – Sodding

32 92 23.10 Sodding Systems

		Crew	Daily Output	Labor-Hours	Unit	Material	Labor	Equipment	Total	Total Incl O&P
0010	**SODDING SYSTEMS**									
0020	Sodding, 1" deep, bluegrass sod, on level ground, over 8 MSF	B-63	22	1.818	M.S.F.	246	63	7.35	316.35	375
0200	4 M.S.F.		17	2.353		273	81.50	9.50	364	435
0300	1000 S.F.		13.50	2.963		297	103	11.95	411.95	495
0500	Sloped ground, over 8 M.S.F.		6	6.667		246	232	27	505	655
0600	4 M.S.F.		5	8		273	278	32.50	583.50	760
0700	1000 S.F.		4	10		297	345	40.50	682.50	900
1000	Bent grass sod, on level ground, over 6 M.S.F.		20	2		525	69.50	8.10	602.60	695
1100	3 M.S.F.		18	2.222		605	77	9	691	795
1200	Sodding 1000 S.F. or less		14	2.857		695	99.50	11.55	806.05	930
1500	Sloped ground, over 6 M.S.F.		15	2.667		525	92.50	10.75	628.25	735
1600	3 M.S.F.		13.50	2.963		605	103	11.95	719.95	835
1700	1000 S.F.		12	3.333		695	116	13.45	824.45	955

32 93 Plants

32 93 10 – General Planting Costs

32 93 10.12 Travel

		Crew	Daily Output	Labor-Hours	Unit	Material	Labor	Equipment	Total	Total Incl O&P
0010	**TRAVEL** to all nursery items, for 10 to 20 miles, add									
0015	10 to 20 miles one way, add				All					5%
0100	30 to 50 miles one way, add				"					10%

32 93 13 – Ground Covers

32 93 13.10 Ground Cover Plants

		Crew	Daily Output	Labor-Hours	Unit	Material	Labor	Equipment	Total	Total Incl O&P
0010	**GROUND COVER PLANTS**									
0012	Plants, pachysandra, in prepared beds	B-1	15	1.600	C	27	54		81	114
0200	Vinca minor, 1 yr, bare root, in prepared beds		12	2	"	40	67.50		107.50	148
0600	Stone chips, in 50 lb. bags, Georgia marble		520	.046	Bag	5.25	1.56		6.81	8.20
0700	Onyx gemstone		260	.092		22.50	3.12		25.62	30
0800	Quartz		260	.092		8.50	3.12		11.62	14.15
0900	Pea gravel, truckload lots		28	.857	Ton	27.50	29		56.50	74.50

32 93 33 – Shrubs

32 93 33.10 Shrubs and Trees

		Crew	Daily Output	Labor-Hours	Unit	Material	Labor	Equipment	Total	Total Incl O&P
0010	**SHRUBS AND TREES**									
0011	Evergreen, in prepared beds, B & B									
0100	Arborvitae pyramidal, 4'-5'	B-17	30	1.067	Ea.	48.50	37.50	24.50	110.50	137
0150	Globe, 12"-15"	B-1	96	.250		21	8.45		29.45	36
0300	Cedar, blue, 8'-10'	B-17	18	1.778		219	62.50	41	322.50	380
0500	Hemlock, canadian, 2-1/2'-3'	B-1	36	.667		33.50	22.50		56	71.50
0550	Holly, Savannah, 8' - 10' H		9.68	2.479		485	83.50		568.50	665

32 93 33 – Shrubs

32 93 33.10 Shrubs and Trees

	32 93 33.10 Shrubs and Trees	Crew	Daily Output	Labor-Hours	Unit	Material	Labor (2010 Bare Costs)	Equipment (2010 Bare Costs)	Total (2010 Bare Costs)	Total Incl O&P
0600	Juniper, andorra, 18"-24"	B-1	80	.300	Ea.	16.80	10.15		26.95	34
0620	Wiltoni, 15"-18"		80	.300		21	10.15		31.15	38.50
0640	Skyrocket, 4-1/2'-5'	B-17	55	.582		56	20.50	13.45	89.95	107
0660	Blue pfitzer, 2'-2-1/2'	B-1	44	.545		28	18.40		46.40	59.50
0680	Ketleerie, 2-1/2'-3'		50	.480		34	16.20		50.20	62
0700	Pine, black, 2-1/2'-3'		50	.480		45.50	16.20		61.70	75
0720	Mugo, 18"-24"		60	.400		36	13.50		49.50	60.50
0740	White, 4'-5'	B-17	75	.427		58.50	15	9.85	83.35	98
0800	Spruce, blue, 18"-24"	B-1	60	.400		42.50	13.50		56	68
0840	Norway, 4'-5'	B-17	75	.427		67	15	9.85	91.85	107
0900	Yew, denisforma, 12"-15"	B-1	60	.400		26	13.50		39.50	49.50
1000	Capitata, 18"-24"		30	.800		24	27		51	67.50
1100	Hicksi, 2'-2-1/2'		30	.800		31.50	27		58.50	76.50

32 93 33.20 Shrubs

	32 93 33.20 Shrubs	Crew	Daily Output	Labor-Hours	Unit	Material	Labor	Equipment	Total	Total Incl O&P
0010	**SHRUBS**									
0011	Broadleaf Evergreen, planted in prepared beds									
0100	Andromeda, 15"-18", container	B-1	96	.250	Ea.	26.50	8.45		34.95	42
0200	Azalea, 15" - 18", container		96	.250		33	8.45		41.45	49
0300	Barberry, 9"-12", container		130	.185		11.75	6.25		18	22.50
0400	Boxwood, 15"-18", B&B		96	.250		30	8.45		38.45	45.50
0500	Euonymus, emerald gaiety, 12" to 15", container		115	.209		20.50	7.05		27.55	33.50
0600	Holly, 15"-18", B & B		96	.250		18.65	8.45		27.10	33.50
0900	Mount laurel, 18" - 24", B & B		80	.300		56.50	10.15		66.65	77.50
1000	Paxistema, 9 – 12" high		130	.185		18.30	6.25		24.55	29.50
1100	Rhododendron, 18"-24", container		48	.500		31.50	16.90		48.40	60.50
1200	Rosemary, 1 gal container		600	.040		62.50	1.35		63.85	71
2000	Deciduous, planted in prepared beds, amelanchier, 2'-3', B & B		57	.421		90.50	14.20		104.70	122
2100	Azalea, 15"-18", B & B		96	.250		24.50	8.45		32.95	40
2300	Bayberry, 2'-3', B & B		57	.421		28	14.20		42.20	52.50
2600	Cotoneaster, 15"-18", B & B		80	.300		16.75	10.15		26.90	34
2800	Dogwood, 3'-4', B & B	B-17	40	.800		26.50	28	18.50	73	92.50
2900	Euonymus, alatus compacta, 15" to 18", container	B-1	80	.300		22.50	10.15		32.65	40
3200	Forsythia, 2'-3', container	"	60	.400		20	13.50		33.50	43
3300	Hibiscus, 3'-4', B & B	B-17	75	.427		15.05	15	9.85	39.90	50.50
3400	Honeysuckle, 3'-4', B & B	B-1	60	.400		22	13.50		35.50	45.50
3500	Hydrangea, 2'-3', B & B	"	57	.421		26	14.20		40.20	51
3600	Lilac, 3'-4', B & B	B-17	40	.800		23	28	18.50	69.50	88.50
3900	Privet, bare root, 18"-24"	B-1	80	.300		12.55	10.15		22.70	29.50
4100	Quince, 2'-3', B & B	"	57	.421		24	14.20		38.20	48.50
4200	Russian olive, 3'-4', B & B	B-17	75	.427		24	15	9.85	48.85	60
4400	Spirea, 3'-4', B & B	B-1	70	.343		16	11.60		27.60	35.50
4500	Viburnum, 3'-4', B & B	B-17	40	.800		29.50	28	18.50	76	96

32 93 43 – Trees

32 93 43.20 Trees

	32 93 43.20 Trees	Crew	Daily Output	Labor-Hours	Unit	Material	Labor	Equipment	Total	Total Incl O&P
0010	**TREES**									
0011	Deciduous, in prep. beds, balled & burlapped (B&B)									
0100	Ash, 2" caliper G	B-17	8	4	Ea.	115	141	92.50	348.50	445
0200	Beech, 5'-6' G		50	.640		223	22.50	14.80	260.30	297
0300	Birch, 6'-8', 3 stems G		20	1.600		114	56.50	37	207.50	252
0500	Crabapple, 6'-8' G		20	1.600		160	56.50	37	253.50	305
0600	Dogwood, 4'-5' G		40	.800		63.50	28	18.50	110	134
0700	Eastern redbud 4'-5' G		40	.800		157	28	18.50	203.50	237

32 93 Plants

32 93 43 – Trees

32 93 43.20 Trees			Crew	Daily Output	Labor-Hours	Unit	Material	2010 Bare Costs Labor	Equipment	Total	Total Incl O&P
0800	Elm, 8′-10′	G	B-17	20	1.600	Ea.	119	56.50	37	212.50	258
0900	Ginkgo, 6′-7′	G		24	1.333		163	47	31	241	286
1000	Hawthorn, 8′-10′, 1″ caliper	G		20	1.600		135	56.50	37	228.50	275
1100	Honeylocust, 10′-12′, 1-1/2″ caliper	G		10	3.200		145	113	74	332	415
1300	Larch, 8′	G		32	1		105	35	23	163	195
1400	Linden, 8′-10′, 1″ caliper	G		20	1.600		113	56.50	37	206.50	251
1500	Magnolia, 4′-5′	G		20	1.600		71.50	56.50	37	165	205
1600	Maple, red, 8′-10′, 1-1/2″ caliper	G		10	3.200		241	113	74	428	520
1700	Mountain ash, 8′-10′, 1″ caliper	G		16	2		182	70.50	46	298.50	360
1800	Oak, 2-1/2″-3″ caliper	G		6	5.333		269	188	123	580	720
2100	Planetree, 9′-11′, 1-1/4″ caliper	G		10	3.200		105	113	74	292	370
2200	Plum, 6′-8′, 1″ caliper	G		20	1.600		95.50	56.50	37	189	232
2300	Poplar, 9′-11′, 1-1/4″ caliper	G		10	3.200		90	113	74	277	355
2500	Sumac, 2′-3′	G		75	.427		23	15	9.85	47.85	59
2700	Tulip, 5′-6′	G		40	.800		53	28	18.50	99.50	122
2800	Willow, 6′-8′, 1″ caliper	G	↓	20	1.600	↓	68.50	56.50	37	162	202

32 94 Planting Accessories

32 94 13 – Landscape Edging

32 94 13.20 Edging		Crew	Daily Output	Labor-Hours	Unit	Material	Labor	Equipment	Total	Total Incl O&P
0010	**EDGING**									
0050	Aluminum alloy, including stakes, 1/8″ x 4″, mill finish	B-1	390	.062	L.F.	3.15	2.08		5.23	6.65
0051	Black paint		390	.062		3.65	2.08		5.73	7.20
0052	Black anodized	↓	390	.062		4.22	2.08		6.30	7.85
0100	Brick, set horizontally, 1-1/2 bricks per L.F.	D-1	370	.043		1.40	1.63		3.03	4
0150	Set vertically, 3 bricks per L.F.	″	135	.119		4.24	4.47		8.71	11.40
0200	Corrugated aluminum, roll, 4″ wide	1 Carp	650	.012		.72	.51		1.23	1.58
0250	6″ wide	″	550	.015		.90	.60		1.50	1.92
0600	Railroad ties, 6″ x 8″	2 Carp	170	.094		3.15	3.91		7.06	9.50
0650	7″ x 9″		136	.118		3.50	4.89		8.39	11.40
0750	Redwood 2″ x 4″	↓	330	.048		2.26	2.01		4.27	5.60
0800	Steel edge strips, incl. stakes, 1/4″ x 5″	B-1	390	.062		4.20	2.08		6.28	7.80
0850	3/16″ x 4″	″	390	.062	↓	3.32	2.08		5.40	6.85

32 94 50 – Tree Guying

32 94 50.10 Tree Guying Systems		Crew	Daily Output	Labor-Hours	Unit	Material	Labor	Equipment	Total	Total Incl O&P
0010	**TREE GUYING SYSTEMS**									
0015	Tree guying Including stakes, guy wire and wrap									
0100	Less than 3″ caliper, 2 stakes	2 Clab	35	.457	Ea.	17.55	15.15		32.70	43
0200	3″ to 4″ caliper, 3 stakes	″	21	.762	″	20	25		45	61
1000	Including arrowhead anchor, cable, turnbuckles and wrap									
1100	Less than 3″ caliper, 3″ anchors	2 Clab	20	.800	Ea.	51	26.50		77.50	97
1200	3″ to 6″ caliper, 4″ anchors		15	1.067		51	35.50		86.50	111
1300	6″ caliper, 6″ anchors		12	1.333		95.50	44		139.50	173
1400	8″ caliper, 8″ anchors	↓	9	1.778	↓	107	59		166	209

32 96 Transplanting

32 96 23 – Plant and Bulb Transplanting

32 96 23.23 Planting		Crew	Daily Output	Labor-Hours	Unit	Material	2010 Bare Costs Labor	Equipment	Total	Total Incl O&P
0010	**PLANTING**									
0012	Moving shrubs on site, 12" ball	B-62	28	.857	Ea.		30.50	5.75	36.25	53.50
0100	24" ball	"	22	1.091	"		39	7.35	46.35	67.50
32 96 23.43 Moving Trees										
0290	**MOVING TREES**, On site									
0300	Moving trees on site, 36" ball	B-6	3.75	6.400	Ea.		229	90	319	450
0400	60" ball	"	1	24	"		860	340	1,200	1,675

Estimating Tips

33 10 00 Water Utilities
33 30 00 Sanitary Sewerage Utilities
33 40 00 Storm Drainage Utilities

- Never assume that the water, sewer, and drainage lines will go in at the early stages of the project. Consider the site access needs before dividing the site in half with open trenches, loose pipe, and machinery obstructions. Always inspect the site to establish that the site drawings are complete. Check off all existing utilities on your drawings as you locate them. Be especially careful with underground because appurtenances are sometimes buried during regrading or repaving operations. If you find any discrepancies, mark up the site plan for further research. Differing site conditions can be very costly if discovered later in the project.
- See also Section 33 01 00 for restoration of pipe where removal/replacement may be undesirable. Use of new types of piping materials can reduce the overall project cost. Owners/design engineers should consider the installing contractor as a valuable source of current information on utility products and local conditions that could lead to significant cost savings.

Reference Numbers

Reference numbers are shown in shaded boxes at the beginning of some major classifications. These numbers refer to related items in the Reference Section. The reference information may be an estimating procedure, an alternate pricing method, or technical information.

Note: Not all subdivisions listed here necessarily appear in this publication.

Note: **Trade Service,** *in part, has been used as a reference source for some of the material prices used in Division 33.*

33 01 Operation and Maintenance of Utilities

33 01 10 – Operation and Maintenance of Water Utilities

33 01 10.10 Corrosion Resistance		Crew	Daily Output	Labor-Hours	Unit	Material	2010 Bare Costs Labor	Equipment	Total	Total Incl O&P
0010	**CORROSION RESISTANCE**									
0012	Wrap & coat, add to pipe, 4" diameter				L.F.	2.02			2.02	2.22
0040	6" diameter					2.10			2.10	2.31
0060	8" diameter					3.25			3.25	3.58
0100	12" diameter					4.86			4.86	5.35
0200	24" diameter					9.20			9.20	10.15
0500	Coating, bituminous, per diameter inch, 1 coat, add					.39			.39	.43
0540	3 coat					.58			.58	.64
0560	Coal tar epoxy, per diameter inch, 1 coat, add					.23			.23	.25
0600	3 coat					.48			.48	.53

33 01 30 – Operation and Maintenance of Sewer Utilities

33 01 30.72 Relining Sewers		Crew	Daily Output	Labor-Hours	Unit	Material	Labor	Equipment	Total	Total Incl O&P
0010	**RELINING SEWERS**									
0011	With cement incl. bypass & cleaning									
0020	Less than 10,000 L.F., urban, 6" to 10"	C-17E	130	.615	L.F.	8.75	26.50	.67	35.92	51
0200	24" to 36"		90	.889		13.95	38	.97	52.92	75.50
0300	48" to 72"		80	1		22	43	1.09	66.09	91.50

33 05 Common Work Results for Utilities

33 05 16 – Utility Structures

33 05 16.13 Utility Boxes		Crew	Daily Output	Labor-Hours	Unit	Material	Labor	Equipment	Total	Total Incl O&P
0010	**UTILITY BOXES** Precast concrete, 6" thick									
0050	5' x 10' x 6' high, I.D.	B-13	2	28	Ea.	2,300	1,000	375	3,675	4,450
0100	6' x 10' x 6' high, I.D.		2	28		2,375	1,000	375	3,750	4,525
0150	5' x 12' x 6' high, I.D.		2	28		2,525	1,000	375	3,900	4,700
0200	6' x 12' x 6' high, I.D.		1.80	31.111		2,825	1,100	415	4,340	5,250
0250	6' x 13' x 6' high, I.D.		1.50	37.333		3,700	1,325	500	5,525	6,675
0300	8' x 14' x 7' high, I.D.		1	56		4,000	2,000	750	6,750	8,275
0350	Hand hole, precast concrete, 1-1/2" thick									
0400	1'-0" x 2'-0" x 1'-9", I.D., light duty	B-1	4	6	Ea.	390	203		593	740
0450	4'-6" x 3'-2" x 2'-0", O.D., heavy duty	B-6	3	8	"	1,200	287	113	1,600	1,875

33 05 23 – Trenchless Utility Installation

33 05 23.19 Microtunneling		Crew	Daily Output	Labor-Hours	Unit	Material	Labor	Equipment	Total	Total Incl O&P
0010	**MICROTUNNELING**									
0011	Not including excavation, backfill, shoring,									
0020	or dewatering, average 50'/day, slurry method									
0100	24" to 48" outside diameter, minimum				L.F.				800	880
0110	Adverse conditions, add				%				50%	50%
1000	Rent microtunneling machine, average monthly lease				Month				90,000	99,000
1010	Operating technician				Day				560	620
1100	Mobilization and demobilization, minimum				Job				40,000	44,000
1110	Maximum				"				410,000	451,000

33 05 23.20 Horizontal Boring		Crew	Daily Output	Labor-Hours	Unit	Material	Labor	Equipment	Total	Total Incl O&P
0010	**HORIZONTAL BORING**									
0011	Casing only, 100' minimum,									
0020	not incl. jacking pits or dewatering									
0100	Roadwork, 1/2" thick wall, 24" diameter casing	B-42	20	3.200	L.F.	117	119	68.50	304.50	390
0200	36" diameter		16	4		186	149	85.50	420.50	530
0300	48" diameter		15	4.267		273	158	91	522	650
0500	Railroad work, 24" diameter		15	4.267		117	158	91	366	475

33 05 Common Work Results for Utilities

33 05 23 – Trenchless Utility Installation

33 05 23.20 Horizontal Boring		Crew	Daily Output	Labor-Hours	Unit	2010 Bare Costs Material	Labor	Equipment	Total	Total Incl O&P
0600	36″ diameter	B-42	14	4.571	L.F.	186	170	97.50	453.50	575
0700	48″ diameter		12	5.333		273	198	114	585	735
0900	For ledge, add									20%

33 05 26 – Utility Line Signs, Markers, and Flags

33 05 26.10 Utility Accessories		Crew	Daily Output	Labor-Hours	Unit	Material	Labor	Equipment	Total	Total Incl O&P
0010	**UTILITY ACCESSORIES**									
0400	Underground tape, detectable, reinforced, alum. foil core, 2″	1 Clab	150	.053	C.L.F.	2.36	1.77		4.13	5.30
0500	6″	″	140	.057	″	7.20	1.89		9.09	10.80

33 11 Water Utility Distribution Piping

33 11 13 – Public Water Utility Distribution Piping

33 11 13.15 Water Supply, Ductile Iron Pipe		Crew	Daily Output	Labor-Hours	Unit	Material	Labor	Equipment	Total	Total Incl O&P
0010	**WATER SUPPLY, DUCTILE IRON PIPE** R331113-80									
0020	Not including excavation or backfill									
2000	Pipe, class 50 water piping, 18′ lengths									
2020	Mechanical joint, 4″ diameter	B-21A	200	.200	L.F.	16.55	8.25	3.28	28.08	34.50
2040	6″ diameter		160	.250		19.20	10.30	4.10	33.60	41
2060	8″ diameter		133.33	.300		21	12.40	4.92	38.32	47.50
2080	10″ diameter		114.29	.350		28.50	14.45	5.75	48.70	60
2100	12″ diameter		105.26	.380		35.50	15.70	6.25	57.45	69.50
2120	14″ diameter		100	.400		45	16.50	6.55	68.05	81.50
2140	16″ diameter		72.73	.550		49	22.50	9	80.50	98.50
2160	18″ diameter		68.97	.580		61.50	24	9.50	95	114
2170	20″ diameter		57.14	.700		72.50	29	11.45	112.95	136
2180	24″ diameter		47.06	.850		92.50	35	13.95	141.45	170
3000	Tyton, push-on joint, 4″ diameter		400	.100		12.40	4.13	1.64	18.17	21.50
3020	6″ diameter		333.33	.120		15.85	4.95	1.97	22.77	27
3040	8″ diameter		200	.200		22	8.25	3.28	33.53	40.50
3060	10″ diameter		181.82	.220		31	9.10	3.61	43.71	51.50
3080	12″ diameter		160	.250		33	10.30	4.10	47.40	56
3100	14″ diameter		133.33	.300		36	12.40	4.92	53.32	63.50
3120	16″ diameter		114.29	.350		58	14.45	5.75	78.20	92.50
3140	18″ diameter		100	.400		64.50	16.50	6.55	87.55	103
3160	20″ diameter		88.89	.450		71	18.55	7.40	96.95	115
3180	24″ diameter		76.92	.520		84.50	21.50	8.50	114.50	134
8000	Fittings, mechanical joint									
8006	90° bend, 4″ diameter	B-20A	16	2	Ea.	230	81		311	375
8020	6″ diameter		12.80	2.500		315	101		416	500
8040	8″ diameter		10.67	2.999		455	121		576	685
8060	10″ diameter	B-21A	11.43	3.500		500	144	57.50	701.50	830
8080	12″ diameter		10.53	3.799		905	157	62.50	1,124.50	1,300
8100	14″ diameter		10	4		1,375	165	65.50	1,605.50	1,825
8120	16″ diameter		7.27	5.502		2,050	227	90	2,367	2,700
8140	18″ diameter		6.90	5.797		3,400	239	95	3,734	4,225
8160	20″ diameter		5.71	7.005		4,575	289	115	4,979	5,575
8180	24″ diameter		4.70	8.511		5,025	350	139	5,514	6,225
8200	Wye or tee, 4″ diameter	B-20A	10.67	2.999		350	121		471	570
8220	6″ diameter		8.53	3.751		465	152		617	745
8240	8″ diameter		7.11	4.501		665	182		847	1,000
8260	10″ diameter	B-21A	7.62	5.249		1,375	217	86	1,678	1,950

33 11 13.15 Water Supply, Ductile Iron Pipe		Crew	Daily Output	Labor-Hours	Unit	2010 Bare Costs Material	Labor	Equipment	Total	Total Incl O&P
8280	12″ diameter	B-21A	7.02	5.698	Ea.	1,750	235	93.50	2,078.50	2,375
8300	14″ diameter		6.67	5.997		1,950	247	98.50	2,295.50	2,625
8320	16″ diameter		4.85	8.247		2,025	340	135	2,500	2,900
8340	18″ diameter		4.60	8.696		5,300	360	143	5,803	6,525
8360	20″ diameter		3.81	10.499		8,275	435	172	8,882	9,950
8380	24″ diameter		3.14	12.739		9,800	525	209	10,534	11,800
8450	Decreaser, 6″ x 4″ diameter	B-20A	14.22	2.250		193	91		284	350
8460	8″ x 6″ diameter	″	11.64	2.749		247	111		358	440
8470	10″ x 6 ″ diameter	B-21A	13.33	3.001		310	124	49	483	585
8480	12″ x 6″ diameter		12.70	3.150		385	130	51.50	566.50	680
8490	16″ x 6″ diameter		10	4		885	165	65.50	1,115.50	1,300
8500	20″ x 6″ diameter		8.42	4.751		2,175	196	78	2,449	2,775
8550	Piping, butterfly valves, cast iron									
8560	4″ diameter	B-20	6	4	Ea.	900	148		1,048	1,225
8570	6″ diameter	″	5	4.800		1,100	177		1,277	1,475
8580	8″ diameter	B-21	4	7		1,500	266	34.50	1,800.50	2,100
8590	10″ diameter		3.50	8		2,000	305	39.50	2,344.50	2,700
8600	12″ diameter		3	9.333		2,850	355	46	3,251	3,725
8610	14″ diameter		2	14		3,475	530	69	4,074	4,725
8620	16″ diameter		2	14		5,375	530	69	5,974	6,825

33 11 13.25 Water Supply, Polyvinyl Chloride Pipe		Crew	Daily Output	Labor-Hours	Unit	Material	Labor	Equipment	Total	Total Incl O&P
0010	**WATER SUPPLY, POLYVINYL CHLORIDE PIPE**									
0020	Not including excavation or backfill, unless specified									
2100	PVC pipe, Class 150, 1-1/2″ diameter	Q-1A	750	.013	L.F.	.30	.70		1	1.38
2120	2″ diameter		686	.015		.46	.76		1.22	1.66
2140	2-1/2″ diameter		500	.020		.68	1.05		1.73	2.32
2160	3″ diameter	B-20	430	.056		1.04	2.06		3.10	4.31
3010	AWWA C905, PR 100, DR 25									
3030	14″ diameter	B-20A	213	.150	L.F.	10	6.10		16.10	20
3040	16″ diameter		200	.160		13	6.50		19.50	24
3050	18″ diameter		160	.200		19.50	8.10		27.60	34
3060	20″ diameter		133	.241		19.50	9.75		29.25	36.50
3070	24″ diameter		107	.299		27.50	12.10		39.60	49
3080	30″ diameter		80	.400		55	16.20		71.20	85
3090	36″ diameter		80	.400		85	16.20		101.20	118
3100	42″ diameter		60	.533		115	21.50		136.50	160
3200	48″ diameter		60	.533		150	21.50		171.50	198
4520	Pressure pipe Class 150, SDR 18, AWWA C900, 4″ diameter		380	.084		2.34	3.41		5.75	7.70
4530	6″ diameter		316	.101		4.63	4.10		8.73	11.30
4540	8″ diameter		264	.121		8.15	4.91		13.06	16.40
4550	10″ diameter		220	.145		12.35	5.90		18.25	22.50
4560	12″ diameter		186	.172		17.45	6.95		24.40	29.50
8000	Fittings with rubber gasket									
8003	Class 150, D.R. 18									
8006	90° Bend , 4″ diameter	B-20	100	.240	Ea.	51.50	8.85		60.35	70
8020	6″ diameter		90	.267		93	9.85		102.85	117
8040	8″ diameter		80	.300		176	11.10		187.10	210
8060	10″ diameter		50	.480		268	17.75		285.75	325
8080	12″ diameter		30	.800		410	29.50		439.50	495
8100	Tee, 4″ diameter		90	.267		54.50	9.85		64.35	75
8120	6″ diameter		80	.300		146	11.10		157.10	178
8140	8″ diameter		70	.343		205	12.65		217.65	245

33 11 Water Utility Distribution Piping

33 11 13 – Public Water Utility Distribution Piping

33 11 13.25 Water Supply, Polyvinyl Chloride Pipe		Crew	Daily Output	Labor-Hours	Unit	Material	Labor	Equipment	Total	Total Incl O&P
						2010 Bare Costs				
8160	10" diameter	B-20	40	.600	Ea.	430	22		452	510
8180	12" diameter		20	1.200		605	44.50		649.50	735
8200	45° Bend, 4" diameter		100	.240		51	8.85		59.85	69.50
8220	6" diameter		90	.267		92	9.85		101.85	116
8240	8" diameter		50	.480		173	17.75		190.75	218
8260	10" diameter		50	.480		273	17.75		290.75	330
8280	12" diameter		30	.800		365	29.50		394.50	445
8300	Reducing tee 6" x 4"		100	.240		120	8.85		128.85	146
8320	8" x 6"		90	.267		290	9.85		299.85	335
8330	10" x 6"		90	.267		335	9.85		344.85	385
8340	10" x 8"		90	.267		360	9.85		369.85	410
8350	12" x 6"		90	.267		450	9.85		459.85	510
8360	12" x 8"		90	.267		500	9.85		509.85	565
8400	Tapped service tee (threaded type) 6" x 6" x 3/4"		100	.240		78	8.85		86.85	99.50
8430	6" x 6" x 1"		90	.267		78	9.85		87.85	101
8440	6" x 6" x 1 1/2"		90	.267		78	9.85		87.85	101
8450	6" x 6" x 2"		90	.267		78	9.85		87.85	101
8460	8" x 8" x 3/4"		90	.267		123	9.85		132.85	151
8470	8" x 8" x 1"		90	.267		123	9.85		132.85	151
8480	8" x 8" x 1 1/2"		90	.267		123	9.85		132.85	151
8490	8" x 8" x 2"		90	.267		123	9.85		132.85	151
8500	Repair coupling 4"		100	.240		108	8.85		116.85	133
8520	6" diameter		90	.267		131	9.85		140.85	159
8540	8" diameter		50	.480		155	17.75		172.75	198
8560	10" diameter		50	.480		198	17.75		215.75	246
8580	12" diameter		50	.480		234	17.75		251.75	285
8600	Plug end 4"		100	.240		38	8.85		46.85	55.50
8620	6" diameter		90	.267		73.50	9.85		83.35	96
8640	8" diameter		50	.480		79	17.75		96.75	115
8660	10" diameter		50	.480		126	17.75		143.75	167
8680	12" diameter	↓	50	.480	↓	156	17.75		173.75	200

33 12 Water Utility Distribution Equipment

33 12 19 – Water Utility Distribution Fire Hydrants

33 12 19.10 Fire Hydrants		Crew	Daily Output	Labor-Hours	Unit	Material	Labor	Equipment	Total	Total Incl O&P
0010	**FIRE HYDRANTS**									
0020	Mechanical joints unless otherwise noted									
1000	Fire hydrants, two way; excavation and backfill not incl.									
1100	4-1/2" valve size, depth 2'-0"	B-21	10	2.800	Ea.	1,250	106	13.80	1,369.80	1,550
1120	2'-6"		10	2.800		1,325	106	13.80	1,444.80	1,625
1140	3'-0"		10	2.800		1,375	106	13.80	1,494.80	1,700
1300	7'-0"	↓	6	4.667		1,575	177	23	1,775	2,050
2400	Lower barrel extensions with stems, 1'-0"	B-20	14	1.714		300	63.50		363.50	430
2480	3'-0"	"	12	2	↓	575	74		649	750

33 16 Water Utility Storage Tanks

33 16 13 – Aboveground Water Utility Storage Tanks

33 16 13.13 Steel Water Storage Tanks		Crew	Daily Output	Labor-Hours	Unit	2010 Bare Costs Material	Labor	Equipment	Total	Total Incl O&P
0010	**STEEL WATER STORAGE TANKS**									
0910	Steel, ground level, ht/diam. less than 1, not incl. fdn., 100,000 gallons				Ea.				168,000	184,000
1000	250,000 gallons								178,500	196,000
1200	500,000 gallons								272,000	299,000
1250	750,000 gallons								324,000	356,000
1300	1,000,000 gallons								494,000	543,000
1500	2,000,000 gallons								797,000	877,000
1600	4,000,000 gallons								1,239,000	1,363,000
1800	6,000,000 gallons								1,719,000	1,891,000
1850	8,000,000 gallons								2,250,000	2,474,000
1910	10,000,000 gallons								3,356,000	3,691,000
2100	Steel standpipes, hgt/diam. more than 1, 100' to overflow, no fdn									
2200	500,000 gallons				Ea.				356,000	392,000
2400	750,000 gallons								435,000	478,000
2500	1,000,000 gallons								529,000	582,000
2700	1,500,000 gallons								739,000	813,000
2800	2,000,000 gallons								908,000	999,000

33 16 13.16 Prestressed Conc. Aboveground Water Utility Storage Tanks

		Crew	Daily Output	Labor-Hours	Unit	Material	Labor	Equipment	Total	Total Incl O&P
0010	**PRESTRESSED CONC. ABOVEGROUND WATER UTILITY STORAGE TANKS**									
0020	Not including fdn., pipe or pumps, 250,000 gallons				Ea.				270,000	297,000
0100	500,000 gallons								412,000	453,000
0300	1,000,000 gallons								597,000	657,000
0400	2,000,000 gallons								906,000	997,000
0600	4,000,000 gallons								1,442,000	1,586,000
0700	6,000,000 gallons								1,916,000	2,107,000
0750	8,000,000 gallons								2,472,000	2,719,000
0800	10,000,000 gallons								2,987,000	3,286,000

33 16 13.23 Plastic-Coated Fabric Pillow Water Tanks

		Crew	Daily Output	Labor-Hours	Unit	Material	Labor	Equipment	Total	Total Incl O&P
0010	**PLASTIC-COATED FABRIC PILLOW WATER TANKS**									
7000	Water tanks, vinyl coated fabric pillow tanks, freestanding, 5,000 gallons	4 Clab	4	8	Ea.	9,500	265		9,765	10,900
7100	Supporting embankment not included, 25,000 gallons	6 Clab	2	24		14,900	795		15,695	17,600
7200	50,000 gallons	8 Clab	1.50	42.667		27,600	1,400		29,000	32,600
7300	100,000 gallons	9 Clab	.90	80		60,000	2,650		62,650	70,000
7400	150,000 gallons		.50	144		82,000	4,775		86,775	97,500
7500	200,000 gallons		.40	180		100,500	5,950		106,450	119,500
7600	250,000 gallons		.30	240		143,000	7,950		150,950	170,000

33 16 19 – Elevated Water Utility Storage Tanks

33 16 19.50 Elevated Water Storage Tanks

		Crew	Daily Output	Labor-Hours	Unit	Material	Labor	Equipment	Total	Total Incl O&P
0010	**ELEVATED WATER STORAGE TANKS**									
0011	Not incl. pipe, pumps or foundation									
3000	Elevated water tanks, 100' to bottom capacity line, incl. painting									
3010	50,000 gallons				Ea.				213,000	233,000
3300	100,000 gallons								297,000	327,000
3400	250,000 gallons								409,000	450,000
3600	500,000 gallons								660,000	726,000
3700	750,000 gallons								912,000	1,003,000
3900	1,000,000 gallons								916,000	1,007,000

33 21 Water Supply Wells

33 21 13 – Public Water Supply Wells

33 21 13.10 Wells and Accessories		Crew	Daily Output	Labor-Hours	Unit	2010 Bare Costs Material	Labor	Equipment	Total	Total Incl O&P
0010	**WELLS & ACCESSORIES**									
0011	Domestic									
0100	Drilled, 4" to 6" diameter	B-23	120	.333	L.F.		11.15	24	35.15	43.50
0200	8" diameter	"	95.20	.420	"		14.10	30.50	44.60	55
0400	Gravel pack well, 40' deep, incl. gravel & casing, complete									
0500	24" diameter casing x 18" diameter screen	B-23	.13	307	Total	31,800	10,300	22,200	64,300	75,500
0600	36" diameter casing x 18" diameter screen		.12	333	"	34,200	11,200	24,100	69,500	81,500
0800	Observation wells, 1-1/4" riser pipe	↓	163	.245	V.L.F.	17.50	8.20	17.70	43.40	51.50
0900	For flush Buffalo roadway box, add	1 Skwk	16.60	.482	Ea.	48	20.50		68.50	84
1200	Test well, 2-1/2" diameter, up to 50' deep (15 to 50 GPM)	B-23	1.51	26.490	"	715	885	1,900	3,500	4,275
1300	Over 50' deep, add	"	121.80	.328	L.F.	19.10	11	23.50	53.60	64
1500	Pumps, installed in wells to 100' deep, 4" submersible									
1510	1/2 H.P.	Q-1	3.22	4.969	Ea.	455	233		688	850
1520	3/4 H.P.		2.66	6.015		620	282		902	1,100
1600	1 H.P.	↓	2.29	6.987		840	325		1,165	1,425
1700	1-1/2 H.P.	Q-22	1.60	10		1,050	470	410	1,930	2,325
1800	2 H.P.		1.33	12.030		1,275	565	495	2,335	2,775
1900	3 H.P.		1.14	14.035		1,725	660	575	2,960	3,525
2000	5 H.P.		1.14	14.035		2,350	660	575	3,585	4,225
3000	Pump, 6" submersible, 25' to 150' deep, 25 H.P., 249 to 297 GPM		.89	17.978		5,275	840	735	6,850	7,875
3100	25' to 500' deep, 30 H.P., 100 to 300 GPM	↓	.73	21.918	↓	5,600	1,025	900	7,525	8,700
8000	Steel well casing	B-23A	3020	.008	Lb.	.84	.29	.90	2.03	2.37
9950	See Div. 31 23 19.40 for wellpoints									
9960	See Div. 31 23 19.30 for drainage wells									

33 21 13.20 Water Supply Wells, Pumps		Crew	Daily Output	Labor-Hours	Unit	Material	Labor	Equipment	Total	Total Incl O&P
0010	**WATER SUPPLY WELLS, PUMPS**									
0011	With pressure control									
1000	Deep well, jet, 42 gal. galvanized tank									
1040	3/4 HP	1 Plum	.80	10	Ea.	1,450	520		1,970	2,375
3000	Shallow well, jet, 30 gal. galvanized tank									
3040	1/2 HP	1 Plum	2	4	Ea.	1,200	208		1,408	1,625

33 31 Sanitary Utility Sewerage Piping

33 31 13 – Public Sanitary Utility Sewerage Piping

33 31 13.15 Sewage Collection, Concrete Pipe		Crew	Daily Output	Labor-Hours	Unit	Material	Labor	Equipment	Total	Total Incl O&P
0010	**SEWAGE COLLECTION, CONCRETE PIPE**									
0020	See Div. 33 41 13.60 for sewage/drainage collection, concrete pipe									

33 31 13.25 Sewage Collection, Polyvinyl Chloride Pipe		Crew	Daily Output	Labor-Hours	Unit	Material	Labor	Equipment	Total	Total Incl O&P
0010	**SEWAGE COLLECTION, POLYVINYL CHLORIDE PIPE**									
0020	Not including excavation or backfill									
2000	20' lengths, S.D.R. 35, B&S, 4" diameter	B-20	375	.064	L.F.	1.68	2.36		4.04	5.50
2040	6" diameter		350	.069		3.64	2.53		6.17	7.90
2080	13' lengths , S.D.R. 35, B&S, 8" diameter	↓	335	.072		7.85	2.65		10.50	12.70
2120	10" diameter	B-21	330	.085		11.90	3.22	.42	15.54	18.50
2160	12" diameter		320	.088		13.60	3.33	.43	17.36	20.50
2200	15" diameter	↓	240	.117		15.90	4.43	.58	20.91	25
4000	Piping, DWV PVC, no exc/bkfill, 10' L, Sch 40, 4" diameter	B-20	375	.064		3.71	2.36		6.07	7.75
4010	6" diameter		350	.069		7.25	2.53		9.78	11.85
4020	8" diameter	↓	335	.072	↓	12.35	2.65		15	17.60

33 36 Utility Septic Tanks

33 36 13 – Utility Septic Tank and Effluent Wet Wells

33 36 13.10 Septic Tanks		Crew	Daily Output	Labor-Hours	Unit	Material	2010 Bare Costs Labor	Equipment	Total	Total Incl O&P
0010	**SEPTIC TANKS**									
0011	Not including excavation or piping									
0015	Septic tanks, precast, 1,000 gal	B-21	8	3.500	Ea.	750	133	17.25	900.25	1,050
0060	1,500 gallon		7	4		1,175	152	19.70	1,346.70	1,525
0100	2,000 gallon		5	5.600		1,675	213	27.50	1,915.50	2,200
0200	5,000 gallon	B-13	3.50	16		7,625	570	214	8,409	9,500
0300	15,000 gallon, 4 piece	B-13B	1.70	32.941		21,100	1,175	710	22,985	25,800
0400	25,000 gallon, 4 piece		1.10	50.909		44,700	1,825	1,100	47,625	53,000
0500	40,000 gallon, 4 piece		.80	70		52,500	2,500	1,500	56,500	63,500
0520	50,000 gallon, 5 piece	B-13C	.60	93.333		60,500	3,325	2,875	66,700	75,000
0540	75,000 gallon, cast in place	C-14C	.25	448		73,500	17,900	121	91,521	109,000
0560	100,000 gallon	"	.15	746		91,000	29,800	202	121,002	147,000
0600	High density polyethylene, 1,000 gallon	B-21	6	4.667		985	177	23	1,185	1,375
0700	1,500 gallon		4	7		1,375	266	34.50	1,675.50	1,975
0900	Galley, 4′ x 4′ x 4′		16	1.750		275	66.50	8.65	350.15	415
1000	Distribution boxes, concrete, 7 outlets	2 Clab	16	1		80	33		113	139
1100	9 outlets	"	8	2		445	66		511	590
1150	Leaching field chambers, 13′ x 3′-7″ x 1′-4″, standard	B-13	16	3.500		525	125	46.50	696.50	825
1200	Heavy duty, 8′ x 4′ x 1′-6″		14	4		275	143	53.50	471.50	580
1300	13′ x 3′-9″ x 1′-6″		12	4.667		1,025	167	62.50	1,254.50	1,450
1350	20′ x 4′ x 1′-6″		5	11.200		1,075	400	150	1,625	1,975
1400	Leaching pit, precast concrete, 3′ diameter, 3′ deep	B-21	8	3.500		550	133	17.25	700.25	830
1500	6′ diameter, 3′ section		4.70	5.957		1,050	226	29.50	1,305.50	1,525
2000	Velocity reducing pit, precast conc., 6′ diameter, 3′ deep		4.70	5.957		1,500	226	29.50	1,755.50	2,025
2200	Excavation for septic tank, 3/4 C.Y. backhoe	B-12F	145	.110	C.Y.		4.28	4.67	8.95	11.65
2400	4′ trench for disposal field, 3/4 C.Y. backhoe	"	335	.048	L.F.		1.85	2.02	3.87	5.05
2600	Gravel fill, run of bank	B-6	150	.160	C.Y.	20	5.75	2.25	28	33
2800	Crushed stone, 3/4″	"	150	.160	"	31	5.75	2.25	39	45
3000	Effluent filter, 4″ dia.	B-21	4.70	5.957	Ea.	44.50	226	29.50	300	425

33 41 Storm Utility Drainage Piping

33 41 13 – Public Storm Utility Drainage Piping

33 41 13.40 Piping, Storm Drainage, Corrugated Metal		Crew	Daily Output	Labor-Hours	Unit	Material	Labor	Equipment	Total	Total Incl O&P
0010	**PIPING, STORM DRAINAGE, CORRUGATED METAL**									
0020	Not including excavation or backfill									
2000	Corrugated metal pipe, galvanized									
2020	Bituminous coated with paved invert, 20′ lengths									
2040	8″ diameter, 16 ga.	B-14	330	.145	L.F.	12.30	5.05	1.02	18.37	22.50
2060	10″ diameter, 16 ga.		260	.185		14.75	6.40	1.30	22.45	27.50
2080	12″ diameter, 16 ga.		210	.229		17.65	7.95	1.61	27.21	33.50
2100	15″ diameter, 16 ga.		200	.240		21.50	8.35	1.69	31.54	38
2120	18″ diameter, 16 ga.		190	.253		27.50	8.80	1.78	38.08	45.50
2140	24″ diameter, 14 ga. **CN**		160	.300		33.50	10.45	2.11	46.06	55.50
2160	30″ diameter, 14 ga.	B-13	120	.467		44.50	16.70	6.25	67.45	81.50
2180	36″ diameter, 12 ga.		120	.467		65	16.70	6.25	87.95	104
2200	48″ diameter, 12 ga.		100	.560		99	20	7.50	126.50	148
2220	60″ diameter, 10 ga.	B-13B	75	.747		128	26.50	16.10	170.60	200
2240	72″ diameter, 8 ga.	"	45	1.244		192	44.50	27	263.50	310
2500	Galvanized, uncoated, 20′ lengths									
2520	8″ diameter, 16 ga.	B-14	355	.135	L.F.	10.05	4.71	.95	15.71	19.30
2540	10″ diameter, 16 ga.		280	.171		11	5.95	1.21	18.16	22.50

33 41 13 – Public Storm Utility Drainage Piping

33 41 13.40 Piping, Storm Drainage, Corrugated Metal		Crew	Daily Output	Labor-Hours	Unit	2010 Bare Costs Material	Labor	Equipment	Total	Total Incl O&P
2560	12″ diameter, 16 ga.	B-14	220	.218	L.F.	12.55	7.60	1.54	21.69	27
2580	15″ diameter, 16 ga.		220	.218		16	7.60	1.54	25.14	31
2600	18″ diameter, 16 ga.		205	.234		21	8.15	1.65	30.80	38
2620	24″ diameter, 14 ga.		175	.274		30.50	9.55	1.93	41.98	51
2640	30″ diameter, 14 ga.	B-13	130	.431		39	15.40	5.75	60.15	73
2660	36″ diameter, 12 ga.		130	.431		64	15.40	5.75	85.15	100
2680	48″ diameter, 12 ga.		110	.509		85.50	18.20	6.80	110.50	130
2690	60″ diameter, 10 ga.	B-13B	78	.718		135	25.50	15.50	176	204
2780	End sections, 8″ diameter	B-14	24	2	Ea.	98	69.50	14.10	181.60	231
2785	10″ diameter		22	2.182		101	76	15.35	192.35	244
2790	12″ diameter		35	1.371		113	47.50	9.65	170.15	209
2800	18″ diameter		30	1.600		135	55.50	11.25	201.75	246
2810	24″ diameter	B-13	25	2.240		196	80	30	306	370
2820	30″ diameter		25	2.240		375	80	30	485	565
2825	36″ diameter		20	2.800		495	100	37.50	632.50	740
2830	48″ diameter		10	5.600		1,025	200	75	1,300	1,525
2835	60″ diameter	B-13B	5	11.200		1,325	400	242	1,967	2,350
2840	72″ diameter	″	4	14		2,400	500	300	3,200	3,750
33 41 13.60 Sewage/Drainage Collection, Concrete Pipe										
0010	**SEWAGE/DRAINAGE COLLECTION, CONCRETE PIPE**									
0020	Not including excavation or backfill									
1000	Non-reinforced pipe, extra strength, B&S or T&G joints									
1010	6″ diameter	B-14	265.04	.181	L.F.	6.25	6.30	1.27	13.82	17.95
1020	8″ diameter		224	.214		6.90	7.45	1.51	15.86	20.50
1030	10″ diameter		216	.222		7.65	7.75	1.56	16.96	22
1040	12″ diameter		200	.240		9.35	8.35	1.69	19.39	25
1050	15″ diameter		180	.267		11.15	9.30	1.88	22.33	28.50
1060	18″ diameter		144	.333		13.15	11.60	2.35	27.10	35
1070	21″ diameter		112	.429		16.20	14.90	3.02	34.12	44
1080	24″ diameter		100	.480		21.50	16.70	3.38	41.58	53
2000	Reinforced culvert, class 3, no gaskets									
2010	12″ diameter	B-14	150	.320	L.F.	10.65	11.15	2.25	24.05	31.50
2020	15″ diameter		150	.320		13.95	11.15	2.25	27.35	35
2030	18″ diameter		132	.364		18.45	12.65	2.56	33.66	42.50
2035	21″ diameter		120	.400		20	13.90	2.82	36.72	46.50
2040	24″ diameter **CN**		100	.480		26.50	16.70	3.38	46.58	58
2045	27″ diameter	B-13	92	.609		33.50	22	8.15	63.65	79
2050	30″ diameter		88	.636		40	22.50	8.50	71	87.50
2060	36″ diameter		72	.778		53.50	28	10.40	91.90	113
2070	42″ diameter	B-13B	72	.778		67.50	28	16.80	112.30	135
2080	48″ diameter		64	.875		82	31.50	18.90	132.40	159
2090	60″ diameter		48	1.167		121	41.50	25	187.50	224
2100	72″ diameter		40	1.400		175	50	30	255	305
2120	84″ diameter		32	1.750		266	62.50	38	366.50	430
2140	96″ diameter		24	2.333		320	83.50	50.50	454	535
2200	With gaskets, class 3, 12″ diameter	B-21	168	.167		11.70	6.35	.82	18.87	23.50
2220	15″ diameter		160	.175		15.35	6.65	.86	22.86	28
2230	18″ diameter		152	.184		20.50	7	.91	28.41	34.50
2240	24″ diameter		136	.206		33.50	7.80	1.02	42.32	50
2260	30″ diameter	B-13	88	.636		48.50	22.50	8.50	79.50	97
2270	36″ diameter	″	72	.778		64	28	10.40	102.40	124
2290	48″ diameter	B-13B	64	.875		98.50	31.50	18.90	148.90	177

33 41 Storm Utility Drainage Piping

33 41 13 – Public Storm Utility Drainage Piping

33 41 13.60 Sewage/Drainage Collection, Concrete Pipe		Crew	Daily Output	Labor-Hours	Unit	Material	2010 Bare Costs Labor	Equipment	Total	Total Incl O&P
2310	72" diameter	B-13B	40	1.400	L.F.	202	50	30	282	330
2330	Flared ends, 6'-1" long, 12" diameter	B-21	190	.147		43.50	5.60	.73	49.83	57
2340	15" diameter		155	.181		48.50	6.85	.89	56.24	64.50
2400	6'-2" long, 18" diameter		122	.230		56.50	8.70	1.13	66.33	77
2420	24" diameter		88	.318		76	12.10	1.57	89.67	104
2440	36" diameter	B-13	60	.933		134	33.50	12.45	179.95	212
3080	Radius pipe, add to pipe prices, 12" to 60" diameter					50%				
3090	Over 60" diameter, add					20%				
3500	Reinforced elliptical, 8' lengths, C507 class 3									
3520	14" x 23" inside, round equivalent 18" diameter	B-21	82	.341	L.F.	52.50	13	1.68	67.18	80
3530	24" x 38" inside, round equivalent 30" diameter	B-13	58	.966		78	34.50	12.90	125.40	152
3540	29" x 45" inside, round equivalent 36" diameter		52	1.077		93.50	38.50	14.40	146.40	178
3550	38" x 60" inside, round equivalent 48" diameter		38	1.474		155	52.50	19.65	227.15	272
3560	48" x 76" inside, round equivalent 60" diameter		26	2.154		196	77	29	302	365
3570	58" x 91" inside, round equivalent 72" diameter		22	2.545		296	91	34	421	500
3780	Concrete slotted pipe, class 4 mortar joint									
3800	12" diameter	B-21	168	.167	L.F.	12	6.35	.82	19.17	24
3840	18" diameter	"	152	.184	"	18	7	.91	25.91	31.50
3900	Class 4 O-ring									
3940	12" diameter	B-21	168	.167	L.F.	11	6.35	.82	18.17	22.50
3960	18" diameter	"	152	.184	"	17.75	7	.91	25.66	31.50

33 42 Culverts

33 42 16 – Concrete Culverts

33 42 16.15 Oval Arch Culverts		Crew	Daily Output	Labor-Hours	Unit	Material	Labor	Equipment	Total	Total Incl O&P
0010	**OVAL ARCH CULVERTS**									
3000	Corrugated galvanized or aluminum, coated & paved									
3020	17" x 13", 16 ga., 15" equivalent	B-14	200	.240	L.F.	38.50	8.35	1.69	48.54	56.50
3040	21" x 15", 16 ga., 18" equivalent		150	.320		49.50	11.15	2.25	62.90	74
3060	28" x 20", 14 ga., 24" equivalent		125	.384		71	13.35	2.70	87.05	101
3080	35" x 24", 14 ga., 30" equivalent		100	.480		86.50	16.70	3.38	106.58	124
3100	42" x 29", 12 ga., 36" equivalent	B-13	100	.560		129	20	7.50	156.50	181
3120	49" x 33", 12 ga., 42" equivalent		90	.622		155	22	8.30	185.30	214
3140	57" x 38", 12 ga., 48" equivalent		75	.747		172	26.50	9.95	208.45	242
3160	Steel, plain oval arch culverts, plain									
3180	17" x 13", 16 ga., 15" equivalent	B-14	225	.213	L.F.	21	7.40	1.50	29.90	36
3200	21" x 15", 16 ga., 18" equivalent		175	.274		24.50	9.55	1.93	35.98	44
3220	28" x 20", 14 ga., 24" equivalent		150	.320		39.50	11.15	2.25	52.90	63
3240	35" x 24", 14 ga., 30" equivalent	B-13	108	.519		49.50	18.55	6.90	74.95	90.50
3260	42" x 29", 12 ga., 36" equivalent		108	.519		82	18.55	6.90	107.45	126
3280	49" x 33", 12 ga., 42" equivalent		92	.609		96.50	22	8.15	126.65	148
3300	57" x 38", 12 ga., 48" equivalent		75	.747		114	26.50	9.95	150.45	177
3320	End sections, 17" x 13"		22	2.545	Ea.	114	91	34	239	300
3340	42" x 29"		17	3.294	"	465	118	44	627	745
3360	Multi-plate arch, steel	B-20	1690	.014	Lb.	1.30	.52		1.82	2.24

33 44 Storm Utility Water Drains

33 44 13 – Utility Area Drains

33 44 13.13 Catchbasins		Crew	Daily Output	Labor-Hours	Unit	2010 Bare Costs Material	Labor	Equipment	Total	Total Incl O&P
0010	**CATCHBASINS**									
0011	Not including footing & excavation									
1600	Frames & covers, C.I., 24" square, 500 lb.	B-6	7.80	3.077	Ea.	300	110	43.50	453.50	545
1700	26" D shape, 600 lb.		7	3.429		515	123	48.50	686.50	810
1800	Light traffic, 18" diameter, 100 lb.		10	2.400		169	86	34	289	355
1900	24" diameter, 300 lb.		8.70	2.759		261	99	39	399	480
2000	36" diameter, 900 lb.		5.80	4.138		525	148	58.50	731.50	865
2100	Heavy traffic, 24" diameter, 400 lb.		7.80	3.077		252	110	43.50	405.50	495
2200	36" diameter, 1150 lb.		3	8		835	287	113	1,235	1,475
2300	Mass. State standard, 26" diameter, 475 lb.		7	3.429		630	123	48.50	801.50	935
2400	30" diameter, 620 lb.		7	3.429		400	123	48.50	571.50	680
2500	Watertight, 24" diameter, 350 lb.		7.80	3.077		430	110	43.50	583.50	685
2600	26" diameter, 500 lb.		7	3.429		405	123	48.50	576.50	685
2700	32" diameter, 575 lb.		6	4		905	143	56.50	1,104.50	1,275
2800	3 piece cover & frame, 10" deep,									
2900	1200 lbs., for heavy equipment	B-6	3	8	Ea.	1,300	287	113	1,700	1,975
3000	Raised for paving 1-1/4" to 2" high									
3100	4 piece expansion ring									
3200	20" to 26" diameter	1 Clab	3	2.667	Ea.	141	88.50		229.50	291
3300	30" to 36" diameter	"	3	2.667	"	197	88.50		285.50	355
3320	Frames and covers, existing, raised for paving, 2", including									
3340	row of brick, concrete collar, up to 12" wide frame	B-6	18	1.333	Ea.	44	48	18.75	110.75	142
3360	20" to 26" wide frame		11	2.182		67	78	30.50	175.50	227
3380	30" to 36" wide frame		9	2.667		82.50	95.50	37.50	215.50	279
3400	Inverts, single channel brick	D-1	3	5.333		94.50	201		295.50	410
3500	Concrete		5	3.200		74	121		195	264
3600	Triple channel, brick		2	8		144	300		444	615
3700	Concrete		3	5.333		127	201		328	445

33 46 Subdrainage

33 46 16 – Subdrainage Piping

33 46 16.20 Piping, Subdrainage, Concrete		Crew	Daily Output	Labor-Hours	Unit	Material	Labor	Equipment	Total	Total Incl O&P
0010	**PIPING, SUBDRAINAGE, CONCRETE**									
0021	Not including excavation and backfill									
3000	Porous wall concrete underdrain, std. strength, 4" diameter	B-20	335	.072	L.F.	3.40	2.65		6.05	7.80
3020	6" diameter	"	315	.076		4.42	2.81		7.23	9.20
3040	8" diameter	B-21	310	.090		5.45	3.43	.45	9.33	11.75
3060	12" diameter		285	.098		11.55	3.73	.48	15.76	18.95
3080	15" diameter		230	.122		13.25	4.63	.60	18.48	22.50
3100	18" diameter		165	.170		17.50	6.45	.84	24.79	30
4000	Extra strength, 6" diameter	B-20	315	.076		4.09	2.81		6.90	8.85
4020	8" diameter	B-21	310	.090		6.15	3.43	.45	10.03	12.50
4040	10" diameter		295	.095		12.25	3.61	.47	16.33	19.55
4060	12" diameter		285	.098		13.30	3.73	.48	17.51	21
4080	15" diameter		230	.122		14.70	4.63	.60	19.93	24
4100	18" diameter		165	.170		21.50	6.45	.84	28.79	34.50

33 46 16.25 Piping, Subdrainage, Corrugated Metal		Crew	Daily Output	Labor-Hours	Unit	Material	Labor	Equipment	Total	Total Incl O&P
0010	**PIPING, SUBDRAINAGE, CORRUGATED METAL**									
0021	Not including excavation and backfill									
2010	Aluminum, perforated									
2020	6" diameter, 18 ga.	B-14	380	.126	L.F.	7.25	4.40	.89	12.54	15.75

33 46 Subdrainage

33 46 16 – Subdrainage Piping

33 46 16.25 Piping, Subdrainage, Corrugated Metal		Crew	Daily Output	Labor-Hours	Unit	2010 Bare Costs Material	Labor	Equipment	Total	Total Incl O&P
2200	8″ diameter, 16 ga.	B-14	370	.130	L.F.	10.50	4.51	.91	15.92	19.45
2220	10″ diameter, 16 ga.		360	.133		13.15	4.64	.94	18.73	22.50
2240	12″ diameter, 16 ga.		285	.168		14.70	5.85	1.19	21.74	26.50
2260	18″ diameter, 16 ga.		205	.234		22	8.15	1.65	31.80	39
3000	Uncoated galvanized, perforated									
3020	6″ diameter, 18 ga.	B-20	380	.063	L.F.	7.25	2.33		9.58	11.60
3200	8″ diameter, 16 ga.	"	370	.065		10.50	2.40		12.90	15.25
3220	10″ diameter, 16 ga.	B-21	360	.078		15.75	2.96	.38	19.09	22.50
3240	12″ diameter, 16 ga.		285	.098		16.50	3.73	.48	20.71	24.50
3260	18″ diameter, 16 ga.		205	.137		25	5.20	.67	30.87	36
4000	Steel, perforated, asphalt coated									
4020	6″ diameter 18 ga.	B-20	380	.063	L.F.	7.25	2.33		9.58	11.60
4030	8″ diameter 18 ga.	"	370	.065		10.50	2.40		12.90	15.25
4040	10″ diameter 16 ga.	B-21	360	.078		12.75	2.96	.38	16.09	19
4050	12″ diameter 16 ga.		285	.098		14.50	3.73	.48	18.71	22
4060	18″ diameter 16 ga.		205	.137		20.50	5.20	.67	26.37	31

33 46 16.40 Piping, Subdrainage, Polyvinyl Chloride

		Crew	Daily Output	Labor-Hours	Unit	Material	Labor	Equipment	Total	Total Incl O&P
0010	**PIPING, SUBDRAINAGE, POLYVINYL CHLORIDE**									
0020	Perforated, price as solid pipe, Division 33 31 13.25									

33 49 Storm Drainage Structures

33 49 13 – Storm Drainage Manholes, Frames, and Covers

33 49 13.10 Storm Drainage Manholes, Frames and Covers

		Crew	Daily Output	Labor-Hours	Unit	Material	Labor	Equipment	Total	Total Incl O&P
0010	**STORM DRAINAGE MANHOLES, FRAMES & COVERS**									
0020	Excludes footing, excavation, backfill (See line items for frame & cover)									
0050	Brick, 4′ inside diameter, 4′ deep	D-1	1	16	Ea.	440	605		1,045	1,400
0100	6′ deep		.70	22.857		625	860		1,485	1,975
0150	8′ deep		.50	32		800	1,200		2,000	2,700
0200	For depths over 8′, add		4	4	V.L.F.	167	151		318	410
0400	Concrete blocks (radial), 4′ I.D., 4′ deep		1.50	10.667	Ea.	365	400		765	1,000
0500	6′ deep		1	16		490	605		1,095	1,450
0600	8′ deep		.70	22.857		615	860		1,475	1,975
0700	For depths over 8′, add		5.50	2.909	V.L.F.	64.50	110		174.50	236
0800	Concrete, cast in place, 4′ x 4′, 8″ thick, 4′ deep	C-14H	2	24	Ea.	475	985	15.10	1,475.10	2,075
0900	6′ deep		1.50	32		690	1,325	20	2,035	2,800
1000	8′ deep		1	48		990	1,975	30	2,995	4,150
1100	For depths over 8′, add		8	6	V.L.F.	108	246	3.78	357.78	505
1110	Precast, 4′ I.D., 4′ deep	B-22	4.10	7.317	Ea.	450	281	50.50	781.50	980
1120	6′ deep		3	10		800	385	69	1,254	1,550
1130	8′ deep		2	15		925	575	104	1,604	2,025
1140	For depths over 8′, add		16	1.875	V.L.F.	75	72	12.95	159.95	207
1150	5′ I.D., 4′ deep	B-6	3	8	Ea.	1,100	287	113	1,500	1,750
1160	6′ deep		2	12		1,200	430	169	1,799	2,175
1170	8′ deep		1.50	16		1,300	575	225	2,100	2,550
1180	For depths over 8′, add		12	2	V.L.F.	184	71.50	28	283.50	340
1190	6′ I.D., 4′ deep		2	12	Ea.	1,825	430	169	2,424	2,850
1200	6′ deep		1.50	16		1,975	575	225	2,775	3,300
1210	8′ deep		1	24		2,700	860	340	3,900	4,650
1220	For depths over 8′, add		8	3	V.L.F.	310	107	42	459	550
1250	Slab tops, precast, 8″ thick									
1300	4′ diameter manhole	B-6	8	3	Ea.	195	107	42	344	425

33 49 Storm Drainage Structures

33 49 13 – Storm Drainage Manholes, Frames, and Covers

33 49 13.10 Storm Drainage Manholes, Frames and Covers		Crew	Daily Output	Labor-Hours	Unit	2010 Bare Costs Material	Labor	Equipment	Total	Total Incl O&P
1400	5′ diameter manhole	B-6	7.50	3.200	Ea.	400	115	45	560	665
1500	6′ diameter manhole		7	3.429		555	123	48.50	726.50	850
3800	Steps, heavyweight cast iron, 7″ x 9″	1 Bric	40	.200		13.10	8.35		21.45	27
3900	8″ x 9″		40	.200		19.60	8.35		27.95	34
3928	12″ x 10-1/2″		40	.200		22.50	8.35		30.85	37
4000	Standard sizes, galvanized steel		40	.200		18.90	8.35		27.25	33.50
4100	Aluminum		40	.200		25.50	8.35		33.85	41

33 51 Natural-Gas Distribution

33 51 13 – Natural-Gas Piping

33 51 13.10 Piping, Gas Service and Distribution, P.E.		Crew	Daily Output	Labor-Hours	Unit	Material	Labor	Equipment	Total	Total Incl O&P
0010	**PIPING, GAS SERVICE AND DISTRIBUTION, POLYETHYLENE**									
0020	Not including excavation or backfill									
1000	60 psi coils, compression coupling @ 100′, 1/2″ diameter, SDR 11	B-20A	608	.053	L.F.	1.66	2.13		3.79	5.05
1010	1″ diameter, SDR 11		544	.059		1.83	2.38		4.21	5.60
1040	1-1/4″ diameter, SDR 11		544	.059		2.57	2.38		4.95	6.45
1100	2″ diameter, SDR 11		488	.066		3.18	2.65		5.83	7.55
1160	3″ diameter, SDR 11		408	.078		6.65	3.17		9.82	12.10
1500	60 PSI 40′ joints with coupling, 3″ diameter, SDR 11	B-21A	408	.098		6.65	4.05	1.61	12.31	15.15
1540	4″ diameter, SDR 11		352	.114		15.20	4.69	1.86	21.75	26
1600	6″ diameter, SDR 11		328	.122		47.50	5.05	2	54.55	62
1640	8″ diameter, SDR 11		272	.147		64.50	6.05	2.41	72.96	83

33 52 Liquid Fuel Distribution

33 52 16 – Gasoline Distribution

33 52 16.13 Gasoline Piping		Crew	Daily Output	Labor-Hours	Unit	Material	Labor	Equipment	Total	Total Incl O&P
0010	**GASOLINE PIPING**									
0020	Primary containment pipe, fiberglass-reinforced									
0030	Plastic pipe 15′ & 30′ lengths									
0040	2″ diameter	Q-6	425	.056	L.F.	5.10	2.73		7.83	9.70
0050	3″ diameter		400	.060		6.65	2.91		9.56	11.70
0060	4″ diameter		375	.064		8.60	3.10		11.70	14.10
0100	Fittings									
0110	Elbows, 90° & 45°, bell-ends, 2″	Q-6	24	1	Ea.	51.50	48.50		100	129
0120	3″ diameter		22	1.091		54	53		107	139
0130	4″ diameter		20	1.200		71.50	58		129.50	166
0200	Tees, bell ends, 2″		21	1.143		62.50	55.50		118	152
0210	3″ diameter		18	1.333		63	64.50		127.50	166
0220	4″ diameter		15	1.600		86	77.50		163.50	211
0230	Flanges bell ends, 2″		24	1		20.50	48.50		69	95
0240	3″ diameter		22	1.091		26	53		79	108
0250	4″ diameter		20	1.200		35.50	58		93.50	126
0260	Sleeve couplings, 2″		21	1.143		13.15	55.50		68.65	97.50
0270	3″ diameter		18	1.333		18.60	64.50		83.10	118
0280	4″ diameter		15	1.600		25.50	77.50		103	145
0290	Threaded adapters 2″		21	1.143		17.25	55.50		72.75	102
0300	3″ diameter		18	1.333		30.50	64.50		95	131
0310	4″ diameter		15	1.600		41	77.50		118.50	161
0320	Reducers, 2″		27	.889		23	43		66	89.50

33 52 Liquid Fuel Distribution

33 52 16 – Gasoline Distribution

33 52 16.13 Gasoline Piping		Crew	Daily Output	Labor-Hours	Unit	Material	2010 Bare Costs Labor	Equipment	Total	Total Incl O&P
0330	3″ diameter	Q-6	22	1.091	Ea.	26.50	53		79.50	108
0340	4″ diameter		20	1.200		34.50	58		92.50	125
1010	Gas station product line for secondary containment (double wall)									
1100	Fiberglass reinforced plastic pipe 25′ lengths									
1120	Pipe, plain end, 3″ diameter	Q-6	375	.064	L.F.	8.75	3.10		11.85	14.25
1130	4″ diameter		350	.069		14.35	3.32		17.67	20.50
1140	5″ diameter		325	.074		18.60	3.58		22.18	26
1150	6″ diameter		300	.080		19.15	3.87		23.02	27
1200	Fittings									
1230	Elbows, 90° & 45°, 3″ diameter	Q-6	18	1.333	Ea.	63	64.50		127.50	166
1240	4″ diameter		16	1.500		110	72.50		182.50	230
1250	5″ diameter		14	1.714		254	83		337	405
1260	6″ diameter		12	2		257	97		354	430
1270	Tees, 3″ diameter		15	1.600		93	77.50		170.50	218
1280	4″ diameter		12	2		137	97		234	296
1290	5″ diameter		9	2.667		277	129		406	500
1300	6″ diameter		6	4		290	194		484	610
1310	Couplings, 3″ diameter		18	1.333		44	64.50		108.50	146
1320	4″ diameter		16	1.500		114	72.50		186.50	234
1330	5″ diameter		14	1.714		237	83		320	385
1340	6″ diameter		12	2		245	97		342	415
1350	Cross-over nipples, 3″ diameter		18	1.333		10.10	64.50		74.60	108
1360	4″ diameter		16	1.500		11.85	72.50		84.35	122
1370	5″ diameter		14	1.714		17.55	83		100.55	143
1380	6″ diameter		12	2		18.40	97		115.40	166
1400	Telescoping, reducers, concentric 4″ x 3″		18	1.333		33.50	64.50		98	134
1410	5″ x 4″		17	1.412		87.50	68.50		156	199
1420	6″ x 5″		16	1.500		210	72.50		282.50	340

33 71 Electrical Utility Transmission and Distribution

33 71 16 – Electrical Utility Poles

33 71 16.33 Wood Electrical Utility Poles		Crew	Daily Output	Labor-Hours	Unit	Material	Labor	Equipment	Total	Total Incl O&P
0010	**WOOD ELECTRICAL UTILITY POLES**									
0011	Excludes excavation, backfill and cast-in-place concrete									
6200	Electric & tel sitework, 20′ high, treated wd, see Div. 26 56 13.10	R-3	3.10	6.452	Ea.	266	310	44.50	620.50	805
6400	25′ high		2.90	6.897		281	335	47.50	663.50	860
6600	30′ high		2.60	7.692		310	370	53	733	955
6800	35′ high		2.40	8.333		405	400	57.50	862.50	1,100
7000	40′ high		2.30	8.696		495	420	60	975	1,225
7200	45′ high		1.70	11.765		605	570	81.50	1,256.50	1,600
7400	Cross arms with hardware & insulators									
7600	4′ long	1 Elec	2.50	3.200	Ea.	126	157		283	370
7800	5′ long		2.40	3.333		146	163		309	405
8000	6′ long		2.20	3.636		169	178		347	450

33 71 19 – Electrical Underground Ducts and Manholes

33 71 19.17 Electric and Telephone Underground		Crew	Daily Output	Labor-Hours	Unit	Material	Labor	Equipment	Total	Total Incl O&P
0010	**ELECTRIC AND TELEPHONE UNDERGROUND**									
0011	Not including excavation									
0200	backfill and cast in place concrete									
0400	Hand holes, precast concrete, with concrete cover									

33 71 Electrical Utility Transmission and Distribution

33 71 19 – Electrical Underground Ducts and Manholes

33 71 19.17 Electric and Telephone Underground		Crew	Daily Output	Labor-Hours	Unit	2010 Bare Costs Material	Labor	Equipment	Total	Total Incl O&P
0600	2′ x 2′ x 3′ deep	R-3	2.40	8.333	Ea.	395	400	57.50	852.50	1,100
0800	3′ x 3′ x 3′ deep		1.90	10.526		510	510	72.50	1,092.50	1,400
1000	4′ x 4′ x 4′ deep	↓	1.40	14.286	↓	1,025	690	98.50	1,813.50	2,250
1200	Manholes, precast with iron racks & pulling irons, C.I. frame									
1400	and cover, 4′ x 6′ x 7′ deep	B-13	2	28	Ea.	1,975	1,000	375	3,350	4,075
1600	6′ x 8′ x 7′ deep		1.90	29.474		2,400	1,050	395	3,845	4,675
1800	6′ x 10′ x 7′ deep	↓	1.80	31.111	↓	2,700	1,100	415	4,215	5,125
4200	Underground duct, banks ready for concrete fill, min. of 7.5″									
4400	between conduits, center to center									
4580	PVC, type EB, 1 @ 2″ diameter	2 Elec	480	.033	L.F.	.81	1.63		2.44	3.32
4600	2 @ 2″ diameter		240	.067		1.62	3.27		4.89	6.65
4800	4 @ 2″ diameter		120	.133		3.24	6.55		9.79	13.25
5000	2 @ 3″ diameter		200	.080		2.17	3.92		6.09	8.25
5200	4 @ 3″ diameter		100	.160		4.35	7.85		12.20	16.45
5400	2 @ 4″ diameter		160	.100		3.26	4.90		8.16	10.90
5600	4 @ 4″ diameter		80	.200		6.50	9.80		16.30	22
5800	6 @ 4″ diameter		54	.296		9.80	14.50		24.30	32.50
6200	Rigid galvanized steel, 2 @ 2″ diameter		180	.089		14.30	4.36		18.66	22
6400	4 @ 2″ diameter		90	.178		28.50	8.70		37.20	44.50
6800	2 @ 3″ diameter		100	.160		34.50	7.85		42.35	49
7000	4 @ 3″ diameter		50	.320		68.50	15.70		84.20	99
7200	2 @ 4″ diameter		70	.229		49.50	11.20		60.70	71
7400	4 @ 4″ diameter		34	.471		99	23		122	144
7600	6 @ 4″ diameter	↓	22	.727	↓	148	35.50		183.50	216

33 81 Communications Structures

33 81 13 – Communications Transmission Towers

33 81 13.10 Radio Towers		Crew	Daily Output	Labor-Hours	Unit	Material	Labor	Equipment	Total	Total Incl O&P
0010	**RADIO TOWERS**									
0020	Guyed, 50′ H, 40 lb. sec., 70 MPH basic wind spd.	2 Sswk	1	16	Ea.	3,225	750		3,975	4,875
0100	Wind load 90 MPH basic wind speed	″	1	16		3,225	750		3,975	4,875
0300	190′ high, 40 lb. section, wind load 70 MPH basic wind speed	K-2	.33	72.727		8,750	3,100	895	12,745	15,900
0400	200′ high, 70 lb. section, wind load 90 MPH basic wind speed		.33	72.727		17,600	3,100	895	21,595	25,700
0600	300′ high, 70 lb. section, wind load 70 MPH basic wind speed		.20	120		25,000	5,125	1,475	31,600	37,800
0700	270′ high, 90 lb. section, wind load 90 MPH basic wind speed		.20	120		28,800	5,125	1,475	35,400	42,000
0800	400′ high, 100 lb. section, wind load 70 MPH basic wind speed		.14	171		42,000	7,325	2,125	51,450	61,000
0900	Self-supporting, 60′ high, wind load 70 MPH basic wind speed		.80	30		6,750	1,275	370	8,395	10,000
0910	60′ high, wind load 90 MPH basic wind speed		.45	53.333		12,100	2,275	660	15,035	18,000
1000	120′ high, wind load 70 MPH basic wind speed		.40	60		16,600	2,550	740	19,890	23,400
1200	190′ high, wind load 90 MPH basic wind speed	↓	.20	120		40,200	5,125	1,475	46,800	54,500
2000	For states west of Rocky Mountains, add for shipping				↓	10%				

Division Notes

	CREW	DAILY OUTPUT	LABOR-HOURS	UNIT	2010 BARE COSTS MAT.	LABOR	EQUIP.	TOTAL	TOTAL INCL O&P

Estimating Tips

34 11 00 Rail Tracks

This subdivision includes items that may involve either repair of existing, or construction of new, railroad tracks. Additional preparation work, such as the roadbed earthwork, would be found in Division 31. Additional new construction siding and turnouts are found in Subdivision 34 72. Maintenance of railroads is found under 34 01 23 Operation and Maintenance of Railways.

34 41 13 Traffic Signals

This subdivision includes traffic signal systems. Other traffic control devices such as traffic signs are found in Subdivision 10 14 53 Traffic Signage.

34 71 13 Vehicle Barriers

This subdivision includes security vehicle barriers, guide and guard rails, crash barriers, and delineators. The actual maintenance and construction of concrete and asphalt pavement is found in Division 32.

Reference Numbers

Reference numbers are shown in shaded boxes at the beginning of some major classifications. These numbers refer to related items in the Reference Section. The reference information may be an estimating procedure, an alternate pricing method, or technical information.

Note: Not all subdivisions listed here necessarily appear in this publication.

34 01 Operation and Maintenance of Transportation

34 01 23 – Operation and Maintenance of Railways

34 01 23.51 Maintenance of Railroads		Crew	Daily Output	Labor-Hours	Unit	Material	2010 Bare Costs Labor	Equipment	Total	Total Incl O&P
0010	**MAINTENANCE OF RAILROADS**									
0400	Resurface and realign existing track	B-14	200	.240	L.F.		8.35	1.69	10.04	14.65
0600	For crushed stone ballast, add	"	500	.096	"	14	3.34	.68	18.02	21

34 11 Rail Tracks

34 11 13 – Track Rails

34 11 13.23 Heavy Rail Track

		Crew	Daily Output	Labor-Hours	Unit	Material	Labor	Equipment	Total	Total Incl O&P
0010	**HEAVY RAIL TRACK** R347216-10									
1000	Rail, 100 lb. prime grade				L.F.	28.50			28.50	31.50
1500	Relay rail				"	14.35			14.35	15.75

34 11 33 – Track Cross Ties

34 11 33.13 Concrete Track Cross Ties

		Crew	Daily Output	Labor-Hours	Unit	Material	Labor	Equipment	Total	Total Incl O&P
0010	**CONCRETE TRACK CROSS TIES**									
1400	Ties, concrete, 8′-6″ long, 30″ O.C.	B-14	80	.600	Ea.	145	21	4.22	170.22	197

34 11 33.16 Timber Track Cross Ties

		Crew	Daily Output	Labor-Hours	Unit	Material	Labor	Equipment	Total	Total Incl O&P
0010	**TIMBER TRACK CROSS TIES**									
1600	Wood, pressure treated, 6″ x 8″ x 8′-6″, C.L. lots	B-14	90	.533	Ea.	42.50	18.55	3.75	64.80	79.50
1700	L.C.L. lots		90	.533		45	18.55	3.75	67.30	82
1900	Heavy duty, 7″ x 9″ x 8′-6″, C.L. lots		70	.686		47	24	4.83	75.83	93.50
2000	L.C.L. lots	↓	70	.686	↓	47	24	4.83	75.83	93.50

34 11 33.17 Timber Switch Ties

		Crew	Daily Output	Labor-Hours	Unit	Material	Labor	Equipment	Total	Total Incl O&P
0010	**TIMBER SWITCH TIES**									
1200	Switch timber, for a #8 switch, pressure treated	B-14	3.70	12.973	M.B.F.	2,550	450	91.50	3,091.50	3,625
1300	Complete set of timbers, 3.7 M.B.F. for #8 switch	"	1	48	Total	9,850	1,675	340	11,865	13,700

34 11 93 – Track Appurtenances and Accessories

34 11 93.50 Track Accessories

		Crew	Daily Output	Labor-Hours	Unit	Material	Labor	Equipment	Total	Total Incl O&P
0010	**TRACK ACCESSORIES**									
0020	Car bumpers, test	B-14	2	24	Ea.	3,150	835	169	4,154	4,925
0100	Heavy duty R347216-20		2	24		5,975	835	169	6,979	8,025
0200	Derails hand throw (sliding)		10	4.800		1,000	167	34	1,201	1,425
0300	Hand throw with standard timbers, open stand & target		8	6	↓	1,100	209	42	1,351	1,575
2400	Wheel stops, fixed		18	2.667	Pr.	790	93	18.75	901.75	1,025
2450	Hinged	↓	14	3.429	"	1,075	119	24	1,218	1,375

34 11 93.60 Track Material

		Crew	Daily Output	Labor-Hours	Unit	Material	Labor	Equipment	Total	Total Incl O&P
0010	**TRACK MATERIAL**									
0020	Track bolts				Ea.	3.31			3.31	3.64
0100	Joint bars				Pr.	76.50			76.50	84
0200	Spikes				Ea.	.77			.77	.85
0300	Tie plates				"	12.80			12.80	14.05

34 41 Roadway Signaling and Control Equipment

34 41 13 – Traffic Signals

34 41 13.10 Traffic Signals Systems		Crew	Daily Output	Labor-Hours	Unit	Material	2010 Bare Costs Labor	Equipment	Total	Total Incl O&P
0010	**TRAFFIC SIGNALS SYSTEMS**									
0020	Mid block pedestrian crosswalk with pushbutton and mast arms	R-11	.30	186	Total	86,000	8,625	2,850	97,475	110,500
0600	Traffic signals, school flashing system, solar powered, remote controlled	"	1	56	Signal	19,700	2,575	860	23,135	26,500
1000	Intersection traffic signals, LED, mast, programmable, no lane control									
1010	Includes all labor, material and equip. for complete installation	R-11	1	56	Signal	200,000	2,575	860	203,435	225,000
1100	Intersection traffic signals, LED, mast, programmable, R/L lane control									
1110	Includes all labor, material and equip. for complete installation	R-11	1	56	Signal	262,000	2,575	860	265,435	293,000
1200	Add protective/permissive left turns to existing traffic light									
1210	Includes all labor, material and equip. for complete installation	R-11	1	56	Signal	80,000	2,575	860	83,435	93,000
1300	Replace existing light heads with LED heads wire hung									
1310	Includes all labor, material and equip. for complete installation	R-11	1	56	Signal	54,000	2,575	860	57,435	64,500
1400	Replace existing light heads with LED heads mast arm hung									
1410	Includes all labor, material and equip. for complete installation	R-11	1	56	Signal	100,000	2,575	860	103,435	115,000

34 71 Roadway Construction

34 71 13 – Vehicle Barriers

34 71 13.17 Security Vehicle Barriers		Crew	Daily Output	Labor-Hours	Unit	Material	Labor	Equipment	Total	Total Incl O&P
0010	**SECURITY VEHICLE BARRIERS**									
0020	Security planters excludes filling material									
0100	Concrete security planter, exposed aggregate 36" dia x 30" high	B-11M	8	2	Ea.	455	76	48.50	579.50	670
0200	48" dia x 36" high		8	2		495	76	48.50	619.50	715
0300	53" dia x 18" high		8	2		585	76	48.50	709.50	815
0400	72" dia x 18" high with seats		8	2		1,050	76	48.50	1,174.50	1,325
0450	84" dia x 36" high		8	2		1,300	76	48.50	1,424.50	1,600
0500	36" x 36" x 24" high square		8	2		385	76	48.50	509.50	590
0600	36" x 36" x 30" high square		8	2		455	76	48.50	579.50	670
0700	48" L x 24" W x 30" H rectangle		8	2		495	76	48.50	619.50	715
0800	72" L x 24" W x 30" H rectangle		8	2		650	76	48.50	774.50	885
0900	96" L x 24" W x 30" H rectangle		8	2		780	76	48.50	904.50	1,025
0950	Decorative geometric concrete barrier, 96" L x 24" W x 36" H		8	2		970	76	48.50	1,094.50	1,250
1000	Concrete security planter, filling with washed sand or gravel °1 CY		8	2		31	76	48.50	155.50	203
1050	Concrete security planter, filling with washed sand or gravel °2 CY	↓	6	2.667		62.50	101	64.50	228	294
1200	Jersey concrete barrier, 10' L x 2' by 0.5' W x 30" H	B-21B	16	2.500		410	89.50	41	540.50	630
1300	10' L x 2' by 0.5' W x 30" H, 10 or more same site		24	1.667		410	59.50	27.50	497	570
1400	Jersey concrete barrier, 10' L x 2' by 0.5' W x 32" H		16	2.500		410	89.50	41	540.50	630
1500	10' L x 2' by 6" W x 32" H, 10 or more same site		24	1.667		410	59.50	27.50	497	570
1600	Jersey concrete barrier, 20' L x 2' by 0.5' W x 30" H		12	3.333		635	119	54.50	808.50	940
1700	20' L x 2' by 0.5' W x 30" H, 10 or more same site		18	2.222		635	79.50	36.50	751	860
1800	Jersey concrete barrier, 20' L x 2' by 0.5' W x 32" H		12	3.333		920	119	54.50	1,093.50	1,250
1900	20' L x 2' by 0.5' W x 32" H, 10 or more same site		18	2.222		920	79.50	36.50	1,036	1,150
2000	GFRC decorative security barrier per 10 feet section including concrete		4	10		2,225	360	164	2,749	3,175
2100	GFRC decorative security barrier per 12 feet section including concrete	↓	4	10	↓	2,675	360	164	3,199	3,650
2210	GFRC decorative security barrier will stop 30 MPH, 4000 lb vehicle									
2300	High security barrier base prep per 12 feet section on bare ground	B-11C	4	4	Ea.	31.50	152	84.50	268	360
2310	GFRC barrier base prep does not include haul away of excavated matl									
2400	GFRC decorative high security barrier per 12 feet section w/concrete	B-21B	3	13.333	Ea.	5,525	475	219	6,219	7,050
2410	GFRC decorative high security barrier will stop 50 MPH, 15000 lb vehl									
2500	GFRC decorative impaler security barrier per 10 feet section w/prep	B-6	4	6	Ea.	2,150	215	84.50	2,449.50	2,800
2600	GFRC decorative impaler security barrier per 12 feet section w/prep	"	4	6	"	2,575	215	84.50	2,874.50	3,275
2610	Impaler barrier should stop 50 MPH, 15000 lb vehl w/some penetr									
2700	Pipe bollards, steel, concrete filled/painted, 8' L x 4' D hole, 8" diam.	B-6	10	2.400	Ea.	945	86	34	1,065	1,225

34 71 Roadway Construction

34 71 13 – Vehicle Barriers

34 71 13.17 Security Vehicle Barriers		Crew	Daily Output	Labor-Hours	Unit	Material	2010 Bare Costs Labor	Equipment	Total	Total Incl O&P
2710	Schedule 80 concrete bollards will stop 4000 lb vehicle @ 30 MPH									
2800	GFRC decorative jersey barrier cover per 10 feet section excludes soil	B-6	8	3	Ea.	905	107	42	1,054	1,200
2900	GFRC decorative jersey barrier cover per 12 feet section excludes soil		8	3		1,075	107	42	1,224	1,400
3000	GFRC decorative 8" diameter bollard cover		12	2		475	71.50	28	574.50	665
3100	GFRC decorative barrier face 10 foot section excludes earth backing		10	2.400		905	86	34	1,025	1,175
3200	Drop arm crash barrier, 15000 lb vehl @ 50 MPH									
3205	Includes all material, labor for complete installation									
3210	12' width				Ea.				67,000	74,000
3310	24' width								89,000	98,000
3410	Wedge crash barrier, 10' width								88,000	97,000
3510	12.5' width								99,000	109,000
3520	15' width								111,000	122,000
3610	Sliding crash barrier, 12' width								143,000	157,000
3710	Sliding roller crash barrier, 20' width								164,000	180,000
3810	Sliding cantilever crash barrier, 20' width								164,000	180,000
3890	Note: Raised bollard crash barriers should be used w/tire shredders									
3910	Raised bollard crash barrier, 10' width				Ea.				36,000	40,000
4010	12' width								42,000	46,000
4110	Raised bollard crash barrier, 10' width, solar powered								44,000	48,000
4210	12' width								49,000	54,000
4310	In ground tire shredder, 16' width								41,000	45,000

34 71 13.26 Vehicle Guide Rails		Crew	Daily Output	Labor-Hours	Unit	Material	Labor	Equipment	Total	Total Incl O&P
0010	**VEHICLE GUIDE RAILS**									
0012	Corrugated stl, galv. stl posts, 6'-3" O.C.	B-80	850	.038	L.F.	21.50	1.33	.88	23.71	27
0200	End sections, galvanized, flared		50	.640	Ea.	90	22.50	14.90	127.40	150
0300	Wrap around end		50	.640	"	135	22.50	14.90	172.40	200
0400	Timber guide rail, 4" x 8" with 6" x 8" wood posts, treated		960	.033	L.F.	25	1.18	.78	26.96	30
0600	Cable guide rail, 3 at 3/4" cables, steel posts, single face		900	.036		9.20	1.26	.83	11.29	12.95
0700	Wood posts		950	.034		12.75	1.19	.78	14.72	16.75
0900	Guide rail, steel box beam, 6" x 6"		120	.267		31.50	9.45	6.20	47.15	56
1100	Median barrier, steel box beam, 6" x 8"		215	.149		41.50	5.25	3.47	50.22	57.50
1400	Resilient guide fence and light shield, 6' high	B-2	130	.308		35	10.30		45.30	54.50
1500	Concrete posts, individual, 6'-5", triangular	B-80	110	.291	Ea.	64.50	10.30	6.80	81.60	94
1550	Square	"	110	.291	"	69	10.30	6.80	86.10	99
2000	Median, precast concrete, 3'-6" high, 2' wide, single face	B-29	380	.147	L.F.	50	5.25	2.14	57.39	65.50
2200	Double face	"	340	.165	"	57.50	5.90	2.39	65.79	74.50
2400	Speed bumps, thermoplastic, 10-1/2" x 2-1/4" x 48" long	B-2	120	.333	Ea.	135	11.15		146.15	166
3030	Impact barrier, UTMCD, barrel type	B-16	30	1.067	"	485	36	22.50	543.50	610

34 71 19 – Vehicle Delineators

34 71 19.13 Fixed Vehicle Delineators		Crew	Daily Output	Labor-Hours	Unit	Material	Labor	Equipment	Total	Total Incl O&P
0010	**FIXED VEHICLE DELINEATORS**									
0020	Crash barriers									
0100	Traffic channelizing pavement markers, layout only	A-7	2000	.012	Ea.		.54	.03	.57	.86
0110	13" x 7-1/2" x 2-1/2" high, non-plowable install	2 Clab	96	.167		26	5.50		31.50	37
0200	8" x 8" x 3-1/4" high, non-plowable, install		96	.167		23.50	5.50		29	34.50
0230	4" x 4" x 3/4" high, non-plowable, install		120	.133		3.52	4.41		7.93	10.65
0240	9-1/4" x 5-7/8" x 1/4" high, plowable, concrete pav't	A-2A	70	.343		20	11.25	5.75	37	45.50
0250	9-1/4" x 5-7/8" x 1/4" high, plowable, asphalt pav't	"	120	.200		4.02	6.55	3.34	13.91	18.20
0300	Barrier and curb delineators, reflectorized, 2" x 4"	2 Clab	150	.107		2.26	3.53		5.79	7.95
0310	3" x 5"	"	150	.107		4.52	3.53		8.05	10.40
0500	Rumble strip, polycarbonate									
0510	24" x 3-1/2" x 1/2" high	2 Clab	50	.320	Ea.	8.05	10.60		18.65	25

34 72 Railway Construction

34 72 16 – Railway Siding

34 72 16.50 Railroad Sidings		Crew	Daily Output	Labor-Hours	Unit	Material	2010 Bare Costs Labor	Equipment	Total	Total Incl O&P
0010	**RAILROAD SIDINGS** R347216-10									
0800	Siding, yard spur, level grade									
0820	100 lb. new rail	B-14	57	.842	L.F.	110	29.50	5.95	145.45	173
1002	Steel ties in concrete, incl. fasteners & plates									
1020	100 lb. new rail	B-14	22	2.182	L.F.	156	76	15.35	247.35	305

34 72 16.60 Railroad Turnouts		Crew	Daily Output	Labor-Hours	Unit	Material	Labor	Equipment	Total	Total Incl O&P
0010	**RAILROAD TURNOUTS**									
2200	Turnout, #8 complete, w/rails, plates, bars, frog, switch point,									
2250	timbers, and ballast to 6" below bottom of ties									
2280	90 lb. rails	B-13	.25	224	Ea.	33,000	8,000	3,000	44,000	52,000
2290	90 lb. relay rails		.25	224		22,200	8,000	3,000	33,200	39,900
2300	100 lb. rails		.25	224		36,800	8,000	3,000	47,800	56,000
2310	100 lb. relay rails		.25	224		24,700	8,000	3,000	35,700	42,700
2320	110 lb. rails		.25	224		40,600	8,000	3,000	51,600	60,000
2330	110 lb. relay rails		.25	224		27,300	8,000	3,000	38,300	45,500
2340	115 lb. rails		.25	224		44,400	8,000	3,000	55,400	64,500
2350	115 lb. relay rails		.25	224		29,800	8,000	3,000	40,800	48,300
2360	132 lb. rails		.25	224		51,000	8,000	3,000	62,000	71,500
2370	132 lb. relay rails	▼	.25	224	▼	33,600	8,000	3,000	44,600	52,500

Division Notes

		CREW	DAILY OUTPUT	LABOR-HOURS	UNIT	2010 BARE COSTS MAT.	LABOR	EQUIP.	TOTAL	TOTAL INCL O&P

Estimating Tips

35 20 16 Hydraulic Gates

This subdivision includes various types of gates that are commonly used in waterway and canal construction. Various earthwork items and structural support is found in Division 31, and concrete work in Division 3.

35 20 23 Dredging

This subdivision includes barge and shore dredging systems for rivers, canals, and channels.

35 31 00 Shoreline Protection

This subdivision includes breakwaters, bulkheads, and revetments for ocean and river inlets. Additional earthwork may be required from Division 31, and concrete work from Division 3.

35 41 00 Levees

Information on levee construction, including estimated cost of clay cone material.

35 49 00 Waterway Structures

This subdivision includes breakwaters and bulkheads for canals.

35 51 00 Floating Construction

This section includes floating piers, docks, and dock accessories. Fixed Pier Timber Construction is found in 06 13 33. Driven piles are found in Division 31, as well as sheet piling, cofferdams, and riprap.

Reference Numbers

Reference numbers are shown in shaded boxes at the beginning of some major classifications. These numbers refer to related items in the Reference Section. The reference information may be an estimating procedure, an alternate pricing method, or technical information.

Note: Not all subdivisions listed here necessarily appear in this publication.

35 20 Waterway and Marine Construction and Equipment

35 20 23 – Dredging

35 20 23.13 Mechanical Dredging		Crew	Daily Output	Labor-Hours	Unit	Material	2010 Bare Costs Labor	Equipment	Total	Total Incl O&P
0010	**MECHANICAL DREDGING**									
0020	Dredging mobilization and demobilization, add to below, minimum	B-8	.53	120	Total		4,400	5,975	10,375	13,300
0100	Maximum	"	.10	640	"		23,300	31,600	54,900	70,500
0300	Barge mounted clamshell excavation into scows									
0310	Dumped 20 miles at sea, minimum	B-57	310	.155	B.C.Y.		5.80	5.50	11.30	14.90
0400	Maximum	"	213	.225	"		8.45	8.05	16.50	21.50
0500	Barge mounted dragline or clamshell, hopper dumped,									
0510	pumped 1000′ to shore dump, minimum	B-57	340	.141	B.C.Y.		5.30	5.05	10.35	13.60
0525	All pumping uses 2000 gallons of water per cubic yard									
0600	Maximum	B-57	243	.198	B.C.Y.		7.40	7.05	14.45	19

35 20 23.23 Hydraulic Dredging		Crew	Daily Output	Labor-Hours	Unit	Material	Labor	Equipment	Total	Total Incl O&P
0010	**HYDRAULIC DREDGING**									
1000	Hydraulic method, pumped 1000′ to shore dump, minimum	B-57	460	.104	B.C.Y.		3.92	3.72	7.64	10.05
1100	Maximum		310	.155			5.80	5.50	11.30	14.90
1400	Into scows dumped 20 miles, minimum		425	.113			4.24	4.03	8.27	10.90
1500	Maximum	↓	243	.198			7.40	7.05	14.45	19
1600	For inland rivers and canals in South, deduct				↓				30%	30%

35 51 Floating Construction

35 51 13 – Floating Piers

35 51 13.23 Floating Wood Piers		Crew	Daily Output	Labor-Hours	Unit	Material	Labor	Equipment	Total	Total Incl O&P
0010	**FLOATING WOOD PIERS**									
0020	Polyethylene encased polystyrene, no pilings included	F-3	330	.121	S.F.	38.50	5.10	1.99	45.59	52
0200	Pile supported, shore constructed, bare, 3″ decking		130	.308		26.50	12.95	5.05	44.50	54.50
0250	4″ decking		120	.333		25.50	14.05	5.45	45	55.50
0400	Floating, small boat, prefab, no shore facilities, minimum		250	.160		13.30	6.75	2.62	22.67	28
0500	Maximum		150	.267	↓	44	11.25	4.37	59.62	70.50
0700	Per slip, minimum (180 S.F. each)		1.59	25.157	Ea.	2,900	1,050	410	4,360	5,250
0800	Maximum	↓	1.40	28.571	"	8,500	1,200	470	10,170	11,700

Estimating Tips

Products such as conveyors, material handling cranes and hoists, as well as other items specified in this division, may require trained installers. The general contractor may not have any choice as to who will perform the installation or when it will be performed. Long lead times are often required for these products, making early decisions in purchasing and scheduling necessary.

The installation of this type of equipment may require the embedment of mounting hardware during construction of floors, structural walls, or interior walls/partitions. Electrical connections will require coordination with the electrical contractor.

Reference Numbers

Reference numbers are shown in shaded boxes at the beginning of some major classifications. These numbers refer to related items in the Reference Section. The reference information may be an estimating procedure, an alternate pricing method, or technical information.

Note: Not all subdivisions listed here necessarily appear in this publication.

41 21 Conveyors

41 21 23 – Piece Material Conveyors

41 21 23.16 Material Handling Conveyors		Crew	Daily Output	Labor-Hours	Unit	2010 Bare Costs Material	Labor	Equipment	Total	Total Incl O&P
0010	**MATERIAL HANDLING CONVEYORS**									
0020	Gravity fed, 2" rollers, 3" O.C.									
0050	10' sections with 2 supports, 600 lb. capacity, 18" wide				Ea.	450			450	495
0100	24" wide					515			515	570
0150	1400 lb. capacity, 18" wide					415			415	455
0200	24" wide					460			460	505
0350	Horizontal belt, center drive and takeup, 60 fpm									
0400	16" belt, 26.5' length	2 Mill	.50	32	Ea.	3,150	1,375		4,525	5,475
0450	24" belt, 41.5' length		.40	40		4,900	1,725		6,625	7,875
0500	61.5' length		.30	53.333		6,525	2,300		8,825	10,500
0600	Inclined belt, 10' rise with horizontal loader and									
0620	End idler assembly, 27.5' length, 18" belt	2 Mill	.30	53.333	Ea.	9,200	2,300		11,500	13,500
0700	24" belt	"	.15	106	"	10,300	4,575		14,875	18,100
3600	Monorail, overhead, manual, channel type									
3700	125 lb. per L.F.	1 Mill	26	.308	L.F.	10.25	13.20		23.45	30.50
3900	500 lb. per L.F.	"	21	.381	"	9.25	16.35		25.60	34
4000	Trolleys for above, 2 wheel, 125 lb. capacity				Ea.	77			77	84.50
4200	4 wheel, 250 lb. capacity					207			207	227
4300	8 wheel, 1,000 lb. capacity					380			380	420

41 22 Cranes and Hoists

41 22 13 – Cranes

41 22 13.10 Crane Rail

		Crew	Daily Output	Labor-Hours	Unit	Material	Labor	Equipment	Total	Total Incl O&P
0010	**CRANE RAIL**									
0020	Box beam bridge, no equipment included	E-4	3400	.009	Lb.	1.19	.45	.04	1.68	2.14
0200	Running track only, 104 lb per yard	"	5600	.006	"	.62	.27	.03	.92	1.19

41 22 13.13 Bridge Cranes

		Crew	Daily Output	Labor-Hours	Unit	Material	Labor	Equipment	Total	Total Incl O&P
0010	**BRIDGE CRANES**									
0100	1 girder, 20' span, 3 ton	M-3	1	34	Ea.	23,800	1,650	156	25,606	28,800
0125	5 ton		1	34		26,000	1,650	156	27,806	31,200
0150	7.5 ton		1	34		31,000	1,650	156	32,806	36,700
0175	10 ton		.80	42.500		41,000	2,050	196	43,246	48,300
0200	15 ton		.80	42.500		52,500	2,050	196	54,746	61,500
0225	30' span, 3 ton		1	34		24,700	1,650	156	26,506	29,800
0250	5 ton		1	34		27,200	1,650	156	29,006	32,500
0275	7.5 ton		1	34		32,600	1,650	156	34,406	38,400
0300	10 ton		.80	42.500		42,400	2,050	196	44,646	49,900
0325	15 ton		.80	42.500		55,000	2,050	196	57,246	64,000
0350	2 girder, 40' span, 3 ton	M-4	.50	72		40,900	3,450	505	44,855	50,500
0375	5 ton		.50	72		42,800	3,450	505	46,755	53,000
0400	7.5 ton		.50	72		46,900	3,450	505	50,855	57,000
0425	10 ton		.40	90		55,000	4,325	635	59,960	67,500
0450	15 ton		.40	90		74,500	4,325	635	79,460	89,000
0475	25 ton		.30	120		88,000	5,750	845	94,595	106,500
0500	50' span, 3 ton		.50	72		46,500	3,450	505	50,455	56,500
0525	5 ton		.50	72		48,400	3,450	505	52,355	58,500
0550	7.5 ton		.50	72		52,000	3,450	505	55,955	62,500
0575	10 ton		.40	90		59,500	4,325	635	64,460	72,500
0600	15 ton		.40	90		78,000	4,325	635	82,960	93,000
0625	25 ton		.30	120		92,500	5,750	845	99,095	111,000

41 22 Cranes and Hoists

41 22 23 – Hoists

41 22 23.10 Material Handling		Crew	Daily Output	Labor-Hours	Unit	Material	2010 Bare Costs Labor	Equipment	Total	Total Incl O&P
0010	**MATERIAL HANDLING**, cranes, hoists and lifts									
1500	Cranes, portable hydraulic, floor type, 2,000 lb capacity				Ea.	2,475			2,475	2,700
1600	4,000 lb capacity					4,150			4,150	4,575
1800	Movable gantry type, 12′ to 15′ range, 2,000 lb capacity					3,600			3,600	3,975
1900	6,000 lb capacity					5,225			5,225	5,750
2100	Hoists, electric overhead, chain, hook hung, 15′ lift, 1 ton cap.					2,250			2,250	2,475
2200	3 ton capacity					2,800			2,800	3,075
2500	5 ton capacity				↓	6,625			6,625	7,275
2600	For hand-pushed trolley, add					15%				
2700	For geared trolley, add					30%				
2800	For motor trolley, add					75%				
3000	For lifts over 15′, 1 ton, add				L.F.	23.50			23.50	25.50
3100	5 ton, add				"	52			52	57
3300	Lifts, scissor type, portable, electric, 36″ high, 2,000 lb				Ea.	4,275			4,275	4,700
3400	48″ high, 4,000 lb				"	4,675			4,675	5,150

Division Notes

	CREW	DAILY OUTPUT	LABOR-HOURS	UNIT	2010 BARE COSTS MAT.	LABOR	EQUIP.	TOTAL	TOTAL INCL O&P

Estimating Tips

There are four basic types of pollution: Air, Noise, Water, and Solid.

These systems may be interrelated and care must be taken that the complete systems are estimated. For example, Air Pollution Equipment may include dust and air-entrained particles that have to be collected. The vacuum systems could be noisy, requiring silencers to reduce noise pollution, and the collected solids have to be disposed of to prevent Solid Pollution.

Water Treatment may be accomplished as comparatively small packaged plants providing Chemical, Biological or Thermal Treatment. These are usually priced by GPH or GPD.

Large plants and municipal treatment facilities are sized MGD. These facilities are usually priced in two steps. First, the large concrete structures used as settling ponds, etc.; and second, the equipment required. Storm water does not need as much treatment as sanitary wastewater.

The total cost of the system is greatly affected by the level of purification required by specifications and codes before release.

An additional cost may be encountered if there is a possibility that the water being treated may contain petroleum products, which have to be isolated and disposed of in an environmentally acceptable manner.

Reference Numbers

Reference numbers are shown in shaded boxes at the beginning of some major classifications. These numbers refer to related items in the Reference Section. The reference information may be an estimating procedure, an alternate pricing method, or technical information.

Note: Not all subdivisions listed here necessarily appear in this publication.

44 11 16 – Industrial Dust Collectors

44 11 16.10 Dust Collection Systems		Crew	Daily Output	Labor-Hours	Unit	Material	2010 Bare Costs Labor	Equipment	Total	Total Incl O&P
0010	**DUST COLLECTION SYSTEMS** Commercial/Industrial									
0120	Central vacuum units									
0130	Includes stand, filters and motorized shaker									
0200	500 CFM, 10" inlet, 2 HP	Q-20	2.40	8.333	Ea.	3,900	375		4,275	4,875
0220	1000 CFM, 10" inlet, 3 HP		2.20	9.091		4,100	410		4,510	5,150
0240	1500 CFM, 10" inlet, 5 HP		2	10		4,325	450		4,775	5,425
0260	3000 CFM, 13" inlet, 10 HP		1.50	13.333		12,200	600		12,800	14,300
0280	5000 CFM, 16" inlet, 2 @ 10 HP		1	20		12,900	905		13,805	15,600
1000	Vacuum tubing, galvanized									
1100	2-1/8" OD, 16 ga.	Q-9	440	.036	L.F.	2.55	1.61		4.16	5.25
1110	2-1/2" OD, 16 ga.		420	.038		2.93	1.68		4.61	5.75
1120	3" OD, 16 ga.		400	.040		3.35	1.77		5.12	6.35
1130	3-1/2" OD, 16 ga.		380	.042		4.86	1.86		6.72	8.15
1140	4" OD, 16 ga.		360	.044		5.45	1.96		7.41	8.95
1150	5" OD, 14 ga.		320	.050		9.80	2.21		12.01	14.10
1160	6" OD, 14 ga.		280	.057		11.30	2.53		13.83	16.25
1170	8" OD, 14 ga.		200	.080		17.05	3.54		20.59	24
1180	10" OD, 12 ga.		160	.100		17.05	4.42		21.47	25.50
1190	12" OD, 12 ga.		120	.133		29.50	5.90		35.40	41.50
1200	14" OD, 12 ga.		80	.200		36	8.85		44.85	53
1940	Hose, flexible wire reinforced rubber									
1956	3" dia.	Q-9	400	.040	L.F.	7.60	1.77		9.37	11.05
1960	4" dia.		360	.044		8.80	1.96		10.76	12.60
1970	5" dia.		320	.050		11.05	2.21		13.26	15.50
1980	6" dia.		280	.057		11.95	2.53		14.48	16.95
2000	90° Elbow, slip fit									
2110	2-1/8" dia.	Q-9	70	.229	Ea.	10.65	10.10		20.75	27
2120	2-1/2" dia.		65	.246		14.90	10.90		25.80	33
2130	3" dia.		60	.267		20	11.80		31.80	40
2140	3-1/2" dia.		55	.291		25	12.85		37.85	47
2150	4" dia.		50	.320		31	14.15		45.15	55.50
2160	5" dia.		45	.356		58.50	15.70		74.20	88.50
2170	6" dia.		40	.400		81	17.70		98.70	116
2180	8" dia.		30	.533		154	23.50		177.50	205
2400	45° Elbow, slip fit									
2410	2-1/8" dia.	Q-9	70	.229	Ea.	9.30	10.10		19.40	25.50
2420	2-1/2" dia.		65	.246		13.70	10.90		24.60	31.50
2430	3" dia.		60	.267		16.55	11.80		28.35	36
2440	3-1/2" dia.		55	.291		21	12.85		33.85	42.50
2450	4" dia.		50	.320		27	14.15		41.15	51
2460	5" dia.		45	.356		45.50	15.70		61.20	74
2470	6" dia.		40	.400		61.50	17.70		79.20	95
2480	8" dia.		35	.457		121	20		141	164
2800	90° TY, slip fit thru 6" dia.									
2810	2-1/8" dia.	Q-9	42	.381	Ea.	21	16.85		37.85	48.50
2820	2-1/2" dia.		39	.410		27	18.15		45.15	57.50
2830	3" dia.		36	.444		35.50	19.65		55.15	69.50
2840	3-1/2" dia.		33	.485		45.50	21.50		67	82.50
2850	4" dia.		30	.533		65.50	23.50		89	108
2860	5" dia.		27	.593		124	26		150	177
2870	6" dia.		24	.667		169	29.50		198.50	231
2880	8" dia., butt end					360			360	395
2890	10" dia., butt end					650			650	715

44 11 Air Pollution Control Equipment

44 11 16 – Industrial Dust Collectors

44 11 16.10 Dust Collection Systems		Crew	Daily Output	Labor-Hours	Unit	Material	Labor	Equipment	Total	Total Incl O&P
						2010 Bare Costs				
2900	12" dia., butt end				Ea.	650			650	715
2910	14" dia., butt end					1,175			1,175	1,300
2920	6" x 4" dia., butt end					103			103	114
2930	8" x 4" dia., butt end					196			196	215
2940	10" x 4" dia., butt end					256			256	282
2950	12" x 4" dia., butt end				↓	294			294	325
3100	90° Elbow, butt end segmented									
3110	8" dia., butt end, segmented				Ea.	154			154	169
3120	10" dia., butt end, segmented					380			380	420
3130	12" dia., butt end, segmented					485			485	530
3140	14" dia., butt end, segmented				↓	605			605	665
3200	45° Elbow, butt end segmented									
3210	8" dia., butt end, segmented				Ea.	121			121	133
3220	10" dia., butt end, segmented					276			276	305
3230	12" dia., butt end, segmented					282			282	310
3240	14" dia., butt end, segmented				↓	345			345	380
3400	All butt end fittings require one coupling per joint.									
3410	Labor for fitting included with couplings.									
3460	Compression coupling, galvanized, neoprene gasket									
3470	2-1/8" dia.	Q-9	44	.364	Ea.	12.35	16.05		28.40	38
3480	2-1/2" dia.		44	.364		12.35	16.05		28.40	38
3490	3" dia.		38	.421		17.80	18.60		36.40	47.50
3500	3-1/2" dia.		35	.457		20	20		40	52.50
3510	4" dia.		33	.485		22	21.50		43.50	56.50
3520	5" dia.		29	.552		25	24.50		49.50	64.50
3530	6" dia.		26	.615		25	27		52	68.50
3540	8" dia.		22	.727		52	32		84	106
3550	10" dia.		20	.800		70.50	35.50		106	131
3560	12" dia.		18	.889		87	39.50		126.50	156
3570	14" dia.	↓	16	1	↓	111	44		155	189
3800	Air gate valves, galvanized									
3810	2-1/8" dia.	Q-9	30	.533	Ea.	180	23.50		203.50	234
3820	2-1/2" dia.		28	.571		199	25.50		224.50	258
3830	3" dia.		26	.615		217	27		244	280
3840	4" dia.		23	.696		273	31		304	345
3850	6" dia.	↓	18	.889	↓	370	39.50		409.50	465

44 41 Packaged Water Treatment

44 41 13 – Packaged Water Treatment Plants

44 41 13.16 Biological Packaged Water Treatment Plants

		Crew	Daily Output	Labor-Hours	Unit	Material	Labor	Equipment	Total	Total Incl O&P
0010	**BIOLOGICAL PACKAGED WATER TREATMENT PLANTS**									
0011	Not including fencing or external piping									
0020	Steel packaged, blown air aeration plants									
0100	1,000 GPD				Gal.				100	115
0200	5,000 GPD								60	69
0300	15,000 GPD								50	57
0400	30,000 GPD								45	52
0500	50,000 GPD								40	46
0600	100,000 GPD								35	40
0700	200,000 GPD								30	34
0800	500,000 GPD				↓				20	23

44 41 Packaged Water Treatment

44 41 13 – Packaged Water Treatment Plants

44 41 13.16 Biological Packaged Water Treatment Plants		Crew	Daily Output	Labor-Hours	Unit	Material	2010 Bare Costs Labor	Equipment	Total	Total Incl O&P
1000	Concrete, extended aeration, primary and secondary treatment									
1010	10,000 GPD				Gal.				50	57
1100	30,000 GPD								45	52
1200	50,000 GPD								40	46
1400	100,000 GPD								35	40
1500	500,000 GPD				↓				20	23
1700	Municipal wastewater treatment facility									
1720	1.0 MGD				Gal.				10	11.50
1740	1.5 MGD								9.50	10.90
1760	2.0 MGD								9	10.30
1780	3.0 MGD								7	8
1800	5.0 MGD				↓				6	6.90
2000	Holding tank system, not incl. excavation or backfill									
2010	Recirculating chemical water closet	2 Plum	4	4	Ea.	945	208		1,153	1,350
2100	For voltage converter, add	"	16	1		250	52		302	355
2200	For high level alarm, add	1 Plum	7.80	1.026	↓	143	53.50		196.50	237

44 41 13.17 Wastewater Treatment System		Crew	Daily Output	Labor-Hours	Unit	Material	Labor	Equipment	Total	Total Incl O&P
0010	**WASTEWATER TREATMENT SYSTEM**									
0020	Fiberglass, 1,000 gallon	B-21	1.29	21.705	Ea.	3,300	825	107	4,232	5,025
0100	1,500 gallon	B-21	1.03	27.184	Ea.	7,200	1,025	134	8,359	9,650

Reference Section

All the reference information is in one section, making it easy to find what you need to know . . . and easy to use the book on a daily basis. This section is visually identified by a vertical gray bar on the page edges.

In this Reference Section, we've included Equipment Rental Costs, a listing of rental and operating costs; Crew Listings, a full listing of all crews and equipment, and their costs; Historical Cost Indexes for cost comparisons over time; City Cost Indexes and Location Factors for adjusting costs to the region you are in; Reference Tables, where you will find explanations, estimating information and procedures, or technical data; Change Orders, information on pricing changes to contract documents; Square Foot Costs that allow you to make a rough estimate for the overall cost of a project; and an explanation of all the Abbreviations in the book.

Table of Contents

Estimating Tips

- This section contains the average costs to rent and operate hundreds of pieces of construction equipment. This is useful information when estimating the time and material requirements of any particular operation in order to establish a unit or total cost. Equipment costs include not only rental, but also operating costs for equipment under normal use.

Rental Costs

- Equipment rental rates are obtained from industry sources throughout North America-contractors, suppliers, dealers, manufacturers, and distributors.
- Rental rates vary throughout the country, with larger cities generally having lower rates. Lease plans for new equipment are available for periods in excess of six months, with a percentage of payments applying toward purchase.
- Monthly rental rates vary from 2% to 5% of the purchase price of the equipment depending on the anticipated life of the equipment and its wearing parts.
- Weekly rental rates are about 1/3 the monthly rates, and daily rental rates are about 1/3 the weekly rate.

Operating Costs

- The operating costs include parts and labor for routine servicing, such as repair and replacement of pumps, filters and worn lines. Normal operating expendables, such as fuel, lubricants, tires and electricity (where applicable), are also included.
- Extraordinary operating expendables with highly variable wear patterns, such as diamond bits and blades, are excluded. These costs can be found as material costs in the Unit Price section.
- The hourly operating costs listed do not include the operator's wages.

Equipment Cost/Day

- Any power equipment required by a crew is shown in the Crew Listings with a daily cost.
- The daily cost of equipment needed by a crew is based on dividing the weekly rental rate by 5 (number of working days in the week), and then adding the hourly operating cost times 8 (the number of hours in a day). This "Equipment Cost/ Day" is shown in the far right column of the Equipment Rental pages.
- If equipment is needed for only one or two days, it is best to develop your own cost by including components for daily rent and hourly operating cost. This is important when the listed Crew for a task does not contain the equipment needed, such as a crane for lifting mechanical heating/cooling equipment up onto a roof.
- If the quantity of work is less than the crew's Daily Output shown for a Unit Price line item that includes a bare unit equipment cost, it is recommended to estimate one day's rental cost and operating cost for equipment shown in the Crew Listing for that line item.

Mobilization/ Demobilization

- The cost to move construction equipment from an equipment yard or rental company to the jobsite and back again is not included in equipment rental costs listed in the Reference section, nor in the bare equipment cost of any Unit Price line item, nor in any equipment costs shown in the Crew listings.
- Mobilization (to the site) and demobilization (from the site) costs can be found in the Unit Price section.
- If a piece of equipment is already at the jobsite, it is not appropriate to utilize mobil./ demob. costs again in an estimate.

01 54 33 | Equipment Rental

		Description	UNIT	HOURLY OPER. COST	RENT PER DAY	RENT PER WEEK	RENT PER MONTH	EQUIPMENT COST/DAY	
10	0010	**CONCRETE EQUIPMENT RENTAL** without operators R015433-10							10
	0200	Bucket, concrete lightweight, 1/2 C.Y.	Ea.	.70	20	60	180	17.60	
	0300	1 C.Y.		.75	24.50	73	219	20.60	
	0400	1-1/2 C.Y.		.95	33	99	297	27.40	
	0500	2 C.Y.		1.05	40	120	360	32.40	
	0580	8 C.Y.		5.65	260	780	2,350	201.20	
	0600	Cart, concrete, self-propelled, operator walking, 10 C.F.		3.05	56.50	170	510	58.40	
	0700	Operator riding, 18 C.F.		5.00	90	270	810	94	
	0800	Conveyer for concrete, portable, gas, 16" wide, 26' long		11.95	123	370	1,100	169.60	
	0900	46' long		12.30	150	450	1,350	188.40	
	1000	56' long		12.45	158	475	1,425	194.60	
	1100	Core drill, electric, 2-1/2 H.P., 1" to 8" bit diameter		1.75	65.50	197	590	53.40	
	1150	11 H.P., 8" to 18" cores		5.60	115	345	1,025	113.80	
	1200	Finisher, concrete floor, gas, riding trowel, 96" wide		12.35	145	435	1,300	185.80	
	1300	Gas, walk-behind, 3 blade, 36" trowel		2.05	19.65	59	177	28.20	
	1400	4 blade, 48" trowel		4.10	27	81	243	49	
	1500	Float, hand-operated (Bull float) 48" wide		.08	13.65	41	123	8.85	
	1570	Curb builder, 14 H.P., gas, single screw		13.55	247	740	2,225	256.40	
	1590	Double screw		14.20	287	860	2,575	285.60	
	1600	Grinder, concrete and terrazzo, electric, floor		2.48	104	312	935	82.25	
	1700	Wall grinder		1.24	52	156	470	41.10	
	1800	Mixer, powered, mortar and concrete, gas, 6 C.F., 18 H.P.		8.20	118	355	1,075	136.60	
	1900	10 C.F., 25 H.P.		10.30	143	430	1,300	168.40	
	2000	16 C.F.		10.60	167	500	1,500	184.80	
	2100	Concrete, stationary, tilt drum, 2 C.Y.		6.30	232	695	2,075	189.40	
	2120	Pump, concrete, truck mounted 4" line 80' boom		23.35	920	2,760	8,275	738.80	
	2140	5" line, 110' boom		30.30	1,225	3,670	11,000	976.40	
	2160	Mud jack, 50 C.F. per hr		6.95	128	385	1,150	132.60	
	2180	225 C.F. per hr		9.00	147	440	1,325	160	
	2190	Shotcrete pump rig, 12 C.Y./hr		14.15	227	680	2,050	249.20	
	2200	35 C.Y./hr		16.75	243	730	2,200	280	
	2600	Saw, concrete, manual, gas, 18 H.P.		6.35	38.50	115	345	73.80	
	2650	Self-propelled, gas, 30 H.P.		12.80	98.50	295	885	161.40	
	2700	Vibrators, concrete, electric, 60 cycle, 2 H.P.		.45	7.35	22	66	8	
	2800	3 H.P.		.67	10.35	31	93	11.55	
	2900	Gas engine, 5 H.P.		1.95	14.35	43	129	24.20	
	3000	8 H.P.		2.65	15.35	46	138	30.40	
	3050	Vibrating screed, gas engine, 8 H.P.		2.72	72.50	217	650	65.15	
	3120	Concrete transit mixer, 6 x 4, 250 H.P., 8 C.Y., rear discharge		59.20	575	1,725	5,175	818.60	
	3200	Front discharge		69.95	710	2,125	6,375	984.60	
	3300	6 x 6, 285 H.P., 12 C.Y., rear discharge		69.00	665	2,000	6,000	952	
	3400	Front discharge		72.20	715	2,150	6,450	1,008	
20	0010	**EARTHWORK EQUIPMENT RENTAL** without operators R015433-10							20
	0040	Aggregate spreader, push type 8' to 12' wide	Ea.	2.90	25.50	77	231	38.60	
	0045	Tailgate type, 8' wide		2.75	33.50	100	300	42	
	0055	Earth auger, truck-mounted, for fence & sign posts, utility poles		18.90	495	1,490	4,475	449.20	
	0060	For borings and monitoring wells		43.85	665	1,995	5,975	749.80	
	0070	Portable, trailer mounted		3.05	25.50	77	231	39.80	
	0075	Truck-mounted, for caissons, water wells		96.05	3,050	9,115	27,300	2,591	
	0080	Horizontal boring machine, 12" to 36" diameter, 45 H.P.		22.80	200	600	1,800	302.40	
	0090	12" to 48" diameter, 65 H.P.		32.65	355	1,060	3,175	473.20	
	0095	Auger, for fence posts, gas engine, hand held		.55	5.65	17	51	7.80	
	0100	Excavator, diesel hydraulic, crawler mounted, 1/2 C.Y. cap.		20.20	375	1,125	3,375	386.60	
	0120	5/8 C.Y. capacity		30.30	535	1,610	4,825	564.40	
	0140	3/4 C.Y. capacity		37.85	625	1,875	5,625	677.80	
	0150	1 C.Y. capacity		47.70	700	2,100	6,300	801.60	
	0200	1-1/2 C.Y. capacity		57.00	935	2,805	8,425	1,017	
	0300	2 C.Y. capacity		76.90	1,175	3,530	10,600	1,321	

	01 54 33 | Equipment Rental		UNIT	HOURLY OPER. COST	RENT PER DAY	RENT PER WEEK	RENT PER MONTH	EQUIPMENT COST/DAY	
20	0320	2-1/2 C.Y. capacity	Ea.	104.45	1,575	4,740	14,200	1,784	20
	0325	3-1/2 C.Y. capacity		138.85	2,075	6,240	18,700	2,359	
	0330	4-1/2 C.Y. capacity		166.15	2,575	7,705	23,100	2,870	
	0335	6 C.Y. capacity		211.30	2,825	8,500	25,500	3,390	
	0340	7 C.Y. capacity		216.05	3,050	9,125	27,400	3,553	
	0342	Excavator attachments, bucket thumbs		2.95	238	715	2,150	166.60	
	0345	Grapples		2.65	207	620	1,850	145.20	
	0347	Hydraulic hammer for boom mounting, 5000 ft lbs		12.45	410	1,225	3,675	344.60	
	0349	11,000 ft lbs		21.05	735	2,200	6,600	608.40	
	0350	Gradall type, truck mounted, 3 ton @ 15' radius, 5/8 C.Y.		38.60	840	2,520	7,550	812.80	
	0370	1 C.Y. capacity		43.55	1,025	3,050	9,150	958.40	
	0400	Backhoe-loader, 40 to 45 H.P., 5/8 C.Y. capacity		12.80	193	580	1,750	218.40	
	0450	45 H.P. to 60 H.P., 3/4 C.Y. capacity		20.35	292	875	2,625	337.80	
	0460	80 H.P., 1-1/4 C.Y. capacity		24.45	320	960	2,875	387.60	
	0470	112 H.P., 1-1/2 C.Y. capacity		38.55	525	1,570	4,700	622.40	
	0482	Backhoe-loader attachment, compactor, 20,000 lb.		5.25	142	425	1,275	127	
	0485	Hydraulic hammer, 750 ft lbs		2.95	91.50	275	825	78.60	
	0486	Hydraulic hammer, 1200 ft lbs		5.40	185	555	1,675	154.20	
	0500	Brush chipper, gas engine, 6" cutter head, 35 H.P.		10.65	108	325	975	150.20	
	0550	12" cutter head, 130 H.P.		18.15	180	540	1,625	253.20	
	0600	15" cutter head, 165 H.P.		27.75	217	650	1,950	352	
	0750	Bucket, clamshell, general purpose, 3/8 C.Y.		1.15	36.50	110	330	31.20	
	0800	1/2 C.Y.		1.25	45	135	405	37	
	0850	3/4 C.Y.		1.40	55	165	495	44.20	
	0900	1 C.Y.		1.45	58.50	175	525	46.60	
	0950	1-1/2 C.Y.		2.35	80	240	720	66.80	
	1000	2 C.Y.		2.45	90	270	810	73.60	
	1010	Bucket, dragline, medium duty, 1/2 C.Y.		.65	23.50	71	213	19.40	
	1020	3/4 C.Y.		.70	24.50	74	222	20.40	
	1030	1 C.Y.		.70	26.50	80	240	21.60	
	1040	1-1/2 C.Y.		1.10	40	120	360	32.80	
	1050	2 C.Y.		1.20	45	135	405	36.60	
	1070	3 C.Y.		1.80	63.50	190	570	52.40	
	1200	Compactor, manually guided 2-drum vibratory smooth roller, 7.5 H.P.		6.20	167	500	1,500	149.60	
	1250	Rammer/tamper, gas, 8"		2.60	43.50	130	390	46.80	
	1260	15"		2.85	48.50	145	435	51.80	
	1300	Vibratory plate, gas, 18" plate, 3000 lb. blow		2.55	23	69	207	34.20	
	1350	21" plate, 5000 lb. blow		3.20	30	90	270	43.60	
	1370	Curb builder/extruder, 14 H.P., gas, single screw		13.55	247	740	2,225	256.40	
	1390	Double screw		14.20	287	860	2,575	285.60	
	1500	Disc harrow attachment, for tractor		.40	67	201	605	43.40	
	1810	Feller buncher, shearing & accumulating trees, 100 H.P.		34.20	560	1,680	5,050	609.60	
	1860	Grader, self-propelled, 25,000 lb.		31.75	475	1,420	4,250	538	
	1910	30,000 lb.		36.65	550	1,645	4,925	622.20	
	1920	40,000 lb.		55.90	1,050	3,120	9,350	1,071	
	1930	55,000 lb.		72.55	1,375	4,155	12,500	1,411	
	1950	Hammer, pavement breaker, self-propelled, diesel, 1000 to 1250 lb.		28.25	345	1,040	3,125	434	
	2000	1300 to 1500 lb.		42.38	695	2,080	6,250	755.05	
	2050	Pile driving hammer, steam or air, 4150 ft lbs @ 225 bpm		9.65	470	1,405	4,225	358.20	
	2100	8750 ft lbs @ 145 bpm		11.65	650	1,945	5,825	482.20	
	2150	15,000 ft lbs @ 60 bpm		13.20	785	2,360	7,075	577.60	
	2200	24,450 ft lbs @ 111 bpm		14.25	880	2,635	7,900	641	
	2250	Leads, 60' high for pile driving hammers up to 20,000 ft lbs		3.00	78	234	700	70.80	
	2300	90' high for hammers over 20,000 ft lbs		4.50	141	422	1,275	120.40	
	2350	Diesel type hammer, 22,400 ft lbs		29.80	635	1,905	5,725	619.40	
	2400	41,300 ft lbs		38.60	690	2,070	6,200	722.80	
	2450	141,000 ft lbs		56.40	1,175	3,540	10,600	1,159	
	2500	Vib. elec. hammer/extractor, 200 kW diesel generator, 34 H.P.		53.70	660	1,980	5,950	825.60	

01 54 33 | Equipment Rental

		01 54 33 | Equipment Rental	UNIT	HOURLY OPER. COST	RENT PER DAY	RENT PER WEEK	RENT PER MONTH	EQUIPMENT COST/DAY
20	2550	80 H.P.	Ea.	99.00	970	2,910	8,725	1,374
	2600	150 H.P.		188.90	1,875	5,650	17,000	2,641
	2700	Extractor, steam or air, 700 ft lbs		17.50	510	1,525	4,575	445
	2750	1000 ft lbs		19.95	625	1,870	5,600	533.60
	2800	Log chipper, up to 22" diameter, 600 H.P.		52.70	445	1,340	4,025	689.60
	2850	Logger, for skidding & stacking logs, 150 H.P.		50.90	835	2,500	7,500	907.20
	2860	Mulcher, diesel powered, trailer mounted		23.50	218	655	1,975	319
	2900	Rake, spring tooth, with tractor		14.34	285	855	2,575	285.70
	3000	Roller, vibratory, tandem, smooth drum, 20 H.P.		8.25	143	430	1,300	152
	3050	35 H.P.		10.40	253	760	2,275	235.20
	3100	Towed type vibratory compactor, smooth drum, 50 H.P.		25.90	380	1,135	3,400	434.20
	3150	Sheepsfoot, 50 H.P.		28.15	440	1,320	3,950	489.20
	3170	Landfill compactor, 220 H.P.		82.85	1,475	4,390	13,200	1,541
	3200	Pneumatic tire roller, 80 H.P.		16.15	345	1,040	3,125	337.20
	3250	120 H.P.		24.20	610	1,835	5,500	560.60
	3300	Sheepsfoot vibratory roller, 240 H.P.		69.85	1,150	3,420	10,300	1,243
	3320	340 H.P.		94.00	1,400	4,220	12,700	1,596
	3350	Smooth drum vibratory roller, 75 H.P.		23.20	550	1,655	4,975	516.60
	3400	125 H.P.		29.90	655	1,970	5,900	633.20
	3410	Rotary mower, brush, 60", with tractor		18.25	260	780	2,350	302
	3420	Rototiller, walk-behind, gas, 5 H.P.		2.12	65.50	196	590	56.15
	3422	8 H.P.		3.30	100	300	900	86.40
	3440	Scrapers, towed type, 7 C.Y. capacity		5.15	107	320	960	105.20
	3450	10 C.Y. capacity		6.45	175	525	1,575	156.60
	3500	15 C.Y. capacity		7.15	192	575	1,725	172.20
	3525	Self-propelled, single engine, 14 C.Y. capacity		103.95	1,475	4,425	13,300	1,717
	3550	Dual engine, 21 C.Y. capacity		153.65	1,700	5,065	15,200	2,242
	3600	31 C.Y. capacity		208.65	2,575	7,750	23,300	3,219
	3640	44 C.Y. capacity		264.30	3,625	10,910	32,700	4,296
	3650	Elevating type, single engine, 11 C.Y. capacity		66.70	1,075	3,255	9,775	1,185
	3700	22 C.Y. capacity		126.05	2,125	6,375	19,100	2,283
	3710	Screening plant 110 H.P. w/5' x 10' screen		36.90	415	1,240	3,725	543.20
	3720	5' x 16' screen		39.05	515	1,550	4,650	622.40
	3850	Shovel, crawler-mounted, front-loading, 7 C.Y. capacity		228.10	2,850	8,555	25,700	3,536
	3855	12 C.Y. capacity		305.30	3,625	10,860	32,600	4,614
	3860	Shovel/backhoe bucket, 1/2 C.Y.		2.20	63.50	190	570	55.60
	3870	3/4 C.Y.		2.25	70	210	630	60
	3880	1 C.Y.		2.35	78.50	235	705	65.80
	3890	1-1/2 C.Y.		2.50	93.50	280	840	76
	3910	3 C.Y.		2.80	127	380	1,150	98.40
	3950	Stump chipper, 18" deep, 30 H.P.		7.56	217	652	1,950	190.90
	4110	Tractor, crawler, with bulldozer, torque converter, diesel 80 H.P.		26.45	375	1,125	3,375	436.60
	4150	105 H.P.		39.30	575	1,730	5,200	660.40
	4200	140 H.P.		42.45	715	2,150	6,450	769.60
	4260	200 H.P.		68.60	1,075	3,215	9,650	1,192
	4310	300 H.P.		82.50	1,550	4,660	14,000	1,592
	4360	410 H.P.		109.70	2,000	6,005	18,000	2,079
	4370	500 H.P.		147.85	2,700	8,135	24,400	2,810
	4380	700 H.P.		245.45	4,400	13,170	39,500	4,598
	4400	Loader, crawler, torque conv., diesel, 1-1/2 C.Y., 80 H.P.		24.10	400	1,205	3,625	433.80
	4450	1-1/2 to 1-3/4 C.Y., 95 H.P.		28.20	515	1,545	4,625	534.60
	4510	1-3/4 to 2-1/4 C.Y., 130 H.P.		42.30	835	2,505	7,525	839.40
	4530	2-1/2 to 3-1/4 C.Y., 190 H.P.		56.50	1,050	3,125	9,375	1,077
	4560	3-1/2 to 5 C.Y., 275 H.P.		81.25	1,500	4,490	13,500	1,548
	4610	Front end loader, 4WD, articulated frame, 1 to 1-1/4 C.Y., 70 H.P.		18.05	208	625	1,875	269.40
	4620	1-1/2 to 1-3/4 C.Y., 95 H.P.		23.95	293	880	2,650	367.60
	4650	1-3/4 to 2 C.Y., 130 H.P.		27.65	345	1,035	3,100	428.20
	4710	2-1/2 to 3-1/2 C.Y., 145 H.P.		26.35	375	1,125	3,375	435.80

01 54 33 | Equipment Rental

			UNIT	HOURLY OPER. COST	RENT PER DAY	RENT PER WEEK	RENT PER MONTH	EQUIPMENT COST/DAY	
20	4730	3 to 4-1/2 C.Y., 185 H.P.	Ea.	35.85	570	1,705	5,125	627.80	20
	4760	5-1/4 to 5-3/4 C.Y., 270 H.P.		56.30	905	2,720	8,150	994.40	
	4810	7 to 9 C.Y., 475 H.P.		97.15	1,625	4,900	14,700	1,757	
	4870	9 - 11 C.Y., 620 H.P.		129.55	2,600	7,825	23,500	2,601	
	4880	Skid steer loader, wheeled, 10 C.F., 30 H.P. gas		9.30	145	435	1,300	161.40	
	4890	1 C.Y., 78 H.P., diesel		19.00	232	695	2,075	291	
	4892	Skid-steer attachment, auger		.55	91.50	275	825	59.40	
	4893	Backhoe		.67	112	336	1,000	72.55	
	4894	Broom		.73	121	364	1,100	78.65	
	4895	Forks		.24	40	120	360	25.90	
	4896	Grapple		.57	95.50	286	860	61.75	
	4897	Concrete hammer		1.06	177	532	1,600	114.90	
	4898	Tree spade		.82	136	408	1,225	88.15	
	4899	Trencher		.79	131	393	1,175	84.90	
	4900	Trencher, chain, boom type, gas, operator walking, 12 H.P.		4.75	46.50	140	420	66	
	4910	Operator riding, 40 H.P.		19.55	297	890	2,675	334.40	
	5000	Wheel type, diesel, 4' deep, 12" wide		82.05	825	2,480	7,450	1,152	
	5100	6' deep, 20" wide		86.95	1,900	5,735	17,200	1,843	
	5150	Chain type, diesel, 5' deep, 8" wide		36.25	580	1,745	5,225	639	
	5200	Diesel, 8' deep, 16" wide		88.95	2,300	6,890	20,700	2,090	
	5202	Rock trencher, wheel type, 6" wide x 18" deep		19.55	315	940	2,825	344.40	
	5206	Chain type, 18" wide x 7' deep		106.25	2,875	8,605	25,800	2,571	
	5210	Tree spade, self-propelled		13.10	267	800	2,400	264.80	
	5250	Truck, dump, 2-axle, 12 ton, 8 C.Y. payload, 220 H.P.		33.35	225	675	2,025	401.80	
	5300	Three axle dump, 16 ton, 12 C.Y. payload, 400 H.P.		59.35	320	965	2,900	667.80	
	5310	Four axle dump, 25 ton, 18 C.Y. payload, 450 H.P.		69.55	470	1,410	4,225	838.40	
	5350	Dump trailer only, rear dump, 16-1/2 C.Y.		4.90	135	405	1,225	120.20	
	5400	20 C.Y.		5.35	153	460	1,375	134.80	
	5450	Flatbed, single axle, 1-1/2 ton rating		24.85	68.50	205	615	239.80	
	5500	3 ton rating		29.75	96.50	290	870	296	
	5550	Off highway rear dump, 25 ton capacity		74.80	1,325	4,005	12,000	1,399	
	5600	35 ton capacity		70.50	1,175	3,515	10,500	1,267	
	5610	50 ton capacity		95.65	1,625	4,895	14,700	1,744	
	5620	65 ton capacity		99.70	1,625	4,860	14,600	1,770	
	5630	100 ton capacity		128.25	2,125	6,350	19,100	2,296	
	6000	Vibratory plow, 25 H.P., walking		8.15	61.50	185	555	102.20	
40	0010	**GENERAL EQUIPMENT RENTAL** without operators R015433 -10							40
	0150	Aerial lift, scissor type, to 15' high, 1000 lb. cap., electric	Ea.	2.80	55	165	495	55.40	
	0160	To 25' high, 2000 lb. capacity		3.20	70	210	630	67.60	
	0170	Telescoping boom to 40' high, 500 lb. capacity, gas		19.30	320	965	2,900	347.40	
	0180	To 45' high, 500 lb. capacity		20.35	370	1,115	3,350	385.80	
	0190	To 60' high, 600 lb. capacity		22.55	490	1,465	4,400	473.40	
	0195	Air compressor, portable, 6.5 cfm, electric		.50	13	39	117	11.80	
	0196	Gasoline		.75	19.65	59	177	17.80	
	0200	Towed type, gas engine, 60 cfm		13.70	48.50	145	435	138.60	
	0300	160 cfm		16.00	50	150	450	158	
	0400	Diesel engine, rotary screw, 250 cfm		16.30	105	315	945	193.40	
	0500	365 cfm		22.20	130	390	1,175	255.60	
	0550	450 cfm		28.25	163	490	1,475	324	
	0600	600 cfm		49.20	228	685	2,050	530.60	
	0700	750 cfm		50.10	237	710	2,125	542.80	
	0800	For silenced models, small sizes, add to rent		3%	5%	5%	5%		
	0900	Large sizes, add to rent		5%	7%	7%	7%		
	0930	Air tools, breaker, pavement, 60 lb.	Ea.	.45	9.65	29	87	9.40	
	0940	80 lb.		.45	10.35	31	93	9.80	
	0950	Drills, hand (jackhammer) 65 lb.		.55	17.35	52	156	14.80	
	0960	Track or wagon, swing boom, 4" drifter		57.35	765	2,295	6,875	917.80	
	0970	5" drifter		71.65	825	2,480	7,450	1,069	

01 54 33 | Equipment Rental

		01 54 33 Equipment Rental	UNIT	HOURLY OPER. COST	RENT PER DAY	RENT PER WEEK	RENT PER MONTH	EQUIPMENT COST/DAY	
40	0975	Track mounted quarry drill, 6" diameter drill	Ea.	96.10	1,275	3,790	11,400	1,527	40
	0980	Dust control per drill		1.04	19	57	171	19.70	
	0990	Hammer, chipping, 12 lb.		.50	23.50	70	210	18	
	1000	Hose, air with couplings, 50' long, 3/4" diameter		.03	4.33	13	39	2.85	
	1100	1" diameter		.04	6.35	19	57	4.10	
	1200	1-1/2" diameter		.06	9.65	29	87	6.30	
	1300	2" diameter		.11	19	57	171	12.30	
	1400	2-1/2" diameter		.12	20	60	180	12.95	
	1410	3" diameter		.15	25	75	225	16.20	
	1450	Drill, steel, 7/8" x 2'		.05	8	24	72	5.20	
	1460	7/8" x 6'		.05	9	27	81	5.80	
	1520	Moil points		.02	4	12	36	2.55	
	1525	Pneumatic nailer w/accessories		.46	30.50	91	273	21.90	
	1530	Sheeting driver for 60 lb. breaker		.04	6	18	54	3.90	
	1540	For 90 lb. breaker		.12	8	24	72	5.75	
	1550	Spade, 25 lb.		.40	6	18	54	6.80	
	1560	Tamper, single, 35 lb.		.53	35	105	315	25.25	
	1570	Triple, 140 lb.		.79	52.50	158	475	37.90	
	1580	Wrenches, impact, air powered, up to 3/4" bolt		.40	9	27	81	8.60	
	1590	Up to 1-1/4" bolt		.50	23	69	207	17.80	
	1600	Barricades, barrels, reflectorized, 1 to 50 barrels		.03	4.60	13.80	41.50	3	
	1610	100 to 200 barrels		.02	3.53	10.60	32	2.30	
	1620	Barrels with flashers, 1 to 50 barrels		.03	5.25	15.80	47.50	3.40	
	1630	100 to 200 barrels		.03	4.20	12.60	38	2.75	
	1640	Barrels with steady burn type C lights		.04	7	21	63	4.50	
	1650	Illuminated board, trailer mounted, with generator		.75	118	355	1,075	77	
	1670	Portable barricade, stock, with flashers, 1 to 6 units		.03	5.25	15.80	47.50	3.40	
	1680	25 to 50 units		.03	4.90	14.70	44	3.20	
	1690	Butt fusion machine, electric		12.80	53.50	160	480	134.40	
	1695	Electro fusion machine		9.45	38.50	115	345	98.60	
	1700	Carts, brick, hand powered, 1000 lb. capacity		.36	60.50	182	545	39.30	
	1800	Gas engine, 1500 lb., 7-1/2' lift		4.01	120	359	1,075	103.90	
	1822	Dehumidifier, medium, 6 lb/hr, 150 cfm		.98	60	180	540	43.85	
	1824	Large, 18 lb/hr, 600 cfm		1.98	122	366	1,100	89.05	
	1830	Distributor, asphalt, trailer mounted, 2000 gal., 38 H.P. diesel		9.40	340	1,015	3,050	278.20	
	1840	3000 gal., 38 H.P. diesel		10.75	370	1,105	3,325	307	
	1850	Drill, rotary hammer, electric		.88	25	75	225	22.05	
	1860	Carbide bit, 1-1/2" diameter, add to electric rotary hammer		.04	7.60	22.75	68.50	4.85	
	1865	Rotary, crawler, 250 H.P.		141.75	2,075	6,245	18,700	2,383	
	1870	Emulsion sprayer, 65 gal., 5 H.P. gas engine		2.83	99	297	890	82.05	
	1880	200 gal., 5 H.P. engine		7.45	165	495	1,475	158.60	
	1930	Floodlight, mercury vapor, or quartz, on tripod, 1000 watt		.36	14	42	126	11.30	
	1940	2000 watt		.63	26	78	234	20.65	
	1950	Floodlights, trailer mounted with generator, 1 - 300 watt light		3.35	71.50	215	645	69.80	
	1960	2 - 1000 watt lights		4.45	90	270	810	89.60	
	2000	4 - 300 watt lights		4.20	88.50	265	795	86.60	
	2005	Foam spray rig, incl. box trailer, compressor, generator, proportioner		32.33	490	1,465	4,400	551.65	
	2020	Forklift, straight mast, 12' lift, 5000 lb., 2 wheel drive, gas		25.20	200	600	1,800	321.60	
	2040	21' lift, 5000 lb., 4 wheel drive, diesel		20.20	247	740	2,225	309.60	
	2050	For rough terrain, 42' lift, 35' reach, 9000 lb., 110 H.P.		26.40	445	1,335	4,000	478.20	
	2060	For plant, 4 ton capacity, 80 H.P., 2 wheel drive, gas		15.50	102	305	915	185	
	2080	10 ton capacity, 120 H.P., 2 wheel drive, diesel		23.40	178	535	1,600	294.20	
	2100	Generator, electric, gas engine, 1.5 kW to 3 kW		3.55	11.35	34	102	35.20	
	2200	5 kW		4.60	15.35	46	138	46	
	2300	10 kW		8.70	36.50	110	330	91.60	
	2400	25 kW		10.20	75	225	675	126.60	
	2500	Diesel engine, 20 kW		11.50	75	225	675	137	
	2600	50 kW	↓	21.55	103	310	930	234.40	

01 54 33 | Equipment Rental

		01 54 33 Equipment Rental	UNIT	HOURLY OPER. COST	RENT PER DAY	RENT PER WEEK	RENT PER MONTH	EQUIPMENT COST/DAY	
40	2700	100 kW	Ea.	42.45	132	395	1,175	418.60	40
	2800	250 kW		82.90	233	700	2,100	803.20	
	2850	Hammer, hydraulic, for mounting on boom, to 500 ft lbs		2.25	68.50	205	615	59	
	2860	1000 ft lbs		3.80	112	335	1,000	97.40	
	2900	Heaters, space, oil or electric, 50 MBH		1.69	7	21	63	17.70	
	3000	100 MBH		3.06	9.35	28	84	30.10	
	3100	300 MBH		9.79	31.50	94	282	97.10	
	3150	500 MBH		19.75	41.50	125	375	183	
	3200	Hose, water, suction with coupling, 20' long, 2" diameter		.02	3	9	27	1.95	
	3210	3" diameter		.03	4.67	14	42	3.05	
	3220	4" diameter		.03	5.35	16	48	3.45	
	3230	6" diameter		.11	18.35	55	165	11.90	
	3240	8" diameter		.20	33.50	100	300	21.60	
	3250	Discharge hose with coupling, 50' long, 2" diameter		.01	1.33	4	12	.90	
	3260	3" diameter		.02	2.67	8	24	1.75	
	3270	4" diameter		.02	3.67	11	33	2.35	
	3280	6" diameter		.06	9.65	29	87	6.30	
	3290	8" diameter		.20	33.50	100	300	21.60	
	3295	Insulation blower		.22	6	18	54	5.35	
	3300	Ladders, extension type, 16' to 36' long		.14	23	69	207	14.90	
	3400	40' to 60' long		.20	33.50	100	300	21.60	
	3405	Lance for cutting concrete		3.07	104	313	940	87.15	
	3407	Lawn mower, rotary, 22", 5 H.P.		2.12	65	195	585	55.95	
	3408	48" self propelled		3.28	114	341	1,025	94.45	
	3410	Level, laser type, for pipe and sewer leveling		1.46	97	291	875	69.90	
	3430	Electronic		.75	50	150	450	36	
	3440	Laser type, rotating beam for grade control		1.17	78	234	700	56.15	
	3460	Builders level with tripod and rod		.09	14.65	44	132	9.50	
	3500	Light towers, towable, with diesel generator, 2000 watt		4.20	88.50	265	795	86.60	
	3600	4000 watt		4.45	98.50	295	885	94.60	
	3700	Mixer, powered, plaster and mortar, 6 C.F., 7 H.P.		2.55	18.65	56	168	31.60	
	3800	10 C.F., 9 H.P.		2.70	30	90	270	39.60	
	3850	Nailer, pneumatic		.46	30.50	91	273	21.90	
	3900	Paint sprayers complete, 8 cfm		.69	45.50	137	410	32.90	
	4000	17 cfm		1.28	85	255	765	61.25	
	4020	Pavers, bituminous, rubber tires, 8' wide, 50 H.P., diesel		39.45	1,050	3,165	9,500	948.60	
	4030	10' wide, 150 H.P.		85.85	1,700	5,070	15,200	1,701	
	4050	Crawler, 8' wide, 100 H.P., diesel		83.90	1,800	5,420	16,300	1,755	
	4060	10' wide, 150 H.P.		96.80	2,125	6,365	19,100	2,047	
	4070	Concrete paver, 12' to 24' wide, 250 H.P.		96.25	1,575	4,750	14,300	1,720	
	4080	Placer-spreader-trimmer, 24' wide, 300 H.P.		140.15	2,600	7,785	23,400	2,678	
	4100	Pump, centrifugal gas pump, 1-1/2" diam., 65 GPM		3.60	45	135	405	55.80	
	4200	2" diameter, 130 GPM		5.00	51.50	155	465	71	
	4300	3" diameter, 250 GPM		5.30	53.50	160	480	74.40	
	4400	6" diameter, 1500 GPM		29.25	177	530	1,600	340	
	4500	Submersible electric pump, 1-1/4" diameter, 55 GPM		.41	16	48	144	12.90	
	4600	1-1/2" diameter, 83 GPM		.44	18.35	55	165	14.50	
	4700	2" diameter, 120 GPM		1.25	23	69	207	23.80	
	4800	3" diameter, 300 GPM		2.15	41.50	125	375	42.20	
	4900	4" diameter, 560 GPM		9.55	163	490	1,475	174.40	
	5000	6" diameter, 1590 GPM		14.05	220	660	1,975	244.40	
	5100	Diaphragm pump, gas, single, 1-1/2" diameter		1.10	48	144	430	37.60	
	5200	2" diameter		4.00	60	180	540	68	
	5300	3" diameter		4.00	60	180	540	68	
	5400	Double, 4" diameter		5.35	80	240	720	90.80	
	5450	Pressure Washer 5 GPM, 3000 psi		4.60	40	120	360	60.80	
	5500	Trash pump, self-priming, gas, 2" diameter		4.35	20.50	62	186	47.20	
	5600	Diesel, 4" diameter	↓	11.90	58.50	175	525	130.20	

01 54 33 | Equipment Rental

			UNIT	HOURLY OPER. COST	RENT PER DAY	RENT PER WEEK	RENT PER MONTH	EQUIPMENT COST/DAY	
40	5650	Diesel, 6" diameter	Ea.	40.15	130	390	1,175	399.20	40
	5655	Grout Pump		18.10	95	285	855	201.80	
	5700	Salamanders, L.P. gas fired, 100,000 Btu		3.10	16	48	144	34.40	
	5705	50,000 Btu		2.31	9.35	28	84	24.10	
	5720	Sandblaster, portable, open top, 3 C.F. capacity		.55	27	81	243	20.60	
	5730	6 C.F. capacity		.85	40	120	360	30.80	
	5740	Accessories for above		.12	20	60	180	12.95	
	5750	Sander, floor		.85	19	57	171	18.20	
	5760	Edger		.81	28	84	252	23.30	
	5800	Saw, chain, gas engine, 18" long		2.15	18.35	55	165	28.20	
	5900	Hydraulic powered, 36" long		.65	56.50	170	510	39.20	
	5950	60" long		.65	56.50	170	510	39.20	
	6000	Masonry, table mounted, 14" diameter, 5 H.P.		1.34	56.50	170	510	44.70	
	6050	Portable cut-off, 8 H.P.		2.40	32.50	98	294	38.80	
	6100	Circular, hand held, electric, 7-1/4" diameter		.21	4.67	14	42	4.50	
	6200	12" diameter		.27	7.65	23	69	6.75	
	6250	Wall saw, w/hydraulic power, 10 H.P		9.30	60	180	540	110.40	
	6275	Shot blaster, walk-behind, 20" wide		4.60	298	895	2,675	215.80	
	6280	Sidewalk broom, walk-behind		2.05	61.50	185	555	53.40	
	6300	Steam cleaner, 100 gallons per hour		3.50	53.50	161	485	60.20	
	6310	200 gallons per hour		5.05	63.50	190	570	78.40	
	6340	Tar Kettle/Pot, 400 gallons		4.68	80	240	720	85.45	
	6350	Torch, cutting, acetylene-oxygen, 150' hose		.50	25	75	225	19	
	6360	Hourly operating cost includes tips and gas		9.90				79.20	
	6410	Toilet, portable chemical		.12	19.35	58	174	12.55	
	6420	Recycle flush type		.14	23.50	70	210	15.10	
	6430	Toilet, fresh water flush, garden hose,		.16	27	81	243	17.50	
	6440	Hoisted, non-flush, for high rise		.14	23	69	207	14.90	
	6450	Toilet, trailers, minimum		.24	40	120	360	25.90	
	6460	Maximum		.73	122	365	1,100	78.85	
	6465	Tractor, farm with attachment		17.10	267	800	2,400	296.80	
	6500	Trailers, platform, flush deck, 2 axle, 25 ton capacity		4.95	110	330	990	105.60	
	6600	40 ton capacity		6.40	153	460	1,375	143.20	
	6700	3 axle, 50 ton capacity		6.95	170	510	1,525	157.60	
	6800	75 ton capacity		8.65	223	670	2,000	203.20	
	6810	Trailer mounted cable reel for high voltage line work		4.94	235	706	2,125	180.70	
	6820	Trailer mounted cable tensioning rig		9.80	465	1,400	4,200	358.40	
	6830	Cable pulling rig		65.52	2,625	7,860	23,600	2,096	
	6900	Water tank trailer, engine driven discharge, 5000 gallons		6.45	147	440	1,325	139.60	
	6925	10,000 gallons		8.80	203	610	1,825	192.40	
	6950	Water truck, off highway, 6000 gallons		84.65	805	2,420	7,250	1,161	
	7010	Tram car for high voltage line work, powered, 2 conductor		6.81	128	383	1,150	131.10	
	7020	Transit (builder's level) with tripod		.09	14.65	44	132	9.50	
	7030	Trench box, 3000 lbs. 6' x 8'		.56	93	279	835	60.30	
	7040	7200 lbs. 6' x 20'		1.05	175	525	1,575	113.40	
	7050	8000 lbs., 8' x 16'		.95	158	475	1,425	102.60	
	7060	9500 lbs., 8' x 20'		1.19	199	596	1,800	128.70	
	7065	11,000 lbs., 8' x 24'		1.27	212	637	1,900	137.55	
	7070	12,000 lbs., 10' x 20'		1.71	285	855	2,575	184.70	
	7100	Truck, pickup, 3/4 ton, 2 wheel drive		13.25	56.50	170	510	140	
	7200	4 wheel drive		13.50	71.50	215	645	151	
	7250	Crew carrier, 9 passenger		18.80	86.50	260	780	202.40	
	7290	Flat bed truck, 20,000 lbs. GVW		21.20	127	380	1,150	245.60	
	7300	Tractor, 4 x 2, 220 H.P.		29.50	197	590	1,775	354	
	7410	330 H.P.		43.80	270	810	2,425	512.40	
	7500	6 x 4, 380 H.P.		50.30	315	950	2,850	592.40	
	7600	450 H.P.		60.85	380	1,145	3,425	715.80	
	7620	Vacuum truck, hazardous material, 2500 gallons	↓	11.15	305	920	2,750	273.20	

01 54 33 | Equipment Rental

				UNIT	HOURLY OPER. COST	RENT PER DAY	RENT PER WEEK	RENT PER MONTH	EQUIPMENT COST/DAY	
40	7625	5,000 gallons		Ea.	13.23	430	1,290	3,875	363.85	40
	7640	Tractor, with A frame, boom and winch, 225 H.P.			32.15	273	820	2,450	421.20	
	7650	Vacuum, HEPA, 16 gallon, wet/dry			.84	18	54	162	17.50	
	7655	55 gallon, wet/dry			.86	27	81	243	23.10	
	7660	Water tank, portable			.17	28.50	85.50	257	18.45	
	7690	Sewer/catch basin vacuum, 14 C.Y., 1500 gallons			17.78	645	1,940	5,825	530.25	
	7700	Welder, electric, 200 amp			3.86	17.35	52	156	41.30	
	7800	300 amp			5.72	21.50	64	192	58.55	
	7900	Gas engine, 200 amp			14.15	25.50	76	228	128.40	
	8000	300 amp			16.20	27	81	243	145.80	
	8100	Wheelbarrow, any size			.07	11.65	35	105	7.55	
	8200	Wrecking ball, 4000 lbs.		↓	2.15	71.50	215	645	60.20	
50	0010	**HIGHWAY EQUIPMENT RENTAL** without operators	R015433 -10							50
	0050	Asphalt batch plant, portable drum mixer, 100 ton/hr		Ea.	66.15	1,450	4,355	13,100	1,400	
	0060	200 ton/hr			73.90	1,525	4,600	13,800	1,511	
	0070	300 ton/hr			85.75	1,800	5,425	16,300	1,771	
	0100	Backhoe attachment, long stick, up to 185 H.P., 10.5' long			.33	21.50	65	195	15.65	
	0140	Up to 250 H.P., 12' long			.35	23.50	70	210	16.80	
	0180	Over 250 H.P., 15' long			.48	31.50	95	285	22.85	
	0200	Special dipper arm, up to 100 H.P., 32' long			.98	65	195	585	46.85	
	0240	Over 100 H.P., 33' long			1.22	81	243	730	58.35	
	0280	Catch basin/sewer cleaning truck, 3 ton, 9 C.Y., 1000 gal			40.90	405	1,210	3,625	569.20	
	0300	Concrete batch plant, portable, electric, 200 C.Y./hr			17.80	515	1,540	4,625	450.40	
	0520	Grader/dozer attachment, ripper/scarifier, rear mounted, up to 135 H.P.			3.20	65	195	585	64.60	
	0540	Up to 180 H.P.			3.75	83.50	250	750	80	
	0580	Up to 250 H.P.			4.15	95	285	855	90.20	
	0700	Pvmt. removal bucket, for hyd. excavator, up to 90 H.P.			1.70	50	150	450	43.60	
	0740	Up to 200 H.P.			1.90	71.50	215	645	58.20	
	0780	Over 200 H.P.			2.05	83.50	250	750	66.40	
	0900	Aggregate spreader, self-propelled, 187 H.P.			50.35	695	2,090	6,275	820.80	
	1000	Chemical spreader, 3 C.Y.			3.05	43.50	130	390	50.40	
	1900	Hammermill, traveling, 250 H.P.			63.25	1,875	5,650	17,000	1,636	
	2000	Horizontal borer, 3" diameter, 13 H.P. gas driven			6.00	56.50	170	510	82	
	2150	Horizontal directional drill, 20,000 lb. thrust, 78 H.P. diesel			29.70	690	2,070	6,200	651.60	
	2160	30,000 lb. thrust, 115 H.P.			37.15	1,050	3,155	9,475	928.20	
	2170	50,000 lb. thrust, 170 H.P.			53.00	1,350	4,040	12,100	1,232	
	2190	Mud trailer for HDD, 1500 gallons, 175 H.P., gas			30.75	153	460	1,375	338	
	2200	Hydromulchers, gas power, 3000 gallons, for truck mounting			22.90	258	775	2,325	338.20	
	2400	Joint & crack cleaner, walk behind, 25 H.P.			3.80	50	150	450	60.40	
	2500	Filler, trailer mounted, 400 gallons, 20 H.P.			9.15	217	650	1,950	203.20	
	3000	Paint striper, self-propelled, double line, 30 H.P.			6.95	162	485	1,450	152.60	
	3200	Post drivers, 6" I-Beam frame, for truck mounting			18.75	425	1,280	3,850	406	
	3400	Road sweeper, self-propelled, 8' wide, 90 H.P.			37.80	580	1,740	5,225	650.40	
	3450	Road sweeper, vacuum assisted, 4 C.Y., 220 gallons			73.55	635	1,905	5,725	969.40	
	4000	Road mixer, self-propelled, 130 H.P.			44.05	775	2,325	6,975	817.40	
	4100	310 H.P.			79.75	2,200	6,585	19,800	1,955	
	4220	Cold mix paver, incl. pug mill and bitumen tank, 165 H.P.			96.80	2,150	6,440	19,300	2,062	
	4250	Paver, asphalt, wheel or crawler, 130 H.P., diesel			95.35	2,050	6,155	18,500	1,994	
	4300	Paver, road widener, gas 1' to 6', 67 H.P.			44.65	835	2,500	7,500	857.20	
	4400	Diesel, 2' to 14', 88 H.P.			59.80	1,075	3,190	9,575	1,116	
	4600	Slipform pavers, curb and gutter, 2 track, 75 H.P.			44.25	785	2,355	7,075	825	
	4700	4 track, 165 H.P.			54.95	825	2,475	7,425	934.60	
	4800	Median barrier, 215 H.P.			55.65	855	2,570	7,700	959.20	
	4901	Trailer, low bed, 75 ton capacity			9.35	218	655	1,975	205.80	
	5000	Road planer, walk behind, 10" cutting width, 10 H.P.			3.45	30.50	91	273	45.80	
	5100	Self-propelled, 12" cutting width, 64 H.P.			10.20	127	380	1,150	157.60	
	5120	Traffic line remover, metal ball blaster, truck mounted, 115 H.P.			52.80	770	2,305	6,925	883.40	
	5140	Grinder, truck mounted, 115 H.P.		↓	56.60	835	2,505	7,525	953.80	

01 54 33 | Equipment Rental

				UNIT	HOURLY OPER. COST	RENT PER DAY	RENT PER WEEK	RENT PER MONTH	EQUIPMENT COST/DAY	
50	5160	Walk-behind, 11 H.P.		Ea.	3.90	46.50	140	420	59.20	50
	5200	Pavement profiler, 4' to 6' wide, 450 H.P.			244.25	3,400	10,220	30,700	3,998	
	5300	8' to 10' wide, 750 H.P.			391.35	4,950	14,865	44,600	6,104	
	5400	Roadway plate, steel, 1" x 8' x 20'			.07	11.65	35	105	7.55	
	5600	Stabilizer, self-propelled, 150 H.P.			47.20	615	1,850	5,550	747.60	
	5700	310 H.P.			81.20	1,350	4,070	12,200	1,464	
	5800	Striper, thermal, truck mounted 120 gallon paint, 150 H.P.			62.30	505	1,520	4,550	802.40	
	6000	Tar kettle, 330 gallon, trailer mounted			4.30	61.50	185	555	71.40	
	7000	Tunnel locomotive, diesel, 8 to 12 ton			30.00	580	1,745	5,225	589	
	7005	Electric, 10 ton			24.00	665	1,995	5,975	591	
	7010	Muck cars, 1/2 C.Y. capacity			1.85	23.50	70	210	28.80	
	7020	1 C.Y. capacity			2.05	31.50	95	285	35.40	
	7030	2 C.Y. capacity			2.15	36.50	110	330	39.20	
	7040	Side dump, 2 C.Y. capacity			2.35	45	135	405	45.80	
	7050	3 C.Y. capacity			3.20	51.50	155	465	56.60	
	7060	5 C.Y. capacity			4.55	65	195	585	75.40	
	7100	Ventilating blower for tunnel, 7-1/2 H.P.			1.39	51.50	155	465	42.10	
	7110	10 H.P.			.83	.52	1.55	4.65	6.95	
	7120	20 H.P.			1.66	.67	2	6	13.70	
	7140	40 H.P.			3.31	.97	2.90	8.70	27.05	
	7160	60 H.P.			7.23	152	455	1,375	148.85	
	7175	75 H.P.			9.22	202	607	1,825	195.15	
	7180	200 H.P.			21.05	305	910	2,725	350.40	
	7800	Windrow loader, elevating		↓	52.00	1,350	4,060	12,200	1,228	
60	0010	**LIFTING AND HOISTING EQUIPMENT RENTAL** without operators	R015433-10							60
	0120	Aerial lift truck, 2 person, to 80'		Ea.	28.05	735	2,205	6,625	665.40	
	0140	Boom work platform, 40' snorkel	R015433-15		17.60	265	795	2,375	299.80	
	0150	Crane, flatbed mounted, 3 ton capacity			15.70	195	585	1,750	242.60	
	0200	Crane, climbing, 106' jib, 6000 lb. capacity, 410 fpm	R312316-45		38.95	1,500	4,490	13,500	1,210	
	0300	101' jib, 10,250 lb. capacity, 270 fpm			44.95	1,900	5,690	17,100	1,498	
	0500	Tower, static, 130' high, 106' jib, 6200 lb. capacity at 400 fpm			42.45	1,725	5,190	15,600	1,378	
	0600	Crawler mounted, lattice boom, 1/2 C.Y., 15 tons at 12' radius			35.47	685	2,050	6,150	693.75	
	0700	3/4 C.Y., 20 tons at 12' radius			47.29	855	2,570	7,700	892.30	
	0800	1 C.Y., 25 tons at 12' radius			63.05	1,150	3,420	10,300	1,188	
	0900	1-1/2 C.Y., 40 tons at 12' radius			63.05	1,150	3,445	10,300	1,193	
	1000	2 C.Y., 50 tons at 12' radius			66.90	1,350	4,055	12,200	1,346	
	1100	3 C.Y., 75 tons at 12' radius			71.80	1,575	4,710	14,100	1,516	
	1200	100 ton capacity, 60' boom			81.40	1,800	5,405	16,200	1,732	
	1300	165 ton capacity, 60' boom			103.75	2,100	6,320	19,000	2,094	
	1400	200 ton capacity, 70' boom			125.80	2,675	7,990	24,000	2,604	
	1500	350 ton capacity, 80' boom			177.05	4,075	12,195	36,600	3,855	
	1600	Truck mounted, lattice boom, 6 x 4, 20 tons at 10' radius			36.33	1,225	3,690	11,100	1,029	
	1700	25 tons at 10' radius			39.34	1,350	4,020	12,100	1,119	
	1800	8 x 4, 30 tons at 10' radius			42.56	1,425	4,280	12,800	1,196	
	1900	40 tons at 12' radius			45.29	1,500	4,470	13,400	1,256	
	2000	60 tons at 15' radius			50.61	1,575	4,730	14,200	1,351	
	2050	82 tons at 15' radius			56.35	1,675	5,050	15,200	1,461	
	2100	90 tons at 15' radius			63.07	1,825	5,510	16,500	1,607	
	2200	115 tons at 15' radius			71.12	2,050	6,160	18,500	1,801	
	2300	150 tons at 18' radius			88.35	2,150	6,480	19,400	2,003	
	2350	165 tons at 18' radius			83.09	2,300	6,870	20,600	2,039	
	2400	Truck mounted, hydraulic, 12 ton capacity			41.45	540	1,620	4,850	655.60	
	2500	25 ton capacity			43.70	665	1,990	5,975	747.60	
	2550	33 ton capacity			44.25	680	2,045	6,125	763	
	2560	40 ton capacity			57.20	800	2,405	7,225	938.60	
	2600	55 ton capacity			75.80	1,000	3,015	9,050	1,209	
	2700	80 ton capacity			97.75	1,275	3,815	11,400	1,545	
	2720	100 ton capacity		↓	91.65	1,525	4,600	13,800	1,653	

01 54 33 | Equipment Rental

		01 54 33 Equipment Rental	UNIT	HOURLY OPER. COST	RENT PER DAY	RENT PER WEEK	RENT PER MONTH	EQUIPMENT COST/DAY	
60	2740	120 ton capacity	Ea.	106.15	1,650	4,920	14,800	1,833	60
	2760	150 ton capacity		124.30	2,250	6,735	20,200	2,341	
	2800	Self-propelled, 4 x 4, with telescoping boom, 5 ton		16.90	235	705	2,125	276.20	
	2900	12-1/2 ton capacity		41.45	540	1,620	4,850	655.60	
	3000	15 ton capacity		39.70	585	1,755	5,275	668.60	
	3050	20 ton capacity		35.20	575	1,720	5,150	625.60	
	3100	25 ton capacity		36.45	605	1,820	5,450	655.60	
	3150	40 ton capacity		57.95	915	2,750	8,250	1,014	
	3200	Derricks, guy, 20 ton capacity, 60' boom, 75' mast		21.70	365	1,100	3,300	393.60	
	3300	100' boom, 115' mast		34.23	630	1,890	5,675	651.85	
	3400	Stiffleg, 20 ton capacity, 70' boom, 37' mast		24.01	475	1,430	4,300	478.10	
	3500	100' boom, 47' mast		36.96	760	2,280	6,850	751.70	
	3550	Helicopter, small, lift to 1250 lbs. maximum, w/pilot		92.16	2,950	8,880	26,600	2,513	
	3600	Hoists, chain type, overhead, manual, 3/4 ton		.10	1	3	9	1.40	
	3900	10 ton		.75	9.35	28	84	11.60	
	4000	Hoist and tower, 5000 lb. cap., portable electric, 40' high		4.82	211	633	1,900	165.15	
	4100	For each added 10' section, add		.10	16.65	50	150	10.80	
	4200	Hoist and single tubular tower, 5000 lb. electric, 100' high		6.48	295	884	2,650	228.65	
	4300	For each added 6'-6" section, add		.17	28.50	85	255	18.35	
	4400	Hoist and double tubular tower, 5000 lb., 100' high		6.93	325	974	2,925	250.25	
	4500	For each added 6'-6" section, add		.19	31.50	95	285	20.50	
	4550	Hoist and tower, mast type, 6000 lb., 100' high		7.53	335	1,010	3,025	262.25	
	4570	For each added 10' section, add		.12	20	60	180	12.95	
	4600	Hoist and tower, personnel, electric, 2000 lb., 100' @ 125 fpm		15.10	895	2,690	8,075	658.80	
	4700	3000 lb., 100' @ 200 fpm		17.31	1,025	3,050	9,150	748.50	
	4800	3000 lb., 150' @ 300 fpm		19.11	1,125	3,410	10,200	834.90	
	4900	4000 lb., 100' @ 300 fpm		19.88	1,150	3,480	10,400	855.05	
	5000	6000 lb., 100' @ 275 fpm	↓	21.55	1,225	3,650	11,000	902.40	
	5100	For added heights up to 500', add	L.F.	.01	1.67	5	15	1.10	
	5200	Jacks, hydraulic, 20 ton	Ea.	.05	2	6	18	1.60	
	5500	100 ton		.35	11.35	34	102	9.60	
	6100	Jacks, hydraulic, climbing w/ 50' jackrods, control console, 30 ton cap.		1.82	121	364	1,100	87.35	
	6150	For each added 10' jackrod section, add		.05	3.33	10	30	2.40	
	6300	50 ton capacity		2.93	195	585	1,750	140.45	
	6350	For each added 10' jackrod section, add		.06	4	12	36	2.90	
	6500	125 ton capacity		7.70	515	1,540	4,625	369.60	
	6550	For each added 10' jackrod section, add		.53	35	105	315	25.25	
	6600	Cable jack, 10 ton capacity with 200' cable		1.53	102	305	915	73.25	
	6650	For each added 50' of cable, add	↓	.18	11.65	35	105	8.45	
70	0010	**WELLPOINT EQUIPMENT RENTAL** without operators (R015433-10)							70
	0020	Based on 2 months rental							
	0100	Combination jetting & wellpoint pump, 60 H.P. diesel	Ea.	14.71	300	902	2,700	298.10	
	0200	High pressure gas jet pump, 200 H.P., 300 psi	"	33.40	257	771	2,325	421.40	
	0300	Discharge pipe, 8" diameter	L.F.	.01	.49	1.47	4.41	.35	
	0350	12" diameter		.01	.72	2.16	6.50	.50	
	0400	Header pipe, flows up to 150 GPM, 4" diameter		.01	.44	1.32	3.96	.35	
	0500	400 GPM, 6" diameter		.01	.52	1.57	4.71	.40	
	0600	800 GPM, 8" diameter		.01	.72	2.16	6.50	.50	
	0700	1500 GPM, 10" diameter		.01	.75	2.26	6.80	.55	
	0800	2500 GPM, 12" diameter		.02	1.43	4.29	12.85	1	
	0900	4500 GPM, 16" diameter		.03	1.83	5.50	16.50	1.35	
	0950	For quick coupling aluminum and plastic pipe, add	↓	.03	1.90	5.69	17.05	1.40	
	1100	Wellpoint, 25' long, with fittings & riser pipe, 1-1/2" or 2" diameter	Ea.	.06	3.78	11.35	34	2.75	
	1200	Wellpoint pump, diesel powered, 4" suction, 20 H.P.		6.44	173	520	1,550	155.50	
	1300	6" suction, 30 H.P.		8.72	215	645	1,925	198.75	
	1400	8" suction, 40 H.P.		11.79	295	884	2,650	271.10	
	1500	10" suction, 75 H.P.	↓	17.73	345	1,033	3,100	348.45	

01 54 33 | Equipment Rental

		01 54 33 | Equipment Rental	UNIT	HOURLY OPER. COST	RENT PER DAY	RENT PER WEEK	RENT PER MONTH	EQUIPMENT COST/DAY	
70	1600	12" suction, 100 H.P.	Ea.	25.55	550	1,650	4,950	534.40	70
	1700	12" suction, 175 H.P.	"	37.17	605	1,810	5,425	659.35	
80	0010	**MARINE EQUIPMENT RENTAL** without operators R015433-10							80
	0200	Barge, 400 Ton, 30' wide x 90' long	Ea.	15.50	1,075	3,220	9,650	768	
	0240	800 Ton, 45' wide x 90' long		18.80	1,300	3,900	11,700	930.40	
	2000	Tugboat, diesel, 100 H.P.		36.45	203	610	1,825	413.60	
	2040	250 H.P.		77.10	375	1,120	3,350	840.80	
	2080	380 H.P.	↓	149.05	1,100	3,325	9,975	1,857	

Crew A-1	Hr.	Daily	Hr.	Daily	Bare Costs	Incl. O&P
1 Building Laborer	$33.10	$264.80	$51.05	$408.40	$33.10	$51.05
1 Concrete saw, gas manual		73.80		81.18	9.22	10.15
8 L.H., Daily Totals		$338.60		$489.58	$42.33	$61.20

Crew A-1A	Hr.	Daily	Hr.	Daily	Bare Costs	Incl. O&P
1 Skilled Worker	$42.60	$340.80	$65.50	$524.00	$42.60	$65.50
1 Shot Blaster, 20"		215.80		237.38	26.98	29.67
8 L.H., Daily Totals		$556.60		$761.38	$69.58	$95.17

Crew A-1B	Hr.	Daily	Hr.	Daily	Bare Costs	Incl. O&P
1 Building Laborer	$33.10	$264.80	$51.05	$408.40	$33.10	$51.05
1 Concrete Saw		161.40		177.54	20.18	22.19
8 L.H., Daily Totals		$426.20		$585.94	$53.27	$73.24

Crew A-1C	Hr.	Daily	Hr.	Daily	Bare Costs	Incl. O&P
1 Building Laborer	$33.10	$264.80	$51.05	$408.40	$33.10	$51.05
1 Chain Saw, gas, 18"		28.20		31.02	3.52	3.88
8 L.H., Daily Totals		$293.00		$439.42	$36.63	$54.93

Crew A-1D	Hr.	Daily	Hr.	Daily	Bare Costs	Incl. O&P
1 Building Laborer	$33.10	$264.80	$51.05	$408.40	$33.10	$51.05
1 Vibrating plate, gas, 18"		34.20		37.62	4.28	4.70
8 L.H., Daily Totals		$299.00		$446.02	$37.38	$55.75

Crew A-1E	Hr.	Daily	Hr.	Daily	Bare Costs	Incl. O&P
1 Building Laborer	$33.10	$264.80	$51.05	$408.40	$33.10	$51.05
1 Vibratory Plate, gas, 21"		43.60		47.96	5.45	6.00
8 L.H., Daily Totals		$308.40		$456.36	$38.55	$57.05

Crew A-1F	Hr.	Daily	Hr.	Daily	Bare Costs	Incl. O&P
1 Building Laborer	$33.10	$264.80	$51.05	$408.40	$33.10	$51.05
1 Rammer/tamper, gas, 8"		46.80		51.48	5.85	6.43
8 L.H., Daily Totals		$311.60		$459.88	$38.95	$57.48

Crew A-1G	Hr.	Daily	Hr.	Daily	Bare Costs	Incl. O&P
1 Building Laborer	$33.10	$264.80	$51.05	$408.40	$33.10	$51.05
1 Rammer/tamper, gas, 15"		51.80		56.98	6.47	7.12
8 L.H., Daily Totals		$316.60		$465.38	$39.58	$58.17

Crew A-1H	Hr.	Daily	Hr.	Daily	Bare Costs	Incl. O&P
1 Building Laborer	$33.10	$264.80	$51.05	$408.40	$33.10	$51.05
1 Exterior Steam Cleaner		60.20		66.22	7.53	8.28
8 L.H., Daily Totals		$325.00		$474.62	$40.63	$59.33

Crew A-1J	Hr.	Daily	Hr.	Daily	Bare Costs	Incl. O&P
1 Building Laborer	$33.10	$264.80	$51.05	$408.40	$33.10	$51.05
1 Cultivator, Walk-Behind, 5 H.P.		56.15		61.77	7.02	7.72
8 L.H., Daily Totals		$320.95		$470.17	$40.12	$58.77

Crew A-1K	Hr.	Daily	Hr.	Daily	Bare Costs	Incl. O&P
1 Building Laborer	$33.10	$264.80	$51.05	$408.40	$33.10	$51.05
1 Cultivator, Walk-Behind, 8 H.P.		86.40		95.04	10.80	11.88
8 L.H., Daily Totals		$351.20		$503.44	$43.90	$62.93

Crew A-1M	Hr.	Daily	Hr.	Daily	Bare Costs	Incl. O&P
1 Building Laborer	$33.10	$264.80	$51.05	$408.40	$33.10	$51.05
1 Snow Blower, Walk-Behind		53.40		58.74	6.67	7.34
8 L.H., Daily Totals		$318.20		$467.14	$39.77	$58.39

Crew A-2	Hr.	Daily	Hr.	Daily	Bare Costs	Incl. O&P
2 Laborers	$33.10	$529.60	$51.05	$816.80	$32.82	$50.43
1 Truck Driver (light)	32.25	258.00	49.20	393.60		
1 Flatbed Truck, Gas, 1.5 Ton		239.80		263.78	9.99	10.99
24 L.H., Daily Totals		$1027.40		$1474.18	$42.81	$61.42

Crew A-2A	Hr.	Daily	Hr.	Daily	Bare Costs	Incl. O&P
2 Laborers	$33.10	$529.60	$51.05	$816.80	$32.82	$50.43
1 Truck Driver (light)	32.25	258.00	49.20	393.60		
1 Flatbed Truck, Gas, 1.5 Ton		239.80		263.78		
1 Concrete Saw		161.40		177.54	16.72	18.39
24 L.H., Daily Totals		$1188.80		$1651.72	$49.53	$68.82

Crew A-2B	Hr.	Daily	Hr.	Daily	Bare Costs	Incl. O&P
1 Truck Driver (light)	$32.25	$258.00	$49.20	$393.60	$32.25	$49.20
1 Flatbed Truck, Gas, 1.5 Ton		239.80		263.78	29.98	32.97
8 L.H., Daily Totals		$497.80		$657.38	$62.23	$82.17

Crew A-3A	Hr.	Daily	Hr.	Daily	Bare Costs	Incl. O&P
1 Truck Driver (light)	$32.25	$258.00	$49.20	$393.60	$32.25	$49.20
1 Pickup truck, 4 x 4, 3/4 ton		151.00		166.10	18.88	20.76
8 L.H., Daily Totals		$409.00		$559.70	$51.13	$69.96

Crew A-3B	Hr.	Daily	Hr.	Daily	Bare Costs	Incl. O&P
1 Equip. Oper. (med.)	$42.95	$343.60	$64.30	$514.40	$38.05	$57.42
1 Truck Driver (heav.)	33.15	265.20	50.55	404.40		
1 Dump Truck, 12 C.Y., 400 H.P.		667.80		734.58		
1 F.E. Loader, W.M.,2.5 C.Y.		435.80		479.38	68.97	75.87
16 L.H., Daily Totals		$1712.40		$2132.76	$107.03	$133.30

Crew A-3C	Hr.	Daily	Hr.	Daily	Bare Costs	Incl. O&P
1 Equip. Oper. (light)	$41.30	$330.40	$61.85	$494.80	$41.30	$61.85
1 Loader, Skid Steer, 78 H.P.		291.00		320.10	36.38	40.01
8 L.H., Daily Totals		$621.40		$814.90	$77.67	$101.86

Crew A-3D	Hr.	Daily	Hr.	Daily	Bare Costs	Incl. O&P
1 Truck Driver, Light	$32.25	$258.00	$49.20	$393.60	$32.25	$49.20
1 Pickup truck, 4 x 4, 3/4 ton		151.00		166.10		
1 Flatbed Trailer, 25 Ton		105.60		116.16	32.08	35.28
8 L.H., Daily Totals		$514.60		$675.86	$64.33	$84.48

Crew A-3E	Hr.	Daily	Hr.	Daily	Bare Costs	Incl. O&P
1 Equip. Oper. (crane)	$44.40	$355.20	$66.45	$531.60	$38.77	$58.50
1 Truck Driver (heavy)	33.15	265.20	50.55	404.40		
1 Pickup truck, 4 x 4, 3/4 ton		151.00		166.10	9.44	10.38
16 L.H., Daily Totals		$771.40		$1102.10	$48.21	$68.88

Crew A-3F	Hr.	Daily	Hr.	Daily	Bare Costs	Incl. O&P
1 Equip. Oper. (crane)	$44.40	$355.20	$66.45	$531.60	$38.77	$58.50
1 Truck Driver (heavy)	33.15	265.20	50.55	404.40		
1 Pickup truck, 4 x 4, 3/4 ton		151.00		166.10		
1 Truck Tractor, 6x4, 380 H.P.		592.40		651.64		
1 Lowbed Trailer, 75 Ton		205.80		226.38	59.33	65.26
16 L.H., Daily Totals		$1569.60		$1980.12	$98.10	$123.76

Crew A-3G	Bare Costs Hr.	Bare Costs Daily	Incl. Subs O&P Hr.	Incl. Subs O&P Daily	Cost Per Labor-Hour Bare Costs	Cost Per Labor-Hour Incl. O&P
1 Equip. Oper. (crane)	$44.40	$355.20	$66.45	$531.60	$38.77	$58.50
1 Truck Driver (heavy)	33.15	265.20	50.55	404.40		
1 Pickup truck, 4 x 4, 3/4 ton		151.00		166.10		
1 Truck Tractor, 6x4, 450 H.P.		715.80		787.38		
1 Lowbed Trailer, 75 Ton		205.80		226.38	67.04	73.74
16 L.H., Daily Totals		$1693.00		$2115.86	$105.81	$132.24

Crew A-3H	Hr.	Daily	Hr.	Daily	Bare Costs	Incl. O&P
1 Equip. Oper. (crane)	$44.40	$355.20	$66.45	$531.60	$44.40	$66.45
1 Hyd. crane, 12 Ton (daily)		871.60		958.76	108.95	119.85
8 L.H., Daily Totals		$1226.80		$1490.36	$153.35	$186.29

Crew A-3I	Hr.	Daily	Hr.	Daily	Bare Costs	Incl. O&P
1 Equip. Oper. (crane)	$44.40	$355.20	$66.45	$531.60	$44.40	$66.45
1 Hyd. crane, 25 Ton (daily)		1015.00		1116.50	126.88	139.56
8 L.H., Daily Totals		$1370.20		$1648.10	$171.28	$206.01

Crew A-3J	Hr.	Daily	Hr.	Daily	Bare Costs	Incl. O&P
1 Equip. Oper. (crane)	$44.40	$355.20	$66.45	$531.60	$44.40	$66.45
1 Hyd. crane, 40 Ton (daily)		1258.00		1383.80	157.25	172.97
8 L.H., Daily Totals		$1613.20		$1915.40	$201.65	$239.43

Crew A-3K	Hr.	Daily	Hr.	Daily	Bare Costs	Incl. O&P
1 Equip. Oper. (crane)	$44.40	$355.20	$66.45	$531.60	$41.35	$61.90
1 Equip. Oper. Oiler	38.30	306.40	57.35	458.80		
1 Hyd. crane, 55 Ton (daily)		1611.00		1772.10		
1 P/U Truck, 3/4 Ton (daily)		161.00		177.10	110.75	121.83
16 L.H., Daily Totals		$2433.60		$2939.60	$152.10	$183.72

Crew A-3L	Hr.	Daily	Hr.	Daily	Bare Costs	Incl. O&P
1 Equip. Oper. (crane)	$44.40	$355.20	$66.45	$531.60	$41.35	$61.90
1 Equip. Oper. Oiler	38.30	306.40	57.35	458.80		
1 Hyd. crane, 80 Ton (daily)		2052.00		2257.20		
1 P/U Truck, 3/4 Ton (daily)		161.00		177.10	138.31	152.14
16 L.H., Daily Totals		$2874.60		$3424.70	$179.66	$214.04

Crew A-3M	Hr.	Daily	Hr.	Daily	Bare Costs	Incl. O&P
1 Equip. Oper. (crane)	$44.40	$355.20	$66.45	$531.60	$41.35	$61.90
1 Equip. Oper. Oiler	38.30	306.40	57.35	458.80		
1 Hyd. crane, 100 Ton (daily)		2268.00		2494.80		
1 P/U Truck, 3/4 Ton (daily)		161.00		177.10	151.81	166.99
16 L.H., Daily Totals		$3090.60		$3662.30	$193.16	$228.89

Crew A-3N	Hr.	Daily	Hr.	Daily	Bare Costs	Incl. O&P
1 Equip. Oper. (crane)	$44.40	$355.20	$66.45	$531.60	$44.40	$66.45
1 Tower crane (monthly)		1049.00		1153.90	131.13	144.24
8 L.H., Daily Totals		$1404.20		$1685.50	$175.53	$210.69

Crew A-3P	Hr.	Daily	Hr.	Daily	Bare Costs	Incl. O&P
1 Equip. Oper., Light	$41.30	$330.40	$61.85	$494.80	$41.30	$61.85
1 A.T. Forklift, 42' lift		478.20		526.02	59.77	65.75
8 L.H., Daily Totals		$808.60		$1020.82	$101.08	$127.60

Crew A-4	Hr.	Daily	Hr.	Daily	Bare Costs	Incl. O&P
2 Carpenters	$41.55	$664.80	$64.05	$1024.80	$39.82	$60.73
1 Painter, Ordinary	36.35	290.80	54.10	432.80		
24 L.H., Daily Totals		$955.60		$1457.60	$39.82	$60.73

Crew A-5	Hr.	Daily	Hr.	Daily	Bare Costs	Incl. O&P
2 Laborers	$33.10	$529.60	$51.05	$816.80	$33.01	$50.84
.25 Truck Driver (light)	32.25	64.50	49.20	98.40		
.25 Flatbed Truck, Gas, 1.5 Ton		59.95		65.94	3.33	3.66
18 L.H., Daily Totals		$654.05		$981.14	$36.34	$54.51

Crew A-6	Hr.	Daily	Hr.	Daily	Bare Costs	Incl. O&P
1 Instrument Man	$42.60	$340.80	$65.50	$524.00	$41.40	$63.02
1 Rodman/Chainman	40.20	321.60	60.55	484.40		
1 Laser Transit/Level		69.90		76.89	4.37	4.81
16 L.H., Daily Totals		$732.30		$1085.29	$45.77	$67.83

Crew A-7	Hr.	Daily	Hr.	Daily	Bare Costs	Incl. O&P
1 Chief Of Party	$52.25	$418.00	$79.90	$639.20	$45.02	$68.65
1 Instrument Man	42.60	340.80	65.50	524.00		
1 Rodman/Chainman	40.20	321.60	60.55	484.40		
1 Laser Transit/Level		69.90		76.89	2.91	3.20
24 L.H., Daily Totals		$1150.30		$1724.49	$47.93	$71.85

Crew A-8	Hr.	Daily	Hr.	Daily	Bare Costs	Incl. O&P
1 Chief of Party	$52.25	$418.00	$79.90	$639.20	$43.81	$66.63
1 Instrument Man	42.60	340.80	65.50	524.00		
2 Rodmen/Chainmen	40.20	643.20	60.55	968.80		
1 Laser Transit/Level		69.90		76.89	2.18	2.40
32 L.H., Daily Totals		$1471.90		$2208.89	$46.00	$69.03

Crew A-9	Hr.	Daily	Hr.	Daily	Bare Costs	Incl. O&P
1 Asbestos Foreman	$46.05	$368.40	$71.55	$572.40	$45.61	$70.89
7 Asbestos Workers	45.55	2550.80	70.80	3964.80		
64 L.H., Daily Totals		$2919.20		$4537.20	$45.61	$70.89

Crew A-10A	Hr.	Daily	Hr.	Daily	Bare Costs	Incl. O&P
1 Asbestos Foreman	$46.05	$368.40	$71.55	$572.40	$45.72	$71.05
2 Asbestos Workers	45.55	728.80	70.80	1132.80		
24 L.H., Daily Totals		$1097.20		$1705.20	$45.72	$71.05

Crew A-10B	Hr.	Daily	Hr.	Daily	Bare Costs	Incl. O&P
1 Asbestos Foreman	$46.05	$368.40	$71.55	$572.40	$45.67	$70.99
3 Asbestos Workers	45.55	1093.20	70.80	1699.20		
32 L.H., Daily Totals		$1461.60		$2271.60	$45.67	$70.99

Crew A-10C	Hr.	Daily	Hr.	Daily	Bare Costs	Incl. O&P
3 Asbestos Workers	$45.55	$1093.20	$70.80	$1699.20	$45.55	$70.80
1 Flatbed Truck, Gas, 1.5 Ton		239.80		263.78	9.99	10.99
24 L.H., Daily Totals		$1333.00		$1962.98	$55.54	$81.79

Crew A-10D	Hr.	Daily	Hr.	Daily	Bare Costs	Incl. O&P
2 Asbestos Workers	$45.55	$728.80	$70.80	$1132.80	$43.45	$66.35
1 Equip. Oper. (crane)	44.40	355.20	66.45	531.60		
1 Equip. Oper. Oiler	38.30	306.40	57.35	458.80		
1 Hydraulic Crane, 33 Ton		763.00		839.30	23.84	26.23
32 L.H., Daily Totals		$2153.40		$2962.50	$67.29	$92.58

Crew A-11	Hr.	Daily	Hr.	Daily	Bare Costs	Incl. O&P
1 Asbestos Foreman	$46.05	$368.40	$71.55	$572.40	$45.61	$70.89
7 Asbestos Workers	45.55	2550.80	70.80	3964.80		
2 Chip. Hammers, 12 Lb., Elec.		36.00		39.60	.56	.62
64 L.H., Daily Totals		$2955.20		$4576.80	$46.17	$71.51

Crew No.	Bare Costs		Incl. Subs O&P		Cost Per Labor-Hour	

Crew A-12	Hr.	Daily	Hr.	Daily	Bare Costs	Incl. O&P
1 Asbestos Foreman	$46.05	$368.40	$71.55	$572.40	$45.61	$70.89
7 Asbestos Workers	45.55	2550.80	70.80	3964.80		
1 Trk-mtd vac, 14 CY, 1500 Gal.		530.25		583.27		
1 Flatbed Truck, 20,000 GVW		245.60		270.16	12.12	13.33
64 L.H., Daily Totals		$3695.05		$5390.64	$57.74	$84.23

Crew A-13	Hr.	Daily	Hr.	Daily	Bare Costs	Incl. O&P
1 Equip. Oper. (light)	$41.30	$330.40	$61.85	$494.80	$41.30	$61.85
1 Trk-mtd vac, 14 CY, 1500 Gal.		530.25		583.27		
1 Flatbed Truck, 20,000 GVW		245.60		270.16	96.98	106.68
8 L.H., Daily Totals		$1106.25		$1348.23	$138.28	$168.53

Crew B-1	Hr.	Daily	Hr.	Daily	Bare Costs	Incl. O&P
1 Labor Foreman (outside)	$35.10	$280.80	$54.10	$432.80	$33.77	$52.07
2 Laborers	33.10	529.60	51.05	816.80		
24 L.H., Daily Totals		$810.40		$1249.60	$33.77	$52.07

Crew B-1A	Hr.	Daily	Hr.	Daily	Bare Costs	Incl. O&P
1 Laborer Foreman	$35.10	$280.80	$54.10	$432.80	$33.77	$52.07
2 Laborers	33.10	529.60	51.05	816.80		
2 Cutting Torches		38.00		41.80		
2 Set of Gases		158.40		174.24	8.18	9.00
24 L.H., Daily Totals		$1006.80		$1465.64	$41.95	$61.07

Crew B-1B	Hr.	Daily	Hr.	Daily	Bare Costs	Incl. O&P
1 Laborer Foreman	$35.10	$280.80	$54.10	$432.80	$36.42	$55.66
2 Laborers	33.10	529.60	51.05	816.80		
1 Equip. Oper. (crane)	44.40	355.20	66.45	531.60		
2 Cutting Torches		38.00		41.80		
2 Set of Gases		158.40		174.24		
1 Hyd. Crane, 12 Ton		655.60		721.16	26.63	29.29
32 L.H., Daily Totals		$2017.60		$2718.40	$63.05	$84.95

Crew B-2	Hr.	Daily	Hr.	Daily	Bare Costs	Incl. O&P
1 Labor Foreman (outside)	$35.10	$280.80	$54.10	$432.80	$33.50	$51.66
4 Laborers	33.10	1059.20	51.05	1633.60		
40 L.H., Daily Totals		$1340.00		$2066.40	$33.50	$51.66

Crew B-3	Hr.	Daily	Hr.	Daily	Bare Costs	Incl. O&P
1 Labor Foreman (outside)	$35.10	$280.80	$54.10	$432.80	$35.09	$53.60
2 Laborers	33.10	529.60	51.05	816.80		
1 Equip. Oper. (med.)	42.95	343.60	64.30	514.40		
2 Truck Drivers (heavy)	33.15	530.40	50.55	808.80		
1 Crawler Loader, 3 C.Y.		1077.00		1184.70		
2 Dump Trucks 12 C.Y., 400 H.P.		1335.60		1469.16	50.26	55.29
48 L.H., Daily Totals		$4097.00		$5226.66	$85.35	$108.89

Crew B-3A	Hr.	Daily	Hr.	Daily	Bare Costs	Incl. O&P
4 Laborers	$33.10	$1059.20	$51.05	$1633.60	$35.07	$53.70
1 Equip. Oper. (med.)	42.95	343.60	64.30	514.40		
1 Hyd. Excavator, 1.5 C.Y.		1017.00		1118.70	25.43	27.97
40 L.H., Daily Totals		$2419.80		$3266.70	$60.49	$81.67

Crew B-3B	Hr.	Daily	Hr.	Daily	Bare Costs	Incl. O&P
2 Laborers	$33.10	$529.60	$51.05	$816.80	$35.58	$54.24
1 Equip. Oper. (med.)	42.95	343.60	64.30	514.40		
1 Truck Driver (heavy)	33.15	265.20	50.55	404.40		
1 Backhoe Loader, 80 H.P.		387.60		426.36		
1 Dump Truck, 12 C.Y., 400 H.P.		667.80		734.58	32.98	36.28
32 L.H., Daily Totals		$2193.80		$2896.54	$68.56	$90.52

Crew No.	Bare Costs		Incl. Subs O&P		Cost Per Labor-Hour	

Crew B-3C	Hr.	Daily	Hr.	Daily	Bare Costs	Incl. O&P
3 Laborers	$33.10	$794.40	$51.05	$1225.20	$35.56	$54.36
1 Equip. Oper. (med.)	42.95	343.60	64.30	514.40		
1 Crawler Loader, 4 C.Y.		1548.00		1702.80	48.38	53.21
32 L.H., Daily Totals		$2686.00		$3442.40	$83.94	$107.58

Crew B-4	Hr.	Daily	Hr.	Daily	Bare Costs	Incl. O&P
1 Labor Foreman (outside)	$35.10	$280.80	$54.10	$432.80	$33.44	$51.48
4 Laborers	33.10	1059.20	51.05	1633.60		
1 Truck Driver (heavy)	33.15	265.20	50.55	404.40		
1 Truck Tractor, 220 H.P.		354.00		389.40		
1 Flatbed Trailer, 40 Ton		143.20		157.52	10.36	11.39
48 L.H., Daily Totals		$2102.40		$3017.72	$43.80	$62.87

Crew B-5	Hr.	Daily	Hr.	Daily	Bare Costs	Incl. O&P
1 Labor Foreman (outside)	$35.10	$280.80	$54.10	$432.80	$36.20	$55.27
4 Laborers	33.10	1059.20	51.05	1633.60		
2 Equip. Oper. (med.)	42.95	687.20	64.30	1028.80		
1 Air Compressor, 250 cfm		193.40		212.74		
2 Breakers, Pavement, 60 lb.		18.80		20.68		
2 -50' Air Hoses, 1.5"		12.60		13.86		
1 Crawler Loader, 3 C.Y.		1077.00		1184.70	23.25	25.57
56 L.H., Daily Totals		$3329.00		$4527.18	$59.45	$80.84

Crew B-5A	Hr.	Daily	Hr.	Daily	Bare Costs	Incl. O&P
1 Foreman	$35.10	$280.80	$54.10	$432.80	$35.60	$54.33
6 Laborers	33.10	1588.80	51.05	2450.40		
2 Equip. Oper. (med.)	42.95	687.20	64.30	1028.80		
1 Equip. Oper. (light)	41.30	330.40	61.85	494.80		
2 Truck Drivers (heavy)	33.15	530.40	50.55	808.80		
1 Air Compressor, 365 cfm		255.60		281.16		
2 Breakers, Pavement, 60 lb.		18.80		20.68		
8 -50' Air Hoses, 1"		32.80		36.08		
2 Dump Trucks, 8 C.Y., 220 H.P.		803.60		883.96	11.57	12.73
96 L.H., Daily Totals		$4528.40		$6437.48	$47.17	$67.06

Crew B-5B	Hr.	Daily	Hr.	Daily	Bare Costs	Incl. O&P
1 Powderman	$42.60	$340.80	$65.50	$524.00	$37.99	$57.63
2 Equip. Oper. (med.)	42.95	687.20	64.30	1028.80		
3 Truck Drivers (heavy)	33.15	795.60	50.55	1213.20		
1 F.E. Loader, W.M.,2.5 C.Y.		435.80		479.38		
3 Dump Trucks, 12 C.Y., 400 H.P.		2003.40		2203.74		
1 Air Compressor, 365 cfm		255.60		281.16	56.14	61.76
48 L.H., Daily Totals		$4518.40		$5730.28	$94.13	$119.38

Crew B-5C	Hr.	Daily	Hr.	Daily	Bare Costs	Incl. O&P
3 Laborers	$33.10	$794.40	$51.05	$1225.20	$36.41	$55.29
1 Equip. Oper. (med.)	42.95	343.60	64.30	514.40		
2 Truck Drivers (heavy)	33.15	530.40	50.55	808.80		
1 Equip. Oper. (crane)	44.40	355.20	66.45	531.60		
1 Equip. Oper. Oiler	38.30	306.40	57.35	458.80		
2 Dump Trucks, 12 C.Y., 400 H.P.		1335.60		1469.16		
1 Crawler Loader, 4 C.Y.		1548.00		1702.80		
1 S.P. Crane, 4x4, 25 Ton		655.60		721.16	55.30	60.83
64 L.H., Daily Totals		$5869.20		$7431.92	$91.71	$116.12

Crew B-6	Hr.	Daily	Hr.	Daily	Bare Costs	Incl. O&P
2 Laborers	$33.10	$529.60	$51.05	$816.80	$35.83	$54.65
1 Equip. Oper. (light)	41.30	330.40	61.85	494.80		
1 Backhoe Loader, 48 H.P.		337.80		371.58	14.07	15.48
24 L.H., Daily Totals		$1197.80		$1683.18	$49.91	$70.13

Crew No.	Bare Costs		Incl. Subs O&P		Cost Per Labor-Hour	

Crew B-6A	Hr.	Daily	Hr.	Daily	Bare Costs	Incl. O&P
.5 Labor Foreman (outside)	$35.10	$140.40	$54.10	$216.40	$37.44	$56.96
1 Laborer	33.10	264.80	51.05	408.40		
1 Equip. Oper. (med.)	42.95	343.60	64.30	514.40		
1 Vacuum Trk.,5000 Gal.		363.85		400.24	18.19	20.01
20 L.H., Daily Totals		$1112.65		$1539.43	$55.63	$76.97

Crew B-6B	Hr.	Daily	Hr.	Daily	Bare Costs	Incl. O&P
2 Labor Foremen (out)	$35.10	$561.60	$54.10	$865.60	$33.77	$52.07
4 Laborers	33.10	1059.20	51.05	1633.60		
1 S.P. Crane, 4x4, 5 Ton		276.20		303.82		
1 Flatbed Truck, Gas, 1.5 Ton		239.80		263.78		
1 Butt Fusion Machine		134.40		147.84	13.55	14.90
48 L.H., Daily Totals		$2271.20		$3214.64	$47.32	$66.97

Crew B-7	Hr.	Daily	Hr.	Daily	Bare Costs	Incl. O&P
1 Labor Foreman (outside)	$35.10	$280.80	$54.10	$432.80	$35.08	$53.77
4 Laborers	33.10	1059.20	51.05	1633.60		
1 Equip. Oper. (med.)	42.95	343.60	64.30	514.40		
1 Brush Chipper, 12", 130 H.P.		253.20		278.52		
1 Crawler Loader, 3 C.Y.		1077.00		1184.70		
2 Chain Saws, gas, 36" Long		78.40		86.24	29.35	32.28
48 L.H., Daily Totals		$3092.20		$4130.26	$64.42	$86.05

Crew B-7A	Hr.	Daily	Hr.	Daily	Bare Costs	Incl. O&P
2 Laborers	$33.10	$529.60	$51.05	$816.80	$35.83	$54.65
1 Equip. Oper. (light)	41.30	330.40	61.85	494.80		
1 Rake w/Tractor		285.70		314.27		
2 Chain Saws, gas, 18"		56.40		62.04	14.25	15.68
24 L.H., Daily Totals		$1202.10		$1687.91	$50.09	$70.33

Crew B-8	Hr.	Daily	Hr.	Daily	Bare Costs	Incl. O&P
1 Labor Foreman (outside)	$35.10	$280.80	$54.10	$432.80	$36.48	$55.41
2 Laborers	33.10	529.60	51.05	816.80		
2 Equip. Oper. (med.)	42.95	687.20	64.30	1028.80		
1 Equip. Oper. Oiler	38.30	306.40	57.35	458.80		
2 Truck Drivers (heavy)	33.15	530.40	50.55	808.80		
1 Hyd. Crane, 25 Ton		747.60		822.36		
1 Crawler Loader, 3 C.Y.		1077.00		1184.70		
2 Dump Trucks, 12 C.Y., 400 H.P.		1335.60		1469.16	49.38	54.32
64 L.H., Daily Totals		$5494.60		$7022.22	$85.85	$109.72

Crew B-9	Hr.	Daily	Hr.	Daily	Bare Costs	Incl. O&P
1 Labor Foreman (outside)	$35.10	$280.80	$54.10	$432.80	$33.50	$51.66
4 Laborers	33.10	1059.20	51.05	1633.60		
1 Air Compressor, 250 cfm		193.40		212.74		
2 Breakers, Pavement, 60 lb.		18.80		20.68		
2 -50' Air Hoses, 1.5"		12.60		13.86	5.62	6.18
40 L.H., Daily Totals		$1564.80		$2313.68	$39.12	$57.84

Crew B-9A	Hr.	Daily	Hr.	Daily	Bare Costs	Incl. O&P
2 Laborers	$33.10	$529.60	$51.05	$816.80	$33.12	$50.88
1 Truck Driver (heavy)	33.15	265.20	50.55	404.40		
1 Water Tank Trailer, 5000 Gal.		139.60		153.56		
1 Truck Tractor, 220 H.P.		354.00		389.40		
2 -50' Discharge Hoses, 3"		3.50		3.85	20.71	22.78
24 L.H., Daily Totals		$1291.90		$1768.01	$53.83	$73.67

Crew B-9B	Hr.	Daily	Hr.	Daily	Bare Costs	Incl. O&P
2 Laborers	$33.10	$529.60	$51.05	$816.80	$33.12	$50.88
1 Truck Driver (heavy)	33.15	265.20	50.55	404.40		
2 -50' Discharge Hoses, 3"		3.50		3.85		
1 Water Tank Trailer, 5000 Gal.		139.60		153.56		
1 Truck Tractor, 220 H.P.		354.00		389.40		
1 Pressure Washer		60.80		66.88	23.25	25.57
24 L.H., Daily Totals		$1352.70		$1834.89	$56.36	$76.45

Crew B-9D	Hr.	Daily	Hr.	Daily	Bare Costs	Incl. O&P
1 Labor Foreman (Outside)	$35.10	$280.80	$54.10	$432.80	$33.50	$51.66
4 Common Laborers	33.10	1059.20	51.05	1633.60		
1 Air Compressor, 250 cfm		193.40		212.74		
2 -50' Air Hoses, 1.5"		12.60		13.86		
2 Air Powered Tampers		50.50		55.55	6.41	7.05
40 L.H., Daily Totals		$1596.50		$2348.55	$39.91	$58.71

Crew B-10	Hr.	Daily	Hr.	Daily	Bare Costs	Incl. O&P
1 Equip. Oper. (med.)	$42.95	$343.60	$64.30	$514.40	$39.67	$59.88
.5 Laborer	33.10	132.40	51.05	204.20		
12 L.H., Daily Totals		$476.00		$718.60	$39.67	$59.88

Crew B-10A	Hr.	Daily	Hr.	Daily	Bare Costs	Incl. O&P
1 Equip. Oper. (med.)	$42.95	$343.60	$64.30	$514.40	$39.67	$59.88
.5 Laborer	33.10	132.40	51.05	204.20		
1 Roller, 2-Drum, W.B., 7.5 H.P.		149.60		164.56	12.47	13.71
12 L.H., Daily Totals		$625.60		$883.16	$52.13	$73.60

Crew B-10B	Hr.	Daily	Hr.	Daily	Bare Costs	Incl. O&P
1 Equip. Oper. (med.)	$42.95	$343.60	$64.30	$514.40	$39.67	$59.88
.5 Laborer	33.10	132.40	51.05	204.20		
1 Dozer, 200 H.P.		1192.00		1311.20	99.33	109.27
12 L.H., Daily Totals		$1668.00		$2029.80	$139.00	$169.15

Crew B-10C	Hr.	Daily	Hr.	Daily	Bare Costs	Incl. O&P
1 Equip. Oper. (med.)	$42.95	$343.60	$64.30	$514.40	$39.67	$59.88
.5 Laborer	33.10	132.40	51.05	204.20		
1 Dozer, 200 H.P.		1192.00		1311.20		
1 Vibratory Roller, Towed, 23 Ton		434.20		477.62	135.52	149.07
12 L.H., Daily Totals		$2102.20		$2507.42	$175.18	$208.95

Crew B-10D	Hr.	Daily	Hr.	Daily	Bare Costs	Incl. O&P
1 Equip. Oper. (med.)	$42.95	$343.60	$64.30	$514.40	$39.67	$59.88
.5 Laborer	33.10	132.40	51.05	204.20		
1 Dozer, 200 H.P.		1192.00		1311.20		
1 Sheepsft. Roller, Towed		489.20		538.12	140.10	154.11
12 L.H., Daily Totals		$2157.20		$2567.92	$179.77	$213.99

Crew B-10E	Hr.	Daily	Hr.	Daily	Bare Costs	Incl. O&P
1 Equip. Oper. (med.)	$42.95	$343.60	$64.30	$514.40	$39.67	$59.88
.5 Laborer	33.10	132.40	51.05	204.20		
1 Tandem Roller, 5 Ton		152.00		167.20	12.67	13.93
12 L.H., Daily Totals		$628.00		$885.80	$52.33	$73.82

Crew B-10F	Hr.	Daily	Hr.	Daily	Bare Costs	Incl. O&P
1 Equip. Oper. (med.)	$42.95	$343.60	$64.30	$514.40	$39.67	$59.88
.5 Laborer	33.10	132.40	51.05	204.20		
1 Tandem Roller, 10 Ton		235.20		258.72	19.60	21.56
12 L.H., Daily Totals		$711.20		$977.32	$59.27	$81.44

Crew B-10G	Bare Costs Hr.	Bare Costs Daily	Incl. Subs O&P Hr.	Incl. Subs O&P Daily	Bare Costs	Incl. O&P
1 Equip. Oper. (med.)	$42.95	$343.60	$64.30	$514.40	$39.67	$59.88
.5 Laborer	33.10	132.40	51.05	204.20		
1 Sheepsft. Roll., 240 H.P.		1243.00		1367.30	103.58	113.94
12 L.H., Daily Totals		$1719.00		$2085.90	$143.25	$173.82

Crew B-10H	Hr.	Daily	Hr.	Daily	Bare Costs	Incl. O&P
1 Equip. Oper. (med.)	$42.95	$343.60	$64.30	$514.40	$39.67	$59.88
.5 Laborer	33.10	132.40	51.05	204.20		
1 Diaphragm Water Pump, 2"		68.00		74.80		
1 -20' Suction Hose, 2"		1.95		2.15		
2 -50' Discharge Hoses, 2"		1.80		1.98	5.98	6.58
12 L.H., Daily Totals		$547.75		$797.52	$45.65	$66.46

Crew B-10I	Hr.	Daily	Hr.	Daily	Bare Costs	Incl. O&P
1 Equip. Oper. (med.)	$42.95	$343.60	$64.30	$514.40	$39.67	$59.88
.5 Laborer	33.10	132.40	51.05	204.20		
1 Diaphragm Water Pump, 4"		90.80		99.88		
1 -20' Suction Hose, 4"		3.45		3.79		
2 -50' Discharge Hoses, 4"		4.70		5.17	8.25	9.07
12 L.H., Daily Totals		$574.95		$827.45	$47.91	$68.95

Crew B-10J	Hr.	Daily	Hr.	Daily	Bare Costs	Incl. O&P
1 Equip. Oper. (med.)	$42.95	$343.60	$64.30	$514.40	$39.67	$59.88
.5 Laborer	33.10	132.40	51.05	204.20		
1 Centrifugal Water Pump, 3"		74.40		81.84		
1 -20' Suction Hose, 3"		3.05		3.36		
2 -50' Discharge Hoses, 3"		3.50		3.85	6.75	7.42
12 L.H., Daily Totals		$556.95		$807.64	$46.41	$67.30

Crew B-10K	Hr.	Daily	Hr.	Daily	Bare Costs	Incl. O&P
1 Equip. Oper. (med.)	$42.95	$343.60	$64.30	$514.40	$39.67	$59.88
.5 Laborer	33.10	132.40	51.05	204.20		
1 Centr. Water Pump, 6"		340.00		374.00		
1 -20' Suction Hose, 6"		11.90		13.09		
2 -50' Discharge Hoses, 6"		12.60		13.86	30.38	33.41
12 L.H., Daily Totals		$840.50		$1119.55	$70.04	$93.30

Crew B-10L	Hr.	Daily	Hr.	Daily	Bare Costs	Incl. O&P
1 Equip. Oper. (med.)	$42.95	$343.60	$64.30	$514.40	$39.67	$59.88
.5 Laborer	33.10	132.40	51.05	204.20		
1 Dozer, 80 H.P.		436.60		480.26	36.38	40.02
12 L.H., Daily Totals		$912.60		$1198.86	$76.05	$99.91

Crew B-10M	Hr.	Daily	Hr.	Daily	Bare Costs	Incl. O&P
1 Equip. Oper. (med.)	$42.95	$343.60	$64.30	$514.40	$39.67	$59.88
.5 Laborer	33.10	132.40	51.05	204.20		
1 Dozer, 300 H.P.		1592.00		1751.20	132.67	145.93
12 L.H., Daily Totals		$2068.00		$2469.80	$172.33	$205.82

Crew B-10N	Hr.	Daily	Hr.	Daily	Bare Costs	Incl. O&P
1 Equip. Oper. (med.)	$42.95	$343.60	$64.30	$514.40	$39.67	$59.88
.5 Laborer	33.10	132.40	51.05	204.20		
1 F.E. Loader, T.M., 1.5 C.Y.		433.80		477.18	36.15	39.77
12 L.H., Daily Totals		$909.80		$1195.78	$75.82	$99.65

Crew B-10O	Hr.	Daily	Hr.	Daily	Bare Costs	Incl. O&P
1 Equip. Oper. (med.)	$42.95	$343.60	$64.30	$514.40	$39.67	$59.88
.5 Laborer	33.10	132.40	51.05	204.20		
1 F.E. Loader, T.M., 2.25 C.Y.		839.40		923.34	69.95	76.94
12 L.H., Daily Totals		$1315.40		$1641.94	$109.62	$136.83

Crew B-10P	Hr.	Daily	Hr.	Daily	Bare Costs	Incl. O&P
1 Equip. Oper. (med.)	$42.95	$343.60	$64.30	$514.40	$39.67	$59.88
.5 Laborer	33.10	132.40	51.05	204.20		
1 Crawler Loader, 3 C.Y.		1077.00		1184.70	89.75	98.72
12 L.H., Daily Totals		$1553.00		$1903.30	$129.42	$158.61

Crew B-10Q	Hr.	Daily	Hr.	Daily	Bare Costs	Incl. O&P
1 Equip. Oper. (med.)	$42.95	$343.60	$64.30	$514.40	$39.67	$59.88
.5 Laborer	33.10	132.40	51.05	204.20		
1 Crawler Loader, 4 C.Y.		1548.00		1702.80	129.00	141.90
12 L.H., Daily Totals		$2024.00		$2421.40	$168.67	$201.78

Crew B-10R	Hr.	Daily	Hr.	Daily	Bare Costs	Incl. O&P
1 Equip. Oper. (med.)	$42.95	$343.60	$64.30	$514.40	$39.67	$59.88
.5 Laborer	33.10	132.40	51.05	204.20		
1 F.E. Loader, W.M., 1 C.Y.		269.40		296.34	22.45	24.70
12 L.H., Daily Totals		$745.40		$1014.94	$62.12	$84.58

Crew B-10S	Hr.	Daily	Hr.	Daily	Bare Costs	Incl. O&P
1 Equip. Oper. (med.)	$42.95	$343.60	$64.30	$514.40	$39.67	$59.88
.5 Laborer	33.10	132.40	51.05	204.20		
1 F.E. Loader, W.M., 1.5 C.Y.		367.60		404.36	30.63	33.70
12 L.H., Daily Totals		$843.60		$1122.96	$70.30	$93.58

Crew B-10T	Hr.	Daily	Hr.	Daily	Bare Costs	Incl. O&P
1 Equip. Oper. (med.)	$42.95	$343.60	$64.30	$514.40	$39.67	$59.88
.5 Laborer	33.10	132.40	51.05	204.20		
1 F.E. Loader, W.M.,2.5 C.Y.		435.80		479.38	36.32	39.95
12 L.H., Daily Totals		$911.80		$1197.98	$75.98	$99.83

Crew B-10U	Hr.	Daily	Hr.	Daily	Bare Costs	Incl. O&P
1 Equip. Oper. (med.)	$42.95	$343.60	$64.30	$514.40	$39.67	$59.88
.5 Laborer	33.10	132.40	51.05	204.20		
1 F.E. Loader, W.M., 5.5 C.Y.		994.40		1093.84	82.87	91.15
12 L.H., Daily Totals		$1470.40		$1812.44	$122.53	$151.04

Crew B-10V	Hr.	Daily	Hr.	Daily	Bare Costs	Incl. O&P
1 Equip. Oper. (med.)	$42.95	$343.60	$64.30	$514.40	$39.67	$59.88
.5 Laborer	33.10	132.40	51.05	204.20		
1 Dozer, 700 H.P.		4598.00		5057.80	383.17	421.48
12 L.H., Daily Totals		$5074.00		$5776.40	$422.83	$481.37

Crew B-10W	Hr.	Daily	Hr.	Daily	Bare Costs	Incl. O&P
1 Equip. Oper. (med.)	$42.95	$343.60	$64.30	$514.40	$39.67	$59.88
.5 Laborer	33.10	132.40	51.05	204.20		
1 Dozer, 105 H.P.		660.40		726.44	55.03	60.54
12 L.H., Daily Totals		$1136.40		$1445.04	$94.70	$120.42

Crew B-10X	Hr.	Daily	Hr.	Daily	Bare Costs	Incl. O&P
1 Equip. Oper. (med.)	$42.95	$343.60	$64.30	$514.40	$39.67	$59.88
.5 Laborer	33.10	132.40	51.05	204.20		
1 Dozer, 410 H.P.		2079.00		2286.90	173.25	190.57
12 L.H., Daily Totals		$2555.00		$3005.50	$212.92	$250.46

Crew B-10Y	Hr.	Daily	Hr.	Daily	Bare Costs	Incl. O&P
1 Equip. Oper. (med.)	$42.95	$343.60	$64.30	$514.40	$39.67	$59.88
.5 Laborer	33.10	132.40	51.05	204.20		
1 Vibr. Roller, Towed, 12 Ton		516.60		568.26	43.05	47.35
12 L.H., Daily Totals		$992.60		$1286.86	$82.72	$107.24

Crew B-11A	Bare Costs Hr.	Bare Costs Daily	Incl. Subs O&P Hr.	Incl. Subs O&P Daily	Bare Costs	Incl. O&P
1 Equipment Oper. (med.)	$42.95	$343.60	$64.30	$514.40	$38.02	$57.67
1 Laborer	33.10	264.80	51.05	408.40		
1 Dozer, 200 H.P.		1192.00		1311.20	74.50	81.95
16 L.H., Daily Totals		$1800.40		$2234.00	$112.53	$139.63

Crew B-11B	Hr.	Daily	Hr.	Daily	Bare Costs	Incl. O&P
1 Equipment Oper. (light)	$41.30	$330.40	$61.85	$494.80	$37.20	$56.45
1 Laborer	33.10	264.80	51.05	408.40		
1 Air Powered Tamper		25.25		27.77		
1 Air Compressor, 365 cfm		255.60		281.16		
2 -50' Air Hoses, 1.5"		12.60		13.86	18.34	20.17
16 L.H., Daily Totals		$888.65		$1225.99	$55.54	$76.62

Crew B-11C	Hr.	Daily	Hr.	Daily	Bare Costs	Incl. O&P
1 Equipment Oper. (med.)	$42.95	$343.60	$64.30	$514.40	$38.02	$57.67
1 Laborer	33.10	264.80	51.05	408.40		
1 Backhoe Loader, 48 H.P.		337.80		371.58	21.11	23.22
16 L.H., Daily Totals		$946.20		$1294.38	$59.14	$80.90

Crew B-11J	Hr.	Daily	Hr.	Daily	Bare Costs	Incl. O&P
1 Equipment Oper. (med.)	$42.95	$343.60	$64.30	$514.40	$38.02	$57.67
1 Laborer	33.10	264.80	51.05	408.40		
1 Grader, 30,000 Lbs.		622.20		684.42		
1 Ripper, beam & 1 shank		80.00		88.00	43.89	48.28
16 L.H., Daily Totals		$1310.60		$1695.22	$81.91	$105.95

Crew B-11K	Hr.	Daily	Hr.	Daily	Bare Costs	Incl. O&P
1 Equipment Oper. (med.)	$42.95	$343.60	$64.30	$514.40	$38.02	$57.67
1 Laborer	33.10	264.80	51.05	408.40		
1 Trencher, Chain Type, 8' D		2090.00		2299.00	130.63	143.69
16 L.H., Daily Totals		$2698.40		$3221.80	$168.65	$201.36

Crew B-11L	Hr.	Daily	Hr.	Daily	Bare Costs	Incl. O&P
1 Equipment Oper. (med.)	$42.95	$343.60	$64.30	$514.40	$38.02	$57.67
1 Laborer	33.10	264.80	51.05	408.40		
1 Grader, 30,000 Lbs.		622.20		684.42	38.89	42.78
16 L.H., Daily Totals		$1230.60		$1607.22	$76.91	$100.45

Crew B-11M	Hr.	Daily	Hr.	Daily	Bare Costs	Incl. O&P
1 Equipment Oper. (med.)	$42.95	$343.60	$64.30	$514.40	$38.02	$57.67
1 Laborer	33.10	264.80	51.05	408.40		
1 Backhoe Loader, 80 H.P.		387.60		426.36	24.23	26.65
16 L.H., Daily Totals		$996.00		$1349.16	$62.25	$84.32

Crew B-11N	Hr.	Daily	Hr.	Daily	Bare Costs	Incl. O&P
1 Labor Foreman	$35.10	$280.80	$54.10	$432.80	$35.54	$54.00
2 Equipment Operators (med.)	42.95	687.20	64.30	1028.80		
6 Truck Drivers (hvy.)	33.15	1591.20	50.55	2426.40		
1 F.E. Loader, W.M., 5.5 C.Y.		994.40		1093.84		
1 Dozer, 410 H.P.		2079.00		2286.90		
6 Dump Trucks, Off Hwy., 50 Ton		10464.00		11510.40	188.02	206.82
72 L.H., Daily Totals		$16096.60		$18779.14	$223.56	$260.82

Crew B-11Q	Hr.	Daily	Hr.	Daily	Bare Costs	Incl. O&P
1 Equipment Operator (med.)	$42.95	$343.60	$64.30	$514.40	$39.67	$59.88
.5 Laborer	33.10	132.40	51.05	204.20		
1 Dozer, 140 H.P.		769.60		846.56	64.13	70.55
12 L.H., Daily Totals		$1245.60		$1565.16	$103.80	$130.43

Crew B-11R	Hr.	Daily	Hr.	Daily	Bare Costs	Incl. O&P
1 Equipment Operator (med.)	$42.95	$343.60	$64.30	$514.40	$39.67	$59.88
.5 Laborer	33.10	132.40	51.05	204.20		
1 Dozer, 200 H.P.		1192.00		1311.20	99.33	109.27
12 L.H., Daily Totals		$1668.00		$2029.80	$139.00	$169.15

Crew B-11S	Hr.	Daily	Hr.	Daily	Bare Costs	Incl. O&P
1 Equipment Operator (med.)	$42.95	$343.60	$64.30	$514.40	$39.67	$59.88
.5 Laborer	33.10	132.40	51.05	204.20		
1 Dozer, 300 H.P.		1592.00		1751.20		
1 Ripper, beam & 1 shank		80.00		88.00	139.33	153.27
12 L.H., Daily Totals		$2148.00		$2557.80	$179.00	$213.15

Crew B-11T	Hr.	Daily	Hr.	Daily	Bare Costs	Incl. O&P
1 Equipment Operator (med.)	$42.95	$343.60	$64.30	$514.40	$39.67	$59.88
.5 Laborer	33.10	132.40	51.05	204.20		
1 Dozer, 410 H.P.		2079.00		2286.90		
1 Ripper, beam & 2 shanks		90.20		99.22	180.77	198.84
12 L.H., Daily Totals		$2645.20		$3104.72	$220.43	$258.73

Crew B-11U	Hr.	Daily	Hr.	Daily	Bare Costs	Incl. O&P
1 Equipment Operator (med.)	$42.95	$343.60	$64.30	$514.40	$39.67	$59.88
.5 Laborer	33.10	132.40	51.05	204.20		
1 Dozer, 520 H.P.		2810.00		3091.00	234.17	257.58
12 L.H., Daily Totals		$3286.00		$3809.60	$273.83	$317.47

Crew B-11V	Hr.	Daily	Hr.	Daily	Bare Costs	Incl. O&P
3 Laborers	$33.10	$794.40	$51.05	$1225.20	$33.10	$51.05
1 Roller, 2-Drum, W.B., 7.5 H.P.		149.60		164.56	6.23	6.86
24 L.H., Daily Totals		$944.00		$1389.76	$39.33	$57.91

Crew B-11W	Hr.	Daily	Hr.	Daily	Bare Costs	Incl. O&P
1 Equipment Operator (med.)	$42.95	$343.60	$64.30	$514.40	$33.96	$51.74
1 Common Laborer	33.10	264.80	51.05	408.40		
10 Truck Drivers (hvy.)	33.15	2652.00	50.55	4044.00		
1 Dozer, 200 H.P.		1192.00		1311.20		
1 Vibratory Roller, Towed, 23 Ton		434.20		477.62		
10 Dump Trucks, 8 C.Y., 220 H.P.		4018.00		4419.80	58.79	64.67
96 L.H., Daily Totals		$8904.60		$11175.42	$92.76	$116.41

Crew B-11Y	Hr.	Daily	Hr.	Daily	Bare Costs	Incl. O&P
1 Labor Foreman (Outside)	$35.10	$280.80	$54.10	$432.80	$36.61	$55.81
5 Common Laborers	33.10	1324.00	51.05	2042.00		
3 Equipment Operators (med.)	42.95	1030.80	64.30	1543.20		
1 Dozer, 80 H.P.		436.60		480.26		
2 Rollers, 2-Drum, W.B., 7.5 H.P.		299.20		329.12		
4 Vibratory Plates, gas, 21"		174.40		191.84	12.64	13.91
72 L.H., Daily Totals		$3545.80		$5019.22	$49.25	$69.71

Crew B-12A	Hr.	Daily	Hr.	Daily	Bare Costs	Incl. O&P
1 Equip. Oper. (crane)	$44.40	$355.20	$66.45	$531.60	$38.75	$58.75
1 Laborer	33.10	264.80	51.05	408.40		
1 Hyd. Excavator, 1 C.Y.		801.60		881.76	50.10	55.11
16 L.H., Daily Totals		$1421.60		$1821.76	$88.85	$113.86

Crew B-12B	Hr.	Daily	Hr.	Daily	Bare Costs	Incl. O&P
1 Equip. Oper. (crane)	$44.40	$355.20	$66.45	$531.60	$38.75	$58.75
1 Laborer	33.10	264.80	51.05	408.40		
1 Hyd. Excavator, 1.5 C.Y.		1017.00		1118.70	63.56	69.92
16 L.H., Daily Totals		$1637.00		$2058.70	$102.31	$128.67

Crew B-12C	Bare Costs Hr.	Bare Costs Daily	Incl. Subs O&P Hr.	Incl. Subs O&P Daily	Cost Per Labor-Hour Bare Costs	Cost Per Labor-Hour Incl. O&P
1 Equip. Oper. (crane)	$44.40	$355.20	$66.45	$531.60	$38.75	$58.75
1 Laborer	33.10	264.80	51.05	408.40		
1 Hyd. Excavator, 2 C.Y.		1321.00		1453.10	82.56	90.82
16 L.H., Daily Totals		$1941.00		$2393.10	$121.31	$149.57

Crew B-12D	Hr.	Daily	Hr.	Daily	Bare Costs	Incl. O&P
1 Equip. Oper. (crane)	$44.40	$355.20	$66.45	$531.60	$38.75	$58.75
1 Laborer	33.10	264.80	51.05	408.40		
1 Hyd. Excavator, 3.5 C.Y.		2359.00		2594.90	147.44	162.18
16 L.H., Daily Totals		$2979.00		$3534.90	$186.19	$220.93

Crew B-12E	Hr.	Daily	Hr.	Daily	Bare Costs	Incl. O&P
1 Equip. Oper. (crane)	$44.40	$355.20	$66.45	$531.60	$38.75	$58.75
1 Laborer	33.10	264.80	51.05	408.40		
1 Hyd. Excavator, .5 C.Y.		386.60		425.26	24.16	26.58
16 L.H., Daily Totals		$1006.60		$1365.26	$62.91	$85.33

Crew B-12F	Hr.	Daily	Hr.	Daily	Bare Costs	Incl. O&P
1 Equip. Oper. (crane)	$44.40	$355.20	$66.45	$531.60	$38.75	$58.75
1 Laborer	33.10	264.80	51.05	408.40		
1 Hyd. Excavator, .75 C.Y.		677.80		745.58	42.36	46.60
16 L.H., Daily Totals		$1297.80		$1685.58	$81.11	$105.35

Crew B-12G	Hr.	Daily	Hr.	Daily	Bare Costs	Incl. O&P
1 Equip. Oper. (crane)	$44.40	$355.20	$66.45	$531.60	$38.75	$58.75
1 Laborer	33.10	264.80	51.05	408.40		
1 Crawler Crane, 15 Ton		693.75		763.13		
1 Clamshell Bucket, .5 C.Y.		37.00		40.70	45.67	50.24
16 L.H., Daily Totals		$1350.75		$1743.83	$84.42	$108.99

Crew B-12H	Hr.	Daily	Hr.	Daily	Bare Costs	Incl. O&P
1 Equip. Oper. (crane)	$44.40	$355.20	$66.45	$531.60	$38.75	$58.75
1 Laborer	33.10	264.80	51.05	408.40		
1 Crawler Crane, 25 Ton		1188.00		1306.80		
1 Clamshell Bucket, 1 C.Y.		46.60		51.26	77.16	84.88
16 L.H., Daily Totals		$1854.60		$2298.06	$115.91	$143.63

Crew B-12I	Hr.	Daily	Hr.	Daily	Bare Costs	Incl. O&P
1 Equip. Oper. (crane)	$44.40	$355.20	$66.45	$531.60	$38.75	$58.75
1 Laborer	33.10	264.80	51.05	408.40		
1 Crawler Crane, 20 Ton		892.30		981.53		
1 Dragline Bucket, .75 C.Y.		20.40		22.44	57.04	62.75
16 L.H., Daily Totals		$1532.70		$1943.97	$95.79	$121.50

Crew B-12J	Hr.	Daily	Hr.	Daily	Bare Costs	Incl. O&P
1 Equip. Oper. (crane)	$44.40	$355.20	$66.45	$531.60	$38.75	$58.75
1 Laborer	33.10	264.80	51.05	408.40		
1 Gradall, 5/8 C.Y.		812.80		894.08	50.80	55.88
16 L.H., Daily Totals		$1432.80		$1834.08	$89.55	$114.63

Crew B-12K	Hr.	Daily	Hr.	Daily	Bare Costs	Incl. O&P
1 Equip. Oper. (crane)	$44.40	$355.20	$66.45	$531.60	$38.75	$58.75
1 Laborer	33.10	264.80	51.05	408.40		
1 Gradall, 3 Ton, 1 C.Y.		958.40		1054.24	59.90	65.89
16 L.H., Daily Totals		$1578.40		$1994.24	$98.65	$124.64

Crew B-12L	Hr.	Daily	Hr.	Daily	Bare Costs	Incl. O&P
1 Equip. Oper. (crane)	$44.40	$355.20	$66.45	$531.60	$38.75	$58.75
1 Laborer	33.10	264.80	51.05	408.40		
1 Crawler Crane, 15 Ton		693.75		763.13		
1 F.E. Attachment, .5 C.Y.		55.60		61.16	46.83	51.52
16 L.H., Daily Totals		$1369.35		$1764.29	$85.58	$110.27

Crew B-12M	Hr.	Daily	Hr.	Daily	Bare Costs	Incl. O&P
1 Equip. Oper. (crane)	$44.40	$355.20	$66.45	$531.60	$38.75	$58.75
1 Laborer	33.10	264.80	51.05	408.40		
1 Crawler Crane, 20 Ton		892.30		981.53		
1 F.E. Attachment, .75 C.Y.		60.00		66.00	59.52	65.47
16 L.H., Daily Totals		$1572.30		$1987.53	$98.27	$124.22

Crew B-12N	Hr.	Daily	Hr.	Daily	Bare Costs	Incl. O&P
1 Equip. Oper. (crane)	$44.40	$355.20	$66.45	$531.60	$38.75	$58.75
1 Laborer	33.10	264.80	51.05	408.40		
1 Crawler Crane, 25 Ton		1188.00		1306.80		
1 F.E. Attachment, 1 C.Y.		65.80		72.38	78.36	86.20
16 L.H., Daily Totals		$1873.80		$2319.18	$117.11	$144.95

Crew B-12O	Hr.	Daily	Hr.	Daily	Bare Costs	Incl. O&P
1 Equip. Oper. (crane)	$44.40	$355.20	$66.45	$531.60	$38.75	$58.75
1 Laborer	33.10	264.80	51.05	408.40		
1 Crawler Crane, 40 Ton		1193.00		1312.30		
1 F.E. Attachment, 1.5 C.Y.		76.00		83.60	79.31	87.24
16 L.H., Daily Totals		$1889.00		$2335.90	$118.06	$145.99

Crew B-12P	Hr.	Daily	Hr.	Daily	Bare Costs	Incl. O&P
1 Equip. Oper. (crane)	$44.40	$355.20	$66.45	$531.60	$38.75	$58.75
1 Laborer	33.10	264.80	51.05	408.40		
1 Crawler Crane, 40 Ton		1193.00		1312.30		
1 Dragline Bucket, 1.5 C.Y.		32.80		36.08	76.61	84.27
16 L.H., Daily Totals		$1845.80		$2288.38	$115.36	$143.02

Crew B-12Q	Hr.	Daily	Hr.	Daily	Bare Costs	Incl. O&P
1 Equip. Oper. (crane)	$44.40	$355.20	$66.45	$531.60	$38.75	$58.75
1 Laborer	33.10	264.80	51.05	408.40		
1 Hyd. Excavator, 5/8 C.Y.		564.40		620.84	35.27	38.80
16 L.H., Daily Totals		$1184.40		$1560.84	$74.03	$97.55

Crew B-12S	Hr.	Daily	Hr.	Daily	Bare Costs	Incl. O&P
1 Equip. Oper. (crane)	$44.40	$355.20	$66.45	$531.60	$38.75	$58.75
1 Laborer	33.10	264.80	51.05	408.40		
1 Hyd. Excavator, 2.5 C.Y.		1784.00		1962.40	111.50	122.65
16 L.H., Daily Totals		$2404.00		$2902.40	$150.25	$181.40

Crew B-12T	Hr.	Daily	Hr.	Daily	Bare Costs	Incl. O&P
1 Equip. Oper. (crane)	$44.40	$355.20	$66.45	$531.60	$38.75	$58.75
1 Laborer	33.10	264.80	51.05	408.40		
1 Crawler Crane, 75 Ton		1516.00		1667.60		
1 F.E. Attachment, 3 C.Y.		98.40		108.24	100.90	110.99
16 L.H., Daily Totals		$2234.40		$2715.84	$139.65	$169.74

Crew B-12V	Hr.	Daily	Hr.	Daily	Bare Costs	Incl. O&P
1 Equip. Oper. (crane)	$44.40	$355.20	$66.45	$531.60	$38.75	$58.75
1 Laborer	33.10	264.80	51.05	408.40		
1 Crawler Crane, 75 Ton		1516.00		1667.60		
1 Dragline Bucket, 3 C.Y.		52.40		57.64	98.03	107.83
16 L.H., Daily Totals		$2188.40		$2665.24	$136.78	$166.58

Crew No.	Bare Costs		Incl. Subs O&P		Cost Per Labor-Hour	
Crew B-13	**Hr.**	**Daily**	**Hr.**	**Daily**	**Bare Costs**	**Incl. O&P**
1 Labor Foreman (outside)	$35.10	$280.80	$54.10	$432.80	$35.74	$54.59
4 Laborers	33.10	1059.20	51.05	1633.60		
1 Equip. Oper. (crane)	44.40	355.20	66.45	531.60		
1 Equip. Oper. Oiler	38.30	306.40	57.35	458.80		
1 Hyd. Crane, 25 Ton		747.60		822.36	13.35	14.69
56 L.H., Daily Totals		$2749.20		$3879.16	$49.09	$69.27
Crew B-13A	**Hr.**	**Daily**	**Hr.**	**Daily**	**Bare Costs**	**Incl. O&P**
1 Foreman	$35.10	$280.80	$54.10	$432.80	$36.21	$55.13
2 Laborers	33.10	529.60	51.05	816.80		
2 Equipment Operators(med.)	42.95	687.20	64.30	1028.80		
2 Truck Drivers (heavy)	33.15	530.40	50.55	808.80		
1 Crawler Crane, 75 Ton		1516.00		1667.60		
1 Crawler Loader, 4 C.Y.		1548.00		1702.80		
2 Dump Trucks, 8 C.Y., 220 H.P.		803.60		883.96	69.06	75.97
56 L.H., Daily Totals		$5895.60		$7341.56	$105.28	$131.10
Crew B-13B	**Hr.**	**Daily**	**Hr.**	**Daily**	**Bare Costs**	**Incl. O&P**
1 Labor Foreman (outside)	$35.10	$280.80	$54.10	$432.80	$35.74	$54.59
4 Laborers	33.10	1059.20	51.05	1633.60		
1 Equip. Oper. (crane)	44.40	355.20	66.45	531.60		
1 Equip. Oper. Oiler	38.30	306.40	57.35	458.80		
1 Hyd. Crane, 55 Ton		1209.00		1329.90	21.59	23.75
56 L.H., Daily Totals		$3210.60		$4386.70	$57.33	$78.33
Crew B-13C	**Hr.**	**Daily**	**Hr.**	**Daily**	**Bare Costs**	**Incl. O&P**
1 Labor Foreman (outside)	$35.10	$280.80	$54.10	$432.80	$35.74	$54.59
4 Laborers	33.10	1059.20	51.05	1633.60		
1 Equip. Oper. (crane)	44.40	355.20	66.45	531.60		
1 Equip. Oper. Oiler	38.30	306.40	57.35	458.80		
1 Crawler Crane, 100 Ton		1732.00		1905.20	30.93	34.02
56 L.H., Daily Totals		$3733.60		$4962.00	$66.67	$88.61
Crew B-13D	**Hr.**	**Daily**	**Hr.**	**Daily**	**Bare Costs**	**Incl. O&P**
1 Laborer	$33.10	$264.80	$51.05	$408.40	$38.75	$58.75
1 Equip. Oper. (crane)	44.40	355.20	66.45	531.60		
1 Hyd. Excavator, 1 C.Y.		801.60		881.76		
1 Trench Box		113.40		124.74	57.19	62.91
16 L.H., Daily Totals		$1535.00		$1946.50	$95.94	$121.66
Crew B-13E	**Hr.**	**Daily**	**Hr.**	**Daily**	**Bare Costs**	**Incl. O&P**
1 Laborer	$33.10	$264.80	$51.05	$408.40	$38.75	$58.75
1 Equip. Oper. (crane)	44.40	355.20	66.45	531.60		
1 Hyd. Excavator, 1.5 C.Y.		1017.00		1118.70		
1 Trench Box		113.40		124.74	70.65	77.72
16 L.H., Daily Totals		$1750.40		$2183.44	$109.40	$136.47
Crew B-13F	**Hr.**	**Daily**	**Hr.**	**Daily**	**Bare Costs**	**Incl. O&P**
1 Laborer	$33.10	$264.80	$51.05	$408.40	$38.75	$58.75
1 Equip. Oper. (crane)	44.40	355.20	66.45	531.60		
1 Hyd. Excavator, 3.5 C.Y.		2359.00		2594.90		
1 Trench Box		113.40		124.74	154.53	169.98
16 L.H., Daily Totals		$3092.40		$3659.64	$193.28	$228.73
Crew B-13G	**Hr.**	**Daily**	**Hr.**	**Daily**	**Bare Costs**	**Incl. O&P**
1 Laborer	$33.10	$264.80	$51.05	$408.40	$38.75	$58.75
1 Equip. Oper. (crane)	44.40	355.20	66.45	531.60		
1 Hyd. Excavator, .75 C.Y.		677.80		745.58		
1 Trench Box		113.40		124.74	49.45	54.40
16 L.H., Daily Totals		$1411.20		$1810.32	$88.20	$113.15

Crew No.	Bare Costs		Incl. Subs O&P		Cost Per Labor-Hour	
Crew B-13H	**Hr.**	**Daily**	**Hr.**	**Daily**	**Bare Costs**	**Incl. O&P**
1 Laborer	$33.10	$264.80	$51.05	$408.40	$38.75	$58.75
1 Equip. Oper. (crane)	44.40	355.20	66.45	531.60		
1 Gradall, 5/8 C.Y.		812.80		894.08		
1 Trench Box		113.40		124.74	57.89	63.68
16 L.H., Daily Totals		$1546.20		$1958.82	$96.64	$122.43
Crew B-13I	**Hr.**	**Daily**	**Hr.**	**Daily**	**Bare Costs**	**Incl. O&P**
1 Laborer	$33.10	$264.80	$51.05	$408.40	$38.75	$58.75
1 Equip. Oper. (crane)	44.40	355.20	66.45	531.60		
1 Gradall, 3 Ton, 1 C.Y.		958.40		1054.24		
1 Trench Box		113.40		124.74	66.99	73.69
16 L.H., Daily Totals		$1691.80		$2118.98	$105.74	$132.44
Crew B-13J	**Hr.**	**Daily**	**Hr.**	**Daily**	**Bare Costs**	**Incl. O&P**
1 Laborer	$33.10	$264.80	$51.05	$408.40	$38.75	$58.75
1 Equip. Oper. (crane)	44.40	355.20	66.45	531.60		
1 Hyd. Excavator, 2.5 C.Y.		1784.00		1962.40		
1 Trench Box		113.40		124.74	118.59	130.45
16 L.H., Daily Totals		$2517.40		$3027.14	$157.34	$189.20
Crew B-14	**Hr.**	**Daily**	**Hr.**	**Daily**	**Bare Costs**	**Incl. O&P**
1 Labor Foreman (outside)	$35.10	$280.80	$54.10	$432.80	$34.80	$53.36
4 Laborers	33.10	1059.20	51.05	1633.60		
1 Equip. Oper. (light)	41.30	330.40	61.85	494.80		
1 Backhoe Loader, 48 H.P.		337.80		371.58	7.04	7.74
48 L.H., Daily Totals		$2008.20		$2932.78	$41.84	$61.10
Crew B-14A	**Hr.**	**Daily**	**Hr.**	**Daily**	**Bare Costs**	**Incl. O&P**
1 Equip. Oper. (crane)	$44.40	$355.20	$66.45	$531.60	$40.63	$61.32
.5 Laborer	33.10	132.40	51.05	204.20		
1 Hyd. Excavator, 4.5 C.Y.		2870.00		3157.00	239.17	263.08
12 L.H., Daily Totals		$3357.60		$3892.80	$279.80	$324.40
Crew B-14B	**Hr.**	**Daily**	**Hr.**	**Daily**	**Bare Costs**	**Incl. O&P**
1 Equip. Oper. (crane)	$44.40	$355.20	$66.45	$531.60	$40.63	$61.32
.5 Laborer	33.10	132.40	51.05	204.20		
1 Hyd. Excavator, 6 C.Y.		3390.00		3729.00	282.50	310.75
12 L.H., Daily Totals		$3877.60		$4464.80	$323.13	$372.07
Crew B-14C	**Hr.**	**Daily**	**Hr.**	**Daily**	**Bare Costs**	**Incl. O&P**
1 Equip. Oper. (crane)	$44.40	$355.20	$66.45	$531.60	$40.63	$61.32
.5 Laborer	33.10	132.40	51.05	204.20		
1 Hyd. Excavator, 7 C.Y.		3553.00		3908.30	296.08	325.69
12 L.H., Daily Totals		$4040.60		$4644.10	$336.72	$387.01
Crew B-14F	**Hr.**	**Daily**	**Hr.**	**Daily**	**Bare Costs**	**Incl. O&P**
1 Equip. Oper. (crane)	$44.40	$355.20	$66.45	$531.60	$40.63	$61.32
.5 Laborer	33.10	132.40	51.05	204.20		
1 Hyd. Shovel, 7 C.Y.		3536.00		3889.60	294.67	324.13
12 L.H., Daily Totals		$4023.60		$4625.40	$335.30	$385.45
Crew B-14G	**Hr.**	**Daily**	**Hr.**	**Daily**	**Bare Costs**	**Incl. O&P**
1 Equip. Oper. (crane)	$44.40	$355.20	$66.45	$531.60	$40.63	$61.32
.5 Laborer	33.10	132.40	51.05	204.20		
1 Hyd. Shovel, 12 C.Y.		4614.00		5075.40	384.50	422.95
12 L.H., Daily Totals		$5101.60		$5811.20	$425.13	$484.27

Crew No.	Bare Costs		Incl. Subs O&P		Cost Per Labor-Hour	
Crew B-14J	**Hr.**	**Daily**	**Hr.**	**Daily**	**Bare Costs**	**Incl. O&P**
1 Equip. Oper. (med.)	$42.95	$343.60	$64.30	$514.40	$39.67	$59.88
.5 Laborer	33.10	132.40	51.05	204.20		
1 F.E. Loader, 8 C.Y.		1757.00		1932.70	146.42	161.06
12 L.H., Daily Totals		$2233.00		$2651.30	$186.08	$220.94
Crew B-14K	**Hr.**	**Daily**	**Hr.**	**Daily**	**Bare Costs**	**Incl. O&P**
1 Equip. Oper. (med.)	$42.95	$343.60	$64.30	$514.40	$39.67	$59.88
.5 Laborer	33.10	132.40	51.05	204.20		
1 F.E. Loader, 10 C.Y.		2601.00		2861.10	216.75	238.43
12 L.H., Daily Totals		$3077.00		$3579.70	$256.42	$298.31
Crew B-15	**Hr.**	**Daily**	**Hr.**	**Daily**	**Bare Costs**	**Incl. O&P**
1 Equipment Oper. (med.)	$42.95	$343.60	$64.30	$514.40	$35.94	$54.55
.5 Laborer	33.10	132.40	51.05	204.20		
2 Truck Drivers (heavy)	33.15	530.40	50.55	808.80		
2 Dump Trucks, 12 C.Y., 400 H.P.		1335.60		1469.16		
1 Dozer, 200 H.P.		1192.00		1311.20	90.27	99.30
28 L.H., Daily Totals		$3534.00		$4307.76	$126.21	$153.85
Crew B-16	**Hr.**	**Daily**	**Hr.**	**Daily**	**Bare Costs**	**Incl. O&P**
1 Labor Foreman (outside)	$35.10	$280.80	$54.10	$432.80	$33.61	$51.69
2 Laborers	33.10	529.60	51.05	816.80		
1 Truck Driver (heavy)	33.15	265.20	50.55	404.40		
1 Dump Truck, 12 C.Y., 400 H.P.		667.80		734.58	20.87	22.96
32 L.H., Daily Totals		$1743.40		$2388.58	$54.48	$74.64
Crew B-17	**Hr.**	**Daily**	**Hr.**	**Daily**	**Bare Costs**	**Incl. O&P**
2 Laborers	$33.10	$529.60	$51.05	$816.80	$35.16	$53.63
1 Equip. Oper. (light)	41.30	330.40	61.85	494.80		
1 Truck Driver (heavy)	33.15	265.20	50.55	404.40		
1 Backhoe Loader, 48 H.P.		337.80		371.58		
1 Dump Truck, 8 C.Y., 220 H.P.		401.80		441.98	23.11	25.42
32 L.H., Daily Totals		$1864.80		$2529.56	$58.27	$79.05
Crew B-17A	**Hr.**	**Daily**	**Hr.**	**Daily**	**Bare Costs**	**Incl. O&P**
2 Laborer Foremen	$35.10	$561.60	$54.10	$865.60	$35.60	$54.85
6 Laborers	33.10	1588.80	51.05	2450.40		
1 Skilled Worker Foreman	44.60	356.80	68.55	548.40		
1 Skilled Worker	42.60	340.80	65.50	524.00		
80 L.H., Daily Totals		$2848.00		$4388.40	$35.60	$54.85
Crew B-18	**Hr.**	**Daily**	**Hr.**	**Daily**	**Bare Costs**	**Incl. O&P**
1 Labor Foreman (outside)	$35.10	$280.80	$54.10	$432.80	$33.77	$52.07
2 Laborers	33.10	529.60	51.05	816.80		
1 Vibratory Plate, gas, 21"		43.60		47.96	1.82	2.00
24 L.H., Daily Totals		$854.00		$1297.56	$35.58	$54.06
Crew B-19	**Hr.**	**Daily**	**Hr.**	**Daily**	**Bare Costs**	**Incl. O&P**
1 Pile Driver Foreman	$42.30	$338.40	$68.00	$544.00	$41.33	$64.68
4 Pile Drivers	40.30	1289.60	64.80	2073.60		
2 Equip. Oper. (crane)	44.40	710.40	66.45	1063.20		
1 Equip. Oper. Oiler	38.30	306.40	57.35	458.80		
1 Crawler Crane, 40 Ton		1193.00		1312.30		
1 Lead, 90' high		120.40		132.44		
1 Hammer, Diesel, 22k ft-lb		619.40		681.34	30.20	33.22
64 L.H., Daily Totals		$4577.60		$6265.68	$71.53	$97.90

Crew No.	Bare Costs		Incl. Subs O&P		Cost Per Labor-Hour	
Crew B-19A	**Hr.**	**Daily**	**Hr.**	**Daily**	**Bare Costs**	**Incl. O&P**
1 Pile Driver Foreman	$42.30	$338.40	$68.00	$544.00	$41.33	$64.68
4 Pile Drivers	40.30	1289.60	64.80	2073.60		
2 Equip. Oper. (crane)	44.40	710.40	66.45	1063.20		
1 Equip. Oper. Oiler	38.30	306.40	57.35	458.80		
1 Crawler Crane, 75 Ton		1516.00		1667.60		
1 Lead, 90' high		120.40		132.44		
1 Hammer, Diesel, 41k ft-lb		722.80		795.08	36.86	40.55
64 L.H., Daily Totals		$5004.00		$6734.72	$78.19	$105.23
Crew B-20	**Hr.**	**Daily**	**Hr.**	**Daily**	**Bare Costs**	**Incl. O&P**
1 Labor Foreman (out)	$35.10	$280.80	$54.10	$432.80	$36.93	$56.88
1 Skilled Worker	42.60	340.80	65.50	524.00		
1 Laborer	33.10	264.80	51.05	408.40		
24 L.H., Daily Totals		$886.40		$1365.20	$36.93	$56.88
Crew B-20A	**Hr.**	**Daily**	**Hr.**	**Daily**	**Bare Costs**	**Incl. O&P**
1 Labor Foreman	$35.10	$280.80	$54.10	$432.80	$40.48	$61.40
1 Laborer	33.10	264.80	51.05	408.40		
1 Plumber	52.05	416.40	78.00	624.00		
1 Plumber Apprentice	41.65	333.20	62.45	499.60		
32 L.H., Daily Totals		$1295.20		$1964.80	$40.48	$61.40
Crew B-21	**Hr.**	**Daily**	**Hr.**	**Daily**	**Bare Costs**	**Incl. O&P**
1 Labor Foreman (out)	$35.10	$280.80	$54.10	$432.80	$38.00	$58.25
1 Skilled Worker	42.60	340.80	65.50	524.00		
1 Laborer	33.10	264.80	51.05	408.40		
.5 Equip. Oper. (crane)	44.40	177.60	66.45	265.80		
.5 S.P. Crane, 4x4, 5 Ton		138.10		151.91	4.93	5.43
28 L.H., Daily Totals		$1202.10		$1782.91	$42.93	$63.68
Crew B-21A	**Hr.**	**Daily**	**Hr.**	**Daily**	**Bare Costs**	**Incl. O&P**
1 Labor Foreman	$35.10	$280.80	$54.10	$432.80	$41.26	$62.41
1 Laborer	33.10	264.80	51.05	408.40		
1 Plumber	52.05	416.40	78.00	624.00		
1 Plumber Apprentice	41.65	333.20	62.45	499.60		
1 Equip. Oper. (crane)	44.40	355.20	66.45	531.60		
1 S.P. Crane, 4x4, 12 Ton		655.60		721.16	16.39	18.03
40 L.H., Daily Totals		$2306.00		$3217.56	$57.65	$80.44
Crew B-21B	**Hr.**	**Daily**	**Hr.**	**Daily**	**Bare Costs**	**Incl. O&P**
1 Laborer Foreman	$35.10	$280.80	$54.10	$432.80	$35.76	$54.74
3 Laborers	33.10	794.40	51.05	1225.20		
1 Equip. Oper. (crane)	44.40	355.20	66.45	531.60		
1 Hyd. Crane, 12 Ton		655.60		721.16	16.39	18.03
40 L.H., Daily Totals		$2086.00		$2910.76	$52.15	$72.77
Crew B-21C	**Hr.**	**Daily**	**Hr.**	**Daily**	**Bare Costs**	**Incl. O&P**
1 Laborer Foreman	$35.10	$280.80	$54.10	$432.80	$35.74	$54.59
4 Laborers	33.10	1059.20	51.05	1633.60		
1 Equip. Oper. (crane)	44.40	355.20	66.45	531.60		
1 Equip. Oper. Oiler	38.30	306.40	57.35	458.80		
2 Cutting Torches		38.00		41.80		
2 Set of Gases		158.40		174.24		
1 Lattice Boom Crane, 90 Ton		1607.00		1767.70	32.20	35.42
56 L.H., Daily Totals		$3805.00		$5040.54	$67.95	$90.01

Crew No.	Bare Costs		Incl. Subs O&P		Cost Per Labor-Hour	
Crew B-22	**Hr.**	**Daily**	**Hr.**	**Daily**	**Bare Costs**	**Incl. O&P**
1 Labor Foreman (out)	$35.10	$280.80	$54.10	$432.80	$38.43	$58.80
1 Skilled Worker	42.60	340.80	65.50	524.00		
1 Laborer	33.10	264.80	51.05	408.40		
.75 Equip. Oper. (crane)	44.40	266.40	66.45	398.70		
.75 S.P. Crane, 4x4, 5 Ton		207.15		227.87	6.91	7.60
30 L.H., Daily Totals		$1359.95		$1991.77	$45.33	$66.39
Crew B-22A	**Hr.**	**Daily**	**Hr.**	**Daily**	**Bare Costs**	**Incl. O&P**
1 Labor Foreman (out)	$35.10	$280.80	$54.10	$432.80	$37.31	$57.17
1 Skilled Worker	42.60	340.80	65.50	524.00		
2 Laborers	33.10	529.60	51.05	816.80		
.75 Equipment Oper. (crane)	44.40	266.40	66.45	398.70		
.75 S.P. Crane, 4x4, 5 Ton		207.15		227.87		
1 Generator, 5 kW		46.00		50.60		
1 Butt Fusion Machine		134.40		147.84	10.20	11.22
38 L.H., Daily Totals		$1805.15		$2598.61	$47.50	$68.38
Crew B-22B	**Hr.**	**Daily**	**Hr.**	**Daily**	**Bare Costs**	**Incl. O&P**
1 Skilled Worker	$42.60	$340.80	$65.50	$524.00	$37.85	$58.27
1 Laborer	33.10	264.80	51.05	408.40		
1 Electro Fusion Machine		98.60		108.46	6.16	6.78
16 L.H., Daily Totals		$704.20		$1040.86	$44.01	$65.05
Crew B-23	**Hr.**	**Daily**	**Hr.**	**Daily**	**Bare Costs**	**Incl. O&P**
1 Labor Foreman (outside)	$35.10	$280.80	$54.10	$432.80	$33.50	$51.66
4 Laborers	33.10	1059.20	51.05	1633.60		
1 Drill Rig, Truck-Mounted		2591.00		2850.10		
1 Flatbed Truck, Gas, 3 Ton		296.00		325.60	72.17	79.39
40 L.H., Daily Totals		$4227.00		$5242.10	$105.68	$131.05
Crew B-23A	**Hr.**	**Daily**	**Hr.**	**Daily**	**Bare Costs**	**Incl. O&P**
1 Labor Foreman (outside)	$35.10	$280.80	$54.10	$432.80	$37.05	$56.48
1 Laborer	33.10	264.80	51.05	408.40		
1 Equip. Operator (med.)	42.95	343.60	64.30	514.40		
1 Drill Rig, Truck-Mounted		2591.00		2850.10		
1 Pickup Truck, 3/4 Ton		140.00		154.00	113.79	125.17
24 L.H., Daily Totals		$3620.20		$4359.70	$150.84	$181.65
Crew B-23B	**Hr.**	**Daily**	**Hr.**	**Daily**	**Bare Costs**	**Incl. O&P**
1 Labor Foreman (outside)	$35.10	$280.80	$54.10	$432.80	$37.05	$56.48
1 Laborer	33.10	264.80	51.05	408.40		
1 Equip. Operator (med.)	42.95	343.60	64.30	514.40		
1 Drill Rig, Truck-Mounted		2591.00		2850.10		
1 Pickup Truck, 3/4 Ton		140.00		154.00		
1 Centr. Water Pump, 6"		340.00		374.00	127.96	140.75
24 L.H., Daily Totals		$3960.20		$4733.70	$165.01	$197.24
Crew B-24	**Hr.**	**Daily**	**Hr.**	**Daily**	**Bare Costs**	**Incl. O&P**
1 Cement Finisher	$39.70	$317.60	$57.95	$463.60	$38.12	$57.68
1 Laborer	33.10	264.80	51.05	408.40		
1 Carpenter	41.55	332.40	64.05	512.40		
24 L.H., Daily Totals		$914.80		$1384.40	$38.12	$57.68
Crew B-25	**Hr.**	**Daily**	**Hr.**	**Daily**	**Bare Costs**	**Incl. O&P**
1 Labor Foreman	$35.10	$280.80	$54.10	$432.80	$35.97	$54.94
7 Laborers	33.10	1853.60	51.05	2858.80		
3 Equip. Oper. (med.)	42.95	1030.80	64.30	1543.20		
1 Asphalt Paver, 130 H.P.		1994.00		2193.40		
1 Tandem Roller, 10 Ton		235.20		258.72		
1 Roller, Pneum. Whl, 12 Ton		337.20		370.92	29.16	32.08
88 L.H., Daily Totals		$5731.60		$7657.84	$65.13	$87.02

Crew No.	Bare Costs		Incl. Subs O&P		Cost Per Labor-Hour	
Crew B-25B	**Hr.**	**Daily**	**Hr.**	**Daily**	**Bare Costs**	**Incl. O&P**
1 Labor Foreman	$35.10	$280.80	$54.10	$432.80	$36.55	$55.72
7 Laborers	33.10	1853.60	51.05	2858.80		
4 Equip. Oper. (med.)	42.95	1374.40	64.30	2057.60		
1 Asphalt Paver, 130 H.P.		1994.00		2193.40		
2 Tandem Rollers, 10 Ton		470.40		517.44		
1 Roller, Pneum. Whl, 12 Ton		337.20		370.92	29.18	32.10
96 L.H., Daily Totals		$6310.40		$8430.96	$65.73	$87.82
Crew B-25C	**Hr.**	**Daily**	**Hr.**	**Daily**	**Bare Costs**	**Incl. O&P**
1 Labor Foreman	$35.10	$280.80	$54.10	$432.80	$36.72	$55.98
3 Laborers	33.10	794.40	51.05	1225.20		
2 Equip. Oper. (med.)	42.95	687.20	64.30	1028.80		
1 Asphalt Paver, 130 H.P.		1994.00		2193.40		
1 Tandem Roller, 10 Ton		235.20		258.72	46.44	51.09
48 L.H., Daily Totals		$3991.60		$5138.92	$83.16	$107.06
Crew B-26	**Hr.**	**Daily**	**Hr.**	**Daily**	**Bare Costs**	**Incl. O&P**
1 Labor Foreman (outside)	$35.10	$280.80	$54.10	$432.80	$36.92	$56.58
6 Laborers	33.10	1588.80	51.05	2450.40		
2 Equip. Oper. (med.)	42.95	687.20	64.30	1028.80		
1 Rodman (reinf.)	46.80	374.40	75.40	603.20		
1 Cement Finisher	39.70	317.60	57.95	463.60		
1 Grader, 30,000 Lbs.		622.20		684.42		
1 Paving Mach. & Equip.		2678.00		2945.80	37.50	41.25
88 L.H., Daily Totals		$6549.00		$8609.02	$74.42	$97.83
Crew B-26A	**Hr.**	**Daily**	**Hr.**	**Daily**	**Bare Costs**	**Incl. O&P**
1 Labor Foreman (outside)	$35.10	$280.80	$54.10	$432.80	$36.92	$56.58
6 Laborers	33.10	1588.80	51.05	2450.40		
2 Equip. Oper. (med.)	42.95	687.20	64.30	1028.80		
1 Rodman (reinf.)	46.80	374.40	75.40	603.20		
1 Cement Finisher	39.70	317.60	57.95	463.60		
1 Grader, 30,000 Lbs.		622.20		684.42		
1 Paving Mach. & Equip.		2678.00		2945.80		
1 Concrete Saw		161.40		177.54	39.34	43.27
88 L.H., Daily Totals		$6710.40		$8786.56	$76.25	$99.85
Crew B-26B	**Hr.**	**Daily**	**Hr.**	**Daily**	**Bare Costs**	**Incl. O&P**
1 Labor Foreman (outside)	$35.10	$280.80	$54.10	$432.80	$37.42	$57.22
6 Laborers	33.10	1588.80	51.05	2450.40		
3 Equip. Oper. (med.)	42.95	1030.80	64.30	1543.20		
1 Rodman (reinf.)	46.80	374.40	75.40	603.20		
1 Cement Finisher	39.70	317.60	57.95	463.60		
1 Grader, 30,000 Lbs.		622.20		684.42		
1 Paving Mach. & Equip.		2678.00		2945.80		
1 Concrete Pump, 110' Boom		976.40		1074.04	44.55	49.00
96 L.H., Daily Totals		$7869.00		$10197.46	$81.97	$106.22
Crew B-27	**Hr.**	**Daily**	**Hr.**	**Daily**	**Bare Costs**	**Incl. O&P**
1 Labor Foreman (outside)	$35.10	$280.80	$54.10	$432.80	$33.60	$51.81
3 Laborers	33.10	794.40	51.05	1225.20		
1 Berm Machine		285.60		314.16	8.93	9.82
32 L.H., Daily Totals		$1360.80		$1972.16	$42.52	$61.63
Crew B-28	**Hr.**	**Daily**	**Hr.**	**Daily**	**Bare Costs**	**Incl. O&P**
2 Carpenters	$41.55	$664.80	$64.05	$1024.80	$38.73	$59.72
1 Laborer	33.10	264.80	51.05	408.40		
24 L.H., Daily Totals		$929.60		$1433.20	$38.73	$59.72

Crew No.	Bare Costs		Incl. Subs O&P		Cost Per Labor-Hour	
Crew B-29	**Hr.**	**Daily**	**Hr.**	**Daily**	**Bare Costs**	**Incl. O&P**
1 Labor Foreman (outside)	$35.10	$280.80	$54.10	$432.80	$35.74	$54.59
4 Laborers	33.10	1059.20	51.05	1633.60		
1 Equip. Oper. (crane)	44.40	355.20	66.45	531.60		
1 Equip. Oper. Oiler	38.30	306.40	57.35	458.80		
1 Gradall, 5/8 C.Y.		812.80		894.08	14.51	15.97
56 L.H., Daily Totals		$2814.40		$3950.88	$50.26	$70.55
Crew B-30	**Hr.**	**Daily**	**Hr.**	**Daily**	**Bare Costs**	**Incl. O&P**
1 Equip. Oper. (med.)	$42.95	$343.60	$64.30	$514.40	$36.42	$55.13
2 Truck Drivers (heavy)	33.15	530.40	50.55	808.80		
1 Hyd. Excavator, 1.5 C.Y.		1017.00		1118.70		
2 Dump Trucks, 12 C.Y., 400 H.P.		1335.60		1469.16	98.03	107.83
24 L.H., Daily Totals		$3226.60		$3911.06	$134.44	$162.96
Crew B-31	**Hr.**	**Daily**	**Hr.**	**Daily**	**Bare Costs**	**Incl. O&P**
1 Labor Foreman (outside)	$35.10	$280.80	$54.10	$432.80	$35.19	$54.26
3 Laborers	33.10	794.40	51.05	1225.20		
1 Carpenter	41.55	332.40	64.05	512.40		
1 Air Compressor, 250 cfm		193.40		212.74		
1 Sheeting Driver		5.75		6.33		
2 -50' Air Hoses, 1.5"		12.60		13.86	5.29	5.82
40 L.H., Daily Totals		$1619.35		$2403.32	$40.48	$60.08
Crew B-32	**Hr.**	**Daily**	**Hr.**	**Daily**	**Bare Costs**	**Incl. O&P**
1 Laborer	$33.10	$264.80	$51.05	$408.40	$40.49	$60.99
3 Equip. Oper. (med.)	42.95	1030.80	64.30	1543.20		
1 Grader, 30,000 Lbs.		622.20		684.42		
1 Tandem Roller, 10 Ton		235.20		258.72		
1 Dozer, 200 H.P.		1192.00		1311.20	64.04	70.45
32 L.H., Daily Totals		$3345.00		$4205.94	$104.53	$131.44
Crew B-32A	**Hr.**	**Daily**	**Hr.**	**Daily**	**Bare Costs**	**Incl. O&P**
1 Laborer	$33.10	$264.80	$51.05	$408.40	$39.67	$59.88
2 Equip. Oper. (med.)	42.95	687.20	64.30	1028.80		
1 Grader, 30,000 Lbs.		622.20		684.42		
1 Roller, Vibratory, 25 Ton		633.20		696.52	52.31	57.54
24 L.H., Daily Totals		$2207.40		$2818.14	$91.97	$117.42
Crew B-32B	**Hr.**	**Daily**	**Hr.**	**Daily**	**Bare Costs**	**Incl. O&P**
1 Laborer	$33.10	$264.80	$51.05	$408.40	$39.67	$59.88
2 Equip. Oper. (med.)	42.95	687.20	64.30	1028.80		
1 Dozer, 200 H.P.		1192.00		1311.20		
1 Roller, Vibratory, 25 Ton		633.20		696.52	76.05	83.66
24 L.H., Daily Totals		$2777.20		$3444.92	$115.72	$143.54
Crew B-32C	**Hr.**	**Daily**	**Hr.**	**Daily**	**Bare Costs**	**Incl. O&P**
1 Labor Foreman	$35.10	$280.80	$54.10	$432.80	$38.36	$58.18
2 Laborers	33.10	529.60	51.05	816.80		
3 Equip. Oper. (med.)	42.95	1030.80	64.30	1543.20		
1 Grader, 30,000 Lbs.		622.20		684.42		
1 Tandem Roller, 10 Ton		235.20		258.72		
1 Dozer, 200 H.P.		1192.00		1311.20	42.70	46.97
48 L.H., Daily Totals		$3890.60		$5047.14	$81.05	$105.15
Crew B-33A	**Hr.**	**Daily**	**Hr.**	**Daily**	**Bare Costs**	**Incl. O&P**
1 Equip. Oper. (med.)	$42.95	$343.60	$64.30	$514.40	$40.14	$60.51
.5 Laborer	33.10	132.40	51.05	204.20		
.25 Equip. Oper. (med.)	42.95	85.90	64.30	128.60		
1 Scraper, Towed, 7 C.Y.		105.20		115.72		
1.25 Dozers, 300 H.P.		1990.00		2189.00	149.66	164.62
14 L.H., Daily Totals		$2657.10		$3151.92	$189.79	$225.14

Crew No.	Bare Costs		Incl. Subs O&P		Cost Per Labor-Hour	
Crew B-33B	**Hr.**	**Daily**	**Hr.**	**Daily**	**Bare Costs**	**Incl. O&P**
1 Equip. Oper. (med.)	$42.95	$343.60	$64.30	$514.40	$40.14	$60.51
.5 Laborer	33.10	132.40	51.05	204.20		
.25 Equip. Oper. (med.)	42.95	85.90	64.30	128.60		
1 Scraper, Towed, 10 C.Y.		156.60		172.26		
1.25 Dozers, 300 H.P.		1990.00		2189.00	153.33	168.66
14 L.H., Daily Totals		$2708.50		$3208.46	$193.46	$229.18
Crew B-33C	**Hr.**	**Daily**	**Hr.**	**Daily**	**Bare Costs**	**Incl. O&P**
1 Equip. Oper. (med.)	$42.95	$343.60	$64.30	$514.40	$40.14	$60.51
.5 Laborer	33.10	132.40	51.05	204.20		
.25 Equip. Oper. (med.)	42.95	85.90	64.30	128.60		
1 Scraper, Towed, 15 C.Y.		172.20		189.42		
1.25 Dozers, 300 H.P.		1990.00		2189.00	154.44	169.89
14 L.H., Daily Totals		$2724.10		$3225.62	$194.58	$230.40
Crew B-33D	**Hr.**	**Daily**	**Hr.**	**Daily**	**Bare Costs**	**Incl. O&P**
1 Equip. Oper. (med.)	$42.95	$343.60	$64.30	$514.40	$40.14	$60.51
.5 Laborer	33.10	132.40	51.05	204.20		
.25 Equip. Oper. (med.)	42.95	85.90	64.30	128.60		
1 S.P. Scraper, 14 C.Y.		1717.00		1888.70		
.25 Dozer, 300 H.P.		398.00		437.80	151.07	166.18
14 L.H., Daily Totals		$2676.90		$3173.70	$191.21	$226.69
Crew B-33E	**Hr.**	**Daily**	**Hr.**	**Daily**	**Bare Costs**	**Incl. O&P**
1 Equip. Oper. (med.)	$42.95	$343.60	$64.30	$514.40	$40.14	$60.51
.5 Laborer	33.10	132.40	51.05	204.20		
.25 Equip. Oper. (med.)	42.95	85.90	64.30	128.60		
1 S.P. Scraper, 21 C.Y.		2242.00		2466.20		
.25 Dozer, 300 H.P.		398.00		437.80	188.57	207.43
14 L.H., Daily Totals		$3201.90		$3751.20	$228.71	$267.94
Crew B-33F	**Hr.**	**Daily**	**Hr.**	**Daily**	**Bare Costs**	**Incl. O&P**
1 Equip. Oper. (med.)	$42.95	$343.60	$64.30	$514.40	$40.14	$60.51
.5 Laborer	33.10	132.40	51.05	204.20		
.25 Equip. Oper. (med.)	42.95	85.90	64.30	128.60		
1 Elev. Scraper, 11 C.Y.		1185.00		1303.50		
.25 Dozer, 300 H.P.		398.00		437.80	113.07	124.38
14 L.H., Daily Totals		$2144.90		$2588.50	$153.21	$184.89
Crew B-33G	**Hr.**	**Daily**	**Hr.**	**Daily**	**Bare Costs**	**Incl. O&P**
1 Equip. Oper. (med.)	$42.95	$343.60	$64.30	$514.40	$40.14	$60.51
.5 Laborer	33.10	132.40	51.05	204.20		
.25 Equip. Oper. (med.)	42.95	85.90	64.30	128.60		
1 Elev. Scraper, 22 C.Y.		2283.00		2511.30		
.25 Dozer, 300 H.P.		398.00		437.80	191.50	210.65
14 L.H., Daily Totals		$3242.90		$3796.30	$231.64	$271.16
Crew B-33H	**Hr.**	**Daily**	**Hr.**	**Daily**	**Bare Costs**	**Incl. O&P**
.5 Laborer	$33.10	$132.40	$51.05	$204.20	$40.14	$60.51
1 Equipment Operator (med.)	42.95	343.60	64.30	514.40		
.25 Equipment Operator (med.)	42.95	85.90	64.30	128.60		
1 S.P. Scraper, 44 C.Y.		4296.00		4725.60		
.25 Dozer, 410 H.P.		519.75		571.73	343.98	378.38
14 L.H., Daily Totals		$5377.65		$6144.52	$384.12	$438.89
Crew B-33J	**Hr.**	**Daily**	**Hr.**	**Daily**	**Bare Costs**	**Incl. O&P**
1 Equipment Operator (med.)	$42.95	$343.60	$64.30	$514.40	$42.95	$64.30
1 S.P. Scraper, 14 C.Y.		1717.00		1888.70	214.63	236.09
8 L.H., Daily Totals		$2060.60		$2403.10	$257.57	$300.39

Crew No.	Bare Costs		Incl. Subs O&P		Cost Per Labor-Hour	
Crew B-33K	**Hr.**	**Daily**	**Hr.**	**Daily**	**Bare Costs**	**Incl. O&P**
1 Equipment Operator (med.)	$42.95	$343.60	$64.30	$514.40	$40.14	$60.51
.25 Equipment Operator (med.)	42.95	85.90	64.30	128.60		
.5 Laborer	33.10	132.40	51.05	204.20		
1 S.P. Scraper, 31 C.Y.		3219.00		3540.90		
.25 Dozer, 410 H.P.		519.75		571.73	267.05	293.76
14 L.H., Daily Totals		$4300.65		$4959.82	$307.19	$354.27
Crew B-34A	**Hr.**	**Daily**	**Hr.**	**Daily**	**Bare Costs**	**Incl. O&P**
1 Truck Driver (heavy)	$33.15	$265.20	$50.55	$404.40	$33.15	$50.55
1 Dump Truck, 8 C.Y., 220 H.P.		401.80		441.98	50.23	55.25
8 L.H., Daily Totals		$667.00		$846.38	$83.38	$105.80
Crew B-34B	**Hr.**	**Daily**	**Hr.**	**Daily**	**Bare Costs**	**Incl. O&P**
1 Truck Driver (heavy)	$33.15	$265.20	$50.55	$404.40	$33.15	$50.55
1 Dump Truck, 12 C.Y., 400 H.P.		667.80		734.58	83.47	91.82
8 L.H., Daily Totals		$933.00		$1138.98	$116.63	$142.37
Crew B-34C	**Hr.**	**Daily**	**Hr.**	**Daily**	**Bare Costs**	**Incl. O&P**
1 Truck Driver (heavy)	$33.15	$265.20	$50.55	$404.40	$33.15	$50.55
1 Truck Tractor, 6x4, 380 H.P.		592.40		651.64		
1 Dump Trailer, 16.5 C.Y.		120.20		132.22	89.08	97.98
8 L.H., Daily Totals		$977.80		$1188.26	$122.22	$148.53
Crew B-34D	**Hr.**	**Daily**	**Hr.**	**Daily**	**Bare Costs**	**Incl. O&P**
1 Truck Driver (heavy)	$33.15	$265.20	$50.55	$404.40	$33.15	$50.55
1 Truck Tractor, 6x4, 380 H.P.		592.40		651.64		
1 Dump Trailer, 20 C.Y.		134.80		148.28	90.90	99.99
8 L.H., Daily Totals		$992.40		$1204.32	$124.05	$150.54
Crew B-34E	**Hr.**	**Daily**	**Hr.**	**Daily**	**Bare Costs**	**Incl. O&P**
1 Truck Driver (heavy)	$33.15	$265.20	$50.55	$404.40	$33.15	$50.55
1 Dump Truck, Off Hwy., 25 Ton		1399.00		1538.90	174.88	192.36
8 L.H., Daily Totals		$1664.20		$1943.30	$208.03	$242.91
Crew B-34F	**Hr.**	**Daily**	**Hr.**	**Daily**	**Bare Costs**	**Incl. O&P**
1 Truck Driver (heavy)	$33.15	$265.20	$50.55	$404.40	$33.15	$50.55
1 Dump Truck, Off Hwy., 35 Ton		1267.00		1393.70	158.38	174.21
8 L.H., Daily Totals		$1532.20		$1798.10	$191.53	$224.76
Crew B-34G	**Hr.**	**Daily**	**Hr.**	**Daily**	**Bare Costs**	**Incl. O&P**
1 Truck Driver (heavy)	$33.15	$265.20	$50.55	$404.40	$33.15	$50.55
1 Dump Truck, Off Hwy., 50 Ton		1744.00		1918.40	218.00	239.80
8 L.H., Daily Totals		$2009.20		$2322.80	$251.15	$290.35
Crew B-34H	**Hr.**	**Daily**	**Hr.**	**Daily**	**Bare Costs**	**Incl. O&P**
1 Truck Driver (heavy)	$33.15	$265.20	$50.55	$404.40	$33.15	$50.55
1 Dump Truck, Off Hwy., 65 Ton		1770.00		1947.00	221.25	243.38
8 L.H., Daily Totals		$2035.20		$2351.40	$254.40	$293.93
Crew B-34I	**Hr.**	**Daily**	**Hr.**	**Daily**	**Bare Costs**	**Incl. O&P**
1 Truck Driver (heavy)	$33.15	$265.20	$50.55	$404.40	$33.15	$50.55
1 Dump Truck, 18 C.Y., 450 H.P.		838.40		922.24	104.80	115.28
8 L.H., Daily Totals		$1103.60		$1326.64	$137.95	$165.83
Crew B-34J	**Hr.**	**Daily**	**Hr.**	**Daily**	**Bare Costs**	**Incl. O&P**
1 Truck Driver (heavy)	$33.15	$265.20	$50.55	$404.40	$33.15	$50.55
1 Dump Truck, Off Hwy., 100 Ton		2296.00		2525.60	287.00	315.70
8 L.H., Daily Totals		$2561.20		$2930.00	$320.15	$366.25

Crew No.	Bare Costs		Incl. Subs O&P		Cost Per Labor-Hour	
Crew B-34K	**Hr.**	**Daily**	**Hr.**	**Daily**	**Bare Costs**	**Incl. O&P**
1 Truck Driver (heavy)	$33.15	$265.20	$50.55	$404.40	$33.15	$50.55
1 Truck Tractor, 6x4, 450 H.P.		715.80		787.38		
1 Lowbed Trailer, 75 Ton		205.80		226.38	115.20	126.72
8 L.H., Daily Totals		$1186.80		$1418.16	$148.35	$177.27
Crew B-34L	**Hr.**	**Daily**	**Hr.**	**Daily**	**Bare Costs**	**Incl. O&P**
1 Equip. Oper. (light)	$41.30	$330.40	$61.85	$494.80	$41.30	$61.85
1 Flatbed Truck, Gas, 1.5 Ton		239.80		263.78	29.98	32.97
8 L.H., Daily Totals		$570.20		$758.58	$71.28	$94.82
Crew B-34N	**Hr.**	**Daily**	**Hr.**	**Daily**	**Bare Costs**	**Incl. O&P**
1 Truck Driver (heavy)	$33.15	$265.20	$50.55	$404.40	$33.15	$50.55
1 Dump Truck, 8 C.Y., 220 H.P.		401.80		441.98		
1 Flatbed Trailer, 40 Ton		143.20		157.52	68.13	74.94
8 L.H., Daily Totals		$810.20		$1003.90	$101.28	$125.49
Crew B-34P	**Hr.**	**Daily**	**Hr.**	**Daily**	**Bare Costs**	**Incl. O&P**
1 Pipe Fitter	$51.90	$415.20	$77.80	$622.40	$42.37	$63.77
1 Truck Driver (light)	32.25	258.00	49.20	393.60		
1 Equip. Oper. (med.)	42.95	343.60	64.30	514.40		
1 Flatbed Truck, Gas, 3 Ton		296.00		325.60		
1 Backhoe Loader, 48 H.P.		337.80		371.58	26.41	29.05
24 L.H., Daily Totals		$1650.60		$2227.58	$68.78	$92.82
Crew B-34Q	**Hr.**	**Daily**	**Hr.**	**Daily**	**Bare Costs**	**Incl. O&P**
1 Pipe Fitter	$51.90	$415.20	$77.80	$622.40	$42.85	$64.48
1 Truck Driver (light)	32.25	258.00	49.20	393.60		
1 Eqip. Oper. (crane)	44.40	355.20	66.45	531.60		
1 Flatbed Trailer, 25 Ton		105.60		116.16		
1 Dump Truck, 8 C.Y., 220 H.P.		401.80		441.98		
1 Hyd. Crane, 25 Ton		747.60		822.36	52.29	57.52
24 L.H., Daily Totals		$2283.40		$2928.10	$95.14	$122.00
Crew B-34R	**Hr.**	**Daily**	**Hr.**	**Daily**	**Bare Costs**	**Incl. O&P**
1 Pipe Fitter	$51.90	$415.20	$77.80	$622.40	$42.85	$64.48
1 Truck Driver (light)	32.25	258.00	49.20	393.60		
1 Eqip. Oper. (crane)	44.40	355.20	66.45	531.60		
1 Flatbed Trailer, 25 Ton		105.60		116.16		
1 Dump Truck, 8 C.Y., 220 H.P.		401.80		441.98		
1 Hyd. Crane, 25 Ton		747.60		822.36		
1 Hyd. Excavator, 1 C.Y.		801.60		881.76	85.69	94.26
24 L.H., Daily Totals		$3085.00		$3809.86	$128.54	$158.74
Crew B-34S	**Hr.**	**Daily**	**Hr.**	**Daily**	**Bare Costs**	**Incl. O&P**
2 Pipe Fitters	$51.90	$830.40	$77.80	$1244.80	$45.34	$68.15
1 Truck Driver (heavy)	33.15	265.20	50.55	404.40		
1 Eqip. Oper. (crane)	44.40	355.20	66.45	531.60		
1 Flatbed Trailer, 40 Ton		143.20		157.52		
1 Truck Tractor, 6x4, 380 H.P.		592.40		651.64		
1 Hyd. Crane, 80 Ton		1545.00		1699.50		
1 Hyd. Excavator, 2 C.Y.		1321.00		1453.10	112.55	123.81
32 L.H., Daily Totals		$5052.40		$6142.56	$157.89	$191.96
Crew B-34T	**Hr.**	**Daily**	**Hr.**	**Daily**	**Bare Costs**	**Incl. O&P**
2 Pipe Fitters	$51.90	$830.40	$77.80	$1244.80	$45.34	$68.15
1 Truck Driver (heavy)	33.15	265.20	50.55	404.40		
1 Eqip. Oper. (crane)	44.40	355.20	66.45	531.60		
1 Flatbed Trailer, 40 Ton		143.20		157.52		
1 Truck Tractor, 6x4, 380 H.P.		592.40		651.64		
1 Hyd. Crane, 80 Ton		1545.00		1699.50	71.27	78.40
32 L.H., Daily Totals		$3731.40		$4689.46	$116.61	$146.55

Crew B-35	Bare Costs Hr.	Bare Costs Daily	Incl. Subs O&P Hr.	Incl. Subs O&P Daily	Bare Costs (Per L.H.)	Incl. O&P (Per L.H.)
1 Laborer Foreman (out)	$35.10	$280.80	$54.10	$432.80	$40.92	$62.08
1 Skilled Worker	42.60	340.80	65.50	524.00		
1 Welder (plumber)	52.05	416.40	78.00	624.00		
1 Laborer	33.10	264.80	51.05	408.40		
1 Equip. Oper. (crane)	44.40	355.20	66.45	531.60		
1 Equip. Oper. Oiler	38.30	306.40	57.35	458.80		
1 Welder, electric, 300 amp		58.55		64.41		
1 Hyd. Excavator, .75 C.Y.		677.80		745.58	15.34	16.87
48 L.H., Daily Totals		$2700.75		$3789.59	$56.27	$78.95

Crew B-35A	Hr.	Daily	Hr.	Daily	Bare Costs	Incl. O&P
1 Laborer Foreman (out)	$35.10	$280.80	$54.10	$432.80	$39.81	$60.50
2 Laborers	33.10	529.60	51.05	816.80		
1 Skilled Worker	42.60	340.80	65.50	524.00		
1 Welder (plumber)	52.05	416.40	78.00	624.00		
1 Equip. Oper. (crane)	44.40	355.20	66.45	531.60		
1 Equip. Oper. Oiler	38.30	306.40	57.35	458.80		
1 Welder, gas engine, 300 amp		145.80		160.38		
1 Crawler Crane, 75 Ton		1516.00		1667.60	29.68	32.64
56 L.H., Daily Totals		$3891.00		$5215.98	$69.48	$93.14

Crew B-36	Hr.	Daily	Hr.	Daily	Bare Costs	Incl. O&P
1 Labor Foreman (outside)	$35.10	$280.80	$54.10	$432.80	$37.44	$56.96
2 Laborers	33.10	529.60	51.05	816.80		
2 Equip. Oper. (med.)	42.95	687.20	64.30	1028.80		
1 Dozer, 200 H.P.		1192.00		1311.20		
1 Aggregate Spreader		38.60		42.46		
1 Tandem Roller, 10 Ton		235.20		258.72	36.65	40.31
40 L.H., Daily Totals		$2963.40		$3890.78	$74.08	$97.27

Crew B-36A	Hr.	Daily	Hr.	Daily	Bare Costs	Incl. O&P
1 Labor Foreman (outside)	$35.10	$280.80	$54.10	$432.80	$39.01	$59.06
2 Laborers	33.10	529.60	51.05	816.80		
4 Equip. Oper. (med.)	42.95	1374.40	64.30	2057.60		
1 Dozer, 200 H.P.		1192.00		1311.20		
1 Aggregate Spreader		38.60		42.46		
1 Tandem Roller, 10 Ton		235.20		258.72		
1 Roller, Pneum. Whl, 12 Ton		337.20		370.92	32.20	35.42
56 L.H., Daily Totals		$3987.80		$5290.50	$71.21	$94.47

Crew B-36B	Hr.	Daily	Hr.	Daily	Bare Costs	Incl. O&P
1 Labor Foreman (outside)	$35.10	$280.80	$54.10	$432.80	$38.28	$57.99
2 Laborers	33.10	529.60	51.05	816.80		
4 Equip. Oper. (med.)	42.95	1374.40	64.30	2057.60		
1 Truck Driver, Heavy	33.15	265.20	50.55	404.40		
1 Grader, 30,000 Lbs.		622.20		684.42		
1 F.E. Loader, crl, 1.5 C.Y.		534.60		588.06		
1 Dozer, 300 H.P.		1592.00		1751.20		
1 Roller, Vibratory, 25 Ton		633.20		696.52		
1 Truck Tractor, 6x4, 450 H.P.		715.80		787.38		
1 Water Tank Trailer, 5000 Gal.		139.60		153.56	66.21	72.83
64 L.H., Daily Totals		$6687.40		$8372.74	$104.49	$130.82

Crew B-36C	Hr.	Daily	Hr.	Daily	Bare Costs	Incl. O&P
1 Labor Foreman (outside)	$35.10	$280.80	$54.10	$432.80	$39.42	$59.51
3 Equip. Oper. (med.)	42.95	1030.80	64.30	1543.20		
1 Truck Driver, Heavy	33.15	265.20	50.55	404.40		
1 Grader, 30,000 Lbs.		622.20		684.42		
1 Dozer, 300 H.P.		1592.00		1751.20		
1 Roller, Vibratory, 25 Ton		633.20		696.52		
1 Truck Tractor, 6x4, 450 H.P.		715.80		787.38		
1 Water Tank Trailer, 5000 Gal.		139.60		153.56	92.57	101.83
40 L.H., Daily Totals		$5279.60		$6453.48	$131.99	$161.34

Crew B-37	Hr.	Daily	Hr.	Daily	Bare Costs	Incl. O&P
1 Labor Foreman (outside)	$35.10	$280.80	$54.10	$432.80	$34.80	$53.36
4 Laborers	33.10	1059.20	51.05	1633.60		
1 Equip. Oper. (light)	41.30	330.40	61.85	494.80		
1 Tandem Roller, 5 Ton		152.00		167.20	3.17	3.48
48 L.H., Daily Totals		$1822.40		$2728.40	$37.97	$56.84

Crew B-38	Hr.	Daily	Hr.	Daily	Bare Costs	Incl. O&P
1 Labor Foreman (outside)	$35.10	$280.80	$54.10	$432.80	$37.11	$56.47
2 Laborers	33.10	529.60	51.05	816.80		
1 Equip. Oper. (light)	41.30	330.40	61.85	494.80		
1 Equip. Oper. (med.)	42.95	343.60	64.30	514.40		
1 Backhoe Loader, 48 H.P.		337.80		371.58		
1 Hyd.Hammer, (1200 lb.)		154.20		169.62		
1 F.E. Loader, W.M., 4 C.Y.		627.80		690.58		
1 Pvmt. Rem. Bucket		58.20		64.02	29.45	32.40
40 L.H., Daily Totals		$2662.40		$3554.60	$66.56	$88.86

Crew B-39	Hr.	Daily	Hr.	Daily	Bare Costs	Incl. O&P
1 Labor Foreman (outside)	$35.10	$280.80	$54.10	$432.80	$34.80	$53.36
4 Laborers	33.10	1059.20	51.05	1633.60		
1 Equip. Oper. (light)	41.30	330.40	61.85	494.80		
1 Air Compressor, 250 cfm		193.40		212.74		
2 Breakers, Pavement, 60 lb.		18.80		20.68		
2 -50' Air Hoses, 1.5"		12.60		13.86	4.68	5.15
48 L.H., Daily Totals		$1895.20		$2808.48	$39.48	$58.51

Crew B-40	Hr.	Daily	Hr.	Daily	Bare Costs	Incl. O&P
1 Pile Driver Foreman (out)	$42.30	$338.40	$68.00	$544.00	$41.33	$64.68
4 Pile Drivers	40.30	1289.60	64.80	2073.60		
2 Equip. Oper. (crane)	44.40	710.40	66.45	1063.20		
1 Equip. Oper. Oiler	38.30	306.40	57.35	458.80		
1 Crawler Crane, 40 Ton		1193.00		1312.30		
1 Vibratory Hammer & Gen.		2641.00		2905.10	59.91	65.90
64 L.H., Daily Totals		$6478.80		$8357.00	$101.23	$130.58

Crew B-40B	Hr.	Daily	Hr.	Daily	Bare Costs	Incl. O&P
1 Laborer Foreman	$35.10	$280.80	$54.10	$432.80	$36.18	$55.17
3 Laborers	33.10	794.40	51.05	1225.20		
1 Equip. Oper. (crane)	44.40	355.20	66.45	531.60		
1 Equip. Oper. Oiler	38.30	306.40	57.35	458.80		
1 Lattice Boom Crane, 40 Ton		1256.00		1381.60	26.17	28.78
48 L.H., Daily Totals		$2992.80		$4030.00	$62.35	$83.96

Crew B-41	Hr.	Daily	Hr.	Daily	Bare Costs	Incl. O&P
1 Labor Foreman (outside)	$35.10	$280.80	$54.10	$432.80	$34.21	$52.59
4 Laborers	33.10	1059.20	51.05	1633.60		
.25 Equip. Oper. (crane)	44.40	88.80	66.45	132.90		
.25 Equip. Oper. Oiler	38.30	76.60	57.35	114.70		
.25 Crawler Crane, 40 Ton		298.25		328.07	6.78	7.46
44 L.H., Daily Totals		$1803.65		$2642.07	$40.99	$60.05

Crew No.	Bare Costs		Incl. Subs O&P		Cost Per Labor-Hour	
Crew B-42	**Hr.**	**Daily**	**Hr.**	**Daily**	**Bare Costs**	**Incl. O&P**
1 Labor Foreman (outside)	$35.10	$280.80	$54.10	$432.80	$37.14	$58.07
4 Laborers	33.10	1059.20	51.05	1633.60		
1 Equip. Oper. (crane)	44.40	355.20	66.45	531.60		
1 Equip. Oper. Oiler	38.30	306.40	57.35	458.80		
1 Welder	46.90	375.20	82.45	659.60		
1 Hyd. Crane, 25 Ton		747.60		822.36		
1 Welder, gas engine, 300 amp		145.80		160.38		
1 Horz. Boring Csg. Mch.		473.20		520.52	21.35	23.49
64 L.H., Daily Totals		$3743.40		$5219.66	$58.49	$81.56
Crew B-43	**Hr.**	**Daily**	**Hr.**	**Daily**	**Bare Costs**	**Incl. O&P**
1 Labor Foreman (outside)	$35.10	$280.80	$54.10	$432.80	$36.18	$55.17
3 Laborers	33.10	794.40	51.05	1225.20		
1 Equip. Oper. (crane)	44.40	355.20	66.45	531.60		
1 Equip. Oper. Oiler	38.30	306.40	57.35	458.80		
1 Drill Rig, Truck-Mounted		2591.00		2850.10	53.98	59.38
48 L.H., Daily Totals		$4327.80		$5498.50	$90.16	$114.55
Crew B-44	**Hr.**	**Daily**	**Hr.**	**Daily**	**Bare Costs**	**Incl. O&P**
1 Pile Driver Foreman	$42.30	$338.40	$68.00	$544.00	$40.67	$63.89
4 Pile Drivers	40.30	1289.60	64.80	2073.60		
2 Equip. Oper. (crane)	44.40	710.40	66.45	1063.20		
1 Laborer	33.10	264.80	51.05	408.40		
1 Crawler Crane, 40 Ton		1193.00		1312.30		
1 Lead, 60' high		70.80		77.88		
1 Hammer, diesel, 15K ft.-lbs.		577.60		635.36	28.77	31.65
64 L.H., Daily Totals		$4444.60		$6114.74	$69.45	$95.54
Crew B-45	**Hr.**	**Daily**	**Hr.**	**Daily**	**Bare Costs**	**Incl. O&P**
1 Equip. Oper. (med.)	$42.95	$343.60	$64.30	$514.40	$38.05	$57.42
1 Truck Driver (heavy)	33.15	265.20	50.55	404.40		
1 Dist. Tanker, 3000 Gallon		307.00		337.70		
1 Truck Tractor, 6x4, 380 H.P.		592.40		651.64	56.21	61.83
16 L.H., Daily Totals		$1508.20		$1908.14	$94.26	$119.26
Crew B-46	**Hr.**	**Daily**	**Hr.**	**Daily**	**Bare Costs**	**Incl. O&P**
1 Pile Driver Foreman	$42.30	$338.40	$68.00	$544.00	$37.03	$58.46
2 Pile Drivers	40.30	644.80	64.80	1036.80		
3 Laborers	33.10	794.40	51.05	1225.20		
1 Chain Saw, gas, 36" Long		39.20		43.12	.82	.90
48 L.H., Daily Totals		$1816.80		$2849.12	$37.85	$59.36
Crew B-47	**Hr.**	**Daily**	**Hr.**	**Daily**	**Bare Costs**	**Incl. O&P**
1 Blast Foreman	$35.10	$280.80	$54.10	$432.80	$36.50	$55.67
1 Driller	33.10	264.80	51.05	408.40		
1 Equip. Oper. (light)	41.30	330.40	61.85	494.80		
1 Air Track Drill, 4"		917.80		1009.58		
1 Air Compressor, 600 cfm		530.60		583.66		
2 -50' Air Hoses, 3"		32.40		35.64	61.70	67.87
24 L.H., Daily Totals		$2356.80		$2964.88	$98.20	$123.54
Crew B-47A	**Hr.**	**Daily**	**Hr.**	**Daily**	**Bare Costs**	**Incl. O&P**
1 Drilling Foreman	$35.10	$280.80	$54.10	$432.80	$39.27	$59.30
1 Equip. Oper. (heavy)	44.40	355.20	66.45	531.60		
1 Oiler	38.30	306.40	57.35	458.80		
1 Air Track Drill, 5"		1069.00		1175.90	44.54	49.00
24 L.H., Daily Totals		$2011.40		$2599.10	$83.81	$108.30

Crew No.	Bare Costs		Incl. Subs O&P		Cost Per Labor-Hour	
Crew B-47C	**Hr.**	**Daily**	**Hr.**	**Daily**	**Bare Costs**	**Incl. O&P**
1 Laborer	$33.10	$264.80	$51.05	$408.40	$37.20	$56.45
1 Equip. Oper. (light)	41.30	330.40	61.85	494.80		
1 Air Compressor, 750 cfm		542.80		597.08		
2 -50' Air Hoses, 3"		32.40		35.64		
1 Air Track Drill, 4"		917.80		1009.58	93.31	102.64
16 L.H., Daily Totals		$2088.20		$2545.50	$130.51	$159.09
Crew B-47E	**Hr.**	**Daily**	**Hr.**	**Daily**	**Bare Costs**	**Incl. O&P**
1 Laborer Foreman	$35.10	$280.80	$54.10	$432.80	$33.60	$51.81
3 Laborers	33.10	794.40	51.05	1225.20		
1 Flatbed Truck, Gas, 3 Ton		296.00		325.60	9.25	10.18
32 L.H., Daily Totals		$1371.20		$1983.60	$42.85	$61.99
Crew B-47G	**Hr.**	**Daily**	**Hr.**	**Daily**	**Bare Costs**	**Incl. O&P**
1 Laborer Foreman	$35.10	$280.80	$54.10	$432.80	$35.65	$54.51
2 Laborers	33.10	529.60	51.05	816.80		
1 Equip. Oper. (light)	41.30	330.40	61.85	494.80		
1 Air Track Drill, 4"		917.80		1009.58		
1 Air Compressor, 600 cfm		530.60		583.66		
2 -50' Air Hoses, 3"		32.40		35.64		
1 Grout Pump		201.80		221.98	52.58	57.84
32 L.H., Daily Totals		$2823.40		$3595.26	$88.23	$112.35
Crew B-47H	**Hr.**	**Daily**	**Hr.**	**Daily**	**Bare Costs**	**Incl. O&P**
1 Skilled Worker Foreman	$44.60	$356.80	$68.55	$548.40	$43.10	$66.26
3 Skilled Workers	42.60	1022.40	65.50	1572.00		
1 Flatbed Truck, Gas, 3 Ton		296.00		325.60	9.25	10.18
32 L.H., Daily Totals		$1675.20		$2446.00	$52.35	$76.44
Crew B-48	**Hr.**	**Daily**	**Hr.**	**Daily**	**Bare Costs**	**Incl. O&P**
1 Labor Foreman (outside)	$35.10	$280.80	$54.10	$432.80	$36.91	$56.13
3 Laborers	33.10	794.40	51.05	1225.20		
1 Equip. Oper. (crane)	44.40	355.20	66.45	531.60		
1 Equip. Oper. Oiler	38.30	306.40	57.35	458.80		
1 Equip. Oper. (light)	41.30	330.40	61.85	494.80		
1 Centr. Water Pump, 6"		340.00		374.00		
1 -20' Suction Hose, 6"		11.90		13.09		
1 -50' Discharge Hose, 6"		6.30		6.93		
1 Drill Rig, Truck-Mounted		2591.00		2850.10	52.66	57.93
56 L.H., Daily Totals		$5016.40		$6387.32	$89.58	$114.06
Crew B-49	**Hr.**	**Daily**	**Hr.**	**Daily**	**Bare Costs**	**Incl. O&P**
1 Labor Foreman (outside)	$35.10	$280.80	$54.10	$432.80	$38.34	$58.75
3 Laborers	33.10	794.40	51.05	1225.20		
2 Equip. Oper. (crane)	44.40	710.40	66.45	1063.20		
2 Equip. Oper. Oilers	38.30	612.80	57.35	917.60		
1 Equip. Oper. (light)	41.30	330.40	61.85	494.80		
2 Pile Drivers	40.30	644.80	64.80	1036.80		
1 Hyd. Crane, 25 Ton		747.60		822.36		
1 Centr. Water Pump, 6"		340.00		374.00		
1 -20' Suction Hose, 6"		11.90		13.09		
1 -50' Discharge Hose, 6"		6.30		6.93		
1 Drill Rig, Truck-Mounted		2591.00		2850.10	42.01	46.21
88 L.H., Daily Totals		$7070.40		$9236.88	$80.35	$104.96

Crew No.	Bare Costs		Incl. Subs O&P		Cost Per Labor-Hour	
Crew B-50	**Hr.**	**Daily**	**Hr.**	**Daily**	**Bare Costs**	**Incl. O&P**
2 Pile Driver Foremen	$42.30	$676.80	$68.00	$1088.00	$39.49	$62.01
6 Pile Drivers	40.30	1934.40	64.80	3110.40		
2 Equip. Oper. (crane)	44.40	710.40	66.45	1063.20		
1 Equip. Oper. Oiler	38.30	306.40	57.35	458.80		
3 Laborers	33.10	794.40	51.05	1225.20		
1 Crawler Crane, 40 Ton		1193.00		1312.30		
1 Lead, 60' high		70.80		77.88		
1 Hammer, diesel, 15K ft.-lbs.		577.60		635.36		
1 Air Compressor, 600 cfm		530.60		583.66		
2 -50' Air Hoses, 3"		32.40		35.64		
1 Chain Saw, gas, 36" Long		39.20		43.12	21.82	24.00
112 L.H., Daily Totals		$6866.00		$9633.56	$61.30	$86.01
Crew B-51	**Hr.**	**Daily**	**Hr.**	**Daily**	**Bare Costs**	**Incl. O&P**
1 Labor Foreman (outside)	$35.10	$280.80	$54.10	$432.80	$33.29	$51.25
4 Laborers	33.10	1059.20	51.05	1633.60		
1 Truck Driver (light)	32.25	258.00	49.20	393.60		
1 Flatbed Truck, Gas, 1.5 Ton		239.80		263.78	5.00	5.50
48 L.H., Daily Totals		$1837.80		$2723.78	$38.29	$56.75
Crew B-52	**Hr.**	**Daily**	**Hr.**	**Daily**	**Bare Costs**	**Incl. O&P**
1 Carpenter Foreman	$43.55	$348.40	$67.15	$537.20	$38.42	$58.88
1 Carpenter	41.55	332.40	64.05	512.40		
3 Laborers	33.10	794.40	51.05	1225.20		
1 Cement Finisher	39.70	317.60	57.95	463.60		
.5 Rodman (reinf.)	46.80	187.20	75.40	301.60		
.5 Equip. Oper. (med.)	42.95	171.80	64.30	257.20		
.5 Crawler Loader, 3 C.Y.		538.50		592.35	9.62	10.58
56 L.H., Daily Totals		$2690.30		$3889.55	$48.04	$69.46
Crew B-53	**Hr.**	**Daily**	**Hr.**	**Daily**	**Bare Costs**	**Incl. O&P**
1 Equip. Oper. (light)	$41.30	$330.40	$61.85	$494.80	$41.30	$61.85
1 Trencher, Chain, 12 H.P.		66.00		72.60	8.25	9.07
8 L.H., Daily Totals		$396.40		$567.40	$49.55	$70.92
Crew B-54	**Hr.**	**Daily**	**Hr.**	**Daily**	**Bare Costs**	**Incl. O&P**
1 Equip. Oper. (light)	$41.30	$330.40	$61.85	$494.80	$41.30	$61.85
1 Trencher, Chain, 40 H.P.		334.40		367.84	41.80	45.98
8 L.H., Daily Totals		$664.80		$862.64	$83.10	$107.83
Crew B-54A	**Hr.**	**Daily**	**Hr.**	**Daily**	**Bare Costs**	**Incl. O&P**
.17 Labor Foreman (outside)	$35.10	$47.74	$54.10	$73.58	$41.81	$62.82
1 Equipment Operator (med.)	42.95	343.60	64.30	514.40		
1 Wheel Trencher, 67 H.P.		1152.00		1267.20	123.08	135.38
9.36 L.H., Daily Totals		$1543.34		$1855.18	$164.89	$198.20
Crew B-54B	**Hr.**	**Daily**	**Hr.**	**Daily**	**Bare Costs**	**Incl. O&P**
.25 Labor Foreman (outside)	$35.10	$70.20	$54.10	$108.20	$41.38	$62.26
1 Equipment Operator (med.)	42.95	343.60	64.30	514.40		
1 Wheel Trencher, 150 H.P.		1843.00		2027.30	184.30	202.73
10 L.H., Daily Totals		$2256.80		$2649.90	$225.68	$264.99
Crew B-54C	**Hr.**	**Daily**	**Hr.**	**Daily**	**Bare Costs**	**Incl. O&P**
1 Laborer	$33.10	$264.80	$51.05	$408.40	$38.02	$57.67
1 Equipment Operator (med.)	42.95	343.60	64.30	514.40		
1 Wheel Trencher, 67 H.P.		1152.00		1267.20	72.00	79.20
16 L.H., Daily Totals		$1760.40		$2190.00	$110.03	$136.88

Crew No.	Bare Costs		Incl. Subs O&P		Cost Per Labor-Hour	
Crew B-54D	**Hr.**	**Daily**	**Hr.**	**Daily**	**Bare Costs**	**Incl. O&P**
1 Laborer	$33.10	$264.80	$51.05	$408.40	$38.02	$57.67
1 Equipment Operator (med.)	42.95	343.60	64.30	514.40		
1 Rock trencher, 6" width		344.40		378.84	21.52	23.68
16 L.H., Daily Totals		$952.80		$1301.64	$59.55	$81.35
Crew B-54E	**Hr.**	**Daily**	**Hr.**	**Daily**	**Bare Costs**	**Incl. O&P**
1 Laborer	$33.10	$264.80	$51.05	$408.40	$38.02	$57.67
1 Equipment Operator (med.)	42.95	343.60	64.30	514.40		
1 Rock trencher, 18" width		2571.00		2828.10	160.69	176.76
16 L.H., Daily Totals		$3179.40		$3750.90	$198.71	$234.43
Crew B-55	**Hr.**	**Daily**	**Hr.**	**Daily**	**Bare Costs**	**Incl. O&P**
2 Laborers	$33.10	$529.60	$51.05	$816.80	$32.82	$50.43
1 Truck Driver (light)	32.25	258.00	49.20	393.60		
1 Truck-mounted earth auger		749.80		824.78		
1 Flatbed Truck, Gas, 3 Ton		296.00		325.60	43.58	47.93
24 L.H., Daily Totals		$1833.40		$2360.78	$76.39	$98.37
Crew B-56	**Hr.**	**Daily**	**Hr.**	**Daily**	**Bare Costs**	**Incl. O&P**
1 Laborer	$33.10	$264.80	$51.05	$408.40	$37.20	$56.45
1 Equip. Oper. (light)	41.30	330.40	61.85	494.80		
1 Air Track Drill, 4"		917.80		1009.58		
1 Air Compressor, 600 cfm		530.60		583.66		
1 -50' Air Hose, 3"		16.20		17.82	91.54	100.69
16 L.H., Daily Totals		$2059.80		$2514.26	$128.74	$157.14
Crew B-57	**Hr.**	**Daily**	**Hr.**	**Daily**	**Bare Costs**	**Incl. O&P**
1 Labor Foreman (outside)	$35.10	$280.80	$54.10	$432.80	$37.55	$56.98
2 Laborers	33.10	529.60	51.05	816.80		
1 Equip. Oper. (crane)	44.40	355.20	66.45	531.60		
1 Equip. Oper. (light)	41.30	330.40	61.85	494.80		
1 Equip. Oper. Oiler	38.30	306.40	57.35	458.80		
1 Crawler Crane, 25 Ton		1188.00		1306.80		
1 Clamshell Bucket, 1 C.Y.		46.60		51.26		
1 Centr. Water Pump, 6"		340.00		374.00		
1 -20' Suction Hose, 6"		11.90		13.09		
20 -50' Discharge Hoses, 6"		126.00		138.60	35.68	39.24
48 L.H., Daily Totals		$3514.90		$4618.55	$73.23	$96.22
Crew B-58	**Hr.**	**Daily**	**Hr.**	**Daily**	**Bare Costs**	**Incl. O&P**
2 Laborers	$33.10	$529.60	$51.05	$816.80	$35.83	$54.65
1 Equip. Oper. (light)	41.30	330.40	61.85	494.80		
1 Backhoe Loader, 48 H.P.		337.80		371.58		
1 Small Helicopter, w/pilot		2513.00		2764.30	118.78	130.66
24 L.H., Daily Totals		$3710.80		$4447.48	$154.62	$185.31
Crew B-59	**Hr.**	**Daily**	**Hr.**	**Daily**	**Bare Costs**	**Incl. O&P**
1 Truck Driver (heavy)	$33.15	$265.20	$50.55	$404.40	$33.15	$50.55
1 Truck Tractor, 220 H.P.		354.00		389.40		
1 Water Tank Trailer, 5000 Gal.		139.60		153.56	61.70	67.87
8 L.H., Daily Totals		$758.80		$947.36	$94.85	$118.42
Crew B-59A	**Hr.**	**Daily**	**Hr.**	**Daily**	**Bare Costs**	**Incl. O&P**
2 Laborers	$33.10	$529.60	$51.05	$816.80	$33.12	$50.88
1 Truck Driver (heavy)	33.15	265.20	50.55	404.40		
1 Water Tank Trailer, 5000 Gal.		139.60		153.56		
1 Truck Tractor, 220 H.P.		354.00		389.40	20.57	22.62
24 L.H., Daily Totals		$1288.40		$1764.16	$53.68	$73.51

Crew B-60	Bare Costs Hr.	Bare Costs Daily	Incl. Subs O&P Hr.	Incl. Subs O&P Daily	Cost Per Labor-Hour Bare Costs	Cost Per Labor-Hour Incl. O&P
1 Labor Foreman (outside)	$35.10	$280.80	$54.10	$432.80	$38.09	$57.67
2 Laborers	33.10	529.60	51.05	816.80		
1 Equip. Oper. (crane)	44.40	355.20	66.45	531.60		
2 Equip. Oper. (light)	41.30	660.80	61.85	989.60		
1 Equip. Oper. Oiler	38.30	306.40	57.35	458.80		
1 Crawler Crane, 40 Ton		1193.00		1312.30		
1 Lead, 60' high		70.80		77.88		
1 Hammer, diesel, 15K ft.-lbs.		577.60		635.36		
1 Backhoe Loader, 48 H.P.		337.80		371.58	38.91	42.81
56 L.H., Daily Totals		$4312.00		$5626.72	$77.00	$100.48

Crew B-61	Hr.	Daily	Hr.	Daily	Bare Costs	Incl. O&P
1 Labor Foreman (outside)	$35.10	$280.80	$54.10	$432.80	$35.14	$53.82
3 Laborers	33.10	794.40	51.05	1225.20		
1 Equip. Oper. (light)	41.30	330.40	61.85	494.80		
1 Cement Mixer, 2 C.Y.		189.40		208.34		
1 Air Compressor, 160 cfm		158.00		173.80	8.69	9.55
40 L.H., Daily Totals		$1753.00		$2534.94	$43.83	$63.37

Crew B-62	Hr.	Daily	Hr.	Daily	Bare Costs	Incl. O&P
2 Laborers	$33.10	$529.60	$51.05	$816.80	$35.83	$54.65
1 Equip. Oper. (light)	41.30	330.40	61.85	494.80		
1 Loader, Skid Steer, 30 H.P., gas		161.40		177.54	6.72	7.40
24 L.H., Daily Totals		$1021.40		$1489.14	$42.56	$62.05

Crew B-63	Hr.	Daily	Hr.	Daily	Bare Costs	Incl. O&P
4 Laborers	$33.10	$1059.20	$51.05	$1633.60	$34.74	$53.21
1 Equip. Oper. (light)	41.30	330.40	61.85	494.80		
1 Loader, Skid Steer, 30 H.P., gas		161.40		177.54	4.04	4.44
40 L.H., Daily Totals		$1551.00		$2305.94	$38.77	$57.65

Crew B-64	Hr.	Daily	Hr.	Daily	Bare Costs	Incl. O&P
1 Laborer	$33.10	$264.80	$51.05	$408.40	$32.67	$50.13
1 Truck Driver (light)	32.25	258.00	49.20	393.60		
1 Power Mulcher (small)		150.20		165.22		
1 Flatbed Truck, Gas, 1.5 Ton		239.80		263.78	24.38	26.81
16 L.H., Daily Totals		$912.80		$1231.00	$57.05	$76.94

Crew B-65	Hr.	Daily	Hr.	Daily	Bare Costs	Incl. O&P
1 Laborer	$33.10	$264.80	$51.05	$408.40	$32.67	$50.13
1 Truck Driver (light)	32.25	258.00	49.20	393.60		
1 Power Mulcher (large)		319.00		350.90		
1 Flatbed Truck, Gas, 1.5 Ton		239.80		263.78	34.92	38.42
16 L.H., Daily Totals		$1081.60		$1416.68	$67.60	$88.54

Crew B-66	Hr.	Daily	Hr.	Daily	Bare Costs	Incl. O&P
1 Equip. Oper. (light)	$41.30	$330.40	$61.85	$494.80	$41.30	$61.85
1 Loader-Backhoe		218.40		240.24	27.30	30.03
8 L.H., Daily Totals		$548.80		$735.04	$68.60	$91.88

Crew B-67	Hr.	Daily	Hr.	Daily	Bare Costs	Incl. O&P
1 Millwright	$42.95	$343.60	$62.70	$501.60	$42.13	$62.27
1 Equip. Oper. (light)	41.30	330.40	61.85	494.80		
1 Forklift, R/T, 4,000 Lb.		309.60		340.56	19.35	21.29
16 L.H., Daily Totals		$983.60		$1336.96	$61.48	$83.56

Crew B-68	Hr.	Daily	Hr.	Daily	Bare Costs	Incl. O&P
2 Millwrights	$42.95	$687.20	$62.70	$1003.20	$42.40	$62.42
1 Equip. Oper. (light)	41.30	330.40	61.85	494.80		
1 Forklift, R/T, 4,000 Lb.		309.60		340.56	12.90	14.19
24 L.H., Daily Totals		$1327.20		$1838.56	$55.30	$76.61

Crew B-69	Hr.	Daily	Hr.	Daily	Bare Costs	Incl. O&P
1 Labor Foreman (outside)	$35.10	$280.80	$54.10	$432.80	$36.18	$55.17
3 Laborers	33.10	794.40	51.05	1225.20		
1 Equip Oper. (crane)	44.40	355.20	66.45	531.60		
1 Equip Oper. Oiler	38.30	306.40	57.35	458.80		
1 Hyd. Crane, 80 Ton		1545.00		1699.50	32.19	35.41
48 L.H., Daily Totals		$3281.80		$4347.90	$68.37	$90.58

Crew B-69A	Hr.	Daily	Hr.	Daily	Bare Costs	Incl. O&P
1 Labor Foreman	$35.10	$280.80	$54.10	$432.80	$36.17	$54.92
3 Laborers	33.10	794.40	51.05	1225.20		
1 Equip. Oper. (med.)	42.95	343.60	64.30	514.40		
1 Concrete Finisher	39.70	317.60	57.95	463.60		
1 Curb/Gutter Paver, 2-Track		825.00		907.50	17.19	18.91
48 L.H., Daily Totals		$2561.40		$3543.50	$53.36	$73.82

Crew B-69B	Hr.	Daily	Hr.	Daily	Bare Costs	Incl. O&P
1 Labor Foreman	$35.10	$280.80	$54.10	$432.80	$36.17	$54.92
3 Laborers	33.10	794.40	51.05	1225.20		
1 Equip. Oper. (med.)	42.95	343.60	64.30	514.40		
1 Cement Finisher	39.70	317.60	57.95	463.60		
1 Curb/Gutter Paver, 4-Track		934.60		1028.06	19.47	21.42
48 L.H., Daily Totals		$2671.00		$3664.06	$55.65	$76.33

Crew B-70	Hr.	Daily	Hr.	Daily	Bare Costs	Incl. O&P
1 Labor Foreman (outside)	$35.10	$280.80	$54.10	$432.80	$37.61	$57.16
3 Laborers	33.10	794.40	51.05	1225.20		
3 Equip. Oper. (med.)	42.95	1030.80	64.30	1543.20		
1 Grader, 30,000 Lbs.		622.20		684.42		
1 Ripper, beam & 1 shank		80.00		88.00		
1 Road Sweeper, S.P., 8' wide		650.40		715.44		
1 F.E. Loader, W.M., 1.5 C.Y.		367.60		404.36	30.72	33.79
56 L.H., Daily Totals		$3826.20		$5093.42	$68.33	$90.95

Crew B-70A	Hr.	Daily	Hr.	Daily	Bare Costs	Incl. O&P
1 Laborer	$33.10	$264.80	$51.05	$408.40	$40.98	$61.65
4 Equip. Oper. (med.)	42.95	1374.40	64.30	2057.60		
1 Grader, 40,000 Lbs.		1071.00		1178.10		
1 F.E. Loader, W.M.,2.5 C.Y.		435.80		479.38		
1 Dozer, 80 H.P.		436.60		480.26		
1 Roller, Pneum. Whl, 12 Ton		337.20		370.92	57.02	62.72
40 L.H., Daily Totals		$3919.80		$4974.66	$98.00	$124.37

Crew B-71	Hr.	Daily	Hr.	Daily	Bare Costs	Incl. O&P
1 Labor Foreman (outside)	$35.10	$280.80	$54.10	$432.80	$37.61	$57.16
3 Laborers	33.10	794.40	51.05	1225.20		
3 Equip. Oper. (med.)	42.95	1030.80	64.30	1543.20		
1 Pvmt. Profiler, 750 H.P.		6104.00		6714.40		
1 Road Sweeper, S.P., 8' wide		650.40		715.44		
1 F.E. Loader, W.M., 1.5 C.Y.		367.60		404.36	127.18	139.90
56 L.H., Daily Totals		$9228.00		$11035.40	$164.79	$197.06

Crew No.	Bare Costs		Incl. Subs O&P		Cost Per Labor-Hour	
Crew B-72	**Hr.**	**Daily**	**Hr.**	**Daily**	**Bare Costs**	**Incl. O&P**
1 Labor Foreman (outside)	$35.10	$280.80	$54.10	$432.80	$38.27	$58.06
3 Laborers	33.10	794.40	51.05	1225.20		
4 Equip. Oper. (med.)	42.95	1374.40	64.30	2057.60		
1 Pvmt. Profiler, 750 H.P.		6104.00		6714.40		
1 Hammermill, 250 H.P.		1636.00		1799.60		
1 Windrow Loader		1228.00		1350.80		
1 Mix Paver 165 H.P.		2062.00		2268.20		
1 Roller, Pneum. Whl, 12 Ton		337.20		370.92	177.61	195.37
64 L.H., Daily Totals		$13816.80		$16219.52	$215.89	$253.43
Crew B-73	**Hr.**	**Daily**	**Hr.**	**Daily**	**Bare Costs**	**Incl. O&P**
1 Labor Foreman (outside)	$35.10	$280.80	$54.10	$432.80	$39.51	$59.71
2 Laborers	33.10	529.60	51.05	816.80		
5 Equip. Oper. (med.)	42.95	1718.00	64.30	2572.00		
1 Road Mixer, 310 H.P.		1955.00		2150.50		
1 Tandem Roller, 10 Ton		235.20		258.72		
1 Hammermill, 250 H.P.		1636.00		1799.60		
1 Grader, 30,000 Lbs.		622.20		684.42		
.5 F.E. Loader, W.M., 1.5 C.Y.		183.80		202.18		
.5 Truck Tractor, 220 H.P.		177.00		194.70		
.5 Water Tank Trailer, 5000 Gal.		69.80		76.78	76.23	83.86
64 L.H., Daily Totals		$7407.40		$9188.50	$115.74	$143.57
Crew B-74	**Hr.**	**Daily**	**Hr.**	**Daily**	**Bare Costs**	**Incl. O&P**
1 Labor Foreman (outside)	$35.10	$280.80	$54.10	$432.80	$38.29	$57.93
1 Laborer	33.10	264.80	51.05	408.40		
4 Equip. Oper. (med.)	42.95	1374.40	64.30	2057.60		
2 Truck Drivers (heavy)	33.15	530.40	50.55	808.80		
1 Grader, 30,000 Lbs.		622.20		684.42		
1 Ripper, beam & 1 shank		80.00		88.00		
2 Stabilizers, 310 H.P.		2928.00		3220.80		
1 Flatbed Truck, Gas, 3 Ton		296.00		325.60		
1 Chem. Spreader, Towed		50.40		55.44		
1 Roller, Vibratory, 25 Ton		633.20		696.52		
1 Water Tank Trailer, 5000 Gal.		139.60		153.56		
1 Truck Tractor, 220 H.P.		354.00		389.40	79.74	87.71
64 L.H., Daily Totals		$7553.80		$9321.34	$118.03	$145.65
Crew B-75	**Hr.**	**Daily**	**Hr.**	**Daily**	**Bare Costs**	**Incl. O&P**
1 Labor Foreman (outside)	$35.10	$280.80	$54.10	$432.80	$39.02	$58.99
1 Laborer	33.10	264.80	51.05	408.40		
4 Equip. Oper. (med.)	42.95	1374.40	64.30	2057.60		
1 Truck Driver (heavy)	33.15	265.20	50.55	404.40		
1 Grader, 30,000 Lbs.		622.20		684.42		
1 Ripper, beam & 1 shank		80.00		88.00		
2 Stabilizers, 310 H.P.		2928.00		3220.80		
1 Dist. Tanker, 3000 Gallon		307.00		337.70		
1 Truck Tractor, 6x4, 380 H.P.		592.40		651.64		
1 Roller, Vibratory, 25 Ton		633.20		696.52	92.19	101.41
56 L.H., Daily Totals		$7348.00		$8982.28	$131.21	$160.40

Crew No.	Bare Costs		Incl. Subs O&P		Cost Per Labor-Hour	
Crew B-76	**Hr.**	**Daily**	**Hr.**	**Daily**	**Bare Costs**	**Incl. O&P**
1 Dock Builder Foreman	$42.30	$338.40	$68.00	$544.00	$41.21	$64.69
5 Dock Builders	40.30	1612.00	64.80	2592.00		
2 Equip. Oper. (crane)	44.40	710.40	66.45	1063.20		
1 Equip. Oper. Oiler	38.30	306.40	57.35	458.80		
1 Crawler Crane, 50 Ton		1346.00		1480.60		
1 Barge, 400 Ton		768.00		844.80		
1 Hammer, diesel, 15K ft.-lbs.		577.60		635.36		
1 Lead, 60' high		70.80		77.88		
1 Air Compressor, 600 cfm		530.60		583.66		
2 -50' Air Hoses, 3"		32.40		35.64	46.19	50.80
72 L.H., Daily Totals		$6292.60		$8315.94	$87.40	$115.50
Crew B-76A	**Hr.**	**Daily**	**Hr.**	**Daily**	**Bare Costs**	**Incl. O&P**
1 Laborer Foreman	$35.10	$280.80	$54.10	$432.80	$35.41	$54.14
5 Laborers	33.10	1324.00	51.05	2042.00		
1 Equip. Oper. (crane)	44.40	355.20	66.45	531.60		
1 Equip. Oper. Oiler	38.30	306.40	57.35	458.80		
1 Crawler Crane, 50 Ton		1346.00		1480.60		
1 Barge, 400 Ton		768.00		844.80	33.03	36.33
64 L.H., Daily Totals		$4380.40		$5790.60	$68.44	$90.48
Crew B-77	**Hr.**	**Daily**	**Hr.**	**Daily**	**Bare Costs**	**Incl. O&P**
1 Labor Foreman	$35.10	$280.80	$54.10	$432.80	$33.33	$51.29
3 Laborers	33.10	794.40	51.05	1225.20		
1 Truck Driver (light)	32.25	258.00	49.20	393.60		
1 Crack Cleaner, 25 H.P.		60.40		66.44		
1 Crack Filler, Trailer Mtd.		203.20		223.52		
1 Flatbed Truck, Gas, 3 Ton		296.00		325.60	13.99	15.39
40 L.H., Daily Totals		$1892.80		$2667.16	$47.32	$66.68
Crew B-78	**Hr.**	**Daily**	**Hr.**	**Daily**	**Bare Costs**	**Incl. O&P**
1 Labor Foreman	$35.10	$280.80	$54.10	$432.80	$33.29	$51.25
4 Laborers	33.10	1059.20	51.05	1633.60		
1 Truck Driver (light)	32.25	258.00	49.20	393.60		
1 Paint Striper, S.P.		152.60		167.86		
1 Flatbed Truck, Gas, 3 Ton		296.00		325.60		
1 Pickup Truck, 3/4 Ton		140.00		154.00	12.26	13.49
48 L.H., Daily Totals		$2186.60		$3107.46	$45.55	$64.74
Crew B-78A	**Hr.**	**Daily**	**Hr.**	**Daily**	**Bare Costs**	**Incl. O&P**
1 Equip. Oper. (light)	$41.30	$330.40	$61.85	$494.80	$41.30	$61.85
1 Line Rem. (metal balls) 115 H.P.		883.40		971.74	110.43	121.47
8 L.H., Daily Totals		$1213.80		$1466.54	$151.72	$183.32
Crew B-78B	**Hr.**	**Daily**	**Hr.**	**Daily**	**Bare Costs**	**Incl. O&P**
2 Laborers	$33.10	$529.60	$51.05	$816.80	$34.01	$52.25
.25 Equip. Oper. (light)	41.30	82.60	61.85	123.70		
1 Pickup Truck, 3/4 Ton		140.00		154.00		
1 Line Rem., 11 H.P., walk behind		59.20		65.12		
.25 Road Sweeper, S.P., 8' wide		162.60		178.86	20.10	22.11
18 L.H., Daily Totals		$974.00		$1338.48	$54.11	$74.36
Crew B-79	**Hr.**	**Daily**	**Hr.**	**Daily**	**Bare Costs**	**Incl. O&P**
1 Labor Foreman	$35.10	$280.80	$54.10	$432.80	$33.33	$51.29
3 Laborers	33.10	794.40	51.05	1225.20		
1 Truck Driver (light)	32.25	258.00	49.20	393.60		
1 Thermo. Striper, T.M.		802.40		882.64		
1 Flatbed Truck, Gas, 3 Ton		296.00		325.60		
2 Pickup Trucks, 3/4 Ton		280.00		308.00	34.46	37.91
40 L.H., Daily Totals		$2711.60		$3567.84	$67.79	$89.20

Crew No.	Bare Costs		Incl. Subs O&P		Cost Per Labor-Hour	
Crew B-79A	**Hr.**	**Daily**	**Hr.**	**Daily**	**Bare Costs**	**Incl. O&P**
1.5 Equip. Oper. (light)	$41.30	$495.60	$61.85	$742.20	$41.30	$61.85
.5 Line Remov. (grinder) 115 H.P.		476.90		524.59		
1 Line Rem. (metal balls) 115 H.P.		883.40		971.74	113.36	124.69
12 L.H., Daily Totals		$1855.90		$2238.53	$154.66	$186.54

Crew B-80	Hr.	Daily	Hr.	Daily	Bare Costs	Incl. O&P
1 Labor Foreman	$35.10	$280.80	$54.10	$432.80	$35.44	$54.05
1 Laborer	33.10	264.80	51.05	408.40		
1 Truck Driver (light)	32.25	258.00	49.20	393.60		
1 Equip. Oper. (light)	41.30	330.40	61.85	494.80		
1 Flatbed Truck, Gas, 3 Ton		296.00		325.60		
1 Earth Auger, Truck-Mtd.		449.20		494.12	23.29	25.62
32 L.H., Daily Totals		$1879.20		$2549.32	$58.73	$79.67

Crew B-80A	Hr.	Daily	Hr.	Daily	Bare Costs	Incl. O&P
3 Laborers	$33.10	$794.40	$51.05	$1225.20	$33.10	$51.05
1 Flatbed Truck, Gas, 3 Ton		296.00		325.60	12.33	13.57
24 L.H., Daily Totals		$1090.40		$1550.80	$45.43	$64.62

Crew B-80B	Hr.	Daily	Hr.	Daily	Bare Costs	Incl. O&P
3 Laborers	$33.10	$794.40	$51.05	$1225.20	$35.15	$53.75
1 Equip. Oper. (light)	41.30	330.40	61.85	494.80		
1 Crane, Flatbed Mounted, 3 Ton		242.60		266.86	7.58	8.34
32 L.H., Daily Totals		$1367.40		$1986.86	$42.73	$62.09

Crew B-80C	Hr.	Daily	Hr.	Daily	Bare Costs	Incl. O&P
2 Laborers	$33.10	$529.60	$51.05	$816.80	$32.82	$50.43
1 Truck Driver (light)	32.25	258.00	49.20	393.60		
1 Flatbed Truck, Gas, 1.5 Ton		239.80		263.78		
1 Manual fence post auger, gas		7.80		8.58	10.32	11.35
24 L.H., Daily Totals		$1035.20		$1482.76	$43.13	$61.78

Crew B-81	Hr.	Daily	Hr.	Daily	Bare Costs	Incl. O&P
1 Laborer	$33.10	$264.80	$51.05	$408.40	$36.40	$55.30
1 Equip. Oper. (med.)	42.95	343.60	64.30	514.40		
1 Truck Driver (heavy)	33.15	265.20	50.55	404.40		
1 Hydromulcher, T.M.		338.20		372.02		
1 Truck Tractor, 220 H.P.		354.00		389.40	28.84	31.73
24 L.H., Daily Totals		$1565.80		$2088.62	$65.24	$87.03

Crew B-82	Hr.	Daily	Hr.	Daily	Bare Costs	Incl. O&P
1 Laborer	$33.10	$264.80	$51.05	$408.40	$37.20	$56.45
1 Equip. Oper. (light)	41.30	330.40	61.85	494.80		
1 Horiz. Borer, 6 H.P.		82.00		90.20	5.13	5.64
16 L.H., Daily Totals		$677.20		$993.40	$42.33	$62.09

Crew B-82A	Hr.	Daily	Hr.	Daily	Bare Costs	Incl. O&P
1 Laborer	$33.10	$264.80	$51.05	$408.40	$37.20	$56.45
1 Equip. Oper. (light)	41.30	330.40	61.85	494.80		
1 Flatbed Truck, Gas, 3 Ton		296.00		325.60		
1 Flatbed Trailer, 25 Ton		105.60		116.16		
1 Horiz. Dir. Drill, 20k lb. thrust		651.60		716.76	65.83	72.41
16 L.H., Daily Totals		$1648.40		$2061.72	$103.03	$128.86

Crew B-82B	Hr.	Daily	Hr.	Daily	Bare Costs	Incl. O&P
2 Laborers	$33.10	$529.60	$51.05	$816.80	$35.83	$54.65
1 Equip. Oper. (light)	41.30	330.40	61.85	494.80		
1 Flatbed Truck, Gas, 3 Ton		296.00		325.60		
1 Flatbed Trailer, 25 Ton		105.60		116.16		
1 Horiz. Dir. Drill, 30k lb. thrust		928.20		1021.02	55.41	60.95
24 L.H., Daily Totals		$2189.80		$2774.38	$91.24	$115.60

Crew B-82C	Hr.	Daily	Hr.	Daily	Bare Costs	Incl. O&P
2 Laborers	$33.10	$529.60	$51.05	$816.80	$35.83	$54.65
1 Equip. Oper. (light)	41.30	330.40	61.85	494.80		
1 Flatbed Truck, Gas, 3 Ton		296.00		325.60		
1 Flatbed Trailer, 25 Ton		105.60		116.16		
1 Horiz. Dir. Drill, 50k lb. thrust		1232.00		1355.20	68.07	74.87
24 L.H., Daily Totals		$2493.60		$3108.56	$103.90	$129.52

Crew B-82D	Hr.	Daily	Hr.	Daily	Bare Costs	Incl. O&P
1 Equip. Oper. (light)	$41.30	$330.40	$61.85	$494.80	$41.30	$61.85
1 Mud Trailer for HDD, 1500 gallon		338.00		371.80	42.25	46.48
8 L.H., Daily Totals		$668.40		$866.60	$83.55	$108.33

Crew B-83	Hr.	Daily	Hr.	Daily	Bare Costs	Incl. O&P
1 Tugboat Captain	$42.95	$343.60	$64.30	$514.40	$38.02	$57.67
1 Tugboat Hand	33.10	264.80	51.05	408.40		
1 Tugboat, 250 H.P.		840.80		924.88	52.55	57.81
16 L.H., Daily Totals		$1449.20		$1847.68	$90.58	$115.48

Crew B-84	Hr.	Daily	Hr.	Daily	Bare Costs	Incl. O&P
1 Equip. Oper. (med.)	$42.95	$343.60	$64.30	$514.40	$42.95	$64.30
1 Rotary Mower/Tractor		302.00		332.20	37.75	41.52
8 L.H., Daily Totals		$645.60		$846.60	$80.70	$105.83

Crew B-85	Hr.	Daily	Hr.	Daily	Bare Costs	Incl. O&P
3 Laborers	$33.10	$794.40	$51.05	$1225.20	$35.08	$53.60
1 Equip. Oper. (med.)	42.95	343.60	64.30	514.40		
1 Truck Driver (heavy)	33.15	265.20	50.55	404.40		
1 Aerial Lift Truck, 80'		665.40		731.94		
1 Brush Chipper, 12", 130 H.P.		253.20		278.52		
1 Pruning Saw, Rotary		6.75		7.42	23.13	25.45
40 L.H., Daily Totals		$2328.55		$3161.89	$58.21	$79.05

Crew B-86	Hr.	Daily	Hr.	Daily	Bare Costs	Incl. O&P
1 Equip. Oper. (med.)	$42.95	$343.60	$64.30	$514.40	$42.95	$64.30
1 Stump Chipper, S.P.		190.90		209.99	23.86	26.25
8 L.H., Daily Totals		$534.50		$724.39	$66.81	$90.55

Crew B-86A	Hr.	Daily	Hr.	Daily	Bare Costs	Incl. O&P
1 Equip. Oper. (med.)	$42.95	$343.60	$64.30	$514.40	$42.95	$64.30
1 Grader, 30,000 Lbs.		622.20		684.42	77.78	85.55
8 L.H., Daily Totals		$965.80		$1198.82	$120.72	$149.85

Crew B-86B	Hr.	Daily	Hr.	Daily	Bare Costs	Incl. O&P
1 Equip. Oper. (med.)	$42.95	$343.60	$64.30	$514.40	$42.95	$64.30
1 Dozer, 200 H.P.		1192.00		1311.20	149.00	163.90
8 L.H., Daily Totals		$1535.60		$1825.60	$191.95	$228.20

Crew B-87	Hr.	Daily	Hr.	Daily	Bare Costs	Incl. O&P
1 Laborer	$33.10	$264.80	$51.05	$408.40	$40.98	$61.65
4 Equip. Oper. (med.)	42.95	1374.40	64.30	2057.60		
2 Feller Bunchers, 100 H.P.		1219.20		1341.12		
1 Log Chipper, 22" Tree		689.60		758.56		
1 Dozer, 105 H.P.		660.40		726.44		
1 Chain Saw, gas, 36" Long		39.20		43.12	65.21	71.73
40 L.H., Daily Totals		$4247.60		$5335.24	$106.19	$133.38

Crew B-88	Hr.	Daily	Hr.	Daily	Bare Costs	Incl. O&P
1 Laborer	$33.10	$264.80	$51.05	$408.40	$41.54	$62.41
6 Equip. Oper. (med.)	42.95	2061.60	64.30	3086.40		
2 Feller Bunchers, 100 H.P.		1219.20		1341.12		
1 Log Chipper, 22" Tree		689.60		758.56		
2 Log Skidders, 50 H.P.		1814.40		1995.84		
1 Dozer, 105 H.P.		660.40		726.44		
1 Chain Saw, gas, 36" Long		39.20		43.12	78.98	86.88
56 L.H., Daily Totals		$6749.20		$8359.88	$120.52	$149.28

Crew B-89	Hr.	Daily	Hr.	Daily	Bare Costs	Incl. O&P
1 Equip. Oper. (light)	$41.30	$330.40	$61.85	$494.80	$36.77	$55.52
1 Truck Driver (light)	32.25	258.00	49.20	393.60		
1 Flatbed Truck, Gas, 3 Ton		296.00		325.60		
1 Concrete Saw		161.40		177.54		
1 Water Tank, 65 Gal.		18.45		20.30	29.74	32.71
16 L.H., Daily Totals		$1064.25		$1411.84	$66.52	$88.24

Crew B-89A	Hr.	Daily	Hr.	Daily	Bare Costs	Incl. O&P
1 Skilled Worker	$42.60	$340.80	$65.50	$524.00	$37.85	$58.27
1 Laborer	33.10	264.80	51.05	408.40		
1 Core Drill (large)		113.80		125.18	7.11	7.82
16 L.H., Daily Totals		$719.40		$1057.58	$44.96	$66.10

Crew B-89B	Hr.	Daily	Hr.	Daily	Bare Costs	Incl. O&P
1 Equip. Oper. (light)	$41.30	$330.40	$61.85	$494.80	$36.77	$55.52
1 Truck Driver, Light	32.25	258.00	49.20	393.60		
1 Wall Saw, Hydraulic, 10 H.P.		110.40		121.44		
1 Generator, Diesel, 100 kW		418.60		460.46		
1 Water Tank, 65 Gal.		18.45		20.30		
1 Flatbed Truck, Gas, 3 Ton		296.00		325.60	52.72	57.99
16 L.H., Daily Totals		$1431.85		$1816.19	$89.49	$113.51

Crew B-90	Hr.	Daily	Hr.	Daily	Bare Costs	Incl. O&P
1 Labor Foreman (outside)	$35.10	$280.80	$54.10	$432.80	$35.41	$54.01
3 Laborers	33.10	794.40	51.05	1225.20		
2 Equip. Oper. (light)	41.30	660.80	61.85	989.60		
2 Truck Drivers (heavy)	33.15	530.40	50.55	808.80		
1 Road Mixer, 310 H.P.		1955.00		2150.50		
1 Dist. Truck, 2000 Gal.		278.20		306.02	34.89	38.38
64 L.H., Daily Totals		$4499.60		$5912.92	$70.31	$92.39

Crew B-90A	Hr.	Daily	Hr.	Daily	Bare Costs	Incl. O&P
1 Labor Foreman	$35.10	$280.80	$54.10	$432.80	$39.01	$59.06
2 Laborers	33.10	529.60	51.05	816.80		
4 Equip. Oper. (med.)	42.95	1374.40	64.30	2057.60		
2 Graders, 30,000 Lbs.		1244.40		1368.84		
1 Tandem Roller, 10 Ton		235.20		258.72		
1 Roller, Pneum. Whl, 12 Ton		337.20		370.92	32.44	35.69
56 L.H., Daily Totals		$4001.60		$5305.68	$71.46	$94.74

Crew B-90B	Hr.	Daily	Hr.	Daily	Bare Costs	Incl. O&P
1 Labor Foreman	$35.10	$280.80	$54.10	$432.80	$38.36	$58.18
2 Laborers	33.10	529.60	51.05	816.80		
3 Equip. Oper. (med.)	42.95	1030.80	64.30	1543.20		
1 Tandem Roller, 10 Ton		235.20		258.72		
1 Roller, Pneum. Whl, 12 Ton		337.20		370.92		
1 Road Mixer, 310 H.P.		1955.00		2150.50	52.65	57.92
48 L.H., Daily Totals		$4368.60		$5572.94	$91.01	$116.10

Crew B-91	Hr.	Daily	Hr.	Daily	Bare Costs	Incl. O&P
1 Labor Foreman (outside)	$35.10	$280.80	$54.10	$432.80	$38.28	$57.99
2 Laborers	33.10	529.60	51.05	816.80		
4 Equip. Oper. (med.)	42.95	1374.40	64.30	2057.60		
1 Truck Driver (heavy)	33.15	265.20	50.55	404.40		
1 Dist. Tanker, 3000 Gallon		307.00		337.70		
1 Truck Tractor, 6x4, 380 H.P.		592.40		651.64		
1 Aggreg. Spreader, S.P.		820.80		902.88		
1 Roller, Pneum. Whl, 12 Ton		337.20		370.92		
1 Tandem Roller, 10 Ton		235.20		258.72	35.82	39.40
64 L.H., Daily Totals		$4742.60		$6233.46	$74.10	$97.40

Crew B-92	Hr.	Daily	Hr.	Daily	Bare Costs	Incl. O&P
1 Labor Foreman (outside)	$35.10	$280.80	$54.10	$432.80	$33.60	$51.81
3 Laborers	33.10	794.40	51.05	1225.20		
1 Crack Cleaner, 25 H.P.		60.40		66.44		
1 Air Compressor, 60 cfm		138.60		152.46		
1 Tar Kettle, T.M.		71.40		78.54		
1 Flatbed Truck, Gas, 3 Ton		296.00		325.60	17.70	19.47
32 L.H., Daily Totals		$1641.60		$2281.04	$51.30	$71.28

Crew B-93	Hr.	Daily	Hr.	Daily	Bare Costs	Incl. O&P
1 Equip. Oper. (med.)	$42.95	$343.60	$64.30	$514.40	$42.95	$64.30
1 Feller Buncher, 100 H.P.		609.60		670.56	76.20	83.82
8 L.H., Daily Totals		$953.20		$1184.96	$119.15	$148.12

Crew B-94A	Hr.	Daily	Hr.	Daily	Bare Costs	Incl. O&P
1 Laborer	$33.10	$264.80	$51.05	$408.40	$33.10	$51.05
1 Diaphragm Water Pump, 2"		68.00		74.80		
1 -20' Suction Hose, 2"		1.95		2.15		
2 -50' Discharge Hoses, 2"		1.80		1.98	8.97	9.87
8 L.H., Daily Totals		$336.55		$487.32	$42.07	$60.92

Crew B-94B	Hr.	Daily	Hr.	Daily	Bare Costs	Incl. O&P
1 Laborer	$33.10	$264.80	$51.05	$408.40	$33.10	$51.05
1 Diaphragm Water Pump, 4"		90.80		99.88		
1 -20' Suction Hose, 4"		3.45		3.79		
2 -50' Discharge Hoses, 4"		4.70		5.17	12.37	13.61
8 L.H., Daily Totals		$363.75		$517.25	$45.47	$64.66

Crew B-94C	Hr.	Daily	Hr.	Daily	Bare Costs	Incl. O&P
1 Laborer	$33.10	$264.80	$51.05	$408.40	$33.10	$51.05
1 Centrifugal Water Pump, 3"		74.40		81.84		
1 -20' Suction Hose, 3"		3.05		3.36		
2 -50' Discharge Hoses, 3"		3.50		3.85	10.12	11.13
8 L.H., Daily Totals		$345.75		$497.44	$43.22	$62.18

Crew B-94D	Hr.	Daily	Hr.	Daily	Bare Costs	Incl. O&P
1 Laborer	$33.10	$264.80	$51.05	$408.40	$33.10	$51.05
1 Centr. Water Pump, 6"		340.00		374.00		
1 -20' Suction Hose, 6"		11.90		13.09		
2 -50' Discharge Hoses, 6"		12.60		13.86	45.56	50.12
8 L.H., Daily Totals		$629.30		$809.35	$78.66	$101.17

Crew No.	Bare Costs		Incl. Subs O&P		Cost Per Labor-Hour	

Crew C-1	Hr.	Daily	Hr.	Daily	Bare Costs	Incl. O&P
3 Carpenters	$41.55	$997.20	$64.05	$1537.20	$39.44	$60.80
1 Laborer	33.10	264.80	51.05	408.40		
32 L.H., Daily Totals		$1262.00		$1945.60	$39.44	$60.80

Crew C-2	Hr.	Daily	Hr.	Daily	Bare Costs	Incl. O&P
1 Carpenter Foreman (out)	$43.55	$348.40	$67.15	$537.20	$40.48	$62.40
4 Carpenters	41.55	1329.60	64.05	2049.60		
1 Laborer	33.10	264.80	51.05	408.40		
48 L.H., Daily Totals		$1942.80		$2995.20	$40.48	$62.40

Crew C-2A	Hr.	Daily	Hr.	Daily	Bare Costs	Incl. O&P
1 Carpenter Foreman (out)	$43.55	$348.40	$67.15	$537.20	$40.17	$61.38
3 Carpenters	41.55	997.20	64.05	1537.20		
1 Cement Finisher	39.70	317.60	57.95	463.60		
1 Laborer	33.10	264.80	51.05	408.40		
48 L.H., Daily Totals		$1928.00		$2946.40	$40.17	$61.38

Crew C-3	Hr.	Daily	Hr.	Daily	Bare Costs	Incl. O&P
1 Rodman Foreman	$48.80	$390.40	$78.60	$628.80	$42.94	$68.02
4 Rodmen (reinf.)	46.80	1497.60	75.40	2412.80		
1 Equip. Oper. (light)	41.30	330.40	61.85	494.80		
2 Laborers	33.10	529.60	51.05	816.80		
3 Stressing Equipment		28.80		31.68		
.5 Grouting Equipment		80.00		88.00	1.70	1.87
64 L.H., Daily Totals		$2856.80		$4472.88	$44.64	$69.89

Crew C-4	Hr.	Daily	Hr.	Daily	Bare Costs	Incl. O&P
1 Rodman Foreman	$48.80	$390.40	$78.60	$628.80	$47.30	$76.20
3 Rodmen (reinf.)	46.80	1123.20	75.40	1809.60		
3 Stressing Equipment		28.80		31.68	.90	.99
32 L.H., Daily Totals		$1542.40		$2470.08	$48.20	$77.19

Crew C-4A	Hr.	Daily	Hr.	Daily	Bare Costs	Incl. O&P
2 Rodmen (reinf.)	$46.80	$748.80	$75.40	$1206.40	$46.80	$75.40
4 Stressing Equipment		38.40		42.24	2.40	2.64
16 L.H., Daily Totals		$787.20		$1248.64	$49.20	$78.04

Crew C-5	Hr.	Daily	Hr.	Daily	Bare Costs	Incl. O&P
1 Rodman Foreman	$48.80	$390.40	$78.60	$628.80	$45.53	$72.00
4 Rodmen (reinf.)	46.80	1497.60	75.40	2412.80		
1 Equip. Oper. (crane)	44.40	355.20	66.45	531.60		
1 Equip. Oper. Oiler	38.30	306.40	57.35	458.80		
1 Hyd. Crane, 25 Ton		747.60		822.36	13.35	14.69
56 L.H., Daily Totals		$3297.20		$4854.36	$58.88	$86.69

Crew C-6	Hr.	Daily	Hr.	Daily	Bare Costs	Incl. O&P
1 Labor Foreman (outside)	$35.10	$280.80	$54.10	$432.80	$34.53	$52.71
4 Laborers	33.10	1059.20	51.05	1633.60		
1 Cement Finisher	39.70	317.60	57.95	463.60		
2 Gas Engine Vibrators		60.80		66.88	1.27	1.39
48 L.H., Daily Totals		$1718.40		$2596.88	$35.80	$54.10

Crew C-7	Hr.	Daily	Hr.	Daily	Bare Costs	Incl. O&P
1 Labor Foreman (outside)	$35.10	$280.80	$54.10	$432.80	$35.73	$54.33
5 Laborers	33.10	1324.00	51.05	2042.00		
1 Cement Finisher	39.70	317.60	57.95	463.60		
1 Equip. Oper. (med.)	42.95	343.60	64.30	514.40		
1 Equip. Oper. (oiler)	38.30	306.40	57.35	458.80		
2 Gas Engine Vibrators		60.80		66.88		
1 Concrete Bucket, 1 C.Y.		20.60		22.66		
1 Hyd. Crane, 55 Ton		1209.00		1329.90	17.92	19.71
72 L.H., Daily Totals		$3862.80		$5331.04	$53.65	$74.04

Crew C-7A	Hr.	Daily	Hr.	Daily	Bare Costs	Incl. O&P
1 Labor Foreman (outside)	$35.10	$280.80	$54.10	$432.80	$33.36	$51.31
5 Laborers	33.10	1324.00	51.05	2042.00		
2 Truck Drivers (Heavy)	33.15	530.40	50.55	808.80		
2 Conc. Transit Mixers		2016.00		2217.60	31.50	34.65
64 L.H., Daily Totals		$4151.20		$5501.20	$64.86	$85.96

Crew C-7B	Hr.	Daily	Hr.	Daily	Bare Costs	Incl. O&P
1 Labor Foreman (outside)	$35.10	$280.80	$54.10	$432.80	$35.41	$54.14
5 Laborers	33.10	1324.00	51.05	2042.00		
1 Equipment Oper. (crane)	44.40	355.20	66.45	531.60		
1 Equipment Oiler	38.30	306.40	57.35	458.80		
1 Conc. Bucket, 2 C.Y.		32.40		35.64		
1 Lattice Boom Crane, 165 Ton		2039.00		2242.90	32.37	35.60
64 L.H., Daily Totals		$4337.80		$5743.74	$67.78	$89.75

Crew C-7C	Hr.	Daily	Hr.	Daily	Bare Costs	Incl. O&P
1 Labor Foreman (outside)	$35.10	$280.80	$54.10	$432.80	$35.81	$54.74
5 Laborers	33.10	1324.00	51.05	2042.00		
2 Equipment Operators (med.)	42.95	687.20	64.30	1028.80		
2 F.E. Loaders, W.M., 4 C.Y.		1255.60		1381.16	19.62	21.58
64 L.H., Daily Totals		$3547.60		$4884.76	$55.43	$76.32

Crew C-7D	Hr.	Daily	Hr.	Daily	Bare Costs	Incl. O&P
1 Labor Foreman (outside)	$35.10	$280.80	$54.10	$432.80	$34.79	$53.38
5 Laborers	33.10	1324.00	51.05	2042.00		
1 Equip. Oper. (med.)	42.95	343.60	64.30	514.40		
1 Concrete Conveyer		194.60		214.06	3.48	3.82
56 L.H., Daily Totals		$2143.00		$3203.26	$38.27	$57.20

Crew C-8	Hr.	Daily	Hr.	Daily	Bare Costs	Incl. O&P
1 Labor Foreman (outside)	$35.10	$280.80	$54.10	$432.80	$36.68	$55.35
3 Laborers	33.10	794.40	51.05	1225.20		
2 Cement Finishers	39.70	635.20	57.95	927.20		
1 Equip. Oper. (med.)	42.95	343.60	64.30	514.40		
1 Concrete Pump (small)		738.80		812.68	13.19	14.51
56 L.H., Daily Totals		$2792.80		$3912.28	$49.87	$69.86

Crew C-8A	Hr.	Daily	Hr.	Daily	Bare Costs	Incl. O&P
1 Labor Foreman (outside)	$35.10	$280.80	$54.10	$432.80	$35.63	$53.86
3 Laborers	33.10	794.40	51.05	1225.20		
2 Cement Finishers	39.70	635.20	57.95	927.20		
48 L.H., Daily Totals		$1710.40		$2585.20	$35.63	$53.86

Crew C-8B	Bare Costs Hr.	Bare Costs Daily	Incl. Subs O&P Hr.	Incl. Subs O&P Daily	Cost Per Labor-Hour Bare Costs	Cost Per Labor-Hour Incl. O&P
1 Labor Foreman (outside)	$35.10	$280.80	$54.10	$432.80	$35.47	$54.31
3 Laborers	33.10	794.40	51.05	1225.20		
1 Equip. Oper. (med.)	42.95	343.60	64.30	514.40		
1 Vibrating Power Screed		65.15		71.67		
1 Roller, Vibratory, 25 Ton		633.20		696.52		
1 Dozer, 200 H.P.		1192.00		1311.20	47.26	51.98
40 L.H., Daily Totals		$3309.15		$4251.78	$82.73	$106.29

Crew C-8C	Hr.	Daily	Hr.	Daily	Bare Costs	Incl. O&P
1 Labor Foreman (outside)	$35.10	$280.80	$54.10	$432.80	$36.17	$54.92
3 Laborers	33.10	794.40	51.05	1225.20		
1 Cement Finisher	39.70	317.60	57.95	463.60		
1 Equip. Oper. (med.)	42.95	343.60	64.30	514.40		
1 Shotcrete Rig, 12 C.Y./hr		249.20		274.12	5.19	5.71
48 L.H., Daily Totals		$1985.60		$2910.12	$41.37	$60.63

Crew C-8D	Hr.	Daily	Hr.	Daily	Bare Costs	Incl. O&P
1 Labor Foreman (outside)	$35.10	$280.80	$54.10	$432.80	$37.30	$56.24
1 Laborer	33.10	264.80	51.05	408.40		
1 Cement Finisher	39.70	317.60	57.95	463.60		
1 Equipment Oper. (light)	41.30	330.40	61.85	494.80		
1 Air Compressor, 250 cfm		193.40		212.74		
2 -50' Air Hoses, 1"		8.20		9.02	6.30	6.93
32 L.H., Daily Totals		$1395.20		$2021.36	$43.60	$63.17

Crew C-8E	Hr.	Daily	Hr.	Daily	Bare Costs	Incl. O&P
1 Labor Foreman (outside)	$35.10	$280.80	$54.10	$432.80	$37.30	$56.24
1 Laborer	33.10	264.80	51.05	408.40		
1 Cement Finisher	39.70	317.60	57.95	463.60		
1 Equipment Oper. (light)	41.30	330.40	61.85	494.80		
1 Air Compressor, 250 cfm		193.40		212.74		
2 -50' Air Hoses, 1"		8.20		9.02		
1 Concrete Pump (small)		738.80		812.68	29.39	32.33
32 L.H., Daily Totals		$2134.00		$2834.04	$66.69	$88.56

Crew C-10	Hr.	Daily	Hr.	Daily	Bare Costs	Incl. O&P
1 Laborer	$33.10	$264.80	$51.05	$408.40	$37.50	$55.65
2 Cement Finishers	39.70	635.20	57.95	927.20		
24 L.H., Daily Totals		$900.00		$1335.60	$37.50	$55.65

Crew C-10B	Hr.	Daily	Hr.	Daily	Bare Costs	Incl. O&P
3 Laborers	$33.10	$794.40	$51.05	$1225.20	$35.74	$53.81
2 Cement Finishers	39.70	635.20	57.95	927.20		
1 Concrete Mixer, 10 C.F.		168.40		185.24		
2 Trowels, 48" Walk-Behind		98.00		107.80	6.66	7.33
40 L.H., Daily Totals		$1696.00		$2445.44	$42.40	$61.14

Crew C-10C	Hr.	Daily	Hr.	Daily	Bare Costs	Incl. O&P
1 Laborer	$33.10	$264.80	$51.05	$408.40	$37.50	$55.65
2 Cement Finishers	39.70	635.20	57.95	927.20		
1 Trowel, 48" Walk-Behind		49.00		53.90	2.04	2.25
24 L.H., Daily Totals		$949.00		$1389.50	$39.54	$57.90

Crew C-10D	Hr.	Daily	Hr.	Daily	Bare Costs	Incl. O&P
1 Laborer	$33.10	$264.80	$51.05	$408.40	$37.50	$55.65
2 Cement Finishers	39.70	635.20	57.95	927.20		
1 Vibrating Power Screed		65.15		71.67		
1 Trowel, 48" Walk-Behind		49.00		53.90	4.76	5.23
24 L.H., Daily Totals		$1014.15		$1461.17	$42.26	$60.88

Crew C-10E	Hr.	Daily	Hr.	Daily	Bare Costs	Incl. O&P
1 Laborer	$33.10	$264.80	$51.05	$408.40	$37.50	$55.65
2 Cement Finishers	39.70	635.20	57.95	927.20		
1 Vibrating Power Screed		65.15		71.67		
1 Cement Trowel, 96" Ride-On		185.80		204.38	10.46	11.50
24 L.H., Daily Totals		$1150.95		$1611.65	$47.96	$67.15

Crew C-11	Hr.	Daily	Hr.	Daily	Bare Costs	Incl. O&P
1 Struc. Steel Foreman	$48.90	$391.20	$85.95	$687.60	$45.89	$78.27
6 Struc. Steel Workers	46.90	2251.20	82.45	3957.60		
1 Equip. Oper. (crane)	44.40	355.20	66.45	531.60		
1 Equip. Oper. Oiler	38.30	306.40	57.35	458.80		
1 Lattice Boom Crane, 150 Ton		2003.00		2203.30	27.82	30.60
72 L.H., Daily Totals		$5307.00		$7838.90	$73.71	$108.87

Crew C-12	Hr.	Daily	Hr.	Daily	Bare Costs	Incl. O&P
1 Carpenter Foreman (out)	$43.55	$348.40	$67.15	$537.20	$40.95	$62.80
3 Carpenters	41.55	997.20	64.05	1537.20		
1 Laborer	33.10	264.80	51.05	408.40		
1 Equip. Oper. (crane)	44.40	355.20	66.45	531.60		
1 Hyd. Crane, 12 Ton		655.60		721.16	13.66	15.02
48 L.H., Daily Totals		$2621.20		$3735.56	$54.61	$77.82

Crew C-13	Hr.	Daily	Hr.	Daily	Bare Costs	Incl. O&P
1 Struc. Steel Worker	$46.90	$375.20	$82.45	$659.60	$45.12	$76.32
1 Welder	46.90	375.20	82.45	659.60		
1 Carpenter	41.55	332.40	64.05	512.40		
1 Welder, gas engine, 300 amp		145.80		160.38	6.08	6.68
24 L.H., Daily Totals		$1228.60		$1991.98	$51.19	$83.00

Crew C-14	Hr.	Daily	Hr.	Daily	Bare Costs	Incl. O&P
1 Carpenter Foreman (out)	$43.55	$348.40	$67.15	$537.20	$40.72	$62.94
5 Carpenters	41.55	1662.00	64.05	2562.00		
4 Laborers	33.10	1059.20	51.05	1633.60		
4 Rodmen (reinf.)	46.80	1497.60	75.40	2412.80		
2 Cement Finishers	39.70	635.20	57.95	927.20		
1 Equip. Oper. (crane)	44.40	355.20	66.45	531.60		
1 Equip. Oper. Oiler	38.30	306.40	57.35	458.80		
1 Hyd. Crane, 80 Ton		1545.00		1699.50	10.73	11.80
144 L.H., Daily Totals		$7409.00		$10762.70	$51.45	$74.74

Crew C-14A	Hr.	Daily	Hr.	Daily	Bare Costs	Incl. O&P
1 Carpenter Foreman (out)	$43.55	$348.40	$67.15	$537.20	$41.78	$64.72
16 Carpenters	41.55	5318.40	64.05	8198.40		
4 Rodmen (reinf.)	46.80	1497.60	75.40	2412.80		
2 Laborers	33.10	529.60	51.05	816.80		
1 Cement Finisher	39.70	317.60	57.95	463.60		
1 Equip. Oper. (med.)	42.95	343.60	64.30	514.40		
1 Gas Engine Vibrator		30.40		33.44		
1 Concrete Pump (small)		738.80		812.68	3.85	4.23
200 L.H., Daily Totals		$9124.40		$13789.32	$45.62	$68.95

Crew C-14B	Hr.	Daily	Hr.	Daily	Bare Costs	Incl. O&P
1 Carpenter Foreman (out)	$43.55	$348.40	$67.15	$537.20	$41.70	$64.46
16 Carpenters	41.55	5318.40	64.05	8198.40		
4 Rodmen (reinf.)	46.80	1497.60	75.40	2412.80		
2 Laborers	33.10	529.60	51.05	816.80		
2 Cement Finishers	39.70	635.20	57.95	927.20		
1 Equip. Oper. (med.)	42.95	343.60	64.30	514.40		
1 Gas Engine Vibrator		30.40		33.44		
1 Concrete Pump (small)		738.80		812.68	3.70	4.07
208 L.H., Daily Totals		$9442.00		$14252.92	$45.39	$68.52

Crew No.	Bare Costs		Incl. Subs O&P		Cost Per Labor-Hour	
Crew C-14C	**Hr.**	**Daily**	**Hr.**	**Daily**	**Bare Costs**	**Incl. O&P**
1 Carpenter Foreman (out)	$43.55	$348.40	$67.15	$537.20	$39.90	$61.74
6 Carpenters	41.55	1994.40	64.05	3074.40		
2 Rodmen (reinf.)	46.80	748.80	75.40	1206.40		
4 Laborers	33.10	1059.20	51.05	1633.60		
1 Cement Finisher	39.70	317.60	57.95	463.60		
1 Gas Engine Vibrator		30.40		33.44	.27	.30
112 L.H., Daily Totals		$4498.80		$6948.64	$40.17	$62.04
Crew C-14D	**Hr.**	**Daily**	**Hr.**	**Daily**	**Bare Costs**	**Incl. O&P**
1 Carpenter Foreman (out)	$43.55	$348.40	$67.15	$537.20	$41.36	$63.81
18 Carpenters	41.55	5983.20	64.05	9223.20		
2 Rodmen (reinf.)	46.80	748.80	75.40	1206.40		
2 Laborers	33.10	529.60	51.05	816.80		
1 Cement Finisher	39.70	317.60	57.95	463.60		
1 Equip. Oper. (med.)	42.95	343.60	64.30	514.40		
1 Gas Engine Vibrator		30.40		33.44		
1 Concrete Pump (small)		738.80		812.68	3.85	4.23
200 L.H., Daily Totals		$9040.40		$13607.72	$45.20	$68.04
Crew C-14E	**Hr.**	**Daily**	**Hr.**	**Daily**	**Bare Costs**	**Incl. O&P**
1 Carpenter Foreman (out)	$43.55	$348.40	$67.15	$537.20	$41.17	$64.36
2 Carpenters	41.55	664.80	64.05	1024.80		
4 Rodmen (reinf.)	46.80	1497.60	75.40	2412.80		
3 Laborers	33.10	794.40	51.05	1225.20		
1 Cement Finisher	39.70	317.60	57.95	463.60		
1 Gas Engine Vibrator		30.40		33.44	.35	.38
88 L.H., Daily Totals		$3653.20		$5697.04	$41.51	$64.74
Crew C-14F	**Hr.**	**Daily**	**Hr.**	**Daily**	**Bare Costs**	**Incl. O&P**
1 Laborer Foreman (out)	$35.10	$280.80	$54.10	$432.80	$37.72	$55.99
2 Laborers	33.10	529.60	51.05	816.80		
6 Cement Finishers	39.70	1905.60	57.95	2781.60		
1 Gas Engine Vibrator		30.40		33.44	.42	.46
72 L.H., Daily Totals		$2746.40		$4064.64	$38.14	$56.45
Crew C-14G	**Hr.**	**Daily**	**Hr.**	**Daily**	**Bare Costs**	**Incl. O&P**
1 Laborer Foreman (out)	$35.10	$280.80	$54.10	$432.80	$37.16	$55.43
2 Laborers	33.10	529.60	51.05	816.80		
4 Cement Finishers	39.70	1270.40	57.95	1854.40		
1 Gas Engine Vibrator		30.40		33.44	.54	.60
56 L.H., Daily Totals		$2111.20		$3137.44	$37.70	$56.03
Crew C-14H	**Hr.**	**Daily**	**Hr.**	**Daily**	**Bare Costs**	**Incl. O&P**
1 Carpenter Foreman (out)	$43.55	$348.40	$67.15	$537.20	$41.04	$63.27
2 Carpenters	41.55	664.80	64.05	1024.80		
1 Rodman (reinf.)	46.80	374.40	75.40	603.20		
1 Laborer	33.10	264.80	51.05	408.40		
1 Cement Finisher	39.70	317.60	57.95	463.60		
1 Gas Engine Vibrator		30.40		33.44	.63	.70
48 L.H., Daily Totals		$2000.40		$3070.64	$41.67	$63.97
Crew C-14L	**Hr.**	**Daily**	**Hr.**	**Daily**	**Bare Costs**	**Incl. O&P**
1 Carpenter Foreman (out)	$43.55	$348.40	$67.15	$537.20	$38.75	$59.47
6 Carpenters	41.55	1994.40	64.05	3074.40		
4 Laborers	33.10	1059.20	51.05	1633.60		
1 Cement Finisher	39.70	317.60	57.95	463.60		
1 Gas Engine Vibrator		30.40		33.44	.32	.35
96 L.H., Daily Totals		$3750.00		$5742.24	$39.06	$59.81

Crew No.	Bare Costs		Incl. Subs O&P		Cost Per Labor-Hour	
Crew C-15	**Hr.**	**Daily**	**Hr.**	**Daily**	**Bare Costs**	**Incl. O&P**
1 Carpenter Foreman (out)	$43.55	$348.40	$67.15	$537.20	$39.13	$59.97
2 Carpenters	41.55	664.80	64.05	1024.80		
3 Laborers	33.10	794.40	51.05	1225.20		
2 Cement Finishers	39.70	635.20	57.95	927.20		
1 Rodman (reinf.)	46.80	374.40	75.40	603.20		
72 L.H., Daily Totals		$2817.20		$4317.60	$39.13	$59.97
Crew C-16	**Hr.**	**Daily**	**Hr.**	**Daily**	**Bare Costs**	**Incl. O&P**
1 Labor Foreman (outside)	$35.10	$280.80	$54.10	$432.80	$38.93	$59.81
3 Laborers	33.10	794.40	51.05	1225.20		
2 Cement Finishers	39.70	635.20	57.95	927.20		
1 Equip. Oper. (med.)	42.95	343.60	64.30	514.40		
2 Rodmen (reinf.)	46.80	748.80	75.40	1206.40		
1 Concrete Pump (small)		738.80		812.68	10.26	11.29
72 L.H., Daily Totals		$3541.60		$5118.68	$49.19	$71.09
Crew C-17	**Hr.**	**Daily**	**Hr.**	**Daily**	**Bare Costs**	**Incl. O&P**
2 Skilled Worker Foremen	$44.60	$713.60	$68.55	$1096.80	$43.00	$66.11
8 Skilled Workers	42.60	2726.40	65.50	4192.00		
80 L.H., Daily Totals		$3440.00		$5288.80	$43.00	$66.11
Crew C-17A	**Hr.**	**Daily**	**Hr.**	**Daily**	**Bare Costs**	**Incl. O&P**
2 Skilled Worker Foremen	$44.60	$713.60	$68.55	$1096.80	$43.02	$66.11
8 Skilled Workers	42.60	2726.40	65.50	4192.00		
.125 Equip. Oper. (crane)	44.40	44.40	66.45	66.45		
.125 Hyd. Crane, 80 Ton		193.13		212.44	2.38	2.62
81 L.H., Daily Totals		$3677.53		$5567.69	$45.40	$68.74
Crew C-17B	**Hr.**	**Daily**	**Hr.**	**Daily**	**Bare Costs**	**Incl. O&P**
2 Skilled Worker Foremen	$44.60	$713.60	$68.55	$1096.80	$43.03	$66.12
8 Skilled Workers	42.60	2726.40	65.50	4192.00		
.25 Equip. Oper. (crane)	44.40	88.80	66.45	132.90		
.25 Hyd. Crane, 80 Ton		386.25		424.88		
.25 Trowel, 48" Walk-Behind		12.25		13.48	4.86	5.35
82 L.H., Daily Totals		$3927.30		$5860.05	$47.89	$71.46
Crew C-17C	**Hr.**	**Daily**	**Hr.**	**Daily**	**Bare Costs**	**Incl. O&P**
2 Skilled Worker Foremen	$44.60	$713.60	$68.55	$1096.80	$43.05	$66.12
8 Skilled Workers	42.60	2726.40	65.50	4192.00		
.375 Equip. Oper. (crane)	44.40	133.20	66.45	199.35		
.375 Hyd. Crane, 80 Ton		579.38		637.31	6.98	7.68
83 L.H., Daily Totals		$4152.57		$6125.46	$50.03	$73.80
Crew C-17D	**Hr.**	**Daily**	**Hr.**	**Daily**	**Bare Costs**	**Incl. O&P**
2 Skilled Worker Foremen	$44.60	$713.60	$68.55	$1096.80	$43.07	$66.13
8 Skilled Workers	42.60	2726.40	65.50	4192.00		
.5 Equip. Oper. (crane)	44.40	177.60	66.45	265.80		
.5 Hyd. Crane, 80 Ton		772.50		849.75	9.20	10.12
84 L.H., Daily Totals		$4390.10		$6404.35	$52.26	$76.24
Crew C-17E	**Hr.**	**Daily**	**Hr.**	**Daily**	**Bare Costs**	**Incl. O&P**
2 Skilled Worker Foremen	$44.60	$713.60	$68.55	$1096.80	$43.00	$66.11
8 Skilled Workers	42.60	2726.40	65.50	4192.00		
1 Hyd. Jack with Rods		87.35		96.08	1.09	1.20
80 L.H., Daily Totals		$3527.35		$5384.89	$44.09	$67.31

Crew C-18	Bare Costs Hr.	Bare Costs Daily	Incl. Subs O&P Hr.	Incl. Subs O&P Daily	Bare Costs	Incl. O&P
.125 Labor Foreman (out)	$35.10	$35.10	$54.10	$54.10	$33.32	$51.39
1 Laborer	33.10	264.80	51.05	408.40		
1 Concrete Cart, 10 C.F.		58.40		64.24	6.49	7.14
9 L.H., Daily Totals		$358.30		$526.74	$39.81	$58.53

Crew C-19	Hr.	Daily	Hr.	Daily	Bare Costs	Incl. O&P
.125 Labor Foreman (out)	$35.10	$35.10	$54.10	$54.10	$33.32	$51.39
1 Laborer	33.10	264.80	51.05	408.40		
1 Concrete Cart, 18 C.F.		94.00		103.40	10.44	11.49
9 L.H., Daily Totals		$393.90		$565.90	$43.77	$62.88

Crew C-20	Hr.	Daily	Hr.	Daily	Bare Costs	Incl. O&P
1 Labor Foreman (outside)	$35.10	$280.80	$54.10	$432.80	$35.41	$53.95
5 Laborers	33.10	1324.00	51.05	2042.00		
1 Cement Finisher	39.70	317.60	57.95	463.60		
1 Equip. Oper. (med.)	42.95	343.60	64.30	514.40		
2 Gas Engine Vibrators		60.80		66.88		
1 Concrete Pump (small)		738.80		812.68	12.49	13.74
64 L.H., Daily Totals		$3065.60		$4332.36	$47.90	$67.69

Crew C-21	Hr.	Daily	Hr.	Daily	Bare Costs	Incl. O&P
1 Labor Foreman (outside)	$35.10	$280.80	$54.10	$432.80	$35.41	$53.95
5 Laborers	33.10	1324.00	51.05	2042.00		
1 Cement Finisher	39.70	317.60	57.95	463.60		
1 Equip. Oper. (med.)	42.95	343.60	64.30	514.40		
2 Gas Engine Vibrators		60.80		66.88		
1 Concrete Conveyer		194.60		214.06	3.99	4.39
64 L.H., Daily Totals		$2521.40		$3733.74	$39.40	$58.34

Crew C-22	Hr.	Daily	Hr.	Daily	Bare Costs	Incl. O&P
1 Rodman Foreman	$48.80	$390.40	$78.60	$628.80	$46.92	$75.37
4 Rodmen (reinf.)	46.80	1497.60	75.40	2412.80		
.125 Equip. Oper. (crane)	44.40	44.40	66.45	66.45		
.125 Equip. Oper. Oiler	38.30	38.30	57.35	57.35		
.125 Hyd. Crane, 25 Ton		93.45		102.80	2.23	2.45
42 L.H., Daily Totals		$2064.15		$3268.20	$49.15	$77.81

Crew C-23	Hr.	Daily	Hr.	Daily	Bare Costs	Incl. O&P
2 Skilled Worker Foremen	$44.60	$713.60	$68.55	$1096.80	$42.75	$65.39
6 Skilled Workers	42.60	2044.80	65.50	3144.00		
1 Equip. Oper. (crane)	44.40	355.20	66.45	531.60		
1 Equip. Oper. Oiler	38.30	306.40	57.35	458.80		
1 Lattice Boom Crane, 90 Ton		1607.00		1767.70	20.09	22.10
80 L.H., Daily Totals		$5027.00		$6998.90	$62.84	$87.49

Crew C-23A	Hr.	Daily	Hr.	Daily	Bare Costs	Incl. O&P
1 Labor Foreman (outside)	$35.10	$280.80	$54.10	$432.80	$36.80	$56.00
2 Laborers	33.10	529.60	51.05	816.80		
1 Equip. Oper. (crane)	44.40	355.20	66.45	531.60		
1 Equip. Oper. Oiler	38.30	306.40	57.35	458.80		
1 Crawler Crane, 100 Ton		1732.00		1905.20		
3 Conc. Buckets, 8 C.Y.		603.60		663.96	58.39	64.23
40 L.H., Daily Totals		$3807.60		$4809.16	$95.19	$120.23

Crew C-24	Hr.	Daily	Hr.	Daily	Bare Costs	Incl. O&P
2 Skilled Worker Foremen	$44.60	$713.60	$68.55	$1096.80	$42.75	$65.39
6 Skilled Workers	42.60	2044.80	65.50	3144.00		
1 Equip. Oper. (crane)	44.40	355.20	66.45	531.60		
1 Equip. Oper. Oiler	38.30	306.40	57.35	458.80		
1 Lattice Boom Crane, 150 Ton		2003.00		2203.30	25.04	27.54
80 L.H., Daily Totals		$5423.00		$7434.50	$67.79	$92.93

Crew C-25	Hr.	Daily	Hr.	Daily	Bare Costs	Incl. O&P
2 Rodmen (reinf.)	$46.80	$748.80	$75.40	$1206.40	$36.50	$59.48
2 Rodmen Helpers	26.20	419.20	43.55	696.80		
32 L.H., Daily Totals		$1168.00		$1903.20	$36.50	$59.48

Crew C-27	Hr.	Daily	Hr.	Daily	Bare Costs	Incl. O&P
2 Cement Finishers	$39.70	$635.20	$57.95	$927.20	$39.70	$57.95
1 Concrete Saw		161.40		177.54	10.09	11.10
16 L.H., Daily Totals		$796.60		$1104.74	$49.79	$69.05

Crew C-28	Hr.	Daily	Hr.	Daily	Bare Costs	Incl. O&P
1 Cement Finisher	$39.70	$317.60	$57.95	$463.60	$39.70	$57.95
1 Portable Air Compressor, Gas		17.80		19.58	2.23	2.45
8 L.H., Daily Totals		$335.40		$483.18	$41.92	$60.40

Crew D-1	Hr.	Daily	Hr.	Daily	Bare Costs	Incl. O&P
1 Bricklayer	$41.75	$334.00	$62.95	$503.60	$37.70	$56.85
1 Bricklayer Helper	33.65	269.20	50.75	406.00		
16 L.H., Daily Totals		$603.20		$909.60	$37.70	$56.85

Crew D-2	Hr.	Daily	Hr.	Daily	Bare Costs	Incl. O&P
3 Bricklayers	$41.75	$1002.00	$62.95	$1510.80	$38.79	$58.61
2 Bricklayer Helpers	33.65	538.40	50.75	812.00		
.5 Carpenter	41.55	166.20	64.05	256.20		
44 L.H., Daily Totals		$1706.60		$2579.00	$38.79	$58.61

Crew D-3	Hr.	Daily	Hr.	Daily	Bare Costs	Incl. O&P
3 Bricklayers	$41.75	$1002.00	$62.95	$1510.80	$38.65	$58.35
2 Bricklayer Helpers	33.65	538.40	50.75	812.00		
.25 Carpenter	41.55	83.10	64.05	128.10		
42 L.H., Daily Totals		$1623.50		$2450.90	$38.65	$58.35

Crew D-4	Hr.	Daily	Hr.	Daily	Bare Costs	Incl. O&P
1 Bricklayer	$41.75	$334.00	$62.95	$503.60	$37.59	$56.58
2 Bricklayer Helpers	33.65	538.40	50.75	812.00		
1 Equip. Oper. (light)	41.30	330.40	61.85	494.80		
1 Grout Pump, 50 C.F./hr.		132.60		145.86	4.14	4.56
32 L.H., Daily Totals		$1335.40		$1956.26	$41.73	$61.13

Crew D-5	Hr.	Daily	Hr.	Daily	Bare Costs	Incl. O&P
1 Bricklayer	$41.75	$334.00	$62.95	$503.60	$41.75	$62.95
8 L.H., Daily Totals		$334.00		$503.60	$41.75	$62.95

Crew D-6	Hr.	Daily	Hr.	Daily	Bare Costs	Incl. O&P
3 Bricklayers	$41.75	$1002.00	$62.95	$1510.80	$37.85	$57.14
3 Bricklayer Helpers	33.65	807.60	50.75	1218.00		
.25 Carpenter	41.55	83.10	64.05	128.10		
50 L.H., Daily Totals		$1892.70		$2856.90	$37.85	$57.14

Crew D-7	Hr.	Daily	Hr.	Daily	Bare Costs	Incl. O&P
1 Tile Layer	$39.35	$314.80	$57.50	$460.00	$35.27	$51.55
1 Tile Layer Helper	31.20	249.60	45.60	364.80		
16 L.H., Daily Totals		$564.40		$824.80	$35.27	$51.55

Crew D-8	Hr.	Daily	Hr.	Daily	Bare Costs	Incl. O&P
3 Bricklayers	$41.75	$1002.00	$62.95	$1510.80	$38.51	$58.07
2 Bricklayer Helpers	33.65	538.40	50.75	812.00		
40 L.H., Daily Totals		$1540.40		$2322.80	$38.51	$58.07

Crew D-9	Hr.	Daily	Hr.	Daily	Bare Costs	Incl. O&P
3 Bricklayers	$41.75	$1002.00	$62.95	$1510.80	$37.70	$56.85
3 Bricklayer Helpers	33.65	807.60	50.75	1218.00		
48 L.H., Daily Totals		$1809.60		$2728.80	$37.70	$56.85

Crew D-10	Hr.	Daily	Hr.	Daily	Bare Costs	Incl. O&P
1 Bricklayer Foreman	$43.75	$350.00	$66.00	$528.00	$40.89	$61.54
1 Bricklayer	41.75	334.00	62.95	503.60		
1 Bricklayer Helper	33.65	269.20	50.75	406.00		
1 Equip. Oper. (crane)	44.40	355.20	66.45	531.60		
1 S.P. Crane, 4x4, 12 Ton		655.60		721.16	20.49	22.54
32 L.H., Daily Totals		$1964.00		$2690.36	$61.38	$84.07

Crew D-11	Hr.	Daily	Hr.	Daily	Bare Costs	Incl. O&P
1 Bricklayer Foreman	$43.75	$350.00	$66.00	$528.00	$39.72	$59.90
1 Bricklayer	41.75	334.00	62.95	503.60		
1 Bricklayer Helper	33.65	269.20	50.75	406.00		
24 L.H., Daily Totals		$953.20		$1437.60	$39.72	$59.90

Crew D-12	Hr.	Daily	Hr.	Daily	Bare Costs	Incl. O&P
1 Bricklayer Foreman	$43.75	$350.00	$66.00	$528.00	$38.20	$57.61
1 Bricklayer	41.75	334.00	62.95	503.60		
2 Bricklayer Helpers	33.65	538.40	50.75	812.00		
32 L.H., Daily Totals		$1222.40		$1843.60	$38.20	$57.61

Crew D-13	Hr.	Daily	Hr.	Daily	Bare Costs	Incl. O&P
1 Bricklayer Foreman	$43.75	$350.00	$66.00	$528.00	$39.79	$60.16
1 Bricklayer	41.75	334.00	62.95	503.60		
2 Bricklayer Helpers	33.65	538.40	50.75	812.00		
1 Carpenter	41.55	332.40	64.05	512.40		
1 Equip. Oper. (crane)	44.40	355.20	66.45	531.60		
1 S.P. Crane, 4x4, 12 Ton		655.60		721.16	13.66	15.02
48 L.H., Daily Totals		$2565.60		$3608.76	$53.45	$75.18

Crew E-1	Hr.	Daily	Hr.	Daily	Bare Costs	Incl. O&P
1 Welder Foreman	$48.90	$391.20	$85.95	$687.60	$45.70	$76.75
1 Welder	46.90	375.20	82.45	659.60		
1 Equip. Oper. (light)	41.30	330.40	61.85	494.80		
1 Welder, gas engine, 300 amp		145.80		160.38	6.08	6.68
24 L.H., Daily Totals		$1242.60		$2002.38	$51.77	$83.43

Crew E-2	Hr.	Daily	Hr.	Daily	Bare Costs	Incl. O&P
1 Struc. Steel Foreman	$48.90	$391.20	$85.95	$687.60	$45.60	$77.08
4 Struc. Steel Workers	46.90	1500.80	82.45	2638.40		
1 Equip. Oper. (crane)	44.40	355.20	66.45	531.60		
1 Equip. Oper. Oiler	38.30	306.40	57.35	458.80		
1 Lattice Boom Crane, 90 Ton		1607.00		1767.70	28.70	31.57
56 L.H., Daily Totals		$4160.60		$6084.10	$74.30	$108.64

Crew E-3	Hr.	Daily	Hr.	Daily	Bare Costs	Incl. O&P
1 Struc. Steel Foreman	$48.90	$391.20	$85.95	$687.60	$47.57	$83.62
1 Struc. Steel Worker	46.90	375.20	82.45	659.60		
1 Welder	46.90	375.20	82.45	659.60		
1 Welder, gas engine, 300 amp		145.80		160.38	6.08	6.68
24 L.H., Daily Totals		$1287.40		$2167.18	$53.64	$90.30

Crew E-4	Hr.	Daily	Hr.	Daily	Bare Costs	Incl. O&P
1 Struc. Steel Foreman	$48.90	$391.20	$85.95	$687.60	$47.40	$83.33
3 Struc. Steel Workers	46.90	1125.60	82.45	1978.80		
1 Welder, gas engine, 300 amp		145.80		160.38	4.56	5.01
32 L.H., Daily Totals		$1662.60		$2826.78	$51.96	$88.34

Crew E-5	Hr.	Daily	Hr.	Daily	Bare Costs	Incl. O&P
2 Struc. Steel Foremen	$48.90	$782.40	$85.95	$1375.20	$46.19	$79.04
5 Struc. Steel Workers	46.90	1876.00	82.45	3298.00		
1 Equip. Oper. (crane)	44.40	355.20	66.45	531.60		
1 Welder	46.90	375.20	82.45	659.60		
1 Equip. Oper. Oiler	38.30	306.40	57.35	458.80		
1 Lattice Boom Crane, 90 Ton		1607.00		1767.70		
1 Welder, gas engine, 300 amp		145.80		160.38	21.91	24.10
80 L.H., Daily Totals		$5448.00		$8251.28	$68.10	$103.14

Crew E-6	Hr.	Daily	Hr.	Daily	Bare Costs	Incl. O&P
3 Struc. Steel Foremen	$48.90	$1173.60	$85.95	$2062.80	$46.23	$79.25
9 Struc. Steel Workers	46.90	3376.80	82.45	5936.40		
1 Equip. Oper. (crane)	44.40	355.20	66.45	531.60		
1 Welder	46.90	375.20	82.45	659.60		
1 Equip. Oper. Oiler	38.30	306.40	57.35	458.80		
1 Equip. Oper. (light)	41.30	330.40	61.85	494.80		
1 Lattice Boom Crane, 90 Ton		1607.00		1767.70		
1 Welder, gas engine, 300 amp		145.80		160.38		
1 Air Compressor, 160 cfm		158.00		173.80		
2 Impact Wrenches		35.60		39.16	15.21	16.73
128 L.H., Daily Totals		$7864.00		$12285.04	$61.44	$95.98

Crew E-7	Hr.	Daily	Hr.	Daily	Bare Costs	Incl. O&P
1 Struc. Steel Foreman	$48.90	$391.20	$85.95	$687.60	$46.19	$79.04
4 Struc. Steel Workers	46.90	1500.80	82.45	2638.40		
1 Equip. Oper. (crane)	44.40	355.20	66.45	531.60		
1 Equip. Oper. Oiler	38.30	306.40	57.35	458.80		
1 Welder Foreman	48.90	391.20	85.95	687.60		
2 Welders	46.90	750.40	82.45	1319.20		
1 Lattice Boom Crane, 90 Ton		1607.00		1767.70		
2 Welders, gas engine, 300 amp		291.60		320.76	23.73	26.11
80 L.H., Daily Totals		$5593.80		$8411.66	$69.92	$105.15

Crew E-8	Hr.	Daily	Hr.	Daily	Bare Costs	Incl. O&P
1 Struc. Steel Foreman	$48.90	$391.20	$85.95	$687.60	$45.92	$78.24
4 Struc. Steel Workers	46.90	1500.80	82.45	2638.40		
1 Welder Foreman	48.90	391.20	85.95	687.60		
4 Welders	46.90	1500.80	82.45	2638.40		
1 Equip. Oper. (crane)	44.40	355.20	66.45	531.60		
1 Equip. Oper. Oiler	38.30	306.40	57.35	458.80		
1 Equip. Oper. (light)	41.30	330.40	61.85	494.80		
1 Lattice Boom Crane, 90 Ton		1607.00		1767.70		
4 Welders, gas engine, 300 amp		583.20		641.52	21.06	23.17
104 L.H., Daily Totals		$6966.20		$10546.42	$66.98	$101.41

Crew E-9	Hr.	Daily	Hr.	Daily	Bare Costs	Incl. O&P
2 Struc. Steel Foremen	$48.90	$782.40	$85.95	$1375.20	$46.23	$79.25
5 Struc. Steel Workers	46.90	1876.00	82.45	3298.00		
1 Welder Foreman	48.90	391.20	85.95	687.60		
5 Welders	46.90	1876.00	82.45	3298.00		
1 Equip. Oper. (crane)	44.40	355.20	66.45	531.60		
1 Equip. Oper. Oiler	38.30	306.40	57.35	458.80		
1 Equip. Oper. (light)	41.30	330.40	61.85	494.80		
1 Lattice Boom Crane, 90 Ton		1607.00		1767.70		
5 Welders, gas engine, 300 amp		729.00		801.90	18.25	20.07
128 L.H., Daily Totals		$8253.60		$12713.60	$64.48	$99.33

Crew No.	Bare Costs		Incl. Subs O&P		Cost Per Labor-Hour	
Crew E-10	**Hr.**	**Daily**	**Hr.**	**Daily**	**Bare Costs**	**Incl. O&P**
1 Welder Foreman	$48.90	$391.20	$85.95	$687.60	$47.90	$84.20
1 Welder	46.90	375.20	82.45	659.60		
1 Welder, gas engine, 300 amp		145.80		160.38		
1 Flatbed Truck, Gas, 3 Ton		296.00		325.60	27.61	30.37
16 L.H., Daily Totals		$1208.20		$1833.18	$75.51	$114.57
Crew E-11	**Hr.**	**Daily**	**Hr.**	**Daily**	**Bare Costs**	**Incl. O&P**
2 Painters, Struc. Steel	$37.40	$598.40	$66.90	$1070.40	$37.30	$61.67
1 Building Laborer	33.10	264.80	51.05	408.40		
1 Equip. Oper. (light)	41.30	330.40	61.85	494.80		
1 Air Compressor, 250 cfm		193.40		212.74		
1 Sandblaster, portable, 3 C.F.		20.60		22.66		
1 Set Sand Blasting Accessories		12.95		14.24	7.09	7.80
32 L.H., Daily Totals		$1420.55		$2223.24	$44.39	$69.48
Crew E-12	**Hr.**	**Daily**	**Hr.**	**Daily**	**Bare Costs**	**Incl. O&P**
1 Welder Foreman	$48.90	$391.20	$85.95	$687.60	$45.10	$73.90
1 Equip. Oper. (light)	41.30	330.40	61.85	494.80		
1 Welder, gas engine, 300 amp		145.80		160.38	9.11	10.02
16 L.H., Daily Totals		$867.40		$1342.78	$54.21	$83.92
Crew E-13	**Hr.**	**Daily**	**Hr.**	**Daily**	**Bare Costs**	**Incl. O&P**
1 Welder Foreman	$48.90	$391.20	$85.95	$687.60	$46.37	$77.92
.5 Equip. Oper. (light)	41.30	165.20	61.85	247.40		
1 Welder, gas engine, 300 amp		145.80		160.38	12.15	13.37
12 L.H., Daily Totals		$702.20		$1095.38	$58.52	$91.28
Crew E-14	**Hr.**	**Daily**	**Hr.**	**Daily**	**Bare Costs**	**Incl. O&P**
1 Welder Foreman	$48.90	$391.20	$85.95	$687.60	$48.90	$85.95
1 Welder, gas engine, 300 amp		145.80		160.38	18.23	20.05
8 L.H., Daily Totals		$537.00		$847.98	$67.13	$106.00
Crew E-16	**Hr.**	**Daily**	**Hr.**	**Daily**	**Bare Costs**	**Incl. O&P**
1 Welder Foreman	$48.90	$391.20	$85.95	$687.60	$47.90	$84.20
1 Welder	46.90	375.20	82.45	659.60		
1 Welder, gas engine, 300 amp		145.80		160.38	9.11	10.02
16 L.H., Daily Totals		$912.20		$1507.58	$57.01	$94.22
Crew E-17	**Hr.**	**Daily**	**Hr.**	**Daily**	**Bare Costs**	**Incl. O&P**
1 Structural Steel Foreman	$48.90	$391.20	$85.95	$687.60	$47.90	$84.20
1 Structural Steel Worker	46.90	375.20	82.45	659.60		
16 L.H., Daily Totals		$766.40		$1347.20	$47.90	$84.20
Crew E-18	**Hr.**	**Daily**	**Hr.**	**Daily**	**Bare Costs**	**Incl. O&P**
1 Structural Steel Foreman	$48.90	$391.20	$85.95	$687.60	$46.51	$79.52
3 Structural Steel Workers	46.90	1125.60	82.45	1978.80		
1 Equipment Operator (med.)	42.95	343.60	64.30	514.40		
1 Lattice Boom Crane, 20 Ton		1029.00		1131.90	25.73	28.30
40 L.H., Daily Totals		$2889.40		$4312.70	$72.23	$107.82
Crew E-19	**Hr.**	**Daily**	**Hr.**	**Daily**	**Bare Costs**	**Incl. O&P**
1 Structural Steel Worker	$46.90	$375.20	$82.45	$659.60	$45.70	$76.75
1 Structural Steel Foreman	48.90	391.20	85.95	687.60		
1 Equip. Oper. (light)	41.30	330.40	61.85	494.80		
1 Lattice Boom Crane, 20 Ton		1029.00		1131.90	42.88	47.16
24 L.H., Daily Totals		$2125.80		$2973.90	$88.58	$123.91

Crew No.	Bare Costs		Incl. Subs O&P		Cost Per Labor-Hour	
Crew E-20	**Hr.**	**Daily**	**Hr.**	**Daily**	**Bare Costs**	**Incl. O&P**
1 Structural Steel Foreman	$48.90	$391.20	$85.95	$687.60	$45.76	$77.75
5 Structural Steel Workers	46.90	1876.00	82.45	3298.00		
1 Equip. Oper. (crane)	44.40	355.20	66.45	531.60		
1 Oiler	38.30	306.40	57.35	458.80		
1 Lattice Boom Crane, 40 Ton		1256.00		1381.60	19.63	21.59
64 L.H., Daily Totals		$4184.80		$6357.60	$65.39	$99.34
Crew E-22	**Hr.**	**Daily**	**Hr.**	**Daily**	**Bare Costs**	**Incl. O&P**
1 Skilled Worker Foreman	$44.60	$356.80	$68.55	$548.40	$43.27	$66.52
2 Skilled Workers	42.60	681.60	65.50	1048.00		
24 L.H., Daily Totals		$1038.40		$1596.40	$43.27	$66.52
Crew E-24	**Hr.**	**Daily**	**Hr.**	**Daily**	**Bare Costs**	**Incl. O&P**
3 Structural Steel Workers	$46.90	$1125.60	$82.45	$1978.80	$45.91	$77.91
1 Equipment Operator (med.)	42.95	343.60	64.30	514.40		
1 Hyd. Crane, 25 Ton		747.60		822.36	23.36	25.70
32 L.H., Daily Totals		$2216.80		$3315.56	$69.28	$103.61
Crew E-25	**Hr.**	**Daily**	**Hr.**	**Daily**	**Bare Costs**	**Incl. O&P**
1 Welder Foreman	$48.90	$391.20	$85.95	$687.60	$48.90	$85.95
1 Cutting Torch		19.00		20.90		
1 Set of Gases		79.20		87.12	12.28	13.50
8 L.H., Daily Totals		$489.40		$795.62	$61.17	$99.45
Crew F-3	**Hr.**	**Daily**	**Hr.**	**Daily**	**Bare Costs**	**Incl. O&P**
4 Carpenters	$41.55	$1329.60	$64.05	$2049.60	$42.12	$64.53
1 Equip. Oper. (crane)	44.40	355.20	66.45	531.60		
1 Hyd. Crane, 12 Ton		655.60		721.16	16.39	18.03
40 L.H., Daily Totals		$2340.40		$3302.36	$58.51	$82.56
Crew F-4	**Hr.**	**Daily**	**Hr.**	**Daily**	**Bare Costs**	**Incl. O&P**
4 Carpenters	$41.55	$1329.60	$64.05	$2049.60	$41.48	$63.33
1 Equip. Oper. (crane)	44.40	355.20	66.45	531.60		
1 Equip. Oper. Oiler	38.30	306.40	57.35	458.80		
1 Hyd. Crane, 55 Ton		1209.00		1329.90	25.19	27.71
48 L.H., Daily Totals		$3200.20		$4369.90	$66.67	$91.04
Crew F-5	**Hr.**	**Daily**	**Hr.**	**Daily**	**Bare Costs**	**Incl. O&P**
1 Carpenter Foreman	$43.55	$348.40	$67.15	$537.20	$42.05	$64.83
3 Carpenters	41.55	997.20	64.05	1537.20		
32 L.H., Daily Totals		$1345.60		$2074.40	$42.05	$64.83
Crew F-6	**Hr.**	**Daily**	**Hr.**	**Daily**	**Bare Costs**	**Incl. O&P**
2 Carpenters	$41.55	$664.80	$64.05	$1024.80	$38.74	$59.33
2 Building Laborers	33.10	529.60	51.05	816.80		
1 Equip. Oper. (crane)	44.40	355.20	66.45	531.60		
1 Hyd. Crane, 12 Ton		655.60		721.16	16.39	18.03
40 L.H., Daily Totals		$2205.20		$3094.36	$55.13	$77.36
Crew F-7	**Hr.**	**Daily**	**Hr.**	**Daily**	**Bare Costs**	**Incl. O&P**
2 Carpenters	$41.55	$664.80	$64.05	$1024.80	$37.33	$57.55
2 Building Laborers	33.10	529.60	51.05	816.80		
32 L.H., Daily Totals		$1194.40		$1841.60	$37.33	$57.55

Crew G-1	Bare Costs Hr.	Bare Costs Daily	Incl. Subs O&P Hr.	Incl. Subs O&P Daily	Cost Per Labor-Hour Bare Costs	Cost Per Labor-Hour Incl. O&P
1 Roofer Foreman	$37.40	$299.20	$62.15	$497.20	$33.06	$54.95
4 Roofers, Composition	35.40	1132.80	58.85	1883.20		
2 Roofer Helpers	26.20	419.20	43.55	696.80		
1 Application Equipment		188.40		207.24		
1 Tar Kettle/Pot		85.45		94.00		
1 Crew Truck		202.40		222.64	8.50	9.35
56 L.H., Daily Totals		$2327.45		$3601.07	$41.56	$64.30

Crew G-2	Hr.	Daily	Hr.	Daily	Bare Costs	Incl. O&P
1 Plasterer	$37.30	$298.40	$55.65	$445.20	$34.77	$52.43
1 Plasterer Helper	33.90	271.20	50.60	404.80		
1 Building Laborer	33.10	264.80	51.05	408.40		
1 Grout Pump, 50 C.F./hr		132.60		145.86	5.53	6.08
24 L.H., Daily Totals		$967.00		$1404.26	$40.29	$58.51

Crew G-2A	Hr.	Daily	Hr.	Daily	Bare Costs	Incl. O&P
1 Roofer, composition	$35.40	$283.20	$58.85	$470.80	$31.57	$51.15
1 Roofer Helper	26.20	209.60	43.55	348.40		
1 Building Laborer	33.10	264.80	51.05	408.40		
1 Foam spray rig, trailer-mtd.		551.65		606.82		
1 Pickup Truck, 3/4 Ton		140.00		154.00	28.82	31.70
24 L.H., Daily Totals		$1449.25		$1988.42	$60.39	$82.85

Crew G-3	Hr.	Daily	Hr.	Daily	Bare Costs	Incl. O&P
2 Sheet Metal Workers	$49.10	$785.60	$74.35	$1189.60	$41.10	$62.70
2 Building Laborers	33.10	529.60	51.05	816.80		
32 L.H., Daily Totals		$1315.20		$2006.40	$41.10	$62.70

Crew G-4	Hr.	Daily	Hr.	Daily	Bare Costs	Incl. O&P
1 Labor Foreman (outside)	$35.10	$280.80	$54.10	$432.80	$33.77	$52.07
2 Building Laborers	33.10	529.60	51.05	816.80		
1 Flatbed Truck, Gas, 1.5 Ton		239.80		263.78		
1 Air Compressor, 160 cfm		158.00		173.80	16.57	18.23
24 L.H., Daily Totals		$1208.20		$1687.18	$50.34	$70.30

Crew G-5	Hr.	Daily	Hr.	Daily	Bare Costs	Incl. O&P
1 Roofer Foreman	$37.40	$299.20	$62.15	$497.20	$32.12	$53.39
2 Roofers, Composition	35.40	566.40	58.85	941.60		
2 Roofer Helpers	26.20	419.20	43.55	696.80		
1 Application Equipment		188.40		207.24	4.71	5.18
40 L.H., Daily Totals		$1473.20		$2342.84	$36.83	$58.57

Crew G-6A	Hr.	Daily	Hr.	Daily	Bare Costs	Incl. O&P
2 Roofers Composition	$35.40	$566.40	$58.85	$941.60	$35.40	$58.85
1 Small Compressor, Electric		11.80		12.98		
2 Pneumatic Nailers		43.80		48.18	3.48	3.82
16 L.H., Daily Totals		$622.00		$1002.76	$38.88	$62.67

Crew G-7	Hr.	Daily	Hr.	Daily	Bare Costs	Incl. O&P
1 Carpenter	$41.55	$332.40	$64.05	$512.40	$41.55	$64.05
1 Small Compressor, Electric		11.80		12.98		
1 Pneumatic Nailer		21.90		24.09	4.21	4.63
8 L.H., Daily Totals		$366.10		$549.47	$45.76	$68.68

Crew H-1	Hr.	Daily	Hr.	Daily	Bare Costs	Incl. O&P
2 Glaziers	$40.20	$643.20	$60.55	$968.80	$43.55	$71.50
2 Struc. Steel Workers	46.90	750.40	82.45	1319.20		
32 L.H., Daily Totals		$1393.60		$2288.00	$43.55	$71.50

Crew H-2	Hr.	Daily	Hr.	Daily	Bare Costs	Incl. O&P
2 Glaziers	$40.20	$643.20	$60.55	$968.80	$37.83	$57.38
1 Building Laborer	33.10	264.80	51.05	408.40		
24 L.H., Daily Totals		$908.00		$1377.20	$37.83	$57.38

Crew H-3	Hr.	Daily	Hr.	Daily	Bare Costs	Incl. O&P
1 Glazier	$40.20	$321.60	$60.55	$484.40	$35.90	$54.50
1 Helper	31.60	252.80	48.45	387.60		
16 L.H., Daily Totals		$574.40		$872.00	$35.90	$54.50

Crew J-1	Hr.	Daily	Hr.	Daily	Bare Costs	Incl. O&P
3 Plasterers	$37.30	$895.20	$55.65	$1335.60	$35.94	$53.63
2 Plasterer Helpers	33.90	542.40	50.60	809.60		
1 Mixing Machine, 6 C.F.		136.60		150.26	3.42	3.76
40 L.H., Daily Totals		$1574.20		$2295.46	$39.35	$57.39

Crew J-2	Hr.	Daily	Hr.	Daily	Bare Costs	Incl. O&P
3 Plasterers	$37.30	$895.20	$55.65	$1335.60	$36.11	$53.75
2 Plasterer Helpers	33.90	542.40	50.60	809.60		
1 Lather	36.95	295.60	54.35	434.80		
1 Mixing Machine, 6 C.F.		136.60		150.26	2.85	3.13
48 L.H., Daily Totals		$1869.80		$2730.26	$38.95	$56.88

Crew J-3	Hr.	Daily	Hr.	Daily	Bare Costs	Incl. O&P
1 Terrazzo Worker	$39.00	$312.00	$57.00	$456.00	$35.45	$51.80
1 Terrazzo Helper	31.90	255.20	46.60	372.80		
1 Terrazzo Grinder, Electric		82.25		90.47		
1 Terrazzo Mixer		184.80		203.28	16.69	18.36
16 L.H., Daily Totals		$834.25		$1122.56	$52.14	$70.16

Crew K-1	Hr.	Daily	Hr.	Daily	Bare Costs	Incl. O&P
1 Carpenter	$41.55	$332.40	$64.05	$512.40	$36.90	$56.63
1 Truck Driver (light)	32.25	258.00	49.20	393.60		
1 Flatbed Truck, Gas, 3 Ton		296.00		325.60	18.50	20.35
16 L.H., Daily Totals		$886.40		$1231.60	$55.40	$76.97

Crew K-2	Hr.	Daily	Hr.	Daily	Bare Costs	Incl. O&P
1 Struc. Steel Foreman	$48.90	$391.20	$85.95	$687.60	$42.68	$72.53
1 Struc. Steel Worker	46.90	375.20	82.45	659.60		
1 Truck Driver (light)	32.25	258.00	49.20	393.60		
1 Flatbed Truck, Gas, 3 Ton		296.00		325.60	12.33	13.57
24 L.H., Daily Totals		$1320.40		$2066.40	$55.02	$86.10

Crew L-1	Hr.	Daily	Hr.	Daily	Bare Costs	Incl. O&P
1 Electrician	$49.00	$392.00	$72.85	$582.80	$50.52	$75.42
1 Plumber	52.05	416.40	78.00	624.00		
16 L.H., Daily Totals		$808.40		$1206.80	$50.52	$75.42

Crew L-2	Hr.	Daily	Hr.	Daily	Bare Costs	Incl. O&P
1 Carpenter	$41.55	$332.40	$64.05	$512.40	$36.58	$56.25
1 Carpenter Helper	31.60	252.80	48.45	387.60		
16 L.H., Daily Totals		$585.20		$900.00	$36.58	$56.25

Crew L-3	Hr.	Daily	Hr.	Daily	Bare Costs	Incl. O&P
1 Carpenter	$41.55	$332.40	$64.05	$512.40	$45.30	$68.83
.5 Electrician	49.00	196.00	72.85	291.40		
.5 Sheet Metal Worker	49.10	196.40	74.35	297.40		
16 L.H., Daily Totals		$724.80		$1101.20	$45.30	$68.83

Crew L-3A	Bare Costs Hr.	Bare Costs Daily	Incl. Subs O&P Hr.	Incl. Subs O&P Daily	Cost Per Labor-Hour Bare Costs	Cost Per Labor-Hour Incl. O&P
1 Carpenter Foreman (outside)	$43.55	$348.40	$67.15	$537.20	$45.40	$69.55
.5 Sheet Metal Worker	49.10	196.40	74.35	297.40		
12 L.H., Daily Totals		$544.80		$834.60	$45.40	$69.55

Crew L-4	Hr.	Daily	Hr.	Daily	Bare Costs	Incl. O&P
2 Skilled Workers	$42.60	$681.60	$65.50	$1048.00	$38.93	$59.82
1 Helper	31.60	252.80	48.45	387.60		
24 L.H., Daily Totals		$934.40		$1435.60	$38.93	$59.82

Crew L-5	Hr.	Daily	Hr.	Daily	Bare Costs	Incl. O&P
1 Struc. Steel Foreman	$48.90	$391.20	$85.95	$687.60	$46.83	$80.66
5 Struc. Steel Workers	46.90	1876.00	82.45	3298.00		
1 Equip. Oper. (crane)	44.40	355.20	66.45	531.60		
1 Hyd. Crane, 25 Ton		747.60		822.36	13.35	14.69
56 L.H., Daily Totals		$3370.00		$5339.56	$60.18	$95.35

Crew L-5A	Hr.	Daily	Hr.	Daily	Bare Costs	Incl. O&P
1 Structural Steel Foreman	$48.90	$391.20	$85.95	$687.60	$46.77	$79.33
2 Structural Steel Workers	46.90	750.40	82.45	1319.20		
1 Equip. Oper. (crane)	44.40	355.20	66.45	531.60		
1 S.P. Crane, 4x4, 25 Ton		655.60		721.16	20.49	22.54
32 L.H., Daily Totals		$2152.40		$3259.56	$67.26	$101.86

Crew L-5B	Hr.	Daily	Hr.	Daily	Bare Costs	Incl. O&P
1 Structural Steel Foreman	$48.90	$391.20	$85.95	$687.60	$47.47	$75.11
2 Structural Steel Workers	46.90	750.40	82.45	1319.20		
2 Electricians	49.00	784.00	72.85	1165.60		
2 Steamfitters/Pipefitters	51.90	830.40	77.80	1244.80		
1 Equip. Oper. (crane)	44.40	355.20	66.45	531.60		
1 Equip. Oper. Oiler	38.30	306.40	57.35	458.80		
1 Hyd. Crane, 80 Ton		1545.00		1699.50	21.46	23.60
72 L.H., Daily Totals		$4962.60		$7107.10	$68.92	$98.71

Crew L-6	Hr.	Daily	Hr.	Daily	Bare Costs	Incl. O&P
1 Plumber	$52.05	$416.40	$78.00	$624.00	$51.03	$76.28
.5 Electrician	49.00	196.00	72.85	291.40		
12 L.H., Daily Totals		$612.40		$915.40	$51.03	$76.28

Crew L-7	Hr.	Daily	Hr.	Daily	Bare Costs	Incl. O&P
2 Carpenters	$41.55	$664.80	$64.05	$1024.80	$40.20	$61.59
1 Building Laborer	33.10	264.80	51.05	408.40		
.5 Electrician	49.00	196.00	72.85	291.40		
28 L.H., Daily Totals		$1125.60		$1724.60	$40.20	$61.59

Crew L-8	Hr.	Daily	Hr.	Daily	Bare Costs	Incl. O&P
2 Carpenters	$41.55	$664.80	$64.05	$1024.80	$43.65	$66.84
.5 Plumber	52.05	208.20	78.00	312.00		
20 L.H., Daily Totals		$873.00		$1336.80	$43.65	$66.84

Crew L-9	Hr.	Daily	Hr.	Daily	Bare Costs	Incl. O&P
1 Labor Foreman (inside)	$33.60	$268.80	$51.80	$414.40	$38.04	$60.62
2 Building Laborers	33.10	529.60	51.05	816.80		
1 Struc. Steel Worker	46.90	375.20	82.45	659.60		
.5 Electrician	49.00	196.00	72.85	291.40		
36 L.H., Daily Totals		$1369.60		$2182.20	$38.04	$60.62

Crew L-10	Hr.	Daily	Hr.	Daily	Bare Costs	Incl. O&P
1 Structural Steel Foreman	$48.90	$391.20	$85.95	$687.60	$46.73	$78.28
1 Structural Steel Worker	46.90	375.20	82.45	659.60		
1 Equip. Oper. (crane)	44.40	355.20	66.45	531.60		
1 Hyd. Crane, 12 Ton		655.60		721.16	27.32	30.05
24 L.H., Daily Totals		$1777.20		$2599.96	$74.05	$108.33

Crew L-11	Hr.	Daily	Hr.	Daily	Bare Costs	Incl. O&P
2 Wreckers	$33.10	$529.60	$55.50	$888.00	$37.98	$59.83
1 Equip. Oper. (crane)	44.40	355.20	66.45	531.60		
1 Equip. Oper. (light)	41.30	330.40	61.85	494.80		
1 Hyd. Excavator, 2.5 C.Y.		1784.00		1962.40		
1 Loader, Skid Steer, 78 H.P.		291.00		320.10	64.84	71.33
32 L.H., Daily Totals		$3290.20		$4196.90	$102.82	$131.15

Crew M-1	Hr.	Daily	Hr.	Daily	Bare Costs	Incl. O&P
3 Elevator Constructors	$61.70	$1480.80	$91.70	$2200.80	$58.61	$87.11
1 Elevator Apprentice	49.35	394.80	73.35	586.80		
5 Hand Tools		58.00		63.80	1.81	1.99
32 L.H., Daily Totals		$1933.60		$2851.40	$60.42	$89.11

Crew M-3	Hr.	Daily	Hr.	Daily	Bare Costs	Incl. O&P
1 Electrician Foreman (out)	$51.00	$408.00	$75.85	$606.80	$48.44	$72.48
1 Common Laborer	33.10	264.80	51.05	408.40		
.25 Equipment Operator, Med.	42.95	85.90	64.30	128.60		
1 Elevator Constructor	61.70	493.60	91.70	733.60		
1 Elevator Apprentice	49.35	394.80	73.35	586.80		
.25 S.P. Crane, 4x4, 20 Ton		156.40		172.04	4.60	5.06
34 L.H., Daily Totals		$1803.50		$2636.24	$53.04	$77.54

Crew M-4	Hr.	Daily	Hr.	Daily	Bare Costs	Incl. O&P
1 Electrician Foreman (out)	$51.00	$408.00	$75.85	$606.80	$47.96	$71.76
1 Common Laborer	33.10	264.80	51.05	408.40		
.25 Equipment Operator, Crane	44.40	88.80	66.45	132.90		
.25 Equipment Operator, Oiler	38.30	76.60	57.35	114.70		
1 Elevator Constructor	61.70	493.60	91.70	733.60		
1 Elevator Apprentice	49.35	394.80	73.35	586.80		
.25 S.P. Crane, 4x4, 40 Ton		253.50		278.85	7.04	7.75
36 L.H., Daily Totals		$1980.10		$2862.05	$55.00	$79.50

Crew Q-1	Hr.	Daily	Hr.	Daily	Bare Costs	Incl. O&P
1 Plumber	$52.05	$416.40	$78.00	$624.00	$46.85	$70.22
1 Plumber Apprentice	41.65	333.20	62.45	499.60		
16 L.H., Daily Totals		$749.60		$1123.60	$46.85	$70.22

Crew Q-1A	Hr.	Daily	Hr.	Daily	Bare Costs	Incl. O&P
.25 Plumber Foreman (out)	$54.05	$108.10	$81.00	$162.00	$52.45	$78.60
1 Plumber	52.05	416.40	78.00	624.00		
10 L.H., Daily Totals		$524.50		$786.00	$52.45	$78.60

Crew Q-1C	Hr.	Daily	Hr.	Daily	Bare Costs	Incl. O&P
1 Plumber	$52.05	$416.40	$78.00	$624.00	$45.55	$68.25
1 Plumber Apprentice	41.65	333.20	62.45	499.60		
1 Equip. Oper. (med.)	42.95	343.60	64.30	514.40		
1 Trencher, Chain Type, 8' D		2090.00		2299.00	87.08	95.79
24 L.H., Daily Totals		$3183.20		$3937.00	$132.63	$164.04

Crew Q-2	Hr.	Daily	Hr.	Daily	Bare Costs	Incl. O&P
2 Plumbers	$52.05	$832.80	$78.00	$1248.00	$48.58	$72.82
1 Plumber Apprentice	41.65	333.20	62.45	499.60		
24 L.H., Daily Totals		$1166.00		$1747.60	$48.58	$72.82

Crew Q-3	Hr.	Daily	Hr.	Daily	Bare Costs	Incl. O&P
1 Plumber Foreman (inside)	$52.55	$420.40	$78.75	$630.00	$49.58	$74.30
2 Plumbers	52.05	832.80	78.00	1248.00		
1 Plumber Apprentice	41.65	333.20	62.45	499.60		
32 L.H., Daily Totals		$1586.40		$2377.60	$49.58	$74.30

Crew Q-4	Hr.	Daily	Hr.	Daily	Bare Costs	Incl. O&P
1 Plumber Foreman (inside)	$52.55	$420.40	$78.75	$630.00	$49.58	$74.30
1 Plumber	52.05	416.40	78.00	624.00		
1 Welder (plumber)	52.05	416.40	78.00	624.00		
1 Plumber Apprentice	41.65	333.20	62.45	499.60		
1 Welder, electric, 300 amp		58.55		64.41	1.83	2.01
32 L.H., Daily Totals		$1644.95		$2442.01	$51.40	$76.31

Crew Q-5	Hr.	Daily	Hr.	Daily	Bare Costs	Incl. O&P
1 Steamfitter	$51.90	$415.20	$77.80	$622.40	$46.70	$70.00
1 Steamfitter Apprentice	41.50	332.00	62.20	497.60		
16 L.H., Daily Totals		$747.20		$1120.00	$46.70	$70.00

Crew Q-6	Hr.	Daily	Hr.	Daily	Bare Costs	Incl. O&P
2 Steamfitters	$51.90	$830.40	$77.80	$1244.80	$48.43	$72.60
1 Steamfitter Apprentice	41.50	332.00	62.20	497.60		
24 L.H., Daily Totals		$1162.40		$1742.40	$48.43	$72.60

Crew Q-7	Hr.	Daily	Hr.	Daily	Bare Costs	Incl. O&P
1 Steamfitter Foreman (inside)	$52.40	$419.20	$78.55	$628.40	$49.42	$74.09
2 Steamfitters	51.90	830.40	77.80	1244.80		
1 Steamfitter Apprentice	41.50	332.00	62.20	497.60		
32 L.H., Daily Totals		$1581.60		$2370.80	$49.42	$74.09

Crew Q-8	Hr.	Daily	Hr.	Daily	Bare Costs	Incl. O&P
1 Steamfitter Foreman (inside)	$52.40	$419.20	$78.55	$628.40	$49.42	$74.09
1 Steamfitter	51.90	415.20	77.80	622.40		
1 Welder (steamfitter)	51.90	415.20	77.80	622.40		
1 Steamfitter Apprentice	41.50	332.00	62.20	497.60		
1 Welder, electric, 300 amp		58.55		64.41	1.83	2.01
32 L.H., Daily Totals		$1640.15		$2435.20	$51.25	$76.10

Crew Q-9	Hr.	Daily	Hr.	Daily	Bare Costs	Incl. O&P
1 Sheet Metal Worker	$49.10	$392.80	$74.35	$594.80	$44.20	$66.92
1 Sheet Metal Apprentice	39.30	314.40	59.50	476.00		
16 L.H., Daily Totals		$707.20		$1070.80	$44.20	$66.92

Crew Q-10	Hr.	Daily	Hr.	Daily	Bare Costs	Incl. O&P
2 Sheet Metal Workers	$49.10	$785.60	$74.35	$1189.60	$45.83	$69.40
1 Sheet Metal Apprentice	39.30	314.40	59.50	476.00		
24 L.H., Daily Totals		$1100.00		$1665.60	$45.83	$69.40

Crew Q-11	Hr.	Daily	Hr.	Daily	Bare Costs	Incl. O&P
1 Sheet Metal Foreman (inside)	$49.60	$396.80	$75.10	$600.80	$46.77	$70.83
2 Sheet Metal Workers	49.10	785.60	74.35	1189.60		
1 Sheet Metal Apprentice	39.30	314.40	59.50	476.00		
32 L.H., Daily Totals		$1496.80		$2266.40	$46.77	$70.83

Crew Q-12	Hr.	Daily	Hr.	Daily	Bare Costs	Incl. O&P
1 Sprinkler Installer	$50.40	$403.20	$75.45	$603.60	$45.35	$67.90
1 Sprinkler Apprentice	40.30	322.40	60.35	482.80		
16 L.H., Daily Totals		$725.60		$1086.40	$45.35	$67.90

Crew Q-13	Hr.	Daily	Hr.	Daily	Bare Costs	Incl. O&P
1 Sprinkler Foreman (inside)	$50.90	$407.20	$76.20	$609.60	$48.00	$71.86
2 Sprinkler Installers	50.40	806.40	75.45	1207.20		
1 Sprinkler Apprentice	40.30	322.40	60.35	482.80		
32 L.H., Daily Totals		$1536.00		$2299.60	$48.00	$71.86

Crew Q-14	Hr.	Daily	Hr.	Daily	Bare Costs	Incl. O&P
1 Asbestos Worker	$45.55	$364.40	$70.80	$566.40	$41.00	$63.73
1 Asbestos Apprentice	36.45	291.60	56.65	453.20		
16 L.H., Daily Totals		$656.00		$1019.60	$41.00	$63.73

Crew Q-15	Hr.	Daily	Hr.	Daily	Bare Costs	Incl. O&P
1 Plumber	$52.05	$416.40	$78.00	$624.00	$46.85	$70.22
1 Plumber Apprentice	41.65	333.20	62.45	499.60		
1 Welder, electric, 300 amp		58.55		64.41	3.66	4.03
16 L.H., Daily Totals		$808.15		$1188.01	$50.51	$74.25

Crew Q-16	Hr.	Daily	Hr.	Daily	Bare Costs	Incl. O&P
2 Plumbers	$52.05	$832.80	$78.00	$1248.00	$48.58	$72.82
1 Plumber Apprentice	41.65	333.20	62.45	499.60		
1 Welder, electric, 300 amp		58.55		64.41	2.44	2.68
24 L.H., Daily Totals		$1224.55		$1812.01	$51.02	$75.50

Crew Q-17	Hr.	Daily	Hr.	Daily	Bare Costs	Incl. O&P
1 Steamfitter	$51.90	$415.20	$77.80	$622.40	$46.70	$70.00
1 Steamfitter Apprentice	41.50	332.00	62.20	497.60		
1 Welder, electric, 300 amp		58.55		64.41	3.66	4.03
16 L.H., Daily Totals		$805.75		$1184.41	$50.36	$74.03

Crew Q-17A	Hr.	Daily	Hr.	Daily	Bare Costs	Incl. O&P
1 Steamfitter	$51.90	$415.20	$77.80	$622.40	$45.93	$68.82
1 Steamfitter Apprentice	41.50	332.00	62.20	497.60		
1 Equip. Oper. (crane)	44.40	355.20	66.45	531.60		
1 Hyd. Crane, 12 Ton		655.60		721.16		
1 Welder, electric, 300 amp		58.55		64.41	29.76	32.73
24 L.H., Daily Totals		$1816.55		$2437.17	$75.69	$101.55

Crew Q-18	Hr.	Daily	Hr.	Daily	Bare Costs	Incl. O&P
2 Steamfitters	$51.90	$830.40	$77.80	$1244.80	$48.43	$72.60
1 Steamfitter Apprentice	41.50	332.00	62.20	497.60		
1 Welder, electric, 300 amp		58.55		64.41	2.44	2.68
24 L.H., Daily Totals		$1220.95		$1806.81	$50.87	$75.28

Crew Q-19	Hr.	Daily	Hr.	Daily	Bare Costs	Incl. O&P
1 Steamfitter	$51.90	$415.20	$77.80	$622.40	$47.47	$70.95
1 Steamfitter Apprentice	41.50	332.00	62.20	497.60		
1 Electrician	49.00	392.00	72.85	582.80		
24 L.H., Daily Totals		$1139.20		$1702.80	$47.47	$70.95

Crew Q-20	Hr.	Daily	Hr.	Daily	Bare Costs	Incl. O&P
1 Sheet Metal Worker	$49.10	$392.80	$74.35	$594.80	$45.16	$68.11
1 Sheet Metal Apprentice	39.30	314.40	59.50	476.00		
.5 Electrician	49.00	196.00	72.85	291.40		
20 L.H., Daily Totals		$903.20		$1362.20	$45.16	$68.11

Crew Q-21	Hr.	Daily	Hr.	Daily	Bare Costs	Incl. O&P
2 Steamfitters	$51.90	$830.40	$77.80	$1244.80	$48.58	$72.66
1 Steamfitter Apprentice	41.50	332.00	62.20	497.60		
1 Electrician	49.00	392.00	72.85	582.80		
32 L.H., Daily Totals		$1554.40		$2325.20	$48.58	$72.66

Crew No.	Bare Costs		Incl. Subs O&P		Cost Per Labor-Hour	
Crew Q-22	**Hr.**	**Daily**	**Hr.**	**Daily**	**Bare Costs**	**Incl. O&P**
1 Plumber	$52.05	$416.40	$78.00	$624.00	$46.85	$70.22
1 Plumber Apprentice	41.65	333.20	62.45	499.60		
1 Hyd. Crane, 12 Ton		655.60		721.16	40.98	45.07
16 L.H., Daily Totals		$1405.20		$1844.76	$87.83	$115.30

Crew Q-22A	Hr.	Daily	Hr.	Daily	Bare Costs	Incl. O&P
1 Plumber	$52.05	$416.40	$78.00	$624.00	$42.80	$64.49
1 Plumber Apprentice	41.65	333.20	62.45	499.60		
1 Laborer	33.10	264.80	51.05	408.40		
1 Equip. Oper. (crane)	44.40	355.20	66.45	531.60		
1 Hyd. Crane, 12 Ton		655.60		721.16	20.49	22.54
32 L.H., Daily Totals		$2025.20		$2784.76	$63.29	$87.02

Crew Q-23	Hr.	Daily	Hr.	Daily	Bare Costs	Incl. O&P
1 Plumber Foreman	$54.05	$432.40	$81.00	$648.00	$49.68	$74.43
1 Plumber	52.05	416.40	78.00	624.00		
1 Equip. Oper. (med.)	42.95	343.60	64.30	514.40		
1 Lattice Boom Crane, 20 Ton		1029.00		1131.90	42.88	47.16
24 L.H., Daily Totals		$2221.40		$2918.30	$92.56	$121.60

Crew R-1	Hr.	Daily	Hr.	Daily	Bare Costs	Incl. O&P
1 Electrician Foreman	$49.50	$396.00	$73.60	$588.80	$43.28	$64.84
3 Electricians	49.00	1176.00	72.85	1748.40		
2 Helpers	31.60	505.60	48.45	775.20		
48 L.H., Daily Totals		$2077.60		$3112.40	$43.28	$64.84

Crew R-1A	Hr.	Daily	Hr.	Daily	Bare Costs	Incl. O&P
1 Electrician	$49.00	$392.00	$72.85	$582.80	$40.30	$60.65
1 Helper	31.60	252.80	48.45	387.60		
16 L.H., Daily Totals		$644.80		$970.40	$40.30	$60.65

Crew R-2	Hr.	Daily	Hr.	Daily	Bare Costs	Incl. O&P
1 Electrician Foreman	$49.50	$396.00	$73.60	$588.80	$43.44	$65.07
3 Electricians	49.00	1176.00	72.85	1748.40		
2 Helpers	31.60	505.60	48.45	775.20		
1 Equip. Oper. (crane)	44.40	355.20	66.45	531.60		
1 S.P. Crane, 4x4, 5 Ton		276.20		303.82	4.93	5.43
56 L.H., Daily Totals		$2709.00		$3947.82	$48.38	$70.50

Crew R-3	Hr.	Daily	Hr.	Daily	Bare Costs	Incl. O&P
1 Electrician Foreman	$49.50	$396.00	$73.60	$588.80	$48.28	$71.87
1 Electrician	49.00	392.00	72.85	582.80		
.5 Equip. Oper. (crane)	44.40	177.60	66.45	265.80		
.5 S.P. Crane, 4x4, 5 Ton		138.10		151.91	6.91	7.60
20 L.H., Daily Totals		$1103.70		$1589.31	$55.19	$79.47

Crew R-4	Hr.	Daily	Hr.	Daily	Bare Costs	Incl. O&P
1 Struc. Steel Foreman	$48.90	$391.20	$85.95	$687.60	$47.72	$81.23
3 Struc. Steel Workers	46.90	1125.60	82.45	1978.80		
1 Electrician	49.00	392.00	72.85	582.80		
1 Welder, gas engine, 300 amp		145.80		160.38	3.65	4.01
40 L.H., Daily Totals		$2054.60		$3409.58	$51.37	$85.24

Crew No.	Bare Costs		Incl. Subs O&P		Cost Per Labor-Hour	
Crew R-5	**Hr.**	**Daily**	**Hr.**	**Daily**	**Bare Costs**	**Incl. O&P**
1 Electrician Foreman	$49.50	$396.00	$73.60	$588.80	$42.72	$64.05
4 Electrician Linemen	49.00	1568.00	72.85	2331.20		
2 Electrician Operators	49.00	784.00	72.85	1165.60		
4 Electrician Groundmen	31.60	1011.20	48.45	1550.40		
1 Crew Truck		202.40		222.64		
1 Flatbed Truck, 20,000 GVW		245.60		270.16		
1 Pickup Truck, 3/4 Ton		140.00		154.00		
.2 Hyd. Crane, 55 Ton		241.80		265.98		
.2 Hyd. Crane, 12 Ton		131.12		144.23		
.2 Earth Auger, Truck-Mtd.		89.84		98.82		
1 Tractor w/Winch		421.20		463.32	16.73	18.40
88 L.H., Daily Totals		$5231.16		$7255.16	$59.45	$82.44

Crew R-6	Hr.	Daily	Hr.	Daily	Bare Costs	Incl. O&P
1 Electrician Foreman	$49.50	$396.00	$73.60	$588.80	$42.72	$64.05
4 Electrician Linemen	49.00	1568.00	72.85	2331.20		
2 Electrician Operators	49.00	784.00	72.85	1165.60		
4 Electrician Groundmen	31.60	1011.20	48.45	1550.40		
1 Crew Truck		202.40		222.64		
1 Flatbed Truck, 20,000 GVW		245.60		270.16		
1 Pickup Truck, 3/4 Ton		140.00		154.00		
.2 Hyd. Crane, 55 Ton		241.80		265.98		
.2 Hyd. Crane, 12 Ton		131.12		144.23		
.2 Earth Auger, Truck-Mtd.		89.84		98.82		
1 Tractor w/Winch		421.20		463.32		
3 Cable Trailers		542.10		596.31		
.5 Tensioning Rig		179.20		197.12		
.5 Cable Pulling Rig		1048.00		1152.80	36.83	40.52
88 L.H., Daily Totals		$7000.46		$9201.39	$79.55	$104.56

Crew R-7	Hr.	Daily	Hr.	Daily	Bare Costs	Incl. O&P
1 Electrician Foreman	$49.50	$396.00	$73.60	$588.80	$34.58	$52.64
5 Electrician Groundmen	31.60	1264.00	48.45	1938.00		
1 Crew Truck		202.40		222.64	4.22	4.64
48 L.H., Daily Totals		$1862.40		$2749.44	$38.80	$57.28

Crew R-8	Hr.	Daily	Hr.	Daily	Bare Costs	Incl. O&P
1 Electrician Foreman	$49.50	$396.00	$73.60	$588.80	$43.28	$64.84
3 Electrician Linemen	49.00	1176.00	72.85	1748.40		
2 Electrician Groundmen	31.60	505.60	48.45	775.20		
1 Pickup Truck, 3/4 Ton		140.00		154.00		
1 Crew Truck		202.40		222.64	7.13	7.85
48 L.H., Daily Totals		$2420.00		$3489.04	$50.42	$72.69

Crew R-9	Hr.	Daily	Hr.	Daily	Bare Costs	Incl. O&P
1 Electrician Foreman	$49.50	$396.00	$73.60	$588.80	$40.36	$60.74
1 Electrician Lineman	49.00	392.00	72.85	582.80		
2 Electrician Operators	49.00	784.00	72.85	1165.60		
4 Electrician Groundmen	31.60	1011.20	48.45	1550.40		
1 Pickup Truck, 3/4 Ton		140.00		154.00		
1 Crew Truck		202.40		222.64	5.35	5.88
64 L.H., Daily Totals		$2925.60		$4264.24	$45.71	$66.63

Crew R-10	Hr.	Daily	Hr.	Daily	Bare Costs	Incl. O&P
1 Electrician Foreman	$49.50	$396.00	$73.60	$588.80	$46.18	$68.91
4 Electrician Linemen	49.00	1568.00	72.85	2331.20		
1 Electrician Groundman	31.60	252.80	48.45	387.60		
1 Crew Truck		202.40		222.64		
3 Tram Cars		393.30		432.63	12.41	13.65
48 L.H., Daily Totals		$2812.50		$3962.87	$58.59	$82.56

Crew No.	Bare Costs		Incl. Subs O&P		Cost Per Labor-Hour	
Crew R-11	**Hr.**	**Daily**	**Hr.**	**Daily**	**Bare Costs**	**Incl. O&P**
1 Electrician Foreman	$49.50	$396.00	$73.60	$588.80	$46.14	$68.93
4 Electricians	49.00	1568.00	72.85	2331.20		
1 Equip. Oper. (crane)	44.40	355.20	66.45	531.60		
1 Common Laborer	33.10	264.80	51.05	408.40		
1 Crew Truck		202.40		222.64		
1 Hyd. Crane, 12 Ton		655.60		721.16	15.32	16.85
56 L.H., Daily Totals		$3442.00		$4803.80	$61.46	$85.78

Crew R-12	**Hr.**	**Daily**	**Hr.**	**Daily**	**Bare Costs**	**Incl. O&P**
1 Carpenter Foreman	$42.05	$336.40	$64.85	$518.80	$39.14	$61.09
4 Carpenters	41.55	1329.60	64.05	2049.60		
4 Common Laborers	33.10	1059.20	51.05	1633.60		
1 Equip. Oper. (med.)	42.95	343.60	64.30	514.40		
1 Steel Worker	46.90	375.20	82.45	659.60		
1 Dozer, 200 H.P.		1192.00		1311.20		
1 Pickup Truck, 3/4 Ton		140.00		154.00	15.14	16.65
88 L.H., Daily Totals		$4776.00		$6841.20	$54.27	$77.74

Crew R-13	**Hr.**	**Daily**	**Hr.**	**Daily**	**Bare Costs**	**Incl. O&P**
1 Electrician Foreman	$49.50	$396.00	$73.60	$588.80	$46.84	$69.74
3 Electricians	49.00	1176.00	72.85	1748.40		
.25 Equip. Oper. (crane)	44.40	88.80	66.45	132.90		
1 Equipment Oiler	38.30	306.40	57.35	458.80		
.25 Hydraulic Crane, 33 Ton		190.75		209.82	4.54	5.00
42 L.H., Daily Totals		$2157.95		$3138.72	$51.38	$74.73

Crew R-15	**Hr.**	**Daily**	**Hr.**	**Daily**	**Bare Costs**	**Incl. O&P**
1 Electrician Foreman	$49.50	$396.00	$73.60	$588.80	$47.80	$71.14
4 Electricians	49.00	1568.00	72.85	2331.20		
1 Equipment Oper. (light)	41.30	330.40	61.85	494.80		
1 Aerial Lift Truck		347.40		382.14	7.24	7.96
48 L.H., Daily Totals		$2641.80		$3796.94	$55.04	$79.10

Crew R-18	**Hr.**	**Daily**	**Hr.**	**Daily**	**Bare Costs**	**Incl. O&P**
.25 Electrician Foreman	$49.50	$99.00	$73.60	$147.20	$38.33	$57.89
1 Electrician	49.00	392.00	72.85	582.80		
2 Helpers	31.60	505.60	48.45	775.20		
26 L.H., Daily Totals		$996.60		$1505.20	$38.33	$57.89

Crew R-19	**Hr.**	**Daily**	**Hr.**	**Daily**	**Bare Costs**	**Incl. O&P**
.5 Electrician Foreman	$49.50	$198.00	$73.60	$294.40	$49.10	$73.00
2 Electricians	49.00	784.00	72.85	1165.60		
20 L.H., Daily Totals		$982.00		$1460.00	$49.10	$73.00

Crew R-21	**Hr.**	**Daily**	**Hr.**	**Daily**	**Bare Costs**	**Incl. O&P**
1 Electrician Foreman	$49.50	$396.00	$73.60	$588.80	$48.97	$72.82
3 Electricians	49.00	1176.00	72.85	1748.40		
.1 Equip. Oper. (med.)	42.95	34.36	64.30	51.44		
.1 S.P. Crane, 4x4, 25 Ton		65.56		72.12	2.00	2.20
32.8 L.H., Daily Totals		$1671.92		$2460.76	$50.97	$75.02

Crew R-22	**Hr.**	**Daily**	**Hr.**	**Daily**	**Bare Costs**	**Incl. O&P**
.66 Electrician Foreman	$49.50	$261.36	$73.60	$388.61	$41.60	$62.48
2 Helpers	31.60	505.60	48.45	775.20		
2 Electricians	49.00	784.00	72.85	1165.60		
37.28 L.H., Daily Totals		$1550.96		$2329.41	$41.60	$62.48

Crew No.	Bare Costs		Incl. Subs O&P		Cost Per Labor-Hour	
Crew R-30	**Hr.**	**Daily**	**Hr.**	**Daily**	**Bare Costs**	**Incl. O&P**
.25 Electrician Foreman (out)	$51.00	$102.00	$75.85	$151.70	$39.37	$59.67
1 Electrician	49.00	392.00	72.85	582.80		
2 Laborers, (Semi-Skilled)	33.10	529.60	51.05	816.80		
26 L.H., Daily Totals		$1023.60		$1551.30	$39.37	$59.67

Crew R-31	**Hr.**	**Daily**	**Hr.**	**Daily**	**Bare Costs**	**Incl. O&P**
1 Electrician	$49.00	$392.00	$72.85	$582.80	$49.00	$72.85
1 Core Drill, Electric, 2.5 H.P.		53.40		58.74	6.67	7.34
8 L.H., Daily Totals		$445.40		$641.54	$55.67	$80.19

Crew W-41E	**Hr.**	**Daily**	**Hr.**	**Daily**	**Bare Costs**	**Incl. O&P**
.5 Plumber Foreman (out)	$54.05	$216.20	$81.00	$324.00	$44.87	$67.82
1 Plumber	52.05	416.40	78.00	624.00		
1 Laborer	33.10	264.80	51.05	408.40		
20 L.H., Daily Totals		$897.40		$1356.40	$44.87	$67.82

Historical Cost Indexes

The table below lists both the RSMeans Historical Cost Index based on Jan. 1, 1993 = 100 as well as the computed value of an index based on Jan. 1, 2010 costs. Since the Jan. 1, 2010 figure is estimated, space is left to write in the actual index figures as they become available through either the quarterly "RSMeans Construction Cost Indexes" or as printed in the "Engineering News-Record." To compute the actual index based on Jan. 1, 2010 = 100, divide the Historical Cost Index for a particular year by the actual Jan. 1, 2010 Construction Cost Index. Space has been left to advance the index figures as the year progresses.

Year	Historical Cost Index Jan. 1, 1993 = 100		Current Index Based on Jan. 1, 2010 = 100		Year	Historical Cost Index Jan. 1, 1993 = 100	Current Index Based on Jan. 1, 2010 = 100		Year	Historical Cost Index Jan. 1, 1993 = 100	Current Index Based on Jan. 1, 2010 = 100	
	Est.	Actual	Est.	Actual		Actual	Est.	Actual		Actual	Est.	Actual
Oct 2010					July 1995	107.6	58.9		July 1977	49.5	27.1	
July 2010					1994	104.4	57.1		1976	46.9	25.7	
April 2010					1993	101.7	55.6		1975	44.8	24.5	
Jan 2010	182.8		100.0	100.0	1992	99.4	54.4		1974	41.4	22.6	
July 2009		180.1	98.5		1991	96.8	53.0		1973	37.7	20.6	
2008		180.4	98.7		1990	94.3	51.6		1972	34.8	19.0	
2007		169.4	92.7		1989	92.1	50.4		1971	32.1	17.6	
2006		162.0	88.6		1988	89.9	49.2		1970	28.7	15.7	
2005		151.6	82.9		1987	87.7	48.0		1969	26.9	14.7	
2004		143.7	78.6		1986	84.2	46.1		1968	24.9	13.6	
2003		132.0	72.2		1985	82.6	45.2		1967	23.5	12.9	
2002		128.7	70.4		1984	82.0	44.8		1966	22.7	12.4	
2001		125.1	68.4		1983	80.2	43.9		1965	21.7	11.9	
2000		120.9	66.1		1982	76.1	41.7		1964	21.2	11.6	
1999		117.6	64.3		1981	70.0	38.3		1963	20.7	11.3	
1998		115.1	63.0		1980	62.9	34.4		1962	20.2	11.1	
1997		112.8	61.7		1979	57.8	31.6		1961	19.8	10.8	
1996		110.2	60.3		1978	53.5	29.3		1960	19.7	10.8	

Adjustments to Costs

The Historical Cost Index can be used to convert National Average building costs at a particular time to the approximate building costs for some other time.

Example:

Estimate and compare construction costs for different years in the same city.

To estimate the National Average construction cost of a building in 1970, knowing that it cost $900,000 in 2010:

INDEX in 1970 = 28.7

INDEX in 2010 = 182.8

Note: The City Cost Indexes for Canada can be used to convert U.S. National averages to local costs in Canadian dollars.

Time Adjustment using the Historical Cost Indexes:

$$\frac{\text{Index for Year A}}{\text{Index for Year B}} \times \text{Cost in Year B} = \text{Cost in Year A}$$

$$\frac{\text{INDEX 1970}}{\text{INDEX 2010}} \times \text{Cost 2010} = \text{Cost 1970}$$

$$\frac{28.7}{182.8} \times \$900{,}000 = .157 \times \$900{,}000 = \$141{,}300$$

The construction cost of the building in 1970 is $141,300.

How to Use the City Cost Indexes

What you should know before you begin

RSMeans City Cost Indexes (CCI) are an extremely useful tool to use when you want to compare costs from city to city and region to region.

This publication contains average construction cost indexes for 731 U.S. and Canadian cities covering over 930 three-digit zip code locations, as listed directly under each city.

Keep in mind that a City Cost Index number is a percentage ratio of a specific city's cost to the national average cost of the same item at a stated time period.

In other words, these index figures represent relative construction factors (or, if you prefer, multipliers) for Material and Installation costs, as well as the weighted average for Total In Place costs for each CSI MasterFormat division. Installation costs include both labor and equipment rental costs. When estimating equipment rental rates only, for a specific location, use 015433 CONTRACTOR EQUIPMENT index.

The 30 City Average Index is the average of 30 major U.S. cities and serves as a National Average.

Index figures for both material and installation are based on the 30 major city average of 100 and represent the cost relationship as of July 1, 2008. The index for each division is computed from representative material and labor quantities for that division. The weighted average for each city is a weighted total of the components listed above it, but does not include relative productivity between trades or cities.

As changes occur in local material prices, labor rates, and equipment rental rates, (including fuel costs) the impact of these changes should be accurately measured by the change in the City Cost Index for each particular city (as compared to the 30 City Average).

Therefore, if you know (or have estimated) building costs in one city today, you can easily convert those costs to expected building costs in another city.

In addition, by using the Historical Cost Index, you can easily convert National Average building costs at a particular time to the approximate building costs for some other time. The City Cost Indexes can then be applied to calculate the costs for a particular city.

Quick Calculations

Location Adjustment Using the City Cost Indexes:

$$\frac{\text{Index for City A}}{\text{Index for City B}} \times \text{Cost in City B} = \text{Cost in City A}$$

Time Adjustment for the National Average Using the Historical Cost Index:

$$\frac{\text{Index for Year A}}{\text{Index for Year B}} \times \text{Cost in Year B} = \text{Cost in Year A}$$

Adjustment from the National Average:

$$\frac{\text{Index for City A}}{100} \times \text{National Average Cost} = \text{Cost in City A}$$

Since each of the other RSMeans publications contains many different items, any *one* item multiplied by the particular city index may give incorrect results. However, the larger the number of items compiled, the closer the results should be to actual costs for that particular city.

The City Cost Indexes for Canadian cities are calculated using Canadian material and equipment prices and labor rates, in Canadian dollars. Therefore, indexes for Canadian cities can be used to convert U.S. National Average prices to local costs in Canadian dollars.

How to use this section

1. Compare costs from city to city.

In using the RSMeans Indexes, remember that an index number is not a fixed number but a ratio: It's a percentage ratio of a building component's cost at any stated time to the National Average cost of that same component at the same time period. Put in the form of an equation:

$$\frac{\text{Specific City Cost}}{\text{National Average Cost}} \times 100 = \text{City Index Number}$$

Therefore, when making cost comparisons between cities, do not subtract one city's index number from the index number of another city and read the result as a percentage difference. Instead, divide one city's index number by that of the other city. The resulting number may then be used as a multiplier to calculate cost differences from city to city.

The formula used to find cost differences between cities for the purpose of comparison is as follows:

$$\frac{\text{City A Index}}{\text{City B Index}} \times \text{City B Cost (Known)} = \text{City A Cost (Unknown)}$$

In addition, you can use RSMeans CCI to calculate and compare costs division by division between cities using the same basic formula. (Just be sure that you're comparing similar divisions.)

2. Compare a specific city's construction costs with the National Average.

When you're studying construction location feasibility, it's advisable to compare a prospective project's cost index with an index of the National Average cost.

For example, divide the weighted average index of construction costs of a specific city by that of the 30 City Average, which = 100.

$$\frac{\text{City Index}}{100} = \text{\% of National Average}$$

As a result, you get a ratio that indicates the relative cost of construction in that city in comparison with the National Average.

3. Convert U.S. National Average to actual costs in Canadian City.

$$\frac{\text{Index for Canadian City}}{100} \times \text{National Average Cost} = \text{Cost in Canadian City in \$ CAN}$$

4. **Adjust construction cost data based on a National Average.**

When you use a source of construction cost data which is based on a National Average (such as RSMeans cost data publications), it is necessary to adjust those costs to a specific location.

$$\frac{\text{City Index}}{100} \times \text{"Book" Cost Based on National Average Costs} = \text{City Cost (Unknown)}$$

5. **When applying the City Cost Indexes to demolition projects, use the appropriate division installation index.** For example, for removal of existing doors and windows, use Division 8 (Openings) index.

What you might like to know about how we developed the Indexes

The information presented in the CCI is organized according to the Construction Specifications Institute (CSI) MasterFormat 2004.

To create a reliable index, RSMeans researched the building type most often constructed in the United States and Canada. Because it was concluded that no one type of building completely represented the building construction industry, nine different types of buildings were combined to create a composite model.

The exact material, labor and equipment quantities are based on detailed analysis of these nine building types, then each quantity is weighted in proportion to expected usage. These various material items, labor hours, and equipment rental rates are thus combined to form a composite building representing as closely as possible the actual usage of materials, labor and equipment used in the North American Building Construction Industry.

The following structures were chosen to make up that composite model:

1. Factory, 1 story
2. Office, 2–4 story
3. Store, Retail
4. Town Hall, 2–3 story
5. High School, 2–3 story
6. Hospital, 4–8 story
7. Garage, Parking
8. Apartment, 1–3 story
9. Hotel/Motel, 2–3 story

For the purposes of ensuring the timeliness of the data, the components of the index for the composite model have been streamlined. They currently consist of:

- specific quantities of 66 commonly used construction materials;
- specific labor-hours for 21 building construction trades; and
- specific days of equipment rental for 6 types of construction equipment (normally used to install the 66 material items by the 21 trades.) Fuel costs and routine maintenance costs are included in the equipment cost.

A sophisticated computer program handles the updating of all costs for each city on a quarterly basis. Material and equipment price quotations are gathered quarterly from 731 cities in the United States and Canada. These prices and the latest negotiated labor wage rates for 21 different building trades are used to compile the quarterly update of the City Cost Index.

The 30 major U.S. cities used to calculate the National Average are:

Atlanta, GA	Memphis, TN
Baltimore, MD	Milwaukee, WI
Boston, MA	Minneapolis, MN
Buffalo, NY	Nashville, TN
Chicago, IL	New Orleans, LA
Cincinnati, OH	New York, NY
Cleveland, OH	Philadelphia, PA
Columbus, OH	Phoenix, AZ
Dallas, TX	Pittsburgh, PA
Denver, CO	St. Louis, MO
Detroit, MI	San Antonio, TX
Houston, TX	San Diego, CA
Indianapolis, IN	San Francisco, CA
Kansas City, MO	Seattle, WA
Los Angeles, CA	Washington, DC

What the CCI does not indicate

The weighted average for each city is a total of the divisional components weighted to reflect typical usage, but it does not include the productivity variations between trades or cities.

In addition, the CCI does not take into consideration factors such as the following:

- managerial efficiency
- competitive conditions
- automation
- restrictive union practices
- unique local requirements
- regional variations due to specific building codes

DIVISION		UNITED STATES 30 CITY AVERAGE			ALABAMA ANNISTON 362			BIRMINGHAM 350 - 352			BUTLER 369			DECATUR 356			DOTHAN 363		
		MAT.	INST.	TOTAL	MAT.	INST.	TOTAL	MAT.	INST.	TOTAL	MAT.	INST.	TOTAL	MAT.	INST.	TOTAL	MAT.	INST.	TOTAL
015433	CONTRACTOR EQUIPMENT		100.0	100.0		100.9	100.9		101.0	101.0		97.8	97.8		100.9	100.9		97.8	97.8
0241, 31 - 34	SITE & INFRASTRUCTURE, DEMOLITION	100.0	100.0	100.0	89.9	93.2	92.2	87.0	93.8	91.8	104.2	87.3	92.4	82.9	92.1	89.3	101.7	87.2	91.6
0310	Concrete Forming & Accessories	100.0	100.0	100.0	90.4	52.3	57.4	90.6	70.2	72.9	87.2	47.4	52.8	94.1	50.3	56.2	95.0	46.9	53.4
0320	Concrete Reinforcing	100.0	100.0	100.0	83.4	87.3	85.4	83.4	87.7	85.6	88.1	51.2	69.3	83.4	76.8	80.0	88.1	50.6	68.9
0330	Cast-in-Place Concrete	100.0	100.0	100.0	94.1	53.2	78.6	100.8	68.8	88.7	91.8	59.3	79.5	93.6	67.2	83.6	91.8	49.6	75.8
03	CONCRETE	100.0	100.0	100.0	93.3	60.8	78.1	92.3	74.1	83.8	94.1	54.0	75.4	88.9	62.6	76.6	93.4	50.4	73.3
04	MASONRY	100.0	100.0	100.0	85.2	74.3	78.6	87.6	83.4	85.0	89.2	54.1	68.1	86.0	57.1	68.6	90.1	44.2	62.5
05	METALS	100.0	100.0	100.0	95.6	93.1	94.7	96.6	94.7	95.9	94.6	80.5	89.8	97.8	87.6	94.3	94.6	79.2	89.4
06	WOOD, PLASTICS & COMPOSITES	100.0	100.0	100.0	90.2	47.7	66.1	92.0	67.1	77.9	85.8	47.3	64.0	94.7	48.0	68.2	96.3	47.3	68.6
07	THERMAL & MOISTURE PROTECTION	100.0	100.0	100.0	91.9	57.5	78.9	94.0	83.9	90.2	91.9	60.9	80.2	91.7	60.4	79.9	91.9	54.2	77.6
08	OPENINGS	100.0	100.0	100.0	93.6	53.8	83.8	97.2	73.0	91.2	93.7	49.3	82.7	97.5	54.1	86.8	93.7	49.3	82.7
0920	Plaster & Gypsum Board	100.0	100.0	100.0	90.8	46.6	60.2	98.5	66.6	76.4	89.1	46.2	59.4	95.3	46.9	61.8	98.1	46.2	62.2
0950, 0980	Ceilings & Acoustic Treatment	100.0	100.0	100.0	91.3	46.6	62.8	94.1	66.6	76.6	91.3	46.2	62.6	96.5	46.9	64.9	91.3	46.2	62.6
0960	Flooring	100.0	100.0	100.0	100.2	36.1	82.5	103.1	64.9	92.6	106.3	60.3	93.6	103.1	54.5	89.7	113.9	29.6	90.6
0970, 0990	Wall Finishes & Painting/Coating	100.0	100.0	100.0	96.1	34.9	59.3	96.1	61.7	75.4	96.1	59.2	73.9	96.1	61.8	75.5	96.1	59.2	73.9
09	FINISHES	100.0	100.0	100.0	94.1	46.4	68.2	96.5	67.6	80.8	97.9	50.6	72.2	96.2	51.6	72.0	101.1	44.4	70.3
COVERS	DIVS. 10 - 14, 25, 28, 41, 43, 44	100.0	100.0	100.0	100.0	80.8	95.9	100.0	85.2	96.9	100.0	48.5	89.1	100.0	50.7	89.5	100.0	48.6	89.1
21, 22, 23	FIRE SUPPRESSION, PLUMBING & HVAC	100.0	100.0	100.0	98.3	62.8	83.7	99.8	66.0	85.9	96.1	37.7	72.1	99.8	46.1	77.7	96.1	37.6	72.1
26, 27, 3370	ELECTRICAL, COMMUNICATIONS & UTIL.	100.0	100.0	100.0	96.7	64.3	79.8	108.0	64.3	85.2	98.8	45.1	70.7	98.5	68.4	82.8	97.6	62.2	79.1
MF2004	WEIGHTED AVERAGE	100.0	100.0	100.0	95.2	67.0	82.8	97.4	75.2	87.6	95.8	53.9	77.3	96.0	61.8	80.9	95.9	53.7	77.3

DIVISION		ALABAMA EVERGREEN 364			GADSDEN 359			HUNTSVILLE 357 - 358			JASPER 355			MOBILE 365 - 366			MONTGOMERY 360 - 361		
		MAT.	INST.	TOTAL	MAT.	INST.	TOTAL	MAT.	INST.	TOTAL	MAT.	INST.	TOTAL	MAT.	INST.	TOTAL	MAT.	INST.	TOTAL
015433	CONTRACTOR EQUIPMENT		97.8	97.8		100.9	100.9		100.9	100.9		100.9	100.9		97.8	97.8		97.8	97.8
0241, 31 - 34	SITE & INFRASTRUCTURE, DEMOLITION	104.7	87.6	92.8	88.4	92.7	91.4	82.7	92.9	89.8	88.0	92.5	91.1	96.1	88.3	90.7	94.1	87.7	89.6
0310	Concrete Forming & Accessories	84.1	49.4	54.1	87.3	45.7	51.4	94.1	67.4	71.0	92.0	38.0	45.3	94.2	59.6	64.3	92.2	47.9	53.9
0320	Concrete Reinforcing	88.2	51.2	69.3	88.4	86.5	87.4	83.4	77.2	80.2	83.4	86.4	84.9	86.0	85.7	85.8	86.0	86.3	86.1
0330	Cast-in-Place Concrete	91.8	62.1	80.6	93.6	52.2	77.9	91.2	65.4	81.4	103.7	46.8	82.2	96.2	72.0	87.1	98.6	49.7	80.1
03	CONCRETE	94.2	55.9	76.3	93.0	57.4	76.4	87.7	69.7	79.3	96.5	52.1	75.8	90.6	70.2	81.1	91.6	57.5	75.7
04	MASONRY	89.2	59.2	71.2	84.6	47.8	62.5	87.1	63.5	72.9	80.9	45.8	59.8	87.8	58.7	70.3	86.2	43.7	60.7
05	METALS	94.6	80.5	89.9	95.6	92.0	94.4	97.8	89.3	94.9	95.6	91.4	94.2	96.7	93.8	95.7	96.5	91.6	94.9
06	WOOD, PLASTICS & COMPOSITES	82.3	47.3	62.5	86.4	43.2	61.9	94.7	68.7	80.0	92.0	35.2	59.8	94.9	59.1	74.6	91.6	47.3	66.5
07	THERMAL & MOISTURE PROTECTION	91.9	63.0	81.0	91.8	67.8	82.7	91.6	75.7	85.6	91.8	50.0	76.0	91.6	75.5	85.5	90.8	66.5	81.6
08	OPENINGS	93.7	49.3	82.7	93.6	52.5	83.5	97.5	65.3	89.5	93.6	51.4	83.2	97.5	64.0	89.2	97.5	57.2	87.5
0920	Plaster & Gypsum Board	87.3	46.2	58.9	88.6	42.0	56.4	95.3	68.2	76.6	91.2	33.7	51.4	95.3	58.4	69.8	98.5	46.2	62.3
0950, 0980	Ceilings & Acoustic Treatment	91.3	46.2	62.6	91.3	42.0	59.9	98.4	68.2	79.2	91.3	33.7	54.6	98.4	58.4	72.9	94.1	46.2	63.6
0960	Flooring	103.6	60.3	91.6	98.2	48.5	84.4	103.1	58.2	90.7	101.2	36.1	83.2	112.8	60.3	98.3	111.1	31.9	89.2
0970, 0990	Wall Finishes & Painting/Coating	96.1	59.2	73.9	96.1	67.4	78.9	96.1	64.9	77.4	96.1	42.8	64.1	100.1	60.3	76.1	96.1	59.2	73.9
09	FINISHES	96.9	51.9	72.5	93.1	47.1	68.1	96.7	65.2	79.6	94.3	36.9	63.1	101.6	58.8	78.4	99.9	44.8	70.0
COVERS	DIVS. 10 - 14, 25, 28, 41, 43, 44	100.0	50.4	89.5	100.0	77.0	95.1	100.0	82.2	96.2	100.0	46.6	88.7	100.0	82.0	96.2	100.0	76.8	95.1
21, 22, 23	FIRE SUPPRESSION, PLUMBING & HVAC	96.1	40.3	73.2	101.5	40.7	76.5	99.7	58.6	82.8	101.5	56.0	82.8	99.7	60.4	83.5	99.8	38.4	74.5
26, 27, 3370	ELECTRICAL, COMMUNICATIONS & UTIL.	96.0	45.1	69.4	98.4	64.3	80.5	99.6	68.4	83.3	98.0	63.3	79.9	99.6	56.2	76.9	102.7	68.2	84.7
MF2004	WEIGHTED AVERAGE	95.4	55.6	77.8	95.9	59.2	79.7	96.0	70.1	84.6	96.3	58.5	79.6	97.1	67.4	84.0	97.1	58.3	80.0

DIVISION		ALABAMA PHENIX CITY 368			SELMA 367			TUSCALOOSA 354			ALASKA ANCHORAGE 995 - 996			FAIRBANKS 997			JUNEAU 998		
		MAT.	INST.	TOTAL	MAT.	INST.	TOTAL	MAT.	INST.	TOTAL	MAT.	INST.	TOTAL	MAT.	INST.	TOTAL	MAT.	INST.	TOTAL
015433	CONTRACTOR EQUIPMENT		97.8	97.8		97.8	97.8		100.9	100.9		116.2	116.2		116.2	116.2		116.2	116.2
0241, 31 - 34	SITE & INFRASTRUCTURE, DEMOLITION	108.8	87.8	94.1	101.5	87.7	91.9	83.1	92.9	89.9	137.3	131.5	133.2	123.9	131.5	129.2	136.4	131.5	132.9
0310	Concrete Forming & Accessories	90.3	42.4	48.9	88.3	47.1	52.7	94.0	51.7	57.4	135.6	118.5	120.8	141.4	119.1	122.1	137.8	118.5	121.1
0320	Concrete Reinforcing	88.1	68.3	78.0	88.1	86.1	87.1	83.4	86.4	84.9	146.4	106.5	126.0	150.6	106.5	128.1	112.1	106.5	109.3
0330	Cast-in-Place Concrete	91.8	50.3	76.1	91.8	49.5	75.8	94.9	52.2	78.8	162.5	114.5	144.4	142.6	117.1	132.9	163.4	114.5	144.9
03	CONCRETE	96.8	51.9	75.8	92.9	57.0	76.1	89.6	60.0	75.8	136.5	114.2	126.1	123.1	115.3	119.5	132.2	114.2	123.8
04	MASONRY	89.2	45.6	63.0	92.6	42.9	62.8	86.2	51.1	65.1	202.7	119.0	152.4	211.4	118.8	155.7	194.2	119.0	149.0
05	METALS	94.6	84.9	91.3	94.6	90.6	93.2	96.9	92.0	95.2	120.2	101.2	113.8	120.3	101.4	113.9	120.2	101.1	113.8
06	WOOD, PLASTICS & COMPOSITES	90.0	39.5	61.4	87.4	47.3	64.7	94.7	49.6	69.2	114.1	118.6	116.7	117.8	119.3	118.7	114.1	118.6	116.7
07	THERMAL & MOISTURE PROTECTION	92.2	66.4	82.5	91.8	55.8	78.2	91.7	69.4	83.3	185.0	114.3	158.3	183.1	114.4	157.2	183.7	114.3	157.5
08	OPENINGS	93.6	48.6	82.5	93.6	57.2	84.6	97.5	63.6	89.1	121.0	113.5	119.1	118.1	113.6	117.0	118.1	113.5	117.0
0920	Plaster & Gypsum Board	92.1	38.2	54.8	90.4	46.2	59.8	95.3	48.6	63.0	137.8	119.0	124.8	157.2	119.7	131.3	137.8	119.0	124.8
0950, 0980	Ceilings & Acoustic Treatment	91.3	38.2	57.5	91.3	46.2	62.6	98.4	48.6	66.7	121.3	119.0	119.8	126.5	119.7	122.2	121.3	119.0	119.8
0960	Flooring	109.2	31.9	87.8	107.1	23.6	84.0	103.1	40.4	85.8	159.3	122.5	149.1	159.3	122.5	149.1	159.3	122.5	149.1
0970, 0990	Wall Finishes & Painting/Coating	96.1	59.2	73.9	96.1	59.2	73.9	96.1	51.0	69.0	154.8	112.8	129.5	154.8	125.7	137.3	154.8	112.8	129.5
09	FINISHES	99.6	40.5	67.5	98.0	43.4	68.3	96.7	48.5	70.5	147.2	119.2	132.0	148.7	121.0	133.7	146.0	119.2	131.4
COVERS	DIVS. 10 - 14, 25, 28, 41, 43, 44	100.0	76.4	95.0	100.0	48.6	89.1	100.0	79.0	95.5	100.0	110.3	102.2	100.0	110.3	102.2	100.0	110.3	102.2
21, 22, 23	FIRE SUPPRESSION, PLUMBING & HVAC	96.1	38.8	72.6	96.1	37.9	72.2	99.8	38.2	74.4	100.4	101.7	100.9	100.3	104.5	102.1	100.4	99.0	99.8
26, 27, 3370	ELECTRICAL, COMMUNICATIONS & UTIL.	98.3	75.5	86.3	97.3	45.1	70.0	99.0	64.3	80.8	141.6	110.0	125.1	147.9	110.0	128.1	142.0	110.0	125.3
MF2004	WEIGHTED AVERAGE	96.4	57.2	79.1	95.6	53.5	77.0	96.0	60.2	80.2	127.4	112.3	120.7	126.3	113.3	120.5	126.0	111.7	119.7

	DIVISION	ALASKA			ARIZONA															
		KETCHIKAN			CHAMBERS			FLAGSTAFF			GLOBE			KINGMAN			MESA/TEMPE		
		999			865			860			855			864			851,852		
		MAT.	INST.	TOTAL	MAT.	INST.	TOTAL	MAT.	INST.	TOTAL	MAT.	INST.	TOTAL	MAT.	INST.	TOTAL	MAT.	INST.	TOTAL
015433	CONTRACTOR EQUIPMENT		116.2	116.2		94.9	94.9		94.9	94.9		98.0	98.0		94.9	94.9		98.0	98.0
0241, 31 - 34	SITE & INFRASTRUCTURE, DEMOLITION	185.0	131.5	147.7	63.0	97.9	87.3	80.1	99.5	93.6	93.0	101.4	98.8	62.9	99.4	88.3	83.9	101.9	96.5
0310	Concrete Forming & Accessories	133.6	118.5	120.5	95.4	47.5	54.0	100.0	63.3	68.3	98.0	47.3	54.2	93.7	60.8	65.3	100.5	57.4	63.2
0320	Concrete Reinforcing	117.0	106.5	111.6	96.8	70.4	83.3	96.6	79.7	88.0	96.6	70.4	83.2	96.9	79.6	88.1	97.2	79.6	88.2
0330	Cast-in-Place Concrete	288.7	114.5	222.8	91.8	56.1	78.3	91.9	75.7	85.8	97.3	56.5	81.8	91.5	57.3	78.6	98.1	62.6	84.6
03	CONCRETE	211.4	114.2	165.9	98.0	55.2	78.0	119.0	70.8	96.4	114.9	55.2	87.0	97.7	63.4	81.6	105.3	63.7	85.8
04	MASONRY	220.1	119.0	159.3	95.5	43.4	64.2	95.2	60.8	74.5	109.8	43.4	69.9	95.4	55.0	71.2	110.0	45.8	71.4
05	METALS	120.4	101.1	113.9	96.7	65.5	86.2	97.3	73.6	89.3	93.4	65.9	84.1	97.4	73.2	89.2	93.7	73.8	87.0
06	WOOD, PLASTICS & COMPOSITES	109.3	118.6	114.6	99.6	45.6	69.1	105.0	62.3	80.8	95.7	45.8	67.4	96.1	62.3	77.0	98.8	62.5	78.2
07	THERMAL & MOISTURE PROTECTION	185.8	114.3	158.8	93.9	53.4	78.6	95.6	66.2	84.5	103.2	51.1	83.5	93.9	59.0	80.7	102.4	56.0	84.9
08	OPENINGS	119.8	113.5	118.3	100.9	50.6	88.5	101.1	67.9	92.8	96.7	50.7	85.3	101.1	64.7	92.1	96.8	64.8	88.9
0920	Plaster & Gypsum Board	143.5	119.0	126.5	92.1	44.0	58.8	95.4	61.2	71.7	100.3	44.0	61.4	87.6	61.2	69.4	102.4	61.2	73.9
0950, 0980	Ceilings & Acoustic Treatment	121.7	119.0	120.0	125.3	44.0	73.5	126.2	61.2	84.8	116.2	44.0	70.2	126.2	61.2	84.8	116.2	61.2	81.2
0960	Flooring	159.3	122.5	149.1	94.9	42.6	80.4	97.4	42.8	82.3	100.7	42.6	84.6	93.9	57.8	83.9	102.1	53.3	88.7
0970, 0990	Wall Finishes & Painting/Coating	154.8	112.8	129.5	90.4	41.0	60.7	90.4	55.1	69.2	95.0	41.0	62.5	90.4	55.1	69.2	95.0	55.1	71.0
09	FINISHES	150.7	119.2	133.6	102.0	44.9	71.0	105.1	57.7	79.4	104.2	45.0	72.0	101.3	58.9	78.3	103.9	55.8	77.8
COVERS	DIVS. 10 - 14, 25, 28, 41, 43, 44	100.0	110.3	102.2	100.0	78.4	95.4	100.0	82.4	96.3	100.0	78.7	95.5	100.0	80.3	95.8	100.0	77.3	95.2
21, 22, 23	FIRE SUPPRESSION, PLUMBING & HVAC	98.8	99.0	98.9	96.0	67.4	84.3	100.2	76.8	90.6	93.5	62.0	80.6	96.0	73.7	86.8	100.3	62.7	84.8
26, 27, 3370	ELECTRICAL, COMMUNICATIONS & UTIL.	147.9	110.0	128.1	102.0	67.1	83.7	100.8	58.9	78.9	97.7	63.8	80.0	102.0	58.9	79.4	94.2	58.9	75.7
MF2004	WEIGHTED AVERAGE	139.3	111.7	127.1	97.3	61.8	81.6	101.6	70.4	87.9	99.5	60.5	82.3	97.2	67.9	84.3	99.4	64.4	83.9

	DIVISION	ARIZONA												ARKANSAS					
		PHOENIX			PRESCOTT			SHOW LOW			TUCSON			BATESVILLE			CAMDEN		
		850,853			863			859			856 - 857			725			717		
		MAT.	INST.	TOTAL	MAT.	INST.	TOTAL	MAT.	INST.	TOTAL	MAT.	INST.	TOTAL	MAT.	INST.	TOTAL	MAT.	INST.	TOTAL
015433	CONTRACTOR EQUIPMENT		98.6	98.6		94.9	94.9		98.0	98.0		98.0	98.0		87.4	87.4		87.4	87.4
0241, 31 - 34	SITE & INFRASTRUCTURE, DEMOLITION	84.4	103.3	97.5	68.9	97.9	89.1	95.0	101.4	99.5	79.9	103.0	96.0	74.4	83.9	81.0	75.2	83.5	81.0
0310	Concrete Forming & Accessories	101.4	71.4	75.4	96.6	47.3	53.9	103.9	47.6	55.2	100.7	70.8	74.9	84.4	49.6	54.3	84.1	32.8	39.8
0320	Concrete Reinforcing	95.6	79.9	87.6	96.6	70.7	83.4	97.2	64.2	80.3	80.4	79.7	80.0	98.3	67.5	82.6	99.6	45.6	72.0
0330	Cast-in-Place Concrete	98.1	76.3	89.8	91.8	56.1	78.3	97.3	56.6	81.9	100.9	76.1	91.5	83.4	48.4	70.2	85.5	39.1	68.0
03	CONCRETE	105.0	74.6	90.8	103.7	55.1	81.0	117.3	54.2	87.8	104.1	74.2	90.2	83.5	53.2	69.3	85.1	38.6	63.4
04	MASONRY	96.6	64.6	77.3	95.5	48.9	67.5	109.8	45.0	70.9	94.6	60.8	74.3	101.9	45.4	67.9	110.6	34.4	64.8
05	METALS	95.4	75.4	88.7	97.3	65.6	86.6	93.1	63.8	83.2	94.6	74.1	87.6	98.4	67.0	87.8	98.4	58.5	85.0
06	WOOD, PLASTICS & COMPOSITES	99.7	72.8	84.4	100.9	45.6	69.6	102.8	45.8	70.5	99.0	72.8	84.2	89.2	50.8	67.5	89.0	33.0	57.3
07	THERMAL & MOISTURE PROTECTION	102.3	68.1	89.4	94.4	52.8	78.7	103.3	53.7	84.6	103.6	63.6	88.5	97.0	48.2	78.5	96.8	34.8	73.4
08	OPENINGS	97.8	73.6	91.8	101.1	50.6	88.6	95.8	49.1	84.2	93.2	73.6	88.3	96.5	49.1	84.8	92.7	38.2	79.2
0920	Plaster & Gypsum Board	105.2	71.9	82.1	92.3	44.0	58.9	104.6	44.0	62.7	108.8	71.9	83.2	78.7	49.9	58.8	78.7	31.6	46.1
0950, 0980	Ceilings & Acoustic Treatment	123.9	71.9	90.8	124.4	44.0	73.2	116.2	44.0	70.2	117.0	71.9	88.3	88.6	49.9	64.0	88.6	31.6	52.3
0960	Flooring	102.4	55.7	89.5	95.8	42.6	81.1	104.0	48.5	88.7	93.0	42.8	79.2	103.2	63.6	92.3	103.1	42.7	86.4
0970, 0990	Wall Finishes & Painting/Coating	95.0	60.9	74.5	90.4	41.0	60.7	95.0	41.0	62.5	95.2	55.1	71.1	98.6	45.7	66.8	98.6	49.9	69.3
09	FINISHES	106.1	66.9	84.8	102.6	44.9	71.3	105.9	46.0	73.4	101.9	63.9	81.3	90.4	51.8	69.4	90.4	36.3	61.0
COVERS	DIVS. 10 - 14, 25, 28, 41, 43, 44	100.0	84.0	96.6	100.0	78.4	95.4	100.0	78.7	95.5	100.0	83.9	96.6	100.0	46.2	88.6	100.0	41.3	87.5
21, 22, 23	FIRE SUPPRESSION, PLUMBING & HVAC	100.3	76.9	90.6	100.2	67.1	86.6	93.5	67.4	82.8	100.2	65.6	86.0	93.4	38.5	70.8	93.4	31.3	67.8
26, 27, 3370	ELECTRICAL, COMMUNICATIONS & UTIL.	101.4	67.0	83.4	100.5	58.8	78.7	94.6	63.8	78.5	96.5	58.4	76.6	99.9	45.8	71.6	96.1	40.8	67.2
MF2004	WEIGHTED AVERAGE	99.9	74.6	88.7	99.1	61.1	82.4	99.6	61.6	82.8	98.2	69.9	85.7	94.2	51.4	75.3	94.1	42.2	71.2

	DIVISION	ARKANSAS																	
		FAYETTEVILLE			FORT SMITH			HARRISON			HOT SPRINGS			JONESBORO			LITTLE ROCK		
		727			729			726			719			724			720 - 722		
		MAT.	INST.	TOTAL	MAT.	INST.	TOTAL	MAT.	INST.	TOTAL	MAT.	INST.	TOTAL	MAT.	INST.	TOTAL	MAT.	INST.	TOTAL
015433	CONTRACTOR EQUIPMENT		87.4	87.4		87.4	87.4		87.4	87.4		87.4	87.4		106.6	106.6		87.4	87.4
0241, 31 - 34	SITE & INFRASTRUCTURE, DEMOLITION	73.8	84.1	81.0	78.6	84.1	82.4	79.3	83.9	82.5	78.8	83.5	82.1	100.1	98.0	98.6	86.5	84.1	84.8
0310	Concrete Forming & Accessories	80.5	41.1	46.4	97.4	39.1	47.0	88.3	49.9	55.0	81.9	38.7	44.5	87.8	53.2	57.9	89.6	65.6	68.9
0320	Concrete Reinforcing	98.3	65.0	81.3	99.4	71.9	85.4	97.8	67.4	82.3	97.7	58.8	77.9	95.0	68.3	81.4	99.6	66.9	82.9
0330	Cast-in-Place Concrete	83.4	45.4	69.1	95.4	64.9	83.9	92.6	46.4	75.1	87.5	40.4	69.7	90.9	59.2	78.9	95.9	65.1	84.2
03	CONCRETE	83.2	48.0	66.8	91.3	55.1	74.4	90.6	52.7	72.9	88.5	44.1	67.7	88.1	59.4	74.7	91.1	66.0	79.3
04	MASONRY	91.9	42.4	62.2	97.8	54.8	71.9	102.3	43.5	66.9	82.7	28.5	50.1	93.3	46.7	65.3	94.6	54.8	70.7
05	METALS	98.4	66.3	87.6	100.8	70.7	90.6	99.5	67.0	88.5	98.4	63.3	86.5	95.2	80.3	90.2	96.9	69.4	87.6
06	WOOD, PLASTICS & COMPOSITES	85.2	39.6	59.4	105.1	34.6	65.2	94.4	50.8	69.7	86.3	39.8	60.0	92.8	54.0	70.8	99.7	70.0	82.9
07	THERMAL & MOISTURE PROTECTION	97.8	47.9	78.9	98.1	54.7	81.7	97.2	46.8	78.2	97.0	38.7	75.0	101.7	54.0	83.7	97.4	58.4	82.7
08	OPENINGS	96.5	46.7	84.2	97.5	43.8	84.2	97.5	49.8	85.6	92.7	43.4	80.5	99.5	57.6	89.1	97.5	63.1	88.9
0920	Plaster & Gypsum Board	76.6	38.3	50.1	84.1	33.2	48.9	82.8	49.9	60.0	77.4	38.6	50.5	94.6	52.8	65.7	79.9	69.6	72.8
0950, 0980	Ceilings & Acoustic Treatment	88.6	38.3	56.6	93.0	33.2	54.9	92.0	49.9	65.2	88.6	38.6	56.7	92.7	52.8	67.3	91.3	69.6	77.5
0960	Flooring	99.8	63.6	89.8	111.6	65.5	98.8	106.2	63.6	94.4	101.6	63.6	91.1	74.7	57.5	70.0	112.9	65.5	99.8
0970, 0990	Wall Finishes & Painting/Coating	98.6	31.9	58.5	98.6	59.8	75.2	98.6	45.7	66.8	98.6	39.6	63.1	87.6	53.7	67.2	98.6	61.3	76.2
09	FINISHES	89.1	43.4	64.3	94.9	45.0	67.8	92.9	51.9	70.6	90.1	43.0	64.6	87.1	53.8	69.0	94.5	65.9	79.0
COVERS	DIVS. 10 - 14, 25, 28, 41, 43, 44	100.0	56.1	90.7	100.0	71.7	94.0	100.0	57.8	91.0	100.0	42.6	87.8	100.0	52.9	90.0	100.0	75.8	94.9
21, 22, 23	FIRE SUPPRESSION, PLUMBING & HVAC	93.4	42.6	72.5	100.2	45.4	77.6	93.4	44.5	73.3	93.4	35.8	69.7	100.4	45.3	77.7	100.1	62.6	84.7
26, 27, 3370	ELECTRICAL, COMMUNICATIONS & UTIL.	93.3	57.4	74.5	97.0	72.4	84.1	98.4	41.6	68.7	98.5	43.0	69.5	104.0	50.2	75.8	102.4	73.6	87.3
MF2004	WEIGHTED AVERAGE	92.8	51.8	74.7	97.3	57.8	79.9	95.5	52.2	76.4	93.3	45.3	72.1	96.9	57.9	79.7	97.2	67.0	83.9

DIVISION		ARKANSAS												CALIFORNIA					
		PINE BLUFF			RUSSELLVILLE			TEXARKANA			WEST MEMPHIS			ALHAMBRA			ANAHEIM		
		716			728			718			723			917 - 918			928		
		MAT.	INST.	TOTAL	MAT.	INST.	TOTAL	MAT.	INST.	TOTAL	MAT.	INST.	TOTAL	MAT.	INST.	TOTAL	MAT.	INST.	TOTAL
015433	CONTRACTOR EQUIPMENT		87.4	87.4		87.4	87.4		88.0	88.0		106.6	106.6		98.6	98.6		101.6	101.6
0241, 31 - 34	SITE & INFRASTRUCTURE, DEMOLITION	80.5	84.1	83.0	75.7	84.0	81.5	94.8	84.6	87.7	107.7	98.1	101.0	102.4	109.4	107.3	100.6	109.4	106.7
0310	Concrete Forming & Accessories	81.5	65.4	67.5	84.9	51.9	56.3	87.4	43.9	49.7	92.6	53.1	58.4	122.6	108.3	110.2	105.8	125.9	123.2
0320	Concrete Reinforcing	99.5	66.9	82.8	99.0	60.6	79.4	99.1	66.3	82.4	95.0	60.3	77.3	117.7	117.8	117.8	100.6	118.0	109.5
0330	Cast-in-Place Concrete	87.5	65.0	79.0	87.4	48.5	72.7	95.4	46.0	76.7	95.3	59.1	81.6	96.3	116.8	104.1	97.5	119.8	105.9
03	CONCRETE	89.1	65.8	78.2	86.5	53.0	70.9	88.1	49.7	70.1	94.9	57.9	77.6	107.8	112.3	109.9	105.2	121.3	112.7
04	MASONRY	117.9	54.8	80.0	97.0	41.6	63.7	97.8	31.9	58.2	80.4	46.7	60.2	120.8	109.6	114.1	86.3	111.0	101.2
05	METALS	99.2	69.1	89.0	98.4	64.7	87.0	92.5	66.0	83.5	94.3	77.7	88.7	86.2	99.9	90.8	103.7	102.1	103.2
06	WOOD, PLASTICS & COMPOSITES	85.8	70.0	76.9	90.5	53.8	69.7	94.0	46.9	67.3	98.4	54.0	73.2	94.4	104.3	100.0	93.6	127.0	112.5
07	THERMAL & MOISTURE PROTECTION	97.0	58.4	82.5	98.0	46.9	78.7	97.8	43.9	77.4	102.1	54.0	83.9	102.5	107.3	104.3	103.4	115.2	107.8
08	OPENINGS	92.7	63.1	85.4	96.5	52.1	85.5	98.0	49.6	86.0	99.5	56.0	88.7	92.4	106.2	95.8	104.4	120.8	108.5
0920	Plaster & Gypsum Board	77.0	69.6	71.9	78.7	53.0	60.9	81.6	45.9	56.8	96.7	52.8	66.3	114.8	104.5	107.7	107.3	127.7	121.4
0950, 0980	Ceilings & Acoustic Treatment	89.6	69.6	76.9	88.6	53.0	65.9	94.7	45.9	63.6	90.7	52.8	66.6	100.7	104.5	103.1	112.2	127.7	122.1
0960	Flooring	101.3	65.5	91.4	102.8	63.6	92.0	104.0	54.9	90.4	77.4	57.5	71.9	102.9	96.0	101.0	114.5	114.5	114.5
0970, 0990	Wall Finishes & Painting/Coating	98.6	61.3	76.2	98.6	38.9	62.7	98.6	31.5	58.3	87.6	53.7	67.2	107.5	110.0	109.0	104.6	102.2	103.2
09	FINISHES	90.3	65.9	77.1	90.5	52.8	70.0	93.4	44.5	66.9	88.4	53.8	69.6	104.3	105.6	105.0	108.4	121.9	115.7
COVERS	DIVS. 10 - 14, 25, 28, 41, 43, 44	100.0	75.8	94.9	100.0	46.6	88.7	100.0	38.0	86.8	100.0	52.9	90.0	100.0	109.4	102.0	100.0	112.6	102.7
21, 22, 23	FIRE SUPPRESSION, PLUMBING & HVAC	100.2	47.8	78.6	93.4	48.4	74.9	100.2	33.4	72.7	93.6	45.2	73.7	93.4	96.3	94.6	100.1	112.9	105.4
26, 27, 3370	ELECTRICAL, COMMUNICATIONS & UTIL.	96.3	73.6	84.4	97.0	50.1	72.5	98.4	40.3	68.0	105.7	50.2	76.7	120.1	100.5	109.8	93.5	101.1	97.5
MF2004	WEIGHTED AVERAGE	96.8	63.9	82.3	94.1	53.8	76.3	96.2	46.5	74.2	95.8	57.4	78.9	100.2	104.2	102.0	101.1	112.6	106.2

DIVISION		CALIFORNIA																	
		BAKERSFIELD			BERKELEY			EUREKA			FRESNO			INGLEWOOD			LONG BEACH		
		932 - 933			947			955			936 - 938			903 - 905			906 - 908		
		MAT.	INST.	TOTAL	MAT.	INST.	TOTAL	MAT.	INST.	TOTAL	MAT.	INST.	TOTAL	MAT.	INST.	TOTAL	MAT.	INST.	TOTAL
015433	CONTRACTOR EQUIPMENT		99.3	99.3		98.7	98.7		99.0	99.0		99.3	99.3		98.9	98.9		98.9	98.9
0241, 31 - 34	SITE & INFRASTRUCTURE, DEMOLITION	106.3	106.7	106.6	127.1	103.2	110.5	114.5	103.9	107.1	106.0	106.3	106.2	91.2	106.5	101.8	98.6	106.5	104.1
0310	Concrete Forming & Accessories	104.3	125.2	122.3	122.1	131.0	129.8	116.2	126.5	125.1	101.9	127.5	124.0	108.0	108.3	108.3	103.4	108.4	107.7
0320	Concrete Reinforcing	113.0	118.2	115.6	109.1	118.4	113.8	109.6	118.2	114.0	96.0	118.2	107.4	111.3	117.8	114.6	110.3	117.8	114.1
0330	Cast-in-Place Concrete	104.3	118.6	109.7	133.3	111.7	125.1	105.8	113.8	108.8	102.8	114.2	107.1	95.3	117.8	103.8	108.6	117.8	112.1
03	CONCRETE	107.7	120.5	113.7	122.6	120.5	121.6	116.6	119.4	117.9	106.9	120.0	113.0	103.2	112.8	107.7	114.1	112.8	113.5
04	MASONRY	107.0	116.1	112.5	145.4	117.2	128.4	112.7	118.6	116.3	110.0	115.8	113.5	78.5	109.7	97.3	87.3	109.7	100.8
05	METALS	99.0	101.4	99.8	104.2	99.1	102.5	103.6	99.5	102.2	103.6	100.4	102.5	94.9	101.7	97.2	94.8	101.8	97.2
06	WOOD, PLASTICS & COMPOSITES	87.9	127.1	110.1	120.0	132.1	126.8	106.6	129.1	119.4	99.4	129.1	116.2	92.3	104.3	99.1	86.4	104.3	96.5
07	THERMAL & MOISTURE PROTECTION	102.0	107.2	104.0	116.5	119.1	117.5	107.5	108.0	107.7	96.9	111.1	102.3	102.2	108.7	104.7	102.4	110.9	105.6
08	OPENINGS	101.9	118.6	106.0	103.4	124.3	108.5	103.2	111.0	105.2	103.7	118.2	107.3	92.4	106.2	95.8	92.4	106.2	95.8
0920	Plaster & Gypsum Board	102.1	127.7	119.8	112.7	132.5	126.4	111.8	129.8	124.3	102.3	129.8	121.3	100.0	104.5	103.1	96.5	104.5	102.0
0950, 0980	Ceilings & Acoustic Treatment	113.6	127.7	122.6	111.4	132.5	124.8	117.4	129.8	125.3	115.1	129.8	124.5	104.3	104.5	104.4	104.3	104.5	104.4
0960	Flooring	114.4	84.4	106.1	124.0	111.6	120.6	119.2	106.5	115.7	121.1	128.5	123.1	104.3	96.0	102.0	100.8	96.0	99.5
0970, 0990	Wall Finishes & Painting/Coating	106.3	98.2	101.4	108.3	123.9	117.7	105.6	51.4	73.0	124.0	97.1	107.8	100.0	110.0	106.0	100.0	110.0	106.0
09	FINISHES	110.0	115.8	113.1	114.9	127.7	121.9	114.6	116.6	115.7	113.9	125.7	120.3	103.6	105.6	104.6	102.6	105.6	104.2
COVERS	DIVS. 10 - 14, 25, 28, 41, 43, 44	100.0	109.6	102.0	100.0	126.3	105.6	100.0	124.5	105.2	100.0	124.6	105.2	100.0	109.5	102.0	100.0	109.5	102.0
21, 22, 23	FIRE SUPPRESSION, PLUMBING & HVAC	100.1	94.2	97.7	93.4	123.2	105.7	93.3	102.4	97.1	100.2	107.8	103.4	93.4	96.3	94.6	93.4	96.3	94.6
26, 27, 3370	ELECTRICAL, COMMUNICATIONS & UTIL.	105.5	93.8	99.4	101.7	125.0	113.9	101.6	86.6	93.7	92.8	90.8	91.8	105.2	100.5	102.7	104.9	100.5	102.6
MF2004	WEIGHTED AVERAGE	102.9	106.6	104.6	107.5	119.2	112.7	104.1	107.3	105.5	102.8	110.6	106.3	96.8	104.2	100.1	98.6	104.3	101.1

DIVISION		CALIFORNIA																	
		LOS ANGELES			MARYSVILLE			MODESTO			MOJAVE			OAKLAND			OXNARD		
		900 - 902			959			953			935			946			930		
		MAT.	INST.	TOTAL	MAT.	INST.	TOTAL	MAT.	INST.	TOTAL	MAT.	INST.	TOTAL	MAT.	INST.	TOTAL	MAT.	INST.	TOTAL
015433	CONTRACTOR EQUIPMENT		100.6	100.6		99.0	99.0		99.0	99.0		99.3	99.3		98.7	98.7		98.2	98.2
0241, 31 - 34	SITE & INFRASTRUCTURE, DEMOLITION	97.8	109.0	105.6	110.6	104.9	106.6	104.4	106.1	105.6	99.3	106.7	104.4	133.5	103.2	112.4	106.4	104.7	105.2
0310	Concrete Forming & Accessories	105.6	126.2	123.4	106.7	127.0	124.3	103.2	127.5	124.2	114.8	107.0	108.1	111.1	143.2	138.8	107.0	125.9	123.3
0320	Concrete Reinforcing	111.3	118.6	115.0	109.6	117.8	113.8	113.5	117.8	115.7	115.1	117.6	116.4	111.5	118.9	115.3	113.0	117.8	115.4
0330	Cast-in-Place Concrete	103.9	118.5	109.4	118.3	113.8	116.6	105.8	114.6	109.1	88.9	118.4	100.1	126.3	117.7	123.1	103.5	119.0	109.4
03	CONCRETE	110.3	121.1	115.3	118.2	119.5	118.8	108.5	120.1	113.9	98.3	112.3	104.8	122.7	128.1	125.2	107.5	120.9	113.8
04	MASONRY	93.1	120.4	109.5	113.7	108.0	110.3	111.3	108.2	109.4	108.1	106.4	107.1	153.5	124.9	136.3	111.8	105.7	108.2
05	METALS	102.2	103.7	102.7	103.1	98.9	101.6	100.2	100.9	100.5	100.6	100.2	100.5	98.2	100.2	98.9	98.8	101.4	99.7
06	WOOD, PLASTICS & COMPOSITES	88.9	126.8	110.4	94.4	129.1	114.1	90.3	129.1	112.3	99.2	103.1	101.4	106.6	147.9	130.0	94.1	127.1	112.8
07	THERMAL & MOISTURE PROTECTION	102.4	118.9	108.7	107.1	108.0	107.4	106.8	109.4	107.8	103.0	101.3	102.3	113.9	122.8	117.3	106.8	111.2	108.5
08	OPENINGS	98.6	123.9	104.9	102.4	119.0	106.5	101.1	120.2	105.9	97.7	103.1	99.1	103.5	132.9	110.7	100.8	120.8	105.8
0920	Plaster & Gypsum Board	100.9	127.7	119.5	105.1	129.8	122.2	107.7	129.8	123.0	114.6	103.0	106.6	106.0	148.9	135.7	107.3	127.7	121.4
0950, 0980	Ceilings & Acoustic Treatment	114.7	127.7	123.0	116.5	129.8	125.0	112.2	129.8	123.4	116.5	103.0	107.9	114.2	148.9	136.3	117.0	127.7	123.8
0960	Flooring	102.4	114.5	105.8	113.8	105.3	111.4	114.4	101.2	110.7	121.8	84.4	111.5	118.2	111.6	116.4	112.9	114.5	113.3
0970, 0990	Wall Finishes & Painting/Coating	100.0	110.0	106.0	105.6	101.2	102.9	105.6	96.5	100.1	105.6	86.9	94.4	108.3	139.6	127.2	105.6	93.2	98.1
09	FINISHES	105.8	122.6	114.9	111.3	122.0	117.1	110.3	120.8	116.0	114.1	100.4	106.7	113.5	138.8	127.3	110.8	120.9	116.3
COVERS	DIVS. 10 - 14, 25, 28, 41, 43, 44	100.0	112.2	102.6	100.0	124.6	105.2	100.0	124.6	105.2	100.0	106.8	101.4	100.0	128.2	106.0	100.0	112.8	102.7
21, 22, 23	FIRE SUPPRESSION, PLUMBING & HVAC	100.2	112.9	105.4	93.3	98.7	95.5	100.1	107.8	103.3	93.4	94.0	93.6	100.2	136.1	115.0	100.1	112.9	105.4
26, 27, 3370	ELECTRICAL, COMMUNICATIONS & UTIL.	101.9	115.8	109.1	97.8	103.4	100.7	100.5	95.0	97.7	91.7	88.0	89.8	100.9	125.0	113.5	97.6	95.6	96.6
MF2004	WEIGHTED AVERAGE	101.7	116.0	108.0	103.3	108.8	105.7	103.0	109.9	106.1	98.9	100.4	99.6	108.5	126.0	116.2	102.5	110.6	106.1

DIVISION		CALIFORNIA																	
		PALM SPRINGS			PALO ALTO			PASADENA			REDDING			RICHMOND			RIVERSIDE		
		922			943			910 - 912			960			948			925		
		MAT.	INST.	TOTAL	MAT.	INST.	TOTAL	MAT.	INST.	TOTAL	MAT.	INST.	TOTAL	MAT.	INST.	TOTAL	MAT.	INST.	TOTAL
015433	CONTRACTOR EQUIPMENT		100.3	100.3		98.7	98.7		98.6	98.6		99.0	99.0		98.7	98.7		100.3	100.3
0241, 31 - 34	SITE & INFRASTRUCTURE, DEMOLITION	91.9	107.2	102.6	122.6	103.2	109.1	99.0	109.4	106.2	110.4	105.0	106.6	133.1	103.2	112.3	98.9	107.3	104.7
0310	Concrete Forming & Accessories	102.3	108.4	107.6	109.1	130.9	128.0	112.5	108.3	108.9	106.5	127.0	124.2	124.7	142.7	140.3	105.9	125.7	123.0
0320	Concrete Reinforcing	115.3	117.7	116.5	109.1	118.4	113.8	118.7	117.8	118.3	109.6	117.8	113.8	109.1	118.4	113.8	112.0	117.8	114.9
0330	Cast-in-Place Concrete	93.0	119.4	103.0	112.6	117.5	114.5	91.4	116.8	101.0	117.3	113.9	116.0	129.6	117.5	125.0	101.1	119.7	108.1
03	CONCRETE	99.6	113.3	106.0	110.4	122.5	116.1	103.1	112.3	107.4	117.1	119.6	118.2	124.8	127.7	126.2	106.1	121.1	113.1
04	MASONRY	84.0	106.6	97.6	120.5	117.2	118.5	104.0	109.6	107.4	113.7	108.0	110.3	145.1	117.2	128.3	85.0	106.6	98.0
05	METALS	104.3	100.9	103.1	95.6	99.1	96.8	86.2	99.9	90.8	103.4	99.2	102.0	95.6	99.2	96.8	103.9	101.6	103.1
06	WOOD, PLASTICS & COMPOSITES	88.7	104.4	97.6	103.9	132.1	119.8	82.0	104.3	94.6	94.1	129.1	113.9	123.8	147.9	137.5	93.6	127.0	112.5
07	THERMAL & MOISTURE PROTECTION	103.3	104.9	103.9	113.5	120.3	116.1	102.2	107.3	104.1	107.3	106.6	107.0	114.1	120.7	116.6	103.6	108.8	105.5
08	OPENINGS	99.9	106.3	101.4	103.5	124.3	108.6	92.4	106.2	95.8	103.5	119.0	107.3	103.5	132.9	110.8	102.9	120.8	107.4
0920	Plaster & Gypsum Board	102.9	104.5	104.0	104.3	132.5	123.8	109.2	104.5	105.9	105.3	129.8	122.3	114.2	148.9	138.2	106.8	127.7	121.3
0950, 0980	Ceilings & Acoustic Treatment	110.5	104.5	106.7	112.3	132.5	125.1	100.7	104.5	103.1	122.2	129.8	127.0	112.3	148.9	135.6	115.3	127.7	123.2
0960	Flooring	116.4	96.0	110.8	117.2	108.8	114.9	96.9	96.0	96.6	113.7	105.3	111.3	126.0	111.6	122.0	118.5	114.5	117.4
0970, 0990	Wall Finishes & Painting/Coating	102.7	123.6	115.3	108.3	123.9	117.7	107.5	110.0	109.0	105.6	101.2	102.9	108.3	123.9	117.7	102.7	102.2	102.4
09	FINISHES	107.3	107.2	107.2	111.8	127.2	120.2	101.6	105.6	103.7	112.5	122.0	117.7	116.5	137.0	127.7	109.9	121.9	116.4
COVERS	DIVS. 10 - 14, 25, 28, 41, 43, 44	100.0	109.9	102.1	100.0	126.3	105.6	100.0	109.4	102.0	100.0	124.6	105.2	100.0	128.2	106.0	100.0	112.5	102.7
21, 22, 23	FIRE SUPPRESSION, PLUMBING & HVAC	93.3	96.3	94.6	93.4	122.3	105.3	93.4	96.3	94.6	100.1	98.7	99.5	93.4	119.7	104.2	100.1	112.9	105.3
26, 27, 3370	ELECTRICAL, COMMUNICATIONS & UTIL.	97.5	92.5	94.9	100.8	123.9	112.9	116.5	100.5	108.1	101.0	94.3	97.5	101.4	119.5	110.9	93.6	92.6	93.1
MF2004	WEIGHTED AVERAGE	98.3	103.0	100.4	102.8	119.2	110.0	98.0	104.2	100.7	105.4	107.6	106.3	106.8	120.6	112.9	101.1	110.6	105.3

DIVISION		CALIFORNIA																	
		SACRAMENTO			SALINAS			SAN BERNARDINO			SAN DIEGO			SAN FRANCISCO			SAN JOSE		
		942,956 - 958			939			923 - 924			919 - 921			940 - 941			951		
		MAT.	INST.	TOTAL	MAT.	INST.	TOTAL	MAT.	INST.	TOTAL	MAT.	INST.	TOTAL	MAT.	INST.	TOTAL	MAT.	INST.	TOTAL
015433	CONTRACTOR EQUIPMENT		98.3	98.3		99.3	99.3		100.3	100.3		98.6	98.6		107.5	107.5		99.8	99.8
0241, 31 - 34	SITE & INFRASTRUCTURE, DEMOLITION	106.4	110.0	108.9	121.0	106.7	111.1	77.4	107.2	98.2	105.1	102.8	103.5	135.8	110.2	118.0	140.9	100.9	113.0
0310	Concrete Forming & Accessories	109.4	128.8	126.2	109.6	130.6	127.8	109.1	108.2	108.4	108.8	110.4	110.2	110.6	145.2	140.5	109.1	144.1	139.4
0320	Concrete Reinforcing	103.4	118.3	111.0	113.5	118.3	115.9	112.0	118.2	115.1	115.0	118.1	116.6	127.0	119.6	123.2	100.0	119.4	109.9
0330	Cast-in-Place Concrete	106.8	115.2	110.0	102.3	114.8	107.0	69.8	119.3	88.5	100.2	106.7	102.7	129.5	120.0	125.9	123.1	118.2	121.2
03	CONCRETE	111.8	120.6	116.0	118.2	121.6	119.8	78.6	113.3	94.8	108.5	110.0	109.2	126.5	130.4	128.3	115.4	129.2	121.9
04	MASONRY	122.8	115.9	118.6	107.7	114.1	111.5	91.5	116.4	106.5	95.4	109.4	103.8	154.1	133.0	141.4	143.9	132.7	137.1
05	METALS	94.0	96.9	95.0	103.2	101.9	102.8	103.9	101.4	103.0	101.8	102.7	102.1	104.8	110.0	106.6	98.6	108.5	102.0
06	WOOD, PLASTICS & COMPOSITES	100.0	130.4	117.2	98.5	132.5	117.8	97.3	104.4	101.3	94.9	108.1	102.4	106.6	148.1	130.1	102.0	147.8	127.9
07	THERMAL & MOISTURE PROTECTION	120.6	112.0	117.4	103.6	113.8	107.4	102.6	108.5	104.8	108.8	104.0	107.0	116.5	136.4	124.0	103.2	135.7	115.5
08	OPENINGS	116.3	120.9	117.4	102.2	124.5	107.7	99.9	106.3	101.5	100.1	109.0	102.3	107.8	136.8	115.0	94.1	137.3	104.8
0920	Plaster & Gypsum Board	100.3	131.0	121.5	108.5	133.3	125.7	108.1	104.5	105.6	115.8	108.1	110.5	108.8	148.9	136.5	102.3	148.9	134.5
0950, 0980	Ceilings & Acoustic Treatment	112.3	131.0	124.2	116.5	133.3	127.2	112.2	104.5	107.3	108.4	108.1	108.2	121.7	148.9	139.0	107.5	148.9	133.8
0960	Flooring	117.6	109.8	115.4	115.2	111.6	114.2	120.9	90.4	112.5	105.7	105.9	105.7	118.2	125.5	120.2	106.4	125.5	111.7
0970, 0990	Wall Finishes & Painting/Coating	105.6	110.5	108.5	106.6	141.3	127.5	102.7	98.7	100.3	104.3	108.1	106.6	108.3	151.6	134.4	106.2	141.3	127.3
09	FINISHES	110.1	124.7	118.1	113.4	129.5	122.2	108.5	103.3	105.7	107.6	109.1	108.4	115.8	143.3	130.7	108.1	141.9	126.4
COVERS	DIVS. 10 - 14, 25, 28, 41, 43, 44	100.0	125.2	105.4	100.0	125.1	105.3	100.0	96.1	99.2	100.0	109.5	102.0	100.0	128.8	106.1	100.0	127.8	105.9
21, 22, 23	FIRE SUPPRESSION, PLUMBING & HVAC	100.1	110.8	104.5	93.4	108.2	99.5	93.3	96.4	94.6	100.2	111.3	104.8	100.2	165.1	126.9	100.1	138.5	115.9
26, 27, 3370	ELECTRICAL, COMMUNICATIONS & UTIL.	96.3	99.7	98.1	92.8	108.6	101.1	97.5	95.0	96.2	101.7	102.5	102.1	101.0	157.9	130.7	104.2	142.1	124.0
MF2004	WEIGHTED AVERAGE	104.9	112.6	108.3	102.9	114.3	107.9	95.7	103.6	99.2	102.4	107.5	104.6	110.7	140.4	123.8	105.8	131.3	117.1

DIVISION		CALIFORNIA																	
		SAN LUIS OBISPO			SAN MATEO			SAN RAFAEL			SANTA ANA			SANTA BARBARA			SANTA CRUZ		
		934			944			949			926 - 927			931			950		
		MAT.	INST.	TOTAL	MAT.	INST.	TOTAL	MAT.	INST.	TOTAL	MAT.	INST.	TOTAL	MAT.	INST.	TOTAL	MAT.	INST.	TOTAL
015433	CONTRACTOR EQUIPMENT		99.3	99.3		98.7	98.7		98.8	98.8		100.3	100.3		99.3	99.3		99.8	99.8
0241, 31 - 34	SITE & INFRASTRUCTURE, DEMOLITION	112.4	106.7	108.4	130.3	103.2	111.4	120.8	108.8	112.5	90.2	107.2	102.1	106.3	106.7	106.6	140.1	100.7	112.7
0310	Concrete Forming & Accessories	116.5	108.1	109.2	115.0	143.1	139.3	117.3	142.5	139.1	109.7	108.4	108.5	108.0	125.6	123.2	109.1	130.8	127.9
0320	Concrete Reinforcing	115.1	117.6	116.4	109.1	118.6	113.9	109.8	118.5	114.3	115.9	118.2	117.0	113.0	118.0	115.5	123.8	118.3	121.0
0330	Cast-in-Place Concrete	109.9	118.4	113.1	125.5	117.7	122.5	146.1	116.3	134.8	89.2	119.4	100.6	103.1	118.8	109.1	122.2	116.8	120.1
03	CONCRETE	115.9	112.8	114.4	120.8	128.0	124.2	145.9	127.0	137.1	97.0	113.3	104.6	107.4	120.7	113.6	118.4	122.6	120.3
04	MASONRY	109.8	106.2	107.6	144.7	120.1	129.9	119.2	120.0	119.7	80.3	116.1	101.8	108.3	110.4	109.5	148.2	114.2	127.8
05	METALS	101.3	100.2	100.9	95.5	99.8	96.9	97.1	95.9	96.7	103.9	101.5	103.1	99.2	101.4	99.9	105.6	105.6	105.6
06	WOOD, PLASTICS & COMPOSITES	101.4	104.5	103.2	111.9	147.9	132.3	108.8	147.7	130.8	98.8	104.4	102.0	94.1	127.1	112.8	102.0	132.6	119.3
07	THERMAL & MOISTURE PROTECTION	103.8	102.0	103.1	113.9	123.1	117.4	118.4	120.0	119.0	103.5	107.4	105.0	103.2	106.2	104.3	103.0	116.5	108.1
08	OPENINGS	100.0	103.9	101.0	103.4	132.9	110.7	113.5	132.7	118.3	99.1	106.3	100.9	101.9	120.8	106.6	95.4	124.6	102.6
0920	Plaster & Gypsum Board	115.9	104.5	108.0	109.0	148.9	136.6	111.5	148.9	137.4	108.6	104.5	105.7	107.3	127.7	121.4	111.3	133.3	126.5
0950, 0980	Ceilings & Acoustic Treatment	116.5	104.5	108.9	112.3	148.9	135.6	120.0	148.9	138.4	112.2	104.5	107.3	117.0	127.7	123.8	113.9	133.3	126.3
0960	Flooring	123.0	84.4	112.3	120.3	111.6	117.9	131.8	108.8	125.5	121.6	96.0	114.5	114.4	84.4	106.1	111.0	111.6	111.2
0970, 0990	Wall Finishes & Painting/Coating	105.6	93.2	98.1	108.3	123.9	117.7	104.5	123.9	116.2	102.7	98.7	100.3	105.6	93.2	98.1	106.5	141.3	127.5
09	FINISHES	115.7	101.5	108.0	113.9	137.0	126.5	117.1	136.4	127.5	109.9	104.4	107.0	111.5	115.7	113.8	112.1	129.6	121.6
COVERS	DIVS. 10 - 14, 25, 28, 41, 43, 44	100.0	107.0	101.5	100.0	128.2	106.0	100.0	127.5	105.8	100.0	109.9	102.1	100.0	112.8	102.7	100.0	125.4	105.4
21, 22, 23	FIRE SUPPRESSION, PLUMBING & HVAC	93.4	96.4	94.6	93.4	137.1	111.4	93.4	132.9	109.6	93.3	96.3	94.6	100.1	112.9	105.4	100.1	108.2	103.4
26, 27, 3370	ELECTRICAL, COMMUNICATIONS & UTIL.	91.7	89.3	90.5	100.8	127.1	114.5	97.3	103.4	100.5	97.5	94.5	95.9	90.6	96.4	93.7	103.0	108.6	105.9
MF2004	WEIGHTED AVERAGE	102.1	101.4	101.8	105.8	125.7	114.6	108.5	121.4	114.2	97.9	104.0	100.6	101.7	110.5	105.6	107.7	114.4	110.7

	DIVISION	CALIFORNIA SANTA ROSA 954			CALIFORNIA STOCKTON 952			CALIFORNIA SUSANVILLE 961			CALIFORNIA VALLEJO 945			CALIFORNIA VAN NUYS 913 - 916			COLORADO ALAMOSA 811		
		MAT.	INST.	TOTAL	MAT.	INST.	TOTAL	MAT.	INST.	TOTAL	MAT.	INST.	TOTAL	MAT.	INST.	TOTAL	MAT.	INST.	TOTAL
015433	CONTRACTOR EQUIPMENT		99.6	99.6		99.0	99.0		99.0	99.0		98.8	98.8		98.6	98.6		96.4	96.4
0241, 31 - 34	SITE & INFRASTRUCTURE, DEMOLITION	105.1	106.1	105.8	104.1	106.1	105.5	117.7	105.0	108.9	106.0	110.0	108.8	117.8	109.4	111.9	124.7	92.6	102.4
0310	Concrete Forming & Accessories	103.3	142.5	137.2	107.1	127.4	124.6	107.8	127.0	124.4	107.8	142.3	137.6	118.4	108.3	109.7	105.4	68.4	73.4
0320	Concrete Reinforcing	110.6	119.3	115.0	113.5	117.8	115.7	109.6	117.8	113.8	111.1	118.6	114.9	118.7	117.8	118.3	104.3	75.0	89.3
0330	Cast-in-Place Concrete	116.2	116.0	116.1	103.1	114.5	107.4	106.7	114.0	109.4	116.3	116.3	116.3	96.4	116.8	104.1	102.4	77.2	92.9
03	CONCRETE	117.9	127.5	122.4	107.4	120.0	113.3	118.8	119.6	119.2	117.5	127.0	122.0	118.3	112.3	115.5	118.2	73.0	97.1
04	MASONRY	111.3	131.7	123.5	111.4	108.2	109.5	112.6	108.0	109.9	87.3	118.9	106.3	120.8	109.6	114.1	128.6	70.9	93.9
05	METALS	104.3	104.3	104.3	100.4	100.8	100.5	102.5	99.2	101.4	97.1	98.0	97.4	85.3	99.9	90.2	98.7	80.0	92.4
06	WOOD, PLASTICS & COMPOSITES	89.9	147.5	122.5	95.7	129.1	114.6	96.0	129.1	114.8	97.3	147.7	125.8	89.6	104.3	97.9	99.4	70.1	82.8
07	THERMAL & MOISTURE PROTECTION	103.9	123.1	111.2	106.8	107.7	107.2	107.8	104.1	106.4	116.2	119.4	117.4	103.1	107.3	104.6	103.1	75.2	92.5
08	OPENINGS	100.9	136.4	109.7	101.1	120.2	105.9	103.3	119.0	107.2	115.9	136.5	121.0	92.2	106.2	95.7	92.4	75.4	88.2
0920	Plaster & Gypsum Board	103.8	148.9	135.0	107.7	129.8	123.0	106.4	129.8	122.6	105.9	148.9	135.6	112.7	104.5	107.0	84.4	68.9	73.7
0950, 0980	Ceilings & Acoustic Treatment	112.2	148.9	135.5	117.0	129.8	125.2	116.5	129.8	125.0	121.9	148.9	139.1	98.1	104.5	102.2	121.0	68.9	87.8
0960	Flooring	117.1	125.5	119.4	114.4	101.2	110.7	114.4	105.3	111.9	126.5	125.5	126.2	100.4	96.0	99.2	114.4	59.0	99.1
0970, 0990	Wall Finishes & Painting/Coating	102.7	123.9	115.4	105.6	101.5	103.2	105.6	114.2	110.8	105.4	123.9	116.5	107.5	110.0	109.0	108.4	26.2	58.9
09	FINISHES	109.0	139.1	125.3	111.4	121.3	116.8	112.5	123.4	118.4	113.6	139.2	127.5	103.8	105.6	104.8	111.4	61.9	84.6
COVERS	DIVS. 10 - 14, 25, 28, 41, 43, 44	100.0	126.6	105.6	100.0	124.7	105.2	100.0	124.6	105.2	100.0	127.0	105.7	100.0	109.4	102.0	100.0	87.5	97.4
21, 22, 23	FIRE SUPPRESSION, PLUMBING & HVAC	93.3	132.1	109.3	100.1	98.8	99.6	93.4	98.7	95.6	100.1	112.0	105.0	93.4	96.3	94.6	93.3	63.8	81.2
26, 27, 3370	ELECTRICAL, COMMUNICATIONS & UTIL.	97.8	103.4	100.8	100.5	107.4	104.1	101.4	94.3	97.7	93.3	111.9	103.0	116.5	100.5	108.1	96.1	74.5	84.8
MF2004	WEIGHTED AVERAGE	102.6	123.5	111.9	103.0	109.7	106.0	104.0	107.7	105.6	103.9	118.8	110.5	101.4	104.2	102.6	102.6	72.5	89.3

	DIVISION	COLORADO BOULDER 803			COLORADO SPRINGS 808 - 809			DENVER 800 - 802			DURANGO 813			FORT COLLINS 805			FORT MORGAN 807		
		MAT.	INST.	TOTAL	MAT.	INST.	TOTAL	MAT.	INST.	TOTAL	MAT.	INST.	TOTAL	MAT.	INST.	TOTAL	MAT.	INST.	TOTAL
015433	CONTRACTOR EQUIPMENT		96.1	96.1		94.7	94.7		99.2	99.2		96.4	96.4		96.1	96.1		96.1	96.1
0241, 31 - 34	SITE & INFRASTRUCTURE, DEMOLITION	88.8	96.9	94.4	90.2	94.7	93.3	89.4	103.2	99.0	118.6	92.6	100.5	100.2	96.6	97.7	91.4	96.7	95.1
0310	Concrete Forming & Accessories	103.3	73.0	77.1	94.9	77.5	79.8	99.9	81.6	84.1	110.6	68.5	74.2	101.0	71.4	75.4	103.7	72.5	76.7
0320	Concrete Reinforcing	102.0	75.1	88.3	101.0	82.2	91.4	101.0	82.3	91.5	104.3	75.0	89.3	102.1	74.0	87.7	102.2	75.0	88.3
0330	Cast-in-Place Concrete	97.2	77.0	89.5	99.8	80.0	92.3	94.5	82.0	89.8	117.8	77.2	102.5	109.7	74.8	96.5	95.3	76.1	88.1
03	CONCRETE	101.1	75.2	89.0	104.6	79.6	92.9	98.8	82.1	91.0	120.5	73.1	98.4	111.6	73.4	93.8	99.7	74.5	87.9
04	MASONRY	94.0	72.4	81.0	97.4	73.1	82.8	99.3	79.1	87.2	116.3	70.9	89.0	109.6	50.7	74.2	109.4	71.1	86.4
05	METALS	102.2	82.5	95.5	105.4	87.0	99.2	108.0	87.4	101.1	98.7	80.1	92.4	103.5	78.4	95.0	101.9	79.8	94.5
06	WOOD, PLASTICS & COMPOSITES	101.3	75.2	86.5	93.1	80.8	86.2	99.7	83.6	90.6	106.5	70.1	85.9	99.5	75.2	85.8	101.7	75.2	86.7
07	THERMAL & MOISTURE PROTECTION	99.0	76.0	90.3	99.6	77.5	91.3	98.5	76.2	90.1	103.0	75.2	92.5	99.4	61.9	85.2	98.9	71.6	88.6
08	OPENINGS	95.5	78.2	91.2	99.7	83.3	95.6	99.5	85.4	96.0	99.7	75.4	93.7	95.4	78.2	91.2	95.4	78.2	91.1
0920	Plaster & Gypsum Board	106.8	74.7	84.6	90.2	80.3	83.3	104.5	83.4	89.9	93.9	68.9	76.6	102.4	74.7	83.3	108.1	74.7	85.0
0950, 0980	Ceilings & Acoustic Treatment	110.4	74.7	87.7	118.1	80.3	94.0	117.3	83.4	95.7	121.0	68.9	87.8	110.4	74.7	87.7	110.4	74.7	87.7
0960	Flooring	112.5	83.3	104.4	103.5	73.5	95.2	108.6	91.3	103.8	119.4	59.0	102.7	108.5	59.0	94.8	113.0	59.0	98.1
0970, 0990	Wall Finishes & Painting/Coating	104.5	74.0	86.1	104.2	45.0	68.6	104.5	74.0	86.1	108.4	26.2	58.9	104.5	44.7	68.5	104.5	59.5	77.4
09	FINISHES	108.2	74.4	89.9	105.3	73.5	88.0	108.2	82.7	94.4	113.3	61.9	85.4	107.0	66.4	85.0	108.5	68.6	86.8
COVERS	DIVS. 10 - 14, 25, 28, 41, 43, 44	100.0	87.1	97.3	100.0	88.1	97.5	100.0	89.6	97.8	100.0	87.5	97.4	100.0	86.3	97.1	100.0	87.1	97.3
21, 22, 23	FIRE SUPPRESSION, PLUMBING & HVAC	93.2	75.7	86.0	100.1	70.8	88.1	100.0	83.2	93.1	93.3	66.2	82.1	100.0	74.5	89.5	93.2	75.6	86.0
26, 27, 3370	ELECTRICAL, COMMUNICATIONS & UTIL.	98.3	82.0	89.8	102.2	80.0	90.6	103.9	85.4	94.2	95.5	75.6	85.1	98.3	82.0	89.8	98.7	82.0	90.0
MF2004	WEIGHTED AVERAGE	98.2	78.9	89.7	101.5	78.7	91.4	101.6	85.1	94.3	102.9	73.2	89.8	102.4	74.4	90.0	98.9	77.5	89.5

	DIVISION	COLORADO GLENWOOD SPRINGS 816			GOLDEN 804			GRAND JUNCTION 815			GREELEY 806			MONTROSE 814			PUEBLO 810		
		MAT.	INST.	TOTAL	MAT.	INST.	TOTAL	MAT.	INST.	TOTAL	MAT.	INST.	TOTAL	MAT.	INST.	TOTAL	MAT.	INST.	TOTAL
015433	CONTRACTOR EQUIPMENT		99.2	99.2		96.1	96.1		99.2	99.2		96.1	96.1		97.8	97.8		96.4	96.4
0241, 31 - 34	SITE & INFRASTRUCTURE, DEMOLITION	133.2	99.7	109.9	101.1	97.0	98.2	118.1	99.3	105.0	87.6	95.4	93.1	127.1	95.8	105.3	110.0	94.5	99.2
0310	Concrete Forming & Accessories	102.9	72.3	76.5	97.1	72.5	75.8	109.5	71.7	76.8	99.2	45.0	52.3	102.5	72.2	76.3	107.1	77.8	81.7
0320	Concrete Reinforcing	103.1	75.0	88.7	102.2	75.0	88.3	103.5	73.9	88.4	102.0	73.5	87.4	103.0	74.9	88.6	99.7	82.1	90.7
0330	Cast-in-Place Concrete	102.4	76.5	92.6	95.4	76.2	88.1	113.4	75.7	99.1	91.7	53.1	77.1	102.4	76.5	92.6	101.6	81.1	93.9
03	CONCRETE	123.8	74.6	100.8	110.2	74.5	93.6	116.7	73.8	96.6	96.4	54.2	76.7	114.1	74.5	95.6	106.3	80.1	94.0
04	MASONRY	104.2	71.1	84.3	112.5	72.1	88.2	136.9	52.7	86.3	104.1	33.6	61.7	108.5	70.9	85.9	100.8	71.5	83.2
05	METALS	98.4	80.4	92.3	102.1	80.1	94.7	100.0	77.6	92.5	103.4	76.9	94.5	97.6	79.5	91.5	101.7	87.8	97.0
06	WOOD, PLASTICS & COMPOSITES	94.5	75.4	83.7	95.0	75.2	83.8	104.0	75.4	87.8	97.2	44.7	67.4	96.0	75.5	84.4	101.5	81.1	90.0
07	THERMAL & MOISTURE PROTECTION	102.9	71.0	90.8	99.8	69.5	88.3	101.9	58.7	85.6	98.8	51.9	81.1	103.1	75.8	92.7	101.5	76.9	92.2
08	OPENINGS	98.9	78.3	93.8	95.4	78.2	91.2	99.7	78.3	94.4	95.4	61.7	87.1	99.9	78.4	94.6	94.2	83.4	91.5
0920	Plaster & Gypsum Board	114.0	74.7	86.8	100.3	74.7	82.6	123.6	74.7	89.8	101.2	43.3	61.1	83.1	74.7	77.3	89.1	80.3	83.0
0950, 0980	Ceilings & Acoustic Treatment	120.1	74.7	91.2	110.4	74.7	87.7	120.1	74.7	91.2	110.4	43.3	67.6	121.0	74.7	91.5	131.3	80.3	98.8
0960	Flooring	113.4	54.3	97.1	106.1	59.0	93.1	118.9	59.0	102.3	107.2	59.0	93.9	116.8	48.9	98.0	115.3	91.3	108.7
0970, 0990	Wall Finishes & Painting/Coating	108.4	59.5	78.9	104.5	74.0	86.1	108.4	74.0	87.7	104.5	27.8	58.4	108.4	26.2	58.9	108.4	41.7	68.3
09	FINISHES	114.9	67.9	89.4	106.7	70.2	86.9	116.0	70.3	91.2	105.7	44.4	72.4	112.0	63.1	85.4	112.9	76.8	93.3
COVERS	DIVS. 10 - 14, 25, 28, 41, 43, 44	100.0	87.4	97.3	100.0	87.1	97.3	100.0	87.3	97.3	100.0	79.5	95.6	100.0	87.7	97.4	100.0	88.8	97.6
21, 22, 23	FIRE SUPPRESSION, PLUMBING & HVAC	93.3	65.9	82.0	93.2	75.4	85.9	100.0	61.4	84.1	100.0	69.8	87.6	93.3	65.9	82.0	100.0	64.1	85.2
26, 27, 3370	ELECTRICAL, COMMUNICATIONS & UTIL.	92.7	75.7	83.8	98.7	82.0	90.0	95.1	59.3	76.4	98.3	82.0	89.8	95.1	59.3	76.4	96.1	74.6	84.9
MF2004	WEIGHTED AVERAGE	102.7	74.8	90.4	100.5	77.8	90.5	105.4	69.3	89.5	99.7	64.5	84.1	101.6	71.7	88.4	101.5	76.9	90.7

	DIVISION	COLORADO			CONNECTICUT															
		SALIDA			BRIDGEPORT			BRISTOL			HARTFORD			MERIDEN			NEW BRITAIN		
		812			066			060			061			064			060		
		MAT.	INST.	TOTAL	MAT.	INST.	TOTAL	MAT.	INST.	TOTAL	MAT.	INST.	TOTAL	MAT.	INST.	TOTAL	MAT.	INST.	TOTAL
015433	CONTRACTOR EQUIPMENT		97.8	97.8		100.6	100.6		100.6	100.6		100.6	100.6		101.0	101.0		100.6	100.6
0241, 31 - 34	SITE & INFRASTRUCTURE, DEMOLITION	118.3	96.1	102.9	97.7	104.4	102.3	96.9	104.3	102.1	94.4	104.3	101.3	95.2	105.1	102.1	97.1	104.3	102.1
0310	Concrete Forming & Accessories	109.4	72.3	77.3	97.7	121.9	118.6	97.7	121.6	118.3	95.6	121.6	118.1	97.4	121.7	118.4	97.9	121.6	118.4
0320	Concrete Reinforcing	102.7	74.9	88.5	110.6	127.2	119.1	110.6	127.1	119.1	110.6	127.1	119.1	110.6	127.2	119.1	110.6	127.1	119.1
0330	Cast-in-Place Concrete	117.3	76.5	101.9	101.0	117.5	107.3	94.7	117.4	103.3	97.1	117.4	104.8	91.1	117.4	101.0	96.2	117.4	104.2
03	CONCRETE	115.2	74.5	96.2	109.2	121.2	114.8	106.1	121.0	113.0	107.1	121.0	113.6	104.3	121.1	112.1	106.8	121.0	113.4
04	MASONRY	138.2	70.9	97.8	107.1	132.0	122.1	99.6	132.0	119.1	99.8	132.0	119.2	99.3	132.0	119.0	101.6	132.0	119.9
05	METALS	97.3	79.6	91.3	100.8	125.0	109.0	100.8	124.7	108.9	105.5	124.7	112.0	98.0	124.9	107.1	97.1	124.7	106.4
06	WOOD, PLASTICS & COMPOSITES	102.6	75.5	87.3	98.8	119.9	110.8	98.8	119.9	110.8	97.4	119.9	110.1	98.8	119.9	110.8	98.8	119.9	110.8
07	THERMAL & MOISTURE PROTECTION	102.0	75.8	92.1	95.9	127.1	107.6	96.0	123.9	106.5	96.7	123.9	106.9	96.0	123.9	106.5	96.0	123.9	106.5
08	OPENINGS	92.5	78.4	89.0	104.3	129.7	110.6	104.3	126.7	109.9	105.0	126.7	110.4	107.1	129.7	112.7	104.3	126.7	109.9
0920	Plaster & Gypsum Board	83.9	74.7	77.6	99.6	119.8	113.6	99.6	119.8	113.6	101.1	119.8	114.1	101.4	119.8	114.1	99.6	119.8	113.6
0950, 0980	Ceilings & Acoustic Treatment	121.0	74.7	91.5	98.7	119.8	112.1	98.7	119.8	112.1	98.7	119.8	112.1	102.8	119.8	113.6	98.7	119.8	112.1
0960	Flooring	122.3	48.9	102.0	97.6	123.1	104.6	97.6	107.5	100.3	97.6	107.5	100.3	97.6	123.1	104.6	97.6	107.5	100.3
0970, 0990	Wall Finishes & Painting/Coating	108.4	26.2	58.9	94.6	118.2	108.8	94.6	118.2	108.8	94.6	118.2	108.8	94.6	118.2	108.8	94.6	118.2	108.8
09	FINISHES	112.8	63.1	85.8	101.1	121.5	112.2	101.1	118.8	110.8	100.8	118.8	110.6	102.3	121.5	112.8	101.1	118.8	110.8
COVERS	DIVS. 10 - 14, 25, 28, 41, 43, 44	100.0	87.8	97.4	100.0	114.1	103.0	100.0	114.1	103.0	100.0	114.1	103.0	100.0	114.1	103.0	100.0	114.1	103.0
21, 22, 23	FIRE SUPPRESSION, PLUMBING & HVAC	93.3	63.8	81.1	100.1	112.8	105.4	100.1	112.8	105.3	100.1	112.8	105.3	93.4	112.8	101.4	100.1	112.8	105.3
26, 27, 3370	ELECTRICAL, COMMUNICATIONS & UTIL.	95.4	74.5	84.5	97.9	109.4	103.9	97.9	109.1	103.7	98.0	110.0	104.3	97.8	109.1	103.7	98.0	109.1	103.8
MF2004	WEIGHTED AVERAGE	102.4	73.4	89.6	101.7	118.2	109.0	100.9	117.5	108.2	101.7	117.6	108.7	99.0	118.1	107.4	100.6	117.5	108.1

	DIVISION	CONNECTICUT																	
		NEW HAVEN			NEW LONDON			NORWALK			STAMFORD			WATERBURY			WILLIMANTIC		
		065			063			068			069			067			062		
		MAT.	INST.	TOTAL	MAT.	INST.	TOTAL	MAT.	INST.	TOTAL	MAT.	INST.	TOTAL	MAT.	INST.	TOTAL	MAT.	INST.	TOTAL
015433	CONTRACTOR EQUIPMENT		101.0	101.0		101.0	101.0		100.6	100.6		100.6	100.6		100.6	100.6		100.6	100.6
0241, 31 - 34	SITE & INFRASTRUCTURE, DEMOLITION	96.9	105.1	102.6	89.9	105.1	100.5	97.5	104.4	102.3	98.1	104.4	102.5	97.3	104.4	102.2	97.5	104.3	102.3
0310	Concrete Forming & Accessories	97.4	121.5	118.3	97.4	121.1	117.9	97.7	122.1	118.8	97.6	122.4	119.1	97.7	121.7	118.5	97.6	121.3	118.1
0320	Concrete Reinforcing	110.6	127.1	119.0	86.8	127.1	107.3	110.6	127.3	119.1	110.6	127.4	119.2	110.6	127.2	119.1	110.6	127.1	119.0
0330	Cast-in-Place Concrete	97.9	108.5	101.9	83.4	108.3	92.8	99.4	128.8	110.5	101.0	129.0	111.6	101.0	117.4	107.2	94.4	98.8	96.0
03	CONCRETE	123.6	117.9	121.0	93.8	117.7	105.0	108.4	125.2	116.2	109.2	125.4	116.8	109.2	121.1	114.8	105.9	114.5	109.9
04	MASONRY	99.8	132.0	119.2	98.2	132.0	118.5	99.3	133.6	119.9	100.1	133.6	120.3	100.1	132.0	119.3	99.5	132.0	119.0
05	METALS	97.3	124.7	106.6	97.0	124.3	106.2	100.8	125.4	109.1	100.8	125.7	109.2	100.8	124.9	108.9	100.6	124.5	108.6
06	WOOD, PLASTICS & COMPOSITES	98.8	119.9	110.8	98.8	119.9	110.8	98.8	119.9	110.8	98.8	119.9	110.8	98.8	119.9	110.8	98.8	119.9	110.8
07	THERMAL & MOISTURE PROTECTION	96.0	122.5	106.0	95.9	122.5	106.0	96.0	129.2	108.6	96.0	129.3	108.5	96.0	123.9	106.5	96.1	121.0	105.5
08	OPENINGS	104.3	129.7	110.6	107.5	117.7	110.0	104.3	129.7	110.6	104.3	129.7	110.6	104.3	129.7	110.6	107.6	129.7	113.1
0920	Plaster & Gypsum Board	99.6	119.8	113.6	99.6	119.8	113.6	99.6	119.8	113.6	99.6	119.8	113.6	99.6	119.8	113.6	99.6	119.8	113.6
0950, 0980	Ceilings & Acoustic Treatment	98.7	119.8	112.1	96.8	119.8	111.4	98.7	119.8	112.1	98.7	119.8	112.1	98.7	119.8	112.1	96.8	119.8	111.4
0960	Flooring	97.6	123.1	104.6	97.6	107.5	100.3	97.6	123.1	104.6	97.6	123.1	104.6	97.6	123.1	104.6	97.6	123.1	104.6
0970, 0990	Wall Finishes & Painting/Coating	94.6	118.2	108.8	94.6	118.2	108.8	94.6	118.2	108.8	94.6	118.2	108.8	94.6	118.2	108.8	94.6	118.2	108.8
09	FINISHES	101.2	121.5	112.2	100.2	118.8	110.3	101.2	121.5	112.2	101.2	121.5	112.2	101.0	121.5	112.2	100.8	121.5	112.1
COVERS	DIVS. 10 - 14, 25, 28, 41, 43, 44	100.0	114.1	103.0	100.0	114.1	103.0	100.0	114.1	103.0	100.0	114.3	103.0	100.0	114.1	103.0	100.0	114.1	103.0
21, 22, 23	FIRE SUPPRESSION, PLUMBING & HVAC	100.1	112.8	105.3	93.4	112.7	101.3	100.1	112.8	105.4	100.1	112.9	105.4	100.1	112.8	105.3	100.1	112.8	105.3
26, 27, 3370	ELECTRICAL, COMMUNICATIONS & UTIL.	97.8	109.1	103.7	94.5	109.1	102.1	97.9	107.8	103.1	97.9	156.0	128.3	97.4	109.4	103.7	97.9	110.0	104.2
MF2004	WEIGHTED AVERAGE	102.6	117.6	109.2	96.9	116.6	105.6	101.2	118.7	109.0	101.4	125.5	112.0	101.3	118.0	108.7	101.2	117.1	108.2

	DIVISION	D.C.			DELAWARE									FLORIDA					
		WASHINGTON			DOVER			NEWARK			WILMINGTON			DAYTONA BEACH			FORT LAUDERDALE		
		200 - 205			199			197			198			321			333		
		MAT.	INST.	TOTAL	MAT.	INST.	TOTAL	MAT.	INST.	TOTAL	MAT.	INST.	TOTAL	MAT.	INST.	TOTAL	MAT.	INST.	TOTAL
015433	CONTRACTOR EQUIPMENT		104.4	104.4		117.7	117.7		117.7	117.7		117.8	117.8		97.8	97.8		90.4	90.4
0241, 31 - 34	SITE & INFRASTRUCTURE, DEMOLITION	112.3	92.8	98.7	101.0	112.0	108.7	100.9	112.0	108.6	90.9	112.3	105.8	116.3	89.2	97.5	101.6	77.3	84.7
0310	Concrete Forming & Accessories	101.3	82.1	84.7	97.8	101.9	101.3	99.9	101.9	101.6	100.6	101.9	101.7	94.8	75.7	78.2	93.5	76.5	78.8
0320	Concrete Reinforcing	103.0	89.0	95.8	93.1	102.4	97.9	93.9	102.4	98.2	93.9	102.4	98.2	86.0	80.7	83.3	86.0	80.4	83.1
0330	Cast-in-Place Concrete	126.8	88.3	112.2	94.4	102.7	97.5	85.6	102.7	92.1	89.5	102.7	94.5	93.4	76.7	87.1	98.0	84.6	92.9
03	CONCRETE	117.1	86.8	102.9	97.6	103.2	100.2	93.6	103.2	98.1	95.5	103.2	99.1	89.3	78.0	84.0	91.5	81.1	86.7
04	MASONRY	104.8	83.0	91.7	104.8	98.7	101.1	103.1	98.7	100.5	108.7	98.7	102.7	86.6	73.1	78.4	86.8	76.2	80.4
05	METALS	94.3	108.4	99.0	107.7	117.1	110.9	107.7	117.1	110.9	104.7	117.1	108.9	98.1	93.8	96.7	98.1	95.7	97.3
06	WOOD, PLASTICS & COMPOSITES	99.9	81.8	89.6	96.3	101.2	99.1	99.0	101.2	100.3	101.1	101.2	101.2	95.6	77.5	85.4	92.0	75.2	82.5
07	THERMAL & MOISTURE PROTECTION	102.4	86.1	96.3	99.1	115.3	105.2	98.8	115.3	105.1	99.4	115.3	105.4	93.9	81.8	89.3	93.9	89.9	92.4
08	OPENINGS	107.8	91.0	103.6	96.2	108.8	99.3	96.2	108.8	99.3	95.9	108.8	99.1	100.1	74.2	93.7	97.5	73.3	91.5
0920	Plaster & Gypsum Board	107.5	81.2	89.3	109.9	101.1	103.8	112.0	101.1	104.5	111.7	101.1	104.4	95.3	77.4	82.9	94.9	75.0	81.1
0950, 0980	Ceilings & Acoustic Treatment	107.9	81.2	90.9	96.6	101.1	99.5	99.2	101.1	100.4	92.2	101.1	97.9	98.4	77.4	85.0	98.4	75.0	83.5
0960	Flooring	127.0	97.0	118.7	83.5	108.2	90.4	83.7	108.2	90.5	83.3	108.2	90.2	116.6	79.3	106.3	116.6	74.7	105.0
0970, 0990	Wall Finishes & Painting/Coating	125.0	84.5	100.6	92.5	102.7	98.7	92.5	102.7	98.7	92.5	102.7	98.7	111.0	82.3	93.7	106.1	58.2	77.3
09	FINISHES	108.9	84.6	95.7	97.9	102.5	100.4	99.0	102.5	100.9	97.3	102.5	100.2	105.7	77.0	90.1	103.1	73.5	87.0
COVERS	DIVS. 10 - 14, 25, 28, 41, 43, 44	100.0	98.3	99.6	100.0	101.2	100.2	100.0	101.2	100.2	100.0	101.2	100.2	100.0	85.7	97.0	100.0	89.1	97.7
21, 22, 23	FIRE SUPPRESSION, PLUMBING & HVAC	100.2	93.1	97.3	100.0	115.7	106.4	100.1	115.7	106.5	100.1	115.7	106.5	99.8	66.8	86.2	99.8	68.7	87.0
26, 27, 3370	ELECTRICAL, COMMUNICATIONS & UTIL.	104.1	99.9	101.9	100.7	109.4	105.3	98.2	109.4	104.1	103.7	109.4	106.7	98.4	61.6	79.1	98.4	76.0	86.7
MF2004	WEIGHTED AVERAGE	104.0	92.2	98.8	100.5	108.7	104.1	99.8	108.7	103.7	100.1	108.7	103.9	98.2	75.3	88.1	97.5	77.4	88.7

DIVISION		FLORIDA																	
		FORT MYERS			GAINESVILLE			JACKSONVILLE			LAKELAND			MELBOURNE			MIAMI		
		339,341			326,344			320,322			338			329			330 - 332,340		
		MAT.	INST.	TOTAL	MAT.	INST.	TOTAL	MAT.	INST.	TOTAL	MAT.	INST.	TOTAL	MAT.	INST.	TOTAL	MAT.	INST.	TOTAL
015433	CONTRACTOR EQUIPMENT		97.8	97.8		97.8	97.8		97.8	97.8		97.8	97.8		97.8	97.8		90.4	90.4
0241, 31 - 34	SITE & INFRASTRUCTURE, DEMOLITION	113.5	88.8	96.3	127.1	88.2	100.0	116.4	88.5	97.0	115.8	88.8	97.0	124.8	88.5	99.5	105.1	77.1	85.6
0310	Concrete Forming & Accessories	89.5	83.5	84.3	90.3	59.9	64.0	94.6	60.1	64.8	86.2	84.2	84.5	91.5	77.5	79.4	97.7	77.0	79.8
0320	Concrete Reinforcing	86.9	103.4	95.3	91.2	62.2	76.4	86.0	62.8	74.1	89.1	104.3	96.9	86.9	80.7	83.8	86.0	80.5	83.2
0330	Cast-in-Place Concrete	102.2	74.5	91.7	107.2	61.0	89.8	94.3	73.8	86.5	104.5	75.8	93.6	112.6	79.3	100.0	106.3	85.7	98.5
03	CONCRETE	92.4	85.1	88.9	100.4	62.2	82.6	89.8	66.7	79.0	94.4	85.9	90.4	101.1	79.7	91.1	95.9	81.7	89.2
04	MASONRY	80.1	69.0	73.4	99.8	58.9	75.2	86.4	68.1	75.4	95.5	84.4	88.8	84.3	77.6	80.3	86.1	80.2	82.6
05	METALS	100.1	103.5	101.3	97.1	84.3	92.8	96.8	85.2	92.8	100.1	103.7	101.3	106.1	93.9	102.0	102.8	94.4	99.9
06	WOOD, PLASTICS & COMPOSITES	89.1	86.8	87.8	90.1	57.9	71.9	95.6	57.9	74.3	85.2	86.8	86.1	91.6	77.5	83.6	97.3	75.2	84.8
07	THERMAL & MOISTURE PROTECTION	93.7	88.9	91.9	94.1	64.9	83.1	94.1	68.1	84.2	93.6	94.1	93.8	94.2	84.4	90.5	102.7	83.6	95.5
08	OPENINGS	98.8	83.3	94.9	98.2	56.5	87.9	100.1	57.2	89.5	98.7	81.2	94.4	99.3	77.9	94.0	97.5	73.2	91.5
0920	Plaster & Gypsum Board	90.4	86.9	88.0	90.8	57.1	67.5	95.3	57.1	68.9	88.2	86.9	87.3	90.8	77.4	81.5	92.8	75.0	80.5
0950, 0980	Ceilings & Acoustic Treatment	91.3	86.9	88.5	91.3	57.1	69.5	98.4	57.1	72.1	91.3	86.9	88.5	93.2	77.4	83.1	94.1	75.0	81.9
0960	Flooring	112.6	47.5	94.6	113.2	41.4	93.3	116.6	69.5	103.6	109.7	60.9	96.2	113.4	79.3	104.0	124.0	78.6	111.5
0970, 0990	Wall Finishes & Painting/Coating	111.0	66.2	84.0	111.0	47.7	72.9	111.0	51.6	75.3	111.0	66.2	84.0	111.0	102.1	105.6	106.1	58.2	77.3
09	FINISHES	102.0	75.1	87.4	103.4	54.2	76.7	105.7	60.3	81.1	100.9	77.9	88.4	103.6	80.4	91.0	104.4	74.7	88.3
COVERS	DIVS. 10 - 14, 25, 28, 41, 43, 44	100.0	85.9	97.0	100.0	83.8	96.6	100.0	81.6	96.1	100.0	86.0	97.0	100.0	87.4	97.3	100.0	89.8	97.8
21, 22, 23	FIRE SUPPRESSION, PLUMBING & HVAC	96.1	61.7	82.0	98.8	67.1	85.8	99.8	49.2	79.0	96.1	91.7	94.3	99.8	69.1	87.2	99.8	74.4	89.3
26, 27, 3370	ELECTRICAL, COMMUNICATIONS & UTIL.	100.8	45.5	71.9	98.7	80.5	89.2	98.1	69.8	83.3	98.7	50.4	73.5	99.6	71.7	85.0	106.3	78.8	91.9
MF2004	WEIGHTED AVERAGE	97.3	73.9	87.0	99.8	69.1	86.2	98.0	66.3	84.0	98.1	83.0	91.4	100.8	78.5	91.0	100.1	79.4	90.9

DIVISION		FLORIDA																	
		ORLANDO			PANAMA CITY			PENSACOLA			SARASOTA			ST. PETERSBURG			TALLAHASSEE		
		327 - 328,347			324			325			342			337			323		
		MAT.	INST.	TOTAL	MAT.	INST.	TOTAL	MAT.	INST.	TOTAL	MAT.	INST.	TOTAL	MAT.	INST.	TOTAL	MAT.	INST.	TOTAL
015433	CONTRACTOR EQUIPMENT		97.8	97.8		97.8	97.8		97.8	97.8		97.8	97.8		97.8	97.8		97.8	97.8
0241, 31 - 34	SITE & INFRASTRUCTURE, DEMOLITION	119.0	88.3	97.6	131.2	86.5	100.1	130.8	87.8	100.9	117.9	88.6	97.5	117.2	88.1	96.9	114.5	87.5	95.7
0310	Concrete Forming & Accessories	94.5	80.2	82.1	94.0	47.4	53.7	92.0	57.4	62.0	94.8	83.6	85.2	93.0	52.2	57.7	96.9	48.7	55.2
0320	Concrete Reinforcing	86.0	77.7	81.7	89.7	58.9	74.0	92.1	58.8	75.1	86.0	104.2	95.3	89.1	66.2	77.4	86.0	62.4	73.9
0330	Cast-in-Place Concrete	123.1	76.5	105.5	99.1	52.8	81.6	122.4	71.3	103.1	106.8	75.6	95.0	105.7	69.5	92.0	103.9	60.5	87.5
03	CONCRETE	105.3	79.4	93.2	98.0	52.7	76.8	108.9	63.9	87.9	96.0	85.6	91.1	95.7	62.4	80.2	94.6	57.1	77.1
04	MASONRY	90.6	73.1	80.1	90.8	52.2	67.6	109.1	60.5	79.9	87.6	84.4	85.7	132.7	54.5	85.7	89.4	56.1	69.4
05	METALS	103.7	92.2	99.8	97.9	71.4	89.0	99.1	83.5	93.8	100.8	103.1	101.6	101.0	85.8	95.9	91.9	84.5	89.4
06	WOOD, PLASTICS & COMPOSITES	95.4	84.3	89.1	94.6	46.6	67.5	93.0	57.6	73.0	95.6	86.8	90.6	93.3	50.7	69.2	93.2	46.1	66.5
07	THERMAL & MOISTURE PROTECTION	95.2	82.5	90.4	94.3	55.2	79.5	94.2	67.2	84.0	93.9	94.1	93.9	93.8	62.4	82.0	99.8	78.4	91.7
08	OPENINGS	100.1	74.7	93.8	97.7	41.2	83.7	97.7	57.9	87.8	100.1	83.0	95.9	98.7	54.9	87.9	100.6	50.5	88.2
0920	Plaster & Gypsum Board	105.8	84.3	90.9	93.4	45.5	60.3	94.3	56.9	68.4	95.3	86.9	89.5	92.5	49.8	62.9	111.1	45.0	65.4
0950, 0980	Ceilings & Acoustic Treatment	98.4	84.3	89.4	93.2	45.5	62.8	93.2	56.9	70.1	96.5	86.9	90.4	93.2	49.8	65.5	98.4	45.0	64.4
0960	Flooring	116.8	79.3	106.4	116.0	46.3	96.8	110.4	67.8	98.6	116.6	56.7	100.1	115.1	59.4	99.7	116.6	60.6	101.2
0970, 0990	Wall Finishes & Painting/Coating	111.0	61.1	81.0	111.0	52.3	75.7	111.0	50.9	74.9	111.0	66.2	84.0	111.0	53.7	76.5	111.0	48.1	73.2
09	FINISHES	106.9	78.6	91.6	105.5	47.5	74.0	103.5	58.1	78.9	105.2	77.1	90.0	103.6	52.8	76.0	107.2	49.6	75.9
COVERS	DIVS. 10 - 14, 25, 28, 41, 43, 44	100.0	86.4	97.1	100.0	54.2	90.3	100.0	54.4	90.3	100.0	85.9	97.0	100.0	61.5	91.8	100.0	77.8	95.3
21, 22, 23	FIRE SUPPRESSION, PLUMBING & HVAC	99.7	64.1	85.0	99.8	42.1	76.1	99.8	59.1	83.0	99.7	62.2	84.3	99.8	57.2	82.2	99.8	44.0	76.8
26, 27, 3370	ELECTRICAL, COMMUNICATIONS & UTIL.	106.5	48.4	76.1	97.0	46.4	70.6	101.1	58.6	78.9	98.4	46.7	71.4	98.7	51.5	74.0	106.4	46.0	74.8
MF2004	WEIGHTED AVERAGE	102.2	73.2	89.4	99.5	53.1	79.0	102.2	64.4	85.5	99.4	76.1	89.1	101.6	61.6	83.9	99.2	57.7	80.9

DIVISION		FLORIDA						GEORGIA											
		TAMPA			WEST PALM BEACH			ALBANY			ATHENS			ATLANTA			AUGUSTA		
		335 - 336,346			334,349			317,398			306			300 - 303,399			308 - 309		
		MAT.	INST.	TOTAL	MAT.	INST.	TOTAL	MAT.	INST.	TOTAL	MAT.	INST.	TOTAL	MAT.	INST.	TOTAL	MAT.	INST.	TOTAL
015433	CONTRACTOR EQUIPMENT		97.8	97.8		90.4	90.4		91.2	91.2		92.6	92.6		93.1	93.1		92.6	92.6
0241, 31 - 34	SITE & INFRASTRUCTURE, DEMOLITION	117.6	88.8	97.5	98.4	77.4	83.7	101.9	78.4	85.6	108.0	92.4	97.2	103.6	94.4	97.2	100.0	92.1	94.5
0310	Concrete Forming & Accessories	95.6	84.3	85.8	96.5	75.9	78.7	94.1	47.7	54.0	93.1	45.5	52.0	96.2	75.0	77.9	94.0	68.3	71.8
0320	Concrete Reinforcing	86.0	104.3	95.3	88.4	80.1	84.1	85.8	87.1	86.5	91.1	85.3	88.2	90.5	88.8	89.6	91.5	75.8	83.5
0330	Cast-in-Place Concrete	103.2	75.8	92.9	93.1	71.1	84.8	99.8	49.0	80.6	110.5	54.5	89.3	110.5	73.0	96.3	104.4	51.2	84.2
03	CONCRETE	94.2	86.0	90.4	88.4	76.2	82.7	92.3	57.2	75.9	102.2	56.9	81.0	99.8	77.0	89.1	95.2	64.1	80.7
04	MASONRY	87.8	84.4	85.8	86.4	74.3	79.1	89.5	43.9	62.1	75.5	59.9	66.2	89.5	71.4	78.6	89.6	43.5	61.9
05	METALS	100.0	103.8	101.3	97.2	94.9	96.4	97.5	88.2	94.3	90.1	76.1	85.3	90.9	81.0	87.6	89.8	72.7	84.0
06	WOOD, PLASTICS & COMPOSITES	96.8	86.8	91.1	96.3	75.2	84.4	94.7	42.4	65.1	90.3	39.2	61.3	94.0	76.7	84.2	91.4	73.8	81.4
07	THERMAL & MOISTURE PROTECTION	94.0	94.1	94.0	93.6	76.2	87.0	91.6	62.7	80.7	91.8	55.4	78.0	91.8	76.4	86.0	91.4	58.5	79.0
08	OPENINGS	100.1	83.8	96.1	96.7	73.2	90.9	97.5	49.0	85.5	92.9	47.2	81.6	98.7	73.8	92.5	93.0	65.6	86.2
0920	Plaster & Gypsum Board	95.3	86.9	89.5	97.7	75.0	82.0	95.1	41.2	57.8	111.9	37.7	60.6	113.6	76.4	87.9	112.3	73.4	85.4
0950, 0980	Ceilings & Acoustic Treatment	98.4	86.9	91.1	91.3	75.0	80.9	97.5	41.2	61.6	106.8	37.7	62.8	106.8	76.4	87.4	107.7	73.4	85.9
0960	Flooring	116.6	60.9	101.2	119.2	71.6	106.0	116.6	33.2	93.6	87.8	58.0	79.6	89.4	70.2	84.1	88.3	42.3	75.6
0970, 0990	Wall Finishes & Painting/Coating	111.0	66.2	84.0	106.1	57.9	77.1	106.1	55.5	75.6	97.5	38.9	62.3	97.5	86.2	90.7	97.5	40.6	63.3
09	FINISHES	105.7	77.9	90.6	102.3	72.8	86.3	102.8	43.9	70.8	97.5	45.5	69.3	97.7	75.2	85.5	97.2	61.1	77.6
COVERS	DIVS. 10 - 14, 25, 28, 41, 43, 44	100.0	86.5	97.1	100.0	89.1	97.7	100.0	77.2	95.2	100.0	45.8	88.5	100.0	81.8	96.1	100.0	58.1	91.1
21, 22, 23	FIRE SUPPRESSION, PLUMBING & HVAC	99.8	91.7	96.4	96.1	71.1	85.8	99.7	69.9	87.4	93.2	74.1	85.3	100.0	76.0	90.1	100.0	69.4	87.4
26, 27, 3370	ELECTRICAL, COMMUNICATIONS & UTIL.	98.4	51.5	73.9	99.6	59.3	78.6	93.8	57.0	74.6	99.0	74.3	86.1	98.4	77.1	87.3	99.7	69.3	83.8
MF2004	WEIGHTED AVERAGE	99.2	83.3	92.2	96.0	74.2	86.4	97.1	61.7	81.5	94.8	65.6	81.9	97.4	77.8	88.8	96.0	66.5	83.0

DIVISION		GEORGIA																	
		COLUMBUS			DALTON			GAINESVILLE			MACON			SAVANNAH			STATESBORO		
		318 - 319			307			305			310 - 312			313 - 314			304		
		MAT.	INST.	TOTAL	MAT.	INST.	TOTAL	MAT.	INST.	TOTAL	MAT.	INST.	TOTAL	MAT.	INST.	TOTAL	MAT.	INST.	TOTAL
015433	CONTRACTOR EQUIPMENT		91.2	91.2		104.8	104.8		92.6	92.6		102.1	102.1		92.2	92.2		92.0	92.0
0241, 31 - 34	SITE & INFRASTRUCTURE, DEMOLITION	101.8	78.7	85.7	108.9	96.4	100.2	107.5	92.4	97.0	102.3	94.2	96.6	104.4	79.3	86.9	109.4	76.2	86.3
0310	Concrete Forming & Accessories	94.0	59.7	64.4	87.1	51.3	56.1	95.9	47.3	53.9	93.3	56.9	61.8	94.1	53.0	58.6	82.1	44.3	49.4
0320	Concrete Reinforcing	86.0	88.1	87.0	90.7	78.9	84.7	90.9	84.4	87.6	87.1	87.4	87.2	86.8	76.0	81.3	90.3	46.0	67.7
0330	Cast-in-Place Concrete	99.3	58.1	83.7	107.3	48.1	84.9	116.1	54.2	92.7	98.0	51.6	80.5	115.9	50.5	91.2	110.3	62.4	92.2
03	CONCRETE	92.1	65.8	79.8	101.2	56.9	80.5	104.1	57.4	82.3	91.6	62.2	77.9	100.4	57.9	80.5	101.7	52.6	78.7
04	MASONRY	88.7	61.6	72.4	76.8	35.7	52.1	84.9	43.6	60.1	101.2	41.6	65.4	88.6	54.3	68.0	78.3	43.8	57.6
05	METALS	96.9	90.6	94.7	90.7	85.7	89.0	89.3	74.4	84.2	93.2	89.8	92.1	93.6	84.7	90.6	94.2	73.1	87.1
06	WOOD, PLASTICS & COMPOSITES	94.7	58.3	74.1	73.8	53.0	62.0	93.8	42.2	64.6	104.5	57.2	77.7	110.4	50.0	76.2	67.7	43.9	54.3
07	THERMAL & MOISTURE PROTECTION	91.5	63.8	81.0	93.6	53.1	78.3	91.7	53.9	77.4	90.1	65.3	80.7	93.9	58.2	80.4	92.4	50.6	76.6
08	OPENINGS	97.5	59.6	88.1	94.6	48.5	83.2	92.9	43.9	80.8	95.7	58.1	86.4	98.7	50.3	86.7	95.4	35.0	80.4
0920	Plaster & Gypsum Board	95.1	57.5	69.1	96.0	52.0	65.6	113.6	40.9	63.3	104.4	56.4	71.2	94.1	49.0	62.9	97.5	42.7	59.6
0950, 0980	Ceilings & Acoustic Treatment	97.5	57.5	72.1	120.1	52.0	76.8	106.8	40.9	64.8	92.6	56.4	69.6	98.4	49.0	66.9	115.6	42.7	69.2
0960	Flooring	116.6	54.3	99.4	88.7	38.6	74.9	89.4	38.6	75.3	90.9	39.1	76.5	117.4	49.9	98.8	104.0	48.3	88.6
0970, 0990	Wall Finishes & Painting/Coating	106.1	67.5	82.9	88.2	59.3	70.8	97.5	38.9	62.3	108.2	55.5	76.5	108.0	59.1	78.6	95.7	36.4	60.0
09	FINISHES	102.8	58.7	78.8	105.8	48.9	74.9	98.0	42.7	68.0	91.0	52.3	70.0	103.5	52.3	75.7	108.5	43.6	73.2
COVERS	DIVS. 10 - 14, 25, 28, 41, 43, 44	100.0	79.2	95.6	100.0	25.1	84.1	100.0	46.1	88.6	100.0	77.0	95.1	100.0	57.3	90.9	100.0	42.5	87.8
21, 22, 23	FIRE SUPPRESSION, PLUMBING & HVAC	99.8	52.6	80.4	93.2	60.9	79.9	93.2	74.1	85.3	99.8	67.7	86.6	99.8	60.9	83.8	93.6	55.7	78.0
26, 27, 3370	ELECTRICAL, COMMUNICATIONS & UTIL.	94.0	74.3	83.7	108.6	75.9	91.5	99.0	74.3	86.1	92.8	63.6	77.5	97.2	58.1	76.8	99.3	58.1	77.8
MF2004	WEIGHTED AVERAGE	97.0	66.2	83.4	96.6	61.8	81.2	95.4	63.4	81.3	95.8	65.8	82.6	98.4	61.4	82.0	96.6	54.7	78.1

DIVISION		GEORGIA						HAWAII									IDAHO		
		VALDOSTA			WAYCROSS			HILO			HONOLULU			STATES & POSS., GUAM			BOISE		
		316			315			967			968			969			836 - 837		
		MAT.	INST.	TOTAL	MAT.	INST.	TOTAL	MAT.	INST.	TOTAL	MAT.	INST.	TOTAL	MAT.	INST.	TOTAL	MAT.	INST.	TOTAL
015433	CONTRACTOR EQUIPMENT		91.2	91.2		91.2	91.2		99.8	99.8		99.8	99.8		166.6	166.6		100.7	100.7
0241, 31 - 34	SITE & INFRASTRUCTURE, DEMOLITION	112.2	78.5	88.7	109.0	77.1	86.8	130.3	106.3	113.6	134.6	106.3	114.9	172.3	105.9	126.0	76.7	99.8	92.8
0310	Concrete Forming & Accessories	84.7	46.8	51.9	86.4	67.7	70.2	109.1	136.8	133.0	116.4	136.8	134.0	111.4	62.8	69.3	99.9	72.2	75.9
0320	Concrete Reinforcing	87.8	73.8	80.7	87.8	81.0	84.3	114.0	107.1	110.5	113.0	107.1	110.0	205.9	33.5	117.9	98.3	69.2	83.5
0330	Cast-in-Place Concrete	97.6	53.1	80.8	110.3	51.8	88.2	197.4	121.2	168.6	168.1	121.2	150.3	171.1	113.1	149.1	92.7	77.9	87.1
03	CONCRETE	96.3	55.7	77.3	100.4	65.5	84.1	154.0	124.6	140.3	139.9	124.6	132.7	154.2	75.1	117.2	102.5	73.8	89.1
04	MASONRY	94.3	49.3	67.3	95.3	43.7	64.3	131.5	124.9	127.6	122.9	124.9	124.1	187.2	48.4	103.8	126.2	62.0	87.6
05	METALS	96.5	83.8	92.2	95.6	81.2	90.7	110.4	102.1	107.6	131.3	102.1	121.5	142.2	80.0	121.2	102.5	72.9	92.5
06	WOOD, PLASTICS & COMPOSITES	83.3	41.3	59.5	85.4	75.2	79.6	97.5	141.5	122.4	108.1	141.5	127.0	108.4	64.2	83.3	93.5	72.9	81.8
07	THERMAL & MOISTURE PROTECTION	91.7	63.5	81.1	91.5	54.1	77.4	106.8	123.2	113.0	116.5	123.2	119.0	122.9	70.2	103.0	93.4	68.3	83.9
08	OPENINGS	93.0	44.7	81.0	93.0	60.7	85.0	100.7	129.9	107.9	106.0	129.9	111.9	105.5	54.3	92.8	93.0	68.3	86.9
0920	Plaster & Gypsum Board	87.3	40.1	54.6	88.6	75.0	79.2	107.8	142.5	131.8	133.6	142.5	139.8	211.9	51.6	100.9	89.8	72.0	77.5
0950, 0980	Ceilings & Acoustic Treatment	93.2	40.1	59.4	91.3	75.0	80.9	113.9	142.5	132.1	116.2	142.5	132.9	233.1	51.6	117.5	125.5	72.0	91.4
0960	Flooring	108.8	41.7	90.3	110.2	32.2	88.6	141.3	128.4	137.7	165.2	128.4	155.0	164.9	50.0	133.2	99.5	68.4	90.9
0970, 0990	Wall Finishes & Painting/Coating	106.1	55.5	75.6	106.1	44.3	68.9	99.8	138.8	123.2	104.1	138.8	124.9	105.6	39.9	66.1	97.9	43.8	65.4
09	FINISHES	99.3	45.1	69.9	99.1	59.5	77.6	120.5	137.1	129.5	132.7	137.1	135.1	197.0	58.7	121.9	104.7	69.2	85.4
COVERS	DIVS. 10 - 14, 25, 28, 41, 43, 44	100.0	56.3	90.7	100.0	46.0	88.5	100.0	115.1	103.2	100.0	115.1	103.2	100.0	86.0	97.0	100.0	67.2	93.0
21, 22, 23	FIRE SUPPRESSION, PLUMBING & HVAC	99.8	69.8	87.4	95.3	54.1	78.4	100.1	107.0	103.0	100.1	107.0	102.9	102.4	42.2	77.7	100.0	65.2	85.7
26, 27, 3370	ELECTRICAL, COMMUNICATIONS & UTIL.	92.0	57.7	74.1	96.5	63.8	79.4	108.4	116.6	112.7	112.8	116.6	114.7	161.4	46.2	101.2	96.1	74.1	84.6
MF2004	WEIGHTED AVERAGE	97.1	61.1	81.2	96.8	61.4	81.2	113.5	117.9	115.4	116.9	117.9	117.3	135.6	61.7	103.0	100.4	71.8	87.7

DIVISION		IDAHO															ILLINOIS		
		COEUR D'ALENE			IDAHO FALLS			LEWISTON			POCATELLO			TWIN FALLS			BLOOMINGTON		
		838			834			835			832			833			617		
		MAT.	INST.	TOTAL	MAT.	INST.	TOTAL	MAT.	INST.	TOTAL	MAT.	INST.	TOTAL	MAT.	INST.	TOTAL	MAT.	INST.	TOTAL
015433	CONTRACTOR EQUIPMENT		94.6	94.6		100.7	100.7		94.6	94.6		100.7	100.7		100.7	100.7		101.2	101.2
0241, 31 - 34	SITE & INFRASTRUCTURE, DEMOLITION	76.8	94.8	89.3	74.6	98.9	91.5	83.2	95.3	91.6	77.6	99.8	93.0	83.9	98.0	93.7	95.7	96.9	96.5
0310	Concrete Forming & Accessories	118.3	59.0	67.0	94.7	43.2	50.1	122.7	60.8	69.1	100.1	71.9	75.7	100.8	30.6	40.0	90.7	117.5	113.9
0320	Concrete Reinforcing	106.5	90.8	98.5	100.4	67.4	83.6	106.5	91.0	98.6	98.7	69.5	83.8	100.7	67.8	83.9	99.9	103.5	101.7
0330	Cast-in-Place Concrete	100.2	85.0	94.4	88.2	53.9	75.2	104.2	86.6	97.5	95.3	77.8	88.7	97.8	40.6	76.1	98.1	109.8	102.5
03	CONCRETE	109.8	74.4	93.3	94.2	52.3	74.7	113.7	75.8	96.0	101.7	73.6	88.6	109.8	42.2	78.2	96.0	112.2	103.6
04	MASONRY	127.7	77.7	97.7	121.5	33.2	68.4	128.1	82.2	100.5	123.5	67.9	90.1	126.3	32.5	69.9	112.8	114.1	113.6
05	METALS	97.0	83.0	92.3	110.4	69.9	96.7	96.4	84.2	92.3	110.4	73.0	97.8	110.5	69.4	96.6	92.0	110.0	98.1
06	WOOD, PLASTICS & COMPOSITES	100.1	53.2	73.5	87.9	41.8	61.8	104.9	53.2	75.6	93.5	72.9	81.8	94.1	29.6	57.6	87.5	116.4	103.9
07	THERMAL & MOISTURE PROTECTION	144.7	71.4	117.0	93.0	45.9	75.2	144.8	75.1	118.5	93.4	63.7	82.2	94.1	39.9	73.7	99.6	112.0	104.3
08	OPENINGS	112.7	57.2	98.9	97.0	45.8	84.3	112.6	60.0	99.6	93.7	62.7	86.0	97.0	36.9	82.1	90.0	105.0	93.7
0920	Plaster & Gypsum Board	156.1	51.8	83.9	83.3	40.0	53.3	157.4	51.8	84.3	87.0	72.0	76.6	86.9	27.4	45.7	92.7	116.8	109.4
0950, 0980	Ceilings & Acoustic Treatment	161.2	51.8	91.5	121.8	40.0	69.7	161.2	51.8	91.5	131.3	72.0	93.5	125.3	27.4	62.9	92.0	116.8	107.8
0960	Flooring	138.6	49.4	114.0	99.1	46.4	84.6	141.8	86.1	126.4	102.7	68.4	93.2	104.0	46.4	88.1	93.5	110.7	98.3
0970, 0990	Wall Finishes & Painting/Coating	118.7	65.2	86.5	97.9	36.0	60.7	118.7	65.2	86.5	97.8	45.6	66.4	97.9	26.9	55.2	96.2	126.2	114.3
09	FINISHES	172.9	56.6	109.7	102.5	42.5	69.9	174.1	64.6	114.7	106.6	69.4	86.4	106.3	32.2	66.1	95.3	117.7	107.5
COVERS	DIVS. 10 - 14, 25, 28, 41, 43, 44	100.0	55.9	90.6	100.0	42.9	87.9	100.0	70.2	93.7	100.0	67.2	93.0	100.0	38.3	86.9	100.0	105.6	101.2
21, 22, 23	FIRE SUPPRESSION, PLUMBING & HVAC	100.0	76.2	90.2	100.8	57.7	83.0	101.0	78.7	91.8	100.0	65.2	85.6	100.0	54.2	81.1	93.2	108.5	99.5
26, 27, 3370	ELECTRICAL, COMMUNICATIONS & UTIL.	88.0	75.0	81.2	87.6	68.9	77.8	85.9	75.0	80.2	93.1	69.9	81.0	89.1	61.1	74.5	93.7	98.7	96.3
MF2004	WEIGHTED AVERAGE	109.4	73.9	93.7	99.7	57.3	81.0	110.1	76.8	95.4	101.2	71.4	88.1	102.5	51.8	80.1	95.2	108.5	101.0

DIVISION		ILLINOIS																	
		CARBONDALE			CENTRALIA			CHAMPAIGN			CHICAGO			DECATUR			EAST ST. LOUIS		
		629			628			618 - 619			606 - 608			625			620 - 622		
		MAT.	INST.	TOTAL	MAT.	INST.	TOTAL	MAT.	INST.	TOTAL	MAT.	INST.	TOTAL	MAT.	INST.	TOTAL	MAT.	INST.	TOTAL
015433	**CONTRACTOR EQUIPMENT**		107.7	107.7		107.7	107.7		101.9	101.9		93.4	93.4		101.9	101.9		107.7	107.7
0241, 31 - 34	**SITE & INFRASTRUCTURE, DEMOLITION**	103.0	97.4	99.1	103.3	98.4	99.9	105.2	97.7	100.0	103.4	93.6	96.6	89.5	97.6	95.1	105.2	98.4	100.5
0310	Concrete Forming & Accessories	99.1	106.1	105.2	101.0	110.6	109.3	96.3	115.3	112.7	101.7	154.8	147.6	100.5	114.4	112.6	97.0	111.0	109.1
0320	Concrete Reinforcing	98.7	88.5	93.5	98.7	105.2	102.0	99.9	100.9	100.4	98.8	151.1	125.5	96.7	104.7	100.8	98.6	105.2	102.0
0330	Cast-in-Place Concrete	91.8	88.1	90.4	92.2	111.9	99.7	113.7	111.5	112.9	109.7	143.5	122.5	100.2	107.4	102.9	93.8	112.2	100.7
03	**CONCRETE**	86.7	97.4	91.7	87.2	110.7	98.2	108.4	111.2	109.7	103.6	149.0	124.8	97.2	110.2	103.3	88.2	111.0	98.9
04	**MASONRY**	75.1	104.5	92.8	75.1	116.4	100.0	136.3	113.7	122.7	97.7	153.4	131.2	71.8	109.2	94.3	75.3	116.5	100.1
05	**METALS**	91.2	109.3	97.4	91.3	118.6	100.5	92.0	107.3	97.1	92.0	130.3	105.0	94.8	108.2	99.3	92.4	119.1	101.4
06	**WOOD, PLASTICS & COMPOSITES**	101.5	105.5	103.8	103.9	108.5	106.5	94.1	114.6	105.7	106.0	154.3	133.3	101.0	114.6	108.7	99.0	108.5	104.4
07	**THERMAL & MOISTURE PROTECTION**	94.7	96.9	95.5	94.7	106.6	99.2	100.1	112.2	104.7	102.2	143.2	117.7	100.2	103.8	101.6	94.7	110.2	100.6
08	**OPENINGS**	86.3	108.1	91.7	86.3	114.1	93.2	90.7	110.9	95.7	101.8	153.2	114.6	96.6	112.0	100.4	86.4	114.1	93.3
0920	Plaster & Gypsum Board	96.6	105.6	102.8	97.8	108.7	105.3	95.7	115.0	109.0	95.0	155.8	137.0	99.6	115.0	110.2	95.7	108.7	104.7
0950, 0980	Ceilings & Acoustic Treatment	92.0	105.6	100.7	92.0	108.7	102.6	92.0	115.0	106.6	108.2	155.8	138.5	98.1	115.0	108.8	92.0	108.7	102.6
0960	Flooring	111.7	101.7	108.9	112.9	118.0	114.3	96.9	107.2	99.8	91.3	152.8	108.3	100.9	106.8	102.6	110.4	118.0	112.5
0970, 0990	Wall Finishes & Painting/Coating	103.9	101.5	102.5	103.9	109.2	107.1	96.2	111.9	105.6	93.0	152.1	128.6	96.2	107.7	103.1	103.9	109.2	107.1
09	**FINISHES**	99.2	104.9	102.3	99.7	111.9	106.3	97.4	114.0	106.4	97.6	155.2	128.9	99.9	113.3	107.1	98.7	112.2	106.1
COVERS	**DIVS. 10 - 14, 25, 28, 41, 43, 44**	100.0	91.9	98.3	100.0	93.7	98.7	100.0	98.3	99.6	100.0	126.3	105.6	100.0	98.1	99.6	100.0	93.7	98.7
21, 22, 23	**FIRE SUPPRESSION, PLUMBING & HVAC**	93.2	98.5	95.4	93.2	94.2	93.6	93.2	106.4	98.6	100.1	134.5	114.2	100.0	100.1	100.0	100.0	95.4	98.1
26, 27, 3370	**ELECTRICAL, COMMUNICATIONS & UTIL.**	92.5	102.2	97.6	94.0	103.5	99.0	97.2	99.0	98.1	95.8	130.5	113.9	96.7	94.3	95.4	93.6	103.5	98.8
MF2004	**WEIGHTED AVERAGE**	91.9	101.4	96.1	92.2	106.1	98.3	98.8	107.3	102.6	99.0	137.6	116.0	96.5	104.5	100.0	94.0	106.6	99.6

DIVISION		ILLINOIS																	
		EFFINGHAM			GALESBURG			JOLIET			KANKAKEE			LA SALLE			NORTH SUBURBAN		
		624			614			604			609			613			600 - 603		
		MAT.	INST.	TOTAL	MAT.	INST.	TOTAL	MAT.	INST.	TOTAL	MAT.	INST.	TOTAL	MAT.	INST.	TOTAL	MAT.	INST.	TOTAL
015433	**CONTRACTOR EQUIPMENT**		101.9	101.9		101.2	101.2		91.4	91.4		91.4	91.4		101.2	101.2		91.4	91.4
0241, 31 - 34	**SITE & INFRASTRUCTURE, DEMOLITION**	96.9	97.2	97.1	98.1	96.8	97.2	103.2	93.1	96.2	97.0	91.9	93.5	97.5	97.9	97.8	102.4	92.5	95.5
0310	Concrete Forming & Accessories	104.7	111.9	110.9	96.2	116.6	113.9	103.9	158.6	151.2	97.9	134.3	129.4	108.6	122.5	120.6	103.3	152.6	146.0
0320	Concrete Reinforcing	99.9	97.3	98.6	99.3	103.4	101.4	98.8	148.6	124.2	99.7	142.6	121.6	99.4	139.7	120.0	98.8	150.7	125.3
0330	Cast-in-Place Concrete	99.9	107.0	102.5	101.0	96.3	99.2	109.6	147.2	123.8	102.2	125.7	111.1	100.8	111.8	105.0	109.7	144.9	123.0
03	**CONCRETE**	98.0	107.5	102.4	98.7	107.2	102.6	103.7	151.3	126.0	97.6	132.0	113.7	99.4	122.2	110.0	103.7	148.3	124.5
04	**MASONRY**	79.0	106.0	95.2	112.9	114.1	113.7	100.7	149.8	130.2	97.2	129.0	116.3	112.9	118.2	116.1	97.7	146.2	126.8
05	**METALS**	91.9	105.3	96.4	92.0	109.1	97.8	89.8	127.1	102.4	89.8	121.5	100.5	92.1	129.8	104.8	91.0	127.4	103.3
06	**WOOD, PLASTICS & COMPOSITES**	103.2	114.6	109.7	93.9	116.5	106.7	107.8	159.9	137.3	100.9	134.7	120.1	107.6	122.0	115.7	106.0	154.2	133.3
07	**THERMAL & MOISTURE PROTECTION**	99.8	104.7	101.7	99.7	105.4	101.9	102.2	141.7	117.1	101.5	129.8	112.2	99.9	113.6	105.0	102.7	136.5	115.5
08	**OPENINGS**	90.5	110.0	95.3	90.0	108.9	94.7	99.8	155.7	113.6	91.9	140.5	103.9	90.1	125.7	98.9	99.8	153.1	113.0
0920	Plaster & Gypsum Board	99.1	115.0	110.1	95.7	116.8	110.3	92.0	161.7	140.2	89.8	135.7	121.6	101.3	122.5	116.0	95.0	155.8	137.0
0950, 0980	Ceilings & Acoustic Treatment	92.0	115.0	106.6	92.0	116.8	107.8	108.2	161.7	142.2	108.2	135.7	125.7	92.0	122.5	111.4	108.2	155.8	138.5
0960	Flooring	102.3	107.2	103.6	96.9	110.7	100.7	90.9	142.0	105.0	87.9	142.0	102.9	104.6	112.1	106.7	91.3	142.0	105.3
0970, 0990	Wall Finishes & Painting/Coating	96.2	106.8	102.6	96.2	99.0	97.9	91.2	142.2	121.9	91.2	126.2	112.3	96.2	126.2	114.3	93.0	142.2	122.6
09	**FINISHES**	98.9	111.5	105.7	96.8	114.6	106.5	97.1	155.5	128.8	95.4	133.3	116.0	99.9	120.3	111.0	97.6	150.9	126.5
COVERS	**DIVS. 10 - 14, 25, 28, 41, 43, 44**	100.0	77.2	95.2	100.0	105.5	101.2	100.0	126.7	105.7	100.0	119.2	104.1	100.0	105.8	101.2	100.0	124.4	105.2
21, 22, 23	**FIRE SUPPRESSION, PLUMBING & HVAC**	93.2	102.1	96.9	93.2	102.4	97.0	100.2	129.5	112.2	93.4	121.1	104.8	93.2	120.7	104.5	100.1	126.9	111.1
26, 27, 3370	**ELECTRICAL, COMMUNICATIONS & UTIL.**	94.4	102.2	98.5	94.7	89.8	92.1	95.1	126.5	111.5	89.7	125.6	108.5	91.6	123.2	108.1	94.8	125.2	110.7
MF2004	**WEIGHTED AVERAGE**	94.2	104.1	98.6	95.8	104.8	99.8	98.5	135.8	115.0	94.3	124.3	107.5	96.0	119.4	106.3	98.5	133.3	113.9

DIVISION		ILLINOIS																	
		PEORIA			QUINCY			ROCK ISLAND			ROCKFORD			SOUTH SUBURBAN			SPRINGFIELD		
		615 - 616			623			612			610 - 611			605			626 - 627		
		MAT.	INST.	TOTAL	MAT.	INST.	TOTAL	MAT.	INST.	TOTAL	MAT.	INST.	TOTAL	MAT.	INST.	TOTAL	MAT.	INST.	TOTAL
015433	**CONTRACTOR EQUIPMENT**		101.2	101.2		101.9	101.9		101.2	101.2		101.2	101.2		91.4	91.4		101.9	101.9
0241, 31 - 34	**SITE & INFRASTRUCTURE, DEMOLITION**	98.2	96.5	97.0	95.7	97.2	96.8	96.2	96.3	96.3	97.4	97.6	97.5	102.4	92.5	95.5	96.6	97.7	97.3
0310	Concrete Forming & Accessories	98.9	114.5	112.4	102.7	111.8	110.6	97.7	103.0	102.3	102.9	132.7	128.7	103.3	152.6	146.0	101.0	112.4	110.8
0320	Concrete Reinforcing	96.7	103.1	100.0	99.4	100.3	99.9	99.3	103.4	101.4	91.5	138.5	115.5	98.8	150.7	125.3	96.7	104.8	100.8
0330	Cast-in-Place Concrete	97.9	112.0	103.2	100.1	100.3	100.1	98.8	99.4	99.0	100.2	118.0	107.0	109.7	144.9	123.0	89.1	106.0	95.5
03	**CONCRETE**	96.0	111.6	103.3	97.6	105.8	101.4	96.8	102.2	99.3	96.6	128.5	111.5	103.7	148.3	124.5	91.7	108.8	99.7
04	**MASONRY**	113.4	108.1	110.2	98.8	101.2	100.2	112.8	98.5	104.2	88.0	132.5	114.7	97.7	146.2	126.8	76.6	110.0	96.7
05	**METALS**	94.8	109.5	99.7	91.9	105.7	96.6	92.0	108.5	97.6	94.7	127.3	105.7	91.0	127.4	103.3	96.9	108.6	100.9
06	**WOOD, PLASTICS & COMPOSITES**	101.0	111.5	106.9	100.8	114.6	108.6	95.4	102.3	99.3	101.0	131.5	118.2	106.0	154.2	133.3	102.3	110.9	107.2
07	**THERMAL & MOISTURE PROTECTION**	100.3	110.2	104.1	99.8	100.3	100.0	99.7	98.9	99.4	102.8	125.8	111.5	102.7	136.5	115.5	100.1	109.2	103.5
08	**OPENINGS**	96.7	110.2	100.1	91.4	110.9	96.2	90.1	101.3	92.9	96.7	131.7	105.4	99.8	153.1	113.0	97.7	110.0	100.8
0920	Plaster & Gypsum Board	99.6	111.7	108.0	97.8	115.0	109.7	95.7	102.3	100.2	99.6	132.3	122.2	95.0	155.8	137.0	95.4	111.1	106.3
0950, 0980	Ceilings & Acoustic Treatment	98.1	111.7	106.8	92.0	115.0	106.6	92.0	102.3	98.6	98.1	132.3	119.9	108.2	155.8	138.5	96.4	111.1	105.8
0960	Flooring	100.9	112.1	104.0	100.9	102.3	101.3	98.2	97.4	98.0	100.9	115.0	104.8	91.3	142.0	105.3	100.9	106.4	102.4
0970, 0990	Wall Finishes & Painting/Coating	96.2	126.2	114.3	96.2	108.9	103.9	96.2	99.0	97.9	96.2	133.7	118.8	93.0	142.2	122.6	96.2	107.3	102.9
09	**FINISHES**	100.0	115.5	108.4	98.3	111.1	105.3	97.2	101.6	99.6	100.0	130.2	116.4	97.6	150.9	126.5	98.8	111.2	105.5
COVERS	**DIVS. 10 - 14, 25, 28, 41, 43, 44**	100.0	105.6	101.2	100.0	80.5	95.9	100.0	99.6	99.9	100.0	101.9	100.4	100.0	124.4	105.2	100.0	97.9	99.6
21, 22, 23	**FIRE SUPPRESSION, PLUMBING & HVAC**	100.0	107.4	103.0	93.2	98.2	95.3	93.2	98.3	95.3	100.1	117.6	107.3	100.1	126.9	111.1	99.9	104.5	101.8
26, 27, 3370	**ELECTRICAL, COMMUNICATIONS & UTIL.**	95.8	98.9	97.4	91.7	83.6	87.5	86.4	94.1	90.4	96.1	123.9	110.7	94.8	125.2	110.7	101.6	96.2	98.8
MF2004	**WEIGHTED AVERAGE**	98.7	107.3	102.5	94.9	100.0	97.2	94.8	99.7	96.9	97.5	122.8	108.7	98.5	133.3	113.9	97.0	105.4	100.7

	DIVISION	INDIANA																	
		ANDERSON			BLOOMINGTON			COLUMBUS			EVANSVILLE			FORT WAYNE			GARY		
		460			474			472			476 - 477			467 - 468			463 - 464		
		MAT.	INST.	TOTAL	MAT.	INST.	TOTAL	MAT.	INST.	TOTAL	MAT.	INST.	TOTAL	MAT.	INST.	TOTAL	MAT.	INST.	TOTAL
015433	CONTRACTOR EQUIPMENT		96.3	96.3		86.8	86.8		86.8	86.8		117.5	117.5		96.3	96.3		96.3	96.3
0241, 31 - 34	SITE & INFRASTRUCTURE, DEMOLITION	87.7	94.6	92.5	76.4	94.7	89.2	73.3	94.5	88.1	81.0	125.6	112.0	88.7	94.6	92.8	88.2	98.0	95.0
0310	Concrete Forming & Accessories	94.7	79.5	81.6	98.6	74.9	78.1	93.7	77.7	79.8	93.4	79.9	81.7	93.5	75.9	78.3	94.8	113.2	110.7
0320	Concrete Reinforcing	95.7	79.9	87.6	91.2	79.7	85.4	91.6	79.7	85.5	100.0	78.0	88.7	95.7	79.0	87.2	95.7	103.7	99.8
0330	Cast-in-Place Concrete	96.2	81.8	90.7	95.2	78.2	88.8	94.8	76.7	87.9	91.0	87.7	89.7	102.2	77.8	93.0	100.5	111.0	104.5
03	CONCRETE	91.6	80.9	86.6	99.0	76.9	88.7	98.5	77.5	88.7	99.5	82.4	91.5	94.5	77.8	86.7	93.7	110.5	101.6
04	MASONRY	82.6	83.2	83.0	86.0	80.9	83.0	85.9	80.9	82.9	82.2	84.6	83.7	86.2	81.2	83.2	83.8	113.0	101.4
05	METALS	96.4	88.0	93.5	94.8	77.0	88.8	94.9	76.2	88.6	88.7	84.9	87.4	96.4	87.9	93.5	96.4	105.2	99.4
06	WOOD, PLASTICS & COMPOSITES	103.7	78.8	89.6	117.9	72.2	92.0	113.3	76.0	92.2	96.7	77.9	86.1	103.4	74.9	87.3	101.7	112.3	107.7
07	THERMAL & MOISTURE PROTECTION	96.5	80.3	90.4	93.3	80.6	88.5	93.0	81.0	88.5	97.3	85.4	92.8	96.3	83.9	91.6	95.4	109.7	100.8
08	OPENINGS	95.7	80.6	91.9	103.5	77.0	96.9	98.8	79.1	94.0	95.7	79.6	91.7	95.7	74.9	90.5	95.7	114.2	100.3
0920	Plaster & Gypsum Board	92.7	78.6	82.9	87.6	72.0	76.8	84.6	76.0	78.6	83.1	76.6	78.6	91.8	74.6	79.9	87.1	113.1	105.1
0950, 0980	Ceilings & Acoustic Treatment	87.5	78.6	81.8	76.7	72.0	73.7	76.7	76.0	76.2	81.4	76.6	78.4	87.5	74.6	79.3	87.5	113.1	103.8
0960	Flooring	93.2	83.6	90.5	104.5	80.8	98.0	99.1	80.8	94.1	98.0	86.9	94.9	93.2	80.7	89.8	93.2	119.0	100.3
0970, 0990	Wall Finishes & Painting/Coating	93.8	73.8	81.8	90.4	87.8	88.8	90.4	87.8	88.8	96.4	87.2	90.9	93.8	78.4	84.5	93.8	115.4	106.8
09	FINISHES	92.0	79.6	85.3	93.5	76.7	84.4	91.5	79.0	84.7	92.0	81.8	86.5	91.8	77.1	83.8	91.3	114.7	104.0
COVERS	DIVS. 10 - 14, 25, 28, 41, 43, 44	100.0	91.4	98.2	100.0	90.2	97.9	100.0	90.7	98.0	100.0	81.3	96.0	100.0	91.9	98.3	100.0	108.7	101.9
21, 22, 23	FIRE SUPPRESSION, PLUMBING & HVAC	99.8	77.0	90.4	99.7	82.0	92.4	92.9	80.4	87.8	99.9	82.9	92.9	99.8	72.5	88.6	99.8	107.6	103.0
26, 27, 3370	ELECTRICAL, COMMUNICATIONS & UTIL.	86.1	90.4	88.4	100.1	86.5	93.0	99.2	91.1	95.0	95.8	83.9	89.6	86.8	76.9	81.6	99.0	103.9	101.5
MF2004	WEIGHTED AVERAGE	94.5	83.5	89.7	97.4	81.6	90.4	94.8	82.4	89.3	95.1	86.6	91.4	95.2	79.6	88.3	96.1	108.3	101.5

	DIVISION	INDIANA																	
		INDIANAPOLIS			KOKOMO			LAFAYETTE			LAWRENCEBURG			MUNCIE			NEW ALBANY		
		461 - 462			469			479			470			473			471		
		MAT.	INST.	TOTAL	MAT.	INST.	TOTAL	MAT.	INST.	TOTAL	MAT.	INST.	TOTAL	MAT.	INST.	TOTAL	MAT.	INST.	TOTAL
015433	CONTRACTOR EQUIPMENT		91.7	91.7		96.3	96.3		86.8	86.8		106.0	106.0		95.8	95.8		95.5	95.5
0241, 31 - 34	SITE & INFRASTRUCTURE, DEMOLITION	87.4	97.9	94.7	84.8	94.5	91.6	74.4	94.4	88.3	72.6	112.3	100.2	76.4	94.8	89.2	69.0	98.0	89.2
0310	Concrete Forming & Accessories	95.4	83.4	85.0	97.3	77.8	80.4	91.8	81.4	82.8	90.5	75.3	77.3	92.2	79.2	81.0	87.1	72.6	74.5
0320	Concrete Reinforcing	95.1	80.0	87.4	86.6	79.8	83.1	91.2	79.8	85.4	90.3	82.1	86.1	101.0	79.9	90.2	91.9	80.2	85.9
0330	Cast-in-Place Concrete	92.5	83.7	89.2	95.2	84.8	91.3	95.3	84.1	91.1	89.1	75.8	84.1	100.1	81.0	92.9	92.0	74.2	85.3
03	CONCRETE	93.6	82.6	88.4	88.6	81.1	85.1	98.7	81.7	90.8	91.9	77.4	85.1	97.9	80.5	89.8	97.1	75.0	86.8
04	MASONRY	89.9	86.0	87.6	82.3	82.5	82.4	90.8	82.9	86.1	71.8	74.3	73.3	88.0	83.2	85.1	78.7	71.4	74.3
05	METALS	98.6	78.7	91.9	92.9	87.7	91.1	93.3	76.3	87.6	90.2	87.6	89.3	96.7	88.1	93.8	92.0	82.8	88.9
06	WOOD, PLASTICS & COMPOSITES	100.5	82.7	90.4	106.4	76.2	89.3	110.7	80.9	93.8	96.0	73.8	83.5	113.3	78.5	93.6	97.3	72.9	83.5
07	THERMAL & MOISTURE PROTECTION	94.7	83.7	90.6	96.2	80.0	90.1	93.0	80.8	88.4	96.4	78.0	89.5	96.0	80.3	90.0	85.0	69.8	79.3
08	OPENINGS	103.3	82.7	98.2	90.1	79.2	87.4	97.3	81.7	93.4	97.1	77.2	92.2	98.5	80.5	94.0	95.3	76.2	90.5
0920	Plaster & Gypsum Board	90.4	82.4	84.9	97.0	75.9	82.4	81.3	81.0	81.1	67.4	73.6	71.7	83.1	78.6	80.0	81.8	72.4	75.3
0950, 0980	Ceilings & Acoustic Treatment	91.0	82.4	85.5	87.5	75.9	80.2	71.5	81.0	77.5	86.6	73.6	78.3	77.5	78.6	78.2	81.4	72.4	75.6
0960	Flooring	93.0	90.0	92.2	97.6	88.1	95.0	98.0	91.6	96.2	72.9	90.0	77.6	97.5	83.6	93.6	96.1	65.9	87.8
0970, 0990	Wall Finishes & Painting/Coating	93.8	87.8	90.2	93.8	79.7	85.3	90.4	84.9	87.1	91.9	70.8	79.2	90.4	73.8	80.4	96.4	71.7	81.6
09	FINISHES	92.9	85.2	88.7	93.7	79.7	86.1	89.5	83.7	86.4	83.4	77.7	80.3	90.7	79.4	84.6	91.5	71.3	80.5
COVERS	DIVS. 10 - 14, 25, 28, 41, 43, 44	100.0	92.8	98.5	100.0	84.5	96.7	100.0	91.3	98.1	100.0	49.3	89.2	100.0	90.8	98.0	100.0	48.4	89.0
21, 22, 23	FIRE SUPPRESSION, PLUMBING & HVAC	99.8	82.9	92.8	93.0	80.5	87.9	92.9	81.6	88.3	93.6	78.3	87.3	99.7	76.9	90.3	93.1	76.1	86.1
26, 27, 3370	ELECTRICAL, COMMUNICATIONS & UTIL.	101.0	90.4	95.5	91.0	81.6	86.1	98.6	85.6	91.8	93.5	77.8	85.3	91.1	76.6	83.5	94.1	74.7	83.9
MF2004	WEIGHTED AVERAGE	97.7	85.7	92.4	92.0	82.7	87.9	94.5	83.4	89.6	91.3	80.4	86.5	96.0	81.5	89.6	92.5	76.1	85.3

	DIVISION	INDIANA									IOWA								
		SOUTH BEND			TERRE HAUTE			WASHINGTON			BURLINGTON			CARROLL			CEDAR RAPIDS		
		465 - 466			478			475			526			514			522 - 524		
		MAT.	INST.	TOTAL	MAT.	INST.	TOTAL	MAT.	INST.	TOTAL	MAT.	INST.	TOTAL	MAT.	INST.	TOTAL	MAT.	INST.	TOTAL
015433	CONTRACTOR EQUIPMENT		105.1	105.1		117.5	117.5		117.5	117.5		100.0	100.0		100.0	100.0		96.5	96.5
0241, 31 - 34	SITE & INFRASTRUCTURE, DEMOLITION	90.5	95.0	93.6	82.9	125.8	112.8	82.9	123.6	111.3	95.7	95.6	95.7	84.4	93.5	90.8	97.7	95.3	96.1
0310	Concrete Forming & Accessories	96.2	79.6	81.8	94.3	81.6	83.3	94.9	78.6	80.8	97.7	65.2	69.6	87.4	43.0	49.0	102.3	73.9	77.7
0320	Concrete Reinforcing	95.7	77.4	86.3	100.0	80.1	89.8	92.4	54.1	72.9	93.1	73.0	82.9	93.9	43.8	68.3	93.8	76.7	85.1
0330	Cast-in-Place Concrete	97.4	81.7	91.5	88.1	90.3	88.9	95.9	89.5	93.5	106.8	48.6	84.8	106.8	52.6	86.3	107.1	75.8	95.3
03	CONCRETE	88.3	81.2	85.0	102.2	84.5	93.9	107.8	77.8	93.8	101.1	62.1	82.9	100.1	47.8	75.6	101.1	75.8	89.3
04	MASONRY	82.9	81.4	82.0	88.9	83.4	85.6	82.4	84.5	83.7	100.5	53.2	72.1	102.9	39.9	65.0	106.9	72.9	86.5
05	METALS	96.4	99.5	97.4	89.3	86.9	88.5	84.1	71.2	79.8	85.9	87.0	86.3	85.9	67.9	79.8	88.3	88.2	88.3
06	WOOD, PLASTICS & COMPOSITES	101.1	79.0	88.6	98.5	80.4	88.3	98.9	77.8	87.0	100.6	71.1	83.9	88.3	44.5	63.5	106.2	73.1	87.5
07	THERMAL & MOISTURE PROTECTION	91.1	85.3	88.9	97.4	84.1	92.4	97.5	82.9	92.0	101.7	57.8	85.1	102.1	46.8	81.2	102.6	74.7	92.0
08	OPENINGS	89.2	79.1	86.7	96.3	81.5	92.6	92.9	69.9	87.2	93.0	65.2	86.1	98.2	40.9	84.0	99.3	76.2	93.6
0920	Plaster & Gypsum Board	84.4	78.8	80.6	83.1	79.2	80.4	82.3	76.5	78.3	97.6	70.3	78.7	92.8	42.9	58.3	101.4	72.6	81.5
0950, 0980	Ceilings & Acoustic Treatment	83.2	78.8	80.4	81.4	79.2	80.0	74.5	76.5	75.8	107.0	70.3	83.6	107.0	42.9	66.2	111.3	72.6	86.6
0960	Flooring	91.6	85.7	90.0	98.0	90.0	95.8	99.1	82.3	94.5	113.3	40.7	93.2	106.2	35.8	86.7	131.3	48.0	108.3
0970, 0990	Wall Finishes & Painting/Coating	93.3	85.1	88.4	96.4	88.4	91.5	96.4	87.8	91.2	102.1	68.2	81.7	102.1	34.5	61.4	105.6	70.8	84.7
09	FINISHES	89.5	81.3	85.0	92.0	83.7	87.5	90.9	80.9	85.5	106.7	61.4	82.1	102.3	40.5	68.8	113.8	68.0	88.9
COVERS	DIVS. 10 - 14, 25, 28, 41, 43, 44	100.0	76.9	95.1	100.0	94.2	98.8	100.0	81.2	96.0	100.0	74.4	94.6	100.0	70.4	93.7	100.0	79.9	95.7
21, 22, 23	FIRE SUPPRESSION, PLUMBING & HVAC	99.7	79.7	91.5	99.9	81.9	92.5	93.1	82.1	88.6	93.3	63.1	80.9	93.3	42.8	72.5	100.1	76.6	90.4
26, 27, 3370	ELECTRICAL, COMMUNICATIONS & UTIL.	106.3	88.8	97.2	93.9	86.6	90.1	94.4	80.3	87.0	97.8	68.7	82.6	98.6	35.3	65.5	95.3	75.5	84.9
MF2004	WEIGHTED AVERAGE	95.1	84.7	90.5	95.9	87.9	92.3	93.5	83.3	89.0	96.1	67.9	83.6	96.0	49.3	75.4	99.5	77.5	89.8

DIVISION		IOWA																	
		COUNCIL BLUFFS			CRESTON			DAVENPORT			DECORAH			DES MOINES			DUBUQUE		
		515			508			527 - 528			521			500 - 503,509			520		
		MAT.	INST.	TOTAL	MAT.	INST.	TOTAL	MAT.	INST.	TOTAL	MAT.	INST.	TOTAL	MAT.	INST.	TOTAL	MAT.	INST.	TOTAL
015433	**CONTRACTOR EQUIPMENT**		96.1	96.1		100.0	100.0		100.0	100.0		100.0	100.0		101.7	101.7		95.2	95.2
0241, 31 - 34	**SITE & INFRASTRUCTURE, DEMOLITION**	105.4	92.3	96.3	84.3	94.4	91.3	95.5	99.3	98.1	94.3	93.9	94.0	87.9	99.6	96.0	95.5	92.2	93.2
0310	Concrete Forming & Accessories	86.8	51.9	56.6	87.9	63.0	66.4	102.0	84.3	86.7	95.5	40.3	47.8	103.2	69.3	73.9	88.0	65.7	68.7
0320	Concrete Reinforcing	95.7	71.8	83.5	93.3	75.2	84.0	93.8	88.1	90.9	93.1	44.0	68.0	93.8	73.4	83.4	92.6	76.6	84.4
0330	Cast-in-Place Concrete	111.4	73.8	97.2	106.5	54.7	86.9	103.2	88.7	97.7	103.9	49.6	83.4	98.9	89.1	95.2	104.9	85.7	97.6
03	**CONCRETE**	103.7	64.5	85.4	99.9	63.4	82.8	99.1	87.2	93.6	99.0	45.7	74.1	95.9	77.8	87.4	98.0	75.6	87.5
04	**MASONRY**	107.7	73.3	87.0	102.9	43.3	67.1	102.8	81.8	90.1	122.0	37.1	70.9	97.9	66.8	79.2	107.9	64.7	81.9
05	**METALS**	92.9	85.6	90.4	85.9	82.3	84.7	88.3	98.2	91.7	86.0	67.3	79.7	88.5	87.8	88.2	86.9	87.7	87.2
06	**WOOD, PLASTICS & COMPOSITES**	87.3	45.6	63.7	88.7	69.0	77.6	106.2	83.0	93.0	97.8	41.3	65.8	106.0	69.0	85.1	88.9	65.5	75.6
07	**THERMAL & MOISTURE PROTECTION**	102.1	61.9	86.9	102.8	50.8	83.2	102.1	83.5	95.1	102.0	41.9	79.3	103.6	71.6	91.5	102.3	65.8	88.5
08	**OPENINGS**	98.2	60.2	88.8	111.2	67.2	100.3	99.3	85.5	95.9	96.5	42.7	83.2	99.3	73.6	92.9	98.3	74.0	92.2
0920	Plaster & Gypsum Board	92.8	44.2	59.2	92.8	68.2	75.8	101.4	82.5	88.3	96.3	39.6	57.1	90.4	68.2	75.0	93.3	64.7	73.5
0950, 0980	Ceilings & Acoustic Treatment	107.0	44.2	67.0	107.0	68.2	82.3	111.3	82.5	93.0	107.0	39.6	64.1	107.0	68.2	82.3	107.0	64.7	80.1
0960	Flooring	105.1	39.2	86.9	106.4	35.8	86.9	116.0	89.4	108.7	112.7	50.8	95.6	115.7	35.8	93.6	120.4	33.4	96.4
0970, 0990	Wall Finishes & Painting/Coating	98.6	67.9	80.1	102.1	42.6	66.3	102.1	85.5	92.1	102.1	36.6	62.7	102.1	73.8	85.1	104.6	70.8	84.3
09	**FINISHES**	103.2	49.2	73.9	102.4	56.7	77.6	109.0	85.1	96.0	106.3	42.0	71.4	105.3	62.8	82.2	108.3	59.3	81.7
COVERS	**DIVS. 10 - 14, 25, 28, 41, 43, 44**	100.0	78.7	95.5	100.0	74.6	94.6	100.0	83.0	96.4	100.0	42.8	87.9	100.0	78.5	95.4	100.0	76.8	95.1
21, 22, 23	**FIRE SUPPRESSION, PLUMBING & HVAC**	100.1	73.3	89.1	93.3	44.6	73.3	100.1	82.9	93.0	93.3	37.1	70.2	100.0	66.4	86.2	100.1	68.7	87.2
26, 27, 3370	**ELECTRICAL, COMMUNICATIONS & UTIL.**	100.9	79.2	89.6	93.0	46.1	68.5	93.3	83.9	88.4	95.3	35.2	63.8	108.1	69.9	88.2	99.4	72.1	85.1
MF2004	**WEIGHTED AVERAGE**	100.2	71.6	87.5	96.7	58.8	80.0	98.4	86.7	93.2	97.0	46.8	74.9	98.7	73.6	87.7	98.6	72.6	87.2

DIVISION		IOWA																	
		FORT DODGE			MASON CITY			OTTUMWA			SHENANDOAH			SIBLEY			SIOUX CITY		
		505			504			525			516			512			510 - 511		
		MAT.	INST.	TOTAL	MAT.	INST.	TOTAL	MAT.	INST.	TOTAL	MAT.	INST.	TOTAL	MAT.	INST.	TOTAL	MAT.	INST.	TOTAL
015433	**CONTRACTOR EQUIPMENT**		100.0	100.0		100.0	100.0		95.2	95.2		96.1	96.1		100.0	100.0		100.0	100.0
0241, 31 - 34	**SITE & INFRASTRUCTURE, DEMOLITION**	91.3	93.7	93.0	91.4	93.9	93.1	96.4	90.6	92.4	103.7	89.5	93.8	105.8	93.5	97.2	106.8	95.5	99.0
0310	Concrete Forming & Accessories	88.4	47.2	52.8	92.3	41.3	48.2	93.9	54.5	59.8	88.3	52.4	57.2	88.6	36.8	43.8	102.3	59.3	65.1
0320	Concrete Reinforcing	93.3	43.9	68.1	93.1	56.4	74.4	93.1	75.6	84.2	95.7	70.0	82.6	95.7	55.8	75.3	93.8	57.8	75.5
0330	Cast-in-Place Concrete	100.0	45.2	79.2	100.0	50.0	81.1	107.6	55.8	88.0	107.7	45.8	84.3	105.5	45.8	82.9	106.2	51.1	85.4
03	**CONCRETE**	95.2	47.2	72.8	95.4	48.6	73.5	100.7	60.2	81.8	101.1	54.6	79.4	100.1	45.1	74.4	100.4	57.4	80.3
04	**MASONRY**	101.2	33.4	60.4	114.6	41.9	70.9	104.3	44.5	68.4	107.3	30.2	61.0	127.3	34.0	71.2	100.3	51.0	70.6
05	**METALS**	85.9	68.2	80.0	86.0	74.0	81.9	85.9	86.1	85.9	91.8	78.0	87.2	86.1	72.3	81.5	88.3	77.8	84.8
06	**WOOD, PLASTICS & COMPOSITES**	89.3	50.5	67.3	93.6	41.3	64.0	95.5	56.4	73.3	88.9	59.0	72.0	89.6	36.8	59.7	106.2	59.6	79.8
07	**THERMAL & MOISTURE PROTECTION**	102.2	34.8	76.7	101.7	43.1	79.6	102.5	52.6	83.6	101.5	39.7	78.1	101.8	44.4	80.1	102.1	50.1	82.5
08	**OPENINGS**	103.6	43.1	88.6	94.0	46.1	82.1	98.2	61.4	89.1	87.8	57.2	80.2	94.3	42.2	81.4	99.3	57.2	88.8
0920	Plaster & Gypsum Board	93.3	49.0	62.7	93.7	39.6	56.3	94.6	55.4	67.4	93.3	58.1	68.9	93.3	35.0	52.9	101.4	58.5	71.7
0950, 0980	Ceilings & Acoustic Treatment	107.0	49.0	70.1	107.0	39.6	64.1	107.0	55.4	74.1	107.0	58.1	75.9	107.0	35.0	61.1	111.3	58.5	77.7
0960	Flooring	108.2	50.8	92.3	110.9	50.8	94.3	124.1	54.0	104.8	106.0	36.9	86.9	107.2	36.6	87.7	116.5	50.6	98.3
0970, 0990	Wall Finishes & Painting/Coating	102.1	18.9	52.0	102.1	33.4	60.8	104.6	76.4	87.6	98.6	23.6	53.5	102.1	34.5	61.4	103.1	58.9	76.5
09	**FINISHES**	104.3	45.3	72.3	105.2	41.7	70.7	109.7	56.0	80.6	103.4	47.4	73.0	105.9	36.2	68.0	110.8	57.6	81.9
COVERS	**DIVS. 10 - 14, 25, 28, 41, 43, 44**	100.0	71.1	93.9	100.0	43.7	88.0	100.0	72.0	94.1	100.0	70.6	93.8	100.0	52.5	89.9	100.0	75.9	94.9
21, 22, 23	**FIRE SUPPRESSION, PLUMBING & HVAC**	93.3	62.7	80.7	93.3	69.0	83.3	93.3	70.0	83.7	93.3	28.4	66.6	93.3	33.8	68.8	100.1	68.1	87.0
26, 27, 3370	**ELECTRICAL, COMMUNICATIONS & UTIL.**	100.5	46.9	72.5	99.4	62.4	80.0	97.7	68.7	82.5	95.3	28.0	60.1	95.3	35.3	63.9	95.3	68.7	81.4
MF2004	**WEIGHTED AVERAGE**	96.4	54.9	78.1	96.1	59.0	79.7	97.0	66.4	83.5	96.4	47.4	74.8	97.5	45.7	74.6	99.1	66.1	84.5

DIVISION		IOWA						KANSAS											
		SPENCER			WATERLOO			BELLEVILLE			COLBY			DODGE CITY			EMPORIA		
		513			506 - 507			669			677			678			668		
		MAT.	INST.	TOTAL	MAT.	INST.	TOTAL	MAT.	INST.	TOTAL	MAT.	INST.	TOTAL	MAT.	INST.	TOTAL	MAT.	INST.	TOTAL
015433	**CONTRACTOR EQUIPMENT**		100.0	100.0		100.0	100.0		102.9	102.9		102.9	102.9		102.9	102.9		101.2	101.2
0241, 31 - 34	**SITE & INFRASTRUCTURE, DEMOLITION**	105.8	93.5	97.2	96.0	94.7	95.1	113.2	92.1	98.5	110.5	92.9	98.2	112.8	92.5	98.7	104.2	90.5	94.7
0310	Concrete Forming & Accessories	93.8	36.8	44.5	102.7	41.6	49.8	98.7	46.9	53.9	104.5	58.4	64.6	98.3	58.3	63.7	91.2	34.4	42.1
0320	Concrete Reinforcing	95.7	35.4	64.9	93.8	76.3	84.9	104.7	51.8	77.7	107.3	51.8	79.0	104.7	52.1	77.9	103.2	52.4	77.3
0330	Cast-in-Place Concrete	105.5	45.8	82.9	107.1	43.1	82.9	118.8	50.4	92.9	114.2	51.2	90.4	116.3	50.9	91.5	114.9	39.8	86.5
03	**CONCRETE**	100.4	41.1	72.7	101.1	50.3	77.4	116.1	50.5	85.4	113.2	56.0	86.5	114.5	55.9	87.1	108.7	41.6	77.3
04	**MASONRY**	127.3	34.0	71.2	101.8	53.5	72.8	97.1	43.4	64.8	98.3	42.3	64.6	105.5	49.4	71.7	102.6	52.3	72.3
05	**METALS**	86.1	63.3	78.4	88.3	86.1	87.6	90.1	73.8	84.6	90.5	76.5	85.8	92.0	76.0	86.6	89.8	77.4	85.6
06	**WOOD, PLASTICS & COMPOSITES**	95.4	36.8	62.2	106.7	41.3	69.7	95.5	49.6	69.5	102.1	64.8	81.0	94.8	64.8	77.8	87.7	33.4	57.0
07	**THERMAL & MOISTURE PROTECTION**	102.6	39.1	78.6	101.9	46.4	80.9	100.9	52.2	82.5	100.9	53.4	83.0	100.9	55.2	83.6	99.7	54.3	82.5
08	**OPENINGS**	107.8	36.3	90.1	94.9	53.6	84.7	96.4	47.0	84.1	96.5	55.3	86.3	96.4	55.3	86.2	94.0	38.2	80.2
0920	Plaster & Gypsum Board	94.6	35.0	53.3	101.4	39.6	58.7	97.4	48.0	63.2	101.3	63.6	75.2	96.6	63.6	73.8	94.0	31.3	50.6
0950, 0980	Ceilings & Acoustic Treatment	107.0	35.0	61.1	111.3	39.6	65.7	92.0	48.0	64.0	92.0	63.6	74.0	92.0	63.6	74.0	92.0	31.3	53.4
0960	Flooring	110.5	36.6	90.1	117.9	50.8	99.3	100.5	42.2	84.4	104.5	42.2	87.3	100.3	42.2	84.2	94.8	39.2	79.5
0970, 0990	Wall Finishes & Painting/Coating	102.1	34.5	61.4	103.1	30.9	59.7	96.2	43.2	64.3	96.2	43.2	64.3	96.2	43.2	64.3	96.2	43.2	64.3
09	**FINISHES**	107.1	36.2	68.6	109.6	41.3	72.6	99.4	45.6	70.2	100.9	54.5	75.7	99.2	54.5	74.9	96.3	34.9	63.0
COVERS	**DIVS. 10 - 14, 25, 28, 41, 43, 44**	100.0	69.5	93.5	100.0	70.2	93.7	100.0	41.0	87.5	100.0	42.7	87.8	100.0	42.8	87.9	100.0	36.1	86.4
21, 22, 23	**FIRE SUPPRESSION, PLUMBING & HVAC**	93.3	33.9	68.9	100.1	37.5	74.4	93.2	62.7	80.7	93.2	56.0	77.9	100.0	55.9	81.9	93.2	61.9	80.3
26, 27, 3370	**ELECTRICAL, COMMUNICATIONS & UTIL.**	97.1	35.3	64.8	95.3	62.4	78.1	99.4	62.4	80.1	99.4	62.4	80.1	96.0	69.0	81.9	96.4	69.3	82.2
MF2004	**WEIGHTED AVERAGE**	99.2	44.5	75.0	98.4	56.0	79.7	98.7	58.7	81.0	98.6	60.0	81.5	100.4	61.5	83.3	96.9	57.3	79.4

	DIVISION	KANSAS																	
		FORT SCOTT			HAYS			HUTCHINSON			INDEPENDENCE			KANSAS CITY			LIBERAL		
		667			676			675			673			660 - 662			679		
		MAT.	INST.	TOTAL	MAT.	INST.	TOTAL	MAT.	INST.	TOTAL	MAT.	INST.	TOTAL	MAT.	INST.	TOTAL	MAT.	INST.	TOTAL
015433	CONTRACTOR EQUIPMENT		102.1	102.1		102.9	102.9		102.9	102.9		102.9	102.9		99.7	99.7		102.9	102.9
0241, 31 - 34	SITE & INFRASTRUCTURE, DEMOLITION	100.6	91.7	94.4	116.7	92.9	100.1	93.3	92.9	93.0	114.3	92.9	99.4	93.0	92.1	92.4	116.5	92.9	100.0
0310	Concrete Forming & Accessories	105.3	73.9	78.1	102.3	58.4	64.3	93.4	56.9	61.8	112.4	51.8	59.9	102.1	98.8	99.2	98.7	57.0	62.6
0320	Concrete Reinforcing	102.6	83.0	92.6	104.7	51.9	77.7	104.7	52.1	77.8	104.1	54.7	78.9	99.5	92.0	95.6	106.2	52.1	78.6
0330	Cast-in-Place Concrete	106.5	54.6	86.9	90.0	51.2	75.4	83.4	38.7	66.5	116.8	39.2	87.4	91.2	98.0	93.8	90.0	38.7	70.6
03	CONCRETE	103.6	69.8	87.8	104.3	56.0	81.7	87.4	51.1	70.4	115.7	49.4	84.7	95.4	97.5	96.4	106.5	51.1	80.6
04	MASONRY	103.2	53.6	73.4	106.2	43.7	68.6	97.6	45.7	66.4	95.6	46.4	66.1	105.5	97.0	100.4	103.3	45.7	68.7
05	METALS	89.7	88.6	89.4	90.0	76.7	85.5	89.8	75.8	85.1	89.7	77.3	85.5	97.7	98.5	98.0	90.3	75.9	85.4
06	WOOD, PLASTICS & COMPOSITES	104.1	82.5	91.8	99.4	64.8	79.8	90.2	64.8	75.8	111.3	56.9	80.5	100.4	100.2	100.3	95.5	64.8	78.1
07	THERMAL & MOISTURE PROTECTION	100.3	67.0	87.7	101.1	53.8	83.2	100.0	52.2	81.9	100.9	60.0	85.4	99.8	97.5	98.9	101.3	52.2	82.7
08	OPENINGS	94.0	76.2	89.6	96.4	55.3	86.2	96.3	55.3	86.2	94.0	51.6	83.5	95.5	92.5	94.7	96.4	55.3	86.2
0920	Plaster & Gypsum Board	99.6	81.8	87.3	99.6	63.6	74.7	95.7	63.6	73.5	106.9	55.5	71.3	92.2	100.1	97.7	97.4	63.6	74.0
0950, 0980	Ceilings & Acoustic Treatment	92.0	81.8	85.5	92.0	63.6	74.0	92.0	63.6	74.0	92.0	55.5	68.8	92.0	100.1	97.2	92.0	63.6	74.0
0960	Flooring	111.4	42.5	92.4	103.2	42.2	86.3	96.9	42.2	81.8	109.6	42.2	91.0	89.8	97.7	92.0	100.5	42.2	84.4
0970, 0990	Wall Finishes & Painting/Coating	98.2	43.2	65.1	96.2	43.2	64.3	96.2	43.2	64.3	96.2	43.2	64.3	103.9	75.2	86.7	96.2	43.2	64.3
09	FINISHES	102.3	66.0	82.6	100.9	54.5	75.7	96.4	53.6	73.2	103.3	49.1	73.9	96.7	95.3	96.0	100.0	53.6	74.9
COVERS	DIVS. 10 - 14, 25, 28, 41, 43, 44	100.0	50.5	89.5	100.0	42.7	87.8	100.0	41.6	87.6	100.0	40.9	87.4	100.0	71.0	93.8	100.0	41.6	87.6
21, 22, 23	FIRE SUPPRESSION, PLUMBING & HVAC	93.2	58.5	79.0	93.2	56.0	77.9	93.2	54.2	77.2	93.2	61.4	80.1	99.9	96.5	98.5	93.2	54.2	77.2
26, 27, 3370	ELECTRICAL, COMMUNICATIONS & UTIL.	95.6	69.3	81.9	98.2	62.4	79.5	92.8	62.4	76.9	95.2	69.3	81.7	101.8	94.9	98.2	96.0	69.0	81.9
MF2004	WEIGHTED AVERAGE	96.7	68.6	84.3	97.9	60.1	81.2	93.6	59.0	78.3	98.3	60.9	81.8	98.6	95.3	97.1	97.7	59.9	81.0

	DIVISION	KANSAS									KENTUCKY								
		SALINA			TOPEKA			WICHITA			ASHLAND			BOWLING GREEN			CAMPTON		
		674			664 - 666			670 - 672			411 - 412			421 - 422			413 - 414		
		MAT.	INST.	TOTAL	MAT.	INST.	TOTAL	MAT.	INST.	TOTAL	MAT.	INST.	TOTAL	MAT.	INST.	TOTAL	MAT.	INST.	TOTAL
015433	CONTRACTOR EQUIPMENT		102.9	102.9		101.2	101.2		102.9	102.9		98.1	98.1		95.5	95.5		102.1	102.1
0241, 31 - 34	SITE & INFRASTRUCTURE, DEMOLITION	102.5	92.9	95.8	96.7	90.5	92.4	95.2	92.7	93.5	103.2	86.7	91.7	69.1	98.7	89.7	77.7	99.2	92.6
0310	Concrete Forming & Accessories	95.2	47.1	53.6	102.4	44.0	51.9	101.3	48.6	55.7	89.1	103.2	101.3	83.8	82.3	82.5	89.1	50.1	55.3
0320	Concrete Reinforcing	104.1	67.2	85.3	96.7	94.1	95.4	96.7	77.2	86.7	93.7	103.9	98.9	90.6	84.0	87.2	91.4	67.0	79.0
0330	Cast-in-Place Concrete	100.9	44.0	79.3	92.3	48.3	75.6	88.0	52.0	74.4	83.7	101.5	90.5	83.1	103.2	90.7	92.7	61.1	80.7
03	CONCRETE	101.5	51.4	78.1	93.4	56.6	76.2	91.2	56.6	75.0	93.2	103.2	97.9	91.3	89.9	90.6	95.1	58.0	77.8
04	MASONRY	120.7	43.3	74.2	100.0	56.6	73.9	93.6	56.0	71.0	87.6	105.1	98.1	89.2	74.0	80.1	87.7	37.5	57.5
05	METALS	91.8	82.5	88.6	96.9	94.6	96.1	96.9	85.3	93.0	91.2	109.5	97.4	92.8	84.2	89.9	92.0	74.9	86.2
06	WOOD, PLASTICS & COMPOSITES	91.6	49.6	67.8	96.7	41.6	65.5	98.1	47.5	69.4	79.0	103.1	92.7	93.0	81.8	86.7	91.8	48.5	67.3
07	THERMAL & MOISTURE PROTECTION	100.5	50.9	81.8	101.7	62.1	86.7	100.1	57.0	83.8	89.7	100.0	93.6	84.9	84.1	84.6	97.6	51.3	80.1
08	OPENINGS	96.3	49.5	84.7	96.7	57.8	87.1	96.7	55.3	86.5	94.0	98.0	95.0	95.3	77.7	90.9	96.9	58.7	87.4
0920	Plaster & Gypsum Board	95.7	48.0	62.7	94.1	39.7	56.5	91.0	45.8	59.7	54.7	103.3	88.3	78.8	81.6	80.7	78.8	46.4	56.3
0950, 0980	Ceilings & Acoustic Treatment	92.0	48.0	64.0	92.0	39.7	58.7	91.2	45.8	62.3	76.6	103.3	93.6	81.4	81.6	81.5	81.4	46.4	59.1
0960	Flooring	98.3	30.8	79.7	101.9	39.2	84.5	100.7	67.9	91.7	78.1	104.3	85.3	93.9	81.6	90.5	95.4	44.4	81.3
0970, 0990	Wall Finishes & Painting/Coating	96.2	43.2	64.3	96.2	58.4	73.4	96.2	53.3	70.4	98.9	93.6	95.7	96.4	71.5	81.4	96.4	38.4	61.5
09	FINISHES	97.7	43.1	68.1	98.1	42.8	68.1	97.2	52.3	72.8	79.9	102.8	92.4	90.3	81.0	85.3	90.9	46.6	66.9
COVERS	DIVS. 10 - 14, 25, 28, 41, 43, 44	100.0	52.5	89.9	100.0	55.5	90.6	100.0	54.7	90.4	100.0	83.6	96.5	100.0	66.1	92.8	100.0	57.4	91.0
21, 22, 23	FIRE SUPPRESSION, PLUMBING & HVAC	100.0	55.9	81.8	100.0	60.3	83.7	99.8	54.2	81.0	93.1	89.9	91.8	99.9	83.1	93.0	93.2	68.5	83.0
26, 27, 3370	ELECTRICAL, COMMUNICATIONS & UTIL.	95.7	69.0	81.7	107.7	69.3	87.6	103.3	69.0	85.3	91.5	98.6	95.2	94.4	76.3	84.9	91.6	74.1	82.5
MF2004	WEIGHTED AVERAGE	99.1	59.2	81.5	99.0	63.8	83.4	97.7	62.7	82.2	92.0	98.2	94.8	93.9	82.6	88.9	93.3	63.7	80.2

	DIVISION	KENTUCKY																	
		CORBIN			COVINGTON			ELIZABETHTOWN			FRANKFORT			HAZARD			HENDERSON		
		407 - 409			410			427			406			417 - 418			424		
		MAT.	INST.	TOTAL	MAT.	INST.	TOTAL	MAT.	INST.	TOTAL	MAT.	INST.	TOTAL	MAT.	INST.	TOTAL	MAT.	INST.	TOTAL
015433	CONTRACTOR EQUIPMENT		102.1	102.1		106.0	106.0		95.5	95.5		102.1	102.1		102.1	102.1		117.5	117.5
0241, 31 - 34	SITE & INFRASTRUCTURE, DEMOLITION	75.3	99.1	91.9	73.8	114.4	102.1	64.5	98.5	88.2	78.5	100.1	93.5	75.7	100.8	93.2	72.4	126.1	109.8
0310	Concrete Forming & Accessories	86.7	49.7	54.7	85.4	101.3	99.1	79.9	80.7	80.6	95.8	60.9	65.6	86.7	43.2	49.0	91.9	79.7	81.3
0320	Concrete Reinforcing	91.0	67.0	78.8	89.9	97.3	93.6	91.0	92.0	91.5	92.4	82.2	87.2	91.9	67.7	79.5	90.7	88.9	89.8
0330	Cast-in-Place Concrete	87.7	61.0	77.6	88.7	94.0	90.7	75.1	75.5	75.3	83.9	68.2	78.0	89.1	89.4	89.2	73.4	91.7	80.3
03	CONCRETE	90.5	57.7	75.2	93.6	97.9	95.6	83.9	81.2	82.6	89.4	68.1	79.4	92.1	64.7	79.3	88.6	85.7	87.3
04	MASONRY	87.5	39.7	58.8	100.6	111.3	107.0	73.8	79.0	77.0	84.9	73.8	78.2	84.7	47.0	62.1	92.9	93.2	93.1
05	METALS	91.9	74.6	86.1	90.2	94.0	91.5	91.9	87.9	90.5	98.8	84.6	94.0	92.0	74.6	86.1	83.8	89.2	85.6
06	WOOD, PLASTICS & COMPOSITES	89.1	48.5	66.2	90.1	94.3	92.5	88.7	81.8	84.8	99.7	58.7	76.5	89.1	35.6	58.8	95.0	76.1	84.3
07	THERMAL & MOISTURE PROTECTION	97.4	52.9	80.6	96.6	100.5	98.1	84.7	78.8	82.4	97.4	71.6	87.7	97.5	50.7	79.8	97.0	93.5	95.7
08	OPENINGS	96.2	51.2	85.0	98.2	91.4	96.5	95.3	84.4	92.6	95.4	67.7	88.6	97.3	42.0	83.6	93.4	80.4	90.2
0920	Plaster & Gypsum Board	77.9	46.4	56.1	64.0	94.7	85.3	77.5	81.6	80.3	84.9	56.8	65.5	77.9	33.0	46.8	79.7	74.7	76.3
0950, 0980	Ceilings & Acoustic Treatment	81.4	46.4	59.1	84.0	94.7	90.8	81.4	81.6	81.5	82.3	56.8	66.1	81.4	33.0	50.6	74.5	74.7	74.6
0960	Flooring	93.5	44.4	79.9	70.4	89.1	75.6	91.1	70.3	85.3	98.4	58.6	87.4	93.5	39.3	78.5	97.2	87.5	94.5
0970, 0990	Wall Finishes & Painting/Coating	96.4	38.4	61.5	91.9	85.7	88.2	96.4	79.0	85.9	96.4	52.2	69.8	96.4	31.5	57.3	96.4	98.8	97.8
09	FINISHES	90.0	46.6	66.5	81.8	97.4	90.3	88.9	78.5	83.3	92.8	58.4	74.1	90.1	39.6	62.7	89.3	82.7	85.7
COVERS	DIVS. 10 - 14, 25, 28, 41, 43, 44	100.0	51.7	89.7	100.0	102.2	100.5	100.0	90.4	98.0	100.0	70.6	93.8	100.0	59.2	91.3	100.0	74.0	94.5
21, 22, 23	FIRE SUPPRESSION, PLUMBING & HVAC	93.1	68.0	82.8	93.7	96.1	94.7	93.6	81.0	88.4	100.0	69.5	87.5	93.2	44.6	73.2	93.6	78.0	87.2
26, 27, 3370	ELECTRICAL, COMMUNICATIONS & UTIL.	91.8	74.1	82.6	95.7	81.1	88.1	91.4	76.3	83.5	103.5	97.7	100.5	91.6	56.3	73.1	93.8	82.3	87.8
MF2004	WEIGHTED AVERAGE	92.4	63.3	79.6	93.2	97.4	95.1	90.0	82.3	86.6	96.3	76.1	87.4	92.6	56.5	76.7	91.3	87.3	89.5

DIVISION		KENTUCKY																	
		LEXINGTON			LOUISVILLE			OWENSBORO			PADUCAH			PIKEVILLE			SOMERSET		
		403 - 405			400 - 402			423			420			415 - 416			425 - 426		
		MAT.	INST.	TOTAL	MAT.	INST.	TOTAL	MAT.	INST.	TOTAL	MAT.	INST.	TOTAL	MAT.	INST.	TOTAL	MAT.	INST.	TOTAL
015433	CONTRACTOR EQUIPMENT		102.1	102.1		95.5	95.5		117.5	117.5		117.5	117.5		98.1	98.1		102.1	102.1
0241, 31 - 34	SITE & INFRASTRUCTURE, DEMOLITION	76.8	101.1	93.7	66.9	98.5	88.9	81.1	125.8	112.2	75.0	125.5	110.2	113.1	85.6	93.9	69.7	99.7	90.6
0310	Concrete Forming & Accessories	96.7	74.5	77.5	91.8	80.7	82.2	90.5	71.9	74.4	88.7	80.9	82.0	96.4	69.2	72.9	87.3	48.1	53.4
0320	Concrete Reinforcing	100.0	91.5	95.6	100.0	92.0	95.9	90.7	90.6	90.6	91.2	85.4	88.3	94.2	90.5	92.3	91.0	81.0	85.9
0330	Cast-in-Place Concrete	89.7	81.6	86.6	93.6	75.5	86.8	85.7	99.4	90.9	78.2	90.0	82.7	92.1	94.3	92.9	73.4	104.1	85.0
03	CONCRETE	93.4	80.5	87.3	95.0	81.2	88.5	100.4	85.2	93.3	93.0	85.0	89.3	106.3	83.1	95.5	79.0	74.4	76.9
04	MASONRY	86.2	65.7	73.9	87.4	79.0	82.4	85.9	71.5	77.2	89.0	90.9	90.2	85.5	67.3	74.5	80.3	56.8	66.2
05	METALS	94.3	88.5	92.3	101.0	87.9	96.6	85.4	88.5	86.4	82.3	87.1	83.9	91.1	97.9	93.4	91.9	81.2	88.3
06	WOOD, PLASTICS & COMPOSITES	102.0	74.5	86.4	99.7	81.8	89.5	93.4	68.2	79.1	91.4	79.9	84.9	87.1	61.5	72.6	89.8	41.1	62.2
07	THERMAL & MOISTURE PROTECTION	97.6	85.9	93.2	91.5	78.8	86.7	97.4	75.7	89.2	97.1	82.8	91.7	90.2	61.2	79.3	97.0	56.7	81.8
08	OPENINGS	96.3	78.5	91.9	96.3	84.4	93.3	93.4	78.6	89.7	92.5	79.2	89.2	94.7	70.6	88.8	96.2	58.2	86.8
0920	Plaster & Gypsum Board	85.9	73.1	77.1	82.0	81.6	81.7	78.9	66.6	70.4	78.0	78.6	78.4	57.8	60.5	59.6	78.3	38.7	50.9
0950, 0980	Ceilings & Acoustic Treatment	86.6	73.1	78.0	82.3	81.6	81.8	74.5	66.6	69.5	74.5	78.6	77.1	76.6	60.5	66.3	81.4	38.7	54.2
0960	Flooring	98.8	59.8	88.0	97.3	70.3	89.8	96.4	81.6	92.3	95.1	63.9	86.5	82.9	52.9	74.6	93.8	44.4	80.1
0970, 0990	Wall Finishes & Painting/Coating	96.4	69.7	80.3	96.4	79.0	85.9	96.4	98.7	97.7	96.4	80.3	86.7	98.9	90.2	93.7	96.4	52.9	70.2
09	FINISHES	93.9	70.6	81.2	91.5	78.5	84.4	89.5	75.7	82.0	88.6	77.5	82.6	82.5	67.3	74.3	89.6	46.1	65.9
COVERS	DIVS. 10 - 14, 25, 28, 41, 43, 44	100.0	90.9	98.1	100.0	90.4	98.0	100.0	84.2	96.6	100.0	70.5	93.7	100.0	62.1	92.0	100.0	61.4	91.8
21, 22, 23	FIRE SUPPRESSION, PLUMBING & HVAC	99.9	65.0	85.5	99.9	81.0	92.1	99.9	68.8	87.1	93.6	81.8	88.7	93.1	78.6	87.1	93.6	64.5	81.6
26, 27, 3370	ELECTRICAL, COMMUNICATIONS & UTIL.	95.0	73.6	83.8	106.2	76.3	90.6	93.9	82.3	87.8	96.5	84.1	90.0	94.8	82.3	88.2	92.0	73.6	82.4
MF2004	WEIGHTED AVERAGE	95.5	76.3	87.0	97.1	82.4	90.6	94.3	81.7	88.8	91.6	86.7	89.4	94.5	78.0	87.2	90.6	67.7	80.5

DIVISION		LOUISIANA																	
		ALEXANDRIA			BATON ROUGE			HAMMOND			LAFAYETTE			LAKE CHARLES			MONROE		
		713 - 714			707 - 708			704			705			706			712		
		MAT.	INST.	TOTAL	MAT.	INST.	TOTAL	MAT.	INST.	TOTAL	MAT.	INST.	TOTAL	MAT.	INST.	TOTAL	MAT.	INST.	TOTAL
015433	CONTRACTOR EQUIPMENT		88.0	88.0		88.0	88.0		88.7	88.7		88.7	88.7		88.0	88.0		88.0	88.0
0241, 31 - 34	SITE & INFRASTRUCTURE, DEMOLITION	101.9	85.4	90.4	116.4	85.1	94.6	116.7	86.5	95.7	117.3	86.3	95.8	118.0	84.9	95.0	101.9	84.7	89.9
0310	Concrete Forming & Accessories	83.5	37.9	44.0	94.9	60.9	65.5	80.3	50.7	54.7	96.1	46.8	53.5	97.8	52.1	58.3	83.0	38.4	44.4
0320	Concrete Reinforcing	101.0	60.2	80.1	104.9	60.3	82.1	103.4	58.1	80.3	104.9	59.9	81.9	104.9	58.0	81.0	99.8	60.1	79.5
0330	Cast-in-Place Concrete	99.7	40.0	77.1	102.9	59.4	86.4	100.2	45.9	79.7	99.7	46.4	79.5	105.0	68.8	91.3	99.7	57.6	83.8
03	CONCRETE	93.4	44.2	70.4	98.0	60.8	80.6	95.7	51.3	74.9	96.5	50.1	74.8	99.2	59.7	80.7	93.2	50.2	73.1
04	MASONRY	115.3	50.4	76.3	106.8	52.2	74.0	106.0	54.0	74.7	106.0	45.4	69.5	104.9	59.8	77.8	110.2	48.3	73.0
05	METALS	91.1	72.8	84.9	103.1	70.0	91.9	98.9	68.9	88.8	97.9	69.3	88.2	97.9	69.6	88.4	91.0	70.6	84.1
06	WOOD, PLASTICS & COMPOSITES	89.1	36.8	59.5	102.7	66.1	82.0	88.7	52.5	68.2	108.9	48.4	74.7	107.1	53.7	76.9	88.4	37.2	59.4
07	THERMAL & MOISTURE PROTECTION	98.2	49.3	79.7	98.0	57.8	82.8	99.9	57.3	83.8	100.4	51.9	82.1	99.6	59.3	84.4	98.2	52.2	80.8
08	OPENINGS	99.6	43.7	85.8	102.6	59.4	91.9	98.1	55.5	87.6	102.6	49.8	89.6	102.6	53.2	90.4	99.6	47.8	86.8
0920	Plaster & Gypsum Board	78.2	35.5	48.7	94.2	65.5	74.4	93.2	51.5	64.3	100.9	47.3	63.8	100.9	52.8	67.6	77.8	35.9	48.8
0950, 0980	Ceilings & Acoustic Treatment	92.2	35.5	56.1	99.9	65.5	78.0	99.8	51.5	69.0	98.1	47.3	65.7	99.1	52.8	69.6	92.2	35.9	56.3
0960	Flooring	101.6	66.0	91.8	107.0	66.0	95.7	96.7	49.9	83.8	107.2	37.4	87.9	107.2	78.5	99.3	101.1	61.5	90.2
0970, 0990	Wall Finishes & Painting/Coating	98.6	35.4	60.6	103.2	40.6	65.5	103.2	53.7	73.4	103.2	40.6	65.5	103.2	37.4	63.6	98.6	42.3	64.7
09	FINISHES	92.2	42.0	64.9	101.7	60.3	79.3	98.3	50.8	72.5	102.1	43.8	70.4	102.3	55.3	76.8	92.0	42.2	65.0
COVERS	DIVS. 10 - 14, 25, 28, 41, 43, 44	100.0	55.5	90.5	100.0	78.6	95.5	100.0	53.2	90.1	100.0	76.2	95.0	100.0	77.2	95.2	100.0	50.8	89.6
21, 22, 23	FIRE SUPPRESSION, PLUMBING & HVAC	100.2	52.1	80.4	99.9	51.2	79.9	93.2	48.4	74.8	100.0	52.1	80.3	100.0	52.7	80.5	100.2	46.2	78.0
26, 27, 3370	ELECTRICAL, COMMUNICATIONS & UTIL.	94.9	53.4	73.2	116.1	60.4	87.0	101.2	50.3	74.6	102.7	58.7	79.7	102.2	58.7	79.5	96.9	54.5	74.8
MF2004	WEIGHTED AVERAGE	97.5	54.0	78.3	103.0	61.2	84.5	98.2	55.8	79.5	100.9	56.1	81.1	101.1	60.7	83.3	97.4	53.4	77.9

DIVISION		LOUISIANA									MAINE								
		NEW ORLEANS			SHREVEPORT			THIBODAUX			AUGUSTA			BANGOR			BATH		
		700 - 701			710 - 711			703			043			044			045		
		MAT.	INST.	TOTAL	MAT.	INST.	TOTAL	MAT.	INST.	TOTAL	MAT.	INST.	TOTAL	MAT.	INST.	TOTAL	MAT.	INST.	TOTAL
015433	CONTRACTOR EQUIPMENT		89.0	89.0		88.0	88.0		88.7	88.7		100.6	100.6		100.6	100.6		100.6	100.6
0241, 31 - 34	SITE & INFRASTRUCTURE, DEMOLITION	121.0	87.7	97.8	103.5	84.6	90.4	119.6	86.4	96.5	81.0	101.1	95.0	80.2	98.8	93.1	78.4	101.1	94.2
0310	Concrete Forming & Accessories	100.1	63.7	68.6	96.9	40.7	48.3	93.0	58.9	63.5	96.5	102.2	101.4	92.3	91.5	91.6	88.7	102.2	100.4
0320	Concrete Reinforcing	104.9	61.8	82.9	99.4	60.5	79.5	103.4	58.1	80.3	90.4	103.1	96.9	90.1	102.8	96.6	89.1	103.0	96.2
0330	Cast-in-Place Concrete	104.0	69.0	90.7	99.1	46.0	79.0	107.7	49.1	85.5	84.3	63.7	76.5	78.3	62.3	72.3	78.3	63.7	72.8
03	CONCRETE	98.8	65.8	83.4	93.1	47.4	71.7	101.0	56.0	80.0	103.0	88.5	96.2	99.3	82.9	91.7	99.4	88.4	94.3
04	MASONRY	106.2	57.5	76.9	101.3	49.9	70.4	133.9	51.1	84.1	98.0	63.2	77.1	112.7	59.0	80.4	119.1	59.9	83.5
05	METALS	111.0	74.0	98.5	87.8	70.7	82.0	98.9	68.9	88.8	88.3	85.1	87.2	87.9	80.1	85.3	86.3	85.0	85.8
06	WOOD, PLASTICS & COMPOSITES	103.7	66.1	82.4	104.4	40.4	68.2	96.9	63.0	77.7	97.5	114.0	106.8	92.7	99.9	96.8	88.2	114.0	102.8
07	THERMAL & MOISTURE PROTECTION	100.4	61.1	85.6	97.5	51.1	79.9	100.1	54.9	83.0	95.8	64.5	83.9	96.1	60.1	82.5	96.1	65.2	84.4
08	OPENINGS	103.4	66.4	94.2	97.5	45.8	84.7	103.7	61.2	93.1	101.1	91.4	98.7	101.0	85.1	97.1	101.0	92.7	99.0
0920	Plaster & Gypsum Board	99.6	65.5	76.0	83.6	39.2	52.9	96.7	62.3	72.9	102.3	113.7	110.2	94.7	99.2	97.8	92.1	113.7	107.0
0950, 0980	Ceilings & Acoustic Treatment	98.1	65.5	77.3	91.3	39.2	58.1	99.8	62.3	75.9	90.1	113.7	105.1	90.9	99.2	96.2	89.0	113.7	104.7
0960	Flooring	107.6	66.0	96.1	111.0	66.0	98.6	103.8	47.5	88.3	97.6	59.5	87.1	95.3	59.2	85.3	93.2	59.5	83.9
0970, 0990	Wall Finishes & Painting/Coating	104.7	64.0	80.2	98.6	35.4	60.6	104.7	53.7	74.0	94.6	38.3	60.7	94.6	35.9	59.3	94.6	38.3	60.7
09	FINISHES	102.6	64.2	81.8	95.8	44.1	67.7	101.2	56.6	77.0	98.1	89.1	93.2	96.7	81.2	88.3	95.3	89.1	91.9
COVERS	DIVS. 10 - 14, 25, 28, 41, 43, 44	100.0	80.5	95.9	100.0	70.9	93.8	100.0	78.8	95.5	100.0	91.1	98.1	100.0	105.3	101.1	100.0	91.1	98.1
21, 22, 23	FIRE SUPPRESSION, PLUMBING & HVAC	100.0	64.1	85.2	100.1	54.8	81.5	93.2	56.0	77.9	100.0	69.0	87.3	100.1	64.7	85.5	93.4	69.1	83.4
26, 27, 3370	ELECTRICAL, COMMUNICATIONS & UTIL.	102.8	71.7	86.6	104.4	68.9	85.8	99.7	61.0	79.5	98.9	81.5	89.8	95.8	80.7	87.9	93.7	81.5	87.4
MF2004	WEIGHTED AVERAGE	103.2	68.2	87.8	97.4	57.7	79.9	101.1	61.0	83.4	97.7	81.5	90.6	97.5	77.5	88.7	95.6	81.3	89.3

	DIVISION	MAINE																	
		HOULTON			KITTERY			LEWISTON			MACHIAS			PORTLAND			ROCKLAND		
		047			039			042			046			040 - 041			048		
		MAT.	INST.	TOTAL	MAT.	INST.	TOTAL	MAT.	INST.	TOTAL	MAT.	INST.	TOTAL	MAT.	INST.	TOTAL	MAT.	INST.	TOTAL
015433	CONTRACTOR EQUIPMENT		100.6	100.6		100.6	100.6		100.6	100.6		100.6	100.6		100.6	100.6		100.6	100.6
0241, 31 - 34	SITE & INFRASTRUCTURE, DEMOLITION	80.3	100.9	94.6	75.1	101.1	93.2	77.8	98.8	92.4	79.6	100.9	94.5	79.3	98.8	92.9	76.5	101.0	93.5
0310	Concrete Forming & Accessories	95.1	101.3	100.5	89.2	102.7	100.9	96.7	91.5	92.2	92.8	101.3	100.2	95.5	91.5	92.0	93.6	101.7	100.6
0320	Concrete Reinforcing	90.1	103.1	96.7	88.5	103.0	95.9	110.6	102.8	106.6	90.1	103.1	96.7	110.6	102.8	106.6	90.1	103.2	96.8
0330	Cast-in-Place Concrete	78.3	61.8	72.1	78.3	64.7	73.1	79.9	62.3	73.3	78.3	61.9	72.1	101.6	62.3	86.8	79.9	63.0	73.5
03	CONCRETE	100.3	87.4	94.3	94.3	89.0	91.8	98.7	82.9	91.4	99.8	87.4	94.0	109.4	82.9	97.0	96.4	88.0	92.5
04	MASONRY	96.3	61.1	75.1	109.7	62.7	81.4	96.9	59.0	74.1	96.3	61.3	75.3	97.2	59.0	74.2	90.5	62.9	73.9
05	METALS	86.5	85.1	86.0	86.2	85.1	85.8	91.1	80.1	87.4	86.5	85.0	86.0	92.3	80.1	88.2	86.4	85.2	86.0
06	WOOD, PLASTICS & COMPOSITES	96.3	114.0	106.3	88.6	114.0	103.0	98.1	99.9	99.2	93.5	114.0	105.1	97.3	99.9	98.8	94.3	114.0	105.5
07	THERMAL & MOISTURE PROTECTION	96.2	63.6	83.9	95.8	66.2	84.6	96.0	60.1	82.4	96.2	63.6	83.9	99.1	60.1	84.4	95.9	64.5	84.0
08	OPENINGS	101.1	91.0	98.6	101.0	92.2	98.8	104.3	85.1	99.6	101.1	89.8	98.3	104.3	85.1	99.6	101.0	89.8	98.2
0920	Plaster & Gypsum Board	96.9	113.7	108.5	92.1	113.7	107.0	100.4	99.2	99.6	95.6	113.7	108.1	106.9	99.2	101.6	95.6	113.7	108.1
0950, 0980	Ceilings & Acoustic Treatment	89.0	113.7	104.7	89.0	113.7	104.7	98.7	99.2	99.0	89.0	113.7	104.7	96.1	99.2	98.1	89.0	113.7	104.7
0960	Flooring	96.5	55.7	85.3	93.5	62.1	84.8	98.3	59.2	87.5	95.5	46.2	81.9	97.6	59.2	87.0	95.8	55.7	84.8
0970, 0990	Wall Finishes & Painting/Coating	94.6	39.8	61.6	94.6	38.3	60.7	94.6	35.9	59.3	94.6	39.8	61.6	94.6	35.9	59.3	94.6	39.8	61.6
09	FINISHES	96.9	88.0	92.1	95.0	89.8	92.2	99.7	81.2	89.7	96.4	86.5	91.0	99.6	81.2	89.6	96.2	88.3	91.9
COVERS	DIVS. 10 - 14, 25, 28, 41, 43, 44	100.0	90.3	97.9	100.0	92.5	98.4	100.0	105.3	101.1	100.0	90.4	98.0	100.0	105.3	101.1	100.0	90.6	98.0
21, 22, 23	FIRE SUPPRESSION, PLUMBING & HVAC	93.4	67.9	82.9	93.4	69.7	83.6	100.1	64.7	85.5	93.4	68.0	82.9	100.0	64.7	85.5	93.4	68.4	83.1
26, 27, 3370	ELECTRICAL, COMMUNICATIONS & UTIL.	98.0	81.5	89.4	93.7	81.5	87.4	98.0	81.3	89.2	98.0	81.5	89.4	104.5	81.3	92.4	97.9	81.5	89.3
MF2004	WEIGHTED AVERAGE	95.2	80.7	88.8	94.3	81.9	88.9	97.8	77.6	88.9	95.1	80.5	88.7	100.1	77.6	90.2	94.2	81.1	88.4

	DIVISION	MAINE			MARYLAND														
		WATERVILLE			ANNAPOLIS			BALTIMORE			COLLEGE PARK			CUMBERLAND			EASTON		
		049			214			210 - 212			207 - 208			215			216		
		MAT.	INST.	TOTAL	MAT.	INST.	TOTAL	MAT.	INST.	TOTAL	MAT.	INST.	TOTAL	MAT.	INST.	TOTAL	MAT.	INST.	TOTAL
015433	CONTRACTOR EQUIPMENT		100.6	100.6		99.2	99.2		102.9	102.9		104.4	104.4		99.2	99.2		99.2	99.2
0241, 31 - 34	SITE & INFRASTRUCTURE, DEMOLITION	80.0	101.1	94.7	97.8	90.3	92.6	95.2	94.4	94.6	109.5	92.0	97.3	87.2	89.9	89.1	93.8	86.1	88.4
0310	Concrete Forming & Accessories	88.3	102.2	100.4	95.3	52.3	58.1	98.3	75.9	78.9	90.5	74.4	76.6	89.9	78.9	80.4	88.1	30.5	38.2
0320	Concrete Reinforcing	90.1	103.1	96.7	109.9	84.7	97.1	109.9	84.8	97.1	103.0	81.3	92.0	97.1	75.8	86.2	96.3	17.8	56.2
0330	Cast-in-Place Concrete	78.3	63.7	72.8	122.0	78.2	105.5	103.9	79.1	94.5	127.9	80.0	109.8	92.8	81.4	88.5	103.0	40.3	79.3
03	CONCRETE	101.1	88.5	95.2	113.8	68.9	92.8	105.1	79.7	93.2	117.2	78.9	99.3	93.8	80.1	87.4	101.9	33.5	69.9
04	MASONRY	105.8	63.2	80.2	100.2	70.5	82.3	101.1	70.5	82.7	118.8	73.1	91.3	98.5	81.3	88.2	113.2	32.3	64.6
05	METALS	86.5	85.2	86.0	92.2	95.6	93.4	93.3	96.4	94.3	82.0	96.9	87.1	89.9	92.5	90.8	90.2	59.8	79.9
06	WOOD, PLASTICS & COMPOSITES	87.7	114.0	102.6	93.0	45.7	66.2	95.9	77.5	85.5	86.2	75.0	79.9	86.6	77.5	81.4	84.3	30.6	53.9
07	THERMAL & MOISTURE PROTECTION	96.2	66.1	84.8	93.9	76.0	87.1	94.4	79.6	88.8	102.5	78.3	93.3	93.9	79.5	88.5	94.0	36.1	72.1
08	OPENINGS	101.1	91.4	98.7	91.1	66.5	85.0	93.1	83.7	90.8	101.9	78.1	96.0	92.2	79.5	89.1	90.4	27.6	74.9
0920	Plaster & Gypsum Board	92.1	113.7	107.0	98.5	44.5	61.1	100.5	77.0	84.2	99.1	74.3	81.9	96.2	77.2	83.1	95.3	28.8	49.3
0950, 0980	Ceilings & Acoustic Treatment	89.0	113.7	104.7	91.2	44.5	61.4	93.8	77.0	83.1	98.4	74.3	83.0	93.8	77.2	83.2	93.8	28.8	52.4
0960	Flooring	93.0	59.5	83.7	89.2	75.1	85.3	89.2	75.1	85.3	117.2	55.6	100.2	85.0	84.5	84.9	84.2	15.9	65.3
0970, 0990	Wall Finishes & Painting/Coating	94.6	38.3	60.7	97.0	81.4	87.6	97.0	81.4	87.6	125.0	81.3	98.7	97.0	81.5	87.7	97.0	22.6	52.2
09	FINISHES	95.4	89.1	92.0	91.9	57.1	73.0	92.7	75.7	83.5	102.6	71.2	85.6	90.4	80.1	84.8	90.4	27.5	56.3
COVERS	DIVS. 10 - 14, 25, 28, 41, 43, 44	100.0	91.1	98.1	100.0	84.0	96.6	100.0	88.1	97.5	100.0	61.0	91.7	100.0	91.1	98.1	100.0	41.2	87.5
21, 22, 23	FIRE SUPPRESSION, PLUMBING & HVAC	93.4	69.1	83.4	99.9	87.7	94.9	100.0	87.7	94.9	93.4	87.0	90.8	93.2	73.6	85.1	93.2	23.8	64.6
26, 27, 3370	ELECTRICAL, COMMUNICATIONS & UTIL.	97.9	81.5	89.4	99.0	92.1	95.4	97.5	95.0	96.2	103.5	96.9	100.1	95.0	79.5	86.9	94.5	64.6	78.9
MF2004	WEIGHTED AVERAGE	95.6	81.6	89.4	98.5	79.3	90.0	97.8	85.2	92.2	100.0	83.7	92.8	93.3	81.0	87.9	95.0	41.8	71.5

	DIVISION	MARYLAND															MASSACHUSETTS		
		ELKTON			HAGERSTOWN			SALISBURY			SILVER SPRING			WALDORF			BOSTON		
		219			217			218			209			206			020 - 022, 024		
		MAT.	INST.	TOTAL	MAT.	INST.	TOTAL	MAT.	INST.	TOTAL	MAT.	INST.	TOTAL	MAT.	INST.	TOTAL	MAT.	INST.	TOTAL
015433	CONTRACTOR EQUIPMENT		99.2	99.2		99.2	99.2		99.2	99.2		96.8	96.8		96.8	96.8		106.5	106.5
0241, 31 - 34	SITE & INFRASTRUCTURE, DEMOLITION	81.8	86.6	85.1	85.4	90.9	89.2	93.7	85.7	88.1	97.1	84.5	88.3	104.7	83.7	90.1	85.3	108.1	101.2
0310	Concrete Forming & Accessories	93.4	82.4	83.9	89.1	78.1	79.5	100.4	28.8	38.4	99.3	53.8	60.0	105.4	47.1	55.0	100.0	139.7	134.4
0320	Concrete Reinforcing	96.3	52.1	73.7	97.1	75.7	86.2	96.3	51.9	73.6	101.8	77.0	89.2	102.5	52.1	76.8	109.4	147.7	129.0
0330	Cast-in-Place Concrete	83.4	45.8	69.2	88.4	58.6	77.1	103.0	39.0	78.8	131.0	61.8	104.8	146.8	55.0	112.1	101.1	150.6	119.8
03	CONCRETE	86.3	64.5	76.1	90.0	71.9	81.6	102.7	38.1	72.5	115.2	62.6	90.6	127.1	52.3	92.1	112.4	144.0	127.2
04	MASONRY	99.2	35.2	60.8	105.0	81.3	90.7	112.6	30.0	62.9	117.5	63.3	84.9	100.5	59.8	76.0	115.7	156.6	140.3
05	METALS	90.2	68.7	82.9	90.1	91.7	90.6	90.2	61.0	80.4	86.0	92.0	88.0	86.0	75.8	82.5	103.0	128.5	111.6
06	WOOD, PLASTICS & COMPOSITES	90.7	97.4	94.5	85.5	77.3	80.9	99.6	27.1	58.5	92.4	52.8	70.0	99.1	47.3	69.8	101.4	138.9	122.7
07	THERMAL & MOISTURE PROTECTION	93.7	47.6	76.2	93.7	69.9	84.7	94.2	54.7	79.3	104.9	76.5	94.2	105.3	71.3	92.5	96.1	150.2	116.5
08	OPENINGS	90.5	69.7	85.3	90.4	77.9	87.3	90.7	23.2	74.0	95.2	65.0	87.7	96.1	46.0	83.7	100.5	142.6	111.0
0920	Plaster & Gypsum Board	97.9	97.7	97.8	95.3	77.0	82.7	102.7	25.3	49.1	104.9	52.3	68.5	108.0	46.6	65.5	103.3	139.5	128.3
0950, 0980	Ceilings & Acoustic Treatment	93.8	97.7	96.3	94.7	77.0	83.4	93.8	25.3	50.2	107.9	52.3	72.5	107.9	46.6	68.8	97.7	139.5	124.3
0960	Flooring	86.7	52.7	77.3	84.7	84.5	84.7	90.7	52.7	80.2	125.2	55.6	106.0	130.2	52.7	108.8	98.0	168.9	117.6
0970, 0990	Wall Finishes & Painting/Coating	97.0	81.3	87.6	97.0	34.5	59.4	97.0	25.5	54.0	131.9	81.3	101.5	131.9	59.5	88.3	97.2	159.0	134.4
09	FINISHES	90.8	79.9	84.9	90.2	74.7	81.8	93.3	31.9	59.9	104.1	55.5	77.7	106.4	48.5	74.9	99.7	147.6	125.7
COVERS	DIVS. 10 - 14, 25, 28, 41, 43, 44	100.0	50.0	89.4	100.0	91.1	98.1	100.0	40.3	87.3	100.0	52.9	90.0	100.0	43.3	88.0	100.0	120.6	104.4
21, 22, 23	FIRE SUPPRESSION, PLUMBING & HVAC	93.2	76.0	86.1	100.0	87.8	94.9	93.2	51.0	75.8	93.4	81.5	88.5	93.4	79.1	87.5	100.1	129.1	112.0
26, 27, 3370	ELECTRICAL, COMMUNICATIONS & UTIL.	96.4	55.3	74.9	94.8	79.5	86.8	93.2	64.6	78.3	100.5	96.9	98.6	97.6	96.8	97.2	97.1	136.8	117.8
MF2004	WEIGHTED AVERAGE	92.3	66.7	81.0	94.5	81.9	88.9	95.4	48.8	74.8	99.2	75.2	88.6	100.0	69.2	86.4	102.0	136.5	117.2

DIVISION		MASSACHUSETTS																	
		BROCKTON			BUZZARDS BAY			FALL RIVER			FITCHBURG			FRAMINGHAM			GREENFIELD		
		023			025			027			014			017			013		
		MAT.	INST.	TOTAL	MAT.	INST.	TOTAL	MAT.	INST.	TOTAL	MAT.	INST.	TOTAL	MAT.	INST.	TOTAL	MAT.	INST.	TOTAL
015433	CONTRACTOR EQUIPMENT		102.5	102.5		102.5	102.5		103.5	103.5		100.6	100.6		101.7	101.7		100.6	100.6
0241, 31 - 34	SITE & INFRASTRUCTURE, DEMOLITION	82.7	104.3	97.8	74.7	104.2	95.2	81.8	104.4	97.5	75.8	104.2	95.6	72.8	103.9	94.4	78.8	102.9	95.5
0310	Concrete Forming & Accessories	100.5	127.1	123.6	98.3	126.5	122.7	100.5	126.8	123.3	93.4	122.3	118.4	99.3	127.1	123.3	91.9	106.0	104.1
0320	Concrete Reinforcing	110.6	147.5	129.5	88.7	123.2	106.3	110.6	123.3	117.1	88.7	133.5	111.6	88.7	147.5	118.7	92.9	115.1	104.2
0330	Cast-in-Place Concrete	96.2	146.0	115.0	79.9	145.8	104.8	93.0	146.4	113.2	83.8	145.8	107.3	83.8	139.5	104.9	86.1	124.8	100.8
03	CONCRETE	109.7	136.7	122.3	92.6	131.7	110.9	108.2	132.0	119.3	90.6	131.6	109.8	93.8	134.4	112.8	94.7	113.6	103.5
04	MASONRY	111.9	149.4	134.4	102.6	149.4	130.7	112.0	149.3	134.4	98.8	149.5	129.2	104.9	149.6	131.8	103.2	125.8	116.8
05	METALS	99.6	125.8	108.5	94.4	115.3	101.5	99.6	115.8	105.1	94.3	117.0	102.0	94.4	125.6	104.9	96.5	104.5	99.2
06	WOOD, PLASTICS & COMPOSITES	100.2	125.9	114.8	97.4	125.9	113.6	100.2	126.2	114.9	94.2	120.3	109.0	99.9	125.6	114.5	92.4	103.2	98.5
07	THERMAL & MOISTURE PROTECTION	96.2	144.2	114.3	95.6	138.8	112.0	96.2	138.1	112.0	95.7	138.5	111.9	95.7	143.5	113.8	95.7	118.7	104.4
08	OPENINGS	98.6	132.7	107.0	94.0	121.9	100.9	98.6	122.1	104.4	103.0	125.9	108.6	94.5	132.5	103.9	103.1	107.8	104.2
0920	Plaster & Gypsum Board	97.6	126.0	117.2	93.1	126.0	115.9	97.6	126.0	117.2	94.7	120.2	112.3	97.3	126.0	117.1	96.1	102.6	100.6
0950, 0980	Ceilings & Acoustic Treatment	99.7	126.0	116.4	90.0	126.0	112.9	99.7	126.0	116.4	89.0	120.2	108.9	89.0	126.0	112.6	96.8	102.6	100.5
0960	Flooring	98.5	168.9	118.0	97.1	168.9	117.0	98.4	168.9	117.8	95.3	168.9	115.6	97.0	168.9	116.9	94.4	138.1	106.5
0970, 0990	Wall Finishes & Painting/Coating	96.7	143.5	124.9	96.7	143.5	124.9	96.7	143.5	124.9	94.6	143.5	124.0	95.7	143.5	124.5	94.6	108.0	102.7
09	FINISHES	99.6	136.7	119.7	95.8	136.7	118.0	99.6	136.9	119.8	95.1	133.4	115.9	95.8	136.5	117.9	97.1	111.9	105.1
COVERS	DIVS. 10 - 14, 25, 28, 41, 43, 44	100.0	116.8	103.6	100.0	116.8	103.6	100.0	117.5	103.7	100.0	96.4	99.2	100.0	116.3	103.5	100.0	106.3	101.3
21, 22, 23	FIRE SUPPRESSION, PLUMBING & HVAC	100.1	111.5	104.8	93.4	111.2	100.7	100.1	111.3	104.7	94.4	111.6	101.5	94.4	124.9	107.0	94.4	102.8	97.9
26, 27, 3370	ELECTRICAL, COMMUNICATIONS & UTIL.	96.2	100.5	98.4	93.3	100.4	97.0	96.1	100.4	98.4	98.4	102.9	100.8	94.8	128.6	112.4	98.4	93.4	95.8
MF2004	WEIGHTED AVERAGE	100.6	123.2	110.6	94.3	120.9	106.0	100.4	121.1	109.5	95.5	120.6	106.5	95.0	129.5	110.2	96.8	107.3	101.4

DIVISION		MASSACHUSETTS																	
		HYANNIS			LAWRENCE			LOWELL			NEW BEDFORD			PITTSFIELD			SPRINGFIELD		
		026			019			018			027			012			010 - 011		
		MAT.	INST.	TOTAL	MAT.	INST.	TOTAL	MAT.	INST.	TOTAL	MAT.	INST.	TOTAL	MAT.	INST.	TOTAL	MAT.	INST.	TOTAL
015433	CONTRACTOR EQUIPMENT		102.5	102.5		102.5	102.5		100.6	100.6		103.5	103.5		100.6	100.6		100.6	100.6
0241, 31 - 34	SITE & INFRASTRUCTURE, DEMOLITION	79.9	104.2	96.8	83.5	104.3	98.0	82.3	104.3	97.6	80.4	104.4	97.1	83.3	102.7	96.8	82.8	102.9	96.8
0310	Concrete Forming & Accessories	93.4	126.5	122.0	100.3	127.3	123.6	97.7	127.5	123.5	100.5	126.8	123.3	97.7	108.4	106.9	97.9	109.4	107.9
0320	Concrete Reinforcing	88.7	123.2	106.3	109.7	137.2	123.7	110.6	137.2	124.2	110.6	123.3	117.1	92.2	109.7	101.1	110.6	115.1	112.9
0330	Cast-in-Place Concrete	87.8	145.8	109.7	97.0	146.1	115.6	88.2	146.2	110.1	82.0	146.4	106.3	96.2	122.7	106.2	91.8	124.2	104.0
03	CONCRETE	99.8	131.7	114.7	109.9	134.8	121.6	100.3	134.7	116.4	102.7	132.0	116.4	101.6	112.9	106.9	102.1	114.9	108.1
04	MASONRY	110.8	149.4	134.0	111.4	149.4	134.2	97.5	149.5	128.7	110.8	149.3	133.9	98.2	122.1	112.5	97.8	124.7	114.0
05	METALS	96.0	115.3	102.5	96.9	121.7	105.3	96.9	119.5	104.5	99.6	115.8	105.1	96.7	102.3	98.6	99.5	104.5	101.2
06	WOOD, PLASTICS & COMPOSITES	91.6	125.9	111.0	100.2	125.9	114.8	99.8	125.9	114.6	100.2	126.2	114.9	99.8	108.5	104.7	99.8	108.5	104.7
07	THERMAL & MOISTURE PROTECTION	95.9	138.8	112.1	96.2	144.2	114.3	96.0	144.2	114.2	96.2	138.1	112.0	96.0	117.6	104.2	96.0	118.7	104.6
08	OPENINGS	94.7	121.9	101.4	98.6	129.9	106.3	104.3	129.9	110.6	98.6	122.1	104.4	104.3	109.2	105.5	104.3	110.6	105.9
0920	Plaster & Gypsum Board	90.1	126.0	114.9	100.4	126.0	118.1	100.4	126.0	118.1	97.6	126.0	117.2	100.4	108.0	105.7	100.4	108.0	105.7
0950, 0980	Ceilings & Acoustic Treatment	91.9	126.0	113.6	98.7	126.0	116.1	98.7	126.0	116.1	99.7	126.0	116.4	98.7	108.0	104.6	98.7	108.0	104.6
0960	Flooring	94.7	168.9	115.2	97.6	168.9	117.3	97.6	168.9	117.3	98.4	168.9	117.8	97.8	138.1	108.9	97.5	138.1	108.7
0970, 0990	Wall Finishes & Painting/Coating	96.7	143.5	124.9	94.7	143.5	124.1	94.6	143.5	124.0	96.7	143.5	124.9	94.6	108.0	102.7	96.1	108.0	103.3
09	FINISHES	95.5	136.7	117.9	99.3	136.7	119.6	99.3	136.7	119.6	99.5	136.9	119.8	99.3	114.1	107.3	99.3	114.7	107.7
COVERS	DIVS. 10 - 14, 25, 28, 41, 43, 44	100.0	116.8	103.6	100.0	116.9	103.6	100.0	116.9	103.6	100.0	117.5	103.7	100.0	105.6	101.2	100.0	106.5	101.4
21, 22, 23	FIRE SUPPRESSION, PLUMBING & HVAC	100.1	111.2	104.7	100.1	114.4	106.0	100.1	124.9	110.3	100.1	111.3	104.7	100.1	98.7	99.6	100.1	101.6	100.7
26, 27, 3370	ELECTRICAL, COMMUNICATIONS & UTIL.	93.8	100.4	97.3	97.4	128.5	113.7	98.0	128.5	113.9	97.0	100.4	98.8	97.9	93.4	95.6	98.0	93.4	95.6
MF2004	WEIGHTED AVERAGE	97.7	120.9	107.9	100.3	127.0	112.1	99.0	129.0	112.2	99.7	121.1	109.1	99.2	106.1	102.3	99.6	107.7	103.2

DIVISION		MASSACHUSETTS			MICHIGAN														
		WORCESTER			ANN ARBOR			BATTLE CREEK			BAY CITY			DEARBORN			DETROIT		
		015 - 016			481			490			487			481			482		
		MAT.	INST.	TOTAL	MAT.	INST.	TOTAL	MAT.	INST.	TOTAL	MAT.	INST.	TOTAL	MAT.	INST.	TOTAL	MAT.	INST.	TOTAL
015433	CONTRACTOR EQUIPMENT		100.6	100.6		110.3	110.3		103.5	103.5		110.3	110.3		110.3	110.3		97.6	97.6
0241, 31 - 34	SITE & INFRASTRUCTURE, DEMOLITION	82.7	104.2	97.7	75.3	95.7	89.5	81.4	87.8	85.9	67.5	94.3	86.2	75.1	95.8	89.5	89.0	97.4	94.9
0310	Concrete Forming & Accessories	98.2	122.6	119.3	95.4	106.7	105.2	97.6	94.1	94.6	95.4	91.5	92.1	95.2	115.6	112.8	96.0	115.7	113.0
0320	Concrete Reinforcing	110.6	147.5	129.4	116.3	126.4	121.5	100.0	88.0	93.9	116.3	125.5	121.0	116.3	126.6	121.6	115.5	126.6	121.2
0330	Cast-in-Place Concrete	91.2	146.0	111.9	80.9	117.1	94.6	89.6	103.3	94.8	77.4	95.7	84.3	79.1	114.8	92.6	86.8	114.8	97.4
03	CONCRETE	101.8	134.5	117.1	91.9	114.4	102.4	93.4	95.4	94.3	90.2	100.1	94.8	91.0	117.6	103.4	94.7	116.4	104.8
04	MASONRY	97.3	149.5	128.7	92.3	113.2	104.9	92.8	90.9	91.6	92.0	91.0	91.4	92.2	117.3	107.3	91.5	117.3	107.0
05	METALS	99.6	123.0	107.5	91.2	122.0	101.6	93.2	86.1	90.8	91.8	118.6	100.9	91.3	122.6	101.9	95.0	104.2	98.1
06	WOOD, PLASTICS & COMPOSITES	100.1	120.3	111.5	98.9	105.0	102.3	98.5	94.5	96.3	98.9	91.0	94.4	98.9	115.7	108.4	98.9	115.7	108.4
07	THERMAL & MOISTURE PROTECTION	96.0	138.5	112.1	102.3	114.5	106.9	93.5	91.2	92.6	100.0	97.5	99.0	100.8	119.5	107.9	99.1	119.5	106.8
08	OPENINGS	104.3	129.6	110.6	94.0	108.4	97.6	90.7	86.5	89.7	94.0	98.5	95.1	94.0	114.2	99.0	95.7	114.2	100.3
0920	Plaster & Gypsum Board	100.4	120.2	114.1	97.2	104.1	101.9	84.6	90.3	88.5	97.2	89.7	92.0	97.2	115.2	109.6	97.2	115.2	109.6
0950, 0980	Ceilings & Acoustic Treatment	98.7	120.2	112.4	91.1	104.1	99.4	86.6	90.3	88.9	92.0	89.7	90.6	91.1	115.2	106.4	92.0	115.2	106.8
0960	Flooring	97.6	168.9	117.3	92.7	120.4	100.4	97.7	99.0	98.0	92.5	80.3	89.1	92.4	121.2	100.3	92.4	121.2	100.4
0970, 0990	Wall Finishes & Painting/Coating	94.6	143.5	124.0	93.5	107.5	101.9	96.4	83.8	88.8	93.5	87.8	90.1	93.5	107.6	102.0	95.0	107.6	102.6
09	FINISHES	99.3	133.4	117.8	92.3	109.1	101.4	93.3	94.1	93.7	92.0	88.4	90.1	92.2	116.0	105.1	93.4	116.0	105.7
COVERS	DIVS. 10 - 14, 25, 28, 41, 43, 44	100.0	110.1	102.2	100.0	107.0	101.5	100.0	109.3	102.0	100.0	98.8	99.7	100.0	108.9	101.9	100.0	108.9	101.9
21, 22, 23	FIRE SUPPRESSION, PLUMBING & HVAC	100.1	111.8	104.9	100.1	100.2	100.1	99.9	92.4	96.8	100.1	88.9	95.5	100.1	110.5	104.3	100.1	112.7	105.2
26, 27, 3370	ELECTRICAL, COMMUNICATIONS & UTIL.	98.0	102.9	100.5	94.6	92.4	93.5	94.5	84.3	89.2	93.6	85.7	89.4	94.6	112.2	103.8	95.7	112.1	104.3
MF2004	WEIGHTED AVERAGE	99.6	122.1	109.5	94.9	106.1	99.9	95.0	91.0	93.2	94.4	94.3	94.3	94.7	113.4	103.0	96.4	112.1	103.3

DIVISION		MICHIGAN																	
		FLINT			GAYLORD			GRAND RAPIDS			IRON MOUNTAIN			JACKSON			KALAMAZOO		
		484 - 485			497			493,495			498 - 499			492			491		
		MAT.	INST.	TOTAL	MAT.	INST.	TOTAL	MAT.	INST.	TOTAL	MAT.	INST.	TOTAL	MAT.	INST.	TOTAL	MAT.	INST.	TOTAL
015433	CONTRACTOR EQUIPMENT		110.3	110.3		105.8	105.8		103.5	103.5		94.1	94.1		105.8	105.8		103.5	103.5
0241, 31 - 34	SITE & INFRASTRUCTURE, DEMOLITION	65.7	95.0	86.1	77.3	86.0	83.3	82.9	87.6	86.2	84.4	94.6	91.5	96.5	87.3	90.1	81.7	87.8	85.9
0310	Concrete Forming & Accessories	97.9	93.4	94.0	94.5	57.8	62.8	97.8	82.8	84.8	86.8	90.4	89.9	92.4	91.0	91.2	97.6	93.7	94.3
0320	Concrete Reinforcing	116.3	125.9	121.2	93.1	97.1	95.2	100.0	85.9	92.8	93.2	96.9	95.1	90.4	125.7	108.4	100.0	88.0	93.8
0330	Cast-in-Place Concrete	81.5	98.5	87.9	89.4	72.8	83.1	90.7	98.6	93.7	105.9	91.7	100.5	89.3	94.4	91.2	91.4	103.1	95.9
03	CONCRETE	92.3	102.0	96.8	89.6	71.9	81.4	93.9	88.3	91.3	99.1	92.1	95.8	84.6	99.4	91.5	96.6	95.1	95.9
04	MASONRY	92.4	100.4	97.2	103.0	83.9	91.5	88.6	56.3	69.2	89.7	91.8	91.0	84.2	97.6	92.3	91.5	90.9	91.1
05	METALS	91.3	119.5	100.8	94.3	103.3	97.3	94.3	81.0	89.8	93.7	92.0	93.1	94.5	117.1	102.1	93.2	85.7	90.7
06	WOOD, PLASTICS & COMPOSITES	101.8	91.3	95.9	90.9	51.4	68.5	95.7	82.7	88.3	85.9	90.8	88.6	89.9	89.6	89.7	98.5	94.5	96.3
07	THERMAL & MOISTURE PROTECTION	100.0	101.0	100.4	91.9	69.9	83.6	95.4	60.7	82.3	95.0	87.7	92.2	91.3	100.1	94.6	93.5	91.2	92.6
08	OPENINGS	94.0	98.7	95.2	92.5	56.8	83.7	94.2	71.2	88.5	97.5	76.8	92.4	91.6	91.2	91.5	90.7	86.5	89.7
0920	Plaster & Gypsum Board	98.9	90.1	92.8	83.8	48.6	59.4	82.8	78.1	79.5	54.1	91.0	79.6	83.1	88.0	86.5	84.6	90.3	88.5
0950, 0980	Ceilings & Acoustic Treatment	91.1	90.1	90.4	84.0	48.6	61.4	84.8	78.1	80.5	84.6	91.0	88.7	85.7	88.0	87.1	86.6	90.3	88.9
0960	Flooring	92.5	84.9	90.4	90.8	51.9	80.1	97.8	36.5	80.9	110.6	96.2	106.7	89.2	71.4	84.3	97.7	78.0	92.2
0970, 0990	Wall Finishes & Painting/Coating	93.5	94.9	94.3	91.9	48.0	65.5	96.4	35.8	60.0	114.2	39.1	69.0	91.9	79.7	84.6	96.4	83.8	88.8
09	FINISHES	91.6	91.1	91.3	93.0	53.7	71.7	92.9	69.0	79.9	96.3	86.3	90.9	94.1	85.4	89.4	93.3	89.9	91.5
COVERS	DIVS. 10 - 14, 25, 28, 41, 43, 44	100.0	100.0	100.0	100.0	96.3	99.2	100.0	107.6	101.6	100.0	99.2	99.8	100.0	103.5	100.7	100.0	109.3	102.0
21, 22, 23	FIRE SUPPRESSION, PLUMBING & HVAC	100.1	98.9	99.6	93.5	83.0	89.2	99.9	83.2	93.0	93.5	85.5	90.2	93.5	93.5	93.5	99.9	81.8	92.4
26, 27, 3370	ELECTRICAL, COMMUNICATIONS & UTIL.	94.6	100.2	97.5	92.2	51.7	71.0	104.6	51.5	76.9	99.1	86.9	92.7	96.5	92.4	94.3	94.3	84.5	89.2
MF2004	WEIGHTED AVERAGE	94.6	100.2	97.1	93.4	74.2	84.9	96.4	74.8	86.8	95.5	88.8	92.5	92.7	95.5	94.0	95.3	88.2	92.2

DIVISION		MICHIGAN															MINNESOTA		
		LANSING			MUSKEGON			ROYAL OAK			SAGINAW			TRAVERSE CITY			BEMIDJI		
		488 - 489			494			480,483			486			496			566		
		MAT.	INST.	TOTAL	MAT.	INST.	TOTAL	MAT.	INST.	TOTAL	MAT.	INST.	TOTAL	MAT.	INST.	TOTAL	MAT.	INST.	TOTAL
015433	CONTRACTOR EQUIPMENT		110.3	110.3		103.5	103.5		94.9	94.9		110.3	110.3		94.1	94.1		98.8	98.8
0241, 31 - 34	SITE & INFRASTRUCTURE, DEMOLITION	83.5	94.5	91.1	79.4	87.7	85.2	79.2	95.1	90.3	68.4	94.3	86.5	71.8	93.9	87.2	93.2	98.6	97.0
0310	Concrete Forming & Accessories	98.7	92.0	92.9	98.1	92.1	92.9	92.1	104.1	102.5	95.4	91.5	92.0	86.8	73.6	75.4	89.2	98.2	97.0
0320	Concrete Reinforcing	116.3	125.7	121.1	100.6	87.0	93.7	105.4	126.4	116.1	116.3	125.4	121.0	94.6	87.0	90.7	95.9	113.8	105.1
0330	Cast-in-Place Concrete	83.1	96.3	88.1	89.3	99.8	93.3	70.8	114.4	87.3	79.9	95.7	85.9	82.6	64.6	75.8	101.4	109.1	104.3
03	CONCRETE	93.2	100.6	96.6	91.7	93.1	92.3	79.3	110.9	94.1	91.4	100.1	95.5	81.4	73.6	77.8	96.2	105.9	100.8
04	MASONRY	86.3	98.1	93.4	90.3	74.2	80.6	87.5	117.0	105.2	93.7	91.0	92.1	87.9	83.8	85.4	101.0	108.6	105.6
05	METALS	89.9	119.0	99.7	90.9	84.0	88.6	94.1	100.4	96.2	91.3	118.5	100.5	93.6	87.6	91.6	88.9	125.5	101.3
06	WOOD, PLASTICS & COMPOSITES	101.1	90.3	95.0	94.6	93.3	93.8	94.4	100.9	98.0	94.3	91.0	92.5	85.9	73.2	78.7	75.4	96.0	87.1
07	THERMAL & MOISTURE PROTECTION	101.3	95.9	99.3	92.7	79.7	87.8	98.7	117.0	105.6	100.8	97.5	99.5	94.2	70.5	85.3	103.2	94.3	99.9
08	OPENINGS	94.0	98.1	95.0	89.9	87.6	89.4	94.0	105.1	96.8	91.9	98.5	93.6	97.5	66.6	89.8	99.0	117.0	103.4
0920	Plaster & Gypsum Board	95.4	89.0	91.0	69.4	89.0	82.9	94.5	99.9	98.2	97.2	89.7	92.0	54.1	72.9	67.1	99.0	96.1	97.0
0950, 0980	Ceilings & Acoustic Treatment	89.4	89.0	89.1	87.4	89.0	88.4	90.3	99.9	96.4	91.1	89.7	90.2	84.6	72.9	77.2	133.6	96.1	109.7
0960	Flooring	102.5	84.9	97.6	96.9	67.2	88.7	90.0	116.7	97.4	92.7	80.3	89.3	110.6	51.9	94.3	109.6	130.1	115.2
0970, 0990	Wall Finishes & Painting/Coating	105.5	88.6	95.3	95.7	68.5	79.3	95.0	107.5	102.5	93.5	87.8	90.1	114.2	48.0	74.4	98.6	101.9	100.6
09	FINISHES	96.3	89.7	92.7	90.8	85.1	87.7	91.2	106.3	99.4	92.0	88.4	90.0	95.3	66.3	79.6	111.3	104.6	107.6
COVERS	DIVS. 10 - 14, 25, 28, 41, 43, 44	100.0	103.1	100.7	100.0	108.9	101.9	100.0	106.7	101.4	100.0	98.8	99.7	100.0	94.8	98.9	100.0	97.8	99.5
21, 22, 23	FIRE SUPPRESSION, PLUMBING & HVAC	100.0	96.9	98.7	99.7	85.9	94.0	93.7	106.0	98.8	100.1	88.6	95.3	93.5	82.5	88.9	93.5	86.6	90.6
26, 27, 3370	ELECTRICAL, COMMUNICATIONS & UTIL.	100.6	98.2	99.3	94.8	81.4	87.8	96.8	99.3	98.1	92.5	93.5	93.0	94.0	80.9	87.2	101.7	85.4	93.2
MF2004	WEIGHTED AVERAGE	95.7	98.7	97.0	93.9	85.6	90.2	92.1	105.7	98.1	94.2	95.3	94.7	92.2	79.8	86.8	97.2	100.1	98.5

DIVISION		MINNESOTA																	
		BRAINERD			DETROIT LAKES			DULUTH			MANKATO			MINNEAPOLIS			ROCHESTER		
		564			565			556 - 558			560			553 - 555			559		
		MAT.	INST.	TOTAL	MAT.	INST.	TOTAL	MAT.	INST.	TOTAL	MAT.	INST.	TOTAL	MAT.	INST.	TOTAL	MAT.	INST.	TOTAL
015433	CONTRACTOR EQUIPMENT		101.7	101.7		98.8	98.8		101.0	101.0		101.6	101.6		104.7	104.7		101.0	101.0
0241, 31 - 34	SITE & INFRASTRUCTURE, DEMOLITION	91.4	103.7	100.0	91.3	99.0	96.6	95.2	102.1	100.0	87.9	104.4	99.4	93.9	107.7	103.5	92.9	101.9	99.2
0310	Concrete Forming & Accessories	93.2	100.8	99.8	86.8	93.5	92.6	100.9	121.7	118.9	99.9	105.8	105.0	102.9	135.3	130.9	103.7	112.4	111.2
0320	Concrete Reinforcing	94.6	114.1	104.6	95.9	113.8	105.0	92.2	114.4	103.5	94.5	125.0	110.1	92.4	125.6	109.4	92.2	125.2	109.0
0330	Cast-in-Place Concrete	110.3	113.4	111.5	98.5	112.6	103.8	108.3	109.5	108.7	101.5	112.5	105.7	112.4	122.9	116.4	110.3	105.0	108.3
03	CONCRETE	100.4	108.6	104.2	93.7	105.0	99.0	101.5	116.6	108.6	95.6	112.6	103.6	105.5	129.4	116.7	102.7	113.1	107.6
04	MASONRY	124.9	119.4	121.6	124.3	115.0	118.7	109.6	122.8	117.5	112.6	116.3	114.8	108.4	133.6	123.5	107.6	120.4	115.3
05	METALS	89.9	126.0	102.1	88.9	125.2	101.2	89.4	128.6	102.7	89.7	132.0	104.0	92.2	136.2	107.1	89.3	134.1	104.4
06	WOOD, PLASTICS & COMPOSITES	97.5	95.7	96.5	72.8	86.7	80.7	110.6	122.8	117.5	105.8	104.1	104.9	113.7	134.9	125.7	113.7	111.0	112.2
07	THERMAL & MOISTURE PROTECTION	102.9	107.8	104.8	103.1	97.1	100.8	103.6	121.5	110.4	103.2	99.4	101.8	103.8	133.1	114.8	104.0	107.9	105.4
08	OPENINGS	86.9	116.9	94.3	99.0	112.0	102.2	95.2	126.7	103.0	91.7	125.5	100.1	98.4	142.2	109.2	95.2	129.3	103.7
0920	Plaster & Gypsum Board	86.8	96.1	93.3	98.1	86.6	90.1	93.2	123.9	114.5	90.7	104.8	100.4	98.4	136.2	124.6	97.9	111.8	107.5
0950, 0980	Ceilings & Acoustic Treatment	69.0	96.1	86.3	133.6	86.6	103.7	92.2	123.9	112.4	69.0	104.8	91.8	95.6	136.2	121.5	93.9	111.8	105.3
0960	Flooring	108.4	130.1	114.4	108.2	130.1	114.3	107.4	131.5	114.0	110.7	125.8	114.9	105.1	130.1	112.0	107.5	93.6	103.7
0970, 0990	Wall Finishes & Painting/Coating	93.4	101.9	98.5	98.6	101.9	100.6	89.6	113.4	103.9	103.7	105.0	104.5	97.3	126.8	115.1	92.5	105.0	100.1
09	FINISHES	94.2	106.0	100.6	110.6	100.8	105.3	98.6	123.4	112.1	95.8	109.7	103.3	99.6	134.1	118.4	99.6	108.9	104.7
COVERS	DIVS. 10 - 14, 25, 28, 41, 43, 44	100.0	99.7	99.9	100.0	98.9	99.8	100.0	106.5	101.4	100.0	89.0	97.7	100.0	111.7	102.5	100.0	104.9	101.0
21, 22, 23	FIRE SUPPRESSION, PLUMBING & HVAC	92.8	89.9	91.6	93.5	89.4	91.8	99.8	100.0	99.9	92.8	94.9	93.6	99.8	124.1	109.8	99.8	99.3	99.6
26, 27, 3370	ELECTRICAL, COMMUNICATIONS & UTIL.	99.2	114.9	107.4	101.5	65.0	82.4	99.7	114.9	107.6	106.2	84.9	95.1	99.2	115.0	107.5	98.2	84.9	91.3
MF2004	WEIGHTED AVERAGE	96.1	107.4	101.1	97.9	97.7	97.8	98.6	114.6	105.6	96.2	105.5	100.3	99.7	126.3	111.5	98.5	107.8	102.6

DIVISION		MINNESOTA															MISSISSIPPI		
		SAINT PAUL			ST. CLOUD			THIEF RIVER FALLS			WILLMAR			WINDOM			BILOXI		
		550 - 551			563			567			562			561			395		
		MAT.	INST.	TOTAL	MAT.	INST.	TOTAL	MAT.	INST.	TOTAL	MAT.	INST.	TOTAL	MAT.	INST.	TOTAL	MAT.	INST.	TOTAL
015433	CONTRACTOR EQUIPMENT		101.0	101.0		101.6	101.6		98.8	98.8		101.7	101.7		101.7	101.7		98.3	98.3
0241, 31 - 34	SITE & INFRASTRUCTURE, DEMOLITION	97.7	102.7	101.2	86.1	105.8	99.8	91.9	98.6	96.6	85.8	99.4	95.3	79.4	98.4	92.6	104.0	88.7	93.4
0310	Concrete Forming & Accessories	98.1	131.8	127.2	90.8	130.6	125.3	89.8	97.8	96.7	90.6	51.6	56.9	94.4	48.1	54.4	93.0	59.1	63.7
0320	Concrete Reinforcing	89.0	125.6	107.7	94.7	125.4	110.4	96.3	113.8	105.2	94.3	124.6	109.7	94.3	123.6	109.2	86.0	60.2	72.8
0330	Cast-in-Place Concrete	114.4	122.2	117.4	97.2	121.9	106.6	100.5	95.1	98.5	98.8	72.3	88.8	85.4	52.6	73.0	108.5	57.2	89.1
03	CONCRETE	108.2	127.7	117.3	91.6	126.9	108.1	94.8	100.9	97.7	91.5	74.5	83.6	82.2	66.0	74.7	96.6	60.2	79.6
04	MASONRY	118.3	133.5	127.5	109.5	131.1	122.5	100.9	108.6	105.6	113.4	91.7	100.4	123.9	66.9	89.6	88.9	48.2	64.4
05	METALS	88.7	135.8	104.6	90.6	134.0	105.3	89.1	125.1	101.2	89.7	123.3	101.0	89.6	121.8	100.5	94.7	84.7	91.3
06	WOOD, PLASTICS & COMPOSITES	106.9	130.3	120.2	95.0	129.1	114.3	76.3	96.0	87.4	94.7	48.4	68.5	98.7	48.4	70.2	95.1	62.8	76.8
07	THERMAL & MOISTURE PROTECTION	105.1	132.4	115.4	103.0	117.5	108.5	104.2	93.6	100.2	102.9	57.2	85.6	102.9	48.0	82.1	91.6	59.7	79.6
08	OPENINGS	92.8	139.7	104.4	91.7	139.0	103.4	99.0	117.0	103.4	89.3	75.5	85.9	93.6	75.5	89.1	97.5	63.2	89.0
0920	Plaster & Gypsum Board	90.5	131.7	119.0	86.3	130.5	116.9	98.5	96.1	96.9	86.3	47.4	59.4	86.8	47.4	59.5	99.2	62.2	73.6
0950, 0980	Ceilings & Acoustic Treatment	90.4	131.7	116.7	69.0	130.5	108.2	133.6	96.1	109.7	69.0	47.4	55.2	69.0	47.4	55.2	98.4	62.2	75.3
0960	Flooring	101.1	130.1	109.1	104.3	130.1	111.5	109.2	130.1	115.0	106.2	117.3	109.3	109.1	117.3	111.3	116.6	56.9	100.1
0970, 0990	Wall Finishes & Painting/Coating	97.3	132.2	118.3	103.7	126.8	117.6	98.6	101.9	100.6	98.6	101.9	100.6	98.6	105.0	102.5	106.1	48.3	71.3
09	FINISHES	96.7	132.0	115.9	93.2	130.7	113.6	111.1	104.6	107.6	93.4	65.8	78.4	93.8	64.2	77.7	103.5	57.7	78.6
COVERS	DIVS. 10 - 14, 25, 28, 41, 43, 44	100.0	110.8	102.3	100.0	110.5	102.2	100.0	97.7	99.5	100.0	48.9	89.2	100.0	46.2	88.6	100.0	65.4	92.6
21, 22, 23	FIRE SUPPRESSION, PLUMBING & HVAC	99.7	115.9	106.4	99.5	124.0	109.6	93.5	86.1	90.4	92.8	99.9	95.7	92.8	77.0	86.3	99.8	53.0	80.5
26, 27, 3370	ELECTRICAL, COMMUNICATIONS & UTIL.	98.1	115.0	106.9	99.2	115.0	107.4	98.6	65.0	81.0	99.2	66.1	81.9	106.2	84.8	95.0	98.9	60.5	78.8
MF2004	WEIGHTED AVERAGE	99.3	123.4	109.9	96.2	124.2	108.6	96.7	96.4	96.6	94.4	84.4	90.0	94.8	77.8	87.3	97.9	62.2	82.1

DIVISION		MISSISSIPPI																	
		CLARKSDALE			COLUMBUS			GREENVILLE			GREENWOOD			JACKSON			LAUREL		
		386			397			387			389			390 - 392			394		
		MAT.	INST.	TOTAL	MAT.	INST.	TOTAL	MAT.	INST.	TOTAL	MAT.	INST.	TOTAL	MAT.	INST.	TOTAL	MAT.	INST.	TOTAL
015433	CONTRACTOR EQUIPMENT		98.3	98.3		98.3	98.3		98.3	98.3		98.3	98.3		98.3	98.3		98.3	98.3
0241, 31 - 34	SITE & INFRASTRUCTURE, DEMOLITION	101.5	88.5	92.5	102.7	88.8	93.0	107.8	88.8	94.6	104.7	88.2	93.2	99.2	88.8	91.9	109.3	88.6	94.9
0310	Concrete Forming & Accessories	84.8	55.2	59.2	83.2	57.4	60.9	81.9	71.4	72.8	92.9	55.1	60.2	92.3	64.3	68.1	83.2	59.3	62.5
0320	Concrete Reinforcing	92.6	46.3	69.0	92.1	47.4	69.3	93.2	67.1	79.9	92.6	58.5	75.2	86.0	59.2	72.3	92.6	43.5	67.6
0330	Cast-in-Place Concrete	105.8	58.1	87.8	110.7	63.8	92.9	109.0	60.5	90.6	113.8	57.7	92.6	100.8	61.3	85.9	108.2	59.9	90.0
03	CONCRETE	94.6	56.1	76.6	98.3	59.3	80.1	99.8	68.1	85.0	100.9	58.1	80.9	92.7	63.7	79.2	100.4	58.1	80.6
04	MASONRY	88.7	50.2	65.5	111.4	57.5	79.0	129.8	53.9	84.2	89.4	50.0	65.7	90.0	53.9	68.3	107.9	53.7	75.3
05	METALS	91.8	77.8	87.0	91.6	78.8	87.3	92.8	87.9	91.2	91.8	79.4	87.6	94.7	84.5	91.2	91.7	77.7	87.0
06	WOOD, PLASTICS & COMPOSITES	84.0	57.3	68.9	82.2	58.4	68.7	80.6	76.3	78.2	95.1	57.3	73.7	94.0	66.7	78.5	82.8	61.1	70.5
07	THERMAL & MOISTURE PROTECTION	91.5	52.1	76.6	91.6	56.9	78.5	91.8	66.1	82.1	91.7	52.0	76.7	93.6	65.1	82.8	91.7	58.7	79.2
08	OPENINGS	96.9	55.1	86.5	96.8	55.3	86.5	96.9	68.6	89.9	96.9	56.6	86.9	97.9	61.2	88.8	93.5	56.2	84.2
0920	Plaster & Gypsum Board	88.6	56.5	66.4	88.2	57.6	67.1	87.3	76.1	79.6	97.7	56.5	69.2	93.4	66.2	74.6	87.3	60.4	68.7
0950, 0980	Ceilings & Acoustic Treatment	91.3	56.5	69.1	91.3	57.6	69.9	93.2	76.1	82.3	91.3	56.5	69.1	94.1	66.2	76.3	91.3	60.4	71.6
0960	Flooring	109.8	56.9	95.2	108.5	63.8	96.1	107.6	56.9	93.6	116.6	56.9	100.1	117.0	56.9	100.4	106.5	56.9	92.8
0970, 0990	Wall Finishes & Painting/Coating	106.1	51.2	73.1	106.1	58.8	77.7	106.1	73.2	86.3	106.1	51.2	73.1	106.1	73.2	86.3	106.1	67.7	83.0
09	FINISHES	98.3	55.3	75.0	98.1	59.1	76.9	98.5	69.8	82.9	101.9	55.3	76.6	102.0	64.2	81.5	97.9	60.3	77.5
COVERS	DIVS. 10 - 14, 25, 28, 41, 43, 44	100.0	65.8	92.7	100.0	67.1	93.0	100.0	69.0	93.4	100.0	65.7	92.7	100.0	67.9	93.2	100.0	43.8	88.1
21, 22, 23	FIRE SUPPRESSION, PLUMBING & HVAC	96.9	46.8	76.3	96.1	42.5	74.1	99.8	54.0	80.9	96.9	44.6	75.4	99.8	59.7	83.3	96.1	52.0	78.0
26, 27, 3370	ELECTRICAL, COMMUNICATIONS & UTIL.	98.2	52.0	74.1	96.2	55.1	74.7	98.2	70.3	83.6	98.2	48.7	72.3	103.3	70.3	86.0	97.8	60.6	78.4
MF2004	WEIGHTED AVERAGE	95.8	57.8	79.0	97.0	59.2	80.4	99.6	67.9	85.6	97.1	57.3	79.5	97.8	67.0	84.2	97.1	60.9	81.2

DIVISION		MISSISSIPPI									MISSOURI								
		MCCOMB			MERIDIAN			TUPELO			BOWLING GREEN			CAPE GIRARDEAU			CHILLICOTHE		
		396			393			388			633			637			646		
		MAT.	INST.	TOTAL	MAT.	INST.	TOTAL	MAT.	INST.	TOTAL	MAT.	INST.	TOTAL	MAT.	INST.	TOTAL	MAT.	INST.	TOTAL
015433	CONTRACTOR EQUIPMENT		98.3	98.3		98.3	98.3		98.3	98.3		106.7	106.7		106.7	106.7		101.9	101.9
0241, 31 - 34	SITE & INFRASTRUCTURE, DEMOLITION	94.9	88.4	90.4	99.0	89.1	92.1	98.9	88.4	91.6	94.2	92.7	93.2	95.3	93.2	93.8	99.3	90.3	93.0
0310	Concrete Forming & Accessories	83.2	57.1	60.6	80.9	70.8	72.2	82.4	57.4	60.8	99.2	83.0	85.2	93.2	70.9	73.9	93.1	83.2	84.6
0320	Concrete Reinforcing	93.2	45.7	68.9	92.1	59.2	75.3	90.7	58.6	74.3	103.3	88.8	95.9	104.5	79.1	91.5	98.9	80.3	89.4
0330	Cast-in-Place Concrete	96.1	57.5	81.5	102.8	63.9	88.1	105.8	59.7	88.4	95.5	76.3	88.2	94.5	80.4	89.2	90.9	64.1	80.8
03	CONCRETE	88.0	56.6	73.3	92.4	67.5	80.8	94.2	59.8	78.1	94.5	83.1	89.2	93.7	77.4	86.1	98.3	76.7	88.2
04	MASONRY	113.2	49.5	74.9	88.5	58.3	70.4	119.9	53.4	79.9	119.2	91.4	102.5	116.7	74.3	91.2	101.4	81.9	89.6
05	METALS	91.8	75.8	86.4	92.8	84.7	90.0	91.7	79.7	87.6	93.0	107.5	97.9	94.1	102.3	96.9	90.1	88.2	89.5
06	WOOD, PLASTICS & COMPOSITES	82.2	60.4	69.8	79.6	73.0	75.9	81.1	58.4	68.3	93.8	83.7	88.1	87.3	69.0	76.9	89.0	89.7	89.4
07	THERMAL & MOISTURE PROTECTION	91.2	51.2	76.1	91.4	67.7	82.4	91.4	58.5	79.0	101.0	92.1	97.6	100.9	75.1	91.1	99.1	76.2	90.4
08	OPENINGS	96.9	57.0	87.0	96.8	70.8	90.4	96.8	56.5	86.8	98.4	79.7	93.7	98.4	69.3	91.2	89.4	76.2	86.1
0920	Plaster & Gypsum Board	88.2	59.7	68.5	87.3	72.7	77.2	87.3	57.6	66.8	102.1	83.2	89.0	99.9	68.0	77.8	94.4	89.3	90.9
0950, 0980	Ceilings & Acoustic Treatment	91.3	59.7	71.2	93.2	72.7	80.1	91.3	57.6	69.9	87.6	83.2	84.8	87.6	68.0	75.1	86.8	89.3	88.4
0960	Flooring	108.5	56.9	94.2	106.4	56.9	92.7	107.9	56.9	93.8	94.0	64.5	85.9	89.9	64.5	82.9	96.6	46.3	82.7
0970, 0990	Wall Finishes & Painting/Coating	106.1	47.0	70.5	106.1	84.6	93.1	106.1	65.2	81.5	101.0	101.6	101.4	101.0	75.2	85.4	96.0	49.8	68.2
09	FINISHES	97.5	56.5	75.3	97.4	70.3	82.7	97.6	58.4	76.3	95.8	81.4	88.0	94.2	69.3	80.7	97.4	75.0	85.2
COVERS	DIVS. 10 - 14, 25, 28, 41, 43, 44	100.0	69.8	93.6	100.0	70.2	93.7	100.0	67.1	93.0	100.0	59.0	91.3	100.0	56.7	90.8	100.0	56.9	90.9
21, 22, 23	FIRE SUPPRESSION, PLUMBING & HVAC	96.1	39.2	72.7	99.8	62.1	84.3	97.3	49.0	77.5	93.0	85.8	90.1	99.8	84.7	93.6	93.4	42.8	72.6
26, 27, 3370	ELECTRICAL, COMMUNICATIONS & UTIL.	94.6	55.5	74.2	97.8	61.3	78.7	97.9	55.2	75.6	100.6	82.0	90.9	100.5	104.1	102.4	92.6	36.9	63.5
MF2004	WEIGHTED AVERAGE	95.5	56.8	78.4	96.1	68.6	84.0	97.3	60.4	81.0	97.0	86.6	92.4	98.4	83.8	92.0	94.7	66.1	82.1

	DIVISION	MISSOURI																	
		COLUMBIA			FLAT RIVER			HANNIBAL			HARRISONVILLE			JEFFERSON CITY			JOPLIN		
		652			636			634			647			650 - 651			648		
		MAT.	INST.	TOTAL	MAT.	INST.	TOTAL	MAT.	INST.	TOTAL	MAT.	INST.	TOTAL	MAT.	INST.	TOTAL	MAT.	INST.	TOTAL
015433	CONTRACTOR EQUIPMENT		107.7	107.7		106.7	106.7		106.7	106.7		101.9	101.9		107.7	107.7		105.3	105.3
0241, 31 - 34	SITE & INFRASTRUCTURE, DEMOLITION	103.4	95.8	98.1	97.0	93.3	94.4	91.5	92.1	91.9	91.0	94.4	93.4	105.2	95.7	98.6	98.7	97.8	98.1
0310	Concrete Forming & Accessories	92.5	75.4	77.7	104.4	79.3	82.7	97.7	60.9	65.8	90.5	94.5	93.9	101.7	57.5	63.5	103.9	66.1	71.2
0320	Concrete Reinforcing	100.4	112.4	106.5	104.5	98.0	101.2	102.7	76.1	89.1	98.5	99.9	99.2	99.1	107.5	103.4	102.1	67.5	84.4
0330	Cast-in-Place Concrete	89.7	85.2	88.0	98.5	76.1	90.0	90.4	69.9	82.6	93.2	94.6	93.7	89.8	85.1	88.0	98.5	66.6	86.4
03	CONCRETE	85.4	87.4	86.3	97.4	83.2	90.7	90.7	68.8	80.4	94.5	96.0	95.2	86.6	78.5	82.8	97.6	67.7	83.6
04	MASONRY	132.4	82.3	102.3	117.3	75.9	92.4	110.3	80.5	92.4	96.0	94.1	94.8	95.8	82.3	87.7	93.8	56.6	71.4
05	METALS	90.0	118.4	99.6	92.9	110.5	98.8	93.0	101.6	95.9	90.5	105.0	95.4	89.4	116.4	98.5	93.0	84.2	90.1
06	WOOD, PLASTICS & COMPOSITES	94.3	72.1	81.7	100.5	78.9	88.3	92.0	60.2	74.0	85.4	94.6	90.6	104.5	47.7	72.3	99.3	65.2	80.0
07	THERMAL & MOISTURE PROTECTION	94.8	80.4	89.4	101.1	87.5	95.9	100.8	80.4	93.1	98.5	95.6	97.4	94.6	78.3	88.4	98.8	62.3	85.0
08	OPENINGS	91.2	92.3	91.5	98.3	79.8	93.7	98.4	63.5	89.7	89.7	96.0	91.3	88.5	77.8	85.8	90.6	67.3	84.8
0920	Plaster & Gypsum Board	94.8	71.2	78.4	106.8	78.2	87.0	100.8	59.0	71.8	87.9	94.2	92.3	93.4	46.1	60.6	100.4	64.0	75.2
0950, 0980	Ceilings & Acoustic Treatment	92.0	71.2	78.7	87.6	78.2	81.6	87.6	59.0	69.4	86.8	94.2	91.5	90.3	46.1	62.1	89.4	64.0	73.2
0960	Flooring	107.3	86.1	101.4	97.5	64.5	88.4	93.0	64.5	85.1	92.0	93.2	92.3	115.6	86.1	107.4	122.0	43.4	100.3
0970, 0990	Wall Finishes & Painting/Coating	103.9	72.9	85.3	101.0	65.9	79.9	101.0	71.0	83.0	100.5	101.6	101.2	103.9	72.9	85.3	95.5	39.0	61.5
09	FINISHES	97.5	74.4	85.0	97.6	74.5	85.1	95.1	61.5	76.9	94.9	95.0	94.9	99.8	60.1	78.2	104.7	59.1	80.0
COVERS	DIVS. 10 - 14, 25, 28, 41, 43, 44	100.0	96.7	99.3	100.0	58.4	91.2	100.0	52.4	89.9	100.0	84.3	96.7	100.0	93.8	98.7	100.0	65.3	92.6
21, 22, 23	FIRE SUPPRESSION, PLUMBING & HVAC	100.0	86.4	94.4	93.0	85.7	90.0	93.0	79.8	87.6	93.3	89.3	91.6	99.8	91.3	96.3	100.1	49.5	79.3
26, 27, 3370	ELECTRICAL, COMMUNICATIONS & UTIL.	95.1	83.7	89.1	105.4	104.1	104.8	99.2	75.4	86.8	99.6	97.2	98.3	101.9	83.7	92.4	90.5	66.1	77.8
MF2004	WEIGHTED AVERAGE	96.8	88.2	93.0	98.0	87.4	93.3	95.8	76.8	87.4	94.2	94.8	94.5	95.6	85.0	90.9	96.8	65.2	82.9

	DIVISION	MISSOURI																	
		KANSAS CITY			KIRKSVILLE			POPLAR BLUFF			ROLLA			SEDALIA			SIKESTON		
		640 - 641			635			639			654 - 655			653			638		
		MAT.	INST.	TOTAL	MAT.	INST.	TOTAL	MAT.	INST.	TOTAL	MAT.	INST.	TOTAL	MAT.	INST.	TOTAL	MAT.	INST.	TOTAL
015433	CONTRACTOR EQUIPMENT		103.2	103.2		98.9	98.9		101.2	101.2		107.7	107.7		99.7	99.7		101.2	101.2
0241, 31 - 34	SITE & INFRASTRUCTURE, DEMOLITION	92.4	97.9	96.2	93.1	88.5	89.9	79.2	92.3	88.3	103.0	93.5	96.4	95.0	90.2	91.7	82.9	92.4	89.5
0310	Concrete Forming & Accessories	102.8	107.2	106.6	90.3	52.2	57.4	90.6	66.5	69.8	99.0	85.0	86.9	96.0	58.3	63.4	91.5	66.6	70.0
0320	Concrete Reinforcing	96.9	113.6	105.4	103.5	75.4	89.2	106.7	72.0	89.0	100.9	49.7	74.8	99.5	91.3	95.3	106.0	64.0	84.5
0330	Cast-in-Place Concrete	91.6	107.5	97.6	98.5	61.1	84.3	75.8	59.1	69.4	91.8	83.6	88.7	95.8	64.8	84.1	81.0	67.0	75.7
03	CONCRETE	93.8	108.8	100.8	107.9	61.2	86.1	83.9	66.2	75.6	87.1	78.9	83.2	99.1	68.2	84.7	87.6	67.5	78.2
04	MASONRY	98.0	108.1	104.1	122.7	68.2	89.9	115.2	54.2	78.5	109.3	61.5	80.6	115.1	51.8	77.0	114.9	59.2	81.4
05	METALS	100.7	113.9	105.1	92.6	91.8	92.3	93.1	89.6	91.9	89.3	82.9	87.2	88.1	98.1	91.4	93.4	85.8	90.8
06	WOOD, PLASTICS & COMPOSITES	98.6	106.5	103.1	80.3	50.7	63.5	79.6	69.0	73.6	101.4	89.2	94.5	94.0	51.3	69.8	80.8	69.0	74.1
07	THERMAL & MOISTURE PROTECTION	98.7	109.8	102.9	106.9	73.5	94.3	105.9	69.2	92.1	95.0	64.3	83.4	100.3	73.9	90.3	106.0	63.5	90.0
08	OPENINGS	96.5	109.0	99.6	101.9	58.6	91.2	103.0	69.8	94.8	91.2	67.1	85.2	94.4	70.0	88.4	103.0	60.8	92.5
0920	Plaster & Gypsum Board	96.5	106.6	103.5	95.3	49.1	63.3	96.0	68.0	76.6	96.6	88.8	91.2	90.0	49.7	62.1	97.3	68.0	77.0
0950, 0980	Ceilings & Acoustic Treatment	95.4	106.6	102.5	85.0	49.1	62.2	87.6	68.0	75.1	92.0	88.8	90.0	92.0	49.7	65.1	87.6	68.0	75.1
0960	Flooring	98.5	107.9	101.1	72.7	47.5	65.7	85.4	53.7	76.7	111.6	40.5	91.9	86.2	93.2	88.2	85.8	43.4	74.1
0970, 0990	Wall Finishes & Painting/Coating	100.5	117.6	110.8	96.4	39.0	61.8	95.9	76.0	84.0	103.9	80.9	90.1	103.9	101.6	102.5	95.9	56.0	71.9
09	FINISHES	99.9	108.2	104.4	93.6	48.4	69.0	94.1	65.3	78.5	99.2	77.8	87.5	95.6	67.6	80.4	94.7	61.1	76.5
COVERS	DIVS. 10 - 14, 25, 28, 41, 43, 44	100.0	101.4	100.3	100.0	48.5	89.1	100.0	51.4	89.7	100.0	60.2	91.6	100.0	76.2	94.9	100.0	52.1	89.8
21, 22, 23	FIRE SUPPRESSION, PLUMBING & HVAC	100.1	112.9	105.3	93.1	78.3	87.0	93.1	68.2	82.8	93.2	70.0	83.7	93.1	84.7	89.6	93.1	69.0	83.2
26, 27, 3370	ELECTRICAL, COMMUNICATIONS & UTIL.	101.4	104.4	103.0	99.3	75.3	86.8	99.6	104.1	102.0	93.4	81.9	87.4	94.6	97.2	95.9	98.6	104.1	101.5
MF2004	WEIGHTED AVERAGE	98.7	108.2	102.9	98.9	70.7	86.5	95.4	74.7	86.3	94.0	75.8	86.0	95.7	78.8	88.2	96.0	74.2	86.4

	DIVISION	MISSOURI									MONTANA								
		SPRINGFIELD			ST. JOSEPH			ST. LOUIS			BILLINGS			BUTTE			GREAT FALLS		
		656 - 658			644 - 645			630 - 631			590 - 591			597			594		
		MAT.	INST.	TOTAL	MAT.	INST.	TOTAL	MAT.	INST.	TOTAL	MAT.	INST.	TOTAL	MAT.	INST.	TOTAL	MAT.	INST.	TOTAL
015433	CONTRACTOR EQUIPMENT		102.1	102.1		101.9	101.9		107.8	107.8		99.0	99.0		98.8	98.8		98.8	98.8
0241, 31 - 34	SITE & INFRASTRUCTURE, DEMOLITION	96.9	93.6	94.6	93.9	93.3	93.5	95.2	97.2	96.6	93.7	96.9	95.9	103.4	96.2	98.4	107.3	96.9	100.0
0310	Concrete Forming & Accessories	103.6	71.2	75.6	102.6	76.0	79.6	104.4	103.9	103.9	99.8	63.2	68.2	89.6	66.8	69.9	99.7	63.9	68.7
0320	Concrete Reinforcing	96.7	112.0	104.5	95.8	99.7	97.8	95.6	108.7	102.3	93.9	67.3	80.3	101.8	67.3	84.2	93.8	67.2	80.2
0330	Cast-in-Place Concrete	97.4	74.2	88.6	91.5	101.4	95.3	94.5	110.2	100.4	110.6	66.9	94.0	121.7	67.8	101.3	128.6	65.4	104.7
03	CONCRETE	96.0	80.9	89.0	93.6	90.2	92.0	93.1	107.8	100.0	103.0	66.3	85.8	106.4	68.2	88.5	111.5	66.0	90.3
04	MASONRY	88.9	81.5	84.5	97.7	86.9	91.2	97.6	110.0	105.0	126.2	69.4	92.0	121.6	71.6	91.6	126.0	70.4	92.6
05	METALS	94.0	104.8	97.6	97.0	104.6	99.6	98.8	119.8	105.9	105.6	84.0	98.3	99.0	84.1	94.0	102.0	83.7	95.8
06	WOOD, PLASTICS & COMPOSITES	101.2	69.5	83.3	99.3	71.4	83.5	100.3	101.8	101.1	101.0	62.2	79.0	90.2	67.0	77.1	102.1	62.2	79.5
07	THERMAL & MOISTURE PROTECTION	99.1	75.1	90.0	99.0	89.1	95.3	100.8	107.3	103.2	102.2	70.2	90.1	102.1	71.0	90.4	102.6	65.6	88.6
08	OPENINGS	96.7	78.5	92.2	95.3	86.5	93.1	97.2	108.8	100.1	97.9	61.0	88.7	95.8	62.9	87.7	99.3	60.2	89.6
0920	Plaster & Gypsum Board	97.8	68.5	77.5	101.9	70.4	80.1	107.9	101.8	103.7	95.8	61.3	71.9	94.0	66.3	74.8	101.4	61.3	73.7
0950, 0980	Ceilings & Acoustic Treatment	92.0	68.5	77.0	94.5	70.4	79.2	92.8	101.8	98.5	104.5	61.3	77.0	109.6	66.3	82.0	111.3	61.3	79.5
0960	Flooring	110.0	43.4	91.6	101.5	93.2	99.2	97.3	98.6	97.6	109.4	75.7	100.1	106.1	79.9	98.9	114.7	79.9	105.1
0970, 0990	Wall Finishes & Painting/Coating	98.2	55.8	72.7	96.0	82.5	87.9	101.0	109.0	105.8	100.7	52.0	71.4	98.6	56.9	73.5	98.6	54.0	71.8
09	FINISHES	101.3	63.7	80.9	101.0	78.4	88.7	98.8	102.7	100.9	104.4	63.8	82.4	104.0	68.1	84.5	108.2	65.3	84.9
COVERS	DIVS. 10 - 14, 25, 28, 41, 43, 44	100.0	92.8	98.5	100.0	94.0	98.7	100.0	104.5	101.0	100.0	81.1	96.0	100.0	81.8	96.1	100.0	81.9	96.2
21, 22, 23	FIRE SUPPRESSION, PLUMBING & HVAC	100.0	67.7	86.7	100.1	85.0	93.9	99.8	105.0	102.0	100.0	72.4	88.6	100.1	66.0	86.1	100.1	70.1	87.8
26, 27, 3370	ELECTRICAL, COMMUNICATIONS & UTIL.	99.3	70.3	84.1	99.6	80.7	89.7	103.3	110.5	107.0	93.9	72.6	82.7	101.8	69.7	85.0	93.9	69.7	81.2
MF2004	WEIGHTED AVERAGE	97.7	77.8	88.9	98.0	87.2	93.3	98.6	107.2	102.4	102.0	72.9	89.2	102.0	72.3	88.9	103.4	72.2	89.6

DIVISION		MONTANA																	
		HAVRE			HELENA			KALISPELL			MILES CITY			MISSOULA			WOLF POINT		
		595			596			599			593			598			592		
		MAT.	INST.	TOTAL	MAT.	INST.	TOTAL	MAT.	INST.	TOTAL	MAT.	INST.	TOTAL	MAT.	INST.	TOTAL	MAT.	INST.	TOTAL
015433	CONTRACTOR EQUIPMENT		98.8	98.8		98.8	98.8		98.8	98.8		98.8	98.8		98.8	98.8		98.8	98.8
0241, 31 - 34	SITE & INFRASTRUCTURE, DEMOLITION	112.5	96.9	101.6	104.4	96.1	98.6	92.9	96.1	95.1	100.2	96.5	97.6	84.1	96.3	92.6	120.5	96.6	103.9
0310	Concrete Forming & Accessories	83.7	60.0	63.2	99.9	63.1	68.1	92.1	57.9	62.5	98.3	58.5	63.9	92.1	65.4	69.0	92.3	53.3	58.5
0320	Concrete Reinforcing	102.6	67.2	84.5	97.1	67.1	81.8	104.5	76.3	90.1	102.2	67.2	84.4	103.5	76.4	89.7	103.7	67.2	85.0
0330	Cast-in-Place Concrete	131.1	65.4	106.2	108.5	63.9	91.6	105.7	62.9	89.5	115.7	65.1	96.6	89.7	65.0	80.3	129.6	65.5	105.4
03	CONCRETE	114.9	64.3	91.3	102.1	65.2	84.9	95.8	64.3	81.1	103.1	63.6	84.6	84.1	68.4	76.7	118.8	61.4	91.9
04	MASONRY	122.6	70.4	91.2	118.2	68.9	88.5	120.6	70.5	90.5	128.0	66.6	91.1	146.2	70.2	100.5	129.2	66.2	91.3
05	METALS	95.2	83.6	91.3	101.1	83.4	95.1	95.1	87.5	92.5	94.3	83.8	90.8	95.6	88.0	93.0	94.4	83.6	90.8
06	WOOD, PLASTICS & COMPOSITES	82.8	57.0	68.2	102.2	62.2	79.5	93.3	55.2	71.7	99.5	57.4	75.7	93.3	65.7	77.7	92.3	50.0	68.4
07	THERMAL & MOISTURE PROTECTION	102.5	65.1	88.3	102.3	70.9	90.5	101.8	69.8	89.7	102.1	65.8	88.4	101.5	70.5	89.8	102.8	64.4	88.3
08	OPENINGS	95.9	57.4	86.4	98.7	59.0	88.9	95.9	59.1	86.8	95.3	57.6	86.0	95.9	64.5	88.1	95.4	53.4	85.0
0920	Plaster & Gypsum Board	89.7	56.0	66.3	93.2	61.3	71.1	94.5	54.1	66.5	100.2	56.4	69.9	94.5	65.0	74.1	95.9	48.8	63.3
0950, 0980	Ceilings & Acoustic Treatment	109.6	56.0	75.4	107.0	61.3	77.9	109.6	54.1	74.2	107.0	56.4	74.8	109.6	65.0	81.2	107.0	48.8	69.9
0960	Flooring	102.8	79.9	96.5	114.7	63.6	100.6	108.3	79.9	100.5	114.5	75.7	103.7	108.3	79.9	100.5	110.0	75.7	100.5
0970, 0990	Wall Finishes & Painting/Coating	98.6	54.0	71.8	98.6	48.0	68.2	98.6	41.5	64.3	98.6	49.9	69.3	98.6	52.9	71.1	98.6	54.0	71.8
09	FINISHES	103.1	62.3	80.9	106.0	61.5	81.8	104.1	59.4	79.8	106.5	60.1	81.3	103.5	66.6	83.5	106.0	56.3	79.0
COVERS	DIVS. 10 - 14, 25, 28, 41, 43, 44	100.0	81.3	96.0	100.0	81.4	96.1	100.0	80.3	95.8	100.0	79.6	95.7	100.0	81.1	96.0	100.0	79.0	95.5
21, 22, 23	FIRE SUPPRESSION, PLUMBING & HVAC	93.3	70.1	83.8	100.0	66.2	86.1	93.3	65.2	81.8	93.3	67.5	82.7	100.1	64.6	85.5	93.3	67.9	82.9
26, 27, 3370	ELECTRICAL, COMMUNICATIONS & UTIL.	93.9	69.7	81.2	99.7	69.7	84.0	98.4	73.7	85.5	93.9	72.6	82.8	99.5	73.7	86.0	93.9	67.6	80.1
MF2004	WEIGHTED AVERAGE	100.3	71.3	87.5	101.9	70.6	88.1	97.8	70.9	86.0	99.0	70.4	86.4	99.2	72.7	87.5	101.5	68.6	87.0

DIVISION		NEBRASKA																	
		ALLIANCE			COLUMBUS			GRAND ISLAND			HASTINGS			LINCOLN			MCCOOK		
		693			686			688			689			683 - 685			690		
		MAT.	INST.	TOTAL	MAT.	INST.	TOTAL	MAT.	INST.	TOTAL	MAT.	INST.	TOTAL	MAT.	INST.	TOTAL	MAT.	INST.	TOTAL
015433	CONTRACTOR EQUIPMENT		96.8	96.8		101.2	101.2		101.2	101.2		101.2	101.2		101.2	101.2		101.2	101.2
0241, 31 - 34	SITE & INFRASTRUCTURE, DEMOLITION	97.3	97.8	97.7	96.4	91.1	92.7	100.9	91.9	94.7	100.0	91.5	94.1	90.6	91.9	91.5	101.0	91.3	94.2
0310	Concrete Forming & Accessories	92.5	59.3	63.8	100.9	62.1	67.4	100.4	79.1	82.0	103.5	78.2	81.6	105.2	62.5	68.3	98.3	59.5	64.7
0320	Concrete Reinforcing	115.6	49.9	82.0	105.9	48.7	76.7	105.3	70.0	87.3	105.3	49.4	76.7	96.7	69.5	82.8	107.3	69.6	88.1
0330	Cast-in-Place Concrete	108.2	60.6	90.2	110.5	57.2	90.3	117.0	71.4	99.7	117.0	66.5	97.9	93.3	74.0	86.0	117.0	60.2	95.5
03	CONCRETE	118.1	58.5	90.2	103.6	59.0	82.7	108.2	75.4	92.8	108.4	69.4	90.1	94.1	68.8	82.3	108.3	62.7	87.0
04	MASONRY	109.8	70.5	86.2	115.0	75.5	91.3	108.1	81.6	92.2	117.1	93.5	102.9	96.9	71.7	81.8	105.0	70.6	84.3
05	METALS	95.1	66.3	85.4	88.1	76.2	84.1	89.9	84.7	88.2	90.4	77.8	86.1	94.8	83.7	91.0	90.6	83.2	88.1
06	WOOD, PLASTICS & COMPOSITES	87.1	58.4	70.8	98.6	63.5	78.7	97.9	81.0	88.4	101.4	81.0	89.9	100.6	58.8	76.9	96.0	58.5	74.7
07	THERMAL & MOISTURE PROTECTION	105.9	71.0	92.7	99.9	67.8	87.8	100.0	77.4	91.5	100.1	91.1	96.7	99.6	72.5	89.3	100.0	68.0	87.9
08	OPENINGS	90.6	54.5	81.6	90.4	59.2	82.7	90.4	74.5	86.5	90.4	68.7	85.0	95.9	58.2	86.6	90.5	59.7	82.8
0920	Plaster & Gypsum Board	87.3	57.1	66.4	96.1	62.4	72.8	95.7	80.3	85.1	97.4	80.3	85.6	91.4	57.4	67.9	95.7	57.1	69.0
0950, 0980	Ceilings & Acoustic Treatment	95.3	57.1	71.0	92.0	62.4	73.1	92.0	80.3	84.6	92.0	80.3	84.6	93.8	57.4	70.6	92.0	57.1	69.8
0960	Flooring	99.4	88.4	96.4	99.4	108.1	101.8	99.2	115.5	103.7	100.8	108.1	102.8	100.9	90.0	97.9	98.1	88.4	95.4
0970, 0990	Wall Finishes & Painting/Coating	168.0	60.8	103.5	96.2	70.8	80.9	96.2	75.1	83.5	96.2	70.8	80.9	96.2	86.5	90.4	96.2	53.0	70.2
09	FINISHES	100.5	64.0	80.7	98.1	70.7	83.2	98.3	85.0	91.1	99.0	82.5	90.0	98.3	69.6	82.7	97.9	63.1	79.0
COVERS	DIVS. 10 - 14, 25, 28, 41, 43, 44	100.0	74.4	94.6	100.0	77.9	95.3	100.0	83.2	96.4	100.0	82.3	96.2	100.0	80.6	95.9	100.0	73.6	94.4
21, 22, 23	FIRE SUPPRESSION, PLUMBING & HVAC	93.4	62.8	80.8	93.2	70.2	83.7	100.0	75.4	89.9	93.2	73.7	85.2	99.9	75.4	89.8	93.2	72.0	84.5
26, 27, 3370	ELECTRICAL, COMMUNICATIONS & UTIL.	91.4	77.8	84.3	91.1	86.4	88.6	89.7	71.1	79.9	89.0	93.7	91.5	103.1	71.1	86.3	93.7	77.8	85.4
MF2004	WEIGHTED AVERAGE	98.7	68.7	85.5	95.8	73.4	85.9	97.9	79.2	89.6	96.8	81.5	90.0	97.8	74.1	87.3	96.5	72.2	85.8

DIVISION		NEBRASKA												NEVADA					
		NORFOLK			NORTH PLATTE			OMAHA			VALENTINE			CARSON CITY			ELKO		
		687			691			680 - 681			692			897			898		
		MAT.	INST.	TOTAL	MAT.	INST.	TOTAL	MAT.	INST.	TOTAL	MAT.	INST.	TOTAL	MAT.	INST.	TOTAL	MAT.	INST.	TOTAL
015433	CONTRACTOR EQUIPMENT		91.8	91.8		101.2	101.2		91.8	91.8		94.7	94.7		100.7	100.7		100.7	100.7
0241, 31 - 34	SITE & INFRASTRUCTURE, DEMOLITION	79.7	90.5	87.2	101.9	91.5	94.7	80.8	90.8	87.7	84.7	94.9	91.8	68.0	101.5	91.3	56.2	101.2	87.5
0310	Concrete Forming & Accessories	87.2	77.3	78.6	100.4	78.0	81.0	95.4	69.2	72.7	88.3	58.9	62.9	100.7	88.9	90.5	107.0	76.2	80.4
0320	Concrete Reinforcing	106.1	51.5	78.2	106.7	70.8	88.4	101.3	69.7	85.1	107.4	48.4	77.3	103.5	107.8	105.7	105.0	96.1	100.5
0330	Cast-in-Place Concrete	111.4	65.7	94.1	117.0	68.0	98.4	100.3	75.9	91.1	103.2	57.9	86.1	104.8	87.7	98.4	98.2	71.8	88.2
03	CONCRETE	102.3	68.7	86.6	108.4	73.8	92.2	96.7	71.9	85.1	105.2	57.2	82.8	107.1	92.2	100.1	100.6	78.9	90.4
04	MASONRY	121.6	73.3	92.6	92.3	93.4	93.0	102.4	77.2	87.2	105.0	70.6	84.3	120.3	75.0	93.1	122.0	67.7	89.4
05	METALS	91.2	69.1	83.7	90.2	85.3	88.5	94.1	75.4	87.8	100.5	67.4	89.3	96.9	98.6	97.5	95.9	91.7	94.5
06	WOOD, PLASTICS & COMPOSITES	81.4	80.5	80.9	97.8	81.0	88.3	91.3	68.2	78.2	81.5	57.9	68.1	93.3	89.7	91.3	101.6	77.2	87.8
07	THERMAL & MOISTURE PROTECTION	99.2	77.3	91.0	100.0	84.6	94.2	95.2	79.8	89.4	99.9	70.8	88.9	97.0	84.4	92.2	98.1	70.5	87.7
08	OPENINGS	92.2	69.3	86.5	89.7	73.8	85.8	98.9	68.0	91.2	92.3	55.7	83.2	93.6	103.8	96.1	96.3	80.2	92.3
0920	Plaster & Gypsum Board	95.6	80.3	85.0	95.7	80.3	85.1	94.1	67.7	75.8	98.2	57.1	69.8	86.8	89.3	88.5	95.1	76.3	82.1
0950, 0980	Ceilings & Acoustic Treatment	106.2	80.3	89.7	92.0	80.3	84.6	107.1	67.7	82.0	109.4	57.1	76.1	121.0	89.3	100.8	125.3	76.3	94.1
0960	Flooring	123.7	108.1	119.4	99.1	108.1	101.6	129.0	52.3	107.8	126.4	86.0	115.2	104.0	55.1	90.5	109.8	55.1	94.7
0970, 0990	Wall Finishes & Painting/Coating	167.0	70.8	109.1	96.2	70.8	80.9	167.0	69.3	108.2	167.0	73.4	110.7	98.9	84.8	90.4	97.9	81.0	87.7
09	FINISHES	118.2	82.5	98.8	98.2	82.5	89.7	119.9	65.4	90.3	120.6	64.4	90.1	104.8	81.6	92.2	107.3	72.9	88.7
COVERS	DIVS. 10 - 14, 25, 28, 41, 43, 44	100.0	80.9	95.9	100.0	77.4	95.2	100.0	80.4	95.8	100.0	71.3	93.9	100.0	103.7	100.8	100.0	70.4	93.7
21, 22, 23	FIRE SUPPRESSION, PLUMBING & HVAC	93.0	73.6	85.0	100.0	73.7	89.2	99.7	75.3	89.7	92.9	60.8	79.7	99.9	82.3	92.7	96.8	67.2	84.6
26, 27, 3370	ELECTRICAL, COMMUNICATIONS & UTIL.	90.0	87.1	88.5	91.8	86.4	89.0	102.9	80.0	90.9	88.9	97.1	93.2	98.1	95.5	96.7	93.9	58.9	75.6
MF2004	WEIGHTED AVERAGE	97.5	77.1	88.5	97.2	81.6	90.4	99.8	75.6	89.2	98.5	70.6	86.2	99.9	89.5	95.3	98.2	74.4	87.7

	DIVISION	NEVADA									NEW HAMPSHIRE								
		ELY			LAS VEGAS			RENO			CHARLESTON			CLAREMONT			CONCORD		
		893			889 - 891			894 - 895			036			037			032 - 033		
		MAT.	INST.	TOTAL	MAT.	INST.	TOTAL	MAT.	INST.	TOTAL	MAT.	INST.	TOTAL	MAT.	INST.	TOTAL	MAT.	INST.	TOTAL
015433	CONTRACTOR EQUIPMENT		100.7	100.7		100.7	100.7		100.7	100.7		100.6	100.6		100.6	100.6		100.6	100.6
0241, 31 - 34	SITE & INFRASTRUCTURE, DEMOLITION	60.9	100.8	88.7	62.6	105.1	92.2	60.8	101.5	89.2	77.2	96.8	90.9	71.7	96.8	89.2	84.3	98.8	94.4
0310	Concrete Forming & Accessories	100.9	61.7	67.0	103.0	116.4	114.6	98.1	89.0	90.2	87.2	45.6	51.2	92.4	45.6	51.9	94.0	82.4	83.9
0320	Concrete Reinforcing	103.8	88.8	96.1	95.5	125.7	110.9	98.2	124.2	111.5	88.5	49.0	68.3	88.5	49.0	68.3	88.5	93.3	90.9
0330	Cast-in-Place Concrete	105.4	69.6	91.8	102.0	116.3	107.4	111.5	87.8	102.5	93.7	51.6	77.8	86.1	51.6	73.0	99.1	90.0	95.6
03	CONCRETE	109.7	70.3	91.3	104.7	117.6	110.8	109.5	95.3	102.9	104.1	49.1	78.4	95.2	49.1	73.7	104.9	87.0	96.5
04	MASONRY	127.2	60.5	87.1	114.6	112.5	113.4	121.1	75.0	93.4	92.4	39.9	60.8	92.6	39.9	60.9	95.3	84.5	88.8
05	METALS	95.8	88.5	93.3	103.2	111.1	105.9	97.3	105.4	100.1	93.7	62.7	83.2	93.7	62.7	83.2	94.4	88.9	92.6
06	WOOD, PLASTICS & COMPOSITES	93.5	61.3	75.3	93.3	115.7	106.0	89.8	89.7	89.8	87.0	47.3	64.5	92.9	47.3	67.1	96.6	83.7	89.3
07	THERMAL & MOISTURE PROTECTION	98.7	65.7	86.2	112.1	104.9	109.4	98.3	84.4	93.0	95.4	43.0	75.6	95.3	43.0	75.5	94.3	86.6	91.4
08	OPENINGS	96.2	69.7	89.6	95.2	122.3	101.9	94.0	103.6	96.4	100.9	42.8	86.5	102.2	42.8	87.5	102.1	78.7	96.3
0920	Plaster & Gypsum Board	91.7	60.0	69.8	93.9	116.0	109.2	87.0	89.3	88.6	92.1	45.0	59.5	93.8	45.0	60.0	96.7	82.5	86.9
0950, 0980	Ceilings & Acoustic Treatment	125.3	60.0	83.7	135.6	116.0	123.1	131.3	89.3	104.5	89.0	45.0	61.0	89.0	45.0	61.0	89.0	82.5	84.9
0960	Flooring	107.2	61.6	94.6	97.9	106.4	100.2	103.6	55.1	90.2	91.8	34.6	76.0	94.6	34.6	78.0	92.0	96.4	93.2
0970, 0990	Wall Finishes & Painting/Coating	97.9	83.3	89.1	100.3	122.4	113.6	97.9	84.8	90.0	94.6	50.8	68.2	94.6	50.8	68.2	94.6	101.8	98.9
09	FINISHES	106.7	63.4	83.2	106.4	115.0	111.1	106.2	81.6	92.8	93.6	44.4	66.9	94.2	44.4	67.2	95.2	87.3	90.9
COVERS	DIVS. 10 - 14, 25, 28, 41, 43, 44	100.0	58.8	91.3	100.0	101.7	100.4	100.0	103.7	100.8	100.0	41.8	87.6	100.0	41.8	87.6	100.0	66.3	92.8
21, 22, 23	FIRE SUPPRESSION, PLUMBING & HVAC	96.8	63.4	83.1	100.1	111.1	104.6	100.0	82.5	92.8	93.4	41.5	72.0	93.4	41.6	72.1	100.0	85.6	94.1
26, 27, 3370	ELECTRICAL, COMMUNICATIONS & UTIL.	94.1	58.9	75.7	98.4	123.0	111.3	94.5	95.5	95.0	95.3	36.0	64.3	95.3	36.0	64.3	97.9	62.7	79.5
MF2004	WEIGHTED AVERAGE	99.6	69.1	86.1	101.0	113.8	106.7	100.0	90.5	95.8	95.8	48.8	75.0	94.7	48.8	74.5	98.5	83.3	91.8

	DIVISION	NEW HAMPSHIRE															NEW JERSEY		
		KEENE			LITTLETON			MANCHESTER			NASHUA			PORTSMOUTH			ATLANTIC CITY		
		034			035			031			030			038			082,084		
		MAT.	INST.	TOTAL	MAT.	INST.	TOTAL	MAT.	INST.	TOTAL	MAT.	INST.	TOTAL	MAT.	INST.	TOTAL	MAT.	INST.	TOTAL
015433	CONTRACTOR EQUIPMENT		100.6	100.6		100.6	100.6		100.6	100.6		100.6	100.6		100.6	100.6		98.7	98.7
0241, 31 - 34	SITE & INFRASTRUCTURE, DEMOLITION	83.5	97.0	92.9	71.9	96.6	89.1	81.9	98.8	93.7	84.4	98.8	94.5	78.8	100.2	93.7	86.3	103.8	98.5
0310	Concrete Forming & Accessories	91.2	43.6	50.0	100.3	48.6	55.6	96.0	83.0	84.7	97.9	83.0	85.0	88.3	87.1	87.2	105.2	118.5	116.7
0320	Concrete Reinforcing	88.5	58.3	73.1	89.3	35.1	61.6	110.6	93.4	101.8	110.6	93.4	101.8	88.5	93.5	91.0	84.3	112.8	98.9
0330	Cast-in-Place Concrete	94.2	53.1	78.6	84.5	53.8	72.9	102.0	113.3	106.3	89.1	113.3	98.3	84.5	119.6	97.8	79.9	131.6	99.5
03	CONCRETE	103.7	50.5	78.9	94.5	48.5	73.0	109.6	95.3	102.9	103.3	95.3	99.6	94.7	99.4	96.9	96.8	121.0	108.1
04	MASONRY	96.2	42.1	63.7	105.1	39.7	65.8	98.3	84.5	90.0	98.1	84.5	89.9	93.7	94.3	94.0	100.0	128.4	117.1
05	METALS	94.4	67.1	85.1	94.4	56.5	81.6	99.6	89.5	96.2	99.5	89.5	96.2	95.9	92.5	94.8	94.1	103.6	97.3
06	WOOD, PLASTICS & COMPOSITES	91.4	42.2	63.5	101.2	48.3	71.2	96.5	83.7	89.3	99.8	83.7	90.7	88.2	83.7	85.7	109.4	115.9	113.1
07	THERMAL & MOISTURE PROTECTION	95.8	45.1	76.7	95.4	46.9	77.0	95.5	90.1	93.5	96.1	90.1	93.9	95.8	114.5	102.9	95.8	124.4	106.6
08	OPENINGS	99.4	47.7	86.6	103.2	40.5	87.7	104.3	78.7	98.0	104.3	78.7	98.0	105.0	73.8	97.3	101.8	116.1	105.3
0920	Plaster & Gypsum Board	93.0	39.7	56.1	105.1	46.0	64.2	104.1	82.5	89.1	100.4	82.5	88.0	92.1	82.5	85.5	103.8	115.7	112.0
0950, 0980	Ceilings & Acoustic Treatment	89.0	39.7	57.6	89.0	46.0	61.6	96.1	82.5	87.4	98.7	82.5	88.4	90.9	82.5	85.6	89.0	115.7	106.0
0960	Flooring	94.0	56.7	83.7	104.0	34.6	84.8	98.6	96.4	98.0	97.6	96.4	97.3	92.4	96.4	93.5	101.6	136.9	111.3
0970, 0990	Wall Finishes & Painting/Coating	94.6	50.8	68.2	94.6	66.6	77.7	94.6	101.8	98.9	94.6	101.8	98.9	94.6	101.8	98.9	94.6	127.6	114.5
09	FINISHES	95.6	46.5	68.9	98.5	47.9	71.0	99.3	87.3	92.8	99.7	87.3	93.0	94.9	89.6	92.0	99.6	122.8	112.2
COVERS	DIVS. 10 - 14, 25, 28, 41, 43, 44	100.0	57.5	91.0	100.0	43.7	88.1	100.0	66.4	92.9	100.0	66.4	92.9	100.0	69.8	93.6	100.0	111.6	102.5
21, 22, 23	FIRE SUPPRESSION, PLUMBING & HVAC	93.4	45.0	73.5	93.4	69.3	83.5	100.0	85.6	94.1	100.1	85.6	94.2	100.1	90.5	96.2	99.6	126.3	110.6
26, 27, 3370	ELECTRICAL, COMMUNICATIONS & UTIL.	95.3	36.0	64.3	96.4	62.6	78.7	97.4	80.3	88.4	97.9	80.3	88.7	95.8	80.3	87.7	92.9	139.6	117.3
MF2004	WEIGHTED AVERAGE	96.3	51.3	76.4	96.1	58.2	79.3	100.4	87.0	94.5	99.9	87.0	94.2	97.3	90.9	94.4	97.6	122.2	108.5

	DIVISION	NEW JERSEY																	
		CAMDEN			DOVER			ELIZABETH			HACKENSACK			JERSEY CITY			LONG BRANCH		
		081			078			072			076			073			077		
		MAT.	INST.	TOTAL	MAT.	INST.	TOTAL	MAT.	INST.	TOTAL	MAT.	INST.	TOTAL	MAT.	INST.	TOTAL	MAT.	INST.	TOTAL
015433	CONTRACTOR EQUIPMENT		98.7	98.7		100.6	100.6		100.6	100.6		100.6	100.6		98.7	98.7		98.2	98.2
0241, 31 - 34	SITE & INFRASTRUCTURE, DEMOLITION	87.3	104.4	99.2	96.0	105.0	102.3	99.9	105.0	103.4	97.1	105.5	102.9	87.3	105.4	99.9	91.8	105.3	101.2
0310	Concrete Forming & Accessories	97.9	118.8	115.9	94.8	119.4	116.1	104.9	119.4	117.4	94.8	119.4	116.1	97.9	119.5	116.6	98.3	119.0	116.2
0320	Concrete Reinforcing	110.6	121.9	116.4	85.2	123.8	104.9	85.2	123.8	104.9	85.2	123.8	104.9	110.6	123.8	117.4	85.2	123.8	104.9
0330	Cast-in-Place Concrete	77.5	133.4	98.7	99.4	130.4	111.1	85.3	130.4	102.4	97.1	130.4	109.7	77.5	130.5	97.5	86.2	131.4	103.3
03	CONCRETE	97.6	123.4	109.7	104.6	123.3	113.3	100.2	123.3	111.0	102.7	123.3	112.3	97.6	123.2	109.6	102.6	123.3	112.3
04	MASONRY	91.7	128.4	113.8	95.4	129.0	115.6	111.4	129.0	122.0	99.3	129.0	117.1	89.2	129.0	113.1	104.0	128.4	118.7
05	METALS	99.4	106.6	101.8	94.1	112.1	100.2	95.6	112.1	101.2	94.2	112.1	100.2	99.5	109.7	102.9	94.2	109.3	99.3
06	WOOD, PLASTICS & COMPOSITES	99.8	115.9	108.9	99.2	115.9	108.6	112.6	115.9	114.5	99.2	115.9	108.6	99.8	115.9	108.9	100.8	115.9	109.3
07	THERMAL & MOISTURE PROTECTION	95.7	125.0	106.8	96.2	128.6	108.4	96.3	128.6	108.5	96.0	125.3	107.0	95.7	128.6	108.1	95.9	126.3	107.4
08	OPENINGS	104.3	118.0	107.7	107.1	119.7	110.2	105.1	119.7	108.7	104.3	119.7	108.1	104.3	119.7	108.1	100.2	119.6	105.0
0920	Plaster & Gypsum Board	100.4	115.7	111.0	96.9	115.7	109.9	104.2	115.7	112.1	96.9	115.7	109.9	100.4	115.7	111.0	99.0	115.7	110.5
0950, 0980	Ceilings & Acoustic Treatment	98.7	115.7	109.5	89.0	115.7	106.0	90.9	115.7	106.7	89.0	115.7	106.0	98.7	115.7	109.5	89.0	115.7	106.0
0960	Flooring	97.6	136.9	108.4	96.4	145.8	110.1	102.1	145.8	114.1	96.4	145.8	110.1	97.6	145.8	110.9	97.8	145.8	111.1
0970, 0990	Wall Finishes & Painting/Coating	94.6	127.6	114.5	94.5	129.7	115.7	94.5	129.7	115.7	94.5	129.7	115.7	94.6	129.7	115.7	94.6	127.6	114.5
09	FINISHES	100.1	122.8	112.4	97.3	124.7	112.2	100.9	124.7	113.8	97.2	124.7	112.2	100.1	124.7	113.5	98.3	124.3	112.4
COVERS	DIVS. 10 - 14, 25, 28, 41, 43, 44	100.0	111.6	102.5	100.0	118.4	103.9	100.0	118.4	103.9	100.0	118.4	103.9	100.0	118.4	103.9	100.0	111.5	102.4
21, 22, 23	FIRE SUPPRESSION, PLUMBING & HVAC	100.1	126.3	110.9	99.7	129.6	112.0	100.1	127.1	111.2	99.7	129.6	112.0	100.1	129.6	112.3	99.7	126.2	110.6
26, 27, 3370	ELECTRICAL, COMMUNICATIONS & UTIL.	98.3	139.6	119.9	94.3	136.2	116.2	95.0	136.2	116.5	94.3	140.0	118.2	99.4	140.0	120.6	94.0	132.1	113.9
MF2004	WEIGHTED AVERAGE	99.0	122.9	109.5	99.1	124.3	110.2	100.1	123.8	110.6	98.8	124.8	110.3	99.0	124.7	110.3	98.5	122.4	109.1

	DIVISION	NEW JERSEY																	
		NEW BRUNSWICK			NEWARK			PATERSON			POINT PLEASANT			SUMMIT			TRENTON		
		088 - 089			070 - 071			074 - 075			087			079			085 - 086		
		MAT.	INST.	TOTAL	MAT.	INST.	TOTAL	MAT.	INST.	TOTAL	MAT.	INST.	TOTAL	MAT.	INST.	TOTAL	MAT.	INST.	TOTAL
015433	**CONTRACTOR EQUIPMENT**		98.2	98.2		100.6	100.6		100.6	100.6		98.2	98.2		100.6	100.6		98.2	98.2
0241, 31 - 34	**SITE & INFRASTRUCTURE, DEMOLITION**	98.5	105.3	103.3	102.6	105.0	104.2	98.5	105.5	103.4	100.3	105.3	103.8	97.7	105.0	102.8	87.1	105.3	99.8
0310	Concrete Forming & Accessories	100.5	119.3	116.7	94.6	119.4	116.0	96.7	119.4	116.3	95.9	118.8	115.7	97.3	119.4	116.4	95.1	119.0	115.7
0320	Concrete Reinforcing	85.2	123.8	104.9	110.6	123.8	117.4	110.6	123.8	117.4	85.2	123.8	104.9	85.2	123.8	104.9	110.6	119.3	115.0
0330	Cast-in-Place Concrete	98.8	131.8	111.2	99.1	130.4	110.9	98.8	130.4	110.8	98.8	131.4	111.1	82.5	130.4	100.6	88.2	131.4	104.6
03	**CONCRETE**	115.3	123.5	119.1	108.1	123.3	115.2	108.0	123.3	115.2	115.0	123.2	118.8	97.0	123.3	109.3	102.7	122.4	111.9
04	**MASONRY**	98.3	128.9	116.7	100.4	129.0	117.6	95.9	129.0	115.8	89.4	128.4	112.9	97.9	129.0	116.6	87.8	128.4	112.2
05	**METALS**	94.2	109.5	99.3	99.4	112.1	103.7	94.7	112.1	100.6	94.2	109.2	99.3	94.1	112.1	100.2	94.6	107.1	98.8
06	**WOOD, PLASTICS & COMPOSITES**	103.6	115.9	110.6	98.5	115.9	108.4	101.5	115.9	109.7	97.5	115.9	107.9	102.6	115.9	110.1	96.9	115.9	107.6
07	**THERMAL & MOISTURE PROTECTION**	96.0	126.5	107.5	95.8	128.6	108.2	96.3	125.3	107.2	96.1	126.3	107.5	96.6	128.6	108.6	91.9	125.4	104.5
08	**OPENINGS**	96.3	119.6	102.1	110.1	119.7	112.5	110.1	119.7	112.5	98.3	119.6	103.6	112.3	119.7	114.2	104.9	118.2	108.2
0920	Plaster & Gypsum Board	100.7	115.7	111.1	104.1	115.7	112.1	100.4	115.7	111.0	96.9	115.7	109.9	99.0	115.7	110.5	104.1	115.7	112.1
0950, 0980	Ceilings & Acoustic Treatment	89.0	115.7	106.0	96.1	115.7	108.6	98.7	115.7	109.5	89.0	115.7	106.0	89.0	115.7	106.0	96.1	115.7	108.6
0960	Flooring	99.2	145.8	112.0	97.8	145.8	111.0	97.6	145.8	110.9	96.4	136.9	107.6	97.9	145.8	111.1	98.0	145.8	111.2
0970, 0990	Wall Finishes & Painting/Coating	94.6	129.7	115.7	94.5	129.7	115.7	94.5	129.7	115.7	94.6	127.6	114.5	94.5	129.7	115.7	94.6	127.6	114.5
09	**FINISHES**	99.7	124.7	113.3	100.2	124.7	113.5	100.3	124.7	113.5	98.4	122.8	111.6	98.3	124.7	112.7	99.9	124.3	113.2
COVERS	**DIVS. 10 - 14, 25, 28, 41, 43, 44**	100.0	118.2	103.9	100.0	118.4	103.9	100.0	118.4	103.9	100.0	111.5	102.4	100.0	118.4	103.9	100.0	111.5	102.4
21, 22, 23	**FIRE SUPPRESSION, PLUMBING & HVAC**	99.6	126.4	110.7	100.2	127.1	111.3	100.1	129.6	112.2	99.6	126.0	110.5	99.7	127.1	111.0	100.2	126.1	110.9
26, 27, 3370	**ELECTRICAL, COMMUNICATIONS & UTIL.**	93.7	137.2	116.4	99.4	140.0	120.7	99.4	136.2	118.6	92.9	132.1	113.4	95.0	136.2	116.5	102.5	142.2	123.3
MF2004	**WEIGHTED AVERAGE**	99.7	123.6	110.2	101.9	124.3	111.8	100.9	124.3	111.2	99.2	122.2	109.3	99.0	123.8	110.0	99.0	123.4	109.8

	DIVISION	NEW JERSEY			NEW MEXICO														
		VINELAND			ALBUQUERQUE			CARRIZOZO			CLOVIS			FARMINGTON			GALLUP		
		080,083			870 - 872			883			881			874			873		
		MAT.	INST.	TOTAL	MAT.	INST.	TOTAL	MAT.	INST.	TOTAL	MAT.	INST.	TOTAL	MAT.	INST.	TOTAL	MAT.	INST.	TOTAL
015433	**CONTRACTOR EQUIPMENT**		98.7	98.7		113.4	113.4		113.4	113.4		113.4	113.4		113.4	113.4		113.4	113.4
0241, 31 - 34	**SITE & INFRASTRUCTURE, DEMOLITION**	90.4	104.4	100.2	77.2	108.6	99.1	97.9	108.6	105.4	86.3	108.6	101.8	83.6	108.6	101.0	91.2	108.6	103.3
0310	Concrete Forming & Accessories	93.7	118.6	115.3	98.8	68.8	72.8	98.9	68.8	72.9	98.9	68.4	72.5	98.9	68.8	72.9	98.9	68.8	72.9
0320	Concrete Reinforcing	84.3	112.8	98.9	98.9	71.5	84.9	106.1	71.5	88.4	107.2	48.5	77.2	108.1	71.5	89.4	103.4	71.5	87.1
0330	Cast-in-Place Concrete	86.2	131.6	103.3	104.2	76.1	93.6	97.0	76.1	89.1	96.9	76.0	89.0	105.2	76.1	94.2	99.0	76.1	90.3
03	**CONCRETE**	102.2	121.1	111.0	106.1	72.8	90.5	122.1	72.8	99.1	109.6	68.2	90.2	109.9	72.8	92.6	116.9	72.8	96.3
04	**MASONRY**	88.1	128.4	112.4	111.2	63.5	82.6	108.4	63.5	81.4	108.4	63.5	81.4	117.6	63.5	85.1	108.5	63.5	81.5
05	**METALS**	94.1	103.7	97.3	103.1	88.8	98.3	100.0	88.8	96.2	99.6	77.4	92.1	100.9	88.8	96.8	100.0	88.8	96.2
06	**WOOD, PLASTICS & COMPOSITES**	94.8	115.9	106.8	93.4	69.7	80.0	93.5	69.7	80.0	93.5	69.7	80.0	93.5	69.7	80.0	93.5	69.7	80.0
07	**THERMAL & MOISTURE PROTECTION**	95.7	124.4	106.5	97.8	72.8	88.4	99.4	72.8	89.4	98.2	72.8	88.6	98.1	72.8	88.5	99.1	72.8	89.2
08	**OPENINGS**	97.8	116.1	102.3	93.8	71.6	88.3	92.6	71.6	87.4	92.8	64.4	85.7	96.6	71.6	90.4	96.6	71.6	90.4
0920	Plaster & Gypsum Board	95.6	115.7	109.5	92.2	68.4	75.7	83.9	68.4	73.2	83.9	68.4	73.2	83.9	68.4	73.2	83.9	68.4	73.2
0950, 0980	Ceilings & Acoustic Treatment	89.0	115.7	106.0	125.7	68.4	89.2	121.0	68.4	87.5	121.0	68.4	87.5	121.0	68.4	87.5	121.0	68.4	87.5
0960	Flooring	95.5	136.9	106.9	101.2	66.3	91.6	103.6	66.3	93.3	103.6	66.3	93.3	103.6	66.3	93.3	103.6	66.3	93.3
0970, 0990	Wall Finishes & Painting/Coating	94.6	127.6	114.5	102.4	53.1	72.7	97.9	53.1	70.9	97.9	53.1	70.9	97.9	53.1	70.9	97.9	53.1	70.9
09	**FINISHES**	97.1	122.8	111.1	105.6	66.7	84.5	105.9	66.7	84.6	104.6	66.7	84.0	104.2	66.7	83.8	105.3	66.7	84.3
COVERS	**DIVS. 10 - 14, 25, 28, 41, 43, 44**	100.0	111.6	102.5	100.0	71.3	93.9	100.0	71.3	93.9	100.0	71.3	93.9	100.0	71.3	93.9	100.0	71.3	93.9
21, 22, 23	**FIRE SUPPRESSION, PLUMBING & HVAC**	99.6	126.3	110.6	100.1	70.8	88.0	96.1	70.8	85.7	96.1	70.3	85.5	100.0	70.8	88.0	95.9	70.8	85.6
26, 27, 3370	**ELECTRICAL, COMMUNICATIONS & UTIL.**	92.9	139.6	117.3	89.5	70.9	79.8	89.5	70.9	79.8	86.9	70.9	78.6	87.6	70.9	78.9	86.8	70.9	78.5
MF2004	**WEIGHTED AVERAGE**	97.1	122.2	108.2	99.8	74.8	88.8	100.8	74.8	89.3	98.5	72.7	87.1	100.4	74.8	89.1	100.0	74.8	88.9

	DIVISION	NEW MEXICO																	
		LAS CRUCES			LAS VEGAS			ROSWELL			SANTA FE			SOCORRO			TRUTH/CONSEQUENCES		
		880			877			882			875			878			879		
		MAT.	INST.	TOTAL	MAT.	INST.	TOTAL	MAT.	INST.	TOTAL	MAT.	INST.	TOTAL	MAT.	INST.	TOTAL	MAT.	INST.	TOTAL
015433	**CONTRACTOR EQUIPMENT**		87.8	87.8		113.4	113.4		113.4	113.4		113.4	113.4		113.4	113.4		87.8	87.8
0241, 31 - 34	**SITE & INFRASTRUCTURE, DEMOLITION**	87.8	86.5	86.9	83.3	108.6	101.0	88.2	108.6	102.4	82.9	108.6	100.8	79.6	108.6	99.8	99.1	86.5	90.3
0310	Concrete Forming & Accessories	95.6	67.1	70.9	98.9	68.8	72.8	98.9	68.5	72.6	98.1	68.8	72.7	98.9	68.8	72.9	95.6	67.2	71.0
0320	Concrete Reinforcing	101.6	48.2	74.3	105.1	71.4	87.9	107.2	48.5	77.2	106.1	71.5	88.4	107.2	71.5	89.0	100.4	48.2	73.7
0330	Cast-in-Place Concrete	91.5	67.9	82.6	102.3	76.1	92.4	96.9	76.0	89.0	120.2	76.1	103.5	100.3	76.1	91.1	110.0	67.9	94.1
03	**CONCRETE**	86.5	64.5	76.2	107.0	72.8	91.0	110.5	68.2	90.7	114.9	72.8	95.2	105.7	72.8	90.3	98.9	64.5	82.8
04	**MASONRY**	104.5	61.1	78.4	108.8	61.4	80.3	118.9	63.5	85.6	113.7	63.5	83.6	108.7	63.5	81.6	106.1	61.1	79.0
05	**METALS**	100.8	70.3	90.5	99.7	88.6	96.0	100.9	77.5	93.0	100.9	88.8	96.8	100.0	88.8	96.2	99.6	70.4	89.7
06	**WOOD, PLASTICS & COMPOSITES**	83.8	68.5	75.1	93.5	69.7	80.0	93.5	69.7	80.0	92.9	69.7	79.8	93.5	69.7	80.0	83.8	68.5	75.1
07	**THERMAL & MOISTURE PROTECTION**	84.5	67.4	78.1	97.8	72.2	88.1	98.3	72.8	88.7	97.7	72.8	88.3	97.7	72.8	88.3	85.0	67.4	78.4
08	**OPENINGS**	86.8	63.7	81.0	92.8	71.6	87.5	92.5	64.4	85.6	92.7	71.6	87.5	92.6	71.6	87.4	86.7	63.7	81.0
0920	Plaster & Gypsum Board	86.5	68.4	73.9	83.9	68.4	73.2	83.9	68.4	73.2	93.7	68.4	76.2	83.9	68.4	73.2	86.5	68.4	73.9
0950, 0980	Ceilings & Acoustic Treatment	113.2	68.4	84.7	121.0	68.4	87.5	121.0	68.4	87.5	121.0	68.4	87.5	121.0	68.4	87.5	113.2	68.4	84.7
0960	Flooring	137.0	62.6	116.4	103.6	66.3	93.3	103.6	66.3	93.3	103.6	66.3	93.3	103.6	66.3	93.3	137.0	62.6	116.4
0970, 0990	Wall Finishes & Painting/Coating	90.4	53.1	67.9	97.9	53.1	70.9	97.9	53.1	70.9	97.9	53.1	70.9	97.9	53.1	70.9	90.4	53.1	67.9
09	**FINISHES**	118.9	65.0	89.7	104.1	66.7	83.8	104.7	66.7	84.1	105.5	66.7	84.4	104.0	66.7	83.7	119.6	65.0	89.9
COVERS	**DIVS. 10 - 14, 25, 28, 41, 43, 44**	100.0	68.1	93.2	100.0	71.3	93.9	100.0	71.3	93.9	100.0	71.3	93.9	100.0	71.3	93.9	100.0	68.2	93.2
21, 22, 23	**FIRE SUPPRESSION, PLUMBING & HVAC**	100.3	70.2	87.9	95.9	70.8	85.6	100.0	70.6	87.9	100.0	70.8	88.0	95.9	70.8	85.6	95.8	70.2	85.3
26, 27, 3370	**ELECTRICAL, COMMUNICATIONS & UTIL.**	88.4	54.9	70.9	89.5	70.9	79.8	88.4	70.9	79.3	100.9	70.9	85.3	87.3	70.9	78.8	91.5	70.9	80.8
MF2004	**WEIGHTED AVERAGE**	96.8	66.7	83.5	98.2	74.6	87.8	100.4	72.8	88.2	101.8	74.8	89.9	97.8	74.8	87.6	97.9	68.9	85.1

DIVISION		NEW MEXICO			NEW YORK														
		TUCUMCARI			ALBANY			BINGHAMTON			BRONX			BROOKLYN			BUFFALO		
		884			120 - 122			137 - 139			104			112			140 - 142		
		MAT.	INST.	TOTAL	MAT.	INST.	TOTAL	MAT.	INST.	TOTAL	MAT.	INST.	TOTAL	MAT.	INST.	TOTAL	MAT.	INST.	TOTAL
015433	CONTRACTOR EQUIPMENT		113.4	113.4		114.6	114.6		114.9	114.9		115.2	115.2		115.8	115.8		96.3	96.3
0241, 31 - 34	SITE & INFRASTRUCTURE, DEMOLITION	85.9	108.6	101.7	73.0	107.5	97.0	95.1	91.5	92.6	118.9	126.1	123.9	121.9	129.8	127.4	97.2	96.7	96.9
0310	Concrete Forming & Accessories	98.9	68.4	72.5	98.1	96.2	96.5	102.4	82.5	85.2	107.4	176.8	167.4	108.6	176.2	167.1	98.3	112.5	110.6
0320	Concrete Reinforcing	105.1	48.5	76.2	94.7	92.9	93.7	93.8	93.1	93.4	96.4	188.0	143.1	95.4	187.9	142.6	93.7	97.5	95.7
0330	Cast-in-Place Concrete	96.9	76.0	89.0	80.9	103.4	89.4	99.8	93.9	97.6	95.6	162.8	121.0	101.7	160.9	124.1	107.6	116.9	111.2
03	CONCRETE	108.8	68.2	89.8	98.7	98.8	98.8	98.5	90.1	94.6	104.0	172.1	135.8	111.9	171.2	139.6	109.2	110.5	109.8
04	MASONRY	121.7	63.5	86.7	92.1	102.0	98.0	107.7	87.3	95.5	96.3	164.8	137.5	117.2	164.8	145.8	107.0	117.0	113.0
05	METALS	99.6	77.4	92.1	99.0	105.6	101.2	95.7	114.8	102.2	101.7	143.2	115.7	103.4	142.5	116.6	99.4	93.2	97.3
06	WOOD, PLASTICS & COMPOSITES	93.5	69.7	80.0	97.5	94.7	95.9	107.3	82.3	93.1	108.3	180.9	149.4	108.6	180.6	149.3	101.2	112.3	107.5
07	THERMAL & MOISTURE PROTECTION	98.2	72.8	88.6	90.2	96.0	92.4	104.9	84.1	97.1	110.6	162.5	130.2	107.8	158.8	127.1	101.7	108.4	104.2
08	OPENINGS	92.5	64.4	85.5	96.1	88.3	94.2	91.0	80.7	88.4	91.8	172.4	111.8	90.5	172.2	110.7	94.5	100.4	96.0
0920	Plaster & Gypsum Board	83.9	68.4	73.2	106.0	94.3	97.9	119.4	81.3	93.0	119.6	182.9	163.4	116.5	182.9	162.5	99.4	112.4	108.4
0950, 0980	Ceilings & Acoustic Treatment	121.0	68.4	87.5	93.6	94.3	94.0	98.7	81.3	87.6	83.7	182.9	146.9	93.7	182.9	150.5	103.9	112.4	109.3
0960	Flooring	103.6	66.3	93.3	85.0	102.6	89.8	95.7	89.1	93.9	92.1	178.4	116.0	102.2	178.4	123.3	91.4	115.0	97.9
0970, 0990	Wall Finishes & Painting/Coating	97.9	53.1	70.9	82.9	82.7	82.8	86.9	86.2	86.5	91.2	149.3	126.1	112.3	149.3	134.5	91.2	111.8	103.6
09	FINISHES	104.6	66.7	84.0	94.4	95.9	95.2	97.6	83.8	90.1	101.3	175.3	141.5	112.3	175.0	146.3	95.0	113.7	105.1
COVERS	DIVS. 10 - 14, 25, 28, 41, 43, 44	100.0	71.3	93.9	100.0	97.9	99.6	100.0	94.1	98.7	100.0	134.7	107.4	100.0	133.8	107.2	100.0	105.2	101.1
21, 22, 23	FIRE SUPPRESSION, PLUMBING & HVAC	96.1	70.3	85.5	100.1	93.1	97.2	100.4	82.6	93.1	100.3	160.9	125.2	99.8	160.8	124.9	100.0	94.7	97.8
26, 27, 3370	ELECTRICAL, COMMUNICATIONS & UTIL.	89.5	70.9	79.8	104.7	93.4	98.8	100.9	91.6	96.0	98.6	178.5	140.4	100.4	178.5	141.2	98.3	95.5	96.8
MF2004	WEIGHTED AVERAGE	99.3	72.7	87.6	97.7	97.6	97.7	98.8	89.5	94.7	100.8	162.5	128.0	103.9	162.4	129.7	100.3	102.8	101.4

DIVISION		NEW YORK																	
		ELMIRA			FAR ROCKAWAY			FLUSHING			GLENS FALLS			HICKSVILLE			JAMAICA		
		148 - 149			116			113			128			115,117,118			114		
		MAT.	INST.	TOTAL	MAT.	INST.	TOTAL	MAT.	INST.	TOTAL	MAT.	INST.	TOTAL	MAT.	INST.	TOTAL	MAT.	INST.	TOTAL
015433	CONTRACTOR EQUIPMENT		116.4	116.4		115.8	115.8		115.8	115.8		114.6	114.6		115.8	115.8		115.8	115.8
0241, 31 - 34	SITE & INFRASTRUCTURE, DEMOLITION	92.9	92.3	92.4	125.9	129.8	128.6	125.9	129.8	128.6	63.4	106.9	93.7	114.2	128.9	124.5	119.6	129.8	126.7
0310	Concrete Forming & Accessories	85.1	76.4	77.5	96.4	176.2	165.4	99.8	176.2	165.9	87.9	80.1	81.1	93.2	146.8	139.6	99.8	176.2	165.9
0320	Concrete Reinforcing	94.8	89.7	92.2	95.4	187.9	142.6	97.0	187.9	143.4	94.3	92.7	93.5	95.4	187.7	142.5	95.4	187.9	142.6
0330	Cast-in-Place Concrete	100.6	88.6	96.1	110.0	160.9	129.3	110.0	160.9	129.3	75.8	100.5	85.2	93.4	156.6	117.3	101.7	160.9	124.1
03	CONCRETE	102.7	85.2	94.5	118.7	171.2	143.2	119.2	171.2	143.5	89.3	90.6	89.9	103.3	156.6	128.2	111.3	171.2	139.3
04	MASONRY	103.3	82.0	90.5	121.3	164.8	147.5	115.0	164.8	144.9	93.1	97.3	95.7	111.4	158.5	139.7	119.2	164.8	146.6
05	METALS	94.7	116.2	102.0	103.4	142.5	116.6	103.4	142.5	116.6	95.1	104.8	98.4	105.0	140.0	116.8	103.4	142.5	116.6
06	WOOD, PLASTICS & COMPOSITES	86.6	75.5	80.3	93.6	180.6	142.8	97.9	180.6	144.7	85.2	76.1	80.0	90.1	145.4	121.4	97.9	180.6	144.7
07	THERMAL & MOISTURE PROTECTION	102.1	78.6	93.2	107.7	158.7	127.0	107.8	158.7	127.0	89.5	91.5	90.3	107.4	150.5	123.7	107.6	158.7	126.9
08	OPENINGS	95.6	76.3	90.8	88.9	172.2	109.5	88.9	172.2	109.5	89.6	78.2	86.8	88.9	153.1	104.8	88.9	172.2	109.5
0920	Plaster & Gypsum Board	94.7	74.5	80.8	106.9	182.9	159.5	109.1	182.9	160.2	103.6	75.1	83.9	106.0	146.7	134.2	109.1	182.9	160.2
0950, 0980	Ceilings & Acoustic Treatment	100.4	74.5	83.9	83.3	182.9	146.8	83.3	182.9	146.8	91.9	75.1	81.2	82.4	146.7	123.3	83.3	182.9	146.8
0960	Flooring	81.4	76.4	80.0	96.6	178.4	119.2	98.2	178.4	120.4	76.5	92.9	81.0	95.4	153.7	111.5	98.2	178.4	120.4
0970, 0990	Wall Finishes & Painting/Coating	89.7	82.0	85.1	112.3	149.3	134.5	112.3	149.3	134.5	82.9	82.7	82.8	112.3	149.3	134.5	112.3	149.3	134.5
09	FINISHES	90.2	76.3	82.6	107.5	175.0	144.2	108.2	175.0	144.5	90.2	81.9	85.7	105.8	147.7	128.5	107.8	175.0	144.3
COVERS	DIVS. 10 - 14, 25, 28, 41, 43, 44	100.0	91.9	98.3	100.0	133.8	107.2	100.0	133.8	107.2	100.0	63.7	92.3	100.0	127.3	105.8	100.0	133.8	107.2
21, 22, 23	FIRE SUPPRESSION, PLUMBING & HVAC	93.4	81.5	88.5	93.0	160.8	120.9	93.0	160.8	120.9	93.6	84.4	89.8	99.8	142.6	117.4	93.0	160.8	120.9
26, 27, 3370	ELECTRICAL, COMMUNICATIONS & UTIL.	96.6	98.6	97.7	108.5	178.5	145.1	108.5	178.5	145.1	95.9	93.4	94.6	99.8	148.8	125.4	98.7	178.5	140.4
MF2004	WEIGHTED AVERAGE	96.3	87.8	92.6	103.6	162.4	129.5	103.4	162.4	129.4	92.2	90.5	91.5	101.6	146.4	121.4	101.5	162.4	128.4

DIVISION		NEW YORK																	
		JAMESTOWN			KINGSTON			LONG ISLAND CITY			MONTICELLO			MOUNT VERNON			NEW ROCHELLE		
		147			124			111			127			105			108		
		MAT.	INST.	TOTAL	MAT.	INST.	TOTAL	MAT.	INST.	TOTAL	MAT.	INST.	TOTAL	MAT.	INST.	TOTAL	MAT.	INST.	TOTAL
015433	CONTRACTOR EQUIPMENT		93.1	93.1		115.8	115.8		115.8	115.8		115.8	115.8		115.2	115.2		115.2	115.2
0241, 31 - 34	SITE & INFRASTRUCTURE, DEMOLITION	94.2	92.9	93.3	121.4	126.5	125.0	123.1	129.8	127.8	116.6	126.5	123.5	127.7	121.0	123.0	126.5	121.0	122.7
0310	Concrete Forming & Accessories	85.1	82.7	83.0	89.7	113.0	109.8	103.7	176.2	166.4	96.0	113.4	111.0	98.3	132.7	128.0	112.2	132.7	129.9
0320	Concrete Reinforcing	95.0	77.1	85.9	94.6	137.7	116.6	95.4	187.9	142.6	93.9	137.8	116.3	94.9	187.1	142.0	94.9	187.1	142.0
0330	Cast-in-Place Concrete	104.4	97.8	101.9	102.4	136.0	115.1	104.9	160.9	126.1	95.9	136.2	111.1	106.8	136.0	117.9	106.8	136.0	117.9
03	CONCRETE	106.1	86.8	97.1	112.5	125.1	118.4	114.7	171.2	141.1	106.9	125.3	115.5	115.0	143.0	128.1	114.0	143.0	127.6
04	MASONRY	112.2	94.0	101.2	108.5	140.6	127.8	114.0	164.8	144.5	101.6	140.6	125.0	102.3	134.2	121.5	102.3	134.2	121.5
05	METALS	92.1	82.6	88.9	103.4	116.8	107.9	103.4	142.5	116.6	103.4	117.1	108.0	101.5	136.9	113.5	101.8	136.9	113.7
06	WOOD, PLASTICS & COMPOSITES	85.2	79.1	81.7	86.4	108.2	98.7	102.9	180.6	146.9	94.1	108.2	102.1	97.7	135.4	119.0	115.7	135.4	126.9
07	THERMAL & MOISTURE PROTECTION	101.7	87.5	96.3	109.4	139.9	120.9	107.7	158.7	127.0	109.0	139.9	120.7	111.5	140.5	122.4	111.5	140.5	122.5
08	OPENINGS	95.4	77.2	90.9	94.2	121.8	101.1	88.9	172.2	109.5	88.9	121.8	97.1	91.8	147.7	105.7	91.9	147.7	105.7
0920	Plaster & Gypsum Board	89.6	78.2	81.7	102.1	108.4	106.5	111.7	182.9	161.0	105.1	108.4	107.4	113.6	136.0	129.1	126.6	136.0	133.1
0950, 0980	Ceilings & Acoustic Treatment	97.0	78.2	85.0	77.3	108.4	97.1	83.3	182.9	146.8	77.3	108.4	97.1	81.1	136.0	116.1	81.1	136.0	116.1
0960	Flooring	84.0	94.9	87.0	89.4	78.1	86.2	99.9	178.4	121.6	92.3	78.1	88.3	84.0	162.9	105.8	92.1	162.9	111.7
0970, 0990	Wall Finishes & Painting/Coating	91.2	82.4	85.9	110.3	108.4	109.1	112.3	149.3	134.5	110.3	107.2	108.5	89.2	149.3	125.3	89.2	149.3	125.3
09	FINISHES	89.8	84.3	86.8	102.8	104.0	103.5	108.8	175.0	144.8	103.7	103.9	103.8	98.2	139.7	120.8	102.0	139.7	122.5
COVERS	DIVS. 10 - 14, 25, 28, 41, 43, 44	100.0	98.4	99.7	100.0	115.5	103.3	100.0	133.8	107.2	100.0	115.5	103.3	100.0	121.3	104.5	100.0	121.3	104.5
21, 22, 23	FIRE SUPPRESSION, PLUMBING & HVAC	93.2	86.7	90.6	93.5	102.3	97.1	99.8	160.8	124.9	93.5	124.0	106.0	93.6	124.5	106.3	93.6	124.5	106.3
26, 27, 3370	ELECTRICAL, COMMUNICATIONS & UTIL.	95.5	84.5	89.7	97.4	122.1	110.3	99.3	178.5	140.7	97.4	122.1	110.3	96.6	152.0	125.5	96.6	152.0	125.5
MF2004	WEIGHTED AVERAGE	96.6	86.9	92.3	101.2	117.9	108.6	103.5	162.4	129.5	99.6	122.5	109.7	100.6	136.0	116.2	100.9	136.0	116.4

DIVISION		NEW YORK																	
		NEW YORK			NIAGARA FALLS			PLATTSBURGH			POUGHKEEPSIE			QUEENS			RIVERHEAD		
		100 - 102			143			129			125 - 126			110			119		
		MAT.	INST.	TOTAL	MAT.	INST.	TOTAL	MAT.	INST.	TOTAL	MAT.	INST.	TOTAL	MAT.	INST.	TOTAL	MAT.	INST.	TOTAL
015433	CONTRACTOR EQUIPMENT		115.5	115.5		93.1	93.1		99.0	99.0		115.8	115.8		115.8	115.8		115.8	115.8
0241, 31 - 34	SITE & INFRASTRUCTURE, DEMOLITION	128.2	126.8	127.2	96.4	93.7	94.5	87.5	102.9	98.2	117.3	126.4	123.6	117.9	129.8	126.2	115.4	128.9	124.8
0310	Concrete Forming & Accessories	111.6	186.1	176.0	85.1	111.5	107.9	91.1	84.9	85.8	89.7	161.7	152.0	93.4	176.2	165.0	97.4	146.7	140.1
0320	Concrete Reinforcing	102.0	188.0	145.9	93.7	95.8	94.8	99.1	92.4	95.7	94.6	138.0	116.8	97.0	187.9	143.4	97.2	184.6	141.8
0330	Cast-in-Place Concrete	107.6	171.3	131.7	108.0	104.7	106.7	92.7	96.4	94.1	99.2	134.5	112.6	96.6	160.9	120.9	94.9	156.5	118.2
03	CONCRETE	115.3	179.1	145.1	108.7	105.4	107.2	103.3	90.0	97.1	109.4	146.3	126.6	106.8	171.2	136.9	103.9	155.9	128.2
04	MASONRY	107.4	170.5	145.3	120.4	115.3	117.3	89.2	91.2	90.4	101.9	136.5	122.7	108.3	164.8	142.3	117.1	158.5	142.0
05	METALS	115.5	143.4	125.0	94.7	90.6	93.3	99.4	84.9	94.5	103.4	118.9	108.7	103.4	142.5	116.6	105.4	138.8	116.7
06	WOOD, PLASTICS & COMPOSITES	112.8	190.4	156.8	85.2	110.9	99.7	91.8	83.1	86.9	86.4	173.8	135.9	90.2	180.6	141.4	95.2	145.4	123.6
07	THERMAL & MOISTURE PROTECTION	110.9	166.2	131.8	101.8	103.6	102.5	102.9	87.2	97.0	109.3	145.4	123.0	107.5	158.7	126.8	108.0	150.5	124.0
08	OPENINGS	97.4	179.4	117.7	95.4	98.5	96.2	98.8	82.0	94.6	94.3	157.3	109.9	88.9	172.2	109.5	88.9	150.8	104.2
0920	Plaster & Gypsum Board	128.6	192.7	172.9	89.6	110.9	104.4	123.3	81.8	94.6	102.1	176.0	153.3	106.0	182.9	159.3	108.3	146.7	134.9
0950, 0980	Ceilings & Acoustic Treatment	105.3	192.7	161.0	97.0	110.9	105.9	103.5	81.8	89.7	77.3	176.0	140.2	83.3	182.9	146.8	84.1	146.7	123.9
0960	Flooring	93.6	178.4	117.1	83.9	109.7	91.0	98.6	102.6	99.7	89.4	149.9	106.1	95.4	178.4	118.3	96.6	153.7	112.4
0970, 0990	Wall Finishes & Painting/Coating	91.2	156.8	130.7	91.2	105.7	99.9	105.4	79.4	89.8	110.3	107.2	108.5	112.3	149.3	134.5	112.3	149.3	134.5
09	FINISHES	108.5	183.1	149.0	89.9	111.5	101.6	100.0	87.7	93.3	102.6	156.2	131.7	106.2	175.0	143.6	106.8	147.7	129.0
COVERS	DIVS. 10 - 14, 25, 28, 41, 43, 44	100.0	138.7	108.2	100.0	106.4	101.4	100.0	54.4	90.3	100.0	121.7	104.6	100.0	133.8	107.2	100.0	127.3	105.8
21, 22, 23	FIRE SUPPRESSION, PLUMBING & HVAC	100.3	163.8	126.4	93.2	95.5	94.2	93.5	84.1	89.6	93.5	123.6	105.9	99.8	160.8	124.9	99.9	141.2	116.9
26, 27, 3370	ELECTRICAL, COMMUNICATIONS & UTIL.	107.2	178.5	144.5	94.0	93.6	93.8	93.7	87.4	90.4	97.4	122.1	110.3	99.8	178.5	140.9	101.6	148.8	126.3
MF2004	WEIGHTED AVERAGE	107.0	166.3	133.2	97.7	100.8	99.1	97.1	87.3	92.8	100.3	134.3	115.3	101.8	162.4	128.6	102.4	145.9	121.6

DIVISION		NEW YORK																	
		ROCHESTER			SCHENECTADY			STATEN ISLAND			SUFFERN			SYRACUSE			UTICA		
		144 - 146			123			103			109			130 - 132			133 - 135		
		MAT.	INST.	TOTAL	MAT.	INST.	TOTAL	MAT.	INST.	TOTAL	MAT.	INST.	TOTAL	MAT.	INST.	TOTAL	MAT.	INST.	TOTAL
015433	CONTRACTOR EQUIPMENT		117.1	117.1		114.6	114.6		115.2	115.2		115.2	115.2		114.6	114.6		114.6	114.6
0241, 31 - 34	SITE & INFRASTRUCTURE, DEMOLITION	73.0	109.1	98.1	72.9	107.5	97.0	132.6	126.1	128.1	122.9	122.1	122.3	94.0	106.5	102.7	71.1	104.6	94.4
0310	Concrete Forming & Accessories	99.6	96.6	97.0	102.1	95.9	96.7	97.7	176.8	166.1	106.3	129.6	126.5	101.1	88.4	90.1	102.8	82.4	85.2
0320	Concrete Reinforcing	94.5	89.8	92.1	93.4	92.9	93.1	96.4	188.0	143.1	94.9	138.0	117.0	94.7	90.5	92.5	94.7	90.8	92.7
0330	Cast-in-Place Concrete	96.8	99.8	97.9	91.6	102.8	95.9	106.9	162.8	128.0	103.6	132.9	114.7	92.6	97.7	94.5	84.7	90.9	87.0
03	CONCRETE	114.6	97.3	106.5	104.1	98.4	101.4	117.1	172.1	142.8	110.8	131.6	120.5	102.6	92.9	98.0	100.9	87.9	94.8
04	MASONRY	103.1	101.6	102.2	91.6	101.0	97.2	110.6	164.8	143.2	102.5	129.7	118.8	98.4	98.9	98.7	90.2	83.5	86.2
05	METALS	96.0	106.4	99.5	99.1	105.6	101.3	99.7	143.2	114.4	99.7	121.4	107.0	99.0	103.0	100.3	96.9	102.7	98.9
06	WOOD, PLASTICS & COMPOSITES	98.1	96.7	97.3	103.4	94.7	98.5	96.1	180.9	144.1	108.0	135.4	123.5	103.4	86.8	94.0	103.4	84.0	92.4
07	THERMAL & MOISTURE PROTECTION	97.2	96.7	97.0	90.6	95.5	92.5	110.9	162.5	130.4	111.4	138.3	121.6	101.2	96.4	99.4	90.4	90.3	90.4
08	OPENINGS	98.3	89.3	96.0	96.1	88.3	94.2	91.8	172.4	111.8	91.9	138.3	103.4	94.1	80.2	90.7	96.1	81.0	92.3
0920	Plaster & Gypsum Board	98.4	96.6	97.2	113.7	94.3	100.2	114.0	182.9	161.7	118.4	136.0	130.6	113.7	86.1	94.6	113.7	83.3	92.6
0950, 0980	Ceilings & Acoustic Treatment	103.1	96.6	99.0	98.7	94.3	95.9	83.7	182.9	146.9	81.1	136.0	116.1	98.7	86.1	90.7	98.7	83.3	88.9
0960	Flooring	81.6	103.5	87.7	84.6	102.6	89.6	87.6	178.4	112.7	87.8	153.7	106.0	86.3	89.1	87.1	84.6	93.0	86.9
0970, 0990	Wall Finishes & Painting/Coating	90.4	96.7	94.2	82.9	82.7	82.8	91.2	149.3	126.1	89.2	115.9	105.2	89.1	93.2	91.5	82.9	82.5	82.7
09	FINISHES	94.6	98.2	96.5	96.2	95.7	95.9	100.4	175.3	141.1	99.5	130.1	116.1	97.8	88.8	92.9	96.2	84.3	89.7
COVERS	DIVS. 10 - 14, 25, 28, 41, 43, 44	100.0	99.7	99.9	100.0	97.6	99.5	100.0	134.7	107.4	100.0	119.6	104.2	100.0	96.5	99.2	100.0	91.7	98.2
21, 22, 23	FIRE SUPPRESSION, PLUMBING & HVAC	100.0	91.9	96.7	100.2	92.6	97.1	100.3	160.9	125.2	93.6	118.2	103.8	100.2	85.9	94.3	100.2	82.8	93.1
26, 27, 3370	ELECTRICAL, COMMUNICATIONS & UTIL.	99.8	90.9	95.1	100.9	93.4	97.0	98.6	178.5	140.4	105.0	130.6	118.4	100.9	91.3	95.9	98.8	90.4	94.4
MF2004	WEIGHTED AVERAGE	99.8	97.4	98.7	98.2	97.3	97.8	103.1	162.5	129.3	100.7	126.7	112.2	99.3	93.0	96.5	97.2	88.9	93.5

DIVISION		NEW YORK									NORTH CAROLINA								
		WATERTOWN			WHITE PLAINS			YONKERS			ASHEVILLE			CHARLOTTE			DURHAM		
		136			106			107			287 - 288			281 - 282			277		
		MAT.	INST.	TOTAL	MAT.	INST.	TOTAL	MAT.	INST.	TOTAL	MAT.	INST.	TOTAL	MAT.	INST.	TOTAL	MAT.	INST.	TOTAL
015433	CONTRACTOR EQUIPMENT		114.6	114.6		115.2	115.2		115.2	115.2		95.2	95.2		95.2	95.2		100.8	100.8
0241, 31 - 34	SITE & INFRASTRUCTURE, DEMOLITION	79.2	107.0	98.5	118.5	121.0	120.3	127.6	120.8	122.9	106.6	75.1	84.6	110.9	75.1	86.0	107.0	84.0	91.0
0310	Concrete Forming & Accessories	89.4	91.7	91.4	111.0	132.7	129.7	111.3	132.7	129.8	95.4	39.3	46.9	99.4	41.3	49.1	97.6	39.9	47.7
0320	Concrete Reinforcing	95.3	90.1	92.6	94.9	187.1	142.0	98.7	187.2	143.9	100.5	43.0	71.2	101.0	40.1	69.9	102.2	50.3	75.7
0330	Cast-in-Place Concrete	98.6	98.4	98.5	94.8	136.0	110.4	106.1	136.0	117.4	106.9	46.7	84.1	120.1	48.7	93.1	108.6	46.9	85.3
03	CONCRETE	114.7	94.5	105.2	103.4	143.0	121.9	114.2	143.0	127.7	106.1	44.5	77.3	112.2	45.5	81.0	106.6	46.2	78.4
04	MASONRY	91.3	97.5	95.0	101.9	134.2	121.3	107.0	134.2	123.4	88.3	40.3	59.4	95.1	44.1	64.5	93.9	36.3	59.2
05	METALS	97.0	102.6	98.9	101.3	136.9	113.3	111.1	137.0	119.9	85.0	74.3	81.4	88.9	73.4	83.6	97.7	77.5	90.9
06	WOOD, PLASTICS & COMPOSITES	86.9	90.6	89.0	114.0	135.4	126.1	113.8	135.4	126.0	95.3	38.8	63.3	100.3	41.0	66.7	97.5	40.3	65.1
07	THERMAL & MOISTURE PROTECTION	90.6	94.6	92.1	111.3	140.5	122.3	111.5	140.5	122.5	99.3	43.3	78.1	97.0	44.9	77.3	100.0	45.4	79.3
08	OPENINGS	96.1	86.1	93.6	91.9	147.7	105.7	95.5	149.5	108.9	94.3	38.6	80.5	98.5	40.1	84.0	98.5	42.7	84.6
0920	Plaster & Gypsum Board	106.3	90.1	95.1	121.9	136.0	131.7	128.1	136.0	133.6	103.3	36.8	57.2	105.1	39.0	59.3	106.6	38.3	59.3
0950, 0980	Ceilings & Acoustic Treatment	98.7	90.1	93.2	81.1	136.0	116.1	103.6	136.0	124.2	92.6	36.8	57.0	95.2	39.0	59.4	96.1	38.3	59.3
0960	Flooring	77.0	93.0	81.5	90.1	162.9	110.2	89.6	162.9	109.9	106.8	44.9	89.7	110.4	45.0	92.3	110.9	44.9	92.7
0970, 0990	Wall Finishes & Painting/Coating	82.9	89.2	86.7	89.2	149.3	125.3	89.2	149.3	125.3	116.0	35.6	67.6	116.0	40.6	70.6	116.0	37.3	68.6
09	FINISHES	93.7	91.8	92.7	100.1	139.7	121.6	106.6	139.7	124.6	99.4	39.7	67.0	101.2	41.7	68.9	101.9	40.1	68.4
COVERS	DIVS. 10 - 14, 25, 28, 41, 43, 44	100.0	90.5	98.0	100.0	121.3	104.5	100.0	122.1	104.7	100.0	70.6	93.8	100.0	71.1	93.9	100.0	75.6	94.8
21, 22, 23	FIRE SUPPRESSION, PLUMBING & HVAC	100.2	78.9	91.5	100.5	124.5	110.4	100.5	124.5	110.4	100.2	39.0	75.1	100.0	39.2	75.0	100.3	37.8	74.6
26, 27, 3370	ELECTRICAL, COMMUNICATIONS & UTIL.	100.9	90.4	95.4	96.6	152.0	125.5	105.1	152.0	129.6	98.3	38.6	67.1	100.5	50.3	74.2	99.0	48.9	72.8
MF2004	WEIGHTED AVERAGE	99.1	91.9	95.9	100.7	136.0	116.3	105.7	136.1	119.2	97.4	47.2	75.3	99.8	49.7	77.7	100.3	49.7	78.0

DIVISION		NORTH CAROLINA																	
		ELIZABETH CITY			FAYETTEVILLE			GASTONIA			GREENSBORO			HICKORY			KINSTON		
		279			283			280			270,272 - 274			286			285		
		MAT.	INST.	TOTAL	MAT.	INST.	TOTAL	MAT.	INST.	TOTAL	MAT.	INST.	TOTAL	MAT.	INST.	TOTAL	MAT.	INST.	TOTAL
015433	CONTRACTOR EQUIPMENT		105.3	105.3		100.8	100.8		95.2	95.2		100.8	100.8		100.8	100.8		100.8	100.8
0241, 31 - 34	SITE & INFRASTRUCTURE, DEMOLITION	112.4	85.2	93.4	105.7	84.1	90.7	106.4	75.0	84.6	106.9	84.1	91.0	105.9	84.1	90.7	104.9	83.9	90.3
0310	Concrete Forming & Accessories	85.3	34.3	41.2	95.0	40.7	48.0	101.3	40.3	48.5	97.4	40.6	48.3	91.8	38.4	45.6	88.7	32.5	40.1
0320	Concrete Reinforcing	100.0	41.8	70.3	104.4	50.4	76.8	101.0	42.9	71.3	101.0	59.0	79.5	100.5	19.5	59.1	100.0	18.9	58.6
0330	Cast-in-Place Concrete	108.9	36.8	81.6	112.1	50.0	88.6	104.5	54.2	85.5	107.8	48.2	85.3	106.8	47.8	84.5	103.1	38.2	78.6
03	CONCRETE	107.1	38.6	75.1	108.1	47.6	79.8	104.6	47.5	77.9	106.0	48.6	79.2	105.9	40.2	75.2	102.6	34.2	70.6
04	MASONRY	106.5	26.7	58.6	92.0	41.3	61.5	93.2	38.1	60.1	91.9	36.9	58.9	77.7	41.5	55.9	84.1	29.1	51.1
05	METALS	87.0	73.5	82.5	101.3	77.6	93.3	85.9	74.3	82.0	92.1	80.9	88.3	85.1	67.4	79.1	84.2	66.3	78.1
06	WOOD, PLASTICS & COMPOSITES	82.5	35.9	56.1	94.2	40.3	63.7	102.9	40.2	67.4	97.2	40.2	64.9	90.2	38.6	61.0	86.8	32.7	56.2
07	THERMAL & MOISTURE PROTECTION	99.3	32.7	74.2	99.2	43.9	78.3	99.5	43.6	78.4	99.8	42.9	78.3	99.5	42.2	77.9	99.4	32.0	74.0
08	OPENINGS	95.3	33.4	80.0	94.4	42.7	81.6	98.5	40.2	84.0	98.5	45.0	85.2	94.3	31.0	78.7	94.4	28.0	78.0
0920	Plaster & Gypsum Board	100.0	33.1	53.7	107.2	38.3	59.5	108.6	38.2	59.9	108.6	38.2	59.9	103.3	36.5	57.1	102.5	30.5	52.7
0950, 0980	Ceilings & Acoustic Treatment	96.1	33.1	55.9	93.5	38.3	58.3	96.1	38.2	59.2	96.1	38.2	59.2	92.6	36.5	56.9	96.1	30.5	54.3
0960	Flooring	100.3	16.0	77.0	106.9	44.9	89.8	110.9	44.9	92.7	110.9	41.5	91.7	106.6	34.6	86.7	103.4	19.8	80.3
0970, 0990	Wall Finishes & Painting/Coating	116.0	46.0	73.9	116.0	35.6	67.6	116.0	40.8	70.8	116.0	34.6	67.0	116.0	34.6	67.0	116.0	31.5	65.2
09	FINISHES	98.2	32.0	62.2	100.1	40.5	67.8	102.1	40.9	68.9	102.2	39.8	68.3	99.5	36.7	65.4	99.1	29.9	61.5
COVERS	DIVS. 10 - 14, 25, 28, 41, 43, 44	100.0	76.2	94.9	100.0	76.4	95.0	100.0	70.6	93.7	100.0	70.8	93.8	100.0	70.5	93.7	100.0	74.4	94.6
21, 22, 23	FIRE SUPPRESSION, PLUMBING & HVAC	93.5	30.4	67.5	100.1	39.0	75.0	100.2	38.6	74.9	100.2	39.1	75.1	93.5	32.1	68.2	93.5	25.9	65.7
26, 27, 3370	ELECTRICAL, COMMUNICATIONS & UTIL.	98.8	30.6	63.2	97.1	44.4	69.6	97.6	35.9	65.4	98.0	38.8	67.1	95.5	32.5	62.6	95.3	50.6	71.9
MF2004	WEIGHTED AVERAGE	97.3	41.6	72.7	100.0	50.1	78.0	98.2	47.2	75.7	99.2	49.2	77.1	94.9	43.9	72.4	94.6	41.6	71.2

DIVISION		NORTH CAROLINA															NORTH DAKOTA		
		MURPHY			RALEIGH			ROCKY MOUNT			WILMINGTON			WINSTON-SALEM			BISMARCK		
		289			275 - 276			278			284			271			585		
		MAT.	INST.	TOTAL	MAT.	INST.	TOTAL	MAT.	INST.	TOTAL	MAT.	INST.	TOTAL	MAT.	INST.	TOTAL	MAT.	INST.	TOTAL
015433	CONTRACTOR EQUIPMENT		95.2	95.2		100.8	100.8		100.8	100.8		95.2	95.2		100.8	100.8		98.8	98.8
0241, 31 - 34	SITE & INFRASTRUCTURE, DEMOLITION	108.4	74.8	85.0	111.1	84.2	92.3	110.2	83.7	91.7	108.0	75.0	85.1	107.2	84.1	91.1	99.3	96.7	97.5
0310	Concrete Forming & Accessories	101.7	24.9	35.3	97.8	41.5	49.1	90.9	29.8	38.0	96.6	39.9	47.5	98.6	39.6	47.6	97.2	43.3	50.6
0320	Concrete Reinforcing	100.0	18.7	58.5	101.0	50.4	75.1	100.0	27.5	63.0	101.3	48.6	74.4	101.0	39.9	69.8	101.8	66.3	83.7
0330	Cast-in-Place Concrete	110.6	35.2	82.1	123.2	53.5	96.8	106.4	33.9	79.0	106.4	47.9	84.3	110.5	45.6	86.0	102.7	50.1	82.8
03	CONCRETE	109.3	29.8	72.1	113.6	49.2	83.5	107.8	33.0	72.8	106.0	46.2	78.0	107.4	43.6	77.6	99.8	51.4	77.1
04	MASONRY	79.9	26.8	48.0	94.2	42.2	62.9	84.5	26.3	49.5	77.6	38.2	53.9	92.1	38.0	59.6	104.8	55.8	75.3
05	METALS	83.2	65.7	77.3	89.9	77.6	85.7	86.2	66.5	79.6	84.9	76.6	82.1	89.9	72.8	84.1	92.5	77.7	87.5
06	WOOD, PLASTICS & COMPOSITES	103.6	23.1	58.0	97.5	40.9	65.5	89.3	30.3	55.9	96.9	39.6	64.4	97.2	39.5	64.6	84.7	38.7	58.7
07	THERMAL & MOISTURE PROTECTION	99.5	30.8	73.5	99.1	44.1	78.3	99.8	28.3	72.8	99.3	43.3	78.1	99.8	42.6	78.1	102.4	48.5	82.0
08	OPENINGS	94.3	25.7	77.3	95.2	42.8	82.2	94.5	26.5	77.6	94.5	41.2	81.3	98.5	39.4	83.8	99.3	44.3	85.7
0920	Plaster & Gypsum Board	107.6	20.6	47.4	105.1	39.0	59.3	102.2	28.0	50.9	104.4	37.6	58.1	108.6	37.5	59.4	99.4	37.2	56.3
0950, 0980	Ceilings & Acoustic Treatment	92.6	20.6	46.7	95.2	39.0	59.4	93.5	28.0	51.8	93.5	37.6	57.9	96.1	37.5	58.8	139.6	37.2	74.4
0960	Flooring	111.3	21.1	86.4	110.9	44.9	92.7	104.9	16.0	80.3	107.7	44.9	90.4	110.9	44.9	92.7	115.2	64.7	101.2
0970, 0990	Wall Finishes & Painting/Coating	116.0	17.2	56.6	116.0	39.2	69.8	116.0	33.4	66.3	116.0	34.5	67.0	116.0	36.3	68.0	98.6	32.7	58.9
09	FINISHES	101.5	22.8	58.8	101.6	41.6	69.0	99.4	27.4	60.3	100.0	39.9	67.4	102.2	40.0	68.4	114.9	44.8	76.8
COVERS	DIVS. 10 - 14, 25, 28, 41, 43, 44	100.0	67.5	93.1	100.0	76.8	95.1	100.0	73.0	94.3	100.0	76.1	94.9	100.0	70.5	93.7	100.0	75.3	94.8
21, 22, 23	FIRE SUPPRESSION, PLUMBING & HVAC	93.5	19.2	62.9	100.1	37.3	74.2	93.5	33.3	68.7	100.2	38.7	74.9	100.2	37.7	74.5	100.1	57.4	82.6
26, 27, 3370	ELECTRICAL, COMMUNICATIONS & UTIL.	99.3	27.1	61.6	104.7	38.5	70.1	101.1	36.9	67.6	98.5	37.3	66.5	98.0	42.3	68.9	101.3	63.7	81.6
MF2004	WEIGHTED AVERAGE	95.8	34.0	68.5	100.3	49.4	77.8	96.3	40.3	71.6	96.9	47.6	75.1	99.1	47.8	76.5	100.4	60.4	82.8

DIVISION		NORTH DAKOTA																	
		DEVILS LAKE			DICKINSON			FARGO			GRAND FORKS			JAMESTOWN			MINOT		
		583			586			580 - 581			582			584			587		
		MAT.	INST.	TOTAL	MAT.	INST.	TOTAL	MAT.	INST.	TOTAL	MAT.	INST.	TOTAL	MAT.	INST.	TOTAL	MAT.	INST.	TOTAL
015433	CONTRACTOR EQUIPMENT		98.8	98.8		98.8	98.8		98.8	98.8		98.8	98.8		98.8	98.8		98.8	98.8
0241, 31 - 34	SITE & INFRASTRUCTURE, DEMOLITION	104.0	95.1	97.8	113.3	94.1	99.9	98.8	96.7	97.3	108.1	94.1	98.3	102.9	94.1	96.7	105.3	96.7	99.3
0310	Concrete Forming & Accessories	100.3	38.2	46.6	91.3	37.7	44.9	98.5	43.6	51.0	94.6	37.5	45.2	92.7	37.4	44.8	91.1	57.8	62.3
0320	Concrete Reinforcing	101.8	66.8	83.9	102.8	43.8	72.7	93.8	65.6	79.4	100.3	66.7	83.1	102.4	52.2	76.8	103.7	66.7	84.8
0330	Cast-in-Place Concrete	117.9	48.9	91.8	106.7	47.7	84.4	98.9	51.9	81.1	106.7	47.6	84.3	116.5	47.5	90.4	106.7	49.0	84.9
03	CONCRETE	110.4	48.7	81.6	110.4	43.6	79.2	107.1	52.0	81.3	107.2	47.8	79.4	109.1	45.0	79.1	105.7	57.5	83.2
04	MASONRY	113.1	63.0	83.0	115.7	57.4	80.7	106.4	41.8	67.6	107.4	62.7	80.5	126.3	36.5	72.3	107.3	62.7	80.5
05	METALS	92.6	77.5	87.5	92.5	65.3	83.3	95.0	76.6	88.8	92.5	73.7	86.1	92.5	66.9	83.9	92.8	78.4	88.0
06	WOOD, PLASTICS & COMPOSITES	88.1	35.6	58.4	77.9	35.6	54.0	84.6	39.3	58.9	81.7	35.6	55.6	79.6	35.6	54.7	77.6	58.1	66.6
07	THERMAL & MOISTURE PROTECTION	103.0	50.5	83.2	103.5	48.2	82.6	100.8	46.3	80.2	103.2	50.7	83.4	102.9	43.8	80.6	103.0	52.2	83.8
08	OPENINGS	99.3	39.0	84.4	99.3	33.6	83.0	99.2	44.6	85.7	99.3	39.0	84.4	99.3	35.9	83.6	99.4	54.8	88.4
0920	Plaster & Gypsum Board	111.4	34.0	57.9	103.2	34.0	55.3	99.4	37.8	56.7	105.0	34.0	55.9	104.5	34.0	55.7	103.2	57.1	71.3
0950, 0980	Ceilings & Acoustic Treatment	142.2	34.0	73.3	142.2	34.0	73.3	139.6	37.8	74.7	142.2	34.0	73.3	142.2	34.0	73.3	142.2	57.1	88.0
0960	Flooring	119.4	38.7	97.1	111.0	38.7	91.0	114.8	38.7	93.7	113.4	38.7	92.8	112.2	38.7	91.9	110.8	77.6	101.6
0970, 0990	Wall Finishes & Painting/Coating	98.6	25.0	54.3	98.6	25.0	54.3	98.6	67.3	79.8	98.6	25.0	54.3	98.6	25.0	54.3	98.6	27.4	55.7
09	FINISHES	118.5	35.6	73.5	115.9	35.6	72.3	114.8	43.7	76.2	116.3	35.6	72.5	115.4	35.6	72.1	115.0	58.2	84.2
COVERS	DIVS. 10 - 14, 25, 28, 41, 43, 44	100.0	37.8	86.8	100.0	37.9	86.8	100.0	75.4	94.8	100.0	37.9	86.8	100.0	73.1	94.3	100.0	77.6	95.2
21, 22, 23	FIRE SUPPRESSION, PLUMBING & HVAC	93.5	57.7	78.8	93.5	67.3	82.7	100.1	60.9	84.0	100.2	37.4	74.4	93.5	37.3	70.4	100.2	56.9	82.4
26, 27, 3370	ELECTRICAL, COMMUNICATIONS & UTIL.	93.3	40.1	65.5	102.6	64.7	82.8	102.8	63.7	82.3	97.0	60.0	77.7	93.3	40.1	65.5	100.5	64.8	81.8
MF2004	WEIGHTED AVERAGE	100.3	55.0	80.3	101.3	57.7	82.0	101.8	59.5	83.2	101.5	52.8	80.0	100.4	47.1	76.9	101.5	64.6	85.2

DIVISION		NORTH DAKOTA			OHIO														
		WILLISTON			AKRON			ATHENS			CANTON			CHILLICOTHE			CINCINNATI		
		588			442 - 443			457			446 - 447			456			451 - 452		
		MAT.	INST.	TOTAL	MAT.	INST.	TOTAL	MAT.	INST.	TOTAL	MAT.	INST.	TOTAL	MAT.	INST.	TOTAL	MAT.	INST.	TOTAL
015433	CONTRACTOR EQUIPMENT		98.8	98.8		96.6	96.6		90.8	90.8		96.6	96.6		101.3	101.3		101.1	101.1
0241, 31 - 34	SITE & INFRASTRUCTURE, DEMOLITION	106.3	94.1	97.8	91.4	104.0	100.2	87.8	92.2	90.9	91.5	103.9	100.1	78.1	106.2	97.7	75.7	105.7	96.6
0310	Concrete Forming & Accessories	96.1	37.7	45.6	99.7	93.6	94.5	94.7	90.4	90.9	99.7	82.1	84.5	95.9	89.1	90.0	97.5	81.9	84.0
0320	Concrete Reinforcing	104.7	43.8	73.6	101.4	91.4	96.3	91.6	85.5	88.5	101.4	78.8	89.9	87.9	84.4	86.1	93.2	84.5	88.7
0330	Cast-in-Place Concrete	106.7	47.7	84.3	98.2	100.6	99.1	96.0	97.1	96.4	99.2	96.5	98.2	87.1	91.9	88.9	80.2	84.9	82.0
03	CONCRETE	107.1	43.6	77.4	96.5	94.8	95.7	96.6	90.7	93.9	97.0	85.9	91.8	91.5	89.2	90.4	86.4	83.6	85.1
04	MASONRY	101.9	57.4	75.1	90.1	97.0	94.2	69.6	92.2	83.2	90.8	86.5	88.2	76.8	96.5	88.6	76.5	89.2	84.1
05	METALS	92.7	65.3	83.4	93.6	80.2	89.1	100.3	70.8	90.3	93.6	74.4	87.1	92.7	86.2	90.5	95.0	86.2	92.0
06	WOOD, PLASTICS & COMPOSITES	82.9	35.6	56.1	94.0	92.8	93.3	82.7	92.1	88.0	94.3	81.9	87.3	95.1	87.8	91.0	97.2	79.3	87.1
07	THERMAL & MOISTURE PROTECTION	103.2	48.2	82.4	105.0	96.7	101.9	101.8	93.2	98.6	106.0	91.1	100.4	103.1	91.5	98.7	101.5	91.8	97.9
08	OPENINGS	99.4	33.6	83.1	107.4	92.8	103.8	99.1	82.6	95.0	101.6	73.1	94.5	90.5	83.8	88.9	98.4	80.0	93.8
0920	Plaster & Gypsum Board	105.0	34.0	55.9	87.3	92.2	90.7	86.6	91.6	90.0	88.2	81.0	83.2	87.1	87.7	87.5	88.6	79.0	82.0
0950, 0980	Ceilings & Acoustic Treatment	142.2	34.0	73.3	86.0	92.2	90.0	96.8	91.6	93.5	86.0	81.0	82.8	91.2	87.7	88.9	92.0	79.0	83.7
0960	Flooring	114.6	38.7	93.6	102.1	84.7	97.3	128.0	96.6	119.3	102.3	77.7	95.5	105.3	100.1	103.9	106.4	92.1	102.5
0970, 0990	Wall Finishes & Painting/Coating	98.6	25.0	54.3	104.2	104.4	104.3	108.1	46.3	70.9	104.2	84.2	92.1	105.2	89.4	95.7	105.2	87.8	94.7
09	FINISHES	116.6	35.6	72.6	98.1	93.2	95.5	100.6	88.1	93.8	98.3	81.5	89.2	97.0	91.5	94.0	97.5	83.9	90.1
COVERS	DIVS. 10 - 14, 25, 28, 41, 43, 44	100.0	37.9	86.8	100.0	85.0	96.8	100.0	60.6	91.6	100.0	64.2	92.4	100.0	80.6	95.9	100.0	88.5	97.6
21, 22, 23	FIRE SUPPRESSION, PLUMBING & HVAC	93.5	67.3	82.7	99.9	91.9	96.6	93.2	57.8	78.6	99.9	77.1	90.5	93.7	84.9	90.1	100.0	84.6	93.6
26, 27, 3370	ELECTRICAL, COMMUNICATIONS & UTIL.	97.5	64.7	80.4	101.3	84.3	92.4	97.0	73.5	84.7	100.4	91.7	95.9	96.5	82.1	89.0	95.3	81.2	87.9
MF2004	WEIGHTED AVERAGE	99.6	57.7	81.1	98.7	91.9	95.7	95.6	78.4	88.0	98.2	83.7	91.8	93.1	89.1	91.3	94.9	86.4	91.1

DIVISION		OHIO																	
		CLEVELAND			COLUMBUS			DAYTON			HAMILTON			LIMA			LORAIN		
		441			430 - 432			453 - 454			450			458			440		
		MAT.	INST.	TOTAL	MAT.	INST.	TOTAL	MAT.	INST.	TOTAL	MAT.	INST.	TOTAL	MAT.	INST.	TOTAL	MAT.	INST.	TOTAL
015433	CONTRACTOR EQUIPMENT		96.9	96.9		95.6	95.6		96.1	96.1		101.3	101.3		93.4	93.4		96.6	96.6
0241, 31 - 34	SITE & INFRASTRUCTURE, DEMOLITION	91.3	104.7	100.6	90.8	100.9	97.8	74.9	105.8	96.4	74.7	105.9	96.4	82.9	93.4	90.2	90.8	103.5	99.6
0310	Concrete Forming & Accessories	99.8	99.2	99.3	97.1	87.0	88.3	97.5	77.7	80.3	97.6	80.8	83.1	94.7	84.0	85.5	99.8	91.6	92.7
0320	Concrete Reinforcing	102.0	91.8	96.8	93.2	84.5	88.8	93.2	76.6	84.7	93.2	84.4	88.7	91.6	76.8	84.0	101.4	91.7	96.4
0330	Cast-in-Place Concrete	96.4	107.8	100.7	85.9	94.9	89.3	74.6	89.1	80.1	80.0	83.5	81.3	88.2	97.9	91.9	93.5	104.3	97.6
03	CONCRETE	95.7	99.9	97.7	89.4	88.9	89.1	83.6	81.2	82.5	86.2	82.6	84.5	90.2	87.2	88.8	94.2	95.2	94.7
04	MASONRY	94.8	105.0	100.9	94.2	98.3	96.7	76.0	86.1	82.1	76.4	86.7	82.6	95.1	84.0	88.4	86.6	99.9	94.6
05	METALS	95.2	83.8	91.4	95.2	81.4	90.5	94.2	75.8	88.0	94.3	86.0	91.5	100.3	79.0	93.1	94.2	81.9	90.0
06	WOOD, PLASTICS & COMPOSITES	93.0	96.7	95.1	103.3	83.4	92.0	98.6	74.8	85.1	97.2	79.3	87.1	82.6	83.7	83.2	94.0	89.9	91.7
07	THERMAL & MOISTURE PROTECTION	103.9	108.6	105.7	100.1	94.9	98.2	106.2	86.6	98.8	103.2	89.5	98.0	101.3	95.8	99.2	106.0	103.7	105.1
08	OPENINGS	97.8	94.9	97.1	100.9	82.1	96.2	98.7	74.8	92.8	96.1	80.0	92.1	99.1	79.0	94.1	101.6	91.3	99.1
0920	Plaster & Gypsum Board	86.4	96.3	93.2	86.4	82.9	84.0	88.6	74.3	78.7	88.6	79.0	82.0	86.6	82.9	84.0	87.3	89.3	88.7
0950, 0980	Ceilings & Acoustic Treatment	84.3	96.3	91.9	83.6	82.9	83.1	93.0	74.3	81.1	92.0	79.0	83.7	95.8	82.9	87.6	86.0	89.3	88.1
0960	Flooring	102.0	102.8	102.2	92.7	87.9	91.3	109.1	82.8	101.9	106.4	92.1	102.5	127.2	89.5	116.8	102.3	107.5	103.8
0970, 0990	Wall Finishes & Painting/Coating	104.2	111.8	108.8	94.2	93.8	93.9	105.2	82.6	91.6	105.2	84.0	92.4	108.2	80.5	91.5	104.2	111.8	108.8
09	FINISHES	97.6	101.0	99.5	93.0	87.4	90.0	98.6	78.6	87.7	97.5	82.8	89.5	99.8	84.4	91.5	98.2	96.6	97.3
COVERS	DIVS. 10 - 14, 25, 28, 41, 43, 44	100.0	97.8	99.5	100.0	94.9	98.9	100.0	87.7	97.4	100.0	87.5	97.4	100.0	82.8	96.4	100.0	95.1	99.0
21, 22, 23	FIRE SUPPRESSION, PLUMBING & HVAC	99.9	99.4	99.7	99.8	90.9	96.1	100.8	81.0	92.6	100.5	83.3	93.4	93.2	79.8	87.7	99.9	79.7	91.6
26, 27, 3370	ELECTRICAL, COMMUNICATIONS & UTIL.	100.9	106.5	103.8	99.4	84.5	91.6	93.9	81.4	87.4	94.2	67.3	80.1	97.3	78.5	87.5	100.6	91.0	95.6
MF2004	WEIGHTED AVERAGE	98.0	100.2	99.0	96.9	89.8	93.7	94.7	82.9	89.5	94.6	83.6	89.7	96.0	83.3	90.4	97.7	91.5	95.0

DIVISION		OHIO																	
		MANSFIELD			MARION			SPRINGFIELD			STEUBENVILLE			TOLEDO			YOUNGSTOWN		
		448 - 449			433			455			439			434 - 436			444 - 445		
		MAT.	INST.	TOTAL	MAT.	INST.	TOTAL	MAT.	INST.	TOTAL	MAT.	INST.	TOTAL	MAT.	INST.	TOTAL	MAT.	INST.	TOTAL
015433	CONTRACTOR EQUIPMENT		96.6	96.6		95.3	95.3		96.1	96.1		100.2	100.2		98.3	98.3		96.6	96.6
0241, 31 - 34	SITE & INFRASTRUCTURE, DEMOLITION	87.8	104.5	99.5	87.3	99.4	95.7	75.1	104.3	95.5	131.1	109.0	115.7	90.0	100.6	97.4	91.2	104.5	100.5
0310	Concrete Forming & Accessories	91.1	90.1	90.3	94.3	75.5	78.1	97.5	83.7	85.6	97.3	86.6	88.0	97.1	99.0	98.7	99.7	86.8	88.5
0320	Concrete Reinforcing	92.3	79.3	85.6	85.9	84.5	85.2	93.2	84.0	88.5	83.3	87.5	85.4	93.2	91.3	92.2	101.4	87.1	94.1
0330	Cast-in-Place Concrete	91.0	93.2	91.8	78.4	88.1	82.1	76.6	89.1	81.4	85.2	94.7	88.8	85.9	103.1	92.4	97.3	100.7	98.6
03	CONCRETE	89.1	88.5	88.8	82.0	81.3	81.7	84.6	85.3	84.9	85.8	88.9	87.2	89.3	98.4	93.6	96.1	91.0	93.7
04	MASONRY	89.2	97.5	94.2	96.0	91.2	93.1	76.2	86.1	82.2	82.0	92.9	88.5	103.1	99.8	101.1	90.4	92.9	91.9
05	METALS	94.4	76.4	88.3	94.4	77.2	88.6	94.2	78.4	88.8	91.1	79.4	87.1	95.0	87.2	92.3	93.6	78.9	88.6
06	WOOD, PLASTICS & COMPOSITES	83.9	89.9	87.3	99.8	75.2	85.9	100.2	83.1	90.5	95.9	85.2	89.8	103.3	99.3	101.0	94.0	84.9	88.9
07	THERMAL & MOISTURE PROTECTION	105.5	93.5	101.0	99.6	82.8	93.3	106.1	87.5	99.1	111.4	92.8	104.3	101.6	105.6	103.1	106.2	93.7	101.5
08	OPENINGS	100.7	82.8	96.3	95.2	75.6	90.4	96.6	81.1	92.7	96.1	83.4	92.9	98.9	94.8	97.9	101.6	86.5	97.9
0920	Plaster & Gypsum Board	82.2	89.3	87.1	84.7	74.5	77.6	88.6	82.9	84.7	85.6	84.2	84.6	86.4	99.3	95.3	87.3	84.2	85.1
0950, 0980	Ceilings & Acoustic Treatment	87.7	89.3	88.7	83.6	74.5	77.8	93.0	82.9	86.6	80.6	84.2	82.9	83.6	99.3	93.6	86.0	84.2	84.8
0960	Flooring	97.3	106.2	99.8	91.0	75.1	86.6	109.1	82.8	101.9	118.3	85.7	109.3	91.9	94.9	92.7	102.3	89.5	98.8
0970, 0990	Wall Finishes & Painting/Coating	104.2	92.7	97.3	94.2	52.8	69.3	105.2	82.6	91.6	108.7	94.3	100.0	94.2	102.8	99.4	104.2	93.8	97.9
09	FINISHES	96.1	93.9	94.9	91.9	72.8	81.5	98.6	83.4	90.4	110.0	86.8	97.4	92.7	98.9	96.1	98.2	87.6	92.4
COVERS	DIVS. 10 - 14, 25, 28, 41, 43, 44	100.0	87.6	97.4	100.0	59.3	91.4	100.0	88.6	97.6	100.0	87.3	97.3	100.0	86.5	97.1	100.0	83.0	96.4
21, 22, 23	FIRE SUPPRESSION, PLUMBING & HVAC	93.1	78.6	87.1	93.0	80.2	87.7	100.8	81.0	92.7	93.2	80.6	88.0	99.8	95.1	97.9	99.9	93.1	97.1
26, 27, 3370	ELECTRICAL, COMMUNICATIONS & UTIL.	97.9	81.5	89.3	93.0	76.8	84.5	93.9	79.0	86.1	87.2	101.5	94.7	99.6	100.9	100.3	100.6	82.6	91.2
MF2004	WEIGHTED AVERAGE	94.9	87.1	91.5	92.9	80.6	87.5	94.6	84.3	90.1	94.9	89.6	92.6	97.1	97.1	97.1	98.1	89.6	94.3

	DIVISION	OHIO			OKLAHOMA															
		ZANESVILLE			ARDMORE			CLINTON			DURANT			ENID			GUYMON			
		437 - 438			734			736			747			737			739			
		MAT.	INST.	TOTAL	MAT.	INST.	TOTAL	MAT.	INST.	TOTAL	MAT.	INST.	TOTAL	MAT.	INST.	TOTAL	MAT.	INST.	TOTAL	
015433	CONTRACTOR EQUIPMENT		95.3	95.3		80.4	80.4		79.5	79.5		79.5	79.5		79.5	79.5		79.5	79.5	
0241, 31 - 34	SITE & INFRASTRUCTURE, DEMOLITION	89.9	100.7	97.4	104.9	90.3	94.7	106.5	88.8	94.2	99.3	88.6	91.9	107.9	88.8	94.6	111.5	87.3	94.6	
0310	Concrete Forming & Accessories	91.7	76.5	78.5	92.2	41.4	48.3	91.1	35.7	43.2	84.7	41.3	47.2	94.2	35.1	43.1	97.6	21.8	32.0	
0320	Concrete Reinforcing	85.4	87.1	86.2	99.3	79.7	89.3	99.8	79.7	89.6	100.2	57.3	78.3	99.2	79.7	89.2	99.8	28.5	63.4	
0330	Cast-in-Place Concrete	82.6	89.1	85.1	101.8	44.0	79.9	98.4	44.0	77.8	95.2	43.9	75.8	98.4	46.5	78.8	98.4	30.0	72.5	
03	CONCRETE	85.4	82.7	84.1	95.2	50.0	74.0	94.4	47.5	72.4	89.8	45.7	69.2	94.9	48.0	73.0	97.4	26.6	64.3	
04	MASONRY	93.1	84.1	87.7	95.1	57.8	72.6	121.7	57.8	83.3	93.6	67.1	77.7	103.7	57.8	76.1	97.9	21.0	51.7	
05	METALS	95.7	80.4	90.5	92.4	69.8	84.8	92.5	69.8	84.8	92.4	60.6	81.7	94.0	69.8	85.8	93.0	39.7	75.0	
06	WOOD, PLASTICS & COMPOSITES	96.5	75.2	84.5	99.6	40.1	65.9	98.7	32.5	61.2	90.8	40.1	62.1	102.4	31.6	62.3	106.1	20.9	57.8	
07	THERMAL & MOISTURE PROTECTION	99.7	88.2	95.4	98.2	60.6	84.0	98.3	59.8	83.7	97.9	59.1	83.3	98.4	59.7	83.7	98.6	26.6	71.4	
08	OPENINGS	95.2	77.7	90.9	95.7	50.8	84.6	95.7	46.6	83.6	95.7	45.0	83.2	95.7	46.0	83.4	95.9	22.4	77.7	
0920	Plaster & Gypsum Board	82.5	74.5	76.9	80.9	39.0	51.9	80.5	31.2	46.4	77.0	39.0	50.7	81.8	30.3	46.1	82.3	19.2	38.6	
0950, 0980	Ceilings & Acoustic Treatment	83.6	74.5	77.8	84.3	39.0	55.5	84.3	31.2	50.5	84.3	39.0	55.5	85.3	30.3	50.2	86.0	19.2	43.5	
0960	Flooring	89.0	79.2	86.3	108.1	46.5	91.1	106.6	44.3	89.4	102.5	63.4	91.7	108.8	44.3	91.0	111.2	26.2	87.7	
0970, 0990	Wall Finishes & Painting/Coating	94.2	75.2	82.7	98.6	56.9	73.6	98.6	56.9	73.6	98.6	56.9	73.6	98.6	56.9	73.6	98.6	16.9	49.5	
09	FINISHES	91.2	76.3	83.1	93.3	42.2	65.6	93.0	37.4	62.8	90.8	45.8	66.3	94.2	36.8	63.0	95.5	21.9	55.5	
COVERS	DIVS. 10 - 14, 25, 28, 41, 43, 44	100.0	76.8	95.1	100.0	73.8	94.4	100.0	72.9	94.2	100.0	73.9	94.5	100.0	72.8	94.2	100.0	69.8	93.6	
21, 22, 23	FIRE SUPPRESSION, PLUMBING & HVAC	93.0	75.6	85.8	93.4	62.0	80.5	93.4	62.0	80.5	93.4	62.0	80.5	100.2	62.0	84.5	93.4	23.1	64.5	
26, 27, 3370	ELECTRICAL, COMMUNICATIONS & UTIL.	93.1	71.8	81.9	95.6	69.9	82.1	96.7	69.9	82.7	98.4	54.5	75.5	96.7	69.9	82.7	98.4	17.3	56.0	
MF2004	WEIGHTED AVERAGE	93.3	80.0	87.5	95.1	61.2	80.2	96.6	59.8	80.3	94.2	58.6	78.5	97.7	59.8	80.9	96.4	30.8	67.4	

	DIVISION	OKLAHOMA																	
		LAWTON			MCALESTER			MIAMI			MUSKOGEE			OKLAHOMA CITY			PONCA CITY		
		735			745			743			744			730 - 731			746		
		MAT.	INST.	TOTAL	MAT.	INST.	TOTAL	MAT.	INST.	TOTAL	MAT.	INST.	TOTAL	MAT.	INST.	TOTAL	MAT.	INST.	TOTAL
015433	CONTRACTOR EQUIPMENT		80.4	80.4		79.5	79.5		88.0	88.0		88.0	88.0		80.7	80.7		79.5	79.5
0241, 31 - 34	SITE & INFRASTRUCTURE, DEMOLITION	103.2	90.3	94.2	91.8	88.8	89.7	93.7	86.1	88.4	93.7	84.7	87.4	102.6	90.8	94.4	100.0	88.8	92.2
0310	Concrete Forming & Accessories	97.3	47.3	54.1	82.9	42.7	48.2	94.4	50.4	56.4	98.4	31.0	40.1	96.4	41.1	48.5	90.4	41.7	48.2
0320	Concrete Reinforcing	99.4	79.7	89.4	99.8	42.8	70.7	98.3	79.7	88.8	99.2	32.6	65.2	99.4	79.7	89.4	99.2	79.7	89.2
0330	Cast-in-Place Concrete	95.2	46.5	76.8	83.4	46.4	69.4	87.4	45.6	71.6	88.5	35.9	68.6	98.0	49.3	79.6	97.7	38.9	75.5
03	CONCRETE	91.2	53.5	73.5	80.7	44.4	63.7	85.0	55.3	71.1	86.6	34.3	62.1	92.5	51.6	73.4	91.8	48.3	71.5
04	MASONRY	97.8	57.8	73.7	111.5	58.3	79.5	96.1	59.0	73.8	113.6	47.0	73.6	97.5	59.6	74.7	88.6	58.3	70.4
05	METALS	97.6	69.8	88.2	92.4	53.3	79.2	92.3	82.2	88.9	93.9	56.2	81.1	99.4	69.8	89.4	92.3	69.4	84.6
06	WOOD, PLASTICS & COMPOSITES	105.1	48.2	72.9	88.4	43.5	63.0	102.3	52.1	73.9	106.6	31.3	63.9	104.3	38.7	67.2	98.0	40.7	65.5
07	THERMAL & MOISTURE PROTECTION	98.2	61.4	84.3	97.6	57.6	82.5	98.0	59.3	83.4	98.0	41.8	76.8	95.0	61.2	82.2	98.1	59.7	83.6
08	OPENINGS	97.5	54.9	86.9	95.7	42.3	82.5	95.7	57.8	86.3	95.7	30.4	79.5	97.5	49.8	85.7	95.7	51.1	84.7
0920	Plaster & Gypsum Board	84.1	47.3	58.6	75.7	42.5	52.8	81.8	51.2	60.6	84.5	29.8	46.6	79.9	37.6	50.6	80.5	39.6	52.2
0950, 0980	Ceilings & Acoustic Treatment	93.0	47.3	63.9	84.3	42.5	57.7	84.3	51.2	63.2	93.0	29.8	52.7	91.3	37.6	57.1	84.3	39.6	55.8
0960	Flooring	111.6	44.3	93.0	101.1	44.3	85.4	109.8	63.4	97.0	112.7	35.7	91.4	111.6	44.3	93.0	106.2	44.3	89.1
0970, 0990	Wall Finishes & Painting/Coating	98.6	56.9	73.6	98.6	42.0	64.5	98.6	73.5	83.5	98.6	30.5	57.6	98.6	56.9	73.6	98.6	56.9	73.6
09	FINISHES	96.5	46.5	69.4	89.6	42.2	63.9	92.9	54.8	72.2	96.2	32.1	61.4	95.4	41.4	66.1	92.4	42.3	65.2
COVERS	DIVS. 10 - 14, 25, 28, 41, 43, 44	100.0	74.7	94.6	100.0	74.3	94.6	100.0	64.0	92.4	100.0	60.2	91.5	100.0	74.3	94.5	100.0	74.0	94.5
21, 22, 23	FIRE SUPPRESSION, PLUMBING & HVAC	100.2	62.0	84.5	93.4	31.8	68.1	93.4	58.3	79.0	100.2	24.2	68.9	100.1	62.9	84.8	93.4	57.9	78.8
26, 27, 3370	ELECTRICAL, COMMUNICATIONS & UTIL.	98.5	64.0	80.5	96.7	61.5	78.3	98.3	61.5	79.1	96.3	29.5	61.4	103.7	69.9	86.0	96.3	63.9	79.4
MF2004	WEIGHTED AVERAGE	97.9	61.7	81.9	93.5	51.0	74.7	93.8	62.6	80.1	96.9	39.6	71.6	98.6	61.7	82.3	94.2	59.2	78.8

	DIVISION	OKLAHOMA												OREGON					
		POTEAU			SHAWNEE			TULSA			WOODWARD			BEND			EUGENE		
		749			748			740 - 741			738			977			974		
		MAT.	INST.	TOTAL	MAT.	INST.	TOTAL	MAT.	INST.	TOTAL	MAT.	INST.	TOTAL	MAT.	INST.	TOTAL	MAT.	INST.	TOTAL
015433	CONTRACTOR EQUIPMENT		87.4	87.4		79.5	79.5		88.0	88.0		79.5	79.5		99.3	99.3		99.3	99.3
0241, 31 - 34	SITE & INFRASTRUCTURE, DEMOLITION	76.8	84.8	82.4	103.3	88.6	93.1	100.3	86.1	90.4	106.9	88.8	94.3	117.3	104.0	108.0	105.3	104.0	104.4
0310	Concrete Forming & Accessories	88.4	44.0	50.0	84.6	40.3	46.3	98.2	40.8	48.6	91.2	35.7	43.2	110.1	95.8	97.8	106.9	95.7	97.2
0320	Concrete Reinforcing	100.3	79.7	89.8	99.2	47.1	72.6	99.4	79.6	89.3	99.2	79.7	89.2	105.4	90.9	98.0	110.1	90.9	100.3
0330	Cast-in-Place Concrete	87.4	45.3	71.5	100.9	43.3	79.1	96.4	45.3	77.1	98.4	44.0	77.8	104.3	99.1	102.3	101.0	99.0	100.3
03	CONCRETE	86.9	52.4	70.8	93.7	43.0	70.0	91.8	51.0	72.7	94.7	47.4	72.6	115.6	95.8	106.3	105.9	95.7	101.1
04	MASONRY	97.0	58.4	73.8	112.8	57.0	79.3	96.9	59.6	74.5	92.0	57.8	71.4	114.2	94.4	102.3	111.6	94.4	101.3
05	METALS	92.4	82.2	88.9	92.3	55.1	79.7	97.2	81.3	91.8	92.5	69.8	84.9	89.6	91.3	90.2	90.3	91.2	90.6
06	WOOD, PLASTICS & COMPOSITES	95.2	43.7	66.0	90.7	39.9	61.9	105.7	39.9	68.4	98.8	32.4	61.2	97.9	95.8	96.7	94.0	95.8	95.0
07	THERMAL & MOISTURE PROTECTION	98.1	57.9	82.9	98.1	58.9	83.3	98.0	56.3	82.3	98.3	59.8	83.8	107.1	91.1	101.1	106.5	88.0	99.5
08	OPENINGS	95.7	53.2	85.2	95.7	39.3	81.7	97.5	50.0	85.7	95.8	46.6	83.6	98.8	96.8	98.3	99.2	96.8	98.6
0920	Plaster & Gypsum Board	79.6	42.5	54.0	77.0	38.8	50.6	84.1	38.7	52.6	81.0	31.1	46.5	105.7	95.5	98.6	104.0	95.5	98.1
0950, 0980	Ceilings & Acoustic Treatment	84.3	42.5	57.7	84.3	38.8	55.3	93.0	38.7	58.4	86.0	31.1	51.0	101.0	95.5	97.5	101.9	95.5	97.8
0960	Flooring	105.4	63.4	93.8	102.3	35.3	83.8	111.4	46.1	93.4	106.7	46.5	90.1	114.6	95.1	109.2	112.6	95.1	107.8
0970, 0990	Wall Finishes & Painting/Coating	98.6	56.9	73.6	98.6	37.9	62.1	98.6	45.0	66.4	98.6	56.9	73.6	110.1	65.5	83.2	110.1	65.5	83.2
09	FINISHES	90.4	47.9	67.3	91.0	37.8	62.1	95.9	41.1	66.2	93.6	37.7	63.2	109.0	92.5	100.0	107.2	92.5	99.2
COVERS	DIVS. 10 - 14, 25, 28, 41, 43, 44	100.0	74.7	94.6	100.0	73.5	94.4	100.0	62.8	92.1	100.0	72.9	94.2	100.0	95.6	99.1	100.0	95.6	99.1
21, 22, 23	FIRE SUPPRESSION, PLUMBING & HVAC	93.4	58.0	78.8	93.4	61.2	80.2	100.2	50.0	79.5	93.4	62.0	80.5	93.3	93.8	93.5	100.1	93.8	97.5
26, 27, 3370	ELECTRICAL, COMMUNICATIONS & UTIL.	96.4	61.5	78.2	98.6	69.9	83.6	98.5	38.5	67.1	98.3	69.9	83.5	102.3	91.4	96.6	100.8	91.4	95.9
MF2004	WEIGHTED AVERAGE	93.2	61.1	79.0	95.9	57.4	78.9	97.7	54.8	78.8	95.3	59.9	79.6	101.2	94.4	98.2	100.9	94.3	98.0

DIVISION		OREGON																	
		KLAMATH FALLS			MEDFORD			PENDLETON			PORTLAND			SALEM			VALE		
		976			975			978			970 - 972			973			979		
		MAT.	INST.	TOTAL	MAT.	INST.	TOTAL	MAT.	INST.	TOTAL	MAT.	INST.	TOTAL	MAT.	INST.	TOTAL	MAT.	INST.	TOTAL
015433	CONTRACTOR EQUIPMENT		99.3	99.3		99.3	99.3		96.7	96.7		99.3	99.3		99.3	99.3		96.7	96.7
0241, 31 - 34	SITE & INFRASTRUCTURE, DEMOLITION	121.9	103.9	109.4	114.2	104.0	107.0	112.4	97.2	101.9	108.2	104.0	105.3	101.9	104.0	103.3	98.7	97.2	97.6
0310	Concrete Forming & Accessories	103.9	95.7	96.8	103.1	95.7	96.7	103.5	95.9	96.9	108.2	96.1	97.7	105.8	95.9	97.2	109.2	95.1	97.0
0320	Concrete Reinforcing	105.4	90.9	98.0	107.2	90.9	98.9	104.4	91.2	97.6	110.9	91.2	100.8	111.2	91.2	101.0	101.8	91.0	96.3
0330	Cast-in-Place Concrete	104.3	97.4	101.7	104.3	99.0	102.3	105.1	98.8	102.7	103.8	99.2	102.1	91.3	99.1	94.2	82.7	98.5	88.7
03	CONCRETE	118.9	95.2	107.8	112.8	95.7	104.8	97.0	95.8	96.5	107.4	96.0	102.1	101.2	95.9	98.7	81.0	95.3	87.7
04	MASONRY	129.6	94.4	108.5	108.7	94.4	100.1	118.6	99.3	107.0	112.5	99.2	104.5	116.8	99.2	106.3	116.8	99.3	106.3
05	METALS	89.6	91.1	90.1	89.9	91.2	90.3	95.4	92.1	94.3	91.1	91.9	91.4	90.7	91.7	91.0	95.3	91.3	94.0
06	WOOD, PLASTICS & COMPOSITES	89.6	95.8	93.1	88.5	95.8	92.6	91.2	95.9	93.9	95.1	95.8	95.5	91.5	95.8	93.9	99.2	95.9	97.4
07	THERMAL & MOISTURE PROTECTION	107.3	90.0	100.8	107.0	90.0	100.6	98.7	91.4	95.9	106.5	96.5	102.7	104.1	92.4	99.7	98.2	89.0	94.7
08	OPENINGS	98.8	96.8	98.3	101.8	96.8	100.5	94.2	96.9	94.9	96.8	96.8	96.8	98.7	96.8	98.2	94.1	90.7	93.3
0920	Plaster & Gypsum Board	102.9	95.5	97.8	102.0	95.5	97.5	82.2	95.5	91.4	107.5	95.5	99.2	96.2	95.5	95.7	87.8	95.5	93.1
0950, 0980	Ceilings & Acoustic Treatment	110.5	95.5	100.9	115.3	95.5	102.7	69.8	95.5	86.2	103.9	95.5	98.5	102.4	95.5	98.0	69.8	95.5	86.2
0960	Flooring	110.9	95.1	106.5	110.1	95.1	105.9	77.8	95.1	82.6	110.0	95.1	105.9	112.8	95.1	107.9	79.9	95.1	84.1
0970, 0990	Wall Finishes & Painting/Coating	110.1	60.0	80.0	110.1	60.0	80.0	96.0	70.3	80.5	109.6	70.3	86.0	110.1	65.5	83.2	96.0	65.5	77.6
09	FINISHES	110.2	91.9	100.3	110.1	91.9	100.2	74.9	93.2	84.8	107.3	93.1	99.6	106.1	92.5	98.7	75.2	92.6	84.6
COVERS	DIVS. 10 - 14, 25, 28, 41, 43, 44	100.0	95.6	99.1	100.0	95.6	99.1	100.0	95.8	99.1	100.0	95.7	99.1	100.0	95.6	99.1	100.0	95.9	99.1
21, 22, 23	FIRE SUPPRESSION, PLUMBING & HVAC	93.3	92.9	93.2	100.1	93.8	97.5	94.7	96.1	95.3	100.1	98.3	99.4	100.0	93.8	97.5	94.7	71.5	85.2
26, 27, 3370	ELECTRICAL, COMMUNICATIONS & UTIL.	100.9	81.7	90.9	104.7	81.7	92.7	92.3	89.2	90.7	101.0	95.9	98.3	108.2	91.4	99.4	92.3	70.0	80.7
MF2004	WEIGHTED AVERAGE	102.4	92.7	98.1	102.7	92.9	98.4	95.5	94.7	95.1	101.1	96.8	99.2	101.0	95.0	98.4	93.1	86.3	90.1

DIVISION		PENNSYLVANIA																	
		ALLENTOWN			ALTOONA			BEDFORD			BRADFORD			BUTLER			CHAMBERSBURG		
		181			166			155			167			160			172		
		MAT.	INST.	TOTAL	MAT.	INST.	TOTAL	MAT.	INST.	TOTAL	MAT.	INST.	TOTAL	MAT.	INST.	TOTAL	MAT.	INST.	TOTAL
015433	CONTRACTOR EQUIPMENT		114.6	114.6		114.6	114.6		111.6	111.6		114.6	114.6		114.6	114.6		113.9	113.9
0241, 31 - 34	SITE & INFRASTRUCTURE, DEMOLITION	92.9	106.4	102.3	97.1	106.2	103.5	110.4	103.1	105.3	92.4	105.7	101.7	88.2	107.7	101.8	86.9	103.8	98.7
0310	Concrete Forming & Accessories	100.5	117.6	115.3	87.5	81.0	81.8	85.9	83.5	83.8	89.2	80.2	81.4	88.9	96.5	95.4	85.7	82.8	83.2
0320	Concrete Reinforcing	94.7	111.1	103.1	91.8	91.2	91.5	93.2	83.1	88.0	93.8	91.4	92.5	92.4	103.4	98.1	92.2	99.6	96.0
0330	Cast-in-Place Concrete	83.9	104.7	91.7	93.5	88.1	91.4	101.0	74.7	91.0	89.3	90.3	89.7	82.6	99.0	88.8	90.7	74.6	84.6
03	CONCRETE	97.1	112.6	104.4	91.7	86.8	89.4	99.4	81.8	91.1	99.1	87.3	93.6	83.2	99.9	91.0	105.6	84.6	95.8
04	MASONRY	95.4	104.3	100.8	97.5	71.8	82.0	99.2	86.2	91.4	94.3	84.8	88.6	99.6	98.8	99.2	95.7	91.3	93.1
05	METALS	99.2	124.6	107.8	93.2	112.0	99.6	92.9	106.3	97.4	97.0	112.1	102.1	92.9	119.9	102.0	97.0	115.0	103.1
06	WOOD, PLASTICS & COMPOSITES	102.8	121.8	113.6	82.6	83.9	83.3	81.6	83.8	82.9	89.2	77.8	82.8	84.1	94.1	89.8	92.5	83.9	87.6
07	THERMAL & MOISTURE PROTECTION	101.2	119.9	108.3	100.6	88.6	96.1	103.4	87.4	97.4	101.2	91.1	97.4	100.4	101.8	100.9	100.2	74.4	90.5
08	OPENINGS	94.1	117.9	100.0	88.6	87.8	88.4	90.2	85.7	89.1	94.5	88.4	93.0	88.5	103.1	92.1	90.9	84.4	89.3
0920	Plaster & Gypsum Board	111.1	122.2	118.8	104.4	83.1	89.7	89.2	83.1	85.0	105.3	76.9	85.6	104.8	93.7	97.1	108.8	83.1	91.0
0950, 0980	Ceilings & Acoustic Treatment	90.1	122.2	110.6	93.6	83.1	86.9	97.1	83.1	88.2	91.9	76.9	82.4	94.5	93.7	94.0	91.9	83.1	86.3
0960	Flooring	86.3	94.9	88.6	80.2	49.1	71.6	90.8	49.9	79.5	79.8	99.2	85.2	81.1	87.2	82.8	83.7	49.7	74.3
0970, 0990	Wall Finishes & Painting/Coating	89.1	78.4	82.6	85.1	105.5	97.4	97.5	75.1	84.0	89.1	92.2	91.0	85.1	105.5	97.4	89.1	64.2	74.1
09	FINISHES	95.5	108.9	102.8	94.0	76.7	84.6	97.6	75.5	85.6	93.3	83.1	87.8	94.0	94.6	94.3	93.9	74.4	83.4
COVERS	DIVS. 10 - 14, 25, 28, 41, 43, 44	100.0	107.5	101.6	100.0	96.1	99.2	100.0	99.1	99.8	100.0	98.8	99.7	100.0	103.8	100.8	100.0	97.2	99.4
21, 22, 23	FIRE SUPPRESSION, PLUMBING & HVAC	100.2	113.3	105.6	99.8	85.5	93.9	93.1	89.9	91.8	93.6	91.6	92.8	93.1	96.9	94.6	93.6	89.8	92.0
26, 27, 3370	ELECTRICAL, COMMUNICATIONS & UTIL.	100.1	98.4	99.2	89.5	102.7	96.4	93.1	102.7	98.1	93.3	102.7	98.2	90.2	102.7	96.7	91.8	85.0	88.3
MF2004	WEIGHTED AVERAGE	98.2	110.5	103.6	94.9	90.3	92.9	95.6	91.0	93.6	95.6	93.8	94.8	92.2	101.6	96.3	95.8	89.7	93.1

DIVISION		PENNSYLVANIA																	
		DOYLESTOWN			DUBOIS			ERIE			GREENSBURG			HARRISBURG			HAZLETON		
		189			158			164 - 165			156			170 - 171			182		
		MAT.	INST.	TOTAL	MAT.	INST.	TOTAL	MAT.	INST.	TOTAL	MAT.	INST.	TOTAL	MAT.	INST.	TOTAL	MAT.	INST.	TOTAL
015433	CONTRACTOR EQUIPMENT		94.4	94.4		111.6	111.6		114.6	114.6		111.6	111.6		113.9	113.9		114.6	114.6
0241, 31 - 34	SITE & INFRASTRUCTURE, DEMOLITION	107.2	91.7	96.4	116.3	103.2	107.2	93.6	106.8	102.8	105.9	105.2	105.4	83.5	105.3	98.7	86.1	106.6	100.4
0310	Concrete Forming & Accessories	86.2	133.9	127.5	85.3	84.7	84.8	100.4	87.4	89.2	91.4	96.4	95.8	93.4	88.5	89.2	84.3	89.7	89.0
0320	Concrete Reinforcing	91.7	139.5	116.1	92.7	91.6	92.1	93.8	93.1	93.4	92.7	103.4	98.2	94.7	101.2	98.0	91.9	107.5	99.9
0330	Cast-in-Place Concrete	79.3	91.8	84.0	97.3	92.8	95.6	91.9	86.2	89.8	93.6	98.6	95.5	93.2	96.3	94.4	79.3	92.8	84.4
03	CONCRETE	93.0	120.2	105.7	100.9	90.1	95.8	90.3	89.4	89.9	94.3	99.6	96.8	99.3	95.0	97.3	89.6	95.5	92.3
04	MASONRY	98.4	133.0	119.2	99.2	87.1	91.9	87.5	90.8	89.5	109.3	98.8	103.0	95.3	92.5	93.6	107.6	99.2	102.6
05	METALS	96.8	129.3	107.8	92.9	111.2	99.1	93.4	112.7	99.9	92.8	118.4	101.4	101.2	119.2	107.3	98.9	121.9	106.7
06	WOOD, PLASTICS & COMPOSITES	84.9	137.5	114.7	80.7	83.8	82.5	98.7	86.3	91.7	88.0	94.1	91.4	97.3	87.7	91.8	83.5	87.7	85.9
07	THERMAL & MOISTURE PROTECTION	99.6	132.1	111.9	103.6	95.7	100.6	101.0	93.4	98.1	103.3	101.4	102.6	107.2	110.6	108.5	100.8	109.6	104.1
08	OPENINGS	95.8	143.2	107.5	90.2	91.6	90.6	88.7	88.8	88.8	90.2	103.0	93.4	94.1	92.8	93.8	94.8	94.6	94.8
0920	Plaster & Gypsum Board	102.8	138.3	127.4	88.3	83.1	84.7	111.1	85.7	93.5	90.7	93.7	92.8	108.4	87.0	93.6	103.3	87.0	92.1
0950, 0980	Ceilings & Acoustic Treatment	89.4	138.3	120.6	97.1	83.1	88.2	90.1	85.7	87.3	96.3	93.7	94.6	91.8	87.0	88.8	91.1	87.0	88.5
0960	Flooring	70.4	105.2	80.0	90.5	99.2	92.9	87.8	73.1	83.7	94.3	73.3	88.5	86.8	86.4	86.7	76.9	89.5	80.4
0970, 0990	Wall Finishes & Painting/Coating	88.6	76.6	81.4	97.5	105.5	102.3	95.1	92.2	93.4	97.5	105.5	102.3	89.1	87.7	88.3	89.1	97.6	94.2
09	FINISHES	86.6	122.6	106.1	97.9	87.4	92.2	96.9	84.6	90.2	98.3	92.8	95.3	94.4	87.7	90.8	91.4	88.6	89.9
COVERS	DIVS. 10 - 14, 25, 28, 41, 43, 44	100.0	80.5	95.9	100.0	99.1	99.8	100.0	100.7	100.2	100.0	103.7	100.8	100.0	98.0	99.6	100.0	101.7	100.4
21, 22, 23	FIRE SUPPRESSION, PLUMBING & HVAC	93.1	129.9	108.2	93.1	90.1	91.9	99.8	93.5	97.2	93.1	95.1	93.9	100.1	91.1	96.4	93.6	93.5	93.5
26, 27, 3370	ELECTRICAL, COMMUNICATIONS & UTIL.	92.7	120.2	107.1	93.7	102.7	98.4	91.4	84.1	87.6	93.7	102.7	98.4	98.8	85.1	91.6	94.2	84.0	88.9
MF2004	WEIGHTED AVERAGE	94.8	122.6	107.1	96.1	94.7	95.5	94.7	93.1	94.0	95.6	100.6	97.8	98.5	95.1	97.0	95.1	96.8	95.9

	DIVISION	PENNSYLVANIA																	
		INDIANA			JOHNSTOWN			KITTANNING			LANCASTER			LEHIGH VALLEY			MONTROSE		
		157			159			162			175 - 176			180			188		
		MAT.	INST.	TOTAL	MAT.	INST.	TOTAL	MAT.	INST.	TOTAL	MAT.	INST.	TOTAL	MAT.	INST.	TOTAL	MAT.	INST.	TOTAL
015433	CONTRACTOR EQUIPMENT		111.6	111.6		111.6	111.6		114.6	114.6		113.9	113.9		114.6	114.6		114.6	114.6
0241, 31 - 34	SITE & INFRASTRUCTURE, DEMOLITION	103.8	104.1	104.0	111.0	104.3	106.4	90.9	107.7	102.6	78.8	105.3	97.2	90.3	106.3	101.5	89.2	104.6	99.9
0310	Concrete Forming & Accessories	86.4	88.6	88.3	85.3	84.9	85.0	88.9	96.6	95.5	87.4	88.0	87.9	94.8	117.6	114.5	85.0	88.9	88.4
0320	Concrete Reinforcing	91.9	103.5	97.8	93.2	103.2	98.3	92.4	103.5	98.1	91.9	101.0	96.6	91.9	110.5	101.4	96.3	107.3	101.9
0330	Cast-in-Place Concrete	91.8	98.5	94.3	101.8	93.1	98.6	85.8	99.1	90.8	77.1	96.4	84.4	85.8	104.6	92.9	84.1	91.2	86.8
03	CONCRETE	92.0	96.1	93.9	100.2	92.5	96.6	85.9	99.9	92.5	92.4	94.7	93.5	96.2	112.5	103.8	94.7	94.3	94.5
04	MASONRY	95.5	100.4	98.5	96.3	86.9	90.6	103.2	100.5	101.5	101.0	93.1	96.2	95.5	104.3	100.8	95.4	98.9	97.5
05	METALS	92.9	118.2	101.5	92.9	116.7	100.9	93.0	120.1	102.2	97.0	118.4	104.2	98.8	124.0	107.3	97.1	116.9	103.8
06	WOOD, PLASTICS & COMPOSITES	82.4	83.8	83.2	80.7	83.8	82.5	84.1	94.1	89.8	95.0	87.7	90.9	95.8	121.8	110.5	84.3	88.0	86.4
07	THERMAL & MOISTURE PROTECTION	103.2	99.3	101.7	103.4	94.3	100.0	100.5	102.2	101.1	99.8	99.3	99.6	101.1	108.1	103.7	100.8	95.0	98.6
08	OPENINGS	90.2	95.2	91.5	90.2	91.3	90.5	88.6	100.8	91.6	90.9	96.5	92.2	94.8	117.8	100.5	91.1	94.8	92.0
0920	Plaster & Gypsum Board	89.6	83.1	85.1	88.1	83.1	84.6	104.8	93.7	97.1	110.9	87.0	94.4	107.2	122.2	117.6	103.6	87.4	92.4
0950, 0980	Ceilings & Acoustic Treatment	97.1	83.1	88.2	96.3	83.1	87.9	94.5	93.7	94.0	91.9	87.0	88.8	91.1	122.2	110.9	91.9	87.4	89.1
0960	Flooring	91.2	99.2	93.4	90.5	62.6	82.8	81.1	99.2	86.1	84.8	90.4	86.3	82.7	94.9	86.0	77.7	62.3	73.4
0970, 0990	Wall Finishes & Painting/Coating	97.5	105.5	102.3	97.5	105.5	102.3	85.1	105.5	97.4	89.1	63.5	73.7	89.1	73.8	79.9	89.1	97.6	94.2
09	FINISHES	97.3	90.7	93.7	97.2	81.8	88.8	94.1	96.8	95.6	94.0	83.9	88.5	94.0	108.5	101.9	92.2	84.5	88.0
COVERS	DIVS. 10 - 14, 25, 28, 41, 43, 44	100.0	102.5	100.5	100.0	99.4	99.9	100.0	103.8	100.8	100.0	98.1	99.6	100.0	107.6	101.6	100.0	101.9	100.4
21, 22, 23	FIRE SUPPRESSION, PLUMBING & HVAC	93.1	95.1	93.9	93.1	90.6	92.1	93.1	99.8	95.8	93.6	91.3	92.7	93.6	113.3	101.7	93.6	101.0	96.6
26, 27, 3370	ELECTRICAL, COMMUNICATIONS & UTIL.	93.7	102.7	98.4	93.7	102.7	98.4	89.5	102.7	96.4	93.3	49.4	70.3	94.2	129.4	112.6	93.3	96.7	95.0
MF2004	WEIGHTED AVERAGE	94.4	99.4	96.6	95.6	94.9	95.3	92.7	102.6	97.1	94.4	89.5	92.2	95.7	114.3	103.9	94.5	98.4	96.2

	DIVISION	PENNSYLVANIA																	
		NEW CASTLE			NORRISTOWN			OIL CITY			PHILADELPHIA			PITTSBURGH			POTTSVILLE		
		161			194			163			190 - 191			150 - 152			179		
		MAT.	INST.	TOTAL	MAT.	INST.	TOTAL	MAT.	INST.	TOTAL	MAT.	INST.	TOTAL	MAT.	INST.	TOTAL	MAT.	INST.	TOTAL
015433	CONTRACTOR EQUIPMENT		114.6	114.6		97.5	97.5		114.6	114.6		96.9	96.9		112.9	112.9		113.9	113.9
0241, 31 - 34	SITE & INFRASTRUCTURE, DEMOLITION	88.6	107.6	101.9	94.1	98.9	97.5	87.1	106.0	100.2	101.0	98.1	98.9	109.4	107.4	108.0	81.8	105.3	98.2
0310	Concrete Forming & Accessories	88.9	95.9	95.0	87.6	134.5	128.1	88.9	81.6	82.6	102.1	138.8	133.8	98.1	96.6	96.8	80.1	89.4	88.1
0320	Concrete Reinforcing	91.3	94.1	92.7	92.6	153.4	123.7	92.4	93.9	93.2	95.6	153.4	125.1	93.7	103.5	98.7	91.2	101.0	96.2
0330	Cast-in-Place Concrete	83.3	98.8	89.2	80.9	133.2	100.7	81.0	95.3	86.4	99.2	133.2	112.1	97.3	98.6	97.8	82.2	80.4	81.5
03	CONCRETE	83.7	97.8	90.3	88.4	137.2	111.2	82.0	90.1	85.8	100.6	139.2	118.6	97.5	99.7	98.5	96.5	89.9	93.4
04	MASONRY	98.6	98.8	98.8	110.5	133.5	124.3	99.2	92.2	95.0	97.1	133.8	119.2	90.3	99.1	95.6	95.2	92.6	93.6
05	METALS	93.0	115.1	100.5	103.4	138.0	115.1	93.0	113.5	99.9	107.5	138.1	117.8	94.2	118.6	102.5	97.2	118.9	104.5
06	WOOD, PLASTICS & COMPOSITES	84.1	94.1	89.8	84.0	133.9	112.2	84.1	77.8	80.5	101.3	139.6	123.0	96.6	94.1	95.2	85.9	87.7	86.9
07	THERMAL & MOISTURE PROTECTION	100.4	98.6	99.7	98.6	137.4	113.3	100.3	94.1	98.0	99.3	138.5	114.1	103.5	101.5	102.7	99.9	108.9	103.3
08	OPENINGS	88.6	97.8	90.8	90.2	145.4	103.9	88.6	80.8	86.6	98.3	148.5	110.7	92.6	103.0	95.2	90.9	92.7	91.3
0920	Plaster & Gypsum Board	104.8	93.7	97.1	95.7	134.7	122.7	104.8	76.9	85.5	103.5	140.6	129.2	96.1	93.7	94.4	105.8	87.0	92.8
0950, 0980	Ceilings & Acoustic Treatment	94.5	93.7	94.0	94.0	134.7	119.9	94.5	76.9	83.3	94.0	140.6	123.7	97.1	93.7	94.9	91.9	87.0	88.8
0960	Flooring	81.1	61.9	75.8	77.6	143.3	95.7	81.1	99.2	86.1	85.6	143.3	101.5	98.4	99.2	98.6	80.6	90.3	83.3
0970, 0990	Wall Finishes & Painting/Coating	85.1	105.5	97.4	94.0	146.9	125.8	85.1	105.5	97.4	94.0	146.9	125.8	97.5	107.2	103.4	89.1	97.6	94.2
09	FINISHES	94.0	89.9	91.8	93.5	137.0	117.1	93.9	85.2	89.1	97.2	140.4	120.7	100.6	97.2	98.8	92.3	90.1	91.1
COVERS	DIVS. 10 - 14, 25, 28, 41, 43, 44	100.0	103.8	100.8	100.0	121.4	104.5	100.0	100.2	100.0	100.0	122.1	104.7	100.0	103.7	100.8	100.0	98.9	99.8
21, 22, 23	FIRE SUPPRESSION, PLUMBING & HVAC	93.1	96.7	94.6	93.2	131.7	109.0	93.1	93.9	93.4	100.0	131.7	113.0	99.9	98.6	99.4	93.6	100.7	96.5
26, 27, 3370	ELECTRICAL, COMMUNICATIONS & UTIL.	90.2	95.3	92.8	92.4	139.3	116.9	92.2	102.7	97.7	97.1	144.4	121.8	96.3	102.7	99.7	91.3	92.5	92.0
MF2004	WEIGHTED AVERAGE	92.2	98.9	95.2	95.4	132.6	111.8	92.2	95.6	93.7	100.3	134.2	115.3	97.7	102.1	99.6	94.3	97.7	95.8

	DIVISION	PENNSYLVANIA																	
		READING			SCRANTON			STATE COLLEGE			STROUDSBURG			SUNBURY			UNIONTOWN		
		195 - 196			184 - 185			168			183			178			154		
		MAT.	INST.	TOTAL	MAT.	INST.	TOTAL	MAT.	INST.	TOTAL	MAT.	INST.	TOTAL	MAT.	INST.	TOTAL	MAT.	INST.	TOTAL
015433	CONTRACTOR EQUIPMENT		117.7	117.7		114.6	114.6		113.9	113.9		114.6	114.6		114.6	114.6		111.6	111.6
0241, 31 - 34	SITE & INFRASTRUCTURE, DEMOLITION	98.4	111.4	107.5	93.3	106.6	102.6	84.0	105.1	98.7	87.8	104.7	99.6	91.9	106.3	102.0	104.5	105.2	105.0
0310	Concrete Forming & Accessories	101.6	89.5	91.1	100.7	90.5	91.9	84.9	80.9	81.5	89.9	90.0	89.9	92.9	88.1	88.7	80.4	96.3	94.1
0320	Concrete Reinforcing	93.9	101.3	97.7	94.7	107.6	101.3	93.0	91.8	92.3	95.0	124.3	109.9	93.8	99.8	96.9	92.7	103.4	98.1
0330	Cast-in-Place Concrete	72.5	95.4	81.2	87.6	93.1	89.7	84.5	70.4	79.2	82.6	76.9	80.4	89.9	96.1	92.2	91.8	98.5	94.4
03	CONCRETE	87.2	95.2	90.9	99.0	96.0	97.6	99.5	80.9	90.8	93.5	93.2	93.3	99.8	94.4	97.3	91.8	99.5	95.4
04	MASONRY	98.7	95.9	97.0	95.8	99.3	97.9	100.9	83.1	90.2	92.2	103.9	99.2	94.3	90.1	91.8	111.7	98.8	104.0
05	METALS	103.7	119.7	109.1	101.3	122.4	108.4	96.9	112.7	102.2	98.9	124.4	107.5	96.9	117.9	104.0	92.6	118.2	101.3
06	WOOD, PLASTICS & COMPOSITES	101.5	87.3	93.5	102.8	88.0	94.4	91.4	83.9	87.1	90.1	88.0	88.9	93.4	87.7	90.2	74.8	94.1	85.7
07	THERMAL & MOISTURE PROTECTION	98.9	112.4	104.0	101.1	100.8	101.0	100.0	90.8	96.5	100.9	85.6	95.1	101.1	106.7	103.2	103.1	101.4	102.5
08	OPENINGS	95.9	96.3	96.0	94.1	94.8	94.3	90.8	87.8	90.1	94.8	92.6	94.3	91.0	92.3	91.3	90.2	103.0	93.4
0920	Plaster & Gypsum Board	107.9	86.8	93.3	113.7	87.4	95.5	107.4	83.1	90.6	105.0	87.4	92.8	106.1	87.0	92.9	85.5	93.7	91.2
0950, 0980	Ceilings & Acoustic Treatment	84.4	86.8	85.9	98.7	87.4	91.5	90.2	83.1	85.7	89.4	87.4	88.1	90.2	87.0	88.2	96.3	93.7	94.6
0960	Flooring	83.7	90.4	85.5	86.3	109.7	92.7	83.2	53.6	75.0	80.3	55.3	73.4	81.6	45.8	71.7	87.5	69.0	82.4
0970, 0990	Wall Finishes & Painting/Coating	92.5	98.6	96.2	89.1	97.6	94.2	89.1	105.5	99.0	89.1	71.7	78.7	89.1	97.6	94.2	97.5	105.5	102.3
09	FINISHES	95.0	89.4	92.0	97.8	94.1	95.8	93.0	77.7	84.7	92.5	81.2	86.4	93.6	81.8	87.2	95.4	91.3	93.2
COVERS	DIVS. 10 - 14, 25, 28, 41, 43, 44	100.0	100.8	100.2	100.0	101.9	100.4	100.0	94.1	98.7	100.0	70.8	93.8	100.0	98.0	99.6	100.0	103.7	100.8
21, 22, 23	FIRE SUPPRESSION, PLUMBING & HVAC	100.1	109.9	104.1	100.2	101.7	100.8	93.6	85.8	90.4	93.6	103.4	97.6	93.6	91.0	92.5	93.1	95.1	93.9
26, 27, 3370	ELECTRICAL, COMMUNICATIONS & UTIL.	99.2	92.6	95.7	100.2	96.7	98.4	92.3	102.7	97.8	94.2	146.2	121.3	91.7	84.7	88.1	90.5	102.7	96.9
MF2004	WEIGHTED AVERAGE	97.9	101.5	99.5	99.0	100.9	99.8	95.2	90.7	93.2	95.0	105.1	99.5	95.2	93.8	94.6	94.6	100.4	97.2

	DIVISION	PENNSYLVANIA																	
		WASHINGTON			WELLSBORO			WESTCHESTER			WILKES-BARRE			WILLIAMSPORT			YORK		
		153			169			193			186 - 187			177			173 - 174		
		MAT.	INST.	TOTAL	MAT.	INST.	TOTAL	MAT.	INST.	TOTAL	MAT.	INST.	TOTAL	MAT.	INST.	TOTAL	MAT.	INST.	TOTAL
015433	CONTRACTOR EQUIPMENT		111.6	111.6		114.6	114.6		97.5	97.5		114.6	114.6		114.6	114.6		113.9	113.9
0241, 31 - 34	SITE & INFRASTRUCTURE, DEMOLITION	104.6	105.2	105.0	96.4	104.5	102.0	100.3	96.7	97.8	85.9	106.6	100.3	83.1	105.1	98.4	82.1	105.3	98.3
0310	Concrete Forming & Accessories	86.5	96.6	95.2	88.7	86.7	87.0	93.6	135.7	130.0	92.2	90.0	90.3	89.9	64.4	67.8	83.1	88.8	88.0
0320	Concrete Reinforcing	92.7	103.5	98.2	93.0	107.2	100.2	91.8	113.4	102.8	93.8	107.5	100.8	93.0	71.6	82.1	93.8	101.2	97.5
0330	Cast-in-Place Concrete	91.8	98.7	94.4	88.5	88.9	88.7	89.5	132.6	105.8	79.3	92.8	84.4	76.0	76.7	76.3	82.6	96.7	87.9
03	CONCRETE	92.1	99.7	95.7	102.0	92.6	97.6	96.1	129.9	111.9	90.3	95.7	92.8	87.0	71.9	79.9	97.7	95.2	96.5
04	MASONRY	95.6	100.4	98.5	100.4	92.8	95.8	104.7	133.5	122.0	108.2	98.9	102.6	86.5	94.5	91.3	95.4	93.1	94.0
05	METALS	92.5	118.7	101.4	96.9	117.0	103.7	103.4	118.6	108.6	97.1	122.2	105.6	97.0	103.6	99.2	98.7	119.2	105.6
06	WOOD, PLASTICS & COMPOSITES	82.4	94.1	89.0	88.7	87.7	88.1	90.9	137.4	117.2	92.5	87.9	89.9	90.1	57.1	71.4	89.5	87.7	88.5
07	THERMAL & MOISTURE PROTECTION	103.2	101.9	102.7	101.4	92.6	98.1	98.9	137.9	113.6	100.8	109.5	104.1	100.6	102.7	101.4	100.0	110.9	104.1
08	OPENINGS	90.2	103.0	93.3	94.5	94.6	94.5	90.2	132.2	100.6	91.0	98.5	92.9	91.0	59.8	83.3	90.9	92.8	91.3
0920	Plaster & Gypsum Board	89.4	93.7	92.4	104.8	87.0	92.5	97.8	138.3	125.9	106.2	87.3	93.1	105.8	55.5	71.0	107.5	87.0	93.3
0950, 0980	Ceilings & Acoustic Treatment	96.3	93.7	94.6	90.2	87.0	88.2	94.0	138.3	122.2	91.9	87.3	89.0	91.9	55.5	68.8	91.0	87.0	88.5
0960	Flooring	91.3	99.2	93.5	79.6	54.7	72.7	81.1	143.3	98.3	81.3	89.5	83.6	80.3	53.6	72.9	82.1	90.4	84.4
0970, 0990	Wall Finishes & Painting/Coating	97.5	105.5	102.3	89.1	97.6	94.2	94.0	146.9	125.8	89.1	97.6	94.2	89.1	97.6	94.2	89.1	87.7	88.3
09	FINISHES	97.1	96.8	96.9	93.1	81.7	86.9	95.4	137.0	118.0	93.4	89.8	91.4	92.8	63.8	77.1	92.8	88.6	90.5
COVERS	DIVS. 10 - 14, 25, 28, 41, 43, 44	100.0	103.7	100.8	100.0	98.0	99.6	100.0	118.1	103.8	100.0	101.7	100.4	100.0	94.7	98.9	100.0	98.2	99.6
21, 22, 23	FIRE SUPPRESSION, PLUMBING & HVAC	93.1	98.6	95.4	93.6	90.4	92.2	93.2	129.8	108.2	93.6	94.4	93.9	93.6	91.1	92.6	100.2	91.4	96.6
26, 27, 3370	ELECTRICAL, COMMUNICATIONS & UTIL.	93.1	102.7	98.1	93.3	82.5	87.7	92.2	127.5	110.7	94.2	84.0	88.9	92.2	52.2	71.3	93.3	85.1	89.0
MF2004	WEIGHTED AVERAGE	94.3	102.1	97.7	96.3	92.8	94.8	96.4	127.0	109.9	94.8	97.4	95.9	92.9	81.1	87.7	96.5	95.4	96.0

	DIVISION	PUERTO RICO			RHODE ISLAND						SOUTH CAROLINA								
		SAN JUAN			NEWPORT			PROVIDENCE			AIKEN			BEAUFORT			CHARLESTON		
		009			028			029			298			299			294		
		MAT.	INST.	TOTAL	MAT.	INST.	TOTAL	MAT.	INST.	TOTAL	MAT.	INST.	TOTAL	MAT.	INST.	TOTAL	MAT.	INST.	TOTAL
015433	CONTRACTOR EQUIPMENT		89.7	89.7		102.2	102.2		102.2	102.2		100.5	100.5		100.5	100.5		100.5	100.5
0241, 31 - 34	SITE & INFRASTRUCTURE, DEMOLITION	119.8	90.8	99.6	77.9	104.0	96.1	81.2	104.0	97.1	124.3	86.3	97.8	118.8	84.6	95.0	101.1	84.2	89.3
0310	Concrete Forming & Accessories	92.6	18.1	28.1	100.5	117.1	114.8	99.1	117.1	114.7	99.1	70.7	74.5	98.0	41.4	49.0	97.2	43.2	50.5
0320	Concrete Reinforcing	198.7	12.8	103.8	110.6	114.7	112.7	110.6	114.7	112.7	102.1	68.8	85.1	101.2	28.6	64.1	101.0	61.8	81.0
0330	Cast-in-Place Concrete	101.1	31.1	74.6	78.2	115.4	92.3	95.1	115.4	102.8	84.3	72.5	79.9	84.3	50.5	71.5	99.0	51.2	80.9
03	CONCRETE	112.2	22.5	70.2	100.9	115.6	107.7	109.1	115.6	112.1	110.7	72.0	92.6	107.6	44.2	77.9	101.6	51.1	78.0
04	MASONRY	87.9	17.1	45.4	104.4	126.4	117.6	108.0	126.4	119.1	82.3	63.4	71.0	98.7	36.7	61.4	100.2	41.0	64.6
05	METALS	113.3	33.7	86.4	99.6	110.3	103.2	99.6	110.3	103.2	83.9	87.0	85.0	83.9	71.1	79.6	85.8	79.2	83.5
06	WOOD, PLASTICS & COMPOSITES	93.8	17.8	50.8	100.1	116.0	109.1	99.3	116.0	108.7	100.3	72.0	84.3	98.4	42.2	66.6	97.2	42.2	66.0
07	THERMAL & MOISTURE PROTECTION	125.2	21.7	86.1	96.1	117.6	104.2	94.0	117.6	102.9	100.4	69.6	88.8	100.2	39.7	77.3	99.4	44.4	78.6
08	OPENINGS	149.0	15.2	115.8	98.6	115.5	102.8	99.2	115.5	103.3	94.3	68.3	87.9	94.4	40.9	81.1	98.5	46.1	85.5
0920	Plaster & Gypsum Board	116.3	15.3	46.4	97.1	115.8	110.0	98.1	115.8	110.3	106.7	71.0	82.0	109.5	40.3	61.6	111.4	40.3	62.2
0950, 0980	Ceilings & Acoustic Treatment	229.1	15.3	92.9	96.0	115.8	108.6	96.2	115.8	108.7	92.6	71.0	78.8	96.1	40.3	60.5	96.1	40.3	60.5
0960	Flooring	202.3	14.1	150.3	98.4	131.7	107.6	98.0	131.7	107.3	109.6	68.8	98.3	111.4	55.2	95.9	110.9	62.7	97.6
0970, 0990	Wall Finishes & Painting/Coating	203.7	13.3	89.1	96.7	124.2	113.2	96.7	124.2	113.2	116.0	74.3	90.9	116.0	38.5	69.4	116.0	43.5	72.4
09	FINISHES	205.4	17.0	103.1	98.7	120.8	110.7	98.8	120.8	110.8	103.5	70.8	85.8	104.8	44.0	71.8	102.9	46.5	72.2
COVERS	DIVS. 10 - 14, 25, 28, 41, 43, 44	100.0	17.9	82.6	100.0	108.7	101.8	100.0	108.7	101.8	100.0	77.2	95.2	100.0	50.9	89.6	100.0	70.6	93.8
21, 22, 23	FIRE SUPPRESSION, PLUMBING & HVAC	103.6	14.0	66.8	100.1	111.5	104.8	100.0	111.5	104.7	93.5	66.3	82.3	93.5	26.8	66.0	100.2	42.1	76.3
26, 27, 3370	ELECTRICAL, COMMUNICATIONS & UTIL.	124.0	13.3	66.1	97.0	100.5	98.8	97.0	100.5	98.8	95.6	69.5	82.0	99.4	34.0	65.2	97.7	99.6	98.7
MF2004	WEIGHTED AVERAGE	121.2	24.5	78.5	99.0	112.8	105.1	100.2	112.8	105.8	96.5	72.0	85.7	97.3	44.1	73.8	98.1	59.7	81.2

	DIVISION	SOUTH CAROLINA															SOUTH DAKOTA		
		COLUMBIA			FLORENCE			GREENVILLE			ROCK HILL			SPARTANBURG			ABERDEEN		
		290 - 292			295			296			297			293			574		
		MAT.	INST.	TOTAL	MAT.	INST.	TOTAL	MAT.	INST.	TOTAL	MAT.	INST.	TOTAL	MAT.	INST.	TOTAL	MAT.	INST.	TOTAL
015433	CONTRACTOR EQUIPMENT		100.5	100.5		100.5	100.5		100.5	100.5		100.5	100.5		100.5	100.5		98.8	98.8
0241, 31 - 34	SITE & INFRASTRUCTURE, DEMOLITION	102.4	84.1	89.7	112.0	84.1	92.6	106.5	83.8	90.7	104.4	83.6	89.9	106.2	83.8	90.6	93.7	94.9	94.5
0310	Concrete Forming & Accessories	98.8	42.2	49.9	86.2	42.4	48.3	96.8	42.2	49.6	95.5	39.7	47.3	99.7	42.2	50.0	98.1	40.9	48.6
0320	Concrete Reinforcing	101.0	61.6	80.8	100.6	61.7	80.7	100.5	49.2	74.3	101.4	49.0	74.6	100.5	49.2	74.3	100.4	42.2	70.7
0330	Cast-in-Place Concrete	106.0	51.8	85.5	84.3	51.0	71.7	84.3	50.9	71.7	84.3	45.6	69.7	84.3	50.9	71.7	101.7	47.1	81.0
03	CONCRETE	105.2	50.8	79.8	101.7	50.6	77.8	99.9	48.2	75.8	97.9	45.3	73.3	100.1	48.2	75.9	99.8	44.7	74.0
04	MASONRY	94.6	38.8	61.1	82.3	41.0	57.5	80.2	41.0	56.6	106.8	36.0	64.2	82.3	41.0	57.5	108.8	58.1	78.3
05	METALS	85.8	78.3	83.3	84.7	78.6	82.7	84.7	74.1	81.1	83.9	73.5	80.4	84.7	74.1	81.1	98.1	67.1	87.6
06	WOOD, PLASTICS & COMPOSITES	99.6	41.7	66.8	83.7	41.7	59.9	97.2	41.7	65.8	95.5	39.6	63.9	101.0	41.7	67.4	100.1	40.1	66.2
07	THERMAL & MOISTURE PROTECTION	98.6	43.1	77.6	99.7	44.4	78.8	99.6	44.4	78.7	99.5	41.9	77.7	99.7	44.4	78.8	102.0	50.8	82.7
08	OPENINGS	98.5	45.8	85.4	94.4	45.8	82.4	94.3	42.9	81.6	94.4	41.9	81.4	94.3	42.9	81.6	95.0	41.0	81.6
0920	Plaster & Gypsum Board	105.1	39.8	59.9	99.2	39.8	58.1	105.0	39.8	59.8	104.1	37.6	58.1	107.1	39.8	60.5	103.5	38.6	58.6
0950, 0980	Ceilings & Acoustic Treatment	95.2	39.8	59.9	93.5	39.8	59.3	92.6	39.8	58.9	92.6	37.6	57.6	92.6	39.8	58.9	110.0	38.6	64.6
0960	Flooring	107.4	47.0	90.7	100.9	47.0	86.0	108.1	61.6	95.3	107.1	47.9	90.7	109.8	61.6	96.5	115.4	54.2	98.5
0970, 0990	Wall Finishes & Painting/Coating	113.5	43.5	71.4	116.0	43.5	72.4	116.0	43.5	72.4	116.0	43.5	72.4	116.0	43.5	72.4	98.6	41.2	64.1
09	FINISHES	100.5	43.0	69.3	98.9	43.5	68.8	101.2	46.0	71.2	100.5	41.3	68.4	102.0	46.0	71.6	108.0	43.0	72.7
COVERS	DIVS. 10 - 14, 25, 28, 41, 43, 44	100.0	70.3	93.7	100.0	70.3	93.7	100.0	70.3	93.7	100.0	69.4	93.5	100.0	70.3	93.7	100.0	47.6	88.9
21, 22, 23	FIRE SUPPRESSION, PLUMBING & HVAC	100.0	37.3	74.2	100.2	37.3	74.4	100.2	35.8	73.7	93.5	34.8	69.3	100.2	35.8	73.7	100.0	45.1	77.4
26, 27, 3370	ELECTRICAL, COMMUNICATIONS & UTIL.	101.0	50.9	74.8	95.5	48.4	70.9	97.8	52.4	74.0	97.8	52.4	74.0	97.8	52.4	74.0	97.6	57.4	76.6
MF2004	WEIGHTED AVERAGE	98.3	51.2	77.5	96.3	51.2	76.4	96.3	50.8	76.2	95.6	48.9	74.9	96.5	50.8	76.3	100.0	54.0	79.7

	DIVISION	SOUTH DAKOTA																	
		MITCHELL			MOBRIDGE			PIERRE			RAPID CITY			SIOUX FALLS			WATERTOWN		
		573			576			575			577			570 - 571			572		
		MAT.	INST.	TOTAL	MAT.	INST.	TOTAL	MAT.	INST.	TOTAL	MAT.	INST.	TOTAL	MAT.	INST.	TOTAL	MAT.	INST.	TOTAL
015433	CONTRACTOR EQUIPMENT		98.8	98.8		98.8	98.8		98.8	98.8		98.8	98.8		99.8	99.8		98.8	98.8
0241, 31 - 34	SITE & INFRASTRUCTURE, DEMOLITION	90.7	94.9	93.7	90.7	94.9	93.6	92.9	95.0	94.3	91.8	95.0	94.0	93.5	96.7	95.7	90.6	94.9	93.6
0310	Concrete Forming & Accessories	97.4	41.2	48.8	89.9	41.0	47.6	96.9	41.8	49.2	103.8	39.6	48.3	97.6	44.5	51.6	87.0	40.8	47.1
0320	Concrete Reinforcing	99.8	51.6	75.2	102.4	42.2	71.7	100.0	53.2	76.1	93.8	53.0	73.0	93.8	53.2	73.1	97.1	42.2	69.1
0330	Cast-in-Place Concrete	98.8	45.7	78.7	98.8	47.1	79.2	93.5	45.2	75.2	98.0	45.8	78.3	93.5	46.2	75.6	98.8	47.0	79.2
03	CONCRETE	97.6	45.9	73.4	97.5	44.7	72.8	95.0	46.5	72.3	96.7	45.7	72.9	92.9	48.1	72.0	96.5	44.6	72.3
04	MASONRY	97.1	57.5	73.3	105.8	58.1	77.1	104.6	57.8	76.4	105.8	53.7	74.5	101.6	55.9	74.1	131.7	57.8	87.3
05	METALS	97.1	67.5	87.1	97.2	67.1	87.0	98.1	72.4	89.4	100.0	72.4	90.7	100.2	72.9	91.0	97.1	67.5	87.1
06	WOOD, PLASTICS & COMPOSITES	99.1	40.7	66.1	89.7	40.4	61.8	98.0	41.0	65.7	103.4	37.2	65.9	98.8	43.9	67.7	86.7	40.1	60.3
07	THERMAL & MOISTURE PROTECTION	101.9	48.9	81.9	101.9	50.9	82.7	102.0	49.2	82.0	102.4	50.7	82.9	104.0	52.5	84.5	101.7	50.8	82.5
08	OPENINGS	93.4	41.6	80.6	96.4	40.5	82.6	98.2	44.8	85.0	99.3	42.8	85.3	99.5	46.4	86.4	93.4	40.7	80.4
0920	Plaster & Gypsum Board	101.9	39.3	58.5	96.6	38.9	56.7	96.5	39.5	57.1	103.2	35.6	56.4	100.2	42.5	60.2	95.0	38.6	56.0
0950, 0980	Ceilings & Acoustic Treatment	107.4	39.3	64.0	110.0	38.9	64.7	104.8	39.5	63.2	113.5	35.6	63.9	107.4	42.5	66.1	107.4	38.6	63.6
0960	Flooring	115.0	54.2	98.2	109.4	54.2	94.1	114.7	38.6	93.7	114.7	68.2	101.8	114.7	73.7	103.4	107.7	54.2	92.9
0970, 0990	Wall Finishes & Painting/Coating	98.6	45.1	66.4	98.6	46.3	67.1	98.6	49.5	69.1	98.6	49.5	69.1	98.6	49.5	69.1	98.6	41.2	64.1
09	FINISHES	106.9	43.8	72.6	105.1	43.7	71.8	105.6	41.3	70.7	108.4	45.3	74.1	106.8	50.1	76.0	103.8	43.0	70.8
COVERS	DIVS. 10 - 14, 25, 28, 41, 43, 44	100.0	44.1	88.1	100.0	47.6	88.9	100.0	71.4	93.9	100.0	71.2	93.9	100.0	71.8	94.0	100.0	47.6	88.9
21, 22, 23	FIRE SUPPRESSION, PLUMBING & HVAC	93.2	44.7	73.3	93.2	45.2	73.5	100.0	44.7	77.2	100.0	45.2	77.5	100.0	43.5	76.7	93.2	45.2	73.5
26, 27, 3370	ELECTRICAL, COMMUNICATIONS & UTIL.	95.8	49.3	71.5	97.6	50.3	72.9	98.7	49.8	73.1	93.9	49.8	70.8	101.1	66.4	83.0	94.9	50.3	71.6
MF2004	WEIGHTED AVERAGE	96.8	53.0	77.4	97.5	53.2	77.9	99.3	54.2	79.4	99.8	54.2	79.7	99.8	57.8	81.2	98.0	53.1	78.2

	DIVISION	TENNESSEE																	
		CHATTANOOGA			COLUMBIA			COOKEVILLE			JACKSON			JOHNSON CITY			KNOXVILLE		
		373 - 374			384			385			383			376			377 - 379		
		MAT.	INST.	TOTAL	MAT.	INST.	TOTAL	MAT.	INST.	TOTAL	MAT.	INST.	TOTAL	MAT.	INST.	TOTAL	MAT.	INST.	TOTAL
015433	CONTRACTOR EQUIPMENT		104.0	104.0		97.8	97.8		97.8	97.8		104.1	104.1		97.7	97.7		97.7	97.7
0241, 31 - 34	SITE & INFRASTRUCTURE, DEMOLITION	102.6	98.7	99.9	92.2	86.4	88.1	98.5	86.4	90.1	100.9	95.7	97.3	111.3	87.1	94.5	88.6	87.8	88.1
0310	Concrete Forming & Accessories	94.0	57.2	62.1	82.4	45.5	50.5	82.5	38.6	44.6	88.4	44.3	50.3	82.9	41.1	46.8	93.0	45.8	52.2
0320	Concrete Reinforcing	79.3	65.2	72.1	81.3	63.1	72.0	81.3	65.1	73.0	81.3	61.1	71.0	79.9	61.5	70.5	79.3	63.1	71.1
0330	Cast-in-Place Concrete	100.7	64.4	87.0	94.3	48.6	77.0	106.9	46.2	84.0	104.4	42.0	80.8	81.0	48.8	68.8	94.6	51.7	78.4
03	CONCRETE	90.9	62.9	77.8	92.2	51.9	73.4	102.1	48.4	77.0	93.7	48.7	72.7	96.3	49.8	74.5	88.1	53.1	71.7
04	MASONRY	96.7	54.2	71.2	110.1	48.4	73.0	105.3	48.7	71.3	110.5	38.2	67.0	108.4	37.2	65.6	74.5	57.0	64.0
05	METALS	100.0	89.4	96.4	92.7	88.7	91.3	92.8	88.4	91.3	95.0	84.1	91.3	97.2	86.4	93.5	100.8	87.9	96.4
06	WOOD, PLASTICS & COMPOSITES	101.7	58.2	77.1	71.0	46.5	57.1	71.1	37.5	52.1	86.8	45.8	63.6	76.1	41.2	56.3	88.4	41.2	61.7
07	THERMAL & MOISTURE PROTECTION	95.0	61.2	82.2	89.3	54.3	76.0	89.6	49.1	74.3	91.4	46.0	74.3	90.4	54.3	76.7	88.6	61.1	78.2
08	OPENINGS	100.4	58.2	89.9	91.9	48.8	81.2	91.9	46.1	80.6	101.0	52.8	89.1	96.1	48.8	84.4	93.0	49.1	82.1
0920	Plaster & Gypsum Board	78.3	57.5	63.9	84.9	45.4	57.6	84.9	36.1	51.1	86.7	44.7	57.6	91.7	40.0	55.9	98.9	40.0	58.1
0950, 0980	Ceilings & Acoustic Treatment	95.7	57.5	71.3	84.4	45.4	59.6	84.4	36.1	53.7	93.7	44.7	62.5	91.9	40.0	58.8	92.8	40.0	59.1
0960	Flooring	98.7	61.0	88.3	90.8	22.1	71.8	91.0	20.6	71.5	91.4	20.6	71.8	93.0	43.8	79.4	98.4	58.2	87.3
0970, 0990	Wall Finishes & Painting/Coating	105.0	58.6	77.1	95.1	37.1	60.2	95.1	40.5	62.3	96.3	55.1	71.5	102.6	46.8	69.0	102.6	46.8	69.0
09	FINISHES	94.0	57.9	74.4	92.6	39.6	63.8	93.2	33.9	61.0	93.6	41.1	65.1	95.7	41.3	66.2	90.2	47.1	66.8
COVERS	DIVS. 10 - 14, 25, 28, 41, 43, 44	100.0	47.7	88.9	100.0	48.9	89.1	100.0	48.0	89.0	100.0	49.2	89.2	100.0	49.4	89.3	100.0	53.6	90.1
21, 22, 23	FIRE SUPPRESSION, PLUMBING & HVAC	99.9	61.9	84.3	96.5	72.8	86.8	96.5	73.0	86.8	99.9	47.1	78.2	99.7	49.2	78.9	99.7	64.1	85.0
26, 27, 3370	ELECTRICAL, COMMUNICATIONS & UTIL.	104.5	70.0	86.4	95.0	64.4	79.0	96.8	63.2	79.3	102.1	65.6	83.0	93.8	52.0	71.9	100.0	57.6	77.8
MF2004	WEIGHTED AVERAGE	98.6	66.9	84.6	94.9	62.3	80.5	96.3	60.5	80.5	98.4	56.0	79.7	97.9	54.2	78.6	94.9	61.8	80.2

	DIVISION	TENNESSEE									TEXAS								
		MCKENZIE			MEMPHIS			NASHVILLE			ABILENE			AMARILLO			AUSTIN		
		382			375,380 - 381			370 - 372			795 - 796			790 - 791			786 - 787		
		MAT.	INST.	TOTAL	MAT.	INST.	TOTAL	MAT.	INST.	TOTAL	MAT.	INST.	TOTAL	MAT.	INST.	TOTAL	MAT.	INST.	TOTAL
015433	CONTRACTOR EQUIPMENT		97.8	97.8		101.6	101.6		105.2	105.2		88.0	88.0		88.0	88.0		87.2	87.2
0241, 31 - 34	SITE & INFRASTRUCTURE, DEMOLITION	98.0	86.2	89.8	91.5	91.9	91.8	97.1	99.7	98.9	100.7	85.6	90.2	101.9	86.8	91.4	91.1	85.8	87.4
0310	Concrete Forming & Accessories	89.3	32.4	40.1	93.4	63.2	67.3	94.0	66.8	70.5	95.8	41.7	49.0	96.4	55.3	60.8	90.4	53.8	58.7
0320	Concrete Reinforcing	81.5	30.0	55.2	84.9	66.3	75.4	84.9	65.5	75.0	96.6	53.9	74.8	96.6	50.3	72.9	93.7	50.2	71.5
0330	Cast-in-Place Concrete	104.6	41.0	80.6	90.7	64.3	80.7	92.6	68.4	83.4	100.0	45.3	79.3	99.1	53.4	81.8	87.9	50.4	73.7
03	CONCRETE	100.5	37.2	70.9	87.0	65.7	77.0	88.4	68.4	79.1	93.1	46.4	71.2	92.7	54.5	74.8	82.1	52.7	68.4
04	MASONRY	108.8	47.7	72.1	87.9	67.0	75.3	81.7	62.1	70.0	100.5	53.7	72.3	100.5	51.1	70.8	92.8	50.2	67.2
05	METALS	92.8	72.6	86.0	101.5	90.0	97.6	102.1	88.3	97.5	100.8	68.7	89.9	100.8	67.5	89.5	96.8	64.9	86.0
06	WOOD, PLASTICS & COMPOSITES	79.2	30.2	51.5	92.8	64.5	76.8	99.3	68.5	81.9	102.1	40.6	67.3	100.6	58.0	76.5	93.8	56.6	72.7
07	THERMAL & MOISTURE PROTECTION	89.6	47.1	73.6	89.2	67.2	80.9	94.3	64.1	82.9	98.1	50.8	80.2	98.4	50.6	80.3	87.6	52.7	74.4
08	OPENINGS	91.9	30.4	76.7	99.2	65.0	90.8	98.4	67.4	90.7	93.7	44.3	81.5	93.7	52.6	83.6	100.5	55.2	89.3
0920	Plaster & Gypsum Board	88.3	28.7	47.0	87.2	63.8	71.0	95.2	68.0	76.3	84.1	39.4	53.2	79.4	57.3	64.1	85.2	55.9	64.9
0950, 0980	Ceilings & Acoustic Treatment	84.4	28.7	48.9	92.9	63.8	74.4	95.4	68.0	77.9	93.0	39.4	58.9	91.3	57.3	69.6	85.7	55.9	66.7
0960	Flooring	94.4	42.4	80.0	93.0	38.9	78.1	101.3	69.2	92.4	111.6	67.5	99.4	111.4	61.7	97.7	94.2	42.6	79.9
0970, 0990	Wall Finishes & Painting/Coating	95.1	47.2	66.3	97.5	58.2	73.9	105.2	72.6	85.6	97.2	54.6	71.6	97.2	43.7	65.0	95.3	44.4	64.7
09	FINISHES	94.5	34.5	62.0	89.6	57.7	72.3	99.9	68.1	82.6	95.9	47.3	69.5	95.2	55.3	73.6	87.6	50.5	67.4
COVERS	DIVS. 10 - 14, 25, 28, 41, 43, 44	100.0	29.3	85.0	100.0	81.0	96.0	100.0	76.6	95.0	100.0	76.2	95.0	100.0	77.7	95.3	100.0	77.9	95.3
21, 22, 23	FIRE SUPPRESSION, PLUMBING & HVAC	96.5	65.6	83.8	99.8	72.5	88.6	99.8	78.4	91.0	100.2	39.6	75.3	100.1	52.4	80.5	100.1	55.6	81.8
26, 27, 3370	ELECTRICAL, COMMUNICATIONS & UTIL.	96.6	70.5	83.0	101.6	68.6	84.4	101.4	64.4	82.0	98.6	52.5	74.5	103.3	70.8	86.3	102.7	65.8	83.4
MF2004	WEIGHTED AVERAGE	96.4	55.6	78.4	96.4	71.6	85.5	97.5	73.9	87.1	98.2	52.8	78.2	98.6	60.5	81.8	95.5	59.4	79.6

		TEXAS																	
	DIVISION	BEAUMONT 776 - 777			BROWNWOOD 768			BRYAN 778			CHILDRESS 792			CORPUS CHRISTI 783 - 784			DALLAS 752 - 753		
		MAT.	INST.	TOTAL	MAT.	INST.	TOTAL	MAT.	INST.	TOTAL	MAT.	INST.	TOTAL	MAT.	INST.	TOTAL	MAT.	INST.	TOTAL
015433	CONTRACTOR EQUIPMENT		89.1	89.1		88.0	88.0		89.1	89.1		88.0	88.0		94.9	94.9		97.9	97.9
0241, 31 - 34	SITE & INFRASTRUCTURE, DEMOLITION	96.9	86.0	89.3	107.6	84.1	91.3	86.9	86.9	86.9	112.9	84.7	93.3	123.4	81.4	94.2	115.9	87.1	95.8
0310	Concrete Forming & Accessories	102.2	52.5	59.2	94.3	29.3	38.1	81.8	43.3	48.5	94.3	53.1	58.7	97.0	39.5	47.3	94.6	58.7	63.6
0320	Concrete Reinforcing	98.3	44.6	70.9	96.9	26.5	60.9	101.0	50.3	75.1	96.7	41.7	68.6	92.9	49.5	70.8	94.6	53.9	73.8
0330	Cast-in-Place Concrete	98.0	54.7	81.6	107.9	38.9	81.8	78.5	68.1	74.6	102.6	48.1	82.0	95.0	47.0	76.8	102.0	57.8	85.3
03	CONCRETE	94.8	52.7	75.1	100.5	33.4	69.1	79.3	54.3	67.6	101.2	50.0	77.3	88.0	45.9	68.3	94.0	59.0	77.6
04	MASONRY	99.6	59.4	75.4	128.1	45.4	78.4	133.5	59.7	89.1	104.9	48.3	70.9	82.9	51.9	64.3	106.1	60.9	79.0
05	METALS	99.5	66.0	88.2	98.4	51.9	82.7	99.1	68.8	88.9	98.1	60.8	85.5	96.4	76.2	89.6	100.4	80.7	93.8
06	WOOD, PLASTICS & COMPOSITES	115.8	52.9	80.2	100.8	28.5	59.8	81.2	36.6	55.9	101.5	58.0	76.9	112.7	39.2	71.1	103.0	59.1	78.1
07	THERMAL & MOISTURE PROTECTION	106.5	58.4	88.3	98.2	40.1	76.3	98.8	58.3	83.5	98.7	45.2	78.5	96.7	47.7	78.2	89.8	63.3	79.7
08	OPENINGS	95.7	49.3	84.2	89.7	29.2	74.7	97.3	44.7	84.3	90.7	49.6	80.5	107.6	40.6	91.0	103.3	54.8	91.3
0920	Plaster & Gypsum Board	98.0	52.1	66.2	83.1	26.9	44.2	87.1	35.3	51.3	83.1	57.3	65.2	89.9	37.8	53.8	89.1	58.3	67.8
0950, 0980	Ceilings & Acoustic Treatment	103.0	52.1	70.6	88.6	26.9	49.3	96.7	35.3	57.6	88.6	57.3	68.7	87.4	37.8	55.8	94.2	58.3	71.3
0960	Flooring	110.7	72.7	100.2	110.4	36.1	89.8	82.4	56.8	75.3	109.3	44.1	91.2	106.6	43.0	89.0	100.1	51.4	86.7
0970, 0990	Wall Finishes & Painting/Coating	94.5	52.9	69.4	97.2	31.8	57.9	89.8	61.5	72.8	97.2	42.0	64.0	108.1	62.8	80.8	104.2	55.5	74.9
09	FINISHES	93.5	56.0	73.2	95.0	30.5	60.0	81.1	45.9	62.0	95.3	50.9	71.2	95.2	41.9	66.3	98.4	56.8	75.8
COVERS	DIVS. 10 - 14, 25, 28, 41, 43, 44	100.0	80.0	95.7	100.0	39.3	87.1	100.0	78.9	95.5	100.0	76.7	95.0	100.0	77.1	95.1	100.0	81.4	96.0
21, 22, 23	FIRE SUPPRESSION, PLUMBING & HVAC	100.1	62.6	84.7	93.4	27.6	66.4	93.3	69.0	83.3	93.4	50.7	75.8	100.1	42.3	76.3	100.0	64.8	85.5
26, 27, 3370	ELECTRICAL, COMMUNICATIONS & UTIL.	95.2	62.5	78.1	94.3	29.6	60.5	93.4	67.6	79.9	98.6	40.4	68.2	96.2	53.3	73.8	97.0	71.4	83.6
MF2004	WEIGHTED AVERAGE	98.1	62.1	82.2	97.9	38.6	71.7	94.4	63.0	80.6	97.4	53.4	78.0	97.8	52.7	77.9	99.6	66.8	85.2

		TEXAS																	
	DIVISION	DEL RIO 788			DENTON 762			EASTLAND 764			EL PASO 798 - 799,885			FORT WORTH 760 - 761			GALVESTON 775		
		MAT.	INST.	TOTAL	MAT.	INST.	TOTAL	MAT.	INST.	TOTAL	MAT.	INST.	TOTAL	MAT.	INST.	TOTAL	MAT.	INST.	TOTAL
015433	CONTRACTOR EQUIPMENT		87.2	87.2		94.3	94.3		88.0	88.0		88.0	88.0		88.0	88.0		98.6	98.6
0241, 31 - 34	SITE & INFRASTRUCTURE, DEMOLITION	110.1	83.7	91.7	107.3	78.2	87.0	110.8	84.4	92.4	98.9	85.1	89.3	103.3	86.0	91.3	122.8	84.8	96.3
0310	Concrete Forming & Accessories	91.4	26.6	35.3	99.5	35.1	43.8	95.0	34.5	42.6	95.1	47.2	53.6	91.9	58.1	62.7	90.0	64.7	68.1
0320	Concrete Reinforcing	93.7	18.7	55.4	98.5	50.2	73.8	97.2	37.3	66.6	96.6	49.4	72.5	96.6	53.8	74.7	100.4	64.6	82.1
0330	Cast-in-Place Concrete	102.2	34.8	76.7	83.4	46.7	69.5	114.1	39.7	85.9	91.8	42.6	73.2	104.3	53.5	85.0	104.3	69.2	91.0
03	CONCRETE	103.7	29.2	68.9	79.3	43.9	62.7	105.1	38.0	73.7	89.0	46.9	69.3	94.9	56.4	76.9	96.5	67.8	83.1
04	MASONRY	96.3	45.3	65.7	136.0	50.7	84.7	99.2	43.4	65.7	98.1	50.5	69.5	95.2	60.9	74.6	94.6	59.8	73.7
05	METALS	95.2	47.5	79.1	98.0	76.2	90.6	98.2	57.1	84.3	100.5	64.1	88.2	97.8	68.9	88.0	100.7	88.5	96.6
06	WOOD, PLASTICS & COMPOSITES	93.1	26.5	55.4	110.6	34.7	67.6	108.0	35.6	67.0	101.7	49.9	72.4	103.5	58.9	78.2	99.2	65.1	79.9
07	THERMAL & MOISTURE PROTECTION	93.3	35.3	71.4	96.8	47.1	78.0	98.6	39.9	76.4	94.4	54.5	79.3	99.1	55.1	82.5	97.9	66.3	86.0
08	OPENINGS	101.3	22.5	81.8	104.4	39.6	88.3	78.5	34.0	67.5	93.7	45.9	81.9	88.2	54.8	79.9	102.4	64.0	92.9
0920	Plaster & Gypsum Board	85.4	24.9	43.5	86.5	33.2	49.6	83.1	34.3	49.3	82.3	49.0	59.3	83.2	58.3	66.0	93.4	64.6	73.5
0950, 0980	Ceilings & Acoustic Treatment	84.0	24.9	46.3	92.9	33.2	54.9	88.6	34.3	54.0	91.3	49.0	64.4	93.0	58.3	70.9	99.3	64.6	77.2
0960	Flooring	91.3	20.2	71.7	102.7	38.4	85.0	143.5	30.3	112.2	111.6	67.2	99.3	144.9	43.8	117.0	96.9	56.8	85.9
0970, 0990	Wall Finishes & Painting/Coating	95.3	28.0	54.8	107.9	32.2	62.3	98.6	33.2	59.3	97.2	38.0	61.6	98.6	55.5	72.7	102.0	51.1	71.4
09	FINISHES	87.8	25.7	54.1	91.6	34.9	60.8	105.7	33.9	66.7	95.2	50.2	70.7	106.6	55.1	78.7	91.4	61.5	75.2
COVERS	DIVS. 10 - 14, 25, 28, 41, 43, 44	100.0	37.5	86.7	100.0	40.9	87.5	100.0	40.3	87.3	100.0	78.1	95.4	100.0	81.0	96.0	100.0	84.2	96.7
21, 22, 23	FIRE SUPPRESSION, PLUMBING & HVAC	93.2	20.1	63.1	93.4	42.0	72.3	93.4	28.5	66.7	100.1	37.4	74.3	100.1	60.0	83.6	93.2	68.4	83.0
26, 27, 3370	ELECTRICAL, COMMUNICATIONS & UTIL.	98.0	21.1	57.8	96.8	48.0	71.3	94.2	32.3	61.9	99.7	54.6	76.1	99.6	63.5	80.7	95.0	68.9	81.3
MF2004	WEIGHTED AVERAGE	96.9	33.7	69.0	97.1	49.2	76.0	96.8	40.8	72.1	97.4	52.7	77.6	98.2	62.7	82.6	97.4	70.1	85.3

		TEXAS																	
	DIVISION	GIDDINGS 789			GREENVILLE 754			HOUSTON 770 - 772			HUNTSVILLE 773			LAREDO 780			LONGVIEW 756		
		MAT.	INST.	TOTAL	MAT.	INST.	TOTAL	MAT.	INST.	TOTAL	MAT.	INST.	TOTAL	MAT.	INST.	TOTAL	MAT.	INST.	TOTAL
015433	CONTRACTOR EQUIPMENT		87.2	87.2		95.4	95.4		98.4	98.4		89.1	89.1		87.2	87.2		89.8	89.8
0241, 31 - 34	SITE & INFRASTRUCTURE, DEMOLITION	96.7	84.1	87.9	108.3	82.0	90.0	120.3	84.5	95.4	106.3	84.5	91.1	90.6	85.5	87.1	104.1	89.4	93.9
0310	Concrete Forming & Accessories	89.8	34.5	42.0	87.3	29.1	37.0	91.8	64.8	68.4	87.9	32.6	40.1	91.5	39.7	46.7	84.5	30.4	37.7
0320	Concrete Reinforcing	94.3	37.1	65.1	95.1	26.8	60.2	100.1	64.7	82.0	101.3	29.3	64.5	93.7	49.4	71.1	93.9	24.5	58.4
0330	Cast-in-Place Concrete	86.6	39.7	68.9	92.9	38.6	72.4	101.1	69.3	89.0	108.0	43.4	83.6	73.1	59.5	68.0	108.0	36.3	80.9
03	CONCRETE	84.4	37.8	62.6	86.9	34.0	62.1	94.2	67.8	81.8	102.3	37.0	71.8	79.5	49.4	65.4	101.8	32.5	69.4
04	MASONRY	103.2	46.1	68.9	165.9	49.4	95.9	94.4	61.0	74.3	133.4	31.4	72.1	90.6	49.7	66.0	161.4	46.1	92.1
05	METALS	94.6	54.7	81.1	97.6	63.9	86.2	103.7	88.6	98.6	99.0	54.7	84.0	97.8	63.9	86.4	91.4	51.0	77.8
06	WOOD, PLASTICS & COMPOSITES	92.2	35.6	60.1	94.6	28.5	57.2	101.4	65.1	80.9	89.4	33.5	57.7	93.1	39.1	62.5	88.4	30.9	55.8
07	THERMAL & MOISTURE PROTECTION	94.0	40.9	73.9	89.6	42.1	71.6	97.8	66.7	86.1	99.6	38.0	76.3	92.4	50.6	76.6	90.5	36.9	70.3
08	OPENINGS	100.5	32.6	83.7	102.0	28.2	83.7	104.7	64.0	94.6	97.3	30.1	80.6	101.2	41.2	86.3	92.6	28.2	76.6
0920	Plaster & Gypsum Board	84.1	34.3	49.6	84.7	26.9	44.7	96.0	64.6	74.3	91.4	32.1	50.4	86.4	37.8	52.8	81.7	29.4	45.5
0950, 0980	Ceilings & Acoustic Treatment	84.0	34.3	52.3	90.8	26.9	50.1	103.6	64.6	78.7	96.7	32.1	55.6	87.4	37.8	55.8	86.5	29.4	50.2
0960	Flooring	91.0	30.3	74.2	95.6	42.1	80.8	98.2	56.8	86.8	86.2	30.3	70.7	91.2	43.0	77.9	101.4	37.1	83.6
0970, 0990	Wall Finishes & Painting/Coating	95.3	33.2	57.9	104.2	27.3	57.9	102.0	61.5	77.6	89.8	34.1	56.3	95.3	52.1	69.3	96.0	31.9	57.4
09	FINISHES	86.7	33.7	57.9	95.2	31.1	60.4	97.8	62.7	78.8	83.9	32.4	56.0	87.3	40.6	62.0	100.0	31.7	63.0
COVERS	DIVS. 10 - 14, 25, 28, 41, 43, 44	100.0	42.4	87.8	100.0	39.0	87.1	100.0	84.2	96.7	100.0	42.0	87.7	100.0	74.8	94.6	100.0	24.5	84.0
21, 22, 23	FIRE SUPPRESSION, PLUMBING & HVAC	93.3	28.9	66.8	93.2	27.1	66.0	100.0	69.2	87.3	93.3	28.0	66.4	100.0	40.9	75.7	93.1	27.2	66.0
26, 27, 3370	ELECTRICAL, COMMUNICATIONS & UTIL.	94.6	36.9	64.5	93.5	29.7	60.1	97.0	67.6	81.7	93.4	36.2	63.5	98.1	62.4	79.4	94.0	48.1	70.0
MF2004	WEIGHTED AVERAGE	93.9	41.6	70.8	98.8	40.0	72.9	100.0	70.3	86.9	98.1	39.3	72.2	94.9	53.0	76.4	98.9	41.0	73.4

	DIVISION	TEXAS																	
		LUBBOCK			LUFKIN			MCALLEN			MCKINNEY			MIDLAND			ODESSA		
		793 - 794			759			785			750			797			797		
		MAT.	INST.	TOTAL	MAT.	INST.	TOTAL	MAT.	INST.	TOTAL	MAT.	INST.	TOTAL	MAT.	INST.	TOTAL	MAT.	INST.	TOTAL
015433	CONTRACTOR EQUIPMENT		97.0	97.0		89.8	89.8		94.9	94.9		95.4	95.4		97.0	97.0		88.0	88.0
0241, 31 - 34	SITE & INFRASTRUCTURE, DEMOLITION	127.9	83.9	97.3	98.3	89.5	92.2	127.7	81.4	95.5	103.9	82.6	89.1	131.9	83.9	98.5	100.9	86.0	90.5
0310	Concrete Forming & Accessories	93.7	44.5	51.1	87.3	39.7	46.2	96.4	37.3	45.3	86.4	39.4	45.7	97.0	41.1	48.7	95.7	41.0	48.4
0320	Concrete Reinforcing	97.9	53.7	75.3	95.5	32.7	63.4	93.1	49.5	70.9	95.1	50.2	72.2	99.1	53.5	75.8	96.6	53.5	74.6
0330	Cast-in-Place Concrete	100.2	50.0	81.2	96.6	41.4	75.7	103.2	45.4	81.3	87.1	47.5	72.1	106.5	47.1	84.0	100.0	46.0	79.6
03	CONCRETE	92.0	50.0	72.3	94.1	40.1	68.9	93.4	44.3	70.5	82.4	46.0	65.4	96.5	47.5	73.6	93.1	46.3	71.2
04	MASONRY	99.8	48.5	69.0	123.7	37.2	71.7	95.9	51.9	69.5	178.5	52.1	102.5	117.8	46.0	74.6	100.5	45.9	67.7
05	METALS	104.1	80.9	96.2	97.7	57.5	84.1	95.1	76.0	88.7	97.5	75.0	89.9	102.2	80.3	94.8	100.0	67.7	89.1
06	WOOD, PLASTICS & COMPOSITES	101.1	45.4	69.6	97.3	40.7	65.2	111.3	36.5	69.0	93.2	39.6	62.8	105.5	41.2	69.1	102.1	41.1	67.6
07	THERMAL & MOISTURE PROTECTION	88.6	50.8	74.3	90.3	40.8	71.6	96.7	46.0	77.5	89.4	49.2	74.2	88.8	46.5	72.8	98.1	45.7	78.3
08	OPENINGS	104.0	45.7	89.5	74.9	37.8	65.7	105.9	39.7	89.5	102.0	42.1	87.2	102.8	43.0	88.0	93.7	42.9	81.2
0920	Plaster & Gypsum Board	84.6	44.2	56.7	81.1	39.5	52.3	89.9	35.1	51.9	83.4	38.3	52.2	86.0	39.9	54.1	84.1	39.9	53.5
0950, 0980	Ceilings & Acoustic Treatment	94.7	44.2	62.6	78.7	39.5	53.8	87.4	35.1	54.1	90.8	38.3	57.4	91.2	39.9	58.5	93.0	39.9	59.2
0960	Flooring	103.6	41.6	86.5	135.7	40.4	109.3	106.0	67.9	95.5	95.2	37.6	79.3	105.2	41.1	87.5	111.6	41.1	92.1
0970, 0990	Wall Finishes & Painting/Coating	108.4	35.5	64.5	96.0	41.1	63.0	108.1	31.4	61.9	104.2	44.8	68.5	108.4	35.0	64.3	97.2	35.5	60.1
09	FINISHES	98.1	42.3	67.8	108.5	39.8	71.2	95.3	41.9	66.4	94.6	39.4	64.6	98.4	39.8	66.5	95.9	39.7	65.4
COVERS	DIVS. 10 - 14, 25, 28, 41, 43, 44	100.0	76.5	95.0	100.0	75.9	94.9	100.0	74.7	94.6	100.0	45.3	88.4	100.0	77.8	95.3	100.0	77.5	95.2
21, 22, 23	FIRE SUPPRESSION, PLUMBING & HVAC	99.7	45.7	77.5	93.1	33.2	68.5	93.3	32.1	68.1	93.2	42.7	72.4	92.9	37.4	70.1	100.2	37.3	74.3
26, 27, 3370	ELECTRICAL, COMMUNICATIONS & UTIL.	97.2	46.9	70.9	95.5	43.5	68.3	95.4	37.5	65.1	93.5	49.5	70.5	97.2	44.2	69.5	98.7	44.1	70.2
MF2004	WEIGHTED AVERAGE	99.9	53.8	79.6	95.8	45.6	73.6	97.2	48.0	75.5	98.8	51.2	77.8	99.6	50.5	77.9	98.1	49.3	76.5

	DIVISION	TEXAS																	
		PALESTINE			SAN ANGELO			SAN ANTONIO			TEMPLE			TEXARKANA			TYLER		
		758			769			781 - 782			765			755			757		
		MAT.	INST.	TOTAL	MAT.	INST.	TOTAL	MAT.	INST.	TOTAL	MAT.	INST.	TOTAL	MAT.	INST.	TOTAL	MAT.	INST.	TOTAL
015433	CONTRACTOR EQUIPMENT		89.8	89.8		88.0	88.0		89.8	89.8		88.0	88.0		89.8	89.8		89.8	89.8
0241, 31 - 34	SITE & INFRASTRUCTURE, DEMOLITION	106.1	89.2	94.3	103.5	86.2	91.5	90.3	89.9	90.0	90.6	84.9	86.6	91.8	90.2	90.7	103.6	90.2	94.3
0310	Concrete Forming & Accessories	79.8	34.4	40.5	94.5	38.9	46.4	91.5	53.3	58.4	97.5	36.8	45.0	93.5	36.7	44.4	88.8	35.5	42.7
0320	Concrete Reinforcing	93.2	37.9	65.0	96.8	53.5	74.7	100.1	48.9	74.0	96.9	51.1	73.6	93.1	53.5	72.9	93.9	53.7	73.4
0330	Cast-in-Place Concrete	88.3	39.0	69.6	101.8	44.3	80.0	71.7	65.8	69.5	83.4	50.2	70.8	89.0	45.4	72.5	106.1	44.3	82.7
03	CONCRETE	94.6	37.7	68.0	95.7	44.6	71.8	79.7	57.6	69.4	81.8	45.3	64.7	87.4	44.1	67.2	101.2	43.2	74.1
04	MASONRY	119.0	43.8	73.8	124.8	45.3	77.0	90.5	58.3	71.1	135.5	51.4	85.0	183.1	48.9	102.4	172.2	51.9	99.9
05	METALS	97.5	54.0	82.8	98.6	65.2	87.3	98.6	66.8	87.8	98.2	64.3	86.8	91.3	67.4	83.2	97.3	67.8	87.4
06	WOOD, PLASTICS & COMPOSITES	88.1	35.6	58.4	101.0	39.3	66.1	93.1	51.6	69.6	110.5	35.6	68.1	99.0	35.1	62.8	98.8	33.1	61.6
07	THERMAL & MOISTURE PROTECTION	90.7	40.7	71.8	98.1	45.3	78.1	92.4	63.3	81.4	97.7	47.0	78.5	90.2	49.7	74.9	90.6	54.3	76.9
08	OPENINGS	74.8	33.3	64.6	89.7	41.9	77.9	103.1	51.7	90.4	76.3	38.1	66.9	92.5	40.5	79.7	74.8	39.4	66.0
0920	Plaster & Gypsum Board	76.8	34.3	47.4	83.1	38.0	51.9	86.4	50.6	61.6	83.5	34.3	49.4	85.3	33.8	49.7	81.1	31.7	46.9
0950, 0980	Ceilings & Acoustic Treatment	78.7	34.3	50.4	88.6	38.0	56.4	87.4	50.6	64.0	88.6	34.3	54.0	85.6	33.8	52.6	78.7	31.7	48.8
0960	Flooring	126.4	30.3	99.9	110.5	36.7	90.1	91.2	68.1	84.8	146.4	44.1	118.2	110.0	62.8	96.9	137.5	39.2	110.4
0970, 0990	Wall Finishes & Painting/Coating	96.0	33.2	58.2	97.2	36.6	60.8	95.3	52.1	69.3	98.6	36.6	61.3	96.0	42.8	64.0	96.0	42.8	64.0
09	FINISHES	105.8	33.7	66.6	94.7	37.5	63.7	87.3	54.8	69.7	105.3	37.4	68.4	102.3	41.3	69.2	109.4	35.3	69.2
COVERS	DIVS. 10 - 14, 25, 28, 41, 43, 44	100.0	39.6	87.2	100.0	74.7	94.6	100.0	79.4	95.6	100.0	40.4	87.4	100.0	74.8	94.6	100.0	74.5	94.6
21, 22, 23	FIRE SUPPRESSION, PLUMBING & HVAC	93.1	28.2	66.4	93.4	38.1	70.6	100.0	65.4	85.8	93.4	35.3	69.5	93.1	36.3	69.8	93.1	60.0	79.5
26, 27, 3370	ELECTRICAL, COMMUNICATIONS & UTIL.	91.2	36.6	62.7	98.6	50.2	73.3	98.5	62.4	79.6	95.5	71.1	82.8	95.3	67.0	80.5	94.0	56.0	74.1
MF2004	WEIGHTED AVERAGE	95.1	41.5	71.4	97.4	49.3	76.2	95.3	63.7	81.3	95.1	51.0	75.6	98.3	52.6	78.1	99.3	55.6	80.0

	DIVISION	TEXAS															UTAH		
		VICTORIA			WACO			WAXAHACKIE			WHARTON			WICHITA FALLS			LOGAN		
		779			766 - 767			751			774			763			843		
		MAT.	INST.	TOTAL	MAT.	INST.	TOTAL	MAT.	INST.	TOTAL	MAT.	INST.	TOTAL	MAT.	INST.	TOTAL	MAT.	INST.	TOTAL
015433	CONTRACTOR EQUIPMENT		97.4	97.4		88.0	88.0		95.4	95.4		98.6	98.6		88.0	88.0		99.9	99.9
0241, 31 - 34	SITE & INFRASTRUCTURE, DEMOLITION	128.8	80.7	95.3	99.5	86.0	90.1	106.2	83.2	90.2	136.2	82.5	98.8	100.3	85.6	90.0	87.6	98.4	95.2
0310	Concrete Forming & Accessories	90.8	38.2	45.3	96.2	42.3	49.6	86.4	47.6	52.9	85.5	34.8	41.7	96.2	40.2	47.8	104.8	52.9	59.9
0320	Concrete Reinforcing	96.4	32.7	63.9	96.6	50.2	72.9	95.1	52.4	73.3	100.3	55.2	77.3	96.6	53.4	74.5	98.7	70.4	84.3
0330	Cast-in-Place Concrete	117.1	42.9	89.1	90.2	54.7	76.8	91.9	43.2	73.5	120.4	47.3	92.7	96.4	46.8	77.6	88.4	68.2	80.7
03	CONCRETE	104.3	41.0	74.7	88.3	49.2	70.0	86.0	48.7	68.6	108.4	45.1	78.8	91.3	46.2	70.2	110.3	62.1	87.8
04	MASONRY	109.6	37.3	66.2	97.4	58.8	74.2	166.6	52.1	97.8	95.9	32.7	57.9	97.9	59.7	74.9	110.5	55.7	77.6
05	METALS	99.2	72.4	90.2	100.7	66.8	89.2	97.6	78.0	91.0	100.7	78.6	93.2	100.6	69.5	90.1	102.7	72.2	92.4
06	WOOD, PLASTICS & COMPOSITES	103.4	39.1	67.0	108.8	38.2	68.8	93.2	50.4	68.9	93.5	35.4	60.6	108.8	38.1	68.8	84.4	49.9	64.9
07	THERMAL & MOISTURE PROTECTION	101.6	42.3	79.2	98.6	50.6	80.4	89.5	47.1	73.5	98.2	46.4	78.6	98.6	53.9	81.7	96.9	61.4	83.5
08	OPENINGS	102.2	38.3	86.4	88.2	39.2	76.0	102.0	48.5	88.8	102.4	40.3	87.0	88.2	43.5	77.1	88.6	52.2	79.6
0920	Plaster & Gypsum Board	92.0	37.7	54.4	84.1	36.9	51.5	83.6	49.4	59.9	90.0	34.0	51.2	84.1	36.8	51.4	84.2	48.2	59.3
0950, 0980	Ceilings & Acoustic Treatment	100.1	37.7	60.4	93.0	36.9	57.3	91.6	49.4	64.7	99.3	34.0	57.7	93.0	36.8	57.2	121.8	48.2	75.0
0960	Flooring	95.8	40.4	80.5	144.9	38.2	115.4	95.2	40.4	80.1	94.1	30.3	76.5	146.1	77.9	127.2	103.6	51.4	89.2
0970, 0990	Wall Finishes & Painting/Coating	102.2	41.1	65.4	98.6	36.6	61.3	104.2	41.1	66.2	102.0	34.1	61.1	102.1	49.8	70.6	97.9	41.7	64.1
09	FINISHES	89.8	39.0	62.2	106.4	39.7	70.2	95.0	45.7	68.3	91.1	33.7	59.9	107.0	47.4	74.6	105.4	50.6	75.6
COVERS	DIVS. 10 - 14, 25, 28, 41, 43, 44	100.0	41.6	87.6	100.0	78.6	95.5	100.0	43.5	88.0	100.0	40.2	87.3	100.0	75.0	94.7	100.0	60.0	91.5
21, 22, 23	FIRE SUPPRESSION, PLUMBING & HVAC	93.3	33.0	68.5	100.2	54.0	81.2	93.2	44.9	73.3	93.2	27.4	66.1	100.2	48.1	78.8	100.0	68.3	86.9
26, 27, 3370	ELECTRICAL, COMMUNICATIONS & UTIL.	100.0	43.5	70.5	98.7	71.3	84.4	93.5	71.4	81.9	98.9	33.3	64.6	100.7	61.1	80.0	94.5	64.2	78.7
MF2004	WEIGHTED AVERAGE	99.6	45.2	75.6	97.8	58.1	80.3	98.7	56.4	80.0	99.6	42.9	74.6	98.4	56.5	79.9	100.3	65.0	84.7

DIVISION		UTAH												VERMONT					
		OGDEN			PRICE			PROVO			SALT LAKE CITY			BELLOWS FALLS			BENNINGTON		
		842,844			845			846 - 847			840 - 841			051			052		
		MAT.	INST.	TOTAL	MAT.	INST.	TOTAL	MAT.	INST.	TOTAL	MAT.	INST.	TOTAL	MAT.	INST.	TOTAL	MAT.	INST.	TOTAL
015433	CONTRACTOR EQUIPMENT		99.9	99.9		98.8	98.8		98.8	98.8		99.9	99.9		100.6	100.6		100.6	100.6
0241, 31 - 34	SITE & INFRASTRUCTURE, DEMOLITION	76.6	98.4	91.8	85.5	95.4	92.4	84.2	96.7	92.9	76.3	98.4	91.7	74.2	97.2	90.2	73.5	97.9	90.5
0310	Concrete Forming & Accessories	104.9	52.9	59.9	106.7	28.4	39.0	106.0	53.0	60.1	107.4	53.0	60.3	97.7	41.8	49.4	95.6	52.4	58.3
0320	Concrete Reinforcing	98.3	70.4	84.1	105.8	65.8	85.4	106.7	70.4	88.2	100.5	70.4	85.1	88.5	66.5	77.3	88.5	66.4	77.2
0330	Cast-in-Place Concrete	89.7	68.2	81.6	88.5	42.5	71.1	88.5	68.2	80.8	98.1	68.2	86.8	90.9	60.0	79.2	90.9	65.6	81.3
03	CONCRETE	99.2	62.1	81.8	111.5	41.6	78.8	109.8	62.1	87.5	119.8	62.1	92.8	99.9	53.6	78.3	99.8	60.1	81.3
04	MASONRY	104.3	55.7	75.1	115.7	35.7	67.6	115.8	55.7	79.7	118.0	55.7	80.6	95.9	33.0	58.1	105.4	49.4	71.7
05	METALS	103.2	72.2	92.7	99.8	68.8	89.4	100.8	72.3	91.2	108.1	72.3	96.0	93.9	70.1	85.9	93.9	69.1	85.5
06	WOOD, PLASTICS & COMPOSITES	84.4	49.9	64.9	86.9	26.5	52.7	85.7	49.9	65.4	86.0	49.9	65.6	99.7	40.3	66.1	96.9	47.2	68.8
07	THERMAL & MOISTURE PROTECTION	95.8	61.4	82.8	98.5	48.3	79.6	98.5	61.4	84.5	102.3	61.4	86.8	95.4	49.7	78.1	95.4	50.4	78.4
08	OPENINGS	88.6	52.2	79.6	92.8	37.5	79.1	92.9	52.2	82.8	90.5	52.2	81.0	101.0	43.2	86.7	101.0	45.7	87.3
0920	Plaster & Gypsum Board	84.2	48.2	59.3	85.9	24.2	43.2	84.6	48.2	59.4	87.0	48.2	60.2	98.1	37.8	56.4	96.0	44.9	60.6
0950, 0980	Ceilings & Acoustic Treatment	121.8	48.2	75.0	121.8	24.2	59.7	121.8	48.2	75.0	116.1	48.2	72.9	89.0	37.8	56.4	89.0	44.9	60.9
0960	Flooring	101.3	51.4	87.5	104.8	38.2	86.4	104.4	51.4	89.7	104.4	51.4	89.8	97.6	32.8	79.7	96.3	32.8	78.8
0970, 0990	Wall Finishes & Painting/Coating	97.9	41.7	64.1	97.9	41.7	64.1	97.9	53.3	71.1	100.6	53.3	72.1	94.6	29.7	55.5	94.6	29.7	55.5
09	FINISHES	103.5	50.6	74.8	106.5	30.3	65.2	106.1	51.9	76.7	104.2	51.9	75.8	95.9	37.3	64.1	95.2	45.4	68.2
COVERS	DIVS. 10 - 14, 25, 28, 41, 43, 44	100.0	60.0	91.5	100.0	44.7	88.3	100.0	60.0	91.5	100.0	60.0	91.5	100.0	40.9	87.5	100.0	47.5	88.8
21, 22, 23	FIRE SUPPRESSION, PLUMBING & HVAC	100.0	68.3	86.9	96.5	33.5	70.6	100.0	68.3	86.9	100.1	68.3	87.0	93.4	87.4	90.9	93.4	90.2	92.1
26, 27, 3370	ELECTRICAL, COMMUNICATIONS & UTIL.	94.9	64.2	78.9	100.4	38.1	67.8	95.2	67.0	80.5	97.9	67.0	81.8	97.7	100.1	99.0	97.7	66.0	81.2
MF2004	WEIGHTED AVERAGE	98.2	65.0	83.6	100.7	44.4	75.8	100.8	65.4	85.2	103.0	65.5	86.5	95.9	67.2	83.2	96.3	67.0	83.4

DIVISION		VERMONT																	
		BRATTLEBORO			BURLINGTON			GUILDHALL			MONTPELIER			RUTLAND			ST. JOHNSBURY		
		053			054			059			056			057			058		
		MAT.	INST.	TOTAL	MAT.	INST.	TOTAL	MAT.	INST.	TOTAL	MAT.	INST.	TOTAL	MAT.	INST.	TOTAL	MAT.	INST.	TOTAL
015433	CONTRACTOR EQUIPMENT		100.6	100.6		100.6	100.6		100.6	100.6		100.6	100.6		100.6	100.6		100.6	100.6
0241, 31 - 34	SITE & INFRASTRUCTURE, DEMOLITION	74.8	98.2	91.1	76.5	97.3	91.0	73.4	96.6	89.6	77.6	97.3	91.3	76.6	97.3	91.1	73.5	96.8	89.7
0310	Concrete Forming & Accessories	97.9	47.8	54.6	95.2	53.4	59.1	95.9	45.8	52.6	99.5	54.0	60.1	98.2	54.0	59.9	94.5	45.9	52.5
0320	Concrete Reinforcing	87.6	66.4	76.8	110.6	83.9	97.0	89.3	39.3	63.8	88.1	83.9	86.0	110.6	83.9	97.0	87.6	44.8	65.8
0330	Cast-in-Place Concrete	93.7	66.0	83.2	95.1	103.3	98.2	88.1	56.5	76.1	100.0	103.5	101.3	89.0	103.5	94.5	88.1	56.8	76.3
03	CONCRETE	102.4	58.3	81.7	106.1	76.8	92.4	97.4	49.0	74.8	105.6	77.1	92.3	103.3	77.1	91.1	97.1	50.1	75.1
04	MASONRY	105.2	49.5	71.7	107.6	51.7	74.0	105.4	37.6	64.7	91.0	51.7	67.4	86.6	51.7	65.6	130.9	37.7	74.9
05	METALS	93.9	69.9	85.8	101.4	78.9	93.8	93.9	58.3	81.9	93.9	79.5	89.0	99.5	79.5	92.8	93.9	58.8	82.1
06	WOOD, PLASTICS & COMPOSITES	100.0	40.3	66.2	95.0	53.4	71.4	96.4	47.2	68.5	99.1	53.4	73.2	100.1	53.4	73.7	92.7	47.2	66.9
07	THERMAL & MOISTURE PROTECTION	95.4	50.5	78.5	95.5	53.5	79.6	95.2	40.7	74.6	94.9	61.6	82.3	95.5	61.6	82.7	95.2	40.7	74.6
08	OPENINGS	101.0	43.2	86.7	104.3	54.6	92.0	101.0	40.1	85.9	101.0	54.6	89.5	104.3	54.6	92.0	101.0	40.1	85.9
0920	Plaster & Gypsum Board	98.1	37.8	56.4	100.2	51.3	66.3	103.3	44.9	62.9	109.9	51.3	69.3	98.9	51.3	65.9	106.4	44.9	63.8
0950, 0980	Ceilings & Acoustic Treatment	89.0	37.8	56.4	92.7	51.3	66.3	89.0	44.9	60.9	89.0	51.3	65.0	93.5	51.3	66.6	89.0	44.9	60.9
0960	Flooring	97.7	32.8	79.8	97.8	59.4	87.2	101.1	32.8	82.2	105.0	59.4	92.4	97.6	59.4	87.0	105.0	32.8	85.0
0970, 0990	Wall Finishes & Painting/Coating	94.6	29.7	55.5	94.6	34.5	58.4	94.6	23.0	51.5	94.6	34.5	58.4	94.6	34.5	58.4	94.6	24.1	52.2
09	FINISHES	96.0	41.4	66.3	97.2	51.6	72.5	97.5	40.8	66.7	99.8	51.6	73.6	97.1	51.6	72.4	99.1	40.9	67.5
COVERS	DIVS. 10 - 14, 25, 28, 41, 43, 44	100.0	46.8	88.7	100.0	77.1	95.1	100.0	41.8	87.6	100.0	77.1	95.1	100.0	77.1	95.1	100.0	41.8	87.6
21, 22, 23	FIRE SUPPRESSION, PLUMBING & HVAC	93.4	93.5	93.4	100.0	65.5	85.8	93.4	63.8	81.2	93.2	65.6	81.9	100.1	65.6	85.9	93.4	63.8	81.2
26, 27, 3370	ELECTRICAL, COMMUNICATIONS & UTIL.	97.7	100.1	99.0	99.7	66.1	82.1	97.7	66.0	81.2	101.2	66.1	82.8	97.8	66.1	81.2	97.7	66.1	81.2
MF2004	WEIGHTED AVERAGE	96.7	71.5	85.6	100.6	67.3	85.9	96.2	56.4	78.6	97.1	67.7	84.1	98.8	67.7	85.0	97.6	56.6	79.5

DIVISION		VERMONT			VIRGINIA														
		WHITE RIVER JCT.			ALEXANDRIA			ARLINGTON			BRISTOL			CHARLOTTESVILLE			CULPEPER		
		050			223			222			242			229			227		
		MAT.	INST.	TOTAL	MAT.	INST.	TOTAL	MAT.	INST.	TOTAL	MAT.	INST.	TOTAL	MAT.	INST.	TOTAL	MAT.	INST.	TOTAL
015433	CONTRACTOR EQUIPMENT		100.6	100.6		102.2	102.2		100.8	100.8		100.8	100.8		105.3	105.3		100.8	100.8
0241, 31 - 34	SITE & INFRASTRUCTURE, DEMOLITION	76.9	96.7	90.7	117.5	88.4	97.2	128.2	85.2	98.3	111.7	84.8	93.0	117.0	86.9	96.1	115.4	85.6	94.6
0310	Concrete Forming & Accessories	93.4	44.1	50.8	92.6	76.2	78.4	92.5	73.8	76.3	88.8	46.4	52.1	87.4	52.9	57.5	85.1	72.5	74.2
0320	Concrete Reinforcing	88.5	39.4	63.4	89.7	81.1	85.3	101.3	75.0	87.9	101.3	69.8	85.2	100.6	72.8	86.4	101.3	68.9	84.8
0330	Cast-in-Place Concrete	93.7	56.2	79.6	110.6	82.3	99.9	107.6	79.0	96.8	107.2	50.8	85.8	111.5	59.4	91.8	110.2	69.3	94.7
03	CONCRETE	104.5	48.2	78.2	108.9	80.3	95.5	113.8	76.5	96.4	109.8	54.1	83.8	110.6	60.5	87.2	107.3	71.6	90.6
04	MASONRY	117.6	39.8	70.8	95.0	72.6	81.5	108.4	68.5	84.4	96.3	43.8	64.8	123.0	50.6	79.5	111.1	65.2	83.5
05	METALS	93.9	58.8	82.0	93.3	95.7	94.1	92.0	84.0	89.3	90.9	87.3	89.6	91.1	90.7	91.0	91.2	83.0	88.4
06	WOOD, PLASTICS & COMPOSITES	94.2	45.4	66.5	96.4	76.1	84.9	93.6	76.1	83.7	86.6	44.6	62.8	84.9	50.9	65.6	83.8	76.1	79.5
07	THERMAL & MOISTURE PROTECTION	95.5	36.5	73.2	98.3	81.7	92.0	100.1	71.2	89.2	99.8	50.6	81.2	99.4	56.4	83.2	99.6	65.1	86.6
08	OPENINGS	101.0	41.5	86.2	98.5	77.4	93.3	96.3	71.8	90.3	99.8	50.5	87.6	97.7	56.1	87.4	98.1	67.2	90.4
0920	Plaster & Gypsum Board	94.7	43.0	58.9	108.6	75.2	85.5	106.1	75.2	84.7	102.0	42.8	61.0	101.1	48.5	64.7	101.8	75.2	83.4
0950, 0980	Ceilings & Acoustic Treatment	89.0	43.0	59.7	96.1	75.2	82.8	93.5	75.2	81.9	92.6	42.8	60.9	92.6	48.5	64.5	93.5	75.2	81.9
0960	Flooring	95.3	32.8	78.0	110.9	86.0	104.0	109.0	59.3	95.2	104.6	40.3	86.8	103.0	47.5	87.6	103.0	59.3	90.9
0970, 0990	Wall Finishes & Painting/Coating	94.6	20.1	49.8	130.0	85.5	103.2	130.0	85.5	103.2	116.0	46.2	74.0	116.0	65.3	85.5	130.0	58.1	86.7
09	FINISHES	95.0	39.2	64.7	103.1	78.7	89.8	102.8	73.0	86.6	99.1	44.4	69.4	98.6	52.3	73.4	99.4	69.8	83.3
COVERS	DIVS. 10 - 14, 25, 28, 41, 43, 44	100.0	41.3	87.5	100.0	88.4	97.5	100.0	76.6	95.0	100.0	73.5	94.4	100.0	80.0	95.8	100.0	65.0	92.6
21, 22, 23	FIRE SUPPRESSION, PLUMBING & HVAC	93.4	63.4	81.0	100.2	87.9	95.2	100.2	85.5	94.2	93.5	56.8	78.4	93.5	60.2	79.8	93.5	66.4	82.3
26, 27, 3370	ELECTRICAL, COMMUNICATIONS & UTIL.	97.7	66.0	81.2	98.1	96.6	97.3	95.4	96.3	95.9	97.7	39.7	67.4	97.6	72.5	84.5	100.4	58.5	78.5
MF2004	WEIGHTED AVERAGE	97.6	56.1	79.3	100.3	85.4	93.7	101.2	81.0	92.3	98.0	56.3	79.6	99.4	65.2	84.3	98.8	69.6	85.9

	DIVISION	VIRGINIA																	
		FAIRFAX			FARMVILLE			FREDERICKSBURG			GRUNDY			HARRISONBURG			LYNCHBURG		
		220 - 221			239			224 - 225			246			228			245		
		MAT.	INST.	TOTAL	MAT.	INST.	TOTAL	MAT.	INST.	TOTAL	MAT.	INST.	TOTAL	MAT.	INST.	TOTAL	MAT.	INST.	TOTAL
015433	CONTRACTOR EQUIPMENT		100.8	100.8		105.3	105.3		100.8	100.8		100.8	100.8		100.8	100.8		100.8	100.8
0241, 31 - 34	SITE & INFRASTRUCTURE, DEMOLITION	127.3	85.8	98.4	112.8	86.2	94.3	114.9	85.8	94.6	109.4	83.6	91.4	124.2	84.9	96.8	110.4	84.9	92.7
0310	Concrete Forming & Accessories	87.5	74.2	76.0	98.0	48.1	54.8	87.5	52.4	57.2	91.5	40.1	47.0	84.1	46.4	51.5	88.9	66.5	69.5
0320	Concrete Reinforcing	101.3	81.0	90.9	101.3	55.5	77.9	102.1	80.0	90.8	99.9	56.1	77.5	101.3	50.5	75.3	100.6	70.5	85.2
0330	Cast-in-Place Concrete	107.6	79.6	97.0	111.5	57.2	91.0	109.2	55.8	89.0	107.2	51.2	86.0	107.6	62.6	90.6	107.2	62.4	90.3
03	CONCRETE	113.5	78.5	97.1	109.4	54.2	83.6	106.9	60.5	85.2	108.4	48.5	80.4	111.2	54.8	84.8	108.4	67.2	89.1
04	MASONRY	108.3	68.6	84.5	106.1	49.9	72.3	109.8	48.5	73.0	100.4	52.8	71.8	107.0	54.9	75.7	115.3	57.2	80.4
05	METALS	91.3	95.8	92.8	91.1	79.7	87.2	91.2	93.1	91.8	90.9	73.2	84.9	91.1	86.3	89.5	91.0	88.7	90.3
06	WOOD, PLASTICS & COMPOSITES	86.6	76.1	80.7	98.3	49.1	70.4	86.6	50.2	66.0	89.4	39.1	60.9	82.8	43.5	60.5	86.6	70.1	77.3
07	THERMAL & MOISTURE PROTECTION	100.1	71.6	89.3	99.5	51.0	81.2	99.6	53.5	82.2	99.7	50.3	81.1	99.9	63.7	86.2	99.6	58.7	84.1
08	OPENINGS	96.3	78.5	91.9	98.1	48.0	85.6	97.7	57.5	87.8	99.8	38.3	84.6	98.1	54.7	87.3	98.0	64.3	89.7
0920	Plaster & Gypsum Board	102.2	75.2	83.5	107.6	46.7	65.4	102.2	48.5	65.0	102.4	37.1	57.2	101.1	41.6	59.9	102.0	69.0	79.2
0950, 0980	Ceilings & Acoustic Treatment	93.5	75.2	81.9	92.6	46.7	63.4	93.5	48.5	64.8	92.6	37.1	57.2	92.6	41.6	60.1	92.6	69.0	77.6
0960	Flooring	104.9	59.3	92.3	110.9	64.9	98.2	104.9	79.7	98.0	106.3	36.3	87.0	102.5	79.7	96.2	104.6	45.0	88.1
0970, 0990	Wall Finishes & Painting/Coating	130.0	85.5	103.2	116.0	43.0	72.1	130.0	71.3	94.7	116.0	39.0	69.7	130.0	56.4	85.7	116.0	46.2	74.0
09	FINISHES	101.1	73.0	85.9	101.5	51.0	74.1	100.0	58.4	77.4	99.6	38.9	66.6	99.6	52.6	74.1	99.0	60.7	78.2
COVERS	DIVS. 10 - 14, 25, 28, 41, 43, 44	100.0	76.6	95.0	100.0	53.5	90.1	100.0	77.7	95.3	100.0	51.3	89.7	100.0	76.5	95.0	100.0	77.2	95.2
21, 22, 23	FIRE SUPPRESSION, PLUMBING & HVAC	93.5	85.8	90.3	93.5	38.7	71.0	93.5	83.1	89.2	93.5	43.5	72.9	93.5	56.8	78.4	93.5	59.6	79.5
26, 27, 3370	ELECTRICAL, COMMUNICATIONS & UTIL.	98.9	96.3	97.6	94.6	52.6	72.7	95.7	96.3	96.0	97.7	50.0	72.8	97.9	87.7	92.5	98.9	59.1	78.1
MF2004	WEIGHTED AVERAGE	99.6	82.8	92.2	98.4	54.5	79.0	98.2	73.9	87.5	98.1	51.7	77.6	99.0	65.7	84.3	98.8	66.1	84.3

	DIVISION	VIRGINIA																	
		NEWPORT NEWS			NORFOLK			PETERSBURG			PORTSMOUTH			PULASKI			RICHMOND		
		236			233 - 235			238			237			243			230 - 232		
		MAT.	INST.	TOTAL	MAT.	INST.	TOTAL	MAT.	INST.	TOTAL	MAT.	INST.	TOTAL	MAT.	INST.	TOTAL	MAT.	INST.	TOTAL
015433	CONTRACTOR EQUIPMENT		105.4	105.4		106.0	106.0		105.3	105.3		105.3	105.3		100.8	100.8		105.3	105.3
0241, 31 - 34	SITE & INFRASTRUCTURE, DEMOLITION	111.0	87.6	94.7	112.1	88.6	95.7	115.4	88.0	96.3	109.4	86.8	93.6	108.6	83.6	91.2	111.9	88.0	95.2
0310	Concrete Forming & Accessories	97.4	69.5	73.3	98.7	69.6	73.6	91.6	62.8	66.7	88.1	57.0	61.2	91.5	40.2	47.2	96.8	62.8	67.4
0320	Concrete Reinforcing	101.0	73.2	86.8	101.0	73.2	86.8	100.6	73.6	86.8	100.6	70.5	85.2	99.9	56.1	77.6	101.0	73.6	87.0
0330	Cast-in-Place Concrete	108.3	62.7	91.1	119.1	62.8	97.8	115.0	57.4	93.2	107.3	62.0	90.1	107.2	51.2	86.0	110.0	57.4	90.1
03	CONCRETE	106.3	69.2	88.9	111.7	69.2	91.8	112.2	64.4	89.8	105.1	62.7	85.3	108.4	48.6	80.5	107.0	64.4	87.1
04	MASONRY	101.1	57.0	74.6	107.0	57.0	77.0	114.5	58.7	81.0	106.4	56.8	76.6	92.7	52.8	68.7	99.3	58.7	74.9
05	METALS	93.3	91.6	92.7	92.3	91.7	92.1	91.2	92.2	91.5	92.3	88.1	90.9	90.9	73.4	85.0	95.5	92.2	94.4
06	WOOD, PLASTICS & COMPOSITES	97.2	74.1	84.1	99.2	74.1	85.0	89.9	65.5	76.0	86.1	57.4	69.9	89.4	39.1	60.9	96.3	65.5	78.8
07	THERMAL & MOISTURE PROTECTION	99.4	59.1	84.2	97.1	59.1	82.7	99.4	59.4	84.3	99.5	56.7	83.3	99.7	50.3	81.1	98.4	59.4	83.7
08	OPENINGS	98.5	67.7	90.8	98.5	67.7	90.8	97.7	64.0	89.4	98.5	58.0	88.5	99.8	38.3	84.6	98.5	64.0	89.9
0920	Plaster & Gypsum Board	108.6	72.4	83.5	105.3	72.4	82.5	102.6	63.5	75.6	101.7	55.3	69.6	102.4	37.1	57.2	114.9	63.5	79.3
0950, 0980	Ceilings & Acoustic Treatment	96.1	72.4	81.0	96.1	72.4	81.0	93.5	63.5	74.4	96.1	55.3	70.1	92.6	37.1	57.2	99.5	63.5	76.6
0960	Flooring	110.9	54.4	95.3	110.4	54.4	94.9	105.9	76.7	97.8	102.0	68.8	92.8	106.3	36.3	87.0	110.4	76.7	101.1
0970, 0990	Wall Finishes & Painting/Coating	116.0	46.7	74.3	116.0	65.7	85.7	116.0	65.3	85.5	116.0	65.7	85.7	116.0	39.0	69.7	116.0	65.3	85.5
09	FINISHES	102.3	64.8	81.9	101.6	66.9	82.7	99.8	65.8	81.3	98.7	59.6	77.5	99.6	38.9	66.6	103.4	65.8	83.0
COVERS	DIVS. 10 - 14, 25, 28, 41, 43, 44	100.0	82.5	96.3	100.0	82.5	96.3	100.0	80.3	95.8	100.0	80.5	95.9	100.0	51.3	89.7	100.0	80.3	95.8
21, 22, 23	FIRE SUPPRESSION, PLUMBING & HVAC	100.2	65.0	85.7	100.1	64.1	85.3	93.5	58.2	78.9	100.2	64.0	85.3	93.5	43.8	73.0	100.1	58.2	82.9
26, 27, 3370	ELECTRICAL, COMMUNICATIONS & UTIL.	97.7	61.3	78.7	104.8	60.9	81.9	97.9	72.5	84.6	95.8	60.9	77.6	97.7	54.0	74.9	104.2	72.5	87.6
MF2004	WEIGHTED AVERAGE	100.0	69.2	86.4	101.4	69.3	87.2	99.3	68.7	85.8	99.4	66.2	84.8	97.7	52.3	77.7	101.0	68.7	86.7

	DIVISION	VIRGINIA									WASHINGTON								
		ROANOKE			STAUNTON			WINCHESTER			CLARKSTON			EVERETT			OLYMPIA		
		240 - 241			244			226			994			982			985		
		MAT.	INST.	TOTAL	MAT.	INST.	TOTAL	MAT.	INST.	TOTAL	MAT.	INST.	TOTAL	MAT.	INST.	TOTAL	MAT.	INST.	TOTAL
015433	CONTRACTOR EQUIPMENT		100.8	100.8		105.3	105.3		100.8	100.8		91.1	91.1		103.9	103.9		103.9	103.9
0241, 31 - 34	SITE & INFRASTRUCTURE, DEMOLITION	108.6	84.9	92.1	113.2	86.8	94.8	122.7	85.0	96.4	101.4	89.2	92.9	100.3	113.4	109.4	103.8	114.3	111.1
0310	Concrete Forming & Accessories	97.1	66.2	70.4	91.2	55.2	60.0	86.2	51.6	56.3	124.0	72.8	79.7	109.4	101.1	102.2	100.2	104.0	103.5
0320	Concrete Reinforcing	101.0	70.5	85.4	100.6	54.4	77.0	100.6	84.3	92.3	111.7	81.9	96.5	107.5	99.3	103.3	108.2	99.4	103.7
0330	Cast-in-Place Concrete	121.8	62.3	99.3	111.5	55.4	90.3	107.6	58.6	89.1	106.9	77.9	95.9	102.6	97.4	100.6	95.9	110.6	101.4
03	CONCRETE	112.9	67.0	91.4	109.6	56.8	84.9	110.4	61.7	87.7	103.5	76.3	90.8	95.9	98.9	97.3	93.2	104.8	98.7
04	MASONRY	101.7	58.9	76.0	112.0	57.3	79.1	104.3	48.2	70.6	112.8	77.3	91.5	133.3	92.7	108.9	123.8	99.5	109.2
05	METALS	93.1	88.5	91.5	91.1	84.9	89.0	91.2	91.6	91.3	89.9	78.2	86.0	106.3	89.6	100.6	101.2	91.4	97.9
06	WOOD, PLASTICS & COMPOSITES	97.2	70.1	81.9	89.4	55.7	70.3	85.2	50.2	65.4	97.4	71.5	82.7	99.5	104.1	102.1	87.4	104.1	96.9
07	THERMAL & MOISTURE PROTECTION	99.4	57.7	83.6	99.4	55.0	82.6	100.0	59.9	84.9	154.7	74.4	124.4	103.6	92.5	99.4	103.4	92.9	99.4
08	OPENINGS	98.5	64.3	90.0	98.1	53.7	87.1	99.8	58.6	89.6	108.0	73.3	99.4	103.3	99.5	102.4	103.1	100.9	102.6
0920	Plaster & Gypsum Board	108.6	69.0	81.2	102.4	53.5	68.5	102.2	48.5	65.0	139.3	70.4	91.6	119.7	104.2	109.0	115.8	104.2	107.8
0950, 0980	Ceilings & Acoustic Treatment	96.1	69.0	78.8	92.6	53.5	67.7	93.5	48.5	64.8	103.6	70.4	82.4	110.4	104.2	106.4	102.5	104.2	103.5
0960	Flooring	110.9	42.2	91.9	105.6	42.5	88.1	104.2	79.7	97.4	107.5	50.8	91.8	121.4	98.6	115.1	113.9	98.8	109.7
0970, 0990	Wall Finishes & Painting/Coating	116.0	46.2	74.0	116.0	32.2	65.6	130.0	49.6	81.6	109.4	65.0	82.7	106.3	78.8	89.8	106.3	88.6	95.7
09	FINISHES	102.1	60.2	79.4	99.3	50.9	73.0	100.4	56.0	76.3	119.7	67.4	91.3	112.2	99.0	105.0	107.8	101.7	104.5
COVERS	DIVS. 10 - 14, 25, 28, 41, 43, 44	100.0	77.2	95.2	100.0	80.2	95.8	100.0	60.2	91.6	100.0	88.3	97.5	100.0	93.5	98.6	100.0	102.8	100.6
21, 22, 23	FIRE SUPPRESSION, PLUMBING & HVAC	100.2	50.1	79.6	93.5	58.5	79.1	93.5	71.2	84.3	93.6	76.7	86.7	100.1	90.9	96.3	100.0	94.5	97.7
26, 27, 3370	ELECTRICAL, COMMUNICATIONS & UTIL.	97.7	42.7	68.9	96.4	46.3	70.3	96.1	48.1	71.0	101.7	87.2	94.1	105.0	86.8	95.5	107.9	95.2	101.3
MF2004	WEIGHTED AVERAGE	100.7	61.9	83.6	98.6	60.7	81.9	98.9	64.2	83.5	102.8	78.3	92.0	104.2	95.1	100.1	102.5	99.4	101.1

	DIVISION	WASHINGTON																	
		RICHLAND			SEATTLE			SPOKANE			TACOMA			VANCOUVER			WENATCHEE		
		993			980 - 981,987			990 - 992			983 - 984			986			988		
		MAT.	INST.	TOTAL	MAT.	INST.	TOTAL	MAT.	INST.	TOTAL	MAT.	INST.	TOTAL	MAT.	INST.	TOTAL	MAT.	INST.	TOTAL
015433	CONTRACTOR EQUIPMENT		91.1	91.1		103.8	103.8		91.1	91.1		103.9	103.9		98.1	98.1		103.9	103.9
0241, 31 - 34	SITE & INFRASTRUCTURE, DEMOLITION	102.4	90.4	94.0	104.6	112.3	110.0	101.7	90.4	93.8	103.7	114.3	111.1	115.8	99.9	104.7	115.6	111.9	113.0
0310	Concrete Forming & Accessories	124.1	77.1	83.4	101.8	104.7	104.3	128.3	77.1	84.0	101.8	104.1	103.8	103.2	87.8	89.9	103.1	72.4	76.6
0320	Concrete Reinforcing	107.2	82.0	94.3	106.0	99.7	102.8	107.9	82.1	94.7	106.0	99.4	102.7	107.1	99.0	102.9	107.0	82.0	94.2
0330	Cast-in-Place Concrete	107.1	83.2	98.1	107.7	110.9	108.9	111.3	83.2	100.7	105.5	110.6	107.5	117.9	94.6	109.0	107.7	79.2	96.9
03	CONCRETE	103.1	80.1	92.3	98.9	105.3	101.9	105.5	80.1	93.6	97.8	104.9	101.1	107.9	92.2	100.6	105.2	76.7	91.9
04	MASONRY	112.6	78.2	91.9	128.4	104.2	113.9	113.3	80.5	93.6	128.3	99.5	110.9	128.7	89.4	105.1	131.7	75.4	97.8
05	METALS	90.3	78.8	86.4	108.1	93.8	103.3	92.7	79.0	88.1	108.1	91.5	102.5	105.3	90.5	100.3	105.5	80.8	97.1
06	WOOD, PLASTICS & COMPOSITES	97.5	76.5	85.6	91.5	104.1	98.7	105.9	76.5	89.2	90.3	104.1	98.1	82.8	87.8	85.6	91.2	71.2	79.9
07	THERMAL & MOISTURE PROTECTION	155.8	76.9	126.0	103.6	98.3	101.6	152.2	78.4	124.3	103.3	94.5	100.0	103.4	83.8	96.0	103.8	73.9	92.5
08	OPENINGS	110.5	71.4	100.8	105.5	100.9	104.3	111.1	71.4	101.3	104.0	100.9	103.2	101.9	89.9	99.0	103.4	68.5	94.8
0920	Plaster & Gypsum Board	139.3	75.6	95.2	114.5	104.2	107.3	136.0	75.6	94.2	117.3	104.2	108.2	115.3	87.6	96.1	118.6	70.3	85.1
0950, 0980	Ceilings & Acoustic Treatment	108.4	75.6	87.5	112.8	104.2	107.3	105.5	75.6	86.4	113.0	104.2	107.4	108.8	87.6	95.3	106.8	70.3	83.5
0960	Flooring	107.7	37.3	88.3	113.8	106.8	111.9	108.2	67.5	96.9	114.3	98.8	110.0	120.1	68.6	105.9	117.5	67.5	103.7
0970, 0990	Wall Finishes & Painting/Coating	109.4	65.0	82.7	106.3	88.6	95.7	109.3	65.0	82.7	106.3	88.6	95.7	112.6	61.9	82.1	106.3	65.0	81.5
09	FINISHES	121.1	67.8	92.2	110.0	103.4	106.4	120.2	74.0	95.1	110.5	101.7	105.7	109.2	81.2	94.0	111.2	70.1	88.9
COVERS	DIVS. 10 - 14, 25, 28, 41, 43, 44	100.0	71.5	93.9	100.0	102.9	100.6	100.0	71.4	93.9	100.0	102.8	100.6	100.0	68.0	93.2	100.0	87.2	97.3
21, 22, 23	FIRE SUPPRESSION, PLUMBING & HVAC	100.5	91.0	96.6	100.1	111.7	104.9	100.4	77.4	90.9	100.1	94.5	97.8	100.2	95.2	98.2	93.4	73.6	85.2
26, 27, 3370	ELECTRICAL, COMMUNICATIONS & UTIL.	99.3	87.2	93.0	104.8	100.4	102.5	96.4	73.4	84.4	104.8	95.2	99.8	110.8	93.8	101.9	105.8	78.7	91.6
MF2004	WEIGHTED AVERAGE	104.6	81.6	94.5	104.6	104.7	104.6	104.9	77.9	93.0	104.3	99.4	102.2	105.8	90.8	99.2	103.9	78.5	92.7

	DIVISION	WASHINGTON			WEST VIRGINIA														
		YAKIMA			BECKLEY			BLUEFIELD			BUCKHANNON			CHARLESTON			CLARKSBURG		
		989			258 - 259			247 - 248			262			250 - 253			263 - 264		
		MAT.	INST.	TOTAL	MAT.	INST.	TOTAL	MAT.	INST.	TOTAL	MAT.	INST.	TOTAL	MAT.	INST.	TOTAL	MAT.	INST.	TOTAL
015433	CONTRACTOR EQUIPMENT		103.9	103.9		100.8	100.8		100.8	100.8		100.8	100.8		100.8	100.8		100.8	100.8
0241, 31 - 34	SITE & INFRASTRUCTURE, DEMOLITION	106.7	112.2	110.5	103.3	86.9	91.9	103.0	86.9	91.7	109.6	86.6	93.6	104.3	88.2	93.1	110.3	86.6	93.8
0310	Concrete Forming & Accessories	102.2	95.5	96.4	87.7	82.6	83.3	88.9	81.7	82.7	88.3	81.2	82.2	102.5	82.5	85.2	86.2	81.1	81.8
0320	Concrete Reinforcing	106.6	81.6	93.8	99.5	85.0	92.1	99.5	67.5	83.1	100.1	84.3	92.0	101.0	85.2	92.9	100.1	85.8	92.8
0330	Cast-in-Place Concrete	112.8	79.1	100.0	104.8	104.2	104.6	104.8	103.8	104.4	104.5	102.4	103.7	102.9	104.2	103.4	114.5	102.3	109.9
03	CONCRETE	102.7	86.9	95.3	103.8	91.2	97.9	103.8	87.5	96.2	107.2	89.9	99.1	103.9	91.3	98.0	111.5	90.1	101.5
04	MASONRY	120.7	61.0	84.8	99.9	90.4	94.2	94.4	90.4	92.0	106.4	85.9	94.1	98.0	92.3	94.5	108.3	85.9	94.9
05	METALS	106.2	81.2	97.7	91.1	96.6	93.0	91.1	89.7	90.6	91.3	97.2	93.3	93.3	97.8	94.8	91.3	97.6	93.4
06	WOOD, PLASTICS & COMPOSITES	90.6	104.1	98.3	86.6	81.9	83.9	88.2	81.9	84.6	87.5	79.3	82.9	103.2	79.3	89.7	84.5	79.3	81.6
07	THERMAL & MOISTURE PROTECTION	103.4	74.9	92.6	99.5	87.6	95.0	99.5	87.6	95.0	99.8	87.0	94.9	96.9	88.1	93.6	99.7	87.0	94.9
08	OPENINGS	103.4	84.2	98.6	99.1	78.6	94.0	100.0	74.5	93.7	100.0	77.0	94.3	99.7	77.2	94.1	100.0	79.7	95.0
0920	Plaster & Gypsum Board	116.9	104.2	108.1	99.5	81.2	86.8	99.9	81.2	86.9	100.4	78.5	85.2	104.8	78.5	86.6	98.7	78.5	84.7
0950, 0980	Ceilings & Acoustic Treatment	108.7	104.2	105.8	90.0	81.2	84.4	90.0	81.2	84.4	91.7	78.5	83.3	94.3	78.5	84.2	91.7	78.5	83.3
0960	Flooring	115.5	56.0	99.1	101.4	108.1	103.2	102.0	108.1	103.7	101.6	102.1	101.8	110.9	108.1	110.1	100.3	102.1	100.8
0970, 0990	Wall Finishes & Painting/Coating	106.3	65.0	81.5	116.0	38.9	69.6	116.0	38.9	69.6	116.0	88.2	99.3	116.0	87.6	98.9	116.0	88.2	99.3
09	FINISHES	110.0	86.1	97.0	96.7	82.9	89.2	96.9	82.9	89.3	97.8	85.3	91.0	101.1	87.3	93.6	97.0	85.3	90.7
COVERS	DIVS. 10 - 14, 25, 28, 41, 43, 44	100.0	96.5	99.3	100.0	60.8	91.7	100.0	60.7	91.7	100.0	98.0	99.6	100.0	99.2	99.8	100.0	98.0	99.6
21, 22, 23	FIRE SUPPRESSION, PLUMBING & HVAC	100.1	87.7	95.0	93.5	83.7	89.5	93.5	57.3	78.6	93.5	89.0	91.6	100.1	85.6	94.1	93.5	82.3	88.9
26, 27, 3370	ELECTRICAL, COMMUNICATIONS & UTIL.	108.1	87.2	97.2	93.6	88.8	91.1	96.5	69.8	82.6	98.1	92.7	95.2	102.1	88.8	95.2	98.1	92.7	95.2
MF2004	WEIGHTED AVERAGE	104.6	86.1	96.4	96.6	86.7	92.2	96.7	77.1	88.1	98.2	89.0	94.1	99.8	89.1	95.0	98.8	87.7	93.9

	DIVISION	WEST VIRGINIA																	
		GASSAWAY			HUNTINGTON			LEWISBURG			MARTINSBURG			MORGANTOWN			PARKERSBURG		
		266			255 - 257			249			254			265			261		
		MAT.	INST.	TOTAL	MAT.	INST.	TOTAL	MAT.	INST.	TOTAL	MAT.	INST.	TOTAL	MAT.	INST.	TOTAL	MAT.	INST.	TOTAL
015433	CONTRACTOR EQUIPMENT		100.8	100.8		100.8	100.8		100.8	100.8		100.8	100.8		100.8	100.8		100.8	100.8
0241, 31 - 34	SITE & INFRASTRUCTURE, DEMOLITION	106.9	87.0	93.1	107.7	89.4	95.0	119.6	86.9	96.8	107.2	84.9	91.7	104.2	86.8	92.1	112.3	88.2	95.5
0310	Concrete Forming & Accessories	87.8	80.2	81.3	98.5	94.0	94.6	86.5	82.2	82.8	87.7	80.6	81.5	86.5	81.0	81.7	90.1	81.1	82.3
0320	Concrete Reinforcing	100.1	85.2	92.5	101.0	87.9	94.3	100.1	67.5	83.5	99.5	74.8	86.9	100.1	85.8	92.8	99.5	84.1	91.6
0330	Cast-in-Place Concrete	109.4	102.4	106.8	114.3	105.8	111.1	104.9	104.0	104.5	109.8	92.6	103.3	104.5	102.3	103.7	106.8	95.2	102.4
03	CONCRETE	107.6	89.7	99.2	109.3	97.4	103.7	114.5	87.8	102.0	107.7	84.2	96.7	103.6	90.0	97.3	109.3	87.4	99.1
04	MASONRY	109.5	89.0	97.2	101.1	85.6	91.8	99.8	90.4	94.2	102.0	74.4	85.4	129.6	85.9	103.3	84.7	86.2	85.6
05	METALS	91.2	97.9	93.5	93.4	98.3	95.0	91.2	90.1	90.8	91.4	85.9	89.6	91.3	97.5	93.4	92.0	96.9	93.6
06	WOOD, PLASTICS & COMPOSITES	86.8	77.8	81.7	97.2	93.4	95.1	84.8	81.9	83.1	86.6	82.4	84.2	84.8	79.3	81.7	87.8	77.8	82.2
07	THERMAL & MOISTURE PROTECTION	99.5	87.6	95.0	99.6	88.6	95.5	100.3	87.6	95.5	99.7	67.7	87.6	99.6	87.0	94.8	99.4	86.5	94.6
08	OPENINGS	97.9	76.4	92.6	98.5	85.6	95.3	100.0	74.5	93.7	101.4	69.9	93.6	101.4	79.7	96.1	99.1	76.2	93.4
0920	Plaster & Gypsum Board	100.0	77.0	84.0	107.6	93.0	97.5	98.7	81.2	86.6	100.0	81.7	87.3	98.7	78.5	84.7	101.7	77.0	84.6
0950, 0980	Ceilings & Acoustic Treatment	91.7	77.0	82.3	92.6	93.0	92.9	91.7	81.2	85.0	91.7	81.7	85.3	91.7	78.5	83.3	91.7	77.0	82.3
0960	Flooring	101.4	108.1	103.3	110.7	89.9	105.0	100.4	108.1	102.5	101.4	57.8	89.3	100.4	102.1	100.9	105.2	95.7	102.6
0970, 0990	Wall Finishes & Painting/Coating	116.0	87.6	98.9	116.0	87.6	98.9	116.0	38.9	69.6	116.0	45.9	73.8	116.0	88.2	99.3	116.0	87.7	99.0
09	FINISHES	97.4	85.6	91.0	101.1	92.6	96.5	98.1	82.9	89.9	97.3	73.5	84.4	96.7	85.3	90.5	99.1	83.8	90.8
COVERS	DIVS. 10 - 14, 25, 28, 41, 43, 44	100.0	97.8	99.5	100.0	101.9	100.4	100.0	60.7	91.7	100.0	60.4	91.6	100.0	70.7	93.8	100.0	98.9	99.8
21, 22, 23	FIRE SUPPRESSION, PLUMBING & HVAC	93.5	91.2	92.5	100.2	83.7	93.4	93.5	83.5	89.4	93.5	87.3	90.9	93.5	88.9	91.6	100.1	86.2	94.4
26, 27, 3370	ELECTRICAL, COMMUNICATIONS & UTIL.	98.1	88.8	93.2	97.7	99.1	98.4	93.6	69.8	81.2	100.0	79.4	89.3	98.3	92.7	95.3	98.2	91.7	94.8
MF2004	WEIGHTED AVERAGE	98.1	89.3	94.2	100.2	91.7	96.4	98.6	82.8	91.6	98.3	80.3	90.3	98.9	88.3	94.2	99.1	87.8	94.1

DIVISION		WEST VIRGINIA									WISCONSIN								
		PETERSBURG			ROMNEY			WHEELING			BELOIT			EAU CLAIRE			GREEN BAY		
		268			267			260			535			547			541 - 543		
		MAT.	INST.	TOTAL	MAT.	INST.	TOTAL	MAT.	INST.	TOTAL	MAT.	INST.	TOTAL	MAT.	INST.	TOTAL	MAT.	INST.	TOTAL
015433	**CONTRACTOR EQUIPMENT**		100.8	100.8		100.8	100.8		100.8	100.8		101.5	101.5		100.9	100.9		98.8	98.8
0241, 31 - 34	**SITE & INFRASTRUCTURE, DEMOLITION**	103.3	87.2	92.1	106.2	87.2	93.0	113.1	87.4	95.2	94.8	106.1	102.7	93.7	102.7	100.0	97.3	99.0	98.4
0310	Concrete Forming & Accessories	89.2	82.5	83.4	85.9	82.1	82.6	91.5	81.3	82.7	100.8	90.2	91.6	100.7	93.8	94.8	104.9	93.1	94.6
0320	Concrete Reinforcing	99.5	75.7	87.3	100.1	71.9	85.7	98.9	85.8	92.2	93.7	117.2	105.7	92.8	95.4	94.1	90.8	81.5	86.1
0330	Cast-in-Place Concrete	104.5	95.2	101.0	109.4	95.1	104.0	106.8	102.4	105.2	102.5	91.9	98.5	99.5	97.3	98.6	102.9	94.9	99.9
03	**CONCRETE**	103.7	86.4	95.6	107.5	85.5	97.2	109.3	90.3	100.4	98.4	96.2	97.4	98.2	95.6	97.0	101.1	91.8	96.8
04	**MASONRY**	102.1	85.9	92.3	100.2	66.6	80.0	109.5	85.9	95.3	105.1	91.3	96.8	93.0	97.5	95.7	125.2	94.7	106.8
05	**METALS**	91.4	93.2	92.0	91.4	90.5	91.1	92.1	97.9	94.1	97.9	107.3	101.1	91.8	99.4	94.3	93.9	93.0	93.6
06	**WOOD, PLASTICS & COMPOSITES**	88.4	82.4	85.0	84.0	82.4	83.1	89.4	79.3	83.7	109.0	90.8	98.7	112.4	93.5	101.7	112.5	93.5	101.7
07	**THERMAL & MOISTURE PROTECTION**	99.6	73.3	89.7	99.7	68.4	87.9	99.8	87.4	95.1	101.1	90.3	97.0	101.2	84.6	94.9	102.5	82.7	95.0
08	**OPENINGS**	101.5	76.7	95.3	101.4	76.1	95.1	100.0	79.7	95.0	103.4	99.1	102.3	100.7	90.8	98.2	99.0	89.1	96.6
0920	Plaster & Gypsum Board	100.4	81.7	87.4	98.2	81.7	86.8	102.1	78.5	85.8	96.6	90.9	92.7	98.6	93.6	95.1	95.2	93.6	94.1
0950, 0980	Ceilings & Acoustic Treatment	91.7	81.7	85.3	91.7	81.7	85.3	91.7	78.5	83.3	84.7	90.9	88.7	99.2	93.6	95.6	91.5	93.6	92.8
0960	Flooring	102.5	102.1	102.4	100.2	65.1	90.5	106.3	102.1	105.2	103.3	107.2	104.4	96.4	102.5	98.1	115.9	102.5	112.2
0970, 0990	Wall Finishes & Painting/Coating	116.0	87.6	98.9	116.0	87.6	98.9	116.0	88.2	99.3	96.0	90.3	92.6	89.6	88.7	89.1	101.1	72.0	83.6
09	**FINISHES**	97.5	87.1	91.9	96.8	79.6	87.4	99.5	85.3	91.8	98.3	93.7	95.8	98.4	95.3	96.7	103.6	93.3	98.0
COVERS	**DIVS. 10 - 14, 25, 28, 41, 43, 44**	100.0	71.0	93.8	100.0	98.3	99.6	100.0	96.1	99.2	100.0	81.0	96.0	100.0	84.0	96.6	100.0	83.4	96.5
21, 22, 23	**FIRE SUPPRESSION, PLUMBING & HVAC**	93.5	88.8	91.6	93.5	88.6	91.4	100.2	88.9	95.6	99.8	83.4	93.1	100.0	77.6	90.8	100.2	81.0	92.3
26, 27, 3370	**ELECTRICAL, COMMUNICATIONS & UTIL.**	101.8	79.5	90.1	101.0	79.5	89.7	95.1	92.7	93.8	97.5	82.5	89.7	101.4	84.2	92.4	95.4	80.1	87.4
MF2004	**WEIGHTED AVERAGE**	97.9	85.4	92.4	98.1	82.7	91.3	100.3	89.2	95.4	99.6	92.1	96.3	98.3	90.5	94.8	100.5	88.5	95.2

DIVISION		WISCONSIN																	
		KENOSHA			LA CROSSE			LANCASTER			MADISON			MILWAUKEE			NEW RICHMOND		
		531			546			538			537			530,532			540		
		MAT.	INST.	TOTAL	MAT.	INST.	TOTAL	MAT.	INST.	TOTAL	MAT.	INST.	TOTAL	MAT.	INST.	TOTAL	MAT.	INST.	TOTAL
015433	**CONTRACTOR EQUIPMENT**		99.5	99.5		100.9	100.9		101.5	101.5		101.5	101.5		89.2	89.2		101.6	101.6
0241, 31 - 34	**SITE & INFRASTRUCTURE, DEMOLITION**	101.0	103.6	102.8	87.1	102.7	98.0	94.4	106.2	102.6	94.5	106.5	102.8	95.5	96.8	96.4	88.6	103.4	98.9
0310	Concrete Forming & Accessories	104.1	100.3	100.8	89.8	93.4	92.9	100.1	90.1	91.5	100.2	94.4	95.2	102.9	115.3	113.6	96.9	99.3	99.0
0320	Concrete Reinforcing	93.5	101.1	97.4	92.5	86.4	89.4	95.2	83.9	89.4	93.7	86.8	90.2	93.7	101.8	97.8	89.8	95.1	92.5
0330	Cast-in-Place Concrete	111.5	97.2	106.1	89.4	91.8	90.3	101.9	94.1	98.9	100.3	97.3	99.1	100.1	109.8	103.8	103.5	83.9	96.1
03	**CONCRETE**	103.1	99.5	101.4	89.6	91.8	90.6	98.2	90.6	94.6	97.3	94.1	95.8	97.5	110.0	103.4	96.0	93.4	94.8
04	**MASONRY**	102.3	103.7	103.1	92.2	97.6	95.5	105.2	96.7	100.1	103.0	101.2	101.9	105.0	119.4	113.7	119.5	91.6	102.8
05	**METALS**	98.6	102.5	99.9	91.7	95.1	92.8	95.4	91.3	94.0	100.7	94.5	98.6	101.5	96.4	99.8	91.7	98.3	93.9
06	**WOOD, PLASTICS & COMPOSITES**	108.8	99.6	103.6	98.9	93.5	95.8	108.3	90.8	98.4	106.1	93.5	99.0	110.8	114.9	113.1	101.6	103.8	102.8
07	**THERMAL & MOISTURE PROTECTION**	100.9	93.8	98.2	100.8	83.4	94.2	100.9	78.2	92.3	97.5	92.0	95.4	99.3	113.4	104.6	103.0	91.4	98.7
08	**OPENINGS**	98.7	103.4	99.9	100.6	81.3	95.8	98.4	80.1	93.9	103.4	93.1	100.8	105.6	111.7	107.1	88.9	96.4	90.7
0920	Plaster & Gypsum Board	84.5	100.0	95.2	93.3	93.6	93.5	95.0	90.9	92.2	92.4	93.6	93.3	97.6	115.6	110.0	86.9	104.4	99.0
0950, 0980	Ceilings & Acoustic Treatment	84.7	100.0	94.4	97.5	93.6	95.0	79.6	90.9	86.8	86.7	93.6	91.1	88.2	115.6	105.6	66.5	104.4	90.6
0960	Flooring	121.4	107.7	117.6	89.0	109.1	94.5	102.5	100.9	102.0	97.6	100.9	98.5	106.1	119.3	109.8	108.3	102.5	106.7
0970, 0990	Wall Finishes & Painting/Coating	105.0	98.4	101.0	89.6	69.1	77.3	96.0	73.8	82.6	96.0	90.3	92.6	98.0	114.6	108.0	103.7	88.7	94.7
09	**FINISHES**	103.2	102.1	102.6	94.7	94.5	94.6	96.6	90.1	93.1	96.3	95.4	95.8	100.2	116.5	109.1	94.1	98.7	96.6
COVERS	**DIVS. 10 - 14, 25, 28, 41, 43, 44**	100.0	100.2	100.0	100.0	84.1	96.6	100.0	53.2	90.1	100.0	83.8	96.6	100.0	104.8	101.0	100.0	83.3	96.5
21, 22, 23	**FIRE SUPPRESSION, PLUMBING & HVAC**	100.0	83.5	93.2	100.0	77.4	90.7	93.0	76.4	86.2	99.6	98.8	99.3	99.8	102.9	101.1	92.8	74.9	85.4
26, 27, 3370	**ELECTRICAL, COMMUNICATIONS & UTIL.**	97.9	99.9	98.9	101.7	84.2	92.6	97.2	84.2	90.4	100.1	85.5	92.5	98.9	100.2	99.6	99.3	84.2	91.4
MF2004	**WEIGHTED AVERAGE**	100.4	97.6	99.1	96.6	89.0	93.2	96.9	86.8	92.4	99.7	95.5	97.8	100.5	106.6	103.2	95.7	89.9	93.2

DIVISION		WISCONSIN																	
		OSHKOSH			PORTAGE			RACINE			RHINELANDER			SUPERIOR			WAUSAU		
		549			539			534			545			548			544		
		MAT.	INST.	TOTAL	MAT.	INST.	TOTAL	MAT.	INST.	TOTAL	MAT.	INST.	TOTAL	MAT.	INST.	TOTAL	MAT.	INST.	TOTAL
015433	**CONTRACTOR EQUIPMENT**		98.8	98.8		101.5	101.5		101.5	101.5		98.8	98.8		101.7	101.7		98.8	98.8
0241, 31 - 34	**SITE & INFRASTRUCTURE, DEMOLITION**	88.9	98.7	95.7	84.5	106.3	99.7	94.6	107.2	103.4	102.2	98.8	99.8	85.2	103.6	98.0	84.5	99.0	94.6
0310	Concrete Forming & Accessories	91.0	89.3	89.6	94.2	91.1	91.5	101.1	100.5	100.6	89.0	89.9	89.7	95.6	90.6	91.3	90.5	92.9	92.6
0320	Concrete Reinforcing	90.9	81.6	86.2	95.3	84.1	89.6	93.7	101.1	97.5	91.1	82.1	86.5	89.8	97.3	93.6	91.1	86.4	88.7
0330	Cast-in-Place Concrete	95.4	92.6	94.3	87.4	96.9	91.0	100.5	97.0	99.2	108.2	91.8	102.0	97.6	92.6	95.7	88.9	94.9	91.2
03	**CONCRETE**	90.9	89.4	90.2	86.1	92.0	88.9	97.5	99.5	98.4	103.3	89.4	96.8	90.4	93.0	91.6	85.9	92.7	89.0
04	**MASONRY**	105.9	92.8	98.0	104.1	97.1	99.9	105.0	103.7	104.2	123.9	91.1	104.2	118.8	100.0	107.5	105.4	96.8	100.2
05	**METALS**	92.0	92.8	92.2	96.1	92.3	94.8	99.7	102.6	100.7	91.9	92.2	92.0	92.7	100.7	95.4	91.7	95.0	92.8
06	**WOOD, PLASTICS & COMPOSITES**	95.1	90.9	92.7	100.3	90.8	94.9	109.2	99.6	103.8	92.7	90.9	91.7	100.0	89.9	94.3	94.6	93.5	94.0
07	**THERMAL & MOISTURE PROTECTION**	101.8	82.3	94.4	100.5	87.8	95.7	101.1	95.0	98.8	102.5	77.3	93.0	102.8	92.9	99.0	101.6	78.1	92.7
08	**OPENINGS**	94.3	87.7	92.7	98.6	87.4	95.8	103.4	103.4	103.4	94.3	81.2	91.1	88.8	92.9	89.8	94.5	90.5	93.5
0920	Plaster & Gypsum Board	85.7	90.9	89.3	90.1	90.9	90.7	96.6	100.0	98.9	85.3	90.9	89.2	86.7	90.1	89.1	85.7	93.6	91.1
0950, 0980	Ceilings & Acoustic Treatment	91.5	90.9	91.1	80.4	90.9	87.1	84.7	100.0	94.4	91.5	90.9	91.1	67.3	90.1	81.9	91.5	93.6	92.8
0960	Flooring	106.2	102.5	105.2	98.5	107.7	101.1	103.3	107.7	104.5	105.3	98.4	103.4	109.7	116.3	111.5	106.0	106.4	106.1
0970, 0990	Wall Finishes & Painting/Coating	98.6	72.0	82.6	96.0	73.8	82.6	96.0	98.4	97.4	98.6	67.0	79.6	93.4	98.2	96.3	98.6	72.0	82.6
09	**FINISHES**	98.7	89.5	93.7	94.4	92.2	93.2	98.3	102.1	100.3	99.5	89.6	94.1	93.7	96.3	95.1	98.4	94.1	96.0
COVERS	**DIVS. 10 - 14, 25, 28, 41, 43, 44**	100.0	70.0	93.6	100.0	69.6	93.6	100.0	100.2	100.0	100.0	59.5	91.4	100.0	82.0	96.2	100.0	83.3	96.5
21, 22, 23	**FIRE SUPPRESSION, PLUMBING & HVAC**	93.5	78.9	87.5	93.0	84.0	89.3	99.8	88.5	95.2	93.5	80.5	88.1	92.8	90.0	91.6	93.5	86.6	90.6
26, 27, 3370	**ELECTRICAL, COMMUNICATIONS & UTIL.**	99.9	80.4	89.7	101.3	88.6	94.7	97.2	101.1	99.2	99.2	74.4	86.2	104.5	84.8	94.2	101.1	74.4	87.2
MF2004	**WEIGHTED AVERAGE**	95.5	86.6	91.6	95.4	90.6	93.3	99.7	99.1	99.5	98.3	85.2	92.5	95.5	93.6	94.7	94.8	89.4	92.4

	DIVISION	WYOMING																	
		CASPER			CHEYENNE			NEWCASTLE			RAWLINS			RIVERTON			ROCK SPRINGS		
		826			820			827			823			825			829 - 831		
		MAT.	INST.	TOTAL	MAT.	INST.	TOTAL	MAT.	INST.	TOTAL	MAT.	INST.	TOTAL	MAT.	INST.	TOTAL	MAT.	INST.	TOTAL
015433	CONTRACTOR EQUIPMENT		100.7	100.7		100.7	100.7		100.7	100.7		100.7	100.7		100.7	100.7		100.7	100.7
0241, 31 - 34	SITE & INFRASTRUCTURE, DEMOLITION	91.7	98.9	96.7	85.1	98.9	94.7	77.9	97.8	91.8	92.0	97.8	96.0	85.4	97.8	94.0	81.5	97.8	92.8
0310	Concrete Forming & Accessories	102.2	42.8	50.8	103.7	58.5	64.6	95.4	38.4	46.1	99.0	38.4	46.6	94.3	38.3	45.9	100.5	38.3	46.7
0320	Concrete Reinforcing	105.2	43.8	73.8	99.4	44.6	71.4	106.7	44.4	74.9	106.3	44.4	74.7	107.3	44.4	75.2	107.3	44.4	75.2
0330	Cast-in-Place Concrete	98.3	72.9	88.7	93.8	73.1	86.0	94.8	53.8	79.3	94.9	53.8	79.3	94.9	53.8	79.3	94.8	53.8	79.3
03	CONCRETE	104.2	54.2	80.9	101.8	61.4	82.9	102.3	45.8	75.9	116.7	45.8	83.6	110.9	45.8	80.5	102.7	45.8	76.1
04	MASONRY	104.8	37.8	64.5	101.5	52.1	71.8	98.3	46.8	67.3	98.3	46.8	67.3	98.4	46.8	67.3	158.7	46.8	91.4
05	METALS	100.4	61.0	87.1	101.7	62.5	88.5	96.6	60.4	84.3	96.6	60.4	84.4	96.7	60.3	84.4	97.5	60.3	84.9
06	WOOD, PLASTICS & COMPOSITES	97.3	40.4	65.1	98.9	60.9	77.4	89.7	37.2	60.0	93.5	37.2	61.6	88.7	37.2	59.5	96.2	37.2	62.8
07	THERMAL & MOISTURE PROTECTION	95.3	49.8	78.1	96.7	55.8	81.2	98.0	47.8	79.1	99.6	47.8	80.1	99.0	47.9	79.7	98.1	47.8	79.1
08	OPENINGS	93.1	39.9	79.9	95.2	51.1	84.3	99.2	38.2	84.1	98.8	38.2	83.8	99.0	38.2	84.0	99.5	38.2	84.3
0920	Plaster & Gypsum Board	102.2	38.4	58.1	90.9	59.6	69.2	85.9	35.1	50.8	86.8	35.1	51.0	85.9	35.1	50.8	93.2	35.1	53.0
0950, 0980	Ceilings & Acoustic Treatment	126.3	38.4	70.3	119.1	59.6	81.2	121.9	35.1	66.7	121.9	35.1	66.7	121.9	35.1	66.7	121.9	35.1	66.7
0960	Flooring	102.6	35.5	84.1	107.0	57.1	93.2	100.6	50.2	86.6	103.6	50.2	88.8	99.9	50.2	86.1	106.0	50.2	90.6
0970, 0990	Wall Finishes & Painting/Coating	97.8	47.8	67.7	101.0	47.8	69.0	97.1	30.6	57.1	97.1	30.6	57.1	97.1	30.6	57.1	97.1	30.6	57.1
09	FINISHES	109.7	41.0	72.4	108.8	57.4	80.9	103.6	38.6	68.3	106.2	38.6	69.5	104.4	38.6	68.6	106.2	38.6	69.5
COVERS	DIVS. 10 - 14, 25, 28, 41, 43, 44	100.0	74.7	94.6	100.0	77.2	95.2	100.0	61.3	91.8	100.0	61.3	91.8	100.0	61.3	91.8	100.0	61.3	91.8
21, 22, 23	FIRE SUPPRESSION, PLUMBING & HVAC	100.1	57.1	82.4	100.0	56.4	82.1	96.5	54.0	79.0	96.5	54.0	79.0	96.5	54.0	79.0	100.0	54.0	81.0
26, 27, 3370	ELECTRICAL, COMMUNICATIONS & UTIL.	100.3	57.0	77.7	97.4	69.9	83.0	95.3	70.5	82.4	95.3	70.5	82.4	95.3	58.0	75.8	93.5	58.0	74.9
MF2004	WEIGHTED AVERAGE	100.5	56.0	80.9	100.0	63.2	83.7	97.8	56.0	79.3	100.3	56.0	80.7	99.2	54.2	79.3	102.2	54.2	81.0

	DIVISION	WYOMING												CANADA					
		SHERIDAN			WHEATLAND			WORLAND			YELLOWSTONE NAT'L PA			BARRIE, ONTARIO			BATHURST, NEW BRUNSWICK		
		828			822			824			821								
		MAT.	INST.	TOTAL	MAT.	INST.	TOTAL	MAT.	INST.	TOTAL	MAT.	INST.	TOTAL	MAT.	INST.	TOTAL	MAT.	INST.	TOTAL
015433	CONTRACTOR EQUIPMENT		100.7	100.7		100.7	100.7		100.7	100.7		100.7	100.7		101.5	101.5		101.2	101.2
0241, 31 - 34	SITE & INFRASTRUCTURE, DEMOLITION	85.6	98.9	94.9	82.3	97.8	93.1	79.8	97.8	92.3	79.9	98.2	92.6	116.5	100.2	105.2	95.9	96.5	96.3
0310	Concrete Forming & Accessories	101.9	43.1	51.0	97.1	38.7	46.6	97.2	38.3	46.3	97.2	40.7	48.3	121.6	83.6	88.8	102.5	59.6	65.4
0320	Concrete Reinforcing	107.3	44.5	75.2	106.7	44.4	74.9	107.3	44.4	75.2	109.2	44.4	76.1	180.4	80.6	129.5	149.6	54.4	101.0
0330	Cast-in-Place Concrete	98.0	72.9	88.5	99.0	54.3	82.1	94.8	53.8	79.3	94.8	57.3	80.6	167.7	83.2	135.7	138.2	58.3	108.0
03	CONCRETE	110.8	54.5	84.5	107.3	46.1	78.7	102.5	45.8	76.0	102.8	48.0	77.2	152.0	83.2	119.8	135.6	58.9	99.8
04	MASONRY	98.6	51.5	70.3	98.7	47.7	68.0	98.4	46.8	67.3	98.4	53.0	71.1	168.7	92.3	122.8	158.9	61.2	100.2
05	METALS	100.4	62.2	87.5	96.5	60.4	84.3	96.7	60.3	84.4	97.4	60.3	84.8	107.4	86.9	100.5	104.7	69.6	92.9
06	WOOD, PLASTICS & COMPOSITES	100.1	40.4	66.3	91.4	37.2	60.7	91.4	37.2	60.7	91.4	37.2	60.7	117.2	82.2	97.4	94.6	59.7	74.8
07	THERMAL & MOISTURE PROTECTION	99.0	53.4	81.8	98.3	48.8	79.6	98.1	47.8	79.1	97.7	51.2	80.1	111.1	84.3	101.0	104.3	59.1	87.2
08	OPENINGS	99.8	40.9	85.2	97.8	38.2	83.0	99.4	38.2	84.2	92.6	38.2	79.2	91.7	80.7	89.0	86.9	52.2	78.3
0920	Plaster & Gypsum Board	112.1	38.4	61.1	85.9	35.1	50.8	85.9	35.1	50.8	86.4	35.1	50.9	155.8	81.6	104.5	162.5	58.4	90.4
0950, 0980	Ceilings & Acoustic Treatment	124.8	38.4	69.8	121.9	35.1	66.7	121.9	35.1	66.7	123.7	35.1	67.3	95.6	81.6	86.7	96.5	58.4	72.2
0960	Flooring	103.3	50.2	88.6	102.1	50.2	87.7	102.1	50.2	87.7	102.1	50.2	87.7	132.2	91.0	120.8	113.9	43.5	94.5
0970, 0990	Wall Finishes & Painting/Coating	99.2	47.8	68.3	97.1	30.6	57.1	97.1	30.6	57.1	97.1	30.6	57.1	110.9	83.9	94.6	110.1	48.2	72.8
09	FINISHES	111.3	43.6	74.6	104.4	38.8	68.8	104.1	38.6	68.5	104.6	40.1	69.6	118.9	85.1	100.6	112.6	55.4	81.5
COVERS	DIVS. 10 - 14, 25, 28, 41, 43, 44	100.0	74.7	94.6	100.0	61.7	91.9	100.0	61.3	91.8	100.0	63.5	92.3	140.0	75.2	126.2	140.0	66.0	124.3
21, 22, 23	FIRE SUPPRESSION, PLUMBING & HVAC	96.5	56.3	80.0	96.5	54.4	79.2	96.5	54.0	79.0	96.5	57.2	80.3	98.8	95.2	97.3	98.7	66.9	85.6
26, 27, 3370	ELECTRICAL, COMMUNICATIONS & UTIL.	97.7	58.0	77.0	95.3	69.8	82.0	95.3	58.0	75.8	94.2	58.0	75.3	120.6	89.1	104.1	122.2	60.9	90.2
MF2004	WEIGHTED AVERAGE	100.7	58.0	81.9	98.5	56.2	79.8	98.0	54.2	78.7	97.4	56.2	79.2	118.0	89.2	105.3	113.2	64.8	91.8

	DIVISION	CANADA																	
		BRANDON, MANITOBA			BRANTFORD, ONTARIO			BRIDGEWATER, NOVA SCOTIA			CALGARY, ALBERTA			CAP-DE-LA-MADELEINE, QUEBEC			CHARLESBOURG, QUEBEC		
		MAT.	INST.	TOTAL	MAT.	INST.	TOTAL	MAT.	INST.	TOTAL	MAT.	INST.	TOTAL	MAT.	INST.	TOTAL	MAT.	INST.	TOTAL
015433	CONTRACTOR EQUIPMENT		103.8	103.8		101.5	101.5		100.9	100.9		106.7	106.7		101.8	101.8		101.8	101.8
0241, 31 - 34	SITE & INFRASTRUCTURE, DEMOLITION	112.0	99.5	103.3	114.2	100.7	104.8	98.8	98.0	98.2	124.3	105.3	111.0	94.2	99.5	97.9	94.2	99.5	97.9
0310	Concrete Forming & Accessories	120.9	67.8	75.0	121.5	90.6	94.8	97.2	67.0	71.1	126.5	86.3	91.7	126.5	82.3	88.3	126.5	82.3	88.3
0320	Concrete Reinforcing	162.4	51.2	105.6	175.4	79.4	126.4	149.6	44.9	96.1	148.3	60.1	103.3	149.6	75.8	111.9	149.6	75.8	111.9
0330	Cast-in-Place Concrete	151.1	72.2	121.3	163.6	103.1	140.7	168.2	67.9	130.2	186.7	97.2	152.8	131.9	91.5	116.6	131.9	91.5	116.6
03	CONCRETE	134.3	67.0	102.8	149.3	92.9	122.9	147.8	64.0	108.6	159.4	85.5	124.8	130.3	84.5	108.9	130.3	84.5	108.9
04	MASONRY	163.1	63.8	103.4	165.0	96.7	123.9	160.0	67.8	104.6	213.9	87.9	138.2	160.0	82.1	113.2	160.0	82.1	113.2
05	METALS	107.6	74.7	96.5	106.8	88.1	100.5	105.9	72.1	94.5	147.7	84.3	126.3	105.0	84.8	98.2	105.0	84.8	98.2
06	WOOD, PLASTICS & COMPOSITES	112.9	68.5	87.8	115.0	89.2	100.4	87.2	66.2	75.3	113.3	85.7	97.7	125.7	82.0	101.0	125.7	82.0	101.0
07	THERMAL & MOISTURE PROTECTION	104.6	68.0	90.8	110.9	89.4	102.8	105.9	66.3	90.9	123.1	84.6	108.5	104.7	84.9	97.2	104.7	84.9	97.2
08	OPENINGS	92.8	60.5	84.8	91.6	86.4	90.3	83.5	60.5	77.8	92.8	76.3	88.7	92.8	75.7	88.5	92.8	75.7	88.5
0920	Plaster & Gypsum Board	126.8	67.2	85.5	155.8	88.9	109.5	156.2	65.2	93.2	172.8	84.8	111.9	184.7	81.3	113.1	184.7	81.3	113.1
0950, 0980	Ceilings & Acoustic Treatment	95.6	67.2	77.5	95.6	88.9	91.3	95.6	65.2	76.2	142.4	84.8	105.7	95.6	81.3	86.5	95.6	81.3	86.5
0960	Flooring	132.1	64.7	113.5	132.1	91.0	120.8	110.0	62.0	96.7	134.5	85.3	120.9	132.1	91.7	120.9	132.1	91.7	120.9
0970, 0990	Wall Finishes & Painting/Coating	110.1	54.8	76.8	110.1	92.1	99.2	110.1	59.8	79.8	110.0	97.3	102.4	110.1	85.6	95.3	110.1	85.6	95.3
09	FINISHES	114.4	66.0	88.2	117.7	91.1	103.2	110.2	65.3	85.8	132.8	87.8	108.4	120.4	84.5	100.9	120.4	84.5	100.9
COVERS	DIVS. 10 - 14, 25, 28, 41, 43, 44	140.0	68.4	124.8	140.0	77.3	126.7	140.0	67.2	124.6	140.0	96.6	130.8	140.0	86.1	128.6	140.0	86.1	128.6
21, 22, 23	FIRE SUPPRESSION, PLUMBING & HVAC	98.7	80.7	91.3	98.7	98.1	98.4	98.7	81.0	91.4	98.2	82.5	91.7	99.0	87.3	94.2	99.0	87.3	94.2
26, 27, 3370	ELECTRICAL, COMMUNICATIONS & UTIL.	122.1	68.6	94.1	120.3	88.6	103.7	126.0	63.4	93.3	126.9	81.4	103.1	119.0	71.5	94.2	119.0	71.5	94.2
MF2004	WEIGHTED AVERAGE	115.0	72.9	96.4	117.1	92.9	106.4	114.8	71.7	95.8	129.4	86.3	110.4	113.8	84.0	100.7	113.8	84.0	100.7

DIVISION		CANADA																	
		CHARLOTTETOWN, PRINCE EDWARD ISLAND			CHICOUTIMI, QUEBEC			CORNER BROOK, NEWFOUNDLAND			CORNWALL, ONTARIO			DALHOUSIE, NEW BRUNSWICK			DARTMOUTH, NOVA SCOTIA		
		MAT.	INST.	TOTAL	MAT.	INST.	TOTAL	MAT.	INST.	TOTAL	MAT.	INST.	TOTAL	MAT.	INST.	TOTAL	MAT.	INST.	TOTAL
015433	CONTRACTOR EQUIPMENT		100.9	100.9		101.8	101.8		102.1	102.1		101.5	101.5		101.2	101.2		100.9	100.9
0241, 31 - 34	SITE & INFRASTRUCTURE, DEMOLITION	106.6	95.7	99.1	93.7	99.1	97.4	114.9	96.9	102.4	112.1	100.3	103.9	95.8	96.5	96.3	106.5	98.0	100.6
0310	Concrete Forming & Accessories	97.5	54.5	60.3	126.5	89.3	94.3	104.1	57.7	64.0	119.5	83.7	88.6	102.5	59.6	65.4	97.2	67.0	71.1
0320	Concrete Reinforcing	159.1	45.0	100.8	112.8	89.4	100.8	149.6	46.7	97.1	175.4	79.1	126.3	149.6	54.4	101.0	155.9	44.9	99.2
0330	Cast-in-Place Concrete	159.4	57.6	120.9	130.2	96.4	117.4	172.4	66.3	132.3	147.1	94.1	127.1	138.2	58.3	108.0	169.9	67.9	131.3
03	CONCRETE	153.8	54.7	107.5	124.3	91.8	109.1	162.6	59.6	114.5	141.1	86.7	115.6	135.6	58.9	99.8	149.6	64.0	109.6
04	MASONRY	160.4	59.1	99.5	160.4	90.0	118.1	160.3	60.5	100.3	163.7	88.7	118.6	156.9	61.2	99.4	171.4	67.8	109.1
05	METALS	105.9	65.8	92.4	104.6	89.9	99.6	108.1	70.8	95.5	106.8	86.7	100.0	104.7	69.6	92.9	108.0	72.1	95.9
06	WOOD, PLASTICS & COMPOSITES	87.5	53.9	68.5	125.7	89.7	105.3	95.6	56.6	73.5	113.8	83.0	96.3	94.6	59.7	74.8	87.2	66.2	75.3
07	THERMAL & MOISTURE PROTECTION	105.2	56.8	86.9	104.7	92.7	100.2	107.8	58.7	89.3	110.7	83.3	100.3	108.4	59.1	89.8	105.9	66.3	90.9
08	OPENINGS	95.0	46.6	83.0	92.3	76.7	88.5	99.0	53.3	87.7	92.8	80.4	89.7	86.9	52.2	78.3	83.5	60.5	77.8
0920	Plaster & Gypsum Board	160.5	52.4	85.7	184.7	89.2	118.6	162.5	55.2	88.3	227.5	82.4	127.1	162.5	58.4	90.4	159.0	65.2	94.1
0950, 0980	Ceilings & Acoustic Treatment	108.5	52.4	72.8	95.6	89.2	91.5	96.5	55.2	70.2	99.1	82.4	88.5	96.5	58.4	72.2	105.1	65.2	79.7
0960	Flooring	110.0	57.9	95.6	132.1	91.7	120.9	114.7	52.7	97.6	132.1	89.7	120.4	113.9	43.5	94.5	110.0	62.0	96.7
0970, 0990	Wall Finishes & Painting/Coating	110.1	40.1	68.0	110.1	100.7	104.4	110.1	57.7	78.5	110.1	85.8	95.5	110.1	48.2	72.8	110.1	59.8	79.8
09	FINISHES	114.6	53.8	81.6	120.4	91.6	104.8	113.8	56.6	82.8	127.2	85.2	104.4	112.6	55.4	81.5	112.7	65.3	87.0
COVERS	DIVS. 10 - 14, 25, 28, 41, 43, 44	140.0	65.5	124.2	140.0	87.9	128.9	140.0	66.3	124.4	140.0	74.8	126.2	140.0	66.0	124.3	140.0	67.2	124.6
21, 22, 23	FIRE SUPPRESSION, PLUMBING & HVAC	98.7	61.2	83.3	98.7	88.8	94.6	98.7	68.9	86.4	99.0	96.1	97.8	98.7	66.9	85.6	98.7	81.0	91.4
26, 27, 3370	ELECTRICAL, COMMUNICATIONS & UTIL.	118.8	51.5	83.6	118.3	82.8	99.7	119.0	57.4	86.8	120.6	89.6	104.4	122.5	60.9	90.3	124.0	63.4	92.3
MF2004	WEIGHTED AVERAGE	116.6	60.5	91.8	112.8	89.4	102.5	118.7	65.0	95.0	116.9	89.5	104.8	113.2	64.8	91.8	116.2	71.7	96.5

DIVISION		CANADA																	
		EDMONTON, ALBERTA			FORT MCMURRAY, ALBERTA			FREDERICTON, NEW BRUNSWICK			GATINEAU, QUEBEC			GRANBY, QUEBEC			HALIFAX, NOVA SCOTIA		
		MAT.	INST.	TOTAL	MAT.	INST.	TOTAL	MAT.	INST.	TOTAL	MAT.	INST.	TOTAL	MAT.	INST.	TOTAL	MAT.	INST.	TOTAL
015433	CONTRACTOR EQUIPMENT		106.7	106.7		103.9	103.9		101.2	101.2		101.8	101.8		101.8	101.8		100.9	100.9
0241, 31 - 34	SITE & INFRASTRUCTURE, DEMOLITION	154.1	105.3	120.1	120.2	102.0	107.5	95.4	96.5	96.2	94.0	99.5	97.8	94.6	99.5	98.0	99.2	97.9	98.3
0310	Concrete Forming & Accessories	129.0	86.2	92.0	119.7	86.3	90.8	119.2	59.8	67.9	126.5	82.2	88.2	126.5	82.1	88.1	97.4	72.9	76.2
0320	Concrete Reinforcing	148.3	60.1	103.3	162.4	60.1	110.2	150.5	54.5	101.5	158.1	75.8	116.1	158.1	75.8	116.1	155.9	58.2	106.0
0330	Cast-in-Place Concrete	208.8	97.2	166.6	217.4	96.4	171.6	131.6	58.4	103.9	130.2	91.5	115.5	134.5	91.4	118.2	171.3	71.9	133.7
03	CONCRETE	170.4	85.5	130.7	173.8	85.3	132.4	133.6	59.1	98.8	130.7	84.5	109.1	132.8	84.4	110.2	150.3	70.4	112.9
04	MASONRY	202.5	87.9	133.6	204.2	87.9	134.3	177.5	62.7	108.5	159.9	82.1	113.1	160.2	82.1	113.3	173.1	79.7	116.9
05	METALS	148.3	84.2	126.6	132.0	84.3	115.9	104.7	70.0	93.0	105.0	84.6	98.2	105.0	84.6	98.1	118.9	76.7	104.7
06	WOOD, PLASTICS & COMPOSITES	110.1	85.7	96.3	110.0	85.6	96.2	115.7	59.7	84.0	125.7	82.0	101.0	125.7	82.0	101.0	87.4	72.1	78.7
07	THERMAL & MOISTURE PROTECTION	124.2	84.6	109.2	118.4	83.9	105.3	108.6	60.1	90.2	104.7	84.9	97.2	104.7	83.4	96.6	106.5	73.2	93.9
08	OPENINGS	92.8	76.3	88.7	92.8	76.2	88.7	86.8	51.1	77.9	92.8	71.3	87.4	92.8	71.3	87.4	83.5	65.7	79.1
0920	Plaster & Gypsum Board	165.3	84.8	109.6	156.2	84.8	106.8	179.0	58.4	95.5	142.1	81.3	100.0	145.0	81.3	100.9	159.5	71.2	98.4
0950, 0980	Ceilings & Acoustic Treatment	157.4	84.8	111.2	105.1	84.8	92.2	102.5	58.4	74.4	95.6	81.3	86.5	95.6	81.3	86.5	105.1	71.2	83.5
0960	Flooring	135.6	85.3	121.7	132.1	85.3	119.2	127.3	43.5	104.2	132.1	91.7	120.9	132.1	91.7	120.9	110.0	62.0	96.7
0970, 0990	Wall Finishes & Painting/Coating	110.1	89.1	97.5	110.1	89.1	97.5	110.1	61.4	80.8	110.1	85.6	95.3	110.1	85.6	95.3	110.1	71.9	87.1
09	FINISHES	137.6	86.9	110.1	121.3	86.8	102.6	120.2	56.8	85.8	115.2	84.5	98.5	115.5	84.5	98.7	112.8	71.0	90.1
COVERS	DIVS. 10 - 14, 25, 28, 41, 43, 44	140.0	96.6	130.8	140.0	96.3	130.7	140.0	66.0	124.3	140.0	86.1	128.6	140.0	86.1	128.6	140.0	69.2	125.0
21, 22, 23	FIRE SUPPRESSION, PLUMBING & HVAC	98.2	82.5	91.8	99.1	94.8	97.3	98.7	76.1	89.4	99.0	87.3	94.2	98.7	87.3	94.0	97.3	71.6	86.8
26, 27, 3370	ELECTRICAL, COMMUNICATIONS & UTIL.	117.8	81.4	98.8	113.2	84.2	98.0	132.1	75.9	102.7	119.0	71.5	94.2	119.8	71.5	94.5	124.6	71.6	96.9
MF2004	WEIGHTED AVERAGE	130.6	86.2	111.0	126.0	88.8	109.6	115.8	69.2	95.2	113.4	83.8	100.3	113.7	83.8	100.5	117.4	74.6	98.5

DIVISION		CANADA																	
		HAMILTON, ONTARIO			HULL, QUEBEC			JOLIETTE, QUEBEC			KAMLOOPS, BRITISH COLUMBIA			KINGSTON, ONTARIO			KITCHENER, ONTARIO		
		MAT.	INST.	TOTAL	MAT.	INST.	TOTAL	MAT.	INST.	TOTAL	MAT.	INST.	TOTAL	MAT.	INST.	TOTAL	MAT.	INST.	TOTAL
015433	CONTRACTOR EQUIPMENT		108.6	108.6		101.8	101.8		101.8	101.8		105.2	105.2		103.9	103.9		103.8	103.8
0241, 31 - 34	SITE & INFRASTRUCTURE, DEMOLITION	115.0	112.9	113.5	94.0	99.5	97.8	94.7	99.5	98.0	117.5	103.0	107.4	112.1	104.1	106.6	101.5	105.0	103.9
0310	Concrete Forming & Accessories	130.1	91.2	96.5	126.5	82.2	88.2	126.5	82.3	88.3	123.4	85.0	90.2	119.6	83.8	88.6	116.3	83.9	88.3
0320	Concrete Reinforcing	161.3	86.0	122.8	158.1	75.8	116.1	149.6	75.8	111.9	117.2	73.9	95.0	175.4	79.1	126.3	110.8	85.9	98.1
0330	Cast-in-Place Concrete	151.9	97.5	131.3	130.2	91.5	115.5	135.6	91.5	118.9	117.3	95.2	108.9	147.1	94.1	127.0	140.8	78.9	117.4
03	CONCRETE	142.1	92.5	118.9	130.7	84.5	109.1	132.1	84.5	109.9	139.6	86.6	114.8	141.1	86.7	115.7	122.0	82.9	103.7
04	MASONRY	180.9	97.7	130.9	159.9	82.1	113.1	160.3	82.1	113.3	166.7	89.5	120.3	170.1	88.7	121.2	161.5	94.0	120.9
05	METALS	125.8	90.3	113.9	105.0	84.6	98.2	105.0	84.8	98.2	107.4	84.8	99.8	108.5	86.7	101.1	115.3	89.9	106.8
06	WOOD, PLASTICS & COMPOSITES	114.3	90.6	100.9	125.7	82.0	101.0	125.7	82.0	101.0	98.4	83.3	89.8	113.8	83.1	96.4	108.7	82.4	93.8
07	THERMAL & MOISTURE PROTECTION	121.6	92.9	110.7	104.7	84.9	97.2	104.7	84.9	97.2	120.2	82.1	105.8	110.7	84.3	100.7	110.5	89.7	102.7
08	OPENINGS	92.8	87.4	91.4	92.8	71.3	87.4	92.8	75.7	88.5	88.6	79.9	86.5	92.8	80.2	89.6	84.0	81.4	83.3
0920	Plaster & Gypsum Board	192.1	90.3	121.6	142.1	81.3	100.0	184.7	81.3	113.1	138.2	82.4	99.6	231.4	82.6	128.4	154.3	81.9	104.2
0950, 0980	Ceilings & Acoustic Treatment	124.8	90.3	102.8	95.6	81.3	86.5	95.6	81.3	86.5	95.6	82.4	87.2	112.0	82.6	93.3	105.1	81.9	90.3
0960	Flooring	134.5	91.0	122.5	132.1	91.7	120.9	132.1	91.7	120.9	129.7	51.2	108.0	132.1	89.7	120.4	127.8	91.0	117.7
0970, 0990	Wall Finishes & Painting/Coating	110.1	97.1	102.3	110.1	85.6	95.3	110.1	85.6	95.3	110.1	78.4	91.0	110.1	79.3	91.5	110.1	86.7	96.0
09	FINISHES	130.1	91.9	109.3	115.2	84.5	98.5	120.4	84.5	100.9	116.9	78.2	95.9	130.6	84.5	105.6	117.6	85.2	100.0
COVERS	DIVS. 10 - 14, 25, 28, 41, 43, 44	140.0	100.6	131.6	140.0	86.1	128.6	140.0	86.1	128.6	140.0	96.6	130.8	140.0	74.8	126.2	140.0	98.7	131.2
21, 22, 23	FIRE SUPPRESSION, PLUMBING & HVAC	98.3	86.6	93.5	98.7	87.3	94.0	98.7	87.3	94.0	98.7	90.5	95.3	99.0	96.2	97.9	97.3	83.7	91.7
26, 27, 3370	ELECTRICAL, COMMUNICATIONS & UTIL.	127.2	92.3	109.0	121.3	71.5	95.3	119.8	71.5	94.5	123.8	80.7	101.3	120.6	88.3	103.7	121.1	89.8	104.7
MF2004	WEIGHTED AVERAGE	121.9	93.2	109.2	113.6	83.8	100.4	114.1	84.0	100.8	116.4	86.9	103.4	117.8	89.6	105.4	113.3	88.5	102.4

DIVISION		CANADA																	
		LAVAL, QUEBEC			LETHBRIDGE, ALBERTA			LLOYDMINSTER, ALBERTA			LONDON, ONTARIO			MEDICINE HAT, ALBERTA			MONCTON, NEW BRUNSWICK		
		MAT.	INST.	TOTAL	MAT.	INST.	TOTAL	MAT.	INST.	TOTAL	MAT.	INST.	TOTAL	MAT.	INST.	TOTAL	MAT.	INST.	TOTAL
015433	CONTRACTOR EQUIPMENT		101.8	101.8		103.9	103.9		103.9	103.9		104.0	104.0		103.9	103.9		101.2	101.2
0241, 31 - 34	SITE & INFRASTRUCTURE, DEMOLITION	94.6	99.5	98.0	112.0	102.0	105.0	111.8	101.6	104.7	114.6	105.2	108.0	110.5	101.6	104.3	95.3	96.7	96.3
0310	Concrete Forming & Accessories	126.7	82.2	88.2	120.8	86.3	91.0	119.2	77.2	82.9	130.2	85.1	91.2	120.8	77.1	83.0	102.5	59.9	65.7
0320	Concrete Reinforcing	158.1	75.8	116.1	162.4	60.1	110.2	162.4	60.1	110.2	131.1	84.6	107.4	162.4	60.1	110.2	149.6	57.3	102.4
0330	Cast-in-Place Concrete	134.5	91.5	118.2	163.0	96.4	137.8	151.1	92.9	129.1	149.3	95.5	129.0	151.1	92.9	129.1	133.1	58.7	105.0
03	CONCRETE	132.8	84.5	110.2	147.1	85.3	118.2	141.2	80.0	112.6	136.5	88.7	114.2	141.3	80.0	112.6	133.1	59.8	98.8
04	MASONRY	160.2	82.1	113.3	178.0	87.9	123.8	160.2	81.8	113.0	182.2	95.3	130.0	160.2	81.8	113.0	158.5	61.3	100.1
05	METALS	105.1	84.6	98.2	125.3	84.3	111.5	107.6	84.1	99.7	123.4	88.2	111.5	107.6	84.1	99.7	104.7	72.7	93.9
06	WOOD, PLASTICS & COMPOSITES	125.9	82.0	101.0	113.0	85.6	97.5	110.0	76.5	91.0	114.3	83.3	96.7	113.0	76.5	92.3	94.6	59.7	74.8
07	THERMAL & MOISTURE PROTECTION	105.3	84.9	97.6	116.0	83.9	103.8	113.1	79.8	100.5	122.4	89.7	110.1	119.2	79.8	104.3	108.4	60.8	90.4
08	OPENINGS	92.8	71.3	87.4	92.8	76.2	88.7	92.8	71.3	87.4	93.9	81.6	90.8	92.8	71.3	87.4	86.9	53.7	78.7
0920	Plaster & Gypsum Board	148.2	81.3	101.9	138.3	84.8	101.3	133.5	75.4	93.3	195.9	82.7	117.6	135.7	75.4	94.0	162.5	58.4	90.4
0950, 0980	Ceilings & Acoustic Treatment	95.6	81.3	86.5	104.2	84.8	91.9	95.6	75.4	82.7	137.7	82.7	102.7	95.6	75.4	82.7	96.5	58.4	72.2
0960	Flooring	132.1	91.7	120.9	132.1	85.3	119.2	132.1	85.3	119.2	134.7	91.0	122.6	132.1	85.3	119.2	113.9	53.5	97.2
0970, 0990	Wall Finishes & Painting/Coating	110.1	85.6	95.3	110.0	97.3	102.4	110.1	75.8	89.5	110.1	93.7	100.2	110.0	75.8	89.4	110.1	48.2	72.8
09	FINISHES	115.9	84.5	98.9	118.0	87.7	101.6	115.5	78.5	95.4	133.6	86.9	108.2	115.6	78.5	95.5	112.6	57.4	82.7
COVERS	DIVS. 10 - 14, 25, 28, 41, 43, 44	140.0	86.1	128.6	140.0	96.3	130.7	140.0	93.1	130.0	140.0	99.2	131.3	140.0	93.1	130.0	140.0	66.0	124.3
21, 22, 23	FIRE SUPPRESSION, PLUMBING & HVAC	97.3	87.3	93.2	98.9	91.6	95.9	99.0	91.7	96.0	98.3	83.8	92.3	98.7	88.4	94.5	98.7	67.0	85.7
26, 27, 3370	ELECTRICAL, COMMUNICATIONS & UTIL.	120.5	71.5	94.9	115.0	84.2	98.9	112.2	84.2	97.6	119.1	89.8	103.8	112.2	84.2	97.6	127.3	60.9	92.6
MF2004	WEIGHTED AVERAGE	113.5	83.8	100.4	119.9	88.3	105.9	115.1	85.2	101.9	120.6	89.6	106.9	115.3	84.5	101.7	113.5	65.6	92.3

DIVISION		CANADA																	
		MONTREAL, QUEBEC			MOOSE JAW, SASKATCHEWAN			NEW GLASGOW, NOVA SCOTIA			NEWCASTLE, NEW BRUNSWICK			NORTH BAY, ONTARIO			OSHAWA, ONTARIO		
		MAT.	INST.	TOTAL	MAT.	INST.	TOTAL	MAT.	INST.	TOTAL	MAT.	INST.	TOTAL	MAT.	INST.	TOTAL	MAT.	INST.	TOTAL
015433	CONTRACTOR EQUIPMENT		103.4	103.4		99.9	99.9		100.9	100.9		101.2	101.2		101.5	101.5		103.8	103.8
0241, 31 - 34	SITE & INFRASTRUCTURE, DEMOLITION	101.2	99.4	99.9	111.7	96.0	100.8	99.0	98.0	98.3	95.9	96.5	96.3	114.2	99.9	104.3	114.1	104.3	107.3
0310	Concrete Forming & Accessories	134.9	89.6	95.7	106.0	56.7	63.3	97.2	67.0	71.1	102.5	59.6	65.4	121.5	81.4	86.8	121.7	83.8	89.0
0320	Concrete Reinforcing	146.6	89.4	117.4	114.6	59.5	86.5	149.6	44.9	96.1	149.6	54.4	101.0	175.4	78.7	126.0	175.4	81.2	127.3
0330	Cast-in-Place Concrete	174.6	97.8	145.6	146.6	66.9	116.5	169.9	67.9	131.3	138.2	58.3	108.0	163.6	81.6	132.6	162.9	83.3	132.8
03	CONCRETE	151.4	92.4	123.8	124.7	61.5	95.1	148.7	64.0	109.1	135.6	58.9	99.8	149.3	81.3	117.5	149.0	83.5	118.3
04	MASONRY	167.0	90.1	120.8	158.7	60.7	99.8	160.1	67.8	104.6	158.9	61.2	100.2	165.0	84.8	116.8	164.6	90.2	119.9
05	METALS	122.1	90.4	111.4	103.9	72.0	93.1	105.9	72.1	94.5	104.7	69.6	92.9	106.8	86.5	99.9	106.9	88.8	100.8
06	WOOD, PLASTICS & COMPOSITES	125.9	89.9	105.5	96.1	55.1	72.9	87.2	66.2	75.3	94.6	59.7	74.8	115.0	81.7	96.1	115.2	82.2	96.5
07	THERMAL & MOISTURE PROTECTION	118.4	93.3	108.9	104.2	60.9	87.8	105.9	66.3	90.9	108.4	59.1	89.8	110.9	81.0	99.6	111.4	83.1	100.7
08	OPENINGS	92.8	78.5	89.2	87.6	53.0	79.1	83.5	60.5	77.8	86.9	52.2	78.3	91.6	78.1	88.3	91.6	82.0	89.2
0920	Plaster & Gypsum Board	165.5	89.2	112.7	131.0	53.8	77.6	156.2	65.2	93.2	162.5	58.4	90.4	155.8	81.1	104.1	157.5	81.6	105.0
0950, 0980	Ceilings & Acoustic Treatment	114.5	89.2	98.4	95.6	53.8	69.0	95.6	65.2	76.2	96.5	58.4	72.2	95.6	81.1	86.4	99.9	81.6	88.3
0960	Flooring	134.0	91.7	122.3	119.8	58.3	102.8	110.0	62.0	96.7	113.9	43.5	94.5	132.1	89.7	120.4	132.1	93.4	121.4
0970, 0990	Wall Finishes & Painting/Coating	110.1	100.7	104.4	110.1	62.4	81.4	110.1	59.8	79.8	110.1	48.2	72.8	110.1	85.1	95.1	110.1	101.4	104.8
09	FINISHES	123.2	91.8	106.1	111.2	57.2	81.9	110.2	65.3	85.8	112.6	55.4	81.5	117.7	83.6	99.2	118.8	87.3	101.7
COVERS	DIVS. 10 - 14, 25, 28, 41, 43, 44	140.0	88.5	129.1	140.0	65.3	124.1	140.0	67.2	124.6	140.0	66.0	124.3	140.0	73.5	125.9	140.0	98.8	131.2
21, 22, 23	FIRE SUPPRESSION, PLUMBING & HVAC	98.3	88.9	94.4	99.0	73.1	88.3	98.7	81.0	91.4	98.7	66.9	85.6	98.7	94.2	96.8	97.3	97.2	97.3
26, 27, 3370	ELECTRICAL, COMMUNICATIONS & UTIL.	127.6	82.8	104.2	122.5	61.8	90.8	119.4	63.4	90.1	121.4	60.9	89.8	120.3	89.5	104.2	122.4	89.1	105.0
MF2004	WEIGHTED AVERAGE	120.8	89.7	107.1	112.2	66.9	92.2	114.2	71.7	95.5	113.2	64.8	91.8	117.1	87.5	104.0	117.0	90.9	105.5

DIVISION		CANADA																	
		OTTAWA, ONTARIO			OWEN SOUND, ONTARIO			PETERBOROUGH, ONTARIO			PORTAGE LA PRAIRIE, MANITOBA			PRINCE ALBERT, SASKATCHEWAN			PRINCE GEORGE, BRITISH COLUMBIA		
		MAT.	INST.	TOTAL	MAT.	INST.	TOTAL	MAT.	INST.	TOTAL	MAT.	INST.	TOTAL	MAT.	INST.	TOTAL	MAT.	INST.	TOTAL
015433	CONTRACTOR EQUIPMENT		103.8	103.8		101.5	101.5		101.5	101.5		103.8	103.8		99.9	99.9		105.2	105.2
0241, 31 - 34	SITE & INFRASTRUCTURE, DEMOLITION	111.9	104.9	107.0	116.5	100.1	105.1	114.2	100.2	104.5	112.4	99.5	103.4	106.1	96.2	99.2	121.3	103.0	108.6
0310	Concrete Forming & Accessories	131.2	87.7	93.6	121.6	80.0	85.6	121.5	82.4	87.7	120.9	67.8	75.0	106.0	56.5	63.2	110.8	85.0	88.5
0320	Concrete Reinforcing	160.0	84.6	121.5	180.4	80.6	129.5	175.4	79.2	126.3	162.4	51.2	105.6	119.7	59.5	89.0	117.2	73.9	95.0
0330	Cast-in-Place Concrete	153.0	94.6	130.9	167.7	77.7	133.6	163.6	83.2	133.2	151.1	72.2	121.3	132.8	66.8	107.9	147.1	95.2	127.5
03	CONCRETE	142.5	89.7	117.8	152.0	79.7	118.2	149.3	82.4	118.0	134.3	66.9	102.8	118.6	61.4	91.9	153.5	86.6	122.2
04	MASONRY	179.7	93.5	127.9	168.7	90.1	121.5	165.0	91.2	120.6	163.1	63.8	103.4	157.7	60.7	99.4	169.1	89.5	121.2
05	METALS	124.3	89.8	112.6	107.4	86.7	100.4	106.8	86.9	100.1	107.6	74.6	96.5	104.0	71.8	93.1	107.4	84.8	99.8
06	WOOD, PLASTICS & COMPOSITES	114.1	87.1	98.8	117.2	78.6	95.3	115.0	80.4	95.4	112.9	68.5	87.8	96.1	55.1	72.9	98.4	83.3	89.8
07	THERMAL & MOISTURE PROTECTION	124.6	89.2	111.3	111.1	81.8	100.0	110.9	86.1	101.5	104.6	68.0	90.8	104.1	59.8	87.4	114.3	82.1	102.1
08	OPENINGS	92.8	84.5	90.7	91.7	77.5	88.2	91.6	79.8	88.7	92.8	60.5	84.8	86.2	53.0	78.0	88.6	79.9	86.5
0920	Plaster & Gypsum Board	234.2	86.7	132.1	155.8	77.9	101.9	155.8	79.8	103.2	126.8	67.2	85.5	131.0	53.8	77.6	135.4	82.4	98.7
0950, 0980	Ceilings & Acoustic Treatment	135.1	86.7	104.3	95.6	77.9	84.3	95.6	79.8	85.5	95.6	67.2	77.5	95.6	53.8	69.0	95.6	82.4	87.2
0960	Flooring	134.5	89.7	122.1	132.2	91.0	120.8	132.1	89.7	120.4	132.1	64.7	113.5	119.8	58.3	102.8	126.6	51.2	105.8
0970, 0990	Wall Finishes & Painting/Coating	110.1	88.4	97.0	110.9	83.9	94.6	110.1	87.3	96.3	110.1	54.8	76.8	110.1	53.2	75.8	110.1	78.4	91.0
09	FINISHES	137.1	88.1	110.5	118.9	82.5	99.1	117.7	84.2	99.5	114.5	66.0	88.2	111.2	56.1	81.3	115.5	78.2	95.3
COVERS	DIVS. 10 - 14, 25, 28, 41, 43, 44	140.0	97.2	130.9	140.0	73.9	126.0	140.0	74.9	126.2	140.0	68.3	124.8	140.0	65.3	124.1	140.0	96.6	130.8
21, 22, 23	FIRE SUPPRESSION, PLUMBING & HVAC	98.3	84.2	92.5	98.8	94.1	96.8	98.7	97.6	98.2	98.7	80.7	91.3	99.0	65.8	85.4	98.7	90.5	95.3
26, 27, 3370	ELECTRICAL, COMMUNICATIONS & UTIL.	118.9	90.7	104.2	122.2	88.2	104.4	120.3	89.1	104.0	122.1	59.1	89.2	122.5	61.8	90.8	120.4	80.7	99.6
MF2004	WEIGHTED AVERAGE	121.5	90.1	107.6	118.1	87.5	104.6	117.1	89.3	104.8	115.0	71.6	95.9	111.1	65.2	90.9	117.6	86.9	104.1

	DIVISION	CANADA																	
		QUEBEC, QUEBEC			RED DEER, ALBERTA			REGINA, SASKATCHEWAN			RIMOUSKI, QUEBEC			ROUYN-NORANDA, QUEBEC			SAINT HYACINTHE, QUEBEC		
		MAT.	INST.	TOTAL	MAT.	INST.	TOTAL	MAT.	INST.	TOTAL	MAT.	INST.	TOTAL	MAT.	INST.	TOTAL	MAT.	INST.	TOTAL
015433	**CONTRACTOR EQUIPMENT**		103.8	103.8		103.9	103.9		99.9	99.9		101.8	101.8		101.8	101.8		101.8	101.8
0241, 31 - 34	**SITE & INFRASTRUCTURE, DEMOLITION**	99.8	99.4	99.6	110.5	101.6	104.3	113.0	96.0	101.2	94.5	99.1	97.7	94.0	99.5	97.8	94.6	99.5	98.0
0310	Concrete Forming & Accessories	136.5	89.8	96.1	139.2	77.1	85.5	106.2	56.7	63.4	126.5	89.3	94.3	126.5	82.2	88.2	126.5	82.2	88.2
0320	Concrete Reinforcing	140.7	89.4	114.5	162.4	60.1	110.2	127.1	59.6	92.6	112.8	89.4	100.8	158.1	75.8	116.1	158.1	75.8	116.1
0330	Cast-in-Place Concrete	148.3	98.2	129.3	151.1	92.9	129.1	156.3	66.9	122.5	136.9	96.4	121.6	130.2	91.5	115.5	134.5	91.5	118.2
03	**CONCRETE**	137.7	92.6	116.7	142.4	80.0	113.2	131.2	61.5	98.6	127.6	91.8	110.9	130.7	84.5	109.1	132.8	84.5	110.2
04	**MASONRY**	174.6	90.1	123.8	160.2	81.8	113.0	165.9	61.1	102.9	159.6	90.0	117.8	159.9	82.1	113.1	160.2	82.1	113.3
05	**METALS**	123.6	90.6	112.5	107.6	84.1	99.7	104.8	72.0	93.7	104.6	89.9	99.6	105.0	84.6	98.2	105.0	84.6	98.2
06	**WOOD, PLASTICS & COMPOSITES**	126.5	90.0	105.8	113.0	76.5	92.3	96.2	55.1	73.0	125.7	89.7	105.3	125.7	82.0	101.0	125.7	82.0	101.0
07	**THERMAL & MOISTURE PROTECTION**	120.6	93.4	110.4	128.9	79.8	110.3	105.9	61.0	88.9	104.7	92.7	100.2	104.7	84.9	97.2	105.0	84.9	97.4
08	**OPENINGS**	92.8	85.6	91.0	92.8	71.3	87.4	87.6	53.0	79.1	92.3	76.7	88.5	92.8	71.3	87.4	92.8	71.3	87.4
0920	Plaster & Gypsum Board	193.0	89.2	121.2	135.7	75.4	94.0	166.7	53.8	88.5	184.2	89.2	118.4	141.6	81.3	99.9	147.3	81.3	101.6
0950, 0980	Ceilings & Acoustic Treatment	120.2	89.2	100.4	95.6	75.4	82.7	118.9	53.8	77.4	93.9	89.2	90.9	93.9	81.3	85.9	93.9	81.3	85.9
0960	Flooring	134.5	91.7	122.7	134.5	85.3	120.9	119.8	58.3	102.8	132.1	91.7	120.9	132.1	91.7	120.9	132.1	91.7	120.9
0970, 0990	Wall Finishes & Painting/Coating	110.5	100.7	104.6	110.0	75.8	89.4	110.1	62.4	81.4	110.1	100.7	104.4	110.1	85.6	95.3	110.1	85.6	95.3
09	**FINISHES**	128.3	91.8	108.5	116.4	78.5	95.8	121.2	57.2	86.4	119.9	91.6	104.6	114.7	84.5	98.3	115.4	84.5	98.6
COVERS	**DIVS. 10 - 14, 25, 28, 41, 43, 44**	140.0	88.7	129.1	140.0	93.1	130.0	140.0	65.3	124.1	140.0	87.9	128.9	140.0	86.1	128.6	140.0	86.1	128.6
21, 22, 23	**FIRE SUPPRESSION, PLUMBING & HVAC**	98.2	88.9	94.4	98.7	88.4	94.5	97.5	73.1	87.4	98.7	88.8	94.6	98.7	87.3	94.0	96.0	87.3	92.4
26, 27, 3370	**ELECTRICAL, COMMUNICATIONS & UTIL.**	118.7	82.8	99.9	112.2	84.2	97.6	123.6	61.8	91.3	119.8	82.8	100.4	119.8	71.5	94.5	120.5	71.5	94.9
MF2004	**WEIGHTED AVERAGE**	119.4	90.1	106.4	115.9	84.5	102.0	114.2	67.0	93.4	113.3	89.4	102.8	113.4	83.8	100.3	113.2	83.8	100.2

	DIVISION	CANADA																	
		SAINT JOHN, NEW BRUNSWICK			SARNIA, ONTARIO			SASKATOON, SASKATCHEWAN			SAULT STE MARIE, ONTARIO			SHERBROOKE, QUEBEC			SOREL, QUEBEC		
		MAT.	INST.	TOTAL	MAT.	INST.	TOTAL	MAT.	INST.	TOTAL	MAT.	INST.	TOTAL	MAT.	INST.	TOTAL	MAT.	INST.	TOTAL
015433	**CONTRACTOR EQUIPMENT**		101.2	101.2		101.5	101.5		99.9	99.9		101.5	101.5		101.8	101.8		101.8	101.8
0241, 31 - 34	**SITE & INFRASTRUCTURE, DEMOLITION**	96.0	97.4	97.0	112.6	100.3	104.0	107.3	96.2	99.5	102.5	99.9	100.7	94.6	99.5	98.0	94.7	99.5	98.0
0310	Concrete Forming & Accessories	119.2	59.2	67.3	120.3	89.1	93.3	106.2	56.6	63.3	110.9	83.3	87.0	126.5	82.2	88.2	126.5	82.3	88.3
0320	Concrete Reinforcing	149.6	57.4	102.5	125.3	80.5	102.4	119.7	59.5	89.0	112.7	79.0	95.5	158.1	75.8	116.1	149.6	75.8	111.9
0330	Cast-in-Place Concrete	136.2	58.7	106.9	150.9	95.5	130.0	142.0	66.8	113.6	135.6	81.8	115.2	134.5	91.5	118.2	135.6	91.5	118.9
03	**CONCRETE**	135.8	59.6	100.1	135.9	89.8	114.3	123.1	61.5	94.3	119.4	82.3	102.0	132.8	84.5	110.2	132.1	84.5	109.9
04	**MASONRY**	176.9	62.9	108.4	175.5	93.7	126.3	165.5	61.1	102.8	161.4	90.2	118.6	160.2	82.1	113.3	160.3	82.1	113.3
05	**METALS**	104.7	73.2	94.0	106.7	87.3	100.2	104.9	71.9	93.7	105.9	88.1	99.9	105.0	84.6	98.2	105.0	84.8	98.2
06	**WOOD, PLASTICS & COMPOSITES**	115.7	58.4	83.3	114.2	88.2	99.5	94.5	55.1	72.2	102.5	83.7	91.8	125.7	82.0	101.0	125.7	82.0	101.0
07	**THERMAL & MOISTURE PROTECTION**	108.6	60.7	90.5	110.9	89.5	102.8	105.0	59.9	88.0	109.9	84.0	100.1	104.7	84.9	97.2	104.7	84.9	97.2
08	**OPENINGS**	86.8	52.0	78.2	93.9	83.4	91.3	86.7	53.0	78.4	84.4	81.0	83.6	92.8	71.3	87.4	92.8	75.7	88.5
0920	Plaster & Gypsum Board	179.0	57.1	94.6	177.8	87.9	115.5	146.9	53.8	82.4	141.9	83.2	101.2	144.4	81.3	100.7	184.2	81.3	113.0
0950, 0980	Ceilings & Acoustic Treatment	102.5	57.1	73.6	99.9	87.9	92.2	118.9	53.8	77.4	95.6	83.2	87.7	93.9	81.3	85.9	93.9	81.3	85.9
0960	Flooring	127.3	53.5	106.9	132.1	94.8	121.8	119.8	58.3	102.8	123.3	89.7	114.0	132.1	91.7	120.9	132.1	91.7	120.9
0970, 0990	Wall Finishes & Painting/Coating	110.1	66.8	84.0	110.1	98.7	103.2	110.1	53.2	75.8	110.1	91.4	98.8	110.1	85.6	95.3	110.1	85.6	95.3
09	**FINISHES**	120.2	58.7	86.8	121.4	91.5	105.1	118.5	56.1	84.7	112.5	85.5	97.8	115.1	84.5	98.5	119.9	84.5	100.7
COVERS	**DIVS. 10 - 14, 25, 28, 41, 43, 44**	140.0	65.8	124.3	140.0	76.4	126.5	140.0	65.3	124.1	140.0	97.7	131.0	140.0	86.1	128.6	140.0	86.1	128.6
21, 22, 23	**FIRE SUPPRESSION, PLUMBING & HVAC**	98.7	72.8	88.0	98.7	103.4	100.6	97.4	73.1	87.4	98.7	91.0	95.5	99.0	87.3	94.2	98.7	87.3	94.0
26, 27, 3370	**ELECTRICAL, COMMUNICATIONS & UTIL.**	130.6	75.8	101.9	123.9	91.3	106.8	123.8	61.8	91.4	122.3	89.5	105.1	119.8	71.5	94.5	119.8	71.5	94.5
MF2004	**WEIGHTED AVERAGE**	115.9	69.2	95.3	116.8	93.5	106.5	112.7	66.8	92.4	111.7	88.8	101.6	113.8	83.8	100.5	114.0	84.0	100.8

	DIVISION	CANADA																	
		ST CATHARINES, ONTARIO			ST JEROME, QUEBEC			ST JOHNS, NEWFOUNDLAND			SUDBURY, ONTARIO			SUMMERSIDE, PRINCE EDWARD ISLAND			SYDNEY, NOVA SCOTIA		
		MAT.	INST.	TOTAL	MAT.	INST.	TOTAL	MAT.	INST.	TOTAL	MAT.	INST.	TOTAL	MAT.	INST.	TOTAL	MAT.	INST.	TOTAL
015433	**CONTRACTOR EQUIPMENT**		101.5	101.5		101.8	101.8		102.1	102.1		101.5	101.5		100.9	100.9		100.9	100.9
0241, 31 - 34	**SITE & INFRASTRUCTURE, DEMOLITION**	102.4	101.5	101.7	94.0	99.5	97.8	115.4	97.8	103.2	102.5	101.0	101.5	106.9	95.7	99.1	94.1	98.0	96.8
0310	Concrete Forming & Accessories	114.5	88.6	92.1	126.5	82.2	88.2	104.1	61.2	67.0	110.9	84.6	88.1	97.5	54.5	60.3	97.2	67.0	71.1
0320	Concrete Reinforcing	111.7	85.9	98.5	158.1	75.8	116.1	167.7	54.9	110.1	112.7	84.2	98.1	149.6	45.0	96.2	149.6	44.9	96.1
0330	Cast-in-Place Concrete	134.5	97.3	120.4	130.2	91.5	115.5	172.4	73.6	135.0	135.6	93.7	119.7	161.2	57.6	122.1	130.9	67.9	107.1
03	**CONCRETE**	118.9	91.2	105.9	130.7	84.5	109.1	165.2	65.1	118.4	119.4	87.9	104.7	153.4	54.7	107.2	129.5	64.0	98.9
04	**MASONRY**	161.0	97.7	122.9	159.9	82.1	113.1	163.3	62.6	102.8	161.1	92.9	120.1	159.4	59.1	99.1	157.0	67.8	103.4
05	**METALS**	106.0	89.8	100.5	105.0	84.6	98.2	108.1	73.8	96.5	105.9	89.6	100.4	105.9	65.8	92.4	105.9	72.1	94.5
06	**WOOD, PLASTICS & COMPOSITES**	106.4	87.6	95.8	125.7	82.0	101.0	95.8	61.3	76.3	102.5	83.7	91.8	87.5	53.9	68.5	87.2	66.2	75.3
07	**THERMAL & MOISTURE PROTECTION**	110.5	89.9	102.7	104.7	84.9	97.2	109.3	60.8	91.0	109.9	87.8	101.5	105.2	56.8	86.9	105.9	66.3	90.9
08	**OPENINGS**	83.4	84.5	83.7	92.8	71.3	87.4	99.0	57.4	88.7	84.4	81.0	83.6	95.0	46.6	83.0	83.5	60.5	77.8
0920	Plaster & Gypsum Board	136.8	87.2	102.5	141.6	81.3	99.9	165.2	60.1	92.4	141.9	83.2	101.2	156.6	52.4	84.5	156.2	65.2	93.2
0950, 0980	Ceilings & Acoustic Treatment	99.9	87.2	91.8	93.9	81.3	85.9	104.2	60.1	76.1	95.6	83.2	87.7	95.6	52.4	68.1	95.6	65.2	76.2
0960	Flooring	126.1	91.0	116.4	132.1	91.7	120.9	114.7	52.7	97.6	123.3	89.7	114.0	110.0	57.9	95.6	110.0	62.0	96.7
0970, 0990	Wall Finishes & Painting/Coating	110.1	97.1	102.3	110.1	85.6	95.3	110.1	62.4	81.4	110.1	87.8	96.7	110.1	40.1	68.0	110.1	59.8	79.8
09	**FINISHES**	113.7	90.1	100.9	114.7	84.5	98.3	116.0	59.7	85.4	112.5	85.7	97.9	111.1	53.8	80.0	110.2	65.3	85.8
COVERS	**DIVS. 10 - 14, 25, 28, 41, 43, 44**	140.0	76.3	126.5	140.0	86.1	128.6	140.0	66.5	124.4	140.0	98.6	131.2	140.0	65.5	124.2	140.0	67.2	124.6
21, 22, 23	**FIRE SUPPRESSION, PLUMBING & HVAC**	97.3	85.3	92.4	98.7	87.3	94.0	97.6	60.4	82.3	98.7	84.6	92.9	98.7	61.2	83.3	98.7	81.0	91.4
26, 27, 3370	**ELECTRICAL, COMMUNICATIONS & UTIL.**	123.0	90.8	106.2	120.5	71.5	94.9	119.2	63.2	89.9	120.3	90.8	104.9	118.2	51.5	83.4	119.4	63.4	90.1
MF2004	**WEIGHTED AVERAGE**	111.4	90.4	102.1	113.4	83.8	100.4	119.2	66.1	95.7	111.4	89.1	101.6	116.1	60.5	91.6	111.6	71.7	94.0

DIVISION		CANADA																
		THUNDER BAY, ONTARIO			TIMMINS, ONTARIO			TORONTO, ONTARIO			TROIS RIVIERES, QUEBEC			TRURO, NOVA SCOTIA			VANCOUVER, BRITISH COLUMBIA		
		MAT.	INST.	TOTAL	MAT.	INST.	TOTAL	MAT.	INST.	TOTAL	MAT.	INST.	TOTAL	MAT.	INST.	TOTAL	MAT.	INST.	TOTAL
015433	**CONTRACTOR EQUIPMENT**		101.5	101.5		101.5	101.5		103.9	103.9		101.8	101.8		100.9	100.9		112.1	112.1
0241, 31 - 34	**SITE & INFRASTRUCTURE, DEMOLITION**	107.8	101.4	103.3	114.2	99.9	104.3	138.6	105.7	115.7	94.7	99.5	98.0	99.0	98.0	98.3	121.7	107.4	111.7
0310	Concrete Forming & Accessories	121.6	88.9	93.3	121.5	81.4	86.8	131.4	96.3	101.1	126.5	82.3	88.3	97.2	67.0	71.1	126.0	74.4	81.4
0320	Concrete Reinforcing	99.9	85.3	92.4	175.4	78.7	126.0	157.1	86.6	121.1	149.6	75.8	111.9	149.6	45.0	96.1	158.3	68.8	112.6
0330	Cast-in-Place Concrete	148.0	95.9	128.3	163.6	81.6	132.6	147.4	105.5	131.5	135.6	91.5	118.9	169.9	67.9	131.3	162.9	88.2	134.7
03	**CONCRETE**	127.8	90.7	110.5	149.3	81.3	117.5	139.3	97.5	119.8	132.1	84.5	109.9	148.7	64.0	109.1	153.1	78.8	118.3
04	**MASONRY**	161.8	97.5	123.1	165.0	84.8	116.8	193.3	102.7	138.8	160.3	82.1	113.3	160.1	67.8	104.6	177.0	78.0	117.5
05	**METALS**	105.8	88.8	100.0	106.8	86.5	99.9	126.0	91.1	114.2	105.0	84.8	98.2	105.9	72.1	94.5	162.4	84.7	136.1
06	**WOOD, PLASTICS & COMPOSITES**	115.2	87.7	99.7	115.0	81.7	96.1	115.2	95.2	103.9	125.7	82.0	101.0	87.2	66.2	75.3	114.9	71.8	90.5
07	**THERMAL & MOISTURE PROTECTION**	110.7	91.3	103.4	110.9	81.0	99.6	122.6	97.4	113.1	104.7	84.9	97.2	105.9	66.3	90.9	133.0	78.9	112.5
08	**OPENINGS**	82.4	84.0	82.8	91.6	78.1	88.3	91.6	91.8	91.6	92.8	75.7	88.5	83.5	60.5	77.8	94.5	72.4	89.0
0920	Plaster & Gypsum Board	170.4	87.4	112.9	155.8	81.1	104.1	179.3	95.1	121.0	184.2	81.3	113.0	156.2	65.2	93.2	155.7	70.4	96.7
0950, 0980	Ceilings & Acoustic Treatment	95.6	87.4	90.4	95.6	81.1	86.4	133.5	95.1	109.0	93.9	81.3	85.9	95.6	65.2	76.2	148.3	70.4	98.7
0960	Flooring	132.1	52.4	110.1	132.1	89.7	120.4	135.6	96.5	124.8	132.1	91.7	120.9	110.0	62.0	96.7	134.9	88.4	122.0
0970, 0990	Wall Finishes & Painting/Coating	110.1	90.1	98.0	110.1	85.1	95.1	112.0	101.4	105.6	110.1	85.6	95.3	110.1	59.8	79.8	110.0	85.7	95.4
09	**FINISHES**	119.1	83.0	99.5	117.7	83.6	99.2	132.8	97.0	113.4	119.9	84.5	100.7	110.2	65.3	85.8	131.9	77.7	102.5
COVERS	**DIVS. 10 - 14, 25, 28, 41, 43, 44**	140.0	76.7	126.6	140.0	73.5	125.9	140.0	102.4	132.0	140.0	86.1	128.6	140.0	67.2	124.6	140.0	94.1	130.3
21, 22, 23	**FIRE SUPPRESSION, PLUMBING & HVAC**	97.3	85.4	92.4	98.7	94.2	96.8	98.2	90.3	95.0	98.7	87.3	94.0	98.7	81.0	91.4	98.2	69.0	86.2
26, 27, 3370	**ELECTRICAL, COMMUNICATIONS & UTIL.**	121.0	89.1	104.4	122.3	89.5	105.1	121.1	93.0	106.4	119.8	71.5	94.5	119.4	63.4	90.1	121.7	72.1	95.8
MF2004	**WEIGHTED AVERAGE**	112.9	89.1	102.4	117.3	87.5	104.1	122.5	95.8	110.7	114.0	84.0	100.8	114.2	71.7	95.5	128.7	78.6	106.6

DIVISION		CANADA																
		VICTORIA, BRITISH COLUMBIA			WHITEHORSE, YUKON			WINDSOR, ONTARIO			WINNIPEG, MANITOBA			YARMOUTH, NOVA SCOTIA			YELLOWKNIFE, NWT		
		MAT.	INST.	TOTAL	MAT.	INST.	TOTAL	MAT.	INST.	TOTAL	MAT.	INST.	TOTAL	MAT.	INST.	TOTAL	MAT.	INST.	TOTAL
015433	**CONTRACTOR EQUIPMENT**		108.5	108.5		101.2	101.2		101.5	101.5		105.9	105.9		100.9	100.9		101.0	101.0
0241, 31 - 34	**SITE & INFRASTRUCTURE, DEMOLITION**	121.3	106.8	111.2	107.9	96.7	100.1	97.2	101.1	99.9	119.3	101.0	106.5	98.8	98.0	98.2	130.4	100.7	109.7
0310	Concrete Forming & Accessories	110.8	73.6	78.6	106.1	55.8	62.6	121.6	86.0	90.8	130.7	62.4	71.7	97.2	67.0	71.1	106.1	76.3	80.3
0320	Concrete Reinforcing	117.2	68.7	92.4	119.7	57.7	88.1	109.5	84.6	96.8	148.3	53.3	99.8	149.6	44.9	96.1	119.7	61.3	89.9
0330	Cast-in-Place Concrete	147.1	87.4	124.5	146.6	66.1	116.2	137.7	96.9	122.3	183.4	68.8	140.1	168.2	67.9	130.2	243.4	83.6	183.0
03	**CONCRETE**	153.5	77.9	118.1	125.4	60.6	95.1	120.6	89.7	106.2	158.0	63.7	114.0	147.8	64.0	108.6	173.1	76.4	127.9
04	**MASONRY**	168.2	78.0	114.0	158.1	59.2	98.7	161.2	96.1	122.0	179.8	59.8	107.7	160.0	67.8	104.6	166.7	71.8	109.7
05	**METALS**	107.4	80.2	98.2	105.2	71.6	93.8	105.8	89.5	100.3	155.8	74.2	128.2	105.9	72.1	94.5	106.0	77.4	96.3
06	**WOOD, PLASTICS & COMPOSITES**	98.4	71.7	83.3	96.1	54.4	72.5	115.2	84.5	97.8	113.2	63.0	84.8	87.2	66.2	75.3	96.1	77.5	85.6
07	**THERMAL & MOISTURE PROTECTION**	114.3	77.3	100.3	104.1	58.9	87.0	110.5	90.1	102.8	115.7	64.6	96.4	105.9	66.3	90.9	107.1	75.2	95.1
08	**OPENINGS**	88.6	68.9	83.7	86.2	51.8	77.7	82.2	82.7	82.3	92.8	58.0	84.1	83.5	60.5	77.8	86.2	65.5	81.1
0920	Plaster & Gypsum Board	135.4	70.4	90.4	116.8	52.9	72.6	159.0	84.0	107.1	163.4	61.4	92.8	156.2	65.2	93.2	169.6	76.8	105.4
0950, 0980	Ceilings & Acoustic Treatment	95.6	70.4	79.6	95.6	52.9	68.4	95.6	84.0	88.2	136.1	61.4	88.5	95.6	65.2	76.2	130.1	76.8	96.1
0960	Flooring	126.6	42.7	103.4	119.8	56.5	102.3	132.1	91.7	121.0	133.4	64.7	114.4	110.0	62.0	96.7	119.8	86.7	110.6
0970, 0990	Wall Finishes & Painting/Coating	110.1	73.1	87.8	110.1	52.3	75.3	110.1	89.7	97.8	110.1	46.4	71.8	110.1	59.8	79.8	110.1	76.2	89.7
09	**FINISHES**	115.5	68.4	89.9	109.5	55.2	80.0	117.3	87.5	101.1	128.9	61.4	92.2	110.2	65.3	85.8	123.9	77.5	98.7
COVERS	**DIVS. 10 - 14, 25, 28, 41, 43, 44**	140.0	70.3	125.2	140.0	64.6	124.0	140.0	75.8	126.4	140.0	67.2	124.5	140.0	67.2	124.6	140.0	68.8	124.9
21, 22, 23	**FIRE SUPPRESSION, PLUMBING & HVAC**	98.7	67.3	85.8	98.7	71.8	87.6	97.3	85.0	92.3	98.3	61.8	83.3	98.7	81.0	91.4	99.0	91.7	96.0
26, 27, 3370	**ELECTRICAL, COMMUNICATIONS & UTIL.**	121.4	71.8	95.5	123.0	61.3	90.7	126.3	90.8	107.7	125.1	64.4	93.4	119.4	63.4	90.1	123.3	85.3	103.5
MF2004	**WEIGHTED AVERAGE**	117.7	75.6	99.1	112.0	65.9	91.7	112.0	89.4	102.0	127.7	66.7	100.8	114.1	71.7	95.4	120.5	82.0	103.5

Location Factors

Costs shown in RSMeans cost data publications are based on national averages for materials and installation. To adjust these costs to a specific location, simply multiply the base cost by the factor and divide by 100 for that city. The data is arranged alphabetically by state and postal zip code numbers. For a city not listed, use the factor for a nearby city with similar economic characteristics.

STATE/ZIP	CITY	MAT.	INST.	TOTAL
ALABAMA				
350-352	Birmingham	97.4	75.2	87.6
354	Tuscaloosa	96.0	60.2	80.2
355	Jasper	96.3	58.5	79.6
356	Decatur	96.0	61.8	80.9
357-358	Huntsville	96.0	70.1	84.6
359	Gadsden	95.9	59.2	79.7
360-361	Montgomery	97.1	58.3	80.0
362	Anniston	95.2	67.0	82.8
363	Dothan	95.9	53.7	77.3
364	Evergreen	95.4	55.6	77.8
365-366	Mobile	97.1	67.4	84.0
367	Selma	95.6	53.5	77.0
368	Phenix City	96.4	57.2	79.1
369	Butler	95.8	53.9	77.3
ALASKA				
995-996	Anchorage	127.4	112.3	120.7
997	Fairbanks	126.3	113.3	120.5
998	Juneau	126.0	111.7	119.7
999	Ketchikan	139.3	111.7	127.1
ARIZONA				
850,853	Phoenix	99.9	74.6	88.7
851,852	Mesa/Tempe	99.4	64.4	83.9
855	Globe	99.5	60.5	82.3
856-857	Tucson	98.2	69.9	85.7
859	Show Low	99.6	61.6	82.8
860	Flagstaff	101.6	70.4	87.9
863	Prescott	99.1	61.1	82.4
864	Kingman	97.2	67.9	84.3
865	Chambers	97.3	61.8	81.6
ARKANSAS				
716	Pine Bluff	96.8	63.9	82.3
717	Camden	94.1	42.2	71.2
718	Texarkana	96.2	46.5	74.2
719	Hot Springs	93.3	45.3	72.1
720-722	Little Rock	97.2	67.0	83.9
723	West Memphis	95.8	57.4	78.9
724	Jonesboro	96.9	57.9	79.7
725	Batesville	94.2	51.4	75.3
726	Harrison	95.5	52.2	76.4
727	Fayetteville	92.8	51.8	74.7
728	Russellville	94.1	53.8	76.3
729	Fort Smith	97.3	57.8	79.9
CALIFORNIA				
900-902	Los Angeles	101.7	116.0	108.0
903-905	Inglewood	96.8	104.2	100.1
906-908	Long Beach	98.6	104.3	101.1
910-912	Pasadena	98.0	104.2	100.7
913-916	Van Nuys	101.4	104.2	102.6
917-918	Alhambra	100.2	104.2	102.0
919-921	San Diego	102.4	107.5	104.6
922	Palm Springs	98.3	103.0	100.4
923-924	San Bernardino	95.7	103.6	99.2
925	Riverside	101.1	110.6	105.3
926-927	Santa Ana	97.9	104.0	100.6
928	Anaheim	101.1	112.6	106.2
930	Oxnard	102.5	110.6	106.1
931	Santa Barbara	101.7	110.5	105.6
932-933	Bakersfield	102.9	106.6	104.6
934	San Luis Obispo	102.1	101.4	101.8
935	Mojave	98.9	100.4	99.6
936-938	Fresno	102.8	110.6	106.3
939	Salinas	102.9	114.3	107.9
940-941	San Francisco	110.7	140.4	123.8
942,956-958	Sacramento	104.9	112.6	108.3
943	Palo Alto	102.8	119.2	110.0
944	San Mateo	105.8	125.7	114.6
945	Vallejo	103.9	118.8	110.5
946	Oakland	108.5	126.0	116.2
947	Berkeley	107.5	119.2	112.7
948	Richmond	106.8	120.6	112.9
949	San Rafael	108.5	121.4	114.2
950	Santa Cruz	107.7	114.4	110.7

STATE/ZIP	CITY	MAT.	INST.	TOTAL
CALIFORNIA (CONT'D)				
951	San Jose	105.8	131.3	117.1
952	Stockton	103.0	109.7	106.0
953	Modesto	103.0	109.9	106.1
954	Santa Rosa	102.6	123.5	111.9
955	Eureka	104.1	107.3	105.5
959	Marysville	103.3	108.8	105.7
960	Redding	105.4	107.6	106.3
961	Susanville	104.0	107.7	105.6
COLORADO				
800-802	Denver	101.6	85.1	94.3
803	Boulder	98.2	78.9	89.7
804	Golden	100.5	77.8	90.5
805	Fort Collins	102.4	74.4	90.0
806	Greeley	99.7	64.5	84.1
807	Fort Morgan	98.9	77.5	89.5
808-809	Colorado Springs	101.5	78.7	91.4
810	Pueblo	101.5	76.9	90.7
811	Alamosa	102.6	72.5	89.3
812	Salida	102.4	73.4	89.6
813	Durango	102.9	73.2	89.8
814	Montrose	101.6	71.7	88.4
815	Grand Junction	105.4	69.3	89.5
816	Glenwood Springs	102.7	74.8	90.4
CONNECTICUT				
060	New Britain	100.6	117.5	108.1
061	Hartford	101.7	117.6	108.7
062	Willimantic	101.2	117.1	108.2
063	New London	96.9	116.6	105.6
064	Meriden	99.0	118.1	107.4
065	New Haven	102.6	117.6	109.2
066	Bridgeport	101.7	118.2	109.0
067	Waterbury	101.3	118.0	108.7
068	Norwalk	101.2	118.7	109.0
069	Stamford	101.4	125.5	112.0
D.C.				
200-205	Washington	104.0	92.2	98.8
DELAWARE				
197	Newark	99.8	108.7	103.7
198	Wilmington	100.1	108.7	103.9
199	Dover	100.5	108.7	104.1
FLORIDA				
320,322	Jacksonville	98.0	66.3	84.0
321	Daytona Beach	98.2	75.3	88.1
323	Tallahassee	99.2	57.7	80.9
324	Panama City	99.5	53.1	79.0
325	Pensacola	102.2	64.4	85.5
326,344	Gainesville	99.8	69.1	86.2
327-328,347	Orlando	102.2	73.2	89.4
329	Melbourne	100.8	78.5	91.0
330-332,340	Miami	100.1	79.4	90.9
333	Fort Lauderdale	97.5	77.4	88.7
334,349	West Palm Beach	96.0	74.2	86.4
335-336,346	Tampa	99.2	83.3	92.2
337	St. Petersburg	101.6	61.6	83.9
338	Lakeland	98.1	83.0	91.4
339,341	Fort Myers	97.3	73.9	87.0
342	Sarasota	99.4	76.1	89.1
GEORGIA				
300-303,399	Atlanta	97.4	77.8	88.8
304	Statesboro	96.6	54.7	78.1
305	Gainesville	95.4	63.4	81.3
306	Athens	94.8	65.6	81.9
307	Dalton	96.6	61.8	81.2
308-309	Augusta	96.0	66.5	83.0
310-312	Macon	95.8	65.8	82.6
313-314	Savannah	98.4	61.4	82.0
315	Waycross	96.8	61.4	81.2
316	Valdosta	97.1	61.1	81.2
317,398	Albany	97.1	61.7	81.5
318-319	Columbus	97.0	66.2	83.4

Location Factors

STATE/ZIP	CITY	MAT.	INST.	TOTAL
HAWAII				
967	Hilo	113.5	117.9	115.4
968	Honolulu	116.9	117.9	117.3
STATES & POSS.				
969	Guam	135.6	61.7	103.0
IDAHO				
832	Pocatello	101.2	71.4	88.1
833	Twin Falls	102.5	51.8	80.1
834	Idaho Falls	99.7	57.3	81.0
835	Lewiston	110.1	76.8	95.4
836-837	Boise	100.4	71.8	87.7
838	Coeur d'Alene	109.4	73.9	93.7
ILLINOIS				
600-603	North Suburban	98.5	133.3	113.9
604	Joliet	98.5	135.8	115.0
605	South Suburban	98.5	133.3	113.9
606-608	Chicago	99.0	137.6	116.0
609	Kankakee	94.3	124.3	107.5
610-611	Rockford	97.5	122.8	108.7
612	Rock Island	94.8	99.7	96.9
613	La Salle	96.0	119.4	106.3
614	Galesburg	95.8	104.8	99.8
615-616	Peoria	98.7	107.3	102.5
617	Bloomington	95.2	108.5	101.0
618-619	Champaign	98.8	107.3	102.6
620-622	East St. Louis	94.0	106.6	99.6
623	Quincy	94.9	100.0	97.2
624	Effingham	94.2	104.1	98.6
625	Decatur	96.5	104.5	100.0
626-627	Springfield	97.0	105.4	100.7
628	Centralia	92.2	106.1	98.3
629	Carbondale	91.9	101.4	96.1
INDIANA				
460	Anderson	94.5	83.5	89.7
461-462	Indianapolis	97.7	85.7	92.4
463-464	Gary	96.1	108.3	101.5
465-466	South Bend	95.1	84.7	90.5
467-468	Fort Wayne	95.2	79.6	88.3
469	Kokomo	92.0	82.7	87.9
470	Lawrenceburg	91.3	80.4	86.5
471	New Albany	92.5	76.1	85.3
472	Columbus	94.8	82.4	89.3
473	Muncie	96.0	81.5	89.6
474	Bloomington	97.4	81.6	90.4
475	Washington	93.5	83.3	89.0
476-477	Evansville	95.1	86.6	91.4
478	Terre Haute	95.9	87.9	92.3
479	Lafayette	94.5	83.4	89.6
IOWA				
500-503,509	Des Moines	98.7	73.6	87.7
504	Mason City	96.1	59.0	79.7
505	Fort Dodge	96.4	54.9	78.1
506-507	Waterloo	98.4	56.0	79.7
508	Creston	96.7	58.8	80.0
510-511	Sioux City	99.1	66.1	84.5
512	Sibley	97.5	45.7	74.6
513	Spencer	99.2	44.5	75.0
514	Carroll	96.0	49.3	75.4
515	Council Bluffs	100.2	71.6	87.5
516	Shenandoah	96.4	47.4	74.8
520	Dubuque	98.6	72.6	87.2
521	Decorah	97.0	46.8	74.9
522-524	Cedar Rapids	99.5	77.5	89.8
525	Ottumwa	97.0	66.4	83.5
526	Burlington	96.1	67.9	83.6
527-528	Davenport	98.4	86.7	93.2
KANSAS				
660-662	Kansas City	98.6	95.3	97.1
664-666	Topeka	99.0	63.8	83.4
667	Fort Scott	96.7	68.6	84.3
668	Emporia	96.9	57.3	79.4
669	Belleville	98.7	58.7	81.0
670-672	Wichita	97.7	62.7	82.2
673	Independence	98.3	60.9	81.8
674	Salina	99.1	59.2	81.5
675	Hutchinson	93.6	59.0	78.3
676	Hays	97.9	60.1	81.2
677	Colby	98.6	60.0	81.5

STATE/ZIP	CITY	MAT.	INST.	TOTAL
KANSAS (CONT'D)				
678	Dodge City	100.4	61.5	83.3
679	Liberal	97.7	59.9	81.0
KENTUCKY				
400-402	Louisville	97.1	82.4	90.6
403-405	Lexington	95.5	76.3	87.0
406	Frankfort	96.3	76.1	87.4
407-409	Corbin	92.4	63.3	79.6
410	Covington	93.2	97.4	95.1
411-412	Ashland	92.0	98.2	94.8
413-414	Campton	93.3	63.7	80.2
415-416	Pikeville	94.5	78.0	87.2
417-418	Hazard	92.6	56.5	76.7
420	Paducah	91.6	86.7	89.4
421-422	Bowling Green	93.9	82.6	88.9
423	Owensboro	94.3	81.7	88.8
424	Henderson	91.3	87.3	89.5
425-426	Somerset	90.6	67.7	80.5
427	Elizabethtown	90.0	82.3	86.6
LOUISIANA				
700-701	New Orleans	103.2	68.2	87.8
703	Thibodaux	101.1	61.0	83.4
704	Hammond	98.2	55.8	79.5
705	Lafayette	100.9	56.1	81.1
706	Lake Charles	101.1	60.7	83.3
707-708	Baton Rouge	103.0	61.2	84.5
710-711	Shreveport	97.4	57.7	79.9
712	Monroe	97.4	53.4	77.9
713-714	Alexandria	97.5	54.0	78.3
MAINE				
039	Kittery	94.3	81.9	88.9
040-041	Portland	100.1	77.6	90.2
042	Lewiston	97.8	77.6	88.9
043	Augusta	97.7	81.5	90.6
044	Bangor	97.5	77.5	88.7
045	Bath	95.6	81.3	89.3
046	Machias	95.1	80.5	88.7
047	Houlton	95.2	80.7	88.8
048	Rockland	94.2	81.1	88.4
049	Waterville	95.6	81.6	89.4
MARYLAND				
206	Waldorf	100.0	69.2	86.4
207-208	College Park	100.0	83.7	92.8
209	Silver Spring	99.2	75.2	88.6
210-212	Baltimore	97.8	85.2	92.2
214	Annapolis	98.5	79.3	90.0
215	Cumberland	93.3	81.0	87.9
216	Easton	95.0	41.8	71.5
217	Hagerstown	94.5	81.9	88.9
218	Salisbury	95.4	48.8	74.8
219	Elkton	92.3	66.7	81.0
MASSACHUSETTS				
010-011	Springfield	99.6	107.7	103.2
012	Pittsfield	99.2	106.1	102.3
013	Greenfield	96.8	107.3	101.4
014	Fitchburg	95.5	120.6	106.5
015-016	Worcester	99.6	122.1	109.5
017	Framingham	95.0	129.5	110.2
018	Lowell	99.0	129.0	112.2
019	Lawrence	100.3	127.0	112.1
020-022, 024	Boston	102.0	136.5	117.2
023	Brockton	100.6	123.2	110.6
025	Buzzards Bay	94.3	120.9	106.0
026	Hyannis	97.7	120.9	107.9
027	New Bedford	99.7	121.1	109.1
MICHIGAN				
480,483	Royal Oak	92.1	105.7	98.1
481	Ann Arbor	94.9	106.1	99.9
482	Detroit	96.4	112.1	103.3
484-485	Flint	94.6	100.2	97.1
486	Saginaw	94.2	95.3	94.7
487	Bay City	94.4	94.3	94.3
488-489	Lansing	95.7	98.7	97.0
490	Battle Creek	95.0	91.0	93.2
491	Kalamazoo	95.3	88.2	92.2
492	Jackson	92.7	95.5	94.0
493,495	Grand Rapids	96.4	74.8	86.8
494	Muskegon	93.9	85.6	90.2

STATE/ZIP	CITY	MAT.	INST.	TOTAL
MICHIGAN (CONT'D)				
496	Traverse City	92.2	79.8	86.8
497	Gaylord	93.4	74.2	84.9
498-499	Iron mountain	95.5	88.8	92.5
MINNESOTA				
550-551	Saint Paul	99.3	123.4	109.9
553-555	Minneapolis	99.7	126.3	111.5
556-558	Duluth	98.6	114.6	105.6
MINNESOTA (CONT'd)				
559	Rochester	98.5	107.8	102.6
560	Mankato	96.2	105.5	100.3
561	Windom	94.8	77.8	87.3
562	Willmar	94.4	84.4	90.0
563	St. Cloud	96.2	124.2	108.6
564	Brainerd	96.1	107.4	101.1
565	Detroit Lakes	97.9	97.7	97.8
566	Bemidji	97.2	100.1	98.5
567	Thief River Falls	96.7	96.4	96.6
MISSISSIPPI				
386	Clarksdale	95.8	57.8	79.0
387	Greenville	99.6	67.9	85.6
388	Tupelo	97.3	60.4	81.0
389	Greenwood	97.1	57.3	79.5
390-392	Jackson	97.8	67.0	84.2
393	Meridian	96.1	68.6	84.0
394	Laurel	97.1	60.9	81.2
395	Biloxi	97.9	62.2	82.1
396	Mccomb	95.5	56.8	78.4
397	Columbus	97.0	59.2	80.4
MISSOURI				
630-631	St. Louis	98.6	107.2	102.4
633	Bowling Green	97.0	86.6	92.4
634	Hannibal	95.8	76.8	87.4
635	Kirksville	98.9	70.7	86.5
636	Flat River	98.0	87.4	93.3
637	Cape Girardeau	98.4	83.8	92.0
638	Sikeston	96.0	74.2	86.4
639	Poplar Bluff	95.4	74.7	86.3
640-641	Kansas City	98.7	108.2	102.9
644-645	St. Joseph	98.0	87.2	93.3
646	Chillicothe	94.7	66.1	82.1
647	Harrisonville	94.2	94.8	94.5
648	Joplin	96.8	65.2	82.9
650-651	Jefferson City	95.6	85.0	90.9
652	Columbia	96.8	88.2	93.0
653	Sedalia	95.7	78.8	88.2
654-655	Rolla	94.0	75.8	86.0
656-658	Springfield	97.7	77.8	88.9
MONTANA				
590-591	Billings	102.0	72.9	89.2
592	Wolf Point	101.5	68.6	87.0
593	Miles City	99.0	70.4	86.4
594	Great Falls	103.4	72.2	89.6
595	Havre	100.3	71.3	87.5
596	Helena	101.9	70.6	88.1
597	Butte	102.0	72.3	88.9
598	Missoula	99.2	72.7	87.5
599	Kalispell	97.8	70.9	86.0
NEBRASKA				
680-681	Omaha	99.8	75.6	89.2
683-685	Lincoln	97.8	74.1	87.3
686	Columbus	95.8	73.4	85.9
687	Norfolk	97.5	77.1	88.5
688	Grand Island	97.9	79.2	89.6
689	Hastings	96.8	81.5	90.0
690	Mccook	96.5	72.2	85.8
691	North Platte	97.2	81.6	90.4
692	Valentine	98.5	70.6	86.2
693	Alliance	98.7	68.7	85.5
NEVADA				
889-891	Las Vegas	101.0	113.8	106.7
893	Ely	99.6	69.1	86.1
894-895	Reno	100.0	90.5	95.8
897	Carson City	99.9	89.5	95.3
898	Elko	98.2	74.4	87.7
NEW HAMPSHIRE				
030	Nashua	99.9	87.0	94.2
031	Manchester	100.4	87.0	94.5

STATE/ZIP	CITY	MAT.	INST.	TOTAL
NEW HAMPSHIRE (CONT'D)				
032-033	Concord	98.5	83.3	91.8
034	Keene	96.3	51.3	76.4
035	Littleton	96.1	58.2	79.3
036	Charleston	95.8	48.8	75.0
037	Claremont	94.7	48.8	74.5
038	Portsmouth	97.3	90.9	94.4
NEW JERSEY				
070-071	Newark	101.9	124.3	111.8
072	Elizabeth	100.1	123.8	110.6
073	Jersey City	99.0	124.7	110.3
074-075	Paterson	100.9	124.3	111.2
076	Hackensack	98.8	124.8	110.3
077	Long Branch	98.5	122.4	109.1
078	Dover	99.1	124.3	110.2
079	Summit	99.0	123.8	110.0
080,083	Vineland	97.1	122.2	108.2
081	Camden	99.0	122.9	109.5
082,084	Atlantic City	97.6	122.2	108.5
085-086	Trenton	99.0	123.4	109.8
087	Point Pleasant	99.2	122.2	109.3
088-089	New Brunswick	99.7	123.6	110.2
NEW MEXICO				
870-872	Albuquerque	99.8	74.8	88.8
873	Gallup	100.0	74.8	88.9
874	Farmington	100.4	74.8	89.1
875	Santa Fe	101.8	74.8	89.9
877	Las Vegas	98.2	74.6	87.8
878	Socorro	97.8	74.8	87.6
879	Truth/Consequences	97.9	68.9	85.1
880	Las Cruces	96.8	66.7	83.5
881	Clovis	98.5	72.7	87.1
882	Roswell	100.4	72.8	88.2
883	Carrizozo	100.8	74.8	89.3
884	Tucumcari	99.3	72.7	87.6
NEW YORK				
100-102	New York	107.0	166.3	133.2
103	Staten Island	103.1	162.5	129.3
104	Bronx	100.8	162.5	128.0
105	Mount Vernon	100.6	136.0	116.2
106	White Plains	100.7	136.0	116.3
107	Yonkers	105.7	136.1	119.2
108	New Rochelle	100.9	136.0	116.4
109	Suffern	100.7	126.7	112.2
110	Queens	101.8	162.4	128.6
111	Long Island City	103.5	162.4	129.5
112	Brooklyn	103.9	162.4	129.7
113	Flushing	103.4	162.4	129.4
114	Jamaica	101.5	162.4	128.4
115,117,118	Hicksville	101.6	146.4	121.4
116	Far Rockaway	103.6	162.4	129.5
119	Riverhead	102.4	145.9	121.6
120-122	Albany	97.7	97.6	97.7
123	Schenectady	98.2	97.3	97.8
124	Kingston	101.2	117.9	108.6
125-126	Poughkeepsie	100.3	134.3	115.3
127	Monticello	99.6	122.5	109.7
128	Glens Falls	92.2	90.5	91.5
129	Plattsburgh	97.1	87.3	92.8
130-132	Syracuse	99.3	93.0	96.5
133-135	Utica	97.2	88.9	93.5
136	Watertown	99.1	91.9	95.9
137-139	Binghamton	98.8	89.5	94.7
140-142	Buffalo	100.3	102.8	101.4
143	Niagara Falls	97.7	100.8	99.1
144-146	Rochester	99.8	97.4	98.7
147	Jamestown	96.6	86.9	92.3
148-149	Elmira	96.3	87.8	92.6
NORTH CAROLINA				
270,272-274	Greensboro	99.2	49.2	77.1
271	Winston-Salem	99.1	47.8	76.5
275-276	Raleigh	100.3	49.4	77.8
277	Durham	100.3	49.7	78.0
278	Rocky Mount	96.3	40.3	71.6
279	Elizabeth City	97.3	41.6	72.7
280	Gastonia	98.2	47.2	75.7
281-282	Charlotte	99.8	49.7	77.7
283	Fayetteville	100.0	50.1	78.0
284	Wilmington	96.9	47.6	75.1
285	Kinston	94.6	41.6	71.2

STATE/ZIP	CITY	MAT.	INST.	TOTAL
NORTH CAROLINA (CONT'D)				
286	Hickory	94.9	43.9	72.4
287-288	Asheville	97.4	47.2	75.3
289	Murphy	95.8	34.0	68.5
NORTH DAKOTA				
580-581	Fargo	101.8	59.5	83.2
582	Grand Forks	101.5	52.8	80.0
583	Devils Lake	100.3	55.0	80.3
584	Jamestown	100.4	47.1	76.9
585	Bismarck	100.4	60.4	82.8
586	Dickinson	101.3	57.7	82.0
587	Minot	101.5	64.6	85.2
588	Williston	99.6	57.7	81.1
OHIO				
430-432	Columbus	96.9	89.8	93.7
433	Marion	92.9	80.6	87.5
434-436	Toledo	97.1	97.1	97.1
437-438	Zanesville	93.3	80.0	87.5
439	Steubenville	94.9	89.6	92.6
440	Lorain	97.7	91.5	95.0
441	Cleveland	98.0	100.2	99.0
442-443	Akron	98.7	91.9	95.7
444-445	Youngstown	98.1	89.6	94.3
446-447	Canton	98.2	83.7	91.8
448-449	Mansfield	94.9	87.1	91.5
450	Hamilton	94.6	83.6	89.7
451-452	Cincinnati	94.9	86.4	91.1
453-454	Dayton	94.7	82.9	89.5
455	Springfield	94.6	84.3	90.1
456	Chillicothe	93.1	89.1	91.3
457	Athens	95.6	78.4	88.0
458	Lima	96.0	83.3	90.4
OKLAHOMA				
730-731	Oklahoma City	98.6	61.7	82.3
734	Ardmore	95.1	61.2	80.2
735	Lawton	97.9	61.7	81.9
736	Clinton	96.6	59.8	80.3
737	Enid	97.7	59.8	80.9
738	Woodward	95.3	59.9	79.6
739	Guymon	96.4	30.8	67.4
740-741	Tulsa	97.7	54.8	78.8
743	Miami	93.8	62.6	80.1
744	Muskogee	96.9	39.6	71.6
745	Mcalester	93.5	51.0	74.7
746	Ponca City	94.2	59.2	78.8
747	Durant	94.2	58.6	78.5
748	Shawnee	95.9	57.4	78.9
749	Poteau	93.2	61.1	79.0
OREGON				
970-972	Portland	101.1	96.8	99.2
973	Salem	101.0	95.0	98.4
974	Eugene	100.9	94.3	98.0
975	Medford	102.7	92.9	98.4
976	Klamath Falls	102.4	92.7	98.1
977	Bend	101.2	94.4	98.2
978	Pendleton	95.5	94.7	95.1
979	Vale	93.1	86.3	90.1
PENNSYLVANIA				
150-152	Pittsburgh	97.7	102.1	99.6
153	Washington	94.3	102.1	97.7
154	Uniontown	94.6	100.4	97.2
155	Bedford	95.6	91.0	93.6
156	Greensburg	95.6	100.6	97.8
157	Indiana	94.4	99.4	96.6
158	Dubois	96.1	94.7	95.5
159	Johnstown	95.6	94.9	95.3
160	Butler	92.2	101.6	96.3
161	New Castle	92.2	98.9	95.2
162	Kittanning	92.7	102.6	97.1
163	Oil City	92.2	95.6	93.7
164-165	Erie	94.7	93.1	94.0
166	Altoona	94.9	90.3	92.9
167	Bradford	95.6	93.8	94.8
168	State College	95.2	90.7	93.2
169	Wellsboro	96.3	92.8	94.8
170-171	Harrisburg	98.5	95.1	97.0
172	Chambersburg	95.8	89.7	93.1
173-174	York	96.5	95.4	96.0
175-176	Lancaster	94.4	89.5	92.2

STATE/ZIP	CITY	MAT.	INST.	TOTAL
PENNSYLVANIA (CONT'D)				
177	Williamsport	92.9	81.1	87.7
178	Sunbury	95.2	93.8	94.6
179	Pottsville	94.3	97.7	95.8
180	Lehigh Valley	95.7	114.3	103.9
181	Allentown	98.2	110.5	103.6
182	Hazleton	95.1	96.8	95.9
183	Stroudsburg	95.0	105.1	99.5
184-185	Scranton	99.0	100.9	99.8
186-187	Wilkes-Barre	94.8	97.4	95.9
188	Montrose	94.5	98.4	96.2
189	Doylestown	94.8	122.6	107.1
190-191	Philadelphia	100.3	134.2	115.3
193	Westchester	96.4	127.0	109.9
194	Norristown	95.4	132.6	111.8
195-196	Reading	97.9	101.5	99.5
PUERTO RICO				
009	San Juan	121.2	24.5	78.5
RHODE ISLAND				
028	Newport	99.0	112.8	105.1
029	Providence	100.2	112.8	105.8
SOUTH CAROLINA				
290-292	Columbia	98.3	51.2	77.5
293	Spartanburg	96.5	50.8	76.3
294	Charleston	98.1	59.7	81.2
295	Florence	96.3	51.2	76.4
296	Greenville	96.3	50.8	76.2
297	Rock Hill	95.6	48.9	74.9
298	Aiken	96.5	72.0	85.7
299	Beaufort	97.3	44.1	73.8
SOUTH DAKOTA				
570-571	Sioux Falls	99.8	57.8	81.2
572	Watertown	98.0	53.1	78.2
573	Mitchell	96.8	53.0	77.4
574	Aberdeen	100.0	54.0	79.7
575	Pierre	99.3	54.2	79.4
576	Mobridge	97.5	53.2	77.9
577	Rapid City	99.8	54.2	79.7
TENNESSEE				
370-372	Nashville	97.5	73.9	87.1
373-374	Chattanooga	98.6	66.9	84.6
375,380-381	Memphis	96.4	71.6	85.5
376	Johnson City	97.9	54.2	78.6
377-379	Knoxville	94.9	61.8	80.2
382	Mckenzie	96.4	55.6	78.4
383	Jackson	98.4	56.0	79.7
384	Columbia	94.9	62.3	80.5
385	Cookeville	96.3	60.5	80.5
TEXAS				
750	Mckinney	98.8	51.2	77.8
751	Waxahackie	98.7	56.4	80.0
752-753	Dallas	99.6	66.8	85.2
754	Greenville	98.8	40.0	72.9
755	Texarkana	98.3	52.6	78.1
756	Longview	98.9	41.0	73.4
757	Tyler	99.3	55.6	80.0
758	Palestine	95.1	41.5	71.4
759	Lufkin	95.8	45.6	73.6
760-761	Fort Worth	98.2	62.7	82.6
762	Denton	97.1	49.2	76.0
763	Wichita Falls	98.4	56.5	79.9
764	Eastland	96.8	40.8	72.1
765	Temple	95.1	51.0	75.6
766-767	Waco	97.8	58.1	80.3
768	Brownwood	97.9	38.6	71.7
769	San Angelo	97.4	49.3	76.2
770-772	Houston	100.0	70.3	86.9
773	Huntsville	98.1	39.3	72.2
774	Wharton	99.6	42.9	74.6
775	Galveston	97.4	70.1	85.3
776-777	Beaumont	98.1	62.1	82.2
778	Bryan	94.4	63.0	80.6
779	Victoria	99.6	45.2	75.6
780	Laredo	94.9	53.0	76.4
781-782	San Antonio	95.3	63.7	81.3
783-784	Corpus Christi	97.8	52.7	77.9
785	Mc Allen	97.2	48.0	75.5
786-787	Austin	95.5	59.4	79.6

STATE/ZIP	CITY	MAT.	INST.	TOTAL
TEXAS (CONT'D)				
788	Del Rio	96.9	33.7	69.0
789	Giddings	93.9	41.6	70.8
790-791	Amarillo	98.6	60.5	81.8
792	Childress	97.4	53.4	78.0
793-794	Lubbock	99.9	53.8	79.6
795-796	Abilene	98.2	52.8	78.2
797	Midland	99.6	50.5	77.9
798-799,885	El Paso	97.4	52.7	77.6
UTAH				
840-841	Salt Lake City	103.0	65.5	86.5
842,844	Ogden	98.2	65.0	83.6
843	Logan	100.3	65.0	84.7
845	Price	100.7	44.4	75.8
846-847	Provo	100.8	65.4	85.2
VERMONT				
050	White River Jct.	97.6	56.1	79.3
051	Bellows Falls	95.9	67.2	83.2
052	Bennington	96.3	67.0	83.4
053	Brattleboro	96.7	71.5	85.6
054	Burlington	100.6	67.3	85.9
056	Montpelier	97.1	67.7	84.1
057	Rutland	98.8	67.7	85.0
058	St. Johnsbury	97.6	56.6	79.5
059	Guildhall	96.2	56.4	78.6
VIRGINIA				
220-221	Fairfax	99.6	82.8	92.2
222	Arlington	101.2	81.0	92.3
223	Alexandria	100.3	85.4	93.7
224-225	Fredericksburg	98.2	73.9	87.5
226	Winchester	98.9	64.2	83.5
227	Culpeper	98.8	69.6	85.9
228	Harrisonburg	99.0	65.7	84.3
229	Charlottesville	99.4	65.2	84.3
230-232	Richmond	101.0	68.7	86.7
233-235	Norfolk	101.4	69.3	87.2
236	Newport News	100.0	69.2	86.4
237	Portsmouth	99.4	66.2	84.8
238	Petersburg	99.3	68.7	85.8
239	Farmville	98.4	54.5	79.0
240-241	Roanoke	100.7	61.9	83.6
242	Bristol	98.0	56.3	79.6
243	Pulaski	97.7	52.3	77.7
244	Staunton	98.6	60.7	81.9
245	Lynchburg	98.8	66.1	84.3
246	Grundy	98.1	51.7	77.6
WASHINGTON				
980-981,987	Seattle	104.6	104.7	104.6
982	Everett	104.2	95.1	100.1
983-984	Tacoma	104.3	99.4	102.2
985	Olympia	102.5	99.4	101.1
986	Vancouver	105.8	90.8	99.2
988	Wenatchee	103.9	78.5	92.7
989	Yakima	104.6	86.1	96.4
990-992	Spokane	104.9	77.9	93.0
993	Richland	104.6	81.6	94.5
994	Clarkston	102.8	78.3	92.0
WEST VIRGINIA				
247-248	Bluefield	96.7	77.1	88.1
249	Lewisburg	98.6	82.8	91.6
250-253	Charleston	99.8	89.1	95.0
254	Martinsburg	98.3	80.3	90.3
255-257	Huntington	100.2	91.7	96.4
258-259	Beckley	96.6	86.7	92.2
260	Wheeling	100.3	89.2	95.4
261	Parkersburg	99.1	87.8	94.1
262	Buckhannon	98.2	89.0	94.1
263-264	Clarksburg	98.8	87.7	93.9
265	Morgantown	98.9	88.3	94.2
266	Gassaway	98.1	89.3	94.2
267	Romney	98.1	82.7	91.3
268	Petersburg	97.9	85.4	92.4
WISCONSIN				
530,532	Milwaukee	100.5	106.6	103.2
531	Kenosha	100.4	97.6	99.1
534	Racine	99.7	99.1	99.5
535	Beloit	99.6	92.1	96.3
537	Madison	99.7	95.5	97.8

STATE/ZIP	CITY	MAT.	INST.	TOTAL
WISCONSIN (CONT'D)				
538	Lancaster	96.9	86.8	92.4
539	Portage	95.4	90.6	93.3
540	New Richmond	95.7	89.9	93.2
541-543	Green Bay	100.5	88.5	95.2
544	Wausau	94.8	89.4	92.4
545	Rhinelander	98.3	85.2	92.5
546	La Crosse	96.6	89.0	93.2
547	Eau Claire	98.3	90.5	94.8
548	Superior	95.5	93.6	94.7
549	Oshkosh	95.5	86.6	91.6
WYOMING				
820	Cheyenne	100.0	63.2	83.7
821	Yellowstone Nat'l Park	97.4	56.2	79.2
822	Wheatland	98.5	56.2	79.8
823	Rawlins	100.3	56.0	80.7
824	Worland	98.0	54.2	78.7
825	Riverton	99.2	54.2	79.3
826	Casper	100.5	56.0	80.9
827	Newcastle	97.8	56.0	79.3
828	Sheridan	100.7	58.0	81.9
829-831	Rock Springs	102.2	54.2	81.0
CANADIAN FACTORS (reflect Canadian currency)				
ALBERTA				
	Calgary	129.4	86.3	110.4
	Edmonton	130.6	86.2	111.0
	Fort McMurray	126.0	88.8	109.6
	Lethbridge	119.9	88.3	105.9
	Lloydminster	115.1	85.2	101.9
	Medicine Hat	115.3	84.5	101.7
	Red Deer	115.9	84.5	102.0
BRITISH COLUMBIA				
	Kamloops	116.4	86.9	103.4
	Prince George	117.6	86.9	104.1
	Vancouver	128.7	78.6	106.6
	Victoria	117.7	75.6	99.1
MANITOBA				
	Brandon	115.0	72.9	96.4
	Portage la Prairie	115.0	71.6	95.9
	Winnipeg	127.7	66.7	100.8
NEW BRUNSWICK				
	Bathurst	113.2	64.8	91.8
	Dalhousie	113.2	64.8	91.8
	Fredericton	115.8	69.2	95.2
	Moncton	113.5	65.6	92.3
	Newcastle	113.2	64.8	91.8
	St. John	115.9	69.2	95.3
NEWFOUNDLAND				
	Corner Brook	118.7	65.0	95.0
	St Johns	119.2	66.1	95.7
NORTHWEST TERRITORIES				
	Yellowknife	120.5	82.0	103.5
NOVA SCOTIA				
	Bridgewater	114.8	71.7	95.8
	Dartmouth	116.2	71.7	96.5
	Halifax	117.4	74.6	98.5
	New Glasgow	114.2	71.7	95.5
	Sydney	111.6	71.7	94.0
	Truro	114.2	71.7	95.5
	Yarmouth	114.1	71.7	95.4
ONTARIO				
	Barrie	118.0	89.2	105.3
	Brantford	117.1	92.9	106.4
	Cornwall	116.9	89.5	104.8
	Hamilton	121.9	93.2	109.2
	Kingston	117.8	89.6	105.4
	Kitchener	113.3	88.5	102.4
	London	120.6	89.6	106.9
	North Bay	117.1	87.5	104.0
	Oshawa	117.0	90.9	105.5
	Ottawa	121.5	90.1	107.6
	Owen Sound	118.1	87.5	104.6
	Peterborough	117.1	89.3	104.8
	Sarnia	116.8	93.5	106.5
	Sault Ste Marie	111.7	88.8	101.6

STATE/ZIP	CITY	MAT.	INST.	TOTAL
	St. Catharines	111.4	90.4	102.1
	Sudbury	111.4	89.1	101.6
	Thunder Bay	112.9	89.1	102.4
	Timmins	117.3	87.5	104.1
	Toronto	122.5	95.8	110.7
	Windsor	112.0	89.4	102.0
PRINCE EDWARD ISLAND				
	Charlottetown	116.6	60.5	91.8
	Summerside	116.1	60.5	91.6
QUEBEC				
	Cap-de-la-Madeleine	113.8	84.0	100.7
	Charlesbourg	113.8	84.0	100.7
	Chicoutimi	112.8	89.4	102.5
	Gatineau	113.4	83.8	100.3
	Granby	113.7	83.8	100.5
	Hull	113.6	83.8	100.4
	Joliette	114.1	84.0	100.8
	Laval	113.5	83.8	100.4
	Montreal	120.8	89.7	107.1
	Quebec	119.4	90.1	106.4
	Rimouski	113.3	89.4	102.8
	Rouyn-Noranda	113.4	83.8	100.3
	Saint Hyacinthe	113.2	83.8	100.2
	Sherbrooke	113.8	83.8	100.5
	Sorel	114.0	84.0	100.8
	St Jerome	113.4	83.8	100.4
	Trois Rivieres	114.0	84.0	100.8
SASKATCHEWAN				
	Moose Jaw	112.2	66.9	92.2
	Prince Albert	111.1	65.2	90.9
	Regina	114.2	67.0	93.4
	Saskatoon	112.7	66.8	92.4
YUKON				
	Whitehorse	112.0	65.9	91.7

R011105-05 Tips for Accurate Estimating

1. Use pre-printed or columnar forms for orderly sequence of dimensions and locations and for recording telephone quotations.
2. Use only the front side of each paper or form except for certain pre-printed summary forms.
3. Be consistent in listing dimensions: For example, length x width x height. This helps in rechecking to ensure that, the total length of partitions is appropriate for the building area.
4. Use printed (rather than measured) dimensions where given.
5. Add up multiple printed dimensions for a single entry where possible.
6. Measure all other dimensions carefully.
7. Use each set of dimensions to calculate multiple related quantities.
8. Convert foot and inch measurements to decimal feet when listing. Memorize decimal equivalents to .01 parts of a foot (1/8″ equals approximately .01′).
9. Do not "round off" quantities until the final summary.
10. Mark drawings with different colors as items are taken off.
11. Keep similar items together, different items separate.
12. Identify location and drawing numbers to aid in future checking for completeness.
13. Measure or list everything on the drawings or mentioned in the specifications.
14. It may be necessary to list items not called for to make the job complete.
15. Be alert for: Notes on plans such as N.T.S. (not to scale); changes in scale throughout the drawings; reduced size drawings; discrepancies between the specifications and the drawings.
16. Develop a consistent pattern of performing an estimate. For example:
    a. Start the quantity takeoff at the lower floor and move to the next higher floor.
    b. Proceed from the main section of the building to the wings.
    c. Proceed from south to north or vice versa, clockwise or counterclockwise.
    d. Take off floor plan quantities first, elevations next, then detail drawings.
17. List all gross dimensions that can be either used again for different quantities, or used as a rough check of other quantities for verification (exterior perimeter, gross floor area, individual floor areas, etc.).
18. Utilize design symmetry or repetition (repetitive floors, repetitive wings, symmetrical design around a center line, similar room layouts, etc.). Note: Extreme caution is needed here so as not to omit or duplicate an area.
19. Do not convert units until the final total is obtained. For instance, when estimating concrete work, keep all units to the nearest cubic foot, then summarize and convert to cubic yards.
20. When figuring alternatives, it is best to total all items involved in the basic system, then total all items involved in the alternates. Therefore you work with positive numbers in all cases. When adds and deducts are used, it is often confusing whether to add or subtract a portion of an item; especially on a complicated or involved alternate.

R011110-10 Architectural Fees

Tabulated below are typical percentage fees by project size, for good professional architectural service. Fees may vary from those listed depending upon degree of design difficulty and economic conditions in any particular area.

Rates can be interpolated horizontally and vertically. Various portions of the same project requiring different rates should be adjusted proportionately. For alterations, add 50% to the fee for the first $500,000 of project cost and add 25% to the fee for project cost over $500,000.

Architectural fees tabulated below include Structural, Mechanical and Electrical Engineering Fees. They do not include the fees for special consultants such as kitchen planning, security, acoustical, interior design, etc.

Civil Engineering fees are included in the Architectural fee for project sites requiring minimal design such as city sites. However, separate Civil Engineering fees must be added when utility connections require design, drainage calculations are needed, stepped foundations are required, or provisions are required to protect adjacent wetlands.

Building Types	Total Project Size in Thousands of Dollars						
	100	250	500	1,000	5,000	10,000	50,000
Factories, garages, warehouses, repetitive housing	9.0%	8.0%	7.0%	6.2%	5.3%	4.9%	4.5%
Apartments, banks, schools, libraries, offices, municipal buildings	12.2	12.3	9.2	8.0	7.0	6.6	6.2
Churches, hospitals, homes, laboratories, museums, research	15.0	13.6	12.7	11.9	9.5	8.8	8.0
Memorials, monumental work, decorative furnishings	—	16.0	14.5	13.1	10.0	9.0	8.3

General Requirements R0111 Summary of Work

R011110-30 Engineering Fees

Typical **Structural Engineering Fees** based on type of construction and total project size. These fees are included in Architectural Fees.

Type of Construction	Total Project Size (in thousands of dollars)			
	$500	$500-$1,000	$1,000-$5,000	Over $5000
Industrial buildings, factories & warehouses	Technical payroll times 2.0 to 2.5	1.60%	1.25%	1.00%
Hotels, apartments, offices, dormitories, hospitals, public buildings, food stores	↓	2.00%	1.70%	1.20%
Museums, banks, churches and cathedrals	↓	2.00%	1.75%	1.25%
Thin shells, prestressed concrete, earthquake resistive	↓	2.00%	1.75%	1.50%
Parking ramps, auditoriums, stadiums, convention halls, hangars & boiler houses	↓	2.50%	2.00%	1.75%
Special buildings, major alterations, underpinning & future expansion	↓	Add to above 0.5%	Add to above 0.5%	Add to above 0.5%

For complex reinforced concrete or unusually complicated structures, add 20% to 50%.

Typical **Mechanical and Electrical Engineering Fees** are based on the size of the subcontract. The fee structure for both are shown below. These fees are included in Architectural Fees.

Type of Construction	Subcontract Size							
	$25,000	$50,000	$100,000	$225,000	$350,000	$500,000	$750,000	$1,000,000
Simple structures	6.4%	5.7%	4.8%	4.5%	4.4%	4.3%	4.2%	4.1%
Intermediate structures	8.0	7.3	6.5	5.6	5.1	5.0	4.9	4.8
Complex structures	10.1	9.0	9.0	8.0	7.5	7.5	7.0	7.0

For renovations, add 15% to 25% to applicable fee.

General Requirements R0121 Allowances

R012157-20 Construction Time Requirements

Table lists average construction time in months for different types, sizes, and values of building projects. Design time runs 25% to 40% of construction time.

Building Type	Size S.F.	Project Value	Construction Duration
Industrial/Warehouse	100,000	$8,000,000	14 Months
	500,000	$32,000,000	19 Months
	1,000,000	$75,000,000	21 Months
Offices/Retail	50,000	$7,000,000	15 Months
	250,000	$28,000,000	23 months
	500,000	$58,000,000	34 months
Institutional/Hospitals/Laboratory	200,000	$45,000,000	31 months
	500,000	$110,000,000	52 months
	750,000	$160,000,000	55 months
	1,000,000	$210,000,000	60 months

R012909-80 Sales Tax by State

State sales tax on materials is tabulated below (5 states have no sales tax). Many states allow local jurisdictions, such as a county or city, to levy additional sales tax.

Some projects may be sales tax exempt, particularly those constructed with public funds.

State	Tax (%)	State	Tax (%)	State	Tax (%)	State	Tax (%)
Alabama	4	Illinois	6.25	Montana	0	Rhode Island	7
Alaska	0	Indiana	7	Nebraska	5.5	South Carolina	6
Arizona	5.6	Iowa	6	Nevada	6.5	South Dakota	4
Arkansas	6	Kansas	5.3	New Hampshire	0	Tennessee	7
California	8.25	Kentucky	6	New Jersey	7	Texas	6.25
Colorado	2.9	Louisiana	4	New Mexico	5	Utah	4.75
Connecticut	6	Maine	5	New York	4	Vermont	6
Delaware	0	Maryland	6	North Carolina	4.5	Virginia	4
District of Columbia	5.75	Massachusetts	6.25	North Dakota	5	Washington	6.5
Florida	6	Michigan	6	Ohio	7	West Virginia	6
Georgia	4	Minnesota	6.5	Oklahoma	4.5	Wisconsin	5
Hawaii	4	Mississippi	7	Oregon	0	Wyoming	4
Idaho	6	Missouri	4.225	Pennsylvania	6	Average	5.01 %

Sales Tax by Province (Canada)

GST - a value-added tax, which the government imposes on most goods and services provided in or imported into Canada. PST - a retail sales tax, which five of the provinces impose on the price of most goods and some services. QST - a value-added tax, similar to the federal GST, which Quebec imposes. HST - Three provinces have combined their retail sales tax with the federal GST into one harmonized tax.

Province	PST (%)	QST (%)	GST(%)	HST(%)
Alberta	0	0	5	0
British Columbia	7	0	5	0
Manitoba	7	0	5	0
New Brunswick	0	0	0	13
Newfoundland	0	0	0	13
Northwest Territories	0	0	5	0
Nova Scotia	0	0	0	13
Ontario	8	0	5	0
Prince Edward Island	10	0	5	0
Quebec	0	7.5	5	0
Saskatchewan	5	0	5	0
Yukon	0	0	5	0

General Requirements R0129 Payment Procedures

R012909-85 Unemployment Taxes and Social Security Taxes

State Unemployment Tax rates vary not only from state to state, but also with the experience rating of the contractor. The Federal Unemployment Tax rate is 6.2% of the first $7,000 of wages. This is reduced by a credit of up to 5.4% for timely payment to the state. The minimum Federal Unemployment Tax is 0.8% after all credits.

Social Security (FICA) for 2010 is estimated at time of publication to be 7.65% of wages up to $106,800.

R012909-90 Overtime

One way to improve the completion date of a project or eliminate negative float from a schedule is to compress activity duration times. This can be achieved by increasing the crew size or working overtime with the proposed crew.

To determine the costs of working overtime to compress activity duration times, consider the following examples. Below is an overtime efficiency and cost chart based on a five, six, or seven day week with an eight through twelve hour day. Payroll percentage increases for time and one half and double time are shown for the various working days.

Days per Week	Hours per Day	Production Efficiency					Payroll Cost Factors	
		1st Week	2nd Week	3rd Week	4th Week	Average 4 Weeks	@ 1-1/2 Times	@ 2 Times
5	8	100%	100%	100%	100%	100 %	100 %	100 %
	9	100	100	95	90	96.25	105.6	111.1
	10	100	95	90	85	91.25	110.0	120.0
	11	95	90	75	65	81.25	113.6	127.3
	12	90	85	70	60	76.25	116.7	133.3
6	8	100	100	95	90	96.25	108.3	116.7
	9	100	95	90	85	92.50	113.0	125.9
	10	95	90	85	80	87.50	116.7	133.3
	11	95	85	70	65	78.75	119.7	139.4
	12	90	80	65	60	73.75	122.2	144.4
7	8	100	95	85	75	88.75	114.3	128.6
	9	95	90	80	70	83.75	118.3	136.5
	10	90	85	75	65	78.75	121.4	142.9
	11	85	80	65	60	72.50	124.0	148.1
	12	85	75	60	55	68.75	126.2	152.4

General Requirements R0131 Project Management & Coordination

R013113-40 Builder's Risk Insurance

Builder's Risk Insurance is insurance on a building during construction. Premiums are paid by the owner or the contractor. Blasting, collapse and underground insurance would raise total insurance costs above those listed. Floater policy for materials delivered to the job runs $.75 to $1.25 per $100 value. Contractor equipment insurance runs $.50 to $1.50 per $100 value. Insurance for miscellaneous tools to $1,500 value runs from $3.00 to $7.50 per $100 value.

Tabulated below are New England Builder's Risk insurance rates in dollars per $100 value for $1,000 deductible. For $25,000 deductible, rates can be reduced 13% to 34%. On contracts over $1,000,000, rates may be lower than those tabulated. Policies are written annually for the total completed value in place. For "all risk" insurance (excluding flood, earthquake and certain other perils) add $.025 to total rates below.

Coverage	Frame Construction (Class 1)		Brick Construction (Class 4)		Fire Resistive (Class 6)	
	Range	Average	Range	Average	Range	Average
Fire Insurance	$.350 to $ .850	$.600	$.158 to $.189	$.174	$.052 to $.080	$.070
Extended Coverage	.115 to .200	.158	.080 to .105	.101	.081 to .105	.100
Vandalism	.012 to .016	.014	.008 to .011	.011	.008 to .011	.010
Total Annual Rate	$.477 to $1.066	$.772	$.246 to $.305	$.286	$.141 to $.196	$.180

R013113-50 General Contractor's Overhead

There are two distinct types of overhead on a construction project: Project Overhead and Main Office Overhead. Project Overhead includes those costs at a construction site not directly associated with the installation of construction materials. Examples of Project Overhead costs include the following:

1. Superintendent
2. Construction office and storage trailers
3. Temporary sanitary facilities
4. Temporary utilities
5. Security fencing
6. Photographs
7. Clean up
8. Performance and payment bonds

The above Project Overhead items are also referred to as General Requirements and therefore are estimated in Division 1. Division 1 is the first division listed in the CSI MasterFormat but it is usually the last division estimated. The sum of the costs in Divisions 1 through 49 is referred to as the sum of the direct costs.

All construction projects also include indirect costs. The primary components of indirect costs are the contractor's Main Office Overhead and profit. The amount of the Main Office Overhead expense varies depending on the following:

1. Owner's compensation
2. Project managers and estimator's wages
3. Clerical support wages
4. Office rent and utilities
5. Corporate legal and accounting costs
6. Advertising
7. Automobile expenses
8. Association dues
9. Travel and entertainment expenses

These costs are usually calculated as a percentage of annual sales volume. This percentage can range from 35% for a small contractor doing less than $500,000 to 5% for a large contractor with sales in excess of $100 million.

R013113-60 Workers' Compensation Insurance Rates by Trade

The table below tabulates the national averages for Workers' Compensation insurance rates by trade and type of building. The average "Insurance Rate" is multiplied by the "% of Building Cost" for each trade. This produces the "Workers' Compensation Cost" by % of total labor cost, to be added for each trade by building type to determine the weighted average Workers' Compensation rate for the building types analyzed.

Trade	Insurance Rate (% Labor Cost) Range		Insurance Rate (% Labor Cost) Average	% of Building Cost Office Bldgs.	% of Building Cost Schools & Apts.	% of Building Cost Mfg.	Workers' Compensation Office Bldgs.	Workers' Compensation Schools & Apts.	Workers' Compensation Mfg.
Excavation, Grading, etc.	4.2 % to	17.5%	9.4%	4.8%	4.9%	4.5%	0.45%	0.46%	0.42%
Piles & Foundations	4.8 to	57.1	18.5	7.1	5.2	8.7	1.31	0.96	1.61
Concrete	4.8 to	29.9	13.7	5.0	14.8	3.7	0.69	2.03	0.51
Masonry	4.6 to	29.8	13.5	6.9	7.5	1.9	0.93	1.01	0.26
Structural Steel	4.8 to	170.8	36.0	10.7	3.9	17.6	3.85	1.40	6.34
Miscellaneous & Ornamental Metals	4 to	20.3	10.1	2.8	4.0	3.6	0.28	0.40	0.36
Carpentry & Millwork	4.8 to	53.2	16.9	3.7	4.0	0.5	0.63	0.68	0.08
Metal or Composition Siding	4.8 to	35.7	15.5	2.3	0.3	4.3	0.36	0.05	0.67
Roofing	4.8 to	77.1	28.9	2.3	2.6	3.1	0.66	0.75	0.90
Doors & Hardware	3.4 to	34.5	11.3	0.9	1.4	0.4	0.10	0.16	0.05
Sash & Glazing	4.8 to	36	13.3	3.5	4.0	1.0	0.47	0.53	0.13
Lath & Plaster	3.1 to	30.3	11.9	3.3	6.9	0.8	0.39	0.82	0.10
Tile, Marble & Floors	3.1 to	16.8	8.8	2.6	3.0	0.5	0.23	0.26	0.04
Acoustical Ceilings	4.4 to	32	9.8	2.4	0.2	0.3	0.24	0.02	0.03
Painting	4.8 to	29.6	11.5	1.5	1.6	1.6	0.17	0.18	0.18
Interior Partitions	4.8 to	53.2	16.9	3.9	4.3	4.4	0.66	0.73	0.74
Miscellaneous Items	2.4 to	134.5	14.9	5.2	3.7	9.7	0.78	0.55	1.45
Elevators	2.6 to	13.5	6.3	2.1	1.1	2.2	0.13	0.07	0.14
Sprinklers	2.5 to	13.6	7.4	0.5	—	2.0	0.04	—	0.15
Plumbing	3.9 to	13.3	7.6	4.9	7.2	5.2	0.37	0.55	0.40
Heat., Vent., Air Conditioning	3.9 to	21.2	9.1	13.5	11.0	12.9	1.23	1.00	1.17
Electrical	2.9 to	15.4	6.4	10.1	8.4	11.1	0.65	0.54	0.71
Total	2.4 % to	170.8%	—	100.0%	100.0%	100.0%	14.62%	13.15%	16.44%
			Overall Weighted Average	14.74%					

Workers' Compensation Insurance Rates by States

The table below lists the weighted average Workers' Compensation base rate for each state with a factor comparing this with the national average of 14.4%.

State	Weighted Average	Factor	State	Weighted Average	Factor	State	Weighted Average	Factor
Alabama	19.8%	138	Kentucky	16.3%	113	North Dakota	12.9%	90
Alaska	19.6	136	Louisiana	28.2	196	Ohio	13.0	90
Arizona	9.7	67	Maine	14.2	99	Oklahoma	14.5	101
Arkansas	10.4	72	Maryland	14.5	101	Oregon	12.4	86
California	21.4	149	Massachusetts	12.1	84	Pennsylvania	14.7	102
Colorado	9.0	63	Michigan	19.0	132	Rhode Island	11.6	81
Connecticut	20.0	139	Minnesota	26.0	181	South Carolina	19.8	138
Delaware	11.5	80	Mississippi	15.3	106	South Dakota	17.2	119
District of Columbia	11.3	78	Missouri	17.5	122	Tennessee	14.2	99
Florida	12.6	88	Montana	15.0	104	Texas	11.2	78
Georgia	26.8	186	Nebraska	21.2	147	Utah	9.1	63
Hawaii	16.9	117	Nevada	10.3	72	Vermont	15.0	104
Idaho	9.9	69	New Hampshire	19.2	133	Virginia	10.9	76
Illinois	23.0	160	New Jersey	13.3	92	Washington	9.8	68
Indiana	7.2	50	New Mexico	18.4	128	West Virginia	12.0	83
Iowa	11.4	79	New York	6.9	48	Wisconsin	15.0	104
Kansas	8.6	60	North Carolina	16.9	117	Wyoming	5.3	37
			Weighted Average for U.S. is	14.7% of payroll = 100%				

Rates in the following table are the base or manual costs per $100 of payroll for Workers' Compensation in each state. Rates are usually applied to straight time wages only and not to premium time wages and bonuses.

The weighted average skilled worker rate for 35 trades is 14.4%. For bidding purposes, apply the full value of Workers' Compensation directly to total labor costs, or if labor is 38%, materials 42% and overhead and profit 20% of total cost, carry 38/80 x 14.4% =6.8% of cost (before overhead and profit) into overhead. Rates vary not only from state to state but also with the experience rating of the contractor.

Rates are the most current available at the time of publication.

R013113-60 Workers' Compensation Insurance Rates by Trade and State (cont.)

State	Carpentry — 3 stories or less 5651	Carpentry — interior cab. work 5437	Carpentry — general 5403	Concrete Work — NOC 5213	Concrete Work — flat (flr., sdwk.) 5221	Electrical Wiring — inside 5190	Excavation — earth NOC 6217	Excavation — rock 6217	Glaziers 5462	Insulation Work 5479	Lathing 5443	Masonry 5022	Painting & Decorating 5474	Pile Driving 6003	Plastering 5480	Plumbing 5183	Roofing 5551	Sheet Metal Work (HVAC) 5538	Steel Erection — door & sash 5102	Steel Erection — inter., ornam. 5102	Steel Erection — structure 5040	Steel Erection — NOC 5057	Tile Work — (interior ceramic) 5348	Waterproofing 9014	Wrecking 5701
AL	28.11	16.91	33.75	15.19	9.51	10.24	10.24	10.24	22.09	16.53	10.78	23.87	21.44	24.16	15.68	10.18	43.60	10.03	8.28	8.28	41.21	18.79	13.84	7.13	41.21
AK	11.41	14.35	15.47	11.20	10.66	8.55	12.54	12.54	35.96	31.60	8.46	29.81	16.71	31.00	30.26	10.01	35.75	8.86	8.04	8.04	29.65	32.73	6.25	6.29	29.65
AZ	11.55	6.95	17.61	9.58	5.16	4.96	5.32	5.32	7.21	14.50	5.56	7.35	6.53	11.22	5.24	5.77	15.08	6.92	9.96	9.96	17.31	14.27	3.56	3.13	17.31
AR	9.83	6.03	13.66	8.16	6.58	4.73	6.47	6.47	7.41	9.31	5.61	7.02	9.22	12.87	8.87	5.43	17.31	5.37	5.41	5.41	25.61	31.50	5.46	3.15	25.61
CA	34.52	34.52	34.52	12.92	12.92	9.65	15.56	15.56	18.87	14.65	12.69	21.99	16.27	16.32	24.72	13.26	47.80	16.92	13.77	13.77	22.36	17.89	8.82	16.27	17.89
CO	10.43	5.56	7.98	7.19	5.24	3.48	7.24	7.24	6.53	11.47	4.36	9.05	5.77	10.08	5.59	4.64	17.78	4.59	6.83	6.83	27.27	10.66	5.43	3.46	10.66
CT	14.94	15.49	24.99	28.69	11.11	8.50	13.15	13.15	20.92	17.38	17.82	23.28	16.39	20.36	16.27	10.08	34.95	12.68	15.41	15.41	41.41	23.21	13.64	5.64	41.41
DE	11.96	11.96	8.80	8.99	9.49	4.26	7.16	7.16	10.17	8.80	10.97	10.39	11.94	14.39	10.17	6.10	21.13	7.75	9.69	9.69	21.36	9.69	7.19	10.39	21.36
DC	10.70	10.28	7.97	9.28	7.58	6.22	9.12	9.12	16.60	6.06	6.77	8.27	6.06	12.61	10.50	8.51	14.67	8.61	11.03	11.03	24.14	13.86	13.64	3.40	24.14
FL	12.62	7.98	13.59	13.18	5.99	15.44	7.01	7.01	8.77	8.12	5.53	9.08	8.76	38.24	19.38	5.47	21.29	7.98	8.12	8.12	22.59	10.37	5.16	4.53	22.59
GA	35.73	21.35	25.92	17.90	15.31	10.45	16.30	16.30	19.14	22.15	14.61	24.16	24.99	26.75	24.86	12.18	63.28	17.45	20.26	20.26	71.30	39.53	13.97	8.92	71.30
HI	17.38	11.62	28.20	13.74	12.27	7.10	7.73	7.73	20.55	21.19	10.94	17.87	11.33	20.15	15.61	6.06	33.24	8.02	11.11	11.11	31.71	21.70	9.78	11.54	31.71
ID	9.41	6.65	10.29	10.24	5.84	3.48	5.84	5.84	8.87	8.06	9.22	7.35	7.88	11.56	5.48	5.01	26.07	6.56	7.00	7.00	24.33	9.22	9.07	4.91	24.33
IL	25.30	15.57	23.56	29.89	12.90	10.43	10.77	10.77	18.01	20.90	17.31	21.46	11.94	31.27	17.49	12.01	33.86	14.59	19.11	19.11	79.28	21.91	16.76	5.22	79.28
IN	11.86	4.44	6.93	5.02	3.12	2.85	4.38	4.38	6.11	7.10	22.54	4.56	4.86	5.84	3.09	12.76	10.06	4.48	3.95	3.95	12.69	6.07	3.12	2.39	12.69
IA	13.14	10.98	12.06	14.63	7.56	4.65	7.06	7.06	9.01	6.62	5.84	9.42	7.53	9.92	7.28	6.74	22.71	6.08	5.56	5.56	30.86	26.76	8.01	3.61	30.86
KS	12.20	7.16	9.07	6.95	5.58	3.39	5.22	5.22	9.12	7.26	4.46	7.09	7.12	9.74	5.27	4.53	12.45	5.13	4.53	4.53	24.78	16.44	6.24	4.25	16.44
KY	14.98	12.29	25.40	10.80	7.22	5.36	9.95	9.95	13.90	14.09	9.23	7.30	11.88	22.46	12.14	5.65	31.57	11.42	9.86	9.86	60.00	18.72	16.40	4.59	60.00
LA	22.02	24.93	53.17	26.43	15.61	9.88	17.49	17.49	20.14	21.43	24.51	28.33	29.58	31.25	22.37	8.64	77.12	21.20	18.98	18.98	51.76	24.21	13.77	13.78	66.41
ME	15.11	10.18	25.32	18.95	9.40	6.61	8.79	8.79	14.50	11.15	7.69	13.82	13.31	15.89	12.02	9.12	18.69	8.40	8.34	8.34	32.51	19.53	6.58	5.84	32.51
MD	14.33	11.61	13.75	15.69	5.67	5.60	7.82	7.82	11.43	13.56	8.01	9.92	7.12	20.15	12.77	5.69	29.53	6.93	8.31	8.31	52.61	26.15	7.99	5.44	26.15
MA	7.50	5.93	11.92	21.45	6.62	3.17	4.17	4.17	8.93	8.94	6.53	11.62	5.01	4.99	4.89	3.88	30.35	5.74	6.76	6.76	44.61	39.97	6.45	2.50	29.46
MI	20.53	10.95	22.88	15.72	9.41	5.27	10.95	10.95	14.98	13.04	13.59	14.45	13.57	57.09	12.36	7.21	32.64	11.89	11.07	11.07	57.09	19.42	10.83	6.26	57.09
MN	22.68	23.13	32.18	10.25	13.85	6.68	11.88	11.88	27.48	16.33	12.68	19.35	14.58	22.85	12.68	11.28	54.88	16.13	11.70	11.70	170.80	13.03	11.23	6.85	13.03
MS	16.49	10.30	17.28	10.71	7.75	6.48	11.17	11.17	10.34	10.46	6.84	13.96	12.78	30.91	19.04	7.94	34.45	7.79	12.45	12.45	33.05	15.81	8.83	3.94	33.05
MO	20.17	12.95	15.43	19.32	10.66	7.57	10.57	10.57	11.63	15.03	9.23	16.87	12.93	18.60	16.85	10.79	34.32	9.57	11.61	11.61	33.95	41.29	11.92	6.56	33.95
MT	13.10	10.20	23.32	12.63	10.84	5.93	14.51	14.51	8.57	24.07	7.98	12.12	8.43	22.50	10.60	9.71	37.56	10.63	9.09	9.09	22.59	11.57	9.08	7.91	11.57
NE	23.73	18.63	20.88	26.18	14.23	9.80	16.00	16.00	16.90	23.33	11.75	21.90	17.52	20.73	17.43	12.53	36.95	12.48	15.00	15.00	39.83	36.08	10.53	6.83	55.60
NV	14.08	7.98	12.08	9.61	8.20	6.12	7.77	7.77	10.06	7.54	4.60	7.35	7.54	11.58	6.73	6.12	12.82	9.04	8.35	8.35	22.13	16.86	5.83	5.36	16.86
NH	26.15	11.92	19.85	24.52	14.29	7.33	13.00	13.00	10.32	5.63	9.25	21.17	21.90	7.88	10.06	10.99	45.41	8.04	15.28	15.28	64.41	23.21	11.64	6.10	64.41
NJ	14.79	8.84	14.79	15.65	9.31	4.38	8.18	8.18	8.97	12.55	10.73	13.44	11.40	16.59	10.73	5.79	37.97	6.78	9.33	9.33	20.26	16.09	7.96	6.20	21.39
NM	20.42	7.36	18.00	14.67	12.11	8.39	9.67	9.67	21.18	13.47	8.06	15.88	11.90	20.37	15.53	8.71	39.03	11.93	13.94	13.94	44.89	47.45	7.36	6.06	44.89
NY	7.96	3.39	7.17	8.25	6.46	3.17	4.63	4.63	5.71	3.73	6.49	8.74	5.50	8.04	4.21	3.98	14.20	5.50	5.44	5.44	13.31	7.90	3.26	2.83	5.51
NC	16.57	12.67	16.19	18.29	8.30	10.73	11.86	11.86	14.23	12.38	11.28	11.33	11.44	17.48	14.68	9.46	26.41	11.87	10.88	10.88	62.71	20.69	9.08	5.75	62.71
ND	10.87	10.87	10.87	6.50	6.50	3.55	5.58	5.58	10.87	10.87	7.25	17.45	6.87	19.11	7.25	5.19	21.35	5.19	19.14	19.14	19.14	19.14	10.87	21.35	10.90
OH	13.61	7.75	11.13	9.75	9.79	5.67	8.20	8.20	8.42	15.84	31.95	11.77	13.74	13.44	4.26	6.07	30.45	6.93	8.35	8.35	20.46	11.61	9.82	6.91	11.61
OK	13.42	9.17	11.76	11.76	6.34	5.73	10.83	10.83	16.07	19.47	8.45	10.02	8.47	20.25	12.19	7.09	21.13	9.05	13.76	13.76	39.72	23.97	6.50	5.71	39.72
OR	16.06	7.50	14.54	11.48	7.70	4.95	9.25	9.25	12.89	11.25	7.90	12.08	9.99	16.97	11.59	5.41	21.99	7.70	6.54	6.54	22.55	21.87	8.85	5.56	22.55
PA	14.68	14.68	12.10	15.30	11.22	6.32	9.13	9.13	12.06	12.10	12.06	13.10	14.00	16.59	12.06	7.77	29.88	8.29	15.11	15.11	22.64	15.11	8.45	13.10	22.64
RI	10.79	8.37	10.96	13.14	10.29	5.15	7.50	7.50	13.33	17.64	6.04	11.62	9.27	27.21	9.50	4.42	14.82	7.75	5.68	5.68	20.92	14.98	5.62	4.72	14.98
SC	21.43	18.13	22.49	16.89	9.59	11.45	11.78	11.78	16.99	18.58	11.06	13.93	15.50	20.61	15.75	12.36	50.08	13.06	13.91	13.91	41.22	32.53	10.50	6.33	41.22
SD	21.47	9.62	21.86	25.23	7.35	5.33	9.54	9.54	15.06	13.87	8.26	11.62	10.45	20.56	11.87	10.92	26.74	12.70	8.86	8.86	59.15	31.01	7.69	5.99	59.15
TN	14.91	11.20	13.03	12.97	6.75	5.79	13.34	13.34	12.18	9.21	8.32	12.35	11.84	28.35	13.71	6.79	24.19	9.81	8.15	8.15	25.81	25.53	8.36	4.08	25.81
TX	9.17	8.39	9.17	9.00	7.06	5.76	7.88	7.88	8.56	11.97	5.74	10.31	8.18	16.38	8.18	5.88	19.42	12.05	7.87	7.87	33.20	11.24	5.53	6.19	8.70
UT	11.93	6.73	9.82	7.43	6.42	3.47	8.78	8.78	11.14	9.47	4.81	8.51	6.34	9.45	5.50	4.68	18.93	6.11	5.43	5.43	19.45	11.11	5.16	3.24	13.92
VT	11.28	10.68	17.87	12.39	7.41	6.20	12.96	12.96	15.29	16.13	7.53	16.28	8.35	14.60	9.04	8.41	26.14	9.59	9.47	9.47	36.84	30.71	9.76	6.97	36.84
VA	10.22	7.61	9.05	11.32	5.11	4.92	7.70	7.70	10.13	7.82	11.06	8.34	9.02	9.49	9.78	5.63	22.11	6.32	6.90	6.90	32.08	20.56	5.08	2.91	32.08
WA	7.94	7.94	7.94	7.29	7.29	3.15	6.88	6.88	12.21	8.54	7.94	9.48	11.84	16.40	10.06	4.30	17.24	3.92	7.61	7.61	7.61	7.61	8.83	17.24	7.61
WV	14.45	9.91	13.88	11.18	5.99	6.38	8.45	8.45	10.70	9.85	8.29	10.83	10.72	16.15	10.77	6.65	24.50	7.80	7.57	7.57	32.81	7.91	7.67	3.60	17.91
WI	10.46	10.33	15.46	11.52	9.18	5.80	8.13	8.13	11.32	12.60	5.84	14.90	14.06	19.42	10.64	6.15	34.47	7.77	9.80	9.80	22.52	48.12	15.46	4.83	22.52
WY	4.77	4.77	4.77	4.77	4.77	4.77	4.77	4.77	4.77	4.77	4.77	4.77	4.77	4.77	4.77	4.77	4.77	4.77	4.77	4.77	4.77	4.77	4.77	4.77	4.77
AVG.	15.47	11.27	16.88	13.72	8.74	6.38	9.42	9.42	13.27	13.11	9.79	13.47	11.46	18.54	11.92	7.62	28.89	9.14	10.07	10.07	35.99	20.79	8.78	6.48	30.31

R013113-60 Workers' Compensation (cont.) (Canada in Canadian dollars)

Province		Alberta	British Columbia	Manitoba	Ontario	New Brunswick	Newfndld. & Labrador	Northwest Territories	Nova Scotia	Prince Edward Island	Quebec	Saskat-chewan	Yukon
Carpentry—3 stories or less	Rate	5.99	3.64	4.17	4.35	5.05	11.00	3.40	9.23	6.23	13.26	4.37	9.42
	Code	42143	721028	40102	723	238130	4226	4-41	4226	401	80110	1226	202
Carpentry—interior cab. work	Rate	1.86	2.62	4.17	4.35	4.59	4.37	3.40	4.33	3.82	13.26	2.57	9.42
	Code	42133	721021	40102	723	238350	4279	4-41	4274	402	80110	B1127	202
CARPENTRY—general	Rate	5.99	3.64	4.17	4.35	5.05	4.37	3.40	9.23	6.23	13.26	4.37	9.42
	Code	42143	721028	40102	723	238130	4299	4-41	4226	401	80110	12.02	202
CONCRETE WORK—NOC	Rate	4.01	3.87	7.01	16.50	5.05	11.00	3.40	5.19	6.23	15.41	5.44	5.23
	Code	42104	721010	40110	748	238110	4224	4-41	4224	401	80100	B1314	203
CONCRETE WORK—flat	Rate	4.01	3.87	7.01	16.50	5.05	11.00	3.40	5.19	6.23	15.41	5.44	5.23
(flr. sidewalk)	Code	42104	721010	40110	748	238110	4224	4-41	4224	401	80100	B1314	203
ELECTRICAL Wiring—inside	Rate	1.95	1.41	1.90	3.25	2.45	2.91	2.76	2.66	3.82	5.92	2.57	5.23
	Code	42124	721019	40203	704	238210	4261	4-46	4261	402	80170	B1105	206
EXCAVATION—earth NOC	Rate	2.34	3.25	4.02	4.68	5.05	3.84	3.57	2.88	3.13	7.55	2.76	5.23
	Code	40604	721031	40706	711	238190	4214	4-43	4214	404	80030	R1106	207
EXCAVATION—rock	Rate	2.34	3.25	4.02	4.68	5.05	3.84	3.57	2.88	3.13	7.55	2.76	5.23
	Code	40604	721031	40706	711	238190	4214	4-43	4214	404	80030	R1106	207
GLAZIERS	Rate	2.56	3.05	4.17	9.25	5.05	6.11	3.40	5.19	3.82	14.17	5.44	5.23
	Code	42121	715020	40109	751	238150	4233	4-41	4233	402	80150	B1304	212
INSULATION WORK	Rate	2.05	6.91	4.17	9.25	4.59	6.11	3.40	5.19	6.23	13.26	4.37	9.42
	Code	42184	721029	40102	751	238310	4234	4-41	4234	401	80110	B1207	202
LATHING	Rate	4.67	6.27	4.17	4.35	4.59	4.37	3.40	4.33	3.82	13.26	5.44	9.42
	Code	42135	721042	40102	723	238390	4279	4-41	4271	402	80110	B1316	202
MASONRY	Rate	4.01	3.50	4.17	11.15	5.05	6.11	3.40	5.19	6.23	15.41	5.44	9.42
	Code	42102	721037	40102	741	238140	4231	4-41	4231	401	80100	B1318	202
PAINTING & DECORATING	Rate	3.30	3.36	3.28	6.75	4.59	4.37	3.40	4.33	3.82	13.26	4.37	9.42
	Code	42111	721041	40105	719	238320	4275	4-41	4275	402	80110	B1201	202
PILE DRIVING	Rate	3.95	3.82	4.02	6.34	5.05	3.84	3.57	5.19	6.23	7.55	5.44	9.42
	Code	42159	722004	40706	732	238190	4129	4-43	4221	401	80030	B1310	202
PLASTERING	Rate	4.67	6.27	4.59	6.75	4.59	4.37	3.40	4.33	3.82	13.26	4.37	9.42
	Code	42135	721042	40108	719	238390	4271	4-41	4271	402	80110	B1221	202
PLUMBING	Rate	1.95	1.90	3.04	3.98	2.45	3.08	2.76	2.52	3.82	6.19	2.57	5.23
	Code	42122	721043	40204	707	238220	4241	4-46	4241	402	80160	B1101	214
ROOFING	Rate	6.41	6.18	7.37	13.30	5.05	11.00	3.40	9.23	6.23	19.71	5.44	9.42
	Code	42118	721036	40403	728	238160	4236	4-41	4236	401	80130	B1320	202
SHEET METAL WORK (HVAC)	Rate	1.95	1.90	7.37	3.98	2.45	3.08	2.76	2.52	3.82	6.19	2.57	5.23
	Code	42117	721043	40402	707	238220	4244	4-46	4244	402	80160	B1107	208
STEEL ERECTION—door & sash	Rate	2.05	12.66	10.05	16.50	5.05	11.00	3.40	5.19	6.23	21.98	5.44	9.42
	Code	42106	722005	40502	748	238120	4227	4-41	4227	401	80080	B1322	202
STEEL ERECTION—inter., ornam.	Rate	2.05	12.66	10.05	16.50	5.05	11.00	3.40	5.19	6.23	21.98	5.44	9.42
	Code	42106	722005	40502	748	238120	4227	4-41	4227	401	80080	B1322	202
STEEL ERECTION—structure	Rate	2.05	12.66	10.05	16.50	5.05	11.00	3.40	5.19	6.23	21.98	5.44	9.42
	Code	42106	722005	40502	748	238120	4227	4-41	4227	401	80080	B1322	202
STEEL ERECTION—NOC	Rate	2.05	12.66	10.05	16.50	5.05	11.00	3.40	5.19	6.23	21.98	5.44	9.42
	Code	42106	722005	40502	748	238120	4227	4-41	4227	401	80080	B1322	202
TILE WORK—inter. (ceramic)	Rate	2.89	3.62	1.95	6.75	4.59	4.37	3.40	4.33	3.82	13.26	5.44	9.42
	Code	42113	721054	40103	719	238340	4276	4-41	4276	402	80110	B1301	202
WATERPROOFING	Rate	3.30	3.35	4.17	4.35	5.05	4.37	3.40	5.19	3.82	19.71	4.37	9.42
	Code	42139	721016	40102	723	238190	4299	4-41	4239	402	80130	B1217	202
WRECKING	Rate	2.34	4.44	6.37	16.50	5.05	3.84	3.57	2.88	6.23	13.26	2.76	9.42
	Code	40604	721005	40106	748	238190	4211	4-43	4211	401	80110	R1125	202

R013113-80 Performance Bond

This table shows the cost of a Performance Bond for a construction job scheduled to be completed in 12 months. Add 1% of the premium cost per month for jobs requiring more than 12 months to complete. The rates are "standard" rates offered to contractors that the bonding company considers financially sound and capable of doing the work. Preferred rates are offered by some bonding companies based upon financial strength of the contractor. Actual rates vary from contractor to contractor and from bonding company to bonding company. Contractors should prequalify through a bonding agency before submitting a bid on a contract that requires a bond.

Contract Amount	Building Construction Class B Projects	Highways & Bridges	
		Class A New Construction	Class A-1 Highway Resurfacing
First $ 100,000 bid	$25.00 per M	$15.00 per M	$9.40 per M
Next 400,000 bid	$ 2,500 plus $15.00 per M	$ 1,500 plus $10.00 per M	$ 940 plus $7.20 per M
Next 2,000,000 bid	8,500 plus 10.00 per M	5,500 plus 7.00 per M	3,820 plus 5.00 per M
Next 2,500,000 bid	28,500 plus 7.50 per M	19,500 plus 5.50 per M	15,820 plus 4.50 per M
Next 2,500,000 bid	47,250 plus 7.00 per M	33,250 plus 5.00 per M	28,320 plus 4.50 per M
Over 7,500,000 bid	64,750 plus 6.00 per M	45,750 plus 4.50 per M	39,570 plus 4.00 per M

General Requirements R0151 Temporary Utilities

R015113-65 Temporary Power Equipment

Cost data for the temporary equipment was developed utilizing the following information.

1) Re-usable material-services, transformers, equipment and cords are based on new purchase and prorated to three projects.
2) PVC feeder includes trench and backfill.
3) Connections include disconnects and fuses.
4) Labor units include an allowance for removal.
5) No utility company charges or fees are included.
6) Concrete pads or vaults are not included.
7) Utility company conduits not included.

R015423-10 Steel Tubular Scaffolding

On new construction, tubular scaffolding is efficient up to 60′ high or five stories. Above this it is usually better to use a hung scaffolding if construction permits. Swing scaffolding operations may interfere with tenants. In this case, the tubular is more practical at all heights.

In repairing or cleaning the front of an existing building the cost of tubular scaffolding per S.F. of building front increases as the height increases above the first tier. The first tier cost is relatively high due to leveling and alignment.

The minimum efficient crew for erecting and dismantling is three workers. They can set up and remove 18 frame sections per day up to 5 stories high. For 6 to 12 stories high, a crew of four is most efficient. Use two or more on top and two on the bottom for handing up or hoisting. They can also set up and remove 18 frame sections per day. At 7' horizontal spacing, this will run about 800 S.F. per day of erecting and dismantling. Time for placing and removing planks must be added to the above. A crew of three can place and remove 72 planks per day up to 5 stories. For over 5 stories, a crew of four can place and remove 80 planks per day.

The table below shows the number of pieces required to erect tubular steel scaffolding for 1000 S.F. of building frontage. This area is made up of a scaffolding system that is 12 frames (11 bays) long by 2 frames high.

For jobs under twenty-five frames, add 50% to rental cost. Rental rates will be lower for jobs over three months duration. Large quantities for long periods can reduce rental rates by 20%.

Description of Component	Number of Pieces for 1000 S.F. of Building Front	Unit
5′ Wide Standard Frame, 6′-4″ High	24	Ea.
Leveling Jack & Plate	24	
Cross Brace	44	
Side Arm Bracket, 21″	12	
Guardrail Post	12	
Guardrail, 7′ section	22	
Stairway Section	2	
Stairway Starter Bar	1	
Stairway Inside Handrail	2	
Stairway Outside Handrail	2	
Walk-Thru Frame Guardrail	2	↓

Scaffolding is often used as falsework over 15′ high during construction of cast-in-place concrete beams and slabs. Two foot wide scaffolding is generally used for heavy beam construction. The span between frames depends upon the load to be carried with a maximum span of 5′.

Heavy duty shoring frames with a capacity of 10,000#/leg can be spaced up to 10′ O.C. depending upon form support design and loading.

Scaffolding used as horizontal shoring requires less than half the material required with conventional shoring.

On new construction, erection is done by carpenters.

Rolling towers supporting horizontal shores can reduce labor and speed the job. For maintenance work, catwalks with spans up to 70′ can be supported by the rolling towers.

R015423-20 Pump Staging

Pump staging is generally not available for rent. The table below shows the number of pieces required to erect pump staging for 2400 S.F. of building frontage. This area is made up of a pump jack system that is 3 poles (2 bays) wide by 2 poles high.

Item	Number of Pieces for 2400 S.F. of Building Front	Unit
Aluminum pole section, 24′ long	6	Ea.
Aluminum splice joint, 6′ long	3	
Aluminum foldable brace	3	
Aluminum pump jack	3	
Aluminum support for workbench/back safety rail	3	
Aluminum scaffold plank/workbench, 14″ wide x 24′ long	4	
Safety net, 22′ long	2	
Aluminum plank end safety rail	2	↓

The cost in place for this 2400 S.F. will depend on how many uses are realized during the life of the equipment.

R015433-10 Contractor Equipment

Rental Rates shown elsewhere in the book pertain to late model high quality machines in excellent working condition, rented from equipment dealers. Rental rates from contractors may be substantially lower than the rental rates from equipment dealers depending upon economic conditions; for older, less productive machines, reduce rates by a maximum of 15%. Any overtime must be added to the base rates. For shift work, rates are lower. Usual rule of thumb is 150% of one shift rate for two shifts; 200% for three shifts.

For periods of less than one week, operated equipment is usually more economical to rent than renting bare equipment and hiring an operator.

Costs to move equipment to a job site (mobilization) or from a job site (demobilization) are not included in rental rates, nor in any Equipment costs on any Unit Price line items or crew listings. These costs can be found elsewhere. If a piece of equipment is already at a job site, it is not appropriate to utilize mob/demob costs in an estimate again.

Rental rates vary throughout the country with larger cities generally having lower rates. Lease plans for new equipment are available for periods in excess of six months with a percentage of payments applying toward purchase.

Monthly rental rates vary from 2% to 5% of the cost of the equipment depending on the anticipated life of the equipment and its wearing parts. Weekly rates are about 1/3 the monthly rates and daily rental rates about 1/3 the weekly rate.

The hourly operating costs for each piece of equipment include costs to the user such as fuel, oil, lubrication, normal expendables for the equipment, and a percentage of mechanic's wages chargeable to maintenance. The hourly operating costs listed do not include the operator's wages.

The daily cost for equipment used in the standard crews is figured by dividing the weekly rate by five, then adding eight times the hourly operating cost to give the total daily equipment cost, not including the operator. This figure is in the right hand column of the Equipment listings under Equipment Cost/Day.

Pile Driving rates shown for pile hammer and extractor do not include leads, crane, boiler or compressor. Vibratory pile driving requires an added field specialist during set-up and pile driving operation for the electric model. The hydraulic model requires a field specialist for set-up only. Up to 125 reuses of sheet piling are possible using vibratory drivers. For normal conditions, crane capacity for hammer type and size are as follows.

Crane Capacity	Hammer Type and Size		
	Air or Steam	Diesel	Vibratory
25 ton	to 8,750 ft.-lb.		70 H.P.
40 ton	15,000 ft.-lb.	to 32,000 ft.-lb.	170 H.P.
60 ton	25,000 ft.-lb.		300 H.P.
100 ton		112,000 ft.-lb.	

Cranes should be specified for the job by size, building and site characteristics, availability, performance characteristics, and duration of time required.

Backhoes & Shovels rent for about the same as equivalent size cranes but maintenance and operating expense is higher. Crane operators rate must be adjusted for high boom heights. Average adjustments: for 150′ boom add 2% per hour; over 185′, add 4% per hour; over 210′, add 6% per hour; over 250′, add 8% per hour and over 295′, add 12% per hour.

Tower Cranes of the climbing or static type have jibs from 50′ to 200′ and capacities at maximum reach range from 4,000 to 14,000 pounds. Lifting capacities increase up to maximum load as the hook radius decreases.

Typical rental rates, based on purchase price are about 2% to 3% per month.

Erection and dismantling runs between 500 and 2000 labor hours. Climbing operation takes 10 labor hours per 20′ climb. Crane dead time is about 5 hours per 40′ climb. If crane is bolted to side of the building add cost of ties and extra mast sections. Climbing cranes have from 80′ to 180′ of mast while static cranes have 80′ to 800′ of mast.

Truck Cranes can be converted to tower cranes by using tower attachments. Mast heights over 400′ have been used.

A single 100′ high material **Hoist and Tower** can be erected and dismantled in about 400 labor hours; a double 100′ high hoist and tower in about 600 labor hours. Erection times for additional heights are 3 and 4 labor hours per vertical foot respectively up to 150′, and 4 to 5 labor hours per vertical foot over 150′ high. A 40′ high portable Buck hoist takes about 160 labor hours to erect and dismantle. Additional heights take 2 labor hours per vertical foot to 80′ and 3 labor hours per vertical foot for the next 100′. Most material hoists do not meet local code requirements for carrying personnel.

A 150′ high **Personnel Hoist** requires about 500 to 800 labor hours to erect and dismantle. Budget erection time at 5 labor hours per vertical foot for all trades. Local code requirements or labor scarcity requiring overtime can add up to 50% to any of the above erection costs.

Earthmoving Equipment: The selection of earthmoving equipment depends upon the type and quantity of material, moisture content, haul distance, haul road, time available, and equipment available. Short haul cut and fill operations may require dozers only, while another operation may require excavators, a fleet of trucks, and spreading and compaction equipment. Stockpiled material and granular material are easily excavated with front end loaders. Scrapers are most economically used with hauls between 300′ and 1-1/2 miles if adequate haul roads can be maintained. Shovels are often used for blasted rock and any material where a vertical face of 8′ or more can be excavated. Special conditions may dictate the use of draglines, clamshells, or backhoes. Spreading and compaction equipment must be matched to the soil characteristics, the compaction required and the rate the fill is being supplied.

R015433-15 Heavy Lifting

Hydraulic Climbing Jacks

The use of hydraulic heavy lift systems is an alternative to conventional type crane equipment. The lifting, lowering, pushing, or pulling mechanism is a hydraulic climbing jack moving on a square steel jackrod from 1-5/8″ to 4″ square, or a steel cable. The jackrod or cable can be vertical or horizontal, stationary or movable, depending on the individual application. When the jackrod is stationary, the climbing jack will climb the rod and push or pull the load along with itself. When the climbing jack is stationary, the jackrod is movable with the load attached to the end and the climbing jack will lift or lower the jackrod with the attached load. The heavy lift system is normally operated by a single control lever located at the hydraulic pump.

The system is flexible in that one or more climbing jacks can be applied wherever a load support point is required, and the rate of lift synchronized.

Economic benefits have been demonstrated on projects such as: erection of ground assembled roofs and floors, complete bridge spans, girders and trusses, towers, chimney liners and steel vessels, storage tanks, and heavy machinery. Other uses are raising and lowering offshore work platforms, caissons, tunnel sections and pipelines.

Existing Conditions — R0241 Demolition

R024119-10 Demolition Defined

Whole Building Demolition - Demolition of the whole building with no concern for any particular building element, component, or material type being demolished. This type of demolition is accomplished with large pieces of construction equipment that break up the structure, load it into trucks and haul it to a disposal site, but disposal or dump fees are not included. Demolition of below-grade foundation elements, such as footings, foundation walls, grade beams, slabs on grade, etc., is not included. Certain mechanical equipment containing flammable liquids or ozone-depleting refrigerants, electric lighting elements, communication equipment components, and other building elements may contain hazardous waste, and must be removed, either selectively or carefully, as hazardous waste before the building can be demolished.

Foundation Demolition - Demolition of below-grade foundation footings, foundation walls, grade beams, and slabs on grade. This type of demolition is accomplished by hand or pneumatic hand tools, and does not include saw cutting, or handling, loading, hauling, or disposal of the debris.

Gutting - Removal of building interior finishes and electrical/mechanical systems down to the load-bearing and sub-floor elements of the rough building frame, with no concern for any particular building element, component, or material type being demolished. This type of demolition is accomplished by hand or pneumatic hand tools, and includes loading into trucks, but not hauling, disposal or dump fees, scaffolding, or shoring. Certain mechanical equipment containing flammable liquids or ozone-depleting refrigerants, electric lighting elements, communication equipment components, and other building elements may contain hazardous waste, and must be removed, either selectively or carefully, as hazardous waste, before the building is gutted.

Selective Demolition - Demolition of a selected building element, component, or finish, with some concern for surrounding or adjacent elements, components, or finishes (see the first Subdivision (s) at the beginning of appropriate Divisions). This type of demolition is accomplished by hand or pneumatic hand tools, and does not include handling, loading, storing, hauling, or disposal of the debris, scaffolding, or shoring. "Gutting" methods may be used in order to save time, but damage that is caused to surrounding or adjacent elements, components, or finishes may have to be repaired at a later time.

Careful Removal - Removal of a piece of service equipment, building element or component, or material type, with great concern for both the removed item and surrounding or adjacent elements, components or finishes. The purpose of careful removal may be to protect the removed item for later re-use, preserve a higher salvage value of the removed item, or replace an item while taking care to protect surrounding or adjacent elements, components, connections, or finishes from cosmetic and/or structural damage. An approximation of the time required to perform this type of removal is 1/3 to 1/2 the time it would take to install a new item of like kind (see Reference Number R220105-10). This type of removal is accomplished by hand or pneumatic hand tools, and does not include loading, hauling, or storing the removed item, scaffolding, shoring, or lifting equipment.

Cutout Demolition - Demolition of a small quantity of floor, wall, roof, or other assembly, with concern for the appearance and structural integrity of the surrounding materials. This type of demolition is accomplished by hand or pneumatic hand tools, and does not include saw cutting, handling, loading, hauling, or disposal of debris, scaffolding, or shoring.

Rubbish Handling - Work activities that involve handling, loading or hauling of debris. Generally, the cost of rubbish handling must be added to the cost of all types of demolition, with the exception of whole building demolition.

Minor Site Demolition - Demolition of site elements outside the footprint of a building. This type of demolition is accomplished by hand or pneumatic hand tools, or with larger pieces of construction equipment, and may include loading a removed item onto a truck (check the Crew for equipment used). It does not include saw cutting, hauling or disposal of debris, and, sometimes, handling or loading.

R024119-20 Dumpsters

Dumpster rental costs on construction sites are presented in two ways.

The cost per week rental includes the delivery of the dumpster; its pulling or emptying once per week, and its final removal. The assumption is made that the dumpster contractor could choose to empty a dumpster by simply bringing in an empty unit and removing the full one. These costs also include the disposal of the materials in the dumpster.

The Alternate Pricing can be used when actual planned conditions are not approximated by the weekly numbers. For example, these lines can be used when a dumpster is needed for 4 weeks and will need to be emptied 2 or 3 times per week. Conversely the Alternate Pricing lines can be used when a dumpster will be rented for several weeks or months but needs to be emptied only a few times over this period.

Existing Conditions — R0265 Underground Storage Tank Removal

R026510-20 Underground Storage Tank Removal

Underground Storage Tank Removal can be divided into two categories: Non-Leaking and Leaking. Prior to removing an underground storage tank, tests should be made, with the proper authorities present, to determine whether a tank has been leaking or the surrounding soil has been contaminated.

To safely remove Liquid Underground Storage Tanks:

1. Excavate to the top of the tank.
2. Disconnect all piping.
3. Open all tank vents and access ports.
4. Remove all liquids and/or sludge.
5. Purge the tank with an inert gas.
6. Provide access to the inside of the tank and clean out the interior using proper personal protective equipment (PPE).
7. Excavate soil surrounding the tank using proper PPE for on-site personnel.
8. Pull and properly dispose of the tank.
9. Clean up the site of all contaminated material.
10. Install new tanks or close the excavation.

R028213-20 Asbestos Removal Process

Asbestos removal is accomplished by a specialty contractor who understands the federal and state regulations regarding the handling and disposal of the material. The process of asbestos removal is divided into many individual steps. An accurate estimate can be calculated only after all the steps have been priced.

The steps are generally as follows:

1. Obtain an asbestos abatement plan from an industrial hygienist.
2. Monitor the air quality in and around the removal area and along the path of travel between the removal area and transport area. This establishes the background contamination.
3. Construct a two part decontamination chamber at entrance to removal area.
4. Install a HEPA filter to create a negative pressure in the removal area.
5. Install wall, floor and ceiling protection as required by the plan, usually 2 layers of fireproof 6 mil polyethylene.
6. Industrial hygienist visually inspects work area to verify compliance with plan.
7. Provide temporary supports for conduit and piping affected by the removal process.
8. Proceed with asbestos removal and bagging process. Monitor air quality as described in Step #2. Discontinue operations when contaminate levels exceed applicable standards.
9. Document the legal disposal of materials in accordance with EPA standards.
10. Thoroughly clean removal area including all ledges, crevices and surfaces.
11. Post abatement inspection by industrial hygienist to verify plan compliance.
12. Provide a certificate from a licensed industrial hygienist attesting that contaminate levels are within acceptable standards before returning area to regular use.

Existing Conditions R0283 Lead Remediation

R028319-60 Lead Paint Remediation Methods

Lead paint remediation can be accomplished by the following methods.

1. Abrasive blast
2. Chemical stripping
3. Power tool cleaning with vacuum collection system
4. Encapsulation
5. Remove and replace
6. Enclosure

Each of these methods has strengths and weakness depending on the specific circumstances of the project. The following is an overview of each method.

1. **Abrasive blasting** is usually accomplished with sand or recyclable metallic blast. Before work can begin, the area must be contained to ensure the blast material with lead does not escape to the atmosphere. The use of vacuum blast greatly reduces the containment requirements. Lead abatement equipment that may be associated with this work includes a negative air machine. In addition, it is necessary to have an industrial hygienist monitor the project on a continual basis. When the work is complete, the spent blast sand with lead must be disposed of as a hazardous material. If metallic shot was used, the lead is separated from the shot and disposed of as hazardous material. Worker protection includes disposable clothing and respiratory protection.
2. **Chemical stripping** requires strong chemicals be applied to the surface to remove the lead paint. Before the work can begin, the area under/adjacent to the work area must be covered to catch the chemical and removed lead. After the chemical is applied to the painted surface it is usually covered with paper. The chemical is left in place for the specified period, then the paper with lead paint is pulled or scraped off. The process may require several chemical applications. The paper with chemicals and lead paint adhered to it, plus the containment and loose scrapings collected by a HEPA (High Efficiency Particulate Air Filter) vac, must be disposed of as a hazardous material. The chemical stripping process usually requires a neutralizing agent and several wash downs after the paint is removed. Worker protection includes a neoprene or other compatible protective clothing and respiratory protection with face shield. An industrial hygienist is required intermittently during the process.
3. **Power tool cleaning** is accomplished using shrouded needle blasting guns. The shrouding with different end configurations is held up against the surface to be cleaned. The area is blasted with hardened needles and the shroud captures the lead with a HEPA vac and deposits it in a holding tank. An industrial hygienist monitors the project, protective clothing and a respirator is required until air samples prove otherwise. When the work is complete the lead must be disposed of as a hazardous material.
4. **Encapsulation** is a method that leaves the well bonded lead paint in place after the peeling paint has been removed. Before the work can begin, the area under/adjacent to the work must be covered to catch the scrapings. The scraped surface is then washed with a detergent and rinsed. The prepared surface is covered with approximately 10 mils of paint. A reinforcing fabric can also be embedded in the paint covering. The scraped paint and containment must be disposed of as a hazardous material. Workers must wear protective clothing and respirators.
5. **Remove and replace** is an effective way to remove lead paint from windows, gypsum walls and concrete masonry surfaces. The painted materials are removed and new materials are installed. Workers should wear a respirator and tyvek suit. The demolished materials must be disposed of as hazardous waste if it fails the TCLP (Toxicity Characteristic Leachate Process) test.
6. **Enclosure** is the process that permanently seals lead painted materials in place. This process has many applications such as covering lead painted drywall with new drywall, covering exterior construction with tyvek paper then residing, or covering lead painted structural members with aluminum or plastic. The seams on all enclosing materials must be securely sealed. An industrial hygienist monitors the project, and protective clothing and a respirator is required until air samples prove otherwise.

All the processes require clearance monitoring and wipe testing as required by the hygienist.

R031113-10 Wall Form Materials

Aluminum Forms

Approximate weight is 3 lbs. per S.F.C.A. Standard widths are available from 4″ to 36″ with 36″ most common. Standard lengths of 2′, 4′, 6′ to 8′ are available. Forms are lightweight and fewer ties are needed with the wider widths. The form face is either smooth or textured.

Metal Framed Plywood Forms

Manufacturers claim over 75 reuses of plywood and over 300 reuses of steel frames. Many specials such as corners, fillers, pilasters, etc. are available. Monthly rental is generally about 15% of purchase price for first month and 9% per month thereafter with 90% of rental applied to purchase for the first month and decreasing percentages thereafter. Aluminum framed forms cost 25% to 30% more than steel framed.

After the first month, extra days may be prorated from the monthly charge. Rental rates do not include ties, accessories, cleaning, loss of hardware or freight in and out. Approximate weight is 5 lbs. per S.F. for steel; 3 lbs. per S.F. for aluminum.

Forms can be rented with option to buy.

Plywood Forms, Job Fabricated

There are two types of plywood used for concrete forms.

1. Exterior plyform which is completely waterproof. This is face oiled to facilitate stripping. Ten reuses can be expected with this type with 25 reuses possible.
2. An overlaid type consists of a resin fiber fused to exterior plyform. No oiling is required except to facilitate cleaning. This is available in both high density (HDO) and medium density overlaid (MDO). Using HDO, 50 reuses can be expected with 200 possible.

Plyform is available in 5/8″ and 3/4″ thickness. High density overlaid is available in 3/8″, 1/2″, 5/8″ and 3/4″ thickness.

5/8″ thick is sufficient for most building forms, while 3/4″ is best on heavy construction.

Plywood Forms, Modular, Prefabricated

There are many plywood forming systems without frames. Most of these are manufactured from 1-1/8″ (HDO) plywood and have some hardware attached. These are used principally for foundation walls 8′ or less high. With care and maintenance, 100 reuses can be attained with decreasing quality of surface finish.

Steel Forms

Approximate weight is 6-1/2 lbs. per S.F.C.A. including accessories. Standard widths are available from 2″ to 24″, with 24″ most common. Standard lengths are from 2′ to 8′, with 4′ the most common. Forms are easily ganged into modular units.

Forms are usually leased for 15% of the purchase price per month prorated daily over 30 days.

Rental may be applied to sale price, and usually rental forms are bought. With careful handling and cleaning 200 to 400 reuses are possible.

Straight wall gang forms up to 12′ x 20′ or 8′ x 30′ can be fabricated. These crane handled forms usually lease for approx. 9% per month.

Individual job analysis is available from the manufacturer at no charge.

R031113-30 Slipforms

The slipform method of forming may be used for forming circular silo and multi-celled storage bin type structures over 30′ high, and building core shear walls over eight stories high. The shear walls, usually enclose elevator shafts, stairwells, mechanical spaces, and toilet rooms. Reuse of the form on duplicate structures will reduce the height necessary and spread the cost of building the form. Slipform systems can be used to cast chimneys, towers, piers, dams, underground shafts or other structures capable of being extruded.

Slipforms are usually 4′ high and are raised semi-continuously by jacks climbing on rods which are embedded in the concrete. The jacks are powered by a hydraulic, pneumatic, or electric source and are available in 3, 6, and 22 ton capacities. Interior work decks and exterior scaffolds must be provided for placing inserts, embedded items, reinforcing steel, and concrete. Scaffolds below the form for finishers may be required. The interior work decks are often used as roof slab forms on silos and bin work. Form raising rates will range from 6″ to 20″ per hour for silos; 6″ to 30″ per hour for buildings; and 6″ to 48″ per hour for shaft work.

Reinforcing bars and stressing strands are usually hoisted by crane or gin pole, and the concrete material can be hoisted by crane, winch-powered skip, or pumps. The slipform system is operated on a continuous 24-hour day when a monolithic structure is desired. For least cost, the system is operated only during normal working hours.

Placing concrete will range from 0.5 to 1.5 labor-hours per C.Y. Bucks, blockouts, keyways, weldplates, etc. are extra.

R031113-40 Forms for Reinforced Concrete

Design Economy

Avoid many sizes in proportioning beams and columns.

From story to story avoid changing column dimensions. Gain strength by adding steel or using a richer mix. If a change in size of column is necessary, vary one dimension only to minimize form alterations. Keep beams and columns the same width.

From floor to floor in a multi-story building vary beam depth, not width, as that will leave slab panel form unchanged. It is cheaper to vary the strength of a beam from floor to floor by means of steel area than by 2″ changes in either width or depth.

Cost Factors

Material includes the cost of lumber, cost of rent for metal pans or forms if used, nails, form ties, form oil, bolts and accessories.

Labor includes the cost of carpenters to make up, erect, remove and repair, plus common labor to clean and move. Having carpenters remove forms minimizes repairs.

Improper alignment and condition of forms will increase finishing cost. When forms are heavily oiled, concrete surfaces must be neutralized before finishing. Special curing compounds will cause spillages to spall off in first frost. Gang forming methods will reduce costs on large projects.

Materials Used

Boards are seldom used unless their architectural finish is required. Generally, steel, fiberglass and plywood are used for contact surfaces. Labor on plywood is 10% less than with boards. The plywood is backed up with 2 x 4's at 12″ to 32″ O.C. Walers are generally 2 - 2 x 4's. Column forms are held together with steel yokes or bands. Shoring is with adjustable shoring or scaffolding for high ceilings.

Reuse

Floor and column forms can be reused four or possibly five times without excessive repair. Remember to allow for 10% waste on each reuse.

When modular sized wall forms are made, up to twenty uses can be expected with exterior plyform.

When forms are reused, the cost to erect, strip, clean and move will not be affected. 10% replacement of lumber should be included and about one hour of carpenter time for repairs on each reuse per 100 S.F.

The reuse cost for certain accessory items normally rented on a monthly basis will be lower than the cost for the first use.

After fifth use, new material required plus time needed for repair prevent form cost from dropping further and it may go up. Much depends on care in stripping, the number of special bays, changes in beam or column sizes and other factors.

Costs for multiple use of formwork may be developed as follows:

2 Uses

$$\frac{\text{(1st Use + Reuse)}}{2} = \text{avg. cost/2 uses}$$

3 Uses

$$\frac{\text{(1st Use + 2 Reuse)}}{3} = \text{avg. cost/3 uses}$$

4 Uses

$$\frac{\text{(1st use + 3 Reuse)}}{4} = \text{avg. cost/4 uses}$$

R031113-60 Formwork Labor-Hours

Item	Hours Required				Total Hours	Multiple Use		
	Unit	Fabricate	Erect & Strip	Clean & Move	1 Use	2 Use	3 Use	4 Use
Beam and Girder, interior beams, 12″ wide	100 S.F.	6.4	8.3	1.3	16.0	13.3	12.4	12.0
Hung from steel beams		5.8	7.7	1.3	14.8	12.4	11.6	11.2
Beam sides only, 36″ high		5.8	7.2	1.3	14.3	11.9	11.1	10.7
Beam bottoms only, 24″ wide		6.6	13.0	1.3	20.9	18.1	17.2	16.7
Box out for openings		9.9	10.0	1.1	21.0	16.6	15.1	14.3
Buttress forms, to 8′ high		6.0	6.5	1.2	13.7	11.2	10.4	10.0
Centering, steel, 3/4″ rib lath			1.0		1.0			
3/8″ rib lath or slab form			0.9		0.9			
Chamfer strip or keyway	100 L.F.		1.5		1.5	1.5	1.5	1.5
Columns, fiber tube 8″ diameter			20.6		20.6			
12″			21.3		21.3			
16″			22.9		22.9			
20″			23.7		23.7			
24″			24.6		24.6			
30″			25.6		25.6			
Columns, round steel, 12″ diameter			22.0		22.0	22.0	22.0	22.0
16″			25.6		25.6	25.6	25.6	25.6
20″			30.5		30.5	30.5	30.5	30.5
24″			37.7		37.7	37.7	37.7	37.7
Columns, plywood 8″ x 8″	100 S.F.	7.0	11.0	1.2	19.2	16.2	15.2	14.7
12″ x 12″		6.0	10.5	1.2	17.7	15.2	14.4	14.0
16″ x 16″		5.9	10.0	1.2	17.1	14.7	13.8	13.4
24″ x 24″		5.8	9.8	1.2	16.8	14.4	13.6	13.2
Columns, steel framed plywood 8″ x 8″			10.0	1.0	11.0	11.0	11.0	11.0
12″ x 12″			9.3	1.0	10.3	10.3	10.3	10.3
16″ x 16″			8.5	1.0	9.5	9.5	9.5	9.5
24″ x 24″			7.8	1.0	8.8	8.8	8.8	8.8
Drop head forms, plywood		9.0	12.5	1.5	23.0	19.0	17.7	17.0
Coping forms		8.5	15.0	1.5	25.0	21.3	20.0	19.4
Culvert, box			14.5	4.3	18.8	18.8	18.8	18.8
Curb forms, 6″ to 12″ high, on grade		5.0	8.5	1.2	14.7	12.7	12.1	11.7
On elevated slabs		6.0	10.8	1.2	18.0	15.5	14.7	14.3
Edge forms to 6″ high, on grade	100 L.F.	2.0	3.5	0.6	6.1	5.6	5.4	5.3
7″ to 12″ high	100 S.F.	2.5	5.0	1.0	8.5	7.8	7.5	7.4
Equipment foundations		10.0	18.0	2.0	30.0	25.5	24.0	23.3
Flat slabs, including drops		3.5	6.0	1.2	10.7	9.5	9.0	8.8
Hung from steel		3.0	5.5	1.2	9.7	8.7	8.4	8.2
Closed deck for domes		3.0	5.8	1.2	10.0	9.0	8.7	8.5
Open deck for pans		2.2	5.3	1.0	8.5	7.9	7.7	7.6
Footings, continuous, 12″ high		3.5	3.5	1.5	8.5	7.3	6.8	6.6
Spread, 12″ high		4.7	4.2	1.6	10.5	8.7	8.0	7.7
Pile caps, square or rectangular		4.5	5.0	1.5	11.0	9.3	8.7	8.4
Grade beams, 24″ deep		2.5	5.3	1.2	9.0	8.3	8.0	7.9
Lintel or Sill forms		8.0	17.0	2.0	27.0	23.5	22.3	21.8
Spandrel beams, 12″ wide		9.0	11.2	1.3	21.5	17.5	16.2	15.5
Stairs			25.0	4.0	29.0	29.0	29.0	29.0
Trench forms in floor		4.5	14.0	1.5	20.0	18.3	17.7	17.4
Walls, Plywood, at grade, to 8′ high		5.0	6.5	1.5	13.0	11.0	9.7	9.5
8′ to 16′		7.5	8.0	1.5	17.0	13.8	12.7	12.1
16′ to 20′		9.0	10.0	1.5	20.5	16.5	15.2	14.5
Foundation walls, to 8′ high		4.5	6.5	1.0	12.0	10.3	9.7	9.4
8′ to 16′ high		5.5	7.5	1.0	14.0	11.8	11.0	10.6
Retaining wall to 12′ high, battered		6.0	8.5	1.5	16.0	13.5	12.7	12.3
Radial walls to 12′ high, smooth		8.0	9.5	2.0	19.5	16.0	14.8	14.3
2′ chords		7.0	8.0	1.5	16.5	13.5	12.5	12.0
Prefabricated modular, to 8′ high		—	4.3	1.0	5.3	5.3	5.3	5.3
Steel, to 8′ high		—	6.8	1.2	8.0	8.0	8.0	8.0
8′ to 16′ high		—	9.1	1.5	10.6	10.3	10.2	10.2
Steel framed plywood to 8′ high		—	6.8	1.2	8.0	7.5	7.3	7.2
8′ to 16′ high		—	9.3	1.2	10.5	9.5	9.2	9.0

R032110-10 Reinforcing Steel Weights and Measures

Bar Designation No.**	Nominal Weight Lb./Ft.	U.S. Customary Units Nominal Dimensions*			SI Units Nominal Dimensions*			
		Diameter in.	Cross Sectional Area, in.^{2}	Perimeter in.	Nominal Weight kg/m	Diameter mm	Cross Sectional Area, cm^2	Perimeter mm
3	.376	.375	.11	1.178	.560	9.52	.71	29.9
4	.668	.500	.20	1.571	.994	12.70	1.29	39.9
5	1.043	.625	.31	1.963	1.552	15.88	2.00	49.9
6	1.502	.750	.44	2.356	2.235	19.05	2.84	59.8
7	2.044	.875	.60	2.749	3.042	22.22	3.87	69.8
8	2.670	1.000	.79	3.142	3.973	25.40	5.10	79.8
9	3.400	1.128	1.00	3.544	5.059	28.65	6.45	90.0
10	4.303	1.270	1.27	3.990	6.403	32.26	8.19	101.4
11	5.313	1.410	1.56	4.430	7.906	35.81	10.06	112.5
14	7.650	1.693	2.25	5.320	11.384	43.00	14.52	135.1
18	13.600	2.257	4.00	7.090	20.238	57.33	25.81	180.1

* The nominal dimensions of a deformed bar are equivalent to those of a plain round bar having the same weight per foot as the deformed bar.
** Bar numbers are based on the number of eighths of an inch included in the nominal diameter of the bars.

R032110-20 Metric Rebar Specification - ASTM A615-81

Grade 300 (300 MPa* = 43,560 psi; +8.7% vs. Grade 40)				
Grade 400 (400 MPa* = 58,000 psi; –3.4% vs. Grade 60)				
Bar No.	Diameter mm	Area mm^2	Equivalent in.^{2}	Comparison with U.S. Customary Bars
10M	11.3	100	.16	Between #3 & #4
15M	16.0	200	.31	#5 (.31 in.^{2})
20M	19.5	300	.47	#6 (.44 in.^{2})
25M	25.2	500	.78	#8 (.79 in.^{2})
30M	29.9	700	1.09	#9 (1.00 in.^{2})
35M	35.7	1000	1.55	#11 (1.56 in.^{2})
45M	43.7	1500	2.33	#14 (2.25 in.^{2})
55M	56.4	2500	3.88	#18 (4.00 in.^{2})

* MPa = megapascals

R032110-40 Weight of Steel Reinforcing Per Square Foot of Wall (PSF)

Reinforced Weights: The table below suggests the weights per square foot for reinforcing steel in walls. Weights are approximate and will be the same for all grades of steel bars. For bars in two directions, add weights for each size and spacing.

C/C Spacing in Inches	Bar Size #3 Wt. (PSF)	#4 Wt. (PSF)	#5 Wt. (PSF)	#6 Wt. (PSF)	#7 Wt. (PSF)	#8 Wt. (PSF)	#9 Wt. (PSF)	#10 Wt. (PSF)	#11 Wt. (PSF)
2″	2.26	4.01	6.26	9.01	12.27				
3″	1.50	2.67	4.17	6.01	8.18	10.68	13.60	17.21	21.25
4″	1.13	2.01	3.13	4.51	6.13	8.10	10.20	12.91	15.94
5″	.90	1.60	2.50	3.60	4.91	6.41	8.16	10.33	12.75
6″	.752	1.34	2.09	3.00	4.09	5.34	6.80	8.61	10.63
8″	.564	1.00	1.57	2.25	3.07	4.01	5.10	6.46	7.97
10″	.451	.802	1.25	1.80	2.45	3.20	4.08	5.16	6.38
12″	.376	.668	1.04	1.50	2.04	2.67	3.40	4.30	5.31
18″	.251	.445	.695	1.00	1.32	1.78	2.27	2.86	3.54
24″	.188	.334	.522	.751	1.02	1.34	1.70	2.15	2.66
30″	.150	.267	.417	.600	.817	1.07	1.36	1.72	2.13
36″	.125	.223	.348	.501	.681	.890	1.13	1.43	1.77
42″	.107	.191	.298	.429	.584	.753	.97	1.17	1.52
48″	.094	.167	.261	.376	.511	.668	.85	1.08	1.33

R032110-50 Minimum Wall Reinforcement Weight (PSF)

This table lists the approximate minimum wall reinforcement weights per S.F. according to the specification of .12% of gross area for vertical bars and .20% of gross area for horizontal bars.

Location	Wall Thickness	Bar Size	Horizontal Steel Spacing C/C	Sq. In. Req'd per S.F.	Total Wt. per S.F.	Bar Size	Vertical Steel Spacing C/C	Sq. In. Req'd per S.F.	Total Wt. per S.F.	Horizontal & Vertical Steel Total Weight per S.F.
Both Faces	10″	#4	18″	.24	.89#	#3	18″	.14	.50#	1.39#
	12″	#4	16″	.29	1.00	#3	16″	.17	.60	1.60
	14″	#4	14″	.34	1.14	#3	13″	.20	.69	1.84
	16″	#4	12″	.38	1.34	#3	11″	.23	.82	2.16
	18″	#5	17″	.43	1.47	#4	18″	.26	.89	2.36
One Face	6″	#3	9″	.15	.50	#3	18″	.09	.25	.75
	8″	#4	12″	.19	.67	#3	11″	.12	.41	1.08
	10″	#5	15″	.24	.83	#4	16″	.14	.50	1.34

R032110-70 Bend, Place and Tie Reinforcing

Placing and tying by rodmen for footings and slabs runs from nine hrs. per ton for heavy bars to fifteen hrs. per ton for light bars. For beams, columns, and walls, production runs from eight hrs. per ton for heavy bars to twenty hrs. per ton for light bars. Overall average for typical reinforced concrete buildings is about fourteen hrs. per ton. These production figures include the time for placing of accessories and usual inserts, but not their material cost (allow 15% of the cost of delivered bent rods). Equipment handling is necessary for the larger-sized bars so that installation costs for the very heavy bars will not decrease proportionately.

Installation costs for splicing reinforcing bars include allowance for equipment to hold the bars in place while splicing as well as necessary scaffolding for iron workers.

R032110-80 Shop-Fabricated Reinforcing Steel

The material prices for reinforcing, shown in the unit cost sections of the book, are for 50 tons or more of shop-fabricated reinforcing steel and include:

1. Mill base price of reinforcing steel
2. Mill grade/size/length extras
3. Mill delivery to the fabrication shop
4. Shop storage and handling
5. Shop drafting/detailing
6. Shop shearing and bending
7. Shop listing
8. Shop delivery to the job site

Both material and installation costs can be considerably higher for small jobs consisting primarily of smaller bars, while material costs may be slightly lower for larger jobs.

R032205-30 Common Stock Styles of Welded Wire Fabric

This table provides some of the basic specifications, sizes, and weights of welded wire fabric used for reinforcing concrete.

New Designation		Old Designation		Steel Area per Foot				Approximate Weight per 100 S.F.	
Spacing — Cross Sectional Area (in.) — (Sq. in. 100)		Spacing — Wire Gauge (in.) — (AS & W)		Longitudinal		Transverse			
				in.	cm	in.	cm	lbs	kg
Rolls	6 x 6 — W1.4 x W1.4	6 x 6 — 10 x 10		.028	.071	.028	.071	21	9.53
	6 x 6 — W2.0 x W2.0	6 x 6 — 8 x 8	1	.040	.102	.040	.102	29	13.15
	6 x 6 — W2.9 x W2.9	6 x 6 — 6 x 6		.058	.147	.058	.147	42	19.05
	6 x 6 — W4.0 x W4.0	6 x 6 — 4 x 4		.080	.203	.080	.203	58	26.91
	4 x 4 — W1.4 x W1.4	4 x 4 — 10 x 10		.042	.107	.042	.107	31	14.06
	4 x 4 — W2.0 x W2.0	4 x 4 — 8 x 8	1	.060	.152	.060	.152	43	19.50
	4 x 4 — W2.9 x W2.9	4 x 4 — 6 x 6		.087	.227	.087	.227	62	28.12
	4 x 4 — W4.0 x W4.0	4 x 4 — 4 x 4		.120	.305	.120	.305	85	38.56
Sheets	6 x 6 — W2.9 x W2.9	6 x 6 — 6 x 6		.058	.147	.058	.147	42	19.05
	6 x 6 — W4.0 x W4.0	6 x 6 — 4 x 4		.080	.203	.080	.203	58	26.31
	6 x 6 — W5.5 x W5.5	6 x 6 — 2 x 2	2	.110	.279	.110	.279	80	36.29
	4 x 4 — W1.4 x W1.4	4 x 4 — 4 x 4		.120	.305	.120	.305	85	38.56

NOTES: 1. Exact W—number size for 8 gauge is W2.1
2. Exact W—number size for 2 gauge is W5.4

R033053-50 Industrial Chimneys

Foundation requirements in C.Y. of concrete for various sized chimneys.

Size Chimney	2 Ton Soil	3 Ton Soil	Size Chimney	2 Ton Soil	3 Ton Soil	Size Chimney	2 Ton Soil	3 Ton Soil
75′ x 3′-0″	13 C.Y.	11 C.Y.	160′ x 6′-6″	86 C.Y.	76 C.Y.	300′ x 10′-0″	325 C.Y.	245 C.Y.
85′ x 5′-6″	19	16	175′ x 7′-0″	108	95	350′ x 12′-0″	422	320
100′ x 5′-0″	24	20	200′ x 6′-0″	125	105	400′ x 14′-0″	520	400
125′ x 5′-6″	43	36	250′ x 8′-0″	230	175	500′ x 18′-0″	725	575

R033105-10 Proportionate Quantities

The tables below show both quantities per S.F. of floor areas as well as form and reinforcing quantities per C.Y. Unusual structural requirements would increase the ratios below. High strength reinforcing would reduce the steel weights. Figures are for 3000 psi concrete and 60,000 psi reinforcing unless specified otherwise.

Type of Construction	Live Load	Span	Per S.F. of Floor Area: Concrete	Forms	Reinf.	Pans	Per C.Y. of Concrete: Forms	Reinf.	Pans
Flat Plate	50 psf	15 Ft.	.46 C.F.	1.06 S.F.	1.71lb.		62 S.F.	101 lb.	
		20	.63	1.02	2.40		44	104	
		25	.79	1.02	3.03		35	104	
	100	15	.46	1.04	2.14		61	126	
		20	.71	1.02	2.72		39	104	
		25	.83	1.01	3.47		33	113	
Flat Plate (waffle construction) 20″ domes	50	20	.43	1.00	2.10	.84 S.F.	63	135	53 S.F.
		25	.52	1.00	2.90	.89	52	150	46
		30	.64	1.00	3.70	.87	42	155	37
	100	20	.51	1.00	2.30	.84	53	125	45
		25	.64	1.00	3.20	.83	42	135	35
		30	.76	1.00	4.40	.81	36	160	29
Waffle Construction 30″ domes	50	25	.69	1.06	1.83	.68	42	72	40
		30	.74	1.06	2.39	.69	39	87	39
		35	.86	1.05	2.71	.69	33	85	39
		40	.78	1.00	4.80	.68	35	165	40
Flat Slab (two way with drop panels)	50	20	.62	1.03	2.34		45	102	
		25	.77	1.03	2.99		36	105	
		30	.95	1.03	4.09		29	116	
	100	20	.64	1.03	2.83		43	119	
		25	.79	1.03	3.88		35	133	
		30	.96	1.03	4.66		29	131	
	200	20	.73	1.03	3.03		38	112	
		25	.86	1.03	4.23		32	133	
		30	1.06	1.03	5.30		26	135	
One Way Joists 20″ Pans	50	15	.36	1.04	1.40	.93	78	105	70
		20	.42	1.05	1.80	.94	67	120	60
		25	.47	1.05	2.60	.94	60	150	54
	100	15	.38	1.07	1.90	.93	77	140	66
		20	.44	1.08	2.40	.94	67	150	58
		25	.52	1.07	3.50	.94	55	185	49
One Way Joists 8″ x 16″ filler blocks	50	15	.34	1.06	1.80	.81 Ea.	84	145	64 Ea.
		20	.40	1.08	2.20	.82	73	145	55
		25	.46	1.07	3.20	.83	63	190	49
	100	15	.39	1.07	1.90	.81	74	130	56
		20	.46	1.09	2.80	.82	64	160	48
		25	.53	1.10	3.60	.83	56	190	42
One Way Beam & Slab	50	15	.42	1.30	1.73		84	111	
		20	.51	1.28	2.61		68	138	
		25	.64	1.25	2.78		53	117	
	100	15	.42	1.30	1.90		84	122	
		20	.54	1.35	2.69		68	154	
		25	.69	1.37	3.93		54	145	
	200	15	.44	1.31	2.24		80	137	
		20	.58	1.40	3.30		65	163	
		25	.69	1.42	4.89		53	183	
Two Way Beam & Slab	100	15	.47	1.20	2.26		69	130	
		20	.63	1.29	3.06		55	131	
		25	.83	1.33	3.79		43	123	
	200	15	.49	1.25	2.70		41	149	
		20	.66	1.32	4.04		54	165	
		25	.88	1.32	6.08		41	187	

R033105-10 Proportionate Quantities (cont.)

4000 psi Concrete and 60,000 psi Reinforcing—Form and Reinforcing Quantities per C.Y.					
Item	**Size**	**Forms**	**Reinforcing**	**Minimum**	**Maximum**
Columns (square tied)	10″ x 10″	130 S.F.C.A.	#5 to #11	220 lbs.	875 lbs.
	12″ x 12″	108	#6 to #14	200	955
	14″ x 14″	92	#7 to #14	190	900
	16″ x 16″	81	#6 to #14	187	1082
	18″ x 18″	72	#6 to #14	170	906
	20″ x 20″	65	#7 to #18	150	1080
	22″ x 22″	59	#8 to #18	153	902
	24″ x 24″	54	#8 to #18	164	884
	26″ x 26″	50	#9 to #18	169	994
	28″ x 28″	46	#9 to #18	147	864
	30″ x 30″	43	#10 to #18	146	983
	32″ x 32″	40	#10 to #18	175	866
	34″ x 34″	38	#10 to #18	157	772
	36″ x 36″	36	#10 to #18	175	852
	38″ x 38″	34	#10 to #18	158	765
	40″ x 40″	32 ▼	#10 to #18	143	692

Item	**Size**	**Form**	**Spiral**	**Reinforcing**	**Minimum**	**Maximum**
Columns (spirally reinforced)	12″ diameter	34.5 L.F.	190 lbs.	#4 to #11	165 lbs.	1505 lb.
		34.5	190	#14 & #18	—	1100
	14″	25	170	#4 to #11	150	970
		25	170	#14 & #18	800	1000
	16″	19	160	#4 to #11	160	950
		19	160	#14 & #18	605	1080
	18″	15	150	#4 to #11	160	915
		15	150	#14 & #18	480	1075
	20″	12	130	#4 to #11	155	865
		12	130	#14 & #18	385	1020
	22″	10	125	#4 to #11	165	775
		10	125	#14 & #18	320	995
	24″	9	120	#4 to #11	195	800
		9	120	#14 & #18	290	1150
	26″	7.3	100	#4 to #11	200	729
		7.3	100	#14 & #18	235	1035
	28″	6.3	95	#4 to #11	175	700
		6.3	95	#14 & #18	200	1075
	30″	5.5	90	#4 to #11	180	670
		5.5	90	#14 & #18	175	1015
	32″	4.8	85	#4 to #11	185	615
		4.8	85	#14 & #18	155	955
	34″	4.3	80	#4 to #11	180	600
		4.3	80	#14 & #18	170	855
	36″	3.8	75	#4 to #11	165	570
		3.8	75	#14 & #18	155	865
	40″	3.0	70	#4 to #11	165	500
		3.0	70	#14 & #18	145	765

R033105-10 Proportionate Quantities (cont.)

3000 psi Concrete and 60,000 psi Reinforcing—Form and Reinforcing Quantities per C.Y.

Item	Type	Loading	Height	C.Y./L.F.	Forms/C.Y.	Reinf./C.Y.
Retaining Walls	Cantilever	Level Backfill	4 Ft.	0.2 C.Y.	49 S.F.	35 lbs.
			8	0.5	42	45
			12	0.8	35	70
			16	1.1	32	85
			20	1.6	28	105
		Highway Surcharge	4	0.3	41	35
			8	0.5	36	55
			12	0.8	33	90
			16	1.2	30	120
			20	1.7	27	155
		Railroad Surcharge	4	0.4	28	45
			8	0.8	25	65
			12	1.3	22	90
			16	1.9	20	100
			20	2.6	18	120
	Gravity, with Vertical Face	Level Backfill	4	0.4	37	None
			7	0.6	27	↓
			10	1.2	20	↓
		Sloping Backfill	4	0.3	31	↓
			7	0.8	21	↓
			10	1.6	15	↓

Item	Span	Live Load in Kips per Linear Foot							
		Under 1 Kip		2 to 3 Kips		4 to 5 Kips		6 to 7 Kips	
		Forms	Reinf.	Forms	Reinf.	Forms	Reinf.	Forms	Reinf.
Beams	10 Ft.	—	—	90 S.F.	170 #	85 S.F.	175 #	75 S.F.	185 #
	16	130 S.F.	165 #	85	180	75	180	65	225
	20	110	170	75	185	62	200	51	200
	26	90	170	65	215	62	215	—	—
	30	85	175	60	200	—	—	—	—

Item	Size	Type	Forms per C.Y.	Reinforcing per C.Y.
Spread Footings	Under 1 C.Y.	1,000 psf soil	24 S.F.	44 lbs.
		5,000	24	42
		10,000	24	52
	1 C.Y. to 5 C.Y.	1,000	14	49
		5,000	14	50
		10,000	14	50
	Over 5 C.Y.	1,000	9	54
		5,000	9	52
		10,000	9	56
Pile Caps (30 Ton Concrete Piles)	Under 5 C.Y.	shallow caps	20	65
		medium	20	50
		deep	20	40
	5 C.Y. to 10 C.Y.	shallow	14	55
		medium	15	45
		deep	15	40
	10 C.Y. to 20 C.Y.	shallow	11	60
		medium	11	45
		deep	12	35
	Over 20 C.Y.	shallow	9	60
		medium	9	45
		deep	10	40

R033105-10 Proportionate Quantities (cont.)

3000 psi Concrete and 60,000 psi Reinforcing — Form and Reinforcing Quantities per C.Y.						
Item	Size	Pile Spacing	50 T Pile	100 T Pile	50 T Pile	100 T Pile
Pile Caps (Steel H Piles)	Under 5 C.Y.	24″ O.C.	24 S.F.	24 S.F.	75 lbs.	90 lbs.
		30″	25	25	80	100
		36″	24	24	80	110
	5 C.Y. to 10 C.Y.	24″	15	15	80	110
		30″	15	15	85	110
		36″	15	15	75	90
	Over 10 C.Y.	24″	13	13	85	90
		30″	11	11	85	95
		36″	10	10	85	90

Item		8″ Thick		10″ Thick		12″ Thick		15″ Thick	
	Height	Forms	Reinf.	Forms	Reinf.	Forms	Reinf.	Forms	Reinf.
Basement Walls	7 Ft.	81 S.F.	44 lbs.	65 S.F.	45 lbs.	54 S.F.	44 lbs.	41 S.F.	43 lbs.
	8	↓	44	↓	45	↓	44	↓	43
	9	↓	46	↓	45	↓	44	↓	43
	10	↓	57	↓	45	↓	44	↓	43
	12	↓	83	↓	50	↓	52	↓	43
	14	↓	116	↓	65	↓	64	↓	51
	16	↓		↓	86	↓	90	↓	65
	18	↓		↓		↓	106	↓	70

R033105-20 Materials for One C.Y. of Concrete

This is an approximate method of figuring quantities of cement, sand and coarse aggregate for a field mix with waste allowance included.

With crushed gravel as coarse aggregate, to determine barrels of cement required, divide 10 by total mix; that is, for 1:2:4 mix, 10 divided by 7 = 1-3/7 barrels.

If the coarse aggregate is crushed stone, use 10-1/2 instead of 10 as given for gravel.

To determine tons of sand required, multiply barrels of cement by parts of sand and then by 0.2; that is, for the 1:2:4 mix, as above, 1-3/7 x 2 x .2 = .57 tons.

Tons of crushed gravel are in the same ratio to tons of sand as parts in the mix, or 4/2 x .57 = 1.14 tons.

1 bag cement = 94#
4 bags = 1 barrel

1 C.Y. sand or crushed gravel = 2700#
1 ton sand or crushed gravel = 20 C.F.

1 C.Y. crushed stone = 2575#
1 ton crushed stone = 21 C.F.

Average carload of cement is 692 bags; of sand or gravel is 56 tons.

Do not stack stored cement over 10 bags high.

R033105-30 Metric Equivalents of Cement Content for Concrete Mixes

94 Pound Bags per Cubic Yard	Kilograms per Cubic Meter	94 Pound Bags per Cubic Yard	Kilograms per Cubic Meter
1.0	55.77	7.0	390.4
1.5	83.65	7.5	418.3
2.0	111.5	8.0	446.2
2.5	139.4	8.5	474.0
3.0	167.3	9.0	501.9
3.5	195.2	9.5	529.8
4.0	223.1	10.0	557.7
4.5	251.0	10.5	585.6
5.0	278.8	11.0	613.5
5.5	306.7	11.5	641.3
6.0	334.6	12.0	669.2
6.5	362.5	12.5	697.1

a. If you know the cement content in pounds per cubic yard, multiply by .5933 to obtain kilograms per cubic meter.

b. If you know the cement content in 94 pound bags per cubic yard, multiply by 55.77 to obtain kilograms per cubic meter.

R033105-40 Metric Equivalents of Common Concrete Strengths (to convert other psi values to megapascals, multiply by 0.006895)

U.S. Values psi	SI Value Megapascals	Non-SI Metric Value kgf/cm²*
2000	14	140
2500	17	175
3000	21	210
3500	24	245
4000	28	280
4500	31	315
5000	34	350
6000	41	420
7000	48	490
8000	55	560
9000	62	630
10,000	69	705

* kilograms force per square centimeter

R033105-50 Quantities of Cement, Sand and Stone for One C.Y. of Concrete per Various Mixes

This table can be used to determine the quantities of the ingredients for smaller quantities of site mixed concrete.

Concrete (C.Y.)	Mix = 1:1:1-3/4			Mix = 1:2:2.25			Mix = 1:2.25:3			Mix = 1:3:4		
	Cement (sacks)	Sand (C.Y.)	Stone (C.Y.)	Cement (sacks)	Sand (C.Y.)	Stone (C.Y.)	Cement (sacks)	Sand (C.Y.)	Stone (C.Y.)	Cement (sacks)	Sand (C.Y.)	Stone (C.Y.)
1	10	.37	.63	7.75	.56	.65	6.25	.52	.70	5	.56	.74
2	20	.74	1.26	15.50	1.12	1.30	12.50	1.04	1.40	10	1.12	1.48
3	30	1.11	1.89	23.25	1.68	1.95	18.75	1.56	2.10	15	1.68	2.22
4	40	1.48	2.52	31.00	2.24	2.60	25.00	2.08	2.80	20	2.24	2.96
5	50	1.85	3.15	38.75	2.80	3.25	31.25	2.60	3.50	25	2.80	3.70
6	60	2.22	3.78	46.50	3.36	3.90	37.50	3.12	4.20	30	3.36	4.44
7	70	2.59	4.41	54.25	3.92	4.55	43.75	3.64	4.90	35	3.92	5.18
8	80	2.96	5.04	62.00	4.48	5.20	50.00	4.16	5.60	40	4.48	5.92
9	90	3.33	5.67	69.75	5.04	5.85	56.25	4.68	6.30	45	5.04	6.66
10	100	3.70	6.30	77.50	5.60	6.50	62.50	5.20	7.00	50	5.60	7.40
11	110	4.07	6.93	85.25	6.16	7.15	68.75	5.72	7.70	55	6.16	8.14
12	120	4.44	7.56	93.00	6.72	7.80	75.00	6.24	8.40	60	6.72	8.88
13	130	4.82	8.20	100.76	7.28	8.46	81.26	6.76	9.10	65	7.28	9.62
14	140	5.18	8.82	108.50	7.84	9.10	87.50	7.28	9.80	70	7.84	10.36
15	150	5.56	9.46	116.26	8.40	9.76	93.76	7.80	10.50	75	8.40	11.10
16	160	5.92	10.08	124.00	8.96	10.40	100.00	8.32	11.20	80	8.96	11.84
17	170	6.30	10.72	131.76	9.52	11.06	106.26	8.84	11.90	85	9.52	12.58
18	180	6.66	11.34	139.50	10.08	11.70	112.50	9.36	12.60	90	10.08	13.32
19	190	7.04	11.98	147.26	10.64	12.36	118.76	9.84	13.30	95	10.64	14.06
20	200	7.40	12.60	155.00	11.20	13.00	125.00	10.40	14.00	100	11.20	14.80
21	210	7.77	13.23	162.75	11.76	13.65	131.25	10.92	14.70	105	11.76	15.54
22	220	8.14	13.86	170.05	12.32	14.30	137.50	11.44	15.40	110	12.32	16.28
23	230	8.51	14.49	178.25	12.88	14.95	143.75	11.96	16.10	115	12.88	17.02
24	240	8.88	15.12	186.00	13.44	15.60	150.00	12.48	16.80	120	13.44	17.76
25	250	9.25	15.75	193.75	14.00	16.25	156.25	13.00	17.50	125	14.00	18.50
26	260	9.64	16.40	201.52	14.56	16.92	162.52	13.52	18.20	130	14.56	19.24
27	270	10.00	17.00	209.26	15.12	17.56	168.76	14.04	18.90	135	15.02	20.00
28	280	10.36	17.64	217.00	15.68	18.20	175.00	14.56	19.60	140	15.68	20.72
29	290	10.74	18.28	224.76	16.24	18.86	181.26	15.08	20.30	145	16.24	21.46

R033105-65 Field-Mix Concrete

Presently most building jobs are built with ready-mixed concrete except at isolated locations and some larger jobs requiring over 10,000 C.Y. where land is readily available for setting up a temporary batch plant.

The most economical mix is a controlled mix using local aggregate proportioned by trial to give the required strength with the least cost of material.

R033105-70 Placing Ready-Mixed Concrete

For ground pours allow for 5% waste when figuring quantities.

Prices in the front of the book assume normal deliveries. If deliveries are made before 8 A.M. or after 5 P.M. or on Saturday afternoons add 30%. Negotiated discounts for large volumes are not included in prices in front of book.

For the lower floors without truck access, concrete may be wheeled in rubber-tired buggies, conveyer handled, crane handled or pumped. Pumping is economical if there is top steel. Conveyers are more efficient for thick slabs.

At higher floors the rubber-tired buggies may be hoisted by a hoisting tower and wheeled to location. Placement by a conveyer is limited to three floors and is best for high-volume pours. Pumped concrete is best when building has no crane access. Concrete may be pumped directly as high as thirty-six stories using special pumping techniques. Normal maximum height is about fifteen stories.

Best pumping aggregate is screened and graded bank gravel rather than crushed stone.

Pumping downward is more difficult than pumping upward. Horizontal distance from pump to pour may increase preparation time prior to pour. Placing by cranes, either mobile, climbing or tower types, continues as the most efficient method for high-rise concrete buildings.

R033105-85 Lift Slabs

The cost advantage of the lift slab method is due to placing all concrete, reinforcing steel, inserts and electrical conduit at ground level and in reduction of formwork. Minimum economical project size is about 30,000 S.F. Slabs may be tilted for parking garage ramps.

It is now used in all types of buildings and has gone up to 22 stories high in apartment buildings. Current trend is to use post-tensioned flat plate slabs with spans from 22′ to 35′. Cylindrical void forms are used when deep slabs are required. One pound of prestressing steel is about equal to seven pounds of conventional reinforcing.

To be considered cured for stressing and lifting, a slab must have attained 75% of design strength. Seven days are usually sufficient with four to five days possible if high early strength cement is used. Slabs can be stacked using two coats of a non-bonding agent to insure that slabs do not stick to each other. Lifting is done by companies specializing in this work. Lift rate is 5′ to 15′ per hour with an average of 10′ per hour. Total areas up to 33,000 S.F. have been lifted at one time. 24 to 36 jacking columns are common. Most economical bay sizes are 24′ to 28′ with four to fourteen stories most efficient. Continuous design reduces reinforcing steel cost. Use of post-tensioned slabs allows larger bay sizes.

R034105-30 Prestressed Precast Concrete Structural Units

Type	Location	Depth	Span in Ft.		Live Load Lb. per S.F.
Double Tee (8′ to 10′)	Floor	28″ to 34″	60 to 80		50 to 80
	Roof	12″ to 24″	30 to 50		40
	Wall	Width 8′	Up to 55′ high		Wind
Multiple Tee (8′)	Roof	8″ to 12″	15 to 40		40
	Floor	8″ to 12″	15 to 30		100
Plank or	Roof or Floor		Roof	Floor	40 for Roof 100 for Floor
		4″	13	12	
		6″	22	18	
		8″	26	25	
		10″	33	29	
		12″	42	32	
Single Tee (8′ to 10′)	Roof	28″	40		40
		32″	80		
		36″	100		
		48″	120		
AASHTO Girder	Bridges	Type 4	100		Highway
		5	110		
		6	125		
Box Beam (4′)	Bridges	15″ 27″ 33″	40 to 100		Highway

The majority of precast projects today utilize double tees rather than single tees because of speed and ease of installation. As a result casting beds at manufacturing plants are normally formed for double tees. Single tee projects will therefore require an initial set up charge to be spread over the individual single tee costs.

For floors, a 2″ to 3″ topping is field cast over the shapes. For roofs, insulating concrete or rigid insulation is placed over the shapes.

Member lengths up to 40′ are standard haul, 40′ to 60′ require special permits and lengths over 60′ must be escorted. Over width and/or over length can add up to 100% on hauling costs.

Large heavy members may require two cranes for lifting which would increase erection costs by about 45%. An eight man crew can install 12 to 20 double tees, or 45 to 70 quad tees or planks per day.

Grouting of connections must also be included.

Several system buildings utilizing precast members are available. Heights can go up to 22 stories for apartment buildings. Optimum design ratio is 3 S.F. of surface to 1 S.F. of floor area.

R034136-90 Prestressed Concrete, Post-Tensioned

In post-tensioned concrete the steel tendons are tensioned after the concrete has reached about 3/4 of its ultimate strength. The cableways are grouted after tensioning to provide bond between the steel and concrete. If bond is to be prevented, the tendons are coated with a corrosion-preventative grease and wrapped with waterproofed paper or plastic. Bonded tendons are usually used when ultimate strength (beams & girders) are controlling factors.

High strength concrete is used to fully utilize the steel, thereby reducing the size and weight of the member. A plasticizing agent may be added to reduce water content. Maximum size aggregate ranges from 1/2″ to 1-1/2″ depending on the spacing of the tendons.

The types of steel commonly used are bars and strands. Job conditions determine which is best suited. Bars are best for vertical prestresses since they are easy to support. The trend is for steel manufacturers to supply a finished package, cut to length, which reduces field preparation to a minimum.

Bars vary from 3/4″ to 1-3/8″ diameter. Table below gives time in labor-hours per tendon for placing, tensioning and grouting (if required) a 75′ beam. Tendons used in buildings are not usually grouted; tendons for bridges usually are grouted. For strands the table indicates the labor-hours per pound for typical prestressed units 100′ long. Simple span beams usually require one-end stressing regardless of lengths. Continuous beams are usually stressed from two ends. Long slabs are poured from the center outward and stressed in 75′ increments after the initial 150′ center pour.

Labor Hours per Tendon and per Pound of Prestressed Steel						
Length	**100′ Beam**		**75′ Beam**		**100′ Slab**	
Type Steel	**Strand**		**Bars**		**Strand**	
Diameter	**0.5″**		**3/4″**	**1-3/8″**	**0.5″**	**0.6″**
Number	**4**	**12**	**1**	**1**	**1**	**1**
Force in Kips	**100**	**300**	**42**	**143**	**25**	**35**
Preparation & Placing Cables	3.6	7.4	0.9	2.9	0.9	1.1
Stressing Cables	2.0	2.4	0.8	1.6	0.5	0.5
Grouting, if required	2.5	3.0	0.6	1.3		
Total Labor Hours	8.1	12.8	2.3	5.8	1.4	1.6
Prestressing Steel Weights (Lbs.)	215	640	115	380	53	74
Labor-hours per Lb. Bonded	0.038	0.020	0.020	0.015		
Non-bonded					0.026	0.022

Flat Slab construction — 4000 psi concrete with span-to-depth ratio between 36 and 44. Two way post-tensioned steel averages 1.0 lb. per S.F. for 24′ to 28′ bays (usually strand) and additional reinforcing steel averages .5 lb. per S.F.

Pan and Joist construction — 4000 psi concrete with span-to-depth ratio 28 to 30. Post-tensioned steel averages .8 lb. per S.F. and reinforcing steel about 1.0 lb. per S.F. Placing and stressing averages 40 hours per ton of total material.

Beam construction — 4000 to 5000 psi concrete. Steel weights vary greatly.

Labor cost per pound goes down as the size and length of the tendon increase. The primary economic consideration is the cost per kip for the member.

Post-tensioning becomes feasible for beams and girders over 30′ long; for continuous two-way slabs over 20′ clear; also in transferring upper building loads over longer spans at lower levels. Post-tension suppliers will provide engineering services at no cost to the user. Substantial economies are possible by using post-tensioned Lift Slabs.

Concrete — R0345 Precast Architectural Concrete

R034513-10 Precast Concrete Wall Panels

Panels are either solid or insulated with plain, colored or textured finishes. Transportation is an important cost factor. Prices shown in the unit cost section of the book are based on delivery within 50 miles of a plant including fabricators' overhead and profit. Engineering data is available from fabricators to assist with construction details. Usual minimum job size for economical use of panels is about 5000 S.F. Small jobs can double the prices shown. For large, highly repetitive jobs, deduct up to 15% from the prices shown.

2″ thick panels cost about the same as 3″ thick panels, and maximum panel size is less. For building panels faced with granite, marble or stone, add the material prices from those unit cost sections to the plain panel price shown. There is a growing trend toward aggregate facings and broken rib finish rather than plain gray concrete panels.

No allowance has been made in the unit cost section for supporting steel framework. On one story buildings, panels may rest on grade beams and require only wind bracing and fasteners. On multi-story buildings panels can span from column to column and floor to floor. Plastic-designed steel-framed structures may have large deflections which slow down erection and raise costs.

Large panels are more economical than small panels on a S.F. basis. When figuring areas include all protrusions, returns, etc. Overhangs can triple erection costs. Panels over 45′ have been produced. Larger flat units should be prestressed. Vacuum lifting of smooth finish panels eliminates inserts and can speed erection.

Concrete — R0347 Site-Cast Concrete

R034713-20 Tilt Up Concrete Panels

The advantage of tilt up construction is in the low cost of forms and the placing of concrete and reinforcing. Panels up to 75′ high and 5-1/2″ thick have been tilted using strongbacks. Tilt up has been used for one to five story buildings and is well-suited for warehouses, stores, offices, schools and residences.

The panels are cast in forms on the floor slab. Most jobs use 5-1/2″ thick solid reinforced concrete panels. Sandwich panels with a layer of insulating materials are also used. Where dampness is a factor, lightweight aggregate is used. Optimum panel size is 300 to 500 S.F.

Slabs are usually poured with 3000 psi concrete which permits tilting seven days after pouring. Slabs may be stacked on top of each other and are separated from each other by either two coats of bond breaker or a film of polyethylene. Use of high early-strength cement allows tilting two days after a pour. Tilting up is done with a roller outrigger crane with a capacity of at least 1-1/2 times the weight of the panel at the required reach. Exterior precast columns can be set at the same time as the panels; interior precast columns can be set first and the panels clipped directly to them. The use of cast-in-place concrete columns is diminishing due to shrinkage problems. Structural steel columns are sometimes used if crane rails are planned. Panels can be clipped to the columns or lowered between the flanges. Steel channels with anchors may be used as edge forms for the slab. When the panels are lifted the channels form an integral steel column to take structural loads. Roof loads can be carried directly by the panels for wall heights to 14′.

Requirements of local building codes may be a limiting factor and should be checked. Building floor slabs should be poured first and should be a minimum of 5″ thick with 100% compaction of soil or 6″ thick with less than 100% compaction.

Setting times as fast as nine minutes per panel have been observed, but a safer expectation would be four panels per hour with a crane and a four-man setting crew. If crane erects from inside building, some provision must be made to get crane out after walls are erected. Good yarding procedure is important to minimize delays. Equalizing three-point lifting beams and self-releasing pick-up hooks speed erection. If panels must be carried to their final location, setting time per panel will be increased and erection costs may approach the erection cost range of architectural precast wall panels. Placing panels into slots formed in continuous footers will speed erection.

Reinforcing should be with #5 bars with vertical bars on the bottom. If surface is to be sandblasted, stainless steel chairs should be used to prevent rust staining.

Use of a broom finish is popular since the unavoidable surface blemishes are concealed.

Precast columns run from three to five times the C.Y. price of the panels only.

Concrete — R0352 Lightweight Concrete Roof Insulation

R035216-10 Lightweight Concrete

Lightweight aggregate concrete is usually purchased ready mixed, but it can also be field mixed.

Vermiculite or Perlite comes in bags of 4 C.F. under various trade names. Weight is about 8 lbs. per C.F. For insulating roof fill use 1:6 mix. For structural deck use 1:4 mix over gypsum boards, steeltex, steel centering, etc., supported by closely spaced joists or bulb trees. For structural slabs use 1:3:2 vermiculite sand concrete over steeltex, metal lath, steel centering, etc., on joists spaced 2′-0″ O.C. for maximum L.L. of 80 P.S.F. Use same mix for slab base fill over steel flooring or regular reinforced concrete slab when tile, terrazzo or other finish is to be laid over.

For slabs on grade use 1:3:2 mix when tile, etc., finish is to be laid over. If radiant heating units are installed use a 1:6 mix for a base. After coils are in place, cover with a regular granolithic finish (mix 1:3:2) to a minimum depth of 1-1/2″ over top of units.

Reinforce all slabs with 6 x 6 or 10 x 10 welded wire mesh.

R040130-10 Cleaning Face Brick

On smooth brick a person can clean 70 S.F. an hour; on rough brick 50 S.F. per hour. Use one gallon muriatic acid to 20 gallons of water for 1000 S.F. Do not use acid solution until wall is at least seven days old, but a mild soap solution may be used after two days.

Time has been allowed for clean-up in brick prices.

R040513-10 Cement Mortar (material only)

Type N - 1:1:6 mix by volume. Use everywhere above grade except as noted below. - 1:3 mix using conventional masonry cement which saves handling two separate bagged materials.

Type M - 1:1/4:3 mix by volume, or 1 part cement, 1/4 (10% by wt.) lime, 3 parts sand. Use for heavy loads and where earthquakes or hurricanes may occur. Also for reinforced brick, sewers, manholes and everywhere below grade.

Mix Proportions by Volume and Compressive Strength of Mortar

Where Used	Allowable Proportions by Volume					Compressive Strength @ 28 days
	Mortar Type	Portland Cement	Masonry Cement	Hydrated Lime	Masonry Sand	
Plain Masonry		1	1	—	6	
	M	1	—	1/4	3	2500 psi
		1/2	1	—	4	
	S	1	—	1/4 to 1/2	4	1800 psi
		—	1	—	3	
	N	1	—	1/2 to 1-1/4	6	750 psi
		—	1	—	3	
	O	1	—	1-1/4 to 2-1/2	9	350 psi
	K	1	—	2-1/2 to 4	12	75 psi
Reinforced Masonry	PM	1	1	—	6	2500 psi
	PL	1	—	1/4 to 1/2	4	2500 psi

Note: The total aggregate should be between 2.25 to 3 times the sum of the cement and lime used.

The labor cost to mix the mortar is included in the productivity and labor cost of unit price lines in unit cost sections for brickwork, blockwork and stonework.

The material cost of mixed mortar is included in the material cost of those same unit price lines and includes the cost of renting and operating a 10 C.F. mixer at the rate of 200 C.F. per day.

There are two types of mortar color used. One type is the inert additive type with about 100 lbs. per M brick as the typical quantity required. These colors are also available in smaller-batch-sized bags (1 lb. to 15 lb.) which can be placed directly into the mixer without measuring. The other type is premixed and replaces the masonry cement. Dark green color has the highest cost.

R040519-50 Masonry Reinforcing

Horizontal joint reinforcing helps prevent wall cracks where wall movement may occur and in many locations is required by code. Horizontal joint reinforcing is generally not considered to be structural reinforcing and an unreinforced wall may still contain joint reinforcing.

Reinforcing strips come in 10′ and 12′ lengths and in truss and ladder shapes, with and without drips. Field labor runs between 2.7 to 5.3 hours per 1000 L.F. for wall thicknesses up to 12″.

The wire meets ASTM A82 for cold drawn steel wire and the typical size is 9 ga. sides and ties with 3/16″ diameter also available. Typical finish is mill galvanized with zinc coating at .10 oz. per S.F. Class I (.40 oz. per S.F.) and Class III (.80 oz per S.F.) are also available, as is hot dipped galvanizing at 1.50 oz. per S.F.

R042110-10 Economy in Bricklaying

Have adequate supervision. Be sure bricklayers are always supplied with materials so there is no waiting. Place best bricklayers at corners and openings.

Use only screened sand for mortar. Otherwise, labor time will be wasted picking out pebbles. Use seamless metal tubs for mortar as they do not leak or catch the trowel. Locate stack and mortar for easy wheeling.

Have brick delivered for stacking. This makes for faster handling, reduces chipping and breakage, and requires less storage space. Many dealers will deliver select common in 2′ x 3′ x 4′ pallets or face brick packaged. This affords quick handling with a crane or forklift and easy tonging in units of ten, which reduces waste.

Use wider bricks for one wythe wall construction. Keep scaffolding away from wall to allow mortar to fall clear and not stain wall.

On large jobs develop specialized crews for each type of masonry unit.

Consider designing for prefabricated panel construction on high rise projects.

Avoid excessive corners or openings. Each opening adds about 50% to labor cost for area of opening.

Bolting stone panels and using window frames as stops reduces labor costs and speeds up erection.

R042110-20 Common and Face Brick

Common building brick manufactured according to ASTM C62 and facing brick manufactured according to ASTM C216 are the two standard bricks available for general building use.

Building brick is made in three grades; SW, where high resistance to damage caused by cyclic freezing is required; MW, where moderate resistance to cyclic freezing is needed; and NW, where little resistance to cyclic freezing is needed. Facing brick is made in only the two grades SW and MW. Additionally, facing brick is available in three types; FBS, for general use; FBX, for general use where a higher degree of precision and lower permissible variation in size than FBS is needed; and FBA, for general use to produce characteristic architectural effects resulting from non-uniformity in size and texture of the units.

In figuring the material cost of brickwork, an allowance of 25% mortar waste and 3% brick breakage was included. If bricks are delivered palletized with 280 to 300 per pallet, or packaged, allow only 1-1/2% for breakage. Packaged or palletized delivery is practical when a job is big enough to have a crane or other equipment available to handle a package of brick. This is so on all industrial work but not always true on small commercial buildings.

The use of buff and gray face is increasing, and there is a continuing trend to the Norman, Roman, Jumbo and SCR brick.

Common red clay brick for backup is not used that often. Concrete block is the most usual backup material with occasional use of sand lime or cement brick. Building brick is commonly used in solid walls for strength and as a fire stop.

Brick panels built on the ground and then crane erected to the upper floors have proven to be economical. This allows the work to be done under cover and without scaffolding.

R042110-50 Brick, Block & Mortar Quantities

Running Bond						For Other Bonds Standard Size Add to S.F. Quantities in Table to Left		
Number of Brick per S.F. of Wall - Single Wythe with 3/8″ Joints				C.F. of Mortar per M Bricks, Waste Included				
Type Brick	Nominal Size (incl. mortar) L H W	Modular Coursing	Number of Brick per S.F.	3/8″ Joint	1/2″ Joint	Bond Type	Description	Factor
Standard	8 x 2-2/3 x 4	3C=8″	6.75	10.3	12.9	Common	full header every fifth course	+20%
Economy	8 x 4 x 4	1C=4″	4.50	11.4	14.6		full header every sixth course	+16.7%
Engineer	8 x 3-1/5 x 4	5C=16″	5.63	10.6	13.6	English	full header every second course	+50%
Fire	9 x 2-1/2 x 4-1/2	2C=5″	6.40	550 # Fireclay	—	Flemish	alternate headers every course	+33.3%
Jumbo	12 x 4 x 6 or 8	1C=4″	3.00	23.8	30.8		every sixth course	+5.6%
Norman	12 x 2-2/3 x 4	3C=8″	4.50	14.0	17.9	Header = W x H exposed		+100%
Norwegian	12 x 3-1/5 x 4	5C=16″	3.75	14.6	18.6	Rowlock = H x W exposed		+100%
Roman	12 x 2 x 4	2C=4″	6.00	13.4	17.0	Rowlock stretcher = L x W exposed		+33.3%
SCR	12 x 2-2/3 x 6	3C=8″	4.50	21.8	28.0	Soldier = H x L exposed		—
Utility	12 x 4 x 4	1C=4″	3.00	15.4	19.6	Sailor = W x L exposed		-33.3%

Concrete Blocks Nominal Size	Approximate Weight per S.F. Standard	Approximate Weight per S.F. Lightweight	Blocks per 100 S.F.	Mortar per M block, waste included Partitions	Mortar per M block, waste included Back up
2″ x 8″ x 16″	20 PSF	15 PSF	113	27 C.F.	36 C.F.
4″ ↓	30 ↓	20 ↓	↓	41 ↓	51 ↓
6″	42	30		56	66
8″	55	38		72	82
10″	70	47		87	97
12″	85	55		102	112

R042210-20 Concrete Block

The material cost of special block such as corner, jamb and head block can be figured at the same price as ordinary block of same size. Labor on specials is about the same as equal-sized regular block.

Bond beam and 16″ high lintel blocks are more expensive than regular units of equal size. Lintel blocks are 8″ long and either 8″ or 16″ high.

Use of motorized mortar spreader box will speed construction of continuous walls.

Hollow non-load-bearing units are made according to ASTM C129 and hollow load-bearing units according to ASTM C90.

R050516-30 Coating Structural Steel

On field-welded jobs, the shop-applied primer coat is necessarily omitted. All painting must be done in the field and usually consists of red oxide rust inhibitive paint or an aluminum paint. The table below shows paint coverage and daily production for field painting.

See Division 09 97 13.23 for hot-dipped galvanizing and for field-applied cold galvanizing and other paints and protective coatings.

See Division 05 01 10.51 for steel surface preparation treatments such as wire brushing, pressure washing and sand blasting.

Type Construction	Surface Area per Ton	Coat	One Gallon Covers		In 8 Hrs. Person Covers		Average per Ton Spray	
			Brush	Spray	Brush	Spray	Gallons	Labor-hours
Light Structural	300 S.F. to 500 S.F.	1st	500 S.F.	455 S.F.	640 S.F.	2000 S.F.	0.9 gals.	1.6 L.H.
		2nd	450	410	800	2400	1.0	1.3
		3rd	450	410	960	3200	1.0	1.0
Medium	150 S.F. to 300 S.F.	All	400	365	1600	3200	0.6	0.6
Heavy Structural	50 S.F. to 150 S.F.	1st	400	365	1920	4000	0.2	0.2
		2nd	400	365	2000	4000	0.2	0.2
		3rd	400	365	2000	4000	0.2	0.2
Weighted Average	225 S.F.	All	400	365	1350	3000	0.6	0.6

R050521-20 Welded Structural Steel

Usual weight reductions with welded design run 10% to 20% compared with bolted or riveted connections. This amounts to about the same total cost compared with bolted structures since field welding is more expensive than bolts. For normal spans of 18′ to 24′ figure 6 to 7 connections per ton.

Trusses — For welded trusses add 4% to weight of main members for connections. Up to 15% less steel can be expected in a welded truss compared to one that is shop bolted. Cost of erection is the same whether shop bolted or welded.

General — Typical electrodes for structural steel welding are E6010, E6011, E60T and E70T. Typical buildings vary between 2# to 8# of weld rod per ton of steel. Buildings utilizing continuous design require about three times as much welding as conventional welded structures. In estimating field erection by welding, it is best to use the average linear feet of weld per ton to arrive at the welding cost per ton. The type, size and position of the weld will have a direct bearing on the cost per linear foot. A typical field welder will deposit 1.8# to 2# of weld rod per hour manually. Using semiautomatic methods can increase production by as much as 50% to 75%.

R050523-10 High Strength Bolts

Common bolts (A307) are usually used in secondary connections (see Division 05 05 23.10).

High strength bolts (A325 and A490) are usually specified for primary connections such as column splices, beam and girder connections to columns, column bracing, connections for supports of operating equipment or of other live loads which produce impact or reversal of stress, and in structures carrying cranes of over 5-ton capacity.

Allow 20 field bolts per ton of steel for a 6 story office building, apartment house or light industrial building. For 6 to 12 stories allow 25 bolts per ton, and above 12 stories, 30 bolts per ton. On power stations, 20 to 25 bolts per ton are needed.

R051223-10 Structural Steel

The bare material prices for structural steel, shown in the unit cost sections of the book, are for 100 tons of shop-fabricated structural steel and include:

1. Mill base price of structural steel
2. Mill scrap/grade/size/length extras
3. Mill delivery to a metals service center (warehouse)
4. Service center storage and handling
5. Service center delivery to a fabrication shop
6. Shop storage and handling
7. Shop drafting/detailing
8. Shop fabrication
9. Shop coat of primer paint
10. Shop listing
11. Shop delivery to the job site

In unit cost sections of the book that contain items for field fabrication of steel components, the bare material cost of steel includes:

1. Mill base price of structural steel
2. Mill scrap/grade/size/length extras
3. Mill delivery to a metals service center (warehouse)
4. Service center storage and handling
5. Service center delivery to the job site

R051223-20 Steel Estimating Quantities

One estimate on erection is that a crane can handle 35 to 60 pieces per day. Say the average is 45. With usual sizes of beams, girders, and columns, this would amount to about 20 tons per day. The type of connection greatly affects the speed of erection. Moment connections for continuous design slow down production and increase erection costs.

Short open web bar joists can be set at the rate of 75 to 80 per day, with 50 per day being the average for setting long span joists.

After main members are calculated, add the following for usual allowances: base plates 2% to 3%; column splices 4% to 5%; and miscellaneous details 4% to 5%, for a total of 10% to 13% in addition to main members.

The ratio of column to beam tonnage varies depending on type of steels used, typical spans, story heights and live loads.

It is more economical to keep the column size constant and to vary the strength of the column by using high strength steels. This also saves floor space. Buildings have recently gone as high as ten stories with 8″ high strength columns. For light columns under W8X31 lb. sections, concrete filled steel columns are economical.

High strength steels may be used in columns and beams to save floor space and to meet head room requirements. High strength steels in some sizes sometimes require long lead times.

Round, square and rectangular columns, both plain and concrete filled, are readily available and save floor area, but are higher in cost per pound than rolled columns. For high unbraced columns, tube columns may be less expensive.

Below are average minimum figures for the weights of the structural steel frame for different types of buildings using A36 steel, rolled shapes and simple joints. For economy in domes, rise to span ratio = .13. Open web joist framing systems will reduce weights by 10% to 40%. Composite design can reduce steel weight by up to 25% but additional concrete floor slab thickness may be required. Continuous design can reduce the weights up to 20%. There are many building codes with different live load requirements and different structural requirements, such as hurricane and earthquake loadings which can alter the figures.

Structural Steel Weights per S.F. of Floor Area

Type of Building	No. of Stories	Avg. Spans	L.L. #/S.F.	Lbs. Per S.F.	Type of Building	No. of Stories	Avg. Spans	L.L. #/S.F.	Lbs. Per S.F.
		20′x20′		8	Apartments	2-8	20′x20′	40	8
Steel Frame Mfg.	1	30′x30′	40	13		9-25			14
		40′x40′		18		to 10			10
Parking garage	4	Various	80	8.5	Office	20	Various	80	18
Domes	1	200′	30	10		30			26
(Schwedler)*		300′		15		over 50			35

R051223-25 Common Structural Steel Specifications

ASTM A992 (formerly A36, then A572 Grade 50) is the all-purpose carbon grade steel widely used in building and bridge construction.

The other high-strength steels listed below may each have certain advantages over ASTM A992 stuctural carbon steel, depending on the application. They have proven to be economical choices where, due to lighter members, the reduction of dead load and the associated savings in shipping cost can be significant.

ASTM A588 atmospheric weathering, high-strength low-alloy steels can be used in the bare (uncoated) condition, where exposure to normal atmosphere causes a tightly adherant oxide to form on the surface protecting the steel from further oxidation. ASTM A242 corrosion-resistant, high-strength low-alloy steels have enhanced atmospheric corrosion resistance of at least two times that of carbon structural steels with copper, or four times that of carbon structural steels without copper. The reduction or elimination of maintenance resulting from the use of these steels often offsets their higher initial cost.

Steel Type	ASTM Designation	Minimum Yield Stress in KSI	Shapes Available
Carbon	A36	36	All structural shape groups, and plates & bars up thru 8″ thick
	A529	50	Structural shape group 1, and plates & bars up thru 2″ thick
High-Strength Low-Alloy Quenched & Self-Tempered	A913	50	All structural shape groups
		60	
		65	
		70	
High-Strength Low-Alloy Columbium-Vanadium	A572	42	All structural shape groups, and plates & bars up thru 6″ thick
		50	All structural shape groups, and plates & bars up thru 4″ thick
		55	Structural shape groups 1 & 2, and plates & bars up thru 2″ thick
		60	Structural shape groups 1 & 2, and plates & bars up thru 1-1/4″ thick
		65	Structural shape group 1, and plates & bars up thru 1-1/4″ thick
High-Strength Low-Alloy Columbium-Vanadium	A992	50	All structural shape groups
Weathering High-Strength Low-Alloy	A242	42	Structural shape groups 4 & 5, and plates & bars over 1-1/2″ up thru 4″ thick
		46	Structural shape group 3, and plates & bars over 3/4″ up thru 1-1/2″ thick
		50	Structural shape groups 1 & 2, and plates & bars up thru 3/4″ thick
Weathering High-Strength Low-Alloy	A588	42	Plates & bars over 5″ up thru 8″ thick
		46	Plates & bars over 4″ up thru 5″ thick
		50	All structural shape groups, and plates & bars up thru 4″ thick
Quenched and Tempered Low-Alloy	A852	70	Plates & bars up thru 4″ thick
Quenched and Tempered Alloy	A514	90	Plates & bars over 2-1/2″ up thru 6″ thick
		100	Plates & bars up thru 2-1/2″ thick

R051223-30 High Strength Steels

The mill price of high strength steels may be higher than A992 carbon steel but their proper use can achieve overall savings thru total reduced weights. For columns with L/r over 100, A992 steel is best; under 100, high strength steels are economical. For heavy columns, high strength steels are economical when cover plates are eliminated. There is no economy using high strength steels for clip angles or supports or for beams where deflection governs. Thinner members are more economical than thick.

The per ton erection and fabricating costs of the high strength steels will be higher than for A992 since the same number of pieces, but less weight, will be installed.

R051223-35 Common Steel Sections

The upper portion of this table shows the name, shape, common designation and basic characteristics of commonly used steel sections. The lower portion explains how to read the designations used for the above illustrated common sections.

Shape & Designation	Name & Characteristics	Shape & Designation	Name & Characteristics
W	W Shape Parallel flange surfaces	MC	Miscellaneous Channel Infrequently rolled by some producers
S	American Standard Beam (I Beam) Sloped inner flange	L	Angle Equal or unequal legs, constant thickness
M	Miscellaneous Beams Cannot be classified as W, HP or S; infrequently rolled by some producers	T	Structural Tee Cut from W, M or S on center of web
C	American Standard Channel Sloped inner flange	HP	Bearing Pile Parallel flanges and equal flange and web thickness

Common drawing designations follow:

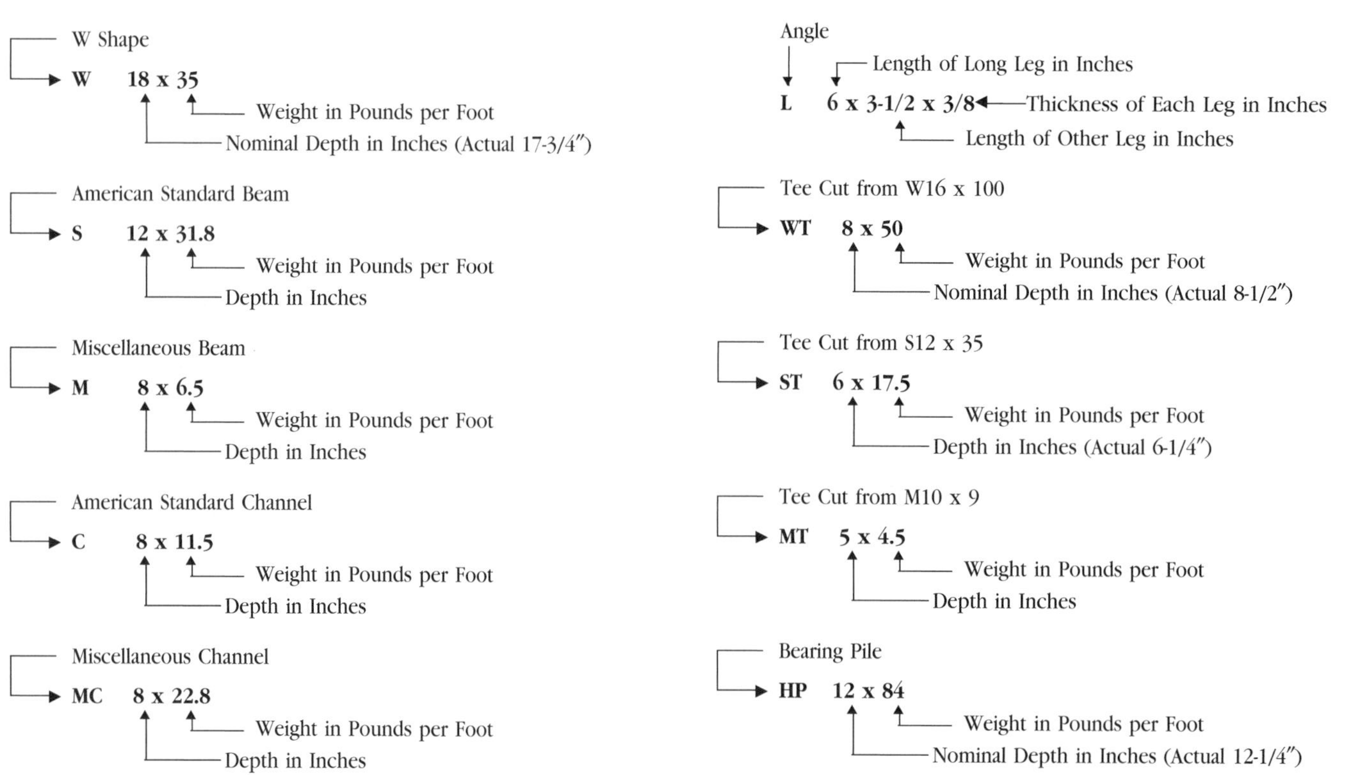

R051223-45 Installation Time for Structural Steel Building Components

The following tables show the expected average installation times for various structural steel shapes. Table A presents installation times for columns, Table B for beams, Table C for light framing and bolts, and Table D for structural steel for various project types.

Table A

Description	Labor-Hours	Unit
Columns		
Steel, Concrete Filled		
3-1/2″ Diameter	.933	Ea.
6-5/8″ Diameter	1.120	Ea.
Steel Pipe		
3″ Diameter	.933	Ea.
8″ Diameter	1.120	Ea.
12″ Diameter	1.244	Ea.
Structural Tubing		
4″ x 4″	.966	Ea.
8″ x 8″	1.120	Ea.
12″ x 8″	1.167	Ea.
W Shape 2 Tier		
W8 x 31	.052	L.F.
W8 x 67	.057	L.F.
W10 x 45	.054	L.F.
W10 x 112	.058	L.F.
W12 x 50	.054	L.F.
W12 x 190	.061	L.F.
W14 x 74	.057	L.F.
W14 x 176	.061	L.F.

Table B

Description	Labor-Hours	Unit	Labor-Hours	Unit
Beams, W Shape				
W6 x 9	.949	Ea.	.093	L.F.
W10 x 22	1.037	Ea.	.085	L.F.
W12 x 26	1.037	Ea.	.064	L.F.
W14 x 34	1.333	Ea.	.069	L.F.
W16 x 31	1.333	Ea.	.062	L.F.
W18 x 50	2.162	Ea.	.088	L.F.
W21 x 62	2.222	Ea.	.077	L.F.
W24 x 76	2.353	Ea.	.072	L.F.
W27 x 94	2.581	Ea.	.067	L.F.
W30 x 108	2.857	Ea.	.067	L.F.
W33 x 130	3.200	Ea.	.071	L.F.
W36 x 300	3.810	Ea.	.077	L.F.

Table C

Description	Labor-Hours	Unit
Light Framing		
Angles 4″ and Larger	.055	lbs.
Less than 4″	.091	lbs.
Channels 8″ and Larger	.048	lbs.
Less than 8″	.072	lbs.
Cross Bracing Angles	.055	lbs.
Rods	.034	lbs.
Hanging Lintels	.069	lbs.
High Strength Bolts in Place		
3/4″ Bolts	.070	Ea.
7/8″ Bolts	.076	Ea.

Table D

Description	Labor-Hours	Unit	Labor-Hours	Unit
Apartments, Nursing Homes, etc.				
1-2 Stories	4.211	Piece	7.767	Ton
3-6 Stories	4.444	Piece	7.921	Ton
7-15 Stories	4.923	Piece	9.014	Ton
Over 15 Stories	5.333	Piece	9.209	Ton
Offices, Hospitals, etc.				
1-2 Stories	4.211	Piece	7.767	Ton
3-6 Stories	4.741	Piece	8.889	Ton
7-15 Stories	4.923	Piece	9.014	Ton
Over 15 Stories	5.120	Piece	9.209	Ton
Industrial Buildings				
1 Story	3.478	Piece	6.202	Ton

R051223-50 Subpurlins

Bulb tee subpurlins are structural members designed to support and reinforce a variety of roof deck systems such as precast cement fiber roof deck tiles, monolithic roof deck systems, and gypsum or lightweight concrete over formboard. Other uses include interstitial service ceiling systems, wall panel systems, and joist anchoring in bond beams. See Unit Price section for pricing on a square foot basis at 32-5/8″ O.C. Maximum span is based on a 3-span condition with a total allowable vertical load of 40 psf.

R051223-80 Dimensions and Weights of Sheet Steel

Gauge No.	Approximate Thickness				Weight		
	Inches (in fractions)	Inches (in decimal parts)		Millimeters			
	Wrought Iron	Wrought Iron	Steel	Steel	per S.F. in Ounces	per S.F. in Lbs.	per Square Meter in Kg.
0000000	1/2″	.5	.4782	12.146	320	20.000	97.650
000000	15/32″	.46875	.4484	11.389	300	18.750	91.550
00000	7/16″	.4375	.4185	10.630	280	17.500	85.440
0000	13/32″	.40625	.3886	9.870	260	16.250	79.330
000	3/8″	.375	.3587	9.111	240	15.000	73.240
00	11/32″	.34375	.3288	8.352	220	13.750	67.130
0	5/16″	.3125	.2989	7.592	200	12.500	61.030
1	9/32″	.28125	.2690	6.833	180	11.250	54.930
2	17/64″	.265625	.2541	6.454	170	10.625	51.880
3	1/4″	.25	.2391	6.073	160	10.000	48.820
4	15/64″	.234375	.2242	5.695	150	9.375	45.770
5	7/32″	.21875	.2092	5.314	140	8.750	42.720
6	13/64″	.203125	.1943	4.935	130	8.125	39.670
7	3/16″	.1875	.1793	4.554	120	7.500	36.320
8	11/64″	.171875	.1644	4.176	110	6.875	33.570
9	5/32″	.15625	.1495	3.797	100	6.250	30.520
10	9/64″	.140625	.1345	3.416	90	5.625	27.460
11	1/8″	.125	.1196	3.038	80	5.000	24.410
12	7/64″	.109375	.1046	2.657	70	4.375	21.360
13	3/32″	.09375	.0897	2.278	60	3.750	18.310
14	5/64″	.078125	.0747	1.897	50	3.125	15.260
15	9/128″	.0713125	.0673	1.709	45	2.813	13.730
16	1/16″	.0625	.0598	1.519	40	2.500	12.210
17	9/160″	.05625	.0538	1.367	36	2.250	10.990
18	1/20″	.05	.0478	1.214	32	2.000	9.765
19	7/160″	.04375	.0418	1.062	28	1.750	8.544
20	3/80″	.0375	.0359	.912	24	1.500	7.324
21	11/320″	.034375	.0329	.836	22	1.375	6.713
22	1/32″	.03125	.0299	.759	20	1.250	6.103
23	9/320″	.028125	.0269	.683	18	1.125	5.490
24	1/40″	.025	.0239	.607	16	1.000	4.882
25	7/320″	.021875	.0209	.531	14	.875	4.272
26	3/160″	.01875	.0179	.455	12	.750	3.662
27	11/640″	.0171875	.0164	.417	11	.688	3.357
28	1/64″	.015625	.0149	.378	10	.625	3.052

R053100-10 Decking Descriptions

General - All Deck Products

Steel deck is made by cold forming structural grade sheet steel into a repeating pattern of parallel ribs. The strength and stiffness of the panels are the result of the ribs and the material properties of the steel. Deck lengths can be varied to suit job conditions, but because of shipping considerations, are usually less than 40 feet. Standard deck width varies with the product used but full sheets are usually 12″, 18″, 24″, 30″, or 36″. Deck is typically furnished in a standard width with the ends cut square. Any cutting for width, such as at openings or for angular fit, is done at the job site.

Deck is typically attached to the building frame with arc puddle welds, self-drilling screws, or powder or pneumatically driven pins. Sheet to sheet fastening is done with screws, button punching (crimping), or welds.

Composite Floor Deck

After installation and adequate fastening, floor deck serves several purposes. It (a) acts as a working platform, (b) stabilizes the frame, (c) serves as a concrete form for the slab, and (d) reinforces the slab to carry the design loads applied during the life of the building. Composite decks are distinguished by the presence of shear connector devices as part of the deck. These devices are designed to mechanically lock the concrete and deck together so that the concrete and the deck work together to carry subsequent floor loads. These shear connector devices can be rolled-in embossments, lugs, holes, or wires welded to the panels. The deck profile can also be used to interlock concrete and steel.

Composite deck finishes are either galvanized (zinc coated) or phosphatized/painted. Galvanized deck has a zinc coating on both the top and bottom surfaces. The phosphatized/painted deck has a bare (phosphatized) top surface that will come into contact with the concrete. This bare top surface can be expected to develop rust before the concrete is placed. The bottom side of the deck has a primer coat of paint.

Composite floor deck is normally installed so the panel ends do not overlap on the supporting beams. Shear lugs or panel profile shape often prevent a tight metal to metal fit if the panel ends overlap; the air gap caused by overlapping will prevent proper fusion with the structural steel supports when the panel end laps are shear stud welded.

Adequate end bearing of the deck must be obtained as shown on the drawings. If bearing is actually less in the field than shown on the drawings, further investigation is required.

Roof Deck

Roof deck is not designed to act compositely with other materials. Roof deck acts alone in transferring horizontal and vertical loads into the building frame. Roof deck rib openings are usually narrower than floor deck rib openings. This provides adequate support of rigid thermal insulation board.

Roof deck is typically installed to endlap approximately 2″ over supports. However, it can be butted (or lapped more than 2″) to solve field fit problems. Since designers frequently use the installed deck system as part of the horizontal bracing system (the deck as a diaphragm), any fastening substitution or change should be approved by the designer. Continuous perimeter support of the deck is necessary to limit edge deflection in the finished roof and may be required for diaphragm shear transfer.

Standard roof deck finishes are galvanized or primer painted. The standard factory applied paint for roof deck is a primer paint and is not intended to weather for extended periods of time. Field painting or touching up of abrasions and deterioration of the primer coat or other protective finishes is the responsibility of the contractor.

Cellular Deck

Cellular deck is made by attaching a bottom steel sheet to a roof deck or composite floor deck panel. Cellular deck can be used in the same manner as floor deck. Electrical, telephone, and data wires are easily run through the chase created between the deck panel and the bottom sheet.

When used as part of the electrical distribution system, the cellular deck must be installed so that the ribs line up and create a smooth cell transition at abutting ends. The joint that occurs at butting cell ends must be taped or otherwise sealed to prevent wet concrete from seeping into the cell. Cell interiors must be free of welding burrs, or other sharp intrusions, to prevent damage to wires.

When used as a roof deck, the bottom flat plate is usually left exposed to view. Care must be maintained during erection to keep good alignment and prevent damage.

Cellular deck is sometimes used with the flat plate on the top side to provide a flat working surface. Installation of the deck for this purpose requires special methods for attachment to the frame because the flat plate, now on the top, can prevent direct access to the deck material that is bearing on the structural steel. It may be advisable to treat the flat top surface to prevent slipping.

Cellular deck is always furnished galvanized or painted over galvanized.

Form Deck

Form deck can be any floor or roof deck product used as a concrete form. Connections to the frame are by the same methods used to anchor floor and roof deck. Welding washers are recommended when welding deck that is less than 20 gauge thickness.

Form deck is furnished galvanized, prime painted, or uncoated. Galvanized deck must be used for those roof deck systems where form deck is used to carry a lightweight insulating concrete fill.

Wood, Plastics & Comp. R0611 Wood Framing

R061110-30 Lumber Product Material Prices

The price of forest products fluctuates widely from location to location and from season to season depending upon economic conditions. The bare material prices in the unit cost sections of the book show the National Average material prices in effect Jan. 1 of this book year. It must be noted that lumber prices in general may change significantly during the year.

Availability of certain items depends upon geographic location and must be checked prior to firm-price bidding.

Wood, Plastics & Comp. R0616 Sheathing

R061636-20 Plywood

There are two types of plywood used in construction: interior, which is moisture-resistant but not waterproofed, and exterior, which is waterproofed.

The grade of the exterior surface of the plywood sheets is designated by the first letter: A, for smooth surface with patches allowed; B, for solid surface with patches and plugs allowed; C, which may be surface plugged or may have knot holes up to 1″ wide; and D, which is used only for interior type plywood and may have knot holes up to 2-1/2″ wide. "Structural Grade" is specifically designed for engineered applications such as box beams. All CC & DD grades have roof and floor spans marked on them.

Underlayment-grade plywood runs from 1/4″ to 1-1/4″ thick. Thicknesses 5/8″ and over have optional tongue and groove joints which eliminate the need for blocking the edges. Underlayment 19/32″ and over may be referred to as Sturd-i-Floor.

The price of plywood can fluctuate widely due to geographic and economic conditions.

Typical uses for various plywood grades are as follows:

AA-AD Interior — cupboards, shelving, paneling, furniture

BB Plyform — concrete form plywood

CDX — wall and roof sheathing

Structural — box beams, girders, stressed skin panels

AA-AC Exterior — fences, signs, siding, soffits, etc.

Underlayment — base for resilient floor coverings

Overlaid HDO — high density for concrete forms & highway signs

Overlaid MDO — medium density for painting, siding, soffits & signs

303 Siding — exterior siding, textured, striated, embossed, etc.

Thermal & Moist. Protec. R0731 Shingles & Shakes

R073126-20 Roof Slate

16″, 18″ and 20″ are standard lengths, and slate usually comes in random widths. For standard 3/16″ thickness use 1-1/2″ copper nails. Allow for 3% breakage.

Thermal & Moist. Protec. R0751 Built-Up Bituminous Roofing

R075113-20 Built-Up Roofing

Asphalt is available in kegs of 100 lbs. each; coal tar pitch in 560 lb. kegs. Prepared roofing felts are available in a wide range of sizes, weights and characteristics. However, the most commonly used are #15 (432 S.F. per roll, 13 lbs. per square) and #30 (216 S.F. per roll, 27 lbs. per square).

Inter-ply bitumen varies from 24 lbs. per sq. (asphalt) to 30 lbs. per sq. (coal tar) per ply, MF4@ 25%. Flood coat bitumen also varies from 60 lbs. per sq. (asphalt) to 75 lbs. per sq. (coal tar), MF4@ 25%. Expendable equipment (mops, brooms, screeds, etc.) runs about 16% of the bitumen cost. For new, inexperienced crews this factor may be much higher.

Rigid insulation board is typically applied in two layers. The first is mechanically attached to nailable decks or spot or solid mopped to non-nailable decks; the second layer is then spot or solid mopped to the first layer. Membrane application follows the insulation, except in protected membrane roofs, where the membrane goes down first and the insulation on top, followed with ballast (stone or concrete pavers). Insulation and related labor costs are NOT included in prices for built-up roofing.

R075213-30 Modified Bitumen Roofing

The cost of modified bitumen roofing is highly dependent on the type of installation that is planned. Installation is based on the type of modifier used in the bitumen. The two most popular modifiers are atactic polypropylene (APP) and styrene butadiene styrene (SBS). The modifiers are added to heated bitumen during the manufacturing process to change its characteristics. A polyethylene, polyester or fiberglass reinforcing sheet is then sandwiched between layers of this bitumen. When completed, the result is a pre-assembled, built-up roof that has increased elasticity and weatherablility. Some manufacturers include a surfacing material such as ceramic or mineral granules, metal particles or sand.

The preferred method of adhering SBS-modified bitumen roofing to the substrate is with hot-mopped asphalt (much the same as built-up roofing). This installation method requires a tar kettle/pot to heat the asphalt, as well as the labor, tools and equipment necessary to distribute and spread the hot asphalt.

The alternative method for applying APP and SBS modified bitumen is as follows. A skilled installer uses a torch to melt a small pool of bitumen off the membrane. This pool must form across the entire roll for proper adhesion. The installer must unroll the roofing at a pace slow enough to melt the bitumen, but fast enough to prevent damage to the rest of the membrane.

Modified bitumen roofing provides the advantages of both built-up and single-ply roofing. Labor costs are reduced over those of built-up roofing because only a single ply is necessary. The elasticity of single-ply roofing is attained with the reinforcing sheet and polymer modifiers. Modifieds have some self-healing characteristics and because of their multi-layer construction, they offer the reliability and safety of built-up roofing.

Thermal & Moist. Protec. R0784 Firestopping

R078413-30 Firestopping

Firestopping is the sealing of structural, mechanical, electrical and other penetrations through fire-rated assemblies. The basic components of firestop systems are safing insulation and firestop sealant on both sides of wall penetrations and the top side of floor penetrations.

Pipe penetrations are assumed to be through concrete, grout, or joint compound and can be sleeved or unsleeved. Costs for the penetrations and sleeves are not included. An annular space of 1″ is assumed. Escutcheons are not included.

Metallic pipe is assumed to be copper, aluminum, cast iron or similar metallic material. Insulated metallic pipe is assumed to be covered with a thermal insulating jacket of varying thickness and materials.

Non-metallic pipe is assumed to be PVC, CPVC, FR Polypropylene or similar plastic piping material. Intumescent firestop sealant or wrap strips are included. Collars on both sides of wall penetrations and a sheet metal plate on the underside of floor penetrations are included.

Ductwork is assumed to be sheet metal, stainless steel or similar metallic material. Duct penetrations are assumed to be through concrete, grout or joint compound. Costs for penetrations and sleeves are not included. An annular space of 1/2″ is assumed.

Multi-trade openings include costs for sheet metal forms, firestop mortar, wrap strips, collars and sealants as necessary.

Structural penetrations joints are assumed to be 1/2″ or less. CMU walls are assumed to be within 1-1/2″ of metal deck. Drywall walls are assumed to be tight to the underside of metal decking.

Metal panel, glass or curtain wall systems include a spandrel area of 5′ filled with mineral wool foil-faced insulation. Fasteners and stiffeners are included.

Openings — R0813 Metal Doors

R081313-20 Steel Door Selection Guide

Standard steel doors are classified into four levels, as recommended by the Steel Door Institute in the chart below. Each of the four levels offers a range of construction models and designs, to meet architectural requirements for preference and appearance, including full flush, seamless, and stile & rail. Recommended minimum gauge requirements are also included.

For complete standard steel door construction specifications and available sizes, refer to the Steel Door Institute Technical Data Series, ANSI A250.8-98 (SDI-100), and ANSI A250.4-94 Test Procedure and Acceptance Criteria for Physical Endurance of Steel Door and Hardware Reinforcements.

Level		Model	Construction	For Full Flush or Seamless		
				Min. Gauge	Thickness (in)	Thickness (mm)
I	Standard Duty	1 2	Full Flush Seamless	20	0.032	0.8
II	Heavy Duty	1 2	Full Flush Seamless	18	0.042	1.0
III	Extra Heavy Duty	1 2 3	Full Flush Seamless *Stile & Rail	16	0.053	1.3
IV	Maximum Duty	1 2	Full Flush Seamless	14	0.067	1.6

*Stiles & rails are 16 gauge; flush panels, when specified, are 18 gauge

Openings — R0851 Metal Windows

R085123-10 Steel Sash

Ironworker crew will erect 25 S.F. or 1.3 sash unit per hour, whichever is less.

Mechanic will point 30 L.F. per hour.

Painter will paint 90 S.F. per coat per hour.

Glazier production depends on light size.

Allow 1 lb. special steel sash putty per 16″ x 20″ light.

Openings — R0852 Wood Windows

R085216-10 Window Estimates

To ensure a complete window estimate, be sure to include the material and labor costs for each window, as well as the material and labor costs for an interior wood trim set.

R087120-10 Hinges

All closer equipped doors should have ball bearing hinges. Lead lined or extremely heavy doors require special strength hinges. Usually 1-1/2 pair of hinges are used per door up to 7′-6″ high openings. Table below shows typical hinge requirements.

Use Frequency	Type Hinge Required	Type of Opening	Type of Structure
High	Heavy weight ball bearing	Entrances Toilet Rooms	Banks, Office buildings, Schools, Stores & Theaters Office buildings and Schools
Average	Standard weight ball bearing	Entrances Corridors Toilet Rooms	Dwellings Office buildings and Schools Stores
Low	Plain bearing	Interior	Dwellings

Door Thickness	Weight of Doors in Pounds per Square Foot				
	White Pine	Oak	Hollow Core	Solid Core	Hollow Metal
1-3/8″	3psf	6psf	1-1/2psf	3-1/2 — 4psf	6-1/2psf
1-3/4″	3-1/2	7	2-1/2	4-1/2 — 5-1/4	6-1/2
2-1/4″	4-1/2	9	—	5-1/2 — 6-3/4	6-1/2

R088110-10 Glazing Productivity

Some glass sizes are estimated by the "united inch" (height + width). The table below shows the number of lights glazed in an eight-hour period by the crew size indicated, for glass up to 1/4″ thick. Square or nearly square lights are more economical on a S.F. basis. Long slender lights will have a high S.F. installation cost. For insulated glass reduce production by 33%. For 1/2″ float glass reduce production by 50%. Production time for glazing with two glaziers per day averages: 1/4″ float glass 120 S.F.; 1/2″ float glass 55 S.F.; 1/2″ insulated glass 95 S.F.; 3/4″ insulated glass 75 S.F.

Glazing Method	United Inches per Light							
	40″	60″	80″	100″	135″	165″	200″	240″
Number of Men in Crew	1	1	1	1	2	3	3	4
Industrial sash, putty	60	45	24	15	18	—	—	—
With stops, putty bed	50	36	21	12	16	8	4	3
Wood stops, rubber	40	27	15	9	11	6	3	2
Metal stops, rubber	30	24	14	9	9	6	3	2
Structural glass	10	7	4	3	—	—	—	—
Corrugated glass	12	9	7	4	4	4	3	—
Storefronts	16	15	13	11	7	6	4	4
Skylights, putty glass	60	36	21	12	16	—	—	—
Thiokol set	15	15	11	9	9	6	3	2
Vinyl set, snap on	18	18	13	12	12	7	5	4
Maximum area per light	2.8 S.F.	6.3 S.F.	11.1 S.F.	17.4 S.F.	31.6 S.F.	47 S.F.	69 S.F.	100 S.F.

R092000-50 Lath, Plaster and Gypsum Board

Gypsum board lath is available in 3/8″ thick x 16″ wide x 4′ long sheets as a base material for multi-layer plaster applications. It is also available as a base for either multi-layer or veneer plaster applications in 1/2″ and 5/8″ thick–4′ wide x 8′, 10′ or 12′ long sheets. Fasteners are screws or blued ring shank nails for wood framing and screws for metal framing.

Metal lath is available in diamond mesh pattern with flat or self-furring profiles. Paper backing is available for applications where excessive plaster waste needs to be avoided. A slotted mesh ribbed lath should be used in areas where the span between structural supports is greater than normal. Most metal lath comes in 27″ x 96″ sheets. Diamond mesh weighs 1.75, 2.5 or 3.4 pounds per square yard, slotted mesh lath weighs 2.75 or 3.4 pounds per square yard. Metal lath can be nailed, screwed or tied in place.

Many **accessories** are available. Corner beads, flat reinforcing strips, casing beads, control and expansion joints, furring brackets and channels are some examples. Note that accessories are not included in plaster or stucco line items.

Plaster is defined as a material or combination of materials that when mixed with a suitable amount of water, forms a plastic mass or paste. When applied to a surface, the paste adheres to it and subsequently hardens, preserving in a rigid state the form or texture imposed during the period of elasticity.

Gypsum plaster is made from ground calcined gypsum. It is mixed with aggregates and water for use as a base coat plaster.

Vermiculite plaster is a fire-retardant plaster covering used on steel beams, concrete slabs and other heavy construction materials. Vermiculite is a group name for certain clay minerals, hydrous silicates or aluminum, magnesium and iron that have been expanded by heat.

Perlite plaster is a plaster using perlite as an aggregate instead of sand. Perlite is a volcanic glass that has been expanded by heat.

Gauging plaster is a mix of gypsum plaster and lime putty that when applied produces a quick drying finish coat.

Veneer plaster is a one or two component gypsum plaster used as a thin finish coat over special gypsum board.

Keenes cement is a white cementitious material manufactured from gypsum that has been burned at a high temperature and ground to a fine powder. Alum is added to accelerate the set. The resulting plaster is hard and strong and accepts and maintains a high polish, hence it is used as a finishing plaster.

Stucco is a Portland cement based plaster used primarily as an exterior finish.

Plaster is used on both interior and exterior surfaces. Generally it is applied in multiple-coat systems. A three-coat system uses the terms scratch, brown and finish to identify each coat. A two-coat system uses base and finish to describe each coat. Each type of plaster and application system has attributes that are chosen by the designer to best fit the intended use.

Gypsum Plaster Quantities for 100 S.Y.	2 Coat, 5/8″ Thick		3 Coat, 3/4″ Thick		
	Base 1:3 Mix	Finish 2:1 Mix	Scratch 1:2 Mix	Brown 1:3 Mix	Finish 2:1 Mix
Gypsum plaster	1,300 lb.		1,350 lb.	650 lb.	
Sand	1.75 C.Y.		1.85 C.Y.	1.35 C.Y.	
Finish hydrated lime		340 lb.			340 lb.
Gauging plaster		170 lb.			170 lb.

Vermiculite or Perlite Plaster Quantities for 100 S.Y.	2 Coat, 5/8″ Thick		3 Coat, 3/4″ Thick		
	Base	Finish	Scratch	Brown	Finish
Gypsum plaster	1,250 lb.		1,450 lb.	800 lb.	
Vermiculite or perlite	7.8 bags		8.0 bags	3.3 bags	
Finish hydrated lime		340 lb.			340 lb.
Gauging plaster		170 lb.			170 lb.

Stucco–Three-Coat System Quantities for 100 S.Y.	On Wood Frame	On Masonry
Portland cement	29 bags	21 bags
Sand	2.6 C.Y.	2.0 C.Y.
Hydrated lime	180 lb.	120 lb.

R092910-10 Levels of Gypsum Drywall Finish

In the past, contract documents often used phrases such as "industry standard" and "workmanlike finish" to specify the expected quality of gypsum board wall and ceiling installations. The vagueness of these descriptions led to unacceptable work and disputes.

In order to resolve this problem, four major trade associations concerned with the manufacture, erection, finish and decoration of gypsum board wall and ceiling systems have developed an industry-wide *Recommended Levels of Gypsum Board Finish.*

The finish of gypsum board walls and ceilings for specific final decoration is dependent on a number of factors. A primary consideration is the location of the surface and the degree of decorative treatment desired. Painted and unpainted surfaces in warehouses and other areas where appearance is normally not critical may simply require the taping of wallboard joints and 'spotting' of fastener heads. Blemish-free, smooth, monolithic surfaces often intended for painted and decorated walls and ceilings in habitated structures, ranging from single-family dwellings through monumental buildings, require additional finishing prior to the application of the final decoration.

Other factors to be considered in determining the level of finish of the gypsum board surface are (1) the type of angle of surface illumination (both natural and artificial lighting), and (2) the paint and method of application or the type and finish of wallcovering specified as the final decoration. Critical lighting conditions, gloss paints, and thin wall coverings require a higher level of gypsum board finish than do heavily textured surfaces which are subsequently painted or surfaces which are to be decorated with heavy grade wall coverings.

The following descriptions were developed jointly by the Association of the Wall and Ceiling Industries-International (AWCI), Ceiling & Interior Systems Construction Association (CISCA), Gypsum Association (GA), and Painting and Decorating Contractors of America (PDCA) as a guide.

Level 0: Used in temporary construction or wherever the final decoration has not been determined. Unfinished. No taping, finishing or corner beads are required. Also could be used where non-predecorated panels will be used in demountable-type partitions that are to be painted as a final finish. .

Level 1 Frequently used in plenum areas above ceilings, in attics, in areas where the assembly would generally be concealed, or in building service corridors and other areas not normally open to public view. Some degree of sound and smoke control is provided; in some geographic areas, this level is referred to as "fire-taping," although this level of finish does not typically meet fire-resistant assembly requirements. Where a fire resistance rating is required for the gypsum board assembly, details of construction should be in accordance with reports of fire tests of assemblies that have met the requirements of the fire rating acceptable.

All joints and interior angles shall have tape embedded in joint compound. Accessories are optional at specifier discretion in corridors and other areas with pedestrian traffic. Tape and fastener heads need not be covered with joint compound. Surface shall be free of excess joint compound. Tool marks and ridges are acceptable.

Level 2 It may be specified for standard gypsum board surfaces in garages, warehouse storage, or other similar areas where surface appearance is not of primary importance.

All joints and interior angles shall have tape embedded in joint compound and shall be immediately wiped with a joint knife or trowel, leaving a thin coating of joint compound over all joints and interior angles. Fastener heads and accessories shall be covered with a coat of joint compound. Surface shall be free of excess joint compound. Tool marks and ridges are acceptable.

Level 3 Typically used in areas that are to receive heavy texture (spray or hand applied) finishes before final painting, or where commercial-grade (heavy duty) wall coverings are to be applied as the final decoration. This level of finish should not be used where smooth painted surfaces or where lighter weight wall coverings are specified. The prepared surface shall be coated with a drywall primer prior to the application of final finishes.

All joints and interior angles shall have tape embedded in joint compound and shall be immediately wiped with a joint knife or trowel, leaving a thin coating of joint compound over all joints and interior angles. One additional coat of joint compound shall be applied over all joints and interior angles. Fastener heads and accessories shall be covered with two separate coats of joint compound. All joint compounds shall be smooth and free of tool marks and ridges. The prepared surface shall be covered with a drywall primer prior to the application of the final decoration.

Level 4 This level should be used where residential grade (light duty) wall coverings, flat paints, or light textures are to be applied. The prepared surface shall be coated with a drywall primer prior to the application of final finishes. Release agents for wall coverings are specifically formulated to minimize damage if coverings are subsequently removed.

The weight, texture, and sheen level of the wall covering material selected should be taken into consideration when specifying wall coverings over this level of drywall treatment. Joints and fasteners must be sufficiently concealed if the wall covering material is lightweight, contains limited pattern, has a glossy finish, or has any combination of these features. In critical lighting areas, flat paints applied over light textures tend to reduce joint photographing. Gloss, semi-gloss, and enamel paints are not recommended over this level of finish.

All joints and interior angles shall have tape embedded in joint compound and shall be immediately wiped with a joint knife or trowel, leaving a thin coating of joint compound over all joints and interior angles. In addition, two separate coats of joint compound shall be applied over all flat joints and one separate coat of joint compound applied over interior angles. Fastener heads and accessories shall be covered with three separate coats of joint compound. All joint compounds shall be smooth and free of tool marks and ridges. The prepared surface shall be covered with a drywall primer like Sheetrock first coat prior to the application of the final decoration.

Level 5 The highest quality finish is the most effective method to provide a uniform surface and minimize the possibility of joint photographing and of fasteners showing through the final decoration. This level of finish is required where gloss, semi-gloss, or enamel is specified; when flat joints are specified over an untextured surface; or where critical lighting conditions occur. The prepared surface shall be coated with a drywall primer prior to the application of final decoration.

All joints and interior angles shall have tape embedded in joint compound and be immediately wiped with a joint knife or trowel, leaving a thin coating of joint compound over all joints and interior angles. Two separate coats of joint compound shall be applied over all flat joints and one separate coat of joint compound applied over interior angles. Fastener heads and accessories shall be covered with three separate coats of joint compound.

A thin skim coat of joint compound shall be trowel applied to the entire surface. Excess compound is immediately troweled off, leaving a film or skim coating of compound completely covering the paper. As an alternative to a skim coat, a material manufactured especially for this purpose may be applied such as Sheetrock Tuff-Hide primer surfacer. The surface must be smooth and free of tool marks and ridges. The prepared surface shall be covered with a drywall primer prior to the application of the final decoration.

Finishes — R0966 Terrazzo Flooring

R096613-10 Terrazzo Floor

The table below lists quantities required for 100 S.F. of 5/8″ terrazzo topping, either bonded or not bonded.

Description	Bonded to Concrete 1-1/8″ Bed, 1:4 Mix	Not Bonded 2-1/8″ Bed and 1/4″ Sand
Portland cement, 94 lb. Bag	6 bags	8 bags
Sand	10 C.F.	20 C.F.
Divider strips, 4′ squares	50 L.F.	50 L.F.
Terrazzo fill, 50 lb. Bag	12 bags	12 bags
15 Lb. tarred felt		1 C.S.F.
Mesh 2 x 2 #14 galvanized		1 C.S.F.
Crew J-3	0.77 days	0.87 days

2′ x 2′ panels require 1.00 L.F. divider strip per S.F.
3′ x 3′ panels require 0.67 L.F. divider strip per S.F.
4′ x 4′ panels require 0.50 L.F. divider strip per S.F.
5′ x 5′ panels require 0.40 L.F. divider strip per S.F.
6′ x 6′ panels require 0.33 L.F. divider strip per S.F.

Finishes — R0972 Wall Coverings

R097223-10 Wall Covering

The table below lists the quantities required for 100 S.F. of wall covering.

Description	Medium-Priced Paper	Expensive Paper
Paper	1.6 dbl. rolls	1.6 dbl. rolls
Wall sizing	0.25 gallon	0.25 gallon
Vinyl wall paste	0.6 gallon	0.6 gallon
Apply sizing	0.3 hour	0.3 hour
Apply paper	1.2 hours	1.5 hours

Most wallpapers now come in double rolls only.
To remove old paper, allow 1.3 hours per 100 S.F.

R099100-10 Painting Estimating Techniques

Proper estimating methodology is needed to obtain an accurate painting estimate. There is no known reliable shortcut or square foot method. The following steps should be followed:

- List all surfaces to be painted, with an accurate quantity (area) of each. Items having similar surface condition, finish, application method and accessibility may be grouped together.
- List all the tasks required for each surface to be painted, including surface preparation, masking, and protection of adjacent surfaces. Surface preparation may include minor repairs, washing, sanding and puttying.
- Select the proper Means line for each task. Review and consider all adjustments to labor and materials for type of paint and location of work. Apply the height adjustment carefully. For instance, when applying the adjustment for work over 8' high to a wall that is 12' high, apply the adjustment only to the area between 8' and 12' high, and not to the entire wall.

When applying more than one percent (%) adjustment, apply each to the base cost of the data, rather than applying one percentage adjustment on top of the other.

When estimating the cost of painting walls and ceilings remember to add the brushwork for all cut-ins at inside corners and around windows and doors as a LF measure. One linear foot of cut-in with brush equals one square foot of painting.

All items for spray painting include the labor for roll-back.

Deduct for openings greater than 100 SF, or openings that extend from floor to ceiling and are greater than 5' wide. Do not deduct small openings.

The cost of brushes, rollers, ladders and spray equipment are considered to be part of a painting contractor's overhead, and should not be added to the estimate. The cost of rented equipment such as scaffolding and swing staging should be added to the estimate.

R099100-20 Painting

Item	Coat	One Gallon Covers			In 8 Hours a Laborer Covers			Labor-Hours per 100 S.F.		
		Brush	Roller	Spray	Brush	Roller	Spray	Brush	Roller	Spray
Paint wood siding	prime	250 S.F.	225 S.F.	290 S.F.	1150 S.F.	1300 S.F.	2275 S.F.	.695	.615	.351
	others	270	250	290	1300	1625	2600	.615	.492	.307
Paint exterior trim	prime	400	—	—	650	—	—	1.230	—	—
	1st	475	—	—	800	—	—	1.000	—	—
	2nd	520	—	—	975	—	—	.820	—	—
Paint shingle siding	prime	270	255	300	650	975	1950	1.230	.820	.410
	others	360	340	380	800	1150	2275	1.000	.695	.351
Stain shingle siding	1st	180	170	200	750	1125	2250	1.068	.711	.355
	2nd	270	250	290	900	1325	2600	.888	.603	.307
Paint brick masonry	prime	180	135	160	750	800	1800	1.066	1.000	.444
	1st	270	225	290	815	975	2275	.981	.820	.351
	2nd	340	305	360	815	1150	2925	.981	.695	.273
Paint interior plaster or drywall	prime	400	380	495	1150	2000	3250	.695	.400	.246
	others	450	425	495	1300	2300	4000	.615	.347	.200
Paint interior doors and windows	prime	400	—	—	650	—	—	1.230	—	—
	1st	425	—	—	800	—	—	1.000	—	—
	2nd	450	—	—	975	—	—	.820	—	—

Special Construction R1311 Swimming Pools

R131113-20 Swimming Pools

Pool prices given per square foot of surface area include pool structure, filter and chlorination equipment, pumps, related piping, ladders/steps, maintenance kit, skimmer and vacuum system. Decks and electrical service to equipment are not included.

Residential in-ground pool construction can be divided into two categories: vinyl lined and gunite. Vinyl lined pool walls are constructed of different materials including wood, concrete, plastic or metal. The bottom is often graded with sand over which the vinyl liner is installed. Vermiculite or soil cement bottoms may be substituted for an added cost.

Gunite pool construction is used both in residential and municipal installations. These structures are steel reinforced for strength and finished with a white cement limestone plaster.

Municipal pools will have a higher cost because plumbing codes require more expensive materials, chlorination equipment and higher filtration rates.

Municipal pools greater than 1,800 S.F. require gutter systems to control waves. This gutter may be formed into the concrete wall. Often a vinyl/stainless steel gutter or gutter/wall system is specified, which will raise the pool cost.

Competition pools usually require tile bottoms and sides with contrasting lane striping, which will also raise the pool cost.

R133113-10 Air Supported Structures

Air supported structures are made from fabrics that can be classified into two groups: temporary and permanent. Temporary fabrics include nylon, woven polyethylene, vinyl film, and vinyl coated dacron. These have lifespans that range from five to fifteen plus years. The cost per square foot includes a fabric shell, tension cables, primary and back-up inflation systems and doors. The lower cost structures are used for construction shelters, bulk storage and pond covers. The more expensive are used for recreational structures and warehouses.

Permanent fabrics are teflon coated fiberglass. The life of this structure is twenty plus years. The high cost limits its application to architectural designed structures which call for a clear span covered area, such as stadiums and convention centers. Both temporary and permanent structures are available in translucent fabrics which eliminates the need for daytime lighting.

Areas to be covered vary from 10,000 S.F. to any area up to 1000 foot wide by any length. Height restrictions range from a maximum of 1/2 of width to a minimum of 1/6 of the width. Erection of even the largest of the temporary structures requires no more than a week.

Centrifugal fans provide the inflation necessary to support the structure during application of live loads. Airlocks are usually used at large entrances to prevent loss of static pressure. Some manufacturers employ propeller fans which generate sufficient airflow (30,000 CFM) to eliminate the need for airlocks. These fans may also be automatically controlled to resist high wind conditions, regulate humidity (air changes), and provide cooling and heat.

Insulation can be provided with the addition of a second or even third interior liner, creating a dead air space with an "R" value of four to nine. Some structures allow for the liner to be collapsed into the outer shell to enable the internal heat to melt accumulated snow. For cooling or air conditioning, the exterior face of the liner can be aluminized to reflect the sun's heat.

R133113-90 Seismic Bracing

Sometimes referred to as anti-sway bracing, this support system is required in earthquake areas. The individual components must be assembled to make a required system.

Example

C-Clamp 3/8″ rod	2 ea.
Rod, continuous thread 3/8″	10 L.F.
Field, weld 1″	2 ea.

Additionally, height factors must be taken into account. Add the following percentages to labor for elevated installations:

15′ to 20′ high	10%
21′ to 25′ high	20%
26′ to 30′ high	30%
31′ to 35′ high	40%
36′ to 40′ high	50%
41′ to 50′ high	60%

R133419-10 Pre-Engineered Steel Buildings

These buildings are manufactured by many companies and normally erected by franchised dealers throughout the U.S. The four basic types are: Rigid Frames, Truss type, Post and Beam and the Sloped Beam type. Most popular roof slope is low pitch of 1″ in 12″. The minimum economical area of these buildings is about 3000 S.F. of floor area. Bay sizes are usually 20′ to 24′ but can go as high as 30′ with heavier girts and purlins. Eave heights are usually 12′ to 24′ with 18′ to 20′ most typical.

Material prices shown in the Unit Price section are bare costs for the building shell only and do not include floors, foundations, anchor bolts, interior finishes or utilities. Costs assume at least three bays of 24′ each, a 1″ in 12″ roof slope, and they are based on 30 psf roof load and 20 psf wind load (wind load is a function of wind speed, building height, and terrain characteristics; this should be determined by a Registered Structural Engineer) and no unusual requirements. Costs include the structural frame, 26 ga. non-insulated colored corrugated or ribbed roofing and siding panels, fasteners, closures, trim and flashing but no allowance for insulation, doors, windows, skylights, gutters or downspouts. Very large projects would generally cost less for materials than the prices shown. For roof panel substitutions and wall panel substitutions, see appropriate Unit Price sections.

Conditions at the site, weather, shape and size of the building, and labor availability will affect the erection cost of the building.

R133423-30 Dome Structures

Steel — The four types are Lamella, Schwedler, Arch and Geodesic. For maximum economy, rise should be about 15 to 20% of diameter. Most common diameters are in the 200′ to 300′ range. Lamella domes weigh about 5 P.S.F. of floor area less than Schwedler domes. Schwedler dome weight in lbs. per S.F. approaches .046 times the diameter. Domes below 125′ diameter weigh .07 times diameter and the cost per ton of steel is higher. See R051223-20 for estimating weight.

Wood — Small domes are of sawn lumber, larger ones are laminated. In larger sizes, triaxial and triangular cost about the same; radial domes cost more. Radial domes are economical in the 60′ to 70′ diameter range. Most economical range of all types is 80′ to 200′ diameters. Diameters can run over 400′. All costs are quoted above the foundation. Prices include 2″ decking and a tension tie ring in place.

Plywood — Stock prefab geodesic domes are available with diameters from 24′ to 60′.

Fiberglass — Aluminum framed translucent sandwich panels with spans from 5′ to 45′ are commercially available.

Aluminum — Stressed skin aluminum panels form geodesic domes with spans ranging from 82′ to 232′. An aluminum space truss, triangulated or nontriangulated, with aluminum or clear acrylic closure panels can be used for clear spans of 40′ to 415′.

R142000-10 Freight Elevators

Capacities run from 2,000 lbs. to over 100,000 lbs. with 3,000 lbs. to 10,000 lbs. most common. Travel speeds are generally lower and control less intricate than on passenger elevators. Costs in the Unit Price sections are for hydraulic and geared elevators.

R142000-20 Elevator Selective Costs See R142000-40 for cost development.

	Passenger		Freight		Hospital	
A. Base Unit	Hydraulic	Electric	Hydraulic	Electric	Hydraulic	Electric
Capacity	1,500 lb.	2,000 lb.	2,000 lb.	4,000 lb.	4,000 lb.	4,000 lb.
Speed	100 F.P.M.	200 F.P.M.	100 F.P.M.	200 F.P.M.	100 F.P.M.	200 F.P.M.
#Stops/Travel Ft.	2/12	4/40	2/20	4/40	2/20	4/40
Push Button Oper.	Yes	Yes	Yes	Yes	Yes	Yes
Telephone Box & Wire	"	"	"	"	"	"
Emergency Lighting	"	"	No	No	"	"
Cab	Plastic Lam. Walls	Plastic Lam. Walls	Painted Steel	Painted Steel	Plastic Lam. Walls	Plastic Lam. Walls
Cove Lighting	Yes	Yes	No	No	Yes	Yes
Floor	V.C.T.	V.C.T.	Wood w/Safety Treads	Wood w/Safety Treads	V.C.T.	V.C.T.
Doors, & Speedside Slide	Yes	Yes	Yes	Yes	Yes	Yes
Gates, Manual	No	No	No	No	No	No
Signals, Lighted Buttons	Car and Hall	Car and Hall	Car and Hall	Car and Hall	Car and Hall	Car and Hall
O.H. Geared Machine	N.A.	Yes	N.A.	Yes	N.A.	Yes
Variable Voltage Contr.	"	"	N.A.	"	"	"
Emergency Alarm	Yes	"	Yes	"	Yes	"
Class "A" Loading	N.A.	N.A.	"	"	N.A.	N.A.

R142000-30 Passenger Elevators

Electric elevators are used generally but hydraulic elevators can be used for lifts up to 70′ and where large capacities are required. Hydraulic speeds are limited to 200 F.P.M. but cars are self leveling at the stops. On low rises, hydraulic installation runs about 15% less than standard electric types but on higher rises this installation cost advantage is reduced. Maintenance of hydraulic elevators is about the same as electric type but underground portion is not included in the maintenance contract.

In electric elevators there are several control systems available, the choice of which will be based upon elevator use, size, speed and cost criteria. The two types of drives are geared for low speeds and gearless for 450 F.P.M. and over.

The tables on the preceding pages illustrate typical installed costs of the various types of elevators available.

R142000-40 Elevator Cost Development

To price a new car or truck from the factory, you must start with the manufacturer's basic model, then add or exchange optional equipment and features. The same is true for pricing elevators.

Requirement: One-passenger elevator, five-story hydraulic, 2,500 lb. capacity, 12′ floor to floor, speed 150 F.P.M., emergency power switching and maintenance contract.

Example:

Description	Adjustment
A. Base Elevator: Hydraulic Passenger, 1500 lb. Capacity, 100 fpm, 2 Stops, Standard Finish	1 Ea.
B. Capacity Adjustment (2,500 lb.)	1 Ea.
C. Excess Travel Adjustment: 48′ Total Travel (4 x 12′) minus 12′ Base Unit Travel =	36 V.L.F.
D. Stops Adjustment: 5 Total Stops minus 2 Stops (Base Unit) =	3 Stops
E. Speed Adjustment (150 F.P.M.)	1 Ea.
F. Options:	
1. Intercom Service	1 Ea.
2. Emergency Power Switching, Automatic	1 Ea.
3. Stainless Steel Entrance Doors	5 Ea.
4. Maintenance Contract (12 Months)	1 Ea.
5. Position Indicator for main floor level (none indicated in Base Unit)	1 Ea.

Conveying Equipment — R1431 Escalators

R143110-10 Escalators

Moving stairs can be used for buildings where 600 or more people are to be carried to the second floor or beyond. Freight cannot be carried on escalators and at least one elevator must be available for this function.

Carrying capacity is 5,000 to 8,000 people per hour. Power requirement is 2 to 3 KW per hour and incline angle is 30°.

Conveying Equipment — R1432 Moving Walks

R143210-20 Moving Ramps and Walks

These are a specialized form of conveyor 3′ to 6′ wide with capacities of 3,600 to 18,000 persons per hour. Maximum speed is 140 F.P.M. and normal incline is 0° to 15°.

Local codes will determine the maximum angle. Outdoor units would require additional weather protection.

Plumbing — R2201 Operation & Maintenance of Plumbing

R220105-10 Demolition (Selective vs. Removal for Replacement)

Demolition can be divided into two basic categories.

One type of demolition involves the removal of material with no concern for its replacement. The labor-hours to estimate this work are found under "Selective Demolition" in the Fire Protection, Plumbing and HVAC Divisions. It is selective in that individual items or all the material installed as a system or trade grouping such as plumbing or heating systems are removed. This may be accomplished by the easiest way possible, such as sawing, torch cutting, or sledge hammer as well as simple unbolting.

The second type of demolition is the removal of some item for repair or replacement. This removal may involve careful draining, opening of unions, disconnecting and tagging of electrical connections, capping of pipes/ducts to prevent entry of debris or leakage of the material contained as well as transport of the item away from its in-place location to a truck/dumpster. An approximation of the time required to accomplish this type of demolition is to use half of the time indicated as necessary to install a new unit. For example; installation of a new pump might be listed as requiring 6 labor-hours so if we had to estimate the removal of the old pump we would allow an additional 3 hours for a total of 9 hours. That is, the complete replacement of a defective pump with a new pump would be estimated to take 9 labor-hours.

Plumbing — R2211 Facility Water Distribution

R221113-50 Pipe Material Considerations

1. Malleable fittings should be used for gas service.
2. Malleable fittings are used where there are stresses/strains due to expansion and vibration.
3. Cast fittings may be broken as an aid to disassembling of heating lines frozen by long use, temperature and minerals.
4. Cast iron pipe is extensively used for underground and submerged service.
5. Type M (light wall) copper tubing is available in hard temper only and is used for nonpressure and less severe applications than K and L.
6. Type L (medium wall) copper tubing, available hard or soft for interior service.
7. Type K (heavy wall) copper tubing, available in hard or soft temper for use where conditions are severe. For underground and interior service.
8. Hard drawn tubing requires fewer hangers or supports but should not be bent. Silver brazed fittings are recommended, however soft solder is normally used.
9. Type DMV (very light wall) copper tubing designed for drainage, waste and vent plus other non-critical pressure services.

Domestic/Imported Pipe and Fittings Cost

The prices shown in this publication for steel/cast iron pipe and steel, cast iron, malleable iron fittings are based on domestic production sold at the normal trade discounts. The above listed items of foreign manufacture may be available at prices of 1/3 to 1/2 those shown. Some imported items after minor machining or finishing operations are being sold as domestic to further complicate the system.

Caution: Most pipe prices in this book also include a coupling and pipe hangers which for the larger sizes can add significantly to the per foot cost and should be taken into account when comparing "book cost" with quoted supplier's cost.

R224000-40 Plumbing Fixture Installation Time

Item	Rough-In	Set	Total Hours	Item	Rough-In	Set	Total Hours
Bathtub	5	5	10	Shower head only	2	1	3
Bathtub and shower, cast iron	6	6	12	Shower drain	3	1	4
Fire hose reel and cabinet	4	2	6	Shower stall, slate		15	15
Floor drain to 4 inch diameter	3	1	4	Slop sink	5	3	8
Grease trap, single, cast iron	5	3	8	Test 6 fixtures			14
Kitchen gas range		4	4	Urinal, wall	6	2	8
Kitchen sink, single	4	4	8	Urinal, pedestal or floor	6	4	10
Kitchen sink, double	6	6	12	Water closet and tank	4	3	7
Laundry tubs	4	2	6	Water closet and tank, wall hung	5	3	8
Lavatory wall hung	5	3	8	Water heater, 45 gals. gas, automatic	5	2	7
Lavatory pedestal	5	3	8	Water heaters, 65 gals. gas, automatic	5	2	7
Shower and stall	6	4	10	Water heaters, electric, plumbing only	4	2	6

Fixture prices in front of book are based on the cost per fixture set in place. The rough-in cost, which must be added for each fixture, includes carrier, if required, some supply, waste and vent pipe connecting fittings and stops. The lengths of rough-in pipe are nominal runs which would connect to the larger runs and stacks. The supply runs and DWV runs and stacks must be accounted for in separate entries. In the eastern half of the United States it is common for the plumber to carry these to a point 5′ outside the building.

R224000-50 Water Cooler Application

Type of Service	Requirement
Office, School or Hospital	12 persons per gallon per hour
Office, Lobby or Department Store	4 or 5 gallons per hour per fountain
Light manufacturing	7 persons per gallon per hour
Heavy manufacturing	5 persons per gallon per hour
Hot heavy manufacturing	4 persons per gallon per hour
Hotel	.08 gallons per hour per room
Theatre	1 gallon per hour per 100 seats

Heating, Ventilating & A.C. R2331 HVAC Ducts & Casings

R233100-20 Ductwork

Duct Weight in Pounds per L.F., Straight Runs

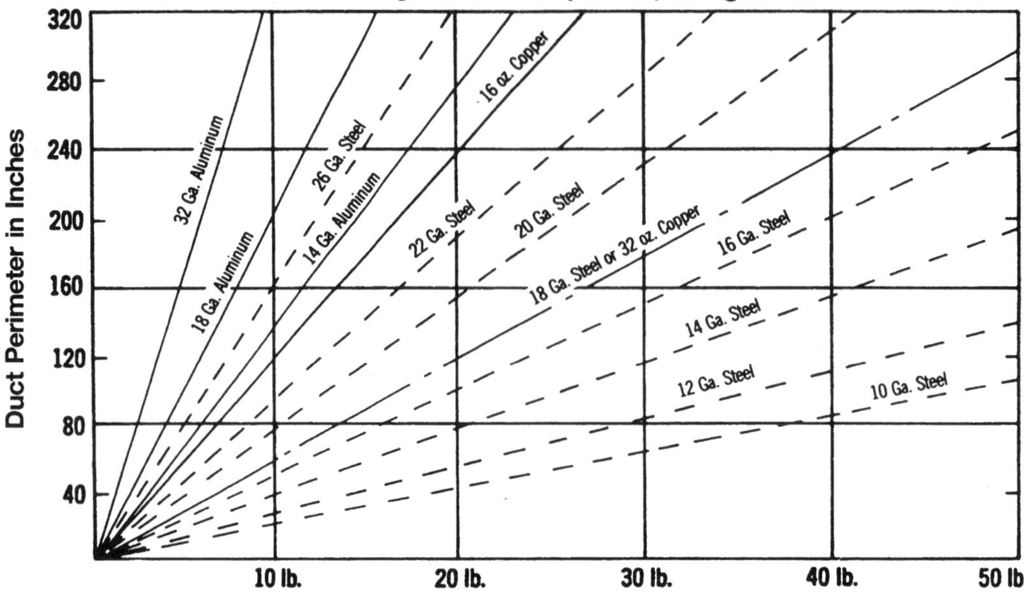

Add to the above for fittings; 90° elbow is 3 L.F.; 45° elbow is 2.5 L.F.; offset is 4 L.F.; transition offset is 6 L.F.; square-to-round transition is 4 L.F.; 90° reducing elbow is 5 L.F. For bracing and waste, add 20% to aluminum and copper, 15% to steel.

Heating, Ventilating & A.C. R2334 HVAC Fans

R233400-10 Recommended Ventilation Air Changes

Table below lists range of time in minutes per change for various types of facilities.

Facility	Minutes	Facility	Minutes	Facility	Minutes
Assembly Halls	2-10	Dance Halls	2-10	Laundries	1-3
Auditoriums	2-10	Dining Rooms	3-10	Markets	2-10
Bakeries	2-3	Dry Cleaners	1-5	Offices	2-10
Banks	3-10	Factories	2-5	Pool Rooms	2-5
Bars	2-5	Garages	2-10	Recreation Rooms	2-10
Beauty Parlors	2-5	Generator Rooms	2-5	Sales Rooms	2-10
Boiler Rooms	1-5	Gymnasiums	2-10	Theaters	2-8
Bowling Alleys	2-10	Kitchens-Hospitals	2-5	Toilets	2-5
Churches	5-10	Kitchens-Restaurant	1-3	Transformer Rooms	1-5

CFM air required for changes = Volume of room in cubic feet ÷ Minutes per change.

Heating, Ventilating & A.C. R2350 Central Heating Equipment

R235000-35 Heating (42° degrees latitude)

$$\text{Approximate S.F. radiation} = \frac{\text{S.F. sash}}{2} + \frac{\text{S.F. wall + roof — sash}}{20} + \frac{\text{C.F. building}}{200}$$

R235000-50 Factor for Determining Heat Loss for Various Types of Buildings

General: While the most accurate estimates of heating requirements would naturally be based on detailed information about the building being considered, it is possible to arrive at a reasonable approximation using the following procedure:

1. Calculate the cubic volume of the room or building.
2. Select the appropriate factor from Table 1 below. Note that the factors apply only to inside temperatures listed in the first column and to 0° F outside temperature.
3. If the building has bad north and west exposures, multiply the heat loss factor by 1.1.
4. If the outside design temperature is other than 0° F, multiply the factor from Table 1 by the factor from Table 2.
5. Multiply the cubic volume by the factor selected from Table 1. This will give the estimated BTUH heat loss which must be made up to maintain inside temperature.

Table 1 — Building Type	Conditions	Qualifications	Loss Factor*
Factories & Industrial Plants General Office Areas at 70° F	One Story	Skylight in Roof	6.2
		No Skylight in Roof	5.7
	Multiple Story	Two Story	4.6
		Three Story	4.3
		Four Story	4.1
		Five Story	3.9
		Six Story	3.6
	All Walls Exposed	Flat Roof	6.9
		Heated Space Above	5.2
	One Long Warm Common Wall	Flat Roof	6.3
		Heated Space Above	4.7
	Warm Common Walls on Both Long Sides	Flat Roof	5.8
		Heated Space Above	4.1
Warehouses at 60° F	All Walls Exposed	Skylights in Roof	5.5
		No Skylight in Roof	5.1
		Heated Space Above	4.0
	One Long Warm Common Wall	Skylight in Roof	5.0
		No Skylight in Roof	4.9
		Heated Space Above	3.4
	Warm Common Walls on Both Long Sides	Skylight in Roof	4.7
		No Skylight in Roof	4.4
		Heated Space Above	3.0

*Note: This table tends to be conservative particularly for new buildings designed for minimum energy consumption.

Table 2 — Outside Design Temperature Correction Factor (for Degrees Fahrenheit)									
Outside Design Temperature	50	40	30	20	10	0	-10	-20	-30
Correction Factor	.29	.43	.57	.72	.86	1.00	1.14	1.28	1.43

Heating, Ventilating & A.C. R2356 Solar Energy Heating Equipment

R235616-60 Solar Heating (Space and Hot Water)

Collectors should face as close to due South as possible, however, variations of up to 20 degrees on either side of true South are acceptable. Local climate and collector type may influence the choice between east or west deviations. Obviously they should be located so they are not shaded from the sun's rays. Incline collectors at a slope of latitude minus 5 degrees for domestic hot water and latitude plus 15 degrees for space heating.

Flat plate collectors consist of a number of components as follows: Insulation to reduce heat loss through the bottom and sides of the collector. The enclosure which contains all the components in this assembly is usually weatherproof and prevents dust, wind and water from coming in contact with the absorber plate. The cover plate usually consists of one or more layers of a variety of glass or plastic and reduces the reradiation by creating an air space which traps the heat between the cover and the absorber plates.

The absorber plate must have a good thermal bond with the fluid passages. The absorber plate is usually metallic and treated with a surface coating which improves absorptivity. Black or dark paints or selective coatings are used for this purpose, and the design of this passage and plate combination helps determine a solar system's effectiveness.

Heat transfer fluid passage tubes are attached above and below or integral with an absorber plate for the purpose of transferring thermal energy from the absorber plate to a heat transfer medium. The heat exchanger is a device for transferring thermal energy from one fluid to another.

Piping and storage tanks should be well insulated to minimize heat losses.

Size domestic water heating storage tanks to hold 20 gallons of water per user, minimum, plus 10 gallons per dishwasher or washing machine. For domestic water heating an optimum collector size is approximately 3/4 square foot of area per gallon of water storage. For space heating of residences and small commercial applications the collector is commonly sized between 30% and 50% of the internal floor area. For space heating of large commercial applications, collector areas less than 30% of the internal floor area can still provide significant heat reductions.

A supplementary heat source is recommended for Northern states for December through February.

The solar energy transmission per square foot of collector surface varies greatly with the material used. Initial cost, heat transmittance and useful life are obviously interrelated.

Heating, Ventilating & A.C. R2360 Central Cooling Equipment

R236000-20 Air Conditioning Requirements

BTUs per hour per S.F. of floor area and S.F. per ton of air conditioning.

Type of Building	BTU/Hr per S.F.	S.F. per Ton	Type of Building	BTU/Hr per S.F.	S.F. per Ton	Type of Building	BTU/Hr per S.F.	S.F. per Ton
Apartments, Individual	26	450	Dormitory, Rooms	40	300	Libraries	50	240
Corridors	22	550	Corridors	30	400	Low Rise Office, Exterior	38	320
Auditoriums & Theaters	40	300/18*	Dress Shops	43	280	Interior	33	360
Banks	50	240	Drug Stores	80	150	Medical Centers	28	425
Barber Shops	48	250	Factories	40	300	Motels	28	425
Bars & Taverns	133	90	High Rise Office—Ext. Rms.	46	263	Office (small suite)	43	280
Beauty Parlors	66	180	Interior Rooms	37	325	Post Office, Individual Office	42	285
Bowling Alleys	68	175	Hospitals, Core	43	280	Central Area	46	260
Churches	36	330/20*	Perimeter	46	260	Residences	20	600
Cocktail Lounges	68	175	Hotel, Guest Rooms	44	275	Restaurants	60	200
Computer Rooms	141	85	Corridors	30	400	Schools & Colleges	46	260
Dental Offices	52	230	Public Spaces	55	220	Shoe Stores	55	220
Dept. Stores, Basement	34	350	Industrial Plants, Offices	38	320	Shop'g. Ctrs., Supermarkets	34	350
Main Floor	40	300	General Offices	34	350	Retail Stores	48	250
Upper Floor	30	400	Plant Areas	40	300	Specialty	60	200

*Persons per ton
12,000 BTU = 1 ton of air conditioning

R260519-92 Minimum Copper and Aluminum Wire Size Allowed for Various Types of Insulation

Minimum Wire Sizes									
	Copper		Aluminum			Copper		Aluminum	
Amperes	THW THWN or XHHW	THHN XHHW *	THW XHHW	THHN XHHW *	Amperes	THW THWN or XHHW	THHN XHHW *	THW XHHW	THHN XHHW *
15A	#14	#14	#12	#12	195	3/0	2/0	250kcmil	4/0
20	#12	#12	#10	#10	200	3/0	3/0	250kcmil	4/0
25	#10	#10	#10	#10	205	4/0	3/0	250kcmil	4/0
30	#10	#10	# 8	# 8	225	4/0	3/0	300kcmil	250kcmil
40	# 8	# 8	# 8	# 8	230	4/0	4/0	300kcmil	250kcmil
45	# 8	# 8	# 6	# 8	250	250kcmil	4/0	350kcmil	300kcmil
50	# 8	# 8	# 6	# 6	255	250kcmil	4/0	400kcmil	300kcmil
55	# 6	# 8	# 4	# 6	260	300kcmil	4/0	400kcmil	350kcmil
60	# 6	# 6	# 4	# 6	270	300kcmil	250kcmil	400kcmil	350kcmil
65	# 6	# 6	# 4	# 4	280	300kcmil	250kcmil	500kcmil	350kcmil
75	# 4	# 6	# 3	# 4	285	300kcmil	250kcmil	500kcmil	400kcmil
85	# 4	# 4	# 2	# 3	290	350kcmil	250kcmil	500kcmil	400kcmil
90	# 3	# 4	# 2	# 2	305	350kcmil	300kcmil	500kcmil	400kcmil
95	# 3	# 4	# 1	# 2	310	350kcmil	300kcmil	500kcmil	500kcmil
100	# 3	# 3	# 1	# 2	320	400kcmil	300kcmil	600kcmil	500kcmil
110	# 2	# 3	1/0	# 1	335	400kcmil	350kcmil	600kcmil	500kcmil
115	# 2	# 2	1/0	# 1	340	500kcmil	350kcmil	600kcmil	500kcmil
120	# 1	# 2	1/0	1/0	350	500kcmil	350kcmil	700kcmil	500kcmil
130	# 1	# 2	2/0	1/0	375	500kcmil	400kcmil	700kcmil	600kcmil
135	1/0	# 1	2/0	1/0	380	500kcmil	400kcmil	750kcmil	600kcmil
150	1/0	# 1	3/0	2/0	385	600kcmil	500kcmil	750kcmil	600kcmil
155	2/0	1/0	3/0	3/0	420	600kcmil	500kcmil		700kcmil
170	2/0	1/0	4/0	3/0	430		500kcmil		750kcmil
175	2/0	2/0	4/0	3/0	435		600kcmil		750kcmil
180	3/0	2/0	4/0	4/0	475		600kcmil		

*Dry Locations Only

Notes:

1. Size #14 to 4/0 is in AWG units (American Wire Gauge).
2. Size 250 to 750 is in kcmil units (Thousand Circular Mils).
3. Use next higher ampere value if exact value is not listed in table.
4. For loads that operate continuously increase ampere value by 25% to obtain proper wire size.
5. Refer to Table R260519-91 for the maximum circuit length for the various size wires.
6. Table R260519-92 has been written for estimating purpose only, based on ambient temperature of 30°C (86° F); for ambient temperature other than 30°C (86° F), ampacity correction factors will be applied.

R260533-22 Conductors in Conduit

Table below lists maximum number of conductors for various sized conduit using THW, TW or THWN insulations.

Copper Wire Size	Conduit Size																										
	½″			¾″			1″			1-¼″			1-½″			2″			2-½″			3″		3-½″		4″	
	TW	THW	THWN	TW	THW	THWN	TW	THW	THWN	TW	THW	THWN	TW	THW	THWN	TW	THW	THWN	TW	THW	THWN	THW	THWN	THW	THWN	THW	THWN
#14	9	6	13	15	10	24	25	16	39	44	29	69	60	40	94	99	65	154	142	93		143		192			
#12	7	4	10	12	8	18	19	13	29	35	24	51	47	32	70	78	53	114	111	76	164	117		157			
#10	5	4	6	9	6	11	15	11	18	26	19	32	36	26	44	60	43	73	85	61	104	95	160	127		163	
#8	2	1	3	4	3	5	7	5	9	12	10	16	17	13	22	28	22	36	40	32	51	49	79	66	106	85	136
#6		1	1		2	4		4	6		7	11		10	15		16	26		23	37	36	57	48	76	62	98
#4		1	1		1	2		3	4		5	7		7	9		12	16		17	22	27	35	36	47	47	60
#3		1	1		1	1		2	3		4	6		6	8		10	13		15	19	23	29	31	39	40	51
#2		1	1		1	1		2	3		4	5		5	7		9	11		13	16	20	25	27	33	34	43
#1					1	1		1	1		3	3		4	5		6	8		9	12	14	18	19	25	25	32
1/0					1	1		1	1		2	3		3	4		5	7		8	10	12	15	16	21	21	27
2/0					1	1		1	1		1	2		3	3		5	6		7	8	10	13	14	17	18	22
3/0					1	1		1	1		1	1		2	3		4	5		6	7	9	11	12	14	15	18
4/0						1		1	1		1	1		1	2		3	4		5	6	7	9	10	12	13	15
250 kcmil								1	1		1	1		1	1		2	3		4	4	6	7	8	10	10	12
300								1	1		1	1		1	1		2	3		3	4	5	6	7	8	9	11
350									1		1	1		1	1		1	2		3	3	4	5	6	7	8	9
400											1	1		1	1		1	1		2	3	4	5	5	6	7	8
500											1	1		1	1		1	1		1	2	3	4	4	5	6	7
600												1		1	1		1	1		1	1	3	3	4	4	5	5
700														1	1		1	1		1	1	2	3	3	4	4	5
750 ▼														1	1		1	1		1	1	2	2	3	3	4	4

R312316-40 Excavating

The selection of equipment used for structural excavation and bulk excavation or for grading is determined by the following factors.

1. Quantity of material
2. Type of material
3. Depth or height of cut
4. Length of haul
5. Condition of haul road
6. Accessibility of site
7. Moisture content and dewatering requirements
8. Availability of excavating and hauling equipment

Some additional costs must be allowed for hand trimming the sides and bottom of concrete pours and other excavation below the general excavation.

When planning excavation and fill, the following should also be considered.

1. Swell factor
2. Compaction factor
3. Moisture content
4. Density requirements

A typical example for scheduling and estimating the cost of excavation of a 15′ deep basement on a dry site when the material must be hauled off the site is outlined below.

Assumptions:

1. Swell factor, 18%
2. No mobilization or demobilization
3. Allowance included for idle time and moving on job
4. No dewatering, sheeting, or bracing
5. No truck spotter or hand trimming

Number of B.C.Y. per truck = 1.5 C.Y. bucket x 8 passes = 12 loose C.Y.

$$= 12 \times \frac{100}{118} = 10.2 \text{ B.C.Y. per truck}$$

Truck Haul Cycle:

Load truck, 8 passes	=	4 minutes
Haul distance, 1 mile	=	9 minutes
Dump time	=	2 minutes
Return, 1 mile	=	7 minutes
Spot under machine	=	1 minute
		23 minute cycle

Fleet Haul Production per day in B.C.Y.

$$4 \text{ trucks} \times \frac{50 \text{ min. hour}}{23 \text{ min. haul cycle}} \times 8 \text{ hrs.} \times 10.2 \text{ B.C.Y.}$$

$$= 4 \times 2.2 \times 8 \times 10.2 = 718 \text{ B.C.Y./day}$$

Add the mobilization and demobilization costs to the total excavation costs. When equipment is rented for more than three days, there is often no mobilization charge by the equipment dealer. On larger jobs outside of urban areas, scrapers can move earth economically provided a dump site or fill area and adequate haul roads are available. Excavation within sheeting bracing or cofferdam bracing is usually done with a clamshell and production is low, since the clamshell may have to be guided by hand between the bracing. When excavating or filling an area enclosed with a wellpoint system, add 10% to 15% to the cost to allow for restricted access. When estimating earth excavation quantities for structures, allow work space outside the building footprint for construction of the foundation and a slope of 1:1 unless sheeting is used.

R312316-45 Excavating Equipment

The table below lists THEORETICAL hourly production in C.Y./hr. bank measure for some typical excavation equipment. Figures assume 50 minute hours, 83% job efficiency, 100% operator efficiency, 90° swing and properly sized hauling units, which must be modified for adverse digging and loading conditions. Actual production costs in the front of the book average about 50% of the theoretical values listed here.

Equipment	Soil Type	B.C.Y. Weight	% Swell	1 C.Y.	1-1/2 C.Y.	2 C.Y.	2-1/2 C.Y.	3 C.Y.	3-1/2 C.Y.	4 C.Y.
Hydraulic Excavator "Backhoe" 15′ Deep Cut	Moist loam, sandy clay	3400 lb.	40%	165	195	200	275	330	385	440
	Sand and gravel	3100	18	140	170	225	240	285	330	380
	Common earth	2800	30	150	180	230	250	300	350	400
	Clay, hard, dense	3000	33	120	140	190	200	240	260	320
Power Shovel Optimum Cut (Ft.)	Moist loam, sandy clay	3400	40	170	245	295	335	385	435	475
				(6.0)	(7.0)	(7.8)	(8.4)	(8.8)	(9.1)	(9.4)
	Sand and gravel	3100	18	165	225	275	325	375	420	460
				(6.0)	(7.0)	(7.8)	(8.4)	(8.8)	(9.1)	(9.4)
	Common earth	2800	30	145	200	250	295	335	375	425
				(7.8)	(9.2)	(10.2)	(11.2)	(12.1)	(13.0)	(13.8)
	Clay, hard, dense	3000	33	120	175	220	255	300	335	375
				(9.0)	(10.7)	(12.2)	(13.3)	(14.2)	(15.1)	(16.0)
Drag Line Optimum Cut (Ft.)	Moist loam, sandy clay	3400	40	130	180	220	250	290	325	385
				(6.6)	(7.4)	(8.0)	(8.5)	(9.0)	(9.5)	(10.0)
	Sand and gravel	3100	18	130	175	210	245	280	315	375
				(6.6)	(7.4)	(8.0)	(8.5)	(9.0)	(9.5)	(10.0)
	Common earth	2800	30	110	160	190	220	250	280	310
				(8.0)	(9.0)	(9.9)	(10.5)	(11.0)	(11.5)	(12.0)
	Clay, hard, dense	3000	33	90	130	160	190	225	250	280
				(9.3)	(10.7)	(11.8)	(12.3)	(12.8)	(13.3)	(12.0)
				Wheel Loaders				Track Loaders		
				3 C.Y.	4 C.Y.	6 C.Y.	8 C.Y.	2-1/4 C.Y.	3 C.Y.	4 C.Y.
Loading Tractors	Moist loam, sandy clay	3400	40	260	340	510	690	135	180	250
	Sand and gravel	3100	18	245	320	480	650	130	170	235
	Common earth	2800	30	230	300	460	620	120	155	220
	Clay, hard, dense	3000	33	200	270	415	560	110	145	200
	Rock, well-blasted	4000	50	180	245	380	520	100	130	180

R312319-90 Wellpoints

A single stage wellpoint system is usually limited to dewatering an average 15′ depth below normal ground water level. Multi-stage systems are employed for greater depth with the pumping equipment installed only at the lowest header level. Ejectors with unlimited lift capacity can be economical when two or more stages of wellpoints can be replaced or when horizontal clearance is restricted, such as in deep trenches or tunneling projects, and where low water flows are expected. Wellpoints are usually spaced on 2-1/2′ to 10′ centers along a header pipe. Wellpoint spacing, header size, and pump size are all determined by the expected flow as dictated by soil conditions.

In almost all soils encountered in wellpoint dewatering, the wellpoints may be jetted into place. Cemented soils and stiff clays may require sand wicks about 12″ in diameter around each wellpoint to increase efficiency and eliminate weeping into the excavation. These sand wicks require 1/2 to 3 C.Y. of washed filter sand and are installed by using a 12″ diameter steel casing and hole puncher jetted into the ground 2′ deeper than the wellpoint. Rock may require predrilled holes.

Labor required for the complete installation and removal of a single stage wellpoint system is in the range of 3/4 to 2 labor-hours per linear foot of header, depending upon jetting conditions, wellpoint spacing, etc.

Continuous pumping is necessary except in some free draining soil where temporary flooding is permissible (as in trenches which are backfilled after each day's work). Good practice requires provision of a stand-by pump during the continuous pumping operation.

Systems for continuous trenching below the water table should be installed three to four times the length of expected daily progress to ensure uninterrupted digging, and header pipe size should not be changed during the job.

For pervious free draining soils, deep wells in place of wellpoints may be economical because of lower installation and maintenance costs. Daily production ranges between two to three wells per day, for 25′ to 40′ depths, to one well per day for depths over 50′.

Detailed analysis and estimating for any dewatering problem is available at no cost from wellpoint manufacturers. Major firms will quote "sufficient equipment" quotes or their affiliates offer lump sum proposals to cover complete dewatering responsibility.

Description for 200′ System with 8″ Header		Quantities
Equipment & Material	Wellpoints 25′ long, 2″ diameter @ 5′ O.C.	40 Each
Equipment & Material	Header pipe, 8″ diameter	200 L.F.
Equipment & Material	Discharge pipe, 8″ diameter	100 L.F.
Equipment & Material	8″ valves	3 Each
Equipment & Material	Combination jetting & wellpoint pump (standby)	1 Each
Equipment & Material	Wellpoint pump, 8″ diameter	1 Each
Equipment & Material	Transportation to and from site	1 Day
Equipment & Material	Fuel for 30 days x 60 gal./day	1800 Gallons
Equipment & Material	Lubricants for 30 days x 16 lbs./day	480 Lbs.
Equipment & Material	Sand for points	40 C.Y.
Labor	Technician to supervise installation	1 Week
Labor	Labor for installation and removal of system	300 Labor-hours
Labor	4 Operators straight time 40 hrs./wk. for 4.33 wks.	693 Hrs.
Labor	4 Operators overtime 2 hrs./wk. for 4.33 wks.	35 Hrs.

R312323-30 Compacting Backfill

Compaction of fill in embankments, around structures, in trenches, and under slabs is important to control settlement. Factors affecting compaction are:

1. Soil gradation
2. Moisture content
3. Equipment used
4. Depth of fill per lift
5. Density required

Production Rate:

$$\frac{1.75' \text{ plate width x 50 F.P.M. x 50 min./hr. x .67' lift}}{27 \text{ C.F. per C.Y.}} = 108.5 \text{ C.Y./hr.}$$

Production Rate for 4 Passes:

$$\frac{108.5 \text{ C.Y.}}{4 \text{ passes}} = 27.125 \text{ C.Y./hr. x 8 hrs.} = 217 \text{ C.Y./day}$$

Example:

Compact granular fill around a building foundation using a 21″ wide x 24″ vibratory plate in 8″ lifts. Operator moves at 50 F.P.M. working a 50 minute hour to develop 95% Modified Proctor Density with 4 passes.

Earthwork — R3141 Shoring

R314116-40 Wood Sheet Piling

Wood sheet piling may be used for depths to 20′ where there is no ground water. If moderate ground water is encountered Tongue & Groove sheeting will help to keep it out. When considerable ground water is present, steel sheeting must be used.

For estimating purposes on trench excavation, sizes are as follows:

Depth	Sheeting	Wales	Braces	B.F. per S.F.
To 8′	3 x 12′s	6 x 8′s, 2 line	6 x 8′s, @ 10′	4.0 @ 8′
8′ x 12′	3 x 12′s	10 x 10′s, 2 line	10 x 10′s, @ 9′	5.0 average
12′ to 20′	3 x 12′s	12 x 12′s, 3 line	12 x 12′s, @ 8′	7.0 average

Sheeting to be toed in at least 2′ depending upon soil conditions. A five person crew with an air compressor and sheeting driver can drive and brace 440 SF/day at 8′ deep, 360 SF/day at 12′ deep, and 320 SF/day at 16′ deep. For normal soils, piling can be pulled in 1/3 the time to install. Pulling difficulty increases with the time in the ground. Production can be increased by high pressure jetting.

R314116-45 Steel Sheet Piling

Limiting weights are 22 to 38#/S.F. of wall surface with 27#/S.F. average for usual types and sizes. (Weights of piles themselves are from 30.7#/L.F. to 57#/L.F. but they are 15″ to 21″ wide.) Lightweight sections 12″ to 28″ wide from 3 ga. to 12 ga. thick are also available for shallow excavations. Piles may be driven two at a time with an impact or vibratory hammer (use vibratory to pull) hung from a crane without leads. A reasonable estimate of the life of steel sheet piling is 10 uses with up to 125 uses possible if a vibratory hammer is used. Used piling costs from 50% to 80% of new piling depending on location and market conditions. Sheet piling and H piles can be rented for about 30% of the delivered mill price for the first month and 5% per month thereafter. Allow 1 labor-hour per pile for cleaning and trimming after driving. These costs increase with depth and hydrostatic head. Vibratory drivers are faster in wet granular soils and are excellent for pile extraction. Pulling difficulty increases with the time in the ground and may cost more than driving. It is often economical to abandon the sheet piling, especially if it can be used as the outer wall form. Allow about 1/3 additional length or more for toeing into ground. Add bracing, waler and strut costs. Waler costs can equal the cost per ton of sheeting.

Earthwork — R3145 Vibroflotation & Densification

R314513-90 Vibroflotation and Vibro Replacement Soil Compaction

Vibroflotation is a proprietary system of compacting sandy soils in place to increase relative density to about 70%. Typical bearing capacities attained will be 6000 psf for saturated sand and 12,000 psf for dry sand. Usual range is 4000 to 8000 psf capacity. Costs in the front of the book are for a vertical foot of compacted cylinder 6′ to 10′ in diameter.

Vibro replacement is a proprietary system of improving cohesive soils in place to increase bearing capacity. Most silts and clays above or below the water table can be strengthened by installation of stone columns.

The process consists of radial displacement of the soil by vibration. The created hole is then backfilled in stages with coarse granular fill which is thoroughly compacted and displaced into the surrounding soil in the form of a column.

The total project cost would depend on the number and depth of the compacted cylinders. The installing company guarantees relative soil density of the sand cylinders after compaction and the bearing capacity of the soil after the replacement process. Detailed estimating information is available from the installer at no cost.

R316326-60 Caissons

The three principal types of cassions are:

(1) Belled Caissons, which except for shallow depths and poor soil conditions, are generally recommended. They provide more bearing than shaft area. Because of its conical shape, no horizontal reinforcement of the bell is required.

(2) Straight Shaft Caissons are used where relatively light loads are to be supported by caissons that rest on high value bearing strata. While the shaft is larger in diameter than for belled types this is more than offset by the saving in time and labor.

(3) Keyed Caissons are used when extremely heavy loads are to be carried. A keyed or socketed caisson transfers its load into rock by a combination of end-bearing and shear reinforcing of the shaft. The most economical shaft often consists of a steel casing, a steel wide flange core and concrete. Allowable compressive stresses of .225 f′c for concrete, 16,000 psi for the wide flange core, and 9,000 psi for the steel casing are commonly used. The usual range of shaft diameter is 18″ to 84″. The number of sizes specified for any one project should be limited due to the problems of casing and auger storage. When hand work is to be performed, shaft diameters should not be less than 32″. When inspection of borings is required a minimum shaft diameter of 30″ is recommended. Concrete caissons are intended to be poured against earth excavation so permanent forms which add to cost should not be used if the excavation is clean and the earth sufficiently impervious to prevent excessive loss of concrete.

Soil Conditions for Belling		
Good	**Requires Handwork**	**Not Recommended**
Clay	Hard Shale	Silt
Sandy Clay	Limestone	Sand
Silty Clay	Sandstone	Gravel
Clayey Silt	Weathered Mica	Igneous Rock
Hard-pan		
Soft Shale		
Decomposed Rock		

R329219-50 Seeding

The type of grass is determined by light, shade and moisture content of soil plus intended use. Fertilizer should be disked 4″ before seeding. For steep slopes disk five tons of mulch and lay two tons of hay or straw on surface per acre after seeding. Surface mulch can be staked, lightly disked or tar emulsion sprayed. Material for mulch can be wood chips, peat moss, partially rotted hay or straw, wood fibers and sprayed emulsions. Hemp seed blankets with fertilizer are also available. For spring seeding, watering is necessary. Late fall seeding may have to be reseeded in the spring. Hydraulic seeding, power mulching, and aerial seeding can be used on large areas.

R331113-80 Piping Designations

There are several systems currently in use to describe pipe and fittings. The following paragraphs will help to identify and clarify classifications of piping systems used for water distribution.

Piping may be classified by schedule. Piping schedules include 5S, 10S, 10, 20, 30, Standard, 40, 60, Extra Strong, 80, 100, 120, 140, 160 and Double Extra Strong. These schedules are dependent upon the pipe wall thickness. The wall thickness of a particular schedule may vary with pipe size.

Ductile iron pipe for water distribution is classified by Pressure Classes such as Class 150, 200, 250, 300 and 350. These classes are actually the rated water working pressure of the pipe in pounds per square inch (psi). The pipe in these pressure classes is designed to withstand the rated water working pressure plus a surge allowance of 100 psi.

The American Water Works Association (AWWA) provides standards for various types of **plastic pipe.** C-900 is the specification for polyvinyl chloride (PVC) piping used for water distribution in sizes ranging from 4″ through 12″. C-901 is the specification for polyethylene (PE) pressure pipe, tubing and fittings used for water distribution in sizes ranging from 1/2″ through 3″. C-905 is the specification for PVC piping sizes 14″ and greater.

PVC pressure-rated pipe is identified using the standard dimensional ratio (SDR) method. This method is defined by the American Society for Testing and Materials (ASTM) Standard D 2241. This pipe is available in SDR numbers 64, 41, 32.5, 26, 21, 17, and 13.5. Pipe with an SDR of 64 will have the thinnest wall while pipe with an SDR of 13.5 will have the thickest wall. When the pressure rating (PR) of a pipe is given in psi, it is based on a line supplying water at 73 degrees F.

The National Sanitation Foundation (NSF) seal of approval is applied to products that can be used with potable water. These products have been tested to ANSI/NSF Standard 14.

Valves and strainers are classified by American National Standards Institute (ANSI) Classes. These Classes are 125, 150, 200, 250, 300, 400, 600, 900, 1500 and 2500. Within each class there is an operating pressure range dependent upon temperature. Design parameters should be compared to the appropriate material dependent, pressure-temperature rating chart for accurate valve selection.

Transportation | R3472 Railway Construction

R347216-10 Single Track R.R. Siding

The costs for a single track RR siding in the Unit Price section include the components shown in the table below.

Description of Component	Qty. per L.F. of Track	Unit
Ballast, 1-1/2″ crushed stone	.667	C.Y.
6″ x 8″ x 8′-6″ Treated timber ties, 22″ O.C.	.545	Ea.
Tie plates, 2 per tie	1.091	Ea.
Track rail	2.000	L.F.
Spikes, 6″, 4 per tie	2.182	Ea.
Splice bars w/ bolts, lock washers & nuts, @ 33′ O.C.	.061	Pair
Crew B-14 @ 57 L.F./Day	.018	Day

R347216-20 Single Track, Steel Ties, Concrete Bed

The costs for a R.R. siding with steel ties and a concrete bed in the Unit Price section include the components shown in the table below.

Description of Component	Qty. per L.F. of Track	Unit
Concrete bed, 9′ wide, 10″ thick	.278	C.Y.
Ties, W6x16 x 6′-6″ long, @ 30″ O.C.	.400	Ea.
Tie plates, 4 per tie	1.600	Ea.
Track rail	2.000	L.F.
Tie plate bolts, 1″, 8 per tie	3.200	Ea.
Splice bars w/bolts, lock washers & nuts, @ 33′ O.C.	.061	Pair
Crew B-14 @ 22 L.F./Day	.045	Day

Change Orders

Change Order Considerations

A Change Order is a written document, usually prepared by the design professional, and signed by the owner, the architect/ engineer and the contractor. A change order states the agreement of the parties to: an addition, deletion, or revision in the work; an adjustment in the contract sum, if any; or an adjustment in the contract time, if any. Change orders, or "extras" in the construction process occur after execution of the construction contract and impact architects/ engineers, contractors and owners.

Change orders that are properly recognized and managed can ensure orderly, professional and profitable progress for all who are involved in the project. There are many causes for change orders and change order requests. In all cases, change orders or change order requests should be addressed promptly and in a precise and prescribed manner. The following paragraphs include information regarding change order pricing and procedures.

The Causes of Change Orders

Reasons for issuing change orders include:

- Unforeseen field conditions that require a change in the work
- Correction of design discrepancies, errors or omissions in the contract documents
- Owner-requested changes, either by design criteria, scope of work, or project objectives
- Completion date changes for reasons unrelated to the construction process
- Changes in building code interpretations, or other public authority requirements that require a change in the work
- Changes in availability of existing or new materials and products

Procedures

Properly written contract documents must include the correct change order procedures for all parties—owners, design professionals and contractors—to follow in order to avoid costly delays and litigation.

Being "in the right" is not always a sufficient or acceptable defense. The contract provisions requiring notification and documentation must be adhered to within a defined or reasonable time frame.

The appropriate method of handling change orders is by a written proposal and acceptance by all parties involved. Prior to starting work on a project, all parties should identify their authorized agents who may sign and accept change orders, as well as any limits placed on their authority.

Time may be a critical factor when the need for a change arises. For such cases, the contractor might be directed to proceed on a "time and materials" basis, rather than wait for all paperwork to be processed—a delay that could impede progress. In this situation, the contractor must still follow the prescribed change order procedures, including but not limited to, notification and documentation.

All forms used for change orders should be dated and signed by the proper authority. Lack of documentation can be very costly, especially if legal judgments are to be made and if certain field personnel are no longer available. For time and material change orders, the contractor should keep accurate daily records of all labor and material allocated to the change. Forms that can be used to document change order work are available in *Means Forms for Building Construction Professionals.*

Owners or awarding authorities who do considerable and continual building construction (such as the federal government) realize the inevitability of change orders for numerous reasons, both predictable and unpredictable. As a result, the federal government, the American Institute of Architects (AIA), the Engineers Joint Contract Documents Committee (EJCDC) and other contractor, legal and technical organizations have developed standards and procedures to be followed by all parties to achieve contract continuance and timely completion, while being financially fair to all concerned.

In addition to the change order standards put forth by industry associations, there are also many books available on the subject.

Pricing Change Orders

When pricing change orders, regardless of their cause, the most significant factor is when the change occurs. The need for a change may be perceived in the field or requested by the architect/engineer *before* any of the actual installation has begun, or may evolve or appear *during* construction when the item of work in question is partially installed. In the latter cases, the original sequence of construction is disrupted, along with all contiguous and supporting systems. Change orders cause the greatest impact when they occur *after* the installation has been completed and must be uncovered, or even replaced. Post-completion changes may be caused by necessary design changes, product failure, or changes in the owner's requirements that are not discovered until the building or the systems begin to function.

Specified procedures of notification and record keeping must be adhered to and enforced regardless of the stage of construction: *before, during,* or *after* installation. Some bidding documents anticipate change orders by requiring that unit prices including overhead and profit percentages—for additional as well as deductible changes—be listed. Generally these unit prices do not fully take into account the ripple effect, or impact on other trades, and should be used for general guidance only.

When pricing change orders, it is important to classify the time frame in which the change occurs. There are two basic time frames for change orders: *pre-installation change orders,* which occur before the start of construction, and *post-installation change orders,* which involve reworking after the original installation. Change orders that occur between these stages may be priced according to the extent of work completed using a combination of techniques developed for pricing *pre-* and *post-installation* changes.

The following factors are the basis for a check list to use when preparing a change order estimate.

Factors To Consider When Pricing Change Orders

As an estimator begins to prepare a change order, the following questions should be reviewed to determine their impact on the final price.

General

- Is the change order work *pre-installation* or *post-installation?*

  Change order work costs vary according to how much of the installation has been completed. Once workers have the project scoped in their mind, even though they have not started, it can be difficult to refocus. Consequently they may spend more than the normal amount of time understanding the change. Also, modifications to work in place such as trimming or refitting usually take more time than was initially estimated. The greater the amount of work in place, the more reluctant workers are to change it. Psychologically they may resent the change and as a result the rework takes longer than normal. Post-installation change order estimates must include demolition of existing work as required to accomplish the change. If the work is performed at a later time, additional obstacles such as building finishes may be present which must be protected. Regardless of whether the change occurs pre-installation or post-installation, attempt to isolate the identifiable factors and price them separately. For example, add shipping costs that may be required pre-installation or any demolition required post-installation. Then analyze the potential impact on productivity of psychological and/or learning curve factors and adjust the output rates accordingly. One approach is to break down the typical workday into segments and quantify the impact on each segment. The following chart may be useful as a guide:

Activities (Productivity) Expressed as Percentages of a Workday			
Task	**Means Mechanical Cost Data (for New Construction)**	**Pre-Installation Change Orders**	**Post-Installation Change Orders**
1. Study plans	3%	6%	6%
2. Material procurement	3%	3%	3%
3. Receiving and storing	3%	3%	3%
4. Mobilization	5%	5%	5%
5. Site movement	5%	5%	8%
6. Layout and marking	8%	10%	12%
7. Actual installation	64%	59%	54%
8. Clean-up	3%	3%	3%
9. Breaks—non-productive	6%	6%	6%
Total	100%	100%	100%

Change Order Installation Efficiency

The labor-hours expressed (for new construction) are based on average installation time, using an efficiency level of approximately 60-65%. For change order situations, adjustments to this efficiency level should reflect the daily labor-hour allocation for that particular occurrence.

If any of the specific percentages expressed in the above chart do not apply to a particular project situation, then those percentage points should be reallocated to the appropriate task(s). Example: Using data for new construction, assume there is no new material being utilized. The percentages for Tasks 2 and 3 would therefore be reallocated to other tasks. If the time required for Tasks 2 and 3 can now be applied to installation, we can add the time allocated for *Material Procurement* and *Receiving and Storing* to the *Actual Installation* time for new construction, thereby increasing the Actual Installation percentage.

This chart shows that, due to reduced productivity, labor costs will be higher than those for new construction by 5% to 15% for pre-installation change orders and by 15% to 25% for post-installation change orders. Each job and change order is unique and must be examined individually. Many factors, covered elsewhere in this section, can each have a significant impact on productivity and change order costs. All such factors should be considered in every case.

- Will the change substantially delay the original completion date?

  A significant change in the project may cause the original completion date to be extended. The extended schedule may subject the contractor to new wage rates dictated by relevant labor contracts. Project supervision and other project overhead must also be extended beyond the original completion date. The schedule extension may also put installation into a new weather season. For example, underground piping scheduled for October installation was delayed until January. As a result, frost penetrated the trench area, thereby changing the degree of difficulty of the task. Changes and delays may have a ripple effect throughout the project. This effect must be analyzed and negotiated with the owner.

- What is the net effect of a deduct change order?

  In most cases, change orders resulting in a deduction or credit reflect only bare costs. The contractor may retain the overhead and profit based on the original bid.

Materials

- Will you have to pay more or less for the new material, required by the change order, than you paid for the original purchase?

  The same material prices or discounts will usually apply to materials purchased for change orders as new construction. In some instances, however, the contractor may forfeit the advantages of competitive pricing for change orders. Consider the following example:

  A contractor purchased over $20,000 worth of fan coil units for an installation, and obtained the maximum discount. Some time later it was determined the project required an additional matching unit. The contractor has to purchase this unit from the original supplier to ensure a match. The supplier at this time may not discount the unit because of the small quantity, and the fact that he is no longer in a competitive situation. The impact of quantity on purchase can add between 0% and 25% to material prices and/or subcontractor quotes.

- If materials have been ordered or delivered to the job site, will they be subject to a cancellation charge or restocking fee?

  Check with the supplier to determine if ordered materials are subject to a cancellation charge. Delivered materials not used as result of a change order may be subject to a restocking fee if returned to the supplier. Common restocking charges run between 20% and 40%. Also, delivery charges to return the goods to the supplier must be added.

Labor

- How efficient is the existing crew at the actual installation?

  Is the same crew that performed the initial work going to do the change order? Possibly the change consists of the installation of a unit identical to one already installed; therefore the change should take less time. Be sure to consider this potential productivity increase and modify the productivity rates accordingly.

- If the crew size is increased, what impact will that have on supervision requirements?

  Under most bargaining agreements or management practices, there is a point at which a working foreman is replaced by a nonworking foreman. This replacement increases project overhead by adding a nonproductive worker. If additional workers are added to accelerate the project or to perform changes while maintaining the schedule, be sure to add additional supervision time if warranted. Calculate the hours involved and the additional cost directly if possible.

- What are the other impacts of increased crew size?

  The larger the crew, the greater the potential for productivity to decrease. Some of the factors that cause this productivity loss are: overcrowding (producing restrictive conditions in the working space), and possibly a shortage of any special tools and equipment required. Such factors affect not only the crew working on the elements directly involved in the change order, but other crews whose movement may also be hampered.

  As the crew increases, check its basic composition for changes by the addition or deletion of apprentices or nonworking foreman and quantify the potential effects of equipment shortages or other logistical factors.

- As new crews, unfamiliar with the project, are brought onto the site, how long will it take them to become oriented to the project requirements?

  The orientation time for a new crew to become 100% effective varies with the site and type of project. Orientation is easiest at a new construction site, and most difficult at existing, very restrictive renovation sites. The type of work also affects orientation time. When all elements of the work are exposed, such as concrete or masonry work, orientation is decreased. When the work is concealed or less visible, such as existing electrical systems, orientation takes longer. Usually orientation can be accomplished in one day or less. Costs for added orientation should be itemized and added to the total estimated cost.

- How much actual production can be gained by working overtime?

  Short term overtime can be used effectively to accomplish more work in a day. However, as overtime is scheduled to run beyond several weeks, studies have shown marked decreases in output. The following chart shows the effect of long term overtime on worker efficiency. If the anticipated change requires extended overtime to keep the job on schedule, these factors can be used as a guide to predict the impact on time and cost. Add project overhead, particularly supervision, that may also be incurred.

Days per Week	Hours per Day	Production Efficiency					Payroll Cost Factors	
		1 Week	2 Weeks	3 Weeks	4 Weeks	Average 4 Weeks	@ 1-1/2 Times	@ 2 Times
5	8	100%	100%	100%	100%	100%	100%	100%
	9	100	100	95	90	96.25	105.6	111.1
	10	100	95	90	85	91.25	110.0	120.0
	11	95	90	75	65	81.25	113.6	127.3
	12	90	85	70	60	76.25	116.7	133.3
6	8	100	100	95	90	96.25	108.3	116.7
	9	100	95	90	85	92.50	113.0	125.9
	10	95	90	85	80	87.50	116.7	133.3
	11	95	85	70	65	78.75	119.7	139.4
	12	90	80	65	60	73.75	122.2	144.4
7	8	100	95	85	75	88.75	114.3	128.6
	9	95	90	80	70	83.75	118.3	136.5
	10	90	85	75	65	78.75	121.4	142.9
	11	85	80	65	60	72.50	124.0	148.1
	12	85	75	60	55	68.75	126.2	152.4

Effects of Overtime

Caution: Under many labor agreements, Sundays and holidays are paid at a higher premium than the normal overtime rate.

The use of long-term overtime is counterproductive on almost any construction job; that is, the longer the period of overtime, the lower the actual production rate. Numerous studies have been conducted, and while they have resulted in slightly different numbers, all reach the same conclusion. The figure above tabulates the effects of overtime work on efficiency.

As illustrated, there can be a difference between the *actual* payroll cost per hour and the *effective* cost per hour for overtime work. This is due to the reduced production efficiency with the increase in weekly hours beyond 40. This difference between actual and effective cost results from overtime work over a prolonged period. Short-term overtime work does not result in as great a reduction in efficiency, and in such cases, effective cost may not vary significantly from the actual payroll cost. As the total hours per week are increased on a regular basis, more time is lost because of fatigue, lowered morale, and an increased accident rate.

As an example, assume a project where workers are working 6 days a week, 10 hours per day. From the figure above (based on productivity studies), the average effective productive hours over a four-week period are:

$$0.875 \text{ x } 60 = 52.5$$

Depending upon the locale and day of week, overtime hours may be paid at time and a half or double time. For time and a half, the overall (average) *actual* payroll cost (including regular and overtime hours) is determined as follows:

$$\frac{40 \text{ reg. hrs.} + (20 \text{ overtime hrs. x } 1.5)}{60 \text{ hrs.}} = 1.167$$

Based on 60 hours, the payroll cost per hour will be 116.7% of the normal rate at 40 hours per week. However, because the effective production (efficiency) for 60 hours is reduced to the equivalent of 52.5 hours, the effective cost of overtime is calculated as follows:

For time and a half:

$$\frac{40 \text{ reg. hrs.} + (20 \text{ overtime hrs. x } 1.5)}{52.5 \text{ hrs.}} = 1.33$$

Installed cost will be 133% of the normal rate (for labor).

Thus, when figuring overtime, the actual cost per unit of work will be higher than the apparent overtime payroll dollar increase, due to the reduced productivity of the longer workweek. These efficiency calculations are true only for those cost factors determined by hours worked. Costs that are applied weekly or monthly, such as equipment rentals, will not be similarly affected.

Equipment

- What equipment is required to complete the change order?

  Change orders may require extending the rental period of equipment already on the job site, or the addition of special equipment brought in to accomplish the change work. In either case, the additional rental charges and operator labor charges must be added.

Summary

The preceding considerations and others you deem appropriate should be analyzed and applied to a change order estimate. The impact of each should be quantified and listed on the estimate to form an audit trail.

Change orders that are properly identified, documented, and managed help to ensure the orderly, professional and profitable progress of the work. They also minimize potential claims or disputes at the end of the project.

Estimating Tips

- The cost figures in this Square Foot Cost section were derived from approximately 11,200 projects contained in the RSMeans database of completed construction projects. They include the contractor's overhead and profit, but do not generally include architectural fees or land costs. The figures have been adjusted to January of the current year. New projects are added to our files each year, and outdated projects are discarded. For this reason, certain costs may not show a uniform annual progression. In no case are all subdivisions of a project listed.
- These projects were located throughout the U.S. and reflect a tremendous variation in square foot (S.F.) and cubic foot (C.F.) costs. This is due to differences, not only in labor and material costs, but also in individual owners' requirements. For instance, a bank in a large city would have different features than one in a rural area. This is true of all the different types of buildings analyzed. Therefore, caution should be exercised when using these Square Foot costs. For example, for court houses, costs in the database are local court house costs and will not apply to the larger, more elaborate federal court houses. As a general rule, the projects in the 1/4 column do not include any site work or equipment, while the projects in the 3/4 column may include both equipment and site work. The median figures do not generally include site work.
- None of the figures "go with" any others. All individual cost items were computed and tabulated separately. Thus, the sum of the median figures for Plumbing, HVAC and Electrical will not normally total up to the total Mechanical and Electrical costs arrived at by separate analysis and tabulation of the projects.
- Each building was analyzed as to total and component costs and percentages. The figures were arranged in ascending order with the results tabulated as shown. The 1/4 column shows that 25% of the projects had lower costs and 75% had higher. The 3/4 column shows that 75% of the projects had lower costs and 25% had higher. The median column shows that 50% of the projects had lower costs and 50% had higher.
- There are two times when square foot costs are useful. The first is in the conceptual stage when no details are available. Then square foot costs make a useful starting point. The second is after the bids are in and the costs can be worked back into their appropriate units for information purposes. As soon as details become available in the project design, the square foot approach should be discontinued and the project priced as to its particular components. When more precision is required, or for estimating the replacement cost of specific buildings, the current edition of *RSMeans Square Foot Costs* should be used.
- In using the figures in this section, it is recommended that the median column be used for preliminary figures if no additional information is available. The median figures, when multiplied by the total city construction cost index figures (see City Cost Indexes) and then multiplied by the project size modifier at the end of this section, should present a fairly accurate base figure, which would then have to be adjusted in view of the estimator's experience, local economic conditions, code requirements, and the owner's particular requirements. There is no need to factor the percentage figures, as these should remain constant from city to city. All tabulations mentioning air conditioning had at least partial air conditioning.
- The editors of this book would greatly appreciate receiving cost figures on one or more of your recent projects, which would then be included in the averages for next year. All cost figures received will be kept confidential, except that they will be averaged with other similar projects to arrive at Square Foot cost figures for next year's book. See the last page of the book for details and the discount available for submitting one or more of your projects.

50 17 00 | S.F. Costs

		50 17 00 | S.F. Costs	UNIT	UNIT COSTS			% OF TOTAL			
				1/4	MEDIAN	3/4	1/4	MEDIAN	3/4	
01	0010	**APARTMENTS Low Rise (1 to 3 story)**	S.F.	66	83.50	111				01
	0020	Total project cost	C.F.	5.95	7.85	9.70				
	0100	Site work	S.F.	5.65	7.70	13.60	6.05%	10.55%	14.05%	
	0500	Masonry		1.30	3.03	5.25	1.54%	3.67%	6.35%	
	1500	Finishes		7	9.65	11.90	9.05%	10.75%	12.85%	
	1800	Equipment		2.16	3.28	4.87	2.73%	4.03%	5.95%	
	2720	Plumbing		5.15	6.60	8.40	6.65%	8.95%	10.05%	
	2770	Heating, ventilating, air conditioning		3.28	4.04	5.95	4.20%	5.60%	7.60%	
	2900	Electrical		3.82	5.10	6.85	5.20%	6.65%	8.40%	
	3100	Total: Mechanical & Electrical	↓	13.25	16.90	21	15.90%	18.05%	23%	
	9000	Per apartment unit, total cost	Apt.	61,500	94,000	138,500				
	9500	Total: Mechanical & Electrical	"	11,600	18,300	23,900				
02	0010	**APARTMENTS Mid Rise (4 to 7 story)**	S.F.	87.50	106	131				02
	0020	Total project costs	C.F.	6.85	9.45	12.90				
	0100	Site work	S.F.	3.51	6.95	12.50	5.25%	6.70%	9.15%	
	0500	Masonry		5.85	8.05	11	5.10%	7.25%	10.50%	
	1500	Finishes		11.05	14.40	18.15	10.55%	13.45%	17.70%	
	1800	Equipment		2.55	3.82	5	2.54%	3.48%	4.31%	
	2500	Conveying equipment		1.89	2.41	2.92	1.94%	2.27%	2.69%	
	2720	Plumbing		5.15	8.25	8.70	5.70%	7.20%	8.95%	
	2900	Electrical		5.80	7.85	9.55	6.65%	7.20%	8.95%	
	3100	Total: Mechanical & Electrical	↓	18.50	23	28	18.50%	21%	23%	
	9000	Per apartment unit, total cost	Apt.	99,000	117,000	193,500				
	9500	Total: Mechanical & Electrical	"	18,700	21,700	27,400				
03	0010	**APARTMENTS High Rise (8 to 24 story)**	S.F.	99.50	115	137				03
	0020	Total project costs	C.F.	9.65	11.25	14.35				
	0100	Site work	S.F.	3.61	5.85	8.15	2.58%	4.84%	6.15%	
	0500	Masonry		5.75	10.45	13	4.74%	9.65%	11.05%	
	1500	Finishes		11.05	13.80	16.30	9.75%	11.80%	13.70%	
	1800	Equipment		3.20	3.93	5.20	2.78%	3.49%	4.35%	
	2500	Conveying equipment		2.26	3.43	4.66	2.23%	2.78%	3.37%	
	2720	Plumbing		7.35	8.65	12.10	6.80%	7.20%	10.45%	
	2900	Electrical		6.85	8.65	11.65	6.45%	7.65%	8.80%	
	3100	Total: Mechanical & Electrical	↓	20.50	26	31.50	17.95%	22.50%	24.50%	
	9000	Per apartment unit, total cost	Apt.	103,500	114,000	158,000				
	9500	Total: Mechanical & Electrical	"	22,400	25,500	27,000				
04	0010	**AUDITORIUMS**	S.F.	103	140	202				04
	0020	Total project costs	C.F.	6.45	9	12.90				
	2720	Plumbing	S.F.	6.25	9.05	11	5.85%	7.20%	8.70%	
	2900	Electrical		8	11.65	18.85	6.80%	9.05%	11.30%	
	3100	Total: Mechanical & Electrical	↓	22.50	45.50	55	24.50%	29%	31.50%	
05	0010	**AUTOMOTIVE SALES**	S.F.	76.50	104	129				05
	0020	Total project costs	C.F.	5.05	6.05	7.85				
	2720	Plumbing	S.F.	3.49	6.05	6.60	2.89%	6.05%	6.50%	
	2770	Heating, ventilating, air conditioning		5.40	8.20	8.90	4.61%	10%	10.35%	
	2900	Electrical		6.15	9.65	13.10	7.25%	9.80%	12.15%	
	3100	Total: Mechanical & Electrical	↓	19.30	27.50	33	19.15%	20.50%	22%	
06	0010	**BANKS**	S.F.	150	187	237				06
	0020	Total project costs	C.F.	10.70	14.55	19.20				
	0100	Site work	S.F.	17.20	26	38	7.85%	12.95%	16.95%	
	0500	Masonry		8.10	15.45	27	3.46%	7.65%	10.10%	
	1500	Finishes		13.90	20.50	25.50	5.85%	8.60%	11.55%	
	1800	Equipment		5.65	12.45	25.50	1%	5.55%	10.50%	
	2720	Plumbing		4.70	6.70	9.80	2.82%	3.90%	4.93%	
	2770	Heating, ventilating, air conditioning		8.95	11.95	15.90	4.86%	7.15%	8.50%	
	2900	Electrical		14.20	18.95	25	8.20%	10.20%	12.20%	
	3100	Total: Mechanical & Electrical	↓	34	46	55	16%	19.40%	23%	
	3500	See also division 11020 & 11030 (MF2004 11 16 00 & 11 17 00)								

	50 17 00 | S.F. Costs		UNIT	UNIT COSTS 1/4	UNIT COSTS MEDIAN	UNIT COSTS 3/4	% OF TOTAL 1/4	% OF TOTAL MEDIAN	% OF TOTAL 3/4	
13	0010	CHURCHES	S.F.	101	128	166				13
	0020	Total project costs	C.F.	6.25	7.90	10.40				
	1800	Equipment	S.F.	1.21	2.89	6.15	.95%	2.11%	4.50%	
	2720	Plumbing		3.93	5.50	8.10	3.51%	4.96%	6.25%	
	2770	Heating, ventilating, air conditioning		9.20	11.95	16.95	7.50%	10%	12%	
	2900	Electrical		8.50	11.65	15.65	7.35%	8.75%	10.95%	
	3100	Total: Mechanical & Electrical	↓	26.50	34	45.50	18.30%	22%	24.50%	
	3500	See also division 11040 (MF2004 11 91 00)								
15	0010	CLUBS, COUNTRY	S.F.	108	131	164				15
	0020	Total project costs	C.F.	8.75	10.65	14.70				
	2720	Plumbing	S.F.	6.55	9.70	22	5.60%	7.90%	10%	
	2900	Electrical		8.50	11.65	15.20	7%	8.95%	11%	
	3100	Total: Mechanical & Electrical	↓	45.50	56.50	59.50	19%	26.50%	29.50%	
17	0010	CLUBS, SOCIAL Fraternal	S.F.	86.50	124	166				17
	0020	Total project costs	C.F.	5.40	8.20	9.75				
	2720	Plumbing	S.F.	5.45	6.75	10.25	5.60%	6.90%	8.55%	
	2770	Heating, ventilating, air conditioning		7.85	9.50	12.20	8.20%	9.25%	14.40%	
	2900	Electrical		6.50	10.70	12.25	6.50%	9.50%	10.55%	
	3100	Total: Mechanical & Electrical	↓	19.20	36.50	46	21%	23%	23.50%	
18	0010	CLUBS, Y.M.C.A.	S.F.	109	141	177				18
	0020	Total project costs	C.F.	5	8.35	12.45				
	2720	Plumbing	S.F.	6.85	13.65	15.30	5.65%	7.60%	10.85%	
	2900	Electrical		8.70	10.85	14.95	6.05%	7.60%	9.25%	
	3100	Total: Mechanical & Electrical	↓	33	37	50.50	18.40%	21.50%	28.50%	
19	0010	COLLEGES Classrooms & Administration	S.F.	110	150	199				19
	0020	Total project costs	C.F.	8	11.65	17.95				
	0500	Masonry	S.F.	8.05	15.15	18.60	5.65%	8.25%	10.50%	
	2720	Plumbing		5.60	11.40	20.50	5.10%	6.60%	8.95%	
	2900	Electrical		9.15	13.85	18.95	7.70%	9.85%	12%	
	3100	Total: Mechanical & Electrical	↓	36.50	51	60.50	24%	28%	31.50%	
21	0010	COLLEGES Science, Engineering, Laboratories	S.F.	204	238	276				21
	0020	Total project costs	C.F.	11.70	17.05	19.35				
	1800	Equipment	S.F.	11.35	25.50	28	2%	6.45%	12.65%	
	2900	Electrical		16.80	23	36.50	7.10%	9.40%	12.10%	
	3100	Total: Mechanical & Electrical	↓	62.50	74	115	28.50%	31.50%	41%	
	3500	See also division 11600 (MF2004 11 53 00)								
23	0010	COLLEGES Student Unions	S.F.	130	176	213				23
	0020	Total project costs	C.F.	7.25	9.50	11.75				
	3100	Total: Mechanical & Electrical	S.F.	49	53	62.50	23.50%	26%	29%	
25	0010	COMMUNITY CENTERS	S.F.	107	132	178				25
	0020	Total project costs	C.F.	7	10	12.95				
	1800	Equipment	S.F.	2.60	4.39	7	1.48%	3.01%	5.45%	
	2720	Plumbing		5.10	8.90	12.15	4.85%	7%	8.95%	
	2770	Heating, ventilating, air conditioning		8.15	11.90	17	6.80%	10.35%	12.90%	
	2900	Electrical		8.80	11.70	16.80	7.30%	9%	10.45%	
	3100	Total: Mechanical & Electrical	↓	30	37.50	53.50	20%	25%	31%	
28	0010	COURT HOUSES	S.F.	155	178	229				28
	0020	Total project costs	C.F.	11.85	14.20	17.90				
	2720	Plumbing	S.F.	7.35	10.30	14.80	5.95%	7.45%	8.20%	
	2900	Electrical		16.45	18.25	27	8.90%	10.45%	11.55%	
	3100	Total: Mechanical & Electrical	↓	41.50	57	63.50	22.50%	27.50%	30.50%	
30	0010	DEPARTMENT STORES	S.F.	57	77.50	98				30
	0020	Total project costs	C.F.	3.07	3.98	5.40				
	2720	Plumbing	S.F.	1.78	2.25	3.42	1.82%	4.21%	5.90%	
	2770	Heating, ventilating, air conditioning	↓	5.20	8.05	12.10	8.20%	9.10%	14.80%	

		50 17 00 | S.F. Costs	UNIT	UNIT COSTS 1/4	UNIT COSTS MEDIAN	UNIT COSTS 3/4	% OF TOTAL 1/4	% OF TOTAL MEDIAN	% OF TOTAL 3/4	
30	2900	Electrical	S.F.	6.55	9	10.65	9.05%	12.15%	14.95%	30
	3100	Total: Mechanical & Electrical		11.55	14.80	26	13.20%	21.50%	50%	
31	0010	**DORMITORIES Low Rise (1 to 3 story)**	S.F.	109	141	188				31
	0020	Total project costs	C.F.	6.15	9.90	14.85				
	2720	Plumbing	S.F.	6.55	8.75	11.05	8.05%	9%	9.65%	
	2770	Heating, ventilating, air conditioning		6.90	8.25	11	4.61%	8.05%	10%	
	2900	Electrical		7.20	10.95	14.95	6.40%	8.65%	9.50%	
	3100	Total: Mechanical & Electrical		37.50	40	62.50	22%	25%	27%	
	9000	Per bed, total cost	Bed	46,000	51,000	109,500				
32	0010	**DORMITORIES Mid Rise (4 to 8 story)**	S.F.	133	174	215				32
	0020	Total project costs	C.F.	14.70	16.15	19.30				
	2900	Electrical	S.F.	14.15	16.10	21.50	8.20%	10.20%	11.95%	
	3100	Total: Mechanical & Electrical	"	39.50	41	80.50	25.50%	34.50%	37.50%	
	9000	Per bed, total cost	Bed	19,000	43,300	109,500				
34	0010	**FACTORIES**	S.F.	50.50	75	116				34
	0020	Total project costs	C.F.	3.24	4.83	8				
	0100	Site work	S.F.	5.75	10.50	16.60	6.95%	11.45%	17.95%	
	2720	Plumbing		2.72	5.05	8.35	3.73%	6.05%	8.10%	
	2770	Heating, ventilating, air conditioning		5.30	7.60	10.25	5.25%	8.45%	11.35%	
	2900	Electrical		6.25	9.90	15.15	8.10%	10.50%	14.20%	
	3100	Total: Mechanical & Electrical		17.90	24	36.50	21%	28.50%	35.50%	
36	0010	**FIRE STATIONS**	S.F.	100	138	185				36
	0020	Total project costs	C.F.	5.85	8.05	10.70				
	0500	Masonry	S.F.	14.80	25	34	8.35%	11.55%	16.15%	
	1140	Roofing		3.26	8.85	10.05	1.90%	4.94%	5.05%	
	1580	Painting		2.53	3.78	3.87	1.37%	1.57%	2.07%	
	1800	Equipment		1.24	2.38	4.41	.62%	1.86%	3.54%	
	2720	Plumbing		5.60	9.05	13.05	5.85%	7.35%	9.45%	
	2770	Heating, ventilating, air conditioning		5.55	9	13.90	5.15%	7.40%	9.40%	
	2900	Electrical		7.15	12.55	17	6.80%	8.60%	10.60%	
	3100	Total: Mechanical & Electrical		36.50	47	53	18.40%	23%	26%	
37	0010	**FRATERNITY HOUSES & Sorority Houses**	S.F.	100	129	176				37
	0020	Total project costs	C.F.	9.95	10.40	12.50				
	2720	Plumbing	S.F.	7.55	8.65	15.85	6.80%	8%	10.85%	
	2900	Electrical		6.60	14.25	17.50	6.60%	9.90%	10.65%	
	3100	Total: Mechanical & Electrical		7.80	25	30		15.10%	15.90%	
38	0010	**FUNERAL HOMES**	S.F.	106	144	261				38
	0020	Total project costs	C.F.	10.75	12	23				
	2900	Electrical	S.F.	4.66	8.55	9.35	3.58%	4.44%	5.95%	
	3100	Total: Mechanical & Electrical		16.50	24	33.50	12.90%	12.90%	12.90%	
39	0010	**GARAGES, COMMERCIAL (Service)**	S.F.	60	92.50	128				39
	0020	Total project costs	C.F.	3.93	5.80	8.45				
	1800	Equipment	S.F.	3.37	7.60	11.80	2.21%	4.62%	6.80%	
	2720	Plumbing		4.14	6.40	11.65	5.45%	7.85%	10.65%	
	2730	Heating & ventilating		5.45	7.20	9.75	5.25%	6.85%	8.20%	
	2900	Electrical		5.70	8.65	12.50	7.15%	9.25%	10.85%	
	3100	Total: Mechanical & Electrical		12.55	24	35.50	12.35%	17.40%	26%	
40	0010	**GARAGES, MUNICIPAL (Repair)**	S.F.	88	117	165				40
	0020	Total project costs	C.F.	5.50	6.95	11.95				
	0500	Masonry	S.F.	8.25	16.10	25	5.60%	9.15%	12.50%	
	2720	Plumbing		3.94	7.55	14.25	3.59%	6.70%	7.95%	
	2730	Heating & ventilating		6.75	9.75	18.85	6.15%	7.45%	13.50%	
	2900	Electrical		6.50	10.20	14.70	6.65%	8.15%	11.15%	
	3100	Total: Mechanical & Electrical		30	41.50	60.50	21.50%	25.50%	28.50%	

50 17 00 | S.F. Costs

		50 17 00 | S.F. Costs	UNIT	UNIT COSTS			% OF TOTAL			
				1/4	MEDIAN	3/4	1/4	MEDIAN	3/4	
41	0010	**GARAGES, PARKING**	S.F.	34	50	85.50				41
	0020	Total project costs	C.F.	3.21	4.36	6.35				
	2720	Plumbing	S.F.	.97	1.50	2.32	1.72%	2.70%	3.85%	
	2900	Electrical		1.87	2.30	3.61	4.33%	5.20%	6.30%	
	3100	Total: Mechanical & Electrical		3.84	5.35	6.65	7%	8.90%	11.05%	
	3200									
	9000	Per car, total cost	Car	14,400	18,100	23,100				
43	0010	**GYMNASIUMS**	S.F.	95.50	127	172				43
	0020	Total project costs	C.F.	4.74	6.45	7.90				
	1800	Equipment	S.F.	2.26	4.24	8.15	1.81%	3.30%	6.70%	
	2720	Plumbing		6	7.15	9.20	4.65%	6.40%	7.75%	
	2770	Heating, ventilating, air conditioning		6.45	9.85	19.80	5.15%	9.05%	11.10%	
	2900	Electrical		7.30	9.95	13.10	6.60%	8.30%	10.30%	
	3100	Total: Mechanical & Electrical		26.50	36	43	19.75%	23.50%	29%	
	3500	See also division 11480 (MF2004 11 67 00)								
46	0010	**HOSPITALS**	S.F.	182	225	310				46
	0020	Total project costs	C.F.	13.85	17.20	24.50				
	1800	Equipment	S.F.	4.63	8.90	15.35	.96%	2.63%	5%	
	2720	Plumbing		15.70	22	28.50	7.60%	9.10%	10.85%	
	2770	Heating, ventilating, air conditioning		23	29	40	7.80%	12.95%	16.65%	
	2900	Electrical		20	27	38	9.90%	11.75%	14%	
	3100	Total: Mechanical & Electrical		56	80.50	122	27%	33.50%	36.50%	
	9000	Per bed or person, total cost	Bed	211,500	291,000	335,500				
	9900	See also division 11700 (MF2004 11 71 00)								
48	0010	**HOUSING For the Elderly**	S.F.	89.50	114	140				48
	0020	Total project costs	C.F.	6.40	8.90	11.35				
	0100	Site work	S.F.	6.25	9.70	14.20	5.05%	7.90%	12.10%	
	0500	Masonry		2.74	10.20	14.95	1.30%	6.05%	11%	
	1800	Equipment		2.17	2.99	4.76	1.88%	3.23%	4.43%	
	2510	Conveying systems		2.19	2.93	3.98	1.78%	2.20%	2.81%	
	2720	Plumbing		6.65	8.50	10.70	8.15%	9.55%	10.50%	
	2730	Heating, ventilating, air conditioning		3.42	4.84	7.25	3.30%	5.60%	7.25%	
	2900	Electrical		6.70	9.10	11.65	7.30%	8.50%	10.25%	
	3100	Total: Mechanical & Electrical		23	27.50	36.50	18.10%	22.50%	29%	
	9000	Per rental unit, total cost	Unit	83,500	97,500	108,500				
	9500	Total: Mechanical & Electrical	″	18,600	21,400	25,000				
50	0010	**HOUSING Public (Low Rise)**	S.F.	75.50	105	136				50
	0020	Total project costs	C.F.	6.70	8.40	10.40				
	0100	Site work	S.F.	9.60	13.85	22.50	8.35%	11.75%	16.50%	
	1800	Equipment		2.05	3.35	5.10	2.26%	3.03%	4.24%	
	2720	Plumbing		5.45	7.20	9.10	7.15%	9.05%	11.60%	
	2730	Heating, ventilating, air conditioning		2.73	5.30	5.80	4.26%	6.05%	6.45%	
	2900	Electrical		4.56	6.80	9.45	5.10%	6.55%	8.25%	
	3100	Total: Mechanical & Electrical		21.50	28	31	14.50%	17.55%	26.50%	
	9000	Per apartment, total cost	Apt.	83,000	94,000	118,500				
	9500	Total: Mechanical & Electrical	″	17,700	21,800	24,100				
51	0010	**ICE SKATING RINKS**	S.F.	64.50	151	166				51
	0020	Total project costs	C.F.	4.74	4.85	5.60				
	2720	Plumbing	S.F.	2.41	4.52	4.62	3.12%	3.23%	5.65%	
	2900	Electrical		6.90	10.60	11.20	6.30%	10.15%	15.05%	
	3100	Total: Mechanical & Electrical		11.45	16.20	20	18.95%	18.95%	18.95%	
52	0010	**JAILS**	S.F.	196	254	325				52
	0020	Total project costs	C.F.	17.70	24.50	29				
	1800	Equipment	S.F.	7.65	22.50	38.50	2.80%	5.55%	10.35%	
	2720	Plumbing		18.90	25.50	33.50	7%	8.90%	13.35%	
	2770	Heating, ventilating, air conditioning		17.70	23.50	45.50	7.50%	9.45%	17.75%	
	2900	Electrical		20.50	27	34.50	8.20%	11.55%	14.95%	

	50 17 00 | S.F. Costs		UNIT	UNIT COSTS 1/4	UNIT COSTS MEDIAN	UNIT COSTS 3/4	% OF TOTAL 1/4	% OF TOTAL MEDIAN	% OF TOTAL 3/4	
52	3100	Total: Mechanical & Electrical	S.F.	53.50	98	116	28%	30%	34%	52
53	0010	**LIBRARIES**	S.F.	124	159	207				53
	0020	Total project costs	C.F.	8.50	10.65	13.60				
	0500	Masonry	S.F.	9.95	17.35	29	5.80%	7.80%	11.80%	
	1800	Equipment		1.70	4.58	6.90	.37%	1.50%	4.07%	
	2720	Plumbing		4.58	6.65	9.05	3.38%	4.60%	5.70%	
	2770	Heating, ventilating, air conditioning		10.15	17.25	22.50	7.80%	10.95%	12.80%	
	2900	Electrical		12.65	16.40	21	8.30%	10.25%	11.95%	
	3100	Total: Mechanical & Electrical	↓	37.50	48	59	20.50%	23%	26.50%	
54	0010	**LIVING, ASSISTED**	S.F.	115	135	159				54
	0020	Total project costs	C.F.	9.65	11.30	12.80				
	0500	Masonry	S.F.	3.38	4.02	4.73	2.37%	3.16%	3.86%	
	1800	Equipment		2.61	3.03	3.89	2.12%	2.45%	2.66%	
	2720	Plumbing		9.60	12.90	13.35	6.05%	8.15%	10.60%	
	2770	Heating, ventilating, air conditioning		11.40	11.95	13.10	7.95%	9.35%	9.70%	
	2900	Electrical		11.25	12.40	14.35	9%	10%	10.70%	
	3100	Total: Mechanical & Electrical	↓	31.50	37	42.50	26%	29%	31.50%	
55	0010	**MEDICAL CLINICS**	S.F.	117	144	183				55
	0020	Total project costs	C.F.	8.55	11.05	14.70				
	1800	Equipment	S.F.	3.16	6.65	10.35	1.05%	2.94%	6.35%	
	2720	Plumbing		7.75	10.95	14.60	6.15%	8.40%	10.10%	
	2770	Heating, ventilating, air conditioning		9.25	12.15	17.85	6.65%	8.85%	11.35%	
	2900	Electrical		10.05	14.25	18.65	8.10%	10%	12.25%	
	3100	Total: Mechanical & Electrical	↓	32	43.50	59.50	22.50%	27%	33.50%	
	3500	See also division 11700 (MF2004 11 71 00)								
57	0010	**MEDICAL OFFICES**	S.F.	110	136	167				57
	0020	Total project costs	C.F.	8.20	11.10	15				
	1800	Equipment	S.F.	3.63	7.15	10.20	.70%	5.10%	7.05%	
	2720	Plumbing		6.05	9.35	12.60	5.60%	6.80%	8.50%	
	2770	Heating, ventilating, air conditioning		7.35	10.60	14	6.10%	8%	9.70%	
	2900	Electrical		8.95	12.80	17.85	7.60%	9.80%	11.70%	
	3100	Total: Mechanical & Electrical	↓	24	34	49.50	19.30%	23%	28.50%	
59	0010	**MOTELS**	S.F.	69.50	100	131				59
	0020	Total project costs	C.F.	6.15	8.25	13.50				
	2720	Plumbing	S.F.	7.05	8.95	10.65	9.45%	10.60%	12.55%	
	2770	Heating, ventilating, air conditioning		4.28	6.40	11.45	5.60%	5.60%	10%	
	2900	Electrical		6.55	8.30	10.30	7.45%	9.05%	10.45%	
	3100	Total: Mechanical & Electrical	↓	21	28	48	18.50%	24%	25.50%	
	5000									
	9000	Per rental unit, total cost	Unit	35,300	67,000	72,500				
	9500	Total: Mechanical & Electrical	"	6,875	10,400	12,100				
60	0010	**NURSING HOMES**	S.F.	109	140	172				60
	0020	Total project costs	C.F.	8.55	10.65	14.60				
	1800	Equipment	S.F.	3.42	4.54	7.55	2.02%	3.62%	4.99%	
	2720	Plumbing		9.30	14.20	17	8.75%	10.10%	12.70%	
	2770	Heating, ventilating, air conditioning		9.80	14.90	19.75	9.70%	11.45%	11.80%	
	2900	Electrical		10.75	13.45	18.30	9.40%	10.55%	12.45%	
	3100	Total: Mechanical & Electrical	↓	25.50	36	60	26%	29.50%	30.50%	
	9000	Per bed or person, total cost	Bed	48,200	60,500	78,000				
61	0010	**OFFICES Low Rise (1 to 4 story)**	S.F.	91.50	119	154				61
	0020	Total project costs	C.F.	6.55	9	11.85				
	0100	Site work	S.F.	7.35	12.70	19.15	6.20%	9.70%	13.60%	
	0500	Masonry		3.54	7.25	13.05	2.62%	5.50%	8.60%	
	1800	Equipment		.97	1.91	5.20	.62%	1.50%	3.51%	
	2720	Plumbing		3.27	5.05	7.40	3.66%	4.50%	6.10%	
	2770	Heating, ventilating, air conditioning		7.25	10.10	14.75	7.20%	10.30%	11.70%	
	2900	Electrical	↓	7.50	10.65	15.10	7.45%	9.60%	11.35%	

50 17 00 | S.F. Costs

		50 17 00 | S.F. Costs	UNIT	UNIT COSTS 1/4	UNIT COSTS MEDIAN	UNIT COSTS 3/4	% OF TOTAL 1/4	% OF TOTAL MEDIAN	% OF TOTAL 3/4	
61	3100	Total: Mechanical & Electrical	S.F.	20.50	28	41	18.15%	22%	27%	61
62	0010	**OFFICES Mid Rise (5 to 10 story)**	S.F.	97	117	155				62
	0020	Total project costs	C.F.	6.85	8.75	12.45				
	2720	Plumbing	S.F.	2.93	4.54	6.55	2.83%	3.74%	4.50%	
	2770	Heating, ventilating, air conditioning		7.35	10.55	16.80	7.65%	9.40%	11%	
	2900	Electrical		7.20	9.35	12.80	6.35%	7.80%	10%	
	3100	Total: Mechanical & Electrical		18.65	24	45.50	19.15%	21.50%	27.50%	
63	0010	**OFFICES High Rise (11 to 20 story)**	S.F.	119	150	185				63
	0020	Total project costs	C.F.	8.30	10.40	14.95				
	2900	Electrical	S.F.	7.25	8.85	13.10	5.80%	7.85%	10.50%	
	3100	Total: Mechanical & Electrical		23.50	31.50	52.50	16.90%	23.50%	34%	
64	0010	**POLICE STATIONS**	S.F.	143	188	237				64
	0020	Total project costs	C.F.	11.35	13.95	19.10				
	0500	Masonry	S.F.	13.40	23.50	29.50	6.70%	9.10%	11.35%	
	1800	Equipment		2.27	9.95	15.80	.98%	3.35%	6.70%	
	2720	Plumbing		8	15.95	19.85	5.65%	6.90%	10.75%	
	2770	Heating, ventilating, air conditioning		12.50	16.60	22.50	5.85%	10.55%	11.70%	
	2900	Electrical		15.60	22.50	29.50	9.80%	11.85%	14.80%	
	3100	Total: Mechanical & Electrical		51	61.50	83.50	25%	31.50%	32.50%	
65	0010	**POST OFFICES**	S.F.	113	139	177				65
	0020	Total project costs	C.F.	6.80	8.60	9.75				
	2720	Plumbing	S.F.	5.10	6.30	7.95	4.24%	5.30%	5.60%	
	2770	Heating, ventilating, air conditioning		7.95	9.80	10.90	6.65%	7.15%	9.35%	
	2900	Electrical		9.30	13.10	15.55	7.25%	9%	11%	
	3100	Total: Mechanical & Electrical		27	35	40	16.25%	18.80%	22%	
66	0010	**POWER PLANTS**	S.F.	780	1,000	1,900				66
	0020	Total project costs	C.F.	21.50	47	100				
	2900	Electrical	S.F.	55.50	117	175	9.30%	12.75%	21.50%	
	8100	Total: Mechanical & Electrical		138	450	1,000	32.50%	32.50%	52.50%	
67	0010	**RELIGIOUS EDUCATION**	S.F.	91	119	146				67
	0020	Total project costs	C.F.	5.05	7.20	9.05				
	2720	Plumbing	S.F.	3.80	5.40	7.65	4.40%	5.30%	7.10%	
	2770	Heating, ventilating, air conditioning		9.60	10.85	15.35	10.05%	11.45%	12.35%	
	2900	Electrical		7.20	10.15	13.45	7.60%	9.05%	10.35%	
	3100	Total: Mechanical & Electrical		29.50	38.50	47	22%	23%	27%	
69	0010	**RESEARCH Laboratories & Facilities**	S.F.	136	195	284				69
	0020	Total project costs	C.F.	10.25	19.80	24				
	1800	Equipment	S.F.	6.05	11.90	29	.94%	4.58%	9.10%	
	2720	Plumbing		13.65	17.45	28	6.15%	8.30%	10.80%	
	2770	Heating, ventilating, air conditioning		12.25	41	48.50	7.25%	16.50%	17.50%	
	2900	Electrical		15.90	26	43	9.45%	11.15%	15.40%	
	3100	Total: Mechanical & Electrical		52	90.50	131	31.50%	37.50%	42%	
70	0010	**RESTAURANTS**	S.F.	132	170	221				70
	0020	Total project costs	C.F.	11.10	14.55	19.10				
	1800	Equipment	S.F.	8.50	21	31.50	6.10%	13%	15.65%	
	2720	Plumbing		10.45	12.65	16.65	6.10%	8.15%	9%	
	2770	Heating, ventilating, air conditioning		13.25	18.40	22.50	9.20%	12%	12.40%	
	2900	Electrical		14	17.20	22.50	8.35%	10.55%	11.55%	
	3100	Total: Mechanical & Electrical		43.50	46.50	60	21%	24.50%	29.50%	
	9000	Per seat unit, total cost	Seat	4,850	6,450	7,625				
	9500	Total: Mechanical & Electrical	"	1,225	1,600	1,900				
72	0010	**RETAIL STORES**	S.F.	61.50	82.50	110				72
	0020	Total project costs	C.F.	4.17	5.95	8.25				
	2720	Plumbing	S.F.	2.23	3.73	6.35	3.26%	4.60%	6.80%	
	2770	Heating, ventilating, air conditioning		4.82	6.60	9.90	6.75%	8.75%	10.15%	
	2900	Electrical		5.55	7.60	10.95	7.25%	9.90%	11.65%	
	3100	Total: Mechanical & Electrical		14.75	18.95	26	17.05%	21%	23.50%	

50 17 00 | S.F. Costs

			UNIT	UNIT COSTS			% OF TOTAL			
				1/4	MEDIAN	3/4	1/4	MEDIAN	3/4	
74	0010	**SCHOOLS Elementary**	S.F.	99.50	123	152				74
	0020	Total project costs	C.F.	6.60	8.40	10.90				
	0500	Masonry	S.F.	9	15.50	23	5.80%	11%	14.95%	
	1800	Equipment		2.75	4.68	8.65	1.89%	3.32%	4.71%	
	2720	Plumbing		5.80	8.15	10.90	5.70%	7.15%	9.35%	
	2730	Heating, ventilating, air conditioning		8.65	13.80	19.25	8.15%	10.80%	14.90%	
	2900	Electrical		9.45	12.50	15.75	8.40%	10.05%	11.70%	
	3100	Total: Mechanical & Electrical	↓	33.50	42.50	52	25%	27.50%	30%	
	9000	Per pupil, total cost	Ea.	11,500	17,100	49,500				
	9500	Total: Mechanical & Electrical	″	3,250	4,125	14,300				
76	0010	**SCHOOLS Junior High & Middle**	S.F.	104	127	155				76
	0020	Total project costs	C.F.	6.60	8.50	9.55				
	0500	Masonry	S.F.	13.25	17.40	20.50	8.60%	11.60%	14.35%	
	1800	Equipment		3.31	5.35	8.05	1.79%	3.09%	4.86%	
	2720	Plumbing		6.05	7.45	9.25	5.30%	6.80%	7.25%	
	2770	Heating, ventilating, air conditioning		12.05	14.65	26	8.90%	11.55%	14.20%	
	2900	Electrical		10.15	12.25	15.80	7.90%	9.35%	10.60%	
	3100	Total: Mechanical & Electrical	↓	33	42.50	52.50	23.50%	27%	29.50%	
	9000	Per pupil, total cost	Ea.	13,100	17,200	23,200				
78	0010	**SCHOOLS Senior High**	S.F.	107	132	165				78
	0020	Total project costs	C.F.	6.50	9.60	15.40				
	1800	Equipment	S.F.	2.83	6.65	9.40	1.86%	2.91%	4.30%	
	2720	Plumbing		6.05	9.05	16.60	5.60%	6.90%	8.30%	
	2770	Heating, ventilating, air conditioning		12.35	14.15	27	8.95%	11.60%	15%	
	2900	Electrical		10.85	14.05	20.50	8.65%	10.15%	12.35%	
	3100	Total: Mechanical & Electrical	↓	36	42	70.50	23.50%	26.50%	29%	
	9000	Per pupil, total cost	Ea.	10,200	20,700	25,900				
80	0010	**SCHOOLS Vocational**	S.F.	87.50	127	157				80
	0020	Total project costs	C.F.	5.40	7.80	10.75				
	0500	Masonry	S.F.	5.10	12.65	19.35	3.53%	6.70%	10.95%	
	1800	Equipment		2.73	6.80	9.45	1.24%	3.10%	4.26%	
	2720	Plumbing		5.60	8.30	12.25	5.40%	6.90%	8.55%	
	2770	Heating, ventilating, air conditioning		7.80	14.55	24.50	8.60%	11.90%	14.65%	
	2900	Electrical		9.10	11.90	16.35	8.45%	10.95%	13.20%	
	3100	Total: Mechanical & Electrical	↓	31.50	35	60	23.50%	27.50%	31%	
	9000	Per pupil, total cost	Ea.	12,200	32,500	48,500				
83	0010	**SPORTS ARENAS**	S.F.	76.50	102	157				83
	0020	Total project costs	C.F.	4.14	7.40	9.55				
	2720	Plumbing	S.F.	4.42	6.70	14.15	4.35%	6.35%	9.40%	
	2770	Heating, ventilating, air conditioning		9.50	11.25	15.65	8.80%	10.20%	13.55%	
	2900	Electrical		7.95	10.80	13.95	8.60%	9.90%	12.25%	
	3100	Total: Mechanical & Electrical	↓	19.75	34.50	46.50	21.50%	25%	27.50%	
85	0010	**SUPERMARKETS**	S.F.	70.50	81.50	95.50				85
	0020	Total project costs	C.F.	3.92	4.74	7.20				
	2720	Plumbing	S.F.	3.93	4.96	5.75	5.40%	6%	7.45%	
	2770	Heating, ventilating, air conditioning		5.80	7.70	9.40	8.60%	8.65%	9.60%	
	2900	Electrical		8.80	10.15	12	10.40%	12.45%	13.60%	
	3100	Total: Mechanical & Electrical	↓	22.50	24.50	32	20.50%	26.50%	31%	
86	0010	**SWIMMING POOLS**	S.F.	114	191	405				86
	0020	Total project costs	C.F.	9.15	11.40	12.45				
	2720	Plumbing	S.F.	10.55	12.05	16.30	4.80%	9.70%	20.50%	
	2900	Electrical		8.55	13.90	20	5.75%	6.95%	7.60%	
	3100	Total: Mechanical & Electrical	↓	21	54	72.50	11.15%	14.10%	23.50%	
87	0010	**TELEPHONE EXCHANGES**	S.F.	151	222	281				87
	0020	Total project costs	C.F.	9.40	15.10	21				
	2720	Plumbing	S.F.	6.40	9.90	14.45	4.52%	5.80%	6.90%	
	2770	Heating, ventilating, air conditioning	↓	14.85	30	37	11.80%	16.05%	18.40%	

	50 17 00 | S.F. Costs		UNIT	UNIT COSTS 1/4	UNIT COSTS MEDIAN	UNIT COSTS 3/4	% OF TOTAL 1/4	% OF TOTAL MEDIAN	% OF TOTAL 3/4	
87	2900	Electrical	S.F.	15.40	24.50	43.50	10.90%	14%	17.85%	87
	3100	Total: Mechanical & Electrical	↓	45.50	86.50	122	29.50%	33.50%	44.50%	
91	0010	**THEATERS**	S.F.	95.50	118	180				91
	0020	Total project costs	C.F.	4.40	6.50	9.55				
	2720	Plumbing	S.F.	3.18	3.44	14.05	2.92%	4.70%	6.80%	
	2770	Heating, ventilating, air conditioning		9.25	11.20	13.85	8%	12.25%	13.40%	
	2900	Electrical		8.35	11.25	23	8.05%	9.35%	12.25%	
	3100	Total: Mechanical & Electrical	↓	21.50	29	34	21.50%	25.50%	27.50%	
94	0010	**TOWN HALLS City Halls & Municipal Buildings**	S.F.	106	135	176				94
	0020	Total project costs	C.F.	9.70	11.65	16.35				
	2720	Plumbing	S.F.	4.44	8.30	15.30	4.31%	5.95%	7.95%	
	2770	Heating, ventilating, air conditioning		8	15.95	23.50	7.05%	9.05%	13.45%	
	2900	Electrical		10.10	14.40	19.65	8.05%	9.45%	11.65%	
	3100	Total: Mechanical & Electrical	↓	35	44.50	68	22%	26.50%	31%	
97	0010	**WAREHOUSES & Storage Buildings**	S.F.	40	59	85				97
	0020	Total project costs	C.F.	2.08	3.25	5.40				
	0100	Site work	S.F.	4.10	8.15	12.30	6.05%	12.95%	19.85%	
	0500	Masonry		2.38	5.65	12.15	3.73%	7.40%	12.30%	
	1800	Equipment		.64	1.37	7.70	.91%	1.82%	5.55%	
	2720	Plumbing		1.32	2.37	4.44	2.90%	4.80%	6.55%	
	2730	Heating, ventilating, air conditioning		1.51	4.26	5.70	2.41%	5%	8.90%	
	2900	Electrical		2.35	4.42	7.30	5.15%	7.20%	10.10%	
	3100	Total: Mechanical & Electrical	↓	6.55	10.05	19.80	12.75%	18.90%	26%	
99	0010	**WAREHOUSE & OFFICES Combination**	S.F.	49	65	89				99
	0020	Total project costs	C.F.	2.50	3.62	5.35				
	1800	Equipment	S.F.	.85	1.64	2.43	.52%	1.20%	2.40%	
	2720	Plumbing		1.89	3.34	4.89	3.74%	4.76%	6.30%	
	2770	Heating, ventilating, air conditioning		2.98	4.65	6.50	5%	5.65%	10.05%	
	2900	Electrical		3.27	4.86	7.65	5.75%	8%	10%	
	3100	Total: Mechanical & Electrical	↓	9.10	14.05	22	14.40%	19.95%	24.50%	

Square Foot Project Size Modifier

One factor that affects the S.F. cost of a particular building is the size. In general, for buildings built to the same specifications in the same locality, the larger building will have the lower S.F. cost. This is due mainly to the decreasing contribution of the exterior walls plus the economy of scale usually achievable in larger buildings. The Area Conversion Scale shown below will give a factor to convert costs for the typical size building to an adjusted cost for the particular project.

The Square Foot Base Size lists the median costs, most typical project size in our accumulated data, and the range in size of the projects.

The Size Factor for your project is determined by dividing your project area in S.F. by the typical project size for the particular Building Type. With this factor, enter the Area Conversion Scale at the appropriate Size Factor and determine the appropriate cost multiplier for your building size.

Example: Determine the cost per S.F. for a 100,000 S.F. Mid-rise apartment building.

$$\frac{\text{Proposed building area} = 100{,}000 \text{ S.F.}}{\text{Typical size from below} = 50{,}000 \text{ S.F.}} = 2.00$$

Enter Area Conversion scale at 2.0, intersect curve, read horizontally the appropriate cost multiplier of .94. Size adjusted cost becomes .94 x $106.00 = $100.00 based on national average costs.

Note: For Size Factors less than .50, the Cost Multiplier is 1.1
For Size Factors greater than 3.5, the Cost Multiplier is .90

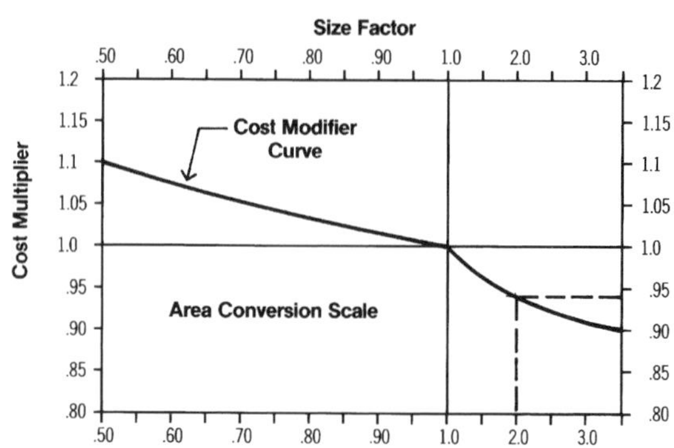

Building Type	Median Cost per S.F.	Typical Size Gross S.F.	Typical Range Gross S.F.	Building Type	Median Cost per S.F.	Typical Size Gross S.F.	Typical Range Gross S.F.
Apartments, Low Rise	$ 83.50	21,000	9,700 - 37,200	Jails	$ 254.00	40,000	5,500 - 145,000
Apartments, Mid Rise	106.00	50,000	32,000 - 100,000	Libraries	159.00	12,000	7,000 - 31,000
Apartments, High Rise	115.00	145,000	95,000 - 600,000	Living, Assisted	135.00	32,300	23,500 - 50,300
Auditoriums	140.00	25,000	7,600 - 39,000	Medical Clinics	144.00	7,200	4,200 - 15,700
Auto Sales	104.00	20,000	10,800 - 28,600	Medical Offices	136.00	6,000	4,000 - 15,000
Banks	187.00	4,200	2,500 - 7,500	Motels	100.00	40,000	15,800 - 120,000
Churches	128.00	17,000	2,000 - 42,000	Nursing Homes	140.00	23,000	15,000 - 37,000
Clubs, Country	131.00	6,500	4,500 - 15,000	Offices, Low Rise	119.00	20,000	5,000 - 80,000
Clubs, Social	124.00	10,000	6,000 - 13,500	Offices, Mid Rise	117.00	120,000	20,000 - 300,000
Clubs, YMCA	141.00	28,300	12,800 - 39,400	Offices, High Rise	150.00	260,000	120,000 - 800,000
Colleges (Class)	150.00	50,000	15,000 - 150,000	Police Stations	188.00	10,500	4,000 - 19,000
Colleges (Science Lab)	238.00	45,600	16,600 - 80,000	Post Offices	139.00	12,400	6,800 - 30,000
College (Student Union)	176.00	33,400	16,000 - 85,000	Power Plants	1000.00	7,500	1,000 - 20,000
Community Center	132.00	9,400	5,300 - 16,700	Religious Education	119.00	9,000	6,000 - 12,000
Court Houses	178.00	32,400	17,800 - 106,000	Research	195.00	19,000	6,300 - 45,000
Dept. Stores	77.50	90,000	44,000 - 122,000	Restaurants	170.00	4,400	2,800 - 6,000
Dormitories, Low Rise	141.00	25,000	10,000 - 95,000	Retail Stores	82.50	7,200	4,000 - 17,600
Dormitories, Mid Rise	174.00	85,000	20,000 - 200,000	Schools, Elementary	123.00	41,000	24,500 - 55,000
Factories	75.00	26,400	12,900 - 50,000	Schools, Jr. High	127.00	92,000	52,000 - 119,000
Fire Stations	138.00	5,800	4,000 - 8,700	Schools, Sr. High	132.00	101,000	50,500 - 175,000
Fraternity Houses	129.00	12,500	8,200 - 14,800	Schools, Vocational	127.00	37,000	20,500 - 82,000
Funeral Homes	144.00	10,000	4,000 - 20,000	Sports Arenas	102.00	15,000	5,000 - 40,000
Garages, Commercial	92.50	9,300	5,000 - 13,600	Supermarkets	81.50	44,000	12,000 - 60,000
Garages, Municipal	117.00	8,300	4,500 - 12,600	Swimming Pools	191.00	20,000	10,000 - 32,000
Garages, Parking	50.00	163,000	76,400 - 225,300	Telephone Exchange	222.00	4,500	1,200 - 10,600
Gymnasiums	127.00	19,200	11,600 - 41,000	Theaters	118.00	10,500	8,800 - 17,500
Hospitals	225.00	55,000	27,200 - 125,000	Town Halls	135.00	10,800	4,800 - 23,400
House (Elderly)	114.00	37,000	21,000 - 66,000	Warehouses	59.00	25,000	8,000 - 72,000
Housing (Public)	105.00	36,000	14,400 - 74,400	Warehouse & Office	65.00	25,000	8,000 - 72,000
Ice Rinks	151.00	29,000	27,200 - 33,600				

Abbreviations

A	Area Square Feet; Ampere
ABS	Acrylonitrile Butadiene Stryrene; Asbestos Bonded Steel
A.C.	Alternating Current; Air-Conditioning; Asbestos Cement; Plywood Grade A & C
A.C.I.	American Concrete Institute
AD	Plywood, Grade A & D
Addit.	Additional
Adj.	Adjustable
af	Audio-frequency
A.G.A.	American Gas Association
Agg.	Aggregate
A.H.	Ampere Hours
A hr.	Ampere-hour
A.H.U.	Air Handling Unit
A.I.A.	American Institute of Architects
AIC	Ampere Interrupting Capacity
Allow.	Allowance
alt.	Altitude
Alum.	Aluminum
a.m.	Ante Meridiem
Amp.	Ampere
Anod.	Anodized
Approx.	Approximate
Apt.	Apartment
Asb.	Asbestos
A.S.B.C.	American Standard Building Code
Asbe.	Asbestos Worker
ASCE.	American Society of Civil Engineers
A.S.H.R.A.E.	American Society of Heating, Refrig. & AC Engineers
A.S.M.E.	American Society of Mechanical Engineers
A.S.T.M.	American Society for Testing and Materials
Attchmt.	Attachment
Avg., Ave.	Average
A.W.G.	American Wire Gauge
AWWA	American Water Works Assoc.
Bbl.	Barrel
B&B	Grade B and Better; Balled & Burlapped
B.&S.	Bell and Spigot
B.&W.	Black and White
b.c.c.	Body-centered Cubic
B.C.Y.	Bank Cubic Yards
BE	Bevel End
B.F.	Board Feet
Bg. cem.	Bag of Cement
BHP	Boiler Horsepower; Brake Horsepower
B.I.	Black Iron
Bit., Bitum.	Bituminous
Bit., Conc.	Bituminous Concrete
Bk.	Backed
Bkrs.	Breakers
Bldg.	Building
Blk.	Block
Bm.	Beam
Boil.	Boilermaker
B.P.M.	Blows per Minute
BR	Bedroom
Brg.	Bearing
Brhe.	Bricklayer Helper
Bric.	Bricklayer
Brk.	Brick
Brng.	Bearing
Brs.	Brass
Brz.	Bronze
Bsn.	Basin
Btr.	Better
Btu	British Thermal Unit
BTUH	BTU per Hour
B.U.R.	Built-up Roofing
BX	Interlocked Armored Cable
°C	degree centegrade
c	Conductivity, Copper Sweat
C	Hundred; Centigrade
C/C	Center to Center, Cedar on Cedar
C-C	Center to Center
Cab.	Cabinet
Cair.	Air Tool Laborer
Calc	Calculated
Cap.	Capacity
Carp.	Carpenter
C.B.	Circuit Breaker
C.C.A.	Chromate Copper Arsenate
C.C.F.	Hundred Cubic Feet
cd	Candela
cd/sf	Candela per Square Foot
CD	Grade of Plywood Face & Back
CDX	Plywood, Grade C & D, exterior glue
Cefi.	Cement Finisher
Cem.	Cement
CF	Hundred Feet
C.F.	Cubic Feet
CFM	Cubic Feet per Minute
c.g.	Center of Gravity
CHW	Chilled Water; Commercial Hot Water
C.I.	Cast Iron
C.I.P.	Cast in Place
Circ.	Circuit
C.L.	Carload Lot
Clab.	Common Laborer
Clam	Common maintenance laborer
C.L.F.	Hundred Linear Feet
CLF	Current Limiting Fuse
CLP	Cross Linked Polyethylene
cm	Centimeter
CMP	Corr. Metal Pipe
C.M.U.	Concrete Masonry Unit
CN	Change Notice
Col.	Column
CO_2	Carbon Dioxide
Comb.	Combination
Compr.	Compressor
Conc.	Concrete
Cont.	Continuous; Continued, Container
Corr.	Corrugated
Cos	Cosine
Cot	Cotangent
Cov.	Cover
C/P	Cedar on Paneling
CPA	Control Point Adjustment
Cplg.	Coupling
C.P.M.	Critical Path Method
CPVC	Chlorinated Polyvinyl Chloride
C.Pr.	Hundred Pair
CRC	Cold Rolled Channel
Creos.	Creosote
Crpt.	Carpet & Linoleum Layer
CRT	Cathode-ray Tube
CS	Carbon Steel, Constant Shear Bar Joist
Csc	Cosecant
C.S.F.	Hundred Square Feet
CSI	Construction Specifications Institute
C.T.	Current Transformer
CTS	Copper Tube Size
Cu	Copper, Cubic
Cu. Ft.	Cubic Foot
cw	Continuous Wave
C.W.	Cool White; Cold Water
Cwt.	100 Pounds
C.W.X.	Cool White Deluxe
C.Y.	Cubic Yard (27 cubic feet)
C.Y./Hr.	Cubic Yard per Hour
Cyl.	Cylinder
d	Penny (nail size)
D	Deep; Depth; Discharge
Dis., Disch.	Discharge
Db.	Decibel
Dbl.	Double
DC	Direct Current
DDC	Direct Digital Control
Demob.	Demobilization
d.f.u.	Drainage Fixture Units
D.H.	Double Hung
DHW	Domestic Hot Water
DI	Ductile Iron
Diag.	Diagonal
Diam., Dia	Diameter
Distrib.	Distribution
Div.	Division
Dk.	Deck
D.L.	Dead Load; Diesel
DLH	Deep Long Span Bar Joist
Do.	Ditto
Dp.	Depth
D.P.S.T.	Double Pole, Single Throw
Dr.	Drive
Drink.	Drinking
D.S.	Double Strength
D.S.A.	Double Strength A Grade
D.S.B.	Double Strength B Grade
Dty.	Duty
DWV	Drain Waste Vent
DX	Deluxe White, Direct Expansion
dyn	Dyne
e	Eccentricity
E	Equipment Only; East
Ea.	Each
E.B.	Encased Burial
Econ.	Economy
E.C.Y	Embankment Cubic Yards
EDP	Electronic Data Processing
EIFS	Exterior Insulation Finish System
E.D.R.	Equiv. Direct Radiation
Eq.	Equation
EL	elevation
Elec.	Electrician; Electrical
Elev.	Elevator; Elevating
EMT	Electrical Metallic Conduit; Thin Wall Conduit
Eng.	Engine, Engineered
EPDM	Ethylene Propylene Diene Monomer
EPS	Expanded Polystyrene
Eqhv.	Equip. Oper., Heavy
Eqlt.	Equip. Oper., Light
Eqmd.	Equip. Oper., Medium
Eqmm.	Equip. Oper., Master Mechanic
Eqol.	Equip. Oper., Oilers
Equip.	Equipment
ERW	Electric Resistance Welded
E.S.	Energy Saver
Est.	Estimated
esu	Electrostatic Units
E.W.	Each Way
EWT	Entering Water Temperature
Excav.	Excavation
Exp.	Expansion, Exposure
Ext.	Exterior
Extru.	Extrusion
f.	Fiber stress
F	Fahrenheit; Female; Fill
Fab.	Fabricated

FBGS	Fiberglass
F.C.	Footcandles
f.c.c.	Face-centered Cubic
f'c.	Compressive Stress in Concrete; Extreme Compressive Stress
F.E.	Front End
FEP	Fluorinated Ethylene Propylene (Teflon)
F.G.	Flat Grain
F.H.A.	Federal Housing Administration
Fig.	Figure
Fin.	Finished
Fixt.	Fixture
Fl. Oz.	Fluid Ounces
Flr.	Floor
F.M.	Frequency Modulation; Factory Mutual
Fmg.	Framing
Fdn.	Foundation
Fori.	Foreman, Inside
Foro.	Foreman, Outside
Fount.	Fountain
fpm	Feet per Minute
FPT	Female Pipe Thread
Fr.	Frame
F.R.	Fire Rating
FRK	Foil Reinforced Kraft
FRP	Fiberglass Reinforced Plastic
FS	Forged Steel
FSC	Cast Body; Cast Switch Box
Ft.	Foot; Feet
Ftng.	Fitting
Ftg.	Footing
Ft lb.	Foot Pound
Furn.	Furniture
FVNR	Full Voltage Non-Reversing
FXM	Female by Male
Fy.	Minimum Yield Stress of Steel
g	Gram
G	Gauss
Ga.	Gauge
Gal., gal.	Gallon
gpm, GPM	Gallon per Minute
Galv.	Galvanized
Gen.	General
G.F.I.	Ground Fault Interrupter
Glaz.	Glazier
GPD	Gallons per Day
GPH	Gallons per Hour
GPM	Gallons per Minute
GR	Grade
Gran.	Granular
Grnd.	Ground
H	High Henry
H.C.	High Capacity
H.D.	Heavy Duty; High Density
H.D.O.	High Density Overlaid
H.D.P.E.	high density polyethelene
Hdr.	Header
Hdwe.	Hardware
Help.	Helper Average
HEPA	High Efficiency Particulate Air Filter
Hg	Mercury
HIC	High Interrupting Capacity
HM	Hollow Metal
HMWPE	high molecular weight polyethylene
H.O.	High Output
Horiz.	Horizontal
H.P.	Horsepower; High Pressure
H.P.F.	High Power Factor
Hr.	Hour
Hrs./Day	Hours per Day
HSC	High Short Circuit
Ht.	Height
Htg.	Heating
Htrs.	Heaters
HVAC	Heating, Ventilation & Air-Conditioning
Hvy.	Heavy
HW	Hot Water
Hyd.;Hydr.	Hydraulic
Hz.	Hertz (cycles)
I.	Moment of Inertia
IBC	International Building Code
I.C.	Interrupting Capacity
ID	Inside Diameter
I.D.	Inside Dimension; Identification
I.F.	Inside Frosted
I.M.C.	Intermediate Metal Conduit
In.	Inch
Incan.	Incandescent
Incl.	Included; Including
Int.	Interior
Inst.	Installation
Insul.	Insulation/Insulated
I.P.	Iron Pipe
I.P.S.	Iron Pipe Size
I.P.T.	Iron Pipe Threaded
I.W.	Indirect Waste
J	Joule
J.I.C.	Joint Industrial Council
K	Thousand; Thousand Pounds; Heavy Wall Copper Tubing, Kelvin
K.A.H.	Thousand Amp. Hours
KCMIL	Thousand Circular Mils
KD	Knock Down
K.D.A.T.	Kiln Dried After Treatment
kg	Kilogram
kG	Kilogauss
kgf	Kilogram Force
kHz	Kilohertz
Kip.	1000 Pounds
KJ	Kiljoule
K.L.	Effective Length Factor
K.L.F.	Kips per Linear Foot
Km	Kilometer
K.S.F.	Kips per Square Foot
K.S.I.	Kips per Square Inch
kV	Kilovolt
kVA	Kilovolt Ampere
K.V.A.R.	Kilovar (Reactance)
KW	Kilowatt
KWh	Kilowatt-hour
L	Labor Only; Length; Long; Medium Wall Copper Tubing
Lab.	Labor
lat	Latitude
Lath.	Lather
Lav.	Lavatory
lb.; #	Pound
L.B.	Load Bearing; L Conduit Body
L. & E.	Labor & Equipment
lb./hr.	Pounds per Hour
lb./L.F.	Pounds per Linear Foot
lbf/sq.in.	Pound-force per Square Inch
L.C.L.	Less than Carload Lot
L.C.Y.	Loose Cubic Yard
Ld.	Load
LE	Lead Equivalent
LED	Light Emitting Diode
L.F.	Linear Foot
L.F. Nose	Linear Foot of Stair Nosing
L.F. Rsr	Linear Foot of Stair Riser
Lg.	Long; Length; Large
L & H	Light and Heat
LH	Long Span Bar Joist
L.H.	Labor Hours
L.L.	Live Load
L.L.D.	Lamp Lumen Depreciation
lm	Lumen
lm/sf	Lumen per Square Foot
lm/W	Lumen per Watt
L.O.A.	Length Over All
log	Logarithm
L-O-L	Lateralolet
long.	longitude
L.P.	Liquefied Petroleum; Low Pressure
L.P.F.	Low Power Factor
LR	Long Radius
L.S.	Lump Sum
Lt.	Light
Lt. Ga.	Light Gauge
L.T.L.	Less than Truckload Lot
Lt. Wt.	Lightweight
L.V.	Low Voltage
M	Thousand; Material; Male; Light Wall Copper Tubing
M^2CA	Meters Squared Contact Area
m/hr.; M.H.	Man-hour
mA	Milliampere
Mach.	Machine
Mag. Str.	Magnetic Starter
Maint.	Maintenance
Marb.	Marble Setter
Mat; Mat'l.	Material
Max.	Maximum
MBF	Thousand Board Feet
MBH	Thousand BTU's per hr.
MC	Metal Clad Cable
M.C.F.	Thousand Cubic Feet
M.C.F.M.	Thousand Cubic Feet per Minute
M.C.M.	Thousand Circular Mils
M.C.P.	Motor Circuit Protector
MD	Medium Duty
M.D.O.	Medium Density Overlaid
Med.	Medium
MF	Thousand Feet
M.F.B.M.	Thousand Feet Board Measure
Mfg.	Manufacturing
Mfrs.	Manufacturers
mg	Milligram
MGD	Million Gallons per Day
MGPH	Thousand Gallons per Hour
MH, M.H.	Manhole; Metal Halide; Man-Hour
MHz	Megahertz
Mi.	Mile
MI	Malleable Iron; Mineral Insulated
mm	Millimeter
Mill.	Millwright
Min., min.	Minimum, minute
Misc.	Miscellaneous
ml	Milliliter, Mainline
M.L.F.	Thousand Linear Feet
Mo.	Month
Mobil.	Mobilization
Mog.	Mogul Base
MPH	Miles per Hour
MPT	Male Pipe Thread

Abbreviations

MRT	Mile Round Trip
ms	Millisecond
M.S.F.	Thousand Square Feet
Mstz.	Mosaic & Terrazzo Worker
M.S.Y.	Thousand Square Yards
Mtd., mtd.	Mounted
Mthe.	Mosaic & Terrazzo Helper
Mtng.	Mounting
Mult.	Multi; Multiply
M.V.A.	Million Volt Amperes
M.V.A.R.	Million Volt Amperes Reactance
MV	Megavolt
MW	Megawatt
MXM	Male by Male
MYD	Thousand Yards
N	Natural; North
nA	Nanoampere
NA	Not Available; Not Applicable
N.B.C.	National Building Code
NC	Normally Closed
N.E.M.A.	National Electrical Manufacturers Assoc.
NEHB	Bolted Circuit Breaker to 600V.
N.L.B.	Non-Load-Bearing
NM	Non-Metallic Cable
nm	Nanometer
No.	Number
NO	Normally Open
N.O.C.	Not Otherwise Classified
Nose.	Nosing
N.P.T.	National Pipe Thread
NQOD	Combination Plug-on/Bolt on Circuit Breaker to 240V.
N.R.C.	Noise Reduction Coefficient/ Nuclear Regulator Commission
N.R.S.	Non Rising Stem
ns	Nanosecond
nW	Nanowatt
OB	Opposing Blade
OC	On Center
OD	Outside Diameter
O.D.	Outside Dimension
ODS	Overhead Distribution System
O.G.	Ogee
O.H.	Overhead
O&P	Overhead and Profit
Oper.	Operator
Opng.	Opening
Orna.	Ornamental
OSB	Oriented Strand Board
O.S.&Y.	Outside Screw and Yoke
Ovhd.	Overhead
OWG	Oil, Water or Gas
Oz.	Ounce
P.	Pole; Applied Load; Projection
p.	Page
Pape.	Paperhanger
P.A.P.R.	Powered Air Purifying Respirator
PAR	Parabolic Reflector
Pc., Pcs.	Piece, Pieces
P.C.	Portland Cement; Power Connector
P.C.F.	Pounds per Cubic Foot
P.C.M.	Phase Contrast Microscopy
P.E.	Professional Engineer; Porcelain Enamel; Polyethylene; Plain End
Perf.	Perforated
PEX	Cross linked polyethylene
Ph.	Phase
P.I.	Pressure Injected
Pile.	Pile Driver
Pkg.	Package
Pl.	Plate
Plah.	Plasterer Helper
Plas.	Plasterer
Pluh.	Plumbers Helper
Plum.	Plumber
Ply.	Plywood
p.m.	Post Meridiem
Pntd.	Painted
Pord.	Painter, Ordinary
pp	Pages
PP, PPL	Polypropylene
P.P.M.	Parts per Million
Pr.	Pair
P.E.S.B.	Pre-engineered Steel Building
Prefab.	Prefabricated
Prefin.	Prefinished
Prop.	Propelled
PSF, psf	Pounds per Square Foot
PSI, psi	Pounds per Square Inch
PSIG	Pounds per Square Inch Gauge
PSP	Plastic Sewer Pipe
Pspr.	Painter, Spray
Psst.	Painter, Structural Steel
P.T.	Potential Transformer
P. & T.	Pressure & Temperature
Ptd.	Painted
Ptns.	Partitions
Pu	Ultimate Load
PVC	Polyvinyl Chloride
Pvmt.	Pavement
Pwr.	Power
Q	Quantity Heat Flow
Qt.	Quart
Quan., Qty.	Quantity
Q.C.	Quick Coupling
r	Radius of Gyration
R	Resistance
R.C.P.	Reinforced Concrete Pipe
Rect.	Rectangle
Reg.	Regular
Reinf.	Reinforced
Req'd.	Required
Res.	Resistant
Resi.	Residential
Rgh.	Rough
RGS	Rigid Galvanized Steel
R.H.W.	Rubber, Heat & Water Resistant; Residential Hot Water
rms	Root Mean Square
Rnd.	Round
Rodm.	Rodman
Rofc.	Roofer, Composition
Rofp.	Roofer, Precast
Rohe.	Roofer Helpers (Composition)
Rots.	Roofer, Tile & Slate
R.O.W.	Right of Way
RPM	Revolutions per Minute
R.S.	Rapid Start
Rsr	Riser
RT	Round Trip
S.	Suction; Single Entrance; South
SC	Screw Cover
SCFM	Standard Cubic Feet per Minute
Scaf.	Scaffold
Sch., Sched.	Schedule
S.C.R.	Modular Brick
S.D.	Sound Deadening
S.D.R.	Standard Dimension Ratio
S.E.	Surfaced Edge
Sel.	Select
S.E.R., S.E.U.	Service Entrance Cable
S.F.	Square Foot
S.F.C.A.	Square Foot Contact Area
S.F. Flr.	Square Foot of Floor
S.F.G.	Square Foot of Ground
S.F. Hor.	Square Foot Horizontal
S.F.R.	Square Feet of Radiation
S.F. Shlf.	Square Foot of Shelf
S4S	Surface 4 Sides
Shee.	Sheet Metal Worker
Sin.	Sine
Skwk.	Skilled Worker
SL	Saran Lined
S.L.	Slimline
Sldr.	Solder
SLH	Super Long Span Bar Joist
S.N.	Solid Neutral
S-O-L	Socketolet
sp	Standpipe
S.P.	Static Pressure; Single Pole; Self-Propelled
Spri.	Sprinkler Installer
spwg	Static Pressure Water Gauge
S.P.D.T.	Single Pole, Double Throw
SPF	Spruce Pine Fir
S.P.S.T.	Single Pole, Single Throw
SPT	Standard Pipe Thread
Sq.	Square; 100 Square Feet
Sq. Hd.	Square Head
Sq. In.	Square Inch
S.S.	Single Strength; Stainless Steel
S.S.B.	Single Strength B Grade
sst, ss	Stainless Steel
Sswk.	Structural Steel Worker
Sswl.	Structural Steel Welder
St.;Stl.	Steel
S.T.C.	Sound Transmission Coefficient
Std.	Standard
Stg.	Staging
STK	Select Tight Knot
STP	Standard Temperature & Pressure
Stpi.	Steamfitter, Pipefitter
Str.	Strength; Starter; Straight
Strd.	Stranded
Struct.	Structural
Sty.	Story
Subj.	Subject
Subs.	Subcontractors
Surf.	Surface
Sw.	Switch
Swbd.	Switchboard
S.Y.	Square Yard
Syn.	Synthetic
S.Y.P.	Southern Yellow Pine
Sys.	System
t.	Thickness
T	Temperature; Ton
Tan	Tangent
T.C.	Terra Cotta
T & C	Threaded and Coupled
T.D.	Temperature Difference
Tdd	Telecommunications Device for the Deaf
T.E.M.	Transmission Electron Microscopy
TFE	Tetrafluoroethylene (Teflon)
T. & G.	Tongue & Groove; Tar & Gravel
Th.,Thk.	Thick
Thn.	Thin
Thrded	Threaded
Tilf.	Tile Layer, Floor
Tilh.	Tile Layer, Helper
THHN	Nylon Jacketed Wire
THW.	Insulated Strand Wire

Abbreviations

THWN	Nylon Jacketed Wire
T.L.	Truckload
T.M.	Track Mounted
Tot.	Total
T-O-L	Threadolet
T.S.	Trigger Start
Tr.	Trade
Transf.	Transformer
Trhv.	Truck Driver, Heavy
Trlr	Trailer
Trlt.	Truck Driver, Light
TTY	Teletypewriter
TV	Television
T.W.	Thermoplastic Water Resistant Wire
UCI	Uniform Construction Index
UF	Underground Feeder
UGND	Underground Feeder
U.H.F.	Ultra High Frequency
U.I.	United Inch
U.L.	Underwriters Laboratory
Uld.	unloading
Unfin.	Unfinished
URD	Underground Residential Distribution
US	United States
USP	United States Primed
UTP	Unshielded Twisted Pair
V	Volt
V.A.	Volt Amperes
V.C.T.	Vinyl Composition Tile
VAV	Variable Air Volume
VC	Veneer Core
Vent.	Ventilation
Vert.	Vertical
V.F.	Vinyl Faced
V.G.	Vertical Grain
V.H.F.	Very High Frequency
VHO	Very High Output
Vib.	Vibrating
V.L.F.	Vertical Linear Foot
Vol.	Volume
VRP	Vinyl Reinforced Polyester
W	Wire; Watt; Wide; West
w/	With
W.C.	Water Column; Water Closet
W.F.	Wide Flange
W.G.	Water Gauge
Wldg.	Welding
W. Mile	Wire Mile
W-O-L	Weldolet
W.R.	Water Resistant
Wrck.	Wrecker
W.S.P.	Water, Steam, Petroleum
WT., Wt.	Weight
WWF	Welded Wire Fabric
XFER	Transfer
XFMR	Transformer
XHD	Extra Heavy Duty
XHHW, XLPE	Cross-Linked Polyethylene Wire Insulation
XLP	Cross-linked Polyethylene
Y	Wye
yd	Yard
yr	Year
Δ	Delta
%	Percent
~	Approximately
Ø	Phase; diameter
@	At
#	Pound; Number
<	Less Than
>	Greater Than
Z	zone

A

B

C

D

E

F

G

H

I

N

O

P

Q

R

S

U

V

W

X

Y

Z

Notes

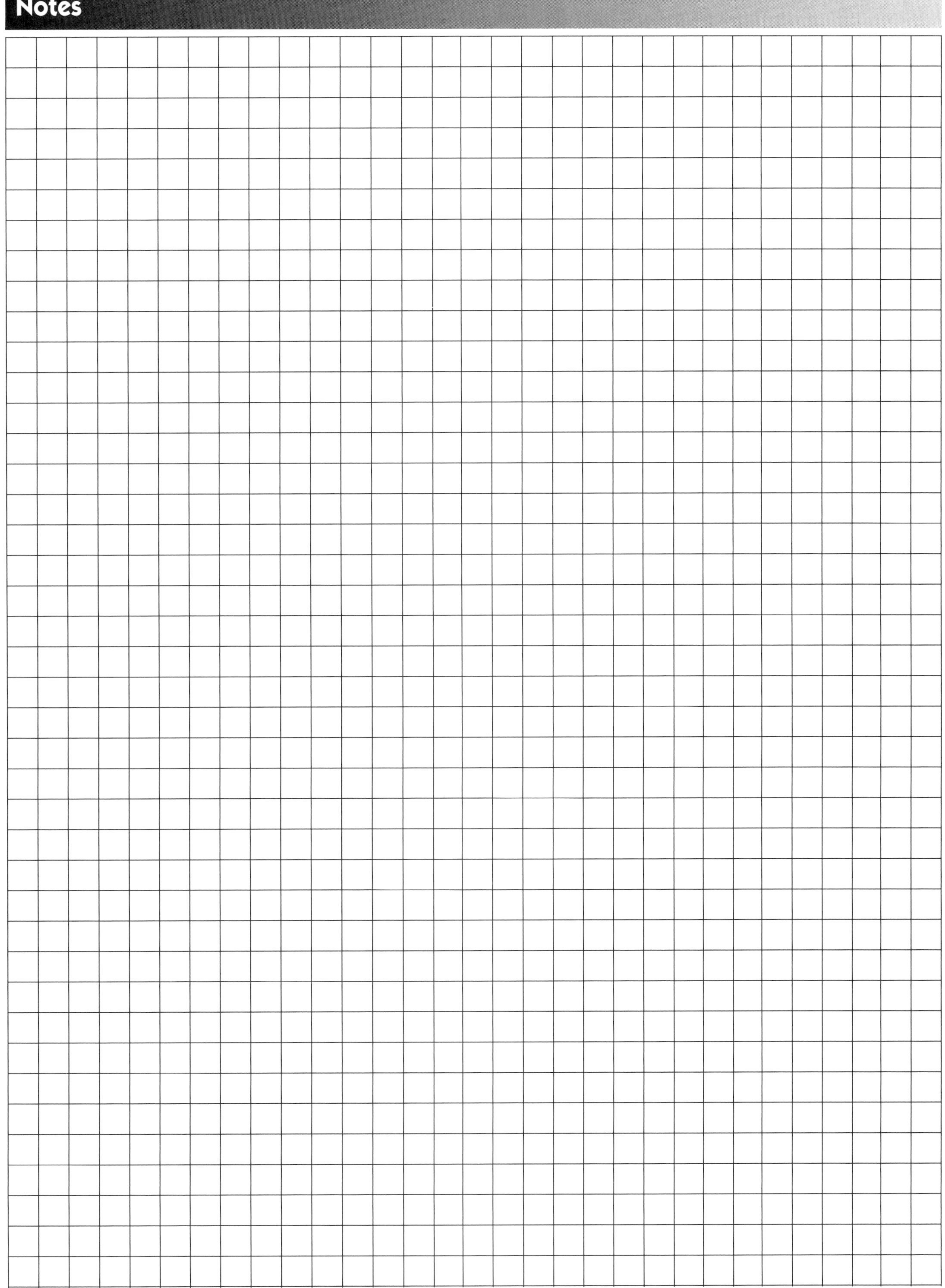

Notes

Notes

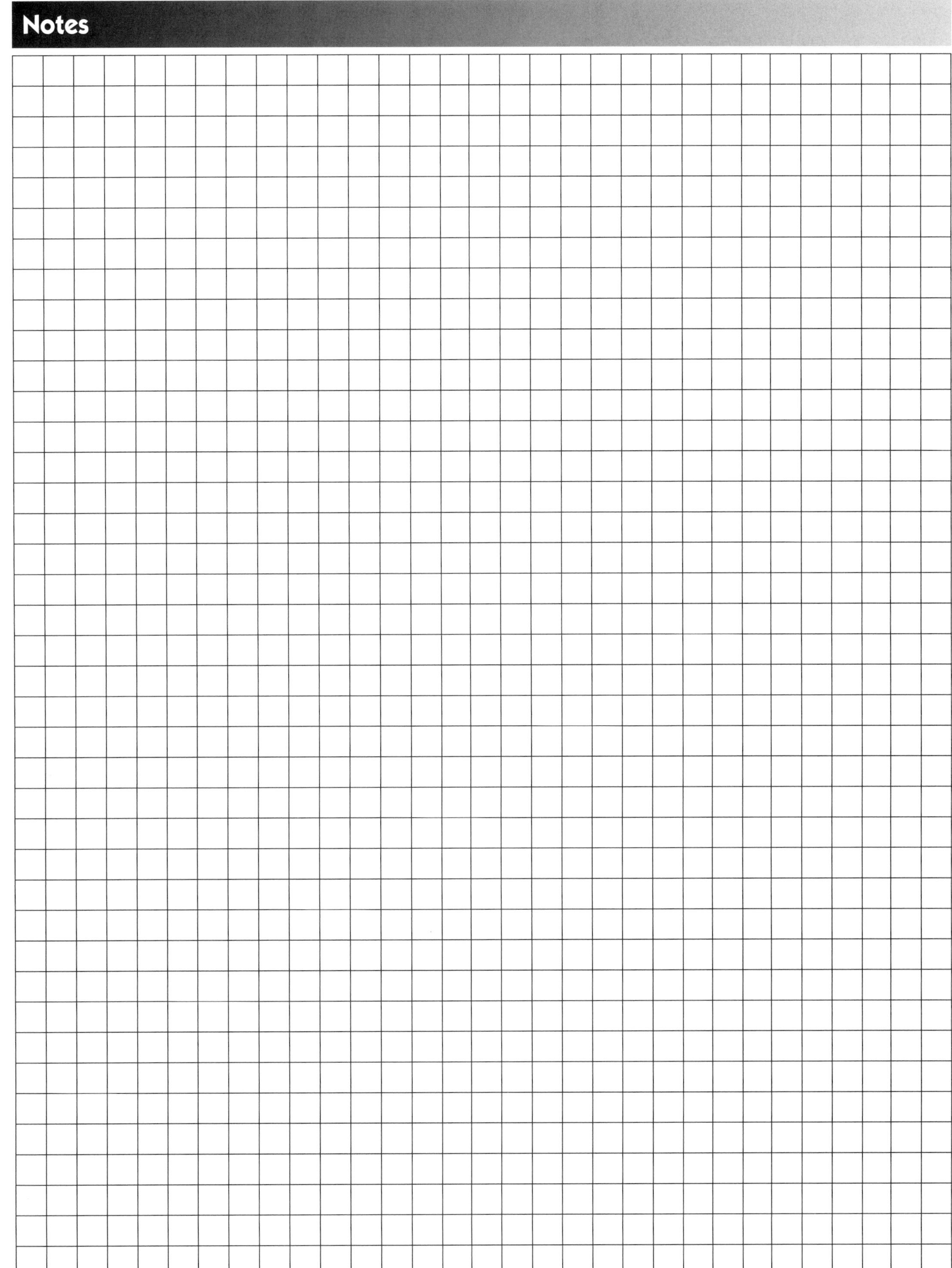

Reed Construction Data/RSMeans—a tradition of excellence in construction cost information and services since 1942

Table of Contents

For more information visit the RSMeans website at www.rsmeans.com

Book Selection Guide

The following table provides definitive information on the content of each cost data publication. The number of lines of data provided in each unit price or assemblies division, as well as the number of reference tables and crews, is listed for each book. The presence of other elements such as equipment rental costs, historical cost indexes, city cost indexes, square foot models, or cross-referenced indexes is also indicated. You can use the table to help select the RSMeans book that has the quantity and type of information you most need in your work.

Unit Cost Divisions	Building Construction Costs	Mechanical	Electrical	Repair & Remodel	Square Foot	Site Work Landsc.	Assemblies	Interior	Concrete Masonry	Open Shop	Heavy Construction	Light Commercial	Facilities Construction	Plumbing	Residential
1	553	372	385	474		489		301	437	552	483	220	967	386	157
2	737	264	85	676		964		358	215	734	706	434	1154	288	242
3	1506	213	201	850		1357		277	1837	1506	1497	347	1422	182	359
4	897	24	0	697		699		572	1105	873	592	489	1145	0	405
5	1851	201	156	958		795		1021	758	1838	1050	826	1841	276	722
6	1841	67	69	1429		449		1160	299	1820	384	1530	1453	22	2043
7	1438	209	125	1462		524		546	466	1436	0	1143	1500	218	905
8	2048	61	10	2118		294		1737	694	1974	0	1527	2320	0	1397
9	1783	72	26	1598		268		1875	367	1737	15	1511	2027	54	1331
10	986	17	10	597		224		800	157	986	29	481	1072	231	214
11	1020	214	163	487		126		865	30	1005	0	231	1034	173	110
12	547	0	2	303		248		1691	12	536	19	349	1736	23	317
13	733	119	111	247		344		255	69	732	256	80	743	74	79
14	280	36	0	229		28		259	0	279	0	12	299	21	6
21	78	0	16	30		0		250	0	78	0	60	368	373	0
22	1148	7031	155	1106		1466		622	20	1103	1648	823	6940	8763	634
23	1185	7032	609	774		158		679	38	1105	110	660	5089	1869	418
26	1246	493	9875	1003		739		1053	55	1234	523	1080	9811	438	559
27	67	0	210	34		9		59	0	67	35	48	212	0	4
28	86	58	107	61		0		68	0	71	0	40	122	44	25
31	1477	735	608	798		2793		7	1208	1421	2865	600	1538	660	606
32	811	114	0	675		3967		359	274	784	1481	373	1443	226	422
33	519	1025	465	201		1967		0	225	518	1890	117	1553	1196	115
34	109	0	20	4		144		0	31	64	122	0	129	0	0
35	18	0	0	0		166		0	0	18	281	0	83	0	0
41	52	0	0	24		7		22	0	52	30	0	59	14	0
44	98	90	0	0		32		0	0	23	32	0	98	105	0
Totals	23114	18447	13408	16835		18257		14836	8297	22546	14048	12981	46158	15636	11070

Assembly Divisions	Building Construction Costs	Mechanical	Elec-trical	Repair & Remodel	Square Foot	Site Work Landscape	Assemblies	Interior	Concrete Masonry	Open Shop	Heavy Construction	Light Commercial	Facilities Construction	Plumbing	Asm Div	Residential
A		15	0	182	150	530	598	0	536		570	154	24	0	1	374
B		0	0	809	2471	0	5577	329	1914		368	2031	143	0	2	217
C		0	0	635	865	0	1235	1567	146		0	765	247	0	3	588
D		1066	811	694	1793	0	2422	767	0		0	1271	1025	1078	4	861
E		0	0	85	260	0	297	5	0		0	257	5	0	5	392
F		0	0	0	123	0	124	0	0		0	123	3	0	6	358
G		527	294	311	202	2366	668	0	535		1194	200	293	685	7	300
															8	760
															9	80
															10	0
															11	0
															12	0
Totals		1608	1105	2716	5864	2896	10921	2668	3131		2132	4801	1740	1763		3932

Reference Section	Building Construction Costs	Mechanical	Electrical	Repair & Remodel	Square Foot	Site Work Landscape	Assemblies	Interior	Concrete Masonry	Open Shop	Heavy Construction	Light Commercial	Facilities Construction	Plumbing	Residential
Reference Tables	yes	yes	yes	yes	no	yes	yes	yes	yes	yes	yes	yes	yes	yes	yes
Models					107							46			28
Crews	487	487	487	467		487		487	487	465	487	465	467	487	465
Equipment Rental Costs	yes	yes	yes	yes		yes		yes	yes	yes	yes	yes	yes	yes	yes
Historical Cost Indexes	yes	yes	yes	yes	yes	yes	yes	yes	yes	yes	yes	yes	yes	yes	no
City Cost Indexes	yes	yes	yes	yes	yes	yes	yes	yes	yes	yes	yes	yes	yes	yes	yes

RSMeans Building Construction Cost Data 2010

Offers you unchallenged unit price reliability in an easy-to-use format. Whether used for verifying complete, finished estimates or for periodic checks, it supplies more cost facts better and faster than any comparable source. Over 21,000 unit prices have been updated for 2010. The City Cost Indexes and Location Factors cover over 930 areas, for indexing to any project location in North America. Order and get *RSMeans Quarterly Update Service* FREE. You'll have year-long access to the RSMeans Estimating **HOTLINE** FREE with your subscription. Expert assistance when using RSMeans data is just a phone call away.

$169.95 per copy
Available Sept. 2009
Catalog no. 60010

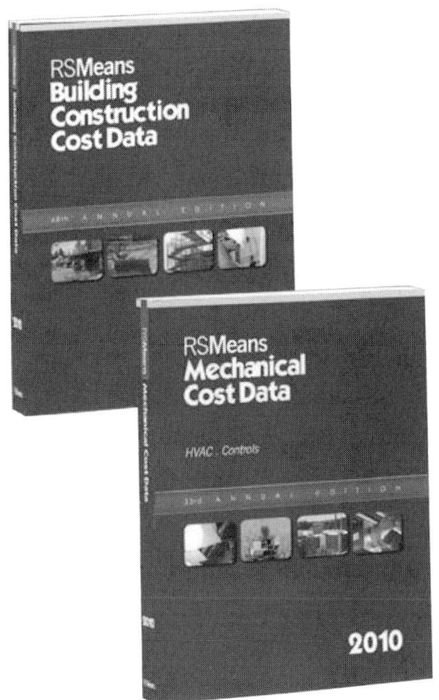

Unit prices organized according to the latest MasterFormat!

RSMeans Mechanical Cost Data 2010

- **HVAC**
- **Controls**

Total unit and systems price guidance for mechanical construction. . . materials, parts, fittings, and complete labor cost information. Includes prices for piping, heating, air conditioning, ventilation, and all related construction.

Plus new 2010 unit costs for:

- Thousands of installed HVAC/controls assemblies components
- "On-site" Location Factors for over 930 cities and towns in the U.S. and Canada
- Crews, labor, and equipment

$169.95 per copy
Available Oct. 2009
Catalog no. 60020

RSMeans Plumbing Cost Data 2010

Comprehensive unit prices and assemblies for plumbing, irrigation systems, commercial and residential fire protection, point-of-use water heaters, and the latest approved materials. This publication and its companion, *RSMeans Mechanical Cost Data*, provide full-range cost estimating coverage for all the mechanical trades.

Now contains updated costs for potable water and radiant heat systems and high efficiency fixtures.

$169.95 per copy
Available Oct. 2009
Catalog no. 60210

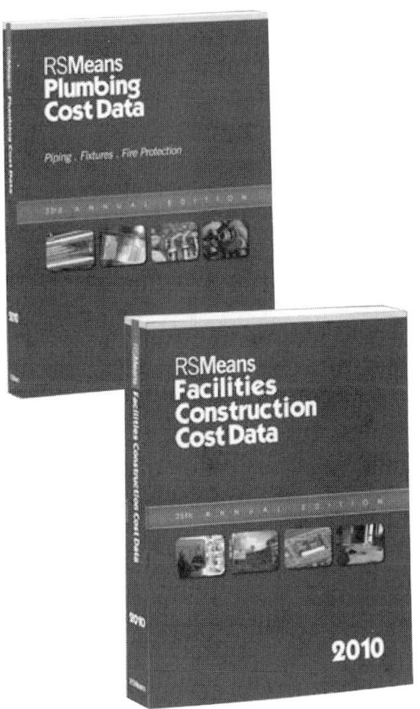

Unit prices organized according to the latest MasterFormat!

RSMeans Facilities Construction Cost Data 2010

For the maintenance and construction of commercial, industrial, municipal, and institutional properties. Costs are shown for new and remodeling construction and are broken down into materials, labor, equipment, and overhead and profit. Special emphasis is given to sections on mechanical, electrical, furnishings, site work, building maintenance, finish work, and demolition.

More than 43,000 unit costs, plus assemblies costs and a comprehensive Reference Section are included.

$419.95 per copy
Available Nov. 2009
Catalog no. 60200

Annual Cost Guides

RSMeans Square Foot Costs 2010

It's Accurate and Easy To Use!

- **Updated price information** based on nationwide figures from suppliers, estimators, labor experts, and contractors
- "How-to-Use" sections, with **clear examples** of commercial, residential, industrial, and institutional structures
- Realistic graphics, offering true-to-life illustrations of building projects
- Extensive information on using square foot cost data, including sample estimates and alternate pricing methods

$184.95 per copy
Available Oct. 2009
Catalog no. 60050

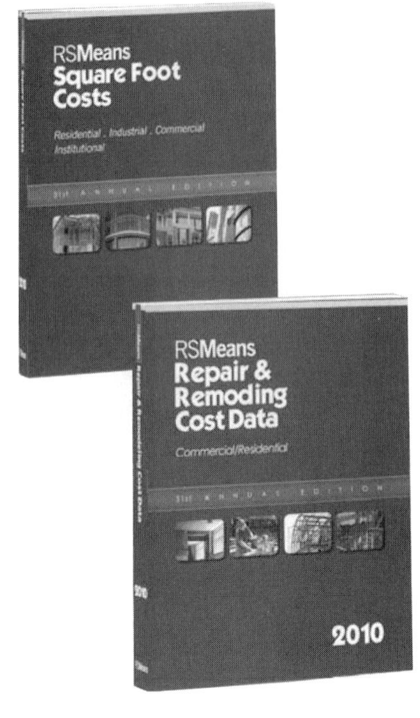

RSMeans Repair & Remodeling Cost Data 2010

Commercial/Residential

Use this valuable tool to estimate commercial and residential renovation and remodeling.

Includes: New costs for hundreds of unique methods, materials, and conditions that only come up in repair and remodeling, PLUS:

- Unit costs for over 15,000 construction components
- Installed costs for over 90 assemblies
- Over 930 "on-site" localization factors for the U.S. and Canada.

Unit prices organized according to the latest MasterFormat!

$144.95 per copy
Available Oct. 2009
Catalog no. 60040

RSMeans Electrical Cost Data 2010

Pricing information for every part of electrical cost planning. More than 13,000 unit and systems costs with design tables; clear specifications and drawings; engineering guides; illustrated estimating procedures; complete labor-hour and materials costs for better scheduling and procurement; and the latest electrical products and construction methods.

- A variety of special electrical systems including cathodic protection
- Costs for maintenance, demolition, HVAC/mechanical, specialties, equipment, and more

$169.95 per copy
Available Oct. 2009
Catalog no. 60030

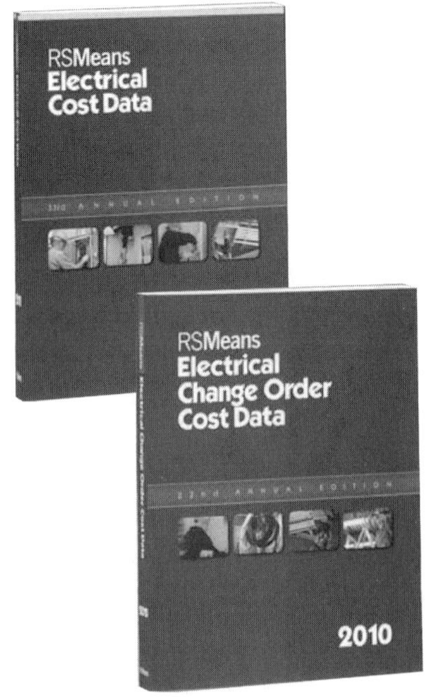

Unit prices organized according to the latest MasterFormat!

RSMeans Electrical Change Order Cost Data 2010

RSMeans Electrical Change Order Cost Data provides you with electrical unit prices exclusively for pricing change orders—based on the recent, direct experience of contractors and suppliers. Analyze and check your own change order estimates against the experience others have had doing the same work. It also covers productivity analysis and change order cost justifications. With useful information for calculating the effects of change orders and dealing with their administration.

$169.95 per copy
Available Dec. 2009
Catalog no. 60230

Annual Cost Guides

For more information visit the RSMeans website at www.rsmeans.com

RSMeans Assemblies Cost Data 2010

RSMeans Assemblies Cost Data takes the guesswork out of preliminary or conceptual estimates. Now you don't have to try to calculate the assembled cost by working up individual component costs. We've done all the work for you.

Presents detailed illustrations, descriptions, specifications, and costs for every conceivable building assembly—240 types in all—arranged in the easy-to-use UNIFORMAT II system. Each illustrated "assembled" cost includes a complete grouping of materials and associated installation costs, including the installing contractor's overhead and profit.

$279.95 per copy
Available Sept. 2009
Catalog no. 60060

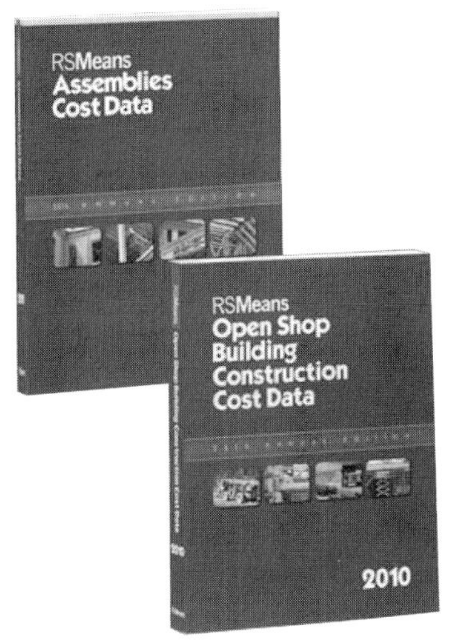

RSMeans Open Shop Building Construction Cost Data 2010

The latest costs for accurate budgeting and estimating of new commercial and residential construction. . . renovation work. . . change orders. . . cost engineering.

RSMeans Open Shop "BCCD" will assist you to:

- Develop benchmark prices for change orders
- Plug gaps in preliminary estimates and budgets
- Estimate complex projects
- Substantiate invoices on contracts
- Price ADA-related renovations

Unit prices organized according to the latest MasterFormat!

$169.95 per copy
Available Dec. 2009
Catalog no. 60150

RSMeans Residential Cost Data 2010

Contains square foot costs for 30 basic home models with the look of today, plus hundreds of custom additions and modifications you can quote right off the page. With costs for the 100 residential systems you're most likely to use in the year ahead. Complete with blank estimating forms, sample estimates, and step-by-step instructions.

Now contains line items for cultured stone and brick, PVC trim lumber, and TPO roofing.

$144.95 per copy
Available Oct. 2009
Catalog no. 60170

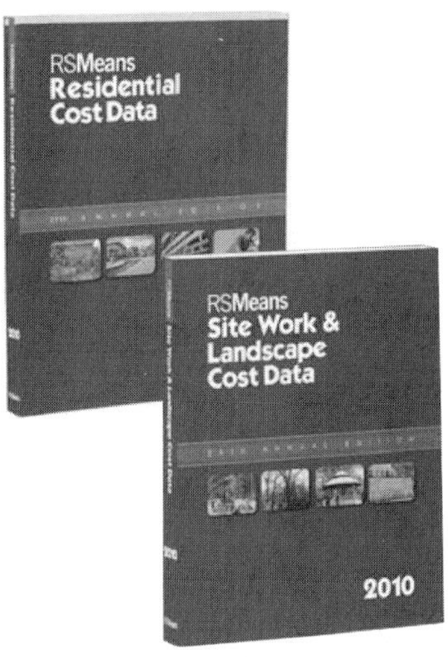

Unit prices organized according to the latest MasterFormat!

RSMeans Site Work & Landscape Cost Data 2010

Includes unit and assemblies costs for earthwork, sewerage, piped utilities, site improvements, drainage, paving, trees & shrubs, street openings/repairs, underground tanks, and more. Contains 57 tables of assemblies costs for accurate conceptual estimates.

Includes:

- Estimating for infrastructure improvements
- Environmentally-oriented construction
- ADA-mandated handicapped access
- Hazardous waste line items

$169.95 per copy
Available Nov. 2009
Catalog no. 60280

RSMeans Facilities Maintenance & Repair Cost Data 2010

RSMeans Facilities Maintenance & Repair Cost Data gives you a complete system to manage and plan your facility repair and maintenance costs and budget efficiently. Guidelines for auditing a facility and developing an annual maintenance plan. Budgeting is included, along with reference tables on cost and management, and information on frequency and productivity of maintenance operations.

The only nationally recognized source of maintenance and repair costs. Developed in cooperation with the Civil Engineering Research Laboratory (CERL) of the Army Corps of Engineers.

$379.95 per copy
Available Nov. 2009
Catalog no. 60300

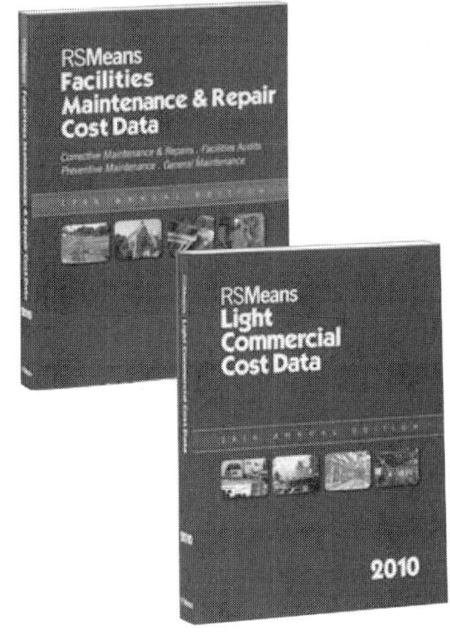

RSMeans Light Commercial Cost Data 2010

Specifically addresses the light commercial market, which is a specialized niche in the construction industry. Aids you, the owner/designer/contractor, in preparing all types of estimates—from budgets to detailed bids. Includes new advances in methods and materials.

Assemblies Section allows you to evaluate alternatives in the early stages of design/planning.

Over 11,000 unit costs ensure that you have the prices you need. . . when you need them.

Unit prices organized according to the latest MasterFormat !

$144.95 per copy
Available Nov. 2009
Catalog no. 60180

Annual Cost Guides

RSMeans Concrete & Masonry Cost Data 2010

Provides you with cost facts for virtually all concrete/masonry estimating needs, from complicated formwork to various sizes and face finishes of brick and block—all in great detail. The comprehensive Unit Price Section contains more than 7,500 selected entries. Also contains an Assemblies [Cost] Section, and a detailed Reference Section that supplements the cost data.

Unit prices organized according to the latest MasterFormat!

$154.95 per copy
Available Dec. 2009
Catalog no. 60110

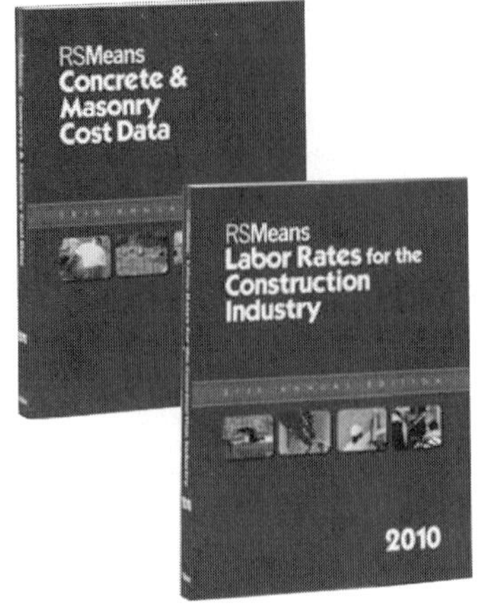

RSMeans Labor Rates for the Construction Industry 2010

Complete information for estimating labor costs, making comparisons, and negotiating wage rates by trade for over 300 U.S. and Canadian cities. With 46 construction trades listed by local union number in each city, and historical wage rates included for comparison. Each city chart lists the county and is alphabetically arranged with handy visual flip tabs for quick reference.

$379.95 per copy
Available Dec. 2009
Catalog no. 60120

RSMeans Construction Cost Indexes 2010

What materials and labor costs will change unexpectedly this year? By how much?

- Breakdowns for 316 major cities
- National averages for 30 key cities
- Expanded five major city indexes
- Historical construction cost indexes

$362.00 per year (subscription)
Catalog no. 50140

$90.50 individual quarters
Catalog no. 60140 A,B,C,D

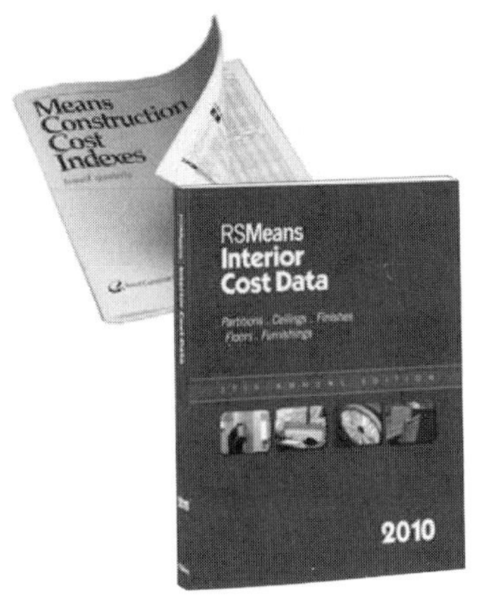

RSMeans Interior Cost Data 2010

Provides you with prices and guidance needed to make accurate interior work estimates. Contains costs on materials, equipment, hardware, custom installations, furnishings, and labor costs. . . for new and remodel commercial and industrial interior construction, including updated information on office furnishings, and reference information.

Unit prices organized according to the latest MasterFormat!

$169.95 per copy
Available Nov. 2009
Catalog no. 60090

RSMeans Heavy Construction Cost Data 2010

A comprehensive guide to heavy construction costs. Includes costs for highly specialized projects such as tunnels, dams, highways, airports, and waterways. Information on labor rates, equipment, and materials costs is included. Features unit price costs, systems costs, and numerous reference tables for costs and design.

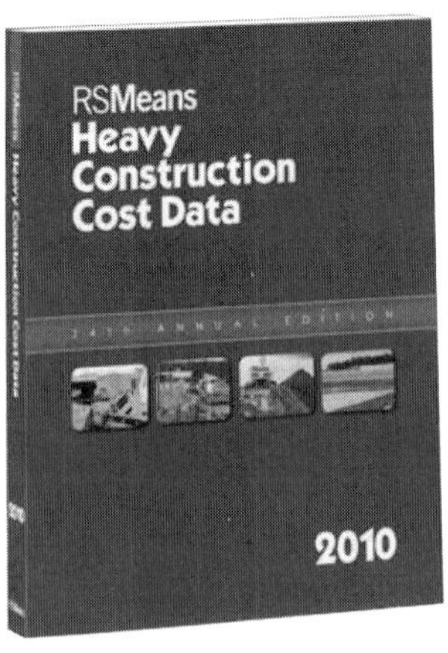

$169.95 per copy
Available Nov. 2009
Catalog no. 60160

Unit prices organized according to the latest MasterFormat!

Value Engineering: Practical Applications

For Design, Construction, Maintenance & Operations

by Alphonse Dell'Isola, PE

A tool for immediate application—for engineers, architects, facility managers, owners, and contractors. Includes making the case for VE—the management briefing; integrating VE into planning, budgeting, and design; conducting life cycle costing; using VE methodology in design review and consultant selection; case studies; VE workbook; and a life cycle costing program on disk.

$79.95 per copy
Over 450 pages, illustrated, softcover
Catalog no. 67319A

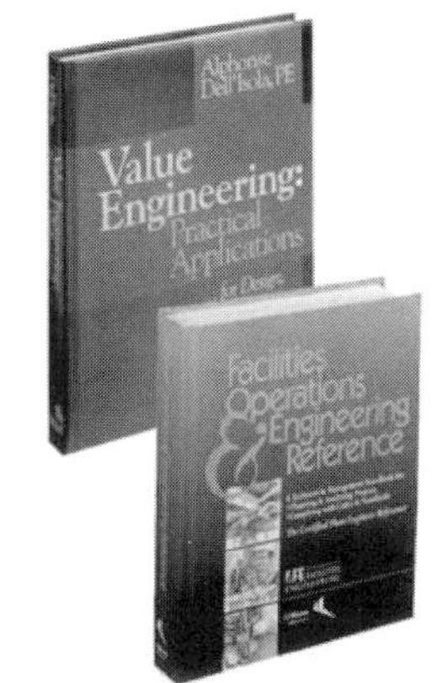

Facilities Operations & Engineering Reference

by the Association for Facilities Engineering and RSMeans

An all-in-one technical reference for planning and managing facility projects and solving day-to-day operations problems. Selected as the official certified plant engineer reference, this handbook covers financial analysis, maintenance, HVAC and energy efficiency, and more.

$54.98 per copy
Over 700 pages, illustrated, hardcover
Catalog no. 67318

The Building Professional's Guide to Contract Documents

3rd Edition

by Waller S. Poage, AIA, CSI, CVS

A comprehensive reference for owners, design professionals, contractors, and students

- Structure your documents for maximum efficiency.
- Effectively communicate construction requirements.
- Understand the roles and responsibilities of construction professionals.
- Improve methods of project delivery.

$64.95 per copy, 400 pages
Diagrams and construction forms, hardcover
Catalog no. 67261A

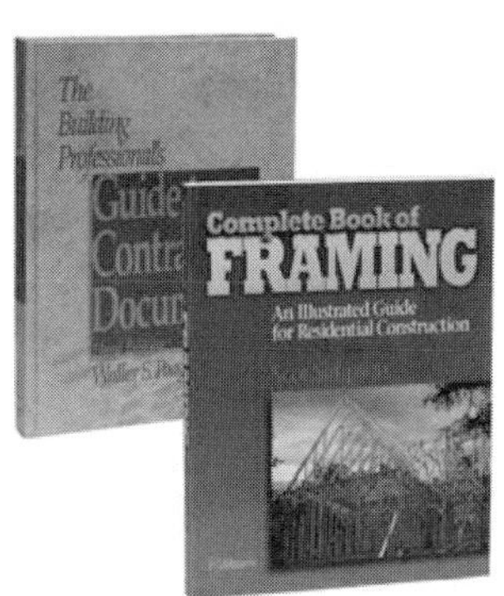

Complete Book of Framing

by Scot Simpson

This straightforward, easy-to-learn method will help framers, carpenters, and handy homeowners build their skills in rough carpentry and framing. Shows how to frame all the parts of a house: floors, walls, roofs, door & window openings, and stairs—with hundreds of color photographs and drawings that show every detail.

The book gives beginners all the basics they need to go from zero framing knowledge to a journeyman level.

$29.95 per copy
352 pages, softcover
Catalog no. 67353

Cost Planning & Estimating for Facilities Maintenance

In this unique book, a team of facilities management authorities shares their expertise on:

- Evaluating and budgeting maintenance operations
- Maintaining and repairing key building components
- Applying *RSMeans Facilities Maintenance & Repair Cost Data* to your estimating

Covers special maintenance requirements of the ten major building types.

$89.95 per copy
Over 475 pages, hardcover
Catalog no. 67314

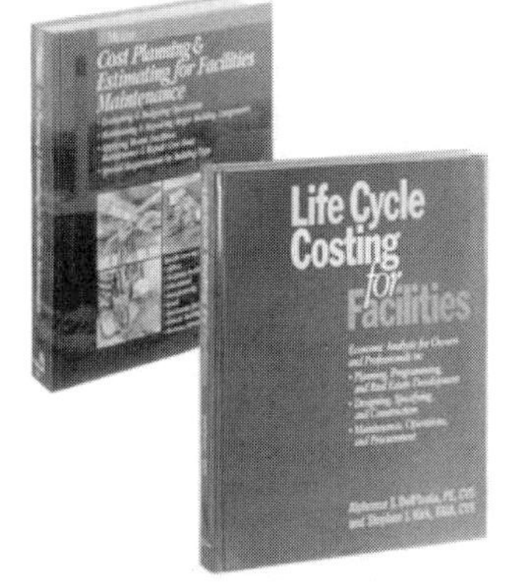

Life Cycle Costing for Facilities

by Alphonse Dell'Isola and Dr. Steven Kirk

Guidance for achieving higher quality design and construction projects at lower costs! Cost-cutting efforts often sacrifice quality to yield the cheapest product. Life cycle costing enables building designers and owners to achieve both. The authors of this book show how LCC can work for a variety of projects — from roads to HVAC upgrades to different types of buildings.

$99.95 per copy
450 pages, hardcover
Catalog no. 67341

Planning & Managing Interior Projects 2nd Edition

by Carol E. Farren, CFM

Expert guidance on managing renovation & relocation projects

This book guides you through every step in relocating to a new space or renovating an old one. From initial meeting through design and construction, to post-project administration, it helps you get the most for your company or client. Includes sample forms, spec lists, agreements, drawings, and much more!

$34.98 per copy
200 pages, softcover
Catalog no. 67245A

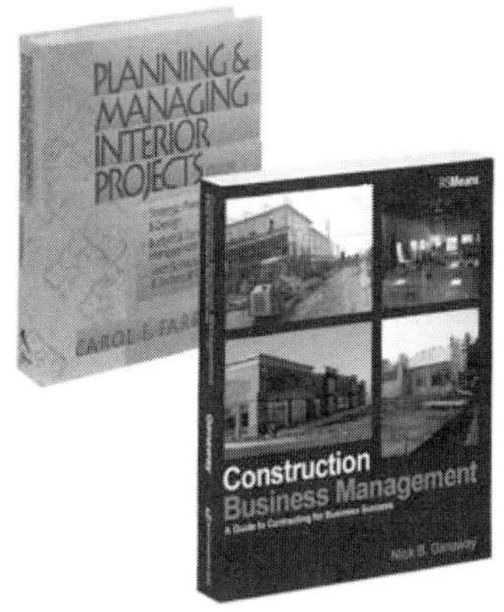

Construction Business Management

by Nick Ganaway

Only 43% of construction firms stay in business after four years. Make sure your company thrives with valuable guidance from a pro with 25 years of success as a commercial contractor. Find out what it takes to build all aspects of a business that is profitable, enjoyable, and enduring. With a bonus chapter on retail construction.

$49.95 per copy
200 pages, softcover
Catalog no. 67352

Reference Books

Interior Home Improvement Costs 9th Edition

Updated estimates for the most popular remodeling and repair projects—from small, do-it-yourself jobs to major renovations and new construction. Includes: kitchens & baths; new living space from your attic, basement, or garage; new floors, paint, and wallpaper; tearing out or building new walls; closets, stairs, and fireplaces; new energy-saving improvements, home theaters, and more!

$24.95 per copy
250 pages, illustrated, softcover
Catalog no. 67308E

Exterior Home Improvement Costs 9th Edition

Updated estimates for the most popular remodeling and repair projects—from small, do-it-yourself jobs, to major renovations and new construction. Includes: curb appeal projects—landscaping, patios, porches, driveways, and walkways; new windows and doors; decks, greenhouses, and sunrooms; room additions and garages; roofing, siding, and painting; "green" improvements to save energy & water.

$24.95 per copy
Over 275 pages, illustrated, softcover
Catalog no. 67309E

Builder's Essentials: Plan Reading & Material Takeoff

For Residential and Light Commercial Construction

by Wayne J. DelPico

A valuable tool for understanding plans and specs, and accurately calculating material quantities. Step-by-step instructions and takeoff procedures based on a full set of working drawings.

$35.95 per copy
Over 420 pages, softcover
Catalog no. 67307

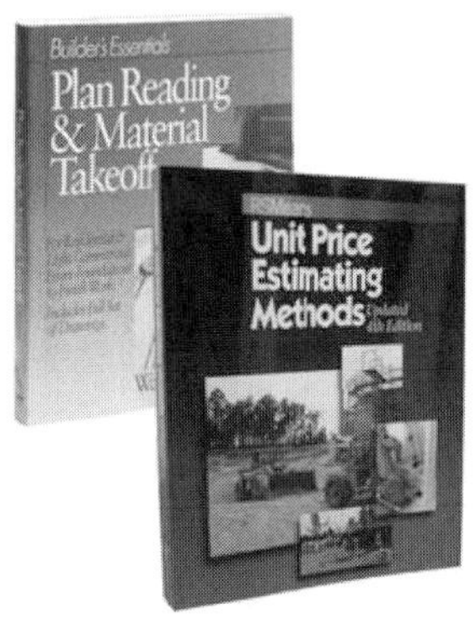

Means Unit Price Estimating Methods

New 4th Edition

This new edition includes up-to-date cost data and estimating examples, updated to reflect changes to the CSI numbering system and new features of RSMeans cost data. It describes the most productive, universally accepted ways to estimate, and uses checklists and forms to illustrate shortcuts and timesavers. A model estimate demonstrates procedures. A new chapter explores computer estimating alternatives.

$59.95 per copy
Over 350 pages, illustrated, softcover
Catalog no. 67303B

Total Productive Facilities Management

by Richard W. Sievert, Jr.

Today, facilities are viewed as strategic resources. . . elevating the facility manager to the role of asset manager supporting the organization's overall business goals. Now, Richard Sievert Jr., in this well-articulated guidebook, sets forth a new operational standard for the facility manager's emerging role. . . a comprehensive program for managing facilities as a true profit center.

$29.98 per copy
275 pages, softcover
Catalog no. 67321

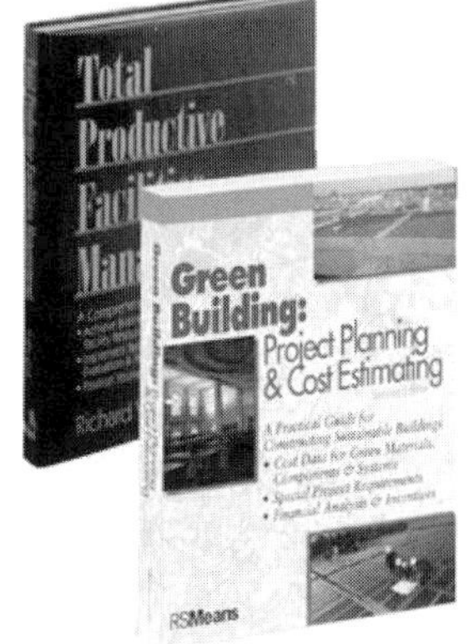

Green Building: Project Planning & Cost Estimating 2nd Edition

This new edition has been completely updated with the latest in green building technologies, design concepts, standards, and costs. Now includes a 2010 Green Building *CostWorks* CD with more than 300 green building assemblies and over 5,000 unit price line items for sustainable building. The new edition is also full-color with all new case studies—plus a new chapter on deconstruction, a key aspect of green building.

$129.95 per copy
350 pages, softcover
Catalog no. 67338A

Concrete Repair and Maintenance Illustrated

by Peter Emmons

Hundreds of illustrations show users how to analyze, repair, clean, and maintain concrete structures for optimal performance and cost effectiveness. From parking garages to roads and bridges to structural concrete, this comprehensive book describes the causes, effects, and remedies for concrete wear and failure. Invaluable for planning jobs, selecting materials, and training employees, this book is a must-have for concrete specialists, general contractors, facility managers, civil and structural engineers, and architects.

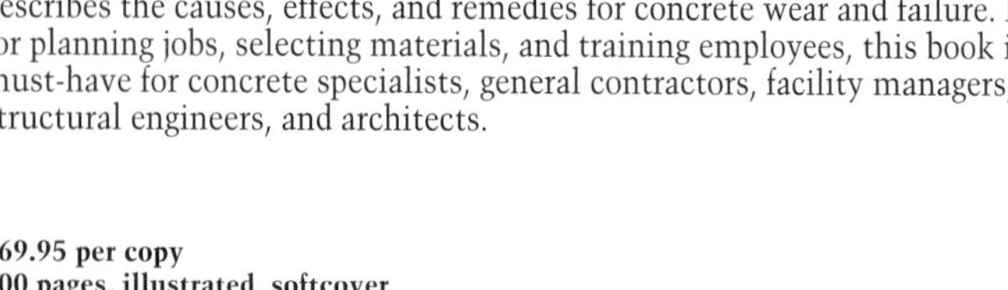

$69.95 per copy
300 pages, illustrated, softcover
Catalog no. 67146

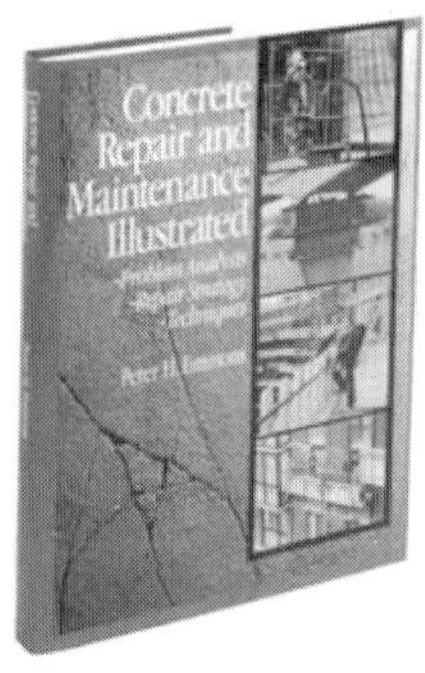

Reference Books

For more information visit the RSMeans website at www.rsmeans.com

Means Illustrated Construction Dictionary Condensed, 2nd Edition

The best portable dictionary for office or field use—an essential tool for contractors, architects, insurance and real estate personnel, facility managers, homeowners, and anyone who needs quick, clear definitions for construction terms. The second edition has been further enhanced with updates and hundreds of new terms and illustrations . . . in keeping with the most recent developments in the construction industry.

Now with a quick-reference Spanish section. Includes tools and equipment, materials, tasks, and more!

$59.95 per copy
Over 500 pages, softcover
Catalog no. 67282A

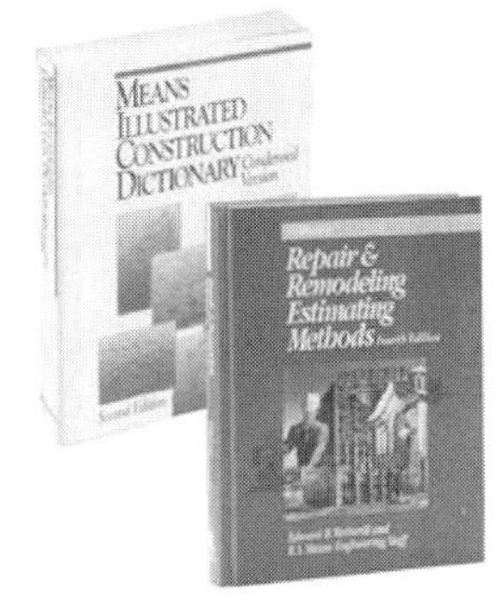

Means Repair & Remodeling Estimating 4th Edition

By Edward B. Wetherill and RSMeans

This important reference focuses on the unique problems of estimating renovations of existing structures, and helps you determine the true costs of remodeling through careful evaluation of architectural details and a site visit.

New section on disaster restoration costs.

$69.95 per copy
Over 450 pages, illustrated, hardcover
Catalog no. 67265B

Facilities Planning & Relocation

by David D. Owen

A complete system for planning space needs and managing relocations. Includes step-by-step manual, over 50 forms, and extensive reference section on materials and furnishings.

New lower price and user-friendly format.

$49.98 per copy
Over 450 pages, softcover
Catalog no. 67301

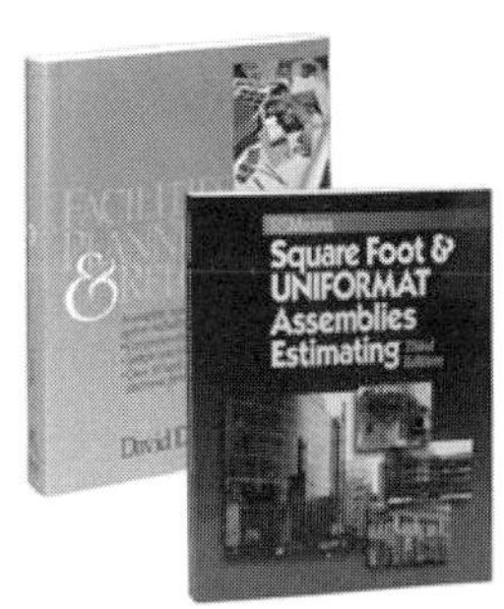

Means Square Foot & UNIFORMAT Assemblies Estimating Methods 3rd Edition

Develop realistic square foot and assemblies costs for budgeting and construction funding. The new edition features updated guidance on square foot and assemblies estimating using UNIFORMAT II. An essential reference for anyone who performs conceptual estimates.

$69.95 per copy
Over 300 pages, illustrated, softcover
Catalog no. 67145B

Means Electrical Estimating Methods 3rd Edition

Expanded edition includes sample estimates and cost information in keeping with the latest version of the CSI MasterFormat and UNIFORMAT II. Complete coverage of fiber optic and uninterruptible power supply electrical systems, broken down by components, and explained in detail. Includes a new chapter on computerized estimating methods. A practical companion to *RSMeans Electrical Cost Data.*

$64.95 per copy
Over 325 pages, hardcover
Catalog no. 67230B

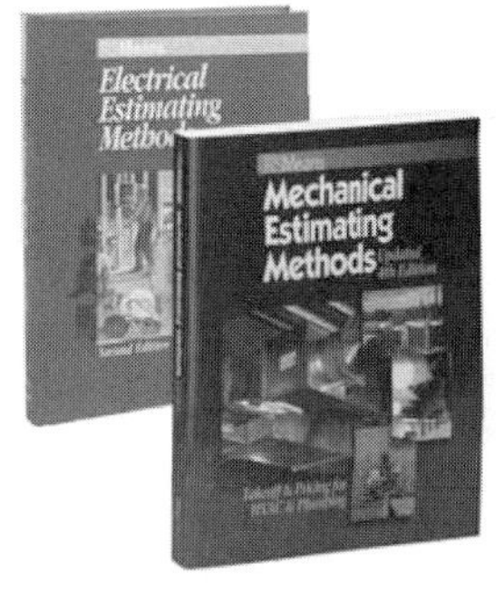

Means Mechanical Estimating Methods 4th Edition

Completely updated, this guide assists you in making a review of plans, specs, and bid packages, with suggestions for takeoff procedures, listings, substitutions, and pre-bid scheduling for all components of HVAC. Includes suggestions for budgeting labor and equipment usage. Compares materials and construction methods to allow you to select the best option.

$64.95 per copy
Over 350 pages, illustrated, softcover
Catalog no. 67294B

Means ADA Compliance Pricing Guide 2nd Edition

by Adaptive Environments and RSMeans

Completely updated and revised to the new 2004 *Americans with Disabilities Act Accessibility Guidelines,* this book features more than 70 of the most commonly needed modifications for ADA compliance. Projects range from installing ramps and walkways, widening doorways and entryways, and installing and refitting elevators, to relocating light switches and signage.

$79.95 per copy
Over 350 pages, illustrated, softcover
Catalog no. 67310A

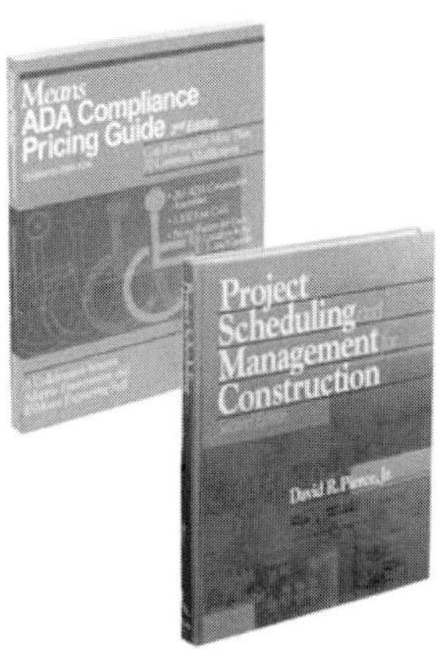

Project Scheduling & Management for Construction 3rd Edition

by David R. Pierce, Jr.

A comprehensive yet easy-to-follow guide to construction project scheduling and control—from vital project management principles through the latest scheduling, tracking, and controlling techniques. The author is a leading authority on scheduling, with years of field and teaching experience at leading academic institutions. Spend a few hours with this book and come away with a solid understanding of this essential management topic.

$64.95 per copy
Over 300 pages, illustrated, hardcover
Catalog no. 67247B

The Practice of Cost Segregation Analysis

by Bruce A. Desrosiers and Wayne J. DelPico

This expert guide walks you through the practice of cost segregation analysis, which enables property owners to defer taxes and benefit from "accelerated cost recovery" through depreciation deductions on assets that are properly identified and classified.

With a glossary of terms, sample cost segregation estimates for various building types, key information resources, and updates via a dedicated website, this book is a critical resource for anyone involved in cost segregation analysis.

$99.95 per copy
Over 225 pages
Catalog no. 67345

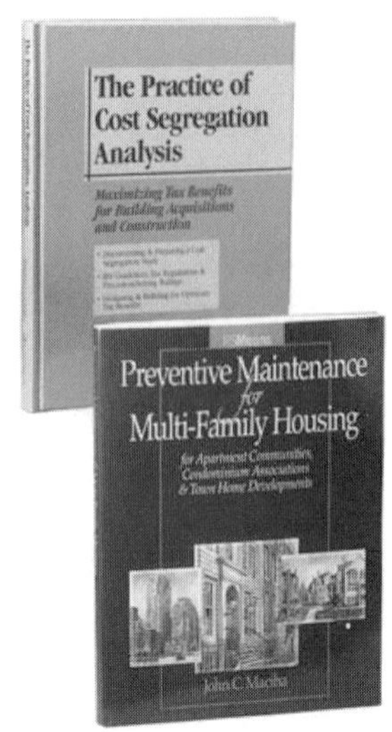

Preventive Maintenance for Multi-Family Housing

by John C. Maciha

Prepared by one of the nation's leading experts on multi-family housing.

This complete PM system for apartment and condominium communities features expert guidance, checklists for buildings and grounds maintenance tasks and their frequencies, a reusable wall chart to track maintenance, and a dedicated website featuring customizable electronic forms. A must-have for anyone involved with multi-family housing maintenance and upkeep.

$89.95 per copy
225 pages
Catalog no. 67346

How to Estimate with Means Data & CostWorks

New 3rd Edition

by RSMeans and Saleh A. Mubarak, Ph.D.

New 3rd Edition—fully updated with new chapters, plus new CD with updated *CostWorks* cost data and MasterFormat organization. Includes all major construction items—with more than 300 exercises and two sets of plans that show how to estimate for a broad range of construction items and systems—including general conditions and equipment costs.

$59.95 per copy
272 pages, softcover
Includes CostWorks CD
Catalog no. 67324B

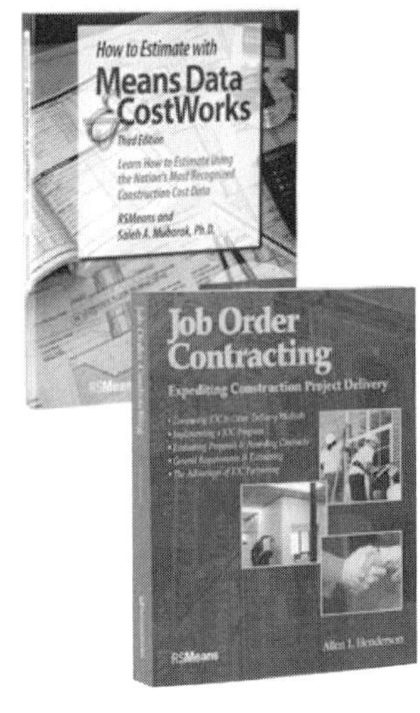

Job Order Contracting

Expediting Construction Project Delivery

by Allen Henderson

Expert guidance to help you implement JOC—fast becoming the preferred project delivery method for repair and renovation, minor new construction, and maintenance projects in the public sector and in many states and municipalities. The author, a leading JOC expert and practitioner, shows how to:

- Establish a JOC program
- Evaluate proposals and award contracts
- Handle general requirements and estimating
- Partner for maximum benefits

$89.95 per copy
192 pages, illustrated, hardcover
Catalog no. 67348

Builder's Essentials: Estimating Building Costs

For the Residential & Light Commercial Contractor

by Wayne J. DelPico

Step-by-step estimating methods for residential and light commercial contractors. Includes a detailed look at every construction specialty—explaining all the components, takeoff units, and labor needed for well-organized, complete estimates. Covers correctly interpreting plans and specifications, and developing accurate and complete labor and material costs.

$29.95 per copy

Over 400 pages, illustrated, softcover
Catalog no. 67343

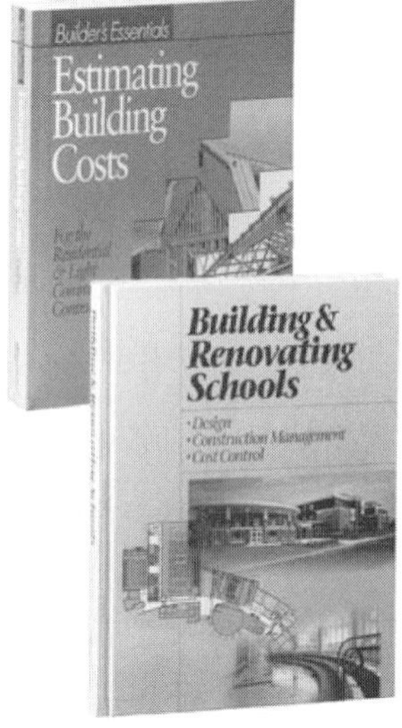

Building & Renovating Schools

This all-inclusive guide covers every step of the school construction process—from initial planning, needs assessment, and design, right through moving into the new facility. A must-have resource for anyone concerned with new school construction or renovation. With square foot cost models for elementary, middle, and high school facilities, and real-life case studies of recently completed school projects.

The contributors to this book—architects, construction project managers, contractors, and estimators who specialize in school construction—provide start-to-finish, expert guidance on the process.

$69.98 per copy
Over 425 pages, hardcover
Catalog no. 67342

Reference Books

For more information visit the RSMeans website at www.rsmeans.com

The Homeowner's Guide to Mold By Michael Pugliese

Expert guidance to protect your health and your home.

Mold, whether caused by leaks, humidity or flooding, is a real health and financial issue—for homeowners and contractors. This full-color book explains:

- Construction and maintenance practices to prevent mold
- How to inspect for and remove mold
- Mold remediation procedures and costs
- What to do after a flood
- How to deal with insurance companies

$21.95 per copy
144 pages, softcover
Catalog no. 67344

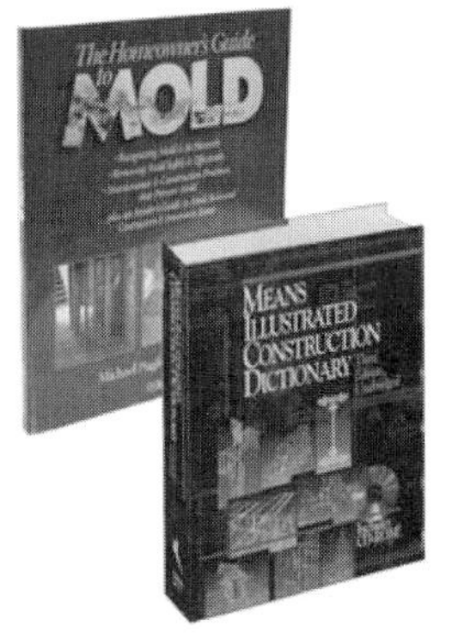

Means Illustrated Construction Dictionary

Unabridged 3rd Edition, with CD-ROM

Long regarded as the industry's finest, *Means Illustrated Construction Dictionary* is now even better. With the addition of over 1,000 new terms and hundreds of new illustrations, it is the clear choice for the most comprehensive and current information. The companion CD-ROM that comes with this new edition adds many extra features: larger graphics, expanded definitions, and links to both CSI MasterFormat numbers and product information.

$99.95 per copy
Over 790 pages, illustrated, hardcover
Catalog no. 67292A

Means Landscape Estimating Methods

New 5th Edition

Answers questions about preparing competitive landscape construction estimates, with up-to-date cost estimates and the new MasterFormat classification system. Expanded and revised to address the latest materials and methods, including new coverage on approaches to green building. Includes:

- Step-by-step explanation of the estimating process
- Sample forms and worksheets that save time and prevent errors

$64.95 per copy
Over 350 pages, softcover
Catalog no. 67295C

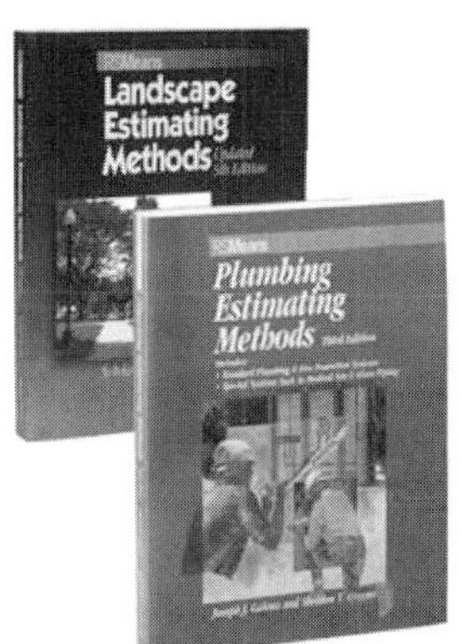

Means Plumbing Estimating Methods 3rd Edition

by Joseph Galeno and Sheldon Greene

Updated and revised! This practical guide walks you through a plumbing estimate, from basic materials and installation methods through change order analysis. *Plumbing Estimating Methods* covers residential, commercial, industrial, and medical systems, and features sample takeoff and estimate forms and detailed illustrations of systems and components.

$29.98 per copy
330+ pages, softcover
Catalog no. 67283B

Understanding & Negotiating Construction Contracts

by Kit Werremeyer

Take advantage of the author's 30 years' experience in small-to-large (including international) construction projects. Learn how to identify, understand, and evaluate high risk terms and conditions typically found in all construction contracts—then negotiate to lower or eliminate the risk, improve terms of payment, and reduce exposure to claims and disputes. The author avoids "legalese" and gives real-life examples from actual projects.

$69.95 per copy
300 pages, softcover
Catalog no. 67350

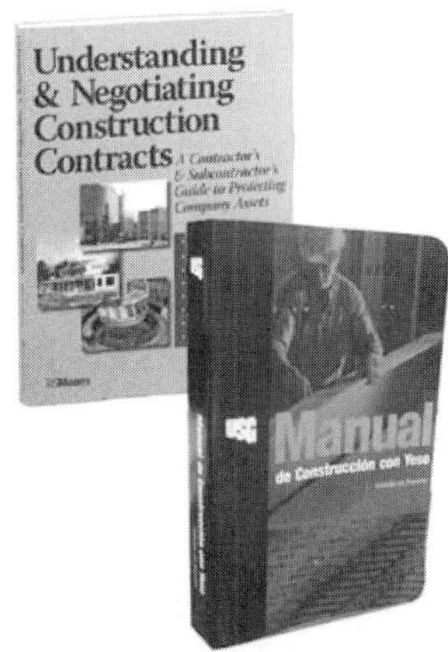

The Gypsum Construction Handbook (Spanish)

The Gypsum Construction Handbook, Spanish edition, is an illustrated, comprehensive, and authoritative guide on all facets of gypsum construction. You'll find comprehensive guidance on available products, installation methods, fire- and sound-rated construction information, illustrated framing-to-finish application instructions, estimating and planning information, and more.

$29.95 per copy
Over 450 pages, softcover
Catalog No. 67357S

Means Spanish/English Construction Dictionary 2nd Edition

by RSMeans and the International Code Council

This expanded edition features thousands of the most common words and useful phrases in the construction industry with easy-to-follow pronunciations in both Spanish and English. Over 800 new terms, phrases, and illustrations have been added. It also features a new stand-alone "Safety & Emergencies" section, with colored pages for quick access.

Unique to this dictionary are the systems illustrations showing the relationship of components in the most common building systems for all major trades.

$23.95 per copy
Over 400 pages
Catalog no. 67327A

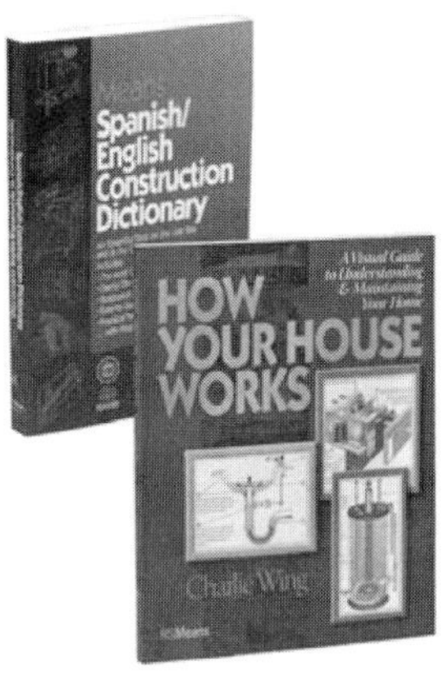

How Your House Works

by Charlie Wing

A must-have reference for every homeowner, handyman, and contractor—for repair, remodeling, and new construction. This book uncovers the mysteries behind just about every major appliance and building element in your house—from electrical, heating and AC, to plumbing, framing, foundations and appliances. Clear, "exploded" drawings show exactly how things should be put together and how they function—what to check if they don't work, and what you can do that might save you having to call in a professional.

$21.95 per copy
160 pages, softcover
Catalog no. 67351

Professional Development

Means CostWorks® CD Training

This one-day course helps users become more familiar with the functionality of *Means CostWorks* program. Each menu, icon, screen, and function found in the program is explained in depth. Time is devoted to hands-on estimating exercises.

Some of what you'll learn:
- Searching the database utilizing all navigation methods
- Exporting RSMeans Data to your preferred spreadsheet format
- Viewing crews, assembly components, and much more!
- Automatically regionalizing the database

You are required to bring your own laptop computer for this course.

When you register for this course you will receive an outline for your laptop requirements.

Also offering web training for CostWorks CD!

Unit Price Estimating

This interactive *two-day* seminar teaches attendees how to interpret project information and process it into final, detailed estimates with the greatest accuracy level.

The most important credential an estimator can take to the job is the ability to visualize construction, and estimate accurately.

Some of what you'll learn:
- Interpreting the design in terms of cost
- The most detailed, time-tested methodology for accurate pricing
- Key cost drivers—material, labor, equipment, staging, and subcontracts
- Understanding direct and indirect costs for accurate job cost accounting and change order management

Who should attend: Corporate and government estimators and purchasers, architects, engineers... and others needing to produce accurate project estimates.

Square Foot & Assemblies Estimating

This *two-day* course teaches attendees how to quickly deliver accurate square foot estimates using limited budget and design information.

Some of what you'll learn:
- How square foot costing gets the estimate done faster
- Taking advantage of a "systems" or "assemblies" format
- The RSMeans "building assemblies/square foot cost approach"
- How to create a reliable preliminary systems estimate using bare-bones design information

Who should attend: Facilities managers, facilities engineers, estimators, planners, developers, construction finance professionals... and others needing to make quick, accurate construction cost estimates at commercial, government, educational, and medical facilities.

Repair & Remodeling Estimating

This *two-day* seminar emphasizes all the underlying considerations unique to repair/remodeling estimating and presents the correct methods for generating accurate, reliable R&R project costs using the unit price and assemblies methods.

Some of what you'll learn:
- Estimating considerations—like labor-hours, building code compliance, working within existing structures, purchasing materials in smaller quantities, unforeseen deficiencies
- Identifying problems and providing solutions to estimating building alterations
- Rules for factoring in minimum labor costs, accurate productivity estimates, and allowances for project contingencies
- R&R estimating examples calculated using unit price and assemblies data

Who should attend: Facilities managers, plant engineers, architects, contractors, estimators, builders... and others who are concerned with the proper preparation and/or evaluation of repair and remodeling estimates.

Mechanical & Electrical Estimating

This *two-day* course teaches attendees how to prepare more accurate and complete mechanical/electrical estimates, avoiding the pitfalls of omission and double-counting, while understanding the composition and rationale within the RSMeans Mechanical/Electrical database.

Some of what you'll learn:
- The unique way mechanical and electrical systems are interrelated
- M&E estimates–conceptual, planning, budgeting, and bidding stages
- Order of magnitude, square foot, assemblies, and unit price estimating
- Comparative cost analysis of equipment and design alternatives

Who should attend: Architects, engineers, facilities managers, mechanical and electrical contractors... and others needing a highly reliable method for developing, understanding, and evaluating mechanical and electrical contracts.

Green Building: Planning & Construction

In this *two-day* course, learn about tailoring a building and its placement on the site to the local climate, site conditions, culture, and community. Includes information to reduce resource consumption and augment resource supply.

Some of what you'll learn:
- Green technologies, materials, systems, and standards
- Modeling tools and how to use them
- Cost vs. value of green products over their life cycle
- Low-cost green strategies, and economic incentives and funding
- Health, comfort, and productivity

Who should attend: Contractors, project managers, building owners, building officials, healthcare and insurance professionals.

Facilities Maintenance & Repair Estimating

This *two-day* course teaches attendees how to plan, budget, and estimate the cost of ongoing and preventive maintenance and repair for existing buildings and grounds.

Some of what you'll learn:
- The most financially favorable maintenance, repair, and replacement scheduling and estimating
- Auditing and value engineering facilities
- Preventive planning and facilities upgrading
- Determining both in-house and contract-out service costs
- Annual, asset-protecting M&R plan

Who should attend: Facility managers, maintenance supervisors, buildings and grounds superintendents, plant managers, planners, estimators... and others involved in facilities planning and budgeting.

Practical Project Management for Construction Professionals

In this *two-day* course, acquire the essential knowledge and develop the skills to effectively and efficiently execute the day-to-day responsibilities of the construction project manager.

Covers:
- General conditions of the construction contract
- Contract modifications: change orders and construction change directives
- Negotiations with subcontractors and vendors
- Effective writing: notification and communications
- Dispute resolution: claims and liens

Who should attend: Architects, engineers, owner's representatives, project managers.

Facilities Construction Estimating

In this *two-day* course, professionals working in facilities management can get help with their daily challenges to establish budgets for all phase of a project

Some of what you'll learn:
- Determine the full scope of a project
- Identify the scope of risks & opportunities
- Creative solutions to estimating issues
- Organizing estimates for presentation & discussion
- Special techniques for repair/remodel and maintenance projects
- Negotiating project change orders

Who should attend: Facility managers, engineers, contractors, facility trades-people, planners & project managers

Professional Development

For more information visit the RSMeans website at www.rsmeans.com

Scheduling and Project Management

This *two-day* course teaches attendees the most current and proven scheduling and management techniques needed to bring projects in on time and on budget.

Some of what you'll learn:
- Crucial phases of planning and scheduling
- How to establish project priorities and develop realistic schedules and management techniques
- Critical Path and Precedence Methods
- Special emphasis on cost control

Who should attend: Construction project managers, supervisors, engineers, estimators, contractors... and others who want to improve their project planning, scheduling, and management skills.

Assessing Scope of Work for Facility Construction Estimating

This *two-day* practical training program addresses the vital importance of understanding the SCOPE of projects in order to produce accurate cost estimates for facility repair and remodeling.

Some of what you'll learn:
- Discussions of site visits, plans/specs, record drawings of facilities, and site-specific lists
- Review of CSI divisions, including means, methods, materials, and the challenges of scoping each topic
- Exercises in SCOPE identification and SCOPE writing for accurate estimating of projects
- Hands-on exercises that require SCOPE, take-off, and pricing

Who should attend: Corporate and government estimators, planners, facility managers, and others who need to produce accurate project estimates.

Site Work & Heavy Construction Estimating

This *two-day* course teaches attendees how to estimate earthwork, site utilities, foundations, and site improvements, using the assemblies and the unit price methods.

Some of what you'll learn:
- Basic site work and heavy construction estimating skills
- Estimating foundations, utilities, earthwork, and site improvements
- Correct equipment usage, quality control, and site investigation for estimating purposes

Who should attend: Project managers, design engineers, estimators, and contractors doing site work and heavy construction.

2010 RSMeans Seminar Schedule

Note: call for exact dates and details.

Location	Dates
Las Vegas, NV	March
Washington, DC	April
Denver, CO	May
San Francisco, CA	June
Washington, DC	September
Dallas, TX	September
Las Vegas, NV	October
Atlantic City, NJ	October
Orlando, FL	November
San Diego, CA	December

2010 Regional Seminar Schedule

Location	Dates
Seattle, WA	January and August
Dallas/Ft. Worth, TX	January
Austin, TX	February
Anchorage, AK	March and September
Oklahoma City, OK	March
Phoenix, AZ	March
Aiea, HI	May
Kansas City, MO	May
Bethesda, MD	June
Columbus, GA	June
El Segundo, CA	August

Location	Dates
St. Louis, MO	August
Houston, TX	October
Colorado Springs, CO	October
Baltimore, MD	November
Sacramento, CA	November
San Antonio, TX	December
Raleigh, NC	December

1-800-334-3509, ext. 5115

Registration Information

Register early... Save up to $100! Register 30 days before the start date of a seminar and save $100 off your total fee. *Note: This discount can be applied only once per order. It cannot be applied to team discount registrations or any other special offer.*

How to register Register by phone today! The RSMeans toll-free number for making reservations is **1-800-334-3509, ext. 5115.**

Individual seminar registration fee - $935. *Means CostWorks®* training registration fee - $375. To register by mail, complete the registration form and return, with your full fee, to: RSMeans Seminars, 63 Smiths Lane, Kingston, MA 02364.

Government pricing All federal government employees save off the regular seminar price. Other promotional discounts cannot be combined with the government discount.

Team discount program for two to four seminar registrations. Call for pricing: 1-800-334-3509, ext. 5115

Multiple course discounts When signing up for two or more courses, call for pricing.

Refund policy Cancellations will be accepted up to ten business days prior to the seminar start. There are no refunds for cancellations received later than ten working days prior to the first day of the seminar. A $150 processing fee will be applied for all cancellations. Written notice of cancellation is required. Substitutions can be made at any time before the session starts. **No-shows are subject to the full seminar fee.**

AACE approved courses Many seminars described and offered here have been approved for 14 hours (1.4 recertification credits) of credit by the AACE International Certification Board toward meeting the continuing education requirements for recertification as a Certified Cost Engineer/Certified Cost Consultant.

AIA Continuing Education We are registered with the AIA Continuing Education System (AIA/CES) and are committed to developing quality learning activities in accordance with the CES criteria. Many seminars meet the AIA/CES criteria for Quality Level 2. AIA members may receive (14) learning units (LUs) for each two-day RSMeans course.

Daily course schedule The first day of each seminar session begins at 8:30 a.m. and ends at 4:30 p.m. The second day begins at 8:00 a.m. and ends at 4:00 p.m. Participants are urged to bring a hand-held calculator, since many actual problems will be worked out in each session.

Continental breakfast Your registration includes the cost of a continental breakfast, and a morning and afternoon refreshment break. These informal segments allow you to discuss topics of mutual interest with other seminar attendees. (You are free to make your own lunch and dinner arrangements.)

Hotel/transportation arrangements RSMeans arranges to hold a block of rooms at most host hotels. To take advantage of special group rates when making your reservation, be sure to mention that you are attending the RSMeans seminar. You are, of course, free to stay at the lodging place of your choice. **(Hotel reservations and transportation arrangements should be made directly by seminar attendees.)**

Important Class sizes are limited, so please register as soon as possible.

Note: *Pricing subject to change.*

Registration Form

ADDS-1000

Call 1-800-334-3509, ext. 5115 to register or FAX this form 1-800-632-6732. Visit our Web site: www.rsmeans.com

Please register the following people for the RSMeans construction seminars as shown here. We understand that we must make our own hotel reservations if overnight stays are necessary.

☐ Full payment of $____________________ enclosed.

☐ Bill me

Please print name of registrant(s)

(To appear on certificate of completion)

____

____

____

____

P.O. #: ____________________________

GOVERNMENT AGENCIES MUST SUPPLY PURCHASE ORDER NUMBER OR TRAINING FORM.

Firm name______________________________

Address______________________________

City/State/Zip______________________________

Telephone no.____________________ Fax no.____________________

E-mail address______________________________

Charge registration(s) to: ☐ MasterCard ☐ VISA ☐ American Express

Account no.____________________ Exp. date____________________

Cardholder's signature______________________________

Seminar name______________________________

Seminar City ______________________________

Please mail check to: RSMeans Seminars, 63 Smiths Lane, P.O. Box 800, Kingston, MA 02364 USA

New Titles

The Gypsum Construction Handbook by USG

An invaluable reference of construction procedures for gypsum drywall, cement board, veneer plaster, and conventional plaster. This new edition includes the newest product developments, installation methods, fire- and sound-rated construction information, illustrated framing-to-finish application instructions, estimating and planning information, and more. Great for architects and engineers; contractors, builders, and dealers; apprentices and training programs; building inspectors and code officials; and anyone interested in gypsum construction. Features information on tools and safety practices, a glossary of construction terms, and a list of important agencies and associations. Also available in Spanish.

$29.95
Over 575 pages, illustrated, softcover
Catalog no. 67357A

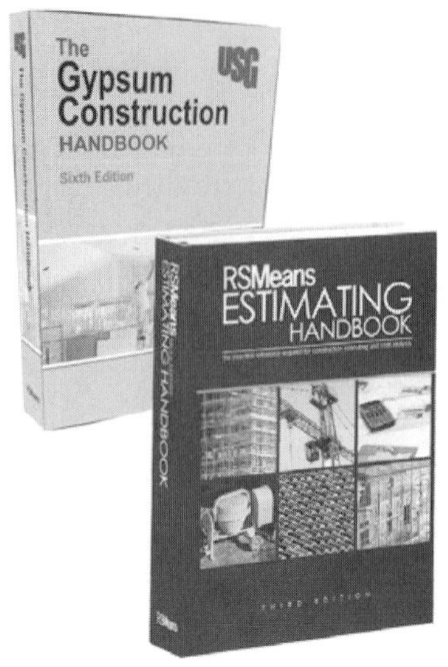

RSMeans Estimating Handbook, 3rd Edition

Widely used in the industry for tasks ranging from routine estimates to special cost analysis projects, this handbook has been completely updated and reorganized with new and expanded technical information.

RSMeans Estimating Handbook will help construction professionals:

- Evaluate architectural plans and specifications
- Prepare accurate quantity takeoffs
- Compare design alternatives and costs
- Perform value engineering
- Double-check estimates and quotes
- Estimate change orders

$99.95 per copy
Over 900 pages, hardcover
Catalog No. 67276B

BIM Handbook
A Guide to Building Information Modeling for Owners, Managers, Designers, Engineers and Contractors
by Chuck Eastman, Paul Teicholz, Rafael Sacks, Kathleen Liston

Building Information Modeling (BIM) is a new approach to design, construction, and facility management where a digital depiction of the building process is used to facilitate the exchange of information in digital format.

The BIM Handbook provides an in-depth understanding of BIM technologies, the business and organizational issues associated with its implementation, and the profound advantages that effective use of BIM can provide to all members of a project team. Also includes a rich set of informative BIM case studies.

$85.00
504 pages
Catalog no. 67356

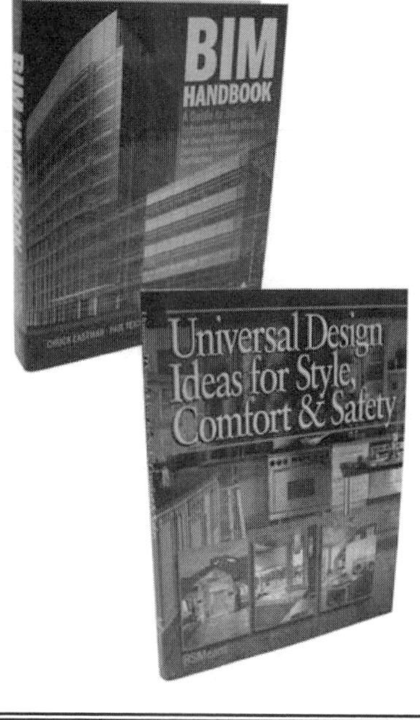

Universal Design Ideas for Style, Comfort & Safety
by RSMeans and Lexicon Consulting, Inc.

Incorporating universal design when building or remodeling helps people of any age and physical ability more fully and safely enjoy their living spaces. This book shows how universal design can be artfully blended into the most attractive homes. It discusses specialized products like adjustable countertops and chair lifts, as well as simple ways to enhance a home's safety and comfort. With color photos and expert guidance, every area of the home is covered. Includes budget estimates that give an idea how much projects will cost.

$21.95
160 pages, illustrated, softcover
Catalog no. 67354

Green Home Improvement
by Daniel D. Chiras, PhD

With energy costs rising and environmental awareness increasing, many people are looking to make their homes greener. This book makes it easy, with 65 projects and actual costs and projected savings that will help homeowners prioritize their green improvements.

Projects range from simple water savers that cost only a few dollars, to bigger-ticket items such as efficient appliances and HVAC systems, green roofing, flooring, landscaping, and much more. With color photos and cost estimates, each project compares options and describes the work involved, the benefits, and the savings.

$24.95
320 pages, illustrated, softcover
Catalog no. 67355

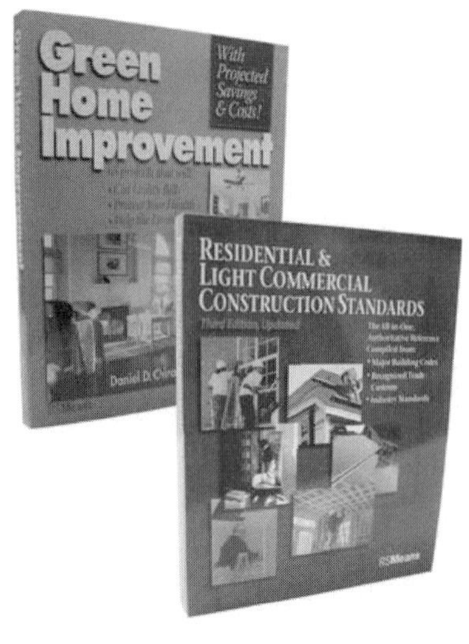

Residential & Light Commercial Construction Standards, 3rd Edition, Updated
by RSMeans and contributing authors

This book provides authoritative requirements and recommendations compiled from leading professional associations, industry publications, and building code organizations. This all-in-one reference helps establish a standard for workmanship, quickly resolve disputes, and avoid defect claims. Updated third edition includes new coverage of green building, seismic, hurricane, and mold-resistant construction.

$59.95
Over 550 pages, illustrated, softcover
Catalog no. 67322B

ORDER TOLL FREE 1-800-334-3509
OR FAX 1-800-632-6732

2010 Order Form

Qty.	Book no.	COST ESTIMATING BOOKS	Unit Price	Total
	60060	Assemblies Cost Data 2010	$279.95	
	60010	Building Construction Cost Data 2010	169.95	
	60110	Concrete & Masonry Cost Data 2010	154.95	
	50140	Construction Cost Indexes 2010 (subscription)	362.00	
	60140A	Construction Cost Index–January 2010	90.50	
	60140B	Construction Cost Index–April 2010	90.50	
	60140C	Construction Cost Index–July 2010	90.50	
	60140D	Construction Cost Index–October 2010	90.50	
	60340	Contr. Pricing Guide: Resid. R & R Costs 2010	39.95	
	60330	Contr. Pricing Guide: Resid. Detailed 2010	39.95	
	60320	Contr. Pricing Guide: Resid. Sq. Ft. 2010	39.95	
	60230	Electrical Change Order Cost Data 2010	169.95	
	60030	Electrical Cost Data 2010	169.95	
	60200	Facilities Construction Cost Data 2010	419.95	
	60300	Facilities Maintenance & Repair Cost Data 2010	379.95	
	60160	Heavy Construction Cost Data 2010	169.95	
	60090	Interior Cost Data 2010	169.95	
	60120	Labor Rates for the Const. Industry 2010	379.95	
	60180	Light Commercial Cost Data 2010	144.95	
	60020	Mechanical Cost Data 2010	169.95	
	60150	Open Shop Building Const. Cost Data 2010	169.95	
	60210	Plumbing Cost Data 2010	169.95	
	60040	Repair & Remodeling Cost Data 2010	144.95	
	60170	Residential Cost Data 2010	144.95	
	60280	Site Work & Landscape Cost Data 2010	169.95	
	60050	Square Foot Costs 2010	184.95	
	62019	Yardsticks for Costing (2009)	154.95	
	62010	Yardsticks for Costing (2010)	169.95	
		REFERENCE BOOKS		
	67310A	ADA Compliance Pricing Guide, 2nd Ed.	79.95	
	67356	BIM Handbook	85.00	
	67330	Bldrs Essentials: Adv. Framing Methods	12.98	
	67329	Bldrs Essentials: Best Bus. Practices for Bldrs	29.95	
	67298A	Bldrs Essentials: Framing/Carpentry 2nd Ed.	15.98	
	67298AS	Bldrs Essentials: Framing/Carpentry Spanish	15.98	
	67307	Bldrs Essentials: Plan Reading & Takeoff	35.95	
	67342	Building & Renovating Schools	69.98	
	67339	Building Security: Strategies & Costs	44.98	
	67353	Complete Book of Framing	29.95	
	67146	Concrete Repair & Maintenance Illustrated	69.95	
	67352	Construction Business Management	49.95	
	67314	Cost Planning & Est. for Facil. Maint.	89.95	
	67328A	Designing & Building with the IBC, 2nd Ed.	99.95	
	67230B	Electrical Estimating Methods, 3rd Ed.	64.95	
	64777A	Environmental Remediation Est. Methods, 2nd Ed.	49.98	
	67343	Estimating Bldg. Costs for Resi. & Lt. Comm.	29.95	
	67276B	Estimating Handbook, 3rd Ed.	$99.95	
	67318	Facilities Operations & Engineering Reference	54.98	
	67301	Facilities Planning & Relocation	49.98	
	67338A	Green Building: Proj. Planning & Cost Est., 2nd Ed.	129.95	
	67355	Green Home Improvement	24.95	

Qty.	Book no.	REFERENCE BOOKS (Cont.)	Unit Price	Total
	67357A	The Gypsum Construction Handbook	29.95	
	67357S	The Gypsum Construction Handbook (Spanish)	29.95	
	67323	Historic Preservation: Proj. Planning & Est.	49.98	
	67308E	Home Improvement Costs–Int. Projects, 9th Ed.	24.95	
	67309E	Home Improvement Costs–Ext. Projects, 9th Ed.	24.95	
	67344	Homeowner's Guide to Mold	21.95	
	67324B	How to Est.w/Means Data & CostWorks, 3rd Ed.	59.95	
	67351	How Your House Works	21.95	
	67282A	Illustrated Const. Dictionary, Condensed, 2nd Ed.	59.95	
	67292A	Illustrated Const. Dictionary, w/CD-ROM, 3rd Ed.	99.95	
	67348	Job Order Contracting	89.95	
	67347	Kitchen & Bath Project Costs	19.98	
	67295C	Landscape Estimating Methods, 5th Ed.	64.95	
	67341	Life Cycle Costing for Facilities	99.95	
	67294B	Mechanical Estimating Methods, 4th Ed.	64.95	
	67245A	Planning & Managing Interior Projects, 2nd Ed.	34.98	
	67283B	Plumbing Estimating Methods, 3rd Ed.	29.98	
	67345	Practice of Cost Segregation Analysis	99.95	
	67337	Preventive Maint. for Higher Education Facilities	149.95	
	67346	Preventive Maint. for Multi-Family Housing	89.95	
	67326	Preventive Maint. Guidelines for School Facil.	149.95	
	67247B	Project Scheduling & Management for Constr. 3rd Ed.	64.95	
	67265B	Repair & Remodeling Estimating Methods, 4th Ed.	69.95	
	67322B	Resi. & Light Commercial Const. Stds., 3rd Ed.	59.95	
	67327A	Spanish/English Construction Dictionary, 2nd Ed.	23.95	
	67145B	Sq. Ft. & Assem. Estimating Methods, 3rd Ed.	69.95	
	67321	Total Productive Facilities Management	29.98	
	67350	Understanding and Negotiating Const. Contracts	69.95	
	67303B	Unit Price Estimating Methods, 4th Ed.	59.95	
	67354	Universal Design	21.95	
	67319A	Value Engineering: Practical Applications	79.95	

MA residents add 6.25% state sales tax	
Shipping & Handling**	
Total (U.S. Funds)*	

Prices are subject to change and are for U.S. delivery only. *Canadian customers may call for current prices. **Shipping & handling charges: Add 7% of total order for check and credit card payments. Add 9% of total order for invoiced orders.

Send order to: **ADDV-1000**

Name (please print) ______________________________

Company ______________________________

☐ **Company**

☐ **Home** **Address** ______________________________

____

City/State/Zip ______________________________

Phone # ____________________ P.O. # ____________________

(Must accompany all orders being billed)

Mail to: RSMeans, P.O. Box 800, Kingston, MA 02364

RSMeans Project Cost Report

By fiiling out this report, your project data will contribute to the database that supports the RSMeans Project Cost Square Foot Data. When you fill out this form, RSMeans will provide a $30 discount off one of the RSMeans products advertised in the preceding pages. Please complete the form including all items where you have cost data, and all the items marked (✓).

$30.00 Discount per product for each report you submit.

Project Description (No remodeling projects, please.)

✓ Building Use (Office, School...) ____________

✓ Address (City, State) ____________

____

✓ Total Building Area (SF) ____________

✓ Ground Floor (SF) ____________

✓ Frame (Wood, Steel...) ____________

✓ Exterior Wall (Brick, Tilt-up...) ____________

✓ Basement: (check one) ☐ Full ☐ Partial ☐ None

✓ Number Stories ____________

✓ Floor-to-Floor Height ____________

✓ Volume (C.F.) ____________

% Air Conditioned ____________ Tons ____________

Total Project Cost $ ____________

Owner ____________

Architect ____________

General Contractor ____________

✓ Bid Date ____________

✓ Typical Bay Size ____________

✓ Capacity ____________

✓ Labor Force: ____ % Union ____ % Non-Union

✓ Project Description (Circle one number in each line)

1. Economy 2. Average 3. Custom 4. Luxury
1. Square 2. Rectangular 3. Irregular 4. Very Irregular

Comments ____________

____

A	✓	General Conditions	$
B	✓	Site Work	$
C	✓	Concrete	$
D	✓	Masonry	$
E	✓	Metals	$
F	✓	Wood & Plastics	$
G	✓	Thermal & Moisture Protection	$
GR		Roofing & Flashing	$
H	✓	Doors and Windows	$
J	✓	Finishes	$
JP		Painting & Wall Covering	$

K	✓	Specialties	$
L	✓	Equipment	$
M	✓	Furnishings	$
N	✓	Special Construction	$
P	✓	Conveying Systems	$
Q	✓	Mechanical	$
QP		Plumbing	$
QB		HVAC	$
R	✓	Electrical	$
S	✓	Mech./Elec. Combined	$

Please specify the RSMeans product you wish to receive. Complete the address information.

Product Name ____________

Product Number ____________

Your Name ____________

Title ____________

Company ____________

☐ Company

☐ Home Street Address ____________

City, State, Zip ____________

Return by mail or Fax 888-492-6770.

Method of Payment:

Credit Card # ____________

Expiration Date ____________

Check ____________

Purchase Order ____________

RSMeans
a division of Reed Construction Data
Square Foot Costs Department
P.O. Box 800
Kingston, MA 02364-3008